An Introduction to the Aquatic Insects of North America

Third Edition

Edited by
Richard W. Merritt
Department of Entomology
Michigan State University

and

Kenneth W. Cummins
Ecosystem Research and Implementation Department
South Florida Water Management District

KENDALL/HUNT PUBLISHING COMPANY
4050 Westmark Drive   Dubuque, Iowa   52002

Cover photograph of mayfly, *Rhithrogena* sp., taken by H. V. Daly
(Univ. of California, Berkeley).

Copyright © 1978, 1984, 1996 by Kendall/Hunt Publishing Company

Library of Congress Catalog Card Number: 95-80182

ISBN 0-7872-3241-6 — Spiral Bound
ISBN 0-7872-3240-8 — Perfect Bound

All rights reserved. No part of this publication may be reproduced,
stored in a retrieval system, or transmitted, in any form or by any
means, electronic, mechanical, photocopying, recording, or otherwise,
without prior written permission of the copyright owner.

Printed in the United States of America
10 9 8 7

Dedication

This book is dedicated to students of aquatic entomology and ecology.

Contents

Preface, vii

Acknowledgments, ix

Acknowledgments to the First Edition, xi

Acknowledgments to the Second Edition, xii

List of Contributors, xiii

1. Introduction, 1
   by R. W. Merritt and K. W. Cummins

2. General Morphology of Aquatic Insects, 5
   by K. W. Cummins and R. W. Merritt

3. Design of Aquatic Insect Studies: Collecting, Sampling and Rearing Procedures, 12
   by R. W. Merritt, V. H. Resh and K. W. Cummins

4. Aquatic Insect Respiration, 29
   by C. H. Eriksen, V. H. Resh, and G. A. Lamberti

5. Habitat, Life History, and Behavioral Adaptations of Aquatic Insects, 41
   by J. B. Wallace and N. H. Anderson

6. Ecology and Distribution of Aquatic Insects, 74
   by K. W. Cummins and R.W. Merritt

7. Use of Aquatic Insects in Biomonitoring, 87
   by D. M. Rosenberg and V. H. Resh

8. Phylogenetic Relationships and Evolutionary Adaptations of Aquatic Insects, 98
   by V. H. Resh and J. O. Solem

9. General Classification and Key to the Orders of Aquatic and Semiaquatic Insects, 108
   by H. V. Daly

10. Aquatic Collembola, 113
    by K. A. Christiansen and R. J. Snider

11. Ephemeroptera, 126
    by G. F. Edmunds, Jr., and R. D. Waltz

12. Odonata, 164
    by M. J. Westfall, Jr., and K. J. Tennessen

13. Semiaquatic Orthoptera, 212
    by I. J. Cantrall and M. A. Brusven

14. Plecoptera, 217
    by K. W. Stewart and P. P. Harper

15. Aquatic and Semiaquatic Hemiptera, 267
    by J. T. Polhemus

16. Megaloptera and Aquatic Neuroptera, 298
    by E. D. Evans and H. H. Neunzig

17. Trichoptera Families, 309
    by G. B. Wiggins

18. Trichoptera Genera, 350
    by J. C. Morse and R. W. Holzenthal

19. Aquatic and Semiaquatic Lepidoptera, 387
    by W. H. Lange

20. Aquatic Coleoptera, 399
    by D. S. White and W. U. Brigham

21. Aquatic Hymenoptera, 474
    by K. S. Hagen

22. Aquatic Diptera, 484
    Part One. Larvae of Aquatic Diptera, 484
    by G. W. Courtney, H. J. Teskey, R. W. Merritt, and B. A. Foote

    Part Two. Pupae and Adults of Aquatic Diptera, 515
    by R. W. Merritt, D. W. Webb, and E. I. Schlinger

23. Tipulidae, 549
    by G. W. Byers

24. Culicidae, 571
    by E. D. Walker and H. D. Newson

25. Simuliidae, 591
    by B. V. Peterson

26. Chironomidae, 635
    by W. P. Coffman and L. C. Ferrington, Jr.

Bibliography, 755

Index, 849
   by J. Novak

Preface

As in the second edition, this revised 3rd edition is intended to serve as a standard guide to the aquatic insects of North America. The revision includes generic keys for each order, the completeness of which varies with order, life stage, and the present state of knowledge. Significant taxonomic changes have been made and additional figures added to several chapters, particularly the Ephemeroptera, Plecoptera, Odonata, Diptera, Coleoptera, and Chironomidae, to mention a few. Chapters dealing with phylogeny, respiration, life history, behavior, sampling and ecology have been expanded, and a new chapter dealing with the use of aquatic insects in biomonitoring has been added. In addition, the ecological tables appearing in each order (or family) chapter have been updated with the appropriate references added. As with any compendium, coverage of the literature is only partial and is a continuing problem as the literature treating aquatic insects has grown exponentially in the past 11 years. In addition to those references supplied by the authors of each chapter, we surveyed numerous literature sources (e.g., NABS bibliographies, *Current Contents*, *Current Adv. Ecol. Environ. Sci.*, etc.) and solicited members of the North American Benthological Society (NABS) to send us reprints on feeding ecology, so that we could update the ecological tables. Despite these efforts, we realize that some pertinent references were missed and we sincerely apologize to those scientists for these omissions. Our coverage has expanded from over 2800 (through 1984) to nearly 5000 references (1995). As before, we strongly encourage all users of the book to continually update material in their own areas of interest. We again hope this new edition will be of even greater use to both professional and lay groups.

Acknowledgments

As with previous editions of this book, many people have contributed in different ways to make the 3rd edition possible. The editors would first like to thank the contributors for their cooperation during the production of this book. At Michigan State University, RWM would like to thank the Chairperson of Entomology, Dr. Mark Scriber, for his continued support of this project and for allowing me sabbatical time to help complete this edition of the book. I would like to acknowledge my current graduate students, John Wallace, Mac Strand, Mike Alexander and Mike Higgins for frequent discussions and assisting with various aspects of editing, and figure and table preparation when needed. We would especially like to thank the following individuals: Bill Morgan for his help on figure format preparation and numerous other computing and editing skills; Teresa Petersen (Ypsilanti, MI) for art work; Becky Blasius for her help, assistance and patience with formatting over 2000 new literature references; and D. J. Nguyen, Kristy Bodary, and Beth Capaldi for library searches and associated work. Thanks go to Roger Wotton (University of London), Ned Walker, Mike Kaufman, Tom Ellis, Bill Cooper (Michigan State University), Doug Craig (University of Alberta), and Larry Sernick (MAC) for their candid opinions and comments at the most opportune times. We would like to thank Dan Lawson (S. C. Johnson and Sons, Wisconsin) for his computer expertise in helping with programs to merge new and old references, followed by table insertion, and Lois Caprio for proofreading. We would like to gratefully acknowledge the help of Jan Eschbach, our expert in word processing and formatting, who was not only a pleasure to work with but was totally responsible for preparing the final manuscript (text, keys, tables, figure legends, and references) before it was given to the publisher. We would like to thank Jim Magovern (Oklahoma City Health Dept.), George Craig (University of Notre Dame), Rosemary Mackay (University of Toronto), Mike Swift (University of Minnesota), and many others over the past 10 years for providing errata for the previous edition. We would like to thank the members of the North American Benthological Society (NABS), too numerous to single out, who sent us reprints to help update the Ecological Tables. We thank Howell Daly (University of California, Berkeley) for use of his picture of the mayfly on the front cover, and John Novak, Dept. of Biology, Colgate University for preparing the index. Finally, I (RWM) would like to thank my son, Scott, for providing me with frequent reprieves from chapter writing or editing with Indiana University soccer games, and my scientist son, Brett, for our late night trivial discussions that usually ended up with the question, why am I doing this book again? Finally, to my wife Pam, for her eternal patience, understanding and optimism, which often balanced my eternal impatience and pessimism. I (KWC) would like to acknowledge the support of the South Florida Water Management District, especially Patrick Gostel, in allowing me time to contribute to the preparation of the 3rd edition. In addition, I acknowledge the continued input of ideas by my wife, Dr. Margaret Wilzbach, throughout preparation of the 3rd edition and to my son Paul for helping me maintain my sanity.

Acknowledgments and credits by contributors and the editors for specific chapters are as follows: *General Morphology*—K. Stewart (North Texas State Univ.) for review; *Use of Aquatic Insects in Biomonitoring*—the authors thank A. Wiens for preparing Figure 1 and D. Laroque for word processing. J. Alba-Tercedor, K. Cummins, I. Davies, J. Davis, W. Davis, and R. Merritt reviewed earlier versions of the manuscript; *Phylogenetic Relationships and Evolutionary Adaptations*—B. Waltz (Indiana Dept. Nat. Res.), K. Stewart (North Texas State Univ.), J. Polhemus (Englewood, CO), J. Morse (Clemson University), G. Courtney (Grand Valley State University), K. Tennessen (T.V.A.), and D. White (Murray State University) for their comments on the genealogical trees prepared for the 3rd edition; R. W. Merritt and W. T. Morgan (M.S.U.) for comments and revised dendrograms. *Ephemeroptera*—W. McCafferty (Purdue University) and I. Campbell (Monash University) for comments and literature information; *Odonata*—D. Johnson (East Tennessee State University) for table information; *Plecoptera*—the Plecoptera larval key was developed, in part, from a study by K. W. Stewart, supported by the National Science Foundation; *Megaloptera and Aquatic Neuroptera*— A. Contreras-Ramos (University of Minnesota) for table information; *Trichoptera*—the editors would like to thank G. Wiggins and J. McNeill (Royal Ontario Museum) for their cooperation on copyright matters; *Aquatic and Semiaquatic Lepidoptera*—D. Habeck (University of Florida) for literature information; *Aquatic Coleoptera*—J. Doyen and G. Ulrich who were involved in earlier editions and provided the strong base for this edition; H. B. Leech and F. N. Young who offered many suggestions for the first two editions; Kathleen Sweeney (University of Michigan), Aleta Holt (INHS), and Celeste Green (University of California) who produced most of the more than 300 illustrations; C. Millenbach, M.L. Giovannini, and E. Rogers for editorial assistance on the second edition and K. Johnston, B. Lee, and B. Sharp (Murray State University) for assistance on this edition; the Center for Reservoir Research (MSU) for research support; *Aquatic Diptera*—the editors would like to thank the following individuals for permission to use figures, reviewing sections of the tables, and/or providing data on certain taxa: S. Murphree (Belmont University), B. Grogan (Salisbury State College),W. Hilsenhoff (University of Wisconsin), J. MacDonald (Purdue University), L. Hribar, J. Linley (Florida Med. Ent. Lab), J. Burger (University of New Hampshire), R. Lane (University of California, Berkeley), G. Byers (University of Kansas), B. Foote (Kent State University), B. Peterson, N. Woodley, C. Thompson and other research scientists (Systematic Ent. Lab, U.S. Nat. Mus., Washington, D.C.), B. Romoser (Ohio University), M. Peters (University of Mass.), W. Mathis and other research scientists (Smithsonian Instit.), J. Cummings and other research scientists (Biosystematics Res. Center, Agric. Canada), B. Sinclair (Carleton University), S. Marshall (University of Guelph), D. Craig (University of Alberta), D. Currie (Royal Ontario Museum), D. Webb (Illinois Nat. Hist. Survey), M. Berg (Loyola University), G. Courtney (Grand Valley State University), W. Coffman (University of Pittsburgh), B. Peckarsky (Cornell University). We would like to acknowledge the Centre for Land and Biological Resources Research, Research Branch, Agriculture and Agri-Food, Canada, for permission to use figures (noted in figure legends) from Vols. 1(1981) and 2(1987), *Manual of Nearctic Diptera*, McAlpine et al. (coords.). Reproduced with the permission of the Minister of Supply and Services Canada, 1994; *Chironomidae*—M. Berg (Loyola University) for ecological table information.

Original research supported, in part, by the Michigan State University Agricultural Experiment Station, National Institutes of Health grant awarded to RWM and E. D. Walker, NE-118 Regional Research Project (RWM), National Science Foundation (KWC, RWM), and the U.S. Dept. of Energy (KWC).

Acknowledgments to the First Edition

We gratefully acknowledge the continual encouragement and support on this project by James E. Bath, Chairman, Department of Entomology, Michigan State University. Original research reported by R. W. Merritt was supported, in part, by the Michigan State University Agricultural Experiment Station, Ecosystem Studies of the National Science Foundation (grant DEB76-20122) and the Office of Water Research and Technology (grant A-085-MICH). Original research reported by K. W. Cummins was supported by the U.S. Energy Research and Development Administration (contract EY-76-S-2002-A001) and Ecosystem Studies of the National Science Foundation (grant GB-36069X).

We would like to acknowledge Drs. Herbert H. Ross (Department of Entomology, University of Georgia) and Carl W. Richards (Rockford, Michigan) for evaluating the pre-publication copy of the manuscript. We are also grateful to Lana Tackett for art work, Paul Mescher and Vivian Napoli for graphics, John E. Sturwold (Graphics Department, Pest Management Curriculum, Michigan State University) for cover design, Josephine Maybee and Charlotte Seeley for manuscript and table preparation, and Doug Ross, Neil Kagan, and Edward F. Gersabeck, Jr., for proofing. We would also like to thank Jane Cummins for bibliographic work and Pam, Brett, and Scott Merritt for their patience and understanding. Others have been acknowledged in specific chapters.

Acknowledgments to the Second Edition

Many people have helped in the preparation of the second edition in many different ways. As in the first edition, we especially acknowledge Dr. Jim Bath, Chairman of the Department of Entomology, Michigan State University, for providing his continual support and optimism during the entire project. We gratefully acknowledge the assistance of the following individuals for either manuscript review, table preparation, reference verification and/or editorial suggestions: Tom Burton, Mike and Sue Kaufman, Dave Cornelius, Dave and Terry Grant, Jim Keller, Dan Lawson, Bill Taft, Kevin Webb, and Peggy Wilzbach. We thank R. Mattingly for assistance in editing. A special thanks goes to Gail Motyka for the extra hours devoted to the book in lieu of finishing her thesis. We are grateful to Lana Tackett for scientific illustration, Josephine Maybee, Susan Battenfield, Patti Vuich, and Carol Asbahr for various aspects of manuscript preparation. Our special thanks go to Ken Dimoff, computer coordinator for the MSU Department of Entomology, for designing the data management program to merge the old and new reference lists. Finally, we would like to again thank Pam, Brett, and Scott Merritt for their patience and understanding, as well as their permission to convert the entire home living room into "book central" for a two-year period. Acknowledgments and credits by contributors and the editors for specific chapters are as follows: *General Morphology*—Roger Akre (Washington State Univ.) and H. V. Daly (Univ. of California) for review; Ken Manuel (Duke Power Co.) for critical comments; *Aquatic Insect Respiration*—Doug Craig (Univ. of Alberta) and Mike Wiley (Illinois Nat. Hist. Surv.) for review; *Habitat, Life History and Behavioral Adaptations*—Chuck Hawkins (Oregon State Univ.) for review; *Ecology and Distribution of Aquatic Insects*—Doug Ross (Biochem Products), Jean Stout, Dan Lawson (Michigan State Univ.) and Peggy Wilzbach (Oregon State Univ.) for review; Al Grigarick (Univ. of California) for critical comments; *Phylogenetic Relationships and Evolutionary Adaptations*—R. Garrison and M. Westfall, Jr. (Odonata), G. Edmunds (Ephemeroptera), P. Harper and K. Stewart (Plecoptera), J. Polhemus (Hemiptera), G. Wiggins and J. Wood (Trichoptera), and E. Schlinger (Diptera) for their comments on the genealogical trees prepared for the second edition; *Plecoptera*—the Plecoptera nymph key was developed, in part, from a study by K. W. Stewart, supported by the National Science Foundation; *Aquatic and Semiaquatic Hemiptera*—Drs. R. F. Denno (Univ. of Maryland) and J. P. Kramer (USDA, USNM) for their suggestions and comments on the semiaquatic Homoptera; Dan A. Polhemus for preparing many original figures; the Entomological Society of Canada for permission to use figures from Brooks and Kelton (1967); the University of California Press for permission to use figures from Usinger (1956a) and Menke et al. (1979); R. Aiken (Erindale College) for critical comments; *Megaloptera and Aquatic Neuroptera*—D. C. Tarter (Marshall Univ.) for providing larval material for study and 0. S. Flint (Smithsonian Institution) for suggestions relative to revising the chapter; *Trichoptera*—Zile Zichmanis for illustrations of adults and pupae; Anker Odum for illustrations of larvae and cases; Toshio Yamamoto for assistance in reviewing characters for the keys; John Morse (Clemson Univ.) for critical comments; *Trichoptera Genera*—0. S. Flint, Jr., S. W. Hamilton, R. W. Kelley, J. S. Weaver, III, and G. B. Wiggins for their cooperation in providing specimens, information, and comments; J. P. Norton for selected illustrations; S. W. Hamilton and W. R. English for editorial tasks; N. H. Brewer for typing the final manuscript; *Aquatic and Semiaquatic Lepidoptera*—Fred Stehr (Michigan State Univ.) and Dale Habeck (Univ. of Florida) for comments and review; *Aquatic Coleoptera*—H. B. Leech who reviewed the chapter for the original edition and F. N. Young, Jr., who has offered many suggestions for this edition; Kathleen Sweeney (Univ. of Michigan), Aleta Holt (INHS), and Celeste Green (Univ. of California) for their artistic abilities in producing most of the more than 300 illustrations; C. Millenbach, M. L. Giovannini, and E. Rogers for their assistance in producing the manuscript; "We gratefully acknowledge G. Ulrich (Univ. of California) for his significant contributions to the original edition." *Aquatic Hymenoptera*—F. E. Skinner and Natalie Vandenberg for several drawings; *Aquatic Diptera*—R. Foote, R. Gagne, C. Thompson, D. D. Wilder, W. Wirth (Systematic Entomology Laboratory, USDA, Washington, D.C.), Lloyd Knutson (IIBIII, USDA/ARS), Wayne Mathis (Smithsonian Institution), Ben Foote (Kent State Univ.), Robert Lane (Univ. of California), W. Turner (Washington State Univ.), J. B. Wallace (Univ. of Georgia), J. Burger (Univ. of New Hampshire) assisted with revisions of the Diptera table; *Chironomidae*—S. Roback (Philadelphia Acad. Sciences) and Don Oliver (Biosystematics Research Instit., Ottawa) for review. Additional credits for figure use are given in specific chapters. Original research supported, in part, by the Michigan State and Oregon State University Agricultural Experiment Stations, Ecosystem Studies of the National Science Foundation (grants DEB-80-22634; DEB-81-12455), and the Department of Energy, Ecological Sciences Division (Contract DE-AT06-79E1004).

List of Contributors

N. H. ANDERSON, Department of Entomology, Oregon State University, Corvallis, Oregon 97331.

W. U. BRIGHAM, Illinois Natural History Survey, 607 East Peabody Drive, Champaign, Illinois 61820.

M. A. BRUSVEN, Department of Entomology, University of Idaho, Moscow, Idaho 83843.

G. W. BYERS, Department of Entomology, University of Kansas, Lawrence, Kansas 66044.

I. J. CANTRALL, Museum of Zoology, University of Michigan, Ann Arbor, Michigan 48104.

G. W. COURTNEY, Department of Biology, Grand Valley State University, Allendale, Michigan 49401.

K. CHRISTIANSEN, Department of Biology, Grinnell College, Grinnell, Iowa 50112.

W. P. COFFMAN, Department of Biological Sciences, University of Pittsburgh, Pittsburgh, Pennsylvania 15213.

K. W. CUMMINS, Ecosystem Research and Implementation Department, South Florida Water Management District, 3301 Gun Club Rd., West Palm Beach, Florida 33416.

H. V. DALY, Department of Environmental Science, Policy and Management, University of California, Berkeley, California 94720.

G. F. EDMUNDS, Jr., Department of Biology, University of Utah, Salt Lake City, Utah 84112.

C. H. ERIKSEN, Joint Sciences Department, The Claremont Colleges, Claremont, California 91711.

E. D. EVANS, Michigan Department of Natural Resources, P.O. Box 30028, Lansing, Michigan 48909.

L. C. FERRINGTON, Jr., State Biological Survey of Kansas, University of Kansas, Lawrence, Kansas 66044.

B. A. FOOTE, Department of Biology, Kent State University, Kent, Ohio 44240.

K. S. HAGEN, Department of Environmental Science, Policy and Management, University of California, Berkeley, California 94720.

P. P. HARPER, Department of Biology, Université de Montréal, Montréal, Quebec, Canada H3C 3J7.

R. W. HOLZENTHAL, Department of Entomology, University of Minnesota, St. Paul, Minnesota 55108.

G. A. LAMBERTI, Department of Biological Sciences, University of Notre Dame, Notre Dame, Indiana 46556.

W. H. LANGE, 1119 Harvard Dr., Davis, California 95616.

R. W. MERRITT, Department of Entomology, Michigan State University, East Lansing, Michigan 48824.

J. C. MORSE, Department of Entomology, Fisheries and Wildlife, Clemson University, Clemson, South Carolina 29631.

H. D. NEWSON, 149 S. Temelec Circle, Sonoma, California 49684.

H. H. NEUNZIG, Department of Entomology, North Carolina State University, Raleigh, North Carolina 27650.

J.A. NOVAK, Department of Biology, Colgate University, Hamilton, NY 13346.

B. V. PETERSON, Systematic Entomology Laboratory, USDA, c/o U.S. National Museum of Natural History, NHB 168, Washington D.C. 20560.

J. T. POLHEMUS, 3115 S. York, Englewood, Colorado 80110.

V. H. RESH, Department of Environmental Science, Policy and Management, University of California, Berkeley, California 94720.

D. M. ROSENBERG, Freshwater Institute, 501 University Crescent, Winnipeg, Manitoba, Canada R3T 2N6.

E. I. SCHLINGER, 1944 Edison St., Santa Ynez, California 93460.

R. I. SNIDER, Department of Zoology, Michigan State University, East Lansing, Michigan 48824.

J. O. SOLEM, The Museum, University of Trondheim, Royal Norwegian Society of Sciences and Letters, Trondheim, Norway.

K. W. STEWART, Department of Biological Sciences, North Texas State University, Denton, Texas 76203.

K. J. TENNESSEN, Tennessee Valley Authority, Muscle Shoals, Alabama 35660.

H. J. TESKEY, Biosystematics Research Institute, Canada Agriculture, Ottawa, Ontario, Canada KIA OC6.

J. B. WALLACE, Department of Entomology, University of Georgia, Athens, Georgia 30602.

E. D. WALKER, Department of Entomology, Michigan State University, East Lansing, Michigan 48824.

R. D. WALTZ, Divsion of Entomology and Plant Pathology, Indiana Department of Natural Resources, Indianapolis, Indiana 46204.

D. W. WEBB, Illinois Natural History Survey, 607 East Peabody Drive, Champaign, Illinois 61820.

M. J. WESTFALL, Jr., Department of Zoology, University of Florida, Gainesville, Florida 32611.

D. S. WHITE, Hancock Biological Station, Murray State University, Murray, Kentucky 42071.

G. B. WIGGINS, Department of Entomology, Royal Ontario Museum, Toronto, Ontario, Canada M5S 2C6.

INTRODUCTION

Richard W. Merritt
Michigan State University, East Lansing

Kenneth W. Cummins
South Florida Water Management District, West Palm Beach

The emphasis on aquatic insect studies, which has expanded exponentially in the last three decades, has been largely ecological. This interest in aquatic insects has grown from early limnological roots (e.g., Forbes 1887) and sport fishery-related investigations of the '30s and '40s (Needham 1934), to the use of aquatic insects as indicators of water quality during the '50s and '60s (e.g., Kuehne 1962; Bartsch and Ingram 1966; Wilhm and Dorris 1968; Warren 1971; Cairns and Pratt 1993). In the '70s and '80s, aquatic insects became the dominant forms used in freshwater investigations of basic ecological questions (e.g., Barnes and Minshall 1983). The work on aquatic insects has embraced most major areas of ecological inquiry (e.g., population dynamics, predator-prey interactions, physiological and trophic ecology, competition; e.g., Resh and Rosenberg 1984; Cummins 1994) and management applications of this basic research (Wright *et al.* 1991). In addition, fly fishermen have enthusiastically sought knowledge about aquatic insects, both as fish foods to be imitated and as interesting cohabitants with their quarry (e.g., Swisher and Richards 1971, 1991; Schweibert 1973; Caucci and Nastasi 1975; Borger 1980; LaFontaine 1981; McCafferty 1981; Whitlock 1982). Initially, the primary justification for this book was that the systematics of aquatic insects has lagged behind the needs of aquatic ecologists and water managers, and the inquisitiveness of anglers. This is still true, but to a lesser extent, although the sophistication of scientists, managers, and anglers perpetuates the need for better taxonomic and ecological treatments.

Aquatic insects also are of concern to those involved in teaching (Resh and Rosenberg 1979; Merritt and Cummins 1984) and in outdoor recreation activities because certain groups (e.g., mosquitoes, black flies, horse flies) are frequently pests of humans and other animals in water-based environments (Merritt and Newson 1978, Kim and Merritt 1987). Identification is the first step toward a basic understanding of the biology and ecology of aquatic insects; this will eventually allow for the development of proper management strategies. The amateur naturalist and primary or secondary school educator also require basic identification as an important initial step in familiarization. Thus, for all concerned, identification and basic ecological and life history information is important for categorizing the aquatic insects collected.

A number of well-known general works (Usinger 1956a; Edmondson 1959; Klots 1966) and specific studies (Ross 1944; Burks 1953) tend to be taxonomically and ecologically, at least partially, out-of-date. More recent comprehensive treatments of the Ephemeroptera (Edmunds *et al.* 1976), Plecoptera (Stewart *et al.* 1988), and Trichoptera (Wiggins 1977, 1995) are currently available, as are some general works (e.g., Stehr 1987, 1991; Pennak 1978; Thorp and Covich 1991). However, at the present time only this 3rd edition gathers together comprehensive, updated, generic treatments of immature and adult stages of aquatic insects of North America. This 3rd edition is intended to serve as a standard reference on the biology and ecology of aquatic insects with keys to separate life stages of all major taxonomic groupings.

The taxonomic coverage is coupled with summaries of related information on aquatic insect phylogeny, ecology, water quality, respiration, sampling, life history, and behavior. Generic keys to immatures and adults are provided for all but a few groups of Diptera. Further, pupal keys are now provided for most holometabolous orders. All the keys have been revised, and very significant revisions have been made to many groups, such as the Ephemeroptera, Odonata, Plecoptera, aquatic Coleoptera and Diptera. Because of the size of the order Diptera, separate chapters have been devoted to individual families (i.e., Chironomidae, Simuliidae, Culicidae).

The distinction between aquatic or semiaquatic and terrestrial insects is arbitrary. In the 3rd edition those orders and families with one or more life stages associated with aquatic habitats and frequently encountered in collections made from aquatic environments are covered. This includes the Collembola, Orthoptera, and Hymenoptera, even though they are only marginally associated with aquatic habitats. Since terrestrial insects frequently become trapped in the surface film of aquatic systems (e.g., Collembola), a wide range of terrestrial species are encountered with varying frequency. A specimen not fitting the keys in this edition probably belongs to a terrestrial taxon and generally can be identified using Borror *et al.* (1989).

An annotated list of general references to works dealing with aquatic insect taxonomy and ecology is given in Table 1A. As indicated above, many of the taxonomic works are outdated; however, they include a great deal of useful biological information on the groups. More specific references are given at the end of the appropriate order (or family) chapter.

Various combinations of taxa can be categorized so as to permit ecological questions to be addressed at the functional

Table 1A. Selected North American literature dealing with general aquatic insect identification and ecology.

Taxonomic treatments Source	Immatures	Adults	GENERAL COVERAGE Keys	Biology	General Comments
Ward and Whipple (1918)	x	x	Generic	x	Contains much information on biology.
Chu (1949)	x		Family	x[†]	Generalized treatment of immatures.
Peterson (1951)	x		Family	x[†]	Limited to holometabolous groups; descriptive in nature.
Usinger (1956a)	x	x[‡]	Primarily generic level, with keys to Calif. species	x	Considerable information on West Coast species.
Edmondson (1959)	x	x[*]	Generic		A standard reference on freshwater invertebrates; some keys outdated.
Eddy and Hodson (1961)	x		Order		Keys to common animals, including water mites, of the North Central states.
Needham and Needham (1962)	x	x[*]	Generic		Keys to many genera; field manual.
Klots (1966)	x	x[*]	Primarily family level, with keys to some genera	x[†]	Field manual; some keys based on ecology and behavior of group.
Borror and White (1970)	x[§]	x	Orders, some keys to families	x	Comprehensive field manual on insects; primarily based on examination of insects in the hand; color plates.
Swisher and Richards (1971)	x	x	Generic (mayflies only)	x	Anglers' guide, primarily to mayflies; color photographs; seasonal data.
Schweibert (1973)	x		None	x	Extensive treatment of immature aquatic insects for anglers; color plates; seasonal and distributional data.
Caucci and Nastasi (1975)	x	x	None	x	Anglers' guide to mayflies and stoneflies; color photographs; seasonal and distributional data.
Parrish (1975)	x		Generic		Keys only to Southeastern United States; limited to water quality indicator organisms.
Smith and Carlton (1975)		x	Primarily family level, with keys to some species	x[†]	Keys only to intertidal insects of the central California coast.
Tarter (1976)	x	x[‡]	Generic	x[†]	Keys only to West Virginia genera and occasionally species.
Merritt and Cummins (1978, 1984)	x	x	Orders, families, and genera of North American aquatic insects	x	Chapters on morphology, ecology, phylogeny, life history, behavior, and sampling; summary tables on ecology and distribution with references.
Pennak (1978, 1989)	x	x[*]	Generic	x	Extensive treatment on biology of many freshwater invertebrates. Last edition (1989) does not treat aquatic insects.
Lehmkuhl (1979a)	x	x[*]	Families	x[†]	Field guide to aquatic insects.
Borger (1980)	x		Orders and families	x[†]	An angler's guide to the major food organisms of trout.
Borror et al. (1981)		x	Families	x[†]	Generalized treatment of adults.
Hilsenhoff (1981)	x	x[*]	Generic	x[†]	Keys only to Wisconsin genera, but generally applicable to Great Lakes region.
McCafferty (1981)	x	x[‡]	Pictorial keys to aquatic insect families	x[†]	A thorough scientific introduction to aquatic insects for the fly fisherman; excellent illustrations.
LaFontaine (1981)	x	x	None	x[†]	An angler's guide to the caddisflies; good biological section on caddis.
Brigham et al. (1982)	x	x[*]	Families and genera for eastern North America; species for the Carolinas	x[†]	A thorough treatment of the aquatic insects and oligochaetes of the Carolinas.

[*] Only adult keys to Hemiptera and Coleoptera.
[†] Contains notes on biology or ecology.
[‡] Covers adults of some groups.
[§] Primarily adult coverage, brief treatment of immatures of some groups.

Table 1A. Continued

Taxonomic treatments Source	Immatures	Adults	GENERAL COVERAGE Keys	Biology	General Comments
Whitlock (1982)	x		None	x[†]	A good practical book on fly-fishing entomology.
Leiser and Boyle (1982)	x		None	x[†]	Good information on stonefly biology and emergence patterns for anglers.
Arbona (1989)	x	x	None	x	Good section on mayfly biology for anglers.
Guthrie (1989)	x	x	Family keys to animals found at the water surface	x[†]	Practical guide to animals found at the surface of freshwaters, including some excellent photos and drawings; Good biological information.
Stehr (1987, 1991)	x		Orders and families of terrestrial and aquatic insect immatures	x[†]	A thorough treatment of terrestrial and aquatic immatures with notes on relationships, diagnosis, and biology. Literature sources provided. Intended as text for classes on immature insects. Excellent illustrations.
Peckarsky et al. (1990)	x	x[*]	Families and genera of immature freshwater macroinvertebrates	x[†]	Good regional treatment of aquatic macroinvertebrates of northeastern North America.
Pobst (1990)	x	x[‡]	None	x[†]	A streamside guide for anglers to the major trout-stream insects with excellent color photos.
Swisher and Richards (1991)	x	x	None		x	A good presentation on the emergence of mayflies, caddisflies, and stoneflies for anglers.
Thorp and Covich (1991)	x	x[*]	Orders and families of aquatic insects and differential taxonomic treatment of other invertebrates	x[†]	Emphasis mainly on freshwater invertebrates, other than insects. Comprehensive treatment.
Clifford (1991)	x	x[*]	Pictorial keys to families and genera of immature aquatic insects and other invertebrates	x[†]	Written for aquatic invertebrates of Alberta, Canada, but much broader coverage. Excellent drawings and color photographs.
Ward and Kondratieff (1992)	x		Orders, families and some genera of selected aquatic insects	x[†]	Very useful guide with illustrated keys to mountain stream insects of Colorado.

Ecological treatments Source	General Coverage
Hynes (1972)	The "classic" on the ecology of running waters. A "must have" for every student and researcher. Wide coverage of topics on biology of rivers and streams with emphasis on aquatic invertebrates.
Ward and Stanford (1979)	Comprehensive treatment on the ecology of stream regulation, with an emphasis on downstream effects on biotic (especially aquatic insects) and abiotic components.
Lock and Williams (1981)	This text was written in honor of the retirement of H.B.N Hynes by former graduate students. Contains chapters on migrations, distributions, hydrodynamics, and ecology of aquatic insects.
Barnes and Minshall (1982)	The first attempt for the application and testing of general ecological theory to stream ecology. Several chapters discuss the ways in which aquatic insects can be used to empirically test ecological theory.
Resh and Rosenberg (1984)	A very good overview of aquatic insect ecology, highlighting research needs and suggested avenues of investigation. Good reviews on several current topics in aquatic ecology. Widely used refernce source.
Williams (1987)	An introduction to the ecology of temporary aquatic habitats, with a discussion of the abiotic features of these environments and the biology of invertebrates colonizing these habitats.

[*] Only adult keys to Hemiptera and Coleoptera.
[†] Contains notes on biology or ecology.
[‡] Covers adults of some groups.

Table 1A. Continued

Ecological treatments Source	General Coverage
Ward (1992)	A treatment of the evolutionary considerations, habitat occurrences of aquatic insect communities, and the relationship of aquatic insects to environmental variables. Good treatment of physical aspects of aquatic insect biology and their habitat.
Williams and Feltmate (1992)	An introductory text to the study of aquatic insects, with background information on the aquatic insect orders and good coverage of life histories, adaptations, population biology, trophic relationships and experimental design and sampling methods.
Rosenberg and Resh (1993)	A very thorough reference source dealing with many different approaches for using benthic macroinvertebrates in biological monitoring programs.
Wotton (1994)	This book takes a functional approach in reviewing the role of particulate and dissolved matter in a wide range of marine and freshwater ecosystems. Specific chapters deal the food of aquatic insects and the manner in which they capture particles in their environment. A good reference source for students and researchers alike.
Allan (1995)	A beginning text in stream ecology, with good overall coverage of biotic and abiotic factors influencing aquatic insect distributions and abundance. Good coverage on subjects such as drift, predation, competition, feeding ecology of fish, and the modification of running waters by humankind.

level. For example, some groups are based on morpho-behavioral adaptations for food gathering, habitat selection, or habits of attachment, concealment, and movement (Chap. 6, and ecological tables at the end of each taxonomic chapter). Different levels of taxonomic identification are required to functionally classify aquatic insects. For example, the ordinal level may be sufficient to define functional trophic relations for the Odonata, but even the generic level may be insufficient in some of the Chironomidae (Diptera). Ecologists also have resorted to "habitat taxonomy" of a single aquatic system (e.g., Coffman *et al.* 1971) or regionalized keys (e.g., Brigham *et al.* 1982; Peckarsky *et al.* 1990; Ward and Kondratieff 1992), where the fauna of a given system or region is studied for an extended period in sufficient detail to permit such system-specific keys to be written. Although this allows for significant simplification in such keys by excluding taxa from other systems or regions, changes in species composition, the key element in disturbance, can be masked by the restricted nature of this approach. This means that such keys must be used *cautiously* and *verified continually* against the full range of taxonomic information available.

The particular emphasis on ecology and field techniques in this 3rd edition reflects our conviction that the most critical task at hand is the integration of taxonomic and ecological approaches in a manner that will permit important questions concerning environmental quality and management to be addressed. The intent in the first two editions was to provide some of the tools and directions for such analyses of aquatic ecosystems. Hopefully, this expanded 3rd edition provides another significant step toward this goal.

GENERAL MORPHOLOGY OF AQUATIC INSECTS

Kenneth W. Cummins
Oregon State University, Corvallis

Richard W. Merritt
Michigan State University, East Lansing

OVERVIEW

A stonefly (order Plecoptera, family Pteronarcyidae) serves to illustrate the general external morphological features of aquatic insects used in taxonomic determinations. This primitive insect exhibits basic morphological features in a relatively unmodified or nonspecialized form. However, modifications of the general morphological plan are found in each insect order having aquatic representatives. These modifications and associated terminology are presented with the introductory material for each group and should be carefully studied before attempting to use the keys in the following chapters.

The insect body represents the fusion and modification of the basic segmentation plan characteristic of the Annelida-Arthropoda evolutionary line (e.g., Snodgrass 1935; Manton and Anderson 1979). Each segment of the body can be compared to a box, with the dorsal (top) portion, the *tergum* or *notum*, joined to the ventral (bottom) portion, the *sternum*, and to the sides or lateral portions, the *pleura*, by membranes. The legs and wings are hinged (articulated) on the pleura of the mid-body region, the *thorax*. The body regions, head, thorax, and abdomen, and associated appendages of a stonefly larva are shown in Figures 2.1 and 2.2. The life cycle of stoneflies is representative of those orders characterized by *simple (incomplete* by some authors) metamorphosis, consisting of egg, larva (immature), and adult stages; more advanced orders exhibit *complete* metamorphosis, consisting of egg, larva (immature), pupa, and adult stages.

Head

The generalized insect head represents the evolutionary fusion of six or seven anterior segments in the ancestral Annelida-Arthropoda line (e.g., Snodgrass 1935; Rempel 1975). Two or three preoral (procephalic) segments, or *somites,* were fused and now bear important sensory structures used by present-day insects to monitor their environment—the compound eyes, light-sensitive ocelli (simple eyes), and the antennae (Figs. 2.1-2.2). The *labrum*, which forms the upper lip, is joined at its base to the *clypeus*, which in turn is fused to the frons, or face. The margins of the clypeus and frons are bounded by the anterior portion of the Y-shaped *epicranial suture* (in Fig. 2.1, the line of joining between the clypeus and frons, termed a *suture* [sulcus], is not externally visible so the structure is referred to as the *frontoclypeus* [Nelson and Hanson 1971]).

Three postoral (gnathocephalic) segments are fused in modern insects to form the posterior portion of the head and bear the remaining structures of the feeding apparatus (Snodgrass 1935). As described above, the labrum forms the upper lip and the paired *mandibles* and *maxillae* form the mouth region laterally (Figs. 2.2-2.3). The bottom of the mouth is set by the *labium* or lower lip (Figs. 2.2-2.3). The maxillae and the labium bear *palps* (palpi), which are sensory in function (Figs. 2.2-2.3). The mandibles are used for chewing or crushing the food or may be modified for piercing (piercing herbivores or predators) or scraping (scraping herbivores that graze on attached algae). The maxillae and labium are variously used for tearing and manipulating food, or they may be highly modified as in the Hemiptera, adult Lepidoptera, Hymenoptera, and Diptera. The *hypopharynx* or insect "tongue," located just anterior to the labium, is a small inconspicuous lobe in some larval forms, but is subject to extreme modification in some orders (e.g., Diptera).

The sides of the head are referred to as *genae* (singular, gena; Fig. 2.2) and the top of the head as the *vertex*. Immediately behind the vertex is a large area called the *occiput* (Fig. 2.1). The head is joined to the thorax by a membranous neck region or *cervix* (Fig. 2.1). If the head is joined to the thorax so that the mouthparts are directed downward (ventrally), the condition is termed *hypognathous* (e.g., many caddisfly larvae). Mouthparts directed forward (anteriorly) are *prognathous* (e.g., beetle larvae) and those directed backward (posteriorly) are *opisthognathous* (e.g., some true bugs).

In aquatic insects that are dorsoventrally flattened, as are some stoneflies and mayflies, the sensory structures (eyes, ocelli, and antennae) are dorsal and the food-gathering apparatus is ventral. These modifications allow certain groups to move through the interstices of coarse sediments and cling to exposed surfaces in rapidly flowing streams.

Thorax

The midregion of the body, or *thorax,* bears the jointed *legs* (Fig. 2.4) and the *wings,* and is divided into three segments (Figs. 2.1-2.2, 2.5-2.6). The *prothorax* bears the forelegs, the *mesothorax* the midlegs and forewings, and the *metathorax* the hindlegs and hindwings (if wings are present).

The jointed legs are five-segmented: the *coxa, trochanter, femur, tibia,* and the three- to five-segmented *tarsus,* which terminates in one or two *tarsal claws* (Fig. 2.4). In aquatic

5

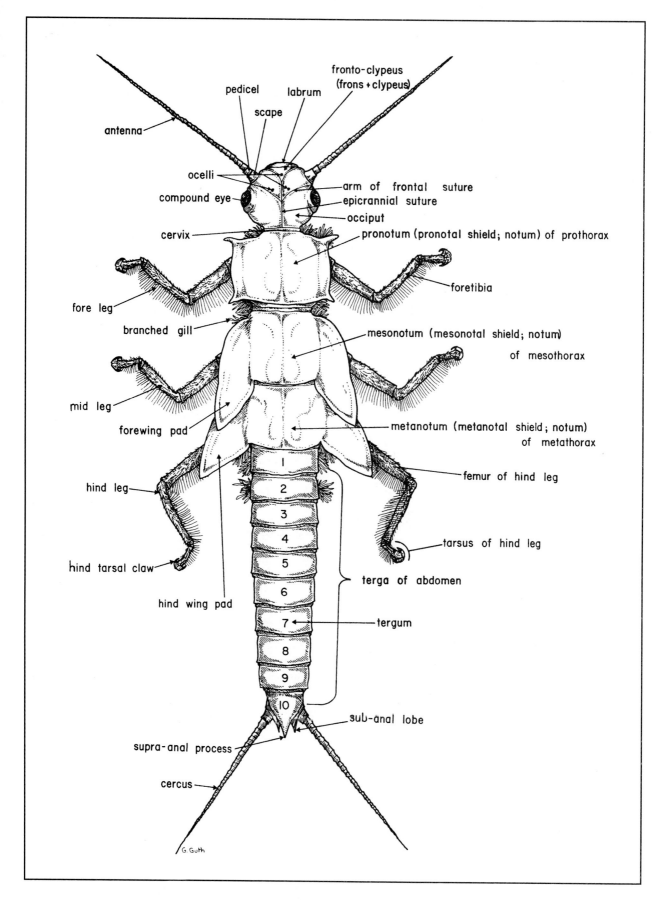

Figure 2.1. Dorsal view of *Pteronarcys* sp. larva (Plecoptera: Pteronarcyidae).

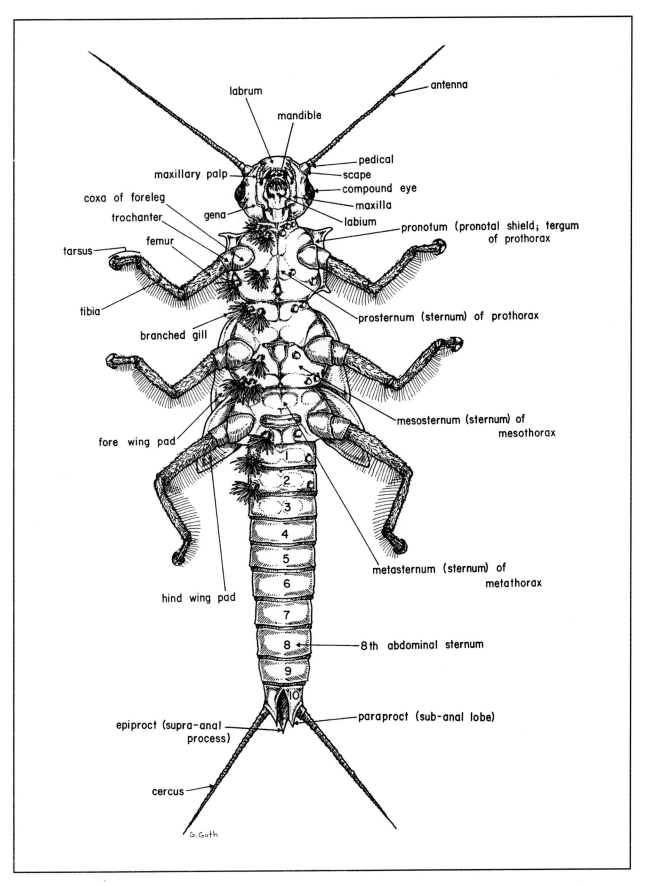

Figure 2.2. Ventral view of *Pteronarcys* sp. larva (Plectoptera: Pteronarcyidae). Gills of left side of thorax and first two abdominal segments removed to show underlying structures.

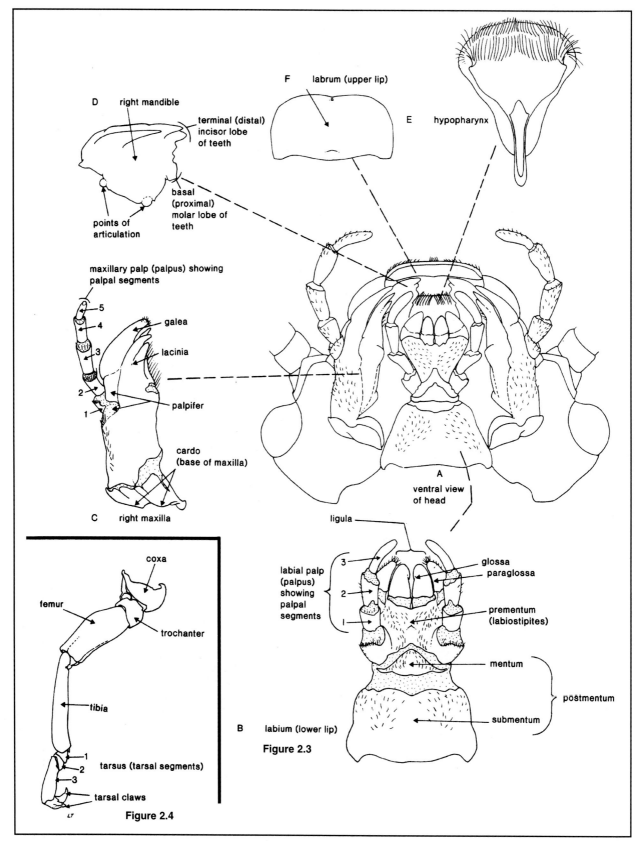

Figure 2.3. Ventral view of head and mouthparts of *Pteronarcys* sp. (Plecoptera: Pteronarcyidae: A, ventral view of head, B, labium, C, right maxilla; D, right mandible; E, hypopharynx; F, labrum.

Figure 2.4. Foreleg of *Pteronarcys* sp. (Plecoptera: Pteronarcyidae) showing segments.

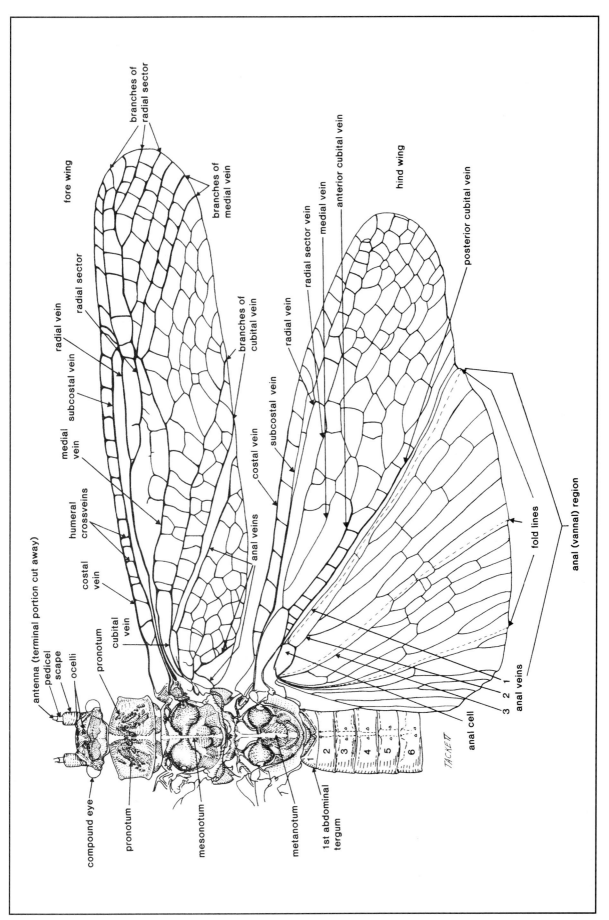

Figure 2.5. Adult *Pteronarcys* sp. (Plecoptera: Pteronarcyidae) showing head, thorax, basal portion of abdomen, and fore and hind wings.

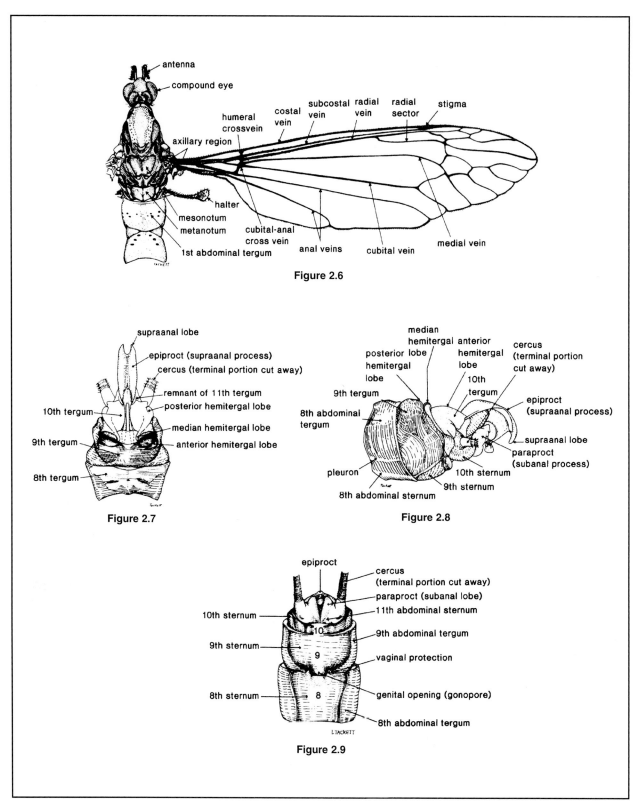

Figure 2.6. Dorsal view of adult *Tipula* sp. (Diptera: Tipulidae) showing head, thorax, basal portion of abdomen, fore wing and halter.

Figure 2.7. Dorsal view of terminal male abdominal segments of *Pteronarcys* sp. (Plecoptera: Pteronarcyidae); terminology after Snodgrass (1935) and Nelson and Hanson (1971).

Figure 2.8. Lateral view of terminal male abdominal segments of *Pteronarcys* sp. (Plecoptera: Pteronarcyidae).

Figure 2.9. Ventral view of terminal female abdominal segments of *Pteronarcys* sp. (Plecoptera: Pteronarcyidae).

insects, modifications of the hindlegs for swimming (e.g., a fringe of tibial hairs) are common in certain adult Coleoptera, some larval and adult Hemiptera, and a few larval Trichoptera. The forelegs are modified for burrowing in Ephemeridae (Ephemeroptera), Gomphidae (Odonata), and some semiaquatic Orthoptera.

Most adult forms of aquatic insects bear two pairs of wings; some mayflies and all Diptera have only one pair. The second pair of wings in Diptera is modified into balancing organs (*halteres,* Fig. 2.6) that function somewhat as gyroscopes. Collembola are wingless (apterous), as are females of certain species of Trichoptera and Diptera.

The structures that extend into the wings are termed *veins.* The form and location of these veins are used extensively in insect taxonomy. Two extreme types of wing venation are shown in Figures 2.5-2.6. The primitive stonefly wings have many branches of the major veins with many crossveins between them. The highly evolved wing of a dipteran Tipulidae (*Tipula* sp.) is characterized by the fusion of veins and the loss of branches and crossveins.

The general venation pattern (Figs. 2.5-2.6) consists of: a *costal vein* (C), the anterior marginal vein; a *subcostal vein* (Sc) just behind the costal vein and often with two branches near the wing tip; a *radial vein* (R), often the heaviest vein of the wing, which forks near the middle of the wing, with the main part forming the *radial sector vein* (Rs), which typically divides into two branches, each of which may divide into two or more branches near the wing margin; a *medial vein* (M) (the fourth major vein), which has a maximum of four major branches (typically two or three); a *cubital vein* (Cu), which has two major branches, the anterior of which usually forks into two branches; and an *anal* (vannal) *vein* (A), which has a maximum of three major branches with considerable secondary branching, particularly in the more primitive forms. Although crossveins are highly variable, certain ones are usually present. There are generally at least one *humeral crossvein* (h) between the base of the wing and the apex (tip) of the subcosta; a *radial crossvein* (r) between the radius and the first branch of the radial sector; a *radial-medial crossvein* (r-m) between the lower first fork of the radial sector and the upper first fork of the medial vein; and a *medial-cubital crossvein* (m-cu) between the lower first fork of the medial and the upper first fork of the cubital (see Snodgrass 1935; Daly *et al.* 1978; Borror *et al.* 1981; and discussion of taxonomically significant wing veins given in the order and family chapters below).

Abdomen

There are primitively eleven abdominal segments, although in most adults fusion of the last two makes them difficult to distinguish. In some immature forms (notably Ephemeroptera and Megaloptera), gills arise from the pleural regions—being extensions of the tracheal system borne in variously shaped plates or filaments (fingerlike gills). In the stonefly shown in Figure 2.2, the branched filamentous gills are attached to the sterna of the thorax and the first two abdominal segments.

The end of the abdomen of paurometabolous and hemimetabolous insects (i.e., Hemiptera, Orthoptera, Ephemeroptera, Odonata, and Plecoptera) bears the reproductive structures (Figs. 2.1-2.2, 2.7-2.9). The terminal segment bears the *anus* at its apex and the *cerci* laterally. The dorsal surface is covered by a triangular or shield-shaped tergal plate, the *epiproct,* and the ventral surface bears two lobes, the *paraprocts.* In males, the ninth sternum often bears two lateral *styli* or *claspers (harpagones).* These accessory structures bound the *phallobase* and *aedeagus* that comprise the main reproductive organ, the *penis* or *phallus.*

The terminal segments of adult females, in addition to the dorsal epiproct and lateral paraprocts below the cerci, generally consist of three pairs of lobes or *valvae* (valves), which form the visible portion of the ovipositor and arise from the eighth and ninth sterna. The bases of the valvae are usually covered by the projecting eighth sternum (Fig. 2.9).

Specific morphological modifications in each of the orders (or families receiving special treatment) are detailed in the introductory material covering the respective groups. The modifications usually represent fusion or specialization of the basic structures discussed above. However, some of the terms used in naming the various segments of the genitalia have restricted meanings, and homology with primitive forms is not always possible (Tuxen 1970; Scudder 1971a). For a more complete treatment of insect morphology the student should consult Snodgrass (1935), DuPorte (1959), Matsuda (1965, 1970, 1976), and Chapman (1982). Consult Torre-Bueno (1937) for further explanation of terms.

DESIGN OF AQUATIC INSECT STUDIES:
Collecting, Sampling and Rearing Procedures

Richard W. Merritt
Michigan State University, East Lansing

Vincent H. Resh
University of California, Berkeley

Kenneth W. Cummins
South Florida Water Management District, West Palm Beach

COLLECTING AND SAMPLING METHODS

Ultimately, the decision to use a specific sampler or collecting device should depend on the objectives of the study (e.g., Andre *et al.* 1981) and a thorough characterization of the habitat to be sampled. Table 3A has been prepared as a guide to appropriate sampling and collecting techniques that have been used in aquatic entomology. The table is not a complete listing of the references on sampling methods, and students and researchers should continually check the current literature for methods suited to their specific objectives. The bibliographies of the North American Benthological Society contain scores of other papers describing devices for collecting aquatic insects, as does the extensive, updated bibliography of Elliott *et al.* (1993). As noted by Cummins (1962), the number of different samplers used for benthic macroinvertebrates is nearly equal to the number of benthic investigations!

The classification system used in Table 3A is based primarily on the habitat and community being sampled. Substrate composition, although not outlined in detail, also is an important consideration when sampling the benthos (e.g., Minshall and Minshall 1977; Rabeni and Minshall 1977; Lamberti and Resh 1978; Reice 1980; Minshall 1984), and the equipment and techniques listed below may require modification depending on the substrate type. For example, an Ekman grab is listed as an appropriate device for littoral benthos (IV. A.1. a. in Table 3A); however, the presence of sticks or even small stones could prevent the jaws from closing, resulting in the collected material being washed through the slightly opened jaws as the sample is pulled to the surface. Resh *et al.* (1990) have prepared a videotape of the operation of twenty of the devices listed in Table 3A (contact VHR to obtain a copy).

Comprehensive reviews and discussions of different collecting and sampling methods for macroinvertebrates have been presented by Welch (1948), Macan (1958), Albrecht (1959), Cummins (1962), Hrbácek (1962), Hynes (1970a), Edmondson and Winberg (1971), Weber (1973), Hellawell (1978, 1986), Southwood (1978), and Downing and Rigler (1984). Additional information on sampling devices is available in Peckarsky (1984), Winterbourn (1985), ASTM (1987), Britton and Greeson (1987), APHA (1989), Klemm *et al.* (1990), Williams and Feltmate (1992), and Cuffney *et al.* (1993a).

A variety of factors may affect benthic sampling devices and result in sampling bias. The factors may be related to characteristics of the sampling devices and the habitat being sampled (Table 3B), or to the organisms themselves (Table 3C). The non-random, clumped (negative binomial) distribution patterns exhibited by most populations of aquatic insects may result from microhabitat preferences for substrate type, current, food, behavioral interactions, etc. (Resh 1979a).

SAMPLING DESIGN AND ANALYSIS

Sampling can be qualitative (a general assessment of the taxa of aquatic insects present, possibly with some observations of their relative abundance) or quantitative (an estimate of the numbers so that a statistical confidence of the estimate can be obtained). A widely used qualitative or semi-quantitative (e.g., when a specific area to be sampled is delineated) method for sampling stream benthos is the kick net (Table 3A; Hynes 1970a), whereby a standard number of kicks dislodge organisms from comparable areas into a collection net. Qualitative sampling is also suitable for determining ratios of various functional feeding groups of aquatic insects, for example, ratios relating the number of invertebrates in the sample that skeletonize leaf litter (shredders) or the number that graze on attached algae (scrapers) to the total numbers, or to other functional feeding groups (See Table 6C, Chap. 6; also Merritt and Cummins 1996).

Quantitative sampling requires not only the selection of the most appropriate collection device, given the habitat type and composition of the macroinvertebrate community (Table 3A), but also consideration of sample site location, sampling frequency, and numbers of sample units (or sample replicates) to be taken on each collection date. Sampling frequency should be related to the type of study. For example, production estimates (Chap. 6) of one (or several) species require a sampling

Table 3A. Collecting and sampling methods for aquatic macroinvertebrates based on the habitat and community being sampled.

Major Sampling Habitat	Subhabitat and Ecological Community	Quantitative Sampler	Figure	Reference(s)	Qualitative or Semi-quantitative Sampler	Figure	References
LOTIC HABITATS **I. Shallow Streams, Rivers, and Springs**	A. Riffles (erosional zones) 1. Benthos a. Sediments	Surber sampler	3.7	636, 1849 2388, 2726 2868, 3903 588, 3868	Aquatic net	3.2	4074, 2413
		Hess or modified Hess sampler	3.6	1704, 1944, 4264, 546, 1084	Hand screen collector	3.3	4074
		T-sampler	3.8	2427	Artificial substrate samplers	3.23	1849, 2388, 2503, 3410, 3430, 3903, 588, 4301, 2230, 556, 1943
		Individual stone, bedrock, or rock-outcrop sampler		3609, 1840, 4575 4148, 2608	Individual stone sampler		929, 4575, 2608
		Wilding or stovepipe sampler; box-type sampler	3.5	66, 3819 4296, 4411	Kick sampling		1321, 1893, 3173, 2413, 3868 2317
		Suction samplers (air-lift and water pump)	3.40	376, 2935, 4523 1497, 444	Recolonization		4437
					Hand collection		2388, 4074
					Leaf packs	3.25	2668, 3109, 3623 377, 921
		Portable invertebrate box sampler	3.9	2107, 133 4148	Photographic methods		787, 788
		Riffle sampler		671	Graded sieves	3.33	176, 2797 4074
	b. Hyporheic area (subterranean)	Implants	3.41	320, 678 1400, 1898, 2765, 3177	Freeze-core samplers (including electro-freezing)		1040, 3852, 404 3223, 403, 2372
		Standpipe corer	3.42	1127, 4429 4436, 4438	Canister sampler	3.41	1400
	c. Plants	Bag sampler		1897, 4296	Needham apron net		4074
		Surber sampler	3.7		Aquatic net Snag sampler	3.2	4074 889
		Stovepipe sampler	3.5	66, 4284			
		Lambourn sampler		1717			
	d. Drift and Neuston (surface)	Drift net	3.10	117, 1048 4257, 4437 4474, 44, 1084	Aquatic net	3.2	4074
		Plankton net	3.11	1048, 1897			
		Cushing-Mundie drift sampler		813, 2795			
		Hardy plankton indicator type sampler		1046, 1897			
		Surface film sampler		670			
		Colonization cages		1659			

Table 3A. Continued

Major Sampling Habitat	Subhabitat and Ecological Community	Quantitative Sampler	Figure	Reference(s)	Qualitative or Semi-quantitative Sampler	Figure	References
	2. Emerging Adults	Surface film sampler		670	Hand screen collector	3.3	4074
		Mundie pyramid trap	3.26	2793	Pan traps		1486, 2208
		Stationary screen trap		1538, 1906, 2511	Window traps		607, 3913
		Enclosed channels		4245	Light traps		3758
		Floating emergence traps		2326, 2390, 2756, 860			
	B. Pools depositional zones) 1. Benthos a. Sediments	Ekman grab with pole	3.16	1897	Aquatic net	3.2	4074
		Wilding or stovepipe sampler	3.5		Graded sieves	3.33	4074
		Single corer with pole	3.12	680, 1340, 4161			
		Suction samplers (air-lift and water-pump)	3.40	376, 444			
	b. Plants	Bag sampler		1897	Needham apron net		4074
	c. Drift and Neuston	See Section I.A.I.C.			See Section I.A.I.C.		
	2. Emerging Adults	See Section I.A.Z.			Section I.A.Z.		
II. Large Rivers	A. Riffles 1. Benthos a. Sediments	Suction samplers (air-lift and water pump)	3.40	1065, 1344, 1499, 4127, 972, 1568, 2935, 973	Basket or cylindrical-type artificial substrate samplers	3.24	78, 101, 216, 763, 1211, 1342, 1536, 1717, 1726, 2505, 2509, 3001, 3432, 4286
					Suction samplers (air-lift and water pump)	3.40	972, 3044, 2424
		SCUBA diving		1342, 1343, 1717, 3236	Single or multiple-plate samplers	3.23	912, 1211 1599, 1705, 3432
		Grab samplers*		37, 1063, 1065	Drag-type samplers		1064, 1065 4079
	b. Drift and Neuston	Floating drift trap (with floats)		1048, 2796			
		Hardy plankton indicator-type sampler		1046, 1897			
	2. Emerging Adults	Insect emergence traps		2244, 2756, 860			
		Floating drift trap (with floats)		2796			

Table 3A. Continued

Major Sampling Habitat	Subhabitat and Ecological Community	Quantitative Sampler	Figure	Reference(s)	Qualitative or Semi-quantitative Sampler	Figure	References
	B. Pools 1. Benthos a. Sediments	Ponar grab	3.17	1063, 1065, 2338, 3193	Basket-type artificial substrate samplers	3.24	1211, 1342, 2505, 3432
		Petersen-type grabs	3.19	1063, 2338, 3125	SCUBA diving		1065, 1897
		Ekman grab	3.16	1045, 1063, 1065, 1897, 2338			
		Core sampler	3.12	1065, 1208, 1340, 1394			
		Allan hand-operated grab	3.18	37			
		Also, see **LENTIC HABITATS**					
	b. Plants	See **Lentic Habitats**, Section III.A.1.b.			See **Lentic Habitats**, Section III.A.1.b		
	c. Drift and Neuston	See Section II.A.1.b.					
	2. Emerging Adults	See Section II.A.Z.					
LENTIC HABITATS **III. Shallow Standing-Water Habitats** (ponds, oxidation lakes, marshes, bogs, rice fields, etc.)	A. Vegetated 1. Benthos a. Sediments	Core sampler with pole	3.12	213, 680, 1340, 4161	Aquatic net	3.2	4074
		Wilding or Hess-type sampler	3.5, 3.6	1704, 1717, 4411	Graded sieves	3.33	2241, 3967, 4074
		Kellen grab	3.13	2064, 4252			
		Water column sampler		2302, 3915			
		Suction sampler	3.40	1943, 444			
	b. Rooted Plants, Periphyton	Macan sampler	3.20	1830, 2037, 2386, 954, 950	Glass slides, plastic squares, and other artificial substrates		2396, 3688, 3689
		Minto sampler		950	Modified KUG sampler		445
		Lambourn sampler		1717			
		Gerking sampler	3.21	1379	Modified sweep net		2046, 4144
		Modified Gerking sampler	3.22	2046, 2723, 1360	Drop traps and pull-up traps		1297, 172
		Quadrat clipping		950			
		Removal of natural substrates		3689			
		Plexiglass or metal tubes		2236, 3689, 1113			
		Douglas method		947			

Table 3A. Continued

Major Sampling Habitat	Subhabitat and Ecological Community	Quantitative Sampler	Figure	Reference(s)	Qualitative or Semi-quantitative Sampler	Figure	References
	c. Neuston	Water column sampler		2302, 240	Subaquatic light traps	3.28	7, 560, 561, 1870, 1879, 4252, 4646
		D-Vac vacuum sampler		1349	Telescope method		1473
					Bottle traps		1753, 1739
		Funnel-trap		1674	Plankton tow net	3.11	
					Hand dipper	3.4	3628, 3758, 3901
					Aquatic net	3.2	4074, 4144, 3901
	2. Emerging Adults	Emergence traps (primarily surface traps which float)	3.27	705, 1111, 1509, 2016, 2091, 2231, 2326, 2345, 2604, 2756, 2757, 2793, 2798, 3210, 3437, 860, 815, 4577, 1084	Subaquatic light traps	3.28	7, 352, 108, 560, 561, 1107, 1870, 1879, 4252, 4646
	B. Nonvegetated 1. Benthos a. Sediments	Ekman with or without pole	3.16	1045, 1830, 2037, 954	Dredges		2037
		Kellen grab	3.13	2064	Aquatic net	3.2	4074
		Petite Ponar grab					
		Suction sampler (air-lift and water pump)	3.40	376, 3866			
		Core sampler with pole	3.12	213, 680, 1340, 2037, 4161			
		Water column sampler		3915			
	b. Neuston	See Section III.A.1.c.			See Section III.A.1.c.		
	2. Emerging Adults	See Section III.A.2.		See Section		III.A.1.c.	
IV. Lakes	A. Littoral 1. Benthos a. Nonvegetated (wave swept)	Wilding or stovepipe sampler	3.5	2037, 4296, 4411	Basket-type artificial substrate samplers*	3.24	102, 118, 216, 1211, 2738, 3432, 4150, 4286
		Suction samplers (air-lift and water pump)	3.40	1190, 1499, 954			
		Core sampler with or without pole	3.12	213, 680, 1208, 1340, 4161, 954	Graded sieves	3.33	4074
		Ekman or similar type grab with pole	3.16	886, 1063, 1828, 2964, 4522, 954			
		Ponar grab	3.17	102, 1063, 1208, 1828, 2037, 3193			
		SCUBA diving with sampling gear		1129, 1341, 954			
		Also, see Section III.B.1.a.					

Table 3A. Continued

Major Sampling Habitat	Subhabitat and Ecological Community	Quantitative Sampler	Figure	Reference(s)	Qualitative or Semi-quantitative Sampler	Figure	References
	b. Vegetated (plant zone)	Wilding or stovepipe sampler	3.5	4284, 4411	Dredges and grabs		954
		Gillespie and Brown sampler		1393	Photographic methods		4284
					Modified KUG sampler		445
		Macan sampler with attachment rod	3.20	2037, 2964			
		Removal of natural substrate by SCUBA		4319			
	2. Emerging Adults			See Section III.A.2.			
	B. Profundal 1. Benthos a. Sediments	Single core sampler	3.13	418, 1208, 1811, 2037, 2040, 2055, 2694, 4296, 4346, 954, 33	Dredges		954
		Multiple core sampler	3.14	1208, 1539, 2037, 2694			
					Photographic methods		2037
		Ekman or modified Ekman grab	3.16	1045, 1208, 2037, 2816, 954			
		Ponar grab	3.17	1208, 2037, 3193, 954			
		Petersen-type grab	3.19	2037, 3125, 3724, 4296			
		SCUBA diving with sampling gear		1129, 2037, 4150			
	2. Emerging Adults	Emergence traps (mainly funnel and submerged traps)	3.27	469, 1144, 3436 1509, 2016, 2756, 2757, 2793, 2798, 3437, 4346, 860,			
TERRESTRIAL HABITATS V. **Stream and Lake Margins** (flood plains, bogs, beach zones, and vegetation mats)							
	A. Adult Aquatic or Semiaquatic Insects	Light traps	3.30, 3.31	1824, 2784, 3628, 3758 3899	Aerial insect net		4074
		Window traps		2022			
					Aspirator		4074
		Malaise traps and other emergence traps	3.29	2451, 3628, 3705, 4031, 4033	Graded sieves		3.33
					Sieve sample splitter		2507, 776
	B. Immatures and Adults	Berlaise or Tullgren funnels	3.32	3758, 4074			
		Behavioral extraction procedure		1133			
		Field washing		3174			
		Elutriation and/or flotation	3.43	2283, 4354, 756, 776, 777, 2271, 4546			

Table 3B. Selected examples of factors that affect benthic sampling devices and may result in sampling bias (modified from Resh 1979a).

Factor	Examples of Samplers Affected	Problems Created	Remedy
A. *Factors related to characteristics of the samplers*			
Backwash created in netted samplers by water not being able to pass through net	Netted and kick samplers	Loss of benthos around sides of sampler	Increase the net's surface area and/or decrease size of net opening; use enclosed double netted sampler (2797); alternatively use a hand-operated Ekman grab or cylinder box sampler (1897)
Washout of surface organisms upon placement of sampler	Hess sampler	Turbulence scours substrate surface	Use permeable sides
Disruption of substrate surface by shockwave when sampler strikes bottom	Corer and Grab samplers	Loss of small organisms and surface dwellers	Modify Ekman grab by removing screens and incorporating heavier materials in design; alternatively use a pneumatic grab (2816) or a modified corer (418, 2036, 2037)
Disturbance of biota	Surber sampler and Allan grab	Underestimation of biota caused by disruption when sampler is set in place	Add screened openings on top
	Shovel sampler	Loss of motile organisms	Add screened openings on top
Variable depth of penetration into substrate by sampler	Grabs	Inconsistent volume of sediment sampled; loss caused by overfilling or incomplete closure	Leave 5-cm space above substrate; use a corer whenever possible (2036)
	Surber	Failure to consider stream hyporheic zone	Use two stage sampling for surface and hyporheic zone
Sampler mesh size too coarse	Netted samplers	Early instars, small and slender organisms missed	Use finer mesh; preferably use a double bag sampler (176, 3869, 3687, 4149, 665)
Sampler mesh too fine	Netted samplers	May cause backwash (see above)	Use coarser mesh as in a double bag sampler (176, 665, 3687)
Sampler dimension too large	All samplers	May increase sorting time; may reduce number of sample units that can be taken	Take smaller samples
Sampler dimension too small	All samplers	Variability increases because of edge effect	Use nested sampler to determine optimal sampler dimension
Operator inconsistency	All samplers	Repeated, systematic error in taking samples	Use a single operator; or develop correction factor for each operator (1112, 662)
B. *Factors related to characteristics of the environment*			
Water depth limitations in lotic environments	Surber and Hess samplers	Because surber sampler limited to <30 cm depth, organisms may drift over sampler top.	From 0.5 to 4 m depth, use an airlift sampler (Fig. 3.40); 0.4 to 10 m deep, use SCUBA and dome suction sampler (1344) or modified Hess sampler (3235, 1084); or use a modified Allan hand-operated grab (37) or artificial substrates (3432)
Substrate - stony	Grab samplers, corers	Grabs may not close; corers cannot penetrate	Substitute airlift or dome suction sampler or use artificial substrates
Substrate - mixed	Grab samplers, corers	Differential penetration into substrates	Use specific specific samplers for different substrate types (1208)
Current too slow	Surber and kick samplers	Organisms do not drift into sampler net	Use an enclosed sampler, such as a Hess or modified-Hess
Current too fast	Netted samplers	Backwash, resulting in a loss of organisms	Substitute a modified sampler with controlled flow (e.g., 2797)
Current fluctuations	All samplers	Rapid change in flow may scour study area	
Low air temperatures	Netted samplers	Samples freeze in net before organisms are removed	Use catch bottle at end of net or a zippered net (4264)
Sampling in water current	Lotic: drift samplers; Lentic: all samplers	Lotic: net clogging and changes in current and flow pattern affect estimation of water volume sampled	Use a waterwheel drift sampler (3045)
		Lentic: sampling a consistent volume of water; scattering of organisms	Use a water column sampler (1674, 2302), or pull-up trap (1297, 172)
Habitat small	All samplers	Destructive sampling destroys habitat and eliminates rare species	Reduce sampler dimension or use artificial substrates, field sorting, and replacement of organisms
Water chemistry	All samplers	Presence of springs, man-made outfalls, and other conditions may influence microhabitat distribution of biota	Reconnaissance of study sites

Table 3C. Factors related to biological characteristics of aquatic insects that may result in sampling bias.

Factor	Samplers affected	Problems created	Remedy
Lithophilic behavior	All samplers	Loss of organisms because of concealment or attachment	Dip rocks in dilute HCl / alcohol; careful examination of substrate
Hyporheic distribution	Surber, Hess samplers	Inconsistent underestimations of population density; failure to find species that are primarily hyporheic	Hyporheic fauna samplers (Figs. 3.41, 3.42; e.g., 403, 404, 3223, 1400)
Motility of organisms	All samplers	Organisms flee when disturbed	
	Artificial substrates	Organisms flee from sampler when artificial substrate is removed	Use a net to retain organisms in artificial substrates
Population movements	All samplers	Sampling regimes and techniques not applicable through all life cycle stages (e.g., planktonic first instars, pre-pupation, and pre-emergence movements)	Use multi-stage sampling approaches involving different sampling techniques
Attractability (or catchability) of certain species	All samplers	Less motile species may appear more abundant than they really are in relation to more motile species	
	Artificial substrates	Numbers and kinds of organisms in artificial substrates differ from that found in benthos	Use representative artificial substrates matching that found in stream bed (3432)
Spatial distribution patterns	All samplers	Nonrandom patterns require greater numbers of samples to estimate density with the same precision as for random distributed populations. Aggregations may result from microhabitat preferences of substrate, current, food source, etc. or behavioral interactions. Benthic populations generally exhibit aggregated spatial distribution patterns	Sampling universe matches population universe through stratification (see text).

regime keyed to specific life cycle patterns, whereas faunal studies require at least seasonal frequency. A useful approach is to sample every 100-300 degree-days (Chap. 6; i.e., cumulated mean daily temperature [e.g., 10 days at a mean temp. of 10°C = 100 degree-days]; Southwood 1978) because most aquatic insect life cycles are under general thermal control. For example, dividing the in-stream portion of the life cycle of a given species requiring 1000 degree-days into four sampling units of 250 degree-days, would likely provide a better coverage of the growth patterns of populations then regularly spaced (e.g., monthly) samples.

A complete discussion of sampling design is beyond the scope of this chapter. Several comprehensive treatments that are related to studies of aquatic insects and other benthic macroinvertebrates include those by Cummins (1975), Elliott (1977), Green (1979), Hellawell (1978, 1986), Resh (1979a), Waters and Resh (1979), Peckarsky (1984), Prepas (1984), Winterbourn (1985), Norris and Georges (1986, 1993), APHA (1989), Voshell *et al.* (1989), Norris *et al.* (1992), Williams and Feldmate (1992), and Cooper and Barmuta (1993).

The number of samples that are taken in aquatic insect studies is often times a choice based on tradition rather than on sound reasoning. The number of sampling units (n) required is a function of: (1) the size of the mean, (2) the degree of aggregation exhibited by the population, and (3) the desired precision of the mean estimate (Resh 1979a). If (2) and (3) are the same for two populations, the one with a higher density will require fewer samples than the one that is less dense. In terms of (2), most ecological studies of aquatic insects have indicated that the majority of species are distributed in a patchy fashion (clumped, non-random, negative binomial, and aggregated distributions are all synonyms for this phenomenon), but interpretations of aggregation may be a function of the size of the sampling unit and the number of samples collected (Elliott 1977). Assuming (1) and (3) are equal, populations with non-random distributions require greater numbers of samples than those that are randomly distributed. The higher the desired precision of the mean estimate, the greater the number of samples required. If the acceptable 95% confidence limits (= two times standard error of the mean) are cut by half (e.g., from $\pm$ 40% to $\pm$ 20%), four times as many samples will be required. The formula for sample size determination is given in Table 3D and an example using a caddisfly population is presented in Table 3E. A graphic (i.e., mean stabilization) technique can also be used to approximate the number of sample units required to adequately sample aquatic insect populations. With this approach, a cumulative plot of mean and total numbers of individuals from a pilot study is made, and the number of samples after which the change in the mean becomes less than 10% as the number of individuals increases is chosen as the sample size. Confidence limits of the mean can then be established based on mean and variance estimates.

The above discussion of sample size is concerned with studies in which an estimate of population density is the objec-

Table 3D. Symbols and formulae for calculating numbers of requisite samples ($n_{est.}$) for a desired level of precision and testing for difference between means (from Elliott 1977, Sokal and Rohlf 1981, and other sources).

Symbol	Formula	Description
n		Total number of sampling units in sample.
$\bar{x}$	$\dfrac{\Sigma x}{n}$	*Mea*n number of individuals ($\bar{x}$) in n sample units in which each x is the number in a given sample unit (Σ = summation).
s^2	$s^2 = \dfrac{\Sigma (x-\bar{x})^2}{n-1}$ or $s^2 = \dfrac{\Sigma (x^2) - \bar{x}\Sigma x}{n-1}$	Variance (s^2) is the sum of the deviations of $\bar{x}$ from each x divided by the degrees of freedom (one less than the number of sampling units). It is also equal to the standard deviation (s) squared.
SE	$SE = \sqrt{\dfrac{s^2}{n}}$	Standard error
D	$D = \dfrac{SE}{\bar{x}}$	Precision (D) is the standard error expressed as a proportion (or, if multiplied by 100, as %) of the mean. The usual range of D in benthic studies is 0.10-0.40, but choice of D should be a function of sampling objectives.
t^2		$t \approx 2$ ($t^2 \approx 4$) if n > 30 for 95% probability level of D in Student's t-distribution.
(1) $n_{est.}$	$n_{est.} = \dfrac{t^2 s^2}{D^2 \bar{x}^2}$	General formula for sample size; if 95% confidence limits of $\pm$ 40% of $n_{est.} = \dfrac{25 S^2}{\bar{x}^2}$
(2) $n_{est.}$	$\geq 2\left(\dfrac{\sigma}{\delta}\right)^2 \{t_{\alpha[v]} + t_{2(1-p)[v]}\}^2$	(Sokal & Rohlf 1981, p. 263).
σ	=	true standard deviation
δ	=	difference between means expressed as a percent of $\bar{Y}$, e.g. = 20 for a 20% difference between means
P	=	desired probability that a difference will be found to be significant
v	=	degrees of freedom of the sample standard deviation with *a* groups and *n* replications per group
$t_{\alpha[v]}$ and $t_{2(1-P)(v)}$	=	values from a *t*–table with v degrees of freedom and corresponding to probabilities of α and 2(1–P), respectively. Note if P = $\dfrac{1}{2}$ then $t_{2=0}$.

Table 3E. Determination of the number of sample units required to estimate age-specific and total population size of the caddisfly *Glossosoma nigrior* Banks (Trichoptera: Glossosomatidae) in two first-order Michigan streams. See Table 3C for sample size formula.

Stream	n	*Glossosoma nigrior* Age Class	$\bar{x}/0.016 \, m^2$	s^2	Sample Sizes (to nearest integer) for 95% Confidence Limits where		
					D=± 40%	± 20%	± 10%
Augusta Creek (August)	62	Instar 1	2.6	16.8	63	249	994
		2	13.4	292.4	41	165	652
		3	7.8	44.9	19	74	296
		4	1.8	4.0	31	124	494
		5	5.6	19.4	16	62	248
		Prepupae	2.2	5.8	30	120	480
		Pupae	4.5	51.8	64	256	1024
		Total	32.8	681.2	16	64	254
Spring Brook (July)	44	Instar 1	0.5	1.0	100	400	1600
		2	4.3	21.2	29	115	459
		3	6.7	37.2	21	83	332
		4	5.3	22.1	20	79	315
		5	1.7	4.0	35	139	554
		Prepupae and pupae	2.1	6.3	36	143	571
		Total	20.6	201.6	12	49	196

tive; however, many sampling programs are concerned with determining whether there is a change in a population or a difference between two populations. A different formula (the last one in Table 3D) is used in determining required sample sizes, and considerations about the level of significance (Type I error) and the power of the test (Type II error) are necessary as well. As a general rule, if the coefficients of variation (s/x) do not overlap, the samples are likely to be statistically different. This topic has received far less attention than population estimates in aquatic insect studies even though it could be used to address many appropriate questions, e.g., changes along a pollution gradient. A detailed discussion of this type of sample-size determination is presented in Norris et al. (1992).

Choice of location where the desired number of sample units will be taken also is an important consideration. Sample units could be collected randomly through the study area or they could be collected randomly only within defined strata (termed stratified random sampling), which could be defined by habitat (e.g., a riffle), certain substrate sizes, depth (usually in lakes), hydraulic features (e.g., see Statzner et al. 1988), or a variety of other factors. In practice, stratification is used to reduce variability or to facilitate comparisons. Concordance of the true population universe (when the organisms occur) with the sampling universe (where samples are taken) is a goal of stratified sampling approaches. However, one problem in using stratified sampling is that extrapolation of trends observed to those expected in other strata is difficult. For more information on stratification, see Norris et al. (1992) and Resh and McElravy (1993). In some studies it may be most appropriate to use a transect technique that acknowledges the existence of known habitat gradients, such as from the margin of a stream or lake to the deepest portion near the center (e.g., Cummins 1975).

If a sample unit contains a very large number of individuals in a given taxon, e.g., midge (Chironomidae) larvae, it may be necessary to subsample to obtain an estimate because the time required for a total count is prohibitive (Waters 1969; Elliott 1977; Wrona et. al. 1982; Fig. 3.1). If the subsample counts fulfill the criteria of randomness, then a single subsample count can be used to estimate the number in the original sample unit. To satisfy randomness, the mean of at least five subsample counts should fall within the 95% confidence interval, that is, a chi-square (X^2) value between the 5% significance levels for n-1 degrees of freedom found in a standard statistical table $\left(\frac{\Sigma(x-\bar{x})^2}{x^2} \right)$ should be obtained. The count should be for the for the category of interest, e.g., all taxa or a particular taxon such as a species, age class, or life stage of a species or functional group. For a further discussion of subsampling, see Elliott (1977).

A similar rationale can be applied to aggregates of entire sample units. Composites of sample units from sites can be used when the objective is to determine if two (or more) sites are different (e.g., distinguishing stream A from stream B, or an upstream site above a pollution source from a site below the source, or changes in a stream site before and after a disturbance) and knowledge of the variance of each site alone is not critical. The strategy is to collect a large number of samples (many more than the normal 3 to 5, e.g., 25-50) in recognition of the natural high variability (e.g., see Table 3E). The samples from each site are composited and subsampled according to the chi-square considerations discussed above. This can be a powerful technique for distinguishing between two sites with differences that would usually be masked by variability (S. Hurlbert, pers. comm.).

Finally, sampling other key structural and functional components (Resh et al. 1988) of aquatic insect populations and communities requires special attention. Detailed discussions on sampling considerations for these components are presented in the following reviews: drift (Waters 1972; Brittain and Eikeland 1988); secondary production (Benke 1984; Rigler and Downing 1984); taxonomic richness (Resh and McElravy 1993); structural, functional, and trophic diversity (Chap. 6; Cummins and Klug 1979); nutrient cycling (Webster and Benfield 1986); life-history patterns (Butler 1984); size spectra (Clifford and Zelt 1972); and biotic interactions (Peckarsky 1984; Powers et al. 1988).

SORTING AND SAMPLE PRESERVATION

Once the sample has been collected, it must be treated according to the nature of the substrate materials and the types of analyses to be made. We have presented a generalized flow diagram summarizing some general procedures that might be used in analyzing stream bottom samples (Fig. 3.1). Sorting can be a time-consuming and, consequently, a costly procedure; the use of elutriation or flotation techniques, sieves, and stains can greatly reduce the time required (see Table 3F, and Resh and McElravy [1993] for details). Sorting must be done with care and the advantages gained from adequate sampling designs and appropriate numbers of samples taken can be obliterated by introduced bias during sorting (Table 3F). Additional information on sorting procedures is available in Weber (1973), Cummins (1975), Hellawell (1978), Downing (1984), Winterbourn (1985), APHA (1989), Cuffney (1993b), and Resh and McElravy (1993).

Considerations about the use of sample and specimen preservatives has undergone major revision because of environmental and individual health concerns. Although formaldehyde was widely used in the past, ethanol now is usually substituted. For sample preservation, 95% ethanol is often recommended (to account for dilution from water in samples); 70-80% ethanol is recommended for specimen preservation. Proper disposal of specimens and preservatives is essential and should be in accordance with your institution's approved hazardous-waste disposal program. In some cases, it might be suitable to store samples frozen. However, careful, slow thawing is necessary to minimize the breakage of specimens that often results from freezing.

REARING METHODS

Most ecological studies of aquatic insects have dealt with their immature stages because it is the larvae that normally occur in aquatic habitats and represent the major portion of insect life cycles. The identification of most aquatic immatures is difficult because: (1) taxonomic names are generally based on characteristics present in the adult stage, (2) for many North American species, immatures and adults have not been associated, and (3) insufficient comparative analysis of congeneric (i.e., in the same genus) larvae has been completed to produce species-level keys.

Table 3F. Selected examples of factors related to benthic sorting procedures that may result in sampling bias (modified from Resh 1979a).

Procedure	Potential problems	Remedy or comment
Preservation	Alcohol preservative may cause a weight loss resulting in erroneous biomass and secondary production estimates (2332, 4370, 936)	Weight loss stabilizes over time so correction factors may be used; alternatively use other preservatives (936, 2332) or other weighing methods (4370).
Live sorting by electroshocking	Selective for large organisms (2332)	Useful technique when large amounts of organic matter present
Sieves too coarse	Underestimation of numbers; misinterpretation of life histories	Use finer mesh
Sieving	Physical damage to specimens; too much material to sieve	Sieve under water, without spraying directly on top; elutriation may be preferable (2283, 4354, 2271); use sieve sample-splitter in field (2507)
Flotation	Flotation time varies with preservative used, taxa present, and instar within same taxa; animals remain in same fraction as organic detritus	Live flotation or formalin and sugar solution maximize flotation time; try phase separation technique (181) or centrifugal flotation (756)
	Selective against case-bearing caddisflies, molluscs and microbenthos	Repeated flotation with freshwater washes and examination of remaining material may be necessary
Subsampling	Rare organisms or life history stages may be missed	Subsample size should be adjusted so that smallest taxon counted > 100; subsample by weight (3621)
Counting	Smaller specimens may be missed; too many specimens to count (3700)	Maximum counting efficiency at 25X magnification with transmitted light; try biovolume calculations
Sampling and/or processing time and costs	Costs prohibitive, takes too much time	Use smaller sampler or subsample (641, 3633); consider redefining study objectives

The required level of identification is essentially a reflection of study objectives (Resh and McElravy 1993). However, species-level identifications are important in ecological studies because congeneric species do not necessarily have identical ecological requirements or water quality tolerances. Also, the inability to distinguish between coexisting species may mask population dynamics or trends, and, without species identification, comparisons with results obtained from other studies (possibly even with related species) are difficult (see individual order chapters for exceptions). The calculation of diversity indices, a technique often criticized but still widely used in aquatic insect community studies, may result in significant underestimates when generic- or family-level identifications, rather than those made at the species level, are used or applied in an uneven fashion to different higher taxa (Resh 1979b).

The taxonomic problems of identifying the immature stages of aquatic insects can be solved either by rearing the larva to the adult stage, or in some groups by collecting associated adult and immature stages. For example, associations can be made by examining mature pupae (Milne 1938) and cast larval skins (e.g., Trichoptera), and by collecting exuviae in organic foam accumulations, drift (streams), or windrows (lakes) (e.g., Chironomidae; Coffman 1973). Rearing techniques range from very simple to highly complex, and no single technique is suitable for all aquatic insects or even all species of a given genus or family. In Table 3G, general references to techniques for obtaining adult stages of immature insects from lotic and lentic habitats are given, as are rearing (i.e., a single generation) and culturing (i.e., rearing through subsequent generations) methods appropriate for the different orders of aquatic insects. In the future, the use of innovative rearing methods and application of molecular and biochemical techniques (such as protein or nucleic acid determinations) could greatly increase the number of species that have associated immature and adult stages.

Published reports of rearing techniques generally fall into three categories: (1) descriptions of various running-water systems (artificial streams; see discussion in Vogel and LaBarbera (1978), (2) methods of maintaining larvae and pupae until emergence occurs, and (3) methods of obtaining eggs from adult females and then rearing the newly hatched larvae as in (2). Because lotic insects are often more difficult to rear than lentic ones, a greater number of techniques has been published on the former. In recent years, culturing methods (i.e., for continuous generations) have been improved, largely in response to the need for maintaining organisms for bioassay tests (e.g., Anderson 1980; Buikema and Voshell 1993). The methods outlined by Lawrence (1981) and references in Table 3G should be consulted for more detailed information on the approaches used. The subject of culturing of invertebrates for bioassays has been covered by Buikema and Cairns (1980).

Field rearings are generally more successful than laboratory rearings, but are often impractical because of the time or frequency required to be on-site. The choice of mass (many species per container) versus individual rearings is based on the degree of similarity among larvae being reared. Collection of adults in the vicinity of the larval aquatic habitat with sweep nets or light traps can give some idea of the presence of systematically related species whose immature stages may not be distinguished easily from those under examination. This information may help avoid the situation in which two or more species of adults emerge from a presumed single-species rearing. An alternative approach is to obtain eggs from known females and rear these to maturity (e.g., Resh 1972).

Table 3G. Selected references on aquatic insect rearing methods.

Order	Immature to Adult Rearing Methods				Laboratory Culture Methods
	Field References	Figure(s)	Laboratory References	Figure(s)	References
Lentic insects in most orders†			118*	3.37	
Rheophilic (current-loving) insects in all orders		3.35	323, 736*, 738, 2270, 1568, 2508, 3583	3.38-3.39	
Collembola					3773
Ephemeroptera	535, 637*, 1782, 1963	3.34	880, 1303, 2387, 3079	3.37-3.39	241, 1303, 1304, 2813, 3775
Odonata	1764	3.35	210*, 524, 2428*, 3073, 4324	3.38	1479, 2196, 2428, 3547, 4324
Plecoptera	805, 1166	3.34	411, 1307, 1308, 1368, 1891*, 2052, 3079, 4578	3.37-3.39	323, 2052
Hemiptera			210*, 2161, 2233, 2634, 2234, 2636, 2637, 2999	3.37	1071, 1794, 1968
Trichoptera	1454	3.34-3.35	85*–87*, 91, 1716, 3130, 3288*, 3712, 4374*	3.37-3.39	89, 323, 3288
Neuroptera			450*, 2968		
Megaloptera			3207*, 3704*		
Lepidoptera		3.35	2242		
Hymenoptera	See methods for rearing specific hosts				
Coleoptera					
Lentic species			12, 80, 210*, 4598		456, 4334
Lotic species			456, 4334*		
Diptera					
Ceratopogonidae			2083, 2354, 3994		1527, 2352
Chironomidae		3.36	210*, 396, 1033	3.36	312, 749, 951, 2651, 4615
Dixidae			1438		
Culicidae			1309, 1377*		1150, 1377, 3703
Sciomyzidae			2871		
Simuliidae	1409		562, 1531, 1614, 3937* 3938, 3988		402*, 1179, 1289, 2782, 2783, 3257, 3659, 3660, 4530, 4531, 281*, 1016*, 4567
Tabanidae			2239, 2471, 3399, 3400, 3596, 3968		
Tipulidae	1953		1663*, 3423		
Parasitic mites on aquatic insects			764, 765, 3304		

* Recommended techniques.
† Hemimetabolous or with aquatic pupal stage.

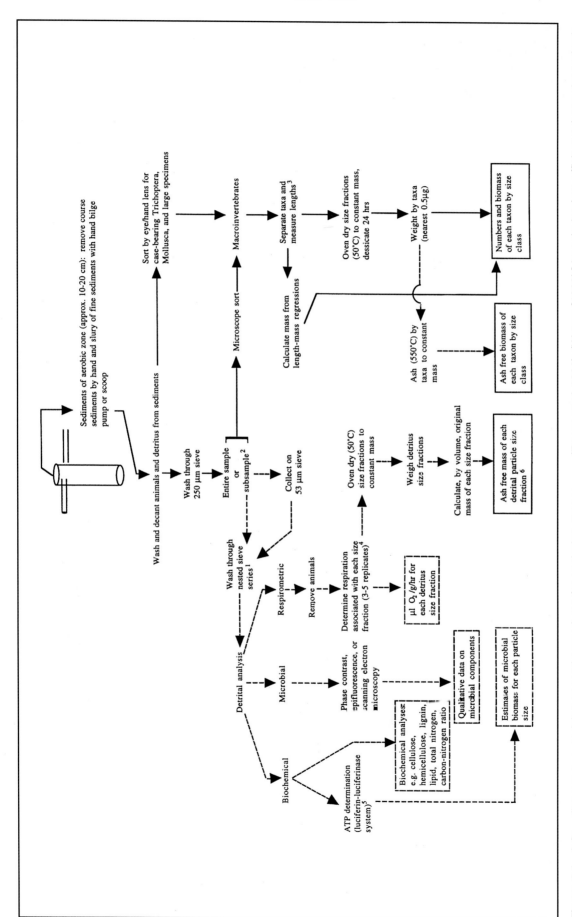

Figure 3.1. A flow diagram summarizing general procedures for analyzing stream bottom samples. *Solid arrows* indicate procedures normally used on all samples, *dashed arrows* are procedures used only on selected samples or subsamples. Notations: *1*, size classes are based on the Wentworth (1922) scale (modified by Cummins *et al.* [1973]); *2*, size fraction and subsample volumes should be based on the nature of the sample, especially the density of macroinvertebrates, see chap. 3; note also that animals can be classed into length or weight groups by sieve size (Reger *et al.* 1982); (Waters [1969b] has devised an efficient subsampler and sub-sampling is discussed by Elliott [1977]); *3*, a number of computer based image analysis systems are available (e.g. Bioquant) that greatly facilitate measuring; *4*, Gilson differential respirometer or oxygen electrode in small circulatory chambers, Bilson (1963); *5*, Asmus (1973); *6*, long ashing times are required to achieve constant weight, particularly with large detritus samples. If sample contains significant clay content, peroxide digestion may be preferable to combustion. Also, see chap. 6.

SAMPLING

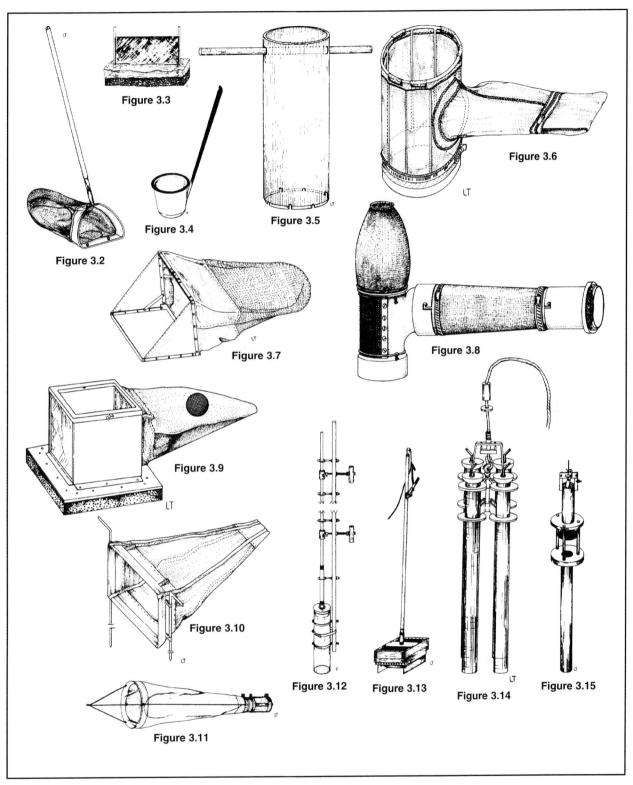

Figure 3.2. D-frame aquatic net.
Figure 3.3. Hand screen collector.
Figure 3.4. Hand dipper.
Figure 3.5. Wilding or stovepipe sampler.
Figure 3.6. Modified Hess sampler.
Figure 3.7. Surber sampler.
Figure 3.8. Stream bottom T-sampler.
Figure 3.9. Portable invertebrate box sampler.
Figure 3.10. Drift net.
Figure 3.11. Plankton tow net.
Figure 3.12. Core sampler with pole.
Figure 3.13. Kellen grab.
Figure 3.14. Multiple core sampler.
Figure 3.15. Single core sampler.

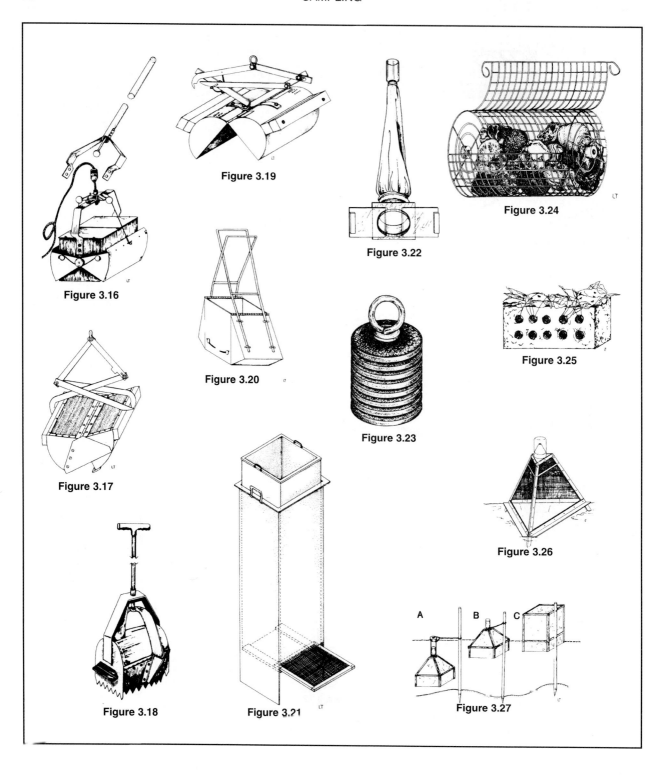

Figure 3.16. Ekman grab (with and without pole attachment).
Figure 3.17. Ponar grab.
Figure 3.18. Allan grab.
Figure 3.19. Petersen grab.
Figure 3.20. Macan sampler.
Figure 3.21. Gerking sampler.
Figure 3.22. Modified Gerking sampler.
Figure 3.23. Multiple-plate artificial substrate sampler.
Figure 3.24. Basket-type artificial substrate sampler.
Figure 3.25. Leaf pack sampler.
Figure 3.26. Mundie pyramid trap.
Figure 3.27. Emergence traps: A, submerged; B, floating pyramid; C, staked box.

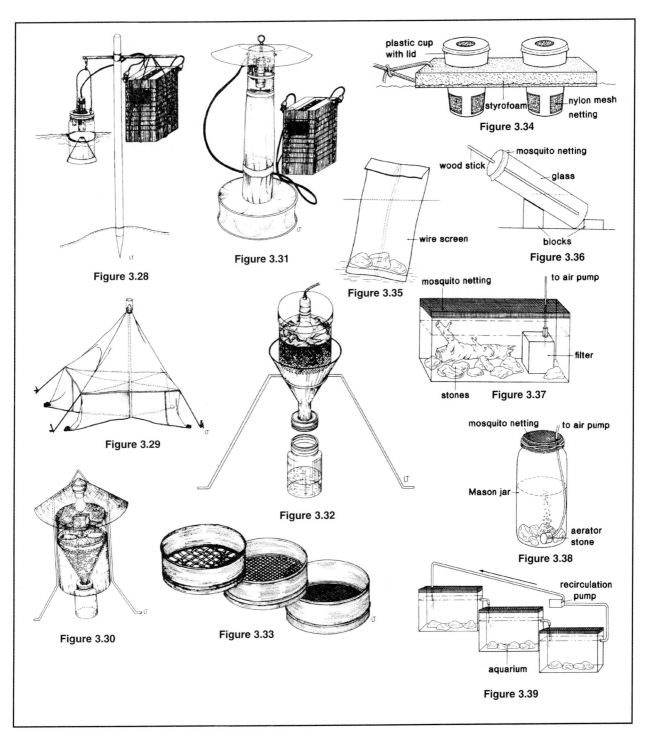

Figure 3.28. Subaquatic light trap.
Figure 3.29. Malaise trap.
Figure 3.30. New Jersey light trap.
Figure 3.31. CDC light trap.
Figure 3.32. Berlese-Tullgren funnel.
Figure 3.33. Graded sieves.
Figure 3.34. Floating cages (drawn after Edmunds et al. [1976]). A mesh lining attached along the inside wall of the cups will allow the subimagos to cling to the side and not slip back into the water.

Figure 3.35. Pillow cage (drawn after Peterson [1934]). The inclusion of larger stones to serve as ballast may prevent the pillow cage from being washed away. The top portion of the cage must be above the water surface.
Figure 3.36. Vial rearings (drawn after Peterson [1934]). Fungal growth will be retarded if distilled water is used and the temperature is kept 16°C or lower.
Figure 3.37. Aquarium rearing method.
Figure 3.38. Quart jar rearing method.
Figure 3.39. Artificial stream rearing design.

One of the most common problems encountered in rearing aquatic insects involves mortality during transport from the field to the laboratory because of inadequate oxygen supply and/or temperature control. Agitation during transport will maintain oxygen levels but may damage delicate specimens. Alternative methods include transporting the animals in damp moss (excess water removed), burlap, or paper towels; using small "bait bucket" aerators; or attaching tubing to an exterior funnel that can pick up a "wind stream" while a vehicle is moving. To maintain cool temperatures, thermos containers or ice coolers should be used.

Laboratory rearings can be maintained at field temperatures using an immersible refrigeration unit or by recirculating water through a cooling reservoir. If laboratory temperatures do not match those in the field, mortality can be reduced by allowing temperatures to equilibrate slowly. To maintain water quality in the laboratory, tap water should be dechlorinated and distilled or spring water added to replace evaporation loss.

Algal and detrital food supplies are often best maintained by periodic replenishment from the field. Detritivores that eat leaf litter (shredders) require conditioned material, i.e., leaves colonized by aquatic hyphomycete fungi and bacteria. Wheat and other grains can be used to supplement the diets of detritivores, and scrapers can frequently be fed on spinach leaves. The addition of enchytraeid worms or wheat grains to a detritus-based diet not only reduced development time but also increased the weight of individuals in a limnephilid caddisfly culture (Anderson 1976; Hanson et al. 1983). Larvae of *Drosophila*, house flies, mosquitoes, and tubificid or enchytraeid worms can serve as food for predators. The most critical problem in rearing aquatic insects that require a highly specific food (e.g., freshwater sponges, bryozoans) may be in the culture of the food source itself. Mortality can be reduced by choosing only larvae close to emergence or those about to pupate. Adjustments of photoperiod (using a light/dark regime similar to that in the field during emergence) or temperature may be required to break the arrested development of species that undergo diapause.

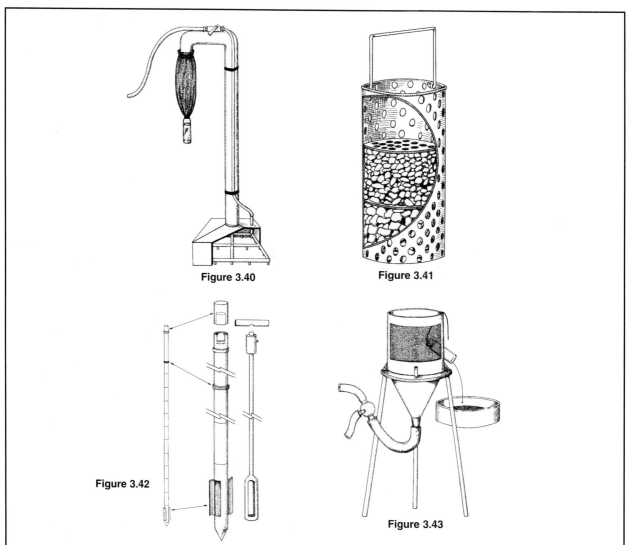

Figure 3.40. Modified air-lift sampler (redrawn from Norris [1980]).

Figure 3.41. Hyporheic cannister (implant) sampler (redrawn from Gilpin and Brusven [1976]).

Figure 3.42. Hyporheic standpipe corer (redrawn from Williams and Hynes [1974]).

Figure 3.43. Benthic invertebrate elutriation apparatus (redrawn from Worswick and Barbour [1974]).

AQUATIC INSECT RESPIRATION

Clyde H. Eriksen
The Claremont Colleges, Claremont, California

Gary A. Lamberti
University of Notre Dame, Notre Dame, Indiana

Vincent H. Resh
University of California, Berkeley, California

INTRODUCTION

A major challenge for any aquatic insect is to obtain sufficient quantities of oxygen (O_2) for its metabolic needs. Aquatic habitats contain much less O_2 than terrestrial environments even under the most favorable conditions. In addition, the aquatic O_2 supply is often highly variable and in some habitats oxygen may be totally lacking, a condition referred to as anaerobic.

Insects originally evolved on land where they developed a gas-filled (*tracheal*) respiratory system (see Chap. 8); as some insects adapted to a water environment, this air-filled tracheal system had to serve as the structural plan for their aquatic respiratory system as well. A number of options exist for obtaining O_2 with a tracheal system even though the habitat is water and, as might be expected, insects have been very successful in taking advantage of those options. Whatever the aquatic habitat or the O_2 supply, insects are normally present.

Why is O_2 so important for aquatic insects? Oxygen is used in cellular respiration where it is the final electron acceptor in a series of mitochondrial reactions that release energy from organic molecules obtained from digested food. This energy is used to do all bodily work. Cellular respiration that uses O_2 releases 19 times more energy than respiration that occurs in its absence. Because the energy needs of most multicellular and all highly active organisms are greater than the energy made available by anaerobic respiration, obtaining O_2 from the environment is absolutely essential.

OXYGEN: FROM SOURCE TO CELL

Oxygen in the Environment

Because the insect respiratory system evolved to obtain and transport the abundant O_2 from the atmosphere, aquatic insects that use air directly do not encounter oxygen limitations. However, insects that use O_2 dissolved in water face a number of difficult problems.

When water and air are in contact, an equilibrium will become established between the gaseous components of each. Gases that make up "air" have different solubilities in water, and their proportions and absolute amounts in water and air are quite different. For example, air contains approximately 21% O_2, but it is over 33% of the gases normally dissolved in water. Carbon dioxide (CO_2) is only 0.03% of air, but almost 3% of water. Nitrogen (N_2) is 78% of air, but is less than 67% of the dissolved gases in water. However, even though O_2 is more soluble in water than some other gases (e.g., N_2), its absolute amount in water is very small compared with that in an equal amount of air. For example, over 200,000 ppm of O_2 are found in air but, at maximum, there are only about 12-15 ppm in saturated, cold water!

Environmental conditions affect gas solubility in water. As temperature, salinity, or altitude increase, the amount of O_2 that can be dissolved decreases. However, as water depth increases, more and more O_2 can be dissolved. Water turbulence (flow in which velocity and direction vary unpredictably over very short distances) increases the exchange of oxygen by increasing the water surface area, forcing aeration, and moving water of lower O_2 concentration to the surface. Once dissolved, if O_2 were distributed by diffusion alone in water devoid of this gas, years would be required for even traces of O_2 to reach several meters of depth. Thus, wind-generated currents and turbulence are vital to the mixing of gases within the water column. In fact, gas exchange and mixing are significantly reduced by anything that inhibits the effect of wind on water, such as limited water surface area, protective vegetation, and ice cover.

Typically, dissolved oxygen levels are higher in flowing water than in still aquatic environments (although large amounts of organic matter or sewage pollution may reduce O_2 concentrations in either). However, more O_2 may be present than the equilibrium amount expected (a situation called *supersaturation*). For example, photosynthesis by phytoplankton, periphyton, and macrophytes may make a warm, algae-rich pond higher in O_2 than a cold, turbulent stream. This would occur only during daylight hours, of course. At night, O_2 levels in the pond may decline drastically because supersaturation persists only when photosynthesis adds oxygen to the water in excess of that lost to community respiration, mixing, and diffusion into the atmosphere. Thus, the normally higher daytime and lower nighttime dissolved oxygen levels clearly reflect the daily cycle in photosynthesis and its relation to the continual

29

respiration of the aquatic community. The ratio of daily gross photosynthesis to daily respiration (P/R) often is used as an index of aquatic community metabolism (Cummins 1974).

Oxygen often is nearly absent in groundwater because of the bacterial respiration that occurs during its slow movement through the soil. Once groundwater comes to the surface, O_2 is replaced at a rate determined by local conditions, especially sediment type, gradient, current, and turbulence. More-detailed treatments of O_2 in aquatic environments are given in Hutchinson (1957), Hynes (1970a), Wetzel (1983), Cole (1983), and Goldman and Horne (1983).

Obtaining Oxygen from the Environment

Whatever the environmental conditions, it is the process of diffusion that ultimately moves gases both to and through respiratory surfaces. The speed of that movement depends on the molecular weight of the gas, the permeability of the medium through which it must pass, the concentration gradient over the distance to be moved, and the distance itself. The interrelationship among these factors determines the diffusion rate for any particular gas and is described by Fick's law: diffusion rate equals a permeability constant, multiplied by the gradient, and then divided by the distance.

Oxygen diffuses rapidly through air but in water its rate of movement is a staggering 324,000 times slower! If that is not problem enough, insect cuticle reduces the O_2 diffusion rate even more, slowing its speed almost 850,000 times over the rate of diffusion of O_2 in air (Miller 1964a). The difference in gas concentration between the respiratory system of the insect and the surrounding medium determines the diffusion gradient. The distance over which that gradient exists includes not only the cuticle-tissue thickness, but also the adjacent layer of water (Fig. 4.1). As water moves over a respiratory surface (or any surface), frictional resistance increasingly slows the adjacent water molecules until flow ceases at the surface. When very slow, water movement changes from turbulent (variable flow and direction) to laminar (slow-flowing sheets of water moving parallel to each other). This region of laminar flow is termed the *boundary layer* and, because this layer lacks turbulence and mixing, gases must move through it by diffusion alone (Ambühl 1959; Feldmeth 1968; Vogel 1988). The thickness of the boundary layer decreases as the rate of flow in the adjacent water increases. The rate of flow is a function of the stream's velocity or, in still water, of an organism's self-generated ventilation currents (e.g., moving its abdomen or gills). Nevertheless, even in the most rapid flows that occur in streams or that are produced by the movement of the organism, a thin boundary layer persists.

Because of their flattened shape or small size, a number of aquatic insects can dwell within the boundary layer that is created as stream water flows over rock surfaces. Here an insect is removed from significant water movement, but it remains close enough to turbulent water flow to obtain adequate O_2 supplies.

Tracheal System and Respiratory Surfaces

In most insects, gas distribution takes place through a network of internal, air-filled tubes known as the *tracheal system* (Chapman 1982). The blood, or hemolymph, plays little or no role in this process. The larger tubes, or *tracheae*, exchange respiratory gases with the atmosphere through segmentally arranged lateral pores called *spiracles*. Tracheae, which are cuticular ingrowths, branch internally from the spiracles and become progressively smaller (to 2-5 µm in diameter). Further branching forms capillaries called *tracheoles*, which are generally less than 1 µm across and end where they contact or indent individual cells. Gases are exchanged by diffusion between tracheoles and cells at these points of contact.

This tracheal respiratory system evolved terrestrially and was undoubtedly the basic design that insects used as they adapted to live in water. Thus, most terrestrial and some aquatic insects have multiple pairs of spiracles (usually 8-10) that open on the body surface (referred to as polypneustic systems; Fig. 4.2A-B). The *oligopneustic* systems developed from this ancestral, polypneustic design and have only one or two pairs of functional spiracles, often located at the posterior end (Fig. 4.2C). Both designs are frequently referred to as *open* tracheal systems because of the presence of functioning spiracles. In contrast, tracheal systems with no functional spiracles are *closed* (referred to as *apneustic*; Fig. 4.2D-F) and, although otherwise complete, lack direct contact with the outside and rely on gases diffusing through the cuticle for respiratory exchange.

The amount of oxygen exchange is partially determined by the amount of surface through which gas molecules can pass. In air, where O_2 concentration is high, open spiracles are sufficient to serve as the respiratory surface; in water, where dissolved O_2 is sparse, larger respiratory surfaces are necessary. The body surface of a small, elongate organism (e.g., an early instar chironomid, Fig. 26.5) is large enough to allow sufficient oxygen diffusion to meet the organism's metabolic needs. However, as that animal increases in size its volume will increase more rapidly than its surface area, and O_2 intake will become surface limited. Additional gas exchange surfaces, which are either large, thin, tracheated body outgrowths called *gills* (Fig. 4.2E-F), or air bubbles (Fig. 4.3), serve to counter this trend. As an insect grows, these surfaces often become very large in order to maintain a permeable surface-to-volume ratio that can meet the insect's respiratory needs.

RESPIRATORY OPTIONS WITH AN OPEN TRACHEAL SYSTEM

Open tracheal systems are characteristic of insects that breathe air. Aquatic insects with open tracheal systems must therefore establish spiracle-air contact, either by connecting directly with a stationary air source (e.g., the atmosphere) or by carrying a store of air with it when it dives (see Table 4A).

Stationary Air Sources

Aquatic insects that connect with a stationary air source have an oligopneustic tracheal system with the functional spiracles located at the end of the abdomen of larvae or on the thorax of pupae. The submerged insect obtains O_2 either by placing its spiracles above the water surface (*atmospheric breathers*) or by forcing them into plant air stores (*plant breathers*).

Atmospheric breathers seldom maintain a continuous connection with their air source. Therefore, spiracles must be adapted to prevent flooding both when the insect submerges

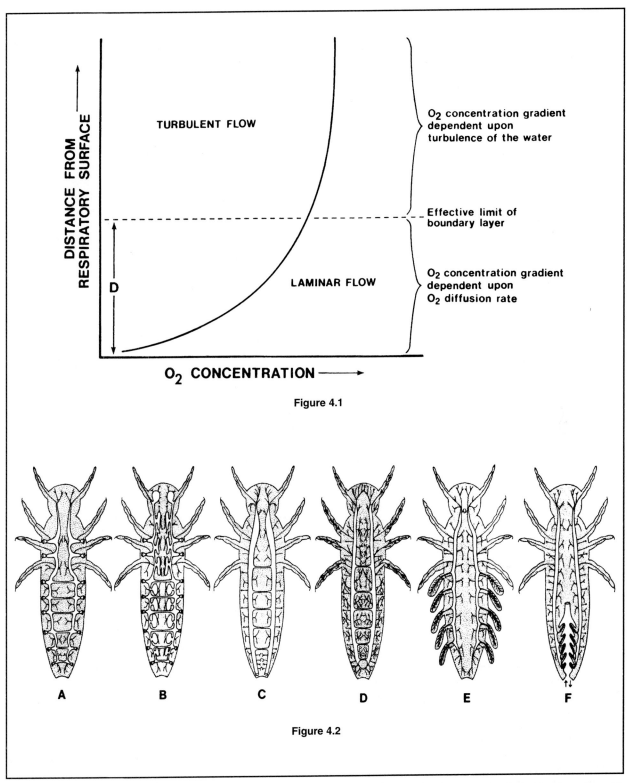

Figure 4.1. The boundary layer and its effect upon O_2 reaching an organism's respiratory surface. The *dashed line* demarcates the boundary layer's outer limit. An O_2 concentration gradient (*solid line*) is established because O_2 is consumed at the organism's surface and is replaced within the boundary layer only by diffusion. The diffusion rate is dependent on the thickness (*D*) of the boundary layer and the steepness of the O_2 gradient (modified from Feldmeth 1968).

Figure 4.2. *A.* Open polypneustic tracheal system; *B.* Polypneustic tracheal system with air-sacs for ventilation; *C.* Oligopneustic tracheal system in which the terminal spiracles alone are functional; *D.* Closed tracheal system allowing cutaneous repiration only; *E.* Closed tracheal system with abdominal tracheal gills; *F.* Closed tracheal system with rectal tracheal gills (modified from Wigglesworth 1972).

Table 4A. Respiratory options with open and closed tracheal systems. (The life stages known or inferred to use a particular respiratory option are indicated by the following: L=larvae; P=pupae; A=adults.)

Respiratory Option	Tracheal System	Oxygen Source	Examples	Selected References
Atmospheric Breathers	open	atmosphere	Diptera: Culicidae (L, P), Dolichopodidae (L), Ephydridae (L, P), Psychodidae (L), Stratiomyidae (L, P), Syrphidae (L, P), Tabanidae (L, P), Tipulidae (L, P), Ptychopteridae (L,P) Coleoptera: Amphizoidae (L), Dytiscidae (L, A), Hydrophilidae (L, A) Hemiptera: Nepidae (L, A)	2063, 2187, 3209
Plant Breathers	open	plants	Coleoptera: Chrysomelidae (L, P, A), Curculionidae (L) Diptera: Culicidae (L, P), Ephydridae (L, P), Syrphidae (L)	1821, 1822, 1823, 4112
Temporary Air Store	open	atmosphere and dissolved	Coleoptera: Dytiscidae (A), Gyrinidae (A), Haliplidae (A), Helodidae (A), Hydraenidae (A), Hydrophilidae (A) Hemiptera: Belostomatidae (L, A), Corixidae (L, A), Naucoridae (L, A), Notonectidae (L, A), Pleidae (L, A)	1041, 3027, 3028, 3179, 3243
Permanent Air Store Plastrons	open	dissolved	Coleoptera: Curculionidae (A), Dryopidae (A), Elmidae (A), Hydraenidae (A), Hydrophilidae (A) Hemiptera: Naucoridae (L, A) Lepidoptera: Pyralidae (L, P)	1590, 1591, 1774, 3999, 4000, 4001, 4002, 4003
Spiracular Gills	open	dissolved	Coleoptera: Hydroscaphidae (L), Psephenidae (P), Sphaeriidae (L), Torridincolidae (L,P) Diptera: Blephariceridae (P), Canacidae (P), Deuterophlebiidae (P), Dolichopodidae (P), Empididae (P), Simuliidae (P), Tanyderidae (P), Tipulidae (P)	1770
Tracheal Gills	closed	dissolved	Ephemeroptera (L), Odonata (L), Plecoptera (L), Megaloptera (L), Neuroptera (Sisyridae) (L), Coleoptera (several families), Diptera (several families), Trichoptera (L), Lepidoptera (Pyralidae) (L)	1092, 1093, 1094, 2242, 2754, 4357, 4481, 1096, 1098, 2934, 3639
Cutaneous	closed	dissolved	Diptera: Ceratopogonidae (L,P), Chaoboridae (L,P), Chironomidae (L, P), Simuliidae (L), Tipulidae (L) Lepidoptera Plecoptera (gill-less L) Trichoptera (gill-less L)	923, 1276, 2242, 3209, 3022, 4235
Hemoglobin	open or closed	atmosphere or dissolved	Hemiptera: Notonectidae (L, A) Diptera: Chironomidae (L, P)	2707, 4207, 4208, 4209

and when it comes to the surface and again contacts the air. Spiracles are commonly surrounded with water-repellent (*hydrofuge*) cuticle or hairs (Figs. 22.47, 22.50). On submergence, flooding is prevented by these hydrofuge hairs, by retractable fleshy lobes that seal the spiracular openings, or by holding an air bubble over the openings.

Many aquatic insect families contain atmospheric breathers (see Table 4A). Among the most conspicuous are the larvae of dytiscid and hydrophilid beetles which have functional spiracles at the end of the abdomen (Figs. 20.68, 20.111). The majority of Diptera larvae also have functional spiracles located posteriorly but they are at the end of a tube called the respiratory siphon (e.g., Figs. 22.20, 22.23, 22.58, 22.63, 22.76, 22.86). In *Eristalis* sp. (Syrphidae), this siphon can extend to six times the body length (Fig. 22.76) and gave rise to the common name of rat-tailed maggot. Larvae of species with shorter siphons often are restricted to shallow seeps (e.g., ephydrids, ptychopterids), to living near the surface in algal mats (e.g., dolichopodids, ptychopterids) or along pond and stream margins (e.g., tabanids), or to swimming short distances away from and back to the water surface (e.g., ephydrids, culicids). Undoubtedly the culicid larvae, commonly called mosquito wrigglers, are far and away the most familiar of the oligopneustic atmospheric breathers. Their common occurrence in pools and puddles has allowed many of us to view them hanging from the water's surface, or wriggling down into their watery home to feed or escape possible surface threat.

Plant breathers (see Table 4A) have spiracles modified to pierce submerged portions of aquatic plants and tap the plants' specialized air channels (*aerenchyma*). Larvae of at least some syrphids (e.g., *Chrysogaster* sp.), ephydrids (e.g., the brine fly *Notiphila* sp.), and curculionids (e.g., the weevil *Lissorhaptrus* sp.) live at the base of aquatic plants in anoxic mud and obtain their O_2 in this manner. Larvae, pupae and, less frequently, adults of a chrysomelid beetle (*Donacia* sp.) use the air source in the roots of such aquatic macrophytes as *Typha* sp. (cattails), *Juncus* sp. (rushes), and *Nymphaea* sp. (water lilies). During winter when the aerial portions of these plants die and the living parts are often completely submerged, O_2 concentrations in the air spaces of the roots decline. Under such conditions, aquatic insects dependent on this air source go into metabolic diapause and thus greatly reduce their oxygen requirements (e.g., *Donacia* sp.; Houlihan 1969b, 1970). The sharp, barbed respiratory siphons of mosquitoes in the genera *Mansonia*, *Coquillettidia*, and *Taeniorhynchus* (Culicidae; Fig. 4.4) can pierce the roots and stems of plants in open water, thus allowing larvae to remain submerged until adult emergence (Keilin 1944). Because these larvae have a thin cuticle and inhabit open water, the O_2 obtained from plant stores can be supplemented by cutaneous respiration. Nonetheless, such larvae appear unable to gain sufficient O_2 for an active existence because they move and feed slowly, and have an extended life cycle (Gillett 1972).

Transportable Air Stores

Aquatic insects that rely solely on stationary oxygen sources (e.g., atmosphere) can leave those sources for only brief periods of time or must remain relatively inactive while separated from them. By contrast, aquatic insects that carry their own air supply can stay submerged longer and be more active. When a transportable air supply is exposed to the water, it can serve not only as an air reserve but also as a *physical gill*. As such, the gas bubble is able to supply more O_2 than it contained originally. Thus, the diving time is extended, sometimes to the extent that the insect never has to come to the surface to renew it.

Temporary Air Stores: When an insect begins its dive with a temporary air store (bubble; Fig. 4.3; see Table 4A), gases in the atmosphere, the bubble, and the water are in equilibrium. As the insect consumes O_2 from its bubble, a like amount of CO_2, produced in cellular respiration, replaces the O_2. However, because CO_2 then diffuses so rapidly out of the bubble and into the surrounding water, it has little effect on the size or gas composition of the temporary air store. As O_2 pressure in the bubble decreases with use, O_2 from the surrounding oxygen-rich water diffuses in. But, because O_2 replacement is not immediate, an increase in the relative amount of N_2 in the bubble causes the outward diffusion of N_2. For this reason, the bubble decreases in size. Still, because O_2 diffuses into the bubble 2-3 times faster than N_2 diffuses out, this temporary physical gill (sometimes called a *compressible gill*; Mill 1973) continues to extract oxygen from the water, supplying about eight times more O_2 than the original air store contained.

The length of time a bubble can act as a gill depends upon the ratio of O_2 consumption to the bubble-water surface area. The larger this ratio, the shorter the lifetime of the gill (Rahn and Paganelli 1968). Large insects that have high O_2 demands must refill their air stores often because they carry bubbles with relatively less surface exposed. For example, so little of the air stores of adult *Belostoma* sp. (Belostomatidae), *Hydrous* sp. (Naucoridae), and *Dytiscus* sp. (Dytiscidae) are exposed that their bubbles serve as effective physical gills only when water temperatures are low (e.g., winter). Under such conditions, more O_2 dissolves in the water, metabolic rates are lower, and thus O_2 consumption by the insects decreases (Ege 1915; de Ruiter *et al.* 1952; Popham 1962). The water scorpion *Ranatra* sp. (Nepidae) does not achieve the physical gill effect because its air store is held completely beneath the hard hemelytra. Other small diving insects that have relatively more bubble area exposed, such as notonectid, pleid, and corixid bugs, use their physical gill continuously (Ege 1915; Gittelman 1975). Consequently, these insects can swim some distance below the surface. As a result, and because they are generally the smallest of the diving Hemiptera, corixids have essentially become bottom dwellers and thereby avoid competition for food at the water surface (Popham 1960).

Several factors reduce the effectiveness of the physical gill and, therefore, increase the surfacing frequency of an insect. Deeper dives increase hydrostatic pressure, which causes N_2 to diffuse out of the bubble faster. Lower O_2 concentrations in surrounding water decrease the diffusion gradient and therefore decrease the inward rate of oxygen diffusion. Finally, increased water temperature causes both lower O_2 concentrations and higher O_2 consumption by the insect. Hutchinson (1981, 1993) has suggested that this relationship between gill efficiency and water temperature may explain the predominance, within the Corixidae and other aquatic groups using temporary air stores, of smaller species in warmer climates.

Permanent Air Stores (Plastrons): A number of aquatic insects have hydrofuge hairs or cuticular meshworks that hold water away from the body surface and thus allow a permanent gas film to form in those areas (see Table 4A). This permanent gas film is called a *plastron* (sometimes known as an *incompressible gill*; Mill 1973) and is composed mainly of N_2. The N_2 can not diffuse out of the plastron and, because the plastron is in contact with open spiracles at the same time it presents a considerable surface to the water, it serves as a physical gill that allows continuous diffusion of O_2 inward.

Many aquatic insects with plastrons never need to surface to replenish their air stores. Oxygen consumption, and thus their metabolic rates, are determined by the rate of O_2 diffusion through the fixed and limited surface area of the plastron. As a consequence, most insects with plastrons are slow moving and are limited to habitats with high dissolved oxygen (e.g., Elmidae, which are commonly called riffle beetles because they occur in fast-flowing streams). Those insects that use plastrons in still water must either be good swimmers (e.g., hydrophilid beetles) or capable of crawling out of the water (e.g., curculionid beetles) to avoid low O_2 when such a condition occurs (Hinton 1976a). The ability of these still-water insects to detect and avoid low oxygen is absolutely essential because O_2 will diffuse away from the insect if the concentration of O_2 is higher in the plastron than in the surrounding water.

Plastrons are quite variable in structure. Hydrofuge hair systems evolved in a wide variety of taxa, including lepidopterans (e.g., *Acentropus* sp.), several kinds of beetles (e.g., the weevil *Phytobius* sp. and the elmid *Stenelmis* sp.), and the true bugs (e.g., *Aphelocheirus* sp.). The latter has one of the most efficient plastrons known, consisting of a dense mat of hydrofuge hairs (4.3×10^6 hairs/mm^2) that are bent at the tips

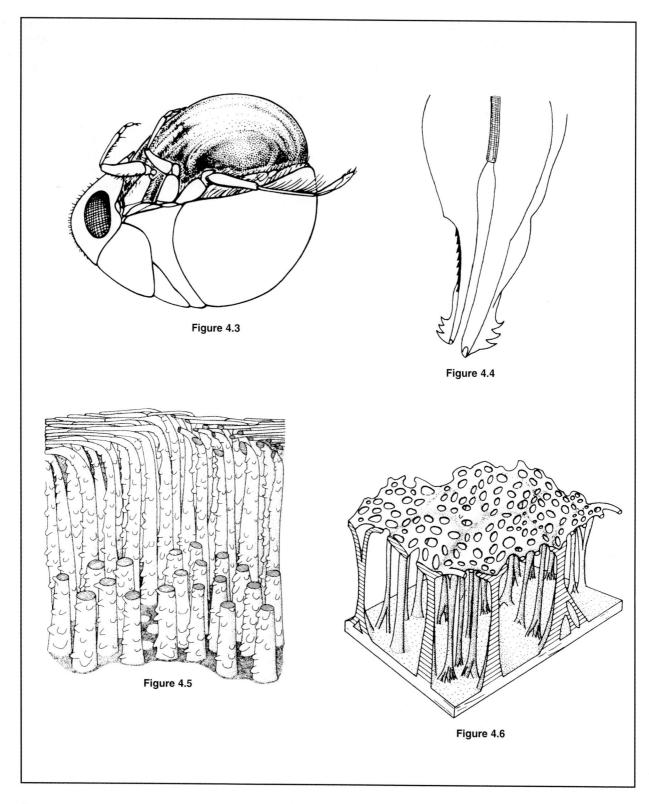

Figure 4.3. Ventral air bubble, which also serves as a temporary physical gill, of a pleid backswimmer *Neoplea* sp. (after Gittelman 1975).

Figure 4.4. Longitudinal section of postabdominal respiratory siphon of a *Taeniorhynchus* sp. (Culicidae) larva. Barbed hooks allow the larva to maintain contact between the spiracle and the plant air stores (modified from Keilin 1944).

Figure 4.5. Hydrofuge hairs comprising the plastron of the aphelocheirid bug *Aphelocheirus* sp. (after Hinton 1976a).

Figure 4.6. Hydrofuge cuticular network found in the spiracular gill of the tipulid *Dicranomyia* sp. (after Hinton 1968).

(Fig. 4.5). This mat covers most of the ventral surface of the insect and, because *Aphelocheirus* is wingless, most of the dorsal surface as well (Hinton 1976a). The extensive nature of its plastron allows this insect to spend its entire life cycle under water. In contrast, elmids have a plastron formed by short, dense hairs located on the lateral and ventrolateral body surfaces and on the dorsum of the thorax; their plastron is overlain by a second, temporary air store, sometimes referred to as a macroplastron, which is formed by longer, less dense hairs. The macroplastron air store is used when the O_2 demands of the beetle are high (Thorpe 1950).

Hydrofuge cuticular networks are always associated with outgrowths of the area around the spiracles. They often rise as columns from the body surface and divide at the top to form an open canopy (Fig. 4.6). An air film held beneath the canopy serves as a physical gill. These *spiracular gills* are found in pupae of many Coleoptera and Diptera, and in larvae of torridincolid, sphaeriid and hydroscaphid beetles (Table 4A; Hinton 1968). Such insects often inhabit streams with highly fluctuating water levels. In these habitats, spiracular gills serve in both O_2 acquisition when the insect is submerged and water retention when the insect is exposed to air. Undoubtedly, the blackflies (Simuliidae) are the most common insects that possess spiracular gills. Their pupae have a filamentous gill tuft on each anteriolateral corner of the thorax (Fig. 25.29). The fact that these gill tufts are associated with spiral-shaped vortices in the flow that moves over them suggests their aquatic respiratory function (Eymann 1991).

RESPIRATORY OPTIONS WITH A CLOSED TRACHEAL SYSTEM

Closed tracheal systems have no functional spiracles, so gas exchange must occur by diffusion through the cuticle. Such cuticle is thin and permeable, and closely below it lies a dense network of tracheoles that provides a large surface area for exchange of gases. Because of either the surface-to-volume considerations mentioned previously or the presence of thick, O_2-impermeable cuticular surfaces, most insects with closed tracheal systems cannot fulfill their O_2 requirements solely by diffusion through the general body surface (i.e., *cutaneous respiration*). Therefore, cutaneous respiration has commonly been supplemented with highly tracheated, thin, body wall outgrowths called *tracheal gills*.

Cutaneous Respiration

Because the amount of gas exchanged is proportional to surface area, only aquatic insects that have a high surface-to-volume ratio rely on cutaneous respiration alone (see Table 4A). Such a situation is demonstrated by the smallest of the aquatic Hemiptera, the non-North American *Idiocoris* sp. and *Paskia* sp. (Helotrephidae), which are the only known adult, free-living, apneustic (no functional spiracles) insects (Esaki and China 1927, cited by Hutchinson 1981). Most of the strictly cutaneously respiring species are small, worm-shaped larvae such as chironomids (Fig. 22.40; Fox 1920; Parkinson and Ring 1982), ceratopogonids (Fig. 22.45; Ward 1991), chaoborids (Fig. 22.30), some tipulids (Pritchard and Stewart 1982), simuliids, and gill-less plecopterans and trichopterans. However, gas exchange in the early life stages of a large number of aquatic insects probably is through cutaneous respiration as well. For example, young Trichoptera larvae respire exclusively through the cuticle; gill filaments develop and become important only in later instars (Wiggins 1977) where their addition presumably maintains a high surface-to-volume ratio. A similar situation is evident in some Plecoptera (Shepard and Stewart 1983). Even for those insects with tracheal gills, cutaneous respiration probably accounts for a significant but variable portion of total O_2 intake. For example, Eriksen and Moeur (1990) demonstrated that even the mayfly *Siphlonurus occidentalis*, with proportionately very large tracheal gills, uses its abdominal surface for about 30% of its O_2 intake. The zygopteran *Lestes disjunctus*, whose gills are "...so large that one may suppose it would be an embarrassment to them if they grew any bigger...." (MacNeill 1960), normally uses cutaneous respiration to meet 70 to 80% of its needs at high O_2 concentrations (Eriksen 1986), with gill surfaces providing the remaining 20 to 30%.

Tracheal Gills

Tracheal gills are present in the immature stages of at least some species in every aquatic insect order except the Hemiptera (see Table 4A). Tracheal gills have been shown to serve in ventilation, protection, hydraulic streamlining, and ion exchange. However, whether or not they are really used in respiration has been the subject of considerable debate. For example, removal of tracheal gills from larvae of the caddisfly *Macronema* sp. resulted in no difference in O_2 intake between normal and gill-less individuals (Morgan and O'Neil 1931); in fact, gill-less larvae generally behaved normally and eventually pupated. However, such experiments were typically conducted under the favorable conditions of high dissolved oxygen concentration and low temperature. When respiratory studies were conducted under varying O_2 and temperature conditions with the mayfly *Cloeon dipterum*, larvae maintained a high metabolic rate down to 1 ppm O_2 when gills were intact, but gill-less individuals experienced oxygen stress below 3 ppm (Wingfield 1939). Gills of the zygopteran *Lestes disjunctus* performed no respiratory role at 7°C and 5.5 ppm O_2; however, they became increasingly important as temperature rose and O_2 decreased until they accounted for up to 80% of total oxygen intake (Eriksen 1986). In general, tracheal gills apparently are not very important as O_2 intake sites under high environmental O_2 conditions but become progressively more important as oxygen concentration decreases or temperature increases.

Segmental pairs of lateral abdominal gills are found on at least some species in a number of groups (e.g., Megaloptera, Coleoptera, Zygoptera, Neuroptera) but they show the greatest structural diversity in the Ephemeroptera. Gills of mayflies vary from leaflike (Fig. 11.57) to two-branched structures with single (e.g., *Paraleptophlebia* sp.; Fig. 11.5) or multiple (e.g., *Habrophlebia* sp.; Fig. 11.26) filaments. Combinations of leaflike and filamentous gills occur in a number of mayfly genera including *Ephoron* (Fig. 11.4) and *Caenis* (Fig. 11.15b). In *Tricorythodes* sp. the first gill pair is enlarged to cover the posterior pairs (Fig. 11.14), presumably to shield them from being covered by fine sediments. In a number of ephemeropterans that occur in rapidly flowing water (e.g., *Iron* sp.) the gills overlap each other and are held against the substrate to provide a flattened shape.

Table 4B. Ventilation methods for insects utilizing dissolved oxygen.

System Ventilated	Ventilation Method	Taxon	References
Cutaneous	Undulation	Chironomidae	2063, 2437, 4209
		Trichoptera (gills lacking)	
		Lepidoptera (gills lacking)	2242
	Swimming	Chaoboridae	*
		Chironomidae	*
	Natural water flow	Trichoptera (caseless, gills lacking)	923, 1441
		Plecoptera (gills lacking)	923, 1441
		Simuliidae	4579
Tracheal Gills	Beating gills	Ephemeroptera	1004, 1005, 1006, 1007, 1092, 1098, 4481,
		Psephenidae	3706
		Corydalidae	2450
		Gyrinidae	2300
	Undulation	Trichoptera	1152, 2288, 3131, 3132, 3133, 3134, 4357
		Lepidoptera	4298
		Chironomidae	4209
	Leg contractions that move body (push-ups)	Plecoptera	242, 1368, 2051, 2118, 2119, 2611, 2836,
		Lestidae	1095, 2837
	Rectal pump	Anisoptera	2053, 2696, 2697
	Natural water flow	Heptageniidae	64
		Plecoptera	923
		Zygoptera	4621
		Trichoptera (caseless)	64
		Trichoptera (with case)	1152, 3133, 3134
		Blephariceridae	*
Temporary and Permanent Air Stores	Leg movements	Notonectidae	883
		Naucoridae	883
		Corixidae	883, 3179
	Swimming	All taxa that swim with exposed air bubble	883
	Natural water flow	Simuliidae (pupae)	1128, 1770
		Dryopidae (adult)	*
		Lepidoptera (larvae and pupae)	263, 264

*Eriksen, C. H. Personal observation.

The abdominal gills of trichopteran larvae appear as scattered single (Fig. 17.77) or clustered (Fig. 17.39) filaments, which may or may not be branched. Their number and size tend to increase with increasing body size. Wichard (1978) reported that gill number in the European species *Molanna angustata* was inversely related to the average environmental O_2 concentration, a phenomenon suggested much earlier for the North American aquatic insects by Dodds and Hisaw (1924b).

Although two families of damselfly larvae (Zygoptera: Odonata) have paired lateral abdominal gills that are used in respiration (Norling 1982), this group is noted for its terminally placed abdominal gills. These *caudal gills* are usually leaf-like structures (Fig. 12.15), with two of them placed laterally and one medially (Fig. 12.12). MacNeill (1960) describes two gill types among Zygoptera: (1) the *simplex* type which increases in size uniformly as the larva grows and is typical of the Lestidae; and (2) the *duplex* gill which consists of a thick proximal area and a thin distal zone that becomes disproportionately larger at each molt. Duplex gills are typical of the Coenagrionidae.

Internal placement of tracheal gills is found in dragonflies (Odonata: Anisoptera), where six longitudinal rows of gills are located in an enlarged, anterior portion of the rectum called the *branchial chamber*. Nymphs respire by exchanging water through their anus, a behavior which ventilates their rectal gills; as a result, those species that burrow in mud must protrude their anus above the sediment surface to prevent fouling of their branchial chamber and gills (Corbet et al., 1960).

Although uncommon, gills (or what appear to be gills) can be found on the head and thorax of some Plecoptera, Diptera, and Lepidoptera. In stoneflies care is required to determine whether or not gills are truly respiratory organs (i.e., tracheate structures). The submental gills (Fig. 14.24) of Perlodidae, the so-called cervical gills (Fig. 14.20) of nemourids, and the coxal gills of *Taeniopteryx* (Taeniopterygidae) are not tracheate but rather are hemolymph-filled evaginations of the membranes between sclerites that function primarily as *osmobranchiae* (Shepard and Stewart 1983). According to these authors there are no tracheal gills on the head or cervical region of stoneflies, and the only thoracic gills in the Plecoptera that serve primarily in oxygen intake are found in the Perlidae and Pteronarcyidae. Thus, Tozer's (1979) suggestion that the cervical gills retained in adult *Zapada cinctipes* (Nemouridae) probably function in respiration when adults enter the water to avoid subzero air temperatures appears to be incorrect.

RESPIRATORY PIGMENTS

One of the most vivid images in aquatic entomology is the discovery of bright red "bloodworms" in dark, anoxic sediments of lakes and ponds. The red color of these chironomids is caused by the respiratory pigment, *hemoglobin*. Although hemoglobin is characteristic of vertebrate blood, it occurs in some species of most animal phyla (Terwilliger 1980). Among insects, hemoglobin is restricted to species of Chironomidae, Notonectidae (see Table 4A), and the terrestrial dipteran family Gasterophilidae (bot flies that parasitize a variety of mammals).

Insect hemoglobin differs from that of vertebrates by containing two (instead of four) heme groups. *Chironomus* possesses a high-affinity hemoglobin, which means that the pigment releases O_2 at low external oxygen pressures, thus benefiting its possessor in the low O_2 concentrations found in water and mud. By contrast, vertebrate and other insect hemoglobins tend to be low-affinity pigments. These hemoglobins release their O_2 in the high oxygen concentrations found in air, thus aiding those that have such pigment in their terrestrial environment.

The *Bohr effect* (the inverse relationship between the CO_2 concentration and the amount of O_2 held by hemoglobin) results in a significant advantage for organisms that obtain O_2 from a high O_2 environment (i.e., lungs) and release it in a high CO_2 environment (i.e., tissues). However, in environments where the oxygen concentration is always low and CO_2 is plentiful (e.g., lake muds), low-affinity hemoglobin would be maladaptive.

When chironomid larvae undulate their bodies in their mud burrows to bring in water of higher oxygen content, their hemoglobin becomes O_2-saturated and apparently performs no function. However, when O_2 is depleted between periods of undulation, hemoglobin releases O_2 to the tissues. If the hemoglobin's approximate nine minute supply of oxygen (as determined by Walshe [1950] for a species of chironomid) is less than is needed in the interval between undulations, anaerobic respiration becomes necessary. When undulations resume, hemoglobin enables a larva to recover rapidly from these anaerobic periods because the pigment takes up O_2 and passes it to the tissues more quickly than is possible by diffusion alone (Walshe 1950).

Two genera of Notonectidae (*Anisops* and *Buenoa*) have a low-affinity hemoglobin that performs a very different function than just described for midges. Miller (1964b, 1966) and Wells et al. (1981) observed that O_2 released from hemoglobin contained in certain richly tracheated abdominal cells markedly reduced the rate of depletion of the temporary external air store during diving. Because these hemoglobin-containing cells provide about 75% of the O_2 used during a dive, the insect can carry a small air bubble, does not have to fight the buoyancy of a larger air store, and can therefore remain submerged longer. By doing so, these notonectids have exploited resources in the sparsely colonized, midwater habitat in ponds.

VENTILATION AND REGULATION

Oxygen ultimately reaches an organism's tissues by diffusion, but the amount obtained can be influenced by structural and behavioral adaptations of the organism that control the diffusion rate. For example, if the thickness of the boundary layer around respiratory surfaces is reduced, the distance over which diffusion must occur is less. *Ventilation*, the flow of air or water by active or passive means over respiratory surfaces or through part of the tracheal system, thins the boundary layer. Ventilation currents may result from abdominal contractions, body undulations, gill beating, swimming through the water, movement to a more favorable microhabitat, utilization of stream flow, or a combination of these mechanisms (Table 4B).

Because the air-filled tracheal system does not collapse and ends blindly among the cells, diffusion is the only way O_2 can be delivered directly to cells. However, some large or highly active insects can advantageously ventilate at least the outer portion of their tracheal system. In such species, contraction of abdominal muscles compresses nonrigid air-sacs that are located along the longitudinal tracheal trunks (Fig. 4.2B). Compression empties the air-sacs and flushes their contents through the tracheal trunks and out the spiracles. When abdominal muscles relax, higher external atmospheric pressure forces air back through the spiracles and tracheal trunks to refill the air-sacs. In diving aquatic insects, air-sac flushing probably occurs only during contact with air. Although air-sacs in most terrestrial insects seldom compress more than 10 to 20% (Miller 1964a), *Dytiscus* sp. (Dytiscidae) and *Eristalis* sp. (Syrphidae) can flush about 65% of their entire system per compression (Krogh 1920, 1943), and *Hydrocyrius* sp. (Belostomatidae) may completely collapse parts of their air-sac system (Miller 1961). High-volume ventilation conveys a substantial advantage to divers that only briefly contact the atmosphere.

As mentioned previously, an organism has a much more difficult time obtaining sufficient O_2 from a dissolved source than it does from air. Not only is there much less O_2 available even under the most favorable conditions, but, because water is so much heavier than air, the boundary layer is more difficult to move and thus to thin by ventilation. Ventilation by an insect generally pushes water posteriorly over the gills and dorsal body surface. Eastham (1934, 1936, 1937, 1939) demonstrated that mayflies beat their gills to create respiratory currents, and the frequency of gill beat increases as O_2 concentration decreases (Eriksen 1963a; Eriksen and Moeur 1990). Riley (1879, as cited by Tracy and Hazelwood 1983) alludes to a similar behavior for *Corydalus* (Megaloptera). Trichopterans, chironomids, and aquatic lepidopterans use body undulations to pump water through their cases or tubes (Welch and Sehon 1928; Walshe 1950; Feldmeth 1970). For some Trichoptera, efficient ventilation apparently depends on the presence of a case that restricts and directs water flow (Williams et al. 1987), as some larvae removed from their cases ultimately die even though they continue to undulate.

Anisoptera larvae possess an especially effective ventilation mechanism for their gills which are located in a blind sac off the rectum. Contraction, mainly of dorsoventral abdominal muscles, increases pressure in the branchial chamber and forces water out the anus. When muscular relaxation occurs, negative pressure in the chamber allows O_2-rich water to return. Ventilations increase in frequency as O_2 decreases and temperature increases (Mantula 1911; Mill and Hughes 1966; Cofrancesco and Howell 1982).

Some organisms, such as plecopterans and lestid zygopterans, perform "push-up" ventilatory movements (Knight and Gaufin 1963; Eriksen 1984). Although helpful, these movements are initiated only during periods of respiratory stress because they are inefficient and merely stir up surrounding water rather than force freshly oxygenated water over the gills. Some insects are unable to accomplish any self-generated ventilation because they have adapted so completely to the ventilation provided by natural water movement (see Table 4B). In the absence of water flow, these insects cannot acquire sufficient O_2 and soon die (Jaag and Ambühl 1964).

Whether or not an insect uses ventilation to compensate for a changing environmental O_2 supply determines whether it is considered a *respiratory regulator* or a *respiratory conformer*. Aquatic insects that use atmospheric air do not encounter variable oxygen concentrations and consequently are respiratory regulators, which means that they control the supply of O_2 to cells by the length of time the spiracles are open, the number of ventilation movements, or the frequency of surfacing. Regulators that use oxygen dissolved in water often face a decreasing O_2 supply. Under such conditions they maintain fairly constant intake of O_2, and thus a fairly constant rate of activity, by increasing the rate of their ventilation movements. However, at some minimum O_2 level, referred to as the *critical point*, ventilation rate reaches its maximum, metabolism is significantly disrupted, and the animal dies if water higher in dissolved oxygen is not soon available.

In contrast to regulators, respiratory conformers are unable to create significant respiratory currents and, as a result, their O_2 intake is proportional to O_2 availability in the microhabitat. Because some rheophilic species use natural water current to ventilate their respiratory surfaces (e.g., rhyacophilid caddisflies, blepharicerid midges), it seems logical that as stream flow slows below some critical speed, so does metabolic rate unless the organism changes position to remain in a favorable ventilatory environment. No studies have tested this latter hypothesis, but Ambühl (1959) has demonstrated that as current speed increases so does respiration (conformity), at least until some plateau (regulation) is reached. Such a response makes sense because many rheophilic insects cannot ventilate for themselves and, therefore, they let natural water flow bathe respiratory surfaces for them. Golubkov et al. (1992) note that some species, which might otherwise be thought of as conformers, demonstrate a constant level of respiration in rapidly flowing water. Feldmeth (1970) found that current speed is related to the intensity of respiration in the caddisflies *Pycnopsyche lepida* and *P. guttifer*, but locomotor behavior, as influenced by current velocity, is most important in setting the respiratory rate.

Studies of aquatic insect respiration have shown a variety of abilities ranging from absolute conformity to strict regulation. However, Eriksen (1963a) and Nagell (1973) have demonstrated that some Ephemeroptera and Plecoptera species can appear to be either respiratory regulators or conformers depending on experimental conditions. In contrast, a variety of experimental conditions did not seem to change a lestid zygopteran from being intermediate between regulation and conformity (Eriksen 1986). Whether or not the array of O_2-conforming and O_2-regulating abilities described to date is real, or caused by conditions of the experiments (i.e., conditions that do not appropriately represent microenvironments in which the insects actually dwell) has yet to be settled.

EFFECTS OF TOXICANTS ON RESPIRATION

Toxic chemicals are a human-induced and pervasive component of aquatic environments that potentially affect the physiology and morphology of aquatic insects. Toxicants can stress respiratory or other systems of aquatic insects and thereby modify respiration rates or inflict physical damage. Doherty and Hummon (1980) warn that although assumptions based on respirometry may indicate physiological stress, they fail to identify the specific toxic mechanism. The respiratory system of aquatic insects may be particularly sensitive to toxicants because tracheal gills not only function as respiratory surfaces, but may also serve as sites for active uptake of ions (e.g., Komnick 1977) and thus possibly for toxic chemicals as well. Most studies of respiratory responses to toxicants have been conducted in the laboratory because of the obvious need for controlled dosages, and necessary restrictions on field releases of chemicals (Johnson et al. 1993). Also, such studies have dealt almost exclusively with aquatic insects that possess tracheal gills as opposed to other respiratory mechanisms.

Aquatic insects have been shown to respond to sublethal concentrations of toxic chemicals with both increased and decreased respiratory rates. Both responses have major physiological implications, because increased respiration has significant metabolic costs and reduced respiration can lead to death. For example, exposure to sublethal concentrations of copper (Kapoor 1976) or Dibrom, an organophosphate pesticide (Maki et al. 1973), increased the oxygen consumption rates of plecopterans and megalopterans and reduced their tolerance to low oxygen concentrations. Similarly, the haloform byproducts of water chlorination consistently increased the respiration rate of dragonfly larvae in laboratory studies (Correa et al. 1985a; Calabrese et al. 1987; Dominguez et al. 1988). In contrast, oxygen consumption by chironomid larvae declined after exposure to naphthalene (a highly toxic polycyclic aromatic hydrocarbon), and hemoglobin-lacking *Tanytarsus dissimilis* larvae were more sensitive to naphthalene than hemoglobin-possessing *Chironomus attenuatus* larvae (Darville and Wilhm 1984). Respiration can also be lowered when metal ions, such as iron, precipitate on gill surfaces under acidic conditions or displace functional cations from the active sites of enzymes and result in respiratory failure (Gerhardt 1992).

The interactive effects of environmental stressors on respiration have received some attention, but results of studies are not consistent. For example, when exposed to low pH and high aluminum concentration a variety of different aquatic insects tested displayed reduced respiration rates (Rockwood et al. 1990), increased respiration (Correa et al. 1985b; Herrmann and Andersson 1986), or showed no effect (Correa et al. 1986). Toxicant interactions undoubtedly are common occurrences in a number of aquatic ecosystems and deserve more study with controlled experiments.

In some cases, respiration may not be affected directly by toxicants but respiratory surfaces may be involved in the absorption of toxic chemicals that ultimately affect other organ systems. Tracheal gills appear to be particularly important in this regard, probably because they are much more permeable than the more heavily sclerotized cuticle. For example, mercury can enter *Hexagenia rigida* (mayfly) larvae by direct absorption across gill lamellae in amounts that exceed those obtained from their diet (Saouter et al. 1991). Once in the mayfly, the metal is dispersed through the body by the tracheal system and hemolymph. The ultrastructure of gill tissue in *Pteronarcys dorsata* (stonefly) larvae can be altered by environmentally extreme acidic (pH <4) or alkaline (pH >10) conditions, thereby resulting in the loss of sodium ions and eventual death (Lechleitner et al. 1985).

Relatively few studies of toxicant effects on aquatic insect respiration have been conducted in natural aquatic ecosystems. Herrmann and Andersson (1986) found that various species of mayflies dominated in natural streams having different pH, a situation which corroborated their laboratory findings that low pH elicited variable levels of respiratory stress in different mayfly species. In streams receiving chlorinated effluents, perlid stoneflies and hydropsychid caddisflies had atrophied or deformed gills in 62-100% of the specimens occurring at the impacted sites. By contrast, non-polluted upstream populations had normal gill structure (Simpson 1980; Camargo 1991). Aquatic insects with physical gills (transportable air stores) probably are less susceptible to damage by chlorine, as suggested by the hemipterans frequently seen occupying swimming pools.

Clearly, respiratory surfaces, especially tracheal gills, are some of the most sensitive to environmental contaminants. Toxicants can result in physical damage of the respiratory structures, as well as pass through respiratory surfaces to affect other organ systems. Whatever their site of effect, toxicants can lead to respiratory stress that in turn can lead to death. The coupling of laboratory experiments with in situ studies is crucial to understanding both the mechanistic effects and ecological consequences of toxicants on aquatic insect respiration.

RESPIRATORY MECHANISMS AND OXYGEN ENVIRONMENTS

The evolution of insects from a terrestrial to an aquatic existence has resulted in a broad range of adaptations of the original, terrestrial-insect respiratory system (see also Chap. 8). Aquatic insects have developed major structural and behavioral adaptations to the various habitats with seemingly endless variations. Pools in flowing-water, and other still-water habitats (Chap. 6) containing submerged and emergent vegetation, possess such a tremendous variety of microenvironments that insects with most of the aquatic respiratory adaptations are found there. The deep parts of lakes lack aquatic insects that use oxygen from the air because of the dual problem of having to travel to the surface to renew their O_2 supplies and losing O_2 through diffusion to the surrounding O_2-poor water when they return to the lake bottom. Those insects that do occur in the depths of lakes commonly possess hemoglobin and have strong ventilatory or migratory abilities. The few species that inhabit the open water (away from the shore and bottom) of ponds and lakes have a high surface-to-volume ratio and consequently use cutaneous respiration. The constantly high-O_2 habitat of rapidly flowing water is suitable for species using plastrons, spiracular gills, cutaneous respiration, or tracheal gills. Those insects that require air contact cannot live in rapid flow because stable connection with the atmosphere is usually impossible given the turbulence of the water. More specific generalizations cannot be made because the major habitats that aquatic insects occupy encompass so many different O_2 microhabitats.

APPENDIX 1: DEMONSTRATIONS OF RESPIRATORY PROCESSES

The respiratory structures and processes that have been described are best understood when they are seen. As a means to that end, simple experiments are summarized in Table 4C that show the structural and behavioral abilities, and also the limitations, of aquatic insects that are subjected to varying environmental conditions. When keeping aquatic insects in the laboratory, or conducting experiments with them, always avoid stressing the organisms unless it is part of the experimental design. Likewise, always provide suitable substrate (pebbles, plastic mesh, glass burrows, etc.) and normal environmental O_2, temperature, and water current conditions (see Rearing Methods, Chap. 3) when holding and using the insects in experiments and as controls. Detailed explanations of additional experiments can be found in Kalmus (1963) and Cummins et al. (1965).

Table 4C. Demonstrations of respiratory processes (*superscripts* refer to Section C. *Useful Equipment*). Table 4B contains relevant literature.

A. *Closed Respiratory System* (larvae only)
   1. Ventilation Methods and Behavior
      a. Beating gills[1,2,6] (e.g., burrowing, climbing, sprawling Ephemeroptera, Corydalidae)
      b. Push-ups[1,6] (e.g., Plecoptera, Lestidae)
      c. Undulation[1,2,3] (e.g., Trichoptera, Lepidoptera, Chironomidae)
      d. Muscular rectal pump[1,6] (Anisoptera)
      e. Swimming[1] (e.g., Chaoboridae, Chironomidae)
      f. None (other than possible position change) (e.g., Blephariceridae, Simuliidae, fast-water Ephemeroptera, Plecoptera)
   2. Respiratory Currents Produced by Insect (Section A.1.a-d)[8]
   3. Micro-areas from which Respiratory Water Obtained (Section A.1.a-d)[8]
   4. Environmental Effects on Ventilation Frequency and Volume of Respiratory Flow (Section A.1.a-d)
      Vary: O_2 concentration[10]
      water current[7]
      temperature[11]

B. *Open Respiratory System* (larvae and adults)
   1. Ventilation Methods and Behavior
      a. Leg movements[4] (e.g., Corixidae, Naucoridae, Notonectidae)
      b. Swimming[5] (any species with exposed air store)
   2. Environmental Effects on Diving Time (any species with temporary air store)
      Vary: dissolved O_2 concentration[5,10]
      dissolved CO_2 concentration[5,10]
      temperature[5,11]
   3. Diving Stimulus (any species with temporary air store)
      Provide: air atmosphere
      O_2 atmosphere[5,10]
      CO_2 atmosphere[5,10]
      N_2 atmosphere[5,10]
   4. Need for Surface Tension to Establish Atmospheric Connection[5,9] (e.g., Culicidae, Tipulidae, Syrphidae, Notonectidae, Dytiscidae)
   5. Plastron (any species using plastron respiration only)
      Vary: dissolved O_2 concentration[1,10]
      current velocity[1,7]
      temperature[1,11]

C. *Useful Equipment*
   1. Narrow (e.g., < 3 cm) plexiglass observation aquarium.
   2. U-shaped glass burrows simulating natural dimensions. Portion restricted with coarse mesh screen for containing animal but allowing current flow (Walshe 1950; Eriksen 1963a).
   3. Artificial Trichoptera case: glass or plastic tubing approximating case interior diameter and length with one end restricted to a 1 mm central pore (Feldmeth 1970).
   4. Vertical, clear "diving tube," 2-3 cm by about 30 cm. Vertical strip of plastic screening near surface simulating vegetation. Horizontal screening just below water level. No bottom substrate.
   5. Vertical, clear "diving tube," 2-3 cm by 100-200 cm. Horizontal screening on bottom as substrate.
   6. Plastic window screen cut to appropriate shapes.
   7. Water current generation:
      — gravitational, from reservoir via appropriate tubing with flow control valves
      — air hose pump
      — magnetic stirrers. Note: these create centrifugal currents; however an organism can be restricted to one area and experience essentially longitudinal current flows (e.g., Philipson 1954; Morris 1963).
      — water current respirometer (e.g., Eriksen and Feldmeth 1967).
   8. Carmine or carbon-black suspension introduced where desired with narrow aperture eyedropper. Observe particle movement.
   9. Detergent or thin oil added to water surface with eyedropper.
   10. Gas concentrations: control concentration of dissolved gases in reservoir with gas mixing valves or a combination of compressed air, O_2, N_2, or CO_2. Monitor with O_2 electrode if available.
   11. Temperature: many heating/cooling devices may be used to adjust reservoir temperature, or use temperature controlled environmental rooms.

HABITAT, LIFE HISTORY, AND BEHAVIORAL ADAPTATIONS OF AQUATIC INSECTS

J. Bruce Wallace
University of Georgia, Athens

N.H. Anderson
Oregon State University, Corvallis

INTRODUCTION

Aquatic insects have received considerable attention within the last decade. Book-length coverage of the subject matter in this chapter is found in: Resh and Rosenberg (eds.) 1984, *The Ecology of Aquatic Insects*; Ward 1992, *Aquatic Insect Ecology*; and Williams and Feltmate 1992, *Aquatic Insects*. Since 1985 the *Annual Review of Entomology* has averaged about an article per year on the ecology or behavior of aquatic insects and the *Journal of the North American Benthological Society* began publication in 1986 and many of the papers pertain to aquatic insects.

Insects are very successful in the freshwater environment. This is demonstrated by their diversity and abundance, broad distribution, and their ability to exploit most types of aquatic habitats. In this chapter, the adaptations that contribute to their success are considered. Examples are given of how some species have adapted to very restricted environments and the life cycle is used to provide a framework for describing different ways that insects cope with the challenges presented by aquatic habitats.

Factors that influence utilization of a particular habitat can be grouped into four broad categories: (1) physiological constraints (e.g., oxygen acquisition, osmoregulation, temperature effects); (2) trophic considerations (e.g., food acquisition); (3) physical constraints (e.g., coping with habitat or habit as given in the ecological tables); and (4) biotic interactions (e.g., predation, competition). However, these categories are so interrelated that detailed analysis of each factor is not appropriate. Respiration is covered in detail in Chapter 4, but it is also included here because activities related to obtaining oxygen are central to behavioral and morphological features associated with most other activities.

The traditional division of freshwater systems into standing (lentic) and running (lotic) waters is useful for indicating physical and biological differences. Most insects are adapted to either a lentic or a lotic habitat, but overlaps are common such as in the floodplains of large rivers. For example, insects inhabiting pools in streams have "lentic" respiratory adaptations, whereas those on wave-swept shores of lakes are similar to stream riffle inhabitants in both oxygen requirements and in clinging adaptations (clingers in ecological tables).

Despite their success in exploiting most types of aquatic environments, insects are only incompletely or secondarily adapted for aquatic life. With very few exceptions, aquatic insects are directly dependent on the terrestrial environment for part of the life cycle. Even Hemiptera and Coleoptera with aquatic adults require access to surface air for respiration. This dependence on the terrestrial environment probably contributes to the prevalence of insects in shallow ponds and streams as compared with deep rivers or lakes and, in part, to their virtual absence from the open sea.

Williams and Feltmate (1992) stated that, "One of the most fascinating questions concerning the distribution of insects is: given the unparalleled success of insects in both terrestrial and freshwater environments, why are they so poorly represented in the sea?" While there are over 30,000 species in fresh water, there are only several hundred that can be called marine. Cheng (1985) records 14 orders and 1400 species of insects in brackish and marine habitats, but only the genus *Halobates* (Hemiptera: Gerridae) in the open seas. One of the most widely accepted theories is that successful resident marine invertebrates evolved long before aquatic insects and occupy many of the same niches as do insects in fresh waters; thus, marine invertebrates have barred many insects from marine habitats by competitive exclusion (e.g., Hynes 1984). Williams and Feltmate offer an interesting alternative viewpoint that since most marine insects are found in "bridging habitats" (e.g., estuaries, salt marshes, the intertidal zone, mangrove swamps), perhaps "...we are currently looking not at the end of an evolutionary pathway leading freshwater insects into marine habitats, but at early steps in the journey." However, this view indicates an extremely slow journey in the last several hundred million years, and the fact remains that the only taxon reported for the open seas, *Halobates*, occurs in a habitat (surface film) occupied by few, if any, marine macroinvertebrates. Some Porifera, Cnidaria, and polychaetes, which are typically considered to be marine, have representatives that have also successfully colonized freshwater habitats (Thorp and Covich 1991).

ADAPTATION TO HABITAT

Osmoregulation

Aquatic insects need to maintain a proper internal salt and water balance. Their body fluids usually contain a much higher salt concentration than does the surrounding water and water tends to pass into the hypertonic (higher osmotic pressure) hemolymph. The insect integument, especially the wax layer of the epicuticle, appears to be especially important in preventing flooding of the tissues (Chapman 1982). Some freshwater insects take in large quantities of water during feeding. Their feces contain more water than the frass of terrestrial counterparts since many aquatic insects excrete nitrogenous wastes as ammonia, which is generally toxic unless diluted with large quantities of water (Chapman 1982). The production of a hypotonic urine (lower osmotic pressure, or more dilute than the body fluids) is an important osmoregulatory mechanism in aquatic insects, accomplished by specialized areas of the hindgut that reabsorb ions before wastes are eliminated.

In contrast, saltwater and terrestrial insects conserve water by producing a rectal fluid that is hypertonic to the hemolymph (Stobbart and Shaw 1974). For example, the dipteran *Ephydra cinerea,* which occurs in Great Salt Lake (salinity > 20% NaCl), maintains water and salt balance by drinking the saline medium and excreting rectal fluid that is more than 20% salt.

Concentrations of freshwater ions vary tremendously. Many insects absorb salts directly from the surrounding water by active transport across specialized regions of the body. Such regions include the rectal gills of the mosquito, *Culex pipiens*; the rectal gills are larger in larvae reared in water with low ionic concentrations, which increases the surface area available for absorption of chloride ions (Wigglesworth 1938). Specialized areas of the integument, including the gills and anal papillae in larvae of Ephemeroptera, Plecoptera, and Trichoptera, facilitate uptake of ions from the hypotonic external media (Wichard and Komnick 1973, 1974; Wichard *et al.* 1975; Wichard 1976, 1978). In the mayfly *Callibaetis* sp. the number of cells involved with chloride uptake decreases as the salinity of the environment increases, an adaptation to the increase in salinity of drying temporary ponds (Wichard and Hauss 1975; Wichard *et al.* 1975).

Temperature

Virtually all facets of life history and distribution of aquatic insects are influenced by temperature (see Sweeney 1984, Table 4.1; Ward 1992). It is an important factor determining the distribution, diversity, and abundance patterns over elevation gradients in lentic and lotic waters (Ward 1992), together with flow regimes that are also very important (Statzner and Higler 1985; Statzner *et al.* 1988). Metabolism, growth, emergence, and reproduction are directly related to temperature, whereas food availability, both quantity and quality, may be indirectly related through associated microbial activity (Anderson and Cummins 1979). The diel cycle of temperature also may be important: for example, the fecundity of *Wyeomyia smithii*, the pitcherplant mosquito, was greater by a factor of seven when exposed to fluctuating as opposed to constant temperatures (Bradshaw 1980).

A few insects occur in waters to almost 50°C, while others can survive frozen in the substrates at the bottom of arctic ponds where temperatures may be as low as -20 to -30°C. In contrast, many taxa inhabiting Alaskan streams cannot survive even moderate freezing temperatures of -1°C; it appears that these species move away from the freezing zone or remain in areas that do not freeze (Irons *et al.* 1992). Chironomids and empidids (Diptera) constitute over 90% of the frozen individuals found in frozen habitats of Alaskan streams (Irons *et al.* 1992).

The thermal death point of most freshwater invertebrates is between 30 and 40°C (Pennak 1978), so species such as the stratiomyiid fly, *Hedriodiscus truquii*, which tolerates temperatures to 47°C in western North America (Stockner 1971), and the ephydrid fly, *Scatella thermarum*, found at 47.7°C in Icelandic hot springs (Tuxen 1944), have developed considerable thermal acclimation. In a general review of thermal spring insect fauna, Pritchard (1991) concluded that no insects live above 50°C, few above 40°C, and both temperature and unusual water chemistry may exclude certain species.

In contrast to the limited number of species found at high temperatures, a diverse fauna exists where water is near freezing, and many lotic species can grow at winter temperatures. Hynes (1963) and Ross (1963) suggest that this is an adaptation to exploit the seasonal pulse of leaf input in the autumn. The ancestral aquatic habitat of many insects is postulated to be cool streams; Hynes (1970a, 1970b) comments that the extant taxa of Plecoptera, Ephemeroptera, Trichoptera, Corydalidae, and nematocerous Diptera occur in cool streams and these are survivors of primitive groups.

Chironomids overwintering as larvae in bottom mud are the dominant taxon in arctic ponds that freeze to the substrate. Danks (1971b) indicates that overwintering is a distinct and important part of the life cycle. The required resistance to the rigors of climate occurs in more than one instar. Larvae exhibit physiological resistance to freezing and many are protected by a winter cocoon. Danks suggests that robust cocoons are important because they can withstand the forces exerted in mud when it freezes and water expands to form ice.

Shallow lentic waters will generally reach higher summer temperatures than do streams of the same surface area, which can result in a greater algal food supply and faster insect growth rates. However, oxygen may become a limiting factor because O_2 concentration is inversely proportional to temperature and high algal respiration during darkness may deplete the available O_2. Thus, a greater proportion of lentic than lotic species utilizes atmospheric O_2 or has developed other more efficient respiratory devices.

Life cycle adaptations have evolved that enable species to utilize favorable periods for growth, coupled with appropriate timing for aerial existence. This may involve diapause or quiescent periods in a resistant life stage for avoiding excessively high or low temperatures. Asynchronous larval cohorts and extended emergence periods occur in springs, where temperature is uniform year around.

Sweeney and Vannote (1978) and Vannote and Sweeney (1980) suggest that an optimal thermal regime exists for a given species and that deviations into warmer (southern) or cooler (northern) waters adversely affect fitness by decreasing body size and fecundity. Alteration of thermal regimes, for example by removal of riparian vegetation or by hypolimnetic release from dams, will obviously affect insect life cycles or species composition (Lehmkuhl 1972a; Ward 1976).

Lotic Habitats

Substrates may range from bedrock and large boulders to fine sediments and be interspersed with large woody debris in a relatively short reach, resulting in a large range of microhabitats. Substratum particle size is influenced by several items, including: geologic structure; the influence of past and present geomorphic processes (e.g., flowing water, glaciation, slope, etc.); and, length of time over which the processes occur. These in turn influence landform which exerts a major influence on various hydrologic characteristics of aquatic habitats (e.g., Newbury 1984). In addition to influencing substratum particle size, velocity of moving water exchanges water surrounding the body and turbulence provides reaeration; thus, dissolved oxygen is rarely limiting to stream inhabitants. Local transport and storage of inorganic and organic materials by the current may be either detrimental (e.g., scouring action) or beneficial (as a food source).

Substratum characteristics are often perceived as a major factor contributing to the distribution of many invertebrates; however, Statzner et al. (1988) suggest that these may be less important than velocity and various key three-dimensional hydraulic characteristics (longitudinal, vertical, and lateral) associated with different substrates. Flow characteristics important to stream ecology have been discussed by Vogel (1981), Newbury (1984), and Gordon et al. (1992). These works are especially useful with respect to stream invertebrates, as the distribution of organisms and their food resources are strongly influenced by local flow conditions.

Friction at the substratum surface results in a velocity gradient or a "boundary layer" adjacent to the substratum surface (Ambühl 1959; Hynes 1970a; Statzner et al. 1988). The boundary layer is regarded as the zone where a sharp velocity reduction prevails close to the substratum (Statzner et al. 1988). Boundary layer thickness is difficult to quantify because of complex substratum profiles, but it generally extends for 1-4 mm above the substratum surface.

Vogel (1981) pointed out that almost all of the information on stream insects prior to the 1980s related to the boundary layer is mainly anecdotal and few, if any, measurements of drag have been made "under reasonable or unreasonable conditions" and the absence of such measurements make conclusions with regard to shape adaptations of torrential fauna somewhat tenuous. Although the shape of dorsoventrally-flattened insects has been proposed as a mechanism to reduce drag and allow them to stay attached in the boundary layer, Vogel (1981) suggests the idea may be an over simplification as a convex upper surface would generate lift. Statzner and Holm (1982, 1989) reached similar conclusions using Laser Doppler Anemometry. Flows in the vicinity of organisms are often characterized by Reynolds number (a measure which relates to laminar and turbulent flows). Undisturbed flows immediately upstream of the organism, the kinematic viscosity, and the shape of the organism or its case determine the organism's Reynolds number and the degree to which various physical forces influence the organism (Vogel 1981). At low Reynolds number flows tend to be laminar and viscous shear stresses predominate. As the Reynolds number increases there is a transition from laminar to turbulent flows and drag forces predominate. Furthermore, the relationship becomes much more complicated for individual organisms since these forces vary as a function of size between small and large individuals of the same species (Statzner et al. 1988). Dorsoventrally-flattened larvae of water pennies (Coleoptera: Psephenidae) have been reported to reduce turbulence over the larva's body by active pumping of water through lateral slots of the carapace, which may decrease drag forces on the larva at high Reynolds numbers (Smith and Dartnall 1980; McShaffrey and McCafferty 1987). However, McShaffrey and McCafferty questioned the significance of the flattened body-shape of *Psephenus herricki* as an adaptation to fast currents, but rather as an adaptation to avoid the current since individuals are often found in crevices and undersides of stones.

In contrast to a flattened body, a fusiform, streamlined, and less flattened shape may be a mechanism to avoid excessive lift, whereas flattening may be viewed as a mechanism of maintaining the body within the boundary layer and increasing the area of contact with the surface substratum (Vogel 1981) or avoiding the current by occupying crevices. Based on hydrodynamic studies of streamlined larvae of the mayfly, *Cloeon dipterum*, Craig (1990) has proposed that streamlining enhances accelerative escape motion and "highly streamlined cylindrical macroinvertebrates could be expected to have accelerative motion in their behavioral repertoire". Considering the diverse array of physical forces acting on insects, it is not surprising that no one taxon would have the "perfect" collection of adaptations to all forces as this would require incredibly different morphologies (Statzner et al. 1988, Statzner and Holm 1989). Wilzbach et al. (1988) related body shape and drag to modes of attachment and movement and to functional feeding modes. Indeed, one wonders to what extent the diverse physical forces influence the array of morphologies found among aquatic insects.

The range of current velocities associated with various substrates also increases habitat diversity, and various taxa are adapted for maintaining position at different velocities. Filter-feeding collectors exploit the current for gathering food with minimal energy expenditure. For example, certain filtering collectors exploit locations where flows converge over and around substrates, thus allowing the animals to exploit sites of greater food delivery (Smith-Cuffney and Wallace 1987; Wetmore et al. 1990). Such areas can be characterized by Froude number, a dimensionless parameter that combines both the depth at which flow is measured and velocity in one term (the ratio of inertial force to the force of gravity; Chow 1959; Newbury 1984). Flows are generally faster and the water column's depth is compressed as Froude number approaches or exceeds 1, whereas Froude numbers much less than 1 are found in deep riffles and pools. In a detailed analysis of the caddisfly, *Brachycentrus occidentalis* Banks, Wetmore et al. (1990) found that Froude number was a much better predictor of microhabitat than individual depth or velocity measurements.

Morphological and Behavioral Adaptations to Current. A general flattening of the body and smooth, streamlined dorsum are typical of many rheophilic (current-loving) insects: e.g., heptageniid mayflies, perlid stoneflies, and some psephenid beetles. Many mayflies and stoneflies have legs that project laterally from the body, thereby reducing drag and simultaneously increasing friction with the substrate. In some caddisflies (e.g., Glossosomatidae), the shape of the case rather than the insect modifies turbulent flow to a laminar sublayer.

True hydraulic suckers are apparently found only in the larvae of the dipteran family Blephariceridae (Fig. 22.11). A V-shaped notch at the anterior edge of each of the six ventral suckers works as a valve out of which water is forced when the sucker is pressed to the substrate. The sucker operates as a piston with the aid of specialized muscles. In addition, a series of small hooks and glands that secrete a sticky substance aid sucker attachment (Brodsky 1980). Blepharicerids move in a "zigzag" fashion, releasing the anterior three suckers, lifting the front portion of the body to a new position, and reattaching the anterior suckers before releasing and moving the posterior ones to a new position. The larvae are commonly found on smooth stones, and Hora (1930) attributes their absence from certain Indian streams to the presence of moss or roughened stones that would interfere with normal sucker function.

Several aquatic insects have structures that simulate the action of suckers. The enlarged gills of some mayflies (e.g., *Epeorus* sp. and *Rhithrogena* sp.) function as a friction pad, and *Drunella doddsi* has a specialized abdominal structure for the same purpose. Brodsky (1980) describes a "pushing-proleg" in some chironomids; it has a circlet of small spines that function as a false sucker when pressed to the substrate. Mountain midge larvae (Deuterophlebiidae) possibly use a similar mechanism to attach their suckerlike prolegs.

Larval black flies (Simuliidae) use a combination of hooks and silk for attachment. The thoracic proleg resembles that of chironomids and deuterophlebiids, described above, and the last abdominal segment bears a circlet of hooks. The larva spreads a web of silk on the substrate to which it attaches either the proleg or posterior hooks. The larva moves forward in an inchworm-like manner, spins silk over the substrate, and attaches the proleg and then the posterior circlet of hooks to the silken web.

Silk is used for attachment by a number of caddisflies (e.g., Hydropsychidae, Philopotamidae, and Psychomyiidae), which build fixed nets and retreats. Some case-making caddisflies (e.g., *Brachycentrus* sp.) also use silk for attaching their cases to the substrate in regions of fairly rapid flow, and free-living caddisflies may use "security threads" as they move over the substrate in fast currents. The silk line is used in combination with their large anal prolegs, which are employed as grapples. Many chironomid larvae construct fixed silken retreats for attachment, and black fly pupae are housed in silken cases that are attached to the substrate. Other morphological adaptations to running water are given in Table 8A and are listed for each genus in the ecological tables.

Despite the fact that unidirectional current is the basic feature of streams, the majority of lotic insects have not adapted to strong currents but instead have developed behavior patterns to *avoid* current. Very few lotic insects are strong swimmers, probably because of the energy expenditure required to swim against a current; downstream transport requires only a movement off the substrate to enter the current. Streamlined forms, such as the mayflies *Baetis* sp., *Isonychia* sp., and *Ameletus* sp., are capable of short rapid bursts of swimming, but most lotic insects move by crawling or passive displacement. The benthic fauna chiefly occurs in cracks and crevices, between or under rocks and gravel, within the boundary layer on surfaces, or in other slack-water regions. Presumably, some portion of the benthic community seeks refuge deeper in the substrates during floods; insects are difficult to find at such times, but normal population levels are found soon after the flows subside (Williams 1984).

The *hyporheic* region is the area below the bed of a stream where interstitial water moves by percolation. In gravelly soils or glacial outwash areas it may also extend laterally from the banks. An extensive fauna occurs down to one meter in such substrates (Williams and Hynes 1974; Williams 1984). Most orders are represented, especially those taxa with slender flexible bodies or small organisms with hard protective exoskeletons. Stanford and Gaufin (1974) report that some stoneflies spend most of their larval period in this subterranean region of a Montana River. They collected these larvae in wells over 4 m deep, located 30-50 m from the river.

Drift. Downstream drift is a characteristic phenomenon of invertebrates in running waters. Despite the adaptations for maintaining their position in the current or avoiding it, occasional individuals could be expected to lose attachment or orientation and be transported downstream. However, the large numbers of some taxa that drift indicate that this is more than a passive activity. Waters (1965) divided drift into three categories: (1) *catastrophic*, resulting from physical disturbance of the bottom fauna, e.g., by floods, high temperatures, and pollutants; (2) *behavioral*, indicated by characteristic behavior patterns resulting in a consistent diel periodicity (usually at night); and (3) *constant*, the continual occurrence of low numbers of most species. Mayflies of the genus *Baetis* consistently exhibit high behavioral drift rates with a night-active periodicity. Other mayflies, stoneflies, caddisflies, black flies, and the amphipod *Gammarus* sp. are frequently abundant in drift. Food resources may also influence drift as many more individuals of *Baetis* sp. (Ephemeroptera) drifted from an unfertilized reach of the Kuparuk River in Alaska, compared to a downstream fertilized reach where food was more abundant (Hershey *et al.* 1993). However, Wilzbach (1990) reported nonconcordance between drift and benthic feeding activity in *Baetis*. Drift is important to stream systems in the recolonization of denuded areas, as a dispersal mechanism, and particularly as a food source for visual predators. Many fish, especially salmonids, select and defend territories best suited for the interception of drift (Waters 1972).

Irrespective of its causes, drift results in a net downstream displacement of some portion of the benthic population. Whether drift losses from upstream areas represent excess production or whether compensatory upstream movements are required is not known. Müller (1954) proposed that upstream flight of adults could be the mechanism to complete the "colonization cycle," and upstream migrations of some mayflies and amphipods in the slow water near shore have been recorded (e.g., Neave 1930; Minckley 1964; Hayden and Clifford 1974). Using stable isotopes, Hershey *et al.* (1993) estimated that one-third to one-half of the adult *Baetis* sp. population in the Kuparuk River flew over 1.6 km upstream from their emergence sites. During three years of seasonal insecticide applications to a headwater stream, colonization by early instars of several taxa showed seasonal occurrences that closely paralleled known life cycles and adult flight periods (Wallace *et al.* 1991). However, under less adverse conditions the relative importance of upstream movement, or even its necessity for most taxa, remains an open question. For example, Wilzbach and Cummins (1989) showed that in the short term, recruitment of immatures made it impossible to detect any population depletion due to drift. Several workers (Bishop and Hynes 1969; Waters 1972; Müller 1974; Williams 1981a; Brittain and Eikeland 1988) have reviewed the extensive literature on the significance of drift to production biology, population dynamics, and life histories.

Unstable Substrates. Sandy substrates of rivers and streams are poor habitats because the shifting nature of the bed affords unsuitable attachment sites and poor food conditions. An extreme example of this instability is the Amazon River, where strong currents move bedload downstream as dunes of coarse sand reaching 8 m in height and up to 180 m in length, largely preventing the establishment of a riverbed fauna (Sioli 1975). Despite substrate instability, some sandy streams are quite productive. Blackwater streams of the Southeast have extensive areas of sand with an average standing stock, primarily small Chironomidae less than 3 mm in length, exceeding $18,000/m^2$. Though their biomass is small, rapid growth rates result in a significant annual production and an important food source for predaceous invertebrates and fish (Benke et al. 1979).

The inhabitants of sandy or silty areas are mostly sprawlers or burrowers (Table 6B), with morphological adaptations to maintain position and to keep respiratory surfaces in contact with oxygenated water. The predaceous mayflies *Pseudiron* sp. and *Analetris* sp. have long, posterior-projecting legs and claws that aid in anchoring the larvae as they face upstream. Some mayflies (e.g., Caenidae, Tricorythidae, and Baetiscidae) have various structures for covering and protecting gills, and others (e.g., Ephemeridae, Behningiidae) have legs and mouthparts adapted for digging. The predaceous mayfly *Dolania* sp. burrows rapidly in sandy substrates of Southeastern streams. The larva utilizes its hairy body and legs to form a cavity underneath the body where the ventral abdominal gills are in contact with oxygenated water.

Many dragonflies (e.g., *Cordulegaster* sp., *Hagenius* sp., Macromiidae, and many Libellulidae) have flattened bodies and long legs for sprawling on sandy and silty substrates. They are camouflaged by dull color patterns and hairy integuments that accumulate a coating of silt. The eyes, which cap the anteriolateral corners of the head, are elevated over the surrounding debris. Many gomphid larvae actually burrow into the sediments using the flattened, wedge-shaped head, and fossorial tibiae. The genus *Aphylla* (Gomphidae) is somewhat unusual in that the last abdominal segment is upturned and elongate, allowing the larvae to respire through rectal gills while buried fairly deep in mucky substrate.

Wood-Associated Insects. Wood debris provides a significant portion of the stable habitat for insects in small streams where water power is insufficient to transport it out of the channel. In addition to the insect component using wood primarily as a substrate, a characteristic xylophilous fauna is associated with particular stages of degradation. These include: chironomid midges and scraping mayflies (*Cinygma* sp. and *Ironodes* sp.) as early colonizers; the elmid beetle, *Lara avara*, and the caddisfly, *Heteroplectron californicum*, as gougers of firm waterlogged wood; chironomids as tunnelers; and the tipulids, *Lipsothrix* spp., in wood in the latest stages of decomposition (Anderson et al. 1978, 1984; Dudley and Anderson 1982; Anderson 1989). Wood debris is most abundant in small forested watersheds, but it is also an important habitat at the margins and on point bars in larger streams with unstable beds. Cudney and Wallace (1980) found that submerged wood in the Coastal Plain region of the Savannah River was the only substrate suitable for net-spinning caddisflies, but that high standing crops and production could be supported in a relatively small space because the caddisflies were exploiting the food resource transported to them by the current. Benke et al. (1979) reported that snags in a small Southeastern blackwater river were highly productive, not only for net-spinning caddisflies, but also for filter-feeding Diptera and other typical "benthic" insects. Wood gouging habits of net-spinning caddisflies were blamed, at least in part, for the 1988 collapse of submerged timber pilings supporting a Maryland bridge (National Transportation Safety Board 1989). Even in the Amazon situation mentioned above, large logs in the lee of dunes are heavily colonized by chironomid midges (Sioli 1975).

Lentic Habitats

Standing-water habitats range from temporary pools to large, deep lakes. They tend to be more closed than are lotic environments, with recycling occurring within the lake basin. The physical environment is governed by the climate and geology of the area and the shape of the basin. The habitats for insects are illustrated in a typical cross section from the surface film, through open waters, to the shallow and deep benthos (Fig. 5.1). Similar habitats occur in lotic situations so many of the insects discussed below can also be found in lotic environments. Merritt et al. (1984) provide additional comparisons of

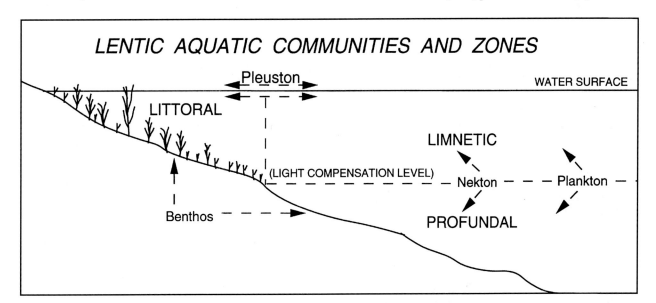

Figure 5.1. Diagram of lentic zones (*capital letters*) and aquatic communities (*lower case letters*).

lotic and lentic ecosystems and nutritional resources available to aquatic insects in each habitat.

Surface Film. The unique properties of the water surface constitute the environment for the *pleuston* community. Water striders (Gerridae), whirligig beetles (Gyrinidae), mosquito larvae (Culicidae), and springtails (Collembola) are common examples. The surface film results from the attractive forces among water molecules. Within the body of water the forces are equal on all sides, but at the surface the attraction is less between air and water than on the other three sides. This results in a slight pull toward the center of the water mass and the surface acts as if it were a stretched elastic membrane. "To an organism of small size, this air-water interface can be an impenetrable barrier, a surface on which to rest, or a ceiling from which to hang suspended" (Usinger 1956a).

Water striders and bugs of the related family Veliidae have preapical claws that enable them to move about without breaking the surface film and a velvety hydrofuge hair pile on the venter that is nonwettable. The skating motion of gerrids is more accurately described as rowing. The weight causes a slight dimpling of the surface film (easily seen in the shadow cast on the bottom of clear shallow waters). The long tarsi of the middle legs are pressed against the depression in the film for propulsion while the hind legs are held outstretched behind as rudders for steering.

The gerrids also exploit the surface film by detecting vibrations from surface ripples with sensors located between the tarsal segments on the meso- and metathoracic legs. The sequence in which these sensors perceive an oncoming ripple tells the insect how much to turn to face the disturbance, caused for example, by a potential prey. The strider turns by moving its rowing legs in opposite directions much as an oarsman would turn a boat. At this juncture, a combination of visual and vibratory information allows the strider to choose between approaching the disturbance or fleeing from it. If the former, it rows forward, pausing between each stroke to evaluate further ripple signals (Milne and Milne 1978). Some gerrids also use vibrations of the surface film for communication in courtship and mating (Wilcox 1979).

Whirligig beetles have divided eyes; the lower part detects events under the water and the upper part detects events on or above the water surface. Glands keep the upper portion of the body greased to repel water, whereas the lower surface and lower eyes are wettable. Whirligig gyrations on the surface are effected by rapid propulsion with paddle-shaped swimming legs. The hindlegs are unique in that the segments can be folded up during the forward stroke and spread apart like a fan for the powerful backward stroke. In addition to propulsion, the swimming activity causes a series of ripples or bow waves that, when reflected back by obstacles, are detected by the antennae touching the surface film, allowing quick course correction even in the dark (Milne and Milne 1978).

Pleuston insects are potentially vulnerable to predation by visual aquatic predators because they are silhouetted against the sky; however, both gerrids and gyrinids seem to be avoided by fish, apparently because secretions produced by repugnatorial glands make them distasteful (e.g., Blum 1981).

Hydrofuge structures are important adaptations for obtaining atmospheric air through the surface film. The openings of terminal spiracles have glands that discharge a waxy secretion on the cuticle; contact with the surface film repels the water exposing the spiracles to the air. The waxy lining of the tracheae prevents water from running into them by capillarity. Some insects (e.g., *Culex* sp., *Limonia* sp., *Stratiomys* sp.) can hang from the surface film supported by a crown of semihydrofuge hairs around the spiracles. The semihydrofuge hair will lie in the water surface, half wetted and half exposed. When a stratiomyiid larva submerges, the crown is lifted off the water and envelops an air bubble that is used as an air store.

Feeding adaptations associated with pleuston specialization are exemplified by mosquito larvae. Though the mouthparts are basically of the primitive chewing type, the larvae have brushes arising from the labrum that sweep floating or suspended material toward the mouth. *Anopheles* sp. larvae lie beneath the surface film supported by tufts of float hairs on each segment. The larva rotates its head so that the mouth brushes are uppermost and sets up a current that draws the microbial-rich layer of water along the underside of the surface film and into the mouth (Merritt *et al.* 1992).

Limnetic Zone. In open waters to the depth of photosynthetically-effective light penetration, a broad distinction is made between *nektonic* and *planktonic* organisms: nekton are swimmers able to navigate at will (e.g., Coleoptera, Hemiptera, some Ephemeroptera), whereas plankton are floating organisms whose horizontal movements are largely dependent on water currents.

The phantom midge *Chaoborus* sp. is the most common insect plankter; it is abundant in many eutrophic (nutrient-rich) ponds, lakes, and some large rivers. *Chaoborus* sp. exhibits vertical migrations, occurring in benthic regions during the day but migrating vertically into the water column at night. These migrations are dependent on light and oxygen concentrations of the water (LaRow 1970). Larvae avoid predation by being almost transparent except for two crescent-shaped air-sacs or buoyancy organs (Fig. 22.30); they lie horizontally in the water, slowly descending or rising by adjusting the volume of the air-sacs. Their prehensile antennae are used as accessory mouthparts to impale zooplankton and deliver them to the mouth.

Many lentic insects are strong swimmers but relatively few are nektonic. They pass through the limnetic zone when surfacing for emergence, but the vast majority of lentic insects occurs in shallow water with emergent plants. As mentioned previously, the scarcity of insects in limnetic areas may be a consequence of the secondary adaptations for aquatic life. There are no resting supports in the limnetic zone so maintaining position requires continuous swimming or neutral buoyancy.

Littoral Zone. The littoral zone, the shallow region with light penetration to the bottom, is typically occupied by macrophytes (macroalgae and rooted vascular plants). It contains a diverse assemblage of insects with representatives of most aquatic orders. Habitats include benthic and plant surfaces, the water column, and the surface film (Table 6A); occupants include burrowers, climbers, sprawlers, clingers, swimmers, and divers (Table 6B). Morphological and behavioral types in the littoral zone or floodplain wetlands of larger rivers are similar to those in slow-moving or backwater regions of lotic habitats. The diversity and abundance of littoral species result in biological factors (e.g., competition and predation) assuming importance in shaping community structure.

The biomass and diversity of invertebrates associated with aquatic macrophytes in lentic or lotic habitats may exceed that of the fauna in the sediments at the same location. The impact of herbivorous insects on many living plants has previously been considered to be low, and it has been suggested that aquatic macrophytes produce secondary plant substances that serve as chemical defenses against herbivores (Otto and Svensson 1981b), or that they may be deficient in some essential amino acids (Smirnov 1962). However, the nitrogen content of macrophytes is similar to that of terrestrial plants and there is now considerable evidence to suggest that herbivory on living macrophytes may be much more important than previously suspected (Lodge 1991; Newman 1991). Herbivore-chewers (shredders-herbivores), miners, and stem borers (see Chap. 6) feed on macrophytes. Floating leaf macrophytes such as water lilies (*Nuphar*) can be heavily consumed by larvae and adults of the chrysomelid, *Galerucella* sp. (Wallace and O'Hop 1985). Some aquatic herbivores include pests of economic importance such as the weevil *Lissorhoptrus oryzophilus* (=*simplex*) (Curculionidae) on rice (Isely and Schwardt 1934), and the caddisfly *Limnephilus lunatus* (Limnephilidae) on watercress (Gower 1967). Newman (1991) found that most herbivory on aquatic macrophytes is usually by specialized oligophagous insects belonging to families and orders that are secondary invaders of aquatic habitats and closely related to pest herbivores in terrestrial habitats (e.g., beetles such as Chrysomelidae and Curculionidae, aquatic Lepidoptera, and specialized Diptera), whereas herbivores from primarily aquatic groups were generalists and also detritivores.

Insects also use macrophytes as substrates for attachment or a surface from which to graze periphyton, rather than directly as food. For example, macrophytes in both lentic and lotic habitats often harbor a number of filter-feeding Ephemeroptera, Trichoptera, and Diptera. Many species found in weed beds are green in color and blend in with their surroundings. Dragonflies of the family Aeshnidae often have contrasting bands of pale and dark green or light brown that add to the effectiveness of the camouflage. Some species in most orders utilize macrophytes as oviposition sites. Larvae of some Coleoptera, e.g., the chrysomelid *Donacia* sp., and Diptera such as the mosquito *Mansonia* sp. and the aquatic syrphid, *Chrysogaster* sp., rely on the intracellular air spaces of aquatic plants for respiration and are thus limited in their distribution by that of their macrophyte host.

Profundal Zone. The number of taxa of aquatic insects that occur below ca. 10 m is limited, but the few species that do occur there can be very abundant. This area includes the sublittoral and profundal regions. The latter is the zone below which light penetration is inadequate for plant growth. The deep water is a stable region because water movement is minimal and temperature varies only slightly between summer and winter. Periodic depletion or absence of dissolved oxygen may occur, especially in eutrophic situations. Substrates are usually soft or flocculent and offer little in the way of habitat diversity or cover. The inhabitants are mostly burrowers that feed on suspended or sedimented materials and are capable of tolerating low dissolved oxygen or even anaerobic conditions. Typical deep-water insects are ephemerid mayflies (e.g., *Hexagenia* sp., *Ephemera* sp.) and many genera of Chironomidae (e.g., *Chironomus, Heterotrissocladius, Tanytarsus*). Predaceous deep-water insects include *Sialis* sp. (Megaloptera) and *Chaoborus* sp. (Diptera).

Dense populations of the midge genus *Chironomus* are characteristic of profundal sediments. The larvae build U-shaped tubes with both openings at the mud-water interface. Body undulations cause a current of water, providing oxygen and particulate food in the form of phytoplankton and fine detritus with its accompanying microbes to be drawn through the tube. Hemoglobin serves as an oxygen store during periods of low dissolved oxygen, but the larvae become quiescent under anaerobic conditions. The life cycle of profundal *Chironomus* typically requires two years, compared with a year or less for the same species in shallow waters. This is due to the slow growth at cold temperatures, low food quality, and extended periods of quiescence when the water is anoxic (Jonasson and Kristiansen 1967; Danks and Oliver 1972).

The profundal chironomid community has been used extensively as an indicator of the nutrient conditions or productivity status of lakes. Saether (1980b) lists about twenty genera of chironomids that are characteristic of particular oxygen-nutrient-substrate combinations, which, in conjunction with other non-insect assemblages, can be used in lake typology classifications. Warwick (1980) documented over 2500 years of land-use practices around Lake Ontario using the subfossil chironomid head capsules in a sediment core; changes in species composition of midges were associated with various periods of eutrophication, deforestation, sedimentation, and contamination.

LIFE CYCLE ADAPTATIONS

Diverse life-history patterns have evolved to enable species to exploit foods that are seasonally available, to time emergence for appropriate environmental conditions, to evade unfavorable physical conditions (e.g., droughts, spates, and lethal temperature), and to minimize repressive biotic interactions such as competition and predation. Typically, a seasonal succession of species can be cataloged at a given location by determining the emergence and flight periods of adults or by studying the larval growth periods. However, comparison of a given species at different sites may indicate considerable flexibility in life histories. Some important adaptations include: (1) responses to temperature and oxygen levels; (2) use of daylength or temperature as environmental cues to synchronize life stages; (3) resting stages to avoid unfavorable conditions; and (4) extended flight, oviposition, or hatching periods to spread the risk in coping with environmental conditions that cannot be avoided entirely. Kohshima (1984) reported an extreme case of adaptation of a *Diamesa* sp. (Chironomidae) inhabiting glacial melt water channels above 5200 m in the Himalayan Mountains. The location of these channels, in which the univoltine species develop, is unpredictable from year to year. Koshima suggested the mated females, which remain active at a temperature of -16°C, overwinter under snow and ice and are ready to oviposit when the new channels develop during the following summer.

All life stages of mayflies exhibit adaptations to warmer waters compared with stoneflies. Egg development in mayflies has higher thermal requirements and is more temperature dependent than in stoneflies (Brittain 1990). A number of mayflies have short life cycles whereas many stoneflies require more than one year to complete their development. Adult Ephemeroptera mate in flight, which requires higher air temperatures than for Plecoptera which generally remain in the vicinity of the larval habitat and mate on solid substratum (Brittain 1990). Such mating behavior may explain, in part, the greater extension of plecopterans to arctic regions, although this involves a trade off between disperal ability and reproductive success in harsh environments. Brittain (1990) has suggested that plecopterans are relatively temperature independent and adapt to alternative food resources, whereas mayflies are more thermally dependent and readily use diatoms, algae, etc. to maintain shorter life cycles in warmer waters. Thus, plecopterans have a competitive advantage at cooler temperatures whereas warmer thermal regimes favor mayflies (Brittain 1990).

The duration of aquatic insect life cycles ranges from less than two weeks (e.g., some Baetidae, Tricorythidae, Culicidae and Chironomidae) to several years (e.g., Elmidae and some Odonata), but in the north temperate zone an annual cycle is most common. Examples of aquatic insect life cycles are given in Table 5B. We have used the terms univoltine, bivoltine, trivoltine, and multivoltine to indicate those taxa reported to complete their life cycles in a year or less; semivoltine for those completing their life cycle in 2 years; and, merovoltine for those that require more than two years from egg to adult. Although many taxa reported in Table 5B are univoltine, larval growth and development may occur over a very short period, e.g., a few weeks or months. For example, a number of univoltine coleopterans require only a month or so to complete their larval development.

Life histories of insects that spend all or a portion of their life history in marine environments are not included in Table 5B. Cheng and Frank (1993) have provided a summary of current information on life history attributes and reproduction in marine insects.

Many species of mayflies maintain univoltine life cycles over broad latitudes in eastern North America. There is a question as to how voltinism and seasonal synchrony are maintained across broad latitudinal ranges and thermal regimes where timing of adult emergence and oviposition may be restricted to one- or two-week periods. To examine this question, Newbold *et al.* (1994) used daily water temperatures and maximum and minimum threshold temperatures for larval development in order to construct a temperature based model for several mayfly species. The model incorporated periods of quiescence associated with both high and low temperatures for 11 river systems ranging from the southeastern United States to Quebec. The model simulations indicate that most mayfly development occurs in the spring and autumn between 34°N and 50°N when thermal regimes are roughly similar at all latitudes (Newbold *et al.* 1994). At lower latitudes, there was a longer period of summer quiescence as the development rate dropped at higher temperatures. The model also predicted the earlier emergence observed at lower latitudes as well as a northward transition to a semivoltine life history for two of the species examined (Newbold *et al.* 1994).

Hynes (1970a) distinguished three main types of life cycles for insects in a temperate stream: slow seasonal, fast seasonal, and nonseasonal cycles. In seasonal cycles, a distinct change of larval size occurs with time, i.e., the progression of growth by cohorts can be discerned by periodic sampling of field populations. In nonseasonal taxa, individuals of several ages are present at all times. The three types are illustrated by three species of glossosomatid caddisflies, all collected from one emergence trap (Fig. 5.2).

Slow-seasonal cycles are common in cool streams and typified by many Plecoptera, some Ephemeroptera, and Trichoptera. Eggs may hatch soon after deposition, and larvae grow slowly, reaching maturity nearly a year later. In many species, an extended hatching period results in recruitment over several months. Larvae grow during winter and most species reach the adult stage early in the year.

Fast-seasonal cycles are those in which growth is rapid after a long egg or larval diapause or after one or more intermediate generations. The caddisfly *Agapetus bifidus* has an egg diapause of 8-9 months and a larval growth period of only 2-3 months (Fig. 5.2). Various fast-seasonal cycles reach full term in spring, early and late summer, and fall. Two or more fast cycles may be exhibited by the same species when rapid generations succeed one another, as in the mayfly *Baetis* sp. and the black fly *Simulium* sp. Individuals that grow rapidly at warm temperatures tend to be much smaller than those of the earlier, slow-growing generation (Hynes 1970b; Ross and Merritt 1978; Sweeney 1978; Cudney and Wallace 1980; Merritt *et al.* 1982; Georgian and Wallace 1983; Freeman and Wallace 1984).

In nonseasonal cycles, individuals of several stages or size classes are present in all seasons. This may be because the life cycle spans more than a year as in some large Plecoptera and Megaloptera, or because a series of overlapping generations occurs, or for other unexplained reasons. Chironomidae often exhibit nonseasonal patterns probably because two or more species have not been distinguished or because short life cycles result in overlapping cohorts.

Poorly synchronized life cycles would be expected to occur in situations where limiting factors for growth or reproduction (e.g., temperature, food, and moisture) are not seasonally dependent. Thus in the tropics the unvarying temperature and light conditions could result in continuous periods of growth and reproduction. It is somewhat surprising that aquatic insects in the southern hemisphere are predominantly of the nonseasonal type in contrast to the situation in the north temperate zone (Winterbourn 1974, 1978; Hynes and Hynes 1975; Corbet 1978; Towns 1981). Hart (1985) suggested that flexible and opportunistic life histories of southern hemisphere stream fauna may be a selective response to unpredictable, or to mild and uniform, environmental conditions.

Habitat Selection

On a gross level, habitat selection is primarily the province of the mated females as deposition of eggs determines where the larvae will initially occur. The chain of behavioral cues leading to oviposition has rarely been elucidated but habitat selection may involve visual, tactile, and chemosensory cues. Coleoptera and Hemiptera detect water while in daytime flight apparently from the reflective surface so they are attracted to ponds, swimming pools, and other shiny surfaces. For most aquatic insects, oviposition occurs near the location from which the adult emerged, but dispersal flights are common especially in species from temporary habitats. Species that overwinter as adults (e.g., Coleoptera, Hemiptera) may have a fall dispersal flight to overwintering sites, and also a spring flight for oviposition.

The widespread tropical dragonfly, *Pantala flavescens*

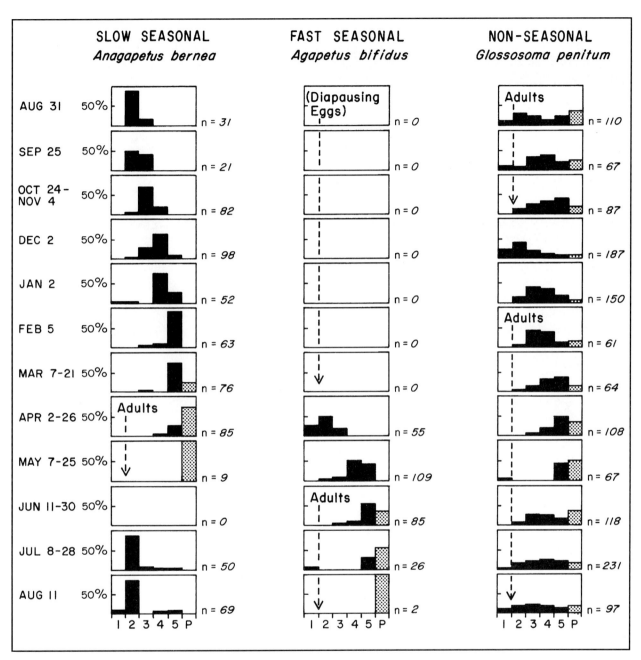

Figure 5.2. Age distribution of three glossosomatid caddisflies, illustrating life cycles. Field data are expressed as percentage composition per month for each instar. There are five larval instars; P = prepupa + pupa; n = number per sample. Flight period of adults is also indicated. (Data from Anderson and Bourne [1974].)

(Libellulidae), an obligate migrant that breeds in temporary ponds, exhibits long-distance dispersal in search of oviposition sites. The adults fly upward at emergence and are then transported by the wind to places where rain will later fall. Wind may transport the adults over 1400 km across the Indian Ocean toward the cold front produced by monsoons (Corbet 1963).

Caddisflies from temporary streams (*Hesperophylax* sp., *Grammotaulius* sp.) and ponds (*Limnephilus indivisus*) (Table 5A) deposit eggs in damp channels or depressions before the habitat fills with autumnal rains or snow. A similar selection of habitat before it is suitable for the larva occurs with snowpool mosquitos (*Aedes* sp.), which deposit diapausing eggs in depressions, and hatching occurs the following spring when eggs are wetted by the melting snow.

Males are responsible for habitat selection in many odonates; they establish territories over ponds or streams, depending on the species, and actively court females to oviposit within that locale. A male in tandem flight may even remain coupled while the female is ovipositing below the surface. Belostomatid bugs, which incubate eggs on the back of the male (Table 5A), are another instance of male influence on habitat selection. Movement by the male through swimming or

Table 5A. Summary of data for oviposition and egg stage of selected aquatic insects.

Taxon	Preoviposition Period	Oviposition Season	Oviposition Site	Oviposition Behavior
EPHEMEROPTERA				
Baetidae				
Baetis spp.	1 day (including subimago)	extended (spring-fall)	Shallow riffles, underside of stones	♀ folds wings along abdomen walks into water and selects site
Callibaetis floridanus	5-7 days (including subimago)	entire year	Lentic, often temporary habitats	♀ lands on water and extrudes milky mass; dies on site
Ephemeridae				
Hexagenia spp.	2-4 days (including subimago)	May-Sept	Lakes, large rivers	♀♀ plummet to water and extrude 2 egg packets; on contact with water, eggs separate and sink. Some ♀♀ can take off and repeat
Leptophlebiidae				
Leptophlebia cupida	2 days (including subimago)	mid May-early July	Mainstream of slow-moving rivers	♀ dips abdomen in water, releases a few eggs at a time; completed in 5 min; flies up or down stream, daytime only
Ephemerellidae				
Ephemerella ignita	2 days (including subimago)	late June-Sept	Turbulent streams, in moss	♀ flies upstream with extruded egg ball; contact with water releases ball, and eggs then separate
Heptageniidae				
Epeorus pleurialis	3 days (including subimago)	Mar-May	Streams, fast water	♀ touches water several times, washes eggs off in batches
ODONATA				
Lestidae				
Lestes congener	3 wk	Aug	Temporary ponds; only in dry stems of *Scirpus* sp., 5-30 cm above water	♂ + ♀ in tandem; eggs inserted singly in incisions 2 cm apart
Archilestes grandis	—	mid June-late July	Lentic; in branches or petioles up to 13 m above water	♂ + ♀ in tandem; oviposition during daytime. Endophytic oviposition; 2 eggs per incision in petioles but about 10 in pithy stems
Aeshnidae				
Anax imperator	11-14 days	early June-early Aug	Ponds; warmest area	♀ alights on floating plants and deposits eggs endophytically; egg may penetrate leaf; 1 egg per incision
Corduliidae				
Tetragoneuria spp.	1 wk	mid May-July (ca. 1 mo at a site) (individuals live 2 wks)	Lotic, ca. 10 or more feet from shore	♀ makes 2 passes to select a site; on 3rd pass she drags abdomen through water trailing egg string until it adheres to solid object. Several ♀♀ add to egg-string aggregations on same day

Table 5A. Continued

Description of Egg or Egg Mass	Number of Eggs	Incubation and Hatching Period	Comments	Geographic Area	Reference
contiguous rows of yellow, spherical eggs, form a flat semicircular plate	*B. rhodani*, winter generation up to 2500; summer generation up to 1200	*B. rhodani*, start of hatching 17 wks @ 3°C, 1 wk @ 22°C. Hatching interval 34 days @ 3°C, 3 days @ 22°C.	Hatching period is variable for species and site. Delayed hatch of some eggs for 40 wk	Europe North America	1050, 1910 661
viscous, reddish brown egg mass on venter of abdomen; mass is released and eggs separate when abdomen touches water	450-500	Hatch in 5-10 sec (ovoviviparous)	Adapted for temporary habitats. ♀ ♀ mate, and embryonic development complete before oviposition	Florida	288, 4049
eggs ellipsoid, .16-.19 X .28-.32 mm; sticky surface results in eggs adhering in clumps	2260-7684; x̄ = 4000 for average-size ♀ ♀ (24-25 mm)	11-26 days; 2 wk @ summer temp	Both low temp. and low D.O. will delay incubation (Fremling 1967)	Midwest	1303, 1871, 1309'
ovoid, .23 X .12 mm; eggs anchored by peg-like structures that spring out after wetting	x̄ (mean-sized ♀), 2959; * Fecundity, F = 2.02 L$^{3.04}$	Hatching started in 10-14 days @ 20°C; 50% emerge on day 1, hatching continues for 43 days	Latitude effect on oviposition period and fecundity. 1072-2065 eggs per ♀ in Penn.	Alberta Pennsylvania	663, 3921
greenish egg ball; eggs have polar anchoring mechanism to attach to substrate	eggs per mass, 156-603; x̄ = 322	112-392 days @ 8.2°C; 56-294 days @ 15.8°C. In stream, 10% hatch by 100 days, 90% by 300 days	Extended hatching period. Temp. > 14.5°C delays hatching; egg diapause in some populations	England, Lake District	1052
cream colored; ellipsoid, .08-.09 X .13-.15 mm; adhere firmly to substrate	2000-6000, x̄ = 4260 (dissected ♀ ♀)	Estimated 7 mo in field @ 11-14°C	Extended hatching period, with peak of small larvae in Feb	Kentucky	2716
elongate, 1.2 mm long; soft and greyish when deposited; cuticle then hardens and darkens	87-297 (x̄=205) mature oocytes; no. of egg clutches per ♀ not determined	Prediapause development for 1 wk; 3 mo diapause; 6-7 wk post-diapause; synchronous hatching (1 wk, in May)	Diapause occurs under snow cover, post-diapause embryogenesis triggered by snowmelt. Hatching threshold is 5°C	Saskatchewan	3549
white with dark anterior end; 1.9 X .3 mm	up to 149/♀; 70-180 (Smith and Pritchard 1956)	15 days; eggs hatch without wetting	At emergence, prolarva "jumps" to water; use of deciduous petioles for egg site possible because eggs do not overwinter	Oklahoma	306
elongate cylinder, .2 X .4 mm, with anterior bladelike projection to anchor egg in leaf	no data; can mature successive batches of eggs	Direct development; at field temp., 22-26 days; in lab, range of 28-51 days @ 15-20°C	Long oviposition period results in many larval size classes overwintering; thus, synchrony required in final instar	Southern England	698, 701, 703
oval, brown eggs, .7 X .4 mm, within a string of gelatinous matrix that swells to 3.5 mm wide X 11 cm long	up to 1000 per string. Egg-string aggregation averages 250 thousand eggs; exceptionally large mass contained over 1 million	Individual strings in lab, hatching began @ 2 wk, 50% complete by 3 wk, continues for 7 wk	Hatching of eggs within large aggregations is retarded and thousands fail to hatch	Michigan	2163

*Fecundity of many mayflies conforms to power law: F = aL^b, where *L* = body length, and *a* and *b* are constants (Elliott 1972; Clifford and Boerger 1974).

Table 5A. Continued

Taxon	Preoviposition Period	Oviposition Season	Oviposition Site	Oviposition Behavior
PLECOPTERA Nemouridae *Nemoura trispinosa*	2-3 wk (feeding required)	mid June-early July	Lotic, midstream	Extruded as mass; flying ♀ dips abdomen into water; mass "explodes" when jelly expands
Amphinemura nigritta	as above	June	As above	As above (?)
Perlodidae *Hydroperla crosbyi*	2-5 days	Feb-Mar	Lotic, head of riffle	♀ alights on water, extrudes egg mass. Mass separates and eggs sink
Perlidae *Paragnetina media*	< 1 day	June (synchronous emergence)	Lotic	Extruded as mass
HEMIPTERA Gerridae *Gerris* spp.	several mo (adults overwinter)	spring and summer	Pools and running water	♀ glues eggs to floating objects or at water's edge, at or below waterline; deposited in parallel rows
Corixidae *Ramphocorixa acuminata*	—	early spring-fall; some eggs overwinter	Ponds, stock watering holes	Preferentially on crayfish (*Cambarus sp.*); also on other smooth surfaces. Several ♀♀ oviposit on one crayfish
Notonectidae *Notonecta undulata*	(*Buenoa sp.*= 16 days; Bare 1926)	early spring through summer	Lentic, widespread (clear pools to slimy ponds)	Eggs glued to submerged plants and other objects; irregular spacing
Belostomatidae *Abedus herberti*	—	nonseasonal	Warm streams and ponds	Starting at apex of wings and moving forward, ♀ deposits a solid mass of eggs on dorsum of ♂
MEGALOPTERA Corydalidae *Orohermes crepusculus*	few days	July-early Sept	Above lotic waters; bridges, trees, rocks	♀ deposits rows of eggs; she may add 3-4 smaller tiers on top of the base layer
Sialidae *Sialis rotunda*	1 day	late Apr-June	On vegetation or other objects overhanging lentic waters	♀ secretes adhesive, then deposits upright rows of eggs forming a tight mass

Table 5A. Continued

Description of Egg or Egg Mass	Number of Eggs	Incubation and Hatching Period	Comments	Geographic Area	Reference
sticky coating attaches eggs to substratum	114-833/batch; x̄ = 514	Immediate development @ 10°C. Incubation minimum of 3 wk, 80% by 6 wk; continues to 12 wk	In field, hatching from mid July to mid Sept.	Ontario	1574
—	90-188/batch; x̄ = 121	4-mo embryonic diapause (germ disk stage)	In lab, hatching began @ 12°C in Sept.; continued to Nov with temp. decrease from 12 to 8°C	Ontario	1574
brown, oval, triangular cross section. Gelatinous coating glues eggs to substratum	172-330 eggs/mass; up to 3 masses/♀; 442-1418, x̄ = 787 (dissected ♀♀)	7 mo egg diapause to Sept–Oct	Synchronous hatching with decreasing temp. from 25 to 19°C	Texas	2953
—	3-7 batches; 1207-2929 egg/♀ (dissected ♀♀)	At 20°C, 32 days; continues for 58 days	Partial hatch in fall, remainder in spring. Some parthenogenesis	Ontario	1573
elongate, cylindrical, 3 X 1 mm; white, turning amber brown before hatching	—	2 wk incubation	♂ of *Rhagadotarsus* calls ♀ with wave patterns; defends oviposition territory (Wilcox 1979)	Kansas	1857
elongate, oval, .9 X .4 mm, with apical nipple. Egg attached by elastic pedestal and disk with strong glue	up to 22 mature eggs in abdomen at one time; successive batches are matured	5-10 day incubation	Continuous reproduction and asynchronous incubation. Synchrony of oviposition induced by freezing or recolonization. Oviposition on crayfish gives protection, aeration, and solid, non-silted, attachment surface	Kansas	1484
elongate, oval, 1.7 X .6 mm; with small tubular micropyle; rough surface with hexagonal sculpturing	—	5-14 day incubation	Continuous oviposition in summer and overlapping broods. Some notonectids insert eggs into plants	Kansas, New York	1857
oval with rounded top; yellow, darkening to tan. Egg swells during development from 3.1 X 1.7 mm to 5.0 X 2.0 mm	one ♀ produced 4 masses in 13 mo, total = 344 eggs	Incubation in lab @ 18°C=21-23 days	♂ aerates eggs by raising and lowering wings. Encumbered ♂♂ occur throughout the year	Arizona	3715
oblong mass, 30 X 25 mm. Eggs, 1.0 X .5 mm, with micropylar projection; greyish yellow, becoming reddish at hatching	1000-1700 eggs/mass x̄ = 1500	In field, 26-63 days, x̄=43 days; in lab (20°C), 25days	Synchronous hatching within a mass	Oregon	1114
cylindrical, rounded at top with micropylar tubercle; .7 X .3 mm; white when laid, turning dark brown before hatching	300-500 eggs/mass; ♀ may deposit a second, smaller mass	8-12 days	Synchronous hatching within a mass, usually at night. Larvae drop from egg mass into the water	Oregon	137, 138

Table 5A. Continued

Taxon	Preoviposition Period	Oviposition Season	Oviposition Site	Oviposition Behavior
TRICHOPTERA Glossosomatidae *Agapetus fuscipes*	few days	Apr-Oct	Chalk streams, slow flow over clean gravel	♀ swims underwater and oviposits on rock, then places "capstone" of small gravel on egg mass before cement is dry; submerged for 15-20 min
Agapetus bifidus	—	July-early Aug	Streams, in crevices on cobble	♀ probably swims to substrate
Hydropsychidae *Hydropsyche* sp. *Cheumatopsyche* sp.	few days	May-Sept	Large rivers, on submerged objects	♀ swims underwater, deposits egg mass on firm surface. Masses concentrated 3-5 ft below surface
Limnephilidae *Limnephilus indivisus*	several wk (ovarial diapause over summer)	early fall	Dry basin of temporary pool	♀ attaches gelatinous egg mass under log or to other protected site
Limnephilus lunatus	variable: 2-3 wk in fall; > 3 mo in spring	fall	Chalk streams, attached to watercress	♀ deposits egg mass on plants above water. Mass absorbs water and swells to 10 mm in diameter
Clistoronia magnifica	2 wk (in lab)	July-Aug	Lentic, attached to submerged logs or plants, or loose in littoral benthos	In lab, ♀ observed to enter water for 5 min and attach egg mass. May also oviposit on surface and egg mass sinks to substrate
Calamoceratidae *Heteroplectron* spp.	few days? (pupae contain mature eggs)	early summer	Small streams; loosely attached masses near waterline; also in drop zones not attached	Some ♀♀ may oviposit underwater, but site of eggs suggests that most oviposit at the waterline
Leptoceridae *Ceraclea* spp.	< 1 day	spring, or spring and mid-late summer	Lentic and lotic	♀ oviposits on surface. Mass floats until it absorbs water, then sinks and adheres to submerged objects
Chathamiidae *Philanisus plebeius*	few wk	summer-fall	Marine; in starfish (*Patiriella*) in tidepools	♀ probably inserts ovipositor through pores on aboral surface of starfish to deposit eggs in coelom

Table 5A. Continued

Description of Egg or Egg Mass	Number of Eggs	Incubation and Hatching Period	Comments	Geographic Area	Reference
round, .2-.25 mm in diameter, creamy white; deposited in single-layered, compact mass	eggs/mass= 12-94, x̄ = 27. In lab, 12 ♀♀ deposited 70 masses	Direct development. At 12°C, hatching starts @ 1 mo, and continues for 3 wk	In cold springbrook, eggs hatched from early May–Oct	Southern England	86
mass, firm matrix, 1.2 X 1.6 mm. Eggs round, .2-.3 mm in diameter	30-100 eggs/mass	7 mo to hatching	Eggs overwinter in obligatory diapause	Oregon	92
concentric rows of closely packed eggs (Badcock 1953)	*Hydropsyche*, dissected ♀♀ =331-465 eggs (Fremling 1960a). 820 deposited in 50 min by a ♀ (Badcock 1953)	In lab, 8-11 days	—	England, Iowa	145, 1302
—	—	Larvae hatch in few wk, but remain in matrix until flooded	Emergence from egg mass dependent on time of flooding; may be in fall or the following spring	Ontario	4381, 4395
dome-shaped mass, opaque yellow when deposited. When swollen, jelly is colorless with eggs in rows. Eggs bluntly elliptical, .4 X .3 mm	270-636/mass	Duration of egg stage (room temp.) averages 17 days	Fall oviposition synchronized by decreasing duration of ovarial diapause as summer progresses	Southern England	1461
spherical colorless mass, up to 3 cm in diameter, takes 8 hr to achieve full size. Eggs greenish, arranged in lines	200-300 eggs per mass; ♀ may deposit a second smaller mass	2-2-1/2 wk @ 16°C; larvae continue to emerge from a mass for 1 wk	Field oviposition period extended by long flight period. May be 2-3 mo ovarial diapause in spring	Oregon, British Columbia	89, 4482
egg mass spherical, 7-15 mm, coated with silt, jelly very fluid; eggs yellowish, arranged in rows	132-348 eggs per mass, x̄ = 209. Dissected ♀♀ contain 300-400 eggs	At 20°C, hatching @ 12 days; 15 days @ 15°C; larvae remain in matrix 2-9 days	Short, synchronous oviposition period. Eggs above waterline require 100% humidity; larvae remain in mass until inundated, then all emerge in ca. 1 min	Pennsylvania; Oregon	3034; R. Wisseman (unpublished)
mass, dark green, ca. 1 mm sphere when deposited. Eggs spherical, green yolk, transparent chorion; eggs swell from .1 mm to .2 mm in diameter by time of hatching	100-300 eggs per mass	In lab, 1-3 wk	Direct embryonic development	Kentucky	3291
eggs spherical, .4 mm, yellow-yellowish gray; single or in small clumps in coelomic cavity of host	dissected ♀♀ contain up to 400 mature eggs, x̄ = ca. 160. Variable numbers in starfish: mostly < 10, but up to 112	At least 5 wk incubation @ 16-18°C	Extended incubation and hatching periods as eggs were found in starfish throughout the year	New Zealand	4487

Table 5A. Continued

Taxon	Preoviposition Period	Oviposition Season	Oviposition Site	Oviposition Behavior
COLEOPTERA Gyrinidae *Dineutus spp.*	—	May-Aug	Ponds, on underside of *Potamogeton sp.* leaves	Deposited in clusters; each egg glued separately to leaf
Haliplidae *Peltodytes sp.,* *Haliplus sp.*	—	May-early July	Lentic, in beds of *Chara* sp. and *Nitella* sp.	*Peltodytes*, eggs glued to macrophytes or algae. *Haliplus*, ♀ chews hole in hollow stems and deposits several eggs within
Dytiscidae *Agabus erichsoni*	—	May-June	Temporary woodland pools	Eggs deposited in clumps of 2-3 among root fibers or moss on bottom of pond
Colymbetes sculptilis	—	late Mar-April	Temporary woodland pools	Eggs firmly attached to submerged vegetation or to edge of rearing container
Hydrophilidae *Hydrophilus triangularis*	—	early-mid summer	Eutrophic ponds, with some vegetation	♀ spins a silken egg case; ellipsoidal shape with elongate "mast." Construction takes > 1 hr
Psephenidae *Psephenus falli*	1 day? (mature ovaries at emergence)	early May-mid Aug	In riffles, under rocks	♀ crawls down a rock and remains submerged for life (1-3 days)
Elmidae *Stenelmis sexlineata*	—	May-Aug	Lotic, in riffles, on sides and bottom of rocks	Submerged ♀ selects depressions or cracks on rocks; deposits group of eggs usually touching each other; each egg pressed against surface for 10-20 sec to glue it down
LEPIDOPTERA Pyralidae *Nymphula sp.*	1 day	July-Aug	Lentic, underside of floating *Potamogeton* sp. leaves	♀ generally does not enter water but extends tip of abdomen to attach egg mass on underside near margin of leaf. Oviposition occurs at night
HYMENOPTERA Agriotypidae *Agriotypus sp.*	few days	May-July	Lentic or lotic; in cases of goerid or odontocerid caddisflies	♀ crawls down a support into water and searches for a host. Eggs only deposited on prepupa or pupa. ♀ may stay underwater for several hr, enveloped in air bubble

Table 5A. Continued

Description of Egg or Egg Mass	Number of Eggs	Incubation and Hatching Period	Comments	Geographic Area	Reference
white, elongated ellipsoid, 1.9 X .6 mm, cluster arranged diagonally @ 45° angle from midrib	7-40 eggs per cluster (Wilson 1923b), 12-17 eggs per ♀ per day (Istock 1966)	5-6 days	Synchronous hatching of clusters, extended oviposition period of the population	Iowa, Michigan	1917, 4464
Peltodytes, oval, with projecting plug, .5 X .3 mm, yellowish brown; *Haliplus,* oval, .4 X .2 mm, whitish	30-40 eggs within a wk	8-10 days @ 21 °C	♀ ♀ live over 1 yr, so several batches of eggs are matured	Michigan	1710, 2515
short oval; pale cream becoming light brown with age; 1.7 X 1.1 mm	♀ ♀ contain 14-31 eggs at one time; ovaries continue to develop eggs	8-9 mo; some embryonic development before pond dries up, then diapause; hatching occurs the following spring	Eggs from dry pond bottom were chilled for 3 mo @ 0°C, then flooded and larvae emerged within a few hr	Ontario	1955
elongate oval, somewhat kidney-shaped; pale yellow with smooth chorion; 1.8 X .7 mm	—	6 days @ 19°C; longer in field as oviposition occurs at < 14°C	Apparently a short incubation and hatching period, as 1st-instar larvae only found for 3 wk in April	Ontario	1955
egg case is yellow, turns brown; eggs, elongate ellipsoid, 4.4 X 1 mm; bright yellow	10-130 eggs per case; ♀ probably matures more than 1 batch	—	Egg case floats and eggs do not hatch if case turns over; mast assumed to aid in stabilizing the case	Iowa	4463
spherical, lemon yellow eggs, deposited in compact, single-layered mass	ca. 500 eggs per ♀; several may oviposit together, forming masses of over 2000 eggs	16-17 days @ 23°C	Apparently synchronous hatching within a mass, but extended oviposition period	Southern California	2821
oblong; whitish-yellow; .55-.62 mm long	—	6-10 days @ 22-25°C	Protracted oviposition period; adults live underwater for > 1 yr	Kentucky	4328
elliptical eggs, .45 X .6 mm; light grey or whitish; about 20 eggs/mass	♀ of *N. badiusalis* laid 441 eggs in one night	6-11 days	Direct development of eggs; synchronous hatching within a mass	Michigan	264
elongate, .9 X .2 mm; tapered to a stalk which is inserted into the host's integument	—	In lab, 5-8 days	Several eggs may be deposited on one host but only one larva can develop per host	Japan, France	649, 1475

Table 5A. Continued

Taxon	Preoviposition Period	Oviposition Season	Oviposition Site	Oviposition Behavior
DIPTERA Tipulidae *Tipula sacra*	< 1 day	June–July	Lentic, in soil or algae mats near shore	♀♀ emerge during the day; mate and begin ovipositing immediately
Lipsothrix nigrilinea	< 12 hr	Mar–Aug peak in May–June	In saturated wood in streams	♀ searches for suitable site on wood near waterline with ovipositor. Deposits egg ca. 1 mm deep in soft wood or cracks; then moves to make another insertion
Ptychopteridae *Ptychoptera lenis*	< 1 day	late May–June	Lentic, stagnant water	Mating and oviposition occur shortly after emergence. Eggs occur loose on substrate, so probably scattered at pond surface and sink to substrate
Simuliidae *Simulium* spp.	variable; blood meal may be required for egg maturation	spring and summer; multivoltine	Lotic; various sites (wetted vegetation, dam faces, debris, etc.)	Variable even within a species; may oviposit in flight, but more commonly on solid surface in masses or strings, at or below waterline
Culicidae *Aedes aegypti*	variable; blood meal required for egg development	nonseasonal	artificial containers: cisterns, cans, old tires	Eggs deposited singly, at or near waterline
Culex pipiens	variable; blood meal required, except in autogenous strains; some overwinter as nulliparous ♀♀	spring–late autumn	Lentic; small catchments and pools with high organic content	♀ lands on water and deposits eggs in raftlike masses. Oviposition usually at night
Chironomidae *Chironomus plumosus*	2-5 days	mid May, July–Sept	Lentic; on water or on flotsam	♀ flies over water (sometimes several mi); extrudes egg mass between hind tibiae and deposits it on first surface that she touches
Tabanidae *Tabanus atratus*	1 wk	June–Oct	On plants, near or over water	♀ faces head downward while depositing egg mass on vertical portion of plant
Ephydridae *Dichaeta sp.* (= *Notiphila*) (Mathis 1979a)	5-15 days	throughout summer	Marshy areas with accumulation of decaying vegetation	♀ scatters eggs along shore or on floating detritus. Eggs not glued to substrate but many in crevices
Sciomyzidae *Sepedon* spp.	4-24 days	—	Lentic; on emergent vegetation, from 5 cm to > 1 m above water	♀ in head downward position, deposits eggs in vertical row

Table 5A. Continued

Description of Egg or Egg Mass	Number of Eggs	Incubation and Hatching Period	Comments	Geographic Area	Reference
shining black, elongate, convex on one side; 1.0 X .4 mm; posterior filament uncoils when wetted as anchoring device	dissected ♀♀, x̄ = 925, range, 500-1600 eggs	In lab, few days; in field, < 1 mo	Direct development of eggs; hatching period from early July-mid Aug	Alberta	3199, 3206
cream colored, elongate, smooth; no anchoring device	dissected ♀♀, x̄ = 185, range, 106-380 eggs	About 3 wk @ 16°C	Direct development, but extended hatching period because of long flight period	Oregon	979
elongate oval; whitish yellow; longitudinal reticulations on chorion; .8-.9 mm long	dissected ♀♀ contain 530-806 eggs	In field, 14-20 days	Egg maturation occurs during pharate adult stage	Alberta	1790
oval to triangular .25 X .14 X .13 mm; whitish, turning brown as they mature.	300-600 eggs per ♀. Eggs may occur in large aggregations (72,000/ft^2)	5 days @ 23°C	Successive generations in summer; overwinter often as diapausing eggs	Ontario	853
elongate oval	average about 140 eggs when fed on humans; may be 2 or more egg cycles	Highly variable; embryonic development completed in 2-4 days after flooding	Direct development in water but eggs withstand desiccation for at least 1 yr	Southeastern U.S.	1618
cylindrical, tapered	100-400 eggs per mass; ♀ lays 2-4 masses	1-3 days	First batch of eggs may mature without a blood meal. Size of later masses depends on blood meals. Several generations per yr	Holarctic	1618
egg mass is dark brown, tear-shaped; swells to 25 X 5 mm. Eggs, cream colored, oval, .5 X .2 mm	eggs per mass: x̄ = 1676, range, 1154-2014	3 days @ 24°C; 14 days @ 9°C	Egg mass floats and larvae remain in it for 1 day after hatching. 2 generations per yr	Wisconsin	1725
egg mass is subconical, oval at base, with 4-5 tiers of eggs, 5-25 mm X 2-10 mm. Eggs white when laid, then darken	500-800 eggs per mass	4-12 days	—	Florida	2018
egg ellipsoidal, convex on venter; longitudinally ridged; white; .9 X .3 mm	—	1-2 days @ 21-25°C	Eggs float when marsh floods and have plastron for underwater respiration	Ohio, Montana	1009
eggs lie horizontal touching preceding one. Egg elongate with coarse, longitudinal striations; white, becoming colored during development	Up to 25 eggs per row; ♀ probably deposits several rows	3-5 days	—	USA	2871

Table 5B. Summary of data providing information on life histories and voltinism of selected aquatic insects.

Taxon	Number of Generations Per Year	Habitat of Study	Location	Reference
EPHEMEROPTERA				
Ameletidae				
Ameletus spp.	univoltine	lotic	Quebec	2414
	univoltine	lotic	North Carolina	1877
Baetidae				
Baetis spp.	uni- to multivoltine	lotic	N. America & Europe	658
Baetis pygmaeus	multivoltine (2-3/yr)	lotic	Quebec	2275
Baetis spp.	multivoltine (~ 10/yr)[†]	lotic	Georgia	1942
Baetis quilleri	multivoltine (18+/yr)[†]	lotic	Arizona	1199, 1469
Callibaetis floridanus	bivoltine	lentic	Virginia	632
Cloeon triangulifer	multivoltine (3/yr)	lotic	Pennsylvania	3923
Heterocloeon curiosum	bivoltine	lotic	Virginia	2156
Isonychiidae				
Isonychia bicolor	bivoltine	lotic	Pennsylvania	3917
Isonychia spp.	bi- or trivoltine(?)	lotic	Georgia	1942
Pseudironidae				
Pseudiron centralis	univoltine	lotic	Alberta	3750
Heptageniidae				
Cinygmula ramaleyi	univoltine	lotic	Alberta	3241
Cinygmula reticulata	univoltine	lotic	Oregon	2312
Leucrocuta sp.	univoltine	lotic	North Carolina	1877
Nixe sp.	univoltine	lotic	North Carolina	1877
Rhithrogena semicolorata	univoltine	lotic	Europe	1648
Stenacron interpunctatum	univoltine to 3 generations in 2 yrs	lotic	Indiana	2591
Stenonema modestum	bivoltine	lotic	Virginia	2155
	bivoltine	lotic	South Carolina	3728
Stenonema vicarium	univoltine	lotic	Minnesota	2193
Ephemerellidae				
Ephemerella (2 spp.)	univoltine	lotic	Oregon	1637
Ephemerella ignita	univoltine	lotic	Europe	1648
Eurylophella funeralis	semivoltine (2 yr)	lotic	Ontario	1387
	univoltine	lotic	Pennsylvania	3922
Eurylophella doris(?)	univoltine	lotic	Georgia	1942
Eurylophella temporalis	univoltine	lotic	South Carolina	3728
Drunella (2 spp.)	univoltine	lotic	Oregon	1637
Drunella doddsi	univoltine	lotic	Alberta	3241
	univoltine	lotic	Oregon	1637
Serratella sp.	univoltine	lotic	North Carolina	1877
Tricorythidae				
Tricorythodes atratus	bivoltine	lotic	Minnesota	1535
Tricorythodes allectus (?)	multivoltine (4 to 5 from May to late Oct.)	lotic	Georgia	1942
Tricorythodes dimorphus	multivoltine (18+/yr)[†]	lotic	Arizona	1199, 1469
Tricorythodes minutus	univoltine	lotic	Alberta	193
Caenidae				
Caenis spp.	uni- to multivoltine		N. America & Europe	658
Caenis amica	bivoltine	lentic	Virginia	632
Baetiscidae				
Baetisca rogersi	univoltine	lotic	Florida	3094
Leptophlebiidae				
Paraleptophlebia (2 spp.)	univoltine	lotic	Oregon	2313
Paraleptophlebia (4 spp.)	univoltine	lotic	Ontario	1387
Paraleptophlebia volitans	univoltine	lotic	South Carolina	3728
Habrophlebia vibrans	semivoltine	lotic	Quebec	2274
	semivoltine	lotic	North Carolina	1877
	semivoltine	lotic	Ontario	1387
Behningiidae				
Dolania americana	semivoltine (ca. 10 mon. as egg)	lotic	South Carolina	1616

[†] based on larval growth rates obtained from *in situ* measurements.

Table 5B. Continued

Taxon	Number of Generations Per Year	Habitat of Study	Location	Reference
Ephemeridae				
Hexagenia bilineata	univoltine	lotic	midwest	1302'
Hexagenia munda	univoltine	lotic	South Carolina	3728
Hexagenia limbata	semivoltine, with variable cohorts	lentic	Manitoba	1656
Ephemera simulans	semivoltine	lentic	Manitoba	1656
Polymitarcyidae				
Ephoron album	univoltine	lotic	Manitoba	1386
Ephoron leukon	univoltine	lotic	Virginia	3738
ODONATA				
Anisoptera				
Gomphidae				
Lanthus vernalis	semivoltine	lotic	North Carolina	1877
Aeshnidae				
Anax junius	univoltine (with 3 & 11-month populations)	lotic	Ontario	4050
Anax junius	uni- & semivoltine	lentic	Indiana	4516
Boyeria vinosa	univoltine	lotic	Virginia	1339
	semivoltine	lotic	South Carolina	3728
Corduliidae				
Neurocordulia molesta	univoltine	lotic	Georgia	258
Libellulidae				
Celithemis fasciata	univoltine	lentic	South Carolina	254
Celithemis elisa	univoltine	lentic	Indiana	4516
Epitheca (2 spp.)	univoltine	lentic	South Carolina	254
Epitheca (2 spp.)	uni- & semivoltine	lentic	Indiana	4516
Epitheca cynosura	uni- & semivoltine	lentic	Tennessee	2494
Ladona deplanata	univoltine	lentic	South Carolina	254
Libellula (3 spp.)	univoltine	lentic	Indiana	4516
Perithemis tenera	univoltine	lentic	Indiana	4516
Zygoptera				
Lestidae				
Archilestes grandis	univoltine	lentic	North Carolina	1913
Lestes (3 spp.)	univoltine	lentic	Saskatchewan	3549
Lestes (2 spp.)	univoltine	lentic	North Carolina	1913
Coenagrionidae				
Enallagma hageni & *E. aspersum*	univoltine uni- & bivoltine	lentic	North Carolina	1914
Telebasis salva	univoltine	lentic	Arizona	3489
PLECOPTERA				
Pteronarcyidae				
Pteronarcys dorsata	univoltine	lotic	Virginia	2292
Pteronarcys proteus	merovoltine (3-4 yrs)	lotic	Massachuchetts	1807
	merovoltine (3 yrs)	lotic	Virginia & West Virginia	4340
Pteronarcys scotti	semivoltine	lotic	Georgia	2611
	semivoltine	lotic	South Carolina	1235
Peltoperlidae				
Tallaperla maria	semivoltine	lotic	Tennessee	1080
			North Carolina	2949
Taeniopterygidae				
Oemopteryx contorta	univoltine	lotic	Tennessee	2875
Strophopteryx limata	univoltine	lotic	Tennessee	2875
Taenionema atlanticum	univoltine	lotic	Quebec	1586
Taeniopteryx nivalis	univoltine	lotic	Ontario	3626
Nemouridae				
Amphinemura (4 spp.)	univoltine	lotic	Ontario, Quebec	1574, 2414
Nemoura trispinosa	univoltine	lotic	Ontario	1574
Ostrocerca albidipennis	univoltine	lotic	Quebec	2414
Prostoia besametsa	univoltine	lotic	New Mexico	1622
Prostoia similis	univoltine	lotic	Minnesota	2192
Shipsa rotunda	univoltine	lotic	Ontario	1574
Soyedina carolinesis	univoltine	lotic	Pennsylvania	3926
Soyedina vallicularia	univoltine	lotic	Quebec	2414
Zapada cinctipes	univoltine	lotic	Oregon	2079
Zapada columbiana	semivoltine	lotic	Alberta	2824

Table 5B. Continued

Taxon	Number of Generations Per Year	Habitat of Study	Location	Reference
Leuctridae				
Leuctra (2 spp.)	uni- & semivoltine	lotic	Ontario	1574
Zealeuctra (2 spp.)	uni- & semivoltine	lotic	Texas	3731
Capniidae				
Allocapnia (4 spp.)	univoltine	lotic	Quebec	1586
Allocapnia nivicola	univoltine	lotic	Quebec	1574
Allocapnia rickeri	univoltine	lotic	Minnesota	2192
Capnia vernalis	univoltine	lotic	Quebec	1586
Paracapnia angulata	univoltine	lotic	Ontario	1583
Perlidae				
Acroneuria evoluta	univoltine	lotic	Oklahoma	1103
Acroneuria lycorias	merovoltine (3 yrs)	lotic	Michigan	1042
Agnetina capitata	semivoltine	lotic	Ontario	1573
	semi- and merovoltine (3 yrs)	lotic	New York	2751
Beloneuria (2 spp.)	semivoltine ?	lotic	North Carolina	1877
Calineuria californica	merovoltine (3 yrs ?)	lotic	Oregon	2079
	semivoltine	lotic	California	3652
Claassenia sabulosa	semi- merovoltine (3yrs)	lotic	Colorado	1331
		lotic	New Mexico	1622
Eccoptura xanthenes	semivoltine	lotic	Kentucky	46
Neoperla clymene	univoltine	lotic	Texas	4119
Paragnetina media	semivoltine	lotic	Kentucky	3946
	semivoltine	lotic	Michigan	1649
	merovoltine (3 yrs)	lotic	Ontario	1573
Perlesta placida	univoltine	lotic	Texas	3732
	univoltine	lotic	Georgia	258
Perlodidae				
Clioperla clio	univoltine	lotic	Kentucky	2719
			Ontario	1573
			Arkansas	1158
Helopicus (2 spp.)	univoltine ?	lotic	Southeastern U.S.	3811
Hydroperla crosbyi	univoltine	lotic	Texas	2953
Isogenoides olivaceus	univoltine	lotic	Minnesota	2192
Isoperla (4 spp.)	univoltine	lotic	Ontario	1573
Isoperla fulva	univoltine	lotic	New Mexico	1622
Isoperla signata	univoltine	lotic	Wisconsin	2026
Chloroperlidae				
Haploperla brevis	univoltine	lotic	Oklahoma	1103
Sweltsa onkos	semivoltine	lotic	Ontario	1573
Sweltsa (2 spp.)	semivoltine	lotic	North Carolina	1877
Triznaka signata	univoltine	lotic	New Mexico	1622
HEMIPTERA				
Hydrometridae				
Hydrometra martini	multivoltine (3-4/yr)	lentic	Michigan & Massachusetts	3780
Gerridae				
Gerris agenticollis	univoltine	lentic	Illinois	2161
Gerris canaliculatus	trivoltine?	lentic	Virginia	346
Gerris remigis	bi- and univoltine	lotic	New York	325
Limnoporus notabilis	univoltine	lentic	British Columbia	1132
Belostomatidae				
Abedus herbti	multivoltine (4+/yr)	lotic	Arizona	1469
Belostoma malkini	multivoltine (5- 6/yr)	lentic	Trinidad	782
Lethocerus americanus	multivoltine ? (33d egg to adult during summer)	lentic	Michigan	3246
Lethocerus maximus	multivoltine (5- 6/yr)	lentic	Trinidad	782
Pleidae				
Neoplea striola	univoltine	lentic	Illinois	2635
Nepidae				
Nepa apiculata	univoltine	lentic	Illinois	2637
Ranatra montezuma	trivoltine	lentic	Arizona	3487

Table 5B. Continued

Taxon	Number of Generations Per Year	Habitat of Study	Location	Reference
Naucoridae				
Cryphocricos hungerfordi	bivoltine	lotic	Texas	3679
Corixidae				
Graptocorixa serrulata	multivoltine (~17/yr)†	lotic	Arizona	1469
Hesperocorixa interrupta	univoltine	lentic & lotic	Virginia	347
Trichocorixa reticulata	bi- and trivoltine	lentic	California	1132
Notonectidae				
Notonecta hoffmanni	univoltine	lotic	California	2633
Notonecta undulata	bivoltine?	lentic	California	4026
MEGALOPTERA				
Sialidae				
Sialis (2 spp.)	univoltine & a few semivoltine	lentic &lotic	Oregon	138
Sialis aequalis	univoltine	lotic	South Carolina	3728
Sialis dreisbachi	univoltine	lotic	Minnesota	2193
Corydalidae				
Corydalus cornutus	univoltine	lotic	Texas	443, 3648
	semivoltine	lotic	West Virginia	3416
Nigronia serricornis	semi- & merovoltine (3 yrs)	lotic	Michigan	3107
NEUROPTERA				
Sisyridae				
Climacia areolaris	trivoltine	lentic	Ohio	450
TRICHOPTERA				
Philopotamidae				
Chimarra (2 spp.)	bivoltine	lotic	Virginia	3019
Chimarra (2 spp.)	bivoltine	lotic	Arkansas	386
Chimarra aterrima	bivoltine	lotic	Quebec	2759
Dolophilodes distinctus	bi -& some trivoltine (?)	lotic	Georgia,N. Carolina	257
	bivoltine	lotic	North Carolina	1878
Wormaldia moestus	bivoltine	lotic	North Carolina	1878
	univoltine	lotic	Arkansas	386
Polycentropodidae				
Neureclipsis bimaculatus	bivoltine	lotic	Alberta	3336
Neureclipsis crepuscularis	bivoltine	lotic	Georgia	775
Polycentropus maculatus	univoltine	lotic	North Carolina	1878
Polycentropus centralis	bivoltine	lotic	Arkansas	386
Psychomyiidae				
Lype diversa	univoltine	lotic	North Carolina	1878
Hydropsychidae				
Arctopsyche grandis	univoltine	lotic	Idaho	778
	semivoltine	lotic	Montana	1628
Arctopsyche irrorata	univoltine	lotic	Georgia, N. Carolina	257
Cheumatopsyche pasella	bivoltine	lotic	Georgia	1628
Cheumatopsyche pettiti	univoltine	lotic	Minnesota	2420
	univoltine		Quebec	2759
Diplectrona modesta	univoltine	lotic	Georgia, N. Carolina	257
Hydropsyche slossonae	univoltine	lotic	Virginia	4458
Hydropsyche (2 spp.)	univoltine some bivoltine?	lotic	Minnesota	2420
Hydropsyche (2 spp.)	univoltine	lotic	Georgia, N. Carolina	257
Hydropsyche (2 spp.)	univoltine	lotic	Quebec	2759
Hydropsyche (5 spp.)	uni- & bivoltine	lotic	Ontario	2418
Hydropsyche (3 spp.)	trivoltine	lotic	Virginia	3019
Hydropsyche (2 spp.)	trivoltine	lotic	Georgia	1299
Hydropsyche (2 spp.)	bi- & trivoltine	lotic	Ontario	2419
Macrostemum carolina	univoltine	lotic	Georgia	775
			South Carolina	3728
Macrostemum zebratum	bivoltine	lotic	Virginia	3019
Parapsyche apicalis	univoltine	lotic	North Carolina	1878, 3445
	univoltine	lotic	Quebec	2414
Parapsyche cardis	univoltine	lotic	Georgia	257

† based on larval growth rates obtained from *in situ* measurements.

Table 5B. Continued

Taxon	Number of Generations Per Year	Habitat of Study	Location	Reference
Rhyacophilidae				
Rhyacophila (5 spp.)	univoltine	lotic	South Carolina	2467
Rhyacophila (2 spp.)	univoltine	lotic	Ontario	2480
Rhyacophila vao	semivoltine	lotic	Alberta	920
Glossosomatidae				
Glossosoma intermedium	bivoltine	lotic	Minnesota	2194
Glossosoma nigrior	bivoltine	lotic	North Carolina	1375
Glossosoma pentium	bivoltine	lotic	Oregon	92
Agapetus sp.	univoltine	lotic	North Carolina	1375
Agapetus illini	univoltine	lotic	Arkansas	386
Agapetus occidentis	univoltine	lotic	Oregon	1375
Anagapetus bernea	univoltine	lotic	Oregon	1375
Hydroptilidae				
Dibusa angata	bivoltine	lotic	Kentucky	3305
Phryganeidae				
Banksiola crotchi	univoltine	lentic	British Columbia	4483
Brachycentridae				
Brachycentrus spinae	univoltine	lotic	North Carolina	3444
Brachycentrus (2 spp.)	univoltine	lotic	Montana	1630
Brachycentrus (2 spp.)	univoltine	lotic	Minnesota	2194
Micrasema kluane	univoltine	lotic	Minnesota	2194
Lepidostomatidae				
Lepidostoma bryanti	univoltine	lotic	Minnesota	2194
Lepidostoma (2 spp.)	univoltine	lotic	Oregon	3112
Lepidostoma (2 spp.)	univoltine	lotic	North Carolina	1878
Limnephilidae				
Clistoronia magnifica	univoltine	lentic	British Columbia	4482
Desmona bethula	univoltine	lotic	California	1101
Dicosmoecus gilvipes	semivoltine	lotic	Montana	1629
Eocosmoecus (2 spp)	semivoltine	lotic	Pacific Northwest	4398
Goerita semata	semivoltine	lotic	North Carolina	1155
Limnephilus indivisus	univoltine	lentic	Ontario	3338
Oligophlebodes zelti	univoltine	lotic	Alberta	2961
Onocosmoecus unicolor	univoltine	lotic	Oregon	4514
Philocasca demita	univoltine	terrestrial (saturated leaf litter)	Oregon	84
Pycnopsyche (3 spp.)	univoltine	lotic	Quebec	2415
Pycnopsyche (2 spp.)	univoltine	lotic	Virginia	3415
Sphagnophylax meiops	univoltine	lentic	N.W.T. Canada	4477
Uenoidae				
Neophylax consimilis	univoltine	lotic	North Carolina	1375
Neophylax mitchelli	univoltine	lotic	North Carolina	1878
Neophylax (5 spp.)	univoltine	lotic	Ontario	217
Neothremma alicia	semivoltine (2 yrs)	lotic	Alberta	2961
Sericostomatidae				
Agarodes libalis	univoltine	lotic	Virginia	3415
Fattigia pele	semivoltine (2 yrs-some univoltine?)	lotic	North Carolina	1878
Molannidae				
Molanna flavicornis	univoltine	lotic	Alberta	3337
Helicopsychidae				
Helicopsyche borealis	univoltine?	lotic	Ontario	4440
	univoltine	lotic	California	1156
Helicopsyche limnella	bivoltine	lotic	Arkansas	386
Helicopsyche mexicana	multivoltine (~ 7/yr)	lotic	Arizona	1469
Calamoceratidae				
Heteroplectron californicum	semivoltine (2 yrs)	lotic	Oregon	93

Table 5B. Continued

Taxon	Number of Generations Per Year	Habitat of Study	Location	Reference
Leptoceridae				
Ceraclea transversa	univoltine	lotic	Kentucky	3292
Ceraclea excisa	univoltine	lotic	Alberta	3337
Mystacides interjecta	univoltine	lotic	Alberta	3337
Nectopsyche albida	univoltine	lentic	Indiana	4037
Oecetis immobilis	univoltine	lotic	Alberta	3337
Oecetis inconspicua	univoltine?	lentic	British Columbia	4482
	univoltine	lotic	Alberta	3337
Triaenodes injusta	univoltine	lotic	Alberta	3337
Ylodes frontalis	univoltine	lotic	Alberta	3337
LEPIDOPTERA				
Pyralidae				
Eoparargyractis plevie	univoltine	lentic	New Hampshire	1178
Petrophila canadensis	bivoltine	lotic	New York	2281
Petrophila confusalis	multivoltine (2-3/yr)	lotic & lentic	California	4060
	univoltine	lotic	Montana	2572
Noctuidae				
Bellura gortynoides	bivoltine	lentic	Indiana	2334
COLEOPTERA				
Adephaga				
Gyrinidae				
Dineutus (2 spp.)	univoltine	lentic	Michigan	1918
Haliplidae				
Haliplus immaculicollis & *Peltodytes* (2 spp.)	uni- & bivoltine(?)	lentic	Michigan	1710
Dytiscidae				
Acilius (3 spp.)	primarily univoltine few bivoltine	lentic	Wisconsin	1744
Agabus spp.	primarily univoltine	lentic	Wisconsin	1746
Cybister fimbriolatus	univoltine	lentic	Wisconsin	1744
Colymbetes spp.	univoltine	lentic	Ontario	1955
	univoltine	lentic	Wisconsin	1745
Copelatus glyphicus	univoltine	lentic	Wisconsin	1745
Coptotomus spp.	univoltine	lentic	Wisconsin	1745
Dytiscus (4 spp.)	univoltine	lentic	Wisconsin	1744
Graphoderus (6 spp.)	primarily univoltine few bivoltine	lentic	Wisconsin	1744
Hydaticus (2 spp.)	primarily univoltine few bivoltine	lentic	Wisconsin	1744
Ilybius spp.	primarily univoltine	lentic	Wisconsin	1746
Laccophilus maculosus	bivoltine	lentic	Wisconsin	1743
Matus bicarinatus	primarily univoltine some bivoltine	lentic	Wisconsin	1745
Rhantus binotatus	primarily univoltine few bivoltine	lentic	Wisconsin	1745
Thermonectus ornaticollis	univoltine	lentic	Wisconsin	1744
Noteridae				
Suphisellus puncticollis	univoltine	lentic	Wisconsin	1743
Polyphaga				
Hydrophilidae				
Tropisternus ellipticus	bi- & trivoltine	lotic	Arizona	1469
Psephenidae				
Psephenus montanus	semivoltine (2 yrs) some univoltine?	lotic	Arizona @elev.= 2,576 m a.s.l.	2819
Scirtidae (= Helodidae)				
Scirtes tibialis	univoltine	lentic	Wisconsin	2175

Table 5B. Continued

Taxon	Number of Generations Per Year	Habitat of Study	Location	Reference
Elmidae				
Ancyronyx variegata	univoltine	lotic	South Carolina	3728
Dubiraphia quadrinotata	univoltine	lotic	South Carolina	3728
Lara avara	merovoltine (4 -6+ yrs)	lotic	Oregon	3824
Macronychus glabratus	univoltine	lotic	South Carolina	3728
Optioservus fastiditus	semivoltine	lotic	Ontario	3956
Promoresia elegans	semivoltine	lotic	Ontario	3956
Stenelmis crenata	merovoltine (3 yrs)	lotic	Quebec	2328
Stenelmis nr. *bicarinata*	univoltine ?	lotic	Ontario	3956
Stenelmis sexlineata	uni- and semivoltine	lotic	Kentucky	4328
Ptilodactylidae				
Anchytarsus bicolor	semivoltine (2-3 yrs)	lotic	Quebec	2329
Chrysomelidae				
Donacia cinerea	uni- and semivoltine	lentic	Sweden	2994
Galerucella nymphaeae	multivoltine (8+/yr)[†]	lotic & lentic	Georgia	4190
G. nymphaeae	bivoltine	lentic	Sweden	2997
Curculionidae				
Lissorhopterus simplex (=*oryzophilus*)	univoltine(?) with multiple cohorts (egg to adult in 36 days) - depending on flooding	lentic	Arkansas	1916
DIPTERA				
Nymphomyiidae				
Nymphomyia spp. = (*Palaeodipteron*)	primarily bivoltine, some trivoltine?	lotic	Eastern N. Amer.	731
Deuterophlebiidae				
Deuterophlebia (4 spp.)	primarily univoltine some bi- & trivoltine	lotic	Western N. America	728
Blephariceridae				
Blepharicera williamsae	univoltine	lotic	North Carolina	1375
Tipulidae				
Antocha saxicola	bivoltine	lotic	Ontario	1326
Dicranota spp.	univoltine	lotic	Minnesota	2193
Dicranota bimaculata	univoltine	lotic	England	1054
Lipsothrix sylvia	univoltine	lotic	Virginia	3426
Lipsothrix (2 spp.)	semi- & merovoltine (3yr)	lotic	Oregon	981
Tipula sacra	semivoltine with some univoltine	lentic	Alberta	3199
Tipula abdominalis	univoltine	lotic	Virginia	3415
Ptychopteridae				
Ptychoptera lenis lenis	univoltine	lentic	Alberta	1790
Culicidae				
Aedes taeniorhynchus	multivoltine (6 –15 days: egg to adult)	salt marsh	Florida	2841
Culex nigripalpus	multivoltine (8 –10 generations/yr)	lentic & containers	Florida	2840
Culex resturans	multivoltine	lentic	Ontario	4442
Chaoboridae				
Chaoborus americanus	univoltine	lentic	British Columbia	1147
C. trivittatus	semivoltine (2 yrs)	lentic	British Columbia	1147
C. punctipennis	bivoltine	lentic	North Carolina	1010
Ceratopogoniidae				
Culicoides variipennis	multivoltine (3/yr ?)	lentic	New York	2787
Sphaeromiini & Palpomyiini	univoltine	lentic	North Carolina	383

[†] based on larval growth rates obtained from *in situ* measurements.

Table 5B. Continued

Taxon	Number of Generations Per Year	Habitat of Study	Location	Reference
Simuliidae				
Prosimulium mixtum/fuscum	univoltine	lotic	Michigan	2677
Simulium jenningsi	multivoltine (3 – 6 / yr)	lotic	West Virginia	4147
Chironomidae				
Chironomus riparius	univoltine	lentic	Alberta	3252
Chironomus (2 spp.)	merovoltine (7 yrs?)	lentic	Alaska	501
Chironomus plumosus	bivoltine	lentic	Wisconsin	1725
Chironomus tenuistylus	merovoltine (3 yrs)	lentic	Wisconsin	502
Cladotantarsus mancus	bivoltine	lotic	Indiana	277
Corynoneura sp.	multivoltine (7-8 /yr)[†]	lotic	North Carolina	1873
Cricotopus bicinctus	trivoltine	lotic	Indiana	277
Cricotopus triannulatus	bivoltine	lotic	Indiana	277
Cricotopus trifascia	bivoltine	lotic	Indiana	277
Cricotopus sp.	multivoltine (~30/yr)[†]	lotic	Arizona	1469
Diamesa incallida	multivoltine (8-10/yr)[†]	lotic	Germany	2932
Diamesa nivoriunda	bivoltine	lotic	Indiana	277
Dicrotendipes sp.	multivoltine (~30/yr)[†]	lotic	Arizona	1469
Einfeldia synchrona	univoltine	lentic	Ontario	837
Eukiefferiella brevinervis	multivoltine (4/yr)	lotic	Indiana	277
Eukiefferiella devonica	trivoltine	lotic	Indiana	277
Glyptotendipes paripes	univoltine	lentic	Alberta	3252
Heleniella sp.	multivoltine (2-3/yr)	lotic	North Carolina	1873
Micropsectra sp.	multivoltine (2 / yr)	lotic	North Carolina	1873
Orthocladius obumbratus	bivoltine	lotic	Indiana	277
Pagastia sp.	bivoltine	lotic	Indiana	277
Parachaetocladius abnobaeus	univoltine	lotic	Ontario	1387
Paratendipes albimanus	univoltine	lotic	Michigan	4229
Pseudorthocladius sp.	univoltine	lotic	North Carolina	1873
Rheosmittia sp.	bivoltine	lotic	Alberta	3749
Rheotanytarsus exiquus	bivoltine	lotic	Indiana	277
Robackia demeijerei	univoltine?	lotic	Alberta	3749
Xylotopus par	univoltine	lotic	Michigan	2059
Tabanidae				
Chrysops (several spp.)	primarily univoltine	lotic & lentic	Florida	2018
Tabanus attratus	univoltine	lotic & lentic	Florida	2018
Tabanus fumipennis	univoltine	lentic	Florida	2018
Athericidae				
Atherix lantha	univoltine	lotic	Quebec	2276
Empididae				
several species	univoltine?	lotic & lentic	Quebec	1575
Hemerodromia empiformis	bivoltine?	lotic	Quebec	1575
Ephydridae				
Dichaeta caudata	multivoltine (4-6/yr)	lentic & riparian	Ohio	1009
Discocerina obscurella	multivoltine (2-5/yr)	lentic & riparian	Ohio	1256

[†] based on larval growth rates obtained from *in situ* measurements.

Table 5B. Continued

Taxon	Number of Generations Per Year	Habitat of Study	Location	Reference
Sciomyzidae				
Elgiva rufa	multivoltine (3-8?/yr)	lentic (littoral)	Northern U.S. & Canada	2135
Pherbellia (13 spp.)	primarily multivoltine (2 - 9 /yr) with some univoltine	lentic (littoral)	Nearctic	398
Sepedon spp.	primarily multivoltine	lentic (littoral)	North America	2871
Tetanocera loewi	univoltine	lentic (littoral)	New York	271
Tetanocera (3 spp.)	trivoltine	lentic (littoral)	Northern U.S.	4046
Scathophagidae				
Cordilura (*Achaetella*) (2 spp.)	bivoltine	(emergent zones)	Virginia & West Virginia	2872
Cordilura (11 spp.)	bivoltine & univoltine	(emergent zones)	Eastern United States	4188
Orthacheta hirtipes	bivoltine	(emergent zones)	Virginia & West Virginia	2873
Muscidae				
Limnophora riparia	bivoltine	lotic	England	2681

overland crawling to nearby pools will determine the site in which the larvae develop.

In general, taxa with specialized habitat requirements might be expected to exhibit the greatest degree of habitat selection by the ovipositing female. Insect parasitism (which is rare in aquatic insects) requires habitat specialization. The Palaearctic wasp, *Agriotypus* sp., crawls into the water enveloped by a film of air and seeks out a mature larva or pupa of the caddisflies *Silo* sp. or *Goera* sp. (Goeridae). An egg is then deposited within the caddisfly case so that the *Agriotypus* sp. larva can develop as a parasite of the pupal stage of the caddisfly (Clausen 1931).

Oviposition and Eggs

The diversity of oviposition strategies is illustrated in Table 5A. Hinton (1981) provides a comprehensive treatment of insect eggs. Much of the life cycle can be deduced from the timing and habits of oviposition and the location of the eggs, but this aspect of natural history frequently receives less attention than it deserves. Elliott and Humpesch (1980) point out that information on fecundity, oviposition behavior, and hatching is essential for: (1) interpretation of life cycles; (2) identification of larval cohorts; (3) study of spatial patterns and movements; (4) construction of life tables; and (5) estimation of growth rates, mortality, and production.

Oviposition behavior has been studied extensively in mosquitoes because disease transmission usually involves an oviposition cycle before pathogen transmission occurs at the next blood meal. Bentley and Day (1989) reviewed the integration of physical and chemical cues involved in the search behavior. Females use vision to identify habitat and site characteristics. Then short-range cues become important; for example, temperature and chemical signals received by contact chemoreceptors (see also Chap. 24).

The number of eggs varies greatly among taxa and among individuals of a species. Within the Ephemeroptera, *Dolania* sp. deposits about 100 eggs (Peters and Peters 1977), whereas the fecundity of *Ecdyonurus* sp. is 5000-8000 eggs per female (Elliott and Humpesch 1980). The number of eggs produced by *Baetis rhodani* ranges from 600 to 2400 per female (Elliott and Humpesch 1980). The within-species variation in fecundity is related to size of the female, which in turn is associated with growth conditions of the larvae. In taxa in which feeding by adults is required for egg maturation, fecundity will depend not only on food availability but also on appropriate oviposition conditions, because eggs may be resorbed if conditions are unsuitable for oviposition.

The eggs of some insects mature in the pupa or last larval stage (e.g., Megaloptera, Ephemeroptera, some Plecoptera) and are ready to be laid soon after emergence and mating. At the other extreme (e.g., Odonata, Hemiptera, some Plecoptera, and Coleoptera), adults emerge with undeveloped ovaries, require a feeding period before oviposition (or overwintering), and deposit eggs over an extended period or in discontinuous clutches. Potential fecundity can be estimated by dissection of newly emerged females if eggs mature all at one time. However, when egg maturation is a continuing process or when successive clutches are produced, total egg production depends on the length of life of individual adults.

The egg is potentially a vulnerable stage because it lacks mobility. However, in many instances the egg is the most resistant stage of the life cycle and spans periods of cold, heat, drought, or food shortages. The hatching period may extend for several months, which spreads the risk over a range of environmental conditions. Extended hatching may be due to extended oviposition periods, irregular rates of development of individual embryos, or to irregular breaking of diapause.

Larval Growth and Feeding

Differences in duration of the larval stage and in growth patterns provide for much of the variation in life histories. The examples presented below illustrate food acquisition strategies and larval growth patterns that adapt species to particular environments.

Though the fauna of temporary habitats is quite limited, most orders of aquatic insects are represented and these exhibit extreme seasonal regulation of life cycles (Williams 1987). Wiggins *et al.* (1980) describe several life-cycle patterns for temporary ponds, depending on the timing of adult colonization and kind of resistant stage. The prime requisite for all species is rapid larval development during the wet phase. Many mosquitoes are temporary-habitat breeders; under optimum conditions some species can complete their growth in five days. Rapid growth implies adaptation to warm waters as even the species that emerge early in snow-melt pools (e.g., limnephilid caddisflies) will be exposed to high temperatures when water volume diminishes. The timing of recruitment in temporary pools is correlated with feeding behavior. The earliest larvae are detritivores, either shredders (e.g., limnephilid caddisflies) or fine-particle feeders (e.g., siphlonurid mayflies, *Aedes* sp. mosquitoes, Chironomidae), then algal feeders (e.g., *Callibaetis* sp. mayflies, haliplid beetles) and finally predators (e.g., Odonata, Hemiptera, dytiscid and gyrinid beetles) are recruited coinciding with abundant prey resources.

Fast development is characteristic of temporary inhabitants of water, but larval diapause or quiescence may extend the life cycle to several years. Corydalidae (Megaloptera) larvae in intermittent streams burrow into the streambed when surface water dries up; growth only occurs during the wet cycle so larval duration may be 3-4 years, depending on the annual duration of stream flow. The extreme example of tolerance of drought is the chironomid, *Polypedilum vanderplanki,* from ephemeral pools in Africa. The larvae can withstand complete dehydration and exist in a state of suspended metabolism, or cryptobiosis. The dehydrated larvae can survive immersion in liquid helium and heating to over 100°C (Hinton 1960). In their natural habitat, they can survive several years in sun-baked mud and rehydrate when wetted (McLachlan and Cantrell 1980). They are normally the first invaders of small pools and can inhabit the shallowest and most ephemeral pools with virtually no competitors.

Many stream insects are adapted to a narrow range of cool temperatures (cold stenothermy). Hynes (1970b) attributes winter growth not only to use of leaf detritus as a food base, but also suggests that this growth pattern may have been selected for because predation by fish would be less at low temperatures. Larval diapause to avoid high summer temperatures occurs in some early-instar Plecoptera (e.g., Capniidae, Taeniopterygidae) and in mature larvae or prepupae of some limnephilid and uenoid caddisflies (e.g., *Dicosmoecus* sp., *Pycnopsyche* sp., *Neophylax* sp.).

Functional feeding groups may facilitate temporal and spatial partitioning of resources. Thus, in addition to shredders, the winter-growing stream species include scrapers (e.g., heptageniid mayflies, glossosomatid caddisflies), filter-feeders (filtering collectors, e.g., Simuliidae), and deposit feeders (gathering collectors, e.g., several mayflies, some Chironomidae). Within a group, coexistence may be based on differential responses to food and temperature. For example, three coexisting species of the caddisfly genus *Pycnopsyche* are all shredders and winter growers, but the food resource is partitioned by the species having different timing of rapid growth intervals and also some differences in microhabitat preferences (Mackay 1972).

Elaborate and specialized feeding adaptations occur in the filter-feeding Ephemeroptera, Trichoptera, and Diptera (Wallace and Merritt 1980). The adaptations include specialized anatomical structures (e.g., leg setae in *Isonychia* spp. mayflies and *Brachycentrus* spp. caddisflies, mouth brushes in mosquitoes, head fans in *Simulium* spp.), and silk nets in many caddisflies and some chironomid midges. Habitat partitioning within the filter-feeding guild occurs along the water velocity gradient. Also, there is some selectivity in feeding habits. Georgian and Wallace (1981) demonstrated that hydropsychid caddisflies that build large-meshed nets filter larger volumes of water than do those species with small nets and mesh size; the former nets will select for animal or algal foods rather than for detritus.

Black fly larvae attach the posterior end of the body to the substratum with the dorsal surface facing upstream and the body rotated 90 - 180 degrees longitudinally. Although the cylindrical body shape reduces drag, lateral deviation (yaw) created by the flowing water generates vortices which pass through the cephalic fans (Craig and Galloway 1986, Currie and Craig 1986). In general, larvae that inhabit fast-flowing water have smaller cephalic fans in proportion to the size of the head, whereas larvae inhabiting slower currents have relatively larger fans (Currie and Craig 1986). Black fly larvae show no selectivity with respect to food quality and will readily ingest inorganic as well as organic materials. Their food includes detritus, bacteria, diatoms, and animal fragments, and particle sizes of ingested material range from colloidal to 350 μm. Because they do not select for food quality, the type of food available will affect growth rates (Wotton 1994). Carlsson *et al.* (1977) found that black fly larvae at lake outfalls occurred in very dense aggregations and had exceptionally high growth rates associated with the availability of colloidal-sized organic material washed into the river at ice melt.

The above examples are primarily winter-growing lotic taxa. Most stream species respond to increasing temperatures in the spring by increasing their rate of growth. Multivoltine stream taxa include *Baetis* sp., *Glossosoma* sp., some hydropsychid caddisflies, and several black fly species. Though typical life cycles are univoltine, the duration and timing are more indeterminate than is usually suggested. Pritchard (1978) cites examples of cohort splitting in which individuals of the same cohort may have 1-, 2-, or even 3-year life cycles, depending on food availability and environmental conditions.

Aquatic insect predators include large conspicuous species with relatively long life cycles compared with those of their prey. Some predaceous beetles belonging to the Adephaga are notable exceptions in that larval development occurs in a few months, or less, and adults are long-lived. Predators have a range of morphological specializations and behavior patterns. Some are ambush or "sit-and-wait" predators, whereas others actively pursue their prey. The Hemiptera and some Coleoptera (e.g., Dytiscidae, Gyrinidae) are aquatic predators both as larvae and adults, whereas all others (e.g., Odonata, Megaloptera, some Diptera) are aquatic only as larvae.

Some of the most specialized predators are larval Odonata; the hinged prehensile labium, or mask, is unique to this order. The mask is projected forward by elevated blood pressure induced by abdominal muscles. Prey are impaled by hooks or setae on the labial palpi. The food is then returned to the mouth when the labium is folded back by adductor muscles. Prey perception involves receptor organs on the antennae and tarsi as well as use of the eyes. Sight is more important in later instars and in climbing species that live on vegetation than in bottom-sprawling or burrowing forms (Corbet 1963). In mature larvae of visual hunters, such as *Anax* sp., prey capture involves a highly integrated binocular vision, resulting from stimulation of certain ommatidia in each eye that enables the distance of the prey to be accurately judged. Feeding behavior of odonates is influenced by factors such as degree of hunger, time since the last molt, and the density, size, and movement of potential prey. As larvae grow, individual prey items become larger and more varied because larger larvae can also consume small prey (Corbet 1980).

Preoral digestion is a feature of hemipteran and some beetle predators. Salivary secretions are injected to immobilize prey and to liquefy tissues with hydrolytic enzymes. In Hemiptera the mouthparts are modified into a 3- or 4-segmented beak that is used to pierce the prey and suck the fluids. Dytiscid beetle larvae have chewing mouthparts, but their long sickle-shaped mandibles are grooved for fluid feeding. Larvae of some hydrophilid beetles (e.g., *Tropisternus* sp. and *Hydrophilus* sp.) are unusual in that the prey is held out of the water. In this position, the prey juices flow down the mandibles and into the oral opening rather than being lost into the surrounding water (Wilson 1923b).

Territoriality and intraspecific competition were shown by Macan (1977) to be important factors affecting growth rate and life cycle of the damselfly *Pyrrhosoma* sp. In years when the larvae were abundant and prey populations were low, two size classes existed at the end of the summer. Macan attributed this to cohort splitting; larvae in superior feeding territories grew rapidly whereas those occupying poorer feeding sites grew slowly and would either require an extra year or die of starvation. Furthermore, Macan suggests that fish predation was selective for the larger specimens, which led to vacancies in the superior feeding sites, and that these were then readily filled by smaller larvae. Thus, predation by fish did not greatly reduce the numbers of damselflies reaching maturity because elimination of large larvae allowed smaller ones to exploit the food resource.

The prevalence of insect predators suggests that predation may be a dominant biotic factor influencing aquatic insects. Anti-predator defenses of aquatic insects include: refugia afforded by the substratum; camouflage including coloration; reduced movement and feeding in the presence of predators; potential detection of predators by chemical cues; and various defensive repertoires used by potential prey (Williams and Feltmate 1992). Defensive repertoires used by aquatic insects include: a) reflex bleeding (Benfield 1974, Moore and Williams 1990) and death feigning in stoneflies; b) deimatic behavior, as displayed by some mayfly prey which display scorpionlike threat postures, or (c) accelerative drift in the presence of predators (Peckarsky 1980; Peckarsky and Dodson 1980).

Other defenses found in aquatic insects include morphological and chemical defenses. For example, hairy bodies that can be effective against other invertebrates such as *Hydra* (Hershey and Dodson 1987). Some chemicals promote rapid escape and return to shore in some semiaquatic beetles and veliid bugs. Glands at the tips of their abdomens discharge secretions that lower the surface tension of water at the rear of the animal so that the animal is being drawn forward by the normal surface tension of water in front (Jenkins 1960; Linsenmair and Jander 1963). This mechanism propels the small insects forward at speeds of 0.4 - 0.75 m/s, and was coined "Entspannungsschwimmen" or expansion swimming by Linsenmair and Jander (1963). There is evidence that some mayflies can detect prey by chemoreception (Peckarsky 1980; Kohler and McPeek 1989); however, in lotic systems with unidirectional or chaotic flow paths the use of chemical cues may be limited (Dodds 1990). Chemical cues may be most effective only in direct contact with predators as this provides precise information about the identity (predator or non-predator) of the intruder; and it avoids the high cost of premature escape in lotic environments (Ode and Wissinger 1993).

Evaluating the impact of predation on aquatic invertebrates, especially in lotic habitats, has been somewhat difficult. Cooper *et al.* (1990) have identified several problems associated with assessing the impact of predators in aquatic ecosystems. These problems involve primarily prey exchange (immigration/emigration) among reaches and may have an overwhelming influence on the perceived effects of predators. In some cases predator/prey relationships have been assessed in cages. Mesh size of cages exerts a strong influence on exchange rates and assessment of impacts of predators. Small meshed cages curtail exhange rates and indicate greater predator effects. At larger spatial scales, drift into and out of study reaches may influence results, and reduced prey mobility increases the probability of a significant predator impact on prey populations. Cooper *et al.* (1990) also suggested that lotic ecosystems may display a lower impact of predators on community structure than that of lentic systems since fast-flowing streams exhibit a much greater exchange rate among habitats than lentic habitats. Sih and Wooster (1994) have published a recent paper that examines some of the assumptions of the above model of Cooper *et al.* 1990 that should be consulted by all students interested in predator-prey relationships.

Metamorphosis and Eclosion

Molting during the larval stages results in a larger insect of essentially the same body form, whereas the final molt produces a more complete change associated with the development of wings and other adult structures. These are considered two distinct types of physiological differentiation, and Chapman (1982) suggests that the term *metamorphosis* be restricted to the latter. Molting and metamorphosis are mediated by hormones. There is no fundamental physiological difference between the metamorphosis of hemimetabolous and holometabolous insects; the difference is a matter of degree rather than kind. The pupal instar may be regarded as the equivalent of the last larval instar of hemimetabolous insects (Gillott 1980).

The primary morphological difference between hemimetabolous (Exopterygota) and holometabolous (Endopterygota) insects is the external development of the wings in the former and the delayed eversion of wings in the latter (Hinton 1963). The gradual development of adult structures is apparent in hemimetabolous insects by the progressive development of external wing pads and rudiments of the genitalia. Changes also occur in the last larval instar of the Holometabola, but these are internal and major differentiation is constrained by lack of space. During the pupal instar, the wings are evaginated to outside the body; this makes room for development of the indirect flight muscles and the reproductive system. The pupal instar is usually of short duration. In the strict sense, the insect becomes an adult immediately after the apolysis (separation) of the pupal cuticle and the formation of the adult epicuticle to which the musculature is now attached (Hinton 1971a). Thus, in most instances, locomotion and mandibular chewing are activities, not of the pupa, but of the *pharate* adult that is enclosed within the pupal exuviae.

Control of molting involves the interaction of a molting hormone (ecdysone) and a juvenile hormone. The latter exerts an influence on development only in the presence of the former. When the concentration of juvenile hormone in the blood is high, the next molt will be larval-larval. At intermediate concentrations, a larval-pupal molt occurs, and when there is little or no circulating juvenile hormone (due to inactivity of the corpora allata of the brain) an adult insect will emerge at the next molt (Gillott 1980).

The final event in metamorphosis is *eclosion,* or emergence, which is the escape of the adult from the cuticle of the pupa or last larval instar. See Chapter 9 for a further discussion of aquatic insect metamorphosis.

The development of a pupal stage has permitted the great divergence of larval and adult forms in the Holometabola. Larvae exploit environments and food resources that result in growth, whereas the activities of the adult center on dispersal and reproduction. If the number of species is considered to be a measure of biological success, complete (holometabolous) metamorphosis is a prime contributor to the success of insects. Hinton (1977) states that this is a focal point of insect evolution because about 88% of the extant insect species belong to the Endopterygota.

Metamorphosis in hemimetabolous aquatic insects differs in detail among the various taxa but the behaviors involved are mostly associated with switching from an aquatic to a terrestrial mode of life (especially respiration and flight) and overcoming the abiotic and biotic hazards during an especially vulnerable stage. Data from Corbet (1963) for Odonata illustrate both morphological and behavioral changes occurring during metamorphosis and eclosion.

The onset of metamorphosis in some dragonflies can be detected several weeks before emergence when the facets of the adult compound eye begin to migrate to the top of the head. Respiration rate also increases prior to metamorphosis associated with increased metabolic requirements. Behavioral changes characteristically involve movement to shallow water or up the stems of plants towards the surface. In *Anax imperator,* histolysis of the labium occurs and feeding stops several days before emergence.

Nearly all Odonata have a diurnal rhythm of emergence. The timing is presumably to restrict emergence to an interval when weather conditions are favorable and when predation is least likely to occur. The major predators are birds or mature adult dragonflies that hunt by sight, so night is the safest time to leave the water. In the tropics, most of the large dragonflies emerge after dusk; they eclose, expand their wings, and are then ready to fly before sunrise. In the temperate regions or at high altitudes where nocturnal temperatures are low, odonates tend to emerge in early morning or during the daytime. Mortality during emergence of *Anax imperator* may amount to 16% of the annual population (Corbet 1963). The individuals are immobile and defenseless for several hours during eclosion and while the wings expand and harden. Cold and wind increase mortality by prolonging ecdysis or by postponing emergence and thereby exposing more individuals to predation. Overcrowding is common in species that have mass emergences; competition for emergence supports is so intense that the first larva to climb a support may be used as a platform by others that follow. Although overcrowding is an ecological disadvantage of mass emergence, this type of synchronization is apparently adaptive, perhaps because it satiates predators and affords the appropriate synchronization of males and females.

Adaptations of pupae pertain to respiration, protection, and emergence from the water. These factors also operate on the pharate adult, so it is convenient to discuss the two stages together. This is usually a quiescent stage, but mosquito pupae (tumblers) are relatively active swimmers, and Hinton (1958a) demonstrated that some black fly pupae both feed and spin a cocoon. Pupation occurs out of water in Megaloptera, Neuroptera, most Coleoptera, and many Diptera. Larvae construct a chamber in the soil (e.g., Megaloptera, Coleoptera) for metamorphosis, and eclosion may occur within the chamber (Coleoptera) or after the pharate adult has worked its way to the soil surface (Megaloptera).

Many nematocerous Diptera have aquatic pupae, some of which are active swimmers (e.g., Culicidae, Dixidae, Chaoboridae, some Chironomidae) throughout the pupal stage, whereas others (e.g., most Chironomidae, some Tipulidae) only swim to the surface for emergence. In taxa in which the pupa is glued to a substrate (e.g., Blephariceridae, Deuterophlebiidae, Simuliidae), the adult emerges underwater and rises to the surface enveloped in a gas bubble. These adults expand their wings and are capable of flying immediately after reaching the surface. Species with swimming pupae will emerge using the exuviae as a platform at the water surface. All Chironomidae emerge in this manner, and the cast skins remain trapped in the surface film for some time. Collections of exuviae are useful for taxonomic purposes and also as a rapid method of sampling the entire chironomid fauna of a water body (Wartinbee and Coffman 1976; also see Chap. 26).

Respiration in most aquatic Diptera pupae differs from that of the larvae in which the terminal abdominal spiracles are generally the most important. The pupae have respiratory horns or other extensions of the prothoracic spiracles. Depending on the species, these may be used for breathing at the surface, in the water column, or even to pierce the tissues of submerged plants.

The case-making caddisflies illustrate several morphological and behavioral adaptations in the pupal stage. The mature larva selects a protected site, such as under a stone or in a crevice in wood, and attaches the case with silk. Then the case is shortened, and the ends are closed with mineral or detrital particles and a silk mesh that allows for water flow. Metamorphosis occurs within the case. The pupa (pharate adult) is active within the case, maintaining a flow of water for respiratory purposes by undulating the body. The pupa has elongate bristles on the labrum and on anal processes that are used for removing debris from the silk grating at either end of the case.

The back-and-forth movement within the case is effected by the dorsal hook-plates on the abdomen. The pharate adult has large mandibles used for cutting an exit hole at emergence. It then swims to the surface or crawls to shore where eclosion occurs. In some limnephilid caddisflies, such as the genus *Pycnopsyche*, the larvae undergo a prepupal inactive period. For example, *P. guttifer* attaches its case to a log or rock and remains sequestered for most of the summer. The digestive tract atrophies, but the legs remain functional so that under conditions of unfavorable habitat changes (e.g., water level) the prepupa can move to a new location.

SURVIVORSHIP

Mortality among most aquatic insects is usually very high; a very small percentage of eggs deposited reach the adult stage. For example, Willis and Hendricks (1992) found that only about 0.5% of the eggs deposited by a caddisfly, *Hydropsyche slossonae*, survived to be adults of the ensuing generation. Sources of mortality include both abiotic (e.g., physical disturbances such as floods and drought) and biotic factors (e.g., predation, disease, parasitism, etc.). Within a given stream reach, downstream drift may also contribute to significant losses (or gains) of individuals through time (e.g., Hershey *et al.* 1993).

Survivorship curves for several aquatic insects are shown in Figures 5.3-5.6. Several of these include only larval stages (*Glossosoma*, *Brachycentrus*, and *Tallaperla*) and were fit by least squares regression beginning with peak abundances of first instar larvae in field samples. These best fit a Type II exponential model of mortality, representing constant deaths through time. However, first instars are generally undersampled (e.g., Benke 1984, Cummins and Wilzbach 1988). If they could be followed in sufficient detail, high mortalities in early instars would shift the curves toward a Type III curve (higher mortality in earlier larval stages). For example Willis and Hendricks (1992) detected no egg mortality in *Hydropsyche slossonae*, exceptionally high mortality (92.5%) among newly hatched larvae, due in part to sibling cannibalism (Fig. 5.5). Cummins and Wilzbach (1988) have proposed that pathogenic mortality in the first two instars might be a major cause of mortality in many cases. There was constant mortality of *H. slossonae* during larval instars II through V, followed by high mortality in the pupal stage. This study indicates that when followed in sufficient detail, Type III survivorship curves may be common in aquatic insects. Following larval mortality from field samples does not give a true picture of mortality over the entire egg to adult cycle.

Eggs, larvae, pupae, and adults of the water lily leaf beetle, *Gallerucella nymphaeae*, occur on the upper surfaces of leaves, and this offers a unique situation in that all stages can be sampled readily. Total number of eggs deposited, larvae reaching each instar, and pupae can be calculated per m^2 of leaf surface area (Otto and Wallace 1989). Comparisons of populations of *G. nymphaeae* from Georgia (U.S.A.) and southern Sweden indicate that both populations display Type II survivorship curves, with constant mortality (Fig. 5.6). However, the Georgia population has much faster growth and shorter generation time, as well as much greater mortality, indicated by the steep slope of the survivorship curve. There were also many other differences between these populations with respect to clutch size, investment in reproduction, larval densities, and terminal weights of larvae and adults. These differences suggest that these populations are subject to very different selection pressures and gene flow, and possibly could be considered different species (Otto and Wallace 1989).

CONCLUDING COMMENTS

The critical importance of systematics to basic life-history studies has been emphasized on numerous occasions (e.g., Wiggins 1966; Ross 1967a; Waters 1979a, 1979b). Excellent progress has been made for several aquatic insect groups in North America (e.g., Edmunds *et al.* 1976; Wiggins 1977; Stewart and Stark 1988) as well as more general works on immatures (e.g., Stehr 1987, 1991); however, eggs and all instars of immatures cannot always be identified. Studies of aquatic insect life histories require that individual investigators develop reliable methodologies for separating early instars of closely related species (e.g., Mackay 1978). It is unfortunate that life histories of so many aquatic insect species remain unknown. In some quarters, the mistaken impression still persists that such efforts are unfashionable and of little value. Ecosystem-level studies are often directed to studies of the processing of organic matter by various groups of animals, and insects are frequently the most abundant group considered. The integration of production, feeding habit, and bioenergetic data can yield a much better understanding of the role of individual species in ecosystems. Benke (1979), Benke *et al.* (1979), and Waters (1979a,b) have emphasized that a knowledge of basic life histories is mandatory for reasonable estimates of production. Waters has pointed out that voltinism and length of aquatic life (=CPI, or Cohort Production Interval [Benke 1979]) are two of the most important life history features influencing secondary production estimates. The "once-per-month" sampling program of most studies is not adequate for many estimates of aquatic insect secondary production, and sampling schedules need to be tailored to the life histories of the organisms being studied (Cummins 1975; Waters 1979a).

Both temperature and food may influence life-history patterns; however, data on the potential combined effects of temperature and food quality on life histories (e.g., Anderson and Cummins 1979) of various groups of aquatic insects are too meager for any broad generalizations as yet. All students are encouraged to develop an appreciation for the importance of systematics, life histories, secondary production, and bioenergetics as interconnecting links toward the basic understanding of the structure and function of aquatic communities.

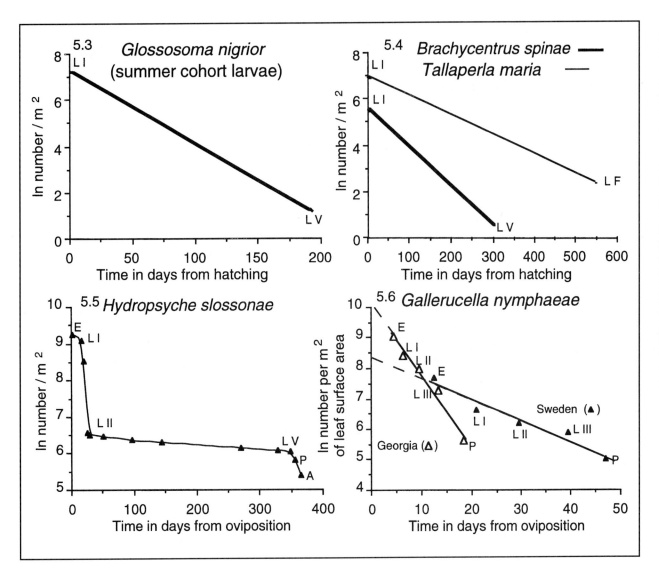

Figure 5.3. Survivorship curves for the summer cohort of the caddisfly *Glossosoma nigrior* (larval instars I [L I] through V [L V] based on instream sampling) (data from Georgian and Wallace [1983]).

Figure 5.4. Survivorship curves for larval stages of the caddisfly *Brachycentrus spinae* (larval instars I [L I] through V [L V]) (data from Ross and Wallace [1981]), and the stonefly *Tallaperla maria* larval instars I (L I) through final instar (L F) (data from O'Hop et al. [1984]).

Figure 5.5. Survivorship curve for the caddisfly *Hydropsyche slossonae* based on egg, larval, pupal, and adult stages. Note the high mortality during the first larval instar (L I), followed by little mortality in later larval stages (L II - L V), and high mortality in pupal to adult stages (data from Willis and Hendricks 1992).

Figure 5.6. Survivorship curve for egg (E), three larval instars (L I - L III), and pupal stage (P) of the chrysomelid beetle, *Gallerucella nymphaeae,* from Georgia (U.S.A.) and Sweden. Note the much shorter life cycle and greater mortality in the Georgia population (data from Otto and Wallace 1989). The equations for the lines are as follows: Georgia, $y = 10.1 - 0.23x$, $r^2 = 0.98$; and, Sweden, $y = 8.4 - 0.07x$, $r^2 = 0.95$.

ECOLOGY AND DISTRIBUTION OF AQUATIC INSECTS

Kenneth W. Cummins
South Florida Water Management District, West Palm Beach

Richard W. Merritt
Michigan State University, East Lansing

INTRODUCTION

As described in the Introduction to this book, aquatic insects have been a major focus of many ecological studies in aquatic habitats. A major reason for this is that aquatic insects, together with other freshwater macroinvertebrates (Cummins 1975; Thorp and Covich 1991), are the the major food items of important commercial and sport fish species. They serve as the link between their food (i.e., detritus plus microorganisms, algae, vascular aquatic plants) and fish, as well as other higher trophic level vertebrates. Furthermore, the distribution of aquatic insects is ubiquitous and they are normally abundant, easy to collect, and of sufficient size to be observed with the unaided eye. Thus, it is not surprising that they have been used widely to evaluate water quality and the food support base for fisheries (Hilsenhoff 1987; Waters 1988; Yount and Niemi 1990; Karr 1991; Rosenberg and Resh 1993; also Chap. 7).

The perpetually imperfect state of our knowledge of aquatic insect species taxonomy, together with pressing environmental and intriguing theoretical problems, continue to provide incentives for understanding aquatic systems by utilizing insect specimens identified to different levels of taxonomic resolution. Although the species may be the appropriate basic unit for many types of ecological questions (e.g., Mayr 1969), such resolution may not always be required in order to achieve significant insights in ecological work, particularly in process-oriented studies. Taxonomic separations ranging from orders to genera, and species in cases requiring a detailed level of taxonomic resolution, can be matched with the requirements of a given study. Stated in this fashion, the goal in each investigation is to maximize ecological information and insight per unit of taxonomic effort. The ecological tables at the end of each taxonomically defined chapter are designed to facilitate this process (the table entries are based on the methods of analysis summarized in Tables 6A-6C)

The major topics covered in this chapter are: (1) organization and functional relationships of aquatic insects with respect to habitat and nutritional resources, and (2) various aspects of the distribution, abundance, and production of aquatic insects. To summarize a large amount of information on the ecology and geographical distribution of aquatic insects, and to better indicate the numerous gaps in our knowledge, a generalized scheme was developed that categorizes aquatic habitats (Table 6A), modes of existence (e.g., habit, locomotion, attachment, concealment; Table 6B), and functional feeding groups (i.e., modes of food acquisition; Table 6C). As indicated above, a designation from each of these three categories is assigned to each genus in the ecological tables at the end of each taxonomic chapter and the definitions of the categories used in these tables appear in Tables 6A-6C. The habitat, habit, and food acquisition classifications are based on our own research (Cummins 1962, 1964, 1972, 1973, 1974, 1975, 1980a, 1980b, 1993; Merritt *et al.* 1978, 1982, 1984, 1992; Cummins and Klug 1979; Merritt and Lawson 1979, 1992; Wallace and Merritt 1980; Merritt and Wallace 1981; Cummins and Wilzbach 1985; Merritt and Cummins 1996), a combination of systems proposed by others (Usinger 1956a; Edmondson 1959; Klots 1966; Pennak 1978), critical reviews of the methods (Simberloff and Dayan 1991; Hawkins and MacMahon 1989), and three decades of teaching aquatic insect-related courses.

HABITAT ORGANIZATION

Aquatic insect adaptations to habitat types are treated in Chapters 5 and 8. The summary classification used here (Table 6A) also stresses the basic distinction between lotic (running waters, i.e., streams-rivers) and lentic (standing waters, i.e., ponds-lakes) habitats. This separation is generally useful in partitioning aquatic insect genera (as well as families, and species in some cases). However, a general separation into erosional and depositional habitats, which has been in the literature for over 80 years (e.g., Shelford 1914; Moon 1939), continues to provide the most all-encompassing conceptual framework. Both stream/river currents and lake shoreline waves create erosional habitats while lake basins, river floodplain pools, and stream/river backwaters provide depositional conditions. In the early phases of flow, intermittent streams may be largely erosional, but become primarily depositional as seasonal flows diminish. Therefore, the key features of habitat organization appear to be the physical-chemical differences among erosional, depositional, and semi-aquatic conditions. Often the same genera (or even species) occur in similar habitats in both lotic and lentic ecosystems. Such correspondence between running and standing water habitats is shown in Figure 6.1. In general, the cobble-riffles of headwater streams are similar to the rocky shores of oligotrophic (clear, low productivity) lakes. The deep pools and floodplain wetlands of larger rivers are similar to

Table 6A. Aquatic habitat classification system.

General Category	Specific Category	Description
Lotic-erosional (running-water riffles)	Sediments	Coarse sediments (cobbles, pebbles, gravel) typical of stream riffles.
	Vascular hydrophytes	Vascular plants growing on (e.g., moss *Fontinalis*) or among (e.g., pondweed *Potamogeton pectinatus*) coarse sediments in riffles.
	Detritus	Leaf packs (accumulations of leaf litter and other coarse particulate detritus at leading edge or behind obstructions such as logs or large cobbles and boulders) and debris (e.g., logs, branches) in riffles.
Lotic-depositional (running-water pools and margins)	Sediments	Fine sediments (sand and silt) typical of stream pools and margins.
	Vascular hydrophytes	Vascular plants growing in fine sediments (e.g., *Elodea*, broad-leaved species of *Potamogeton*, *Ranunculus*).
	Detritus	Leaf litter and other particulate detritus in pools and alcoves (backwaters).
Lentic-limnetic (standing water)	Open water	On the surface or in the water column of lakes, bogs, ponds.
Lentic-littoral (standing water, shallow shore area)	Erosional	Wave-swept shore area of coarse (cobbles, pebbles, gravel) sediments.
	Vascular hydrophytes	Rooted or floating (e.g., duckweed *Lemna*) aquatic vascular plants (usually with associated macroscopic filamentous algae).
	Emergent zone	Plants of the immediate shore area, e.g., *Typha* (cattail), with most of the leaves above water.
	Floating zone	Rooted plants with large floating leaves, e.g., *Nymphaea* (pond lily), and non-rooted plants (e.g., *Lemna*).
	Submerged zone	Rooted plants with most leaves beneath the surface.
	Sediments	Fine sediments (sand and silt) of the vascular plant beds.
Lentic-profundal (standing water, basin)	Sediments	Fine sediments (fine sand, silt and clay) mixed with organic matter of the deeper basins of lakes. (This is the only category of "lentic-profundal.)
Beach zone	Freshwater lakes	Moist sand beach areas of large lakes.
	Marine intertidal	Rocks, sand, and mud flats of the intertidal zone.

Table 6B. Categorization of aquatic insect habits, that is, mode of existence.

Category	Description
Skaters	Adapted for "skating" on the surface where they feed as scavengers on organisms trapped in the surface film (example: Hemiptera, Gerridae-water striders).
Planktonic	Inhabiting the open water limnetic zone of standing waters (lentic; lakes, bogs, ponds). Representatives may float and swim about in the open water, but usually exhibit a diurnal vertical migration pattern (example: Diptera, Chaoboridae-phantom midges) or float at the surface to obtain oxygen and food, diving when alarmed (example: Diptera: Culicidae-mosquitoes).
Divers	Adapted for swimming by "rowing" with the hind legs in lentic habitats and lotic pools. Representatives come to the surface to obtain oxygen, dive and swim when feeding or alarmed; may cling to or crawl on submerged objects such as vascular plants (examples: Hemiptera, Corixidae-water boatmen; Coleoptera, adult Dytiscidae-predaceous diving beetles).
Swimmers	Adapted for "fishlike" swimming in lotic or lentic habitats. Individuals usually cling to submerged objects, such as rocks (lotic riffles) or vascular plants (lentic), between short bursts of swimming (examples: Ephemeroptera in the families Siphlonuridae, Leptophlebiidae).
Clingers	Representatives have behavioral (e.g., fixed retreat construction) and morphological (e.g., long, curved tarsal claws, dorsoventral flattening, and ventral gills arranged as a sucker) adaptations for attachment to surfaces in stream riffles and wave-swept rocky littoral zones of lakes (examples: Ephemeoptera, Heptageniidae; Trichoptera, Hydropsychidae; Diptera, Blephariceridae).
Sprawlers	Inhabiting the surface of floating leaves of vascular hydrophytes or fine sediments, usually with modifications for staying on top of the substrate and maintaining the respiratory surfaces free of silt (examples: Ephemeroptera, Caenidae; Odonata, Libellulidae).
Climbers	Adapted for living on vascular hydrophytes or detrital debris (e.g., overhanging branches, roots and vegetation along streams, and submerged brush in lakes) with modifications for moving vertically on stem-type surfaces (examples: Odonata, Aeshnidae).
Burrowers	Inhabiting the fine sediments of streams (pools) and lakes. Some construct discrete burrows which may have sand grain tubes extending above the surface of the substrate or the individuals may ingest their way through the sediments (examples: Ephemeroptera, Ephemeridae-burrowing mayflies; Diptera, most Chironominae, Chironomini-"blood worm" midges). Some burrow (tunnel) into plant stems, leaves, or roots (miners).

Table 6C. General classification system for aquatic insect trophic relations. (Applicable only to immature and adult stages that occur in the water.)

Functional Group (General category based on feeding mechanism)	Subdivision of Functional Group		General Particle Size Range of Food (microns)	Orders Having Dominant Representatives											
	Dominant Food	Feeding Mechanism		Collembola	Plecoptera	Odonata	Ephemeroptera	Hemiptera	Megaloptera	Neuroptera	Trichoptera	Lepidoptera	Coleoptera	Hymenoptera	Diptera
Shredders	Living vascular hydrophyte plant tissue	Herbivores-chewers and miners of live Macrophytes	>10^3								+	+	+		+
	Decomposing vascular plant tissue-coarse particulate organic matter (CPOM)	Detritivores-chewers of CPOM			+		(+)				+		+		+
	Wood	Gougers-excavate and gallery, wood									+		+		+
Collectors	Decomposing fine particulate organic matter (FPOM)	Detritivores-filterers or suspension feeders	<10^3				+				+				+
		Detritivores-gathers or deposit (sediment) feeders (includes feeders on loose surface films)		+			+	+			+		+		+
Scrapers	Periphyton-attached algae and associated material	Herbivores-grazing scrapers of mineral and organic surfaces	<10^3				+	+			+	+	+		+
Macrophyte Piercers	Living vascular hydrophyte cell and tissue fluids or fimamentons (macroscopic) algal cell fluids	Herbivores-pierce tissues or cells and suck fluids	>10^2->10^3								+	+			
Predators	Living animal tissue	Engulfers-carnivores, attack prey and ingest whole animals or parts	>10^3		+	+	(+)		+		+		+	+	+
		Piercers-carnivores, attack prey, pierce tissues and cells, and suck fluids						+		+			+	+	+
Parasites	Living animal tissue	Internal parasites of eggs, larvae and pupae. External parasites of larvae, prepupae and pupae in cocoons, pupal cases or mines. Also, external parasites of adult spiders.	>10^3											+	+

lentic marshes and eutrophic (low transparency, high productivity) ponds and lakes. In addition, there usually is little difference in the aquatic insect fauna between lakes and artificial impoundments (reservoirs) on large rivers. A well known example of lotic and lentic co-occurrence is the North American snail-case caddisfly *Helicopsyche borealis* (Trichoptera; Helicopsychidae), which inhabits erosional habitats of both lakes and streams. Other examples of co-occurrence in these lotic and lentic erosional habitats are species of ephemerid and heptageniid mayflies and net-spinning hydropsychid caddisflies.

Within a given habitat (Table 6A), the modes by which individuals maintain their location, for example clingers on surfaces in fast-flowing water or burrowers in soft sediments, or move about, such as swimmers or surface skaters, have been categorized in Table 6B. The distribution pattern resulting from habitat selection by a given aquatic insect species reflects the optimal overlap between habit and physical environmental conditions that comprise the habitat: substrate, flow, turbulence, etc. Because food is almost always distributed in a patchy fashion, certain locations within a species' general distribution where the habitat-habit match is maximized will con-

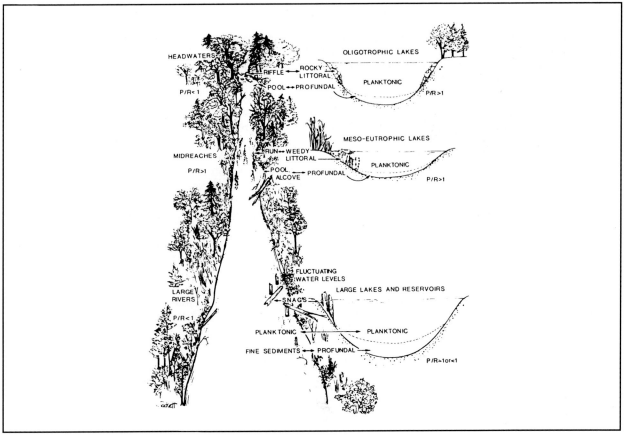

Figure 6.1. Comparison of running (lotic - streams, rivers) and standing (lentic - ponds, lakes) aquatic ecosystems showing similarities. Emphasis is on similarities of shore line and channel vs basin habitats and on nutritional resource categories used by aquatic insects in the two types of systems. Autotrophy to heterotrophy (P/R) index is the ratio of gross primary production to community respiration averaged over an annual cycle. After Merritt et al. (1984).

stitute areas with the highest probability of occurrence of the species. Thus, the discontinuous (= patchy, contagious, negative binomial) distribution pattern of an aquatic insect population is the result of interplay among habitat (e.g., sediment particle size, large wood debris), habit (i.e., mode of existence), and the availability of food resources (Cummins 1964, 1993; Lauff and Cummins 1964; Anderson et al. 1978; Reice 1980; Williams 1981a; Resh and Rosenberg 1984; Pringle et al. 1988).

The habit (mode of locomotion, attachment, or concealment) of a given taxon exerts a strong influence on the propensity and frequency of movement within the habitat. The most significant example is the portion of stream drift that follows a regular diel pattern (Waters 1972; Allan and Russek 1985; Kohler 1985; Allan et al. 1986; Wilzbach et al. 1988; Wilzbach and Cummins 1989; Wilzbach 1990; Hershey et al. 1993). Individuals with a regular drift pattern, that is usually more at night than in the day and with peaks at dusk and dawn, are termed behavioral drifters. Those that show no diel pattern are usually termed accidental drifters. There is an ongoing debate about the mechanisms leading to behavioral drift and adaptations that would favor avoidance of accidental displacement into the drift (e.g., Corkum 1978; Wilzbach et al. 1988). Stream insects are clearly adapted for responses to current, for example through hydrodynamic shapes that facilitate swimming in flowing water, sessile behavior, or avoiding the main thrust of the current by way of a dorsoventrally flattened body shape. There are some studies that support the idea that the latter mechanism is the most significant (Statzner et al. 1988). Behavioral drift might be triggered by depletion of local food supplies. In addition, Wilzbach (1990) using fiberoptics in field and laboratory observations proposed a role for pathogens in weakening individuals. Such weakened individuals are more likely to appear in the drift (Cummins and Wilzbach 1988; Wilzbach et al. 1988).

Functional classifications of habit and feeding groups integrated with drift-related morphology appear in Figure 6.2. Drift propensity is related to habit, which is reflected by body shape (expressed as height to width ratios) and also varies with functional feeding group. For example, many of the active (behavioral) drifters are in the swimming habit group and are usually generalized gathering-collectors, swimming frequently between food patches. They have a cylindrical, streamlined body shape with a height to width ratio near 1. Almost all of the filtering-collectors and scrapers are clingers (an exception would be the swimming, filtering-collector *Isonychia*). Scrapers and predators that are clingers (e.g., Heptageniidae, Psephenidae) have a flattened body (low height to width ratio) that avoids the main thrust of the current. These are normally accidental drifters that are dislodged into the current from their high risk locations on exposed, periphyton-rich surfaces. Burrowers generally have a cylindrical body shape (high ratio), such as many shredders (*Tipula*, and pteronarcid and nemourid stoneflies), that live concealed in their plant litter food resource and are accidentally introduced into the drift when the litter accumulations they inhabit are disturbed. Sprawlers may have low height to width ratios, but the adaptation is largely for wide bodies to maintain position on top of flocculant substrates in depositional areas. Their appearance in the drift is normally accidental at times when depositional areas are scoured. Some shredders (e.g., a number of limnephilid genera) are sprawlers living on litter accumulations or woody debris that constitutes their food (Fig. 6.2, Wilzbach *et al.* 1988).

TROPHIC ORGANIZATION AND FUNCTION

A major observation stemming from insect feeding studies is that, based on food ingested, essentially all aquatic insects are omnivorous, at least in their early instars (Cummins and Klug 1979; Clifford and Hamilton 1987). For instance, species that chew leaf litter (shredders), particularly large species in their later instars, ingest not only the leaf tissue (and associated microbiota: fungi, bacteria, protozoans, and microarthropods), but also diatoms and other algae that may be attached to the leaf surface, as well as very small macroinvertebrates such as first-instar larval midges. For this reason, the trophic level analysis pioneered by Lindeman (1942) and used extensively in investigations of trophic relationships in marine and terrestrial communities usually does not yield consistent information about linkages between insects and their food resources or the manner in which food resources might be partitioned among aquatic insect assemblages (Figs. 6.2, 6.3). Coffman *et al.* (1971) attempted to compensate for this omnivory found in most species in a study of a Pennsylvania woodland stream. They assigned aquatic invertebrate biomass to trophic levels (herbivore, detritivore, carnivore) based on the percent caloric intake by each age class of each species of three food categories — algae, detritus, and animal prey. The analysis, which was extremely time consuming, indicated that, on this basis, the invertebrate community was highly dependent on detritus, and more so than if only the obvious ingesters of dead organic matter were placed in the detrital trophic level. Clifford and Hamilton (1987) reported a similar dominance of detritus (90%) in insects from some Canadian streams.

A far less time consuming approach is a classification that is based on morpho-behavioral mechanisms of food acquisition, namely functional feeding groups. The same general morpho-behavioral mechanisms in different species can result in the ingestion of a wide range of food items, the intake of which constitutes herbivory (living plants), detritivory (dead organic

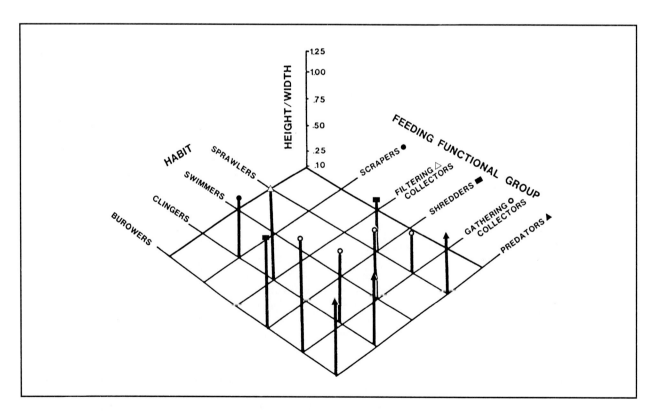

Figure 6.2. A three dimensional matrix relating aquatic insect habit (i. e. mode of locomotion, attachment, concealment) and functional feeding group (scrapers, filtering collectors, etc.) categories to an index of body shape (ratio of maximum body height to width) which provides a prediction of drift propensity. Height to width ratios represent means of measurements made on animals from several taxa belonging to various combinations of habit and feeding groups. After Wilzbach *et al.* (1988).

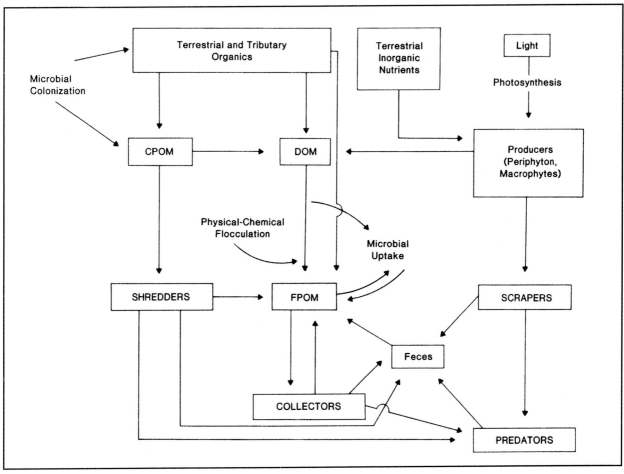

Figure 6.3. Nutritional resource categories and invertebrate functional feeding group categories in lotic ecosystems.

Table 6D. Comparison of functional feeding groups and trophic levels.

Functional Feeding Group	Trophic Level Based on Ingestion*
Shredders (live plant or dead plant)	Detritivores Herbivores (Carnivores)
Collectors	Detritivores Herbivores (Carnivores)
Filtering (suspension feeders)	Detritivores Herbivores (Carnivores)
Gathering (deposit feeders)	
Scrapers (grazers)	Herbivores Detritivores
Predators (engulfers)	Carnivores (Detritivores)
Piercers (plant or animal)	Unrecognizable fluids

*The occasional or minor component (on a biomass basis) of a trophic classification is shown in parentheses.

matter), or carnivory (live animal prey) (Table 6D). Although the type of food intake would be expected to change from season to season, from habitat to habitat, and with growth stage (larval instar, aquatic adult), limitations in food acquisition mechanisms have been shaped over evolutionary time and these are relatively more fixed (Arens 1989, 1990). An example of similar morphological structures that enable species to scrape attached periphyton can be found in the very similar mandibles

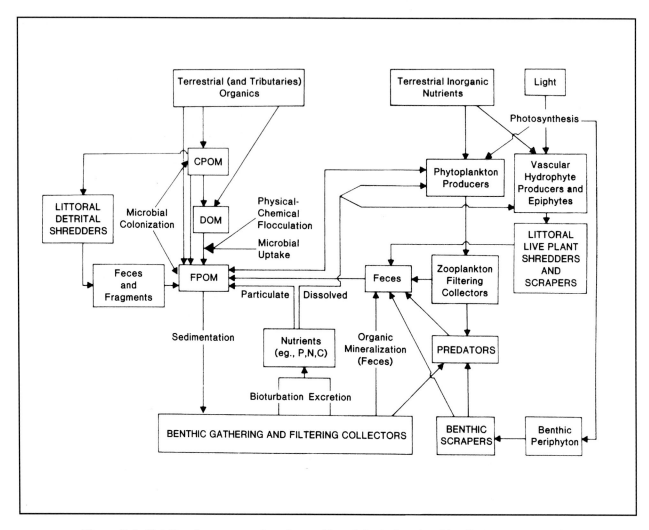

Figure 6.4. Nutritional resource categories and invertebrate functional feeding group categories in lentic ecosystems.

of the Trichoptera and Coleoptera shown in Figure 6.5. The similarity in structure of the scraping mandibles in four families representing two orders is a striking example of convergent evolution. Each cupped mandible has a blade-like inner margin which, when moved across a relatively smooth substrate surface, scrapes off attached algae. Brushes at the base of the inner margin of the mandibles aid in retaining the detached algae and associated material and moving it into the mouth.

Mouthparts, legs, and other morphological structures or constructed devices (e.g., silk nets), together with the associated feeding behavior, may change as part of the larval developmental process. Many species may initiate growth as members of the most generalized food acquisition group — the gathering collectors. This would be a logical consequence of eggs generally hatching in protected locations where fine particulate detritus should be an omnipresent, abundant food resource. The early developmental food acquisition process would then consist of moving the tiny particles, in which the animals are immersed, into the mouth. For example, this was evident in the early instars of the predaceous megalopteran *Nigronia* studied by Petersen (1974). As individuals grow, feeding morphology would be expected to become more restrictive. In many instances, the morphology obviously limits the food type that can be acquired, although the corresponding behavior that manipulates the structures may not be as apparent.

Food resource categories (Table 6A; Cummins 1973, 1974; Cummins and Klug 1979; Merritt and Cummins 1984) relating to food acquisition mechanisms have been chosen on the basis of: (1) size range (e.g., coarse and fine) of the material and (2) general location (e.g., attached to surfaces [periphyton], suspended in the water column, deposited in the sediments, in litter accumulations, or in the form of live invertebrates). This categorization also reflects biochemical differences in the nutritional resources, such as presence of living chlorophyll or microbial-substrate interactions, and the major source of the resource, either from within the aquatic system (autochthonous organic matter) or from the shoreline (riparian) terrestrial areas (allochthonous organic matter).

This classification of feeding adaptations is a "functional group" classification in that it distinguishes insect taxa that perform different functions within aquatic ecosystems with respect to processing of nutritional resource categories (Figs. 6.3 and 6.4). The functional approach reflects both convergent (i.e., inter-ordinal, -familial, or -generic) and parallel evolution lead-

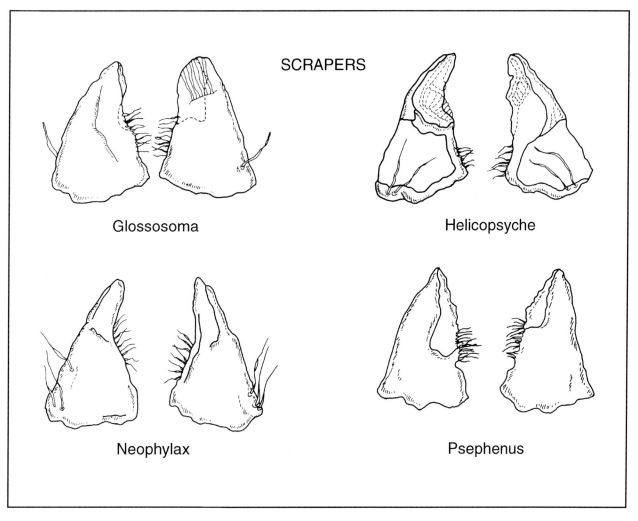

Figure 6.5. Mandible structure in scraper species representing four families and two orders (Trichoptera; Glossosomatidae, Heliocopsychidae, Linnephilidae; Coleoptera: Psephenidae).

ing to functionally similar organisms (e.g., MacMahon et al. 1981). As an example, Wiggins and Mackay (1978) presented an analysis of Trichoptera that indicated a reasonable consistency of functional group at the generic level. The functional groups described in our classification are analogous to "guilds" or sets of organisms using a particular resource class (Root 1973; MacMahon et al. 1981; Georgian and Wallace 1983; Hawkins and MacMahon 1989; Simberloff and Dayan 1991), in as much as function is defined as the use of similar resource classes.

The functional feeding group approach is advantageous in that it enables an assessment, numerically, by standing crop biomass, or by production, of the degree to which the invertebrate biota of a given aquatic system is dependent upon a particular food (nutritional) resource. It also makes more apparent the linkages that exist between food sources and insect morpho-behavioral adaptations (Cummins 1973, 1974; Cummins and Klug 1979; Merritt and Cummins 1996; Figs. 6.3 and 6.4). In an increasing number of examples, freshwater invertebrate assemblages have been analyzed according to functional feeding group categories, expressed as relative abundances (e.g., Vannote et al. 1980; Cummins et al. 1981, 1984, 1989; Hawkins and Sedell 1981; Minshall et al. 1983; Petersen et al. 1989). Functional feeding groups were used as invertebrate categories in the stream ecosystem simulation model of McIntire (1983). These functional feeding group analyses support the notion that linkages exist in riparian-dominated headwater streams between coarse particulate organic matter (CPOM) and shredders, and fine particle organic matter (FPOM) and collectors, and between primary production (e.g., periphyton in mid-sized rivers) and scrapers. The feeding of shredders on riparian litter impacts detrital processing in aquatic systems. About 30% of the conversion of CPOM leaf litter to FPOM has been attributed to shredder feeding (Petersen and Cummins 1974; Cuffney et al. 1990), and this can affect the growth of FPOM-feeding collectors (Short and Maslin 1977). In addition, shredder feeding enhances the release of dissolved organic matter (DOM; Meyer and O'Hop 1983). Some examples of aquatic insect functional feeding group ratios that reflect process-level aquatic ecosystem attributes are summarized in Table 6E. Such analyses link the balance between food resource categories and the predictable response of aquatic insect assemblages.

The efficiency with which ingested food is converted to growth of aquatic insects (ECI, Waldbauer 1968) is dependent

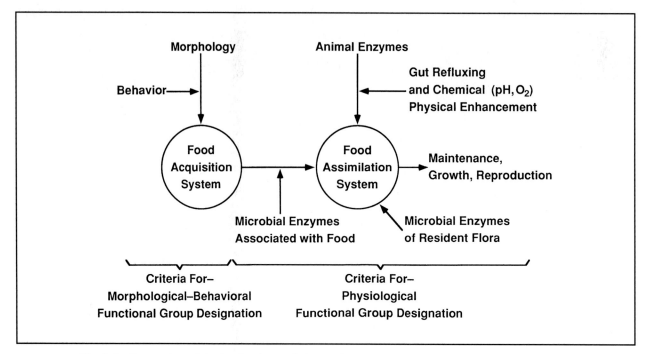

Fig. 6.6. Separation of the basis of aquatic insect functional feeding group classification into an acquisition (ingestion) and an assimilation (digestion absorption) system. The most efficient acquisition is achieved with the best match up between the morphological-behavioral adaptations and the food resource to be harvested. Not all food resources acquired are assimilated with equal efficiency (see Fig. 6.8). Maximum growth per unit of ingested material would occur with the best correspondence between the food acquired and the capability for enzymatic (endogenous and exogenous) digestion and absorption. After Cummins (1988).

on the assimilation system (Fig. 6.6). In general, aquatic insects would be expected to ingest foods that are most efficiently assimilated. However, the food acquisition system might be expected to be less selective than the biochemically-based assimilation system, particularly if resident microorganisms are involved in the digestion process (Fig. 6.7). Within a food resource category, such as CPOM or FPOM, there may be selection for quality of food (Cummins and Klug 1979; Wotton 1994a). Examples would be feeding by shredders in which they select the leaf tissue that is most densely colonized by microorganisms, especially aquatic hyphomycetes (Cummins and Klug 1979; Suberkropp et al. 1983; Arsuffi and Suberkropp 1984), and has concentrations of particular polyunsaturated lipids (Cargill et al. 1983, 1984). Also, selection of food by particle size, somewhat independent of quality, is exhibited by many collectors (Chance 1970; Cummins and Klug 1979, Wallace and Merritt 1980; Merritt et al. 1992; Wotton 1994b). One of the most intriguing features of food assimilation by aquatic insects involves the roles played by microorganisms associated with food and as gut symbionts, which likely produce enzymes and/or metabolites that can be used by the insects (Martin et al. 1980, 1981a,b; Cummins and Wilzbach 1988; Fig. 6.7). The relationships between the shredder functional group and the microbial communities of the leaf substrate and the shredders' digestive tract are exemplified by *Pycnopsyche guttifer*, an eastern limnephilid caddisfly. Figure 6.7 depicts the sequence from the introduction of plant litter (e.g., a red maple leaf) from the riparian zone along a woodland stream (Cummins 1988), to entrainment of the litter by an obstruction in the current, such as large woody debris, and colonization of the leaves by stream microbes, especially aquatic hyphomycete fungi (Suberkropp et al. 1983). This is followed by ingestion of the microbially-conditioned leaf substrate by a shredder, such as *Pycnopsyche*, and the influence on the digestion of the ingested plant substrate plus microbe biomass by resident (and transient) gut microbes in the mid- and hindgut of the larval caddisfly. Enzymes capable of digesting plant tissue ingested by the shredder are likely derived from one or a combination of the following: (1) the extra cellular release from live microbial cells, (2) the lysed microbial cells passing through the shredder gut, or (3) the resident microbes concentrated in the hindgut. Enzymes and digested material refluxed forward to the midgut from the hindgut would be of significant benefit to the shredder (Fig. 6.7).

The assimilation system, and thus insect metabolism and growth, are generally under direct thermal control (Sweeney and Vannote 1978; Vannote and Sweeney 1980; Ward and Stanford 1982). However, some species exhibit temperature compensation as evidenced by growth of many fall-winter species, such as the shredder *Pycnopsyche* discussed above, at temperatures below 5°C (Anderson and Cummins 1979; Cummins et al. 1989). Based on this temperature compensation observed in some taxa, temperate zone shredders can be classified into fall-winter and spring-summer groups. The fall-winter species (e.g., Plecoptera, Nemouridae, *Zapada*; Trichoptera, Limnephilidae, *Pycnopsyche*; Diptera, Tipulidae, *Tipula*) utilize

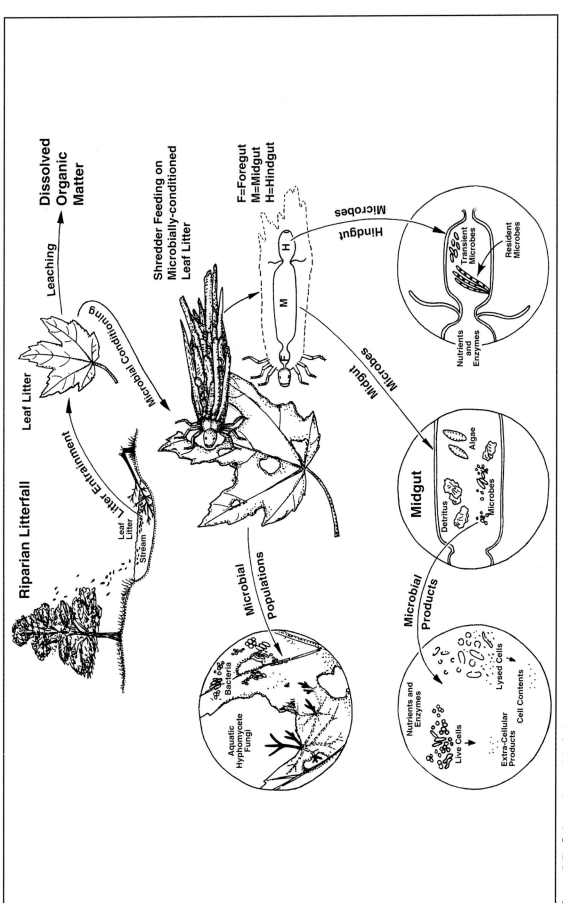

Figure 6.7. Schematic model of shredder invertebrate-microbial interactions in a stream ecosystem. Litter inputs from the riparian zone become entrained in the receiving stream against an obstruction such as large woody debris. After dissolved organic matter (DOM) is leached from the litter, it is colonized by microorganisms, especially the aquatic hyphomycete fungi which are keystone species in the sequence between litter and shredders. Once the litter is appropriately conditioned (softened and with significant microbial biomass accumulation) by the microbes, it is selectively fed upon by shredders. The microbial flora on the litter, and the physical and biochemical changes they bring about, are the basis of feeding selection by the shredders. These microbes are also the source of colonization by gut microbes. The transient and resident mid- and hindgut microbes potentially provide a nutritional base (living cell products and lysed cell constituents) for the shredder or, under certain conditions, they may play a pathogenic role. After Cummins (1995).

autumn-shed litter from plant species such as alder that are more rapidly conditioned by microorganisms and therefore more quickly suitable for shredder ingestion (termed "fast" litter). In contrast, spring-summer shredders (e.g., some lepidostomatid trichopterans) primarily utilize "slow" litter (conifers and oaks) that requires longer conditioning times (Cummins *et al.* 1989).

In addition, temperature exerts a strong effect on food quality. The growth rate of detrital-associated microbes increases with temperature, improving food quality for shredders by virtue of the increased biomass and activity of the associated microbes (Anderson and Cummins 1979; Cummins and Klug 1979). Sweeney *et al.* (1986) observed that diet overrode temperature as a factor in determining the final adult mass of a leptophlebiid mayfly.

Because the efficiency of food conversion to growth reflects the degree of correspondence between the acquisition and assimilation systems operating together (Fig. 6.6) and the food being consumed, obligate specialists within a functional group should have maximum loss of efficiency if the match between food and function is poor (e.g., a shredder scraping algae or feeding on FPOM; Fig. 6.8). Conversely, facultative generalists should be more plastic in their ability to acquire and assimilate a wider range of food resources, but presumably not with the same efficiency as a specialist, for any particular resource (e.g., a collector such as the midge *Stictochironomus*, which is capable of some scraping, or the shredder *Paraleptophlebia* which is capable of collecting; Fig 6.8). A survival advantage for generalists may accrue during unpredictable changes in environmental conditions when particular resources are unavailable. Predictably, such generalists are often pioneer colonists performing better than specialists in newly available aquatic habitats following disturbance, but being outperformed by obligate specialists in mature systems.

Thus, the functional feeding group method of analysis avoids the relatively non-informative necessity to classify the majority of aquatic insect taxa as omnivores and it establishes linkages to basic aquatic food resource categories (CPOM, FPOM, periphyton, and prey) requiring different adaptations for their exploitation. The critical point is that the food acqui-

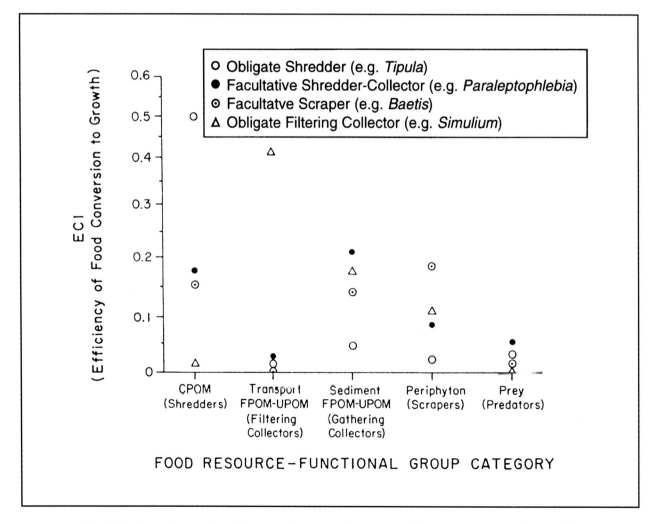

Fig. 6.8. General expected differences between obligate (specialist) and facultative (generalist) members of various functional feeding groups. The range of values for the efficiency of conversion of ingested food to growth (ECI, Waldbauer 1968) would vary with the specific nature of the food in a given resource category and with temperature. After Cummins and Klug (1979).

Table 6E. Example of the use of selected aquatic insect functional feeding group ratios to predict aquatic ecosystem attributes. P/R is the autotrophy to heterotrophy ratio (= gross primary production/total community metabolism; P/R > 1 indicates an autotrophic system). POM Storage/Transport is the ratio of the total particulate organic matter in the sediments (storage) to that in the water column (transport). CPOM/FPOM is the ratio of either benthic storage (i.e., on and in the sediments) or transport of coarse particulate organic matter to fine particulate organic matter (CPOM = particle sizes > 1 mm, Cummins 1974, Vannote et al. 1980). SH = Shredders, TOT COL = total collectors, SC = scrapers, GCOL = gathering collectors, FCOL = filtering collectors. Data from Cummins et al. (1981).

Ecosystem Attributes and Functional Group Ratios	Data by Stream Order/Stream Width			
Stream Width (m)	1	5	10	45
Stream Order	1	2	3	5
P/R	0.47	1.13	0.90	1.23
Storage CPOM/FPOM	0.36	0.11	0.15	0.10
Transport CPOM/FPOM	0.022	0.016	0.019	0.022
SC/SH + TOT COL (P/R)	0.08	0.24	0.22	0.05
SH/TOT COL (Storage)	0.220	0.003	0.002	0.001
SH + SC + GCOL/FCOL (Transport)	3.53	0.46	0.46	0.69

sition system (i.e., morpho-behavioral adaptations for feeding, Fig. 6.6) is less variable than the ingested food and, therefore, gut contents are not a reliable predictor of feeding group. The adaptations can be very specific, especially in the later instars of obligate (specialist) taxa. These taxa would be expected to exhibit the highest efficiency on the matching resource (e.g., shredders on conditioned leaf litter), but very low efficiency on other food categories (Fig. 6.8). As shown by Anderson and Cummins (1979) for populations of the obligate scraper caddisfly, *Glossosoma nigrior*, the consequences of this can be differential growth rates on matching and non-matching food resources. *G. nigrior* showed better growth (prepupae of greater mass) from streams with significant non-filamentous algal periphyton (P/R > 1; Table 6E) than streams with detritus and little algae on the tops of rocks (P/R < 1) where the larvae feed. Facultative forms would be expected to exhibit a much wider range of food resource use but less efficiency than shown by the specialists for any specific food category. For example, *Paraleptophlebia* (Mattingly 1986, 1987) and *Leptophlebia* (Sweeney et al. 1986) have been reported to function as either shredders or as scrapers (Scrimgeour et al. 1991), but comparisons of the conversion of ingestion to growth on the different diets (Fig. 6.8) have not been made to determine if there are significant differences. In general, in the entries in the ecological tables at the end of each taxonomic chapter, multiple functional categories indicate facultative taxa. However, until the critical comparative studies of ECI (Fig. 6.6) are done, and the classifications are based on examination of morphological adaptations and feeding observations rather than on gut contents, many true obligate forms may as yet be undetected.

DISTRIBUTION, ABUNDANCE, AND PRODUCTION

Distribution of an aquatic insect population is ultimately set by the physical-chemical tolerance of the individuals in the population to an array of environmental factors. Within its range of occurrence, e.g., as determined by thermal requirements (Vannote and Sweeney 1980; Rader and Ward 1990; Ward 1992, 1994), population abundance is regulated through interactions between habitat and food suitability and availability. Vannote and Sweeney (1980) proposed the Thermal Equilibrium Hypothesis that relates maximized fecundity of a species, predicted from female body size, to optimal temperature regime along a thermal gradient (related to latitude, longitude, or altitude). Fecundity, interpreted as "fitness" of a species population, is lower in temperature regimes above or below the thermal optimum. In a test of the hypothesis, Rader and Ward (1990) found that fecundity and mean body size were predicted by thermal regime better than were density or biomass. Experimental studies suggest that, at least under certain circumstances, competitors and/or predators may significantly influence or modify habitat or food associations and local distribution patterns of a population (Hart and Resh 1980; Peckarsky 1980; Peckarsky and Dodson 1980; Allan 1982, Power et al. 1988).

Inventory methods for qualitative and quantitative assessments of aquatic insect populations are reviewed in Chapter 3. In lotic waters and fluctuating lentic systems, such as reservoirs (Ward and Stanford 1979), the expansion and contraction of overall habitat with changing flow regimes alters aquatic insect distributions seasonally. This may give the appearance of differences in absolute abundance, expressed as number of organisms per unit area, if care is not taken when analyzing distribution patterns (Elliott 1977; Chap. 3).

In addition to distribution and abundance, often the objective of an aquatic insect study is to estimate production of one or more key species, or in some cases the entire community. Gross production rate (gross productivity) is defined as the elaboration of new biomass per unit area per unit time, regardless of its fate, which may include death by predation or adult emergence. Net production rate (net productivity) of a particular generation of a species can be measured as the biomass of emerging adults from a unit area per unit of time (e.g., with emergence traps see Chap. 3). Techniques used to calculate aquatic insect production rates differ depending on the portion of the aquatic insect community that is covered and the resolution of the data available on species composition, size classes, and seasonal patterns (Cummins 1975; Benke 1977, 1979, 1984; Waters 1977, 1979a, 1979b; Calow and Petts 1993). Most techniques require the same basic field and laboratory data. Several size-specific (usually instar or length-class) estimates of a species' density (e.g., number/m^2) are made over a period of time (usually one year). These densities are converted to size-specific standing stock biomass (e.g., g/m^2). Biomass can be determined by directly weighing specimens on a microbalance or estimated from measurements of length that are converted to mass (or ash-free mass) using size-mass regressions (Smock 1980). Mass also can be estimated from volume displacement by assuming a specific gravity of the specimens of slightly greater than 1 (Stanford 1972), or by using volume-mass regressions. If fresh specimens are not available for weighing, preserved material can be used. However, the results may be highly variable because the amount of leaching of soluble fractions from the specimens depends on the type of preservative, length of preservation, and characteristics of the insect's cuticle. However determined, size-specific biomasses are then used to estimate production.

In a typical study, production estimates of a single species or a few individual species are made (Huryn and Wallace 1988; Benke and Parsons 1990; Huryn 1990; Lugthart et al. 1991), and several techniques are applicable for such estimates (Waters 1977, 1979a; Benke 1984). Sometimes it may be desirable to estimate production of a group of related species (e.g., a taxonomic family or functional group) or the entire benthic insect community (e.g., Benke et al. 1984; Smock et al. 1985). The size-frequency method can be used to estimate both "group" and single species production (Waters 1977; Waters and Hokenstrom 1980). When using this method for a group of species, the researcher must be aware of complications caused by differences in the life cycles, growth rates, and maximum weights of the species in the group as discussed in the literature (e.g., Waters 1977, 1979a; Benke et al. 1984).

Size-frequency production estimates are greatly influenced by a species' life cycle. If the insect has any prolonged non-growing stages, such as eggs, adults, or diapausing larvae, the size-frequency calculations will produce an underestimate of annual production (Benke 1979, 1984). Therefore, size-frequency estimates must be corrected using the cohort production interval of the insect (CPI; Benke 1979, 1984). CPI is the time in days from hatching to the largest larval instar or larval size. The production estimate is multiplied by 365/CPI to correct for non-growing stages. Krueger and Martin (1980) developed a formula to calculate confidence intervals for size-frequency production estimates that makes possible critical comparisons of production among species from the same habitat and different locations.

Aquatic insect growth approaches a logarithmic function in most cases, so instantaneous growth can be approximated as $G = (\ln \bar{x}$ final or maximum mass $- \ln \bar{x}$ initial or minimum mass$) \times (1/\text{time interval})$ (Waters 1977, 1979). If the initial mass is at hatching and the final at the last larval stadium or at pupation, the estimate is for life cycle instantaneous growth rate. The growth rate of an average individual in the population can also be expressed as relative growth rate (Waldbauer 1968): RGR = ($\bar{x}$ final mass $- \bar{x}$ initial mass$) \times (1/\text{median mass}) \times (1/\text{time interval})$, where median mass is (initial mass + final mass)/2. The resulting expression multiplied by 100 is growth as percent body weight/day and is essentially equal to instantaneous growth rate for short time intervals, especially during periods of rapid growth (Cummins et al. 1973).

Another concept involved in production estimation is the expression P/B, where P = annual (or cohort) production and B = mean annual (or cohort) standing stock biomass (Waters 1969a, 1977, 1979). This expression describes the rate at which the biomass of a species is turning over (replacing itself), i.e., how many units of new biomass are produced per unit standing stock biomass per unit area per unit time. This P/B turnover ratio ranges from 3 to 6, with a mean of about five for most univoltine species (Waters 1969a, 1979a; Waters and Crawford 1973; Waters and Hokenstom 1980; Benke 1984; Calow and Petts 1993). Bivoltine and multivoltine species have turnover ratios > 6 (Waters 1969a, 1979; Benke et al. 1984), and for species with rapid turnover populations the ratio might approach 100 (Hauer and Benke 1991). The P/B ratio of a species approximates the instantaneous growth rate G (see above) calculated over the entire life cycle, which is also the turnover ratio (TR). Assuming an average TR of 5 for most univoltine species, this allows an estimate of production rate if the mean annual standing stock biomass (B) is known. A common method of production estimation is to measure G in enclosures in the laboratory (e.g., Cummins et al. 1973) or in the field (Hauer and Benke 1991), as $G = (\ln B_t/B_i)t$, where B_i and B_t are average initial and final biomasses of the enclosed population respectively, and t is the time interval over which the increase occurs. Production then is estimated by multiplying this growth rate (G) by the numerical standing stock of the field population.

In order to allow for comparisons of growth or production rates between locations or seasons that differ in temperature, it is useful to normalize the data for temperature by reporting intervals as temperature-time or degree-days (i.e., the median daily temperature summed over the time interval; Andrewartha and Birch 1954).

USE OF AQUATIC INSECTS IN BIOMONITORING

David M. Rosenberg
Freshwater Institute,
Winnipeg, Manitoba, Canada

Vincent H. Resh
University of California,
Berkeley, California

INTRODUCTION

Biological monitoring, or biomonitoring, is the systematic use of living organisms or their responses to determine the quality of the aquatic environment. Biomonitoring can be separated into two main types: (1) surveys before and after an impact (e.g., a project is built or a potential toxicant is released) to determine effects of the activity or action (e.g., Van Urk *et al.* 1993); and (2) regular sampling or toxicity testing to measure compliance with legally mandated water quality standards (e.g., Humphrey and Dostine 1994). Recent programs launched by U.S. government agencies are attempting another type of biomonitoring - the use of surveys to determine "reference conditions" that could apply to fresh waters over large geographical areas (e.g., Hughes *et al.* 1986). Of the above approaches, surveys before and after an impact are the most commonly used biomonitoring approach involving aquatic insects.

In this chapter, we present an overview of the use of aquatic insects for biomonitoring in North America; for a more detailed treatment of the topics discussed here, consult the recent text of Rosenberg and Resh (1993a). Other chapters in this book also present topics of relevance to biomonitoring: sampling (Chap. 3), physiological processes (Chap. 4), and ecology (Chap. 6).

Fish, algae, protozoans, and other groups of organisms have been recommended for use in water quality assessment but macroinvertebrates, which oftentimes consist mainly of aquatic insects, are the group most frequently used (Hellawell 1986). Rosenberg and Resh (1993b) concluded that this was because: (1) they are ubiquitous and, consequently, are affected by perturbations in many different aquatic habitats; (2) the large number of species exhibit a range of responses to environmental stress; (3) their sedentary nature, relative to other aquatic organisms such as fish, permits effective determination of the spatial extent of perturbations; and (4) their long life cycles, relative to most other groups of organisms, allow temporal changes in characteristics such as abundance and age structure to be examined. However, some characteristics of aquatic insects can hinder their effective use in biomonitoring activities and require special consideration: (1) they do not respond directly to all types of impacts (e.g., herbicides; see Hawkes 1979); (2) their distribution and abundance can be affected by factors other than water quality (e.g., current velocity, type of substrate); (3) their abundance and distribution vary seasonally; and (4) dispersal abilities may carry aquatic insects into and out of areas in which they normally do not occur. Moreover, some groups of aquatic insects lack identification keys.

Biomonitoring differs from the traditional physical and chemical approaches that also are used in water quality assessment. Physical and chemical measurements are analogous to photographs; they are instantaneous and describe conditions that exist when the sample is collected. In contrast, the reliance on organisms present, which is the basis for biomonitoring, is more like using a movie or a video; a temporal component is added to the still photograph because organisms such as aquatic insects are exposed to past conditions as well. However, physical/chemical measurements and biomonitoring are not mutually exclusive; an optimal water quality monitoring program involves both approaches (Rosenberg and Resh 1993b).

HISTORY OF BIOMONITORING

The use of macroinvertebrates to assess water quality began in Germany in the early part of the 20th century. Kolkwitz and Marsson (1908, 1909) developed the idea of *saprobity* (the degree of pollution) in rivers as a measure of the extent of contamination by sewage, which results in decreases in dissolved oxygen, and the effect that this oxygen decrease has on life in rivers (Cairns and Pratt 1993). Observations on the restricted occurrence of certain taxa in response to environmental conditions led to the development of lists of *indicator organisms*. For example, tubificid worms were considered to be pollution tolerant, whereas caddisflies were considered to be pollution intolerant. Many features of these ideas still form the philosophical basis of biological assessment approaches used in Europe and North America, although the concept has been broadened to include *indicator communities*. Today, most practitioners of biomonitoring acknowledge that the original indicator organism approach is overly simplistic. Nevertheless, this concept still remains the basis for the biotic indices and several other biomonitoring approaches that are discussed below.

The use of macroinvertebrates in North American biomonitoring programs followed the European tradition of qualitative evaluations quite closely until the 1970s; at this time, the emphasis in North America shifted to approaches involving quantitative sampling and analysis (Resh and Jackson 1993). These quantitative approaches typically included formal hypothesis testing that required replicate sample units and detailed statistical analysis. However, recent trends indicate a renewed interest in the adoption of some qualitative approaches (perhaps to avoid the high cost and time-consuming nature of quantitative approaches), which have been incorporated in new approaches generally referred to as "rapid assessment procedures". More detailed coverage of the historical development of biomonitoring can be found in Metcalfe (1989), Cairns and Pratt (1993), and Davis (1995).

SCALES OF BIOMONITORING

Biomonitoring approaches using aquatic insects cover different ranges along a spatial-temporal continuum, depending on the hierarchical level used (Fig. 7.1). Hierarchical levels range from the biochemical (subcellular) to the ecosystem; biomonitoring examples within each level are given in Table 7A. Resh and Rosenberg (1989) provide a further discussion of spatial and temporal scales in aquatic insect research.

Biochemical and Physiological Levels: Obviously, responses to environmental stress begin at subcellular levels of the organism, and are reflected in *biochemical* and *physiological* responses; as a result, these subcellular levels should be sensitive early-warning indicators of adverse effects (Johnson *et al.* 1993). These levels occur at the smallest spatial and temporal scales (Fig. 7.1). Examples include the measurement of changes in enzyme activity (biochemical) and in respiratory metabolism (physiological) as a result of exposure to a toxicant (Table 7A). Unfortunately, the use of biochemical and physiological approaches is limited because of a lack of basic knowledge of these processes in most aquatic insects. However, these suborganismal indicators may be crucial to our understanding of modes of action and consequent ecological results. Use of these approaches will probably increase in the future as more basic knowledge is gained and, within the next few decades, biochemical and physiological indicators for aquatic insects may become routine parts of water quality assessment.

Individual Level: Biomonitoring conducted at the level of the individual organism is characterized by analyses at intermediate spatial and temporal scales (Fig. 7.1). The techniques involved (Table 7A) currently are more widely used in biomonitoring than are methods involving the suborganismal biochemical- and physiological-levels. The *morphological deformities* found in individual organisms may reflect developmental anomalies linked to biochemical and physiological processes, and these have been reported from several groups of insects occurring in many different parts of the world (see table 4.8 in Johnson *et al.* 1993). At present, the use of morphological deformities to indicate contaminant stress is mainly qualitative, and routine use of these measures will require a quantitative orientation (Rosenberg 1992; e.g., see Clarke 1993).

Deviations from normal *behavior* in response to a specific contaminant probably result from physiological disorders, so these deviations should be sensitive measures of contaminant stress. Moreover, benthic (bottom-dwelling) aquatic insects may be ideal organisms for linking behavioral responses to toxicity assessments because sediments are often sinks for toxic compounds. In general, aquatic insects have not been as widely used for this kind of biomonitoring as have fish, but efforts to do so have increased substantially in the last decade (Johnson *et al.* 1993).

Life-history endpoints such as survival, growth, and reproduction ultimately determine the success of a species. Therefore, measurements of the effects of contaminants on these endpoints are an important component of biomonitoring, and are now commonly used measures of stress in contaminant-assessment studies (see table 4.9 of Johnson *et al.* 1993). Recent approaches in the use of such endpoints have attempted to integrate field and laboratory studies (e.g., France and LaZerte 1987).

Finally, the use of *sentinel organisms* as bioaccumulators of metals and organic contaminants such as insecticides and polychlorinated biphenyls (PCBs) provides a number of advantages over the direct, chemical analysis of contaminants in water or sediments. For example, this approach can provide a time-integrated indication that the contaminant is "bioavailable", and can warn that other parts of the food web and the ecosystem may be affected (Johnson *et al.* 1993). Aquatic insects have been used as sentinel organisms but other, larger freshwater macroinvertebrates are more commonly selected. The key to the sentinel approach lies in understanding the factors influencing contaminant levels (e.g., Hare 1992; Phillips and Rainbow 1993); optimal use is achieved when unexplained variance between the concentration in the organism and the concentration in the environment is minimized. The next step in advancing the use of sentinel organisms will be to relate contaminant levels in an organism to effects of these contaminants on life-history and population characteristics (Reynoldson and Metcalfe-Smith 1992).

Population and Species Assemblage Level: Like at the individual level, the population level is characterized by analyses at intermediate spatial and temporal scales (Fig. 7.1). However, populations and species assemblages of aquatic insects have been used more frequently and widely than individuals in assessing water quality (Table 7A). For example, populations have been used as indicator taxa (usually in *biotic indices*) to classify the degree of pollution in an aquatic ecosystem by determining the tolerance or sensitivity of species to a given pollutant (see below).

Biotic indices are essentially a univariate approach to environmental analysis, i.e., they are used when species abundance appears to vary uniformly with one feature of the environment (e.g., dissolved oxygen). However, *multivariate analysis* is used when species experience the effects of more than one environmental variable simultaneously (Johnson *et al.* 1993; Norris and Georges 1993). Multivariate approaches are generally used at the community level (see below) but also may be used for populations and species assemblages (Table 7A).

Community Level: Work at this level is done at the second largest spatial and temporal scales (Fig. 7.1). In contrast to subcellular-, individual-, and population-level analyses involving aquatic insects, community measures attempt to summarize the magnitude, ecological consequences, or significance of a particular stress on the system being examined (Johnson *et al.* 1993). For this reason, the analysis of macroinvertebrate communities, including aquatic insects, has received more detailed attention than any other level. Community-level biomonitoring involves both traditional quantitative approaches, which typi-

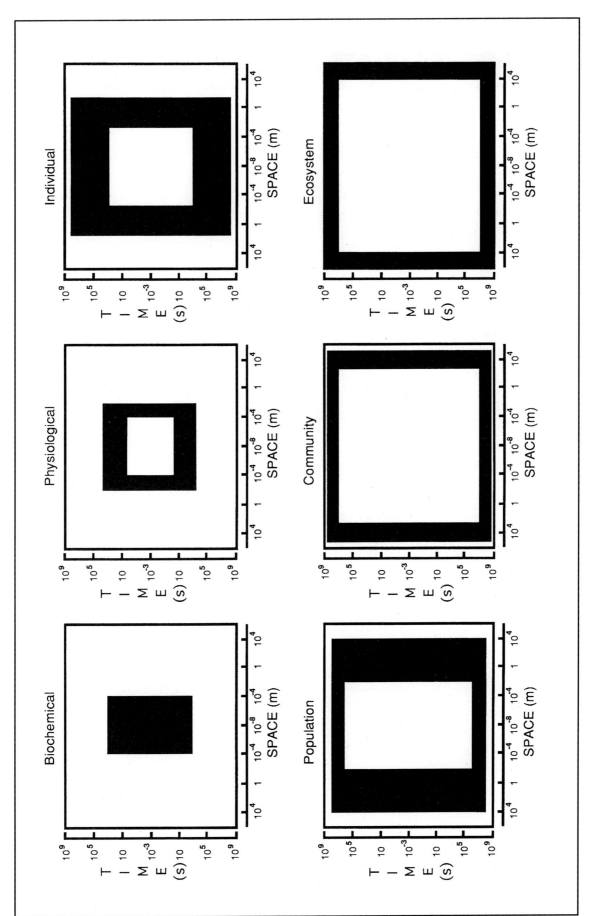

Figure 7.1. Spatial and temporal scales of hierarchical levels used in examining aquatic insects in biomonitoring programs. The shaded part within each box represents the spatial and temporal scales encompassed by that hierarchical level. Adapted from Cooper and Barmuta (1993). s = seconds, m = meters.

Table 7A. Examples of how aquatic insects and other freshwater macroinvertebrates are used in biomonitoring at different spatial and temporal scales. Johnson *et al.* (1993) and Resh and Jackson (1993) provide many additional examples.

Hierarchical level			Examples	Reference(s)
Biochemical	Enzyme activities	-	monitor responses of isozymes of *Argia vivida* larvae (damselfly) to a thermal gradient	3585
		-	monitor responses of acetylcholinesterase activity in *Acroneuria lycorias* and *A. abnormis* larvae (stoneflies) to the pesticide fenitrothion	1210
	RNA, DNA, amino acids, protein	-	monitor responses of concentrations of these constituents in *Neochetina eichhorniae* adults (water hyacinth weevil) to metals	3500, 3501
	Ion regulation	-	monitors responses of Na, Cl, and Ca concentrations in *Limnephilus pallens* larvae (caddisfly), *Chironomus riparius* larvae (midge), and *Orthocladius consobrinus* larvae (midge) to low pH	1631
Physiological	Respiratory metabolism	-	monitors O_2 consumption rates of *Aeshna umbrosa* larvae (dragonfly) in response to di-and trichloroacetic acids and of oligochaete worms (two freshwater species) in response to Cd, Hg, and sodium pentachlorophenol	932, 417
	Scope for growth	-	monitors how various components of the energy budget of individual *Gammarus pulex* (freshwater shrimp) respond to stress	2842
Individual	Morphology	-	uses deformities to indicate contaminant stress in field situations; reports concern aquatic insects (mainly Chironomidae) and Oligochaeta; head-capsule deformities (Chironomidae) are most frequently used	4249, 4248, 915, 2693
	Behavior	-	uses deviations from normal behavior in response to a specific pollutant; examples include altered net-spinning behavior of *Hydropsyche angustipennis* larvae (caddisfly) in response to kraft pulp mill effluent and 4,5,6-trichloroguaiacol, altered feeding behavior of *Glyptotendipes pallens* larvae (midge) in response to cadmium, and altered competition behavior of *Hydropsyche contubernalis* in response to cadmium	3106, 1652, 4154
	Life-history	-	uses life-history endpoints such as survival, growth, and emergence to measure stress; examples include effects of low pH and nickel on survival and growth of *Clistoronia magnifica* larvae (caddisfly), effects of tannery pollution on growth and emergence of two species of mayflies, and effects of cadmium on emergence and oviposition of *Polypedilum nubifer* adults (midges)	4103, 1542, 1623
	Bioaccumulation	-	uses "sentinel organisms" that accumulate contaminants from food or water to provide an indirect estimate of environmental concentrations of such contaminants; the best-known example is the "mussel watch" of marine and estuarine systems (Bayne 1989); examples of freshwater equivalents include the use of *Elliptio complanata* and *Lampsilis radiata radiata* (mussels) to monitor organic contaminants in the St. Lawrence River, Canada, and the use of aquatic insects to monitor metal contamination in two U.S.A. streams	2683, 511, see also table 4.10 in Ref. 2015 for summary
Populations and species assemblages	Biotic indices	-	classify the degree of pollution by determining the tolerance or sensitivity of taxonomically homogeneous organisms (e.g., Chironomidae, Oligochaeta) or several groups of indicator organisms to a given pollutant	
			e.g., organic enrichment/eutrophication	
			- Chironomidae	4365, 3513, 4467
			- Oligochaeta:Chironomidae	4366
			- six groups of key organisms	4538
			acidification - Chironomidae	3238
			- several taxa of indicator organisms	3237, 1207
	Multivariate approaches	-	ordinate indicator taxa or species assemblages along environmental gradients e.g., direct ordination (Canonical Correspondence Analysis) of profundal chironomid taxa from Swedish lakes along environmental gradients	2012

Table 7A. Continued

Hierarchical level			Examples	Reference(s)
Community	Taxa richness	-	describes number of taxa in a community	3303, 2107
		-	EPT richness - number of taxa of Ephemeroptera (mayflies), Trichoptera (caddisflies), and Plecoptera (stoneflies)	2317, 3148
	Enumerations	-	range from counts of all organisms collected to relative abundances of different taxonomic groups (e.g., orders, families, numerically dominant taxa)	3303, 2107, 3148
	Diversity indices	-	combine taxa richness and abundances in a summary statistic (e.g., Shannon's Index); generally require species-level identifications	4250
	Similarity indices	-	compare community structure in space (i.e., different sites) or time (i.e. different years)	
			e.g., Coefficient of Community Loss	4174
			Pinkham-Pearson Index	3146
	Biotic indices	-	use preestablished water-quality tolerance values for taxa (species, genera, or families) (see "Population and species assemblages", above); relative abundances of taxa weighted by tolerance values are sometimes included in the calculation of a biotic index	
			e.g., Saprobic Index	4645, 2682
			Belgian Biotic Index	882
			Biotic Index	637, 1731, 1734, 1738, 2318
			Biological Monitoring Working Party (BMWP) Score	4568
			Biological Condition Index	4479
	Functional feeding-group measures	-	compare the proportion of individuals that are evolutionarily adapted to using certain food resources (see Chap. 6) in a sample or within specific food sources to the resources available (e.g., ratio of scrapers to collector—filterers; proportion of shredders in a leaf pack)	799, 3148, 2667
	Combination indices	-	use more than one measurement (such as those listed above)	
			e.g., Invertebrate Community Index	2962
			Biological Condition Score	3148
	Multivariate approaches	-	account for observed variation in the abundance or presence/absence of species in a community when that community is affected by several environmental variables simultaneously; also used in predictive modeling	
			e.g., direct gradient analysis	2013
			indirect gradient analysis	1337, 4570
			predictive modeling	4568
Ecosystem	Structure of food webs	-	monitors the effects of acidification on an oligotrophic, Canadian Shield lake	3570, 861
	Productivity	-	monitors the effects of nutrient addition on an oligotrophic, Canadian Shield lake	3568, 3569, 859
		-	monitors the effects of nutrient addition on an Alaskan, tundra stream	3114, 3115, 1701, 1754
	Decomposition	-	monitors the effects of methoxychlor insecticide on particulate organic matter dynamics in a small Applachian stream	4196, 4184, 4181, 4182, 780, 779
		-	monitors the effects of acidification on leaves in streams (numerous studies)	1714, 2423, 3993, 1482, 45
		-	monitors the effects of chlorine and ammonia on litter in artificial streams	2904, 3087, 2903
	Chemical cycling	-	monitors the effects of acidification on fluxes of dissolved ions, drifting coarse particulate organic matter (CPOM), and drifting invertebrates in a New Hampshire stream	1533

cally include formal hypothesis-testing using replicated sample units and detailed statistical analyses (see below), and rapid assessment procedures, which reduce the intensity of site-specific study required relative to traditional quantitative approaches and enable a greater number of sites to be examined. However, in actual practice, the distinction between procedures used in quantitative and qualitative approaches is not always clear.

Rapid assessment approaches attempt to provide cost-effective tools for evaluating the biological status of the aquatic habitats under study. They achieve cost-effectiveness by limiting the number of habitats examined, the number of samples collected, the amount of sample-sorting time, and the number and/or level of taxonomic identifications made. Summary scores are used so that site surveys can be understood not only by biologists, but also by managers, decision makers, and the general public. Rapid assessment approaches are an outgrowth of the Index of Biotic Integrity (IBI) approach of Karr (1991), which was originally developed for fish. An alternative approach, which combines many of the rapid assessment objectives with multivariate statistics, is used in the United Kingdom (Wright *et al.* 1984, 1988).

Rapid assessment approaches are analogous to using thermometers to assess human health (Resh and Jackson 1993); easily obtained values are compared to a threshold that is considered to be normal (e.g., the human body temperature of 98.6°F or 37°C). The normal threshold is based on values obtained from control sites or regionally based reference conditions. Differences between observed values and those expected at control sites or from reference conditions are a measure of the degree of environmental ill-health (e.g., Plafkin *et al.* 1989). A disadvantage of most water quality evaluation programs that are based on rapid assessment approaches is the lack of sample replication (and consequent awareness or documentation of variability). Resh and Jackson (1993) and Resh *et al.* (1995) discuss the use of rapid assessment approaches in detail.

Quantitative approaches to community-level analysis (see Chap. 3) and rapid assessment approaches generally include one or more of the measures described below. In rapid assessment approaches, these measures are often called "metrics".

The most commonly used measure to describe macroinvertebrate communities is *taxa richness*, which is the number of taxa in a community (Table 7A). This measure is based on the premise that the number of taxa decreases as water quality declines. Although all organisms present would optimally be identified to the species level, measurement of species richness tends to be inexact because of the lack of species-level identification keys for the immature stages (the stages most commonly encountered in freshwater studies) of many groups of aquatic insects. As a result, measures of taxa richness usually involve resolution to genus, and perhaps the designation of "morphospecies" within a genus, based on different morphological characteristics (e.g., sp. 1, sp. 2, etc.). This approach probably underestimates species richness because more than one species in a genus often co-occur and different species are more likely to be combined than separated as morphospecies. Some studies avoid this problem by reporting taxa richness as number of genera or families. It is important to note, though, that it is difficult to compare taxonomic richness among sites if different levels of taxonomic resolution are used at different sites.

EPT richness [i.e., richness of Ephemeroptera (mayflies), Plecoptera (stoneflies), and Trichoptera (caddisflies)] is a recent, commonly used variation on taxa richness. This measure is based on the idea that most taxa in these three orders are pollution sensitive (Resh 1993, which is not always true, e.g., see Resh and Unzicker 1975). EPT richness has become popular probably because of a perception (which is sometimes true) that these groups are easier to identify or to separate into species groups than many other aquatic insect taxa.

Enumerations involve counts such as the total number of individuals, ratio of EPT abundance to Chironomidae (midge larvae) abundance, and ratio of individuals in numerically dominant taxa to total number of individuals (% dominant taxa). The idea behind the use of enumerations is that certain stresses increase or decrease total numbers of individuals. Many types of enumerations require minimal taxonomic effort, so the taxonomic considerations that apply to taxa richness are mostly circumvented. However, high intersample variability may occur (see Chap. 3 for further discussion of this problem), which makes detection of trends from enumerations difficult (especially in rapid assessment programs involving single samples).

A large number of *diversity indices* have been proposed for use in water quality monitoring. Their use relates to the idea that a high diversity value indicates a balanced, stable community. Rather than considering the pollution tolerances of macroinvertebrates (as in biotic indices; see below), the richness of the sample and the numbers of individuals of each species (the evenness) are considered in calculating the index. Reliance on diversity indices in biomonitoring programs has been strongly, and rightly, criticized (Washington 1984; Norris and Georges 1993). Many of the arguments in favor of using diversity indices are not theoretically valid, the biological or ecological meaning of what they claim to be measuring is poorly understood, and many shortcomings are evident in their application (Norris and Georges 1993). However, diversity indices are used routinely in water quality monitoring programs, and the requirement to use this flawed approach has, in some cases, even been codified through environmental legislation.

Most *similarity indices* compare community structure at different sites, with the rationale being that communities at disturbed and undisturbed sites will become more dissimilar as stress increases. These indices use either taxa richness (e.g., Coefficient of Community Loss; Courtemanch and Davies 1987) or both taxa richness and abundance (e.g., Pinkham-Pearson Index; Pinkham and Pearson 1976) (Table 7A). The use of similarity indices requires similar levels of taxonomic discernment among the communities being compared.

Examples of *biotic indices* that use overall community composition appear in Table 7A. Biotic indices are popular because they provide an easily understood numerical expression of a biological response. In calculating a biotic-index score, organisms in a sample are assigned values according to their tolerance or intolerance, ideally to the pollutant in question but values are often based on responses to organic (sewage) pollution. The sum of individual values (usually multiplied by the abundance of each taxon considered) provides a measure by which the pollution status of a site can be determined from pre-established classification levels. Of course, this technique depends on an accurate assessment of pollution tolerances for different taxa. Biotic indices were originally devised to deal with organic pollution; the Saprobien system of Kolkwitz and Marsson (1909), which is the basis for water quality monitoring programs used in Germany, is an example of this approach, as is the Trent Biotic Index of Woodiwiss (1964), which is the basis for many other European programs.

The application of tolerance values and classification levels from organic to other forms of pollution or to sewage that contains metals and other contaminants is problematic. Recently, a few biotic indices have been developed to deal with acidification (Table 7A). In the future, biotic indices may be developed specifically to detect the effects of other types of pollution such as insecticides, polychlorinated biphenyls (PCBs), or various land uses (e.g., agriculture, forestry). A combination of laboratory-derived tolerance data and field observations would help in this development. Besides being specific to a type of pollution, tolerance values used in biotic indices are usually also specific to the geographic area for which they were developed. Resh et al. (1996) developed a teaching exercise involving collection of macroinvertebrates and calculation of biotic indices.

Functional feeding-group measures differ from the community measures previously described because they measure an aspect of the *functioning* rather than just the *structure* of the macroinvertebrate community. This approach is based on the idea that organisms evolved certain morphological-behavioral food-gathering mechanisms or locomotion-attachment adaptations, and can be placed into particular groups (i.e., functional feeding or habit groups) based on these mechanisms. Consequently, each of these groups are expected to occur in proportionately higher abundances in accumulations of particular food sources or associated with particular habitat types(see also Chap. 6); deviations from expected abundances of functional feeding groups in particular food sources, or habit groups in particular habitats, could indicate perturbation of the community. For example, examination of the occurrence of functional-feeding groups within leaf packs should indicate the presence of organisms that evolved specialized mouthparts and feeding behavior for shredding leaves (described as "shredders"); lower-than-expected proportions of shredders in leaf packs may indicate exposure to a contaminant or disturbance of the riparian zone. Mouthpart morphology and means of food acquisition for each taxon collected should be the basis for functional-feeding group assignment (Cummins and Klug 1979; Cummins and Wilzbach 1985; Cummins 1988; Palmer et al. 1993a,b), but often the ecological data are based on food ingested, not mode of food acquisition. An example would be scrapers that have gut contents containing either detritus (detritivores) or algae (herbivores), but both types of food resources were acquired by the same mechanism. Care should be taken in making this distinction when the tables in this book are used. The topic is considered in more detail in Chapter 6. Merritt and Cummins (1996) developed a teaching exercise for students involving the use of functional feeding group ratios as a tool for assessing the ecological state of running water communities.

Combination indices are designed to spread the risk of making incorrect assessments by using a variety of the types of measures described above. Incorrect assessments are a hazard when only a single measurement is used as the basis for evaluation. These indices use various measures of richness, enumerations, tolerance values, functional-feeding group designations, differences between taxa present at control and treatment sites, and a variety of measures often unique to each index (e.g., percentage non-dipterans) to determine a score for a site. Reliance on multiple measures (often called "multimetrics") either by using a combination index or by analyzing several measures independently has become standard practice in biomonitoring programs devised by U.S. state and federal agencies (e.g., Plafkin et al. 1989). Scoring criteria are applicable to the geographic region in which they were developed but extrapolation to other regions requires recalibration of these indices.

Multivariate approaches are routinely used for analysis of communities, and they have been applied to macroinvertebrates for water quality assessments. Four approaches are common (from ter Braak and Prentice 1988): (1) direct gradient analysis; (2) inference; (3) indirect gradient analysis; and (4) constrained ordination (Johnson et al. 1993). In direct gradient analysis, taxa abundance or probability of occurrence is related to environmental variables, whereas in inference the species composition of the community is used to infer environmental variables. Indirect gradient analysis searches for major gradients in the species data, which are subsequently interpreted in terms of environmental gradients. In constrained ordination, axes of variation in community composition are constructed to optimize their fit to the environmental data used. The success of multivariate approaches in biomonitoring programs in the United Kingdom (e.g., Wright et al. 1984, 1988) has led many specialists in the use of macroinvertebrates for water quality monitoring to predict that the future of community-level biomonitoring lies in multivariate approaches (see Norris and Georges 1993 and Chevenet et al. 1994 for a further discussion of this topic).

Ecosystem Level: Biomonitoring at this level uses analyses at the largest spatial and temporal scales (Fig. 7.1). An ecosystem perspective is also the most inclusive hierarchical level because it extends the focus from individuals and small-scale phenomena to populations, communities, and large-scale phenomena. This approach shifts from the structural and functional emphases seen at the community level to processes operating at the ecosystem level as it attempts to detect and understand the effects of disturbance in natural systems (Reice and Wohlenberg 1993; see also Chap. 6). Aquatic insects are important in ecosystem-level biomonitoring both because they are an essential component of most ecosystems and they are sensitive indicators of ecosystem deterioration (Reice and Wohlenberg 1993).

In summary, given ideal conditions biomonitoring would include all of the hierarchical levels discussed above (e.g., see Krantzberg 1992). Community- and ecosystem-level studies can demonstrate that a change has occurred, but subcellular-, individual-, and population-level studies are needed to establish the underlying causes of the observed changes. The addition of toxicity testing and experimental approaches, which are discussed next, can also provide explanations of underlying causes.

OTHER BIOMONITORING APPROACHES

In addition to the laboratory and field approaches already discussed, aquatic insects are used in other, often supplemental ways in biological monitoring programs. These include paleolimnological investigations, toxicity testing, and field experimentation. These approaches involve quantitative methods and are used at a variety of hierarchical levels.

Paleolimnological Techniques: Paleolimnological techniques involve the reconstruction of the history of aquatic systems that have been exposed to impacts such as eutrophication (Warwick 1980), acidification (Uutala 1986, 1990), industrial pollution (Kansanen 1985, 1986; Kansanen and Jaakola 1985), or climate change (Walker et al. 1991a,b). Paleolimnology offers one major advantage over other approaches to biomonitoring: no pre-impact investigation is necessary because a pre-impact record is already preserved in the sediments (Walker 1993).

Paleolimnology involves the study of the many different constituents of sediment, which include the remains of aquatic and terrestrial organisms. As with the use of living aquatic insects in biomonitoring, each species of fossil organism is characteristic of certain environmental conditions present during the period in which it lived. The most common aquatic insect fossils found in the sediments of lakes and rivers are fragments of the dipteran families Chaoboridae (phantom midges) and Chironomidae. The remains of other aquatic insects such as Ephemeroptera, Plecoptera, Trichoptera, and Heteroptera (true bugs) are less abundant but they are equally informative about past environmental conditions (e.g., Klink 1989).

The methods of paleolimnology are highly specialized, and involve techniques for the removal of cores (vertical sections of sediment), considerations of adequate sample storage and subsampling, and techniques for sediment dating (essential in accurately interpreting historical change), statistical interpretation, and inference (Walker 1993). The methods for statistical interpretation and inference overlap broadly with biomonitoring of living populations of aquatic macroinvertebrates.

Paleolimnological studies offer biologists a way to reconstruct long-past biotic responses to anthropogenic environmental change, but given suitable conditions they also offer means for detecting very recent limnological change (i.e., events of the past 10 years). Moreover, because of the importance currently being placed on rehabilitation of aquatic ecosystems, paleolimnology should play an increasingly important role in defining what natural conditions existed, and thereby serve as a realistic goal for rehabilitation efforts.

Paleolimnological techniques, like many other areas of biomonitoring using aquatic insects, depend heavily on the quality of taxonomic identifications; taxonomic quality assurance and quality control are essential. As a final point, it should be noted that paleolimnological studies have mainly involved lakes, techniques available do not apply to streams (because of substrate disturbance) to the same extent as to ponds or lakes. However, former lotic habitats such as reservoirs and beaver ponds are virtually unexploited and deserve future attention from a paleolimnological perspective.

Toxicity Testing: Historically, toxicity tests precede all other biomonitoring approaches. The first toxicity experiments involving freshwater macroinvertebrates were published in 1816 (see Buikema and Voshell 1993), and long before that Aristotle conceivably was conducting a bioassay when he placed freshwater fish in seawater to observe their reactions!

The use of acute (i.e., short-term) toxicity tests for predicting potential environmental damage began in earnest in the 1940s and 1950s. However, even though the use and types of toxicity tests have expanded, little has been added to the conceptual framework of acute toxicity testing since then. The 1970s and 1980s marked the rise of toxicity tests to manage pollution by using them to: (1) regulate discharges; (2) compare toxicants, animal sensitivities, or factors affecting toxicity; and (3) predict environmental effects.

Toxicity tests form a continuum of complexity from single-species, laboratory-based acute tests to those conducted on multispecies subsets of natural ecosystems (Buikema and Voshell 1993). Aquatic insects are used in three basic types of tests: acute single-species tests (e.g., Julin and Sanders 1978; Kosalwat and Knight 1987); chronic single-species tests (e.g., Thornton and Wilhm 1974; Dauble et al. 1982; Reynoldson et al. 1994); and multispecies tests (e.g., Hansen and Garton 1982; Zischke et al. 1983) (Table 7B). Toxicity tests extend from short term (e.g., 24 h) to long term in duration (e.g., weeks to months) and range from being environmentally unrealistic (i.e., done in a laboratory beaker) to large-scale tests that accurately reflect natural conditions. Unfortunately, single-species tests in small containers are by far the most common (and environmentally unrealistic) type used. Toxicity-test endpoints measured range from simple mortality to effects on ecological processes such as secondary production (Table 7B). All toxicity tests obviously require a great deal of basic information about the biology of the populations being tested.

Multispecies toxicity tests can be arbitrarily divided into two size categories: *microcosms* (small systems, generally <10 m^3) and *mesocosms* (large systems, generally >10 m^3) (Buikema and Voshell 1993); however, definitions of the size of microcosms and mesocosms are highly variable (Cooper and Barmuta 1993). Biotic colonization in microcosms is controlled entirely or is restricted to a subset of the natural assemblage of organisms found in larger ecosystems (Hansen and Garton 1982; Selby et al. 1985). *Mesocosms* are outdoor experimental systems that resemble natural ecosystems more closely than do microcosms (Zischke et al. 1983; Cushman and Goyert 1984). Aquatic insects usually are important elements of both microcosm and mesocosm tests (e.g., Buikema and Voshell 1993; Cooper and Barmuta 1993).

Environmental protection has centered around the use of existing standardized tests, or refinements of them, to rank the biological effects of chemicals and effluents. These standardized tests should involve more than the small percentage of species of aquatic insects that are currently used. However, environmental protection also requires the maintenance of ecosystem integrity, which is not currently done with toxicity testing and is a major future challenge in its use. Aquatic insects must play a role in this regard in future studies.

Field Experiments in Biomonitoring: Field experimentation directed at basic ecological questions has resulted in important insights into the structure and functioning of ecological systems. This approach has recently become an integral part of biomonitoring because field experiments can assist in: (1) interpreting observed responses of aquatic insects to environmental change; (2) calibrating biomonitoring programs; (3) identifying sensitive (and possibly indicator) species; (4) predicting responses of aquatic insects to potential perturbations;

Table 7B. Types of toxicity tests using aquatic insects. Adapted from Buikema and Voshell (1993).

Types of tests	Duration	Endpoints
Single species: acute	48 to 96 hours	Mortality, growth, behavior
Single species: chronic		
Life cycle	One generation	Mortality, growth, reproduction
Sub-life cycle	Egg through early life stages	Mortality, growth, behavior, metamorphosis
Multispecies		
Artificial assemblage	Days to weeks to months	Mortality, growth, competition, predator-prey interactions, biomass, density, diversity, taxa richness, community metabolism
Natural assemblage	Days to weeks to months	Mortality, growth, competition, predator-prey interactions, biomass, density, diversity, taxa richness, community metabolism, secondary production, behavior, functional feeding groups

(5) distinguishing between direct and indirect effects of environmental perturbations; and (6) determining direct and interactive effects of different variables on an ecosystem (Cooper and Barmuta 1993). Thus, *in situ* controlled experiments circumvent many of the design problems inherent in field-survey approaches and laboratory toxicity tests because researchers can manipulate environmental variables of interest and then, with appropriate controls, examine responses of aquatic organisms to those manipulations.

Field experiments range in size from microcosms (Hopkins *et al.* 1989; Barmuta *et al.* 1990) to mesocosms (Rosenberg and Wiens 1978) to whole-lake (Schindler 1974; Schindler and Fee 1974; Schindler *et al.* 1985) and whole-stream (Hall *et al.* 1980) manipulations. Treatment durations range from single-application short-term experiments to multiple-application experiments conducted over several years.

Field experiments can be complemented with information from field surveys and laboratory studies, but it is essential that field experiments be rigorously conducted to allow unambiguous interpretation of results. Much of the information necessary for the design, analysis, and interpretation of field experiments is discussed in Cooper and Barmuta (1993), and Chapter 3 provides a further consideration of some of the elements of field experiments.

DESIGN OF BIOMONITORING SURVEYS

The design of any aquatic-insect biomonitoring survey involves making decisions about the use of control or reference sites, the duration of study, the frequency of sampling, the number of sample replicates taken, whether sampling will be confined to specific habitats (i.e., habitat stratification), the level of taxonomic identifications, and the measures and analysis techniques to be used. Selected examples of reviews and textbooks dealing with various aspects of the design and analysis of biomonitoring using aquatic macroinvertebrates are summarized in Table 7C (see also Chap. 3 for a detailed consideration of sampling).

In conducting biomonitoring surveys, four types of studies should be distinguished: (1) pilot or reconnaissance studies; (2) studies to describe population or community characteristics; (3) studies to detect differences in populations or communities between or among sites; and (4) studies based on standardized, rapid assessment procedures. Pilot or reconnaissance studies are essential to the successful completion of any biomonitoring study that will involve more than a standardized approach (as in 4). These studies reveal what organisms occur, their approximate densities, and their spatial aggregation, all of which are essential elements in deciding how many samples to take and whether to stratify the sampling according to specific habitats.

A description of population or community characteristics requires a different study design than detection of differences in populations or communities between or among sites. Studies that describe population characteristics such as the mean density of populations have received the most attention in discussions of the numbers of required samples and, inevitably, large numbers of samples are necessary for a precise description of density (Chap. 3). Sometimes, precise estimates of density are an appropriate goal, as for example in a study to establish the present density of a potentially endangered species. However, for studies of perturbation or disturbance, it is whether a *difference* exists between mean densities of populations or communities over time or between different sites that is the measure of interest. Study types (2) and (3) require different analyses and numbers of samples. This topic is considered in detail in Chapter 3.

Rapid assessment procedures, which are designed to reduce costs of environmental assessment at a site or group of sites, have both strong supporters and strong detractors. Differences in opinion between these two groups often center on how results from rapid assessment are used (Resh *et al.* 1995). The majority of both groups would probably support their use as pilot studies or for surveys in areas where taxonomic keys and logistic support are difficult to obtain. For example, use of measures such as the Family Biotic Index (Hilsenhoff 1988), a common component of programs that use rapid assessment

Table 7C. Selected reports dealing with the design and analysis of benthic research and biomonitoring. An asterisk indicates detailed treatment. Adapted from Resh and McElravy (1993). See also discussion and references in Chapter 3.

Author	Lentic	Lotic	Survey design-stratification/sampling considerations	Study design/frequency of studies	Sampling devices	Sorting procedures	Data storage and retrieval/quality control
APHA (1989)	X	X	X		X	X	
Cooper and Barmuta (1993)	X	X	X	X			
Cummins (1975)		X	X			X	
Dawson and Hellenthal (1986)	X	X					X*
Downing (1984)	X				X*	X	
Elliott (1977)	X	X	X				
Green (1979)	X	X	X	X			
Hellawell (1978)		X	X		X*	X	
Hellawell (1986)	X	X	X		X		
Klemm et al. (1990)	X	X	X		X	X	
Norris and Georges (1986)		X	X				X
Norris et al. (1992)		X	X	X			
Ortal and Ritman (1985)		X					X*
Peckarsky (1984)		X	X		X		
Prepas (1984)	X	X	X				
Resh (1979)		X	X		X	X	
Resh et al. (1990)	X	X			X		
Resh and McElravy (1993)	X	X	X	X*	X	X	
Southwood (1978)	X	X	X		X		
Voshell et al. (1989)		X	X	X	X		
Winterbourn (1985)		X	X		X	X	

procedures, or the Sequential Comparison Index (Cairns et al. 1968) can roughly characterize the pollution status of a site or group of sites, and can be used by inexperienced investigators (see also Resh 1995). Likewise, few biologists would claim that rapid assessment procedures always could substitute for studies based on hypothesis testing, replicate samples, and statistical analysis. The area between these extremes is the area of disagreement as to whether the results of rapid assessment approaches are an endpoint or an indication of the need for further studies. Whatever the use, the reliability of any rapid assessment program depends on its ability to detect environmental perturbation outside the range of natural variability. Clearly, cost savings achieved in a biomonitoring program are of little value if the technique does not accurately characterize site conditions.

However, the main difficulty in using rapid assessment procedures is determining, in the absence of a priori statistical objectives, what deviation from normal constitutes impact. (Think back to the thermometer analogy used above: how many degrees deviation from normal temperature is a signal of ill health?). Also, rapid assessment procedures have been developed mainly for small streams, for only a few kinds of pollution (see comments under biotic indices above), and for limited geographical areas. Protocols need to be expanded to large rivers and lakes, tolerance scores need to be expanded and calibrated for different regions of North America, and methods must extend to many different kinds of pollution, especially multiple-pollution inputs.

CONCLUSION

Aquatic insects offer an excellent way to examine biological aspects of water quality; their use in this fashion is an idea whose time has certainly come. Scientists in many countries have realized this and are increasingly using (and in many cases emphasizing) water quality criteria based on macroinvertebrates (and among them, mainly aquatic insects).

A number of present needs in biomonitoring can be identified. First, the importance of accuracy in taxonomic identification, a central message of this book, cannot be overestimated. Second, multivariate approaches may enhance our understanding of pollution stress because, typically, many confounding factors are associated with pollution. The use of new multivariate techniques and the predictive modelling they allow can be a benefit to biomonitoring programs in that these methods introduce objectivity into data handling and analyses, thus avoiding inter-investigator subjectivity. Third, given the pressing need for toxicity testing to address ecosystem concerns, mesocosms will be increasingly required in risk assessment for pollutants, especially pesticides. Fourth, the ecosystem approach offers an unparalleled opportunity to study the effects of various perturbations on the structure and functioning of whole ecosystems. However, this level of experimentation is expensive and will require extraordinary cooperation among researchers in universities, government agencies, and the private sector.

A number of future needs in biomonitoring also should be identified. First, an improved understanding of the natural variability in response to pollution at all spatial and temporal scales would help in the assessment of anthropogenically generated impact. Second, a critical need in field experimentation is an identification of the degree to which experimental results at one scale can be extrapolated to other spatial and temporal scales (e.g., see Fee and Hecky 1992; Fee *et al*. 1992). Third, traditional biomonitoring deals with point sources of contaminants (i.e., investigations upstream and downstream of an effluent that is suspected of causing impact). An urgent need exists to apply these techniques, or develop modifications of them, to non point-source situations. Finally, "... although the need for applied freshwater ecology is obvious, it is impossible to apply knowledge that one does not have. A sound understanding of macroinvertebrate ecology is a prerequisite to the implementation of a biological approach to ecosystem management" (Johnson *et al*. 1993, p. 104). This is a contribution that all students of aquatic entomology can make.

PHYLOGENETIC RELATIONSHIPS AND EVOLUTIONARY ADAPTATIONS OF AQUATIC INSECTS

Vincent H. Resh
University of California, Berkeley

John O. Solem
The Royal Norwegian Society of Sciences and Letters, Trondheim, Norway

Evolutionary biologists generally accept the hypothesis that life originated in the sea and from there all species of plants and animals evolved. Of the animals, the insects have perhaps undergone the most remarkable adaptive radiation, and today representatives of this arthropod class can be found in almost every conceivable terrestrial and aquatic habitat.

The origin of the insects has been the subject of numerous discussions in the entomological literature and some controversy still exists as to whether or not insects are primarily or secondarily adapted to aquatic environments. That is, did insects originally evolve in fresh waters, or did they first evolve in terrestrial habitats and then secondarily occupy habitats in freshwater environments?

Ross (1965a) assumes that the pre-terrestrial progenitor of the myriapod-insect group (millipedes, centipedes, insects) lived in leaf litter areas along margins of pondlike environments. In this moist terrestrial environment, many present-day myriapods (centipedes and millipedes) are found along with many primitive species of insects. However, Riek (1971) postulates that the original insects were aquatic and spent their whole lives in water. These primitive insects lacked the tracheal respiratory system that is typical of the modern-day terrestrial forms, and their descendants developed a tracheal system when they left the water for part of their life cycle. After tracheae developed in the terrestrial adult stage, they eventually became incorporated in the immature stages. Currently, the majority of the evidence seems to favor the terrestrial origin of both aquatic and terrestrial insects (Wootton 1972; Pritchard *et al.* 1993).

The Odonata and the Ephemeroptera are the only Paleoptera (primitive orders of insects whose adults cannot fold their wings) that are known to have had aquatic juveniles. The Plecoptera originated from a complex of Paleozoic protorthopteroid groups (precursors of the grasshoppers, cockroaches, crickets, etc.), many of which are known to have been terrestrial. However, even though the aquatic insects may have originated from terrestrial species, their survival and success through geologic time probably resulted from the exploitation of freshwater environments by their immature stages (Fig. 8.1).

The understanding of aquatic insect evolution and phylogeny has been hampered by the relatively poor fossil record of freshwater animals; Wootton (1988) has reviewed what is known about fossils of aquatic insects. Marine animals with calcareous ($CaCO_3$) exoskeletons tend to be preserved as fossils in fairly good condition and, as a result, our knowledge of the evolution of marine organisms is far more complete than our knowledge of the evolution of freshwater groups. One reason that few freshwater fossils exist is that the deposition of sediment occurs at a slower rate in freshwater than in marine environments. Because the chitin in the exoskeleton of aquatic insects slowly dissolves in water, insects usually decompose too rapidly for preservation as fossils (Corbet 1960).

The fossil record can be valuable in reconstructing evolutionary patterns by indicating: (1) intermediates between two higher taxonomic categories (such as families); (2) the direction of the evolutionary trend within a group; (3) the time of occurrence of evolutionary grades within phylogenetic lines; and (4) distributions of phylogenetic lines in geographical regions from which they are now extinct (Edmunds 1972). Interpretations of the fossil record must be made with great caution. For example, fossils used in evaluating the terrestrial/aquatic origin of insects were found to be not primitive insects at all but merely fossilized segments of crustaceans! With so few insect fossils available and fossils absent from critical geologic periods, it is difficult to base evolutionary trends in any of the insect orders solely on the fossil record.

The lack of a complete fossil record does not preclude systematic studies of aquatic insects or the construction of phylogenetic trees. Systematists use a wide variety of evidence, including the determination of ancestral and derived character states (often referred to as *plesiomorphic* and *apomorphic* characters, respectively). The interpretation and use of these characteristics vary depending on the approach of the systematist. There are many sound arguments in favor of each of the three main approaches to systematic research: the classical phylogenetic method (in which taxonomic groupings are deduced by similarity of characteristics; e.g., Fig. 8.2); the numerical taxonomic methods (in which the arrangement is based on overall similarity of all available characters); and the now widely used cladistic approach (in which the emphasis is on recency of common ancestry; e.g., Fig. 8.2). Many systematists use combinations of these techniques, and lively debates about interpretations can result (e.g., Weaver 1992a, 1992b; Wiggins 1992). For further discussions on systematic methods, the texts

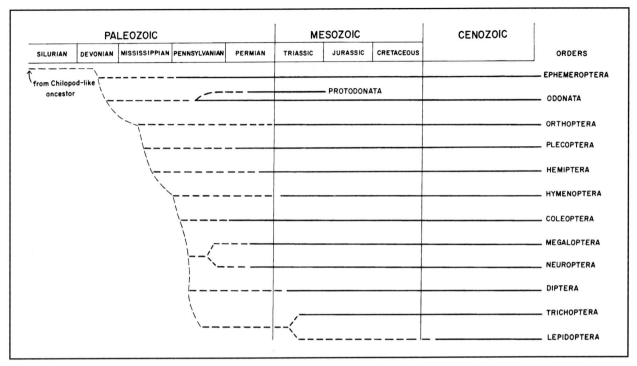

Figure 8.1. Phylogeny of the insect orders containing aquatic representatives plotted against geologic time (after Ross 1965).

of Hennig (1966), Sneath and Sokal (1973), Ross (1974), Wiley (1981), Hillis and Moritz (1990), Mayr and Ashlock (1991), Wiley et al. (1991), and Forey et al. (1992) should be consulted.

The Protodonata are presumed to have been carnivores as are their immediate successors, the Odonata. The largest Protodonata, with a wingspan of 75 cm, probably had larvae that required a number of years to reach maturity; in their late instars these larvae may well have eaten fish (Hutchinson 1981).

The fossil record of damselflies and dragonflies (Fig. 8.3) is more complete than the records of other aquatic insect orders and is mainly composed of wing fragment impressions. Fraser (1957) concluded that the suborder Zygoptera (in particular the family Coenagrionidae) represented the most primitive origins of all the present-day odonates. From this relationship, the evolutionary role of the Carboniferous dragonflies in the family Meganeuridae, whose large size initially gave the false impression that the Anisoptera were the most primitive, is now placed in a new perspective.

The order Odonata has retained a recognizable individuality for at least 200 million years. How has this order survived for such a vast period of time without appreciable change? Corbet (1960) suggests that part of the explanation lies in the ecological niche that the adults occupy, which places a premium on aerial agility and visual acuity but not the specialization of feeding. By retaining generalized feeding habits and living in an environment where overcrowding and competition for food are virtually absent, the adult odonates have avoided two of the most severe selective forces with which other animals must contend. Odonate larvae may have a limited food supply or live in a habitat with severe environmental restrictions, but it is also the larval stage that exhibits the greatest diversity of form and habit.

Edmunds (1972, 1975) has reviewed evolutionary trends within the Ephemeroptera, an aquatic order with a poor fossil record. Mayflies may be regarded as "flying Thysanura," because they are almost certainly derived from lepismatoid (bristletails) origins and are similar in having three caudal filaments (Edmunds and Traver 1954). The oldest known fossil mayflies in North America are those from the lower Permian, and these were found in Kansas and Oklahoma (Peters 1988). Studies of mayfly evolution have had a unique orientation when compared with those of the other insect orders in that both immature and adult stages have been considered in the development of phylogenetic relationships (Fig. 8.3); phylogenies of other orders are now following this example. The reviews of Edmunds (1972, 1975) on the evolution of mayflies and of Edmunds and McCafferty (1988) on the mayfly subimago (i.e., the winged but sexually immature adults that must molt to the adult stage before reproducing) are recommended not only as excellent studies of mayflies but also for their thought-provoking comments on systematic methodology. More recently, McCafferty (1991) proposed a classification of this order based on cladistic analysis (Fig. 8.2; see also Chapter 11).

Illies (1965) reviews the evolutionary history of the Plecoptera (Fig. 8.4), and discusses their ancestral relationship to the Orthoptera and the Blattoidea (which includes the cockroaches). The stonefly fossil record includes over thirty species that have been described from the different strata of the Permian to the middle Tertiary periods (Wootten 1972). Extant stonefly families are restricted to either the northern or southern hemispheres, and there is a pattern indicating that the more primitive families occur in the southern hemisphere (Illies 1965). The present distribution of stoneflies can be explained by continental drift patterns. Illies' (1965) study of Plecoptera describes worldwide distribution patterns and Hynes (1988)

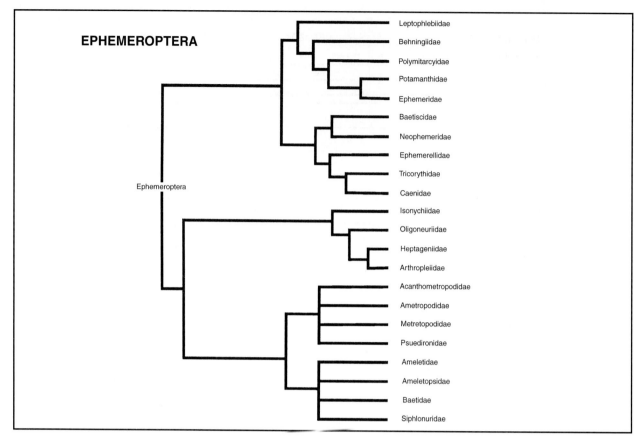

Figure 8.2. Phylogenetic relationships of North American Ephemeroptera based on cladistic data (after McCafferty 1991a, b, and McCafferty and Wang 1994a).

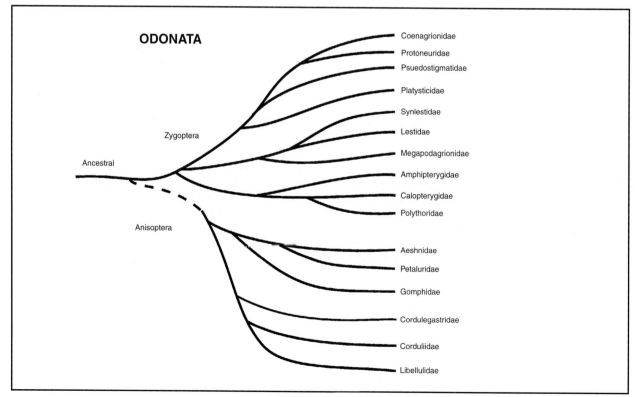

Figure 8.3. Genealogical tree of Odonata of North America (modified from Fraser 1957, Pfau 1991, and Nel *et al.* 1993 by K. J. Tennessen).

concentrates on those patterns for North American species. The study of Ross and Ricker (1971) on the winter-stonefly genus *Allocapnia* provides interesting information on the distribution and dispersal routes used during the Pleistocene glaciation.

The order Trichoptera appears to have arisen from a megalopteran-like ancestor with a larval stage much like that of the present-day caddisfly family Philopotamidae (Ross 1956). The evolution of the approximately thirty families of Trichoptera is more apparent in the structure and habits of the immature stages (Fig. 8.5), whereas adult structures are valuable in elucidating the evolution of genera within these families (Ross 1956). Both the Trichoptera and Lepidoptera are closely related to a third group, the Zeugloptera (considered a suborder by some authors to include those Lepidoptera with functional mandibles). In fact, certain authors have suggested that these three groups should be considered as distinct suborders within one single order (Hinton 1958b). Kristensen (1975, 1981) recently concluded that the monophyly of the Trichoptera is generally accepted and substantiated, although the Zeugloptera is cladistically a member of the Lepidoptera. Wiggins and Mackay (1979) suggested that ecological diversification in Trichoptera can be attributed to the many ways in which species of this order use silk for food gathering, and Wiggins and Wichard (1989) emphasized the importance of how silk is used in preparing for the pupal stage.

Several orders in which most species are terrestrial also play important roles in aquatic systems. The commonly encountered aquatic hemipterans include the water striders of the family Gerridae. These are among the most conspicuous inhabitants of ponds or streams, and this group has been extensively studied by systematists, behaviorists, and ecologists (Spence and Anderson 1994). This family also includes the sea skaters, *Halobates*, which are the only insects known to occur in the open oceans (Cheng 1985). The Chironomidae of the Diptera is remarkable for both the diversity of habitats its species can occupy and the wide range of environmental conditions many species can tolerate (Pinder 1986). Systematically, chironomids have contributed much to our knowledge of zoogeography (e.g., Brundin 1966). The families of Coleoptera that are known collectively as "riffle beetles" (the Elmidae, Dryopidae, some Lutrochidae, and the Psephenidae) have closely related species that occur in regions as widely separated as Australia and South America. However, through modern concepts of plate tectonics and continental drift, such patterns are explainable and the antiquity of these insects is emphasized (Brown 1987).

The Hemiptera (Fig. 8.6), Diptera (Fig. 8.7), and Coleoptera are large orders that contain representative aquatic species in several different families. In all these groups the invasion into aquatic habitats probably occurred independently several different times. This polyphyletic pattern is especially true in the Coleoptera, the largest order of insects.

The phylogenetic studies of the aquatic insect orders mentioned above (Fraser 1957; Edmunds 1972, 1975; etc.), along with studies of more specialized groups such as Brundin's (1966) study of the Chironomidae, have greatly increased our knowledge of animal dispersal patterns. Three general affinities are evident for the North American insect fauna (e.g., Noonan 1986) and several zoogeographic studies of aquatic insects (e.g., Hynes 1988; Peters 1988; Stewart and Stark 1988) have supported this pattern: (1) Asia-North America, which were intermittently joined as Beringia from the Cretaceous to the Pleistocene (100-70 mya, i.e., million years ago); (2) Europe-eastern North America, which were joined until the opening of the North Atlantic 35-50 mya; and (3) South America-North America, which were joined from 4 mya to the present. Human activities have also influenced the present distribution of aquatic insects; the Asian tiger mosquito (*Aedes albopictus*, Rai 1991) and the fauna of the Hawaiian Islands (Howarth and Polhemus 1991; Resh and de Szalay 1995) are examples of such influences.

In discussing the present-day distribution of all these groups, the question of dispersal mechanisms is logically raised. Active dispersal is accomplished by adult females, although species can also be dispersed passively as windblown eggs or by birds. The short adult life span of mayflies, stoneflies, and caddisflies would prevent long dispersal flights, but a windblown egg mass that could withstand desiccation for long periods would certainly aid in widespread distribution. Examination of the present-day distribution of aquatic insect species suggests that the question of how the species initially arrived in the habitat is less interesting than what are the ecological conditions that have enabled it to survive there.

There are several excellent reviews covering a variety of aspects of the evolutionary biology of aquatic insects in the orders Odonata (Corbet 1980), Ephemeroptera (Brittain 1982), Plecoptera (Hynes 1976), and (Mackay and Wiggins 1979), the families Chironomidae (Pinder 1986) and Gerridae (Spence and Anderson 1994), and the several families collectively referred to as the riffle beetles (Brown 1987).

PHYSIOLOGICAL AND MORPHOLOGICAL ADAPTATIONS

During their evolution and adaptation to aquatic environments, insects have solved the problems of respiration in many different ways. These include the use of air-tubes to obtain atmospheric oxygen, cutaneous and gill respiration, the extraction of air from plants, hemoglobin pigments, air bubbles, and plastrons. Air-tubes, which tend to restrict activity to the water surface, have evolved in the Hemiptera (Nepidae, Belostomatidae) and the Diptera (*Aedes*, *Culex*, and *Eristalis*). Cutaneous and gill respiration is widespread in the immature stages of most of the aquatic insect orders and these mechanisms enable the submerged species to occupy habitats below the water surface and within the substrate. Most of the species that rely on this type of respiration require well-oxygenated water, although certain species of the Chironominae that are found in the profundal regions of eutrophic lakes and that normally rely on cutaneous respiratory mechanisms may survive periods of oxygen depletion through the use of hemoglobin pigments that aid in oxygen transfer. Respiration by adult aquatic insects such as the beetles and true bugs is often facilitated by the use of an air bubble, although certain species have evolved a more advanced respiratory mechanism, the plastron (a system of microhairs or papillae that hold an air film, Fig. 4.5). A plastron enables the adult to stay submerged for far longer periods than would be possible if an air bubble mechanism were used (Thorpe 1950). (For a detailed description of aquatic insect respiration see Chap. 4.)

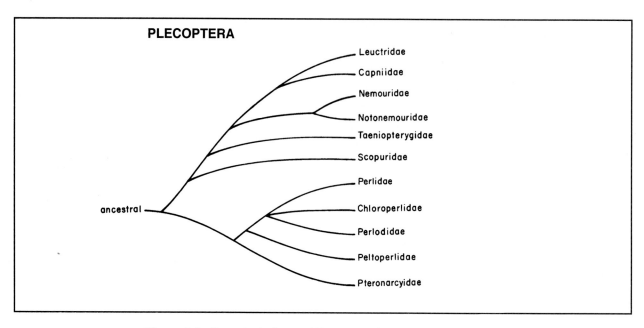

Figure 8.4. Genealogical tree of Plecoptera (after Illies 1965).

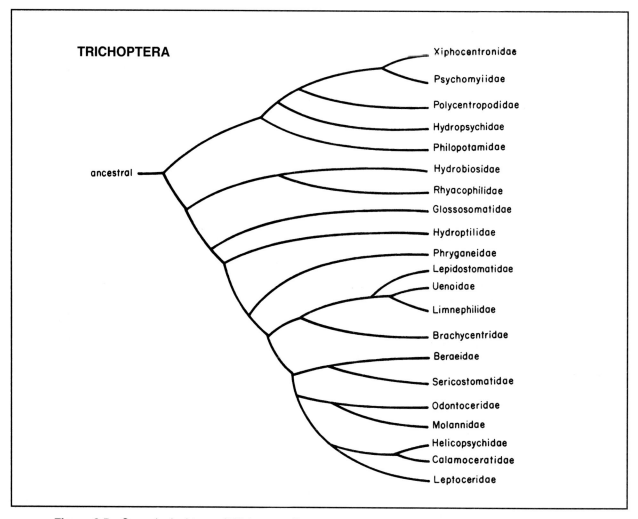

Figure 8.5. Genealogical tree of Trichoptera (from Ross 1967b). For other phylogenies, see Weaver (1983, 1984) and Wiggins and Wichard (1989).

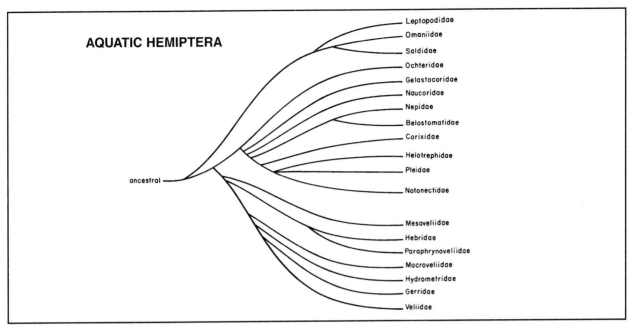

Figure 8.6. Phylogeny of aquatic Hemiptera (modified from China 1955 and Usinger 1956b by J. T. Polhemus).

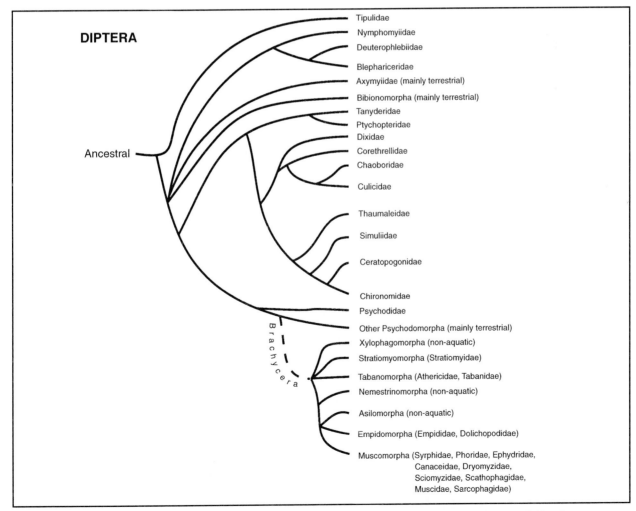

Figure 8.7. Phylogeny of Diptera families and infraorders (after Wood and Borkent 1989 and Woodley 1989). All taxa from Tipulidae through other Psychodomorpha are Nematocerous Diptera.

The distribution of aquatic insects in the wide variety of habitats present in freshwater environments has led to the evolution of many types of adaptations. Recently, it has been demonstrated that aquatic insects having the same species traits use particular habitats (Usseglio-Polatera 1994).

Adaptations occur in all stages of the insect life cycle, and are characterized by their great flexibility in terms of the evolution of several different mechanisms in response to a specific selective pressure. For example, many adaptations in the egg stage are found in species that occupy temporary pool habitats. The adult females of these species may deposit egg masses close to the ground on the underside of pieces of wood or bark, thereby gaining the advantage of increased humidity and protection from the sun and wind. Certain caddisfly species that are primarily adapted for life in temporary pools have evolved mechanisms that delay or suspend development of the immature stages until soil moisture or the surface water in the basin is sufficient to sustain the newly hatched larvae (Wiggins 1973a). In addition, modifications of the gelatinous egg-mass matrix protect the eggs and larvae from desiccation and freezing for periods up to seven months (Wiggins 1973a). Other mechanisms used by species inhabiting temporary pools include delaying oviposition until the pools contain water, followed by the prompt hatching and development of the eggs (Corbet 1964; Wiggins et al. 1980).

The time spent in the egg stage varies considerably from species to species, but the time may also vary within a single species population. Whereas all eggs of *Lestes sponsa* (Odonata) will diapause through the late summer, autumn, and winter (Macan 1973), both diapausing and nondiapausing eggs will be found in the same batch of *Diura bicaudata* (Plecoptera) eggs (Hynes 1970a). These two egg types of this latter species imply that not all larvae hatch at the same time. Similarly, eggs of *Baetis rhodani* (Ephemeroptera) hatch at different intervals (Macan 1973). A diapause during the egg stage may enable insects to survive unfavorable periods. A staggered hatching pattern may also prevent overcrowding of newly hatched larvae in an area that may have limited food resources. The exploding egg masses in the stonefly family Nemouridae, which disperse eggs over a wide area, and the swimming ability of newly hatched caddisfly larvae in genera such as *Phryganea, Agrypnia,* and *Triaenodes* may also aid in enhancing dispersal and preventing overcrowding.

Morphological adaptations of larvae of freshwater insects (Table 8A, based on the discussion of Hynes 1970a, 1970b; also see Hora 1930; Nielsen 1951b) are closely followed by behavioral adaptations. In running-water environments, many of the adaptations can be directly related to hydraulic stress and the continuous struggle of the organisms to remain on the substrate (Stazner et al. 1988). There are, however, species that actively exhibit what Waters (1972) has termed *behavioral drift,* in which individuals enter into the water column and move downstream from their original points of attachment during certain well-defined periods of their daily cycle (for a recent review on drift, see Brittain and Eikeland (1988)). In standing water, current does not force animals into the open water. However, insects such as *Chaoborus* sp. (Diptera) and many Coleoptera and Hemiptera actively move through the water column in these environments. (For a description of aquatic insect behavior, see Chap 5.)

Of the holometabolous insects, aquatic pupae are found in nearly all species of Trichoptera (cf. Anderson 1967) and aquatic Diptera. Some of the aquatic Lepidoptera (e.g., *Petrophila confusalis*) that pupate underwater do so in air-filled cocoons. Terrestrial pupae are found in all Megaloptera and aquatic Neuroptera. Although the aquatic Coleoptera have species of all the above types, a terrestrial pupa is by far the most common condition among the aquatic beetles. Aquatic pupae are found only in *Donacia, Neohaemonia,* and *Macroplea* of the Chrysomelidae, *Lissorhoptrus* of the Curculionidae, *Psephenoides* of the Psephenidae, and *Noterus* of the Noteridae. Of all aquatic beetles, the pupae of species in the European genus *Psephenoides* are the only ones surrounded with water and that could be described as truly aquatic; the other taxa mentioned above pupate in air-filled cocoons (Leech and Chandler 1956).

The different climatic zones in which aquatic insects exist expose these species to a variety of abiotic factors, the most pronounced being temperature, which varies considerably on a yearly basis (although generally less than terrestrial temperatures) within the different geographical regions of the world. Temperature fluctuations greatly affect the poikilothermous (cold-blooded) insects (Sweeney 1984). Possibly because of the relatively constant temperature, more types of emergence patterns are seen in the tropical regions of the world than in either the arctic or temperate zones. In permanent water bodies of the tropics, many insects have continuous emergence throughout the year (Corbet 1964; McElravy et al. 1982). This occurs primarily in areas that are located near the equator and undergo small fluctuations in temperature. However, continuous emergence has recently been demonstrated in constant low-temperature mountain brooks where two species of caddisflies, *Philopotamus ludificatus* and *Wormaldia copiosa* (Philopotamidae), have acyclic development and show continuous emergence (Malicky 1980).

Several species also show a rhythmic emergence pattern, and lunar emergence rhythms have been recognized in several aquatic insects that live in tropical climates (Tjonneland 1960; Corbet 1964). The peak in their emergence pattern coincides with different lunar phases, depending on the species. The chironomid *Tanytarsus balteatus* has an emergence pattern that follows the phases of the new moon (Corbet 1964), whereas other species may have two emergence peaks where the minimum activity occurs during the new and full moon phases (Tjonneland 1960).

In temperate regions, the lunar periodicity in emergence patterns is exemplified by the chironomid *Clunio marinus* (Neumann 1976). This midge larva lives in the intertidal zone of sandy seashores and emergence is restricted to a few days at the time of the new and full moon (Caspers 1951). A sporadic pattern, in which emergence appears to occur at irregular intervals and seemingly without any environmental cues, seems to be present in only a few aquatic insects (Corbet 1964).

Distinct seasonal emergence is typical of insects living at high latitudes where regular changes in temperature occur. However, wet and dry season climatic patterns may also lead to seasonal emergence patterns as has been demonstrated in the tropics (Corbet 1964). As one moves from the tropics to higher latitudes, emergence periods become increasingly shorter, with the extreme condition occurring in the arctic regions where many species have adapted to cycles in which emergence is limited to a few days out of the entire year.

The most widespread rhythm exhibited by aquatic insects is the diel pattern (Remmert 1962), which may influence initial hatching from the egg, feeding behavior of the immature stages, emergence, flight activity, oviposition, and other life history features (Resh and Rosenberg 1989). The significance of diel emergence patterns is that in a short-lived adult insect, such as a mayfly or caddisfly, simultaneous emergence and subsequent swarming by males help to ensure the continuity of the population through the next generation. Periodicity in mate attraction by caddisflies that use sex pheromones also may be an especially important adaptation in this regard (Jackson and Resh 1991).

LIFE CYCLE ADAPTATIONS

Of the major groups of freshwater insects, almost all are regularly represented throughout the world. However, several groups only occur in running-water habitats and many others reach their maximum development and diversity there. Hynes (1970a) suggests that this may be a consequence of the permanence of streams and rivers when compared with the longevity of most lake and pond habitats. Many river systems have been in continuous existence for long periods of geologic time, whereas lakes persist for relatively short periods and have had little opportunity to develop a purely lacustrine fauna. Lakes are certainly capable of producing evolutionary change, as shown in the endemic species that occupy ancient lakes such as Lake Baikal in Russia. In time, however, all lakes fill in and disappear; their faunas perish. River faunas have a much greater chance for continuity and development because although rivers and river systems may change, they rarely disappear entirely. They are not "evolutionary traps" and, because of this, many of the most primitive species of freshwater organisms are primarily found in lotic environments (Hynes 1970a).

In addition, most primitive species of aquatic insects are found in lotic habitats because many groups probably first became aquatic by using lotic habitats where respiration conditions for larvae and resistance to desiccation for adults were most propitious. These factors seem to apply for the Plecoptera, Ephemeroptera, Odonata, Trichoptera, and Chironomidae. Thus, the absence of primitive members of these groups from lentic habitats could be explained by the hypothesis that they never lived there. Wiggins *et al.* (1980) give support to this speculation in that insects that occur in temporary pools are primarily derivative groups because the specialized features required to sustain dry periods arose from permanent lentic-dwelling species and these, in turn, arose from lotic species.

Aquatic insects exhibit the same versatility as terrestrial insects in their ability to use a wide variety of food sources (see Chap. 6). In pre-Tertiary times, the main food items available to the primitive aquatic insects probably were floating and encrusting algae, plant debris (with its associated decomposing microorganisms), and animals, including smaller insects, other invertebrates, and fish fry (Wootton 1972). As small predators feeding on primary consumers, these early aquatic insects probably had few competitors. In time, they developed morphological and behavioral adaptations in response to increased competition and other selective pressures. Eventually, these adaptations were such that the aquatic insects were able to exploit a wider range of food sources and develop into the variety of macro- and micro-feeders that operate from deep within the substrate to the top of the surface film (Wootton 1972).

Very few aquatic insects have adapted to a completely submerged life cycle (but see Jewett 1963, for a Plecoptera exception). At one time or another, nearly all species spend a period in the terrestrial habitat. A major problem in being submerged for even part of the life cycle is respiration because, in order to respire while submerged, the insect must receive oxygen from the surrounding aquatic environment (see Chap. 4). Many species have evolved respiratory systems that function in well-aerated water but have not developed survival mechanisms for low oxygen concentrations. In regard to the latter, there is a major difference between running- and standing-water environments. Normally, streams have higher oxygen concentrations (because of turbulence) than do either ponds or lakes. This is certainly a factor in the distribution of Plecoptera and most species of Ephemeroptera and Trichoptera, which are groups that have their maximum development in running water. Oxygen saturation and temperature are integrally related, and the cooler temperatures that often prevail in running water aid in survival. In high-altitude (or latitude) lakes where the water is cold and highly oxygenated at all times, the distinction between standing- and running-water faunas becomes less clear.

In addition to the adaptations in the immature and adult stages of aquatic insects, the life cycles often exhibit unique phenological patterns. Aquatic insect populations may produce single or multiple generations during a year or, in some species, greater than a year. Life-cycle completion time may vary greatly throughout the range of a species, between populations of the same species (e.g., from more than one generation in a year in the warmer areas to more than one year for each generation in the colder areas), or even between the upper and lower reaches of the same stream (e.g., Resh and Rosenberg 1989).

The presence of a diapause in the egg stage, or the formation of a quiescent prepupa, is an important modification that enables the insect to conserve energy or survive unfavorable conditions. Certain life history patterns in aquatic insects may have a definite selective advantage, particularly in maximizing the efficient use of food sources that have only seasonal availability. An example of this can be seen in the leptocerid caddisflies of the genus *Ceraclea* that feed on freshwater sponge. Because the sponge is available only certain times of the year (only the nonedible gemmules are present during the winter months), life cycles have been modified and alternative food sources are used (Resh 1976a). The significance of life cycle flexibility in the presence of coexisting, systematically related species has yet to be fully understood, but the ability of these species to share available resources presents interesting implications for habitat partitioning and community evolution. This and many other fascinating aspects of aquatic insect evolution provide fertile areas for future research.

PROSPECTS FOR FUTURE RESEARCH

As Ross (1974) points out, the future of systematic research can best be predicted by examining the future of biology. This is especially true in considering evolutionary studies of aquatic insects. Thousands of species of aquatic insects have been described; however, in all groups, and particularly in the holometabolous orders, the immature stages are poorly known.

Table 8A. Examples of morphological adaptations of aquatic insects to running-water environments.

Adaptation	Significance	Representative Groups and Structures	Exceptions or Comments
Flattening of body surface	Allows species to live on top of flattened stones and allows them to crawl through closely compacted substrate	Psephenidae (Coleoptera); *Epeorus, Rhithrogena* (Ephemeroptera); Gomphidae, Libellulidae (Odonata); *Molanna, Glossosoma, Ceraclea ancylus* (Trichoptera)	Many stream inhabitants that do not live on exposed surfaces are also flattened
Streamlining	A fusiform body offers least resistance to fluids	*Baetis, Centroptilum* (Ephemeroptera); *Simulium* (Diptera); Dytiscidae (Coleoptera)	Except in the Coleoptera, this body shape is relatively rare
Reduction of projecting structures	Projecting structures increase water resistance	Gills of *Baetis* and loss of central cerci in many mayflies (Ephemeroptera)	*Atherix* (Diptera), Corydalidae (Megaloptera), and Gyrinidae have large lateral projections; gills of *Baetis* may reflect respiratory physiology
Suckers	Provide attachment to smooth surfaces	Blephariceridae (Diptera); some Dytiscidae (Coleoptera)	Rare in stream animals; if surfaces are irregular because of moss, algae, etc., suckers are inefficient
Friction pads and marginal contact with substrate	Close contact with substrate increases frictional resistance and reduces chances of being dislodged by current	*Psephenus* (Coleoptera); *Dicercomyzon, Drunella doddsi, Rhithrogena* (Ephemeroptera)	Marginal contact devices are not confined to insects living in the torrential parts of streams
Hooks and grapples	Attachment to rough areas of substrate	Elmidae, Dryopidae (Coleoptera); Rhyacophilidae (Trichoptera); Corydalidae (Megaloptera)	Modified structures include tarsal claws, clawlike legs, and posterior prolegs
Small size	Small sizes permit them to crawl through closely compacted substrates.	Water mites (Hydracarina) in streams are never as large as still-water Hydracarina and lack the swimming hairs that are found in these latter water mites; large hydrophilids and dytiscids (Coleoptera) inhabit still water; nearly all beetles in fast water are small	Stream animals in other groups are not noticeably smaller than still-water relatives
Silk and sticky secretions	Allows attachment to stones in swift current	*Rheotanytarsus*, Orthocladiinae, *Simulium* (Diptera); Psychomyiidae, Hydropsychidae, Leptoceridae (Trichoptera); *Paragyractis* (Lepidoptera); Plecoptera and Ephemeroptera eggs	Even free-living caddisflies use silk attachments as do the cased species when they molt and pupate
Ballast	Incorporating large stones in cases makes the insects heavier and less easily swept away	*Goera, Stenophylax* (Trichoptera)	This is at least partially caused by these sized particles being the only ones available for case building
Attachment claws and dorsal processes	Stout claws aid in attachment and fixation to plants	Calopterygidae (Odonata); some *Taeniopteryx* (Plecoptera); *Ephemerellidae* (Ephemeroptera)	Most animals that live in vegetation show no particular adaptations that distinguish them from still-water species in vegetation
Reduction in powers of flight	A loss of individuals from a restricted habitat may result in smaller populations in following generations	Reduced wings of stoneflies *Allocapnia* (Plecoptera); loss of hind wings in Elmidae (Coleoptera); wingless females of *Dolophiloides* (Trichoptera)	Reduced flight powers may be a disadvantage because it reduces dispersal ability
Hairy bodies	Keeps sand and soil particles away while burrowing in substrate	*Hexagenia, Potamanthus* (Ephemeroptera)	Permits open spaces for water to flow over body

Because it is the larval stages that are usually collected by aquatic entomologists, the lack of association of the immature and the taxonomically named adult stage has limited the precision of many studies in aquatic ecology. Associations between larval and adult stages are needed, and they can be made by a variety of rearing techniques (see Chap. 3, especially Table 3F for rearing method references). Association of the egg and pupal stages with the adult is also important in elucidating phylogenetic relationships (Wiggins 1966; Koss 1970).

The association of immature and adult stages is an area of research in which all students of aquatic entomology can make valuable contributions because rearing techniques do not require elaborate equipment and they can provide associations that eventually can be used in constructing taxonomic keys. In the future, we also will likely see the application of molecular-systematic methods (e.g., DNA sequencing) in associating immature and adult stages. These techniques are likely to be more powerful than currently used electrophoretic approaches; however, even using these older techniques, advances in the separation of morphologically similar early instars (Zloty *et al.* 1993) and cryptic species (Jackson and Resh 1992) of aquatic insects have been accomplished.

Studies of systematically related, coexisting species often reveal both obvious and subtle mechanisms of resource partitioning and ecological segregation (e.g., Cummins 1964; Grant and Mackay 1969; Mackay 1972; Butler 1984). Similarly, studies of water quality tolerances of congeneric species have been useful in developing the important concept of biological indicators of environmental quality (e.g., Resh and Unzicker 1975). Oftentimes congeneric species co-occur, providing ample opportunity for such comparative studies.

In almost all cases, evolutionary relationships of aquatic insects have been based on studies of morphological structures, and this has proven to be a useful way of analyzing systematically related groups. However, species are often assigned to different higher taxa because of conflicting opinions on the validity of the various morphological characteristics. In these cases, behavioral studies may be valuable. For instance, behavior patterns of larvae of two European species of the mayfly genus *Leptophlebia* (Solem 1973) agree with that reported for a North American *Leptophlebia* (Hayden and Clifford 1974). If the diversity of such rhythms is comparable at the generic level, as has been proposed for trophic status by Wiggins and Mackay (1979) (but see Hawkins and MacMahon (1989) for other considerations in this approach), behavioral studies could provide valuable information on the higher classification of aquatic insects. By combining behavior and other alternative approaches to descriptive systematics, classification and taxonomy may soon be used extensively in predicting ecological features of systematically related species.

GENERAL CLASSIFICATION AND KEY TO THE ORDERS OF AQUATIC AND SEMIAQUATIC INSECTS

Howell V. Daly
University of California, Berkeley

GENERAL CLASSIFICATION

Although only three percent of the insect species are aquatic or semiaquatic, that is, with one or more life stages living in or closely associated with aquatic habitats, representatives may be found in about half (13) of all the insect orders. Since many species are aquatic only during their immature stages, the study of aquatic insects involves both terrestrial and aquatic or semiaquatic life forms.

Insects are named according to strict international rules of nomenclature that apply to all animals. The hierarchical system is exemplified by the classification of a species of burrowing mayfly below:

Taxon	*Name*
Kingdom	Animalia
Phylum	Arthropoda
Class	Insecta
Subclass	Pterygota
Infraclass	Paleoptera
Order	Ephemeroptera
Superfamily	Ephemeroidea
Family	Ephemeridae
Subfamily	Ephemerinae
Genus	*Hexagenia*
Subgenus	(*Hexagenia*)
Species	*limbata*
Author	(Jean G. A. Serville)

Each species is assigned a specific name and placed in a genus with related species. The combination of Latin or Latinized generic and specific names is unique for each species of insect and is called the *scientific name*. The scientific name is printed in italics with the generic name capitalized, e.g., *Hexagenia limbata* (Serville). The name of the first person to publish a description of and assign a specific name to the insect is cited after the scientific name. If, as in this case, the author initially placed the species in another genus, the author's name is enclosed in parentheses. A more detailed coverage of the principles of classification is given in Mayr and Ashlock (1991).

A named group of insects at any level is referred to as a *taxon* (plural, taxa)—a species or group of species, a genus or group of genera, etc. One or more genera that share certain common characteristics are placed in a *tribe* designated by the ending -ini. Tribes are grouped in *subfamilies* (-inae), subfamilies in *families* (-idae) and phylogenetically related families are grouped in *superfamilies* (-oidea). Related superfamilies and their families are included in an *order*. The names of most insect orders have the ending "-ptera," from the Greek word meaning wing, but a few such as Collembola (springtails) and Odonata (dragonflies and damselflies) simply end with "-a". Finally, all insect orders are included in the class Insecta, sometimes referred to as Hexapoda (six legs).

Although scientific names of taxa may be hard to remember and difficult to pronounce, they eliminate the confusion that often surrounds common names. The mayfly, *Hexagenia limbata*, is also commonly known as the Michigan caddis, fishfly, sandfly, and Great Olive-Winged Drake. Entomologists and knowledgeable anglers use caddis or caddisflies to refer to the order Trichoptera, fishflies for certain Megaloptera, and sandflies for certain Diptera. Even the angler's "Drake" applies to mayflies belonging to at least six genera and three families.

Below is a list of orders that have at least some aquatic or semiaquatic species. Orders with almost all species having one or more aquatic stages are marked by an asterisk (*), followed by the common names where appropriate. The Subclass Apterygota is reserved for insects that do not have wings at any stage of the life history and, more importantly, are believed to have ancestors that were wingless. The Subclass Pterygota includes winged insects and those whose ancestors were winged. Among these, two major categories can be recognized: (1) the Paleoptera or "old" type of wing that can not twist at the base, and (2) the Neoptera or "new" type of wing that can be twisted at the base so the wings can be folded over the back. Within the Neoptera are two Divisions: (1) the Exopterygota, with wings developing externally as wing pads in the immatures, and (2) the Endopterygota, with wings developing internally.

Class Insecta (Insects)
  Subclass Apterygota
    Order Collembola (Springtails)
  Subclass Pterygota
    Infraclass Paleoptera
      Order Ephemeroptera* (Mayflies)
      Order Odonata* (Dragonflies and damselflies)
    Infraclass Neoptera
      Division Exopterygota
        Order Orthoptera (Grasshoppers and their allies)

Order Plecoptera* (Stoneflies)
Order Hemiptera (True bugs, leaf hoppers, etc.)
Division Endopterygota
Order Neuroptera (Spongillaflies)
Order Megaloptera* (Dobsonflies, fishflies, and alderflies)
Order Trichoptera* (Caddisflies)
Order Lepidoptera (Moths)
Order Coleoptera (Beetles)
Order Diptera (Flies)
Order Hymenoptera (Wasps)

INSECT LIFE HISTORIES

Identification of insects is complicated by the existence of several distinctive forms during the life cycle. The change in structure and form during the life of an insect is called *metamorphosis*.

To increase in size or change form, an insect periodically sheds (or casts) its rigid exoskeleton, a process called *molting*. A portion of the old cuticle is digested and absorbed, while a new cuticle is deposited beneath. The outer part of the old cuticle is loosened and finally split, beginning at the top of the head and thorax. The brief act of casting or shedding the old skin is termed *ecdysis*. The discarded skin, termed the *exuviae*, is known as the "shuck" by anglers.

The act of ecdysis provides a useful means for subdividing an insect's life history. The form of an insect between ecdyses is called an *instar*. After hatching from the egg, the insect is a first instar and after the initial ecdysis it is the second instar, etc. Most insects have four to six instars, but fifteen to thirty or more instars are known in some mayflies, dragonflies, and stoneflies. It is now common to refer to all active immature insects without wings as *larvae*. The adult insect or *imago* is the imaginal instar, is distinguished by functional external reproductive organs, and is usually winged. The life histories of insects can be divided into three main kinds depending on the degree of change between the immature instars and the imago: *ametabolous* metamorphosis (meaning without change), *hemimetabolous* metamorphosis (incomplete change, Fig. 9.1B), and *holometabolous* metamorphosis (complete change, Fig. 9.1A)

An example of ametabolous metamorphosis is the Collembola. This is an order of tiny, primitive, wingless insects that live in moist litter or soil. Some species are semiaquatic. They change little in form after hatching except to become larger as they pass through the larval instars. After the reproductive organs become functional, the adults may continue to molt.

All species of the orders of the Paleoptera and, among the Neoptera, the Division Exopterygota have hemimetabolous metamorphosis. The rudimentary wings of the larvae are recognizable as stiff, immovable pads on the thorax, and well-developed legs, antennae, and compound eyes are visible. Wing development of the larva culminates when the wings become functional after the last ecdysis. In older textbooks larvae of hemimetabolous insects were sometimes called *nymphs* whether they lived in the water or not. Larvae of Orthoptera and Hemiptera generally resemble the adults, except that adults are sexually active and usually have wings. However, larvae of Odonata, Ephemeroptera, and Plecoptera are all aquatic and highly modified for life under water. In the older textbooks they were sometimes called *naiads*. Again, the adults are sexually active and have wings except for one plecopteran species that is aquatic. Anglers use the word nymph for a wide variety of aquatic insects regardless of the scientific classification or kind of metamorphosis.

The Ephemeroptera are unique among all winged insects in having two winged instars. The first winged instar, called the *subimago*, or dun by the angler, emerges from the water. Usually within 24 hours, the insect molts again to the *imago*, or spinner in angler terminology. Reproduction is usually restricted to the imaginal or last instar.

The last Division, the Endopterygota, have holometabolous metamorphosis and include the largest number of aquatic or semiaquatic species. After hatching, the larval instars are totally unlike the adult. There is no external evidence of the wings, which develop internally as invaginated buds of tissue. The legs and antennae may be reduced or missing entirely, and the eyes are simple ocelli as opposed to the large compound eyes of adults. The larva is often wormlike and distinctive forms are called a caterpillar (Lepidoptera), maggot (Diptera) or wiggler (mosquito). Once fully grown, the larva transforms to a nonfeeding and usually less active instar, the *pupa*. The wings, legs, antennae, and compound eyes of the adult are formed during the

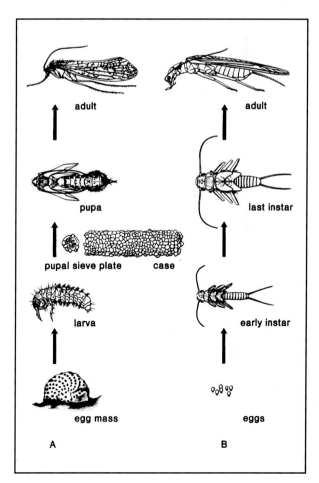

Figure 9.1. Examples of life stages in aquatic insects: *A.* holometabolous metamorphosis of a caddisfly (Trichoptera); *B.* hemimetabolous metamorphosis of a stonefly (Plecoptera). (Figure by Bonnie Hall.)

pupal stage, but are not functional because further development is necessary before the pupa transforms into the winged adult.

Seven orders of endopterygotes have at least some aquatic or semiaquatic species, primarily during the immature stages: Megaloptera, Neuroptera (only the family Sisyridae or spongillaflies are aquatic), Trichoptera, Lepidoptera (certain moth families), Hymenoptera (some parasitic wasps have aquatic hosts), Coleoptera (many families), and Diptera (many families). The last instar of aquatic larvae of Megaloptera, Neuroptera, and most Coleoptera crawl out of the water and pupate in terrestrial sites whereas the other orders remain in the water. The pupae of Trichoptera, Megaloptera, Neuroptera (Sisyridae), and the dipterans, Culicidae (mosquitoes) and Chironomidae (midges), are fairly active. Adults of all orders return near to, or actually reenter, the water to lay their eggs. Only species of adult Coleoptera regularly spend their adult life in water. An oddity among aquatic beetles is the family Dryopidae where the immature stages are terrestrial and only the adults are aquatic.

Taxonomic Keys

A key is simply an aid for identifying different kinds of organisms. Once you are familiar with the higher taxa, most can be recognized at a glance. A dichotomous key consists of numbered, paired statements of contrasting characteristics. In using a key, one reads the first couplet and decides which statement best describes the specimen at hand. Each statement indicates the number of the next couplet to be read. Couplet by couplet, the number of possible choices is reduced until the search terminates with the name of a taxon.

If the specimen does not exactly match either statement, try both possible routes to an identification. It usually becomes evident that a wrong choice was made when the specimen clearly lacks the features mentioned in both statements of subsequent couplets. To confirm a tentative identification, consult the following chapters or other texts describing or illustrating the taxon. Of course, if possible, specimens already identified should be examined. (Locations of good reference collections of aquatic specimens are listed at the end of the chapter.)

KEY TO THE ORDERS OF AQUATIC INSECTS

The following key is designed to identify the orders of all stages of insects treated in this text. The key is a modification of several previous keys, especially that of Bentinck (1956). In the great majority of cases, insects normally found living on the water surface or submerged are included in the key. Insects found at the water's edge, flying near water, or trapped in the surface film may be mainly terrestrial in habit. If the characters of such insects do not correspond with the choices in the key, then a key to all orders of insects should be consulted (e.g., Daly *et al.* 1978; Hill *et al.* 1987; Borror *et al.* 1989).

Immatures and Adults

1.	Wings or wing pads present, fore wings sometimes hard and shell-like, concealing hind wings; legs present (Figs. 2.5, 11.1, 11.94, 121, 15.4, 20.97, 22.215)	2
1'.	Wings or wing pads entirely absent; legs present or absent (Figs. 16.5, 17.23, 19.4, 22.1)	27
2(1).	Wings fully developed, usually conspicuous and movable (Figs. 11.94, 15.1, 16.8-16.9, 22.215). Adults	3
2'.	Wings developing in fixed wing pads (Figs. 11.1, 12.12, 14.1, 15.4). Larvae and pupae	15
3(2).	Fore wings leathery or hard, at least in the basal (nearest the body) half (Figs. 13.3, 15.1, 20.97)	4
3'.	Wings entirely membranous (Figs. 11.94, 16.8, 17.34, 19.36)	6
4(3).	Chewing mouthparts, mandibles visible (Fig. 20.229)	5
4'.	Sucking mouthparts united in a jointed beak, mandibles concealed (Figs. 15.1-15.3)	*Hemiptera*
5(4).	Fore wings leathery, veins distinct; femora of hind legs greatly enlarged, suited for jumping (Fig. 13.3)	*Orthoptera*
5'.	Fore wings hard (called elytra), veins indistinct (Figs. 20.97, 20.98); hind legs suited for walking or swimming (Figs. 20.99-20.101)	*Coleoptera*
6(3').	One pair of wings (Figs. 22.215-22.242)	7
6'.	Two pairs of wings (Figs. 11.94, 16.8, 17.34, 19.38)	8
7(6).	Abdomen ending in 2 or 3 long filaments; mouthparts inconspicuous, thorax without halteres (Fig. 11.94)	*Ephemeroptera* (in part)
7'.	Abdomen without conspicuous filaments (Figs. 22.215-22.242); mouthparts well developed, forming a proboscis (Figs. 21.224, 22.226); thorax with halteres (Fig. 22.216)	*Diptera*
8(6').	Wings covered with scales or hairs, obscuring venation (Figs.9.1A, 17.112, 19.38-19.43)	9
8'.	Wings bare or with minute hairs, venation clearly visible (Figs. 11.94, 12.1, 12.2, 16.8, 16.9)	10
9(8).	Wings with scales (Figs. 19.38-19.43); mouthparts usually fitted with a coiled sucking tube (Fig. 19.28)	*Lepidoptera*
9'.	Wings with hairs (Figs. 9.1A, 17.112); mouthparts without a coiled sucking tube (Fig. 17.32)	*Trichoptera*

10(8').	Antennae short, bristlelike, and inconspicuous (Figs. 11.94, 12.3, 12.4)	11
10'.	Antennae of various shapes, conspicuous, not bristlelike (Figs. 14.1, 16.8, 16.52, 21.20)	12
11(10).	Abdomen ending in 2 or 3 long filaments; hind wings much smaller than fore wings (Fig. 11.94)	***Ephemeroptera*** (in part)
11'.	Abdomen without long filaments; wings about equal in size (Figs. 12.1, 12.2)	***Odonata***
12(10').	Tarsi 2- or 3-segmented; abdomen ending with 2 conspicuous cerci (reduced in some adult Nemouridae) (Fig. 14.1)	***Plecoptera***
12'.	Tarsi 5-segmented (except in a few Hymenoptera, with 3 [Fig. 21.11]); abdomen without conspicuous appendages (Figs. 16.8, 16.9, 21.13)	13
13(12').	Abdomen broadly joined to thorax (Figs. 16.8, 16.9); front margin of fore wing in basal half with many small veins perpendicular to edge; wings with more than 20 closed cells (Figs. 16.32-16.35, 16.48-16.49)	14
13'.	Abdomen with narrow constriction at junction with the thorax (Fig. 21.13); marginal veins in basal half of fore wing parallel to leading edge; wings with fewer than 20 closed cells (Figs. 21.19-21.20); aquatic forms are very small, usually less than 3 mm long (Figs. 21.13, 21.16)	***Hymenoptera***
14(13).	Hind wings folded or pleated lengthwise (Figs. 16.8, 16.9)	***Megaloptera***
14'.	Hind wings not folded (Fig. 16.52)	***Neuroptera***
15(2').	Active insects with legs freely movable; not in cocoons or capsulelike cases (Figs. 11.1, 12.12, 12.21, 14.1, 15.4). Larvae	16
15'.	Usually inactive insects, "mummylike" with appendages drawn up and free or fused to body (Figs. 16.6, 16.7, 17.30, 21.10); sometimes in cocoons or sealed in capsulelike cases or puparia (Figs. 19.7, 25.27). Pupae	20
16(15) .	Chewing mouthparts with mandibles distinct (Figs. 11.2, 12.3)	17
16'.	Sucking mouthparts united in a jointed beak with mandibles concealed (Fig. 15.3)	***Hemiptera***
17(16).	Hind legs suited for jumping, hind femora greatly enlarged; abdomen without long cerci; found in moist places and only temporarily in water (Fig. 13.2)	***Orthoptera***
17'.	Hind legs suited for crawling, hind femora not greatly enlarged, approximately the same size as front and middle femora (Figs. 11.1, 12.12); abdomen with or without conspicuous terminal appendages; usually submerged, truly aquatic	18
18(17')	Labium (lower lip) masklike, extendable into a scooplike structure longer than head (Figs. 12.12, 12.34, 12.21)	***Odonata***
18'.	Labium normal, smaller than head, not large and masklike (Figs. 2.3, 11.2, 14.8)	19
19(18').	Tarsi with one claw (Fig. 11.1); abdomen ending in 3 long filaments, less commonly with 2 filaments; gills located on sides of abdomen, may be platelike, filamentous, or feathery (Figs. 11.1, 11.26-11.30)	***Ephemeroptera***
19'.	Tarsi with 2 claws (Fig. 13.1); abdomen ending in only 2 filaments; gills present, fingerlike and located at base of mouthparts (inconspicuous), head, or legs, or on thorax or abdomen (Figs. 2.1, 14.1, 14.4a-14.4d)	***Plecoptera***
20(15').	Appendages free, distinct, not fused to body (termed exarate pupae) (Figs. 16.6, 17.30, 21.10, 24.2, 25.27)	21
20'.	Appendages fused to body or concealed in hardened capsule (termed obtect and coarctate pupae, respectively) (Fig. 19.7)	26
21(20) .	Abdomen broadly joined to thorax (Figs. 16.6, 16.7, 17.30)	22
21'.	Abdomen with constriction where joined to thorax (Fig. 21.10)	***Hymenoptera***
22(21).	One pair of wing pads (Figs. 24.2, 25.27, 25.28)	***Diptera***
22'.	Two pairs of wing pads (Figs. 16.6-16.7, 17.30)	23
23(22')	Pads of fore wings thickened; antennae usually 11-segmented or less	***Coleoptera***
23'.	Pads of fore wings not thickened; antennae of 12 or more segments (Figs. 16.6, 16.7, 17.30)	24
24(23')	Mandibles stout, not crossing each other; pupae terrestrial (near water's edge), not normally submerged (Figs. 16.6, 16.7, 16.47)	25
24'.	Mandibles curved, projecting forward and usually crossing each other (Figs. 17.89, 16.91); pupae usually submerged in water (may be in damp areas of overhanging stream banks); always in cases (Figs. 9.1A, 17.30)	***Trichoptera***
25(24).	Smaller, 10 mm or less in length; pupae in double-walled, meshlike cocoons in sheltered places (Fig. 16.47)	***Neuroptera***
25'.	Larger, 12 mm or more; pupae in chambers in soil or rotten wood, without cocoons (Figs. 16.6-16.7)	***Megaloptera***
26(20').	Appendages visible on surface of pupa; without obvious breathing tubes or gills; two pairs of wing pads present (hind wings mostly concealed beneath fore wings) (Figs. 19.1d, 19.7)	***Lepidoptera***

26'.	Appendages visible or entirely concealed in barrel-shaped capsule; if appendages visible, then usually with projecting respiratory organs (Figs. 25.27, 25.28) or paired, dorsal, prothoracic breathing tubes (Fig. 24.2); sometimes with gills at the abdominal tip; one pair of wing pads	*Diptera*
27(1').	Abdomen with 6 or fewer segments and with a ventral tube; minute, 5 mm or less (Fig. 10.2)	*Collembola*
27'.	Abdomen with more than 6 segments and without a ventral tube; usually larger than 5 mm (Figs. 16.5, 19.4, 20.71, 22.1)	28
28(27').	Three pairs of jointed legs present on thorax (Figs. 16.5, 19.4, 20.74)	29
28'.	True legs absent; fleshy, leglike protuberances or prolegs may be present on thorax, but fewer than three pairs and not jointed (Figs. 22.1-22.3, 22.38-22.46)	34
29(28).	Middle and hind legs long and slender, extending considerably beyond the abdomen; compound eyes present (wingless Gerridae)	*Hemiptera*
29'.	Legs not longer than the abdomen; compound eyes absent (Figs. 16.5, 19.4, 20.195)	30
30(29')	Abdomen with at least two pairs of ventral, fleshy, leglike protuberances tipped with tiny hooks (prolegs with crochets) (Figs. 19.4-19.6)	*Lepidoptera*
30'.	Abdomen without leglike protuberances, or, if present, not tipped with tiny hooks (Figs. 16.5, 17.23, 17.39, 20.231)	31
31(30').	Last abdominal segment with lateral appendages bearing hooks (anal hooks) (Figs. 17.39, 17.42), antennae 1-segmented, inconspicuous; gills, if present, seldom confined to lateral margins of body; larvae free-living or in cases made of sand grains and/or bits of plant matter (Figs. 17.1-17.22)	*Trichoptera*
31'.	Last abdominal segment without anal hooks; or, if anal hooks present, antennae of more than 1 segment, and gill insertions lateral (Figs. 16.22, 16.23, 20.71, 20.131)	32
32(31')	Mandibles and maxillae united at each side to form long, straight or slightly recurved, threadlike suctorial tubes; laterally inserted, segmented gills folded beneath abdomen; small, 10 mm or less, and found in or on freshwater sponges (Fig. 16.43)	*Neuroptera*
32'.	Mandibles not united with maxillae (Figs. 16.17, 20.133); if mandibles suctorial, then strongly curved; gills seldom segmented and not folded beneath abdomen (Figs. 16.22, 16.23, 20.46, 20.52); not associated with sponges	33
33(32').	Abdomen with 7 or 8 pairs of lateral filaments or gills, arranged 1 pair on each segment (Figs. 16.5, 16.17, 16.22, 16.23); segment 9 with hooked lateral appendages (anal hooks) (Figs. 16.22, 16.23) or a single, medial, caudal filament (Fig. 16.5)	*Megaloptera*
33'.	Abdomen usually without lateral gills; if segmental gills present, then (a) anal hooks absent, or (b) segment 10 with 4 gills, or (c) caudal appendage paired or absent, never single (Figs. 20.46, 20.52, 20.71, 20.183, 20.195, 20.196)	*Coleoptera*
34(28').	Head capsule distinct, partly or entirely hardened, and usually pigmented (Figs. 20.15, 22.10-22.37), but may be deeply withdrawn in prothorax (Figs.22.1-22.3)	35
34'.	Head capsule absent, not distinct, hardened, or pigmented, often consisting of a few pale rods (Figs. 22.66-22.88)	36
35(34).	Posterior end of body with at least one or a combination of gills, hair brushes, a sucker, or breathing tube (Figs. 22.1-22.37)	*Diptera* (in part)
35'.	Posterior end of body simple or with small processes or isolated hairs, but without gills, brushes, suckers, or breathing tubes (Fig. 20.15)	(Curculionidae) *Coleoptera*
36(34').	Body usually 5 mm or larger, elongate, somewhat cylindrical, spindle-shaped or maggotlike; mouthparts may be reduced to a pair of retractile mouth hooks that move vertically (Figs. 22.66-22.88)	*Diptera* (in part)
36'.	Body usually 5 mm or less; mouthparts may be reduced to a pair of opposable, acute mandibles that move horizontally; parasitoids on or inside insect hosts (Fig. 21.9)	*Hymenoptera*

Collections of identified aquatic immatures are maintained at several North American universities and museums, such as: Illinois Natural History Survey, Urbana; Philadelphia Academy of Natural Sciences, Pennsylvania; American Museum of Natural History, New York City; California Academy of Sciences, San Francisco; United States National Museum of Natural History and Smithsonian Institution, Washington, D.C.; Ohio State University, Columbus; University of Wisconsin, Madison; Cornell University, Ithaca, New York; Oregon State University, Corvallis; Canadian National Collection of Insects, Ottawa; and the Royal Ontario Museum, Toronto.

AQUATIC COLLEMBOLA

Kenneth A. Christiansen
Grinnell College, Iowa

Richard J. Snider
Michigan State University, East Lansing

INTRODUCTION

Collembola are primarily inhabitants of soil, litter, and moist vegetation. Their ubiquitous distribution and small size suggest that almost any species may accidentally be found on the water surface. Their size and hydrophobic (water-repelling) integument keep them afloat on the surface film. However, a number of specialized aquatic species form part of the freshwater neuston and others are regularly found in the intertidal zone (Christiansen 1964, Christiansen and Bellinger 1988). A variety of individual species from different genera are inhabitants of one or another of these habitats.

Rapoport and Sanchez (1963) recognized two types of neustonic conditions occupied by Collembola: *transitory*, formed by rain into temporary pools, and *permanent*, typically lakes, lagoons, or ponds. They also concluded that rivers, streams, and floods act as dispersal mechanisms. Waltz and McCafferty (1979) ranked the Collembola into categories based on the relative degree of association and adaptation to the aquatic environment. First, the *primary aquatic associates* are those found exclusively in aquatic habitats; the *secondary aquatic associates* may be found in and around aquatic habitats, but may also inhabit other areas where high humidity exists; and the *tertiary aquatic associates* are those typical (in part) of the transitory neuston of Rapoport and Sanchez (1963) and having the least apparent adaptation to aquatic habitats. Waltz and McCafferty (1979) were able to classify some 50 species representing 29 genera most commonly associated with freshwater habitats. The great majority of Nearctic neustonic Collembola are either *Podura aquatica* (Fig. 10.5) or members of the genus *Sminthurides* (Fig. 10.47).

Among intertidal species the most common is *Anurida maritima* (Fig. 10.8), which is frequently found in North Atlantic tidal marshes, rocky coasts, and sometimes sandy beaches (Joosse 1966). Commonly, it clusters together among shells during low tide before inundation occurs. This species is replaced in the southern part of the U.S. by the sibling species *Anurida ashbyae* (Christiansen and Bellinger 1988). Short body setae hold an air bubble that functions as a physical gill. When currents are not too strong, clustering individuals form "nests" in rock cracks or under shells where the animals seek refuge and deposit eggs. In northern European regions, the late fall population dies out, leaving diapausing eggs until spring (Joosse 1966). *Anurida maritima* is widespread; however, *A. ashbyae* has so far only been identified from Florida and Bermuda and many Nearctic specimens identified as the widespread *maritima* are probably *ashbyae*. *Archisotoma besselsi* and *Onyciurus dentatus* are also widespread (see Fig. 10.1); however, other littoral species appear to be much more limited in distribution.

Collembolans have also been reported from two specialized aquatic habitats, interstitial (water-filled spaces between sand grains) littoral habitats (Delamare-Deboutteville 1960) and the surface film in cave habitats (Vandel 1964). No Nearctic species are presently known to be limited to such habitats, although some members of the genus *Arrhopalites* appear to be largely inhabitants of the latter. Because the interstitial habitat intergrades with the true littoral zone, some Nearctic species are found in both environments.

The Collembola are generally detritus feeders capable of consuming a wide variety of dead plant material and microflora; however, most species show a strong preference for particular foods (Christiansen 1964; Christiansen and Bellinger 1988). Neustonic forms feed primarily on diatoms, unicellular algae, and plankton trapped in the surface film. In a number of species, the highly modified mouthparts (Fig. 10.10) suggest fluid or bacterial diets. This postulate agrees with the work of Rapoport and Sanchez (1963) who reported fluid intestinal contents and the observation of biting the surface film by some species. They concluded that the springtails fed on a lipoprotein layer of the film or bacterial populations beneath it. This would also agree with the observation of Christiansen and Bellinger (1988) that specimens of *Anurida* and *Pseudanurida* generally had no differentiable gut contents.

Collembola display a number of unusual features including a great divergence in total instar number (2-50+) and the absence of a definitive adult form in many species (Christiansen and Bellinger 1980-81). Although the great majority of Collembola show little sexual differentiation, the aquatic genera *Sminthurides* and *Bourletiella* have strong dimorphism associated with an elaborate behavior pattern during reproduction. The neustonic *Hydroisotoma schaefferi* shows sexual dimorphism only in certain localities.

EXTERNAL MORPHOLOGY

The Collembola are apterygote (wingless) insects characterized by the presence of a *collophore* or ventral tube (respiratory, adhesive, osmoregulatory organ) on the venter of the first

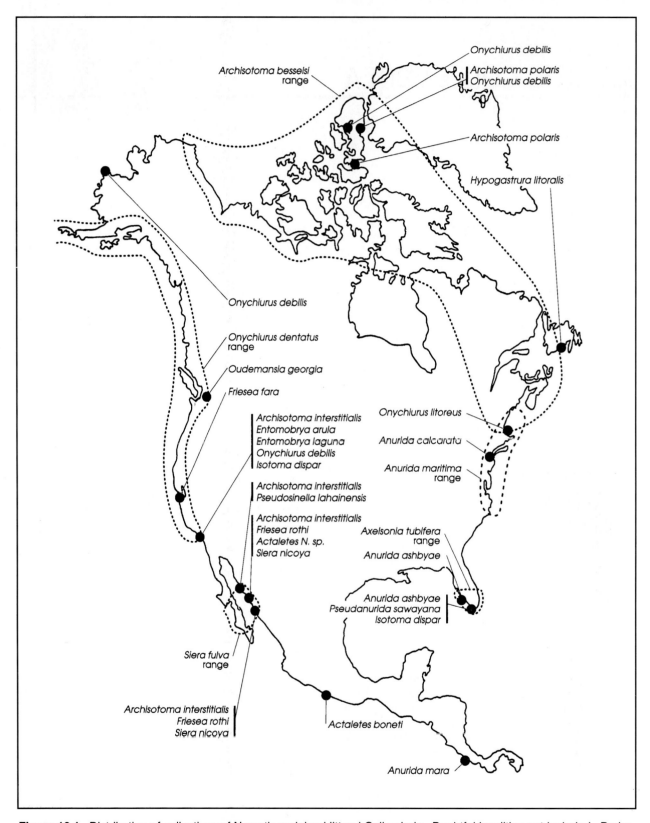

Figure 10.1. Distribution of collections of Nearctic mainland littoral Collembola. Doubtful localities not included. Probable continuous distributions of five species. Letters = localities with one species. Numbers = localities with 2+ species. *Hypogastrura littoralis*—A., *Friesea fara*—B., *Oudemansia georgia*—C., *Anurida calcarata*—D., *Onychiurus debilis*—E and 6., *Onychiurus litoreus*—F., *Anurida ashbyae*—G and 5., *Actaletes boneti*—H., *Anurida mara*—J., *Actaletes polaris*—K and 6., *Archisotoma interstitialis* 1, 2, 3, and 4. *Isotoma marisca*—*Entomobrya arula*—1., *Entomobryna laguna*—1., *Pseudosinella lahainesis*—2., *Friesea rothi*—3 and 4., *Actaletes* N. sp—3., *Siera nicoya*—3 and 4., *Pseudanurida sawayana*—5., *Isotoma dispar*—5.

abdominal segment (Fig. 10.2). Primitive forms are equipped on the venter of the fourth abdominal segment with a peculiar bifurcate jumping organ or *furcula* (Fig. 10.2), which is held in place by the *tenaculum* (Fig. 10.2) when the animal is at rest. Eight eyes on two generally trapezoidal, pigmented eye patches are located on each side of the head (Fig. 10.2). Both furcula and eye patches are often reduced or absent in specialized forms. Mouthparts vary from primitive chewing forms to complex filtering types, or have specialized piercing and sucking structures (Fig. 10.10). Primitive forms have three thoracic and six abdominal segments, which may be fused in various ways. Each tarsus is typically equipped with a single *unguis* and a small opposable *unguiculus* (Fig. 10.2).

The body is clothed with various types of setae and sometimes with complex scales. The antennae are always present and generally four-segmented. Cerci are absent, although anal horns or anal appendages of various sorts may be present (Fig. 10.2). There is no metamorphosis (ametabolous) and molting continues after the adult stage is reached. All aquatic species are small, being rarely larger than 3 mm.

KEY TO FAMILIES AND GENERA OF COLLEMBOLA

1.	Body linear (Fig. 10.2)	2
1'.	Body subglobular (Figs. 10.46, 10.47) or broadly oval (Fig. 10.44)	79
2(1).	First thoracic segment dorsally visible, with dorsal setae (Fig. 10.3) (Poduromorpha)	3
2'.	First thoracic segment without dorsal setae and frequently not visible dorsally (Fig. 10.4) (Entomobryomorpha)	32
3(2).	Dentes more than 3 times as long as manubrium (Fig. 10.5)	PODURIDAE—*Podura*†
3'.	Dentes absent or less than 2.5 times as long as manubrium	4
4(3').	Pseudocelli present (Figs. 10.6, 10.7); eyes always absent	ONYCHIURIDAE 5
4'.	Pseudocelli absent; eyes present or (rarely) absent (Fig. 10.8)	HYPOGASTRURIDAE 8
5(4).	Furcula with dens and mucro present; tenaculum present (Fig. 10.2)	*Lophognathella*
5'.	Furcula of paired knobs, single fold, or absent; mucro and tenaculum absent	6
6(5').	Last two antennal segments completely fused	*Sensiphorura*
6'.	Last two antennal segments distinctly separated	7
7(6').	Apical sense organ of 3rd antennal segment with 2 sense clubs behind 4-6 integumentary papillae (Fig. 10.9A)	*Onychiurus* s.l.†
	(including: ***Archaphorura*** and ***Protaphorura***)	
7'.	Apical sense organ of 3rd antennal segment with 1-3 exposed sense clubs (Fig.10.9B)	*Tullbergia* s.l.*
	(including: ***Metaphorura*** and ***Neonaphorura***)	
8(4').	Mandible with basal molar plate (Fig. 10.10A)	(HYPOGASTRURINAE) 9
8'.	Mandible absent or without a molar plate (Figs. 10.10B-C)	(NEANURINAE) 20
9(8).	PAO (postantennal organ) absent (Fig. 10.3)	10
9'.	PAO (postantennal organ) present (Fig. 10.2)	11
10(9).	Four or 5 eyes per side	*Xenylla**
10'.	Eyes absent	*Acherontiella*
11(9').	Anal spines 3; 9 or more dorsal dental setae	*Triacanthella*
11'.	Anal spines 0 or 2 (Fig. 10.2); 7 or fewer dorsal dental setae	12
12(11').	Eyes 8 + 8	13
12'.	Eyes 6 + 6 or fewer	15
13(12).	Furcula absent	*Knowltonella*
13'.	Furcula present	14
14(13').	Mandible with apical teeth (Fig. 10.11A); PAO (postantennal organ) lobed	*Hypogastrura**
	(including: ***Ceratophysella***, ***Mitchellania*** and ***Schoetella***)	
14'.	Mandible without apical teeth (Fig. 10.11B); PAO (postantennal organ) simple	*Stenogastrura*
15(12').	Furcula absent (Fig. 10.8)	16
15'.	Furcula present (Fig. 10.2)	18
16(15).	PAO (postantennal organ) elongate with 2 rows of oblong tubercles (Fig. 10.12A)	ONYCHIURIDAE—*Sensiphorura*
16'.	PAO (postantennal organ) circular with triangular tubercles (Fig. 10.12B)	17
17(16').	Unguis strongly toothed; 3 clavate tenent hairs (Fig. 10.1)	*Tafallia*
17'.	Unguis without tooth; at most 1 clavate tenent hair	*Willemia*
18(15').	More than 1 clavate (clubbed) tenent hair (Fig. 10.2) per foot	*Mesachorutes*
18'.	One acuminate (tapering) or clavate tenent hair per foot	19

* These genera are often found on water surface. Those marked † have truly aquatic species. The remainder are terrestrial or rarely associated with water but are included to ensure reliable determination of the primarily aquatic or semiaquatic genera.

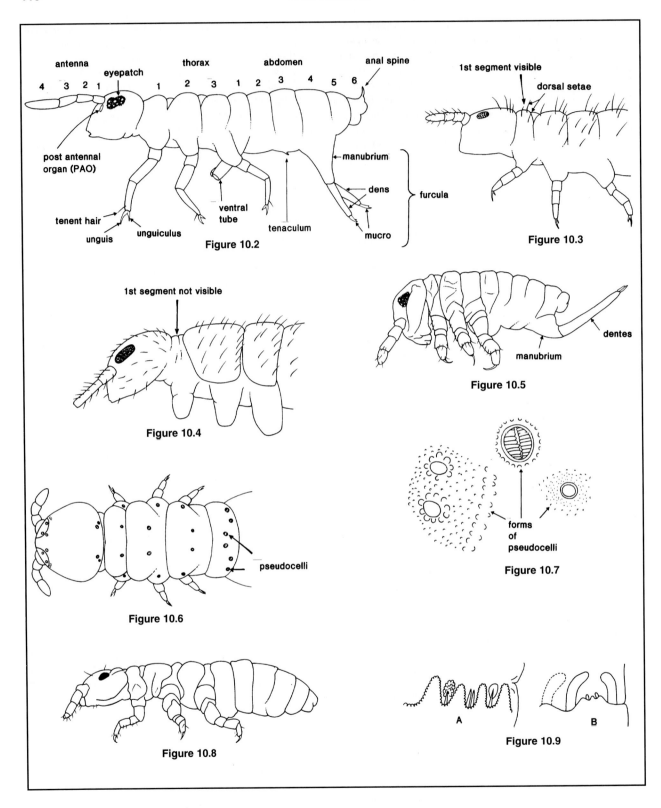

Figure 10.2. Generalized diagram of collembolan anatomy.

Figure 10.3. Lateral view of podurimorph thorax and head.

Figure 10.4. Lateral view of entomobryomorph thorax and head.

Figure 10.5. *Podura aquatica* (Poduridae).

Figure 10.6. Onychiurid thorax and first abdominal segment showing pseudocelli.

Figure 10.7. Various forms of pseudocelli.

Figure 10.8. *Anurida maritima* (Hypogastruidae).

Figure 10.9. Typical apical organs of third antennal segment: A, *Onychiurus* sp. and B, *Tullbergia* sp. (Onychiuridae).

19(18').	Mandible with strong molar plate and several apical teeth (Fig. 10.11A)	*Schaefferia**
	(including: **Bonetograstrura** and **Typhlogastrura**)	
19'.	Mandible with weak molar plate and at most 1 apical tooth (Fig. 10.13)	*Microgastrura*
20(8').	Mucro trilamellate (Fig. 10.14)	*Odontella* (*Odontella*)
20'.	Mucro not trilamellate; frequently reduced or absent	21
21(20').	Unguiculus present (Fig. 10.2); PAO (postantennal organ) triradiate	*Odontella* (*Xenyllodes*)
21'.	Unguiculus absent; PAO (postantennal organ), when present, lobed (Fig. 10.12B)	22
22(21').	Maxilla sickle-shaped (Fig. 10.15A); 3 or more anal spines or spine-like setae, or 2 such setae and mucro fused to dens	*Friesea*†
22'.	Maxilla quadrate (Fig. 10.15B), narrow and lamellate (Fig. 10.15C), or needle-like (Fig. 10.15D); anal spines absent, or 2 anal spines and mucro separate from dens	23
23(22').	Furcula present with all segments distinct (Fig. 10.2)	24
23'.	Furcula reduced or absent; mucro always absent	28
24(23).	Head of maxilla quadrate (Fig. 10.15B)	*Brachystomella*†
24'.	Maxilla otherwise (Figs. 10.15C-D)	25
25(24').	Mandible absent	*Rapoportella*
25'.	Mandible present	26
26(25').	PAO (postantennal organ) present; anal spines absent	*Pseudachorutes**
26'.	PAO (postantennal organ) absent; 2 anal spines	27
27(26').	Dens more than 4 times as long as mucro	*Pseudanurida*†
27'.	Dens less than 3 times as long as mucro	*Oudemansia*†
28(23').	Fourth antennal segment with one blunt seta much larger than others (Fig.10.16A)	*Sensillanura*
28'.	Fourth antennal segment with blunt setae similar in size (Fig. 10.16B)	29
29(28').	Last abdominal segment bilobed; body usually with conspicuous tubercles (Fig. 10.17A)	30
29'.	Last abdominal segment rounded; body without conspicuous tubercles (Fig. 10.17B)	31
30(29).	PAO (postantennal organ) present	*Morulina*
30'.	PAO (postantennal organ) absent	*Neanura* s.l.*
	(Including: **Christobella**, **Deutonura**, **Endonura**, **Morulodes**, etc.)	
31(29').	PAO (postantennal organ) present (Fig. 10.2)	*Anurida* s.l.†
	(Including: **Micranurida**, **Anuridella**, **Protachoutes**, **Pratanurida**, and **Rusekella**)	
31'.	PAO (postantennal organ) absent	*Paranura*
32(2').	Dentes with spines (Fig. 10.19)	33
32'.	Dentes without spines	36
33(32).	Mucro bidentate (with 2 teeth) and short (Fig. 10.19C)	ISOTOMIDAE—*Semicerura**
33'.	Mucro elongate, 3 or more teeth (Figs. 10.19 A-B)	ENTOMOBRYIDAE (in part) 34
34(33').	Dental spines relatively small, on basal portion of dens only (Fig. 10.19A) (TOMOCERINAE)	*Tomocerus* s.l.*
	(Including: **Lethemurus**, **Pogonognathellus**, **Plutomurus**, **Tomocerina**, and **Tomolonus**)	
34'.	Dental spines relatively large, most conspicuous toward apex of dens (Fig. 10.19B) (ONCOPODURINAE)	35
35(34').	Eyes and pigment present	*Harlomillsia**
35'.	Eyes and pigment absent	*Oncopodura**
36(32').	Body with scales	ENTOMOBRYIDAE (in part) 37
36'.	Body without scales	43
37(36).	Mucro elongate (Fig. 10.19A) (CYPHODERINAE)	*Cyphoderus*
37'.	Mucro short (Fig. 10.20) (ENTOMOBRYINAE)	38
38(37').	Mucro falcate (sickle-shaped) (Fig. 10.20A), with eyes	*Seira*†
38'.	Mucro bidentate (Fig. 10.20B)	39
39(38').	Dentes with scales on ventral surface (Fig. 10.21)	40
39'.	Dentes without scales	42
40(39).	Fourth abdominal segment at midline more than twice as long as 3rd	41
40'.	Fourth abdominal segment at midline less than twice as long as 3rd	*Heteromurus*
41(40).	Eyes 8 + 8 (Fig. 10.22)	*Lepidocyrtus**
41'.	Eyes 6 + 6 or fewer	*Pseudosinella*†
42(39').	Body scales narrow, without clear markings (Fig. 10.23A)	*Americabrya*

* These genera are often found on water surface. Those marked † have truly aquatic species. The remainder are terrestrial or rarely associated with water but are included to ensure reliable determination of the primarily aquatic or semiaquatic genera.

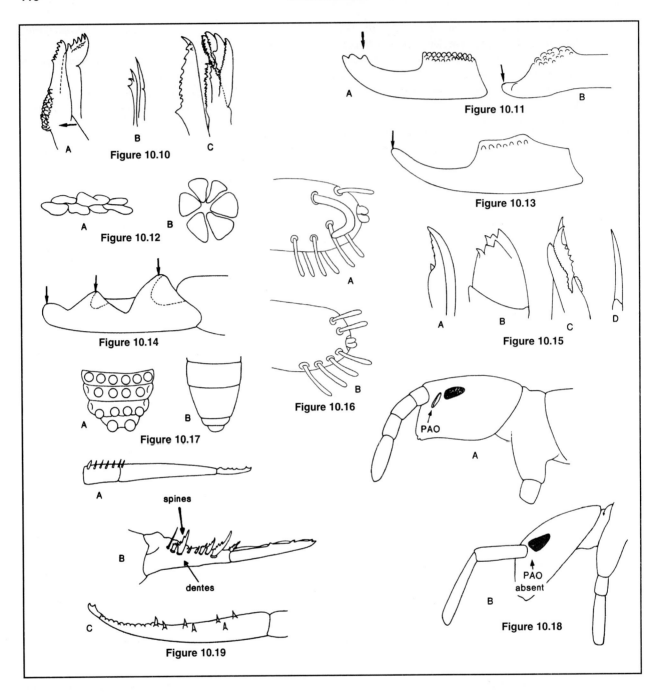

Figure 10.10. Mouthparts: *A*, chewing grinding sort typical of Hypogasturinae; *B* and *C*, piercing and sucking sorts typical of Neanurinae.

Figure 10.11. Mandible typical of *A*, *Hypogastrura* sp. and *B*, *Stenogastura* sp. (Hypogasturidae).

Figure 10.12. Postantennal organ typical of *A*, *Sensiphonrura* sp. (Onychiuridae) and *B*, *Tafallia* sp. (Hypogastruridae) and *Willemia* sp. (Hypogastruridae).

Figure 10.13. Mandible of *Microgastrura* sp. (Hypogastruridae).

Figure 10.14. Mucro typical of *Odontella* sp. (Hypogastruridae).

Figure 10.15. Maxillae typical of *A*, *Friesea* sp.; *B*, *Brachystomella* sp,; *C*, *Anurida* sp.; and *D*, *Neanura* sp. (Hypogastruridae).

Figure 10.16. Fourth antennal segment. *A*, large, blunt seta; *B*, blunt setae of similar size.

Figure 10.17. Diagram of dorsal view of the end of the abdomen; *A*, *Morulina* sp. and *Neanura* sp. (Hypogastruridae), and *B*, other genera of Neanurinae (Hypogastruridae).

Figure 10.18. Diagrammatic lateral view of entomobryomorph heads; *A*, with postantennal organs and *B*, without postantennal organs.

Figure 10.19. Mucro and dens showing spines; *A*, *Tomocerus* sp.; *B*, *Oncopodura* sp. (Entomobryidae); and *C*, *Semicerura* sp. (Isotomidae).

42'.	Body scales broad, and clearly striate (Fig. 10.23B)...*Willowsia**	
43(36').	PAO (postantennal organ) present (Fig. 10.18A) or (3 species) absent; setae at most bilaterally ciliate...**ISOTOMIDAE** (in part)................48	
43'.	PAO (postantennal organ) absent (Fig. 10.18B); some setae multilaterally ciliate...**ENTOMOBRYIDAE** (in part)................44	
44(43').	Eyes 6 + 6 or fewer ....*Sinella*	
44'.	Eyes 8 + 8 45	
45(44').	Fourth abdominal segment at midline less than 3 times as long as 3rd..................................47	
45'.	Fourth abdominal segment at midline more than 3 times as long as 3rd.................................46	
46(44').	Dens dorsally crenulate (with small scallops) and curving upward, basally in line with manubrium (Figs. 10.21, 10.24A)..(ENTOMOBRYINAE) *Entomobrya*†	
46'.	Dens not crenulate, straight and usually forming a basal angle with manubrium (Fig. 10.24B)...(PARONELLINAE) *Salina*	
47(45).	Unguis with 2 unpaired and 2 paired inner teeth (Fig. 10.26A); 1st and 2nd antennal segments subsegmented...*Orchesella*	
47'.	Unguis with 3 unpaired inner teeth (Fig. 10.26B); antennal segments not subsegmented..*Corynothrix**	
48(43).	Anal spines 1-5, terminal at end of abdomen, at least some on projecting papillae (Fig. 10.27A); furcula never reaching end of abdomen...49	
48'.	Abdominal spines absent, or subterminal (Fig. 10.27B), or not on papillae (Fig. 10.27C); furcula sometimes extending beyond end of abdomen.....................52	
49(48).	More than two anal spines 50	
49'.	Two anal spines...*Uzelia*	
50(49).	Five anal spines (Fig. 10.35A)..*Pentacanthella*	
50'.	Four anal spines (Figs. 10.35B,C) 51	
51(50').	Anal spines in a single transverse row (Fig. 10.35B)..*Blissia*	
51'.	Lateral anal spines medial to interior ones (Fig. 10.35C).....................................*Tetracanthella*	
52(48').	Abdominal segments 4-6 fused (no clear nonsetaceous bands separating these segments) (Fig. 10.28)..53	
52'.	At least 4th and 5th abdominal segments separated by clear nonsetaceous band or suture...54	
53(52).	Mucro falcate (Fig. 10.20A)...*Folsomina*	
53'.	Mucro bidentate (Fig. 10.20B) ....*Folsomia**	
54(52').	Furcula well developed; mucro with 2 (Fig. 10.29A) or more teeth, or lamellate (Fig. 10.29B)...60	
54'.	Furcula reduced or absent; mucro absent or unidentate and hook-like (Fig. 10.29C)...55	
55(54').	Furcula bifurcate (Fig. 10.30A); tenaculum present..56	
55'.	Furcula without paired distal projections (Fig. 10.30B) or absent; tenaculum absent............57	
56(55).	Integument coarsely granulate (Fig. 10.31)..*Coloburella (Coloburella)*	
56'.	Coarse integumentary granulations absent ....*Stachanorema*	
57(55).	Eyes 2 + 2 or more; antenna with apical retractile papilla.................*Anurophorus (Anurophorus)*	
57'.	Eyes 1 + 1 or absent; no apical retractile antennal papilla...58	
58(57').	Rudimentary furcula present (Fig. 10.29C); integument coarsely granulate (Fig. 10.31)...*Coloburella (Paranurophorus)*	
58'.	Furcula completely absent; integument finely granulate..59	
59(58').	Fifth and 6th abdominal segments fused..*Micranurophorus*	
59'.	Fifth and 6th abdominal segments separate..................................*Anurophorus (Pseudanurophorus)**	
60(54).	Abdomen with 2-6 subterminal spines set in strongly rugose (wrinkled) surface ....*Isotoma* ("Spinisotoma" ecomorphs)	
60'.	Abdominal spines absent or not set on rugose surface..61	
61(60').	Head wider than 2nd thoracic segment; antennae shorter than head (Fig. 10.32A)..*Metisotoma*	
61'.	Head subequal to or narrower than 2nd thoracic segment; antennae longer than head (Fig. 10.32B) 62	
62(61').	PAO (postantennal organ) absent 63	
62'.	PAO (postantennal organ) present 64	

* These genera are often found on water surface. Those marked † have truly aquatic species. The remainder are terrestrial or rarely associated with water but are included to ensure reliable determination of the primarily aquatic or semiaquatic genera.

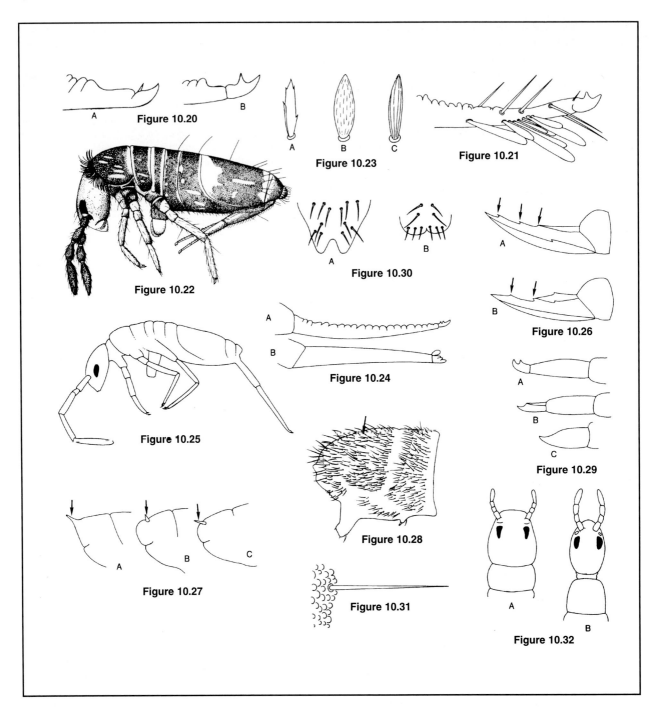

Figure 10.20. Apex of dens and mucro showing; A, falcate and B, bidentate conditions.

Figure 10.21. Apex of dens and mucro of *Lepidocyrtus* sp. (Entomobryidae) showing ventral scales.

Figure 10.22. *Lepidocyrtus* sp. (Entomobryidae) (after Christiansen and Bellinger 1980-81).

Figure 10.23. Scales typical of A, *Janetschekbrya* sp., B and C, *Willowsia* sp. (Entomobryidae).

Figure 10.24. Dens and mucro typical of A, Entomobryinae and B, Paronellinae (Entomobryidae).

Figure 10.25. *Entomobrya laguna* (Entomobryidae).

Figure 10.26. Ungues, characteristic of A, *Orchesella* sp. and B, *Corynothrix* sp. (Entomobryidae).

Figure 10.27. Lateral view of abdomen showing: A, terminal anal spines; and B and C, subterminal anal spines.

Figure 10.28. Posterior 4 abdominal segments of typical *Folsomia* sp. (Isotomidae). The *arrow* shows the location of the unsetaceous band in other genera.

Figure 10.29. Dentes and murcones showing: A, bidentate; B, lamellate; and C, fused conditions.

Figure 10.30. Reduced furculae showing A, presence and B, absence of distal projections.

Figure 10.31. Coarsely granulate integument.

Figure 10.32. Dorsal views of anterior thorax and head of: A, *Metisotoma* sp. and B, a characteristic Isotomidae.

63(62).	Eyes and pigment absent; 3rd antennal segment with 2 to 3 blunt setae (Fig. 10.33B)	*Isotomiella**
63'.	Eyes and pigment present; 3rd antennal segment with 10 or more blunt setae (Fig. 10.33A)	*Axelsonia*†
64(62').	Ventral manubrial setae 10 or more (usually 14 or more)	65
64'.	Ventral manubrial setae 9 or fewer (usually 6 or fewer)	69
65(64').	Dentes dorsally smooth (Fig. 10.36)	*Hydroisotoma*†
65'.	Dentes dorsally crenulate or tuberculate (Fig. 10.39)	66
66(65').	Abdominal macrochaetae usually coarsely, multilaterally ciliate (Fig. 10.37A); 2nd to 4th abdominal segments with bothriotricha (long, slender, fringed hairs) (Fig. 10.37A)	*Isotomurus*†
66'.	All setae smooth or unilaterally ciliate (Fig. 10.37B); bothriotricha absent	67
67(66').	Dentes dorsally tuberculate (tubercle-shaped) (Fig. 10.39A)	68
67'.	Dentes dorsally crenulate (with transverse ridges) (Fig. 10.39B)	*Isotoma**†
	(Including: ***Desoria****, **Granisotoma**, **Panchaetoma**, **Halisotoma**† and **Psammisotoma**†)	
68(67).	Subapical dental seta greatly exceed apex of mucro (Fig. 10.39A, 10.40)	*Agrenia*†
68'.	No dental setae exceed apex of mucro (Fig. 10.39B)	*Isotoma* (Panchaetoma)
69(64').	Cephalic integument reticulate (Fig. 10.41A); 8 anal spines present on papillae	*Weberacantha*
69'.	Cephalic integument granulate (Fig. 10.41C), tuberculate (Fig. 10.41B), or smooth; anal spines present or absent, *rarely* on papillae	70
70(69').	Manubrium ventrally without setae	71
70'.	Manubrium ventrally with setae	75
71(70).	Mucro with 2 basal flaps (Fig. 10.42)	*Archisotoma* †
71'.	Mucro without basal flaps	72
72(71').	Dorsal dental setae 7 or more	73
72'.	Dorsal dental setae 3 or fewer	74
73(72).	Dentes without ventrolateral setae (Fig. 10.43A)	*Bonetrura*
73'.	Dentes with 1 or more ventrolateral setae (Fig. 10.43B)	*Proisotoma* (in part)*
74(72').	Abdomen with spine-like setae on the 6th segment; dens with 2 setae in ventral half	*Isotomodes**
74'.	Abdomen without posterior spine-like setae; dens with 1 or no setae in ventral half	*Folsomides**
75(70').	PAO (postantennal organ) with margin divided into 4 quadrants (Fig. 10.45A); some macrochaetae conspicuously ciliate	*Micrisotoma*
75'.	PAO (postantennal organ) margin not divided into quadrants (Fig. 10.45B); no clearly ciliate setae	76
76(75').	Mucro with 3 or more teeth	*Proisotoma* (in part)*
76'.	Mucro bidentate	77
77(76').	Fifth and 6th abdominal segments separated by clear nonsetaceous band or suture	*Proisotoma* (in part)*
77'.	Fifth and 6th abdominal segments fused	78
78(77').	Dens with 6 or fewer setae in ventral half	*Dagamaea**
78'.	Dens with 7 or more setae in ventral half	*Cryptopygus**
79(1').	Antennae shorter than head, eyes absent (Fig. 10.46)	NEELIDAE—*Neelus*
79'.	Antennae longer than head or eyes present	80
80(79').	Body elongate oval, dens with 3 setae (Fig. 10.44)	MACKENZIELLIDAE-*Mackenziella*
80'.	Body globular (Fig. 10.47); dens with numerous setae	SMINTHURIDAE 81
81(80').	Fourth antennal segment less than half as long as 3rd	*Dicyrtoma* s.l.*
	(Including: ***Bothriovulsus***, ***Dicyrtomina*** and ***Ptenothrix***)	
81'.	Fourth antennal segment more than half as long as 3rd	82
82(81').	Antennae shorter than head	*Sminthurides* (Denisiella)†
82'.	Antennae longer than head	83
83(82').	With 2 or more clavate tenent hairs (Fig. 10.49)	84
83'.	Without clavate tenent hairs	87
84(83).	Body with heavy, outstanding serrate setae on greater abdomen (Fig. 10.48)	*Vesicephalus*
84'.	Body without such setae	85

* These genera are often found on water surface. Those marked † have truly aquatic species. The remainder are terrestrial or rarely associated with water but are included to ensure reliable determination of the primarily aquatic or semiaquatic genera.

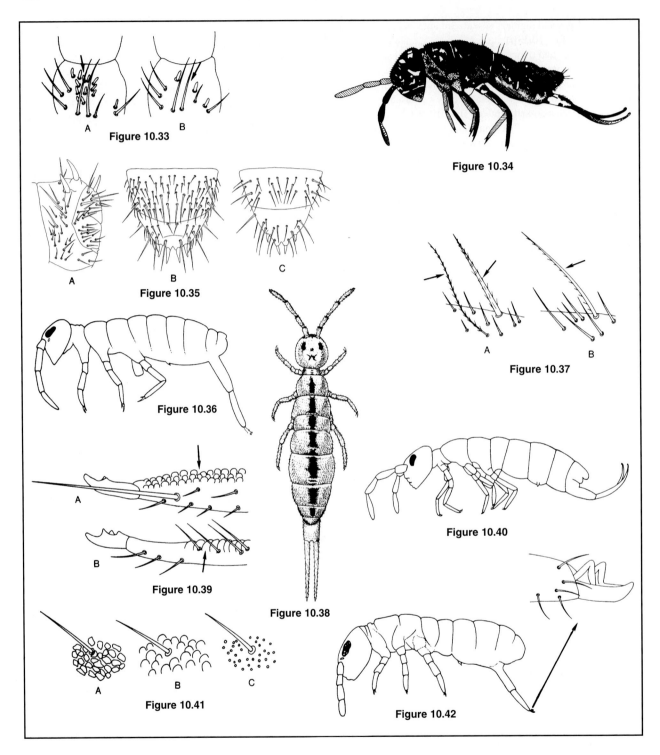

Figure 10.33. Apex of 3rd antennal segment of A, Axelsonia sp. and B, a typical isotomid.

Figure 10.34. Axelsonia sp. (Isotomidae) (after Christiansen and Bellinger 1980-81).

Figure 10.35. Anal segments: A, Pentacanthella sp., lateral view; B, Tetracanthella sp., dorsal view; C, Tetracanthella sp., dorsal view.

Figure 10.36. Isotoma schaefferi (Isotomidae).

Figure 10.37. Abdominal setae of A, Isotomurus sp. showing large ciliate setae and slender bothriotrichium and B, Isotoma sp.

Figure 10.38. Isotomurus tricolor (Isotomidae) seen from above (after Folsom 1937).

Figure 10.39. Apex of dens and mucro typical of A, Agrenia sp. and B, Isotoma sp. (Isotomidae).

Figure 10.40. Agrenia bidenticulata (Isotomidae).

Figure 10.41. Integumentary types of the family Isotomidae; A, reticulate; B, coarsely granulate; and C, finely granulate.

Figure 10.42. Archistoma besselsi (Isotomidae) with enlargement of mucro.

85(84').	Tenent hairs 2-3, parallel to long axis of tibiotarsus, very strongly clavate (Fig. 10.49A)	*Bourletiella*†
85'.	Tenent hairs often more than 3, outstanding, less strongly clavate (Fig. 10.49B)	86
86(85').	Trochanteral organ (posterior distal seta in pit) on hind trochanter (Fig. 10.50A); 4 or more clavate tenent hairs, or 4th antennal segment simple	*Sminthurinus**
86'.	Trochanteral organ absent (hind trochanter without modified external seta) (Fig. 10.50B); 3 clavate tenent hairs or fewer; 4th antennal segment subsegmented	*Sminthurus**
87(83').	Eyes 2 + 2 or less	*Arrhopalites**
87'.	Eyes 4 + 4 or more	88
88(87').	Fourth antennal segment with 8 or more subsegments or annulations	89
88'.	Fourth antennal segment with 7 or fewer subsegments	91
89(88).	Metathoracic trochanter with inner spine (Fig. 10.51A)	90
89'.	Metathoracic trochanter without inner spine (Fig. 10.51B)	*Sminthurus* s. str.*
90(89).	Female subanal appendage palmate (Fig. 10.52)	*Sminthurus* (*Allacma*)
90'.	Female subanal appendage truncate or acuminate	*Sphyrotheca*
91(89').	Body with heavy, denticulate setae (Fig. 10.53)	*Neosminthurus*
91'.	Body without such setae	92
92(91').	Eyes at least 6 + 6 (Fig. 10.47)	*Sminthurides*†
92'.	Eyes 4 + 4	*Collophora*

ADDITIONAL TAXONOMIC REFERENCES

General
Heymons and Heymons (1909); Rimski-Korsakov (1940); Phillips (1955); Gisin (1960); Salmon (1964); Pedigo (1970); Martynova (1972); Joosse (1976); Pennak (1978); Waltz and McCafferty (1979); Betsch (1980).

Regional faunas
Alaska: Fjellberg (1985).
California: Wilkey (1959).
Canada: James (1933).
Colorado: Fjellberg (1983).
Indiana: Hart (1970, 1971, 1973, 1974).
Louisiana: Hepburn and Woodring (1963).
Michigan: Snider (1967).
Pacific Coast: Scott and Yosii (1972).
West Virginia: Lippert and Butler (1976).
North and Central America littoral: Christiansen and Bellinger (1988)

Taxonomic treatments at the family level
General: Guthrie (1903); Mills (1934); Maynard (1951); Hammer (1953); Uchida (1971, 1972a, 1972b); Ellis and Bellinger (1973); Christiansen and Bellinger (1980-81).
Entomobryidae: Stach (1960, 1963); Pedigo (1968).
Hypogastruridae: Folsom (1916); Stach (1949a, 1949b, 1951); Yosii (1960).
Isotomidae: Folsom (1937); Stach (1947).
Onychiuridae: Folsom (1917); Stach (1954).
Sminthuridae: Folsom and Mills (1938); Stach (1956, 1957); Betsch (1980).

---

* These genera are often found on water surface. Those marked † have truly aquatic species. The remainder are terrestrial or rarely associated with water but are included to ensure reliable determination of the primarily aquatic or semiaquatic genera.

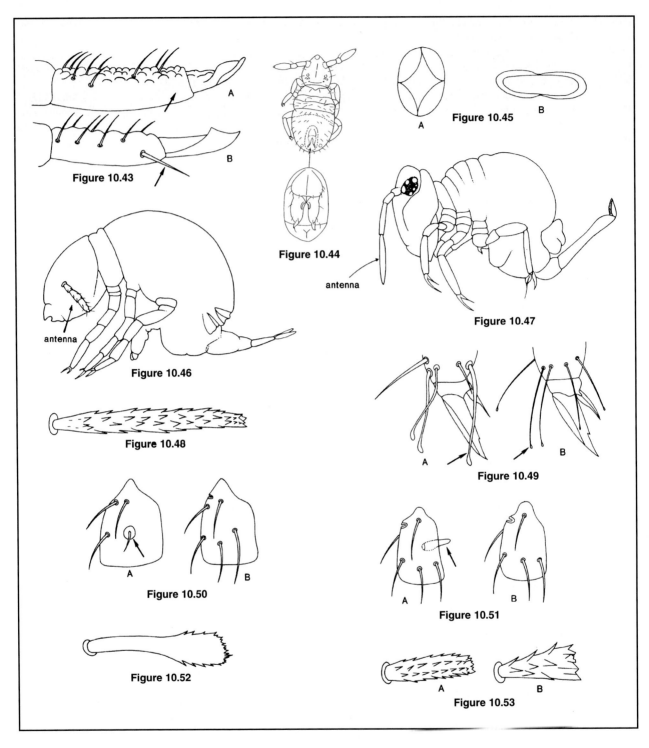

Figure 10.43. Dens and mucro of A, *Bonetrura* sp. and B, *Proisotoma* sp. (Isotomidae).

Figure 10.44. *Mackenziella psocoides*, habitus and furcula detail.

Figure 10.45. Postantennal organ typical of A, *Microisotoma* sp. and B, other isotomid Collembola.

Figure 10.46. Typical Neelidae form.

Figure 10.47. *Sminthurides aquaticus* (Sminthuridae).

Figure 10.48. Large body seta characteristic of *Vesicephalus* sp. (Sminthuridae).

Figure 10.49. Foot complex typical of A, *Bourletiella* sp. and B, *Sminthurinus* sp. (Sminthuridae).

Figure 10.50. Trochanter typical of A, *Sminthurinus* sp. and B, some species of *Sminthurus* (Sminthuridae).

Figure 10.51. Trochanter characteristic of A, *Sminthurus* sp. (Allacma) and *Sphyrotheca* sp., and B, most species of *Sminthurus* (Sminthuridae).

Figure 10.52. Female subanal appendage of *Allacma* sp. (Sminthuridae).

Figure 10.53. Heavy body setae (A and B) characteristic of *Neosminthurus* sp. (Sminthuridae).

Table 10A. Summary of ecological and distributional data for *Semiaquatic Collembola (springtails)*. (For definition of terms see Tables 6A-6C; table prepared by K. Christiansen, R. J. Snider, and K. W. Cummins; only genera with normally aquatic or semiaquatic species are included.)

Taxa (number of species in parentheses)	Habitat	Habit	Trophic Relationships	North American Distribution	Ecological References
Poduridae(1) *Podura*(1)	Lotic—margins (on surface film)	"Skaters" on water surface, "sprawlers" (at margins)	Collectors—gatherers (scavengers)	Widespread	850, 1239, 3556 3572, 3604
Onychiuridae(3) *Onychiurus*(3)	Littoral fringe, intertidal - marine	Sprawlers	Collectors—gatherers	Widespread, Northern half	630
Hypogastruridae(9)					209, 466, 887, 888, 3000, 3076, 3249, 3556, 3595, 3604, 3882, 4107
Oudemansia(1)	Beach zone—marine (intertidal)	"Sprawlers" (active at low tide, hide in protected microhabitats when submerged)	Collectors gatherers (scavengers?)	Washington coast	3604
Brachystomella(1)	Tidal margin—marine	Sprawlers	Collectors (scavengers?)	Florida	630
Friesea(3)	Beach zone—marine (intertidal)	Sprawlers, active at low tide	Collectors—gatherers (scavengers)	California, Northern Mexico	630
Anurida(1)	Beach zone—marine (intertidal), estaurine marshes	"Sprawlers" (active at low tide, aggregate at high tide in protected microhabitats)	Collectors—gatherers (scavengers on dead invertebrates, especially Molluska)	East Coast, Gulf Coast	630
Pseudanurida(1)	Beach, mangroves, (intertidal)	"Sprawlers" active at low tide	Probably feed on bacteria and diatoms	Florida	630
Anuridella*(1)				New York coast	
Isotomidae(6)	Generally marine—intertidal; lotic—margins	Generally "skaters" or "sprawlers" active at low tide.			209, 466, 600, 3000, 3076, 3249, 3556, 3595, 3604
Axelsonia(1)	Beach zone—marine intertidal (on surface film)	"Sprawlers" active at low tide	Collectors—gatherers (scavengers)	Florida coast	630
Archisotoma(1)	Beach zone—marine intertidal (on surface film)	"Sprawlers" active at low tide	Collectors—gatherers (scavengers)	Widespread	630
Agrenia(1)	Lotic margins (on surface film)	"Skaters"	Collectors—gatherers (scavengers)	Widespread	630
Isotoma(2)	Marine intertidal	"Sprawlers" active at low tide	Collectors—gatherers (scavengers)	Widespread	630
Hydroisotoma(1)	Lotic—margins (on surface film)	"Skaters"	Collectors—gatherers (scavengers)	Widespread	3595
Entomobryidae(4)					3000, 3076, 3249, 3556, 3604
Entomobrya(2)	Beach zone—marine	"Sprawlers" (active at low tide)	Collectors—gatherers (scavengers)	Southern California coast	
Pseudosinella(1)	Intertidal—marine	"Sprawlers" (active at low tide)	Collectors—gatherers	Northern Mexico and Hawaii	630
Salina(1)	Lotic—margins	"Sprawlers"	Collectors—gatherers	Eastern United States	3071
Sminthuridae(17)	Generally lotic—margins (on surface film)	Generally "skaters"	Generally collectors gatherers (scavengers), shredders—herbivores (chewers)		629, 3000, 3076, 3249, 3556, 3604
Sminthurides(14)	Lotic margins (on surface film)	"Skaters," "Clingers-sprawlers" (on *Lemna*)	Collectors gatherers (scavengers), shredders—herbivores (chewers of *Lemna*)	Widespread	1135, 1239, 3076, 3249, 3601, 3604
Bourletiella Pseudobourletiella*(2)	Lotic—margins (on surface film)	"Skaters"	Collectors—gatherers (scavengers)	Eastern United States	
Sminthurus(1)	Lotic—margins (on surface film)	"Skaters" on (on submerged vegetation)	Collectors—gatherers	Florida	3734

*Subgenera

EPHEMEROPTERA

G. F. Edmunds, Jr.
University of Utah, Salt Lake City

R. D. Waltz
Indiana Department of Natural Resources, Indianapolis

INTRODUCTION

Mayflies occur in an extremely wide variety of standing- and running-water habitats, the greatest diversity being found in rocky-bottomed, second- and third-order, headwater streams. Most of the taxonomic and biological studies on mayflies have been centered on the larvae. This is primarily a result of the abbreviated adult life and the limited range of adult activities, viz., mating and egg laying. Adult mayflies can be so elusive that adults of some genera remain unknown and others are rarely seen except by specialists. Unless adults come to lights or form conspicuous swarms during daylight hours they are rarely encountered.

The eggs of mayflies are usually deposited at the water surface, a few at a time or all in one or two clusters. In at least some species of *Baetis*, the female crawls beneath the water and lays rows of eggs on the substrate. A few adults drop clusters of eggs from the air. Eggs of most mayflies have sticky coverings, frequently with specialized anchoring devices (Koss and Edmunds 1974; Kopelke and Müller-Liebenau 1981; Provonsha 1990). Egg structure is useful in taxonomic and phylogenetic analyses. Embryonic development usually takes a few weeks. Although diapause has been confirmed experimentally in only a few species, it is almost certainly common in many temperate zone mayflies. In most species, egg hatching may be delayed three to nine months and eggs of *Parameletus* may remain dormant for over 11 months. In others development proceeds directly. A number of species are known to depend on specific hatching temperatures. The combination of specific cool periods to break diapause and specific warm temperatures for hatching appears to play a significant role in the distribution of certain species.

The larvae undergo numerous molts; examples of reported numbers of molts are 12 for *Baetisca rogersi* Berner, 27 for *Baetis tricaudatus* Dodds, and 40-45 for *Stenacron interpunctatum* (Walker). Because more mortality tends to occur at than between molts, reduced numbers of instars should be adaptive. Measures of the number of molts are difficult to obtain and some totals reported may be overestimated. Length of larval life varies with temperature and is usually three to six months. The period can be as short as 10-14 days in some Baetidae, Tricorythidae and Caenidae or as long as two years in *Hexagenia limbata* Serville in some cold habitats. Development of *Hexagenia limbata* varies from one to two years in Lake Winnipeg in northern Canada but the same species may develop in less than 17 weeks in warm canals in Utah and can be laboratory reared in 13 weeks. Some mayfly species may produce several broods a year. Such species develop much faster in summer than do their overwintering broods.

Most mayfly larvae are collectors or scrapers (Chap. 6) and feed on a variety of detritus and algae, and some macrophyte and animal material. A few species are true carnivores (Table 11A). Frequently food habits vary during the growth period. Newly hatched larvae tend to feed largely on fine particle detritus. Many shift to ingestion of algae, and some eventually increase the amount of animal material eaten as they increase in size.

The subimago (a winged, but usually sexually immature stage) in the life cycle of mayflies is unique among living insect orders. The Ephemeroptera have two winged-instars, the subimago and the adult (or imago). Several extinct orders of insects had two or more winged instars (Kukalova-Peck 1978). Subimagos usually perch on shoreline vegetation, but many short-lived ones alight for only one to two minutes or shed the subimaginal exoskeleton from all but the wings without alighting.

The length of the life of the subimago instar tends to be correlated with the length of life of the adult. It may be as short as a few minutes in species whose winged life is about four hours or less (e.g., *Ephoron, Tortopus, Caenis*) or as long as 24 to 48 hours depending on temperature. In warm climates the subimago often emerges at dusk and transforms into an adult before dawn. There is a tendency for the male subimagos to emerge from the larvae prior to the females. This tendency is greatest in some *Tricorythodes*. The males of *Tricorythodes* emerge after dusk and the females emerge a few hours before dawn; both sexes transform to adults by dawn. Mating swarms of *Tricorythodes* usually occur from early full sunlight until near noon.

Subimagos are covered on their bodies, wings, and appendages with a rather dense layer of microtrichia (Edmunds and McCafferty 1988). These microtrichia are the cause for the subimago's lack of luster when compared to the glossy adults. More importantly, they cause the subimago to be relatively waterproof. Edmunds and McCafferty (1988) noted that nearly identical microtrichia occur on the wings of adult caddisflies, true flies, and other adult aquatic insects. The persistence of the subimago in extant mayflies may be largely explained by the reduced chance of being trapped in the water's surface as the subimago escapes from its larval exuviae at the water-air interface. Important changes in the relative proportion of the legs and male genitalia occur at the subimago to adult molt. The males of all mayfly species transform from the subimago to the adult stage. The females of a few North American species mate and lay eggs as a subimago (e.g., *Tortopus, Campsurus* and *Ephoron*).

Females of an unusual *Serratella* (Ephemerellidae) species from Colorado received by the first author demonstrate an apparently novel reproductive behavior as subimagos. These subimago females do not molt to the adult stage and do not fly from the stream surface. The females emerge from their larval exuviae only to spin on the water surface. The eggs of this species issue from the dorsum of the abdomen by a dorsal midline rupture of abdominal segments 1-3, and 4 or 5. No males are known. The species is parthenogenic.

Adults of some species of *Baetis* emerge throughout the year even in strongly seasonal climates. Long emergence periods for mayflies are characteristically found where winters are relatively mild, such as the Pacific Coast and in the Southeast. Adults of most species live two hours to three days, but some live less than 90 minutes. The females of some genera may live for weeks, especially genera that hold the eggs until they are ready to hatch, e.g., *Callibaetis, Cloeon* (Oehme 1972).

EXTERNAL MORPHOLOGY

The following account of the external morphology of mayflies will facilitate use of the keys.

Mature Larvae

Head (Fig. 11.1): The shape of the larval head is variable and the head may possess a variety of processes, projections, and armature. The eyes are usually moderately large and are situated laterally or dorsally near the posterolateral margin. The antennae usually arise anterior or ventral to the eyes. The antennae may vary in length from less than the width of the head to more than twice the width of the head. The mouthparts are illustrated in Figure 11.2.

Thorax (Fig. 11.1): Developing wing pads are found on the meso- and metathorax, although metathoracic wing pads may be absent in some species. Each of the three thoracic segments bears a pair of legs. The legs of some genera are modified for such special functions as burrowing, filtering food, grooming, and gill protection.

Abdomen (Fig. 11.1): All mayflies have a 10-segmented abdomen, although some of the segments may be concealed beneath the mesonotum. The terga may have spines and/or tubercles and the shape of the posterolateral corners can be used as taxonomic characters. To determine the segment number *always count forward from segment 10*. The gills, which are the most variable larval structures, are usually located entirely on the abdomen, although in some species they also occur at the bases of the coxae (*Isonychia, Camelobaetidius*), or at the bases of the maxillae (*Isonychia*). Gill position on the abdomen is variable: they may be ventral, lateral, or dorsal, and occur on abdominal segments 1 through 7, or they may be absent from one or more segments in various combinations. In referring to gill position, gill 4, for example, refers to the gills on segment 4 whether or not there are gills on the first three segments.

Most species have three caudal filaments, composed of a terminal filament and two cerci (Fig. 11.5). In others, only the cerci are well developed, the terminal filament being represented by a short rudiment or entirely absent. When present, the terminal filament may vary in length and thickness relative to the cerci. Length of the caudal filaments varies from shorter than the body to two or three times its length.

Adults

Head (Fig. 11.94): The eyes are usually sexually dimorphic, those of the male being larger than those of the female. In males, the eye facets may be uniform in size or the upper facets may be larger and in the Baetidae the upper facets are turbinate, i.e., raised on a stalklike portion (Fig. 11.125). The mouthparts are vestigial and nonfunctional in the adult.

Thorax (Figs. 11.94-11.95): The thorax consists of three regions, each with one or two pairs of appendages; the prothorax with the fore legs, the mesothorax with the middle pair of legs and the fore wings, and the metathorax with the hind legs and hind wings (which may be absent in some species). The fore legs of most mayflies show sexual dimorphism, with those of the male having very long tibiae and tarsi and usually being much longer than the middle and hind legs and often as long as or longer than the body. In the Polymitarcyidae, the middle and hind legs of the male and all legs of the female are much reduced and presumably nonfunctional. In *Dolania* (Behningiidae), all the legs of both sexes are much reduced but apparently are somewhat functional. Most mayflies have two pairs of wings; the somewhat triangular fore wings and the much smaller hind wings. In the families Caenidae, Tricorythidae, and Baetidae the hind wings have become greatly reduced or completely lost in one or both sexes. The three wing margins are known as the costal, outer, and hind margin (Figs. 11.94-11.95). The surface of the wing has a regular series of corrugations or fluting with the longitudinal veins lying either on a ridge (indicated in the figures by +) or a furrow (indicated by -). The primitive complex venation of giant Carboniferous fossil mayflies (wingspread up to 45 cm) is presented by Kukalova-Peck (1985). The system of venational nomenclature proposed by Tillyard (1932) as discussed by Edmunds and Traver (1954) has been followed herein. Abbreviations designating the major longitudinal veins and their convexity (+) or concavity (-) relative to wing fluting as viewed dorsally are as follows:

C (+)	=	costa
Sc (-)	=	subcosta
R_1 (+), R_2 (-), R_3 (-), R_{4+5} (-)	=	radius$_1$, radius$_2$, etc.
MA_1 (+), MA_2 (+)	=	medius anterior$_1$, etc.
MP_1 (-), MP_2 (-)	=	medius posterior$_1$, etc.
CuA (+)	=	cubitus anterior
CuP (-)	=	cubitus posterior
A_1 (+)	=	anal$_1$

R_2 through R_5 are often referred to as the radial sector (Rs). Intercalary veins lie between the principal veins. When intercalaries are long, they lie opposite the principal veins on either side; for example, IMA (intercalary medius anterior) is a

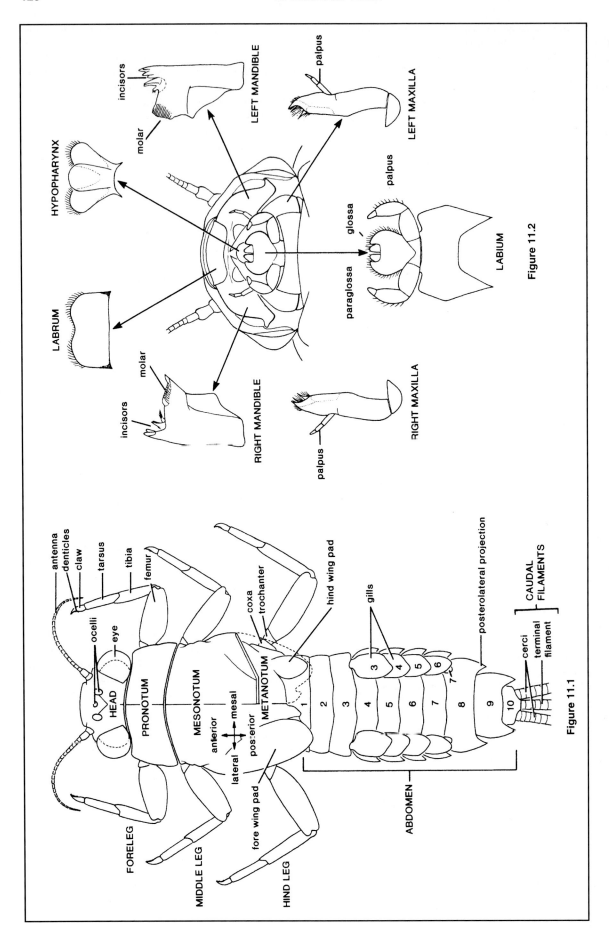

Figure 11.1. Dorsal view of *Ephemerella* sp. larva (Ephemerellidae).

Figure 11.2. Ventral view of mouthparts of *Ephemerella* sp. larva (Ephemerellidae).

furrow (-) vein lying between the two ridge (+) branches of MA (MA$_1$ and MA$_2$). Conversely, IMP is a ridge vein (+) in between the furrow veins of MP. The longer veins alternate as ridges and furrows at the wing margin. Two ridge veins of the fore wing, MA and CuA, are important landmarks in locating and identifying the entire venation. Learning which veins are ridge veins (+) and which are furrow veins (-) is important for efficient use of the keys.

Abdomen (Figs. 11.94, 11.96-11.97): The abdomen is composed of ten segments. Each segment is ring shaped and consists of a dorsal tergum and a ventral sternum. The posterior portion of sternum 9 of the female is referred to as the *subanal plate* and in males (Fig. 11.96) as the *subgenital plate* or *styliger plate*. The posterior margin of the subgenital plate, which is variable in shape, gives rise to a pair of slender and usually segmented appendages called the *forceps* (or claspers). Dorsal to the subgenital plate are the paired *penes* (Fig. 11.96). In most mayflies the penes are well developed and sclerotized, but in the Baetidae they are membranous and extrudable. Arising from the posterior portion of tergum 10 in both sexes are the *caudal filaments* (Figs. 11.94, 11.97). Most species have two caudal filaments, the *cerci* and a vestige of a median *terminal filament*; others have three, the cerci and a median terminal filament.

The above account of the *Ephemeroptera* and the keys below were originally rewritten from Edmunds *et al.* (1976). The keys now are changed considerably. Many of the figures are modified or taken directly from Edmunds *et al.* (1976). We are grateful to the University of Minnesota Press and the co-authors for permission to use this material and the figures. A number of the figures have been taken from work co-authored by one of the authors elsewhere: Figs. 11.71-11.72, 11.155 (Allen and Edmunds 1962); Figs. 11.156-11.158 (Allen and Edmunds 1965); Figs. 11.159-160 (Allen and Edmunds 1963a); Fig. 11.161 (Allen and Edmunds 1963b); Fig. 11.63 (Bednarik and Edmunds 1980); Fig. 11.67 (Traver and Edmunds 1967); Figs. 11.81, 11.170 (Waltz and McCafferty 1989); Fig. 11.175 (Waltz and McCafferty 1987a); Figs. 11.83, 11.84 (Waltz *et al.* 1985). The late Dr. R. K. Allen allowed us to republish Figures 11.68 (Allen 1974), 11.78, 11.80 (Allen 1973), 11.79 (Allen 1967), and 11.161 (Allen 1965). Figures 11.129-11.131, 11.141-11.144, 11.148, 11.162, 11.177 and 11.178 are from Burks (1953) and are published with permission of the Illinois Natural History Survey. Figure 11.41 is from Needham *et al.* (1935) and is published courtesy of Cornell University Press. Figure 11.176 is from Provonsha and McCafferty (1982), courtesy of the authors. Figures 11.90, 11.179 and 11.180 are from Davis (1987). Figures 11.91-11.92 are from Peters (1971) with permission of the author. Figures 11.57-11.59 are from Bednarik and McCafferty (1979), courtesy of the Canadian Bulletin of Fisheries and Aquatic Science.

KEYS TO THE FAMILIES AND GENERA OF EPHEMEROPTERA

The keys are to families and genera; subfamilies are not keyed. The classification used herein is that of McCafferty (1991a,b; McCafferty and Wang 1994a,b) with which the senior author has some differences of opinion. The use of cladistically driven classifications places several genera into families other than those assigned in previous classifications. For example, the placement of some former genera of Siphlonuridae into monobasic families, i.e., Ameletidae (*Ameletus*). *Isonychia*, formerly in a subfamily of Oligoneuriidae is most recently assigned to its own family, Isonychiidae. Evidence has accumulated that *Pseudiron* is a very distinct lineage and it now is in its own family (Pseudironidae) (Landa and Soldán 1985). A cladistic evaluation of the Ephemeroidea has resulted in the reassignment of *Pentagenia* to the Ephemeridae (subfamily Palingeniinae) (McCafferty 1991a).

Several rare genera have been incorporated into this key that were not previously included. The inclusion of rare genera in the keys is essential to discovering more about such forms. Most rare genera occur in large rivers where specialized collecting techniques are required, although some genera, e.g., *Caurinella* (Ephemerellidae), are apparently restricted to small streams.

The adult of the genus *Acanthametropus* (Acanthametropodidae) has been recently associated with its larva by workers in Russia, but it is inadequately described as of this writing. The adult of *Acanthametropus* will apparently key to *Analetris* in the keys provided. McCafferty and Wang (1994a) reunited *Analetris* and *Acanthametropus* in the family Acanthametropodidae based on a revised cladistic analysis.

The generic classification of the Baetidae has undergone a much needed revolutionary revision. The studies affecting the genera in this edition are those of Waltz and McCafferty (1987a,b,c,d; 1989); Waltz *et al.* 1985, 1994; McCafferty and Waltz (1990, 1995); and Waltz (1994). Although knowledge regarding larval stage characters useful for identifying genera and species of Baetidae has increased substantially, less is known about adult characters. For this reason, couplets separating adult Baetidae presented herein should be regarded as somewhat provisional.

In Heptageniidae, the larva formerly referred to *Anepeorus* has been reared and the adult is conspecific with "*Heptagenia cruentata*," now placed as the genus *Raptoheptagenia* (Whiting and Lehmkuhl 1987a). The true larva of *Anepeorus simplex*, which was previously known as *Spinadis*, was associated by McCafferty and Provonsha (1984). A closely related rare larva from the Athabaca River was described and named *Acanthamola pubescens* by Whiting and Lehmkuhl (1987b). The larva of the type species of *Anepeorus*, *A. rusticus*, is unknown. Adults of *A. rusticus* are known from Saskatchewan and from the Green River of Utah, whose fauna reflects its former flow into the Missouri River system through the North Platte River drainage. *Acanthamola* possibly could be the larva of *A. rusticus* (Whiting and Lehmkuhl 1987b; McCafferty and Provonsha 1988) but its rarity in the Athabaca and the failure to collect the larva where adults of *A. rusticus* are known makes the solution to the problem difficult. We continue to regard *Iron* and *Ironopsis* as subgenera of *Epeorus*.

Within Leptophlebiidae, *Neochoroterpes* has been elevated to the rank of genus (McCafferty *et al.* 1993; Henry 1993). The rare genus *Farrodes* is keyed as both larvae and adults.

In Potamanthidae, Bae and McCafferty (1991) have shown that the North American species formerly assigned to *Potamanthus* are not congeneric with the Palearctic-Oriental *Potamanthus* and have erected the new genus *Anthopotamus* for them (McCafferty and Bae 1990).

In Ephemerellidae, the genus *Caurinella* (Allen 1984) is keyed although it is rare and known only as larvae from Idaho. The previously recognized genera *Dannella* and *Dentatella* have been relegated to subgenera of *Timpanoga* and *Eurylophella* respectively (McCafferty and Wang 1994b), although the relationships of *Eurylophella* and *Dentatella* may remain somewhat open until adults and eggs of *E.(D.) bartoni* are associated with the larva (e.g., see Funk and Sweeney 1995). Other changes in generic classifications were withheld in this revision pending further cladistic analysis. Some European workers refer the species *Ephemerella aurivillii* to *Chitonophora*. Such a separation can readily be applied to *E. aurivillii* (a Holarctic species) and *E. septentrionalis* but other species are not easily placed into one genus or another. Some of these species may require new generic names as more is known of their phylogenetic position. For the present, we prefer not to apply the name *Chitonophora* in North America.

All known distributions are tentative and should be regarded as incomplete. Such terms as abundant, common, uncommon, or rare are all relative to place, time, and collector. Rare can be interpreted as hard to collect, active at the wrong time of day, or perhaps, in some cases, truly at low population levels. In any case, a collector using the best technique at the right places at an appropriate time may collect large numbers of a "rare" genus or species.

KEYS TO THE FAMILIES AND GENERA OF EPHEMEROPTERA

Mature Larvae

1. Thoracic notum enlarged to form a shield extended to abdominal segment 6, gills enclosed beneath shield (Fig. 11.3) .... **BAETISCIDAE**—*Baetisca*
1'. Thoracic notum not enlarged as above; at least anterior abdominal gills exposed 2
2(1'). Gills on abdominal segments 2-7 forked and elongate-lanceolate, with margins fringed (Fig. 11.4); most with mandibular tusks projected forward and visible from above head (Figs. 11.6-11.7); if tusks absent, head and thorax with pads of long spines (Fig. 11.8) 3
2'. Gills on abdominal segments 2-7 variable; if gills forked and elongate-lanceolate, margins not fringed (Fig. 11.5); tusks rarely present on mandibles 6
3(2). Head and prothorax with dorsal pad of long spines on each side (Fig. 11.8); without mandibular tusks; gills ventral .... **BEHNINGIIDAE**—*Dolania*
3'. Head and prothorax without pads of spines; mandibular tusks present (Figs. 11.6-11.7); gills lateral or dorsal 4
4(3'). Fore tibiae more or less modified, either expanded or with tubercles, adapted for burrowing (Fig. 11.6); abdominal gills held dorsally 5
4'. Legs unmodified (Fig. 11.7); abdominal gills held laterally .... **POTAMANTHIDAE**—*Anthopotamus*
5(4). Mandibular tusks curved upward apically as viewed laterally (Fig. 11.9); ventral apex of hind tibiae projected into distinct acute point (Fig. 11.11) .... **EPHEMERIDAE** 80
5'. Mandibular tusks not curved upward apically as viewed laterally (Fig. 11.10); ventral apex of hind tibiae rounded (Fig. 11.12) .... **POLYMITARCYIDAE** 83
6(2'). A double row of long setae on inner margins of femora and tibiae of fore legs (Fig. 11.13) 39
6'. Long setae absent on fore legs, or not arranged as above 7
7(6'). Gills on abdominal segment 2 operculate or semioperculate (covering succeeding pairs) (Fig. 11.14) 8
7'. Gills on abdominal segment 2 neither operculate nor semioperculate, either similar to those on succeeding segments or absent 10
8(7). Gills on abdominal segment 2 triangular, subtriangular, or oval, not meeting medially (Fig. 11.14); gill lamellae on segments 3-6 simple or bilobed, without fringed margins .... **TRICORYTHIDAE** 76
8'. Gills on abdominal segment 2 quadrate, meeting or almost meeting medially (Fig. 11.15a); gill lamellae on segments 3-6 with fringed margins (Fig. 11.15b) 9
9(8'). Mesonotum with distinct rounded lobe on anterolateral corners (Fig. 11.16); operculate gills fused medially; developing hind wing pads present; East, mostly Southeast .... **NEOEPHEMERIDAE**—*Neoephemera*
9'. Mesonotum without anterolateral lobes (Fig. 11.17); operculate gills not fused medially; without developing hind wing pads .... **CAENIDAE** 77
10(7'). Gills *absent* on abdominal segment 2, rudimentary or absent on segment 1, and present or absent on segment 3 (Figs. 11.18, 11.41); gills on segments 3-7 or 4-7 consist of anterior (dorsal) oval lamella and posterior (ventral) lamella with numerous lobes (Fig. 11.19); paired tubercles often present on abdominal terga .... **EPHEMERELLIDAE** 67

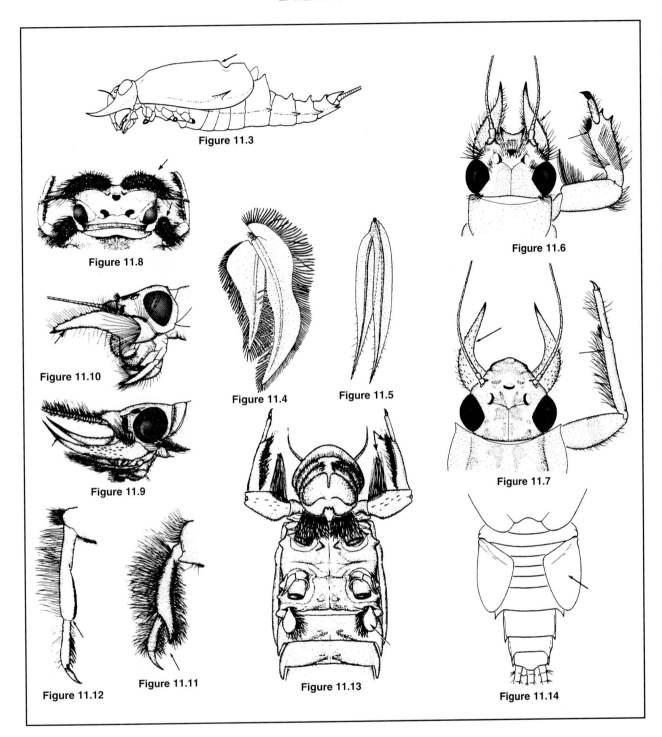

Figure 11.3. Lateral view of *Baetisca* sp. larva (Baetiscidae).

Figure 11.4. Gill 4 of *Ephoron* sp. larva (Polymitarcyidae).

Figure 11.5. Gill 4 of *Paraleptophlebia* sp. larva (Leptophlebiidae).

Figure 11.6. Dorsal view of head and foreleg of *Pentagenia* sp. larva (Ephemeridae).

Figure 11.7. Dorsal view of head and foreleg of *Anthopotamus* sp. larva (Potamanthidae).

Figure 11.8. Dorsal view of head and prothorax of *Dolania* sp. larva (Behningiidae).

Figure 11.9. Lateral view of head of *Ephemera* sp. larva (Ephemeridae).

Figure 11.10. Lateral view of head of *Ephoron* sp. larva (Polymitarcyidae).

Figure 11.11. Hind leg of *Hexagenia* sp. larva (Ephemeridae).

Figure 11.12. Hind leg of *Ephoron* sp. larva (Polymitarcyidae).

Figure 11.13. Ventral view of anterior portion of *Lachlania* sp. larva (Oligoneuriidae).

Figure 11.14. Doral view of abdomen of *Tricorythodes* sp. larva (Tricorythidae).

10'.	Gills present on abdominal segments 1-5, 1-7, or 2-7; paired tubercles rarely present on abdominal terga	11
11(10').	Body distinctly flattened; head flattened; eyes and antennae dorsal; mandibles not visible in dorsal view (Fig. 11.20)	41
11'.	Body not flattened (Figs. 11.21-11.22) or, if flattened, mandibles visible and forming part of the flattened dorsal surface of head (Fig. 11.23)	12
12(11').	Claws of fore legs differ in structure from those on middle and hind legs (Figs. 11.24-11.25); claws of middle and hind legs long and slender, about as long as tibiae (Figs. 11.24-11.25)	13
12'.	Claws of all legs similar in structure, usually sharply pointed, rarely spatulate; claws variable in length, if those of middle and hind legs long and slender, then usually shorter than tibiae (longer than tibiae in three rare genera)	14
13(12).	Claws on fore legs simple, with long slender denticles; spinous pad present on fore coxae (Fig. 11.24a); from Michigan west	AMETROPODIDAE—*Ametropus*
13'.	Claws on fore legs bifid (Fig. 11.25a); without spinous pad on fore coxae	METRETOPODIDAE ... 57
14(12').	Abdominal gills on segments 2-7 either forked (Fig. 11.5), in tufts (Fig. 11.26), with all margins fringed (Fig. 11.27), or with double lamellae terminated in filaments or points (Figs. 11.28-11.30); apicolateral margin of maxillae with a dense brush of hairs (Fig. 11.31)	LEPTOPHLEBIIDAE ... 58
14'.	Abdominal gills not as above; gills either more or less oval or heart-shaped; lamellae either single, double, or triple folded (Figs. 11.33-11.36); fringed on inner margin in one rare genus (Fig. 11.37); never terminating in filaments or points; apicolateral margin of maxillae variable, never with dense brush of hairs (Figs. 11.32, 11.40)	15
15(14').	Labrum with a median notch on distal margin (in Nearctic: absent only in *Apobaetis*) terminal filament variable, may be shorter than tergum 10 or subequal to cerci; antennae long, two or three times or more width of head (Fig. 11.39) (including *Apobaetis*), or, antennae shorter than twice width of head, but labrum with notched distal margin and maxilla with pectinate spines	BAETIDAE ... 22
15'.	Labrum without a median notch on distal margin; terminal filament subequal to cerci; antennae shorter than twice width of head (Fig. 11.38), or labrum with notched distal margin and maxilla with pectinate spines (Fig. 11.40)	16
16(15').	Tibiae and tarsi bowed; claws very long and slender, claws of hind legs about as long as tarsi (similar to Fig. 11.24b); rare; large rivers; scattered distribution	ACANTHAMETROPODIDAE ... 17
16'.	Tibiae and tarsi not bowed; claws usually not long and slender	18
17(16).	Dorsum of abdomen with median hook-like tubercles; hook-like tubercle also present on each thoracic sternum (adult uncharacterized)	*Acanthametropus*
17'.	Dorsum of abdomen without median hook-like tubercles; hook-like tubercles absent on thoracic sterna; rare, scattered distribution	*Analetris*
18(16').	Maxillae with crown of pectinate spines (Fig. 11.40); gills with single lamellae, more or less oval with a sclerotized band along lateral margin and usually with a similar sclerotized band on (Fig. 11.33) or near mesal margin (Fig. 11.34)	AMELETIDAE—*Ameletus*
18'.	Maxillae without pectinate spines; gills variable	SIPHLONURIDAE ... 19
19(18').	Abdominal gills with double lamellae on segments 1-2 (in some, double also on segments 3-7) (Fig. 11.35)	20
19'.	Gills on all abdominal segments with single lamellae, more or less oval (Figs. 11.33-11.34) or heart-shaped	21
20(19).	Gills on abdominal segments 1-2 subtriangular, broadest near apex (Fig. 11.35); claws of middle and hind legs slightly longer than those of fore legs; widespread	*Siphlonurus*
20'.	Gills on abdominal segments 1-2 oval; claws of middle and hind legs distinctly longer than those of fore legs; rare, California	*Edmundsius*
21(19').	Sterna of mesothorax and metathorax each with a median tubercle; abdominal segments 5-9 greatly expanded laterally; rare, New York to Labrador	*Siphlonisca*
21'.	Sterna of thorax without tubercles; abdominal segments 5-9 not greatly expanded; uncommon; Canada or West, mountains	*Parameletus*
22(15).	Claws distinctly spatulate with large apical denticles, tarsi distinctly bowed (Fig. 11.42); uncommon; medium to large rivers; Southwest and Central	*Camelobaetidius*
22'.	Claws sharply pointed; denticles, if present, smaller and ventral (Figs. 11.43-11.45)	23

EPHEMEROPTERA

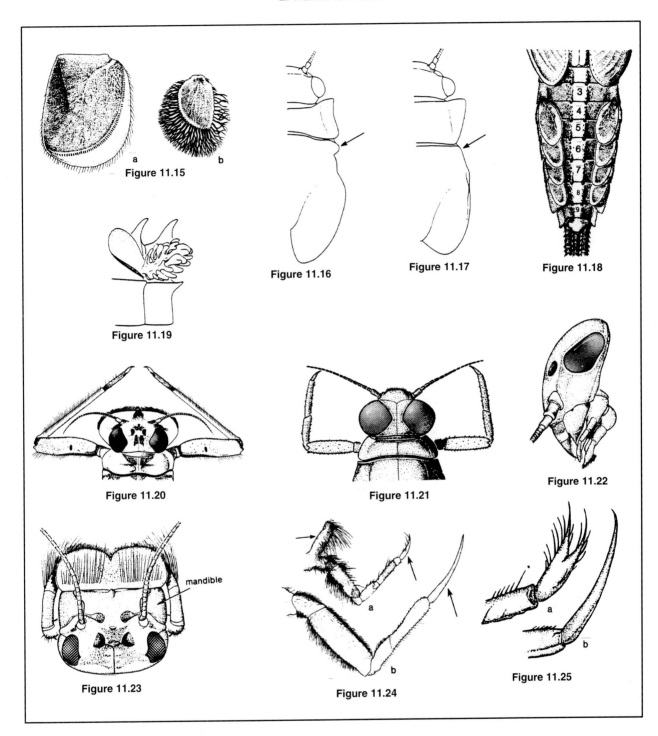

Figure 11.15. Gills 2 (*a*) and 4 (*b*) of *Caenis* sp. larva (Caenidae).

Figure 11.16. Dorsal view of head and thorax of *Neoephemera* sp. larva (Neoephemeridae).

Figure 11.17. Dorsal view of head and thorax of *Caenis* sp. larva (Caenidae).

Figure 11.18. Dorsal view of abdomen of *Serratella* sp. larva (Ephemerellidae).

Figure 11.19. Gill of segment 3 (lamella raised) of *Drunella* sp. larva (Ephemerelidae).

Figure 11.20. Dorsal view of head and prothorax of *Epeorus* (*Iron*) sp. larva (Heptageniidae).

Figure 11.21. Dorsal view of head and prothorax of *Parameletus* sp. larva (Siphlonuridae).

Figure 11.22. Lateral view of head of *Baetis* sp. larva (Baetidae).

Figure 11.23. Dorsal view of *Traverella* sp. larva (Leptophlebiidae).

Figure 11.24. Foreleg (*a*) and hind leg (*b*) of *Ametropus* sp. (Ametropodidae).

Figure 11.25. Claws of foreleg (*a*) and hind leg (*b*) of *Siphloplecton* sp. larva (Metretopodidae).

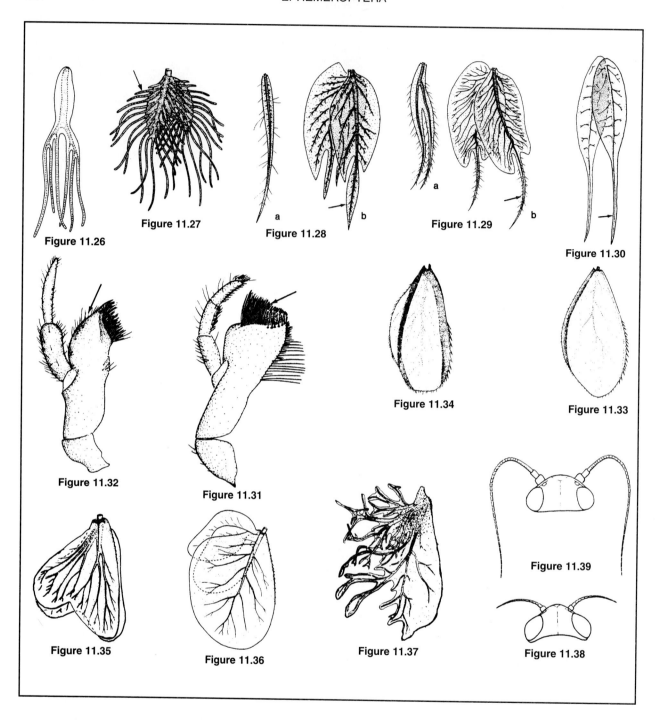

Figure 11.26. Gill 5 of *Habrophlebia* sp. larva (Leptophlebiidae).

Figure 11.27. Gill 4 of *Traverella* larva (Leptophlebidae).

Figure 11.28. Gills 1 (*a*) and 4 (*b*) of *Choroterpes* sp. larva (Leptophlebiidae).

Figure 11.29. Gills 1 (*a*) and 4 (*b*) of *Leptophlebia* sp. larva (Leptophlebiidae).

Figure 11.30. Gill 4 of *Leptophlebia* larva (Leptophlebiidae).

Figure 11.31. Right maxilla of *Leptophlebia* sp. larva (Leptophlebiidae).

Figure 11.32. Right maxilla of *Callibaetis* sp. larva (Baetidae).

Figure 11.33. Gill 4 of *Ameletus* sp. larva (Ameletidae).

Figure 11.34. Gill 4 of *Ameletus* sp. larva (Ameletidae).

Figure 11.35. Gill 4 of *Siphlonurus* sp. larva (Siphlonuridae).

Figure 11.36. Gill 4 of *Analetris* sp. larva (Acanthametropodidae).

Figure 11.37. Gill 4 of *Acanthametropus* sp. larva (Acanthametropodidae).

Figure 11.38. Dorsal view of head of *Siphlonurus* sp. larva (Siphlonuridae).

Figure 11.39. Dorsal view of head of *Baetis* sp. larva (Baetidae).

23(22').	Abdominal gills present on segments 1-5 only, extended ventrally from the pleura; median tubercle present on anterior abdominal terga; Southwest, Texas, Utah..................................***Baetodes***
23'.	Abdominal gills present on segments 1-7 or 2-7, held laterally or dorsally; no tubercles on abdomen..................................24
24(23').	Apex of labial palpi simple and truncate (Fig. 11.46); trachea of gills palmate (Fig. 11.47) or asymmetrical with most branches on median side (Fig. 11.48); tails with distinct dark band every third to fifth segment (Fig. 11.49a,b); gills simple or with dorsal flaps on one or more segments..................................25
24'.	Apex of labial palpi variable, not truncate as above; trachea variable; tails rarely banded as above; gills simple or large and compound..................................27
25(24).	Labrum (Fig. 11.81) with deep triangular median notch at distal margin; paraglossae broad, longer than glossae; claws longer than respective tarsi; gills simple; Midwest..................***Pseudocentroptiloides***
25'.	Labrum (as in Fig. 11.82) with small median notch along anterior margin; paraglossae subequal to glossae; claws usually shorter than respective tarsi; gills simple or with dorsal flaps..................................26
26(25').	Mandible incisors deeply cleft to base or united less than one-half height of incisors; length of maxillary palp segment 3 subequal to segment 2; lateral bristles of caudal filaments present on proximal three-fourths, apices of filaments without bristles; gills simple; widespread..................................***Centroptilum***
26'.	Mandible incisors united above base or fused to apex; length of maxillary palp segment 3, when present, shorter than segment 2; lateral bristles of caudal filaments present to apices; gills simple, or with dorsal flap at least on abdominal segment 1; widespread..................................***Procloeon***
27(24').	Gills simple (Fig. 11.51)..................................29
27'.	Gills on one or more segments compound, with distinct large flaps (Figs. 11.47, 11.50)..................................28
28(27').	Labial palpi apparently two-segmented, blunt apically; hind wing pads present; gills with recurved flaps folded ventrally (Fig. 11.50); widespread..................................***Callibaetis***
28'.	Labial palpi three-segmented, obliquely truncate apically; hind wing pads absent; gills with recurved flaps folded dorsally (as in Fig. 11.47); East, Midwest..................................***Cloeon***
29(27).	Claws clearly less than half as long as respective tarsi (Fig. 11.43)..................................31
29'.	Claws one-half or more as long as respective tarsi (Figs. 11.44-11.45)..................................30
30(29').	Claws almost as long as respective tarsi (Fig. 11.45); labrum without notch on distal margin; rare; West, Southeast..................................***Apobaetis***
30'.	Claws about one-half as long as respective tarsi (Fig. 11.44); labrum with notch on distal margin; widespread..................................***Paracloeodes***
31(29).	Antennae subequal in length to head capsule (Fig. 11.83) and body pale, white/cream colored with contrasting dark bands of pigment completely encircling some abdominal segments (Fig. 11.84); East, Midwest, Colorado..................................***Barbaetis***
31'.	Antennae longer than head capsule, subequal to two or three times width of head, often extending to middle of thorax or beyond; body coloration never as above..................................32
32(31').	Villopore present (Fig. 11.85); terminal filament subequal to cerci or reduced..................................33
32'.	Villopore absent; terminal filament subequal to cerci..................................36
33(32).	Fore coxae each with a single filamentous gill; hind wing pads extremely small; terminal filament reduced to fewer than five segments; claws with two rows of denticles (visible under 400x magnification); widespread..................................***Heterocloeon***
33'.	Fore coxae without gills; hind wing pads variable, present or absent; terminal filament variable..................................34
34(33').	Antennal scapes with distal lobe (Fig. 11.86); maxillary palpi with subapical excavation (Fig. 11.87); segment 2 of labial palpi usually greatly expanded medially; paraglossae much broader than glossae; widespread..................................***Labiobaetis***
34'.	Antennal scapes without distal lobe; maxillary palpi without subapical excavation; segment 2 of labial palpi variable, generally not greatly expanded medially; glossae and paraglossae usually (note couplet 37) subequal in width..................................35
35(34').	Abdominal terga without scales; body usually dorsoventrally compressed; terminal filament reduced to fewer than five segments; prothorax with two pairs of oblique stripes; abdominal terga with paired mid dorsal spots; labium compact, palpi shortened, truncated, apices of palpi attaining apices of paraglossae (slightly exceed in *A. ampla*); femora, tibiae, and tarsi usually with row of long, fine setae (Fig. 11.88); widespread..................................***Acentrella***
35'.	Abdominal terga usually with scales; body usually cylindrical; terminal filament variable; prothorax never patterned as above; paired tergal spots present or absent (usually absent); labium not compact (may be confused only in *B. fuscatus* group but differs from above by shape of palpi); tibiae and tarsi never with distinct row of long setae; widespread..................................***Baetis***

36(32').	Claws without denticles; long, very fine setae present on tibiae and tarsi; tufts of setae present on sternites 2-6 (visible under high magnification); palpi of maxillae reduced; prostheca of right mandible bifid; Arizona, New Mexico	*Cloeodes*
36'.	Claws with denticles; setae on legs short, usually blade-like; tufts of setae on sternites never present; palpi of maxillae variable, not reduced as above; prostheca of right mandible variable	37
37(36').	Labial palpi (Fig. 11.89) with segment 2 bearing median digitate lobe, projecting anteriorly; tuft of setae between incisors and molar area of mandible present; gill 7 slender, — apically pointed (in Nearctic species; may be rounded in Neotropical species); width of paraglossae subequal to glossae (paraglossae wider than glossae in Neotropical species); widespread	*Acerpenna*
37'.	Labial palpi segment 2 weakly expanded; tuft of setae between incisors and molar region of mandible present on at least one mandible; width of paraglossae subequal to glossae	38
38(37').	Gills of abdominal segment 1 absent, gills present on abdominal segments 2-7; prostheca of right mandible bifid, pectinate; widespread	*Diphetor*
38'.	Gills on abdominal segment 1 present, gills present on abdominal segments 1-7; prostheca of right mandible digitate; Southwest, Central, Southeast	*Fallceon*
39(6).	Gills on segment 1 dorsolateral, similar in position and structure to other gills; fibrils shorter than gill plates; widespread and common; larvae minnowlike	ISONYCHIIDAE—*Isonychia*
39'.	Gills on segment 1 ventral (Fig. 11.13); fibrils longer than plate or gill plate absent	OLIGONEURIIDAE 40
40(39').	Gill plates oval on segments 2-7, similar to abdominal gill 1 (Fig. 11.13); claws present on fore legs	*Lachlania*
40'.	Gill plates slender and elongate on segments 2-7; claws absent on fore legs	*Homoeoneuria*
41(11).	Claws as long as or longer than tarsi; tibiae and tarsi bowed; large sandy rivers	PSEUDIRONIDAE—*Pseudiron*
41'.	Claws much shorter than tarsi; tibiae and tarsi straight	HEPTAGENIIDAE 42
42(41').	All gills ventral; gill lamellae slender; numerous gill filaments radiating out from a central plate; rare; large rivers; West, Midwest, Southeast and Canada	*Raptoheptagenia*
42'.	Gills lateral in position at least on abdominal segments 4 to 6; lamellae usually broad; gill filaments variable	43
43(42').	With only two well developed caudal filaments; terminal filament vestigial or absent	44
43'.	With three well developed caudal filaments	48
44(43).	Gills 1 and 2 inserted ventrally, gills on 3 inserted ventrally or ventrolaterally; mouthparts adapted for predation; claws with a single tooth and no denticles (as in Fig. 11.61)	45
44'.	Gills inserted laterally on segments 2-6; may extend ventrally on segments 1 and 7 (Fig. 11.53); claws without a tooth, but with three or more subapical denticles	46
45(44).	Head and thorax without paired dorsal tubercles; abdomen with very small single dorsal tubercles medially; femora broad, flattened, subequal to 2.0-2.5 times width of respective tibiae (adults unknown)	*Acanthamola*
45'.	Head and each thoracic segment with paired dorsal tubercles; abdomen with single dorsal tubercles medially; femora relatively narrow, subequal to 1.0-1.5 times width of respective tibiae; rare; big rivers; Midwest, Southeast, west to Utah and Saskatchewan	*Anepeorus*
46(44').	Well-developed paired tubercles present on hind margin of abdominal terga 1-9; uncommon; Northwest	*Ironodes*
46'.	Abdominal terga without paired tubercles	*Epeorus* 47
47(46').	Terga with a conspicuous dense median row of setae; a median anterior emargination on head; gills purplish brown; Pacific Coast east to western Nevada, Colorado, Utah, Montana	*Epeorus (Ironopsis)*
47'.	Terga may possess a row of setae, but not conspicuous and dense; gills pale or tan; widespread	*Epeorus (Iron)*
48(43').	Second segment of maxillary palpi longer than head is broad, conspicuous at side or behind head (Fig. 11.52); uncommon; in pools, swamps, and small, slow streams; Midwest or Northeast	*Arthroplea**
48'.	Second segment of maxillary palpi not greatly elongated, inconspicuous; in lotic waters	49
49(48').	Gills on abdominal segments 1 and 7 enlarged and meet or almost meet beneath abdomen to form ventral disk (Fig. 11.53)	*Rhithrogena*
49'.	Gills on abdominal segments 1 and 7 do not meet beneath abdomen and usually smaller than intermediate pairs	50

* Family Arthropleidae

EPHEMEROPTERA

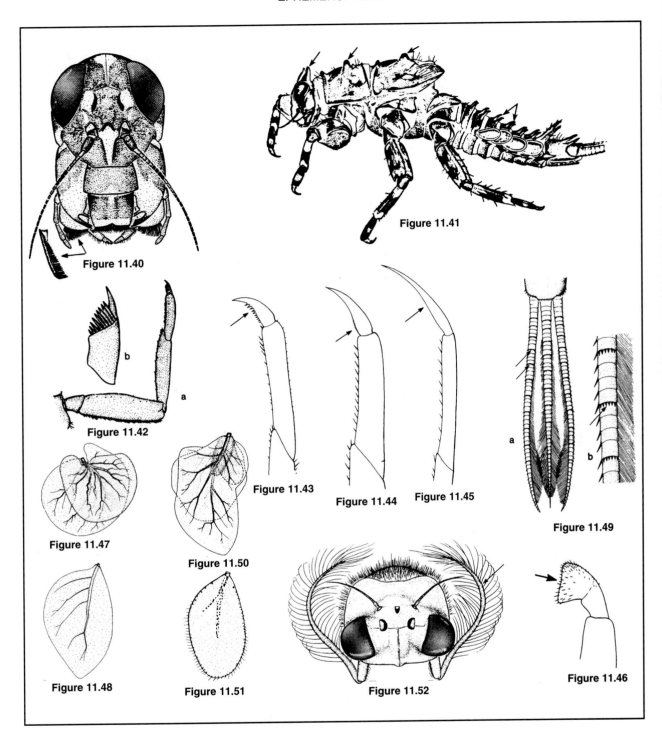

Figure 11.40. Anterior view of head and enlarged pectinate spines of *Ameletus* sp. larva (Ameletidae).

Figure 11.41. Lateral view of *Drunella grandis* larva (Ephemerellidae).

Figure 11.42. Leg (*a*) and claw (*b*) of *Camelobaetidius* sp. larva (Baetidae).

Figure 11.43. Tarsus and claw of *Baetis* sp. larva (Baetidae).

Figure 11.44. Tarsus and claw of *Paracloeodes* sp. larva (Baetidae).

Figure 11.45. Tarsus and claw of *Apobaetis* sp. larva (Baetidae).

Figure 11.46. Labial palpus of *Centroptilum* sp. larva (Baetidae).

Figure 11.47. Gill 1 of *Cloeon* sp. larva (Baetidae).

Figure 11.48. Gill 4 of *Centroptilum* sp. larva (Baetidae).

Figure 11.49. Caudal filaments (*a*) and detail of cercus (*b*) of *Procloeon* sp. larva (Baetidae).

Figure 11.50. Gill 2 of *Callibaetis* sp. larva (Baetidae).

Figure 11.51. Gill 4 of *Baetis* sp. larva (Baetidae).

Figure 11.52. Dorsal view of head of *Arthroplea* sp. (Arthropleidae).

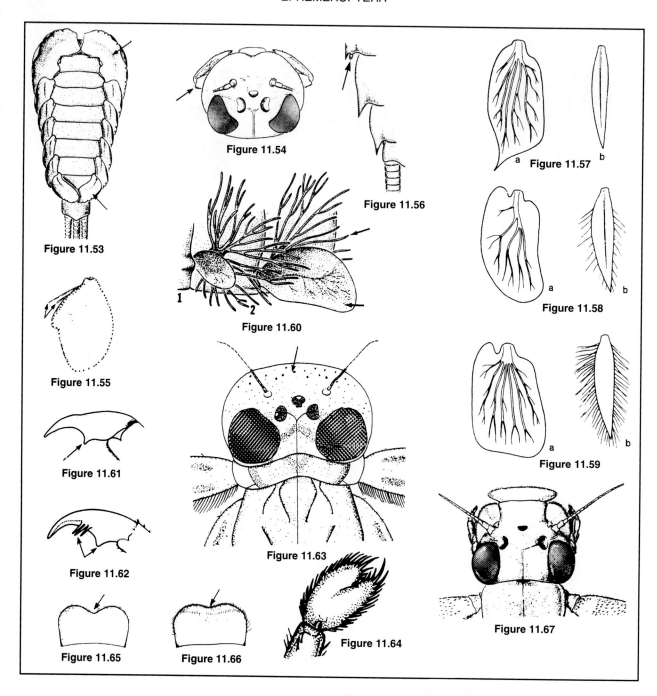

Figure 11.53. Ventral view of abdomen of *Rhithrogena* sp. larva (Heptageniidae).

Figure 11.54. Dorsal view of head of *Cinygmula* sp. larva (Heptageniidae).

Figure 11.55. Gill 4 of *Cinygmula* sp. larva (Heptageniidae).

Figure 11.56. Abdominal segments 7 - 10 of *Macdunnoa* sp. larva (Heptageniidae).

Figure 11.57. Gills 4 (*a*) and 7 (*b*) of *Stenacron* sp. larva (Heptageniidae).

Figure 11.58. Gills 4 (*a*) and 7 (*b*) of *Stenonema* sp. larva (Heptageniidae).

Figure 11.59. Gills 4 (*a*) and 7 (*b*) of *Stenonema* sp. larva (Heptageniidae).

Figure 11.60. Gills of segments 1 and 2 of *Cinygma* sp. larva (Heptageniidae).

Figure 11.61. Claw of *Heptagenia* sp. larva (Heptageniidae).

Figure 11.62. Claw of *Nixe* sp. larva (Heptageniidae).

Figure 11.63. Head and part of thorax of *Leucrocuta* sp. larva (Heptageniidae).

Figure 11.64. Claw of foreleg of *Metretopus alter* (Metretopodidae).

Figure 11.65. Labrum of *Habrophlebiodes* sp. larva (Leptophlebiidae).

Figure 11.66. Labrum of *Paraleptophlebia* sp. larva (Leptophlebiidae).

Figure 11.67. Head of *Thraulodes* sp. larva (Leptophlebiidae).

50(49').	Maxillary palpi protrude at sides of head (Fig. 11.54); fibrilliform portion of gills 2-6 absent or with only a few filaments (Fig. 11.55)	*Cinygmula*
50'.	Maxillary palpi rarely protrude at sides of head; fibrilliform portion of gills 2-6 with many fibrils	51
51(50').	Gills on segment 7 minute, no longer than the posterolateral projections of that segment (Fig. 11.56); rare	*Macdunnoa*
51'.	Gills on segment 7 much larger than above	52
52(51').	Gills on segment 7 reduced to slender filaments; trachea absent or with few or no lateral branches (Figs. 11.57b-11.59b)	53
52'.	Gills on segment 7 similar to preceding pairs but smaller; trachea of gill 7 with lateral branches	54
53(52).	Gills on abdominal segments 1-6 with apex pointed (Fig. 11.57a)	*Stenacron*
53'.	Gills on abdominal segments 1-6 with apex rounded (Fig. 11.58a) or truncate (Fig. 11.59a)	*Stenonema*
54(52').	Gill lamellae on segment 1 less than one-half as long as those on segment 2, with fibrilliform portion of gill 1 much longer than lamella (Fig. 11.60); labrum small, narrower at apex than at base; Northwest	*Cinygma*
54'.	Gill lamellae on segment 1 two-thirds as long as those on segment 2; fibrilliform portion of gill 1 usually subequal to or shorter than lamella; labrum never more narrow at apex than at base	55
55(54').	Gills on segment 7 with fibrilliform portion present and with numerous fibrils; claws without denticles but with one basal tooth (Fig. 11.61)	*Heptagenia*
55'.	Gills on segment 7 without fibrilliform portion; claws with denticles (Fig. 11.62)	56
56(55').	Head large, wider than pronotum, many dark spots near anterior margin (Fig. 11.63)	*Leucrocuta*
56'.	Head not as wide as pronotum, without numerous dark spots	*Nixe*
57(13').	Margins of claws of fore legs densely covered with spines (more than thirty) that are only slightly smaller than the short terminal spines (Fig. 11.64); outer margin of gill 4 with two to seven stout spines in addition to small setae; rare; Canada, Alaska, Michigan to Maine	*Metretopus*
57'.	Margins of claws of fore legs sparsely covered with spines (less than twenty) that are more slender than the long terminal spines (Fig. 11.25); Canada, Midwest and East	*Siphloplecton*
58(14).	Labrum as wide as or wider than head capsule (Fig. 11.23); abdominal gills oval with fringed margins (Fig. 11.27); widespread, in large or medium rivers	*Traverella*
58'.	Labrum much narrower than head capsule (Fig. 11.67)	59
59(58').	Abdominal gills 2-7 consist of clusters of slender filaments (Fig. 11.26); uncommon; eastern	*Habrophlebia*
59'.	Abdominal gills 2-7 forked or bilamellate, not as above (Figs. 11.28-11.30)	60
60(59').	Abdominal gill 1 of single linear lamella (Fig. 11.28a) or asymmetrically forked (Fig. 11.68a,b); gills 2-7 broadly bilamellate with apex three lobed	61
60'.	Abdominal gill 1 forked (Fig. 11.29a) or bilamellate, either similar to or different from those on succeeding segments (Figs. 11.29b, 11.30)	62
61(60).	Abdominal gills 2-7 with dorsal lamellae larger than ventral lamellae and terminated in three lobes with the median lobe longer and often wider than lateral lobes (Fig. 11.28b); dorsal setae of labrum in two transverse rows; setae on lateral margin of mandibles limited to middle	*Choroterpes*
61'.	Abdominal gills 2-7 with dorsal and ventral lamellae similar and terminating in three slender, subequal processes (Fig. 11.68a,b); dorsal setae of labrum scattered; setae on lateral margin of mandibles in distal half	*Neochoroterpes*
62(60').	Gills on abdominal segment 1 much narrower than those on segments 2-7 (Fig. 11.29a); each lamella of gills on segments 2-7 terminated in single slender filament that may (Fig. 11.29b) or may not (Fig. 11.30) be flanked by one or two blunt lobes	*Leptophlebia*
62'.	Gills on abdominal segment 1 not conspicuously narrower than those on segments 2-7; gills on 2-7 forked or bilamellate	63
63(62').	Gills on middle abdominal segments shaped as in Figure 11.69; lateral tracheal branches conspicuous	64
63'.	Gills on middle abdominal segments forked or bilamellate, but either broader, narrower, more deeply forked, or with lateral tracheal branches less conspicuous than Fig. 11.69	65
64(63).	Labrum with moderately deep V-shaped median emargination (Fig. 11.65); small row of spinules present on posterior margins of abdominal terga 6-10 or 7-10 only (Fig. 11.70); uncommon; Midwest or East	*Habrophlebiodes*

64'.	Labrum with shallow, variably shaped median emargination (Fig. 11.66); small rows of spinules present on posterior margins of abdominal terga 1-10; widespread	*Paraleptophlebia* (in part)
65(63').	Labrum at least 55% as wide as head anterior (ventrally) to the eyes (Fig. 11.90); labrum wider than apex of the clypeus; head prognathous; distribution principally in Texas, New Mexico and Arizona but also known from Green River in Utah	66
65'.	Labrum less than one-half as wide as head anterior (ventrally) to the eyes; labrum about as wide as apex of clypeus; head hypognathous; widespread	*Paraleptophlebia* (in part)
66(65).	Gills forked with each lamella slender, gills on segment 4 about 14 times as long as wide; notch on anterior margin of labrum with five small denticles (Fig. 11.91); claws with 14 or more denticles, the apical denticle much the larger (Fig. 11.92); rare; Texas, in small warm streams with rocky riffles	*Farrodes*
66'.	Gills variable, appear as two lamellae, one dorsal and one ventral or forked; gills on segment 4 range from 3.0 to 10.0 times as long as wide (*Farrodes* larva illustrated by Allen and Brusca 1978 as *Thraulodes* sp. F.); apical denticles of claw may be slightly longer but not as in Fig. 11.92; uncommon; Texas, New Mexico, Arizona, Utah	*Thraulodes*
67(10).	Lamellate gills present on abdominal segments 3-7 (Figs. 11.18, 11.41)	68
67'.	Lamellate gills present on abdominal segments 4-7 (Fig. 11.72)	72
68(67).	Terminal filament at least 1.3 times as long as lateral cerci; Northwest	*Caudatella*
68'.	Caudal filaments subequal in length	69
69(68').	Leading margin of fore femora usually armed with conspicuous tubercles (Fig. 11.71); if not (some western species), thorax, head, *and* abdomen with large paired dorsal tubercles (Fig. 11.41) *or* abdominal sterna with attachment disk of long hair	*Drunella*
69'.	Leading margin of fore femora without such tubercles, without large tubercles on dorsum of head, thorax, *and* abdomen, abdominal sterna without disk of long hair	70
70(69').	Caudal filaments with whorls of spines at apex of each segment, otherwise with only sparse setae or none (Fig. 11.76); maxillary palpi reduced in size (Figs. 11.73, 11.74) or absent	71
70'.	Caudal filaments with or without whorls of spines at apex of each segment, apical half of caudal filaments with long intersegmental setae extending laterally (Fig. 11.77); maxillary palpi well developed (Fig. 11.75); widespread	*Ephemerella*
71(70).	Posterolateral extensions on tergum 9 extend beyond segment 10 a distance equal to tergum 10 at the midline and curved strongly upward so that apices reach up to the level of tergum 9 at midline (Fig. 11.93); without paired dorsal tubercles on posterior margin of abdominal terga; rare, Idaho (adult unknown)	*Caurinella*
71'.	Posterolateral extensions on tergum 9 rarely extend beyond segment 10, but less than the length of the midline of segment 9; with paired dorsal tubercles on the posterior margin of terga 4-7 or more; widespread	*Serratella*
72(67').	Gills on tergum 4 operculate, largely covering those on segments 5-7, only about one-third of any following gill visible (Fig. 11.72) and reaching to about posterior margin of abdominal tergum 7	73
72'.	Gills on tergum 4 not operculate, apical half of gills 5 and 6 visible and not reaching posterior margin of abdominal tergum 7	*Attenella*
73(72).	Claws without denticles; maxillae with palpi; West, Midwest, East	*Timpanoga* 74
73'.	Claws with denticles; maxillae without palpi; Midwest, East, Pacific Coast	*Eurylophella* 75
74(73).	Filamentous gills on abdominal segment 1 originate at lateral margin of tergum; posterolateral abdominal processes not extremely developed; abdominal terga without paired tubercles on or near posterior margin; Midwest, East	*Timpanoga* (*Dannella*)
74'.	Filamentous gills on abdominal segment 1 originate sublaterally on tergum; posterolateral abdominal processes extremely developed; abdominal terga with paired tubercles or rarely absent; Midwest, East, Pacific Coast	*Timpanoga* (s.str.)
75(73').	Abdominal tergum 9 approximately 1.4 times midlength of terga 8 or 10; operculate gills relatively narrow, ovate and broadly rounded distally; paired tubercles present on abdominal terga 1 through 7	*Eurylophella* (s.str.)
75'.	Abdomen tergum 9 subequal to midlength of terga 8 or 10; operculate gills broad, but more narrowly rounded distally; paired tubercles present on abdominal terga 5 through 7; Midwest	*Eurylophella* (*Dentatella*)
76(8).	Fore femora with transverse row of long setae (Fig. 11.79); operculate abdominal gill 2 triangular or subtriangular in shape, over two-thirds as wide as long (Fig. 11.14); abundant; widespread	*Tricorythodes*

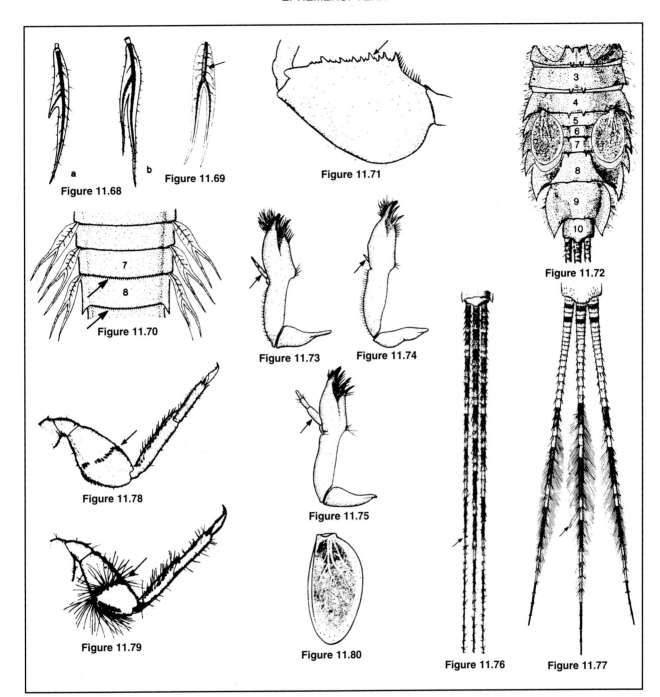

Figure 11.68. Variations (*a* and *b*) of gill 1 of *Neochoroterpes* sp. larva (Leptophlebiidae).

Figure 11.69. Gill 4 of *Habrophlebiodes* sp. larva (Leptophlebiidae).

Figure 11.70. Abdominal segments 5-8 of *Habrophlebiodes* sp. larva (Leptophlebiidae).

Figure 11.71. Forefemur of *Drunella* sp. larva (Ephemerellidae).

Figure 11.72. Abdominal terga of *Eurylophella* sp. larva (Ephemerellidae).

Figure 11.73. Maxilla of *Serratella* sp. larva (Ephemerellidae).

Figure 11.74. Maxilla of *Serratella* sp. larva (Ephemerellidae).

Figure 11.75. Maxilla of *Ephemerella* sp. larva (Ephemerellidae).

Figure 11.76. Caudal filaments of *Serratella* sp. larva (Ephemerellidae).

Figure 11.77. Caudal filaments of *Ephemerella* sp. larva (Ephemerellidae).

Figure 11.78. Foreleg of *Leptohyphes* sp. larva (Tricorythidae).

Figure 11.79. Foreleg of *Tricorythodes* sp. larva (Tricorythidae).

Figure 11.80. Gill 2 of *Leptohyphes* sp. larva (Tricorythidae).

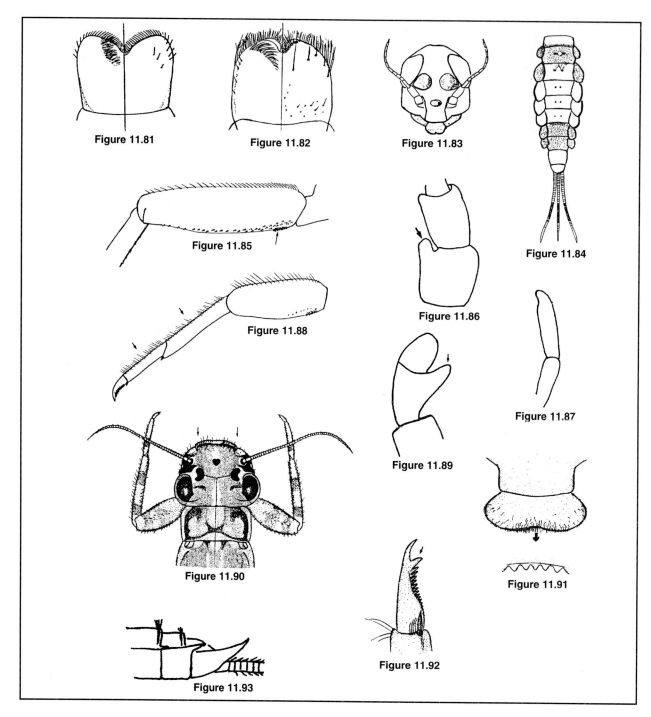

Figure 11.81. Labrum of *Pseudocentroptiloides usa* larva (Baetidae) (left, ventral view; right, dorsal view).

Figure 11.82. Labrum of *Procloeon* sp. larva (Baetidae) (left, ventral view; right, dorsal view).

Figure 11.83. Head capsule of *Barbaetis* sp. larva (Baetidae).

Figure 11.84. Abdomen (dorsal) of *Barbaetis benfieldi* larva (Baetidae).

Figure 11.85. Villopore of *Baetis* sp. larva (Baetidae).

Figure 11.86. Antennal scape of *Labiobaetis* sp. larva (Baetidae).

Figure 11.87. Maxillary palp of *Labiobaetis* sp. larva (Baetidae).

Figure 11.88. Leg of *Acentrella* sp. larva (Baetidae).

Figure 11.89. Labial palp of *Acerpenna* sp. larva (Baetidae).

Figure 11.90. Head capsule of *Farrodes texanus* larva (Leptophlebiidae).

Figure 11.91. Labrum of *Farrodes* sp. larva (Leptophlebiidae).

Figure 11.92. Tarsal claw of *Farrodes* sp. larva (Leptophlebiidae).

Figure 11.93. Apex of abdomen of *Caurinella* sp. larva (Ephemerellidae).

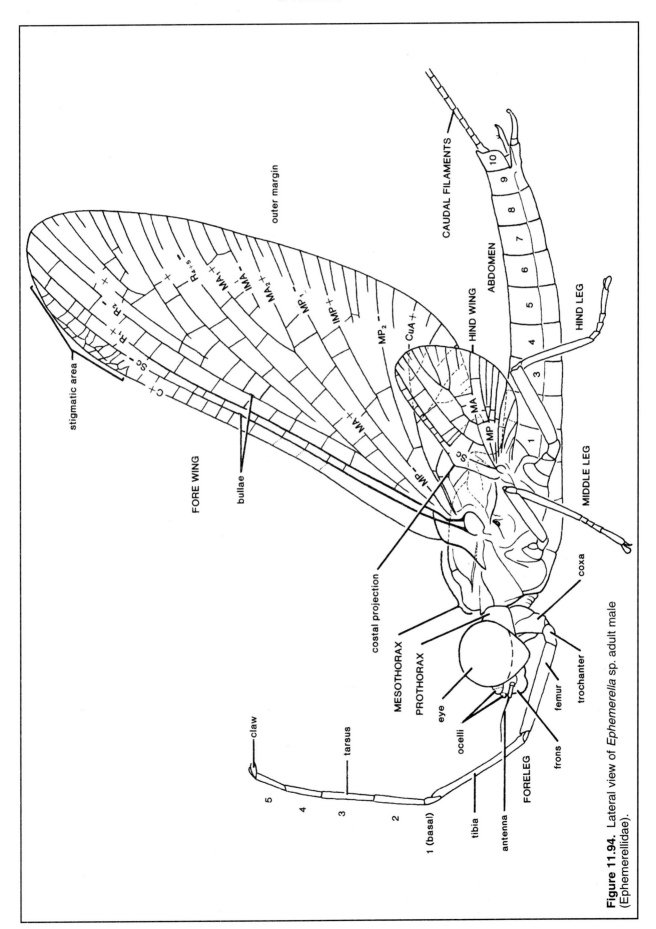

Figure 11.94. Lateral view of *Ephemerella* sp. adult male (Ephemerellidae).

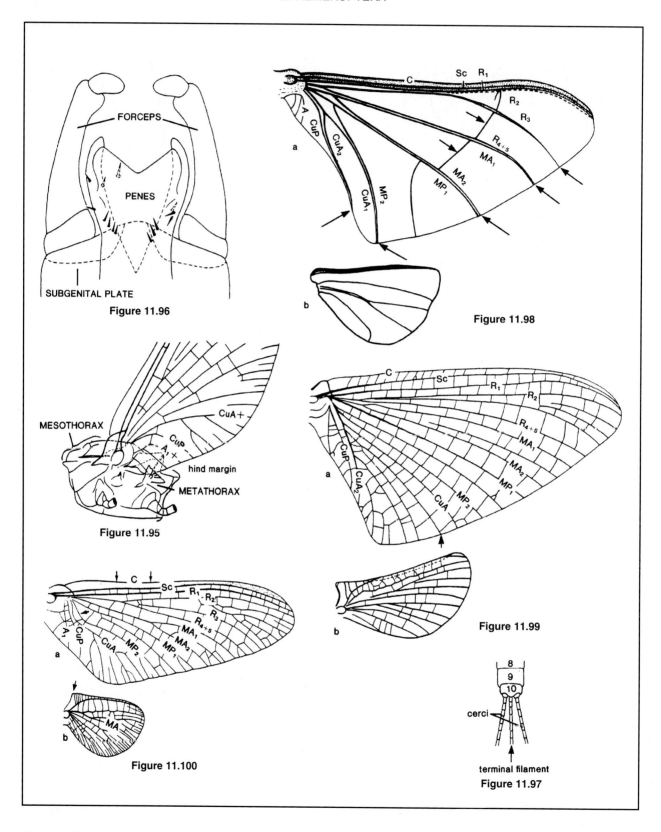

Figure 11.95. Lateral view of thorax and wing of *Ephemerella* sp. (Ephemereliidae).

Figure 11.96. Ventral view of genitalia of *Ephemerella* sp. adult male (Ephemerellidae).

Figure 11.97. Dorsal view of apex of abdomen of *Ephemerella* sp. adult female (Ephemerellidae).

Figure 11.98. Fore wing (*a*) and hind wing (*b*) of *Lachlania* sp. (Oligoneuriidae).

Figure 11.99. Fore wing (*a*) and hind wing (*b*) of *Dolania* sp. (Behningiidae).

Figure 11.100. Fore wing (*a*) and hind wing (*b*) of *Neoephemera* sp. (Neoephemeridae).

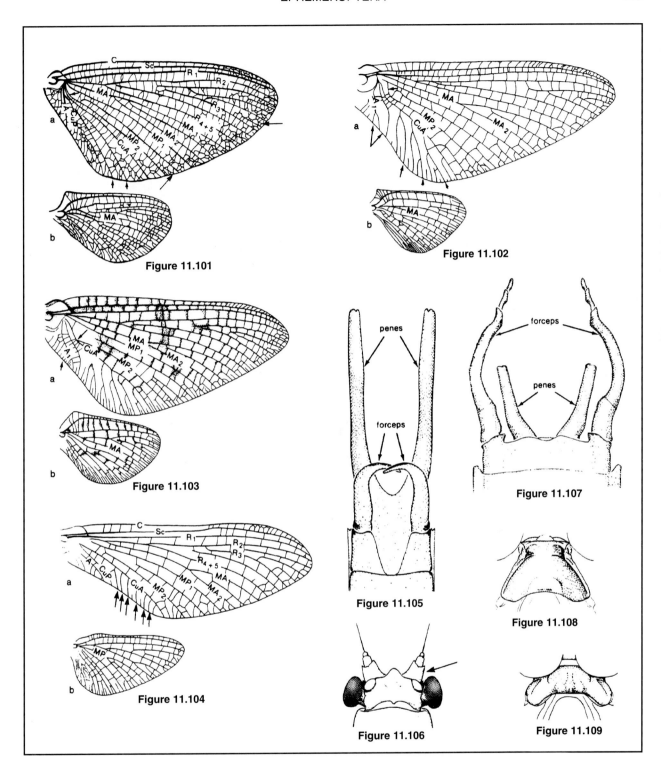

Figure 11.101. Fore wing (*a*) and hind wing (*b*) of *Ephoron* sp. (Polymitarcyidae).

Figure 11.102. Fore wing (*a*) and hind wing (*b*) of *Anthopotamus* sp. (Potamanthidae).

Figure 11.103. Fore wing (*a*) and hind wing (*b*) of *Ephemera* sp. (Ephemeridae).

Figure 11.104. Fore wing (*a*) and hind wing (*b*) of *Siphlonurus* sp. (Siphlonuridae).

Figure 11.105. Ventral view of male genitalia of *Dolania* sp. (Behningiidae).

Figure 11.106. Dorsal view of head of *Dolania* sp. adult female (Behningiidae).

Figure 11.107. Ventral view of genitalia of *Pentagenia* sp. adult male (Ephemeridae).

Figure 11.108. Dorsal view of pronotum of *Ephemera* sp. adult male (Ephemeridae).

Figure 11.109. Dorsal view of pronotum of *Pentagenia* sp. adult male (Ephemeridae).

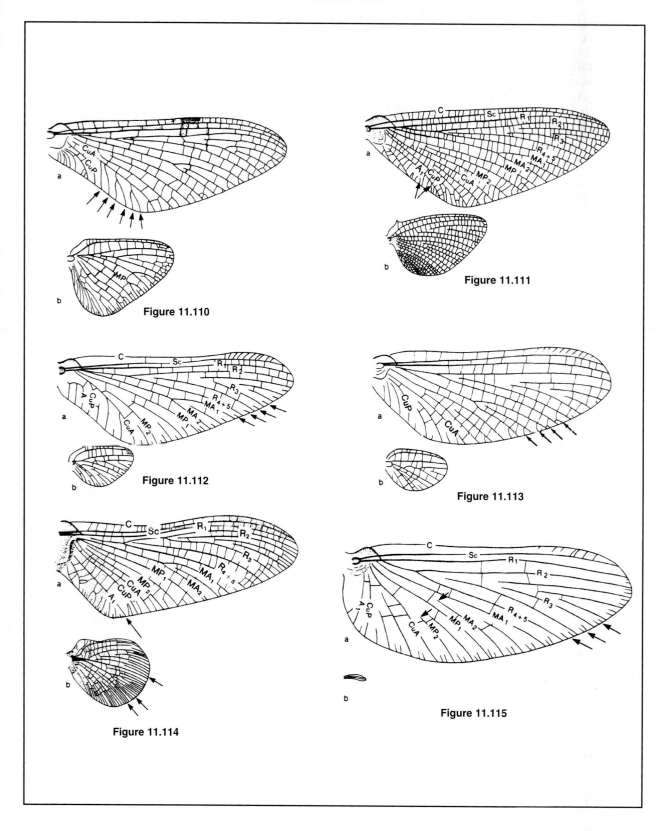

Figure 11.110. Fore wing (*a*) and hind wing (*b*) of *Isonychia* sp. (Isonychiidae).

Figure 11.111. Fore wing (*a*) and hind wing (*b*) of *Ametropus* sp. (Ametropodidae).

Figure 11.112. Fore wing (*a*) and hind wing (*b*) of *Ephemerella* sp. (Ephemerellidae).

Figure 11.113. Fore wing (*a*) and hind wing (*b*) of *Paraleptophlebia* sp. (Leptophlebiidae).

Figure 11.114. Fore wing (*a*) and hind wing (*b*) of *Baetisca rogersi* (Baetiscidae).

Figure 11.115. Fore wing (*a*) and hind wing (*b*) of *Acentrella* sp. (Baetidae).

76'.	Fore femora with transverse row of spines (Fig. 11.78); operculate abdominal gill 2 oval in shape, less than two-thirds as wide as long (Fig. 11.80); uncommon; Southwest, Utah	***Leptohyphes***
77(9').	Head with three prominent ocellar tubercles; maxillary and labial palpi two segmented; fore legs distinctly shorter than middle and hind legs	78
77'.	Head without ocellar tubercles; maxillary and labial palpi three segmented; fore legs subequal in length to middle and hind legs	79
78(77).	Anterior margin of mesosternum flat, without bristles; spines on abdominal segment 6 not curved medially	***Brachycercus***
78'.	Anterior margin of mesosternum convex, with numerous long bristles; spines on abdominal segment 6 strongly curved medially; western and eastern	***Cercobrachys***
79(77').	Inner margin of fore tibiae and tarsi with row of stout spines; segment 3 of labial palpi subequal to or shorter than segment 2, with few long setae and spines; outer margin of operculate gills fringed with long setae; widespread	***Caenis***
79'.	Inner margin of fore tibiae and tarsi densely covered with long setae; segment 3 of labial palpi twice as long as segment 2, densely covered with long setae; outer margin of operculate gills fringed with short bifurcate setae; Southeast, Midwest, Northeast	***Amercaenis***
80(5).	Mandibular tusks with a distinct lateral keel that is more or less toothed and with some spurs; infrequently collected; large rivers (Fig. 11.6)	***Pentagenia***
80'.	Mandibular tusks without a distinct toothed keel (Fig. 11.9)	81
81(80').	Antennae with whorls of long setae on most segments (Fig. 11.9); the small unfringed gills on segment 1 forked	82
81'.	Antennae with only short setae; the small gills on segment 1 single; uncommon; East and Midwest	***Litobrancha***
82(81).	Frontal process of head forked (similar to Fig. 11.6)	***Ephemera***
82'.	Frontal process of head rounded, not forked; abundant; widespread	***Hexagenia***
83(5').	Mandibular tusks with numerous tubercles on upper surface (Fig. 11.10); fore tarsi rounded and clearly demarcated from fore tibiae; abundant; widespread	***Ephoron***
83'.	Mandibular tusks with one to three tubercles on *inner margin,* fore tarsi flattened and broadly fused with tibiae	84
84(83').	Tubercles present almost at apex of inner margin of tusks; abundant; Midwest, Southeast, Texas	***Tortopus***
84'.	Tubercles present in basal half of tusks only; uncommon; Texas	***Campsurus***

Adults

1.	Fore wing venation greatly reduced, apparently only three or four longitudinal veins behind R_1 (Fig. 11.98); body dark .... OLIGONEURIIDAE	41
1'.	Fore wing venation complete or moderately reduced, numerous longitudinal veins present behind R_1 (Figs. 11.99-11.100, 11.120); body variable in color	2
2(1').	Penes of male longer than forceps (Fig. 11.105); antennae of female inserted on prominent anterolateral projections (Fig. 11.106); intercalary behind CuA of fore wing subparallel to and almost as long as CuA; attached to CuA by seven or more unbranched crossveins (Fig. 11.99)	**BEHNINGIIDAE—*Dolania***
2'.	Penes of male shorter than forceps (Fig. 11.107); antennae of female not inserted as above; intercalaries behind CuA variable; intercalary behind CuA much shorter, with few or no crossveins attaching it to CuA (Figs. 11.100, 11.102-11.104, 11.110-11.120) or crossveins anastomosed, especially near wing margin (Fig. 11.101)	3
3(2').	In fore wings, base of veins MP_2 and CuA strongly divergent from base of vein MP_1; MP_2 strongly bent towards CuA basally (Figs. 11.100-11.103) and sometimes fused at base with CuA; hind wings with numerous veins and crossveins; vein MA of hind wings unforked (Figs. 11.100-11.103)	4
3'.	In fore wings, base of veins MP_2 and CuA little divergent from vein MP_1 (vein MP_2 only may diverge from MP_1), fork of MP usually more symmetrical (Figs. 11.104, 11.110-11.120); hind wings variable, may be reduced or absent; if hind wings present, vein MA variable	7
4(3).	Costal angulation of hind wings acute or at a right angle (Fig. 11.100b); fore wing vein A_1 unforked; costal crossveins basal of bullae of fore wings weak or atrophied (Fig. 11.100a); East, mostly Southeast	**NEOEPHEMERIDAE—*Neoephemera***
4'.	Costal angulation of hind wings usually rounded (Figs. 11.101-11.103); if nearly acute or at right angles (Fig. 11.102), fore wing vein A_1 forked near margin (Fig. 11.102); costal crossveins basal of bullae of fore wings well developed (Fig. 11.102)	5

5(4').	Middle and hind legs of male and all legs of female feeble and nonfunctional; color usually pale; wings often somewhat translucent and colorless or with gray or purplish gray shading ....**POLYMITARCYIDAE** 77	
5'.	All legs of both sexes well developed and functional; color variable 6	
6(5').	Fore wing vein A_1 forked near wing margin (Fig. 11.102); abdomen usually yellowish, in some species with reddish lateral stripes or spots on terga ....***POTAMANTHIDAE—Anthopotamus***	
6'.	Fore wing vein A_1 unforked, attached to hind margin by three or more veinlets (Fig. 11.103); abdomen of most species with striking dark pattern on terga and sterna ....**EPHEMERIDAE** 74	
7(3').	Cubital intercalaries of fore wing consist of a series of veinlets, often forking or sinuate, attaching vein CuA to hind margin (Figs. 11.104, 11.110)..8	
7'.	Cubital intercalaries of fore wing variable but not as above (Figs. 11.111-11.113, 11.116-11.117); sometimes absent (Fig. 11.114) 9	
8(7).	Remnants of gill tufts (often purplish colored) present at sides of vestigial maxillae and bases of fore coxae; fore legs largely or entirely dark and middle and hind legs pale; vein MP of hind wing forked near margin (Fig. 11.110b); terminal filament vestigial; widespread and common ....***ISONYCHIIDAE—Isonychia***	
8'.	Remnants of gill tufts not present on vestigial maxillae and fore coxae; leg color not as above; vein MP of hind wing forked near base to mid-length; terminal filament variable 18	
9(7').	Three well-developed caudal filaments (the cerci and terminal filament) present..10	
9'.	Two well-developed caudal filaments (the cerci) present, terminal filament rudimentary or absent..13	
10(9).	Hind wings present and usually relatively large with one or more veins forked; costal projection shorter than wing width (Figs. 11.111-11.113)..11	
10'.	Hind wings small and with two or three simple veins only or absent (Figs. 11.119, 11.120); if hind wing present, costal projection long (1.5 times to 3.0 times width of wing) and straight, or recurved (Fig. 11.118)..17	
11(10).	Vein A_1 of fore wings attached to hind margin by a series of veinlets (Fig. 11.111), and with two pairs of cubital intercalaries present, anterior pair long, posterior pair very short; from Michigan, west ....***AMETROPODIDAE—Ametropus***	
11'.	Vein A_1 not attached to hind margin as above (Figs. 11.112, 11.113); cubital intercalaries not as above 12	
12(11').	Short, basally detached marginal intercalaries present between veins along entire outer margin of wings (Fig. 11.112); genital forceps of male with one short terminal segment (Figs. 11.155-11.161)..**EPHEMERELLIDAE** 65	
12'.	No true basally detached marginal intercalaries in positions indicated above, usually absent along entire outer margin of wings (Fig. 11.113); genital forceps of male with two or three short terminal segments (Fig. 11.154)..**LEPTOPHLEBIIDAE** 57	
13(9').	Hind wings with numerous, long, free marginal intercalaries (Fig. 11.114); cubital intercalaries absent in fore wings with vein A_1 terminating in outer margin of wings (Fig. 11.114) ....***BAETISCIDAE—Baetisca***	
13'.	Hind wings not as above, sometimes absent; cubital intercalaries present in fore wings with vein A_1 terminating in hind margin of wings (Figs. 11.115-11.117) 14	
14(13').	Short, basally detached, single or double marginal intercalaries present in each interspace of fore wings; and veins MA_2 and MP_2 detached basally from their respective stems (Fig. 11.115), hind wings small or absent; penes of male membranous; upper portion of eyes of male turbinate (raised on a stalk like portion; Fig. 11.125) ....**BAETIDAE** 23	
14'.	Marginal intercalaries attached basally to other veins; MA_2 and MP_2 attached basally (Figs. 11.117-11.118); hind wings relatively large; penes of male well developed; eyes of male not turbinate 15	
15(14').	Hind tarsi apparently four segmented, basal segment fused or partially fused to tibiae (Figs. 11.122-11.123), the fusion line being in basal half of tibiae plus tarsi; cubital intercalaries consist of one or two pair (Figs. 11.117-11.118) 16	
15'.	Hind tarsi distinctly five segmented (Fig. 11.124); tarsi shorter than tibiae; cubital intercalaries consist of two pairs similar to Figure 11.116 ....**HEPTAGENIIDAE** 42	
16(15).	Eyes of male contiguous (touching) or nearly contiguous dorsally (similar to Fig. 11.138); fore tarsi three times length of fore tibiae; abdomen of female with apical and basal segments subequal to middle segments in length and width; subanal plate evenly convex ....**METRETOPODIDAE** 56	

EPHEMEROPTERA

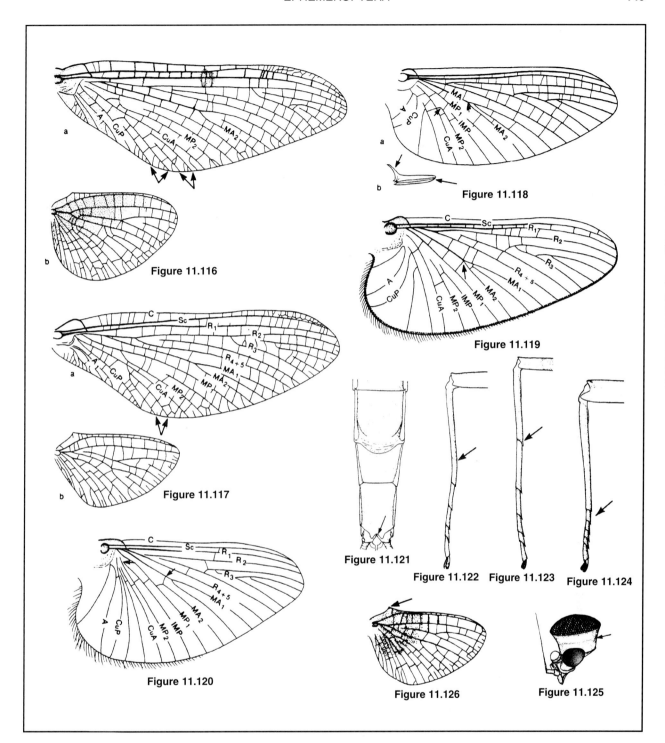

Figure 11.116. Fore wing (*a*) and hind wing (*b*) of *Siphloplecton* sp. (Metretopodidae).

Figure 11.117. Fore wing (*a*) and hind wing (*b*) of *Metretopus* sp. (Metretopodidae).

Figure 11.118. Fore wing (*a*) and hind wing (*b*) of *Leptohyphes* sp. (Tricorythidae).

Figure 11.119. Wing of *Tricorythodes* sp. male (Tricorythidae).

Figure 11.120. Wing of *Caenis* sp. (Caenidae).

Figure 11.121. Ventral view of apex of *Pseudiron centralis* female adult (Pseudironidae).

Figure 11.122. Hind tibia and tarsus of *Pseudiron centralis* adult (Pseudironidae).

Figure 11.123. Hind tibia and tarsus of *Siphloplecton* sp. adult (Metretopodidae).

Figure 11.124. Hind tibia and tarsus of *Heptagenia* sp. adult (Heptageniidae).

Figure 11.125. Lateral view of head of *Baetis* sp. male (Baetidae).

Figure 11.126. Hind wing of *Ameletus* sp. adult (Ameletidae).

16'.	Eyes of male separated dorsally by twice width of median ocellus; fore tarsi two times length of fore tibiae; abdomen of female long and slender, apical segments distinctly more elongate and slender than basal segments; subanal plate with medial emargination (Fig. 11.121); adults rarely collected; large sandy rivers	***PSEUDIRONIDAE—Pseudiron***
17(10').	Vein MA of fore wings forming a more or less symmetrical fork, and veins MP_2 and IMP extend less than three-fourths of distance to base of vein MP (Figs. 11.118-11.119); genital forceps of male two or three segmented; thorax usually black or gray	**TRICORYTHIDAE** 71
17'.	Vein MA of fore wing not as above, MA_2 attached basally by a crossvein; wing veins MP_2 and IMP almost as long as vein MP, and extend nearly to base (Fig. 11.120); genital forceps of male one segmented; thorax usually brown	**CAENIDAE** 72
18(8').	Three caudal filaments; terminal filament distinctly longer than tergum 10; hind wings one-half or more as long as fore wings; rare; scattered distribution	***ACANTHAMETROPODIDAE—Analetris***
18'.	Two apparent caudal filaments; terminal filament vestigial; hind wings less than one-half as long as fore wings	19
19(18').	Claws of each pair dissimilar (1 sharp, 1 blunt) (Fig. 11.127); costal projection of hind wings acute (Fig. 11.126)	***AMELETIDAE—Ameletus***
19'.	Claws of each pair similar, sharp (Fig. 11.128); costal projection of hind wings obtuse or weak (Fig. 11.104b)	**SIPHLONURIDAE** 20
20(19').	Abdominal segments 5-9 of male greatly expanded laterally; tubercles present on sterna of mesothorax and metathorax; rare; New York to Labrador	*Siphlonisca*
20'.	Abdominal segments 5-9 not expanded laterally; without tubercles on sterna of thorax	21
21(20').	Vein MP of hind wings simple, unforked; uncommon; Canada or West, mountains	*Parameletus*
21'.	Vein MP of hind wings forked (Fig. 11.104b)	22
22(21').	Abdominal sterna 3-6 pale with a brown transverse bar; rare; California	*Edmundsius*
22'.	Ventral markings not as above, usually with dark spots, longitudinal stripes, oblique stripes, U-shaped marks, or mostly dark; abundant; widespread	*Siphlonurus*
23(14).	Hind wings present and with blunt costal projection and ten or more crossveins; abdomen with distinct dark speckles; fore wings of most species patterned; widespread	*Callibaetis*
23'.	Hind wings present or absent, if present, then usually with less than six crossveins; costal projection of hind wings never blunt, either pointed (Fig. 11.129), hooked (Fig. 11.130), or absent (Fig. 11.131); abdomen variable, rarely with speckles; fore wings of most species without patterns	24
24(23').	Marginal intercalaries of fore wings occur singly; hind wings present or absent, if present, hind wings with costal projection hooked or recurved (similar to Fig. 11.130)	25
24'.	Marginal intercalaries of fore wings occur in pairs as in Figure 11.115a; hind wings present or absent, if present, costal projection of hind wings variable, rarely hooked	28
25(24).	Spine-like process (Fig. 11.169) present between forceps bases of most species; terminal segment of forceps elongate; widespread	*Centroptilum*
25'.	Spine-like process between forceps bases never present; terminal segment of forceps variable	26
26(25').	Hind wings absent and male genitalia with distinct coniform penes cover between forceps bases; terminal segment of forceps small, rounded; females with brownish pigment at pterostigma; East and Midwest	*Cloeon*
26'.	Hind wings present or absent and male genitalia with rounded or rectangular penes cover between forceps bases; terminal segment of forceps usually elongate; females without brown pigment at pterostigma of fore wings	27
27(26').	Segment 2 of male forceps (Fig. 11.170) with distinct inner swelling or bulge; Midwest	*Pseudocentroptiloides*
27'.	Segment 2 of male forceps without distinct inner swelling or bulge; widespread	*Procloeon*
28(24').	Costal projection of hind wings broadly based, acute at apex, basal anterior margin of hind wing forming a straight line from point of attachment to apex of costal process, giving hind wings characteristic shape (Fig. 11.132a,b); uncommon; medium to large rivers; West and Central	*Camelobaetidius*
28'.	Costal projection of hind wings present or absent (if present, variable in shape); if similar to above, costal projection not broadly based so that hind wings are not shaped as above (Fig. 11.133a,b)	29
29(28').	Anterior process of mesothorax coniform, projecting dorsally (Fig. 11.171); hind wings present or absent, if present, hind wings without costal process	30
29'.	Anterior process of mesothorax not coniform and not projecting dorsally; hind wings, when present, with or without a costal process	31

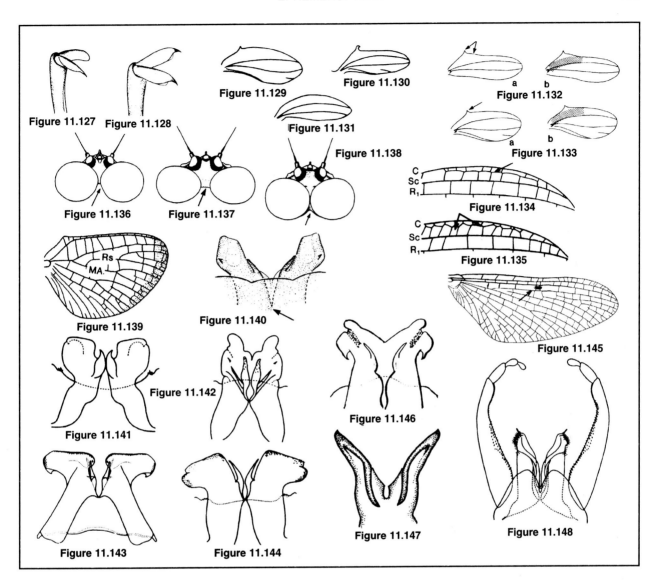

Figure 11.127. Claws of adult *Ameletus* sp. (Ameletidae).

Figure 11.128. Claws of adult *Siphlonurus* sp. (Siphlonuridae).

Figure 11.129. Hind wing of *Baetis* sp. adult (Baetidae).

Figure 11.130. Hind wing of *Fallceon quilleri* adult (Baetidae).

Figure 11.131. Hind wing of *Labiobaetis* sp. adult (Baetidae).

Figure 11.132. Hind wing (a,b) of *Camelobaetidius* sp. adult (Baetidae).

Figure 11.133. Hind wing (a,b) of *Baetis* sp. adult (Baetidae). Shaded arc on Figures 11.132b and 11.133b to emphasize shape of costal projection.

Figure 11.134. Stigmatic area at apex of *Cinygma* sp. adult fore wing (Heptageniidae).

Figure 11.135. Stigmatic area at apex of *Rhithrogena* sp. adult fore wing (Heptageniidae).

Figure 11.136. Dorsal view of head of *Epeorus* (*Ironopsis*) sp. adult male (Heptageniidae).

Figure 11.137. Dorsal view of head of *Ironodes* sp. adult male (Heptageniidae).

Figure 11.138. Dorsal view of head of *Epeorus* (*Iron*) sp. adult male (Heptageniidae).

Figure 11.139. Hind wing of *Stenonema* sp. adult (Heptageniidae).

Figure 11.140. Penes of *Cinygmula* sp. adult male (Heptageniidae).

Figure 11.141. Penes of *Macdunnoa* sp. adult male (Heptageniidae).

Figure 11.142. Penes of *Nixe* sp. adult male (Heptageniidae).

Figure 11.143. Penes of *Stenonema* sp. adult male (Heptageniidae).

Figure 11.144. Penes of *Stenonema* sp. adult male (Heptageniidae).

Figure 11.145. Fore wing of *Stenacron* sp. adult male (Heptageniidae).

Figure 11.146. Penes of *Epeorus* (*Iron*) sp. adult male (Heptageniidae).

Figure 11.147. Penes of *Leptophlebia* sp. adult male (Leptophlebiidae).

Figure 11.148. Penes of *Leptophlebia* sp. adult male (Leptophlebiidae).

30(29).	Hind wings minute in male and female; hind wings without veins (similar to Fig. 11.115b, but without veins); widespread	**Heterocloeon**
30'.	Hind wings present or absent; when present, hind wings with 2 longitudinal veins (Fig. 11.115b) (*Heterocloeon frivolum* (McD) will also key here); widespread	**Acentrella**
31(29').	Hind wings without costal process (Fig. 11.131) or with costal process greatly reduced (hind wings are present in all known North American species); distinct plate present between forceps bases; terminal segment of forceps small and rounded; widespread	**Labiobaetis**
31'.	Hind wings present or absent; if present, hind wings with costal process	32
32(31').	Hind wings with costal process hooked (Fig. 11.130) (some Neotropical adventive species, e.g., *F. eatoni*, with straight costal processes may be found eventually in the Southwest); terminal segment of male forceps elongate in North American species; bilobed process present between forceps bases; Southwest, Central, Southeast	**Fallceon**
32'.	Hind wings present or absent, if present, with costal process rarely appearing hooked, terminal segment of male forceps variable; bilobed process between forceps bases not present	33
33(32').	Prominent penes cover present between forceps bases; bases of compound eyes nearly contiguous posteriorly and widely divergent anteriorly; hind wings absent in all known species; rare; West and Southeast	**Apobaetis**
33'.	Penes cover between forceps bases reduced or absent; eyes variable; hind wings present or absent	34
34(33').	Hind wings (Fig. 11.172) with second longitudinal vein forked, vertex of fork proximad and toward wing base; terminal segment of male forceps elongate; widespread	**Diphetor**
34'.	Hind wings present or absent, if present, hind wings with second vein simple, or if forked, vertex of fork occurs distally beyond middle of hind wing (some individuals of *Baetis magnus* and *B. diablis* may key here, with small round terminal segments of the forceps)	35
35(34').	Costal process of hind wing acute, prominent (similar to Fig. 11.132a,b but with curved anterior margin preceding costal process); beyond costal process anterior margin of hind wing undulate; terminal segment of male forceps approximately four times as long as wide; widespread	**Acerpenna**
35'.	Hind wings present or absent, if present, hind wing costal process variable, not as acute or prominent as above; terminal segment of male forceps variable	36
36(35').	Male genitalia with a mesal projection on segment 1 of forceps (Fig. 11.173); metanotal scutellum reduced or absent; male fore leg tibiae 1.5-2.0 times length of their respective femora; Southwest, Texas, Utah	**Baetodes**
36'.	Male genitalia without mesal projection on segment 1 of forceps; metanotal scutellum not reduced; fore legs of male variable	37
37(36').	Hind wings present	**Baetis** (in part), **Barbaetis** (in part)
37'.	Hind wings absent	38
38(37').	Very small species, fore wing length 2.5 mm-4.5mm; terminal segment of forceps (Fig. 11.174) elongate, segment length subequal to 2.5 to 3.0 times width; without small rectangular plate between forceps bases; widespread	**Paracloeodes**
38'.	Larger species, fore wing length generally greater than 4.0 mm; terminal segment of male forceps usually about as long as wide	39
39(38').	Segment 2 of male forceps with distinct inner bulge (Fig. 11.175); Arizona, New Mexico	**Cloeodes**
39'.	Segment 2 of male forceps lacking distinct inner bulge	40
40(39').	Forceps of male strongly curved (Fig. 11.176); abdominal terga 3 and 4 strongly pigmented in red medially, terga 5 and 6 washed in red; rectangular plate present between forceps bases; East, Midwest, Colorado	**Barbaetis** (in part)
40'.	Forceps of male not strongly curved as above; abdominal terga not pigmented as above; rectangular plate between forceps bases usually absent (present in *fuscatus* group species); widespread	**Baetis** (in part)
41(1).	Three caudal filaments present	**Homoeoneuria**
41'	Two caudal filaments (cerci) present	**Lachlania**
42(15').	Stigmatic area of fore wing divided by a fine, more or less straight vein into two series of cellules (Fig. 11.134); Northwest	**Cinygma**
42'.	Stigmatic area with simple or anastomosed crossveins, not as above	43
43(42').	Vein MA of hind wing simple, unforked; uncommon; Midwest or Northeast	**Arthroplea***
43'.	Vein MA of hind wing forked (Fig. 11.139) (females not keyed beyond this couplet)	44
44(43').	Male fore tarsi three-fourths or less length of fore tibiae; male genitalia as in Figure 11.177; rare; Midwest, Southeast, west to Utah and Saskatchewan	**Anepeorus**

*Family Arthropleidae

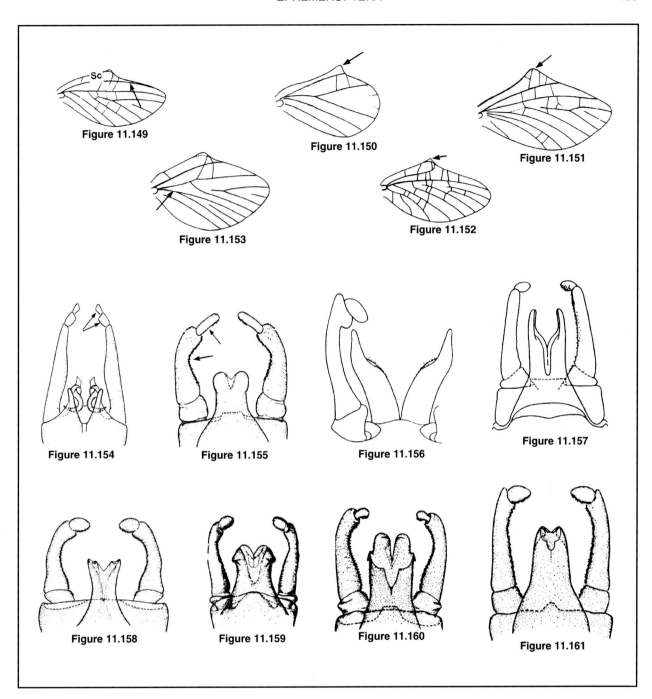

Figure 11.149. Hind wing of *Habrophlebia* sp. adult (Leptophlebiidae).

Figure 11.150. Hind wing of *Habrophlebiodes* sp. adult (Leptophlebiidae).

Figure 11.151. Hind wing of *Choroterpes* sp. adult (Leptophlebiidae).

Figure 11.152. Hind wing of *Traverella* sp. adult (Leptophlebiidae).

Figure 11.153. Hind wing of *Thraulodes* sp. adult (Leptophlebiidae).

Figure 11.154. Genitalia of *Paraleptophlebia* sp. adult male (Leptophlebiidae).

Figure 11.155. Genitalia of *Drunella* sp. adult male (Ephemerellidae).

Figure 11.156. Genitalia of *Ephemerella* sp. adult male (Ephemerellidae).

Figure 11.157. Genitalia of *Ephemerella* sp. adult male (Ephemerellidae).

Figure 11.158. Genitalia of *Ephemerella* sp. adult male (Ephemerellidae).

Figure 11.159. Genitalia of *Serratella* sp. adult male (Ephemerellidae).

Figure 11.160. Genitalia of *Serratella* sp. adult male (Ephemerellidae).

Figure 11.161. Genitalia of *Eurylophella* sp. adult male (Ephemerellidae).

44'.	Male fore tarsi longer than fore tibiae; male genitalia not as above	45
45(44').	Penes as in Figure 11.178; longitudinal veins yellowish, crossveins dark; rare; West, Midwest, Southeast, and Canada	***Raptoheptagenia***
45'.	Penes not as above; venation variable; widespread	46
46(45').	Stigmatic area of wing with from two to many anastomoses of crossveins (Fig. 11.135), and basal segment of fore tarsi one-third or less than one-third length of segment 2; femora usually with dark longitudinal streak near middle	***Rhithrogena***
46'.	Stigmatic area of wing with or without anastomosed crossveins; if crossveins anastomosed (similar to Fig. 11.135) basal segment of fore tarsi half or more than half length of segment 2; femora usually without dark longitudinal streak	47
47(46').	Basal segment of fore tarsi equal to or slightly longer than segment 2	48
47'.	Basal segment of fore tarsi four-fifths or less than four-fifths length of segment 2	50
48(47).	Eyes separated dorsally by more than width of median ocellus (Fig. 11.137); basal costal crossveins strongly developed, attached anteriorly; uncommon; Northwest	***Ironodes***
48'.	Eyes contiguous (touching) (Fig. 11.138) or separated by no more than width of median ocellus (Fig. 11.136); basal costal crossveins weakly developed, apparently detached anteriorly; common; widespread	***Epeorus*** 49
49(48').	Penes with median spines minute or absent; stigmatic crossveins of fore wings usually anastomosed (similar to Fig. 11.135); Pacific Coast east to western Nevada, Colorado, Utah, and Montana	***Epeorus (Ironopsis)***
49'.	Penes with well developed median spines (Fig. 11.146); stigmatic crossveins of fore wings not anastomosed; widespread	***Epeorus (Iron)***
50(46').	Penis lobes separated or appearing separated to near base medially (Fig. 11.140); basal segment of fore tarsi usually two-thirds or more than two-thirds length of segment 2	***Cinygmula***
50'.	Penis lobes fused medially at least in basal half (Figs. 11.141-11.144); basal segment of fore tarsi two-thirds or less than two-thirds length of segment 2, usually less than half length of segment 2	51
51(50').	Wings with two or three crossveins below bullae between veins R_1 and R_2 connected or nearly connected by dark pigmentation (Fig. 11.145), rarely only a dark spot; basal crossveins between R_1 and R_2 dark margined	***Stenacron***
51'.	Wings may have crossveins below bullae clouded but never as above; crossveins between R_1 and R_2 rarely dark margined	52
52(51').	Penes distinctly L-shaped (Figs. 11.143-11.144); basal segment of fore tarsi usually one-third to two-thirds length of segment 2	***Stenonema***
52'.	Penes not distinctly L-shaped as above (Figs. 11.141-11.142); basal segment of fore tarsi usually one-fifth to one-half length of segment 2	53
53(52').	Eyes of male contiguous (touching) on vertex (similar to Fig. 11.138) or separated by less than the diameter of *median* ocellus; penes with dorsolateral spines minute, discal spines present or absent (Fig. 11.142); basal costal veins weak	***Nixe***
53'.	Eyes of male separated at least by width of a *lateral* ocellus; dorsolateral spines of penes present or absent	54
54(53').	Eyes of male separated by little more than diameter of a *lateral* ocellus (see Fig. 11.137); penes with large dorsolateral spines, discal spines absent	***Heptagenia***
54'.	Eyes of male separated at vertex by 3-5 times width of *median* ocellus	55
55(54').	Eyes of male separated at vertex by approximately the width of one compound eye; median titillators of male genitalia slender; penes with large dorsolateral spines, discal spines present; basal costal crossveins well-developed; crossveins behind costa and subcosta usually margined in brown	***Leucrocuta***
55'.	Eyes of male separated by width of 4-5 times width of *median* ocellus; median titillators of male genitalia thick and abruptly narrowed at apex; penes without discal spines or ridges (Fig. 11.141); basal costal crossveins well-developed; anterior crossviens often tan to black; rare	***Macdunnoa***
56(16).	One pair of cubital intercalaries present (Fig. 11.117); rare; Canada, Alaska, Michigan to Maine	***Metretopus***
56'.	Two pairs of cubital intercalaries present (Fig. 11.116); Canada, Midwest and East	***Siphloplecton***
57(12').	Hind wings without costal projection (Fig. 11.113b)	58
57'.	Hind wings with distinct costal projection (Figs. 11.149-11.153)	59
58(57).	Penes of male with long, decurrent, median appendages (Figs. 11.147-11.148); terminal filament often shorter and thinner than lateral cerci	***Leptophlebia***
58'.	Penes of male variable (e.g., Fig. 11.154), not as in Figures 11.147-11.148; three subequal caudal filaments; widespread	***Paraleptophlebia***

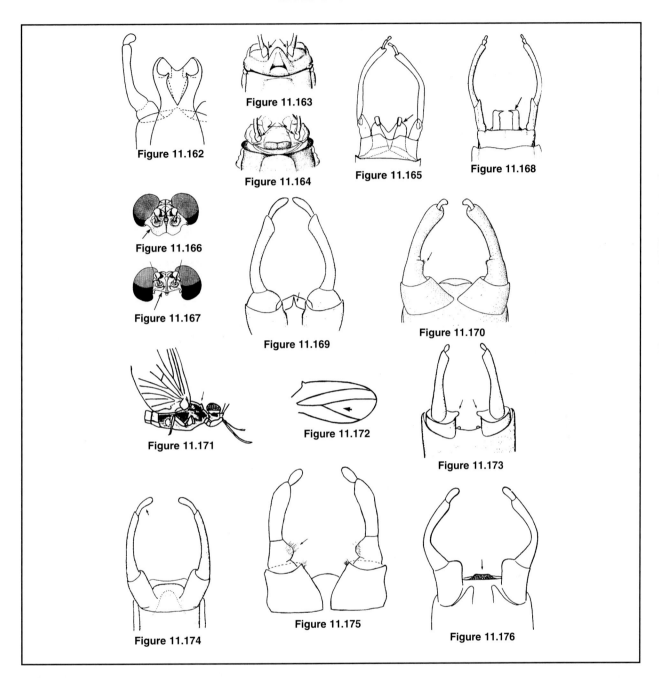

Figure 11.162. Genitalia of *Timpanoga* (*Dannella*) sp. adult male (Ephemerellidae).

Figure 11.163. Ventral view of pronotum of *Caenis* sp. adult (Caenidae).

Figure 11.164. Ventral view of pronotum of *Brachycercus* sp. adult (Caenidae).

Figure 11.165. Genitalia of *Litobrancha* sp. adult male (Ephemeridae).

Figure 11.166. Anterior view of head of *Litobrancha* sp. adult male (Ephemeridae).

Figure 11.167. Anterior view of head of *Hexagenia* sp. adult male (Ephemeridae).

Figure 11.168. Genitalia of *Hexagenia* sp. adult male (Ephemeridae).

Figure 11.169. Ventral view of adult male genitalia, *Centroptilum* sp. (Baetidae).

Figure 11.170. Ventral view of adult male genitalia, *Pseudocentroptiloides usa* (Baetidae).

Figure 11.171. Right lateral view of head and thorax of *Acentrella* sp. adult male (Baetidae). Arrow indicates anterior process of mesonotum.

Figure 11.172. Hind wing of *Diphetor* sp. adult (Baetidae).

Figure 11.173. Ventral view of adult male genitalia, *Baetodes* sp. (Baetidae).

Figure 11.174. Ventral view of adult male genitalia, *Paracloeodes minutus* (Baetidae).

Figure 11.175. Ventral view of adult male genitalia, *Cloeodes* sp. (Baetidae).

Figure 11.176. Ventral view of adult male genitalia, *Barbaetis cestus* (Baetidae).

59(57').	Costal projection of hind wings in distal half of wing (Fig. 11.150); uncommon; Midwest or East	*Habrophlebiodes*
59'.	Costal projection of hind wing near midpoint of length of wings (Figs. 11.149, 11.151-11.153)	60
60(59').	Vein Sc of hind wings extends well beyond costal projection (Fig. 11.149); uncommon; eastern	*Habrophlebia*
60'.	Vein Sc of hind wings ends at or slightly beyond costal projection (Figs. 11.151, 11.152b, 11.153)	61
61(60').	Vein MP of hind wing forked (Fig. 11.153); uncommon; Utah, Texas, New Mexico, Arizona	*Thraulodes*
61'.	Vein MP of hind wing simple, unforked (Figs. 11.151-11.152)	62
62(61').	Costal projection of hind wings rounded (Fig. 11.151)	63
62'.	Costal projection of hind wings acute (Figs. 11.152, 11.190)	64
63(62).	Posterior margin of female 9th abdominal segment sternum with a very shallow indentation	*Choroterpes*
63'.	Posterior margin of female 9th abdominal segment sternum with distinct indentation	*Neochoroterpes*
64(62').	Hind wings (Fig. 11.152) with eight or more longitudinal veins and twelve or more crossveins; males with process near apex of penes directed toward base of penes; widespread	*Traverella*
64'.	Hind wings (Fig. 11.179); with seven or less longitudinal veins and about three crossveins; males with process near apex of penes projecting laterally (Fig. 11.180); rare; Texas	*Farrodes*
65(12).	Adult and subimago with vestiges of gill sockets on abdominal segment 3; eggs with one polar cap and somewhat smooth chorion	66
65'.	Adult and subimago without vestiges of gill sockets on abdominal segment 3; eggs without polar caps, eggs with two polar caps, *or* if eggs with one polar cap then with reticulate chorion	69
66(65).	Cerci one-fourth to three-fourths as long as terminal filament; Northwest	*Caudatella*
66'.	Cerci and terminal filament subequal in length (females not keyed beyond this couplet)	67
67(66').	Terminal segment of male genital forceps more than twice as long as broad; inner margin of long second segment distinctly incurved (Fig. 11.155) or strongly bowed	*Drunella*
67'.	Terminal segment of male forceps less than twice as long as broad (Fig. 11.161); inner margin of long second segment variable but not strongly bowed	68
68(67').	Penes with dorsal and/or ventral spines and shaped similar to Figure 11.96, *or* with long anterolateral lobes (Figs. 11.156-11.157), *or* (one western species) as in Figure 11.158; widespread	*Ephemerella*
68'.	Penes different than above, with lateral subapical projections as in Figures 11.159-11.160; widespread	*Serratella*
69(65').	Abdominal segments 6 and 7 of adult and subimago with poorly developed fingerlike remnants of posterolateral projections; male adult with terminal segments of genital forceps relatively elongate (approximately six times as long as broad); eggs without polar caps and having somewhat smooth chorion	*Attenella*
69'.	Abdominal segment 7 and, often, preceding segments also, of adult and subimago with well-developed fingerlike remnants of posterolateral projections; gill socket formation of abdominal segment 4 relatively well developed; male adult with terminal segment of genital forceps relatively short (length always less than three times width); egg with reticulate chorion and zero, one, or two, polar caps	70
70(69').	Male adult of known species with penes broadest at base, narrowing apically (Fig. 11.161); eggs of known species, without polar caps; Midwest, East, Pacific Coast	*Eurylophella*
70'.	Male adult with penes laterally expanded apically, somewhat narrower basally (Fig. 11.162); eggs with polar caps; West, Midwest, East	*Timpanoga*
71(17).	Wings of male greatly expanded in cubitoanal areas; vein CuP evenly recurved in male and female (Fig. 11.119); hind wings absent; abundant; widespread	*Tricorythodes*
71'.	Wings not expanded in cubitoanal area, broadest near midpoint of length; vein CuP abruptly recurved (Fig. 11.118a); hind wings of male with long costal projection (Fig. 11.118b); hind wings absent in female; uncommon; Southwest, Utah	*Leptohyphes*
72(17').	Prosternum half as long as broad, rectangular in shape; fore coxae widely separated on venter (Fig. 11.164); uncommon	73
72'.	Prosternum two to three times as long as broad, triangular in shape; fore coxae close together on venter (Fig. 11.163); widespread	*Amercaenis, Caenis*
73(72).	Spines on abdominal segment 6 considerably bent dorsad; pedicel only slightly longer than scape or equal in length; eggs spindle shaped, four to six times longer than broad; western and eastern	*Cercobrachys*
73'.	Spines on abdominal segment 6 slightly bent dorsally; pedicel at least 1.5-2.0 times longer than scape; eggs oval, two to three times longer than broad	*Brachycercus*
74(6').	Pronotum of male well developed, no more than twice as wide as long (Fig. 11.108); penes variable, not long and tubular (Figs. 11.165, 11.168); caudal filaments of female longer than body	75

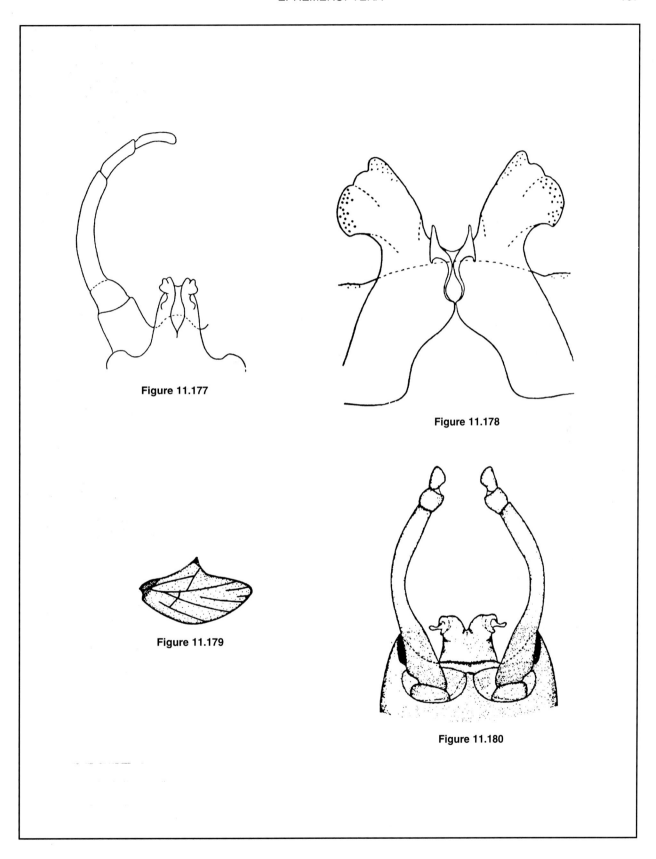

Figure 11.177. Ventral view of adult male genitalia, *Anepeorus simplex* (Heptageniidae).

Figure 11.178. Penes of adult male, *Raptoheptagenia cruentata* (Heptageniidae).

Figure 11.179. Hind wing of *Farrodes texanus* adult (Leptophlebiidae).

Figure 11.180. Ventral view of adult male genitalia, *Farrodes texanus* (Leptophlebiidae).

74'.	Pronotum of male shortened, about three times as wide as long (Fig. 11.109); penes long and tubular (Fig. 11.107); caudal filaments of female shorter than body; large rivers	*Pentagenia*
75(74).	Crossveins of wings crowded together near bullae; fore wings with distinct pattern of dark markings (Fig. 11.103); terminal filament as long as cerci	*Ephemera*
75'.	Crossveins of wings not crowded near bullae; wings without pattern of dark markings although crossveins may be darkened; terminal filament vestigial or distinctly shorter than cerci	76
76(75').	Head with frons greatly expanded below eyes (Fig. 11.166); penes of male recurved ventrally (Fig. 11.165); uncommon; East, Midwest	*Litobrancha*
76'.	Head with frons not extending below eyes (Fig. 11.167); penes of male not recurved (e.g., Fig. 11.168); abundant; widespread	*Hexagenia*
77(5).	Outer margin of fore wings with dense network of reticulate veinlets (Fig. 11.101); abundant, widespread	*Ephoron*
77'.	Outer margin of fore wings with few or no veinlets	78
78(77').	Middle and hind legs with all segments present, but reduced and shriveled; abundant; Midwest, Southeast, Texas	*Tortopus*
78'.	Middle and hind legs greatly reduced, terminating in blade like femora; uncommon; Texas	*Campsurus*

ADDITIONAL TAXONOMIC REFERENCES

General
Tillyard (1932); Needham *et al.* (1935); Edmunds and Traver (1954); Day (1956); Edmunds (1959); Koss (1968); Koss and Edmunds (1974); Edmunds *et al.* (1976); Pennak (1978); McCafferty and Edmunds (1979); McCafferty (1981, 1990, 1991a,b); Edmunds and McCafferty (1988); McCafferty *et al.* (1990).

Regional faunas
Florida: Berner and Pescador (1988)
Illinois and region: Burks (1953).
Michigan: Leonard and Leonard (1962).
North and South Carolina: Unzicker and Carlson (1982).
Saskatchewan: Lehmkuhl (1976a).
Wisconsin: Hilsenhoff (1981).

Taxonomic treatments at the family and generic levels
Acanthametropodidae: McCafferty (1991c); McCafferty and Wang (1994a).
Ametropodidae: Allen and Edmunds (1976).
Baetidae: Cohen and Allen (1978); Morihara and McCafferty (1979a,b,c); Waltz and McCafferty (1985, 1987a,b,c,d, 1989, 1996); Waltz *et al.* (1985, 1994); Lowen and Flannagan (1990a,b, 1991); McCafferty and Waltz (1990, 1995); McCafferty *et al.* (1994); Waltz (1994).
Baetiscidae: Berner (1955); Berner and Pescador (1980); Pescador and Berner (1981).
Behningiidae: Edmunds and Traver (1959); McCafferty (1975); Hubbard (1994).
Caenidae: Provonsha and McCafferty (1985); Soldàn (1986); Provonsha (1990).
Ephemerellidae: Allen and Edmunds (1965; obtain citations for Parts I-VII from this reference); Allen (1980, 1984); McCafferty (1977, 1985, 1992a); Funk and Sweeney (1995); Funk *et al.* (1988); McCafferty and Wang (1994b).
Ephemeridae: Spieth (1941); McCafferty (1975, 1994).
Heptageniidae: Edmunds and Allen (1964); Lewis (1974); Flowers and Hilsenhoff (1975); Bednarik and McCafferty (1979); Lehmkuhl (1979b); Bednarik and Edmunds (1980); Flowers (1980,1982); McCafferty and Provonsha (1984,1985,1988); Whiting and Lehmkuhl (1987a,b); McCafferty (1990, 1991a,b).
Isonychiidae: Kondratieff and Voshell (1983, 1984); McCafferty (1989).
Leptophlebiidae: Traver and Edmunds (1967), Berner (1975); Peters (1980); McCafferty (1992b); Henry (1993); McCafferty *et al.* (1993).
Metretopodidae: Berner (1978).
Neoephemeridae: Berner (1956).
Oligoneuriidae: Edmunds *et. al.* (1958); Edmunds (1961); Pescador and Peters (1980).
Polymitarcyidae: McCafferty (1975).
Potamanthidae: McCafferty (1975); McCafferty and Bae (1990, 1992); Bae and McCafferty (1991).
Pseudironidae: Pescador (1985).
Tricorythidae: Allen (1967).

Table 11A. Summary of ecological and distributional data for *Ephemeroptera (mayflies)*. (For definition of terms see Tables 6A-6C; table prepared by K.W. Cummins, and R.W. Merritt.)

Taxa (number of species in parentheses)	Habitat	Habit	Trophic	North American Relationships	Ecological Distribution
Acanthametropodidae (1)					
Acanthametropus(1)	Lotic depositional (large rivers)	Swimmers, clingers	Predators (engulfers)	Illinois (Rock River), South Carolina, Georgia (Savanna River), Wisconsin	489, 1029
Analetris(1)	Lotic depositional (large rivers)	Swimmers, clingers	Predators (engulfers)	Wyoming, Utah, Saskatchewan	1028, 1029, 2309
Ameletidae(33)					
Ameletus(33)	Lotic—erosional and depositional	Swimmers, clingers	Scrapers, collectors—gatherers (detritus, diatoms)	Northeast, West and Southeast mountains, Illinois	111, 112, 880, 890, 1399, 1721, 2553, 2830, 2866, 2909, 3078, 3919, 4234, 4241
Siphlonuridae (25)	Generally lentic	Swimmers, climbers	Collectors—gatherers		880, 1019, 1020, 1023, 1028, 2112, 2323, 2866, 3076
Edmundsius(1)	Lotic—erosional			California mountains	1028
Parameletus(4)	Lentic—vascular hydrophytes	Swimmers, climbers		West, Ontario, Quebec	1028, 3740
Siphlonisca(1)	Lentic—littoral		Predators (engulfers)	Northeast	1383, 1384
Siphlonurus(19)	Lentic—littoral (vascular hydrophytes or sediments), lotic—depositional	Swimmers, climbers	Collectors—gatherers, scrapers, predators (engulfers; especially *Tanytarsus* in pools), shredders—herbivores?	Widespread	423, 1021, 1028, 1399, 2021, 2752, 2866, 2995, 3629, 4146
Metretopodidae(8)	Generally lotic–erosional	Generally swimmers, clingers	Generally predators (engulfers), collectors—gatherers		880, 1020, 1023, 1028 2323, 2866, 3076
Metretopus(1)	Lotic—erosional		Swimmers, clingers	Canada, Alaska, upper Midwest, Northeast	1028, 1504
Siphloplecton(7)	Lotic—erosional and depositional, lentic—margins	Swimmers, clingers	Collectors—gatherers	Canada, upper Midwest, Southeast	288, 657
Baetidae(121)	Generally lotic erosional and depositional	Generally swimmers, clingers	Generally collectors—gatherers, scrapers		111, 533, 880, 1020, 1023, 1028, 1310, 2112, 2323, 2866, 2939, 3076, 3196, 4268,
Acentrella(5)	Lotic—erosional	Swimmers, clingers	Collectors—gatherers	Widespread	
Acerpenna(4)	Lotic—erosional	Swimmers, clingers	Collectors—gatherers	Widespread	288
Apobaetis(2)	Lotic—depositional (sand)	Swimmers, clingers		California, Colorado Georgia, Kansas, Texas	1028, 2446
Baetis(42)	Lotic—erosional and depositional, lentic—littoral, vascular hydrophytes	Swimmers, climbers, clingers	Collectors—gatherers (detritus, diatoms), scrapers	Widespread	147, 423, 603, 657, 713, 985, 1399, 1852, 2021, 2168, 2439, 2712, 2718, 2746, 2752, 2830, 3584, 3647, 4262, 146, 1330, 2316, 4233, 733, 1330, 2143, 3729, 162, 1430, 2275, 2469, 1504, 3326, 3328, 3608, 734, 784, 925, 1703, 111, 112, 3078, 4439, 890, 1877, 2140, 3012, 3136, 3239, 3607, 4234
Baetodes(5)	Lotic—erosional	Clingers	Scrapers	Texas, Arizona, Oklahoma, Utah	1028
Barbaetis(2)	Lotic—erosional	Swimmers, clingers	Collectors—gatherers	Primarily eastern US, Colorado	

* Emphasis on trophic relationships.
** Subgenera.
*** Unpublished data, K.W. Cummins, Kellogg Biological Station: Oregon State University, University of Maryland.

Table 11A. Continued

Taxa (number of species in parentheses)	Habitat	Habit	Trophic	North American Relationships	Ecological Distribution
Callibaetis(11)	Lentic—vascular hydrophytes	Swimmers, clingers	Collectors—gatherers (filamentous algae)	Widespread	288, 632, 814, 1028, 2866
Camelobaetidius(6) (=Dactylobaetis)	Lotic—erosional and depositional	Swimmers, clingers		West, Northwest, Midwest, Texas	1028
Centroptilum(6)	Lotic—depositional	Swimmers, clingers	Collectors—gatherers (detritus, diatoms), scrapers	Widespread	671, 1023, 1399, 2718, 1310, 2226, 3011, ***
Cloeon(1)	Lentic—vascular hydrophytes	Swimmers, clingers	Collectors—gatherers	Midwest, Northeast	447, 448, 638, 639, 1505, 3012, 3923 2900, 2955, 784
Cloeodes(2)	Lentic and Lotic depositional	Swimmers, clingers	Collectors—gatherers	Arizona, New Mexico	784
Diphetor(2)	Lotic—erosional and depositional	Swimmers, clingers		Widespread	890
Fallceon(1)	Lotic—erosional and depositional	Swimmers, clingers		Widespread	1469, 2218
Heterocloeon(4)	Lotic—erosional	Swimmers,clingers	Scrapers	Midwest, Southeast, Ontario, Colorado	1028, 2156, 2791
Labiobaetis (6)	Lotic —depositional	Swimmers, clingers		Widespread	1942, 2218
Paracloeodes(1)	Lotic—depositional (sand)	Swimmers	Scrapers	West, Midwest, South	1028
Procloeon(29)	Lotic depositional and Lentic—littoral	Swimmers, clingers	Collectors—gatherers (detritus, diatoms) scrapers	Widespread	288
Pseudocentroptiloides(1)	Lotic—-depositional	Swimmers, clingers	Collectors—gatherers	Midwest	
Ametropodidae(3)					
Ametropus(3)	Lotic—erosional (gravel-sand) and depositional (sand) (large rivers)	Burrowers (shallow, with eyes and gills exposed)	Collectors—filterers (using vortices),—gatherers (diatoms)	West, Northwest, Michigan	660, 880, 1020, 1023, 1028, 2866, 3076, 3752
Oligoneuriidae(7)	Generally lotic	Clingers and Burrowers			660
Lachlania(3)	Lotic—erosional and depositional	Clingers	Collectors—filterers Filamentous algae from current	West	880, 1020, 1023, 1028, 1895, 2866, 3076
Homoeoneuria (4) (=*Oligoneuria* of Spieth and Burks)	Lotic depositional (large rivers)	Burrowers	Collectors—filterers	Southwest (north to Utah), lower Mid-west (north to Indiana, Nebraska), Southeast	288, 1028, 3095
Isonychiidae(17)					
Isonychia(17)	Lotic—erosional	Swimmers, clingers	Collectors—filterers (very fine particles), predators (engulfers)	Widespread	288, 489, 653, 671, 1467, 1942, 2460, 3136, 3629, 3917, 4189, ***
Pseudironidae (1)					
Pseudiron (1)	Lotic—depositional (large rivers on sand)	Sprawlers	Predators (engulfers) (uses vortices to search)	West, Midwest Southeast	288, 873, 1019, 1028, 2583, 3092, 3750, 3751, 3753, 4055
Heptageniidae(128**)	Generally lotic and lentic—erosional	Generally clingers	Generally scrapers, collectors—gatherers		284, 533, 880, 922, 1020, 1023, 1028, 1310, 1877, 2112, 2323, 2446, 2866, 3076, 4268
Acanthamola(1)	Lotic- erosional and depositional (large rivers)	Clingers?	Predators?	Saskatchewan, Alberta	

* Emphasis on trophic relationships.
** Subgenera.
*** Unpublished data, K,W. Cummins, Kellogg Biological Station: Oregon State University, University of Maryland.

Table 11A. Continued

Taxa (number of species in parentheses)	Habitat	Habit	Trophic	North American Relationships	Ecological Distribution
Anepeorus (2) (= Spinadis)	Lotic—erosional and depositional	Sprawlers?	Predators (engulfers, primarily midges)	West, Midwest Southeast	873, 1027, 1028, 2218
Cinygma(3)	Lotic—erosional; wood substrate	Clingers	Scrapers, collectors—gatherers	West	96, 1028
Cinygmula(11)	Lotic—erosional	Clingers	Scrapers (diatoms), collectors—gatherers (detritus, diatoms)	West, Northeast, Southeast	603, 890, 1028, 1399, 2746, 4234
Epeorus (19) (=*Iron* + *Ironopsis*)	Lotic —erosional	Clingers	Collectors—gatherers, scrapers	Widespread	96, 111, 112, 603, 1399, 1806, 2716, 2717, 2727, 2316, 3629, 4234, 4588, 2057
Heptagenia(13)	Lotic—erosional	Clingers (swimmers)	Scrapers, collectors gatherers	Widespread	288, 681, 1657, 1942, 2021, 2753, 2829, 2830, 4234, 4242
Ironodes(6)	Lotic—erosional	Clingers	Scrapers, collectors—gatherers	West	1028, 2316, 2866, 3081
Leucrocuta(10)	Lotic—erosional and depositional (slow-flowing warm waters)	Clingers	Scrapers, collectors—gatherers	Widespread	1228, 2752
Macdunnoa(3)	Lotic—erosional and depositional	Clingers	Scrapers, collectors	Midwest, Southeast	288, 2311
Nixe(13)	Lotic—erosional and depositional	Clingers	Scrapers, collectors—gatherers	Widespread	1228, 1399, 2830
Raptoheptagenia (1)	Lotic—erosional	Swimmers, clingers	Predators (engulfers)	Arkansas, Illinois, Indiana, Montana, Nebraska, Ohio, Tennessee, Manitoba, Saskatchewan	2218
Rhithrogena(22)	Lotic—erosional	Clingers	Collectors—gatherers (detritus, diatoms), scrapers	Widespread	111, 112, 1028, 1399, 1430, 2021, 2638, 3629, 4234, 2641
Stenacron(6)	Lotic and lentic erosional	Clingers	Collectors—gatherers, facultative scrapers	East and Central (to Arkansas, Minnesota)	288, 671, 1028, 2232, *** 2638, 2639, 4521,
Stenonema(16)	Lotic and lentic erosional (depositional)	Clingers (under loose cobbles and boulders)	Scrapers, collectors—gatherers	Widespread	317, 671, 1028, 1036, 1942, 2142, 2155, 2232, 2275, 2712, 2752, 2753, 2763, 3341, 3629, 3729, 3777, 3136, 4039, 4283, ***
Arthropleidae (1)					
Arthroplea(1)	Lentic—littoral, lotic depositional	Swimmers, clingers	Collectors—filterers (using vortices)	Northeast (to Wisconsin and Ohio)	425, 1028
Ephemerellidae(90)	Lotic—erosional, some depositional and in macrophytes, a few lentic vascular hydrophytes	Generally clingers, some sprawlers and swimmers	Generally collectors—gatherers (detritus, algae), some scrapers, few shredders (detritivores and herbivores), 1 predator (engulfer)	Widespread	95, 284, 603, 671, 1020, 1023, 1028, 1310, 2323, 2712, 2866, 3076, 3629, 4537
Attenella(4)	Lotic—erosional and depositional	Clingers (detritus, algae)	Collectors—gatherers	West, East	288, 1028, 1635, 4234
Caudatella (5)	Lotic—erosional and depositional	Clingers	Collectors—gatherers (detritus, algae), scrapers?	West	657, 1028, 1635, 2316
Caurinella(1)	Lotic —erosional	Clingers	Collectors—gatherers?	Idaho	

* Emphasis on trophic relationships.
** Subgenera.
*** Unpublished data, K.W. Cummins, Kellogg Biological Station: Oregon State University, University of Maryland.

Table 11A. Continued

Taxa (number of species in parentheses)	Habitat	Habit	Trophic	North American Relationships	Ecological Distribution
Drunella(15)		Clingers, sprawlers	Scrapers, (*D. spinifera* and possibly *D. doddsi* and *D. coloradensis*, predators [engulfers])	Widespread	1635, 2316, 3088, 3239, 4234, 4478, 1029, 1399, 2463, 2464
Ephemerella(32)	Lotic—erosional and depositional	Clingers, some swimmers	Collectors—gatherers, scrapers (*E. infrequens*, a shredder)	Widespread	288, 315, 317, 1028, 1634, 1635, 1636, 1942, 2213, 2275, 2638, 2642, 2643, 2792, 2830, 3080, 2021, 3239, 3752, 3440, 3919, 3922, 4233, 3439, 3608, 4459
Eurylophella(15) (=*Dentatella*)**	Lotic erosional and depositional; Lentic—vascular hydrophytes	Clingers, sprawlers	Collectors gatherers, (*E. funeralis*, a shredder)	West, East	288, 1028, 1541, 1942, 1335, 3136, 3919
Serratella(14)	Lotic—erosional and depositional	Clingers	Collectors—gatherers (detritus)	Widespread	288, 1028, 1635, 2752, 1028, 3088, 4234
Timpanoga(4) (=*Dannella*)**	Lotic—erosional and depositional	Clingers, sprawlers	Collectors—gatherers	West, East	
Tricorythidae(24)	Generally lotic—depositional and lentic—littoral	Generally clingers, sprawlers	Generally collectors—gatherers		880, 1020, 1023, 1028, 1310, 2323, 2866, 3076
Tricorythodes(14)	Lotic—depositional, lentic—littoral (sediments)	Sprawlers, clingers	Collectors—gatherers	Widespread	1028, 1399, 1532, 2168, 2439, 2610, 2897, 288, 890, 925, 3629, 712, 1469, 1942, 3136, 4478, 4480, ***
Leptohyphes(10)	Lotic—depositional	Clingers		Utah, Texas, Maryland	1028, 1469
Neoephemeridae(4)					880, 1020, 1023, 1028, 2866, 3076
Neoephemera(4)	Lotic depositional	Sprawlers, clingers	Collectors—gatherers?	East	288, 1028
Caenidae(26)					880, 1020, 1023, 1028, 3011
Amercaenis (1)	Lotic depositional	Sprawlers	Collectors—filterers	Midwest, Southeast	2112, 2323, 2866, 3076
Brachycercus(12)	Lotic depositional	Sprawlers	Collectors—gatherers	East, Central, Southwest, Idaho and Wyoming	288, 1028
Caenis(11)	Lotic depositional, lentic—littoral (sediments)	Sprawlers, climbers	Collectors—gatherers, scrapers	Widespread	632, 671, 2439, 2739, 3584, 712, 2034, 3212, 1942, 3136, ***
Cercobrachys (2)	Lotic depositional	Sprawlers	Collectors—gatherers	West, Southeast	288
Baetiscidae(12)					880, 1020, 1023, 1028, 2112, 3071, 2323, 3076, 3127
Baetisca(12)	Lotic depositional (sand, with detritus)	Sprawlers, clingers	Collectors—gatherers, scrapers	Widespread	282, 589, 1028, 2307, 3093, 3094, 3629
Leptophlebiidae(72)	Generally lotic—erosional	Generally swimmers, clingers	Generally collectors—gatherers, scrapers		284, 288, 880, 1020, 1023, 1028, 1904, 2112, 2323, 2866, 3010, 3076, 1310, 4268
Choroterpes(11)	Lotic erosional and depositional, lentic littoral (sediments)	Clingers, sprawlers	Collectors—gatherers, scrapers	Widespread	671, 2446, 2606, 2866, 3011, 3012, 4234
Farrodes(1)	Lotic	Clingers, sprawlers		Texas	1028, 873

* Emphasis on trophic relationships.
** Subgenera.
*** Unpublished data, K.W. Cummins, Kellogg Biological Station: Oregon State University, University of Maryland.

Table 11A. Continued

Taxa (number of species in parentheses)	Habitat	Habit	Trophic	North American Relationships	Ecological Distribution
Habrophlebia(1)	Lotic—depositional and erosional			East	288, 1028, 2274, 2275, ***
Habrophlebiodes(4)	Lotic—erosional and depositional	Swimmers, clingers, sprawlers	Scrapers, collectors—gatherers	East, Midwest	288
Leptophlebia(10)	Lotic—erosional (sediments and detritus)	Swimmers, clingers, sprawlers	Collectors—gatherers (fine particles)	Northwest, Midwest, East	285, 422, 423, 661, 663, 711, 1640, 1905, 2739, 2746, 2843, 248, 288, 2275, 3546, 3629, 3136, 3239, 3919, 3925, ***
Neochoroterpes (3)	Lotic—erosional and depositional, lentic, littoral (sediments)	Clingers, sprawlers	Collectors—gatherers, scrapers (fine particles)	Southwest	
Paraleptophlebia (35)	Lotic—erosional (sediments and detritus)	Swimmers, clingers, sprawlers	Collectors—gatherers (coarse detritus, diatoms), faculatative shredders—detritivores	Widespread	288, 603, 711, 1028, 1399, 1430, 1903, 3629, 638, 2556, 3608, 1810, 2316, 3078, 1636
Thraulodes(5)	Lotic—erosional	Clingers, sprawlers		Southwest	1028, 4040
Traverella(4)	Lotic—erosional and depositional	Clingers	Collectors—filterers	West, Southwest, Indiana, and North Dakota	1028, 4242
Behningiidae(1)					880, 1020, 1023, 1028, 1032, 1310, 2866, 3076, 3101, 1180, 1181, 3919, ***
Dolania(1)	Lotic–depositional (sand)	Burrowers	Predators (engulfers, Chironomidae)	Southeast	1616, 1028, 4055, 288
Potamanthidae (4)					2112, 2866, 3076
Anthopotamus(4)	Lotic—depositional	Burrowers	Collectors—filterers	East, Midwest	1028, 1905, 2652, 1078, 2586, 3629, 192, 2587, 2799
Ephemeridae(15)					284, 880, 1020, 1023, 1028, 2112, 2323, 2351, 1310, 2866, 3076
Ephemera(7)	Lotic and lentic depositional (sand-gravel)	Burrowers	Collectors—gatherers, predators (engulfers), filterers	Widespread	135, 419, 671, 2687, 2813, 1657, 2995, 3629, 2213, 2216, 3136, 3919, ***
Hexagenia(5)	Lentic and lotic depositional (sand-silt)	Burrowers	Collectors—gatherers (fine particles, possibly also filter at mouth of burrow)	Widespread	1550, 1818, 1819, 1871, 1567, 2752, 2866, 3629, 1657, 4641, 4642, 4643, 288, 692, 2069, 3254, 1388, 3013, 3729, 4353, 424, 1302, 1303, 1304
Litobrancha(1)	Lotic depositional (small streams and rivers)	Burrowers		East	584, 1028
Pentagenia(2)	Lotic depositional (hard clay banks, large rivers)	Burrowers	Collectors—gatherers, passive filterers	Central, Southeast	660, 1028, 2069
Polymitarcyidae(6)					1020, 1023, 1025, 2687, 2866, 3076
Ephoron(2)	Lotic—erosional and depositional, lentic—	Burrowers	Collectors—gatherers, filterers	Widespread	419, 1028, 1905, 3629, 1386, 3136
Campsurus(1)		Burrowers		Texas	1028
Tortopus(3)	Lotic depositional (hard clay banks, large rivers)	Burrowers		Midwest (north to Manitoba), Southeast	288, 660, 2124, 3606

* Emphasis on trophic relationships.
** Subgenera.
*** Unpublished data, K,W. Cummins, Kellogg Biological Station: Oregon State University, University of Maryland.

ODONATA

Minter J. Westfall, Jr.
University of Florida, Gainesville

Kenneth J. Tennessen
Florence, Alabama

INTRODUCTION

Adult Odonata, known as dragonflies and damselflies, fly conspicuously near almost any body of fresh water on sunny, warm days. The larvae occupy a great diversity of aquatic habitats, although in general the order is most speciose and abundant in lowland streams and ponds. A few species tolerate brackish water. Because the conspicuous adults are colorful, long-bodied, and agile, they have been given imaginative names such as "mosquito hawks", "devil's darning needles", and "snake doctors", along with many interesting superstitions. Some people are still fearful that they can sting, though they are incapable of this. Mosquitoes comprise an important part of the diet of adults and larvae of many odonates, but other insects are preyed upon also, many of which are considered pests. Odonata are thus regarded as beneficial predators. Adult Odonata capture flying insects with their spiny legs or with their mouthparts. They have especially keen eyesight, the compound eyes being unusually large and many-faceted. The larva thrusts out its long, hinged labium to grasp its prey. The larvae are an important part of aquatic food webs involving invertebrates, fish and other aquatic vertebrates. Dragonflies are useful subjects for research on the effects of stream pollution and in certain medical fields. The worldwide number of described species of Odonata is about 5,500, with approximately 650 species occurring in North America.

The scientific name, Odonata, is derived from the Greek "odon", meaning tooth, and apparently refers to the sharp teeth on the mouthparts. Dragonflies were once classified in the order Neuroptera because of the superficial resemblance in wing venation. However, the Odonata are hemimetabolous insects, and the wings are unique in venational characteristics. Wing structure is extremely important in odonate maneuverability, as much of their time is spent in aerial pursuit of food, habitat, or mates, and in escape from predators; some species oviposit while flying. Flight speeds have been estimated to be between 25 and 35 km per hour, possibly as high as 56 km per hour.

Flight habits vary with the species. Some rarely alight during hours of flight ("fliers"), whereas others occupy certain perch sites from which they go forth to capture food or defend an area from other dragonflies ("perchers"). Males of some species (e.g., *Epitheca*) defend a territory over a stretch of water where eggs will be laid, and often patrol back and forth across this area. Many species fly only in bright sunlight and quickly perch when the sun is obscured by clouds. A preponderance of cloudy days can discourage flights for food to the extent that individuals can be decimated before sunny weather returns. Many species of Gomphidae fly for only a few hours at a particular time of the day, such as from mid-morning until noon, and then again from mid-afternoon until late afternoon. A few species (e.g., *Neurocordulia* spp., which feed upon mayflies, *Gynacantha nervosa* Rambur, and *Enallagma vesperum* Calvert) fly briefly only at dawn or dusk. Some species will fly in light rain, such as *Enallagma pollutum* (Hagen) and *Aphylla williamsoni* (Gloyd). *Tauriphila argo* (Hagen) sometimes flies tree-top high, defying the collector. Migrations of dragonflies have been observed but rarely recorded in detail. Longfield (1948) described the appearance of thousands of individuals of *Sympetrum* sp. on the south coast of Ireland over a period of six weeks, probably traveling 500 miles from the coast of Spain. Federly (1908) reported migrations of *Libellula quadrimaculata* Linnaeus in Finland. Flight seasons are also variable in onset of emergence and duration. Some temperate species have an explosive early emergence (in April or May) and soon disappear, whereas others continue to emerge throughout the summer. In subtropical areas, some species emerge throughout the year. The complete flight period of most individuals lasts only a few weeks, although some may live for up to three months (e.g., *Hagenius brevistylus* Selys).

Mating takes place in flight or while perched, over water, near water, or some distance from the water, and at various times of day. Courtship has been observed in two North American families, the Calopterygidae and Libellulidae. In other families, males pursue and quickly grasp females without any apparent behavioral or visual display. Prior to mating, the male must transfer sperm from the sperm duct openings on the venter of segment nine to the penis on segment two. This transfer is performed either before or right after a female is taken into tandem. The male usually grasps the dorsum of a female with his legs and then clasps her by the head (Anisoptera) or prothorax (Zygoptera) with his caudal appendages (modified cerci and epiproct or paraprocts). The female responds by swinging the tip of her abdomen forward to engage the male's accessory genitalia on segment two, where sperm is deposited in the spermatheca. The pair forms a lopsided, heart-shaped loop. Details of the mating behavior of many species are still unknown.

Oviposition is performed by females in tandem with males or by females alone. Tandem pairs may fly over and tap the water releasing eggs, or the pair may perch on the substrate and the female insert the eggs. In some species, the pair may fly over the water where the male releases the female to fly down and release a small number of eggs, after which she flies upward and is again taken into tandem. In many species of Libellulidae, males energetically guard their mates while she lays eggs. Because eggs are not exposed to sperm until release from the female body, a male that continues to clasp his mate in tandem or guards her from a distance during oviposition is effectively ensuring that she lays eggs fertilized by his sperm. Males of many species are capable of removing or packing sperm from a previous mating stored within the spermatheca; this act is a form of sperm competition.

The number of eggs laid by a female in one egg-laying bout varies from a few hundred to a few thousand, the maximum recorded being 5,200. A few corduliid species which lay eggs in gelatinous strings will deposit their string where a string has already been laid, creating masses of hundreds of thousands of eggs that have been credited erroneously to a single female. Zygoptera, Aeshnidae and Petaluridae have sharp ovipositors that insert eggs into plant tissue above or below the water (Fig. 12.8), presumably an adaptation to avoid predation and mediate temperature extremes. On occasion, this endophytic trait has resulted in excessive damage to plant tissue by large numbers of females, especially of the family Lestidae. Needham (1900) reported considerable damage to stems of "blue flag" (*Iris versicolor* Linnaeus) by *Lestes* females. Some species, such as *Calopteryx aequabilis* Say and *Enallagma aspersum* (Hagen), submerge entirely, sometimes descending to a depth of 30 cm or more to oviposit in plant stems, remaining submerged for as long as 30 min. before floating to the surface. The male may descend and remain with the female or rise to the surface soon after submerging. They are able to derive oxygen from the air adhering to the hairs of the body surface and that which is trapped by the wings. Cordulegastridae have elongate, spike-like ovipositors used to insert eggs into silt and sand of shallow streams, whereas Gomphidae, most Corduliidae and Libellulidae have reduced ovipositors (vulvar laminae) to release or scatter eggs in water or mud, or in some *Sympetrum* sp. to drop them from the air into the water or onto moist soil.

Odonata have an incomplete metamorphosis with the immature stage (nymph, naiad, larva) going through 10-15 instars. Wing pads appear after the sixth or seventh molt, becoming swollen late in the final instar, a sign of imminent emergence. Larvae in the last few days of the final instar eat little or no food and are sluggish, while great internal changes are taking place in preparation for emergence to the adult stage. Snodgrass (1954) likens this period to the pupa of holometabolous insects. The larval stage lasts from a few weeks in some species of Zygoptera and Libellulidae to about five years in the Petaluridae. Some univoltine species have a synchronized emergence, and all adults disappear after a few weeks; others continue to emerge throughout the warmer seasons, or even throughout the year farther south. The most favorable time for emergence appears to be a dry, sunny period following wet, cold weather. Most Odonata larvae are found in permanent bodies of water, although some have adapted to temporary aquatic habitats. These species (e.g., *Lestes*) usually have short life cycles. A few odonates are semi-aquatic, living in bog moss or under damp leaves in seepage areas. In tropical regions, several species are adapted for life in water that collects in bromeliads or in tree holes, whereas others are able to withstand long periods of desiccation.

Most odonate eggs hatch within 8-30 days, depending upon weather conditions. Eggs laid in plant tissue above a drying temporary pond do not hatch until the pond fills again. In the temperate zone, some species go through a winter diapause in the egg stage. Upon hatching, the unique *pronymph* (usually defined as instar one) molts almost immediately to produce the second instar larva. These small, mobile larvae are nourished for a few days by yolk retained in the gut, but very soon begin to feed on protozoa or other minute aquatic animals. Dragonfly larvae are generalized carnivores, feeding on any aquatic animal of an appropriate size that they can capture. Feeding behavior has been described in three phases: (1) detecting prey and adjusting position in relation to it; (2) ejecting the labium to grasp the prey; (3) employing the mandibles to devour the prey. The labium, when extended, is approximately as long as the fore legs, but at rest may mask the face up to the eyes. It is designed for grasping and holding prey. It is protruded toward prey and drawn back so swiftly that an observer can not follow it. The sharp hooks and spines of the palpal lobes, situated at the front of the labium, clutch the prey; the maxillae and mandibles cut and manipulate it for ingestion. Any loose fragments are retained in the prementum and also consumed. The labial mechanism, though efficient, requires enough room to reach out for prey and is occasionally entangled in threads of filamentous algae and other plants, probable reason for the scarcity of larvae in dense mats of such plants.

Larvae usually conceal themselves by either burrowing in substrate, sprawling amongst fine sediment and detritus, or climbing on vascular plants. With a few exceptions, species of Petaluridae and Gomphidae burrow in mud, coarse sand, or light silt, and characteristically have a flat, wedge-shaped head and short antennae. The fore and middle legs are much shorter than the hind legs and are modified in some species with stout, curved hooks for shallow burrowing. The end of the abdomen generally is upturned slightly, allowing it to protrude above the mud or silt into the water for respiration. The genus *Aphylla* (Fig. 12.96), which burrows more deeply, has a cylindrical abdomen with an elongated tenth segment forming a siphon. In rare instances, some Libellulidae have been found buried beneath the surface of sand, possibly to escape predators.

Sprawlers are typically more active hunters (with some exceptions, e.g., the sluggish *Hagenius brevistylus* and *Macromia* spp.), having longer, often spider like legs. Numerous setae give them a hairy appearance, and particles of mud and silt adhere to and camouflage the larvae. The setae are probably also tactile organs that assist in prey detection. Color is advantageously protective in patterns of mottled greens and browns. The body is generally flatter than that of the burrowers. The Cordulegastridae includes species intermediate between the sprawlers and burrowers. They are found in clean sand or silt, buried by sand swept over the subcylindrical bodies by the lateral kicking movements of the legs, which are slender and without burrowing hooks. When deep enough, sand covers them except for the tops of the eyes and the tip of the abdomen.

Most Zygoptera and Aeshnidae are classified as climbers, lurking in vegetation or resting on stems of aquatic plants. These larvae stalk their prey, halting their pursuit when the prey stops moving. Some weed dwellers have developed specialized forms of protective coloration that may match their immediate background. *Aeshna* larvae are apparently able to distinguish different colors (Koehler 1924). Larvae of *A. grandis* Linnaeus are reported to become green in summer and brown in winter, and *Anax* spp. possess a highly characteristic type of juvenile coloration. Many Zygoptera position their long bodies parallel to the long axis of a stem, and can move quickly around or along it to avoid predators. A few species of Protoneuridae, Coenagrionidae, Aeshnidae, and Corduliidae (especially *Neurocordulia*), are clingers, holding tenaciously to submerged rocks, sticks, and roots.

Adult Odonata respire through an open tracheal system, i.e., directly through the thoracic and abdominal spiracles. The larvae have a closed respiratory system, spiracles persisting but not functional. In Anisoptera larvae, respiration is chiefly rectal. The expansion and contraction of the rectal walls forces an inflow and outflow of water through the anus, with gas exchange occurring through the thin cuticle and epithelial syncytium of the gills. Respiration in Zygoptera is chiefly cutaneous, the caudal lamellae (gills) serving as supplementary respiratory structures. In later instars, oxygen is absorbed also through the surface of the wing pads. Another function of the rectal pump system in Anisoptera nymphs is locomotion, achieved by a sudden expulsion of water through the anus, useful for escape or for quickly burrowing into sand (e.g., *Progomphus*).

Predators of larvae of Odonata include aquatic birds, fish (e.g., Percidae, Centrarchidae, Salmonidae), and large predaceous insects. Predators of adult Odonata include many species of birds (e.g., falcons, kingbirds, kingfishers, herons, terns, gulls, sandpipers, blackbirds, swallows, swifts, martins, grackles, and red-winged blackbirds), amphibians, bats, spiders, wasps, other dragonflies, and other insects. Larvae of *Aeshna cyanea* Müller have been observed climbing up plant stems to grasp and devour teneral adults of their own species. Because of the predation activity during daylight hours, emergence most often occurs at night. Weather is also responsible for considerable mortality. Cloudy, rainy days may shorten feeding hours, and severe windstorms may drive adults onto large lakes or to sea, where they drown and are washed ashore in large numbers.

Immature stages of water mites (*Hydracarina* spp.) may parasitize lentic odonate larvae and prey on odonate eggs. Several species of Hymenoptera parasitize eggs, especially those of species that oviposit endophytically (Aeshnidae and Coenagrionidae). The adults of Ceratopogonidae (biting midges) and some internal parasites are also causes of mortality. Comprehensive accounts of the biology and ecology of the Odonata are given by Tillyard (1917), Walker (1953), and Corbet (1963, 1980).

EXTERNAL MORPHOLOGY

Larvae

Although Odonata larvae appear different in form from the adults, the body plan is basically the same and body parts mostly bear the same names. Larvae are cryptically colored with body shapes and structures adapted for different aquatic microhabitats. Compared with the adult, the head is smaller and not so freely movable. The three segments of the thorax are more aligned and equal in size than in the adult, and the legs are farther apart and are better adapted for walking. Larvae of Zygoptera are more slender than those of the Anisoptera, the head is wider than the thorax or abdomen, and the antennae are relatively longer than in Anisoptera.

Head: The most unique structure of the larval head is the *labium* or lower lip. It is folded upon itself at midlength and turned backward (caudally) beneath the front legs (Figs. 12.33-12.34). At the front are a pair of strong, hinged *palpal lobes* or labial palpi (formerly called *lateral lobes*) armed with hooks, spines, teeth, and raptorial setae that vary with family and genus (Figs. 12.22, 12.26). Each palpal lobe is two-segmented: the larger first segment articulates with the prementum and is prolonged apically to meet the one on the opposite side. The palpal lobe usually bears strong, raptorial *palpal setae* (formerly called *lateral setae*) on its dorsal surface (Fig. 12.27). The lobe often terminates in an apical projection, the *end hook* (Fig. 12.22). The second segment is a slender, curved or sharp-pointed *movable hook,* which may bear strong, raptorial setae and is articulated at a movable joint with the first segment on its outer margin; it usually overlaps the hook of the opposite side (Figs. 12.13, 12.16, 12.28-12.29). The central portion of the front margin of the prementum may have a projection known as the *ligula* (Figs. 12.22, 12.26). On each side of the midline prominent *premental setae* may be present (formerly called *mental setae*), and their number and placement are used taxonomically (Figs. 12.13, 12.26, 12.30). The antennae are usually composed of six or seven segments, but in the Gomphidae the number is reduced to three or four, the fourth usually being vestigial (Figs. 12.90-12.94).

Thorax: The prothorax is freely movable as in the adult, although it is larger in proportion to the rest of the thorax and less contracted. The propleura may be useful taxonomically when produced as a pair of prothoracic *supracoxal processes.* The mesothorax and metathorax are solidly fused to form the *synthorax.* The wing cases are moved more caudad than in the adult, and the metepisterna meet in front of them. The legs are still in a position to permit walking, not forced so far forward as in the adult. The legs are usually shorter and more robust than in the adult, although in some groups, such as Macromiinae, they are quite long. Setae and spines, so characteristic of adult legs, are missing. The first two pairs of legs of some Anisoptera (especially Gomphidae) may have burrowing hooks (Fig. 12.21). In late instars the two inner wing cases enclose the front wings and the two outer cases enclose the hind wings; thus the ventral side of the hind wings is exposed. The longitudinal veins are clearly seen in the swollen wing cases prior to emergence, and in some species, before the wings are crumpled and swollen, triangles, crossveins, and other features also may be seen clearly. If the wing cases are removed from fresh, mature larvae, slit open, and placed into a weak potassium hydroxide solution, the wings will expand to their full size and adult venational characters can be observed.

Abdomen: The larval abdomen of Anisoptera is always much shorter and stouter than the adult's, whereas in Zygoptera it is slightly shorter and stouter. The abdomen may be long, slender, and tapering to the end, as in Aeshnidae and the Zygoptera, or blunt-tipped, even broad and subcircular, as in some Gomphidae (*Hagenius* sp.) (Fig. 12.101) and in the Macromiinae. In most Anisoptera but very few Zygoptera, the lateral abdominal carina ends in a sharp projection, the *lateral spine,* usually longest on segment 9 (Fig. 12.23). There may be variously shaped, middorsal projections or humps (*dorsal hooks*) on the posterior margin of the segments (Figs. 12.23, 12.25). On the sternum of segment 2 of the male (even in exuviae), the rudiments of the adult genitalia often may be seen as swellings, and the developing ovipositor of the female is visible in the sternal region of segment 9. The terminal anal pyramid of Anisoptera is composed of a dorsal *epiproct* (formerly called *superior appendage*) and two ventral *paraprocts* (formerly called *inferior appendages*), which surround the anal opening from the rectal chamber (Fig. 12.23). Two *cerci* (formerly called *lateral appendages*) develop later on each side of the epiproct. In the Zygoptera, the epiproct and paraprocts are specialized to form the three *gills* (caudal lamellae), one median and two lateral (Fig. 12.12), the shape, tracheation, and markings of which are important taxonomic characters. In the Coenagrionidae, the gills usually have a nodus or point on the edge of the gill where there is a change in the marginal setae, often marked by a notch. The antenodal setae are short and stiff, whereas the postnodal setae may be only fine hairs. The setae are more numerous and stronger on the dorsal margin of the median gill and on the ventral margins of the lateral gills. The cerci are quite small, but in some Zygoptera the shape may be used to separate two species that otherwise appear almost identical. All North American species have easily-shed gills that regenerate when broken off. Because these regenerated gills may be much smaller and quite different from the original ones, gills must be used with caution in making taxonomic identifications.

Adults

Detailed accounts of adult odonate structure are given in Walker (1953) and Needham and Westfall (1955), but structures used in the keys are discussed below.

Head: The huge, many-faceted compound eyes, which occupy a large portion of the head, are more widely separated in Zygoptera (Fig. 12.3) than in Anisoptera (Fig. 12.4). The labium or lower lip has a median cleft in some families (Fig. 12.42), but not in others (Fig. 12.43). The short and bristle-like antennae are of little use in taxonomy (Fig. 12.4). The top of the head between the compound eyes, the *vertex,* bears three *ocelli*. Behind the vertex and fused with it is the *occiput,* forming the rear of the head. In some Anisoptera, as the Aeshnidae (Fig. 12.4), the compound eyes meet and separate the vertex from the occiput. Many Zygoptera have conspicuous, pale *postocular spots* behind the ocelli.

Thorax: The head is freely movable, attached to the small prothorax by a very slender neck. The dorsal sclerite, or *pronotum,* is divided into three lobes. In most Zygoptera during mating, either the middle or posterior lobe of the female's pronotum is grasped by the inferior appendages of the male, and it may bear structural modifications matching the specific form of the male appendages. The meso- and metathorax of Odonata are fused into a *pterothorax* (or synthorax). The dorsal sclerites or plates (*nota*) and the ventral ones (*sterna*) are much reduced. The side sclerites (*pleura*) slope backward and upward, with the legs displaced forward and the wings backward. Immediately anterior to the first abdominal sternum is the *intersternum,* which is sometimes raised into a distinct tubercle. The mesopleural suture is termed the *humeral suture.* The intersegmental or interpleural suture (between meso- and metathorax) is called the first *lateral,* and the metapleural suture, the *second lateral.* The middorsal carina extends the length of the dorsum. There is often an *antehumeral stripe* between the middorsal carina and the humeral suture, and stripes are often present on and between the thoracic sutures.

Legs: Each of the six legs is composed of a basal *coxa* (Fig. 12.118), followed by a thinner *trochanter.* The *femur* and *tibia* (Fig. 12.118) are long and armed with various spines and hairs. The *tarsus* is composed of three segments, increasing in size from first to third. The last segment ends in a pair of *tarsal claws* that may have hooks on the ventral margin. Some species of Corduliidae possess conspicuous *tibial keels.*

Wings: The two pairs of wings of Odonata are well developed. In Anisoptera the wings are held horizontally when at rest, whereas in most Zygoptera they are held together above the body. In some Zygoptera genera, such as *Lestes* and *Chromagrion,* the wings are held partly spread, whereas some Libellulidae may flex the wings downward below horizontal. In the Zygoptera, the front and hind wings are about equal in length and similar in form (Fig. 12.1). Both wings are narrowed toward the base and usually stalked or petiolate, i.e., each is paddlelike with a proximally narrowed portion and an expanded distal portion on the posterior side (Fig. 12.68). In the Anisoptera (Fig. 12.2) the hind wing is much broader at the base than the front wing and has a more or less distinct *anal margin,* which may meet the hind margin of the wing at a distinct angle, the *anal angle,* as in males of most families. The anal margin may meet the hind margin of the wing in a broad curve, as in the females of all families and males of the Libellulidae (Fig. 12.137). An opaque *membranule* borders the anal margin of the hind wing of Anisoptera (Fig. 12.83). Most Odonata have an anterior, enlarged, opaque cell, the *pterostigma* (Fig. 12.2), near the apex of each wing. It is often wider than the neighboring cells and pigmented.

Wing venation: The most stable and reliable characters for identifying the major groups of Odonata to genus are found in the wing venation (Needham 1903, 1951; Tillyard 1917; Munz 1919). The veins are either longitudinal or crossveins. The longitudinal veins appear early in the wing pads of the nymphs, generally along the path of the tracheae. The crossveins, which appear later, are more variable than the longitudinal veins.

The following account, based on Needham (1903b), applies chiefly to the Anisoptera, but many structures are common to both suborders. The *costa* (Fig. 12.2) comprises the front border of the wing. The notch near the middle of that border indicates the position of the thickened crossvein called the *nodus.* The *subcosta,* immediately behind the costa, ends at the nodus. The *radius* (R) is a strong vein parallel to the costa. At its base it is fused with the *media* (M) as far as the *arculus,* and at the nodus it gives off a strong branch posteriorly, the *radial sector* (Rs). The radial sector descends by way of the subnodus and the *oblique vein* (Fig. 12.83) to its position behind the first two branches of media (M_1 and M_2). The media is composed of four branches, and as noted, it is fused at the base with the radius. At the arculus the media descends to meet a crossvein and then bends sharply outward toward the wing tip. At the arculus it gives off a branch rearward (M_4), another (M_3) farther distad, and still another at the subnodus (M_2). From the terminal branch, M_1 runs parallel to the main radial stem (R_1), whereas M_2 parallels the radial sector. The two branches at the first forking are called the *sectors of the arculus.* The *cubitus* (Cu) is a two-branched vein (Fig. 12.83). It is straight to the

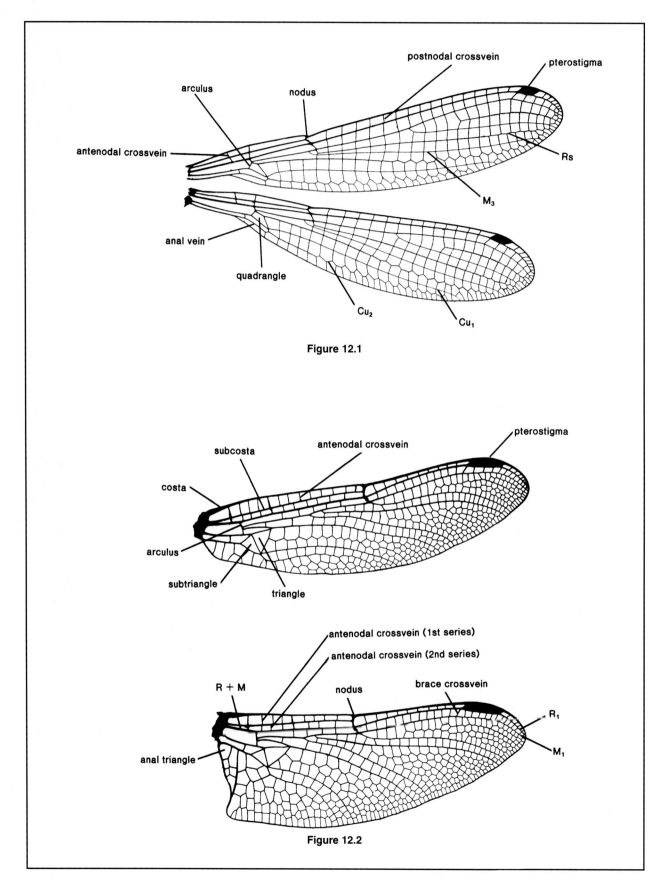

Figure 12.1. Wings of male *Argia moesta* (Hagen) (Coenagrionidae).

Figure 12.2. Wings of male *Ophiogomphus carolus* Needham (Gomphidae).

arculus, but soon turns sharply rearward, forming the inner (proximal) side of the *triangle* (Fig. 12.83). At the hind angle of the triangle, the cubitus is forked and its branches (Cu_1 and Cu_2) run more or less parallel to the hind margin. The *anal vein* (A), when present, lies posterior to the cubitus, and has three branches, A_1-A_3. The path of all these longitudinal veins and their degree of development varies in different groups. At the proximal end of the pterostigma, a crossvein, the *brace vein*, may be present (Figs. 12.82-12.83). Two crossveins between M and Cu complete the triangle and form the front (appears to be a part of Cu) and outer sides. A *cubito-anal crossvein* forms the front side of the *subtriangle* and added cubito-anal crossveins are present in some genera (Fig. 12.82). Behind the costa, as far as the nodus, are two series of *antenodal crossveins*. The first and sixth hind wing crossveins of the first series, anterior to subcosta (Fig. 12.83), are thickened and usually they are the only ones in this series that are continuous from the costa to R + M. This arrangement occurs only in some groups; in other families and genera, the antenodals are not thickened and the first and second series are aligned. The single series of *postnodal crossveins* extends from the nodus to the pterostigma between the costa and R_1. Bordering the triangle on its front side is the *supertriangle* (or supratriangle; Fig. 12.82). All triangles may or may not contain crossveins. A stout, perpendicular *anal crossing* (Ac) connects Cu and A near the wing base; this is especially important in Zygoptera (Figs. 12.52, 12.72). In both wings the cells behind the anal vein, between the hind angle of the triangle and the wing base, are the *paranal cells* (Fig. 12.82). In the male hind wing of some genera, the basal one or two paranals may lie in a strongly bordered *male anal triangle* (Figs. 12.82, 12.86-12.87). Below the radial sector may be a *radial planate* (Figs. 12.82-12.83), formed by a number of cells lined up evenly and subtending one or more rows of cells. Below M_4 may be a *median planate*, and below M_1 toward the apex of the wing, an *apical planate* (Fig. 12.140). In some genera (e.g., *Hagenius*), a *trigonal planate* runs from the triangle toward the wing tip (Fig. 12.113). Between M_1 and M_2 near the nodus, the Libellulidae have a *reverse vein*, which may be strongly slanted as compared with the adjacent crossveins (Fig. 12.138). An *anal loop*, which is markedly foot-shaped in the Libellulidae may be formed between A_1 and A_2 of the hind wing. It has a thick midrib between A_1 and A_2, which may be strongly angulated at the heel (Fig. 12.139) or more nearly straight (Fig. 12.138). Proximal to the arculus is a clear space (with or without crossveins), the *midbasal space* (median space), bounded by R + M in front and Cu behind.

Figures of the wings of Zygoptera show readily apparent differences from the Anisoptera. For example, a *quadrangle* replaces the triangle of the Anisoptera, and there is no subtriangle, supertriangle, or anal loop (Fig. 12.1).

Abdomen: The abdomen consists of ten complete segments, which are usually subcylindrical, and a portion of an eleventh segment bearing terminal appendages (Fig. 12.5). Segments 1 and 10 are shorter than the intervening ones. In Anisoptera, segments 2 and 3 are swollen, and in the male these bear on the ventral side the unique copulatory organs in the *genital pocket* (genital fossa). The anterior margin of the genital pocket, the *anterior lamina*, may be flat, as in *Stylurus* sp., or project above the rim of the genital pocket. It also may be bilobed and armed with spines or denticles. In the pocket is located the penis and accessory genitalia, including two pairs of *hamules* (anterior and posterior) that assist in copulation and often constitute good diagnostic characters. In some families of Anisoptera the males possess earlike projections (auricles) on the sides of segment 2, which may bear scattered, stout, black spinules. The anal appendages of Anisoptera males consist of two dorsal *cerci* (formerly termed superior appendages) and a ventral *epiproct* (formerly termed inferior appendage) (Figs. 12.5, 12.9); males of Zygoptera also have two dorsal cerci, and two ventral *paraprocts* (Fig. 12.7). The genital aperture of the male is on the sternum of segment 9. In Aeshnidae and the Zygoptera, the ovipositor of the female is composed of two pairs of sharp, pointed stylets and a pair of covering genital valves, tipped with a slender stylus on each side (Fig. 12.8).

KEY TO THE SUBORDERS OF ODONATA

Larvae

1. Larvae slender, head wider than thorax and abdomen; 3 long, caudal tracheal gills at tip of abdomen (Fig. 12.12) ....***Zygoptera*** (p. 171)
1'. Larvae stout, head usually narrower than thorax and abdomen; 5 short, stiff, pointed appendages at tip of abdomen (Figs. 12.21, 12.24) ....***Anisoptera*** (p. 171)

Adults

1. Front and hind wings similar in size and shape, each with a quadrangle instead of a triangle and subtriangle (Fig. 12.1); eyes separated by more than their own width (Fig. 12.3); males with 2 inferior appendages (paraprocts, Fig. 12.7); females with a fully developed ovipositor, bearing styli (Fig. 12.8); when perched, wings held together above abdomen or only partly spread ....***Zygoptera*** (p. 171)
1'. Front and hind wings dissimilar in size and shape, the hind wing considerably wider at base than the front wing, each having a triangle and subtriangle (Fig. 12.2); eyes meeting middorsally or not separated by a space greater than their own width (Fig. 12.4, 12.41); males with a single inferior appendage (epiproct, Fig. 12.9), which may be bifid (Fig. 12.115); females with or without an ovipositor; when perched, wings held horizontally ....***Anisoptera*** (p. 174)

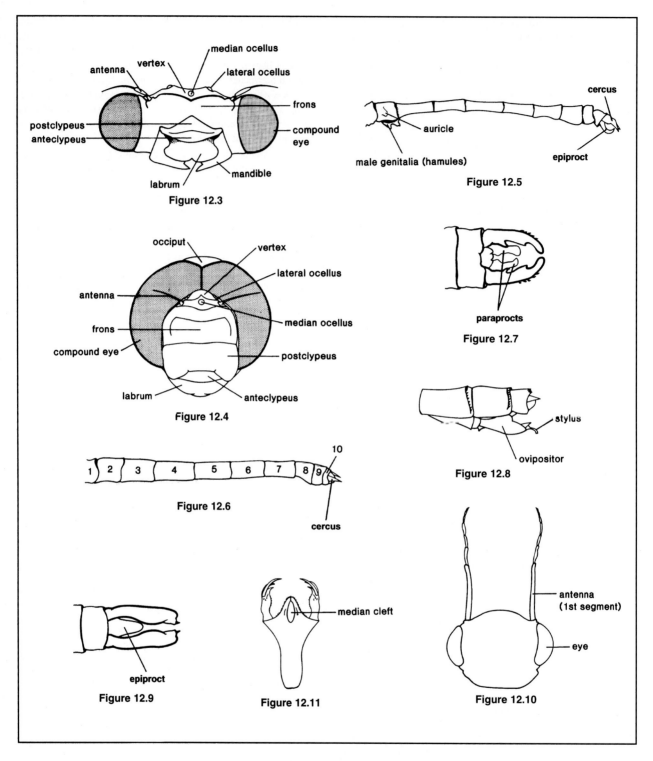

Figure 12.3. Anterior view of head of adult *Calopteryx* sp. (Calopterygidae).

Figure 12.4. Anterior view of head of adult *Aeshna* sp. (Aeshnidae).

Figure 12.5. Lateral view of male abdomen of adult *Erpetogomphus* sp. (Gomphidae).

Figure 12.6. Lateral view of female abdomen of adult *Erpetogomphus* sp. (Gomphidae).

Figure 12.7. Dorsal view of anal appendages of adult male *Lestes* sp. (Lestidae).

Figure 12.8. Lateral view of abdominal segments 8-10 of adult female *Lestes* sp. (Lestidae).

Figure 12.9. Dorsal view of anal appendages of adult male *Aeshna* sp. (Aeshnidae).

Figure 12.10. Dorsal view of head of larva *Calopteryx* sp. (Calopterygidae).

Figure 12.11. Prementum of labium of larva *Hetaerina* sp. (Calopterygidae).

KEY TO THE FAMILIES OF ZYGOPTERA

Larvae

1. First antennal segment greatly elongate, as long as the combined length of the remaining segments (Fig. 12.10); prementum with deep median cleft (Fig. 12.11); lateral caudal gills triangular in cross section ....*CALOPTERYGIDAE* (p. 176)
1'. First antennal segment not so elongate, distinctly less than the combined length of the remaining segments (Fig. 12.12); prementum with small closed cleft or none (Fig. 12.13); lateral caudal gills flat or somewhat inflated 2
2(1'). Prementum distinctly petiolate (stalked) and spoon-shaped, the narrow proximal part (nearest the body) as long as or longer than the expanded distal part (farthest from the body) (Fig. 12.16); movable hook of each palpal lobe with 2 or 3 setae; usually 5-8 dorsal premental setae on each side of median line (when only 4 or 5 prementals present, then with only 3 palpal setae) (Fig. 12.16) ....*LESTIDAE* (p. 176)
2'. Prementum not distinctly petiolate, more or less triangular or subquadrate in shape (Fig. 12.13); movable hook of each palpal lobe without setae; dorsal premental setae 0-3 on each side of median line (some Coenagrionidae with 4 or 5 premental setae have 5-6 palpal setae (Fig. 12.13) 3
3(2'). One dorsal premental seta on each side of median line; palpal setae 3-5; proximal portion of gills thickened and darkened, apical portion thinner and more lightly pigmented, thus nodus very distinctly delineated across entire width of gill (Fig. 12.17) (Texas)..*PROTONEURIDAE* (p. 183)
3'. Dorsal premental setae usually 3-5 on each side of median line or absent (if only 1 present, gills not as above and usually 6 palpal setae present) (Fig. 12.13); palpal setae 0-6; proximal portion of gills usually not differing markedly from distal portion, thus nodus not delineated across entire width of gill *(Nehalennia* with gills an exception to this, have 3-4 dorsal premental setae on each side of median line) (Fig. 12.15) (widespread)..*COENAGRIONIDAE* (p. 176)

Adults

1. Antenodal crossveins numerous (Fig. 12.18); postnodal crossveins not in line with the veins below them; anal vein at its base separate from posterior border of wing; quadrangle with several crossveins (Fig. 12.18) ....*CALOPTERYGIDAE* (p. 176)
1'. Only 2 antenodals present (Fig. 12.19); postnodals in line with the veins below them; anal vein joined with wing margin for a distance from the wing base; quadrangle never with crossveins (Fig. 12.19) 2
2(1'). Veins M_3 and Rs arising nearer the arculus than the nodus (Fig. 12.19) ....*LESTIDAE* (p. 176)
2'. Veins M_3 and Rs arising nearer the nodus than the arculus (Fig. 12.20) 3
3(2'). Anal vein absent or greatly reduced; Cu_2 absent or at most only as long as 1 cell; quadrangle somewhat rectangular (Fig. 12.20) (Texas)............................*PROTONEURIDAE* (p. 185)
3'. Anal vein and Cu_2 of normal length (Fig. 12.1); quadrangle distinctly trapezoidal (Fig. 12.1) (widespread)..*COENAGRIONIDAE* (p. 181)

KEY TO THE FAMILIES OF ANISOPTERA

Larvae

1. Prementum and palpal lobes of labium flat or nearly so, without dorsal premental setae and usually without palpal setae (Fig. 12.22) 2
1'. Prementum and palpal lobes of labium forming a spoon-shaped structure, usually with dorsal premental and always with palpal setae (Figs. 12.26-12.27) 4
2(1). Antennae 4-segmented; fore- and middle tarsi 2-segmented; ligula without a median cleft (Figs. 12.21-12.22) ....*GOMPHIDAE* (p. 192)
2'. Antennae 6- and 7-segmented; fore and middle tarsi 3-segmented; ligula with a median cleft 3
3(2'). Segments of antennae short, thick, hairy (Fig. 12.91); prementum with sides subparallel in distal three-fifths, abruptly narrowed near base (Fig. 12.28); each palpal lobe bearing a strong dorsolateral spur at base of movable hook (Fig. 12.28) ....*PETALURIDAE* (p. 188)

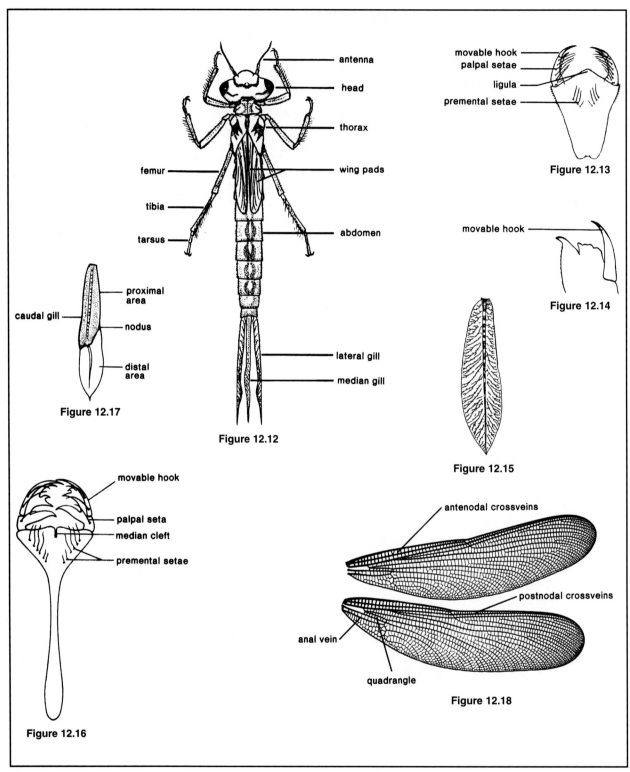

Figure 12.12. Dorsal view of larva *Enallagma civile* (Hagen) (Coenagrionidae) (drawn by K.J. Tennessen).

Figure 12.13. Prementum of labium of larva *Enallagma civile* (Hagen) (Coenagrionidae) (drawn by K.J. Tennessen).

Figure 12.14. Tip of palpal lobe of prementum of larva *Enallagma civile* (Hagen) (Coenagrionidae) (drawn by K.J. Tennessen).

Figure 12.15. Left lateral caudal gill of larva *Enallagma civile* (Hagen) (Coenagrionidae) (drawn by K.J. Tennessen).

Figure 12.16. Prementum of labium of larva *Lestes* sp. (Lestidae).

Figure 12.17. Left lateral caudal gill of larva *Protoneura viridis* Westfall (Protoneuridae).

Figure 12.18. Wings of male *Calopteryx angustipennis* (Selys) (Calopterygidae).

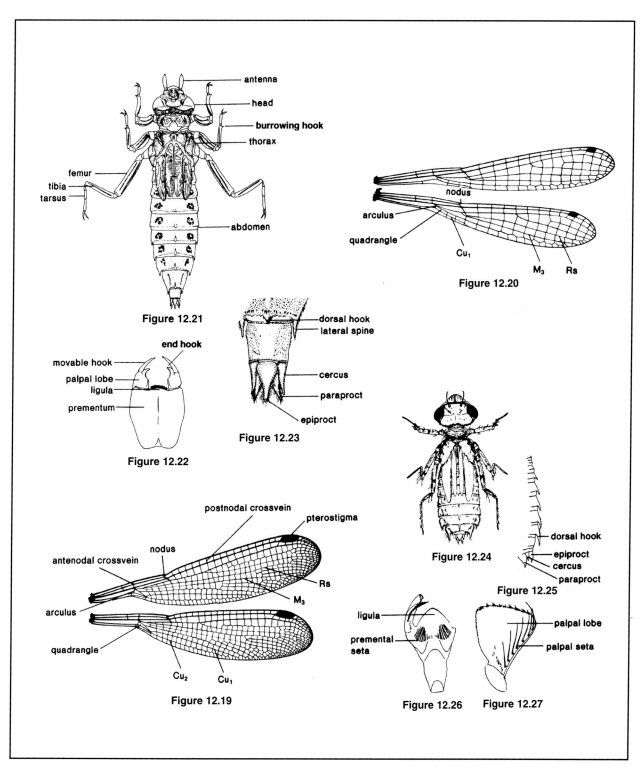

Figure 12.19. Wings of *Lestes vigilax* Hagen (Lestidae).

Figure 12.20. Wings of *Protoneura capillaris* (Rambur) (Protoneuridae).

Figure 12.21. Dorsal view of larva *Gomphus hodgesi* Needham (Gomphidae).

Figure 12.22. Prementum of labium of larva *Gomphus hodgesi* Needham (Gomphidae).

Figure 12.23. Tip of abdomen of larva *Gomphus hodgesi* Needham (Gomphidae).

Figure 12.24. Dorsal view of larva *Miathyria marcella* Selys (Libellulidae).

Figure 12.25. Lateral view of abdomen of larva *Miathyria marcella* Selys (Libellulidae).

Figure 12.26. Prementum of labium of larva with one palpal lobe attached, *Miathyria marcella* Selys (Libellulidae).

Figure 12.27. Palpal lobe of prementum of larva *Miathyria marcella* Selys (Libellulidae).

3'. Segments of antennae usually slender, bristlelike (Fig. 12.90); prementum widest in distal half, much narrower in basal half or more (Fig. 12.29); no spur at base of movable hook (Fig. 12.29) ....AESHNIDAE (p. 185)

4(1'). Distal margin of each palpal lobe cut into large and irregular dentations (toothed projections) without associated setae (Fig. 12.30); ligula represented by a toothlike process cleft in the middle (Fig. 12.30) ....CORDULEGASTRIDAE—*Cordulegaster*

4'. Distal margin of each palpal lobe smooth or with smaller and more regular dentations (Figs. 12.27, 12.31-12.32), each crenation (rounded projection) bearing 1 or more setae (Fig. 12.27); ligula not as above 5

5(4'). Crenulations (rounded projections) on distal margins of palpal lobes of labium separated by deep notches, crenations usually one-fourth to one-half as high as they are broad (as in Fig. 12.135); cerci generally more than one-half as long as paraprocts; lateral spines of abdominal segment 9 usually longer than its middorsal length; middorsal hooks on abdomen often cultriform (sicklelike) ....CORDULIIDAE* (p. 196)

5'. Crenulations on distal margins of palpal lobes of labium generally separated by shallow notches, crenations usually one-tenth to one-sixth as high as they are broad (Fig. 12.136); cerci generally not more than one-half as long as paraprocts; lateral spines of abdominal segment 9 usually shorter than its middorsal length but, if longer, then middorsal hooks on abdomen are not cultriform, but more spinelike, stubby, or absent ....LIBELLULIDAE* (p. 198)

Adults

1. Triangles of fore wings less than twice as far from the arculus as those of hind wings (Fig. 12.37); triangles of both wings similar in shape, generally elongated in long axis of wings; 2 antenodal crossveins thickened, most of the other antenodals of the 1st series not aligned with those of the 2nd series (Fig. 12.37) 2

1'. Triangles of fore wings at least twice as far from the arculus as those of hind wings (Fig. 12.38); triangles of the fore wings generally elongated transversely, those of hind wings longitudinally; no thickened antenodal crossveins, those of the 1st series aligned with those of the 2nd series (Fig. 12.38) 5

2(1). Eyes meeting for a considerable distance (Fig. 12.4); pterostigma supported by an oblique brace crossvein at or very near its inner end (Fig. 12.2); epiproct of male generally triangular, rarely bifurcate (forked) or truncate (cut off squarely at tip) (Fig. 12.9); ovipositor well developed, both genital valves with a stylus (as in Fig. 12.8).........AESHNIDAE (p. 188)

2'. Eyes wide apart or barely meeting (Fig. 12.41); pterostigma with or without a brace crossvein; epiproct of male usually bifurcate or truncate, never triangular; ovipositor small or vestigial (Fig. 12.44), rarely long as in figure 12.45; genital valves, if present, usually without stylus (Fig. 12.45) 3

3(2'). Eyes close together or barely meeting; labium with a median cleft (Fig. 12.42); pterostigma without a brace crossvein; ovipositor extending beyond tip of abdomen ....CORDULEGASTRIDAE—*Cordulegaster*

3'. Eyes wide apart (Fig. 12.41); labium with or without a median cleft (Fig. 12.43); pterostigma with a brace crossvein (Fig. 12.2); ovipositor never reaching tip of abdomen 4

4(3'). Front margin of labium with a median cleft (Fig. 12.42); pterostigma longer than one-fourth the distance from nodus to distal end of R_1; subtriangle of fore wing generally divided into 2 or more cells; last segments of abdomen not enlarged; ovipositor small but complete, both genital valves with a stylus ....PETALURIDAE (p. 188)

4'. Front margin of labium entire (Fig. 12.43); pterostigma shorter than one-fourth the distance from nodus to distal end of R_1 (Fig. 12.2); subtriangle of fore wing generally single-celled (Fig. 12.37); last segments of abdomen usually enlarged; ovipositor represented by a bifid vulvar lamina (Fig. 12.44) ....GOMPHIDAE (p. 194)

5(1'). Anal loop generally foot-shaped, with well-developed toe (Fig. 12.38); males without anal triangle, and anal margin of hind wings rounded as in females (Fig. 12.38); no tubercle on rear margin of eye; coloration varied, but not metallic (in North American species); males without anterior hamules or auricles on abdominal segment 2; legs without tibial keels (elevated ridges) ....LIBELLULIDAE (p. 202)

*No single character will reliably separate all larvae of Corduliidae from those of Libellulidae. For this reason, some specialists have recognized only one family, Libellulidae, and three subfamilies, Libellulinae, Corduliinae, and Macromiinae. Specimens difficult to place in couplet 5 need to be studied through keys to genera of both families and compared with figures.

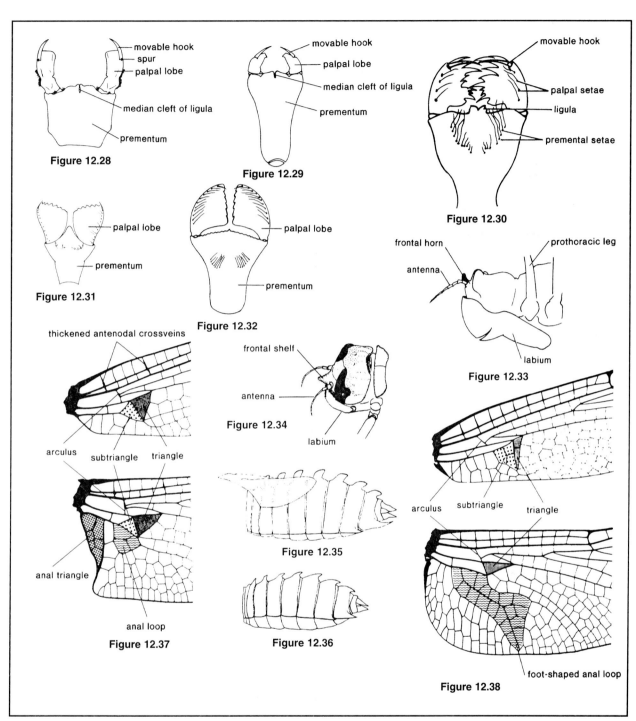

Figure 12.28. Prementum of labium of larva *Tanypteryx* sp. (Petaluridae).

Figure 12.29. Prementum of labium of larva *Aeshna* sp. (Aeshnidae).

Figure 12.30. Prementum of labium of larva *Cordulegaster* sp. (Cordulegastridae).

Figure 12.31. Prementum of labium of larva *Paltothemis* sp. (Libellulidae).

Figure 12.32. Prementum of labium of larva *Plathemis* sp. (Libellulidae).

Figure 12.33. Lateral view of head of larva *Macromia* sp. (Corduliidae, Macromiinae).

Figure 12.34. Dorsolateral view of head of larva *Neurocordulia molesta* (Walsh) (Corduliidae) (drawn by K. J. Tennessen).

Figure 12.35. Lateral view of abdomen of larva *Neurocordulia alabamensis* Hodges (Corduliidae) (drawn by K. J. Tennessen).

Figure 12.36. Lateral view of abdomen of larva *Perithemis* sp. (Libellulidae) (drawn by K. J. Tennessen).

Figure 12.37. Bases of wings of male *Ophiogomphus carolus* Needham (Gomphidae).

Figure 12.38. Bases of wings of male *Erythemis simplicicollis* (Say) (Libellulidae).

5'. Anal loop somewhat foot-shaped, but with little development of the toe (Fig. 12.40); hind wing of males with an anal triangle and anal margin angulate (Fig. 12.37); usually with a low tubercle on rear margin of each compound eye; coloration usually metallic; males with small anterior hamules, and an auricle on each side of abdominal segment 2 (Fig. 12.5); tibial keels on some legs ....*CORDULIIDAE* (p. 197)

KEYS TO THE GENERA OF ZYGOPTERA

Calopterygidae

Larvae

1. Prementum cleft nearly halfway to its base (Fig. 12.46); posterolateral margins of abdominal segments 9 and 10 without spines; lateral caudal gills flat or not strongly triquetral in cross section ....*Calopteryx*

    Prementum cleft only to base of palpal lobes (Fig. 12.47); posterolateral margins of abdominal segments 9 and 10 with small, distinct spines; lateral caudal gills strongly triquetral in cross section ....*Hetaerina*

Adults

1. Space proximal to arculus (midbasal or median space) without crossveins; arculus bent where its sectors (M_3 and M_4) arise, the sectors not proximally curved; anterior margin of quadrangle straight; anal branch of Cu_2 forming somewhat of a loop (Fig. 12.50) ....*Calopteryx*

1'. Space proximal to arculus with several crossveins; arculus not bent where its sectors arise, the sectors proximally curved; anterior margin of quadrangle convex; anal branch of Cu_2 not evident (Fig. 12.51) ....*Hetaerina*

Lestidae

Larvae

1. Distal margin of each palpal lobe divided into 3 sharp processes, the outermost one much shorter than the movable hook (Fig. 12.48); caudal gills with 2 well-defined, dark crossbands ....*Archilestes*

1'. Distal margin of each palpal lobe divided into 4 processes—3 sharp hooks and a short, serrate, truncate projection between the 2 outer hooks, the outermost hook reaching almost as far distad as the tip of the movable hook (Fig. 12.49); caudal gills never with 2 distinct and complete dark crossbands ....*Lestes*

Adults

1. The proximal side (inner end) of the quadrangle of the fore wing almost one-half the length of the inferior side; in species in the United States, vein M_2 arises about one cell beyond the nodus (Fig. 12.52) ....*Archilestes*

1'. The proximal side (inner end) of the quadrangle of the fore wing one-third or less of the length of the inferior side; in species in the United States, vein M_2 arises several cells beyond the nodus (Fig. 12.53) ....*Lestes*

Coenagrionidae

Larvae

1. Dorsal premental setae absent; palpal lobes with 2 distal, pointed hooks (Fig. 12.54); palpal setae 0-3 (rarely 4-5); body form usually short and stout; caudal gills of some species, in dorsal view, quite thick or triquetral (Fig. 12.55) ....*Argia*

1'. Dorsal premental setae present; palpal lobes with 1 distal, pointed hook and a truncate, denticulate lobe (Fig. 12.56); palpal setae 3-7; body form usually longer and more slender; caudal gills in dorsal view never thick or triquetral (Fig. 12.12) 2

2(1'). Lateral caudal gills 2.5–2.6 times longer than wide, apexes rounded, no trace of a pointed tip (Fig. 12.57); total length, including gills, about 19 mm ....*Hesperagrion*

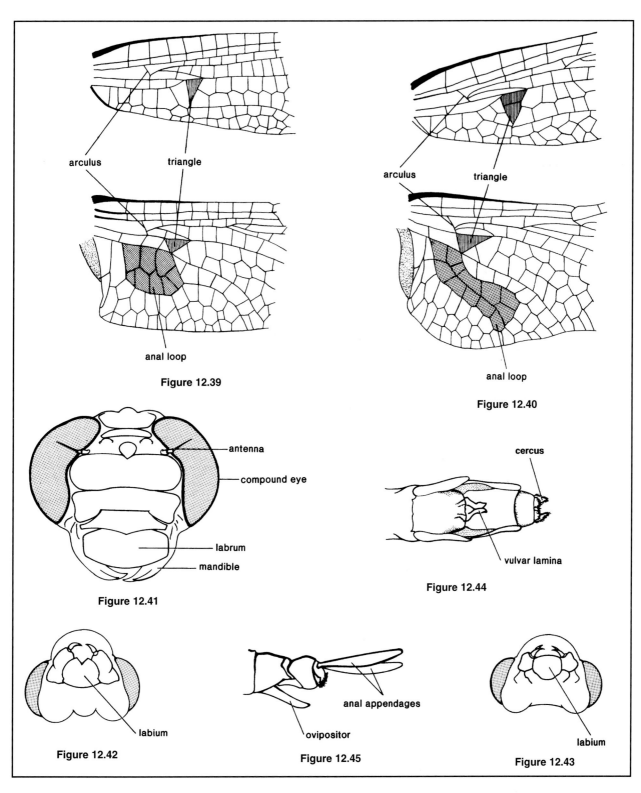

Figure 12.39. Base of wings of male *Didymops* sp. (Corduliidae, Macromiinae).

Figure 12.40. Base of wings of male *Tetragoneuria* sp. (Corduliidae).

Figure 12.41. Anterior view of head of adult *Gomphus* sp. (Gomphidae).

Figure 12.42. Ventral view of head of adult *Cordulegaster maculata* Selys (Cordulegastridae).

Figure 12.43. Ventral view of head of adult *Gomphus dilatatus* Rambur (Gomphidae).

Figure 12.44. Ventral view of abdominal segments 8-10 of adult female *Gomphus* sp. (Gomphidae).

Figure 12.45. Lateral view of abdominal segments 9-10 of adult female *Somatochlora* sp. (Corduliidae).

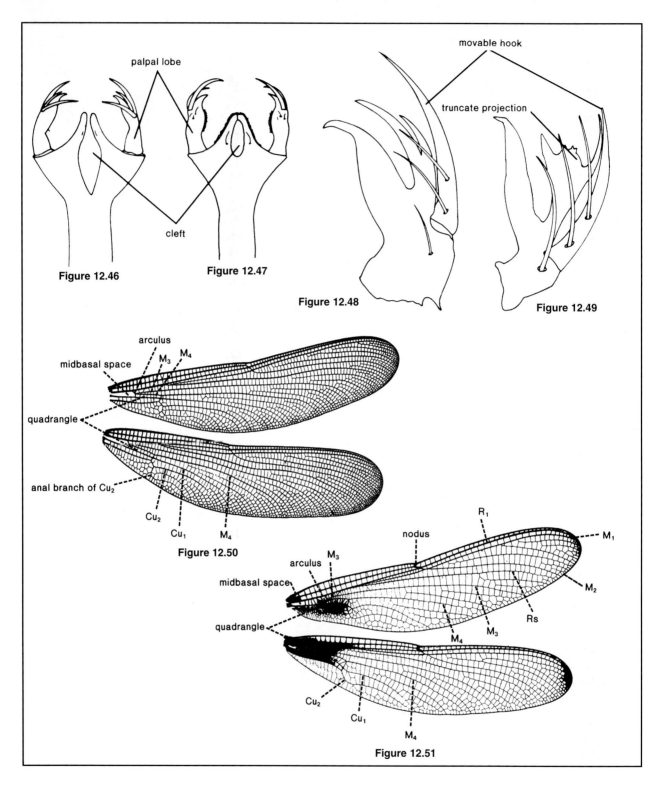

Figure 12.46. Prementum of labium of larva, dorsal view, *Calopteryx maculata* (Beauvois) (Calopterydigae) (drawn by M. S. Westfall).

Figure 12.47. Prementum of labium of larva, doral view, *Hetaerina titia* (Drury) (Calopterygidae) (drawn by M. S. Westfall).

Figure 12.48. Right palpal lobe of prementum of larva, dorsal view, *Archilestes grandis* (Rambur) (Lestidae) (drawn by M. S. Westfall).

Figure 12.49. Right palpal lobe of prementum of larva, dorsal view, *Lestes inaequalis* Walsh (Lestidae) (drawn by M. S. Westfall).

Figure 12.50. Wings of *Calopteryx angustipennis* (Selys) (Calopterygidae). (From *Manual of the Damselflies of North America*, M. J. Westfall, in prep.).

Figure 12.51. Wings of *Hetaerina titia* (Drury) (Calopterygidae) (Westfall manuscript).

2'.	Lateral caudal gills at least 3 times longer than wide, apexes usually acute with a sharply pointed tip (Fig. 12.58) (if apexes rounded, total length including gills less than 16 mm)	3
3(2').	Posterolateral margins of head greatly produced and sharply angulate (Fig. 12.59)	4
3'.	Posterolateral margins of head not so greatly produced, broadly rounded (Fig. 12.60)	5
4(3).	Apexes of caudal gills with acute tip long and sharply pointed; gills about one-sixth as broad as long, margins with widely separated setae; antennae 7-segmented	**Chromagrion**
4'.	Apexes of caudal gills with tip not so long and acute; gills about one-third as broad as long, margins thickly and closely beset with setae (Fig. 12.58); antennae 5- or 6-segmented	**Amphiagrion**
5(3').	Apexes of lateral caudal gills with long, tapering, almost filamentlike tips, thus apical 6th with a terminal angle of less than 20° (Fig. 12.61)	6
5'.	Apexes of lateral caudal gills with short, tapering tips, apical 6th with a terminal angle of 25° or more (Fig. 12.64 a, b)	7
6(5).	Proximal portion of caudal gills conspicuously thicker and darker than apical portion, thus nodus conspicuous; caudal gills with extratracheal dark pigment in form of marginal spots (Fig. 12.61); eyes unpatterned; dorsum of abdomen with a row of small, dark, paired spots (Florida)	**Nehalennia** (in part)
6'.	Proximal portion of caudal gills not conspicuously thicker or darker than apical portion, nodus not so conspicuous; caudal gills without extratracheal dark pigment (tracheae in widely separated clusters) (Fig. 12.62); eyes with a checkered pattern; abdomen without such a row of spots (California)	**Zoniagrion**
7(5').	Lateral caudal gills one-third as broad as long, apical 6th of each with a terminal angle of 70° or more	8
7'.	Lateral caudal gills one-fourth (or less) as broad as long, apical 6th of each with a terminal angle of about 60° or less	9
8(7).	Dorsal premental setae 5 each side of median line; apical half of lateral caudal gills, in dorsal view, with outer margin convex, and with tips incurved (forcepslike) (Fig. 12.63) (Arizona)	**Apanisagrion**
8'.	Dorsal premental setae 1-3 each side of median line; apical half of lateral caudal gills, in dorsal view, not conspicuously convex, and with tips straight (as in Fig. 12.12); gills in lateral view broad and often with marginal dark spots (Fig. 12 .64)	**Telebasis**
9(7').	Lateral caudal gills four-fifths the length of the body from anterior margin of head to posterior tip of abdomen; caudal gills with algalike patches of pigmented tracheae (Fig. 12.65); dorsal premental setae 3 each side of median line; palpal setae 4 (southern Texas)	**Acanthagrion**
9'.	Lateral caudal gills not more than two-thirds length of body; caudal gills usually without such algalike patches of tracheae; almost always a different combination of premental and palpal setae than above	10
10(9').	One dorsal premental seta of normal length present on each side of median line, although 1-3 small setae may be present on mesal side of principal seta	11
10'.	At least 2 dorsal premental setae of normal length present on each side of median line	12
11(10).	Palpal setae 3-4; numerous long, stiff setae on lateral carinae of all abdominal segments beyond 1st; length of final-instar larva 17 mm or more	**Enallagma** (in part)
11'.	Palpal setae 5-6; no long, stiff setae present on lateral carinae of anterior abdominal segments, although often present on posterior segments; length of final-instar larva 15 mm or less	**Nehalennia** (in part)
12(10').	Lateral carinae of abdominal segments 2-8, in dorsal view, slightly concave, with apexes prominent and bearing 2 or more stout, curved setae; usually 3 dorsal premental setae each side of median line; eyes with dark pattern of spots, or lines, forming hexagonal-shaped cells (Fig. 12.66)	13
12'.	Lateral carinae of abdominal segments 2-8 with margins straight or only slightly convex, apexes not prominent, and with apical setae, if present, not noticeably larger than the preceding setae; 2-5 dorsal premental setae each side of median line; eyes, if patterned, usually not as above	14
13(12).	Venter of abdominal segments 2-4 (often 2-6) with a more or less transverse, apical group of stiff, conspicuous setae, or nearly all segments devoid of conspicuous ventral setae, at most with tiny brown specks; lateral carinae of abdominal segment 9 less prominent than those of preceding segments and with no stout setae; usually 4 palpal setae	**Enallagma** (in part)

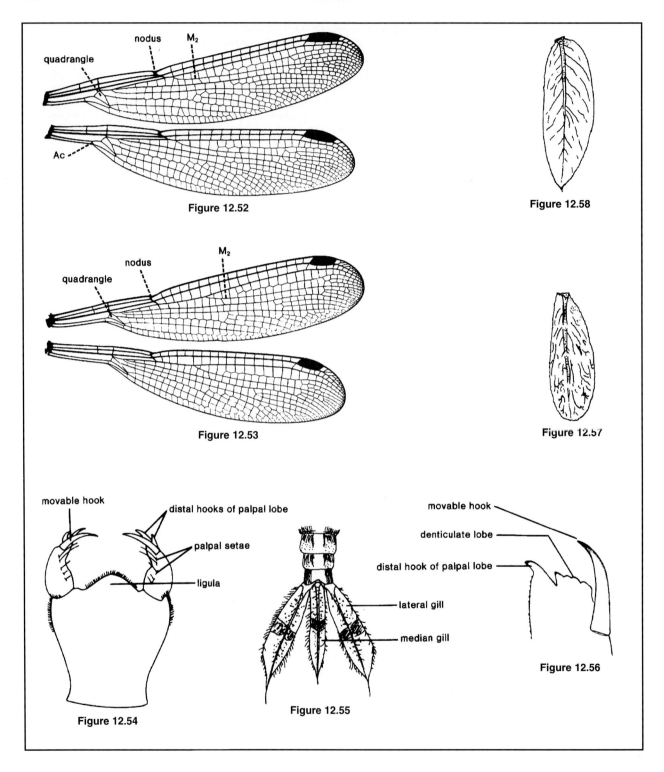

Figure 12.52. Wings of *Archilestes grandis* (Rambur) (Lestidae) (Westfall manuscript).

Figure 12.53. Wings of *Letes vigilax* Hagen (Lestidae) (Westfall manuscript).

Figure 12.54. Prementum of labium of larva, dorsal view, *Argia extranea* (Hagen) (Coenagrionidae) (Westfall manuscript).

Figure 12.55. Dorsal view, apex of larval abdomen with caudal gills, *Argia extranea* (Hagen) (Coenagrionidae) (Westfall manuscript).

Figure 12.56. Tip of palpal lobe of prementum of larva, *Enallagma civile* (Hagen) (Coenagrionidae) (drawn by K. J. Tennessen).

Figure 12.57. Right lateral caudal gill of larva, *Hesperagrion heterodoxum* (Selys) (Coenagrionidae) (Westfall manuscript).

Figure 12.58. Right lateral caudal gill of larva, *Amphiagrion saucium* (Burmeister) (Coenagrionidae) (Westfall manuscript).

13'.	Venter of all abdominal segments without such a conspicuous, transverse, apical group of setae; instead, setae of equal size and evenly scattered; lateral carinae of abdominal segment 9 nearly as prominent as those of segment 8; segments 8 and 9 bearing 1 stout apical seta; usually 5 palpal setae (southern Florida) .... ***Neoerythromma***
14(12').	Eyes usually with a pattern of lateral, alternating pale and dark bands following curvature of eye (Fig. 12.67); antennae usually with 7 distinct segments; lateral carinae of abdominal segments 2-7 with numerous small setae usually not arranged in a single row .... ***Ischnura***
14'.	Eyes solid dark (no pattern), or with dark spots, or with dark lines separating pale spots (Fig. 12.66); antennae usually 6-segmented, although apical 6th segment often with a discernible transverse line that is dissimilar to joints dividing preceding segments; lateral carinae of abdominal segments 2-8 usually with a single row of stout setae .... ***Coenagrion, Enallagma*** (in part)

Adults

1.	Tibial spines twice the length of intervening spaces, at least proximally 2
1'.	Tibial spines at most barely longer than the intervening spaces 3
2(1).	Vein M_{1a} extending at least the length of 8 cells; M_2 arising near 6th or 7th postnodal crossvein in fore wing, near 5th or 6th in hind wing; wings stalked (narrowed) only to level of 1st antenodal crossvein (Fig. 12.68) .... ***Argia***
2'.	Vein M_{1a} extending at most the length of 6 cells; M_2 arising near 4th or 5th postnodal crossvein in fore wing, near 4th in hind wing; wings stalked to level beyond 1st antenodal crossvein (Fig. 12.69) .... ***Nehalennia*** (in part)
3(1').	Intersternum with a prominent, moundlike tubercle bearing numerous long, stiff setae; wings nearly equal in length to abdomen; stocky, red or reddish brown and black species .... ***Amphiagrion***
3'.	No prominent intersternal tubercle present; wings at most three-fourths the length of the abdomen; usually not stocky, red and black species 4
4(3').	Dorsum of thorax and abdomen mostly metallic green; length of abdomen 19-25 mm; prothoracic femora with 2 distinct external black stripes, 1 at base of spines .... ***Nehalennia*** (in part)
4'.	Dorsum of thorax and abdomen usually not metallic green; length of abdomen variable; prothoracic femora without black or with one black stripe which may cover entire external surface 5
5(4').	Postocular area entirely pale or with pale spots ranging from narrow linear areas to large round spots, sometimes confluent with each other or with pale crest of occiput 6
5'.	Postocular area dark, without pale spots, although crest of occiput may be pale 20
6(5).	Vein M_2 arising proximal to or near 4th postnodal crossvein in fore wing, and proximal to or near 3rd postnodal in hind wing 7
6'.	Vein M_2 arising near 5th postnodal crossvein or beyond in fore wing, near 4th or beyond in hind wing (Figs. 12.70-12.71) 9
7(6).	Black humeral stripe divided along its entire length by a narrow, pale stripe; prothorax with a pale dorsomedial spot .... ***Enallagma*** (in part)
7'.	Black humeral stripe entire along its length or lacking; prothorax without a pale, dorsomedial spot or entirely pale 8
8(7').	Anterior margin of fore wing quadrangle less than one-half as long as distal margin; femora with a black apical band; costal margin of pterostigma usually twice as long as proximal margin .... ***Neoerythromma*** (in part)
8'.	Anterior margin of fore wing quadrangle nearly as long as distal margin; femora without a dark apical band; costal margin of pterostigma at most slightly longer than proximal margin .... ***Ischnura*** (in part)
9(6').	Vein Cu_1 of fore wing extends well beyond level of origin of M_{1a}, whereas Cu_1 of hind wing usually extends only to level of origin of M_{1a}; costal margin of fore wing pterostigma (and often hind wing pterostigma) may be shorter than proximal margin (Fig. 12.70) 10
9'.	Vein Cu_1 similar in fore and hind wings, either extending well beyond level of origin of M_{1a} (usually), or nearly to this level; costal margin of pterostigma as long as or longer than proximal margin in both pairs of wings 11
10(9).	Hind wings of male with dense, dark venation at apex (Fig 12.71); abdominal segment 8 of female without a ventroapical spine (Arizona) .... ***Apanisagrion***
10'.	Hind wings of male with normal venation at apex (Fig. 12.70); abdominal segment 8 of female with a distinct, ventroapical spine .... ***Hesperagrion*** (in part)

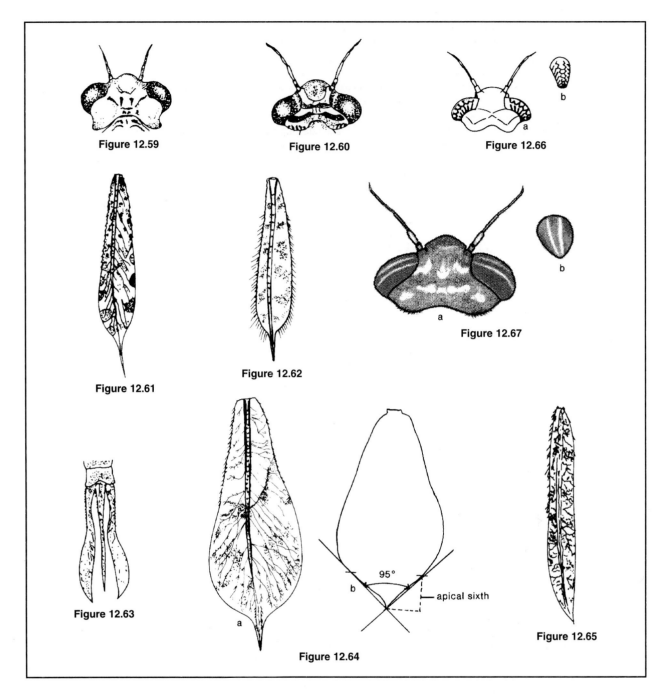

Figure 12.59. Dorsal view of head of larva, *Amphiagrion saucium* (Burmeister) (Coenagrionidae) (Westfall manuscript).

Figure 12.60. Dorsal view of head of larva, *Nehalennia minuta* (Selys) (Coenagrionidae) (Cuba) (Westfall manuscript).

Figure 12.61. Left lateral caudal gill of larva, *Nehalennia minuta* (Selys) (Coenagrionidae) (Cuba) (Westfall manuscript).

Figure 12.62. Left lateral caudal gill of larva, *Zoniagrion exclamationis* (Selys) (Coenagrionidae) (Westfall manuscript).

Figure 12.63. Apex of abdomen of larva, with caudal gills, dorsal view, *Apanisagrion lais* (Selys) (Coenagrionidae) (Westfall manuscript).

Figure 12.64. *a*, Left lateral caudal gill of larva, *Telebasis salva* (Hagen) (Coenagrionidae); *b*, outline of lateral caudal gill, *Telebasis byersi* Westfall (Coenagrionidae) illustrating measurement of angle of apical sixth (drawn by M. S. Westfall).

Figure 12.65. Left lateral caudal gill of larva, *Acanthagrion quadratum* Selys (Coenagrionidae) (Westfall manuscript).

Figure 12.66. *a*, Dorsal view of head of larva, *Enallagma divagans* Selys (Coenagrionidae); *b*, anterolateral view of eye showing pattern (Westfall manuscript).

Figure 12.67. Dorsal view of head of larva, *Ischnura posita* (Hagen) (Coenagrionidae) (drawn by K. J. Tennessen).

11(9').	Anal crossing linking Cu to posterior border in fore and hind wings	12
11'.	Anal crossing linking Cu to A in fore and usually hind wings, not reaching posterior border (as in Fig. 12.70)	14
12(11).	Abdominal color of male predominantly red, sometimes with black; abdominal segment 8 of female without a ventroapical spine	***Telebasis*** (in part)
12'.	Abdominal colors of male blue and black; abdominal segment 8 of female with a ventroapical spine	13
13(12').	Abdominal segment 10 of male in lateral view two and one-half times higher than long, the dorsum projecting dorsoposteriorly; female with a pit on each side of the synthoracic middorsal carina, and cerci approximated at base (southern Texas)	***Acanthagrion***
13'.	Abdominal segment 10 of male in lateral view less than twice as high as long, the dorsum flat, directed straight rearward; female without pits near middorsal carina, and cerci separated at base, usually by width of one cercus	***Enallagma*** (in part)
14(11').	Males	15
14'.	Females	17
15(14).	Pterostigma of fore wing different in shape, color, or size from pterostigma of hind wing; segment 10 with a posterodorsally projecting bifid process at least one-half as long as the segment	***Ischnura*** (in part)
15'.	Pterostigma of fore and hind wings similar in color, shape, and size; segment 10 with at most a very low, widely bifid prominence one-fourth as long as the segment	16
16(15').	Abdominal segment 10 with a low, bifid prominence raised slightly above the dorsum of the segment; penis with a transverse row of short spines at base of spatulate lobe; pale antehumeral stripe interrupted posteriorly by black cross-stripe (California)	***Zoniagrion***
16'.	Abdominal segment 10 without bifid prominence, although posterior margin may be notched; penis without a transverse row of spines at base of spatulate lobe; pale antehumeral stripe usually entire (except *Coenagrion interrogatum*)	***Coenagrion, Enallagma*** (in part)
17(14').	Humeral suture usually pale, but, if black stripe present, then no apical spine on venter of abdominal segment 8	***Ischnura*** (in part)
17'.	Humeral suture usually with a black stripe, but, if pale, an apical spine is present on venter of abdominal segment 8	18
18(17').	Abdominal segment 8 without a ventroapical spine	***Coenagrion***
18'.	Abdominal segment 8 with a ventroapical spine	19
19(18').	Pale antehumeral stripe constricted by black at three-fourths its length; metapleural suture black along entire length; outer surface of all femora wholly black (California)	***Zoniagrion***
19'.	Pale antehumeral stripe usually not constricted posteriorly; metapleural suture pale anteriorly; outer surface of all femora not wholly black	***Enallagma*** (in part)
20(5').	Costal margin of pterostigma of fore wing (and usually hind wing) shorter than proximal margin (Fig. 12.70); female with ventroapical spine on abdominal segment 8; thoracic middorsal black marking reaching humeral suture, thus pale antehumeral stripe split into anterior and posterior spots (young specimens with thorax entirely pale orange to yellow) (Arizona)	***Hesperagrion*** (in part)
20'.	Costal margin of pterostigma much longer than length of proximal margin (as in Fig. 12.68); female without a ventroapical spine on abdominal segment 8; thoracic markings not as above	21
21(20').	M_2 of hind wing arising near 3rd postnodal crossvein; humeral stripe black and of even width along entire length (southern Florida)	***Neoerythromma*** (in part)
21'.	M_2 of hind wing nearly always arising near 4th postnodal crossvein; humeral stripe pale and of various widths (extremely narrow in some *Telebasis*)	22
22(21').	M_{1a} extending for the length of 7-8 cells; metepimeron largely bright yellow; male predominantly blue and green	***Chromagrion***
22'.	M_{1a} extending for the length of 4-5 cells; metepimeron pale, but not bright yellow; males of species recorded in United States predominantly red or reddish brown	***Telebasis*** (in part)

Protoneuridae

Larvae

1.	Wing pads transparent before adult wings form prior to emergence, so that underlying abdominal segments are visible; pterostigma of developing wings conspicuous; ventral margin of apical portion of lateral caudal gills evenly and slightly convex	***Protoneura***

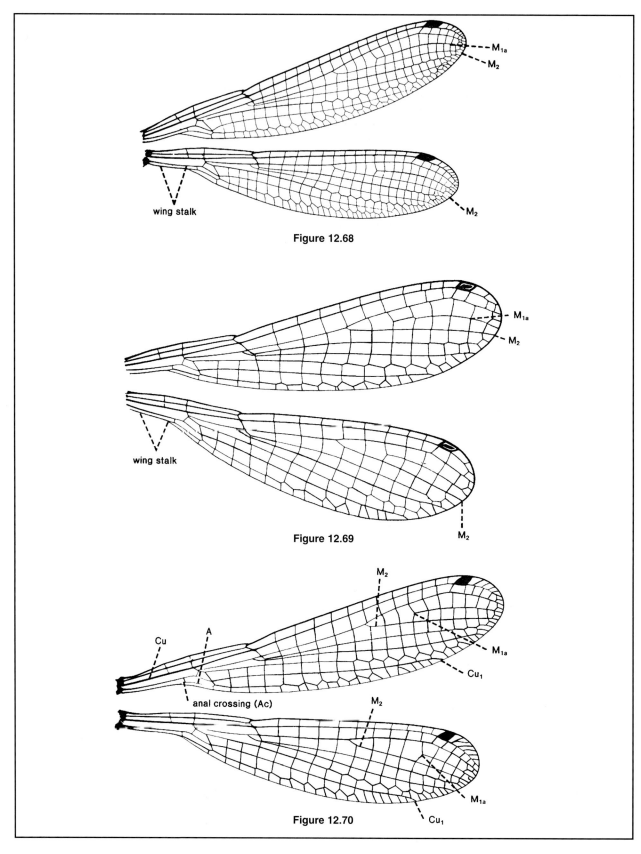

Figure 12.68. Wings of *Argia moesta* (Hagen) (Coenagrionidae) (Westfall manuscript).

Figure 12.69. Wings of *Nehalennia irene* (Hagen) (Coenagrionidae) (Westfall manuscript).

Figure 12.70. Wings of *Hesperagrion heterodoxum* (Selys) (Coenagrionidae) (Westfall manuscript).

1'. Wing pads opaque, so that underlying abdominal segments are not visible; pterostigma of developing wings not conspicuous; dorsal or ventral margin (or both) of apical portion of lateral caudal gills undulate and not smoothly convex............................*Neoneura*

Adults

1. Anal vein (A) present in both pairs of wings for length of one cell (Fig. 12.72); M_2 arising near 4th postnodal crossvein in fore wing, near 3rd in hind wing (Fig. 12.72); M_{1a} of fore wing usually arising only 1-2 cells beyond origin of M_2 (Fig. 12.72) ....*Neoneura*
1'. Anal vein (A) absent in both pairs of wings (Fig. 12.73); M_2 arising near 5th postnodal crossvein or beyond in fore wing, near 4th or beyond in hind wing (Fig. 12.73); M_{1a} of fore wing arising 3-4 cells beyond origin of M_2 (Fig. 12.73) ....*Protoneura*

KEYS TO THE GENERA OF ANISOPTERA

Aeshnidae

Larvae

1. Palpal lobes of labium with stout raptorial setae on dorsal surface (Fig. 12.74) 2
1'. Palpal lobes of labium without stout raptorial setae on dorsal surface (Fig. 12.75) 3
2(1). Palpal setae about 9, nearly uniform in length, palpal lobe without end hook (Fig. 12.76); total length of last instar larva less than 40 mm ....*Triacanthagyna*
2'. Palpal setae about 12-14, less robust, very unequal in length, diminishing to very small ones at proximal end of row, palpal lobe with pronounced end hook (Fig. 12.74); total length of last instar larva more than 40 mm ....*Gynacantha*
3(1'). Posterolateral margins of head decidedly angulate (Fig. 12.79) 4
3'. Posterolateral margins of head rounded (sometimes bluntly angular in *Aeshna*) (Fig. 12.80) 8
4(3). Head flattened and elongate, rectangular in dorsal view; low, broadly rounded eyes directed laterally (Fig. 12.79); ligula usually with a long marginal spine each side of the median cleft and adjacent to it (Fig. 12.75) ....*Coryphaeschna*
4'. Head roughly trapezoidal, strongly narrowed posteriorly; ligula without a long marginal spine each side of median cleft, at most a small tubercle present 5
5(4'). Dorsum of abdomen broadly rounded 6
5'. Dorsum of abdomen with a low median ridge 7
6(5). Palpal lobes of labium squarely truncated at distal ends (Fig. 12.77); paraprocts about as long as abdominal segments 9 + 10; epiproct tapering to apex that may be pointed or cleft; usually with conspicuous pale spot on dorsum of abdominal segment 8 ....*Boyeria*
6'. Palpal lobes of labium tapering to a point (Fig. 12.78); paraprocts longer than abdominal segments 9 + 10; epiproct parallel-sided in distal half, ending in a cleft tip; no conspicuous pale spot on dorsum of abdominal segment 8...................*Basiaeschna*
7(5'). Median ridge of abdomen with blunt dorsal hooks on segments 7-9 ....*Nasiaeschna*
7'. Median ridge of abdomen without dorsal hooks ....*Epiaeschna*
8(3'). Tips of paraprocts strongly incurved (Fig. 12.81); lateral spines well developed on abdominal segments 5-9 (Arizona) ....*Oplonaeschna*
8'. Tips of paraprocts usually quite straight; lateral spines usually on abdominal segments 6 or 7-9 (rarely on 5-9); widely distributed 9
9(8'). Antennae longer than distance from their base to rear of head (Fig. 12.80); distal margin of ligula deeply bilobed, with a V-shaped notch ....*Gomphaeschna*
9'. Antennae about half as long as distance from their base to rear of head (as in Fig. 12.79); distal margin of ligula obtuseangulate, at most very slightly bilobed, with the notch closed 10
10(9'). Compound eyes as long as their greatest width; lateral spines only on abdominal segments 7-9; paraprocts about as long as abdominal segments 8 + 9 ....*Anax*
10'. Compound eyes much shorter than their greatest width; lateral spines usually on abdominal segments 6-9 (rarely on 5-9 or 7-9 only); paraprocts shorter than abdominal segments 8 and 9 (usually about equal to 9 + 10) ....*Aeshna*

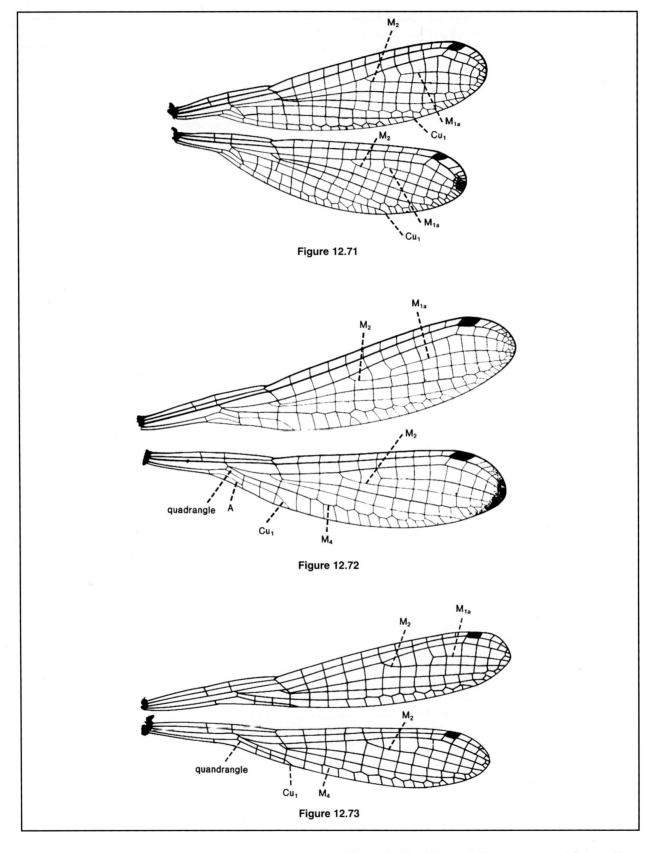

Figure 12.71. Wings of *Apanisagrion lais* (Selys) (Coenagrionidae) (Westfall manuscript).

Figure 12.72. Wings of *Neoneura aaroni* Calvert (Protoneuridae) (Westfall manuscript).

Figure 12.73. Wings of *Protoneura capillaris* (Rambur) (Protoneuridae) (Westfall manuscript).

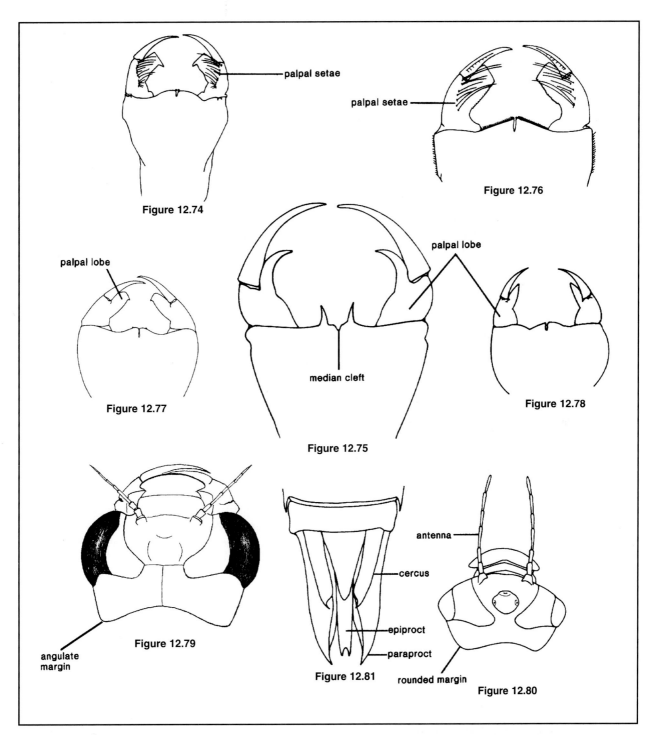

Figure 12.74. Prementum of labium of larva, dorsal view, *Gynacantha nervosa* Rambur (Aeshnidae) (drawn by M. S. Westfall).

Figure 12.75. Prementum of labium of larva, dorsal view, *Coryphaeschna ingens* (Rambur) (Aeshnidae) (drawn by M. S. Westfall).

Figure 12.76. Prementum of labium of larva, dorsal view, *Triacanthagyna trifida* (Rambur) (Aeshnidae) (drawn by M. S. Westfall).

Figure 12.77. Prementum of labium of larva, dorsal view, *Boyeria vinosa* (Say) (Aeshnidae) (drawn by M. S. Westfall).

Figure 12.78. Prementum of labium of larva, dorsal view, *Basiaeschna janata* (Say) (Aeshnidae) (drawn by M. S. Westfall).

Figure 12.79. Dorsal view of head of larva, *Coryphaeschna ingens* (Rambur) (Aeshnidae) (drawn by K. J. Tennessen).

Figure 12.80. Dorsal view of head of larva, *Gomphaeschna furcillata* (Say) (Aeshnidae) (drawn by M. S. Westfall).

Figure 12.81. Dorsal view, tip of abdomen of larva, *Oplonaeschna armata* (Hagen) (Aeshnidae) (drawn by M. S. Westfall).

Adults

1. Midbasal (median) space with more than one crossvein; sides of thorax in species in the United States with 2 rounded, pale spots .... *Boyeria*
1'. Midbasal (median) space with not more than 1 crossvein (as in Fig. 12.82); sides of thorax variously marked but never with only 2 rounded pale spots 2
2(1'). Sectors of arculus arising from its upper end (Fig. 12.83); thorax uniform green; anal border of hind wing rounded in both sexes; auricles absent .... *Anax*
2'. Sectors of arculus arising near its middle (Fig. 12.82); thorax usually brown, marked with blue, green, or yellow; male with anal border of hind wing angulate and with auricles on abdominal segment 2 (as in Fig. 12.5) 3
3(2'). Vein Rs forked (Fig. 12.82) 4
3'. Vein Rs not forked (Fig. 12.83) 9
4(3). Stalk of Rs straight and fork symmetrical (Fig. 12.82) 5
4'. Stalk of Rs bending forward to an asymmetrical fork (Fig. 12.84) 6
5(4). Radial planate subtending one row of cells (Fig. 12.82); viewed from the side, frons prominent and face flat; cerci of male about as long as abdominal segment 9, cerci of female much shorter; anterior lamina of male without a spine .... *Nasiaeschna*
5'. Radial planate subtending more than one row of cells (as in Fig. 12.83); viewed from the side, frons not noticeably prominent and face slightly convex; cerci of male about as long as abdominal segments 9 + 10, cerci of female somewhat longer; anterior lamina of male bearing a curved spine on each side .... *Epiaeschna*
6(4'). Vein Rs forked under the pterostigma (Fig. 12.84); two rows of cells in fork of Rs (Fig. 12.84) .... *Coryphaeschna*
6'. Vein Rs forked proximal to the pterostigma (as in Fig. 12.82); usually more than two rows of cells in fork of Rs (as in Fig. 12.82) 7
7(6'). Supertriangle as long as, or shorter than, midbasal space (as in Fig. 12.82) .... *Aeshna*
7'. Supertriangle distinctly longer than midbasal space (as in Fig. 12.83) 8
8(7'). Hind wing with two rows of cells between M_1 and M_2 beginning distad to base of pterostigma; female with 3 pronged process on venter of abdominal segment 10 (Fig. 12.88) .... *Triacanthagyna*
8'. Hind wing with 2 rows of cells between M_1 and M_2 beginning at base of pterostigma or proximal to it; female with 2-pronged process on venter of abdominal segment 10 (Fig. 12.89) .... *Gynacantha*
9(3'). Two cubito-anal crossveins on both fore and hind wings (Fig. 12.85); pterostigma surmounting 1 crossvein, not counting brace vein (Fig. 12.85); supertriangle without crossveins (Fig. 12.85); epiproct of male deeply forked .... *Gomphaeschna*
9'. Three or more cubito-anal crossveins (Fig. 12.86); pterostigma surmounting 2 or more crossveins (Figs. 12.86-12.87); supertriangle with crossveins (Fig. 12.87); epiproct of male triangular (as in Fig. 12.9) 10
10(9'). Anal triangle of male 2-celled (Fig. 12.87); base of wings with large brown spot; one row of cells between Cu_1 and Cu_2 beginning at the triangle (Fig. 12.87) .... *Basiaeschna*
10'. Anal triangle of male 3-celled (Fig. 12.86); base of wings hyaline; hind wing with two rows of cells between Cu_1 and Cu_2 beginning at the triangle (Fig. 12.86) .... *Oplonaeschna*

Petaluridae

Larvae

1. Antennae 6-jointed; abdomen more slender, lateral margins of segments not expanded and with spines inconspicuous .... *Tanypteryx*
1'. Antennae 7-jointed (Fig. 12.91); abdomen broader, lateral margins of segments expanded and with conspicuous spines on segments 3-9 (Fig. 12.98) .... *Tachopteryx*

Adults

1. Thorax black, spotted on front and sides with yellow; metathorax bearing a round hairy tubercle beneath; 1st and 4th or 5th antenodals thickened; anal loop poorly developed and very variable; subtriangle of fore wing 1- or 2-celled; anal angle of hind wing produced and rather sharply angulated in male (West Coast) .... *Tanypteryx*
1'. Thorax grayish in front, with carina narrowly black; sides lighter gray, with 2 oblique black stripes; metathorax without a round, hairy tubercle beneath; 1st and 6th or 7th antenodals thickened; subtriangle of fore wing usually 3-celled; anal angle of hind wing very obtusely angulated in male (eastern United States) .... *Tachopteryx*

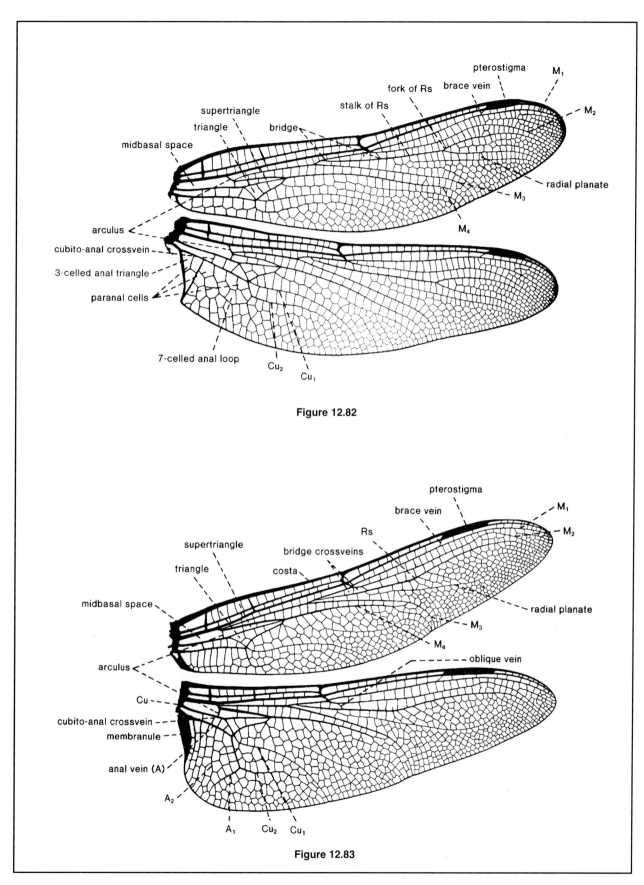

Figure 12.82. Wings of *Nasiaeschna pentacantha* (Rambur) (Aeshnidae) (Needham and Westfall 1955).

Figure 12.83. Wings of *Anax junius* (Drury) (Aeshnidae) (Needham and Westfall 1955).

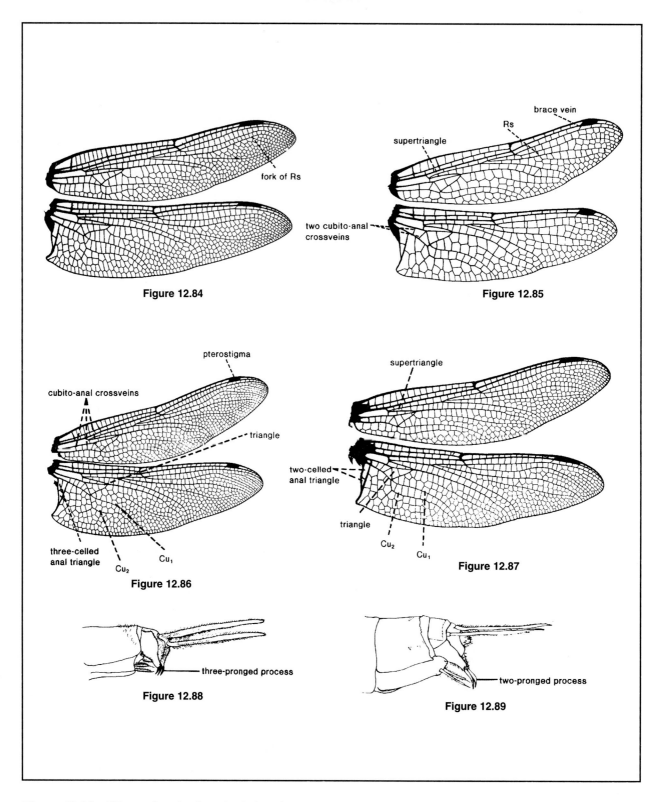

Figure 12.84. Wings of male *Coryphaeschna ingens* (Rambur) (Aeshnidae) (Needham and Westfall 1955).

Figure 12.85. Wings of male *Gomphaeschna furcillata* (Say) (Aeshnidae) (Needham and Westfall 1955).

Figure 12.86. Wings of male *Oplonaeschna armata* (Hagen) (Aeshnidae) (Needham and Westfall 1955).

Figure 12.87. Wings of male *Basiaeschna janata* (Say) (Aeshnidae) (Needham and Westfall 1955).

Figure 12.88. Lateral view, tip of abdomen of adult female, *Triacanthagyna trifida* (Rambur) (Aeshnidae) (drawn by M. S. Westfall).

Figure 12.89. Lateral view, tip of abdomen of adult female, *Gynacantha nervosa* Rambur (Aeshnidae) (drawn by M. S. Westfall).

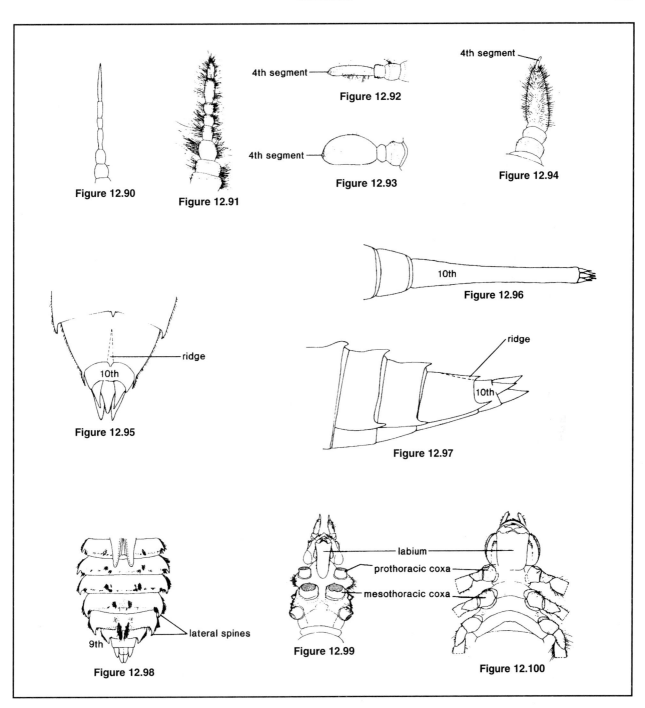

Figure 12.90. Antenna of larva, *Aeshna* sp. (Aeshnidae) (Wright and Peterson 1944).

Figure 12.91. Antenna of larva, *Tachopteryx thoreyi* (Hagen) (Petaluridae) (Wright and Peterson 1944).

Figure 12.92. Antenna of larva, *Dromogomphus* (Gomphidae) (Wright and Peterson 1944).

Figure 12.93. Antenna of larva, *Octogomphus specularis* (Hagen) (Gomphidae) (Wright and Peterson 1944).

Figure 12.94. Antenna of larva, *Progomphus* sp. (Gomphidae) (Wright and Peterson 1944).

Figure 12.95. Dorsal view, tip of abdomen of larva, *Dromogomphus* sp. (Gomphidae) (drawn by M. S. Westfall).

Figure 12.96. Dorsal view, tip of abdomen of larva, *Aphylla williamsoni* (Gloyd) (Gomphidae) (Wright and Peterson 1944).

Figure 12.97. Lateral view, tip of abdomen of larva, *Dromogomphus spinosus* Selys (Gomphidae) (drawn by M. S. Westfall).

Figure 12.98. Dorsal view, abdomen of larva, *Tachopteryx thoreyi* (Hagen) (Petaluridae) (Wright and Peterson 1944).

Figure 12.99. Ventral view, head and thorax of larva, *Progomphus obscurus* (Rambur) (Gomphidae) (Wright and Peterson 1944).

Figure 12.100. Ventral view, head and thorax of larva, *Gomphus* sp. (Gomphidae) (Wright and Peterson 1944).

Gomphidae

Larvae

1. Abdominal segment 10 cylindrical, more than half as long as all the other abdominal segments combined (Fig. 12.96) .... *Aphylla*
1'. Abdominal segment 10 shorter than abdominal segments 8 and 9 combined (Fig. 12.97) 2
2(1'). Middle (mesothoracic) legs closer together at base than fore (prothoracic) legs (Fig. 12.99); 4th antennal segment in species in the United States elongate, about one-fourth as long as the hairy 3rd antennal segment (Fig. 12.94) .... *Progomphus*
2'. Middle (mesothoracic) legs not closer together at base than (prothoracic) fore legs (Fig. 12.100); 4th antennal segment vestigial or merely a small, rounded knob (Figs. 12.92-12.93) 3
3(2'). Abdomen subcircular, very flat (Fig. 12.101); head behind the eyes bearing 2 large and 2 smaller spines (Fig. 12.101); fore and middle legs without tibial burrowing hooks .... *Hagenius*
3'. Abdomen much longer than broad, not so flat (Fig. 12.102); head behind the eyes without spines (Fig. 12.102); fore and middle legs usually with at least vestigial burrowing hooks on tibiae 4
4(3'). Palpal lobes of labium rounded, without an end hook, and usually bearing only microscopic teeth on the inner margin (Fig. 12.104); wing cases may be divergent 5
4'. Palpal lobes of labium usually with recurved end hook at apex (Fig. 12.105), but, if without, then either a sharp dorsal ridge on abdominal segment 9 that ends in a long, sharp spine (Figs. 12.95, 12.97), or the inner margins of the palpal lobes bear coarse teeth that may be irregular and obliquely truncated (Fig. 12.106); wing cases not divergent (Fig. 12.21) 9
5(4). Wing cases divergent (Fig. 12.102); some abdominal segments with middorsal hooks or knobs (Fig. 12.102) 6
5'. Wing cases parallel along midline (Fig. 12.21); no abdominal segment with a middorsal hook or knob 7
6(5). Cerci in full-grown larvae usually about three-fourths or less as long as paraprocts; dorsal hooks almost always well developed on abdominal segments 2- or 3-9 (Fig. 12.102); lateral spines usually absent on abdominal segment 6, but, if present, then species is far Western .... *Ophiogomphus*
6'. Cerci in full-grown larvae about as long as paraprocts; dorsal hooks usually wanting or rudimentary on some posterior abdominal segments, but, if well developed, then dorsal surface of abdomen very granular .... *Erpetogomphus*
7(5'). Short lateral spines on abdominal segments 7-9; 3rd segment of antennae about half as wide as long, widest in distal half (Fig. 12.93); far Western .... *Octogomphus*
7'. Very short lateral spines only on abdominal segments 8 and 9; third segment of antennae about three-fourths as wide as long, widest in proximal half (Fig. 12.103) (Eastern) 8
8(7'). Wide 3rd antennal segment 3 not equilateral, straight on its inner margin (Fig. 12.103); denticles on front border of prementum usually 3; sides of prementum somewhat convergent toward base; lateral spines of abdominal segment 9 wide, short, and stout .... *Stylogomphus*
8'. Wide 3rd antennal segment 3 nearly oval, its inner margin convex; denticles on front border of prementum usually 4; sides of prementum parallel toward base; lateral spines of abdominal segment 9 more slender, longer, and more clawlike .... *Lanthus*
9(4'). Abdominal segment 9 with an acute middorsal ridge usually bearing a long, sharp dorsal hook at its apex (seen best in lateral view; Figs. 12.95, 12.97); segment 9 never as long as wide at its base 10
9'. Abdominal segment 9 quite rounded dorsally and without such a long, sharp dorsal hook at its apex (Figs. 12.108-12.109); abdominal segment 9 may be longer than wide at its base 11
10(9). Prementum moderately produced in a low, rounded spinulose border; abdominal segment 10 a little longer than segment 9; dorsal hook of abdominal segment 9 elevated at apex, seen in lateral view (Western) .... *Phyllogomphoides*
10'. Prementum with straight front border; abdominal segment 10 shorter than abdominal segment 9 (Fig. 12.95); dorsal hook of abdominal segment 9 straight to apex and not elevated when seen in lateral view (Fig. 12.97) .... *Dromogomphus*

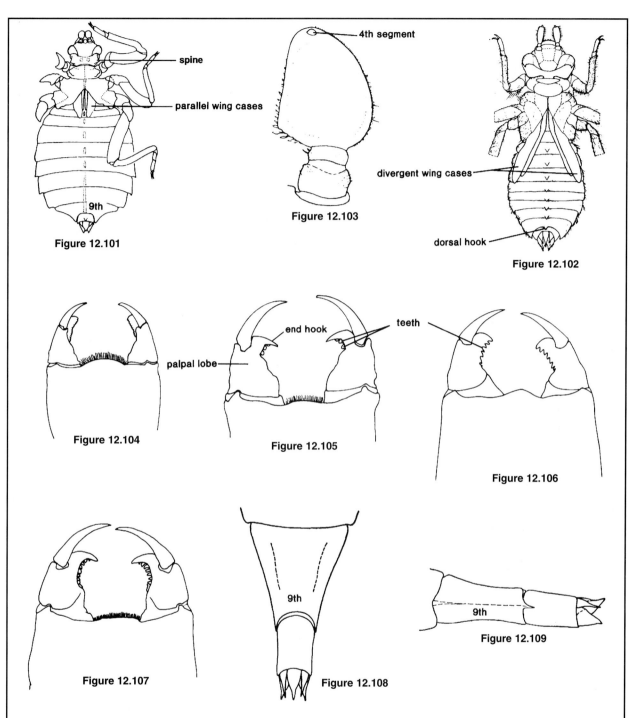

Figure 12.101. Dorsal view, larva of *Hagenius brevistylus* Selys (Gomphidae) (Wright and Peterson 1944).

Figure 12.102. Dorsal view, larva of *Ophiogomphus* (Gomphidae) (Wright and Peterson 1944).

Figure 12.103. Antenna of *Stylogomphus albistylus* (Hagen) (Gomphidae) (Wright and Peterson 1944).

Figure 12.104. Prementum of labium of larva, dorsal view, *Erpetogomphus designatus* Hagen (Gomphidae) (drawn by M. S. Westfall).

Figure 12.105. Prementum of labium of larva, dorsal view, *Stylurus laurae* Williamson (Gomphidae) (drawn by M. S. Westfall).

Figure 12.106. Prementum of labium of larva, dorsal view, *Arigomphus pallidus* (Rambur) (Gomphidae) (drawn by M. S. Westfall).

Figure 12.107. Prementum of labium of larva, dorsal view, *Gomphus lividus* Selys (Gomphidae) (drawn by M. S. Westfall).

Figure 12.108. Dorsal view, tip of abdomen of larva, *Gomphus* sp. (Gomphidae) (drawn by M. S. Westfall).

Figure 12.109. Lateral view, tip of abdomen of larva, *Gomphus* sp. (Gomphidae) (Wright and Peterson 1944).

11(9').	Tibial burrowing hooks absent or vestigial; abdomen elongate, regularly tapering all the way rearward; lateral spines of abdominal segment 9 not noticeably flattened dorsoventrally and with only hairs on the outer margin; each palpal lobe of labium ending in a strong, recurved end hook, followed by 1-4 usually shallow teeth (Fig. 12.105) ....***Stylurus***
11'.	Tibial burrowing hooks well developed (Fig. 12.21); abdomen usually not so elongate, and not tapering regularly all the way rearward; lateral spines of abdominal segment 9 may be noticeably flattened dorsoventrally and with the outer margins spinulose-serrate; palpal lobe of labium usually ending in less of a recurved end hook, followed by more than 4 (sometimes deeply cut) teeth (Figs. 12.106-12.107) 12
12(11').	Abdomen acuminate, rapidly narrowing on abdominal segment 8; abdominal segment 10 is longer than wide; lateral spines of abdominal segment 9 laterally flattened and closely appressed to the sides of 10; palpal lobes of labium without a well-developed end hook, but with inner margins bearing 5-8 coarse teeth that are irregular and obliquely truncate (Fig. 12.106) ....***Arigomphus***
12'.	Abdomen not or scarcely acuminate; segment 10 is usually wider than long; lateral spines of abdominal segment 9 usually neither laterally flattened nor closely appressed to the sides of 10; palpal lobes of labium usually with a well-developed end hook and the inner margins with a series of 4-10 rather regular teeth (Fig. 12.107) ....***Gomphus*** (including subgenera *Gomphus, Gomphurus,* and *Hylogomphus*)

Adults

1.	Basal subcostal crossvein present (Fig. 12.112); at least 1 crossvein in fore wing subtriangle (Fig. 12.112) 2
1'.	Basal subcostal crossvein absent (Fig. 12.113); no crossvein in fore wing subtriangle (Fig. 12.113) 4
2(1).	Supertriangles with 1 or more crossveins (Fig. 12.112) 3
2'.	Supertriangles without crossveins ....***Progomphus***
3(2).	Hind wing subtriangle of 2 or more cells (Fig. 12.112); anal loop of 3-5 cells formed by convergence of veins A_1 and A_2 (Fig. 12.112) ....***Phyllogomphoides***
3'.	Hind wing subtriangle usually 1-celled; veins A_1 and A_2 run directly to hind wing margin and form no anal loop ....***Aphylla***
4(1').	Triangles with a crossvein (Fig. 12.113); each triangle with a supplementary longitudinal vein (trigonal planate) arising from its distal side (Fig. 12.113); hind femora reaching a little beyond the base of abdominal segment 3; very large insects, generally more than 70 mm in length ....***Hagenius***
4'.	Triangles without crossveins, and without a supplementary longitudinal vein arising from distal side; hind femora usually not reaching beyond middle of abdominal segment 2; smaller insects, rarely 69 mm in length 5
5(4').	Hind wing with semicircular anal loop typically of 3 cells; epiproct of male with the 2 branches separated either by a mere slit with the apexes contiguous or by a U-shaped or V-shaped notch (Figs. 12.114-12.115); vulvar lamina nearly as long as sternum of segment 9 ....***Ophiogomphus***
5'.	Hind wing with anal loop absent or of 1-2 weakly bordered cells; epiproct of male usually with more or less widely separated divergent branches (except in *Erpetogomphus*); vulvar lamina usually less than half as long as sternum of abdominal segment 9 6
6(5').	Hind femur long, reaching the base of abdominal segment 3 and bearing 4-7 long ventral spines in addition to the usual short ones (Fig. 12.118) ....***Dromogomphus***
6'.	Hind femur not extending beyond the middle of abdominal segment 2 and usually bearing only the usual numerous, short, ventral spines 7
7(6').	Pterostigma of fore wing short and thick, at its widest about twice as long as wide; hind wing with 5 paranal cells (see Fig. 12.82); branches of epiproct of male long, parallel full length, and strongly hooked upward (Figs. 12.116-12.117) ....***Erpetogomphus***
7'.	Pterostigma of fore wing usually more elongate, about 3 times as long as wide; hind wing with 4-5 paranal cells; branches of epiproct of male shorter and divergent 8
8(7').	Dorsum of thorax pale between broad dorsolateral dark stripes; pterostigma of fore wing scarcely 3 times as long as wide and more than twice as wide as the space immediately behind its middle; epiproct of male 4-branched (far Western) ....***Octogomphus***

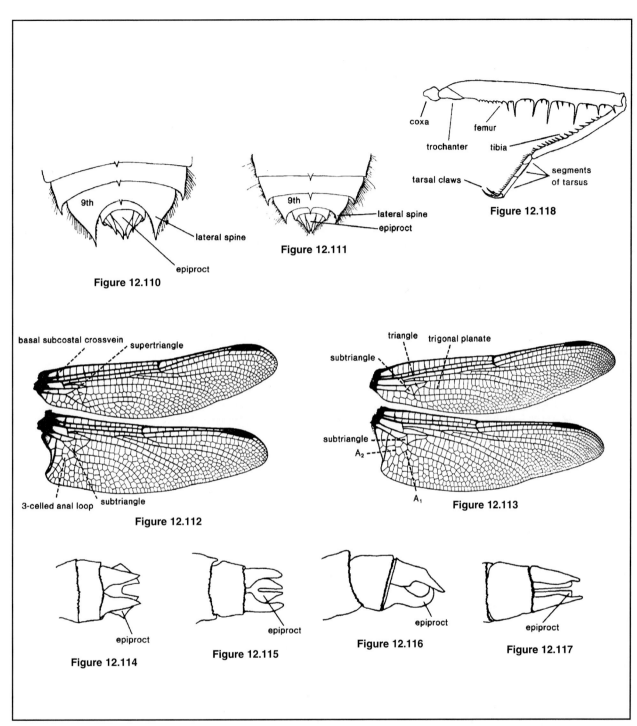

Figure 12.110. Dorsal view, tip of abdomen of larva, *Didymops* sp. (Corduliidae, Macromiinae) (Wright and Peterson 1944).

Figure 12.111. Dorsal view, tip of abdomen of larva, *Macromia* sp. (Corduliidae, Macromiinae) (Wright and Peterson 1944).

Figure 12.112. Wings of *Phyllogomphoides stigmatus* (Say) (Gomphidae) (Needham and Westfall 1955).

Figure 12.113. Wings of *Hagenius brevistylus* Selys (Gomphidae) (Needham and Westfall 1955).

Figure 12.114. Dorsal view, tip of abdomen, adult male, *Ophiogomphus carolus* Needham (Gomphidae) (Needham and Westfall 1955).

Figure 12.115. Dorsal view, tip of abdomen, adult male, *Ophiogomphus colubrinus* Selys (Gomphidae) (Needham and Westfall 1955).

Figure 12.116. Lateral view, tip of abdomen, adult male, *Erpetogomphus designatus* Hagen (Gomphidae) (Needham and Westfall 1955).

Figure 12.117. Dorsal view, tip of abdomen, adult male, *Erpetogomphus designatus* Hagen (Gomphidae) (Needham and Westfall 1955).

Figure 12.118. Hind (metathoracic) leg of adult, *Dromogomphus spinosus* Selys (Gomphidae) (drawn by K. J. Tennessen).

8'.	Dorsum of thorax usually mesally dark between pale dorsolateral stripes; pterostigma more than 3 times as long as wide and usually less than twice as wide as the space behind its middle; epiproct of male 2-branched 9
9(8').	Pterostigma less than 4 times as long as wide, more than twice as wide as the space behind its middle; epiproct of male with its lateral edges subparallel, its 2 branches shorter than the median length of the appendage; size small, less than 40 mm 10
9'.	Pterostigma rarely less than 4 times as long as wide, less than twice as wide as the space behind its middle; epiproct of male with lateral edges widely divergent, its 2 branches longer than the median length of the appendage; size larger, usually more than 40 mm 11
10(9).	Outer side of fore wing triangle slightly angulated near middle; anteclypeus pale; cerci of male pale yellow or white, laterally angulate with apexes very slender and curved upward ....*Stylogomphus*
10'.	Outer side of fore wing triangle straight or nearly so; anteclypeus blackish; cerci of male black, sides not angulate, and apexes straight and not very slender ....*Lanthus*
11(9').	Usually pale species with darker stripe each side of pale middorsal carina of thorax faint or absent; thoracic side stripes much reduced; male with end hook of posterior hamules long and falciform (Fig. 12.119); the abdomen usually constricted at segment 9 and the apexes of cerci usually convergent and very slender ....*Arigomphus*
11'.	Usually darker species, with conspicuous dark stripes on the thorax; male posterior hamule often with a sharp shoulder, but the end hook is not so long or falciform (Fig. 12.120); the abdomen usually not constricted at segment 9, and apexes of cerci variable, but usually not convergent or very slender 12
12(11').	Top of frons 4 times as wide as long; typically long, slender species in which the male cerci are simple, horizontal, without angular processes and only rarely with lateral margins angular; branches of male epiproct diverging at about same angle as cerci, so they are usually not seen from above; anterior hamules vestigial, slender, unarmed, and difficult to see, whereas posterior hamules are usually upright or leaning forward (Fig. 12.120); anterior lamina of male flat, not projecting above rim of genital pocket; vulvar lamina of female very short, sometimes vestigial ....*Stylurus*
12'.	Top of frons about 3 times as wide as long; form varied, male cerci usually with sharply angular processes; branches of male epiproct often diverge more strongly than cerci so that their apexes are conspicuous when seen from above; anterior hamules not vestigial and always bearing at least 1 hook or spine; posterior hamules upright or leaning backward; anterior lamina of male variable, but usually projecting above rim of genital pocket; vulvar lamina of female variable, often half as long as segment 9 (Fig. 12.44) ....*Gomphus* (including subgenera *Gomphus*, *Gomphurus*, and *Hylogomphus*)

Corduliidae

Larvae

1.	Head with a prominent, almost erect frontal horn between bases of antennae (Fig. 12.33); legs very long, the hind femur extending to or beyond the hind margin of abdominal segment 8; metasternum with a broad mesal tubercle near posterior margin; abdomen strongly depressed, almost circular in outline when in dorsal view ....*Macromiinae* 2
1'.	Head without a prominent, almost erect frontal horn (*Neurocordulia molesta* has a triangular, almost flat frontal shelf —Fig. 12.34); legs shorter, the hind femur usually not extending to the hind margin of abdominal segment 8 (Fig. 12.24); metasternum without a tubercle near posterior margin; abdomen less depressed, more cylindrical in outline (Figs. 12.24, 12.35-12.36) ....*Corduliinae* 3
2(1).	Lateral spines of abdominal segment 9 reaching posteriorly to tip of epiproct or beyond (Fig. 12.110); width of head across the eyes about equal to the greatest width behind the eyes; abdominal segment 10 without a middorsal hook ....*Didymops*
2'.	Lateral spines of abdominal segment 9 not reaching posteriorly to tip of epiproct (Fig. 12.111); width of head greater across the eyes than behind the eyes; abdominal segment 10 with a small middorsal hook ....*Macromia*

3(1').	Middorsal hooks present and well developed on some abdominal segments	4
3'.	Middorsal hooks absent or reduced to low knobs	9
4(3).	Without lateral spines on abdominal segment 8, but present on abdominal segment 9	***Williamsonia***
4'.	With lateral spines on abdominal segments 8 and 9	5
5(4').	Crenulations (rounded projections) on distal margin of palpal lobes nearly semicircular or even more deeply cut; lateral spines of abdominal segments 8 and 9 about equal in length, those on 8 divergent (Figs. 12.121-12.122)	***Neurocordulia***
5'.	Crenulations on distal margin of palpal lobes much shallower than a semicircle; lateral spines on abdominal segment 9 about equal in length to those on abdominal segment 8 or much longer	6
6(5').	Lateral spines on abdominal segment 9 at least twice as long as those on abdominal segment 8, at least as long as the middorsal length of abdominal segment 9	7
6'.	Lateral spines on abdominal segment 9 about equal in length to those on abdominal segment 8, less than the middorsal length of abdominal segment 9	8
7(6).	Distal half of dorsal surface of prementum heavily setose (Fig. 12.123); palpal setae usually 4, rarely 5	***Epitheca*** (subgenus *Epicordulia*)
7'.	Distal half of dorsal surface of prementum with few very small setae; palpal setae usually 6-7, rarely 8	***Epitheca*** (subgenus *Tetragoneuria*)
8(6').	Middorsal hooks on abdominal segments 6-9 only, minute on abdominal segment 6, on the other segments as long as the abdominal segment or longer	***Helocordulia***
8'.	Middorsal hooks on abdominal segments 3 or 4 to 9 spine-like and usually curved	***Somatochlora*** (in part)
9(3').	Sides of thorax uniformly colored	***Somatochlora*** (in part)
9'.	Sides of thorax with a broad, dark, longitudinal stripe	10
10(9').	Middorsal hooks absent or vestigial, but, if present, then highest on segment 8; lateral spines of abdominal segment 9 about one-fifth the length of lateral margin of that segment	***Cordulia***
10'.	Middorsal hooks distinct, stubby, and the highest on segment 5; lateral spines of abdominal segment 9 about two-fifths the length of lateral margin of that segment	***Dorocordulia***

Adults

1.	Anal loop about as broad as long, without a midrib (Fig. 12.39); triangle of hind wing remote from arculus (Fig. 12.39); hooks of tarsal claws nearly equal in size, the ventral one usually longer; no ventrolateral carinae (ridges) on abdomen; anterior hamules relatively well developed (Fig. 12.5)	***Macromiinae*** 2
1'.	Anal loop, if present, long and narrow, with 2 rows of cells divided by a midrib, somewhat foot-shaped (Fig. 12.40); triangle of hind wing opposite the arculus (Fig. 12.40); hooks of tarsal claws unequal, the ventral one much shorter; ventrolateral carinae on abdominal segments 3 or 4 to 8 or 9; anterior hamules hardly, if at all, discernible	***Corduliinae*** 3
2(1).	Nodus of fore wing about midway between base and apex of wing; vertex simple, rounded, smaller than occiput, which has a bulbous swelling behind; coloration light brown and yellow, nonmetallic	***Didymops***
2'.	Nodus of fore wing distinctly beyond middle of wing; vertex bilobed, larger than occiput, which lacks a bulbous swelling behind; coloration dark, with a metallic lustre and bright yellow markings	***Macromia***
3(1').	Veins M_4 and Cu_1 in fore wing diverging towards wing margin	4
3'.	Veins M_4 and Cu_1 converging towards wing margin, often being parallel for most of the distance	5
4(3).	Small dragonflies (abdomen and hind wing each less than 25 mm); top of head metallic green; triangles and subtriangles of fore wing and usually of hind wing of a single cell; hind wing without a basal spot	***Williamsonia***
4'.	Medium-sized dragonflies (abdomen and hind wing each more than 30 mm) without metallic lustre; fore wing triangle divided into 2-3 cells; hind wing may have a basal amber spot	***Neurocordulia***
5(3').	Wings usually with large brown spots at nodus and wing tip, especially in Southern specimens, and always with large basal spot in hind wing; length of hind wing 38 mm or more	***Epitheca*** (subgenus *Epicordulia*)
5'.	Wings never with spots at nodus or wing tip, and usually with no more than a trace of color at base of hind wing; length of hind wing much less than 38 mm	6

6(5'). Body without or almost without metallic lustre; tibiae of males with well developed keels on all 3 pairs of legs; vulvar lamina elongate, deeply bilobed and generally flexible, usually longer than sternum of abdominal segment 9 7

6'. Body usually with a metallic green or blue lustre; mesotibiae of males without a keel or with only a vestige at the distal end; vulvar lamina bilobed or entire, sometimes shorter than sternum of abdominal segment 9, and, when as long, more or less rigid, forming an ovipositor (Fig. 12.45) 8

7(6). Abdomen of male widest at segments 6-7; cerci of male separated at base by a space often no wider than one cercus; vulvar lamina bilobed at the base, the lobes longer than sternum of segment 9; wings with one crossvein behind pterostigma ....*Epitheca* (subgenus *Tetragoneuria*)

7'. Abdomen of male widest at segment 8; cerci of male separated at base by a space at least twice the width of one cercus; vulvar lamina bilobed but not to base, the lobes usually shorter than sternum of segment 9; wings usually with 2 crossveins behind pterostigma ....*Helocordulia*

8(6'). Hind wing with 2nd cubito-anal crossvein forming a subtriangle; mesotibiae of males without a keel; epiproct of male normally triangular, rarely bifid.............................*Somatochlora*

8'. Hind wing with only 1 cubito-anal crossvein; mesotibiae of males with a vestigial keel at the distal end; epiproct of male triangular or strongly bifurcate 9

9(8'). Fore wing triangle without a crossvein; each cercus of male tapering to an upturned point; epiproct of male triangular ....*Dorocordulia*

9'. Fore wing triangle with a crossvein; each cercus of male cylindrical, without apex upturned or pointed; epiproct of male strongly bifurcate, each ramus bifid ....*Cordulia*

Libellulidae

Larvae

1. Eyes confined to anterior fourth of head, capping anterolateral part of head and raised dorsally above contour of head capsule (Fig. 12.124); abdomen tapering to an acute point 2

1'. Eyes usually occupying about half the length of head, more broadly rounded and lateral in position, not protruding above contour of head capsule, though sometimes protruding laterally (Fig. 12.125); abdomen variable, usually ending more bluntly 5

2(1). Anterior margin of prementum straight on each side of median line, finely crenulate (rounded) (seen only at magnifications above 15 X), and bases of spinulose setae not in notches between crenulations (Fig. 12.126)....................................*Libellula*

2'. Anterior margin of prementum convex on each side of median line, noticeably crenulate (observable at 5 X), and base of each spinulose seta within notch between two crenulations (Figs. 12.127) 3

3(2'). Dorsal hooks absent on all abdominal segments; eyes protruding dorsally so that a line joining their tips will not touch another part of the head ....*Orthemis*

3'. Dorsal hooks present on middle abdominal segments (at least segments 5 and 6); eyes variable, usually lower 4

4(3'). Dorsal hook absent on abdominal segments 7 and 8, present only on segments 3-6; epiproct longer than middorsal length of abdominal segments 8 + 9; prementum with 8-11 dorsal setae on each side of median line ....*Plathemis*

4'. Dorsal hook present on abdominal segments 4 - 8; epiproct shorter than middorsal length of abdominal segments 8 + 9; prementum with 0-3 dorsal setae on each side of median line, short and inconspicuous when present ....*Ladona*

5(1'). Paraprocts and cerci strongly decurved at tip, best seen in lateral view (Fig. 12.128).........................*Erythemis*

5'. Paraprocts and cerci straight or only slightly decurved in lateral view 6

6(5'). Dorsal hooks, spines, or knobs present on some of abdominal segments 5-9, the low knobs each bearing a conspicuous tuft of hair 7

6'. Dorsal hooks, spines, or knobs absent on abdominal segments 5-9 21

7(6). Dorsal hook, spine, or knob present on abdominal segment 9 8

7'. Dorsal hook, spine, or knob absent on abdominal segment 9 13

8(7'). Dorsal hooks of abdomen cultriform, the series in lateral view like teeth of a circular saw (Fig. 12.36); crenulations (rounded projections) of distal margin of each palpal lobe deep; palpal setae 5-6 ....*Perithemis*

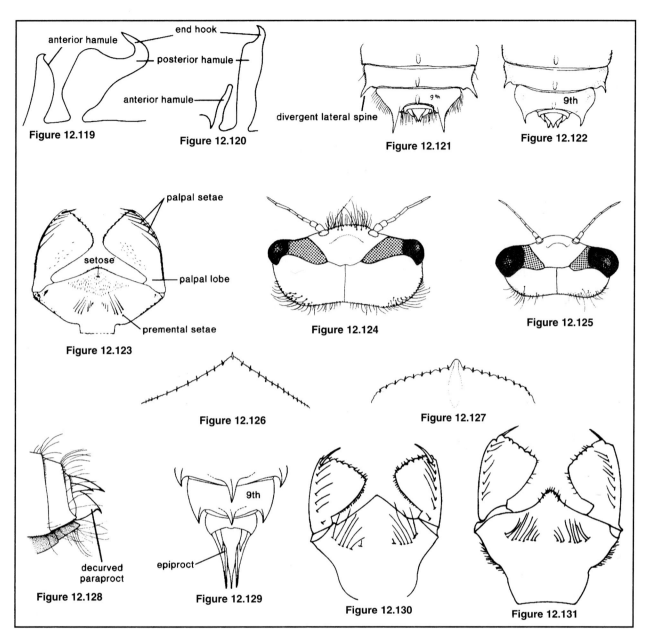

Figure 12.119. Lateral view, hamules of adult male, *Arigomphus villosipes* (Selys) (Gomphidae) (drawn from Needham and Westfall 1955).

Figure 12.120. Lateral view, hamules of adult male, *Stylurus ivae* Williamson (Gomphidae) (drawn from Needham and Westfall 1955).

Figure 12.121. Dorsal view, tip of abdomen of larva, *Neurocordulia obsoleta* (Say) (Corduliidae) (drawn by K. J. Tennessen).

Figure 12.122. Dorsal view, tip of abdomen of larva, *Neurocordulia yamakanensis* (Provacher) (Corduliidae) (Say) (Corduliidae) (Wright and Peterson 1944).

Figure 12.123. Dorsal view, labium of larva, *Epicordulia princeps* (Hagen) (Cordullidae) (Wright and Peterson 1944).

Figure 12.124. Dorsal view, head of larva, *Libellula incesta* (Libellulidae) (drawn by K. J. Tennessen).

Figure 12.125. Dorsal view, head of larva, *Celithemis elisa* (Hagen) (Libellulidae) (drawn by K. J. Tennessen).

Figure 12.126. Anterior margin of prementum of larva, *Libellula incesta* Hagen (Libellulidae) (drawn by K. J. Tennessen).

Figure 12.127. Anterior margin of prementum of larva, *Plathemis lydia* (Drury) (Libellulidae) (drawn by K. J. Tennessen).

Figure 12.128. Lateral view, tip of abdomen of larva, *Erythemis simplicicollis* (Say) (Libellulidae) (drawn by K. J. Tennessen).

Figure 12.129. Dorsal view, tip of abdomen of larva, *Brachymesia gravida* (Calvert) (Libellulidae) (Wright and Peterson 1944).

Figure 12.130. Dorsal view, labium of larva, *Macrothemis celeno* (Selys) (Libellulidae) (Cuba) (drawn by M. S. Westfall).

Figure 12.131. Dorsal view, labium of larva, *Brechmorhoga mendax* (Hagen) (Libellulidae) (drawn by M. S. Westfall).

8'.	Dorsal hooks of abdomen more upright, spinelike, or low and blunt; crenulations of palpal lobes not deep; palpal setae 6-10	9
9(8').	Epiproct about twice as long as its basal width, and as long as or longer than dorsal length of abdominal segments 8 + 9 (Fig. 12.129)	10
9'.	Epiproct as long as, or only slightly longer than, its basal width, and much shorter than dorsal length of abdominal segments 8 + 9	11
10(9)	Cerci about one-third to one-half as long as epiproct in final instar; lateral spine of abdominal segment 9 about one-third as long as lateral margin of which it is a part; palpal setae 6-10	*Brachymesia*
10'.	Cerci about one-fifth as long as epiproct in final instar; lateral spine of abdominal segment 9 about one-half as long as lateral margin of which it is a part; palpal setae 6	*Idiataphe*
11(9').	Crenulations of distal margin of each palpal lobe obsolete (as in Fig. 12.27); surface of larva smooth; palpal setae 7-10	*Dythemis*
11'.	Crenulations of distal margin of each palpal lobe evident (Fig. 12.31); surface of larva smooth or granulose; palpal setae 6-9	12
12(11').	Palpal setae 6; dorsal premental setae in North American species almost in even, smoothly curved line (Fig. 12.130); surface of larva smooth	*Macrothemis*
12'.	Palpal setae 7-9; 2 outermost premental setae set apart from the others and almost at a right angle to them (Fig. 12.131); surface of larva granulose or rough	*Brechmorhoga*
13(7').	Dorsal hook on abdominal segment 8	14
13'.	No dorsal hook on abdominal segment 8	18
14(13).	Palpal setae 7; 9-11 dorsal premental setae on each side of median line (Figs. 12.26-12.27)	*Miathyria*
14.	Palpal setae 9-13; 10-21 dorsal premental setae on each side of median line	15
15(14').	Epiproct shorter than paraprocts, from 0.4 to 0.8 as long; no dorsal hook on abdominal segment 3	16
15'.	Epiproct as long or nearly as long (0.9) as paraprocts; dorsal hook present on abdominal segment 3	17
16(15).	Dorsal hook on abdominal segment 8 extends posteriorly beyond posterior margin of abdominal segment 9; epiproct less than half as long as paraprocts	*Tauriphila*
16'.	Dorsal hook on abdominal segment 8 extends posteriorly only to middle of abdominal segment 9; epiproct 0.6 to 0.8 as long as paraprocts	*Sympetrum* (in part)
17(15').	Dorsal hook on abdominal segment 8 not reaching posterior margin of segment 9; epiproct slightly shorter than paraprocts; 10-15 dorsal premental setae on each side of median line	*Leucorrhinia* (in part)
17'.	Dorsal hook on abdominal segment 8 extending to or beyond posterior margin of abdominal segment 9; epiproct equal to or longer than paraprocts; 16-17 dorsal premental setae on each side of median line	*Macrodiplax*
18(13').	Lateral spines of abdominal segment 9 long and straight, longer than middorsal length of abdominal segment and extending about to tips of paraprocts	*Celithemis*
18'.	Lateral spines of abdominal segment 9 much shorter than middorsal length of segment and not extending to tips of paraprocts	19
19(18').	Distal margin of each palpal lobe with crenulations large, as long as wide; median lobe of prementum markedly angulate (Fig. 12.31)	*Paltothemis*
19'.	Distal margin of each palpal lobe with shallow crenulations, not as deep as width of teeth; median lobe of prementum more rounded (Fig. 12.27)	20
20(19').	Abdomen with low, blunt middorsal prominences on abdominal segments 4-9, each bearing a conspicuous tuft of hairs	*Erythrodiplax* (in part)
20'.	Abdomen with definite middorsal hooks on some abdominal segments, with no conspicuous tufts of hairs on low prominences	*Sympetrum* (in part)
21(6').	Lateral spines of abdominal segment 9 not longer than its middorsal length; no dark ridge running mesad from mesoposterior part of eye; antennae not light with dark bands	22
21'.	Lateral spines of abdominal segment 9 much longer than its middorsal length, or, if equal, with a dark ridge running mesad from mesoposterior part of eye, and antennae light with dark bands	28
22(21).	Small species: length of final instar 10 mm; palpal setae 6; end of abdomen truncated and very hairy	*Nannothemis*
22'.	Larger species: length of final instar 12-20 mm; palpal setae more than 6	23

ODONATA

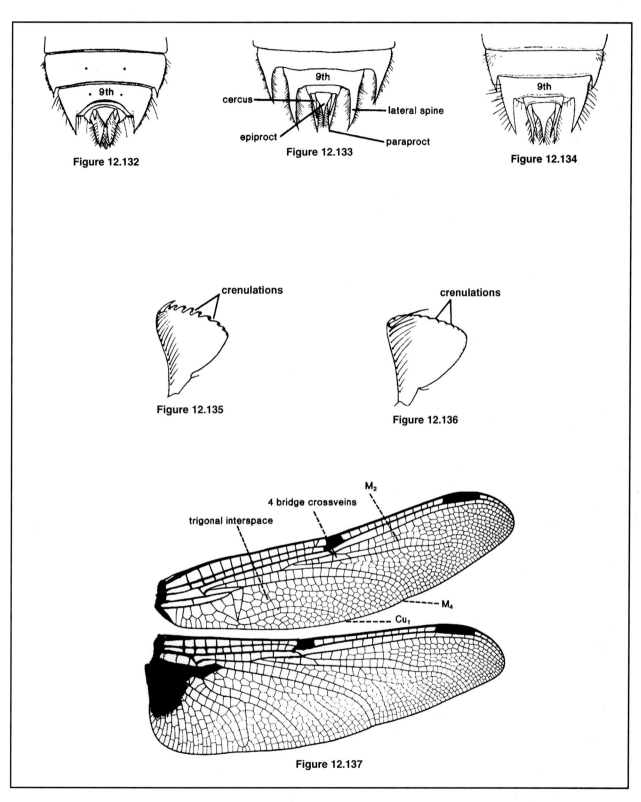

Figure 12.132. Dorsal view, tip of abdomen of larva *Pachydiplax longipennis* (Burmeister) (Libellulidae) (Wright and Peterson 1944).

Figure 12.133. Dorsal view, tip of abdomen of larva, *Tramea* sp. (Libellulidae) (Wright and Peterson 1944).

Figure 12.134. Dorsal view, tip of abdomen of larva, *Pantala flavenscens* (Fabricius) (Libellulidae) (Wright and Peterson 1944).

Figure 12.135. Dorsal view, left palpal lobe of larva, *Pantala falvescens* (Fabricius) (Libellulidae) (Wright and Peterson 1944).

Figure 12.136. Dorsal view, left palpal lobe of larva, *Tramea* sp. (Libellulidae) (Wright and Peterson 1944).

Figure 12.137. Wings of male, *Libellula quadrimaculata* Linnaeus (Libellulidae) (Needham and Westfall 1955).

23(22').	Dorsum of abdominal segments 1-5 conspicuously pale, segments 6-10 much darker	24
23'.	Dorsum of abdominal segments 1-5 not conspicuously paler than segments 6-10	25
24(23).	Lateral margins of abdominal segments 8 and 9 somewhat concave, each bearing 15-16 stout spinules; legs faintly banded; maximum width of abdomen of final instar larva about 6.3 mm (southern Florida)	*Crocothemis servilia*
24'.	Lateral margins of abdominal segments 8 and 9 straight, each bearing 5-6 stout spinules, interspersed with long, thin hairs; legs strongly banded; maximum width of abdomen of final instar larva about 5 mm (southwestern United States)	*Micrathyria hageni*
25(23').	Three wide longitudinal dark bands on underside of abdomen	*Leucorrhinia* (in part)
25'.	No distinct longitudinal band on underside of abdomen	26
26(25').	Abdomen rather abruptly rounded to tip; abdominal segment 6 is about one-fifth as long as wide measured ventrally; epiproct as long as paraprocts; lateral spines of abdominal segments 8 and 9 very small, that of 9 no more than one-fifth of the middorsal length of that segment	*Pseudoleon*
26'.	Abdomen not so abruptly rounded to tip; abdominal segment 6 is about one-fourth as long as wide measured ventrally; epiproct no more than nine-tenths as long as paraprocts; lateral spines of abdominal segments 8 and 9 larger, that of 9 at least two-fifths the middorsal length of that segment	27
27(26').	Epiproct about nine-tenths as long as paraprocts; cerci usually no more than one-half as long as paraprocts in full-grown larvae; lateral spine of abdominal segment 8 at least three-tenths the length of the lateral margin of which it is a part	*Erythrodiplax* (in part)
27'.	Epiproct ususally no more than four-fifths as long as paraprocts; cerci at least three-fifths as long as paraprocts in full-grown larvae, if only one-half, then lateral spines of abdominal segment 8 are minute or absent	*Sympetrum* (in part)
28(21').	Epiproct not more than two-thirds length of paraprocts, nearly equal to middorsal length of segment 9 (Fig. 12.132); a dark line running mesad from mesoposterior part of eye; antennal segment 6 markedly darker than segment 5	*Pachydiplax*
28'.	Epiproct four-fifths or more the length of paraprocts, longer than middorsal length of segment 9 (Figs. 12.133,12.134); no dark line running mesad from mesoposterior part of eye; antennal segment 6 not markedly darker than segment 5	29
29(28').	Dorsal hook present on abdominal segments 2-4; epiproct as long as or longer than paraprocts; lateral spines on abdominal segment 8 less than two-thirds as long as those on segment 9; crenulations of distal margin of palpal lobe about as long as wide (Fig. 12.135)	*Pantala*
29.	No dorsal hook on abdominal segments 2-4; epiproct shorter than paraprocts; lateral spines on abdominal segment 8 about nine-tenths as long as those on segment 9; crenulations of distal margin of palpal lobe wider than long (Fig. 12.136)	*Tramea*

Adults

1.	Fore wing triangle with anterior side bent, thus making the space it encloses a quadrangle; anal loop of hind wing incomplete, open rearward; hind wing 15 mm or less in length	*Nannothemis*
1'.	Fore wing triangle usually normal, 3-sided; anal loop usually complete, and more or less foot-shaped; hind wing longer than 16 mm	2
2(1').	Antenodal crossveins of both wings with row of roundish brown to black spots and distinctive pattern	*Pseudoleon*
2'.	Crossveins with no such coloration	3
3(2').	Mature males brilliant red; front surface of frons with 2 conspicuous flat areas which are roughly triangular; golden basal wing spots; fore wing triangle 2-celled; radial planate subtending one row of cells; single bridge crossvein; Cu_1 in hind wing separated from anal angle of triangle (southern Florida)	*Crocothemis*
3'.	No such combination of characters	4
4(3').	Vein M_2 waved (undulation slight in *Brechmorhoga* and *Macrothemis* (Fig.12.137)	5
4'.	Vein M_2 smoothly curved (Fig. 12.138)	13
5(4).	Wings with several bridge crossveins (except in many *Libellula semifasciata*, which have large spots at nodus and pterostigma but clear wing tips) (Fig. 12.137)	6
5'.	Wings with a single bridge crossvein (Figs. 12.138-140)	8

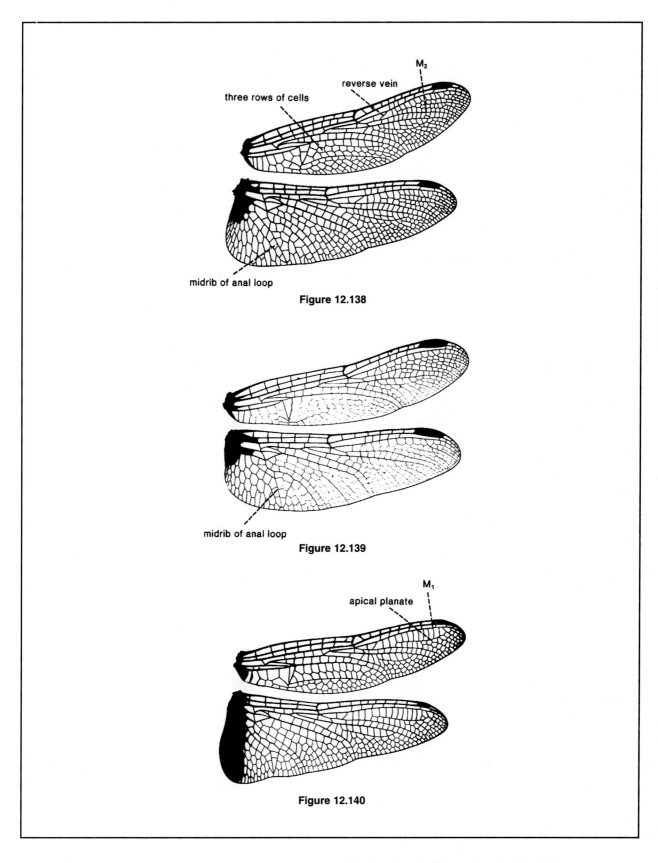

Figure 12.138. Wings of male, *Celithemis verna* Pritchard (Libellulidae) (Needham and Westfall 1955).

Figure 12.139. Wings of male, *Pachydiplax longipennis* (Burmeister) (Libellulidae) (Needham and Westfall 1955).

Figure 12.140. Wings of male, *Miathyria marcella* (Selys) (Libellulidae) (Needham and Westfall 1955).

6(5).	Fore wing triangle of 2 cells	*Ladona*
6'.	Fore wing triangle of 3 or more cells	7
7(6').	First abdominal sternum of male bearing a pair of large, conspicuous processes; abdomen of female with side margins of middle abdominal segments parallel; wings of male and female very differently marked	*Plathemis*
7'.	First abdominal sternum of male bare; abdomen of female slowly tapered rearward on side margins of middle segments; wing markings of male and female similar	*Libellula*
8(5').	Pterostigma very long, surmounting 5-6 crossveins	*Orthemis*
8'.	Pterostigma moderately long, surmounting 1-4 crossveins	9
9(8').	Hind wing with 2 cubito-anal crossveins	*Pantala*
9'.	Hind wing with a single cubito-anal crossvein	10
10(9').	Fore wing with 2 rows of cells beyond triangle	11
10'.	Fore wing with 3 rows of cells beyond triangle	12
11(10).	Fore wing subtriangle with 1-2 cells in species in the United States; median planate absent in fore wing; inner tooth of tarsal claw elongated in species in the United States	*Macrothemis*
11'.	Fore wing subtriangle with 3 cells; median planate present in fore wing; inner tooth of tarsal claw not elongated, not reaching tip of claw	*Brechmorhoga*
12(10').	Two to 4 parallel rows of cells between vein A_2 and marginal row at hind angle of hind wing	*Dythemis*
12'.	Four or 5 very irregular rows of cells between vein A_2 and marginal row at hind angle of hind wing	*Paltothemis*
13(4').	Midrib of anal loop nearly straight or very slightly bent at ankle (Fig. 12.138); reverse vein, postnodal crossveins of 2nd series, and 1st crossvein under pterostigma usually strongly aslant	14
13'.	Midrib of anal loop more angulated (Fig. 12.139); reverse vein and crossveins much less, if at all, aslant	17
14(13).	Fore wing triangle of 2-4 cells, and usually with 3-4 rows of cells beyond in trigonal interspace; radial planate often subtends 2 rows of cells	*Celithemis*
14'.	Fore wing triangle of 1 cell (except in some *Perithemis*), and with 2 rows of cells beyond in trigonal interspace; radial planate subtends a single row of cells	15
15(14').	Fore wing triangle with inner side about as long as front side; more than a single bridge crossvein	*Perithemis*
15'.	Fore wing triangle with inner side much longer than front side; usually with a single bridge crossvein	16
16(15').	Fore wing subtriangle of 1 cell; 2 paranal cells before anal loop	*Idiataphe*
16'.	Fore wing subtriangle of 2 cells; 3 paranal cells before anal loop	*Macrodiplax*
17(13').	Wings with more than a single bridge crossvein	*Micrathyria*
17'.	Wings with 1 bridge crossvein only	18
18(17').	Wings with triple-length vacant space before single crossvein that is either under distal end of pterostigma or just beyond it (Fig. 12.139)	*Pachydiplax*
18'.	Wings with one or more crossveins under pterostigma, with no triple-length space adjacent to them	19
19(18').	Wings with a single crossvein under the pterostigma; without a dark band across entire base of hind wing	*Sympetrum*
19'.	Wings usually with 2 or more crossveins under pterostigma; there may be a dark band across entire base of hind wing	20
20(19').	Fore wing triangle with inner side less than twice as long as front side; pterostigma short and thick, about twice as long as wide, at least in male; face white	*Leucorrhinia*
20'.	Fore wing triangle with inner side more than twice as long as front side; pterostigma about 3 times as long as wide; face not white	21
21(20').	Hind wing with 2 paranal cells before anal loop	*Brachymesia*
21'.	Hind wing with 3 paranal cells before anal loop	22
22(21').	Wings with pterostigma trapezoidal, front side distinctly longer than rear; some double-length cells above apical planate, reaching from planate to M_1 (Fig. 12.140)	23
22'.	Wings with front and rear sides of pterostigma about equal in length; apical planate poorly developed and with no double-length cells above it	25
23(22).	All cells above apical planate double-length and in a single row (Fig. 12.140); one crossvein under pterostigma (Fig. 12.140)	*Miathyria*
23'.	About half of these cells in a single row, then a double row; 2 crossveins under pterostigma	24

24(23'). Fore wing with 3 rows of cells in trigonal interspace (distal to triangle, between veins M_4 and Cu_1 (as in Fig. 12.138) .... *Tauriphila*
24'. Fore wing with 4 rows of cells in trigonal interspace .... *Tramea*
25(22'). Spines on outer angle of hind femur gradually increasing in length distally; fore wing with 5 paranal cells before subtriangle .... *Erythrodiplax*
25'. Spines on basal half or two-thirds of outer angle of hind femur short and of about equal length, with 2-4 large spines on distal half or third .... *Erythemis*

ADDITIONAL TAXONOMIC REFERENCES

General
Muttkowski (1910); Munz (1919); Needham and Heywood (1929); Wright and Peterson (1944); Snodgrass (1954); Needham and Westfall (1955); Gloyd and Wright (1959); Smith and Pritchard (1956); Pennak (1978); Corbet (1980); McCafferty (1981); Bick (1984a,b); Davies and Tobin (1984, 1985); Garrison (1991); Pfau (1991); Tsuda (1991); Bridges (1993); Westfall and May (1996).*

Regional faunas
Alaska: Walker (1953, 1958); Walker and Corbet (1975).
British Columbia: Whitehouse (1941); Scudder et al. (1976); Cannings and Stuart (1977).
California: Kennedy (1917); Smith and Pritchard (1956).
Canada: Walker (1953, 1958); Walker and Corbet (1975).
Connecticut: Garman (1927).
Florida: Byers (1930); Paulson (1966); Johnson and Westfall (1970); Dunkle (1989, 1990, 1992).
Illinois: Needham and Hart (1901); Garman (1917).
Indiana: Williamson (1900).
Massachusetts: Carpenter (1991).
Nevada: Kennedy (1917).
New England: Howe (1917-1923).
New York: Needham (1901, 1903).
North Carolina: Huggins and Brigham (1982).
Oregon: Kennedy (1915).
Philadelphia and vicinity: Calvert (1893).
Quebec: Robert (1963).
South Carolina: Huggins and Brigham (1982).
Southeastern United States: Louton (1982).
Texas: Johnson (1972).
Utah: Musser (1962).
Washington: Kennedy (1915).
Wisconsin: Muttkowski (1908).
Wyoming: Molnar and Lavigne (1979).

Regional species lists
Alabama: Bick and Bick (1983); Tennessen, et al. (1995).**
Alberta: Acorn (1983); Pritchard (1988)
Arizona: Donnelly (1961b).
Arkansas: Bick (1959); Harp and Rickett (1977).
Bahamas: Paulson (1966); Dunkle (1985).
Bermuda: Dunkle (1988).
British Columbia: Whitehouse (1941); Scudder et al. (1976); Cannings, S.G. (1981); Cannings, R.A. (1982a).
Canada: Corbet (1979).
Colorado: Bick and Hornuff (1974); Evans (1988).
Florida: Pearse (1932); Westfall (1953); Dunkle and Westfall (1982); Daigle (1991, 1992).
Illinois: Cashatt (1987).
Indiana: Williamson (1917); Montgomery (1941, 1947, 1948).
Iowa: Miller (1906); Hummel and Haman (1975).
Kansas: Huggins et al. (1976); Huggins (1983).
Kentucky: Garman (1924).
Louisiana: Bick (1957a).
Maine: Borror (1944); White (1989).
Manitoba: Conroy and Kuhn (1977).
Maryland: Fisher (1940).
Massachusetts: Gibbs and Gibbs (1954).
Michigan: Byers (1927); Kormondy (1958).
Minnesota: Whedon (1914); Hamrum et al. (1971).
Mississippi: Bick (1950); Westfall (1952); Lago et al. (1979).
Missouri: Williamson (1932).
Montana: Bick and Hornuff (1974); Miller, and Gustafson (1996).***
Nebraska: Pruess (1968); Bick and Hornuff (1972).
Nevada: LaRivers (1940).
New Hampshire: White and Morse (1973).
New Jersey: Gillespie (1941); Soltesz (1991).
New York: Needham (1901b, 1928); Donnelly (1992).
North Carolina: Brimley (1938); Westfall (1942); Wray (1967); Paulson and Jenner (1971); Cuyler (1984).
North Central United States: Montgomery (1967).
North Dakota: Bick et al. (1977).
Ohio: Borror (1937); Perry (1983); Alrutz (1992); Glotzhober (1995).****
Oklahoma: Bird (1932); Bick (1957b); Bick and Bick (1958).
Pacific Coast States: Paulson and Garrison (1977).
Pennsylvania: Beatty and Beatty (1968); Beatty et al. (1969); Opler (1985).
Prince Edward Island: Hilton (1990).
Quebec: Hutchinson and Larochell (1977); Pilon et al. (1990, 1991); Hilton (1992).
South Carolina: Montgomery (1940); Cross (1955); White et al. (1980, 1983).
South Dakota: Bick and Hornuff (1972); Bick et al. (1977).
Southeastern United States: Cuyler (1968); Krotzer and Krotzer (1992).
Tennessee: Williamson (1903); Wright (1938); Wasscher (1991).
Texas: Ferguson (1940); Gloyd (1958); Donnelly (1978); Young and Bayer (1979); Williams (1982).
Utah: Brown (1934).
Virginia: Byers (1951); Gloyd (1951); Matta (1978); Voshell and Simmons (1978); Carle (1979b).
Washington: Paulson (1970).
Washington, D.C. and vicinity: Donnelly (1961a).
West Virginia: Cruden (1962); Harwood (1971).
Western United States: Montgomery (1968).
Wisconsin: Ries (1967, 1969); Hilsenhoff (1981); Smith et al. (1993).
Wyoming: Bick and Hornuff (1972).
Yukon: Cannings et al. (1991).

Taxonomic treatments at the suborder, family, and generic levels (L = larvae; A = adults)
Aeshnidae: Martin (1908, 1909a, 1909b)-A; Walker (1912)-L, A; Calvert (1934)-L; Dunkle (1977b)-L.
Anisoptera: Daigle (1992)-L.
Calopterygidae: Johnson (1974)-A; Tennessen (1984)-L; Garrison (1990)-A.
Coenagrionidae: Garrison (1984)-L, A.
Cordulegastridae: Fraser (1929)-A; Carle (1983)-A; Lohmann (1992, 1993)-A.
Corduliidae: Martin (1907)-A; Walker (1925)-L, A; Davis (1933)-A; Byers (1937)-L, A; Kormondy (1959)-L,A; Dunkle (1977a)-L.
Gomphidae: Walker (1928)-L, A; Walker (1933)-L; Byers (1939)-A; Needham (1941)-L; Belle (1973)-A; Westfall and Tennessen (1979)-L, A; Carle (1979a, 1980, 1986, 1992)-L, A.
Libellulidae: Ris (1909-1916)-A; Ris (1930)-A; Davis (1933)-A; Kormondy (1959)-L, A; Bennefield (1965)-A.; May (1992)-A.
Odonata: Carpenter (1992)-A (fossil).
Petaluridae: Fraser (1933)-A.
Zygoptera: Munz (1919)-A; Daigle (1991)-L.

---

* Westfall, M.J., Jr. and M.L. May. 1996. Damselflies of North America. Scien. Publ., Gainesville, FLA. 649 p.
** Tennessen, K.J., J.D. Harper, and S. Krotzer. 1995. The distribution of Odonata in Alabama. Bull. Amer. Odonatol. 3: 49-74.
*** Miller, K.B., and D.L. Gustafson. 1996. Distribution records of the Odonata of Montana. Bull. Amer. Odonatol. 3: 75-88.
**** Glotzhober, R.C. 1995. The Odonata of Ohio - a preliminary report. Bull. Amer. Odonatol. 3: 1-30.

Table 12A. Summary of ecological and distributional data for *Odonata (dragonflies and damselflies)*. (For definition of terms see Tables 6A-6C; table prepared by K. W. Cummins, R. W. Merritt, D. M. Johnson, and K. J. Tennessen.)

Taxa (number of species in parentheses)	Habitat	Habit	Trophic Relationships	North American Distribution	Ecological References†
Anisoptera (dragonflies)	Wide variety of lentic and lotic habitats	Sprawlers, climbers, burrowers	Predators (engulfers)		504, 536, 703, 708, 855, 1354, 2002, 2865, 3076, 3391, 4009, 4151, 4171, 4172, 4342
Petaluridae (2)	Lotic—depositional, lentic—littoral	Primarily burrowers	Predators (engulfers)		2865, 3713
Tachopteryx (1) (*thoreyi*)	Lotic—depositional (margins and moss of spring streams; upper edges of hillside seepages in deciduous forests), lentic littoral (bogs)	Sprawlers	Predators (engulfers)	East (particularly mountains)	991, 2072, 2865, 3713
Tanypteryx (1) (*hageni*)	Bogs	Burrowers	Predators (engulfers)	West (mountains)	535, 2865, 3713, 3914, 4171
Gomphidae (97)	Generally lotic—depositional, lentic—littoral (sediments)	Primarily burrowers	Predators (engulfers, generally await prey)		1429, 2112, 2865, 3076, 3713
Aphylla (3)	Lotic—depositional, lentic—littoral	Burrowers	Predators (engulfers)	Widespread (primarily South)	304, 1429, 2865
Arigomphus (7) (= *Orcus*)	Lentic—littoral, some lotic—depositional (sediments, primarily silt)	Burrowers	Predators (engulfers)	East, Southwest	1429, 2865, 4185
Dromogomphus (3)	Lotic—erosional and depositional (detritus, larger rivers)	Burrowers	Predators (engulfers)	East	1429, 2865, 4171, 4311, 2441
Erpetogomphus (5)	Lotic—depositional	Burrowers	Predators (engulfers)	Widespread (except North)	2865, 3713
Gomphus (38)	Lotic—depositional, lentic—littoral (sediments, primarily silt)	Burrowers	Predators (engulfers)	Widespread	2848, 2863, 3713, 4171, 4314
Hagenius (1) (*brevistylus*)	Lotic—erosional and depositional (detritus, at margins), lentic—littoral (at margins)	Sprawlers	Predators (engulfers)	East, Southwest	1429, 2865, 4171
Lanthus (2)	Lotic—erosional and depositional (sand and detritus in spring streams)	Burrowers (active)	Predators (engulfers)	East, Southeast	554, 1429, 2849, 4171, 4180
Octogomphus (1) (*specularis*)	Lotic—depositional (detritus)	Burrowers	Predators (engulfers)	West Coast	554, 2865, 3713, 4171
Ophiogomphus (18)	Lotic—erosional and depositional (sand) of small cold streams	Burrowers	Predators (engulfers)	Widespread (particularly mountains and high latitudes)	555, 556, 987, 1429, 2076, 2168, 2849, 2865, 852, 3873, 4169, 4171
Phyllogomphoides (2)	Lotic—depositional	Burrowers	Predators (engulfers)	Texas	1428, 1429
Progomphus (4)	Lotic—erosional and depositional (sand including intermittent streams), lentic—littoral (sand)	Burrowers	Predators (engulfers)	Widespread (except North)	1428, 2126, 2854, 2863, 3713, 4055
Slylogomphus (1) (*albistylus*)	Lotic—erosional and depositional (sand and detritus in spring streams)	Burrowers (active)	Predators (engulfers)	East, Southeast	1429

* includes subgenera *Gomphus* (17), *Gomphurus* (15), *Hylogomphus* (6).
† Emphasis on trophic relationships.

Table 12A. Continued

Taxa (number of species in parentheses)	Habitat	Habit	Trophic Relationships	North American Distribution	Ecological References†
Stylurus 12)	Lotic—depositional, lentic—littoral (sediments, primarily silt)	Burrowers	Predators (engulfers)	Widespread	1422, 1429, 2861, 2865, 4171
Aeshnidae (38) (darners) (= Aeschnidae)	Primarily lentic—littoral (especially vascular hydrophytes)	Generally climbers	Predators (engulfers, usually stalk prey)		139, 1429, 2112, 2865, 3076
Aeshna (20)	Lentic—vascular hydrophytes (lakes, ponds, marshes, and bogs)	Climbers	Predators (engulfers, stalk prey; Diptera, Coleoptera, Trichoptera, Ephemeroptera)	Widespread	340, 536, 537, 937, 2865, 1993, 3198, 3713, 4171, 4093, 1990, 2009, 4094, 1992, 1994
Anax (4)	Lentic—vascular hydrophytes	Climbers	Predators (engulfers, stalk prey)	Widespread (especially South and except far North)	523, 701, 1240, 2429, 2850, 2951, 3713, 4171, 278, 340, 341, 342, 1242, 3404, 4091, 4518
Basiaeschna (1) (*janata*)	Lotic—depositional (detritus and sediments, under rocks)	Climbers—sprawlers, clingers	Predators (engulfers)	East, Southwest	2863, 4171
Boyeria (2) (=*Fonscolombia*)	Lotic—erosional and depositional (detritus, vascular hydrophytes)	Climbers—sprawlers (clingers under rocks)	Predators (engulfers)	East	1339, 2863, 4171, 4490
Coryphaeschna (4)	Lentic—vascular hydrophytes	Climbers	Predators (engulfers)	Southeast, Arizona	1429, 2865
Epiaeschna (l) (*heros*)	Lentic—littoral (detritus—woodland ponds) and vascular hydrophytes (marshes)	Climbers—sprawlers	Predators (engulfers)	East, Southwest	1429, 2865, 4171, 4455
Gomphaeschna (2)	Lentic—vascular hydrophytes (bogs)	Climbers—clingers	Predators (engulfers)	East, Central	989, 1429, 2865, 4171
Gynacanrha (1) (*nervosa*)	Lentic—vascular hydrophytes (temporary ponds)	Climbers—sprawlers	Predators (engulfers)	Florida, Georgia, California, and Oklahoma	2865, 3713
Nasiaeschna (1) (*pentacantha*)	Lotic—erosional (detritus and debris jams) and depositional (detritus)	Climbers—clingers	Predators (engulfers)	East, Southwest	1429, 2865, 4171
Oplonaeschna (1) (*armata*)	Lotic—erosional (detritus and sediments under rocks)	Clingers—climbers	Predators (engulfers)	Arizona, New Mexico	1425, 1995, 2865
Triacanthagyna (1) (*trifida*)	Lentic—vascular hydrophytes (temporary ponds)	Climbers—sprawlers	Predators (engulfers)	Southeast, California	1429, 2865
Cordulegastridae (8) (biddies, flying adders)	Lotic—depositional	Burrowers	Predators (engulfers)		1429, 2112, 2865, 3076, 3713
Cordulegaster (8) (=*Taeniogaster, Thecaphora, Zoraena*)	Lotic—depositional (headwater streams, sand, silt, detritus)	Burrowers	Predators (engulfers, await prey)	East, West	2072, 2865, 3713, 4171, 2359, 2367
Corduliidae (59)	Primarily lotic—depositional (small streams, large rivers) and lentic—littoral (bogs, sediments)	Both sprawlers and climbers	Predators (engulfers)		1425, 2865, 3713
Macromiinae *Didymops* (2)	Lentic—littoral (sand and silt), lotic—depositional	Sprawlers	Predators (engulfers, await prey)	East, Central, Southeast	2865, 3713

† Emphasis on trophic relationships.

Table 12A. Continued

Taxa (number of species in parentheses)	Habitat	Habit	Trophic Relationships	North American Distribution	Ecological References†
Macromia (8)	Same as *Didymops*	Sprawlers	Predators (engulfers, await prey)	East, West Coast	2071
Corduliinae					
Cordulia (1) (*shurtleffi*)	Lentic—littoral (detritus at margins of bogs and ponds)	Sprawlers	Predators (engulfers; Diptera, amphipods)	North	2865, 3198, 3713, 4172, 1990, 1991, 1992, 4066
Dorocordulia (2)	Lentic—littoral (bogs)	Sprawlers—climbers?	Predators (engulfers)	North	1429, 2865
‡*Epitheca* (10) (includes subgenera *Epicordulia* and *Tetragoneuria*)	Lotic—depositional, lentic—littoral (pond margins in detritus), lotic—depositional (detritus)	Climbers, sprawlers	Predators (engulfers)	Widespread	2004, 3138, 249, 251, 768, 769, 770, 1429, 2381, 2383, 2865, 3713, 2000, 2003, 2163, 2164, 2494, 2660, 4172, 4518
Helocordulia (2)	Lotic—depositional	Sprawlers	Predators (engulfers)	East	1429
Neurocordulia (6)	Lotic—depositional, lentic—littoral	Climbers, clingers on rocks, roots, and sticks	Predators (engulfers)	Widespread (particularly Southeast)	1429, 2865, 4422, 4424, 4426
Somatochlora (26)	Lentic—littoral (bogs), lotic depositional (springs)	Sprawlers	Predators (engulfers)	North	939, 988, 2865, 3713, 3961, 4172, 4421
Williamsonia (2)	Lentic—littoral (bogs)	Sprawlers	Predators (engulfers)	Northeast	2865, 4337
Libellulidae (97)	Generally lentic—vascular hydrophytes	Primarily sprawlers	Predators (engulfers)		367, 1429, 2860, 2865, 3713
Brachymesia (3) (= *Cannacria*)	Lentic—littoral (brackish)	Sprawlers	Predators (engulfers)	South	1429, 2865
Brechmorhoga (1) (*mendax*)	Lotic—depositional (sediments in pools of torrential streams)	Sprawlers	Predators (engulfers)	Southwest	2865, 3713
Cannaphila (1) (*insularis*)			Predators (engulfers)	Texas	1429, 2865
Celithemis (8)	Lentic—vascular hydrophytes	Climbers	Predators (engulfers)	East	250, 251, 254, 1429, 2003, 2660, 2865, 3996, 4518
Dythemis (3)	Lotic—erosional and depositional (sand)	Sprawlers (active)	Predators (engulfers, stalk prey)	Far South (particularly Southwest)	2865, 3713
Erythemis (4) (= *Lepthemis*, *Mesothemis*)	Lentic—littoral (silt in ponds)	Sprawlers	Predators (engulfers)	Widespread (except far North)	296, 1429, 2865, 3404, 3713, 4517, 4518, 4519
Erythrodiplax (4)	Lentic—vascular hydrophytes	Climbers	Predators (engulfers)	South, East	999, 2865, 3713
Idiataphe (1) (*cubensis*) (= *Ephidatia*)	Lentic—vascular hydrophytes (emergent zone—brackish water)	Sprawlers	Predators (engulfers)	Florida	1429, 2865
Ladona (3)	Lentic—littoral (sediments in ponds)	Sprawlers	Predators (engulfers)	Widespread	249, 250, 251, 254, 1429, 2865, 3138
Leucorrhinia (7)	Lentic—vascular hydrophytes (including bogs)	Climbers	Predators (engulfers, await prey; Diptera, Coleoptera, Trichoptera, Ephemeroptera)	Widespread in North	2865, 2951, 3198, 3211, 1672, 3713, 3741, 4172, 1990, 1991, 1992
Libellula (18) (=*Belonia*)	Lentic—littoral (silt and detritus, vascular hydrophytes), lotic—depositional (sediments)	Sprawlers	Predators (engulfers)	Widespread	249, 340, 2737, 2865, 3211, 3404, 3713, 4423, 4517, 4518

† Emphasis on trophic relationships.
‡ includes subgenera *Epicordulia* and *Tetragoneuria*, sometimes considered genera.

Table 12A. Continued

Taxa (number of species in parentheses)	Habitat	Habit	Trophic Relationships	North American Distribution	Ecological References†
Macrodiplax (1) (*balteata*)	Lentic—littoral (brackish water)	Sprawlers	Predators (engulfers)	South (particularly Southeast)	1429, 2865
Macrothemis (2)	Lentic—littoral	Sprawlers	Predators (engulfers)	Arizona, Texas	1429, 2865
Miathyria (1) (*marcella*) (=*Nothifixis*)	Lentic—littoral (in water hyacinth roots)	Climbers	Predators (engulfers)	Extreme South	1429, 2865
Micrathyria (4)	Lentic—littoral (vascular hydrophytes	Sprawlers	Predators (engulfers)	Texas, Arkansas	1429, 2865
Nannothemis (1) (*bella*) (=*Aino*)	Lentic—vascular hydrophytes (emergent zone in small puddles away from water's edge)	Sprawlers—climbers	Predators (engulfers)	East	1429, 2865
Orthemis (1) (*ferruginea*) (=*Neocysta*)	Lentic—littoral	Sprawlers	Predators (engulfers)	South	2865, 3713
Pachydiplax (1) (*longipennis*)	Lentic—littoral (detritus and silt), lotic—depositional (detritus)	Sprawlers	Predators (engulfers)	Widespread (except far North)	986, 2760, 2865, 3713, 3404, 4090, 4518
Paltothemis (1) (*lineatipes*)	Lotic—erosional (among rocks)	Sprawlers	Predators (engulfers)	Southwest	990, 2865, 3713
Pantala (2)	Lentic—littoral (sediments and macroalgae in temporary ponds)	Sprawlers (active foragers)	Predators (engulfers; Chironomidae)	Widespread	212, 507, 2865, 3713
Perithemis (3)	Lotic—depositional (margins)	Sprawlers (active)	Predators (engulfers)	Widespread (except North)	1429, 2760, 2865, 3404 4518
Plathemis (2)	Lentic—littoral	Sprawlers	Predators (engulfers)	Widespread	2865, 3404, 3713
Pseudoleon (1) (*superbus*)	Lotic—depositional	Sprawlers	Predators (engulfers)	Southwest	3713
Sympetrum (15) (=*Diplax*, =*Tarnetrum*)	Lentic—littoral (detritus and vascular hydrophytes in ponds)	Sprawlers—climbers	Predators (engulfers)	Widespread	699, 2865, 2893, 3198 2660, 3138, 3713, 4572, 4518
Tauriphila (2)	Lentic—littoral	Sprawlers	Predators (engulfers)	Florida, Texas	2865
Tramea (7) (=*Trapezostigma*)	Lentic—littoral (silt and detritus, vascular hydrophytes, and macroalgae)	Sprawlers	Predators (engulfers)	Widespread	298, 881, 2294, 2865, 3713, 4091, 4092, 4517, 4518, 4519
Zygoptera (damselflies)	Both lentic and lotic habitats	Generally climbers	Predators (engulfers)		11, 367, 767, 855, 1429, 2002, 2112, 2286, 2287, 3042, 3076, 3391, 3713, 4151, 4170, 4342
Calopterygidae (8) (=Agrionidae, = Agriidae)	Generally lotic—erosional (margins and detritus) and depositional (detritus)	Generally climbers	Predators (engulfers)		1429, 2112, 3076, 3713, 4170
Calopteryx (5) (=*Agrion*)	Lotic—erosional and depositional (margins and detritus)	Climbers	Predators (engulfers)	Widespread	529, 2493, 3199, 3713 4170
Hetaerina (3)	Lotic—erosional and depositional (margins and detritus)	Climbers—clingers	Predators (engulfers)	Widespread	3713, 4170
Lestidae (18)	Lotic—depositional and lentic—littoral	Climbers	Predators (engulfers)		1429, 2112, 3076, 3713 4170

† Emphasis on trophic relationships.

Table 12A. Continued

Taxa (number of species in parentheses)	Habitat	Habit	Trophic Relationships	North American Distribution	Ecological References†
Archilestes(2)	Lotic—depositional (detritus and vascular hydrophytes), lentic—vascular hydrophytes	Climbers	Predators (engulfers)	West	1913, 3713
Lestes(16)	Lentic—vascular hydrophytes, lotic—depositional (vascular hydrophytes)	Climbers—swimmers	Predators (engulfers)	Widespread	539, 699, 1194, 1351 1913, 2382, 2849, 3548 1632, 3713, 4170, 4317, 155, 2009, 2959
Protoneuridae(2)	Lotic—erosional and depositional	Generally climbers—clingers	Predators (engulfers)		1429, 2112, 3076, 3713 4170
Neoneura (1) (*aaroni*)	Lotic—erosional (on rocks)	Climbers—clingers	Predators (engulfers)	Texas	1429
Protoneura (1) (*cara*)	Lotic—depositional (vascular hydrophytes)	Climbers—clingers (on floating leaves)	Predators (engulfers)	Texas	
Coenagrionidae (96) (=Agrionidae)	Wide range of lentic and lotic habitats	Generally climbers	Predators (engulfers)		1429, 2112, 3076, 3713 4170
Acanthagrion (1) (*quadratum*)	Lentic—vascular hydrophytes	Climbers	Predators (engulfers)	Texas	1370
Amphiagrion (2)	Lotic—depositional (vascular hydrophytes), lentic—vascular hydrophytes (emergent zone) including bogs	Climbers	Predators (engulfers)	Widespread (particularly North)	3713, 4170, 4345
Apanisagrion (1) (*lais*)	Lotic—depositional (spring fed regions)	Sprawlers?	Predators (engulfers)	Arizona	
Argia (30) (=*Hyponeura*)	Lotic—erosional (sediments and detritus) and depositional, lentic—erosional and littoral (sediments)	Clingers, climbers—sprawlers	Predators (engulfers)	Widespread	363, 852, 1426, 1427, 2071, 2168, 3202, 3713, 4170
Chromagrion (1) (*conditum*)	Lotic—erosional (detritus) and depositional (detritus and vascular hydrophytes in small spring streams)	Climbers	Predators (engulfers)	East	1353, 1429, 2861, 4170
Coenagrion (3) (=*Agrion*)	Lotic—erosional and depositional (emergent vascular hydrophytes at margin), lentic—vascular hydrophytes (marshes and bogs)	Climbers	Predators (engulfers)	North, West	154, 156, 542, 3023, 3713, 155, 1990, 1991, 4170, 1992, 1994, 2959
Enallagma(35) (=*Teleallagma*)	Lentic—vascular hydrophytes (including brackish and alkaline waters), lotic depositional (vascular hydrophytes)	Climbers	Predators (engulfers; Cladocera)	Widespread	1353, 1843, 1844, 1914, 2072, 2168, 2850, 2861, 3211, 3713, 3963, 4170, 626, 627, 628, 852, 105, 158, 342, 4096, 2004, 2631, 3139
Hesperagrion (1) (*heterodoxum*)	Lentic—vascular hydrophytes	Climbers	Predators (engulfers)	Southwest	1429
Ischnura (14) (including *Ischnuridia, Celaenura, Anomalura Anomalagrion*)	Lentic—vascular hydrophytes, lotic depositional (vascular hydrophytes), small spring seeps	Climbers	Predators (engulfers; Cladocera, Chironomidae)	Widespread	538, 940, 1355, 1359, 1429, 11, 1480, 1999, 2001, 2071, 2849, 2915, 3023, 3713, 1699, 3990, 3991, 4170, 158, 160, 161, 1477, 1644, 2632, 2959, 3403

† Emphasis on trophic relationships.

Table 12A. Continued

Taxa (number of species in parentheses)	Habitat	Habit	Trophic Relationships	North American Distribution	Ecological References†
Nehalennia (4) (=*Argiallagma*)	Lentic—vascular hydrophytes (including edges of bog mats)	Climbers	Predators (engulfers)	East, North	1350, 1370, 1429, 4170
Neoerythromma (1) (*cultellatum*)	Lentic—vascular hydrophytes	Climbers	Predators (engulfers)	Florida	1350
Telebasis (2)	Lentic—vascular hydrophytes	Climbers	Predators (engulfers)	Widespread in South, Southwest	3489, 3713, 4313
Zoniagrion (1) (*exclamationis*)	Lotic—depositional (vascular hydrophytes *Sparganium*)	Climbers (bases of *Sparganium*)	Predators (engulfers)	California	2072, 3713

† Emphasis on trophic relationships.

SEMIAQUATIC ORTHOPTERA

Irving J. Cantrall
University of Michigan, Ann Arbor

Merlyn A. Brusven
University of Idaho, Moscow

INTRODUCTION

Orthopterans are generally terrestrial, yet numerous forms are hydrophilous to some extent, and a few are adapted to live on emergent aquatic vegetation. Most species have one-year life cycles, although some may require up to three years. Larvae undergo gradual metamorphosis and adults lay their eggs in loose soil, plant tissue, or burrows. Most Orthoptera shred live plant tissue; however, some tettigoniids are predators and some gryllids are collectors adapted to feed on detrital particles. Many of the semiaquatic forms have distinctive morphological and behavioral adaptations.

Numerous species of Tetrigidae (grouse or pygmy locusts; Fig. 13.1) live in moist areas along edges of aquatic habitats, and several species of Tettigoniidae (katydids; Figs. 13.7, 13.8) in the genera *Conocephalus* and *Orchelimum* commonly inhabit marshes and the marginal vegetation of freshwater environments. The Tetrigidae have no special adaptations for swimming, but can move readily through or on water. The semiaquatic tetrigids and the tettigoniid *Orchelimum bradleyi* have been observed to dive and swim to submerged objects.

Two species of Tridactylidae (pygmy mole crickets; Fig. 13.2) inhabit wet shores of streams and back bays (minimal wave action) of small lakes where they dig shallow burrows in moist sand. Pairs of long slender plates on the hind tibiae aid in swimming. Of the Acrididae (grasshoppers; Fig. 13.3) known from North America, seven species in four genera commonly occur on emergent aquatic plants, herbaceous riparian vegetation along streams and lakes, or in boggy meadows and swamps. They sometimes dive and cling to submerged vegetation. The hind tibiae aid in swimming; they are widened distally and have a flattened upper surface with laminate edges bearing fine hairs between the spines.

Approximately 10% of the nearly 100 North American species of Gryllidae (crickets; Fig. 13.5) occur on the ground or in foliage along margins of aquatic habitats. They belong to the subfamilies Nemobiinae, Eneopterinae, Mogoplistinae, and especially Trigonidiinae.

Two species in the genus *Neocurtilla* of the Gryllotalpidae (mole crickets; Fig. 13.6) are semiaquatic. They burrow with flattened front tibiae in moist or muddy sand along margins of freshwater habitats. A body covering of short, fine hairs traps air and increases buoyancy.

EXTERNAL MORPHOLOGY

Orthoptera (grasshoppers, pygmy locusts, katydids, crickets) are distinctive insects having generally four wings, chewing mouthparts, and enlarged hind femura for jumping. The front wings, when present, vary from being fully developed to being reduced to small scalelike structures (Tetrigidae). The fore wings (tegmina) are thickened, leathery, or parchmentlike in texture. The hind wings, when present, are membranous, broad, and folded fanlike beneath the front wings. Many orthopterans produce sound with a stridulating apparatus at the base of the tegmina (Figs. 13.7, 13.8).

KEY TO FAMILIES OF ORTHOPTERA CONTAINING ONE OR MORE SEMIAQUATIC SPECIES

1.	Front and middle tarsi 2-segmented	2
1'.	Front and middle tarsi 3- or 4-segmented	3
2(1).	Pronotum prolonged posteriorly to or surpassing apex of abdomen (Fig. 13.1) (pygmy or grouse locusts)	*TETRIGIDAE*
2'.	Pronotum not prolonged posteriorly (Fig. 13.2) (pygmy mole crickets)	*TRIDACTYLIDAE* (p. 214)
3(1').	All tarsi 3-segmented; antennae longer or shorter than body	4
3'.	All tarsi 4-segmented; antennae longer than body (Figs. 13.7, 13.8) (katydids)	*TETTIGONIIDAE*
4(3).	Front legs similar to middle legs; body not velvety in appearance	5
4'.	Front legs strongly flattened, front tibiae enlarged and fitted for digging; body covered with very short hairs, giving a velvety appearance; antennae shorter than body length (Fig. 13.6) (mole crickets)	*GRYLLOTALPIDAE* (p. 214)

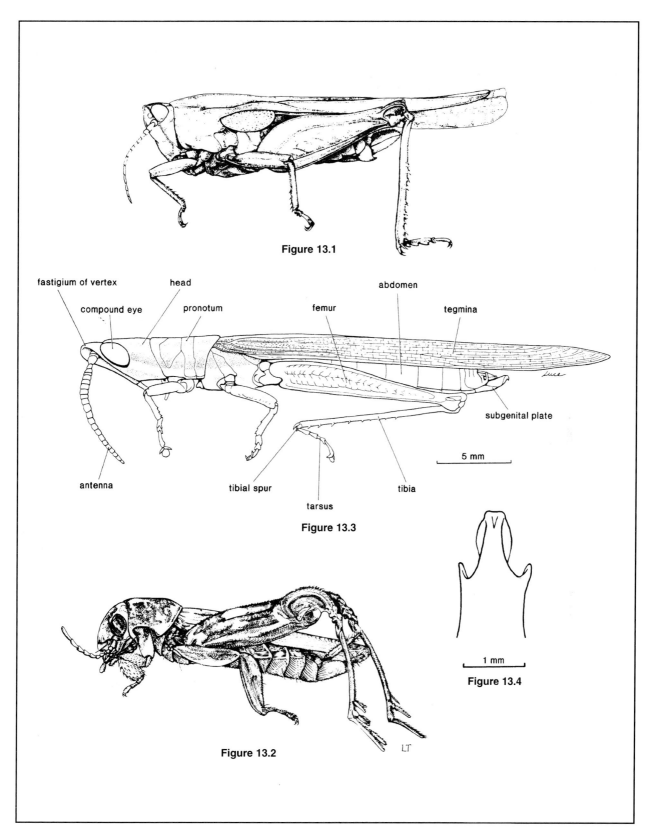

Figure 13.1. Adult female grouse locust, *Textrix subulata* (Linnaeus) (Tetrigidae).

Figure 13.2. Adult male pygmy mole cricket, *Ellipes minuta* (Scudder) (Tridactylidae).

Figure 13.3. Adult male *Leptysma marginicollis* (Serville) (Acrididae).

Figure 13.4. Subgenital plate of adult male *Stenacris vitreipennis* (Marschall) showing well-developed lateral processes.

5(4).	Antennae longer than body; tegmina flat above, bent rather abruptly downward at sides; male with stridulatory apparatus at the base of the tegmina; female with an elongate and cylindrical ovipositor (Fig. 13.5) (crickets)	*GRYLLIDAE*
5'.	Antennae shorter than body; tegmina flattened above and reflexed downward, but not so strongly in either case; tegmina without a stridulatory mechanism at the base; female with ovipositor consisting of 4 short, horny pieces projecting from the tip of the abdomen (Fig. 13.3) (grasshoppers)	*ACRIDIDAE* (p. 214)

KEYS TO SELECTED GENERA BY FAMILY

Tridactylidae

1.	Tarsus of hind leg almost as long as large tibial spur; hind tibiae with 4 pairs of slender plates used in swimming; body more than 5.5 mm in length (one species only, *apicialis* [Say])	*Neotridactylus*
1'.	Tarsus of hind leg absent; hind tibiae with one pair of long slender plates used in swimming (Fig. 13.2); body less than 5.5 mm in length (one species only, *minuta* [Scudder])	*Ellipes*

Acrididae

1.	Head moderately slanted and broadly convex in profile; form moderately robust (boggy meadows, swamps, edges of marshes and lakes; primarily Canadian provinces and north central and northeastern USA) (three species, *lineata* [Scudder], *gracile* [Scudder], *celata* Otte)	*Stethophyma*
1'.	Head strongly slanted, nearly straight to slightly concave in profile; form slender (Fig. 13.3)	2
2.	Tegmina with apex broadly angulate; form moderately slender (mostly eastern USA; sedges and grasses along lakes, ponds, streams, and salt and freshwater marshes) (one species, *brevicornis* [Johannson])	*Metaleptea*
2'.	Tegmina slender and acutely pointed at tip; form slender	3
3.	Head as long as or longer than pronotum; fastigium of vertex as long as eyes and with an obvious median groove; male subgenital plate without a lateral process on each side (Fig. 13.3)	*Leptysma*
3'.	Head shorter than pronotum; fastigium of vertex shorter than eyes and without a median groove; male subgenital plate with well-developed lateral process on each side (Fig. 13.4) (Coastal area from North Carolina into Texas) (only one species, *vitreipennis* [Marschall])	*Stenacris*

Gryllotalpidae

1.	Front tibiae with 2 dactyls (fingerlike structures); hind femora usually longer than pronotum	*Scapteriscus*
1'.	Front tibiae with 4 dactyls; hind femora shorter than pronotum	2
2.	Apical half of hind tibiae armed above with 3-4 long spines on inner margins	*Gryllotalpa*
2'.	Apical half of hind tibiae spined only at apices (Fig. 13.6)	*Neocurtilla*

ADDITIONAL TAXONOMIC REFERENCES

General
Blatchley (1920); Rehn and Grant (1961); Borror *et al.* (1981).

Taxonomic treatments at the family and generic levels
Acrididae: Helfer (1953); Otte (1981); Rehn and Eades (1961).
Gryllidae: Blatchley (1920); Vickery and Johnstone (1970).
Gryllotalpidae: Blatchley (1920).
Tetrigidae: Rehn and Grant (1961).
Tettigoniidae: Rehn and Hebard (1915a,b); Thomas and Alexander (1962); Walker (1971).
Tridactylidae: Günther (1975).

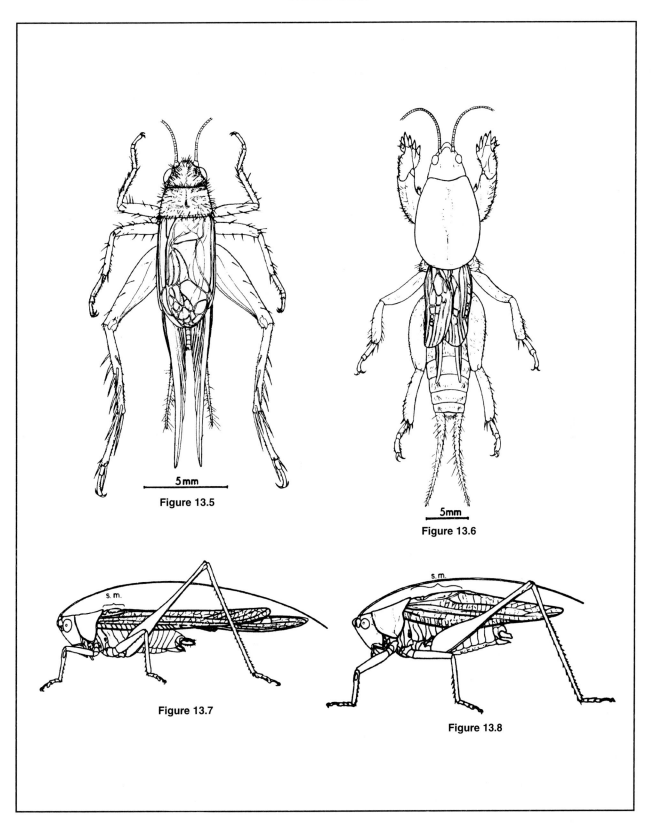

Figure 13.5. Adult male of a small field cricket, *Allonemobius allardi* (Alexander and Thomas) (Gryllidae).

Figure 13.6. Adult male mole cricket, *Neocurtilla hexadactyla* (Perty) (Gryllotalpidae).

Figure 13.7. Adult male *Conocephalus fasciatus* (De Geer) (Tettigoniidae) (Courtesy Michigan Entomological Society), *s.m.*, stridulating mechanism.

Figure 13.8. Adult male *Orchelium vulgare* Harris (Tettigoniidae) (Courtesy Michigan Entomological Society), *s.m.*, stridulating mechanism.

Table 13A. Summary of ecological and distributional data for *Semiaquatic Orthoptera (grasshoppers, crickets, etc.)* (For definition of terms see Tables 6A-6C; table prepared by I. J. Cantrall, K. W. Cummins and M. A. Brusven.)

Taxa (number of species in parentheses)	Habitat	Habit	Trophic Relationships	North American Distribution	Ecological References*
Tetrigidae (?) (grouse or pygmy locusts)	Lentic—vascular hydrophytes (emergent zone, margins)	Sprawlers (semiaquatic)	Generally shredders herbivores (chewers), collectors—gatherers	Widespread	330, 3264
Tridactylidae (2) (pygmy mole crickets)	Lentic and lotic margins (near quiet water away from wave or splash effects)	Burrowers; skaters, "swimmers" (semiaquatic)	Shredders—herbivores (chewers)		330, 1503
Neotridactylus (1)				Widespread	1503
Ellipes (1)				Widespread	1503
Acrididae (3) (short-horned grasshoppers)	Lentic—vascular hydrophytes (emergent zone, margins)	Skaters, "swimmers," climbers (semiaquatic)	Shredders—herbivores (chewers)		330, 3263
Stenacris (1)				East Coast, Gulf	
Leprysma (2)				East to edge of Great Plains, Southwest	
Stethophyma (3)				Northcentral, Northeastern	2993
Metaleptea (1)				Eastern, Gulf Region	1659, 2993
Tettigoniidae (12) (katydids)			Generally shredders—herbivores (chewers), some predators (engulfers)		
Conocephalinae (12)					
Orchelimum (2)	Lentic (and lotic)—vascular hydrophytes (emergent zone, margins)	Climbers, "swimmers" (semiaquatic)	Shredders—herbivores (chewers), predators (engulfers)	Northern Midwest to edge of Great Plains, East Coast	330, 3265, 3987 4176
Conocephalus (10)	Lentic and lotic—margins (and vascular hydrophytes, emergent zone)	Climbers (semiaquatic)	Shredders—herbivores	Widespread	3266
Gryllidae (?) (crickets)	Generally lentic—vascular hydrophytes (emergent zone, margins)	Sprawlers (semiaquatic)	Generally shredders—herbivores (chewers), collectors—gatherers, some predators (engulfers)	Widespread	330, 4128
Eneopterinae Trigonidiinae Mogoplistinae Nemobiinae					
Gryllotalpidae (2) (mole crickets)	Lentic and lotic—margins (near quiet water away from wave or splash effects)	Burrowers (semiaquatic)	Collectors—gatherers, predators (engulfers)	East, Midwest	330, 975
Neocurtilla (2)					

*Emphasis on general ecology, life history.

PLECOPTERA

K. W. Stewart
University of North Texas, Denton

P. P. Harper
Université de Montréal, Canada

INTRODUCTION

Plecoptera (stoneflies) are primarily associated with clean, cool running waters, although a number of species are adapted to life in large oligotrophic, alpine, and boreal lakes. Several species inhabit streams that warm or dry up in summer or that are organically enriched. Eggs and larvae of all North American species are aquatic, and, with the exception of an *Utacapnia* species living in the depths of Lake Tahoe, adults are terrestrial. The following account of the biology and ecology of stoneflies can be supplemented with the comprehensive works of Hitchcock (1974), Hynes (1976) and Stewart and Stark (1988).

Stonefly larvae tend to have specific water temperature, substrate type, and stream size requirements reflected in their distribution and succession along the course of streams and rivers. In addition, a distinct microdistribution of species and of size classes of individual species is frequently observed within a particular stream reach. Microhabitats include boulder surfaces, cobble and gravel interstices, debris accumulations and leaf packs, as well as the hyporheal (*Isocapnia* sp., *Paraperla* spp., some European *Leuctra* spp., and early instars of many other species). Stonefly adults live on riparian vegetation or among rocks or debris.

Food ingested by larvae may vary depending on species, developmental stage, or time of day (Stewart and Stark 1988). Some species are, for example, shredders or predators throughout development; however, others may change their feeding habit as development proceeds. Gut content analyses indicate that shifts from herbivory-detritivory in the earliest instars to omnivory-carnivory in later periods are common. Predators usually ingest their prey whole. Stoneflies may either be opportunistic or select for particular food items in low abundance. Details on individual families are given in Table 14B. Adults of Taeniopterygidae, Nemouridae, Leuctridae, Capniidae, Chloroperlidae, and some Perlodidae feed on epiphytic algae or young leaves and buds of riparian vegetation.

Plecoptera species emerge in clear succession throughout the year, except at latitudes or altitudes experiencing seasons of drought or frost. In general, species with predaceous larvae have an emergence period restricted to the spring and summer months (Table 14A). Adults live about 1-4 weeks, although winter species often have greater longevity.

In many stonefly species, male and female adults are attracted by drumming (Rupprecht 1967; Zeigler and Stewart 1977; Stewart and Maketon 1991), which is a tapping upon the substrate with the tip of the abdomen. Male drummers continue this activity throughout their adult life, but only virgin females respond to males. Males initiate the communication, and females answer either during or immediately after the male drumming call. Male and female signals are species specific, and dialects have been detected in European *Diura* spp. (Rupprecht 1972) and North American *Pteronarcella* spp. (Stewart et al. 1982).

Little courtship has been observed in stoneflies prior to mating, although some Peltoperlidae do a dance involving considerable wing lifting during drumming. During the intersexual duetting, males actively search with triangulation or other patterns (Abbott and Stewart 1993) while females remain stationary, until contact is made. Males mount females, curve the abdomen to the left or right side, and, using structures such as the hooks of the subanal lobes or the epiproct, gain apposition with the subgenital plate (Chap. 2). The aedeagus is everted from beneath the ninth sternum and curves upward between the male cerci into the female genital opening. Sperm are either conveyed internally directly from the aedeagus (Stewart et al. 1969) or through a specially grooved epiproct (Brinck 1956), or introduced externally into a pocket just outside the genital opening, after which the female aspirates the sperm into the bursa (Stewart and Stark 1977). The gravid female releases eggs over the stream surface or deposits them in the water. Eggs are attached to the substrate by either a sticky gelatinous covering or specialized anchoring devices.

In most species, embryonic development proceeds directly and is complete within 3-4 weeks; hatchlings then undergo gradual development requiring some 12-24 months, depending on the species, sex, and environmental conditions. In most cases, growth requires 10-11 months (annual life cycle, univoltine), although growth rates are often low during winter months. Semivoltine (2-3 year) cycles are known or suspected in many Pteronarcyidae and Perlidae, as well as in a few species of Nemouridae, Leuctridae, and Perlodidae (Table 14A). Such cycles are termed *slow seasonal* in Hynes' (1961) classification.

In other species, particularly those inhabiting intermittent aquatic habitats or streams subjected to extremes in temperature, embryonic development may be arrested for 3-6 months and hatching delayed until environmental conditions are more favorable (Table 14A). Hatchlings must then develop rapidly in order to complete a univoltine cycle in the remaining 6-8 months. A variation of this strategy is found in some Taeniopterygidae, Capniidae, and Leuctridae in which diapause occurs in the early larval instars (Harper and Hynes 1970). This type of cycle is called *fast seasonal*. Under drought conditions, species of *Zealeuctra* are able to extend the embryonic diapause for an extra year (Snellen and Stewart 1979a). These observations and Hynes and Hynes' (1975) studies on Australian stoneflies suggest considerable flexibility in stonefly life cycles. Patterns of egg development, life histories and life cycles, and larval growth are reviewed by Stewart and Stark (1988).

EXTERNAL MORPHOLOGY

Larvae and Adults

Larval and adult stoneflies fit the general morphological pattern of a primitive insect (see Chap. 2). Metamorphosis is hemimetabolous, and the main characters used for identifying adults are the wings and the external genitalia. The more difficult key characters are explained below.

Mouthparts: The mouthparts are of the primitive mandibular type, and the maxillae (Figs. 14.39-14.40, 14.72, 14.79-14.81) and labium (Figs. 14.7-14.8, 14.46-14.47, 14.76) are of considerable taxonomic importance. The relative positions of the glossae and paraglossae (Figs. 14.7-14.8) are used extensively both in larval and adult determinations. The larval mouthparts are of two basic types: Type I (herbivorous - detritivorous) and Type II (carnivorous) (Stewart and Stark 1988). The maxillae and mandibles generally reflect a scraping-grinding function in herbivores and stabbing-grasping function in carnivores. The mouthparts of nonfeeding adults remain typical, even though they are membranous and somewhat degenerate.

Gills: Stonefly larvae lack gills or possess gills of various shapes and number at various body locations (Shepard and Stewart 1983; Stewart and Stark 1988). Locations include: the submentum (Figs. 14.8, 14.24, 14.80, 14.76), neck (Figs. 14.18-14.20), thorax (Figs. 14.5-14.6, 14.14, 14.76), abdomen (Figs. 14.5, 14.77), anal region (Figs. 14.56-14.57), and bases of legs (Fig. 14.9). These gills are of two main types: simple flat or cylindrical gills set individually (Figs. 14.6, 14.8-14.9, 14.19, 14.76-14.78) and tufts of gills set on a common stock ("branched gills"; Figs. 14.5, 14.14, 14.18, 14.20, 14.24, 14.56-14.57). Gills may persist in adult stages, but tend to shrivel and sometimes leave scars (Fig. 14.124) or stubs (Figs. 14.114, 14.122, 14.137).

Cerci: As a rule, cerci of both adults and larvae are long whiplike appendages divided into numerous segments (Figs. 14.1-14.2); however, in a few families, cerci of the adults of one or both sexes are reduced to the basal segment (Figs. 14.138-14.139, 14.145-14.150, 14.174-14.177), and, in some cases, are modified as accessory male copulatory organs (Figs. 14.146, 14.174-14.175).

Wings: The wings appear in half-grown larvae as tiny buds on the thorax. They grow with each successive instar and wing pads reach their full development in the terminal instar (Figs. 14.1, 14.36, 14.38). Most species are winged; however, some may be apterous or brachypterous (shortened wings; Fig. 14.37), often depending on sex. Venation is simple and veins are easily recognized (Fig. 14.2).

External genitalia: Adult males possess various types of external genitalia. Accessory copulatory structures are formed from one or more of the following: epiproct (Figs. 14.115, 14.123, 14.127, 14.142, 14.146), paraprocts (Figs. 14.138, 14.142, 14.187), cerci (Figs. 14.146, 14.174-14.175, 14.205), abdominal sternites (Figs. 14.129-14.130), and abdominal tergites (Figs. 14.147, 14.174, 14.186, 14.199, 14.208). Adult females lack extensive external genitalia. In most instances, sternum 8 is modified into a subgenital plate covering the gonopore (Fig. 14.4).

KEY TO FAMILIES AND GENERA OF NORTH AMERICAN PLECOPTERA NYMPHS

The gill systems of Ricker (1959b) and Shepard and Stewart (1983) are used in the following key. Mature larvae can usually be sexed by the developing genitalia. In addition, males are smaller than females and the latter have a subgenital plate, usually manifested by the absence of mesal hairs of the Ab_8 posterior sternal margin. Determination of mouthpart characters may require extraction, mounting, and observation at a magnification of 100-400×. This key is adapted from Stewart and Stark (1988); users are referred to that work for more extensive illustrations. Definitive separation of the capniid genera *Capnia*, *Capnura*, *Mesocapnia* and *Utacapnia* is not presently possible.

Table 14A. Life history data for North American stonefly families. (Only about 15% of the species have been studied in this respect.)

Family	Emergence	Life Cycle	Diapause
Pteronarcyidae	Early summer	1-3 years	none
Peltoperlidae	Early summer	1-2 years	none
Taeniopterygidae	Winter and spring	Univoltine	larva (egg?) in summer
Nemouridae	Succession of spp. throughout the year	1-2 years	egg in summer or in winter, in some spp.
Leuctridae	Succession of spp.	1-2 years throughout the year	egg diapause in a few spp.
Capniidae	Winter and spring	Univoltine	larva in summer
Perlidae	Summer	1-3 years	eggs in summer, some eggs in winter
Perlodidae	Spring-fall	Univoltine	egg (1 species)
Chloroperlidae	Spring-fall	1-2 years	none

KEY TO THE FAMILIES AND GENERA OF NORTH AMERICAN PLECOPTERA LARVAE

Larvae

1.	Ventral tufts of gills present on Th (Thorax) and Ab_{1-2} (Abdomen) or Ab_{1-3} (Fig. 14.5)...***PTERONARCYIDAE*** (Figs. 2.1, 2.2, 14.4a) 9	
1'.	Without ventral gill tufts on Ab_{1-2} or Ab_{1-3} 2	
2(1').	Single, double, or forked conical gills at least behind $coxae_{2-3}$ (Fig. 14.6); Th sterna posteriorly produced, overlapping succeeding segment; some genera roachlike...***PELTOPERLIDAE*** (Fig. 14.4b)................10	
2'.	Gills, if present, not conical; Th sterna not overlapping 3	
3(2').	Paraglossae and glossae produced forward about the same distance (Fig. 14.7)..4	
3'.	Paraglossae much longer than glossae (Fig. 14.8) 7	
4(3).	First and 2nd tarsal segments ca. equal in length (Fig. 14.9); midline of wing pads strongly divergent from body axis ....***TAENIOPTERYGIDAE*** (Fig. 14.4c)................15	
4'.	Second tarsal segment much shorter than first (Fig. 14.10); midline of wing pads parallel or divergent..5	
5(4').	Small, robust, hairy larvae, less than 12 mm in body length; extended hind legs reach about tip of Ab; midline of metathoracic wing pads strongly divergent from body axis; cervical gills present (Figs. 14.18-14.20) or absent ....***NEMOURIDAE*** (Fig. 14.4d)................22	
5'.	Elongate larvae, extended hind legs reach far short of Ab tip; midline of metathoracic wing pads if present about parallel with body axis; no cervical gills 6	
6(5').	Terga and sterna of Ab_{1-9} separated by membranous pleural fold (Fig. 14.11); Ab terga with posterior setal fringe (Fig. 14.12); Ab segments widest posteriorly; hind wing pads about as broad as long, reduced, or absent ....***CAPNIIDAE*** (Fig. 14.4e)................33	
6'.	At most Ab_{1-7} separated by membranous pleural fold (Fig. 14.13); Ab terga of some genera without posterior setal fringe (*Despaxia*, *Perlomyia*, *Paraleuctra*, *Zealeuctra*) (Figs. 14.49, 14.51); Ab segments cylindrical; hind wing pads usually longer than wide....................***LEUCTRIDAE*** (Fig. 14.4f)................39	
7(3').	Highly branched filamentous gills on sides and venter of Th (Fig. 14.14). ....***PERLIDAE*** (Fig. 14.4g)................45	
7'.	Branched, filamentous gills absent from Th 8	
8(7').	Cerci 3/4 or less length of Ab; head, Th terga usually without distinct pigment pattern .... ***CHLOROPERLIDAE*** (Fig. 14.4h) 59	
8'.	Cerci as long or longer than Ab; head, Th terga usually with distinct pigment pattern ....***PERLODIDAE*** (Fig. 14.1) 71	
9(1).	Ab_3 with gills similar to those on Ab_{1-2} ....***Pteronarcella*** Banks (Fig. 14.4a)	
9'.	Ab_3 without gills (Eastern species, previously in *Allonarcys*, with lateral spinelike processes on Ab) ....***Pteronarcys*** Newman (Figs. 2.1, 2.2)	
10(2).	Posterior supracoxal gills on Th_1 (PSC_1) present (Fig. 14.6) 14	
10'.	Gill PSC_1 absent 11	
11(10').	Gill posterior $supracoxal_{2-3}$ single; gill (posterior Th_3) present or absent 12	
11'.	Gill PSC_{2-3} double; gill PT_3 (posterior Th_3) present 13	
12(11).	Gill posterior Th_3 absent... ***Soliperla*** Ricker	
12'.	Gill posterior Th_3 present, single.. ***Viehoperla*** Ricker	
13(11).	Four paired, light-colored round spots on pronotum; paired dark pigment spots on meso- and metanota...***Peltoperla*** Needham	
13'.	Pronotum without paired, light round spots; prominent dark spots absent from meso- and metanota..***Tallaperla*** Stark and Stewart	
14(10).	Apex Ab rounded; dorsal femora with transverse rows of long setae; close-set, median transverse row of spinules and a double posterior row of stout setae on all Th nota.. ***Yoraperla*** Ricker (Fig. 14.4b)	
14'.	Apex Ab with prominent spinelike process (Fig. 14.15); no transverse rows of spinules on femora or Th nota; Western..***Sierraperla*** Jewett	
15(4).	Single, fingerlike, segmented gill on each coxa (Fig. 14.9) ....***Taeniopteryx*** Pictet	
15'.	Coxae without gills 16	
16(15').	Some basal cercal segments or entire cercus with dorsal, silky hair fringe (Fig. 14.16)..17	

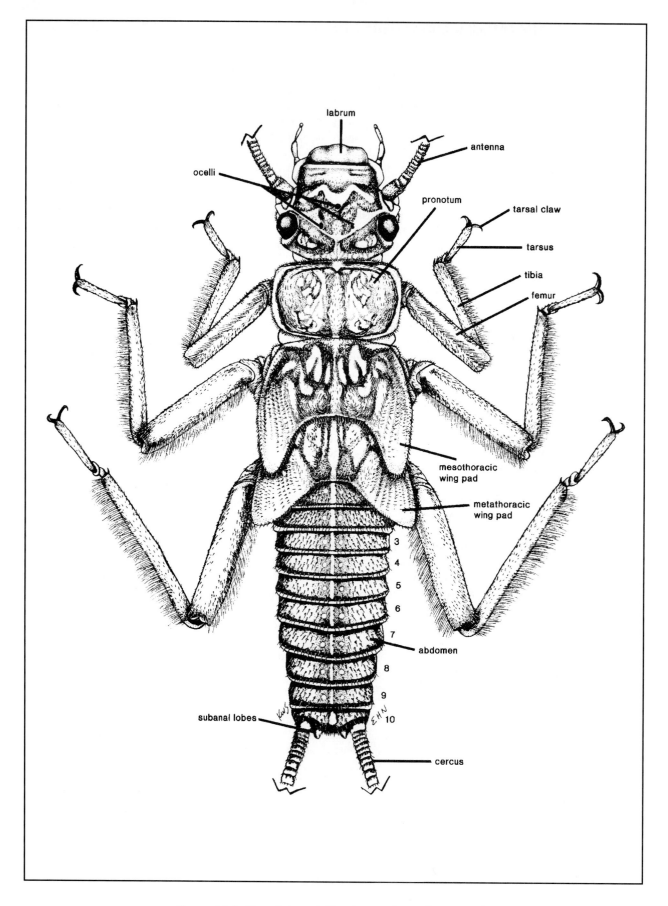

Figure 14.1. Dorsal view of *Skwala americana* (Perlodidae).

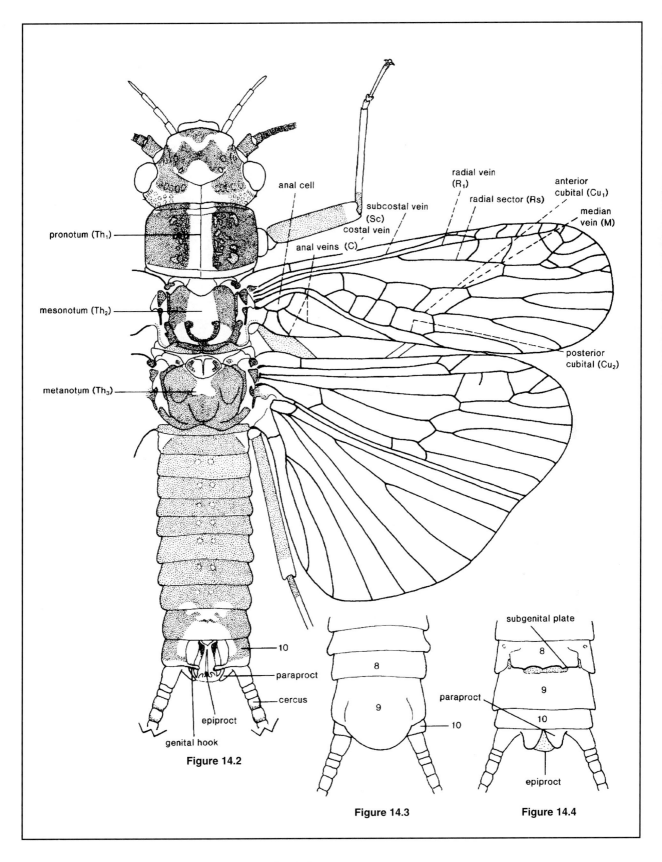

Figure 14.2. Adult male of *Skwala americana* (Perlodidae).

Figure 14.3. Ventral view of terminal abdominal segments of male *Skwala americana* (Perlodidae).

Figure 14.4. Ventral view of terminal abdominal segments of female *Skwala americana* (Perlodidae).

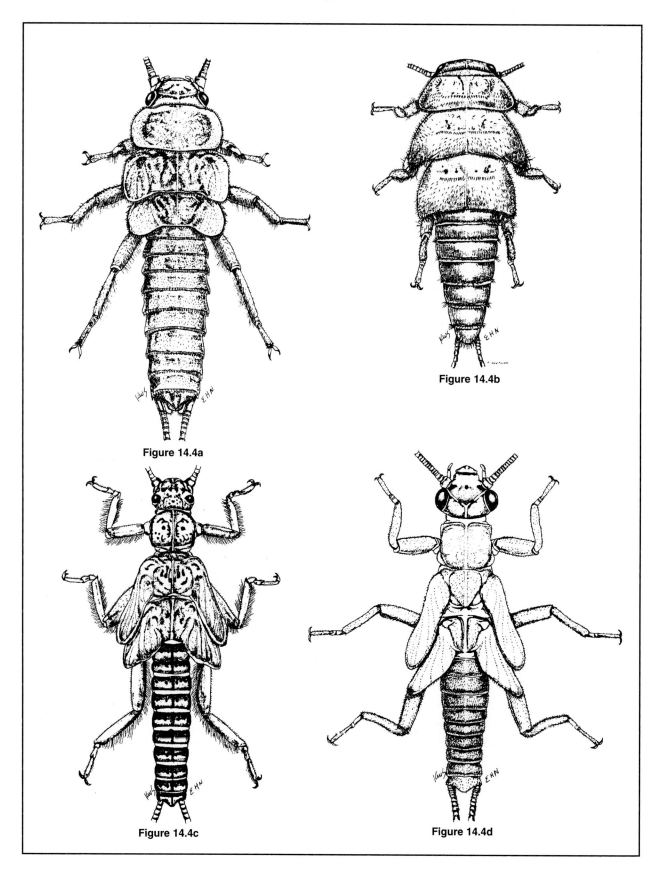

Figure 14.4a. *Pteronarcella badia* (Pteronarcyidae) larval habitus.

Figure 14.4b. *Yoraperla* sp. (Peltoperlidae) larval habitus.

Figure 14.4c. *Strophopteryx fasciata* (Taeniopterygidae) larval habitus.

Figure 14.4d. *Shipsa rotunda* (Nemouridae) larval habitus.

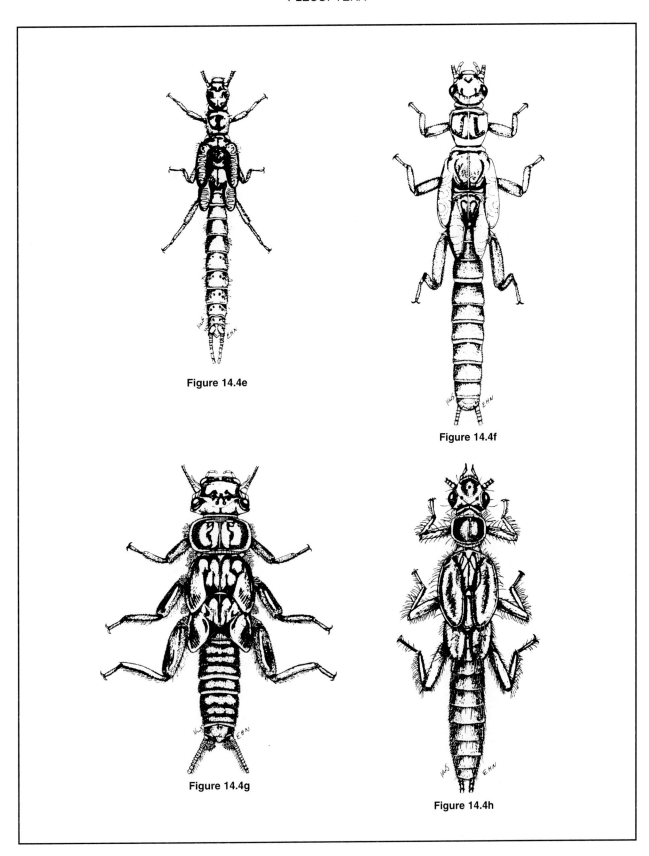

Figure 14.4e. *Eucapnosis brevicauda* (Capniidae) larval habitus.

Figure 14.4f. *Leuctra* sp. (Leuctridae) larval habitus.

Figure 14.4g. *Eccoptura xanthenes* (Perlidae) larval habitus.

Figure 14.4h. *Haploperla brevis* (Chloroperlidae) larval habitus.

16'.	No dorsal silky hair fringe on cercus	20
17(16).	Entire cercus with distinct dorsal hair fringe; Western	***Doddsia*** Needham and Claassen
17'.	Dorsal silky hairs present only on basal few cercal segments (Fig. 14.17)	18
18(17').	Single, short dorsal fringe hair on first 3-6 cercal segments (Fig. 14.17) and single outer hair fringe on tibiae (Fig. 14.9); Eastern	***Strophopteryx*** Frison (in part) (all species except *S. fasciata*)
18'.	First 10-12 cercal segments with multiple short fringe hairs	19
19(18').	Mesodorsal fringe of fine hairs on basal few antennal segments; Eastern	***Bolotoperla*** Ricker
19'.	No fringe of fine hairs on basal antennal segments	***Oemopteryx*** Klapálek (in part) (all eastern species except *O. contorta* that has no cercal or antennal fringe)
20(16').	Body and legs brown, with indistinct pattern; prothorax usually subtriangulate (wider posteriorly, Western, except *T. atlanticum* and Eastern populations of *T. pacificum*)	***Taenionema*** Banks
20'.	Body and legs lighter with distinct pattern; prothorax usually with parallel sides; (Fig. 14.4c); Eastern	21
21(20').	Tibiae without ventral hair fringe (Fig. 14.9)	***Oemopteryx*** Klapálek (in part)
21'.	Tibiae with both dorsal and ventral hair fringe (Fig. 14.4c)	***Strophopteryx*** (in part) (*S. fasciata*)
22(5).	A pair of cervical gills present, each with 5 or more branches arising from the cervical sclerites on each side (Fig. 14.18); submental gills absent; no dorsal transverse row or whorl of femoral spines on fore legs	23
22'.	Cervical gills usually absent or if present each of the 4 gills is simple or 2-4 branched and submental gills are absent (Fig. 14.19), or there are 2 short cervical knobs and branched submental gills are present (Fig. 14.24)	24
23(22).	Gill filaments of different length, arranged palmately from base (Fig. 14.18)	***Malenka*** Ricker
23'.	Gill filaments approximately equal in length emerging equally from base (Fig. 14.20)	***Amphinemura*** Ris
24(22').	Distinct, single lateral fringe of spines on pronotum, sometimes longer on posterior corners (Fig. 14.21); lateral spine fringes on wing pads	25
24'.	No lateral fringe of pronotal spines, or, if present, not arranged in single row, or of unequal lengths; meso- and metanotum without lateral fringes of spines or with sparse fringes near posterolateral corners	27
25(24).	Pair of simple or 2-4 branched cervical gills present on each side of midline (Fig. 14.19); transverse row of spines on all femora	***Zapada*** Ricker
25'.	Cervical gills absent; no transverse femoral spine whorls	26
26(25).	Pronotum with round corners, without a definite lateral notch (Fig. 14.22)	***Nemoura*** Latreille
26'.	Pronotum angular at corners, with a definite lateral notch (Fig. 14.23)	***Soyedina*** Ricker
27(24').	Submental gills absent; fore tibia with outer row of spines and 2 to many fine hairs, or fine hairs only	28
27'.	Branched submental gills present (Fig. 14.24) and fore tibia with outer row of spines and fringe of fine hairs *or* submental gills absent and fore tibia without fine hair fringe	31
28(27).	Pronotum with irregular fringe of moderate to large spines (Fig. 14.25)	***Ostrocerca*** Ricker
28'.	Pronotum without distinct fringe of heavy spines (Fig. 14.26)	29
29(28').	Fore tibia with outer fringe of long hairs, with or without stout outer spines	30
29'.	Fore tibia without outer fringe of long hairs, but with occasional single hairs (Fig. 14.27)	***Podmosta*** Ricker
30(29).	Fore tibia with outer row of stout spines, in addition to hairs (Fig. 14.28)	***Prostoia*** Ricker
30'.	Fore tibia with outer hair fringe only; no stout spines (Fig. 14.29)	***Shipsa*** Ricker (Fig. 14.4d)
31(27').	Branched submental gills present (Fig. 14.24); fore tibia with outer row of long stout black spines and fringe of long hairs	***Visoka*** Ricker
31'.	Submental and cervical gills absent; fore tibia with only occasional or no longer outer hairs (Fig. 14.30)	32
32(31').	Female larvae with distinct posterior vesiclelike lobe on Ab_8 sternum (Fig. 14.31); Glacier Nat. Park, Montana	***Lednia*** Ricker
32'.	Female larvae without a distinct posterior vesiclelike lobe on Ab_8 sternum; Eastern	***Paranemoura*** Needham and Claassen
	(Recurved epiprocts of mature male *Lednia* and *Paranemoura* larvae visible under Ab_{10} cuticle)	
33(6).	Basal and/or apical cercal segments with multiple long, fine hairs forming a prominent vertical fringe (Figs. 14.32-14.33)	34
33'.	Long cercal hairs restricted to apical segmental whorls; no prominent vertical fringe	35
34(33).	Cerci with about 20 segments (Fig. 14.32), cercal length about equal to length of Ab; Western	***Isocapnia*** Banks

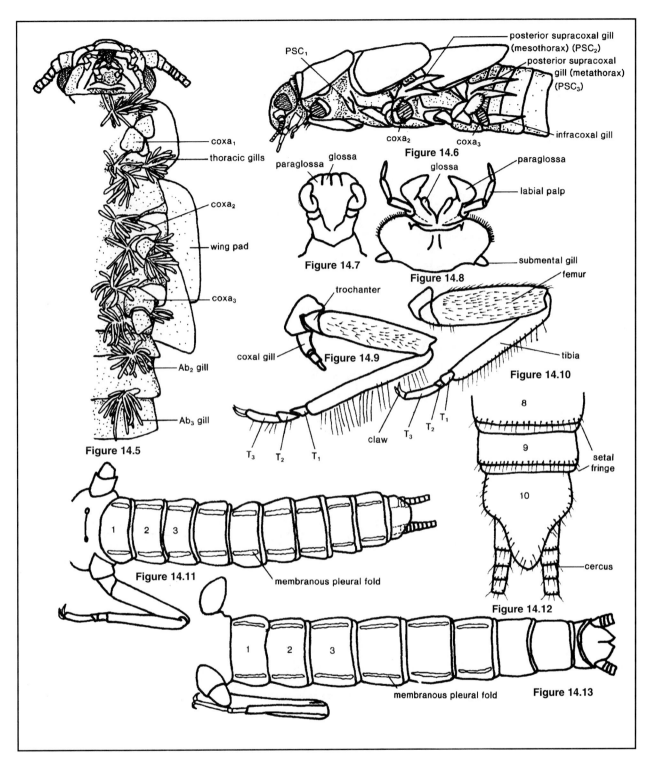

Figure 14.5. *Pteronarcella badia* (Pteronarcyidae) venter with gills.
Figure 14.6. *Peltoperla arcuata* (Peltoperlidae) left side with gills.
Figure 14.7. *Taeniopteryx maura* (Taeniopterygidae) labium (ventral).
Figure 14.8. *Setvena bradleyi* (Perlodidae) labium (ventral).
Figure 14.9. *Taeniopteryx maura* (Taeniopterygidae) left rear leg with gill.
Figure 14.10. *Utacapnia lemoniana* (Capniidae) left rear leg.
Figure 14.11. *Allocapnia granulata* (Capniidae) abdominal venter.
Figure 14.12. *Allocapnia granulata* (Capniidae) terminal abdomen (dorsal).
Figure 14.13. *Zealeuctra claasseni* (Leuctridae) abdominal venter.

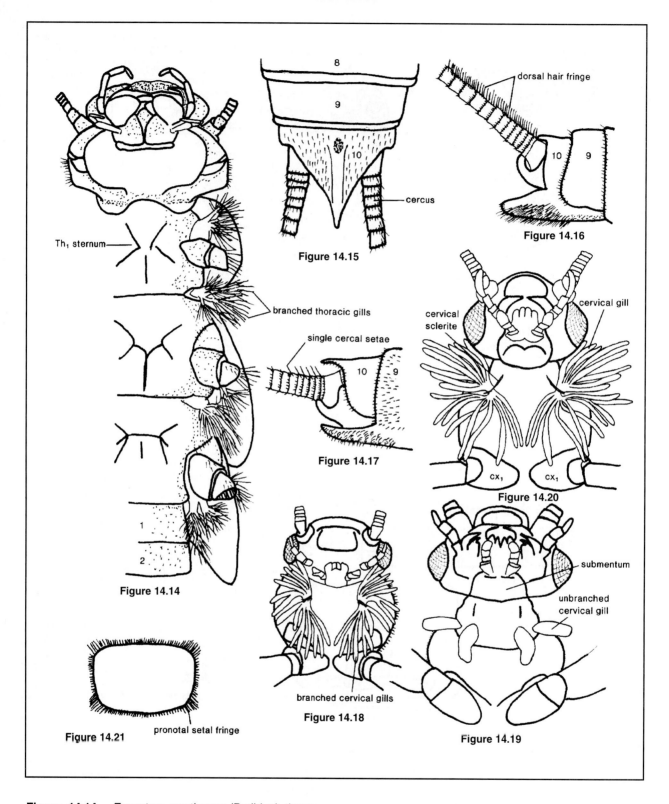

Figure 14.14. *Eccoptura xanthenes* (Perlidae) thorax venter with gills.

Figure 14.15. *Sierraperla cora* (Peltoperlidae) terminal abdomen (dorsal).

Figure 14.16. *Doddsia occidentalis* (Taeniopterygidae) right cercus (lateral).

Figure 14.17. *Oemopteryx glacialis* (Taeniopterygidae) right cercus (lateral).

Figure 14.18. *Malenka californica* (Nemouridae) cervical gills (ventral).

Figure 14.19. *Zapada haysi* (Nemouridae) cervical gills (ventral).

Figure 14.20. *Amphinemura banksi* (Nemouridae) cervical gills (ventral).

Figure 14.21. *Zapada haysi* (Nemouridae) pronotum.

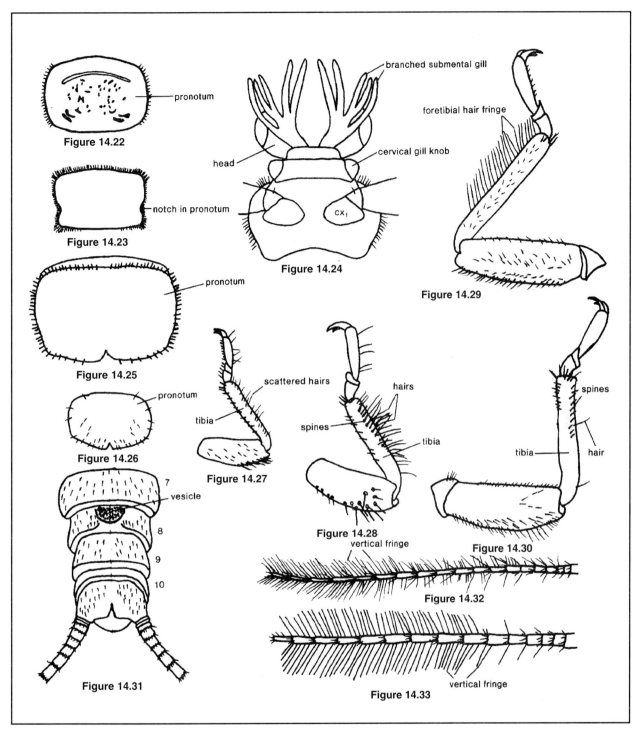

Figure 14.22. *Nemoura arctica* (Nemouridae) pronotum.
Figure 14.23. *Soyedina vallicularia* (Nemouridae) pronotum.
Figure 14.24. *Visoka cataractae* (Nemouridae) submental gills.
Figure 14.25. *Ostrocerca truncata* (Nemouridae) pronotum.
Figure 14.26. *Podmosta decepta* (Nemouridae) pronotum.
Figure 14.27. *Podmosta decepta* (Nemouridae) right front leg.
Figure 14.28. *Prostoia* sp. right front leg.
Figure 14.29. *Shipsa rotunda* (Nemouridae) left front leg.
Figure 14.30. *Paranemoura perfecta* (Nemouridae) right front leg.
Figure 14.31. *Lednia tumana* (Nemouridae) abdominal ventrum.
Figure 14.32. *Isocapnia integra* (Capniidae) left cercus (inside view).
Figure 14.33. *Nemocapnia carolina* (Capniidae) right cercus (outer view).

34'.	Cerci with about 15 segments (Fig. 14.33), length equal to last 7 Ab segments together; Eastern	***Nemocapnia*** Banks
35(33').	Body and appendages densely clothed with stout hairs; Ab terga with posterior fringe of long bristles (Fig. 14.34)	***Paracapnia*** Hanson
35'.	Bristles on body and appendages few and inconspicuous, or absent; Ab terga with posterior fringe of short setae and scattered intercalary setae (Fig. 14.35)	36
36(35').	Inner margin of hind wing pad unnotched or notched close to the tip (Fig. 14.36) (hind wing pad sometimes much reduced as in Fig. 14.37)	***Allocapnia*** Claassen
36'.	Inner margin of hind wing pad notched at about mid length (Fig. 14.38) (notch weak in *Bolshecapnia*)	37
37(36').	Lacinia with single, weakly sclerotized, emarginate cusp; galea longer than lacinia (Fig. 14.39); cercal segments about 17	***Eucapnopsis*** Okamoto (Fig. 14.4e)
37'.	Lacinia with 2 sharp cusps; galea shorter than lacinia (Fig. 14.40); cercal segments 25 or more	38
38(37').	Mature male larvae with developing vesicle (dark lobe) visible through larval skin at base of Ab_9 sternum (Fig. 14.41); body length 8-10 mm	***Bolshecapnia*** Ricker
38'.	No developing vesicle evident at base of Ab_9 sternum of male larvae; body length variable, 5-10 mm	***Capnia*** Pictet, ***Capnura*** Banks, ***Mesocapnia*** Rauser, ***Utacapnia*** Nebeker and Gaufin
39(6').	Ab_{1-7} divided ventrolaterally by membranous pleural fold (Fig. 14.42)	40*
39'.	Fewer than 7 Ab segments divided by membrane (Fig. 14.43)	41
40(39).	Body robust, length less than 8 times width; body conspicuously clothed with hairs about 1/5 the length of middle Ab segment; subanal lobes of mature male a fused strongly keeled plate, much produced with no posterior notch (Fig. 14.44)	***Megaleuctra*** Neave
40'.	Body more elongate, fine hair pile inconspicuous, appearing naked; subanal lobes of mature male larva fused 1/2-2/3 length, leaving a notch at tip; paraprocts bare ventrally	***Perlomyia*** Banks
41(39').	Surface of Ab and Th nearly naked of hairs; Ab_{1-5} divided by membrane (Fig. 14.45);	***Despaxia*** Ricker
41'.	Ab and Th clothed with sparse or numerous hairs; Ab with 1-4 or 1-6 segments ventrolaterally divided by pleural membrane	42
42(41').	Ab_{1-4} divided ventrolaterally by membrane (Fig. 14.43); labial palps long, apical segment fully reaching beyond tips of paraglossae when palp is appressed (Fig. 14.46); clothing of stout hairs on apical Ab segments and subanal lobes; subanal lobes with 2 long innerapical hairs	***Leuctra*** Stephens (Fig. 14.4f)
42'.	Ab_{1-6} divided ventrolaterally; labial palps shorter, tips barely reaching near or beyond tips of paraglossae at most, when appressed (Fig. 14.47)	43
43(42').	Body nearly naked; subanal lobes of male larvae half fused (Fig. 14.48); developing V- or U-shaped cleft sometimes evident in Ab_9 tergum of mature male (Fig. 14.49); paraprocts with fine hairs ventrally	***Zealeuctra*** Ricker
43'.	Body clothed with distinct short hairs and/or bristles; subanal lobes of male larvae separate (Fig. 14.50)	44
44(43').	Body covered with long bristles, many whose length is subequal to greatest width of femora	***Moselia*** Ricker
44'.	Pronotum with a few long bristles only on corners; wing pads and Ab with sparse, scattered bristles (Fig. 14.51), but covered with tiny clothing hairs	***Paraleuctra*** Hanson
45(7).	Occiput with transverse row of regularly spaced spinules (Fig. 14.52), or distinctly elevated ridge	46
45'.	Occiput without spinules, except possibly laterally near the eyes (Figs. 14.53, 14.62), *or* with a sinuate, irregularly spaced spinule row (Fig. 14.54)	49
46(45).	Two ocelli	***Neoperla*** Needham
46'.	Three ocelli	47
47(46').	Ab terga with more than 5 intercalary spinules (Fig. 14.55)	***Claassenia*** Wu
47'.	Ab terga with no more than 4 intercalary spinules	48
48(47').	Posterior spinule fringe of Ab_7 sternum complete (Fig. 14.56); cerci without a long setal fringe	***Agnetina*** Klapálek
48'.	Posterior spinule fringe of Ab sternum incomplete (Fig. 14.57); cerci with at least a few long silky setae	***Paragnetina*** Klapálek
49(45').	Occipital spinules in a sinuate, irregularly spaced row, more or less complete behind ocelli (Fig. 14.54)	50
49'.	No distinct occipital spinule row; a few scattered spinules may be present near the postocular setal fringe (Fig. 14.53)	54
50(49).	Ab terga with fewer than 5 or no intercalary spinules	***Hesperoperla*** Banks

* This character also sends you to couplet 43 (*Zealeuctra*).

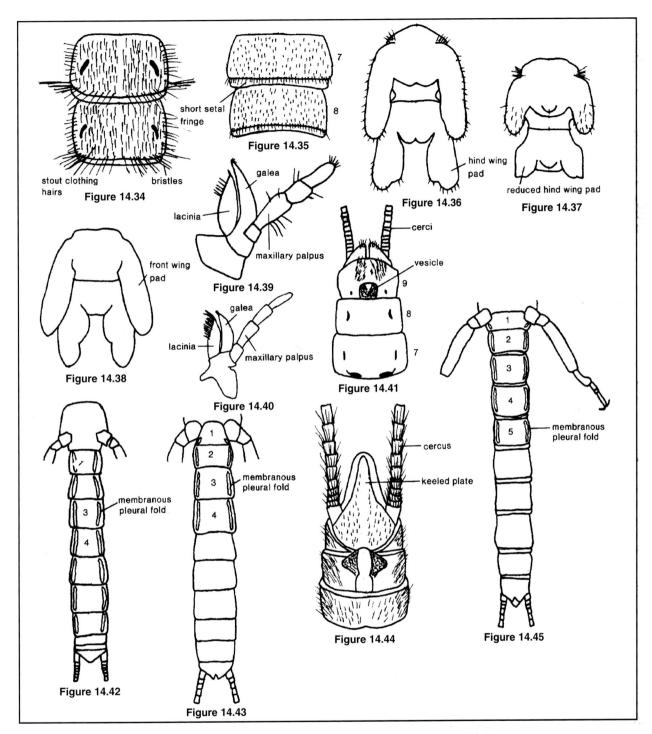

Figure 14.34. *Paracapnia angulata* (Capniidae) abdominal terga.

Figure 14.35. *Capnia vernalis* (Capniidae) abdominal terga.

Figure 14.36. *Allocapnia* sp. (Capniidae) wing pads.

Figure 14.37. *Allocapnia granulata* (Capniidae) male larva wing pads.

Figure 14.38. *Capnia vernalis* (Capniidae) wing pads.

Figure 14.39. *Eucapnopsis brevicauda* (Capniidae) right maxilla (doral).

Figure 14.40. *Utacapnia lemoniana* (Capniidae) right maxilla (dorsal).

Figure 14.41. *Bolshecapnia spenceri* (Capniidae) male larva abdominal venter.

Figure 14.42. *Perlomyia utahensis* (Leuctridae) abdominal venter.

Figure 14.43. *Leuctra sibleyi* (Leuctridae) abdominal venter.

Figure 14.44. *Megaleuctra kincaidi* (Leuctridae) abdominal venter.

Figure 14.45. *Despaxia augusta* (Leuctridae) abdominal venter.

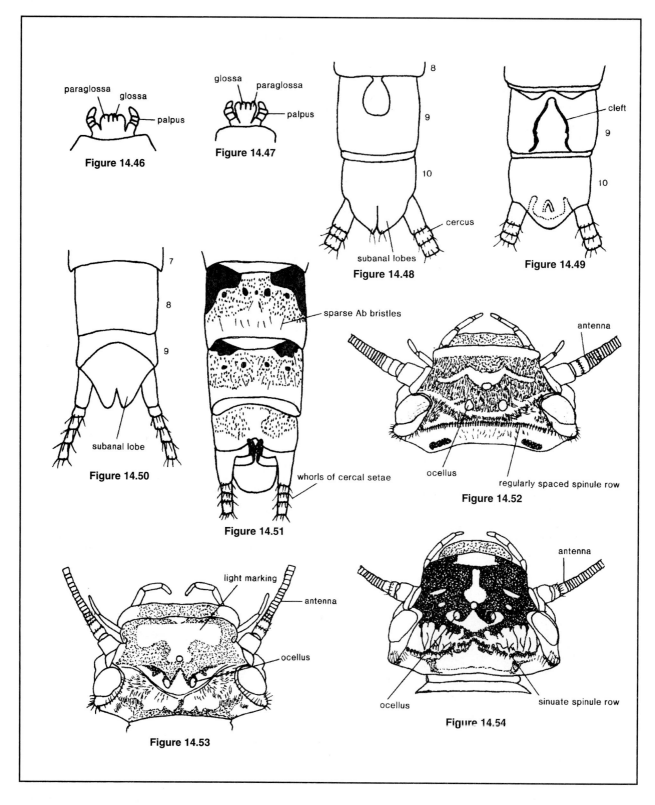

Figure 14.46. *Leuctra sibelyi* (Leuctridae) labium.
Figure 14.47. *Paraleuctra vershina* (Leuctridae) labium.
Figure 14.48. *Zealeuctra claasseni* (Leuctridae) abdominal venter.
Figure 14.49. *Zealeuctra claasseni* (Leuctridae) male larva, abdominal dorsum.
Figure 14.50. *Paraleuctra vershina* (Leuctridae) abdominal dorsum.
Figure 14.51. *Paraleuctra vershina* (Leuctridae) abdominal dorsum.
Figure 14.52. *Claassenia sabulosa* (Perlidae) head.
Figure 14.53. *Eccoptura xanthenes* (Perlidae) head.
Figure 14.54. *Hesperoperla pacifica* (Perlidae) head.

50'.	Ab terga with more than 5 intercalary spinules	51
51(50').	Pronotum laterally fringed with a complete, close-set row of long setae (Fig. 14.58); posterior fringe of Ab terga with numerous long setae whose length is three-fourths or more the length of Ab segments	***Attaneuria*** Ricker
51'.	Pronotum fringed laterally with short setae, not so closely set (Fig. 14.59); posterior fringe of Ab terga mostly of short setae whose length is about one-fourth the length of Ab segments	52
52(51').	Cerci without a dorsal fringe of long silky hairs; Ab of most species speckled with dark pigment at bases of intercalary setae	***Perlesta*** Banks
52'.	Cerci with prominent dorsal fringe of long silky hairs (Fig. 14.60);Ab not speckled	53
53(52').	Dorsum of Th and Ab with a mesal, longitudinal row of long, fine, silky hairs (Fig. 14.61) (best seen in lateral view); Ab_7 sternum usually with incomplete posterior fringe	***Doroneuria*** Needham and Claassen
53'.	No mesal longitudinal row of silky hairs on Th-Ab dorsum; Ab_7 sternum usually with a complete posterior fringe	***Calineuria*** Ricker
54(49').	Postocular fringe reduced to 1-3 long setae (Fig. 14.62); eyes set forward on head; pronotal fringe of 2-3 setae at corners	55
54'.	Postocular fringe with a close-set row of several thick spinules (Fig. 14.53); pronotal fringe well developed, consisting of a close-set row of spinules or setae, occasionally incomplete laterally	56
55(54).	Femora and tibia with dorsal (outer) and ventral (inner) fringes of long silky setae	***Perlinella*** Banks
55'.	Femora and tibia with only dorsal (outer) fringe of long, silky setae	***Hansonoperla*** Nelson
56(54').	Two ocelli (Fig. 14.63); lateral pronotal fringe complete	***Anacroneuria*** Klapálek
56'.	Three ocelli (Figs. 14.52-54); lateral pronotal fringe incomplete	57
57(56').	Head with large area of yellow in front of median ocellus (Fig. 14.53)	***Eccoptura*** Klapálek (Fig. 14.4g)
57'.	Head mostly brown (Fig. 14.64), often with yellow, flat, M-shaped mark in front of median ocellus	58
58(57').	Cerci with fringe of long, silky setae, at least on basal segment	***Acroneuria*** Pictet
58'.	Cerci without basal fringe of silky setae	***Beloneuria*** Needham and Claassen
59(8).	Eyes set far forward, head margin behind eyes forms a right angle (Fig. 14.65); body elongate (except ***Utaperla*** that will fit Couplet 59' except that ecdysial suture truncate; Fig. 14.67)	60
59'.	Eyes set midlaterally, posterolateral angles of head convex (Fig. 14.66); ecdysial suture Y-shaped; body less elongate (Fig. 14.4h)	62
60(59).	Ecdysial suture truncate (Fig. 14.67)	***Utaperla*** Ricker
60'.	Ecdysial suture Y-shaped (Fig. 14.66)	61
61(60').	Head longer than wide (Fig. 14.65); lacinia semiquadrate in shape (Fig. 14.72)	***Kathroperla*** Banks
61'.	Head about as wide as long; lacinia triangular in shape (Fig. 14.73)	***Paraperla*** Banks
62(59').	Apical segments of cerci with vertical, feathery fringe of intercalary hairs (Fig. 14.74a); pronotal setae largely restricted to corners (Fig. 14.70)	***Alloperla*** Banks
62'.	Cercal segments without vertical fringe of intercalary hairs; setae inserted only at apex of segments (Fig. 14.74b); pronotum with variable setation, but always with some setae along hind margin (Fig. 14.71)	63
63(62').	Thick, depressed black hairs present inside coxae on thoracic sternae (Fig. 14.75a); both pairs of wing pads divergent from body axis (Fig. 14.69); long pronotal fringe hairs except on sides (Fig. 14.71)	***Sweltsa*** Ricker
63'.	No thick black hairs inside coxae on thoracic sternae (Fig. 14.75b); wing pads variable in divergence	64
64(63').	Longest apical hairs of distal cercal segments shorter than their following segment (Fig. 14.74b)	65
64'.	Longest apical hairs of distal cercal segments as long or longer than their following segment (Fig. 14.75c)	66
65(64).	Lateral hairs absent from pronotum (Fig. 14.75d)	***Neaviperla*** Ricker
65'.	One or two long, lateral hairs on each side of pronotum (Fig. 14.75e)	***Suwallia*** Ricker
66(64').	Distal segments of cerci each with a single dorsal and ventral long apical hair, in addition to short circlet hairs (Fig. 14.75c)	67
66'.	Distal segments or cerci each with 4 or more long apical hairs in addition to short and medium circlet hairs (Fig. 14.75f)	68
67(66).	Lateral fringe complete around sides of pronotum (Fig. 14.75g); dark pigment forming a striped or checkered pattern on dorsal abdomen	***Triznaka*** Ricker

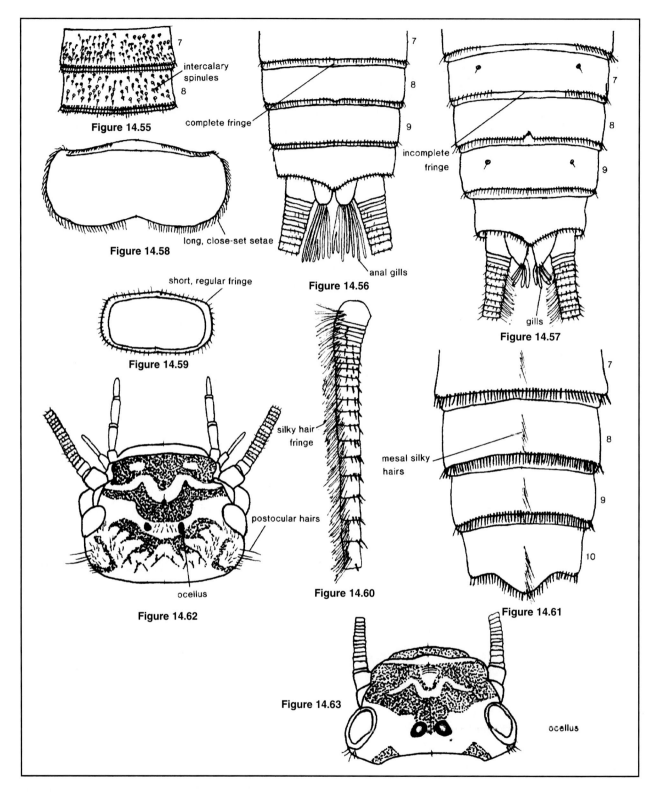

Figure 14.55. *Claassenia sabulosa* (Perlidae) abdominal terga.

Figure 14.56. *Agnetina capitata* (Perlidae) abdominal ventrum.

Figure 14.57. *Paragnetina fumosa* (Perlidae) abdominal ventrum.

Figure 14.58. *Attaneuria ruralis* (Perlidae) pronotum.

Figure 14.59. *Perlesta placida* (Perlidae) pronotum.

Figure 14.60. *Doroneuria baumanni* (Perlidae) basal right cercus (dorsal).

Figure 14.61. *Doroneuria baumanni* (Perlidae) abdominal dorsum.

Figure 14.62. *Perlinella drymo* (Perlidae) head.

Figure 14.63. *Anacroneuria* sp. (Perlidae) head.

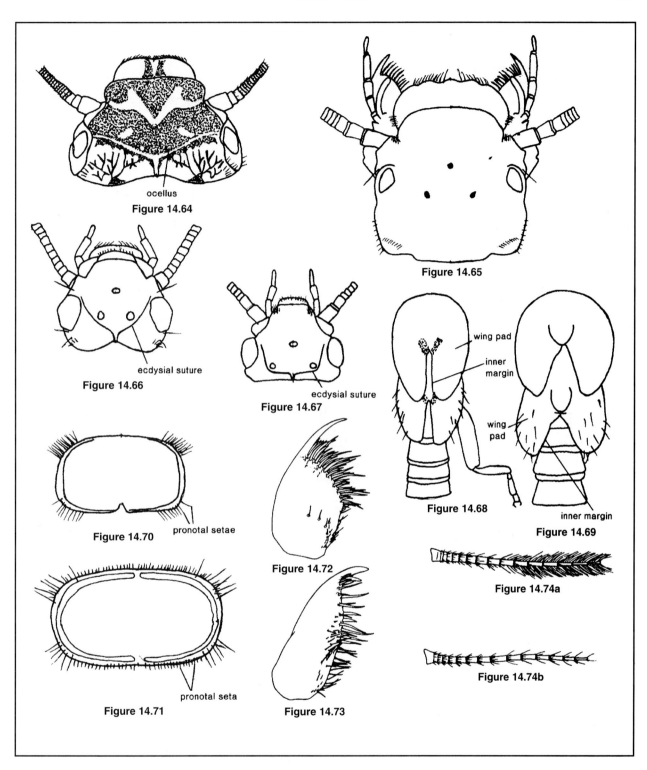

Figure 14.64. *Beloneuria stewarti* (Perlidae) head.
Figure 14.65. *Kathroperla perdita* (Chloroperlidae) head.
Figure 14.66. *Alloperla imbecilla* (Chloroperlidae) head.
Figure 14.67. *Utaperla sopladora* (Chloroperlidae) head.
Figure 14.68. *Haploperla brevis* (Chloroperlidae) wing pads.
Figure 14.69. *Suwallia pallidula* (Chloroperlidae) wing pads.
Figure 14.70. *Alloperla* sp. (Chloroperlidae) pronotum.
Figure 14.71. *Sweltsa fidelis* (Chloroperlidae) pronotum.
Figure 14.72. *Kathroperla perdita* (Chloroperlidae) right lacinia (ventral).
Figure 14.73. *Paraperla frontalis* sp. (Chloroperlidae) right lacinia (ventral).
Figure 14.74a. *Alloperla imbecilla* (Chloroperlidae) right cercus (lateral).
Figure 14.74b. *Suwallia pallidula* (Chloroperlidae) right cercus (lateral).

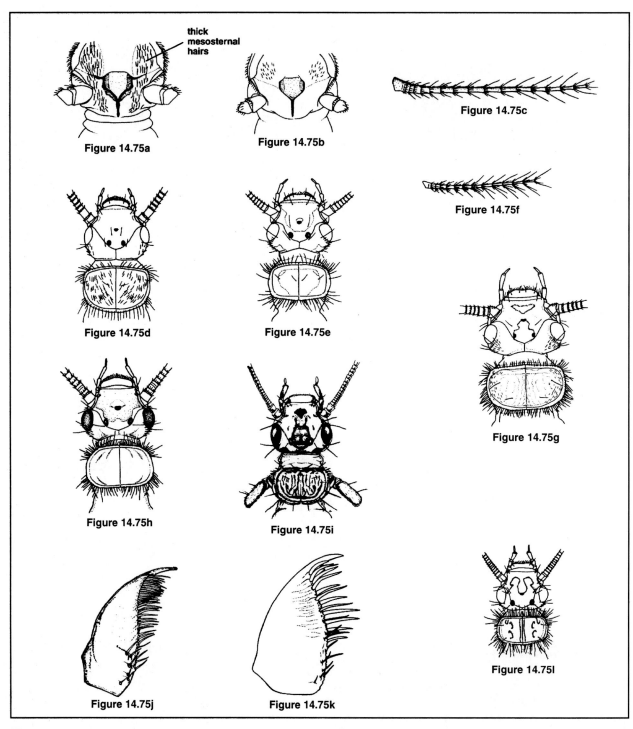

Figure 14.75a. *Sweltsa oregonensis* (Chloroperlidae) mesosternum (ventral).

Figure 14.75b. *Suwallia pallidula* (Chloroperlidae) mesosternum (ventral).

Figure 14.75c. *Plumiperla diversa* (Chloroperlidae) right cercus (lateral).

Figure 14.75d. *Neaviperla forcipata* (Chloroperlidae) head and pronotum (dorsal).

Figure 14.75e. *Suwallia pallidula* (Chloroperlidae) head and pronotum (dorsal).

Figure 14.75f. *Rasvena terna* (Chloroperlidae) right cercus (lateral).

Figure 14.75g. *Triznaka pintada* (Chloroperlidae) head and pronotum (dorsal).

Figure 14.75h. *Plumiperla diversa* (Chloroperlidae) head and pronotum (dorsal).

Figure 14.75i. *Bisancora rutriformis* (Chloroperlidae) head and pronotum (dorsal).

Figure 14.75j. *Bisancora rutriformis* (Chloroperlidae) right lacinia (ventral).

Figure 14.75k. *Haploperla brevis* (Chloroperlidae) right lacinia (ventral).

Figure 14.75l. *Rasvena terna* (Chloroperlidae) head and pronotum (dorsal).

67'.	Lateral hairs of pronotum sparse, with a distinct gap where they are absent on sides of pronotum (Fig. 14.75h); dorsal abdomen concolorous ....***Plumiperla*** Surdick
68(66').	Posterior margin of pronotum with close set fringe of over 40 long hairs (Fig. 14.71); head with U-shaped dark pattern; endemic to Alaska and adjacent Canadian territories ....***Alaskaperla*** Stewart and DeWalt
68'.	Posterior margin of pronotum with fewer than 24 long fringe hairs (Figs. 14.4h, 14.75i); head with variable pattern, not U-shaped; Eastern or Western 69
69(68').	Posterior margin of pronotum with a sparse fringe of about 8 long fringe hairs (Fig. 14.75i); lacinia with a close-set comb-like row of setae just below apical tooth (Fig. 14.75j); endemic to southern California ....***Bisancora*** Surdick
69'.	Posterior-margin of pronotum with 12-24 long fringe hairs alternating with a few short ones (Fig. 14.75h); lacinia without a distinct comb-like row in its marginal setae (Fig. 14.75k) 70
70(69').	Abdomen patterned above with dark spots, forming 4 incomplete longitudinal stripes; body length of mature larvae less than 5.5mm; 20 -24 long fringe hairs on posterior pronotal margin (Fig. 14.75l); Eastern ....***Rasvena*** Ricker
70'.	Abdomen concolorous; mature body length over 6mm; inner margins of both pairs of wing pads nearly parallel to body axis (Fig. 14.68); 12 - 18 long fringe hairs on posterior pronotal margin (Fig. 14.4h); common species *brevis* is Eastern; rare species in West ....***Haploperla*** Navás
71(8').	Gills present on 1 or more thoracic segments (Fig. 14.76) 72
71'.	Gills absent from thoracic segments..76
72(71).	Prothoracic gills present (Fig. 14.76) 73
72'.	Prothoracic gills absent; Western ....***Setvena*** Illies
73(72).	Lateral abdominal gills present (Fig. 14.77); Western ....***Oroperla*** Needham
73'.	Lateral abdominal gills absent..74
74(73').	Cervical gills present (Fig. 14.76); Western ....***Perlinodes*** Needham and Claassen
74'.	Cervical gills absent..75
75(74').	Prothoracic gills reduced to nipplelike stubs; meso- and metathoracic gills forked (Fig. 14.78); Western ....***Salmoperla*** Baumann and Lauck
75'.	Prothoracic gills well developed; meso- and metathoracic gills simple (Fig. 14.76); Western ....***Megarcys*** Klapálek
76(71').	Lacinia unidentate (Fig. 14.79)..77
76'.	Lacinia bidentate (Fig. 14.80)...80
77(76).	Abdomen with dark longitudinal pigment bands ....***Isoperla*** Banks (in part)
77'.	Abdomen without dark longitudinal pigment bands 78
78(77').	Lacinia broad basally, abruptly narrowed into a long terminal spine (Fig. 14.81); Eastern ....***Remenus*** Ricker
78'.	Lacinia gradually narrowed from base to terminal spine (Fig. 14.79).......................................79
79(78').	Mesosternal furcal pits connected by a transverse anterior furrow (Fig. 14.82); Western ....***Rickera*** Jewett
79'.	Mesosternum without transverse anterior furrow; Western ....***Kogotus*** Ricker
80(76').	Abdomen with dark longitudinal pigment bands 81
80'.	Abdomen without dark longitudinal pigment bands 89
81(80).	Abdomen with pale, median, longitudinal pigment band 82
81'.	Abdomen with dark, median, longitudinal pigment band..84
82(81).	Apical lacinial tooth about as long as rest of lacinia (Fig. 14.83); Western ....***Osobenus*** Ricker
82'.	Apical lacinial tooth much shorter than rest of lacinia (Fig. 14.84) 83
83(82').	Inner lacinial margin with row of at least 4-5 long setae; no prominent knob bearing pegs below subapical lacinial tooth (Fig. 14.85); widely distributed ....***Isoperla*** Banks (in part)
83'.	Inner lacinial margin lacking a row of long setae; prominent knob below subapical lacinial tooth bearing 3-4 stout peg-like setae (may appear as a 3rd tooth under low-power magnification; Fig. 14.84); endemic to California ....***Susulus*** Bottorff and Stewart
84(81').	Lacinia quadrate (Fig. 14.86) 85
84'.	Lacinia triangular or subquadrate (Fig. 14.83)...88
85(84).	Anterior lacinial margin with dense brush of stout setae; inner lacinial margin glabrous (Fig. 14.86)..86
85'.	Anterior lacinial margin with row of stout setae; inner lacinial margin setose 87
86(85).	Apical cercal segments fringed with fine hairs along both margins; Western ....***Cascadoperla*** Szczytko and Stewart

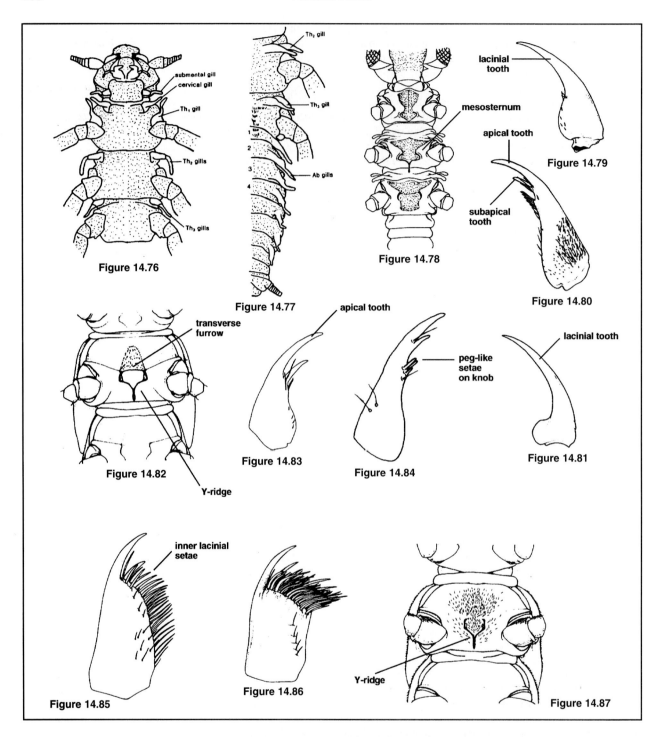

Figure 14.76. *Perlinodes aurea* (Perlodidae) ventral head, thorax.

Figure 14.77. *Oroperla barbara* (Perlodidae) left abdomen (ventral).

Figure 14.78. *Salmoperla sylvanica* (Perlodidae) thorax (ventral).

Figure 14.79. *Kogotus nonus* (Perlodidae) left lacinia (ventral).

Figure 14.80. *Marlirekus hastatus* (Perlodidae) left lacinia (ventral).

Figure 14.81. *Remenus bilobatus* (Perlodidae) left lacinia (ventral).

Figure 14.82. *Rickera sorpta* (Perlodidae) mesothorax (ventral).

Figure 14.83. *Osobenus yakamae* (Perlodidae) right lacinia (ventral).

Figure 14.84. *Susulus venustus* (Perlodidae) right lacinia (ventral).

Figure 14.85. *Isoperla bilineata* (Perlodidae) right lacinia (ventral).

Figure 14.86. *Cascadoperla trictura* (Perlodidae) right lacinia (ventral).

Figure 14.87. *Calliperla luctuosa* (Perlodidae) mesosternum (ventral).

86'.	Apical cercal segments fringed along inner margins or not at all	*Isoperla* Banks (in part)
87(85').	Cerci without marginal fringe; mesosternal Y-arms enclosing an area of abundant clothing hairs (Fig. 14.87); Western	*Calliperla* Banks
87'.	Cerci with marginal fringe on apical segments; mesosternal Y-arms enclosing an area with few or no clothing hairs	*Isoperla* Banks (in part)
88(84').	Mesosternal Y-arms sinuate; cerci not fringed with fine hairs; Western	*Cosumnoperla* Szczytko and Bottorff
88'.	Mesosternal Y-arms straight; cerci with a fringe of fine hairs	*Isoperla* Banks (in part)
89(80').	Mesosternal Y-arms meet or approach anterior corners of furcal pits (Fig. 14.89); mandibles deeply cleft, separating teeth into two major cusps	90
89'.	Mesosternal Y-arms meet or approach posterior corners of furcal pits (Fig. 14.82); mandibles not deeply cleft	92
90(89).	Apical lacinial tooth short, much less than one-third the outer lacinial length (Fig. 14.90); ventral submarginal lacinial setae extending well onto apical tooth as a closely set row; Western	*Frisonia* Ricker
90'.	Apical lacinial tooth long, greater than one-third the outer lacinial length; ventral submarginal lacinial setae end at inner base of apical tooth	91
91(90').	Outermost cusp of both mandibles serrate; mesal tufts of silky setae on occiput; Ab_{1-2} divided by pleural fold; Western	*Skwala* Ricker
91'.	Outermost cusp of both mandibles unserrated, or with indistinct serrations on left mandible only; occiput without mesal tuft of silky setae; Ab_{1-3} divided by pleural fold	*Arcynopteryx* Klapálek
92(89').	Femora and tibiae without long setal fringe; posterolateral margins of pronotum notched; Eastern	*Oconoperla* Stark and Stewart
92'.	Femora or tibiae or both with long setal fringe; posterolateral margins of pronotum smoothly rounded	93
93(92').	Mesosternum with median longitudinal suture connecting fork of Y-arms with transverse suture (Fig. 14.91)	*Isogenoides* Klapálek
93'.	Mesosternum without median longitudinal suture	94
94(93').	Submental gills conspicuous, projecting portion usually 2 times or more as long as basal diameter (Fig. 14.92)	95
94'.	Submental gills absent or barely projecting beyond submentum	98
95(94).	Apical lacinial tooth about half the total outer lacinial length; Western	*Pictetiella* Illies
95'.	Apical lacinial tooth about one-third or less the total outer lacinial length (Fig. 14.80)	96
96(95').	Ventral lacinial surface with basal patch of about fifty dark clothing hairs (Fig. 14.80); Eastern	*Malirekus* Ricker (in part)
96'.	Ventral lacinial surface without dark clothing hairs, a patch of about 10 setae may be present (Fig. 14.85)	97
97(96').	Transverse dark pigment band of frons lateral to median ocellus interrupted by circular yellow areas; ventral lacinial surface with outer patch of about 10 setae; right mandible with 4 teeth; Eastern	*Hydroperla* Frison
97'.	Transverse dark pigment band of frons uninterrupted by enclosed yellow areas (Fig. 14.93); ventral lacinial surface without outer patch of setae; right mandible with 5 teeth; Eastern	*Helopicus* Ricker
98(94').	Inner lacinial margin with a low knob below subapical tooth (Fig. 14.80)	99
98'.	Inner lacinial margin without low knob below subapical tooth (Fig. 14.85)	101
99(98).	Outer ventral lacinial surface with basal patch of about 50 dark clothing hairs (Fig. 14.80); Eastern	*Malirekus* Ricker (in part)
99'.	Outer ventral lacinial surface with fewer than 50 basal clothing hairs (Fig. 14.94)	100
100(99').	Marginal lacinial setal row extending from near subapical tooth to near base (Fig. 14.94); labrum with yellow longitudinal mesal band; submental gills very short, if present	*Yugus* Ricker
100'.	Marginal lacinial setal row restricted to apical half; labrum without longitudinal band; submental gills absent	*Diura* Billberg (in part)
101(98').	Occiput or anterolateral prothoracic margins or both with a row of short stout setae	102
101'.	Occiput and anterolateral prothoracic margins without rows of short, stout setae, a few long setae may be present	106
102(101).	Ocellar region transversed by dark band connecting lateral ocelli and eyes but not covering anterior ocellus (Fig. 14.95)	*Clioperla* Needham and Claassen
102'.	Ocellar triangle more or less completely covered by dark pigment	103

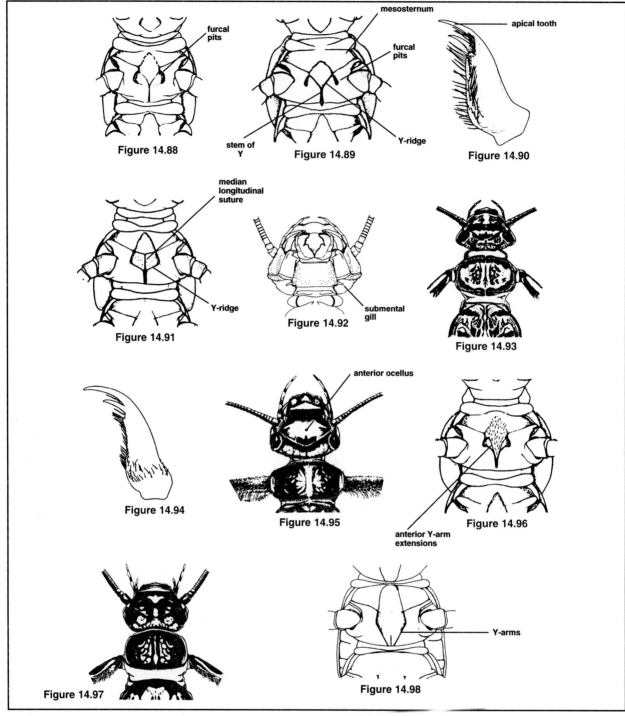

Figure 14.88. *Frisonia pictipes* (Perlodidae) mesosternum (ventral).

Figure 14.89. *Skwala americana* (Perlodidae) mesosternum (ventral).

Figure 14.90. *Frisonia pictipes* (Perlodidae) left lacinia (ventral).

Figure 14.91. *Isogenoides zionensis* (Perlodidae) mesosternum (ventral).

Figure 14.92. *Pictetiella expansa* (Perlodidae) head (ventral).

Figure 14.93. *Helopicus nalatus* (Perlodidae) head and pronotum (dorsal).

Figure 14.94. *Yugus bulbosus* (Perlodidae) left lacinia (ventral).

Figure 14.95. *Clioperla clio* (Perlodidae) head and pronotum (dorsal).

Figure 14.96. *Chernokrilus misnomus* (Perlodidae) mesosternum (ventral).

Figure 14.97. *Diura knowltoni* (Perlodidae) head and pronotum (dorsal).

Figure 14.98. *Diploperla duplicata* (Perlodidae) mesosternum (ventral).

103(102')	Major mandibular cusp serrate; abdominal terga with narrow dark basal bands and conspicuous transverse row of dark spots	***Baumannella*** Stark and Stewart
103'.	Major mandibular cusp without serrations; abdominal terga with variable color pattern	104
104(103').	Mesosternal Y with indistinct anterior extensions which approach furcal pits (Fig. 14.96); Western	***Chernokrilus*** Ricker
104'.	Mesosternal Y without anterior extensions	105
105(104').	Occipital area with a pair of pale oval markings adjacent to compound eyes; posterior margins of ovals marked by dense irregular rows of short, thick setae (Fig. 14.97)	***Diura*** Billberg (in part)
105'.	Occipital area without pale oval markings margined by short, thick setae	***Isoperla*** Banks (in part)
106(101').	Mesosternal Y with basal stem and fork (Fig. 14.89)	***Cultus*** Ricker
106'.	Mesosternal Y without stem and fork (Fig. 14.98); Eastern	***Diploperla*** Needham and Claassen

KEY TO THE FAMILIES AND GENERA OF NORTH AMERICAN PLECOPTERA ADULTS

1.	Glossae and paraglossae of approximately same length and size and set at same level on labium (Figs. 14.7, 14.109, 14.110, 14.135)	2
1'.	Labium with glossae much reduced and set below paraglossae (Figs. 14.8, 14.185)	7
2(1).	First segment of each tarsus short (Fig. 14.99; gill remnants present on the sides of thorax (Figs. 14.5, 14.102, 14.103, 14.114, 14.122)	3
2'.	First segment of each tarsus longer (Figs. 14.100-14.101); no gill remnants on sides of thorax	4
3(2).	Three ocelli; gill remnants on the sides of Ab_{1-2} or Ab_{1-3} (Figs. 14.102, 14.103); 2 rows of anal crossveins in forewing (Fig. 14.104)	**PTERONARCYIDAE** 9
3'.	Two ocelli; no gill remnants on sides of abdomen; no rows of anal crossveins (Fig. 14.105)	**PELTOPERLIDAE** 10
4(2').	Second segment of tarsus of approximately same length as first (Fig. 14.100)	**TAENIOPTERYGIDAE** 15
4'.	Second segment of tarsus much shorter than 1st (Fig. 14.101)	5
5(4').	Cerci with many segments; in fore wing, 2A simple and unforked; intercubital crossveins few, usually 1-2 (Fig. 14.106)	**CAPNIIDAE** 45
5'.	Cerci 1-segmented; in fore wing, 2A forked; intercubital crossveins numerous, usually 5 or more (Figs. 14.107, 14.108)	6
6(5').	At rest, wings folded flat over back; terminal segment of labial palpus circular and larger than preceding segment (Fig. 14.109); a crossvein in costal space just beyond cord, forming with adjoining veins a X-pattern (Fig. 14.107)	**NEMOURIDAE** 27
6'.	At rest, wings rolled, covering both back and sides of abdomen, giving insect a sticklike appearance; terminal segment of labial palpus more elongate, not much larger than preceding segment (Fig. 14.110); no crossvein in costal space beyond cord (Fig. 14.108)	**LEUCTRIDAE** 58
7(1').	Remnants of branched gills on sides of thorax (Fig. 14.14); cubito-anal crossvein of fore wing usually reaching anal cell or not distant from it by more than its own length (Fig. 14.111)	**PERLIDAE** 126
7'.	Remnants of branched gills absent; in a few Perlodidae, remnants of single unbranched or forked gills occur on the thorax or elsewhere (Figs. 14.180, 14.185); cubito-anal crossvein of fore wing, if present, usually removed from anal cell by at least its own length (Figs. 14.112, 14.113)	8
8(7').	Fork of 2A of fore wing set in anal cell; hind wing with 5-10 anal veins (Fig. 14.112)	**PERLODIDAE** 64
8'.	Fork of 2A of fore wing set beyond anal cell or absent; hind wing usually with 1-4 anal veins (Fig. 14.113) (if 5 or more, either 2A of fore wing forked or hind corners of head angulate; Fig. 14.214)	**CHLOROPERLIDAE** 109
9(3).	Gill remnants on Ab_{1-3} (Fig. 14.102); body length 12-20 mm	***Pteronarcella*** Banks
9'.	Gill remnants on Ab_{1-2} (Fig. 14.103); body length 23-40 mm	***Pteronarcys*** Newman
10(3').	Gill remnants present on cervix and prothorax (Fig. 14.114); male epiproct reduced (Fig. 14.115) or absent, and vesicle on sternum 9 wide and oval and set at middle of sternum (Fig. 14.116); female with eggs flat or hemispherical	11
10'.	No gill remnants on cervix and prothorax; male epiproct developed into a rod or plate (Figs. 14.117, 14.123); if undeveloped, vesicle round or quadrate and set at base of sternum (Figs. 14.118-14.119); female with eggs spherical	12
11(10).	Male cerci with 7 segments; female sternum 9 with distinct diagonal sclerotized bars (Fig. 14.120)	***Yoraperla*** Ricker
11'.	Male cerci with 10 segments; female sternum 9 without such bars	***Sierraperla*** Jewett
12(10').	Remnants of supracoxal gills on meso- and metathorax simple; male sternum 9 with vesicle 2 or 3 times as wide as long (Fig. 14.116); vagina of female without large central sclerite, though a pair of weak sclerites are present near vaginal opening, and sometimes another near junction of spermathecal duct	13

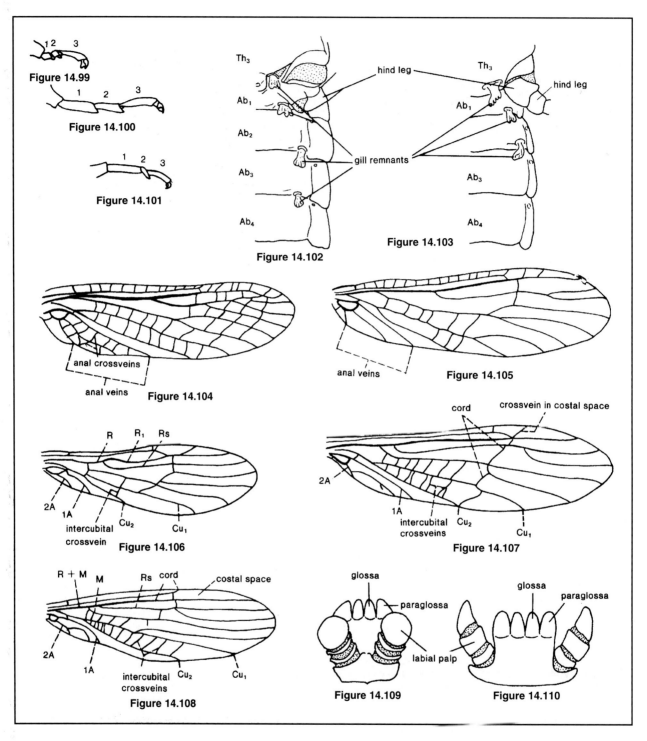

Figure 14.99. Tarsus of *Peltoperla* sp. (Peltoperlidae).
Figure 14.100. Tarsus of *Taeniopteryx* sp. (Taeniopterygidae).
Figure 14.101. Tarsus of *Paracapnia* sp. (Capniidae).
Figure 14.102. Ventral view of metasternum (Th_3) and abdominal sterna 1-4 of *Pteronarcella regularis* (Pteronarcyidae).
Figure 14.103. Ventral view of metasternum (Th_3) and abdominal sterna 1-4 of *Pteronarcys biloba* (Pteronarcyidae).
Figure 14.104. Fore wing of *Pteronarcella badia* (Pteronarcyidae).
Figure 14.105. Fore wing of *Tallaperla maria* (Peltoperlidae).
Figure 14.106. Fore wing of *Capnia* sp. (Capniidae).
Figure 14.107. Fore wing of *Nemoura trispinosa* (Nemouridae).
Figure 14.108. Fore wing of *Leuctra tenuis* (Leuctridae).
Figure 14.109. Labium of *Nemoura trispinosa* (Nemouridae).
Figure 14.110. Labium of *Leuctra tenuis* (Leuctridae).

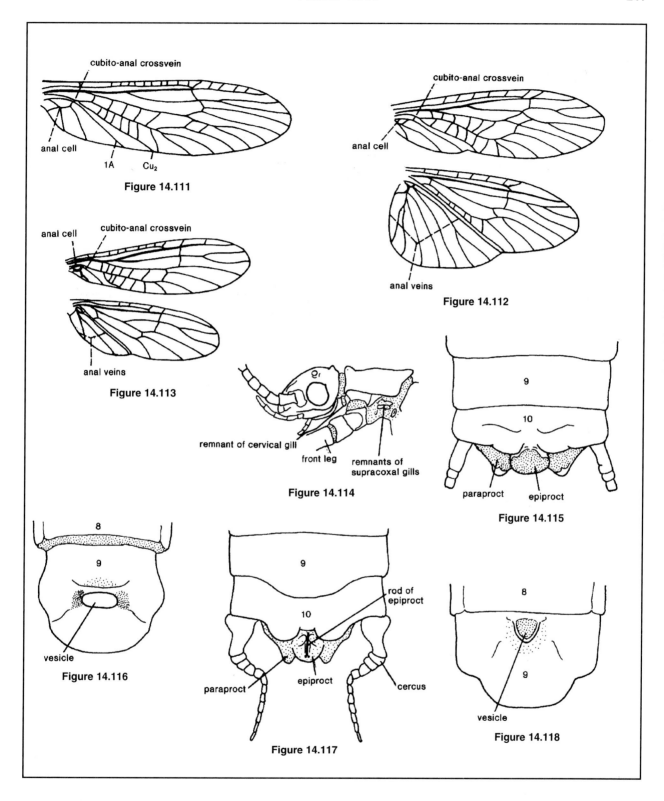

Figure 14.111. Fore wing of *Paragnetina media* (Perlidae).

Figure 14.112. Wings of *Isoperla frisoni* (Perlodidae).

Figure 14.113. Wings of *Sweltsa onkos* (Chloroperlidae).

Figure 14.114. Side view of head and prothorax of *Yoraperla brevis* (Peltoperlidae).

Figure 14.115. Dorsal view of terminal abdominal segments of male *Yoraperla brevis* (Peltoperlidae).

Figure 14.116. Ventral view of abdominal sternum 9 of male *Yoraperla brevis* (Peltoperlidae).

Figure 14.117. Dorsal view of terminal abdominal segments of *Peltoperla arcuata* (Peltoperlidae).

Figure 14.118. Abdominal sternum 9 of male *Peltoperla arcuata* (Peltoperlidae).

12'.	Remnants of supracoxal gills on meso- and metathorax double (Fig. 14.122); male sternum 9 with vesicle about as wide as long (Figs. 14.118-14.119); vagina of female with a well-developed central sclerite (Fig. 14.121)	14
13(12).	Eastern species; male epiproct long and pointed; subgenital plate of female reaching about middle of sternum 9; remnants of an infracoxal gill on metathorax (Fig. 14.122)	*Viehoperla* Ricker
13'.	Western species; epiproct of male a large crenulate plate (Fig. 14.123); subgenital plate of female large, reaching tip of sternum 9; no infracoxal gill	*Soliperla* Ricker
14(12').	Male epiproct developed into a slender rod (Fig. 14.117); subgenital plate of female long and notched	*Peltoperla* Needham
14'.	Male epiproct undeveloped (as in Fig. 14.115); female subgenital plate short and entire	*Tallaperla* Stark and Stewart
15(4).	Gill scar on inner surface of each coxa (Fig. 14.124)	*Taeniopteryx* Pictet
15'.	No such scar	16
16(15').	Males (epiproct and paraprocts complex) (Figs. 14.127, 14.129-14.130)	17
16'.	Females (epiproct and paraprocts unmodified) (Fig. 14.133)	21
17(16).	Hind margin of sternum 9 asymmetrical (Fig. 14.126)	*Bolotoperla* Ricker and Ross
17'.	Sternum 9 symmetrical	18
18(17').	Costal crossveins absent (Fig. 14.125); epiproct with 2 prongs (Fig. 14.127)	*Oemopteryx* Klapálek
18'.	One to 5 costal crossveins (Figs. 14.128, 14.131); epiproct with one prong (Figs. 14.129-14.130)	19
19(18').	Rs with 3 branches, Cu$_1$ with 4-5 branches (Fig. 14.131); tip of epiproct asymmetrical	*Doddsia* Needham and Claassen
19'.	Rs with 2 branches, Cu$_1$ with 2-3 branches (Fig. 14.128); tip of epiproct symmetrical (Figs. 14.129, 14.130)	20
20(19').	Hind corners of sternum 9 elevated (Fig. 14.129)	*Taenionema* Banks
20'.	Middle of hind margin of sternum 9 elevated, corners low (Fig. 14.130)	*Strophopteryx* Frison
21(16').	Costal crossveins absent (Figs. 14.125, 14.132)	22
21'.	Costal crossveins present (Figs. 14.128, 14.131)	24
22(21).	Crossvein c-r present (Fig. 14.132)	23
22'.	Crossvein c-r absent (Fig. 14.125)	*Oemopteryx* Klapálek (in part)
23(22).	Western species (California)	*Oemopteryx* Klapálek (in part)
23'.	Eastern species	*Bolotoperla* Ricker and Ross
24(21').	Rs with 3 branches, Cu$_1$ with 4-5 branches (Fig. 14.131)	*Doddsia* Needham and Claassen
24'.	Rs with 2 branches, Cu$_1$ with 2-3 branches (Fig. 14.132)	25
25(24').	Western species	*Taenionema* Banks (in part)
25'.	Eastern species	26
26(25').	Sternum 9 subtriangular or narrower (Fig. 14.134)	*Strophopteryx* Frison
26'.	Sternum 9 wider, more rounded (Fig. 14.133)	*Taenionema* Banks (in part)
27(6).	Gill remnants on submentum or cervix (Figs. 14.18-14.20, 14.24, 14.135-14.137)	28
27'.	No gill remnants	32
27'.	Gill remnants on cervix (Figs. 14.18-14.20, 14.136-14.137	29
28'.	Gill remnants on submentum (Figs. 14.24, 14.135)	*Visoka* Ricker
29(28).	Each gill unbranched (Figs. 14.19, 14.136)	*Zapada* Ricker (in part)
29'.	Each gill divided into 3 or more branches (Figs. 14.18, 14.20, 14.137)	30
30(29').	Each gill composed of 3-4 branches; male paraprocts with 2 lobes (Fig. 14.138); female sternum 7 covering most of sternum 8 (Fig. 14.150)	*Zapada* Ricker (in part)
30'.	Each gill composed of 5-18 branches (Figs. 14.18, 14.20, 14.137); male paraprocts with 3 lobes (Fig. 14.139); female sternum 7 shorter (Figs. 14.140-14.141)	31
31(30').	All gills divided down to gill base (Fig. 14.20); male cerci unmodified; female sternum 7 broadly rounded (Fig. 14.140)	*Amphinemura* Ris
31'.	Gill branches arising at different levels (Fig. 14.18); male cerci with dorsal lobe at base (Fig. 14.142); female sternum 7 forming a nipplelike projection (Fig. 14.141)	*Malenka* Ricker
32(27').	Veins A1 and A2 joined near margin in fore wing (Fig. 14.143)	*Soyedina* Ricker
32'.	Veins A1 and A2 separate (Figs. 14.107, 14.143)	33
33(32').	Terminal costal crossvein joining Sc (Fig. 14.143)	*Paranemoura* Needham and Claassen
33'.	Terminal crossvein joining R (Figs. 14.107, 14.144)	34
34(33').	Males (epiproct and paraprocts modified) (Figs. 14.145-14.149)	35
34'.	Females (epiproct and paraprocts unmodified) (Fig. 14.150)	40
35(34).	Cerci sclerotized or enlarged (Figs. 14.145-14.146)	36
35'.	Cerci membranous, unmodified (Figs. 14.147-14.149)	37
36(35).	Cerci with subterminal spines (Fig. 14.145)	*Nemoura* Latreille
36'.	Cerci without spines, but elongated and curved (Fig. 14.146)	*Ostrocerca* Ricker

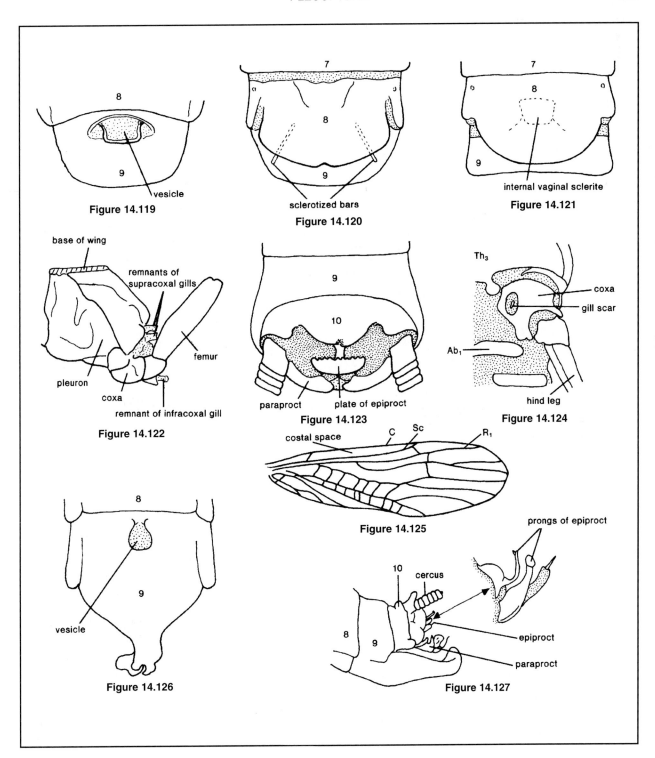

Figure 14.119. Abdominal sternum 9 of male *Tallaperla maria* (Peltoperlidae).

Figure 14.120. Abdominal sterna 8 and 9 of female *Yoraperla brevis* (Peltoperlidae).

Figure 14.121. Subgenital plate of female *Peltoperla arcuata* (Peltoperlidae).

Figure 14.122. Side view of Th_3 of *Tallaperla maria* (Peltoperlidae).

Figure 14.123. Dorsal view of terminal abdominal segments of *Soliperla* sp. (Peltoperlidae).

Figure 14.124. Ventral view of Th_3 of *Taeniopteryx* sp. (Taeniopterygidae).

Figure 14.125. Fore wing of *Oemopteryx glacialis* (Taeniopterygidae).

Figure 14.126. Abdominal sternum 9 of male *Bolotoperla rossi* (Taeniopterygidae).

Figure 14.127. Side view of terminal abdominal segments of *Oemopteryx glacialis* (Taeniopterygidae). Details of the epiproct.

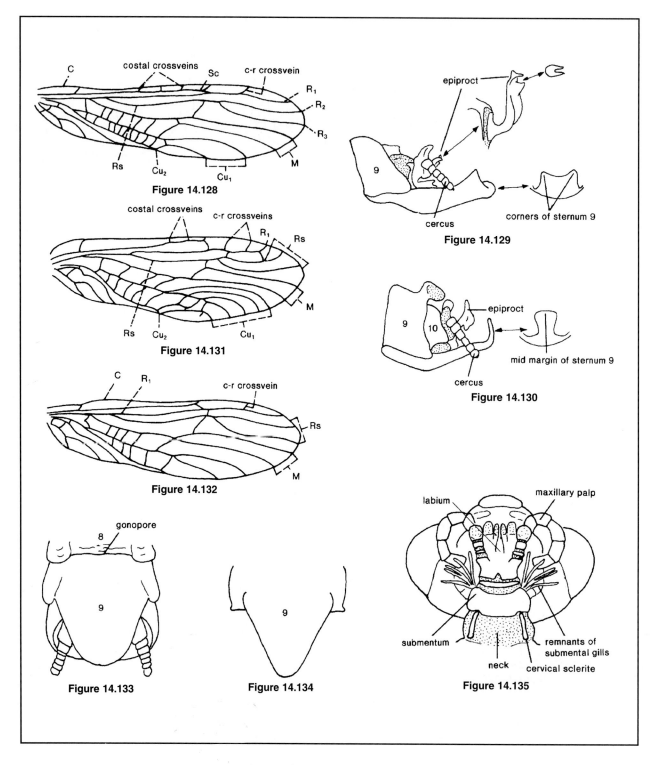

Figure 14.128. Fore wing of *Taenionema oregonense* (Taeniopterygidae).

Figure 14.129. Side view of terminal abdominal segments of *Taenionema oregonense* (Taeniopterygidae). Details of epiproct; hind view of sternum 9.

Figure 14.130. Side view of terminal abdominal segments of *Strophopteryx fasciata* (Taeniopterygidae). Details of epiproct; hind view of sternum 9.

Figure 14.131. Fore wing of *Doddsia occidentalis* (Taeniopterygidae).

Figure 14.132. Fore wing of *Bolotoperla rossi* (Taeniopterygidae).

Figure 14.133. Sternum 9 of female *Taenionema pacificum* (Taeniopterygidae).

Figure 14.134. Sternum 9 of female *Strophopteryx* sp. (Taeniopterygidae).

Figure 14.135. Ventral view of the head of *Visoka cataractae* (Nemouridae).

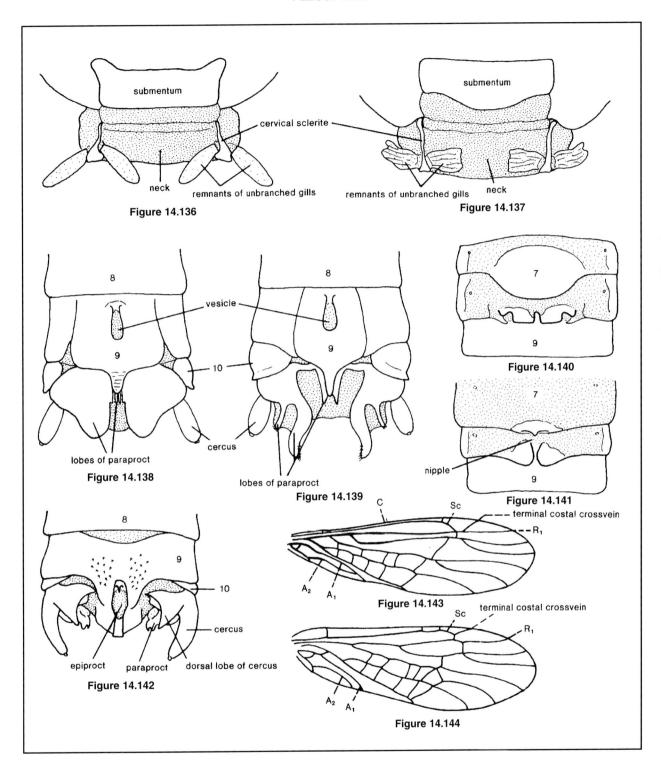

Figure 14.136. Ventral view of neck region of *Zapada* sp. (Nemouridae).

Figure 14.137. Ventral view of neck region of *Amphinemura delosa* (Nemouridae).

Figure 14.138. Ventral view of terminal abdominal segments of male *Zapada* sp. (Nemouridae).

Figure 14.139. Ventral view of terminal abdominal segments of male *Amphinemura delosa* (Nemouridae).

Figure 14.140. Subgenital plate of female *Amphinemura delosa* (Nemouridae).

Figure 14.141. Subgenital plate of female *Malenka* sp. (Nemouridae).

Figure 14.142. Dorsal view of terminal abdominal segments of male *Malenka* sp. (Nemouridae).

Figure 14.143. Fore wing of *Soyedina producta* (Nemouridae).

Figure 14.144. Fore wing of *Paranemoura perfecta* (Nemouridae).

37(35').	Sternum 9 without vesicle	***Lednia*** Ricker
37'.	Sternum 9 with vesicle (Figs. 14.138-14.139, 14.147-14.149)	38
38(37').	Tergum 10 produced into 2 large lobes covering cerci (Figs. 14.147)	***Shipsa*** Ricker
38'.	Lobes of tergum 10, if produced, not so large (Figs. 14.148, 14.149)	39
39(38').	Epiproct long and simple (Fig. 14.149)	***Prostoia*** Ricker
39'.	Epiproct short and complex (Fig. 14.148)	***Podmosta*** Ricker
40(34').	Sternum 7 large and extending over sternum 8 (Fig. 14.150)	***Nemoura*** Latreille
40'.	Sternum 7 short or unmodified	41
41(40').	Sternum 8 extending over sternum 9 as a distinct subgenital plate (Figs. 14.151-14.152)	42
41'.	Sternum 8 not extending over sternum 9, though its margin may be sclerotized (Figs. 14.153-14.155)	43
42(41).	Sternum 8 small (Fig. 14.151)	***Shipsa*** Ricker
42'.	Sternum 8 large; sternum 7 often with a small nipplelike projection (Fig. 14.152)	***Ostrocerca*** Ricker
43(41').	Sternum 8 with a distinct median sclerotized patch (Figs. 14.154-14.155); wings unmarked	44
43'.	Sternum 8 unsclerotized except at margin (Fig. 14.153); wings distinctly banded	***Prostoia*** Ricker
44(43).	Sternum 8 with lateral sclerotized patches in addition to the median sclerotization (Fig. 14.154)	***Lednia*** Ricker
44'.	Sternum 8 with only median sclerotization (Fig. 14.155)	***Podmosta*** Ricker
45(5).	Wings present	49
45'.	Wings absent	46
46(45').	Eastern species	***Allocapnia*** Claassen
46'.	Western species	47
47(46').	Postfurcasternal plates of mesosternum fused to furcasternum (Fig. 14.156)	48
47'.	Postfurcasternal plates free (Fig. 14.157)	50
48(47).	Prothoracic presternum fused to basisternum; males with vesicle at base of sternum 9 (Fig. 14.159)	***Isocapnia*** Banks
48'.	Prothoracic presternum detached from basisternum; no vesicle on male sternum 9 (Figs. 14.160, 14.162)	***Paracapnia*** Hanson
49(45).	R_1 of fore wing curved cephalad just beyond origin of R_s (Fig. 14.106)	50
49'.	R_1 of fore wing straight at this point (Fig. 14.158)	54
50(47', 50').	Sternum 9 of male with vesicle (Fig. 14.159); male epiproct with lateral hooks (Fig. 14.159); sternum 8 of female often produced into a subgenital plate which extends over sternum 9 (Fig. 14.165); body length of female 9 mm or more	***Bolshecapnia*** Ricker
50'.	Sternum 9 of male without vesicle; if vesicle present, male epiproct without lateral hooks; sternum 8 of female usually not produced over sternum 9; if produced, marked with a distinct dark pattern (Fig. 14.162); size of female smaller	51
51(50').	Epiproct of male divided into 2 slender processes, upper process forked at tip (Fig. 14.161); female sternum 8 with a striking color pattern, its margin usually notched (Fig. 14.162)	***Utacapnia*** Nebeker and Gaufin
51'.	Epiproct of male usually simple (Fig. 14.163); if divided, processes are short, or tip of upper process not forked (Fig. 14.160); female sternum 8 not as above	52
52(51').	Epiproct of male simple, terminated by a slender spine, abdominal terga unmodified (Fig. 14.163); female sternum 8 little produced or produced into a short point (Fig. 14.164)	***Mesocapnia*** Rauser
52'.	Epiproct of male variable; if simple, either not ending in a slender spine, or 1 or more abdominal terga with knobs or humps; female sternum 8 variable (Note: females not satisfactorily separable from *Mesocapnia*)	53
53 (52').	Epiproct of male divided into widely separated upper and lower processes (the lower process often reduced and difficult to recognize) (Fig. 14.160); some females with a Y-shaped sclerite on subgenital plate and without a sclerotized bridge uniting sterna 7 and 8 (other females undistinguishable from those of *Capnia*)	***Capnura*** Banks
53'.	Epiproct of male simple or divided; if divided, processes short and not widely separated; female subgenital plate variable; if Y-shaped sclerite present, there is a sclerotized bridge uniting sterna 7 and 8	***Capnia*** Pictet
54(49').	Cerci with fewer than 11 segments	55
54'.	Cerci with more than 11 segments	56
55(54).	Eastern species; 1A of fore wing sharply bent just beyond anal cell (Fig. 14.166)	***Nemocapnia*** Banks
55'.	Western species; 1A of fore wing straight (Fig. 14.158)	***Eucapnopsis*** Okamoto
56(54').	Anal lobe of hind wing nearly as large as rest of wing (Fig. 14.172)	***Allocapnia*** Claassen
56'.	Anal lobe much smaller	57
57(56').	Zero to 3 costal crossveins in fore wing (Fig. 14.167)	***Paracapnia*** Hanson
57'.	Five to 6 costal crossveins in fore wing (Fig. 14.168)	***Isocapnia*** Banks
58(6').	Hind wing with 6 anal veins (Fig. 14.169)	***Megaleuctra*** Neave

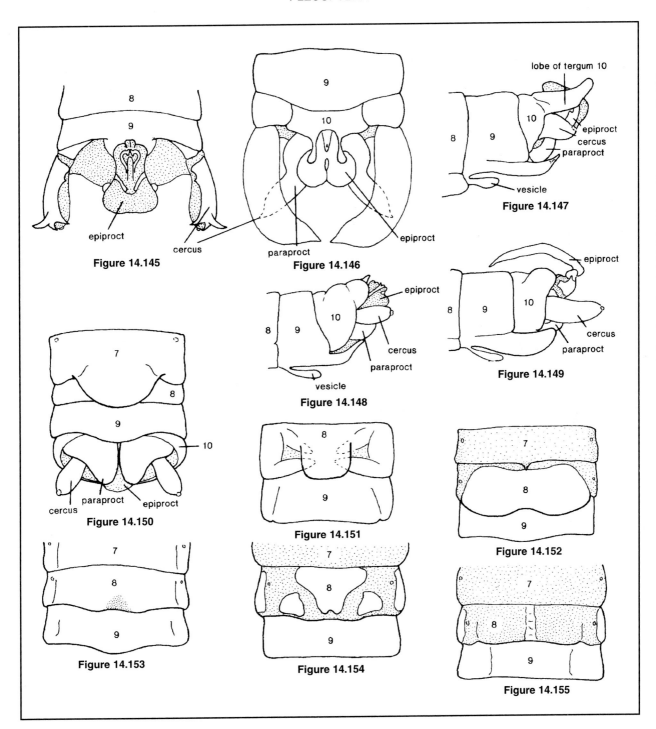

Figure 14.145. Dorsal view of terminal abdominal segments of male *Nemoura trispinosa* (Nemouridae).

Figure 14.146. Dorsal view of terminal abdominal segments of male *Ostrocerca dimicki* (Nemouridae).

Figure 14.147. Side view of terminal abdominal segments of *Shipsa rotunda* (Nemouridae).

Figure 14.148. Side view of terminal abdominal segments of male *Podmosta macdunnoughi* (Nemouridae).

Figure 14.149. Side view of terminal abdominal segments of male *Prostoia besametsa* (Nemouridae).

Figure 14.150. Ventral view of terminal abdominal segments of female *Nemoura trispinosa* (Nemouridae).

Figure 14.151. Subgenital plate of female *Shipsa rotunda* (Nemouridae).

Figure 14.152. Subgenital plate of female *Ostrocerca albidipennis* (Nemouridae).

Figure 14.153. Subgenital plate of female *Prostoia besametsa* (Nemouridae).

Figure 14.154. Subgenital plate of female *Lednia tumana* (Nemouridae).

Figure 14.155. Subgenital plate of female *Podmosta macdunnoughi* (Nemouridae).

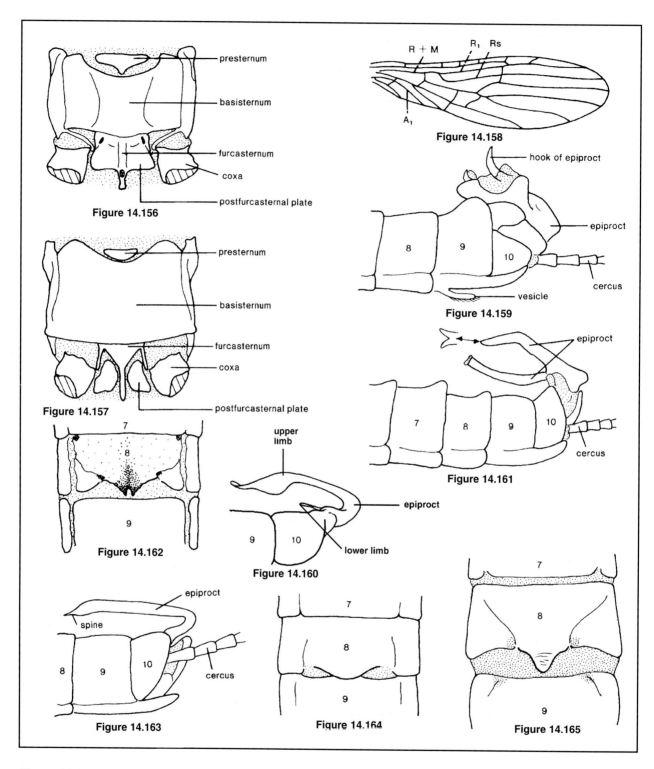

Figure 14.156. Ventral view of mesosternum of *Isocapnia agassizi* (Capniidae).

Figure 14.157. Ventral view of mesosternum of *Capnia* sp. (Capniidae).

Figure 14.158. Fore wing of *Eucapnopsis brevicauda* (Capniidae).

Figure 14.159. Side view of terminal abdominal segments of male *Bolshecapnia spenceri* (Capniidae).

Figure 14.160. Side view of epiproct of *Capnura venosa* (Capniidae).

Figure 14.161. Side view of terminal abdominal segments of male *Utacapnia labradora* (Capniidae).

Figure 14.162. Subgenital plate of female *Utacapnia labradora* (Capniidae).

Figure 14.163. Side view of terminal abdominal segments of male *Mesocapnia frisoni* (Capniidae).

Figure 14.164. Subgenital plate of female *Mesocapnia frisoni* (Capniidae).

Figure 14.165. Subgenital plate of female *Bolshecapnia spenceri* (Capniidae).

58'.	Hind wing with 3 anal veins (Fig. 14.170-14.171)	59
59(58').	R_s and M arising at same point or close together on R in fore wing (Fig. 14.173)	*Perlomyia* Banks
59'.	R_s and M arising separately from R in fore wing (Fig. 14.108)	60
60(59').	Crossvein m-cu joining Cu beyond fork of Cu in hind wing (Fig. 14.170)	61
60'.	Crossvein m-cu joining Cu before fork in hind wing (Fig. 14.171)	62
61(60).	Male tergum 9 entire (Fig. 14.175); female abdomen sclerotized dorsally	*Paraleuctra* Hanson
61'.	Male tergum 9 deeply cleft (Fig. 14.174); female abdomen membranous dorsally	*Zealeuctra* Ricker
62(60').	Cerci of male long; 2 hooklike processes arising from sides of epiproct (Fig. 14.176); female abdomen with sclerotized mid-dorsal stripe	*Moselia* Ricker
62'.	Cerci of male short, no hooklike processes on epiproct (Fig. 14.177); female abdomen membranous dorsally	63
63(62').	Eastern species; sternum 9 of male with vesicle, as in Fig. 14.176; female sternum 8 produced into a bilobed plate reaching over sternum 9 (Fig. 14.178)	*Leuctra* Stephens
63'.	Western species; sternum 9 of male without vesicle (Fig. 14.177); female sternum 8 not produced posteriorly, shallowly excavated (Fig. 14.179)	*Despaxia* Ricker
64(8).	Male tergum 10 usually deeply cleft (Figs. 14.186-14.187,14.189-14.192); when entire, paraprocts extended as long, erect half cylinders (Fig. 14.188); often with gill remnants on submentum, cervix, thorax, or abdomen (Figs. 14.180, 14.185, 14.204); arms of mesoscutal ridge leading to front end (Fig. 14.181) or hind end of furcal pits (Figs. 14.182-14.183), or absent (Fig. 14.184); sometimes with supplementary crossveins in R_1-R_s space (Fig. 14.2); pronotum with narrow, median, light stripe, margins of prontal disk (plate) usually dark (Fig. 14.2)	**Perlodinae** 65
64'.	Male tergum 10 entire, paraprocts unmodified or transformed into recurved hooks or knobs (Fig. 14.200); cleft, if present, shallow or narrow (Fig. 14.199); no gill remnants; arms of mesoscutal ridge leading to hind end of furcal pits (Fig. 14.182); no supplementary veins in R_1-R_s space (Fig. 14.112), though rarely in Rs-M space; pronotal light stripe wider, margins of prontal disk usually light	**Isoperlinae** 101
65(64).	Gill remnants on abdominal segments 1-7 (Fig. 14.77)	*Oroperla* Needham
65'.	No such gill remnants on abdomen	66
66(65').	Gill remnants on the thorax (Figs. 14.76, 14.78, 14-180)	67
66'.	No gill remnants on the thorax	70
67(66).	Gill remnants on all 3 thoracic segments (Figs. 14.76, 14.78, 14.180)	68
67'.	Gill remnants only on meso- and metathorax	*Setvena* Illies
68(67).	Gill remnants on the cervix (Figs. 14.76, 14.180)	*Perlinodes* Needham and Claassen
68'.	No gill remnants on the cervix	69
69(68').	Meso- and metathoracic gills double (forked) (Figs. 14.78, 14.204)	*Salmoperla* Baumann and Lauck
69'.	Meso- and metathoracic gills simple	*Megarcys* Klapálek
70(66').	Arms of mesosternal ridge lead to front end of furcal pits (Figs. 14.88, 14.89, 14.181)	71
70'.	Arms of mesosternal ridge lead to hind end of furcal pits (Figs. 14.182-14.183), or arms absent (Fig. 14.184)	75
71(70).	An irregular network of supplementary crossveins in apical area of wings (Fig. 14.2)	72
71'.	This network absent (Fig. 14.112)	74
72(71).	Transverse ridge present on mesosternum (Figs. 14.182-14.183)	*Frisonia* Ricker
72'.	Transverse ridge absent (Fig. 14.181)	73
73(72').	Male epiproct blunt (Fig. 14.2); lateral stylets present; inhabits temperate zone	*Skwala* Ricker
73'.	Male epiproct a long needlelike structure; inhabits arctic and alpine zones	*Arcynopteryx* Klapálek
74(71').	Wings dark	*Chernokrilus* Ricker (in part)
74'.	Wings pale	*Osobenus* Ricker
75(70').	Median ridge present between arms of mesosternal ridge (Figs. 14.91, 14.183)	*Isogenoides* Klapálek
75'.	Median ridge absent (Figs. 14.182, 14.184)	76
76(75').	Submental gills long, at least twice as long as wide (Fig. 14.185)	77
76'.	Submental gills shorter (Fig. 14.92) or absent	79
77 (76).	Western species	*Susulus* Bottorf and Stewart
77'.	Eastern species	78
78(77').	Male epiproct with lateral stylets (Fig. 14.186); no lobes (vesicles) on sternum 7 and sternum 8; female subgenital plate reaching at most middle of sternum 9	*Hydroperla* Frison
78'.	Male epiproct without lateral stylets (Fig. 14.187); lobes on sternum 7 and sternum 8 marked at least with light pigment; female subgenital plate reaching beyond the middle of sternum 9	*Helopicus* Ricker
79(76').	Lateral arms of mesosternal ridge absent (Fig. 14.184)	*Diploperla* Needham and Claassen
79'.	Lateral arms of mesosternal ridge present (Fig. 14.182)	80
80(79').	Males	81

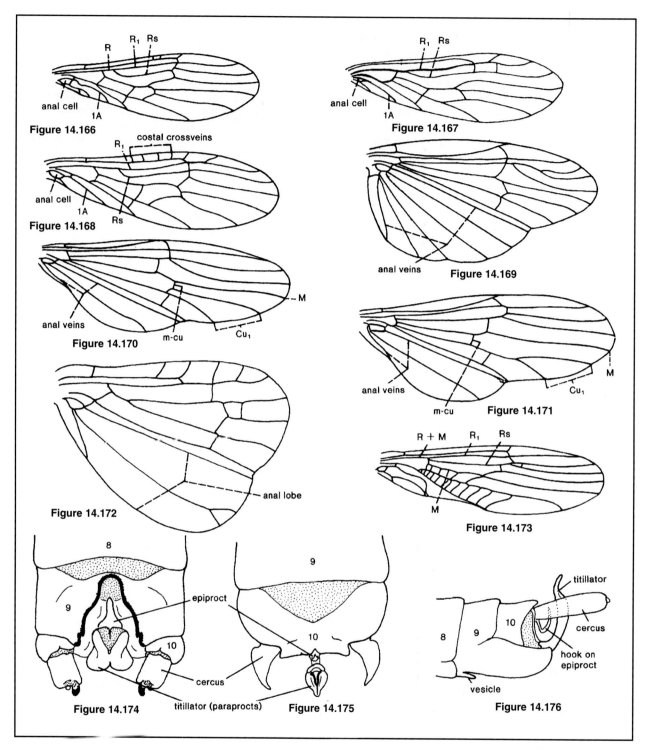

Figure 14.166. Fore wing of *Nemocapnia carolina* (Capniidae).

Figure 14.167. Fore wing of *Paracapnia opis* (Capniidae).

Figure 14.168. Fore wing of *Isocapnia agassizi* (Capniidae).

Figure 14.169. Hind wing of *Megaleuctra* sp. (Leuctridae).

Figure 14.170. Hind wing of *Paraleuctra* sp. (Leuctridae).

Figure 14.171. Hind wing of *Leuctra* sp. (Leuctridae).

Figure 14.172. Hind wing of *Allocapnia granulata* (Capniidae).

Figure 14.173. Fore wing of *Perlomyia collaris* (Leuctridae).

Figure 14.174. Dorsal view of terminal abdominal segments of male *Zealeuctra hitei* (Leuctridae).

Figure 14.175. Dorsal view of terminal abdominal segments of male *Paraleuctra occidentalis* (Leuctridae).

Figure 14.176. Side view of terminal abdominal segments of male *Moselia infuscata* (Leuctridae).

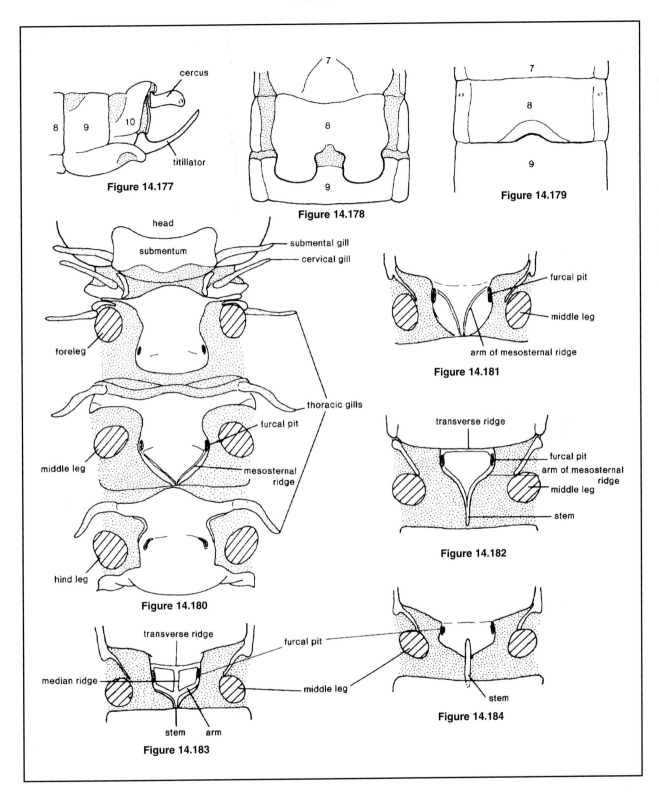

Figure 14.177. Side view of terminal abdominal segments of male *Despaxia augusta* (Leutridae).

Figure 14.178. Subgenital plate of female *Leuctra tenuis* (Leuctridae).

Figure 14.179. Subgenital plate of female *Despaxia augusta* (Leuctridae).

Figure 14.180. Ventral view of neck and thorax of *Perlinodes aurea* (Perlodidae).

Figure 14.181. Mesosternum of *Skwala curvata* (Perlodidae).

Figure 14.182. Mesosternum of *Clioperla clio* (Perlodidae).

Figure 14.183. Mesosternum of *Isogenoides doratus* (Perlodidae).

Figure 14.184. Mesosternum of *Diploperla duplicata* (Perlodidae).

80'.	Females	90
81(80).	Tergum 10 cleft; epiproct prominent (Figs. 14.189-14.192)	82
81'.	Tergum 10 entire; epiproct undeveloped (Fig. 14.188)	*Diura* Billberg
82(81).	Lateral stylets present on epiproct (Figs. 14.190-14.191)	83
82'.	No lateral stylets on epiproct (Figs. 14.189, 14.192)	86
83(82).	Epiproct covered laterally and ventrally with scale-like setae (Fig. 14.205)	*Oconoperla* Stark and Stewart
83'.	Epiproct without such spines	84
84(83).	Lateral stylets of epiproct slender and acute (Fig. 14.190)	*Cultus* Ricker
84'.	Lateral stylets hooked (Fig. 14.191) or with a rounded tip	85
85(84').	Lateral stylets hooked at tip (Fig. 14.191)	*Malirekus* Ricker
85'.	Lateral stylets with rounded tips	*Chernokrilus* Ricker (in part)
86(82').	Tip of epiproct with coiled inner band (Fig. 14.189)	87
86'.	Epiproct without coiled inner band	88
87(86).	Peg-like spinules on abdominal terga 9 and 10 (Fig. 14.189)	*Kogotus* Ricker
87'.	Peg-like spinules on abdominal terga 10 only	*Baumannella* Stark and Stewart
88(86').	No crossvein in costal space beyond cord; epiproct of male ending in a long reversible lash (Fig. 14.192); lobes (vesicles) on sternum 7 and sternum 8 equally well developed	*Remenus* Ricker
88'.	One to 3 crossveins in costal space beyond cord (Fig. 14.194); epiproct without lash; lobe on sternum 8 much smaller than on sternum 7 or absent	89
89(88').	Eastern species (Appalachian)	*Yugus* Ricker
89'.	Western species	*Pictetiella* Illies
90(80').	Eastern species	96
90'.	Western species	91
91(90').	Sternum 8 forming a short subgenital plate (Fig. 14.193)	*Diura* Billberg
91'.	Subgenital plate larger (Figs. 14.195-14.197)	92
92(91').	Subgenital plate large covering most of sternum 9 (Fig. 14.195)	*Pictetiella* Illies
92'.	Subgenital plate smaller (Figs. 14.196-14.197)	93
93(92').	Color pale, yellowish or brownish; no crossveins in costal space beyond cord (Fig. 14.112); wings pale; submental gills absent	94
93'.	Color dark; a few crossveins in costal space beyond cord (Fig. 14.194); wings dark; submental gills present, albeit small	*Chernokrilus* Ricker (in part)
94(93).	Subgenital plate nearly as wide as segment at base (Fig. 14.196)	*Kogotus* Ricker
94'.	Subgenital plate only about 3/4 the width of sternum 8 at base (Fig. 14.197)	95
95(94').	Subgenital plate with a small notch (Fig. 14.206); egg biconcave	*Baumannella* Stark and Stewart
95'.	Subgenital plate entire or notched (Fig. 14.197); egg not biconcave	*Cultus* Ricker
96(90).	Subgenital plate on sternum 8 short and truncate (Fig. 14.193)	97
96'.	Subgenital plate larger and not truncate (Fig. 14.198)	98
97(96).	Boreal or alpine species; margin of subgenital plate straight (Fig. 14.193)	*Diura* Billberg
97'.	Southern Appalachian species; margin of subgenital plate distinctly emarginate (Fig. 14.207)	*Oconoperla* Stark and Stewart
98(96').	Submental gills present, albeit small (Fig. 14.92)	99
98'.	Submental gills absent	100
99(98).	Subgenital plate evenly rounded; small species, less than 15 mm	*Remenus* Ricker
99'.	Subgenital plate broadly excavated (Fig. 14.198); larger species, more than 20 mm	*Malirekus* Ricker
100(98').	No crossvein in costal space beyond cord (Fig. 14.112)	*Cultus* Ricker
100'.	One to 3 crossveins in costal space beyond cord (Fig. 14.194)	*Yugus* Ricker
101(64').	Males	102
101'.	Females	107
102(101).	Tergum 10 partially cleft or slightly excavated with small hooks on each side of the cleft (Fig. 14.199)	103
102'.	Tergum 10 usually entire, no hooks developed from hind margin of tergum (Figs. 14.200, 14.201)	104
103(102).	Ventral lobe present on sternum 8 (Fig. 14.213); epiproct present but largely membranous	*Calliperla* Banks
103'.	No ventral lobe on abdominal sternites; epiproct undeveloped (Fig. 14.199)	*Cascadoperla* Szczytko and Stewart
104(102').	Tergum 10 bearing a large elevated triangular notched process (Fig. 14.208); no ventral lobe on abdominal sterna	*Consumnoperla* Szcytko and Bottorf
104'.	Tergum 10 without a large elevated process; ventral lobe usually present on sternum 7 or 8 or both	105
105(104').	Ventral lobe on sternum 7 only	*Rickera* Jewett
105'.	Ventral lobe present usually only on sternum 8 (Fig. 14.213); in some species, there is no lobe, and in one, there is an extra lobe on sternum 7	106

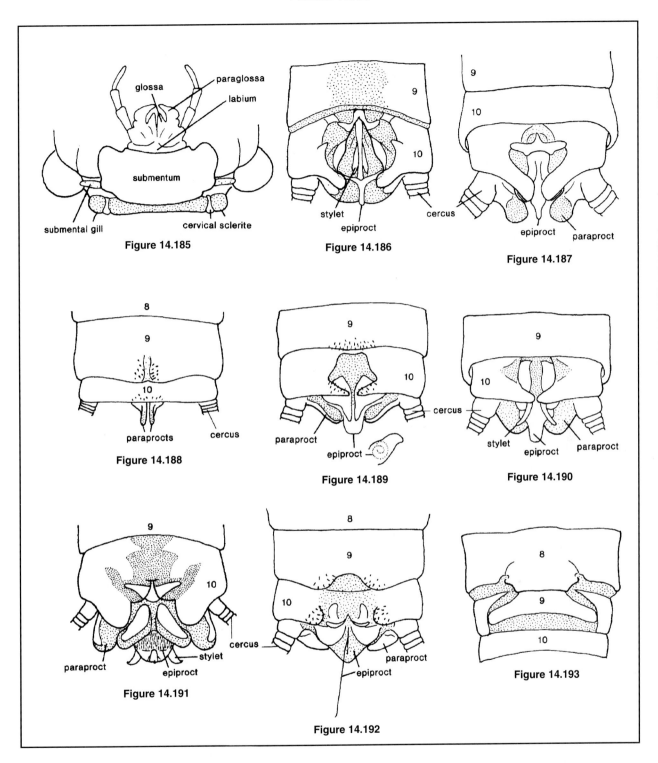

Figure 14.185. Submentum and labium of *Helopicus subvarians* (Perlodidae).

Figure 14.186. Dorsal view of terminal abdominal segments of male *Hydroperla crosbyi* (Perlodidae).

Figure 14.187. Dorsal view of terminal abdominal segments of male *Helopicus subvarians* (Perlodidae).

Figure 14.188. Dorsal view of terminal abdominal segments of male *Diura knowltoni* (Perlodidae).

Figure 14.189. Dorsal view of terminal abdominal segments of male *Kogotus modestus* (Perlodidae).

Figure 14.190. Dorsal view of terminal abdominal segments of male *Cultus decisus* (Perlodidae).

Figure 14.191. Dorsal view of terminal abdominal segments of male *Malirekus iroquois* (Perlodidae).

Figure 14.192. Dorsal view of terminal abdominal segments of male *Remenus bilobatus* (Perlodidae).

Figure 14.193. Subgenital plate of female *Diura knowltoni* (Perlodidae).

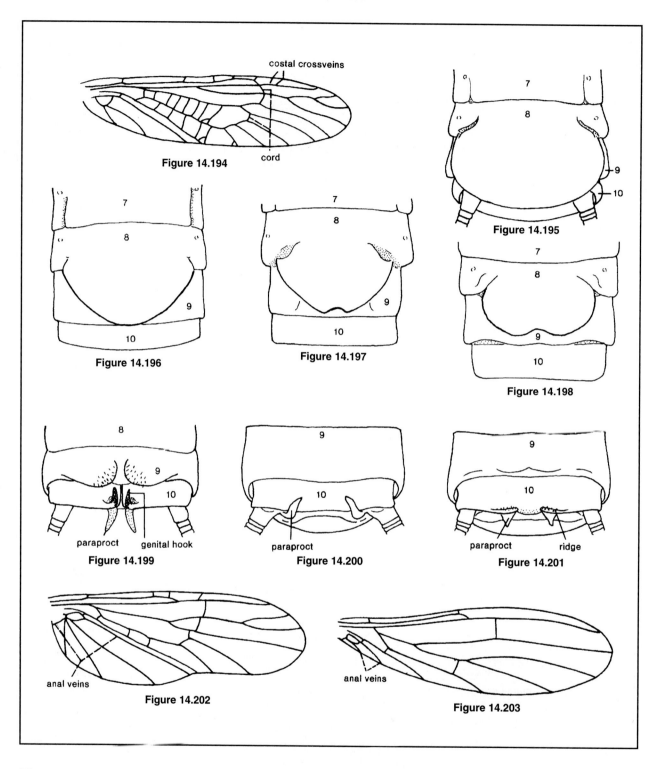

Figure 14.194. Fore wing of *Helopicus subvarians* (Perlodidae).

Figure 14.195. Subgenital plate of female *Pictetiella expansa* (Perlodidae).

Figure 14.196. Subgenital plate of female *Kogotus modestus* (Perlodidae).

Figure 14.197. Subgenital plate of female *Cultus decisus* (Perlodidae).

Figure 14.198. Subgenital plate of female *Malirekus hastatus* (Perlodidae).

Figure 14.199. Dorsal view of terminal abdominal segments of male *Cascadoperla trictura* (Perlodidae).

Figure 14.200. Dorsal view of terminal abdominal segments of male *Isoperla* sp. (Perlodidae).

Figure 14.201. Dorsal view of terminal abdominal segments of male *Clioperla clio* (Perlodidae).

Figure 14.202. Hind wing of *Utaperla sopladora* (Chloroperlidae).

Figure 14.203. Hind wing of *Haploperla brevis* (Chloroperlidae).

106(105').	Tergum 10 with elevated ridges on its hind margin (Figs. 14.201)	***Clioperla*** Needham and Claassen
106'.	Tergum 10 without elevated ridges (Fig. 14.200)	***Isoperla*** Banks
107(101').	Subgenital plate produced only as a small nipple on hind margin of sternum 8	***Cascadoperla*** Szczytko and Stewart
107'.	Subgenital plate larger	108
108(107').	Subgenital plate large covering most of sternum 9	***Rickera*** Jewett
108'.	Subgenital plate otherwise, smaller	***Isoperla*** Banks, ***Calliperla*** Banks, ***Clioperla*** Needham and Claassen, ***Cosumnoperla*** Szczytko and Bottorf
109(8').	Eyes set far forward on the sides of the head; hind corners of head rather angulate (Figs. 14.65, 14.214)	110
109'.	Eyes set in normal position, hind corners of head rounded (Fig. 14.2)	111
110(109).	Head longer than wide (Fig. 14.214)	***Kathroperla*** Banks
110'.	Head as long as wide	***Paraperla*** Banks
111(109').	Hind wing with 5 or more anal veins (Fig. 14.202)	***Utaperla*** Ricker
111'.	Hind wing with 4 or fewer anal veins (Figs. 14.113, 14.203)	112
112(111').	Folded anal area of hind wing absent (Fig. 14.203)	***Haploperla*** Navas
112'.	Folded anal area of hind wing present (Fig. 14.113)	113
113(112').	Basal segment of cercus more than 3 times as long as wide, and curved (in male, it bears knobs and a basal spine) (Fig. 14.215)	***Neaviperla*** Ricker
113'.	Basal segment of cercus shorter, straight	114
114(113').	Body concolorous (uniform in colour) without dark markings, generally green in life (yellow in alcohol)	***Alloperla*** Banks
114'.	Body with dark markings on head, thoracic nota, or abdomen	115
115(114').	2A in fore wing simple, unforked (Fig. 14.210)	***Alaskaperla*** Stewart and DeWalt
115'.	2A in fore wing forked	116
116(115').	Anal lobe of hind wing small (1/2 as large as in Fig. 14.113)	***Rasvena*** Ricker
116'.	Anal lobe of hind wing larger (as in Fig. 14.113)	117
117(116').	Males	118
117'.	Females (following key tentative)	122
118(117).	Fingerlike process at base of cercus (Fig. 14.216)	***Suwallia*** Ricker
118'.	No process at base of cercus	119
119(118').	Epiproct large, its tip hinged and elaborate; epiproct set in a deep groove in tergum 10 (Fig. 14.217); brushes of close-set setae on lateral margins of terminal abdominal segments (Fig. 14.217)	120
119'.	Epiproct small, about as wide as long, tergum 10 entire or just slightly depressed; no lateral brushes of setae on terminal abdominal segments (Fig. 14.212)	121
120(119).	Small species (4-6 mm in length); epiproct flattened, curled, its anchor double (Fig. 14.209)	***Bisancora*** Surdick
120'.	Larger species (8-18 mm in length); epiproct more globular or if flattened, not curled, its anchor simple (Fig. 14.217)	***Sweltsa*** Ricker
121(119').	Color dark, heavily patterned; epiproct rounded in side view; small hammer present on margin of sternum 7	***Triznaka*** Ricker
121'.	Color pale, unpatterned; epiproct pointed in side view (Fig. 14.212); no hammer on sternum 7; 2 plume-like appendages at tip of aedeagus	***Plumiperla*** Surdick
122(117').	Meso- and metanota marked as in Figure 14.218	123
122'.	Nota marked differently, median bar absent or faint (Fig. 14.221)	124
123(122')	Color dark, heavily patterned	***Triznaka*** Ricker
123'.	Color pale, unpatterned	***Plumiperla*** Surdick
124(122').	Sternum 8 forming a large subgenital plate covering most of sternum 9; if plate is shorter, pronotum pale with only narrow dark lateral margins; no brushes of setae on lateral abdominal margins	***Suwallia*** Ricker
124'.	Subgenital plate variable, shorter; disk of pronotum with a dark median line, wide dark borders or with a dark reticulate pattern; brushes of close-set setae on lateral margins of terminal abdominal segments (Fig. 14.217)	125
125(124').	Size small (4-6 mm); mesosternal Y-ridge with median branch extending cephalad (Fig. 14.211); subgenital plate scalloped with long hairs restricted to scalloped margins	***Bisancora*** Surdick
125'.	Size larger (8-18 mm); mesosternal Y-ridge unmodified; subgenital plate variable, if scalloped, hairs arranged otherwise	***Sweltsa*** Ricker
126(7).	Two ocelli	127
126'.	Three ocelli	129
127(126).	Male tergum 10 cleft and forming genital hooks (Fig. 14.219); subgenital plate scarcely developed on female sternum 8 (Fig. 14.220)	***Neoperla*** Needham

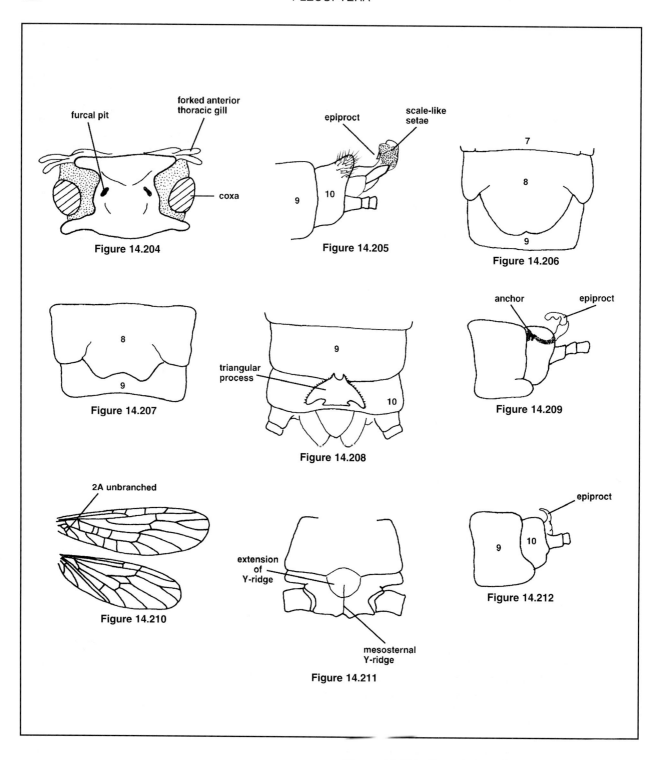

Figure 14.204. Ventral view of metasternum of *Salmoperla sylvanica* (Perlodidae).

Figure 14.205. Side view of terminal abdominal segments of male *Oconoperla weaveri* (Perlodidae).

Figure 14.206. Subgenital plate of *Baumannella alameda* (Perlodidae).

Figure 14.207. Subgenital plate of *Oconoperla weaveri* (Perlodidae).

Figure 14.208. Dorsal view of terminal abdominal segments of male of *Consumnoperla hypocrena* (Perlodidae).

Figure 14.209. Side view of terminal abdominal segments of male *Bisancora* sp. (Chloroperlidae).

Figure 14.210. Wings of *Alaskaperla ovibovis* (Chloroperlidae).

Figure 14.211. Ventral view of mesosternum of *Bisancora* sp. (Chloroperlidae).

Figure 14.212. Side view of terminal abdominal segments of *Plumiperla* sp. (Chloroperlidae).

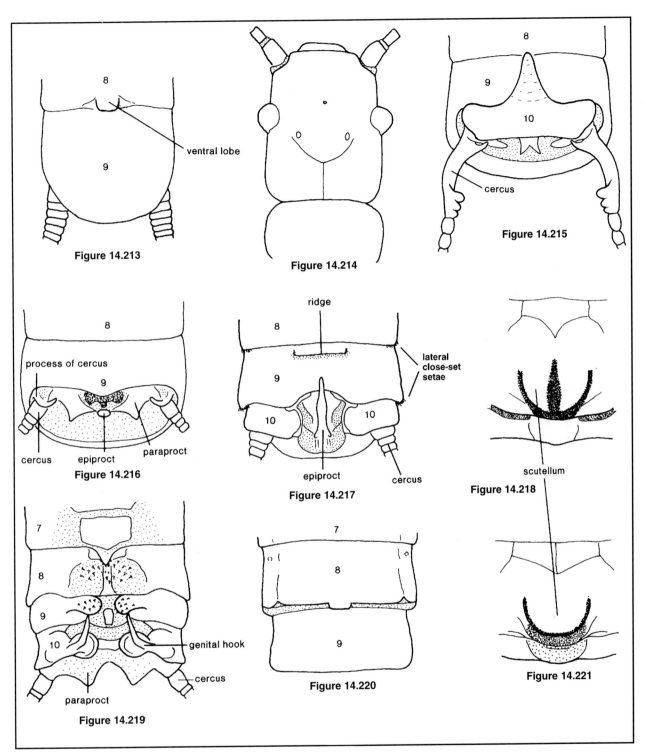

Figure 14.213. Ventral view of terminal abdominal segments of male *Isoperla* sp. (Perlodidae).

Figure 14.214. Outline of head and pronotum of *Kathroperla perdita* (Chloroperlidae).

Figure 14.215. Dorsal view of terminal abdominal segments of male *Neaviperla forcipata* (Chloroperlidae).

Figure 14.216. Dorsal view of terminal abdominal segments of male *Suwallia marginata* (Chloroperlidae).

Figure 14.217. Dorsal view of terminal abdominal segments of male *Sweltsa* sp. (Chloroperlidae).

Figure 14.218. Meso- and metanotum of *Triznaka* sp. (Chloroperlidae).

Figure 14.219. Dorsal view of terminal abdominal segments of male *Neoperla* sp. (Perlidae).

Figure 14.220. Subgenital plate of female *Neoperla* sp. (Perlidae).

Figure 14.221. Mesonotum of *Suwallia* sp. (Chloroperlidae).

127'. Male tergum 10 entire, genital hooks developed from paraprocts (Fig. 14.227); subgenital plate of female either very large or notched (Figs. 14.222-14.223) 128
128(127'). Male sternum 9 with small hammer; female sternum 8 forming a large subgenital plate; sternum 9 with spicule patch (Fig. 14.222) ....*Anacroneuria* Klapálek
128'. Male sternum 9 with a large hammer (Fig. 14.224); female sternum 8 forming a short subgenital plate bearing a large quadrate notch (Fig. 14.223) ....*Perlinella* Banks (in part)
129(126'). Males..................130
129'. Females..................142
130(129). Genital hooks arising from tergum 10 (Figs. 14.225-14.226)..................131
130'. Genital hooks arising from paraprocts (Figs. 14.227-14.228)..................133
131(130). Hammer (vesicle) present on sternum 9..................*Claassenia* Wu
131'. No hammer on sternum 9..................132
132(131'). Genital hooks long, reaching forward to midline of tergum 8 (Fig. 14.225)..................*Agnetina* Klapálek
132'. Genital hooks shorter, not reaching beyond the midline of tergum 9 (Fig. 14.226)..................*Paragnetina* Klapálek
133(130'). Mesal tergite differentiated on tergum 10 (Fig. 14.228)..................134
133'. Tergum 10 without differentiated mesal tergite (Fig. 14.227)..................135
134(133). Hammer on sternum 9 twice as wide as long (Fig. 14.229)..................*Hesperoperla* Banks
134'. Hammer on sternum 9 twice as long as wide (Fig. 14.230)..................*Calineuria* Ricker
135(133'). No hammer on sternum 9..................*Perlesta* Banks
135'. Hammer present on sternum 9..................136
136(135'). Hammer on sternum 9 a small raised knob..................*Eccoptura* Klapálek
136'. Hammer not raised in side view..................137
137 (136'). Hammer oval (Fig. 14.224)..................138
137'. Hammer rectangular or triangular (Figs. 14.229, 14.231)..................140
138(137). Small species (fore wing less than 13 mm); aedeagus with large lateral sclerites..................*Perlinella* Banks (in part)
138'. Larger species (fore wing over 15 mm); aedeagus without such sclerites..................139
139(138'). Spinule patches on both tergum 9 and tergum 10 (Fig. 14.227)..................*Acroneuria* Pictet
139'. Spinule patches absent, a few spinules on tergum 10..................*Attaneuria* Ricker
140(137'). Hammer rectangular (Fig. 14.229)..................*Doroneuria* Needham and Claassen
140'. Hammer triangular (Fig. 14.231)..................141
141(140'). Crossveins present between 1A and 2A, and 2A and 3A (as in Fig. 14.237, but with additional crossveins between 2A and 3A)..................*Hansonoperla* Nelson
141'. No such crossveins..................*Beloneuria* Needham and Claassen
142(129'). Subgenital plate of sternum 8 not produced over sternum 9 (Fig. 14.232)..................143
142'. Subgenital plate extending over sternum 9 (Figs. 14.236, 14.238-14.241)..................147
143(142). Subgenital plate notched (Fig. 14.232)..................144
143'. Subgenital plate entire..................145
144(143). Notch of subgenital plate bordered by a membranous area (Fig. 14.232)..................*Calineuria* Ricker
144'. Notch bordered by a dark, sclerotized area..................*Hansonoperla* Nelson
145(143'). Vagina with accessory glands; spermatheca with few or no glands..................146
145'. Vagina without accessory glands; spermatheca with numerous glands (Fig. 14.233)..................*Acroneuria* Pictet (in part)
146(145). Vagina covered internally with golden brown spinules..................*Doroneuria* Needham and Claassen
146'. Vagina without such spinules..................*Claassenia* Wu
147(142'). Egg collar stalked (Fig. 14.234)..................148
147'. Egg collar not stalked (Fig. 14.235) or absent..................151
148(147). A squarish dark area on margin of subgenital plate; apex of egg flanged around collar (Fig. 14.234)..................*Hesperoperla* Banks
148'. No such dark area; egg without apical flange..................149
149(148'). Fore wing 15 mm or more in length..................*Paragnetina* Klapálek
149'. Fore wing 14 mm or less in length..................150
150(149'). Crossveins present in fore wing between 1A and 2A (Fig. 14.237)..................*Perlinella* Banks (in part)
150'. Crossveins absent between 1A and 2A..................*Perlesta* Banks (in part)
151(147'). Egg collar developed as a small button (Fig. 14.235)..................152
151'. Egg collar absent..................155
152(151). Vagina with sclerites (Fig. 14.236)..................*Agnetina* Klapálek
152'. Vagina without sclerites..................153
153(152'). Vagina covered internally with spinules..................*Acroneuria* Pictet (in part)
153'. Vagina membranous, without spinules..................154
154(153'). Subgenital plate large, reaching middle of sternum 9 (Fig. 14.238)..................*Eccoptura* Klapálek
154'. Subgenital plate shorter, reaching about one-third of length of sternum 9, notched (Fig. 14.239)..................*Perlesta* Banks (in part)

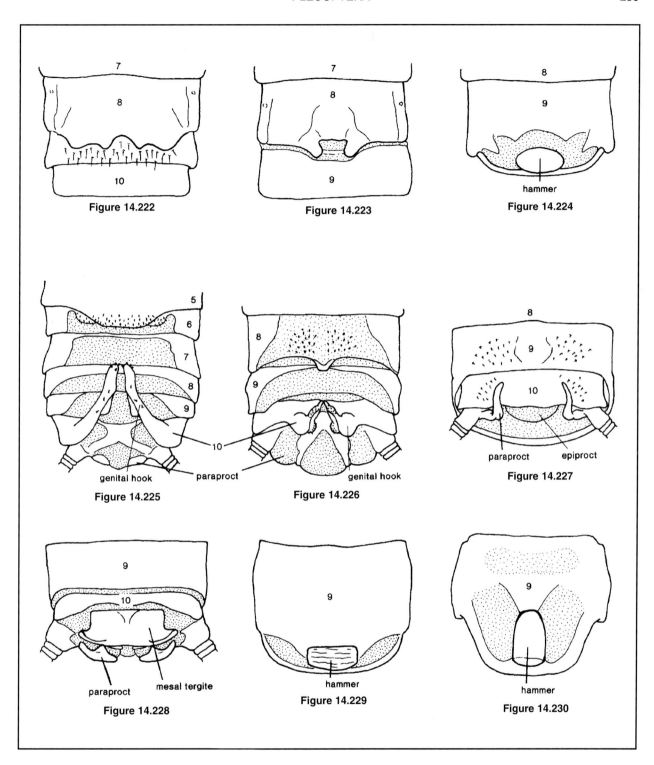

Figure 14.222. Subgenital plate of female *Anacroneuria* sp. (Perlidae).

Figure 14.223. Subgenital plate of female *Perlinella* sp. (Perlidae).

Figure 14.224. Sternum 9 of male *Perlinella* sp. (Perlidae).

Figure 14.225. Dorsal view of terminal abdominal segments of male *Agnetina capitata* (Perlidae).

Figure 14.226. Dorsal view of terminal abdominal segments of male *Paragnetina immarginata* (Perlidae).

Figure 14.227. Dorsal view of terminal abdominal segments of male *Hesperoperla pacifica* (Perlidae).

Figure 14.228. Dorsal view of terminal abdominal segments of male *Acroneuria* sp. (Perlidae).

Figure 14.229. Sternum 9 of male *Hesperoperla pacifica* (Perlidae).

Figure 14.230. Sternum 9 of male *Calineuria californica* (Perlidae).

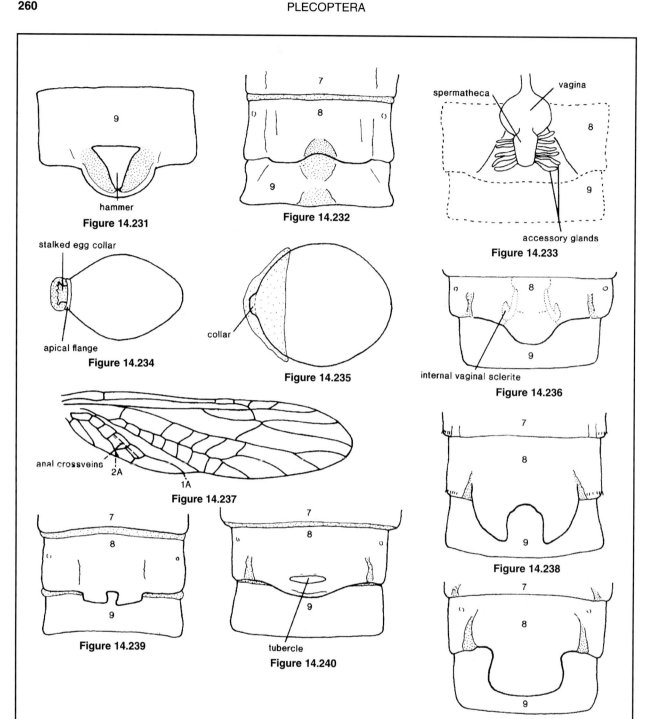

Figure 14.231. Sternum 9 of male *Beloneuria georgiana* (Perlidae).

Figure 14.232. Subgenital plate of female *Calineuria californica* (Perlidae).

Figure 14.233. Vagina and spermatheca of female *Acroneuria abnormis* (Perlidae).

Figure 14.234. Egg of *Hesperoperla pacifica* (Perlidae).

Figure 14.235. Egg of *Agnetina capitata* (Perlidae).

Figure 14.236. Subgenital plate of female *Phasganophora capitata* (Perlidae).

Figure 14.237. Fore wing of *Perlinella* sp. (Perlidae).

Figure 14.238. Subgenital plate of female *Eccoptura xanthenes* (Perlidae).

Figure 14.239. Subgenital plate of female *Perlesta placida* (Perlidae).

Figure 14.240. Subgenital plate of female *Attaneuria ruralis* (Perlidae).

Figure 14.241. Subgenital plate of female *Acroneuria evoluta* (Perlidae).

155(151'). Subgenital plate with submarginal tubercle (Fig. 14.240) .... ***Attaneuria*** Ricker
155'. Subgenital plate without such a tubercle 156
156(155'). Subgenital plate triangular and notched .... ***Beloneuria*** Needham and Claassen
156'. Subgenital plate large, expanding distally, margin entire or shallowly notched
 (Fig. 14.241) .... ***Acroneuria*** Pictet (in part)

ADDITIONAL TAXONOMIC REFERENCES

General
Needham and Claassen (1925); Claassen (1931); Ricker (1952); Stewart and Stark (1988).

Regional faunas
British Columbia: Ricker (1943); Ricker and Scudder (1975).
California: Jewett (1956, 1960).
Connecticut: Hitchcock (1974).
Florida: Stark and Gaufin (1979).
Illinois: Frison (1935).
Louisiana: Stewart et al. (1976).
Minnesota: Harden and Mickel (1952).
Montana: Gaufin et al. (1972); Gaufin and Ricker (1974).
Ozark and Ouachita Mountains: Poulton and Stewart (1991).
Pacific Northwest: Jewett (1959).
Pennsylvania: Surdick and Kim (1976).
Rocky Mountains: Baumann et al. (1977).
Saskatchewan: Dosdall and Lehmkuhl (1979).
Texas: Szczytko and Stewart (1977).
Utah: Gaufin et al. (1966); Baumann et al. (1977); Baumann and Unzicker (1981).
Wisconsin: Hilsenhoff (1981).

Regional species lists
Alabama: Stark and Harris (1986).
Alaska: Jewett (1971): Ellis (1975).
California (*Capniidae*): Nelson and Baumann (1987a).
Canadian Far North: Ricker (1944).
Canadian Maritime Provinces: Ricker (1947); Brinck (1958).
Canadian Prairie Provinces: Ricker (1946).
Colorado: Stark et al. (1973).
Delaware: Lake (1980).
Florida: Stark and Gaufin (1979)
Idaho: Newell and Minshall (1976).
Illinois: Harris and Webb (1994).*
Indiana: Ricker (1945); Bednarik and McCafferty (1977).
Kansas: Stewart and Huggins (1977).
Kentucky: Tarter et al. (1984).
Minnesota: Lager et al. (1979).
Mississippi: Stark (1979).
Nevada: Cather et al. (1975).
New Mexico: Stark et al. (1975).
Ohio: Gaufin (1956); Tkac and Foote (1978).
Oklahoma: Stark and Stewart (1973).
Ozark and Ouachita Mountains (*Neoperla*): Ernst et al. (1986).
Quebec: Ricker et al. (1968).
South Carolina: McCaskill and Prins (1968).
Southwestern United States: Stewart et al. (1974).
Virginia: Kondratieff and Voshell (1979); Kondratieff and Kirchner (1987).
Western Intermountain Area: Gaufin (1964); Logan and Smith (1966).

Taxonomic treatments at the family and generic levels
(L = larvae; A = adults)
Capniidae: Ricker (1959a)-A; Nebeker and Gaufin (1965)-A; Baumann and Gaufin (1970)-A; Harper and Hynes (1971b)-L; Ross and Ricker (1971)-A; Nelson and Baumann (1987b) (*Capnura*)-A; Nelson and Baumann (1989) (*Capnia*)-A; Zenger and Baumann (1995) (*Isocapnia*).
Chloroperlidae: Gaufin (1964)-A: Hitchcock (1968)-A; Fiance (1977)-L; Surdick (1985)-A, L.
Leuctridae: Ricker (1943)-A; Ricker and Ross (1969)-A; Harper and Hynes (1971a)-L; Nelson and Hanson (1973)-A.
Nemouridae: Ricker (1952)-A; Harper and Hynes (1971d)-L; Baumann (1975)-L, A.
Peltoperlidae: Stark and Stewart (1981)-L, A.; Stark and Nelson (1994) (*Yoroperla*)-A.
Perlidae: Stark and Gaufin (1974a, 1976a,b)-L, A; Stark and Szczytko (1976)-A; Stark and Baumann (1978)-A; Ernst et al. (1986)(*Neoperla*)-A; Kondratieff et al. (1988) (*Perlinella*)-A, L; Stark (1986) (*Agnetina*)-A, L; Stark (1989) (*Perlesta*)-A.
Perlodidae: Ricker (1952)-L, A; Hilsenhoff and Billmyer (1973)-L; Stark and Gaufin (1974b)-A; Szczytko and Stewart (1979)-L, A.
Pteronarcyidae: Nelson and Hanson (1971)-A; Nelson et al. (1977)-L, A.
Taeniopterygidae: Ricker and Ross (1968, 1975)-A; Harper and Hynes (1971c)-L; Fullington and Stewart (1980)-L; Stanger and Baumann (1993) (*Taenionema*)-A.

---

*Harris, M. A. and D. W. Webb. 1994. The stoneflies (Plecoptera) of Illinois revisited. J. Kans. Ent. Soc. 67:340-346.

Table 14B. Summary of ecological and distributional data for *Plecoptera (stoneflies)*. (For definition of terms see Tables 6A-6C; table prepared by K. W. Cummins, R. W. Merritt, K. W. Stewart, and P. P. Harper.)

Taxa (number of species in parentheses)	Habitat	Habit	Trophic Relationships	North American Distribution	Ecological References*
Pteronarcyidae (10)	Generally lotic—erosional and depositional (debris jams, leaf packs)	Generally clingers—sprawlers	Generally shredders—detritivores, some facultative scrapers		790, 803, 1308, 1462, 1563, 1783, 1891, 1899, 1908, 1976, 2830, 3076, 3340, 3352, 3629, 3841
Pteronarcella (2)	Lotic—erosional and depositional (logs, leaf litter)	Clingers—sprawlers	Shredders—detritivores and herbivores (macroalgae), facultative predators (engulfers)	West, Southwest	41, 397, 1331, 1333, 3063, 3089, 3841, 3843, 2077, 4234
Pteronarcys (8)	Lotic—erosional and depositional (logs, leaf litter)	Clingers—sprawlers	Shredders—detritivores and herbivores (macroalgae), facultative predators (engulfers), facultative scrapers	Widespread	193, 397, 803, 1331, 1333, 1563, 1807, 1808, 2486, 2611, 2698, 2844, 2859, 2316, 3340, 3629, 3841, 852, 1235, 3675, 478, 1301, 2749, 1559, 2774, 3091, 3740
Peltoperlidae (18)	Generally lotic—erosional and depositional (leaf litter)	Generally clingers—sprawlers	Generally shredders—detritivores		1563, 1783, 1899, 1976, 1877, 2112, 3076, 3352, 1310
Peltoperla (2)	Lotic—erosional and depositional (leaf litter)	Clingers—sprawlers	Shredders—detritivores (leaf litter)	East	1783, 2701, 3484, 3841, 1235, 1482, 2949
Sierraperla (1)				West (Nevada, California)	3841
Soliperla (6)				Pacific Northwest	3841
Tallaperla (6)	Lotic—erosional and depositional (leaf litter)	Clingers—sprawlers	Shredders—detritivores (leaf litter)	East	1080, 1783, 4198, 4537, 1586, 1587, 2949, 3841, 3870
Viehoperla (1)				Appalachians	3841
Yoroperla (2)			Shredders—detritivores, facultative—scrapers	West (mountain, intermountain)	603, 3809, 3841
Taeniopterygidae (35)	Generally lotic—erosional (coarse sediments, debris jams, leaf packs) and depositional at margins	Generally sprawlers, clingers	Generally shredders—detritivores, facultative collectors—gatherers and scrapers		1307, 1563, 1783, 1899, 1976, 2830, 3076, 3352, 111, 112, 1310
Taeniopteryginae (11)					
Taeniopteryx (11)	Lotic—erosional (coarse sediments, wood, leaf packs) and depositional at margins	Sprawlers—clingers	Shredders—detritivores, facultative collectors—gatherers (scrapers)	East (1 in Northwest; Texas, Colorado, New Mexico)	193, 246, 671, 679, 1185, 1307, 1334, 1583, 2122, 1586, 3340, 4212, 4341, 3626, 3841, 4431, 2343
Brachypterinae (24)			Generally scrapers (some shredders)		1891, 2435, 3340, 3841
Bolotoperla (1)				East	3841
Doddsia (1)				West	3841
Oemopteryx (4)				East, Central, California	1586, 1587, 2875, 3626, 3841
Strophopteryx (6)				East	1583, 1587, 1877, 2875, 3841
Taenionema (12)				West, Northwest	1587, 3089, 3841, 3836, 4234
Nemouridae (65)	Generally lotic—erosional (coarse sediments, wood, leaf packs) and depositional (leaf litter), lentic—erosional	Generally sprawlers, clingers	Generally shredders—facultative detritivores, collectors—gatherers	Widespread	420, 603, 1574, 2436, 1060, 2718, 2859, 4578, 2477, 3841, 4650, 96, 1381, 1482, 1310, 1665

* Emphasis on trophic relationships.

Table 14B. Continued

Taxa (number of species in parentheses)	Habitat	Habit	Trophic Relationships	North American Distribution	Ecological References*
Amphinemurinae (24)					
Amphinemura (13)	Lotic—erosional and depositional (in detritus)	Sprawlers—clingers	Shredders-detritivores, facultative collectors—gatherers	Widespread	679, 1574, 2414, 2435, 890, 2436, 3629, 4578, 1104, 1587, 4242, 2025, 2340, 3841
Malenka (11)				West	935, 2316, 3334, 4234
Nemourinae (41)					
Lednia (1)				Montana	205, 3841
Nemoura (5)	Lotic—erosional detritus)	Sprawlers—clingers	Shredders-detritivores (herbivores [macroalgae])	West, North, Northeast	243, 247, 1574, 3629, 2341, 4341, 3836, 3841, 4435, 1103, 2343
Ostrocerca (6)				East, Northwest	1576, 1587, 2414, 3841
Paranemoura (1)				Northeast	3841
Podmosta (5)				West, Northeast	1588, 2305, 3841
Prostoia (4)	Lotic—erosional and depositional	Shredders-detritivores (facultative scrapers)		Northwest, Northeast	679, 1574, 1905, 3629, 1104, 1587, 1622, 2025, 3841, 4234
Shipsa (1)	Lotic—erosional and depositional	Shredders-detritivores, (scrapers)		North, Northeast	193, 1574, 3629, 3841
Soyedina (7)	Spring outflows			Widespread	1587, 2079, 2414, 4578, 3841, 3924, 3926, 3919, 3841
Visoka (1)				West	
Zapada (10)	Lotic—erosional (detritus)	Sprawlers—clingers	Shredders-detritivores (leaf litter)-herbivores (moss)	West, East	581, 1613, 2079, 3241, 3647, 2825, 3334, 3650, 3841, 3335, 3836, 4234
Leuctridae (53)	Generally lotic erosional (coarse sediments, debris jams, leaf packs) and depositional	Generally sprawlers—clingers	Generally shredders—detritivores		1482, 1574, 2021, 3493, 1310
Megaleuctrinae (6)					
Megaleuctra (6)	Springs, seeps			Northwest, Appalachians	3841
Leuctrinae (47)					
Despaxia (1)	Spring outflows			West	3335, 3841
Leuctra (26)	Lotic—erosional and depositional		Shredders—detritivores	East, North Central	243, 1574, 3629, 4343, 2343, 3795, 1104, 1576, 2025, 4630, 2339, 2342, 3841, 4488, 4651, 4652, 1665, 3841, 1057, 1877
Moselia (1)				Pacific Northwest	3841
Paraleuctra (9)				Widespread	3841, 4234
Perlomyia (2)				West	3841
Zealeuctra (8)	Small streams (some intermittent)			East, Central, Texas	3731, 3841
Capniidae (151)	Generally lotic—erosional (coarse sediments, wood, leaf packs) and depositional	Generally sprawlers—clingers	Generally shredders—detritivores		1307, 1310, 1583
Allocapnia (41)		Clingers	Shredders-detritivores	East	246, 679, 1185, 1186, 1188, 1189, 1583, 2414, 3222
Bolshecapnia (6)	Small streams, alpine lakes			West	

* Emphasis on trophic relationships.

Table 14B. Continued

Taxa (number of species in parentheses)	Habitat	Habit	Trophic Relationships	North American Distribution	Ecological References*
Capnia (53)			Shredders-detritivores	West, North, Northeast	147, 935, 2079, 2088, 2305, 3647, 3650, 3986, 4234, 427, 2343, 3078, 1586, 2340, 3841
Capnura (7)			Shredders-detritivores	West, East (1)	1583
Eucapnopsis (1)				West	935, 2079, 3841, 4234
Isocapnia (11)	Hyporheal			West	3793, 3794, 3841
Mesocapnia (15)				West	2880, 3841
Nemocapnia (1)				East	1469, 2880, 3841
Paracapnia (5)				West, East	1483, 1583, 1586, 1589, 890, 3841
Utacapnia (11)	Lotic (1 pelagic sp. in Lake Tahoe)			West, East	3841, 4234
Perlidae (65)	Generally lotic and lentic—erosional	Clingers	Predators (engulfers)		411, 790, 1308, 1462, 1563, 1783, 1891, 1899, 1908, 1976, 2112, 2830, 3076, 533, 3352, 4433, 1310, 3841, 4268
Perlinae (22)					
Agnetina (=*Phasganophora*) (3)	Lotic—erosional	Clingers	Predators (engulfers; Chironomidae, Ephemeroptera, Trichoptera)	East	671, 1573, 2007, 2174 3629, 1104, 2025, 3841, 1325, 4431, 1322, 3069, 1332,†
Claassenia (1)		Clingers	Predators (engulfers; Trichoptera, Ephemeroptera, Chironomidae, Simuliidae)	West, North	38, 193, 1331, 1333, 3841, 640, 1622, 3607, 4234
Neoperla (13)	Lotic—erosional	Clingers	Predators (engulfers)	East, Southwest	1102, 1104, 3833, 4119, 2025, 3808, 3841
Paragnetina (5)	Lotic—erosional	Clingers	Predators (engulfers; Diptera, Ephemeroptera, Hydropsychidae)	East	671, 1573, 1649, 2007, 3629, 1325, 3841, 3946, 852, 1322, 4565, 1153, 3176, 4490, 4491, †
Acroneuriinae (43)					
Acroneuria (14)	Lotic and lentic—erosional	Clingers	Predators (engulfers; Chironomidae, Trichoptera, Ephemeroptera, Plecoptera)	Widespread	671, 1650, 3049, 3063, 1698, 2007, 4212, 4213, 2956, 4490, 4491, 4565, 1042, 1469, 3069, 1104, 2025, 3841
Anacroneuria (2)				Texas, Arizona	3833, 3841
Attaneuria (1)				East, Central	3049, 3841
Beloneuria (3)				Southern Appalachian	1877, 3841
Calineuria (1)	Lotic—erosional	Clingers	Predators (engulfers; Chironomidae, Trichoptera, Ephemeroptera)	West	603, 2079, 2830, 3630, 3632, 2316, 3651, 3841
Doroneuria (2)				West	2316, 3841
Eccoptura (1)				East, Central	46, 3049, 3841
Hansonoperla (1)				Appalachian	3841
Hesperoperla (2)			Predators (engulfers; Chironomidae, Trichoptera, Ephemeroptera)	West	1061, 1333, 1334, 3841 42, 43, 2728, 2008, 2727, 3401

*Emphasis on trophic relationships.
†Unpublished data, K. W. Cummins, Oregon State University.

Table 14B. Continued

Taxa (number of species in parentheses)	Habitat	Habit	Trophic Relationships	North American Distribution	Ecological References*
Perlesta (13)	Lotic—erosional and depositional (in detritus)	Clingers	Predators (engulfers; Chironomidae, Simuliidae, Ephemeroptera, Trichoptera), facultative collectors—gatherers (especially early instars)	East, Central	3629, 3732, 3841, †, 1158
Perlinella (3)				East	2153, 3841
Perlodidae (103+)	Generally lotic and lentic—erosional	Generally clingers	Generally predators (engulfers) (some scrapers, collectors—gatherers)		1976, 3076, 3352, 3598, 1310, 3841, 4268, 4433, 4159
Perlodinae					
Arcynopteryx (1)			Predators (engulfers) (collectors—gatherers?)	Arctic, alpine	2278, 2746, 3631, 3841, 2340, 3836
Baumannella (1)				West	3841
Chernokrilus (3)				West	3841
Cultus (6)	Lotic—erosional	Clingers	Predators (engulfers; Chironomidae, Simuliidae)	West, East	1331, 1587, 2718, 2719, 852, 3817, 3841, 4234
Diploperla (4)			Predators (engulfers)	East	130, 3841
Diura (3)			Scrapers, predators (engulfers)	West, North, East	147, 3492, 3598, 3631, 4070, 1158, 2460, 2806, 3841
Frisonla (1)				West	1491, 3841
Helopicus (4)			Predators (engulfers)	East	2712, 3841
Hydroperla (3)	Lotic—depositional and erosional	Clingers	Predators (engulfers; Simuliidae, Chironomidae, Emphemeroptera)	East	2761, 2953, 3841
Isogenoides (9)	Lotic erosional and depositional		Predators, (engulfers; Diptera, especially Chironomidae)	East, North, West	193, 3629, 3841
Kogotus (2)	Lotic—erosional	Clingers	Predators (engulfers; Ephemeroptera, Trichoptera, Diptera—Simuliidae, Chironomidae) (some scrapers)	West	38, 3634, 3841, 4007, 4156, 3050, 3053, 3054, 3056, 1104, 3058, 3059, 3060, 3061, 3067, 3068, 3070, 4158, 3069, 4157, 4159
Malirekus (2)				East	3841
Megarcys (5)	Lotic—erosional	Clingers	Predators (engulfers; Ephemeroptera, Trichoptera, Diptera—Simuliidae, Chironomidae)	West	38, 580, 3063, 3841, 4156, 696, 2728, 3050, 3053, 3054, 3055, 3056, 3057, 1104, 3058, 3059, 3060, 2952, 3061, 3066, 3068, 2727, 3062, 3069, 4234
Oconoperla (1)				East	3814, 3841
Oroperla (1)				West (California)	3841
Osobenus (1)				West	3241, 3841
Perlinodes (1)				West	3241, 3634
Pictetiella (1)				West (intermountain)	3841
Remenus (1)				East	3841
Salmoperla (1)				West (California)	1586, 3841
Susulus (1)			Predators (engulfers)	West (California)	2339, 3841
Setvena (3)			Predators (engulfers) (scrapers?)	West	3841, 4007
Skwala (2)			Predators (engulfers)	West	1622, 3634, 3649, 3841, 2008

*Emphasis on trophic relationships.
†Unpublished data, K. W. Cummins, Oregon State University.

Table 14B. Continued

Taxa (number of species in parentheses)	Habitat	Habit	Trophic Relationships	North American Distribution	Ecological References*
Yugus (2)				East (Appalachian)	3841, 4537
Isoperlinae (62)	Lotic—erosional and depositional	Generally clingers, sprawlers	Predators, (engulfers) (collectors gatherers)		
Calliperla (1)				West	3841
Cascadoperla (1)		Clingers, sprawlers		West	3841
Clioperla (1)				East	1158, 1573, 2718, 3841
Cosumnoperla (1)				West	3930
Isoperla (57+)	Lotic—erosional and depositional, large cold lake	Clingers sprawlers	Predators (engulfers; Chironomidae, Simuliidae, Ephemeroptera, Plecoptera), facultative collectors—gatherers	Widespread	193, 1131, 1308, 1331, 1470, 2414, 2718, 2719, 4565, 1061, 1103, 3598, 3629, 2343, 2460, 4431, 852, 1587, 1622, 2026, 2963, 3836, 3841, 1158, 2176, 2340, 1104, 2461, 2792, 1573, 2462, 1588, 4490
Rickera (1)	Lotic—erosional	Clingers	Predators (engulfers)	West	3634, 3841, 4007
Chloroperlidae (77)	Generally lotic—erosional	Generally clingers	Generally predators (engulfers), scrapers, collectors—gatherers		671, 1074, 1131, 1308, 1331, 1573, 2906, 3841, 1310, 1333, 4268, 1059
Paraperlinae (6)					
Kathroperla (2)			Collectors—gatherers, scrapers	West	3841
Paraperla (2)	Hyporheal			West	3793, 3841, 4234, 3784
Utaperla (2)				East, West	3841
Chloroperlinae (71)			Generally collectors—gatherers, engulfers (predators)		
Alaskaperla (1)	Lotic—erosional		Predator (engulfers)	West (Alaska)	3834
Alloperla (29)				East, West	3841
Bisancora (2)					3841
Haploperla (4)			Predators (engulfers; Chironomidae); facultative collectors-gatherers and scrapers	East, West	193, 1589, 2025, 3629 3841
Neaviperla (1)				West	3841
Plumiperla (2)				West	3841
Rasvena (1)				East	3841
Suwallia (6)			Predators (engulfers; Chironomidae, Simuliidae)	East, West	1333, 3841, 4234
Sweltsa (23)			Predators, (engulfers; Chironomidae, Simuliidae)	East, West	816, 935, 1333, 1573, 2414, 1877, 3841, 3078, 4234, 1877
Triznaka (2)			Predators (engulfers; Chironomidae, Simuliidae)	West	1333, 1622, 3841, 4234

*Emphasis on tropic relationships.

AQUATIC AND SEMIAQUATIC HEMIPTERA

John T. Polhemus
Englewood, Colorado

Heteroptera

INTRODUCTION

Fifteen of the 17 major families of Heteroptera associated with the aquatic habitat are represented in the North American insect fauna, with only the pantropical Helotrephidae and Old World Aphelocheiridae missing. Six families are totally aquatic; they leave the water only to migrate. Three families of the water striders live on the surface film and adjacent banks, and the remaining six groups live at aquatic margins although some are often found on the water surface. All of these families belong to three suborders of the order Heteroptera: Leptopodomorpha, Gerromorpha, and Nepomorpha. [The Heteroptera are considered to be an order by Henry and Froeschner (1988) in their catalog of North American fauna, but as a suborder by some authors.]

Several other families occur in North America whose members are essentially terrestrial, but they have species that are occasionally found near water. They are the introduced Leptopodidae (related to Saldidae) and Ceratocombidae, Dipsocoridae and Schizopteridae (infraorder Dipsocoromorpha). These are not treated in this work except to list ecological and taxonomic references for the Dipsocoromorpha that provide an entry to the literature.

About 3,800 species of aquatic and semiaquatic Heteroptera occur worldwide; Jaczewski and Kostrowicki (1969) give an underestimate of 2,900, excluding the Saldidae and recently described species. Presently 68 genera and 412 species are recognized from North America. The Nearctic fauna has been enriched by the addition of tropical forms, whereas transverse mountain ranges have blocked such invasions into the Palaearctic, which has a comparatively poor fauna. Holarctic species are known only in the cold-adapted Corixidae, Gerridae and Saldidae, the latter being the most numerous with 14 species. A discussion of the world distribution of water bugs is given by Hungerford (1958) and of Gerromorpha by Andersen (1982). The zoogeography of North American water bugs has been discussed by Menke (1979; California) and Slater (1974; Connecticut). Dispersal of most aquatic Hemiptera occurs by flight, but there is also evidence for hurricane transport (Herring 1958). A habitat key and other significant data on biology are furnished by Hungerford (1920) in his classic work on water bugs; this has been updated by Usinger (1956b) and further by Menke et al. (1979). The characterization of a habitat for each group is only a rough generalization due to the many exceptions (Table 15A).

The aquatic and semiaquatic Heteroptera are remarkable for their diversity of form, reflecting adaptions to a wide variety of niches. They occupy many varied habitats, including saline ponds, high mountain lakes, hot springs, and large rivers. Basically they are predators; many species seem to be relatively resistant to predation, which is often attributed to the possession of characteristic heteropteran scent glands. Only the Corixidae differ significantly: many genera are primarily collectors, feeding on detritus, and are heavily preyed upon (Table 15A).

Some aquatic Heteroptera are of recognized economic importance, and the role of additional forms is presently being investigated. Biting genera such as *Notonecta, Belostoma,* and *Lethocerus* are attracted to lights and may be a nuisance to swimming pools. *Lethocerus,* a large belostomatid, can also be a nuisance at fish hatcheries (Wilson 1958). Corixidae have long been a relished food item in Mexico under the name "Ahuautle" and are also used extensively as food for pet fish and turtles; Belostomatidae are considered a delicacy in Asia. Corixidae have been shown to be good indicators of lentic water quality (Jansson 1977). Both surface and aquatic bugs can be important predators of mosquito larvae and adults (Jenkins 1964; Collins and Washino 1985; Table 15A). *Notonecta undulata* Say prefers mosquito larvae over other food (Ellis and Borden 1970; Toth and Chew 1972b), and these authors and Laird (1956) urge further study of aquatic Hemiptera as biological control agents.

Although there are exceptions, most aquatic Heteroptera lay their eggs in the spring, develop during the warmer months, overwinter as adults, and repeat the cycle. Some saldids overwinter as eggs, many gerrids are bivoltine, and in southern regions a number of species breed throughout the year. Eggs are of various forms and are laid in a wide variety of places, either glued to a substrate or inserted in the earth or plants. They have a wide variety of shapes: spindle-shaped, oval, or occasionally stalked. The chorion is tough, often hexagonally reticulate and successful hatching usually takes place submerged or in damp habitats (see Chap. 5, Table 5A). A splendid review of heteropteran eggs and embryology is given by Cobben (1968). All but a few species have five larval instars; the exceptions have four and include some *Mesovelia* sp., *Microvelia* sp., and *Nepa* sp.

Some true water bugs (Pleidae, Notonectidae, Naucoridae, Corixidae) swim with synchronous oarlike strokes of the hind legs, and other families (i.e., Nepidae, Belostomatidae), by synchronous strokes of the middle and hind pair of legs, those on each side working alternately. When swimming vigorously the latter families stroke both pairs in unison. The belostomatids are the strongest, the nepids the weakest, and the corixids the most agile swimmers.

The water striders (Gerromorpha) are supported by surface tension of the water and the unwettable hydrofuge pile of their tarsi and sometimes their tibiae. With this capability, species of the gerrid genus *Halobates* have colonized the open ocean, the only insects to do so (Cheng 1985, Cheng and Frank 1993). The wettable claws can be retracted in the most specialized families (Gerridae and Veliidae). These bugs are able to easily glide over the water because of the low resistance and use their wettable claws or other pretarsal structures to penetrate the surface for "traction." Forward thrust in the gerrid and veliid species by rowing is caused by the surface film "packing up" behind the tarsi during the thrust. These animals usually steer with the hind legs and unequal strokes of the middle legs (Menke 1979; Andersen 1982).

Most other surface-dwelling Gerromorpha walk over the surface film with tripodal locomotion, each leg alternating movement with its opposite. Some veliids are also able to move very rapidly when alarmed by "expansion skating," being carried forward on a contracting surface film caused by lowering of the surface tension by saliva discharged from the beak (review in Andersen 1982). Most of the true water bugs (Nepomorpha) breathe by means of an air store usually carried dorsally between the wings and abdomen plus an exposed thin bubble (physical gill) on the ventral surface held in place by hydrofuge hairs (Chap. 4). Means of air store replenishment are often distinctive for each family, e.g., through tubes (Nepidae), air straps (Belostomatidae), the pronotum (Corixidae), or the tip of the abdomen (Notonectidae, most Naucoridae). A few naucorid bugs can remain submerged indefinitely utilizing plastron respiration (Chap. 4; Menke 1979).

Sound production in the Gerromorpha has long been reported for the Veliidae (Leston and Pringle 1963), however, a few Gerridae also possess stridulatory mechanisms. Many Saldidae and most Leptopodidae (Leptopodomorpha) possess evident stridulatory mechanisms (Polhemus 1985; Pericart and Polhemus 1990), as do some, if not most, of the families of Nepomorpha. Stridulatory mechanisms have been described in the Old World Helotrephidae, the New World Naucoridae and Nepidae, and cosmopolitan Gelastocoridae, Notonectidae, and Corixidae (Polhemus, 1994b); "songs" of the latter family have been studied in detail (Jansson 1976).

EXTERNAL MORPHOLOGY

Larvae and Adults

The larvae of aquatic and semiaquatic Heteroptera have one-segmented tarsi, a useful characteristic in separating them from adults, especially the apterous gerromorphans (water striders), which always have two tarsal segments, at least on some legs. Larvae of Hemiptera resemble the adults but the body parts have different proportions and developing wings are present as wing pads (Fig. 15.4) in the ultimate and penultimate instars. Keys to the larvae of North American families and subfamilies of Heteroptera based in a large part on the *trichoboth-*

ria (hair-bearing spots on abdomen) and dorsal abdominal scent gland openings (Fig. 15.4) have been published (DeCoursey 1971; Herring and Ashlock 1971; Yonke 1991). A key and synopsis of the families and genera of adult North American Heteroptera has been published by Slater and Baranowski (1978), but differs somewhat in arrangement from this work [e.g., genera *Aquarius* and *Limnoporus* included in *Gerris* (Gerridae); family Macroveliidae included in Mesoveliidae; genera *Platyvelia* and *Steinovelia* included in *Paravelia* (Veliidae)].

With the exception of a few families with head and thorax joined (e.g., Pleidae, Naucoridae), the head, thorax, and abdomen are generally well defined in aquatic and semiaquatic Hemiptera.

Head: The eyes are usually prominent and well developed. Ocelli may be present but are lacking in many aquatic families and ocelli are present only in winged forms of some semiaquatic species (e.g., *Mesovelia*). Antennae are three-, four-, or five-segmented and are ordinarily quite conspicuous in semiaquatic bugs, but hidden in the true aquatics. Two semiaquatic families (Ochteridae, Gelastocoridae), which inhabit the margins of fresh waters and are closely allied to the true aquatics, have short antennae that are largely or entirely hidden. The *rostrum* or beak is segmented, varies in length, and has either three or four visible segments (except for Corixidae; Fig. 15.17). It attaches to the apex of the head and is directed posteriorly underneath (Cobben 1978). The ventral region between the base of the beak and the collar, called the *gula* (Fig. 15.100), is one of the primary characters used in separating the Hemiptera-Heteroptera from the Homoptera (leaf hoppers; see Homoptera section).

Thorax: The three-segmented thorax bears the legs and wings and, due to fusions or extra sutures, the segments are difficult to identify except in the wingless forms (Fig. 15.145). The metasternum usually bears one or more scent glands (Fig. 15.81) and sometimes lateral scent channels (Fig. 15.145). The legs often represent striking adaptations of aquatic Heteroptera to their environment (Figs. 15.80, 15.143, 15.146). Leg segments are variable in length; however, each leg always consists of a coxa articulating with the body followed by a trochanter joining the coxa and femur. The femur and tibia are usually the longest segments, and the tarsi are one-, two-, or three-jointed and bear the claws.

Alary (wing) polymorphism in water bugs is a common phenomenon, but a few groups, such as the marine water strider *Halobates* sp. and the macroveliid *Oravelia* sp., are known only in the wingless state. The hind wings of Heteroptera are membranous, whereas the fore wings or *hemelytra* have a leathery *clavus* and *corium* plus a thin membrane portion (Fig. 15.2), which may appear to be lacking in some aquatic species (e.g., Pleidae). Winged forms are more common in southern regions and wingless or short winged forms are more common in the groups of surface bugs. The most extensive studies of alary polymorphism and its mechanisms have concerned the Gerridae (Brinkhurst 1959, 1960; Vepsalainen 1971a, b, 1974; Andersen 1973).

Abdomen: The abdomen bears the spiracles and genitalia. The first visible segment ventrally is actually the second, and the first seven segments are similar. The eighth through tenth segments form the genitalia and may or may not be distinguishable. In some families of Nepomorpha (e.g., Ochteridae, Gelastocoridae, and Corixidae) the last few abdominal segments are asymmetrical in males but symmetrical or nearly so in females.

Homoptera

INTRODUCTION

The Homoptera have not adapted to truly aquatic life as have some Heteroptera. However, some species of Homoptera can be considered marginally semiaquatic as they have a more or less permanent association with the margins of both the intertidal zone and fresh water, where they feed on aquatic plants.

Homopteran species commonly found in semiaquatic habitats belong to a half dozen or more families. *Draeculacephala* spp. (Cicadellidae) are found along stream margins, and the transcontinental *Helochara communis* Fitch (Cicadellidae) is apparently always associated with low marshy grasses. Adults and larvae of *Megamelus davisi* Van Duzee (Delphacidae) feed on all emergent parts of water lilies *(Nuphar* sp.) (Wilson and McPherson 1981).

Although few, if any, of the homopterans are subjected to significant aquatic inundation, the eggs of *Prokelisia marginata* (Van Duzee) (Delphacidae) are deposited in *Spartina* sp., an intertidal grass, which is sometimes completely submerged. Many intertidal Homoptera (Delphacidae, Cicadellidae, Issidae) inhabit salt marsh vegetation (Denno 1976) that is occasionally inundated, but the adults of these species move up the culms of vegetation to escape rising tides (Davis and Gray 1966). Conversely, Cameron (1976) has suggested that certain Homoptera can withstand long periods of submergence in California marshes, which typically have a steeper littoral gradient than those studied by Davis and Gray (1966) in North Carolina. Certain of the Auchenhorrhyncha may locate in an air bubble trapped in leaf or blade axils during inundation; the bubble could function as a physical gill.

EXTERNAL MORPHOLOGY

Although the inclusion of Homoptera in a work on aquatic insects is problematic, collectors of semiaquatic insects will find them and have the need for identifications. Homoptera differ morphologically from Heteroptera primarily in having the posteriorly directed beak apparently arise from the underside of the thorax rather than the head (Fig. 15.1), and sometimes in having a uniformly leathery, hyaline, or membranous corium rather than the differentiated heteropteran wing described above (see Figs. 15.2, 15.100, and 15.116).

Within the Sternorrhyncha, *Haliaspis spartinae* (Comstock) (Diaspididae), *Eriococcus* sp. (Eriococcidae), and various pseudococcids inhabit intertidal vegetation and are at least occasionally inundated by tides. In Britain, the aphid *Pemphigus treheranei* Foster occurs on the roots of an intertidal aster (Foster 1975).

For the most part, the Homoptera have not evolved the sophisticated respiratory mechanisms found in aquatic Heteroptera. They can be considered no more than semiaquatic, although some scale insects and mealybugs approach an aquatic existence. The respiratory adaptations of salt marsh homopterans have been reviewed by Foster and Treherne (1976; see Chap. 4 also).

KEY TO THE FAMILIES OF AQUATIC AND SEMIAQUATIC HEMIPTERA

1.	Head without a gula; gular region hidden by posteriorly directed head (Fig. 15.1)	(Suborder Homoptera)
1'.	Head with a gula; gular region not hidden (Fig. 15.100)	(Suborder Heteroptera) 2
2(1').	Antennae shorter than head, inserted beneath eyes, not plainly visible from above (Fig. 15.100) except in Ochteridae (Fig. 15.5); aquatic or semiaquatic bugs (at margins of standing- or running-water habitats)	3
2'.	Antennae longer than head, inserted forward of eyes, plainly visible from above (Figs. 15.2-15.3); bugs on surface or at aquatic margins	10
3(2).	Beak triangular, very short, unsegmented (sometimes transversely striated), appearing as apex of head (Fig. 15.15); front tarsus with a single segment, scooplike, fringed with stiff setae forming a rake (Fig. 15.22)	***CORIXIDAE*** (p. 271)
3'.	Beak cylindrical, short to long, 3- or 4-segmented (Fig. 15.100); front tarsus not scooplike or fringed with stiff setae	4
4(3').	Apex of abdomen with respiratory appendages (Figs. 15.12-15.111)	5
4'.	Apex of abdomen without respiratory appendages	6
5(4).	Apex of abdomen with a pair of flat, retractile air straps (Fig. 15.12)	***BELOSTOMATIDAE*** (p. 271)
5'.	Apex of abdomen with cylindrical breathing tube (siphon) composed of 2 slender, nonretractile filaments (Fig. 15.111)	***NEPIDAE*** (p. 278)
6(4').	Middle and hind legs without fringelike swimming hairs; ocelli present (Fig. 15.64) except in *Nerthra rugosa* Desjardins; semiaquatic bugs at margins of aquatic habitats	7
6'.	Middle and hind legs with fringelike swimming hairs (Fig. 15.115); ocelli absent; aquatic bugs	8
7(6).	Front legs raptorial (grasping), femora broad (Figs. 15.60-15.61); rostrum short, not reaching hind coxae; antennae not visible from above (Fig. 15.64)	***GELASTOCORIDAE*** (p. 275)
7'.	Front legs not raptorial, femora not broad (Fig. 15.6); rostrum long, reaching or extending beyond hind coxae; tips of antennae usually visible from above (Figs. 15.5, 15.69)	***OCHTERIDAE***—*Ochterus*

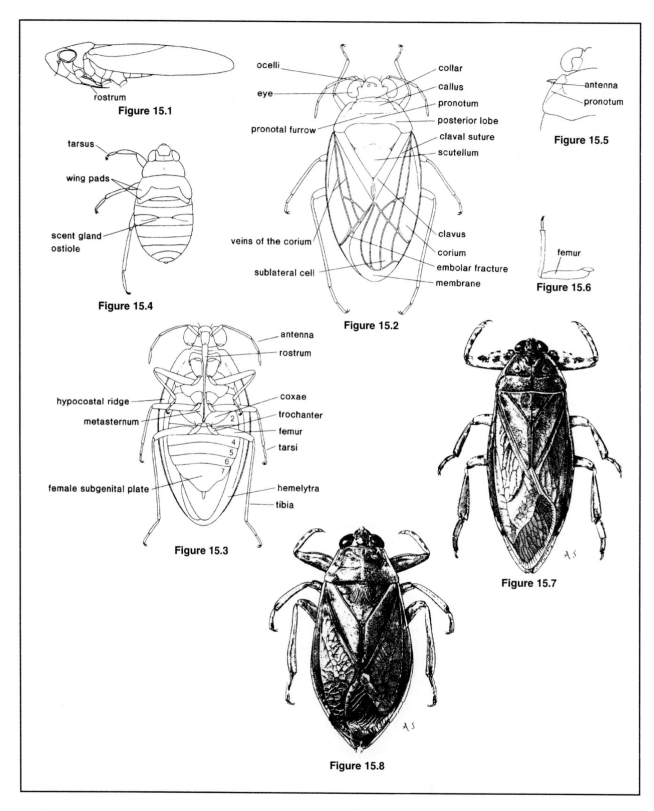

Figure 15.1. Lateral view of adult Cicadellidae (Homoptera).

Figure 15.2. Dorsal view of adult *Saldula* sp. (Saldidae).

Figure 15.3. Ventral view of adult *Saldula* sp. (Saldidae).

Figure 15.4. Dorsal view of Saldidae nymph.

Figure 15.5. Dorsal view of head and pronotum of Ochteridae.

Figure 15.6. Leg of Ochteridae.

Figure 15.7. *Lethocerus americanus* (Leidy), dorsal view (Belostomatidae; from Usinger 1956).

Figure 15.8. *Belostoma bakeri* Montandon, dorsal view (Belostomatidae; from Usinger 1956).

HEMIPTERA

8(6').	Front legs raptorial, femora broad; body dorsoventrally flattened (Fig. 15.100)	*NAUCORIDAE* (p. 278)
8'.	Front legs slender, femora not broad; body strongly convex dorsally (Figs. 15.95-15.116)	9
9(8').	Body form ovoid; 3 mm or less in length (Fig. 15.95); all legs similar; hind tarsus with 2 well-developed claws (Fig. 15.97)	*PLEIDAE* (p. 284)
9'.	Body form elongate, 5 mm or more in length (Fig. 15.116); hind legs long, oarlike; claws of hind tarsus inconspicuous (Fig. 15.115)	*NOTONECTIDAE* (p. 278)
10(2').	Membrane of wing with 4 or 5 distinct similar cells (Fig. 15.2); hind coxae large, transverse, with broad coxal cavity (Fig. 15.3)	*SALDIDAE* (p. 284)
10'.	Membrane of wing without distinct similar cells (Fig. 15.139); hind coxae small, cylindrical, or conical; coxal cavity socketlike (Fig. 15.145)	11
11(10').	Claws of at least front tarsi inserted before apex (Fig. 15.144)	12
11'.	Claws of all legs inserted at tips of tarsi (Fig. 15.92)	13
12(11).	Hind femur short, distally either scarcely or not surpassing apex of abdomen; metasternum with a pair of lateral scent grooves terminating on pleura in front of hind coxae (Fig. 15.145); dorsum of head usually with median longitudinal sulcus or glabrous (smooth) stripe; mid legs inserted about midway between front and hind legs, except *Trochopus* and *Rhagovelia*, which have featherlike structures on the middle tarsus (Fig. 15.146), and *Husseyella*, which has bladelike structures instead of claws on middle tarsus (Fig. 15.143)	*VELIIDAE* (p. 288)
12'.	Hind femur long, distally greatly exceeding apex of abdomen (scarcely in *Rheumatobates* females); metasternal region with single median scent gland opening (omphalium; Fig. 15.81), lateral scent grooves absent; dorsum of head without median groove or line except in *Rheumatobates;* mid legs inserted closer to hind legs than fore legs	*GERRIDAE* (p. 275)
13(11').	Body long, slender; head as long as or longer than combined length of pronotum and scutellum (Fig. 15.94)	*HYDROMETRIDAE—Hydrometra*
13'.	Body stout; head length not greater than combined length of pronotum and scutellum, or pronotum alone in wingless forms (Fig. 15.90)	14
14(13').	Tarsi two-segmented; head ventrally with deep longitudinal channel for reception of rostrum (Fig. 15.86)	*HEBRIDAE* (p. 278)
14'.	Tarsi 3-segmented; head ventrally without longitudinal channel (Fig. 15.91)	15
15(14').	Inner margins of eyes converging anteriorly; femora with at least 1 or 2 black spines on dorsum distally; winged forms with exposed bilobed scutellum (Fig. 15.90)	*MESOVELIIDAE—Mesovelia*
15'.	Inner margins of eyes arcuate, not converging anteriorly (Fig. 15.88); femora without black spines; winged forms with scutellum concealed by pronotum (Fig. 15.88)	*MACROVELIIDAE* (p. 278)

KEYS TO THE GENERA OF AQUATIC AND SEMIAQUATIC HEMIPTERA

Belostomatidae

1.	Tibia and tarsus of hind leg strongly compressed, thin, much broader than middle tibia and tarsus (Figs. 15.7, 15.11); basal segment of beak about half length of 2nd; length 40 mm or more	(Lethocerinae) **Lethocerus** Mayr
1'.	Tibia and tarsus of middle and hind leg similar (Figs. 15.8-15.10); basal segment of beak subequal to 2nd; length 37 mm or less	(Belostomatinae) 2
2(1').	Membrane of hemelytron reduced (Figs. 15.9, 15.12)	**Abedus** Stål
2'.	Membrane of hemelytron not reduced (Figs. 15.8, 15.13); length 26 mm or less	**Belostoma** Latreille

Corixidae

1.	Scutellum exposed, covered by pronotum only at anterior angles (Fig. 15.21)	(Micronectinae) **Tenagobia** Bergroth
1'.	Scutellum concealed (Fig. 15.28)	2
2(1').	Rostrum (beak) without transverse striations; nodal furrow absent (Fig. 15.16)	(Cymatiinae) **Cymatia** Flor
2'.	Rostrum with transverse striations (Figs. 15.15, 15.17); nodal furrow present (Fig. 15.19)	(Corixinae) 3
3(2').	Fore tarsus with rather thick, well-developed apical claw; pala of both sexes narrowly digitiform (fingerlike) (Figs. 15.24, 15.28)	(Graptocorixini) 4
3'.	Fore tarsus with spinelike apical claw usually resembling spines along lower margin of palm; pala not digitiform (Figs. 15.22, 15.35)	5

4(3).	Tergal lobes on left side of male abdomen produced posteriorly (i.e., abdomen sinistral) (Fig. 15.23); strigil absent; female abdomen slightly asymmetrical; female face slightly concave, densely pilose (hairy)	***Neocorixa*** Hungerford
4'.	Tergal lobes on right side of male abdomen produced posteriorly (i.e., abdomen dextral) (Fig. 15.27); strigil present on right side; female abdomen asymmetrical; female face not concave, not densely pilose	***Graptocorixa*** Hungerford
5(3').	Eyes protuberant (produced above surface) with inner anterior angles broadly rounded. Face depressed in both sexes, with dense hair covering; postocular space broad, head transversely depressed behind eyes (Fig. 15.14)	(Glaenocorisini) 6
5'.	Eyes not protuberant, inner anterior angles acute (Fig. 15.15); face of females usually not densely hairy or depressed; if postocular space is broad, then head is not transversely depressed behind eyes	(Corixini) 7
6(5).	Pronotum and clavus strongly rastrate (with longitudinal scratches); metaxyphus (Fig. 15.17) broadly triangular (Fig. 15.26)	***Glaenocorisa*** Thomson
6'.	Pronotum and clavus not strongly rastrate; metaxyphus (Fig. 15.17) narrowly triangular (Fig. 15.25)	***Dasycorixa*** Hungerford
7(5').	Apices of hemelytral clavi not or scarcely surpassing a line drawn through the costal margins at the nodal furrows; abdominal asymmetry of males sinistral, strigil on left; foretibia of male produced over base of pala (Figs. 15.29, 15.34); species smaller than 5.6 mm in length	***Trichocorixa*** Kirkaldy
7'.	Apices of hemelytral clavi clearly surpassing a line drawn through the costal margins at the nodal furrows (Figs. 15.19, 15.33); abdominal asymmetry of males dextral, strigil on right (Fig. 15.27); foretibia of male not produced over base of pala (greatly expanded distally in *Centrocorisa* sp. but not over base of pala; Fig. 15.37); species longer than 5.9 mm	8
8(7').	Pruinose (frosted) area at base of claval suture (clavopruina) short and broadly rounded, usually about one-half to two-thirds as long as postnodal pruinose area (postnodal pruina) (Figs. 15.30, 15.32); clavus rastrate	***Hesperocorixa*** Kirkaldy
8'.	Pruinose area at base of claval suture (clavopruina) either broadly rounded, narrowly rounded or pointed distally, subequal to or longer than postnodal pruinose area (postnodal pruina) (Fig. 15.33) (except two-thirds in *Corisella*, which has smooth clavus)	9
9(8').	Body short, broad, more than one-third as broad as long (width measured across pronotum)	10
9'.	Body elongate, distinctly less than one-third as broad as long	11
10(9).	Pronotum and clavus only faintly rugulose (wrinkled); male foretarsi expanded distally (Fig. 15.37); male without strigil or pedicel (Fig. 15.38); none of the species have a longitudinal groove on the ventral surface of middle femora	***Centrocorisa*** Lundblad
10'.	Pronotum and clavus distinctly rastrate; male fore tarsi not expanded distally in United States species (Fig. 15.35); male with strigil or at least pedicel; some species with a longitudinal groove on the ventral surface of middle femora of both sexes (Fig. 15.31)	***Morphocorixa*** Jaczewski
11(9').	Palar claw of both sexes minutely serrate (sawlike) at base (Fig. 15.46); upper surface of male pala deeply incised (Fig. 15.45); vertex of male usually acuminate (tapering to a point) (Fig. 15.42); usually with hemelytral pattern indistinct or obscure	***Ramphocorixa*** Abbott
11'.	Palar claw not serrate; upper surface of male pala not deeply incised; vertex of male not acuminate (Fig. 15.43); hemelytral pattern usually distinct	12
12(11').	Posterior margin of head sharply curved, embracing very short pronotum (Fig. 15.43); interocular space much narrower than width of an eye (Fig. 15.20); median lobe of male abdominal tergite VII with a hooklike projection (Fig. 15.44)	***Palmacorixa*** Abbott
12'.	Posterior margin of head not sharply curved, pronotum longer (Fig. 15.40); interocular space at least subequal to width of an eye; median lobe of male abdominal tergite VII without a hooklike projection	13
13(12').	Pronotum and clavus smooth and shining, at most faintly rugulose (wrinkled) (Fig. 15.39)	***Corisella*** Lundblad
13'.	Part or all of pronotum and clavus rough, either rastrate (with longitudinal scratches) or rugulose, or both (Fig. 15.40)	14
14(13').	Markings on clavus and corium narrow and broken, usually open reticulate (covered with network of fine lines) with much anastomosing (merging) (Fig. 15.55); pronotal carina (ridge) distinct on at least anterior one-third (Fig. 15.41)	15

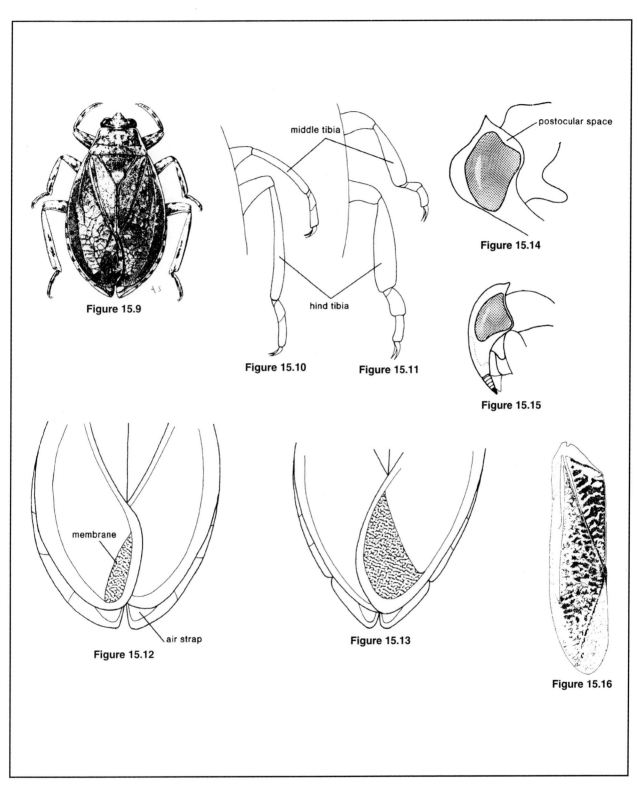

Figure 15.9. *Abedus indentatus* (Haldeman), dorsal view (Belostomatidae; from Usinger 1956).

Figure 15.10. Middle and hind legs of *Belostoma* sp. (Belostomatidae).

Figure 15.11. Middle and hind legs of *Lethocerus* sp. (Belostomatidae).

Figure 15.12. Hemelytra of *Abedus* sp. (Belostomatidae).

Figure 15.13. Hemelytra of *Belostoma* sp. (Belostomatidae).

Figure 15.14. Side view of head of *Glaenocorisa* sp. (Corixidae).

Figure 15.15. Side view of head of *Hesperocorixa* sp. (Corixidae).

Figure 15.16. Left hemelytron of *Cymatia* sp. (Corixidae; from Brooks and Kelton 1967).

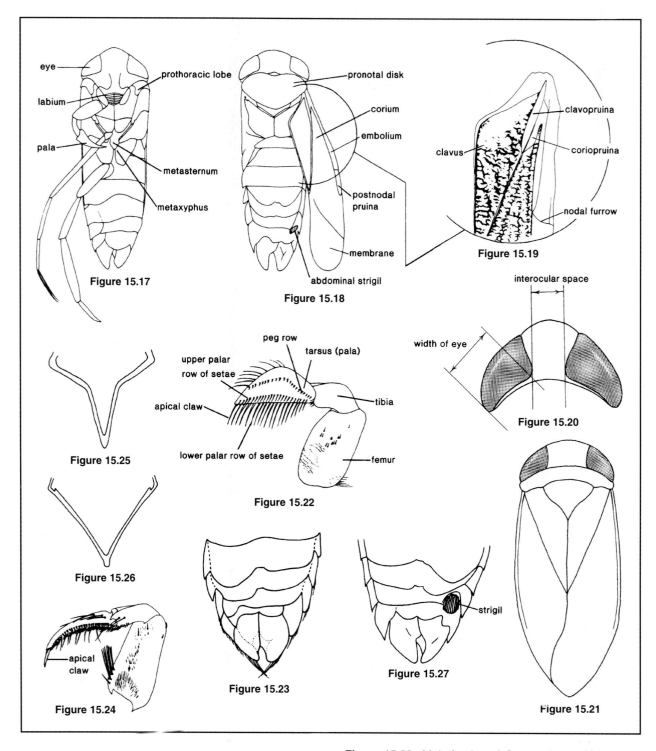

Figure 15.17. Ventral view of *Corisella* sp. (Corixidae; from Menke *et al.* 1979).

Figure 15.18. Dorsal view of *Corisella* sp. (Corixidae; from Menke *et al.* 1979).

Figure 15.19. Enlarged portion of hemelytron of *Corisella* sp. (Corixidae) showing pruinose areas (from Menke *et al.* 1979).

Figure 15.20. Dorsal view of head of *Palmacorixa* sp. (Corixidae).

Figure 15.21. Dorsal view of *Tenagobia* sp. (Corixidae).

Figure 15.22. Male foreleg of *Cenocorixa* sp. (Corixidae; from Menke *et al.* 1979).

Figure 15.23. Abdominal tergites of *Neocorixa* sp. (Corixidae).

Figure 15.24. Male foreleg of *Graptocorixa* sp. (Corixidae; from Menke *et al.* 1979).

Figure 15.25. Metaxyphus of *Dasycorixa* sp. (Corixidae).

Figure 15.26. Metaxyphus of *Glaenocorisa* sp. (Corixidae).

Figure 15.27. Abdominal tergites of *Graptocorixa* sp. (Corixidae; from Menke *et al.* 1979).

14'.	Markings on clavus transverse, those of corium transverse, longitudinal or reticulate (Figs. 15.56-15.59); pronotal carina absent or faintly expressed on anterior border at most (Figs. 15.40, 15.54) 16
15(14).	Median carina of pronotum well defined on anterior two-thirds or more; male fore pala usually digitiform with rather evenly curved row of pegs (Fig. 15.47); peg row interrupted and pala broadened basally only in *A. chancea* Hungerford (Fig. 15.48) ....***Arctocorisa*** Wallengren
15'.	Median carina of pronotum well defined only on anterior one-third (Fig. 15.41); male fore pala broadened medially, peg row sharply curved (Fig. 15.22) or disjunct medially (Fig. 15.51) ....***Cenocorixa*** Hungerford
16(14').	Male strigil present; palar pegs usually in 1 row (Fig. 15.50) with notable exceptions (Fig. 15.49); ground color usually yellowish, dark pattern on hemelytra strongly contrasting (Fig. 15.54, 15.56-15.58) ....***Sigara*** Fabricius
16'.	Male strigil absent; palar pegs in 2 rows (Fig. 15.52); ground color greenish yellow, dark pattern on hemelytra weakly contrasting (Figs. 15.40, 15.59). In all except *C. audeni*, posterior first tarsal segment with black spot (Fig. 15.53) .... ***Callicorixa*** White

Gelastocoridae

1.	Fore tarsus articulating with tibia, 1-segmented with 2 claws in nymphs and adults; fore femur only moderately enlarged at base, twice as long as basal width, not subtriangular (Fig. 15.61); closing face of fore femur flat and bordered by 2 rows of short spines; beak clearly arising at front of head, directed posteriorly; dorsal aspect of body as in Figure 15.62 .... (Gelastocorinae) ***Gelastocoris*** Kirkaldy
1'.	Fore tarsus fused with tibia and terminated by a single claw (adults) or 2 claws (nymphs); fore femur very broad at base, about as long as broad, subtriangular (Fig. 15.60); closing face of fore femur with a dorsal flangelike extension that projects over tibia when it is closed against femur; beak appearing to arise from the back of the head, L-shaped; dorsal aspect of body as in Figure 15.63 .... (Nerthrinae) ***Nerthra*** Say

Gerridae

1.	Inner margins of eyes sinuate or concave behind the middle (Figs. 15.75-15.77); body comparatively long and narrow (Fig. 15.79) ....(Gerrinae) 2
1'.	Inner margins of eyes convex (Fig. 15.78), body comparatively short and broad (Figs. 15.66-15.68) 6
2(1).	Pronotum shiny 3
2'.	Pronotum dull 4
3(2).	Fore lobe of pronotum with pair of long, pale lines (Fig. 15.76) ....***Limnogonus*** Stål
3'.	Fore lobe of pronotum with single, median, pale spot (Fig. 15.77) ....***Neogerris*** Matsumura
4(2').	Antennal segment I not less than nine-tenths of the combined lengths of II and III (Fig. 15.71)..5
4'.	Antennal segment I not more than eight-tenths of the combined lengths of II and III (Fig. 15.70) ....***Limnoporus*** Stål
5(4).	Hind tibia at least 4.0 times as long as first tarsal segment; larger species, with total length at least 11 mm., with prominent connexival spines (Fig. 15.151)..***Aquarius*** Schellenberg
5'	Hind tibia not over 3.2 times as long as first tarsal segment; usually smaller species, with total length less than 10 mm., without prominent connexival spines (Fig. 15.152)..***Gerris*** Fabricius
6(1').	Tibia and first tarsal segment of middle leg with fringe of long hairs (Fig. 15.80); always apterous (without trace of wing pads); the meso- and metanotum fused, without trace of a dividing suture ....(Halobatinae)........ ***Halobates*** Eschscholtz
6'.	Tibia and first tarsal segment of middle leg without a fringe of long hairs; dimorphic (have both apterous and long-winged forms); the meso- and metanotum of apterous forms with a distinct dividing suture 7
7(6').	Antennal segment III with several stiff bristles that are at least as long as diameter of segment (Fig. 15.74); length of antennal segment I much shorter than remaining 3 taken together; abdomen as long as remainder of body (Fig. 15.67) ....***Rheumatobates*** Bergroth
7'.	Antennal segment III with fine pubescence or tuft of short, stiff bristles, but these not as long as diameter of segment; abdomen shorter than remainder of body (Fig. 15.68), or if subequal then length of antennal segment I about equal to remaining 3 taken together (Fig. 15.73) ....(Trepobatinae) 8

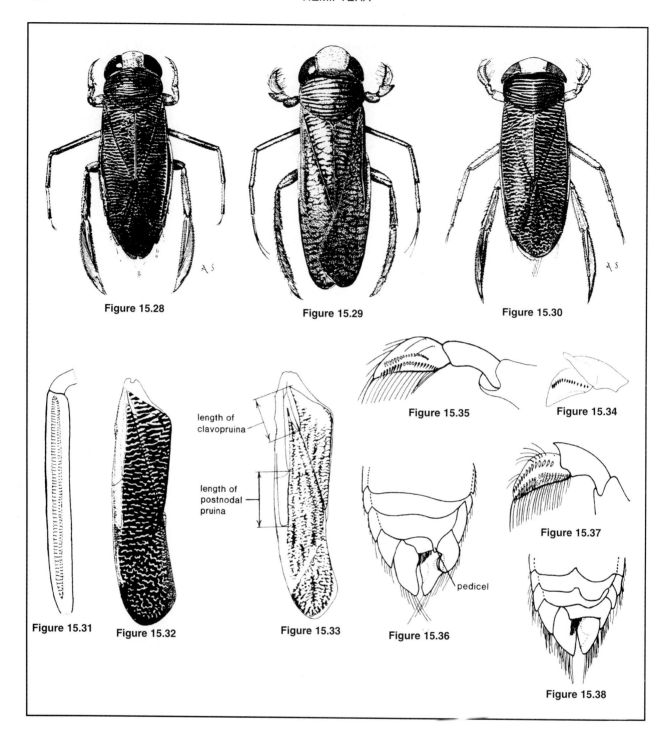

Figure 15.28. *Graptocorixa californica* (Hungerford) (Corixidae; from Usinger 1956).

Figure 15.29. *Trichocorixa reticulata* (Guerin-Meneville) (Corixidae; from Usinger 1956).

Figure 15.30. *Hesperocorixa vulgaris* (Hungerford) (Corixidae; from Usinger 1956).

Figure 15.31. Dorsal view of middle femur of *Morphocorixa* sp. (Corixidae).

Figure 15.32. Left hemelytron of *Hesperocorixa* sp. (Corixidae; from Brooks and Kelton 1967).

Figure 15.33. Left hemelytron of *Corisella* sp. (Corixidae; from Menke et al. 1979).

Figure 15.34. Male foreleg of *Trichocorixa* sp. (Corixidae; from Brooks and Kelton 1967).

Figure 15.35. Male foreleg of *Morphocorixa* sp. (Corixidae).

Figure 15.36. Abdominal tergites of *Morphocorixa* sp. (Corixidae).

Figure 15.37. Male foreleg of *Centrocorisa* sp. (Corixidae).

Figure 15.38. Abdominal tergites of *Centrocorisa* sp. (Corixidae).

HEMIPTERA

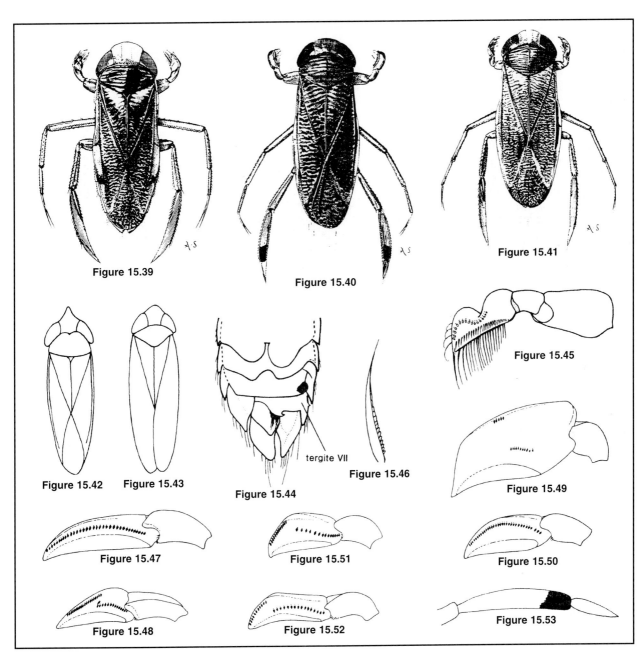

Figure 15.39. *Corisella decolor* (Uhler) (Corixidae; from Usinger 1956b).

Figure 15.40. *Callicorixa vulnerata* (Uhler) (Corixidae; from Usinger 1956b).

Figure 15.41. *Cenocorixa kuiterti* Hungerford (Corixidae; from Usinger 1956b).

Figure 15.42. Dorsal view of *Ramphocorixa* sp. (Corixidae).

Figure 15.43. Dorsal view of *Palmacorixa* sp. (Corixidae).

Figure 15.44. Abdominal tergites of *Palmacorixa* sp. (Corixidae).

Figure 15.45. Male foreleg of *Ramphocorixa* sp. (Corixidae).

Figure 15.46. Palar claw of *Ramphocorixa* sp. (Corixidae).

Figure 15.47. Male foreleg of *Arctocorisa sutilis* (Uhler), semidiagrammatic view (from Brooks and Kelton 1967.

Figure 15.48. Male foreleg of *Arctocorisa chanceae* (Hungerford), semidiagrammatic view (from Brooks and Kelton 1967).

Figure 15.49. Male foreleg of *Sigara fallenoidea* (Hungerford), semidiagrammatic view (from Brooks and Kelton 1967).

Figure 15.50. Male foreleg of *Sigara mathesoni* (Hungerford), semidiagrammatic view (from Brooks and Kelton 1967).

Figure 15.51. Male foreleg of *Cenocorixa* sp. (Corixidae), semidiagrammatic view (from Brooks and Kelton 1967).

Figure 15.52. Male foreleg of *Callicorixa* sp. (Corixidae), semidiagrammatic view (from Brooks and Kelton 1967).

Figure 15.53. Posterior tarsus of *Callicorixa* sp. (Corixidae).

8(7').   Length of antennal segment I subequal to remaining 3 taken together (Fig. 15.73) .... ***Metrobates*** Uhler
8'.   Length of antennal segment I much shorter than remaining 3 taken together (Fig. 15.72) .... ***Trepobates*** Uhler

Hebridae

1.   Antennae distinctly shorter than greatest width of pronotum; antennal segments stout, 4th segment subequal in length to 1st segment (Fig. 15.83) .... ***Merragata*** White
1'.   Antennae distinctly longer than greatest width of pronotum; antennal segments slender, 4th segment much longer than 1st segment (Figs. 15.84-15.85) 2
2(1').   Antennal segment IV without a constriction (false joint structure) in the middle (Fig. 15.84) .... ***Lipogomphus*** Berg
2'.   Antennal segment IV with a constriction (false joint structure) in the middle, appearing 5-segmented (Figs. 15.82, 15.85) .... ***Hebrus*** Curtis

Macroveliidae

1.   Ocelli absent; apterous; posterior margin of pronotum arcuate (arched), scutellum exposed (Fig. 15.89); antennal segments I-III each longer than head width across eyes .... ***Oravelia*** Drake and Chapman
1'.   Ocelli present, well developed (Fig. 15.88); macropterous or brachypterous; posterior margin of pronotum angular, concealing scutellum (Fig. 15.88); antennal segments I-III each shorter than head width across eyes (Fig. 15.87) .... ***Macrovelia*** Uhler

Naucoridae

1.   Anterior margin of pronotum straight or slightly concave behind interocular space (Figs. 15.103, 15.105-15.106) 2
1'.   Anterior margin of pronotum deeply concave behind interocular space (Fig. 15.104) 3
2(1).   Inner margins of eyes diverging anteriorly (Figs. 15.103, 15.106); meso- and metasterna bearing prominent longitudinal carinae (keels) that are broad and foveate (with a deep impression) along middle; body broadly oval, sub-flattened. Embolium may be dilated (Fig. 15.103), or produced outward and backward as an arcuate (arched), acute spine (Fig. 15.106) .... (Limnocorinae) ***Limnocoris*** Stål
2'.   Inner margins of eyes converging anteriorly (Fig. 15.105); meso- and metasterna without longitudinal carinae at middle; body strongly convex above, the embolium rounded. Embolium not dilated, or produced as an acute spine .... (Naucorinae) ***Pelocoris*** Stål
3(1').   Posterior part of prosternum covered by platelike extensions of propleura which are nearly contiguous at midline (Fig. 15.102); abdominal venter densely pubescent (hairy), except glabrous (shining) around spiracles, each spiracle also with a transverse row of small glabrous areas behind; macropterous (Fig. 15.104) .... (Ambrysinae) ***Ambrysus*** Stål
3'.   Prosternum completely exposed, separated from flattened pleura by simple sutures (Fig. 15.101); abdominal venter bare and with a perforated disklike area near each spiracle; dimorphic, the brachypterous forms with hemelytra truncate (shortened and squared-off) at apices, about half as long as abdomen .... (Cryphocricinae) ***Cryphocricos*** Signoret

Nepidae

1.   Anterior lobe of pronotum not wider than head; body long, slender, cylindrical (Fig. 15.107); abdominal sterna of adult undivided; adult female subgenital plate laterally compressed, keel-like .... (Ranatrinae) ***Ranatra*** Fabricius
1'.   Anterior lobe of pronotum wider than head (Fig. 15.110); body flattened; abdominal sterna of adult divided longitudinally into median and parasternites (Figs. 15.108-15.109); adult female subgenital plate broad, flattened .... (Nepinae) 2
2(1').   Median length of 6th sternite twice median length of 5th (Fig. 15.109) .... (Nepini) .... ***Nepa*** Linnaeus
2'.   Median length of 6th sternite about equal to length of 5th (Fig. 15.108) .... (Curictini) ....***Curicta*** Stål

Notonectidae

1.   Hemelytral commissure with a definite hair-lined pit at anterior end (Figs. 15.113, 15.117); antennae 3-segmented .... ***Buenoa*** Kirkaldy

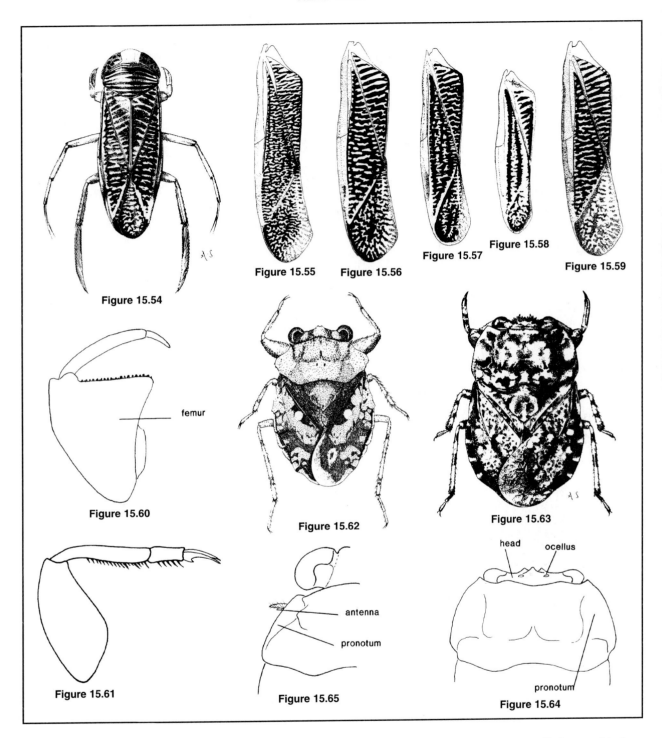

Figure 15.54. Dorsal view of *Sigara mckinstryi* Hungerford (Corixidae; from Usinger 1956b).

Figure 15.55. Left hemelytron of *Arctocorisa* sp. (Corixidae; from Brooks and Kelton 1967).

Figure 15.56. Left hemelytron of *Sigara decoratella* (Hungerford) (Corixidae; from Brooks and Kelton 1967).

Figure 15.57. Left hemelytron of *Sigara mullettensis* (Hungerford) (Corixidae; from Brooks and Kelton 1967).

Figure 15.58. Left hemelytron of *Sigara lineata* (Forster) (Corixidae; from Brooks and Kelton 1967).

Figure 15.59. Left hemelytron of *Callicorixa audeni* Hungerford (Corixidae; from Brooks and Kelton 1967).

Figure 15.60. Foreleg of *Nerthra* sp. (Gelastocoridae).

Figure 15.61. Foreleg of *Gelastocoris* sp. (Gelastocoridae).

Figure 15.62. Dorsal view of *Gelastocoris oculatus* (Fabricius) (Gelastocoridae; from Brooks and Kelton 1967).

Figure 15.63. Dorsal view of *Nerthra martini* Todd (Gelastocoridae; from Usinger 1956).

Figure 15.64. Dorsal view of head and pronotum of *Nerthra* sp. (Gelastocoridae).

Figure 15.65. Dorsal view of head and protonum of *Ochterus* sp. (Ochteridae).

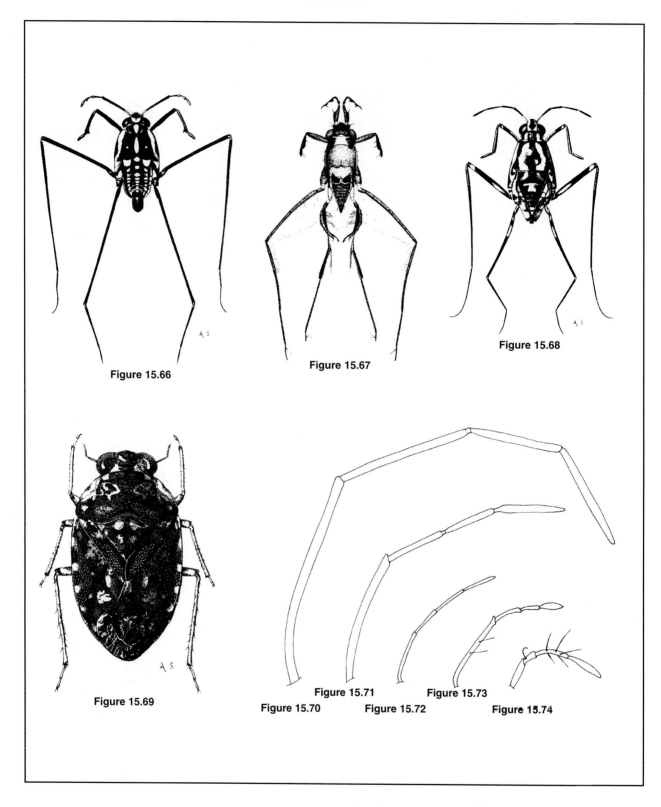

Figure 15.66. Dorsal view of *Metrobates trux infuscatus* Usinger (Gerridae; from Usinger 1956b).

Figure 15.67. Dorsal view of *Rheumatobates rileyi* Bergroth (Gerridae; from Brooks and Kelton 1967).

Figure 15.68. Dorsal view of *Trepobates becki* Drake and Harris (Gerridae; from Usinger 1956b).

Figure 15.69. Dorsal view of *Ochterus barberi* Schell (Ochteridae; from Usinger 1956b).

Figure 15.70. Antenna of *Limnoporus* sp. (Gerridae).

Figure 15.71. Antenna of *Aquarius* sp. (Gerridae).

Figure 15.72. Antenna of *Trepobates* sp. (Gerridae).

Figure 15.73. Antenna of *Metrobates* sp. (Gerridae).

Figure 15.74. Antenna of *Rheumatobates* sp. (Gerridae).

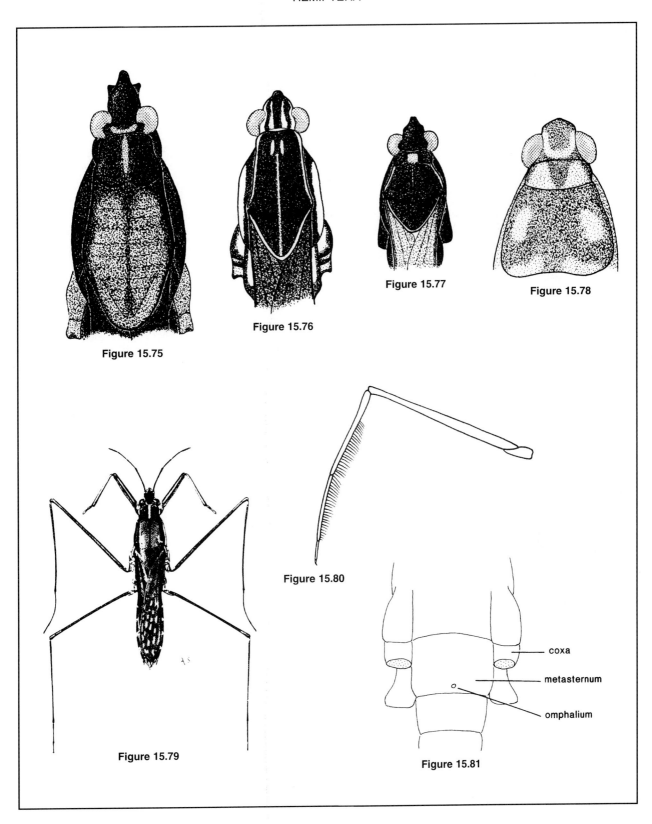

Figure 15.75. Dorsum of head and thorax of *Aquarius* sp. (Gerridae).

Figure 15.76. Dorsum of head and thorax of *Limnogonus* sp. (Gerridae).

Figure 15.77. Dorsum of head and thorax of *Neogerris* sp. (Gerridae).

Figure 15.78. Dorsum of head and thorax of *Trepobates* sp. (Gerridae).

Figure 15.79. Dorsal view of *Aquarius remigis* (Say) (Gerridae; from Usinger 1956b).

Figure 15.80. Hind leg of *Halobates* sp. (Gerridae).

Figure 15.81. Ventral view of thorax of Gerridae.

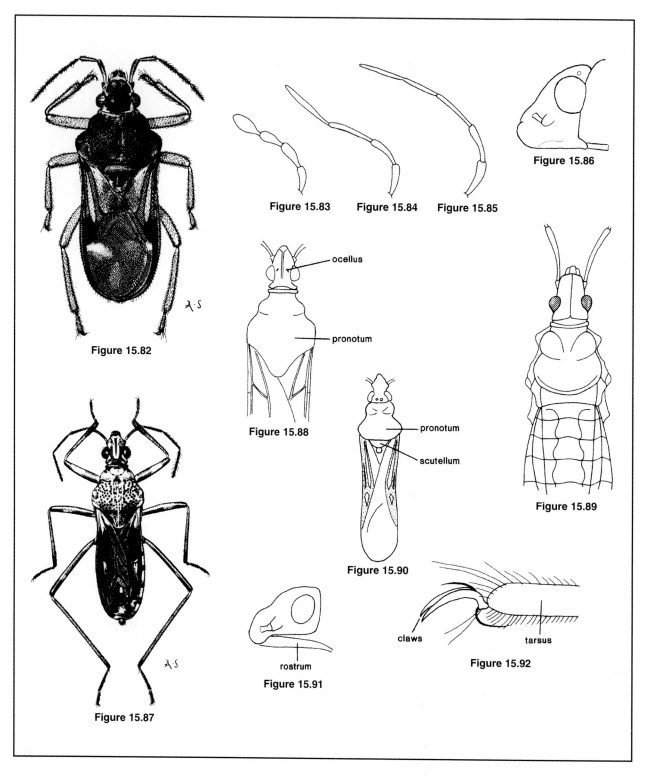

Figure 15.82. Dorsal view of *Hebrus sobrinus* Uhler (Hebridae; from Usinger 1956b).
Figure 15.83. Antenna of *Merragata* sp. (Hebridae).
Figure 15.84. Antenna of *Lipogomphus* sp. (Hebridae).
Figure 15.85. Antenna of *Hebrus* sp. (Hebridae).
Figure 15.86. Lateral view of head of Hebridae.
Figure 15.87. Dorsal view of *Macrovelia hornii* Uhler (Macroveliidae; from Usinger 1956b).
Figure 15.88. Dorsum of head and pronotum of *Macrovelia* sp. (Macroveliidae).
Figure 15.89. Dorsal view of *Orovelia pege* Drake and Chapman (Macroveliidae).
Figure 15.90. Dorsal view of *Mesovelia* sp. (Mesoveliidae).
Figure 15.91. Lateral view of head of *Mesovelia* sp. (Mesoveliidae).
Figure 15.92. Tarsus and claws of Macroveliidae.

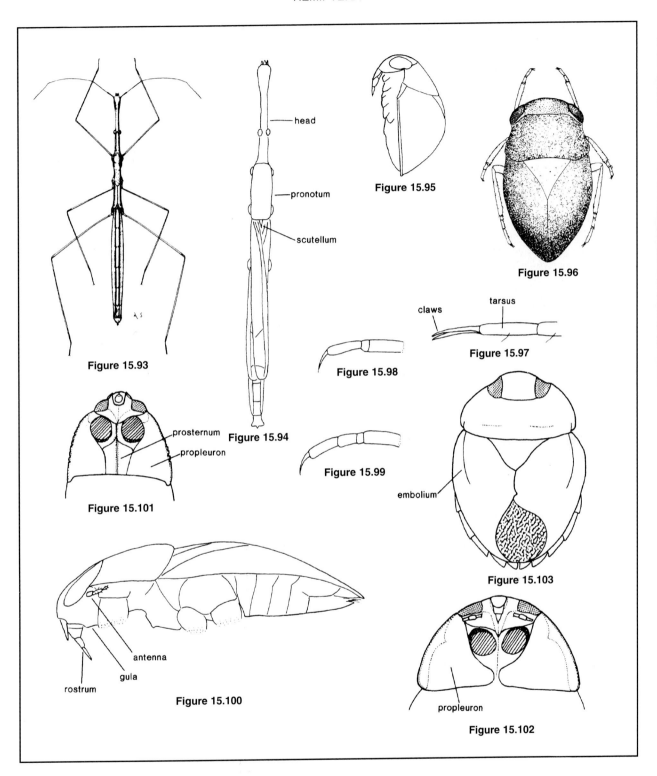

Figure 15.93. Dorsal view of *Hydrometra australis* Say (Hydrometridae; from Usinger 1956b).

Figure 15.94. Dorsal view of *Hydrometra* (Hydrometridae).

Figure 15.95. Lateral view of adult Pleidae.

Figure 15.96. Dorsal view of *Neoplea striola* (Fieber) (Pleidae; from Brooks and Kelton 1967).

Figure 15.97. Hind tarsus of Pleidae.

Figure 15.98. Foretarsus of *Paraplea* sp. (Pleidae).

Figure 15.99. Foretarsus of *Neoplea* sp. (Pleidae).

Figure 15.100. Lateral view of adult Naucoridae.

Figure 15.101. Ventral view of head and thorax of *Cryphocricos* sp. (Naucoridae).

Figure 15.102. Ventral view of head and thorax of *Ambrysus* sp. (Naucoridae).

Figure 15.103. Dorsal view of *Limnocoris* sp. (Naucoridae).

1'. Hemelytral commissure without a definite hair-lined pit at anterior end (Figs. 15.112, 15.114); antennae 4-segmented 2
2(1'). Eyes not holoptic (touching), separated dorsally (Fig. 15.114); intermediate femur with anteapical (before apex) pointed protuberance; anterolateral margins of prothorax not foveate (with a deep impression) .... *Notonecta* Linnaeus
2'. Eyes holoptic, contiguous dorsally (Fig. 15.112); intermediate femur without anteapical pointed protuberance; anterolateral margins of prothorax foveate .... *Martarega* White

Pleidae

1. Anterior tarsi each with 2 segments (Fig. 15.98); abdominal carinae (keels) on ventrites 2-6 .... *Paraplea* Esaki and China
1'. Anterior tarsi each with 3 segments (Fig. 15.99); abdominal carinae on ventrites 2-5 ....*Neoplea* Esaki and China

Saldidae

1. Hemelytra with long embolar fracture reaching forward at least to level of posterior end of claval suture (Fig. 15.121)... (Chiloxanthinae)................ 2
1'. Hemelytra with short embolar fracture, not reaching forward more than half-way from beginning of fracture on costal margin to level of posterior end of claval suture (Figs. 15.2, 15.128) ....(Saldinae) 3
2(1). Sublateral cell of membrane short, only half as long as lateral cell (Fig. 15.127) .... *Chiloxanthus* Reuter
2'. Sublateral cell of membrane subequal in length to lateral cell (Figs. 15.121, 15.133) .... *Pentacora* Reuter
3(1'). Pronotum with 2 prominent conical tubercles on anterior lobe (Fig. 15.126) .... *Saldoida* Osborn
3'. Pronotum without prominent tubercles on anterior lobe (Fig. 15.2) 4
4(3'). Lateral margins of pronotum concave, humeral (posterolateral) angles produced (Fig. 15.132) .... *Lampracanthia* Reuter
4'. Lateral margins of pronotum straight or convex, humeral angles rounded (Fig. 15.131) 5
5(4'). Hypocostal ridge simple, secondary hypocostal ridge absent (Fig. 15.137) 6
5'. Hypocostal ridge complex, secondary hypocostal ridge present (Figs. 15.134-15.136) 7
6(5). Innermost cell of membrane produced anteriorly one-half its length beyond base of 2nd cell (Fig. 15.120); first and 2nd antennal segments of male flattened, oval in cross section, the flattened sides glabrous (shiny) ....*Calacanthia* Reuter
6'. Innermost cell of membrane produced anteriorly only slightly, not more than one-third its length beyond base of 2nd cell (Fig. 15.118); first and 2nd antennal segments of male not flattened, round in cross section, evenly pubescent (hairy) or pilose over entire surface... *Rupisalda* Polhemus
7(5'). Second segment of tarsi usually nearly half again as long as 3rd (Fig. 15.129); innermost cell of membrane short, usually reaching only four-fifths the distance to apex of adjacent cell; outer corium with large, pale spots; clavus with yellow spot on each side in velvety black area (Fig. 15.131); secondary hypocostal ridge present, oblique, meets or projects to costal margin .... *Teloleuca* Reuter
7'. Second segment of tarsi subequal or slightly longer than 3rd segment (Fig. 15.130); innermost cell of membrane long, usually reaching almost to apex of adjacent cell (Figs. 15.119, 15.128); outer corium with or without pale spots; clavus with or without yellow spot on each side in velvety black area; secondary hypocostal ridge present, may or may not meet or project to costal margin 8
8(7'). Males longer than 5.5 mm, females longer than 6 mm or, *if* shorter, then innermost cell of membrane produced two-fifths to one-half its length anteriorly beyond base of 2nd (Fig. 15.119), and dorsal surface unicolorous or with a few small, pale spots on corium and membrane (Fig. 15.125); secondary hypocostal ridge does not meet or project to costal margin (Fig. 15.134) .... *Salda* Fabricius
8'. Males shorter than 5.5 mm, females shorter than 6 mm; innermost cell of membrane produced anteriorly only slightly beyond base of 2nd (Fig. 15.128) *or*, if inner cell is produced more strongly anteriorly, then dorsum has more or less extensive pale markings; secondary hypocostal ridge meets or projects to costal margin (Figs. 15.35-15.36) 9
9(8'). Antennae relatively thick, the 3rd and 4th segments thicker than the distal end of the 2nd segment (Fig. 15.123); secondary hypocostal ridge (hrs) meets or projects to costal margin at approximately three-fourths distance from base to embolar fracture; Strigil (file) present on hrs (Fig. 15.136), plectrum (rasp) on distal portion of hind femur .... *Ioscytus* Reuter

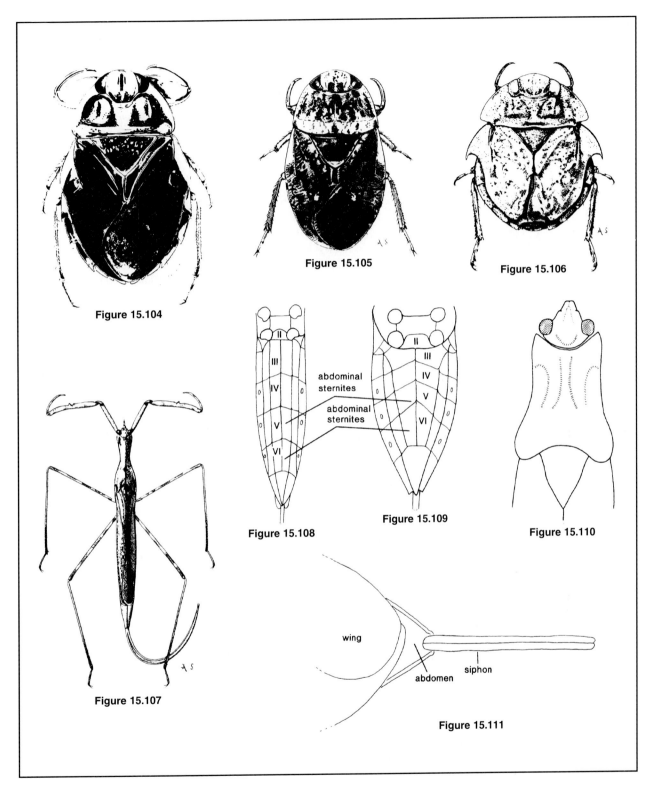

Figure 15.104. Dorsal view of *Ambrysus mormon* Montandon (Naucoridae; from Menke *et al.* 1979).

Figure 15.105. Dorsal view of *Pelocoris shoshone* LaRivers (Naucoridae; from Usinger 1956).

Figure 15.106. Dorsal view of *Limnocoris moapensis* (LaRivers) (Naucoridae; from Usinger 1956b).

Figure 15.107. Dorsal view of *Ranatra brevicollis* Montandon (Nepidae; from Usinger 1956b).

Figure 15.108. Ventral view of abdomen of *Curicta* sp. (Nepidae).

Figure 15.109. Ventral view of abdomen of *Nepa* sp. (Nepidae).

Figure 15.110. Dorsal view of head and pronotum of *Curicta* sp. (Nepidae).

Figure 15.111. Dorsal view of breathing tube of *Nepa* sp. (Nepidae).

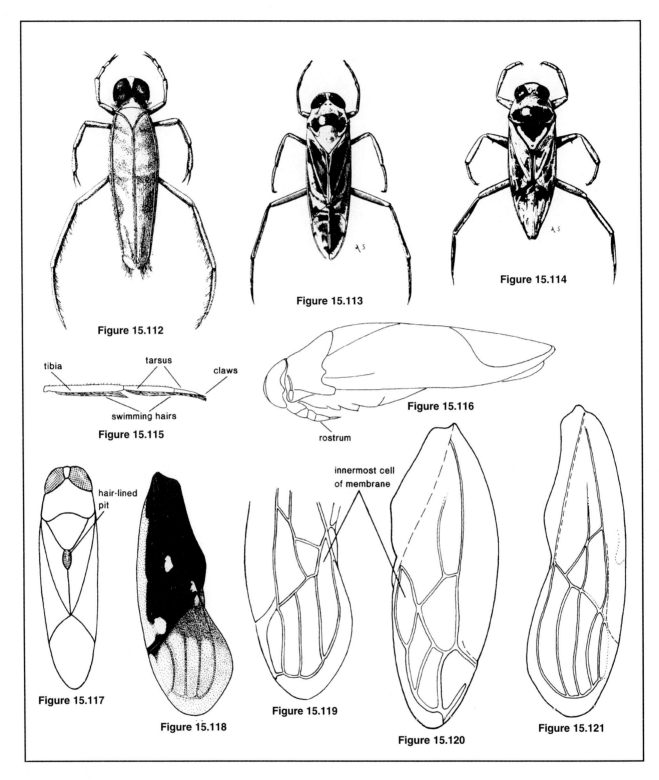

Figure 15.112. Dorsal view of *Martarega mexicana* Truxal (Notonectidae; from Menke *et al.* 1979).

Figure 15.113. Dorsal view of *Buenoa scimitra* Bare (Notonectidae; from Usinger 1956b).

Figure 15.114. Dorsal view of *Notonecta unifasciata* Guerin-Meneville (Notonectidae; from Usinger 1956b).

Figure 15.115. Hind leg of Notonectidae.

Figure 15.116. Lateral view of adult Notonectidae.

Figure 15.117. Dorsal view of *Buenoa* sp. (Notonectidae).

Figure 15.118. Left hemelytron of *Rupisalda* sp. (Saldidae).

Figure 15.119. Left hemelytron of *Salda* sp. (Saldidae).

Figure 15.120. Right hemelytron of *Calacanthia* sp. (Saldidae).

Figure 15.121. Right hemelytron of *Pentacora* sp. (Saldidae).

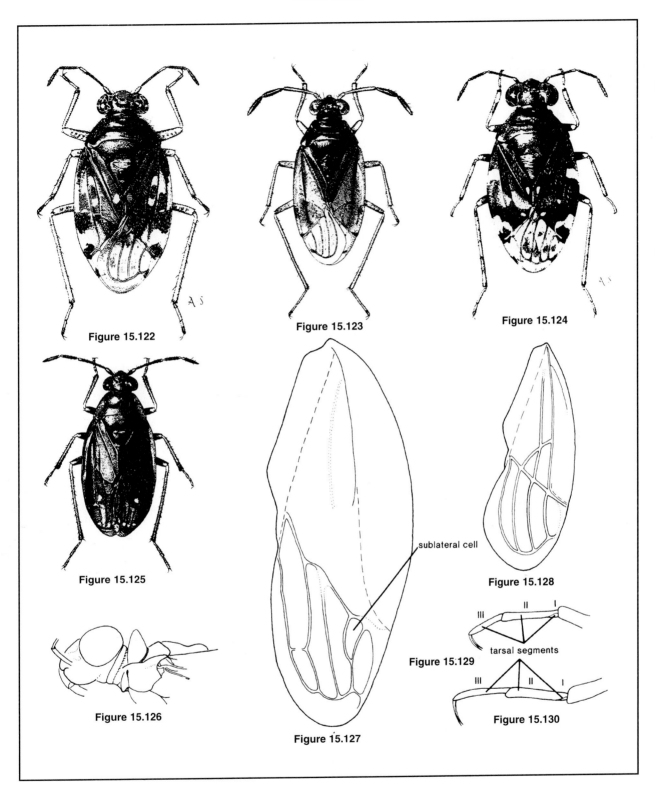

Figure 15.122. Dorsal view of *Saldula pexa* Drake (Saldidae; from Usinger 1956b).

Figure 15.123. Dorsal view of *Ioscytus politus* (Uhler) (Saldidae; from Usinger 1956b).

Figure 15.124. Dorsal view of *Micracanthia quadrimaculata* (Champion) (Saldidae; from Usinger 1956b).

Figure 15.125. Dorsal view of *Salda buenoi* (McDunnough) (Saldidae; from Usinger 1956b).

Figure 15.126. Lateral view of head and thorax of *Saldoida* sp. (Saldidae).

Figure 15.127. Right hemelytron of *Chiloxanthus* sp. (Saldidae).

Figure 15.128. Right hemelytron of *Saldula* sp. (Saldidae).

Figure 15.129. Tarsus of *Teloleuca* sp. (Saldidae).

Figure 15.130. Tarsus of *Salda* sp. (Saldidae).

9'.	Antennae relatively slender, the third and fourth segments not thicker than the distal end of the second segment (Figs. 15.122, 15.124); secondary hypocostal ridge meets or projects to costal margin at three-fifths or less distance from base to embolar fracture (Fig. 15.135); strigil and plectrum absent	10
10(9').	Veins of corium more or less distinct (Fig. 15.122); body usually more than 3.5 mm long; *if* less, then anterior margin of pronotum wider than collar	***Saldula*** Van Duzee
10'.	Veins of corium obsolete (Fig. 15.124); body usually less than 3.5 mm long; anterior margin of pronotum usually narrower than collar	***Micracanthia*** Reuter

Veliidae

1.	Middle tarsi deeply cleft, with leaflike claws and plumose (plumelike) hairs arising from base of cleft (Fig. 15.146)	(Rhagoveliinae) ...2
1'.	Middle tarsi not deeply cleft and without plumose hairs arising from base of cleft	3
2(1).	Hind tarsi 2-segmented, the basal segment very short (Fig. 15.141); apterous; marine	***Trochopus*** Carpenter
2'.	Hind tarsi 3-segmented, the basal segment very short (Fig. 15.144); apterous or macropterous; riffles of streams and rivers or (rarely) lakes	***Rhagovelia*** Mayr
3(1').	Tarsal formula 1:2:2 (Fig. 15.142)	(Microveliinae)...4
3'.	Tarsal formula 3:3:3	(Veliinae)...5
4(3).	Middle tarsi with 4 leaflike blades arising from cleft (Fig. 15.143)	***Husseyella*** Herring
4'.	Middle tarsi with narrow claws arising from cleft (Fig. 15.142)	***Microvelia*** Westwood
5(3').	Body broad (Fig. 15.147); metasternum with lateral tubercles meeting mesoacetabulae (Fig. 15.149)	***Platyvelia*** Polhemus and Polhemus
5'.	Body narrow (Fig. 15.148); metasternum with lateral tubercles meeting mesocoxae (Fig. 15.150)	***Steinovelia*** Polhemus and Polhemus

ADDITIONAL TAXONOMIC REFERENCES

General
VanDuzee (1917); Hungerford (1920, 1958, 1959); Parshley (1925); Blatchley (1926); China (1955); Usinger (1956a); China and Miller (1959); Lawson (1959); Polhemus (1966, 1973); Brooks and Kelton (1967); Cobben (1968); Ruhoff (1968); Jaczewski and Kostrowicki (1969); DeCoursey (1971); Herring and Ashlock (1971); Miller (1971); Bobb (1974); Stys and Kerzhner (1975); Cheng (1976); Slater and Baranowski (1978); Andersen (1981a, 1982); Henry and Froeschner (1988); Stys and Jansson (1988); Yonke (1991); Spence and Andersen (1994).

Regional faunas
Arizona: Polhemus and Polhemus (1976); Blinn and Sanderson (1989).
Arkansas: Kittle (1980); Farris and Harp (1982).
British Columbia: Scudder (1977).
California: Usinger (1956a); Menke *et al.* (1979).
Central Canada: Strickland (1953); Brooks and Kelton (1967).
Connecticut: Britton (1923).
Florida: Herring (1950, 1951a); Chapman (1958).
Idaho: Harris and Shull (1944).
Illinois: Lauck (1959).
Kansas: Slater (1981).
Louisiana: Ellis (1952); Gonsoulin (1973a,b,c, 1974, 1975).
Minnesota: Bennett and Cook (1981).
Mississippi: Wilson (1958); Lago and Testa (1989).
Missouri: Froeschner (1949, 1962).
Montana: Roemhild (1976).
New Jersey: Chapman (1959).
North Carolina: Sanderson (1982a).
Oklahoma: Schaefer (1966); Schaefer and Drew (1964, 1968).
Quebec: Chagnon and Fournier (1948).
Rhode Island: Reichart (1976, 1977, 1978).
South Carolina: Sanderson (1982a).
South Dakota: Harris (1937).
Texas: Millspaugh (1939).
Virginia: Bobb (1974).
Wisconsin: Hilsenhoff (1981, 1984, 1986).

Taxonomic treatments at the family and generic levels
Belostomatidae: Menke (1958, 1960, 1963); Lauck and Menke (1961); Lauck (1963, 1964).
Corixidae: Hungerford (1948); Sailer (1948); Lansbury (1960); Hilsenhoff (1970); Applegate (1973); Scudder (1976); Nieser (1977); Jansson (1978, 1981); Dunn (1979); Jansson and Polhemus (1987).
Dipsocoridae: McAtee and Malloch (1925); Stys (1970).
Gelastocoridae: Martin (1928); Todd (1955, 1961); Polhemus and Lindskog (1994).
Gerridae: Anderson (1932); Drake and Harris (1932, 1934); Deay and Gould (1936); Kuitert (1942); Hussey and Herring (1949); Hungerford (1954); Hungerford and Matsuda (1960); Matsuda (1960); Herring (1961); Cheng and Fernando (1970); Scudder (1971b); Calabrese (1974); Kittle (1977b,c, 1982); Stonedahl and Lattin (1982); Cheng (1985); Spangler, Froeschner and Polhemus (1985); Polhemus and Spangler (1989); Andersen (1990, 1991); Andersen and Spence (1992); Polhemus and Polhemus (1993b, 1995).
Hebridae: Porter (1950); Drake and Chapman (1954, 1958a); Polhemus and Chapman (1966); Andersen (1981b); Polhemus and McKinnon (1983).
Hydrometridae: Torre-Bueno (1926); Hungerford and Evans (1934); Drake and Lauck (1959).
Leptopodidae: Schuh, Galil and Polhemus (1987).
Macroveliidae: McKinstry (1942).
Mesoveliidae: Jaczewski (1930); Andersen and Polhemus (1980).
Naucoridae: Usinger (1941); LaRivers (1949, 1951, 1971, 1974, 1976); Polhemus and Polhemus (1994).
Nepidae: Hungerford (1922c); Polhemus (1976b); Sites and Polhemus (1994).
Notonectidae: Hungerford (1933); Hutchinson (1945); Truxal (1949, 1953); Scudder (1965); Reichart (1971); Voigt and Garcia (1976); Zalom (1977).
Ochteridae: Schell (1943); Polhemus and Polhemus (1976).
Pleidae: Drake and Chapman (1953); Drake and Maldonado (1956).
Saldidae: Hodgden (1949a,b); Drake (1950, 1952); Drake and Hoberlandt (1950); Drake and Hottes (1950); Drake and Chapman (1958b); Chapman (1962); Polhemus (1967, 1976c, 1985, 1994); Schuh (1967); McKinnon and Polhemus (1986); Schuh, Galil and Polhemus (1987); Lindskog and Polhemus (1992).
Schizopteridae: Emsley (1969).
Veliidae: Drake and Hussey (1955); Bacon (1956); Polhemus (1974, 1976a); Smith and Polhemus (1978); Smith (1980); Polhemus and Polhemus (1993a).

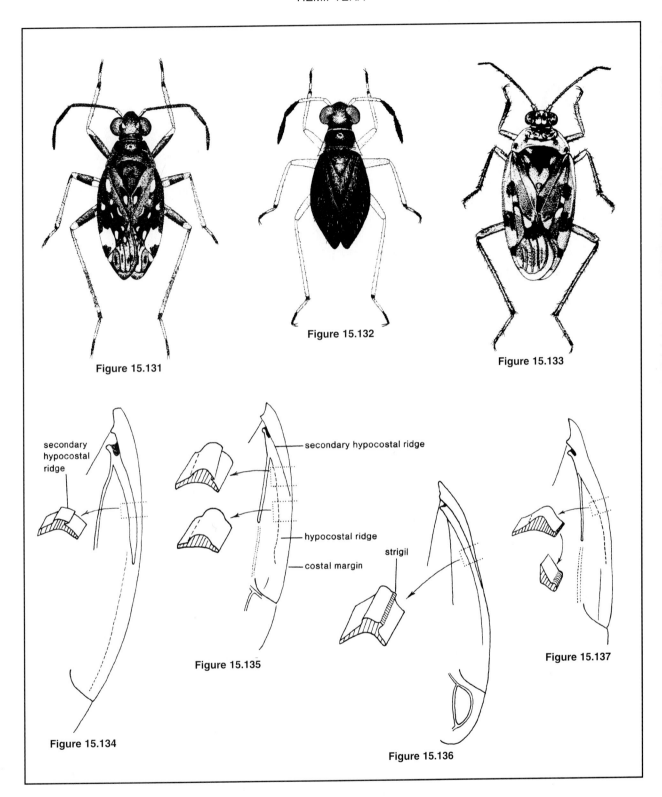

Figure 15.131. Dorsal view of *Teloleuca bifasciata* Thomson (Saldidae; from Brooks and Kelton 1967).

Figure 15.132. Dorsal view of *Lampracanthia crassicornis* (Uhler) (Saldidae; from Brooks and Kelton 1956b).

Figure 15.133. Dorsal view of *Pentacora signoreti* (Guerin-Meneville) (Saldidae; from Usinger 1956b).

Figure 15.134. Ventral view of hemelytron of *Salda* sp. (Saldidae).

Figure 15.135. Ventral view of hemelytron of *Saldula* sp. (Saldidae).

Figure 15.136. Ventral view of hemelytron of *Ioscytus* sp. (Saldidae).

Figure 15.137. Ventral view of hemelytron of *Rupisalda* sp. (Saldidae).

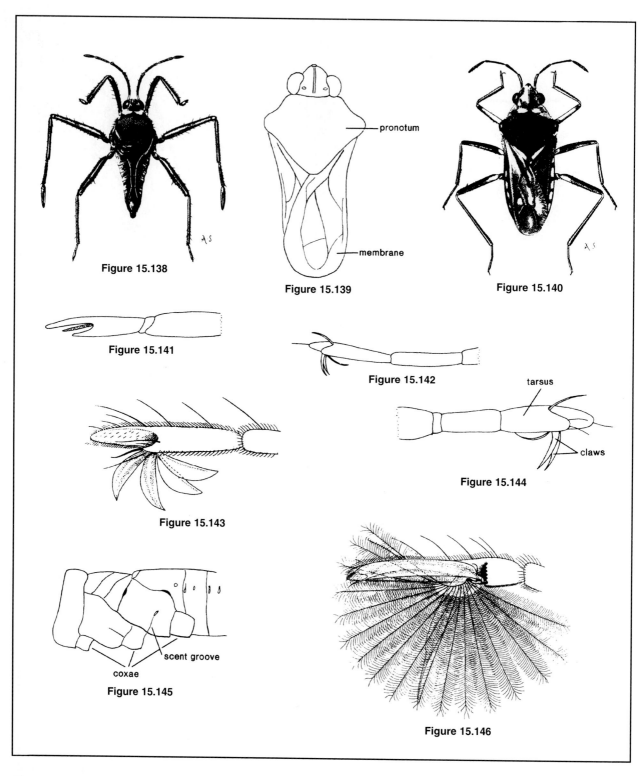

Figure 15.138. Dorsal view of *Rhagovelia distincta* Champion (Veliidae; from Usinger 1956b).

Figure 15.139. Dorsal view of *Microvelia* sp. (Veliidae).

Figure 15.140. Dorsal view of *Microvelia beameri* McKinstry (Veliidae; from Usinger 1956b).

Figure 15.141. Dorsal view of hind tarsus of *Trochopus* sp. (Veliidae).

Figure 15.142. Lateral view of middle tarsus of *Microvelia* sp. (Veliidae).

Figure 15.143. Lateral view of middle tarsus of *Husseyella* sp. (Veliidae).

Figure 15.144. Lateral view of hind tarsus of *Rhagovelia* sp. (Veliidae).

Figure 15.145. Lateral view of thorax of *Microvelia* sp. (Veliidae).

Figure 15.146. Middle tarsus of *Trochopus* sp. (Veliidae) showing swimming plume.

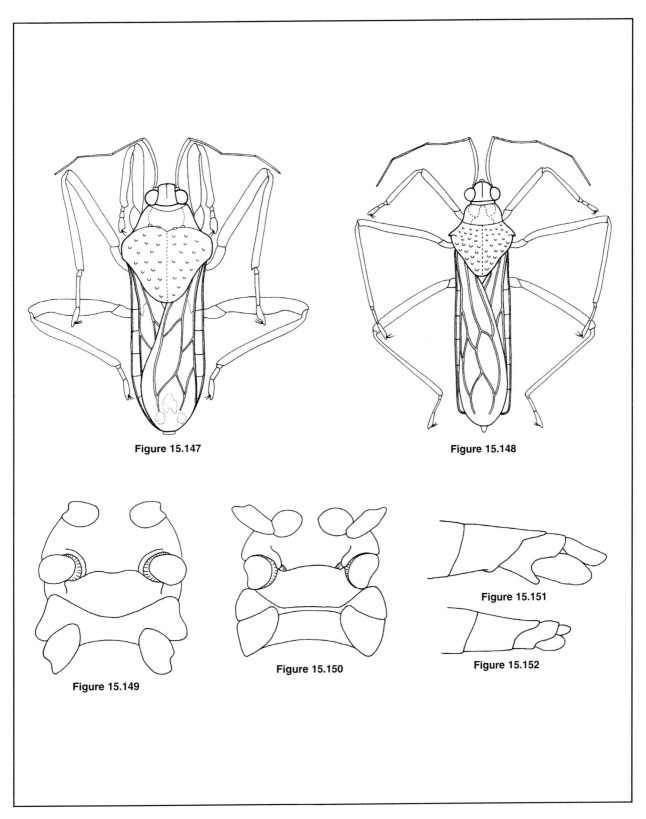

Figure 15.147. *Platyvelia brachialis* (Stål), dorsal view (Veliidae).

Figure 15.148. *Steinovelia* sp., dorsal view (Veliidae).

Figure 15.149. Ventral view of thorax of *Steinovelia* sp. (Veliidae).

Figure 15.150. Ventral view of thorax of *Platyvelia* sp. (Veliidae).

Figure 15.151. Lateral view of abdominal terminalia of *Aquarius* sp. (Gerridae).

Figure 15.152. Lateral view of abdominal terminalia of *Gerris* sp. (Gerridae).

Table 15A. Summary of ecological and distributional data for *Hemiptera (aquatic and semiaquatic bugs)*. (For definition of terms see Tables 6A-6C; table prepared by J. T. Polhemus, K. W. Cummins and R. W. Merritt.)

Taxa (number of species in parentheses)	Habitat	Habit	Trophic Relationships	North American Distribution	Ecological References*
Hemiptera					
Hydrometridae (9) (water measurers)					72, 331, 348, 1867, 1892, 2112, 2222, 2247, 3076, 3534, 4075, 4306
Hydrometra (9)	Lentic—limnetic, surface (in littoral zone), lotic—margins	"Skaters" (slow walkers on surface film)	Predators (piercers) (adult arthropods, dead or live, especially mosquito larvae and pupae, and ostracods)	Widespread	913, 1750, 1857, 2112, 2443, 2654, 2770, 3033, 2234, 3287, 3780
Macroveliidae (2)	Lotic—margins (semiaquatic)	Climbers—sprawlers (rarely in or on the water)	Predators (piercers)		72, 331, 1867, 2112, 2625, 3076, 4075
Macrovelia (1)	Lotic—erosional and depositional margins (in protected areas)	Climbers—sprawlers	Predators (piercers)	West	2625, 2654, 4075
Oravelia (1)	Lotic—margins	Climbers—sprawlers	Predators (piercers)	West	2654
Veliidae (35) (broad-shouldered water striders)	Generally lentic—limnetic surface, lotic surface	Skaters	Generally predators (piercers) (live and dead arthropods)		72, 76, 331, 348, 1689, 1750, 1857, 1867, 1892, 2112, 2222, 2247, 2654, 3076, 3287, 3534, 4075, 439, 4306
Husseyella (1)	Lotic and lentic—surface (brackish water)	Skaters	Predators (piercers)	Southern Florida	76, 2654
Microvelia (20)	Lentic limnetic and lotic—depositional surface	Skaters	Predators (piercers)	Widespread	913, 934, 1306, 2654, 2770, 3276, 3278, 4021, 1678, 1679, 4023, 4075, 1680, 2725, 2838, 4041
Platyvelia (3)	Lentic—limnetic and lotic—depositional surface	Skaters	Predators (piercers)	South, Southwest, East, Midwest	913, 2654, 4075
Rhagovelia (9)	Lotic—erosional surface	Skaters	Predators (piercers) (scavengers)	Widespread	616, 910, 913, 2654, 2770, 3033, 3278, 3389, 4465
Steinovelia (1)	Lentic—limnetic and lotic—depositional surface	Skaters	Predators (piercers)	South, East, Midwest	913, 1798, 2654, 4075
Trochopus (1)	Lentic—limnetic (saltwater bays)	Skaters	Predators (piercers)	Southern Florida coast	76, 2654
Ceratocombidae (4)					
Ceratocombus (3)	Lentic			East, California, New Mexico	1081
Leptonannus (1)				New Mexico	
Dipsocoridae (4) (= Cryptostemmatidae)	Generally lotic—erosional margins	Generally burrowers		West, Southeast	3534, 4075
Cryptostemma (3)	Lotic—erosional margins (under stones at stream edge)	Burrowers (active)		California, Georgia	4075, 4077
Schizopteridae (3)					1081
Corixidea (1)				Tennessee, Louisiana	1081
Nannocoris (1)				Florida	
Schizoptera (1)				Florida	

*Emphasis on trophic relationships.

Table 15A. Continued

Taxa (number of species in parentheses)	Habitat	Habit	Trophic Relationships	North American Distribution	Ecological References*
Gerridae (44) (water striders)	Generally lentic—limnetic surface, lotic surface	Skaters	Generally predators (piercers) (scavengers)		72, 76, 331, 348, 1689, 1750, 1857, 1867, 1892, 2112, 2222, 2247, 2654, 3076, 3269, 3365, 3389, 3534, 3771, 3772, 4075, 4306
Gerrinae (20) *Aquarius* (4)	Lotic-depositional surface, lentic—limnetic and littoral surface	Skaters	Predators (piercers) (scavengers)	Widespread	467, 515, 516, 1338, 1959, 1960, 2179, 2654, 2770, 2811, 3033, 3278, 2724, 4075, 4113, 4407, 2605, 3007, 4512
Gerris (10)	Lotic-depositional surface, lentic—limnetic and littoral surface	Skaters	Predators (piercers) (scavengers)	Widespread	515, 516, 520, 1959, 1960, 2654, 2770, 3033, 3287, 3766, 4075, 4113, 2035, 2161, 2945, 3768, 3770, 4126, 612, 2654
Limnogonus (1)	Lentic—limnetic	Skaters	Predators (piercers)	South (primarily tropical)	612, 1892, 2654, 58
Limnoporus (4)	Lentic—limnetic	Skaters	Predators (piercers)	Widespread	346, 515, 516, 913, 1959, 1960, 2770, 3033, 4113, 4126
Neogerris (1)	Lentic—limnetic	Skaters	Predators (piercers)	South, East	
Halobatinae (2)					
Halobates (2)	Marine (open ocean and protected reefs)	Skaters	Predators (piercers) (scavengers)	East and West Coasts	76, 1691, 2654, 4075, 1677
Rhagadotarsinae (8)					
Rheumatobates (8)	Lotic and lentic—surfaces	Skaters	Predators (piercers)	East, Central, South	76, 2654, 3033, 3657
Trepobatinae (14)					
Metrobates (6)	Lotic—erosional surface (usually large rivers)	Skaters	Predators (piercers)	Widespread (especially East)	2654, 2770, 3033, 4075
Trepobates (8)	Lentic—limnetic surface, lotic depositional surface	Skaters	Predators (piercers)	East, West, South	2654, 3033, 4075
Belostomatidae (19) (giant water bugs)	Generally lotic—depositional—vascular hydrophytes and detritus, lentic—littoral	Climbers—swimmers	Predators (piercers)		331, 348, 786, 1689, 1750, 1857, 1867, 1892, 2112, 2222, 2654, 2658, 2770, 3076, 3269, 3534, 977, 3897, 4075, 4306, 3525, 3959, 4121, 4129
Abedus (6)	Lotic—depositional	Climbers—swimmers	Predators (piercers)	Southwest, Southeast	1881, 2654, 3715, 3717, 4075, 4251
Belostoma (8)	Lotic—depositional, lentic—littoral	Climbers—swimmers	Predators (piercers)	Widespread	782, 933, 1444, 2160, 2654, 3033, 3716, 3827, 693, 4075, 4251, 766, 2062, 2080, 3537, 496
Lethocerus (5)	Lotic—depositional, lentic—littoral	Climbers—swimmers	Predators (piercers)	Widespread	782, 913, 1892, 2654, 1459, 3246, 3389, 4075, 1758, 3719
Nepidae (13) (water scorpions)	Generally lotic—depositional—vascular hydrophytes and detritus, lentic—littoral	Climbers (poor swimmers)	Predators (piercers)		331, 348, 1689, 1857, 1860, 1867, 1892, 3076, 2112, 2222, 2247, 2654, 3269, 3534, 4075
Curicta (2)	Lentic—littoral	Climbers (poor swimmers)	Predators (piercers)	Southwest, South	2654, 3076, 4415

*Emphasis on trophic relationships.

Table 15A. Continued

Taxa (number of species in parentheses)	Habitat	Habit	Trophic Relationships	North American Distribution	Ecological References*
Nepa (1)	Lentic—vascular hydrophytes	Climbers (poor swimmers)	Predators (piercers)	Central, East	2654, 4075
Ranatra (10)	Lotic—depositional vascular hydrophytes, lentic—littoral	Climbers (poor swimmers)	Predators (piercers)	Widespread	337, 866, 869, 913, 1348, 1750, 2654, 2770, 2982, 3033, 3242, 3389, 4020, 666, 1273, 2724, 4306, 3248, 3487, 3957, 338
Pleidae (5) (pigmy back swimmers)					2222, 2247, 3534
Neoplea (3)	Lentic—vascular hydrophytes (especially dense stands)	Swimmers—climbers	Predators (piercers) (especially microcrustacea)	Widespread	178, 212, 331, 348, 913, 1069, 1410, 1412, 1572, 1857, 1867, 2112, 2200, 3076, 683, 3278, 3325, 4075, 3935
Paraplea (2)	Lentic—vascular hydrophytes (especially dense stands)	Swimmers—climbers	Predators (piercers) (especially microcrustacea)	Southeast	961
Naucoridae (22) (creeping water bugs)	Generally lotic—erosional, lentic—littoral	Generally clingers, swimmers	Predators (piercers)		331, 499, 1689, 1857, 1867, 1896, 2112, 2247, 2654, 3076, 3256, 3534, 4075, 4306
Ambrysus (15)	Lotic and lentic—erosional (sediments and vascular hydrophytes)	Clingers—swimmers	Predators (piercers)	West, Southwest	174, 467, 2654, 4078, 3879, 4075, 4465, 3678, 3681
Cryphocricos (1)	Lotic—erosional (sediments)	Clingers	Predators (piercers)	Texas	3029, 3030, 3872, 4075, 3679, 3681, 3879, 4465
Limnocoris (2) (= *Usingerina*)	Lotic—erosional (in sediments, warm springs)	Clingers	Predators (piercers)	Texas, Nevada	2654, 3029, 3872, 4075, 3681, 3879, 4465
Pelocoris (4)	Lentic—vascular hydrophytes	Climbers—swimmers	Predators (piercers) (Diptera, Belostomatidae)	East, Central, Southwest	348, 913, 1572, 1862, 2250, 2473, 2654, 2770, 3033, 3389, 4019, 4075
Corixidae (129) (water boatmen)	Generally lentic—vascular hydrophytes, lotic—depositional (vascular hydrophytes)	Generally swimmers	Generally piercers—herbivores and some predators (engulfers and piercers) or scrapers		331, 348, 467, 499, 755, 1409, 1484, 1689, 1750, 1854, 1857, 1864, 1867, 1892, 1919, 1964, 1965, 2112, 2222, 2245, 2247, 2385, 2389, 2391, 2654, 2770, 3003, 3076, 3103, 3220, 3269, 3365, 3389, 3534, 3614, 3839, 3897, 3912, 4022, 4075, 4306, 1643, 4594
Arctocorisa (5)	Lentic			North	3002, 3003, 3005, 3006, 3007, 3008, 3827, 3247
Callicorixa (6)	Lentic		Predators (piercers)	North	2654, 3002, 3003, 3005, 3006, 3007, 3008, 3528, 3247
Cenocorixa (8)	Lentic			Northwest	1969, 3320, 3321
Centrocorisa (1)			Predators (piercers)	Florida, Texas	
Corisella (4)	Lentic—littoral (including brackish water and saline lakes)		Predators (piercers)	West, North	2654, 2770, 4075, 4251
Cymatia (1)				North	1973, 2246, 3275
Dasycorixa (3)				North	
Glaenocorisa (1)				North	1099, 1673, 2989

*Emphasis on trophic relationships.

Table 15A. Continued

Taxa (number of species in parentheses)	Habitat	Habit	Trophic Relationships	North American Distribution	Ecological References*
Graptocorixa (6)			Predators (piercers)	Southwest, West	1278, 2654
Hesperocorixa (19)	Lotic—depositional	Swimmers, climbers	Piercers—herbivores	Widespread	347, 913, 2654, 2770
Krizousacorixa (3) (=*Ahautlea*)				Extreme Southwest	2654, 3102
Morphocorixa (3) (= *Pseudocorixa*)			Predators (piercers)—herbivores	Southwest	
Neocorixa (1)				Southwest	
Palmacorixa (4)				Widespread (except extreme North)	2770
Ramphocorixa (2)	Lentic—littoral		Predators (engulfers)	Widespread (except West)	1484, 2654, 2770
Sigara (50)	Lotic—depositional	Swimmers, climbers	Piercers—herbivores, collectors—gatherers	Widespread	913, 1855, 1857, 1864, 2168, 2654, 2770, 3278, 2989, 3920
Tenagobia (1)	Lotic—depositional			Southwest, Florida (primarily tropical)	
Trichocorixa (11)	Lentic—littoral (freshwater, some brackish or saltwater including intertidal pools and offshore sea)	Swimmers, climbers	Predators (piercers) (especially chironomid larvae, Oligochaeta; some collectors—gatherers as early instars)	Widespread	532, 871, 872, 1502, 1892, 2070, 2654, 2770, 3033, 3534, 4075, 4580, 2989, 3278
Notonectidae (32) (back swimmers)					212, 331, 348, 1689, 1757, 1857, 1858, 1863, 1892, 2112, 2222, 2247, 2294, 2391, 3076, 3269, 3325, 3534, 3897, 4075, 3051, 3958, 4306
Buenoa (14)	Lentic—littoral, lotic—depositional	Swimmers (rest submerged in hydrostatic balance)	Predators (piercers)	Widespread	913, 1413, 1892, 2654, 2770, 3839, 4027, 4075, 693, 695, 4251, 4623
Martarega (1)	Lotic—depositional	Swimmers (rest at surface in open water)	Predators (piercers)	Southwest	2654, 1892
Notonecta (17)	Lentic—littoral, lotic—depositional	Swimmers—climbers (rest submerged or at surface)	Predators (piercers) (including cannibalism)	Widespread	178, 467, 869, 913, 1072, 1108, 1278, 1279, 1392, 1409, 1412, 1750, 2179, 2200, 2633, 2654, 3033, 3325, 3389, 3653, 3827, 3876, 3881, 3960, 4026, 4075, 1680, 3162, 4251, 4623, 620, 683, 619, 693, 695, 757, 831, 928, 1273, 1390, 1643, 1673, 1780, 1973, 2724, 2807, 2808, 2809, 2985, 3654, 3683, 3877, 3878, 3879, 3992, 3880
Mesoveliidae (3) (water treaders)					
Mesovelia (3)	Lentic—vascular hydrophytes (emergent and floating zone including salt marshes)	Skaters—climbers (or sprawlers at water's edge, semiaquatic)	Predators (piercers) (scavengers)	Widespread	72, 331, 348, 913, 1793, 1856, 1857, 1861, 1867, 1892, 2070, 2112, 2222, 2654, 2770, 3033, 3076, 3389, 3692, 3827, 4075, 1679, 4306
Hebridae (15) (velvet water bugs)	Generally lentic—littoral vascular hydrophytes (emergent zone and detritus)	Generally climbers (at shore, semiaquatic)	Generally predators (piercers)		72, 331, 348, 1857, 1867, 2112, 2654, 2658, 3076, 3184, 3534, 4075, 4306

*Emphasis on trophic relationships.

Table 15A. Continued

Taxa (number of species in parentheses)	Habitat	Habit	Trophic Relationships	North American Distribution	Ecological References*
Hebrus (12)	Lentic—littoral—vascular hydrophytes (emergent zone and sediments at water's edge)	Climbers—"burrowers" (under stones at water's edge)	Predators (piercers)	Widespread	1857, 2654, 4075
Merragata (2)	Lentic—littoral (on mats of floating algae)	Skaters—climbers	Predators (piercers)	Widespread	958, 1572, 2654, 3076
Lipogomphus (1)	Lentic—littoral vascular hydrophytes (emergent zone and sediments at water's edge)	Climbers—"burrowers" (under stones at water's edge)	Predators (piercers)	South	
Saldidae (69) (shore bugs)	Generally lentic—vascular hydrophytes (emergent zone), lotic—depositional, beach zone—freshwater	Generally climbers (at shore, semiaquatic)	Predators (piercers) (scavengers, especially Diptera larvae)		148, 331, 348, 1857, 1867, 2654, 3076, 3160, 3161, 3534, 4075
Calacanthia (1)	Lentic—vascular hydrophytes (emergent zone)	Climbers (at shore, semiaquatic)	Predators (piercers)	Alaska, northern Canada	3161
Chiloxanthus (2)	Lentic and lotic—margins (tundra zone)	Climbers (at shore, semiaquatic)	Predators (piercers)	Alaska, northern Canada	3160, 3161
Isocytus (7)	Lentic (including alkaline marshes) and lotic—margins	Climbers (at shore, semiaquatic)	Predators (piercers) (scavengers)	California, Southwest	2654, 3160, 3161
Lampracanthia (1)	Lentic (marshy meadows)	Climbers (semiaquatic)	Predators (piercers)	Arctic (widespread in boreal zone)	2654, 3161
Micracanthia (10)	Lentic (marshy meadows)	Climbers (at shore, semiaquatic)	Predators (piercers)	Widespread	2654, 3161
Pentacora (6)	Beaches—marine and freshwater, lotic—depositional	Climbers (semiaquatic)	Predators (piercers) (scavengers)	Widespread	2654, 3160, 3161
Rupisalda (3)	Lentic and lotic—margins	Clingers (on wet or dry vertical rock surfaces)	Piercers—carnivores	Arizona, Idaho	3161
Salda (8)	Beach zone—freshwater, lentic (marshy meadows)	Climbers (semiaquatic)	Predators (piercers) (scavengers)	Widespread	2654, 3160, 3161, 4414
Saldoida (3)	Lentic (marshy meadows)	Climbers (semiaquatic)	Predators (piercers) (scavengers)	East, South, Texas, Michigan, Kansas	3160, 3161
Saldula (26)	Lentic—littoral and lotic—shorelines (also salt marshes)	Climbers (at shore, semiaquatic)	Predators (piercers) (scavengers)	Widespread	1857, 2349, 2654, 3160, 3161, 3389, 3839, 3851, 1478, 3318, 4075, 4414
Teloleuca (2)	Lotic—margins	Climbers (at shore, semiaquatic)	Predators (piercers) (scavengers)	Arctic, midlatitude mountains	2654, 3161
Gelastocoridae (7) (toad bugs)	Generally lentic—vascular hydrophytes (emergent zone), lotic depositional beaches—freshwater	Generally sprawlers at shore, semiaquatic)	Generally predators (piercers)		331, 1857, 1867, 2112, 2654, 3076, 3534, 4075, 4306
Gelastocoris (12)	Lentic—littoral (water's edge) and beaches—freshwater	Sprawlers (jumpers)	Predators (piercers)	Widespread	348, 907, 1859, 2426, 2654, 3033, 3276, 3389, 4075
Nerthra (=*Mononyx*) (5)	Lentic—littoral (water's edge) and beaches—freshwater and marine	"Burrowers" (in mud at water's edge and terrestrial under logs, etc.)	Predators (piercers)	Southwest, Southeast	2654, 3160, 4075

*Emphasis on trophic relationships.

Table 15A. Continued

Taxa (number of species in parentheses)	Habitat	Habit	Trophic Relationships	North American Distribution	Ecological References*
Ochteridae (6) *Ochterus* (6)	Lentic—vascular hydrophytes, lotic—margins and seeps on rock surfaces faces)	Climbers (at shore, semiaquatic), clingers (on seeping vertical rock)	Predators (piercers)	Widespread, particularly South and Southwest, North to Nebraska and Great Lakes	345, 348, 2654, 3160, 3167, 3534, 3565, 4075

*Emphasis on trophic relationships.

MEGALOPTERA AND AQUATIC NEUROPTERA

Elwin D. Evans
Michigan Department of Natural Resources, Lansing

H. H. Neunzig
North Carolina State University, Raleigh

INTRODUCTION

The Megaloptera (alderflies, dobsonflies, fishflies, hellgrammites) and aquatic Neuroptera (spongillaflies) constitute a small worldwide fauna of probably less than 300 species, representing three families (Sialidae, Corydalidae, and Sisyridae). This group of holometabolous aquatic insects contains some of the largest and most spectacular species. The aquatic larvae are predaceous and inhabit both lotic and lentic environments in tropical and temperate climates; however, all eggs, pupae, and adults are terrestrial. Large numbers of adults are seldom seen in nature because they are short-lived, secretive, and many species are nocturnal.

Larval sialids are usually abundant in streams, rivers, or lakes where the substrate is soft and detritus is abundant. Larvae usually burrow into the substrate and feed nonselectively on small animals, such as insect larvae, annelids, crustaceans, and mollusks, in the habitat. Sialids pass through as many as 10 instars during a one- to two-year life cycle. Prior to pupation, larvae leave the stream, river, or lake and pupate in an unlined chamber dug 1-10 cm deep in shoreline soil and litter. Adults (alderflies) usually emerge from late spring to early summer and are active during warm midday hours. Flight is brief and infrequent, and most individuals stay in the same general area where the larvae occur. Apparently, the adults do not feed. Eggs are laid in masses primarily on leaves or branches overhanging the aquatic habitat, on large rocks overhanging or projecting from the water, or on bridge abutments.

Larval corydalids (sometimes called hellgrammites) occur in a wide variety of habitats including spring seeps, streams, rivers, lakes, ponds, swamps, and even temporarily dry streambeds. The life cycle is 2-5 years long with the larvae passing through 10-12 instars. As with sialids, corydalid larvae feed on a wide variety of small aquatic invertebrates. Pupation occurs mostly in chambers in the soil adjacent to the larval habitat. However, some species pupate in dry streambeds, and others prefer soft, rotting shoreline logs or stumps. Adults (dobsonflies, fishflies) emerge from late spring to midsummer. Most species of adult corydalids are nocturnal and may fly considerable distances; some are attracted to lights. Diurnal species are usually found resting or flying near the larval habitats. Oviposition habits are similar to those of adult sialids.

Larvae of the sisyrids are usually found associated with freshwater sponges. They occur on the surface or in the cavities of the host, and pierce the sponge cells and suck the fluids with their elongated mouthparts. Larvae pass through three instars, and some species have several generations each year. Just before pupation, the larvae leave the water, climb onto shoreline plants or other objects, and spin a silk cocoon in which to pupate. Sites chosen for pupation are frequently somewhat secluded and usually near shore, but larvae may migrate inland up to 20 m before pupating. The cocoon is usually double walled, i.e., composed of an inner, close-meshed enclosure and an additional, more loosely constructed outer envelope. No feces are voided by the larvae prior to pupation; this is typical of many neuropterans but unique among aquatic insects. Adults (spongillaflies) appear to be primarily nocturnal.

EXTERNAL MORPHOLOGY

Megalopteran eggs are quite distinct and can be separated by the size and appearance of the egg mass, and the size, color, sculpturing, and shape of the micropylar process of individual eggs. The egg-burster, left with the hatched egg by a newly emerged larva, is also diagnostic. Identification of larval Megaloptera is based primarily on the number of abdominal filaments, the presence or absence of ventral abdominal gill tufts, and the appearance and location of the eighth abdominal spiracles. Setae and color patterns are also used, particularly for separating species. Pupal identifications of sialids and corydalids can be made based on size and color patterns. Familial and generic identification of adult Megaloptera relies mainly on wing venation; species identification is based primarily on male and female genitalic characters.

Eggs of the two genera of aquatic Neuroptera have not been studied adequately to provide diagnostic characters. Larval sisyrids are best separated at the generic level by the presence or absence and the location of certain setae, and by the presence or absence of spines associated with setae. The labium is useful in separating pupal sisyrids, and there are possibly some differences in the appearance of the cocoons of the two genera. Adult aquatic Neuroptera are identified mainly by wing venation and genitalia.

Megaloptera (Sialidae, Corydalidae)

Eggs: Sialid eggs are laid in even rows of about 200-900 eggs, in compact, more or less quadrangular masses with the eggs vertical (Fig. 16.1) or horizontal (Fig. 16.2) to the substrate. Each egg is cylindrical, about 0.2 by 0.6 mm in size, with rounded ends; the outer surface is partially or completely covered with small, very short, shield-shaped projections. The micropylar process is cylindrical or slightly fusiform (Fig. 16.13). The egg-burster is V-shaped and sharply toothed (Fig. 16.3).

Corydalid egg masses of 300-3,000 eggs are compact, rounded to quadrangular in shape, and have 1-5 layers (Figs. 16.12, 16.15, 16.16); sometimes the eggs have a white or brown protective covering (Fig. 16.12). Individual eggs, in general, appear similar to sialid eggs; sometimes the eggs are covered with shield-shaped processes (Fig. 16.14), but usually they are relatively smooth. The size of each egg is about 0.5 by 1.5 mm. The position of the eggs relative to the substrate is as in the Sialidae, and the micropylar process is apically enlarged. Egg-bursters are elongate, rounded, or ridged apically, and toothed (Fig. 16.4).

Larvae: Terminal instar sialids (Fig. 16.5) reach a maximum length of approximately 25 mm, including the unsegmented, median caudal filament. Mouthparts consist of a labrum, two well-developed mandibles (for grasping and engulfing the prey), two maxillae, and a labium. Antennae are four-segmented. The quadrate head is patterned, as is the 10-segmented abdomen, which ranges in color from purplish or reddish brown to yellow. Thoracic legs have two claws. Abdominal segments 1-7 bear four- or five-segmented lateral filaments (Fig. 16.5).

Corydalid larvae (Figs. 16.17, 16.22-16.23) are larger than sialids, reaching 30-65 mm or more in length when full grown. Mouthparts are similar to those of sialid larvae but the mandibles are usually more robust. Antennae are four- or five-segmented. Thoracic legs have two claws. The head and thorax may be of a uniform color or patterned. Abdominal segments 1-8 bear two-segmented (a short, basal segment and a long, distal segment) lateral filaments (Figs. 16.17, 16.22-16.23) and the abdomen terminates in a pair of anal prolegs. Each proleg bears paired claws and a dorsal filament (Figs. 16.17-16.23). The last pair of spiracles on the abdomen (segment 8) are sometimes modified with regard to size and location.

Pupae: Both sialid (Fig. 16.6) and corydalid (Fig. 16.7) pupae are exarate (appendages free, not fastened to body), and range in length from 10 to 12 mm and from 30 to 60 mm, respectively.

Adults: Adults of Sialidae are approximately 10-15 mm in length. Their bodies are black, brown, or yellowish orange with similarly colored wings (Fig. 16.8). The head lacks ocelli, and the fourth tarsal segment is dilated (Fig. 16.11).

Corydalid adults are 40-75 mm long, with black, brown, or gray bodies; many species have pale smoky wings mottled with brown (Fig. 16.9). Some species (*Nigronia* spp.) have darker, almost black, wings with white markings (Fig. 16.36). The head has three ocelli and males of some species (*Corydalus* spp.) have very long mandibles (Figs. 16.39-16.40). All tarsal segments are simple (Fig. 16.10).

Aquatic Neuroptera (Sisyridae)

Eggs: Masses are of 2-5, or occasionally as many as 20, oval, whitish to yellowish eggs covered with a web of white silk. Each egg is about 0.1 by 0.3 mm in size with a short micropylar process (Fig. 16.42). The egg-burster is elongate (Fig. 16.41).

Larvae: Terminal instars (Fig. 16.43) are small (4-8 mm in length), stout, and with conspicuous setae. Body color varies from yellowish brown to dark green. Mouthparts are modified into elongate, unsegmented stylets (usually separated in preserved specimens). Antennae are relatively long and legs are slender and bear a single claw. Second and third instars bear two- or three-segmented, transparent ventral gills, which are folded medially and posteriorly on abdominal segments 1-7.

Pupae: All species are exarate (exposed) and housed in hemispherical, usually double-walled, silken cocoons (Fig. 16.47).

Adults: Spongillaflies lack ocelli, have brown bodies with brown wings (Fig. 16.52), and are 6-8 mm in length.

KEYS TO THE FAMILIES AND GENERA OF MEGALOPTERA

1. *Eggs* in masses of approximately 15 mm diameter (Figs. 16.1-16.2); egg-burster V-shaped (Fig. 16.3).
Larvae with 7 pairs of 4 to 5-segmented lateral filaments on abdominal segments 1-7 and a single long caudal filament (Fig. 16.5); 25 mm or less when full grown.
Pupae 10-12+ mm (Fig. 16.6).
Adults less than 25 mm in length (many 10-15 mm) (Fig. 16.8); ocelli absent; 4th tarsal segment dilated (Fig. 16.11) .... *SIALIDAE-Sialis* Latreille

1'. *Eggs* in masses of 20+ mm diameter (Figs. 16.12, 16.15-16.16); egg-bursters elongate, apically rounded, or ridgelike and toothed (Fig. 16.4).
Larvae with 8 pairs of 2-segmented lateral filaments on abdominal segments 1-8, and a pair of 1-segmented filaments on abdominal segment 10 (Fig. 16.17); apex of abdomen with 2 anal prolegs, each bearing a pair of claws; 30-65 mm when full grown.
Pupae greater than 30 mm in length (Fig. 16.7).
Adults over 25 mm in length (Fig. 16.9); ocelli present; 4th tarsal segment simple (Fig. 16.10) .... *CORYDALIDAE*

Corydalidae

Eggs

1.	Egg mass 3-1ayered with a thick, white chalky covering (Fig. 16.12)	***Corydalus*** Latreille
1'.	Egg mass not as above	2
2(1').	Egg mass single layered, may have a thin coating	3
2'.	Egg mass with more than one layer	4
3(2).	Egg chorion with peltate (shield-shaped) processes on dorsum (Fig. 16.14)	***Chauliodes*** Latreille
3'.	Egg chorion smooth; a thin coating may cover egg mass	***Neohermes*** Banks, ***Protochauliodes*** Weele
4(2').	Egg mass 3-5 layered; western United States and Canada (Fig. 16.16)	5
4'.	Egg mass with up to 50 eggs in a second layer; eastern and central United States	***Nigronia*** Banks
5(4).	Egg approximately 2.0 mm long	***Dysmicohermes*** Munroe
5'.	Egg approximately 1.0 mm long	***Orohermes*** Evans

Larvae

1.	Abdominal segments 1-7 with ventral gill tufts at base of lateral filaments (Fig. 16.17)	***Corydalus*** Latreille
1'.	Abdominal ventral gill tufts absent	2
2(1').	Last pair of abdominal spiracles (segment 8) at the apex of 2 long dorsal respiratory tubes extending beyond prolegs (Fig. 16.18)	***Chauliodes*** Latreille
2'.	Last pair of abdominal spiracles not at apex of long respiratory tubes	3
3(2').	Larval head not conspicuously patterned	4
3'.	Larval head with a conspicuous pattern (Figs. 16.22-16.23)	6
4(3).	West coast of United States and Canada; last pair of abdominal spiracles either dorsal, large, and raised on short tubes (Figs. 16.19, 16.30-16.31), or lateral and about same size as elsewhere on abdomen (Figs. 16.20, 16.29)	5
4'.	East and central United States and Canada; last pair of abdominal spiracles dorsal, similar in size to other abdominal spiracles, and borne at apex of short respiratory tubes (Figs. 16.21, 16.27-16.28)	***Nigronia*** Banks
5(4).	Last pair of spiracles dorsal, large, and raised on short tubes (Figs. 16.19, 16.30-16.31)	***Dysmicohermes*** Munroe
5'.	Last pair of spiracles lateral, similar in size to other abdominal spiracles and sessile (Figs. 16.20, 16.29)	***Orohermes*** Evans
6(3').	Spiracles on abdominal segment 8 distinct, associated with raised areas of the integument (Figs. 16.25-16.26); eastern and western United States and Canada	***Neohermes*** Banks
6'.	Spiracles of abdominal segment 8 less conspicuous, integument not raised around spiracle (Fig. 16.24); west coast of United States and Canada	***Protochauliodes*** Weele

Adults

1.	Fore wing with white spots in many cells; 20 veins or more reaching wing margin posteriad of R_1; M vein with 3 branches reaching wing margin (Fig. 16.32)	***Corydalus*** Latreille
1'.	Fore wing lacking white spots (Fig. 16.9), or white spots less widely distributed (Fig. 16.36); less than 20 veins reaching wing margin posteriad of R_1; M vein with 2 branches reaching wing margin (Figs. 16.33-16.36)	2
2(1').	Posterior branch of Rs forked in both pairs of wings (Fig. 16.33)	3
2'.	Posterior branch of Rs simple in both pairs of wings (Fig. 16.34)	4
3(2).	Hind wing with posterior branch of M forked (Fig. 16.33)	***Dysmicohermes*** Munroe
3'.	Hind wing with posterior branch of M simple	***Orohermes*** Evans
4(2').	M vein of hind wing with 3 branches reaching wing margin (Fig. 16.34)	5
4'.	M vein of hind wing with 2 branches reaching wing margin (Fig. 16.35)	6
5(4).	Crossvein present between R_3 and R_4 in fore wing (Fig. 16.34); antennae of male elongate, moniliform (beadlike), setigerous (bearing setae) (Fig. 16.37); apical papilla of gonapophysis lateralis absent in female (compare with Fig. 16.38)	***Neohermes*** Banks

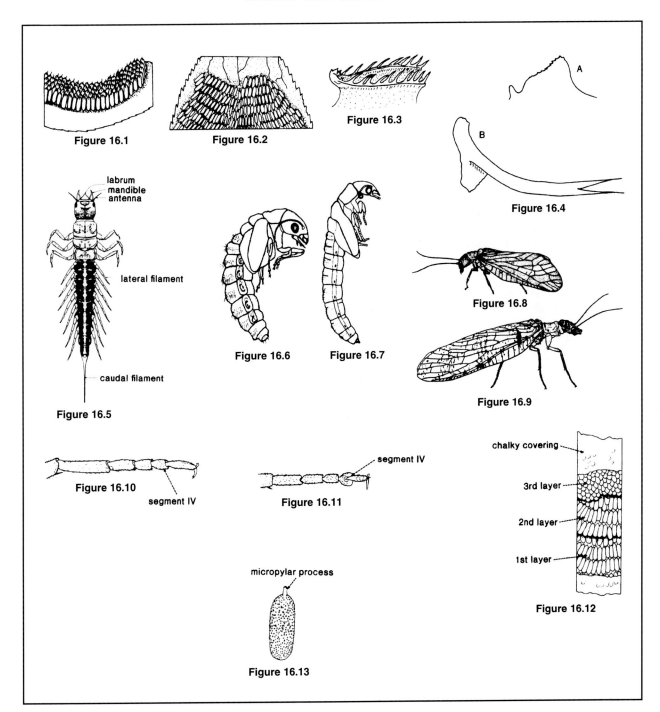

Figure 16.1. Eggs of *Sialis rotunda* Ross (Sialidae).

Figure 16.2. Eggs of *Sialis hamata* Ross (Sialidae).

Figure 16.3. Lateral view of V-shaped egg-burster of *Sialis* sp. (Sialidae) (length 90 µm).

Figure 16.4. Lateral view of corydalid egg-bursters (Corydalidae); a, *Orohermes crepusculus* (Chandler) (length, 115 µm); b, *Corydalus* sp. (length, 160 µm).

Figure 16.5. Dorsal view of larva of *Sialis rotunda* Ross (Sialidae).

Figure 16.6. Lateral view of pupa of *Sialis cornuta* Ross (Sialidae) (after Leischner and Pritchard 1973).

Figure 16.7. Lateral view of *Neohermes* sp. pupa (Corydalidae); length 30+ mm.

Figure 16.8. Adult of *Sialis californica* Ross (Sialidae).

Figure 16.9. Adult of *Orohermes crepusculus* (Chandler) (Corydalidae).

Figure 16.10. Distal part of tibia and the tarsus of *Orohermes crepusculus* (Chandler) with a simple 4th tarsal segment (Corydalidae).

Figure 16.11. Distal part of tibia and the tarsus of *Sialis* sp. with a dilated 4th tarsal segment (Sialidae).

Figure 16.12. Section of *Corydalus* egg mass (Corydalidae); covering partially removed.

Figure 16.13. Egg of *Sialis hasta* Ross (Sialidae) (length, 0.6 mm).

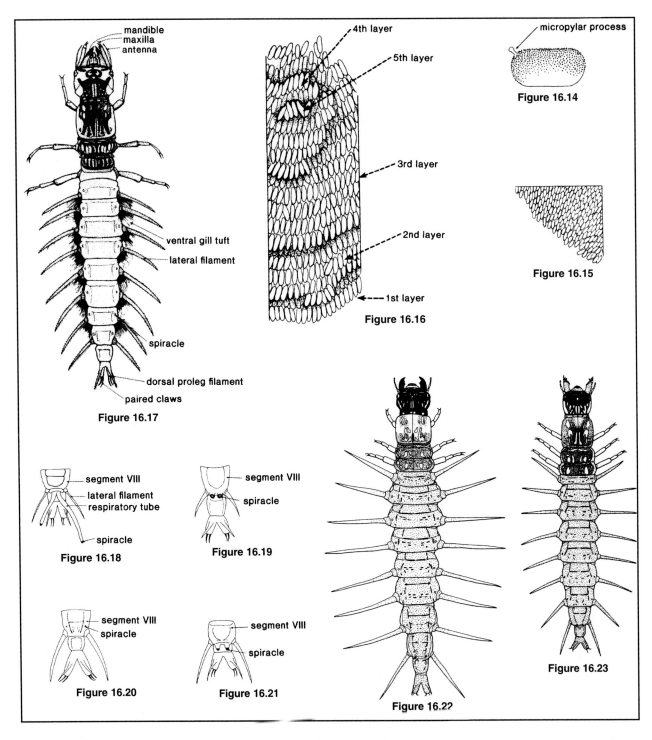

Figure 16.14. Egg of *Chauliodes pectinicornis* (L.) (Corydalidae) (length 1.0 mm).

Figure 16.15. Section of a *Neohermes* sp. egg mass (Corydalidae).

Figure 16.16. Section of a mutilayered egg mass of *Orohermes crepusculus* (Chandler) (Corydalidae).

Figure 16.17. Dorsal view of larva of *Corydalus* sp. (Corydalidae).

Figure 16.18. Dorsal view of caudal segments, respiratory tubes and spiracles of segment VIII of *Chauliodes* sp. larva (Corydalidae).

Figure 16.19. Dorsal view of caudal segments and spiracles of segment VIII of *Dysmicohermes ingens* Chandler larva (Corydalidae).

Figure 16.20. Dorsal view of caudal segments and spiracles of segment VIII of *Orohermes crepusculus* (Chandler) larva (Corydalidae).

Figure 16.21. Dorsal view of caudal segments and spiracles of segment VIII of *Nigronia serricornis* (Say) larva (Corydalidae).

Figure 16.22. Dorsal view of larva of *Neohermes filicornis* (Banks) (Corydalidae).

Figure 16.23. Dorsal view of larva of *Protochauliodes spenceri* Munroe (Corydalidae).

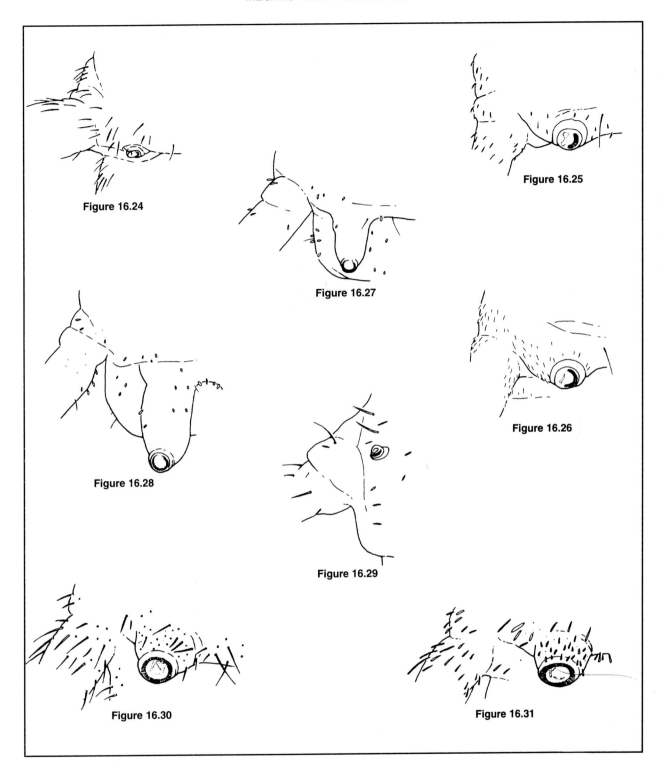

Figure 16.24. Dorsal view of left spiracle on segment VIII of *Protochauliodes spenceri* Munroe larva (Corydalidae).

Figure 16.25. Dorsal view of left spiracle on segment VIII of *Neohermes concolor* (Davis) larva (Corydalidae).

Figure 16.26. Dorsal view of left spiracle on segment VIII of *Neohermes filicornis* (Banks) larva (Corydalidae).

Figure 16.27. Dorsal view of left spiracle on segment VIII of *Nigronia serricornis* (Say) larva (Corydalidae).

Figure 16.28. Dorsal view of left spiracle on segment VIII of *Nigronia fasciatus* (Walker) larva (Corydalidae).

Figure 16.29. Dorsal view of left spiracle on segment VIII of *Orohermes crepusculus* (Chandler) larva (Corydalidae).

Figure 16.30. Dorsal view of left spiracle on segment VIII of *Dysmicohermes ingens* Chandler larva (Corydalidae).

Figure 16.31. Dorsal view of left spiracle on segment VIII of *Dysmicohermes disjunctus* (Walker) larva (Corydalidae).

5'.	Crossvein absent between R_3 and R_4 in fore wing; antennae filiform (threadlike): apical papilla of gonapophysis lateralis present in female (Fig. 16.38)	***Protochauliodes*** Weele
6(4').	Wings dark with white spots and patches (Fig. 16.36)	***Nigronia*** Banks
6'.	Wings pale gray-brown, mottled (Fig. 16.35)	***Chauliodes*** Latreille

KEYS TO THE GENERA OF AQUATIC NEUROPTERA

Sisyridae

Larvae (after Poirrier and Arceneaux [1972])

1.	Pair of dorsal setae present on abdominal segment 8 (Fig. 16.43); ventral pair of medial setae on abdominal segment 8 raised on tubercles and only slightly closer together than those on segment 9 (Fig. 16.46); small acute spines at bases of thoracic setae (Fig. 16.44) (not present in *Climacia californica*)	***Climacia*** McLachlan
1'.	Pair of dorsal setae absent on abdominal segment 8; pair of ventral medial setae on abdominal segment 8 sessile and distinctly closer together than those on segment 9 (Fig. 16.45); small acute spines at bases of thoracic setae absent	***Sisyra*** Burmeister

Pupae

1.	Segments of labial palp similar (Fig. 16.51)	***Climacia*** McLachlan
1'.	Last palpal segment of labium greatly enlarged and triangular (Fig. 16.50)	***Sisyra*** Burmeister

Adults

1.	Rs of fore wing with one fork before pterostigmata (Fig. 16.48), fore wing with brown markings; last segment of labial palp similar in size and shape to other segments (Fig. 16.51)	***Climacia*** McLachlan
1'.	Rs of fore wing with more than one fork before pterostigmata (Fig. 16.49), fore wing uniformly brown; last segment of labial palp greatly enlarged and triangular in shape (Fig. 16.50)	***Sisyra*** Burmeister

ADDITIONAL TAXONOMIC REFERENCES

General
Chandler (1956a,b); Gurney and Parfin (1959); Pennak (1978); McCafferty (1981); Evans and Neunzig (1984); Peckarsky *et al.* (1990).

Regional faunas
Arkansas: Bowles (1989).
California: Chandler (1954, 1956a,b); Whiting (1991c).
Canada: Kevan (1979).
Colorado: Herrmann and Davis (1991).
Kansas: Huggins (1980); Roble (1984).
Minnesota: Parfin (1952).
Mississippi: Poirrier and Holzenthal (1980); Stark and Lago (1980, 1983).
New York: Needham and Betten (1901a).
North Carolina: Cuyler (1956); Brigham *et al.* (1982).
Oklahoma: Bowles (1989).
Pacific Coastal Region: Evans (1972).
South Carolina: Brigham *et al.* (1982).
West Virginia: Watkins *et al.* (1975); Tarter (1976, 1988).

Taxonomic treatments at the family and generic levels
Corydalidae: Munroe (1951b, 1953); Cuyler (1958); Hazard (1960); Flint (1965); Neunzig (1966); Baker and Neunzig (1968); Glorioso (1981); Evans (1984); Contreras-Ramos (1990); Neunzig and Baker (1991).
Sialidae: Davis (1903); Ross (1937); Cuyler (1956); Canterbury (1978); Canterbury and Neff (1980); Neunzig and Baker (1991); Whiting (1991a, 1991b).
Sisyridae: Parfin and Gurney (1956); Poirrier and Arceneaux (1972); Pupedis (1980); Tauber (1991).

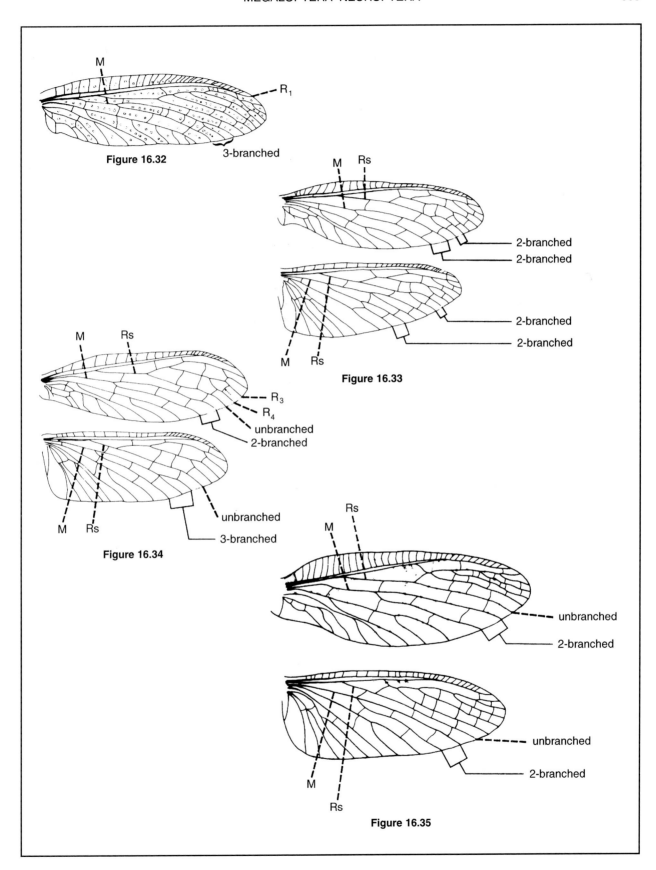

Figure 16.32. Fore wing of *Corydalus* sp. (Corydalidae).
Figure 16.33. Wings of *Dysmicohermes disjunctus* (Walker) (Corydalidae).
Figure 16.34. Wings of *Neohermes* sp. (Corydalidae).
Figure 16.35. Wings of *Chauliodes* sp. (Corydalidae).

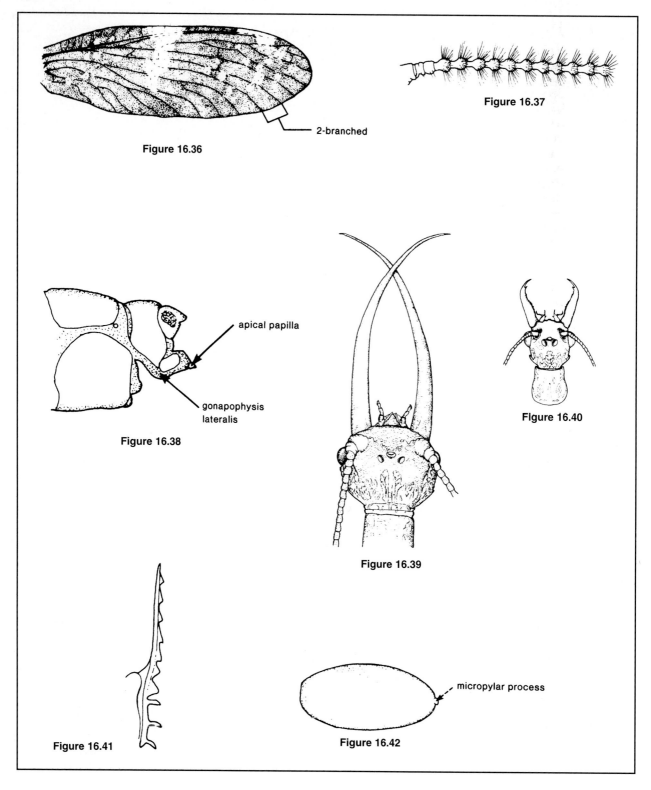

Figure 16.36. Fore wing of *Nigronia* sp. (Corydalidae).

Figure 16.37. Basal part of antenna of male *Neohermes* sp. (Corydalidae).

Figure 16.38. Lateral view of female genitalia of *Protochauliodes spenceri* Munroe (Corydalidae).

Figure 16.39. Dorsal view of head and part of prothorax of male *Corydalus* sp. (eastern North America) (Corydalidae).

Figure 16.40. Dorsal view of head and prothorax of male *Corydalus* sp. (western North America) (Corydalidae).

Figure 16.41. Lateral view of egg-burster of *Climacia* sp. (Sisyridae) (after Brown 1952).

Figure 16.42. Egg of *Climacia* sp. (Sisyridae) (after Brown 1952).

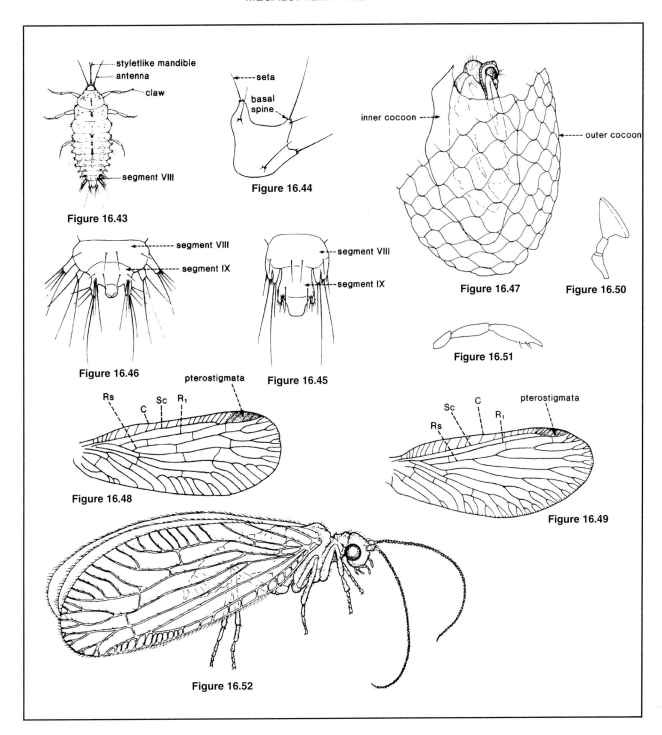

Figure 16.43. Dorsal view of larva of *Climacia* sp. (Sisyridae) (after Brown 1952).

Figure 16.44. Right dorsal plate of pronotum of *Climacia* sp. (Sisyridae) (after Parfin and Gurney 1956).

Figure 16.45. Ventral view of caudal abdominal segments of *Sisyra vicaria* (Walker) (after Parfin and Gurney 1956).

Figure 16.46. Ventral view of caudal abdominal segments of *Climacia areolaris* (Hagen) (after Parfin and Gurney 1956).

Figure 16.47. Lateral view of pupa and cocoons of *Sisyra* sp. (Sisyridae).

Figure 16.48. Fore wing of *Climacia* sp. (Sisyridae) (after Parfin and Gurney 1956).

Figure 16.49. Fore wing of *Sisyra* sp. (Sisyridae) (after Parfin and Gurney 1956).

Figure 16.50. Labial palp of *Sisyra* sp. (Sisyridae) (after Chandler 1956a).

Figure 16.51. Labial palp of *Climacia* sp. (Sisyridae) (after Chandler 1956a).

Figure 16.52. Female *Climacia* sp. (Sisyridae) (after Brown 1952).

Table 16A. Summary of ecological and distributional data for the *Megaloptera (alderflies, dobsonflies, hellgrammites)* and the *Aquatic Neuroptera (spongillaflies).* (For definition of terms see Tables 6A-6C; table prepared by K. W. Cummins, R. W. Merritt, E. D. Evans, and H. H. Neunzig.)

Taxa (number of species in parentheses)	Habitat	Habit	Trophic Relationships	North American Distribution	Ecological References*
Megaloptera					599, 688, 1506, 3076, 1116, 2886, 3069'
Sialidae (alderflies) (24)					1116, 2886
Sialis (24)	Lotic—erosional and depositional (detritus, sediments), lentic—erosional (sediments)	Burrowers—climbers—clingers	Predators (engulfers) (one species reported to be a collector-gatherer)	Widespread	137, 138, 440, 544, 599, 1114, 1369, 1382, 1506, 1713, 2033, 2314, 2344, 2857, 3014, 3076, 3207, 3446, 3929, 3948, 4035, 2514, 2959, 4204, 4540, 1715, 2193
Corydalidae (22)	Generally lotic erosional (lentic—littoral)	Generally clingers—climbers	Predators—(engulfers)		153, 599, 1114, 1506, 3014, 1116, 2886, 3076
Corydalinae (dobsonflies, hellgrammites) (4)					
Corydalus (4)	Lotic—erosional and depositional (sediments, detritus)	Clingers—climbers (swimmers)	Predators (engulfers)	Widespread	153, 443, 599, 1114, 1506, 2120, 3014, 3364, 385, 573, 852, 3835, 688, 1091, 2193, 2438, 2552, 2987, 3416, 3648, †
Chauliodinae (fishflies) (18)					
Chauliodes (2)	Lentic—littoral (sediments, detritus, logs)	Clingers—climbers—burrowers	Predators—(engulfers)	East, Central	153, 820, 874, 1641, 931, 3014, 3364, 3947, 3807
Dysmicohermes (2)	Lotic—erosional and depositional (sediments, detritus)	Clingers—climbers	Predators—(engulfers)	West, Texas	597, 689, 1114, 2802, 2803
Neohermes (5)	Lotic—erosional and spring seeps (sediments, vascular hydrophytes, leaf detritus)	Clingers—climbers	Predators—(engulfers)	West, East	1114, 1219, 3074, 3704, 389, 688, 3406, 3949
Nigronia (2)	Lotic—erosional and depositional (coarse sediments, detritus, especially under bark and in crevices of woody debris)	Clingers—climbers—burrowers (in crevices and under bark)	Predators—(engulfers)	East, Central	153, 518, 669, 671, 821, 2121, 2718, 2885, 3107, 852, 3947, 3950, 4057, 519, 688, 1332, 2680, 1322, 1325, †
Orohermes (1)	Lotic—erosional	Clingers—climbers	Predators—(engulfers)	West	597, 1114, 1115
Protochauliodes (6)	Lotic—erosional (sediments, detritus), lotic—depositional (detritus) (including intermittent habitats)	Clingers—climbers	Predators—(engulfers)	West	597, 1114, 2434, 2802, 2803
Neuroptera					
Sisyridae (6) (spongillaflies)	Generally lotic—erosional (on sponges), lentic—littoral (on alpine sponges)	Generally climbers clingers, burrowers (live in or on freshwater sponges)	Predators—(piercers of *Spongilla, Ephydatia*, etc.)		598, 1506, 2968, 3014, 3015, 3076, 3151, 3152, 1116, 2552, 3290, 3225, 3849, 3955
Climacia (3)				Widespread (2 Central and Eastern, 1 Western)	598, 1506, 2552, 3290, 4327, 645, 3849, 3955, 3069'
Sisyra (3)				Widespread	598, 645, 1488, 3014, 3290, 3849, 3955

*Emphasis on trophic relationships.
† Unpublished data, K W. Cummins, Kellogg Biological Station.

TRICHOPTERA FAMILIES

Glenn B. Wiggins
*Department of Entomology, Royal Ontario Museum and
Department of Zoology, University of Toronto, Canada*

INTRODUCTION

The Trichoptera, or caddisflies, one of the largest groups of aquatic insects, are closely related to the Lepidoptera. They are holometabolous and, except for a few land-dwelling species (Flint 1958; Anderson 1967) that are secondarily adapted to life out of water, are aquatic in the immature stages. Respiration is independent of the surface and of atmospheric oxygen. Adults of almost all species are active, winged insects, although females of at least one North American species are wingless (see Ross 1944, Fig. 171). More than 1,350 species are now known in North America north of the Rio Grande, and these are currently assigned to approximately 147 genera in 22 families. This taxonomic richness is a consequence of the broad ecological diversity of the order (Wiggins and Mackay 1978). Caddisflies occur in most types of freshwater habitats: spring streams and seepage areas, rivers, lakes, marshes, and temporary pools. They have been particularly successful in subdividing resources within these habitats. General summaries of information on the biology of Trichoptera are available in several references: Betten (1934), Balduf (1939), Lepneva (1964), Malicky (1973), Wiggins (1977), Mackay and Wiggins (1979), and Table 17A.

All North American families are represented in cool, lotic waters, and they have also been successful, to varying degrees, in exploiting freshwater habitats that are larger, warmer, and more lentic. Most larvae eat plant materials in one form or another—algae, especially diatoms on rocks, decaying vascular plant tissue, and the associated microorganisms—but living vascular plant tissue evidently is seldom ingested; some larvae are mainly predaceous. Generally, larval Trichoptera show little selectivity of food, but they are highly and diversely specialized for food acquisition (Table 17A).

Caddisfly larvae are perhaps best known for the remarkable nets, retreats, and portable cases they construct. Silk, emitted through an opening at the tip of the labium (Fig. 17.26), is used either by itself (e.g., Figs. 17.1-17.2, 17.6) or to fasten together rock fragments and pieces of plant materials (e.g., Figs. 17.7-17.22). Retreats and cases differ widely in design, materials, and function, but on the whole are consistent at the generic level. Construction behavior coincides so closely with the diverse ecological roles the larvae fill that North American families are usually categorized into five groups on this basis:

1. *Free-living forms*. Rhyacophilidae, Hydrobiosidae (superfamily Rhyacophiloidea, in part). Larvae move actively about and do not construct a retreat or case of any kind until just before pupation when a crude cell of rock fragments is fastened to some substrate, usually a large rock. Within the cell the larva spins a characteristic ovoid, closed cocoon of tough, brown, silk and undergoes metamorphosis inside the cocoon. Larvae are predators for the most part, although some species feed on algae and vascular plant tissue; most inhabit cool running waters, some occur in transient streams. The Rhyacophilidae (Figs. 17.48-17.49) are generally regarded as the most primitive living family in the Trichoptera. The genus *Rhyacophila*, alone comprising more than 500 species in the Northern Hemisphere and the largest genus in the order, is abundant and highly diverse in streams especially in western North America. The Hydrobiosidae (Fig. 17.50), a family mainly of the Southern Hemisphere, extend into North America only in the Southwest.

2. *Saddle-case makers*. Glossosomatidae (superfamily Rhyacophiloidea, in part; Figs. 17.21, 17.40, 17.47). Using rock fragments, larvae construct portable cases resembling the shell of a tortoise. They live in running waters and occasionally along the wave-swept shores of lakes, grazing on diatoms and fine particulate organic matter from the upper exposed surfaces of rocks; larvae are entirely covered by their dome-shaped cases. Freshly aerated water for respiration enters the case through spaces between rock pieces. In preparation for pupation, the larva removes the ventral strap and fastens the dome firmly to a large rock with silk; a closed, brown, silken cocoon is spun within the case as in the Rhyacophilidae.

3. *Purse-case makers*. Hydroptilidae (superfamily Rhyacophiloidea, in part; Figs. 17.22, 17.37, 17.42). Larvae are extremely small, and are free living until the final instar when they construct purse-shaped or barrel-shaped cases, which are portable in most genera. The first four larval instars, completed within three weeks in some species, differ considerably in morphology from the fifth instar, and represent the only example of larval heteromorphosis known in the Trichoptera. Larvae live in all types of permanent habitats, including springs, streams, rivers, and lakes. Their primary food is algae, mainly the cellular contents of filamentous forms, but diatoms are ingested by species in some genera. One of the few examples

in the Trichoptera of specific association with a plant occurs in the genus *Dibusa* where final instar larvae feed only on the freshwater red alga *Lemanea* and use the same plant in constructing their cases (Resh and Houp 1986). Although fifth instar larvae in most genera construct portable cases, those in the tribe Leucotrichiini are sedentary, fixing flattened silken cases resembling the egg capsules of leeches to rocks in running waters; the head and thorax are extended through a small opening at either end to graze periphyton and particulate matter from the area surrounding the case. Food reserves of fifth instars cause the abdomen of hydroptilids to become disproportionately large—depressed in the Leucotrichiini and Ptilocolepinae but compressed in most others.

4. *Net-spinners or retreat-makers*. Philopotamidae, Psychomyiidae, Ecnomidae, Xiphocentronidae, Polycentropodidae, Hydropsychidae (superfamily Hydropsychoidea). Most of these families are sedentary and construct fixed retreats, often with capture nets, to strain food particles from the current. Most are dependent on currents of running water to carry food to their retreats, although some also live along wave-washed shorelines of lakes. Larvae of the Philopotamidae (Figs. 17.52-17.53) live in elongate, fine-meshed nets (Fig. 17.6) in reduced currents on the underside of rocks, where they filter particles smaller than those filtered by other Trichoptera (Wallace and Malas 1976a); the specialized membranous labrum (Fig. 17.52) serves to clear accumulated particles from the net.

Larval Hydropsychidae (Figs. 17.38-17.39) construct retreats of organic and mineral fragments (Fig. 17.4) with a silken sieve net placed adjacent to the anterior entrance to filter particles from the current. Mesh size of the filter net differs: larvae in the Arctopsychinae, living in cold upstream sites with strong currents, spin the largest meshes and feed mainly on other insects; larvae in the Macronematinae, living in downstream sites with slow currents, spin the smallest mesh size and filter small particles; and those in the Hydropsychinae, occupying sites intermediate between these two, spin sieve nets with meshes in a range of intermediate sizes (Wallace 1975a,b; Hauer and Stanford 1981). By rubbing the femur across ridges on the underside of the head, *Hydropsyche* larvae produce sound (Jansson and Vuoristo 1979). Evidently the sound is a defensive behavior of larvae in protecting their retreat against other hydropsychids, but whether this protective role extends to predators generally is not known.

Larvae in the Polycentropodidae (Figs. 17.54, 17.56) construct shelters of several types. Predaceous genera such as *Nyctiophylax* make a flattened tube of silk (Fig. 17.1) in depressions on rocks or logs; from concealment within the tube, the larva darts out to capture prey that strike the silk threads emanating from each opening. Larvae in other genera such as *Neureclipsis* construct a funnel-shaped filter net of silk (Fig. 17.2) in slow currents and rest in the narrowed base; a single net may be 12 cm long. Larvae of the genus *Phylocentropus* (subfamily Dipseudopsinae, treated as a family by some authors) fashion branching tubes of silk and sand in loose sediments (Fig. 17.3), with the ends of the tubes protruding above the sediment; water with food particles in suspension enters the upstream tube, passes through a filter of silk threads that retains the particles, and out of the downstream tube (Wallace et al. 1976).

Although larvae of the Psychomyiidae and Ecnomidae (Fig. 17.55) live mainly in running waters, they do not filter food from the current, but graze on the periphyton and fine particulate matter around the ends of their tubelike retreats. Constructed of fine sand and organic material over a silken lining (Fig. 17.5), the retreats are fastened to rocks and logs. Larvae of the Xiphocentronidae (Fig. 17.57) construct similar tubes (Edwards 1961).

5. *Tube-case makers*. Phryganeidae, Brachycentridae, Limnephilidae, Uenoidae, Lepidostomatidae, Beraeidae, Sericostomatidae, Odontoceridae, Molannidae, Helicopsychidae, Calamoceratidae, Leptoceridae (superfamily Limnephiloidea). Larvae of these families construct portable cases, essentially tubular in form, of various shapes and materials. While serving as shelters, portable cases enable larvae to move from place to place seeking food. Moreover, respiratory dependence on natural currents is moderated because undulating movements by the larva cause a current of water to move through the tubular case, bathing the tracheal gills. Experimental work has shown that larvae remove more oxygen from water and thereby survive longer at low oxygen levels when in their cases than when deprived of them (Jaag and Ambuhl 1964); the rate of ventilation increases at lower oxygen levels (Van Dam 1938; Fox and Sidney 1953). Portable cases seem, therefore, to have released some groups from respiratory dependence on stream currents and to have been an asset in the exploitation of the resources of lentic habitats. Most larvae of the tube-case families are detritivorous shredders; some are scrapers, and a few are collector-gatherers or predators (Table 17A).

The dominant family is the Limnephilidae (Figs. 17.41, 17.59, 17.63, 17.69 17.70, 17.77, 17.79) with nearly 300 species and 50 genera in North America alone; the genera are highly diverse in case-making behavior (e.g., Figs. 17.7-17.11), habitat, and food (Table 17A). Most larvae in subfamilies Dicosmoecinae, Limnephilinae, and Pseudostenophylacinae are detritivorous or omnivorous and have toothed mandibles (as in Fig. 17.26); larvae in the Apataniinae and Goerinae have specialized mandibles without teeth (Fig. 17.27) and feed mainly by scraping exposed rock surfaces for diatoms and fine organic particles. The Goerinae (Fig. 17.79) are often treated as a separate family. Larval Phryganeidae (Figs. 17.15-17.16, 17.45-17.46) are large, up to 40 mm in length, and usually have conspicuous yellow and black markings on the head and pronotum. Pupae in several phryganeid genera are unusual in having pupal mandibles reduced to membranous lobes, a specialized feature correlated with the fact that they do not close the anterior end of the pupal case with a silken sieve membrane prior to metamorphosis as do other larvae (Wiggins 1960b). Leptoceridae (Figs. 17.43-17.44) are biologically diverse and some are able to swim with their cases; larvae in several genera are predaceous, some in *Ceraclea* feeding on sponges (Resh et al. 1976). Larvae of the Odontoceridae (Figs. 17.73, 17.75) and Sericostomatidae (Figs. 17.71-17.72, 17.74) are primarily burrowers in loose sediments; some odontocerids (*Psilotreta*) pupate in dense clusters on rocks. Larval Brachycentridae (Figs. 17.12-17.13, 17.58, 17.60) are confined to running waters, and those of *Brachycentrus* are unusual among the tube-case groups in filtering food from the current with their outstretched legs. The flattened and dorsally cowled larval cases of Molannidae (Figs. 17.18, 17.61-17.62) are unusually cryptic; even more so are those of some Calamoceratidae (Figs. 17.17, 17.65), particularly *Heteroplectron* species using hollow twigs or pieces of bark as cases. Many larvae of the Lepidostomatidae (Fig. 17.76), important components of the shredder

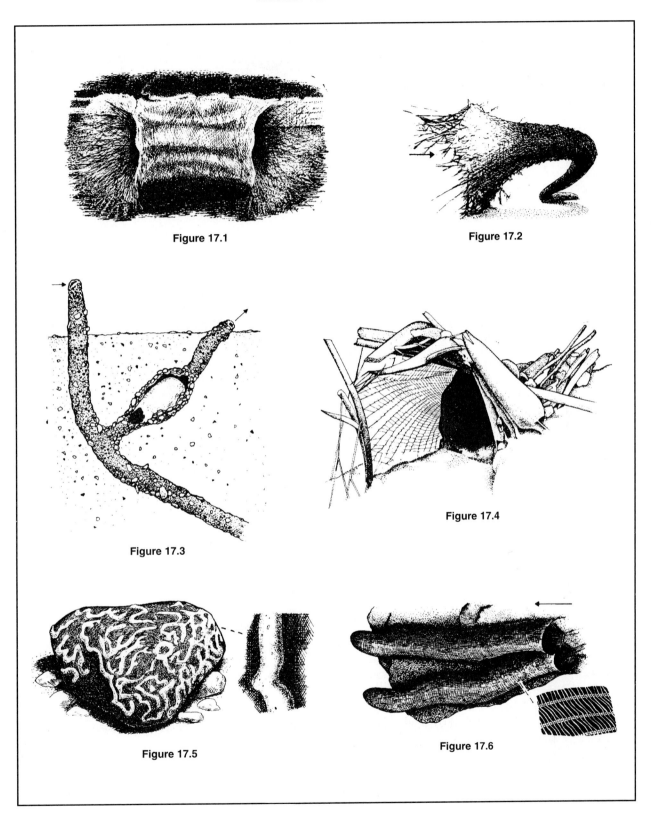

Figure 17.1. *Paranyctiophylax* sp. (Polycentropodidae) retreat.
Figure 17.2. *Neureclipsis* sp. (Polycentropodidae) net.
Figure 17.3. *Phylocentropus* sp. (Polycentropodidae) buried branching tube.
Figure 17.4. *Hydropsyche* sp. (Hydropsychidae) net and retreat.
Figure 17.5. *Psychomyia* sp. (Psychomyiidae) tubular retreats on rock, detail of tube.
Figure 17.6. *Dolophilodes* sp. (Philopotamidae) nets, detail of mesh.

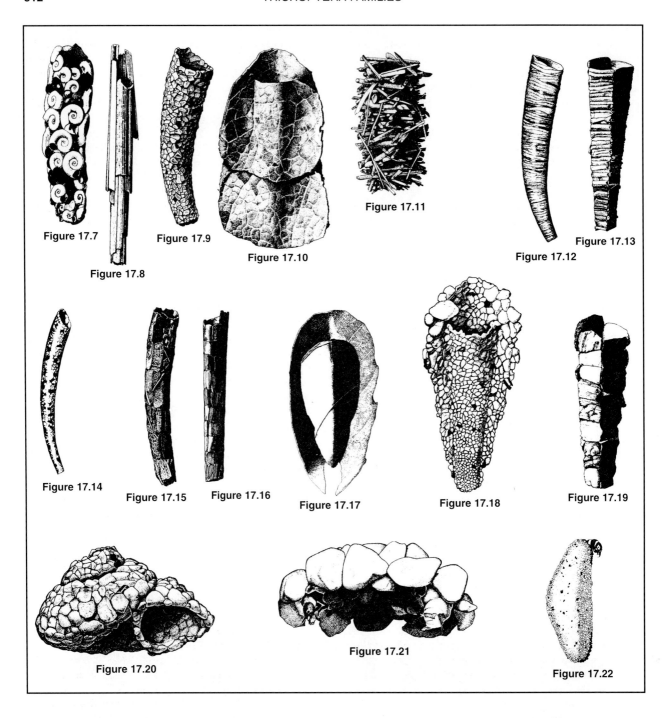

Figure 17.7. *Philarctus* sp. (Limnephilidae) portable case.
Figure 17.8. *Arctopora* sp. (Limnephilidae) portable case.
Figure 17.9. *Pseudostenophylax* sp. (Limnephilidae) portable case.
Figure 17.10. *Clostoeca* sp. (Limnephilidae) portable case.
Figure 17.11. *Platycentropus* sp. (Limnephilidae) portable case.
Figure 17.12. *Micrasema* sp. (Brachycentridae) portable case.
Figure 17.13. *Brachycentrus* sp. (Brachycentridae) portable case.
Figure 17.14. *Neothremma* sp. (Uenoidae) portable case.
Figure 17.15. *Oligostomis* sp. (Phryganeidae) portable case.
Figure 17.16. *Agrypnia* sp. (Phryganeidae) portable case.
Figure 17.17. *Anisocentropus* sp. (Calamoceratidae) portable case.
Figure 17.18. *Molanna* sp. (Molannidae) portable case.
Figure 17.19. *Lepidostoma* sp. (Lepidostomatidae) portable case.
Figure 17.20. *Helicopsyche* sp. (Helicopsychidae) portable case.
Figure 17.21. *Glossosoma* sp. (Glossosomatidae) portable case.
Figure 17.22. *Hydroptila* sp. (Hydroptilidae) portable case.

community in cool streams, construct cases of sand grains or silk in early instars, later changing to four-sided cases of leaf and bark pieces (Fig. 17.19). Larvae of Uenoidae (Figs. 17.14, 17.78) live in running waters and often occur in aggregations on rocks. Beraeidae (Figs. 17.51, 17.67-17.68) are extremely localized in North America and have been recorded only from the water-saturated muck of spring seepage areas in the East. Perhaps the most unusual larvae of all belong to the Helicopsychidae (Figs. 17.20, 17.35), which construct cases coiled like the shell of a snail; these larvae graze diatoms and fine particulate matter from exposed surfaces of rocks in rivers and along wave-swept shorelines of lakes.

Although most species of North American caddisflies are univoltine, some require two years for development, and others less than a year. Larvae of most species have five instars (a few, up to 7 or even 13), after which they fasten the case with silk to a solid substrate, sealing off the ends. In the Hydropsychoidea and Limnephiloidea the ends of the cocoon are perforated, allowing water to pass over the pupa directly (Wiggins and Wichard 1989); pupation in the Rhyacophiloidea occurs in a closed silken cocoon where respiration is sustained solely by diffusion through the semipermeable silk wall of the cocoon. The actual pupal stage lasts from two to three weeks, although in some groups it is preceded by a prepupal phase of up to several weeks' duration when the larva is in diapause (Wiggins 1977). When metamorphosis is complete, the pharate adult within the pupal cuticle (Figs. 17.29-17.30) leaves the pupal case and swims to the surface. Eclosion occurs either on the water surface (e.g., in large lakes) or on some emergent object.

Adults of most species are quiescent during the day, but fly actively during evening and night hours. Some are known to feed on plant nectar (Crichton 1957). Although most adult caddisflies probably live less than one month, adults of the Limnephilidae whose larvae inhabit temporary pools, live for at least three months; their reproductive maturity is delayed by diapause until late summer and early autumn when drought conditions are waning (Novak and Sehnal 1963, Wiggins 1973a). Diapause intervenes to suspend development at various points in the life cycles of other species. Eggs are deposited in water in most families, but above water in some groups of the Limnephilidae, and entirely in the absence of surface water by some species inhabiting temporary vernal pools. A further specialization for drought occurs in some Polycentropodidae whose eggs remain in the dry basin of temporary pools and do not hatch until water is replenished the following spring (Wiggins *et al.* 1980). Eggs are enclosed in a matrix of spumaline (a polysaccharide complex; Hinton 1981), which, in the Limnephiloidea, becomes greatly enlarged as water is absorbed.

EXTERNAL MORPHOLOGY

Larvae

Viewed dorsally (Fig. 17.28), the head capsule is subdivided into three parts by Y-shaped dorsal *ecdysial lines* (sutures); the *frontoclypeal apotome* is separated on each side by *frontoclypeal sutures* from the rounded *parietals*. Posteromesally, the parietals meet along the coronal suture. Ventrally, they come together along the ventral ecdysial line, but frequently the parietals are partially or completely separated mesally by the *ventral apotome*. Peglike antennae (Fig. 17.28) are visible on larvae in families constructing portable cases, but are not apparent in most other groups. The eyes are groups of *stemmata*. Cutting edges of the mandibles are of two basic types correlated with the method of feeding—a series of separate points or teeth (Fig. 17.26) or an entire scraping edge (Fig. 17.27); silk is emitted from an opening at the tip of the labium (Fig. 17.26).

The *pronotum* is covered by a heavily sclerotized plate subdivided by a mid-dorsal ecdysial line (Fig. 17.28); the prosternum sometimes bears small sclerites and, in certain families, a membranous *prosternal horn* (Fig. 17.23). The *trochantin* is a derivative of the prothoracic pleuron. The *mesonotum* may bear sclerotized plates (Fig. 17.28), small sclerites (Fig. 17.46), or be entirely membranous (Fig. 17.52); the *metanotum* (Fig. 17.28) in most families is largely membranous. *Notal setae* on the last two thoracic segments may be single or grouped on sclerites, but their basic arrangement into three setal areas—*sa1*, *sa2* and *sa3*—is usually apparent (Figs. 17.25, 17.28). In families with ambulatory larvae, the middle and hind pairs of legs are substantially longer than the first (Fig. 17.23).

The abdomen (Fig. 17.23) consists of ten segments, usually entirely membranous except for a dorsomedian sclerite on segment IX in some families. This sclerite is not always pigmented and may be difficult to distinguish in certain groups. Segment I of the abdomen in families constructing portable tube-cases usually bears prominent *humps*—one on each side and one dorsally; these humps, which serve as "spacers" facilitating the movement of the respiratory current of water through the case, are retractile and may not be prominent in preserved specimens. *Tracheal gills* are filamentous extensions of the body wall that contain fine tracheoles in the epithelium; some larvae lack gills entirely. Gills may be single (Fig. 17.23) or branched (e.g., Fig. 17.39) and are generally arranged in dorsal, lateral, and ventral pairs of gills on each side of a segment. Lightly sclerotized ovoid rings are discernible ventrally on most abdominal segments in larvae of the Limnephilidae (Fig. 17.23) and dorsally and laterally in certain limnephilid subgroups; these rings enclose chloride epithelia that are areas of the epidermis specialized for ionic transport in osmoregulation (Wichard and Komnick 1973). A lateral fringe of slender, bifid, hollow filaments extends along the mid-lateral surface of most larvae in the Limnephiloidea. Usually closely associated with the lateral fringe are small bifurcate *lateral tubercles*. The function of both structures is unknown; they appear to be correlated in some way with living in portable cases, but do not occur in all of the Limnephiloidea (Kerr and Wiggins 1995). *Anal prolegs* terminating the abdomen bear a pointed *anal claw* that sometimes has a small accessory hook. The prolegs are short and laterally directed (Fig. 17.23) in larvae with portable cases, enabling them to hold fast to the silken lining of the case, but are longer in net-spinning larvae (Fig. 17.24).

Pupae

Heavily sclerotized *pupal mandibles* (Fig. 17.29) serve to cut an opening in the pupal case through which the insect escapes to swim to the surface for emergence (Wiggins 1960b). Stout single setae occur on various parts of the head, those on the labrum are hooked apically in some groups; dorsal and ventrolateral tufts of setae often occur near the bases of the antennae (Fig. 17.29).

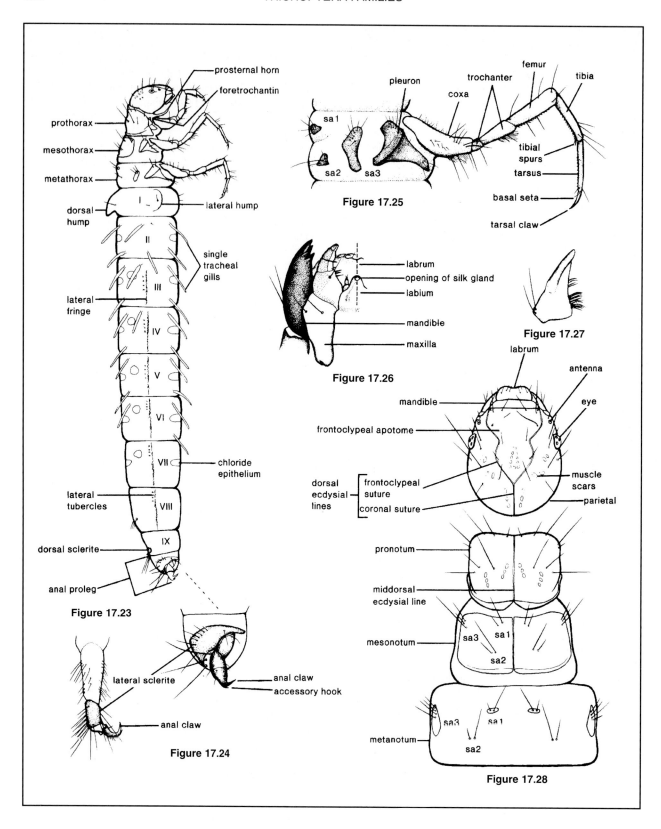

Figure 17.23. Lateral view of larva (Limnephilidae), abdominal segments numbered, detail of anal proleg.

Figure 17.24. Lateral view of anal proleg of larva (Polycentropodidae).

Figure 17.25. Lateral view of metathorax and leg of larva (Limnephilidae).

Figure 17.26. Ventral view of larval mouthparts (Leptoceridae: *Mystacides* sp.), mandible with pointed teeth.

Figure 17.27. Ventral view of larval mandible (Uenoidae: *Neophylax* sp.) with uniform scraping edge.

Figure 17.28. Dorsal view of larval head and thorax (Limnephilidae).

The compacted wings conform tightly to the body, and the legs are folded ventrolaterally. The middle tarsi bear a dense fringe of setae (Fig. 17.30), rendering the leg more effective for swimming from the pupal case to the surface.

Several of the abdominal segments bear paired dorsal sclerites—the *hookplates* (Fig. 17.30); hookplates are designated as anterior (a) or posterior (p) in accordance with their position on a particular segment, with at least segment V usually having both pairs. Hooks on anterior plates are directed posteriorly, and those on posterior plates directed anteriorly. In several families, segment I bears a rough spined ridge. These abdominal sclerites engage with the silken lining of the pupal case, enabling the insect to move within — especially important when it is ready to vacate the case for emergence. Pupal gills generally coincide with larval gills. A lateral fringe of slender filaments is variously developed in a number of families, absent in others; when present, the lateral fringe extends along the side of several segments (Fig. 17.30), turning ventrad on segment VIII (Fig. 17.31). At the apex of the abdomen in many families is a pair of *anal processes* (Fig. 17.31); these are often elongate, but may also be short and lobate (e.g., Fig. 17.106).

Adults

Structural characters of adult Trichoptera are identified in Figures 17.32-17.34. *Setal warts* are widely used in the key as diagnostic characters for families; usually clearly delineated by color and texture, the warts are somewhat dome-shaped in profile and bear setae, only the basal pits of which are illustrated in the figures. Setal warts of the head in most families are reduced in various ways from the generalized condition of Figure 17.33. The three *ocelli* occurring on the head in some families can be distinguished by their rounded, beadlike appearance (Fig. 17.33). The number and relative length of segments in the *maxillary palp* is an important diagnostic character; segments are numbered 1-5 from base to apex (Fig. 17.32). The terminal segment (no. 5) of the maxillary palp in families of the Hydropsychoidea is flexible, and numerous transverse striations can often be seen in the cuticle (Fig. 17.133). Sexual dimorphism occurs in the maxillary palpi of some families of the Trichoptera: palpi of all females are five-segmented and largely unmodified, but those of males are reduced in the number of segments — Phryganeidae, 4; Limnephilidae, Brachycentridae, and Helicopsychidae, 3. Maxillary palpi and basal segments of the antennae may be markedly enlarged and distorted in males of the Lepidostomatidae and Sericostomatidae.

Tibial spurs of the legs are large, modified setae occurring in pairs at the apex of the tibia, and singly or paired in a preapical position (Fig. 17.32). The full complement of spurs is expressed in an abbreviated form, e.g., 3, 4, 4 for Figure 17.32, giving the respective number of spurs on the fore, middle, and hind legs. Wings of caddisflies (Figs. 17.34, 17.128) provide a wealth of taxonomic characters involving shape, venation, and other aspects.

Segment I of the abdomen is reduced ventrally, an important point when locating a particular segment (Fig. 17.32). A pair of glands, probably pheromonal, open on the sternum of segment V, often in a raised ovoid sclerotized area (Fig. 17.150) or in a slender filament. These are present in at least some representatives of the Rhyacophilidae, Glossosomatidae, Hydroptilidae, Hydropsychidae, Psychomyiidae, Polycentropodidae, Phryganeidae, Brachycentridae, Limnephilidae, Beraeidae, and Molannidae (Schmid 1980). Males are readily distinguished from females (Nielsen 1980) by the paired forceps-like *claspers* or *inferior appendages*, and generally more complex genitalic structures terminating the abdomen (Fig. 17.32). (See Ross [1944] for examples of genitalic structures in all families.)

An illustration of the head and first two thoracic segments is provided for each family as a means of confirming identifications based on other characters.

KEYS TO THE FAMILIES OF TRICHOPTERA

Generic keys to larvae of the Nearctic fauna are given in Wiggins (1977, 1995) and in Chapter 18 of this volume. Diagnostic characters to pupae of some genera are given by Ross (1944) but are not yet available for most genera. Keys to adults of many genera are given by Ross (1944) and Schmid (1980).

Larvae

1. Anal claw comb-shaped (Fig. 17.35); larva constructing portable case of sand grains or small rock fragments, coiled to resemble a snail shell (Fig. 17.20). Widespread in rivers, streams, and wave-swept shorelines of lakes .... *HELICOPSYCHIDAE* ...... *Helicopsyche*
1'. Anal claw hook-shaped (Figs. 17.36, 17.39); larval case straight or nearly so, not resembling a snail shell, or larva not constructing a portable case 2
2(1'). Dorsum of each thoracic segment covered by sclerites, usually closely appressed along the mid-dorsal line (Fig. 17.37), sometimes subdivided with thin transverse sutures (Fig. 17.38), or some sclerites undivided 3
2'. Metanotum and sometimes mesonotum entirely membranous, or largely so and bearing several pairs of smaller sclerites (Figs. 17.40-17.41) 4
3(2). Abdomen with ventrolateral rows of branched gills, and with prominent brush of long hairs at base of anal claw (Fig. 17.39); posterior margin of meso- and metanotal plates lobate (Fig. 17.38); larvae construct fixed retreats (e.g., Fig. 17.4). Widespread in rivers and streams, occasionally along rocky shores of lakes .... *HYDROPSYCHIDAE* (p. 356)

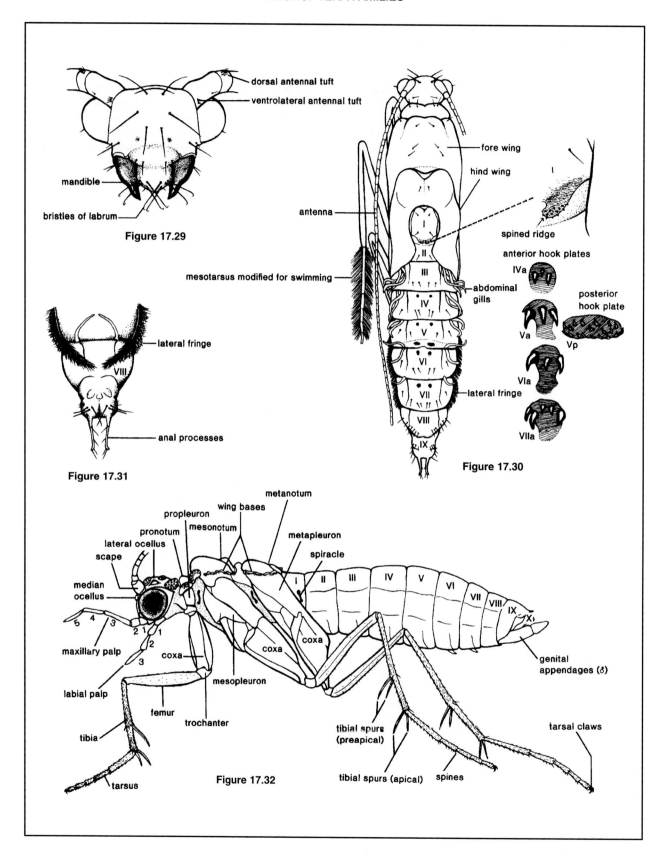

Figure 17.29. Frontal view of pupal head (Limnephilidae).

Figure 17.30. Dorsal view of pupa (Limnephilidae), abdominal segments numbered, detail of hook plates.

Figure 17.31. Ventral view of pupal abdominal segments VIII and IX with anal processes (Limnephilidae).

Figure 17.32. Lateral view of adult (Rhyacophilidae), wings not shown.

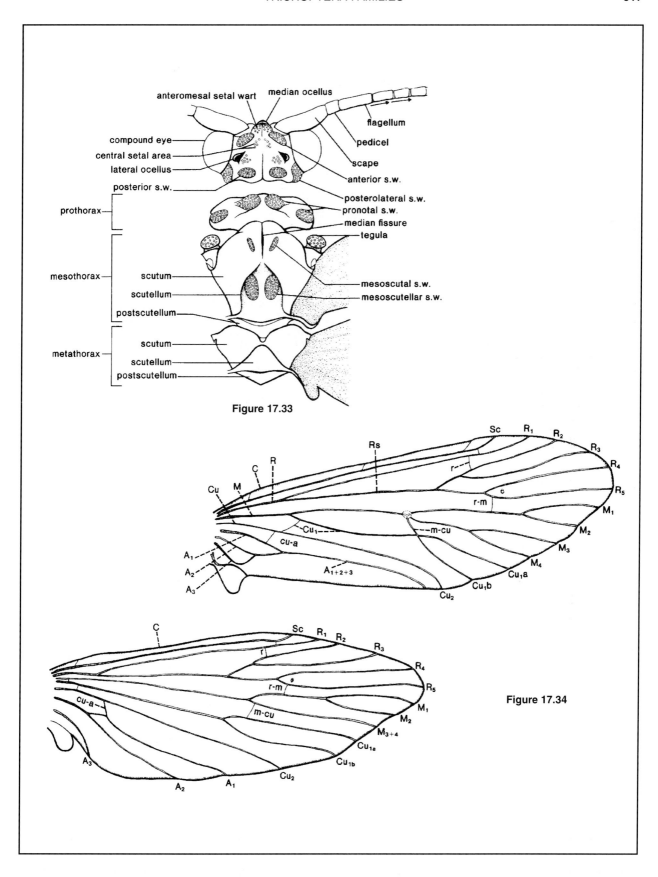

Figure 17.33. Dorsal view of head and thorax of adult trichopteran, generalized, *s.w.*, setal wart.

Figure 17.34. Wing venation of adult (Rhyacophilidae). Major longitudinal veins R_1, M, etc.; crossveins connecting major veins r-m, m-cu, etc.

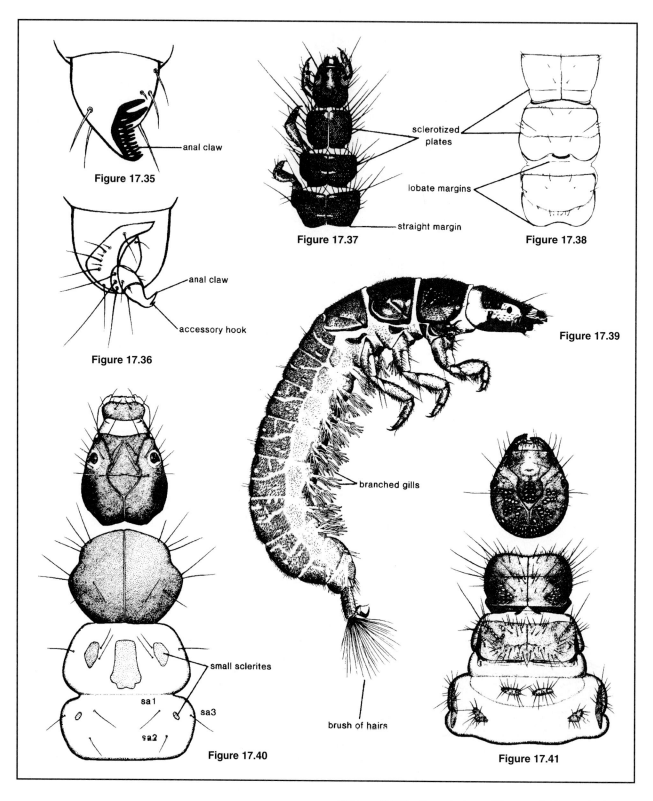

Figure 17.35. Lateral view of anal proleg of *Helicopsyche* sp. (Helicopsychidae) larva.

Figure 17.36. Lateral view of larval anal proleg (Limnephilidae).

Figure 17.37. Dorsal view of head and thorax of *Ochrotrichia* sp. (Hydroptilidae) larva.

Figure 17.38. Dorsal view of thorax of *Homoplectra* sp. (Hydropsychidae) larva.

Figure 17.39. Lateral view of *Hydropsyche* sp. (Hydropsychidae) larva.

Figure 17.40. Dorsal view of head and thorax of *Matrioptila* sp. (Glossosomatidae) larva.

Figure 17.41. Dorsal view of head and thorax of *Homophylax* sp. (Limnephilidae) larva.

3'.	Abdomen lacking ventrolateral gills, and with only 2 or 3 hairs at base of anal claw (Fig. 17.42); posterior margin of meso- and metanotal plates usually straight (Fig. 17.37); minute larvae usually less than 6 mm long, constructing portable cases of sand (e.g., Fig. 17.22), algae, or fixed cases of silk. Widespread in rivers, streams, and lakes	**HYDROPTILIDAE*** (p. 359)
4(2').	Antennae very long and prominent, at least 6 times as long as wide (Fig. 17.43); and/or sclerites on mesonotum lightly pigmented except for a pair of dark curved lines on posterior half (Fig. 17.44); larvae construct portable cases of various materials. Widespread in lakes and rivers	**LEPTOCERIDAE** (p. 362)
4'.	Antennae of normal length, no more than 3 times as long as wide (Fig. 17.45), or not apparent; mesonotum never with a pair of dark curved lines as above	5
5(4').	Mesonotum largely or entirely membranous (Fig. 17.45), or with small sclerites covering not more than half of notum (Fig. 17.46); pronotum never with an anterolateral lobe (Fig. 17.46)	6
5'.	Mesonotum largely covered by sclerotized plates, variously subdivided and usually pigmented (Fig. 17.41), although sometimes lightly; pronotum sometimes with a transverse carina terminating in prominent anterolateral lobes (Fig. 17.51)	13
6(5).	Abdominal segment IX with sclerite on dorsum (Fig. 17.48)	7
6'.	Abdominal segment IX with dorsum entirely membranous (Fig. 17.53)	10
7(6).	Metanotal sa3 usually consisting of a cluster of setae arising from a small rounded sclerite (Figs. 17.45-17.46); prosternal horn present (Fig. 17.45); larvae construct tubular portable cases, mainly of plant materials (e.g., Figs. 17.15-17.16). Widespread in lakes, ponds, and slow streams	**PHRYGANEIDAE** (p. 377)
7'.	Metanotal sa3 consisting of a single seta not arising from a sclerite (Fig. 17.40); prosternal horn absent; larvae either constructing a tortoiselike case of stones or free-living	8
8(7').	Basal half of anal proleg broadly joined with segment IX, anal claw with at least one dorsal accessory hook (Fig. 17.47); larvae construct tortoiselike portable cases of small stones (Fig. 17.21). Widespread in rivers and streams	**GLOSSOSOMATIDAE** (p. 351)
8'.	Most of anal proleg free from segment IX, anal claw without dorsal accessory hooks (Fig. 17.48); larvae free-living without cases or fixed retreats until pupation	9
9(8').	Tibia, tarsus, and claw of fore leg articulating against ventral lobe of femur to form a chelate leg (Fig. 17.50). Southwest, running waters	**HYDROBIOSIDAE** ...*Atopsyche*
9'.	Fore leg normal, not chelate as above (Fig. 17.49). Widespread in running waters	**RHYACOPHILIDAE** (p. 383)
10(6').	Labrum membranous and T-shaped (Fig. 17.52), often withdrawn from view in preserved specimens; larvae construct fixed sac-shaped nets of silk (Fig. 17.6). Widespread in rivers and streams	**PHILOPOTAMIDAE** (p. 377)
10'.	Labrum sclerotized, rounded and articulated in normal way (Fig. 17.54)	11
11(10').	Mesopleuron extended anteriorly as a lobate process, tibiae and tarsi fused together on all legs (Fig. 17.57); larvae construct fixed tubes of sand in small streams. Southern Texas	**XIPHOCENTRONIDAE** ....*Xiphocentron*
11'.	Mesopleuron not extended anteriorly, tibiae and tarsi separate on all legs (Fig. 17.55)	12
12(11').	Trochantin of prothoracic leg with apex acute, fused completely with episternum without separating suture (Fig. 17.56); larvae construct exposed funnel-shaped capture nets (Fig. 17.2), flattened retreats (Fig. 17.1), or tubes buried in loose sediments (Fig. 17.3). Widespread in most types of aquatic habitats	**POLYCENTROPODIDAE **** (p. 380)
12'.	Trochantin of prothoracic leg broad and hatchet-shaped, separated from episternum by dark suture line (Fig. 17.55); larvae construct tubular retreats on rocks and logs (e.g., Fig. 17.5). Widespread in running waters	**PSYCHOMYIIDAE** (p. 383)
13 (5').	Abdominal segment I lacking both dorsal and lateral humps (Fig. 17.60); each metanotal sa1 usually lacking entirely (Fig. 17.58), or, if represented only by a single seta without a sclerite, mesonotal sclerites subdivided similarly to Figure 17.58; larvae construct portable cases of various materials and arrangements (e.g., Figs. 17.12-17.13). Widespread in running waters	**BRACHYCENTRIDAE** (p. 350)

* The larva of Ecnomidae, recently described from Texas (Waltz and McCafferty 1983), keys to Hydroptilidae, but it is 5–7 mm long when mature, its abdominal tergum 9 is entirely membranous, and it never lives in a portable case.

** Includes the family Dipseudopsidae (genus *Phylocentropus*), which may be distinguished by the characters in the key to Polycentropodidae genera, couplet #1, on page 380.

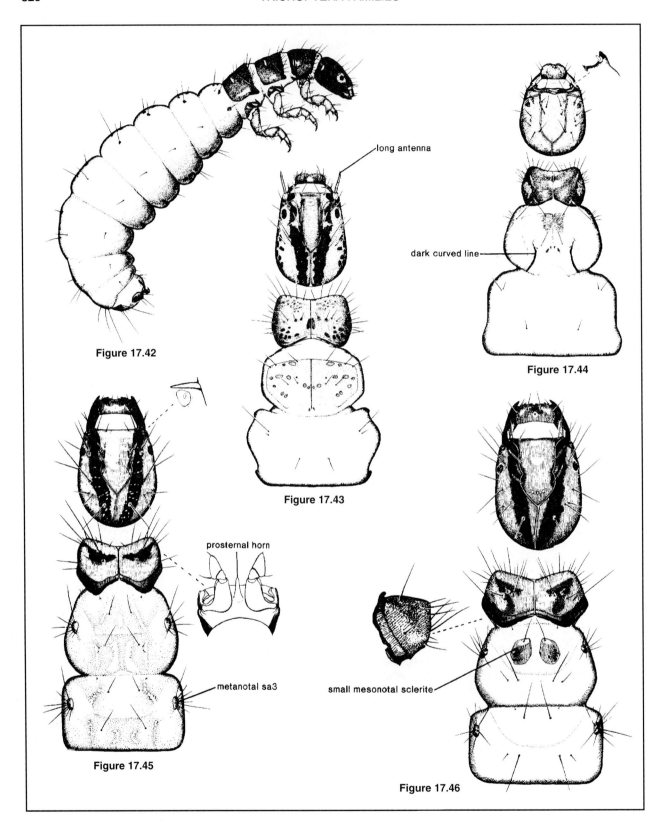

Figure 17.42. Lateral view of *Ochrotrichia* sp. (Hydroptilidae) larva.

Figure 17.43. Dorsal view of head and thorax of *Triaenodes* sp. (Leptoceridae) larva.

Figure 17.44. Dorsal view of head and thorax of *Ceraclea* sp. (Leptoceridae) larva, detail of antenna.

Figure 17.45. Dorsal view of head and thorax of *Ptilostomis* sp. (Phryganeidae) larva, details of antenna and prosternum.

Figure 17.46. Dorsal view of head and thorax of *Oligostomis* sp. (Phryganeidae) larva, detail of pronotum (lateral).

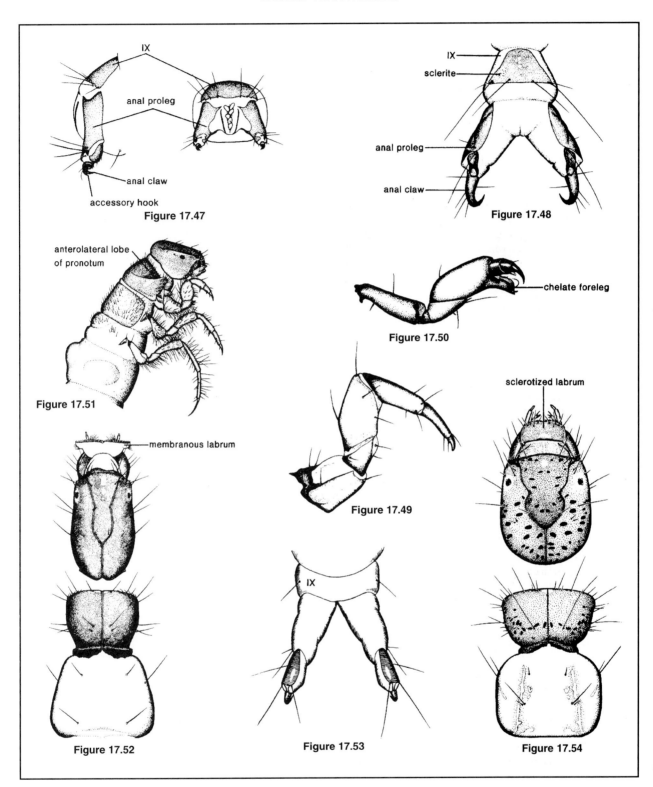

Figure 17.47. Lateral and caudal view of segment IX and anal prolegs of *Glossosoma* sp. (Glossosomatidae) larva.

Figure 17.48. Dorsal view of segment IX and anal prolegs of *Rhyacophila* sp. (Rhyacophilidae) larva.

Figure 17.49. Lateral view of foreleg of *Rhyacophila* sp. (Rhyacophilidae) larva.

Figure 17.50. Lateral view of foreleg of *Atopsyche* sp. (Hydrobiosidae) larva.

Figure 17.51. Lateral view of head, thorax, and abdominal segment I of *Beraea* sp. (Beraeidae) larva.

Figure 17.52. Dorsal view of head and thorax of *Dolophilodes* sp. (Philopotamidae) larva.

Figure 17.53. Dorsal view of segment IX and anal prolegs of *Chimarra* sp. (Philopotamidae) larva.

Figure 17.54. Dorsal view of head and thorax of *Polycentropus* sp. (Polycentropodidae) larva.

13'.	Abdominal segment I always with a lateral hump on each side although not always prominent, and with (Fig. 17.63) or without (Fig. 17.76) a median dorsal hump; metanotal sa1 always present, usually represented by a sclerite bearing several setae (Fig. 17.59) but with at least a single seta; larvae construct portable cases of various materials and arrangements 14
14(13').	Tarsal claw of hind leg modified to form a short setose stub (Fig. 17.61), or a slender filament (Fig. 17.62); larval case of sand grains with a dorsal cowl and lateral flanges (Fig. 17.18). Transcontinental through Canada to Alaska, and eastern, in lakes and large rivers, less commonly in spring streams ....*MOLANNIDAE* (p. 377)
14'.	Tarsal claws of hind legs no different in structure from those of other legs (Fig. 17.64)................................15
15(14').	Labrum with transverse row of approximately 16 long setae across central part (Fig. 17.65); larval case a hollowed twig, or of leaves (Fig. 17.17) and bark variously arranged. Eastern and western streams ....*CALAMOCERATIDAE* (p. 351)
15'.	Labrum with no more than 6 long setae across central part (Fig. 17.66) 16
16(15').	Anal proleg with lateral sclerite much reduced in size and produced posteriorly as a lobe from which a stout apical seta arises (Fig. 17.67); base of anal claw with ventromesal membranous surface bearing a prominent brush of 25-30 fine setae (Fig. 17.68); larval case of sand grains. Eastern, exceedingly local, in wet muck of spring seepage areas ....*BERAEIDAE* ....*Beraea*
16'.	Anal proleg with lateral sclerite not produced posteriorly as a lobe around base of apical seta (Fig. 17.69); base of anal claw with ventromesal surface lacking prominent brush of fine setae (Fig. 17.70) although setae may be present dorsally (Fig. 17.72) 17
17(16').	Antenna situated at or very close to the anterior margin of the head capsule (Fig. 17.71); prosternal horn lacking (Fig. 17.74); larval cases mainly of rock fragments 18
17'.	Antenna removed from the anterior margin of the head capsule and approaching the eye (Figs. 17.76-17.77); prosternal horn present although sometimes short (Fig. 17.77); larval cases of rock fragments or of plant materials 19
18(17).	Anal proleg with dorsal cluster of approximately 30 or more setae posteromesad of lateral sclerite (Fig. 17.72); fore trochantin relatively large, the apex hook-shaped (Fig. 17.74); larval case mainly of sand. Widespread in running waters and along lake shorelines ....*SERICOSTOMATIDAE* (p. 383)
18'.	Anal proleg with no more than 3-5 dorsal setae posteromesad of lateral sclerite, sometimes short spines (Fig. 17.73); fore trochantin small, the apex not hook-shaped (Fig. 17.75); larval case mainly of small rock fragments. Widespread in running waters ....*ODONTOCERIDAE* (p. 377)
19(17').	Antenna situated close to the anterior margin of the eye, median dorsal hump of segment I lacking (Fig. 17.76); larval cases of various materials and arrangements, frequently 4-sided (e.g., Fig. 17.19). Widespread, mainly in lotic habitats ....*LEPIDOSTOMATIDAE* (p. 362)
19'.	Antenna situated approximately halfway between the anterior margin of the head capsule and the eye; median dorsal hump of segment I almost always present (Fig. 17.77) 20
20(19').	Mesopleuron modified, usually extended anteriorly as a prominent process (Fig. 17.79), mesopleuron occasionally only spinose........................ *LIMNEPHILIDAE* (*Goerinae*)* (p. 365)
20'.	Mesopleuron unmodified (Fig. 17.77)................................21
21(20').	Mesonotum with anteromesal border emarginate (e.g., Fig. 17.78), the two primary sclerites closely aligned on the mid-dorsal line........................*UENOIDAE* (p. 383)
21'.	Mesonotum with anteromesal border not emarginate (Fig. 17.59), occasionally the two primary sclerites widely separated on the mid-dorsal line (*Rossiana*) ....*LIMNEPHILIDAE* (excl. *Goerinae*)** (p. 365)

* Now considered a separate family Goeridae. The genus *Goereilla* of Rossianidae also keys to Goeridae, but may be distinguished by the characters provided in couplet #5 on p. 365, in the key to Limnephilidae genera.
** The family Apataniidae and the genus *Rossiana* of Rossianidae also key to Limnephilidae. These may be distinguished in the key to Limnephilidae genera on p. 365.

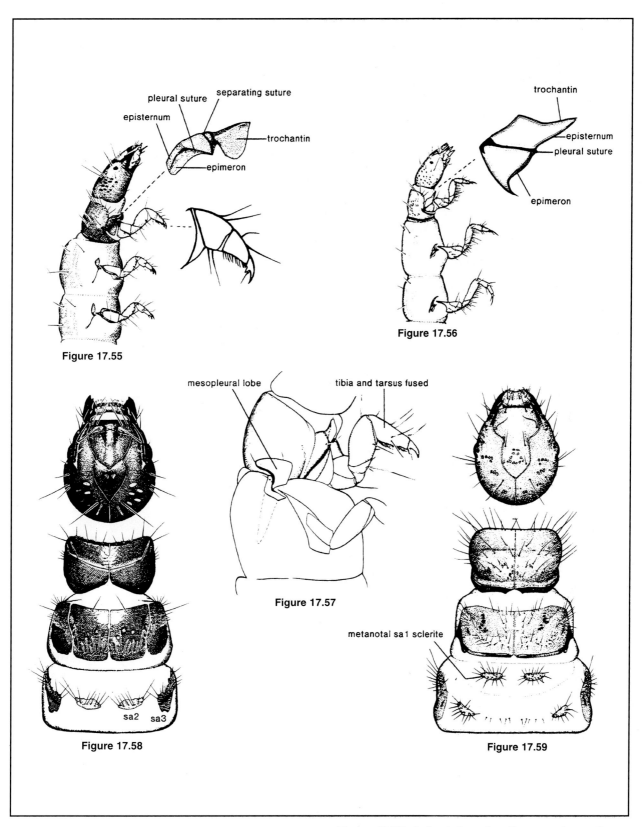

Figure 17.55. Lateral view of head and thorax of *Tinodes* sp. (Psychomyiidae) larva, details of trochantin and foreleg.

Figure 17.56. Lateral view of head and thorax of *Neureclipsis* sp. (Polycentropodidae) larva, detail of trochantin.

Figure 17.57. Lateral view of prothorax and mesothorax of *Xiphocentron* sp. (Xiphocentronidae) larva.

Figure 17.58. Dorsal view of head and thorax of *Brachycentrus* sp. (Brachycentridae) larva.

Figure 17.59. Dorsal view of head and thorax of *Pseudostenophylax* sp. (Limnephilidae) larva.

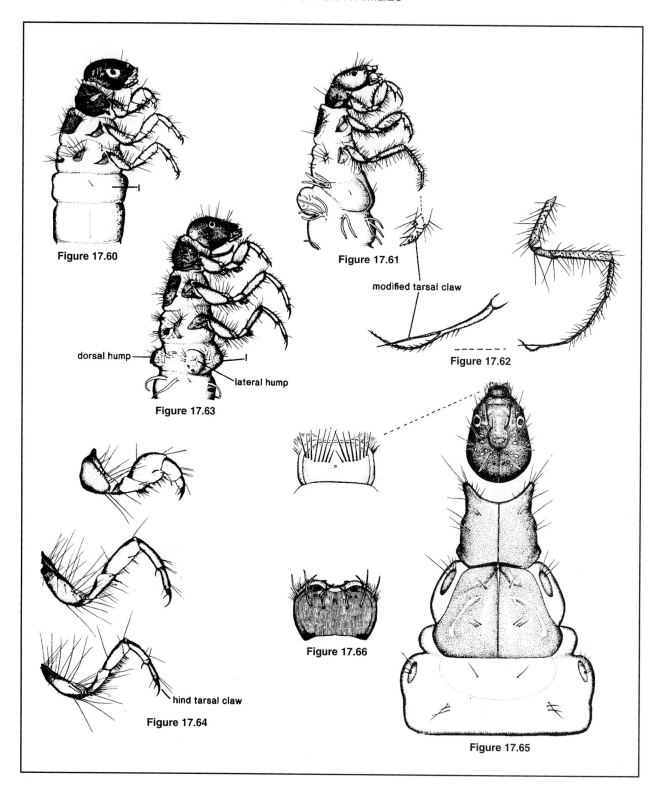

Figure 17.60. Lateral view of *Micrasema* sp. (Brachycentridae) larva.

Figure 17.61. Lateral view of *Molanna* sp. (Molannidae) larva, detail of hind tarsal claw.

Figure 17.62. Lateral view of hind leg of *Molannodes* sp. (Molannidae) larva, detail of tarsal claw.

Figure 17.63. Lateral view of *Desmona* sp. (Limnephilidae) larva.

Figure 17.64. Lateral view of legs of *Eobrachycentrus* sp. (Brachycentridae) larva.

Figure 17.65. Dorsal view of head and thorax of *Anisocentropus* sp. (Calamoceratidae) larva, detail of labrum.

Figure 17.66. Dorsal view of labrum of *Homophylax* sp. (Limnephilidae) larva.

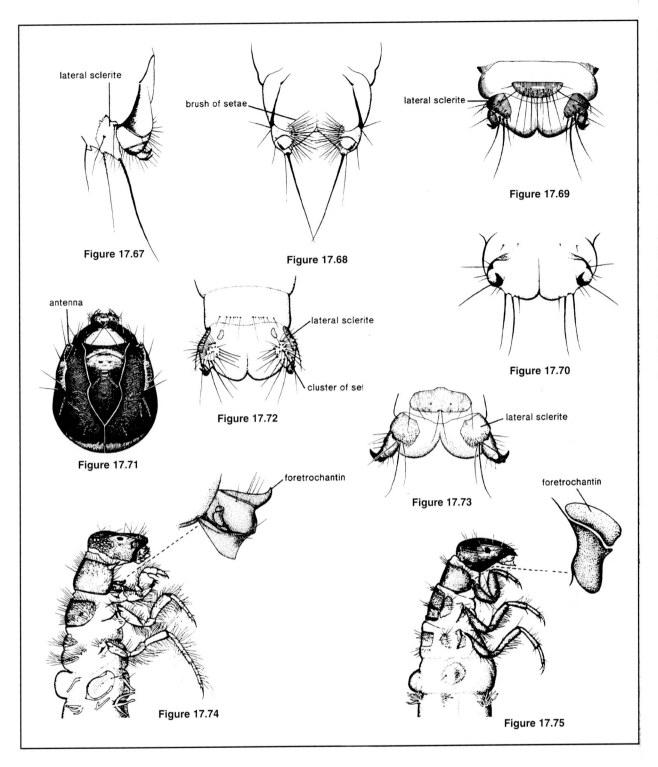

Figure 17.67. Lateral view of anal proleg of *Beraea* sp. (Beraeidae) larva.

Figure 17.68. Ventral view of anal prolegs of *Beraea* sp. (Beraeidae) larva.

Figure 17.69. Dorsal view of segment IX and anal prolegs of *Homophylax* sp. (Limnephilidae) larva.

Figure 17.70. Ventral view of anal prolegs of *Homophylax* sp. (Limnephilidae) larva.

Figure 17.71. Dorsal view of head of *Agarodes* sp. (Sericostomatidae) larva.

Figure 17.72. Dorsal view of segment IX and anal prolegs of *Agarodes* sp. (Sericostomatidae) larva.

Figure 17.73. Dorsal view of segment IX and anal prolegs of *Pseudogoera* sp. (Odontoceridae) larva.

Figure 17.74. Lateral view of *Fattigia* sp. (Sericostomatidae) larva, detail of trochantin.

Figure 17.75. Lateral view of *Marilia* sp. (Odontoceridae) larva, detail of trochantin.

Pupae

1.	Two pairs of hook plates present on each of segments III, IV, and V (Figs. 17.80, 17.95)	2
1'.	Two pairs of hook plates never present together on each of segments III, IV, and V	3
2(1).	Insects very small, less than 5-6 mm long	***HYDROPTILIDAE***
2'.	Insects larger, more than 6 mm long	12
3(1').	Abdomen lacking lateral fringe (Figs. 17.81, 17.84), although isolated setal tufts sometimes present (Figs. 17.85, 17.100)	4
3'.	Abdomen with lateral fringe, the fringe continuous where present (Figs. 17.30-17.31)	16
4(3).	Mandibles with only a single apical point (Fig. 17.99), although this is sometimes extended as a slender filament (Fig. 17.83)	5
4'.	Mandibles with at least one subapical toothlike point more prominent than the others in addition to an apical point (Figs. 17.89, 17.94)	12
5(4).	Hook plates on segment VII, frequently also on segment VIII (Fig. 17.84)	6
5'.	Hook plates not extending posteriorly beyond segment VI (Fig. 17.87)	10
6(5).	Hook plate Vp with 6 or more hooks (Fig. 17.81)	7
6'.	Hook plate Vp with only 2 or 3 hooks (Fig. 17.100)	20
7(6).	Hook plates present on segment II (Fig. 17.81)	8
7'.	Hook plates absent from segment II (Fig. 17.84)	9
8(7).	Hooks of plates on segments II, III, and IV arranged in an arc curved concavely anteriorly (Fig. 17.81)	***PSYCHOMYIIDAE****
8'.	Hooks of plates on segments II, III, and IV arranged in an arc curved concavely posteriorly (Fig. 17.82)	***XIPHOCENTRONIDAE***
9(7').	Anal processes rounded lobes, pair of hook plates usually present on segment VIII (Fig. 17.84)	***POLYCENTROPODIDAE***
9'.	Anal processes very slender and pointed, frequently hooked apically, segment VIII lacking hook plates (Fig. 17.86)	***UENOIDAE***
10(5').	Antennae little if any longer than body, not coiled around anal processes (Fig. 17.30)	11
10'.	Antennae much longer than body, coiled around anal processes (Fig. 17.109)	25
11(10).	Anal processes simple lobes closely appressed along midline, with stout apical bristles longer than processes themselves (Fig. 17.87)	***HELICOPSYCHIDAE***
11'.	Anal processes divergent in dorsal aspect, each one forked apically in lateral aspect, and lacking apical bristles (Figs. 17.102-17.103)	***BERAEIDAE***
12(2', 4').	Abdominal gills present, 2 pairs of hook plates present on segment III (Fig. 17.104)	***HYDROPSYCHIDAE***
12'.	Abdominal gills absent, 1 or 2 pairs of hook plates on segment III (Figs. 17.88, 17.95)	13
13(12').	One pair of hook plates on segment IV (Fig. 17.88)	***PHILOPOTAMIDAE***
13'.	Two pairs of hook plates on segment IV (Fig. 17.95)	14
14(13').	Segments VIII and/or IX with a pair of small hook plates, and/or 2 patches of long setae at apex of abdomen (Fig. 17.93)	***GLOSSOSOMATIDAE***
14'.	Segments VIII and IX lacking hook plates (Fig. 17.90)	15
15(14').	Apex of abdomen bearing 2 patches of setae (Fig. 17.90)	***HYDROBIOSIDAE***
15'.	Apex of abdomen lacking patches of setae (Fig. 17.95)	***RHYACOPHILIDAE***
16(3').	Lateral fringe of abdomen extending anteriorly from segments VII or VIII to V or VI (Fig. 17.96)	17
16'.	Lateral fringe of abdomen extending anteriorly from segments VII or VIII to III or IV (Fig. 17.106)	22
17(16).	Anal processes short or lobate, less than 5 times longer than wide (Figs. 17.105-17.106)	18
17'.	Anal processes long and slender, more than 5 times longer than wide (Fig. 17.100)	19
18(17).	Anterior hook plates longer than wide or both dimensions approximately the same, and most plates with 2-4 hooks (Fig. 17.30)	20
18'.	Anterior hook plates wider than long (Fig. 17.106) and usually each with 5 or more hooks (Fig. 17.96)	23
19(17').	Anterior hook plates wider than long (Fig. 17.96)	***BRACHYCENTRIDAE***
19'.	Anterior hook plates longer than wide, or the two dimensions approximately equal (Fig. 17.30)	20
20(6', 18, 19').	Bristles of labrum usually hooked apically, antennae usually with dorsal tuft (Fig. 17.29); hook plate Vp usually with more than 3 hooks (Fig. 17.30)	***LIMNEPHILIDAE*****

* The pupa of Ecnomidae, recently described from Texas (Waltz and McCafferty 1983), keys to Psychomyiidae, but its mandibles are elongate, curved without mesal serrations, and without a long, thin, filiform apical process.

** The families Apataniidae, Goeridae, and Rossianidae also key to Limnephilidae.

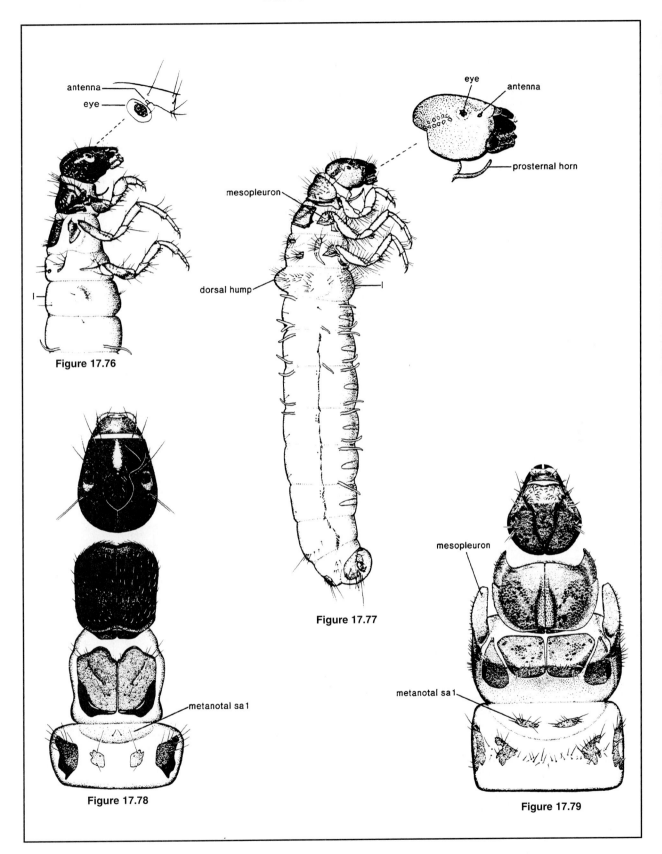

Figure 17.76. Lateral view of *Lepidostoma* sp. (Lepidostomatidae) larva, detail of antenna.

Figure 17.77. Lateral view of *Pseudostenophylax* sp. (Limnephilidae) larva, detail of head.

Figure 17.78. Dorsal view of head and thorax of *Neothremma* sp. (Uenoidae) larva.

Figure 17.79. Dorsal view of head and thorax of *Goera* sp. (Limnephilidae: Goerinae) larva.

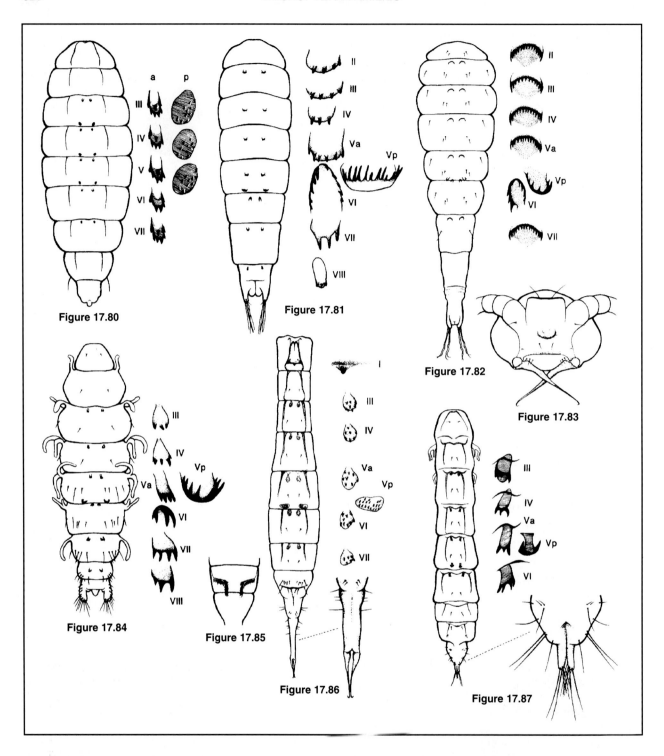

Figure 17.80. Dorsal view of abdomen of *Agraylea* sp. (Hydroptilidae) pupa, detail of hook plates: *a*, anterior, *p*, posterior.

Figure 17.81. Dorsal view of abdomen of *Psychomyia* sp. (Psychomyiidae) pupa, detail of hook plates.

Figure 17.82. Dorsal view of abdomen of *Xiphocentron* sp. (Xiphocentronidae) pupa, detail of hook plates.

Figure 17.83. Frontal view of head of *Xiphocentron* sp. (Xiphocentronidae) pupa.

Figure 17.84. Dorsal view of abdomen of *Polycentropus* sp. (Polycentropodidae) pupa, detail of hook plates.

Figure 17.85. Ventral view of abdominal segments VIII and IX of *Neothremma* sp. (Uenoidae) pupa.

Figure 17.86. Dorsal view of abdomen of *Neothremma* sp. (Uenoidae) pupa, detail of hook plates and anal processes.

Figure 17.87. Dorsal view of abdomen of *Helicopsyche* sp. (Helicopsychidae) pupa, detail of hook plates and anal processes.

20'.	Bristles of labrum never hooked apically, antennae usually with ventrolateral tuft (Fig. 17.99); hook plate Vp with 2-3 hooks (Fig. 17.97)	21
21(20').	Mesal margin of mandible concave in outline, apex pointed (Fig. 17.99); anterior hook plates with 2 or 3 hooks (Fig. 17.97)	**SERICOSTOMATIDAE**
21'.	Mesal margin of mandible straight in outline, apex usually attenuate (Fig. 17.101); most anterior hook plates with 1 hook (Fig. 17.100)	**ODONTOCERIDAE**
22(16').	Anal processes short and quadrate (Fig. 17.106) or triangular (Fig. 17.105) in dorsal aspect	23
22'.	Anal processes long and slender (Fig. 17.108)	24
23(18', 22).	Segment I with dorsomesal spined lobe, anal processes in dorsal aspect short and roughly quadrate (Fig. 17.106)	**PHRYGANEIDAE**
23'.	Segment I lacking dorsomesal spined lobe, anal processes more elongate than above in dorsal aspect and roughly triangular (Fig. 17.105)	**LEPIDOSTOMATIDAE**
24(22').	Dorsal abdominal setae arising in clusters (Fig. 17.107)	**CALAMOCERATIDAE**
24'.	Dorsal abdominal setae arising singly (Fig. 17.108)	25
25(10', 24').	Anal processes with apical bristles approximately one-half as long as processes themselves; antennae little longer than body and not coiled apically (Fig. 17.108)	**MOLANNIDAE**
25'.	Anal processes with much shorter apical bristles (Fig. 17.110) or with none; antennae much longer than body and coiled around base of anal processes (Fig. 17.109)	**LEPTOCERIDAE**

Adults

1.	Small insects, usually 5 mm or less in length; mesoscutum lacking setal warts, mesoscutellar setal warts transverse and meeting mesally to form an angulate ridge (Fig. 17.111); hind wings narrow and apically acute (Fig. 17.112), often with a posterior fringe of long hairs, the longest approximating the width of the hind wing	**HYDROPTILIDAE**
1'.	Insects usually more than 5 mm long; mesoscutum frequently with setal warts (Fig. 17.117), mesoscutellar setal warts usually rounded or elongate (Figs. 17.115, 17.120); hind wings usually broader and rounded apically (Fig. 17.34), posterior fringe, when present, of relatively shorter hairs	2
2(1').	Dorsum of head with 3 ocelli (Fig. 17.120)	3
2'.	Dorsum of head lacking ocelli (Fig. 17.129)	10
3(2).	Maxillary palp of 5 segments, terminal segment 5 flexible, usually at least twice as long as preceding segment 4 (Fig. 17.114)	**PHILOPOTAMIDAE**
3'.	Maxillary palp of 3, 4, or 5 segments, terminal segment similar to others in structure, usually approximately same length as preceding segment (Figs. 17.118, 17.123)	4
4(3').	Maxillary palp of 5 segments, segment 2 short, often rounded, and approximately same length as segment 1 (Fig. 17.118)	5
4'.	Maxillary palp of 3, 4, or 5 segments, segment 2 slender and longer than segment 1 (Figs. 17.123, 17.125)	8
5(4).	Maxillary palp with 2nd segment rounded and globose (Fig. 17.118)	6
5'.	Maxillary palp with 2nd segment not globose, but of same general cylindrical shape as 1st (Fig. 17.121)	**HYDROBIOSIDAE**
6(5).	Fore tibia with a preapical spur (Fig. 17.116)	**RHYACOPHILIDAE**
6'.	Fore tibia lacking a preapical spur (Fig. 17.119)	7
7(6').	Mesal setal warts of pronotum widely spaced (Fig. 17.117)	**GLOSSOSOMATIDAE**
7'.	Mesal setal warts of pronotum close together (as in Fig. 17.111, although not in contact	**HYDROPTILIDAE** (Ptilocolepinae)
8(4').	Middle tibia with 2 preapical spurs (Fig. 17.136); spurs usually 2, 4, 4	**PHRYGANEIDAE**
8'.	Middle tibia with 1 or no preapical spurs; spurs usually 1, 2-3, 4	9
9(8').	Anterior edge of hind wing with row of stout, hooked setae (Fig. 17.128)	**UENOIDAE**
9'.	Anterior edge of hind wing lacking row of stout hooked setae, although straight or slightly curved normal setae present; or if hooked setae present, vein R_1 of hind wing abbreviated and not reaching wing margin (*Lepania*)	**LIMNEPHILIDAE**
	(also includes ***Apataniidae, Rossianidae,*** and ***Lepania*** of ***Goeridae***)	
10(2').	Maxillary palp with 5 or more segments (Fig. 17.133)	11
10'.	Maxillary palp with fewer than 5 segments	15
11(10).	Terminal segment of maxillary palp flexible, with numerous cross-striae and different in structure from preceding segments; usually at least twice as long as preceding segment (Fig. 17.131)	12

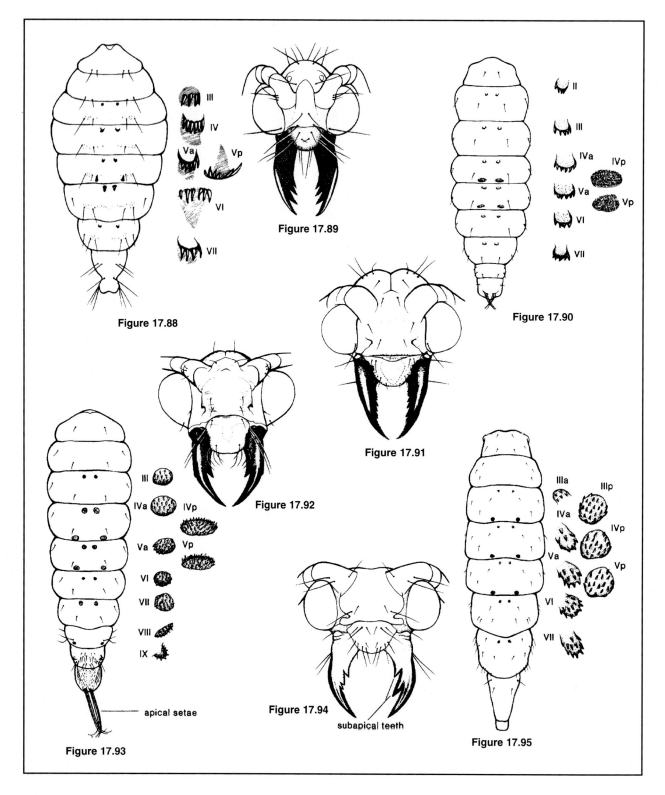

Figure 17.88. Dorsal view of abdomen of *Dolophilodes* sp. (Philopotamidae) pupa, detail of hook plates.

Figure 17.89. Frontal view of head of *Dolophilodes* sp. (Philopotamidae) pupa.

Figure 17.90. Dorsal view of abdomen of *Atopsyche* sp. (Hydrobiosidae) pupa, detail of hook plates.

Figure 17.91. Frontal view of head of *Atopsyche* sp. (Hydrobiosidae) pupa.

Figure 17.92. Frontal view of head of *Glossosoma* sp. (Glossosomatidae) pupa.

Figure 17.93. Dorsal view of abdomen of *Glossosoma* sp. (Glossosomatidae) pupa, detail of hook plates.

Figure 17.94. Frontal view of head of *Rhyacophila* sp. (Rhyacophilidae) pupa.

Figure 17.95. Dorsal view of abdomen of *Rhyacophila* sp. (Rhyacophilidae) pupa, detail of hook plates.

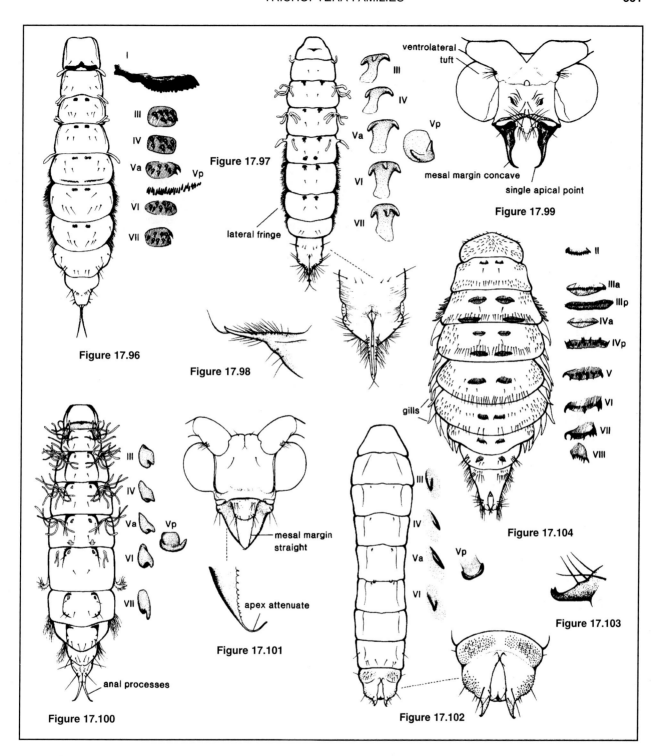

Figure 17.96. Dorsal view of abdomen of *Brachycentrus* sp. (Brachycentridae) pupa, detail of hook plates.

Figure 17.97. Dorsal view of abdomen of *Fattigia* sp. (Sericostomatidae) pupa, detail of hook plates and anal processes.

Figure 17.98. Lateral view of anal process of *Fattigia* sp. (Sericostomatidae) pupa.

Figure 17.99. Frontal view of head of *Fattigia* sp. (Sericostomatidae) pupa.

Figure 17.100. Dorsal view of abdomen of *Psilotreta* sp. (Odontoceridae) pupa, detail of hook plates.

Figure 17.101. Frontal view of head of *Psilotreta* sp. (Odontoceridae) pupa, detail of mandible.

Figure 17.102. Dorsal view of abdomen of *Beraea* sp. (Beraeidae) pupa, detail of hook plates and anal processes.

Figure 17.103. Lateral view of anal process of *Beraea* sp. (Beraeidae) pupa.

Figure 17.104. Dorsal view of abdomen of *Hydropsyche* sp. (Hydropsychidae) pupa, detail of hook plates.

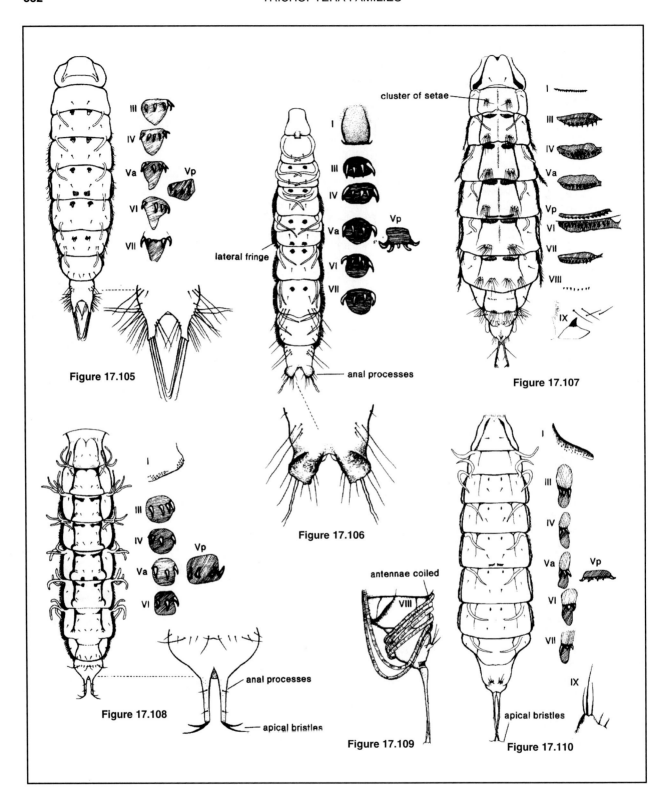

Figure 17.105. Dorsal view of abdomen of *Lepidostoma* sp. (Lepidostomatidae) pupa, detail of hook plates and anal processes.

Figure 17.106. Dorsal view of abdomen of *Banksiola* sp. (Phryganeidae) pupa, detail of hook plates and anal processes.

Figure 17.107. Dorsal view of abdomen of *Heteroplectron* sp. (Calamoceratidae) pupa, detail of hook plates.

Figure 17.108. Dorsal view of abdomen of *Molanna* sp. (Molannidae) pupa, detail of hook plates and anal processes.

Figure 17.109. Lateral view of terminal abdominal segments and antennae of *Oecetis* sp. (Leptoceridae) pupa.

Figure 17.110. Dorsal view of abdomen of *Oecetis* sp. (Leptoceridae) pupa, detail of hook plates.

11'.	Terminal segment of maxillary palp similar to others in structure and usually of approximately the same length as preceding segment (Fig. 17.121), or some segments with long hair brushes (Fig. 17.147)	15
12(11).	Mesoscutum lacking setal warts or setae (Fig. 17.152)	**HYDROPSYCHIDAE**
12'.	Mesoscutum with setal warts (Fig. 17.129)	13
13(12').	Mesoscutal setal warts quadrate and appressed along the median line over a large area approximately the size of the entire mesoscutellum (Fig. 17.137)	**XIPHOCENTRONIDAE**
13'.	Mesoscutal setal warts circular, sometimes touching at the median line, but much smaller than the mesoscutellum (Fig. 17.129)	14
14(13').	Fore tibia with a preapical spur, or if spur absent *(Cernotina)*, length of basal segment of tarsus less than twice the length of the longer apical spur (Fig. 17.130)	**POLYCENTROPODIDAE**[†]
14'.	Fore tibia lacking a preapical spur, and length of basal segment of tarsus at least twice the length of the longer apical spur (Fig. 17.134)	**PSYCHOMYIIDAE**
15(10', 11')	Mesoscutum lacking setal warts and setae (Fig. 17.138); tarsal segments, excepting basal one, with spines only around apex (Fig. 17.139)	**BERAEIDAE**
15'.	Mesoscutum with setal warts (Fig. 17.148) or setal area (Fig. 17.141); tarsal segments with spines usually arranged irregularly (Fig. 17.124)	16
16(15').	Mesoscutal setae arising in diffuse area over almost entire length of mesoscutum (Fig. 17.141)	17
16'.	Mesoscutal setae largely confined to pair of small discrete warts (Fig. 17.145)	19
17(16).	Antennae with scape at most twice as long as pedicel, dorsum of head usually with posteromesal ridge (Fig. 17.140)	**CALAMOCERATIDAE**
17'.	Antennae with scape at least 3 times longer than pedicel, dorsum of head lacking posteromesal ridge (Fig. 17.141)	18
18(17').	Antennae much longer than body, middle tibia lacking preapical spurs (Fig. 17.142)	**LEPTOCERIDAE**
18'.	Antennae little if any longer than body, middle tibia with 2 preapical spurs (Fig. 17.144)	**MOLANNIDAE**
19(16').	Dorsum of head with posterior setal warts very large, extending from mesal margin of eye to mid-dorsal line and anteriorly to middle of head (Fig. 17.145); antennae never longer than fore wing	**HELICOPSYCHIDAE**
19'.	Dorsum of head with posterior setal warts relatively smaller than above (e.g., Fig. 17.146); or antennae 1 1/2 times longer than fore wing	20
20(19').	Mesoscutellum with 1 mesal setal wart (Fig. 17.146)	21
20'.	Mesoscutellum with a pair of setal warts (Fig. 17.148), although sometimes touching along the mid-dorsal line (Fig. 17.153)	22
21(20).	Mesoscutellum almost entirely covered by a single setal wart, setae arising over most of wart (Fig. 17.146); maxillary palpi always 5-segmented	**ODONTOCERIDAE**
21'.	Mesoscutellum with setal wart narrower, setae largely confined to periphery (Fig. 17.122); maxillary palpi 5-segmented in females, but 3-segmented in males	**LIMNEPHILIDAE** (Goerini)[*]
22(20').	Pronotum with 1 pair of setal warts, median fissure of mesoscutum deep (Fig. 17.151)	**SERICOSTOMATIDAE**
22'.	Pronotum with 2 pairs of setal warts, median fissure of mesoscutum not as deep as above (Fig. 17.148)	23
23(22').	Middle tibia with 1 or 2 preapical spurs arising at a point about one-third distance from apex of tibia (Fig. 17.149) or without preapical spurs; abdomen with openings of glands on venter V in a pair of rounded sclerotized lobes (Fig. 17.150)	**BRACHYCENTRIDAE**
23'.	Middle tibia with 2 preapical spurs arising from approximately midpoint of tibia (Fig. 17.154); abdomen with glands on venter V not apparent	**LEPIDOSTOMATIDAE**

[†] The adult of Ecnomidae, recently described from Texas (Waltz and McCafferty 1983), keys to Polycentropodidae, but its fore wing has vein R_1 branched.

[*] Now considered a separate family Goeridae. The Goeridae genus *Lepania* keys to Limnephilidae, couplet #9, p. 329, but is distinguished from Limnephilidae in that couplet.

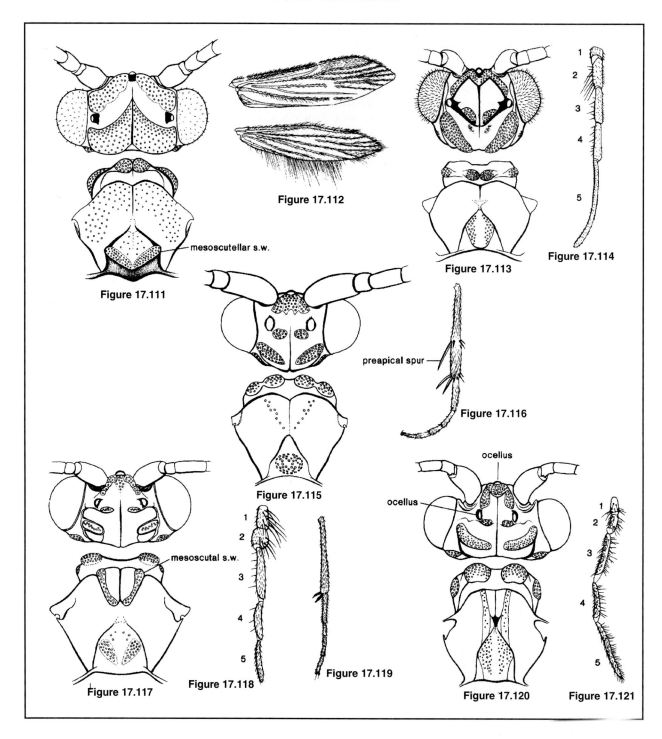

Figure 17.111. Doral view of head, pro-, and mesonotum of *Agraylea* sp. (Hydroptilidae) adult.

Figure 17.112. Dorsal view of wings of *Agraylea* sp. (Hydroptilidae).

Figure 17.113. Dorsal view of head, pro-, and mesonotum of *Dolophilodes* sp. (Philopotamidae) adult.

Figure 17.114. Maxillary palp of *Dolophilodes* sp. (Philopotamidae) adult.

Figure 17.115. Dorsal view of head, pro-, and mesonotum of *Rhyacophila* sp. (Rhyacophilidae) adult.

Figure 17.116. Tibia and tarsus of foreleg of *Rhyacophila* sp. (Rhyacophilidae) adult.

Figure 17.117. Dorsal view of head, pro-, and mesonotum of *Glossosoma* sp. (Glossosomatidae) adult.

Figure 17.118. Maxillary palp of *Glossosoma* sp. (Glossosomatidae) adult.

Figure 17.119. Tibia and tarsus of foreleg of *Glossosoma* sp. (Glossosomatidae) adult.

Figure 17.120. Dorsal view of head, pro-, and mesonotum of *Atopsyche* sp. (Hydrobiosidae) adult.

Figure 17.121. Maxillary palp of *Atopsyche* sp. (Hydrobiosidae) adult.

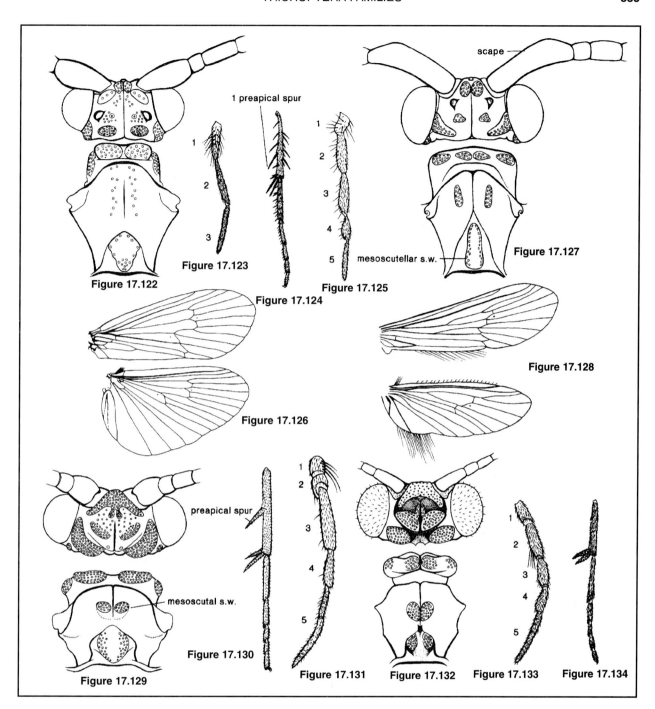

Figure 17.122. Dorsal view of head, pro-, and mesonotum of *Limnephilius* sp. (Limnephilidae) adult.

Figure 17.123. Maxillary palp of *Limnephilus* sp. (Limnephilidae) male.

Figure 17.124. Middle tibia and tarsus of *Limnephilus* sp. (Limnephilidae) adult.

Figure 17.125. Maxilliary palp of *Limnephilus* sp. (Limnephilidae) female.

Figure 17.126. Wings of *Dicosmoecus* sp. (Limnephilidae).

Figure 17.127. Dorsal view of head, pro-, and mesonotum of *Neothremma* sp. (Uenoidae) adult.

Figure 17.128. Wings of *Farula* sp. (Uenoidae).

Figure 17.129. Dorsal view of head, pro-, and mesonotum of *Polycentropus* sp. (Polycentropodidae) adult.

Figure 17.130. Foretibia and tarsus of *Polycentropus* sp. (Polycentropodidae) adult.

Figure 17.131. Maxillary palp of *Polycentropus* sp. (Polycentropodidae) adult.

Figure 17.132. Dorsal view of head, pro-, and mesonotum of *Psychomyia* sp. (Psychomyiidae) adult.

Figure 17.133. Maxillary palp of *Psychomyia* sp. (Psychomylidae) adult.

Figure 17.134. Foretibia and tarsus of *Psychomyia* sp. (Psychomylidae) adult.

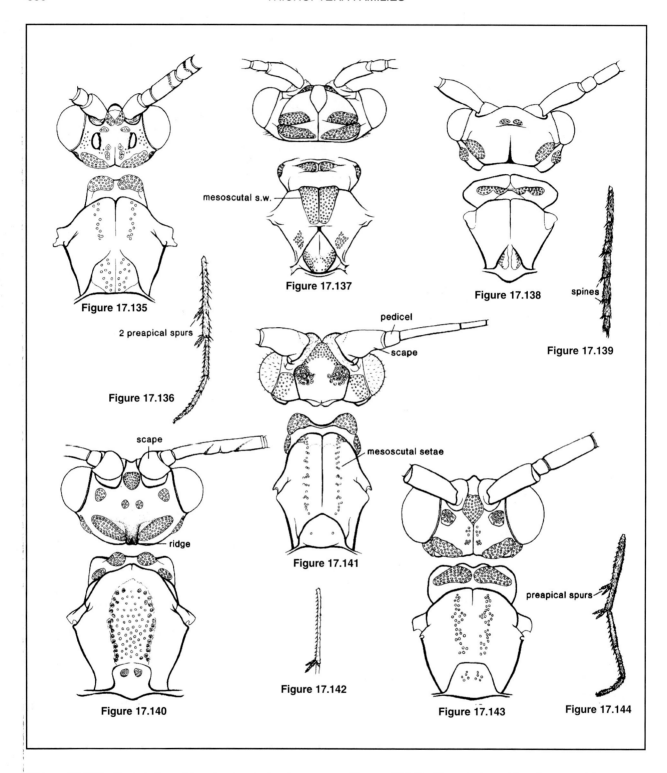

Figure 17.135. Dorsal view of head, pro-, and mesonotum of *Banksiola* sp. (Phryganeidae) adult.

Figure 17.136. Middle tibia and tarsus of *Banksiola* sp. (Phryganeidae) adult.

Figure 17.137. Dorsal view of head, pro-, and mesonotum of *Xiphocentron* sp. (Xiphocentronidae) adult.

Figure 17.138. Dorsal view of head, pro-, and mesonotum of *Beraea* sp. (Beraeidae) adult.

Figure 17.139. Middle tarsus of *Beraea* sp. (Beraeidae) adult.

Figure 17.140. Dorsal view of head, pro-, and mesonotum of *Heteroplectron* sp. (Calamoceratidae) adult.

Figure 17.141. Dorsal view of head, pro-, and mesonotum of *Oecetis* sp. (Leptoceridae) adult.

Figure 17.142. Middle tibia of *Oecetis* sp. (Leptoceridae) adult.

Figure 17.143. Dorsal view of head, pro-, and mesonotum of *Molanna* sp. (Molannidae) adult.

Figure 17.144. Middle tibia and tarsus of *Molanna* sp. (Molannidae) adult.

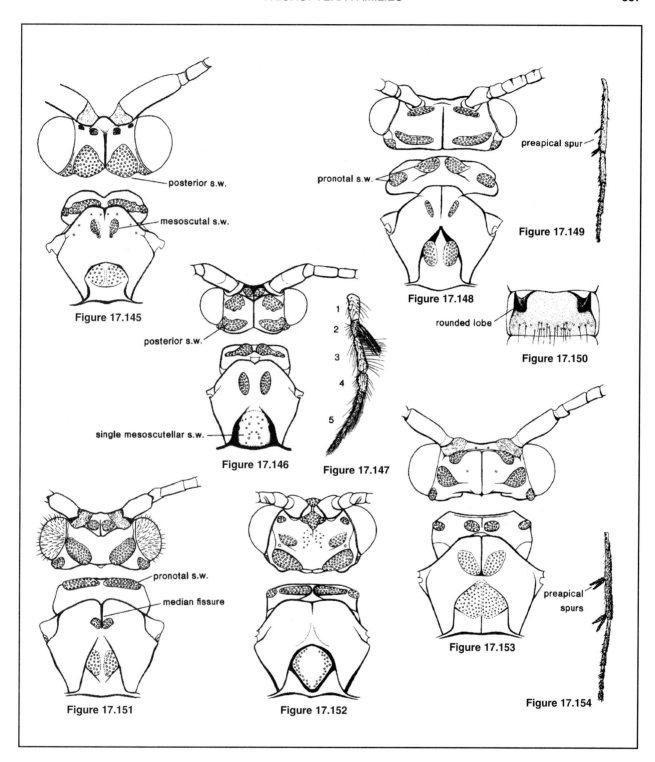

Figure 17.145. Dorsal view of head, pro-, and mesonotum of *Helicopsyche* sp. (Helicopsychidae) adult.

Figure 17.146. Dorsal view of head, pro-, and mesonotum of *Psilotreta* sp. (Odontoceridae) adult.

Figure 17.147. Maxillary palp of *Psilotreta* sp. (Odontoceridae) adult.

Figure 17.148. Dorsal view of head, pro-, and mesonotum of *Brachycentrus* sp. (Brachycentridae) adult.

Figure 17.149. Middle tibia and tarsus of *Brachycentrus* sp. (Brachycentridae) adult.

Figure 17.150. Ventral view of abdominal segment V of *Brachycentrus* sp. (Brachycentridae) adult.

Figure 17.151. Dorsal view of head, pro-, and mesonotum of *Agarodes* sp. (Sericostomatidae) adult.

Figure 17.152. Dorsal view of head, pro-, and mesonotum of *Hydropsyche* sp. (Hydropsychidae) adult.

Figure 17.153. Dorsal view of head, pro-, and mesonotum of *Lepidostoma* sp. (Lepidostomatidae) adult.

Figure 17.154. Middle tibia and tarsus of *Lepidostoma* sp. (Lepidostomatidae) adult.

ADDITIONAL TAXONOMIC REFERENCES
(L = larvae; P = pupae; A = adults)

General
Ross (1944)-L,P,A, (1959)-L; Wiggins (1977, 1995)-L, (1982); (1982)-L, A; Schmid (1980)-A.

Regional faunas
Alberta: Nimmo (1971, 1974, 1977a,b)-A.
California: Denning (1956)-A.
Illinois: Ross (1944)-L,P,A.
New York: Betten (1934)-A.
North and South Carolina: Unzicker et al. (1982)-L.
Wisconsin: Hilsenhoff (1981)-L.

Regional species lists
Alabama: Harris et al. (1991).
Alaska: Nimmo (1986b).
Arkansas: Unzicker et al. (1970); Bowles and Mathis (1989).
British Columbia: Nimmo and Scudder (1978, 1983).
Colorado: Hermann et al. (1986).
Delaware: Lake (1984).
Florida: Blickle (1962); Harris et al. (1982b).
Idaho: Smith (1965).
Indiana: Waltz and McCafferty (1983).
Kansas: Hamilton and Schuster (1978,1979,1980); Hamilton et al. (1983).
Kentucky: Resh (1975).
Louisiana: Harris et al. (1982a); Holzenthal et al. (1982); Lago et al. (1982).
Maine: Blickle (1964); Blickle and Morse (1966).
Manitoba: Flannagan and Flannagan (1982).
Massachusetts: Neves (1979).
Michigan: Leonard and Leonard (1949); Ellis (1962).
Minnesota: Etnier (1965), Lager et al. (1979).
Missouri: Mathis and Bowles (1992).
Mississippi: Harris et al. (1982a); Holzenthal et al. (1982); Lago et al. (1982).
Montana: Roemhild (1982).
New Hampshire: Morse and Blickle (1953, 1957).
Newfoundland: Marshall and Larson (1982).
North Carolina: Denning (1950).
North Dakota: Harris et al. (1980).
Ohio: Huryn and Foote (1983).
Oklahoma: Bowles and Mathis (1992).
Oregon: Anderson (1976b).
Pennsylvania: Masteller and Flint (1992).
Quebec: Roy and Harper (1979).
South Carolina: Morse et al. (1980).
Tennessee: Etnier and Schuster (1979).
Texas: Edwards (1973).
Utah: Baumann and Unzicker (1981).
Virginia: Parker and Voshell (1981).
West Virginia: Tarter (1990).
Wisconsin: Longridge and Hilsenhoff (1975).
Wyoming: Ruiter and Lavigne (1985).
Yukon: Nimmo and Wickstrom (1984).

Taxonomic treatments at the familial and generic levels
Beraeidae: Wiggins (1954)-L, P, A; Hamilton (1985)-L,P,A.
Brachycentridae: Wiggins (1965)-L; Flint (1984)-L,A.
Glossosomatidae: Ross (1956)-A.
Hydrobiosidae: Ross and King (1952)-A.
Hydropsychidae: Denning (1943)-A; Flint (1961, 1974)-L,A; (1974)-L, A; Schmid (1968)-A; Smith (1968a)-L, A; Gordon (1974)-A; Ross and Unzicker (1977)-A; Schuster and Etnier (1978)-L; Flint et al. (1979)-A; Givens and Smith (1980)-L, A; Nielsen (1981)-A; Schefter and Wiggins (1986)-A; Schefter et al. (1986)-A; Nimmo (1987)-A.
Hydroptilidae: Flint (1970)-L, A; Denning and Blickle (1972)-A; Blickle (1979)-A; Marshall (1979)-L, A; Kelley (1982)-A; Kelley and Morse (1982)-A; Kelley (1984)-A.
Lepidostomatidae: Ross (1946)-A; Flint and Wiggins (1961)-A; Wallace and Sherberger (1972)-A; Weaver (1988)-L, A.
Leptoceridae: Yamamoto and Wiggins (1964)-L, A; Yamamoto and Ross (1966)-A; Morse (1975, 1981)-A; Resh (1976b)-L, P; Haddock (1977)-L, A; Holzenthal (1982)-A; Manuel and Nimmo (1984)-L, A; Glover (1993)-L.
Limnephilidae: Ross and Merkley (1952)-A; Schmid (1955, includes references to several generic revisions)-A; Flint (1956, 1960)-L; Denning (1964, 1970, 1975)-A; Wiggins and Anderson (1968)-L, A; Wiggins (1973b,c)-L, A; Wiggins and Richardson (1982, 1987, 1989)-L, A; Wiggins and Winchester (1984)-A; Parker and Wiggins (1985)-L, A; .
Molannidae: Sherberger and Wallace (1971)-L; Roy and Harper (1980)-A.
Odontoceridae: Parker and Wiggins (1987)-L, A.
Philopotamidae: Ross (1956)-A; Armitage (1991)-A.
Phryganeidae: Wiggins (1956)-A, (1960a)-L, (1962)-L, P, A; Wiggins and Larson (1989)-L, A.
Polycentropodidae: Flint (1964b)-L, A; Ross (1965b)-A; Morse (1972)-A; Hudson et al. (1981)-L; Armitage and Hamilton (1990)-A; Schuster and Hamilton (1984)-A; Nimmo (1986a)-A.
Psychomyiidae: Ross and Merkley (1950)-A; Flint (1964b)-L, A.
Rhyacophilidae: Ross (1956)-A; Flint (1962)-L; Smith (1968b)-L; Schmid (1970, 1981)-A; Peck and Smith (1978)-L, A; Weaver and Sykora (1979)-L.
Sericostomatidae: Ross (1948)-A; Ross and Scott (1974)-A; Ross and Wallace (1974)-L, A.
Uenoidae (North American spp. previously included in Limnephilidae): Denning (1958,1975)-A; Wiggins et al. (1985)-L, A; Wiggins and Erman (1987)-A; Wiggins and Wisseman (1990, 1992)-A; Vineyard and Wiggins (1987, 1988)-L, A.
Xiphocentronidae: Ross (1949)-A; Edwards (1961)-L, P; Schmid (1982)-A; Armitage and Hamilton (1990)-A .

Table 17A. Summary of ecological and distributional data for *Trichoptera (caddisflies)*. (For definition of terms see Tables 6A-6C; table prepared by K. W. Cummins, J. R. Wallace, R. W. Merritt, G. B. Wiggins, and J. C. Morse.)

Taxa (number of species in parentheses)	Habitat	Habit	Trophic Relationships	North American Distribution	Ecological References*
Hydropsychoidea					144, 603, 1893, 2112, 4185, 4186, 4197, 4386, 3327, 4268, 4393
Philopotamidae (42)	Generally lotic—erosional	Clingers (sac-like silk net makers)	Generally collectors—filterers		
Chimarra (20)	Lotic—erosional (warmer rivers)	Clingers (saclike silk nets)	Collectors—filterers	Widespread	144, 669, 671, 2944, 3629, 4186, 4197, 4386, 1510, 3020, 4449, †
Dolophilodes (8) (= *Trentonius* =*Sortosa*)	Lotic—erosional (headwater streams and mossy seeps on rock faces)	Clingers (saclike silk nets)	Collectors—filterers	Widespread	88, 257, 2452, 2649, 3629, 4185, 4197, 4439, 4444, †
Wormaldia (14) (=*Dolophilus*)	Lotic—erosional	Clingers (saclike silk nets)	Collectors—filterers	Widespread	2461, 3541, 3674, 4186
Psychomyiidae (14)	Generally lotic—erosional	Clingers (silk-tube retreat makers)	Generally collectors—gatherers		603, 895, 2112, 3076, 3447, 3454, 4386, 4393
Paduniellinae *Paduniella*(1)				Arkansas	
Psychomyiinae *Lype* (1)	Lotic—erosional		Scrapers	East, North Central	2421, 4386, 4393,
Tinodes (9)	Lotic—erosional		Scrapers, collectors—gatherers	West	198, 835, 3080, 4386, 227, 982
Psychomyia (3)	Lotic—erosional	Clingers (silk tube retreats)	Collectors—gatherers scrapers	Widespread	88, 669, 671, 3080, 3629, 4386, 4393, †, 1607
Xiphocentronidae (1) *Xiphocentron* (1)	Lotic—erosional		Collectors—gatherers	Southern Texas	1037, 3887, 4393
Ecnomidae (1) *Austrotinodes* (1)	Lotic—erosional	Clingers (probably silk-tube retreat makers)	Collectors—filterers?	Mexico, Texas	1222, 4216, 4355
Polycentropodidae (76)	Generally lotic—erosional	Clingers (net-spinning retreat makers)	Generally collectors—filterers, some predators (engulfers)		2021, 2112, 2944, 3080, 3134, 3656, 4386, 4393, 978, 2806, 2959
Polycentropodinae					
Cernotina (7)	Lotic and lentic	Clingers (silk tube retreats)	Predators (engulfers)	Central, East	1841
Cyrnellus (1)	Lotic and lentic	Clingers (silk tube retreats)	Collectors—filterers	East, Central	4393
Neureclipsis (6)	Lotic—erosional (on vascular hydrophytes or other supports in water column)	Clingers (trumpet-shaped silk nets)	Collectors—filterers, shredders—herbivores, engulfers (predators)	East, Central	2618, 2944, 3182, 3629, 3333, 3336, 4386, 3108, 3332
Paranyctiophylax (10)	Lotic—erosional and depositional, lentic—littoral	Clingers (silk tube retreats)	Predators (engulfers), collectors—filterers, shredders—herbivores	Widespread, except Southwest	669, 671, 1218, 2618, 3728, 4386
Polycentropus (45)	Lotic—erosional, lentic—littoral	Clingers (silk tube retreats)	Predators (engulfers), collectors—filterers, shredders—herbivores	Widespread	88, 1131, 2618, 2944, 3629, 3933, 3934, 4386, 4322, 4482
Polyplectropus (2)	Lotic—erosional	Clingers (silk tube retreats)		Southwest	

* Emphasis on trophic relationships.
† Unpublished data, K. W. Cummins, Kellogg Biological Station and Oregon State University.

Table 17A. Continued

Taxa (number of species in parentheses)	Habitat	Habit	Trophic Relationships	North American Distribution	Ecological References*
Dipseudopsinae**					
Phylocentropus (5)	Lotic—depositional (sand, headwater streams)	Burrowers (branched, silk, buried tubes)	Collectors—filterers	East, Central	2414, 2421, 4199, 4386, 3728
Hydropsychidae (144)	Generally lotic—erosional, some lentic—erosional	Clingers (net-spinning fixed-retreat makers)	Generally collectors—filterers, some predators (engulfers)		603, 758, 895, 1014, 1707, 1712, 2112, 2831, 2944, 3076, 3447, 3454, 526, 4197, 4386, 4393, 1373, 4234, 4588, 318, 3728, 4439
Arctopsychinae					
Arctopsyche (4)	Lotic—erosional (cool streams)	Clingers (net spinners, fixed retreats)	Collectors—filterers (coarse particles, animal and plant)	Widespread	60, 257, 1628, 2649, 3721, 4178, 4186, 4197, 2316, 4137, 4234
Parapsyche (7)	Lotic—erosional	Clingers (net spinners, fixed retreats)	Collectors—filterers (coarse particles)	Widespread	60, 257, 2414, 2452, 3629, 3721, 4186, 4197, 978, 4537
Diplectroninae					
Diplectrona (4)	Lotic—erosional (headwater streams)	Clingers (net spinners, fixed retreats)	Collectors—filterers (coarse particles, especially detritus)	East, West	816, 1712, 2452, 2718, 2992, 4186, 4197, 4537, †
Homoplectra (11)	Lotic—erosional (rock face springs and seeps)	Clingers (net spinners, fixed retreats)	Collectors—filters	East, West	88, 4386, 4393
Oropsyche (1)	Lotic			North Carolina	
Hydropsychinae					
Ceratopsyche (30)‡ (=Symphitopsyche)	Lotic—erosional, some lentic—erosional	Clingers (net spinners, fixed retreats)	Collectors—filterers (particles include diatoms, detritus, animals)	Widespread	60, 3593, 4197, 4491
Cheumatopsyche (40)	Lotic—erosional (especially warmer streams and rivers)	Clingers (net spinners, fixed retreats)	Collectors—filterers (particles include algae, detritus, some animals)	Widespread	60, 88, 669, 1162, 1302, 2712, 2991, 2992, 3324, 3012, 3108, 4179, 3729, †
Hydropsyche (43)	Lotic—erosional	Clingers (net spinners, fixed retreats)	Collectors—filterers (particles include diatoms, algae, detritus, animals)	Widespread	60, 88, 144, 257, 669, 671, 758, 1131, 1327, 1328, 1329, 1418, 1470, 1711, 3674, 1712, 2021, 2168, 2178, 3105, 1324, 2418, 2609, 2648, 2649, 1323, 2712, 2830, 2882, 2944, 2461, 2991, 2992, 3080, 3133, 3020, 3182, 3324, 3603, 3629, 3656, 3661, 3686, 4179, 4581, 852, 1085, 4186, 4197, 4305, 982, 2316, 3088, 3729, †
Potamyia (1)	Lotic—erosional (larger rivers)	Clingers (net spinners, fixed retreats)	Collectors—filterers (detritus, diatoms)	East, South	1302, 3629
Smicridea (4)	Lotic—erosional	Clingers (net spinners, fixed retreats)	Collectors—filterers	West, Southwest	4386, 4393
Macronematinae					
Leptonema (1)	Lotic—erosional	Clingers (net spinners, fixed retreats)	Collectors—filterers	South Texas, Mexico	1217, 1220, 4386
Macrostemum (3) (=*Macronema* =*Macronemum*)	Lotic—erosional (larger rivers)	Clingers (net spinners, fixed retreats)	Collectors—filterers (fine particles)	East	1225, 3543, 4178, 4179, 4186, 4194, 4195, 4197, 3012, 3728, 3729

* Emphasis on trophic relationships.
** Now considered a distinct family, Dipseudopsidae.
† Unpublished data, K. W. Cummins, Kellogg Biological Station and Oregon State University.
‡ Although some systematists follow Ross and Unzicker (1977) and Nielsen (1981) in recognizing the *morosa-bifida* group as genus *Ceratopsyche,* sound reasons for retaining *Hydropsyche sens. lat.* were given by Schmid (1979).

Table 17A. Continued

Taxa (number of species in parentheses)	Habitat	Habit	Trophic Relationships	North American Distribution	Ecological References*
Rhyacophiloidea Rhyacophilidae (127)	Lotic—erosional	Clingers (free ranging)	Generally predators (engulfers)		895, 2112, 3076, 3447, 3454, 978, 4386, 4393, 4565
Himalopsyche (1)	Lotic—erosional	Clingers (free ranging) (alpine)	Predators (engulfers), scrapers	West	4386
Rhyacophila (126)	Lotic—erosional	Clingers (free ranging)	Predators (engulfers), a few scrapers, collectors—gatherers, shredders—herbivores (chewers)	Widespread (except Great Plains)	493, 2363, 2414, 2421, 2467, 2649, 2907, 3080, 3603, 3629, 3656, 3686, 852, 4006, 4386, 4565, 982, 983, 4234, 978, 2481, 2225, 2316, 2461, 2952, 2806
Hydrobiosidae (3) *Atopsyche* (3)	Lotic—erosional	Clingers (free ranging)	Predators (engulfers)	Southwest	4386, 4393
Glossosomatidae (76)	Lotic—erosional	Clingers (saddle- or turtle-shell-case makers)	Generally scrapers		895, 1707, 2112, 3076, 3447, 3454, 4386, 4393, 2066
Glossosomatinae *Anagapetus* (6)	Lotic—erosional (including small alpine springs)	Clingers (turtle shell case, mineral)	Scrapers	Rocky Mountains	88, 92, 4393
Glossosoma (22)	Lotic—erosional (including large alpine rivers)	Clingers (turtle shell case, mineral)	Scrapers	Widespread	88, 92, 144, 603, 790, 895, 1470, 2021, 2112, 2173, 2363, 2718, 2907, 3076, 3080, 3447, 3454, 3603, 3919, 4012, 4386, 2066, 2957, 3150, 2143, 2144, 4417, 111, 2142, 2956, 146, 2570, 2958, 112
Agapetinae *Agapetus* (30)	Lotic—erosional	Clingers (laterally compressed turtle shell case, mineral)	Scrapers, collectors—gatherers	Widespread	92, 111, 2649, 3656, 112, 1206, 3150, 4234
Protoptilinae *Culoptila* (4)	Lotic—erosional	Clingers (turtle shell case, mineral)	Scrapers	Widespread (primarily Southwest)	4386, 4393
Matrioptila (1)	Lotic—erosional	Clingers (depressed turtle shell case)	Scrapers	Southeast	4386, 4393
Protoptila (13)	Lotic—erosional	Clingers (laterally compressed turtle shell case, mineral)	Scrapers	Widespread	3629, 4386,, †
Hydroptilidae (230) (microcaddis)	Generally lotic and lentic—erosional, lentic—littoral	Generally clingers or climbers (may fasten case down; purse- or barrel-case makers). First 4 instars free living, 5th instar case builders	Generally piercers herbivores, scrapers, collectors—gatherers		895, 1707, 2112, 2908, 3076, 3447, 3454, 4386, 4393
Ptilocolepinae *Palaeagapetus* (2)	Lotic—erosional (cold springs and seeps)	Sprawlers (purse-type case of small leaf fragments, especially liverwort)	Shredders—detritivores	West, East	88, 4393

* Emphasis on trophic relationships.
† Unpublished data, K. W. Cummins, Kellogg Biological Station and Oregon State University.

Table 17A. Continued

Taxa (number of species in parentheses)	Habitat	Habit	Trophic Relationships	North American Distribution	Ecological References*
Hydroptilinae					
Hydroptilini					
Agraylea (4)	Lentic—vascular—hydrophytes (with filamentous algae), lotic—erosional (vascular hydrophytes)	Climbers (purse-type case of silk and algal and plant stem strands)	Piercers—herbivores (filamentous algae), collectors—gatherers	Widespread	88, 183, 2712, 2882, 2908, 3470, 3629, 3656, †
Dibusa (1)	Lotic—erosional	Clingers (purse-type case of silk and *Lemanea*)	Scrapers (red alga, *Lemanea*)?	East	3305, 4386
Hydroptila (100+)	Lotic—erosional and depositional (including seeps)	Clingers (purse-type case of silk and fine mineral)	Piercers—herbivores, scrapers	Widespread	88, 671, 2908, 3080, 982, 4386, †
Ochrotrichia (50+)	Lotic—erosional and depositional	Clingers (purse-type case of silk and fine mineral)	Collectors—gatherers, piercers—herbivores	Widespread	982, 4234, 4386, 4537
Oxyethira (40+)	Lentic—vascular hydrophytes (with filamentous algae), lotic—erosional and depositional (vascular—hydrophytes)	Climbers (purse-type case, flat, flask-shaped, open at back, primarily of silk)	Piercers—herbivores collectors—gatherers, scrapers (?)	Widespread	2908, 4386, †
Paucicalcaria (1)				Arkansas	1598
Stactobiini					
Stactobiella (3) (= *Tascobia*)	Lotic—erosional and depositional (small rapid streams)	Clingers? (purse-type case of silk and algal and plant stem strands)	Shredders	Widespread	88, 4386, 4393
Leucotrichiini					
Leucotrichia (3)	Lotic—erosional	Clingers (purse-type case, fixed)	Scrapers, collectors gatherers	Widespread (especially South)	88, 1221, 3629, 4386, 1604, 2569, 2571, 1603, 1607, 2066, †
Zumatrichia (1)	Lotic—erosional	Clingers (purse-type case, fixed)	Scrapers, collectors gatherers	Montana	4386
Orthotrichiini					
Ithytrichia (2)	Lotic—erosional (on rocks)	Clingers (purse-type case of silk)	Scrapers	Widespread	4386
Orthotrichia (6)	Lentic—vascular hydrophytes (with filamentous algae)	Clingers? (purse-type case of silk)	Piercers—herbivores	Widespread	4386
Neotrichiini					
Neotrichia (16)	Lotic—erosional	Clingers (case, a tube of fine mineral)	Scrapers	Widespread	4386, 4393
Mayatrichia (3)		Clingers? (case of silk, rigid)	Scrapers	Widespread	4386, 4393
Uncertain relationships (within Hydroptilidae)					
Alisotrichia (–*Rioptila*)(1)				Arizona, Utah	334, 335
Limnephiloidea					
Phryganeidae (27)	Generally lentic—littoral, lotic—depositional	Generally climbers (tube-case makers cylinders open at both ends)	Generally shredders—herbivores, predators (engulfers)		895, 2112, 3076, 3447, 3454, 4386, 4393
Yphriinae					
Yphria (1)	Lotic—depositional and erosional	Clingers—sprawlers (case a curved cylinder of mineral and wood pieces)	Predators (engulfers)	West	88, 4378, 4386

* Emphasis on trophic relationships.
† Unpublished data, K. W. Cummins, Kellogg Biological Station and Oregon State University.

Table 17A. Continued

Taxa (number of species in parentheses)	Habitat	Habit	Trophic Relationships	North American Distribution	Ecological References*
Phryganeinae *Agrypnia* (9)	Lentic—littoral (ponds), lotic—depositional	Climbers (case a spirally arranged, tapered cylinder of leaf pieces)	Shredders—detritivores and herbivores (chewers), scrapers(?)	Widespread	88, 4375, 4386, †
Banksiola (5)	Lentic—vascular hydrophytes, lotic depositional	Climbers (case similar to *Agrypnia* but with some irregular strands of vegetation)	Shredders—herbivores (chewers; early instars, filamentous green algae); last two instars predators (engulfers)	Widespread (except western Alaska)	88, 2618, 4375, 4386, 3871, 4482, 4483
Beothukus (1)	Lentic—sphagnum bog pools	Climbers (case similar to *Oligostomis*)		Northeast, Central	4392
Fabria (1)	Lentic—vascular hydrophytes	Climbers (case a rough,"Christmas tree-shaped" tube of long vegetation strands)	Shredders—herbivores	Northeast, Central	4386
Hagenella (1)		Climbers (case a slightly curved cylinder of leaf pieces)		Northeast	4386
Oligostomis (2)	Lotic—erosional and depositional (detritus and vascular hydrophytes)	Climbers (case similar to *Hagenella*)	Predators (engulfers) shredders—herbivores and detritivores (chewers)	East	2145, 3699, 4386
Oligotricha (1)	Lentic—littoral	Climbers (case similar to *Agrypnia*)	Predators (engulfers)	Western Alaska	4386
Phryganea (2)	Lotic—depositional and lentic (detritus and vascular hydrophytes)	Climbers (case a spirally arranged, tapered cylinder constructed of leaf pieces)	Shredders—herbivores and detritivores (chewers), predators (engulfers)	Widespread	293, 1361, 1511, 2145, 2363, 3656, 4375, 2902, 2959, 4386, †
Ptilostomis (4)	Lotic—erosional and depositional (detritus and vascular hydrophytes)	Climbers (case similar to *Hagenella*)	Shredders—herbivores and detritivores (chewers), predators (engulfers)	Widespread	2363, 4375, 4386, 4482, †
Brachycentridae (36)	Generally lotic—erosional	Generally clingers, climbers (tube-case makers, tapered, may be square in cross section)	Generally collectors filterers and gatherers, shredders—herbivores		895, 2112, 3076, 3447, 3454, 4386, 4393
Adicrophleps (1)	Lotic—erosional (in moss)	Clingers, climbers (case tapered, square in cross section)	Shredders?	East	4386
Amiocentrus (1)	Lotic—erosional (rocks and vascular hydrophytes)	Clingers, climbers (case a straight, tapered tube mostly of silk)	Collectors—gatherers	West	88, 4379, 4386
Brachycentrus (13) (*Oligoplectrum* a subgenus)	Lotic—erosional (on logs, branches, or vascular hydrophytes)	Clingers (case tapered, smooth, square in cross section; *Oligoplectrum* case round, of coarse mineral)	Collectors—filterers (particles include algae, detritus, animals), scrapers	Widespread	1224, 1345, 1346, 1347, 1551, 2363, 2648, 2650, 2812, 925, 2830, 3629, 4145, 4379, 1630, 4137, 4386, 4393, 1700
Eobrachycentrus (1)	Lotic—erosional (cold streams in moss)	Clingers (case similar to *Adicrophelps*)	Shredders—herbivores?	West (Mt. Hood area)	4379, 4386
Micrasema (20)	Lotic—erosional (on logs, branches, or vascular hydrophytes)	Clingers—sprawlers (case tapered, sometimes curved, of silk and strands of vegetation)	Shredders—herbivores (chewers), collectors gatherers	Widespread	88, 603, 2421, 3629, 227, 984, 4386, 982, 4234

* Emphasis on trophic relationships.
† Unpublished data, K. W. Cummins, Kellogg Biological Station and Oregon State University.

Table 17A. Continued

Taxa (number of species in parentheses)	Habitat	Habit	Trophic Relationships	North American Distribution	Ecological References*
Lepidostomatidae (80+)	Generally lotic—erosional and depositional (detritus)	Generally climbers sprawlers—clingers (tube-case makers; tapered, often square in cross section)	Shredders—detritivores (chewers)		895, 2112, 3076, 3447, 3010, 3454, 4386
Lepidostoma (75+)	Lotic—erosional and depositional (detritus) (headwater streams and springs)	Climbers—sprawlers—clingers (case square or "rough log cabin" type of leaf and bark fragments, or cylindrical, of sand or silk)	Shredders—detritivores (chewers) (also reported as scavengers)	Widespread	88, 94, 95, 603, 1463, 1553, 2414, 3623, 3629, 3647, 4450, 4482, 235, 3335, 4326, †
Theliopsyche (5)	Lotic—erosional (in gravel)	Climbers—sprawlers (case cylindrical, of sand)	Shredders—detritivores	East	4386
Limnephilidae (300+)	All types of lotic and lentic habitats	Climbers—sprawlers—clingers (tube case makers of great variety)	Generally shredders—detritivores (chewers), collectors—gatherers, scrapers		603, 895, 1214, 1707, 2112, 3076, 3447, 3454, 3728, 4386, 4393
Dicosmoecinae					
Allocosmoecus (1)	Lotic—erosional	Sprawlers (case curved, flattened, rough mineral)	Scrapers, shredders?	West	88, 4386
Amphicosmoecus (1)	Lotic—erosional and depositional, lentic littoral	Sprawlers (case a hollowed twig with anterior bark pieces or entire case of wood pieces)	Shredders	West	2920, 4386, 4393
Cryptochia (7)	Lotic—depositional (detritus; small, cool streams, springs and seeps)	Sprawlers (case flat, tapered, of wood and bark)	Scrapers, shredders	West	88, 4384, 4386, 4393, 2519
Dicosmoecus (4)	Lotic—erosional	Sprawlers (case curved, flattened, rough mineral)	Scrapers, shredders detritivores, engulfers (predators) (also reported as scavengers)	West	88, 474, 1606, 4386, 24, 1458, 4396, 1629, 1946, 2227, 892, 2226, 2229
Ecclisocosmoecus (1)	Lotic—depositional (sand)	Burrowers (case tapered, curved, smooth mineral)	Scrapers, shredders	Northwest	3335, 4386, 4393
Ecclisomyia (3)	Lotic—erosional (cold alpine streams)	Clingers (case a mineral tube, may have long plant pieces)	Collectors—gatherers, scrapers	West	88, 2649, 2920, 3452, 3335, 4386
Eocosmoecus (2)	Lotic—erosional (small spring streams)		Shredders		4339, 4398
Ironoquia (4) (=*Caborius*)	Lotic—depositional, lentic—littoral (temporary streams and ponds)	Sprawlers (case a curved tube of leaf or bark pieces, or mineral)	Shredders	East	179, 1213, 2414, 3447, 4381, 4386, 4443
Onocosmoecus (2)	Lotic—depositional (detritus), lentic—littoral (detritus)	Sprawlers (case of wood, bark, or mineral)	Shredders	Widespread	2920, 4386, 4393, 4450, 4397, 4482, 4514
Apataniinae**					
Apatania (17)	Lotic—erosional (especially springs), lentic—littoral (oligotrophic lakes)	Clingers—climbers sprawlers (case tapered, strongly curved, dorsal projection, mineral)	Scrapers, collectors gatherers	Widespread (primarily North and higher elevations)	88, 1049, 1214, 4386

* Emphasis on trophic relationships.
† Unpublished data, K. W. Cummins, Kellogg Biological Station and Oregon State University.
** Now considered a separate family Apataniidae.

Table 17A. Continued

Taxa (number of species in parentheses)	Habitat	Habit	Trophic Relationships	North American Distribution	Ecological References*
Pseudostenophylacinae					
Pseudostenophylax (3) (=*Drusinus*)	Lotic—erosional (detritus) and depositional (including seeps and temporary streams)	Sprawlers (case tapered, curved, smooth, mineral)	Shredders—detritivores (chewers), collectors—gatherers	East, West	85, 88, 94, 1214, 2414, 4386, 4390
Limnephilinae					
Anabolia (5)	Lentic—vascular hydrophytes, lotic—depositional (including temporary ponds)	Climbers—sprawlers (case a rough tube of leaf and wood pieces; may be three-sided)	Shredders—detritivores (chewers), collectors gatherers	East, North	1551, 2021, 3686, 4386, 289
Arctopora (3) (=*Lenarchulus*)	Lentic—littoral (including temporary ponds)	Climbers—sprawlers (case a smooth tube of long leaf pieces)		Northeast, Idaho	4386
Asynarchus (10)	Lentic—littoral (including temporary ponds), lotic—depositional	Climbers (case variable tube of mineral and plant pieces)		North, Rocky Mountains	88, 1214, 2920, 4386, 317, 4450, 4520
Chilostigma (1)				Minnesota	4384, 4386
Chilostigmodes (1)				Widespread (North)	2183, 4386
Chyranda (1)	Lotic—depositional (detritus)	Sprawlers (case a flat tube of bark and leaf pieces)	Shredders—detritivores	North, Western mountains	88, 4386, 4393, 4450
Clistoronia (4)	Lentic—littoral (sediments and detritus)	Sprawlers (case a rough tube of twig and bark pieces arranged longitudinally)	Collectors—gatherers, shredders—detritivores (chewers)	West	87, 88, 3255, 4386, 551, 4482
Clostoeca (1)	Lotic—depositional (detritus, spring seepage)	Sprawlers (case a flattened tube of large leaf pieces with flanges)	Shredders—detritivores	Northwest	4386, 4393
Desmona (2)	Lotic—depositional (sand)	Sprawlers—burrowers (case a tube of mixed mineral and wood fragments)	Shredders—detritivores and herbivores (chewers)	West	1101, 4386, 4393, 4402
Frenesia (2)	Lotic—erosional and depositional (detritus; including springs)	Sprawlers (case a smooth tube of mineral and wood pieces)	Shredders—detritivores (chewers)	Northeast	2363, 3454, 4386
Glyphopsyche (2)	Lentic—littoral (detritus), lotic—depositional (detritus)	Sprawlers (case a smooth tube of twig and bark pieces)		North, West, Missouri	4386
Grammotaulius (5)	Lentic—littoral and lotic—depositional (vascular hydrophytes), including temporary streams	Climbers (case a tube of long leaf pieces)		North	88, 1214, 1723, 4386
Grensia (1)	Lentic—littoral(?) (tundra lakes)	Sprawlers? (case a curved tube of plant fragments and mineral)		North (tundra zone)	4386
Halesochila (1)	Lentic—littoral (sediments)	Sprawlers (case a rough tube of leaf and wood pieces)	Collectors—gatherers, shredders—detritivores (chewers)	Northwest	1391, 4386, 4482
Hesperophylax (7) (=*Platyphylax*)	Lotic—erosional and depositional (detritus), including temporary streams	Sprawlers (case a slightly curved, slightly rough, coarse mineral tube)	Shredders—detritivores and herbivores (chewers) (scrapers, collectors—gatherers)	Widespread	88, 1470, 2363, 2495, 3629, 4386, 4450, 317, 2077, 3016
Homophylax (10)	Lotic—erosional (sediments and detritus)	Clingers—sprawlers (case a smooth tube of bark pieces)	Shredders—detritivores (chewers)	West	4386

* Emphasis on trophic relationships.

Table 17A. Continued

Taxa (number of species in parentheses)	Habitat	Habit	Trophic Relationships	North American Distribution	Ecological References*
Hydatophylax (4) (=*Astenophylax*)	Lotic—depositional (detritus)	Sprawlers—climbers (case a rough cylinder of wood, bark, mineral, with balance sticks)	Shredders—detritivores (chewers), collectors—gatherers	East, Southeast, Northwest	88, 95, 1214, 2363, 4145, 4386, †
Lenarchus (9)	Lentic—littoral (including temporary ponds)	Sprawlers—climbers (case a tube of long leaf pieces or bark and leaf pieces)	Collectors—gatherers	North, West	4386, 4450, 4482
Leptophylax (1)				Northeast	4386
Limnephilus (95-100)	All types of lotic and lentic habitats (including temporary ponds and streams)	Climbers, sprawlers, clingers (case of stick, leaf and/or sand construction, variable)	Shredders—detritivores and herbivores (chewers), collectors—gatherers (and probably others)	Widespread (especially North and mountains of West)	88, 179, 1214, 1551, 1552, 2168, 2363, 2618, 2691, 3542, 3629, 3686, 4145, 4381, 4386, 4450, 289, 2959, 4520, 1406, 2902, †
Nemotaulius (1) (=*Glyphotaelius*)	Lentic—littoral (detritus), lotic—depositional (detritus)	Sprawlers (case a flat "log cabin" type, of leaf pieces)	Shredders—detritivores (chewers)	Widespread in North	1214, 2363, 3447, 3629, 289, 3686, 4386
Phanocelia (1)				North	1134
Philarctus (1)	Lentic—littoral, lotic—depositional	Sprawlers (case a tube with small shells, seeds, and leaf pieces)		West, Northwest	4386
Philocasca (7)	Lotic—erosional (sediments and detritus) and depositional (detritus-semi-terrestrial)	Clingers—sprawlers (case rough, curved, coarse mineral)	Probably shredders—detritivores (chewers)	West mountains	84, 88, 4386, 4390
Platycentropus (3)	Lentic—littoral, lotic depositional (vascular hydrophytes, detritus)	Climbers (case a rough log cabin type of plant stems and other fine pieces)	Shredders—detritivores (and herbivores) (chewers)	East	1214, 4386, 4450
Psychoglypha (14)	Lotic—erosional (detritus) and depositional	Sprawlers—clingers (case a type of mixed mineral and bark and wood pieces)	Collectors—gatherers, shredders—detritivores (chewers) (including scavengers)	North, West	88, 474, 1214, 4386, 4482
Psychoronia (2)	Lotic	Sprawlers? (case a rough mineral tube)		Rocky Mountains	4386
Pycnopsyche (15)	Lotic—erosional and depositional, lentic—littoral (detritus)	Sprawlers—climbers of clingers (case smooth mineral, or like *Hydatophylax*)	Shredders—detritivores (chewers), scrapers (last instar in some species)	East, North to Rocky Mountains	669, 671, 788, 1152, 1468, 1824, 2363, 2415, 2416, 2422, 2712, 4386, 4445, 316, 317, 3728, 2482, 4450, 3728, 2482, 4057, 4200, 806, 2770, 3675, 3870, 2774
Sphagnophylax (1)	Lentic—transient tundra pools	Sprawlers	Shredders	Arctic	4401, 4476, 4477
Goerinae**					
Goera (6)	Lotic—erosional	Clingers (case mineral with two lateral ballast stones on each side)	Scrapers	East, West	669, 671, 4386, †
Goeraceu (2)	Lotic—erosional (including seeps)	Clingers (case like *Goera* with more than two ballast stones)	Scrapers	West	88, 4382, 4386
Goereilla (1)***	Lotic—depositional (springs)	Clingers (case tapered, curved, mineral)	Collectors—gatherers	West (Montana, Idaho)	4385, 4386
Goerita (2)	Lotic—erosional	Clingers (case tapered, curved, mineral)	Scrapers	Southeast	1155, 4382, 4386
Lepania (1)	Lotic—depositional (springs and seeps)	Sprawlers (case tapered, curved, mineral)	Collectors—gatherers	Oregon, Washington	4382, 4385, 4386

* Emphasis on trophic relationships.
† Unpublished data, K. W. Cummins, Kellogg Biological Station and Oregon State University.
** Now considered a separate family Goeridae.
*** Now included in a separate family Rossianidae with *Rossiana*.

Table 17A. Continued

Taxa (number of species in parentheses)	Habitat	Habit	Trophic Relationships	North American Distribution	Ecological References*
Uncertain relationships (within Limnephilidae)					
Allomyia (11)	Lotic—erosional	Clingers (case tapered, curved, mineral, may have ballast stones)	Shredders—herbivores (chewers, mosses?), scrapers	West	3577, 4383, 4386
Madeophylax (1)	Lotic	Clingers (case mineral)	Scrapers	East	1875
Manophylax (1)	Lotic—erosional (torrential)	Clingers (case tapered, curved, mineral and plant materials)	Scrapers	Idaho Mountains	4383, 4386, 4393
Moselyana (1)	Lotic—depositional (springs and seeps)	Sprawlers (case tapered, curved, fine mineral)	Collectors—gatherers?	Oregon and Washington	4383, 4386
Pedomoecus (1)	Lotic—erosional	Clingers (case tapered, curved, rough mineral)	Scrapers	West	4386, 4393
Rossiana (1)**	Lotic—erosional and depositional (especially in moss)	Clingers (case tapered, curved, rough minerals)	Probably scrapers and shredders—herbivores (chewers)	Northwest	4386
Uenoidae (49)	Lotic—erosional	Clingers (case tapered, slender, curved, mineral)	Scrapers, collectors—gatherers	West	88, 4386, 4399
Farula (10)	Lotic—erosional (on rocks, including seeps)	Clingers (case tapered, curved, long, very slender, mineral)	Scrapers, collectors—gatherers	West	88, 4386
Neothremma (7)	Lotic—erosional (on rocks)	Clingers (case tapered, slightly curved, long, slender mineral)	Scrapers, collectors—gatherers	West	88, 1214, 2649, 4386 2961
Neophylax (21)	Lotic—erosional	Clingers (case tapered, slightly curved, mineral with ballast stones on each side)	Scrapers	East, West	88, 669, 671, 2414, 2718, 3622, 3629, 4386, 217, 1720, 2560, 983, 1722, 2482, 1721
Oligophlebodes (7)	Lotic—erosional	Clingers (case strongly tapered and curved, rough mineral)	Scrapers, collectors—gatherers	West	2961, 3045, 4386
Sericostriata (1)	Lotic—erosional (on rocks)	Clingers (case tapered, slightly curved, long, slender, mineral)	Scrapers, collectors—gatherers	West	4399
Beraeidae (3)					
Beraea (3)	Lotic—depositional (detritus, springs)	Sprawlers (case curved, smooth, fine mineral)	Probably collectors—gatherers	Northeast, Georgia	1543, 4372, 4386
Sericostomatidae (13)	Generally lotic	Generally sprawlers	Generally shredders		895, 1923, 1924, 2021, 2112, 2882, 3447, 3454, 1305, 4004, 4386
Agarodes (10)	Lotic—erosional (detritus) and depositional (detritus)	Sprawlers (case tapered, curved, short, smooth, coarse mineral and wood fragments)	Shredders—detritivores (chewers), collectors—gatherers	East	3629, 4386
Fattigia (1)	Lotic	Sprawlers? (case curved, short, smooth, coarse mineral)	Shredders?	Southeast mountains	4386
Gumaga (2)	Lotic—erosional (spring brooks)	Sprawlers (case tapered, curved, long, smooth, fine mineral)	Shredders	West	3302, 3454, 4386, 4535, 3296, 3297, 4534, 279, 1157, 1876, 1155

* Emphasis on trophic relationships.
** Now included in a separate family Rossianidae with *Goereilla*.

Table 17A. Continued

Taxa (number of species in parentheses)	Habitat	Habit	Trophic Relationships	North American Distribution	Ecological References*
Odontoceridae (13)	Generally lotic—erosional and depositional (detritus)	Generally sprawlers?	Generally shredders (and scavengers)		4386
Marilia (2)	Lotic	Sprawlers? (case curved, smooth, coarse mineral)	Shredders	Southwest, Ontario	2421
Namamyia (1)	Lotic—erosional and depositional	Sprawlers? (case curved, smooth, coarse mineral)	Collectors—gatherers?	Oregon, California	4386
Nerophilus (1)	Lotic—depositional	Sprawlers? (case curved, slightly rough, coarse mineral)	Shredders—detritivores	West	4386
Parthina (2)	Lotic—erosional and depositional (small streams, springs, and seeps)	Sprawlers? (case curved, tapered, smooth, fine mineral and wood fragments)	Shredders—detritivores (chewers)?	Western mountains	88, 2421, 4386
Psilotreta (6)	Lotic—erosional (gravel, detritus) and depositional (detritus), (lentic—erosional)	Sprawlers (case curved, smooth, coarse mineral)	Scrapers, collectors gatherers	East	669, 671, 2414, 2421, 2057, 3017, 4386, †
Pseudogoera (1)	Lotic	Sprawlers? (case curved, tapered, rough, coarse mineral)	Predators (engulfers)	Southeastern mountains	4191, 4386
Molannidae (6)	Generally lentic—erosional, lotic—depositional (especially springs)	Generally sprawlers—clingers (case flat with dorsal projection and flanges)	Generally scrapers, collectors—gatherers, predators (engulfers)		895, 1462, 2112, 3076, 3447, 3454, 3686, 4386, †
Molanna (5)	Lentic—erosional, lotic—depositional (headwater streams and springs)	Sprawlers—clingers (case flat with dorsal projection and lateral ballast flanges, coarse to fine mineral)	Scrapers, collectors—gatherers, predators (engulfers)	Widespread	3686, 4386, †
Molannodes (1)	Lotic depositional (headwater streams and springs)	Sprawlers (case flat with dorsal projection and lateral flanges of wood fragments)	Collectors—gatherers?	Alaska, Yukon	4386
Helicopsychidae (4)					895, 2112, 3076, 3447, 2226, 3454, 3728, 4386, 2229, 3309
Helicopsyche (4)	Lotic and lentic—erosional (including thermal springs)	Clingers (case snail shell shaped, fine mineral)	Scrapers	Widespread	669, 671, 2648, 2649, 1156, 1230, 3629, 4386, 1876, 3308, 4427, 1157, 4114, 4115, 3309, 4116, 4118
Calamoceratidae (5)	Lotic	Sprawlers	Generally shredders detritivores and scrapers		895, 2112, 3076, 3447, 3454, 4386
Anisocentropus (1)	Lotic—depositional (detritus)	Sprawlers? (case flat of large leaf pieces with dorsal projection)	Generally shredders detritivores	Southeast	4192, 4386
Heteroplectron (2)	Lotic—erosional (detritus) and depositional (detritus)	Sprawlers (case a hollowed out stick or piece of bark)	Shredders—detritivores (chewers, of leaf litter and gougers of wood), scrapers?	East, West	88, 95, 2363, 3034, 96, 4386, 4482, 3919

* Emphasis on trophic relationships.
† Unpublished data, K. W. Cummins, Kellogg Biological Station and Oregon State University.

Table 17A. Continued

Taxa (number of species in parentheses)	Habitat	Habit	Trophic Relationships	North American Distribution	Ecological References*
Phylloicus (2)	Lotic	Sprawlers? (case a flat tube with large leaf and bark pieces and dorsal projection)	Shredders—detritivores	Southwest	2077, 4386
Leptoceridae (107)	All types of lotic and lentic habitats (including limnetic)	Climbers, sprawlers, clingers, swimmers (tube case makers of wide variety)	Collectors—gatherers, shredders—herbivores (chewers), predators (engulfers)		895, 2112, 3076, 3447, 1282, 3454, 3908, 4386
Ceraclea (36) (=*Athripsodes*)	Lotic and lentic (some in sponges)	Sprawlers, climbers (case a fine mineral or silk tube, may have dorsal projection and flanges)	Collectors—gatherers, shredders—herbivores (chewers), predators (engulfers of sponge)	Widespread	2618, 3288, 3290, 3291, 3292, 3293, 3311, 3447, 1685, 3629, 4386
Leptocerus (1)	Lentic—vascular—hydrophytes	Swimmers—climbers (case long and slender, primarily of silk)	Shredders—herbivores (chewers)	Northcentral, Northeast to South Carolina	2421, 2902, 4386
Mystacides (3)	Lotic—depositional, lentic—vascular hydrophytes	Sprawlers—climbers (case a rough tube of mineral and vegetation pieces may have balance sticks)	Collectors—gatherers, shredders—herbivores (chewers)	Widespread	88, 1723, 2363, 2618, 4386, 4582, †
Nectopsyche (12) (=*Leptocella*)	Lentic—vascular hydrophytes, lotic—erosional and depositional (vascular hydrophytes)	Climbers—swimmers (case long, slender of mineral and vegetation pieces, may have long balance sticks)	Shredders—herbivores (chewers), collectors gatherers (predators [engulfers])	Widespread	263, 1707, 2618, 2648, 2649, 2882, 3447, 3629, 2902, 4234, 4386
Oecetis (20)	Lotic—erosional and depositional, lentic littoral	Clingers—sprawlers, climbers (case a curved tube, often tapered, of coarse mineral or plant fragments)	Predators (engulfers), shredders—herbivores (chewers)?	Widespread	293, 2618, 3447, 3629, 4234
Setodes (8)	Lotic—erosional and depositional, lentic littoral	Sprawlers—clingers, burrowers (case curved, stout or slender, mineral)	Collectors—gatherers (particles include algae and animals), predators (engulfers)	East	2421, 2659, 4386
Triaenodes (23)	Lentic—littoral, lotic depositional (vascular hydrophytes)	Swimmers—climbers (case long, tapered of spirally arranged leaf and stem fragments)	Shredders—herbivores (chewers)	Widespread	263, 2421, 2618, 4011, 1421, 4386, †
Ylodes (4)	Lentic—littoral, lotic—depositional (vascular hydrophyles)	Swimmers—climbers burrowers (case long, tapered, of spirally arranged leaf and stem fragments	Shredders—herbivores (chewers)	Northern, Western	1421

* Emphasis on trophic relationships.
† Unpublished data, K. W. Cummins, Kellogg Biological Station and Oregon State University.

TRICHOPTERA GENERA

John C. Morse
Clemson University, South Carolina

Ralph W. Holzenthal
University of Minnesota, St. Paul

INTRODUCTION

Through the efforts of various recent workers, especially H. H. Ross, O. S. Flint, and G. B. Wiggins, larvae of most North American caddisfly genera have been associated and described. Exceptions noted below are rare and not likely to be encountered by the general collector. Pupae for the more common genera have been described by Ross (1944). Most pupae may be identified, however, by reference to last instar larval sclerites retained in the pupal case (except Leptoceridae, which rid their pupal cases of these sclerites) and/or to structures of the pharate adult within the pupal cuticle. Adults of North American caddisfly genera may be identified through use of the keys and references provided by Betten (1934), Ross (1944), Schmid (1980), Armitage (1991), and Armitage and Hamilton (1990).

For the most part, the following larval keys are modified from the work by Wiggins (1977). Indeed, use of the keys and corresponding figures in this section should be viewed only as a preliminary step toward assuring accurate larval determinations. The excellent illustrations and biological discussions in the more comprehensive treatises by Wiggins (1977, 1995) should be considered companion material for reliable identification of larval Trichoptera specimens to the generic level.

As the sequel to the preceding chapter, this section assumes the user's familiarity with the information given by Wiggins concerning ecology and life history of caddisflies and especially general trichopteran morphology. Structural features mentioned for the first time in this section are labelled in the figures. Even more so than in the chapter on family-level identification, accurate generic larval determinations are most likely obtained when last instar specimens are examined.

Of the 149 North American genera recognized in this chapter, the larvae of seven have not yet been described, including three genera of Limnephilidae (*Chilostigma*, *Chilostigmodes* and *Leptophylax*), one of Hydropsychidae (*Oropsyche*), one of Psychomyiidae *(Paduniella)* and two of Hydroptilidae (*Alisotrichia* and *Paucicalcaria*). Five families—Beraeidae, Ecnomidae, Helicopsychidae, Hydrobiosidae, and Xiphocentronidae—each contain only one North American genus (*Beraea*, *Austrotinodes*, *Helicopsyche*, *Atopsyche*, and *Xiphocentron*, respectively) and thus are not treated further.

KEYS TO THE GENERA OF TRICHOPTERA LARVAE

Brachycentridae

(modified from Flint 1984)

1. Meso- and metathoracic legs long, their femora about as long as head capsule, their tibiae each produced distally into prominent process from which stout spur arises (Fig. 18.1); case usually square in cross section, composed of small pieces of plant materials fastened transversely (Figs. 17.13, 18.16), although case sometimes cylindrical and largely of silken secretion, or occasionally of small rock fragments............................***Brachycentrus***
1'. Meso- and metathoracic legs shorter, their femora much shorter than head capsule (Fig. 17.60), each tibia not produced distally into prominent process, although spur arises from about same point on unmodified tibia (Fig. 18.2) 2
2(1'). Ventral apotome of head longer than wide, narrowed somewhat posteriorly, rudimentary prosternal horn present on anterior part of prosternum (Fig. 18.6); case 4-sided and tapered, composed of short pieces of plant material placed crosswise with loose ends often protruding (Fig. 18.17) 3

2'.	Ventral apotome of head usually wider than long (Fig. 18.7), sometimes squarish (Fig. 18.8); prosternal horn absent; case cylindrical, tapered, straight or curved, composed of lengths of plant material wound around the circumference (Figs. 17.12, 18.18) or of silk or silk and rock material (Fig. 18.19)	4
3(2).	Each half of mesonotum largely entire, lateral quarter partially delineated by variable suture, posterior margin raised and colored dark brown (Fig. 18.9)	*Eobrachycentrus*
3'.	Each half of mesonotum divided into 3 separate sclerites, posterior margin not conspicuously raised or colored (Fig. 18.10)	*Adicrophleps*
4(2').	Transverse pronotal groove curving anteriorly and usually meeting anterior margin (Fig. 18.11), sometimes forming a rounded lateral lobe; brown, sclerotized band on either side of anus (Fig. 18.13); mesonotal sa1 with multiple setae (Fig. 18.14) or solitary seta	*Micrasema*
4'.	Pronotal groove not curving anteriorly and never reaching anterior margin (Fig. 18.12); no brown, sclerotized bands near anus; mesonotal sa1 with solitary seta (Fig. 18.15)	*Amiocentrus*

Calamoceratidae

1.	Anterolateral corners of pronotum produced into prominent lobes (Figs. 17.65, 18.21-18.22); gills with 2 or 3 branches (Fig. 18.24)	2
1'.	Anterolateral corners of pronotum somewhat extended (Fig. 18.23), but much less than above; gill filaments single (Fig. 18.25); case a hollowed-out twig (Fig. 18.26)	*Heteroplectron*
2(1).	Metathoracic legs about as long as mesothoracic legs; anterolateral corners of pronotum pointed (Fig. 18.22); case of pieces of bark and leaves (Fig. 18.27)	*Phylloicus*
2'.	Metathoracic legs about twice as long as mesothoracic legs; anterolateral corners of pronotum rounded (Fig. 18.21); case of 2 leaf pieces, dorsal piece overlapping ventral one (Fig. 17.17)	*Anisocentropus*

Glossosomatidae

1.	Mesonotum with 2 or 3 sclerites (Figs. 18.28, 18.29); head with ventromesial margins of genae not thickened, posterior median ventral ecdysial line about 1.5 times as long as each anterior divergent branch (Figs. 18.31,18.32); anal opening without dark, sclerotized line on each side	2
1'.	Mesonotum without sclerites; head with ventromesal margins of genae thickened, posterior median ventral ecdysial line about as long as each anterior divergent branch (Fig. 18.30); anal opening with dark, sclerotized line on each side (Fig.18.33) ....(subfamily Glossosomatinae)	5
2(1).	Mesonotum with 3 sclerites (Fig. 18.29); ventral apotome of head a slender, V-shaped sclerite (Fig. 18.32) ....(subfamily Protoptilinae)	3
2'.	Mesonotum with 2 sclerites (Fig. 18.28); ventral apotome of head not as slender (Fig. 18.31) ....(subfamily Agapetinae)	*Agapetus*
3(2).	Each tarsal claw apparently trifid, the three points subequal in length (Fig. 18.34)	*Matrioptila*
3'.	Each tarsal claw with normal single point, basal seta and basal process much smaller (Figs. 18.35-18.36)	4
4(3').	Basal seta of each tarsal claw long, thin, and arising from side of stout basal process (Fig. 18.36); case constructed of relatively large stones (Fig. 18.39)	*Protoptila*
4'.	Basal seta of each tarsal claw short and stout, larger than basal process (Fig. 18.35); case constructed of uniformly small stones (Fig. 18.40)	*Culoptila*
5(1').	Pronotum excised one-third anterolaterally to accommodate coxae (Fig. 18.37)	*Glossosoma*
5'.	Pronotum excised two-thirds anterolaterally to accommodate coxae (Fig. 18.38)	*Anagapetus*

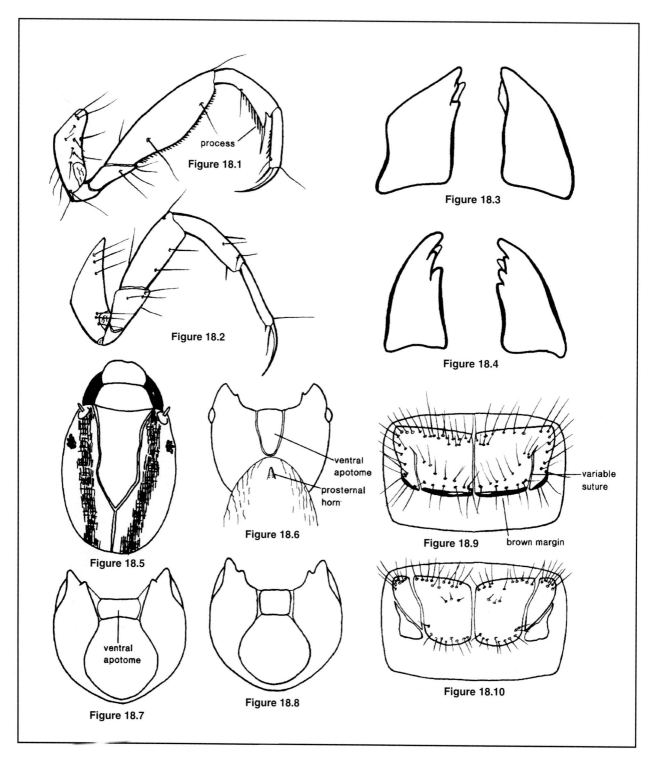

Figure 18.1. *Brachycentrus* sp. (Brachycentridae); lateral view of right metathoracic leg.

Figure 18.2. *Micrasema wataga* Ross (Brachycentridae); lateral view of right metathoracic leg.

Figure 18.3. *Triaenodes* sp. (Leptoceridae); ventral view of mandibles (after Glover 1993).

Figure 18.4. *Mystacides* sp. (Leptoceridae); ventral view of mandibles (after Glover 1993).

Figure 18.5. *Beothukus complicatus* (Banks) (Phryganeidae); dorsal view of head (after Wiggins and Larson 1989).

Figure 18.6. *Adicrophleps hitchcocki* Flint (Brachycentridae); ventral view of head and prosternum.

Figure 18.7. *Amiocentrus aspilus* (Ross) (Brachycentridae); ventral view of head.

Figure 18.8. *Micrasema wataga* Ross (Brachycentridae); ventral view of head.

Figure 18.9. *Eobrachycentrus gelidae* Wiggins (Brachycentridae); dorsal view of mesonotum.

Figure 18.10. *Adicrophleps hitchcocki* Flint (Brachycentridae); dorsal view of mesonotum.

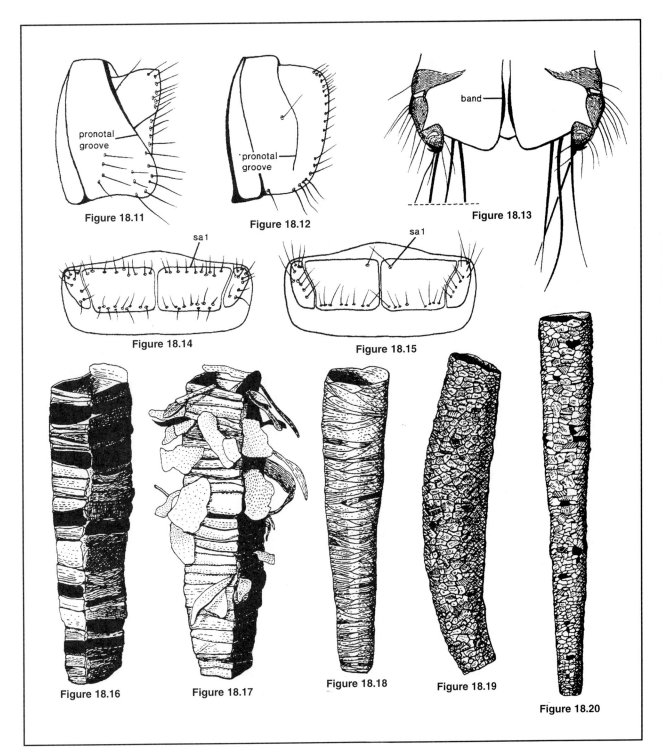

Figure 18.11. *Micrasema wataga* Ross (Brachycentridae); right lateral view of pronotum.

Figure 18.12. *Amiocentrus aspilus* (Ross) (Brachycentridae); right lateral view of pronotum.

Figure 18.13. *Micrasema wataga* Ross (Brachycentridae); posterior view of anus.

Figure 18.14. *Micrasema wataga* Ross (Brachycentridae); dorsal view of mesonotum.

Figure 18.15. *Amiocentrus aspilus* (Ross) (Brachycentridae); dorsal view of mesonotum.

Figure 18.16. *Brachycentrus* sp. (Brachycentridae); larval case.

Figure 18.17. *Adicrophleps hitchcocki* Flint (Brachycentridae); larval case.

Figure 18.18. *Micrasema wataga* Ross (Brachycentridae); larval case.

Figure 18.19. *Micrasema rusticum* (Hagen) (Brachycentridae); larval case.

Figure 18.20. *Brachycentrus echo* (Ross) (Brachycentridae); larval case.

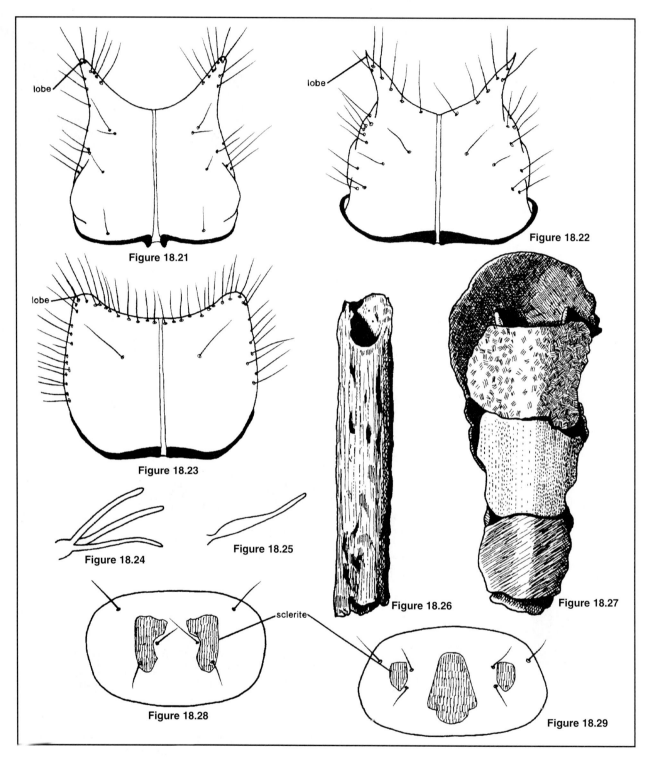

Figure 18.21. *Anisocentropus pyraloides* (Walker) (Calamoceratidae); dorsal view of pronotum.

Figure 18.22. *Phylloicus aeneus* (Banks) (Calamoceratidae); dorsal view of pronotum.

Figure 18.23. *Heteroplectron americanum* (Walker) (Calamoceratidae); dorsal view of pronotum.

Figure 18.24. *Anisocentropus pyraloides* (Walker) (Calamoceratidae); larval gill.

Figure 18.25. *Heteroplectron americanum* (Walker) (Calamoceratidae); larval gill.

Figure 18.26. *Heteroplectron americanum* (Walker) (Calamoceratidae); larval case.

Figure 18.27. *Phylloicus aeneus* (Banks) (Calamoceratidae); larval case.

Figure 18.28. *Agapetus* sp. (Glossosomatidae); dorsal view of mesonotum.

Figure 18.29. *Matrioptila jeanae* (Ross) (Glossosomatidae); dorsal view of mesonotum.

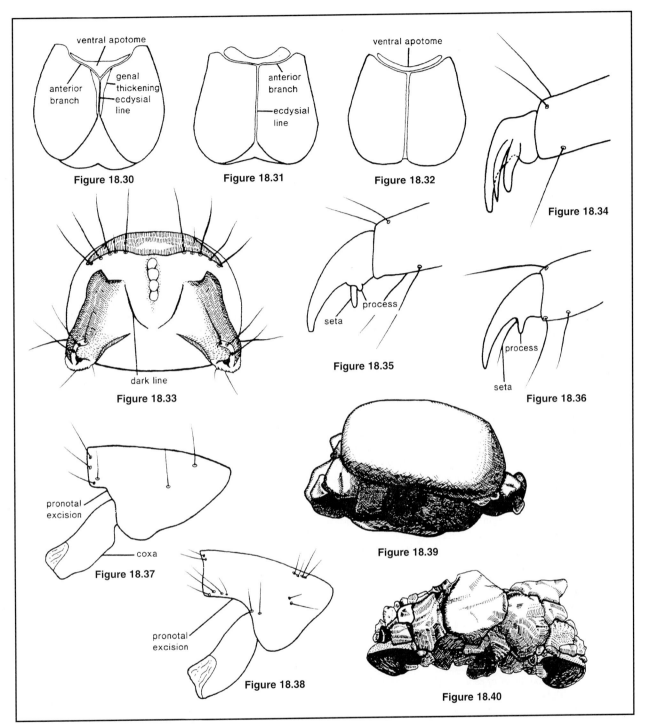

Figure 18.30. *Glossosoma nigrior* Banks (Glossosomatidae); ventral view of head.

Figure 18.31. *Agapetus* sp. (Glossosomatidae); ventral view of head.

Figure 18.32. *Matrioptila jeanae* (Ross) (Glossosomatidae); ventral view of head.

Figure 18.33. *Glossosoma nigrior* Banks (Glossosomatidae); posterior view of anus.

Figure 18.34. *Matrioptila jeanae* (Ross) (Glossosomatidae); lateral view of left mesothoracic tarsus.

Figure 18.35. *Culoptila moselyi* Denning (Glossosomatidae); lateral view of left mesothoracic tarsus.

Figure 18.36. *Protoptila* sp. (Glossosomatidae); lateral view of left mesothoracic tarsus.

Figure 18.37. *Glossosoma nigrior* Banks (Glossosomatidae); left lateral view of pronotum.

Figure 18.38. *Anagapetus* sp. (Glossosomatidae); left lateral view of pronotum.

Figure 18.39. *Protoptila* sp. (Glossosomatidae); larval case.

Figure 18.40. *Culoptila moselyi* Denning (Glossosomatidae); larval case.

Hydropsychidae[1]

(modified from Schuster and Etnier 1978)

1. Genae of head capsule completely separated by single ventral apotome (Figs. 18.41, 18.42) ....(subfamily Arctopsychinae)................4
1'. Genae touching ventrally, separating ventral apotome into anterior and posterior parts (Figs. 18.43, 18.44, 18.46) or posterior part inconspicuous (Fig. 18.45) 2
2(1'). Posterior ventral apotome at least one-half as long as median ecdysial line where genae touch (Fig. 18.43) ....(subfamily Diplectroninae)..................5
2'. Posterior ventral apotome much less than one-half as long as median ecdysial line (Figs. 18.44, 18.46) or inconspicuous (Fig. 18.45) 3
3(2'). Abdominal gills with up to 40 filaments arising fairly uniformly along central stalk (Fig. 18.47); fore trochantin never forked (Figs. 18.70-18.71) ....(subfamily Macronematinae)..................10
3'. Abdominal gills with up to 10 filaments arising mostly near the apex of the central stalk (Figs. 17.39, 18.48); fore trochantin usually forked (Figs. 17.39, 18.50), sometimes not (Fig. 18.49) ....(subfamily Hydropsychinae)..................6
4(1). Most abdominal segments dorsally with tuft of long setae and/or scale hairs on sa2 and sa3 positions (Fig. 18.53); ventral apotome of head usually nearly rectangular (Fig. 18.42) ....*Parapsyche*
4'. Most abdominal segments with single long seta in sa2 and sa3 positions, frequently with 1 or 2 shorter setae, but not a tuft (Fig. 18.54); ventral apotome narrowed posteriorly (Fig. 18.41) ....*Arctopsyche*
5(2). Pronotum with transverse furrow separating narrower posterior one-third from broader anterior two-thirds ....*Homoplectra*
5'. Pronotum without transverse furrow; constricted only slightly at posterior border (Fig. 18.55) ....*Diplectrona*
6(3'). Abdominal sternum VIII with single median sclerite (Fig. 18.62); submentum entire apically (Fig. 18.45) ....*Smicridea*
6'. Abdominal sternum VIII with pair of sclerites (Figs. 18.61, 18.72-18.73); submentum notched apically (Fig. 18.44) 7
7(6'). Prosternum with pair of large sclerites in intersegmental fold posterior to prosternal plate (Fig. 18.66); frontoclypeus entire (Fig. 18.63) 8
7'. Prosternum with pair of usually small sclerites posterior to prosternal plate (Fig. 18.67); if sclerites large, frontoclypeus with shallow mesal excision (Fig. 18.65) 9
8(7). Dorsum of abdomen with numerous plain hairs in addition to minute spines on at least first 3 segments, scale hairs present on at least last 3 segments (Fig. 18.68); club hairs absent ....*Hydropsyche*
8'. Dorsum of abdomen with plain hairs and club hairs only (Fig. 18.69); minute spines and scale hairs absent ....*Ceratopsyche*[2]
9(7'). Anterior ventral apotome of head with prominent anteromedian projection (Fig. 18.46); posterior margin of each sclerite on abdominal sternum IX entire (Fig. 18.73); lateral border of each mandible flanged (Fig. 18.51); fore trochantin forked or not (Fig. 18.49) ....*Potamyia*
9'. Anterior ventral apotome without anteromedian projection; posterior margin of each sclerite on abdominal sternum IX notched (Fig. 18.72); mandibles not flanged (Fig. 18.52); fore trochantin forked (Fig. 18.50) ....*Cheumatopsyche*
10(3). Tibia and tarsus of each prothoracic leg with dense, dorsal setal fringe (Fig. 18.71); dorsum of head flattened and margined with sharp carina (Fig. 18.60) ....*Macrostemum*
10'. Tibia and tarsus of each prothoracic leg lacking dense, dorsal setal fringe (Fig. 18.70); dorsum of head convex and without carina (Fig. 18.64) ....*Leptonema*

[1] The larva of *Oropsyche* is unknown.

[2] This group of species is recognized in recent publications by different systematists as the *Hydropsyche morosa* Species Group (Ross 1944, as *H. bifida* Group, see Schefter and Unzicker 1984), or as the subgenus *Ceratopsyche* of the genus *Symphitopsyche* (Ross and Unzicker 1977; Schuster and Etnier 1978), or as a separate genus *Ceratopsyche* (Nielsen 1981; Schuster 1984). Schuster (1984) presented evidence for the monophyly of each of the three species groups, *Hydropsyche s.s.*, *Symphitopsyche*, and *Ceratopsyche*, as well as for their distinctive ecological roles (Mackay and Wiggins 1979). We think the last mentioned position, recognizing these as distinct genera, ultimately will prove to be in the best interest of science.

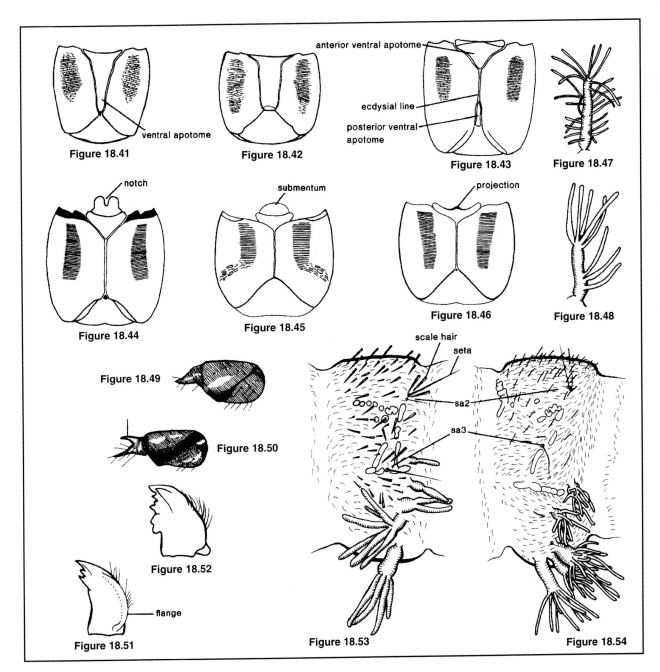

Figure 18.41. *Arctopsyche irrorata* Banks (Hydropsychidae); ventral view of head.

Figure 18.42. *Parapsyche cardis* Ross (Hydropsychidae); ventral view of head.

Figure 18.43. *Diplectrona modesta* Banks (Hydropsychidae); ventral view of head.

Figure 18.44. *Hydropsyche betteni* Ross (Hydropsychidae); ventral view of head (including submentum).

Figure 18.45. *Smicridea fasciatella* Mac Lachlan (Hydropsychidae); ventral view of head (including submentum).

Figure 18.46. *Potamyia flava* (Hagen) (Hydropsychidae); ventral view of head.

Figure 18.47. *Macrostemum carolina* (Banks) ((Hydropsychidae); left lateral view of gill of larval abdominal segment IV.

Figure 18.48. *Hydropsyche betteni* Ross (Hydropsychidae); left lateral view of gill of larval abdominal segment IV.

Figure 18.49. *Potamyia flava* (Hagen) (Hydropsychidae); lateral view of left foretrochantin.

Figure 18.50. *Cheumatopsyche* sp. (Hydropsychidae); lateral view of left foretrochantin.

Figure 18.51. *Potamyia flava* (Hagen) (Hydropsychidae); dorsal view of right mandible.

Figure 18.52. *Cheumatopsyche* sp. (Hydropsychidae); dorsal view of right mandible.

Figure 18.53. *Parapsyche cardis* Ross (Hydropsychidae); left lateral view of abdominal segment IV.

Figure 18.54. *Arctopsyche irrorata* Banks (Hydropsychidae); left lateral view of abdominal segment IV.

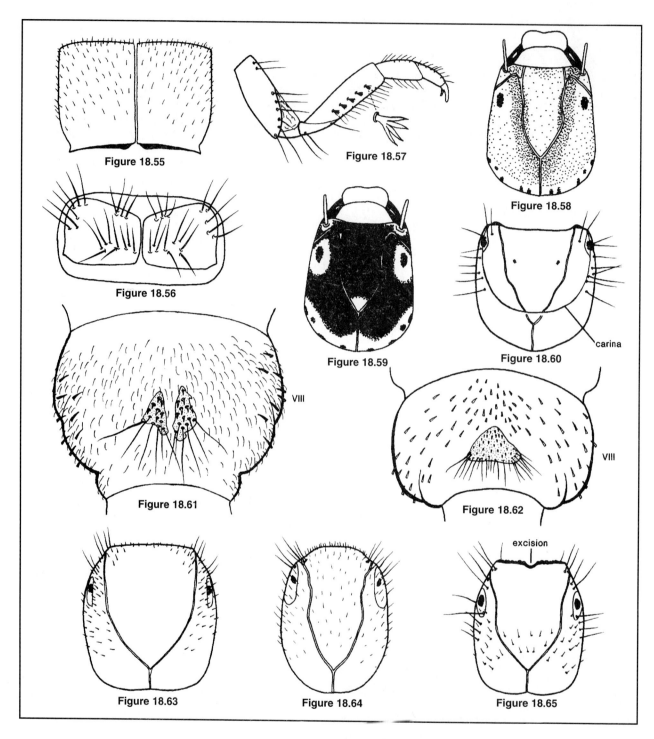

Figure 18.55. *Diplectrona modesta* Banks (Hydropsychidae); dorsal view of pronotum.

Figure 18.56. *Phanocelia canadensis* (Banks) (Limnephilidae); dorsal view of mesonotum (after Fairchild and Wiggins 1989).

Figure 18.57. *Homoplectra* sp. (Hydropsychidae); mesal view of mesothoracic leg.

Figure 18.58. *Ylodes frontalis* (Banks) (Leptoceridae); dorsal view of head (after Glover 1993).

Figure 18.59. *Triaenodes perna* Ross (Leptoceridae); dorsal view of head (after Glover 1993).

Figure 18.60. *Macrostemum carolina* (Banks) (Hydropsychidae); dorsal view of head.

Figure 18.61. *Hydropsyche betteni* Ross (Hydropsychidae); ventral view of abdominal sternum VIII.

Figure 18.62. *Smicridea fasciatella* Mac Lachlan (Hydropsychidae); ventral view of abdominal sternum VIII.

Figure 18.63. *Hydropsyche betteni* Ross (Hydropsychidae); dorsal view of head.

Figure 18.64. *Leptonema* sp. (Hydropsychidae); dorsal view of head.

Figure 18.65. *Cheumatopsyche* sp. (Hydropsychidae); dorsal view of head.

Hydroptilidae[3]

(final larval instar only)

1. Middorsal ecdysial line present on pronotum but usually lacking on meso- and metanota (Figs. 18.76-18.77); tarsal claws short and stout, tarsi about twice as long as claws (Fig. 18.74); abdominal segments V and VI usually abruptly broader than others in dorsal aspect (Figs. 18.76-18.77); flat, oval, silken case, with small circular opening near each end, fastened to rock (Fig. 18.97).................................(subfamily Leucotrichiinae) 2

1'. Middorsal ecdysial line present on all 3 thoracic nota (Fig. 17.37); tarsal claws variable, but, if short and stout, tarsi much less than twice as long as claws (Fig. 18.75); abdominal segments V and VI never abruptly broader than others in dorsal aspect 3

2(1). Sclerite on abdominal tergum IX with scattered short, stout setae (Fig. 18.76 inset); abdominal tergites II-VII with a pair of small circular punctures near the midline (Fig. 18.76 inset)..*Zumatrichia*

2'. Sclerite on abdominal tergum IX usually without short, stout setae, but, if present, in transverse bands (Fig. 18.77 inset); abdominal tergites II-VII solid, without punctures (Fig. 18.77 inset) ....*Leucotrichia*

3(1'). Metathoracic tarsi usually short and thick, each about as long as its claw (Fig. 18.79) or shorter; if metathoracic tarsi slender and slightly longer than claws (Fig. 18.78), prothoracic tibiae each with prominent posteroventral lobe (Fig. 18.81).............................4

3'. Metathoracic tarsi slender, each twice as long as its claw (Fig. 18.80) or longer; if metathoracic tarsi only slightly longer than claws, prothoracic tibiae each without prominent lobe 10

4(3). Larva and its case dorsoventrally depressed; each abdominal segment I-VIII with truncate, fleshy tubercle on each side (Fig. 18.82); case flat, elliptical valves covered with pieces of liverwort (Fig. 18.93)..........................(subfamily Ptilocolepinae).......*Palaeagapetus*

4'. Larva and its case laterally compressed; abdominal segments without lateral tubercles (Figs. 18.83-18.84) ....(subfamily Hydroptilinae) 5

5(4'). Tarsal claws stout and abruptly curved, each with thick, blunt basal spur (Fig. 18.75)...6

5'. Tarsal claws slender, gradually curved, each with thin, pointed basal spur (Figs. 18.78-18.79, 18.81) 7

6(5). Dorsal abdominal setae stout, each arising from small sclerite, dorsal rings distinct (Fig. 18.83); 2 long, parallel-sided valves of case incorporate red algae on which species live (Fig. 18.94) ....*Dibusa*

6'. Dorsal abdominal setae thin and without basal sclerites, dorsal rings indistinct (Fig. 18.84); 2 elliptical valves of silk case with few or no inclusions (Fig. 18.95) ....*Stactobiella*

7(5'). Meso- and metathoracic legs about 2.5 times as long as prothoracic legs or longer; case entirely of silk, shaped like a flask, open posteriorly (Fig. 18.96) ....*Oxyethira*

7'. Meso- and metathoracic legs not more than 1.5 times as long as prothoracic legs (Fig. 17.42).........8

8(7'). Meso- and metathoracic tibiae each about 1-2 times as long as broad (Fig. 18.79); case covered usually with sand (Fig. 17.22), occasionally with diatoms or filamentous algae; if case with algae, meso- and metathoracic tibiae each about as long as broad 9

8'. Meso- and metathoracic tibiae each about 2-3 times as long as broad (Fig. 18.88); case incorporating filamentous algae in concentric circles (Fig. 18.102) ....*Agraylea*

9(8). Anteroventral corners of metanotum each nearly 90° angle, with 1 or more setae (Fig. 18.86); 3 filamentous gills posteriorly: 1 dorsomesally behind abdominal tergite IX, other 2 mesad of lateral sclerites of anal prolegs (Fig. 18.85); case of 2 silk valves covered with sand or occasionally diatoms (Fig. 17.22) ....*Hydroptila**

9'. Anteroventral corners of metanotum each somewhat lobate, without setae (Fig. 18.87); no filamentous gills posteriorly; case of 2 silk valves covered with sand (similar to Fig. 17.22) or occasionally filamentous algae; sometimes one valve carried like a tortoise shell, the other a flat ventral sheet of silk ....*Ochrotrichia***

10(3'). Anal prolegs long and cylindrical, conspicuously projecting from body (Figs. 18.89, 18.90) 11

10'. Anal prolegs short, not especially projecting from body (Fig. 18.91) 12

---

* The larva of the single species of *Paucicalcaria* (Arkansas) keys here to *Hydroptila*. It may be distinguished by the base of each of the middle and hind tarsal claws large, quadrate, forming an angle with the inner margin of the rest of the claw.

** The genus *Matrichia* is sometimes considered a subgenus of *Ochrotrichia*. Larvae of *Metrichia* lack a basal seta on the tarsal claws of each of the middle and hind legs.

3. The larva of North American *Alisotrichia* species (Arizona, Utah) apparently is caseless and possibly can be further distinguished by the dorsoventrally depressed abdominal segments without lateral fleshy tubercles and segments V and VI not abruptly wider than segment IV.

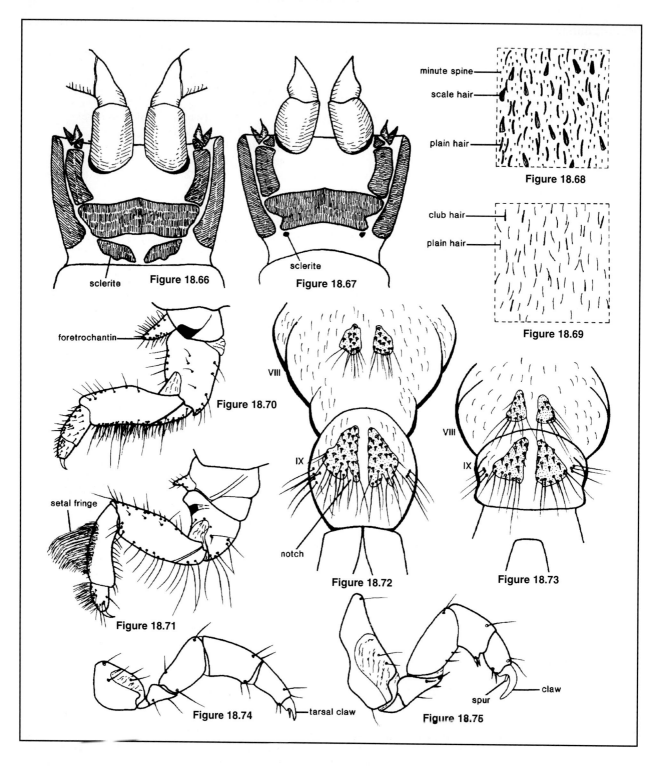

Figure 18.66. *Hydropsyche betteni* Ross (Hydropsychidae); ventral view of prosternum.

Figure 18.67. *Cheumatopsyche* sp. (Hydropsychidae); ventral view of prosternum.

Figure 18.68. *Hydropsyche betteni* (Ross) (Hydropsychidae); detail of setation on abdominal tergum VIII.

Figure 18.69. *Ceratopsyche ventura* (Ross) (Hydropsychidae); detail of setation on abdominal tergum VIII.

Figure 18.70. *Leptonema* sp. (Hydropsychidae); lateral view of left prothoracic leg.

Figure 18.71. *Macrostemum carolina* (Banks) (Hydropsychidae); lateral view of left prothoracic leg.

Figure 18.72. *Cheumatopsyche* sp. (Hydropsychidae); ventral view of abdominal sternum IX.

Figure 18.73. *Potamyia flava* (Hagen) (Hydropsychidae); lateral view of abdominal sternum IX.

Figure 18.74. *Leucotrichia* sp. (Hydropsychidae); lateral view of right metathoracic leg.

Figure 18.75. *Dibusa angata* Ross (Hydroptilidae): lateral view of right metathoracic leg.

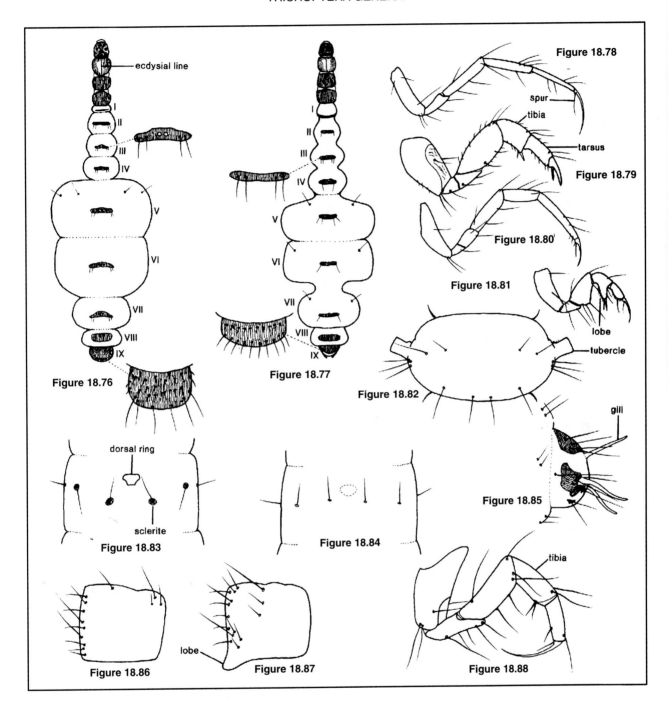

Figure 18.76. *Zumatrichia notosa* (Ross) (Hydroptilidae); dorsal view of larva; *insets*, sclerites of abdominal terga III and IX.

Figure 18.77. *Leucotrichia* sp. (Hydroptilidae); dorsal view of larva: *insets*, sclerites of abdominal terga III and IX.

Figure 18.78. *Oxyethira* sp. (Hydroptilidae); lateral view of right metathoracic leg.

Figure 18.79. *Hydroptila* sp. (Hydroptilidae); lateral view of right metathoracic leg.

Figure 18.80. *Ithytrichia* sp. (Hydroptilidae); lateral view of right metathoracic leg.

Figure 18.81. *Oxyethira* sp. (Hydroptilidae); lateral view of right prothoracic leg.

Figure 18.82. *Palaeagapetus celsus* (Ross) (Hydroptilidae); dorsal view of abdominal segment III.

Figure 18.83. *Dibusa angata* Ross (Hydroptilidae); dorsal view of abdominal segment III.

Figure 18.84. *Stactobiella delira* (Ross) (Hydroptilidae): dorsal view of abdominal segment III.

Figure 18.85. *Hydroptila* sp. (Hydroptilidae); left lateral view of posterior end of larval abdomen.

Figure 18.86. *Hydroptila* sp. (Hydroptilidae); left lateral view of metanotum.

Figure 18.87. *Ochrotrichia* sp. (Hydroptilidae); left lateral view of metanotum.

Figure 18.88. *Agraylea* sp. (Hydroptilidae); lateral view of right metathoracic leg.

11(10).	Abdomen somewhat depressed dorsoventrally, with prominent intersegmental grooves and lateral fringes of hairs (Fig. 18.89); cylindrical case of fine sand (Fig. 18.98)	*Neotrichia*
11'.	Abdomen more inflated, with shallow intersegmental grooves and no lateral fringes of hairs (Fig. 18.90); case of silk, cylindrical, but usually with transverse or longitudinal ridges (Fig. 18.99)	*Mayatrichia*
12(10').	Most abdominal segments with prominent, pointed, dorsal and ventral projections (Fig. 18.92); flat silk case open posteriorly, reduced to small circular opening anteriorly (Fig. 18.100)	*Ithytrichia*
12'.	Abdominal segments without dorsal and ventral projections (Fig. 18.91); silk case with longitudinal ridges (Fig. 18.101)	*Orthotrichia*

Lepidostomatidae

1.	Ventral apotome of head as long as, or longer than, median ecdysial line (Fig. 18.103); case usually 4-sided, of quadrate pieces of leaves or bark (Fig. 17.19), but pieces may be arranged irregularly, transversely, or spirally, or case may be of sand grains	*Lepidostoma*
1'.	Ventral apotome of head shorter than median ecdysial line (Fig. 18.104); case of sand grains (similar to Fig. 18.132)	*Theliopsyche*

Leptoceridae

1.	Tarsal claw of each mesothoracic leg hooked and stout; tarsus curved (Fig. 18.105); slender case of transparent silk (Fig. 18.119)	*Leptocerus*
1'.	Tarsal claw of each mesothoracic leg slightly curved and slender; tarsus straight (Fig. 18.106)	2
2(1').	Sclerotized, concave plate with marginal spines on each side of anal opening and extending onto ventral lobe (Fig. 18.107); cylindrical case of stones (Fig. 18.120)	*Setodes*
2'.	Sclerotized, spiny plates absent, although patches of spines or setae may be present (Fig. 18.108)	3
3(2').	Maxillary palpi extending far beyond labrum; mandibles long and bladelike, with sharp apical tooth separated from remainder of teeth (Fig. 18.110 and inset); cases of various types and materials	*Oecetis*
3'.	Maxillary palpi extending little, if any, beyond labrum; mandibles short, wide, with teeth grouped close to apex around central concavity (Fig. 18.111 and inset)	4
4(3').	Mesonotum with pair of dark, curved bars on weakly sclerotized plates (Fig. 17.44); abdomen broad basally, tapering posteriorly, with gills usually in clusters of 2 or more (Fig. 18.112); cases of various shapes and materials, sometimes including spicules and pieces of freshwater sponges	*Ceraclea*
4'.	Mesonotum without pair of dark bars (Fig. 17.43); abdominal segments I-VII more slender, nearly parallel-sided, with gills single (Fig. 18.113) or absent	5
5(4').	Ventral apotome of head triangular (Fig. 18.114); tibia of each hind leg usually without apparent constriction (Fig. 18.116); pair of ventral bands of uniformly small spines beside anal opening (Fig. 18.109) or spines absent in this position, but no lateral patches of longer spines; slender case of plant fragments, fine sand, and/or diatoms with usually 1 twig or conifer needle extending length of case and beyond 1 or both ends (Fig. 18.121)	*Nectopsyche*
5'.	Ventral apotome of head rectangular (Fig. 18.115), if triangular, case a spiral of plant pieces; tibia of each hind leg with translucent constriction, apparently dividing it into 2 subequal parts (Figs. 18.117-18.118); patch of longer spines laterad of each band of short anal spines (Fig. 18.108)	6
6(5').	Mandibles strongly asymmetrical (Fig. 18.3); metathoracic legs each with close-set fringe of long hairs (Fig. 18.117); slender case a spiral of plant pieces (Fig. 18.122)	7
6'.	Mandibles only slightly asymmetrical (Fig. 18.4); metathoracic legs with only few, scattered, long hairs (Fig. 18.118); irregular case of plant and mineral materials, with twigs or conifer needles extending beyond ends (Fig. 18.123)	*Mystacides*
7(6).	Most of head in dorsal view with light brown pigmentation ending before occipital foramen; only small amount of cream ground color visible as thin, smooth crescent between anterior head pigmentation and occipital foramen (Fig. 18.58)	*Ylodes*
7'.	Most of head in dorsal view not light brown; sometimes almost completely dark, posteriorly with irregularly shaped, broad, cream crescent projecting anteriorly in middle (Fig. 18.59)	*Triaenodes*

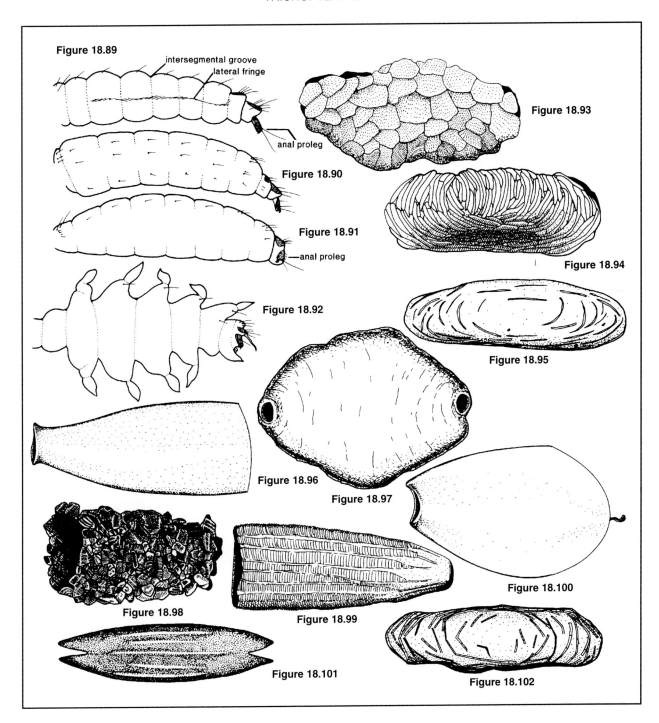

Figure 18.89. *Neotrichia* sp. (Hydroptilidae); left lateral view of abdomen.

Figure 18.90. *Mayatrichia ayama* Mosely (Hydroptilidae); left lateral view of abdomen.

Figure 18.91. *Orthotrichia* sp. (Hydroptilidae); left lateral view of abdomen.

Figure 18.92. *Ithytrichia* sp. (Hydroptilidae); left lateral view of abdomen.

Figure 18.93. *Palaeagapetus celsus* (Ross) (Hydroptilidae); larval case.

Figure 18.94. *Dibusa angata* Ross (Hydroptilidae); larval case.

Figure 18.95. *Stactobiella delira* (Ross) (Hydroptilidae); larval case.

Figure 18.96. *Oxyethira* sp. (Hydroptilidae); larval case.

Figure 18.97. *Leucotrichia* sp. (Hydroptilidae); larval case.

Figure 18.98. *Neotrichia* sp. (Hydroptilidae); larval case.

Figure 18.99. *Mayatrichia ayama* Mosely (Hydroptilidae); larval case.

Figure 18.100. *Ithytrichia* sp. (Hydroptilidae); larval case.

Figure 18.101. *Orthotrichia* sp. (Hydroptilidae); larval case.

Figure 18.102. *Agraylea* sp. (Hydroptilidae); larval case.

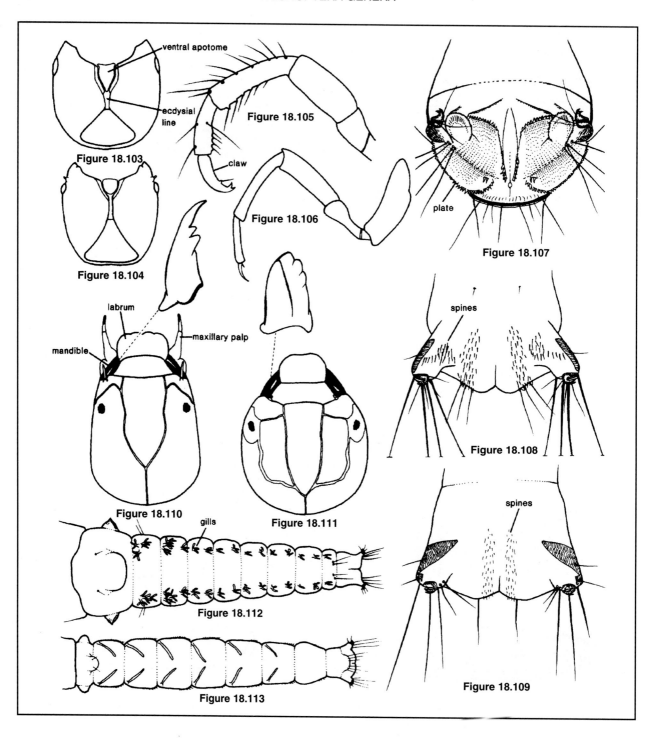

Figure 18.103. *Lepidostoma* sp. (Lepidostomatidae); ventral view of head.

Figure 18.104. *Theliopsyche* sp. (Lepidostomatidae); ventral view of head.

Figure 18.105. *Leptocerus americanus* (Banks) (Leptoceridae); lateral view of left mesothoracic leg.

Figure 18.106. *Oecetis* sp. (Leptoceridae); lateral view of left mesothoracic leg.

Figure 18.107. *Setodes incertus* (Walker) (Leptoceridae); ventral view of posterior end.

Figure 18.108. *Triaenodes tardus* Milne (Leptoceridae); ventral view of posterior end.

Figure 18.109. *Nectopsyche* sp. (Leptoceridae); ventral view of posterior end.

Figure 18.110. *Oecetis* sp. (Leptoceridae); dorsal view of head: *inset*, left mandible.

Figure 18.111. *Ceraclea maculata* (Banks) (Leptoceridae); dorsal view of head, *inset*, left mandible.

Figure 18.112. *Ceraclea* sp. (Leptoceridae); dorsal view of abdomen.

Figure 18.113. *Triaenodes tardus* Milne (Leptoceridae); dorsal view of abdomen.

Limnephilidae[4]

1.	Mesonotum with setal areas on 2 or 3 pairs of sclerites (Figs. 17.79, 18.124-18.126)	2
1'.	Mesonotum with 1 pair of sclerites closely contiguous on mid-dorsal line (Figs. 17.28, 17.41, 17.59)	8
2(1).	Mesepisternum enlarged anteriorly, either as sharp projection (Figs. 17.79, 18.124-18.125) or as short, rounded, spiny prominence (Fig. 18.126) (families Goeridae and, in part, Rossianidae)	3
2'.	Mesepisternum not enlarged anteriorly	7
3(2).	Gills mostly 3-branched, usually on abdominal segments II-VII; case of small rocks, usually with 2 pairs of larger pebbles on sides (Fig. 18.127)	*Goera**
3'.	Gills single, sometimes restricted to abdominal segments III-IV or III, or gills absent	4
4(3').	Gills absent; smooth, tapered, slightly curved case of small rocks (Fig. 18.128)	*Goerita**
4'.	Gills present	5
5(4').	Gills on abdominal segments II-VIII; mesepisternum short, rounded, spiny prominence (Fig. 18.126)	*Goereilla***
5'.	Gills on abdominal segments III or III-V; mesepisternum sharp projection (Figs. 18.124-18.125)	6
6(5').	Pronotum flat or with pair of concavities, mesepisternum laterally compressed, each metanotal sa1 with few (1-6) setae on very small, oval sclerite (Fig. 18.124) or without sclerite; case of rocks with several larger pebbles along each side (Fig. 18.129)	*Goeracea**
6'.	Pronotum convex, mesepisternum dorsoventrally depressed, each metanotal sa1 with several setae on distinct triangular sclerite (Fig. 18.125); tapered, slightly curved case of small rocks without larger lateral pebbles (Fig. 18.130)	*Lepania**
7(2').	Dorsum of head with many unusually stout and prominent primary setae, head without posterolateral flanges (Fig. 18.134); pronotum convex; smooth, tapered, slightly curved case of rocks (Fig. 18.131)	*Pedomoecus****
7'.	Dorsum of head with most primary setae unmodified, with posterolateral flanges (Fig. 18.133); pronotum with pair of concavities (Fig. 18.136); case as above except with rough outline and not as tapered (Fig. 18.132)	*Rossiana***
8(1').	Basal seta of each tarsal claw extending to, or nearly to, tip of claw (Fig. 18.138); mandibles each with apical edge entire, without teeth (Fig. 17.27)	9
8'.	Basal seta of each tarsal claw extending far short of tip of claw (Fig. 18.139); mandibles each with 2 or more teeth (Fig. 18.137)	12
9(8).	Metanotal sa1 sclerites absent, numerous sa1 setae in transverse band (Fig. 18.135); strongly tapered case of rocks, anterior opening usually oblique with dorsal edge extending beyond ventral edge for final instar (Fig. 18.153) (subfamily Apataniinae)	*Apatania****
9'.	Metanotal sa1 sclerites present (Fig. 17.59) or, if absent, setae not numerous and arranged in continuous transverse band (similar to Fig. 18.140)	10
10(9').	Dorsum of head evenly convex, without carina; abdominal sternite I present with small central membranous space (Fig. 18.142)	11
10'.	Dorsum of head flat, often with prominent carina laterally and anteriorly (Fig. 18.143; abdominal sternite I absent; tapered, curved case of rocks, sometimes with larger pebbles laterally (Fig. 18.157)	*Allomyia****
11(10).	Gills absent and lateral fringe of setae absent from abdomen except few sparce setae on abdominal segments II and III and dorsum of abdomen heavily pigmented, slate-colored (Fig. 18.268); anal claw without accessory hook (Fig. 18.270); case of rock fragments, often with plant materials dorsally (Fig. 18.272)	*Madeophylax****†
11'.	Gills present on abdominal segments I-VI, lateral fringe of setae and spicules present on abdominal segments III-VIII, and dorsum of abdomen of same texture and color as venter (Fig. 18.269); anal claw with accessory hook (Fig. 18.271); case of rocks with plant material attached (Fig. 18.156)	*Manophylax****
12(8').	Abdominal gills absent; tapered, curved case of fine rocks with shiny silk covering (Fig. 18.158)	*Moselyana****
12'.	Abdominal gills present	13

* Family Goeridae.
** Family Rossianidae.
*** Family Apataniidae.
[4.] Larvae are unknown for 2 genera: *Chilostigmodes* (Alaska to Labrador), and *Leptophylax* (North Central states).
† Now a synonym of *Manophylax*.

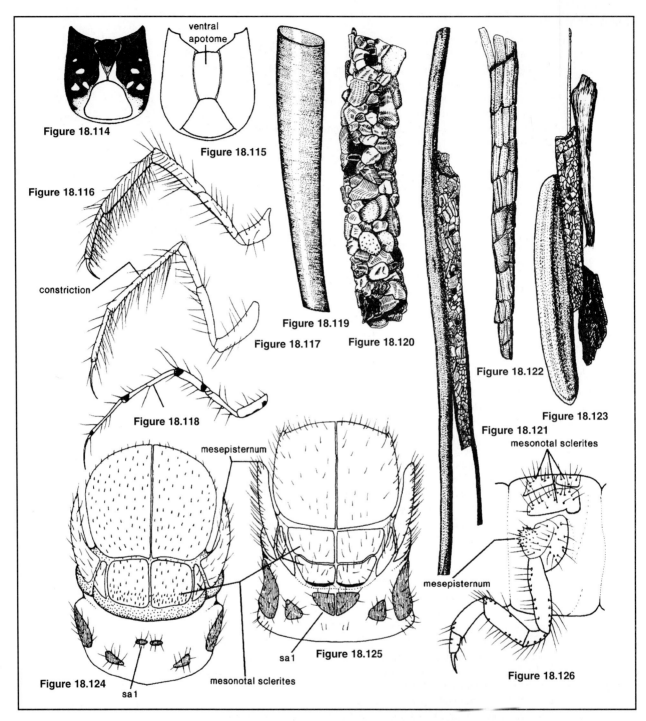

Figure 18.114. *Nectopsyche* sp. (Leptoceridae); ventral view of head.

Figure 18.115. *Triaenodes tardus* Milne (Leptoceridae); ventral view of head.

Figure 18.116. *Nectopsyche* sp. (Leptoceridae); lateral view of left metathoracic leg.

Figure 18.117. *Triaenodes tardus* Milne (Leptoceridae); lateral view of left metathoracic leg.

Figure 18.118. *Mystacides* sp. (Leptoceridae); lateral view of left metathoracic leg.

Figure 18.119. *Leptocerus americanus* (Banks) (Leptoceridae); larval case.

Figure 18.120. *Setodes incertus* (Walker) (Leptoceridae); larval case.

Figure 18.121. *Nectopsyche* sp. (Leptoceridae); larval case.

Figure 18.122. *Triaenodes tardus* Milne (Leptoceridae); larval case.

Figure 18.123. *Mystacides* sp. (Leptoceridae); larval case.

Figure 18.124. *Goeracea* poss. *oregona* Denning (Limnephilidae); dorsal view of thorax.

Figure 18.125. *Lepania cascada* Ross (Limnephilidae); dorsal view of thorax.

Figure 18.126. *Goereilla baumanni* Denning (Limnephilidae); left lateral view of mesothorax.

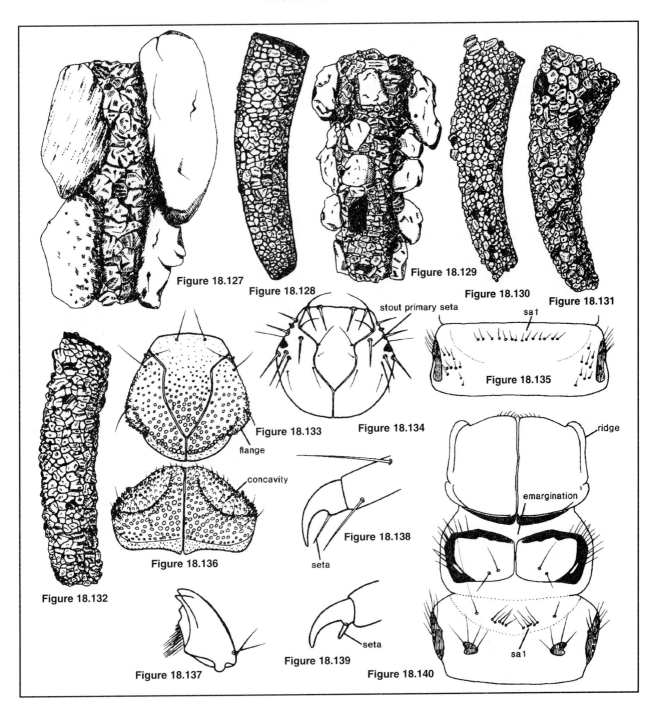

Figure 18.127. *Goera fuscula* Banks (Limnephilidae); larval case.

Figure 18.128. *Goerita semata* Ross (Limnephilidae); larval case.

Figure 18.129. *Goeracea* poss. *oregona* Denning (Limnephilidae); larval case.

Figure 18.130. *Lepania cascada* Ross (Limnephilidae); larval case.

Figure 18.131. *Pedomoecus sierra* Ross (Limnephilidae); larval case.

Figure 18.132. *Rossiana montana* Denning (Limnephilidae); larval case.

Figure 18.133. *Rossiana montana* Denning (Limnephilidae); dorsal view of head.

Figure 18.134. *Pedomoecus sierra* Ross (Limnephilidae); dorsal view of head.

Figure 18.135. *Apatania arizona* Wiggins (Limnephilidae); dorsal view of metanotum (after Wiggins 1977).

Figure 18.136. *Rossiana montana* Denning (Limnephiildae); dorsal view of pronotum.

Figure 18.137. *Limnephilus* sp. (Limnephilidae); ventral view of left mandible.

Figure 18.138. *Neophylax* sp. (Uenoidae); lateral view of left metathoracic claw.

Figure 18.139. *Dicosmoecus* sp. (Limnephilidae); lateral view of left metathoracic claw.

Figure 18.140. *Oligophlebodes* sp. (Uenoidae); dorsal view of thorax.

13(12').	Anterior margin of pronotum densely fringed with long hairs; dorsum of head flat and with 2 bands of dense scale hairs (Fig. 18.144); tapered, depressed case of transverse bits of wood and bark (Fig. 18.159)..........................(subfamily Dicosmoecinae, in part) ....***Cryptochia***	
13'.	Anterior margin of pronotum and dorsum of head without dense hairs (Figs. 17.41, 17.59) or hairs not of type or arrangement described above (Figs. 18.165-18.166)..14	
14(13').	Most abdominal gills single..15	
14'.	Most dorsal and ventral gills multiple, lateral gills sometimes single 27	
15(14).	Metanotal sa1 and sa2 sclerites large, distance beween sa2 sclerites no more than twice maximum dimension of single sa2 sclerite (Fig. 18.145); slender, straight, scarcely tapered tube case of coarse rocks often with long plant material attached (Fig. 18.160) ....(subfamily Dicosmoecinae, in part)........***Ecclisomyia***	
15'.	Metanotal sa1 and sa2 sclerites small, distance between sa2 sclerites more than twice maximum dimension of single sa2 sclerite, sa1 sclerites sometimes fused (Figs. 18.146-18.147)..16	
16(15').	One or 2 sclerites adjacent to base of each lateral hump of abdominal segment I (Figs. 18.148-18.151; sclerites often only lightly pigmented but distinguishable by relatively shinier surfaces)..17	
16'.	No sclerites adjacent to lateral humps of abdominal segment I..22	
17(16).	Large single sclerite at base of each lateral hump of abdominal segment I enclosing posterior half of hump and extending posterodorsad as irregular lobe (Fig. 18.149); case of leaves or bark formed into flattened tube with seams along narrow lateral flanges (Fig. 18.161) ....(subfamily Limnephilinae, tribe Stenophylacini, in part) ....***Chyranda***	
17'.	One or 2 small sclerites at base of each lateral hump of abdominal segment I (Figs. 17.63, 18.150-18.151) 18	
18(17').	Two sclerites at base of each lateral hump of abdominal segment I (Figs. 17.63, 18.150)............................19	
18'.	One long sclerite at posterior edge of base of each lateral hump of abdominal segment I (Figs. 18.148, 18.151)..20	
19(18).	Two small ring sclerites posterodorsally at base of each lateral hump of abdominal segment I (Fig. 17.63); tubular, slightly curved and tapered case largely of rocks with some small pieces of wood incorporated (similar to Fig. 18.132)..(subfamily Limnephilinae, tribe Stenophylacini, in part) ....***Desmona***	
19'.	One rounded posterior sclerite, one long dorsal sclerite at base of each lateral hump of abdominal segment I (Fig. 18.150); rough, straight, untapered tube case of rocks and wood fragments (similar to Fig. 18.162) ....(subfamily Limnephilinae, tribe Chilostigmini, in part).....***Psychoglypha***	
20(18').	Sclerite at base of each lateral hump of abdominal segment I only half as long as basal width of hump (Fig. 18.151); smooth, thin-walled, nearly straight, little-tapered tube base of irregularly arranged bark pieces or occasional flat rocks (Fig. 18.162), sometimes 3-sided ....(subfamily Limnephilinae, tribe Chilostigmini, in part)..............***Homophylax***	
20'.	Sclerite at base of each lateral hump of abdominal segment I nearly as long as basal width of hump (Fig. 18.148).....................(subfamily Limnephilinae, tribe Stenophylacini, in part).........................21	
21(20').	Metanotal sa1 sclerites fused (Fig. 18.146); abdominal sternum II with chloride epithelium (in which case abdominal segment IX with only single seta on each side of dorsal sclerite, Fig. 18.152, *Hydatophylax argus*, eastern North America) or without chloride epithelium (in which case abdominal segment IX with tuft of 3-6 setae, *H. hesperus*, western North America); case of wood or leaves in irregular outline or case of leaves flattened (Fig. 18.163) ....***Hydatophylax***	
21'.	Metanotal sa1 sclerites not fused although often continguous (Fig. 18.147), abdominal sternum II without chloride epithelium and abdominal segment IX with only single seta on each side of dorsal sclerite (similar to Fig. 18.152); case of twigs, gravel, or leaves, variously shaped (similar to Figs. 18.132, 18.163, or 18.199, occasionally 3-sided) ....***Pycnopsyche***	
22(16').	Anterior margin of pronotum with flat scale hairs, dorsum of head flat (Fig. 18.165); coarse mineral case slightly curved (similar to Fig. 18.132)..***Philocasca***	
22'.	Anterior margin of pronotum without flat scale hairs, setae normal; dorsum of head usually convex (flat only in western *Pseudostenophylax edwardsi*; Fig. 18.167)............................23	
23(22').	Mesonotal sa1 and sa2 distinct, separated by gap free of setae (Figs. 18.168-18.169)..................................24	

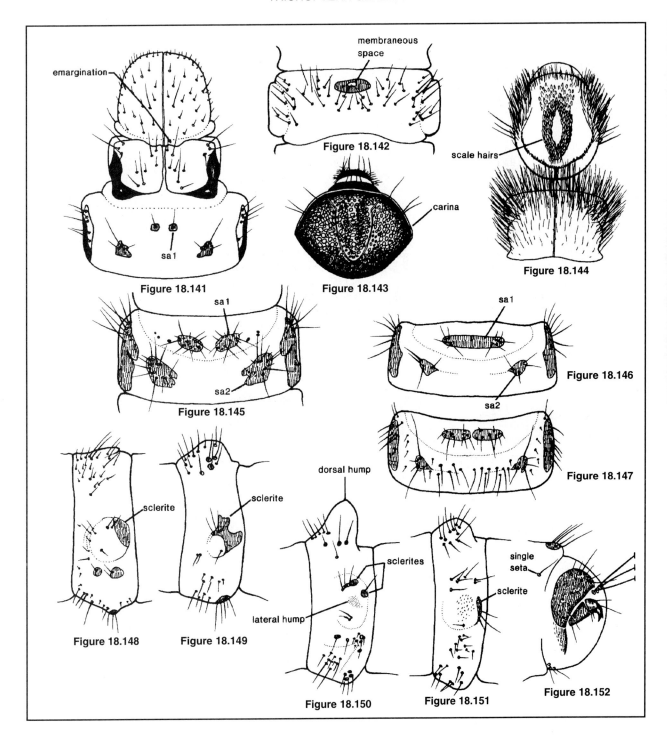

Figure 18.141. *Neophylax* sp. (Uenoidae); dorsal view of thorax.

Figure 18.142. *Manophylax annulatus* Wiggins (Limnephilidae); ventral view of abdominal sternum I.

Figure 18.143. *Allomyia scotti* (Wiggins) (Limnephilidae); dorsal view of larval head.

Figure 18.144. *Cryptochia* sp. (Limnephilidae); dorsal view of head and pronotum.

Figure 18.145. *Ecclisomyia* sp. (Limnephilidae); dorsal view of metanotum.

Figure 18.146. *Hydatophylax* sp. (Limnephilidae); dorsal view of metanotum.

Figure 18.147. *Pycnopsyche* sp. (Limnephilidae); dorsal view of metanotum.

Figure 18.148. *Pycnopsyche* sp. (Limnephilidae); left lateral view of abdominal segment I.

Figure 18.149. *Chyranda centralis* (Banks) (Limnephilidae); left lateral view of abdominal segment I.

Figure 18.150. *Psychoglypha* sp. (Limnephilidae); left lateral view of abdominal segment I.

Figure 18.151. *Homophylax* sp. (Limnephilidae); left lateral view of abdominal segment I.

Figure 18.152. *Hydatophylax argus* (Harris) (Limnephilidae); left lateral view of abdominal segment IX.

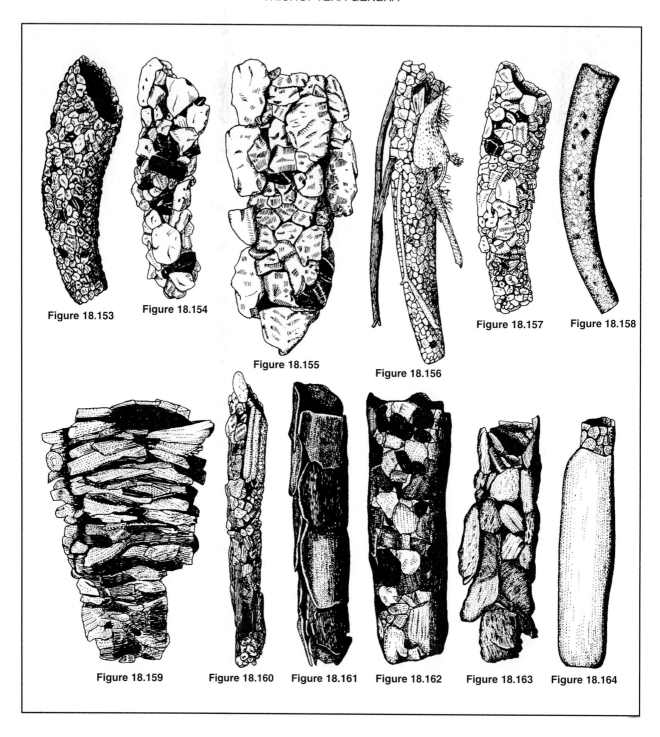

Figure 18.153. *Apatania* sp. (Limnephilidae); larval case.
Figure 18.154. *Oligophlebodes* sp. (Uenoidae); larval case.
Figure 18.155. *Neophylax* sp. (Uenoidae); larval case.
Figure 18.156. *Manophylax annulatus* Wiggins (Limnephilidae); larval case.
Figure 18.157. *Allomyia scotti* (Wiggins) (Limnephilidae); larval case.
Figure 18.158. *Moselyana comosa* Denning (Limnephildae); larval case.
Figure 18.159. *Cryptochia* sp. (Limnephilidae); larval case.
Figure 18.160. *Ecclisomyia* sp. (Limnephilidae); larval case.
Figure 18.161. *Chyranda centralis* (Banks) (Limnephilidae); larval case.
Figure 18.162. *Homophylax* sp. (Limnephilidae); larval case.
Figure 18.163. *Hydatophylax argus* (Harris) (Limnephilidae); larval case.
Figure 18.164. *Amphicosmoecus canax* (Ross) (Limnephilidae); larval case (after Wiggins 1977).

23'.	Mesonotal sa1 and sa2 connected by continuous longitudinal band of setae on each side of meson (Fig. 17.59)	26
24(23).	Pronotum covered with fine spines (Fig. 18.168 inset); case of plant and rock fragments and apparent snail opercula (similar to Fig. 18.162, but smaller diameter)..........................(subfamily Limnephilinae, tribe Chilostigmini, in part)	***Grensia***
24'.	Pronotum smooth and shiny, without fine spines (Fig. 18.169)	25
25(24').	Mesonotal sclerites each wider than long, much shorter mesally than laterally (Fig. 18.56); case of short pieces of *Sphagnum* moss laid transversely (Fig. 18.278)	***Phanocelia*****
25'.	Mesonotal sclerites each about as long as wide and nearly as long mesally as laterally (Fig. 18.169); case of leaf pieces with wide flanges at each side of depressed tube (Fig. 17.10).......................(subfamily Limnephilinae, tribe Stenophylacini, in part)	***Clostoeca*****
26(23').	Head and pronotum strongly inflated and with pebbled texture (Fig. 18.166); smooth, tapered, curved case of small rocks (similar to Fig. 17.9)...(subfamily Dicosmoecinae, in part)	***Ecclisocosmoecus***
26'.	Head and pronotum not unusually inflated (Figs. 17.59, 17.77), although dorsum of head flat in Western species (Fig. 18.167), sclerotized areas not pebbled; smooth, tapered, curved case of rocks (Fig. 17.9)............................(subfamily Pseudostenophylacinae)	***Pseudostenophylax***
27(14').	Most gills with 3 branches, none with more than 4	28
27'.	At least some gills with more than 4 branches	45
28(27).	Dorsum of head with 2 bands of contrasting color extending from coronal suture to bases of mandibles (Fig. 18.170) and/or narrowed posterior portion of frontoclypeal apotome with 3 light areas: 1 along each side and 1 at posterior extremity (Fig. 18.173)..(subfamiy Limnephilinae, tribe Limnephilini, in part)	29
28'.	Dorsum of head lacking bands or other well-defined, contrasting areas, usually uniform in color or with prominent light or dark spots only at points of muscle attachment (Fig. 18.193)	32
29(28).	Abdominal sternum I usually with more than 100 setae overall (Fig. 18.171); small spines on head and pronotum (Fig. 18.173); short, stout setae on lateral sclerite of each anal proleg (Fig. 18.172); cylindrical case usually of irregular pieces of twigs and bark, sometimes small rocks (similar to Fig. 18.163)	***Clistoronia***
29'.	Abdominal sternum I with fewer than 100 setae overall; usually without spines on head and pronotum; usually without short, stout setae on lateral sclerite of anal proleg	30
30(29').	Mesonotal sa1 consisting of single seta (Fig. 18.174)	31
30'.	Mesonotal sa1 consisting of more than 1 seta (Fig. 18.175)	43
31(30).	Chloride epithelia dorsally and ventrally on several abdominal segments (Fig. 18.176); rough, tubular case of wood or leaf fragments (similar to Fig. 18.163), sometimes 3-sided, changed to fine gravel before pupation	***Halesochila***
31'.	Chloride epithelia absent dorsally; case of leaf pieces arranged transversely or longitudinally (similar to Fig. 18.27)	***Nemotaulius***
32(28').	Femur of metathoracic leg with more than 2 major setae on ventral edge (Fig. 18.181) ....(subfamily Dicosmoecinae, in part)	33
32'.	Femur of metathoracic leg with 2 major setae on ventral edge (Fig. 18.180)	37
33(32).	Metanotal sa1 sclerites fused (Fig. 18.178); case a hollow twig with ring of bark pieces anteriorly (Fig. 18.164) or case entirely of wood fragments	***Amphicosmoecus***
33'.	Metanotal sa1 sclerites clearly separate (Fig. 18.179)	34
34(33').	Abdominal tergum I with transverse row of setae posterior to median dorsal hump (Fig. 18.179); scale hairs on dorsum of head (Fig. 18.177); abdominal sternum II with 2 chloride epithelia (Fig. 18.183); irregularly outlined case of small pebbles arranged into slightly curved and flattened cylinder (Fig. 18.196)	***Allocosmoecus***
34'.	Abdominal tergum I usually lacking setae posterior to median hump; dorsum of head without scale hairs; abdominal sternum II with single chloride epithelium (Fig. 18.184), 3 epithelia (smaller epithelia laterally), or without epithelia	35
35(34').	Tibiae each with several pairs of stout spur-like setae (Fig. 18.181); anterolateral margin of pronotum often with stout, dense spines (Fig. 18.273); case of final instar of fine gravel, slightly depressed (Fig. 18.197), case of younger larva with plant materials	***Dicosmoecus***
35'.	Tibiae each with no more than one pair of stout spurs (Fig. 18.182); case of bark, wood, or gravel, but never depressed	36

* It is likely that *Chilostigma* (Minnesota) runs to *Phanocelia* or *Clostoeca* in this key. A presumed specimen of *Chilostigma* differs from *Clostoeca* by having mesonotal sclerites broader than long and from *Phanocelia* by having mesonotal sclerites about as long mesally as laterally.

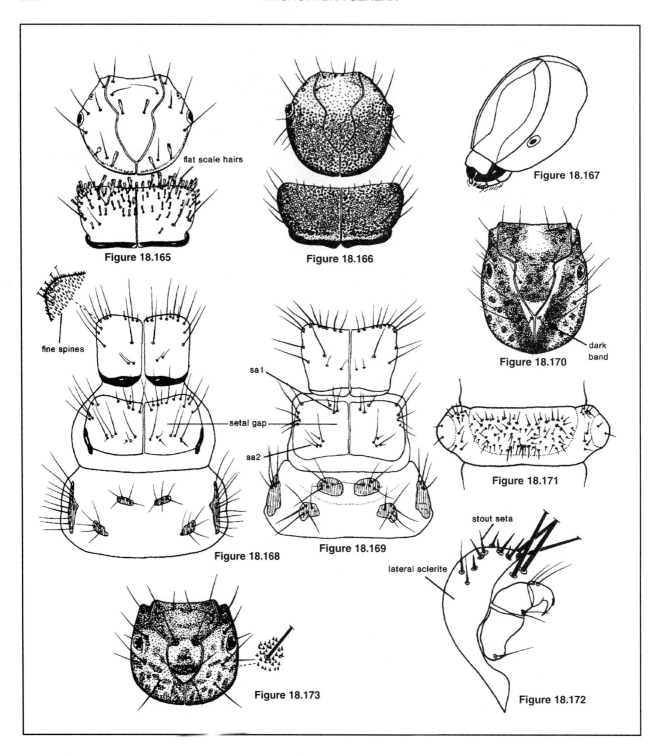

Figure 18.165. *Philocasca rivularis* Wiggins (Limnephilidae); dorsal view of head and pronotum.

Figure 18.166. *Ecclisocosmoecus scylla* (Milne) (Limnephildae); dorsal view of head and pronotum.

Figure 18.167. *Pseudostenophylax edwardsi* (Banks) (Limnephilidae); left dorsolateral oblique view of head (after Wiggins, 1977).

Figure 18.168. *Grensia praeterita* (Walker) (Limnephilidae); dorsal view of thorax, *inset*, microspines enlarged.

Figure 18.169. *Clostoeca disjuncta* (Banks) (Limnephilidae); dorsal view of thorax.

Figure 18.170. *Halesochila taylori* (Banks) (Limnephilidae); dorsal view of head.

Figure 18.171. *Clistoronia magnifica* (Banks) (Limnephilidae); ventral view of abdominal sternum I.

Figure 18.172. *Clistoronia magnifica* (Banks) (Limnephilidae); lateral view of left anal proleg.

Figure 18.173. *Clistoronia magnifica* (Banks) (Limnephildae); dorsal view of head; *inset*, microspines enlarged.

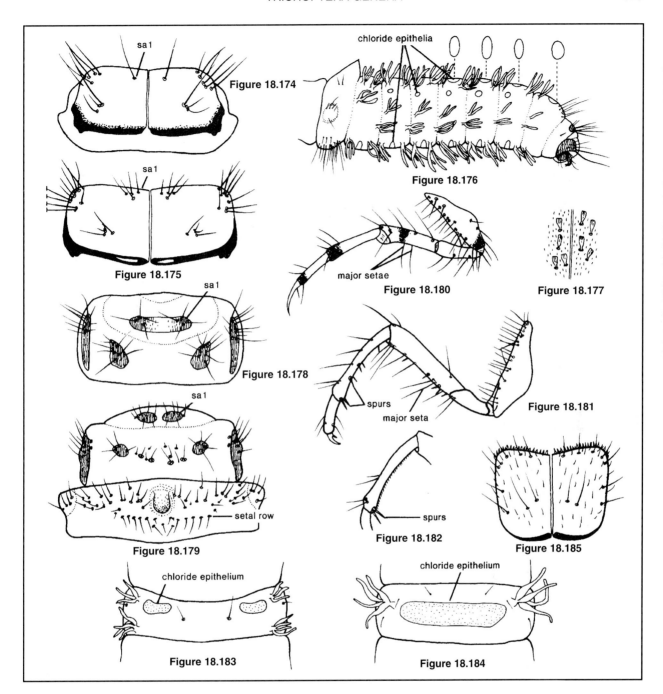

Figure 18.174. *Halesochila taylori* (Banks) (Limnephilidae); dorsal view of mesonotum.

Figure 18.175. *Asynarchus* sp. (Limnephilidae); dorsal view of mesonotum.

Figure 18.176. *Halesochila taylori* (Banks) (Limnephilidae); left lateral view of abdomen: *insets*, dorsal chloride epithelia.

Figure 18.177. *Allocosmoecus partitus* Banks (Limnephilidae); dorsal view of scale hairs on larval head.

Figure 18.178. *Amphicosmoecus canax* (Ross) (Limnephilidae); dorsal view of metanotum (after Wiggins 1977).

Figure 18.179. *Allocosmoecus partitus* Banks (Limnephilidae); dorsal view of metanotum and abdominal tergum I.

Figure 18.180. *Glyphopsyche irrorata* (Fabricius) (Limnephilidae); lateral view of left metathoracic leg.

Figure 18.181. *Dicosmoecus* sp. (Limnephilidae); lateral view of left metathoracic leg.

Figure 18.182. *Onocosmoecus* sp. (Limnephildae); lateral view of left metathoracic tibia.

Figure 18.183. *Allocosmoecus partitus* Banks (Limnephilidae); ventral view of abdominal sternum II.

Figure 18.184. *Dicosmoecus* sp. (Limnephilidae); ventral view of abdominal sternum II.

Figure 18.185. *Frenesia missa* (Milne) (Limnephilidae); dorsal view of pronotum.

36(35').   Pronotum with submarginal setae widely spaced, especially mesally, and with some lateral setae much shorter and thicker than others (Fig. 18.274); dorsum of abdominal segment VII with single long setae on either side of midline posterodorsally, sometimes with one or two much smaller setae (Fig. 18.276); lateral abdominal gills usually lacking from segment V, sometimes from IV or terminating with anterior position of segment IV; case of final instar of stout pieces of wood (Fig. 18.279) or of fine rock fragments (Fig. 18.280), cases of earlier instars of pliable plant materials..*Eocosmoecus*

36'.   Pronotum with submarginal setae more closely spaced, especially submesally, and all of nearly the same length and thickness (Fig. 18.275); dorsum of abdominal segment VII with 1-5 long setae on either side of midline posterodorsally (Fig. 18.277); lateral abdominal gills terminating with anterior position of segment V; case of pieces of wood and bark (similar to Fig. 18.200) ....*Onocosmoecus*

37(32').   Abdominal tergum VIII with median gap in transverse band of setae (Fig. 18.186); cylindrical case of longitudinally arranged sedge or similar leaves (Fig. 18.198) ....subfamily Limnephilinae, tribe Limnephilini, in part)......*Grammotaulius*

37'.   Abdominal tergum VIII with continuous transverse band of setae (Fig. 18.187)...38

38(37').   Pronotum, especially anterior margin (Fig. 18.185), and lateral sclerite of anal proleg (Fig. 18.172) both with short, stout setae ....(subfamily Limnephilinae, tribe Chilostigmini, in part) 39

38'.   Pronotum usually, lateral sclerite of anal proleg always, without short, stout setae ....(subfamily Limnephilinae, tribe Limnephilini, in part) 40

39(38).   Tibiae and tarsi of all legs each with dark, contrasting band (Fig. 18.180); cylindrical case of sand, twigs, and bark (similar to Fig. 18.160) ....*Glyphopsyche*

39'.   Tibiae and tarsi each lacking dark band; smooth, cylindrical case mostly of small stones with wood fragments (similar to Fig. 18.132) ....*Frenesia*

40(38').   Metanotal sa2 with few setae, usually 2, and no sclerite (Fig. 18.188); smooth, cylindrical case of long leaf pieces (Fig. 17.8)...*Arctopora*

40'.   Metanotal sa2 with more than 2 setae and with sclerite (Fig. 18.193) 41

41(40').   Dorsum of head with numerous large spots coalescing into diffuse blotches in some places, especially on frontoclypeal apotome (Fig. 18.193); anterolateral corner of pronotum with small patch of spines (Fig. 18.193); cylindrical case of plant materials (Fig. 18.199), sometimes 3-sided ....*Anabolia*

41'.   Dorsum of head with various markings, including spots, but not coalescing as above; anterolateral corner of pronotum usually lacking patch of spines...42

42(41').   Prosternal horn extending beyond head capsule to mentum of labium (Fig. 18.191); case of transverse, narrow, projecting pieces of plant material (Fig. 17.11) ....*Platycentropus*

42'.   Prosternal horn extending only to distal edge of head capsule (Fig. 18.192) 43

43(30', 42').   Chloride epithelia present dorsally, laterally, and ventrally on most abdominal segments (similar to Fig. 18.176); case of plant and rock materials (similar to Fig. 18.154)..*Asynarchus*

43'.   Chloride epithelia absent dorsally, may be present laterally but always present ventrally on most abdominal segments 44

44(43').   Chloride epithelia present laterally and ventrally on most abdominal segments; cases of wide range of shapes and materials (e.g., resembling Figs. 17.7-17.8, 17.11, 18.132, 18.200) ....*Limnephilus, Philarctus*[5]

44'.   Chloride epithelia absent laterally, but present ventrally on abdominal segments II-VIII; case of pieces of sedge leaves arranged lengthwise and irregularly (Fig. 18.281)..*Sphagnophylax*

45(27').   Femora of meso- and metathoracic legs each with approximately 5 major setae along ventral edge (Fig. 18.189), curved, scarcely tapered case of bark and leaves (Fig. 18.200) or of sand ....(subfamily Dicosmoecinae, in part)...*Ironoquia*

45'.   Femora of meso- and usually metathoracic legs each with 2 major setae along ventral edge (Fig. 18.190) ....(subfamily Limnephilinae, tribe Limnephilini, in part) 46

46(45').   Metanotum with all setae confined to primary sclerites (Fig. 18.194); cylindrical case of longitudinally arranged lengths of sedge leaves (similar to Fig. 17.8) or of fragments of bark and leaves ....*Lenarchus*

46'.   Metanotum with at least a few setae between primary sclerites (Fig. 18.195)....................................47

47(46').   Lateral series of abdominal gills single-branched and on segment II only, occasionally also III; case mostly of coarse, small rocks (similar to Fig. 18.196) ....*Psychoronia*

---

[5] "Because of the close morphological similarity between larvae in *Philarctus* and *Limnephilus* and of the large number of *Limnephilus* spp. for which larvae are not known (about 65% unassociated), separation between the two genera is illusory at this stage in our knowledge" (Wiggins 1977).

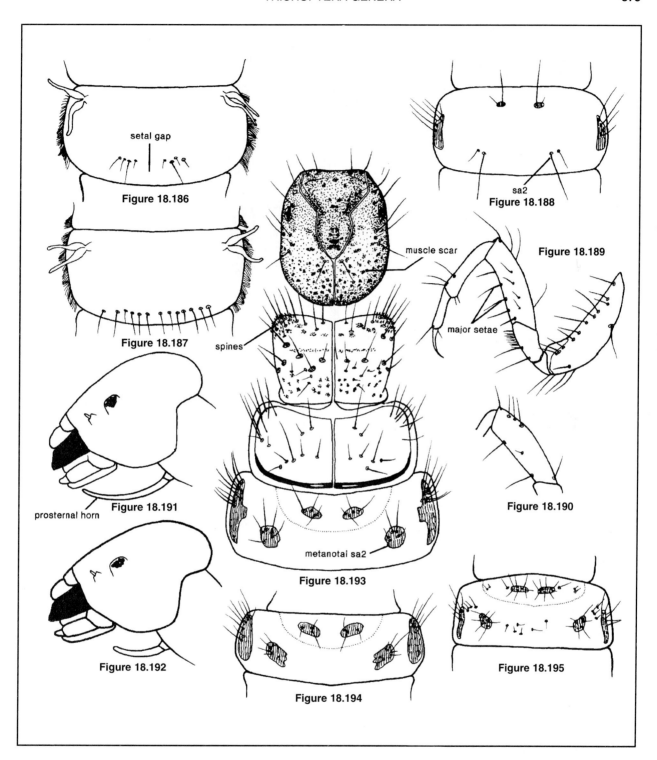

Figure 18.186. *Grammotaulius* sp. (Limnephilidae); dorsal view of abdominal tergum VIII.

Figure 18.187. *Glyphopsyche irrorata* (Fabricius) (Limnephilidae); dorsal view of abdominal tergum VIII.

Figure 18.188. *Arctopora* sp. (Limnephilidae); dorsal view of metanotum.

Figure 18.189. *Ironoquia* sp. (Limnephilidae); lateral view of left metathoracic leg.

Figure 18.190. *Hesperophylax* sp. (Limnephildae); lateral view of left metathoracic femur.

Figure 18.191. *Platycentropus radiatus* (Say) (Limnephilidae); left lateral view of head and prosternal horn.

Figure 18.192. *Limnephilus* sp. (Limnephilidae); left lateral view of head and prosternal horn.

Figure 18.193. *Anabolia bimaculata* (Walker) (Limnephilidae); dorsal view of head and thorax.

Figure 18.194. *Lenarchus* sp. (Limnephilidae); dorsal view of metanotum.

Figure 18.195. *Hesperophylax* sp. (Limnephilidae); dorsal view of metanotum.

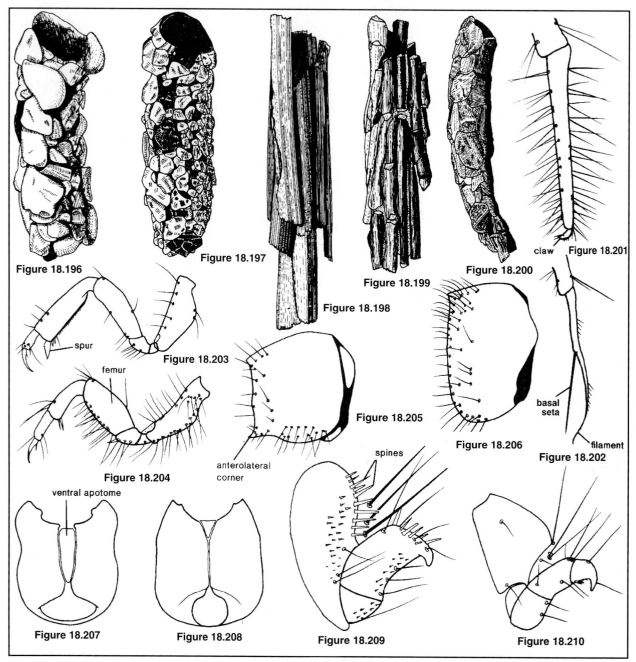

Figure 18.196. *Allocosmoecus partitus* Banks (Limnephilidae); larval case.

Figure 18.197. *Dicosmoecus* sp. (Limnephilidae); larval case.

Figure 18.198. *Grammotaulius* sp. (Limnephilidae); larval case.

Figure 18.199. *Anabolia bimaculata* (Walker) (Limnephilidae); larval case.

Figure 18.200. *Ironoquia* sp. (Limnephilidae); larval case.

Figure 18.201. *Molanna tryphena* Betten (Molannidae); lateral view of left metathoracic tarsas and claw.

Figure 18.202. *Molannodes tinctus* (Zetterstedt) (Molannidae); lateral view of left metathoracic tarsal claw.

Figure 18.203. *Pseudogoera singularis* Carpenter (Odontoceridae); lateral view of left prothoracic leg.

Figure 18.204. *Psilotreta* sp. (Odontoceridae); lateral view of left prothoracic leg.

Figure 18.205. *Psilotreta* sp. (Odontoceridae); left lateral view of pronotum.

Fiture 18.206. *Namamyia* sp. (Odontoceridae); left lateral view of pronotum.

Figure 18.207. *Parthina* prob. *vierra* Denning (Odonotoceridae); ventral view of head.

Figure 18.208. *Psilotreta* sp. (Odontoceridae); ventral view of head.

Figure 18.209. *Parthina* prob. *vierra* Denning (Odontoceridae); lateral view of left anal claw.

Figure 18.210. *Psilotreta* sp. (Odontoceridae); lateral view of left anal claw.

47'. Lateral series of abdominal gills on segments II-V, some with more than 1 branch; case mostly of small rocks (similar to Fig. 18.132), sometimes also with plant pieces .... *Hesperophylax*

Molannidae

1. Tarsal claw of each metathoracic leg curved, broad, setose, much shorter than tarsus (Fig. 18.201) .... *Molanna*
1'. Tarsal claw of each metathoracic leg forming slender filament as long as tarsus (Fig. 18.202) .... *Molannodes*

Odontoceridae

1. Prothoracic femur about as broad as its tibia, prothoracic tibia about 4 times as long as its tarsus, single apical spur of prothoracic tibia broad, clasplike (Fig. 18.203); case of rock fragments, curved and tapered (Fig. 18.215) .... (subfamily Pseudogoerinae) .... *Pseudogoera*
1'. Prothoracic femur distinctly broader than its tibia, prothoracic tibia as long as its tarsus, both apical spurs of prothoracic tibia slender (Fig. 18.204) .... (subfamily Odontocerinae) 2
2(1'). Anterolateral corner of pronotum produced, sharply pointed (Fig. 18.205) 3
2'. Anterolateral corner of pronotum not produced, rounded (Fig. 18.206); sand case (similar to Fig. 17.9) 4
3(2). Ventral apotome of head long, completely separating genae (Fig. 18.207); claw of anal proleg stout, both claw and lateral sclerite with straight spines as well as setae (Fig. 18.209); case of fine sand grains with silken exterior (similar to Fig. 18.158) .... *Parthina*
3'. Ventral apotome of head short, separating genae only in anterior one-third (Fig. 18.208); claw of anal proleg more slender, claw and lateral sclerite with setae only (Fig. 18.210); case of coarse and fine rock fragments, very sturdy (Fig. 18.216) .... *Psilotreta*
4(2'). Mesonotal plates each subdivided into 3 sclerites, metanotal sa1 sclerites large, subrectangular, contiguous mesally (Fig. 18.211) .... *Marilia*
4'. Mesonotal plates undivided, metanotal sa1 sclerites small, oval, separated by a distance equal to, or greater than, the greatest width of one of them (Fig. 18.212) 5
5(4'). Abdominal sternum I with 2 clusters of gill filaments, 2 pairs of setae (Fig. 18.213) .... *Nerophilus*
5'. Abdominal sternum I without gills, with many setae (Fig. 18.214) .... *Namamyia*

Philopotamidae

(modified from Weaver *et al.* 1981)
1. Anterior margin of frontoclypeus with prominent notch asymmetrically right of midline (Fig. 18.219); fore trochantin small, scarcely projecting (Fig. 18.221) and prothoracic coxae with long, slender, subapical seta-bearing process; head with seta no. 18 at level of posterior point of ventral apotome (Fig. 18.218) .... *Chimarra*
1'. Anterior margin of frontoclypeus without prominent notch as above, or only slightly sinuous (Fig. 17.52), or completely symmetrical (Fig. 18.220); fore trochantin small as above or elongate, fingerlike (Fig. 18.222); prothoracic coxae without long subapical process (Fig. 18.222); head with seta no. 18 approximately halfway between posterior edge of ventral apotome and occipital foramen (Fig. 18.217) 2
2(1'). Anterior margin of frontoclypeus slightly (Fig. 17.52) to markedly (as in Fig. 18.219) asymmetrical; fore trochantin projecting, fingerlike (Fig. 18.222) .... *Dolophilodes*
2'. Anterior margin of frontoclypeus evenly convex, symmetrical (Fig. 18.220); fore trochantin small, scarcely projecting (as in Fig. 18.221) .... *Wormaldia*

Phryganeidae

1. Genae of head almost completely separated by ventral apotome (Fig. 18.223); case entirely of plant fragments .... (subfamily Phryganeinae) 2
1'. Genae of head mostly contiguous ventrally, separated anteriorly by tiny ventral apotome (Fig. 18.224); case of plant and mineral fragments, mineral fragments mostly anterior and ventral (Fig. 18.231), pupal case entirely of micalike fragments .... (subfamily Yphriinae) .... *Yphria*
2(1). Mesonotal sa1 sclerite several times larger than sa3 sclerite, sa1 seta near its anterior edge (Fig. 17.46) 9
2'. Mesonotal sa1 sclerite absent (Fig. 17.45) or much smaller than sa3 sclerite (Fig. 18.230) and with its sa1 seta centrally located; case generally of ring or spiral (Fig. 17.16) construction, slightly curved or straight, respectively 3

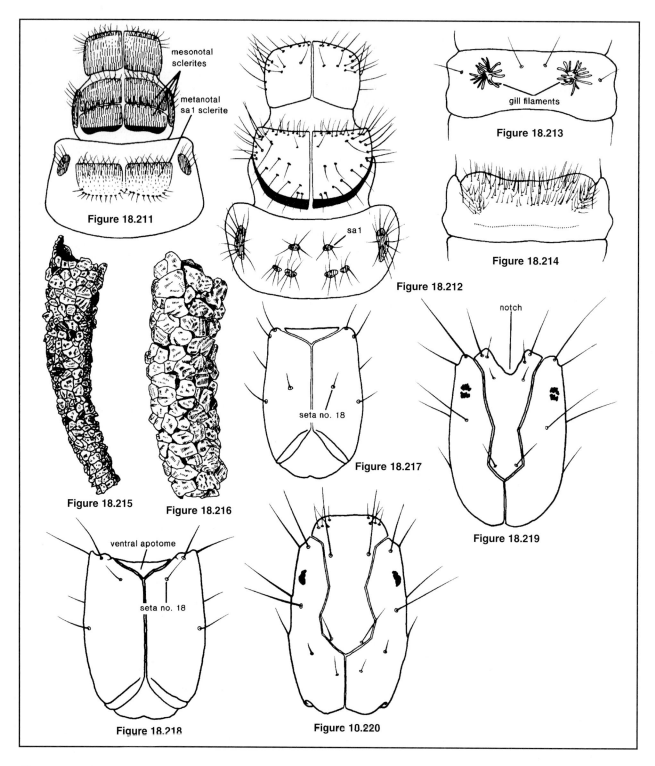

Figure 18.211. *Marilia* sp. (Odontoceridae); dorsal view of thorax.

Figure 18.212. *Nerophilus californicus* (Hagen) (Odontoceridae); dorsal view of thorax.

Figure 18.213. *Nerophilus californicus* (Hagen) (Odontoceridae); ventral view of abdominal sternum I.

Figure 18.214. *Namamyia* sp. (Odontoceridae); ventral view of abdominal sternum I.

Figure 18.215. *Pseudogoera singularis* Carpenter (Odontoceridae); larval case.

Figure 18.216. *Psilotreta* sp. (Odontoceridae); larval case.

Figure 18.217. *Dolophilodes* sp. (Philopotamidae); ventral view of head.

Figure 18.218. *Chimarra* sp. (Philopotamidae); ventral view of head.

Figure 18.219. *Chimarra* sp. (Philopotamidae); dorsal view of head.

Figure 18.220. *Wormaldia* sp. (Philopotamidae); dorsal view of head.

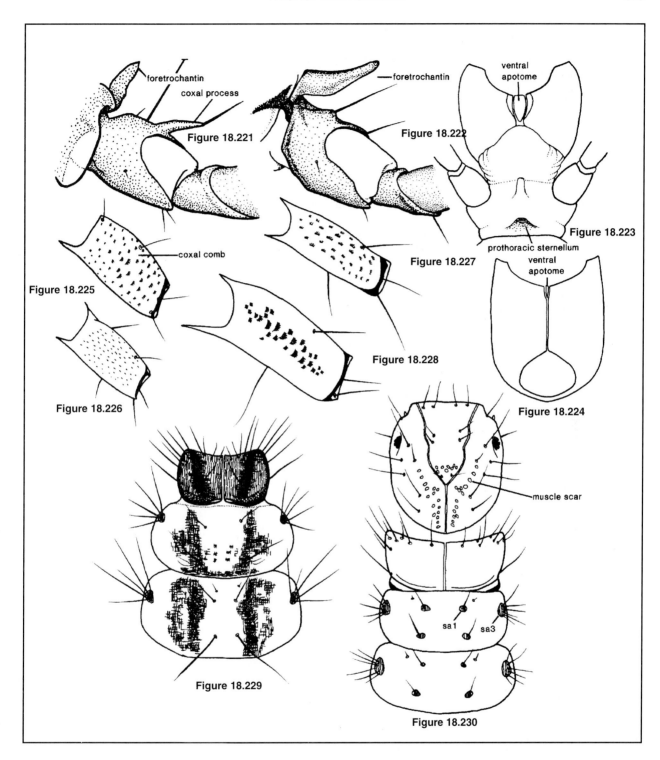

Figure 18.221. *Chimarra* sp. (Philopotamidae); lateral view of right foretrochantin and prothoracic coxa.

Figure 18.222. *Dolophilodes* sp. (Philopotamidae); lateral view of right foretrochantin and prothoracic coxa.

Figure 18.223. *Agrypnia vestita* (Walker) (Phryganeidae); ventral view of head and prosternum.

Figure 18.224. *Yphria californica* (Banks) (Phryganeidae); ventral view of head.

Figure 18.225. *Phryganea* sp. (Phryganeidae); ventral view of left prothoracic coxa.

Figure 18.226. *Banksiola dossuaria* (Say) (Phryganeidae); ventral view of left prothoracic coxa.

Figure 18.227. *Agrypnia vestita* (Walker) (Phryganeidae); ventral view of left mesothoracic coxa.

Figure 18.228. *Phryganea* sp. (Phryganeidae); ventral view of left mesothoracic coxa.

Figure 18.229. *Banksiola dossuaria* (Say) (Phryganeidae); dorsal view of thorax.

Figure 18.230. Prob. *Hagenella canadensis* (Banks) (Phryganeidae); dorsal view of head and thorax.

3(2').	Head and pronotom uniformly light brown except for darker muscle scars on head (Fig. 18.230); case of ring construction (similar to Fig. 17.15) ....probably ***Hagenella***
3'.	Head and pronotum with distinct, dark bands (Fig. 17.45) 4
4(3').	Ventral combs of prothoracic coxae conspicuous, their teeth evident at a magnification of 50X (Fig. 18.225) 5
4.	Ventral combs of prothoracic coxae small, each comb appearing as a tiny raised point at 50X magnification (Fig. 18.226) 6
5(4).	Prothoracic sternellum usually present (Fig. 18.223); ventral combs of mesothoracic coxae with basal axes both transverse and parallel to long axis of coxa (Fig. 18.227); case usually of spiral construction (Fig. 17.16) ....***Agrypnia***
5'.	Prothoracic sternellum absent; ventral combs of mesothoracic coxae with basal axes only transverse to long axis of coxa (Fig. 18.228); case of spiral construction (Fig. 18.232) ....***Phryganea***
6(4').	Meso- and metanota with pair of longitudinal, irregular, dark bands (Fig. 18.229); case of spiral construction 7
6'.	Meso- and metanota nearly uniform in color (Fig. 17.45); case variously constructed 8
7(6).	Abdominal segments VI and VII with anterodorsal gills, segment VII without posteroventral gills (Fig. 18.237); case often with pieces of plant material trailing posteriorly (Fig. 18.233)..........................***Banksiola***
7'.	Abdominal segments VI and VII without anterodorsal gills, segment VII with posteroventral gills (Fig. 18.238); case as in Figure 18.232, without trailing ends..***Oligotricha***
8(6').	Pronotum with dark line along anterior margin, without dark, central, transverse markings (Fig. 18.236); case of spiral construction but with trailing ends of plant fragments giving a bushy appearance (Fig. 18.234) ....***Fabria***
8'.	Pronotum without dark line along anterior margin, with dark transverse markings near center of each sclerite (Fig. 17.45); case of ring construction without trailing ends (Fig. 18.235) ....***Ptilostomis***
9(2).	Antennae each as long as width of pigmented area of clustered stemmata (Fig. 18.5); case straight, with plant materials arranged in discrete rings or bands (Fig. 18.282)...***Beothukus***
9'.	Antennae much shorter than width of pigmented area of clustered stemmata (Fig. 17.46); case slightly curved, with plant materials arranged in discrete rings or bands (Fig. 17.15).........................***Oligostomis***

Polycentropodidae

(modified from Hudson *et al.* 1981)

1.	Tarsi all broader than their tibiae and flat (Fig. 18.239); larva in branching, sand-covered silken tube in stream sand deposits (Fig. 17.3)................................(subfamily Dipseudopsinae).........***Phylocentropus***
1'.	Tarsi all narrower than their tibiae and more nearly cylindrical (Figs. 18.240-18.241); larva in unbranched silken tube on substrate surfaces (Figs. 17.1, 17.2) ....(subfamily Polycentropodinae) 2
2(1').	Anal claw with 6 or fewer conspicuous teeth along ventral, concave margin (Figs. 18.242-18.243)................3
2'.	Anal claw without ventral teeth (Figs. 18.246-18.249) or with 10 or more tiny spines (Fig. 18.244) along ventral, concave margin 4
3(2).	Teeth on anal claw much shorter than apical hook, dorsal accessory spine present (Fig. 18.243); pronotum with short, stout bristle near each lateral margin (Fig. 18.245).........................***Paranyctiophylax***
3'.	Teeth on anal claw almost as long as apical hook, dorsal accessory spine absent (Fig. 18.242); pronotum without short, lateral bristles ....***Polyplectropus***
4(2').	Basal segment of anal proleg about as long as distal segment and with only 2 or 3 apicoventral setae (Fig. 18.244); anal claw with many tiny spines along ventral, concave margin (Fig. 18.244) ....***Neureclipsis***
4'.	Basal segment of anal proleg obviously longer than distal segment in mature specimens and with many setae scattered over most of its surface (Figs. 18.246-18.247); anal claw without tiny ventral spines (Figs. 18.246-18.249) 5
5(4').	Dorsal region between anal claw and sclerite of distal segment of anal proleg with 2 dark bands contiguous mesally (Fig. 18.246, inset); meso- and metanotal sa1 setae short, not more than one-third as long as longest sa2 setae (Fig. 18.250) 6
5'.	Dorsal region between anal claw and sclerite of distal segment of anal proleg with 2 dark bands completely separated mesally (Fig. 18.247, inset); meso- and metanotal sa1 setae about as long as longest sa2 setae (Fig. 18.253) ....***Cyrnellus***
6(5).	Prothoracic tarsi broad and only one-half as long as prothoracic tibiae (Fig. 18.240), *or* anal claw obtusely curved (Fig. 18.248), *or* anal claw with 2 or 3 dorsal accessory spines (Fig. 18.249)...............***Polycentropus***
6'.	Prothoracic tarsi narrow and at least two-thirds as long as prothoracic tibiae (Fig. 18.241), *and* anal claw curved approximately 90° (Fig. 18.246), *and* anal claw with only 1 dorsal accessory spine (Fig. 18.246) ....***Cernotina***

* *Phylocentropus* now in a distinct family, Dipseudopsidae.

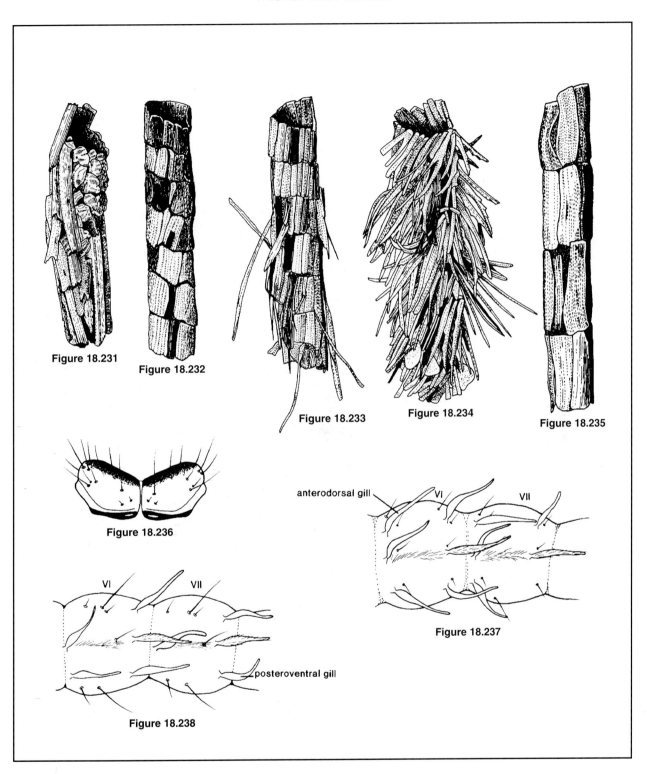

Figure 18.231. *Yphria californica* (Banks) (Phryganidae); larval case.
Figure 18.232. *Phryganea* sp. (Phryganeidae); larval case.
Figure 18.233. *Banksiola dossuaria* (Say) (Phryganeidae); larval case.
Figure 18.234. *Fabria inornata* (Banks) (Phryganeidae); larval case.
Figure 18.235. *Ptilostomis* sp. (Phryganeidae); larval case.
Figure 18.236. *Fabria inornata* (Banks) (Phryganeidae); dorsal view of pronotum.
Figure 18.237. *Banksiola dossuaria* (Say) (Phryganeidae); left lateral view of abdominal segments VI and VII.
Figure 18.238. *Oligotricha lapponica* (Hagen) (Phryganeidae); left lateral view of larval abdominal segments VI and VII.

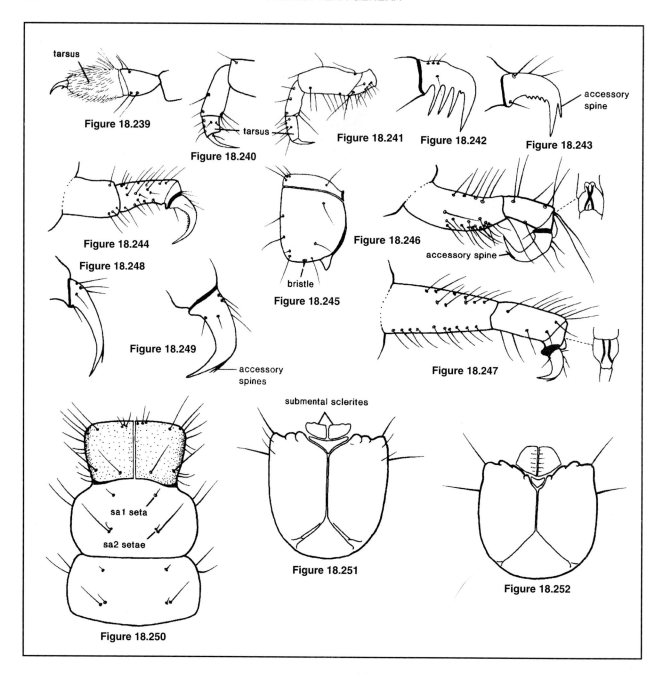

Figure 18.239. *Phylocentropus* sp. (Polycentropodidae); lateral view of left prothoracic tibia, tarsus, and claw.

Figure 18.240. *Polycentropus* (*P.*) sp. (Polycentropodidae); lateral view of left prothoracic tibia, tarsus, and claw.

Figure 18.241. *Cernotina spicata* Ross (Polycentropodidae); lateral view of left prothoracic leg beyond coxa.

Figure 18.242. *Polyplectropus* sp. (Polycentropodidae); lateral view of left anal claw.

Figure 18.243. *Paranyctiophylax* sp. (Polycentropodidae); lateral view of left anal claw.

Figure 18.244. *Neureclipsis* sp. (Polycentropodidae); lateral view of left anal proleg and claw.

Figure 18.245. *Paranyctiophylax* sp. (Polycentropodidae); left dorsolateral oblique view of pronotum.

Figure 18.246. *Cernotina spicata* Ross (Polycentropodidae); lateral view of anal proleg and claw: *inset*, dorsal region at base of anal claw.

Figure 18.247. *Cyrnellus* sp. (Polycentropodidae); lateral view of left anal proleg and claw: *inset*, dorsal region at base of anal claw.

Figure 18.248. *Polycentropus* (*Plectrocnemia*) sp. (Polycentropodidae); lateral view of left anal claw.

Figure 18.249. *Polycentropus* (*Holocentropus*) sp. (Polycentropodidae); lateral view of left anal claw.

Figure 18.250. *Polycentropus* sp. (Polycentropodidae); dorsal view of larval thorax.

Figure 18.251. *Tinodes* sp. (Psychomiidae); ventral view of head.

Figure 18.252. *Psychomyia flavida* Hagen (Psychomiidae); ventral view of head.

Psychomyiidae[6]

1. Anal claw with 3 or 4 conspicuous teeth along ventral, concave margin (Fig. 18.256); paired submental sclerites on ventral surface of labium longer than broad (Fig. 18.252) .... *Psychomyia*
1'. Anal claw without teeth on ventral, concave margin; paired submental sclerites on ventral surface of labium broader than long (Fig. 18.251) 2
2(1'). Dorsolateral edge of each mandible without protuberance, lateral setae about one-third of distance from base (Fig. 18.254) .... *Lype*
2'. Dorsolateral edge of each mandible with rounded protuberance, lateral setae about midway (Fig. 18.255) .... *Tinodes*

Rhyacophilidae

1. Dense tuft of stout gills on sides of meso- and metathorax and abdominal segments I-VII (Fig. 18.257); final instar larva up to 32 mm long .... *Himalopsyche*
1'. Tufts of gills absent or not as dense or not on as many abdominal segments as above; final instar larva less than 30 mm long .... *Rhyacophila*

Sericostomatidae

1. Anterolateral corners of pronotum each acute, projecting (Fig. 18.258); each metanotal sa2 with many setae on transverse sclerite (Fig. 18.260); sand case similar to Figure 17.9 2
1'. Anterolateral corners of pronotum each rounded, not projecting (Fig. 18.259); each metanotal sa2 with single seta and no sclerite (Fig. 18.261); case of small sand grains, frequently long, slender (Fig. 18.262) .... *Gumaga*
2(1). Dorsum of abdominal segment IX with about 40 setae; head flat dorsally with lateral carinae prominent (Fig. 18.263). Higher elevations in southern Appalachian Mountains .... *Fattigia*
2'. Dorsum of abdominal segment IX with about 15 setae; head rounded dorsally with lateral carinae not as prominent (Fig. 18.264). Eastern North America, middle and lower elevations in Southeast .... *Agarodes*

Uenoidae

(modified from Wiggins *et al.* 1985)

1. Pronotum with prominent, lateral, longitudinal ridges; anterior edge of mesonotum with only shallow triangular emargination in middle (Fig. 18.140); case strongly tapered and curved, composed of rocks without larger lateral pebbles (Fig. 18.154) .... *Oligophlebodes*
1'. Pronotum without prominent longitudinal ridges; anterior edge of mesonotum with deep truncate or triangular emargination in middle (Figs. 18.141, 18.284-18.286) 2
2(1'). Mesonotal sclerites with anterior margins straight except for truncate emargination in middle (Fig. 18.141). 3
2'. Mesonotal sclerites with anterior margins convex, with triangular emargination in middle (Figs. 18.284, 18.286) 4
3(2). Pronotum with anterior margin straight and anterolateral corner angulate (Fig. 18.285); case of dark, tough silk, tapered and slightly curved, dorsal surface usually with parallel longitudinal ridges formed of silk, often with slight spiral (Fig. 18.283) .... *Sericostriata*
3'. Pronotum with anterior margin and anterolateral corners uniformly convex (Fig. 18.141); case of coarse rocks, several larger pebbles laterally (Fig. 18.155) .... *Neophylax*
4(2'). Darkened posterolateral corner of each mesonotal sclerite extending anteriad on lateral margin approximately to middle of sclerite (Fig. 18.286); filaments of abdominal lateral fringe arising along half or more of most segments (Fig. 18.265); case of sand grains with thin, silken lining over interior and exterior surfaces (Fig. 17.14) .... *Neothremma*
4'. Darkened posterolateral corner of each mesonotal sclerite extending along posterior one-third of lateral margin (Fig. 18.284); filaments of abdominal lateral fringe scattered and arising along less than half of each segment, but with prominent and discrete tuft of filaments on anterior edge of segment II (Fig. 18.266); case as above, but more slender .... *Farula*

---

6. The larva of *Paduniella* (Arkansas) has 4 conspicuous teeth on each anal claw (similar to Fig. 18.256), and paired submental sclerites on ventral surface of labium small, broader than long (similar to Fig. 18.251).

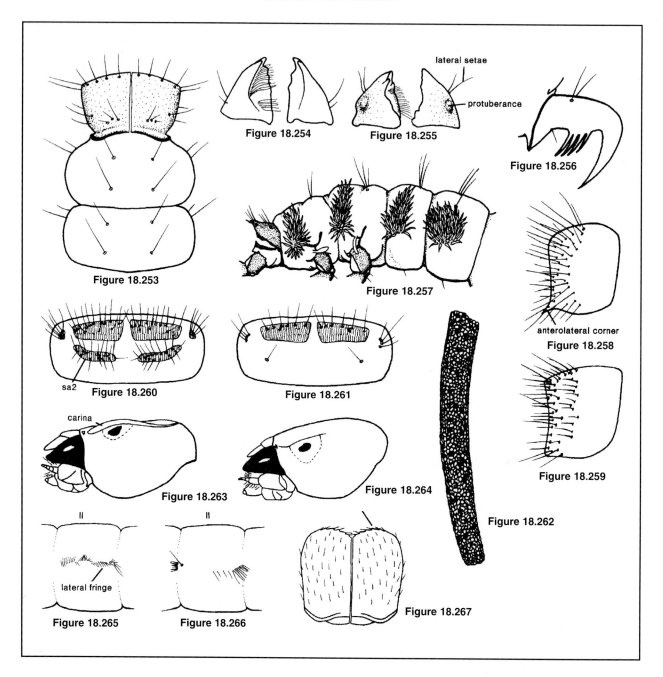

Figure 18.253. *Cyrnellus fraternus* (Banks) (Polycentropodidae); dorsal view of thorax.

Figure 18.254. *Lype diversa* (Banks) (Psychomyiidae); dorsal view of mandibles.

Figure 18.255. *Tinodes* sp. (Psychomyiidae); dorsal view of mandibles.

Figure 18.256. *Psychomyia flavida* Hagen (Psychomyiidae); lateral view of left anal claw.

Figure 18.257. *Himalopsyche phryganea* (Ross) (Rhyacophilidae); left lateral view of larval thorax and abdominal segments I and II (distal portions of thoracic legs omitted).

Figure 18.258. *Agarodes libalis* Ross and Scott (Sericostomatidae); left lateral view of pronotum.

Figure 18.259. *Gumaga* sp. (Sericostomatidae); left lateral view of pronotum.

Figure 18.260. *Agarodes libalis* Ross and Scott (Sericostomatidae); dorsal view of metathorax.

Figure 18.261. *Gumaga* sp. (Sericostomatidae); dorsal view of metathorax.

Figure 18.262. *Gumaga* sp. (Sericostomatidae); larval case.

Figure 18.263. *Fattigia pele* (Ross) (Sericostomatidae); left lateral view of head.

Figure 18.264. *Agarodes libalis* Ross and Scott (Sericostomatidae); left lateral view of head.

Figure 18.265. *Neothremma* sp. (Uenoidae); left lateral view of abdominal segment II.

Figure 18.266. *Farula* sp. (Uenoidae); left lateral view of abdominal segment II.

Figure 18.267. *Neothremma* sp. (Uenoidae); dorsal view of pronotum.

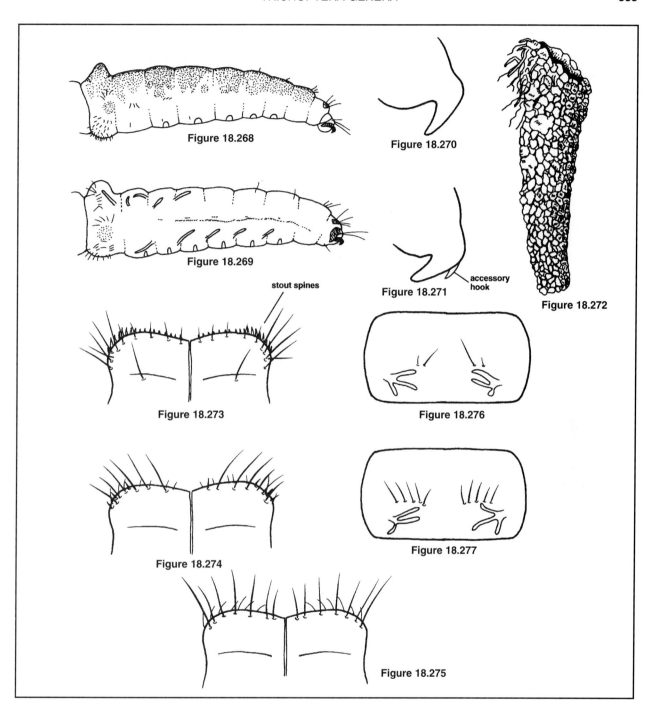

Figure 18.268. *Madeophylax altus* Huryn and Wallace (Limnephilidae); lateral view of abdomen (after Huryn and Wallace 1984).

Figure 18.269. *Manophylax annulatus* Wiggins (Limnephilidae); lateral view of abdomen (after Wiggins 1977).

Figure 18.270. *Madeophylax altus* Huryn and Wallace (Limnephilidae); lateral view of claw of anal proleg (after Huryn and Wallace 1984).

Figure 18.271. *Manophylax annulatus* Wiggins (Limnephilidae); lateral view of claw of anal proleg (after Wiggins 1977).

Figure 18.272. *Madeophylax altus* Huryn and Wallace (Limnephilidae); larval case (after Huryn and Wallace 1984).

Figure 18.273. *Dicosmoecus gilvipes* (Hagen) (Limnephilidae); dorsal view of pronotum (after Wiggins 1977).

Figure 18.274. *Eocosmoecus frontalis* (Banks) (Limnephilidae); dorsal view of pronotum (after Wiggins and Richardson 1989).

Figure 18.275. *Onocosmoecus unicolor* (Banks) (Limnephilidae); dorsal view of pronotum (after Wiggins and Richardson 1989).

Figure 18.276. *Eocosmoecus schmidi* (Wiggins) (Limnephilidae); dorsal view of abdominal segment VII (after Wiggins and Richardson 1989).

Figure 18.277. *Onocosmoecus unicolor* (Banks) (Limnephilidae); dorsal view of abdominal segment VII (after Wiggins and Richardson 1986).

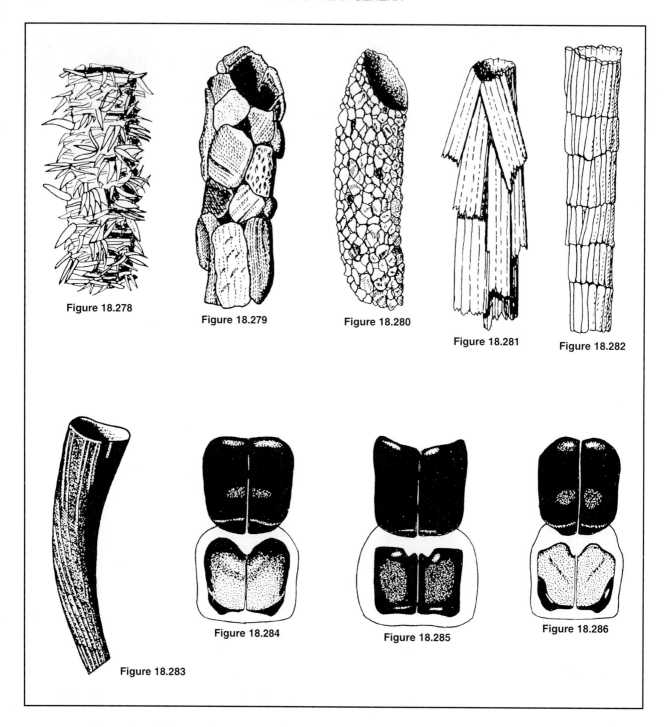

Figure 18.278. *Phanocelia canadensis* (Banks) (Limnephilidae); larval case (after Fairchild and Wiggins 1989).

Figure 18.279. *Eocosmoecus frontalis* (Banks) (Limnephilidae); larval case (after Wiggins and Richardson 1989).

Figure 18.280. *Eocosmoecus schmidi* (Wiggins) (Limnephilidae); larval case (after Wiggins and Richardson 1989).

Figure 18.281. *Sphagnophylax meiops* Wiggins and Winchester (Limnephilidae); larval case (after Winchester et al. 1993).

Figure 18.282. *Beothukus complicatus* (Banks) (Phryganeidae); larval case (after Wiggins and Larson 1989).

Figure 18.283. *Sericostriata surdickae* Wiggins, Weaver and Unzicker (Uenoidae); larval case (after Wiggins et al. 1985).

Figure 18.284. *Farula jewetti* Denning (Uenoidae); dorsal view of pro- and mesonota (after Wiggins et al. 1985).

Figure 18.285. *Sericostriata surdickae* Wiggins, Weaver and Unzicker (Uenoidae); dorsal view of pro- and mesonota (after Wiggins et al. 1985).

Figure 18.286. *Neothremma alicia* Dodds and Hisaw (Uenoidae); dorsal view of pro- and mesonota (after Wiggins et al. 1985).

AQUATIC AND SEMIAQUATIC LEPIDOPTERA

W. H. Lange
University of California, Davis

INTRODUCTION

Adaptation to an aquatic existence in the Lepidoptera is best exemplified by certain members of the family Pyralidae where the immature stages (egg, larva, and pupa) undergo their entire development in the water. In one instance, the brachypterous form of the female of *Acentria nivea* (Olivier) spends her adult life in the water (Berg 1942; Buckingham and Ross 1981). Larvae of the tropical genera *Palustra* and *Maenas* (Arctiidae) have been reported to feed underwater on aquatic plants (Wesenberg-Lund 1943), and some members of the families Nepticulidae, Coleophoridae, Cosmopterigidae, Gelechiidae, Tortricidae, Olethreutidae, Noctuidae, Arctiidae, Cossidae, and Sphingidae are also associated with aquatic or semiaquatic plants. Larval habits include leaf mining, stem or root boring, foliage feeding, and feeding on flower or seed structures. Larvae of some species of the Hawaiian genus *Hyposmocoma* (Cosmopterigidae) construct Trichoptera-like cases (Fig. 19.2) and feed on lichens, algae, or mosses growing on the emergent portions of rocks in fast-flowing streams. The larvae are able to survive periodic submergence.

Among the pyralids, members of the subfamily Nymphulinae are the best suited for an aquatic existence. In the tribe Argyractini, the larvae and pupae occur in a variety of aquatic habitats, including lakes (*Eoparargyractis* sp.), deep and fast-flowing streams, hot springs, intermittent streams, and those receiving organic enrichment (*Petrophila jaliscalis* [Schaus]) (Lange 1956a; Tuskes 1977, 1981). Most of the species in the genus *Petrophila* are scrapers feeding on algae and diatoms on the surfaces of rocks or other submerged objects, but at least one species, *P. drumalis* (Dyar), feeds on the rootlets of water lettuce (*Pistia*) (Dray *et al.* 1989). The larvae of *Neargyractis slossonalis* (Dyar) feed on the roots of *Pistia* and other submerged plants (Habeck 1988). Forno (1983) reports the feeding of *Argyractis subornata* (Dyar) on the roots of water hyacinth (*Eichhornia crassipes*). An increased knowledge of the biology and ecology of plant-feeding nymphuline and related pyralid moths is being used as a basis for the suppression of several aquatic (submergent) weeds. In addition to water hyacinth by *Argyractis subornata* (Ref. 1271), we can include water lettuce by *Sameodes albiguttalis* (Refs. 587, 975), and *Hydrilla vericillata* by *Paraponyx diminutalis* (Refs. 163, 164, 197, 477).

A western species (*Petrophila confusalis* [Walker] [=*truckeealis* Dyar]), is found in well-oxygenated water of streams and lakes. The adult female usually crawls down a rock into the water and deposits groups of eggs on the underside (Fig. 19.1). After hatching, the gilled larva constructs a silken tent on the rock, under which it feeds on diatoms and other algae. The pupal cocoon is made of tightly woven silk with holes along the periphery that allow for water circulation (Fig. 19.1F). The pupa is located in an inner cocoon, and prior to pupation the larva cuts an "escape slit" to assist in adult emergence. The emerging adult swims (using the mid and hind legs, and wings) or floats to the surface from the underwater cocoon.

In northern California two or three generations of *Petrophila confusalis* occur a year. In western Montana McAuliffe and Williams (1983) report a single generation a year for *P. confusalis*. Tuskes (1981) reports that the larvae are most abundant in lakes and streams where the water velocity is between 0.4 and 1.4 m/sec. Other factors influencing numbers and distribution include water temperature, concentration of dissolved oxygen, substrate texture, and algal growth. Although *P. jaliscalis* (Schaus) occurs sympatrically in northern California with *P. confusalis*, the overlap in distribution is limited to streams possessing physical factors conducive to the survival of both species. *Petrophila jaliscalis* can tolerate low oxygen concentration, relatively high temperatures, reduced water velocity, and eutrophication; this allows it to have a more southern distribution and an association with intermittent streams.

In addition to the rock-dwelling species of Nymphulinae (Argyractini), the Nymphulini are associated with vascular hydrophytes and also adapted to fresh water (Fig. 19.3). Many first larval instars are nongilled. Following a molt to the second instar, members of the genus *Parapoynx* (=*Paraponyx*) possess tracheal gills, whereas all instars of the other genera (*Munroessa, Synclita, Neocataclysta*) lack gills (Berg 1949, 1950a). In the absence of gills, a plastron type of respiration has been suggested for *Munroessa icciusalis* (Walker) (Berg 1949, 1950a; Thorpe 1950). The first two larval instars of *Parapoynx maculalis* (Clemens) occur on the bottom and feed on submerged leaves of *Nymphaea*, whereas older larvae generally become surface feeders (Welch 1924). Periodic vibratory movements of *P. maculalis* larvae aid in replenishing oxygen (Welch and Sehon 1928). One species, *Nymphuliella daeckealis* (Haimbach) (=*broweri* Heinrich), feeds on *Cephalozia* sp. in water holes in the mat of sphagnum bogs (Heinrich 1940).

Studies of the aquatic adaptations of *Pyrausta penitalis* Grote (Pyralidae) have revealed that young larvae avoid dis-

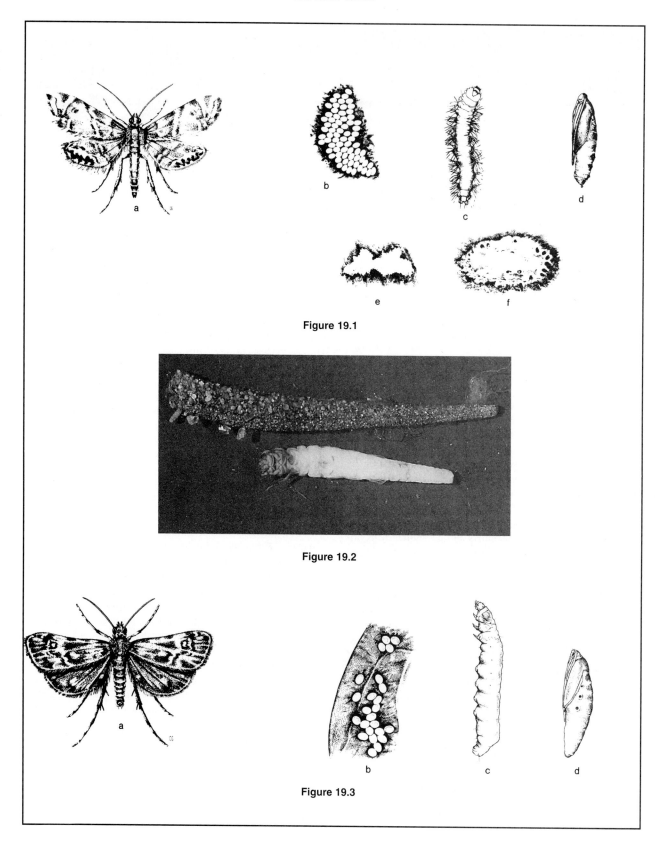

Figure 19.1. Life stages of *Petrophila* (*Parargyractis*) *confusalis* (Walker) (Pyrelidae) (after Lange 1956a). *a*, adult male; *b*, eggs; *c*, mature larva; *d*, pupa, *e*, larval web; *f*, cocoon.

Figure 19.2. Case and larva of *Hyposmocoma* sp. (Cosmopterigidae) (Oahu, Hawaii).

Figure 19.3. Life stages of *Snyclita occidentalis* Lange (Pyralidae) (after Lange 1956a). *a*, adult male, *b*, eggs; *c*, mature larva; *d*, pupa.

placement by feeding under webs on the surface of *Nelumbo lutea* leaves. The older larvae tunnel into the upper end of the petioles and, if dislodged, can swim to other plants. The pupa in a *N. lutea* stem burrow is protected by a woven silk cap, or plug, made by the larva (Welch 1919; Ainslie and Cartwright 1922).

Most of the other lepidopterous larvae associated with aquatic and semiaquatic ecosystems feed on cattail (*Typha* spp.), bulrush (*Scirpus* spp.), or other vascular hydrophytes. These include genera of several pyralids (*Crambus* and *Chilo*), a nepticulid (*Nepticula*), several coleophorids (*Coleophora*) (Ellison 1991), several cosmopterigids (*Cosmopteryx* and *Lymnaecia*), noctuid borers (*Bellura*, *Oligia*, *Neocrastria*, *Archanara*, and *Simyra*), and a tortricid (*Archips*) (Claassen 1921). A few records are available for the aquatic habits of sphingid larvae (Hagen 1880), and recent papers have discussed the biology and ecology of several genera of noctuid borers (*Bellura*, *Apamea*) (MacKay and Rockburne 1958; Vogel and Oliver 1969; Levine 1974; Levine and Chandler 1976).

EXTERNAL MORPHOLOGY

Eggs

The eggs of Lepidoptera are laid singly or in clusters or groups. In many pyralids, the eggs are often greatly flattened and deposited in overlapping layers; in others groups eggs are of the upright type. In some Noctuidae, eggs are characteristically ribbed with a depressed micropyle (minute opening) and are laid singly. In the genus *Spodoptera* (Noctuidae), eggs are deposited in groups that the female covers with hairs from the abdomen, and in some families eggs are inserted into leaves or flower structures.

Larvae

Although many modifications occur, lepidopterous larvae are characterized by: (1) the presence of a distinct head with a ring of ocelli (*stemmata* or simple eyes); (2) a thorax composed of pro-, meso-, and metathoracic segments, each with a pair of segmented legs; (3) 10 abdominal segments with prolegs on segments 3, 4, 5, 6, and 10 (anal prolegs); and (4) spiracles on the prothoracic segment and abdominal segments 1-8 (Fig. 19.4).

Head: The head capsule can be vertical (*hypognathous*) with the mouthparts directed downward (ventrally) (Fig. 19.4) or horizontal (*prognathous*) with mouthparts directed forward (Fig. 19.5). The head of leaf-mining or wood-boring species is often prognathous. An inverted *epicranial suture* and narrow *adfrontal sclerites* are distinct larval features (Fig. 19.11). The *clypeus* and *labrum* are ventrad or anterad to the frons. The antennae are usually short and bear several sensillae. A distinct *spinneret* (silk-producing structure) is present and mouthparts are the chewing type with opposable mandibles. The *ocelli* are usually six in number but can be reduced, modified, or lost (Fig. 19.4).

Thorax: The thorax usually bears a distinct, often sclerotized, *cervical (prothoracic) shield* (Fig. 19.4). The true legs, one pair on each thoracic segment, are five-segmented with terminal claws but can be aborted or missing. Gills occur on some aquatic species.

Abdomen: In most Lepidoptera the abdomen consists of 10 segments, with pairs of fleshy prolegs on segments 3, 4, 5, 6, and 10 (Fig. 19.4); however, some variations occur. In Nepticulidae prolegs occur on segments 2-7, and in the semiloopers (Noctuidae) on segments 5, 6, and 10. The truncate end of each proleg, the *planta*, bears curved hooks called *crochets* (Fig. 19.4), the pattern and arrangement of which are used extensively in larval classification. If the crochets are all the same length, they are termed *uniordinal;* if two or three lengths, *bi-* or *triordinal*. If they arise from a given line, they are referred to as *uniserial;* if arising from three or more lines, *multiserial*. Crochets can be in a complete circle, in an interrupted circle (*penellipse*), in single bands extending longitudinally on the mesal side of a proleg (*mesoseries*), or in a combination of lateroseries and mesoseries. In aquatic forms, *blood gills* or *tracheal gills* may be present or absent.

Chaetotaxy: Larvae of Lepidoptera bear setae or punctures commonly used in their classification. *Primary setae* on the head, thorax, and abdomen have a definite distribution and have a system of nomenclature utilizing Greek letters, Roman numerals, or other identifying symbols (Fracker 1915; Heinrich 1916; Hinton 1946; Common 1970). Some *subprimary setae* do not occur in the first instar but appear in the second. *Secondary setae* occur randomly and are not specifically named. A few microscopic primary setae, which occur at the anterior margin of each segment or on the prothorax, are believed to be proprioceptors (sense organs; Hinton 1946). Setae often arise from a flattened, pigmented area, a *pinaculum*, which if elevated is called a *chalaza*. In hairy caterpillars, such as the Arctiidae, a wartlike elevation bearing several to many setae pointing in different directions is called *verruca*. The Sphingidae have a characteristic *caudal horn* or scar on the dorsum of the eighth abdominal segment, and *anal combs*, or *forks*, located ventrad of the suranal plate (Fig. 19.4), are found in some Tortricidae and Olethreutidae. The cuticle of many larvae has ornamentations of granules, spicules, microspines, or pigmented areas that are sometimes useful in classification.

Pupae

The typical lepidopteran pupa is of the *obtect* type (appendages and body compactly united) and may or may not be enclosed in a cocoon (Fig. 19.7). In many of the truly aquatic forms, the external spiracular openings on abdominal segments 3 and 4 are greatly enlarged and the *cremaster* (apex of the last segment of the abdomen) is hooklike and adapted to firmly anchor the pupa to its silken case (Fig. 19.7).

Adults

Lepidoptera are characterized by the presence of overlapping scales on the two pairs of wings, body, and legs. Aquatic Lepidoptera usually possess a sucking proboscis or haustellum formed from the modified galeae of the maxillae.

Head: The compound eyes are usually prominent and one ocellus may be present over each eye. A pair of sensory organs, *chaetosemata* (Jordan's organs), is found in some aquatic Nymphulinae (Fig. 19.28). A pair of *antennae* are located anterior to the ocelli, and the *scape* (basal segment) sometimes has a tuft of scales (*eye-cap*) over the compound eye. The antennae can be lamellate (leaflike), ciliated, clubbed, moniliform (beadlike), or serrate (sawlike). *Labial palpi* are usually present but *maxillary palpi* may be well-developed or absent.

Thorax: The small *prothorax* possesses a pair of overlapping plates, the *patagia*, and the well-developed *mesothorax* has a pair of laterally placed *tegulae* (articular sclerites of insect wing). The *metathorax* is usually inconspicuous, and the legs are well-developed with 5-segmented tarsi. In some aquatic Nymphulinae, the hind legs possess an oarlike fringe of hairs used in swimming.

Wings: Both surfaces of the wings are usually clothed with overlapping, broad scales. Wing colors and patterns are helpful in determining species, and wing venation is also used. The generalized venation is similar to primitive Trichoptera, with a tendency toward vein reduction in the higher groups. Some families may have *brachypterous* forms in one or both sexes, and in a few species (e.g., *Acentria nivea*) both winged and brachypterous females occur.

Abdomen: The 10-segmented abdomen has the terminal segments modified into genitalia. *Tympanal organs* (hearing) occur at the base of the abdomen in some groups. The *genital* structures are helpful in characterizing taxa at the generic and specific levels, particularly in males in which modification of the *valvae* (claspers), *aedeagus*, *juxta*, *tegumen*, and *uncus* are commonly used. Females have an egg canal and a copulatory opening, the *ductus bursae* (ostium bursae). The sclerotized portions of the ductus bursae and associated structures are particularly diagnostic in certain aquatic forms.

KEYS TO THE FAMILIES OF AQUATIC AND SEMIAQUATIC LEPIDOPTERA

Larvae

1.	Thoracic legs reduced to fleshy swellings; prolegs usually on abdominal segments 2-7 (Fig. 19.8); crochets (curved hooks) lacking; small, leaf-mining species	**NEPTICULIDAE**
1'.	Thoracic legs present, segmented; crochets present (Fig. 19.4)	2
2(1').	Thorax and abdomen with filamentous gills (Figs. 19.5, 19.9)	3
2'.	Thorax and abdomen without filamentous gills	5
3(2).	Head flattened dorsoventrally (Fig. 19.11); prognathous (head horizontal and mouthparts directed forward); thorax and abdomen with numerous blood gills (Figs. 19.1c, 19.5; 19.9); mandibles prominent, adapted to scrape algae and diatoms from rocks in streams, lakes, and springs	**PYRALIDAE**-*Petrophila*
3'.	Head round, typical type; hypognathous (head vertical and mouthparts directed ventrally), or sometimes prognathous; feed on higher aquatic plants	4
4(3').	Thorax and abdomen with branched or unbranched subspiracular and supraspiracular gills (Fig. 19.10); living in cases cut from leaves of aquatic plants	**PYRALIDAE**-*Parapoynx*
4'.	Thorax and abdomen with gills usually arising separately, not branched; a conspicuous tuft of gills on abdominal segments 8-10 (Fig. 19.12)	**PYRALIDAE** (Nymphulinae, genus unknown)
5(2').	Larvae living in cases (Fig. 19.13)	6
5'.	Larvae not living in cases; stem-borers, leaf-miners, foliage feeders; some use plant materials to build retreats (Fig. 19.14)	7
6(5).	Cases made of silk and sand particles, sometimes with operculum (lid) present; semiaquatic trichopteroid type living in conical, thornlike or saclike cases on emergent rocks in swiftly running streams; prolegs absent; crochets present with thoracic legs enlarged (Figs. 19.2, 19.15); endemic to Hawaii	**COSMOPTERIGIDAE** (Cosmopteriginae)-*Hyposmocoma*
6'.	Case made of leaves or plant material, usually associated with lakes, ponds, or quiet water (Fig. 19.3)	**PYRALIDAE**-*Munroessa, Synclita, Neocataclysta*
7(5').	Terminal segment of abdomen with anal fork (Fig. 19.16); feeds on leaves and heads of aquatic plants	**TORTRICIDAE**-*Archips*
7'.	Terminal segment of abdomen without anal fork	8
8(7').	Body with prominent verrucae (elevated portions of cuticle bearing tufts of long setae) (Fig. 19.17)	9
8'.	Body without prominent verrucae	10
9(8).	Crochets heteroideous (middle crochets abruptly longer on each proleg); head of *Estigmene* largely black; setae dark, long, and setal tufts dense (Fig. 19.18)	**ARCTIIDAE**-*Estigmene*
9'.	Crochets homoideous (no abrupt change in crochet length); head of *Simyra* black and white mottled; setae light colored, bristlelike, sparsely distributed (Fig. 19.19)	**NOCTUIDAE**-*Simyra*
10(8').	Anal horn on 8th abdominal segment present (Fig. 19.20) or reduced to prominent dark spot	**SPHINGIDAE***
10'.	Anal horn on 8th abdominal segment absent	11
11(10').	Large larvae, 50-70 mm long at maturity	12
11'.	Small larvae, less than 20 mm long at maturity	13
12(11).	Crochets uniordinal and arranged in a transverse band (*Archanara*) (Fig. 19.21), or arranged as in Fig. 19.22 (*Bellura*); spiracles narrow, height at least 3 times width (Fig. 19.23); burrowers in *Typha, Nuphar, Pontederia,* and *Nymphaea* stems	**NOCTUIDAE**

*Included in the key since some species are associated with aquatic habitats

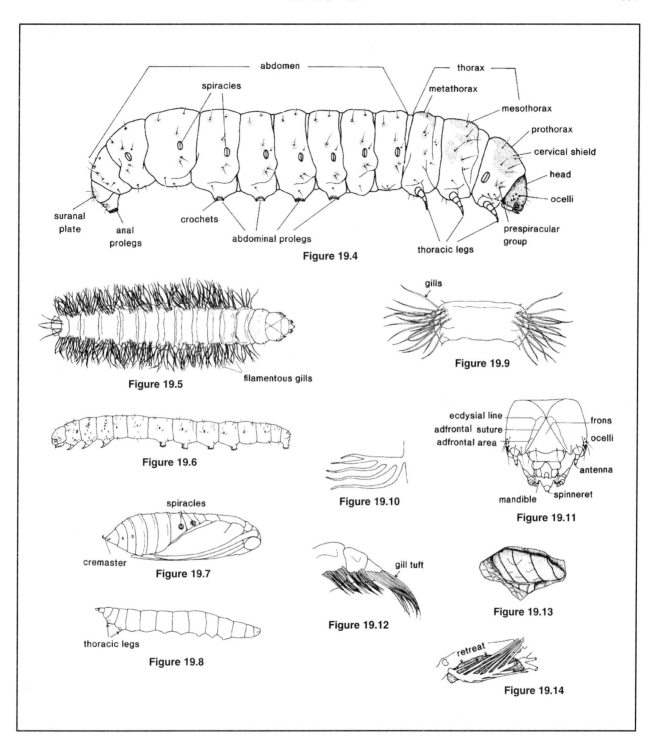

Figure 19.4. Lateral view of larva of wood-boring type, *Prionoxystus robiniae* (Peck) (Cossidae).

Figure 19.5. Dorsal view of gilled larva of rock-dwelling type, *Petrophila confusalis* (Walker) (Pyralidae).

Figure 19.6. Lateral view of stem-boring larva of *Archanara oblonga* (Grote) (Noctuidae).

Figure 19.7. Lateral view of pupa of *Petrophila confusalis* (Walker) (Pyralidae) showing enlarged spiracular openings on specific abdominal segments.

Figure 19.8. Lateral view of larva of *Nepticula* sp. (Nepticulidae).

Figure 19.9. Gills of *Petrophila confusalis* (Walker) (Pyralidae).

Figure 19.10. Branched gill of *Parapoynx* sp. (Pyralidae).

Figure 19.11. Dorsal view of head of *Petrophila confusalis* (Walker) (Pyralidae).

Figure 19.12. Gill tufts on posterior segments of a pyralid (Nymphulinae).

Figure 19.13. Larval case of *Snyclita occidentalis* Lange (Pyralidae).

Figure 19.14. Larval retreat of *Acentria* sp. (Pyralidae) on *Ceratophyllum* sp.

12'.	Crochets triordinal to multiordinal (3 or more different lengths) (Fig. 19.24); prothoracic spiracle enlarged; cervical shield smooth (Fig. 19.4); body white with brown pinacula (seta-bearing papillae) (Fig. 19.25); borer in *Populus* and *Salix* (Fig. 19.4) ....**COSSIDAE**-*Prionoxystus**
13(11').	Crochets biordinal, in an incomplete ellipse with almost parallel sides and open on mesal side (Fig. 19.26); body surface covered with small granules; pinacula (seta-bearing papillae) inconspicuous; in retreats made by fastening together *Ceratophyllum* or other aquatic plants ....**PYRALIDAE**-*Acentria*
13'.	Crochets uniordinal or biordinal, in circles; body with or without granules or spicules; pinacula variable 14
14(13').	Pinacula conspicuous (Fig. 19.25); body covered with granulations; crochets biordinal and arranged in a circle ....**PYRALIDAE**-*Crambus*
14'.	Pinacula inconspicuous; body smooth; crochets uniordinal and arranged in a circle.......................15
15(14').	Body covered with minute spicules (Fig. 19.27); primary setae short; on nutgrass ....**TORTRICIDAE** (Oleuthreutinae)-*Bactra*
15'.	Body smooth, spicules lacking; pinacula not distinct; setae long; on *Helianthus* heads...**COSMOPTERIGIDAE**-*Pyroderces*

Adults

1.	Haustellum (proboscis) short, rudimentary, or missing...2
1'.	Haustellum well developed (Fig. 19.28) 5
2(1).	Wings usually fully developed in males and females 3
2'.	Species with brachypterous wings in some females; male with normal wings ....**PYRALIDAE**-*Acentria*
3(2).	Small species with slender, aculeate, or lanceolate wings (Fig. 19.29); body slender 4
3'.	Larger species (greater than 20 mm); wings of normal width; body stout................**COSSIDAE***
4(3).	Head with ascending (recurved) labial palpi (Fig. 19.30); species over 3 mm in length; eye-cap (greatly enlarged basal segment of antenna with tuft of scales over the eye) absent ....**GELECHOIDEA, (GELECHIIDAE)-(COSMOPTERIGIDAE)**
4'.	Head with drooping labial palpi; very small species (3 mm); eye-cap present............**NEPTICULIDAE**
5(1').	Small species, wing expanse under 30 mm...6
5'.	Larger species, wing expanse 35-115 mm 8
6(5).	Fore wings somewhat square-tipped (Figs. 19.31-19.32) 7
6'.	Fore wings not usually square-tipped, outer margin with projecting apex and often sloping inwardly toward body (Figs. 19.33, 19.38-19.41) .... **PYRALIDAE**
7(6).	Costal margin of fore wings arched (Fig. 19.31) .... **TORTRICIDAE**
7'.	Costal margin of fore wings not arched (Fig. 19.32)..**OLETHREUTIDAE**
8(5').	Hind wing small relative to fore wing (about one-half length of fore wing); body stout, fusiform (spindle-shaped); antennae hooked at tip (Fig. 19.34) ....**SPHINGIDAE***
8'.	Hind wing approaching length of fore wing; body not usually fusiform; antennae not hooked at tip (Fig. 19.35) 9
9(8').	Mostly somber or dark-colored species (Fig. 19.42); abdomen usually lacking dark lateral median markings; veins Sc and R in hind wing fused for only a short distance at base of discal cell (Fig. 19.36) ....**NOCTUIDAE**
9'.	Light-colored species or with combinations of brightly colored or banded patterns (Fig. 19.43); abdomen usually with dorsal and/or lateral dark markings; Sc and R in hind wing fused to middle of discal cell (Fig. 19.37) ....**ARCTIIDAE**

ADDITIONAL REFERENCES

General taxonomic
Müller (1892); Quail (1904); Portier (1911, 1949); Tsou (1914); Fracker (1915); Heinrich (1916); Mosher (1916); Gerasimov (1937); Frohne (1939b); Wesenberg-Lund (1943); Hinton (1946, 1948, 1956); Peterson (1948); Chu (1949, 1956); Patocka (1955); Capps (1956); Lange (1956a,b); Haggett (1955-1961); Speyer (1958); Welch (1959); Okumura (1961); MacKay (1963a, 1964, 1972); Klots (1966); Common (1970); Pennak (1978); Lehmkuhl (1979a); Brigham and Herlong (1982); Scoble (1992).

Regional faunas
California: Lange (1956a); Okumura (1961); Powell (1964); Jackson and Resh (1989).
Florida: Heppner (1976); Minno (1992).
Indiana and adjacent areas: McCafferty and Minno (1979).
New York: Forbes (1923, 1954, 1960).

Taxonomic treatments at the family and generic levels
(L =larvae; A =adults)
Cosmopterigidae: Hodges (1962)-A.
Cosmopterigidae (Cosmopteriginae): Walsingham (1907)-A; Hodges (1962)-A; Zimmerman (1978)-A.
Nepticulidae: Braun (1917)-L, A.
Noctuidae: Crumb (1929, 1956)-L; Forbes (1954)-L, A; MacKay and Rockburne (1958)-L; Beck (1960)-L.
Olethreutidae: MacKay and Rockburne (1958)-L; MacKay (1959, 1962a)-L; Beck (1960)-L.
Pyralidae: Forbes (1910)-L, A; Heinrich (1940)-A; Lange (1956a,b)-L, A; Hasenfuss (1960)-L; Munroe (1972-1973)-A; Heppner (1976)-L, A.
Tortricidae: MacDunnough (1933)-L; Swatchek (1958)-L; MacKay (1926b,1963b)-L; Powell (1964)-L, A.

*Included in the key since some species are associated with aquatic habitats

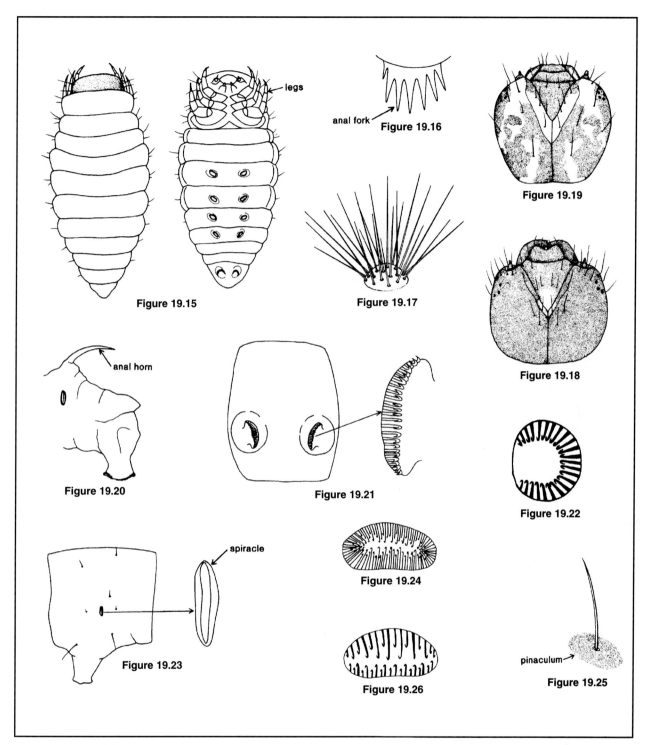

Figure 19.15. Dorsal and ventral view of *Hyposmocoma* sp. (Cosmopterigidae).

Figure 19.16. Anal fork in *Archips* sp. (Tortricidae).

Figure 19.17. Verruca with tuft of setae.

Fiture 19.18. Dorsal view of larval head of *Estigmene* sp. (Arctiidae).

Figure 19.19. Dorsal view of larval head of *Simyra* sp. (Noctuidae).

Figure 19.20. Lateral view of abdomen showing anal horn of Sphingidae.

Figure 19.21. Crochets of *Archanara oblonga* (Grote) (Noctuidae) (on abdominal segments 3-6).

Figure 19.22. Crochets of *Bellura* sp. (Noctuidae) (on abdominal segments 3-6).

Figure 19.23. Lateral view of larval abdominal segment of *Archanara* sp. (Noctuidae) showing spiracle.

Figure 19.24. Crochets of *Prionoxystus* sp. (Cossidae).

Figure 19.25. Pinaculum of *Crambus* sp. (Pyralidae).

Figure 19.26. Crochets of *Acentria* sp. (Pyralidae).

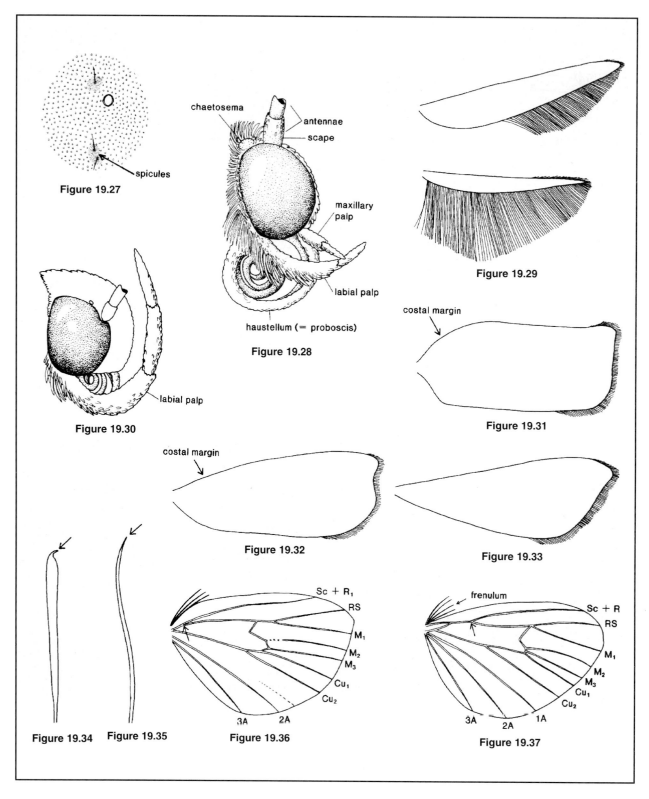

Figure 19.27. Spicules of Olethreutidae.
Figure 19.28. Lateral view of adult Lepidoptera head showing diagnostic characters.
Figure 19.29. Front and hind wing of Cosmopterigidae.
Figure 19.30. Lateral view of adult Lepidoptera head showing recurved labial palpi (Cosmopterigidae).
Figure 19.31. Fore wing of Tortricidae.
Figure 19.32. Fore wing of Olethreutidae.
Figure 19.33. Fore wing of Pyralidae.
Figure 19.34. Antenna of Sphingidae.
Figure 19.35. Antenna of Noctuidae.
Figure 19.36. Hind wing of Noctuidae.
Figure 19.37. Hind wing of *Estigmene acrea* Dru. (Arctiidae).

Figure 19.38

Figure 19.40

Figure 19.39

Figure 19.41

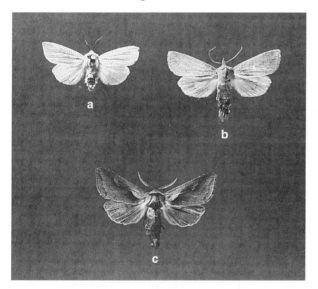

Figure 19.42

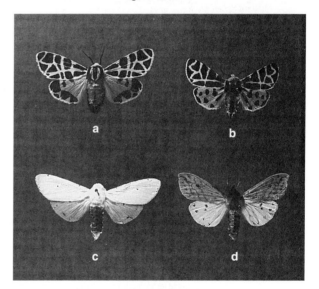

Figure 19.43

Figure 19.38. Adult male of *Parapoynx maculalis* (Clemens) (Pyralidae).

Figure 19.39. Adult male of *Neocataclysta magnificalis* (Hubner) (Pyralidae).

Figure 19.40. Adult male of *Petrophila* (*Parargyractis*) *jaliscalis* (Schaus) (Pyralidae).

Figure 19.41. Adult male of *Eoparargyractis plevie* (Dyar) (Pyralidae).

Figure 19.42. Adults of Noctuidae. *a*, *Simyra henrici* (Grt.); *b*. *Archanara oblonga* (Grt.); *c*. *Arzama obliqua* Wlk.

Figure 19.43. Adults of Arctiidae, *a*, *Apantesis proxima* Guer.; *b*, *A. ornata* Pack.; *c*, *Estigmene acrea* Dru.; *d*, *Isia isabella* A. and S.

Table 19A. Summary of ecological and distributional data for Lepidoptera. (For definition of terms see Tables 6A-6C; table prepared by K. W. Cummins, R. W. Merritt, and W. H. Lange.)

Taxa (number of species in parentheses)	Habitat	Habit	Trophic Relationships	North American Distribution	Ecological References*
Pyralidae(148) (=Pyraustidae)	Generally lentic—vascular hydrophytes burrowers (miners)	Generally climbers, climbers—swimmers, and miners, stem borers	Generally shredders—herbivores (leaf eaters)		9, 2056, 2112, 2243, 3076, 3185, 3186, 3999, 587, 4056, 4297, 4306, 2715, 407, 1518, 2242
Nymphulinae(50) (=Hydrocampinae)	Generally lentic—vascular hydrophytes	Generally climbers, climbers—swimmers	Shredders—herbivores		1683, 2056, 2067, 2112, 2243, 2279, 2280, 2804, 4056, 4297
Ambiini(3)					
Undulambia(3)	Lentic—margins (on ferns)	Climbers	Shredders—herbivores	Southern Florida	1683, 2804
Nymphulini(27)					
Chrysendeton(3)				Southeastern United States	1271, 1842, 2243, 2804
Contiger(1)				Southeastern United States	2243, 2804
Langessa(1)	Lentic—vascular hydrophytes (*Nymphaea, Nymphoides*)	Climbers	Shredders—herbivores	Widespread (common in Florida)	1683, 2243, 2804
Munroessa(4)	Lentic and lotic—vascular hydrophytes (*Potamogeton, Vallisneria, Nymphaea, Ludwigia, Lemna, Brasenia*, etc.)	Climbers	Shredders—herbivores	Widespread	1683, 2243, 2592, 2618, 2804
Neocataclysta(1)	Lentic and lotic—vascular hydrophytes (floating zone: *Lemna*)	Climbers—swimmers (case constructed of *Lemna*)	Shredders—herbivores (on *Lemna*)	Nova Scotia and southern Quebec to Florida	1261, 2243, 2804, 4056, 1842, 2592
Nymphula(1) (=*Hydrocampa* of some authors)	Lentic—vascular hydrophytes	Climbers—swimmers (case constructed of sedges, Cyperaceae)	Shredders—herbivores (chewers and miners)	Northeastern United States and southeastern Canada	177, 263, 264, 1262, 1313, 1620, 1972, 2243, 2618, 2619, 2801, 2804, 3601, 3677, 4306, 4420, ‡
Nymphuliella(1)	Lentic—bog pools	Climbers (case constructed of *Cephalozia fruitans*)	Shredders	Northeastern United States	1655, 2243, 2804
Oligostigmoides(1)				Texas	2243, 2804
Parapoynx(7) (=*Paraponyx* of some authors)	Lentic—vascular hydrophytes (*Nuphar, Nymphaea, Brasenia, Potamogeton, Bacopa, Myriophyllum, Vallisneria, Nymphoides, Orortium*, etc.)	Climbers—swimmers (cases of a wide variety of aquatic plants)	Shredders—herbivores	Widespread	163, 264, 1260, 1600, 1683, 2592, 2618, 2619, 3527, 4291, 4294, 4295, 4297, 975, 164, 477, 1513, 2777, 2998
Synclita(4)	Lentic—vascular hydrophytes (*Nymphaea, Potamogeton, Lemna, Brasenia*, etc.)	Climbers—swimmers (cases constructed of pieces of leaves or *Lemna*)	Shredders—herbivores (chewers)	Widespread (introduced into Hawaii)	1262, 1600, 1683, 2056, 2243, 2592, 2804, 4448, 975, 164, 1517, 2096, 1682
Argyactini(23)					
Eoparargyractis(3)	Lentic—littoral?			East (Canada to Florida)	1178, 2243, 2804
Neargyractis(1)	Lentic and Lotic—streams and lakes	Climbers—swimmers	Shredders—herbivores scrapers—periphytes	Florida and southeastern United States	1516, 2115, 2804
Petrophila(16) (=*Parargyractis*)†	Lotic—erosional, lentic—erosional (algae, diatoms, roots of submerged plants)	Clingers (silk-retreat makers, shelters of plant material)	Scrapers and herbivores	Widespread	2243, 2279, 2280, 2281, 2360, 2592, 2804, 1271, 4056, 4060, 4061, 783, 974, 975, 3307, 258, 2564, 4234, ∫, ¶

*Emphasis on trophic relationships
†See "Checklist of the Lepidoptera north of Mexico," London, 1983.
‡*Nymphula ekthlipsis* (Grote) is the only species known in the genus from North America; some of the references refer to species now placed in related genera.
∫Unpublished data, K. W. Cummins, Oregon State University.
¶Unpublished data, W. H. Lange, Department of Entomology, University of California.

Table 19A. Continued

Taxa (number of species in parentheses)	Habitat	Habit	Trophic Relationships	North American Distribution	Ecological References*
Oxyelophila(1)				Texas	2243, 2804
Usingeriessa(2)	Lotic—erosional?			California, Arizona, Texas	2243, 2804
Schoenobiinae(26)	Generally lentic—vascular hydrophytes (emergent zone)	Generally burrowers (miners—stem borers)	Generally shredders—herbivores		2804
Acentria(1) (=*Acentropus*)	Lentic—vascular hydrophytes (*Ceratophyllum, Potamogeton, Elodea, Hydrilla, Myriophyllum, Elatine*, etc.)	Climbers—swimmers, (early instars, stem borers; later instars, in retreats made of plant material)	Shredders—herbivores	Canada and eastern United States	197, 273, 476, 1263, 2028, 2243, 2592, 2800, 2804, 2918, 4042, 4043, 752, 753
Schoenobius(12)	Lentic—vascular hydrophytes (emergent zone; *Eleocharis, Carex, Scirpus*; semiaquatic)	Burrowers (miners—stem borers below water)	Shredders—herbivores (miners)	Widespread	1313, 1316, 2592, 4290
Crambinae(8)	Lentic—vascular hydrophytes (emergent zone)	Burrowrs (miners—stem borers below water)	Shredders—herbivores (miners)		1316, 2592, 2804
Chilo(3)	Lentic—vascular hydrophytes (emergent zone; *Scirpus, Juncus, Eleocharis, Oryza*)	Burrowers (miners—stem borers)	Shredders—herbivores (miners)	Widespread	1313, 1315, 2592
Acigona(1) (=*Occidentalia*)	Lentic—vascular hydrophytes (emergent zone; *Scirpus*)	Burrowers (miners—stem borers)	Shredders—herbivores (miners)	Eastern United States	1316, 2592
Pyraustinae(80)					2125
Ostrinia(4) (=*Pyrausta*)	Lentic—vascular hydrophytes (*Potamogeton penitalis*) (emergent and floating zone; *Eupatorium, Polygonum*, and Nymphaceae)	Burrowers (in leaves—stem borers) (adapted to aquatic locomotion)	Shredders—herbivores	Widespread	9, 1682, 2592, 4293
Nepticulidae(70) (=Stigmellidae)					
Stigmella (=*Nepticula*)	Lentic—vascular hydrophytes (emergent zone); *Scirpus, Eleocharis*	Burrowers (miners—(stem borers)	Shredders—herbivores (miners)	Widespread	1313, 1316
Cosmopterigidae(371+) (=Lavernidae)	Generally lentic—vascular hydrophytes	Generally burrowers (miners)	Shredders—herbivores		642, 1313, 1518
Cosmopteryx(30)	Lentic—vascular hydrophytes (emergent zone)	Burrowers (miners—stem borers)	Shredders—herbivores (miners)	Widespread	642
Lymnaecia(1)	Lentic—vascular hydrophytes (emergent and floating zones; *Typha*)	Burrowers (in heads, seeds, or stems)	Shredders—herbivores	Widespread	642
Cosmopteriginae(350+?)					
Hyposmocoma(30?)	Lentic—margins (on lichens, algae, and mosses) (unknown number of species semiaquatic)	Climbers (in cases of sand particles, silk, and debris on emergent portion of basalt rocks)	Shredders—herbivores (chewers and scrapers)	Endemic to Hawaii	4211, 4631, ¶

*Emphasis on trophic relationships.
¶Unpublished data, W. H. Lange, Department of Entomology, University of California.

Table 19A. Continued

Taxa (number of species in parentheses)	Habitat	Habit	Trophic Relationships	North American Distribution	Ecological References*
Noctuidae(11) (=Phalaenidae)	Generally lentic—vascular hydrophytes	Generally burrowers and climbers	Shredders—herbivores		1518, 2412
Archanara(4)	Lentic—vascular hydrophytes (emergent and floating zones; *Typha, Scirpus, Juncus, Sparganium*)	Burrowers (miners—stem borers)	Shredders—herbivores	Widespread	774, 1313, 2243, 2592
Bellura(4) (=*Arzama*)	Lentic—vascular hydrophytes (emergent and floating zones; *Typha, Nuphar, Pontederia, Eichornia, Nelumbo, Symplocarpus, Sagittaria, Spanganium*)	Burrowers—leaf miners (early instars); petiole and stem borers (later instars)	Shredders—herbivores (miners)	Central, East, South	264, 642, 774, 2333, 2592, 2618, 2619, 4138, 1512, 1658, 4288, 4547
Neoerastria(1?)	Lentic—vascular hydrophytes (emergent zone)	Burrowers	Shredders—herbivores	Midwest	1514, 1682, 2592
Oligia(1?)	Lentic—vascular hydrophytes (emergent zone; *Scirpus*)	Burrowers—leaf miners (early instars), stem borers (later instars)	Shredders—herbivores	Midwest	2592
Simyra(1)	Lentic—vascular hydrophytes (emergent and floating zones; *Typha, Polygonum, Salix*, Graminaceae)	Climbers	Shredders—herbivores (chewers)	Widespread	578, 774, ¶
Tortricidae(37)					
Archips(1) (=*Choristoneura*)	Lentic—vascular hydrophytes (emergent and floating zones; *Typha* and other plants)	Burrowers—climbers	Shredders—herbivores	Widespread	642, 1682
Coleophoridae(145)					
Colephora(3)	Lentic—salt marsh; *Juncus, Salticornia*	Burrowers—seed capsules	Shredders—herbivores	Widespread; New England	1077, 1682

*Emphasis on trophic relationships.
¶Unpublished data, W. H. Lange, Department of Entomology, University of California.

AQUATIC COLEOPTERA

David S. White
*Hancock Biological Station,
Murray, Kentucky*

Warren U. Brigham
*Illinois Natural History Survey,
Champaign, Illinois*

INTRODUCTION

Coleoptera comprise the largest order of insects and, with approximately 5,000 aquatic members, ranks as one of the major groups of freshwater arthropods. Moreover, beetles occupy a broad spectrum of aquatic habitats, including such systems as cold, rapid mountain streams, brackish, stagnant waters of estuaries and salt marshes, and the intertidal zone of rocky seashores. Beetles are important in some aquatic food chains, and several species are consumed by fish and waterfowl. Few species, however, reach the high population densities or biomass levels exhibited by Ephemeroptera, Trichoptera, or Diptera, and beetles are not the dominant organisms in most lotic aquatic habitats.

The general biology and ecology of Coleoptera are summarized by Crowson (1981). The best general identification manuals for adult North American Coleoptera are Blatchley (1910) and Arnett (1960), and Bosquet (1991) gives a very useful checklist to the species of Canada and Alaska. Leech and Chandler (1956), with emphasis on California, and Leech and Sanderson (1959) provide excellent discussions of the families of aquatic Coleoptera, with keys to adults and larvae of most North American genera. Brigham *et al.* (1982) give keys to genera and species of the aquatic eastern fauna along with extensive summaries of ecology, and Hilsenhoff (1981) provides keys to Wisconsin taxa that is useful for much of the Midwest. Intertidal species are treated by Doyen (1975, 1976). The worldwide work by Bertrand (1972) is the most comprehensive available on aquatic immatures. General treatments of aquatic Coleoptera, as well as references to more detailed taxonomic works, are given in Table 20A.

It is difficult to make broad generalizations about the life histories and ecology of aquatic Coleoptera, even at the family level, because the order appears to have invaded the aquatic habitat many times and in many ways. Several families of the more "primitive" suborder Adephaga (e.g., Dytiscidae, Haliplidae, Gyrinidae) are wholly aquatic, leading some to speculate that the hardened elytra and heavy sclerotization may have evolved as a mechanism to keep water out. The suborder Myxophaga (e.g., Hydroscaphidae) is primarily aquatic living in madicolus habitats (rapidly flowing thin sheets of water). Only a few of the numerous families in the suborder Polyphaga are wholly aquatic, and adaptations to aquatic existence are quite varied. For example, within the superfamily Dryopoidea, Heteroceridae exist only at the water's edge; Lutrochidae and Psephenidae adults are terrestrial while the larvae are aquatic; Dryopidae adults are aquatic while the larvae are primarily terrestrial; and both the adults and larvae of most Elmidae are aquatic. Therefore the introductory materials given below are quite broad and general, and more specific information for individual families has been included within the keys.

Most aquatic beetles are substrate dwellers. Some notable exceptions, including adult Dytiscidae and Hydrophilidae, are efficient swimmers but must return to the surface periodically to renew their air supply (see Chap. 4). Numerous species, including virtually all marine forms, inhabit cracks, crevices, or self-constructed burrows and seldom, if ever, venture into open water. Adults of most species leave the water temporarily on dispersal flights, which may occur once (Elmidae) or repeatedly (many Dytiscidae and Hydrophilidae). There are a few taxa known only from subterranean habitats (e.g., *Stygoparnus*, *Haideoporus*, *Stygoporus*), and undoubtedly more genera and species will be described as aquifers and hyporheic zones (White 1993) are better studied.

Most Adephaga are predators either engulfing their prey or injecting digestive enzymes through piercing mouth parts (especially larvae of Dytiscidae). Some large dytiscids attack small fish or tadpoles. The Myxophaga are scrapers, and hydroscaphids (*Hydroscapha*) are unusual in that their diet consists largely of bluegreen algae. Feeding habits of aquatic Polyphaga are extremely variable (Table 20A) and include all the categories described in Chapter 6.

Respiration in aquatic beetles conforms to four major modes (Chap. 4): (1) reliance on self-contained air reserves (e.g., Dytiscidae, Hydrophilidae, Hydraenidae); (2) transcuticular respiration, with or without tracheal gills (larvae of most families); (3) plastron respiration (adult Dryopidae, Elmidae, etc.); and (4) piercing plant tissues (e.g., larval Chrysomelidae). In most species having an air reservoir, it occupies the space beneath the elytra, and these beetles must regularly return to the surface to renew depleted air supplies. Beetles using plastron respiration mostly occupy fast-moving, well-oxygenated water and are quite sensitive to pollutants that act as wetting agents. Representatives of types (2), (3), and (4) may remain submerged indefinitely.

Adults of most aquatic beetles probably survive a single season or part of a season, but some dytiscids, hydrophilids, and elmids have been maintained for years in aquariums. Species with terrestrial adults (Psephenidae, Scirtidae, Ptilodactylidae) frequently reproduce and die after a short time, sometimes without feeding.

Although *Gyrinus* species form schools of thousands of individuals, aquatic beetles do not form mating aggregations. However, species with short-lived adults often emerge synchronously and may be very abundant locally. Copulation most often occurs with the male mounted dorsally on the female and may be preceded by stroking, stridulation, or other courtship behaviors. The eggs, numbering from one (Hydroscaphidae) to hundreds (Psephenidae), are deposited singly or in masses in diverse situations. Most species with truly aquatic larvae oviposit underwater, but scirtids apparently oviposit in the damp marginal zone, which is also used by georyssids, heterocerids, and other families with subaquatic larvae. Some dytiscids insert the eggs into plant tissue, and a few hydrophilid females carry the egg mass beneath the abdomen. Eggs are generally simple ovoids without a thickened or sculptured chorion. Most are laid naked, but those of hydrophilids and some hydraenids are enclosed in silken cases. Eggs usually hatch in about one to two weeks, but eclosion may be delayed for many months in exceptional cases.

The life histories of aquatic Coleoptera are quite variable. The larvae of some families (e.g., Psephenidae) can be collected the year round, but the adults are present for only a short period of time during summer. Both adults and larvae of Elmidae can be collected together most times of the year. In other families (e.g., Hydrophilidae) it is the adult that is most often collected, while the larvae exist for only a few weeks in the summer. Many aquatic Coleoptera pass through from three to eight larval instars, requiring an average of six to eight months to develop, with a single generation per year in temperate regions. In almost all species pupation is terrestrial, usually in cells excavated by the larvae under stones, logs, or other objects or occasionally in mud cells on aquatic vegetation (some Gyrinidae). Noteridae, Curculionidae, Chrysomelidae, and possibly some Hydrophilidae spin silk cocoons enclosing a bubble of air, within which they pupate while submerged. Psephenidae pupae are pharate, forming beneath the cuticle of the last larval instar, usually under stones at the water's edge. Just prior to pupation, the last larval instar becomes a quiescent prepupa, when diapause may occur. Few Coleoptera diapause as pupae and generally transform to adults in about two to three weeks.

LITTORAL COLEOPTERA

Of all the orders, except perhaps Diptera and Hemiptera, Coleoptera is unique in including many semiaquatic species that inhabit littoral zone environments. Fresh, brackish, and saline waters all support littoral zone faunas, similar in family representation but differing in species composition. Aquatic marine Coleoptera are almost entirely members of families that commonly frequent the littoral zone and are discussed separately below. In fresh waters, however, families well represented in the littoral zone are almost entirely absent from truly aquatic habitats. For example, the large families Staphylinidae and Carabidae contain hundreds of species that occur regularly, if not exclusively, around water. Yet only a few Carabidae could be called aquatic, even in the loosest sense. Species of *Nebria* may submerge temporarily to escape predators but require terrestrial situations to survive and reproduce. In general, littoral zone existence does not appear to have been an evolutionary pathway to aquatic habits.

The littoral zone comprises several distinct habitats. The interstices in gravel or coarse sand substrates alongside bodies of water support a fauna of minute beetles such as Microsporidae, Hydraenidae, and Hydrophiloidea. Many of these species are actually aquatic, living in the water held in interstitial spaces in the substrate. A much larger number of species frequents the surfaces of damp sand or mud adjacent to standing or running water. Adult and larval Carabidae and Staphylinidae are the most obvious and diverse element in this situation, often very dense on mud or sandbars, especially around drying ponds or intermittent streams. Limnichidae and Georyssidae also frequent this zone and may be locally abundant. Adult and larval Heteroceridae inhabit burrows they excavate beside streams. Recently emergent sandbars and banks are favorite locations, but some species dig in mud, and a few live on intertidal mud flats. Larval *Helichus* and *Postelichus* (Dryopidae) burrow through moist sand adjacent to streams and are most likely submerged during spring floods. Early instars of *Acneus* (Psephenidae) are apparently aquatic, like the other members of the family. Later instars cling to moist stones just above the waterline and drown if submerged (G. Ulrich, pers. comm.). Undersides of stones and logs, including those partially submerged, provide shelter for many inhabitants of the littoral zone, which may emerge nocturnally to forage. Accumulations of water-borne organic debris also shelter many littoral zone species. Sampling with Berlaise funnels is effective, especially in obtaining minute forms such as Ptiliidae.

In their peripheral areas, littoral zone habitats merge with purely terrestrial ones. Along steep-banked streams the transition may be abrupt, but along swampy streams and ponds it is almost imperceptible. In addition to the littoral zone inhabitants, many terrestrial taxa are often regular members of such intermediate communities. In particular, members of the terrestrial soil fauna (Pselaphidae, Ptiliidae, Leiodidae, Scydmaenidae, etc.) are likely to occur along with Lampyridae, Cantharidae, and others.

MARINE COLEOPTERA

Although numerous species representing most families of Coleoptera inhabit maritime environments, very few are marine in the sense of regularly occupying aquatic situations. Moreover, the marine representatives are almost exclusively from families found in littoral zone habitats rather than those that are truly aquatic. Thus, Staphylinidae and Carabidae are the dominant marine Coleoptera in numbers of individuals and species. Other primarily terrestrial families with marine representatives include Melyridae, Limnichidae, Salpingidae, and Curculionidae. In contrast, marine representatives of freshwater families include only a few Hydrophiloidea and Hydraenidae. Nearly all marine forms in North America occur on the Pacific coast with a small number of Staphylinidae recorded from the Atlantic.

With the exception of a few Hydrophiloidea genera *(Berosus, Enochrus,* etc.) that swim in brackish estuarine waters, marine Coleoptera are substrate dwellers. Some forms inhabit air-filled crevices in rocks. This group includes most Carabidae, many Staphylinidae, and *Aegialites* (Salpingidae). Many Staphylinidae (and a few exotic Carabidae) inhabit self-constructed burrows in sandy beaches. In most cases these burrows probably entrap air. Thus, most marine Coleoptera are not in direct contact with the water and have no special respiratory mechanisms. Exceptions include several genera of Hydraenidae, which breathe with a "gill" of air held by the hydrofuge ventral cuticle.

Life histories of North American species are largely unknown, but studies of European relatives suggest that adults and larvae have similar habits and requirements. Various instars typically occur together in colonies. Dispersal is apparently by ocean currents, because most submarine forms lack wings and have fused elytra. In contrast, beetles inhabiting the supratidal zones of beaches are commonly active, powerful fliers.

DEFENSE MECHANISMS

Defense mechanisms against predators include behavior patterns, cryptic coloration, speed, concealment, holdfasts, hardened bodies, spines and claws, and distasteful chemicals. Most Coleoptera larvae rely on some form of concealment, residing in cracks and crevices, in debris, or remaining buried in bottom or shoreline sediments, particularly during daylight hours. Larval Psephenidae adhere tightly to rocks and other smooth surfaces and are difficult to dislodge.

Coleoptera and Hemiptera are the only two major orders where most of the adults are both long-lived and also aquatic, and both orders appear to rely heavily on chemical defense mechanisms (reviews in Dettner 1987, Scrimshaw and Kerfoot 1987, White 1989). In the suborder Adephaga, several families have specific glands that produce distasteful and irritating compounds. Dytiscidae and Hygrobiidae possess both pygidial glands that secrete aromatic compounds and prothoracic glands that secrete steroids, giving them a distinctive odor and making them unpalatable. Defensive chemicals in Gyrinidae, Carabidae, Haliplidae, Noteridae and possibly Amphizoidae are produced primarily in the pygidial glands. Among these aquatic families, the chemistry of gyrinid secretions is best known and consists primarily of terpenes that are aromatic and extremely distasteful to most predators.

Chemical defenses of the aquatic families in the suborder Polyphaga are less well understood, although some Chrysomelidae are known to possess defensive glands and secretions. No glands or chemicals have been identified, but adult Elmidae appear to be distasteful and are rejected by fish and other predators. One South American elmid species apparently is used by people as a spice. Short lived aerial adults, such as limnichids and psephenids, seem to have no chemical defenses but rely on rapid movements and will enter the water when disturbed. Although dytiscids often are distasteful to predators, hydrophilids apparently are not, which seems to have led to mimicry, which is particularly common in lentic species. Some adult Hydrophilidae are rejected by fish, but this may be a result of the hard exoskeleton rather than chemicals.

In addition to chemical defenses, adult Dytiscidae, Haliplidae, Gyrinidae and Hydrophilidae are fast swimmers while terrestrial Staphylinidae, Carabidae, and Chrysomelidae are swift fliers and runners. Adult Amphizoidae, some Carabidae, Noteridae, Dryopidae, Elmidae, Microsporidae, Melyridae, Salpingidae, Hydraenidae, and Curculionidae live concealed within crevices, sediments, or vegetation. Further, most adult Coleoptera are hard bodied with sharp claws and many have legs with stiff spines.

EXTERNAL MORPHOLOGY

Adults

Most adult beetles, including nearly all aquatic members, are characterized by a heavily sclerotized, usually compact body. The combination of elytra (see below) and antennae with 11 or fewer segments will separate nearly all beetles from other insects with only cursory examination.

Head and Mouthparts: Coleoptera are mandibulate insects, and the mouthparts are usually visible without dissection. Mandibular structure is broadly indicative of feeding habits. In herbivore scrapers (Dryopidae, Elmidae), the mandible bears a basal flattened molar or grinding lobe as well as a sharp, anterior incisor lobe. In predaceous forms the molar lobe is usually absent.

Posterad of the labium the head capsule consists of a gula, delimited by paired gular sutures, except in Curculionidae, where the sutures are coalesced (Fig. 20.309). Configuration of the antennae is used extensively in identification at all levels; important types are illustrated in the keys.

Thorax and Legs: In most beetles the pronotum has encroached ventrally to the region of the procoxae where notum and sternum are separated by a suture (Fig. 20.150). In Dytiscidae and related families the lateral prothoracic region is occupied by the pleuron so that both notopleural and sternopleural sutures are present (Fig. 20.99).

Dorsally, much of the thorax and abdomen are concealed by the elytra in most beetles. Morphologically the elytra represent heavily sclerotized fore wings without crossveins. Flying wings (hind wings), present in most aquatic beetles, are folded beneath the elytra when the adults are at rest.

The number of tarsal segments on each leg is critical in identification to family. By convention, these numbers are designated by a three-digit tarsal formula (e.g., 5-5-5), indicating the number of tarsal segments (tarsomeres) on the anterior, middle, and posterior legs, respectively. The claws are not counted as separate segments. The tibiae and/or tarsi may bear fringes of long, slender swimming hairs (Figs. 20.101, 20.150), which adhere to the leg in dried specimens and may not be visible without wetting.

Abdomen: Normally, only the abdominal sternites are visible. In Dytiscidae and related families the posterior coxae are greatly enlarged, dividing the basal sternite into two separate sclerites (Fig. 20.99). In counting the number of abdominal sternites, those segments associated with the genitalia and normally drawn into the abdomen should not be tallied. They almost always differ markedly in sculpturing, color, and degree of sclerotization from the external sternites.

Larvae

Larval Coleoptera are exceedingly diverse. Among aquatic forms, the presence of a distinct, sclerotized head capsule with mandibles, maxillae, labium, and two- or three-segmented antennae (except in Scirtidae, which have long, multisegmented antennae) is a reliable distinguishing feature. All but curculionids and a few hydrophiloids have three pairs of thoracic legs bearing one or two apical claws. Beetle pupae are exarate (appendages not fused to body) and bear a general similarity to adults.

Cranial Structures: Mouthparts are extremely important in larval classification and identification, and careful dissection may be necessary to view critical structures. As in adults, the mandibles may possess or lack a molar lobe. In some larvae a fleshy cushion-like or digitate prostheca is inserted on the mandible just anterior to the molar lobe (Fig. 20.168). The maxilla may bear a separate galea and lacinea or these may be united as a single lobe, the mala. The maxillary palp is inserted on the palpiger, a projection of the stipes, which is greatly enlarged in Hydrophiloidea, appearing as a separate segment of the palp (Fig. 20.122).

Abdominal Structures: A variety of gill-like appendages may be present laterally or ventrally on any segment (Figs. 20.46, 20.131, 20.213). "Gills" may be simple or branched, dispersed or clustered, unsegmented or articulated. In some families the gills are concealed in a pocket beneath the terminal abdominal sternite (Fig. 20.262), which forms a lid or operculum, and dissection is frequently necessary for examination.

Diverse unsegmented, immovable dorsal or dorsolateral appendages occur on the apical or preapical abdominal segment (Figs. 20.19, 20.21). These terminal appendages are usually relatively short and strongly sclerotized. They are termed *urogomphi* (singular, urogomphus) regardless of position or origin. Movable appendages, sometimes segmented and often elongate and filamentous (Figs. 20.71, 20.167), may be present also and presumably are homologous to the cerci of many other insects.

KEYS TO THE FAMILIES OF AQUATIC COLEOPTERA

Aquatic Coleoptera include members of the three suborders Adephaga, Myxophaga, and Polyphaga and many subfamilies. The following keys are designed primarily for aquatic beetles, including marine forms. Families that regularly inhabit the littoral zone, although not strictly aquatic, are included because they frequently appear in aquatic samples. Many members of families that are not aquatic may occasionally appear around water, and those that have merely fallen into the water or have been caught in floods cannot be identified with the keys. If one is unsure if a specimen is aquatic or not, it is wise to consult some of the general references listed above. With minor exceptions there are no species level keys to aquatic Coleoptera larvae, and rearing often is necessary for species level identifications.

Taxonomic usage generally follows that in Arnett (1960) and Bosquet (1991). Hydrochidae, Sperchiidae, Epimetopidae, and Helophoridae, however, are no longer treated as subfamilies of Hydrophilidae (Hansen 1991). We have used the superfamily Hydrophiloidea as the end point in the keys to families and then annotate each of the five families separately in the generic keys and ecological table. We are somewhat conservative by combining some closely related taxa into families. For example, the Psephenoididae, Eubriidae, and Psephenidae of Bertrand (1972) are considered here as subfamilies of Psephenidae. The relationships of the genera included here in Ptilodactylidae are poorly understood. The genus *Stenocolus* is morphologically divergent from other Ptilodactylidae and probably will be placed in a separate family, Eulichadidae. The genus *Araeopidius* is also divergent and probably will be removed from Ptilodactylidae. While we have placed the genus *Lutrochus* in Lutrochidae, we continue to treat Limnichidae and Lutrochidae together in the keys until more is known about the larvae. Helodidae (commonly accepted in North America) and Cyphonidae long have been used synonymously with Scirtidae. Because Scirtidae is the oldest name, it is nomenclaturally correct (Pope 1975) and is used here. Sphaeriidae is replaced with Microsporidae as the name Sphaeriidae has been reserved for the family of bivalves commonly known as fingernail clams.

1. Mesothoracic wings (elytra) present, usually covering entire abdomen (Fig. 20.30), sometimes only its base (Figs. 20.32, 20.33); antennae with at least 4 segments, usually 6 or more (Figs. 20.28, 20.35); tarsi with at least 3 segments (Figs. 20.30, 20.33)[a] ....*ADULTS*
1'. Mesothoracic wings absent (Figs. 20.1-20.2); antennae with 3 or fewer segments[b]; tarsi with a single segment (Figs. 20.4, 20.12) ....*LARVAE*

Larvae

1. Legs absent (Fig. 20.15) 10
1'. Legs sometimes small, but always with 3-6 clearly defined segments (Figs. 20.12, 20.53, 20.55, 20.74, 20.134) 2
2(1'). Legs (excluding claws) with 5 segments; tarsi with 2 claws (Figs. 20.4, 20.74; exception: HALIPLIDAE with single claw) 3
2'. Legs with 3-4 segments (Figs. 20.12, 20.134); tarsi with single claw (Fig. 20.12, 20.134) 10
3(2). Abdomen with 2 pairs of stout, terminal hooks on segment 10 (Fig. 20.47); abdominal segments 1-9 bearing lateral gills; Fig. 20.46) ....***GYRINIDAE*** (p. 411)
3'. Abdomen without hooks on terminal segment; abdominal segments usually without lateral gills, occasionally with ventral gills (Fig. 20.213) 4

[a] Lepiceridae, not occurring in United States, have a single tarsomere.
[b] Larvae of Scirtidae have numerous, filamentous antennal segments.

4(3').	Abdomen with 8 segments	6
4'.	Abdomen with 9 or 10 segments	5
5(4').	Tarsi with single claw (Figs. 20.53, 20.55); mandibles grooved internally (Fig. 20.57); at least last larval instar with erect, dorsal projections from thoracic and abdominal tergites (Figs. 20.52, 20.54)	**HALIPLIDAE** (p. 413)
5'.	Tarsi with 2 claws (as in Fig. 20.4); mandibles not grooved; tergites without projections (Fig. 20.1)	**CARABIDAE**[1,2,3,6]
6(4).	Abdominal segment 8 with a pair of large terminal spiracles	7
6'.	Abdominal segment 8 without spiracles	**HYGROBIIDAE**[4,6]
7(6).	Cerci slender, longer (usually much longer) than abdominal segment 1 (Fig. 20.71)	**DYTISCIDAE** (in part) (p. 415)
7'.	Cerci stout, shorter than abdominal segment 1 (Fig. 20.2) or rudimentary or absent	8
8(7').	Thorax and abdomen strongly flattened; tergites expanded laterally as thin, flat projections (Fig. 20.2); gular sutures single (Fig. 20.3)	**AMPHIZOIDAE**............*Amphizoa* (p. 411)
8'.	Thorax and abdomen round or subcylindrical in cross section; tergites not expanded as flat, platelike projections; gular sutures double	9
9(8').	Legs short, stout, adapted for digging (Fig. 20.104); mandibles with enlarged molar portion (Fig. 20.105)	**NOTERIDAE** (p. 425)
9'.	Legs long, slender, adapted for swimming (Fig. 20.74); mandibles falcate (sickle-shaped), without enlarged molar portion (Figs. 20.72-20.73)	**DYTISCIDAE** (in part) (p. 415)
10(1, 2').	Labrum separated from clypeus by distinct suture	14
10'.	Labrum not represented as separate sclerite (Fig. 20.130) (the labium may be visible dorsally)	11
11(10').	Body round or subcylindrical in cross section; head projecting anteriorly from prothorax and visible from above (Fig. 20.131); movable cerci often visible (Fig. 20.167)	12
11'.	Body dorsoventrally flattened with large, transverse thoracic and abdominal tergites; pronotum expanded anteriorly, usually concealing head from above (Figs. 20.5, 20.6)	**LAMPYRIDAE**[1,6]
12(11).	Maxilla with palpifer appearing as a segment of palp (Figs. 20.122, 20.133); spiracles biforous (having 2 openings; Fig. 20.111)	13
12'.	Maxilla with palpifer appearing as part of stipes (Fig. 20.7); spiracles annular (ring-shaped)	**STAPHYLINIDAE**[1,2,6] (in part) (p. 436)
13(12).	Abdomen with 10 segments (Fig. 20.10); cerci very short, palpilliform (Fig. 20.11); legs short, 3-segmented (Fig. 20.12)	**GEORYSSIDAE**[1] ....*Georyssus*
13'.	Abdomen with 8 segments, or rarely if with 10 segments, then cerci long, 2 to 3 segmented, and legs long, 5-segmented (Figs. 20.112, 20.120, 20.131)	**HYDROPHILOIDEA**[5] (p. 427)
14(10).	Thorax and abdomen short, obese, without distinct sclerites (Figs. 20.13-20.15); legs reduced (Fig. 20.13) or absent (Fig. 20.15)	15
14'.	Thorax and abdomen cylindrical, flattened, or fusiform (spindle-shaped), but not markedly obese; thoracic and abdominal tergites clearly defined; legs adapted for walking	16
15(14).	Legs very small but complete and visible (Fig. 20.13); spiracles on abdominal segment 8 forming large, sclerotized dorsal hooks (Fig. 20.14)	**CHRYSOMELIDAE**[6] (p. 455)
15'.	Legs entirely absent (Fig. 20.15); spiracles sometimes set on tubercles, but abdominal segment 8 never with sclerotized dorsal hooks	**CURCULIONIDAE**[2,6] (p. 458)
16(14').	At least abdominal tergite 8 bearing pairs of fleshy, articulated, fingerlike lobes (Fig. 20.16); antennae very short, 2-segmented (Fig. 20.17); minute larvae, less than 2 mm long	17
16'.	Abdominal tergites without dorsal lobes; antennae 3-segmented or more	19
17(16).	Fingerlike articulated lobes present on abdominal segments 1-8	18
17'.	Fingerlike lobes present only on abdominal segments 1 and 8 (Fig. 20.16)	**HYDROSCAPHIDAE**........*Hydroscapha* (p. 427)
18(17).	Fingerlike lobes on abdominal segments 1-8 approximately 1.5 times as long as wide; antenna with 2nd segment about 2-3 times as long as broad and bearing minute, lateral appendage (Fig. 20.17)	**MICROSPORIDAE**[1] *Sphaerius*
18'.	Fingerlike lobes on abdominal segments 1-8 more than 10 times as long as wide; antenna with 2nd segment about 4-5 times as long as broad, lacking appendage	**TORRIDINICOLIDAE**[4,6]

[1] Subaquatic or inhabiting littoral region.
[2] Includes intertidal species in North America.
[3] Larvae of *Brachinus* spp. (Carabidae), which are ectoparasitic on pupae of Hydrophilidae, have 3 leg segments and a single claw.
[4] Family does not occur in United States.
[5] Key to families and genera of Hydrophiloidea.
[6] Keys to genera not given.

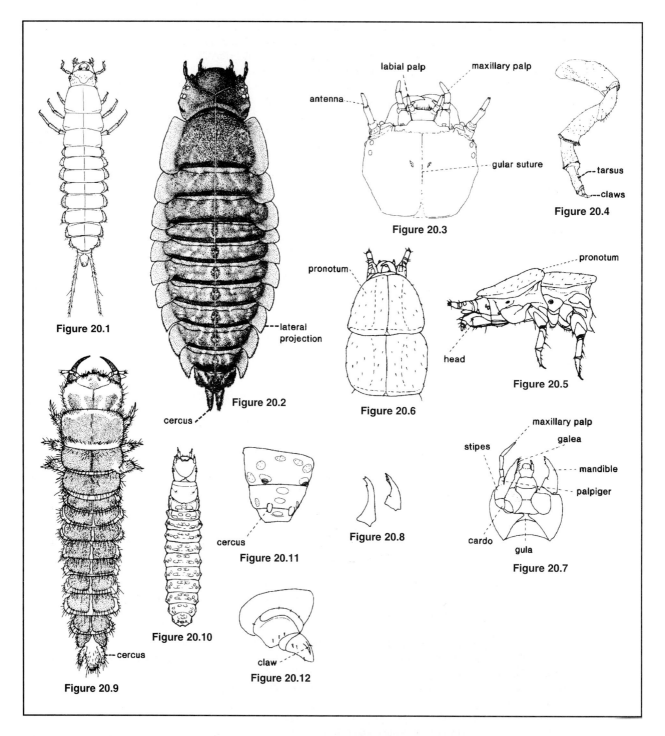

Figure 20.1. *Chlaenius* sp. (Carabidae) larva, dorsal aspect.

Figure 20.2. *Amphizoa* sp. (Amphizoidae) larva, dorsal aspect.

Figure 20.3. *Amphizoa* sp. (Amphizoidae) larva, ventral aspect of head.

Figure 20.4. *Amphizoa* sp. (Amphizoidae) larva, metathoracic leg.

Figure 20.5. Lampyridae larva, lateral aspect of head and thorax.

Figure 20.6. Lampyridae larva, dorsal aspect of head and thorax.

Figure 20.7. *Piestus* sp. (Staphylinidae) larva, ventral aspect of head.

Figure 20.8. *Oxytelus* sp. (left) and *Piestus* sp. (right) larvae (Staphylinidae), mandibles.

Figure 20.9. *Thinopinus* sp. (Staphylinidae) larva, dorsal aspect.

Figure 20.10. *Georyssus* sp. (Georyssidae) larva, dorsal aspect (after Van Emden 1956).

Figure 20.11. *Georyssus* sp. (Georyssidae) larva, abdominal apex.

Figure 20.12. *Georyssus* sp. (Georyssidae) larva, leg (after Van Emden 1956).

19(16').	Abdomen with 10 segments; segment 9 bearing articulated, 1- or 2-jointed cerci (Fig. 20.167)	20
19'.	Abdomen with 9 segments; segment 8 or 9 sometimes bearing immovable urogomphi (Figs. 20.19, 20.21), but articulated cerci never present	22
20(19).	Mandibles with large, asperate (roughened) molar lobe (Fig. 20.168)	21
20'.	Mandibles falcate (sickle-shaped), without molar lobe (Fig. 20.8)	***STAPHYLINIDAE***[1,2,6] (in part) (p. 436)
21(20).	Abdominal segment 10 with pair of recurved ventral hooks (Fig. 20.169); cerci with 2 segments (Fig. 20.167)	***HYDRAENIDAE***[1,6] (p. 434)
21'.	Abdominal segment 10 without hooks; cerci with single segment (as in Fig. 20.9)	***PTILIIDAE***[1,2,6]
22(19').	Antennae much longer than head (Fig. 20.183), multiarticulate (many jointed)	***SCIRTIDAE*** (p. 438)
22'.	Antennae short, with 2 to 3 segments	23
23(22').	Body cylindrical, subcylindrical, or fusiform; head and legs visible in dorsal aspect	24
23'.	Body extremely flattened, with thoracic and abdominal tergites expanded laterally as thin laminae concealing head and legs from above (Figs. 20.195-20.196)	***PSEPHENIDAE*** (p. 440)
24(23).	Abdominal segment 9 with a lidlike operculum covering the anal region ventrally (Figs. 20.221, 20.262); abdominal sternites 1-8 never bearing gills	25
24'.	Abdominal segment 9 without operculum; abdominal sternites 1-8 sometimes bearing fasciculate (clustered) gills (Fig. 20.213)	27
25(24).	Terminal abdominal segment rounded posteriorly (Fig. 20.221); head capsule with groups of 6 ocelli, 5 lateral and 1 ventral, or eyes absent	26
25'.	Terminal abdominal segment bifid or slightly emarginate (notched) posteriorly and with lateral ridges (Fig. 20.231); head capsule with groups of 5 lateral ocelli	***ELMIDAE*** (p. 446)
26(25).	Opercular chamber containing 2 retractile hooks and 3 tufts of retractile gills (Fig. 20.262); mandibles with prostheca (Fig. 20.234)	***LIMNICHIDAE and LUTROCHIDAE*** (p. 444)
26'.	Opercular chamber without hooks or gills; mandibles without prostheca (Fig. 20.222)	***DRYOPIDAE***[1] (p. 444)
27(24').	Abdomen with distinct tufts of gills, either restricted to anal region (as in Fig. 20.262) or present on segments 1-7 (Fig. 20.213)	***PTILODACTYLIDAE*** (p. 441)
27'.	Abdomen without gills	28
28(27').	Abdomen bearing prominent, spinelike urogomphi on terminal segment (Figs. 20.19, 20.21)	29
28'.	Urogomphi absent; abdominal segment 9 rounded	31
29(28).	Urogomphi bifid (Fig. 20.19); spiracles raised on tubercles (Fig. 20.18)	***SALPINGIDAE***[2,6]
29'.	Urogomphi with single points (Fig. 20.21); spiracles not elevated	30
30(29').	Epicranial sutures lyre-shaped (Fig. 20.20)	***ANTHICIDAE***[1,6]
30'.	Epicranial sutures Y-shaped (Fig. 20.22)	***MELYRIDAE***[1,2,6]
31(28').	Mouthparts prognathous; median epicranial suture absent (Fig. 20.23)	***HETEROCERIDAE***[1,6]
31'.	Mouthparts hypognathous; median epicranial suture present (Fig. 20.24)	***TENEBRIONIDAE***[1,2,6]

Adults

1.	Compound eyes divided into separate dorsal and ventral segments (Figs. 20.48-20.50); antennae short, clavate (clubbed) (Fig. 20.51)	***GYRINIDAE*** (p. 411)
1'.	Compound eyes undivided (Fig. 20.98); antennae variable but not short and clavate as in Fig. 20.193	2
2(1').	Head usually produced anteriorly as a rostrum (Figs. 20.309-20.310); anterior tibiae without large teeth; tarsal formula 4-4-4	***CURCULIONIDAE*** (p. 458)
2'.	Head not produced as a rostrum	3
3(2').	Elytra covering entire abdomen or exposing only part of 1 abdominal tergite	6
3'.	Elytra truncate, exposing at least 2 entire abdominal tergites (Figs. 20.26, 20.32, 20.172-20.175)	4
4(3').	Antennae with 11 (rarely 10) segments, terminal segment no longer than combined length of 2 preceding segments (Figs. 20.193, 20.211, 20.216, 20.298)	5
4'.	Antennae with 8 segments, terminal segment as long as combined length of 4 preceding segments (Fig. 20.28); minute beetles, less than 1.5 mm long; hind coxal plates widely separated (Fig. 20.25; adult as in Fig. 20.26)	***HYDROSCAPHIDAE***......***Hydroscapha*** (p. 427)
5(4).	Antennae inserted at base of mandibles distant from eyes (Fig. 20.31); abdomen and thorax with membranous yellow or orange protrusible (extendible) vesicles (Fig. 20.32; most apparent in living beetles); adult as in Fig. 20.33	***MELYRIDAE***[1,2,6]

---

[1] Subaquatic or inhabiting littoral region.
[2] Includes intertidal species in North America.
[6] Keys to genera not given.

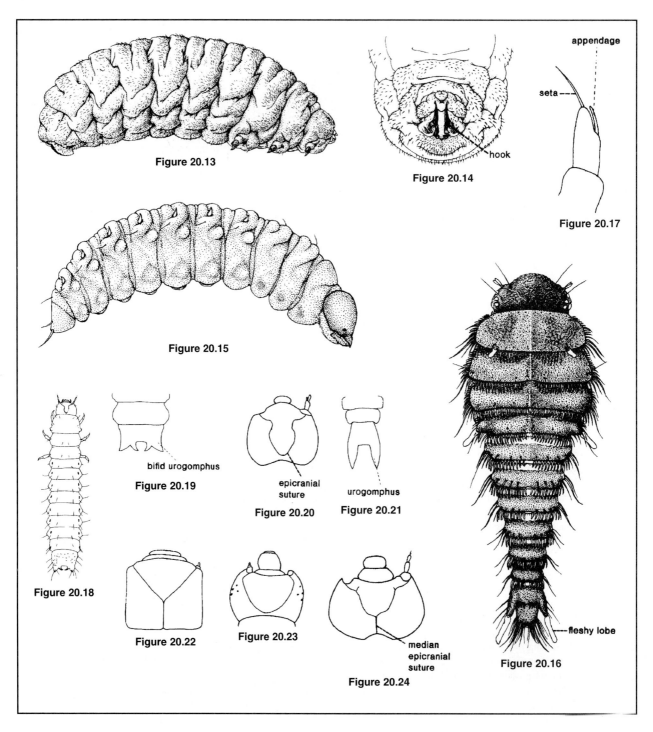

Figure 20.13. *Donacia* sp. (Chrysomelidae) larva, lateral aspect.
Figure 20.14. *Donacia* sp. (Chrysomelidae) larva, ventral aspect of terminal abdominal segment.
Figure 20.15. Curculionidae larva, lateral aspect.
Figure 20.16. *Hydroscapha* sp. (Hydroscaphidae) larva, dorsal aspect.
Figure 20.17. *Sphaerius* sp. (Microsporidae) larva, antenna.
Figure 20.18. *Aegialites* sp. (Salpingidae) larva, dorsal aspect.
Figure 20.19. *Aegialites* sp. (Salpingidae) larva, abdominal apex.
Figure 20.20. *Anthicus* sp. (Anthicidae) larva, dorsal aspect of cranium.
Figure 20.21. *Endodes* sp. (Melyridae) larva, abdominal apex.
Figure 20.22. *Collops* sp. (Melyridae) larva, dorsal aspect of cranium.
Figure 20.23. Heteroceridae larva, dorsal aspect of cranium.
Figure 20.24. Tenebrionidae larva, dorsal aspect of cranium.

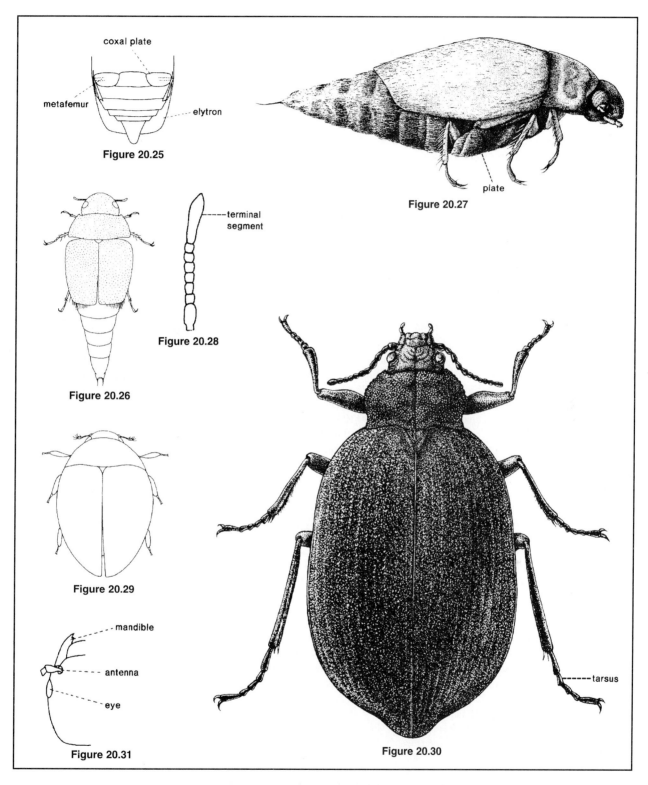

Figure 20.25. *Hydroscapha* sp. (Hydroscaphidae) adult, ventral aspect of abdomen and metathorax.

Figure 20.26. *Hydroscapha* sp. (Hydroscaphidae) adult, dorsal aspect.

Figure 20.27. *Hydroscapha* sp. (Hydroscaphidae) adult, lateral aspect.

Figure 20.28. *Hydroscapha* sp. (Hydroscaphidae) adult, antenna.

Figure 20.29. *Sphaerius* sp. (Microsporidae) adult, dorsal aspect.

Figure 20.30. *Amphizoa* sp. (Amphizoidae) adult, dorsal aspect.

Figure 20.31. *Endeodes* sp. (Melyridae) adult, dorsal aspect of head (after Chamberlin and Ferris 1929).

5'.	Antennae inserted laterally on frons close to eyes or between eyes; protrusible vesicles absent	**STAPHYLINIDAE**[1,2] (p. 436)
6(3').	Hind coxae expanded as broad, flattened plates covering all or part of abdominal sternites 1-3 (Figs. 20.27, 20.63)	7
6'.	Hind coxae sometimes extending posteriorly along midline (Figs. 20.34, 20.99), but never as broad plates	8
7(6).	Hind coxal plates completely covering 2 or 3 basal abdominal segments (Fig. 20.63) and concealing all but apices of hind femora; beetles more than 3 mm long; adult as in Figure 20.64	**HALIPLIDAE** (p. 413)
7'.	Hind coxal plates exposing abdominal segments laterally; bases of hind femora exposed; highly convex beetles less than 1.5 mm long; adult as in Figure 20.29	**MICROSPORIDAE**..........*Sphaerius*
8(6').	Hind coxae with medial portion extending posteriorly to divide abdominal sternite 1 into lateral sclerites (Figs. 20.34, 20.99); prothorax with distinct notopleural sutures (Fig. 20.99)	9
8'.	Hind coxae not extending posteriorly to divide abdominal sternite 1 (Figs. 20.38, 20.150); notopleural sutures almost always absent	14
9(8).	Hind tarsi and usually tibiae flattened, streamlined, and bearing long, stiff swimming bristles (Fig. 20.101)	11
9'.	Hind tibiae and tarsi cylindrical or subcylindrical in cross section, without long, stiff swimming bristles	10
10(9').	Hind coxae extending laterally to margins of elytra; elytra at most with very short hairs; adult as in Figure 20.30	**AMPHIZOIDAE**..............*Amphizoa* (p. 411)
10'.	Hind coxae not extending laterally as far as elytra (Fig. 20.34); elytra bearing several long, slender, erect sensory hairs; representative genera shown in Figures 20.34a-d	**CARABIDAE**[1,2,6]
11(9).	Metasternum with transverse suture (as in Fig. 20.34); eyes strongly protruberant	**HYGROBIIDAE**[4,6]
11'.	Metasternum without transverse suture (Fig. 20.99); eyes streamlined (Figs. 20.97, 20.100)	12
12(11').	Fore and middle tarsi with 5 segments, segment 4 similar in size to segment 3 (Fig. 20.75); scutellum concealed (Figs. 20.97, 20.100) or exposed (Fig. 20.78)	13
12'.	Fore and middle tarsi with 4 segments (Fig. 20.100) or with segment 4 very small, concealed between lobes of segment 3 (Fig. 20.76); scutellum concealed (Figs. 20.97, 20.100) (exception: scutellum exposed in *Celina*)	**DYTISCIDAE** (in part) (p. 419)
13(12).	Hind tarsi with 2 similar claws (Figs. 20.106-20.107); scutellum concealed (as in Figs. 20.97, 20.100)	**NOTERIDAE** (p. 425)
13'.	Hind tarsi with a single claw (Fig. 20.101); if 2 claws are present, scutellum is large, exposed	**DYTISCIDAE** (in part) (p. 419)
14(8').	Antennae with terminal segment as long as combined length of 3-4 preceding segments (Fig. 20.28)	16
14'.	Antennae with terminal segment no longer than combined length of 2 preceding segments (Fig. 20.298); terminal segments may be fused into a globular or elongate club	15
15(14').	Antennae terminating in abrupt, globular (Figs. 20.35, 20.311) or elongate (Fig. 20.170) club	17
15'.	Antennae slender, elongate (Fig. 20.216) or very short, thick, with basal segment enlarged (Fig. 20.227)	21
16(14).	Antennae with 9 segments; minute flattened beetles	**TORRIDINICOLIDAE**[4,6]
16'.	Antennae with 4 segments; minute, coarsely sculptured, globular beetles	**LEPICERIDAE**[4,6]
17(15).	Antennae elbowed, with 2nd segment attached medially on elongate 1st segment; antennal club consisting of several compactly fused segments (as in Figs. 20.320, 20.324)	**HISTERIDAE**[1,6]
17'.	Antennae with 2nd segment attached apically on 1st (Fig. 20.227); antennal club consisting of 2-5 articulated segments (Fig. 20.170)	18
18(17').	Antennae about as long as head (Figs. 20.147, 20.150); club with 3-5 segments (Fig. 20.170); tarsal formula 4-4-4, 5-5-5, or 5-4-4	19
18'.	Antennae much longer than head, length approaching half that of body (Fig. 20.36); club with 2-3 loosely articulated segments (Fig. 20.37); tarsal formula 3-3-3; adult as in Figure 20.36	**PTILIIDAE**[1,2,6] (in part)
19(18).	Abdomen with 5 visible sternites; antennal club with 3 segments (not including segment 6)	20
19'.	Abdomen with 6-7 visible sternites; antennal club with 5 segments (Fig. 20.170)	**HYDRAENIDAE** (p. 434)

[1] Subaquatic or inhabiting littoral region.
[2] Includes intertidal species in North America.
[4] Family does not occur in United States.
[6] Keys to genera not given.

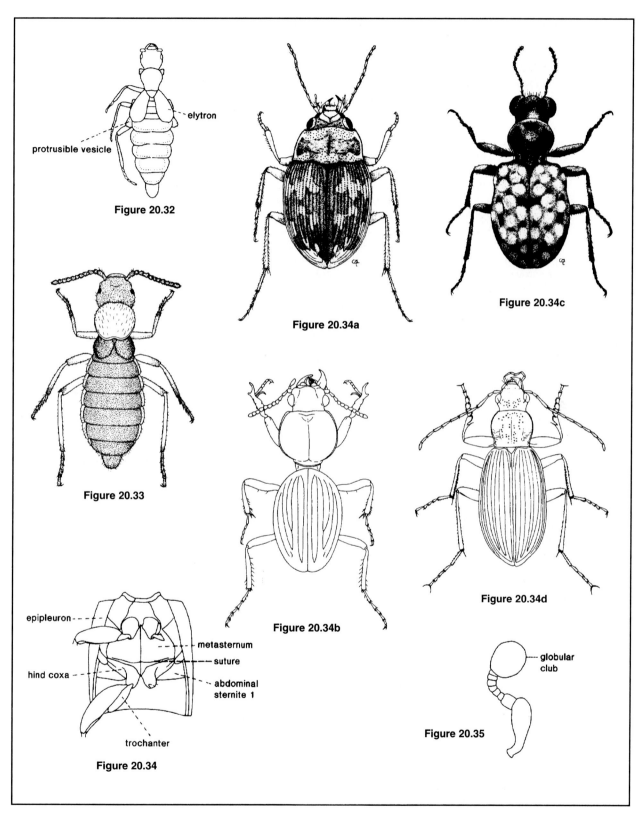

Figure 20.32. *Melyridae* adult, dorsal aspect.

Figure 20.33. *Endeodes* sp. (Melyridae) adult, dorsal aspect.

Figure 20.34. *Harpalus* sp. (Carbaidae) adult, ventral aspect of thoracic region (after Larson 1975). (Figs. 20.34a-d give the dorsal aspect of several representative Carabidae often found around water: 20.34a, Omophroninae, *Omophron* sp.; 20.34b, Scaratinae, *Dyschirius* sp.; 20.34c, Elaphrinae, *Elaphrus* sp.; 20.34d, Chaeninae, *Chlaenius* sp.).

Figure 20.35. *Neopachylopus* sp. (Histeridae) adult, antenna.

20(19).	Fore tarsi with 4 segments; metacoxae widely separated, intercoxal process broadly truncate (Fig. 20.38); adult as in Figs. 20.39, 20.40	***GEORYSSIDAE* Georyssus**
20'.	Fore tarsi with 5 segments; metacoxae nearly contiguous, intercoxal process narrow (Fig. 20.150)	***HYDROPHILOIDEA*[5]** (p. 432)
21(15').	Tarsal formula 5-5-5	22
21'.	Tarsal formula 5-5-4 or less	31
22(21).	Abdomen with 5-6 visible segments	23
22'.	Abdomen with 7-8 visible segments	***LAMPYRIDAE*[1,6]**
23(22).	Prosternum expanded anteriorly as prominent lobe beneath head, head usually contracted into thorax concealing antennae and eyes (Figs. 20.228-20.229, 20.297)	24
23'.	Prosternum not markedly expanded anteriorly beneath head, antennae clearly visible (Fig. 20.219)	27
24(23).	Antennae usually thick, with enlarged basal segment (Figs. 20.227, 20.229), about as long as head	25
24'.	Antennae filiform or serrate (Figs. 20.193, 20.216, 20.298), much longer than head	26
25(24).	Antennae with 10 or fewer segments; hind coxae contiguous	***LIMNICHIDAE* and *LUTROCHIDAE*[1]** (p. 444)
25'.	Antennae with 11 segments; hind coxae separated (Fig. 20.226)	***DRYOPIDAE*** (in part) (p. 444)
26(24').	Anterior coxae transverse with trochantin visible (Fig. 20.229); antennae (if visible) very short, thick, with enlarged basal segment (Fig. 20.227)	***DRYOPIDAE*** (in part) (p. 444)
26'.	Anterior coxae round, trochantin concealed (Figs. 20.269, 20.274), antennae slender, filiform (Fig. 20.298)	***ELMIDAE*** (in part) (p. 450)
27(23').	Tarsi with 4th segment deeply bilobed (Fig. 20.192)	***SCIRTIDAE*[1]** (p. 438)
27'.	Tarsi usually filiform, 4th segment not bilobed	28
28(27').	Antennae serrate or pectinate (Figs. 20.202, 20.211, 20.216), never concealed	29
28'.	Antennae filiform or clavate (Figs. 20.193, 20.298), or partly concealed within prosternum	***ELMIDAE*** (in part) (p. 450)
29(28).	Abdomen with 6 or 7 segments; maxillary palp with 2nd segment longer than next 2 combined	***PSEPHENIDAE*** (in part) (p. 440)
29'.	Abdomen with 5 segments; maxillary palp with 2nd segment much shorter than next 2 combined	30
30(29').	Head with antennae inserted close together between eyes constricting clypeus from frons (Fig. 20.210)	***PSEPHENIDAE*** (in part) (p. 440)
30'.	Head with antennae inserted below eyes, not constricting clypeus (Fig. 20.217)	***PTILODACTYLIDAE*[1]** (p. 441)
31(21').	Tarsal formula apparently 4-4-4 or 3-3-3; hind coxae contiguous or nearly so (as in Fig. 20.34)	34
31'.	Tarsal formula 5-5-4; hind coxae usually separated	32
32(31').	Hind coxae separated by less than coxal width; basal 2 abdominal sternites separated by suture	33
32'.	Hind coxae separated by much more than coxal width; basal 2 abdominal segments fused; adult as in Figure 20.41	***SALPINGIDAE*[2,6]**
33(32).	Eyes emarginate (notched) anteriorly; procoxal cavities enclosed behind by prothorax	***TENEBRIONIDAE*[1,6]**
33'.	Eyes oval or round; procoxal cavities enclosed behind by mesothorax (as in Fig. 20.229)	***ANTHICIDAE*[1,6]**
34(31).	Tarsal formula apparently 4-4-4; beetles larger than 2 mm	35
34'.	Tarsal formula 3-3-3; minute beetles less than 2 mm long; adult as in Figure 20.36	***PTILIIDAE*[1,2,6]** (in part)
35(34).	Antennae thickened apically, shorter than head and thorax; mandibles long, projecting horizontally before head; adult as in Figure 20.42	***HETEROCERIDAE*[1,6]**
35'.	Antennae slender, longer than head and thorax (Fig. 20.301); mandibles small, directed ventrally	***CHRYSOMELIDAE*** (p. 455)

[1] Subaquatic or inhabiting littoral region.
[2] Includes intertidal species in North America.
[5] Key to families and genera of Hydrophiloidea.
[6] Keys to genera not given.

KEYS TO THE GENERA OF AQUATIC COLEOPTERA

Amphizoidae (Trout-Stream Beetles)

The family Amphizoidae contains a single genus: *Amphizoa* LeConte. Presently, three species are recognized, all from northwestern North America where the larvae and adults live in mountain streams. They are especially abundant on driftwood and trash floating in frothy eddies, along undercut banks among roots, or among accumulations of submerged pine needles. Although often rare in collections, large numbers can be collected once their habitat is recognized.

Eggs of amphizoids have been found in cracks on the undersurface of driftwood, although oviposition sites may occur more typically in high humidity or splash zones in partly submerged brush piles. Hatching takes place during middle to late August, and the larvae usually reach the second instar by winter. Larvae (Fig. 20.2) are predaceous and seem to restrict their diet to Plecoptera larvae. They usually are found crawling on twigs or other wood; mature larvae are almost invariably entirely out of, but near, the water. Such larvae readily enter the water to seize prey, but quickly return to a twig or other support to eat the victim.

When dislodged into relatively quiet water, amphizoid larvae assume a characteristic posture with the abdomen at and horizontal to the water's surface (spiracles of the eighth abdominal tergite at the surface), and the thorax and head are folded under the abdomen so that the mandibles lie beneath its apex. This posture permits the larva to respire while afloat and enables it to quickly capture any prey that comes into reach or to grasp any solid object it touches.

Mature larvae have been observed crawling from the water in late July and early August, and two have been found in protective cases lodged in debris-filled crevices between logs; however, pupation has not been observed. Newly emerged adults frequently are mud-covered which suggests that pupation may occur in muddy creek banks.

Adult amphizoids (Fig. 20.30) are poor swimmers usually found crawling on twigs or other submerged plant material. They are predaceous and, like the larvae, show a preference for Plecoptera larvae. Earlier reports that these beetles were scavengers are incorrect.

Adult amphizoids have been observed surfacing briefly, then carrying a bubble of air beneath and surrounding the elytral apices while submerged. The highly oxygenated habitat may enable the air bubble to serve as a physical gill, precluding the need for surfacing or greatly extending the time submerged.

Gyrinidae (Whirligig Beetles)

Gyrinidae is a family of aquatic insects with approximately 700 species. Nearly 60 species and subspecies are recorded for the United States and Canada. Whirligig beetles often are a familiar sight on freshwater ponds, lake margins, open flowing streams, quiet stream margins, bog pools, swamps, and roadside ditches where they may form large aggregations or schools in late summer and autumn. These aggregations may contain a single species or as many as 13. In autumn, Wisconsin pond-inhabiting gyrinids fly to large streams and lakes to overwinter.

Copulation occurs on the water surface, and the female lays her eggs on stems of emergent vegetation a few centimeters below the surface of the water. After hatching in 1-2 weeks, larvae (Fig. 20.46) pass through three instars. They crawl about on submerged objects using their characteristic apical abdominal hooks (Fig. 20.47), and feed on small aquatic organisms. They can swim in an undulating fashion by using the abdominal gills, possibly as an escape mechanism. Pupation takes place on shore above water level.

Adults are unique in having eyes divided into two portions (Figs. 20.48-20.49). The lower portion remains completely submerged surveying the aquatic habitat; the upper portion views the above water habitat. Divided vision and quick swimming movements allow them to avoid predators from above or below. They are predominantly scavengers, feeding upon live or dead insects trapped or floating on the water surface.

Gyrinidae

Larvae[1]

1.	Head suborbicular with collum (neck) narrow and distinct (Fig. 20.43); mandible without retinaculum (tooth) on inner margin; nasale with median produced lobe, which may or may not be emarginate, and with a lower tooth on each side (Fig. 20.43); adult as in Fig. 20.46	***Dineutus***
1'.	Head elongate, with collum not distinct, nearly as wide as remainder of head (Figs. 20.44-20.45); mandible with retinaculum or without	2
2(1').	Nasale with 2-4 teeth in a transverse row (Fig. 20.44)	***Gyrinus***
2'.	Nasale without teeth (Fig. 20.45) (based upon early instar larvae from Missouri)	***Gyretes***

Adults

1.	Dorsal and ventral compound eyes in contact on lateral margin of head, separated only by a narrow ridge (Fig. 20.48); meso- and metatarsal segments as long as or longer than broad; length less than 3 mm	***Spanglerogyrus***
1'.	Dorsal and ventral compound eyes divided, upper eyes inset from lateral margin of head for a distance of at least half the width of an eye (Fig. 20.49); meso- and metatarsal segments 2, 3, and 4 much broader than long; total length 3-15 mm	2

---

[1] The larva of *Spanglerogyrus* is unknown.

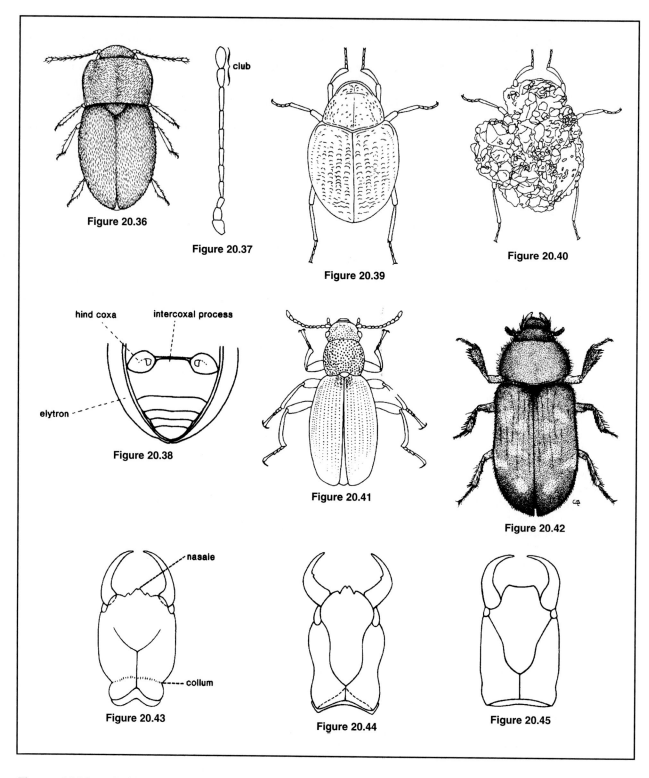

Figure 20.36. *Actidium* sp. (Ptiliidae) adult, dorsal aspect.

Figure 20.37. *Actidium* sp. (Ptiliidae) adult, antenna.

Figure 20.38. *Georyssus* sp. (Georyssidae) adult, ventral aspect of abdomen and metathorax.

Figure 20.39. *Georyssus* sp. (Georyssidae) adult, dorsal aspect with camouflage of sand grains removed.

Figure 20.40. *Georyssus* sp. (Georyssidae) adult, dorsal aspect with camouflage of sand grains.

Figure 20.41. *Aegilites* sp. (Salpingidae) adult, dorsal aspect.

Figure 20.42. Heteroceridae adult, dorsal aspect.

Figure 20.43. *Dineutus* sp. (Gyrinidae) larva, dorsal aspect of head (after Sanderson 1982).

Figure 20.44. *Gyrinus* sp. (Gyrinidae) larva, dorsal aspect of head (after Sanderson 1982).

Figure 20.45. *Gyretes* sp. (Gyrinidae) larva, dorsal aspect of head (after Sanderson 1982).

2(1').	Lateral margins of pronotum and elytron pubescent (Fig. 20.50); elytron without striae; apical abdominal sternites with median longitudinal row of long hairs; scutellum concealed; total length 3-5 mm	*Gyretes*
2'.	Pronotum and elytron entirely glabrous (without pubescense); apical abdominal sternites without longitudinal row of hairs	3
3(2').	Larger species, 9-15 mm in length; elytron smooth or with indistinct striae; scutellum concealed	*Dineutus*
3'.	Smaller species, 4-7 mm in length; elytron with 11 distinct striae; scutellum exposed (as in Fig. 20.48)	*Gyrinus*

Haliplidae (Crawling Water Beetles)

The Haliplidae is a relatively small family of slightly more than 200 species with about 70 species representing four genera from the United States and Canada. Females of the genus *Haliplus* have been observed to cut a hole with their mandibles in the side of a filament of the aquatic macrophytes *Ceratophyllum* and *Nitella* and deposit several eggs within the plant cell. Eggs of the genus *Peltodytes* are deposited on the leaves and stems of aquatic plants where hatching occurs within 8-14 days.

Haliplid larvae pass through three instars and are herbivorous. Pupation occurs from 20-25 days after hatching in a spherical pupal chamber constructed by the larva in rather dry mud. The mature larva remains in the pupal chamber from 4-6 days before transformation.

Numerous investigators have discussed the feeding habits of adult haliplids. Early studies reported that the adults were carnivorous but later studies have shown them to be herbivorous.

The greatly expanded hind coxal plates of Haliplidae (Fig. 20.63) are unique among aquatic Coleoptera. The air store retained by these plates is taken in by way of the tip of the abdomen, retained under the elytra, and acts as both a supplemental air store and in hydrostatic functions.

Haliplidae

Larvae

1.	Body segments each with 2 or more erect, segmented, hollow, spine-tipped filaments, each filament half as long as body (Fig. 20.52); fore legs chelate, 4th segment produced apically and edged with a solid row of small teeth so that 5th segment and claw can be closed on it (Fig. 20.53)	*Peltodytes*
1'.	Body spines, except in 1st instar, never stalked or much longer than length of 1st body segment (Fig. 20.54); apical abdominal segment produced posteriorly in a forked or unforked horn; fore legs, if chelate, with 4th segment less produced and without solid row of small teeth (Fig. 20.55)	2
2(1').	Third antennal segment shorter than 2nd; fore legs moderately chelate, but 3rd instead of 4th segment produced, edged with 2 blunt teeth; apical abdominal segment unforked, strongly curved ventrally (Fig. 20.56); body without conspicuous spines	*Brychius*
2'.	Third antennal segment 2-3 times as long as 2nd; body with or without conspicuous spines	3
3(2').	Fore leg with 3rd segment produced and edged by 2 blunt teeth; apical abdominal segment unforked (except in 1st instar); body with conspicuous spines only on lateral margins	*Apteraliplus*
3'.	Fore leg weakly to moderately chelate, 4th segment more or less produced, usually bearing 2-3 spines (Fig. 20.55); body with or without conspicuous spines	*Haliplus*

Adults

1.	Pronotum rounded with distinct black blotch on each side of middle near posterior margin (Fig. 20.58); last segment of both labial and maxillary palpi cone-shaped, as long as or longer than next to last; hind coxal plates large, only last abdominal sternite completely exposed; elytron with fine sutural stria in at least apical half	*Peltodytes*
1'.	Pronotum immaculate or with median blotch anteriorly (Fig. 20.60), posteriorly, or both or with 2 indistinct blotches (Fig. 20.59); last segment of both labial and maxillary palpi sublate, shorter than next to last; hind coxal plates smaller, leaving last 3 abdominal sternites exposed; elytron without fine sutural stria	2
2(1').	Pronotum with sides of basal two-thirds nearly parallel (Fig. 20.59); epipleuron broad, extending almost to tip of elytron, which is never truncate; metasternum reaching epipleuron	*Brychius*
2'.	Pronotum with sides widest at base, convergent anteriorly (Fig. 20.60); epipleuron evenly narrowed, usually ending near base of last abdominal sternite, never reaching elytral apex; episternum completely separating metasternum from epipleuron	3
3(2').	Median part of prosternum and base of prosternal process forming a plateaulike elevation, at least in part angularly separated from sides of prosternum (Fig. 20.62)	*Haliplus*
3'.	Prosternum evenly rounded from side to side, process raised above base (Fig. 20.61); tiny, length 1.5-2.5 mm; California	*Apteraliplus*

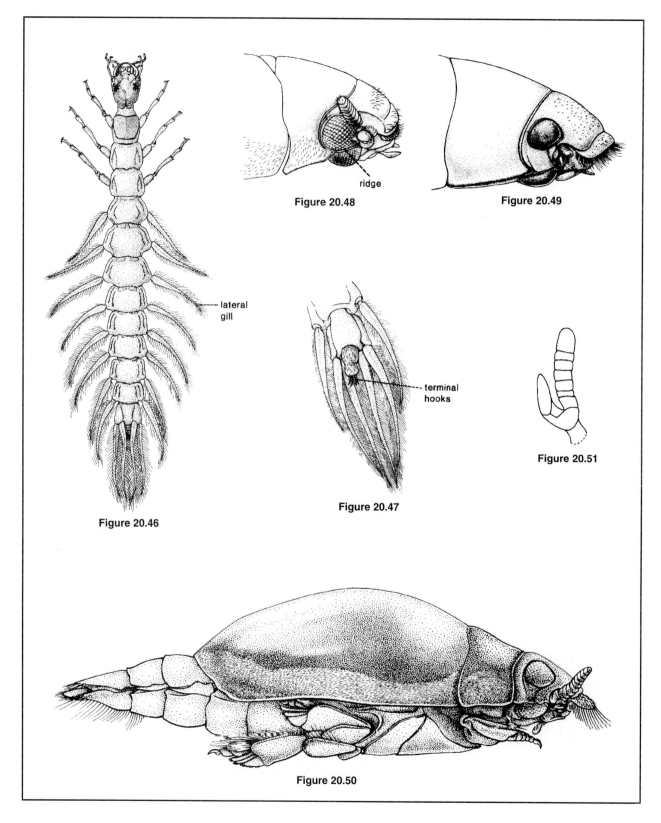

Figure 20.46. *Dineutus* sp. (Gyrinidae) larva, dorsal aspect.
Figure 20.47. *Dineutus* sp. (Gyrinidae) larva, ventral aspect of last abdominal segments.
Figure 20.48. *Spanglerogyrus* sp. (Gyrinidae) adult, lateral aspect of head and pronotum.
Figure 20.49. *Dineutus* sp. (Gyrinidae) adult, lateral aspect of head and pronotum.
Figure 20.50. *Gyretes* sp. (Gyrinidae) adult, lateral aspect.
Figure 20.51. *Gyrinus* sp. (Gyrinidae) adult, antenna.

Dytiscidae (Predaceous Diving Beetles)

The Dytiscidae are among the best adapted insects for aquatic existence and the most diverse of the water beetles with 2,500 described species, more than 500 of which occur in North America.

Despite the size and importance of this group among water beetles, surprisingly little is known of the life history of North American dytiscids. Mating occurs from early spring through autumn. Oviposition sites of dytiscids appear to correlate well with structural modifications of the ovipositor. Species having a long, flexible ovipositor (some *Acilius*) place their eggs loosely, usually 30-50 in a mass, above water in moist soil among grass roots or under organic debris. Most species possessing a cutting ovipositor (some *Agabus, Coptotomus, Cybister, Dytiscus, Hydraticus, Ilybius, Laccophilus,* and *Thermonectus*) insert their eggs into parts of living plants. A third group places its eggs on plant surfaces or inserts them, at most, halfway into the plant tissues (some *Agabus, Colymbetes, Hydraporus,* and *Rhatus*).

The larval stage of the Dytiscidae consists of three instars and requires from several weeks to several months, depending mainly upon season and the availability of food. Most dytiscid larvae rise to the surface and take in air through the large terminal spiracles. Cuticular respiration, however, appears to be common among first-instar larvae of many species. *Agabus, Ilybius,* and some *Hydroporus* have extensive networks of tracheae near the ventral cuticle and are believed to exchange gases through this structure. Larvae of the genus *Coptotomus* possess lateral gills and can remain beneath the surface continuously. All mature dytiscid larvae leave the water to prepare a pupal cell on land near the water's edge.

As the common name for the group implies, larval dytiscids are predaceous. Their selection of food appears to be governed by their ability to catch and overcome their prey. Although most adult dytiscids are active predators, many also are scavengers.

Adult dytiscids readily leave the water and fly. They are unable to take off directly from the water but first crawl onto an object above the water level. This behavior appears to be linked to the shift of the respiratory function; the large thoracic air sacs connected with the first pair of spiracles that are used in flight remain collapsed while the insect is in the water. When returning, adult dytiscids splash directly into the water because the extreme modification of the legs for swimming makes them useless for alighting on surfaces. Impact usually carries them through the surface film, but frequently they return to the surface after a few seconds to fill the subelytral air chamber. Presumably, the tracheal system undergoes a reversal of the process of preparation for flight.

The taxonomy of the Hydroporinae remains as unsettled as it was 35 years ago (Leech and Chandler 1956). We have taken a conservative approach regarding most Hydroporini and followed previous works on Dytiscidae of the United States. A new subterranean dytiscid genus and species (*Stygoporus oregonensis*; Larson and Labonte 1994) was described as this chapter was going to press and is not included in the keys.

Dytiscidae

Mature Larvae[1]

1.	Head with a frontal projection (Figs. 20.65-20.67); maxillary palp usually 3-segmented (4-segmented in *Celina*)	2
1'.	Head without a frontal projection (Fig. 20.68); maxillary palp with 4 or more segments	14
2(1).	Last abdominal segment with recurved extension of lateral tracheal trunks reaching beyond apex of abdomen (Fig. 20.69) ....Methlinae .....Celinini	***Celina***
2'.	Lateral tracheal trunks not extending beyond apex of last abdominal segment, terminating on apex ....Hydroporinae	3
3(2').	Frontal projection of head elongate, constricted near base and broadly spatulate at apex, a lateral branch arising from each side near base (Figs. 20.65-20.66)	4
3'.	Frontal projection usually broadly triangular, not constricted near base and without a lateral branch on each side, although a notch may be present on each side (Fig. 20.67)	6
4(3).	Lateral branch of frontal projection long and curved medially, extending well beyond midpoint of frontal projection anterior to base of lateral branch (Fig. 20.65). ....Vatellini	***Derovatellus***
4'.	Lateral branch of frontal projection short, not reaching midpoint of frontal projection anterior to base of lateral branch (Fig. 20.66) ....Hyphydrini	5
5(4').	Apical segment of antenna bifurcate	***Desmopachria***
5'.	Apical segment of antenna simple	***Pachydrus***
6(3').	Cercus with more than 7 setae ....Hydroporini (in part)	7
6'.	Cercus with only 7 setae	8
7(6).	Cercus short	***Nebrioporus*** and ***Stictotarsus***
7'.	Cercus long or very long	***Oreodytes***
8(6').	Cercus very short, about one-half length of last abdominal segment ....Hydroporini (in part)	***Laccornis***
8'.	Cercus distinctly longer than last abdominal segment	9
9(8').	Frontal projection with a notch at each side (Fig. 20.67) ....Hydroporini (in part)	10

---

[1] The larvae of *Agaporomorphus, Anodocheilus, Bidessonotus, Brachyvatus, Carrhydrus, Laccodytes, Lioporeus, Megadytes,* and *Neobidessus* are unknown or undescribed. In addition, the generic pairs *Nebrioporus* and *Stictotarsus,* and *Liodessus* and *Neoclypeodytes* are incompletely separated.

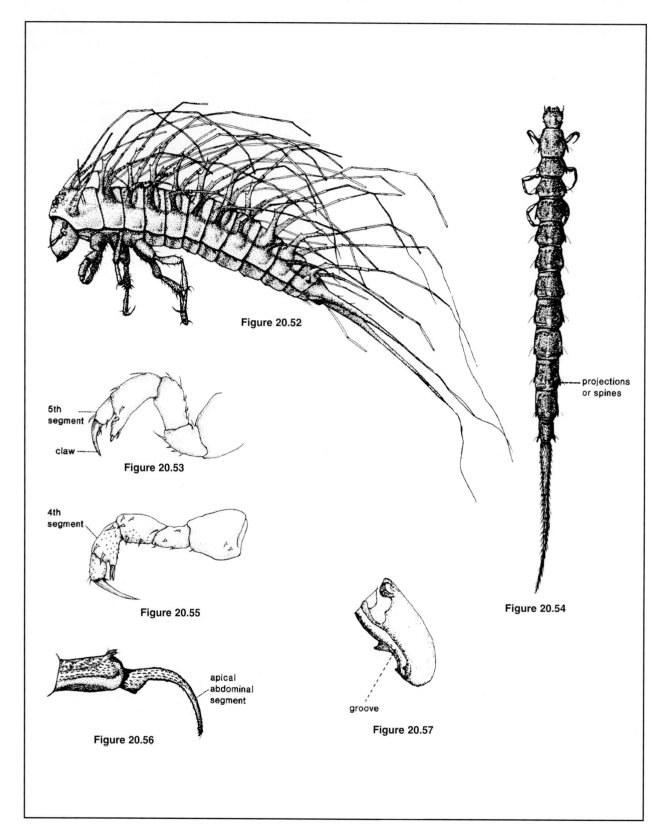

Figure 20.52. *Peltodytes* sp. (Haliplidae) larva, lateral aspect.
Figure 20.53. *Peltodytes* sp. (Haliplidae) larva, fore leg.
Figure 20.54. *Haliplus* sp. (Haliplidae) larva, dorsal aspect.
Figure 20.55. *Haliplus* sp. (Haliplidae) larva, fore leg.
Figure 20.56. *Brychius* sp. (Haliplidae) larva, lateral aspect of apex of abdomen.
Figure 20.57. *Peltodytes* sp. (Haliplidae) larva, mandible.

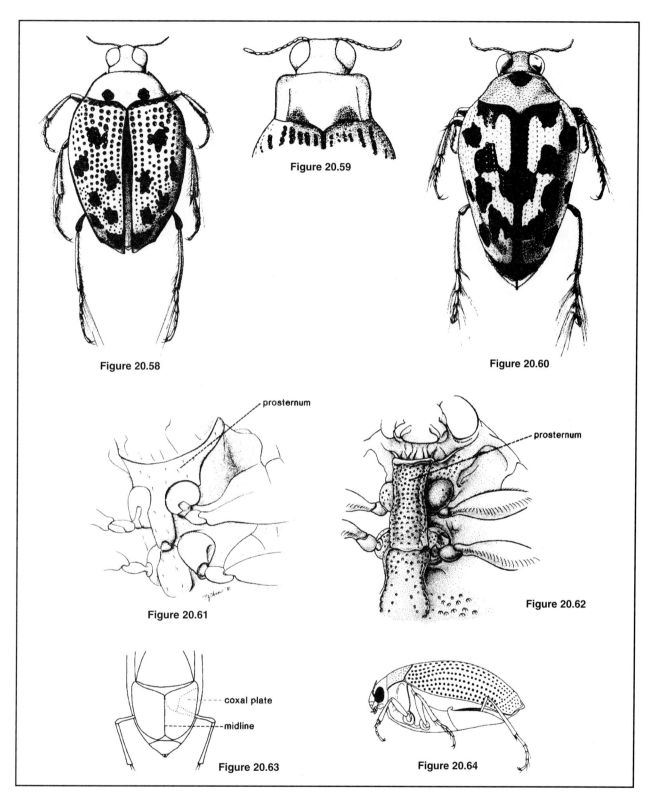

Figure 20.58. *Peltodytes* sp. (Halipidae) adult, dorsal aspect.

Figure 20.59. *Brychius* sp. (Halipidae) adult, dorsal aspect of head and pronotum.

Figure 20.60. *Haliplus* sp. (Haliplidae) adult, dorsal aspect.

Figure 20.61. *Apteraliplus* sp. (Haliplidae) adult, prosternum.

Figure 20.62. *Haliplus* sp. (Haliplidae) adult, prosternum.

Figure 20.63. Haliplidae adult, ventral aspect of metathorax and abdomen.

Figure 20.64. *Peltodytes* sp. (Haliplidae) adult, lateral aspect.

9'.	Frontal projection without notches	12
10(9).	Ocelli absent; cavernicolous, known only from artesian wells in Texas	***Haideoporus***
10'.	Ocelli present; surface dwelling, not normally confined to artesian waters	11
11(10').	Head narrow, maximum width nearly equal to length from base to antenna; frontal projection with a deep notch and a denticle behind it	***Hygrotus***
11'.	Head broad, maximum width greater than length from base to antenna; notch on frontal projection weak, denticle weak or absent	***Hydroporus***
12(9').	Larva greatly widened in middle, greatest width approximately 0.25 of total length	Hydrovatini ... ***Hydrovatus***
12'.	Larva not greatly widened in middle, greatest width 0.20 or less of total length	Bidessini ... 13
13(12').	Basal segment of cercus shorter than last abdominal segment including siphon; siphon equal in length or longer than the base of last abdominal segment	***Uvarus***
13'.	Basal segment of cercus longer than last abdominal segment including siphon; siphon shorter than base of segment	***Liodessus*** and ***Neoclypeodytes***
14(1').	Maxillary stipes broad, suboval, usually with 1-2 strong inner marginal spines or hooks	15
14'.	Maxillary stipes long and slender, usually without inner marginal hooks	31
15(14).	Abdominal segments 7 and/or 8 without lateral fringe of long swimming hairs	16
15'.	Abdominal segments 7 and 8 with swimming hairs	27
16(15).	Last (4th) segment of antenna less than two-thirds the length of the 3rd segment	20
16'.	Last segment of antenna more than two-thirds the length of the 3rd segment	Colymbetinae (in part) ... Colymbetini ... 17
17(16').	Tarsal claw with small spines on lower margin in basal half	19
17'.	Tarsal claw without small spines on lower margin in basal half; labium deeply emarginate apically	18
18(17').	Basal segment of labial palp much shorter than maxillary stipes; southern United States	***Hoperius***
18'.	Basal segment of labial palp much longer than maxillary stipes; boreal United States and Canada	***Neoscutopterus***
19(17).	Mandible almost 3 times as long as broad (Fig. 20.71)	***Rhantus***
19'.	Mandible short, slightly more than twice as long as broad	***Colymbetes***
20(16).	Last antennal segment double, although the lesser lobe may be represented only as a stout seta arising from the apex of the 3rd segment beside the greater lobe	Colymbetinae (in part) ... 21
20'.	Last antennal segment simple; mandible not toothed	22
21(20).	Last antennal segment consisting of 2 unequal lobes; mandible toothed along inner edge	Copelatini ... ***Copelatus***
21'.	Last antennal segment consisting of a short lobe about one-sixth as long as 3rd segment and a stout seta, each arising from the apex of the 3rd segment; mandible falciform, stout at base, slender and tapering to sharp apex, deeply grooved along inner surface, bearing a cluster of short setae ventrobasally	Agabetini ... ***Agabetes***
22(20').	Fore and middle legs chelate (Fig. 20.70) ... Colymbetinae (in part) ... Matini	***Matus***
22'.	Fore and middle legs simple	23
23(22').	Cercus with numerous secondary hairs ... Laccophilinae ... Laccophilini	***Laccophilus***
23'.	Cercus usually with 7 primary hairs in 2 whorls, 3 near middle, 4 apically	Colymbetinae (in part) ... Agabini. ... 24
24(23').	Thorax as wide as long, about equal in length to abdomen; sides of head nearly round, neck area not set off by shallow groove	***Hydrotrupes***
24'.	Thorax longer than wide, never as long as abdomen; sides of head round or squarish, but with neck area set off by occipital suture or shallow groove	25
25(24').	Tibia and tarsus with conspicuous spines confined to apical half, mostly terminal; dorsal margin of middle and hind femora without row of spines; tergites from 2nd thoracic segment to 8th abdominal segment each bearing 3 or more pairs of long, conspicuous setae; sternites of abdominal segments 4 through 6 each with a similar pair of setae	***Agabinus***
25'.	Tibia and tarsus with spines not confined to apical half; dorsal margin of middle and hind femora each with row of spines; tergites and sternites (see above) usually without conspicuous setae, or, if present, they are much smaller than those of 8th and 9th tergites	26
26(25').	Lateral margin of head more or less compressed or keeled, temporal spines on a line that would intersect ocelli or just pass below them	***Ilybius***
26'.	Lateral margin of head not keeled, temporal spines on a line that would run well below ocelli	***Agabus***
27(15').	Anterior 6 abdominal segments each with a pair of long lateral gills	Coptotomini ... ***Coptotomus***
27'.	Abdominal segments without gills	Dytiscinae (in part) ... 28
28(27').	Ligula very short, armed with 4 spines	Eretini ... ***Eretes***
28'.	Ligula long, simple or bifid, but without 4 spines	Thermonectini ... 29
29(28').	Ligula bifid apically	***Acilius***

29'.	Ligula simple	30
30(29').	Ligula not as long as 1st segment of labial palp	***Thermonectus***
30'.	Ligula nearly equal to or exceeding length of 1st segment of labial palp	***Graphoderus***
31(14').	Head dentate anteriorly; ligula long; cerci absent .... Cybistrinae	***Cybister***
31'.	Head lacking dentation anteriorly; ligula absent, or low and bilobed; cerci present (Fig. 20.68) .... Dytiscinae (in part)	32
32(31').	Cercus with lateral fringe; labium without projecting lobes .... Dytiscini	***Dytiscus***
32'.	Cercus without fringe; labium with 2 projecting lobes .... Hydaticini	***Hydaticus***

Adults[1]

1.	Fore and middle tarsi distinctly 5-segmented, 4th segment approximately as long as 3rd (Fig. 20.75)	2
1'.	Fore and middle tarsi each 4-segmented or with 4th segment small and concealed between lobes of 3rd so that 5th segment appears to be 4th (Fig. 20.76) (except in *Bidessonotus*, which has 4th segment small, but not concealed by lobes of 3rd [Fig. 20.77])	5
2(1).	Scutellum entirely visible (Fig. 20.78)	3
2'.	Scutellum covered by pronotum, or rarely a small tip visible; hind tarsus with a single straight claw .... Laccophilinae .... Laccophilini	6
3(2).	Eye emarginate above base of antenna (Fig. 20.79); basal 3 segments of fore tarsus of male widened and with adhesion disks, but never together forming an oval or nearly round plate ..... Colymbetinae	26
3'.	Eye not emarginate above base of antenna; first 3 segments of fore tarsus of male greatly broadened, forming a nearly round or oval plate with adhesion disks	4
4(3').	Inferior spur at apex of hind tibia dilated, much broader than the other large spur (Fig. 20.80); basal 3 segments of male fore tarsus forming an oval plate; large beetles, 20-32 mm long .... Cybistrinae	39
4'.	Inferior spur not as broad as, or but little broader than, other spur (Fig. 20.81); basal 3 segments of male fore tarsus forming a nearly round plate; medium to large beetles, 8-38 mm long .... Dytiscinae	40
5(1').	Scutellum fully visible .... Methlinae .... Celinini	***Celina***
5'.	Scutellum covered by pronotum .... Hydroporinae	7
6(2').	Spines of hind tibia notched or bifid apically; apical third of prosternal process lanceolate, only moderately broad; larger species, 2.5-6.5 mm in length	***Laccophilus***
6'.	Spines of hind tibia simple, acute apically; apical third of prosternal process somewhat diamond-shaped	***Laccodytes***
7(5').	Metepisternum not reaching mesocoxal cavity, excluded by mesepimeron; prosternal process short, broad, not reaching metasternum, its tip ending at front of the contiguous middle coxae .... Vatellini	***Derovatellus***
7'.	Metepisternum reaching mesocoxal cavity; apex of prosternal process reaching metasternum	8
8(7').	Apices of hind coxal processes broad, conjointly divided into 3 parts: 2 widely separated narrow lateral lobes and a broad, depressed middle region (Fig. 20.82); small, broadly ovate beetles about 2.5 mm long .... Hydrovatini	***Hydrovatus***
8'.	Hind coxal processes not conjointly divided into 3 parts as described above, either without lateral lobes (Fig. 20.83) or with these lobes covering bases of hind trochanters (Fig. 20.84)	9
9(8').	Hind coxal process without lateral lobe, base of hind trochanter entirely free (Fig. 20.83)	10
9'.	Sides of hind coxal processes diverging posteriorly, more or less produced into lobes that cover bases of hind trochanters (Fig. 20.84) .... Hydroporini	18
10(9).	Hind tibia straight, of almost uniform width from near base to apex; hind tarsal claws unequal; epipleuron with diagonal carina crossing near base; shining, ventrally convex beetles .... Hyphydrini	11
10'.	Hind tibia slightly arcuate, narrow at base, gradually widening to apex; hind tarsal claws equal; epipleuron without diagonal carina crossing near base (except in *Brachyvatus*, which has a carina) .... Bidessini	12
11(10).	Middle coxae separated by about width of a middle coxa; prosternal process short and broad, apex obtuse	***Pachydrus***
11'.	Middle coxae separated by only one-half width of a middle coxa; prosternal process rhomboid, apex acute	***Desmopachria***
12(10').	Head with a transverse cervical carina or suture behind eyes (Fig. 20.85)	13
12'.	Head without a cervical carina	***Uvarus***
13(12).	Pronotum and elytron plicate or with impressed carinae (Figs. 20.86-20.88)	14
13'.	Basal pronotal striole distinct, elytron without basal carina or sutural or accessory striae; Figure 20.85	***Brachyvatus***

---

[1] The generic pair *Nebrioporus* and *Stictotarsus* is incompletely separated.

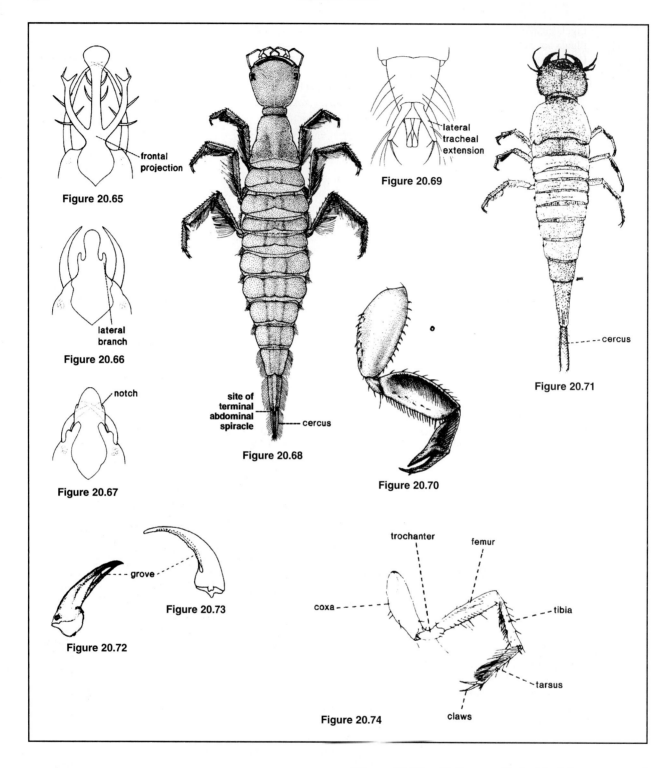

Figure 20.65. *Derovatellus* sp. (Dytiscidae) larva, dorsal aspect of frontal projection of head (after Spangler and Folkerts 1973).

Figure 20.66. *Pachydrus* sp. (Dytiscidae) larva, dorsal aspect of frontal projection of head (after Spangler and Folkerts 1973).

Figure 20.67. *Hydroporus* sp. (Dytiscidae) larva, dorsal aspect of frontal projection of head.

Figure 20.68. *Cybister* sp. (Dytiscidae) larva, dorsal aspect.

Figure 20.69. *Celina* sp. (Dytiscidae) larva, apex of abdomen (after Spangler 1973).

Figure 20.70. *Matus* sp. (Dytiscidae) larva, mesothoracic leg.

Figure 20.71. *Rhantus* sp. (Dytiscidae) larva, dorsal aspect.

Figure 20.72. *Rhantus* sp. (Dytiscidae) larva, mandible.

Figure 20.73. *Copelatus* sp. (Dytiscidae) larva, mandible.

Figure 20.74. *Rhantus* sp. (Dytiscidae) larva, metathoracic leg.

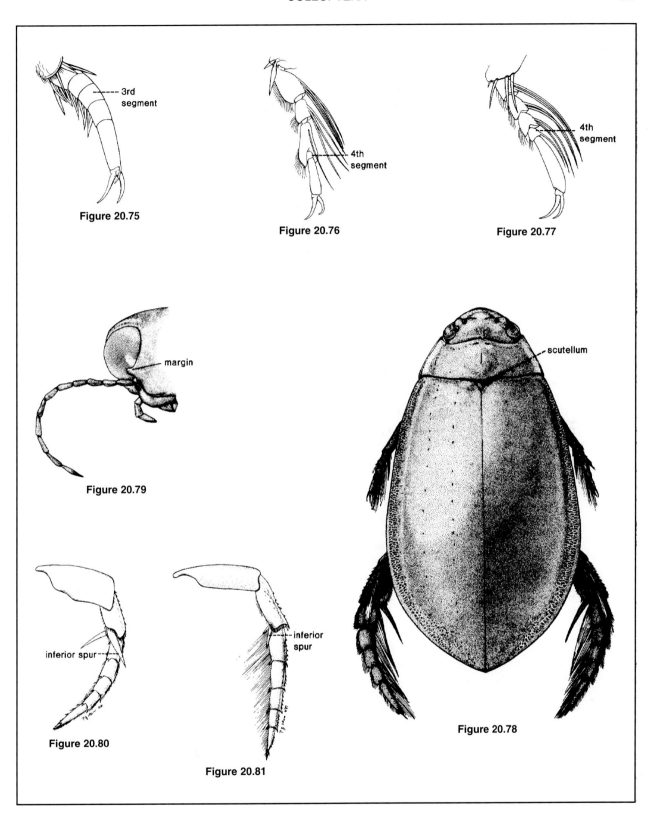

Figure 20.75. *Coptotomus* sp. (Dytiscidae) adult, tarsus of fore leg.

Figure 20.76. *Hydroporus* sp. (Dytiscidae) adult, tarsus of fore leg.

Figure 20.77. *Bidessonotus* sp. (Dytiscidae) adult, tarsus of fore leg.

Figure 20.78. *Cybister* sp. (Dytiscidae) adult, dorsal aspect.

Figure 20.79. *Coptotomus* sp. (Dytiscidae) adult, anterior aspect of head.

Figure 20.80. *Cybister* sp. (Dytiscidae) adult, hind leg.

Figure 20.81. *Acilius* sp. (Dytiscidae) adult, hind leg.

14(13).	Clypeal margin thickened anteriorly, upturned, tuberculate, or distinctly rimmed 15
14'.	Clypeal margin not thickened or only feebly so, not margined, upturned, tuberculate, or rimmed 16
15(14).	Basal pronotal plicae connected by transverse depression; elytron with suture not thickened and without sutural carina (Fig. 20.86) ....*Anodocheilus*
15'.	Basal pronotal plicae distinct, but area between them essentially flat; elytron with suture thickened, sutural carina more or less distinct ....*Neoclypeodytes*
16(14').	Tarsi clearly 5-segmented, the 4th small, but not concealed by lobes of 3rd (Fig. 20.77) ....*Bidessonotus*
16'.	Tarsi each with 4th segment small and concealed between lobes of 3rd so that 5th segment appears to be 4th (as in Fig. 20.76) 17
17(16').	Elytron without sutural stria, but with accessory discal stria of impressed punctures between suture and basal striole (Fig. 20.87); apical sternite broad, without lateral impressions ....*Neobidessus*
17'.	Elytron with sutural stria, but without accessory discal striae of impressed punctures (Fig. 20.88); apical sternite narrow, impressed on either side of middle ....*Liodessus*
18(9').	Cavernicolous, known only from artesian wells in Texas; eyes minute and apparently nonfunctional; body pigmentation reduced; fore and middle coxae enlarged; prosternal process short, not reaching metasternum between middle coxae ....*Haideoporus*
18'.	Surface-dwelling, not normally confined to artesian waters; eyes normal; body pigmentation normal; fore and middle coxae not enlarged; prosternal process reaching metasternum between middle coxae 19
19(18').	Base of hind femur contacting hind coxal lobe ....*Laccornis*
19'.	Hind femur separated from hind coxal lobe by basal part of trochanter 20
20(19').	Epipleuron with diagonal carina crossing near base (Fig. 20.89); fore and middle tarsi 4-segmented ....*Hygrotus*
20'.	Epipleuron without diagonal carina crossing near base; fore and middle tarsi 5-segmented, with 4th segment small and concealed between lobes of 3rd so that 5th segment appears to be 4th (Fig. 20.76) 21
21(20').	Hind margin of conjoined hind coxal processes slightly to deeply, and more or less triangularly, incised medially, lateral lobes more produced posteriorly 22
21'.	Hind margin of conjoined hind coxal processes not in the least abbreviated behind, the apex either truncate or more or less angularly prominent at the middle (Fig. 20.84) 23
22(21).	Pronotum with a longitudinally impressed line or crease on each side, and usually near base; with a shallow transverse impression near base or an impression on each side near base; hind femur with median row of setigerous punctures, otherwise sparsely punctate or nearly smooth; venter densely and finely punctate, with scattered or numerous coarser punctures ....*Oreodytes*
22'.	Pronotum without sublateral impressed lines, usually without basal impression; hind femur usually densely punctate over entire surface; venter densely and finely punctate or somewhat granulate, usually lacking scattered large punctures ....*Nebrioporus* and *Stictotarsus* (in part)
23(21').	Hind margin of conjoined hind coxal processes either truncate or obtusely angulate medially (Fig. 20.84) ....*Hydroporus* (in part)
23'.	Hind margin of conjoined hind coxal processes sinuate and somewhat angularly prominent medially (Fig. 20.90) 24
24(23').	Prosternal process not protuberant; male with 4th or 4th and 5th antennal segments enlarged; basal segment of male protarsus with a ventral cupule; length greater than 3.3 mm ....*Lioporeus*
24'.	Prosternal process usually protuberant (if not protuberant, then length less than 3.3 mm); male with 4th and 5th antennal segments not enlarged; male with protarsal cupule absent 25
25(24').	Hind angle of pronotum rectangular or obtuse ....*Hydroporus* (in part)
25'.	Hind angle of pronotum acute ....*Nebrioporus* and *Stictotarsus* (in part)
26(3).	Inner side of hind femur with a more or less thick group of cilia arising from a linear depression on the inner half of the inner apical angle (Fig. 20.91) ....Agabini 27
26'.	Hind femur without such cilia 31
27(26).	Hind coxal processes parallel-sided, lateral margins straight to apex ....*Agabinus*
27'.	Hind coxal processes not parallel-sided, lateral margins each forming a rounded lobe laterally 28
28(27').	Hind tarsal claws of equal length, if slightly unequal, then both very small, only one-third length of 5th tarsal segment 29
28'.	Hind tarsal claws obviously unequal, outer one of each pair two-thirds or less length of inner claw 30
29(28).	Labial palp very short, apical segment quadrate ....*Hydrotrupes*

29'.	Labial palp approximately as long as maxillary palp, apical segment cylindrical	***Agabus***
30(28').	Labial palp with penultimate segment triangular in cross section, the faces concave and unequal (Fig. 20.92); genital valves of female dorsoventrally compressed, not armed with teeth	***Carrhydrus***
30'.	Labial palp with penultimate segment cylindrical, not enlarged and triangular; genital valves of female laterally compressed, sawlike, with series of sharp teeth along dorsal edge	***Ilybius***
31(26').	Prosternum with median longitudinal furrow; basal 4 hind tarsal segments lobate, each distinctly produced at upper (inner) posterior corner	Matini .... ***Matus***
31'.	Prosternum convex or keeled, without such a furrow; basal 4 hind tarsal segments not lobate at upper posterior corners	32
32(31').	Hind coxal lines divergent anteriorly, coming so close together posteriorly as to almost touch median line, then turning outward almost at right angles onto hind coxal processes (Fig. 20.93); hind tarsal claws equal; pronotum narrowly, but clearly margined laterally	33
32'.	Hind coxal lines never almost touching median line; hind tarsal claws equal or not; pronotum margined or not	34
33(32).	Curved strigae present on venter, particularly on basal abdominal sternites; medium-sized beetles, generally over 3 mm long; widespread	Copelatini .... ***Copelatus***
33'.	Curved strigae absent from venter; very small beetles, 2.3-2.65 mm long; presently known in the United States only from Florida	***Agaporomorphus***
34(32').	Hind tarsal claws of equal length or virtually so; smaller species, 6-9 mm in length	35
34'.	Hind tarsal claws obviously unequal, outer claw only from one-third to two-thirds length of inner claw; larger species, 9-20 mm in length	Colymbetini 36
35(34).	Apical segment of palpi notched or emarginate at tip; pronotum clearly though narrowly margined laterally	Coptotomini ... ***Coptotomus***
35'.	Apical segment of palpi entire, not notched or emarginate at tip; pronotum with an exceedingly fine line along lateral edge, but not margined	Agabetini .... ***Agabetes***
36(34').	Anterior tip of metasternum (between middle coxae) clearly triangularly split to receive apex of prosternal process, the triangular channel usually deep with its apex about in line with posterior margins of middle coxae; pronotum usually margined laterally	37
36'.	Anterior tip of metasternum depressed, with shallow pit or broad notch to receive apex of prosternal process, never with sharply outlined triangular excavation; pronotum not margined	38
37(36).	Prosternal process flat; dorsum unusually flat; pronotum widely margined laterally; elytron lightly reticulate throughout, the meshes rather coarse, unequal, and irregular in shape	***Hoperius***
37'.	Prosternal process convex or carinate; elytron with reticulation lightly impressed, meshes of unequal sizes and shapes, but very small (a superimposed secondary reticulation of deeply impressed lines occurs over parts of the elytron of some females)	***Rhantus***
38(36').	Elytral sculpture consisting of numerous parallel transverse grooves	***Colymbetes***
38'.	Elytron coarsely reticulate, without transverse grooves	***Neoscutopterus***
39(4).	Hind tarsus of male with 2 claws, of female with longer outer claw and rudimentary inner claw	***Megadytes***
39'.	Hind tarsus of male always, of female usually, with only 1 claw (Fig. 20.80)	***Cybister***
40(4').	Large beetles, 20-38 mm long; posterior margins of first 4 hind tarsal segments bare	Dytiscini .... ***Dytiscus***
40'.	Smaller beetles, 8-15 mm long; posterior margins of first 4 hind tarsal segments with a dense fringe of flat, golden cilia	41
41(40').	Posterolateral margin of elytron edged with row of small spines (Figs. 20.94-20.95); apex of prosternal process sharply pointed; pronotum margined laterally; dorsal surface of hind tarsus punctate, bearing fine appressed hairs	Eretini .... ***Eretes***
41'.	Elytron without spines on posterolateral margin; apex of prosternal process rounded; pronotum not margined laterally; dorsal surface of hind tarsus bare, except for marginal cilia	42
42(41').	Outer (shorter) spur at apex of hind tibia acute	Hydaticini .... ***Hydaticus***
42'.	Outer spur at apex of hind tibia blunt, more or less emarginate	Thermonectini 43
43(42').	Elytron densely punctate, also usually fluted and hairy in female	***Acilius***

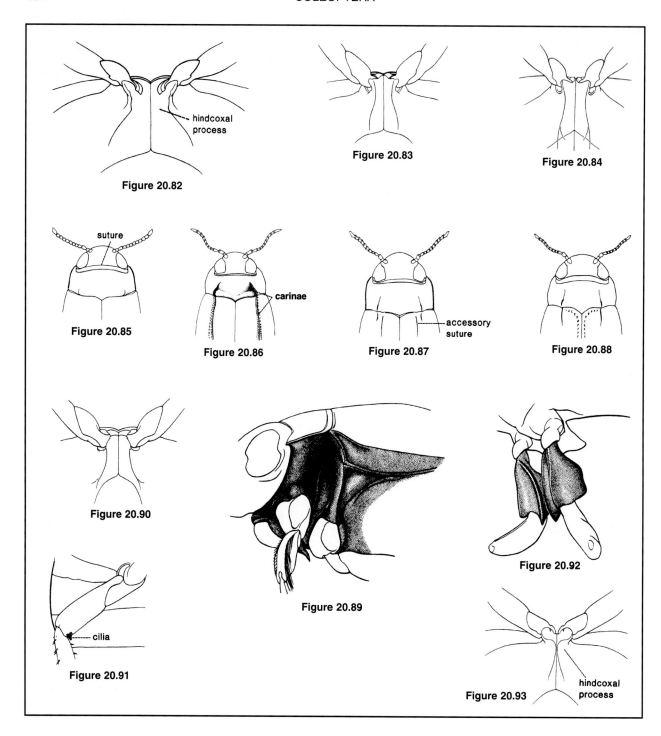

Figure 20.82. *Hydrovatus* sp. (Dytiscidae) adult, region of hind coxal processes.

Figure 20.83. *Hydroporus (Neoporus)* sp. (Dytiscidae) adult, region of hind coxal processes.

Figure 20.84. *Hydroporus* sp. (Dytiscidae) adult, region of hind coxal processes.

Figure 20.85. *Brachyvatus* sp. (Dytiscidae) adult, head, pronotum, and bases of elytra.

Figure 20.86. *Anodocheilus* sp. (Dytiscidae) adult, head, pronotum, and bases of elytra.

Figure 20.87. *Neobidessus* sp. (Dytiscidae) adult, head, pronotum, and bases of elytra.

Figure 20.88. *Liodessus* sp. (Dytiscidae) adult, head, pronotum, and bases of elytra.

Figure 20.89. *Hygrotus* sp. (Dytiscidae) adult, ventrolateral aspect of epipleuron.

Figure 20.90. *Lioporius* sp. (Dytiscidae) adult, region of hind coxal processes (after Wolfe and Matta 1981).

Figure 20.91. *Agabus* sp. (Dytiscidae) adult, hind femur.

Figure 20.92. *Carrhydrus* sp. (Dytiscidae) adult, labial palp.

Figure 20.93. *Copelatus* sp. (Dytiscidae) adult, region of hind coxal processes.

43'.	Elytron with extremely fine punctation or none; some females with superimposed sculpture of elongate grooves, or granulate	44
44(43').	Elytron basically yellowish, uniformly speckled or vermiculate with black; hind margin of middle femur with series of stiff setae which are only about one-half as long as femur is wide	***Graphoderus***
44'.	Elytron black with yellow maculae or transverse bands, or yellow with black spots, or irrorate (freckled); hind margin of middle femur with series of stiff setae which are as long as or longer than the femur is wide (Fig. 20.96) (Caution: these setae are brittle and may be broken in old or roughly handled specimens)	***Thermonectus***

Noteridae (Burrowing Water Beetles)

The family Noteridae is distributed widely throughout the tropical regions of both hemispheres with only a few genera and species reaching the temperate zone. The family is represented in North America north of Mexico by five genera and about 15 species.

The Palaearctic species *Noterus capricornis* Herbst has been investigated thoroughly, and its burrowing habits are the source of the family's common name. Observations of noterid larvae in eastern North America indicate that most, if not all, of these species do not share the strict burrowing habits of their European relatives. In fact, they lack the chitinous point at the apex of the abdomen with which *N. capricornis* is presumed to pierce plant roots for the purpose of obtaining intercellular air.

Food habits of the larvae are unknown. Larval *N. capricornis* have been observed working their mandibles upon the surface of plant roots without appearing to remove any tissue. Noterid larvae also readily attack dead chironomid larvae. The morphology of the mandible suggests an omnivorous diet; adults are predaceous.

In shallow water, larvae of *N. capricornis* renew their air supplies by bringing the apex of the abdomen to the surface. However, in the typical method of obtaining air, the apical spine of the abdomen is used to pierce plant tissue in order to tap intercellular air in the manner of chrysomelid beetles of the genus *Donacia*.

Pupation of *N. capricornis* is similar to that of *Donacia*. The larva constructs a cocoon from small pieces of vegetable material mixed with mud particles on the roots of various species of aquatic plants. Larvae chew or pierce the plant root at the point of attachment of the cocoon. Air escaping from the lacerated tissue is caught in the cocoon as it is being constructed and sealed within when the larva closes the distal end. Whether North American larvae of Noteridae behave in a similar manner to *N. capricornis* is unknown. The length of the pupal period is unknown, but presumably lasts no more than a few weeks.

Noteridae

Larvae[1]

1.	Third antennal segment not longer than 4th; mandible with stout preapical tooth	***Suphisellus***
1'.	Third antennal segment more than twice as long as 4th; mandible not strongly toothed	2
2(1').	Body globular (Fig. 20.102); 3rd antennal segment about 12 times as long as 4th; mandible serrulate	***Suphis***
2'.	Body cylindriform, not globular (Fig. 20.103); 3rd antennal segment about 3 times longer than 4th; mandible simple	***Hydrocanthus***

Adults

1.	Apex of fore tibia with curved hook or spur (Fig. 20.106); length usually over 2.0 mm	2
1'.	Apex of fore tibia without curved hook or spur (Fig. 20.107); length usually less than 1.5 mm	***Notomicrus***
2(1).	Fore tibial hooks strong, curved, and conspicuous (Fig. 20.106); hind femur with angular cilia; prosternal process truncate posteriorly, or if rounded in the male, form is very broad, almost hemispherical	3
2'.	Fore tibial hooks weak and inconspicuous; hind femur usually without angular cilia; prosternal process rounded posteriorly in both sexes	***Pronoterus***
3(2).	Body form very broad, almost hemispherical (Fig. 20.108); color opaque black with irregular reddish marks on each elytron	***Suphis***
3'.	Body form elongate, not hemispherical (Fig. 20.109); elytron uniformly black, reddish brown, or yellowish brown, without markings or with an oblique yellowish crossbar just behind the middle, never with reddish marks	4
4(3').	Length usually less than 3 mm; apical segment of maxillary palpus emarginate at apex; prosternal process not broader than long, apex at least twice its breadth between the anterior coxae	***Suphisellus***
4'.	Length usually over 4 mm; apical segment of maxillary palpus truncate at apex; prosternal process broader than long, apex very broad, at least 2.5-3 times its breadth between the anterior coxae; adult as in Figure 20.109	***Hydrocanthus***

[1] The larvae of *Notomicrus* and *Pronoterus* are unknown.

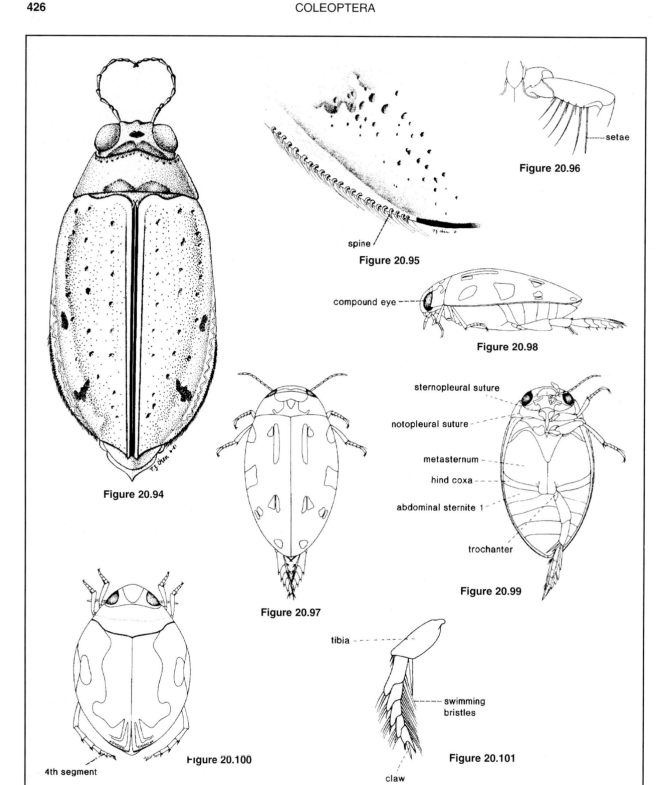

Figure 20.94. *Eretes* sp. (Dytiscidae) adult, dorsal aspect.

Figure 20.95. *Eretes* sp. (Dyticidae) adult, margin of elytron.

Figure 20.96. *Thermonectus* sp. (Dytiscidae) adult, middle femur.

Figure 20.97. *Laccophilus* sp. (Dytiscidae) adult, dorsal aspect.

Figure 20.98. *Laccophilus* sp. (Dytiscidae) adult, lateral aspect.

Figure 20.99. *Laccophilus* sp. (Dytiscidae) adult, ventral aspect.

Figure 20.100. *Desmopachria* sp. (Dysticidae) adult, dorsal aspect.

Figure 20.101. *Laccophilus* sp. (Dyticidae) adult, hind tibia and tarsus.

Hydroscaphidae (Skiff Beetles)

The Hydroscaphidae is a small family of worldwide distribution, although mainly tropical. Only *Hydroscapha* LeConte is known from the United States and is represented by *H. natans* LeConte, which extends northward from Mexico through the western states to Idaho.

Hydroscaphids are rare in collections probably because most collecting is done with nets having too coarse a mesh to retain them. Once located, hydroscaphids frequently are very abundant and hundreds may be collected in a rather short time. Although adults may be found under stones as much as a meter below the surface of fast-flowing streams, larvae and adults occur most commonly on algae over which a thin film of water is flowing. Often the water is so shallow that the adults are only partially submerged. *Hydroscapha natans* occurs over a wide range of temperatures, from hot springs (46°C) to mountain streams that freeze nightly throughout the summer.

Only one, remarkably large egg develops at a time, occupying a fourth of the female's abdomen. Larvae (Fig. 20.16) and adults (Fig. 20.26) remain in and feed upon algae. Pupation occurs at the edge of the water film or even among algae over which the water film is flowing. Pupae lie partially within the last larval exuviae.

In larval *H. natans,* only the first pair of thoracic spiracles and those of the first and eighth abdominal segments are functional. All three pairs are balloon-like, more than twice as long as wide. Presumably gas exchange takes place across the walls of these "balloons" providing a water-air interface available for respiration that is about an order of magnitude greater per milligram wet body weight than that across the spiracles of terrestrial insects.

Adult *Hydroscapha* apparently breathe by means of an air bubble carried beneath the elytra (physical gills, see Chap. 4). A fringe of cilia on the hind wings and setae on the dorsum covered by the elytra serve to retain this bubble.

Hydrophilidae (Water Scavenger Beetles)

With more than 230 species, Hydrophilidae ranks second in abundance to Dytiscidae, and these two taxa most often come to mind when one considers aquatic Coleoptera. In a recent revision by Hansen (1991), four of the aquatic subfamilies of "Hydrophilidae" have been elevated to family status: Helophoidae, *Helophorus*; Epimetopidae, *Epimetopus*; Hydrochidae, *Hydrochus*; and Sperchiidae, *Sperchius* (*Sperchius* is not found in North America). The remaining genera are retained in the family Hydrophilidae, and, except where noted, all discussions refer to hydrophilid genera. The common name is not accurate because most hydrophilid larvae are predators, most adults are omnivores, consuming both living and dead materials, and the subfamily Sphaeridiinae is not aquatic.

Eggs are deposited underwater in silk cases that may contain more than 100 eggs. Larvae go through three instars rapidly in one to several months. Even though a few species possess lateral gills allowing them to occupy deeper habitats, most larvae must obtain oxygen at the waterline through terminal abdominal spiracles. The larvae are poor swimmers and tend to lie in wait for prey. The genera *Hydrophilus, Tropisternus* and *Hydrobius* often consume their prey out of water. When mature, larvae leave the water to construct pupal chambers in moist soil, under rocks, and in organic debris. Some species may pupate in emergent vegetation and floating algal mats some distance from shore.

Adults generally are good swimmers but not as active as many of the Dytiscidae. They must return to the surface to renew their air supply. Typically they break the surface film with the antennae and side of the head; this allows gas exchange along the plastron and air passage on the ventral surface of the thorax. Some gas exchange, providing an auxiliary oxygen source, may occur via the plastron while submerged. Adults are active flyers and have been shown capable of leaving the water many times. Mass emergence and flight periods are not uncommon, and large numbers may be attracted to lights.

Hydrophilidae are most common in small pools and ponds with emergent vegetation and few predators. Adults can be collected throughout most of the year. With the exception of a few of the more common genera, larvae are rarely collected, perhaps because of the relatively short larval cycle, compared with most other families of aquatic Coleoptera.

Hydrophiloidea

Larvae[1]

1.	Nine complete abdominal segments, 10th reduced but distinct (Fig. 20.110); integument noticeably chitinized	**HELOPHORIDAE** ... *Helophorus*
1'.	Eight complete abdominal segments, segments 9 and 10 reduced and forming a stigmatic atrium or cavity (atrium absent in *Berosus*)	2
2(1').	Antennae with points of insertion nearer anterolateral angles of head than are insertion points of mandibles; labium and maxillae inserted in furrow beneath head	**HYDROCHIDAE** ... *Hydrochus*
2'.	Antennae with points of insertion further from anterolateral angles than those of mandibles; labium and maxillae inserted at anterior margin of ventral side of head	**HYDROPHILIDAE** ... 3
3(2').	Abdominal segments with broad (Fig. 20.129) or long (Fig. 20.112) lateral projections	4
3'.	Lateral projections usually absent; if projections present (Fig. 20.128), not nearly so prominent	5
4(3).	First 7 abdominal segments with long, lateral tracheal gills (Fig. 20.112); mandibles as in Figure 20.113, clypeus as in Figure 20.114	*Berosus*
4'.	Abdominal segments and posterior thoracic segments with broad, lateral projections (Fig. 20.129)	*Crenitis*

[1] The larvae of *Crenitulus, Dibolocelus,* and *Epimetopus* either are not well known or are not yet described for North America.

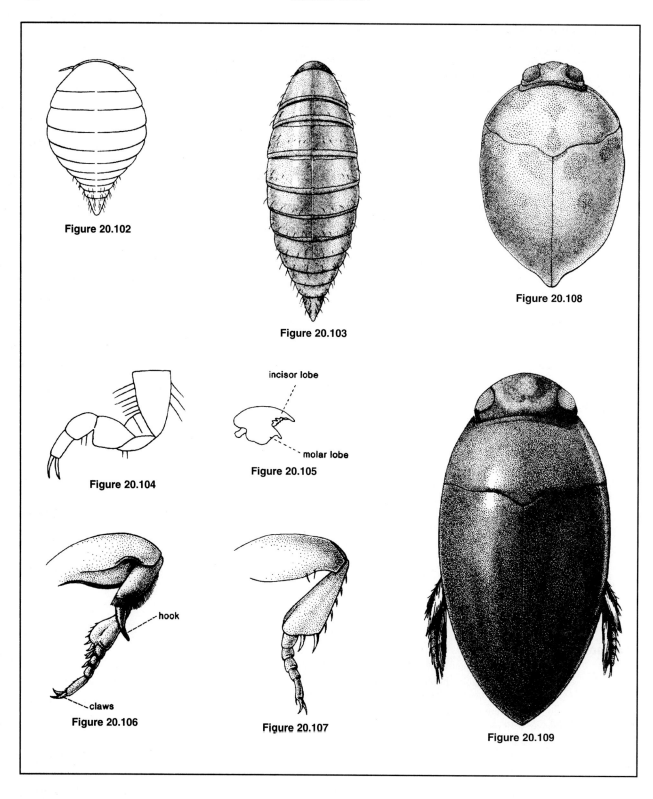

Figure 20.102. *Suphis* sp. (Noteridae) larva, dorsal aspect (after Spangler and Folkerts 1973).

Figure 20.103. *Hydropcanthus* sp. (Noteridae) larva, dorsal aspect.

Figure 20.104. *Noterus* sp. (Noteridae) larva, fore leg (not in key) (after Bertrand 1972).

Figure 20.105. *Noterus* sp. (Noteridae) larva, mandible (not in key) (after Bertrand 1972).

Figure 20.106. *Hydrocanthus* sp. (Noteridae) adult, fore leg.

Figure 20.107. *Notomicrus* sp. (Noteridae) adult, fore leg.

Figure 20.108. *Suphis* sp. (Noteridae) adult, dorsal aspect.

Figure 20.109. *Hydrocanthus* sp. (Noteridae) adult, dorsal aspect.

5(3').	Antennae biramal, terminal articles accompanied by a fingerlike antennal appendage	6
5'.	Antennae not biramal, fingerlike antennal appendages absent (Fig. 20.132)	16
6(5).	Legs completely invisible from above and very reduced, without claws; ligula present	***Chaetarthria***
6'.	Legs sometimes very small and not visible from above, but always complete, with claws (Fig. 20.134); ligula present or absent	7
7(6').	Frontal sutures parallel (Fig. 20.115) and not uniting to form an epicranial suture; ligula absent	***Laccobius***
7'.	Frontal sutures not parallel (as in Fig. 20.142); epicranial suture present or not; ligula present	8
8(7').	Ligula shorter than segment 1 of labial palp	***Helobata***
8'.	Ligula longer than segment 1 of labial palp	9
9(8').	Antenna short; epicranial suture absent; legs reduced	10
9'.	Antenna longer; epicranial suture present, but usually short; legs fairly long, not reduced	11
10(9).	Frons truncate behind; clypeus as in Figure 20.117; mandible with 2 inner teeth (Fig. 20.116); ligula about as long as palp, apparently 2-segmented; anterior margin of pronotum without a fringe of stout setae; legs not visible from above	***Paracymus***
10'.	Frons rounded behind; clypeus with 4 teeth (Fig. 20.119); mandible with 3 inner teeth (Fig. 20.118); ligula not as long as palp, 1-segmented; anterior margin of pronotum with a fringe of stout setae; legs barely visible from above	***Anacaena***
11(9').	Mandibles asymmetrical, the right with 2 inner teeth, the left with only 1 (Fig. 20.121); abdomen with prolegs on segments 3 through 7; larva as in Figure 20.120	***Enochrus***
11'.	Mandibles symmetrical, each with 2-3 teeth (Fig. 20.122); abdomen without prolegs	12
12(11').	Clypeus with 4 (sometimes 5) large and distinct teeth (Fig. 20.123)	***Ametor***
12'.	Clypeus with 5 or more distinct teeth	13
13(12').	Clypeus with 5 distinct teeth though middle tooth may be small (Fig. 20.140)	20
13'.	Clypeus with 6 or more distinct teeth; mandible with 2 inner teeth	14
14(13').	Clypeus with 6 distinct teeth, placed in 2 groups, 2 on the left and 4 on the right (Fig. 20.125); larva as in Figure 20.124	***Helochares***
14'.	Clypeus with more than 6 teeth, those toward right not as clearly defined and with several smaller teeth (Fig. 20.126)	15
15(14').	Clypeus with tooth on left side smaller than tooth on right side (Fig. 20.126)	***Cymbiodyta***
15'.	Clypeus with left and right side teeth equal in size	***Helocombus***
16(5').	Mesothorax, metathorax, and abdominal segment 1 each with 3 setiferous, lateral gills (Fig. 20.127); abdominal segments 2 through 6 each with 4 moderately long, setiferous, lateral gills (Fig. 20.128); femur without fringe of long swimming hairs	***Derallus***
16'.	Gills absent or, if present, with only a single lateral gill on each side of abdominal segments (Fig. 20.131); femur with fringe of long swimming hairs	17
17(16').	Head subspherical; mandibles not symmetrical, left mandible very robust, right mandible much more slender (Fig. 20.130); pronotum not entirely sclerotized	***Hydrophilus***
17'.	Head subquadrangular or subrectangular (Figs. 20.132, 20.142), mandibles symmetrical or not (Figs. 20.132, 20.139), pronotum entirely sclerotized	18
18(17').	Lateral abdominal gills well developed and pubescent (Fig. 20.131); mentum convex toward basal half, anterolateral angles less prominent (Figs. 20.132-20.133)	***Hydrochara***
18'.	Lateral abdominal gills rudimentary, but indicated by tubercular projections, each with several terminal setae (Fig. 20.135); mentum with sides almost straight, anterolateral angles very prominent	19
19(18').	Mesonotal and metanotal sclerites much reduced, triangular, hind margins very narrow, almost pedunculate (Fig. 20.135); apex of ligula not bifid; lateral abdominal gills of segment 9 short (Fig. 20.136)	***Tropisternus***
19'.	Mesonotal and metanotal sclerites not much reduced, trapezoidal hind margins almost as wide as anterior margin; apex of ligula shallowly bifid; lateral abdominal gills of segment 9 very long, conspicuous; clypeus with numerous irregular teeth (Fig. 20.137)	***Hydrobiomorpha***
20(13).	Middle tooth on clypeus smaller than the others (Figs. 20.139-20.140); prosternum entire; larva as in Figure 20.138	***Sperchopsis***
20'.	All teeth on clypeus subequal, outer left tooth usually a little distant from the rest (Fig. 20.142); prosternum with a mesal fracture; larva as in Figure 20.141	***Hydrobius***

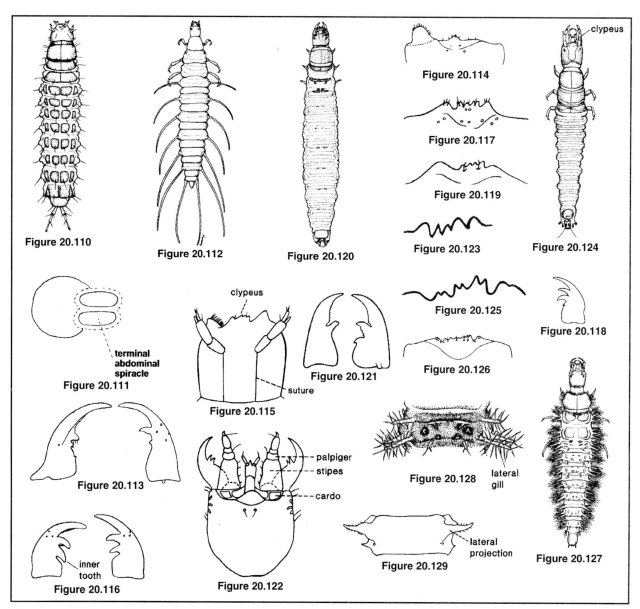

Figure 20.110. *Helophorus* sp. (Helophoridae) larva, dorsal aspect.

Figure 20.111. Hydrophilidae larva, terminal abdominal spiracle.

Figure 20.112. *Berosus* sp. (Hydrophilidae) larva, dorsal aspect.

Figure 20.113. *Berosus* sp. (Hydrophilidae) larva, dorsal aspect of mandibles.

Figure 20.114. *Berosus* sp. (Hydrophilidae) larva, clypeus.

Figure 20.115. *Laccobius* sp. (Hydrophilidae) larva, dorsal aspect of head.

Figure 20.116. *Paracymus* sp. (Hydrophilidae) larva, dorsal aspect of mandibles (after Richmond 1920).

Figure 20.117. *Paracymus* sp. (Hydrophilidae) larva, clypeus (after Richmond 1920).

Figure 20.118. *Anacaena* sp. (Hydrophilidae) larva, dorsal aspect of right mandible (after Bertrand 1972).

Figuree 20.119. *Anacaena* sp. (Hydrophilidae) larva, clypeus (after Brigham 1982).

Figure 20.120. *Enochrus* sp. (Hydrophilidae) larva, dorsal aspect.

Figure 20.121. *Enochrus* sp. (Hydrophilidae) larva, dorsal aspect of mandibles.

Figure 20.122. *Ametor* sp. (Hydrophilidae) larva, ventral aspect of head (after Bertrans 1972).

Figure 20.123. *Ametor* sp. (Hydrophilidae) larva, clypeus (after Bertrand 1972)

Figure 20.124. *Helochares* sp. (Hydrophilidae) larva, dorsal aspect.

Figure 20.125. *Helochares* sp. (Hydrophilidae) larva, clypeus.

Figure 20.126. *Cymbiodyta* sp. (Hydrophilidae) larva, clypeus (after Richmond 1920).

Figure 20.127. *Derallus* sp. (Hydrophilidae) larva, dorsal aspect.

Figure 20.128. *Derallus* sp. (Hydrophilidae) larva, dorsal aspect of 2nd abdominal segment.

Figure 20.129. *Crenitus* sp. (Hydrophilidae) larva, dorsal aspect of 2nd abdominal segment (after Bertrand 1972).

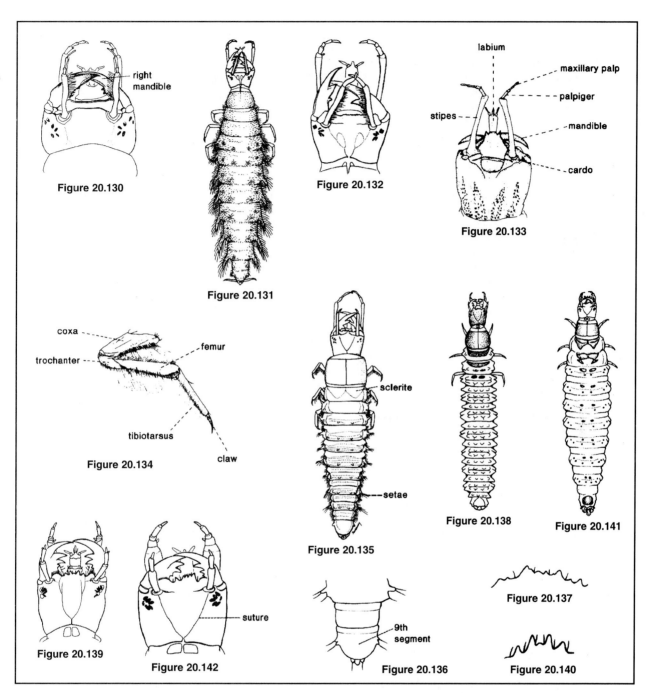

Figure 20.130. *Hydrophilus* sp. (Hydrophilidae) larva, dorsal aspect of head.

Figure 20.131. *Hydrochara* sp. (Hydrophilidae) larva, dorsal aspect.

Figure 20.132. *Hydrochara* sp. (Hydrophilidae) larva, dorsal aspect of head.

Figure 20.133. *Hydrochara* sp. (Hydrophilidae) larva, ventral aspect of head.

Figure 20.134. *Hydrochara* sp. (Hydrophilidae) larva, leg.

Figure 20.135. *Tropisternus* sp. (Hydrophilidae) larva, dorsal aspect.

Figure 20.136. *Tropisternus* sp. (Hydrophilidae) larva, dorsal aspect of last abdominal segments.

Figure 20.137. *Hydrobiomorpha* sp. (Hydrophilidae) larva, clypeus (after Bertrand 1972).

Figure 20.138. *Sperchopsis* sp. (Hydrophilidae) larva, dorsal aspect.

Figure 20.139. *Sperchopsis* sp. (Hydrophilidae) larva, dorsal aspect of head.

Figure 20.140. *Sperchopsis* sp. (Hydrophilidae) larva, clypeus.

Figure 20.141. *Hydrobius* sp. (Hydrophilidae) larva, dorsal aspect.

Figure 20.142. *Hydrobius* sp. (Hydrophilidae) larva, dorsal aspect of head.

Adults

1. Pronotum with 5 longitudinal grooves (Fig. 20.143) or expanded anteriorly (as in Fig. 20.144) 2
1'. Pronotum not as above 3
2(1). Pronotum with 5 longitudinal grooves (Fig. 20.143) .... **HELOPHORIDAE** .... *Helophorus*
2'. Pronotum lacking grooves, expanded anteriorly at middle to cover much of head
   (Fig. 20.144) .... **EPIMETOPIDAE** .... *Epimetopus*
3(1'). Eyes protruding, pronotum distinctly narrower than elytra bases (Fig. 20.145);
   scutellum small; antenna with 3 or fewer segments before cupule .... **HYDROCHIDAE** .... *Hydrochus*
3'. Eyes prominent or not; pronotum not distinctly narrower than elytra (some
   Berosini may have narrow pronotum but scutellum is long and triangular
   as in Fig. 20.149); antenna often having 5 segments before cupule .... **HYDROPHILIDAE** 4
4(3'). First segment of hind tarsi shorter than 2nd; antenna shorter or only about same
   length as maxillary palp; last glabrous segment of antenna asymmetrical and
   often cuplike, embracing 1st segment of pubescent, triarticulate club; 2nd
   segment of maxillary palpi similar in thickness to 3rd or 4th .... Hydrophilinae 5
4'. First segment of hind tarsi longer than 2nd; antennae usually longer than
   maxillary palpi, last glabrous segment of antennae obconic (form of a reversed
   cone) and fitted tightly against pubescent club; 2nd segment of maxillary palpi
   noticeably thicker than 3rd or 4th .... terrestrial subfamily Sphaeridiinae not considered here
5(4). Mesosternum and metasternum with a continuous median longitudinal keel which
   is prolonged posteriorly into a spine between the hind coxae (Fig. 20.146) 6
5'. Mesosternum and metasternum without a continuous spine (Fig. 20.160) 10
6(5). Size smaller, length usually less than 16 mm 7
6'. Size larger, length usually more than 25 mm 9
7(6). Length rarely exceeding 12 mm; prosternum sulcate; metasternal spine long (Fig. 20.146) .... *Tropisternus*
7'. Length usually greater than 13 mm; prosternum carinate; metasternal spine short
   (Fig. 20.150) 8
8(7'). Clypeus broadly emarginate on its anterior border (Fig. 20.148), exposing
   articulation of the labrum; adult as in Figure 20.147 .... *Hydrobiomorpha*
8'. Clypeus truncate; adult as in Figures 20.149-20.151 .... *Hydrochara*
9(6'). Prosternal process closed in front, hood-shaped (Fig. 20.153) .... *Hydrophilus*
9'. Prosternal process not closed in front, bifurcate (Fig. 20.156) .... *Dibolocelus*
10(5'). Basal 2 abdominal sternites with a common excavation on each side that is covered
   with a bilobed ciliated plate .... *Chaetarthria*
10'. Basal abdominal sternites without such an excavation on each side and without a bilobed plate 11
11(10'). Middle and hind tibiae fringed on the inner side with long swimming hairs;
   pronotum not continuous in outline with elytra (Fig. 20.155) 12
11'. Middle and hind tibiae without swimming hairs; pronotum and elytra continuous
   in outline (Fig. 20.160) 13
12(11). Smaller species, 1.5-2 mm long; black, very convex beetles (Fig. 20.154) .... *Derallus*
12'. Larger species, 2-6 mm long, usually greater than 3 mm long; brown to yellowish
   beetles, usually with a highly patterned elytra and dark pronotal markings (Fig. 20.155) .... *Berosus*
13(11'). Elytron without striae, including the sutural stria (Fig. 20.157) .... *Laccobius*
13'. Elytron with at least a sutural stria (Fig. 20.158) 14
14(13'). Maxillary palp short and stout, about the same length as antenna, last segment as
   long as or longer than penultimate segment (Fig. 20.159) 15
14'. Maxillary palp slender, much longer than antenna; last segment usually shorter
   than penultimate segment (Fig. 20.163) 21
15(14). Larger species, at least 4.5 mm long 16
15'. Smaller species, not more than 3 mm long 18
16(15). Lateral margins of elytron and pronotum smooth .... *Hydrobius*
16'. Lateral margins of elytron and pronotum serrate or emarginate (Fig. 20.158) 17
17(16'). Body shape strongly convex, clypeus deeply emarginate; eastern North America;
   adult as in Figures 20.158-20.159 .... *Sperchopsis*
17'. Body oval; clypeus shallowly emarginate; western North America .... *Ametor*
18(15'). Hind tarsal segments united longer than tibia; elytra narrowed posteriorly almost
   from the humerus .... *Crenitulus*
18'. Hind tarsus at most no longer than tibia; body shape short and convex 19
19(18'). Prosternum longitudinally carinate (Fig. 20.161); hind femur at most sparsely
   pubescent (Fig. 20.160) .... *Paracymus*
19'. Prosternum not carinate; hind femur densely pubescent 20

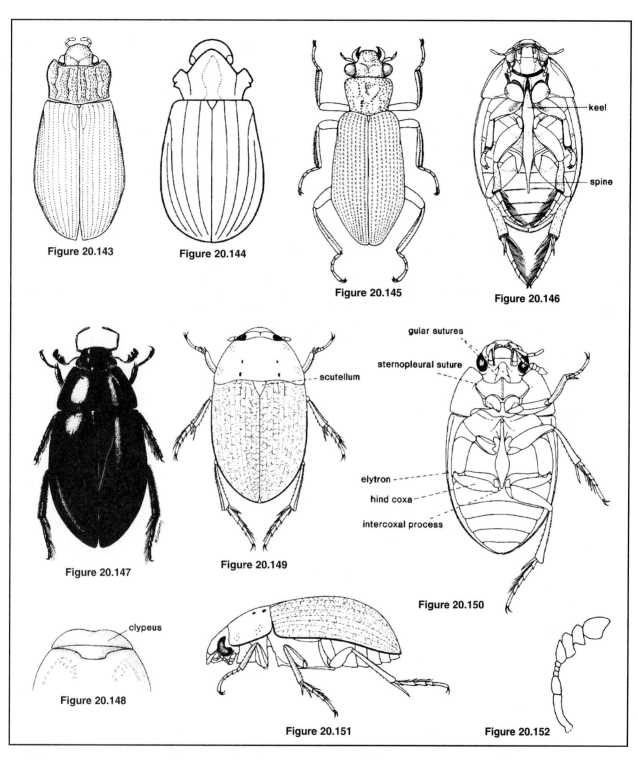

Figure 20.143. *Helophorus* sp. (Helophoridae) adult, dorsal aspect.

Figure 20.144. *Epimetopus* sp. (Epimetopidae) adult, dorsal aspect.

Figure 20.145. *Hydrochus* sp. (Hydrochidae) adult, dorsal aspect.

Figure 20.146. *Tropisternus* sp. (Hydrophilidae) adult, ventral aspect.

Figure 20.147. *Hydrobiomorpha* sp. (Hydrophilidae) adult, dorsal aspect.

Figure 20.148. *Hydrobiomorpha* sp. (Hydrophilidae) adult, clypeus.

Figure 20.149. *Hydrochara* sp. (Hydrophildae) adult, dorsal aspect.

Figure 20.150. *Hydrochara* sp. (Hydrophilidae) adult, ventral aspect.

Figure 20.151. *Hydrochara* sp. (Hydrophilidae) adult, lateral aspect.

Figure 20.152. *Hydrochara* sp. (Hydrophilidae) adult, antenna.

20(19').	Mesosternum simple, not carinate or with a small transverse protuberance before middle coxae	*Crenitis*
20'.	Mesosternum with a prominent angularly elevated or dentiform protuberance before middle coxae (Fig. 20.162)	*Anacaena*
21(14').	All tarsi 5-segmented, basal segment may be very small and difficult to see	22
21'.	Middle and hind tarsi 4-segmented	24
22(21).	Curved pseudobasal segment (actual basal segment very small) of maxillary palp convex anteriorly (Fig. 20.163)	*Enochrus*
22'.	Curved pseudobasal segment of maxillary palp convex posteriorly	23
23(22').	Labrum visible	*Helochares*
23'.	Labrum concealed beneath clypeus that projects laterally in front of eyes (Fig. 20.164)	*Helobata*
24(21').	Mesosternum with a compressed conical process (Fig. 20.165); maxillary palp long and slender	*Helocombus*
24'.	Mesosternal carina transverse or elevated at the middle forming a pyramidal or dentiform protuberance (Fig. 20.166); maxillary palp shorter and stouter	*Cymbiodyta*

Hydraenidae (Minute Moss Beetles)

As both adults and larvae, the Hydraenidae are small (1-2 mm), inconspicuous, and semiaquatic. More than 80 species presently are recognized for the United States and Canada, most of which have been described in the past few years. Most adults and larvae occur along a narrow band from the waterline to less than 8 cm above it. They also may be found as much as 2-3 cm deep in the littoral zone substrata. Hydraenids may commonly occur in lotic systems but, because of their habitat, are not often collected by usual stream sampling methods.

Eggs are deposited singly on wood and stones just beneath the waterline. Newly hatched larvae leave the water once the cuticle has hardened and begin feeding on protozoans and periphyton from leaves and twigs trapped above the water's edge. After larvae mature, in one year or less, pupal chambers are constructed in sand along the littoral zone. Little is known of adult life histories, although they appear to be long lived, feeding on bacteria and fungus from decaying leaves.

Hydraenidae

Larvae

1.	Setae on clypeus not placed at anterior margin and 2 median ones distant from each other; lacinia mobilis narrower; inner lobe of maxillae not distinctly divided apically; cerci nearly contiguous proximally and divergent (Fig. 20.167)	*Ochthebius*
1'.	Setae on clypeus placed at anterior margin and equidistant; lacinia mobilis broader; inner maxillary lobe distinctly divided apically; cerci widely separated proximally and nearly parallel	2
2(1').	Third antennal segment without inner swellings, segment 2 with a single antennal appendage; a pair of pectinate setae at anterior margin of labrum	*Hydraena*
2'.	Third antennal segment with an inner swelling, segment 2 with 2 slender antennal appendages; no pectinate setae at anterior labral margin; inner maxillary lobe strongly divided	*Limnebius*

Adults

1.	Second segment of hind tarsi elongate; longer than the 3rd segment; pronotum as broad basally as base of elytra, smooth, not coarsely punctate or sculptured, sides evenly arcuate; black or rufescent beetles, about 1 mm long ....Limnebiinae	*Limnebius*
1'.	Second segment of hind tarsi short, about as long as 3rd; pronotum at base slightly or decidedly narrower than base of elytra, surface uneven, coarsely punctate or with a transparent lateral border, sides sinuate or irregular; black, reddish, or aenescent (brass-colored) beetles, 1-2 mm long ....Hydraeninae	2
2(1').	Maxillary palpi very long, much longer than antennae (antenna as in Fig. 20.170); pronotum coarsely, closely punctate, sides without a transparent border	*Hydraena*
2'.	Maxillary palpi shorter than antennae; pronotum variously sculptured, often with deep fossae and grooves, almost always with a transparent border in at least basal half; adult as in Figure 20.171	*Ochthebius*

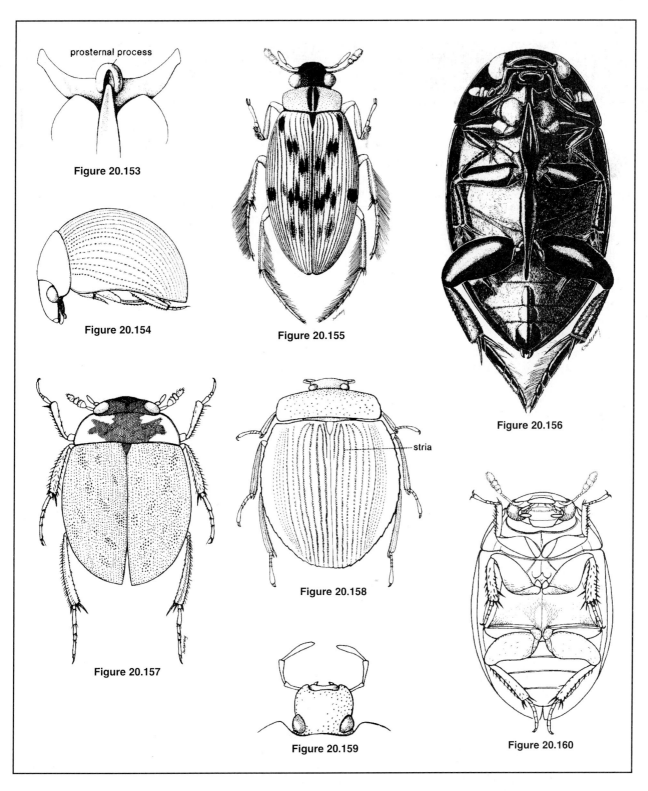

Figure 20.153. *Hydrophilus* sp. (Hydrophilidae) adult, prosternal process.

Figure 20.154. *Deralus* sp. (Hydrophilidae) adult, lateral aspect.

Figure 20.155. *Berosus* sp. (Hydrophilidae) adult, dorsal aspect.

Figure 20.156. *Dibolocelus* sp. (Hydrophildae) adult, ventral aspect.

Figure 20.157. *Laccobius* sp. (Hydrophilidae) adult, dorsal aspect.

Figure 20.158. *Sperchopsis* sp. (Hydrophilidae) adult, dorsal aspect.

Figure 20.159. *Sperchopsis* sp. (Hydrophilidae) adult, dorsal aspect of head.

Figure 20.160. *Paracymus* sp. (Hydrophilidae) adult, ventral aspect.

Staphylinidae (Rove Beetles)

The abbreviated elytra (Figs. 20.172-20.175) of most of these beetles (and the Melyridae) differentiate them from all other aquatic Coleoptera. Staphylinidae are extremely common in damp habitats, including the littoral zone of fresh and saline waters; however, few are truly aquatic. For example, in fresh water, genera such as *Bledius, Stenus,* and *Carpelimus* are abundant in streamside and lakeside environments, but none live beneath the water surface. In contrast, Staphylinidae are the most common Coleoptera in North American marine habitats with at least 12 genera occupying intertidal habitats. A host of other genera frequent decaying wrack (or marine vegetation) on supertidal beaches but are only rarely submerged by excessively high tides and cannot be considered aquatic. Many of the intertidal species inhabit air spaces in crevices in rocks. This group includes the genera *Diaulota, Diglotta, Bryothinusa, Amblopusa,* and *Liparocephalus.* Other species occupy burrows they excavate in sandy beaches (*Thinopinus, Bledius, Micralymma, Diglotta, Pontamalota, Thinusa*), and a few are found in *Salicornia* (glasswort) flats (*Bledius, Carpelimus, Thinobius*).

Biology of most aquatic Staphylinidae is poorly known. Most species are likely to be predaceous, but little precise information is available. Larval staphylinids (Fig. 20.9) generally resemble the adults in body shape and, at least in the case of known marine species, occupy similar niches.

Staphylinidae is one of the largest and most poorly known families of Coleoptera. The subfamily Aleocharinae, which includes most of the marine forms, is in drastic need of comprehensive study. The following key includes only adults of taxa believed to have aquatic members. Larvae, which are incompletely studied, are not included here. The large number of semiaquatic and littoral genera are best identified with general taxonomic works (Arnett 1960; Moore and Legner 1974). Maritime species are treated by Moore and Legner (1976).

Staphylinidae

Adults

1.	Antennae inserted on side margins of head, between eyes and mandibles (Figs. 20.172-20.173)	2
1'.	Antennae inserted on vertex between inner margins of eyes; adult as in Figures 20.174-20.175	8
2(1).	Ocelli absent	4
2'.	Two ocelli present between eyes	Omaliinae ... 3
3(2').	Elytra much longer than wide, covering all but 3-4 terminal abdominal segments	*Psephidonus*
3'.	Elytra about as long as wide, exposing 5-6 abdominal segments	*Micralymma*
4(2).	Abdomen with 7 visible sternites	5
4'.	Abdomen with 6 visible sternites (1st and 2nd sternites coalesced); adult as in Figure 20.173	*Thinopinus*
5(4).	Anterior tibiae bearing 2 rows of stiff, projecting spines (Fig. 20.176)	6
5'.	Anterior tibiae pubescent, lacking large, stiff spines	7
6(5).	Tarsal segmentation 4-4-4	*Bledius*
6'.	Tarsal segmentation 3-3-3	*Psamathobledius, Microbledius*
7(5').	Scutellum concealed; tarsi with 3 segments	*Carpelimus*
7'.	Scutellum visible; tarsi with 2 segments	*Thinobius*
8(1').	Hind coxae expanded laterally (Fig. 20.177)	Aleocharinae ... 9
8'.	Hind coxae conical, not extending laterally (Fig. 20.178)	Steninae ... *Stenus*
9(8).	Maxilla with both lacinia and galea sclerotized, hooked and pointed apically, and lacking brushes of setae (as in Fig. 20.179)	*Bryothinusa*
9'.	Maxilla with lacinia (and usually the galea) membranous apically and bearing brushes of setae (Fig. 20.180)	10
10(9').	Hind tarsi with 5 segments	11
10'.	Hind tarsi with 4 segments	*Diglotta*
11(10).	Middle tarsi with 4 segments	Bolitocharini ... 12
11'.	Middle tarsi with 5 segments	*Pontamalota*
12(11).	Anterior and middle tibiae set with stiff spines	13
12'.	Anterior and middle tibiae pubescent or glabrous, lacking stiff spines	14
13(12).	Abdominal tergite 5 with transverse impression near base; adult as in Figure 20.172	*Phytosus*
13'.	Abdominal tergite 5 without impression near base; adult as in Figure 20.174	*Thinusa*
14(12').	Abdomen with at least 3 tergites impressed at base	15
14'.	Abdomen without impressed tergites (Fig. 20.174)	*Liparocephalus*
15(14).	Abdomen with 5th tergite impressed at base	*Amblopusa*
15'.	Abdomen with 5th tergite not impressed at base	*Diaulota*

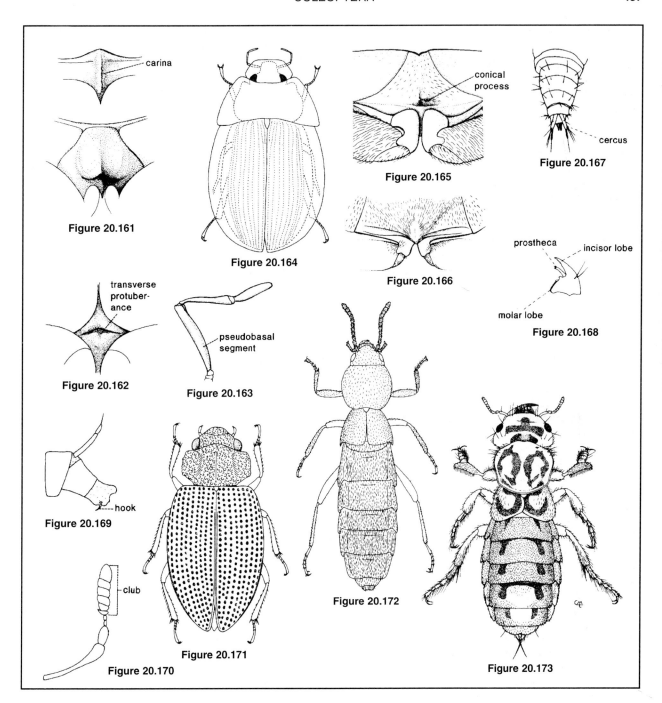

Figure 20.161. *Paracymus* sp. (Hydrophilidae) adult, pro- and mesosternum.

Figure 20.162. *Anacaena* sp. (Hydrophilidae) adult, mesosternum.

Figure 20.163. *Enochrus* sp. (Hydrophilidae) adult, maxillary palp.

Figure 20.164. *Helobata* sp. (Hydrophildae) adult, dorsal aspect.

Figure 20.165. *Helocombus* sp. (Hydrophilidae) adult, mesosternum.

Figure 20.166. *Cymbiodyta* sp. (Hydrophilidae) adult, mesosternum.

Figure 20.167. *Ochthebius* sp. (Hydraenidae) larva, dorsal aspect of last abdominal segments.

Figure 20.168. *Ochthebius* sp. (Hydraenidae) larva, mandible.

Figure 20.169. *Limnebius* sp. (Hydraenidae) larva, lateral view of last abdominal segments.

Figure 20.170. *Hydraena* sp. (Hydraenidae) adult, antenna.

Figure 20.171. *Ochthebius* sp. (Hydraenidae) adult, dorsal aspect.

Figure 20.172. *Pontamalota* sp. (Staphylinidae) adult, dorsal aspect.

Figure 20.173. *Thinopinus* sp. (Staphylinidae) adult, dorsal aspect.

Scirtidae (Marsh Beetles)

The Scirtidae (=Helodidae) are more common in other parts of the world, but at least 30 species occur in North America. In general, the taxonomy and ecology of the group are poorly known even though the aquatic larvae are easily recognized by their long, many segmented antennae.

Larvae go through 4-8 instars and mature within one year, grazing on rotting leaves and decaying vegetation. They tend to remain just beneath the surface film and obtain oxygen from the air through a modified spiracle on segment 8. The retractile trachael gills may provide some auxiliary oxygen supply when the larvae are submerged.

Pupae of some *Cyphon* and *Scirtes* hang freely from submerged objects, but most species leave the water to construct pupal chambers in wood, mud, moss, or sand. The terrestrial adults are short lived, and it is not known if they feed. Little is known of mating or egg deposition.

Scirtidae

Larvae[1]

1.	Head with 3 ocelli on each side; anterior margin of hypopharynx with a central cone bearing 1 pair of flat, usually serrate spines (Fig. 20.181)	***Elodes***
1'.	Head with 1-2 ocelli on each side; cone of hypopharynx with 2 pairs of spines (Fig. 20.182)	2
2(1').	Abdominal segments 3-6 with a regular series of short, flattened, differentiated setae along lateral margin	4
2'.	Sides of abdominal segments with only scattered, thin setae, like those of dorsum (Fig. 20.183)	3
3(2').	Terminal segment of maxillary palpi short (Fig. 20.184); head with only scattered, long setae, no spines along margins	***Cyphon***
3'.	Terminal segment of maxillary palpi very large (Fig. 20.185); setae reduced on head but with marginal spines	***Microcara***
4(2).	Anterior angle of labrum bent under, inner margin projecting from under transverse front margin (Fig. 20.186); adult as in Figure 20.187	***Prionocyphon***
4'.	Anterior angle of labrum on same plane as rest, labrum thus simply emarginate (Fig. 20.188)	***Scirtes***

Adults

1.	Hind femur similar to front and middle femora, not greatly enlarged for jumping	2
1'.	Hind femur greatly enlarged, broadened, modified for jumping (Fig. 20.189)	6
2(1).	Meso- and metasternal processes in contact between middle coxae, separating them	3
2'.	Mesosternal process short, narrow, not contacting metasternum; middle coxae contiguous in apical half	5
3(2).	First antennal segment large, fully twice as broad as any of those following, expanded anteriorly; 2nd segment arising from posterior apical angle of 1st and from under a slight margin (Fig. 20.190)	***Prionocyphon***
3'.	Antennae not as above (Fig. 20.193)	4
4(3').	Labial palpi with 3rd segment arising from side of 2nd (as in Fig. 20.191)	***Microcara***
4'.	Labial palpi with 3rd segment arising from tip of 2nd	***Cyphon***
5(2').	First segment of hind tarsus flattened above, finely margined laterally, 2nd with part of hind margin prolonged, concealing basal portion of 3rd; labial palp with 3rd segment arising from side of 2nd near the midpoint (as in Fig. 20.191)	***Elodes***
5'.	First segment of hind tarsus rounded, not margined laterally, 2nd not produced posteriorly, not concealing base of 3rd; labial palp with 3rd segment arising from apex of 2nd	***Sarabandus***
6(1').	Hind coxae meeting along full length of midline (Fig. 20.189)	***Scirtes***
6'.	Hind coxae touching each other only anteriorly (Fig. 20.192)	***Ora***

---

[1] The larvae of *Ora* and *Sarabandus* are unknown.

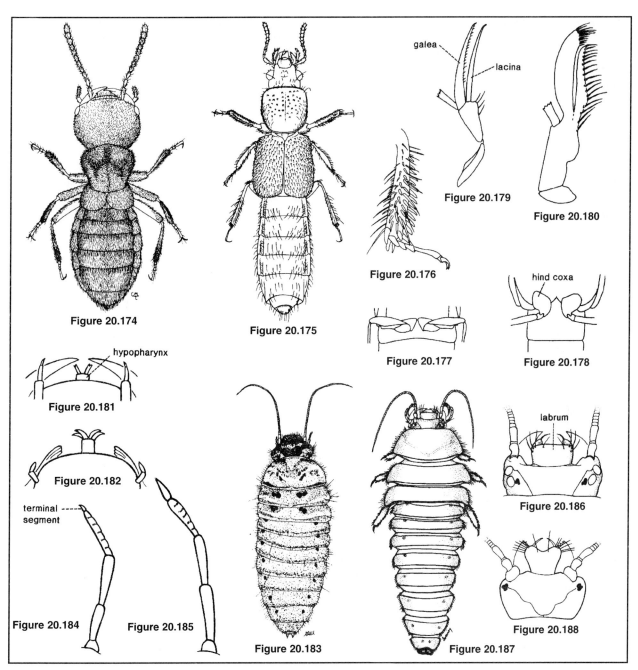

Figure 20.174. Liparocephalus sp. (Staphylinidae) adult, dorsal aspect.

Figure 20.175. Bledius sp. (Staphylinidae) adult, dorsal aspect.

Figure 20.176. Bledius sp. (Staphylinidae) adult, anterior tibia.

Figure 20.177. Aleochara sp. (Staphylinidae) adult (not in key), hind coxae and legs.

Figure 20.178. Stenus sp. (Staphylinidae) adult, hind coxae and legs.

Figure 20.179. Myllaena sp. (Staphylinidae) adult (not in key), maxilla.

Figure 20.180. Liparocephalus sp. (Staphylinidae) adult, maxilla.

Figure 20.181. Elodes sp. (Scirtidae) larva, dorsal aspect of anterior portion of head, (after Hilsenhoff 1981).

Figure 20.182. Microcara sp. (Scirtidae) larva, dorsal aspect of anterior portion of head (after Bertrand 1972).

Figure 20.183. Elodes sp. (Scirtidae) larva, dorsal aspect.

Figure 20.184. Cyphon sp. (Scirtidae) larva, maxillary palp.

Figure 20.185. Microcara sp. (Scirtidae) larva, maxillary palp.

Figure 20.186. Prionocyphon sp. (Scirtidae) larva, dorsal aspect of head.

Figure 20.187. Prionocyphon sp. (Scirtidae) larva, dorsal aspect.

Figure 20.188. Scirtes sp. (Scirtidae) larva, dorsal aspect of head.

Psephenidae (Water Pennies)

For convenience, we take a conservative approach to the systematics of Psephenidae, placing the North American genera and species in one family, although they have been placed variously in Psephenidae, Dascillidae, and Eubriidae. Morphologically there may be good reason to separate the false water pennies (Eubriinae), which possess a ventral operculum, from those without it (the genera *Eubrianax* and *Psephenus*; true water pennies).

The life history of only *Psephenus herricki* has been studied in detail. The bright yellow eggs laid in masses of several thousand can be found on submerged and emergent objects in the swifter parts of riffles. Larval growth occurs in the warmer months and maturity may not be reached until the second year. Mature larvae either leave the water in early summer to construct pupal chambers in soft, moist soils, or they pupate underwater in air-filled chambers (some western species). In both cases, pupation occurs in the last larval skin. The adult stage is very short with little to no feeding.

With the exception of the genus *Psephenus*, information on psephenids is scant. Adults inhabit the moist undersurfaces of logs and rocks overhanging stream habitats. Presumably, these individuals are near oviposition sites, but neither oviposition nor eggs have been observed for most species. Eubriinae apparently do not exhibit the active "play" behavior associated with *Psephenus* mating.

Larvae exhibit positive rheotaxis and thigmokinesis and negative phototaxis. Presumably, all genera are scrapers utilizing similar food sources, primarily periphyton. As with *Psephenus*, larval *Ectopria* appear to be gregarious. Larvae of the genera *Acneus* and *Dicranopselaphus* are extremely rare and usually have been taken singly. Mature larvae undoubtedly leave the water or are stranded by receding water prior to pupation as in *Psephenus*. Larvae previously identified as *Alabameubria* apparently are immatures of *Dicranopselaphus* and *Alabameubria* should be considered a synonym (Barr and Spangler 1995).

Psephenidae

Larvae

1. Abdominal segment 9 with a ventral operculum closing a caudal chamber (Fig. 20.195) containing 3 tufts of retractile filamentous gills; without gills on other parts of abdomen; expanded lateral portions of abdominal segments separated (Figs. 20.195, 20.197, 20.199) 2
1'. Abdominal segment 9 without ventral operculum; with pairs of ventral tufts of filamentous gills on 4-5 abdominal segments; expanded lateral portions of abdominal segments fitting tightly together at margin (Fig. 20.201) 4
2(1). Apex of abdominal segment 9 with a distinct median notch (Figs. 20.194-20.197).................................3
2'. Apex of abdominal segment 9 truncate or arcuate, without a median notch (Fig. 20.198) ....*Ectopria*
3(2). Notch on apex of abdominal segment 9 very shallow, apices broad and flat; lateral (pleural) extensions of thoracic and abdominal segments 1-7 broad and flattened (Figs. 20.194-20.196) ....*Acneus*
3'. Notch on apex of abdominal segment 9 very deep (at least in specimens from the United States, see also Fig. 20.199), about half length of segment, apices acute; pleural extensions of thoracic and abdominal segments very slender and curved posteriorly (Fig. 20.197) ....*Dicranopselaphus*
4(1'). Abdominal segment 8 with lateral expansions; abdomen with 4 pairs of gills (Fig. 20.200) ....*Eubrianax*
4'. Abdominal segment 8 without lateral expansions; abdomen with 5 pairs of gills (Fig. 20.201) ......*Psephenus*

Adults

1. Posterior margin of pronotum crenulate or finely beaded; males with at least the anterior claw on each tarsus forked at apex (Fig. 20.205); adults not aquatic 2
1'. Posterior margin of pronotum smooth (Fig. 20.209) 4
2(1). Prosternum narrow, depressed between coxae; antenna with 3rd segment at least as long as either first 2 segments or next 3 segments combined; male with flabellate antenna (Fig. 20.202) and tarsal claws toothed at base; female larger than male, with serrate antenna and tarsal claws not toothed at base ....*Acneus*
2'. Prosternum of moderate width, not depressed between coxae; tarsal claws of both sexes each with basal tooth (Fig. 20.203); antenna of male not flabellate (Fig. 20.206) 3
3(2'). Tibia sinuate, length approximately equal to length of tarsus, femur slender, narrowing gradually toward apices (Fig. 20.207); tarsus slender, 5th segment 3.8 times as long as its maximum width, 4th segment smaller than 3rd and not prolonged beneath 5th (Figs. 20.207-20.208) ....*Ectopria*
3'. Tibia straight, length approximately 1.25 times length of tarsus, femur stout, narrow toward apex, but much less than in *Ectopria* (Fig. 20.204); tarsus slightly dilated, 2nd, 3rd, and 4th segments feebly emarginate (Fig. 20.205), the 4th slightly prolonged beneath 5th (Fig. 20.204) ....*Dicranopselaphus*

4(1').	Head usually hidden beneath broadly expanded pronotum (Fig. 20.209); base of claws with a membranous appendage nearly reaching to tip of claw; antenna of male pectinate (Fig. 20.211); that of female serrate (Fig. 20.209)	*Eubrianax*
4'.	Head visible from above (Fig. 20.212); base of claws without membranous appendage; antenna of female moniliform, that of male subserrate to serrate	*Psephenus*

Ptilodactylidae (Toed-Winged Beetles)

At present we are considering only three North American genera, each with a single species, to be aquatic. The one species of the western *Areopidius* and one of the central *Odontonyx* most likely are aquatic as well, but descriptions of the larvae have not been published. Adults of these genera can be identified using Arnett (1960) under the family Dascillidae. It is likely that *Stenocolus* will be moved to its own family. Ptilodactylids are widespread but only locally common.

Larvae of *Anchytarsus* occur in small, cool, clear streams, often along with Elmidae and Dryopidae. The number of instars is unknown but may be as many as 10, and the life cycle may take up to three years. Mature larvae leave the water to construct puparia in leaf litter where, on rare occasions, adults have been found along streams. Little is known of adult ecology, mating, or egg deposition.

Ptilodactylidae[1]

Larvae

1.	Abdominal segments 1-7 each with 2 ventral tufts of filamentous gills (Fig. 20.213); submentum not divided; abdominal segment 9 without prehensile appendages bearing hooks	*Stenocolus*
1'.	Abdominal segments 1-7 without gill tufts; submentum divided longitudinally into 3 parts; anal region of abdominal segment 9 with 2 curved prehensile appendages covered with short spines (Fig. 20.214)	2
2(1').	Abdominal segment 9 with numerous fingerlike anal gills (Fig. 20.214), apex without projection	*Anchytarsus*
2'.	Abdominal segment 9 with 3 median anal gills and 1 gill lateral to each prehensile appendage; dorsal, flattened apex of segment 9 with small raised projection	*Anchycteis*

Adults

1.	Mandibles prominent, acutely margined above, rectangularly flexed at tip; head not retracted, moderately deflexed; 14-22 mm long	*Stenocolus*
1'.	Mandibles not prominent, arcuate at tip, not acutely margined above; head strongly deflexed; less than 12 mm long	2
2(1').	Antennae serrate in female, pectinate in male (Fig. 20.216); middle coxae twice as widely separated as anterior coxae; margin of pronotum obtusely rounded laterally (Fig. 20.215); 10 mm long	*Anchycteis*
2'.	Antennae slender; middle coxae no more widely separated than anterior coxae; pronotum obtusely margined laterally; 5-6 mm long	*Anchytarsus*

Limnichidae and Lutrochidae (Marsh-Loving Beetles)

Relationships among the families and genera are not well understood. Although we have included only *Lutrochus* (Lutrochidae) and *Throscinus* (Limnichidae) here, adults of *Limnichus* (Limnichidae) also may be found around streams.

Throscinus is marine, occurring along west coast beaches. The larvae are undescribed but probably occur with the adults as in most littoral zone taxa. The two eastern species of *Lutrochus* are widely distributed in small streams although they are only locally common. Larvae of *Lutrochus* are morphologically similar to elmids and often share the same habitat among rocks and larger organic debris of small, cool, fast-flowing streams. Adults cling to emergent objects at the same locations one finds the larvae. If disturbed, they either fly or enter the water carrying a film of air trapped in the hairs covering the body. It is assumed that adults are short-lived, and it is unknown if they feed. Little has been published on any portion of the life cycle of larvae and adults, mating, oviposition, or eggs.

---

[1] *Araeopidius* and *Odontonyx*, not included here, most likely are aquatic but habits have not been described.

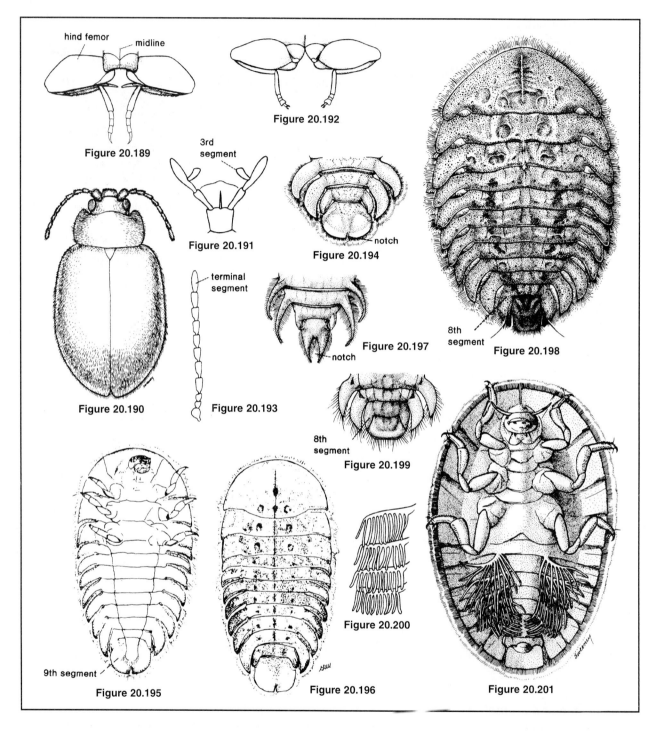

Figure 20.189. *Scirtes* sp. (Scirtidae) adult, hind legs.

Figure 20.190. *Prinocyphon* sp. (Scritidae) adult, dorsal aspect.

Figure 20.191. *Elodes* sp. (Scritidae) adult, labial palpi.

Figure 20.192. *Ora* sp. (Scirtidae) adult, hind legs.

Figure 20.193. *Cyphon* sp. (Scirtidae) adult, antenna.

Figure 20.194. *Acneus* sp. (Psephenidae) larva, dorsal aspect of last abdominal segment.

Figure 20.195. *Acneus* sp. (Psephenidae) larva, ventral aspect.

Figure 20.196. *Acneus* sp. (Psephenidae) larva, dorsal aspect.

Figure 20.197. *Dicranopselaphos* sp. (Psephenidae) larva, dorsal aspect of last abdominal segment.

Figure 20.198. *Ectopria* sp. (Psephenidae) larva, dorsal aspect.

Figure 20.199. *Dicranopselasphus* sp. (Psephenidae) larva, dorsal aspect of last abdominal segment.

Figure 20.200. *Eubrianix* sp. (Psephenidae) larva, abdominal gills.

Figure 20.201. *Psephenus* sp. (Psephenidae) larva, ventral aspect.

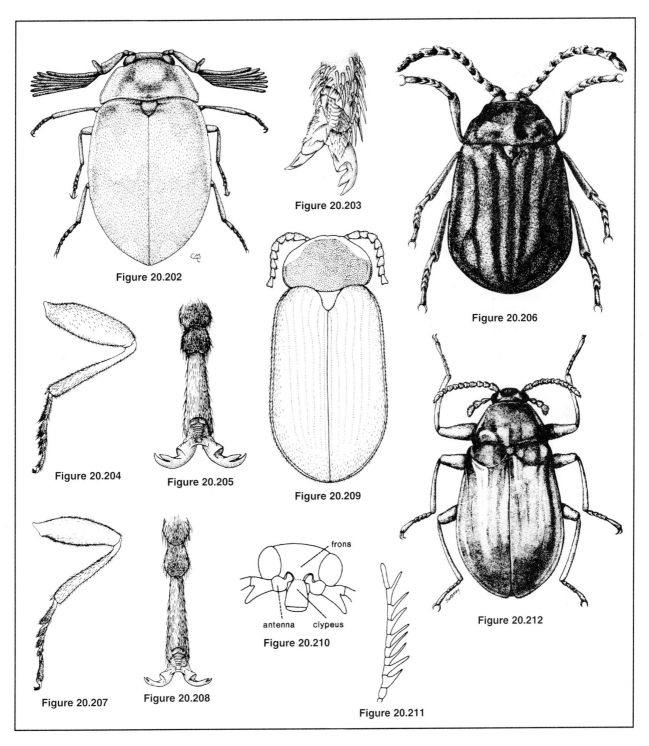

Figure 20.202. *Acneus* sp. (Psephenidae) adult, dorsal aspect.

Figure 20.203. *Dicranopselaphus* sp. (Psephenidae) adult, apex of tarsus.

Figure 20.204. *Dicranopselaphus* sp. (Psephenidae) adult, leg.

Figure 20.205. *Dicranopselaphus* sp. (Psephenidae) adult, 4th and 5th tarsal segments.

Figure 20.206. *Ectopria* sp. (Psephenidae) adult, dorsal aspect.

Figure 20.207. *Ectopria* sp. (Psephenidae) adult, leg.

Figure 20.208. *Ectopria* sp. (Psephenidae) adult, 4th and 5th tarsal segments.

Figure 20.209. *Eubrianix* sp. (Psephenidae) adult, dorsal aspect.

Figure 20.210. *Eubrianix* sp. (Psephenidae) adult, frontal aspect of head.

Figure 20.211. *Eubrianix* sp. (Psephenidae) adult, antenna.

Figure 20.212. *Psephenus* sp. (Psephenidae) adult, dorsal aspect.

Limnichidae and Lutrochidae

Larvae

1. Larvae of *Limnichus* and *Throscinus* (Limnichidae) have not been described. *Limnichus* may be terrestrial, and *Throscinus* most likely will be found with the adults on mudflats and beaches. It is not expected that either will be similar to *Lutrochus* (Lutrochidae). The following description should separate *Lutrochus,* which is truly aquatic.
1'. Abdomen with pleurites on abdominal segments 1-4; each undivided thoracic pleurite with erect hairs along medial margin; thoracic sternites membranous or absent; apex of abdominal segment 9 evenly rounded; each eyespot with 5 ocelli and with 1 ventral ocellus below base of antenna; head often retracted within body; larva as in Figure 20.218 ....***Lutrochus***

Adults

1. Pronotal hypomeron with a transverse or oblique ridge; body plump and convex; adult as in Figure 20.219; near streams ....Limnichinae 2
1'. Pronotal hypomeron without a ridge; body more elongate; on ocean mudflats or beaches ....Cephalobyrrhinae ....***Throscinus***
2(1). Small, less than 2 mm long; antenna slender, with 10 segments; 1st abdominal sternite with grooves for reception of folded hind legs (generally terrestrial) ....***Limnichus***
2'. Larger, at least 2.5 mm long; antenna with 11 segments, the 1st two enlarged and the remaining 9 subpectinate; 1st abdominal sternite without grooves for the reception of folded legs; adult as in Figure 20.219 ....***Lutrochus***

Dryopidae (Long-Toed Water Beetles)

Fourteen species of this primarily tropical family occur in America north of Mexico and represent a variety of associations with the aquatic environment. Only the genus *Helichus* is common and widespread. *Postelichus* is common in the Southwest. *Stygoparnus* is known only from Comal Springs in the Edwards Aquifer of Texas. The adults of *Helichus* and *Postelichus* are found in flowing waters along with the Elmidae, but the larvae appear to be subaquatic or terrestrial, a condition unique among the aquatic insects. *Pelonomus* adults and possibly larvae may be subaquatic to aquatic, having been recorded from small woodland pools with accumulated leaf litter. *Dryops* is rare, but both adults and larvae appear to be at least marginally aquatic. Beyond the above, little is known of life cycles of the North American Dryopidae.

Helichus and *Postelichus* larvae have been reared and described along with their pupae, but at this time cannot be separated from *Pelonomus* or *Stygoparnus* larvae. We have included some previously published drawings of *Helichus* (Figs. 20.220-20.223) that seem to be almost identical to known associated specimens of *Pelonomus*.

Dryopidae

Larvae[1]

1. Tergites with anterior margins smooth ....***Stygoparnus, Dryops***
1'. Tergites (except pronotum) with numerous longitudinal carinae arising near each anterior margin; gular sutures obliterated; Figures 20.220-20.223 ....***Pelonomus, Helichus, Postelichus***

Adults

1. Eyes vestigial, antennae with 8 segments ....***Stygoparnus***
1'. Eyes normal 2
2(1'). Pronotum on each side with a conspicuous, complete sublateral longitudinal sulcus; pubescent (Fig. 20.224) ....***Dryops***
2'. Pronotum without such a sublateral sulcus (Fig. 20.230) 3
3(2'). Second segment of antenna not enlarged; antennae pubescent (Fig. 20.225); bases of antennae very close together; both 3rd and 4th segments of maxillary palp very elongate; without tomentum but extremely hairy; adult as in Figure 20.226 ....***Pelonomus***

---

[1] Separation of Dryopidae larvae remains difficult because so few specimens have been collected or examined in detail. *Postelichus* may be separated from at least *Helichus* in having numerous longitudinal carinae arising near both anterior and posterior margins of each tergite.

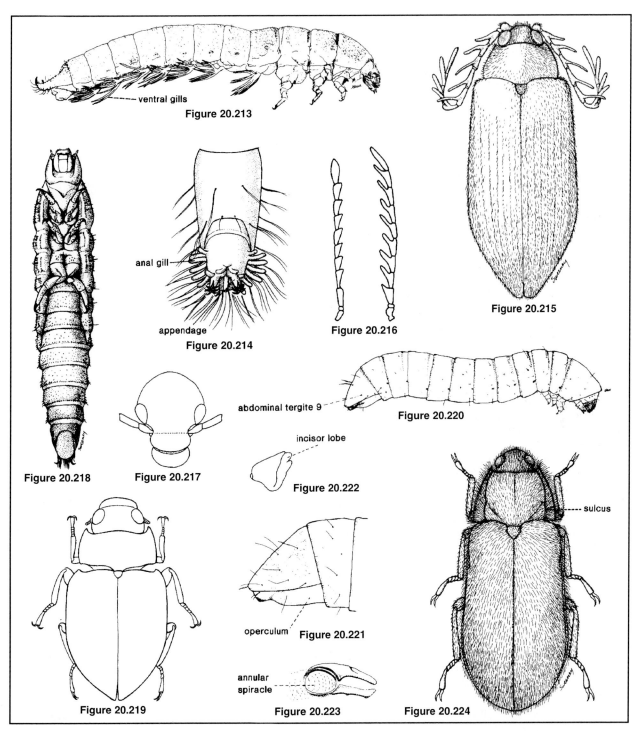

Figure 20.213. *Stenocolus* sp. (Ptilodactylidae) larva, lateral aspect.

Figure 20.214. *Anchydarsus* sp. (Ptilodactylidae) larva, 9th abdominal segment.

Figure 20.215. *Anchyctesis* sp. (Ptilodactylidae) adult, dorsal aspect.

Figure 20.216. *Anchycteis* sp. (Ptilodactylidae) adult, male and female antennae.

Figure 20.217. *Anchycteis* sp. (Ptilodactylidae) adult, frontal aspect of head.

Figure 20.218. *Lutrochus* sp. (Limnichidae) larva, ventral aspect.

Figure 20.219. *Lutrochus* sp. (Limnichidae) adult, dorsal aspect.

Figure 20.220. *Helichus* sp. (Dryopidae) larva, lateral aspect.

Figure 20.221. *Helichus* sp. (Dryopidae) larva, lateral aspect of abdominal apex.

Figure 20.222. *Helichus* sp. (Dryopidae) larva, mandible.

Figure 20.223. *Helichus* sp. (Dryopidae) larva, spiracle and closing apparatus.

Figure 20.224. *Dryops* sp. (Dryopidae) adult, dorsal aspect.

3'.	First, and, even more, 2nd segment of antenna enlarged and heavily sclerotized, forming a shield beneath which remaining segments may be retracted and protected (Fig. 20.227); bases of antennae widely separated; parts of body and legs with tomentum; Figures 20.228-20.230	4
4(3').	Pubescence of last abdominal sternite different from that of preceding sternites, often appearing bare	***Helichus***
4'.	All abdominal sternites similarly and densely pubescent	***Postelichus***

Elmidae (Riffle Beetles)

The nearly 100 species and 27 genera present in North America are typical inhabitants of the swifter portions of streams and rivers. Only the genera *Stenelmis, Optioservus,* and *Dubiraphia,* with a combined total of more than 50 species, are common and widespread. The remaining genera occur primarily in localized populations in eastern, central, western and southwestern mountain ranges. One species of *Xenelmis* has been recorded for Arizona and New Mexico, but it has yet to be worked into the North American keys.

The terminal gills of the larvae and the highly efficient plastron of the adults make the elmids among the most truly aquatic insects. The gills of the larvae, which are retractile and protected by an operculum, can be rhythmically expanded and contracted to increase ventilation when oxygen levels are low. The adult plastron is an extremely fine hydrofuge pile with hairs as dense as several million per mm^2. Although efficient, this adaptation limits adult distribution to habitats rich in oxygen, such as riffles.

Eggs are deposited singly or in small groups on submerged rocks, organic debris, and vegetation. Larvae undergo 6-8 instars, which may take three years or more. Terrestrial pupal chambers are constructed in moist soils, under rocks, or in rotting wood. Newly emerged adults undergo a short flight period, but, after entering the water, most are incapable of further aerial activity. Adult life spans are unknown, although many species may not reach maturity until the second year, and some individuals have been maintained in the laboratory for years. Adults occupy habitats similar to the larvae with which they often are collected.

Elmidae

Larvae

1.	Abdomen with pleura on first 8 segments (Fig. 20.232)	2
1'.	Abdomen with pleura on first 5-7 segments (Figs. 20.253, 20.256)	7
2(1).	Body rather broad, lateral margins expanded (Fig. 20.233)	3
2'.	Body slender, elongate, cylindrical or hemicylindrical (as in Fig. 20.243)	4
3(2).	With coarse, prominent spines along lateral margins (Fig. 20.232); dorsal surface ridged on each side; last segment rather square-sided and flat dorsally (Fig. 20.231); procoxal cavities open behind	***Lara***
3'.	Without marginal spines; body quite flattened and with rather smooth surface; testaceous to brown, somewhat translucent; procoxal cavities closed behind; larva as in Figure 20.233	***Phanocerus***
4(2').	Last abdominal segment very long and slender (at least 4 times as long as wide), operculum confined to posterior third of segment (Fig. 20.235)	***Dubiraphia***
4'.	Last abdominal segment not conspicuously long or slender (less than 4 times as long as wide), operculum not confined to apical third	5
5(4').	Head tuberculate, with suberect spines; anterior margin of head without a prominent frontal tooth on each side (Fig. 20.236; note spines on Figs. 20.246, 20.248); body subcylindrical, yellowish; often more than 8 mm long	***Nurpus***
5'.	Head without suberect spines, anterior margin with a prominent frontal tooth on each side (Fig. 20.248)	6
6(5').	Body cylindrical (Fig. 20.237); pleural sutures extend to basal half of abdominal segment 9; procoxal cavities closed behind	***Cylloepus***
6'.	Body hemicylindrical (Fig. 20.238); pleural sutures not extending onto abdominal segment 9; procoxal cavities open behind	***Rhizelmis***
7(1').	Prothorax with a posterior sternum so procoxal cavities are closed behind (Fig. 20.249)	8
7'.	Prothorax without posterior sternum; procoxal cavities open behind (Fig. 20.251)	17
8(7).	Posterolateral margins of abdominal segments 1-8 produced into spine-like processes; body rather robust (Figs. 20.239-20.240)	***Ancyronyx***
8'.	Margins of abdominal segments not produced into spines; body elongate	9
9(8').	Dorsum of all but last segment bearing spatulate tubercles or short spines arranged in about 10 conspicuous longitudinal or diagonal rows (Fig. 20.241); last segment with a middorsal longitudinal ridge and lateral margins bearing spatulate tubercles	***Heterelmis***

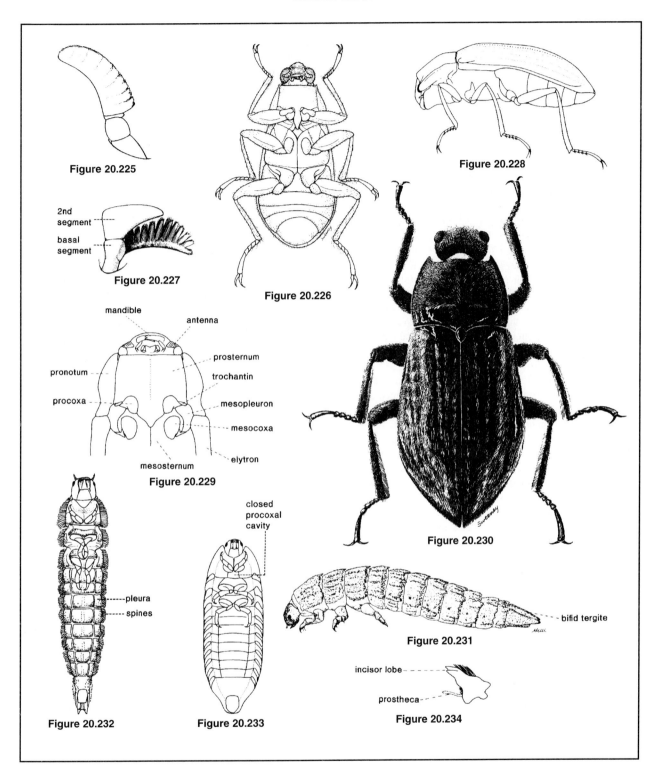

Figure 20.225. *Pelonomus* sp. (Dryopidae) adult, antenna.

Figure 20.226. *Pelonomus* sp. (Dryopidae) adult, ventral aspect.

Figure 20.227. *Helichus* sp. (Dryopidae) adult, antenna.

Figure 20.228. *Helichus* sp. (Dryopidae) adult, lateral aspect.

Figure 20.229. *Helichus* sp. (Dryopidae) adult, ventral aspect of head.

Figure 20.230. *Helichus* sp. (Dryopidae) adult, dorsal aspect.

Figure 20.231. *Lara* sp. (Elmidae) larva, lateral aspect.

Figure 20.232. *Lara* sp. (Elmidae) larva, ventral aspect.

Figure 20.233. *Phanocerus* sp. (Elimdae) larva, ventral aspect.

Figure 20.234. *Elmidae* larva, mandible.

9'.	Dorsum without such spiny tubercles, although there may be rows of small, flat tubercles	10
10(9').	Anterior margin of head on each side with a distinct frontal tooth (Figs. 20.246, 20.248)	13
10'.	Anterior margin of head without distinct frontal tooth (Fig. 20.244)	11
11(10').	Dorsum with relatively conspicuous, flattened tubercles often arranged in longitudinal rows (Fig. 20.243); abdominal tergites often with middorsal pale spots; last segment with a weak middorsal longitudinal ridge	12
11'.	Tubercles of dorsum inconspicuous, not arranged in longitudinal rows; without middorsal pale spots; last segment convex dorsally, without median ridge	*Neoelmis*
12(11).	Last abdominal segment conspicuously long and slender (3 times longer than wide; Fig. 20.242); middorsal spots widest near middle of each segment; dorsal tubercles not arranged in parallel longitudinal rows	*Hexacylloepus*
12'.	Last segment not unusually long or slender; middorsal spots widest near posterior of segments; dorsal tubercles partially arranged in parallel longitudinal rows (Fig. 20.243)	*Microcylloepus*
13(10).	Tergite of last abdominal segment with prominent median and sublateral, longitudinal, carinate ridges (Fig. 20.245)	*Neocylloepus*
13'.	Dorsum of last abdominal segment not carinate or prominently ridged	14
14(13').	Second segment of antenna more than twice as long as 1st (Fig. 20.246); prosternum with anterior suture obliterated; no suture extending from procoxal cavity to lateral margin of pronotum	*Ordobrevia*
14'.	Second segment of antenna less than twice as long as 1st; prosternum with anterior median suture; suture from procoxal cavity to lateral margin may or may not be visible	15
15(14').	Suture from procoxal cavity to lateral margin distinct (Fig. 20.247); body large and rather flattened	*Macrelmis*
15'.	Suture from procoxal cavity to lateral margin indistinct or absent (Fig. 20.249); body more convex and elongate, smaller	16
16(15').	Pleuron on each side of prothorax divided into pre- and postpleurites, division may be faint; Arizona and Southwest	*Huleechius*
16'.	Pleuron not divided (Fig. 20.249); from Texas and Oregon eastward	*Stenelmis*
17(7').	Postpleurite composed of 1 part (Fig. 20.251)	18
17'.	Postpleurite composed of 2 parts (Fig. 20.254)	20
18(17).	Body robust, broad, subtriangular in cross section; with spatulate spines along lateral margins (Fig. 20.251) and middorsal line (Fig. 20.250)	*Ampumixis*
18'.	Body long and slender, hemicylindrical; without prominent clusters of spines	19
19(18').	Abdominal segments 7 and 8 with no longitudinal sutures, segments 1-6 with pleurites; posterior border of nota and terga with tubercles having elongate setae well sclerotized	*Atractelmis*
19'.	Abdominal segments 7 and/or 8 divided by tergosternal sutures, pleurites on segments 1-6 or 1-7; posterior border of tubercles with setae not or poorly sclerotized	*Cleptelmis*
20(17').	Mesopleuron composed of 1 part (Fig. 20.255)	21
20'.	Mesopleuron composed of 2 parts (Fig. 20.257)	22
21(20).	Dorsum of each segment with median and sublateral humps (Fig. 20.252)	*Promoresia*
21'.	Dorsum without such humps (also see Figs. 20.253-20.255)	*Optioservus*
22(20').	First 5 or 6 abdominal segments with pleura; last segment with 2 long, acute, narrowly separated apical processes (Fig. 20.256)	*Macronychus*
22'.	First 7 abdominal segments with pleura	23
23(22').	Body long, slender, and hemicylindrical; apex of last segment rather deeply emarginate, the lateral angles produced and acute (Fig. 20.258)	*Zaitzevia*
23'.	Body usually less elongate, subtriangular in cross section; apex of last segment shallowly emarginate, lateral angles less acute (Fig. 20.261)	24
24(23').	Abdominal segments with middorsal humps that are especially prominent toward the rear (Fig. 20.259), each hump bearing conspicuous scale-like hairs; dorsum of each thoracic segment with 2 longitudinal dark spots on each side	*Gonielmis*
24'.	Abdominal segments without middorsal humps; thorax usually without dark markings	25
25(24').	Western; tubercles of dorsum relatively dense, separated by less than their own widths, crowded along posterior margins of segments; mesothorax with anterior portion of pleuron much smaller than posterior portion; mature larva 4-5 mm long; larva as in Figure 20.260	*Heterlimnius*
25'.	Eastern; tubercles of dorsum sparse, separated by more than their own widths except along middorsal line, marginal tubercles separated by their own widths; mesothorax with anterior portion of pleuron subequal to posterior portion; mature larva not over 3 mm long; larva as in Figure 20.261	*Oulimnius*

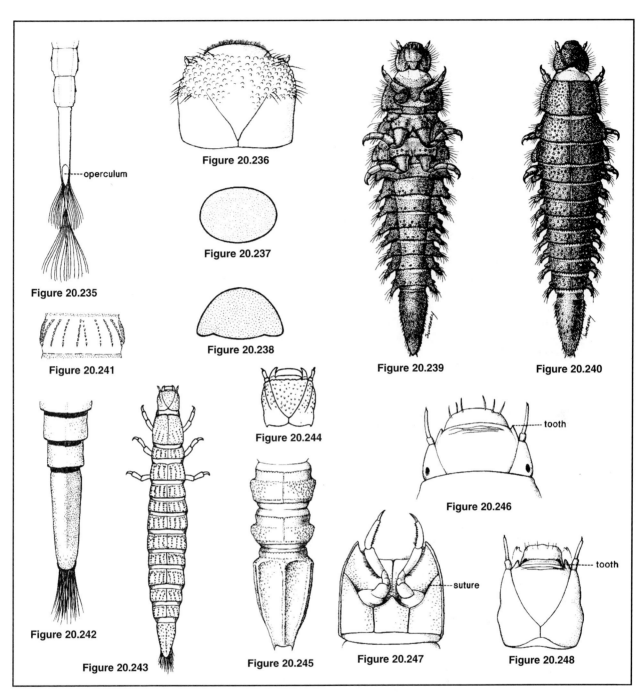

Figure 20.235. *Dubiraphia* sp. (Elmidae) larva, ventral aspect of last abdominal segments.

Figure 20.236. *Narpus* sp. (Elmidae) larva, dorsal aspect of head.

Figure 20.237. *Cylloepus* sp. (Elmidae) larva, shape of abdominal cross section.

Figure 20.238. *Rhizelmis* (Elmidae) larva, shape of abdominal cross section.

Figure 20.239. *Ancyronyx* sp. (Elmidae) larva, ventral aspect.

Figure 20.240. *Ancyronyx* sp. (Elmidae) larva, dorsal aspect.

Figure 20.241. *Heterelmis* sp. (Elmidae) larva, dorsal aspect of an abdominal segment.

Figure 20.242. *Hexacylloepus* sp. (Elmidae) larva, dorsal aspect of last abdominal segments.

Figure 20.243. *Microcylloepus* sp. (Elmidae) larva, dorsal aspect.

Figure 20.244. *Microcylloepus* sp. (Elmidae) larva, dorsal aspect of head.

Figure 20.245. *Neocylloepus* sp. (Elmidae) larva, dorsal aspect of last abdominal segments.

Figure 20.246. *Ordobrevia* sp. (Elmidae) larva, dorsal aspect of head.

Figure 20.247. *Macrelmis* sp. (Elmidae) larva, ventral aspect of prothorax segment.

Figure 20.248. *Stenelmis* sp. (Elmidae) larva, dorsal aspect of head.

Adults

1. Riparian, usually not underwater; agile fliers; rather soft-bodied; pubescent, but without tomentum; procoxae transverse and with trochantin exposed (Fig. 20.265) 2
1'. Aquatic; typically slow moving, clinging to submerged objects; rarely flying except at night; hard-bodied; with tomentum on various ventral parts; procoxae rounded and trochantin concealed 3
2(1). Less than 4 mm long; antennae clubbed (Fig. 20.263); pronotum with sublateral sulci (Fig. 20.264) .... *Phanocerus*
2'. More than 5 mm long; black, antennae not clubbed; pronotum without sublateral sulci (Fig. 20.266) .... *Lara*
3(1'). Hind coxae globular and about same size as other coxae (Fig. 20.269); posterior margin of prosternal process almost as wide as head 4
3'. Hind coxae transverse and larger than other coxae (Figs. 20.274, 20.275); posterior margin of prosternal process much narrower than width of head 5
4(3). Black; elytra with sublateral carinae (Fig. 20.268); antennae with 7 segments, enlarged at apex (Fig. 20.267); pronotum without transverse impressions .... *Macronychus*
4'. Conspicuously colored with black and yellow or orange; elytra and pronotum without sublateral carinae (Fig. 20.270); antennae with 11 segments, filiform; pronotum with oblique transverse impressions at apical third; tarsal claw with a basal tooth .... *Ancyronyx*
5(3'). Antennae with 8 segments, the apical one being enlarged (Fig. 20.271); pronotum with median longitudinal groove; elytra with 3 sublateral carinae .... *Zaitzevia*
5'. Antennae with 10-11 segments (Fig. 20.298), usually filiform 6
6(5'). Anterior tibia with fringe of tomentum (Fig. 20.299) 8
6'. Anterior tibia without fringe of tomentum 7
7(6'). Elytron with an accessory stria (sutural stria confluent with 2nd stria at about 5th puncture, Fig. 20.272); granules of head and legs elongate .... *Ordobrevia*
7'. Elytron without such an accessory stria (Fig. 20.273); granules of head and legs round .... *Stenelmis*
8(6). Lateral margin of 4th or 5th abdominal sternite produced as a prominent lobe or tooth, which is bent upward to clasp the epipleuron (Fig. 20.300); epipleuron widened to receive tooth, then usually narrowing abruptly toward apex 14
8'. Lateral margin of abdominal sternites not produced into a prominent upturned tooth; epipleuron usually tapering uniformly toward apex (Fig. 20.278) 9
9(8'). Pronotum with sublateral carinae (Fig. 20.279) 10
9'. Pronotum smooth, without sublateral carinae (Fig. 20.284) 13
10(9). Prosternum projecting beneath head (Fig. 20.276); epipleuron extending to middle of 5th abdominal segment; black; 2.5-2.6 mm long .... *Rhizelmis*
10'. Prosternum not projecting beneath head (Fig. 20.278); epipleuron ending at base of 5th abdominal segment; less than 2.3 mm long 11
11(10'). Pronotal carinae forked at base (Fig. 20.279) .... *Cleptelmis*
11'. Pronotal carinae not forked 12
12(11'). Sides of pronotum converging anteriorly from base (Fig. 20.280); body rather spindle-shaped; black, each elytron with a broad humeral and an oblique, narrow, subapical spot; tarsi and claws prominent .... *Atractelmis*
12'. Sides of pronotum parallel or divergent anteriorly at base (Fig. 20.281), strongly convergent apically; hump-backed; black, elytra black to red, uniformly colored or with basal half red, with or without broad apical red spots; tarsi and claws not unusually prominent .... *Ampumixis*
13(9'). Maxillary palpi 3-segmented (Fig. 20.282); markings, if present, transverse .... *Narpus*
13'. Maxillary palpi 4-segmented (Fig. 20.283); markings, if present, longitudinal .... *Dubiraphia*
14(8). Tooth that clasps epipleuron arising from lateral margin of 5th abdominal sternite 15
14'. Tooth that clasps epipleuron arising from apical (posterior) lateral margin of 4th abdominal sternite (Fig. 20.300) 22
15(14). Elytron at base with a short accessory stria between sutural and 2nd major stria (Fig. 20.285) .... *Macrelmis*
15'. Elytron without such an accessory stria (Fig. 20.286); testis usually bilobate 16
16(15'). Elytron with 1 sublateral carina; pronotum without oblique sculpturing (Fig. 19.286) 17
16'. Elytron with 2 sublateral carinae (rarely only 1 in *Microcylloepus,* which has oblique sculpturing on posterior half of pronotum) 18

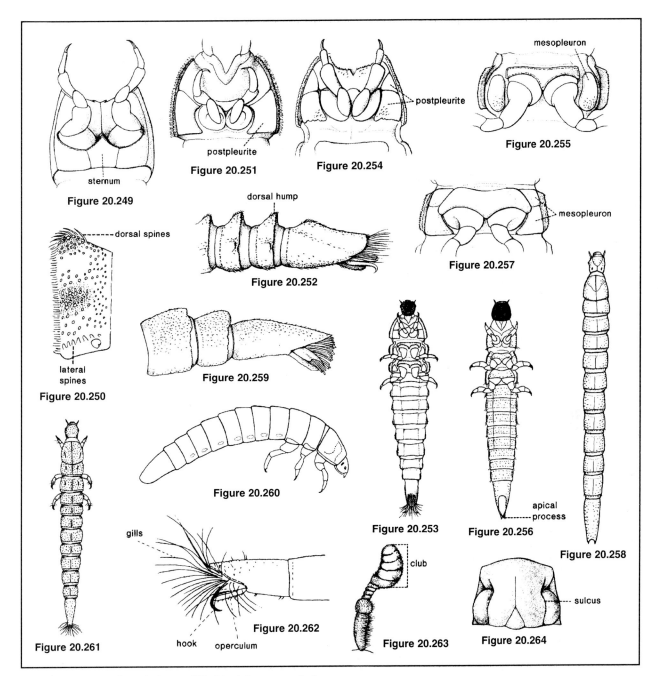

Figure 20.249. *Stenelmis* sp. (Elmidae) larva, ventral aspect of prothorax segment.

Figure 20.250. *Ampumixis* sp. (Elmidae) larva, lateral aspect of an abdominal segment.

Figure 20.251. *Ampumixis* sp. (Elmidae) larva, ventral aspect of prothorax segment.

Figure 20.252. *Promoresia* sp. (Elmidae) larva, ventral aspect of prothorax segment.

Figure 20.253. *Optioservus* sp. (Elmidae) larva, ventral aspect.

Figure 20.254. *Optioservus* sp. (Elmidae) larva, ventral aspect of prothorax.

Figure 20.255. *Optioservus* sp. (Elmidae) larva, ventral aspect of mesothorax.

Figure 20.256. *Macronychus* sp. (Elmidae) larva, ventral aspect.

Figure 20.257. *Macronychus* sp. (Elmidae) larva, ventral aspect of mesothorax.

Figure 20.258. *Zaitzevia* sp. (Elmidae) larva, dorsal aspect.

Figure 20.259. *Gonielmis* sp. (Elmidae) larva, lateral aspect of last abdominal segments.

Figure 20.260. *Heterlimnius* sp. (Elmidae) larva, lateral aspect.

Figure 20.261. *Oulimnius* sp. (Elmidae) larva, dorsal aspect.

Figure 20.262. *Cleptelmis* sp. (Elmidae) larva, lateral aspect of last abdominal segments.

Figure 20.263. *Phanocerus* sp. (Elmidae) adult, antenna.

Figure 20.264. *Phanocerus* sp. (Elmidae) adult, pronotum.

17(16). Posterior half of pronotum divided by a conspicuous median longitudinal impression, with a transverse impression slightly anterior to middle; adult as in Figure 20.286; brown to black .... *Neocylloepus*
17'. Pronotum undivided except by transverse impression at anterior two-fifths (Fig. 20.287); testaceous .... *Neoelmis*
18(16'). Hypomeron of pronotum with a belt of tomentum extending from coxa to lateral margin; pronotum with a shallow, median, longitudinal impression, but with no transverse impressions; testaceous to black .... *Hexacylloepus*
18'. Hypomeron with or without tomentum, but if present it does not reach lateral margin 19
19(18'). Prosternal process broad and truncate; pronotum without median longitudinal impression, usually with transverse impression at middle (Fig. 20.288); pronotal hypomeron with tomentum near coxa; body usually plump; mandible with a lateral lobe .... *Heterelmis*
19'. Prosternal process relatively narrow, elongate with apex tapering or rounded; pronotum with median longitudinal impression (Figs. 20.289, 20.291); hypomeron without tomentum; body not plump 20
20(19'). Pronotum with a transverse impression at anterior two-fifths (Fig. 20.289); mandible with a lateral lobe; epipleuron without tomentum; small, less than 2.3 mm long .... *Microcylloepus*
20'. Pronotum without such a transverse impression; epipleuron with tomentum 21
21(20'). Gula distinctly narrower than submentum or mentum (Fig. 20.290) .... *Huleechius*
21'. Gula not distinctly narrower than submentum or mentum; adult as in Figure 20.291 .... *Cylloepus*
22(14'). Pronotum with sublateral carinae extending from base to anterior margin; elytron with 3 sublateral carinae; adult as in Figure 20.292; brown to black; 1.25-1.6 mm long .... *Oulimnius*
22'. Pronotum with sublateral carinae absent or not extending beyond about middle; elytron without sublateral carinae; larger species, longer than 1.6 mm 23
23(22'). Pronotum smooth, without or with only a trace of carinae; body elongate and spindle-shaped; black, elytron with 2 oblique yellowish spots (Fig. 20.293); legs long, claws prominent and recurved; 2-2.6 mm long .... *Gonielmis*
23'. Pronotum with short, sublateral carinae (Fig. 20.294) 24
24(23'). Body rather elongate; tarsi and claws long and prominent; lateral and posterior margins of pronotum smooth; eastern; adult as in Figure 20.294 .... *Promoresia*
24'. Body plump; tarsi and claws not conspicuously enlarged; lateral margin of pronotum usually slightly serrate, posterior margin usually with many small, closely placed teeth 25
25(24'). Convex, giving a rather hump-backed appearance, with sutural intervals slightly raised; with 3rd or 4th elytral stria converging and merging with 2nd or 3rd stria at about apical third; major striae entire, extending to apex of elytron; antennae with 10-11 segments, last 3 somewhat enlarged; apex of 5th abdominal sternite usually somewhat truncate or emarginate; tarsal claws relatively slender; in western mountains; adult as in Figure 20.295 .... *Heterlimnius*
25'. Less convex; sutural interval usually not raised; elytral striae not ordinarily merging as described above, either being entire or becoming obsolete in posterior portion of elytron; antennae with 11 segments, the last 3 less enlarged; apex of 5th abdominal sternite usually evenly rounded; claws somewhat larger and more curved; adult as in Figure 20.296 .... *Optioservus*

Chrysomelidae (Leaf Beetles)

The Chrysomelidae is a large family of more than 14,000 species, but only a few genera have aquatic species. The genus *Donacia* Fabricius is the largest aquatic or semiaquatic chrysomelid genus, with 50 North American species. Some authors give generic status to *Plateumaris* Thomson, but it is treated here as a subgenus of *Donacia*, *Neohaemonia* Szekessy contains three species. Most existing works include *Galerucella* Crotch (Galerucellinae); however, *Pyrrhalta* Joannis has priority. At least two genera and five species of Chrysomelinae are known to contain aquatic or semiaquatic species in North America: *Prasocuris* Latreille, with a single species, and *Hydrothassa* Thomson with four species.

A number of flea beetles (Alticinae) in the large genus *Disonycha* Chevrolat are known to feed upon aquatic plants, and this number is growing, principally from work on the biological control of alligatorweed (*Alternanthera philoxeroides*). The host plants of many species remain unknown, but undoubtedly several more chrysomelid genera will be added to the aquatic list over the next few years. Two examples not included in the key are *Lysathia ludoviciana* on parrotfeather (*Myriophyllum aquaticum*) (Habeck and Wilderson 1980) and *Pseudolampsis guttata* on watervelvet (*Azola caroliniana*) (Habeck 1979).

Mating of aquatic chrysomelids occurs above water, with the flowers of water lilies (*Nuphar, Nymphaea*) being preferred mating sites. North American Donaciinae usually deposit their eggs in masses, with the structure and location of the egg mass distinct for each known species. Upon hatching, larval aquatic chrysomelids must locate the correct host plant. Field studies indicate that most Donaciinae have at least a two-year life cycle.

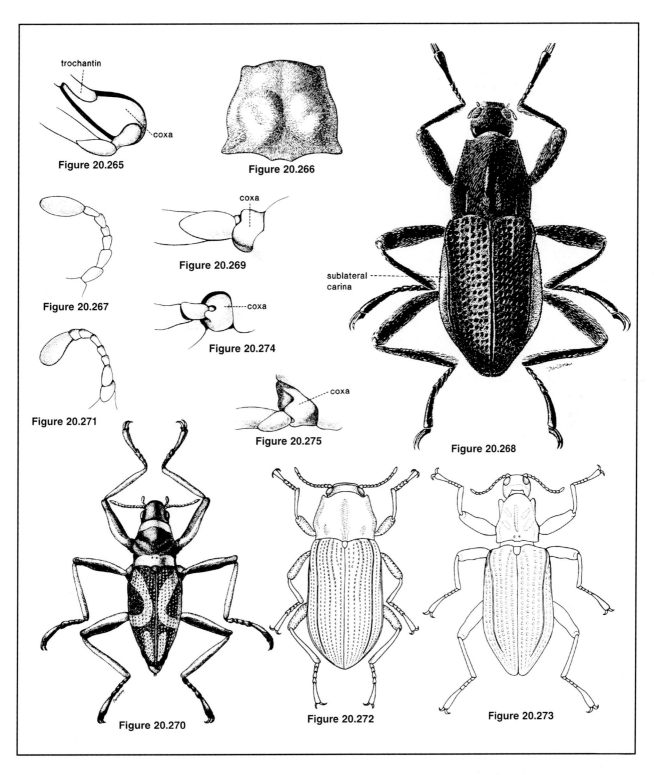

Figure 20.265. *Lara* sp. (Elmidae) adult, procoxa.
Figure 20.266. *Lara* sp. (Elmidae) adult, pronotum.
Figure 20.267. *Macronychus* sp. (Elmidae) adult, antenna.
Figure 20.268. *Macronychus* sp. (Elmidae) adult, dorsal aspect.
Figure 20.269. *Ancyronyx* sp. (Elmidae) adult, hind coxa.
Figure 20.270. *Ancyronyx* sp. (Elmidae) adult, dorsal aspect.
Figure 20.271. *Zaitzevia* sp. (Elmidae) adult, antenna.
Figure 20.272. *Ordobrevia* sp. (Elmidae) adult, dorsal aspect.
Figure 20.273. *Stenelmis* sp. (Elmidae) adult, dorsal aspect.
Figure 20.274. *Stenelmis* sp. (Elmidae) adult, procoxa.
Figure 20.275. *Stenelmis* sp. (Elmidae) adult, hind coxa.

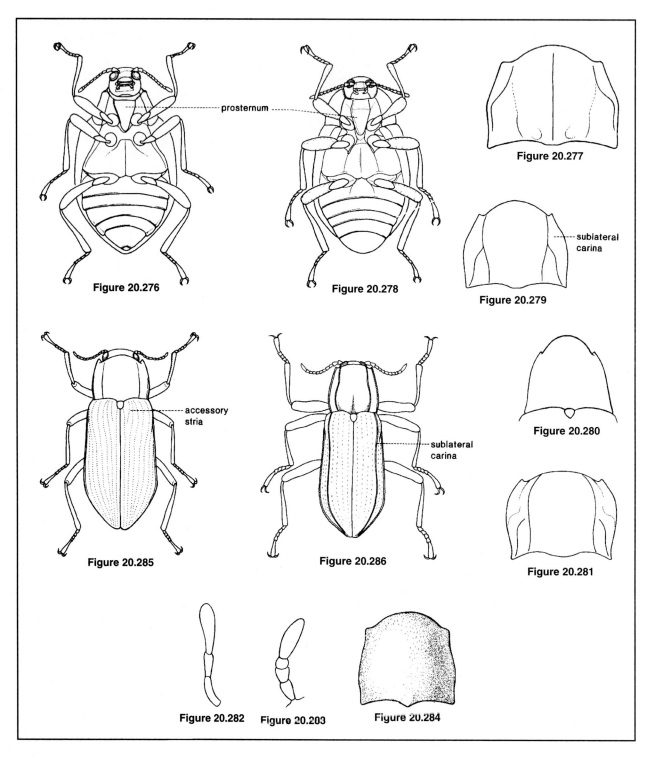

Figure 20.276. *Rhizelmis* sp. (Elmidae) adult, ventral aspect.
Figure 20.277. *Rhizelmis* sp. (Elmidae) adult, pronotum.
Figure 20.278. *Cleptelmis* sp. (Elmidae) adult, ventral aspect.
Figure 20.279. *Cleptelmis* sp. (Elmidae) adult, pronotum.
Figure 20.280. *Actractelmis* sp. (Elmidae) adult, pronotum (after Brown 1972b).
Figure 20.281. *Ampumixis* sp. (Elmidae) adult, pronotum.
Figure 20.282. *Narpus* sp. (Elmidae) adult, maxillary palp.
Figure 20.283. *Dubiraphia* sp. (Elmidae) adult, maxillary palp.
Figure 20.284. *Dubiraphia* sp. (Elmidae) adult, pronotum.
Figure 20.285. *Macrelmis* sp. (Elmidae) adult, dorsal aspect.
Figure 20.286. *Neocylloepus* sp. (Elmidae) adult, dorsal aspect.

Larvae of most species form their cocoons in the same positions on the plants in which the larvae occurred. Cocoons, spun of silk from glands opening in the mouth, are both water and airtight and are filled with intercellular air obtained from the host plant. *Agasicles* pupates within the stem cavities of the host plant.

Larvae and adults of aquatic and semiaquatic Chrysomelidae are live vascular plant shredders and, as far as is known, have a definite preference for a single-host plant taxon or, at most, several species of plants that occupy the same habitat. Although the larvae of Donaciinae feed almost exclusively upon the submerged roots, leaves, and petioles of aquatic plants, the adults may feed upon pollen. *Pyrrhalta nymphaeae* is unique among the Galerucinae in that both larvae and adults feed on the upper surface of floating leaves. Rather than mining, they chew away the epidermis of the leaf, producing an irregular trench that gives the leaf a characteristic brownish, etched appearance.

The larvae of *Donacia* pass their lives underwater even though they do not possess any of the usual morphological adaptations for an aquatic existence. They are equipped with a pair of caudal spines that, at least in some instances, may be used to penetrate the tissues of aquatic plants. It has long been suspected that the caudal spines permit the larva to use oxygen contained in air spaces within the plants. Some larvae, however, can live submerged for as long as 21 days without penetrating plants, so it is likely that they obtain oxygen by other means.

Except for some groups that contain important economic pests, the immature stages of the Chrysomelidae (Fig. 20.13) are poorly known. As a result, it is impossible to provide a key for the separation of aquatic and semiaquatic genera. In fact, although the key to families of larvae provided here will separate chrysomelids from the larvae of other aquatic and semiaquatic families, the key is not adequate for separating aquatic and semiaquatic chrysomelids from the great number of other grub-like larvae that may fall into the water. The inexperienced coleopterist would do well to look for the caudal spines diagnostic for the Donaciinae (Fig. 20.14) and to consider all other identifications as tentative unless supported by reared and associated adult material.

Chrysomelidae

Adults

1.	Antennal insertions nearly contiguous (Fig. 20.301); 1st visible abdominal sternite as long as all others combined ....Donaciinae 2	
1'.	Antennal insertions separated by entire width of frons (Fig. 20.302); if nearly contiguous, 1st visible abdominal sternite no longer than 2nd and 3rd combined 3	
2(1).	Elytron with outer apical angle produced into a strong spine (Fig. 20.303) ....***Neohaemonia***	
2'.	Elytron without spine at outer apical angle (Fig. 20.301) ....***Donacia***	
3(1').	Antennal insertions separated by entire width of frons (Fig. 20.302) ....Chrysomelinae ....Prasocurini 4	
3'.	Antennae inserted closely on front of head (Fig. 20.304) 5	
4(3).	Basal margin of pronotum with a fine elevated bead (Fig. 20.302); form very elongate length, at least 2.5 times width; sutural dark stripe not much wider around scutellum (Fig. 20.305) ....***Prasocuris***	
4'.	Basal margin of pronotum without such a bead; form broader, length less than 2.5 times width; sutural dark stripe abruptly widened around scutellum (Fig. 20.306) ....***Hydrothassa***	
5(3').	Posterior femur greatly enlarged, adapted for jumping (Fig. 20.307) ....Alticinae 6	
5'.	Posterior femur not greatly enlarged, slender, adapted for walking (Fig. 20.308) ....Galerucinae....***Pyrrhalta***	
6(5).	Pronotum narrow, its width approximately two-thirds that of bases of elytra; pronotum unicolorous, black ....***Agasicles***	
6'.	Pronotum wide, its width approximately three-fourths that of bases of elytra; pronotum unicolorous, yellow, or yellow with 2 or 3 medial black spots (Fig. 20.304) ....***Disonycha***	

Curculionidae (Weevils)

The Curculionidae is one of the largest families of animals. The total world fauna through 1971 approached 45,000 described species. About one-tenth of this number is known to occur within North America. For such a large and diverse group, it is surprising that only a few species have invaded the aquatic environment. More than 50 species are known from genera in which at least one species is aquatic.

Lissorhoptrus simplex Say is a serious pest of cultivated rice. Eggs of this species are deposited in roots of rice plants, and other species are known to deposit eggs in and on leaves of aquatic plants. Early instar *L. simplex* larvae feed within the roots, but later instars move to the mud around the roots and continue to feed upon them. Pupation occurs within an air-filled, watertight mud cocoon located on the roots.

The respiratory mechanism of *L. simplex* is unique. Abdominal segments 2-7 each bear a pair of dorsal hooks that connect by large tracheae to the main longitudinal tracheal trunks. The hooks appear to be modified spiracles. The larvae pierce air cells within the rice roots and obtain air from them. Denied access to living roots, submerged larvae, as a rule, die within 24 hours.

Larval *Bagous americanus* mine the floating leaves of *Nymphaea* spp. for approximately three larval instars. These mines remain air-filled. *Bagous* larvae have the abdomen truncate behind, with a pair of enlarged spiracles on the truncated

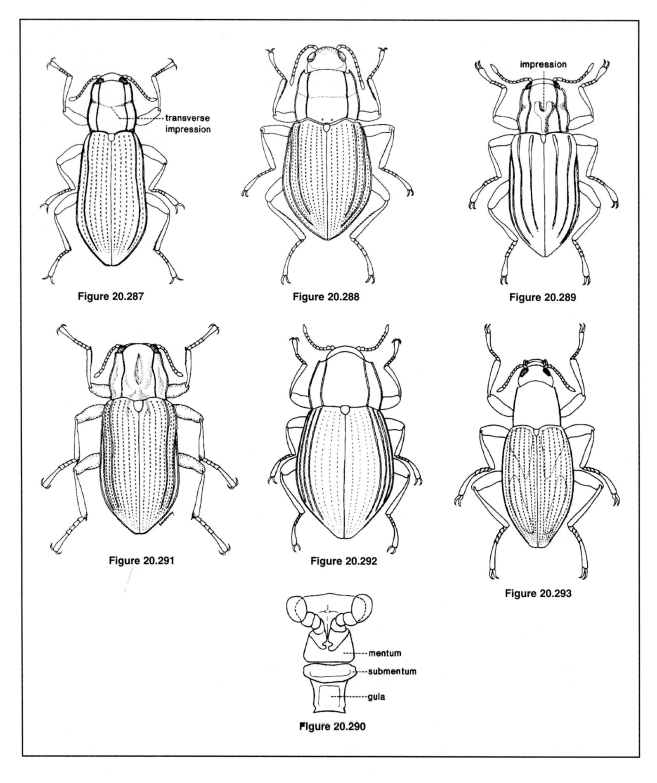

Figure 20.287. *Neoelmis* sp. (Elmidae) adult, dorsal aspect.

Figure 20.288. *Heterelmis* sp. (Elmidae) adult, dorsal aspect.

Figure 20.289. *Microcylloepus* sp. (Elmidae) adult, dorsal aspect.

Figure 20.290. *Huleechius* sp. (Elmidae) adult, ventral aspect of gula, mentum, and submentum (after Brown 1981).

Figure 20.291. *Cylloepus* sp. (Elmidae) adult, dorsal aspect.

Figure 20.292. *Oulimnius* sp. (Elmidae) adult, dorsal aspect.

Figure 20.293. *Gonielmis* sp. (Elmidae) adult, dorsal aspect.

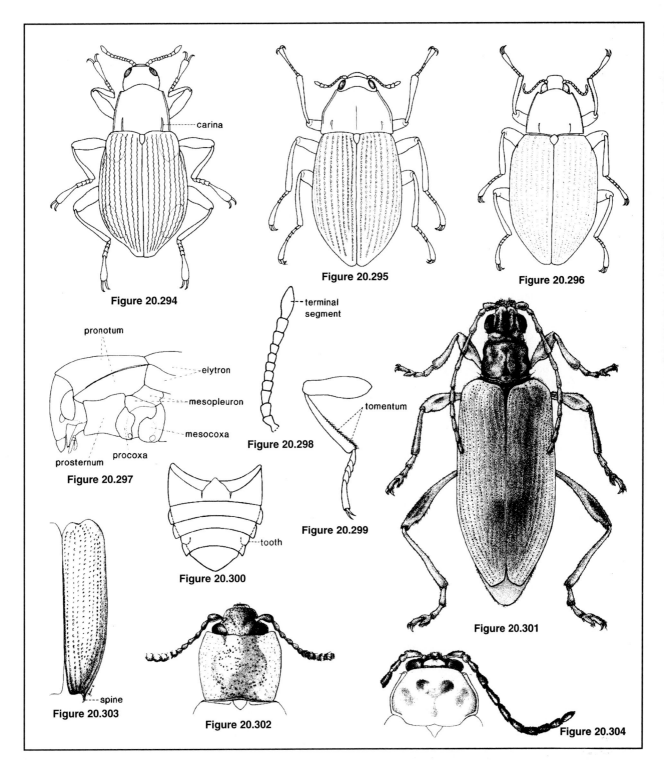

Figure 20.294. *Promoresia* sp. (Elmidae) adult, dorsal aspect.
Figure 20.295. *Heterlimnius* sp. (Elmidae) adult, dorsal aspect.
Figure 20.296. *Optioservus* sp. (Elmidae) adult, dorsal aspect.
Figure 20.297. *Optioservus* sp. (Elmidae) adult, lateral aspect of head.
Figure 20.298. *Optioservus* sp. (Elmidae) adult, antenna.
Figure 20.299. *Optioservus* sp. (Elmidae) adult, foreleg.
Figure 20.300. *Optioservus* sp. (Elmidae) adult, ventral aspect of abdomen.
Figure 20.301. *Donacia* sp. (Chrysomelidae) adult, dorsal aspect.
Figure 20.302. *Prasocuris* sp. (Chrysomelidae) adult, dorsal aspect of head and prontotum.
Figure 20.303. *Neohaemonia* sp. (Chrysomelidae) adult, elytron.
Figure 20.304. *Disonycha* sp. (Chrysomelidae) adult, dorsal aspect of head and pronotum.

surface and the other abdominal spiracles reduced or absent. Thus, larvae can utilize the air-filled mines behind them for respiration. Larvae of *B. longirostris* have been found as deep as 2 m below the water in air-filled mines in petioles of *Nymphaea*.

Pupation of aquatic weevils occurs on or within leaf petioles. The pupal chamber remains dry. *L. simplex* in the laboratory required 32-77 days from deposition of egg to transformation to the adult.

Several curculionid genera and species have been purposefully added to the North American fauna as biological control agents of introduced nuisance aquatic plants. Plants where some control has been achieved include *Eichhornia* and *Salvalina* (see Table 20A). More recently the weevil *Neohydronomus affinis* has shown promise in controling waterlettuce (*Pistia stratiotes*) (Thompson and Habeck 1989). We have not included this genus in the key because of its extremely limited distribution.

Larvae of Curculionidae (Fig. 20.15) may be separated readily from larvae of other families of aquatic Coleoptera because they do not possess legs. However, as the larvae of many genera and most species of aquatic Curculionidae are unknown or have not been described, it is impossible to provide larval keys at this time. Association of adults (which can be keyed) and larvae on host plants is a method of tentative identification of immature weevils.

Curculionidae

Adults

1.	Rostrum free, not received in prosternum	2
1'.	Rostrum received in prosternum in repose; upper surface of body very often uneven	6
2(1).	Lateral angle of 1st sternite covered by elytron	3
2'.	Lateral angle of 1st sternite exposed	*Lixus*
3(2).	Tibiae fossorial (Fig. 20.312); mandible biemarginate, tridentate at apex; procoxae contiguous	*Emphyastes*
3'.	Tibiae not fossorial; mandible emarginate or biemarginate; procoxae contiguous or separated	4
4(3').	Mandible usually emarginate and bidentate at apex	*Listronotus*
4'.	Mandible biemarginate, tridentate at apex	5
5(4').	Tibiae each with inner apical angle drawn into a strong, curved apical claw (Fig. 20.313); size large, middorsal length (excluding head) usually greater than 7 mm	*Steremnius*
5'.	Tibiae without strong apical claw, truncate at apex, if claw present (*Bagous* and *Pnigodes*), then size small, middorsal length (excluding head) usually less than 5 mm	12
6(1').	Procoxae contiguous	12
6'.	Procoxae separated (except in *Listronotus*)	7
7(6').	Second abdominal segment prolonged at sides, cutting off 3rd, which fails to reach margins (Fig. 20.314); pectoral groove extending behind procoxae into mesosternum	*Auleutes*
7'.	Second abdominal segment not prolonged at sides; pectoral groove not extending behind procoxae, sometimes absent	8
8(7').	Third tarsal segment narrow, not or scarcely bilobed; 4th tarsal segment as long as preceding segments united	9
8'.	Third tarsal segment bilobed, 4th tarsal segment shorter than the 2 preceding segments united	10
9(8).	Base of thorax prolonged posteriorly as an acute triangular process	*Phytobius* (in part)
9'.	Base of thorax not prolonged posteriorly	*Perenthis*
10(8').	Prosternum deeply sulcate before coxae, sulcus more or less evidently marked by carinae or bounded by very strongly formed carinae	11
10'.	Prosternum not or feebly sulcate before coxae, lateral margin of sulcus not defined by carinae	*Phytobius* (in part)
11(10).	Eye with distinct supraorbital ridge (Fig. 20.315); funicle of antenna 6-segmented; front coxae separated by one-third the distance of middle coxae	*Pelenomus*
11'.	Eye without supraorbital ridge; funicle of antenna 7-segmented; front coxae separated by one-half the distance of middle coxae	*Rhinoncus*
12(5', 6).	Metasternum one-third length of 1st sternite	*Thalasselephas*
12'.	Metasternum as long as 1st sternite	13
13(12').	Rostrum very short and broad, not longer than head (Fig. 20.316); 3rd segment of hind tarsus deeply emarginate, bilobed	*Stenopelmus*
13'.	Rostrum cylindrical, much longer than head	14
14(13').	Prosternum broadly and deeply excavated in front of coxae to receive rostrum (Fig. 20.317); club of antenna entirely pubescent; tibia slender, curved, armed with a distal hooked spine	15
14'.	Not fitting above description exactly	17
15(14).	Funicle of antenna 6-segmented	*Neochetina*
15'.	Funicle of antenna 7-segmented	16
16(15').	Pronotum feebly constricted anteriorly (Fig. 20.318)	*Bagous*
16'.	Pronotum very strongly constricted anteriorly (Fig. 20.319)	*Pnigodes*

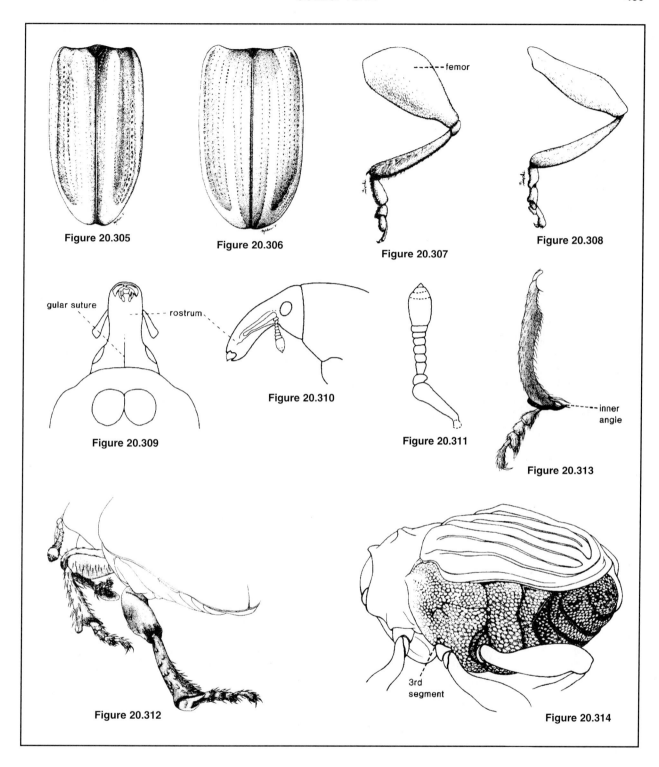

Figure 20.305. *Prasocuris* sp. (Chrysomelidae) adult, scutellum and elytra.

Figure 20.306. *Hydrothassa* sp. (Chrysomelidae) adult, scutellum and elytra.

Figure 20.307. *Disonycha* sp. (Chrysomelidae) adult, hind leg.

Figure 20.308. *Pyrrhalta* sp. (Chrysomelidae) adult, hind leg.

Figure 20.309. *Emphyastes* sp. (Curculionidae) adult, ventral aspect of head and prothorax.

Figure 20.310. *Emphyastes* sp. (Curculionidae) adult, lateral aspect of head and prothorax.

Figure 20.311. *Emphyastes* sp. (Curculionidae) adult, antenna.

Figure 20.312. *Emphyastes* sp. (Curculionidae) adult, lateral oblique aspect.

Figure 20.313. *Steremnius* sp. (Curculionidae) adult, dorsal aspect of hind leg.

Figure 20.314. *Auleutes* sp. (Curculionidae) adult, lateral oblique aspect.

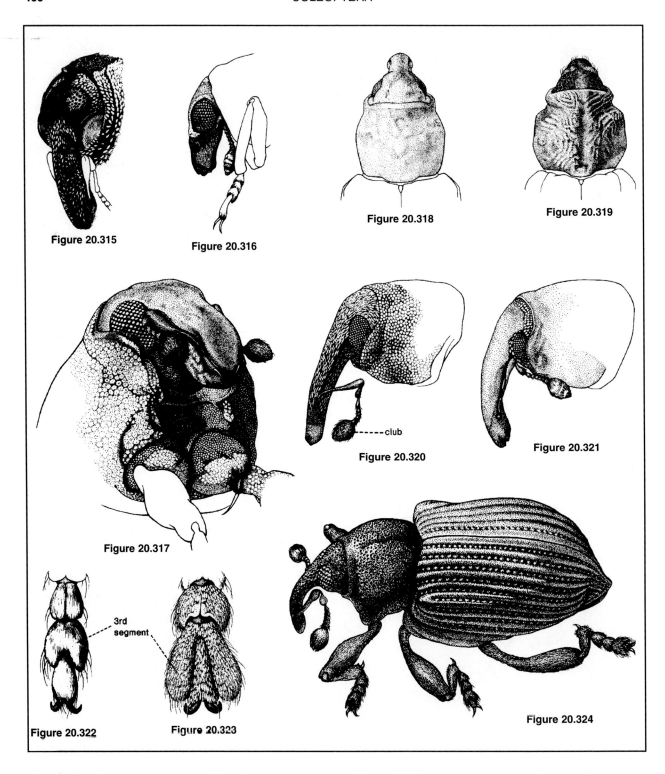

Figure 20.315. *Mecopeltus* sp. (Curculionidae) adult, anterior oblique aspect of head.

Figure 20.316. *Stenopelmus* sp. (Curculionidae) adult, lateral aspect of head.

Figure 20.317. *Bagous* sp. (Curculionidae) adult, ventrolateral aspect of head and prothorax.

Figure 20.318. *Bagous* sp. (Curculionidae) adult, dorsal aspect of head and pronotum.

Figure 20.319. *Pnigodes* sp. (Curculionidae) adult, dorsal aspect of head and pronotum.

Figure 20.320. *Anchodemus* sp. (Curculionidae) adult, lateral aspect of head and pronotum.

Figure 20.321. *Brachybamus* sp. (Curculionidae) adult, lateral aspect of head and pronotum.

Figure 20.322. *Onychylis* sp. (Curculionidae) adult, lateral aspect of apical segments of tarsus.

Figure 20.323. *Endalus* sp. (Curculionidae) adult, lateral aspect of apical segments of tarsus.

Figure 20.324. *Tanysphyrus* sp. (Curculionidae) adult, dorsolateral aspect.

17(14').	Third segment of hind tarsus emarginate or bilobed	18
17'.	Third segment of hind tarsus simple; legs long and slender	***Lissorhoptrus***
18(17).	Rostrum straight or only slightly curved (Fig. 20.320)	***Lixellus***
18'.	Rostrum more curved (Fig. 20.321)	19
19(18').	Tarsi each with a single claw	***Brachybamus***
19'.	Tarsi each with 2 claws	20
20(19').	Last segment of tarsus projecting beyond lobes of 3rd segment (Fig. 20.322)	***Onychylis***
20'.	Last segment of tarsus not projecting beyond lobes of 3rd segment (Fig. 20.323)	21
21(20').	Length usually 2 mm or more; elytra slightly if any wider than thorax	***Notiodes***
21'.	Length less than 1.5 mm; elytra much wider than thorax (Fig. 20.324)	***Tanysphyrus***

SELECTED ADDITIONAL TAXONOMIC REFERENCES

General

Blatchley (1910); Boving and Craighead (1930); Leech and Chandler (1956); Leech and Sanderson (1959); Arnett (1960); Klots (1966); Bertrand (1972); Doyen (1976); Pennak (1978); Lehmkuhl (1979); Crowson (1981); Hilsenhoff (1981); McCafferty (1981); Bosquet (1991); Williams and Felmate (1992).

Regional faunas

California: Leech and Chandler (1956).
Canada (Southhampton Island): Young (1960).
Florida: Young (1954).
Georgia: Turnbow and Smith (1983).
Maine: Malcolm (1971).
Maryland: Staines (1986a,b).
Minnesota: Daussin (1979).
Newfoundland: Bistrom (1978).
North and South Carolina: Brigham (1982).
North Dakota: Gordon and Post (1965).
Ohio: Melin and Graves (1971).
Pacific Northwest: Hatch (1962, 1965).
Wisconsin: Hilsenhoff (1981); Hilsenhoff and Schmude (1992).

Taxonomic treatments at the family and generic levels
(several are regional in scope)

Amphizoidae: Edwards (1951).
Chrysomelidae: Schaeffer (1925, 1928); Mead (1938); Marx (1957); Wilcox (1965); Balsbaugh and Hays (1972); Bayer and Brochmann (1975); Askevold (1987).
Curculionidae: Tanner (1943); Kissinger (1964); Bayer and Brochmann (1975); O'Brien (1975); O'Brien and Marshall (1979).
Dryopidae: Musgrave (1935); Brown (1972b); Ulrich (1986); Barr and Chapin (1988); Nelson (1989); Hilsenhoff and Schmude (1992).
Dytiscidae: Fall (1919, 1922a, 1923); Wallis (1939a, b); Leech (1938, 1940, 1942a); Young (1953, 1956, 1963, 1967, 1974, 1979b, 1981a,b); Anderson, R. D. (1962, 1971, 1976, 1983); McWilliams (1969); Zimmerman (1970, 1980a,b, 1981, 1982, 1985); Hilsenhoff (1975, 1980, 1992, 1993a,b,c); Larson (1975, 1987, 1989); Zimmerman and Smith (1975a,b); Michael and Matta (1977); Matta (1979, 1986); Gordon (1981); Matta and Wolfe (1981); Roughley and Pengelly (1981); Wolfe and Matta (1981); Wolfe and Spangler (1985); Larson and Roughley (1990); Roughley (1990); Wolfe and Roughley (1990); Nilsson and Angus (1992).
Elmidae: Sanderson (1938, 1953, 1954); LaRivers (1950a); Brown and Shoemake (1964); Brown (1970a, 1972a,b, 1981c); Collier (1970); Hilsenhoff (1973); Finni and Skinner (1975); Brown and White (1978); White (1978b, 1982); Barr and Chapin (1988).
Gyrinidae: Roberts (1895); Fall (1922b); Hatch (1930); Wood (1962); Morrissetta (1979); Ferkinhoff and Gunderson (1983); Hilsenhoff (1990, 1992, 1993a,b,c).
Haliplidae: Matheson (1912); Roberts (1913); Wallis (1933); Matta (1976); Hilsenhoff and Brigham (1978); Gundersen and Otremba (1988).
Hydraenidae: Perkins (1975, 1980); Hanson (1991).
Hydrophilidae: Horn (1873); Richmond (1920); LaRivers (1954); Spangler (1960); Miller (1964, 1974); Wooldridge (1965, 1966, 1967); McCorkle (1967); Van Tassell (1966); Wilson (1967); Cheary (1971); Matta (1974, 1983); Smetana (1974, 1980, 1985a,b, 1988); Gundersen (1978); Malcolm (1979); Testa and Lago (1994).
Limnichidae and Lutrochidae: Brown (1972b); Hilsenhoff and Schmude (1992).
Melyridae: Moore and Legner (1975).
Microsporidae: Britton (1966).
Noteridae: Spangler and Folkerts (1973); Young (1978, 1979a, 1985); Hilsenhoff (1992, 1993a, b, c).
Psephenidae: Brown (1972b); Brown and Murvosh (1974); Brigham (1981); Barr and Chapin (1988); Hilsenhoff and Schmude (1992).
Ptilodactylidae: Brown (1972b); Stribling (1986); Hilsenhoff and Schmude (1992).
Salpingidae: Spilman (1967).
Scirtidae: Tetrault (1967); Young and Stribling (1990).
Staphylinidae: Herman (1972); Moore and Legner (1974, 1976).

TABLE 20A. Summary of ecological and distributional data for *aquatic Coleoptera (beetles)*. (For definition of terms see Tables 6A-6C; table prepared by W. U. Brigham, D. S. White, R. W. Merritt and K. W. Cummins.)

Taxa (number of species in parentheses)	Habitat*	Habit*	Trophic Relationships*	North American Distribution	Ecological References**
Adephaga					1034
Amphizoidae (3) (trout-stream beetles)					
Amphizoa (3) Larvae and adults	Lotic—erosional (especially logs)	Clingers	Predators (engulfers; especially on stoneflies)	West, Northwest	1035, 2112, 2300, 2301, 3076
Gyrinidae (60) (whirligig beetles)					1233, 1234, 1624, 1653 2112, 2300, 2301, 3076 3535, 4152, 4153
Larvae	Lentic and lotic—vascular hydrophytes	Generally climbers and swimmers	Generally predators (engulfers)		
Adults	Lentic and lotic—depositional	Generally surface swimmers, divers	Generally predators (engulfers) and surface film (neuston) scavengers		
Dineutus (14)		Surface swimmers, divers		Widespread	244, 867, 1917, 2148 1204, 1300
Cyretes (4)		Surface swimmers, divers		West, South	2300, 4596
Gyrinus (41)		Surface swimmers, divers	Predators (engulfers) and neuston scavengers	Widespread	497, 1917
Spanglerogyrus (1)	Lotic—depositional	Surface swimmers, divers, climbers		South, East	1232, 3830
Carabidae (2) (Predaceous ground beetles)					
Thalassotrechus (2)					
Larvae and adults	Marine rocky coasts	Clingers (rock crevice dwellers)	Predators (engulfers)	Pacific Coast	956, 957, 4101
Haliplidae (70) (crawling water beetles)					1709, 2112, 2300, 2301, 3076, 3394
Larvae	Generally lentic—littoral, vascular hydrophytes and some lotic—depositional	Generally climbers	Generally shredders—herbivores (chewers), piercers—hervibores, some predators (engulfers)		
Adults	Generally same as larvae	Swimmers, climbers	Generally same as larvae		
Apteraliplus (1)				California	2300
Larvae	Lentic	Clingers	Shredders—herbivores (chewers), piercers—herbivores, predators (engulfers)		
Adults	Same as larvae	Swimmers, climbers	Shredders and piercers—herbivores		
Brychius (4)					
Larvae and adults	Lotic and lentic—erosional	Clingers	Scrapers, piercers—herbivores	West, north central	2300, 3875
Haliplus (47)				Widespread	1708, 1710, 2301, 1501, 4234
Larvae	Lentic—vascular hydrophytes	Climbers	Piercers and shredders—herbivores		

* Data listed are for adults unless noted.
** Emphasis on trophic relationships.

TABLE 20A. Continued

Taxa (number of species in parentheses)	Habitat*	Habit*	Trophic Relationships*	North American Distribution	Ecological References**
Adults	Same	Swimmers, climbers	Same		
Peltodytes (18)				Widespread	1708, 1710, 2301, 4602, 1501
Larvae	Lentic—vascular hydrophytes, some lotic—erosional	Climbers, clingers	Piercers and shredders—herbivores, predators (engulfers)		
Adults	Same	Swimmers, climbers, clingers	Same		
Dytiscidae (552) (predaceous diving beetles)					125, 165, 1234, 1789, 2112, 2263, 2300, 2301, 2687, 3076, 4596, 4600, 4624, 750, 1736, 2197
Larvae	Generally lentic—vascular hydrophytes, some lotic—depositional	Generally climbers, swimmers	Generally predators (engulfers)		
Adults	Same as larvae	Generally divers, swimmers	Same as larvae		
Acilius (5)	Lentic—vascular hydrophytes	Swimmers, divers	Predators (piercers)	Widespread	1730, 1746, 2300, 2687
Agabetes (1)	Lentic—littoral	Swimmers, divers	Predators (piercers)	East	3641, 4596
Agabinus (2)	Lotic—erosional and depositional, lentic—littoral	Swimmers, divers	Predators (piercers)	West	2300
Agabus (93)	Lotic—erosional and depositional, lentic—vascular hydrophytes	Swimmers, divers	Predators (piercers)	Widespread	1729, 1929, 1932, 1952, 1736, 2299, 2687, 4596
Agaporomorphus (1)				Florida	4612
Anodocheilus (1)	Lentic—littoral, lotic—depositional (vascular hydrophytes)	Swimmers, climbers	Predators (piercers)	Southeast	4596
Bidessonotus (3)	Lentic—littoral, lotic—depositional	Swimmers, climbers	Predators (piercers)	East, espcially Southeast	4596
Brachyvatus (1)	Lentic—littoral, lotic—depositional	Swimmers, climbers	Predators (piercers)	Widespread	4596
Carrhydrus (1)	Lentic—vascular hydrophytes	Swimmers, divers	Predators (piercers)	Far North	2259
Celina (9)	Lentic—vascular hydrophytes	Swimmers, climbers	Predators (piercers)	Widespread	2300, 4596
Colymbetes (7)	Lotic—depositional	Swimmers, divers	Predators (piercers)	Northern	165, 2300
Copelatus (7)	Lotic—depositional, lentic—vascular hydrophytes	Swimmers, divers	Piercers—carnivores	Widespread	2300
Coptotomus (8)	Lentic—vascular hydrophytes, lotic—depositional	Swimmers, divers, climbers	Predators (piercers)	Widespread	165, 2259, 2300, 2301, 4464, 4596
Cybister (5)	Lentic—vascular hydrophytes	Swimmers, divers	Predators (piercers)	Widespread	165, 1231, 2005, 2006, 370, 2300
Derovatellus (1)	Lentic—littoral	Swimmer, climbers	Predators (piercers)	Florida	4596
Desmopachria (9)	Lotic—depositional, lentic—vascular hydrophytes	Swimmers, climbers	Predators (piercers)	Widespread	2259, 2301, 4464, 4596

* Data listed are for adults unless noted.
** Emphasis on trophic relationships.

TABLE 20A. Continued

Taxa (number of species in parentheses)	Habitat*	Habit*	Trophic Relationships*	North American Distribution	Ecological References**
Dytiscus (14)	Lentic—vascular hydrophytes, lotic—depositional	Swimmers, divers	Predators (piercers)	Widespread	165, 2300, 3566, 4591, 460, 1270, 1809, 2293, 1268, 1269, 1989
Eretes (1)	Lentic—vascular hydrophytes	Swimmers, divers		West, Southwest	2095, 2300
Graphoderus (6)	Lentic—vascular hydrophytes	Swimmers, divers	Predators (piercers)	Widespread	2300
Haideoporus (1)	Subterranean			Texas	3242, 4613
Hoperius (1)		Swimmers, divers, climbers	Predators (piercers)	Arkansas	3762
Hydaticus (4)	Lentic—vascular hydrophytes, lotic—depositional	Swimmers, divers, climbers	Predators (piercers)	Widespread	2300, 2301, 4464
Hydroporus (182)	Lotic—depositional, lentic—vascular hydrophytes	Swimmers, climbers	Predators (piercers)	Widespread	165, 2259, 2300, 2516, 4596
Hydrotrupes (1)	Lotic—erosional and depositional	Clingers, climbers	Predators (piercers)	West Coast	2300
Hydrovatus (8)	Lotic—depositional		Predators (piercers)	Widespread	2300, 4596
Hygrotus (54)	Lotic—depositional, lentic—vascular hydrophytes	Swimmers, climbers	Predators (piercers)	Widespread	2259, 2300
Ilybius (15)	Lotic—depositional, lentic—vascular hydrophytes	Swimmers, divers	Predators (piercers)	East of Rocky Mountains	2259, 2300, 4596 1736
Laccodytes (1)	Lentic—littoral		Predators (piercers)	Florida	4596
Laccophilus (10)	Lotic—depositional, lentic—vascular hydrophytes	Swimmers, divers, climbers		Widespread	1953, 1954, 2300 3395, 4596, 4632
Laccornis (5)	Lentic—vascular hydrophytes	Swimmers, climbers	Predators (piercers)	East	2259, 2297
Liodessus (6)	Lotic—erosional and depositional, lentic—vascular hydrophytes	Swimmers, climbers	Predators (piercers)	Widespread	2300, 4596
Lioporius (2)	Lotic—erosional and depositional, lentic—vascular hydrophytes	Swimmers, climbers	Predators (piercers)	East	2300, 4596
Matus (4)	Lotic—depositional, lentic—vascular hydrophytes	Swimmers, divers	Predators (piercers)	East, South	2300, 4596
Megadytes (2)	Lentic—vascular hydrophytes	Swimmers, climbers	Predators (piercers)	Florida	4596
Nebrioporus (3) (=*Deronectes* in part)	Lotic—erosional and depositional, lentic—vascular hydrophytes	Swimmers, climbers	Predators (piercers)	Widespread	2300
Neobidessus (1)	Lotic erosional and depositional, lentic—vascular hydrophytes	Swimmers, climbers	Predators (piercers)	Florida, Louisiana	2300, 4596
Neoclypeodytes (8)	Lotic—erosional and depositional, lentic—littoral	Swimmers, climbers	Predators (piercers)	West, Southwest	2300
Neoscutopterus (2)	Lentic—vascular hydrophytes			Extreme North	2300

* Data listed are for adults unless noted.
** Emphasis on trophic relationships.

COLEOPTERA

TABLE 20A. Continued

Taxa (number of species in parentheses)	Habitat*	Habit*	Trophic Relationships*	North American Distribution	Ecological References**
Oreodytes (17)	Lotic—erosional and depositional, lentic—littoral	Swimmers, climbers	Predators (piercers)	Widespread (primarily West)	2300
Pachydrus (1)	Lentic—littoral, lotic—depositional	Swimmers, climbers		Florida	4596
Rhantus (11)	Lentic—vascular hydrophytes, lotic—depositional	Swimmers, divers	Predators (piercers)	Widespread	2259, 2300, 2301, 3714, 4464
Stictotarsus (23) (=*Deronectes* in part)	Lotic—erosional and depositional, lentic—vascular hydrophytes	Swimmers, climbers	Predators (piercers)	Widespread	2300
Thermonectus (6)	Lentic—vascular hydrophytes	Swimmers, divers, climbers	Predators (piercers)	Southwest	2300
Uvarus (9)	Lotic—erosional and depositional, lentic—vascular hydrophytes	Swimmers, climbers	Predators (piercers)	Widespread	2300, 4596
Noteridae (18) (burrowing water beetles)					168, 169, 2112, 2300, 2301, 3076, 4596
Larvae	Generally lentic—vascular hydrophytes	Generally burrowers, climbers	Predators (engulfers), collectors gatherers		
Adults	Same as larvae	Generally swimmers, climbers, burrowers	Predators (engulfers)		
Hydrocanthus (6)	Lentic—vascular hydrophytes	Climbers		East, South	4596
Notomicrus (2)	Lotic—depositional, lentic—littoral	Burrowers		South	3641, 4596
Pronoterus (2)	Lentic—vascular hydrophytes	Climbers		East, South	4596
Suphis (1)	Lentic—vascular hydrophytes	Climbers		East, South	4596
Suphisellus (7)	Lotic—depositional, lentic—vascular hydrophytes	Climbers		Widespread	4596
Myxophaga					
Hydroscaphidae (1) (skiff beetles)					1769, 1771, 2112, 2252, 2300, 2301, 3076, 3597
Hydroscapha (1)					
Larvae and adults	Lotic—erosional and margins (including thermal springs)	Clingers	Scrapers (bluegreen algae)	West (especially Southwest)	4234
Microsporidae (3) (=Sphaeriidae) (minute bog beetles)					429, 2300
Sphaerius (3)					
Larvae and adults	Lentic and lotic margins (semiaquatic)	Burrowers, climbers	Scrapers?	Texas, Southern California, Washington	
Polyphaga					
Epimetopidae (2)					see Hydrophilidae
Epimetopus (2)				Southwest	4596

* Data listed are for adults unless noted.
** Emphasis on trophic relationships.

TABLE 20A. Continued

Taxa (number of species in parentheses)	Habitat*	Habit*	Trophic Relationships*	North American Distribution	Ecological References**
Helophoridae (41)					see Hydrophilidae
Helophorus (41)					
Larvae and adults	Lentic and lotic erosional, especially margins	Climbers	Shredders—herbivores	Widespread	1083, 2236, 2237, 3414, 4234
Hydrochidae (19)					see Hydrophilidae
Hydrochus (19)					
Larvae and adults	Lentic and lotic—erosional, especially margins	Climbers	Shredders—herbivores	Widespread	2300
Hydrophilidae (192) (water scavenger beetles)					212, 1829, 2112, 2300, 2301, 3076, 3978
Larvae	Lentic—vascular hydrophytes, lotic—depositional	Generally climbers	Generally predators (engulfers)		
Adults	Same as larvae	Generally divers, swimmers	Generally collectors—gatherers		
Ametor (2)	Lentic—littoral, lotic—depositional	Clingers		West	3761
Anacaena (4)	Lentic—littoral, lotic—depositional (detritus, fine sediments)	Burrowers (silt)		East and West Coasts, Illinois	2236, 2237, 2300
Berosus (26)	Lentic—littoral, lotic—depositional	Swimmers, divers, climbers	Piercers—herbivores, collectors—gatherers, shredders	Widespread	165, 3343, 4596
Chaetarthria (32)	Lentic—littoral, lotic—depositional	Climbers?		Widespread	3085
Crenitis (11)	Lentic—littoral, lotic—depositional	Burrowers (sand and gravel)		East	2300
Cymbiodyta (29)	Lentic—littoral, lotic—depositional	Burrowers (sand and gravel)		Widespread	2300
Derallus (1)	Lentic—littoral, lotic—depositional	Swimmers, divers, climbers		East and Gulf Coasts	2300
Dibolocelus (2)	Lentic—littoral, lotic—depositional	Swimmers, divers		East	4596
Enochrus (25)				Widespread	165, 2236, 2237, 3343
Larvae	Lentic—littoral	Burrowers—sprawlers	Collectors—gatherers		
Adults	Same	Same	Piercers—herbivores		
Helobata (1)				Florida, Louisiana	3763
Helochares (3)				East, Southwest	2300
Helocombus (1)				East	2300
Hydrobiomorpha (1)	Lentic—littoral, lotic—depositional	Swimmers, divers, climbers		South	4596
Hydrobius (3)	Lentic—littoral	Climbers, clingers, sprawlers		East	167, 2236, 2237
Hydrochara (9)	Lentic—littoral, lotic—depositional	Swimmers, divers, climbers		East, South, California	2548, 4596

* Data listed are for adults unless noted.
** Emphasis on trophic relationships.

TABLE 20A. Continued

Taxa (number of species in parentheses)	Habitat*	Habit*	Trophic Relationships*	North American Distribution	Ecological References**
Hydrophilus (3)	Lentic—littoral, lotic—depositional	Swimmers, divers, climbers		Widespread	165, 2687, 4120, 4463, 4625
Larvae			Predators (engulfers)		
Adults			Collectors—gatherers, piercers— herbivores		
Lacobius (9)				Widespread	165, 3085
Larvae					
Adults			Piercers— herbivores		
Paracymus (14)	Lentic—littoral, lotic—depositional	Burrowers (sand and gravel)		Widespread (primarily South)	2300, 4234
Sperchopsis (1)	Lentic—littoral, lotic— depositional (wood)	Clingers		East, South	3760
Tropisternus (15)	Lentic—littoral, lotic—depositional			Widespread	165, 1820, 3343, 4598, 4234
Larvae		Climbers	Predators (engulfers)		3496, 3497, 3498, 3499
Adults		Swimmers, divers, climbers	Collectors—gatherers, piercers—herbivores		
Staphylinidae (rove beetles)	Generally shorelines and beaches freshwater lotic and lentic—littoral, marine intertidal and rocky coasts	Generally clingers and climbers, burrowers	Predators (engulfers)		565, 2300, 2745, 3545, 3329
Amblopusa (2)	Marine coasts (rock crevices)	Clingers		West Coast (Alaska to California)	
Bledius (100)***	Lotic and lentic—littoral	Clingers		Widespread	
Bryothinusa (1)	Marine coasts (rock crevices)	Clingers		California coast	
Carpelimus (78)	Lotic and lentic—littoral (leaf litter)	Clingers		Widespread	
Diaulota (4)	Marine coasts (rock crevices)	Clingers		West Coast (Alaska to Baja California)	
Diglotta (1)	Beaches, marine (sand)	Burrowers		California coast	
Liparocephalus (2)	Marine coasts (rock crevices)	Clingers		West Coast (Alaska to California)	
Micralymma (2)	Marine coasts (rock crevices)	Clingers		Coastal, northern	
Phytosus (1)				New Jersey	
Pontomalota (5)	Beaches marine (sand)	Burrowers		West Coast (British Columbia to California)	
Psephidonus (1)	Lotic—erosional	Climbers (runners)	Predators (engulfers of Simuliidae)	East	852, 3975
Thinobius (3)	Lotic and lentic— littoral	Climbers		Widespread	
Thinopinus (1)	Beaches, marine (sand)	Burrowers	Predators (engulfers of amphipods)	West coast, California to British Columbia	741, 3329, 3330

* Data listed are for adults unless noted.
** Emphasis on trophic relationships.
*** Not all species in the genus are aquatic.

TABLE 20A. Continued

Taxa (number of species in parentheses)	Habitat*	Habit*	Trophic Relationships*	North American Distribution	Ecological References**
Thinusa (2)	Beaches, marine (sand)	Burrowers?	Predators (engulfers)	West Coast (Alaska to California)	
Stenus (159)***	Lotic and lentic—littoral surface	Skaters (eject oils)		Widespread	
Melyridae (5) (soft-winged flower beetles)					324, 2300
Endeodes (5)				Pacific Coast	
Larvae and adults	Beaches, marine intertidal, marine rocky coasts (semiaquatic)	Clingers (rock crevice dwellers), beaches (supratidal)	Predators (engulfers)		
Salpingidae (5) (= Eurystethidae) (narrow-waisted bark beetles)					2300, 3779, 3900, 4101
Aegialites (5)	Marine rocky coasts	Clingers (rock crevice dwellers)	Predators (engulfers, especially mites), shredders—herbivores	Pacific Coast	
Hydraenidae (83) (minute moss beetles)					1619, 2112, 2300, 2301, 3076, 3084, 3085, 3086
Larvae	Lotic—erosional (or stream margins), lentic—vascular hydrophytes (emergent zone)	Generally clingers, swimmers (runners) (primary semiaquatic)	Predators (engulfers)		
Adults	Lotic—erosional (especially margins)	Generally clingers (logs, cobbles, rock crevices)	Scrapers, collectors—gatherers		
Hydraena (25)	Lotic—erosional, lentic—littoral (especially margins)	Clingers, climbers		Widespread	111, 3084, 3085, 3086
Limnebius (14)	Lotic—erosional, lentic—littoral (especially margins)	Clingers, climbers		Widespread	3084, 3085, 3086
Ochthebius (43)	Lotic—erosional and margins (1 intertidal species)	Clingers		Widespread	1619, 3084, 3085, 3086
Neochthebius (1)					
Psephenidae (14) (water pennies)	Generally lotic and lentic—erosional	Clingers	Scrapers		463, 1762, 1768, 2112, 2300, 2301, 2820, 3076, 461
Larvae	Lotic and lentic—erosional	Clingers	Scrapers		3706
Adults	(Females enter water to oviposit)	Ovipositing females are clingers	Non-feeding		
Eubrianax (1)	Lotic—erosional	Clingers	Scrapers	West	646, 2818
Psephenus (7)	Lotic—erosional	Clingers	Scrapers	*East (1 sp.), West	451, 671, 870, 2817, 2821, 3555, 4307, 2640, 2819
Eubriinae (false water pennies)					
Acneus (2)	Lotic—erosional	Clingers	Scrapers	West Coast (south to San Francisco Bay)	

* Data listed are for adults unless noted.
** Emphasis on trophic relationships.
*** Not all species in the genus are aquatic.

COLEOPTERA

TABLE 20A. Continued

Taxa (number of species in parentheses)	Habitat*	Habit*	Trophic Relationships*	North American Distribution	Ecological References**
Dicranopselaphus (1)	Lotic—erosional	Clingers	Scrapers	Widespread	455
Ectopria (3)	Lotic and lentic—erosional	Clingers	Scrapers	East	408, 409, 671, 3555, 4596
Dryopidae (14) (long-toed water beetles)					2112, 2300, 2301, 3076
Larvae	Terrestrial	Burrowers	Generally shredders—herbivores?		
Adults (all entries)	Generally lentic—littoral, lotic—erosional	Clingers, climbers	Generally scrapers, collectors—gatherers?		
Dryops (2)	Lentic—vascular hydrophytes (emergent zone)	Climbers		Southwest, Northeast	1759
Helichus (7)	Lotic—erosional	Clingers		Widespread	1590, 4071
Pelonomus (1)	Lentic—vascular hydrophytes (emergent zone)	Climbers		Texas to Florida, Illinois	455, 4596
Postelichus (3)	Lotic—erosional	Clingers		Southwest	1590, 4071
Stygoparnus (1)	Lotic, cave springs	Clingers		Comal Springs, Texas	3638
Scirtidae (35) (= Helodidae) (marsh beetles)					2112, 2300, 2301, 3076 3909, 4045
Larvae (all entries)	Generally lentic—vascular hydrophytes (emergent zone)	Generally climbers, sprawlers	Generally scrapers, collectors—gatherers, shredders—herbivores, piercers—herbivores		
Adults	Terrestrial (possibly a few semiaquatic)				
Cyphon (14)	Lentic—littoral (sediments, leaf litter; including marshes and bogs), tree holes			Widespread	
Elodes (4)	Tree holes, seeps (including mineral springs)			Widespread	2300
Flavohelodes (3)	Tree holes, seeps			Widespread	2300
Microcara (1)				Michigan	
Ora (1)				Florida, Texas	
Prionocyphon (2)	Tree holes			East	1447, 2300
Sarabandus (3)				Northeast	
Scirtes (7)	Lentic—vascular hydrophytes (floating zone)	Climbers	Shredders—herbivores, piercers—herbivores	Widespread	234, 2175
Elmidae (101) (riffle beetles)					456, 457, 464, 465, 488, 2112, 2300, 2301, 2714, 461, 3076, 3532, 4330, 112, 2901, 4057, 111, 4234
Larvae and adults	Generally lotic and lentic—erosional, few lentic vascular hydrophytes	Clingers, few climbers	Generally collectors, gatherers, scrapers		
Ampumixis (1)	Lotic—erosional	Clingers, burrowers		West	

* Data listed are for adults unless noted.
** Emphasis on trophic relationships.

TABLE 20A. Continued

Taxa (number of species in parentheses)	Habitat*	Habit*	Trophic Relationships*	North American Distribution	Ecological References**
Ancyronyx (1)	Lotic—erosional and depositional (detritus)	Clingers, sprawlers	Generally collectors, scrapers	East, Southeast	3729
Atractelmis (1)	Lotic—erosional	Clingers		California	3638
Cleptelmis (2)	Lotic—erosional (cobbles and submerged roots)	Clingers		West	473
Cylloepus (2)	Lotic—erosional	Clingers		Southwest	
Dubiraphia (11)	Lentic and lotic—erosional (submerged macrophytes)	Clingers, climbers		Widespread	671, 3729
Macrelmis (3)	Lotic—erosional (cobbles and gravel)	Clingers		Southwest	
Gonielmis (1)	Lotic—erosional and depositional (wood debris and submerged roots)	Clingers, climbers		East, Southeast	
Heterelmis (4)	Lotic—erosional (cobbles and wood)	Clingers		Southwest	454, 473
Huleechius (1)	Lotic—erosional	Clingers		Arizona	458
Heterlimnius (2)	Lotic—erosional (cobbles and gravel)	Clingers		West	4234, 4242
Hexacylloepus (1)	Lotic—erosional (cobbles, gravel, and wood)	Clingers		Southwest	
Lara (2)	Lotic—erosional (wood debris, including semiaquatic)	Clingers, burrowers (in wood)	Shredders, detritivores (wood)	West (montane)	95, 96, 473, 3824
Macronychus (1)	Lotic—erosional and depositional (wood debris)	Clingers	Collectors, detritivores	East	876, 3729
Microcylloepus (12)	Lotic—erosional and depositional	Clingers, climbers, burrowers		Widespread	3637
Narpus (4)	Lotic—erosional (cobbles and gravel)	Clingers		West	473, 4234
Neocylloepus (1)	Lotic—erosional (cobbles and gravel)	Clingers		Texas, Arizona	453, 455
Neoelmis (1)	Lotic—erosional (cobbles and gravel)	Clingers		Southwest	
Optioservus (13)	Lotic—erosional and depositional (sediments and detritus)	Clingers	Scrapers, collectors—gatherers	Widespread	170, 473, 603, 671, 2168 3618, 3619, 4329, 4234
Ordobrevia (1)	Lotic—erosional (cobbles and gravel)	Clingers		West	
Oulimnius (1)	Lotic—erosional (cobbles and gravel)	Clingers		East, Southeast	455
Phanocerus (1)				Texas	
Larvae	Lotic—erosional	Clingers, climbers			
Adults	Lotic margins (semiaquatic)	Climbers			
Promoresia (2)	Lotic—erosional (cobbles and gravel)	Clingers		East, Southeast	465
Rhizelmis (1)	Lotic—erosional	Clingers		California	

* Data listed are for adults unless noted.
** Emphasis on trophic relationships.

COLEOPTERA

TABLE 20A. Continued

Taxa (number of species in parentheses)	Habitat*	Habit*	Trophic Relationships*	North American Distribution	Ecological References**
Stenelmis (28)	Lotic—erosional (coarse sediments and detritus)	Clingers	Scrapers, collectors—gatherers	Widespread	455, 671, 1141, 3617, 3618, 3619
Xenelmis (1)				Arizona	460
Zaitzevia (2)	Lotic—erosional (cobbles and gravel)	Clingers		West	4234
Ptilodactylidae (3) (toed-winged beetles)					2300, 2301, 3778
Larvae (all entries)	Generally lotic—erosional and depositional	Generally clingers, burrowers	Generally shredders—detritivores and herbivores		
Adults	Terrestrial (near lotic margins in leaf litter)				
Anchycteis (1)	Lotic—erosional and depositional	Burrowers	Shredders—herbivores (chewers)	California	2300, 2329, 4596
Anchytarsus (1)	Lotic—erosional and depositional	Clingers	Shredders—detritivores (rotting wood)	East	
Stenocolus (1)	Lotic—erosional and depositional	Burrowers	Shredders—herbivores (chewers)	California	
Limnichidae (3) (marsh-loving beetles)		Generally clingers or burrowers	Generally collectors—gatherers?		461, 1136, 2300, 2301, 3636
Throscinus (3)					
Adults and larvae	Beach zone marine intertidal	Burrowers	Collectors—gatherers?	Texas, southern California	2300
Lutrochidae (3) (marsh-loving beetles)					461, 1136, 2300, 2301, 3636
Lutrochus (3)					
Larvae	Lotic—erosional (especially mineral springs)	Clingers		East, Southwest	455, 462
Adults	Mostly terrestrial				
Chrysomelidae (leaf beetles)	Generally lentic—vascular hydrophytes	Generally clingers, sprawlers	Shredders—herbivores		2112, 2300, 2301
Larvae	Lentic—vascular hydrophytes (below surface in floating and submerged zones)	Clingers (obtain air directly from vascular hydrophyte tissues)	Shredders—herbivores (chewers)		
Adults	Lentic—vascular hydrophytes (generally surface of floating leaves)	Sprawlers (on floating leaf surfaces)	Shredders—herbivores (chewers)		
Agasicles (1)	On *Alternanthera*			South	
Disonycha (33)***	On Amaranthaceae			Widespread	
Donacia (50)	On a wide variety of vascular hydrophytes			Widespread	381, 1795, 1796, 2618
Hydrothassa (4)	On *Ranunculus*			North, East	
Neohaemonia (4)	On a wide variety of vascular hydrophytes			East of Rocky Mountains	381, 1795
Prasocuris (1)	On a wide variety of vascular hydrophytes			Northeast	

* Data listed are for adults unless noted.
** Emphasis on trophic relationships.
*** Not all species in the genus are aquatic.

TABLE 20A. Continued

Taxa (number of species in parentheses)	Habitat*	Habit*	Trophic Relationships*	North American Distribution	Ecological References**
Pyrrhalta (5)*** (=*Galerucella*)	On Nymphaeaceae			East	4190
Curculionidae (weevils)					2300, 2301, 2618, 3936, 975, 3989
Larvae	Generally lentic—vascular hydrophytes (most semiaquatic)	Generally clingers, climbers, sprawlers (on leaf surfaces), burrowers (in stems)	Shredders—herbivores (chewers and miners)		
Adults	Same as larvae	Generally clingers and sprawlers	Shredders—herbivores (chewers)		
Auleutes (12)***	Lentic—vascular hydrophytes (emergent and floating zones)			Widespread	
Bagous (30)	Lentic—vascular hydrophytes (floating zone)	Sprawlers, clingers		Widespread	2618
Brachybamus (1)	Lentic—vascular hydrophytes (emergent zone: probably *Sagittaria*)			East	3936
Emphyastes (1)	Beach zone marine (sand, under macroalgae)		Shredders—detritivores (decaying *Fucus*)	Pacific Coast	
Euhrychiopsis	Lentic—vascular hydrophytes (emergent and floating zones: *Myriophyllum* and probably others)	Climbers, clingers	Shredders—herbivores (meristem, leaves and stem)	East, Northern	750, 751, 753 752
Lissorhoptrus (2)	Lentic—vascular hydrophytes (submerged zone: rice)	Climbers, clingers	Shredders—herbivores (roots and submerged foliage of rice)	East	2115, 4058
Listronotus (70+)***	Lentic—vascular hydrophytes (emergent and floating zones)			Widespread	
Lixellus (3)	Lentic—vascular hydrophytes (emergent zone: *Sagittaria*)			East	2618
Lixus (69)***	Lentic—vascular hydrophytes (emergent and floating zones)			Widespread	
Neochetina (1)	Lentic—vascular hydrophytes (emergent zone: *Eichhornia*)			South	
Notiodes (8)***	Lentic—vascular hydrophytes (emergent zone: *Scirpus* and probably others)	Climbers, clingers		Widespread	
Onychylis (3)	Lentic—vascular hydrophytes (floating zone: *Sagittaria*, *Nymphaea*)			East	
Perenthis (1)				East	
Pelenomus (6)***	Lentic—vascular hydrophytes (emergent and floating zones)			East, West	

* Data listed are for adults unless noted.
** Emphasis on trophic relationships.
*** Not all species in the genus are aquatic.

TABLE 20A. Continued

Taxa (number of species in parentheses)	Habitat*	Habit*	Trophic Relationships*	North American Distribution	Ecological References**
Phytobius (8)***	Lentic—vascular hydrophytes (emergent and floating zones)			Widespread	751
Pnigodes (3)	Lentic—vascular hydrophytes (on *Lepidium, Ptilimnium*)			South, West	3936
Rhinoncus (5)				Widespread	
Stenopelmus (1)				Widespread	3936
Steremnius (4)				East, West	
Tanysphyrus (2)	Lentic—vascular hydrophytes (floating zone: duckweeds)		Shredders—herbivores (*Lemna*)	East	
Thalasselephas (1)	Beach zone marine (sand and rock crevices)			Southern California	

* Data listed are for adults unless noted.
** Emphasis on trophic relationships.
*** Not all species in the genus are aquatic.

AQUATIC HYMENOPTERA

Kenneth S. Hagen
University of California, Berkeley

INTRODUCTION

Although the Hymenoptera considered aquatic in this review are all parasites (parasitoids), the group is difficult to define. In the narrowest sense they are wasps that enter the water as adults to contact their aquatic hosts, and in the broadest sense (e.g., Hedqvist's [1967] European list) they are parasitoids of aquatic invertebrates, but do not necessarily parasitize the aquatic stage of the host.

In an earlier review of North American aquatic Hymenoptera (Hagen 1956), only species known or suspected to either dive or crawl beneath the water surface to parasitize or to obtain their hosts were included. The present coverage follows Burghele's (1959) definition of aquatic Hymenoptera as all species parasitizing aquatic stages of insects. Therefore, all known North American hymenopteran families and genera that conform to this definition, plus some selected families and genera that attack semiaquatic species and/or life stages, have been included. Thus, caution is advised if the reader attempts to identify hymenopterans reared from terrestrial (or marginally semiaquatic) stages.

Aquatic Hymenoptera are members of the suborder Apocrita, all families falling under the Section Parasitica. The one known exception is the pompilid *Anoplius depressipes* Banks, which is a member of the section Aculeata. This wasp attacks pisaurid spiders (genus *Dolomedes*) that run over and dive under the surface of the water and may stay submerged for some time. Since *A. depressipes* can crawl into the water and run on the bottom, it may sting the spiders under water before transporting them to nests made in the bank (Evans 1949; Evans and Yoshimoto 1962; Roble 1985).

The superfamilies of Parasitica—Ichneumonoidea, Proctotrupoidea, Cynipoidea, and Chalcidoidea contain a few species that can be considered aquatic. Most are internal parasitoids of aquatic immatures that are usually found in plant tissues. The few external parasitoid species oviposit on larvae, prepupae, and pupae that are either in terrestrial cocoons, puparia, or pupal cases, or in plant mines.

Among the Ichneumonoidea, most aquatic braconids are internal parasitoids of leaf miners and enter the water as adults to oviposit. *Chaenusa* oviposits into ephydrid eggs, whereas other genera *(Ademon, Chorebus, Chorebidea, Chorebidella, Dacnusa,* and *Opius)* oviposit into larvae, but all emerge from ephydrid puparia. Species of these genera have been reported to parasitize 60-90% of *Hydrellia* sp. populations in some instances (Grigarick 1959b; Deonier 1971). Aquatic ichneumonids (*Apsilops* sp. and perhaps *Cremastus* sp.) are solitary internal parasitoids of stem-mining Lepidoptera, and a *Tanychela* sp. is a solitary endoparasitoid of a pyralid and emerges from pupae within underwater cocoons attached to rocks (Resh and Jamieson 1988). *Medophron* spp. are ectoparasitic on *Agabus* pupae in terrestrial pupal cells (R. Garcia, 1993, pers. comm.). The Agriotypidae (often considered a subfamily of the Ichneumonidae), are ectoparasitic as larvae upon the prepupae and pupae of caddisflies (Clausen 1950; Elliott 1983).

Proctotrupoidea associated with aquatic insects include two families, the Scelionidae and Diapriidae. All scelionids are internal parasitoids of insect eggs; and the diapriid genus *Trichopria* oviposits in dipteran puparia (Clausen 1940). Among the Cynipoidea, only the family Eucoilidae is aquatic, attacking Diptera larvae and emerging from pupae; *Hexacola* is the only genus reported from North America thus far that attacks aquatic insects.

In the Chalcidoidea, the tiny mymarid wasps (fairyflies) are all internal parasitoids of insect eggs. The biology of *Caraphractus cinctus* Walker, which is probably the most completely known of any aquatic hymenopteran, is summarized in Figures 21.1-21.10. After mating underwater, on the surface film, or on emergent plants, this species oviposits underwater in dytiscid beetle eggs, either exposed or in plant tissues (Fig. 21.1). The number of eggs deposited depends upon the size of the host egg, but up to 55 progeny have been obtained from a single dytiscid egg. In such cases the adults are often micropterous (with reduced wings). After the egg (Fig. 21.2) is deposited, it enlarges (Figs. 21.3-21.5) and the transparent first larval instar (Fig. 21.6) hatches. The larva feeds by sucking in yolk spheres from the host egg, which can be seen in the gut of the second instar (Fig. 21.7). As the yolk cells are digested, the gut of the third instar becomes paler (Fig. 21.8). The gregarious larvae do not attack each other and face in different directions if they are in the same host egg (Figs. 21.8-21.9). Opaque white areas, which probably represent excretory products, can be seen in the terminal larval instar (Fig. 21.9). Wastes are stored as a single mass in the pupal stage (Fig. 21.10). The adults emerge under water, respire cutaneously, and swim with their wings (Jackson 1958c, 1961b). The female of *C. cinctus* probes each egg of *Agabus* sp. with her ovipositor and usually rejects eggs already parasitized.

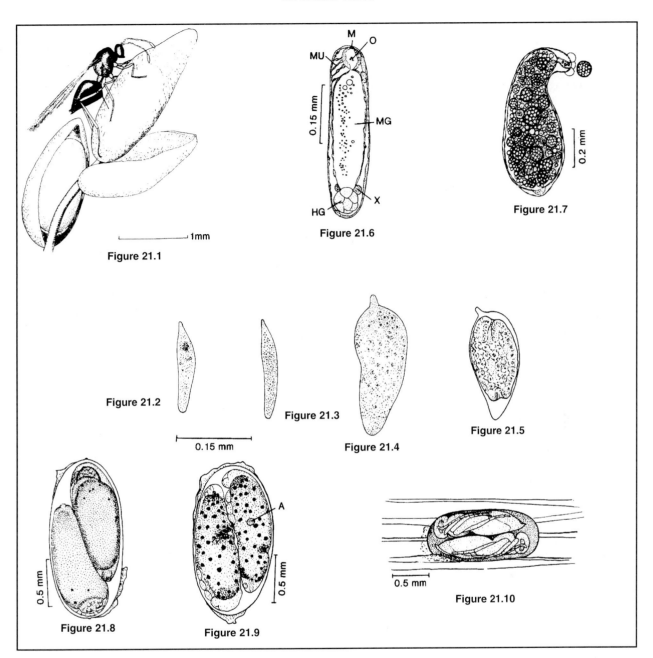

Figure 21.1. Female *Caraphractus cinctus* Walker (Mymaridae) ovipositing underwater in an egg of *Agabus bipustulatus* L. (Dytiscidae) laid in a sphagnum leaf. One other host egg is shown attached to the leaf with gelatinous cement (from Jackson 1958b).

Figure 21.2. *Caraphractus cinctus* ovarian eggs (from Jackson 1961b).

Figure 21.3. *Caraphractus cinctus* egg dissected from host 20 minutes after laying (from Jackson 1961b).

Figure 21.4. *Caraphractus cinctus* egg (laid by same female as egg in fig. 21.3) dissected from another host about 72 hours after laying (from Jackson 1961b).

Figure 21.5. *Caraphractus cinctus* egg of another female about 72 hours after laying; more advanced, showing reduced size and developing embryo.

Figure 21.6. *Caraphractus cinctus* first larval instar dissected from host about 70 hours after laying (from Jackson 1961b): *HG*, hindgut; *M*, mouth; *MG*, midgut; *MU*, muscle cells extending from body wall to esophagus; *O*, esophagus; *X*, probably sex cells.

Figure 21.7. *Caraphractus cinctus* second larval instar (from Jackson 1961b).

Figure 21.8. Egg of *Agabus bipustulatus* L. with two-third stage larvae of *Caraphractus cinctus*, about 9 days after laying (from Jackson 1961b).

Figure 21.9. Two full-grown parasitic larvae of *Caraphractus cinctus* in host egg. *A*, discolored spot around the oviposition puncture in shell of *Agabus bipustulatus* (from Jackson 1961b).

Figure 21.10. Eggs of *Agabus bipustulatus*, on leaf of *Juncus* sp., containing two newly formed pupae of *Caraphractus cinctus*; the one above is a male, the one below, a female (from Jackson 1961b).

Arrhenotokous parthenogenesis occurs as in most parasitic Hymenoptera, where an unmated female will produce only male progeny and a mated female can deposit both fertilized and unfertilized eggs. When *C. cinctus* oviposits in small host eggs like *Agabus bipustulatus,* the proportion of fertilized eggs usually increases when hosts are offered in quick succession; the number of eggs deposited in each host is then reduced to two or one with a resulting sex ratio of about 17% males. In host eggs offered to a female at long intervals, three eggs are deposited, and the resulting sex ratio is usually one male and two females. Under high competition between female parasitoids for a few host eggs, about 47% male progeny results (Jackson 1966).

Jackson (1961a) was able to obtain four or five generations of *C. cinctus* from *A. bipustulatus* reared in an unheated room near a window in Scotland, but by the end of October all full-grown larvae (prepupae) entered a diapause state. These diapausing prepupae pupated in the spring and emerged as adults from March to May. Interestingly, dytiscid eggs do not diapause but develop throughout the year when the temperature is suitable. Jackson (1961a) experimentally varied the photoperiod and found that 9 hours of complete darkness induced diapause in the *C. cinctus* larvae (which were sensitive to extremely low light intensity); she concluded that in ponds, direct development of the parasitoid would occur with long daylight hours combined with faint light at night (twilight) or with a few hours of darkness. Temperature was not the deciding factor in the induction of diapause.

Aquatic Trichogrammatidae are all internal parasitoids of insect eggs. Although some species of the large genus *Trichogramma* can swim with their legs, remain under water up to five days, and have been reported to mate within the submerged host egg, most "aquatic" species will undoubtedly be reared from terrestrial eggs of aquatic insect species. Thus far, no *Trichogramma* species have been reported as parasitoids of submerged host eggs. However, an undescribed species of *Lathromeroidea* is known to use its wings to swim if accidentally submerged, but does not swim under water in search of eggs (Henriquez and Spence 1993). *Trichogramma* spp. enter a diapause state if the host eggs undergo diapause.

Species of Eulophidae and Pteromalidae have diverse habits, including hyperparasitism, a condition in which one parasitoid is in turn parasitized by another. None have been reported to invade water to find their host; however, several species are associated with the semiaquatic or terrestrial stages of aquatic insects.

EXTERNAL MORPHOLOGY

Generally, Hymenoptera that parasitize aquatic insects show no external morphological adaptations in either the adult or larval stages compared with those parasitizing terrestrial hosts. The larvae of internal parasitoids are already modified to live in a liquid environment (host hemolymph), and no special morphological modifications seem to have evolved in the adults that dive beneath water to reach their hosts. However, several recently-described eucoilid species in the endemic Hawaiian genus *Aspidogyrus* exhibit specialized morphological adaptations, particularly in the females (Beardsley 1992). They have elongated tarsi and modified enlarged divided claws (unique in the family Eucoilidae) which enable them to move over rock surfaces under swiftly moving streams, facilitating the parasitization of their dipteran hosts (larval canacaeids and ephydrids) and the subsequent escape of the adults emerging from the submerged fly puparia. The only two known species of the ichneumonid genus *Tanychela* also have elongated tarsi and claws (Dasch 1979). *T. pilosa* Dasch is a parasitoid of moth pupae which are found in silken cocoons attached to underwater rocks in running brooks (Resh and Jamieson 1988). The adult wasp of this species emerges by chewing a hole in its cocoon and the silk tent; a thick covering of hair on the wasp's body causes a bubble of air to form around it as it swims to the surface and begins its terrestrial life (Jamieson and Resh unpublished data). Larvae of Agriotypidae are the only external parasitoids of hosts that live under water and they have a special morphological adaptation: a ribbonlike respiratory filament formed by a last larval instar that remains functional through the pupal stage.

Suborder Apocrita

Section Parasitica

Larvae

Larvae of internal parasites are usually hypermetamorphic, that is, the first instars are distinct from later instars, which are legless and grublike. Maxillary palpi rarely project from the head surface, and the labial palpi are disklike and nonprojecting. There are rarely more than nine spiracles in mature larvae, and they are lacking in early instars of internal parasitoids.

Adults

Hind wings are without an anal lobe (Figs. 21.11, 21.13, 21.16A) and species are rarely wingless. Hind tibiae have one or two spurs (Fig. 21.20), but none are modified for preening.

KEY TO THE FAMILIES OF "AQUATIC" HYMENOPTERA

All known hymenopteran families that contain species that enter the water as adults are included in the key, as well as most families having species that parasitize terrestrial stages of aquatic insects. Since many of the hymenopteran larvae that parasitize aquatic insects cannot be separated satisfactorily on a morphological basis (particularly the Chalcidoidea and Proctotrupoidea), a key to the larvae is not presented. Rather, some general descriptive information for the larvae of each family is given below, followed by a key to adults.

Suborder Apocrita

Section Parasitica

Larvae

1. **Ichneumonoidea:** First instars of solitary internal parasites with large mandibles, no open spiracles and body usually ending in a long tail. First instar of external parasites segmented and with 9 spiracles; mandibles not conspicuous.

Mature larvae segmented, with spiracles; head parts usually outlined by sclerotic rods.

    a. *Braconidae:* First instars with large mandibles and usually without long, pointed tails *(Opius, Chorebus).* Mature larvae usually with spiracles on pro- and metathorax and on first 7 abdominal segments.

    b. *Ichneumonidae:* First instars of solitary internal parasites often have large mandibles and a long, pointed tail. Mature larvae similar to Braconidae.

  2. **Chalcidoidea:** First instar of internal parasites variable, usually greatly different than mature larvae. Mature larvae either saclike with no spiracles or one pair in most egg parasites. In other hosts, segmented grublike larvae with 9, rarely 10, open spiracles; one pair on meso- and metathorax and first 7 abdominal segments.

    a. *Eulophidae:* First instar is usually 13-segmented, with mandibles. Mature larva similar, with no functional spiracles in *Mestocharis* and maxillae with 2-segmented palpi. In most mature chalcidoid larvae, maxilliary palpi are disklike and not projecting.

    b. *Mymaridae:* First instar *(Caraphractus)* elongate, cylindrical with no mandibles or outgrowths and no visible segmentation (Fig. 21.6); first instar *(Anagrus)* saclike, head lobate, separated from body by a constriction. Mature larvae saclike, without mouth hooklets in *Caraphractus* and with hooklets in *Anagrus*.

    c. *Pteromalidae:* First instar 13-segmented; mandibles present. External parasites with 4 pairs of spiracles situated in mesothorax and first 3 abdominal segments; internal parasites lack functional spiracles. Mature larvae similar to first instar but with 9 pairs of spiracles; a pair on meso- and metathorax and first 7 abdominal segments; in *Sisridivora,* 10 spiracles with the additional pair on the 8th abdominal segment. Mandibles sclerotized and supported laterally and above by sclerotic rods.

    d. *Trichogrammatidae:* First instar larvae saclike, with minute mandibles. Mature larvae robust and segmented, without spines and setae; mandibles are elongated, extruded, and lie parallel to each other.

  3. **Proctotrupoidea:** First instar variable. Mature larvae segmented, with mandibles and with less than 9 pairs of spiracles.

    a. *Diapriidae:* First instar clearly segmented, solitary parasites with large mandibles and often with short, forked tail and no spiracles. In gregarious types (many larval parasites in one host; *(Trichopria),* first instar with mouthparts small and indistinct, but with a pair of distinct projecting papillae arising above mouth opening. Two tubercles are present on the last abdominal segment and each bear several teeth. Mature larvae *(Trichopria)* 13-segmented; papillae above mouthparts absent; mandibles sclerotized with or without teeth; spiracles on body segments 3, 4, and 5; posterior abdominal tubercles present or absent.

    b. *Scelionidae:* First instar with complete lack of segmentation, but body divided by a sharp constriction into somewhat equal portions; mandibles large, external, sharply pointed, widely separated, and usually curving ventrally; abdomen terminates in a horn or tail. A transverse row of spines usually around abdomen; no spiracles present. Mature larvae elongate-ovate, mandibles (oral hooks) slender and threadlike with broad base; abdomen without spines and tail.

  4. **Cynipoidea:** First instar segmented, with no spiracles, variable from an unusually elongate body with nearly every segment bearing a ventral appendage and last abdominal segment prolonged into a dorsally curved tail equal in length to the 4 preceding segments, to a body bearing no ventral appendages and a tail nearly as long as the body. Mature larvae segmented, ovoid to spindle-shaped with 6-10 spiracles; mandibles bi- or tridentate.

    Eucoilidae: First instar elongate, bearing long, paired, fleshy ventral processes on the thoracic segments and a long, spined, tapering tail with a ventral projecting appendage near the base. Mature larva with 9 pairs of spiracles on the last 2 thoracic and first 7 abdominal segments; mandibles bidentate.

Section Aculeata

  5. **Vespoidea** *Pompilidae:* Body with head and 13 segments, legless and grublike. Head with conspicuous lobes, mandibles large and toothed, maxillary and labial palpi project as conical tubercles; maxillary palpi and galeae about equal in length. Body with 10 spiracles, but second pair on thorax vestigial.

Adults

1.	Wing venation reduced; front wings usually with no enclosed cells, but, if present, with less than 5 enclosed cells; wings usually with long marginal fringe; hind wings veinless or with one vein and vein-enclosed cells absent; rarely wingless (Figs. 21.11-21.13, 21.16)	2
1'.	Wing venation well developed; front wings with more than 5 enclosed cells and without long marginal fringe, at least on anterior margins; hind wings with more than 2 veins and with at least one enclosed cell (Figs. 21.19-21.22)	8
2(1).	Tarsi 4- or 5-segmented	3
2'.	Tarsi 3-segmented (Fig. 21.11); parasites of insect eggs	**TRICHOGRAMMATIDAE**
3(2).	Tarsi 4-segmented (Fig. 21.13)	4
3'.	Tarsi 5-segmented (Figs. 21.16, 21.20)	5
4(3).	Marginal vein short, terminating within the first third of wing's length; stigmal vein absent (Fig. 21.12); body not metallic; parasites of insect eggs	**MYMARIDAE**
4'.	Marginal vein long, extending beyond one-half of wing's length; stigmal vein present (Fig. 21.13); body with metallic reflections or highly colored; parasites of dytiscid eggs, psephenid prepupae and pupae, leaf-miners, and mymarids	**EULOPHIDAE**

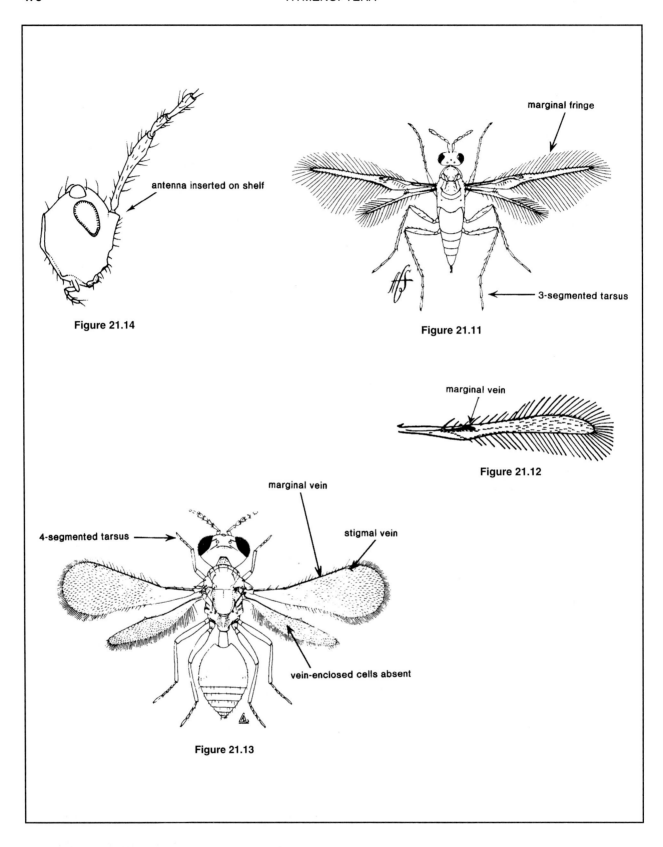

Figure 21.11. Female of *Hydrophylita aquivolans* Matheson and Crosby (Trichogrammatidae) (from Matheson and Crosby 1912).

Figure 21.12. Front wing of *Patasson gerrisophaga* (Doutt) (Mymaridae) (from Doutt 1949).

Figure 21.13. Female *Psephenivorus mexicanus* Burks (Eulophidae) (from Burks 1968).

Figure 21.14. Lateral view of a *Trichopria* sp. head (Diapriidae) (drawn by N. Vandenberg).

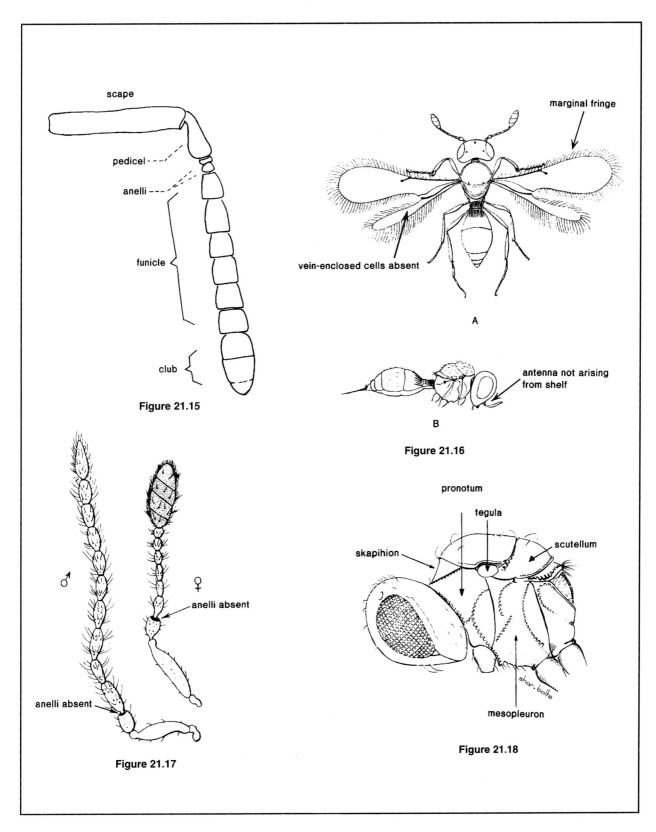

Figure 21.15. Pteromalid (*Eupteromalus* sp.) antenna (drawn by N. Vandenberg).

Figure 21.16. Female of *Tiphodytes gerriphagus* (Marchal) (Scelionidae). *A*, dorsal view; *B*, lateral view (from Marchal 1900).

Figure 21.17. Male and female antennae of *Tiphyodytes gerriphagus* (Marchal) (Scelionidae) (from Marchal 1900).

Figure 21.18. Lateral view of head and thorax of *Tiphyodytes gerriphagus* (Marchal) (Scelionidae) (from Masner 1972).

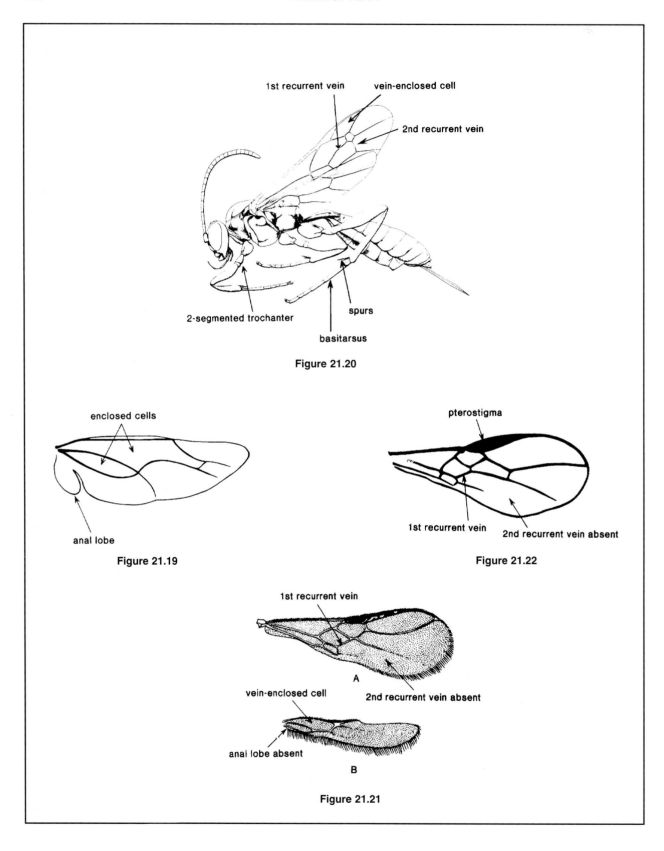

Figure 21.19. Hind wing of pompilid wasp (drawn by N. Vandenberg).

Figure 21.20. Lateral view of *Apsilops hirtifrons* (Ashmead) (Ichneumonidae) (from Townes 1970).

Figure 21.21. Front (*A*) and hind (*B*) wing of *Chorebus aquaticus* Muesebeck (Braconidae) (from Muesebeck 1950).

Figure 21.22. Front wing of an *Opius* sp. (Braconidae) (drawn by N. Vandenberg).

5(3').	Antennae inserted on shelf at middle of face (Fig. 21.14); parasites of dipterous puparia and psephenid pupae .... ***DIAPRIIDAE***
5'.	Antennae not arising from shelf at middle of face (Fig. 21.16B) 6
6(5').	Antennae with 2 distinctly smaller segments (anelli) between pedicel and first funicle segment; together the anelli thinner and usually shorter than first funicle segment (Fig. 21.15); parasites of gyrinid and sisyrid larvae and pupae in pupal cases or cocoons, or emerging from dipterous puparia .... ***PTEROMALIDAE***
6'.	Antennae without 2 distinctly smaller segments (anelli) between pedicel and first funicle segment (Fig. 21.17) 7
7(6').	Scutellum with 2 pits at base and an elevated pit ("cup") on disk; fore wing without stigmal vein; abdomen compressed; parasites of Diptera larvae and pupae .... ***EUCOILIDAE***
7'.	Scutellum without an elevated cup (Fig. 21.18) fore wing with stigmal vein (Figs. 21.13, 21.16A); abdomen depressed; parasites of eggs of Hemiptera, Lepidoptera, and Tabanidae .... ***SCELIONIDAE***
8(1').	Hind wings without an anal lobe (Figs. 21.20, 21.22); hind trochanters 2-segmented (Fig. 21.20); antennae with more than 15 segments; parasites of Diptera and Lepidoptera larvae and pupae 9
8'.	Hind wings with an anal lobe (Fig. 21.19); hind trochanters l-segmented; antennae with less than 15 segments and curled apically; parasites of spiders .... ***POMPILIDAE***
9(8).	Fore wing with 2 recurrent veins (second recurrent vein is a crossvein in apical lower half of wing disk) (Fig. 21.20); parasites of Lepidoptera larvae and pupae and Coleoptera pupae .... ***ICHNEUMONIDAE***
9'.	Fore wing with one recurrent vein (no crossvein in apical lower half of wing) (Figs. 21.21-21.22); parasites of Diptera larvae and puparia and Lepidoptera larvae .... ***BRACONIDAE***

ADDITIONAL TAXONOMIC (AND MORPHOLOGICAL) REFERENCES

General
Matheson and Crosby (1912); Clausen (1940); Muesebeck *et al.* (1951); Short (1952); Michener (1953); Hagen (1956, 1964); Richards (1956); Hedqvist (1967); Krombein and Burks (1967); Finlayson and Hagen (1979); Krombein *et al.* (1979); Evans *et al.* (1987); Grissell and Schauff (1990).

Taxonomic treatments at the family and generic levels
(L = larvae; P = pupae; A = adults)

Agriotypidae: Clausen (1950)-L; Mason (1971)-L, A; Chao and Zhang (1981)-L, A; Elliott (1982)-L, P.

Braconidae: Muesebeck (1950)-A; Burghele (1959)-A; Fischer (1964, 1971)-A; Capek (1970)-L.; Marsh *et al.* (1987)-A; Finlayson (1987b)-L.

Diapriidae: Kieffer (1916)-A; Muesebeck (1949, 1972)-A; Sundholm (1960)- A; O'Neill (1973)-L.

Eucoilidae: Weld (1952)-A; Quinlan (1967)-A.; Eskafi and Legner (1974)-L; Evans (1987a)-L; Beardsley (1992)-A.

Eulophidae: Parker (1924)-L; Graham (1959)-A; Boucek *et al.* (1963)-A; Jackson (1964)-L; Burks (1968)-A.

Ichneumonidae: Short (1959)-L; Mason (1968)-A; Townes (1969, 1970, 1971)-A; Finlayson (1987a)-L.

Mymaridae: Doutt (1949)-A; Annecke and Doutt (1961)-A; Jackson (1961 a)-L.; Schauff (1984)-A; Yoshimoto (1990)-A.

Pompilidae: Evans (1950, 1951)-A; Townes (1957)-A; Evans (1959, 1987c)-L; Krombein *et al.* (1979)-A.

Pteromalidae: Parker (1924)-L; Brown (1951)-L; Graham (1969)-A.

Scelionidae: Marchal (1900)-L, A; Martin (1927)-L; Masner (1972, 1976, 1980)-A; Evans (1987b)-L.

Trichogrammatidae: Parker (1924)-L; Martin (1927)-L; Doutt and Viggiani (1968)-A; Nagaraja and Nagarkatti (1973)-A.

Table 21A. Ecological (host-parasite) relationships and distribution (of host) of aquatic *Hymenoptera (wasps)*. (Table prepared by K. Hagen, K. W. Cummins, and R. W. Merritt.)

Parasite Taxa	Order	Host Family	Genus	Aquatic Life Stage of Host That Are Parasitized†	North American Distribution	Ecological References
Vespoidea						
Pompilidae (1)						
Anoplius (1) (= *Priocnemus*)	Araenea	Pisauridae	*Dolomedes*	Adults	Northeast, Northcentral	290, 585, 1117, 1124, 1523 1269
Proctotrupoidea						
Scelionidae (4)						
Pseudanteris (1)	Hemiptera	Gerridae?		Eggs	East	2499
Telenomus (3)	Lepidoptera	Pyralidae	*Chilo, Occidentalia*	Eggs	East	1317
	Diptera	Tabanidae	*Tabanus, Chrysops*	Eggs	Widespread	2190
Thoron (1) (= *Teleas, Anreris*)	Hemiptera	Nepidae	*Nepa?, Ranatra?*	Eggs	Northeast	2499
Tiphodytes (1) (=*Limnodytes, Hungaroscelio*)	Hemiptera	Gerridae	*Gerris, Trepobates, Limnoporus*	Eggs	Widespread	1523, 2478, 2499, 2502, 2517, 3767
Diapriidae (3)						
Trichopria (3) (= *Phaenopria*)	Diptera	Ephydridae	*Hydrellia*	Pupae	Widespread	263, 903, 1487, 1523
		Sciomyzidae	*Atrichomelina, Dictya, Elgiva, Sepedon*	Pupae	East	2950
	Coleoptera	Psephenidae	*Psephenus*	Prepupae, pupae	Texas	451
Ichneumonoidea						
Ichneumonidae (14)						
Apsilops (3) (=*Neostricklandia, Trichocryptus, Trichestema*)	Lepidoptera	Pyralidae	*Occidentalia, Chilo, Schoenobius, Nymphula*	Larvae	Widespread	263, 590, 1317, 1523
Bathythrix (2) (=*Hemiteles*)	Coleoptera	Gyrinidae	*Gyrinus*	Larvae in pupal cases	Iowa, Michigan	498, 2190
Cremastus (2?)	Lepidoptera	Pyralidae	*Nymphula, Occidentalia, Chilo*	Larvae, pupae	East	263, 1317, 1523
Medophron (2)	Coleoptera	Dytiscidae	*Agabus, Dytiscus*	Pupae	California, East	2504, *
Mesoleptus (2)	Diptera	Sciomyzidae	*Antichaeta*	Pupae	East	2134
Oecotelma (1) (= *Hemiteles*)	Coleoptera	Gyrinidae	*Gyrinus*	Larvae in pupal cases	Michigan	498, 2190
Pleurogyrus (2) (= *Hemiteles*)	Coleoptera	Gyrinidae	*Gyrinus*	Larvae in pupal cases	Michigan	498, 2190
Phygadeuon (1)	Diptera	Sciomyzidae	*Antichaeta*	Pupae	East	2134
Sulcarius	Trichoptera	Limnephilidae	*Limnephilus*	Larvae in cases	Minnesota	2691
Tanychela (2)	Lepidoptera	Pyralidae	*Petrophila*	Pupae	West	3307, 847
Therascopus (1) (=*Eriplanus*)	Coleoptera	Gyrinidae	*Gyrinus*	Larvae in pupal cases	Michigan	498
Braconidae (11)						
Ademon (2)	Diptera	Ephydridae	*Hydrellia*	Larvae-pupae	Widespread	263, 487, 903, 1523
Aphanta (1)	Diptera	Ephydridae	*Hydrellia*	Pupae	East	903
Asobara (1)	Diptera	Ephydridae	*Parydra*	Larvae—pupae	East	909
Bracon (1) (= *Microbracon*)	Lepidoptera	Noctuidae	*Archanara*	Larvae	East	1317

† Emphasis on trophic relationships.
* R. Garcia, pers. comm.

Table 21A. Continued

Parasite Taxa	Order	Host Family	Genus	Aquatic Life Stage of Host That Are Parasitized[†]	North American Distribution	Ecological References
Chaenusa (1 ?)	Diptera	Ephydridae	*Hydrellia*	Eggs—pupae	Widespread	487, 903, 1487,
Chorebidea (1)	Diptera	Ephydridae	*Hydrellia*	Pupae	Michigan	263, 487, 909, 1523
Chorebidella (1)	Diptera	Ephydridae	*Hydrellia*	Pupae	East, Alaska	263, 903, 1523,
Chorebus (1)	Diptera	Ephydridae	*Hydrellia*	Pupae	Widespread	263, 487, 909, 1487
Dacnusa (1 ?)	Diptera	Ephydridae	*Hydrellia*	Pupae	East	263, 909, 1523
Opius (1 ?)	Diptera	Ephydridae	*Hydrellia*	Larvae—pupae	Widespread	487, 909, 1487
Phaenocarpa (1)	Diptera	Sciomyzidae	*Antichaeta*	Pupae	East	2134
Chalcidoidea						
Mymaridae (3)						
Anagrus (1)	Hemiptera	Gyrinidae	*Gyrinus*	Eggs	Kansas	650, 1523, 1792
Caraphractus (1)	Hemiptera	Notonectidae	*Notonecta*	Eggs	East	1523, 2517
	Hemiptera	Gerridae	*Gerris*	Eggs	East	1857
	Odonata	Calopterygidae	*Calopteryx*	Eggs	East	
	Coleoptera	Dytiscidae	*Dytiscus, Agabus*	Eggs	East	1523, 1930, 1931, 2517, 3366
Patasson (1)	Hemiptera	Gerridae	*Gerris*	Eggs	California	948, 1523, 3557
Polynema (1)	Odonata	Lestidae	*Lestes*	Eggs	Illinois	1402
Trichogrammatidae (9)						
Hydrophylita (1)	Odonata	Coenagrionidae	*Ischnura*	Eggs	East	868, 1523, 2517
Lathromeroidea (2)	Odonata	Lestidae	*Lestes*	Eggs	East	949, 2781
(= *Centrobiopsis*)						
	Hemiptera	Gerridae	*Gerris, Limnoporus*	Eggs	Alberta	1675
				Eggs		
Paracentrobia (1)	Odonata	Lestidae	*Lestes*	Eggs	Widespread	2190
Prestwichia (1)	Hemiptera	Nepidae	*Ranatra, Nepa*	Eggs	East?	650, 1317, 1671, 3494
	Coleoptera	Dytiscidae	*Dytiscus*	Eggs		650, 1523, 1930
Trichogramma ?(4)	Hydracarina			Eggs	West?	939, 1097
	Megaloptera	Sialidae	*Sialis*	Eggs	West	138, 2478, 3207
	Megaloptera	Corydalidae	*Chauliodes*	Eggs	East	1403
	Diptera	Sciomyzidae	*Sepedon*	Eggs	East	2030, 2031
	Diptera	Sciomyzidae	*Elgiva*	Eggs	East	2030
	Diptera	Sciomyzidae	*Tetanocera*	Eggs	East	2030
Eulophidae (3)						
Aprostocetus (1)	Odonata	Lestidae	*Lestes*	Eggs (hyperparasite)	East	490
Mestocharis (2)	Coleoptera	Dytiscidae	*Dytiscus?*	Eggs	East	1930, 1931, 1936
Psephenivorus (1)	Coleoptera	Psephenidae	*Psephenus*	Prepupae, pupae	Texas	451, 491
Tetrastichus (1)	Odonata	Lestidae	*Lestes*	Eggs and hyperparasite	East	2190
Pteromalidae (5)						
Eupteromalus (2)	Diptera	Ephydridae	*Hydrellia*	Pupae	California	1487
Gyrinophagus (1)	Coleoptera	Gyrinidae	*Gyrinus, Dineutes*	Larvae in pupal cases	East, Midwest	498, 4362
Halticoptera (11)	Diptera	Ephydridae	*Hydrellia*	Larvae	California	1487
Sisridivora (1)	Neuroptera	Sisyridae	*Climacia*	Larvae, pupae in cocoons	Widespread	449, 1488, 3224
Cynipoidea						
Eucoilidae						
Hexacola (2)	Diptera	Ephydridae	*Parydra*	Larvae—pupae	East	909
	Diptera	Chloropidae	*Hippelates*	Larvae—pupae	California	1106

[†] Emphasis on trophic relationships.

AQUATIC DIPTERA
Part One. Larvae of Aquatic Diptera

G. W. Courtney
*Grand Valley State University,
Allendale, Michigan*

R. W. Merritt
*Michigan State University,
East Lansing*

H. J. Teskey
*Biosystematics Research Institute,
Agriculture Canada, Ottawa*

B. A. Foote
*Kent State University,
Kent, Ohio*

INTRODUCTION

Diptera or true flies are one of the most diverse insect orders, with almost 120,000 described species. The order contains many common and familiar insects, including several important aquatic groups such as mosquitoes (Chap. 24), black flies (Chap. 25), and midges (Chap. 26). Diptera are diverse in not only number of species, but with respect to morphological and ecological characteristics. Although the adults of a few Diptera are briefly aquatic (e.g., while ovipositing or emerging from submerged pupae), only the immature stages of certain species spend extended periods of time partially or fully submerged. Larvae of most species can be considered aquatic in the broadest sense since for survival they must be surrounded by a moist to wet environment, i.e., within tissues of living plants, amid decaying organic materials, as parasites of other animals, or in association with bodies of water. Only the latter habitat is normally considered to encompass aquatic species.

The Diptera have successfully colonized nearly every aquatic habitat except the open ocean, and dispersing adults of some families (e.g., Ephydridae) may even be captured above the latter habitat. Diptera larvae occur in coastal marine, saline and estuarine waters; shallow and deep lakes; ponds; cold and hot springs; natural seeps of crude petroleum; stagnant waters of ground-pools, plant cavities (phytotelmata) and artificial containers; slow to torrential streams; and groundwater zones. Larvae of most groups are free-living and actively crawl or swim in the habitat (e.g., Deuterophlebiidae, Simuliidae and Culicidae); others may be buried in sediments or beneath stones (e.g., Ptychopteridae and Tabanidae), within saturated wood (e.g., the chironomids *Symposiocladius*, *Xylotopus*, and the tipulid *Lipsothrix*), or inside silken tubes attached to rocks, plants and debris (e.g., the tipulid *Antocha* and many Chironomidae). Aquatic habits are most prevalent in larvae of nematocerous Diptera, including all or most species of the Blephariceridae, Chaoboridae, Chironomidae, Culicidae, Deuterophlebiidae, Dixidae, Ptychopteridae and Simuliidae. Among brachycerous flies, aquatic habits are perhaps most common in the Athericidae and Ephydridae. In some families, such as the Muscidae and Scathophagidae, only a few species are aquatic.

Lotic habitats for Diptera larvae range from slow, silty rivers to clear, torrential streams. The larvae of the Blephariceridae, Deuterophlebiidae and Simuliidae are among the most specialized inhabitants of flowing waters; all lack spiracles and rely on exchange of oxygen directly through their skin, and all have structural modifications that permit survival on current-exposed substrates. The larvae of Blephariceridae (net-winged midges) usually inhabit streams where current velocities exceed 2 m/sec, and these flies show perhaps the greatest morphological specialization. Larvae have a compact, somewhat flattened body composed of six primary divisions, the first consisting of the *cephalothorax*, a fused head, thorax and 1st abdominal segment (Fig. 22.10). This division and each of the remaining five have a highly specialized, ventral suctorial disc used to adhere to smooth rocks (Fig. 22.11). The Deuterophlebiidae (mountain midges) occur in similar habitats, often on the same rocks as Blephariceridae. Deuterophlebiid larvae are dorsoventrally flattened and also have unusual attachment devices in the form of elongate lateral prolegs that bear apical rows of *crochets* (small, curved hooks; Fig. 22.18). Another family whose larvae frequent current-exposed habitats is the Simuliidae, a group that can be numerically and economically important. Larvae of these flies use a crochet-tipped posterior proleg and a silk pad to adhere to submerged rocks or vegetation, where they extend their head into the current and strain microscopic food particles with their labral fans (Chap. 25). Other specialized flowing-water habitats include seepages on cliff faces and waterfall splash-zones, habitats frequented by larval Thaumaleidae and many Chironomidae, Tipulidae and Stratiomyidae. The larvae of many other lotic Diptera occur in the sheltered habitats beneath rocks (e.g., many Tipulidae and Chironomidae), inside moss (Nymphomyiidae and the muscid *Limnophora*) or attached to other vegetation (e.g., many Empididae), and among the silt, sand, and gravel of the streambed (e.g., Tanyderidae). Subterranean channels are another potentially important but largely unstudied habitat for larval Diptera, particularly for the Chironomidae.

Lentic habitats for Diptera larvae are equally diverse, including lakes, springs, temporary pools, and tree holes. Stream depositional zones are in some respects similar to these lentic habitats. Larvae of the Culicidae, Chaoboridae and Dixidae are among the lentic groups that lead active and predominantly exposed lives, usually at or near the surface. Although these families include proficient swimmers that can travel to considerable depths, their larvae generally remain near the water surface because of a dependence on atmospheric respiration. The culicid *Coquillettidia* is an exception to this pattern, as its larvae use a specialized siphon to obtain oxygen from submerged vegetation. Free-swimming larvae of Chironomidae and Ceratopogonidae do not depend on atmospheric respiration and can colonize larger and deeper bodies of water. Some chironomids survive at great depths, with one species known from below 1300 m in Lake Baikal. As in lotic habitats, most lentic Diptera are associated with benthic substrates. The larvae of many families that respire through spiracles and inhabit benthic sediments (e.g., Tipulidae, Tabanidae, Stratiomyidae, and Dolichopodidae) must stay within easy migratory distance of the air-water interface. The upper two inches of damp-to-saturated sediments are among the favored habitats of most representatives of these families. This respiratory problem has been alleviated for the larvae of Ptychopteridae and some Syrphidae by the presence of a slender terminal respiratory siphon extensile up to two or three times the body length. Even the pupae of Ptychopteridae have an elongate respiratory horn (Fig. 22.175) that allows them to remain submerged in saturated muck soils, which are the favored habitat.

Insects in general are not considered a major component of marine and brackish-water environments, but fly larvae can in fact be abundant in some habitats. Among the groups that have successfully invaded these zones (e.g., intertidal pools, estuarine marshes, etc.) are members of the Ceratopogonidae, Chironomidae, Tipulidae, Tabanidae, Canacidae, and Ephydridae. The Ephydridae are particularly well adapted to brackish-water habitats, including maritime marshes, tidal salt pools, and alkaline lakes of arid regions. The larvae of many other families (e.g., Coelopidae and Sphaeroceridae) frequent piles of decaying seaweed on the beach, but because these habitats are essentially semiaquatic, these families are not covered in this chapter.

The taxonomic and ecological diversity of Diptera is reflected in the wide range of larval feeding habits, which encompass nearly every trophic group (Table 22B). Shredders, which consume live plants or decomposing plant fragments, are particularly well represented by larvae of the Tipulidae and Ephydridae. These families also contain many collector-gatherers, feeding on decaying, fine organic matter and associated microorganisms; the larvae of several other families (e.g., Psychodidae, Ptychopteridae and Syrphidae) belong to this functional group. Most Culicidae and Simuliidae consume organic matter of comparable size and quality, but these collector-filterers use modified mouth-brushes or labral fans (Figs. 22.20, 22.35) to extract particles from the water. Other groups (e.g., Stratiomyidae, Syrphidae) lack external filtering devices, but accomplish the same task with a filter-like pharynx (Fig. 22.77). Grazers, which feed on the thin periphytic film of algae and organic matter on rocks and other substrates, include the highly specialized families Blephariceridae, Deuterophlebiidae and Thaumaleidae, and various members of the Psychodidae (e.g., *Maruina*), Simuliidae (e.g., *Gymnopais*) and Ephydridae (e.g., *Parydra*). Many families (e.g., Chironomidae, Culicidae, Tipulidae and Ephydridae) contain a few predaceous species, whereas the larvae of some other groups (e.g., Ceratopogonidae, Athericidae, Dolichopodidae and Tabanidae) feed primarily or exclusively on other animals. All of these functional groups are represented in the diverse families Chironomidae, Tipulidae and Ephydridae. Parasitoid habits are rare among aquatic dipteran larvae, but examples include the Sciomyzidae, whose hosts are freshwater snails and fingernail clams, and *Oedoparena* sp. (Dryomyzidae), which feed on barnacles.

Their trophic diversity and numerical abundance make the Diptera an important component in aquatic ecosystems, both as primary consumers and as a food resource for other invertebrates, fish, amphibians, reptiles, birds and mammals. The Chironomidae, which are often the most abundant organism in both number and biomass, can be especially significant in ecosystem functioning. Other families, especially the Simuliidae and Culicidae, are, because of their role as vectors of disease and as pests of humans, livestock and other animals, of great medical and economic importance. The practical significance of aquatic Diptera extends to allergies (e.g., from wind-borne setae and other body parts of certain Chironomidae), aquaculture programs (e.g., as fish food), and water-supply and sewage-treatment facilities (e.g., trickle filters are often colonized by larvae of certain Psychodidae). Aquatic Diptera can serve other significant roles in the fields of water quality, biomonitoring, conservation biology, and in scientific research on the structure and function of aquatic ecosystems, structural and ecological adaptations to aquatic environments, and patterns and mechanisms of evolutionary divergence.

As a holometabolous insect, or one that undergoes complete metamorphosis, the dipteran life-cycle includes a series of distinct stages or instars. A typical life-cycle consists of a brief egg stage (usually a few days or weeks, but sometimes much longer), 3-4 larval instars (typically 3 in Brachycera, 4 in nematocerous Diptera, more in Simuliidae, Tabanidae and a few others), a pupal stage of varying length, and an adult stage that lasts from less than 2 hours (Deuterophlebiidae) to several weeks or even months. The eggs of aquatic Diptera are usually laid singly, in small clusters, or in loose or compact masses in or near the water and attached to rocks or vegetation. In some groups (e.g., Deuterophlebiidae and Nymphomyiidae), the female crawls beneath the water, losing its wings in the process, to select oviposition sites. This behavior ensures that eggs are placed in a habitat that is suitable for larval development. For a particular species, all larval instars usually occur in the same habitat, though exceptions include lentic Chironomidae with planktonic first-instars and benthic later-instars. In general, the duration of the first larval-stage is shortest, whereas that of the last instar is much greater, often several weeks or even months.

EXTERNAL MORPHOLOGY

The larvae of aquatic Diptera can be distinguished from those of most other aquatic insects by the lack of jointed thoracic legs, but otherwise larval dipterans show tremendous structural variation. The larval head may be extensively modified from the complete, fully exposed and heavily sclerotized head capsule of most nematocerous Diptera, through reduction to forms with slender *tentorial arms* and *cephalic rods* that are partly retracted into the thorax (e.g., orthorrhaphous Brachycera; Fig. 22.59). Larval Tipulidae represent a special case among nematocerous flies, as the head capsule is often fully retracted into the thorax and the posterior cranial margin may bear small to extensive *longitudinal incisions* (Figs. 22.7-22.9). The culmination of cranial reduction is in cyclorrhaphous brachycerans, whose heads consist of an external membranous segment and a distinctively shaped *cephalopharyngeal skeleton* that is fully retracted into the thorax (Figs. 22.91, 22.100). The cephalopharyngeal skeleton represents the remnants of internal cranial sclerites (tentorium) and various mouthparts. Cranial modifications are accompanied by general changes in the shape and rotation of mandibles and other mouthparts. The mandible of larval nematocerous Diptera usually consists of a stout, toothed structure that moves in a horizontal or oblique plane and operates as a biting and chewing organ (Figs. 22.8, 22.9). The brachyceran larval mandible is, in contrast, more claw-like, has fewer teeth along the inner surface, moves in a vertical plane and operates as a piercing or slashing organ (Figs. 22.59, 22.60). In most Diptera larvae, the thorax and abdomen are soft, flexible and only occasionally provided with sclerotized plates. The thorax usually consists of three distinct segments and the abdomen usually 8 or 9 segments. Larvae of the Culicidae, Chaoboridae and Corethrellidae are unusual in that the thoracic segments are indistinctly differentiated and form a single large segment that is wider than the rest of the body (Figs. 22.20, 22.21, 22.30, 22.31). In many Diptera larvae the body is covered with dense pubescence (short, fine hairs) or scattered hairs of various size and structure. Tubercles (small, fleshy processes) may be present on one or more segments, particularly on the last abdominal segment and near the posterior spiracles (e.g., Tipulidae). These processes may serve various sensory and/or locomotory purposes. Because Diptera larvae lack jointed thoracic legs, these tubercles and other fleshy organs may be particularly useful for locomotion. Locomotory appendages operate through a combination of turgor pressure and muscle action, and include three basic types: creeping welts, prolegs and suctorial discs. *Creeping welts* are transverse, swollen areas (ridges) that bear one to several modified setae or spines; creeping welts are characteristic of certain Tipulidae, Tabanidae, Empididae and Dolichopodidae (Figs. 22.3, 22.56, 22.66, 22.73). *Prolegs* usually are paired, round, elongate, fleshy, retractable processes that bear apical spines or crochets; prolegs come in a diversity of shapes, sizes and positions, and are typical of larval Chironomidae, Deuterophlebiidae, Nymphomyiidae, Simuliidae, Thaumaleidae, Athericidae, and various members of other groups (e.g., Figs. 22.35, 22.38, 22.43). *Suctorial discs* are true suction devices on the ventral body surface of certain larvae from torrential streams, specifically the Blephariceridae (Fig. 22.11). The presence of these different locomotory appendages, listed in more or less increasing order of specialization, have greatly facilitated the colonization of aquatic habitats by Diptera larvae.

Diptera larvae show a variety of respiratory adaptations. Respiration may be directly from the atmosphere (e.g., Dixidae, most Culicidae), from plant tissues (e.g., some Culicidae), or from the surrounding water (e.g., Blephariceridae, Simuliidae). The presence of haemoglobin in the blood of some Chironomidae may assist the absorption of oxygen. Many aquatic larvae, particularly those from lotic habitats (e.g., Blephariceridae, Deuterophlebiidae, Nymphomyiidae and Simuliidae) are *apneustic* (lack spiracles; e.g., Figs. 22.10, 22.11, 22.16, 22.35) and absorb oxygen directly through the skin. Some families (e.g., Psychodidae, Tanyderidae, Thaumaleidae, and Canacidae) are *amphipneustic*, possessing spiracles on the prothorax and last abdominal segment (e.g., Figs. 22.36, 22.79), or next to last abdominal segment (Fig. 22.17), while others (e.g., Culicidae, Dolichopodidae and Tabanidae) are *metapneustic*, with spiracles on only the last segment (e.g., Figs. 22.20, 22.56, 22.57, 22.70). In several groups (e.g., Ptychopteridae, Ephydridae and Syrphidae), the spiracles are located at the end of a retractile respiratory siphon (= respiratory tube or breathing tube; e.g., Figs. 22.23, 22.76, 22.86), which further assists larvae to colonize aquatic habitats. Further information on the morphology of Diptera larvae is provided by Teskey (1981a).

The following key has been substantially expanded over the one given in the second edition to include the genera, where sufficiently well known, of families that are not treated in separate chapters. These additions have been taken, with various changes and/or modifications, from chapters in *The Manual of Nearctic Diptera*, Vols. 1 and 2 (1981, 1987) (Coords. McAlpine *et al.* 1981, 1987), by C. L. Hogue (Blephariceridae), C. P. Alexander (Ptychopteridae), E. F. Cook (Chaoboridae), T. M. Peters (Dixidae), M. T. James (Stratiomyidae), G. C. Steyskal and L. V. Knutson (Empididae, Sciomyzidae), and W. W. Wirth (Canacidae). The key to ceratopogonid larvae was modified from Peckarsky *et al.* (1990).

KEY TO THE FAMILIES AND SELECTED GENERA OF AQUATIC DIPTERA

Larvae

1. Mandibles moving against one another in a horizontal or oblique plane (Figs. 22.8, 22.9); head capsule usually complete and fully exposed, except retracted and reduced in Tipulidae ....NEMATOCERA 2

1'. Mandibles or mouth hooks moving parallel to one another in a vertical plane (Figs. 22.59, 22.60, 22.64, 22.71, 22.82); head capsule variously reduced posteriorly; partially or almost completely retracted within thorax even if such retracted portions comprise only a few slender rods ....BRACHYCERA 42

2(1).	Head capsule partially to fully retracted within thorax (Figs. 22.1-22.3), usually with longitudinal incisions of varying depths dorsolaterally (Figs. 22.7-22.9); in extreme cases head consisting only of several slender rods (Fig. 22.8); respiratory system metapneustic or apneustic; posterior spiracles usually bordered by 1-3 or 5-7 pairs of short lobes that are often fringed with short to very long hairs (Figs. 22.1-22.6) ....***TIPULIDAE*** (Chap. 23)
2'.	Head capsule complete, usually without longitudinal incisions dorsolaterally, completely exserted (Figs. 22.16-22.21, 22.29); respiratory system amphipneustic, metapneustic, or apneustic; posterior spiracles usually without bordering fringed lobes 3
3(2').	Head not distinctly separated from thorax; body divided into 7 major divisions, the 1st comprising the fused head, thorax, and 1st abdominal segment (Fig. 22.10); each of the 1st 6 divisions with a median suctorial disk ventrally (Fig. 22.11) ....***BLEPHARICERIDAE***................4
3'.	Head showing a distinct constrictive separation from the thorax (Figs. 22.21, 22.29-22.37); suctorial disks absent (except in *Maruina*, Psychodidae) 8
4(3).	Dorsal proleg absent, at least from anal division of abdomen (Fig. 22.15) ....***Blepharicera*** Macquart
4'.	Dorsal proleg present (Fig. 22.10) 5
5(4').	Ventral gill tufts composed of 6 filaments, arranged in a semirosette pattern in the same plane (Fig. 22.12) ....***Bibiocephala*** Osten Sacken
5'.	Ventral gill tufts composed of 3-5 or 7 filaments, spreading and all directed generally anterolaterally (Fig. 22.11) 6
6(5').	Dorsal sclerotized plates or tubercles usually present on abdominal segments (Figs. 22.10, 22.13-22.14); if absent, at least a small dorsolateral tubercle present above dorsal proleg on abdominal segment 1 7
6'.	Dorsal sclerotized processes absent; if conical processes present, 2 transverse series of minute plates also present across thoracic region of anterior body division ....***Agathon*** Röder (***aylmeri*** group)
7(6).	Dorsal proleg double, with a subequal, elongate dorsal branch (Fig. 22.13); or ventral gill tufts with 3 filaments (Fig. 22.11) ....***Philorus*** Kellogg
7'.	Dorsal proleg single, with only a small proximodorsal mammillate (nipplelike) process or setal swelling (Fig. 22.14); ventral gill tufts with 5-7 filaments ....***Agathon*** Röder (in part)
8(3').	A pair of elongate prolegs present on each of 7 or 8 abdominal segments (Fig. 22.16) 9
8'.	Prolegs usually absent, but, if present, on no more than 3 abdominal segments 10
9(8).	Body with 8 pairs of slender, ventrally projecting, abdominal prolegs, each terminating in a comb of clawlike spines; antennae simple, shorter than head length (Fig. 22.16) ....***NYMPHOMYIIDAE***—***Nymphomyia*** Tokunaga
9'.	Seven pairs of rather broad abdominal prolegs projecting ventrolaterally; their apices encircled by several transverse rows of hooked spinules; antennae forked, longer than length of head (Figs. 22.18-22.19) ....***DEUTEROPHLEBIIDAE***—***Deuterophlebia*** Edwards
10(8').	Abdomen terminating in a long, slender, telescopic respiratory siphon (Fig. 22.23); body segments with multiple transverse ridges or rows of small setae or setiferous (setae-bearing) papillae (soft projections); first 3 abdominal segments with a pair of ventral prolegs, sometimes very small, bearing a single, slender, curved claw (Figs. 22.23-22.24) ....***PTYCHOPTERIDAE***.......11
10'.	Not having above characters 13
11(10).	Hypostoma apparently fused with hypostomal bridge that closes head capsule ventrally, its anterior margin bilobed (Fig. 22.25); mandible with a single outer tooth; prolegs prominent ventrally on first 3 abdominal segments, each with a conspicuous curved claw (Fig. 22.24); coloration rusty red or black 12
11'.	Hypostoma separated from hypostomal bridge, its anterior margin multitoothed (Fig. 22.26); mandible with 3 large outer teeth; prolegs and apical curved claws small and inconspicuous (Fig. 22.23); coloration pale yellow or brown ....***Ptychoptera*** Meigen
12(11).	Predominantly blackish in color; respiratory siphon light yellow, entirely retractile; very long projections encased in a black horny substance, covering entire body; mandible with an inner comb of teeth ....***Bittacomorphella*** Alexander
12'.	Rusty red in color; body tapering gradually to a long, slender, partly retractile respiratory siphon; transverse rows of shorter stellate tubercles covering entire body (Fig. 22.24); mandible without an inner comb of teeth ....***Bittacomorpha*** Westwood
13(10').	Thoracic segments fused and indistinctly differentiated, forming a single segment that is wider than any of abdominal segments (Figs. 22.21, 22.30-22.31, 22.33-22.34); thoracic and abdominal segments with prominent, lateral fanlike tufts of long setae, and/or terminal segment with an anal setal fan (Figs. 22.20-22.22, 22.30-22.31, 22.33-22.34) 14

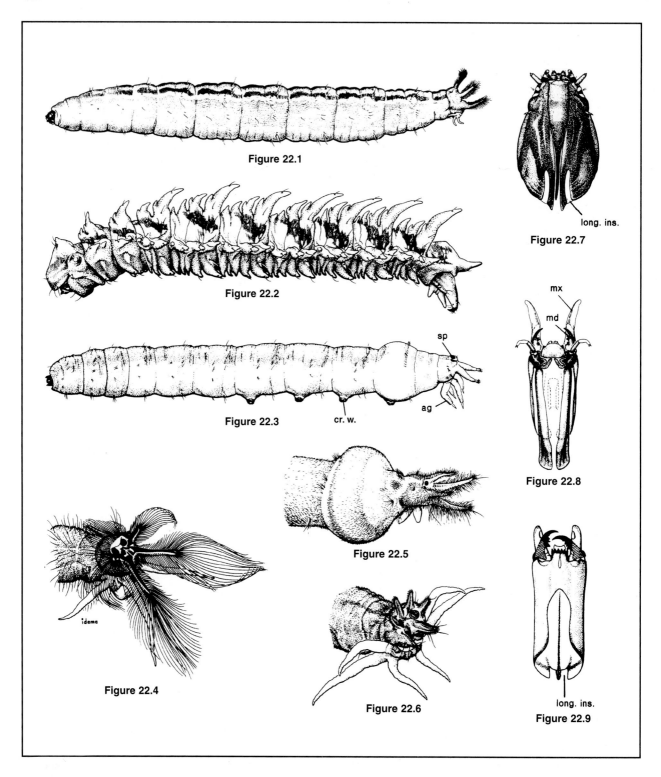

Figure 22.1. Lateral view of *Prionocera sp.* (Tipulidae).

Figure 22.2. Lateral view of *Liogma nodicornis* (Osten Sacken) (Tipulidae).

Figure 22.3. Lateral view of *Pedicia sp.* (Tipulidae); *ag,* anal gill; *cr. w.,* creeping welt; *sp,* spiracle.

Figure 22.4. Caudal segments of *Pseudolimnophila inornata* (Osten Sacken) (Tipulidae).

Figure 22.5. Caudal segments of *Limnophila sp.* (Tipulidae).

Figure 22.6. Caudal segments of *Tipula strepens* Loew (Tipulidae).

Figure 22.7. Dorsal view of head capsule of *Prionocera sp.* (Tipulidae); *long. ins.,* longitudinal incision.

Figure 22.8. Dorsal view of head capsule of *Limnophila sp.* (Tipulidae); *md,* mandible; *mx,* maxilla.

Figure 22.9. Ventral view of head capsule of *Dicranota sp.* (Tipulidae); *long. ins.,* longitudinal incision.

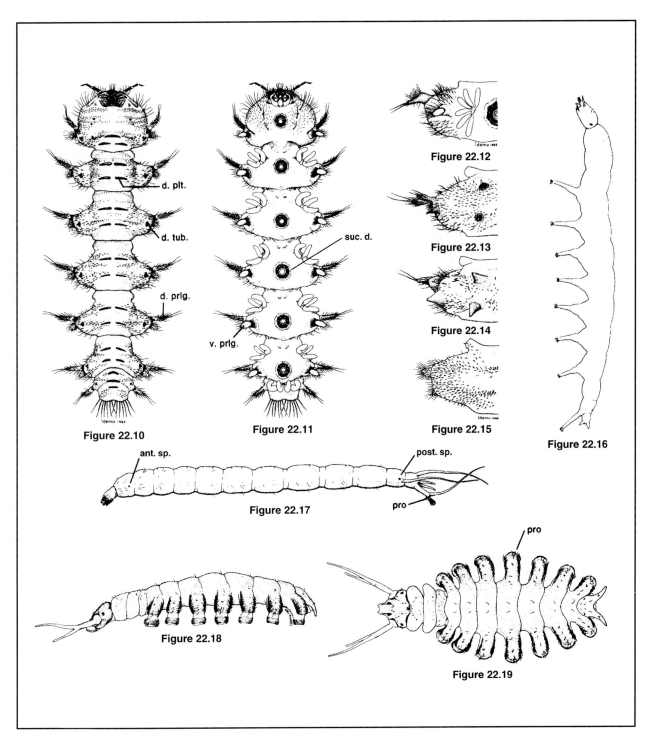

Figure 22.10. Dorsal view of *Philorus californicus* Hogue (Blephariceridae); *d. plt.*, dorsal plate; *d. tub.*, dorsal tubercle; *d. prlg.*, dorsal proleg.

Figure 22.11. Ventral view of *Philorus californicus* Hogue (Blephariceridae); *suc. d.*, suctorial disc; *v. prlg.*, ventral proleg.

Figure 22.12. Ventral view of first abdominal division of *Bibiocephala grandis* Osten Sacken (Blephariceridae).

Figure 22.13. Dorsal view of first abdominal division of *Philorus yosemite* (Osten Sacken) (Blephariceridae).

Figure 22.14. Dorsal view of first abdominal division of *Agathon elegantulus* Röder (Blephariceridae).

Figure 22.15. Dorsal view of first abdominal division of *Blepharicera tenuipes* (Walker) (Blephariceridae).

Figure 22.16. Lateral view of *Nymphomyia walkeri* Ide (Nymphomyiidae); *pro*, proleg.

Figure 22.17. Lateral view of *Protoplasa fitchii* Osten Sacken (Tanyderidae); *ant. sp.*, anterior spiracle; *pro*, proleg.

Figure 22.18. Lateral view of *Deuterophlebia nielsoni* Kennedy (Deuterophlebiidae).

Figure 22.19. Dorsal view of *Deuterophlebia nielsoni* Kennedy (Deuterophlebiidae); *pro*, proleg.

490 DIPTERA

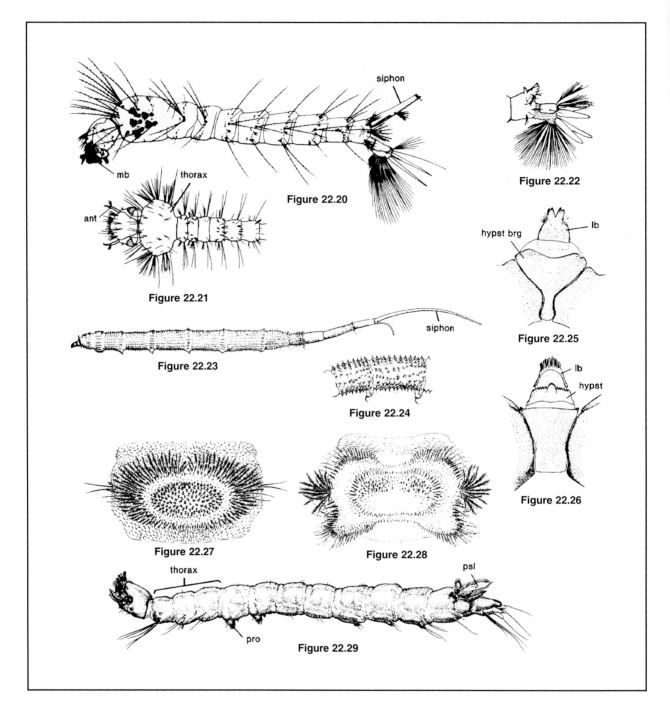

Figure 22.20. Lateral view of *Aedes fitchii* Felt and Young (Culicidae); *mb*, mouth brush.

Figure 22.21. Dorsal view of head, thorax, and six abdominal segments of *Aedes sp.* (Culicidae); *ant*, antenna.

Figure 22.22. Lateral view of caudal segment of *Anopheles sp.* (Culicidae) showing short respiratory siphon.

Figure 22.23. Lateral view of *Ptychoptera sp.* (Ptychopteridae).

Figure 22.24. Lateral view of first two abdominal segments of *Bittacomorpha clavipes* (Fabricius) (Ptychopteridae).

Figure 22.25. Ventral view of central portion of head capsule of *Bittacomorpha clavipes* (Fabricius) (Ptychopteridae) showing absence of a separate hypostoma; *hypst brg*, hypostomal bridge; *lb*, labium.

Figure 22.26. Ventral view of central portion of head capsule of *Ptychoptera sp.* (Ptychopteridae) showing a separate hypostoma; *hypst*, hypostoma; *lb*, labium.

Figure 22.27. Dorsal view of an abdominal segment of *Dixa sp.* (Dixidae) showing dorsal corona.

Figure 22.28. Dorsal view of an abdominal segment of *Meringodixa chalonensis* Nowell (Dixidae) showing dorsal corona and lateral tufts of long hairs.

Figure 22.29. Dorsolateral view of *Dixella sp.* (Dixidae); *pro*, proleg; *psl*, postspiracular lobe.

13'.	Thoracic segments usually individually distinguishable; thorax and abdomen about equal in diameter or abdomen wider (Fig. 22.35); setae on thoracic and abdominal segments not tufted and anal fan of terminal segment absent	18
14(13).	Antennae prehensile (grasping), with long apical setae; mouth brushes absent (Figs. 22.30-22.34)	15
14'.	Antennae not prehensile and with only short apical setae; prominent mouth brushes present on either side of labrum (Figs. 22.20-22.21)	***CULICIDAE*** (Chap. 24)
15(14).	Antennae inserted close together; a transverse row of spiniform setae on each side of head (Fig. 22.32); terminal segment with a tuft of long setae ventrally instead of a fan	***CORETHRELLIDAE***—*Corethrella* Coquillett[*]
15'.	Antennae inserted far apart; head without a transverse row of setae laterally; terminal segment with longitudinal fanlike row of setae (Fig. 22.34)	***CHAOBORIDAE*** 16
16(15').	Abdominal segment 8 with an elongate dorsal respiratory siphon (Fig. 22.34)	***Mochlonyx*** Loew
16'.	Abdominal segment 8 without a long respiratory siphon (Figs. 22.30-22.31)	17
17(16').	Thorax rounded laterally; hydrostatic air-sacs present in thorax; respiratory siphon absent (Fig. 22.30)	***Chaoborus*** Lichtenstein
17'.	Thorax diamond-shaped in dorsal view (Fig. 22.33); hydrostatic air-sacs absent; respiratory siphon on abdominal segment 8 short, stout, terminating in conspicuous flat spiracular apparatus (Fig. 22.31)	***Eucorethra*** Underwood
18(13').	Paired crochet-bearing prolegs ventrally on 1st and usually 2nd abdominal segments (Fig. 22.29); abdomen posteriorly with 2 flattened dorsolateral postspiracular lobes having setose (hair-covered) margins projecting above a conical, dorsally sclerotized segment bearing the terminal anus and anal papillae (Fig. 22.29)	***DIXIDAE*** 19
18'.	Abdominal segments without prolegs, except on anal segment; posterior body segment without flattened, fringed postspiracular lobes and a conical, dorsally sclerotized anal segment	21
19(18).	Abdominal segments 2-7 each with dorsal corona-like fringe of plumose hairs (Figs. 22.27-22.28)	20
19'.	Abdominal segments 2-7 without dorsal coronae (Fig. 22.29)	***Dixella*** Dyar and Shannon
20(19).	Pair of prolegs ventrally on only abdominal segment 1; tuft of longer hairs on each side of abdominal segments 2-6 below coronae (Fig. 22.28)	***Meringodixa*** Nowell
20'.	Pair of prolegs on each of first 2 abdominal segments; lateral tufts of hairs on abdominal segments absent (Fig. 22.27)	***Dixa*** Meigen
21(18').	Prothorax with 1 proleg or a pair of prolegs ventrally (Figs. 22.35, 22.38, 22.40-22.43)	22
21'.	Prothorax lacking prolegs	25
22(21).	Head capsule usually with a pair of conspicuous, folding labral fans dorsolaterally (Fig. 22.35); abdominal segments 5-8 swollen, posterior segment terminating in a ring or circlet of numerous radiating rows of minute hooks (Fig. 22.35)	***SIMULIIDAE*** (Chap. 25)
22'.	Head capsule lacking labral fans; posterior abdominal segments not conspicuously swollen nor with radiating rows of hooks terminally, although anal proleg(s) bearing crochets may be present	23
23(22').	Respiratory system amphipneustic (Fig. 22.38); anterior spiracles on short stalks, and posterior spiracles opening into a transverse cleft between fingerlike processes on abdominal segment 8 (Fig. 22.39); prothoracic and anal prolegs unpaired	***THAUMALEIDAE***
23'.	Respiratory system apneustic (Figs. 22.40-22.46); prothoracic or anal prolegs usually paired even if distinction is only a slight separation of the apical spines (Fig. 22.43)	24
24(23').	All body segments dorsally with prominent tubercles (elevated fleshy processes) and/or setae (Fig. 22.42)	***CERATOPOGONIDAE*** (in part) 29
24'.	Body segments lacking prominent dorsal tubercles and setae (Figs. 22.40-22.41, 22.43)	***CHIRONOMIDAE*** (Chap. 26)
25(21').	Last 2 abdominal segments with long filamentous processes, pairs of such processes arising laterally on the next to last segment, dorsolaterally on the terminal segment, and from near the apex of 2 elongate cylindrical prolegs that project posteroventrally from the terminal segment (Fig. 22.17)	***TANYDERIDAE***

[*] Larvae of *Lutzomiops* Lane unknown.

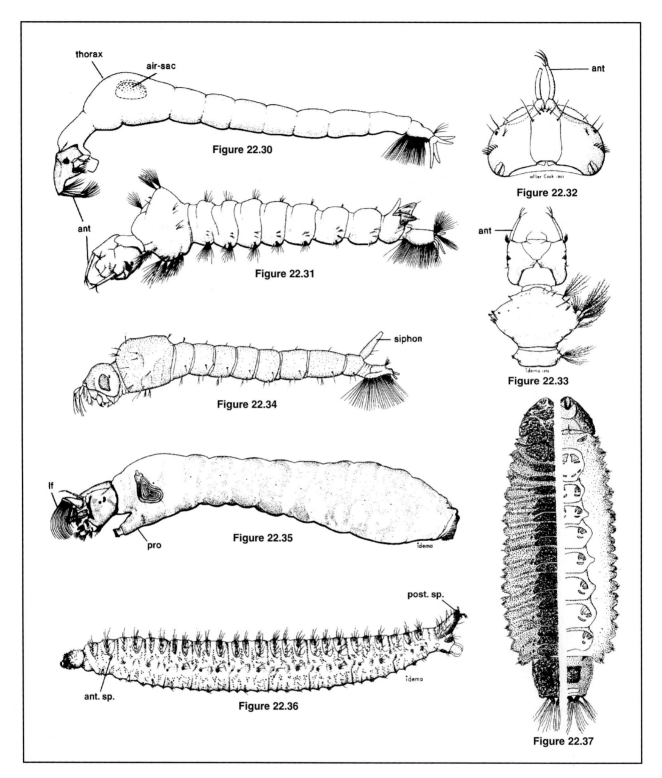

Figure 22.30. Lateral view of *Chaoborus americanus* (Johannsen) (Chaoboridae); *ant*, antenna.

Figure 22.31. Dorsolateral view of *Eucorethra underwoodi* Underwood (Chaoboridae); *ant,* antenna.

Figure 22.32. Dorsal view of head capsule of *Corethrella brakeleyi* (Coquillett) (Corethrellidae); *ant,* antenna.

Figure 22.33. Dorsal view of head, thorax, and first abdominal segment of *Eucorethra underwoodi* Underwood (Chaoboridae); *ant,* antenna.

Figure 22.34. Lateral view of *Mochlonyx sp.* (Chaoboridae).

Figure 22.35. Lateral view of *Simulium venustum* Say (Simuliidae); *lf*, labral fan; *pro*, proleg.

Figure 22.36. Lateral view of *Pericoma* sp. (Psychodidae); *ant. sp.,* anterior spiracle; *post. sp.,* posterior spiracle.

Figure 22.37. Half dorsal, half ventral views of *Maruina sp.* (Psychodidae).

25'.	Posterior abdominal segments without long filamentous processes; at most, only a single anal proleg present	26
26(25').	All body segments secondarily divided into 2 or 3 subdivisions with some or all of these subdivisions bearing dorsal sclerotized plates (Figs. 22.36-22.37); remainder of integument with numerous dark spots that together with the dorsal plates impart a greyish brown coloration to larva; respiratory system amphipneustic; posterior spiracles usually at apex of a relatively short, conical respiratory tube (Fig. 22.36) ......**PSYCHODIDAE**	27
26'.	Body segments usually not secondarily divided; integument smooth, shiny, and creamy white, lacking all surface features except a few setae that may be noticeable at tip of terminal segment (Figs. 22.45-22.46), and sometimes a retractile anal proleg bearing a few crochets (curved hooks; Fig. 22.44); larvae apneustic ......**CERATOPOGONIDAE** (in part)	29
27(26).	Larva distinctly flattened, with a median row of 8 suctorial disks ventrally (Fig. 22.37) ......***Maruina*** Muller	
27'.	Larva more or less cylindrical and lacking ventral suctorial disks	28
28(27').	Preanal plate present and tergal plates on body segments rather uniformly sized, 2 such plates on each of thoracic and 1st abdominal segments and 3 plates on each of abdominal segments 2-7, 26 plates in all (Fig. 22.36)......***Pericoma*** Walker, ***Telmatoscopus*** Eaton	
28'.	Preanal plate absent or tergal body plates generally reduced in size or only on some segments, or fewer than 26 plates present ......***Psychoda*** Latreille	
29(24, 26').	Segments of body with conspicuous dorsal tubercules and/or setae (Figs. 22.42, 22.154)	30
29'.	Segments of body without dorsal tubercles and setae	31
30(29).	Lateral processes on body segments at least as long as segments (Fig. 22.42); cross section of body oval......***Atrichopogon*** Kieffer	
30'.	Lateral processes on body segments absent or less than half length of segments (Fig. 22.154); cross section of body circular......***Forcipomyia*** Meigen	
31(29').	Head capsule sclerotized	32
31'.	Head capsule not sclerotized ......***Leptoconops*** Skuse	
32(31).	Last abdominal segment lacking proleg	33
32'.	Last abdominal segment bearing proleg (may be withdrawn and noticeable only as terminal hooks on segment; Fig. 22.44) ......***Dasyhelea*** Kieffer	
33(32).	Head more than twice as long as broad (Fig. 22.158) or posterior collar with a triangular or hemispherical lobe ventrally (Figs. 22.157-22.158)	34
33'.	Head less than twice as long as broad (Fig. 22.161) and with posterior collar of uniform width ventrally (Fig. 22.161)	37
34(33).	Stout setae on anal segment about one third length of segment; ventral side of posterior collar of head with triangular or hemispherical expansion (Figs. 22.157, 22.158)	35
34'.	Stout setae on anal segment at least one half length of segment (Figs. 22.45, 22.155); posterior collar of head of uniform width ventrally, or if expansion present, head less than 1.7 times as long as wide (Fig. 22.156) ......***Bezzia*** Kieffer, ***Palpomyia*** Meigen	
35(34).	Head more than 1.8 times as long as wide; ventral collar of head with hemispherical expansion (Fig. 22.158)	36
35'.	Head less than 1.8 times as long as wide; ventral collar of head with triangular expansion (Fig. 22.157) ......***Sphaeromias*** Kieffer	
36(35).	Head at least twice as long as wide, anterior end distinctly narrowed (Fig. 22.158)......***Probezzia*** Kieffer	
36'.	Head less than twice as long as broad, anterior end not markedly narrowed ......***Mallochohelea*** Wirth	
37(33').	Head much smaller than body, about one-third length of anal segment (Fig. 22.159) ......***Serromyia*** Meigen	
37'.	Head larger, more than one-third length of anal segment	38
38(37').	Antennae distinctly protruding from head ......***Stilobezzia*** Kieffer	
38'.	Antennae not protruding from head	39
39(38').	Long dark setae on anal segment equalling length of segment (Fig. 22.160)......***Alluaudomyia*** Kieffer	
39'.	Setae of anal segment less than one-half length of segment	40
40(39').	Setae of anal segment thick, black ......***Monohelea*** Kieffer	
40'.	Setae of anal segment short, pale	41
41(40').	Ventral side of head with long narrow sclerotized suture (Fig. 22.161) ......***Ceratopogon*** Meigen	
41'.	Ventral side of head lacking narrow sclerotized suture (Fig. 22.162); larva as in Fig. 22.46 ......***Culicoides*** Latreille	

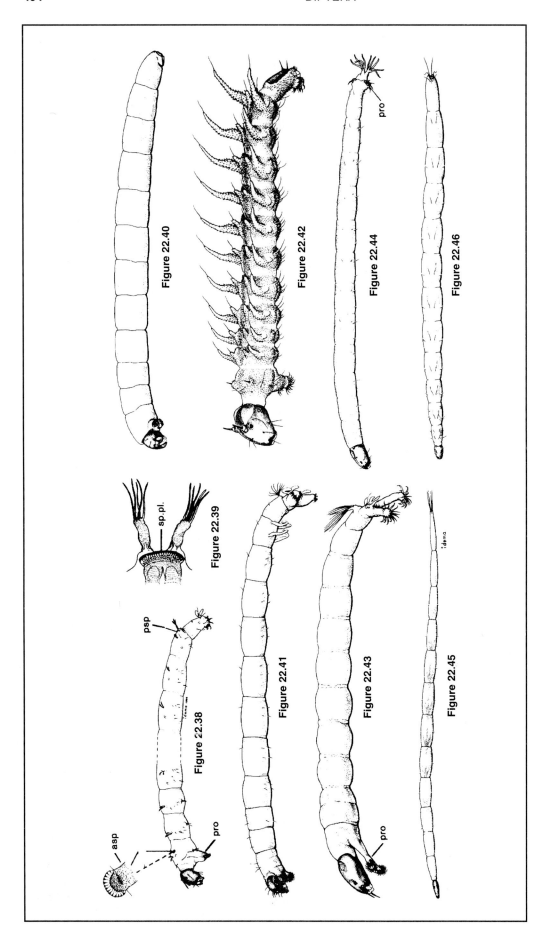

Figure 22.38. Lateral view of *Thaumalea sp.* (Thaumaleidae); *asp*, anterior spiracle; *pro*, proleg; *psp*, posterior spiracle.
Figure 22.39. Dorsal view of posterior spiracle and associated lobes of *Thaumalea sp.* (Thaumaleidae); *sp. pl.*, spiracular plate.
Figure 22.40. Lateral view of *Pseudosmittia sp.* (Chironomidae).
Figure 22.41. Lateral view of *Chironomus sp.* (Chironomidae).
Figure 22.42. Dorsolateral view of *Atrichopogon sp.* (Ceratopogonidae).
Figure 22.43. Lateral view of *Ablabesmyia sp.* (Chironomidae); *pro*, proleg.
Figure 22.44. Lateral view of *Dasyhelea sp.* (Ceratopogonidae); *pro*, proleg.
Figure 22.45. Lateral view of *Bezzia sp.* (Ceratopogonidae).
Figure 22.46. Lateral view of *Culicoides sp.* (Ceratopogonidae).

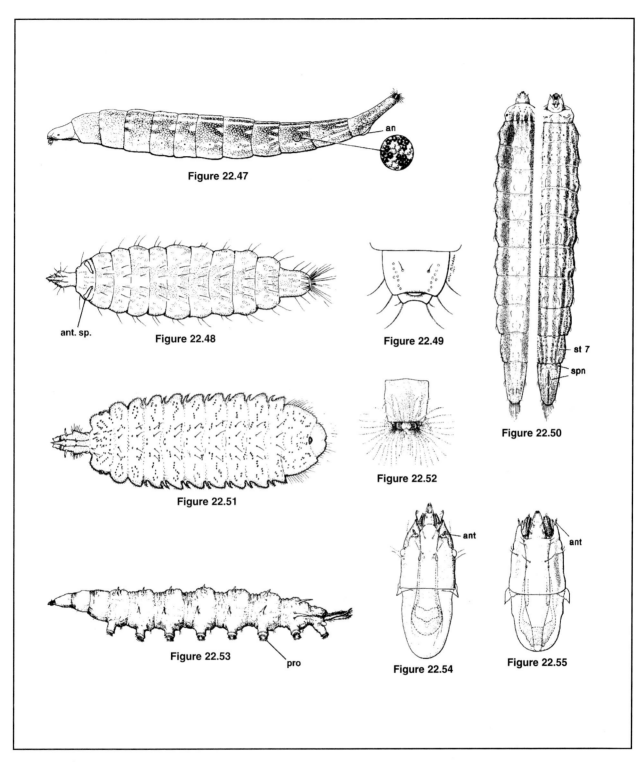

Figure 22.47. Lateral view of *Stratiomys sp.* (Stratiomyidae); *an*, anus.

Figure 22.48. Dorsal view of *Euparyphus sp.* (Stratiomyidae); *ant. sp.*, anterior spiracle.

Figure 22.49. Dorsal view of terminal segment of *Nemotelus kansensis* Adams (Stratiomyidae).

Figure 22.50. Part dorsal and part ventral view of *Odontomyia cincta* Olivier (Stratiomyidae); *spn*, spine; *st 7*, sternite 7.

Figure 22.51. Dorsal view of *Allognosta sp.* (Stratiomyidae).

Figure 22.52. Dorsal view of terminal segment of *Myxosargus nigricornis* Greene (Stratiomyidae).

Figure 22.53. Lateral view of *Atherix sp.* (Athericidae); *pro*, proleg.

Figure 22.54. Dorsal view of head capsule of *Caloparyphus major* (Hine) (Stratiomyidae); *ant*, antenna.

Figure 22.55. Dorsal view of head capsule of *Odontomyia cincta* Olivier (Stratiomyidae); *ant*, antenna.

42(1').	Sclerotized portions of head capsule exposed externally although sometimes greatly reduced, in which case slender tentorial and metacephalic rods prominent internally (Figs. 22.54, 22.59, 22.64, 22.71) ..........................ORTHORRHAPHOUS—BRACHYCERA.	43
42'.	External sclerotized portions of head capsule absent; head reduced to an internal cephalopharyngeal skeleton of rather characteristic form (Figs. 22.77, 22.81-22.82, 22.89) ..........................CYCLORRHAPHOUS—BRACHYCERA	77
43(42).	Body somewhat depressed (Figs. 22.47-22.48, 22.50-22.51); integument toughened and leathery from calcium deposits that are evident as numerous small reticulately arranged facets (Fig. 22.47); head capsule capable of only slight independent movement; usually with distinctive lateral eye prominences (Figs. 22.54-22.55)..........................STRATIOMYIDAE	44
43'.	Larva usually not conspicuously depressed nor with a toughened integument bearing a network of facets; head capsule capable of extensive independent movement; without distinctive eye prominences	53
44(43).	Spiracular cleft at apex of terminal abdominal segment with long marginal hydrofuge (water-repelling) setae (Figs. 22.47-22.48, 22.50, 22.52)	46
44'.	Spiracular cleft situated dorsally near apex of terminal abdominal segment and, if with bordering hydrofuge setae, very short and inconspicuous (Figs. 22.49, 22.51)	45
45(44').	Abdominal segments with lateral margins thinly compressed and bilobate; anterior lobe small and sharply pointed; posterior margin of terminal segment rounded (Fig. 22.51) ....***Allognosta*** Osten Sacken	
45'.	Abdominal segments with lateral margins not strongly compressed or appearing bilobate, anterior lobe blunt and not sharply pointed; posterior margin of terminal segment concave (Fig. 22.49) ....***Nemotelus*** Geoffroy	
46(44).	Antenna situated anterolaterally on head capsule and directed anteriorly or somewhat ventrally (Fig. 22.55)	47
46'.	Antenna situated dorsolaterally on head capsule and directed dorsally (Fig. 22.54)	50
47(46).	Terminal abdominal segment long and slender; apical portion with lateral margins nearly parallel (Fig. 22.47); anus located anteriorly on segment and swollen..........................***Stratiomys*** Geoffroy	
47'.	Terminal abdominal segment rarely over twice as long as broad, uniformly tapered posteriorly (Fig. 22.50); anus not conspicuously swollen	48
48(47').	Abdominal sternite 7 lacking sclerotized hooklike spines on posterior margin; integument covered with minute peltate scale-like pubescence ....***Odontomyia*** (*Catatasina* Enderlein)	
48'.	Abdominal sternite 7 with hooklike spines on posterior margin (Fig. 22.50); integumental pubescence variable	49
49(48').	Only abdominal sternite 7 with hooked spines..........................***Odontomyia*** (*Odontomyiina* Enderlein)	
49'.	Abdominal sternites 6 and 7 with hooked spines (Fig. 22.50)..........................***Odontomyia*** (*Odontomyia* Meigen), ***Hedriodiscus*** Enderlein	
50(46').	Hydrofuge setae arising from 2 lobate structures on lower lip of spiracular cleft (Fig. 22.52); anterior spiracles located at anterior corner of prothorax ....***Myxosargus*** Brauer	
50'.	Hydrofuge setae arising from straight edge of lower lip; spiracular cleft lacking strong lobate structures; anterior spiracles located laterally near middle of prothorax (Fig. 22.48)	51
51(50').	Terminal abdominal segment almost rectangular in outline; posterior margin more or less indented or notched sublaterally ....***Oxycera*** Meigen	
51'.	Terminal segment subconical with lateral and posterior margins rounded	52
52(51').	Anterior spiracle on short stalk or nearly sessile ....***Caloparyphus*** James	
52'.	Anterior spiracle on long stalk (Fig. 22.48) ....***Euparyphus*** Gerstacker	
53(43').	Head capsule well developed dorsally, closed ventrally by a submental plate (Figs. 22.59, 22.64); tentorial rods solidly fused with head capsule internally; a brush of backwardly curved bristles usually present on each side of clypeus above and near base of each mandible (Figs. 22.59, 22.60, 22.64)	54
53'.	Head capsule reduced to a pair of slender metacephalic rods; they and tentorial rods flexibly articulate with anterior cephalic sclerites (Figs. 22.67, 22.71); submental plate and brushes of bristles above mandibles absent	67
54(53).	Posterior spiracles present, opening within slits on either side of a vertically linear stigmatal bar (Fig. 22.58) or a retractile, laterally compressed spine (Figs. 22.61, 22.63); body integument with longitudinal striations, except in some species where integument totally covered by short, velvety pubescence; abdominal segments 1-7 girdled by 3 or 4 pairs of fleshy creeping welts (setae-bearing swellings) or prolegs, these being the only projections from the segments (Figs. 22.56-22.57)..........................TABANIDAE	56

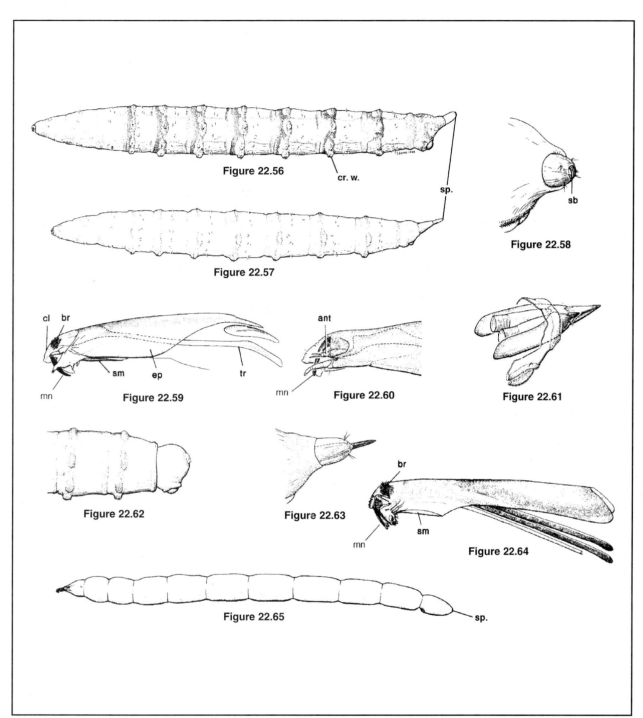

Figure 22.56. Lateral view of *Tabanus reinwardtii* Wiedemann (Tabanidae); *cr. w.*, creeping welt; *sp.*, spiracle.

Figure 22.57. Lateral view of *Chrysops furcatus* Walker (Tabanidae); *sp.*, spiracle.

Figure 22.58. Respiratory siphon of *Tabanus sp.* (Tabanidae); *sb*, stigmatal bar.

Figure 22.59. Lateral view of head capsule of *Tabanus reinwardtii* Wiedemann (Tabanidae); *br*, mandibular bristles; *cl*, clypeus; *ep*, epicranium; *mn*, mandible; *sm*, submentum; *tr*, tentorial rod.

Figure 22.60. Lateral view of head capsule of *Chrysops excitans* Walker (Tabanidae); *ant*, antenna; *mn*, mandible.

Figure 22.61. Respiratory siphon and spine of *Merycomyia whitneyi* (Johnson) (Tabanidae).

Figure 22.62. Terminal segments of *Leucotabanus annulatus* (Say) (Tabanidae).

Figure 22.63. Respiratory siphon and spine of *Chrysops cincticornis* Walker (Tabanidae).

Figure 22.64. Lateral view of head capsule of *Glutops rossi* Pechuman (Pelecorhynchidae); abbreviations as in Figure 22.59.

Figure 22.65. Lateral view of *Glutops rossi* Pechuman (Pelecorhynchidae); *sp.*, spiracle.

54'.	Posterior spiracles greatly reduced or situated within a small terminal cavity (Fig. 22.65); integument without striations or extensive covering of pubescence; prolegs, if present, limited to 1 ventral pair on each abdominal segment 55
55(54').	Larva slightly flattened dorsoventrally; slender tubercles of progressively increasing size situated laterally and dorsolaterally on abdominal segments 1-7; 2 longer caudal tubercles fringed with hairs on terminal segment; all abdominal segments with a ventral pair of prolegs bearing crochets (Fig. 22.53) ....**ATHERICIDAE**—*Atherix* Meigen, **Suragina** Walker (Texas only)
55'.	Larva cylindrical, with smooth shiny integument and with segmentation beadlike; lacking tubercles and prolegs (Fig. 22.65) ....**PELECORHYNCHIDAE**—*Glutops* Burgess
56(54).	Prolegs present only on abdominal segments 1-5; body widest in region of mesothoracic and metathoracic segments; integument of intersegmental regions and posterior borders of prolegs with a reticulate, scalelike pattern; terminal segment hemispherical; posterior spiracle sessile, or nearly sessile, on surface of terminal segment; mandible straight and bladelike ....***Apatolestes*** Williston
56'.	Prolegs present on each of abdominal segments 1-7 (Figs. 22.56-22.57); body normally widest near middle; integument lacking reticulate, fish scale-like patterned areas, instead often with microtrichial pubescence intersegmentally and bordering prolegs; posterior spiracle at least slightly elevated; mandible curved.....................57
57(56').	Three pairs of prolegs situated dorsally, laterally, and ventrally on each of abdominal segments 1-7 (Fig. 22.57) 58
57'.	Four pairs of prolegs, including an additional ventrolateral pair, on abdominal segments 1-7 (Fig. 22.56) 61
58(57).	Body surface, except respiratory siphon, completely clothed with a dense covering of short pubescence 59
58'.	Pubescence restricted to anterior or posterior margin or posterior border of prolegs of 1 or more segments 60
59(58).	Pubescent integumental covering conspicuously mottled with dark and paler areas; 3rd antennal segment shorter than 2nd; respiratory siphon equal to or only slightly longer than its basal diameter ....***Chlorotabanus*** Lutz
59'.	Pubescence not conspicuously mottled; 3rd antennal segment longer than 2nd segment (as in Fig. 22.60); respiratory siphon length about twice its basal diameter ....***Diachlorus*** Osten Sacken
60(58').	Third antennal segment about half the length of 2nd segment; respiratory siphon shorter than its basal diameter, lacking a respiratory spine; prolegs with prominent hooklike crochets ....***Silvius*** (*Zeuximyia* Philip)
60'.	Third antennal segment longer than 2nd segment (Fig. 22.60); respiratory siphon either longer than its basal diameter (Fig. 22.57) or with a respiratory spine (Fig. 22.63) ....***Chrysops*** Meigen
61(57').	Respiratory siphon comprising distal ends of 2 opposed sclerotized plates between which tracheal trunks terminate in an exsertile spicular spine (Fig. 22.61); inconspicuous and incomplete striations present only laterally on segments ....***Merycomyia*** Hine
61'.	Although tracheal trunks sometimes terminating in a spicular spine, respiratory siphon always membranous and lacking sclerotized plates; striations on abdominal segments present or absent 62
62(61').	Respiratory siphon shorter than its basal diameter (as in Fig. 22.62); integumental striations extremely fine on all aspects of body, and usually visible only under high magnification; striations spaced at approximately 5 μm ....***Haematopota*** Meigen
62'.	If respiratory siphon shorter than its basal diameter, striations more coarsely spaced at usually more than 20 μm 63
63(62').	Respiratory siphon very short, projecting no more than half its diameter; terminal abdominal segment usually shorter than greatest diameter, hemispherical (Fig. 22.62); striations uniformly spaced on all aspects of body 64
63'.	Respiratory siphon length ranging from slightly shorter to about 4 times longer than its basal diameter; terminal abdominal segment usually somewhat tapering posteriorly toward respiratory siphon (Fig. 22.56); striations normally absent from dorsal and ventral surfaces of at least the prothorax, and more widely spaced dorsally and ventrally than laterally on other segments 66
64(63).	Pubescence encircling anterior three-fourths of prothorax and broadly encircling posterior half of terminal abdominal segment so that anal lobes and base of respiratory siphon are covered by enlarged pubescent area ....***Silvius*** (*Griseosilvius* Philip)
64'.	Pubescence encircling little more than anterior one-fourth of prothorax, and on terminal abdominal segment restricted to narrow annulus around base of respiratory siphon, and on anal lobes so that pubescence on anal lobes is separated from that encircling base of respiratory siphon 65
65(64').	Terminal abdominal segment two-thirds length of penultimate segment (Fig. 22.62); larva inhabiting decaying wood and tree holes ....***Leucotabanus*** Lutz
64'.	Terminal abdominal segment less than half the length of penultimate segment; larva inhabiting damp sand on coastal beaches ....***Stenotabanus*** (*Aegialomyia* Philip)
66(63').	Median lateral surfaces of terminal segment usually lacking pubescent markings; striations present on dorsal and ventral surfaces of all abdominal segments or, if absent, pubescence restricted, at most, to a prothoracic annulus (ring) and anal ridges...***Hybomitra*** Enderlein

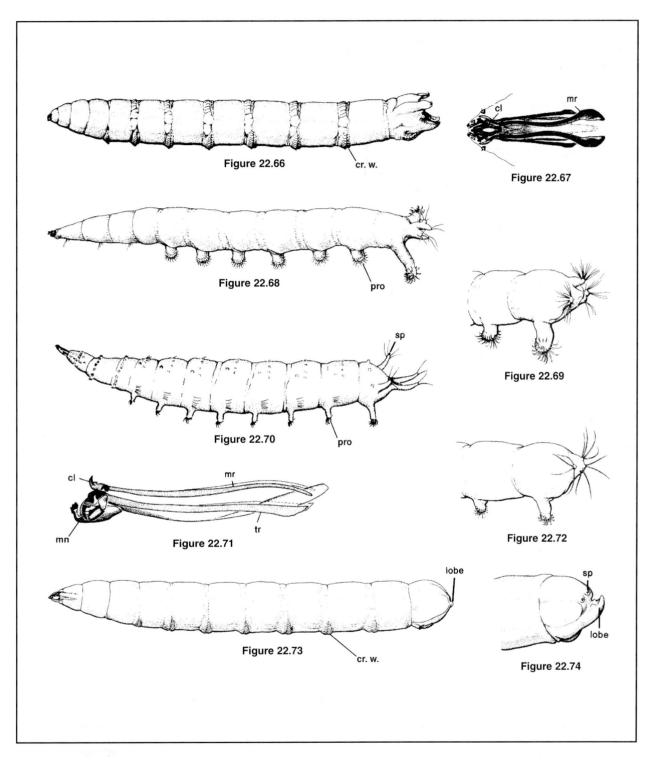

Figure 22.66. Dorsolateral view of *Rhaphium campestre* Curran (Dolichopodidae); *cr. w.,* creeping welt.

Figure 22.67. Dorsal view of head skeleton of *Rhaphium sp.* (Dolichopodidae); *cl,* clypeus; *mr,* metacephalic rod.

Figure 22.68. Dorsolateral view of *Hemerodromia sp.* (Empididae); *pro,* proleg.

Figure 22.69. Dorsolateral view of terminal segments of *Clinocera sp.* (Empididae).

Figure 22.70. Dorsolateral view of *Oreogeton sp.* (Empididae); *pro,* proleg.; *sp,* spiracle.

Figure 22.71. Lateral view of head skeleton of *Hemerodromia sp..* (Empididae); *cl,* clypeus; *mn,* mandible; *mr,* metacephalic rod; *tr,* tentorial rod.

Figure 22.72. Dorsolateral view of terminal segments of *Chelifera sp.* (Empididae).

Figure 22.73. Lateral view of *Rhamphomyia sp.* (Empididae); *cr. w.,* creeping welt.

Figure 22.74. Dorsolateral view of terminal segment of *Phyllodromia sp.* (Empididae); *sp,* spiracle.

66'.	Either median lateral surfaces of anal segment with pubescent markings (Fig. 22.56), or striations absent or poorly developed on dorsal or ventral surface or both surfaces of abdominal segments ....***Tabanus*** Linnaeus, ***Whitneyomyia*** Bequaert, ***Atylotus*** Osten Sacken
67(53').	Larva metapneustic; posterior spiracles situated at the base of upper 2 of 4 smooth primary lobes of last abdominal segment (Fig. 22.66); transverse ventral creeping welts present on abdominal segments; metacephalic rods expanded posteriorly (Fig. 22.67) .... **DOLICHOPODIDAE**
67'.	Larva usually apneustic, terminal abdominal segment with 1-4 rounded lobes bearing apical setae, and abdominal segments bearing paired prolegs with apical crochets (Figs. 22.68-22.70, 22.72); if metapneustic, then posterior segment with only a single lobe below spiracles, and abdominal segments with ventral creeping welts (Figs. 22.73-22.74); metacephalic rods slender posteriorly (Fig. 22.71) ....**EMPIDIDAE**......68
68(67').	Respiratory system usually apneustic (posterior spiracles present in *Roederiodes* and *Oreogeton*); 7 or 8 pairs of abdominal prolegs bearing apical hooked spines or crochets; terminal segment with elongate lobes or short tubercles ending in long setae (Figs. 22.68-22.70, 22.72), rarely ending in only a tuft of apical setae; aquatic or semiaquatic 69
68'.	Respiratory system amphipneustic; abdominal prolegs absent; terminal segment rounded, with a single, bare, mid-ventral lobe (Figs. 22.73-22.74); usually terrestrial, found in soil, leaf litter, and rotting wood, but some in wet soil. ....Several genera (for the most part inseparable)
69(68).	Terminal segment rounded posteriorly, at most with small dorsal and apical tubercles; each tubercle with 1-3 pairs of long setae (Fig. 22.72); 7 pairs of abdominal prolegs ....***Chelifera*** Macquart
69'.	Terminal segment with prominent caudal lobes (Figs. 22.68-22.70); 7 or 8 pairs of abdominal prolegs 70
70(69').	Terminal segment with a single, more or less medially divided, setose caudal lobe (Fig. 22.68); 7 pairs of abdominal prolegs ....***Hemerodromia*** Meigen
70'.	Terminal segment with dorsal and apical caudal lobes; 8 pairs of abdominal prolegs......................................71
71(70').	Terminal segment with 2 dorsolateral lobes and with one more or less divided apical lobe (Fig. 22.69) 72
71'.	Terminal segment with 2 pairs of lobes (Fig. 22.70) 73
72(71).	Dorsolateral lobes more than twice their width..***Trichoclinocera*** Collin
72'.	Dorsolateral lobes equal to or less than twice their width ....***Clinocera*** (*Hydromia* Macquart)
73(71').	Prolegs and caudal lobes rounded, very short ....***Dolichocephala*** Macquart
73'.	Prolegs and caudal lobes elongate, longer 74
74(73').	Terminal segment with apical lobe short and dorsolateral lobes long 75
74'.	Terminal segment with lobes of equal length (Fig. 22.70) 76
75(74).	Posterior spiracles present ....***Roederiodes*** Coquillett
75'.	Posterior spiracles absent ....***Clinocera*** (*Clinocera* Meigen)
76(74').	Posterior spiracles present on dorsolateral lobes; abdominal segments 1-7 each with a group of 6 setae on either side; last thoracic segment with row of brushy protuberances dorsally (Fig. 22.70) ....***Oreogeton*** Schiner
76'.	Posterior spiracles absent; abdominal segments without setae laterally; last thoracic segment without brushy protuberances ....***Wiedemannia*** Zetterstedt
77(42').	Posterior spiracular plates fused or very closely approximated (Fig. 22.80), usually on apex of a telescopic respiratory tube (Figs. 22.76, 22.78-22.79) 78
77'.	Posterior spiracular plates always distinctly separated whether mounted on a telescopic respiratory tube or not 80
78(77).	Prothoracic spiracles with stigmatal openings on branching papillae or, if sessile, arranged along apical half of a long spiracular stalk (Fig. 22.79); cephalopharyngeal skeleton with well-developed mouth hooks (Fig. 22.81) ....**CANACIDAE**.....79
78'.	Prothoracic spiracles, if present, with stigmatal openings near apex of a simple stalk (Fig. 22.75); cephalopharyngeal skeleton lacking mouth hooks in aquatic larvae of family, a ribbed filter chamber in area normally occupied by mouth hooks (Fig. 22.77) ....**SYRPHIDAE***
79(78).	Openings on anterior prothoracic spiracles sessile (Fig. 22.79); distinct creeping welts ventrally; perianal pad greatly reduced, with single spine ....***Canace*** Haliday
79'.	Openings on anterior prothoracic spiracles at tips of elongate papillae (Fig. 22.153); ventral creeping welts absent; perianal pad with numerous claw-like hooks (Fig. 22.153) ....***Canaceoides*** Cresson
80(77').	Anterior spiracles simple, each with 1 to several openings arranged peripherally at apex of short projection; body often dorsoventrally flattened and bearing a series of spicules or tubercles (Fig. 22.151); posterior spiracles with openings arranged in 2 pairs placed one behind the other (Fig. 22.152)...**PHORIDAE**

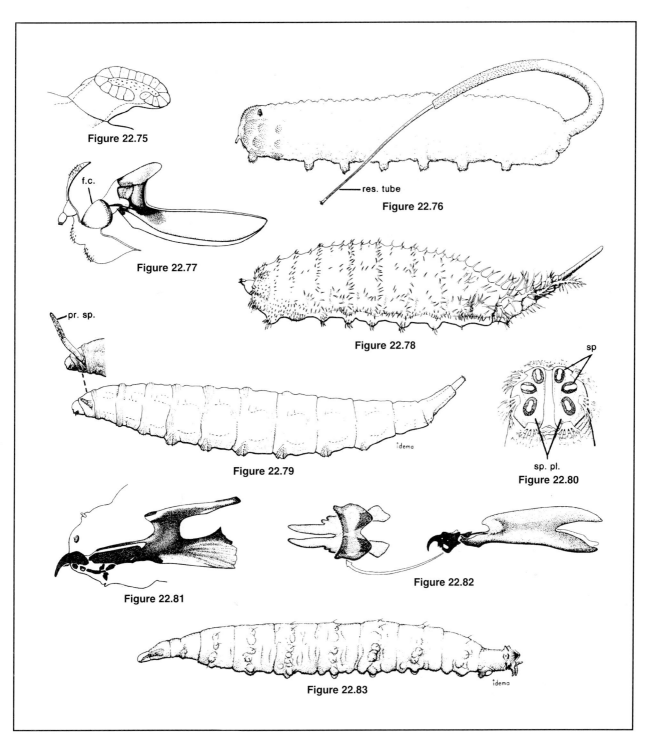

Figure 22.75. Anterior spiracle of *Eristalis tenax* (L.) (Syrphidae).

Figure 22.76. Lateral view of *Eristalis tenax* (L.) (Syrphidae); *res.*, respiratory.

Figure 22.77. Lateral view of a cephalopharyngeal skeleton of *Eristalis tenax* (Syrphidae); *f.c.*, filter chamber.

Figure 22.78. Lateral view of *Chrysogaster sp.* (Syrphidae).

Figure 22.79. Lateral view of *Canace macateei* Malloch (Canacidae) showing enlargement of prothoracic spiracle; *pr. sp.*, prothoracic spiracle.

Figure 22.80. Posterior spiracles of *Canace macateei* Malloch (Canacidae); *sp*, spiracular openings; *sp. pl.*, spiracular plate.

Figure 22.81. Lateral view of cephalopharyngeal skeleton of *Canace macateei* (Canacidae).

Figure 22.82. Lateral view of cephalopharyngeal skeleton of *Sepedon sp.* (Sciomyzidae) with enlarged ventral view of mouth hooks and serrated ventral arch.

Figure 22.83. Lateral view of *Sepedon sp.* (Sciomyzidae).

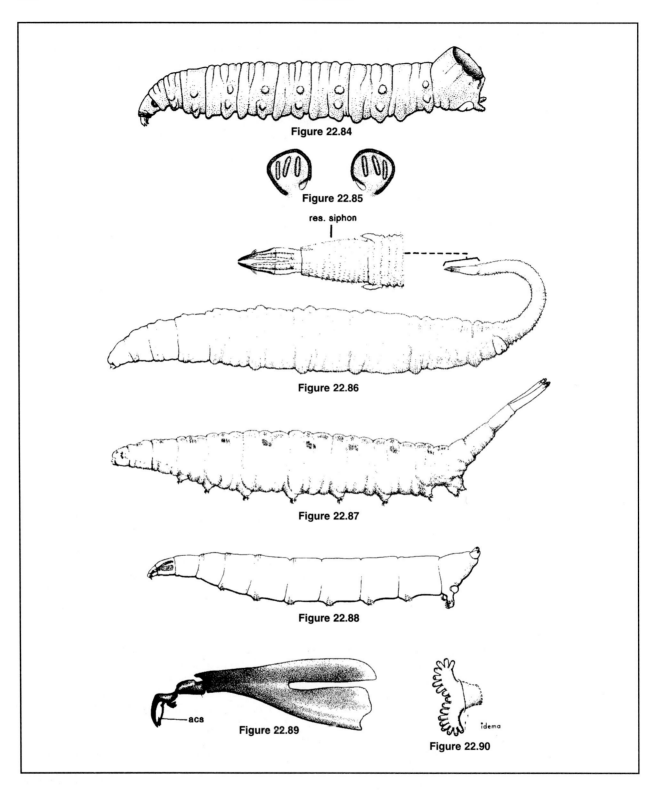

Figure 22.84. Lateral view of *Blaesoxipha fletcheri* (Aldrich) (Sarcophagidae).

Figure 22.85. Posterior spiracles of *Blaesoxipha fletcheri* (Aldrich) (Sarcophagidae).

Figure 22.86. Lateral view of *Notiphila sp.* (Ephydridae) with enlargement of apex of respiratory siphon; *res.*, respiratory.

Figure 22.87. Lateral view of *Ephydra sp.* (Ephydridae).

Figure 22.88. Lateral view of *Limnophora sp.* (Muscidae).

Figure 22.89. Lateral view of cephalopharyngeal skeleton of *Limnophora sp.* (Muscidae); *acs,* accessory oral sclerite.

Figure 22.90. Typical bicornuate anterior spiracle of Scathophagidae.

80'.	Anterior spiracles either absent or bearing 2 or more short or branched papillae (Fig. 22.90); posterior spiracles with openings usually arranged in parallel or radiating pattern (Figs. 22.80, 22.85)	81
81(80').	Cephalopharyngeal skeleton with a sclerotized ventral arch below base of mouth hooks, its anterior margin usually toothed (Fig. 22.82); body segments often extensively covered with short, fine hairs; posterior segment often somewhat tapered, its apex with tubercles surrounding posterior spiracles that are only slightly elevated (Fig. 22.83)	SCIOMYZIDAE.....82
81'.	Cephalopharyngeal skeleton lacking a ventral arch; if body extensively covered with short, fine hairs then a respiratory siphon present, or each spiracle situated on a short tubular projection on posterior segment (Figs. 22.84-22.87)	88
82(81).	Hypopharyngeal and tentoropharyngeal sclerites fused (Figs. 22.91, 22.93); indentation index about 50 (Fig. 22.91)	83
82'	Hypopharyngeal and tentoropharyngeal sclerites separate (Fig. 22.100); indentation index 34-46 (Fig. 22.100)	84
83(82).	Body with ventral spinule patches and dense coat of transparent spinules (Fig. 22.92); accessory teeth and dorsal bridge present (Fig. 22.91); larvae feeding on snail eggs	*Antichaeta* Haliday
83'.	Body without ventral spinule patches and transparent spinules; accessory teeth and dorsal bridge absent (Fig. 22.93); larvae feeding on fingernail clams (Sphaeriidae)	*Renocera* Hendel
84(82').	Ventral arch (Fig. 22.101) triangular to quadrate, with posterior projection (Fig. 22.94); larva as in Fig. 22.95	*Dictya* Meigen
84'.	Ventral arch (Fig. 22.101) bilobed, with posterior indentation (Fig. 22.96)	85
85(84').	Posterior spiracular disc with 10 lobes (Fig. 22.97); ventral arch with posterolateral projections (Figs. 22.82, 22.96); larva as in Fig. 22.83	*Sepedon* Latreille, *Sepedomerus* Steyskal
85'.	Posterior spiracular disc with 8 lobes (Fig. 22.104); ventral arch without posterolateral projections (Fig. 22.98)	86
86(85').	Postanal portion of abdominal segment 8 much longer than wide (Fig. 22.99); 5-7 accessory teeth on mandibles (Fig. 22.100)	*Elgiva* Meigen
86'.	Postanal portion of abdominal segment 8 not much longer than wide (Fig. 22.105); 2-7 accessory teeth on mandibles	87
87(86').	Lateral lobes of posterior spiracular disc nearly as long as ventrolateral and ventral lobes (Fig. 22.102); larva as in Fig. 22.103	*Hedria* Steyskal
87'.	Lateral lobes of posterior spiracular disc much shorter than ventrolateral and ventral lobes (Fig. 22.104); larva as in Fig. 22.106	*Tetanocera* Dumeril
88(81').	Abdominal segments with short, stout, conical projections laterally; prothoracic spiracles absent; parasite of barnacles	DRYOMYZIDAE—*Oedoparena* Curran
88'.	Abdominal segments usually without lateral projections, but, if present, slender; prothoracic spiracles usually present	89
89(88').	Posterior spiracles located in a deep spiracular cavity; spiracular slits inclined more or less vertically (Figs. 22.84-22.85); inhabitants of pitcher plants	SARCOPHAGIDAE—*Blaesoxipha* Loew
89'.	Posterior spiracles not located in a deep cavity but are well exposed terminally; one or more of the spiracular slits nearly horizontal	90
90(89').	Posterior abdominal segment somewhat tapered, sometimes ending in a retractile respiratory tube (Figs. 22.86-22.87); integument of posterior abdominal segments covered with setae (Fig. 22.86) or spinules, or with setaceous (setae-bearing) tubercles on some segments	EPHYDRIDAE*.....91
90'.	Posterior abdominal segment rather truncate (cut off squarely) and/or integument with setae only on intersegmental areas (Fig. 22.88); tubercles, if present, restricted to posterior abdominal segment	114
91(90).	Posterior spiracles borne on tips of short to elongate breathing tubes (Figs. 22.86, 22.87, 22.107, 22.108).	93
91'.	Posterior spiracles sessile on last body segment (Figs. 22.109, 22.110)	92
92(91').	Body with numerous tubercles (Figs. 22.110, 22.111); anterior spiracles (Fig. 22.111) with 8 marginal papillae	*Allotrichoma* Becker
92'.	Body lacking tubercles (Fig. 22.109); anterior spiracles with 5-6 marginal papillae (Fig. 22.112)	*Discocerina* Macquart
93(91).	Posterior spiracles borne on sharply pointed spines (Figs. 22.86, 22.115, 22.116)	94
93'.	Posterior spiracles not borne on sharply pointed spines (Fig. 22.121)	97
94(93).	Body length greater than 8.0 mm; larva detritivorous in bottom sediments, associated with roots of aquatic macrophytes	95
94'.	Body length less than than 6.0 mm; larva mining leaves or stems of wetland plants	96
95(94).	Anterior spiracles absent (Fig. 22.86); widespread	*Notiphila* (*Notiphila* Fallen)
95'.	Anterior spiracles present (Figs. 22.113, 22.114); larva as in Fig. 22.115; restricted to coastal habitats in eastern states	*Dimecoenia* Cresson (in part)

* The following key includes some of the more common shore-fly genera; other genera are not included.

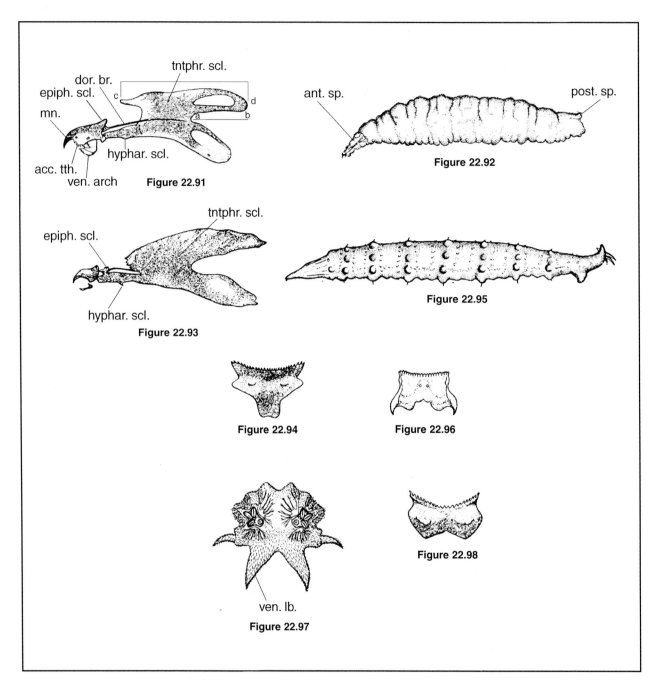

Figure 22.91. Lateral view of cephalopharyngeal skeleton of larva of *Anticheta testacea* Melander (Sciomyzidae) (from Manual of Nearctic Diptera, Vol. 2); acc. tth., accessory teeth; dor. br., dorsal bridge; epiph. scl., epipharyngeal sclerite; hyphar. scl., hypopharyngeal sclerite; mn., mandible; tntphr. scl., tentoropharyngeal sclerite; ven., ventral. Indentation index = length of ab ÷ length of cd X 100.

Figure 22.92. Lateral view of larva of *Antichaeta borealis* Foote (Sciomyzidae) (from Robinson and Foote 1978); ant. sp., anterior spiracle; post. sp., posterior spiracle.

Figure 22.93. Lateral view of cephalopharyngeal skeleton of larva of *Renocera* sp. (Sciomyzidae) (from Manual of Nearctic Diptera, Vol. 2); epiph. scl., epipharyngeal sclerite; hyphar. scl., hypopharyngeal sclerite; tntphr. scl., tentoropharyngeal sclerite.

Figure 22.94. Ventral arch of larva of *Dictya expansa* Steyskal (Sciomyzidae) (from Manual of Nearctic Diptera, Vol. 2).

Figure 22.95. Lateral view of larva of *Dictya pictipes* (Sciomyzidae) (from Johannsen 1935).

Figure 22.96. Ventral arch of larva of *Sepedon tenuicornis* Cresson (Sciomyzidae) (from Manual of Nearctic Diptera, Vol. 2).

Figure 22.97. Posterior spiracular disc of larva of *Sepedon tenuicornis* Cresson (Sciomyzidae) (from Manual of Nearctic Diptera, Vol. 2); ven. lb., ventral lobe.

Figure 22.98. Ventral arch of larva of *Elgiva solicita* (Harris) (Sciomyzidae) (from Manual of Nearctic Diptera, Vol. 2).

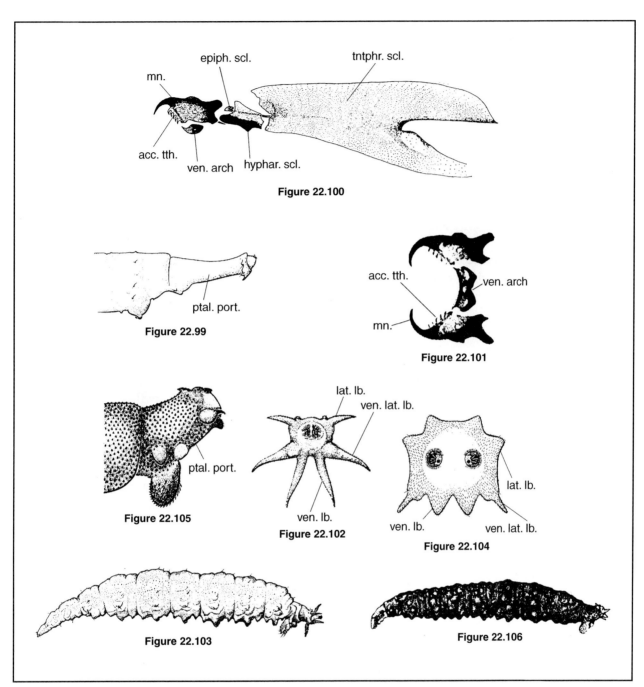

Figure 22.99. Abdominal segments 7 and 8 of larva of *Elgiva solicita* (Harris) (Sciomyzidae) (from Manual of Nearctic Diptera, Vol. 2); ptal. port., postanal portion of segment 8.

Figure 22.100. Cephalopharyngeal skeleton of larva of *Elgiva solicita* (Harris) (Sciomyzidae) (from Stehr 1991); epiph. scl., epipharyngeal sclerite; hyphar. scl., hypopharyngeal sclerite; mn., mandible; tntphr. scl., tentoropharyngeal sclerite; ven., ventral.

Figure 22.101. Mandibles and ventral arch of larva of *Elgiva solicita* (Harris) (Sciomyzidae) (from Stehr 1992); acc. tth, accessory teeth; mn., mandible; ven., ventral.

Figure 22.102. Posterior spiracular disc of larva of *Hedria mixta* Steyskal (Sciomyzidae) (from Manual of Nearctic Diptera, Vol. 2); lat. lb., lateral lobe; ven. lat. lb., ventral lateral lobe; ven. lb., ventral lobe.

Figure 22.103. Lateral view of larva of *Hedria mixta* Steyskal (Sciomyzidae) (from Foote 1971).

Figure 22.104. Posterior spiracular disc of larva of *Tetanocera montana* Day (Sciomyzidae) (from Manual of Nearctic Diptera, Vol. 2); lat. lb., lateral lobe; ven. lat. lb., ventral lateral lobe; ven. lb., ventral lobe.

Figure 22.105. Abdominal segments 7 and 8 of larva of *Tetanocera montana* Day (Sciomyzidae) (from Manual of Nearctic Diptera, Vol. 2); ptal. port., postanal portion of segment 8.

Figure 22.106. Lateral view of larva of *Tetanocera vicina* Macquart (Sciomyzidae).

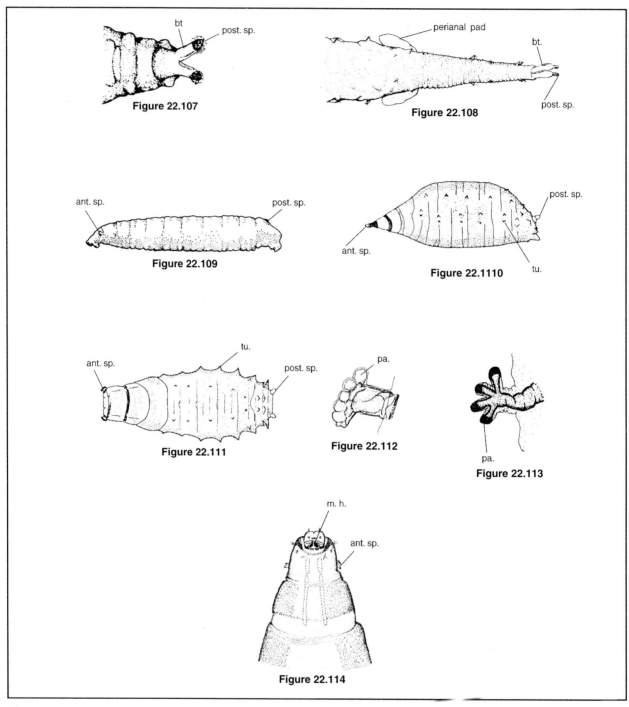

Figure 22.107. Posterior end of larva of *Scatella picea* (Walker) (Ephydridae) (from Eastin 1970); bt., breathing tube; post. sp., posterior spiracle.

Figure 22.108. Dorsal view of posterior end of larva of *Dichaeta caudata* (Fallén) (Ephydridae) (from Eastin and Foote 1971) bt., breathing tube; post. sp., posterior spiracle.

Figure 22.109. Lateral view of larva of *Discocerina obscurella* (Fallén) (Ephydridae) from Foote and Eastin 1974; ant. sp., anterior spiracle; post. sp., posterior spiracle.

Figure 22.110. Lateral view of larva of *Allotrichoma simplex* (Loew) (Ephydridae) (from Runyon and Deonier 1979); ant. sp., anterior spiracle; post. sp., posterior spiracle; tu., tubercle.

Figure 22.111. Dorsal view of larva of *Allotrichoma simplex* (Loew) (Ephydridae) (from Runyon and Deonier 1979); ant. sp., anterior spiracle; post. sp., posterior spiracle; tu., tubercle.

Figure 22.112. Anterior spiracle of larva of *Discocerina obscurella* (Fallén) (Ephydridae) (from Foote and Eastin 1974); pa., papilla.

Figure 22.113. Anterior spiracle of larva of *Dimecoenia spinosa* (Loew) (Ephydridae) (from Mathis and Simpson 1981); pa., papilla.

Figure 22.114. Anteroventral aspect of larva of *Dimecoenia spinosa* (Loew) (Ephydridae) (from Mathis and Simpson 1981); ant. sp., anterior spiracle; m.h., mouth hooks.

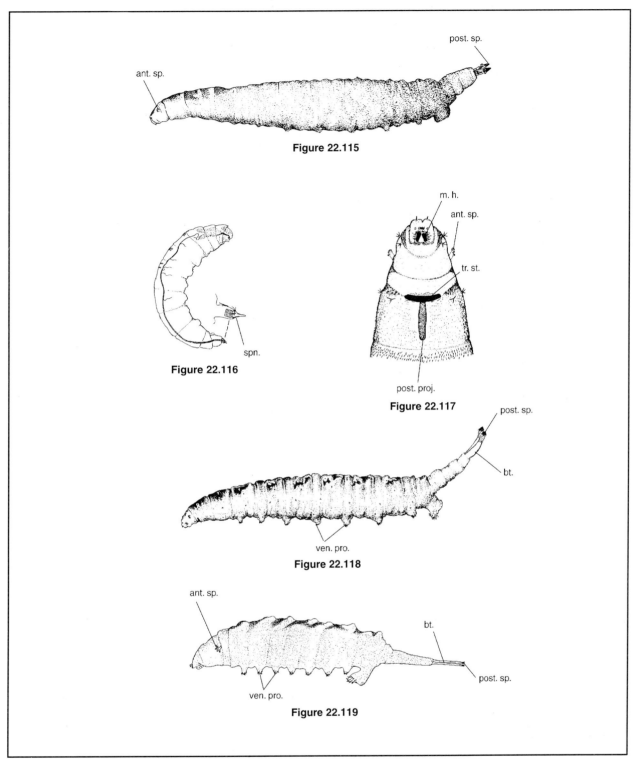

Figure 22.115. Lateral view of larva of *Dimecoenia spinosa* (Loew) (Ephydridae) (from Mathis and Simpson 1981); ant. sp., anterior spiracle; post. sp., posterior spiracle.

Figure 22.116. Lateral view of larva of *Hydrellia williamsi* Cresson showing caudal processes (Ephydridae) (from Williams 1938); spn., spines.

Figure 22.117. Anteroventral view of larva of *Cirrula gigantea* Cresson (Ephydridae) (from Mathis and Simpson 1981); ant. sp., anterior spiracle; m.h., mouth hooks; post. proj., posterior projection; tr. st., transverse strap.

Figure 22.118. Lateral view of larva of *Cirrula austrina* (Coquillett) (Ephydridae) (from Mathis and Simpson 1981); post. sp., posterior spiracle; bt., breathing tube; ven. pro., ventral prolegs.

Figure 22.119. Lateral view of larva of *Ephydra subopaca* Loew (Ephydridae) (from Johannsen 1935); ant. sp., anterior spiracle; post. sp., posterior spiracle; bt., breathing tube; ven. pro., ventral prolegs.

96(94').	Body length less than 2.5 mm; larva mining thalli of duckweed (*Lemna*)	***Lemnaphila*** Cresson
96'.	Body length more than 3.0 mm; larva as in Fig. 22.116; larva mining leaves of many wetland plants	***Hydrellia*** Robineau-Desvoidy
97(93').	Larva living in pools of petroleum	***Helaeomyia*** Cresson
97'.	Larva not living in pools of petroleum	98
98(97').	Larva with ventral prolegs (Figs. 22.87, 22.118-22.120, 22.123)	99
98'.	Larva without ventral prolegs	103
99(98).	Venter of segment three with transverse strap (Fig. 22.117)	100
99'.	Venter of segment three without transverse strap (Fig. 22.114)	101
100(99).	Transverse strap on venter of segment 3 with pale posterior projection (Fig. 22.117); larva as in Fig. 22.118	***Cirrula*** Cresson
100'.	Transverse strap on venter of segment 3 lacking pale projection; larva as in Figs. 22.87, 22.119	***Ephydra*** Fallen (in part)
101(99').	Breathing tube bearing pair of filaments near base (Fig. 22.120)	***Ephydra*** Fallen (in part)
101'.	Breathing tube lacking pair of filaments at base	102
102(101').	Dorsum of larva with distinct pattern composed of darkened scales and spinules (Fig. 22.122); larva as in Fig. 22.123	***Ephydra*** Fallen (in part)
102'.	Dorsum of larva lacking distinct pattern, nearly uniform in color; larva as in Fig. 22.124	***Setacera*** Cresson
103(98').	Breathing tubes short, arising separately on last body segment (Fig. 22.129)	104
103'.	Breathing tubes fused at base to form elongate, apically branched structure (Fig. 22.142)	109
104(103).	Anterior spiracles unbranched, either narrow and elongate (Fig. 22.125) or broader with marginal papillae (Fig. 22.127); larvae detritivores or algivores	105
104'.	Anterior spiracles distinctly 2-branched at base (Fig. 22.132); mandibles slender and hook-like (Fig. 22.133); larvae associated with blue-green algae	107
105(104).	Anterior spiracles narrow and elongate, frequently retracted (Fig. 22.125); mandibles broad, with apical teeth (Fig. 22.126)	***Brachydeutera*** Loew
105'.	Anterior spiracles broader, with marginal papillae (Fig. 22.127); mandibles slender and hook-like (Fig. 22.133)	106
106(105').	Posterior spiracles with 8-9 spiracular openings, without intraspiracular processes (float hairs) (Fig. 22.128); larva as in Fig. 22.129	***Scatophila*** Becker
106'.	Posterior spiracles with 4 spiracular openings, with 4 highly branched intraspiracular processes (float hairs; Fig. 22.130); larva as in Fig. 22.131	***Scatella*** Robineau-Desvoidy
107(104').	Dorsum of larva with distinct pattern formed by blackened scales and spinules (Fig. 22.135)	***Pelina*** Haliday
107'.	Dorsum of larva unpatterned, uniformly colored	108
108(107').	Length of body over 3.0 mm; facial mask with comb-like structures (Fig. 22.134); breathing tubes elongate, at least 1/4 length of body (Fig. 22.136)	***Lytogaster*** Becker
108'.	Length of body less than 2.5 mm; facial mask lacking comb-like structures (Fig. 22.137); breathing tubes short, less than 1/4 length of body (Fig. 22.138)	***Nostima*** Coquillett
109(103').	Facial mask lacking comb-like structures	110
109'.	Facial mask with comb-like structures (Fig. 22.148)	111
110(109).	Anterior spiracles 4-branched (Fig. 22.139); larva as in Fig. 22.140	***Parydra*** Stenhammar
110.	Anterior spiracles unbranched, with 4-7 marginal papillae (Fig. 22.141); larva as in Fig. 22.142	***Ochthera*** Latreille
111(109').	Larva less than 5.0 mm in length; perianal pad transversely elongate (Fig. 22.143); anterior spiracles somewhat hand-shaped, with 4-5 papillae apically (Fig. 22.144)	***Coenia*** Robineau-Desvoidy
111'.	Larva exceeding 8.0 mm in length; perianal pad variable	112
112(111').	Perianal pad nearly circular; anterior spiracles elongate, with 2-4 marginal papillae apically	113
112'	Perianal pad transversely elongate (Fig. 22.145); anterior spiracles broadened apically, with 7-9 marginal papillae (Fig. 22.146); larva as in Fig. 22.86	***Notiphila*** (*Dichaeta* Meigen)
113(112).	Anterior spiracles deeply bilobed (Fig. 22.147); larva as in Fig. 22.149; widespread	***Paracoenia*** Cresson
113'.	Anterior spiracles not bilobed, with 3-4 short marginal papillae (Fig. 22.150); eastern coastal habitats	***Dimecoenia*** (in part)
114(90').	Posterior abdominal segment lacking tubercles other than those bearing spiracles (Fig. 22.88); prothoracic spiracles, if present, fan-shaped and usually with fewer than 10 papillae; accessory oral sclerite present below mouth hooks (Fig. 22.89)	**MUSCIDAE**
114'.	Posterior abdominal segment often with several pairs of tubercles surrounding spiracles; prothoracic spiracle a cribriform (sievelike) plate, or with many papillae arranged in a bicornuate (two-horned) fan (Fig. 22.90); accessory oral sclerite absent	**SCATHOPHAGIDAE**

ADDITIONAL TAXONOMIC REFERENCES
(Larvae and Pupae)*

General
Johannsen (1934, 1935); Hennig (1948, 1950, 1952); Chu (1949); Peterson (1951); Bertrand (1954); Wirth and Stone (1956); Pennak (1978); McAlpine *et al.* (1981, 1987); McCafferty (1981); Ferrar (1987); Foote (1987); Hilsenhoff (1991); Teskey (1991).

Regional faunas
North and South Carolina: Webb and Brigham (1982).
Northeastern U.S.: Peckarsky *et al.* (1990).
Wisconsin: Hilsenhoff (1981).

Taxonomic treatments at the family level
Athericidae: Webb (1981, 1995); Nagatomi (1961a,b, 1962).
Blephariceridae: Hogue (1973b, 1981, 1987).
Ceratopogonidae: Thomsen (1937); Glukhova (1977, 1979); Downes and Wirth (1981); Hribar and Mullen (1991a).
Chaoboridae: Cook (1956, 1981).
Deuterophlebiidae: Kennedy (1981); Courtney (1989, 1990b).
Dixidae: Nowell (1953); Peters (1981).
Dryomyzidae: Ferrar (1987).
Empididae: Steyskal and Knutson (1981).
Ephydridae: Ferrar (1987).
Nymphomyiidae: Kevan and Cutten (1981); Courtney (1994a).
Pelecorhynchidae: Teskey (1981b).
Phoridae: Disney (1991).
Psychodidae: Quate (1955); Vaillant (1971); Hogue (1973a).
Ptychopteridae: Alexander (1981a).
Sciomyzidae: Ferrar (1987); Knutson (1987).
Stratiomyidae: McFadden (1967); James (1981); Roskosny (1982, 1983).
Syrphidae: Hartley (1961); Ferrar (1987).
Tabanidae: Teskey (1969); Pechuman and Teskey (1981); Pechuman *et al.* (1983).
Tanyderidae: Alexander (1981b).
Thaumaleidae: Stone and Peterson (1981).

* Exclusive of Chironomidae, Culicidae, Simuliidae, and Tipulidae.

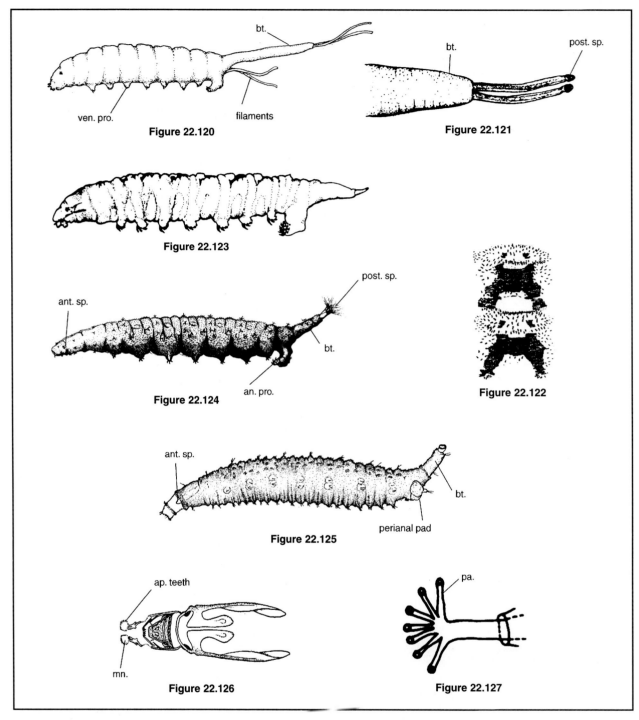

Figure 22.120. Lateral view of larva of *Ephydra cinerea* Jones (Ephydridae) (from Johannsen 1935); bt., breathing tube; ven. pro., ventral prolegs.

Figure 22.121. Posterior end (dorsal view) of larva of *Ephydra riparia* Fallén (Ephydridae) (from Eastin 1970); bt., breathing tube, post. sp., posterior spiracle.

Figure 22.122. Dorsum of abdominal segments 5 and 6 of larva of *Ephydra riparia* Fallén (Ephydridae) (from Eastin 1970).

Figure 22.123. Lateral view of larva of *Ephydra riparia* Fallén (Ephydridae) (from Eastin 1970).

Figure 22.124. Lateral view of *Setacera atrovirens* (Loew) (Ephydridae) (from Foote 1982); ant. sp., anterior spiracle; an. pro., anal proleg; post. sp., posterior spiracle; bt., breathing tube.

Figure 22.125. Lateral view of larva of *Brachydeutera argentata* (Walker) (Ephydridae) (from Johannsen 1935); ant. sp., anterior spiracle; bt., breathing tube.

Figure 22.126. Dorsal view of pharyngeal skeleton of larva of *Brachydeutera argentata* (Walker) (Ephydridae) (from Johannsen 1935); ap., apical; mn., mandible.

Figure 22.127. Anterior spiracle of larva of *Scatophila iowana* Wheeler (Ephydridae) (from Deonier 1974); pa., papilla.

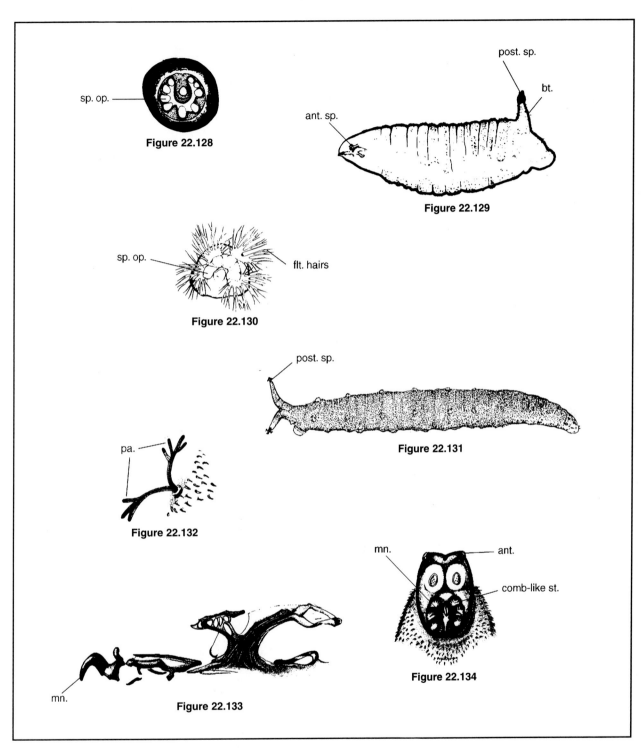

Figure 22.128. Posterior spiracle of larva of *Scatophila iowana* Wheeler (Ephydridae) (from Deonier 1974); sp. op., spiracular opening.

Figure 22.129. Lateral view of larva of *Scatophila iowana* Wheeler (Ephydridae) (from Deonier 1974); ant. sp., anterior spiracle; post. sp., posterior spiracle; bt., breathing tube.

Figure 22.130. Posterior spiracle of larva of *Scatella picea* (Walker) (Ephydridae) (from Eastin 1970); flt., float hairs; sp. op., spiracular opening.

Figure 22.131. Lateral view of larva of *Scatella hawaiiensis* Grimshaw (Ephydridae) (from Williams 1938); post. sp., posterior spiracle.

Figure 22.132. Anterior spiracle of larva of *Pelina truncatula* Loew (Ephydridae) (from Foote 1981b); pa., papilla.

Figure 22.133. Lateral view of cephalopharyngeal skeleton of larva of *Pelina truncatula* Loew (Ephydridae) (from Foote 1981b); mn., mandible.

Figure 22.134. Facial mask of larva of *Lytogaster excavata* (Sturtevant and Wheeler) (Ephydridae) (from Foote 1981a); mn., mandible; st., comb-like structure.

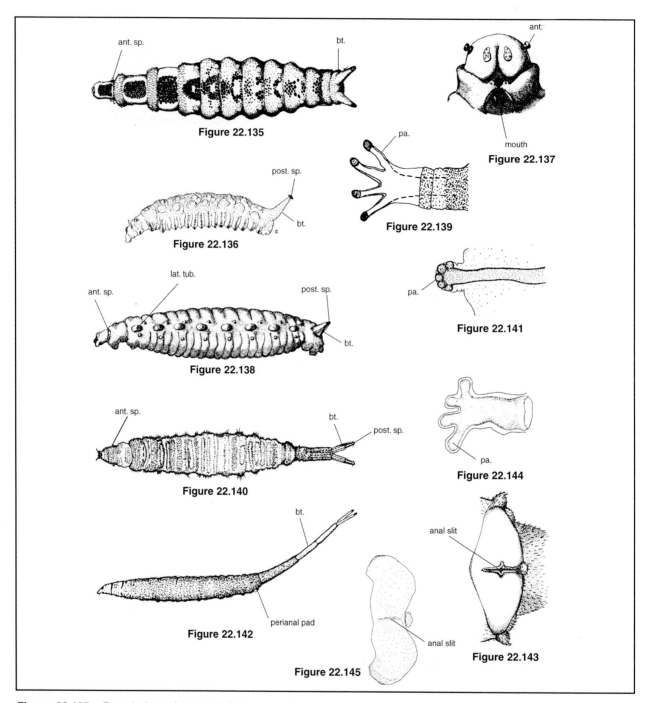

Figure 22.135. Dorsal view of larva of *Pelina truncatula* Loew (Ephydridae) (from Foote 1981b); ant. sp., anterior spiracle; bt., breathing tube.

Figure 22.136. Lateral view of larva of *Lytogaster excavata* (Sturtevant and Wheeler) (Ephydridae) (from Stehr 1991); post. sp., posterior spiracle; bt., breathing tube.

Figure 22.137. Facial mask of larva of *Nostima approximata* (Sturtevant and Wheeler) (Ephydridae) (from Foote 1983); ant., antenna.

Figure 22.138. Lateral view of larva of *Nostima approximata* (Sturtevant and Wheeler) (Ephydridae) (from Foote 1983); ant. sp., anterior spiracle; lat. tub., lateral tubercles; post. sp., posterior spiracle; bt., breathing tube.

Figure 22.139. Anterior spiracle of larva of *Parydra quadrituberculata* Loew (Ephydridae) (from Deonier and Regenburg 1978); pa., papilla.

Figure 22.140. Dorsal view of larva of *Parydra quadrituberculata* Loew (Ephydridae) (from Deonier and Regensburg 1978); ant. sp., anterior spiracle; post. sp., posterior spiracle; bt., breathing tube.

Figure 22.141. Anterior spiracle of larva of *Ochthera mantis* (De Geer) (Ephydridae) (from Simpson 1975); pa., papilla.

Figure 22.142. Lateral view of larva of *Ochthera mantis* (De Geer) (Ephydridae) (from Simpson 1975); bt., breathing tube.

Figure 22.143. Perianal pad of larva of *Coenia curvicauda* (Meigen) (Ephydridae) (from Foote 1990).

Figure 22.144. Anterior spiracle of larva of *Coenia curvicauda* (Meigen) (Ephydridae) (from Foote 1990); pa., papilla.

Figure 22.145. Perianal pad of larva of *Dichaeta caudata* (Fallén) (Ephydridae) (from Eastin and Foote 1971).

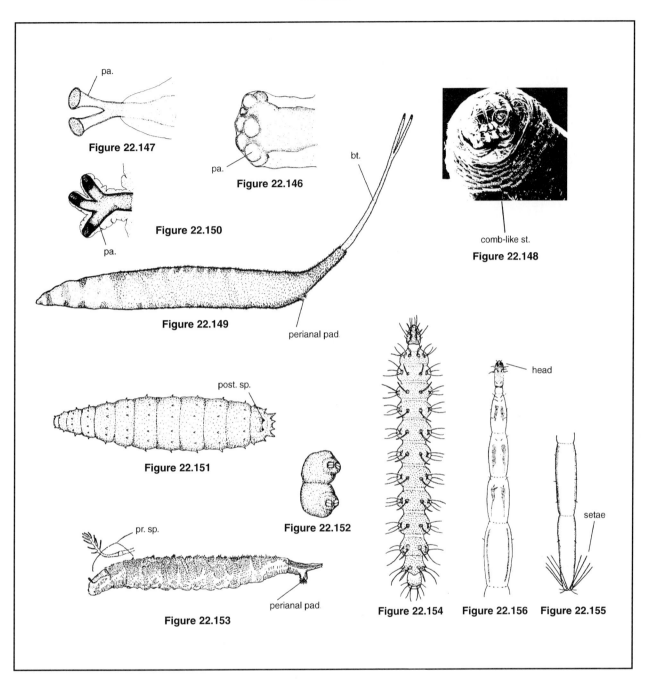

Figure 22.146. Anterior spiracle of larva of *Dichaeta caudata* (Fallén) (Ephydridae) (from Eastin and Foote 1971); pa., papilla.

Figure 22.147. Anterior spiracle of larva of *Paracoenia bisetosa* (Coquillett) (Ephydridae) (from Zack 1983); pa., papilla.

Figure 22.148. Head and anterior segments of larva of *Paracoenia bisetosa* (Coquillett) (Ephydridae) showing comb-like structures. (from Zack 1983); st., comb-like structure.

Figure 22.149. Lateral view of larva of *Paracoenia bisetosa* (Coquillett) (Ephydridae) (from Zack 1983); bt., breathing tube.

Figure 22.150. Anterior spiracle of larva of *Dimecoenia fuscifemur* Steyskal (Ephydridae) (from Mathis and Simpson 1981); pa., papilla.

Figure 22.151. Dorsal view of larva of *Megaselia* sp. (Phoridae).

Figure 22.152. Dorsal view of posterior spiracles of *Megaselia* sp. (Phoridae).

Figure 22.153. Lateral view of larva of *Canaceoides nudata* (Cresson) (Canacidae) (from Williams 1938); pr. sp., prothoracic spiracle.

Figure 22.154. Dorsal view of larva of *Forcipomyia brevipennis* (Macquart) (Ceratopogonidae) (from Stehr 1991).

Figure 22.155. Posterior segments of larva of *Palpomyia pruinescens* Thomsen (Ceratopogonidae) (from Thomsen 1937).

Figure 22.156. Head and anterior segments of larva of *Palpomyia pruinescens* Thomsen (Ceratopogonidae) (from Thomsen 1937).

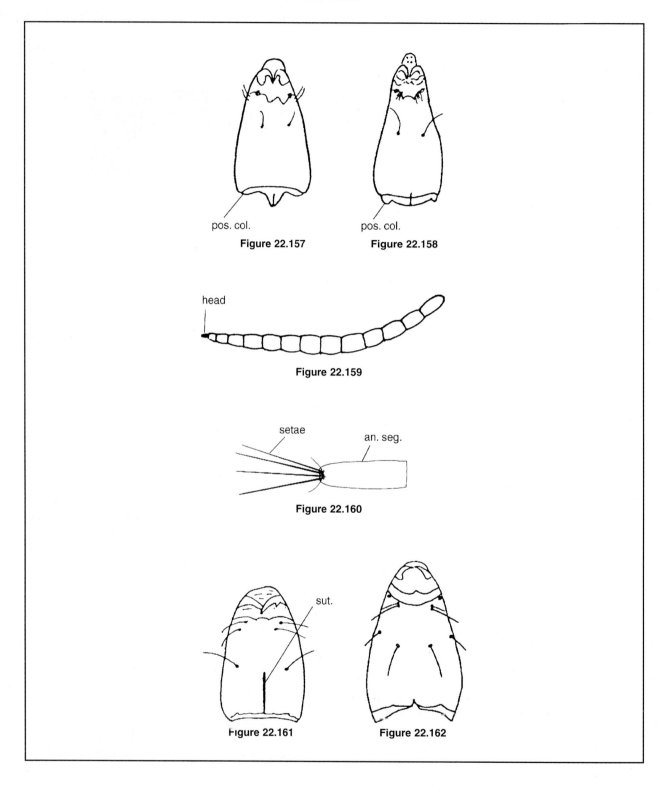

Figure 22.157. Ventral view of head capsule of larva of *Sphaeromias* sp. (Ceratopogonidae) (from Hilsenhoff 1981); pos. col., posterior collar.

Figure 22.158. Ventral view of head capsule of larva of *Probezzia* sp. (Ceratopogonidae) (from Hilsenhoff 1981); pos. col., posterior collar.

Figure 22.159. Lateral view of larva of *Serromyia* sp. (Ceratopogonidae) (from Hilsenhoff 1981).

Figure 22.160. Terminal segment of larva of *Alluaudomyia paraspina* Wirth (Ceratopogonidae) (from Grogan and Messersmith 1976); an. seg., anal segment.

Figure 22.161. Ventral view of head capsule of larva of *Ceratopogon* sp. (Ceratopogonidae) (from Hilsenhoff 1981).

Figure 22.162. Ventral view of head capsule of larva of *Culicoides variipennis* (Coquillett) (Ceratopogonidae) (modified from Murphree and Mullen 1991).

AQUATIC DIPTERA
Part Two. Pupae and Adults of Aquatic Diptera

Richard W. Merritt
Michigan State University, East Lansing

Donald W. Webb
Illinois Natural History Survey, Champaign

Evert I. Schlinger
University of California, Berkeley

INTRODUCTION

The pupae of aquatic and semiaquatic Diptera, or true flies, occur in a variety of aquatic and semiaquatic habitats throughout the world and generally fall into four categories: (1) those that are free-swimming with the pupa rising to the surface before the adult emerges (e.g., Culicidae, Chaoboridae, Dixidae, Chironomidae, etc.); (2) those that are attached to the bottom substrate in a silken cocoon (e.g., Simuliidae) or fully exposed (e.g., Blephariceridae, Deuterophlebiidae, and some Psychodidae), with the adults emerging from below the water surface; (3) those that burrow into the shoreline substrate along a stream, lake or pond (e.g., Athericidae, Tabanidae, Tipulidae, etc.); and (4) those that pupate within the last larval integument (Stratiomyidae, Syrphidae, Sciomyzidae, Ephydridae, etc.).

Adults of most aquatic and semiaquatic flies are found around or very near water, but those families whose species require a blood meal for ovarian development can occur miles away from the nearest water source (e.g., Ceratopogonidae, Culicidae, Simuliidae, Tabanidae). Some adult Diptera spend most of their lives "walking" on or skimming over water (e.g., some Chironomidae, Empididae, Dolichopodidae, Tipulidae); however, most groups spend little time directly associated with water. Species that oviposit directly in or on water may do so without much water contact, whereas others touch the water gently during oviposition. Collection of some of the smaller aquatic families may prove difficult; for example, one of the best places to find adult deuterophlebiids is in orb-weaving spider webs along streams (Courtney 1991). A summary of some common habits and habitats of adult aquatic Diptera is presented in Table 22A.

The ability of adults of some families to emerge from pupae submerged in swiftly flowing water has often been cited as an extreme adaptation to an aquatic habitat. For example, in the Blephariceridae the wings expand to full size during growth within the pupal case. Thus, at emergence the adult needs only to unfold them to render them functional (Hogue 1973b). In the Nymphomyiidae, adults locate a mate soon after emergence; they then couple, crawl beneath the water *in copula*, select an oviposition site, and the female lays a rosette of eggs around the coupled bodies. Finally, the adults die *in copula* under water (Courtney 1944a).

Some aquatic Diptera pass through one (univoltine) or two (bivoltine) generations a year, and adults are usually present during the spring or summer (e.g., some Tipulidae, Tanyderidae, Blephariceridae, and Stratiomyidae). Other groups may undergo several generations a year (multivoltine) and adults may be found during all seasons (e.g., some Culicidae, Ceratopogonidae, and Ephydridae). The latter often occur in warmer, more southern areas of the United States.

The order Diptera contains two suborders, the Nematocera and Brachycera (McAlpine *et al.* 1981). The Nematocera contains six infraorders and 25 families (the family Corethrellidae was added recently). The Brachycera contains 3 infraorders and 84 families. The aquatic Diptera belong mainly to the suborder Nematocera, where 15 of the 25 families contain aquatic species. The Brachycera has 15 of 84 families with aquatic or semiaquatic species. Although only 30 of the 109 families have aquatic or semiaquatic species, approximately 50% of all dipterous species are reported to be aquatic (Peterson 1951).

EXTERNAL MORPHOLOGY

Pupae

The morphology of the Diptera pupa is relatively simple and the structures discussed below are some of the more important characteristics utilized in the key to the families of aquatic Diptera pupae.

Head: The head of most aquatic and semiaquatic Diptera pupae is closely fused to the thorax, although separated from the thorax by the *cephalothoracic suture*. The head is differentiated into a frontal plate which is enclosed by the arms of the *epicranial suture*, along which the fracture occurs that liberates the adult. The size and shape of the *antennal sheath* (Figs. 22.173-22.177) and the sheath covering the maxillary palpus

Table 22A. Some common habits and habitats of adult aquatic Diptera.[1]

	On Rocks in Splash Zone or Under Waterfalls	Flying in Mating Swarms Above Vegetation or Aggregated on Rocks	Predatory on Other Invertebrates	Feeding on Nectar or Pollen at Flowers, or on Insect Secretions on Leaves	Resting on Vegetation or Ground Near Water Source	On Host (females of blood-sucking species)
Tipulidae	(some marine spp.)	x			x	
Tanyderidae		x				
Nymphomyiidae		x				
Psychodidae	x	(aggregated)			x	
Blephariceridae	x	(aggregated)	(females of some spp. feed on insects)	x?	x	
Deuterophlebiidae		x				
Thaumaleidae					x	
Dixidae		(aggregated)			x	
Culicidae		x		x		x
Ceratopogonidae	x	x	(some spp. feed on insects)	x	x	x
Chironomidae	x	x		x	x	x
Simuliidae				x	x	x
Tabanidae		x		x	x	
Stratiomyidae	(some stream spp.)			x	x	
Empididae	x	x	(insects)	x	x	
Dolichopodidae	(some stream spp.)		(insects)	x	x	
Syrphidae				x	x	
Dryomyzidae	(some stream spp.)		(barnacles)			
Sciomyzidae			(snails, slugs)		x	
Canaceidae	(some stream spp.)				x	
Ephydridae	x				x	
Scathophagidae			(some spp. feed on insects)	x	x	
Muscidae	x		(some spp. feed on insects)	x	x	

1. This table includes habits and habitats of exemplar species only and does not intend to show all habitats of species within each family.

are diagnostic characteristics, as well as the position of various setae, calli, and tubercles. The *vertex*, with its various tubercles and setae, comprises the remaining portion of the head. In the families Blephariceridae, Deuterophlebiidae, and some Psychodidae, the head is covered by the thorax (Figs. 22.173, 22.174). In those families of aquatic and semiaquatic Diptera utilizing a *puparium*, the diagnostic characteristics are essentially those of the last larval stage (Figs. 22.207-22.214).

Thorax: The thorax in free-swimming aquatic Diptera is considerably enlarged and thicker than the abdomen (Figs. 22.182, 22.184, 22.188). In those families that adhere to the bottom substrate (e.g., Blephariceridae and Deuterophlebiidae), or pupate in the substrate (Athericidae, Tabanidae, Tipulidae, etc.), the thorax is subequal to or narrower than the abdomen (Figs. 22.171, 22.173, 22.174, 22.196, 22.199). The thorax is characterized by the size and shape of the *thoracic spiracle*, the position of various setae, ridges, and tubercles, and the size and shape of the *leg* and *wing sheaths*.

Abdomen: The abdomen generally consists of eight or nine segments, separated into a dorsal *tergite*, a ventral *sclerite* and two lateral *pleurites*. The lateral pleurites may or may not possess *spiracles*. In the Blephariceridae and Deuterophlebiidae, the pleurites are extremely reduced (Figs. 22.173B, 22.174B). The tergites, sclerites and pleurites in those pupae burrowing into shoreline substrates often possess transverse rows of spines (Figs. 22.196, 22.199-22.201). Abdominal spines generally are absent or few in free-swimming pupae. The terminal segment of the pupa often possesses several *tubercles*, *projections*, or *paddles* (Figs. 22.176, 22.177, 22.179, 22.182, 22.196, 22.201).

Adults

The morphology of adult Diptera is complex and a given structure in one family may not be the same (homologous) as in another family. For this reason, only structures thought to be homologous have been used in construction of the key. The structures discussed below should be carefully understood before attempting to use the key.

Head and Mouthparts: The *antennae* are usually inserted between the eyes and consist of three to numerous (20-30) segments.* When there are only three segments, the terminal segment may have an additional structure, a *style* (Fig. 22.163) or

---

* There are actually only three true segments to the antenna; the third may be divided into many flagellomeres. However, for the purpose of this key, we will use the word "segment" instead of flagellomere.

an *arista* (Fig. 22.165), or this segment may be annulated, consisting of incomplete, segmentlike rings (Fig. 22.164). In some families with three-segmented antennae, the second segment may have a dorsal, seamlike line, termed the *longitudinal seam* (Fig. 22.165A). When counting the number of antennal segments, *do not* count the nonarticulating or nonmovable antennal tubercle as the first segment. The *ptilinal* (*frontal*) *suture* is a crescent-shaped groove, situated on the lower part of the frons between the bases of the antennae and the eyes (Fig. 22.166A), and usually extends ventrally into the facial area. The *proboscis* is a sucking, tubelike mouthpart that is often capable of being extended and may appear rigid, as well as varying in length (Figs. 22.224, 22.226).

Thorax: The *mesonotum* of Diptera is strongly developed and composed of a very narrow *prescutum* and the larger *scutum, scutellum,* and *postnotum* (Fig. 22.228). Several characters of the mesonotum are important in separating families of Diptera. For example, the posteriorly directed V-shaped suture signifies the family Tipulidae (Fig. 22.215). Also, the absence or presence of a transverse suture, complete or incomplete, separates several families (Fig. 22.167). The postnotum is usually round, swollen, and smooth, but in the Chironomidae a longitudinal groove divides this structure (Fig. 22.228).

Wings: These organs are probably the most important structures used to separate families of Diptera. A drawing of a generalized wing (Fig. 22.169) shows the terminology of wing veins and wing cells used in the key below. When examining the wing of a fly under the microscope, the viewer must move the specimen *slowly* to obtain the proper lighting for viewing the insertion or ending of any particular vein. This is especially true when examining the subcostal vein, since this vein is often hidden by the radial vein (R_1), and is even more critical when looking for fractures in the costal vein (Figs. 22.239-22.240). Because venation varies considerably from family to family, several text figures are given with the veins and cells labeled. Also, the venation within a family may be quite different from that shown in a text figure; therefore, learn the location of the veins, rather than the shape of cells enclosed by the veins.

Halteres: These small, knoblike structures are remnants of the second pair of wings, and serve as a significant character to identify the order Diptera. In one family, the Ptychopteridae, an additional structure termed the *prehalter* (Fig. 22.216) is present.

Calypteres: These scalelike structures may appear to be absent, or when fully developed may cover the halteres. The calypter is often called a *squama*. There may be one or two calypteres on each side of the thorax below the wing base. The lower calypter is attached to the thorax, and the upper calypter is attached to the wing base. One calypter may be longer than the other (Figs. 22.241-22.242).

Legs: The apex of the leg often contains several characters useful for separating families of Diptera. Between the tarsal claws are two padlike structures termed *pulvilli* (Fig. 22.168). A hairlike or padlike structure may occur between the pulvilli, and is termed the *empodium* (Fig. 22.168A, C).

KEYS TO THE FAMILIES OF AQUATIC DIPTERA

Pupae*

1.	Pupa free, not covered by last larval skin, although cocoon may be present (Figs. 22.170-22.205, 25.196)	2
1'.	Pupa contained within last larval skin, which is heavily sclerotized (Figs. 22.206-22.214)	24
2(1).	Antennal sheaths short, extending anteriorly; respiratory organs absent; pupa very slender and less than 2.0 mm in length; shape of pupa as in Fig. 22.180	**NYMPHOMYIIDAE**
2'.	Antennal sheaths directed posteriorly or laterally; respiratory organs present or absent, but pupa not as in Fig. 22.180	3
3(2').	Antennal sheaths generally elongate, lying over eyes and extending to or beyond bases of wing sheaths (Figs. 22.175, 22.177, 22.188); prothoracic respiratory organs or spiracles usually conspicuous (Figs. 22.173-22.179)	4
3'.	Antennal sheaths short, directed posteriorly and laterally; not lying over eyes and not reaching beyond wing bases (Figs. 22.199, 22.201); prothoracic respiratory organs or spiracles rudimentary or absent in most cases (Figs. 22.197, 22.198, 22.201, 22.203)	19
4(3).	Pupa usually enclosed in a fibrous, cone-shaped or slipper-like cocoon; respiratory organs consisting of numerous filaments projecting from open end of cocoon (Figs. 22.170, 25.28)	**SIMULIIDAE** (Chap. 25)
4'.	Pupa not enclosed in cocoon, although it may be enclosed in silken tube or case made out of various materials; respiratory organs not as above	5
5(4').	Pupa generally convex in shape; attached limpet-like to rocks in bed of stream (Figs. 22.171, 22.173, 22.174)	6
5'.	Pupa variable in shape, but not attached to rocks in bed of stream	8
6(5).	Pupa flattened, shield-shaped; respiratory organs simple, cylindrical or ovoid in shape (Fig. 22.171)	**PSYCHODIDAE** (in part)
6'.	Pupa more convex; respiratory organs with 3-4 filaments	7

---

* The key to aquatic Diptera pupae (Nematocera and Brachycera) presented here was modified from Johannsen (1934-1937) and Usinger (1956) by R. W. Merritt, B. A. Foote, and D. W. Webb.

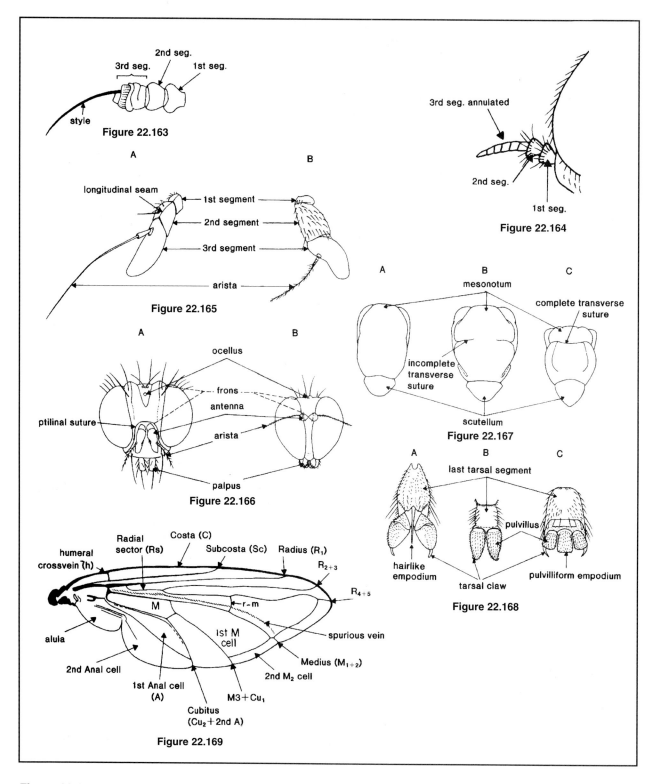

Figure 22.163. Lateral view of antenna of *Microchrysa* sp. (Stratiomyidae).

Figure 22.164. Lateral view of antenna of *Odontomyia* sp. (Stratiomyidae).

Figure 22.165. Antenna of (*A*) *Limnophora* sp. (Muscidae), and (*B*) *Dictya* sp. (Sciomyzidae).

Figure 22.166. Frontal view of head of (*A*) *Cordilura* sp. (Scathophagidae), and (*B*) *Dolichopus* sp. (Dolichopodidae).

Figure 22.167. Dorsal view of thorax of (*A*) *Dictya* sp. (Sciomyzidae), (*B*) *Cordilura* sp. (Scathophagidae), and (*C*) *Limnophora* sp. (Muscidae).

Figure 22.168. Ventral view of last tarsal segment of (*A*) *Tubifera* sp. (Syrphidae), (*B*) *Limnophora* sp. (Muscidae), and (*C*) *Tabanus* sp. (Tabanidae).

Figure 22.169. Generalized wing of Diptera showing veins and cells (Syrphidae).

7(6').	Respiratory organs usually consisting of 4 simple lamellae (thin plates) (Fig. 22.173)	**BLEPHARICERIDAE**
7'.	Respiratory organs consisting of 3-4 slender, crooked projections (Fig. 22.174)	**DEUTEROPHLEBIIDAE**
8(5').	Leg sheaths straight and projecting beyond ends of wing sheaths (Figs. 22.175-22.177); caudal (anal) end of pupa not paddle-shaped, often forked, lobed and/or spined, never cylindrical with 3 terminal setae	9
8'.	Leg sheaths often curved or folded and projecting little, if any, beyond ends of wing sheaths (Figs. 22.172, 22.186-22.188), but if straight and projecting beyond ends of wing sheaths, then caudal lobes cylindrical with 3 terminal setae (Fig. 25.501); otherwise, caudal end of pupa terminating in paddle-like structure; pupae commonly found in open water (Figs. 22.186-22.188)	11
9(8).	One prothoracic respiratory organ longer than body of pupa, other organ short (Fig. 22.175)	**PTYCHOPTERIDAE**
9'.	Both prothoracic respiratory horns distinctly shorter than length of pupa, usually subequal in length (Figs. 22.176-22.179)	10
10(9').	Respiratory organs small and gradually narrowed to terminal point; head with noticeable two-spined crest (Fig. 22.176)	**TANYDERIDAE**
10'.	Without above combination of characters (Figs. 22.177-22.179)	**TIPULIDAE** (Chap. 23)
11(8').	Tergites of abdomen with acute posterolateral angles (Fig. 22.181C); thorax formed into numerous ridges and hollows (Fig. 22.181B); pupa dorsoventrally flattened and less than 4 mm in length (Fig. 22.181A)	**THAUMALEIDAE**
11'.	Tergites of abdomen without acute posterolateral angles; thorax not strongly ridged; combination of characters not as above	12
12(11').	Wing sheaths extending to nearly mid-length of pupa; leg sheaths short, straight, and superimposed (Fig. 22.172); robust pupa, under 4 mm in length	**PSYCHODIDAE** (in part)
12'.	Wing sheaths ending distinctly before mid-length of pupa (Fig. 22.182A); body elongate, length variable	13
13(12').	Caudal (anal) end of pupa with 2 paddles, each paddle with distinct midrib (Fig. 22.182A, B)	14
13'.	Caudal end of pupa either lacking paddles, or if paddles are present, midrib is lacking (Figs. 22.186-22.188)	15
14(13).	Respiratory horn swollen, spindle-shaped, usually pointed, with a reticulate (meshed or netted) surface and nearly closed at tip (Figs. 22.182A, C)	**CHAOBORIDAE**
14'.	Respiratory horn (=trumpet) open at tip (Figs. 24.2, 24.44, 24.47, 24.49, 24.51)	**CULICIDAE** (Chap. 24)
15(13').	Paddles present, consisting of 2 long pointed caudal lobes that are immovable, narrow and fused basally, never with 2 lateral setae (Figs. 22.183-22.185)	16
15'.	Paddles absent or if present, not fused basally; when present and shaped as above, then paddles always with 2 strong lateral setae in addition to fringe of smaller setae that may be present (Figs. 26.286, 26.288, 26.293, 26.300)	17
16(15).	Caudal (anal) paddles without long hairs or spines (Figs. 22.184, 22.185)	**DIXIDAE**
16'.	Caudal paddles with lateral spines and stout spines at tips (Fig. 22.183)	**CORETHRELLIDAE**
17(15').	Prothoracic respiratory organs either without openings or a plastron (Figs. 22.188, 26.368, 26.392, 26.424), or composed of numerous filaments (Fig. 22.187) or absent	**CHIRONOMIDAE** (in part) (Chap. 26)
17'.	Prothoracic respiratory organs with multiple openings or a plastron (Figs. 22.186, 26.283, 26.303, 26.309)	18
18(17').	Prothoracic respiratory organ with a plastron (Figs. 22.186A, 26.283, 26.287, 26.303); anal segment usually terminating in 2-lobed swimming paddle (Figs. 22.186B, 26.304, 26.308)	**CHIRONOMIDAE** (in part) (Chap. 26)
18'.	Prothoracic respiratory organ with multiple small openings, often lined up along one margin or around apex; anal segment ending in pair of pointed processes (Figs. 22.189-22.193)	**CERATOPOGONIDAE**
19(3').	Pupa short and robust; prothoracic respiratory organs or horns greatly elongated (Figs. 22.194-22.196)	20
19'.	Pupa more elongate (Figs. 22.199, 22.201, 22.203); prothoracic respiratory organs shorter and less conspicuous (Figs. 22.201, 22.203)	21
20(19).	Abdomen with elongate lateral spiracular processes (Figs. 22.196); pair of diverging sutures absent on ventral side of head	**EMPIDIDAE** (in part)

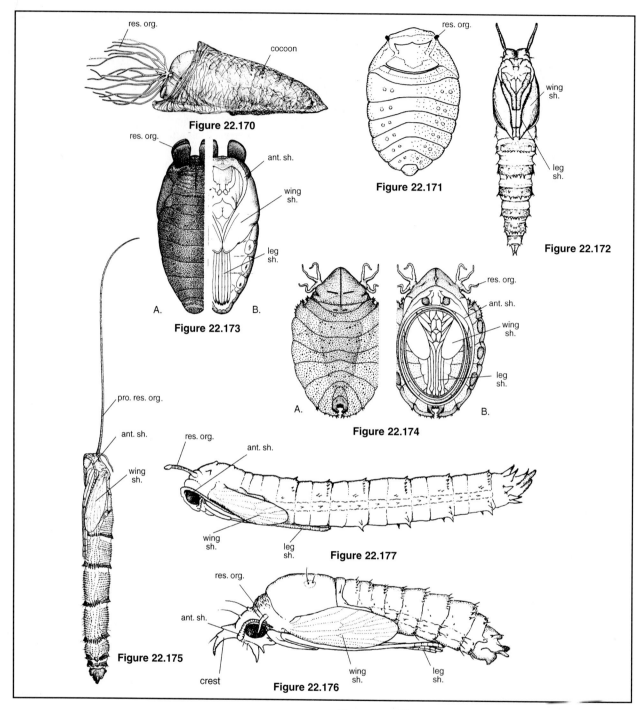

Figure 22.170. Lateral view of pupa of *Simulium vittatum* Zetterstedt (Simuliidae); res. org., respiratory organ.

Figure 22.171. Dorsal view of pupa of *Maruina lanceolata* (Kincaid) (Psychodidae) (redrawn from Johannsen 1934); res. org., respiratory organ.

Figure 22.172. Ventral view of pupa of *Psychoda alternata* Say (Psychodidae) (from Johannsen 1934); sh., sheath.

Figure 22.173. Three-fifths dorsal (A) and three-fifths ventral (B) views of pupa of *Agathon comstocki* (Kellogg) (Blephariceridae) (from Manual of Nearctic Diptera, Vol. 1); res. org., respiratory organ; ant., antennal; sh., sheath.

Figure 22.174. Dorsal (A) and ventral (B) view of pupa of male *Deuterophlebia nielsoni* Kennedy (Deuterophlebiidae) (from Manual of Nearctic Diptera, Vol. 1); res. org., respiratory organ; ant., antennal; sh., sheath.

Figure 22.175. Lateral view of pupa of *Ptychoptera lenis* Osten Sacken (Ptychopteridae) (from Manual of Nearctic Diptera, Vol. 1); pro. res. org., prothoracic respiratory organ; ant., antennal; sh., sheath.

Figure 22.176. Lateral view of male pupa of *Protoplasa fitchii* Osten Sacken (Tanyderidae) (modified from Alexander by Byers); res. org., respiratory organ; ant., antennal; sh., sheath.

Figure 22.177. Lateral view of female pupa of *Tipula* (*Yamatotipula*) *eluta* Loew (Tipulidae) (modified from Malloch by Byers); res. org., respiratory organ; ant., antennal; sh., sheath.

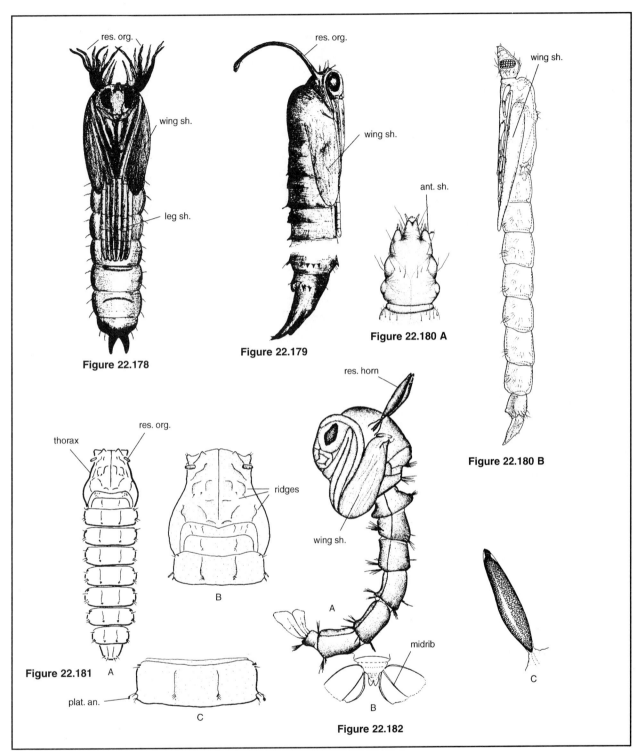

Figure 22.178. Ventral view of pupa of *Antocha saxicola* Osten Sacken (Tipulidae) (from Johannsen 1934); res. org., respiratory organ; sh., sheath.

Figure 22.179. Lateral view of pupa of *Ulomorpha pilosella* (Osten Sacken) (Tipulidae) (from Johannsen 1934); res. org., respiratory organ; sh., sheath.

Figure 22.180. *A*) Dorsal view of pupal head of *Nymphomyia dolichopera* Courtney (Nymphomyiidae) (from Courtney 1994a); ant. sh., antennal sheath; and *B*) Lateral view of pupa of *Nymphomyia walkeri* (Ide) (Nymphomyiidae) (from Manual of Nearctic Diptera, Vol. 1); ant., antennal; sh., sheath.

Figure 22.181. *A*) Dorsal view of pupa of *Trichothaumalea elakalensis* Sinclair (Thaumaleidae); res. org., respiratory organ; *B*) dorsal view of thorax of *T. elakalensis* pupa; *C*) abdominal tergite of *T. elakalensis* pupa (from Sinclair 1992); plat. an., posterolateral angle.

Figure 22.182. *A*) Lateral view of pupa of *Chaoborus americanus* (Johannsen) (Chaoboridae); res., respiratory; sh., sheath; B) caudal paddle of *C. americanus* pupa (from Johannsen 1934); C) respiratory horn of *C. americanus* pupa (from Manual of Nearctic Diptera, Vol. 1).

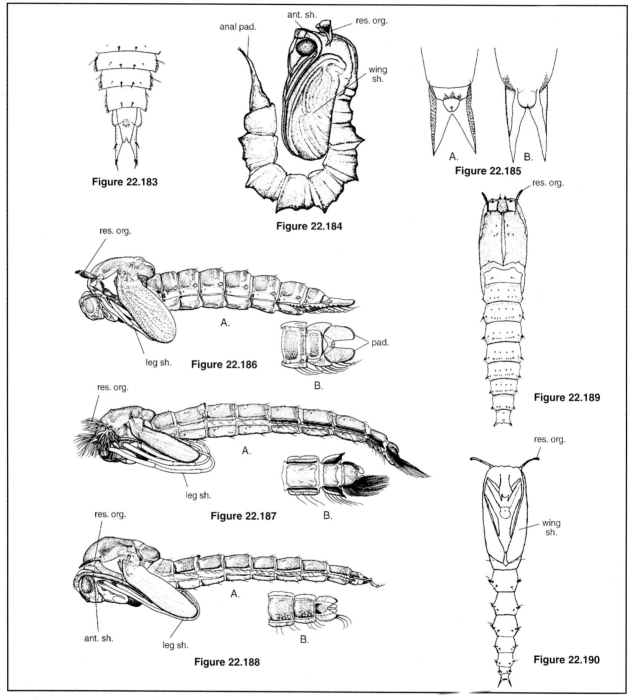

Figure 22.183. Dorsal view of abdomen of pupa of *Corethrella brakeleyi* (Coquillett) (Corethrellidae) (from Johannsen 1934).

Figure 22.184. Lateral view of pupa of *Dixella aliciae* (Johannsen) (Dixidae) (from Johannsen 1934); pad., paddle, res. org., respiratory organ; ant., antennal; sh., sheath.

Figure 22.185. A) Ventral view of caudal lobes of pupa of *Dixella* sp. (Dixidae); B) Ventral view of caudal lobes of pupa of *Dixa* (redrawn from Disney 1975).

Figure 22.186. A) Lateral view of pupa of *Procladius* sp. (Chironomidae); res. org., respiratory organ; sh., sheath; B) terminal segments of pupa of *Procladius* sp. (from Manual of Nearctic Diptera, Vol. 1); pad., paddle.

Figure 22.187. A) Lateral view of pupa of *Chironomus* sp. (Chironomidae); res. org., respiratory organ; sh., sheath; B) terminal segments of pupa of *Chironomus* sp. (from Manual of Nearctic Diptera, Vol. 1).

Figure 22.188. A) Lateral view of pupa of *Orthocladius* sp. (Chironomidae); res. org., respiratory organ; ant., antennal; sh., sheath; B) terminal segments of pupa of *Orthocladius* sp. (from Manual of Nearctic Diptera, Vol. 1).

Figure 22.189. Dorsal view of pupa of *Stilobezzia bulla* Thomsen (Ceratopogonidae)(from Johannsen 1937); res. org., respiratory organ.

Figure 22.190. Ventral view of pupa of *Bezzia varicolor* (Coquillett) (Ceratopogonidae) (from Johannsen 1937); res. org., respiratory organ; sh., sheath.

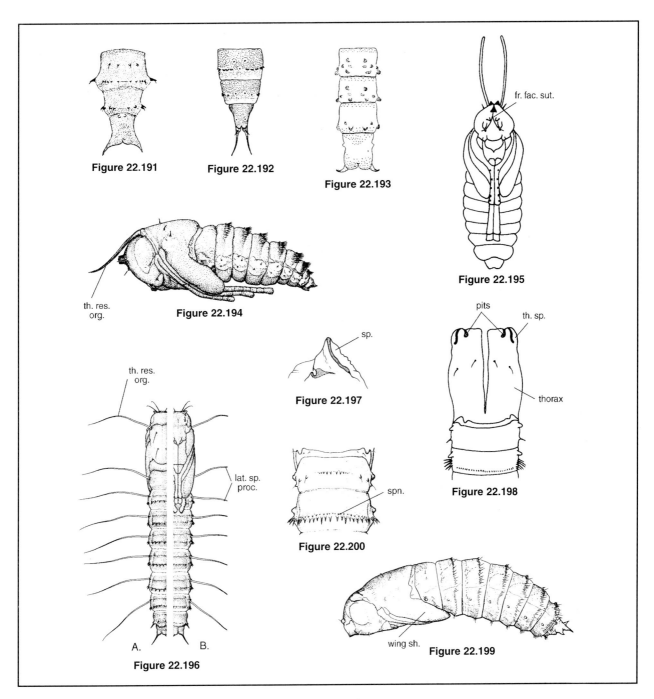

Figure 22.191. Posterior abdominal segments of pupa of *Culicoides venustus* Hoffman (Ceratopogonidae) (from Johannsen 1937).

Figure 22.192. Posterior abdominal segments of pupa of *Dasyhelea traverae* Thomsen (Ceratopogonidae) (from Johannsen 1937).

Figure 22.193. Posterior abdominal segments of pupa of *Forcipomyia johannseni* Thomsen (Ceratopogonidae) (from Johannsen 1937).

Figure 22.194. Lateral view of pupa of *Rhaphium slossonae* (Johnson) (Dolichopodidae) (from Manual of Nearctic Diptera, Vol. 1); th. res. org., thoracic respiratory organ.

Figure 22.195. Ventral view of male pupa of *Hydrophorus oceanus* (Macquart) (Dolichopodidae) (Redrawn from Dyte 1967); fr. fac. sut., fronto-facial suture.

Figure 22.196. Half dorsal (A) and half ventral (B) views of pupa of *Hemerodromia* sp. (Empididae) (from Manual of Nearctic Diptera, Vol. 1); th. res. org., thoracic respiratory organ; lat. sp. proc., lateral spiracular processes.

Figure 22.197. Thoracic spiracle of pupa of *Chrysops* sp. (Tabanidae); sp., spiracle.

Figure 22.198. Dorsal view of thorax of pupa of *Hamatabanus carolinensis* (Macquart) (Tabanidae); th. sp., thoracic spiracle.

Figure 22.199. Lateral view of pupa of *Hybomitra epistates* (Osten Sacken) (Tabanidae) (from Manual of Nearctic Diptera, Vol. 1); sh., sheath.

Figure 22.200. Dorsal view of tergites 1 and 2 of *Chlorotabanus crepuscularis* (Bequaert) (Tabanidae) (from Manual of Nearctic Diptera, Vol. 1); spn., spines.

20'. Abdomen without elongate lateral spiracular processes (Figs. 22.194, 22.195); pair of sutures present on ventral side of head running from apical cephalic tubercle, then diverging and extending to behind the base of antennae on each side (Fig. 22.195) ....*DOLICHOPODIDAE*

21(19'). Dorsal opening of thoracic spiracle generally directed along anterior-posterior axis of pupa (Fig. 22.197, 22.198); prominent pit mediolateral to each spiracle (Fig. 22.198); abdominal segments 2-7 with encircling fringe of spines arranged in 1 or 2 series, the anterior series usually distinctly shorter (and sometimes stouter) than the posterior series (Figs. 22.199, 22.200) ....*TABANIDAE*

21'. Dorsal opening of thoracic spiracle directed transversely (left to right) to axis of pupa and without a pit mediolateral to each spiracle (Figs. 22.201, 22.202); other characters not as above 22

22(21'). Dorsal opening of thoracic spiracle directed transversely to axis of pupa (Figs. 22.201, 22.202)..*ATHERICIDAE*

22'. Dorsal opening of thoracic spiracle circular, generally on short or moderately long tubercle (Figs. 22.203-22.205)..23

23(22'). Thorax with anterior transverse row of dark colored pits (Fig. 22.203); dorsal opening of thoracic spiracle at end of short tubercle (Fig. 22.203)...........................*PELECORHYNCHIDAE*

23'. Thorax without anterior transverse row of dark colored pits; dorsal opening of thoracic spiracle circular, at end of short or moderately long tubercle (Figs. 22.204, 22.205)..*EMPIDIDAE* (in part)

24(1'). Pupa enclosed in unmodified last larval skin; head capsule distinct (Figs. 22.47, 22.48, 22.50, 22.51, 22.206)...*STRATIOMYIDAE*

24'. Pupa enclosed in last larval skin that is modified to form a puparium; no head capsule; variable shapes (Figs. 22.207-22.214)...**(CYCLORRHAPHOUS—BRACHYCERA)**

Adults*

1. Antennae generally longer than thorax, composed of 6 or more freely articulated segments (Nematocera) (Figs. 22.215-22.229)..2

1'. Antennae composed of 5 or fewer (usually 3) freely articulated segments, the last segment often annulated (Fig. 22.164) or bearing a style or arista (Figs. 22.163, 22.165) ....(Cyclorrhaphous-Brachycera) 16

2(1). Mesonotum with an entire V-shaped suture or impression (Fig. 22.215) 3

2'. Mesonotal suture transverse, not V-shaped, or absent (Fig. 22.167) 5

3(2). Wings with 2 anal veins reaching the margin (Fig. 22.215); legs long and slender (crane flies) ....*TIPULIDAE* (Chap. 23)

3'. Wings with a single anal vein reaching the margin (Figs. 22.216-22.217) 4

4(3'). Halteres with a conspicuous process ("prehalter") anterior to their base (Fig. 22.216); wings with 4 branches of radius reaching the margin (Fig. 22.216) (phantom crane flies) ....*PTYCHOPTERIDAE*

4'. Halteres without a conspicuous process anterior to their base; wings with 5 branches of radius reaching the margin (Fig. 22.217) (primitive crane flies) ....*TANYDERIDAE*

5(2'). Wings, if present, with a marginal fringe of long hairs (Fig. 22.218), very narrow and elongate with only rudiments of wing veins present anteriorly (Fig. 22.218); small flies, less than 3 mm (nymphomyiid flies) ....*NYMPHOMYIIDAE*

5'. Wings without marginal fringe of long hairs, broader, not narrow and elongated 6

6(5'). M_3 vein of wing completely detached (Fig. 22.219); eyes divided by an unfacetted stripe; wing venation sometimes accompanied by a fine network of folds and creases (Fig. 22.219) (net-winged midges) ....*BLEPHARICERIDAE*

6'. M_3 vein of wing not detached (Figs. 22.221-22.224); eyes not divided by unfacetted stripe..7

7(6'). Wings large, broadest in basal fourth and densely covered with microtrichia (fine hairs) on anterior half, less dense on posterior half; wings generally lacking most of true venation (Fig. 22.220); mouthparts absent; terminal antennal segment in male may be 3-4 times body length (Fig. 22.220); (mountain midges) ....*DEUTEROPHLEBIIDAE*

7'. Wings not broadest in basal fourth and not densely covered with microtrichia; wing veins well developed, at least anteriorly; mouthparts usually present 8

---

* Key modified from Wirth and Stone (1956) and Cole (1969). This key is for only those members of the included families whose larvae or pupae are aquatic or whose adults are found near, in, or on water.

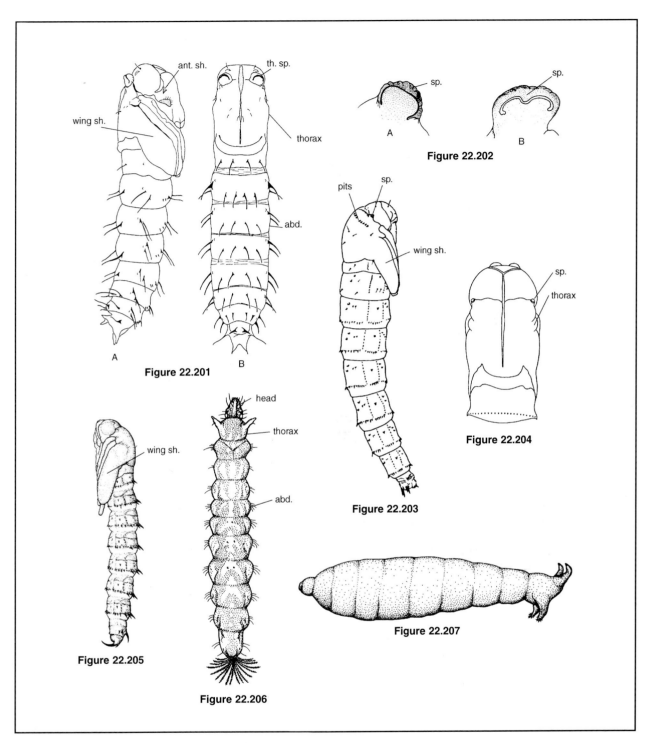

Figure 22.201. Lateral view (*A*) and dorsal view (*B*) of pupa of *Suragina concinna* (Williston) (Athericidae); abd., abdomen; ant., antennal; sh., sheath; th. sp., thoracic spiracle.

Figure 22.202. Thoracic spiracles of pupae of *A*) *Suragina concinna* (Williston) and *B*) *Atherix pachypus* Bigot (Athericidae); sp., spiracle.

Figure 22.203. Lateral view of pupa of *Glutops rossi* Pechuman (Pelecorhynchidae) (from Teskey 1970); sh., sheath; sp., spiracle.

Figure 22.204. Dorsal view of thorax of pupa of *Clinocera* sp. (Empididae); sp., spiracles.

Figure 22.205. Lateral view of pupa of *Clinocera stagnalis* (Haliday) (Empididae) (from Manual of Nearctic Diptera, Vol. 1); sh., sheath.

Figure 22.206. Dorsal view of pupa of *Caloparyphus greylockensis* (Johnson) (Stratiomyidae) (modified from Johannsen 1935); abd., abdomen.

Figure 22.207. Lateral view of puparium of *Limnophora riparia* Fallén (Muscidae).

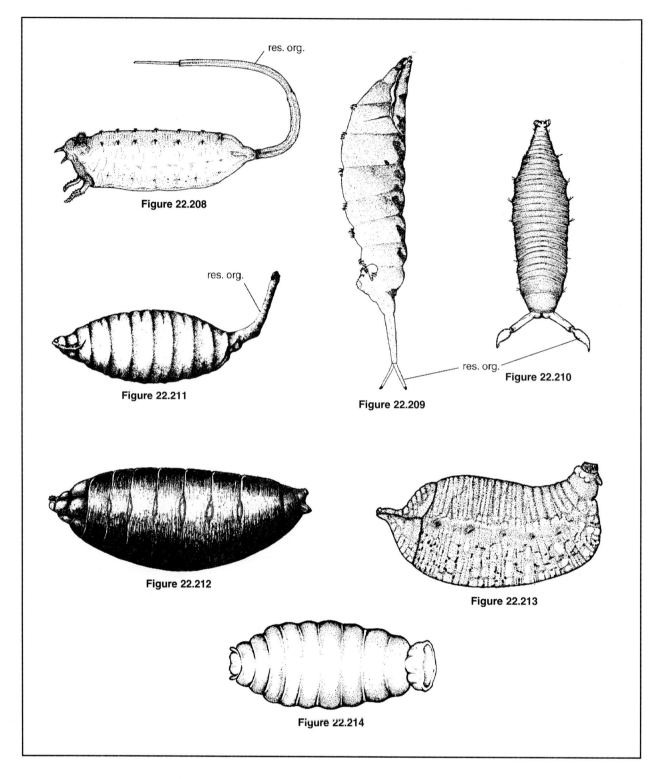

Figure 22.208. Lateral view of puparium of *Eristalis* sp. (Syrphidae); res. org., respiratory organ.

Figure 22.209. Lateral view of puparium of *Ephydra macellaria* Egger (Ephydridae) (from Johannsen 1935); res. org., respiratory organ.

Figure 22.210. Dorsal view of puparium of *Brachydeutera argentata* (Walker) (Ephydridae) (from Johannsen 1935); res. org., respiratory organ.

Figure 22.211. Lateral view of puparium of *Notiphila aenigma* Cresson (Ephydridae) (from Busacca and Foote 1978); res. org., respiratory organ.

Figure 22.212. Lateral view of puparium of *Pherbellia quadrata* Steyskal (Sciomyzidae) (from Bratt et al., 1969).

Figure 22.213. Lateral view of puparium of *Sepedon fuscipennis* Loew (Sciomyzidae) (from Neff and Berg 1966).

Figure 22.214. Dorsal view of puparium of *Fletcherimyia* sp. (Sarcophagidae) (from Johannsen 1935).

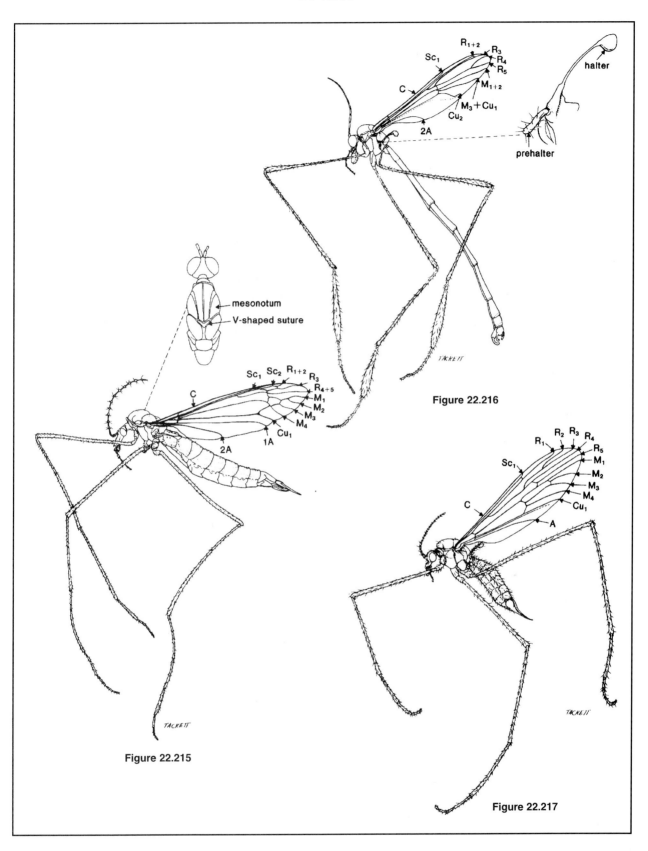

Figure 22.215. Lateral view of Tipulidae showing V-shaped mesonotal suture.

Figure 22.216. Lateral view of *Bittacomorpha* sp. (Ptychopteridae) showing prehalter.

Figure 22.217. Lateral view of *Protoplasa* sp. (Tanyderidae).

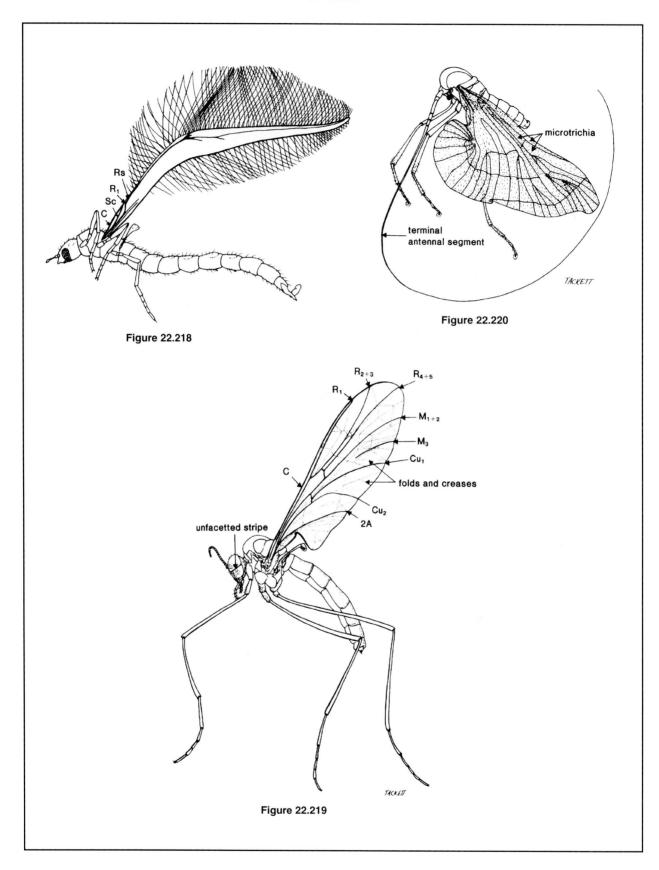

Figure 22.218. Lateral view of Nymphomyiidae.
Figure 22.219. Lateral view of *Agathon* sp. (Blephariceridae).
Figure 22.220. Lateral view of male *Deuterophlebia* sp. (Deuterophlebiidae).

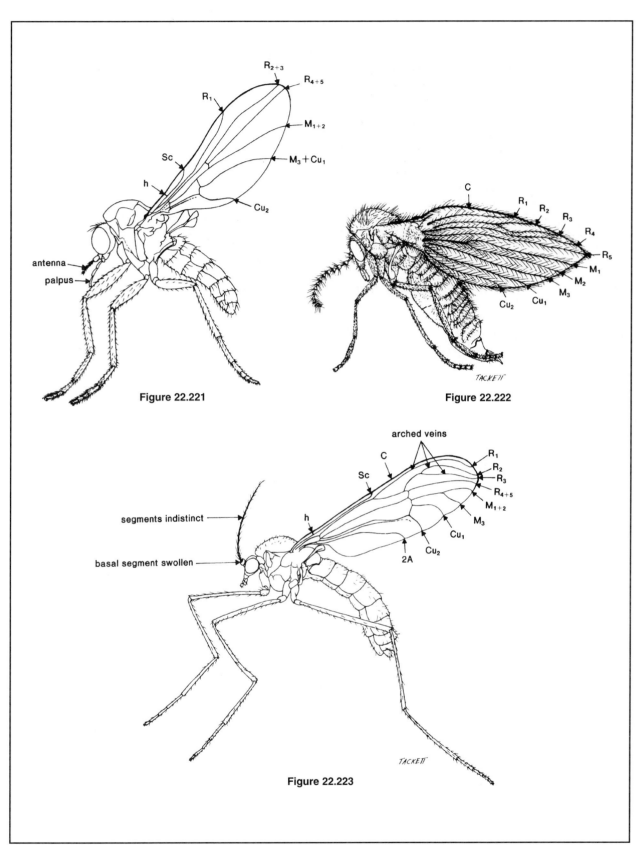

Figure 22.221. Lateral view of *Thaumalea* sp. (Thaumaleidae).

Figure 22.222. Lateral view of *Psychoda* sp. (Psychodidae).

Figure 22.223. Lateral view of *Dixa* sp. (Dixidae).

8(7').	Costa extending around margin of wing, though weaker beyond apex (Figs. 22.221-22.224, 22.226-22.227)	9
8'.	Costa ending at or near the apex of the wing (Figs. 22.225, 22.228-22.231)	14
9(8).	Wings with 7 longitudinal veins reaching margin (Fig. 22.221); antennae subequal to head length and shorter than palpi, with basal segments of flagellum thickened and clavate, the remaining segments filiform (Fig. 22.221); small flies (3-4 mm) (solitary midges)	*THAUMALEIDAE*
9'.	Wings with at least 9 longitudinal veins reaching margin (Figs. 22.222-22.226); flagellum of antennae uniformly filiform or beadlike	10
10(9').	Wings short and broad, pointed apically, usually densely hairy (mothlike) and held rooflike over the body when at rest; crossveins absent except near base of wing (Fig. 22.222) (moth flies)	*PSYCHODIDAE*
10'.	Wings long, or, if broad, the apex very broadly rounded (Figs. 22.223-22.226); wings held flat over the abdomen when at rest and not densely hairy (although scales may be present along wing veins or wing margins); crossveins present	11
11(10').	Proboscis elongated, extending far beyond the clypeus (Fig. 22.224); scales present on wing margins and veins and usually on the body (Fig. 22.224) (mosquitoes)	*CULICIDAE* (Chap. 24)
11'.	Proboscis short, extending little beyond the clypeus (Fig. 22.226); scales absent on wing veins (Fig. 22.226)	12
12(11').	Apical wing veins (R_1-R_3) strongly arched (Fig. 22.223) (dixid flies)	*DIXIDAE*
12'.	Apical wing veins (R_1-R_3) straight or nearly so (Figs. 22.226-22.227)	13
13(12').	R_1 of wing ending in costa closer to Sc than R_2 (Fig. 22.227)	*CORETHRELLIDAE*
13'.	R_1 of wing ending in costa closer to R_2 than Sc (Fig. 22.226) (phantom midges)	*CHAOBORIDAE*
14(8').	Wings broad, anterior veins strong with posterior veins weak and poorly developed (Fig. 22.225); antennae shorter than thorax, never plumose; dark or variously colored flies with humpbacked appearance, rarely over 6 mm in length (Fig. 22.225) (black flies)	*SIMULIIDAE* (Chap. 25)
14'.	Wings narrower and long, the posterior veins usually strong; antennae generally longer than thorax, often plumose (Fig. 22.228)	15
15(14').	Postnotum elongate with a median longitudinal groove or furrow (Fig. 22.228); M vein of wing not forked; femora not swollen; antennae of male plumose (Fig. 22.228); mouthparts not adapted for piercing (non-biting midges)	*CHIRONOMIDAE* (Chap. 26)
15'.	Postnotum rounded without a median longitudinal groove or furrow; M vein of wing nearly always forked (Fig. 22.229); femora often swollen; mouthparts sclerotized and adapted for piercing (Fig. 22.229) (biting midges)	*CERATOPOGONIDAE*
16(1').	Tarsi with 3 nearly equal pads under tarsal claws (empodium pulvilliform) (Fig. 22.168C)	17
16'.	Tarsi with 2 nearly equal pads under tarsal claws (empodium hairlike or absent) (Fig. 22.168A, B)	20
17(16).	Costa ending at or near wing tip (Fig. 22.231); tibiae without spurs (Fig. 22.231) (soldier flies)	*STRATIOMYIDAE*
17'.	Costa continuing well past wing tip, often around entire posterior margin of wing (Fig. 22.232); at least middle tibiae with spurs (Fig. 22.232)	18
18(17').	Third antennal segment compact, not annulated (Fig. 22.230)	*ATHERICIDAE**
18'.	Third antennal segment annulated with 3-8 apparent segments (Figs. 22.164, 22.231-22.232)	19
19(18').	Upper and lower calypteres large, subequal in size; abdominal tergite 1 with deep medial notch in posterior margin (Fig. 22.232) (horse and deer flies)	*TABANIDAE*
19'.	Lower calypter scarcely developed; abdominal tergite 1 without medial notch; uncommon.	*PELECORHYNCHIDAE*
20(16').	Anal cell long, closed just before margin of wing, therefore petiolate; a spurious (false) vein present running obliquely between veins R_{4+5} and M_{1+2} (Figs. 22.169, 22.233) (flower flies)	*SYRPHIDAE*
20'.	Anal cell short, transverse, oblique, or convex apically, or rarely absent (Figs. 22.234-22.236); no spurious vein	21
21(20').	Ptilinal (frontal) suture entirely absent (Fig. 22.166B); frons uniformly sclerotized; alula weak (Fig. 22.234)	22
21'.	Ptilinal (frontal) suture present (Fig. 22.166A); frons divided into distinct regions by ptilinal suture; alula well developed (Fig. 22.242)	23

* Some Rhagionidae may key out here, but they often have an elongated style.

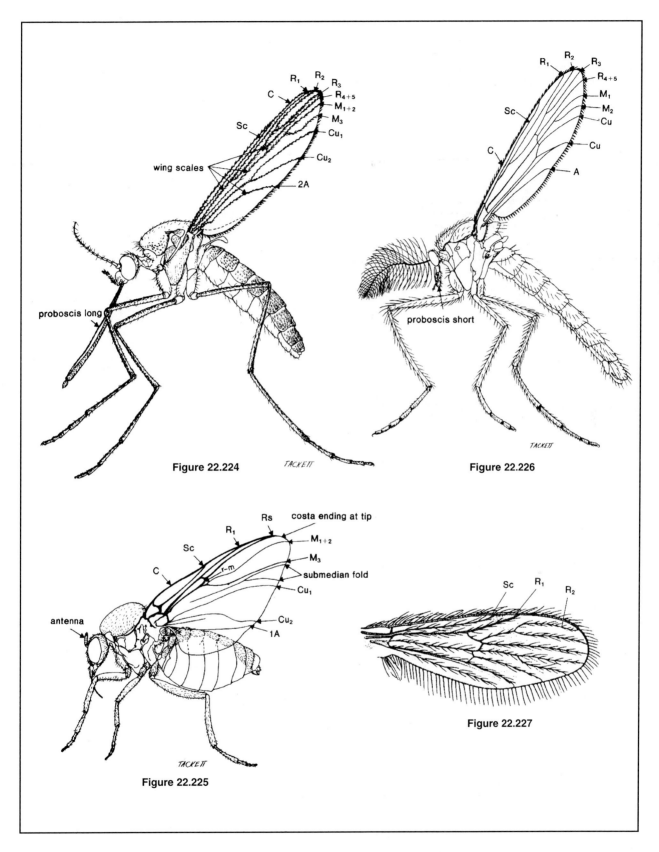

Figure 22.224. Lateral view of female *Aedes* sp. (Culicidae).

Figure 22.225. Lateral view of *Cnephia* sp. (Simuliidae).

Figure 22.226. Lateral view of male *Chaoborus* sp. (Chaoboridae).

Figure 22.227. Wing of *Corethrella* sp. (Cothrerellidae) showing veins.

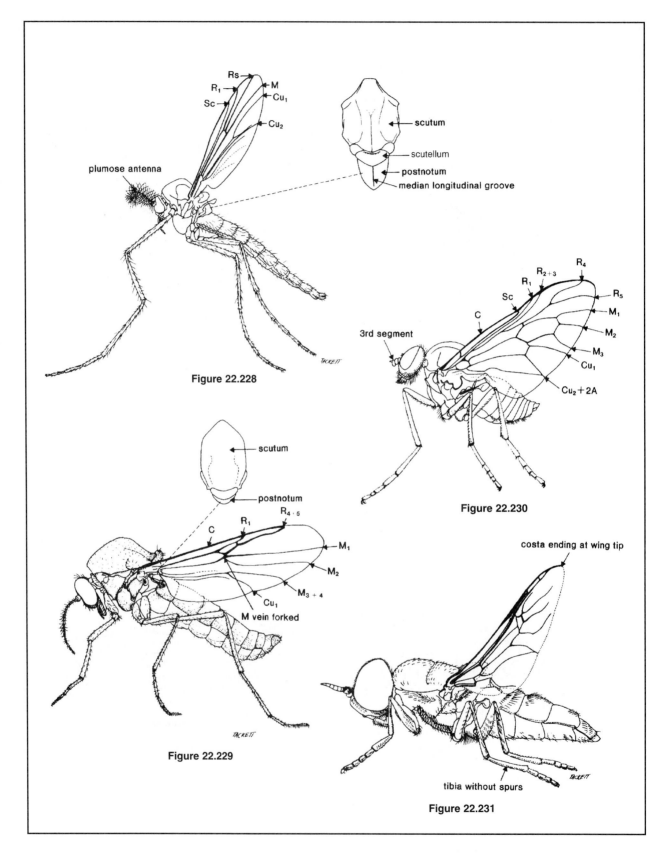

Figure 22.228. Lateral view of male Chironomini (Chironomidae) showing postnotal longitudinal groove.

Figure 22.229. Lateral view of male *Johannsenomyia* sp. (Ceratopogonidae).

Figure 22.230. Lateral view of *Atherix* sp. (Athericeridae).

Figure 22.231. Lateral view of *Odontomyia* sp. (Stratiomyidae).

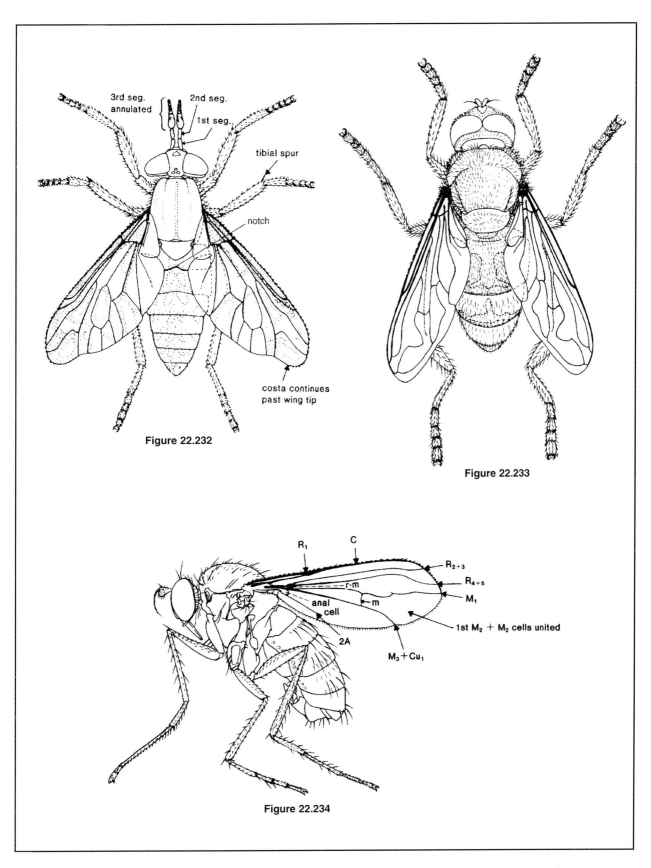

Figure 22.232. Dorsal view of *Chrysops sp.* (Tabanidae).

Figure 22.233. Dorsal view of *Eristalis* sp. (Syrphidae).

Figure 22.234. Lateral view of *Dolichopus* sp. (Dolichopodidae).

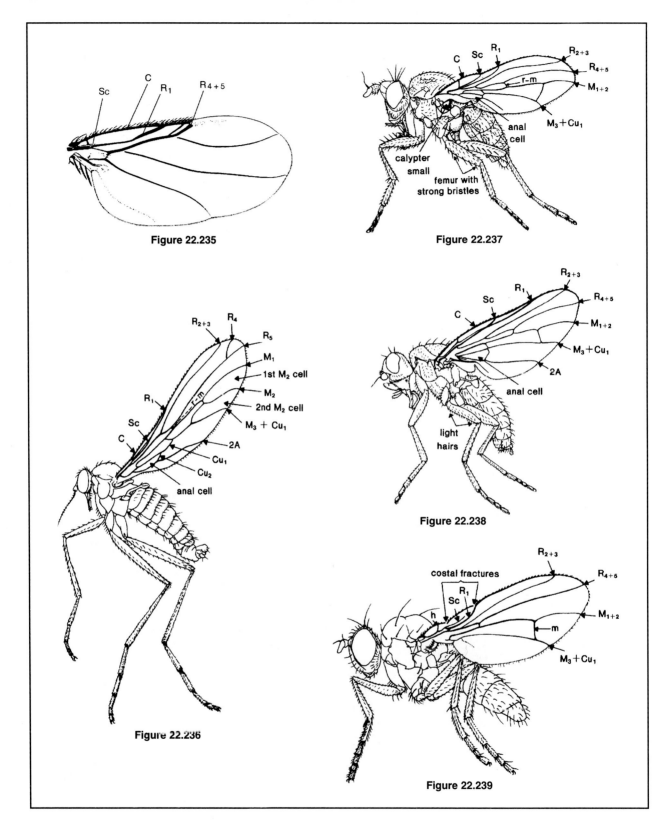

Figure 22.235. Wing of *Dohrniphora cornuta* (Bigot) (Phoridae) showing veins (from Manual of Nearctic Diptera, Vol. 2).

Figure 22.236. Lateral view of Empididae.

Figure 22.237. Lateral view of *Dictya* sp. (Sciomyzidae).

Figure 22.238. Lateral view of *Oedoparena* sp. (Dryomyzidae).

Figure 22.239. Lateral view of *Notiphila* sp. (Ephydridae).

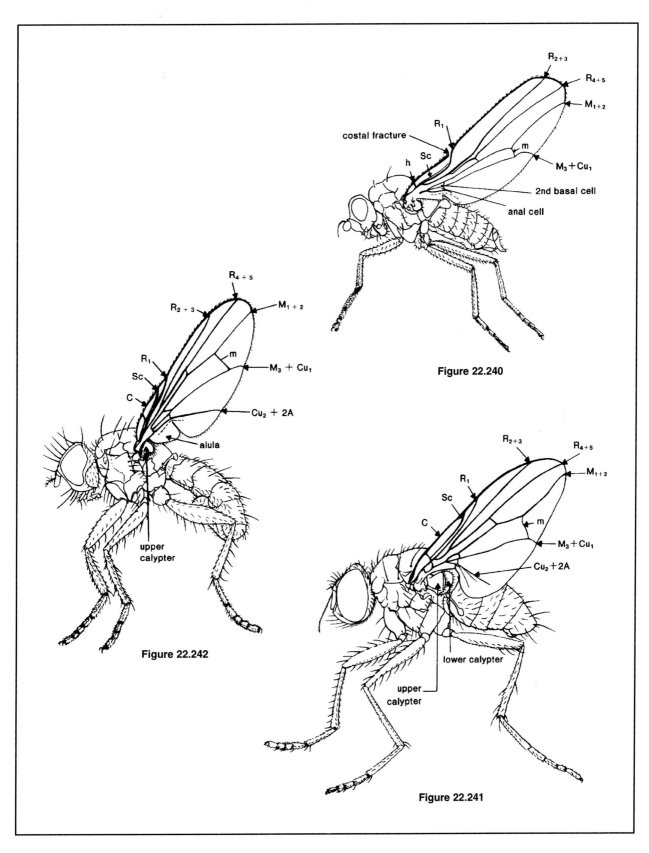

Figure 22.240. Lateral view of *Canaceoides* sp. (Canacidae).

Figure 22.241. Lateral view of *Limnophora* sp. (Muscidae).

Figure 22.242. Lateral view of *Cordilura* sp. (Scathophagidae).

22(21).	Anterior crossvein (r-m) situated at or before the basal fourth of wing (Fig. 22.234); 1st M_2 and 2nd M_2 cells united to form one long cell; vein R_{4+5} not branched (Fig. 22.234)	DOLICHOPODIDAE
22'.	Anterior crossvein (r-m) situated far beyond basal fourth of wing (Fig. 22.236); 1st M_2 and 2nd M_2 cells separated; vein R_{4+5} usually branched (Fig. 22.236) (dance flies)	EMPIDIDAE
23(21').	Second antennal segment without a longitudinal seam on its upper outer edge (Fig. 22.165B); mesonotum without a complete transverse suture (Fig. 22.167A, B); calypter (squama) small (Fig. 22.237)	(Acalypterate flies) 24
23'.	Second antennal segment with a longitudinal seam on its upper outer edge (Fig. 22.165A); mesonotum usually with complete transverse suture (Fig. 22.167C); calypter (squama) large (Fig. 22.241)	(Calypterate flies) 28
24(23).	Costa ending near middle of wing (Fig. 22.235); radial veins strongly thickened and crowded near anterior margin of wing (Fig. 22.235)	PHORIDAE
24'.	Costa extending at least to apex of wing; radial veins not thickened and crowded anteriorly (Figs. 22.237-22.240).	25
25(24').	Costa entire (not fractured); subcosta complete, ends in costa (Fig. 22.237)	26
25'.	Costa fractured just before end of subcosta (Fig. 22.240), and sometimes also fractured at humeral crossvein (Fig. 22.239); subcosta usually incomplete, not reaching costa (Figs. 22.239-22.240)	27
26(25).	Vein R_1 ending at middle of costa (Fig. 22.237); femora with strong bristles (Fig. 22.237) (marsh flies)	SCIOMYZIDAE
26'.	Vein R_1 ending past middle of costa (Fig. 22.238); femora with hairs but without strong bristles (Fig. 22.238) (barnacle flies)	DRYOMYZIDAE
27(25').	Costa fractured twice, once at junction of subcosta and once at humeral crossvein (Fig. 22.239); anal cell absent (Fig. 22.239) (shore and brine flies)	EPHYDRIDAE
27'.	Costa fractured only once, just before junction of subcosta (Fig. 22.240); anal cell present (Fig. 22.240) (beach flies)	CANACIDAE
28(23').	Mesonotum with 3 complete longitudinal stripes; venation similar to Figure 22.242 except vein M_{1+2} nearly joins vein R_{4+5} at apex (*Blaesoxipha* only) (flesh flies)	SARCOPHAGIDAE
28'.	Mesonotum without stripes or with 4 incomplete stripes	29
29(28').	Lower calypter longer than upper calypter (Fig. 22.241); transverse suture complete (Fig. 22.167C)	MUSCIDAE
29'.	Lower calypter not longer than upper calypter (Fig. 22.242); transverse suture incomplete to nearly complete (Fig. 22.167A, B) (Anthomyiidae of some authors)	SCATHOPHAGIDAE

ADDITIONAL TAXONOMIC REFERENCES
(Adults)*

General
Crampton (1942); Brues *et al.* (1954); Wirth and Stone (1956); Cole (1969); Griffiths (1972); Hennig (1973); Rohdendorf (1974); Borror *et al.* (1981); McAlpine *et al.* (1981, 1987); McCafferty (1981).

Taxonomic treatments at the family level
Athericidae: Stuckenberg (1973); Webb (1977, 1981, 1995).
Blephariceridae: Alexander (1958); Hogue (1973b, 1981, 1987).
Canacidae: Wirth (1951b, 1987); Mathis (1988a, 1992).
Ceratopogonidae: Wirth (1952a); Wirth *et al.* (1977); Downes and Wirth (1981); Wirth and Grogan (1988).
Chaoboridae: Cook (1956, 1981); Saether (1970).
Corethrellidae: Cook (1956, 1981).
Deuterophlebiidae: Kennedy (1958, 1981); Courtney (1989, 1990b, 1994b).
Dixidae: Peters and Cook (1966); Peters (1981).
Dolichopodidae: Wirth and Stone (1956); Robinson and Vockeroth (1981).
Dryomyzidae: Steyskal (1957, 1987).
Empididae: Wirth and Stone (1956); Steyskal and Knutson (1981); Wilder (1981a,b,c).
Ephydridae: Sturtevent and Wheeler (1954); Wirth and Stone (1956); Wirth *et al.* (1987); Zatwarnicki (1992).
Muscidae: Wirth and Stone (1956); Huckett and Vockeroth (1987).
Nymphomyiidae: Cutten and Kevan (1970); Kevan and Cutten (1981); Courtney (1994a).
Pelecorhynchidae: Krivosheina (1971); Teskey (1981b).
Phoridae: Disney (1991).
Psychodidae: Quate (1955); Quate and Vockeroth (1981).
Ptychopteridae: Alexander (1942, 1981a).
Sarcophagidae: Aldrich (1916); Shewell (1987).
Scathophagidae: Wirth and Stone (1956); Vockeroth (1987).
Sciomyzidae: Wirth and Stone (1956); Knutson (1987).
Stratiomyidae: Wirth and Stone (1956); James (1981).
Syrphidae: Wirth and Stone (1956); Vockeroth and Thompson (1987).
Tabanidae: Wirth and Stone (1956); Pechuman and Teskey (1981); Pechuman *et al.* (1983); Teskey (1990).
Tanyderidae: Alexander (1942, 1981b).
Thaumaleidae: Wirth and Stone (1956); Stone (1964b); Stone and Peterson (1981); Arnaud and Boussy (1994).

---

* Exclusive of Chironomidae, Culicidae, Simuliidae, and Tipulidae.

Table 22B. Summary of ecological and distributional data for larval aquatic *Diptera*. (For definition of terms see Tables 6A-6C; table prepared by G. W. Courtney, R. W. Merritt, K. W. Cummins, B. A. Foote, D. W. Webb.)

Taxa (number of species in parentheses)	Habitat	Habit	Trophic Relationships	North American Distribution	Ecological References*
Nematocera					
Blephariceridae (28) (net-winged midges)	Lotic—erosional	Clingers (ventral suckers)	Generally scrapers		1802, 1957, 1984, 2112, 1804, 3076, 4511, 4649, 111, 112, 3175
Agathon (8)	Lotic—erosional	Clingers	Scrapers	West mountains	4234
Bibiocephala (1)	Lotic—erosional	Clingers	Scrapers	West mountains	4234
Blepharicera (15)	Lotic—erosional	Clingers	Scrapers	Widespread	1375, 1805
Philorus (4)	Lotic—erosional	Clingers	Scrapers	West mountains	
Ceratopogonidae (501) (= Heleidae; biting midges, "no-see-ums")	Generally lentic—littoral (including tree holes and small temporary ponds and pools), lotic depositional (margins and detritus)	Generally sprawlers, burrowers, or planktonic (swimmers)	Generally predators (engulfers), collectors—gatherers		1957, 2112, 2355, 3076, 3994, 4496, 4510, 4511, 1831, 1835, 2081, 2786, 1832, 1836, 3183, 2116, 2959
Dasyheleinae (37)					
Dasyhelea (37)	Lentic—tree holes, rock pools, littoral (in algal mats, mosses, aquatic plants), lotic—depositional (margins)	Sprawlers—climbers (semiaquatic)	Collectors—gatherers, scrapers	Widespread	594, 3994, 4511
Forcipomyiinae (103)	Generally lotic—erosional	Generally sprawlers and clingers	Collectors—gatherers and scrapers(?)		
Atrichopogon (27)	Lotic—erosional (margins, debris jams), lentic—littoral (algal mats, mosses, leaf litter, aquatic plants)	Sprawlers—clingers		Widespread	353, 2910, 3994, 4511, 3965
Forcipomyia (83)	Lotic—erosional (margins, including warm springs); lentic—littoral and margins (semiaquatic); mosses, leaf litter, aquatic plants	Sprawlers	Scrapers(?)	Widespread	509, 3544, 4496, 4511, 4098, 4506, 4508
Ceratopogoninae (331)	Wide variety of lentic and lotic habitats	Generally burrowers, occasionally planktonic	Generally predators—(engulfers), a few collectors—gatherers		
Alluaudomyia (8)	Lentic—littoral (detritus and algal mats), occasionally limnetic	Burrowers, occasionally planktonic (swimmers)	Predators (engulfers)	Widespread	1493, 1494, 3994, 4511
Bezzia (42)	Lentic—littoral, profundal, and occasionally limnetic; lotic—hot springs (algal mats)	Burrowers, occasionally planktonic (swimmers)	Predators (engulfers)	Widespread	1835, 3994, 4496, 4511, 3078, 4057, 4234
Brachypogon (5)				Widespread	
Ceratopogon (20)				Widespread	
Clinohelea (7)	Lotic—stream margins; lentic—lake bottoms	Burrowers, occasionally swim	Predators (engulfers)	Widespread	2116, 2786
Culicoides (136)	Lentic—littoral (margins), tree holes and temporary pools, brackish marshes, marine shores; lotic—margins	Burrowers, occasionally planktonic (swimmers)	Predators (engulfers), collectors—gatherers	Widespread	199, 229, 328, 1292, 2082, 2289, 2290, 4451, 2054, 2087, 2108, 4511, 1833, 2109, 2810, 134, 2356
Jenkinshelea (4)	Lotic—depositional; lentic—littoral (margins)	Burrowers	Predators (engulfers)	East?	2116, 2786

* Emphasis on trophic relationships.

Table 22B. Continued

Taxa (number of species in parentheses)	Habitat	Habit	Trophic Relationships	North American Distribution	Ecological References*
Johannsenomyia (2)	Lentic—littoral, limnetic	Burrowers, occasionally planktonic	Predators (engulfers)	East	4511
Mallochohelea (11)	Lentic—littoral (margins); lotic	Burrowers; planktonic	Predators (engulfers)	Widespread	2116, 2786, 3994
Monohelea (15)				Widespread	
Nilobezzia (4)	Lentic and lotic; saline pools	Burrowers	Probably predators (engulfers)	Widespread	2116, 2786
Palpomyia (31)	Lotic—erosional and depositional (detritus), lentic—littoral, profundal, and occasionally limnetic	Burrowers, occasionally planktonic (swimmers)	Predators (engulfers), collectors—gatherers	Widespread	669, 671, 1495, 1496, 1751, 2786, 3994, 4511, 3078, 4234
Probezzia (19)	Lentic—littoral and limnetic	Burrowers, occasionally planktonic (swimmers)	Predators (engulfers)	Widespread	4511
Serromyia (7)	Lentic—littoral (margins)	Burrowers	Probably predators (engulfers)	Widespread	2129, 3994, 4511
Sphaeromias (2)	Lentic—littoral (algal mats), occasionally limnetic	Burrowers, occasionally planktonic (swimmers)	Predators (engulfers) collectors—gatherers	Widespread	3994, 4511
Stilobezzia (18)	Lentic—littoral (margins)	Burrowers, occasionally swimmers	Predators (engulfers)	Widespread	2786, 3994, 4511
Leptoconopinae (23)					
Leptoconops (23)	Lentic—littoral (margins), beaches—marine; seepage areas	Burrowers		Widespread (primarily Western)	648, 2352, 2353, 3708, 4505, 4511
Chaoboridae (14) (phantom midges)	Generally lentic—littoral, profundal, and limnetic; lotic—depositional (pools)	Generally sprawlers (day), planktonic (night)	Predators (piercers) and engulfers		690, 927, 1957, 2112, 2687, 3076, 3506, 4511, 4548
Chaoborus (10) (=*Corethra*)	Lentic—limnetic, profundal, and littoral (including temporary ponds and bogs)	Sprawlers (day), planktonic (night)	Predators (piercers), predators (engulfers), primarily microcrustacea	Widespread	272, 362, 390, 1785, 2041, 2829, 2869, 3471, 1440, 3506, 3586, 4536, 692, 2750, 4528
Eucorethra (1)	Lotic—depositional (pools of cold springs)	Planktonic—swimmers	Predators (piercers, feed at surface)	North and West	852, 1785, 3219
Mochlonyx (3)	Lentic—limnetic, profundal and littoral (especially bogs and small ponds)		Predators (engulfers)	Widespread	1785, 1951, 3219
Chironomidae	See Table 26A.				
Corethrellidae (5)					
Corethrella (4)	Lentic—limnetic and margins, lotic—hyporheic	Sprawlers	Predators (engulfers)	South	1805, 1070, 2521, 2621, 2622, 2629, 4528
Luizomiops (1)				Texas	
Culicidae	See Table 24A.				
Deuterophlebiidae (6) (mountain midges)	Lotic—erosional	Clingers (hooks on lateral abdominal prolegs)	Scrapers		27, 1957, 2073, 2074, 2112, 3076, 4494, 4511
Deuterophlebia (6)	Lotic—erosional	Clingers	Scrapers	West mountains	725, 726, 727, 728, 3075, 4234

* Emphasis on trophic relationships.

Table 22B. Continued

Taxa (number of species in parentheses)	Habitat	Habit	Trophic Relationships	North American Distribution	Ecological References*
Dixidae (42) (dixid midges)	Generally lotic—erosional (margins) and depositional, lentic—margins	Swimmers—climbers	Collectors—gatherers (possibly also filterers)		
Dixa (22)	Lotic—erosional (protected areas) and depositional (detritus)	Swimmers—climbers over damp surfaces	Collectors—gatherers	Widespread	603, 2942, 3672, 4511, 4234
Dixella (19)	Lentic—margins	Swimmers—climbers	Collectors—gatherers	Widespread	1438, 3041, 3096
Meringodixa (1)	Lotic—erosional (margins) and depositional	Swimmers—climbers	Collectors—gatherers	West	
Nymphomyiidae (2)					
Nymphomyia (2) (= Palaeodipteron)	Lotic—erosional (aquatic mosses)	Clingers (elongate abdominal prolegs)	Scrapers, collectors—gatherers	East	141, 818, 1907, 2085, 593, 731, 1585, 2449, 2448, 3711
Psychodidae (67) (moth flies)	Generally lotic—depositional, lentic—littoral (detritus)	Generally burrowers	Generally collectors—gatherers		1957, 1984, 2112, 2687, 3076, 3229, 4511
Maruina (3)	Lotic—erosional	Clingers	Scrapers, collectors—gatherers	West	1801, 3231, 4511
Pericoma (24)	Lotic depositional (margins), lentic—littoral (detritus)	Burrowers	Collectors—gatherers	Widespread	669, 3540, 3672, 4511, 2316, 4234, 4236, †
Philosepedon (5)	Lotic depositional (margins), lentic—littoral (detritus)	Burrowers	Collectors—gatherers	Widespread	4087
Psychoda (21)	Lentic—littoral (detritus), lotic—depositional, beaches—marine	Burrowers	Collectors—gatherers	Widespread	669, 2111, 2661, 3539, 4511
Telmatoscopus (17)	Lentic—littoral (including tree holes), lotic depositional	Burrowers	Collectors—gatherers	Widespread	1984, 4088, 4511, †
Threticus (2)	Lentic—littoral (including tree holes)	Burrowers	Collectors—gatherers	East	3672, 4088
Ptychopteridae (16) (=Liriopeidae) (phantom crane flies)	Generally lotic—depositional (including springs), lentic—littoral (sediments, detritus)	Generally burrowers	Generally collectors—gatherers		1957, 1984, 2112, 3076, 4511
Bittacomorpha (2)	Lotic depositional, lentic—vascular hydrophytes (emergent zone)	Burrowers	Collectors—gatherers	Widespread	18, 1594, 3422
Bittacomorphella (4)	Lotic—depositional	Burrowers	Collectors—gatherers	Widespread	18
Ptychoptera (10) (=Liriope)	Lotic depositional, lentic—vascular hydrophytes (bogs)	Burrowers	Collectors—gatherers, shredders (chewers—Sphagnum)	Widespread	18, 1594, 1790, 2557, 3422, 2555
Simuliidae	See Table 25A.				
Tanyderidae (4) (primitive crane flies)	Lotic—erosional (sediments)	Sprawlers—burrowers			
Protanyderus (3)	Lotic—erosional	Sprawlers—burrowers		Western	1126, 2117, 3429, 4528
Protoplasa (1)	Lotic—erosional	Sprawlers—burrowers		Eastern	20

* Emphasis on trophic relationships.
† Unpublished data, K. W. Cummins, Kellogg Biological Station.

Table 22B. Continued

Taxa (number of species in parentheses)	Habitat	Habit	Trophic Relationships	North American Distribution	Ecological References*
Thaumaleidae (7) (solitary midges)	Lotic—madicolous	Clingers	Scrapers		4511
Thaumalea (6)	Lotic—madicolous	Clingers	Scrapers	Widespread	1396, 1984, 3672, 3858, 4511
Trichothaumalea (2)	Lotic—madicolous	Clingers	Scrapers	Western and East	3670, 4511
Tipulidae	See Table 23A.				
Brachycera-Orthorrhapha					
Athericidae (4) (= Rhagionidae, in part)					2835, 3985
Atherix (3)	Lotic—erosional and depositional	Sprawlers—burrowers	Predators (piercers)	Widespread	1957, 1984, 2112, 2891, 4491, 2892, 3076, 3984, 4234, 4278, 1872, 2276
Suragina (1)	Lotic—erosional and depositional	Sprawlers—burrowers	Predators (piercers)	Southwest Texas, Mexico	2833, 2835, 4278, 4279 2834, 4280
Pelecorhynchidae (7)					2185
Glutops (7)	Lotic—depositional	Sprawlers—burrowers	Predators (piercers), shredders—herbivores?	Widespread	3969, 3971
Dolichopodidae (624)	Generally lentic and lotic margins (semiaquatic)	Generally sprawlers—burrowers	Generally predators (engulfers)		714, 1775, 1984, 2687, 577, 1003, 4447, 4511
Aphrosylus (6)	Beach zone—marine (intertidal, on rocks)	Clingers (in algae on rocks)	Predators (engulfers)	West Coast	1775, 3571, 4086, 4511
Argyra (46)	Lentic—littoral (margins and beach zone)	Sprawlers—burrowers	Predators (engulfers)	Widespread	290, 611, 1775, 4086, 4511
Asyndetus (21)	Beach zone—lakes	Sprawlers—burrowers	Predators (engulfers)	Widespread	1002, 1775, 3710
Campsicnemus (21)	Lentic—littoral (margins)	Sprawlers—burrowers	Collectors—gatherers (?)	Widespread	4086, 4511
Dolichopus (311)	Lentic—littoral (margins), estuaries, lotic—margins	Sprawlers—burrowers	Predators (engulfers)	Widespread	1002, 1775, 4086, 4511
Hercostomus (26)	Lotic—depositional (moss) lentic—tree holes	Sprawlers (damp moss)	Predators (engulfers)	Widespread	611, 1002, 3710, 4086, 4511
Hydrophorus (54)	Lentic—littoral (margins), estuaries	Sprawlers—burrowers (in fine detritus)	Predators (engulfers) (especially midges)	Widespread	611, 1002, 3710, 4086, 4511
Hypocharassus (2)	Beach zone—marine	Sprawlers—burrowers	Predators (engulfers)	East Coast	1775, 3710, 4511
Liancalus (5)	Lotic—seeps (algal mats)	Sprawlers—burrowers	Predators (engulfers)	Widespread	716, 290, 3672, 4086
Melanderia (3)	Beach zone—marine (rocky shore line)	Clingers	Predators (engulfers)	West Coast	3571, 4511
Pelastoneurus (29)	Lotic—seeps (algal mats)	Sprawlers—burrowers	Predators (engulfers)	Widespread	715, 717
Sympycnus (32)				Widespread	4086
Systenus (6)	Lentic—tree holes	Sprawlers—burrowers	Predators (engulfers) (especially *Dasyhelea*)	East and South	4497, 4511
Tachytrechus (33)	Lentic—tree holes, beaches; lotic—margins	Sprawlers—burrowers	Predators (engulfers)	Widespread	718, 2199, 4086, 4511
Telmaturgus (1)	Lotic depositional			East	4086
Teuchophorus (4)				Widespread	4086
Thinophilus (24)	Beach zone—lakes	Sprawlers—burrowers	Predators (engulfers)	Widespread	3710
Empididae (265) (dance flies)	Generally lotic—erosional and depositional (detritus), lentic—littoral	Generally sprawlers—burrowers	Generally predators (engulfers), some collectors—gatherers		1957, 1984, 2112, 2136, 2137, 3076, 3848, 4511, 577, 1003, 3909

* Emphasis on trophic relationships.

Table 22B. Continued

Taxa (number of species in parentheses)	Habitat	Habit	Trophic Relationships	North American Distribution	Ecological References*
Chelifera (22)	Lotic–depositional lentic—littoral (bogs)	Sprawlers—burrowers		Widespread	414, 2238, 2399, 2402, 4511, 3078, 4234
Chelipoda (7)	Possibly lentic and semi-permanent lotic			Widespread	1575, 2401, 4044
Clinocera (19)	Lotic—erosional lentic—littoral (bogs)	Clingers		Widespread	414, 2238, 2916, 4085, 4511, 3672, 4234
Dolichocephala (2)	Lotic—erosional	Clingers		Widespread	4084
Heleodromia (1)				New Mexico	
Hemerodromia (22)	Lotic—erosional and depositional (detritus)	Sprawlers—burrowers	Predators (engulfers) collectors—gatherers	Widespread	290, 669, 671, 1575, 1984, 2136, 2238, 4057, 4085, 4511 2399, 2403
Metachela (3)	Lotic—erosional and depositional (also in moss mats at water level, or just above on banks)	Sprawlers—burrowers	Predators (engulfers)?	Widespread	1575, 2238, 2399, 2400
Neoplasta (12)	Lotic—erosional and depositional (also in moss mats at water level, or just above on banks)	Sprawlers—burrowers	Predators (engulfers)?	Widespread	1575, 2238, 2399, 2404, 3563
Oreogeton (8)	Lotic—erosional (moss)	Sprawlers—burrowers	Predators (engulfers) of Simuliidae, caddisflies	West	3755
Oreothalia (5)				Widespread	
Proclinopyga (5)	Lotic—margins (?)	Sprawlers—burrowers	Predators (engulfers)	Widespread	
Rhamphomyia (150)	Lentic—littoral	Sprawlers—burrowers		Widespread	4086
Roederiodes (7)	Lotic—erosional	Clingers	Predators (engulfers) of pupal Simuliidae	Widespread	290, 622, 2238, 2858, 4511
Stilpon (13)	Lotic—erosional	Clingers	Predators (engulfers)	East, Central	785, 4086
Trichoclinocera (16)	Lotic—erosional	Clingers	Predators (engulfers)	Widespread	4086, 3671, 4086
Wiedemannia (2)	Lotic—erosional	Clingers	Predators (engulfers)	West	4086
Stratiomyiidae (178) (soldier flies)	Generally lentic—littoral	Generally sprawlers—swimmers	Generally collectors—gatherers		1957, 1983, 1984, 2112, 2616, 2617, 3076, 3480, 3481, 4511
Anoplodonta (1)	Lotic—erosional and depositional (margins), lentic—littoral	Sprawlers—burrowers	Collectors—gatherers	West	±
Caloparyphus (12)	Lentic—vascular hydrophytes (emergent zone); lotic—erosional	Sprawlers	Collectors—gatherers	Widespread	2616, 3669, 3672
Euparyphus (22)	Lotic—erosional and depositional (margins)	Sprawlers	Collectors—gatherers, scrapers	Widespread	2616, 3278, 3669, 3672
Hedriodiscus (7)	Lotic depositional (vascular hydrophytes)	Climbers	Scrapers	Widespread	2616, 3853
Hoplitimyia (4)	Lotic—erosional and depositional (margins),	Sprawlers—burrowers	Collectors—gatherers	West	±
Labostigmina (20)	Lotic—erosional and depositional (margins),	Sprawlers—burrowers	Collectors—gatherers	Widespread	±
Nemotelus (38)	Lentic—littoral, beaches (saline pools, marshes) lotic—margins	Swimmers, sprawlers	Collectors—gatherers	Widespread	290, 1984, 2616, 3480, 4511
Odontomyia (30) (=*Eulalia*)	Lentic—vascular hydrophytes (emergent zone)	Sprawlers	Collectors—gatherers, scrapers?	Widespread	290, 1983, 2616, 3480, 3672
Oxycera (7) (= *Hermione*)	Lotic—erosional and depositional (margins)	Sprawlers—burrowers	Scrapers	East	112, 2616, 3480, 4511, 111

* Emphasis on trophic relationships.
± N. E. Woodley, pers. comm.

Table 22B. Continued

Taxa (number of species in parentheses)	Habitat	Habit	Trophic Relationships	North American Distribution	Ecological References*
Sargus (6)	Mostly terrestrial, some lentic—vascular hydrophytes (emergent zone)	Climbers	Collectors—gatherers, scrapers	Widespread	2617, 3480
Stratiomys (31) (=*Stratiomyia*)	Lotic—erosional and depositional (margins), lentic—littoral (including saline pools)	Sprawlers—burrowers	Collectors—gatherers (and filterers)	Widespread	1145, 1983, 2172, 2188, 2616, 3480, 4028, 4511
Tabanidae (332) (horse and deer flies)	Generally lotic—depositional, lentic—littoral (margin, sediments, and detritus)	Generally sprawlers burrowers	Generally predators (piercers)		136, 528, 1957, 1984, 2112, 3076, 3400, 3968, 731, 2681, 4282, 4511, 2240, 3975
Apatolestes (11)	Beach zone—marine	Sprawlers—burrowers	Predators (piercers)	California coast; West	2295
Atylotus (14)	Lentic—littoral (sediments and mosses)	Sprawlers—burrowers	Predators (piercers)	Widespread	528, 3968, 4282, 731, 3972, 3973
Agkistrocerus (2)	Lotic—depositional, lentic—littoral	Sprawlers—burrowers	Predators (piecers)	South	485
Bolbodimyia (1)>	Lotic—erosional	Sprawlers	Predators (piercers)	Southwest	481
Brennania (2)	Lentic—coastal dunes	Sprawlers—burrowers	Predators (piercers)	California coast	2692
Chlorotabanus (1)	Lentic—littoral (sediments)	Sprawlers—burrowers	Predators (piercers)	South, East Coast	1449
Chrysops (83)	Lentic—littoral, beaches—marine and estuaries, lotic—depositional	Sprawlers—burrowers	Predators (piercers)	Widespread	290, 484, 3129, 3968, 731, 2440, 2623, 2681, 1298, 3563, 3975
Haematopota (5)	Lentic—littoral (sediments)	Sprawlers—burrowers		Widespread	731, 3480, 528
Hybomitra (55)	Lentic—littoral, lotic—depositional	Sprawlers—burrowers	Predators (piercers)	Widespread	731, 3968, 3975
Leucotabanus (2)	Lentic—tree holes	Sprawlers—burrowers	Predators (piercers)	South	1449
Merycomyia (2)	Lentic—littoral, lotic—depositional	Sprawlers—burrowers		East	1450, 3968
Silvius (11)	Lentic—littoral, lotic—margins	Sprawlers—burrowers		Central, West	731, 2239, 2681, 3965
Stenotabanus (6)	Beach zone—marine	Sprawlers—burrowers	Predators (piercers)	South	1451
Tabanus (108)	Lentic—littoral, lotic—erosional, depositional	Sprawlers—burrowers	Predators (piercers)	Widespread	731, 290, 669, 671, 2681, 2440, 3398, 3411, 3975, 1298, 1452, 1928, 3563
Brachycera-Cyclorrhapha					
Canacidae (14) (beach flies)	Generally beaches—marine intertidal	Generally burrowers	Generally scrapers		1775, 4446, 4495, 4511, 2531, 4504
Canace (2)	Beach zone—marine intertidal	Burrowers		East and West Coasts of United States	1775, 3977
Canaceoides (7)	Beach zone—marine intertidal	Burrowers		California coast	1775, 4446
Nocticanace (3)	Beach zone—marine intertidal	Burrowers	Scrapers (graze algae)	West and South Coasts	1775, 3408, 3409
Paracanace (1)	Beach zone—marine intertidal	Burrowers	Scrapers (graze algae)	Florida	
Procanace (1)	Beach zone—marine intertidal	Burrowers	Scrapers (graze algae)	Southeast Coast	2528
Dryomyzidae (2)					
Oedoparena (2)	Beach zone—marine intertidal	Burrowers—inside barnacles	Predators (engulfers) (barnacles)	West Coast	482, 1161, 2128, 3571

* Emphasis on trophic relationships.

Table 22B. Continued

Taxa (number of species in parentheses)	Habitat	Habit	Trophic Relationships	North American Distribution	Ecological References*
Ephydridae (445) (shore and brine flies)	Generally lentic—littoral (margins and vascular hydrophytes)	Generally burrowers, sprawlers	Generally collectors—gatherers, shredders—herbivores (miners), scrapers, predators (engulfers)		263, 902, 1666, 1957, 1984, 2112, 3076, 3561, 3564, 3663, 4446, 4511, 906, 1161, 1248, 3983, 4509, 1008, 1255
Discomyzinae (32)	Generally lentic—littoral; marine shores, lotic—margins	Generally burrowers (some climbers, planktonic)	Generally collectors—gatherers, some shredders—herbivores, predators (piercers)		
Ceropsilopa (7)	Beach zone—marine			West and South Coasts	
Clanoneurum (1)	Lentic—littoral, esp. saline, brackish water; Marine—vascular hydrophytes, emergent zone (mangrove)	Burrowers—miners	Shredders—herbivores (miners)	West and East Coasts	3663, 4511
Clasiopella (1)	Beach zone—marine			Florida	2534
Cressonomyia (4)	Lentic—littoral and margins			East and South Coasts	
Discomyza (2)	Lentic—littoral and margins (detritus)	Burrowers	Parasites of Mollusca, scavengers of carrion	West and South Coasts	290, 3663, 4511
Helaeomyia (1) (petroleum flies)	Lentic—pools of crude petroleum and waste oil	Sprawlers—burrowers, or large "planktonic" masses of larvae	Collectors—gatherers (organics associated with petroleum)	California, Gulf area of United States	4511
Leptopsilopa (3)	Lentic—littoral	Burrowers—in decaying vegetation	Collectors—gatherers	Widespread	3564, 3831
Guttipsilopa (2)	Beach zone—marine			Florida	
Paratissa (1)	Marine—shores	Burrowers—in decaying vegetation?	Collectors—gatherers	East and West Coasts	2532, 3564
Psilopa (7)	Lentic—littoral; vascular hydrophytes	"Planktonic", burrowers	Collectors—gatherers, shredders—herbivores (miners)	Widespread	290, 1984, 4511
Rhysophora (1)	Lentic—littoral; vascular hydrophytes	Burrowers?	Shredders—herbivores (miners)?	Eastern and southern United States	975, 230, 2521, 3571
Trimerina (1)	Lentic—vascular hydrophytes (emergent zone)	Climbers	Predators (piercers) (spider egg masses)	North, West	
Trimerinoides (1)				West	
Ephydrinae (145)	Generally lentic—littoral lotic—depositional (thermal springs), marine—shores and salt marshes	Generally sprawlers—burrowers or climbers	Generally shredders—herbivores, collectors—gatherers, scrapers		1008
Brachydeutera (4)	Lentic—littoral	Sprawlers—burrowers; swimmers	Collectors—gatherers	Widespread	1984, 3663, 4446, 4500, 1565, 2535, 2538, 4511, 975, 4122
Callinapaea (2)				Widespread	
Calocoenia (2)				North and West	
Cirrula (4) (=Hydropyrus)	Marine—salt marshes Lentic—littoral (alkaline lakes)	Burrowers (in floating algal mats)	Shredders—herbivores (macroalgae), collectors—gatherers	Widespread	14, 1984, 2537, 3663, 4502
Coenia (2)	Lentic—margins	Sprawlers, burrowers	Collectors—gatherers	Widespread	1254, 2520, 3564, 3663
Dimecoenia (2)	Marine—salt marshes	Burrowers (in floating algal mats)	Shredders—herbivores (macroalgae)	East and West Coasts	2537, 3663
Ephydra (13)	Lentic—littoral (in algal mats, including alkaline lakes), lotic—depositional (including thermal springs)	Sprawlers—burrowers	Shredders—herbivores (macroalgae), collectors—gatherers	Widespread	14, 295, 433, 434, 684, 1984, 3145, 3564, 3663, 1008, 4502, 4511, 184

* Emphasis on trophic relationships.

Table 22B. Continued

Taxa (number of species in parentheses)	Habitat	Habit	Trophic Relationships	North American Distribution	Ecological References*
Eutaenionotum (1)				North	
Lamproscatella (12)	Marine—shores and salt marshes, lentic—alkaline ponds			Widespread	2524, 3564, 3663
Limnellia (11)	Lentic—littoral (?)			Widespread	2522
Paracoenia (7)	Lotic—alkaline thermal springs (in algal mats), lentic—littoral	Sprawlers	Scrapers (graze blue-green algae), collectors—gatherers of fecal material	Widespread	433, 3104, 4617, 184
Parydra (34)	Lentic—littoral, lotic—depositional and margins	Burrowers	Scrapers	Widespread	904, 909, 3564, 3663
Philotelma (1)	Lotic—depositional			West	
Rhinonapaea (1)				North	
Setacera (8)	Lentic—littoral (algal mats), including saline pools; lotic—margins	Sprawlers	Shredders—herbivores, scrapers (graze algae)	Widespread	1247, 1251, 1984, 3564, 2525, 4511, 4618, 184
Scatella (17)	Lentic—vascular hydrophytes and algal mats (including thermal springs) marine—shores and marshes, lotic—margins	Sprawlers—climbers	Collectors—gatherers, scrapers (graze algae)	Widespread	686, 904, 1247, 2537, 3663, 4062, 4500, 4619, 1008, 3672
Scatophila (24)	Lentic—littoral, marine—salt marshes, lotic—margins	Sprawlers	Scrapers	Widespread	905, 3663, 4511
Gymnomyzinae (81)	Generally lentic—littoral; marine shores, lotic—margins	Generally burrowers (some climbers, planktonic)	Generally collectors—gatherers, some shredders—herbivores, predators		
Allotrichoma (7)	Lentic—temporary puddles (near dung), lotic—margins	Burrowers (in dung, temporary puddles and decaying snails)	Collectors—gatherers (dung and decaying snails)	Widespread	354, 902, 3564, 4511, 3490
Athyroglossa (8)	Lentic—littoral, vascular hydrophytes; decaying organic material	Burrowers	Collectors—gatherers	Widespread	1489, 2184, 3571
Diclasiopa (1)	Lentic—littoral, lotic—depositional			Widespread	3564
Diphuia (2)	Marine—shores			East Coast	2530, 3663
Discocerina (8)	Lentic and lotic margins (primarily terrestrial)	Burrowers (in moss and algae)	Collectors—gatherers	Widespread	904, 1256, 3564, 4511
Ditrichophora (15)	Lentic—littoral			Widespread	3564
Gastrops (2)	Lentic—vascular hydrophytes (frog eggs)	Burrowers	Predators (engulfers) (frog eggs)	East, South	356, 4130
Glenanthe (3)	Marine—salt marshes			East and West Coasts	3663
Hecamede (1)	Beach zone—marine (semiaquatic)	Burrowers	Probably collectors—gatherers (scavengers)	East Coast	2533, 3663, 4511
Hecamedoides (1)	Lentic—littoral (?)			Widespread	
Hydrochasma (4)	Lentic—littoral, lotic—margins, marine—shores			Widespread	3663
Lipochaeta (1)	Beaches—marine; lentic—saline			East and West Coasts	3562, 3663
Mosillus (3)	Marine—shores, lentic—littoral (incl. alkaline lakes and ponds)			Widespread	2539, 3562, 3663
Ochthera (13)	Lentic—littoral, lotic—depositional and margins	Burrowers	Predators (engulfers) (midge larvae)	Widespread	904, 3662, 3663, 4511
Paraglenanthe (1)	Beach zone—marine			Florida	

* Emphasis on trophic relationships.

Table 22B. Continued

Taxa (number of species in parentheses)	Habitat	Habit	Trophic Relationships	North American Distribution	Ecological References*
Placopsidella (1)	Beach zone—marine	Sprawlers	Shredders—herbivores, predators or scavengers of small molluscs	East Coast	2527, 2529
Platygymnopa (1)	Lotic and lentic—margins	Burrowers (in decomposing snails)	Collectors—gatherers (decomposing snails)	Northern United States	4501
Polytrichophora (4)	Lentic—littoral, marine—salt marshes; lotic—margins			Widespread	3663
Pseudohecamede (5)	Marine—shores, lentic—littoral	Burrowers	Collectors—gatherers (some in dung)	East and West Coasts (occasionally inland)	902, 3490, 3564, 3663
Hydrelliinae (135)	Generally lentic—littoral (vascular hydrophytes), lotic—depositional, marine—shores	Generally burrowers, sprawlers	Generally collectors—gatherers		
Asmeringa (1)	Lentic—saline	Burrowers	Collectors—gatherers	California	2526, 3663
Atissa (5)	Beach zone—marine, lentic—littoral (saline lakes and pools)			Widespread	3564, 3663
Hydrellia (59)	Lentic—vascular hydrophytes, lotic—erosional and depositional (vascular hydrophytes—watercress), marine—shores	Burrowers—miners (especially *Potamogeton*)	Shredders—herbivores (miners)	Widespread	266, 903, 1487, 1984, 2618, 3076, 3663, 4446, 975, 4511
Lemnaphila (1)	Lentic—vascular hydrophytes (floating zone—duckweed)	Burrowers—miners (of duckweed)	Shredders—herbivores (miners of duckweeds)	East	3076, 3601, 4511
Notiphila (52)	Lentic—littoral (detritus, vascular hydrophytes, bogs), lotic—depositional	Burrowers (attached to roots by respiratory spine)	Collectors—gatherers and filterers	Widespread	266, 290, 495, 908, 1666, 1821, 2523, 2618, 3663, 4097, 4112, 4511, 3563
Oedenops (1)	Lentic—littoral, marine—shores			West and South Coasts	3663
Paralimna (5)	Lentic—littoral (margins in sediments near dung)	Sprawlers	Collectors—gatherers	Widespread	3564, 4511
Ptilomyia (6)	Marine—shores, lentic—littoral			Widespread	3564, 3663
Schema (1)	Marine—shores				
Typopsilopa (4)	Lentic—littoral (detritus at margins in emergent vascular hydrophyte beds)	Sprawlers	Collectors—gatherers	Widespread	3564
Ilytheinae (52)	Generally lentic—littoral, lotic—depositional	Generally burrowers, sprawlers	Generally collectors—gatherers		
Axysta (2)	Lentic—littoral	Sprawlers	Scrapers (graze blue-green algae)	North, East	1247, 3564
Hyadina (8)	Lotic and lentic—margins	Sprawlers	Scrapers (graze blue-green algae)	Widespread	1247, 3564
Ilythea (3)	Lentic—littoral lotic—depositional (sediments)	Sprawlers	Collectors—gatherers	West, South	3564
Lytogaster (8)	Lotic and lentic—margins	Sprawlers	Scrapers (graze blue-green algae)	Widespread	1247, 1249, 3564
Microlytogaster (1)				Widespread	
Nostima (9)	Lotic and lentic—margins	Sprawlers	Scrapers (graze blue-green algae)	Widespread	1252
Pelina (10)	Lotic and lentic—margins	Sprawlers	Scrapers (graze blue-green algae)	West, Midwest	1247, 1250, 3564

* Emphasis on trophic relationships.

Table 22B. Continued

Taxa (number of species in parentheses)	Habitat	Habit	Trophic Relationships	North American Distribution	Ecological References*
Philygria (5)	Lentic—vascular hydrophytes	Sprawlers	Scrapers (graze blue-green algae)	Widespread	3564, 4325
Pseudohyadina (1)				East	
Zeros (5)	Lotic—depositional (sediments)	Sprawlers	Collectors—gatherers	East, South	3564
Muscidae (271) (=Anthomyiidae in part, by some authors)	Generally lentic—littoral, lotic—depositional, erosional	General sprawlers	Generally predators (piercers)		1957, 1984, 2112, 3076, 1161, 3684, 3909, 4511
Caricea (10) (=*Lispocephala*)	Lotic—erosional(?)		Predators (piercers)	Widespread	1161, 4446, 4511
Graphomya (9)	Lentic—depositional (small enriched ponds)	Sprawlers	Predators (piercers)	Widespread	1161, 3685
Limnophora (10)	Lotic—erosional (especially mosses)	Burrowers	Predators (piercers) on Oligochaeta, Simuliidae, Tipulidae	Widespread	1470, 1669, 1984, 2472, 1161, 2681, 3397, 4511, 3685, 4564
Lispe (25)	Lentic—littoral, lotic—depositional (margins), marine shores (algae)	Sprawlers	Predators (piercers)	Widespread	1646, 1669, 1984, 4446, 1161, 3685, 4511
Lispoides (1)	Lotic—erosional	Sprawlers?		Widespread	1646, 3672
Phaonia (81)	Lentic—tree holes		Predators (piercers) on mosquito larvae	Widespread	1161, 1669, 3954, 4511
Spilogona (135)	Lotic—depositional, lentic—littoral			Widespread	1161, 1669, 3672, 3685
Phoridae (2)					
Dohrniphora (1)	Lotic—erosional (trickling filter beds)	Burrowers	Collectors—gatherers, predators (engulfers) of Psychodidae	Widespread	918, 2111
Megaselia (1 aquatic species)	Lentic—littoral (sediments and detritus), trickling filter beds	Burrowers	Collectors—gatherers	Widespread	618, 918
Sarcophagidae (4)	Semiaquatic	Burrowers—miners	Collectors—gatherers		
Fletcherimyia (4) (=*Blaesoxipha*, in part)		Burrowers—miners (bases of pitcher plants)	Collectors—gatherers (scavengers on trapped invertebrates)	South, Northeast	1195, 1196, 1272
Scathophagidae (53) (dung flies) (=Scatophagidae = Scopeumatidae = Cordiluridae = Anthomyiidae, in part)	Generally lentic—vascular hydrophytes, lotic depositional	Generally burrowers—miners (in plant stems), sprawlers	Generally shredders—herbivores (miners); predators (engulfers)		1957, 1984, 2112, 3076, 1161, 4133, 4511
Acanthocnema (4)	Lotic—depositional			East, West	
Cordilura (44)	Lentic—vascular hydrophytes (emergent zone)	Burrowers—miners (plant stems)	Shredders—herbivores (miners *Scirpus*, *Juncus*, and *Carex*), predators (engulfers) of Ceratopogonidae	Widespread	1313, 1314, 1984, 2872, 4133, 4188, 4511, 4644
Hydromyza (1)	Lentic—vascular hydrophytes (submerged and floating zones)	Burrowers—miners (plant stems and roots)	Shredders—herbivores (miners in petioles of *Nuphar*, roots of *Potamogeton*)	Northeast	266, 2618, 2852, 3076, 2534, 3885, 4289, 4292
Orthacheta (3)	Lentic—vascular hydrophytes (emergent zone)	Burrowers—miners (plant stems)	Predators (engulfers)	Widespread	2873

* Emphasis on trophic relationships.

Table 22B. Continued

Taxa (number of species in parentheses)	Habitat	Habit	Trophic Relationships	North American Distribution	Ecological References*
Spaziphora (1)	Lentic—littoral (sewage beds in oxidation ponds)	Sprawlers	Scrapers, collectors—gatherers	Widespread	1161, 2029
Sciomyzidae (175) (marsh flies) (=Tetanoceridae = Tetanoceratidae)	Generally lentic—littoral, lotic depositional in snails	Generally burrowers, inside snails	Generally predators (engulfers) or "parasites" of snails		1161, 2132
Antichaeta (8)	Lentic—littoral (marshes and temporary ponds)	Burrowers—in snail egg masses and snails	Predators (engulfers) or "parasites" of snail eggs and snails	North, West	267, 268, 269, 270, 271, 1200, 1957, 1984, 2112
Atrichomelina (1)	Lentic—littoral	Burrowers, inside snails	Predators (engulfers) or "parasites" of snails, some scavengers on decaying snails	Widespread	1258, 2130, 2134
Colobaea (1)	Lentic—littoral			Eastern United States	
Dictya (25)	Lentic—vascular hydrophytes (emergent zone)	Burrowers—inside snails	Predators (engulfers) or "parasites" of snails	Widespread	271, 4089
Dictyacium (2)				North, Midwest	
Elgiva (2)	Lentic—vascular hydrophytes (emergent zone), lotic—margins	Burrowers—inside snails	Predators (engulfers) or "parasites" of snails	North	267, 2135, 2988, 4511
Euthycera (1)				East, Midwest	
Hedria (1)	Lentic—vascular hydrophytes (emergent zone), lotic—margins	Burrowers—inside snails	Predators (engulfers) or "parasites" of snails	North	267, 268, 1245
Hoplodictya (5)	Lentic—littoral, marine—salt marshes	Burrowers—inside snails	Predators (engulfers) or "parasites" of snails; 1 sp. in *Littorina*	Widespread	2870
Limnia (17)	Lentic—vascular hydrophytes (emergent zone)	Burrowers—inside snails	Predators (engulfers) or "parasites" of snails	Widespread	3847
Oedematops (1)				East	
Pherbecta (1)	Lentic—littoral (bogs)			North	2131
Pherbellia (40)	Lentic—littoral (margins) marine—mud flats	Burrowers—inside snails	Predators (engulfers) or "parasites" of snails	Widespread	398, 1244, 2133
Poecilographa (1)	Lentic—littoral (bogs)—semiaquatic			Midwest, East	1984, 4511
Pteromicra (14)		Burrowers—inside snails	Predators (engulfers) or "parasites" of snails	Widespread	1984, 3483, 4511
Renocera (4)	Lentic—littoral (marshes and temporary ponds)	Burrowers—inside fingernail clams	Predators (engulfers) or "parasites" in fingernail clams (Sphaeriidae)	North	267, 1246, 1257
Sepedomerus (1)	Lentic—vascular hydrophytes (emergent zone)	Burrowers—inside snails	Predators (engulfers) or "parasites" of snails	Southwest	268
Sepedon (16)	Lentic—littoral (including temporary ponds)	Burrowers—inside snails	Predators (engulfers) or "parasites" of snails	Widespread	267, 2871
Tetanocera (30)	Lentic—vascular hydrophytes (emergent zone)	Burrowers—inside snails	Predators (engulfers) or "parasites" of snails (and slugs)	Widespread	267, 268, 1984, 3479, 2129, 4046, 4511
Syrphidae (125) (flower flies, rattail maggots)					1615, 1957, 1984, 2112, 1161, 3076, 3396, 4511, 2445, 4135
Blera (16)	Lentic—tree holes			Widespread	
Callicera (4)	Lentic—tree holes	Burrowers	Collectors—gatherers	East, Southwest	1161, 3472
Ceriana (5)	Lentic—tree holes			Widespread	

* Emphasis on trophic relationships.

Table 22B. Continued

Taxa (number of species in parentheses)	Habitat	Habit	Trophic Relationships	North American Distribution	Ecological References*
Chalcosyrphus (27)	Lentic—tree holes	Burrowers	Collectors—gatherers	Widespread	1161
Chrysogaster (10)	Lentic—littoral (pond margins and vascular hydrophytes)	Burrowers	Collectors—gatherers	Widespread	1615, 1984, 2277, 4511
Eristalinus (1)	Lentic—littoral (sediments and detritus)	Burrowers	Collectors—gatherers	Widespread	
Eristalis (18)	Lentic—littoral (sediments and detritus), lotic—depositional	Burrowers (especially low O_2 organic substrates)	Collectors—gatherers	Widespread	1472, 1984, 2687, 4447, 4511
Helophilus (10)	Lentic—littoral (detritus and organic sediments)	Burrowers	Collectors—gatherers	Widespread	290, 1615, 1984, 4511, 1161
Mallota (9)	Lentic—tree holes		Collectors—gatherers	Widespread	290, 1161, 1615, 1984, 4511, 2444
Myolepta (7)	Lentic—tree holes		Collectors—gatherers	Widespread	290, 1472, 1615, 4511
Neoascia (10)	Lentic—littoral (margins, marshes)		Predators—engulfers	Widespread	290, 2380, 2442
Orthonevra (16)	Lentic—tree holes			Widespread	1161
Palpada (10)	Lentic—littoral (sediments and detritus)	Burrowers	Collectors—gatherers	South	4511
Sericomyia (11)	Lentic—littoral (bog mat pools)		Collectors—gatherers	Widespread	290, 1472, 1615, 4511
Spilomyia (11)	Lentic—tree holes			Widespread	1161

* Emphasis on trophic relationships.

TIPULIDAE

George W. Byers
University of Kansas, Lawrence

INTRODUCTION

The Tipulidae or crane flies, as far as is known, comprise the largest family of Diptera. Stone *et al.* (1965) recorded 1,458 species in America north of Mexico, and about 15,000 species are already known in the world fauna. The diversity of form among the North American tipulids is indicated by their taxonomic distribution among 3 subfamilies (Tipulinae, Cylindrotominae, Limoniinae), 5 tribes (Tipulini, Limoniini, Pediciini, Hexatomini, Eriopterini), and 64 genera.

Most crane flies are associated with moist environments, the adults usually being found in low, leafy vegetation along streams or around ponds in wooded areas. Larval tipulid abundance and distribution in woodland floodplains appear to be influenced by high soil moisture and organic content (Merritt and Lawson 1981). There are, however, several species inhabiting open fields, relatively dry rangelands, and even desert environments. The chief economic importance of the Tipulidae involves feeding by the larvae on roots of forage crops or seedling field crops. More significant, however, is the ecological importance of larvae in aquatic and semiaquatic habitats as detrital feeders and of both larval and adult tipulids as food for other invertebrates, birds, mammals, fishes, amphibians, and reptiles.

Tipulidae have successfully exploited nearly every kind of aquatic environment except the open oceans and large, deep freshwater lakes. Larvae of only a few genera, e.g., *Antocha* and *Hesperoconopa,* are truly aquatic in the sense of being apneustic (without spiracles) and able to extract oxygen from the water. Lotic waters ranging from slow, silt-laden rivers to clear, torrential mountain streams are habitats for a great number of species. Slowly flowing water, from seepages on rock cliffs, the splash zone around waterfalls, etc., commonly supports a growth of algae in which larvae of several crane fly species live. Lentic habitats include margins of ponds and lakes, freshwater and brackish marshes, and standing waters in tree-holes or axils of large leaves. A few littoral species inhabit the marine intertidal zone. To these may be added the vast numbers of species that are semiaquatic, spending their larval life in saturated plant debris, mud, or sand near the water's edge, or in wet to saturated bryophytes or sodden, partially submerged, decayed wood (Table 23A).

The life cycle characteristically consists of a brief egg stage (a few days to two weeks), four larval stadia (of which the first three are relatively short as compared with the fourth), a pupal stage of 5-12 days, and an adult stage of a few days. The entire cycle may be as short as six weeks or as long as five years, depending on the species and environmental conditions, particularly temperature and moisture. The cycle of many species having aquatic larvae requires about a year; however, some species have two generations a year at temperate latitudes, with the first in May to early June and the second in August to early September. In the Arctic, some aquatic species may require four or five years to mature because of the rigorous climate (MacLean 1973). Pritchard and Hall (1971) and Pritchard (1976, 1980a) have conducted the most comprehensive study on the life cycle of a North American aquatic crane fly, *Tipula sacra* Alexander.

Only the larva can be described as aquatic, and many "aquatic" crane fly larvae depend on atmospheric oxygen and thus cannot venture far from the water surface. The extent of cutaneous respiration in aquatic tipulids has not been investigated in detail. Metapneustic larvae of several species are capable of remaining submerged for prolonged periods of time. Moreover, the large posterior spiracles of certain species are known to be closed (Pritchard and Stewart 1982; Chap. 4). Elongate, membranous anal gills in some aquatic larvae, such as those of *Pseudolimnophila* sp. (Fig. 23.28) or subgenus *Yamatotipula* of *Tipula* (Fig. 23.4), are probably functional, but these structures are only weakly developed in some other aquatic tipulids. The pupae are also air-breathers (again with rare exceptions, such as *Antocha* spp.), and aquatic larvae leave the water to pupate in nearby soil, moss, or litter. Mating and oviposition are the primary activities of the adult flies, and when not engaged in this activity they usually rest by standing upon or suspending themselves from vegetation, rocks, or other objects in their habitat.

As indicated in Table 23A, many aquatic tipulid larvae feed on organic detritus, such as decaying leaves, plant fragments, and associated microorganisms, that accumulates on pond bottoms or in backwaters of streams. Among the Limoniinae, larvae of *Hexatoma* sp., *Pedicia* sp., *Dicranota* sp., and *Limnophila* sp. are active predators. Larval diet has not been extensively studied, but Martin *et al.* (1980) recently examined the digestive mechanisms of *Tipula (Nippotipula) abdominalis* Say, a stream shredder.

Only about 10% of the North American tipulid species have been reared to the extent that larvae and pupae can definitely be associated with adults. Crane fly biology is a vast and only slightly explored field, much in need of interested students.

EXTERNAL MORPHOLOGY

Larvae

In most tipulid larvae, the body is elongate, generally subcylindrical, somewhat tapered toward the head, and less so posteriorly. The head is strongly sclerotized anteriorly, with well-developed mouthparts, and has sclerotized plates or rods extending backward into the thorax and separated by deep incisions (Fig. 23.21). The skin is attached far forward, only a little behind the mouthparts and antennae, and the entire head can readily be withdrawn into the central thoracic region. There are three thoracic segments and eight evident abdominal segments. The spiracular disk is interpreted as the main part of the ninth abdominal segment, and the anal lobes or *anal gills* (Figs. 23.4, 23.8) as the tenth segment. In the unmodified abdominal segments, a basal anterior ring and a wider (i.e., longer) posterior ring can usually be differentiated (Fig. 23.2).

Larvae of Tipulidae have been given the common name "leather jackets," which applies most aptly to the larger, tough-skinned tipulines. Many of the limoniine larvae have thin, nearly transparent skin. Some aspects of the integument have been applied to the problem of identification in some species, notably the transverse, somewhat swollen zones on the basal rings of certain abdominal segments, termed *creeping welts* (Fig. 23.2). These usually bear dense hairs or small spicules, or denticles, often arranged in rows or ridges.

Behind the eighth abdominal segment is a smooth posterodorsal surface bearing the two large spiracles of the dorsal tracheal trunks. This is the *spiracular disk,* which is surrounded by a variable number of *lobes* and is sclerotized to varying degrees according to the genus (Figs. 23.4-23.5). This structure is a great aid in the recognition of larvae. In the Tipulinae, there are usually four subconical lobes above the spiracular disk and two blunt lobes below the disk (Figs. 23.4, 23.8). Larval Limoniinae usually have only three lobes above the disk and two below (Fig. 23.5), but these may be reduced in prominence, less often in number (Figs. 23.16, 23.31).

Although mouthparts have been used extensively by other authors in differentiating tipulid larvae, this has been avoided where possible, because of the inconvenience of dissecting out these small parts. A mid-ventral structure, termed the *mentum,* fused *mental plates,* or *maxillary plates* by earlier authors, is usually densely sclerotized and visible without dissection. It appears to be derived from ventral extensions of the most posterior sclerotized part of the head bordering the occipital foramen (postocciput) and accordingly is called the *hypostomal bridge* (Figs. 23.21-23.24). In the Hexatomini and Pediciini, the maxillae are conspicuously prolonged anteriorly (Fig. 23.20).

Adults

Typically, adult crane flies possess an elongate, slender body, long and rather narrow wings, and very long, thin legs and tarsi. Tipulidae can be differentiated from other similar nematocerous flies by their V-shaped transverse mesonotal suture (Fig. 22.215) and by the absence of ocelli.

Since classification at the generic level is based almost wholly on wing venation, an understanding of the nomenclature of veins and cells is necessary in order to use the accompanying key to adults. The venational nomenclature adopted here conforms to that in the *Manual of Nearctic Diptera* (McAlpine et al. 1981), which differs in the interpretation of the branching of the media (M) and cubitus (Cu) from virtually all earlier references on Tipulidae (e.g., the works of C. P. Alexander). In Diptera, M is thought to have only three branches; the apparent fourth is regarded as the first branch of the anterior cubitus (CuA_1). The apparent m-cu crossvein becomes the transverse basal section of the CuA_1 (labelled *bscu*). Vein M_3, the posterior branch of M, joins CuA_1 for a short distance, in most species forming the posterior border of the discal cell (cell 1st M_2, labelled *d*). Veins and cells according to this system are shown in Figures 23.36 and 23.37.

Detailed structure of the male hypopygium (generally the ninth abdominal segment) is employed extensively in the differentiation of tipulid species and occasionally genera. Shapes of the ninth tergum and sternum, and also the eighth sternum, are of great taxonomic utility. The paired *basistyles* of the ninth segment, apparently of pleural origin, bear at their apices the genital claspers, or *dististyles,* which vary greatly in shape according to genus and species. There may be a single dististyle on each basistyle, or two of quite different structure (Figs. 23.64-23.65).

KEY TO THE GENERA OF TIPULIDAE

In offering the keys that follow, I acknowledge a great debt to the late Prof. Charles P. Alexander. His extensive studies of the world's Tipulidae, spanning more than 70 years, provide much of the basis for the present keys. I have tried to make the keys conform to his concepts of genera.

KEY TO THE GENERA OF TIPULIDAE

Larvae

1. Both dorsal and lateral longitudinal rows of conspicuous, usually elongate, fleshy projections on thoracic and abdominal segments (Figs. 22.3, 23.2) 2
1'. Thoracic and abdominal segments without dorsal longitudinal rows of conspicuous projections (Figs. 23.1-23.2); lateral abdominal projections, if present, blunt, shorter than their basal diameter 3
2(1). Dorsal projections mostly long, slender; those of thoracic segments simple, posterior ones on most abdominal segments either deeply bifurcate (forked) or, if simple, approximately 10 times as long as basal diameter (Fig. 23.3); in aquatic or semiaquatic mosses ....*Phalacrocera*
2'. Dorsal projections shorter, length 1-3 times basal diameter (Fig. 22.2); posterior pair of dorsal projections on abdominal segments 1-7 with 3 or 4 serrations, not deeply divided; body color brownish; in semiaquatic mosses ....*Triogma*

3(1').	Spiracular disk bordered by 6 (rarely 8) usually subconical lobes, ordinarily 2 dorsal, 2 dorsolateral, 2 below spiracles; lobes sometimes short and blunt (Figs. 22.6, 23.4) ....(subfamily Tipulinae).................4
3'.	Spiracular disk bordered by 5 (rarely 7) or fewer lobes, often 1 dorsomedial, 2 lateral, 2 below spiracles; lobes of variable shape (Fig. 23.5); or spiracles absent ....(subfamily Limoniinae) 9
4(3).	Anal gills palmately or pinnately branched (Fig. 23.6); dorsal lobes of spiracular disk short, bluntly rounded, lower lobes more than twice as long as their basal diameter; aquatic or semiaquatic in small streams, eastern North America ....(subgenus *Longurio*)..........**Leptotarsus**
4'.	Anal gills not pinnately branched; lobes of spiracular disk variable 5
5(4').	All lobes of spiracular disk elongate; lateral and ventral lobes 3-4 times as long as their basal width; lobes bordered by numerous long hairs, outer hairs 2-3 times as long as width of lobe at point of attachment of hair (Figs. 23.7, 23.10) 6
5'.	Lobes of spiracular disk not all elongate, longest ones rarely more than twice their basal width; bordering hairs usually sparse, if numerous not long (Figs. 22.6, 23.8, 23.9) 7
6(5).	Two pairs of elongate, retractile anal gills; lobes of spiracular disk darkened along margins, pale medially (Fig. 23.7); larvae in open-ended tubes of floating vegetation ....**Megistocera**
6'.	Three pairs of elongate anal gills (as in Figs. 22.1, 23.4); lobes of spiracular disk darkened along margins but each with a thin, submedian dark line (Fig. 23.10); larvae not in tubes of vegetation (included here are larvae of species of subgenus *Angarotipula* of *Tipula*, formerly assigned on basis of adult structures to *Prionocera*) ....**Prionocera**
7(5').	Abdominal segments and posterior ring of metathorax covered with dense pilosity (fine, long hairs) giving larva a woolly appearance; thoracic segments otherwise with only short pubescence (microsetae), nearly bare by contrast; spiracular disk relatively small; spiracles subquadrate; dorsal lobes of disk low, inconspicuous (Fig. 23.11); lateral lobes only about as long as diameter of spiracle, with bluntly rounded apex; ventral lobes darkened on discal face, with subapical translucent spot; larvae in dark, organic soil or mud along small streams, in seepage areas, etc........................**Brachypremna**
7'.	Pilosity not dense on abdomen and contrastingly absent on thorax; spiracular disk not relatively small; ventral lobes of disk not constricted near midlength (Figs. 22.6, 23.8, 23.9) 8
8(7).	Spiracles large, separated by less than diameter of a spiracle; lobes of disk less than twice as long as basal width, fringed with long, dark hairs; a thin, black median line on discal face of each lobe (Fig. 23.8); larvae attaining more than 50 mm length in 4th instar; in organic debris and mud at edges of ponds or small streams ....**Holorusia**
8'.	Spiracles generally separated by more than diameter of a spiracle (Figs. 22.6, 23.9); lobes of disk highly variable, from short and rounded to elongate and subconical, ventral pair rarely divided (Figs. 22.6, 23.1, 23.9); large genus with larvae in a variety of aquatic and terrestrial habitats ....**Tipula**
9(3').	Spiracles absent, tracheal system closed; dorsal and lateral lobes of 9th abdominal segment absent or extremely reduced (Fig. 23.2) 10
9'.	Spiracles present, usually conspicuous (may be concealed if lobes of disk are infolded); dorsal and lateral lobes usually present, but absent in some species (Fig. 23.12) 11
10(9).	Ventral lobes of 9th abdominal segment elongate, deeply separated, slightly divergent, with a few tufts of hairs; anal gills elongate; dorsal and ventral creeping welts conspicuous on abdominal segments 2-7 (Fig. 23.2); larvae in silken tubes on stones in swift, well-oxygenated water ....**Antocha**
10'.	Eighth and 9th abdominal segments covered with dense, long pilosity; 9th segment elongate, tapering, shallowly bifurcate (divided) at apex; anal gills short, not extending beneath 9th segment; no conspicuous creeping welts; larvae in sandy bottoms of cold, clear, rapid streams chiefly of Pacific drainage ....**Hesperoconopa**
11(9').	Dorsal and lateral lobes of spiracular disk absent or extremely reduced, ventral lobes elongate (Fig. 23.12) (larvae of *Ornithodes* and *Nasiternella*, at present unknown, may key out here) 12

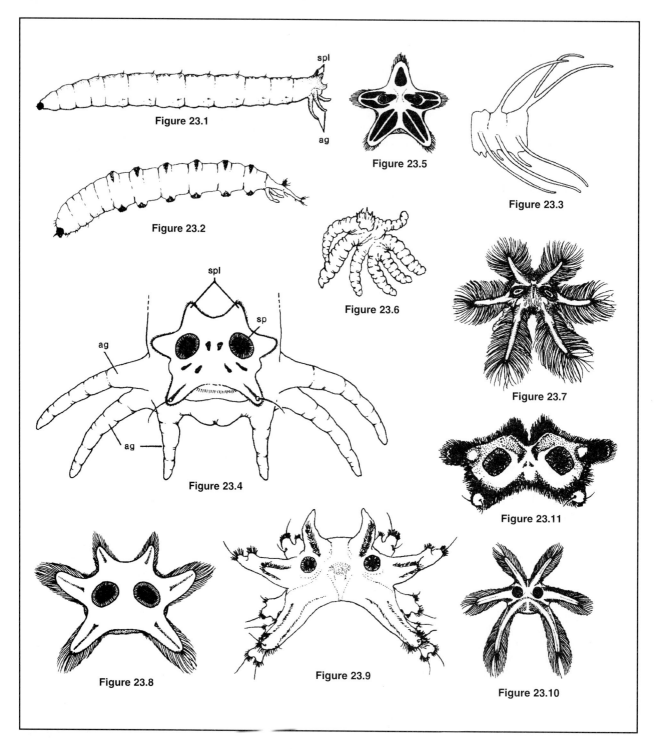

Figure 23.1. Lateral view of larva of *Tipula* (*Yamatotipula*) *furca* Walker, *ag*, anal gills; *spl*, spiracular lobes.

Figure 23.2. Lateral view of larva of *Antocha* sp.

Figure 23.3. A typical abdominal segment of *Phalacrocera* sp. (anterior at left).

Figure 23.4. Posterior end of larva of *Tipula* (*Yamatotipula*) *furca* Walker, *ag*, anal gill; *sp*, spiracle; *spl*, spiracular lobes.

Figure 23.5. Spiracular disk of *Molophilus hirtipennis* (O.S.).

Figure 23.6. Anal gill of larva of *Leptotarsus* (*Longurio*) *testaceus* (Loew).

Figure 23.7. Spiracular disk of *Megistocera longipennis* (Macquart) (after Rogers 1949).

Figure 23.8. Spiracular disk of *Holorusia hespera* Arnaud and Byers.

Figure 23.9. Spiracular disk of *Tipula* (*Nippotipula*) *abdominalis* (Say) (after Gelhaus 1986).

Figure 23.10. Spiracular disk of *Prionocera* sp.

Figure 23.11. Spiracular disk of *Brachypremna dispellens* (Walker) (after Gelhaus and Young 1991).

11'.	Median dorsal and/or lateral lobes of disk well developed if ventral lobes are elongate; ventral lobes usually short, less often absent (Fig. 23.13)	14
12(11).	Paired prolegs (pseudopods) with sclerotized apical crochets (curved hooks) on venter of abdominal segments 3-7 (Fig. 23.14); larvae in wet to saturated soil along streams	most *Dicranota*
12'.	Abdomen without prolegs; roughened creeping welts or broad tubercles on basal annulus of segments 4-7 (Figs. 22.3, 23.15)	13
13(12').	Creeping welts on both dorsum and venter, bearing microscopic spicules (subgenus *Raphidolabina*)	*Dicranota*
13'.	Creeping welts or broad tubercles on venter only, without spicules but with microscopic roughened surface (Figs. 22.3, 23.15); larvae in wet soil of swampy woods, springs, and seepage areas	*Pedicia*
14(11').	Spiracular disk surrounded by 7 lobes, 1 dorsomedial and 1 each dorsolateral, lateral, and ventral on each side; spiracles small, widely separated, at bases of lateral lobes of disk (Fig. 23.13); larvae in organic silt in small streams of Pacific drainage	*Gonomyodes*
14'.	Spiracular disk with 5 or fewer peripheral lobes, or without distinct lobes	15
15(14').	Spiracular disk with 4 or 5 peripheral lobes (Figs. 23.16, 23.17)	16
15'.	Spiracular disk with only 3 lobes, or without distinct lobes (Fig. 23.18)	43
16(15).	Internal portion of head extensively sclerotized dorsally and laterally, with shallow posterior incisions (can be determined by cutting prothoracic skin at one side, or often can be seen through skin) (Fig. 23.19)	17
16'.	Internal portion of head divided by deep posterior incisions into elongate, slender, rodlike to spatulate sclerites, or if sclerites are platelike, they are darkly sclerotized only along margins, giving appearance of separate rods (Figs. 23.20, 23.21)	37
17(16).	Hypostomal bridge (mentum or maxillary plates according to some authors) interrupted medially by membranous area (Figs. 23.21, 23.22) (hypostomal plates in contact though not fused in *Pseudolimnophila* (Fig. 23.23)); abdominal segments without creeping welts	18
17'.	Hypostomal bridge complete, although may be deeply incised posteriorly (Fig. 23.24); creeping welts present on basal ring of abdominal segments, or abdominal segments with transverse bands or patches of dense pilosity on both basal and apical rings	31
18(17).	Spiracular disk surrounded by 5 lobes in form of blackened, spatulate plates with finely toothed margins; larvae in marshy soil (subgenus *Scleroprocta*)	*Ormosia*
18'.	Spiracular disk surrounded by 4 or 5 lobes of rounded or subconical form (Figs. 23.25, 23.26)	19
19(18').	Plane of spiracular disk roughly perpendicular to long axis of body (Fig. 23.25); disk surrounded by 5 lobes (Fig. 23.26)	20
19'.	Plane of spiracular disk diagonal to long axis of body (Fig. 23.27); disk surrounded by 4 peripheral lobes (Figs. 22.4, 23.34)	29
20(19).	Hypostomal prolongations expanded into sclerotized plates with anterior margins toothed (Fig. 23.22)	21
20'.	Hypostomal prolongations, if expanded, not sclerotized or not toothed anteriorly (Fig. 23.21)	22
21(20).	Hypostomal plates with 4 teeth (Fig. 23.22); spiracular disk extensively blackened; black spots on dorsolateral and ventral lobes divided medially by pale line; spot on dorsal lobe nearly always undivided; larvae in wet, humic soil (larvae of *Tasiocera*, at present unknown, may key out here)	*Molophilus*
21'.	Hypostomal plates with 5-8 teeth; spiracular disk small, without extensive blackened areas; larvae in organic mud	some *Erioptera*
22(20').	Posterior faces of all 5 lobes of spiracular disk bearing solidly blackened spot (Fig. 23.29)	23
22'.	Spots on some or all lobes of spiracular disk divided medially by pale line or wider pale zone (Fig. 23.26)	24
23(22).	Blackened areas of dorsolateral lobes continued between spiracles; larvae in organic mud near water	some *Ormosia*
23'.	No blackened areas between spiracles (Fig. 23.29); larvae in muddy stream banks (subgenus *Trimicra*)	*Erioptera*
24(22').	Median dorsal lobe of disk bearing densely sclerotized, hornlike projection with apex bent downward over disk (Fig. 23.26); black, wedge-shaped spots at periphery of disk between ventral lobes, between ventral and lateral lobes, and between lateral lobes and median dorsal lobe (Fig. 23.26); larvae in fine sand, silt, and organic debris at margins of clear streams, Pacific and Arctic drainage	*Arctoconopa*

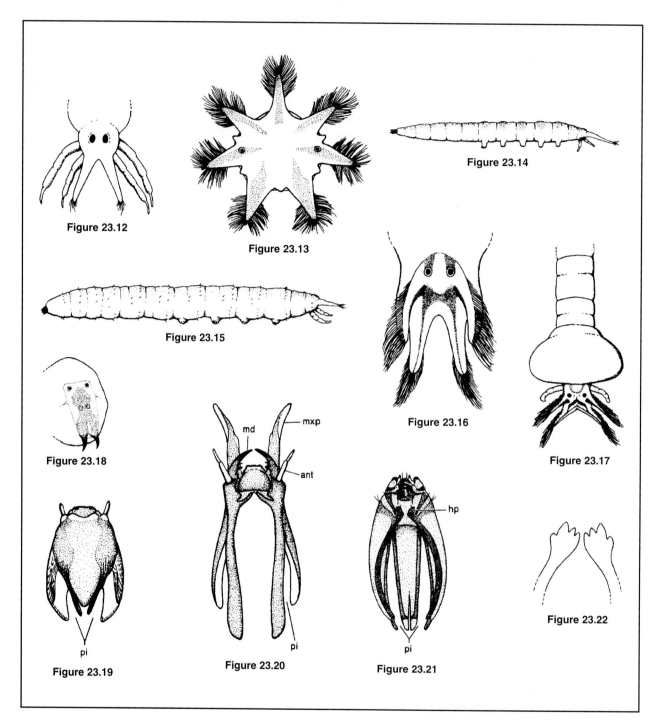

Figure 23.12. Dorsal view of posterior end of larval abdomen of *Pedicia* (*Tricyphona*) *inconstans* (O.S.).

Figure 23.13. Spiracular disk of *Gonomyodes tacoma* (Alex.) (after Hynes 1969).

Figure 23.14. Lateral view of larva of *Dicranota* sp.

Figure 23.15. Lateral view of larva of *Pedicia albivitta* Walker.

Figure 23.16. Spiracular disk of *Hexatoma* (*Hexatoma*) *megacera* (O.S.).

Figure 23.17. Posterior end of larval abdomen of *Hexatoma* (*Eriocera*) *spinosa* (O.S.), showing expanded seventh segment.

Figure 23.18. Posterolateral view of spiracular disk of *Rhabdomastix setigera* (Alex.) (after Hynes 1969).

Figure 23.19. Dorsal view of larval head capsule of *Limonia* (*Geranomyia*) *rostrata* (Say). *pi*, posterior incisions.

Figure 23.20. Dorsal view of larval head capsule of *Limnophila* sp. *ant*, antenna; *md*, mandible; *mxp*, maxillary palp; *pi*, posterior incision.

Figure 23.21. Ventral view of larval head capsule of *Ormosia* sp. *hp*, hypostomal plate (half of incomplete hypostomal bridge); *pi*, posterior incisions.

Figure 23.22. Ventral view of incomplete hypostomal bridge of *Molophilus* sp.

24'.	Median dorsal lobe of disk without sclerotized, hornlike projection; no wedges of black pigmentation between lobes of disk	25
25(24').	Dorsolateral lobes with solidly blackened spots continuous around spiracles and extending to midline or nearly so; spots on ventral lobes divided; larvae in moist earth or sand, usually near water	some *Gonomyia*
25'.	Dorsolateral lobes and ventral lobes with spots divided (Fig. 23.30); if spots of dorsolateral lobes are more completely darkened, 4-6 small, dark spots on central disk	26
26(25').	Peripheral lobes of disk short, blunt; blackened areas of dorsolateral lobes continuous around spiracles, fading toward midline	some *Gonomyia*
26'.	Peripheral lobes of disk nearly as long as their width at base, or longer; blackened areas of dorsolateral lobes not continuous around spiracles (Fig. 23.30)	27
27(26').	Blackened area of median dorsal lobe not divided (Fig. 23.30); larvae in organic mud	some *Ormosia*
27'.	Blackened area of median dorsal lobe (and all others) divided medially by pale line	28
28(27').	Area between spiracles generally unpigmented, not blackened; larvae in organic mud	some *Ormosia*
28'.	Two rounded spots between spiracles; spiracular disk small in proportion to body size; larvae in organic mud	some *Erioptera*
29(19').	Ventral lobes of disk not darkly pigmented on upper surface, not fringed with long hairs; spiracles pale (Fig. 23.27); hypostome reduced to small, longitudinal rod below maxilla on each side; larvae in sandy bottoms of clear, cold streams	*Cryptolabis*
29'.	Ventral lobes of disk darkly pigmented on upper surface, fringed with long hairs (longer than the lobes) (Figs. 22.4, 23.28); spiracles dark; hypostome in form of a toothed plate at each side (Fig. 23.23)	30
30(29').	Hypostomal plates (mental plates or maxillary plates of other authors) each bearing 4 anterior teeth; larvae in organic mud in wet woodlands	*Paradelphomyia*
30'.	Hypostomal plates each bearing 7 or 8 anterior teeth (Fig. 23.23); larvae in thin organic mud in swampy woods, pond margins, marshy areas	*Pseudolimnophila*
31(17').	Spiracular disk with 5 peripheral lobes	32
31'.	Spiracular disk with 4 peripheral lobes (if vestigial median dorsal lobe is present, it is unpigmented) (Fig. 23.28)	33
32(31).	Abdominal segments 2-7 with both dorsal and ventral creeping welts on basal ring; lobes of spiracular disk wider than long, broadly rounded, unpigmented or with only limited darkened spots; large genus with larvae in numerous aquatic and terrestrial habitats	some *Limonia*
32'.	Abdominal segments 2-7 with ventral creeping welts only; ventral lobes of spiracular disk longer than their width at base, darkened at margins with broad median pale zone on each; hypostomal bridge with 5 teeth; larvae brownish with long, appressed (closely applied to body) pubescence; occurring in marsh borders in decomposing aquatic vegetation, or in marshy areas in woods	*Helius*
33(31').	Abdominal segments 2-7 without distinct creeping welts; all segments with transverse bands or patches of dense pilosity; lateral lobes of disk broadly pigmented from spiracles outward, broadly pigmented faces of ventral lobes narrowly connected across lower part of disk; larvae in thin mosses and algal mats on wet, rocky cliffs, rarely in soil (Fig. 23.32a)	*Dactylolabis*
33'.	Abdominal segments 2-7 with distinct creeping welts, without transverse bands of dense pilosity	34
34(33').	Ventral lobes of spiracular disk longer than their width at base, tapering to subacute apex, fringed with long hairs (Figs. 22.4, 23.28)	35
34'.	Ventral lobes of disk shorter than their width at base, broadly rounded and without long marginal hairs (Fig. 23.31)	36
35(34).	Body wide, flattened; ventral creeping welts without minute spines; spiracles dorsoventrally elongated; larvae semiaquatic in indistinct tunnels beneath algal mats on wet cliffs, beside waterfalls, etc.	*Elliptera*
35'.	Body nearly terete (cylindrical), only slightly flattened; ventral creeping welts with numerous rows of minute spines; spiracles transversely elliptical; lobes of spiracular disk narrowly darkened at margins; larvae in sodden, decayed wood, at or just below water level	*Lipsothrix*
36(34').	Nearly entire spiracular disk except spiracles and outer margins of lobes dark reddish brown; spiracles horizontally elongated; larvae in wet, extremely decayed, pulpy wood	(subgenus *Diotrepha*) *Orimarga*

36'.	Spiracular disk with only isolated spots of dark pigmentation, generally pale; spiracles oval, inclined together dorsally (Fig. 23.31); large genus in a variety of habitats ....some *Limonia*
37(16').	Maxillae not prolonged forward, inconspicuous in dorsal aspect; plane of spiracular disk roughly perpendicular to long axis of body (Fig. 23.25) ....(tribe Eriopterini, in part)...........return to 18
37'.	Maxillae prolonged forward, each as a dorsoventrally flattened, tapering (less often subconical) blade, together appearing as divergently curved tusks with apices visible even when head is withdrawn into thoracic segments (Fig. 23.20) 38
38(37').	Mandibles complex, jointed near midlength (Fig. 23.32); maxillae and labrum-epipharynx densely fringed with long yellowish to golden hairs; dorsal plates of head fused into spatulate plate widest posteriorly; spiracular disk small, its upper lobes often infolded, concealing spiracles, marginal hairs protruding from cavity formed by infolding (Fig. 23.33) 39
38'.	Mandibles not jointed near midlength; maxillae and labrum-epipharynx with mostly short pilosity; dorsal plates of head not fused, although each may be widest posteriorly 40
39(38).	Pigmentation of ventral lobes of spiracular disk discontinuous, either as transverse striations near base of lobe, more continuous coloration toward apex, or reduced to short darkened median line; all (4, lateral pair sometimes reduced) lobes fringed with long golden hairs (Fig. 23.33); basal tooth or teeth of apical portion of mandible much less than half as long as main outer tooth (Fig. 23.32); larvae in moist to wet humic soil or decomposing vegetation in swampy woodlands ....*Pilaria*
39'.	Pigmentation of ventral lobes of disk more evenly distributed but more intense toward apex of lobe; all 4 lobes fringed with long hairs; basal tooth of apical portion of mandible about half as long as main outer tooth; larvae in organic mud in swampy woodlands ....*Ulomorpha*
40(38').	Spiracular disk surrounded by 5 short, bluntly rounded lobes; ventral lobes not fringed with long hairs; in some species a densely sclerotized, hornlike projection near apex of median dorsal lobe and lateral lobes; larvae in sandy bottoms and margins of clear streams ....some *Rhabdomastix*
40'.	Lobes of spiracular disk (usually 4) not all short and bluntly rounded, ventral ones usually elongate; ventral lobes fringed with long hairs; no sclerotized, hornlike projections from upper lobes (Fig. 23.17) 41
41(40').	Midventral region of head before line of attachment of skin entirely membranous, without darkened transverse bar just beneath surface; larvae in sand or gravel near margins of clear, cool brooks and streams. (Note: In this genus especially, but also in some others in similar habitats, larvae may be found with the 7th abdominal segment much swollen, possibly as an aid in locomotion or anchorage. This swelling may persist in preserved specimens; Fig. 23.17) ....*Hexatoma*
41'.	Midventral region of head before line of attachment of skin membranous, with darkened, narrow transverse bar (part of hypopharynx) visible just beneath surface 42
42(41').	Lateral lobes of spiracular disk unpigmented on posterior face; mandible with long outer tooth and two smaller teeth of similar size and shape near midlength of inner margin; maxillary projections subconical; larvae in wet organic debris ....*Polymera*
42'.	Lateral lobes of disk pigmented at least along one margin, usually much more extensively (Figs. 22.5, 23.34); mandible without 2 small, similar teeth near midlength of inner margin (more or fewer dissimilar teeth); maxillary projections flattened; large genus with carnivorous, aquatic larvae found usually in organic mud in swampy woods, pond margins, etc., less often in bottom mud or sand of small streams ....*Limnophila*
43(15').	Spiracular disk broadly emarginate (indented) dorsally; larvae in a hardened, flattened, elliptical case, in marshy soil near small streams or springs, Pacific drainage (based on a European species) ....*Thaumastoptera*
43'.	Spiracular disk not broadly emarginate dorsally (Figs. 23.18, 23.31); larvae not in a hardened case 44
44(43').	Internal portion of head divided by deep posterior incisions into elongate, slender, or spatulate sclerites (can be determined by cutting prothoracic skin at one side, or often can be seen through skin); spiracular disk lightly pigmented, vertically subrectangular, with 2 clawlike projections at ventral margin (Fig. 23.18); spiracles minute, pale, separated by about 3 times the diameter of a spiracle (Fig. 23.18); larvae yellowish, aquatic, in sandy bottoms and margins of clear streams ....some *Rhabdomastix*

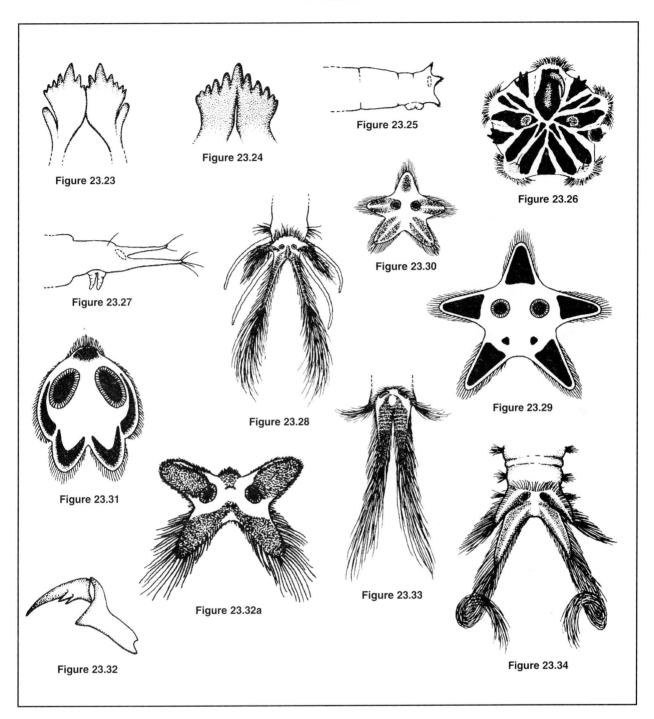

Figure 23.23. Ventral view of hypostomal bridge of *Pseudolimnophila inornata* (O.S.).

Figure 23.24. Ventral view of hypostomal bridge of *Limonia* sp.

Figure 23.25. Lateral view of caudal end of larva of *Molophilus* sp. (position of spiracle indicated by dashed circle).

Figure 23.26. Spiracular disk of *Arctoconopa carbonipes* (Alex.) (after Hynes 1969).

Figure 23.27. Lateral view of caudal end of larva of *Cryptolabis* sp. (position of spiracle indicated by dashed circle).

Figure 23.28. Posterior end of larval abdomen of *Pseudolimnophila inornata* (O.S.).

Figure 23.29. Spiracular disk of *Erioptera* (*Trimicra*) *pilipes* (O.S.).

Figure 23.30. Spiracular disk of *Ormosia meigenii* (O.S.).

Figure 23.31. Spiracular disk of *Limonia* (*Dicranomyia*) *humidicola* (O.S.).

Figure 23.32. Mandible of *Pilaria* sp.

Figure 23.32a. Spiracular disk of *Dactylolabis* sp.

Figure 23.33. Spiracular disk of *Pilaria* sp.

Figure 23.34. Spiracular disk of *Limnophila macrocera* (Say) (after Alexander 1920).

44'.	Internal portion of head extensively sclerotized dorsally and laterally; sclerities platelike, with shallow posterior incisions (Fig. 23.19); abdominal segments 2-7 with both dorsal and ventral creeping welts (of differing structure in some species) on basal ring; spiracular disk roughly circular or broadly oval to transversely subrectangular; spiracles often large, oval, inclined together dorsally (Fig. 23.31); large genus with larvae in a variety of aquatic and terrestrial habitats ....some **Limonia**

Adults*

1.	Terminal segment of maxillary palp elongate, longer than previous 2 segments combined (Fig. 23.35); antennae usually 13-segmented (11 flagellomeres); rostral nasus usually present (Fig. 23.35); vein CuA with slight deflection at bscu (Fig. 23.36) ....(subfamily Tipulinae) 2
1'.	Terminal segment of maxillary palp short, its length less than that of previous 2 segments combined; antennae usually of 14 or 16 segments (12 or 14 flagellomeres); rostral nasus absent; vein CuA not deflected at bscu (Fig. 23.37) 10
2(1).	Tarsi filiform (threadlike), each tarsus longer than corresponding femur and tibia combined; legs long and slender 3
2'.	Length of tarsus not exceeding combined length of corresponding femur and tibia 5
3(2).	Antennae with only 8 segments, unusually short in both sexes (in Florida only) ....**Megistocera***
3'.	Antennae with 12 or 13 segments 4
4(3').	Wing vein R_{1+2} absent; Rs short, subequal to bscu, only slightly curved (Fig. 23.38) ....**Dolichopeza**
4'.	Vein R_{1+2} present; Rs much longer than bscu, strongly curved near base (Fig. 23.39) (in eastern United States) ....**Brachypremna***
5(2').	Flagellomeres 2-9 each bearing 3 or 4 pectinations (making antenna comblike) in male; antennae of female serrate (sawlike), 11-segmented (Fig. 23.40) ....**Ctenophora**
5'.	Flagellomeres 2-9 not branched in male; antennae of female 1-3-segmented 6
6(5').	Vein R_3 deflected toward R_{4+5} near midlength, then forward toward costa, resulting in constriction of cell R_3 near its midlength (Fig. 23.41) (in western states and British Columbia) ....**Holorusia***
6'.	Cell R_3 not strongly constricted by deflection of vein R_3 7
7(6').	Antennal flagellomeres without verticils (strong hairs near base), expanded subapically to give serrate appearance to antenna (Fig. 23.42) ....**Prionocera***
7'.	Flagellomeres usually verticillate (Fig. 23.35), if verticils absent, flagellomeres wider near their bases than apically 8
8(7').	Abdomen in both sexes greatly elongated, more than 12 times average diameter; verticils of outer flagellomeres conspicuously elongated (in eastern United States) ....(subgenus *Longurio*) ....**Leptotarsus**
8'.	Abdomen not greatly elongated, or if so (females of a few species), flagellar verticils not conspicuously elongated 9
9(8').	Rs shorter than bscu; cell M_1 either joining cell 1st M_2 (discal cell) or separated from it by only short petiole (Fig. 23.43) ....**Nephrotoma**
9'.	Rs longer than bscu; cell M, separated from cell 1st M_2 (discal cell) by distinct petiole (Fig. 23.36) ....**Tipula***
10(1').	Short terminal segment of Sc_2 present between R_3 and C; R_{1+2} absent, R_2 curving back to join R_{2+3} at a sharp angle (Fig. 23.44) ....(most Cylindrotominae) 11
10'.	Terminal segment of Sc_2 usually absent (Fig. 23.48) (present in some Limoniini; see couplet 15); R_{1+2} present or absent, if absent R_1 curved back to join R_{2+3} at a blunt angle 14
11(10).	Dorsum of head and prescutum (Fig. 23.35) deeply pitted; prescutum with median longitudinal groove ....**Triogma***
11'.	Dorsum of head and prescutum without deep pits; no median longitudinal groove in prescutum 12
12(11').	Vein M with 3 branches (Fig. 23.44) ....**Cylindrotoma**
12'.	Vein M with only 2 branches (Fig. 23.45) 13
13(12').	Distal end of cell 1st M_2 closed by single crossvein; short r-m crossvein usually present (Fig. 23.45) (see also couplet 16) ....**Phalacrocera** (in part)*
13'.	Distal end of cell 1st M_2 closed by two short crossveins (m and base of M_3) (Fig. 23.46); antennae nodose (swelling on each segment) to bluntly serrate (sawlike) ....**Liogma**

* Adults of almost all genera of Tipulidae are likely to be found near standing or flowing water. Accordingly, the adult key identifies all North American genera, and those whose larvae are aquatic are indicated (*).

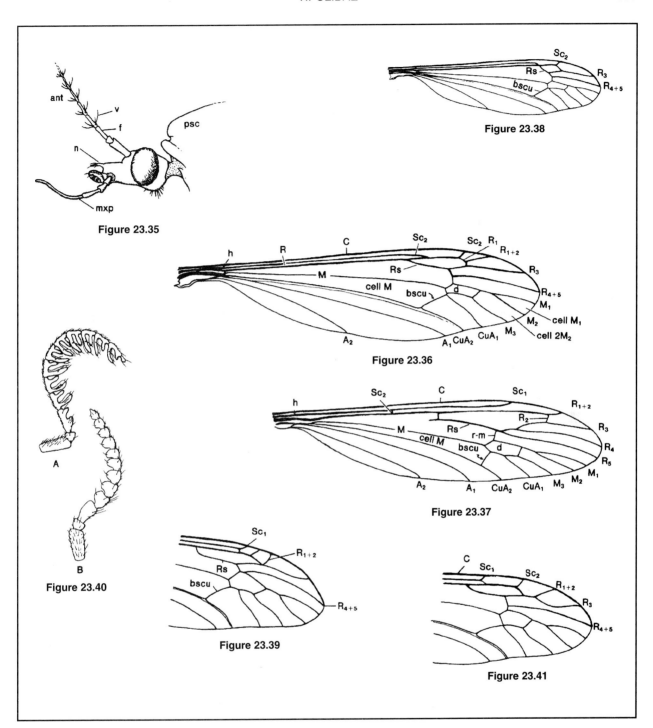

Figure 23.35. Lateral view of head and anterior thorax of *Tipula* (*Yamatotipula*) *caloptera* Loew, *ant*, antenna; *mxp*, maxillary palp; *n*, nasus; *psc*, prescutum of mesonotum; *v*, verticil; *f*, flagellomere.

Figure 23.36. Wing of *Tipula* (*Yamatotipula*) *caloptera* Loew., showing nomenclature of veins and cells. *A*, anal veins; *bscu*, basal section of anterior cubitus; *C*, costa; *Cu*, cubitus; *d*, discal cell; *h*, humeral crossvein; *M*, media; *R*, radius; *Rs*, radial sector; *Sc*, subcosta. Cells are named from the vein (or posteriormost component of composite veins) forming their anterior border; thus, the discal cell (d) is cell first M_2 as there is a second cell $2M_2$ beyond it.

Figure 23.37. Wing of *Pedicia* (*Tricyphona*) *inconstans* (O.S.). Abbreviations as in Figure 23.36.

Figure 23.38. Wing of *Dolichopeza* (*Oropeza*) *subalbipes* (Johnson).

Figure 23.39. Wing apex of *Brachypremna dispellens* (Walker).

Figure 23.40. Antenna of (*A*) male, and (*B*) female *Ctenophora* (*Tanyptera*) *dorsalis* (Walker).

Figure 23.41. Wing apex of *Holorusia hespera* Arnaud and Byers.

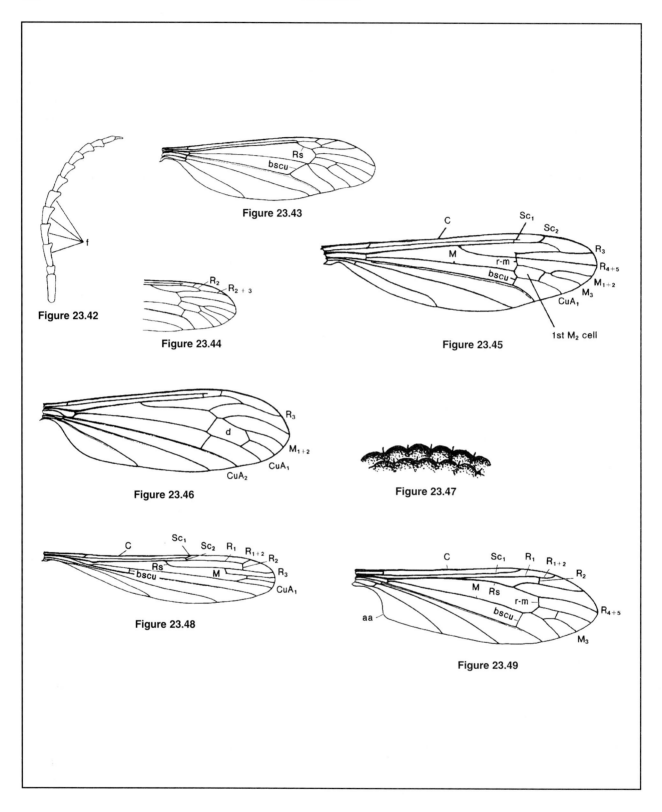

Figure 23.42. Antenna of *Prionocera* sp. *f*, flagellomere.

Figure 23.43. Wing of *Nephrotoma ferruginea* (Fabr.). *d*, discal cell.

Figure 23.44. Wing apex of *Cylindrotoma* sp.

Figure 23.45. Wing of *Phalacrocera tipulina* O.S.

Figure 23.46. Wing of *Liogma nodicornis* (O.S.).

Figure 23.47. Portion of compound eye of *Pedicia* sp., in profile, to show short hairs among ommatidia.

Figure 23.48. Wing of *Orimarga* (*Diotrepha*) *mirabilis* (O.S.).

Figure 23.49. Wing of *Antocha opalizans* O.S. *aa*, anal angle of wing.

14(10').	Eyes bare, no short hairs interspersed among ommatidia (units composing compound eye); vein Sc_1 usually short (not extending far beyond Sc_2), if long, Sc_2 is beyond level of origin of R_s (Fig. 23.48)	15
14'.	Eyes with short hairs interspersed among ommatidia (Fig. 23.47); vein Sc_1 long, Sc_2 before level of origin of Rs	(Pediciini) 25
15(14).	Both short terminal segment of Sc_2 and R_{1+2} present between R_1 and C (Fig. 23.36)	16
15'.	Terminal segment of Sc_2 or R_{1+2} present, but not both (Fig. 23.48)	17
16(15).	Vein bscu beyond first fork of M; discal cell (1st M_2) present (see also couplet 13) (Fig. 23.45)	*Phalacrocera* (in part)*
16'.	Vein bscu before first fork of M; no closed discal cell (Fig. 23.48)	*Orimarga* (in part)*
17(15').	Terminal segment of Sc_2 often present; veins R_4 and R_5 entirely fused; only 2 branches of R_5 present (Fig. 23.50)	(Limoniini) 18
17'.	Terminal segment of Sc_2 absent; veins R_4 and R_5 separate but R_4 usually merged with R_{2+3}; usually 3 branches of R present (Fig. 23.55)	29
18(17).	Short section of R_2 present between R_1 and anterior branch of Rs (Fig. 23.48)	19
18'.	Short section of R_2 absent from between R_1 and anterior branch of Rs (Fig. 23.51)	24
19(18).	Vein bscu twice its length or more before level of 1st fork of M (Fig. 23.48)	20
19'.	Vein bscu near or beyond level of 1st fork of M (Fig. 23.49)	21
20(19).	Vein bscu near base of wing, far before level of origin of Rs (Fig. 23.48)	*Orimarga* (in part)*
20'.	Vein bscu 2 or 3 times its length before 1st fork of M (in California only)	*Thaumastoptera**
21(19').	Vein R_2 far distad (toward wing apex) of level of discal cell (cell 1st M_2); bscu beyond 1st fork of M (Fig. 23.37); pale longitudinal line in outer part of cell Cu	*Dicranoptycha*
21'.	Vein R_2 in nearly transverse alignment with crossvein r-m (near proximal end of discal cell), or rarely beyond; no pale line in cell Cu (Fig. 23.49)	22
22(21').	Antennae with 14 segments; Rs curved	*Limonia**
22'.	Antennae with 16 segments; Rs elongate, straight, or nearly so (Fig. 23.49)	23
23(22').	Anal margin of wing protruding, almost angular; Sc closely paralleling R; Sc_2 absent (Fig. 23.49)	*Antocha**
23'.	Anal margin of wing not protruding, smoothly curved; Sc distinctly separated from R_1; Sc_2 present but far from end of Sc (Fig. 23.50)	*Elliptera* (in part)*
24(18').	Rostrum (snout) much shorter than rest of head; Sc_2 far from end of Sc (Fig. 23.50)	*Elliptera* (in part)*
24'.	Rostrum approximately as long as rest of head, or longer; Sc_2 near tip of Sc (Fig. 23.51)(in eastern North America)	*Helius**
25(14').	Membrane of wings covered with macrotrichia (large microscopic hairs)	*Ula*
25'.	Membrane of wings without macrotrichia	26
26(25').	Antennae with 13 or 15 segments (11 or 13 flagellomeres); wings of male usually 7 mm long or less; usually no closed discal cell (Fig. 23.52)	*Dicranota**
26'.	Antennae with 14 or 16 segments; wing of male more than 7 mm long; closed discal cell usually present	27
27(26').	Rostrum (snout) prolonged to about half length of rest of head (in western North America)	*Ornithodes**
27'.	Rostrum not prolonged, less than half length of rest of head	28
28(27').	A crossvein in cell M	*Nasiternella**
28'.	No crossvein in cell M (Fig. 23.37)	*Pedicia**
29(17').	Tibial spurs (enlarged spines) present on distal end of tibiae (Fig. 23.53)	(Hexatomini) 30
29'.	Tibial spurs absent	(Eriopterini) 44
30(29).	Antennae with no more than 12 segments (10 flagellomeres), although they may be very long, particularly in males	*Hexatoma**
30'.	Antennae with more than 13 segments (11 or more flagellomeres)	31
31(30').	Rostrum (snout) greatly elongated, more than half length of entire body	*Elephantomyia*
31'.	Rostrum short, less than length of rest of head	32
32(31').	Only two branches of R_s entending to margin of wing (Fig. 23.54)	*Atarba*
32'.	Three branches of R_s extending to margin of wing	33
33(32').	Macrotrichia (large microscopic hairs) present in apical cells of wings (Fig. 23.55)	34
33'.	No macrotrichia in apical cells, confined to stigmal area of costal margin if present	35

* Adults of almost all genera of Tipulidae are likely to be found near standing or flowing water. Accordingly, the adult key identifies all North American genera, and those whose larvae are aquatic are indicated (*).

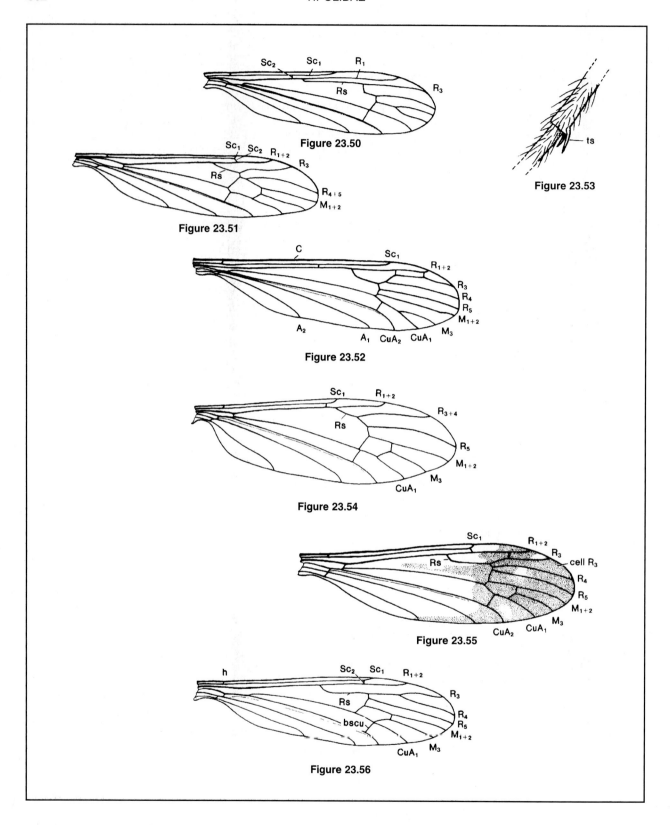

Figure 23.50. Wing of *Elliptera illini* Alex.
Figure 23.51. Wing of *Helius flavipes* (Macq.).
Figure 23.52. Wing of *Dicranota* (*Paradicranota*) *iowa* Alex.
Figure 23.53. End of tibia and part of basitarsus, tribe Hexatomini; *ts*, tibial spur.
Figure 23.54. Wing of *Atarba* (*Atarba*) *picticornis* O.S.
Figure 23.55. Wing of *Ulomorpha pilosella* (O.S.). Stippling = macrotrichia.
Figure 23.56. Wing of *Phyllolabis encausta* O.S.

34(33).	Macrotrichia in all cells, on all parts of membrane except near wing base; petiole of cell R_3 very short or absent (Fig. 23.55)	*Ulomorpha**
34'.	Macrotrichia in apical cells only; cell R_3 distinctly petiolate on vein R_{2+3+4}	*Paradelphomyia*
35(33').	Crossvein in cell C between humeral crossvein and Sc_1	*Epiphragma*
35'.	No crossvein in cell C beyond humeral crossvein (Fig. 23.56)	36
36(35').	Vein R_2 absent (as crossvein between R_1 and anterior branch of Rs); vein bscu near branching of vein M_3 and CuA_1 (Fig. 23.56)	*Phyllolabis*
36'.	Vein R_2 present as crossvein between R_1 and anterior branch of Rs; bscu at or before midlength of vein $M_3 + CuA_1$ (Fig. 23.59)	37
37(36').	Anterior arculus present between R and base of M (Fig. 23.57)	38
37'.	Anterior arculus absent (Fig. 23.58)	41
38(37).	Sc short, Sc_1 joining C before level of fork of Rs (Fig. 23.59)	39
38'.	Sc longer, Sc_1 joining C at or beyond level of fork of Rs	40
39(38).	Antennal verticils (whorls of hair) (Fig. 23.35) conspicuously long; Rs longer than vein R_3 (Fig. 23.59)	*Pilaria**
39'.	Antennal verticils short; Rs short and abruptly curved or angulate near origin	*Shannonomyia*
40(38').	Head prolonged and narrowed backward between eyes and neck (Fig. 23.60)	*Pseudolimnophila**
40'.	Head not prolonged and narrowed backward behind eyes	*Limnophila**
41(37').	No closed discal cell in wing; antennae of male elongate, most flagellomeres constricted near midlength, binodose (2 swellings), producing appearance of 2 segments each (coastal plain of southeastern United States)	*Polymera**
41'.	Closed discal cell (cell 1st M_2) present (Fig. 23.61); antennae of male without binodose flagellomeres	42
42(41').	Vein bscu at or only slightly beyond fork of M (Fig. 23.61)	*Dactylolabis**
42'.	Vein bscu distinctly beyond fork of M_1 usually from one-third to one-half length of discal cell (Fig. 23.62)	43
43(42').	Discal cell (cell 1st M_2) elongate, its proximal end (nearest wing base), acutely angulate and extending much farther toward wing base than level of r-m crossvein (Fig. 23.62)	*Prolimnophila*
43'.	Discal cell not conspicuously elongate, its proximal end obtuse and at about level of r-m	*Austrolimnophila*
44(29').	Wings reduced to tiny vestiges; body and legs stout, giving fly a spiderlike appearance (winter species)	*Chionea*
44'.	Wings well developed; general appearance not spiderlike	45
45(44').	Vein M_{1+2} forked, cell M_1 present (Fig. 23.63)	46
45'.	Vein M_{1+2} not forked, cell M_1 absent	48
46(45).	Cell R_3 shorter than its petiole (R_{2+3+4}) (Fig. 23.63)	*Neolimnophila*
46'.	Cell R_3 3-4 times as long as its petiole	47
47(46').	Two dististyles on each basistyle in male (Fig. 23.64)	*Neoclatura*
47'.	Only 1 stout dististyle on each basistyle in male (Fig. 23.65)	*Cladura*
48(45').	Rostrum elongate, about as long as rest of head and thorax combined	*Toxorhina*
48'.	Rostrum not elongate, shorter than rest of head	49
49(48').	Vein R_{3+4} unbranched, only 2 branches of Rs reaching wing margin (Fig. 23.66)	50
49'.	Vein R_{3+4} branched, 3 branches of Rs reaching wing margin	51
50(49).	Vein R_2 absent between R_1 and anterior branch of Rs; Sc short, Sc_1 ending before origin of short Rs; no closed discal cell (Fig. 23.66)	*Gonomyia* (in part)
50'.	Vein R_2 present, near fork of Rs; Sc_1 beyond origin of R_5; R_{2+3+4} very short, Rs long; discal cell closed	*Teucholabis*
51(49').	Rs branching into R_{2+3} and R_{4+5}; cell R_3 sessile on R_5 (Fig. 23.67)	52
51'.	Rs branching into R_{2+3+4} and R_5; cell R_3 petiolate (Fig. 23.68)	53
52(51).	Cell A_2 elongate and wide; vein bscu beyond fork of M (Fig. 23.67)	*Molophilus**
52'.	Cell A_2 short and narrow (i.e., vein A_2 short, close to hind margin of wing); bscu before fork of M	*Tasiocera*
53(51').	R_5 short and straight, cell R_1 (between R_1 and R_5) nearly an equilateral triangle (Fig. 23.68)	*Cryptolabis**
53'.	R_5 long, cell R_1 either long-triangular or not triangular	54
54(53').	Cell R_3 elongate, 3 or more times the length of its petiole (R_{2+3+4}) (Figs. 23.70-23.71)	55

* Adults of almost all genera of Tipulidae are likely to be found near standing or flowing water. Accordingly, the adult key identifies all North American genera, and those whose larvae are aquatic are indicated (*).

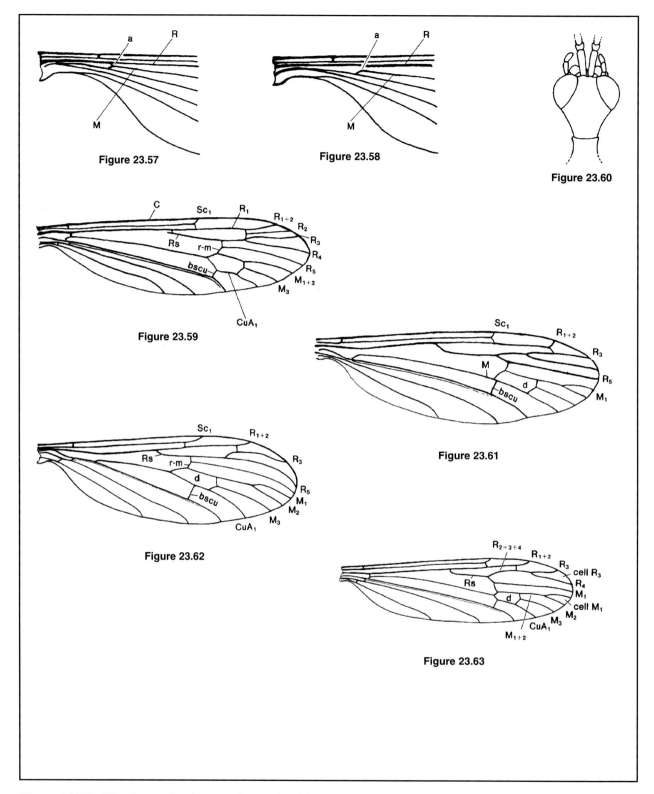

Figure 23.57. Wing base, showing anterior arculus (*a*) present between R and base of M (also see fig. 23.36.

Figure 23.58. Wing base, showing absence of anterior arculus. *a*, position of anterior arculus when present; *M*, medial vein; *R*, radial vein.

Figure 23.59. Wing of *Pilaria quadrata* (O.S.).

Figure 23.60. Dorsal view of head of *Pseudolimnophila luteipennis* (O.S.), showing narrowing behind eyes.

Figure 23.61. Wing of *Dactylolabis montana* (O.S.). *di*, discal cell.

Figure 23.62. Wing of *Prolimnophila areolata* (O.S.). *d*, discal cell.

Figure 23.63. Wing of *Neolinmnophila ultima* (O.S.). *d*, discal cell.

54'.	Cell R_3 short, vein R_3 not more than twice the length of petiole of cell R_3 (Fig. 23.75)	62
55(54).	Coxae of middle and hind legs close together, separated by narrow meron (lateral aspect)	56
55'.	Coxae of middle and hind legs separated by enlarged mesothoracic meron (lateral aspect) (Fig. 23.69)	59
56(55).	Veins Sc, and Sc_2 both short, approximately of equal length (Fig. 22.70)	57
56'.	Vein Sc_1 much longer than Sc_2 (Fig. 23.71)	58
57(56).	Vein R_{2+3+4} longer than vein bscu; veins R_2 and R_{1+2} approximately of equal length (Fig. 23.70)	***Lipsothrix****
57'.	Vein R_{2+3+4} about the same length as bscu, or shorter; vein R_2 much shorter than R_{1+2} (southwestern United States)	***Eugnophomyia***
58(56').	Body black; wings darkly tinged with brownish black; legs black, without scales	***Cnophomyia***
58'.	Body brown; wings yellowish; legs with numerous elongate scales in addition to setae (southwestern United States)	***Idiognophomyia***
59(55').	Abundant macrotrichia in all cells of wings (Fig. 23.71)	***Ormosia****
59'.	No macrotrichia in wing cells, or only a few scattered ones in outermost cells	60
60(59').	Vein Sc_1 shorter than bscu or of equal length (as in Fig. 23.70) (western North America)	***Hesperoconopa****
60'.	Vein Sc_1 long, 2 or 3 times length of bscu (Fig. 23.72)	61
61(60').	Cell 1st M_2 (discal cell) open by absence of crossvein m; vein A_2 nearly straight, not curved toward A_1 in apical portion (Fig. 23.73); vein R_3 much longer than petiole of cell R_3; aedeagus simple, not divided	***Arctoconopa****
61'.	Cell 1st M_2 usually closed by crossvein m between M_{1+2} and M_3; if discal cell open by absence of m, vein A_2 curves toward A_1 apically (Fig. 23.72); aedeagus bifid (forked) (Fig. 23.74)	***Erioptera****
62(54').	Vein R_2 present between R_1 and anterior branch of Rs (Fig. 23.75)	63
62'.	Vein R_2 absent	68
63(62).	Vein R_2 immediately before point at which R_3 and R_4 diverge (Fig. 23.75)	***Gonomyia*** (in part)
63'.	Vein R_2 well before divergence of R_3 and R_4; R_{3+4} longer than R_2 (Fig. 23.76)	64
64(63').	Vein R_4 short, subequal in length to R_{3+4}, curved caudad (backward) (Fig. 23.76)	***Rhabdomastix*** (in part)*
64'.	Vein R_4 longer than R_{3+4}, straight or nearly so (Fig. 23.77)	65
65(64').	Vein Sc short, Sc_1 joining C before level of midlength of Rs (Fig. 23.77)	***Cheilotrichia***
65'.	Vein Sc_1 joining C at level of outer one-quarter of Rs or beyond fork of Rs (Fig. 23.71)	66
66(65').	Vein Sc_1 joining C at level of R_2, well beyond fork of Rs (as in Fig. 23.71)	***Gonempeda***
66'.	Vein Sc_1 joining C opposite or before level of fork of Rs	67
67(66').	Abundant macrotrichia in area surrounding stigma (outer costal margin) in male	***Empedomorpha***
67'.	No macrotrichia in area surrounding stigma	***Gonomyodes****
68(62').	Vein Sc ending before to only slightly beyond origin of Rs (Fig. 23.66)	***Gonomyia*** (in part)
68'.	Vein Sc ending opposite midlength of Rs, or more distally	69
69(68').	Vein bscu well before fork of M	***Gonomyia*** (in part)
69'.	Vein bscu beyond fork of M (Fig. 23.76)	***Rhabdomastix*** (in part)*

ADDITIONAL TAXONOMIC REFERENCES

General
Alexander (1919, 1920, 1927, 1931, 1934, 1942, 1965); Johannsen (1934); Rees and Ferris (1939); Crampton (1942); Wirth and Stone (1956); Frommer (1963); Alexander and Byers (1981); Oosterbroek and Theowald (1991).

Regional faunas
California: Alexander (1967).
Florida (northern): Rogers (1933).
Kansas: Young (1978).
Michigan: Rogers (1942).
Northeastern United States: Alexander (1942).
Northwestern United States: Alexander (1949, 1954).
Tennessee: Rogers (1930).

Taxonomic treatments at the generic level
(L = larvae; P = pupae; A = adults)
Brachypremna: Gelhaus and Young (1991)–L, P.
Holorusia: Arnaud and Byers (1990)–A.
Phalacrocera, Triogma: Brodo (1967)–L, P, A.
Prionocera: Brodo (1987)–A.
Tipula: Gelhaus (1986)–L

---

* Adults of almost all genera of Tipulidae are likely to be found near standing or flowing water. Accordingly, the adult key identifies all North American genera, and those whose larvae are aquatic are indicated (*).

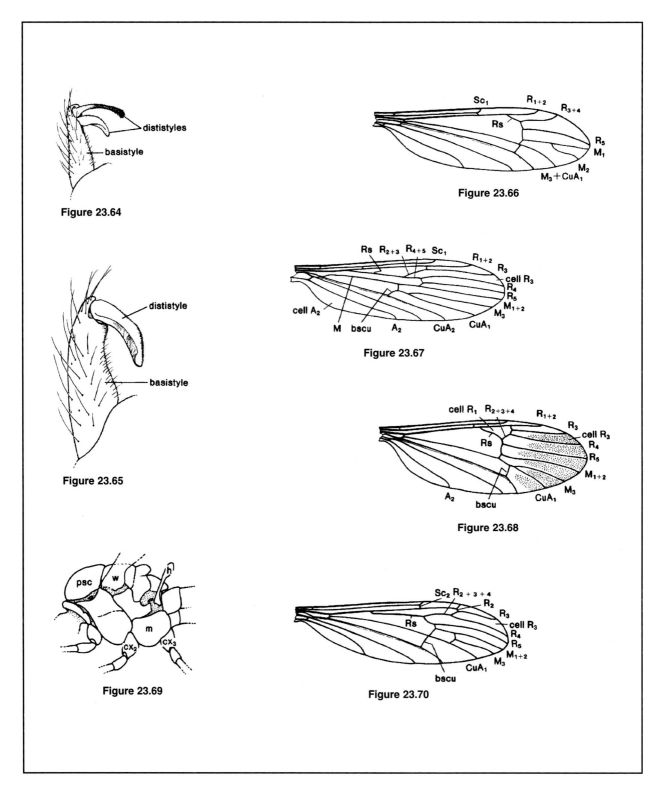

Figure 23.64. Dorsal view of right basistyle and dististyles of male *Neocladura delicatula* Alex.

Figure 23.65. Dorsal view of right basistyle and dististyle of male *Cladura flavoferruginea* O.S.

Figure 23.66. Wing of *Gonomyia* (*Neolipophleps*) *cinerea* (Doane).

Figure 23.67. Wing of *Molophilus* (*Molophilus*) *nitidus* Coq.

Figure 23.68. Wing of *Cryptolabis* (*Cryptolabis*) *paradoxa* (O.S.). Stippling = macrotrichia.

Figure 23.69. Lateral view of thorax of *Erioptera* (*Erioptera*) *chlorophylla* O.S. *cx*, coxa; *h*, halter; *m*, meron; *psc*, prescutum; w, base of wing.

Figure 23.70. Wing of *Lipsothrix sylvia* (Alex.)

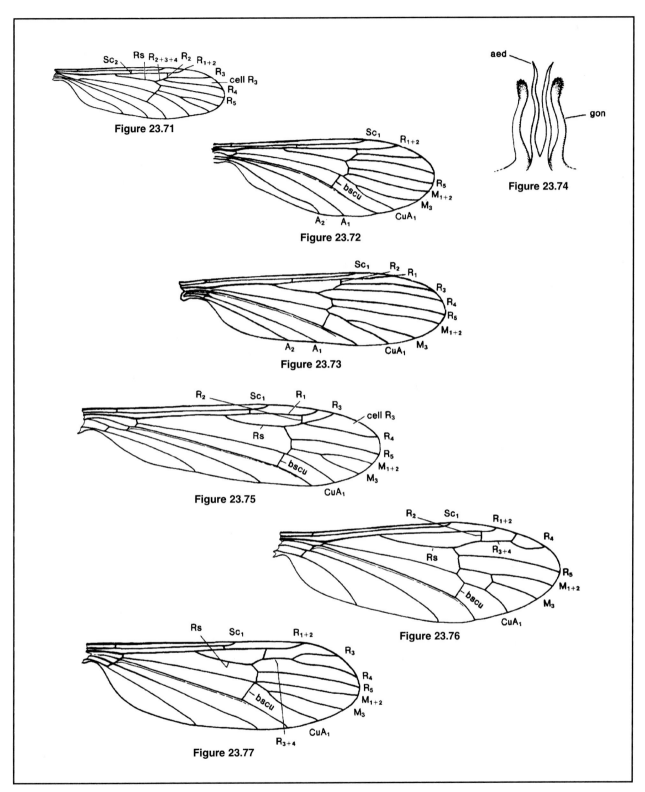

Figure 23.71. Wing of *Ormosia monticola* Alex.
Figure 23.72. Wing of *Erioptera* (*Erioptera*) *chrysocoma* O.S.
Figure 23.73. Wing of *Arctoconopa kluane* Alex.
Figure 23.74. Male aedeagus (*aed*) and gonapophyses (*gon*) of *Erioptera* (*Erioptera*) *subchlorophylla* Alex.
Figure 23.75. Wing of *Gonomyia* (*Progonomyia*) *hesperia* Alex.
Figure 23.76. Wing of *Rhabdomastix* (*Sacandaga*) *monticola* Alex.
Figure 23.77. Wing of *Cheilotrichia* (*Empeda*) *stigmatica* (O.S.).

Table 23A. Summary of ecological and distributional data for *Tipulidae (Diptera)*. (For definition of terms see Tables 6A-6C; table prepared by K. W. Cummins, R. W. Merritt, and G. W. Byers.)

Taxa (number of species in parentheses)	Habitat	Habit	Trophic Relationships	North American Distribution	Ecological§ References
Tipulidae (573+) Crane flies)	Generally lentic—littoral, lotic—erosional and depositional (detritus)	Generally burrowers (and sprawlers)	Generally shredders—detritivores, collectors—gatherers	Widespread	18, 1957, 1984, 2990, 1371, 1584, 3076, 4511
Tipulinae(53+)	Generally lotic—depositional and lenticlittoral	Generally burrowers	Generally shedders?	Widespread	18, 1296, 1957, 1984 3076, 4511
Brachypremna(1)	Lotic—depositional margins (in fine sediments of springs and small streams)	Burrowers		Eastern United States	1372, 3563
Holorusia(1)	Lotic—depositional and lentic margins (in detritus, moss, and fine sediments)	Burrowers	Shredders—detritivores	West	806
Leptotarsus (*Longurio*)(4)	Lotic—depositional (springs and small streams)	Burrowers?		Eastern United States	
Megistocera(1)	Lentic—littoral (vascular hydrophytes—floating zone)	(Form part of the neuston living in surface film in open-ended tubes of floating vegetation)	Shredders—herbivores?	Florida to eastern Texas	3425
Prionocera(16)	Lentic—littoral	Burrowers (in detritus)		Widespread	
Tipula(30+)	Lotic—erosional and depositional (detritus), lentic—littoral (detritus)	Burrowers (in detritus)	Shredders—detritivores and herbivores, collectors—gatherers, possibly some scrapers, predators (engulfers) with animals taken in the process of shredding?	Widespread	185, 576, 669, 671, 722, 1044, 1470, 1530, 1663, 2113, 2272, 2273, 2712, 2990, 3199, 3203, 3206, 1482, 1584, 3563, 3919, 4234, 1371, 806, 4057, 4110, 2077, 3675, 3870, †, ‡, 3286, 3422, 4338, 4511, *, 2671, 2673
Cylindrotominae(5)	Generally lentic and lotic—margins (in moss)	Generally burrowers—sprawlers	Generally shredders—herbivores	North	18, 24, 1228, 1957, 1984, 3076, 4511
Phalacrocera(4)	Lentic—vascular hydrophytes (moss in marsh areas)	Burrowers—sprawlers	Shredders—herbivores (chewers)	Northern United States and southern Canada	2688
Triogma(1)	Lotic and lentic—margins (in moss)	Burrowers—sprawlers (semiaquatic)		Northern United States	
Limoniinae(515+)	Generally lotic—depositional, erosional, and margins, lentic—littoral and margins	Generally—burrowers	Shredders, collectors, and predators (engulfers)	Widespread	18, 24, 1957
Antocha(7)	Lotic—erosional (in fast water on rocks and logs)	Clingers (in silk tube)	Collectors—gatherers	Widespread	669, 671, 2316, 1326, 4234, ‡
Arctoconopa(9)	Lotic—erosional and depositional, margins (clear streams)	Burrowers		Widespread	
Cryptolabis(12)	Lotic—depositional (sand in clear, cold streams)	Burrowers		Widespread	2316

* Includes some important references on related terrestrial species.
† Unpublished data, K. W. Cummins and M. J. Klug, Kellogg Biological Station.
‡ Unpublished data, K. W. Cummins, Kellogg Biological Station.
§ Emphasis on trophic relationships.

Table 23A. Continued

Taxa (number of species in parentheses)	Habitat	Habit	Trophic Relationships	North American Distribution	Ecological§ References
Dactylolabis(18)	Lotic—erosional and depositional (in moss and algal mats, especially rocky seeps)	Burrowers—clingers		Widespread	3668
Dicranota(55)	Lotic—erosional and depositional (detritus), lentic—littoral (detritus); lotic and lentic margins (in damp soil)	Sprawlers—burrowers	Predators (engulfers)	Widespread	669, 2686, 2687, 2193, 4234, ‡
Elliptera(5)	Lotic—burrowers erosional (in moss and algal mats on wet rock faces)	Burrowers—clingers		Widespread	
Erioptera(35+) (includes *Trimicra* as a subgenus)	Lotic—erosional and depositional (sand, detritus, and margins), lentic margins	Burrowers (semiaquatic)	Collectors—gatherers	Widespread	3422
Gonomyia(15+)	Lotic and lentic—margins	Burrowers (semiaquatic)		Widespread	3419, 4234
Gonomyodes(4)	Lotic—depositional	Burrowers		West	
Helius(2)	Lentic and lotic—margins (in detritus in wet woodlands or marshes)	Sprawlers—burrowers		Widespread	3422
Hesperoconopa(5)	Lotic—depositional (sand in cold, clear streams)	Burrowers		West	4234
Hexatoma(34) (includes *Eriocera* as a subgenus)	Lotic—erosional and depositional (detritus and moss) lentic—littoral (detritus)	Burrowers—sprawlers, clingers	Predators (engulfers, Oligochaeta, Diptera)	Widespread	669, 3422, 4511 3563, 4234
Limnophila(40+)	Lotic and lentic margins, lotic—depositional (in fine sediments and detritus)	Burrowers	Predators (engulfers, Oligochaeta, Diptera)	Widespread	3422, 3563, 3078
Limonia(95+) (includes *Geranomyia* as a subgenus)	Lotic and lentic—margins (on exposed objects—rocks and logs; in fine organic sediments and detritus)	Burrowers—sprawlers (semiaquatic)	Shredders—herbivores (chewers—macroalgae)	Widespread	3409, 3420, 3422, 3909
Lipsothrix(4)	Lotic—depositional (soft, saturated wood)	Burrowers	Shredders (decaying wood)	Widespread	96, 979, 3426, 981
Molophilus(15+)	Lotic and lentic margins	Burrowers (semiaquatic)		Widespread	1521, 3422, 3563
Orimarga(5)	Lotic and lentic margins (wet, soft, decaying wood)	Burrowers	Shredders?	Southern United States	
Ormosia(40+)	Lotic and lentic margins (in detritus and fine organic sediments)	Burrowers (semiaquatic)	Collectors—gatherers?	Widespread	4234
Paradelphomyia(8)	Lentic and lotic—margins (in organic sediments—wet woodlands)	Burrowers		Widespread	

‡ Unpublished data, K. W. Cummins, Kellogg Biological Station.
§ Emphasis on trophic relationships.

Table 23A. Continued

Taxa (number of species in parentheses)	Habitat	Habit	Trophic Relationships	North American Distribution	Ecological§ References
Pedicia(57)	Lotic and lentic—margins (detritus in springs, seeps, and swampy woods)	Burrowers (semi-aquatic)	Predators (engulfers)	Widespread	2430, 4234
Pilaria(12)	Lentic—littoral (wet detritus and fine sediments at margin)	Burrowers	Predators (engulfers)?	Widespread	
Polymera(2)	Lotic—depositional (wet detritus at margin)	Burrowers		Southeastern United States	
Pseudolimnophila(5)	Lentic and lotic—margins (in organic sediments—wet woodlands and marshes)	Burrowers		Widespread	
Rhabdomastix(26)	Lotic—depositional (sand in clear streams and at margin)	Burrowers		Widespread	4234
Thaumastoptera(1)	Lotic—depositional (fine sediments of small streams, springs, and seeps)	Burrowers		California	
Ulomorpha(8)	Lentic and lotic—margins (in organic sediments of wet woodlands)	Burrowers	Predators (engulfers)?	Widespread	

§ Emphasis on trophic relationships.

CULICIDAE

E. D. Walker
Michigan State University, East Lansing

H. D. Newson
Sonoma, California

INTRODUCTION

The larvae and pupae of mosquitoes are exclusively aquatic. The eggs are always associated with water, but may occur in dry areas that will eventually be flooded. Mosquito development involves metamorphosis from the egg, through the larval stage, to the motile pupa, and finally to the adult. Depending upon the species of mosquito and the way in which females oviposit, the eggs occur in soil that may be flooded, above the water-line in natural or artificial containers, or on the surface of the water. Eggs laid on the water surface often have morphological characteristics that influence buoyancy and orientation of the eggs. Oviposition behavior is highly variable among genera. *Anopheles* females generally lay eggs singly on the surface of still waters. These floating eggs may aggregate into star-shaped groups with the ends of the eggs touching each other. *Psorophora* spp. and many species of *Aedes* lay eggs in the moist soil of low-lying areas; their eggs may remain dormant for months or even years until flooded by rains or water from other sources such as irrigation or flooding. Many *Aedes* whose larvae occur in containers of water, such as tree holes, rock holes, or tires, lay their eggs on the inside of the containers above the water line, where they will be flooded after rain. Oftentimes, the eggs of those species which are laid in soil or containers are the overwintering stage. Females of the genera *Culex, Mansonia, Culiseta,* and *Uranotaenia* cluster their eggs together into rafts, which float on the surface of the water. Larvae hatch from these eggs after a brief period of embryonation; these eggs do not persist for long periods in the unhatched condition.

After the eggs hatch, larvae (first instars) emerge and feed. The larvae molt and develop through three additional instars, thus mosquitoes have four instars during larval development. The rate of larval development is temperature-dependent, and may also be affected by other density-independent or density-dependent factors such as crowding or food limitation. Many species of mosquitoes complete larval development in 7-10 days if conditions are favorable, but others may require weeks to several months, depending upon life history. Mosquito larvae are active swimmers, showing a characteristic wriggling motion as they move about underwater or at the water surface. Larvae that are disturbed by agitation in the water or shadows above them show an alarm reaction, in which they quickly swim away from the source of the agitation and often down into the water column. The final larval molt produces the pupa, a nonfeeding stage that usually floats directly underneath the water surface. When disturbed by mechanical agitation or sudden changes in light intensity, the pupa also shows an alarm reaction, and dives with a tumbling motion into the water column. Sometimes after diving, a pupa will remain still at the bottom of its habitat before returning by swimming or passively floating to the surface. The buoyancy of the pupa results from an air cavity between the wings of the adult developing within the pupal exoskeleton. The pupal stage usually lasts only three or four days, but in some species may persist for two weeks or more. At the end of this stage, the pupa extends its abdomen nearly parallel to the surface of the water prior to adult emergence. The adult mosquito ingests air that is present between the pupal exoskeleton and the pharate adult exoskeleton, thus creating pressure that splits the dorsal cephalothorax of the pupa and allows the adult to ecdyse passively from the pupal skin. After emerging, the adult uses the cast pupal skin as a float until its body and wings dry and harden.

All mosquito larvae, except members of the genus *Mansonia,* come to the water surface at frequent intervals to obtain oxygen. Except when feeding or disturbed, they usually are found suspended in various positions from the water surface. The respiratory apparatus is located posteriorly and dorsally as paired spiracles either on the eighth abdominal segment (*Anopheles* larvae) or at the end of a siphon in other genera. When larvae (except *Mansonia*) are breathing, the spiracles must be in the plane of the water surface. *Mansonia* larvae attach their modified siphon into the aerenchymous tissue of submerged roots of emergent plants (such as cattails and arrowhead) and obtain their oxygen from these plant tissues. Under conditions of sufficient oxygen concentration, larvae of some species may respire cutaneously. Pupal mosquitoes have a pair of large respiratory trumpets located dorsally on the cephalothorax which enable pupae to exchange air from the water surface. In pupae of *Mansonia,* the trumpets are modified for penetration and attachment to submerged vegetation.

Larval feeding behavior is variable among the different species; most larvae eat microorganisms, detritus, zooplankton, and even inert particles from the water column by filtering, gathering, and collecting. Larvae of some species may collect similar materials from surfaces such as leaves by brushing along these surfaces, and may gnaw at substrates as well. Larvae may move about the habitat to forage different areas for food, or may station themselves in one zone of the water column for feeding. Larvae of the genus *Toxorhynchites*, and some species of *Psorophora*, are predaceous and often feed upon other species of mosquito larvae. The mouthparts of other mosquito larvae are highly complex and consist of setae and spicules configured into brushes, combs and sweepers, often referred to as mouth brushes. The *lateral palatal brushes* are conspicuous collecting organs borne on the *labrum*. The individual *setae* or *filaments* of these brushes extend and flex rapidly. The combined action of the lateral palatal brushes, pumping of the pharynx, and other mouthpart movements creates a system of flows and counterflows that entrain particles of food and draw them toward the head; this entire process has been termed suspension feeding. These entrained particles concentrate into the pharynx, where a food bolus is formed and swallowed at regular intervals. The length and dimensions of the lateral palatal brushes affect the distance in the water from which particles may be gathered. Larvae with longer lateral palatal brushes may entrain particles from as far as 40 mm away from the larval head. Larval mosquitoes ingest a range of particles of different sizes, with older instars eating larger particles. In general, uniform particle sizes range from less than 1 µm to 50 µm, however larger particles of nonuniform dimensions and flexible texture (such as filamentous algae) can be ingested. Larvae may also reduce large particles down to smaller, ingestible size by mastication. Additionally, larvae may ingest colloidal materials, and dissolved matter possibly by drinking. Estimates of filtration rates of water have ranged to as high as 0.69 ml per hour by individual fourth instars.

The variation in larval habitats of the different species of mosquitoes, and the different schemes proposed to classify them, have been reviewed by Laird (1988), who proposed a standard classification consisting of above-ground waters, and subterranean waters (both natural and artificial). The above-ground waters were subdivided into the following groups, based upon morphology of the aquatic system: (1) flowing streams; (2) ponded streams; (3) lake edges; (4) swamps and marshes; (5) shallow, permanent ponds; (6) shallow, temporary pools; (7) intermittent, ephemeral puddles; (8) natural containers; and (9) artificial containers. The term "phytotelmata" has been applied to those larval habitats occurring naturally in plants, such as leaf axils, bamboo internodes, coconut husks, tree holes, and pitcher plants. There has apparently been no attempt to link quantitative, wetland classification systems such as that of Cowardin *et al.* (1979) to classification of larval mosquito habitats.

Adult, female mosquitoes are well known for their blood feeding habits and as vectors of a great number of human and animal pathogens. Females of most mosquito species require a blood meal for egg development and are termed *anautogenous*. Some females are *autogenous*, that is, they can produce eggs in at least their first gonotrophic cycle without having had a previous blood meal, apparently carrying over sufficient nutrients from the larval stage. Both male and female mosquitoes obtain carbohydrates by feeding on plant juices and nectar. The females of a great majority of mosquito species are zoophilous, that is, they blood feed in nature on vertebrate animals other than humans. Feeding preferences vary from one species to another and may include cold-blooded vertebrates, birds, and a variety of mammals, including humans. Although some species are host-group specific in their blood feeding, feeding on only birds, mammals, or amphibians, others are much less discriminating in their choice of hosts.

Longevity of adult mosquitoes varies greatly in nature. Male mosquitoes are thought to be less long-lived than females, although some studies based upon analysis of mark-recapture data show males to be as long-lived as females. Field studies have also shown that females of some species survive for three or more months but the usual life span of most is from two to four weeks. In some species (many *Culex* and *Anopheles*), the inseminated female is the overwintering stage in North America, and these are very long-lived. Both longevity and blood feeding preferences figure critically in the potential of mosquitoes to transmit pathogens of humans or animals.

There is considerable variation in the mating behavior of mosquitoes. For most species, mating occurs when males form a crepuscular (twilight) swarm and the females fly into it. There is evidence to indicate that the sounds formed by the swarming males is species specific and is a major factor in attracting females to the swarm. Some species mate readily within the confines of underground animal burrows, others while swarming in a small space, and still others have larger space requirements and will mate only while swarming several meters above ground. A few species mate after male courtship behaviors are displayed. Reviews of the natural history, physiology, ecology, behavior, or sampling of mosquitoes include Bates (1949), Carpenter and LaCasse (1955), Christophers (1960), Clements (1963, 1992), Gillett (1972), Wood *et al.* (1979), Frank and Lounibos (1983), Lounibos *et al.* (1985), Laird (1988), Merritt *et al.* (1992), and Service (1992).

EXTERNAL MORPHOLOGY

Larvae

Mosquito larvae have three well-differentiated body regions: the *head*, *thorax*, and *abdomen*. Each of these has structures that are used in classification and identification (Fig. 24.1). The head, which is flattened dorsoventrally, is formed of three large sclerites. The mouthparts are ventral, but some portions extend anteriorly and are visible in a dorsal view (Figs. 24.6-24.7). The thorax is composed of the fused pro-, meso-, and metathorax, distinguished only by the hair group present on each segment (Fig. 24.1). The first seven of the nine readily visible abdominal segments are somewhat similar and unmodified. The eighth segment bears the breathing apparatus posterodorsally (Fig. 24.5) and the various modifications of this structure provide useful taxonomic characters. The terminal segment also has several structures of taxonomic importance (Fig. 24.1). Larvae have two posterior *spiracles* situated in a spiracular apparatus located at the tip of the *siphon*, an elongate and sclerotized tube that is part of the eighth abdominal segment. Anopheline larvae lack this siphon and instead have a spiracular apparatus similar to that of dixid larvae.

Pupae

Mosquito pupae have only two clearly defined body regions: an enlarged anterior *cephalothorax* and a segmented abdomen (Fig. 24.2). Two respiratory structures, the *trumpets*, are located on the dorsal portion of the cephalothorax. The abdomen is slender and terminates in a pair of flattened *pad-*

dles. Variations in the *chaetotaxy* (arrangement of setae), the paddles, and the trumpets are important diagnostic pupal features.

Adults

The body of an adult mosquito is comprised of three distinct regions, each with important characters that can be used in identification and classification (Fig. 24.3). The large compound eyes occupy most of the lateral part of the head and the remaining portions are covered with scales of various shapes and colors. Mouthparts are elongate and highly modified for piercing and ingesting liquid foods. Long, slender, 15-segmented antennae arise between the eyes and a five-segmented *maxillary palpus* is located on each side of the *proboscis*. The thorax is composed of three fused segments, the *pro-*, *meso-*, and *metathorax*. The prothorax bears the front pair of legs; the mesothorax, the wings and middle pair of legs; and the metathorax, the halteres and hind pair of legs. The thoracic segments and legs are covered with various types of setae and colored *scales* that provide useful taxonomic characters. Indeed, the Culicidae are one of the few insect taxa outside of the Lepidoptera that have true scales. The wing veins and margins also are densely covered with scales of various shapes and colors. These are diagnostic features for some species, as are the configurations of certain wing veins.

For additional information on morphology of mosquitoes, the reader is referred to Harbach and Knight (1980).

KEYS TO THE GENERA OF CULICIDAE

These keys were designed to include only the species in North America that occur north of Mexico. Characteristics given in the adult key are suitable for identifying both males and females to genus. Carpenter and LaCasse (1955) provided complete descriptions, figures, and keys for the larvae and adults of all species then known to occur in North America. Since then one additional genus, *Haemagogus*, represented by only one species, has been found to occur in the extreme southern part of Texas (Breland 1958). Darsie and Ward (1981, 1989) provided identification keys and the known distribution ranges of all species currently reported from America north of Mexico, with additional corrections and additions given by Ward and Darsie (1982), Moore *et al.* (1988), Barr and Guptavanij (1989), Darsie (1992), and Seawright *et al.* (1992).

The genus *Coquillettidia* was elevated from subgeneric rank from within *Mansonia*, but Wood *et al.* (1979) considered this change ill-advised and retained *Mansonia* for the single species in North America that Darsie and Ward (1981) and others refer to as *Coquillettidia perturbans*. This species is widespread in North America. Here, we use *Mansonia* as the genus for this species, following the argument given by Wood *et al.* (1979). The other two species of *Mansonia* in North America, *M. titillans* and *M. dyari*, occur mainly in certain Gulf Coast states.

The key for pupae utilizes chaetotaxic characteristics to a major extent. Generally, pupal identification is more difficult than adult or larval identification for that reason. The setal numbering system is that of Mattingly (1973) in which the individual setae on the cephalothorax and each abdominal segment are designated by sequential Arabic numbers, and abdominal segments are numbered from anterior to posterior using Roman numerals I through X. Thus, in the key, 9-C represents seta number 9 on the cephalothorax and 9-VII represents seta number 9 on the seventh abdominal segment. Segments I through VIII are distinct and clearly separated, whereas segments IX and X appear more as lobelike appendages.

Larvae (Fourth Instar)

1.	Eighth abdominal segment without an elongate dorsal siphon (Fig. 24.4); abdomen with palmate setae (Fig. 24.4)	***Anopheles***
1'.	Eighth abdominal segment with an elongate dorsal siphon that is as long as or longer than wide (Fig. 24.5)	2
2(1').	Lateral palatal brushes prehensile (adapted for grasping), each composed of about 10 stout, curved rods (Fig. 24.6); larvae very large (up to 15 mm long) and predaceous	***Toxorhynchites***
2'.	Lateral palatal brushes composed of 30 or more hairs (Fig. 24.7)	3
3(2').	Distal half of siphon strongly attenuated (tapered), adapted for piercing roots of aquatic plants (Fig. 24.8); larvae normally attached to submerged vegetation, rarely at water surface (includes *Coquillettidia* of some authors)	***Mansonia***
3'.	Siphon cylindrical or fusiform (spindle-shaped) (Fig. 24.9); not adapted for piercing aquatic plants	4
4(3').	Siphon with a pecten (Fig. 24.10)	5
4'.	Siphon without a pecten (Fig. 24.11)	11
5(4).	Head longer than wide (Fig. 24.12); 8th abdominal segment with a sclerotized comb plate bearing the comb scales on its posterior border (Fig. 24.13)	***Uranotaenia***
5'.	Head at least as wide as long (Fig. 24.14); 8th abdominal segment without a comb plate (Fig. 24.15) (small plate present in some *Psorophora* species)	6
6(5').	Head with prominent triangular pouch on each side (Fig. 24.16); anal segment with divided dorsal and ventral sclerotic plates, membranous laterally (Fig. 24.17)	***Deinocerites***
6'.	Head without a prominent triangular pouch on each side (Fig. 24.18); anal segment without divided dorsal and ventral sclerotic plates (Fig. 24.19)	7

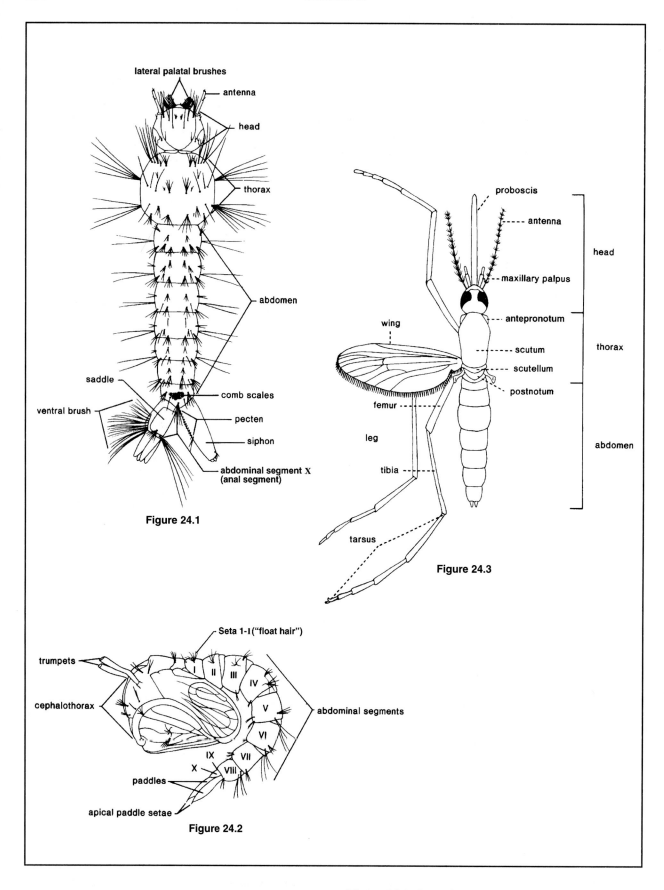

Figure 24.1. Dorsal view of mosquito larva.
Figure 24.2. Lateral view of mosquito pupa.
Figure 24.3. Dorsal view of mosquito adult.

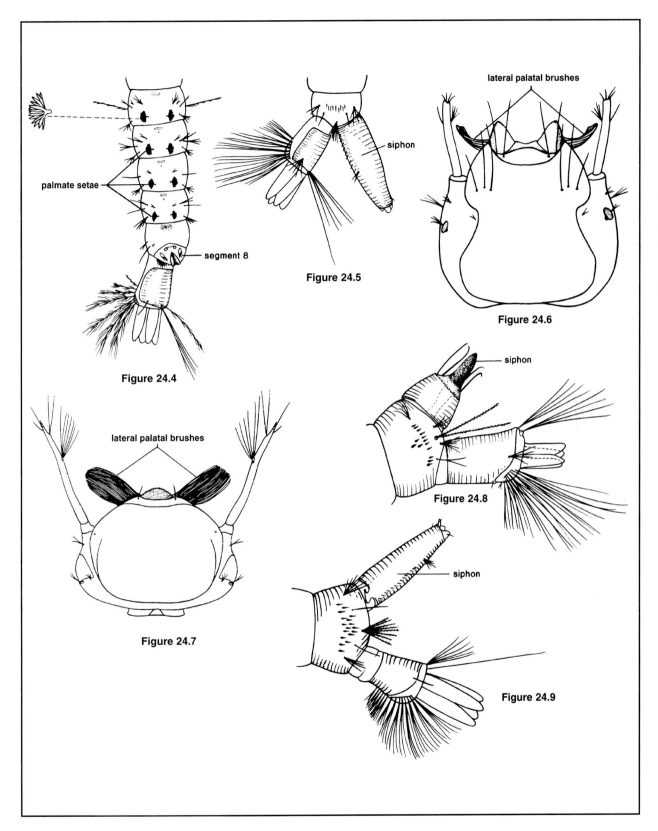

Figure 24.4. Dorsal view of *Anopheles* sp. abdomen.

Figure 24.5. Terminal abdominal segments of typical culicine larva.

Figure 24.6. Dorsal view of *Toxorhynchites* sp. mouth brushes.

Figure 24.7. Dorsal view of typical culicine mouth brushes.

Figure 24.8. Lateral view of *Mansonia* sp. siphon.

Figure 24.9. Lateral view of typically shaped culicine siphon.

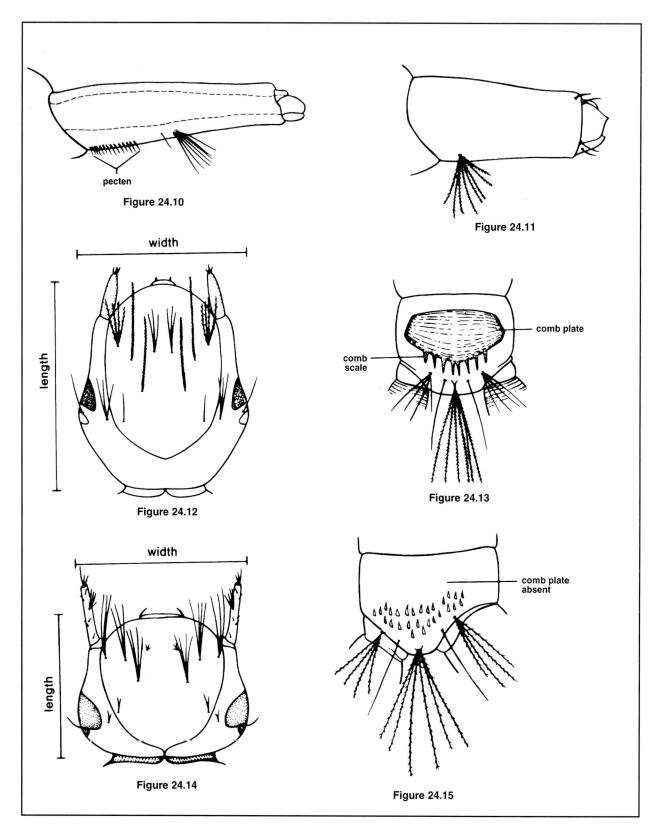

Figure 24.10. Lateral view of culicine siphon.
Figure 24.11. Lateral view of culicine siphon.
Figure 24.12. Dorsal view of *Uranotaenia* sp. head.
Figure 24.13. Lateral view of eighth abdominal segment of *Uranotaenia* sp.
Figure 24.14. Dorsal view of typical culicine larval head.
Figure 24.15. Lateral view of culicine eighth abdominal segment.

7(6').	Siphon with a pair of large basal tufts (Fig. 24.20); or, if tufts small, the comb scales are in a single row and have a barlike arrangement ....***Culiseta***
7'.	Siphon without a pair of basal tufts (Fig. 24.21) 8
8(7').	Siphon with several pairs of siphonal tufts or single hairs (Fig. 24.22) ....***Culex***
8'.	Siphon with one pair of median or subapical siphonal tufts (sometimes vestigial), or one pair of single hairs (Fig. 24.23) 9
9(8').	Anal segment completely ringed by the dorsal saddle and pierced on the midventral line by tufts of the ventral brush (Fig. 24.24) ....***Psorophora***
9'.	Anal segment not completely ringed by the dorsal saddle (Fig. 24.25); or, if ringed, not pierced on the midventral line by tufts of the ventral brush (Fig. 24.26) 10
10(9').	Dorsal saddle with a prominent group of long spines on posterior margin; comb scales in a single row (Fig. 24.27) ....***Haemagogus***
10'.	Dorsal saddle without a prominent group of long spines on the posterior margin (Fig. 24.28); or, if long spines are present on the posterior margin, comb scales are in a patch (Fig. 24.29) ....***Aedes***
11(4').	Anal segment with a prominent median ventral brush consisting of a close-set row of tufts; comb with a double row of pointed scales (Fig. 24.30) ....***Orthopodomyia***
11'.	Anal segment without a median ventral brush, but with a pair of ventrolateral tufts; comb scales arranged in a single row (Fig. 24.31) ....***Wyeomyia***

Pupae

1.	Segment X with a conspicuous branched hair; seta 9 on segment VIII greatly reduced; paddles without apical seta (Fig. 24.32) ....***Toxorhynchites***
1'.	Segment X without branched hair; seta 9-VIII and paddles variable 2
2(1').	Paddles nearly always with an accessory seta arising anterior to, and in line with, the apical seta; if not, then seta 9 on segments IV-VII in the form of a short, stout, dark spine arising from the extreme posterior corner of the segment (Fig. 24.33) and trumpets short, flared, and split nearly to base ....***Anopheles***
2'.	Paddles with accessory seta absent; or, if present, arising level with and laterad of the apical seta; seta 9 and trumpets usually different from above 3
3(2').	Trumpets modified for insertion into aquatic plant tissues (Fig. 24.34); float hairs suppressed ....***Mansonia***
	(includes *Coquillettidia* of some authors)
3'.	Trumpets otherwise, and/or float hair well developed (Fig. 24.35) 4
4(3').	Paddles small, somewhat pointed, and without apical seta (Fig. 24.36) ....***Wyeomyia***
4'.	Paddles not pointed, apical seta present (Fig. 24.37) 5
5(4').	Apex of paddle convex with both borders smooth; apical seta on paddle at least two-thirds as long as paddle; seta 9-VIII long, single, and simple (Fig. 24.38) ....***Deinocerites***
5'.	Without this combination of characters 6
6(5').	Paddles with inner half deeply excavated toward base and usually much broader on outer half; segment IX usually with a pair of small setulae (stiff bristles); paddles fringed or toothed on both borders (Fig. 24.39) ....***Uranotaenia***
6'.	Without this combination of characters 7
7(6').	Paddles somewhat rectangular, with the thickened basal portion of outer edge sometimes spiculate (having minute, pointed spines), but remaining otherwise smooth and hyaline (Fig. 24.40); accessory paddle seta absent; apical paddle seta very short; seta 9 on segments VI-VIII long, stout, plumose, and on segment VIII about half or more the length of paddle (Fig. 24.40) ....***Orthopodomyia***
7'.	Without this combination of characters 8
8(7').	Trumpets tubular for most of their length, either with or without rudimentary basal tracheation (Fig. 24.41); seta 9-VIII arising from the posterolateral corner of the segment with little or no anterior displacement (Fig. 24.42); and, with one of the following characteristics: posterior corner of abdominal segment IV toothed (Fig. 24.43); prominent ventral lobes present on posterior border of segment VIII, partly covering the bases of the paddles (Fig. 24.42); or, accessory paddle seta present (Fig. 24.42) ....***Psorophora***
8'.	Without this combination of characters 9
9(8').	Seta 8-C arising laterad to, and approximately even with, the base of the trumpet, but very much anterior to 9-C (Fig. 24.44); tracheation of trumpet absent or, at most, rudimentary (Fig. 24.44); seta 9-VIII rarely arising cephalad of the posterior border of the segment (Fig. 24.45) 10

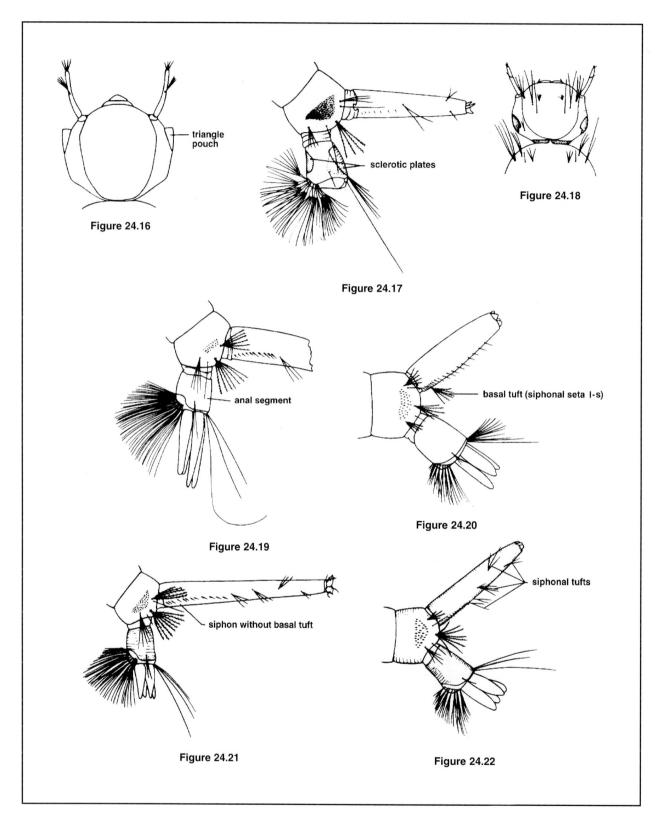

Figure 24.16. Dorsal view of *Deinocerites* sp. head.
Figure 24.17. Lateral view of anal segment of *Deinocerites* sp.
Figure 24.18. Dorsal view of culicine head.
Figure 24.19. Lateral view of culicine anal segment.
Figure 24.20. Lateral view of *Culiseta* sp. siphon and anal segment.
Figure 24.21. Lateral view of culicine siphon and anal segment.
Figure 24.22. Lateral view of culicine siphon and anal segment.

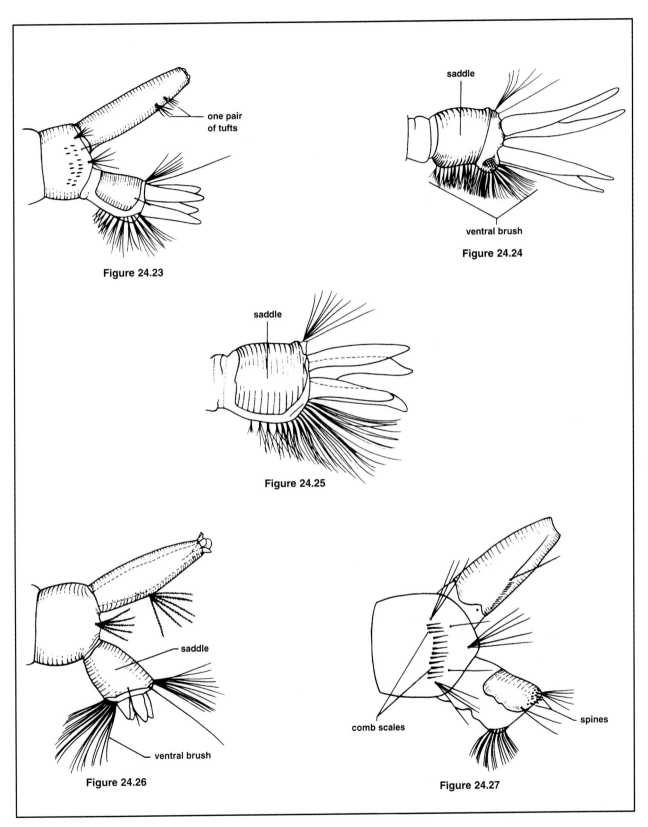

Figure 24.23. Lateral view of culicine siphon and anal segment.

Figure 24.24. Lateral view of *Psorphora* sp. anal segment.

Figure 24.25. Lateral view of culicine anal segment.

Figure 24.26. Lateral view of culicine anal segment and siphon.

Figure 24.27. Lateral view of terminal abdominal segments of *Haemogogus* sp. showing dorsal saddle and comb scales.

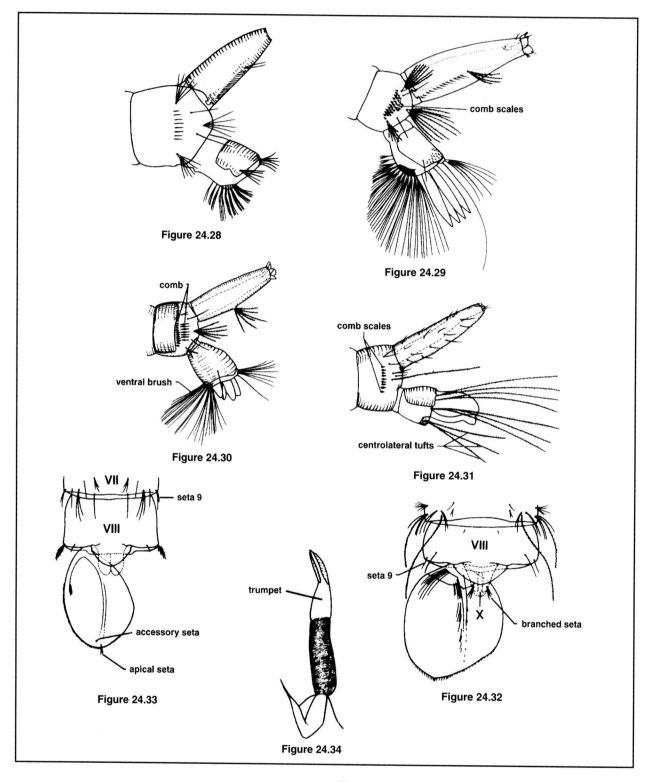

Figure 24.28. Lateral view of terminal abdominal segments of *Aedes* sp. showing dorsal saddle.

Figure 24.29. Lateral view of terminal abdominal segments of *Aedes* sp. showing comb scales.

Figure 24.30. Lateral view of terminal abdominal segments of *Orthopodomyia* sp. showing ventral brush and comb scales.

Figure 24.31. Lateral view of terminal abdominal segments of *Wyeomyia* sp. showing anal segment and comb scales.

Figure 24.32. Dorsal view of terminal abdominal segments of *Toxorhynchites* sp. showing paddle and terminal segments.

Figure 24.33. Dorsal view of paddle and terminal segments of *Anopheles* sp.

Figure 24.34. *Mansonia* sp. trumpet.

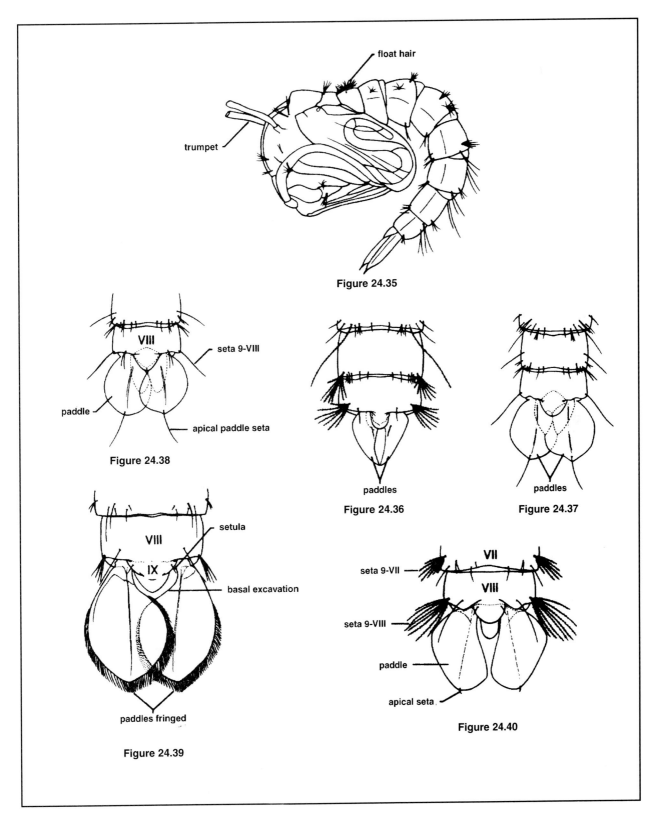

Figure 24.35. Lateral view of typical culicine pupa showing trumpets and float hairs.
Figure 24.36. Dorsal view of *Wyeomyia* sp. paddles.
Figure 24.37. Dorsal view of typical culicine paddles.
Figure 24.38. Dorsal view of *Deinocerites* sp. paddles and terminal segments.
Figure 24.39. Dorsal view of *Uranotaenia* sp. paddles and terminal segments.
Figure 24.40. Dorsal view of *Orthopodomyia* sp. paddles and terminal segments.

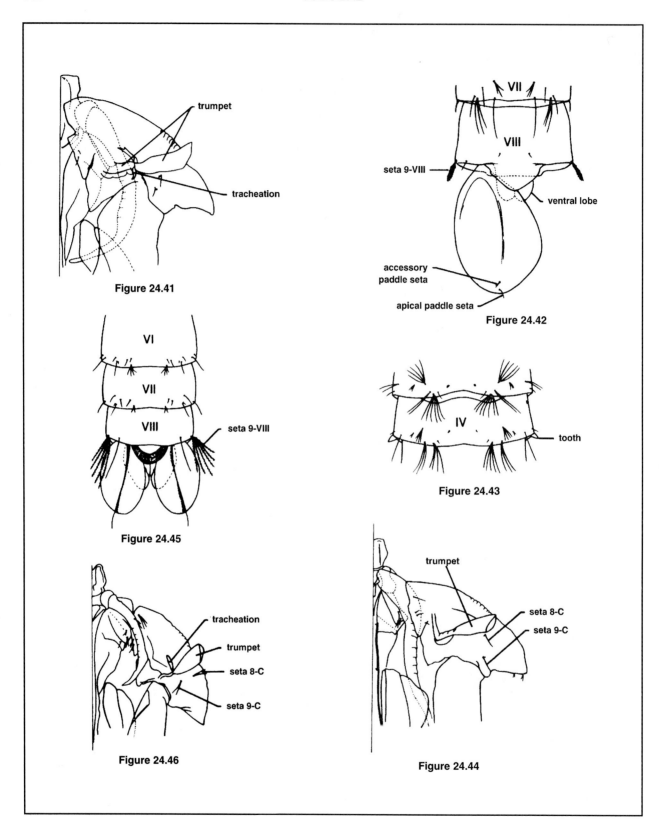

Figure 24.41. Lateral view of *Psorophora* sp. trumpets.
Figure 24.42. Dorsal view of *Psorophora* sp. paddles and terminal segments.
Figure 24.43. Dorsal view of abdominal segment IV of *Psorophora* sp.
Figure 24.44. Lateral view of culicine trumpet.
Figure 24.45. Dorsal view of culicine pupal terminal abdominal segments.
Figure 24.46. Lateral view of culicine trumpet.

9'.	Seta 8-C arising even with, or posterior to, the base of the trumpet and more nearly level with 9-C (Fig. 24.46); trumpets frequently with extensive subbasal tracheation (Fig. 24.46); position of seta 9-VIII variable 11
10(9).	Setae 8-C and 9-C poorly developed (Fig. 24.47); setae 5-II and 5-III not extending to, or barely reaching, the next posterior abdominal segment; *either* with seta 5-VIII as long as or longer than segment VI, *or* with midrib of paddle deeply pigmented, and seta 5 on segments IV-VI each shorter than the following segment; seta 9-VIII with 4 or more branches as long as or longer than one-half the paddle length (Fig. 24.48) ....*Haemagogus*
10'.	Without this combination of characters ....*Aedes*
11(9').	Trumpets with well-developed subbasal tracheation (Fig. 24.49), or seta 9-VIII arising well cephalad of the posterior border of the segment (Fig. 24.50) ....*Culex*
11'.	Trumpets with, at most, rudimentary basal tracheation (Fig. 24.51); seta 9-VIII always arising from the posterior border of the segment (Fig. 24.52) ....*Culiseta*

Adults

1.	Proboscis rigid and stout on basal half, apical half tapered toward apex and strongly curved downward (Fig. 24.53). Large species (up to 15 mm long), with broad, metallic colored scales on head, thorax, abdomen, and legs ....*Toxorhynchites*
1'.	Proboscis slender, never strongly curved downward on apical half, and of nearly uniform thickness (Fig. 24.54) 2
2(1').	Scutellum evenly rounded on posterior margin (Fig. 24.55); palpi of female nearly as long as proboscis (Fig. 24.56); abdomen bare of scales or only sparsely scaled ....*Anopheles*
2'.	Scutellum trilobed on posterior margin (Fig. 24.57); palpi less than one-half as long as proboscis (Fig. 24.58); abdomen densely scaled both dorsally and ventrally 3
3(2').	Second marginal cell of wing short, less than half as long as its petiole (Fig. 24.59) ....*Uranotaenia*
3'.	Second marginal cell of wing as long as or longer than its petiole (Fig. 24.60) 4
4(3').	Postnotum with a tuft of setae (Fig. 24.61); wing squama without fringe of hair (Fig. 24.62) ....*Wyeomyia*
4'.	Postnotum bare (Fig. 24.63); wing squama with fringe of hair (Fig. 24.64) 5
5(4').	Spiracular bristles present (Fig. 24.65) 6
5'.	Spiracular bristles absent (Fig. 24.66) 7
6(5).	Postspiracular bristles present (Fig. 24.67); tip of abdomen pointed (Fig. 24.68) ....*Psorophora*
6'.	Postspiracular bristles absent (Fig. 24.69); tip of abdomen blunt (Fig. 24.70) ....*Culiseta*
7(5').	Anterior pronotal lobes large, collarlike, almost joining dorsally (Fig. 24.71); abdominal scales bright metallic violet or silver ....*Haemagogus*
7'.	Anterior pronotal lobes small, widely separated dorsally (Fig. 24.72); abdomen without bright metallic scales 8
8(7').	Postspiracular bristles present (Fig. 24.67) 9
8'.	Postspiracular bristles absent (Fig. 24.69) 10
9(8).	Wing scales very broad, mixed brown and white (Fig. 24.73); tip of abdomen blunt (Fig. 24.70) ....*Mansonia* (in part)
9'.	Wing scales narrow (rarely moderately broad) (Fig. 24.74); tip of abdomen pointed (Fig. 24.68) ....*Aedes*
10(8').	Antenna much longer than proboscis, 1st flagellar segment longer than the next 2 segments combined (Fig. 24.75) ....*Deinocerites*
10'.	Antenna not longer than proboscis, or only slightly so, 1st flagellar segment about as long as each succeeding segment (Fig. 24.76) 11
11(10').	Scutum bicolored with narrow longitudinal lines of white scales (Fig. 24.77); penultimate (next to the last) segment of front tarsi very short, only about half as long as wide (Fig. 24.78) ....*Orthopodomyia*
11'.	Scutum without longitudinal lines of white scales (Fig. 24.79); penultimate segment of front tarsi much longer than wide (Fig. 24.80) 12
12(11').	Wing scales very broad, mixed brown and white (Fig. 24.73) ....*Mansonia* (in part—includes *Coquillettidia* of some authors)
12'.	Wing scales narrow and uniformly dark (Fig. 24.74) ....*Culex*

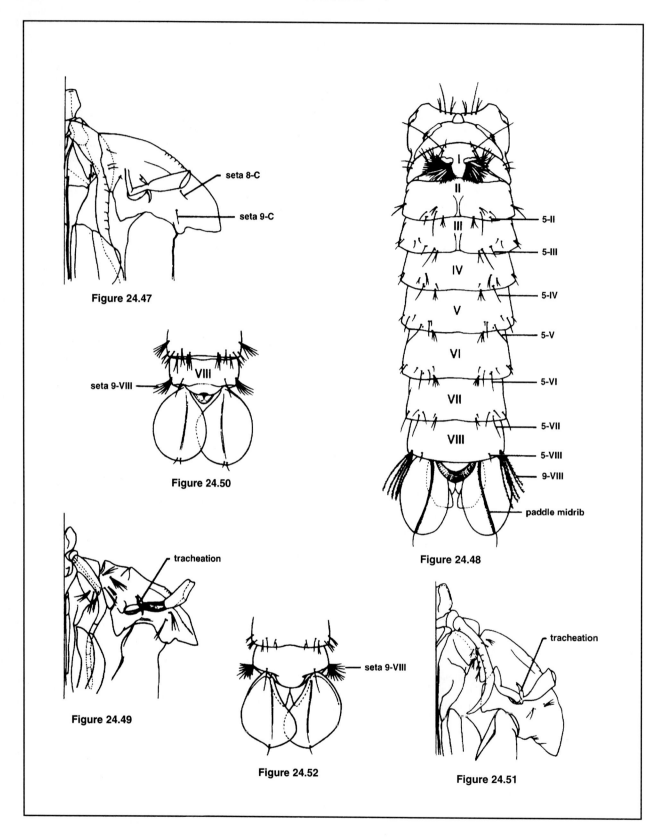

Figure 24.47. Lateral view of culicine trumpet.
Figure 24.48. Dorsal view of *Haemogogus* sp. paddles and abdominal segments.
Figure 24.49. Lateral view of *Culex* sp. trumpet.
Figure 24.50. Dorsal view of segment VIII and paddles of *Culex*.
Figure 24.51. Lateral view of *Culiseta* sp. trumpet.
Figure 24.52. Dorsal view of segment VIII and paddles of *Culiseta* sp.

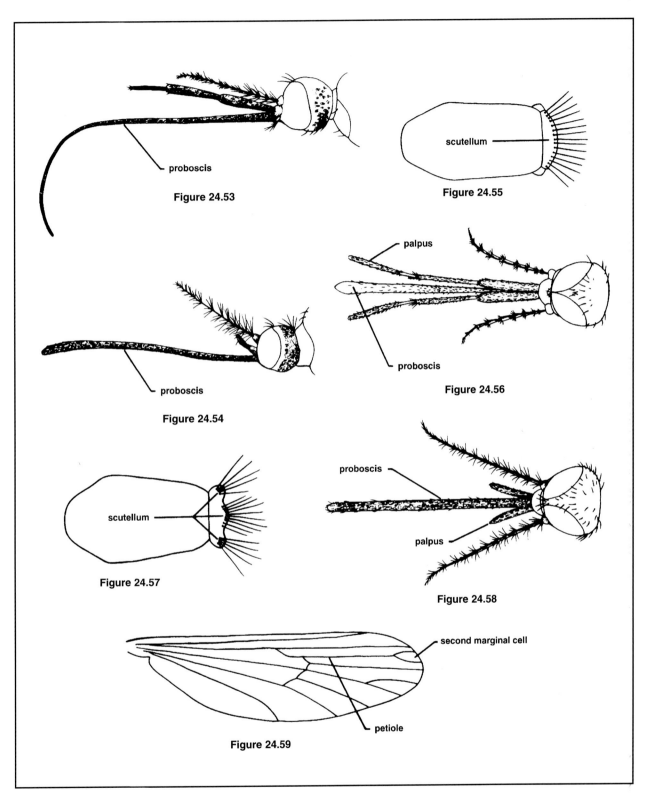

Figure 24.53. Lateral view of head showing *Toxorhynchites* sp. proboscis.

Figure 24.54. Lateral view of head showing typical culicine proboscis.

Figure 24.55. Dorsal view of thorax showing *Anopheles* sp. scutellum.

Figure 24.56. Dorsal view of head showing *Anopheles* sp. palpi.

Figure 24.57. Dorsal view of thorax showing typical culicine scutellum.

Figure 24.58. Dorsal view of head showing typical culicine palpi.

Figure 24.59. *Uranotaenia* sp. wing.

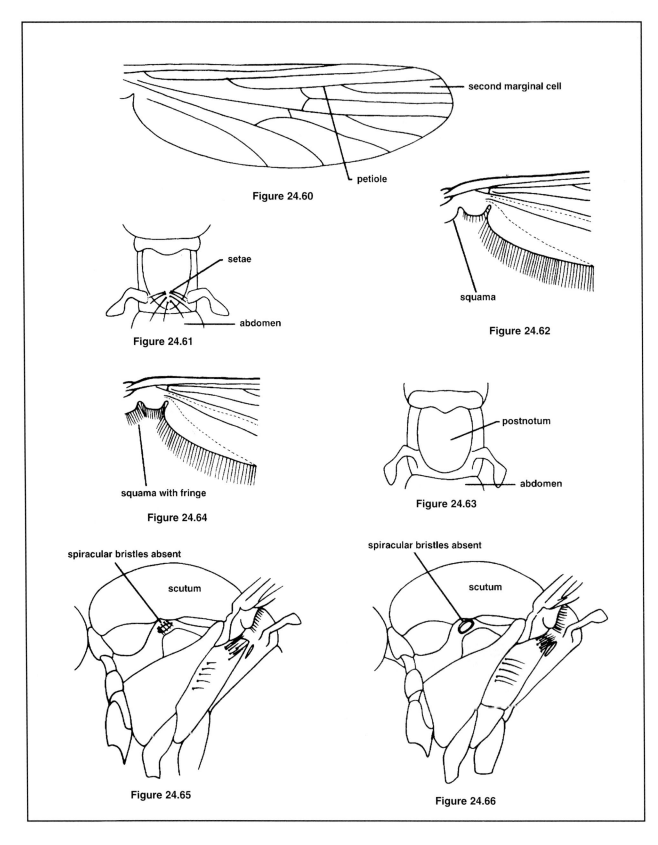

Figure 24.60. Typical culicine wing.
Figure 24.61. Dorsal view of *Wyeomyia* sp. postnotum.
Figure 24.62. *Wyeomyia* sp. squama.
Figure 24.63. Dorsal view of typical culicine postnotum.
Figure 24.64. Typical culicine squama.
Figure 24.65. Lateral view of culicine thorax.
Figure 24.66. Lateral view of culicine thorax.

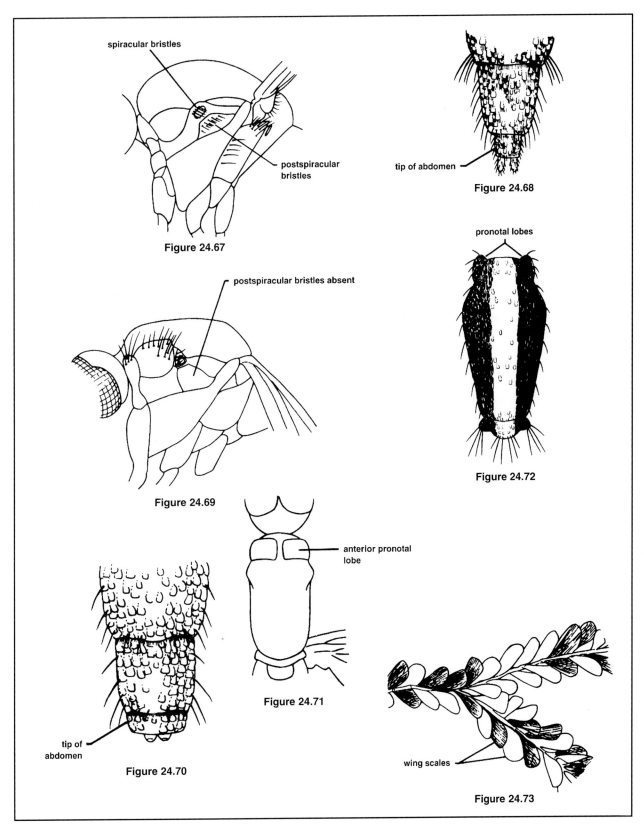

Figure 24.67. Lateral view of culicine thorax.
Figure 24.68. Dorsal view of pointed culicine abdomen.
Figure 24.69. Lateral view of culicine thorax.
Figure 24.70. Dorsal view of blunt culicine abdomen.
Figure 24.71. Dorsal view of pronotal lobes of *Haemogogus* sp.
Figure 24.72. Dorsal view of culicine thorax.
Figure 24.73. Enlarged view of *Mansonia* sp. broad wing scale, mixed brown and white.

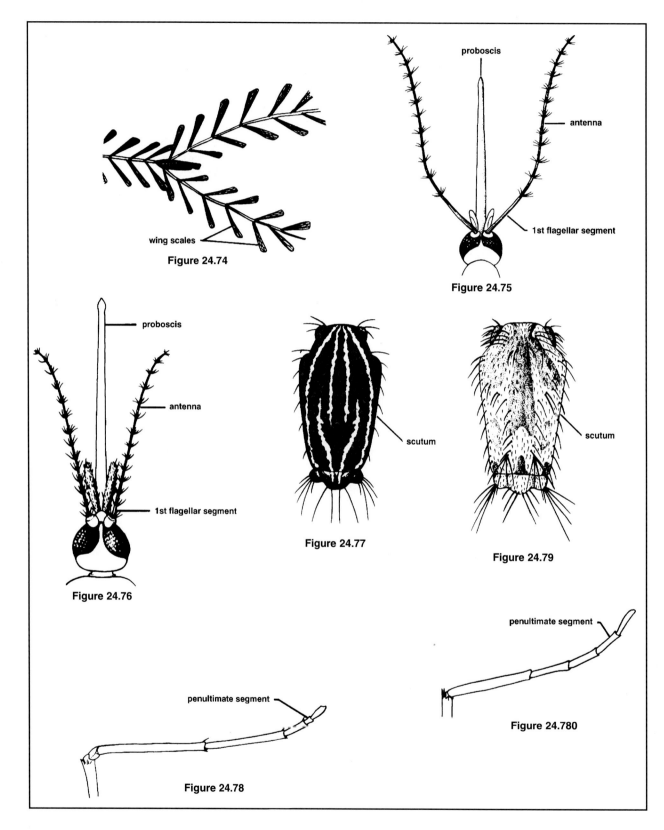

Figure 24.74. Enlarged view of culicine narrow wing scales.
Figure 24.75. Dorsal view of head showing *Deinocerites* sp. antenna.
Figure 24.76. Dorsal view of culicine head showing typical mosquito antenna.
Figure 24.77. Dorsal view of *Orthopodomyia* sp. scutum.
Figure 24.78. *Orthopodomyia* sp. tarsus.
Figure 24.79. Dorsal view of culicine scutum.
Figure 24.80. Typical culicine front tarsus.

ADDITIONAL TAXONOMIC REFERENCES

General
Carpenter and LaCasse (1955); Breland (1958); Darsie and Ward (1981, 1989); Stone (1981).

Regional faunas
Alaska: Gjullin *et al.* (1961).
Arizona: McDonald *et al.* (1973).
Arkansas: Carpenter (1941).
British Columbia: Curtis (1967).
California: Bohart and Washino (1978); Meyer and Durso (1993).
Canada: Wood *et al.* (1979).
Colorado: Harmston and Lawson (1967).
Florida: Breeland and Loyless (1982).
Illinois: Ross and Horsfall (1965).
Indiana: Siverly (1972).
Iowa: Knight and Wonio (1969).
Michigan: Wilmot *et al.* (1990).
Minnesota: Barr (1958).
Montana: Quickenden and Jamison (1979).
New Jersey: Headlee (1945).
New York: Means (1979, 1987).
North Carolina: Slaff and Apperson (1989).
Northwestern United States: Gjullin and Eddy (1972).
Oklahoma: Rozeboom (1942).
Southeastern United States: King *et al.* (1960).
Texas: Randolph and O'Neill (1944).
Utah: Nielsen and Rees (1961).
Virginia: Gladney and Turner (1969).
Wisconsin: Dickinson (1944).

Regional species lists
Colorado: Harmston (1949).
Delaware: Darsie *et al.* (1951).
District of Columbia: Good (1945).
Idaho: Brothers (1971).
Kentucky: Quinby *et al.* (1944).
Manitoba: Trimble (1972).
Maryland: Bickley *et al.* (1971).
Michigan: Cassani and Newson (1980).
Mississippi: Harden *et al.* (1967).
Missouri: Smith and Enns (1968).
New Mexico: Sublette and Sublette (1970).
New York: Jamnback (1969).
North Dakota: Darsie and Anderson (1985).
Northeastern United States: Stojanovich (1961).
Ohio: Parsons *et al.* (1972).
Oklahoma: Parsons and Howell (1971).
Ontario: James *et al.* (1969).
Pennsylvania: Wilson *et al.* (1946).
Southeastern United States: Stojanovich (1961).
Texas: Sublette and Sublette (1970).
Utah: Nielsen (1968).
Vermont: Graham *et al.* (1991).
Virginia: Dorer *et al.* (1944).
West Virginia: Amrine and Butler (1978); Joy *et al.* (1994).

Table 24A. Summary of ecological and distributional data for *Culicidae (Diptera)*. (For definition of terms see Tables 6A-6C; table prepared by E.D. Walker, R.W. Merritt, K.W. Cummins, H. D. Newson.)

Taxa (number of species in parentheses)	Habitat	Habit	Trophic Relationships	North American Distribution	Ecological References*
Culicidae(172) (mosquitoes)	Generally lentic—littoral (limnetic), lotic—depositional (pools and backwaters)	Generally planktonic—swimmers	Generally collectors—filterers and gatherers	Widespread	569, 1415, 1561, 1957, 2112, 3076, 3910, 4187, 3628, 4364, 4529, 655 830, 4165, 4363, 4, 2741, 4162, 1284, 633, 2223, 2373, 2663, 2670
Aedes(80)	Lentic (temporary ponds and pools)	Swimmers (divers, feeding in many habitat zones)	Collectors—gatherers and filterers	Widespread	68, 569, 652, 919, 1415, 1785, 2094, 2674, 3219, 2139, 3910, 4164, 3628, 4529, 633, 65
Anopheles(17)	Lentic—littoral (limnetic), lotic—depositional	Neustonic/planktonic—swimmers	Collectors—filterers	Widespread	196, 236, 569, 654, 1415, 1827, 2094, 3219, 3285, 2664, 3910, 4471, 3628, 4529, 63
Culex(29)	Lentic—limnetic (lakes, ponds, ditches, and ground pools)	Planktonic—swimmers	Collectors—filterers	Widespread	569, 824, 825, 826, 827, 1415, 1526, 2094, 3219, 3624, 3628, 3910, 4529
Culiseta(8) (=Theobaldia)	Lentic—limnetic (ponds and ground pools)	Planktonic—swimmers, divers (feeding at bottom)	Collectors—gatherers and filterers	Widespread	569, 1415, 1785, 2094, 3209, 3219, 3628, 4529
Deinocerites(3)	Beach zone—marine intertidal	Planktonic (crab holes)	Collectors—filterers	Extreme South	569, 1415, 2094, 3628
Haemagogus(1)	Lentic (tree holes)	Planktonic	Collectors—filterers	Southern Texas	401
Mansonia(3)	Lentic—vascular hydrophytes	Clingers (stems and roots of plants, piercing respiratory siphon)	Collectors—gatherers and filterers	Widespread	569, 1415, 2094, 3910, 3628, 4511, 4529
Orthopodomyia(3)	Lentic (tree holes and wooden artificial water containers)	Planktonic	Collectors—filterers	East, South	569, 1415, 2094, 4529
Psorophora(15)	Lentic (temporary ponds and pools)	Planktonic—swimmers	Predators (piercers), collectors—filterers	Widespread	569, 1415, 2094, 3910, 3628, 4511, 4529
Toxorhynchites(2)	Lentic (in tree and rock holes, artificial water containers, other phytotelmata)	Planktonic	Predators (engulfers)	East, South	569, 1415, 2094, 3910, 3628, 4511
Uranotaenia(4)	Lentic—limnetic (ponds and ground pools)	Planktonic—swimmers	Collectors—filterers	Widespread (except for West and intermountain)	569, 1415, 2094, 3910, 3628, 4529
Wyeomyia(4)	Lentic (bog mats in pitcher plants and small pools and water in living and dead plants)	Planktonic (in microhabitats in living and dead plants)	Collectors—filterers	East, Southeast	569, 1196, 1415, 1920, 4, 2094, 3624, 1283, 3628, 4529

* Emphasis on trophic relationships.

SIMULIIDAE

B. V. Peterson[1]
Systematic Entomology Laboratory, USDA
Washington, D.C.

INTRODUCTION

Biting female black flies are among the most important insect pests of humans and animals in many parts of the world because of their role as vectors of certain parasitic disease organisms and the irritation, toxic and allergic manifestations of their bites. Consequently, black flies have a marked impact on the economy of many regions of the world. Not all female black flies bite humans; some species feed strictly on birds or other animals, some are polyphagous, and some species do not feed on blood at all. Adults of both sexes imbibe nectar as a source of energy.

The greatest public health importance of black flies lies in their roles as vectors of the filarial nematode *Onchocerca volvulus* Leuckart, the causative organism of human onchocerciasis ("river blindness"), in tropical Africa, Central America, northern South America, and Yemen where 20-30 million people are infected. Some of the primary vectors, as now known, include several species of the *Simulium damnosum* Theobald complex in Africa, and *S. exiguum* Roubaud complex, *S. guianense* Wise, *S. metallicum* Bellardi complex, *S. ochraceum* Walker complex, and *S. oyapockense* Flock and Abonnenc in the New World tropics. Other species in these areas serve as secondary vectors of *Onchocerca volvulus*, and still other species serve as vectors of other filarial nematodes of humans, cattle, ducks, loons, deer and other cervids (relatives of the deer). In addition to filarial nematodes, some black flies are known to transmit various avian blood protozoans including the genus *Trypanosoma* Gruby and several species of *Leucocytozoon* Ziemann as well as certain viruses, such as vesicular stomatitis New Jersey serotype, that may be pathogenic to various animals. Crosskey (1973) tabulated our knowledge of the role of black flies as vectors of pathogenic organisms, and, in 1990, listed the main pest and vector species affecting avian, mammalian and human hosts.

Black flies occur in all regions of the world except Antarctica and desert areas where there is no running water. They occur from sea level to altitudes over 14,000 feet where the snow and ice melt enough to form streams of flowing water of sufficient temperature and with ample food supply to allow complete development. Although the general life cycle for most black fly species is similar, much remains to be learned about individual species. The female is ready to mate and oviposit usually after egg maturation, for which a blood meal is often required. However, mating can occur shortly after emergence or just before oviposition and takes place in flight or while landed. In some species the male transfers the sperm to the female in a spermatophore (Davies 1965a), and in the recently described *Piezosimulium jeanninae* Peterson (Peterson 1989), the male has what appears to be a functional sperm pump.

After mating, females lay their eggs in a variety of lotic environments ranging from the smallest trickles to the largest rivers; the choice of habitat varies with the species. Most females average 200-500 eggs in a single gonotrophic cycle. Various species oviposit in different manners, ranging from the free distribution of eggs while the female taps her abdomen on the water surface during flight, or ovipositing while landed on wet surfaces such as grass trailing in the water, to crawling underwater to deposit the eggs. Carlsson (1962) reported that pupae of the northern species, *Prosimulium ursinum* (Edwards), produce mature eggs. When climatic conditions prevent the females from fully developing and emerging, the pupae eventually rupture and release eggs that develop without fertilization (parthenogenesis) and later hatch. Apparently, these eggs produce only triploid females (individuals with three sets of single chromosomes).

Depending on the species and environmental factors such as water temperature, incubation time varies from two to 30 days and even longer in those species whose eggs pass through diapause. The egg burster on the larval head can be seen through the egg shell just prior to hatching. Larvae may remain at the site of hatching if the substrate and food supply are adequate, or they may drift downstream on 'silken' threads to suitable sites. Larvae pass through a series of 4-9 molts (usually seven). Growth rates fluctuate with water temperatures and the quantity and quality of available food (Ross and Merritt 1987). The larval period varies from a minimum of about four days to six or seven months in overwintering species. Some larvae are filter feeders (both suspension collectors and bottom-feeding

---

[1] New Address: Monte L. Bean Life Science Museum, Brigham Young University, Provo, Utah 84602

collectors), whereas those without labral fans are collectors (grazers and scrapers) of organic detritus around their attachment sites (Table 25A). Last stage larvae (pharate pupae) spin variously shaped cocoons that serve to anchor and protect the developing pupae.

The duration of the pupal stage also varies according to water temperature, but usually lasts about 2-7 days and, under certain circumstances, can take as long as two or three weeks. When the adult is ready to emerge, it pulls itself out of the pupal skin through a T-shaped slit originating at the back of the head and continuing along the median longitudinal line of the thorax. As the adult emerges its wings expand and the fly rises to the water surface in a bubble of air. In most instances, the adult flies immediately to a nearby support to rest and allow its cuticle to harden. After mating, and a suitable blood meal, if required, the life cycle begins again. The number of annual generations varies with the species and is strongly influenced by geographic and climatic factors. In tropical areas there may be as many as 16 or more generations; in temperate regions there may be one to six generations; and in more arctic habitats there is usually only one or possibly two generations.

EXTERNAL MORPHOLOGY

The family Simuliidae is one of the most homogeneous and easily recognized families in the order Diptera. However, generic and especially specific determinations are often difficult to make because of this homogeneity. Another factor that makes the taxonomy of the Simuliidae unusually difficult is the presence of sibling species (often called cryptic- or cytospecies) that differ biologically and chromosomally but look alike morphologically. What may appear to be a clearly defined species based on morphological features (morphospecies) could, in reality, consist of two or more cytologically (chromosomally) and biologically distinguishable entities. Such morphospecies are referred to as species complexes. Twelve genera and 19 subgenera (see footnote to adult key) have been recognized in North America with about 260 species. The family is briefly characterized as follows.

Eggs

The eggs are small, and range from about 0.15 to 0.54 mm in length. They generally are ovoid, but vary from ellipsoid to subspherical, and may appear triangular or kidney-shaped in one aspect to more oval in another. The anterior end is narrowly rounded and the posterior end more broadly rounded, and the dorsum is strongly convex and the ventral surface more flattened. The egg surface is smooth without obvious features, but under scanning electron microscope magnifications eggs of some species show distinctive sculpturing and pitting. Eggs are pale white when laid and gradually darken to brown as they develop.

Larvae

The larvae are slender, somewhat cylindrical, and vary from about 3.5 to 15.0 mm long. They have an apneustic (without spiracles) respiratory system, and vary from pale gray to pale yellowish brown to nearly black in color (Fig. 25.1). They may be tinged with irregular splotches of reddish, orange, greenish or purplish color. The head is directed forward (*prognathous*), well developed, and heavily sclerotized, with a variable pattern of pale or dark spots, especially dorsally (Figs. 25.2-25.6). There is usually a pair of dorsal *labral fans* used for straining food particles from flowing water (Fig. 25.1); these are reduced or absent in a few taxa (Fig. 25.6). The antennae are slender, three-segmented, and transparent to variously pigmented or patterned.

The head capsule has an anteroventral wedge-shaped or subtriangular *hypostoma* (Figs. 25.10-25.12; 25.19-25.27) that bears a series of heavily *sclerotized teeth* along the anterior margin, and has smooth or variously serrated lateral margins. Usually the head capsule's posteroventral margin has a median incision or *hypostomal cleft* that is variable in size and shape (Figs. 25.10-25.12). The mouthparts are well developed with the *labrum* forming an enlarged lobe or beaklike structure that lies anteroventral to, and is continuous medially with, the ventral surface of the stalks of the *labral fans* (Figs. 25.4-25.5). The *mandibles* (Figs. 25.5, 25.7-25.8) are heavily sclerotized, broadly rectangular, somewhat laterally flattened, and have 3-6 large apical teeth and two series of smaller teeth or serrations plus eight main groups of *setal brushes*. The *maxillae* (Fig. 25.5), which are ventral to the mandibles and dorsolateral to the *labiohypopharyngeal apparatus*, consist of large lobate structures with a single segmented, digitiform (finger–shaped) *palpus*.

The thoracic segments are stout and indistinctly delineated. The anterior segment bears a ventral eversible *proleg* having an apical ring of minute *hooks* arranged in rows, and a lateral sclerite on each side that varies in size and shape (Fig. 25.1). Developing *histoblasts* of legs, wings and halteres, and *respiratory filaments* are visible under the epidermis of later instars. The abdomen has eight poorly defined segments with the anterior segments slender and the posterior segments variously enlarged. A pair of conical *tubercles* or a single *midventral bulge* is sometimes present on segment 8 (Figs. 25.13-25.14). A ring of many rows of minute hooks is present posteriorly, and an *X-shaped* (sometimes subquadrate, Y-shaped or absent) *anal sclerite* is generally present anterodorsal to the ring of hooks (Figs. 25.15-25.18). The rectum has colorless, extrusible *anal papillae* that are apparently osmoregulatory in function. They are composed of three simple digitiform *lobes* (Fig. 25.17) that sometimes bear varying numbers of secondary *lobules* (a condition often referred to as compound; Fig. 25.18). The cuticle of the abdomen usually is smooth but may be roughened or have a microsculpture, and may be ornamented with fine setae or scales, especially posterodorsally.

Pupae

Typical pupae are shown in Figures 25.28-25.29. The head is anteroventral to and flexed beneath the thorax. The frons of the female is shorter and broader than in the male. The female *antennal sheaths* usually reach the hind margin of the head, or slightly beyond, while in the male they extend about one-half to three-quarters of that distance. The thorax is enlarged, strongly arched dorsally, and usually has 2-5 dorsal and 1-2 lateral pairs of specialized setae (*trichomes*) on each half. The pupal cuticle is smooth with a faint reticulate (meshed) or densely rugose (wrinkled) pattern, and sometimes has a series or pattern of flattened to raised granules of variable size and shape. A respiratory organ (*gill*) present at each anterolateral corner of the thorax usually consists of one to many short or long slender filaments; it may be enlarged, clublike or antlerlike, or exhibit various combinations of these forms (Figs. 25.28-25.61). The abdominal segments are beset with a series of setae and small hooks, and the last segment often bears a dorsal pair of short to long terminal spines. The pupae are covered by cocoons that are attached to the substrate, and may consist of a few irregular 'silken' threads, or may be a more dense shapeless sleevelike structure of variable extent, or a woven structure of specific shape and texture.

Adults

Black fly adults are small, stout flies that range from 1.2 to 5.5 mm in length and are usually dark brown to black but may be reddish brown, gray, orange, or yellow in ground color. The females have separated eyes that have facets subequal in size (Figs. 25.74-25.76), whereas in males the eyes usually touch along the midline above the antennae and the facets of the upper half of each eye are usually distinctly larger than those of the lower half (Fig. 25.73). However, there are a few species in which the male eyes nearly meet along the midline below the antennae (Fig. 25.72), or are much like those of the female. There are no ocelli. The antennae are short, rather stout, and erect, with 7-9 *flagellomeres* (Fig. 25.74). The proboscis is short and the mouthparts of the female are usually strong and adapted for sucking blood. Both the mandibles and the laciniae of the maxillae are serrated or toothed for cutting skin; however, in all males and the females of some species the mouthparts are weakly developed and are suitable only for taking in nectar. The thorax is usually high and strongly arched. The legs are short and moderately stout with elongated basal tarsomeres *(basitarsi)*. On the hind leg the basal tarsomere often has an anterior process *(calcipala;* Fig. 25.82) on the inner apical surface. The second hind tarsomere frequently has a variably distinct notch *(pedisulcus)* in the posterior margin (Fig. 25.82). The tarsal claws are short and simple or have a basal or subbasal tooth (Figs. 25.81-25.85). The wings are broad, and have strong anterior and weak posterior veins (Figs. 25.62-25.68). The abdomen has the first tergite modified as a collarlike basal scale that bears a fringe of long setae (Figs. 25.87-25.88). Female and male terminalia vary in size and shape and often are of great value in dinstinguishing species. Female terminalia are generally as in Figures 25.86, 25.89-25.105, and terminalia of the males generally as in Figures 25.106-25.134. For a more detailed morphological description see Peterson (1981), and Crosskey (1990).

The following keys place the larvae and pupae to the generic level, and the adults to the generic and subgeneric levels. In the adult key, characters not marked as either male or female apply to both sexes. Although keys can be constructed to separate many species in the larval and pupal stages, keys for separating the pupae to genera and subgenera are not completely satisfactory. These keys are based largely on those of Peterson (1981) and Peterson and Kondratieff (1995).

KEY TO GENERA OF SIMULIIDAE

Larvae

1.	Labral fans absent; lateral margins of head capsule strongly convex (Fig. 25.6); anal sclerite Y-shaped (Fig. 25.15)	2
1'.	Labral fans present; lateral margins of head capsule less convex (Figs. 25.1-25.3); anal sclerite X-shaped (Figs. 25.17-25.18), nearly rectangular (Fig. 25.16), or absent	3
2(1).	Outer (ventral) apical surface of mandible with a series of short fine teeth and about 10 rows of distinct comblike scales (Fig. 25.8)	**Gymnopais** Stone
2'.	Outer (ventral) apical surface of mandible without a series of short fine teeth, and without comblike scales (Fig. 25.7)	**Twinnia** Stone and Jamnback
3(1').	Postocciput nearly complete dorsally, enclosing cervical sclerites (Fig. 25.3); median tooth of hypostoma compound, with two or three basal or subbasal denticles (Figs. 25.19, 25.25); tips of secondary fan rays, when expanded, arranged in a straight line	4
3'.	Postocciput with a distinct and usually wide gap dorsally, not enclosing cervical sclerites (Fig. 25.2); median tooth of hypostoma single, without basal or subbasal denticles (Fig. 25.20); tips of secondary fan rays, when expanded, variable but usually arranged in an arc	5
4(3).	Head capsule distinctly pigmented and often with a distinct pattern of head spots, however pigment and/or pattern may be pale; hypostomal cleft variable in shape and usually easily discernible; basal two segments of antenna pale, strongly contrasting with darkly pigmented third segment (Fig. 25.3), length of segments variable but not as below; median tooth of hypostoma distinctly and symmetrically trifid (Fig. 25.19), remaining teeth variable but not arranged as in Fig. 25.25; abdominal segment 8 without a midventral bulge	**Prosimulium** Roubaud
4'.	Head capsule largely unpigmented except for a few heavily sclerotized areas; hypostomal cleft either absent or extremely shallow; antenna uniformly pale, basal two segments short, distal segment comprising about 80% of antennal length; median tooth of hypostoma with two lateral tines basally and one lateral tine more distally, remaining teeth arranged on two long lateral lobes as in Fig. 25.25; abdominal segment 8 with a midventral bulge (Fig. 25.14)	**Parasimulium** Malloch

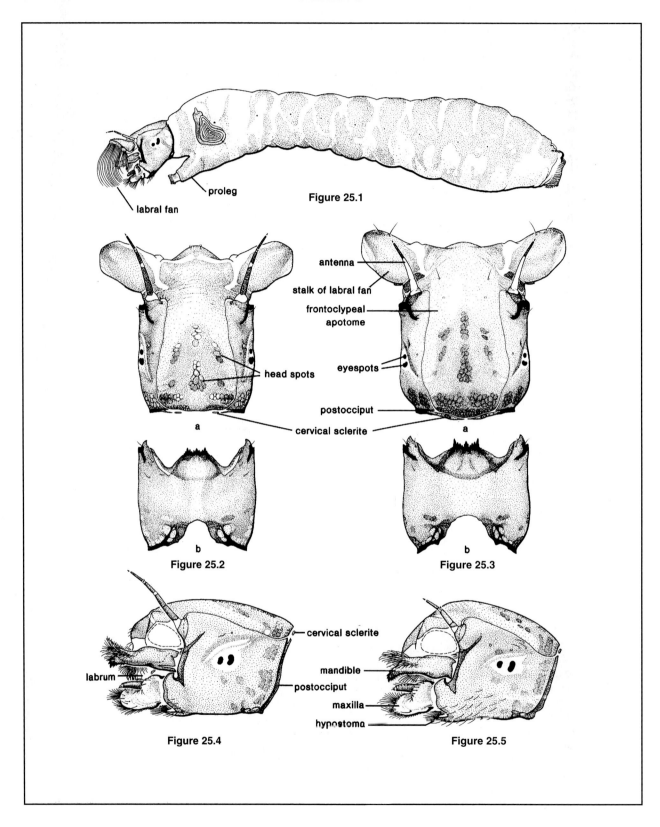

Figure 25.1. Lateral view of larva of *Simulium (Simulium) venustum*.

Figure 25.2. Dorsal (*a*) and ventral (*b*) views of larval head capsule of *Simulium (Psilozia) vittatum*.

Figure 25.3. Dorsal (*a*) and ventral (*b*) views of larval head capsule of *Prosimulium (Prosimulium) ursinum*.

Figure 25.4. Lateral view of larval head capsule of *Simulium (Psilozia) vittatum*.

Figure 25.5. Lateral view of larval head capsule of *Cnephia dacotensis*.

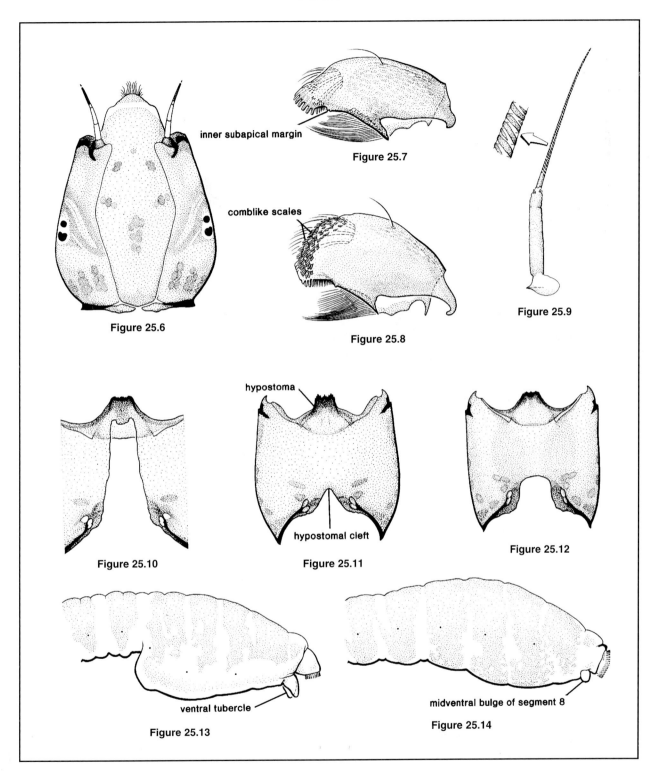

Figure 25.6. Dorsal view of larval head capsule of *Twinnia tibblesi*.

Figure 25.7. Ventrolateral view of larval mandible of *Twinnia* sp.

Figure 25.8. Ventrolateral view of larval mandible of *Gymnopais* sp.

Figure 25.9. Larval antenna of *Greniera* sp.

Figure 25.10. Ventral view of larval head capsule of *Metacnephia* sp.

Figure 25.11. Ventral view of larval head capsule of *Stegopterna mutata*.

Figure 25.12. Ventral view of larval head capsule of *Cnephia dacotensis*.

Figure 25.13. Lateral view of larval abdomen of *Ectemnia invenusta*.

Figure 25.14. Larval view of larval abdomen of *Stegopterna mutata*.

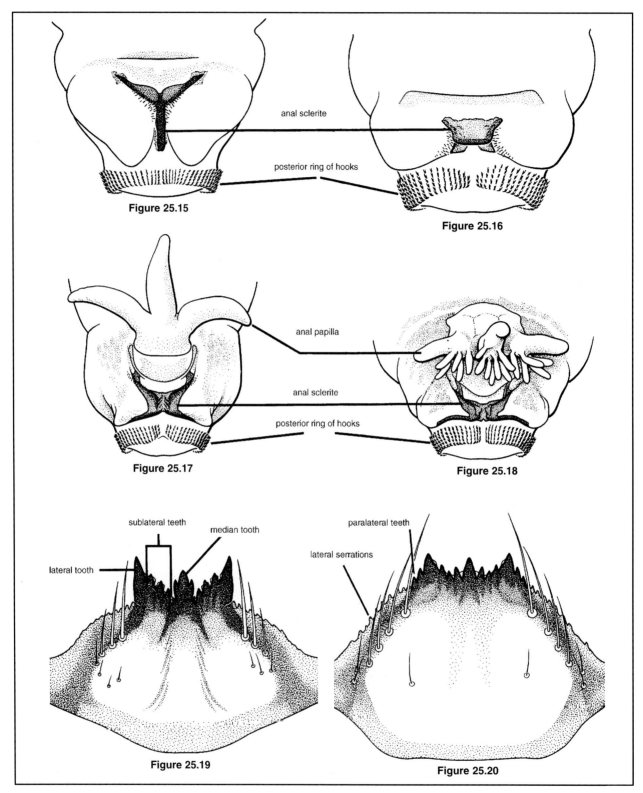

Figure 25.15. Dorsal view of posterior portion of larval abdomen of *Gymnopais* sp.

Figure 25.16. Dorsal view of posterior portion of larval abdomen of *Prosimulium* (*Parahelodon*) *decemarticulatum*.

Figure 25.17. Dorsal view of posterior portion of larval abdomen of *Simulium* (*Psilozia*) *vittatum*.

Figure 25.18. Dorsal view of posterior portion of larval abdomen of *Simulium* (*Hearlea*) *canadense*.

Figure 25.19. Ventral view of anterior portion of larval head capsule of *Prosimulium* (*Prosimulium*) *ursinum* showing hypostomal teeth.

Figure 25.20. Ventral view of anterior portion of larval head capsule of *Simulium* (*Prosimulium*) *vittatum* showing hypostomal teeth.

5(3').	Hypostoma with median and outer lateral (corner) teeth of each side moderately large and subequal in height, and with three variably smaller but nearly equal sublateral (intermediate) teeth on each side (Fig. 25.20); anterodorsal portion of frontoclypeal apotome in lateral view not noticeably arched nor strongly convex (Fig. 25.4); anal papillae usually consisting of three compound lobes, but sometimes lobes simple (Figs. 25.17-25.18) ....*Simulium* Latreille
5'.	Hypostoma either with uniformly small teeth (Figs. 25.10, 25.24), with teeth on each side of median tooth confined to one large lobe (Fig. 25.21), or with teeth clustered in three prominent groups (Fig. 25.22); anterodorsal portion of frontoclypeal apotome in lateral view often strongly convex (Fig. 25.5); anal papillae consisting of three simple lobes (Fig. 25.17) 6
6(5').	Hypostomal cleft extended anteriorly to or slightly beyond posterior margin of hypostoma, and of nearly uniform width throughout, with apex rather truncate; hypostomal teeth all very small (Fig. 25.10) ....*Metacnephia* Crosskey
6'.	Hypostomal cleft not extended anteriorly much more than about one-half the distance to posterior margin of hypostoma, often much less, and usually shaped like an inverted V or U (Figs. 25.11-25.12); hypostomal teeth variable, but often large and distinct (Figs. 25.26-25.27) 7
7(6').	Abdominal segment 8 with two ventral cone-shaped tubercles (Fig. 25.13) 8
7'.	Abdominal segment 8 without two ventral cone-shaped tubercles (Fig. 25.1), but sometimes with a single transverse midventral bulge (Fig. 25.14) 11
8(7).	Abdomen abruptly and greatly expanded at segment 5, with its lateral margins projected ventrally beyond central portion of segment (Fig. 25.13); anal sclerite absent; anterior margin of hypostoma with small indistinct teeth (Fig. 25.23) ....*Ectemnia* Enderlein
8'.	Abdomen of normal shape, not abruptly or greatly expanded at segment 5; anal sclerite present; anterior margin of hypostoma with minute lateral teeth borne on two large nearly parallel-sided lobes that are much longer than or subequal in length to median tooth (Fig. 25.21) 9
9(8').	Third (apical) antennal segment much shorter than second segment, and without spiralike microannulations ....*Greniera* Doby and David (in part) *denaria* (Davies, Peterson and Wood) group
9'.	Third (apical) antennal segment longer than combined lengths of basal two segments, and with spiralike microannulations (Fig. 25.9) 10
10(9').	Median tooth of hypostoma slender and much shorter than large lateral toothed lobes so that hypostoma appears bidentate (Fig. 25.21) .... *Greniera* Doby and David (in part) *abdita* (Peterson) group
10'.	Median tooth of hypostoma rather stout, subequal in length to large lateral toothed lobes so that hypostoma appears tridentate (Fig. 25.26) .... *Greniera* Doby and David (in part)
11(7').	Hypostomal cleft moderately deep with its anterior margin rounded (Fig. 25.12); hypostomal teeth uniformly small (Fig. 25.24) ....*Cnephia* Enderlein
11'.	Hypostomal cleft narrow, shallow and acutely pointed or narrowly rounded, shaped much like an inverted V (Fig. 25.11); hypostomal teeth larger and strong (Fig. 25.22) 12
12(11').	Abdominal segment 8 with a single transverse midventral bulge (Fig. 25.14); hypostomal teeth as in Fig. 25.22; outside margin of lateral cluster of teeth usually sloped inwardly; labral fan with 40-60 primary rays ....*Stegopterna* Enderlein
12'.	Abdominal segment 8 simple, without a transverse midventral bulge (Fig. 25.1); hypostomal teeth as in Fig. 25.27; outside margin of lateral cluster of teeth parallel or slightly sloped outwardly; labral fan with 22-26 rays ....*Mayacnephia* Wygodzinsky and Coscarón

Pupae

1.	Cocoon an irregular, shapeless, sleeve-like structure without a well-defined anterior margin, covering variable portions of pupa (sometimes which seems absent) (Fig. 25.28); terminal abdominal segment of pupa with two long or short, dorsal spines, or none 2
1'.	Cocoon well developed, variously shaped, with a well defined anterior margin, and covering most or all of pupa (Fig. 25.29); terminal abdominal segment of pupa with two short, dorsal spines, or none 11

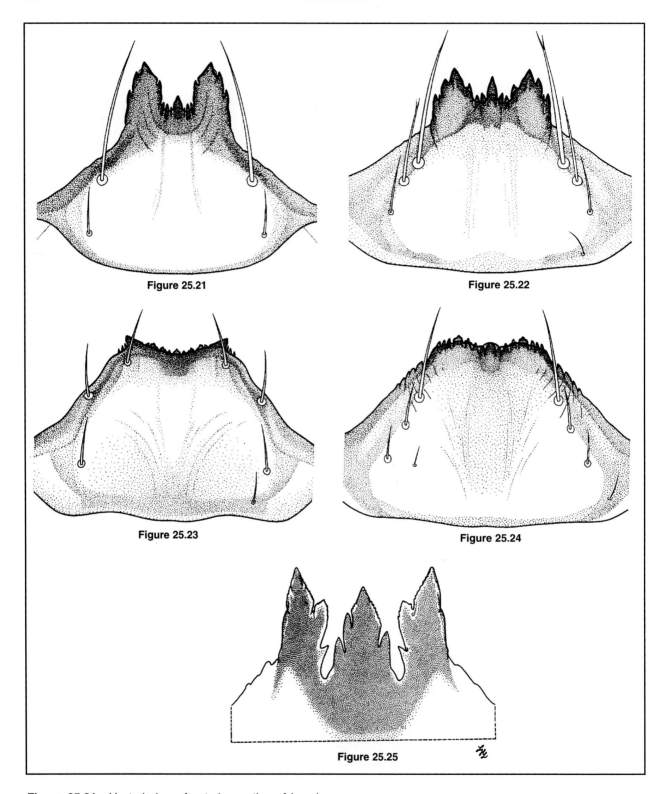

Figure 25.21. Ventral view of anterior portion of larval head capsule of *Greniera abdita* showing hypostomal teeth.

Figure 25.22. Ventral view of anterior portion of larval head capsule of *Stegopterna mutata* showing hypostomal teeth.

Figure 25.23. Ventral view of anterior portion of larval head capsule of *Ectemnia invenusta* showing hypostomal teeth.

Figure 25.24. Ventral view of anterior portion of larval head capsule of *Cnephia dacotensis* showing hypostomal teeth.

Figure 25.25. Ventral view of anterior portion of larval head capsule of *Parasimulium* (*Parasimulium*) *crosskeyi* showing hypostomal teeth (redrawn from Courtney 1986).

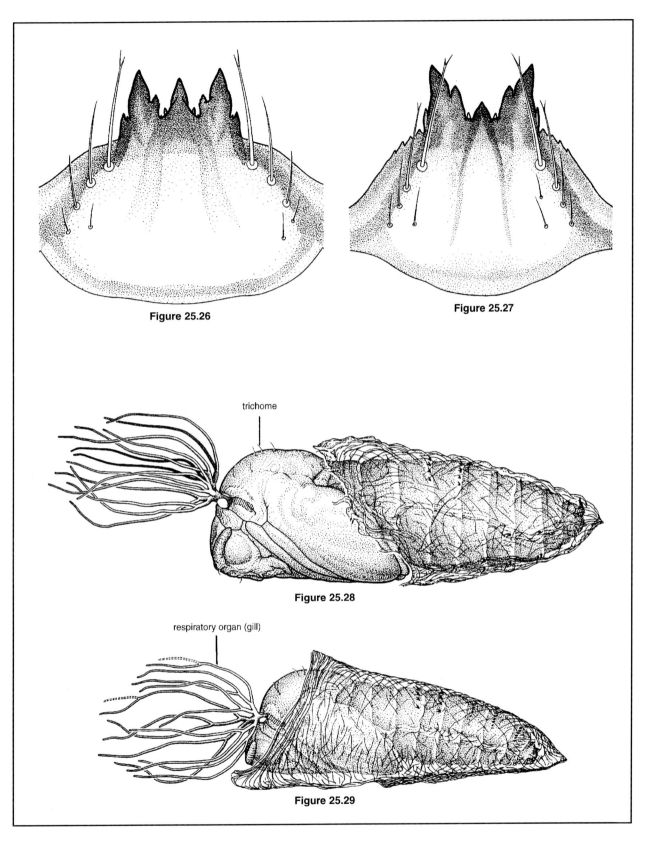

Figure 25.26. Ventral view of anterior portion of larval head capsule of *Greniera* sp. showing hypostomal teeth.

Figure 25.27. Ventral view of anterior portion of larval head capsule of *Mayacnephia* sp. showing hypostomal teeth.

Figure 25.28. Lateral view of pupa of *Prosimulium* (*Prosimulium*) *ursinum*.

Figure 25.29. Lateral view of pupa of *Simulium* (*Psilozia*) *vittatum*.

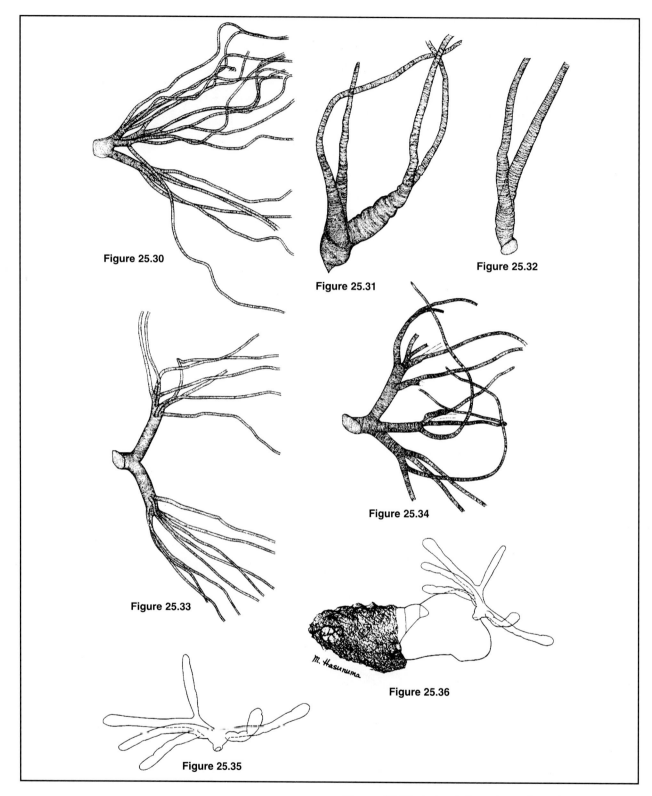

Figure 25.30. Lateral view of pupal respiratory organ of *Greniera abdita*.

Figure 25.31. Lateral view of pupal respiratory organ of *Gymnopais* sp.

Figure 25.32. Lateral view of pupal respiratory organ of *Gymnopais holopticus*.

Figure 25.33. Lateral view of pupal respiratory organ of *Twinnia nova*.

Figure 25.34. Lateral view of pupal respiratory organ of *Twinnia tibblesi*.

Figure 25.35. Lateral view of pupal respiratory organ of *Mayacnephia stewarti*.

Figure 25.36. Lateral view of pupa of *Mayacnephia stewarti*.

2(1).	Abdominal tergites 5-8 without spines, tergite 9 with long or short terminal spines or none 3
2'.	Abdominal tergites 5-8 each usually with an anterior transverse row of posteriorly directed spines as well as terminal spines on tergite 9 5
3(2).	Respiratory organ (gill) with 15-27, long, slender, threadlike filaments arranged in four short petiolate groups (Fig. 25.30); cocoon fragile, loosely woven, non-slimy, readily visible and usually consisting of only a few threads that seldom occur on more than posterior two-thirds of abdomen ....***Greniera*** Doby and David (in part) ....***abdita*** (Peterson) group
3'.	Respiratory organ (gill) with 2-16 filaments that may be longer or shorter and stouter than above, and with a different branching pattern; cocoon a gelatinous, transparent, slimy mass of variable extent and not readily discernible except by attached debris 4
4(3').	Respiratory organ with 2-4 (rarely five) filaments, 2-3 of which may be digitiform and noticeably shorter and stouter than the other one or two longer filaments (Figs. 25.31-25.32); terminal spines of tergite 9 small or absent; cocoon sloughs off, leaving pupa naked, except for a small ventral padlike patch of silk ....***Gymnopais*** Stone
4'.	Respiratory organ with filaments arising from two or three somewhat inflated, non-annulate trunks that may be clublike and with a total of 16 long slender filaments (Figs. 25.33-25.34); terminal spines of tergite 9 long and slender; cocoon more extensive and with at least a few silken strands over abdomen ....***Twinnia*** Stone and Jamnback
5(2').	Respiratory organ with either six long swollen tubular filaments that are rounded distally, of which anterior two are single and posterior two are branched into two petiolate pairs (Figs. 25.35-25.36), or with eight filaments that are somewhat inflated and arise in four main groups - (2 dorsal, 3 dorsolateral, 2 ventrolateral, and 1 ventral); cocoon a loosely woven saclike structure without any definite shape and often covered with detritus ....***Mayacnephia*** Wygodzinsky and Coscarón
5'.	Respiratory organ variable, if with six or eight filaments then with different branching pattern; cocoon variable...6
6(5').	Pupal integument transparent; respiratory organ with an elongate base and two strongly divergent primary trunks, with dorsal trunk simple and ventral trunk branching horizontally into two filaments of subequal length, the three filaments translucent, somewhat inflated and with shallow annulations; cocoon a thinly woven saclike structure, poorly defined anteriorly, and covering abdomen and thorax ....***Parasimulium*** Malloch
6'.	Pupal integument variable but usually well-pigmented; if respiratory organ with three filaments, then with a different branching pattern; cocoon variable 7
7(6').	Respiratory filaments variable in number but usually 30 or more arranged in 6-7 groups that arise from a rounded knob on a short petiole (Fig. 25.43)...***Cnephia*** Enderlein
7'.	Respiratory organ variable in shape; filaments variable in number but do not arise from a rounded knob on a short petiole 8
8(7').	Respiratory organ with 12 slender filaments that arise from two main trunks that diverge from each other; the dorsal trunk divides into two branches with three and four filaments, respectively; the ventral trunk with five filaments (Fig. 25.45); cocoon a loosely woven sac covering most of abdomen ....***Stegopterna*** Enderlein
8'.	Respiratory organ with more or less than 12 filaments, if with 12 filaments, then with a different branching pattern; cocoon variable 9
9(8').	Respiratory organ with 22 filaments that arise from three main trunks that branch into 10 dorsal, four lateral, and eight ventral filaments, most branching of filaments occurs at a considerable distance from base ....***Greniera*** Doby and David (in part) ....***denaria*** (Davies, Peterson and Wood) group
9'.	Respiratory organ with a variable number of filaments arranged in varying patterns, or if with 22 filaments then filaments not arranged as above, branching often occurring close to base 10
10(9').	Respiratory organ with 14 filaments that arise from three, short, non-swollen, primary trunks; ventromedial and mediolateral trunks each with four sessile filaments (Fig. 25.44) ....***Piezosimulium*** Peterson

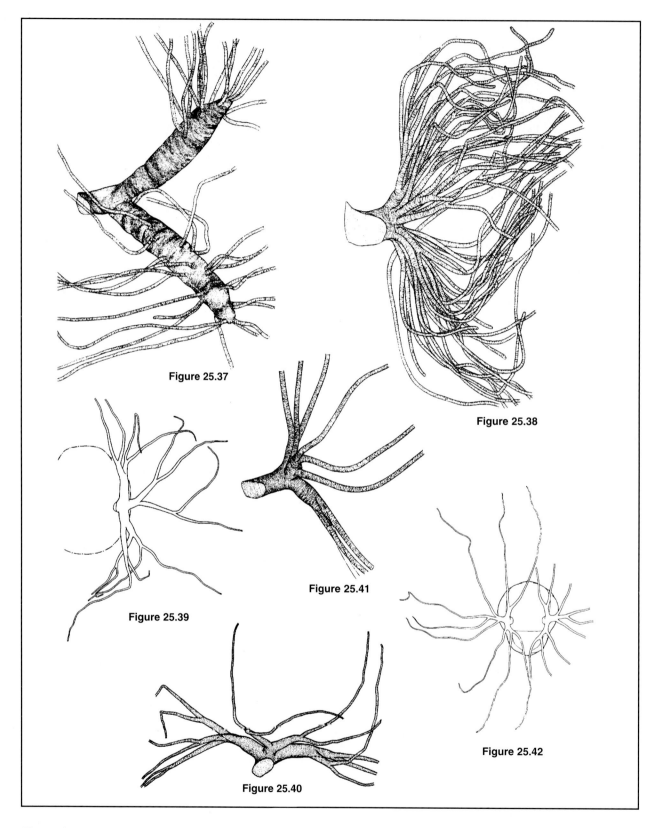

Figure 25.37. Lateral view of pupal respiratory organ of *Prosimulium (Helodon) onychodactylum*.

Figure 25.38. Lateral view of pupal respiratory organ of *Prosimulium (Prosimulium) exigens*.

Figure 25.39. End view of pupal respiratory organ of *Prosimulium (Prosimulium) frohnei*.

Figure 25.40. Lateral view of pupal respiratory organ of *Prosimulium (Prosimulium) frohnei*.

Figure 25.41. Lateral view of pupal respiratory organ of *Prosimulium (Parahelodon) decemarticulatum*.

Figure 25.42. End view of pupa and pupal respiratory organs of *Prosimulium (Parahelodon) decemarticulatum*.

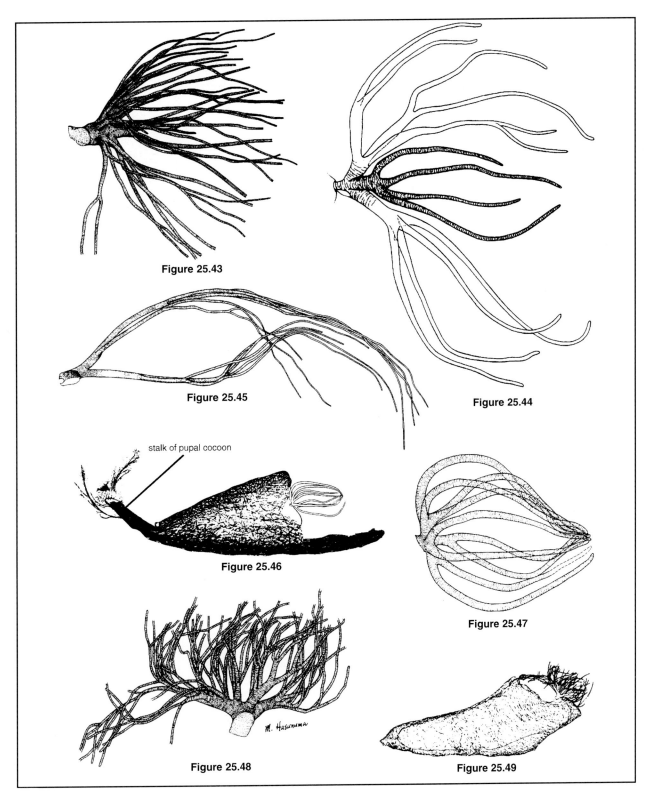

Figure 25.43. Lateral view of pupal respiratory organ of *Cnephia dacotensis*.

Figure 25.44. Lateral view of pupal respiratory organ of *Piezosimulium jeanninae*.

Figure 25.45. Lateral view of pupal respiratory organ of *Stegopterna mutata*.

Figure 25.46. Lateral view of pupa of *Ectemnia invenusta* showing stalk of cocoon.

Figure 25.47. Lateral view of pupal respiratory organ of *Ectemnia taeniatifrons*.

Figure 25.48. Lateral view of pupal respiratory organ of *Metacnephia jeanae*.

Figure 25.49. Lateral view of pupa of *Metacnephia jeanae*.

10'.	Respiratory organ varying in shape from a single swollen club, to two divergent, clublike structures each bearing slender filaments, or consisting of a highly variable number (5-100+) of long or short, usually slender filaments that arise from a short base (Figs. 25.37-25.42); or if with 14 filaments then ventrolateral and mediolateral trunks each with four filaments on short to moderately long petioles (Figs. 25.39-25.40) ....*Prosimulium* Roubaud
11(1').	Respiratory organ with 8 or 10 stout unbranched filaments that arise from base and converge distally toward a common point (Fig. 25.47); cocoon with a short or long stalk, anterior margin of cocoon not well defined (Fig. 25.46)..................*Ectemnia* Enderlein
11'.	Respiratory organ with a variable number of filaments, if with 8 or 10 filaments then filaments do not converge distally toward a common point or at least some filaments branching basally or away from base; cocoon without a stalk and anterior margin of cocoon usually well defined..................12
12(11').	Cocoon rather loosely or finely woven, often transparent, and with a distinctly raised anterodorsal collar of varying length whose ventral margin is set at an angle to the surface on which the cocoon is placed so that it appears vaguely boot-shaped; opening of cocoon dorsal or anterodorsal to position of pupal head, and collar without festoons or windowlike openings (Fig. 25.49); respiratory organ variable in shape, with about 17-100+ filaments (Fig. 25.48) ....*Metacnephia* Crosskey
12'.	Cocoon variable but often rather tightly woven and variably thickened; slipper-shaped or without a prominent raised anterodorsal collar or with, at most, a weak anteroventral connection so opening of cocoon is horizontal to or anterior to position of pupal head; if cocoon boot-shaped then opening with festoons, or two or more windowlike openings dorsolaterally, or both; respiratory organ variously shaped and with 2-100+ filaments (Figs. 25.50- 25.61) ....*Simulium* Latreille

Adults*

1.	R_1 joining C slightly beyond middle of wing; branches of Rs conspicuously separated by membrane, with posterior branch ending well before terminus of C; C, Sc, and branches of R with moderately long setae on dorsal and ventral surfaces; false vein not forked apically; cell bm absent (Figs. 25.62-25.63). Calcipala and pedisulcus absent. Mesepimeral tuft absent (Fig. 25.80). **Male** (in known species) with facets of eye all similar in size except for a few larger facets near middle of anterior margin; eyes broadly separated by frons dorsally, touching or nearly so below antennae (Fig. 25.72); gonostylus without an apical spinule, but with internal apical margin sometimes sclerotized (Figs. 25.106-25.109); a sclerotized, somewhat conical 'inner gonostylus' may be present dorsally between gonocoxite and gonostylus ..................PARASIMULIINAE..................*Parasimulium* Malloch 2
1'.	R_1 joining C well beyond middle of wing (Fig. 25.66); if Rs forked, then branches lying close together, with posterior branch ending near terminus of C; C, Sc, and branches of R with short setae sometimes present on one surface only; false vein distinctly forked apically; cell bm present or absent (Figs. 25.64-25.69); Calcipala and pedisulcus present or absent. Mesepimeral tuft present (Figs. 25.79-25.80). **Male** with facets of dorsal one-half of eye usually conspicuously larger than those of ventral one-half; if facets all of similar size then eyes not touching below antennae, but usually touching or nearly so at middle of head above antennae, with frons narrow (Fig. 25.73). Gonostylus with or without an apical spinule; 'inner gonostylus' absent..................SIMULIINAE 3

* Wygodzinsky (1973) reported the capture of two females of the subgenus *Ectemnaspis* Enderlein, genus *Simulium*, from the Catskill Mountains, Ulster Co., New York. This subgenus is not included in the present key because of the doubtful nature of this record.

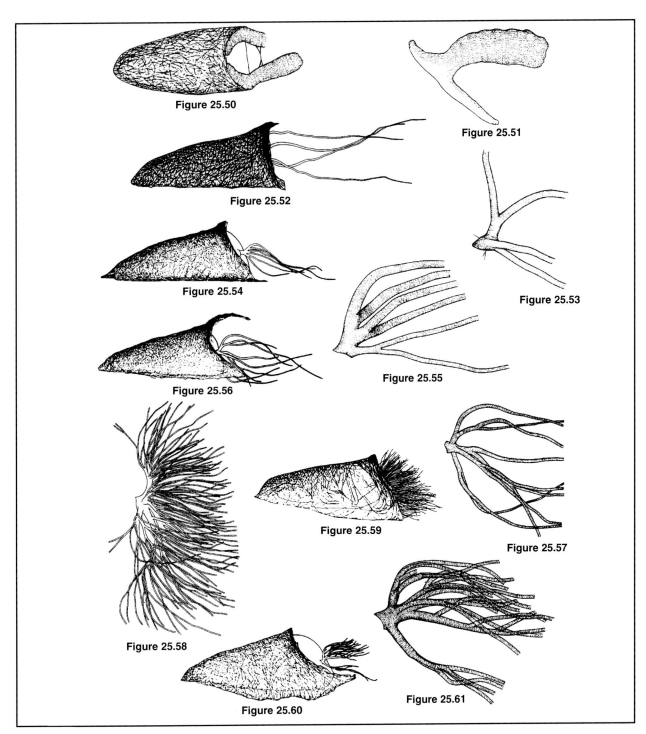

Figure 25.50. Dorsal view of pupa of *Simulium* (*Hearlea*) *canadense*.

Figure 25.51. Lateral view of pupal respiratory organ of *Simulium* (*Hearlea*) *canadense*.

Figure 25.52. Lateral view of pupa of *Simulium* (*Eusimulium*) *aureum*.

Figure 25.53. Lateral view of pupal respiratory organ of *Simulium* (*Eusimulium*) *aureum*.

Figure 25.54. Lateral view of pupa of *Simulium* (*Simulium*) *tuberosum*.

Figure 25.55. Lateral view of pupal respiratory organ of *Simulium* (*Simulium*) *tuberosum*.

Figure 25.56. Lateral view of pupa of *Simulium* (*Simulium*) *piperi*.

Figure 25.57. Lateral view of pupal respiratory organ of *Simulium* (*Simulium*) *piperi*.

Figure 25.58. Lateral view of pupal respiratory organ of *Simulium* (*Simulium*) *hunteri*.

Figure 25.59. Lateral view of pupa of *Simulium* (*Simulium*) *hunteri*.

Figure 25.60. Lateral view of pupa of *Simulium* (*Byssodon*) *meridionale*.

Figure 25.61. Lateral view of pupal respiratory organ of *Simulium* (*Byssodon*) *meridionale*.

2(1). Stem of Rs shorter than or at most subequal to posterior branch (R_{4+5}) of Rs; false vein faint, becoming evanescent at about level of distal end of anterior branch (R_{2+3}) of Rs; A_2 absent or faint (Fig. 25.62). Halter entirely yellow. **Male** gonostylus yellow, narrowed distally and somewhat rounded, or pointed, or with inner distal corner produced as a short rounded upturned lobe (Figs. 25.106-25.108); ventral plate of aedeagus, in ventral view, broad, dorsoventrally flattened, and somewhat H-shaped (Fig. 25.106). **Female** genital fork (sternite nine) with a slender stem that is about twice as long as arm, each arm slender basally, expanded distally into a variously enlarged plate with a slender posteromedial process; anal lobe undivided medially ....Subgenus *Parasimulium* Malloch

2'. Stem of Rs distinctly longer than posterior branch (R_{4+5}) of Rs; false vein weak but extended more nearly to wing margin; A_2 present and distinct (Fig. 25.63). Halter entirely brownish black. **Male** gonostylus black, broad, with apex broadly rounded, and with a short median sclerotization (Fig. 25.109); ventral plate of aedeagus, in ventral view, a narrow subquadrate structure with a short bulbous apical liplike projection, and with basal arms extended laterally at right angles beyond its margins (Fig. 25.109). **Female** unknown..Subgenus *Astoneomyia* Peterson

3(1'). Rs with a long, distinct fork that is conspicuously longer than its petiole; C with fine setae only, not interspersed with spinules; basal section of R always setose (Fig. 25.69). Calcipala and pedisulcus absent 4

3'. Rs unforked, or with a short, usually obscure apical fork that is conspicuously shorter than its petiole; C often with spinules interspersed among its fine setae; basal section of R setose or bare. Calcipala and pedisulcus present or absent (Figs. 25.81-25.85)..10

4(3). **Male** terminalia with ventral plate of aedeagus somewhat M-shaped in ventral view; gonocoxite with a prominent ventromedial lobe that articulates with a small but conspicuous sclerotized, setose plate situated anteriorly between the lobes of the two gonocoxites; sperm pump present (Fig. 25.110). **Female** unknown .... *Piezosimulium* Peterson

4'. **Male** terminalia with ventral plate of aedeagus other than M-shaped in ventral view; gonocoxite with, at most, a small ventromedial lobe; without a sclerotized, setose plate anteroventrally between gonocoxites; sperm pump absent. **Female** variable; claw simple or with a variably sized basal or subbasal tooth 5

5(4'). Wing fumose (smoky), sometimes nearly opaque; petiole of M at least twice as long as Rs from its base to r-m (Fig. 25.64). Head, body, coxae, and femora sparsely covered with rather coarse semierect to erect setae, without fine recumbent setae (Fig. 25.80); clypeus bare except for a few erect setae near lateral margins. Postnotum rather small and strongly arched, sometimes with a variably conspicuous median longitudinal ridge or line; anepisternal membrane usually with some setae dorsally. **Male** gonostylus usually with two or more minute apical spinules that cannot easily be seen under a dissecting microscope. **Female** anal lobe and cercus fused into a single heavily sclerotized piece; spermatheca globular, sclerotized, with a long neck (Fig. 25.89)...*Gymnopais* Stone

5'. Wing usually hyaline (clear), but if slightly fumose or tinted then distinctly transparent; petiole of M less than twice as long as Rs from its base to crossvein r-m (Figs. 25.65-25.66). Head, body, coxae, and femora entirely covered with rather dense fine recumbent setae (Fig. 25.79), with a few erect setae sometimes evident especially posteromedially on scutum; clypeus entirely covered with setae. Postnotum larger, rather evenly arched, without a median longitudinal ridge or line; anepisternal membrane setose or bare. **Male** gonostylus with a variable number of apical spinules that are usually visible under a dissecting microscope. **Female** anal lobe and cercus clearly separated by membrane, usually lightly to moderately sclerotized (Fig. 25.86); spermatheca of various shapes, but if sclerotized then with, at most, a short sclerotized neck 6

6(5'). Antenna with seven flagellomeres, rarely with eight. Posterior margin of eye near middle with a slightly raised but conspicuous shiny stemmatic bulla (Fig. 25.74). **Male** gonostylus with a single (rarely two) apical spinule; lateral margins of ventral plate of aedeagus strongly emarginate near junction with basal arms (Fig. 25.112). **Female** claws simple. Hypogynial valve rather truncate, short, not extended to anal lobe; spermatheca short, broader than long, with a large differentiated area at junction with spermathecal duct (Fig. 25.86) ....*Twinnia* Stone and Jamnback

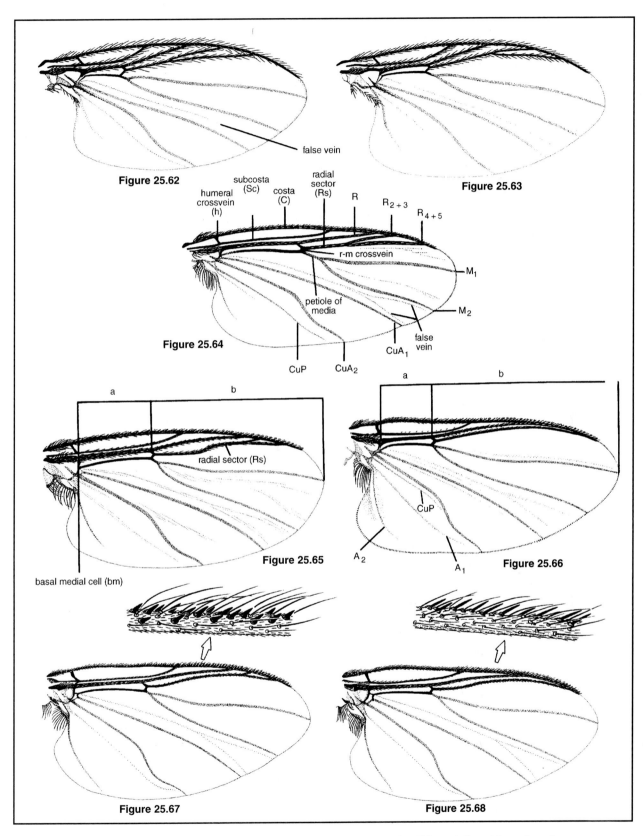

Figure 25.62. Wing of *Parasimulium (Parasimulium) stonei* male.
Figure 25.63. Wing of *Parasimulium (Astoneomyia) melanderi* male.
Figure 25.64. Wing of *Gymnopais holopticus* female.
Figure 25.65. Wing of *Cnephia* sp. female.
Figure 25.66. Wing of *Simulium (Simulium) venustum* female.
Figure 25.67. Wing of *Mayacnephia* sp. female.
Figure 25.68. Wing of *Greniera* sp. female.

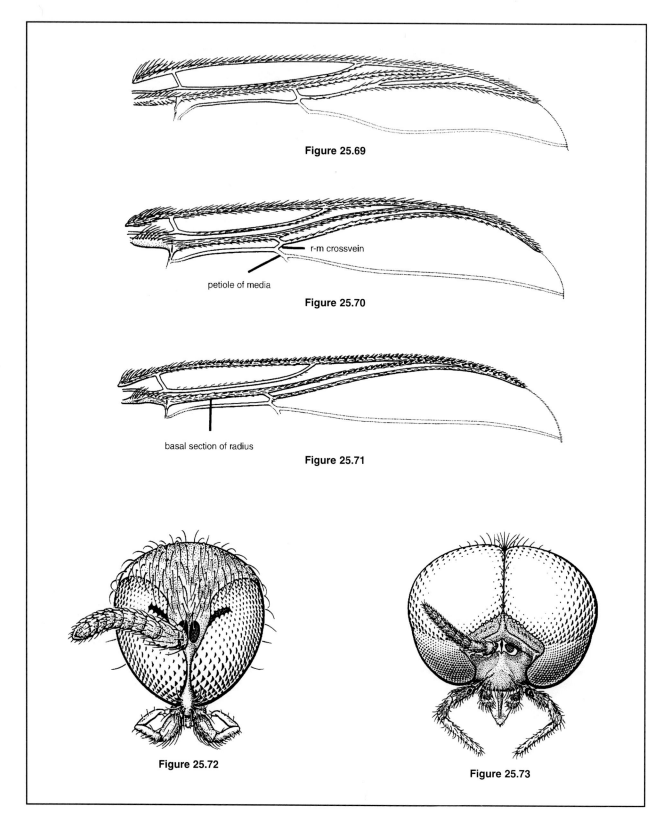

Figure 25.69. Anterior portion of wing of *Prosimulium (Prosimulium) ursinum* female.

Figure 25.70. Anterior portion of wing of *Greniera abdita* female.

Figure 25.71. Anterior portion of wing of *Metacnephia* sp. female.

Figure 25.72. Anterior view of adult head of *Parasimulium (Parasimulium) stonei* male.

Figure 25.73. Anterior view of adult head of *Simulium (Simulium) decorum* male.

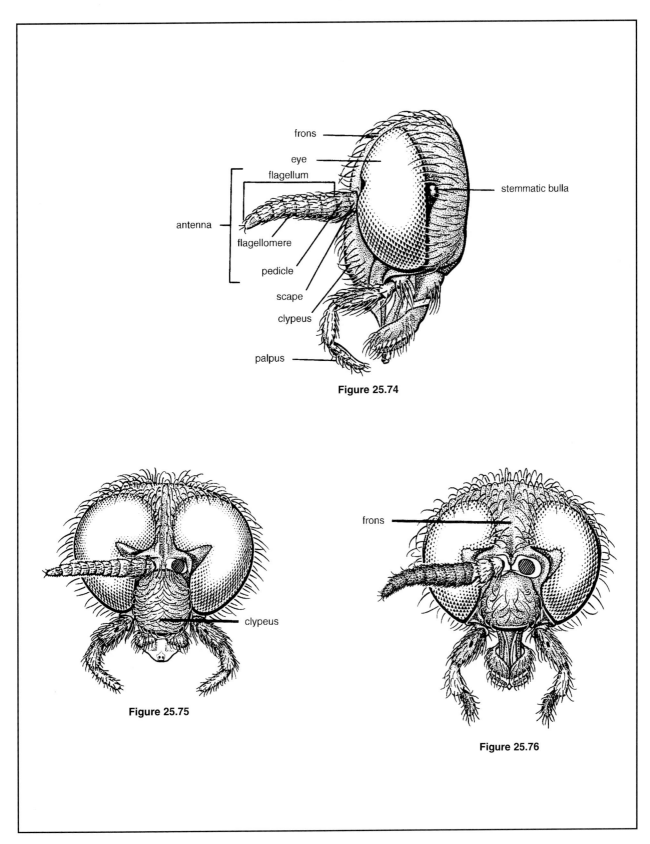

Figure 25.74. Lateral view of adult head of *Twinnia* sp. female.

Figure 25.75. Anterior view of adult head of *Ectemnia taeniatifrons* female.

Figure 25.76. Anterior view of adult head of *Cnephia dacotensis* female.

6'. Antenna usually with nine flagellomeres (Figs. 25.75- 25.76), but sometimes with seven or eight. Posterior margin of eye without a conspicuous stemmatic bulla, but some species with a weak indication of a bulla. **Male** gonostylus often with more than one spinule; ventral plate of aedeagus variable, but not exactly as above. **Female** claws variable. Hypogynial valve short or elongate, but if valve short and truncate then claws with a variably sized but usually conspicuous basal or subbasal tooth; spermatheca variable ....***Prosimulium*** Roubaud 7

7(6'). **Male** ventral plate of aedeagus variably broad or narrow, but neither noticeably compressed dorsoventrally or somewhat H-shaped, dorsal surface not convex but conspicuously concave at least proximally; apical margin usually with a prominent ventrally directed lip or emargination (Fig. 25.113). **Female** hypogynial valve long, narrowly rounded or pointed distally, extended nearly to or beyond anal lobe so that abdomen appears rather pointed posteriorly; anteromedial corner of each valve produced nipplelike. Spermatheca with a large differentiated circular area at junction with spermathecal duct (Fig. 25.90). Claw simple, or at most with a small inconspicuous subbasal tooth ....Subgenus ***Prosimulium*** Roubaud

7'. **Male** ventral plate of aedeagus variably broad, somewhat H-shaped, or noticeably flattened, or if somewhat compressed dorsoventrally then with dorsal surface broadly convex; distal margin deeply emarginate or without a prominent ventrally directed lip although a short ventrally directed lip may be present (Figs. 25.114-25.117). **Female** hypogynial valve short, broadly rounded or truncate distally, not extended to or nearly to anal lobe so that abdomen appears rounded or truncate posteriorly; anteromedial corner of each valve not produced nipplelike. Spermatheca with only a small differentiated circular area, or none, at junction with spermathecal duct (Figs. 25.91-25.93). Claw with a variably sized but usually conspicuous basal or subbasal tooth 8

8(7'). Anepisternum with a small patch of fine setae both anterior and posterior to anepisternal membrane. **Male** ventral plate of aedeagus deeply cleft and basal arms long so that entire structure appears H-shaped; median sclerite Y-shaped, with its stem and basal portion of arms heavily sclerotized, and remainder of arms membranous and projected posteriorly and ventrally beyond margin of ventral plate. Gonostylus thin, pointed, bent at nearly a right angle, with one subterminal and two terminal spinules; paramere with a strong spiniform process that curves ventrally and posteriorly to meet tip of gonostylus (Figs. 25.114-25.116). **Female** with two sclerotized plates, bearing short setae, present posterior to hypogynial valves, and continuous laterally with a prominent protuberance on each side of segment 9 (Fig. 25.91). Spermatheca a greatly enlarged, rounded, thin, delicate bag, neither pigmented nor patterned (often difficult to see, and therefore sometimes seemingly absent) (Fig. 25.91)..Subgenus ***Distosimulium*** Peterson

8'. Anepisternum without a small patch of setae on both sides of anepisternal membrane (*P. neomacropyga* Peterson sometimes has a few setae present anterior to ventral edge of anepisternal membrane). **Male** ventral plate of aedeagus not deeply cleft or H-shaped; median sclerite variable but arms not prolonged. Gonostylus variable; paramere expanded platelike distally but not as a spiniform process curved to meet tip of gonostylus. **Female** without sclerotized plates posterior to hypogynial valves, and without a prominent lateral protuberance on each side of segment 9. Spermatheca not greatly enlarged, at least lightly pigmented, often patterned 9

9(8'). Antenna with 7-9 flagellomeres. **Male** with ventral plate of aedeagus broad, flattened dorsoventrally, without a ventrally directed lip; basal arms short, slender, pointed; proximal margin of dorsal surface produced medially beyond tips of basal arms as a long, slender, hollow, tubelike process or a similar, somewhat flattened, grooved process with which base of median sclerite is fused; median sclerite not Y-shaped; paramere free, not connected to basal arm of ventral plate (Fig. 25.118); gonostylus with a single apical spinule. **Female** spermatheca subcircular to pear-shaped, heavily sclerotized. Arm of genital fork slender at base, abruptly expanded distally into a weakly sclerotized terminal plate that is sometimes wrinkled or denticulate and sometimes bears a basal denticulate median process (Fig. 25.93) ....Subgenus ***Parahelodon*** Peterson

9'. Antenna with nine flagellomeres. **Male** with ventral plate of aedeagus broad, compressed dorsoventrally but dorsal surface broadly convex with either a shallow dorsomedial depression, or furrow, or a median dorsal convexity; apical margin with, at most, a short narrow ventrally directed lip, or none; proximal margin of dorsal surface not produced medially beyond tips of basal arms as a long, slender, hollow, tubelike process or a similar, somewhat flattened, grooved process; median sclerite Y-shaped; paramere connected to basal arm of ventral plate; gonostylus rounded or pointed apically, with two to five apical spinules (Fig. 25.117). **Female** spermatheca elongate, lightly pigmented; arm of genital fork thicker at base, more gradually expanded into a large, heavily sclerotized plate which may have a sharply pointed posteromedian process but never a wrinkled or denticulate basal median process (Fig. 25.92).................Subgenus ***Helodon*** Enderlein

10(3'). Length of basal section of R (*a*, Fig. 25.65) rarely less than one-third distance from base of Rs to apex of wing (*b*, Fig. 25.65); R setose dorsally; cell bm present. Calcipala present but sometimes reduced; pedisulcus absent or, if present, usually very shallow. Anepisternal membrane with or without a tuft of setae dorsally 11

10'. Length of basal section of R (*a*, Fig. 25.66) equal to much less than one-third distance from base of Rs to apex of wing (*b*, Fig. 25.66); R with or without setae dorsally; cell bm absent or greatly reduced. Calcipala usually well developed (Fig. 25.82) although sometimes reduced (Fig. 25.85); pedisulcus present, usually deep and distinct (Fig. 25.82), rarely a shallow depression. Anepisternal membrane bare .... ***Simulium*** Latreille 22

11(10). C with fine uniformly colored setae, some of which are short and stiff but not spiniform or darker in color (Figs. 25.68–25.70).......................***Greniera*** Doby and David 12

11'. C with short black spinules interspersed among longer finer and paler setae (Figs. 25.66–25.67) 13

12(11). Antenna with eight flagellomeres .... ***Greniera*** Doby and David (in part)
....***denaria*** (Davies, Peterson and Wood) group

12'. Antenna with nine flagellomeres .... ***Greniera*** Doby and David (in part)
....***abdita*** (Peterson) group

13(11'). Rs with a short but distinct apical fork whose branches are narrowly separated by membrane (Fig. 25.67). **Male** Sc bare ventrally .... ***Mayacnephia*** Wygodzinsky and Coscarón

13'. Rs simple, or with a short indistinct apical fork whose branches are closely appressed and scarcely or not separated by membrane; **Male** Sc bare or setose ventrally 14

14(13'). Calcipala large and prominent, lamellate, rounded apically, in posterior view overlaps and sometimes conceals base of second tarsomere (Fig. 25.83). Anepisternal membrane bare. **Male** Sc setose ventrally. **Female** claw simple (Fig. 25.83) .... ***Stegopterna*** Enderlein

14'. Calcipala absent, or if present, usually small and bluntly pointed, in posterior view does not conceal base of second tarsomere. Anepisternal membrane bare or with a tuft of short setae dorsally. **Male** Sc bare or setose ventrally. **Female** claw with a small subbasal tooth or a large basal thumblike projection...15

15(14'). **Male**..16
15'. **Female** 19

16(15). R_1 dorsally with scattered black spinules on distal two-thirds or more; these more numerous apically and as stout as spinules on C..17

16'. R_1 dorsally usually with setae only; if spinules present on R_1, these confined to distal one-half or less and not as stout as spinules on C...18

17(16). Sc setose ventrally. Anepisternal membrane with tuft of pale setae dorsally (Fig. 25.77) (this tuft absent in one species)..***Metacnephia*** Crosskey

17'. Sc bare ventrally. Anepisternal membrane bare..***Ectemnia*** Enderlein

18(16'). Halter brown, with knob distinctly paler than stem. Fine setae on pleural membrane of abdominal segments 3 and 4 not conspicuously long (Fig. 25.87) ....***Cnephia*** Enderlein

18'. Halter entirely black. Fine setae on pleural membrane of abdominal segments 3 and 4 long and erect (Fig. 25.88)..***Greniera*** Doby and David

19(15'). R_1 dorsally with scattered black spinules on distal one-half (Fig. 25.71). Anepisternal membrane usually with tuft of pale setae dorsally (Fig. 25.77) (this tuft absent in one species)..***Metacnephia*** Crosskey

19'. R_1 dorsally with setae only on distal one-half. If a few spinules present on R_1, then anepisternal membrane bare..20

20(19'). Frons narrow, four times longer than wide, of nearly uniform width (Fig. 25.75)..***Ectemnia*** Enderlein

20'. Frons wider, not more than twice as long as wide, or if narrower then noticeably widening above (Fig. 25.76)...21

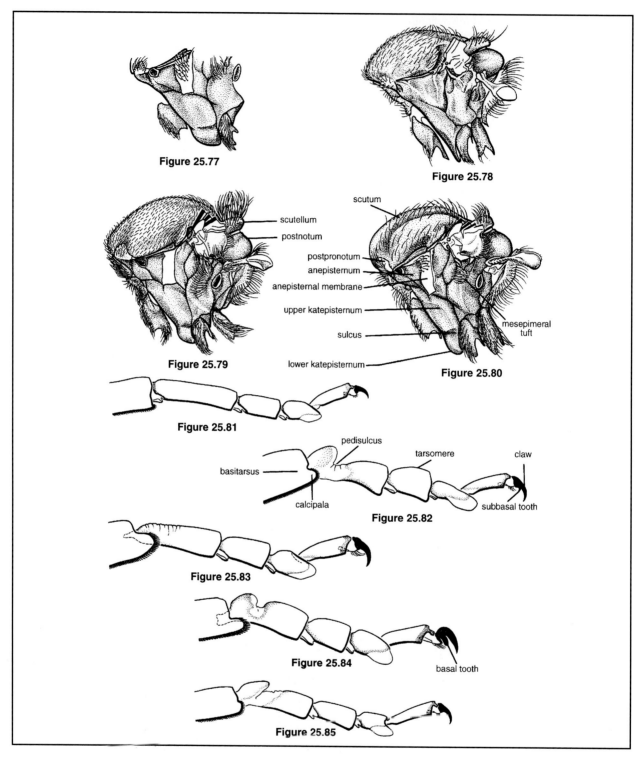

Figure 25.77. Lateral view of anterior portion of adult thorax of *Metacnephia jeanae* female.

Female 25.78. Lateral view of adult thorax of *Parasimulium* (*Parasimulium*) *stonei* male.

Figure 25.79. Lateral view of adult thorax of *Prosimulium* (*Prosimulium*) *mixtum* female.

Figure 25.80. Lateral view of adult thorax of *Gymnopais holopticus* female.

Figure 25.81. Hind tarsomeres of adult *Prosimulium* (*Prosimulium*) *ursinum* female.

Figure 25.82. Hind tarsomeres of adult *Simulium* (*Simulium*) *arcticum* female.

Figure 25.83. Hind tarsomeres of adult *Stegopterna mutata* female.

Figure 25.84. Hind tarsomeres of adult *Simulium* (*Byssodon*) *meridionale* female.

Figure 25.85. Hind tarsomeres of adult *Simulium* (*Psilozia*) *vittatum* female.

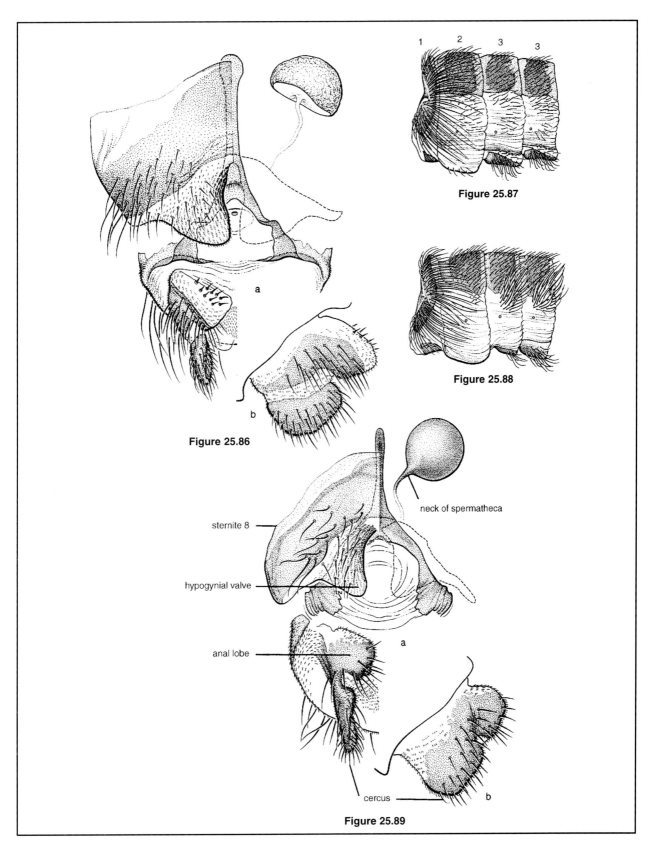

Figure 25.86. Female terminalia of adult *Twinnia tibblesi*: (*a*), ventral view; (*b*), lateral view.

Figure 25.87. Abdominal segments 1-4 of adult *Cnephia dacotensis* male.

Figure 25.88. Abdominal segments 1-4 of adult *Greniera* sp. male.

Figure 25.89. Female terminalia of adult *Gymnopais holopticus*: (*a*), ventral view; (*b*), lateral view.

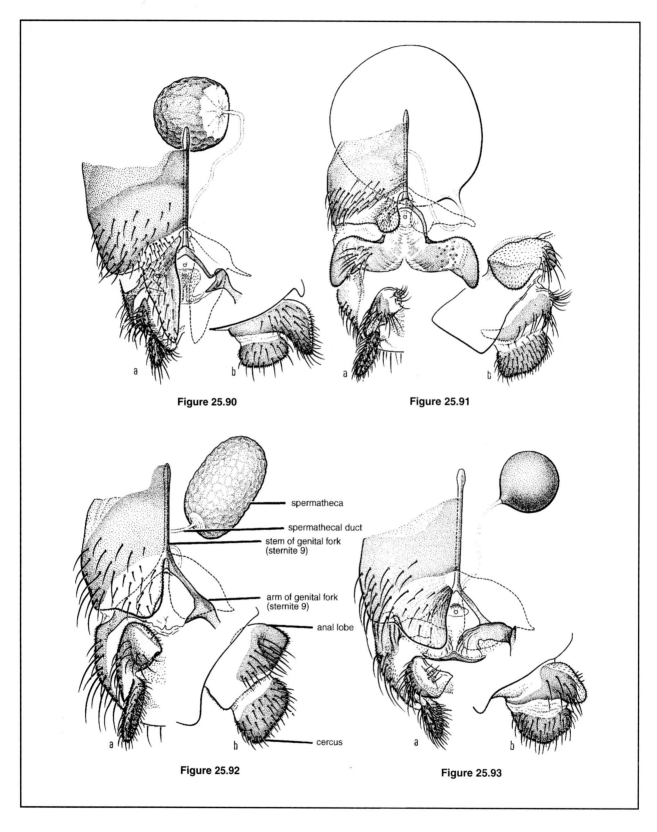

Figure 25.90. Female terminalia of adult *Prosimulium (Prosimulium) mixtum*: (*a*), ventral view; (*b*), lateral view.

Figure 25.91. Female terminalia of adult *Prosimulium (Distosimulium) pleurale*: (*a*), ventral view; (*b*), lateral view.

Figure 25.92. Female terminalia of adult *Prosimulium (Helodon) onychodactylum*: (*a*), ventral view; (*b*), lateral view.

Figure 25.93. Female terminalia of adult *Prosimulium (Parahelodon) decemarticulatum*: (*a*), ventral view; (*b*), lateral view.

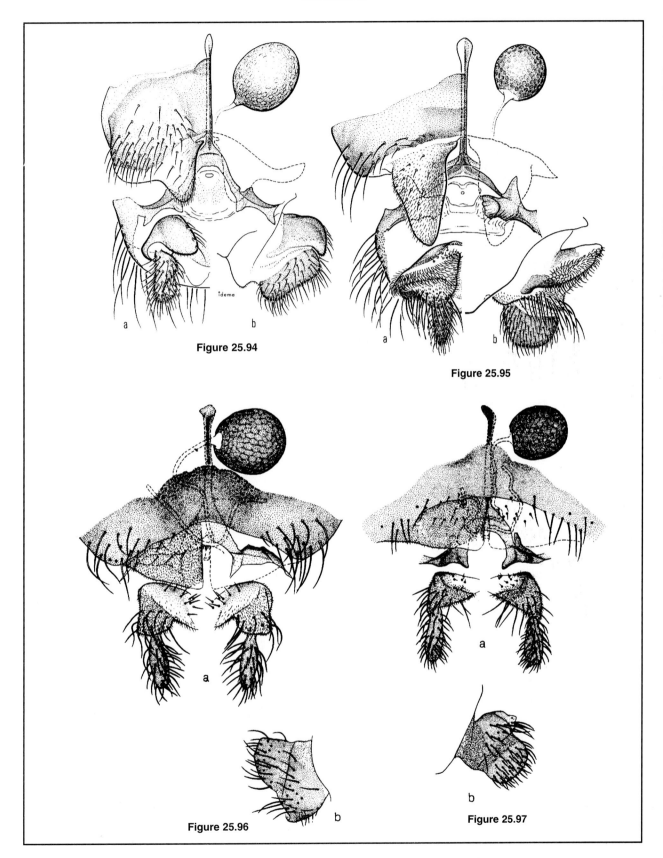

Figure 25.94. Female terminalia of adult *Simulium (Eusimulium) aureum*: (*a*), ventral view; (*b*), lateral view.

Figure 25.95. Female terminalia of adult *Simulium (Hemicnetha) virgatum*: (*a*), ventral view; (*b*), lateral view.

Figure 25.96. Female terminalia of adult *Simulium (Nevermannia) vernum*: (*a*), ventral view; (*b*), lateral view.

Figure 25.97. Female terminalia of adult *Simulium (Hellichiella) canonicola*: (*a*), ventral view; (*b*), lateral view.

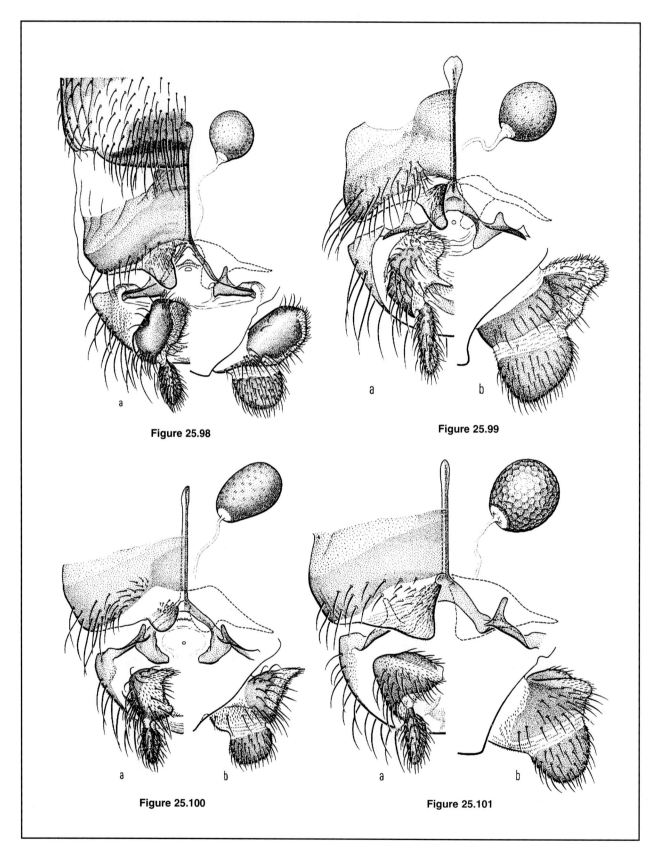

Figure 25.98. Female terminalia of adult *Simulium (Shewellomyia) pictipes*: (*a*), ventral view; (*b*), lateral view.

Figure 25.99. Female terminalia of adult *Simulium (Psilozia) vittatum*: (*a*), ventral view; (*b*), lateral view.

Figure 25.100. Female terminalia of adult *Simulium (Psilopelmia) bivittatum*: (*a*), ventral view; (*b*), lateral view.

Figure 25.101. Female terminalia of adult *Simulium (Byssodon) meridionale*: (*a*), ventral view; (*b*), lateral view.

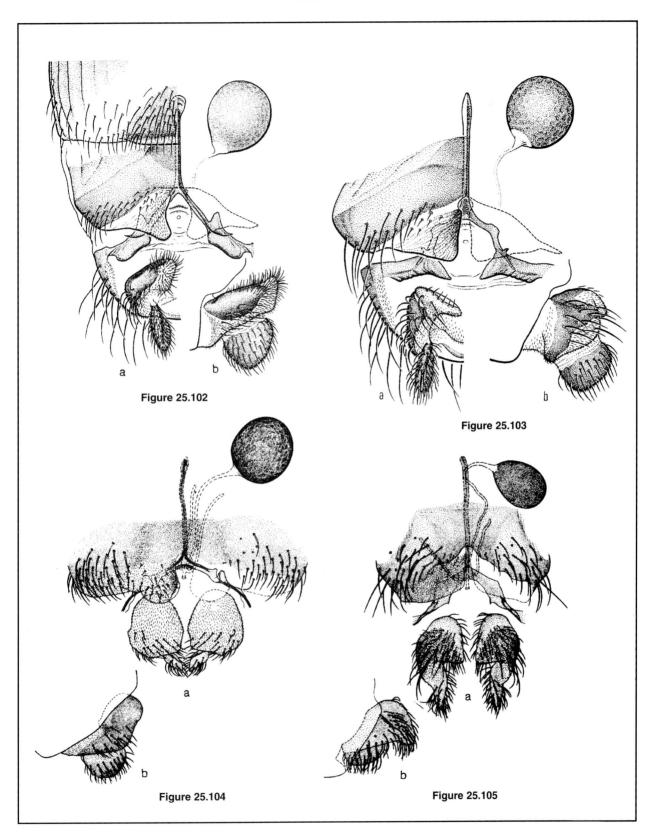

Figure 25.102. Female terminalia of adult *Simulium (Hearlea) canadense*: (*a*), ventral view; (*b*), lateral view.

Figure 25.103. Female terminalia of adult *Simulium (Parabyssodon) transiens*: (*a*), ventral view; (*b*), lateral view.

Figure 25.104. Female terminalia of adult *Simulium (Simulium) arcticum*: (*a*), ventral view; (*b*), lateral view.

Figure 25.105. Female terminalia of adult *Simulium (Simulium) hunteri*: (*a*), ventral view; (*b*), lateral view.

21(20').	Halter pale brown, with knob paler than stem. R_1 dorsally with a few spinules on distal one-half	***Cnephia*** Enderlein
21'.	Halter entirely black. R_1 dorsally with setae only on distal one-half	***Greniera*** Doby and David
22(10').	Basal section of R setose dorsally	23
22'.	Basal section of R bare dorsally	25
23(22).	Postnotum with appressed, yellow setae. Pedisulcus short, deep and distinct, its depth about one-half or more the width of the segment; legs bicolored. **Male** with gonostylus bent at nearly a right angle near base; ventral plate of aedeagus compressed laterally with a somewhat slender, keellike median lobe that is more than twice as long as broad, and with widely divergent basal arms (Fig. 25.119). **Female** terminalia as in Fig. 25.94	Subgenus ***Eusimulium*** Roubaud
23'.	Postnotum bare, without appressed yellow setae. Pedisulcus variable; legs rather uniformly yellowish brown to black, not distinctly bicolored. **Male** with gonostylus straighter, not bent near base at nearly a right angle; ventral plate of aedeagus not keellike but broad and somewhat flattened dorsoventrally. **Female** terminalia not as in Fig. 25.94	24
24(23').	**Male** gonostylus tapered to a pointed apex, with a small, apical spine, and with a slight but distinct concavity on outside margin of apical one-half; ventral plate of aedeagus somewhat quadrate, shallowly concave on distal margin, basal arms slightly convergent (Fig. 25.120). **Female** terminalia as in Fig. 25.97. Pedisulcus long, shallow and indistinct, its depth less than one-third the width of the segment	Subgenus ***Hellichiella*** Rivosecchi and Cardinali
24'.	**Male** gonostylus broad apically, not pointed but with a flattened, triangular flange medially, and a small apical spine directed anteromedially (Fig. 25.121). Ventral plate not as above. **Female** pedisulcus short and deep, its depth usually one-half or more the width of the segment	Subgenus ***Nevermannia*** Enderlein
25(22').	**Male** gonostylus broad, flattened, and with sinuous lateral margins; ventral plate of aedeagus broad, with a strong, slender, median projection that may be one-fourth to one-third as long as gonostylus (Fig. 25.122), or with a prominent, laterally compressed, broadly rounded, keellike lobe. **Female** hypogynial valve elongate, usually reaching or extended slightly beyond anterior margin of cercus; anal lobe extended well below cercus, but scarcely produced posteriorly (Fig. 25.95)	Subgenus ***Hemicnetha*** Enderlein
25'.	**Male** gonostylus variable in width but not unusually broad, and with lateral margins more regular or not strikingly sinuous; ventral plate of aedeagus without a strong median projection or prominent, laterally compressed, broadly rounded keellike lobe. **Female** hypogynial valve short, rarely reaching anterior margin of cercus; anal lobe variable	26
26(25')	**Male** ventral plate, in ventral view, with a deep median cleft so that entire structure appears H-shaped, without denticles on margins of ventral lip; gonostylus without an apical spinule (Fig. 25.123). **Female** with posterior margin of sternite 7 with a conspicuous fringe of long curved setae; ventral portion of anal lobe greatly expanded, much wider than dorsal portion, subquadrate, highly polished (Fig. 25.98)	Subgenus ***Shewellomyia*** Peterson
26'.	**Male** ventral plate of aedeagus variable, but not H-shaped nor with a deep median cleft, and with or without denticles on margins of ventral lip; gonostylus usually with one or more strong apical spinules. **Female** with posterior margin of sternite 7 often setose but without a conspicuous fringe of long curved setae; anal lobe variable in shape, usually without highly polished lateral surface, but if lateral surface polished then shaped as in Fig. 25.102	27
27(26').	**Male** gonostylus shorter than gonocoxite, flattened, subquadrate with inner distal angle prolonged and bearing a single spinule (Fig. 25.124). **Female** anal lobe narrowly angulate ventrally, or attenuate and extended well below cercus (Fig. 25.100)	Subgenus ***Psilopelmia*** Enderlein
27'.	**Male** gonostylus variable in length, more or less cylindrical, or if flattened, usually much longer than greatest width at base; number of apical spinules variable. **Female** anal lobe, if extended well below cercus then more uniformly broadened and with a more broadly rounded ventral margin	28

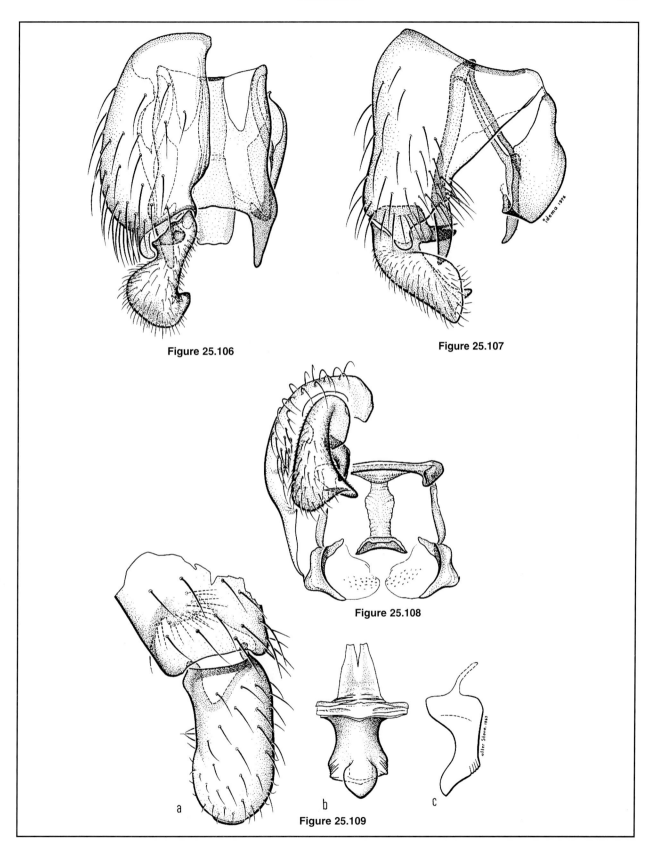

Figure 25.106. Male terminalia of adult *Parasimulium (Parasimulium) furcatum*, ventral view.

Figure 25.107. Male terminalia of adult *Parasimulium (Parasimulium) furcatum*, lateral view.

Figure 25.108. Male terminalia of adult *Parasimulium (Parasimulium) furcatum*, terminal (end) view.

Figure 25.109. Male terminalia of adult *Parasimulium (Astoneomyia) melanderi*, (*a*), ventral view; (*b*), ventral view; (*c*), lateral view.

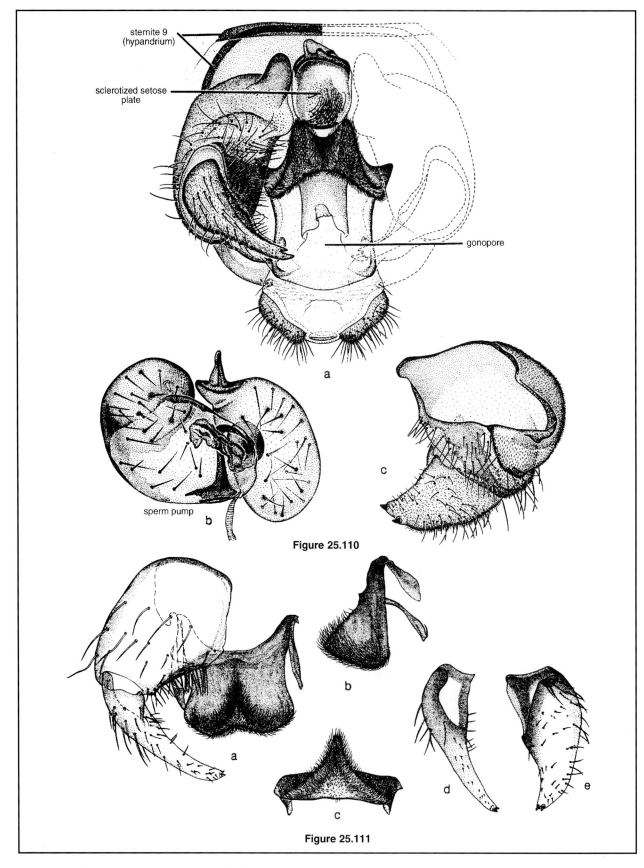

Figure 25.110. Male terminalia of adult *Piezosimulium jeanninae*: (*a*), ventral view; (*b*), sperm pump; (*c*), dorsal view of gonostylus.

Figure 25.111. Male terminalia of adult *Gymnopais* sp.: (*a*), ventral view; (*b*), lateral view; (*c*), terminal (end) view; (*d*), dorsal view of gonostylus; (*e*), lateral view of gonostylus.

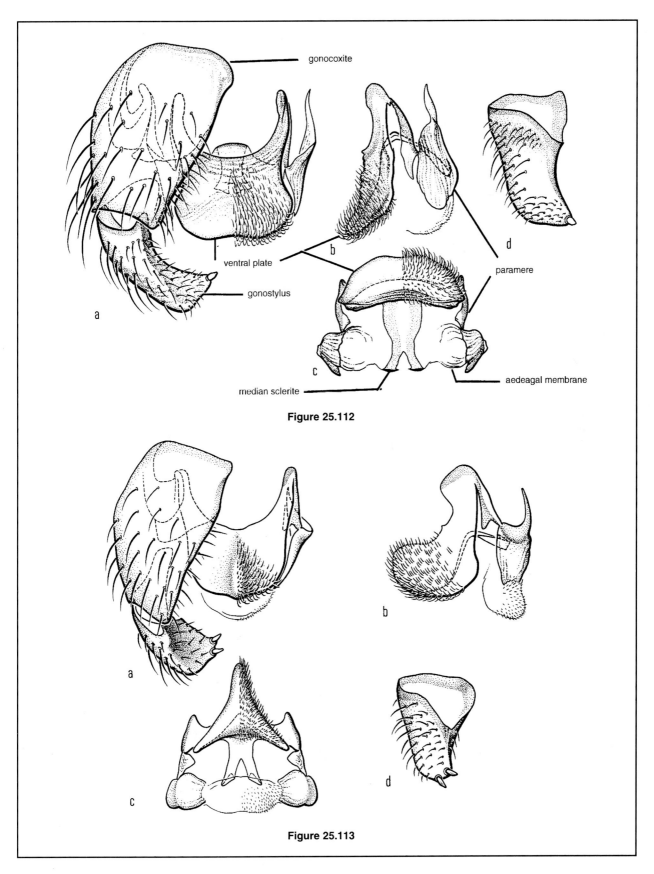

Figure 25.112. Male terminalia of adult *Twinnia tibblesi*: (*a*), ventral view; (*b*), lateral view; (*c*), terminal (end) view; (*d*), dorsal view of gonostylus.

Figure 25.113. Male terminalia of adult *Prosimulium (Prosimulium) mixtum*: (*a*), ventral view; (*b*), lateral view; (*c*), terminal (end) view; (*d*), dorsal view of gonostylus.

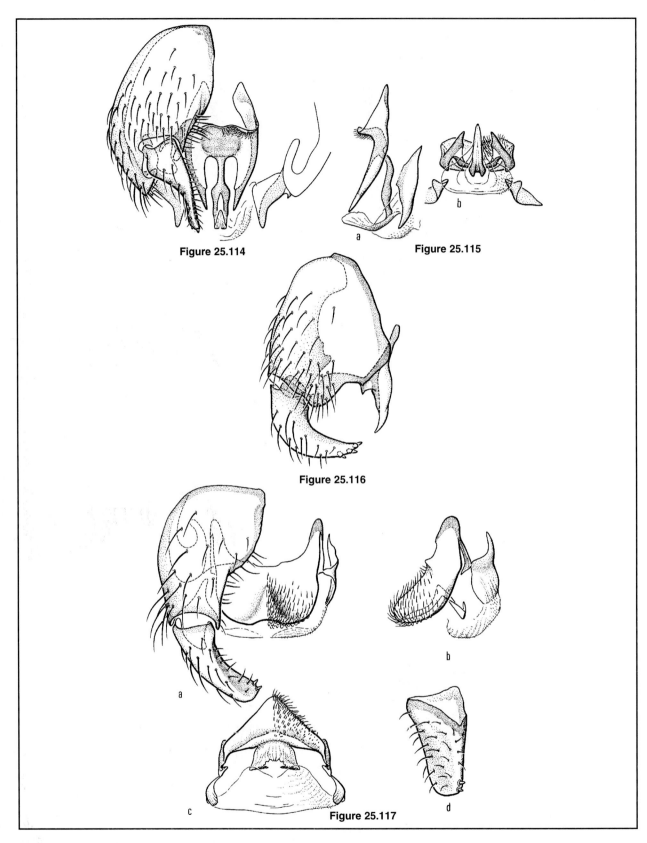

Figure 25.114. Male terminalia of adult *Prosimulium (Distosimulium) pleurale*, ventral view.

Figure 25.115. Male terminalia of adult *Prosimulium (Distosimulium) pleurale*: (*a*), lateral view; (*b*), terminal (end) view.

Figure 25.116. Male terminalia of adult *Prosimulium (Distosimulium) pleurale*, lateral view.

Figure 25.117. Male terminalia of adult *Prosimulium (Helodon) onychodactylum*: (*a*), ventral view; (*b*), lateral view; (*c*), terminal (end) view; (*d*), dorsal view of gonostylus.

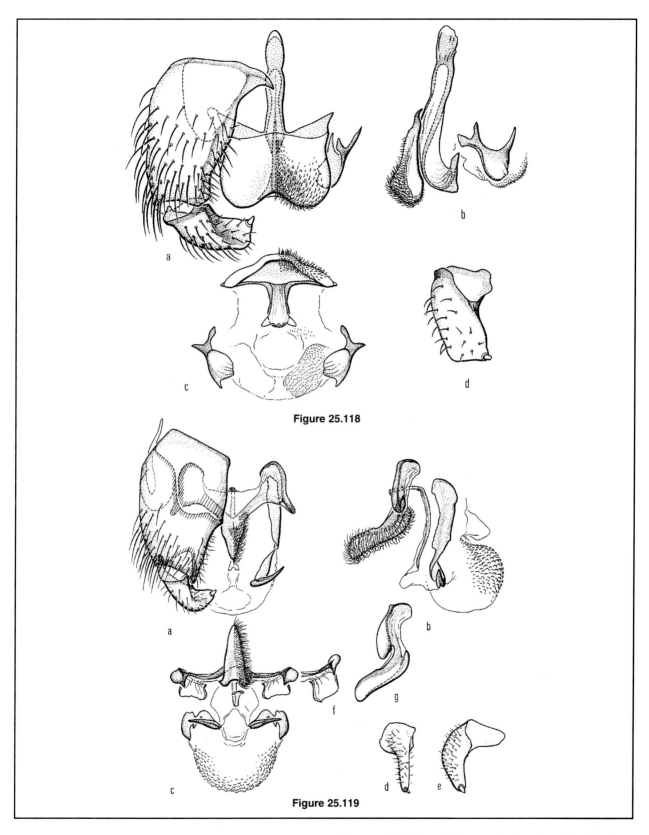

Figure 25.118. Male terminalia of adult *Prosimulium (Parahelodon) decemarticulatum*: (*a*), ventral view; (*b*), lateral view; (*c*), terminal (end) view; (*d*), dorsal view of gonostylus.

Figure 25.119. Male terminalia of adult *Simulium (Eusimulium) aureum*: (*a*), ventral view; (*b*), lateral view; (*c*), terminal (end) view; (*d*), ventral view of gonostylus; (*e*), lateral view of inner surface of gonostylus; (*f*), lateral arm of ventral plate showing variation in development; (*g*), lateral view of ventral plate showing variation in development.

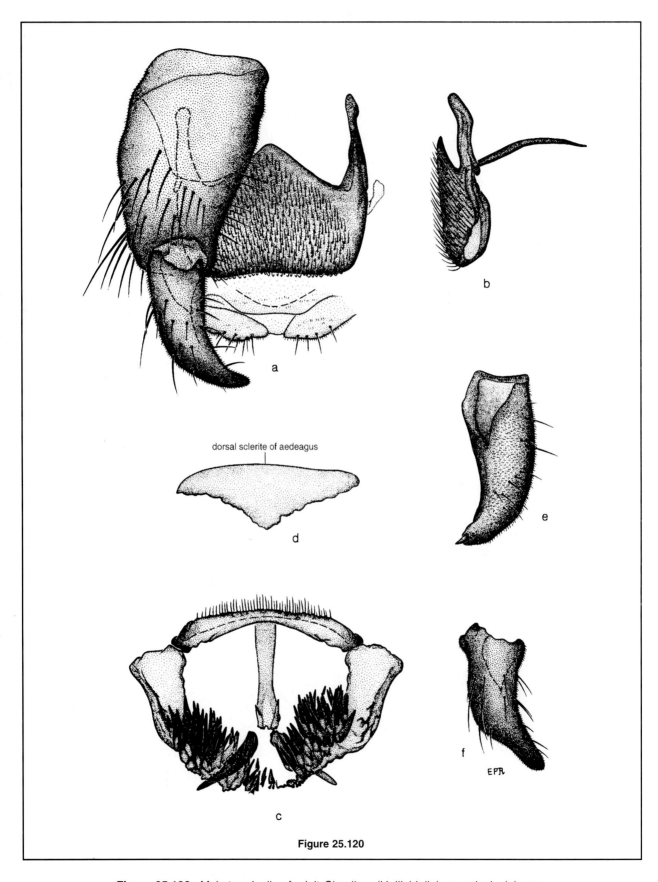

Figure 25.120. Male terminalia of adult *Simulium (Hellichiella) canonicola*: (*a*), ventral view; (*b*), lateral view; (*c*), terminal (end) view; (*d*), dorsal sclerite of aedeagus; (*e*), dorsal view of gonostylus; (*f*), ventral view of gonostylus.

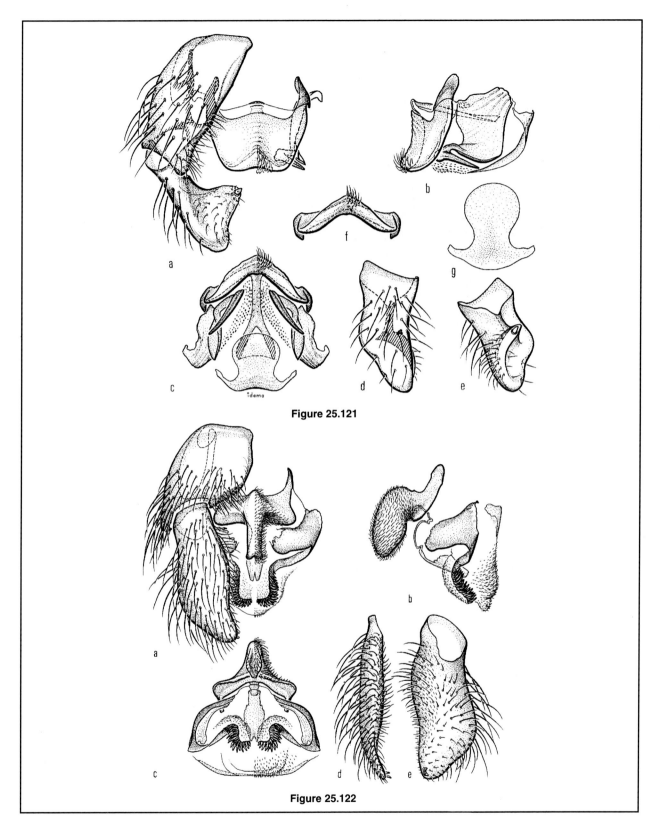

Figure 25.121. Male terminalia of adult *Simulium (Nevermannia) vernum*: (*a*), ventral view; (*b*), lateral view; (*c*), terminal (end) view; (*d*), ventral view of gonostylus; (*e*), dorsal view of gonostylus; (*f*), terminal view of ventral plate showing variation in development; (*g*), dorsal view of dorsal sclerite of aedeagus.

Figure 25.122. Male terminalia of adult *Simulium (Hemicnetha) virgatum*: (*a*), ventral view; (*b*), lateral view; (*c*), terminal (end) view; (*d*), lateral view of gonostylus; (*e*), dorsal view of gonostylus.

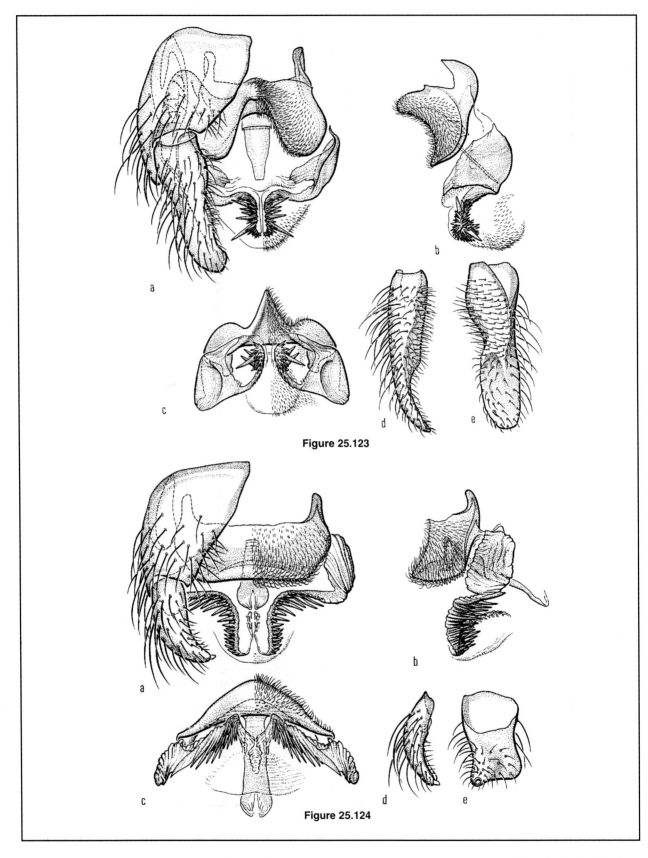

Figure 25.123. Male terminalia of adult *Simulium* (*Shewellomyia*) *pictipes*: (*a*), ventral view; (*b*), lateral view; (*c*), terminal (end) view; (*d*), lateral view of gonostylus; (*e*), dorsal view of gonostylus.

Figure 25.124. Male terminalia of adult *Simulium* (*Psilopelmia*) *bivittatum*: (*a*), ventral view; (*b*), lateral view; (*c*), terminal (end) view; (*d*), lateral view of gonostylus; (*e*), dorsal view of gonostylus.

28(27').	Calcipala small, ending well before pedisulcus (Fig. 25.85). **Male** thorax velvety black, with two submedian stripes of varying length on scutum; gonostylus short, stout, somewhat flattened, subconical to subquadrate, with 2-5 apical spinules; ventral plate of aedeagus, in ventral view, broadly triangular, with short basal arms (Fig. 25.125). **Female** black, with densely ashy gray pruinescence (a minute dust), with five dark stripes on scutum and with a distinct black and gray pattern on abdomen; anal lobe extended well below cercus, broadly triangular in shape (Fig. 25.99) ....Subgenus ***Psilozia*** Enderlein
28'.	Calcipala large, distinct, ending near dorsal margin of pedisulcus (Fig. 25.82). **Male** variable in color, with or without two submedian stripes on scutum; gonostylus long, more or less cylindrical, with a variable number of spinules; ventral plate of aedeagus, in ventral view, variable in shape, but if somewhat triangular then narrower and with long basal arms. **Female**, if black and gray pruinose, with 0-3 stripes on scutum and without a distinct black and gray pattern on abdomen; anal lobe usually not extended well below cercus, but if somewhat extended then not broadly triangular..29
29(28').	**Male** scutum without anterolateral white or silvery spots. Ventral plate of aedeagus broad, lamellate, with distal margin roundly truncate, and lacking marginal denticles (Fig. 25.127). **Female** fore coxa dark; fore tibia entirely brown to black, without a bright white patch anteriorly; claw with a large basal, thumblike projection (Fig. 25.84); anal lobe shortly acuminate ventrally (Fig. 25.101) ....Subgenus ***Byssodon*** Enderlein
29'.	**Male** scutum usually with variably distinct anterolateral silvery or white spots, or with submedian stripes; ventral plate of aedeagus variable, if broad and lamellate then with a median notch and lip, and often with marginal denticles. **Female** fore tibia usually with a bright white patch anteriorly; if fore tibia entirely brown or black, then anal lobe variable but not acuminate ventrally, at most with a short rounded point.............................30
30(29').	**Male** gonostylus long and slender, narrowest at about midlength, with a small, internal, setose basal lobe and a single apical spinule; ventral plate of aedeagus, in ventral view, with a rounded distal margin, a median notch, and a short ventral lip without marginal denticles (Fig. 25.126). **Female** tarsomeres of fore leg slender; claw with a small subbasal tooth (Fig. 25.82); scutum with two distinct stripes; outer surface of anal lobe mostly bare and polished; posterior and inner margins of hypogynial valve shallowly emarginate (Fig. 25.102) ....Subgenus ***Hearlea*** Vargas, Martínez Palacios and Díaz Nájera
30'.	**Male** gonostylus variable, but if long then more uniformly widened; ventral plate of aedeagus, in ventral view, variable, but if somewhat rounded on distal margin then margins of ventral lip with denticles. **Female** tarsomeres of fore leg variable but often noticeably widened and laterally flattened; claw variable; scutum patterned or unpatterned; anal lobe variable, but outer surface usually setose and not polished; hypogynial valves short, their posterior and inner margins nearly straight to rounded 31
31(30').	**Male** scutum anteriorly with a V-shaped pruinose mark that is sometimes continuous anterolaterally with pruinose border that encircles lateral and posterior margins (Fig. 25.129). Ventral plate of aedeagus with apicolateral corners produced shoulderlike, and with a broad somewhat dorsoventrally flattened ventral lip with variably distinct denticles basally, and with a medially notched apex; gonostylus with a densely spinulose internal lobe basally, and with a tiny subapical spinule (Fig. 25.128). **Female** scutum semishiny, covered with a light but even pruinescence that allows underlying integument to show through; claw with a large basal thumblike projection; terminalia as in Fig. 25.103 ....Subgenus ***Parabyssodon*** Rubtsov
31'.	**Male** scutum unpatterned, or if patterned, then without an anterior V-shaped mark although anterolateral spots sometimes obliquely oriented; ventral plate of aedeagus variable but ventral lip, if present, not dorsoventrally flattened, and without distinct denticles on margins; gonostylus variable, but not exactly as above (Figs 25.129-25.134). **Female** scutum variable, but usually distinctly shiny or more densely pruinose; claw usually simple or with a small subbasal tooth, rarely (two species) with a large basal thumblike projection; terminalia not as in Fig. 25.103 above (Figs. 25.104-25.105) ....Subgenus ***Simulium*** Latreille

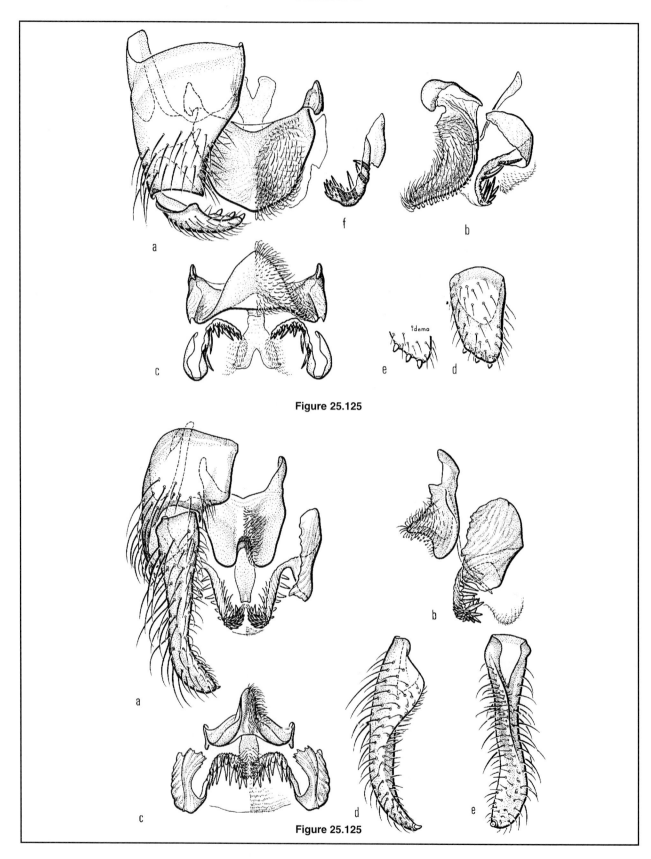

Figure 25.125. Male terminalia of adult *Simulium* (*Psilozia*) *vittatum*: (*a*), ventral view; (*b*), lateral view; (*c*), terminal (end) view; (*d*), ventral view of gonostylus; (*e*), tip of dorsal surface of gonostylus; (*f*), lateral view of paramere.

Figure 25.126. Male terminalia of adult *Simulium* (*Hearlea*) *canadense*: (*a*), ventral view; (*b*), lateral view; (*c*), terminal (end) view; (*d*), lateral view of gonostylus; (*e*), dorsal view of gonostylus.

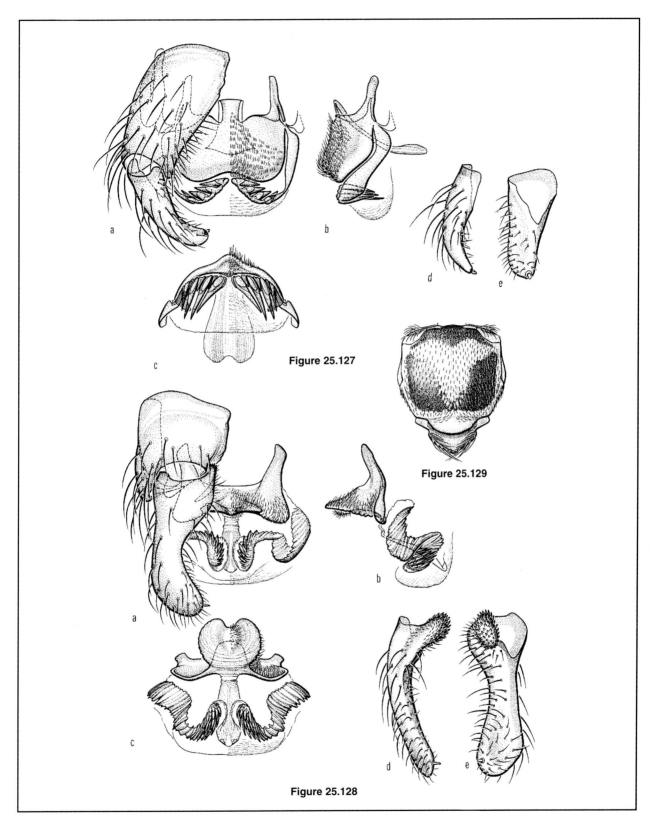

Figure 25.127. Male terminalia of adult *Simulium* (*Byssodon*) *meridionale*: (*a*), ventral view; (*b*), lateral view; (*c*), terminal (end) view; (*d*), lateral view of gonostylus; (*e*), dorsal view of gonostylus.

Figure 25.128. Male terminalia of adult *Simulium* (*Parabyssodon*) *transiens*: (*a*), ventral view; (*b*), lateral view; (*c*), terminal (end) view; (*d*), lateral view of gonostylus; (*e*), dorsal view of gonostylus.

Figure 25.129. Male thorax of adult *Simulium* (*Parabyssodon*) *transiens*, dorsal view.

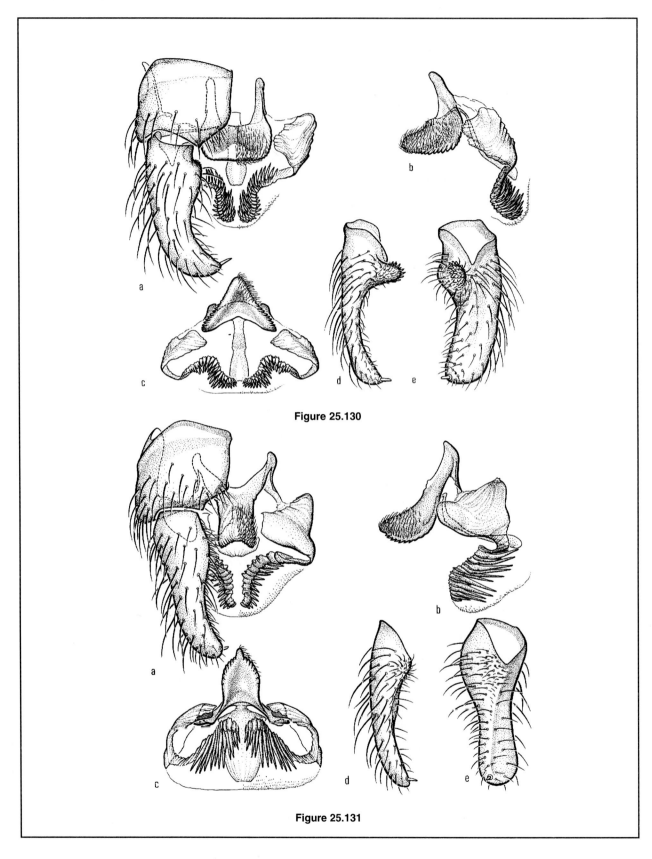

Figure 25.130. Male terminalia of adult *Simulium* (*Simulium*) *tuberosum*: (*a*), ventral view; (*b*), lateral view; (*c*), terminal (end) view; (*d*), lateral view of gonostylus; (*e*), dorsal view of gonostylus.

Figure 25.131. Male terminalia of adult *Simulium* (*Simulium*) *jenningsi*: (*a*), ventral view; (*b*), lateral view; (*c*), terminal (end) view; (*d*), lateral view of gonostylus; (*e*), dorsal view of gonostylus.

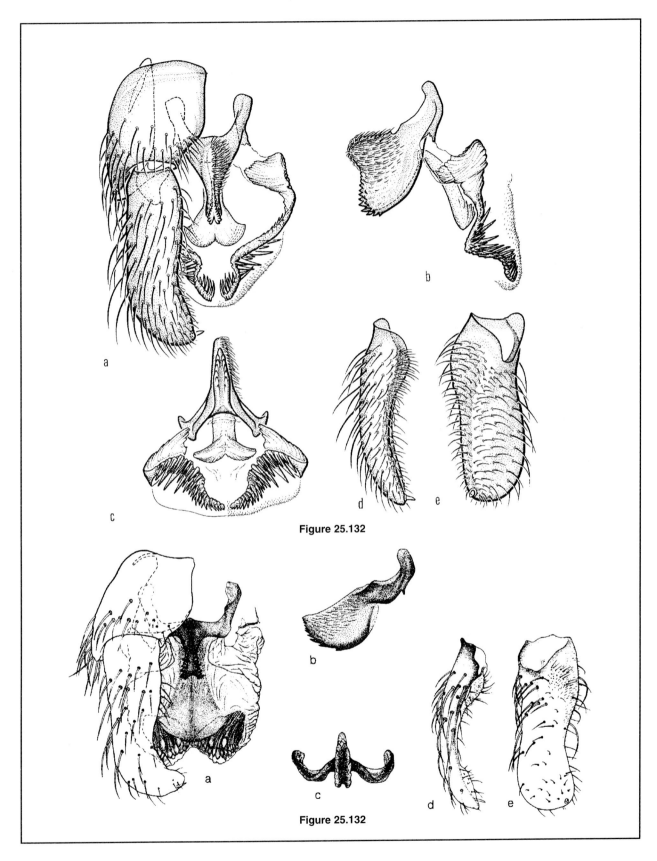

Figure 25.132. Male terminalia of adult *Simulium* (*Simulium*) *decorum*: (*a*), ventral view; (*b*), lateral view; (*c*), terminal (end) view; (*d*), lateral view of gonostylus; (*e*), dorsal view of gonostylus.

Figure 25.133. Male terminalia of adult *Simulium* (*Simulium*) *arcticum*: (*a*), ventral view; (*b*), lateral view; (*c*), terminal (end) view; (*d*), lateral view of gonostylus; (*e*), dorsal view of gonostylus.

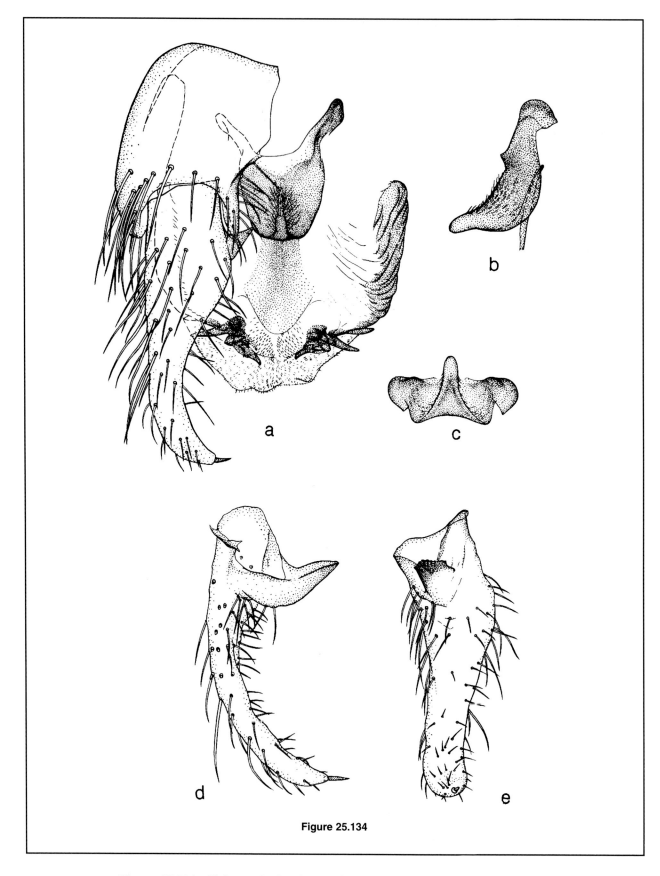

Figure 25.134. Male terminalia of adult *Simulium* (*Simulium*) *hunteri*: (*a*), ventral view; (*b*), lateral view; (*c*), terminal (end) view; (*d*), lateral view of gonostylus; (*e*), dorsal view of gonostylus.

ADDITIONAL TAXONOMIC REFERENCES

General
Coquillett (1898); Malloch (1914); Johannsen (1903, 1934) Dyar and Shannon (1927); Vargas (1945); Nicholson and Mickel (1950); Stone and Jamnback (1955); Wirth and Stone (1956); Carlsson (1962); Davies et al. (1962); Wood et al. (1963); Stone (1964a, 1965); Stone and Snoddy (1969); Crosskey (1973, 1988, 1990); Peterson (1981); Adler and Kim (1986); Peterson and Kondratieff (1995).

Regional faunas
Alabama: Stone and Snoddy (1969).
Alaska: Stone (1952); Sommerman (1953); Peterson (1970); Hershey et al. (1995).
Alberta: Fredeen (1958); Abdelnur (1968); Currie (1986).
British Columbia: Hearle (1932).
California: Wirth and Stone (1956); Hall (1972, 1974); Tietze and Mulla (1989).
Canada: Shewell (1958); Peterson (1970).
Colorado: Ward and Kondratieff (1992); Peterson and Kondratieff (1995)
Connecticut: Stone (1964a).
Delaware: Sutherland and Darsie (1960).
Eastern Canada: Twinn (1936).
Florida: Pinkovsky and Butler (1978).
Idaho: Twinn (1938).
Kansas: Snyder and Huggins (1980).
Maine: Bauer and Granett (1979).
Manitoba: Fredeen (1958).
Maritime Provinces: Lewis and Bennett (1979).
Michigan: Gill and West (1955); Merritt et al. (1978).
Minnesota: Nicholson and Mickel (1950).
Montana: Newell (1970).
Nebraska: Pruess and Peterson (1987).
Newfoundland: Lewis and Bennett (1973).
New Jersey: Crans and McCuiston (1970).*
New York: Stone and Jamnback (1955).
Northeastern United States: Cupp and Gordon (1983).
Ontario: Davies et al. (1962); Wood et al. (1963).
Pennsylvania: Frost (1949); Eckhart and Snetsinger (1969); Adler and Kim (1986).
Quebec: Back and Harper (1978).
Rhode Island: Dimond and Hart (1953).
Saskatchewan: Fredeen (1958, 1981, 1985).
South Carolina: Arnold (1974); Noblet et al. (1979)
Southeastern United States: Snoddy and Noblet (1976).
Utah: Twinn (1938); Peterson (1955, 1960).
Washington: Bacon and McCauley (1959).
Western United States: Stains and Knowlton (1943).
Wisconsin: Anderson (1960).

Taxonomic treatments at the generic level
(L = larvae; P = pupae; A = adults)
Gymnopais: Wood (1978)-L, P, A.
Parasimulium: Peterson (1977)-A.
Piezosimulium: Peterson (1989)-P, A.
Prosimulium: Peterson (1970)-L, P, A.
Simulium: Nicholson and Mickel (1950)-A; Stone and Jamnback (1955)-L, P, A; Peterson (1960, 1993)-L, P, A; Davies et al. (1962)-P, A; Wood et al. (1963)-L; Stone (1964a)-L, P, A; Stone and Snoddy (1969)-L, P, A; Currie (1986)-L, P; Peterson and Kondratieff (1995)-L, P, A. (Also see references under headings **General**, and **Regional faunas**, above.)
Twinnia: Wood (1978)-L, P, A.

---

*Crans, W. J., and L. G. McCuiston. 1970. A checklist of the blackflies of New Jersey (Diptera: Simuliidae). Mosq. News 30:654-55.

Table 25A. Summary of ecological and distributional data for *Simuliidae (Diptera)*. (For definition of terms see Tables 6A-6C; table prepared by R. W. Merritt, K. W. Cummins, and B. V. Peterson.)

Taxa (number of species in parentheses)	Habitat	Habit	Trophic Relationships	North American Distribution	Ecological References*
Simuliidae(143) (black flies, buffalo gnats)	Generally lotic—erosional	Generally clingers (ring of minute hooks on terminal end of abdomen and proleg; silk threads)	Generally collectors—filterers		1957, 1984, 3121, 3126, 596, 3441, 3589, 4187, 809, 3339, 4558, 3443, 4555, 4560, 2090
Cnephia(4)	Lotic—erosional	Clingers	Collectors—filterers	Widespread	82, 595, 1380, 2206, 2207, 2677, 3441, 3442, 2460, 3940, 4550
Ectemnia(2)	Lotic—erosional	Clingers	Collectors—filterers	Widespread	3642
Greniera(4)	Lotic—erosional	Clingers	Collectors—filterers	Northeastern United States and eastern Canada, British Columbia	854
Gymnopais(5)	Lotic—erosional	Clingers	Scrapers	Northwestern North America, from British Columbia to Alaska	864, 3642, 4511, 4527, 4236
Mayacnephia(4)	Lotic—erosional	Clingers	Collectors—filterers	Western North America	
Metacnephia(7)	Lotic—erosional	Clingers	Collectors—filterers	Western United States and Canada; Alaska	563, 3642, 4553, 4554
Parasimulium(4)	Lotic—erosional hyporheic zone	Clingers	Collectors—filterers	Pacific Northwest: California, Oregon, Washington, British Columbia	724, 3119, 3856, 730
Prosimulium(42)	Lotic—erosional	Clingers	Collectors—filterers	Widespread	82, 563, 595, 862, 1380, 1420, 1462, 2206, 2207, 2677, 2678, 3120, 3441, 3442, 3443, 4236, 1697, 2485, 3589, 1605, 4234
Simulium (66)	Lotic and lentic erosional	Clingers	Collectors—filterers	Widespread	82, 152, 226, 492, 563, 595, 603, 671, 674, 675, 737, 1100, 1289, 1290, 1291, 1557, 2078, 2168, 2206, 2207, 2209, 2211, 2212, 2214, 2215, 2447, 2609, 2674, 2677, 2746, 2785, 2830, 2888, 2889, 2890, 3117, 3121, 3227, 3277, 3441, 3442, 3635, 3642, 3940, 4452, 4453, 4549, 4550, 4551, 4552, 4554, 393, 4556, 4579, 1664, 394, 395, 1405, 4620, 1407, 1408, 2488, 4234, 2210, 3443, 4147, 4236, 983, 4557, 4558, 4559, 1702, 3588, 4561, 1324, 2460, 4205
Stegopterna(2)	Lotic—erosional	Clingers	Collectors—filterers	Widespread	2677, 2678, 3441, 3442, 3642
Twinnia(3)	Lotic—erosional	Clingers	Scrapers	Western and northeast- United States and Canada	595, 864, 1462, 4527

* Emphasis on trophic relationships.

CHIRONOMIDAE

W. P. Coffman
*University of Pittsburgh,
Pennsylvania*

L. C. Ferrington, Jr.
*University of Kansas,
Lawrence, Kansas*

INTRODUCTION

The family Chironomidae is an ecologically important group of aquatic insects often occurring in high densities and diversity. The relatively short life cycles and the large total biomass of the numerous larvae confer ecological energetic significance on this taxon (as consumers and prey) and the partitioning of ecological resources by a large number of species presumably enhances the biotic stability of aquatic ecosystems.

Chironomid larvae are known to feed on a great variety of organic substrates: (1) coarse detrital particles (leaf- and wood-shredders); (2) medium detrital particles deposited in or on sediments (gatherers and scrapers); (3) fine detrital particles in suspension, transport, or deposited (filter-feeders, gatherers, and scrapers); (4) algae; benthic, planktonic, or in transport (scrapers, gatherers, and to a lesser extent filter-feeders); (5) vascular plants (miners); (6) fungal spores and hyphae (gatherers); (7) animals (as simple predators, often preying on other chironomid larvae, or as parasites on a variety of taxa, although the latter may most often be commensal relationships).

Most aquatic predators feed extensively on chironomids (larvae, pupae, and/or adults) at some point in their life cycles. As young-of-the-year of some predaceous fish species increase in size, they may rely less on chironomids.

Chironomids occur in most types of aquatic ecosystems, as well as moist soils, tree holes, pitcher plants and dung. The range of conditions under which chironomids are found is more extensive than that of any other group of aquatic insects. Almost the complete range of gradients of temperature, pH, salinity, oxygen concentration, current velocity, depth, productivity, altitude, latitude, and others have been exploited, at least by some chironomid species. The wide ecological amplitude displayed by this family is related to the very extensive array of morphological, physiological, and behavioral adaptations found among the members of the family. Ecologists have utilized the partitioning of gradients by various midge species to characterize the overall ecological conditions of lentic and lotic systems.

Presumably, the great species diversity in this family, with probably more than 2,000 species in the Nearctic Region and, perhaps, 20,000 species worldwide, is the product of its antiquity, relatively low vagility (instances of geographic isolation appear to be common), and evolutionary plasticity. The overall diversity of the family is also reflected in the rich chironomid faunas of many aquatic ecosystems. The number of chironomid species present in most systems often accounts for at least 50% of the total macroinvertebrate species diversity (richness component). Natural lakes, ponds, and streams usually have at least 50 and often more than 100 species. The actual number of species present in a system is the result of the complex of physical, chemical, biological and biogeographic conditions. Obviously, the least diverse faunas are correlated with extreme conditions.

As holometabolous insects, chironomids have four distinct life stages: egg, larva, pupa, and adult (imago). Duration of the larval stage, with four instars, may last from less than two weeks to several years, depending on species and environmental conditions. In general, warm water, high-quality food, and small species correlate with shorter life cycles. Average-sized chironomids (fourth instar about 5 mm long) in temperate systems usually have from one to two generations per year. First larval instars may be planktonic, are often difficult to sample, and may have a unique morphology not treated in keys that are designed for fourth instar larvae. Later instars are usually benthic. Toward the end of the fourth instar, the larval thoracic region begins to swell with the formation of the pupal integument and adult tissues. The pupal stage begins with apolysis, the separation (not shedding) of the larval from the underlying pupal integument. After ecdysis, the pupa usually remains hidden in debris until it swims to the surface where eclosion (adult emergence) takes place. Technically, the adult stage begins with the pupal-adult apolysis (pharate adult), which occurs a short time before eclosion. Chironomid adults usually live a few days, although some species survive for several weeks. The adult stage performs the functions of reproduction and dispersal. As a rule, chironomid adults do not need to feed, as reflected by the usual condition of reduced mouthparts and atrophied gut. However, many species (perhaps most) will take liquid and semiliquid carbohydrate sources such as aphid honeydew and flower nectars. The consumption of these energy rich substances presumably maximizes the potential for the completion of additional ovarian cycles.

Mating takes place in aerial swarms, on the water surface in skating swarms, or on solid substrates. Females may broadcast the eggs at the water surface or, more frequently, deposit gelatinous egg masses on the open water or on emergent vegetation. Egg or larval development may be arrested under unfavorable environmental conditions. Some species are facultatively or obligatorily parthenogenetic, and larvae parasitized by nematodes or nematomorphs may produce intersexual adults. Excellent reviews of chironomid biology may be found in Oliver (1971), Davies (1976), and Pinder (1986).

In most ecological studies the larva is the life stage that is most frequently encountered. However, the quantitative collection of early instars is often difficult (Chap. 3). Even when larval populations can be "adequately" sampled (Chap. 3), sorting, particularly retrieval of early instars, can require prodigious amounts of time (densities of 50,000 larvae per m^2 are not uncommon). After sorting, the identification of the larvae also proves to be difficult because: (1) only a small percentage of the larvae of Nearctic species have been described; (2) larvae of several important groups of genera are essentially inseparable; (3) keys usually will not separate early instars; and (4) slide preparation of specimens is usually required, and even then diagnostic features are often difficult to see.

The collection of pupal exuviae has only recently become a widely used method for the investigation of chironomids (e.g., Coffman 1973; Wilson and Bright 1973), although the idea is not new (Thienemann 1910; Lenz 1955). The advantages of the technique are numerous. Most generic and species-level taxa are easily separated in the pupal stage, even if a generic or species name cannot be assigned. Large numbers of specimens can usually be collected in a short time. Collection takes advantage of the fact that pupal exuviae are not wettable for a period of time and float on the surface where they can be collected at natural or artificial blockades in streams and along the windward shore in lentic systems. Quantitative collections can be made using enclosures because all known species rise to the surface for eclosion. Since the collections are made from the surface or the water column, the substrates are not disturbed and most emerging species, regardless of larval microhabitat, are collected. Once a voucher slide series for species present in a system has been identified, most can be recognized with a dissecting microscope. Because the pupal exuviae have no tissues (except in the thoracic horns), no clearing is required. All the diagnostic features are usually discernible on every specimen because the specimen is depressed by the cover glass into two dimensions. As there is no ambiguity about the age of pupal exuviae, the size of a species is determined. On occasion, the larval and pupal exuviae as well as the adult of a specimen are collected entangled with each other providing ideal material for association. Studies for which the collection of pupal exuviae is well suited include diversity (richness and actual taxonomic composition of faunas), phenology (diel and annual), biogeography, size distribution of species, sex ratios, and production of adults.

EXTERNAL MORPHOLOGY

Larvae

Mature (fourth instar) chironomid larvae range in size from about 2 to 30 mm. There are three main body divisions: head, thorax, and abdomen, all of which have structures that are utilized for generic-level diagnosis. These divisions may appear as pale yellow or white in preserved larvae or may be variously pigmented or patterned. Common pigmentation includes dark yellow, brown, black, or yellow with a black posterior margin in the case of the head capsule and yellow, green, blue, violet, rose, orange, or brown in the case of the thorax and abdomen. In addition the presence of hemoglobin in certain larvae produces a strong "blood red" color in living specimens, which often changes to a more diffuse red, red-orange, or brown color when preserved in ethanol.

Head: The head is in the form of a completely sclerotized capsule, and is never fully retractile. Plates that have characters of diagnostic importance associated with them are the *genae* or lateral sclerites, the *frontoclypeal apotome* or dorsal sclerite, the *labrum* or anterodorsal sclerite (Figs. 26.140-26.141), and the *mentum* or medioventral sclerite (Figs. 26.1-26.4).

Sensory structures situated on the dorsal aspect of the head capsule are the eyespots, antennae, and numerous setae, scales, and lamellae. Eyespots (Figs. 26.1-26.4) occur on the dorsal or dorsolateral aspects of the genae. Antennae (Figs. 26.1-26.3) also arise from the genae but occur anterior to the eyespots, except in Tanypodinae larvae which have antennae that are retractile into the head capsule (Fig. 26.4). Setae originate from all head capsule sclerites, but in dorsal aspect only the S setae of the labrum are of diagnostic importance. The dorsal S setae (Figs. 26.1-26.2, 26.3) are paired medial setae of the labrum and are referred to as SI through SIV setae. The anteriormost setal pair are the SI setae; setae SII through SIV originate in sequence progressing posteriorly. Various scales or lamellae (Figs. 26.239, 26.241) may surround or originate in close proximity to the bases of the S setae, particularly SI.

Structures associated with feeding show extreme variation among the genera of Chironomidae. In general these structures are concentrated on the anteroventral aspect of the head capsule including the region of the mentum and the ventral surface of the labrum. Included in this category are the pecten epipharyngis, premandibles, mandibles, maxillae, prementohypopharyngeal complex, and the mentum. (See Figures 26.1-26.4 for the ventral perspective and orientation of these structures.)

The pecten epipharyngis and premandibles originate from the ventral surface of the labrum. The pecten epipharyngis generally consists of three scales that may be simple or have numerous apical teeth. In some genera the scales may be fused and the pecten epipharyngis thus appears as a single plate or bar. The premandibles are movable ancillary feeding structures and may have a blunt apical cusp or, in the case of some predatory genera, may have from one to several mesally projecting teeth (Fig. 26.140). Premandibles are vestigial or lacking in Tanypodinae and Podonominae larvae.

Paired mandibles and maxillae occur ventral to the premandibles. The mandibles may have strong but blunt apical and lateral teeth (Figs. 26.297-26.299) or may have an elongate and sharply pointed apical tooth (Fig. 26.14); the former is typical of nonpredatory species and the latter, of predatory species. The maxilla, which is usually lightly sclerotized and difficult to discern, bears a maxillary palp which is generally detectable on its anteroventral surface. In Tanypodinae larvae and in some Chironomini species (specifically "*Harnischia* Complex" genera such as *Robackia* and *Beckidia*) the maxillary palp is more elongate and often multisegmented.

The prementohypopharyngeal complex and the mentum occur on the medioventral aspect of the head capsule. In Tanypodinae larvae the prementohypopharyngeal complex is armed

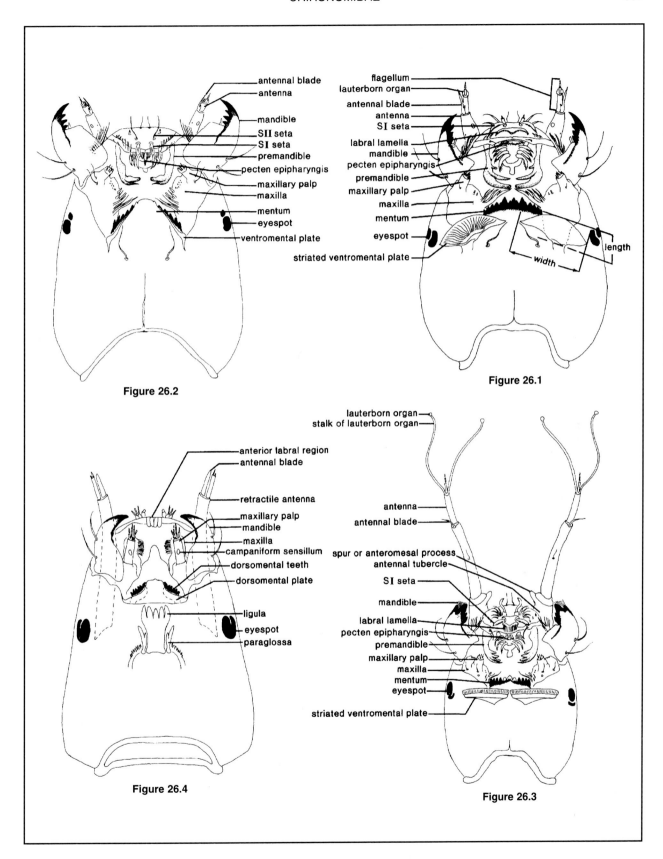

Figure 26.1. Generalized Chironomini head capsule (ventral view).

Figure 26.2. Generalized Orthocladiinae head capsule (ventral view).

Figure 26.3. Generalized Tanytarsini head capsule (ventral view).

Figure 26.4. Generalized Tanypodinae head capsule (ventral view).

apically with a well-developed, readily seen ligula. In most other chironomid larvae the mentum may obscure the greater part of the prementohypopharyngeal complex and, as no well-developed structure homologous to the tanypod ligula is usually present, this structure may be difficult to locate.

The current interpretation of the larval mentum holds that this structure is a double-walled, medioventral plate consisting of a dorsomentum and a ventromentum. In Chironomini, Pseudochironomini, and Tanytarsini the ventromentum is greatly expanded laterally into striated ventromental (paralabial) plates which are quite conspicuous. With the exception of the Tanypodinae, the ventromentum may be only slightly to moderately expanded laterally in the remaining chironomid genera. When only slightly expanded, the ventromental plates may appear to be absent or vestigial. In some genera the plates may be quite obvious and may even have a cardinal beard (i.e., elongate setae arising in definite rows) originating from the dorsal surface.

In most chironomid genera, again excluding Tanypodinae, the dorsomental plate is usually well sclerotized and has conspicuous teeth. In these cases the term "mentum" is simply used to refer to the toothed margin of the dorsomental plate. Notable exceptions occur within certain genera such as *Acamptocladius* and *Protanypus* in which the dorsomental plates do not meet on the ventral midline of the head capsule and the ventromental plate covers all (in the case of *Acamptocladius*) or most (in the case of *Protanypus*) of the dorsomental teeth. A more detailed discussion of the structure of the mentum is provided by Saether (1971a).

In Tanypodinae larvae the ventromental plate appears to be membranous or lightly sclerotized and is difficult to see. The dorsomental plate is very reduced in Pentaneurini genera and is also difficult to discern. In the remaining Tanypodinae genera, the lateral aspect of the dorsomental plate is well developed and gives the appearance of paired, heavily sclerotized, platelike structures armed with teeth along the anterior margin. Medially the dorsomental plate has reduced sclerotization and often is not readily distinguishable from the ventromental plate.

Thorax: The larval thorax, consisting of the first three segments behind the head capsule, is only clearly demarcated from the remainder of the body segments in the late fourth instar. At this time the larva is often referred to as a prepupa since the thorax is swollen by developing pupal structures. Often structures of the pupa, such as the respiratory organ or pronounced setae, are visible, thus allowing association of the larva with the pupal stage. Ventrally, on the first thoracic segment, a pair of prolegs typically arise from a common base and bifurcate apically. However, in some genera no bifurcation may occur and a single proleg appears to exist, or the prolegs may be reduced or vestigial. The prolegs are generally armed apically with sclerotized claws, the size and structure of which may be of specific importance. Numerous setae of species-level importance may arise from the thoracic segments, most notably a distinct ring of strong setae in some case-building Tanytarsini larvae and some *Eukiefferiella* larvae.

Abdomen: The abdomen of the larva consists of all post-thoracic segments. Structures of the abdomen used for generic identification include setation, prolegs, anal tubules, procerci, ventral tubules, and supraanal setae.

In all Tanypodinae larvae except Pentaneurini genera, the lateral margins of at least the anterior five abdominal segments are equipped with a dense row of setae termed a *lateral hair fringe* (Fig. 26.5). In *Natarsia* sp. the fringe is somewhat less dense but still distinguishable in most preserved specimens. A distinct lateral hair fringe is also present in the Orthocladiinae genera *Stackelbergina* and *Xylotopus* (Fig. 26.290). Most other abdominal setae patterns appear to have only species-level significance.

As is the case with the prolegs of the thorax (anterior prolegs), those of the abdomen (posterior prolegs) show quite a bit of variation in structure. The posterior prolegs usually occur as paired structures with apical claws (Figs. 26.5, 26.291, 26.293), but they may occasionally be fused or extremely reduced (Figs. 26.294-26.296).

Anal tubules occur on the anal segment in most chironomid genera and may originate above, between, or occasionally along the basal margin of the posterior prolegs. Anal tubules generally occur as two paired structures, but occasionally may consist of one or three pairs, or rarely be absent.

On the dorsum of the preanal segment, a pair of fleshy tubercles, or procerci, may occur (Figs. 26.339-26.340). One to several strong setae originate from the apex of the procerci and generally 0-2 weak lateral setae are present. Occasionally the procerci may have basal or lateral regions of sclerotization or pigmentation; basal spurs may also be present. Rarely the procerci are vestigial or lacking (Figs. 26.293-26.296).

In some genera of Chironomini the lateral margins of the antepenultimate abdominal segment are produced into fleshy protuberances that are called *ventral tubules*. When present, these occur as one or two pairs. In some species of the genus *Goeldichironomus* the anteriormost pair bifurcates just beyond the base (Fig. 26.177).

Supraanal setae occur on the anal segment, usually just dorsal to the anal tubules (Figs. 26.26, 26.339-26.340). Although these setae are generally only of species-level diagnostic importance, in the Tanypodinae genus *Pentaneura* and in the Podonominae genera *Boreochlus* and *Paraboreochlus* they appear to have generic diagnostic importance when used in concert with other body characters.

GLOSSARY OF MORPHOLOGICAL TERMS FOR KEYS TO LARVAE

ACCESSORY BLADE OF MANDIBLE—bladelike structure arising on inner margin of mandible of some Tanypodinae larvae (e.g., *Hudsonimyia karelena* Roback; Fig. 26.55) and oriented along the inner edge.

ANAL TUBULES—appendages of the anal segment above, between, or at the basal margin of the posterior prolegs (e.g., Figs. 26.26, 26.291, 26.293, 26.295); normally two pairs, occasionally one or three pairs, rarely absent.

ANNULATIONS—areas of reduced sclerotization giving the appearance of concentric rings or annuli on certain antennal segments of most Diamesinae and Podonominae (e.g., Figs. 26.323-26.324); also present on segments of one genus of Tanytarsini (Fig. 26.120) and on the stalks of the Lauterborn organs in another genus of Tanytarsini (Fig. 26.119).

ANTENNA (plural=antennae)—paired, segmented sensory organs, one on each side of the head capsule, directed anteriorly (e.g., Figs. 26.1-26.3, 26.320); retractile in Tanypodinae (e.g., Figs. 26.4, 26.6-26.9); usually with conspicuous Lauterborn organs in Tanytarsini (e.g., Figs.

26.103-26.114) and sometimes in Chironomini (e.g., Figs. 26.138-26.139), variously modified in Orthocladiinae (e.g., Figs. 26.219-26.235, 26.237-26.238).

ANTENNAL BLADE—bladelike structure usually originating at the apex of the first antennal segments (e.g., Figs. 26.1-26.3), occasionally originating from the margin of the second segment (e.g., Figs. 26.145, 26.323) or third segment (Fig. 26.146).

ANTENNAL TUBERCLE—pronounced tubercle from which the antenna originates in all Tanytarsini and in the Orthocladiinae genus *Abiskomyia* (Fig. 26.223); may be simple (i.e., lacking armature; e.g., Figs. 26.110, 26.112-26.113), with anteromesal spur (e.g., Figs. 26.114, 26.116, 26.118), elongate spine (e.g., Figs. 26.103-26.104, 26.106-26.107, 26.117), or armature in the form of a palmate process (Fig. 26.105).

ANTERIOR LABRAL REGION—anterior portion of the labrum (e.g., Figs. 26.4, 26.140-26.141); armed with setae and peglike sensillae in most genera; occasionally with additional scales (Fig. 26.322) and other armature.

ANTERIOR PROLEGS—locomotory structures originating from ventral surface of first thoracic segment (Fig. 26.291) usually armed with claws; may be paired, with a common base, singular or vestigial.

APICAL DISSECTIONS—divided distally into several branches or strands; refers to structure of S setae (Fig. 26.213).

BIFID—cleft or divided into two parts, forked; refers to structure of S setae (Fig. 26.209).

CAMPANIFORM SENSILLUM—sense organ of the antenna or maxillary palp having the appearance of a circular structure with a central area of reduced sclerotization (e.g., Figs. 26.4, 26.23-26.24, 26.29, 26.223-26.227).

CARDINAL BEARD—setae appearing to arise from underneath the ventromental plates in some Orthocladiinae (e.g., Figs. 26.278-26.289) and Prodiamesinae (e.g., Figs. 26.315, 26.317).

CARDO—base of maxilla (Figs. 26.199-26.200).

CRENULATE—with a series of indentations giving the appearance of small, blunt or evenly rounded, convex bumps on the anterior margin of the ventromental plates (e.g., Figs. 26.149, 26.180-26.182); see *scalloped* for contrasting pattern of indentations.

DISTAL—near or toward the free end of any appendage; that part of a structure farthest from the body.

DORSOMENTAL PLATE—heavily sclerotized platelike structure (Fig. 26.4), conspicuous in Tanypodini (Fig. 26.11), Procladiini (Fig. 26.12), Macropelopiini (e.g., Figs. 26.18-26.20, 26.25, 26.60, 26.62, 26.64-26.66), somewhat reduced in Natarsiini (Fig. 26.17) and very reduced or vestigial in Pentaneurini.

DORSOMENTAL TEETH—teeth along the anterior margin of the dorsomental plates in Tanypodini (Fig. 26.11), Procladiini (Fig. 26.12), Macropelopiini (e.g., Figs. 26.18-26.20, 26.25, 26.60, 26.62, 26.64-26.66) and Natarsiini (Fig. 26.17); arranged in longitudinal rows in Coelotanypodini (e.g., Figs. 26.10, 26.13).

EYESPOT—dorsal or dorsolateral pigmentation patches of the head capsule used for detecting light (e.g., Figs. 26.1-26.4); may consist of a single pair, two pairs, or, rarely, three pairs of pigment patches.

FLAGELLUM—the second through terminal antennal segments (Fig. 26.1); used to calculate the antennal ratio, which is a ratio of the combined lengths of these segments relative to the length of the first antennal segment.

FRONTOCLYPEAL APOTOME—dorsal sclerite of the head capsule.

HEAD CAPSULE—fused sclerites of the head region, forming a hard, compact case (e.g., Figs. 26.1-26.4).

HEAD RATIO—length of the head capsule divided by the maximum width of the head capsule.

LABRAL LAMELLA (plural = labral lamellae)—one or more smooth to apically pectinate, scalelike or leaflike, thin plates surrounding or anterior to the bases of the SI setae (e.g., Figs. 26.239, 26.241).

LABRAL SENSILLA (plural = labral sensillae)—labral seta SIV; in some Chironomini seta SIV is greatly enlarged and two- (Fig. 26.141) or three-segmented (Fig. 26.140).

LATERAL HAIR FRINGE—a more or less continuous and conspicuous row of setae originating from the lateral margin of at least the five anterior abdominal segments in all Tanypodinae larvae except Pentaneurini genera (Fig. 26.5), very reduced in Natarsiini larvae; also present in the Orthocladiinae genera *Xylotopus* Oliver and *Stackelbergina* Shilova and Zelentsov (Fig. 26.290).

LAUTERBORN ORGAN—compound sensory organ usually originating at the apex of the second antennal segment and usually having a teardrop shape; most pronounced in Tanytarsini where they may occur at the end of elongate stalks (e.g., Figs. 26.3, 26.113-26.114, 26.119-26.120), arise alternately (e.g., Figs. 26.103-26.104) or oppositely (e.g., Figs. 26.105-26.114, 26.119-26.120), be conspicuously striated (e.g., Figs. 26.103-26.106, 26.108-26.110) or simple (e.g., Figs. 26.107, 26.111, 26.113-26.114, 26.119-26.120); less pronounced in most Chironomini, but occasionally well developed in certain genera (e.g., Figs. 26.138-26.139); poorly developed and often inconspicuous in most Orthocladiinae; lacking, vestigial, or solidly fused to apex of segment 2 (Fig. 26.72) in Tanypodinae.

LAUTERBORN ORGAN STALK—a sclerotized structure basal to the Lauterborn organ and serving as an attachment to the antenna; sometimes elongate and delicate in some Tanytarsini genera (e.g., Figs. 26.3, 26.113-26.114) or short and straplike (e.g., Figs. 26.103-26.106, 26.108-26.109, 26.111-26.112) or vestigial (e.g., Figs. 26.107, 26.110); usually not apparent in most Chironomini and Orthocladiinae.

LIGULA—a well-sclerotized, apically toothed, internal plate of the prementohypopharyngeal complex in Tanypodinae (e.g., Figs. 26.4, 26.13, 26.15-26.16, 26.41-26.53, 26.63, 26.86); reduced, not discernible, or absent in other subfamilies.

M-APPENDAGE—medioventral appendage of prementum; well developed in Tanypodinae and consisting of a hyaline triangular plate often with lateral vesicles and a well to poorly developed pseudoradula (Figs. 26.60, 26.62, 26.64-26.66, 26.69, 26.71, 26.89-26.91).

MANDIBLE—stout, well-sclerotized, chewing or grasping structures (e.g., Figs. 26.1-26.4); usually with well-developed dents (or "teeth") along the inner margin (e.g., Figs. 26.14, 26.56-26.59, 26.142, 26.168, 26.297-26.303, 26.318-26.319), occasionally with dents reduced or vestigial (e.g., Figs. 26.4, 26.54-26.55, 26.84, 26.92, 26.115).

MAXILLA (plural = maxillae)—in the strictest sense the second pair of jaws or eating structures; usually poorly sclerotized and difficult to discern; located ventral and slightly posterior to the mandibles in most larvae (e.g., Figs. 26.1-26.4).

MAXILLARY PALP—palp originating from the ventral margin of the maxilla and armed apically with numerous sensillae and other sensory structures; usually small in most Orthocladiinae and Tanytarsini larvae (e.g., Figs. 26.2, 26.3); occasionally very elongate and/or with numerous segments in Chironomini (e.g., Figs. 26.166-26.167); usually conspicuous and occasionally segmented in Tanypodinae (e.g., Figs. 26.4, 26.23-26.24, 26.27-26.29); synonyms—maxillary palpus, palpus.

MENTUM—usually toothed and well-sclerotized medioventral plate of head capsule (e.g., Figs. 26.1-26.3, 26.98-26.102, 26.130-26.136, 26.143, 26.148-26.165, 26.169-26.176, 26.178-26.184, 26.242-26.289, 26.315-26.317, 26.321, 26.325-26.336, 26.338); reduced or membranous in Tanypodinae.

PALMATE—resembling the structure of a hand; with a palm-like basal portion and fingerlike apical processes; usually referring to structure of the SI seta (Fig. 26.211) or armature of antennal tubercle in Tanytarsini (Fig. 26.105).

PALMATE PROCESS OF ANTENNAL TUBERCLE—anteromesally projecting process of antennal tubercle in some Tanytarsini species (Fig. 26.105), produced mesally to resemble a palmate structure.

PARAGLOSSA—(plural = paraglossae)—curved, bifid, serrated or pectinate rod or plate originating from opposite sides at base of and running parallel to the ligula in Tanypodinae (e.g., Figs. 26.4, 26.30-26.34, 26.76).

PECTINATE—comblike or coarsely featherlike; usually referring to the structure of the SI seta.

PECTEN EPIPHARYNGIS—sclerotized scale or scales originating just ventral to the apex of the labral margin (e.g., Figs. 26.1-26.3, 26.124-26.125, 26.202-26.205, 26.236, 26.240); may be smooth, spinelike, pectinate, serrate, or fused.

PECTEN HYPOPHARYNGIS—small sclerites on mediobasal portion of hypopharynx; in Tanypodinae they occur near the base or sides of the ligula, are well developed, and have transverse rows of spines along anterior margin (Figs. 26.79-26.80, 26.88).

PLUMOSE—very finely featherlike; usually referring to the structure of the SI seta (e.g., Figs. 26.239, 26.241).

POSTERIOR PROLEGS—locomotory structures originating from the ventrolateral margin of the terminal abdominal segment; usually paired and with apical claws (e.g., Figs. 26.5, 26.291, 26.293), occasionally fused or extremely reduced (e.g., Figs. 26.294-26.296).

PREANAL SEGMENT—the penultimate abdominal segment.

PREMANDIBLE—paired, movable, ancillary feeding appendages originating from the ventral surface of the labrum in the vicinity of the pecten epipharyngis; usually with an apical cusp or one to several mesally projecting teeth (e.g., Figs. 26.1-26.3, 26.126-26.127, 26.140); vestigial or lacking in Podonominae and Tanypodinae.

PREMENTOHYPOPHARYNGEAL COMPLEX—a double lobed structure ventral to the esophagus and dorsal to the mentum; usually covered by the mentum in slide mounted specimens or with only the apical portion visible; ligula arises from this structure in Tanypodinae.

PROCERCUS (plural = procerci)—fleshy tubercle originating from dorsal surface of preanal segment (e.g., Figs. 26.194-26.195, 26.339-26.340); usually paired and carrying several strong apical and two weak lateral setae; occasionally with lateral sclerotization, basal spurs, triangular lobes (Fig. 26.314) or pigmentation patches; rarely reduced, vestigial, or lacking (e.g., Figs. 26.293-26.296).

PSEUDORADULA—longitudinal band of fine to minute spinules on dorsal surface of m-appendage in Tanypodinae (Figs. 26.60, 26.62, 26.64-26.66, 26.69, 26.71, 26.89-26.91).

SI SETA (plural = SI setae)—anteriormost dorsolabral seta; occurring as paired setae (e.g., Figs. 26.1-26.3, 26.206-26.207, 26.208-26.218, 26.236, 26.239, 26.241).

SII SETA (plural = SII setae)—dorsolabral seta immediately posterior to SI seta; occurring as paired seta (e.g., Figs. 26.2, 26.216).

SCALE—small, sclerotized, flat evagination of a head capsule plate.

SCALLOPED—with a series of concave indentations giving the appearance of evenly rounded depressions on the anterior margin of the ventromental plates (e.g., Figs. 26.164-26.165); see *crenulate* for contrasting pattern of indentations.

SPIRACULAR RING—lightly sclerotized or differentially pigmented margin of spiracle on the dorsum of the antepenultimate abdominal segment in some Podonominae genera (Fig. 26.340).

STRIATIONS—a series of fine, longitudinal impression lines; referring to the pattern of dorsal striae on the ventromental plates of Chironomini, Tanytarsini, and Pseudochironomini (e.g., Figs. 26.1, 26.3, 26.98-26.102, 26.130-26.136, 26.143, 26.148-26.162, 26.164-26.165, 26.169-26.176, 26.178-26.184, 26.333-26.335).

SUPRAANAL SETA (plural = supraanal setae)—seta on anal segment just dorsal to anal tubules; occurring as paired setae (e.g., Figs. 26.26, 26.339-26.340).

THORAX—first three body segments posterior to the head capsule.

VENTRAL TUBULE—fleshy protuberences originating on the ventrolateral margin of the antepenultimate abdominal segment of some genera (Figs. 26.177, 26.194-26.195); if present usually occurring as one or two pairs; occasionally with the anterior pair bifurcating beyond base.

VENTROMENTAL PLATE—plate occurring ventral to the mentum and extending laterally; usually occurring as well-separated, paired plates with conspicuous striations in Chironomini (e.g., Figs. 26.1, 26.130-26.134, 26.136, 26.143, 26.148-26.162, 26.164-26.165, 26.169-26.174, 26.180-26.184), or without striations (Fig. 26.163) or striated but closely appressed medially (e.g., Figs. 26.175-26.176, 26.178-26.179); usually striated and closely appressed medially in Tanytarsini (e.g., Figs. 26.3, 26.98, 26.100-26.102), less commonly widely separated (Fig. 26.99); usually nonstriated, narrow, inconspicuous, or very reduced in Orthocladiinae (e.g., Figs. 26.2, 26.242-26.243, 26.245-26.247, 26.251-26.254, 26.256-26.265, 26.267-26.270, 26.272-26.273, 26.275-26.277), occasionally somewhat conspicuous, (e.g., Figs. 26.244-26.245, 26.248, 26.250, 26.254, 26.256, 26.267, 26.270, 26.273) or with a cardinal beard (e.g., Figs. 26.278-26.289); vestigial or large and covering the lateral mental teeth in Diamesinae (e.g., Figs. 26.321, 26.325-26.328); vestigial in Podonominae (e.g., Figs. 26.336, 26.338); large and with (e.g., Figs. 26.315, 26.317) or without (Fig. 26.316) a cardinal beard in Prodiamesinae; large, striated, and closely appressed medially in Pseudochironomini (e.g., Figs. 26.333-26.335).

Pupae

Chironomid pupae range in length from about 1.5 mm to 20 mm. There are three main body divisions: head, thorax, and abdomen. The morphology of the pupa is almost competely external, since it is considered to be a modified larval integument within which the adult develops (Figs. 26.341-26.343). The structures of the pupa are best seen in the cast skins (pupal exuviae), and the general description below and keys to follow are based on exuviae. (For reference to particular structures see the glossary for pupal morphology.)

Head: The head region of chironomid pupal exuviae consists primarily of eye, antenna, and mouthpart sheaths. The area of integument covering the vertex of the pharate adult head is the frontal apotome (Fig. 26.341).

Thorax: The thoracic region of chironomid pupal exuviae bears the leg, wing, and halter sheaths (Figs. 26.344-26.346). The thorax also has several groups of setae and the taxonomically very important thoracic horn (Fig. 26.342).

Abdomen: The abdomen of chironomid pupal exuviae consists of eight similar segments and one or more additional segments modified into anal lobes and genital sheaths (Fig. 26.343). The terga (and sometimes sterna) often bear distinctive groupings of spines, recurved hooks and shagreen (Fig. 26.343). The anal lobes often bear a fringe of setae and/or two or more spinelike to hairlike macrosetae (Fig. 26.343).

The pigmentation of pupal exuviae is not as varied as that of larvae. However, most species have a characteristic pigmentation even if it is the absence or near absence of pigment. In others the thorax and abdomen are yellow, golden yellow, yellow brown, brown, dark brown, or gray. Many have such pigment in a distinct pattern, especially on the abdomen. Together with size and shape, the pigmentation of pupal exuviae may be an important aid in sorting and identification.

GLOSSARY OF MORPHOLOGICAL TERMS FOR KEYS TO PUPAE

ANAL LOBE—usually broad "swimming fins" located at tip of abdomen; with a fringe of setae (e.g., Figs. 26.343, 26.379, 26.393, 26.415, 26.417, 26.460, 26.608, 26.704, 26.709) or without (e.g., Figs. 26.347, 26.360, 26.471, 26.474, 26.501, 26.531, 26.572, 26.659); may be atypical in form (e.g., Figs. 26.347, 26.404, 26.476, 26.497, 26.507, 26.582); often with one or more macrosetae.

ANAL LOBE FRINGE—a fringe of setae along outer and sometimes inner margins of anal lobes; may be sparse (e.g., Figs. 26.424-26.425, 26.455) or dense (e.g., Figs. 26.343, 26.393, 26.460, 26.466).

ANAL LOBE MACROSETAE—usually large lateral or terminal setae on anal lobes; two in Tanypodinae, 0-12 in other taxa (e.g., Figs. 26.343, 26.347, 26.349, 26.357, 26.360, 26.377, 26.413, 26.444, 26.467-26.468, 26.497, 26.504); may be very small in some Orthocladiinae (e.g., Figs. 26.481, 26.485, 26.496, 26.526, 26.528); absent in some Orthocladiinae and all Chironominae.

CAUDOLATERAL ARMATURE ON ABDOMINAL SEGMENT 8—a single spine (e.g., Figs. 26.627, 26.634, 26.668, 26.687), a comb (e.g., Figs. 26.641, 26.646, 26.655, 26.705) or a compound spur (Figs. 26.697-26.698) on each caudolateral margin of abdominal segment 8; present in many Chironominae and a very few Orthocladiinae (however, these are probably not homologous to those in Chironominae).

CEPHALIC TUBERCLES—enlarged, sometimes ornate areas of the frontal apotome upon which the frontal setae are located (e.g., Figs. 26.661, 26.663, 26.670-26.671, 26.680); present in many Chironominae and a few Orthocladiinae (not to be confused with frontal warts [Figs. 26.443, 26.595] or frontal tubercles [Figs. 26.678, 26.680] that may be present in addition to simple frontal setae or cephalic tubercles).

CONJUNCTIVA—weakly chitinized connecting areas between abdominal segments (e.g., Fig. 26.710).

DORSAL CENTRAL SETAE—usually 3 or 4 small to large setae located on the central area of the thorax, often close to the thoracic suture and arranged in pairs (e.g., Fig. 26.342).

DORSAL SETAE—a. setae located on abdominal terga; often difficult to see, but they may be large, spinelike or branched (e.g., Figs. 26.385, 26.393); b. one or two setae (usually lamellar) on the dorsal surface of each anal lobe (e.g., Figs. 26.608, 26.620, 26.630, 26.641, 26.647).

FELT CHAMBER—the internal component of the thoracic horn which is apparent as the continuation of the connection with the tracheal system (e.g., Figs. 26.348, 26.361, 26.378, 26.390, 26.398-26.399).

FRONTAL APOTOME—the chitinous region between and posterior (dorsal) to the bases of the antennae; often with frontal setae (e.g., Figs. 26.341, 26.443, 26.603, 26.670, 26.679).

FRONTAL SETAE—a pair of setae usually located on the frontal apotome, often on cephalic tubercles (e.g., Figs. 26.341, 26.443, 26.603, 26.670, 26.679); rarely on prefrons (Fig. 26.602); often absent.

FRONTAL TUBERCLES—a pair of tubercles on frontal apotome of some Chironominae in addition to simple frontal setae or cephalic tubercles (Figs. 26.678, 26.680)

FRONTAL WARTS—a pair of tubercles on the frontal apotome of some Orthocladiinae (Figs. 26.443, 26.595).

GENITAL SHEATHS—usually weakly chitinized sacs that enclose the male and female genitalia; most conspicuous in male (e.g., Figs. 26.347, 26.349, 26.353, 26.367, 26.528).

LAMELLAR LATERAL SETAE—large, broad setae on pleura of abdominal segments 5-8 (e.g., Figs. 26.360, 26.455, 26.638, 26.688); often absent.

LATERAL ABDOMINAL FRINGE—dense rows of setae along pleura of at least some abdominal segments (Figs. 26.357, 26.404, 26.461).

LEG SHEATHS—cuticular covering of pharate adult legs (Figs. 26.344-26.346).

MULTISERIAL FRINGE—a fringe of anal lobe setae consisting of 2 or more rows (e.g., Figs. 26.393, 26.466, 26.715).

"NASE"—a small to moderate "nose-shaped" tubercle near tip of wing sheath (Fig. 26.656); present on most Tanytarsini.

PEARL ROWS—one or more rows of small, rounded projections along the distal margins of the wing sheaths (Figs. 26.414, 26.422).

PLEURA—lateral components of each abdominal segment that connect the terga and sterna.

POSTERIOR SPINE ROW(S)—spines along the posterior margins of at least some terga (e.g., Figs. 26.343, 26.428, 26.442, 26.453, 26.462, 26.480, 26.498, 26.540, 26.548).

PREFRONS—the cuticular region anterior (dorsal) to the frontal apotome (Fig. 26.341).

PRECORNEAL SETAE—usually 3 setae close to the base of the thoracic horn (e.g., Figs. 26.342, 26.420, 26.432, 26.500, 26.537, 26.550, 26.640, 26.642).

PSB II, III—lateral protuberances on abdominal segment 2 or segments 2 and 3 (e.g., Figs. 26.343, 26.423, 26.436, 26.463, 26.487, 26.672).

RECURVED HOOKS (HOOKLETS)—a. one or more rows of anteriorly directed hooks on the posterior margin of abdominal segment 2 (e.g., Figs. 26.343, 26.436, 26.439, 26.592, 26.596, 26.662, 26.672); b. one row or partial row of large anteriorly directed hooks on the posterior margins (or conjunctiva) of abdominal segments 3-5; present on a few Orthocladiinae (e.g., Figs. 26.554, 26.559, 26.561, 26.563, 26.565, 26.576).

SHAGREEN—complete or partial fields of small spinules on terga, pleura, or sterna (e.g., Figs. 26.343, 26.451, 26.482, 26.559, 26.651, 26.654, 26.703).

SPINE GROUPS—groups of spines (long or short) on terga—usually near the center but including those along the posterior margin (e.g., Figs. 26.343, 26.437, 26.440, 26.442, 26.472, 26.609, 26.615, 26.645); the spines of such groups are always at least slightly larger than any shagreen that may be present.

STERNA—the ventral plates of the abdominal segments.

TERGA—the dorsal plates of the abdominal segments.

THORACIC COMB—a row of spines near the base of the thoracic horn of some Tanypodinae (Figs. 26.363, 26.365, 26.370).

THORACIC HORN—a structure of the anterior thorax which has been assumed to have respiratory functions; may be with a plastron (e.g., Figs. 26.348, 26.359, 26.380), with a surface meshwork (e.g., Figs. 26.396, 26.405), simple (e.g., Figs. 26.408, 26.420, 26.427, 26.435, 26.448, 26.465, 26.479, 26.492, 26.500, 26.535, 26.555, 26.585, 26.600), with "hairs" (e.g., Figs. 26.606, 26.618, 26.642), or branched (e.g., Figs. 26.667, 26.696, 26.699, 26.713).

UNISERIAL FRINGE—a fringe of the anal lobes in which the setae are in a single row (e.g., Figs. 26.429, 26.608, 26.627, 26.704, 26.709).

WING SHEATHS—cuticular covering of the pharate adult wings (Figs. 26.342, 26.344).

Adults

Adult chironomids range in size from about 1.5 mm to 20 mm. There are three main body divisions: head, thorax, and abdomen (Fig. 26.761).

Head: The most conspicuous features of the head are the eyes and antennae. The antennae of most male chironomids are plumose (Figs. 26.761, 26.794). The antennae of the females are not plumose and are usually shorter than those of the males and with fewer flagellomeres. However, some male antennae are not plumose and have a reduced number of flagellomeres (e.g., Fig. 26.795).

Thorax: The adult thorax bears the legs, wings, and halteres. The legs of chironomids are moderately long, especially the first pair, and consist of a femur, tibia, and five tarsomeres. The distal ends of the tibiae often carry one or two spines that are of taxonomic importance (Figs. 26.766, 26.768). The fourth and fifth tarsomeres are infrequently cordiform (heart-shaped, Fig. 26.767) or trilobed (Fig. 26.796). The wings of chironomids have a moderately complex venation, but with few crossveins (Figs. 26.762-26.765, 26.770-26.772, 26.782-26.785). The thorax may bear groups of setae and is usually patterned with brown, black, yellow, or green.

Abdomen: The abdomen of adult chironomids consists of eight segments plus several terminal segments modified as genitalia. Groups of setae are usually present on terga, sterna, and pleura and the abdomen may be patterned in brown, black, yellow, blue, or green. The genitalia (termed *hypopygium* in the male) often have a complex external and internal morphology and in the male have a large gonocoxite and a terminal (usually) gonostylus (Figs. 26.761, 26.773-26.781).

NOTES ON PREPARATION OF SPECIMENS AND USE OF KEYS

General

1. The keys below include: freshwater Nearctic larvae (to subfamily and genus), freshwater Nearctic pupae (to subfamily and genus), freshwater Nearctic adults (to tribe), and marine Nearctic larvae, pupae, and adults (to tribe). The freshwater keys to genus include all known Nearctic genera for which larvae and/or pupae have been described. Additionally, the key to pupae includes some taxa that cannot, at present, be assigned to any known genus. Such taxa are designated as numbered genera in the keys to pupae; however, in some cases these numbered genera may simply be the undescribed pupae of named genera known only as adults, or aberrant pupae belonging to established genera. Some of the numbered genera of the 2nd edition of this book have now been assigned names and are so identified in this edition. To avoid confusion, those which are still identified only by number have the same number assigned to them as in the 2nd edition even though this has produced gaps in the numbered genera sequence. The reader must keep in mind that the number of undiscovered immature stages is large-based on the number of species known only from the adult stage and the frequency with which we encounter new forms of immatures. It is to be expected that large collections and/or collections from unusual habitats or geographical

regions will contain species that do not fit the generic concepts used in these keys. The keys do not include genera belonging to subfamilies and tribes which have not yet been recorded from North America.

2. Collections from freshwater and brackish water should be keyed in the main generic keys for larvae and pupae. Marine specimens are treated (usually above the level of genus) in separate keys.

3. The Orthocladiinae, which is the most diverse of the chironomid subfamilies, has not yet been satisfactorily divided into tribes. Tentative names for a few of the orthoclad tribes are indicated by quotation marks.

4. Slides prepared for identifications may be temporary mounts with water or, preferably, permanent mounts with any suitable medium such as Euparal or Canada balsam.

Larvae

1. When possible, use the fourth larval instar. These may often be recognized by the swollen thorax of the prepupa.

2. Sever the head and mount it with the ventral side facing upward. The head must often be gently (sometimes not so gently) depressed to expose the mouthparts. Depending on the thickness of the abdomen, the head and abdomen should be placed under the same cover glass. This is usually not possible for large larvae since the abdomen prevents the depression of the head. In such cases, mount the abdomen under a separate cover glass, but on the same slide.

Pupae

1. Pupae are best determined from pupal exuviae. The identification of pupae (not exuviae) is not easy since the structures are obscured by tissues. If possible, remove pharate adults from their pupal exuviae.

2. Before dissecting pupae or pupal exuviae examine the leg sheath arrangement and type of thoracic horn. Preparation of exuviae often disrupts the leg sheath pattern, and the thoracic horn (particularly of the Chironominae) is often extremely difficult to see.

3. Pupal exuviae are best prepared as follows (these steps should be carried out with the specimen in a drop of mounting medium on a microscope slide):
   a. separate the head and thorax from the abdomen.
   b. split the thorax along the middorsal suture and open the thorax so that the two edges of the suture are on opposite sides of the specimen.
   c. turn the thorax so that the outer side of the integument is facing upwards and arrange on the slide above the abdomen.
   d. move the abdomen, dorsal side up (usually), to a position just below the thorax.
   e. place a cover glass over the specimen, depressing it just enough to cause the exuviae to flatten but not distort.

Adults

1. The keys below are designed for adult males. Males can be separated from females by their generally more slender abdomen, a rather conspicuous set of genital appendages, and, in most species, characteristic plumose antennae.

2. Most adults can be identified to subfamily and many to tribe with the use of a dissecting microscope. To be certain, however, it is best to prepare the specimen for examination with a compound microscope. This is sometimes a tedious and involved process, but has been greatly simplified here for the purposes of a key to tribes.
   a. Remove and mount:
      (1) third (hind) pair of legs
      (2) wings
   b. Remove the abdomen and heat it cautiously in 10% KOH to remove the soft obscuring tissues (alternatively, the abdomen will clear standing for 12-24 hours in 10% KOH); mount the abdomen dorsal side up, being particularly careful to orient the genitalia dorsal side up.

KEY TO TRIBES OR SUBFAMILIES OF NORTH AMERICAN FRESHWATER CHIRONOMIDAE LARVAE

1.	Antennae retractile (Fig. 26.4); prementohypopharyngeal complex with a 4-8 toothed ligula (Figs. 26.4, 26.13, 26.15, 26.16, 26.41-26.53, 26.63, 26.86); mentum almost entirely membranous or with dorsomental teeth arranged in conspicuous plates (Figs. 26.11-26.12, 26.17-26.20, 26.25, 26.60, 26.62-26.64, 26.66) or longitudinal rows (Figs. 26.10, 26.13)	*Tanypodinae* (p. 644)
1'.	Antennae nonretractile (Figs. 26.1-26.3); prementohypopharyngeal complex without a toothed ligula. Mentum usually entirely toothed (Figs. 26.98-26.102, 26.128, 26.130-26.136, 26.143, 26.148-26.165, 26.169-26.176, 26.178-26.187, 26.190-26.193, 26.199-26.200, 26.242-26.289, 26.304, 26.306-26.309, 26.312-26.313, 26.315-26.316, 26.325, 26.329-26.332, 26.333-26.335), or occasionally with weakly sclerotized or translucent portions (Figs. 26.244-26.245, 26.317, 26.321, 26.327), but never membranous	2
2(1').	Procerci long, at least 5 times as long as wide; premandibles absent	*Podonominae* (p. 690)
2'.	Procerci variable, rarely more than 4 times longer than wide; premandibles present (Figs. 26.1-26.3, 26.140, 26.322), usually well developed and conspicuous	3
3(2').	Ventromental plates well developed and with conspicuous striations throughout more than one-half their width (Figs. 26.98-26.102, 26.130-26.136, 26.143, 26.148-26.165, 26.169-26.176, 26.178-26.187, 26.190-26.193, 26.199-26.200, 26.333-26.335)	*Chironominae*[*] 4

[*]The Chironominae genera *Stenochironomus* and *Xestochironomus* violates the couplet in that they do not have ventromental plates with conspicuous striations. Larvae of these genera, however, can be easily recognized by the unusual structure of the mentum and the mandible (see couplet no. 24 of the key to genera of Chironomini larvae and Figs. 26.163, 26.187 and 26.168).

3'. Ventromental plates vestigial to well developed (Figs. 26.242-26.277, 26.304, 26.308-26.309, 26.315-26.317, 26.321, 26.325-26.327, 26.329-26.332), when well developed never with striations although occasionally with setae underneath (Figs. 26.278-26.289, 26.306-26.307) 6

4(3). Antenna arising from distinct tubercle (Figs. 26.103-26.107, 26.113-26.114, 26.116-26.118); first antennal segment elongate and usually at least slightly curved (Figs. 26.103-26.114); lauterborn organs very large and conspicuous (Figs. 26.103-26.112, 26.123) or occurring at the apex of elongated stalks (Figs. 26.113-26.114, 26.119-26.120) .... *Tanytarsini* (p. 653)

4'. Antenna not arising from a distinct tubercle; if first antennal segment elongate then not curved as in Figs. 26.103-26.114, and lauterborn organs not large or occurring at the apex of elongated stalks 5

5(4'). Outermost lateral teeth of mentum rounded and directed laterally; mentum as in Figs. 26.333-26.335 .... *Pseudochironomini**

5'. Outermost lateral teeth usually pointed and directed anteriorly (Figs. 26.130-26.136, 26.143, 26.148-26.165, 26.169-26.176, 26.178-26.187, 26.190-26.193, 26.199-26.200); mentum never as in Figures 26.333-26.335 .... *Chironomini* (p. 661)

6(3'). Mentum as in Figure 26.315, 26.316, or 26.317 .... **Prodiamesinae** (p. 690)

6'. Mentum not as in Figure 26.315, 26.316, or 26.317 7

7(6'). Third antennal segment with areas of reduced sclerotization which give the appearance of annulations (Figs. 26.323-26.324), or mentum similar to Figure 26.321 and anterior labral region with large scales similar to Figure 26.322 .... **Diamesinae** (p. 687)

7'. Third antennal segment occasionally very small (Figs. 26.221, 26.228, 26.231-26.233, 26.237) but never with reduced sclerotization causing an annulated appearance; mentum and anterior labral region never as in Figure 26.321 or 26.322 .... **Orthocladiinae** (p. 673)

KEY TO THE GENERA OF NORTH AMERICAN FRESHWATER CHIRONOMIDAE LARVAE

Tanypodinae

1. Abdominal segments with a lateral hair fringe (Fig. 26.5); dorsomental teeth present in well defined plates (Figs. 26.11-26.12), or arranged in longitudinal rows (Figs. 26.10, 26.13); head ratio 1.0 to 1.5 2

1'. Abdominal segments lacking well defined lateral hair fringe; dorsomental teeth absent or extremely reduced; head ratio 1.5 or greater .... Pentaneurini 16

2(1). Dorsomental teeth arranged in longitudinal rows (Figs. 26.10, 26.13); ligula with 6-7 teeth (Figs. 26.15, 26.16); head capsule with pronounced anterior taper (Fig. 26.6) .... Coelotanypodini 3

2'. Dorsomental teeth present in well defined plates (Figs. 26.11-26.12); ligula with 4-5 teeth (Figs. 26.43, 26.45, 26.53); head capsule more rounded anteriorly (Fig. 26.7) 4

3(2). Ligula with 6 teeth (Fig. 26.15); mandible strongly hooked (Fig. 26.14); ratio of head length to antennal length approximately 1.6 .... *Clinotanypus* Kieffer

3'. Ligula with 7 teeth (Fig. 26.16); mandible not strongly hooked; ratio of head length to antennal length approximately 2.6 .... *Coelotanypus* Kieffer

4(2'). Dorsomental plates each with 13-15 teeth .... **Anatopyniini**[†]

4'. Dorsomental plates with 9-7 or fewer teeth 5

5(4'). Dorsomental plates with 2-3 teeth (Fig. 26.17); antennae about one-third length of head capsule and ratio of length of first antennal segment to length of remaining segments less than 3.5 .... Natarsiini—*Natarsia* Fittkau

5'. Dorsomental plates with more than 3 conspicuous teeth (Figs. 26.18-26.20, 25.25); antennae shorter than one-third the head capsule length, or, if longer than one-third the length of head capsule, the ratio of the length of the first antennal segment to length of remaining segments always 3.5 or greater 6

6(5'). Mandible with bulbous base and very minute lateral teeth (Fig. 26.54); ligula with 5 pale yellow to light brown teeth; teeth of ligula forming a convex arch or (less commonly) of equal lengths (Fig. 26.44) .... Tanypodini—*Tanypus* Meigen

6'. Mandible without conspicuous bulbous base; ligula with 4-5 teeth (Figs. 26.41, 26.42, 26.43, 26.45, 26.53); when 5 pale yellow or light brown teeth are present on ligula the teeth form a concave arch (Figs. 26.41-26.42) 7

---

* In North America the tribe Pseudochironomini is represented by a single genus, *Pseudochironomus*. The characters used in this key to identify specimens to the tribe level thus also serve to identify to the generic level; no additional key to the genus is provided for this tribe.

[†] Not recorded in North America.

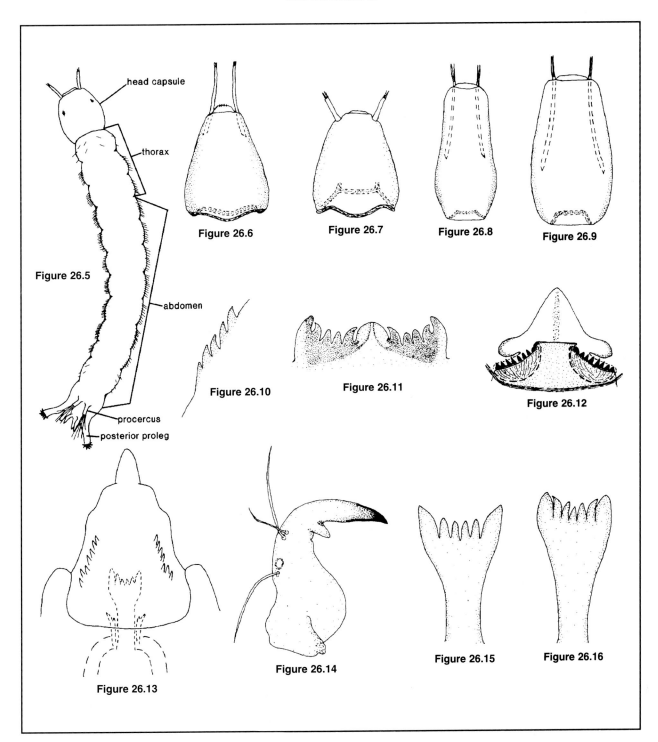

Figure 26.5. Generalized Tanypodinae larva (dorsal view) showing lateral hair fringe.

Figure 26.6. Dorsal view of *Clinotanypus* sp. head capsule.

Figure 26.7. Dorsal view of *Procladius* sp. head capsule.

Figure 26.8. Dorsal view of *Nilotanypus* sp. head capsule.

Figure 26.9. Dorsal view of *Conchapelopia* sp. head capsule.

Figure 26.10. Longitudinal arrangement of dorsomental teeth in *Coelotanypus concinnus* (Coquillett).

Figure 26.11. Dorsomental teeth of *Tanypus* sp.

Figure 26.12. Dorsomental teeth, ventromentum and prementohypopharyngeal element (in ventral view) of *Procladius* sp. (redrawn from Roback [1977]).

Figure 26.13. Longitudinal arrangement of dorsomental teeth on mental region of *Clinotanypus* sp. (ventral view).

Figure 26.14. Mandible of *Clinotanypus* sp.

Figure 26.15. Ligula of *Clinotanypus* sp.

Figure 26.16. Ligula of *Coelotanypus* sp.

7(6').	Ligula with 5 or (less commonly) 4 black teeth (Figs. 26.53, 26.45); paraglossae with 1 main tooth and 1-7 accessory teeth on each side (Figs. 26.30, 26.31) .... Procladiini	8
7'.	Ligula with 5 or 4 light yellow to brown teeth (Figs. 26.41-26.43); paraglossae pectinate (Fig. 26.34); unevenly bifid (Fig. 26.33), or with only accessory teeth on outer margin (Fig. 26.32) .... Macropelopiini	9
8(7).	Antennal blade more than twice as long as the combined lengths of antennal segments 2 through 4 (Fig. 26.22); ligula usually with 4 teeth (Fig. 26.45) .... ***Djalmabatista*** Fittkau	
8'.	Antennal blade subequal in length to the combined lengths of antennal segments 2 through 4 (Fig. 26.21); ligula normally with 5 teeth (Fig. 26.53) .... ***Procladius*** Skuse	
9(7').	Ligula with 4 teeth (Fig. 26.43); paraglossae pectinate (Fig. 26.34); mandible with row of 4 or more inner teeth (Figs. 26.58, 26.59)	10
9'.	Ligula with 5 teeth (Figs. 26.41, 26.42); paraglossae not pectinate (Figs. 26.32, 26.33); mandible not as in Figures 26.58-26.59	11
10(9).	Lateral dorsomental teeth closely appressed (Fig. 26.25); small claws of posterior prolegs as in Figure 26.40 or simple .... ***Derotanypus*** Roback	
10'.	Lateral dorsomental teeth not closely appressed (Fig. 26.19); small claws of posterior prolegs basally ovoid (Fig. 26.39); or simple .... ***Psectrotanypus*** Kieffer	
11(9').	Inner margin of dorsomental plate produced medially into one (Fig. 26.18) or two points (Fig. 26.60)	30
11'.	Inner margins of dorsomental plate rounded (Fig. 26.20)	12
12(11').	Mandible with 2 or 3 complete or partial rows of inner teeth (Fig. 26.61); dorsomental plates not interrupted in the middle (Fig. 26.62); 3 middle teeth of ligula all about the same length (Fig. 26.63) .... ***Fittkauimyia*** Karunakaran	
12'.	Mandible with 4 or fewer inner teeth, not forming multiple rows; dorsomental plates separated in the middle; 3 middle teeth of ligula differing in length	13
13(12').	Pseudoradula widest near base of m-appendage and tapering in width to apex, granules more coarse in proximal half of pseudoradula (Fig. 26.64) .... ***Radotanypus*** Fittkau and Murray	
13'.	Pseudoradula narrowest near median edges of dorsomental plates and expanded apically (Fig. 26.65) or pseudoradula of uniform width but granules weaker near base (Fig. 26.66)	14
14(13').	All teeth of ligula directed anteriorly (Fig. 26.42); length of palpus at least 4 times greater than width at midlength and campaniform sensillum occurring in basal one-third of palpus (Fig. 26.29); pseudoradula of uniform width but granules weaker near base (Fig. 26.66) .... ***Macropelopia*** Thienemann	
14'.	First lateral teeth of ligula outcurved (Fig. 26.41); length of palpus 2.9-4.1 times the width at midlength; campaniform sensillum located near middle of palpus (Figs. 26.23, 26.24); pseudoradula narrowest near median edges of dorsomental plates (Fig. 26.65)	15
15(14').	Dorsomental plate with 7 teeth; palpus about 4.1 times longer than wide at midlength (Fig. 26.24); 2nd antennal segment 4.2-4.7 times longer than maximum width; all ventrolateral setae of mandible simple (Fig. 26.67) .... ***Alotanypus*** Roback	
15'.	Dorsomental plate with 4-5 teeth (Fig. 26.20); palpus 2.9-3.2 times longer than width at midlength (Fig. 26.23); 2nd antennal segment 2.1-2.5 times longer than maximum width; ventrolateral seta 1 of mandible simple, setae 2 and 3 multibranched (Fig. 26.68) .... ***Apsectrotanypus*** Fittkau	
16(1').	Maxillary palp with 2 or more basal segments (Figs. 26.27-26.28)	17
16'.	Maxillary palp with only one basal segment	18
17(16).	Maxillary palp with 2 basal segments and campaniform sensillum very large, about as large as width of palp, and located between first and second segments (Fig. 26.70); pseudoradula broad, widened toward base of m-appendage (Fig. 26.69); dark claws of posterior prolegs, if present, pectinate or bifid (Fig. 26.35) .... ***Paramerina*** Fittkau (in part)	
17'.	Maxillary palp with more than two basal segments, or when only two basal segments then campaniform sensillum much smaller than width of palp; pseudoradula narrow, widest near middle and with granules in parallel, longitudinal rows (Fig. 25.71); dark claws of posterior prolegs, if present, all simple .... ***Ablabesmyia*** Johannsen	
18(16').	Lauterborn organs at tip of second antennal segment well sclerotized and solidly fused with segment so that second segment appears to envelope or flank most of the third segment (Fig. 26.72); third and fourth segments of antenna about equal in length	31
18'.	Lauterborn organs small or vestigial, less than 1/3 the length of segment 3, or not fused to apex of segment 2; Fourth segment of antenna variable in length, but usually substantially shorter than third segment	19

(**Caution**—Sometimes it is difficult to see the structure of the lauterborn organs and the lengths of the distal segments of the antenna. If these structures cannot be seen continue to couplet 19)

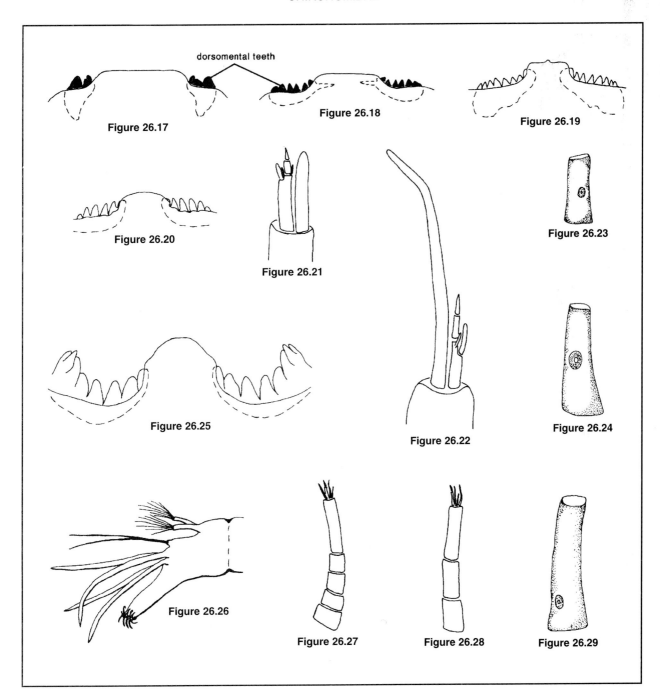

Figure 26.17. Arrangement of dorsomental teeth in *Natarsia* sp.
Figure 26.18. Arrangement of dorsomental teeth in *Brundiniella* sp.
Figure 26.19. Arrangement of dorsomental teeth in *Psectrotanypus dyari* (Coquillett).
Figure 26.20. Arrangement of dorsomental teeth in *Apsectrotanypus* sp.
Figure 26.21. Apex of antenna of *Procladius* sp.
Figure 26.22. Apex of antenna of *Djamabatista pulcher* (Johannsen).
Figure 26.23. Basal segment of maxillary palpus of *Apsectrotanypus* sp.
Figure 26.24. Basal segment of maxillary palpus of *Alotanypus venustus* (Coquillett).
Figure 26.25. Arrangement of dorsomental teeth in *Derotanypus alaskensis* (Malloch).
Figure 26.26. Proceri, supraanal setae, anal tubules, and posterior prolegs of *Pentaneura* sp.
Figure 26.27. Maxillary palpus of *Ablabesmyia* (*Ablabesmyia*) *mallochi* (Walley).
Figure 26.28. Maxillary palpus of *Ablabesmyia* (*Ablabesmyia*) *parajanta* Roback.
Figure 26.29. Basal segment of maxillary palpus of *Macropelopia* sp.

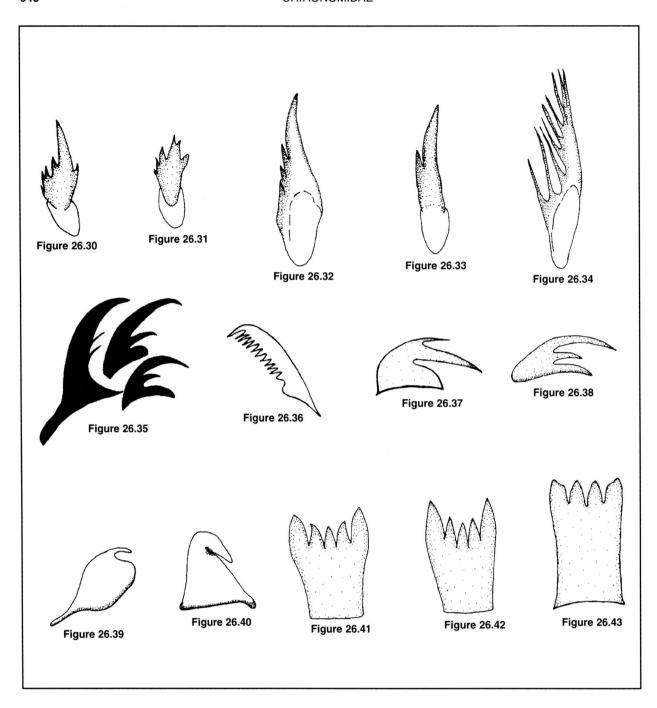

Figure 26.30. Paraglossa of *Procladius* sp.

Figure 26.31. Paraglossa of *Djalmabatista pulcher* (Johannsen).

Figure 26.32. Paraglossa of *Alotanypus venustus* (Coquillett) (redrawn from Roback [1978]).

Figure 26.33. Paraglossa of *Macropelopia decedens* (Walker) (redrawn from Roback [1978]).

Figure 26.34. Paraglossa of *Psectrotanypus dyari* (Coquillett).

Figure 26.35. Pigmented claws of posterior prolegs of *Paramerina smithae* (Sublette) (redrawn with modification from Roback [1972]).

Figure 26.36. Pectinate claw of posterior proleg of *Nilotanypus* sp.

Figure 26.37. Bifid claw of posterior proleg of *Labrundinia* sp.

Figure 26.38. Bifid claw of posterior proleg of *Zavrelimyia* sp.

Figure 26.39. Basally ovoid claw of posterior proleg of *Psectrotanypus dyari* (Coquillett).

Figure 26.40. Smallest claw of posterior proleg of *Derotanypus alaskensis* (Malloch) (redrawn from Roback [1978]).

Figure 26.41. Ligula of *Alotanypus venustus* (Coquillett).

Figure 26.42. Ligula of *Macropelopia* sp.

Figure 26.43. Ligula of *Psectrotanypus dyari* (Coquillett).

19(18').	Anterior margin of ligula straight (i.e., all teeth ending at more or less the same point) (Figs. 26.50-26.51)	20
19'.	Anterior margin of ligula with median tooth distinctly longer than or shorter than the 1st lateral and/or 2nd lateral teeth (Figs. 26.46-26.49, 26.52)	22
20(19).	Posterior proleg with 1 claw darkly pigmented; supraanal seta strong, and originating from distinct papillae; anal tubules longer than posterior prolegs (Fig.26.26)	**Pentaneura** Philippi
20'.	All claws of posterior prolegs unicolorous; supraanal seta weak, or, if conspicuous, not originating from distinct papillae; anal tubules shorter than posterior prolegs	21
21(20').	One claw of posterior proleg bifid (Fig. 26.38); head capsule unicolorous; ratio of length of 1st antennal segments to combined lengths of remaining antennal segments about 3.1	**Zavrelimyia** Fittkau
21'.	All claws of posterior prolegs simple; posterior 1/4 of head capsule with conspicuous blackish brown pigmentation; ratio of length of 1st antennal segment to combined lengths of remaining antennal segments about 2.5	**Paramerina** Fittkau (in part)
22(19').	Median tooth of ligula longer than 1st lateral teeth (Figs. 26.48-26.49)	23
22'.	Median tooth of ligula shorter than (Figs. 26.46-26.47) or equal in length to 1st lateral teeth; when equal in length to 1st lateral teeth, then all 3 median teeth shorter than 2nd lateral teeth (Fig. 26.52)	24
23(22).	Posterior prolegs with 1 bifid claw (Fig. 26.37); 2nd antennal segment usually distinctly more darkly pigmented than 1st segment; ligula as in Figure 26.48	**Labrundinia** Fittkau
23'.	Posterior prolegs with 1 pectinate claw (Fig. 26.36); all antennal segments unicolorous; ligula as in Figure 26.49	**Nilotanypus** Kieffer
24(22').	Second antennal segment with distinct dark brown pigmentation	**Monopelopia** Fittkau (in part)
24'.	All antennal segments unicolorous	25
25(24').	Head capsule with granular appearance visible 100x magnification; body with undulate wrinkles; ratio of length of 1st antennal segment to combined length of remaining segments 6.0-7.5	**Guttipelopia** Fittkau
25'.	Head capsule lacking granular appearance; body smooth; ratio of length of 1st antennal segment to combined lengths of remaining segments variable, but rarely greater than 5.5	26
26(25').	Mandible with at least 1 large conspicuous tooth along inner margin (Figs. 26.56-26.57)	27
26'.	Inner margin of mandible smooth or with only very small teeth present	28
27(26).	Mandible with 1 large, pointed tooth and 1-2 smaller teeth present on inner margin (Fig. 26.56)	**Larsia** Fittkau
27'.	Mandible with 1 large blunt tooth and 1 small tooth present on inner margin (Fig. 26.57)	**Krenopelopia** Fittkau (in part)
28(26').	Head with posterior half pigmented dark brown; mandible with 1 small tooth and well-developed accessory blade (Fig. 26.55); larvae apparently restricted to mats of blue-green algae occurring on steep rock outcrops over which small volumes of water trickle	**Hudsonimyia** Roback
28'.	Head capsule unicolorous; mandible not as in Fig. 26.55; larvae occurring in a wide variety of habitats	29
29(28').	Mandible with small but distinct teeth along inner edge; ratio of length of 1st antennal segment to combined lengths of remaining segments 3.4-3.8; campaniform sensillum in basal 1/3 of maxillary palp; ventrolateral seta 1 of mandible reduced and in small pit, setae 2 and 3 long and simple (Fig. 26.73)	**Trissopelopia** Kieffer
29'.	Teeth of inner margin of mandible very minute, indistinct, or lacking; ratio of length of 1st antennal segment to combined lengths of remaining segments variable but usually 3.8-5.3; campaniform sensillum of maxillary palp variable, but usually in apical 1/3; at least ventrolateral seta 3 of mandible bifid or multibranched (Fig. 26.75)	**Thienemannimyia** group[*] 36

[*] The **Thienemannimyia** group consists of the genera **Arctopelopia**, **Conchapelopia**, **Hayesomyia**, **Helopelopia**, **Meropelopia**, **Rheopelopia**, **Telopelopia** and **Thienemannimyia**. Larvae of these genera are very difficult to identify to genus. Characters given in this key are based on fourth instar larvae, and probably will not work for earlier instars. Earlier instars should only be identified to "**Thienemannimyia group sp.**". Identifications should be confirmed by rearing larvae to pupal or adult stages.

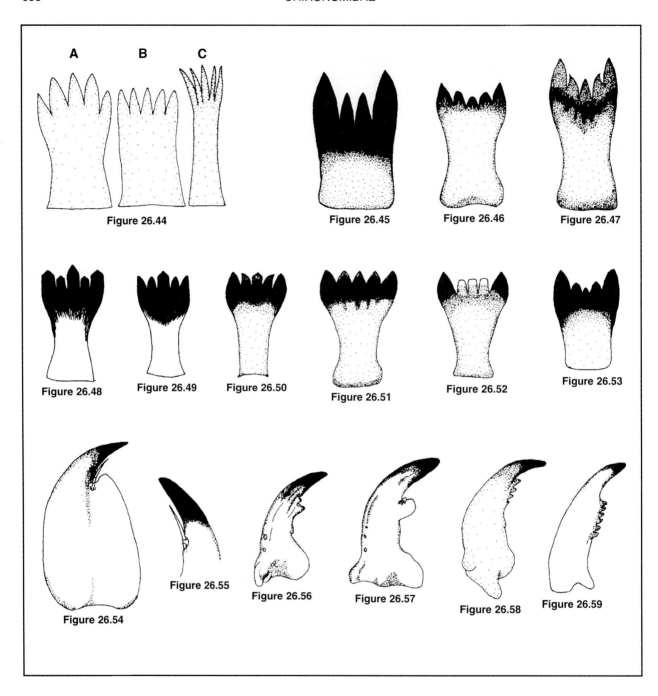

Figure 26.44. Three types of Ligula of *Tanypus* species: (a) *Tanypus* (*Apelopia*) *neopunctipennis* Sublette; (b) *Tanypus* (*Tanypus*) *punctipennis* Meigen; (c) *Tanypus* (*Tanypus*) possibly *concavus* Roback. (*Tanypus* (*Tanypus*) possibly *concavus* Roback redrawn from Roback [1977]).

Figure 26.45. Ligula of *Djalmabatista pulcher* (Johannsen).

Figure 26.46. Ligula of *Larsia* sp.

Figure 26.47. Ligula of *Conchapelopia* sp.

Figure 26.48. Ligula of *Labrundinia* sp.

Figure 26.49. Ligula of *Nilotanypus* sp.

Figure 26.50. Ligula of *Pentaneura* sp.

Figure 26.51. Ligula of *Zavrelimyia* sp.

Figure 26.52. Ligula of *Ablabesmyia annulata* (Say).

Figure 26.53. Ligula of *Procladius* sp.

Figure 26.54. Mandible of *Tanypus* sp.

Figure 26.55. Apex of mandible of *Hudsonimyia karelena* Roback (redrawn with modification from Roback [1979]).

Figure 26.56. Mandible of *Larsia* sp.

Figure 26.57. Mandible of probable *Krenopelopia* sp.

Figure 26.58. Mandible of *Derotanypus*.

Figure 26.59. Mandible of *Psectrotanypus* sp.

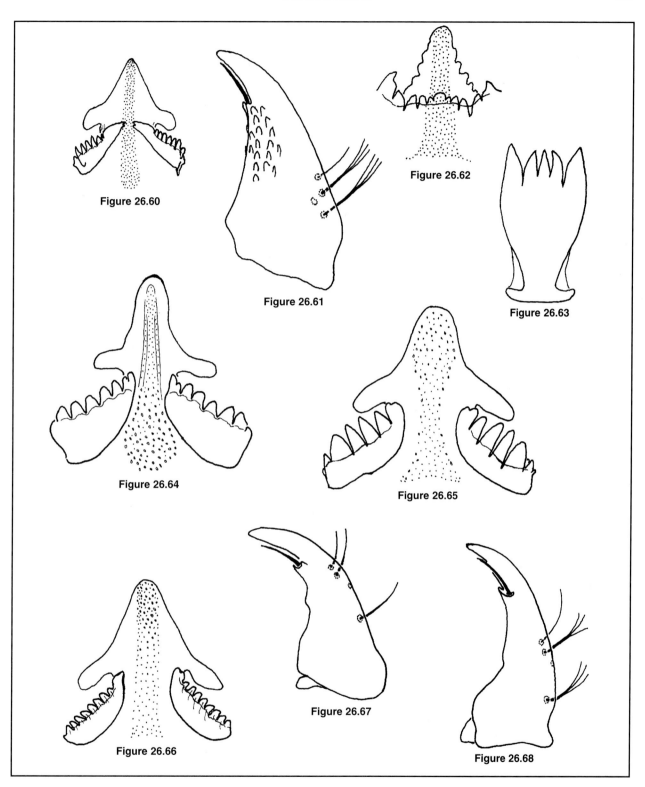

Figure 26.60. Dorsomental plates and m-appendage of *Bethbilbeckia* sp.

Figure 26.61. Mandible of *Fittkauimyia* sp.

Figure 26.62. Dorsomental plates and m-appendage of *Fittkauimyia* sp.

Figure 26.63. Ligula of *Fittkauimyia* sp.

Figure 26.64. Dorsomental plates and m-appendage of *Radotanypus* sp.

Figure 26.65. Dorsomental plates and m-appendage of *Apsectrotanypus* sp.

Figure 26.66. Dorsomental plates and m-appendage of *Macropelopia* sp.

Figure 26.67. Mandible of *Alotanypus* sp.

Figure 26.68. Mandible of *Apsectrotanypus* sp.

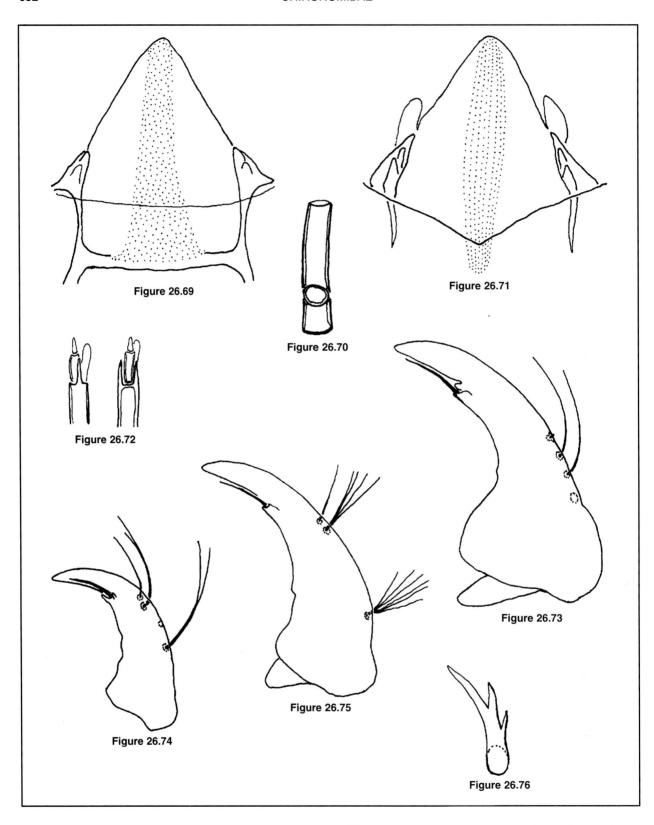

Figure 26.69. Dorsomentum, m-appendage and pseudoradula of *Paramerina* sp.

Figure 26.70. Maxillary palp of *Paramerina* sp.

Figure 26.71. Dorsomentum, m-appendage and pseudoradula of *Ablabesmyia* sp.

Figure 26.72. Lateral and ventral views of apex of second antennal segment of *Monopelopia* sp.

Figure 26.73. Mandible of *Trissopelopia* sp.

Figure 26.74. Mandible of *Bethbilbeckia* sp.

Figure 26.75. Mandible of *Brundiniella* sp.

Figure 26.76. Paraglossa of *Denopelopia* sp.

30(11).	Ventrolateral setae 1 and 3 of mandible simple, seta 2 bifid (Fig. 25.74); campaniform sensillum of maxillary palp located in basal 1/3	***Bethbilbeckia*** Fittkau and Murray
30'.	Ventrolateral seta 1 simple, setae 2 and 3 multibranched (Fig. 26.75); campaniform sensillum of maxillary palp located in middle 1/3	***Brundiniella*** Roback
31(18).	Anterior margin of ligula straight (i.e., all teeth ending at more or less the same point) or slightly convex	32
31'.	Anterior margin of ligula concave	34
32(31).	Paraglossae with two inner teeth (Fig. 26.76); two smallest claws of posterior prolegs each with one large inner tooth and some spines on outer margin, three slightly larger claws with a single spine-like inner tooth, and nine largest claws with only fine spines on inner margin (Fig. 26.77)	***Denopelopia*** Roback and Rutter
32'.	Paraglossae with one inner tooth (Fig. 26.33); at least smaller claws of posterior prolegs with several large inner teeth (Fig. 26.78 or 26.81)	33
33(32').	Second antennal segment brown; campaniform sensillum in apical 1/2 of first antennal segment; some claws of posterior prolegs dark	***Xenopelopia*** Fittkau
33'.	Second antennal segment not dark; campaniform sensillum in basal 1/2 of first antennal segment; all claws of posterior prolegs pale	***Telmatopelopia*** Fittkau
34(31').	Pecten hypopharyngis with fewer than 13 unequal teeth (Fig. 26.79); teeth of ligula deeply concave; campaniform sensillum in apical 1/2 of first antennal segment	35
34'.	Pecten hypopharyngis with 15 teeth of approximately equal lengths (Fig. 26.80); teeth of ligula only moderately concave; campaniform sensillum in basal 1/2 of first antennal segment	***Pentaneurella*** Fittkau and Murray
35(34).	Second antennal segment brown; one claw of posterior prolegs dark, dark claw and some smaller pale claws with large inner teeth (Fig. 26.81)	***Monopelopia*** Fittkau (in part)
35'.	Second antennal segment pale, same color as basal segment; all claws of posterior prolegs unicolorous, pale and without large inner teeth	***Krenopelopia*** Fittkau (in part)
36(29').	B-seta of maxillary palp 2-segmented (Fig. 26.82)	39
36'.	B-seta of maxillary palp 3-segmented (Fig. 26.83)	37
37(36').	Mandible strongly curved in apical 1/2, no teeth or only one very small inner tooth visible on mandible (Fig. 26.84)	***Rheopelopia*** Fittkau
37'.	Mandible uniformly weakly curved, 2 small inner teeth visible on mandible (Fig. 26.85)	38
38(37').	Ventrolateral seta 3 and sensillum minusculum clearly located in basal 1/3 of mandible (Fig. 26.85); middle tooth of ligula about as long as width at base (Fig. 26.86)	***Helopelopia*** Roback
38'.	Ventrolateral seta 3 and sensillum minusculum occurring more distally along outer margin of mandible, generally located in middle 1/3 of mandible (Fig. 26.87); middle tooth of ligula clearly longer than width at base (Fig. 26.47)	***Conchapelopia*** Fittkau
39(36).	Pecten hypopharyngis with about 25 teeth (Fig. 26.88); pseudoradula strongly widened in middle (Fig. 26.89)	***Arctopelopia*** Fittkau
39'.	Pecten hypopharyngis with about 20-22 teeth; pseudoradula either strongly widened proximally (Fig. 26.90) or uniform in width (Fig. 26.91)	40
40(39').	Pseudoradula strongly widened proximally (Fig. 26.90)	41
40'.	Pseudoradula uniform in width (Fig. 26.91)	42
41(40).	Two inner teeth of mandible about same size (Fig. 26.92)	***Hayesomyia*** Murray and Fittkau
41'.	Inner basal tooth of mandible larger than distal tooth (Fig. 26.93)	***Telopelopia*** Roback
42(40').	Distance between origin of ventrolateral setae 2 and 3 of mandible about 1.5 times the distance between ventrolateral setae 1 and 2 (Fig. 26.94)	***Thienemannimyia*** Fittkau
42'.	Distance between origin of ventrolateral setae 2 and 3 of mandible about 3.0 times the distance between ventrolateral setae 1 and 2 (Fig. 26.95)	***Meropelopia*** Roback

Tanytarsini

1.	Ventromental plates well separated, pointed at anteromedial edge (Fig. 26.99); larvae construct portable cases made of sand grains and/or detritus (Figs. 26.96-26.97)	2
1'.	Ventromental plates almost meeting at ventral midline of head capsule, anteromedial edge generally rounded or truncate (Figs. 26.98, 26.100-26.102); larvae may construct tubes or filtering structures but never portable cases	6
2(1).	Lauterborn organs originating alternately at different heights on 2nd antennal segment (Figs. 26.103-26.104); antennal segment 2 subequal to or longer than the combined lengths of segments 3-5; larval case always straight (Fig. 26.96)	3

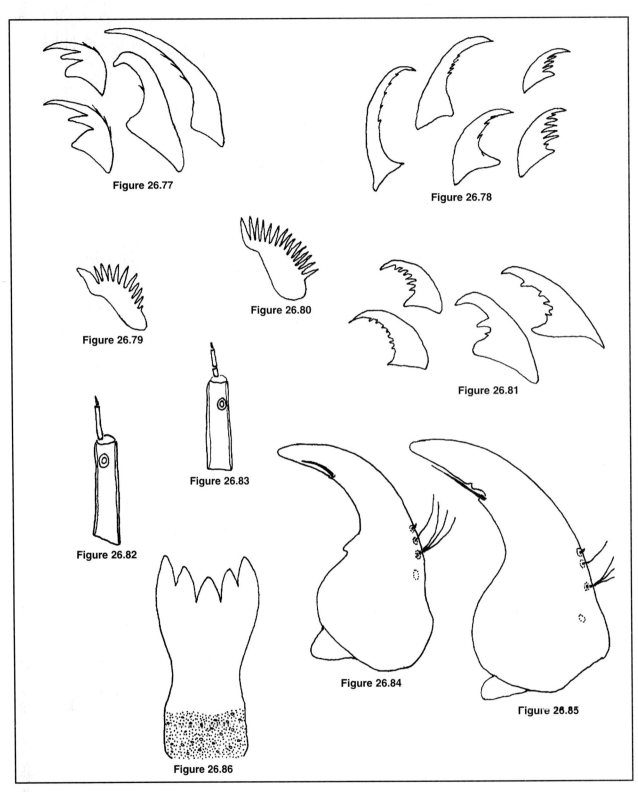

Figure 26.77. Two smallest claws, one larger claw and one largest claw of posterior prolegs of *Denopelopia* sp.

Figure 26.78. Claws of posterior prolegs of *Telmatopelopia* sp.

Figure 26.79. Pecten hypopharyngis of *Krenopelopia* sp.

Figure 26.80. Pecten hypopharyngis of *Pentaneurella* sp.

Figure 26.81. Claws of posterior prolegs of *Monopelopia* sp.

Figure 26.82. Maxillary palp and B-seta of *Arctopelopia* sp.

Figure 26.83. Maxillary palp and B-seta of *Conchapelopia* sp.

Figure 26.84. Mandible of *Rheopelopia* sp.

Figure 26.85. Mandible of *Helopelopia* sp.

Figure 26.86. Ligula of *Helopelopia* sp.

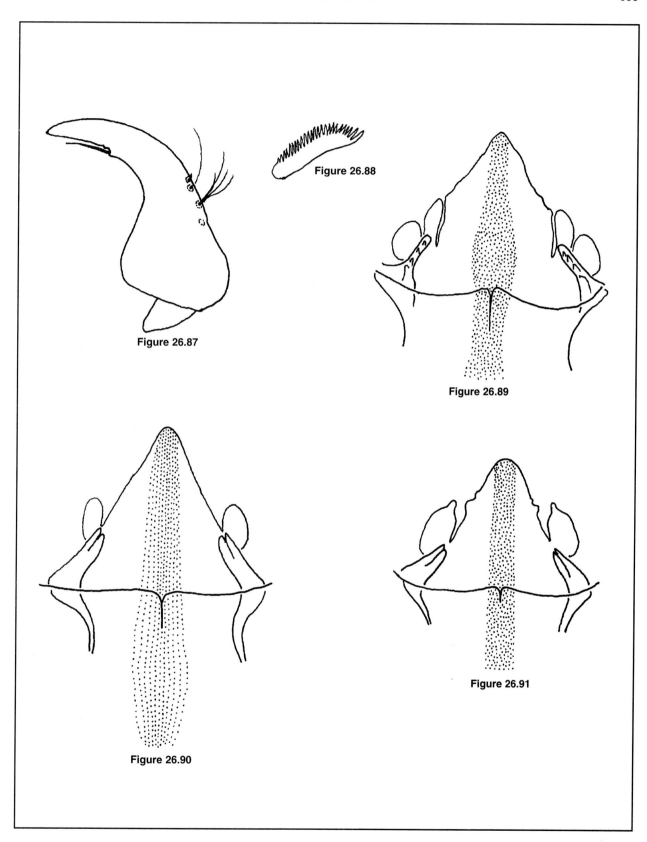

Figure 26.87. Mandible of *Conchapelopia* sp.
Figure 26.88. Pecten hypopharyngis of *Arctopelopia* sp.
Figure 26.89. Dorsomentum, m-appendage and pseudoradula of *Arctopelopia* sp.
Figure 26.90. Dorsomentum, m-appendage and pseudoradula of *Hayesomyia* sp.
Figure 26.91. Dorsomentum, m-appendage and pseudoradula of *Thienemannimyia* sp.

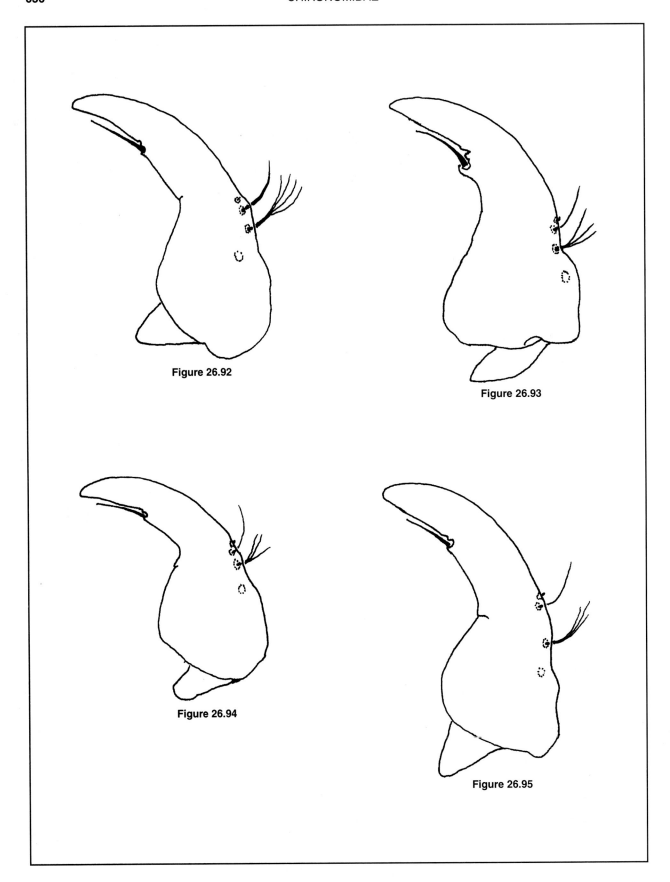

Figure 26.92. Mandible of *Hayesomyia senata*.
Figure 26.93. Mandible of *Telopelopia okoboji*.
Figure 26.94. Mandible of *Thienemannimyia* sp.
Figure 26.95. Mandible of *Meropelopia* sp.

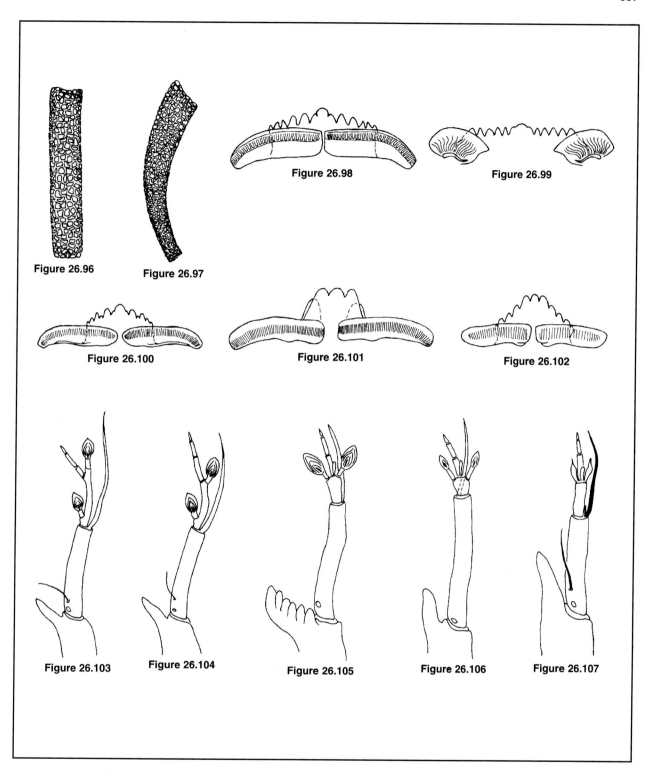

Figure 26.96. Portable sand case of *Stempellinella* sp.
Figure 26.97. Portable sand case of *Constempellina* sp.
Figure 26.98. Mentum of *Rheotanytarsus* sp.
Figure 26.99. Mentum of *Constempellina* sp.
Figure 26.100. Mentum of *Cladotanytarsus (Lenziella)* sp.
Figure 26.101. Mentum of *Corynocera* sp. (redrawn from Hirvenoja [1961]).
Figure 26.102. Mentum of *Sublettea* sp.
Figure 26.103. Antenna of *Stempellinella* sp.
Figure 26.104. Antenna of *Zavrelia* sp. (redrawn from modification from Bause [1913]).
Figure 26.105. Antenna of *Stempellina* sp.
Figure 26.106. Antenna of *Constempellina* sp.
Figure 26.107. Antenna of *Thienemanniola* sp. (redrawn with modification from Lehmann [1973]).

2'.	Lauterborn organs originating on opposite sides at apex of 2nd antennal segment (Figs. 26.105-26.107); antennal segment 2 distinctly shorter than the combined lengths of segments 3-5; larval case straight or curved (Figs. 26.96-26.97)	4
3(2).	Distal lauterborn organ arising preapically on 2nd antennal segment (Fig. 26.104); antennal segment 2 subequal in length to the combined lengths of segments 3-5 (Fig. 26.104)	***Zavrelia*** Kieffer
3'.	Distal lauterborn organ arising from apex of 2nd antennal segment (Fig. 26.103); antennal segment 2 distinctly longer than the combined lengths of segments 3-5 (Fig. 26.103)	***Stempellinella*** Brundin
4(2').	Antennal tubercle with a conspicuous anteromesally projecting palmate process (Fig. 26.105); larval case curved (Fig. 26.97)	***Stempellina*** Bause
4'.	Antennal tubercle lacking palmate process but with a single strong anteromesally directed spine (Figs. 26.106-26.107); larval case curved or straight	5
5(4').	Lauterborn organs on distinct stalks, stalks subequal to or greater in length than antennal segment 3 (Fig. 26.106); larval case curved (Fig. 26.97)	***Constempellina*** Brundin
5'.	Lauterborn organs not conspicuously stalked (Fig. 28.107), or if stalks are apparent then much shorter in length than antennal segment 3; larval case straight (Fig. 26.96)	***Thienemanniola*** Kieffer
6(1').	Mentum with only 3 distinct teeth (Fig. 26.101); mandible blunt distally, lacking apical and lateral dents (teeth) (Fig. 26.115)	***Corynocera*** Zetterstedt
6'.	Mentum with more than 3 distinct teeth; mandible with at least 3 teeth	7
7(6').	Stalks of lauterborn organs appearing annulated along three-fourths of their length (Fig. 26.119)	***Nimbocera*** Reiss
7'.	Stalks of lauterborn organs never appearing annulated	8
8(7').	Second antennal segment annulated (Fig. 26.120)	Tanytarsini-**Genus "A"** (see Roback 1966)
8'.	Second antennal segment not annulated	9
9(8').	Stalks of lauterborn organs less than 1.2 times combined lengths of antennal segments 3-5 (Figs. 26.108-26.112)	11
9'.	Stalks of lauterborn organs greater than 1.25 times (i.e., distinctly longer than) the combined lengths of antennal segments 3-5 (Figs. 26.113-26.114)	10
10(9').	Antennal tubercle with a straight or anteromedially curved spur originating from medial edge (Figs. 26.114, 26.116-26.118)	17
10'.	Antennal tubercle without a straight or anteromedially curved spur (Fig. 26.113)	19
11(9).	Lauterborn organs less than 0.40 times the length of their stalks (Fig. 26.112)	14
11'.	Lauterborn organs about as long as or longer than the length of their stalk (Figs. 26.108-26.111)	12
12(11').	Lauterborn organs about as long as their stalk; lauterborn organs broad and usually with visible longitudinal striations (Figs. 26.108-26.109)	13
12'.	Lauterborn organs at least one-third longer than their stalk (Fig. 26.110); or if subequal in length then organs oval and lacking longitudinal striations (Fig. 26.111); stalks sometimes vestigial or absent (Fig. 26.123)	16
13(12).	Mentum with 2nd lateral tooth smaller than both 1st and 3rd lateral teeth (Fig. 26.100); 2nd antennal segment membranous in apical half and subequal in length to 3rd antennal segment (Fig. 26.108)	***Cladotanytarsus*** (*Lenziella*)
13'.	Mentum with all lateral teeth decreasing in size in an orderly manner, or if 2nd lateral teeth are reduced then 3rd antennal segment distinctly longer than 2nd segment (Fig. 26.109), and 2nd segment not membranous throughout the apical one half	***Cladotanytarsus*** Kieffer
14(11).	Mentum with 5 pairs of lateral teeth	15
14'.	Mentum with 4 pairs of lateral teeth; combined length of antennal segments 2-5 subequal in length to segment 1 in 4th instar specimens	***Neozavrelia*** Goetghebuer
15(14).	Distal portion of 2nd antennal segment greatly expanded (Fig. 26.122); mentum strongly arched (Fig. 26.102)	***Sublettea*** Roback
15'.	Distal portion of 2nd antennal segment only moderately expanded (Fig. 26.121); mentum not strongly arched (Fig. 26.98)	***Rheotanytarsus*** Bause
16(12').	Larvae marine; first antennal segment short, about as long as the combined lengths of the remaining segments (Fig. 26.123); pecten epipharyngis consisting of 3 lobes with apical dissections (Fig. 26.124)	***Pontomyia*** Edwards
16'.	Larvae occurring in a wide variety of aquatic habitats; first antennal segment much longer than the combined lengths of the remaining segments (Figs. 26.110-26.111); pecten epipharyngis consisting of 3-5 scale-like lobes without apical dissections (Fig. 26.125)	***Paratanytarsus*** Bause

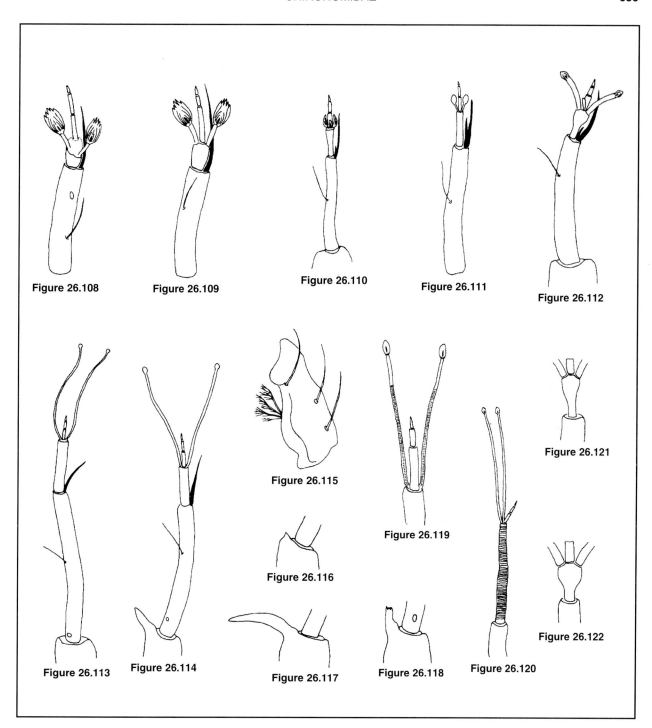

Figure 26.108. Antenna of *Cladotanytarsus (Lenziella)* sp.
Figure 26.109. Antenna of *Cladotanytarsus* sp.
Figure 26.110. Antenna of *Paratanytarsus* sp.
Figure 26.111. Antenna of *Paratanytarsus* sp.
Figure 26.112. Antenna of *Rheotanytarsus* sp.
Figure 26.113. Antenna of *Tanytarsus* sp.
Figure 26.114. Antenna of *Microspectra* sp.
Figure 26.115. Mandible of *Corynocera* sp. (redrawn with modification from Hirvenoja [1961]).
Figure 26.116. Spur on apex of antennal tubercle of *Micropsectra* sp. or *Tanytarsus* sp.
Figure 26.117. Spur on apex of antennal tubercle of *Micropsectra* sp.
Figure 26.118. Spur on apex of antennal tubercle of *Micropsectra* sp. or *Tanytarsus* sp.
Figure 26.119. Apex of antenna of *Nimbocera* sp.
Figure 26.120. Apex of antenna of Tanytarsini Genus "A" (redrawn from Roback [1966]).
Figure 26.121. Second antennal segment of *Rheotanytarsus* sp.
Figure 26.122. Second antennal segment of *Sublettea* sp.

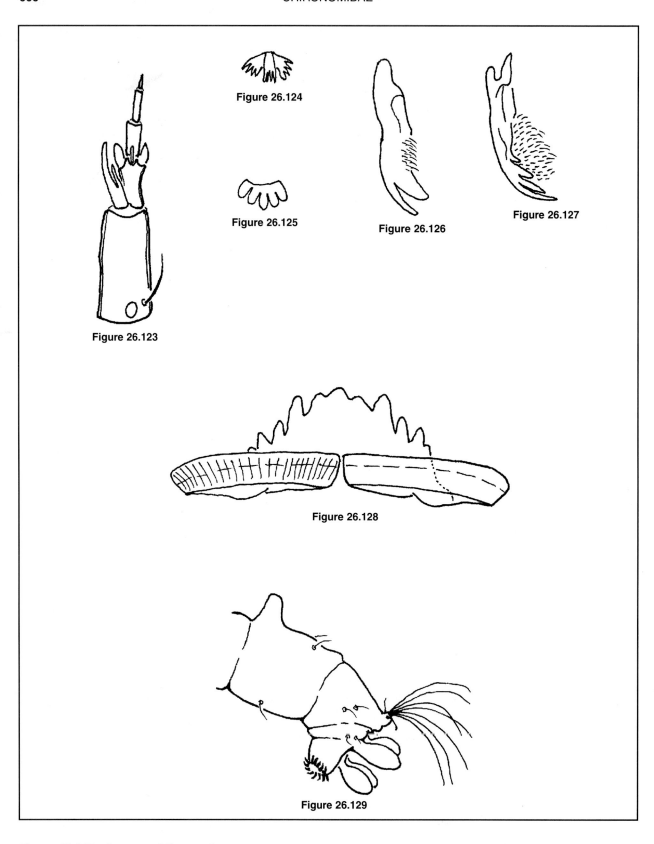

Figure 26.123. Antenna of *Pontomyia* sp.
Figure 26.124. Pecten epipharyngis of *Pontomyia* sp.
Figure 26.125. Pecten epipharyngis of *Paratanytarsus* sp.
Figure 26.126. Premandible of *Micropsectra* sp.
Figure 26.127. Premandible of *Tanytarsus* sp.
Figure 26.128. Mentum and ventromental plates of *Parapsectra* sp.
Figure 26.129. Lateral view of posterior abdominal region of *Krenopsectra* sp.

17(10).	Premandible with 3-4 slender, pointed teeth (Fig. 26.127)	***Tanytarsus*** van der Wulp (in part)
17'.	Premandible with only 2 broader teeth (Fig. 26.126)	18
18(17').	Antennal blade short, subequal in length to antennal segment 3; mentum strongly arched, median 3 teeth set off from remaining 4 lateral teeth (Fig. 26.128)	***Parapsectra*** Reiss
18'.	Antennal blade longer than antennal segment 3, often subequal in length to segment 2; mentum not strongly arched, median tooth often appearing tripartite but well separated from remaining 5 lateral teeth	***Micropsectra*** Kieffer (in part)
19(10').	Premandible with 3-4 slender, pointed teeth (Fig. 26.127)	***Tanytarsus*** van der Wulp (in part)
19'	Premandible with only 2 broader teeth (Fig. 26.126)	20
20(19')	Eighth abdominal segment with well-developed dorsal projection (Fig. 26.129)	***Krenopsectra*** Reiss
20'	Eighth abdominal segment lacking dorsal projection or with only small round bulge [only one species, ***M. roseiventris*** (Kieffer), has this bulge but it is not known from North America]	***Micropsectra*** Kieffer (in part)

Chironomini

1.	Seven anterior abdominal segments subdivided, giving the appearance of a 20-segmented body; larva oligocheatelike	***Chernovskiia*** Saether
1'.	Seven anterior abdominal segments not subdivided, not giving the appearance of a 20-segmented body	2
2(1').	Antenna with 6 segments, lauterborn organs large and alternate at apices of segments 2 and 3 (Figs. 26.138-26.139)	3
2'.	Antenna usually with 5, 7, or 8 segments, (Figs. 26.145-26.147) or if 6 antennal segments are present then lauterborn organs not alternate at apices of 2nd and 3rd segments	8
3(2).	Mentum with a single broad (Fig. 26.134) or narrow (Fig. 26.191) light colored tooth and 5 or 6 pairs of more darkly pigmented lateral teeth	46
3'.	Mentum with paired median teeth	4
4(3').	Median teeth of mentum distinctly lighter in color than outer lateral teeth (Figs. 26.130, 26.132, Fig. 26.185)	5
4'.	Median teeth of mentum not less darkly pigmented than outer lateral teeth (Figs. 26.131, 26.133, 26.135-26.136)	6
5(4).	Two median teeth of mentum distinctly lighter in color than outer lateral teeth (Fig. 26.132, fig 26.185)	47
5'.	Four median teeth of mentum distinctly lighter in color than outer lateral teeth (Fig. 26.130) or light median teeth tiered	***Paratendipes*** Kieffer
6(4').	Antenna elongate and arising from distinct tubercle (Fig. 26.139); anterior edge of ventromental plates straight; mentum as in Figure 26.135 or Figure 26.186;	49
6'.	Not with the above combination of characters	7
7(6').	First lateral teeth closely appressed to 2nd lateral teeth (Fig. 26.133); antenna distinctly longer than mandible	***Omisus*** Townes
7'.	First lateral teeth not closely appressed to 2nd lateral teeth, much longer than median or 2nd lateral teeth (Fig. 26.131); antenna subequal in length to mandible	***Stictochironomus*** Kieffer
8(2').	First antennal segment curved and with a dorsal projection originating near base (Fig. 26.137); mentum as in Figure 26.136	***Pagastiella*** Brundin
8'.	Not with the above combination of characters	9
9(8').	Mentum with a dome-shaped median tooth, median tooth pale yellow in middle, lateral edges and remaining lateral teeth darkly pigmented; lateral teeth longer than the median tooth, giving the mentum an overall concave appearance (Fig. 26.143)	10
9'.	Mentum without a dome-shaped median tooth (Figs. 26.150, 26.169, 26.184), or, if median tooth is dome shaped, then lateral teeth small and not giving the impression of an overall concave mentum (Figs. 26.148, 26.152, 26.157-26.158, 26.165)	13
10(9).	Mentum with 5 lateral teeth (Fig. 26.143)	***Cryptochironomus*** Kieffer (in part)
10'.	Mentum with 7 lateral teeth	11
11(10').	Antenna with 7 segments; blade originating at apex of 3rd segment (Fig. 26.146)	***Demicryptochironomus*** Lenz
11'.	Antenna with 5 segments; blade originating near apex of 2nd segment (Fig. 26.145)	12
12(11').	Mentum with 7 sharp and free lateral teeth; S II bladelike, all other S setae reduced	***Gillotia*** Kieffer

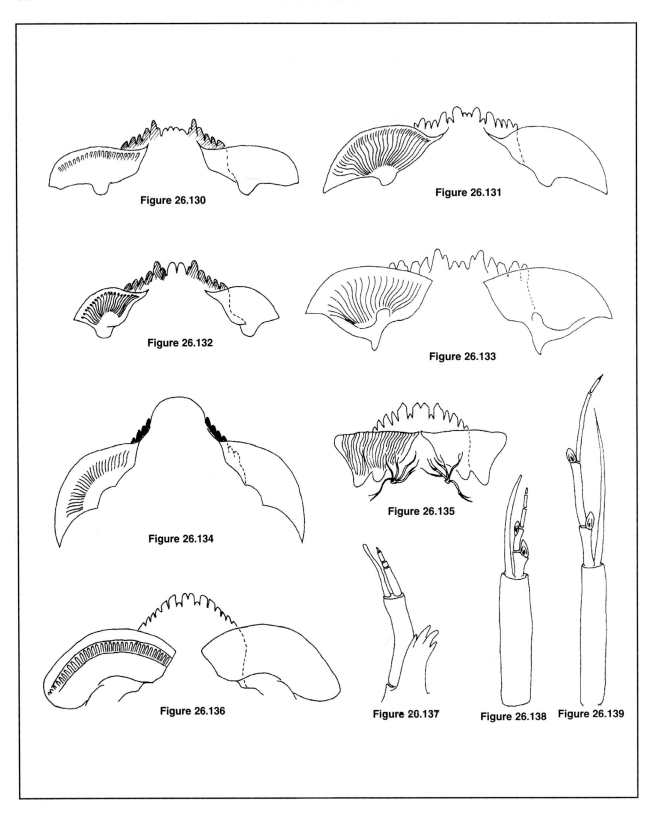

Figure 26.130. Mentum of *Paratendipes* sp.
Figure 26.131. Mentum of *Stictochironomus* sp.
Figure 26.132. Mentum of *Microtendipes* sp.
Figure 26.133. Mentum of *Omisus* sp. (redrawn with modification from Beck and Beck [1970]).
Figure 26.134. Mentum of *Paralauterborniella* sp.
Figure 26.135. Mentum of *Lauterborniella* sp.
Figure 26.136. Mentum of *Pagastiella* sp.
Figure 26.137. Antenna of *Pagastiella* sp.
Figure 26.138. Antenna of *Microtendipes* sp.
Figure 26.139. Antenna of *Lauterborniella* sp.

12'.	Mentum with 7 lateral teeth, however, 1st lateral teeth incompletely separated from the median tooth (Fig. 26.144), and the outermost 2 lateral teeth fused throughout most of their base; median tooth often with 2 small spines in the center (Fig. 26.144) ....***Cryptochironomus*** Kieffer (in part)
13(9').	Two outermost lateral teeth of mentum distinctly enlarged as in Figures 26.149-26.151, and labral sensilla 2-segmented (Fig. 26.141); mentum similar to Figures 26.149-26.151 14
13'.	Two outermost lateral teeth of mentum usually not enlarged, if somewhat enlarged then labral sensilla 3-segmented (Fig. 26.140); mentum not as in Figures 26.149-26.151 16
14(13).	Median tooth of mentum trifid (Fig. 26.151) and antennal blade longer than the combined lengths of segments 2 through 5 ....***Microchironomus*** Pagast
14'.	Median tooth of mentum broadly rounded, medially notched and often giving the appearance of paired teeth (Fig. 26.150), or with lateral notches (Fig. 26.149); antennal blade shorter than flagellum 15
15(14').	Median tooth of mentum medially notched (Fig. 26.150) ....***Cladopelma*** Kieffer
15'.	Median tooth of mentum broadly rounded or with lateral notches (Fig. 26.149)..***Cryptotendipes*** Lenz
16(13').	Median portion of mentum lacking distinct teeth (Figs. 26.155, 26.157), with a large dome-shaped tooth (Figs. 26.152, 26.158), or with 1 to several notches so that the median portion forms a broad convex structure (Figs. 26.153-26.154, 26.156, 26.159) 17
16'.	Median portion of mentum with distinct teeth (Figs. 26.160, 26.163, 26.171, 26.180) 23
17(16).	Antenna more than one-third as long as head capsule; mentum and ventromental plates as in Figures 26.154-26.158 ....***Paracladopelma*** Harnish (in part)
17'.	Antenna not more than one-third as long as head capsule; mentum and ventromental plates variable 18
18(17').	Antennal blade longer than flagellum; median tooth of mentum pointed or subtriangular (Fig. 26.159); anal tubules vestigial ....***Acalcarella*** Shilova
18'.	Antennal blade shorter than flagellum; median tooth of mentum not pointed or subtriangular 19
19(18').	Mandible with an elongate apical tooth and 4 smaller inner teeth that originate from a common base (Fig. 26.142); mentum as in Figure 26.148 ....***Nilothauma*** Kieffer
19'.	Mandible not as in Figure 26.142 and mentum not as in Figure 26.148 20
20(19').	Second antennal segment about as long as 3rd segment and antenna 5-segmented; basal segment of maxillary palp about 4 times as long as wide...............***Harnischia*** Kieffer
20'.	Second antennal segment distinctly longer than 3rd segment or antenna 6-segmented and 2nd segment much shorter than 3rd segment; basal segment of maxillary palp at most 3 times as long as broad 21
21(20').	Antenna 6-segmented; mentum as in Figure 26.152 or 26.153...***Saetheria*** Jackson
21'.	Antenna 5-segmented; mentum not as in Figure 26.152 or 26.153 22
22(21').	Second antennal segment unsclerotized in basal two-thirds (Fig. 26.147) ....***Cyphomella*** Saether
22'.	Second antennal segment fully sclerotized; mentum as in Figures 26.154-26.158 ....***Paracladopelma*** Harnish (in part)
23(16').	Mentum with an even number of teeth 24
23'.	Mentum with an odd number of teeth 33
24(23).	Mentum concave and consisting of 8 or 10 darkly pigmented teeth (Figs. 26.163, 26.187); mandible short and robust (Fig. 26.168) 48
24'.	Mentum and mandibles not as in Figures 26.163, 26.187 and 26.168 25
25(24').	Pecten epipharyngis composed of 3 blunt teeth; mentum as in Figure 26.171; larvae occurring in stems and petioles of aquatic vascular plants ....***Hyporhygma*** Reiss
25'.	Pecten epipharyngis either a single individual plate or, more commonly, consisting of 1-3 plates, each with 2 or more teeth; mentum not as in Figure 26.171 26
26(25').	Apical segment of maxillary palp elongate and well sclerotized (Fig. 26.166); striations of ventromental plates very coarse; mentum as in Figure 26.160 or 26.161 ....***Robackia*** Saether
26'.	Apical segment of maxillary palp not elongate, rarely well sclerotized; ventromental plates with finer striations and mentum not as in Fig. 26.160 or 26.161 27

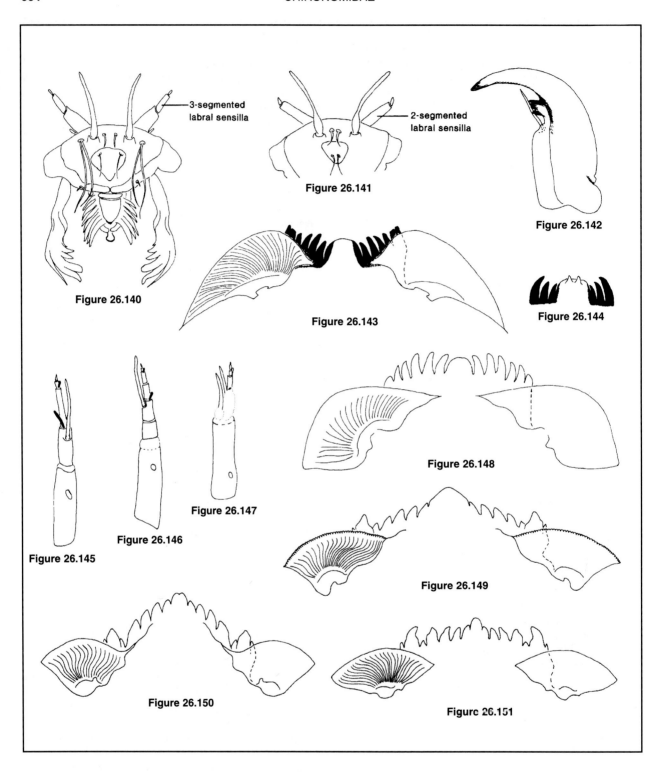

Figure 26.140. Anterior view of labral region of *Paracladopelma* sp. showing position of three-segmented labral sensillae in relation to other labral armature.

Figure 26.141. Anterior view of labral region of *Cladopelma* sp. showing two-segmented labral sensillae.

Figure 26.142. Mandible of *Nilothauma* sp.

Figure 26.143. Mentum of *Crypotchironomus* sp.

Figure 26.144. Detail of median portion of mentum of *Cryptochironomus blarina* Townes.

Figure 26.145. Antenna of *Cryptochironomus* sp.

Figure 26.146. Antenna of *Demicryptochironomus* sp.

Figure 26.147. Antenna of *Cyphomella* sp. (redrawn with modification from Saether [1977]).

Figure 26.148. Mentum of *Nilothauma* sp.

Figure 26.149. Mentum of *Cryptotendipes* sp.

Figure 26.150. Mentum of *Cladopelma* sp.

Figure 26.151. Mentum of *Microchironomus* sp. (redrawn with modification from Kugler [1971]).

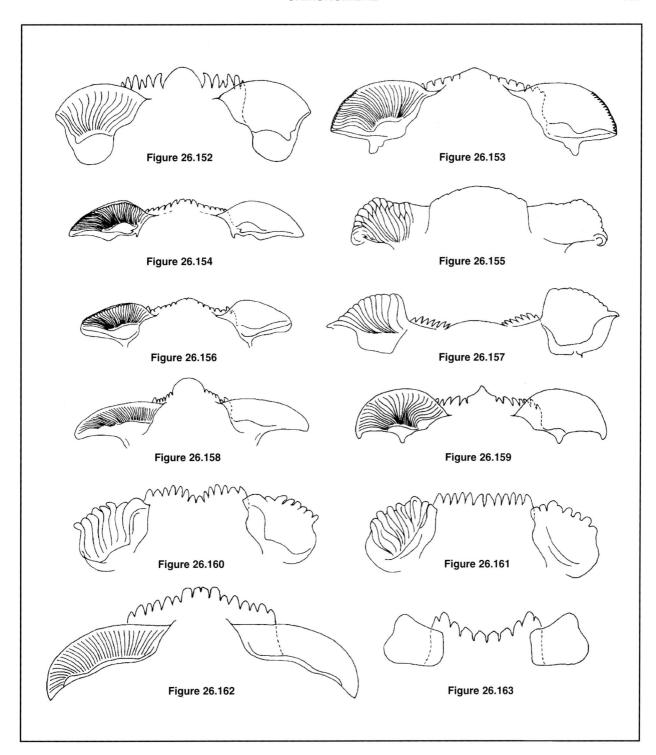

Figure 26.152. Mentum of *Saetheria tylus* (Townes).

Figure 26.153. Mentum of *Saetheria* sp. (redrawn from Jackson [1977]).

Figure 26.154. Mentum of *Paracladopelma galaptera* (Townes) (redrawn from Jackson [1977]).

Figure 26.155. Mentum of *Paracladopelma rolli* (Kirp.) (redrawn with modification from Saether [1977]).

Figure 26.156. Mentum of *Paracladopelma undine* (Townes) (redrawn from Jackson [1977]).

Figure 26.157. Mentum of *Paracladopelma doris* (Townes) (redrawn with modification from Saether [1977]).

Figure 26.158. Mentum of *Paracladopelma longanae* Beck and Beck (redrawn from Jackson [1977]).

Figure 26.159. Mentum of *Acalcarella* sp.

Figure 26.160. Mentum of *Robackia claviger* (Townes).

Figure 26.161. Mentum of *Robackia demeijerei* (Kruseman).

Figure 26.162. Mentum of *Endochironomus* sp.

Figure 26.163. Mentum of *Stenochironomus* sp.

27(26').	Median pair of mental teeth partially fused and distinctly wider than each of the remaining lateral teeth (Fig. 26.164); anterior margin of ventromental plates coarsely scalloped (Fig. 26.164)	***Parachironomus*** Lenz (in part)
27'.	Median pair of mental teeth not partially fused, or if partially fused then not distinctly wider than each of the remaining lateral teeth; anterior margin of ventromental plates not coarsely scalloped	28
28(27').	First lateral teeth of mentum much shorter than median and 2nd lateral teeth (Fig. 26.169)	***Polypedilum*** Kieffer (in part)
28'.	First lateral teeth of mentum subequal to or longer than the median and/or 2nd lateral teeth	29
29(28').	Lateral teeth of mentum gradually decreasing in size and length so that anterior margin of mentum appears to be broadly convex (Figs. 26.170, 26.172)	32
29'.	Second lateral teeth recessed and smaller in size than 1st lateral and 3rd lateral teeth (Fig. 26.174), or median teeth partially fused (Fig. 26.162)	30
30(29').	Median teeth partially fused (Fig. 26.162); ventromental plates 3-4 times wider than their length	53
30'.	Median teeth not partially fused (Fig. 26.174); ventromental plates not more than 3 times as wide as their maximum length	31
31(30').	Mola of mandible without serrations, or with only a simple serration at base of seta subdentalis (Fig. 26.188); distance from basal notch of innermost mandibular tooth to insertion of seta subdentalis usually at least 3/4 the distance from basal notch to apical notch of the apical inner tooth	54
31'	Mola of mandible with 1-3 serrations distinctly separated from seta subdentalis (Fig. 26.189); distance from basal notch of innermost mandibular tooth to insertion of seta subdentalis less than 3/4 the distance from basal notch to apical notch of apical inner tooth	***Tribelos*** Townes
32(29).	Ventromental plates more than 3 times wider than their maximum length and with lateral corners rounded (Fig. 26.172)	***Asheum*** Sublette and Sublette
32'.	Ventromental plates usually less than 3 times wider than their maximum length, lateral corners never rounded (Fig. 26.170)	***Polypedilum*** Kieffer (in part)
33(23').	Anterior margin of ventromental plates coarsely scalloped; median tooth of mentum about 2 times as wide as 1st lateral teeth; all lateral teeth pointed and gradually diminishing in size laterally (Fig. 26.165)	***Parachironomus*** Lenz (in part)
33'.	Anterior margin of ventromental plates smooth (Figs. 26.176, 26.183) or only finely crenulate (Figs. 26.180-26.182); median tooth and lateral teeth of mentum variable, but not as in Figure 26.165	34
34(33').	Median apices of ventromental plates touching or separated from one another by a distance less than the width of the median tooth (Figs. 26.175-26.176, 26.178-26.179, 26.190)	35
34'.	Median apices of ventromental plates separated by a distance equal to or greater than the width of the median tooth (Figs. 26.180, 26.184)	38
35(34).	Ventromental plates touching on midline; mentum as in Figures 26.179 or 26.190	50
35'.	Ventromental plates not touching on midline; mentum not as in Figure 26.179 or 26.190	36
36(35').	Mentum strongly arched, and with alternating large and small teeth (Fig. 26.178); larvae found in freshwater sponges	***Xenochironomus*** Kieffer
36'.	Mentum not strongly arched, not as in Figure 26.178; larvae found in various habitats, but apparently more common in subtropical regions of the United States	37
37(36').	Mentum with distinctly overlapping lateral teeth (Fig. 26.175); abdominal segment 8 with a bifurcate anterior ventral tubule (Fig. 26.177)	***Goeldichironomus*** Fittkau (in part)
37'.	Lateral teeth of mentum not overlapping (Fig. 26.176) abdominal segment 8 with a nonbifurcate anterior ventral tubule	***Goeldichironomus*** Fittkau (in part)
38(34').	Some lateral teeth of mentum projecting at least as far anterior as median tooth, giving the mentum an overall flat (Fig. 26.173) or concave appearance (Fig. 26.192), or median tooth deeply recessed (Fig. 26.191)	51
38'.	Outermost teeth of mentum decreasing in size or length, giving the mentum an overall convex appearance (Figs. 26.181, 26.183-26.184)	39
39(38').	Median tooth of mentum broadly rounded (Figs. 26.180, 26.183) or pointed (Figs. 26.181, 26.182), but lacking lateral notches	40
39'.	Median tooth with lateral notches that give it a trifid appearance (Fig. 26.184)	42
40(39).	Ventromental plate distinctly less than 2 times as wide as long and usually with small crenulations along anterior margin; median tooth and 1st lateral teeth enlarged and somewhat pointed; 1st laterals with (Fig. 26.181) or without lateral notches (Fig. 26.182, 26.193)	55

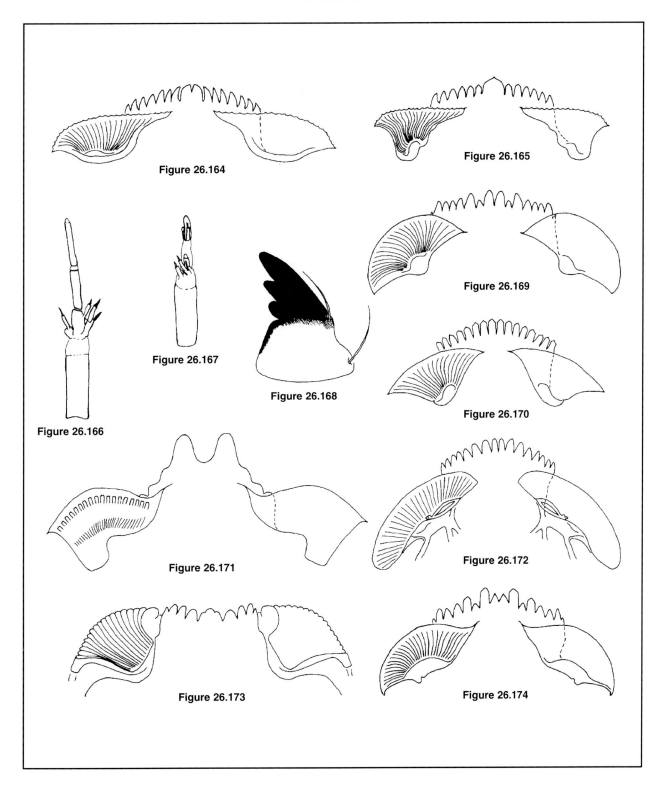

Figure 26.164. Mentum of *Parachironomus* cf. *frequens* (Johannsen).

Figure 26.165. Mentum of *Parachironomus* cf. *abortivus* (Malloch).

Figure 26.166. Maxillary palpus of *Robackia demeijerei* (Kruseman).

Figure 26.167. Maxillary palpus of *Beckidia tethys* (Townes) (redrawn with modification from Saether [1977]).

Figure 26.168. Mandible of *Stenochironomus* sp.

Figure 26.169. Mentum of *Polypedilum* sp.

Figure 26.170. Mentum of *Polypedilum* sp.

Figure 26.171. Mentum of *Hyporhygma* sp.

Figure 26.172. Mentum of *Asheum beckae* Sublette.

Figure 26.173. Mentum of *Beckidia tethys* (Townes) (redrawn with modification from Saether [1977]).

Figure 26.174. Mentum of *Phaenosectra* sp.

40'.	Ventromental plates at least twice as wide as long and anterior margin with or without crenulations; median tooth rounded; 1st lateral teeth variable but never as in Figures 26.181, 26.182 or 26.193 41
41(40').	Median tooth extending anterior to 1st lateral teeth; 4th lateral teeth reduced (Fig. 26.183) ....*Einfeldia* Kieffer (in part)
41'.	Median tooth recessed within 1st lateral teeth (Fig. 26.180), or very broad and extending about as far anterior as 1st lateral teeth, or if projecting farther anterior than 1st lateral teeth then 4th lateral teeth not conspicuously reduced..*Glyptotendipes* Kieffer
42(39').	Premandible with 3 or more teeth ....*Kiefferulus (Kiefferulus)* Goetghebuer (in part)
42'.	Premandible at most bifid 43
43(42').	Frontal apotome with an oval depression just posterior to antennal bases and 8th abdominal segment with 1 pair of ventral tubules ....*Einfeldia* Kieffer (in part)
43'.	Not with the above combination of characters 44
44(43').	Eighth abdominal segment without ventral tubules and pecten epipharyngis with 3-7 teeth ....*Einfeldia* Kieffer (in part)
44'.	Not with the above combination of characters 45
45(44').	Eighth abdominal segment lacking ventral tubules, with 1 pair of ventral tubules, or with 2 pairs of ventral tubules; pecten epipharyngis with 10 or more teeth 56
45'.	Eighth abdominal segment with ventral tubules and pecten epipharyngis with 9 teeth; known only from California...*Kiefferulus (Wirthiella)* Sublette (in part)
46(3).	Median tooth of mentum broad and with 6 pairs of more darkly pigmented lateral teeth (Fig. 26.134) ....*Paralauterborniella* Lenz
46'.	Median tooth of mentum narrow, recessed and with 5 pairs of more darkly pigmented lateral teeth (Fig. 26.191)..*Beardius* Reiss and Sublette (in part)
47(5).	S-I setae well separated, sockets not fused (Fig. 26.206) ....*Microtendipes* Kieffer
47'.	S-I setae originating very close to each other, sockets fused medially (Fig. 26.207) ....*Apedilum* Townes
48(24).	Mentum with 10 darkly pigmented teeth (Fig. 26.163)...*Stenochironomus* Kieffer
48'.	Mentum with 8 darkly pigmented teeth (Fig. 26.187) ....*Xestochironomus* Sublette and Wirth
49(6).	Seta posterior to ventromental plates plumose (Fig. 26.135); abdominal segment eight with posteriorly projecting dorsal hump (Fig. 26.194) ....*Lauterborniella* Bause
49'.	Seta posterior to ventromental plates simple (Fig. 26.186); abdominal segment eight with anteriorly projecting dorsal hump (Fig. 26.195) ....*Zavreliella* Kieffer
50(35).	Inner teeth of mandible sharply pointed (Fig. 26.196); head capsule with well developed tubercles lateral to bases of antennae (Fig. 26.197)...........................*Lipiniella* Shilova
50'.	Inner teeth of mandible low, blunt (Fig. 26.198); head capsule without tubercles lateral to bases of antennae..*Axarus* Roback
51(38).	Maxillary palp more than 4 times longer than wide, one apical segment elongate (Fig. 26.167); mentum as in Figure 26.173 ....*Beckidia* Saether
51'.	Maxillary palp not more than 3 times longer than wide, apical segment not as in Figure 26.167; mentum as in Figure 26.191 or Figure 26.192 52
52(51').	Mentum as in Figure 26.192 ....*Stelechomyia* Reiss
52'.	Mentum as in Figure 26.191...*Beardius* Reiss and Sublette (in part)
53(30).	Oral margin of cardo tuberculate (Fig. 26.199) ....*Endochironomus* Kieffer
53'.	Oral margin of cardo not tuberculate (Fig. 26.200)..*Synendotendipes* Grodhaus
54(31).	Mandible with 4 inner teeth ....*Sergentia* Kieffer
54'.	Mandible with 3 inner teeth..*Phaenopsectra* Kieffer
55(40)	Pecten epipharyngis consisting of 3 plates each with 4-5 apical teeth (Fig. 26.202) and mandible and mentum as in Figures 26.201 and 26.193 ....*Endotribelos* Grodhaus
55'	Pecten epipharyngis consisting of one plate with 3-6 strong or blunt teeth (Fig. 26.203) ....*Dicrotendipes* Kieffer

(**Note**- Some species of *Dicrotendipes* have a mandible similar to Figure 26.201, but the mentum of these species is similar to Figures 26.181 or 26.182)

56(45)	Pecten epipharyngis with numerous fine teeth in 2-3 partial or complete rows (Fig. 26.204) ....*Einfeldia* Kieffer (in part)
56'	Pecten epipharyngis with stronger teeth in only one row (Fig. 26.205) ....*Chironomus* Meigen (in part)

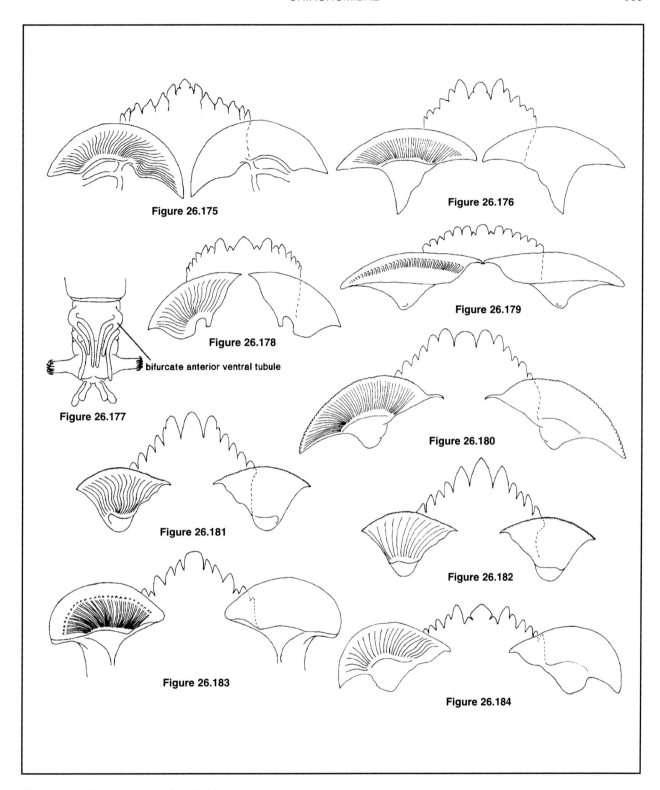

Figure 26.175. Mentum of *Goeldichironomus* sp.

Figure 26.176. Mentum of *Goeldichironomus devineyae* (Beck) (redrawn with modification from Beck and Beck [1970]).

Figure 26.177. Posterior abdominal segments and ventral tubules of *Goeldichironomus* sp. (redrawn with modification from Fittkau [1965]).

Figure 26.178. Mentum of *Xenochironomus xenolabis* (Kieffer).

Figure 26.179. Mentum of *Axarus festivus* (Say).

Figure 26.180. Mentum of *Glyptotendipes* sp.

Figure 26.181. Mentum of *Dicrotendipes* sp.

Figure 26.182. Mentum of *Dicrotendipes* sp.

Figure 26.183. Mentum of *Einfeldia natchitocheae* (Sublette) (redrawn with modification from Sublette [1964]).

Figure 26.184. Mentum of *Chironomus* sp.

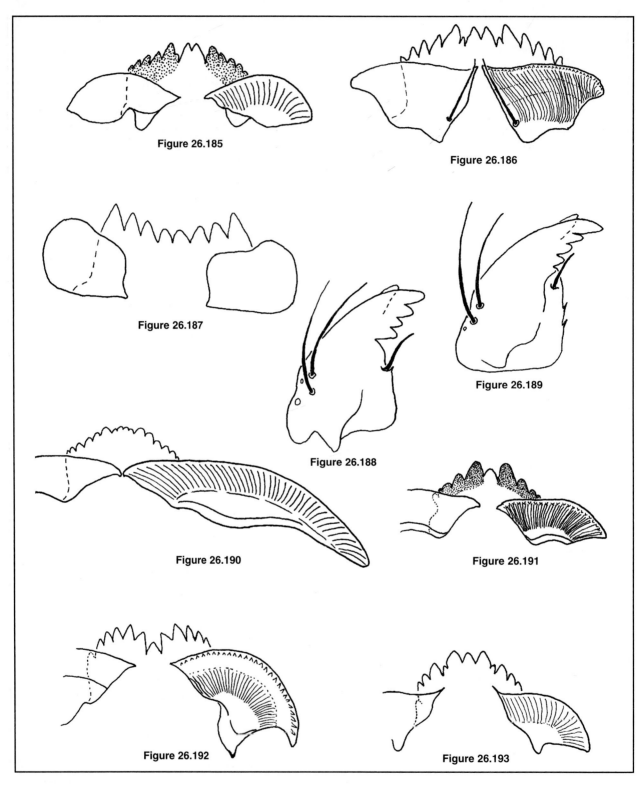

Figure 26.185. Mentum and ventromental plates of *Apedilum* sp.

Figure 26.186. Mentum and ventromental plates of *Zavreliella* sp.

Figure 26.187. Mentum and ventromental plates of *Xestochironomus* sp.

Figure 26.188. Mandible of *Phaenopsectra* sp.

Figure 26.189. Mandible of *Tribelos* sp.

Figure 26.190. Mentum and ventromental plates of *Lipiniella* sp.

Figure 26.191. Mentum and ventromental plates of *Beardius* sp.

Figure 26.192. Mentum and ventromental plates of *Stelechomyia* sp.

Figure 26.193. Mentum and ventromental plates of *Endotribelos* sp.

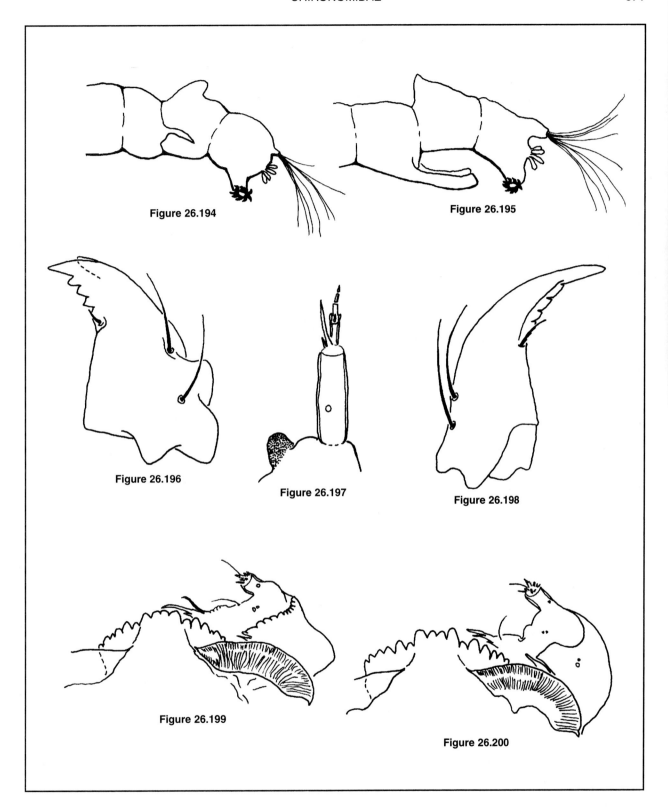

Figure 26.194. Lateral view of apical segments of abdomen in *Lauterborniella* sp.

Figure 26.195. Lateral view of apical segments of abdomen in *Zavreliella* sp.

Figure 26.196. Mandible of *Lipiniella* sp.

Figure 26.197. Antenna and lateral tubercle of *Lipiniella* sp.

Figure 26.198. Mandible of *Axarus* sp.

Figure 26.199. Mentum, ventromental plates, maxilla, maxillary palp and cardo of *Endochironomus* sp.

Figure 26.200. Mentum, ventromental plates, maxilla, maxillary palp and cardo of *Synendotendipes* sp.

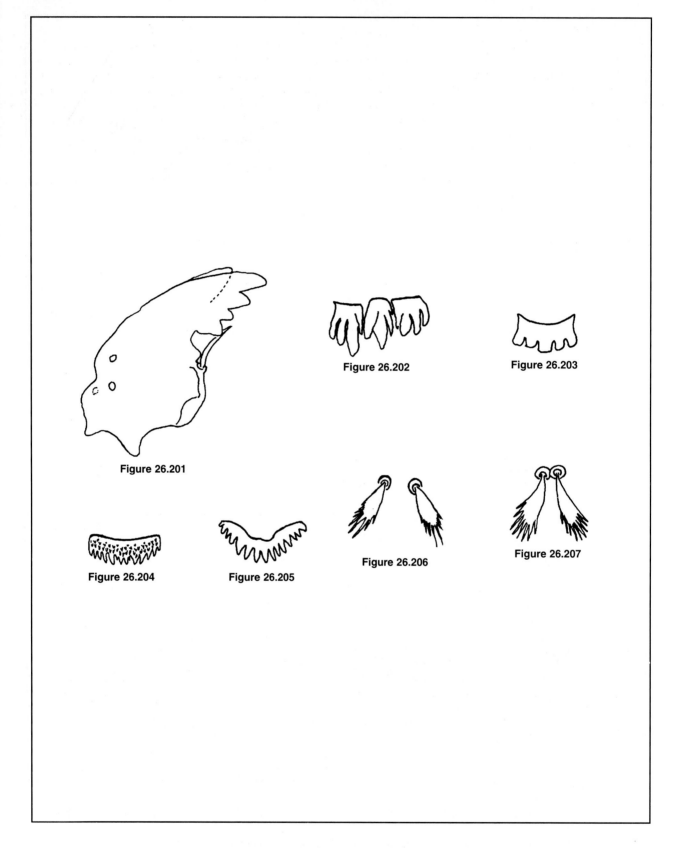

Figure 26.201. Mandible of *Endotribelos* sp.
Figure 26.202. Pecten epipharyngis of *Endotribelos* sp.
Figure 26.203. Pecten epipharyngis of *Dicrotendipes* sp.
Figure 26.204. Pecten epipharyngis of *Einfeldia* sp.
Figure 26.205. Pecten epipharyngis of *Chironomus* sp.
Figure 26.206. SI setae of *Microtendipes* sp.
Figure 26.207. SI seta of *Apedilum* sp.

Orthocladiinae

1.	Antenna elongate, more than two-thirds the length of the head capsule	2
1'.	Antenna not elongate, equal to or less than two-thirds the length of the head capsule	6
2(1).	Second antennal segment with extensive dark brown or golden brown pigmentation (Figs. 26.219-26.220)	3
2'.	Second antennal segment concolorous with remaining segments, lacking distinct dark brown or golden brown pigmentation	4
3(2).	Antenna longer than head capsule, 4-segmented, and usually with some darker pigmentation occurring on the 3rd segment (Fig. 26.219)	***Corynoneura*** Winnertz
3'.	Antenna shorter than head capsule, 5-segmented, and never with dark pigmentation occurring on 3rd segment (Fig. 26.220)	***Thienemanniella*** Kieffer
4(2').	Median portion of mentum concave (Fig. 26.248); antenna with large lauterborn organs originating at different levels on 2nd antennal segment (Fig. 26.226)	***Heterotanytarsus*** Sparck
4'.	Median portion of mentum not concave; antenna not as in Figure 26.226	5
5(4').	Procercus well developed; 1 procercal seta at least twice as long as the length of the posterior prolegs; antenna with an elongate blade and whiplike apical segment (Fig. 26.221); body white when preserved	***Lopescladius*** Oliveira
5'.	Procercus less well developed; all procercal setae less than twice as long as posterior prolegs; apical segment of antenna not whiplike (Fig. 26.222); body often with distinctive purple or brown pigmented areas when preserved	***Rheosmittia*** Brundin
6(1').	Antenna arising from tubercle that has a large anteromedially directed spur (Fig. 26.223); thoracic segments with conspicuous setae, many of which are apically dissected; larvae that build portable sand cases	***Abiskomyia*** Edwards
6'.	Not with the above combination of characters	7
7(6').	One procercal setae elongate, at least one-quarter the length of the larval body (Fig. 26.291); remaining procercal setae vestigial or absent	8
7'.	All procercal setae less than one-fifth the length of the larval body	10
8(7).	Mentum consisting of a single broad, dome shaped median tooth with a central cusp and 6 pairs of narrow pointed lateral teeth (Fig. 26.242); apical tooth of mandible longer than the combined widths of the lateral teeth (Fig. 26.303)	***Krenosmittia*** Thienemann
8'.	Mentum consisting of paired median teeth and 5 paired lateral teeth, the outermost 2 teeth fused throughout most of their length (Fig. 26.243); apical tooth of mandible shorter than combined widths of the lateral teeth (Figs. 26.298-26.299)	9
9(8').	Mandible with 3 inner teeth (Fig. 26.299)	***Pseudorthocladius*** Goetghebuer
9'.	Mandible with 2 inner teeth (Fig. 26.298)	***Parachaetocladius*** Wulker
10(7').	Ventromental plates well developed and covering all lateral teeth of mentum; 3 small median teeth visible in median concavity of mentum (Fig. 26.244); larvae apparently occurring only within colonies of blue-green algae	***Acamptocladius*** Brundin
10'.	Mentum not as in Figure 26.244; larvae living in a variety of habitats	11
11(10').	Cardinal beard present beneath the ventromental plates (Figs. 26.278-26.289, 26.306-26.307)	12
11'.	Cardinal beard not present beneath the ventromental plates (Figs. 26.245-26.277)	21
12(11).	SI seta simple (Fig. 26.208), bifid (Fig. 26.209) or bifid with secondary feathering (Fig. 26.210)	13
12'.	SI seta palmate (Fig. 26.211), plumose (Fig. 26.212), or with multiple apical dissections (Fig. 26.213, 26.217-26.218)	18
13(12).	SI setae simple (Fig. 26.208)	14
13'.	SI setae bifid or bifid with secondary feathering (Figs. 26.209-26.210)	15
14(13)	Cardinal setae very weak and mostly covered by lateral edges of ventromental plates (Fig. 26.306) or occurring in a single row (Fig. 26.307)	62
14'	Cardinal setae elongate, bristlelike (Fig. 26.288) branched or stellate (Fig. 26.285), and extending well beyond lateral edges of ventromental plates	63
15(13').	SI setae bifid with secondary feathering; mentum as in Figure 26.280	***Acricotopus*** Kieffer
15'.	SI setae bifid; mentum not as in Figure 26.280	16
16(15').	Mentum with paired median teeth (Fig. 26.284)	***Rheocricotopus*** Thienemann and Harnisch
16'.	Mentum with an unpaired median tooth (Figs. 26.278, 26.282)	17
17(16').	Median tooth of mentum broad, dome-shaped, and distinctly lighter in color than remaining lateral teeth (Fig. 26.282)	***Paracladius*** Hirvenoja
17'.	Median tooth of mentum narrower, not dome-shaped, and usually concolorous with 1st lateral teeth (Fig. 26.278)	***Halocladius*** (*Halocladius*) Hirvenoja

18(12').	SI seta palmate (Fig. 26.211); procerci with at least 1 basal spur; apical tooth of mandible longer than the combined widths of the lateral teeth (Fig. 26.300); mentum usually with a broad unpaired but peaked median tooth (Fig. 26.283) or with broad paired median teeth (Fig. 26.281)	***Psectrocladius*** Kieffer
18'.	SI seta plumose (Fig. 26.212) or with multiple apical dissections (Fig. 26.213), but not with the remaining combination of characters	19
19(18').	Mentum with a broad, truncated median tooth (Fig. 26.279); larvae occurring in marine littoral zones, not presently recorded from North America	***Halocladius*** *(Psammocladius)* Hirvenoja
19'.	Mentum with rounded or pointed median teeth (Figs. 26.286-26.287, 26.289); larvae occurring in a variety of freshwater habitats	20
20(19').	Mandible with 4 inner teeth; mentum as in Figure 26.286	***Diplocladius*** Kieffer
20'.	Mandible with 3 inner teeth; mentum either with 2-4 pale median teeth (Fig. 26.289) or similar to Figure 26.287	***Zalutschia*** Lipina
21(11').	Procerci present	27
21'.	Procerci absent	22
22(21').	Preanal segment strongly bent ventrally so that the anal segment and posterior prolegs are orientated at a right angle to the long axis of the body (Figs. 26.294-26.295); anal segment and posterior prolegs retractile into preanal segment	23
22'.	Preanal segment at most declivent; anal segment and posterior prolegs not oriented at right angle to main axis of body, and not retractile into preanal segment	24
23(22).	Posterior prolegs each subdivided, the anterior portion bearing a semicircle of claws (Fig. 26.294); no anal tubules present; antenna as in Figure 26.229	***Gymnometriocnemus*** Geotghebuer
23'.	Posterior prolegs not subdivided (Fig. 26.295); 4 short anal tubules present; antenna similar to Figure 26.230	***Bryophaenocladius*** Thienemann
24(22').	Larva ectoparasitic on mayflies; mouthparts reduced, mentum as in Figure 26.245; body robust, especially thoracic segments, giving an overall swollen appearance (Fig. 26.296)	***Symbiocladius*** Kieffer and Zavrel
24'.	Larva not ectoparasitic; mouthparts, mentum, and body not as in Figure 26.245 or 26.296	25
25(24').	SI seta and SII seta bifid and both well developed (Fig. 26.216)	***Pseudosmittia*** Goetghebuer
25'.	SII seta never bifid	26
26(25').	SI seta strong and simple (Fig. 26.214) or with fine apical and lateral serrations (Fig. 26.215); antenna as in Figures 26.225 or Figure 26.310; anal tubules elongate and with numerous constrictions throughout their length (Fig. 26.293, 26.314); mentum as in Figures 26.246, 26.309 or 26.313	65
26'.	SI seta not strong and simple or with fine apical and lateral serrations, but usually with multiple apical dissections (Fig. 26.217) or nearly palmate; antenna similar to Figure 26.231, 26.232 or 26.311; anal tubules not elongate, lacking constrictions; mentum not as in Figure 26.246	67
27(21).	Antenna 7-segmented, 3rd segment shorter than 4th segment and 7th segment hairlike or vestigial (Fig. 26.227); mentum with 1 (*maeaeri* group, Fig. 26.254) or 2 median teeth and 5 pairs of lateral teeth (Fig. 26.256); labral lamella present and divided at least anteriorly (Fig. 26.241); SI seta plumose	***Heterotrissocladius*** Sparck
27'.	Not with the above combination of characters	28
28(27').	Larva with a dense lateral fringe of setae on at least abdominal segments 1 through 5 (Fig. 26.290); mentum as in Figure 26.252 or 26.308	64
28'.	Larva lacking a dense fringe of setae on abdominal segments; mentum not as in Figure 26.252 or 26.308	29
29(28').	Mentum strongly arched and appearing truncated at apex, only 6 anterior teeth readily discernible when head capsule not strongly depressed (Figs. 26.249-26.250); larvae phoretic on Ephemeroptera nymphs or living within the gills of bivalve mollusks	30
29'.	Mentum, if strongly arched, not as in Figures 26.249 and 26.250; larvae living in a variety of habitats	31
30(29).	Body with conspicuous brown setae; mentum as in Figure 26.250; antenna as in Figure 26.233; premandible with 2 apical teeth; larvae phoretic on Ephemeroptera	***Epoicocladius*** Zavrel
30'.	Body lacking conspicuous brown setae; mentum as in Figure 26.249; antenna as in Figure 26.224; premandible with 6 apical teeth; larvae living within gills of bivalve mollusks	***Baeoctenus*** Saether

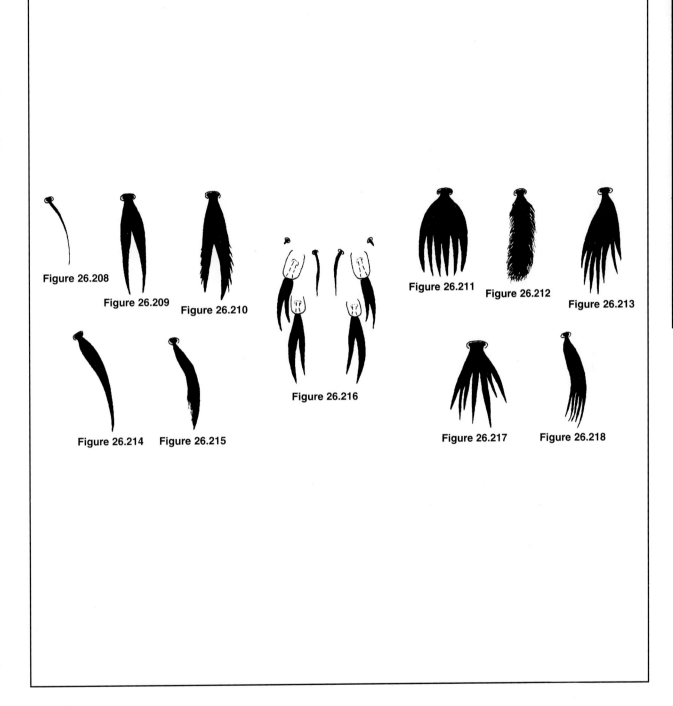

Figure 26.208. Simple SI seta of *Eukiefferiella* sp.
Figure 26.209. Bifid SI seta of *Cricotopus* sp.
Figure 26.210. Bidif SI seta with secondary feathering in *Acricotopus* sp.
Figure 26.211. Palmate SI seta of *Psectrocladius* sp.
Figure 26.212. Plumose SI seta of *Parametriocnemus* sp.
Figure 26.213. SI seta of *Hydrobaenus* sp. with strong apical dissections.

Figure 26.214. Strong and simple SI seta of *Georthocladius* sp.
Figure 26.215. SI seta with fine apical dissections.
Figure 26.216. Bifid SI and SII setal arrangement in *Pseudosmittia* sp.
Figure 26.217. SI seta with strong apical dissections in *Smittia* sp.
Figure 26.218. SI seta with weak apical dissections in *Paracricotopus* sp.

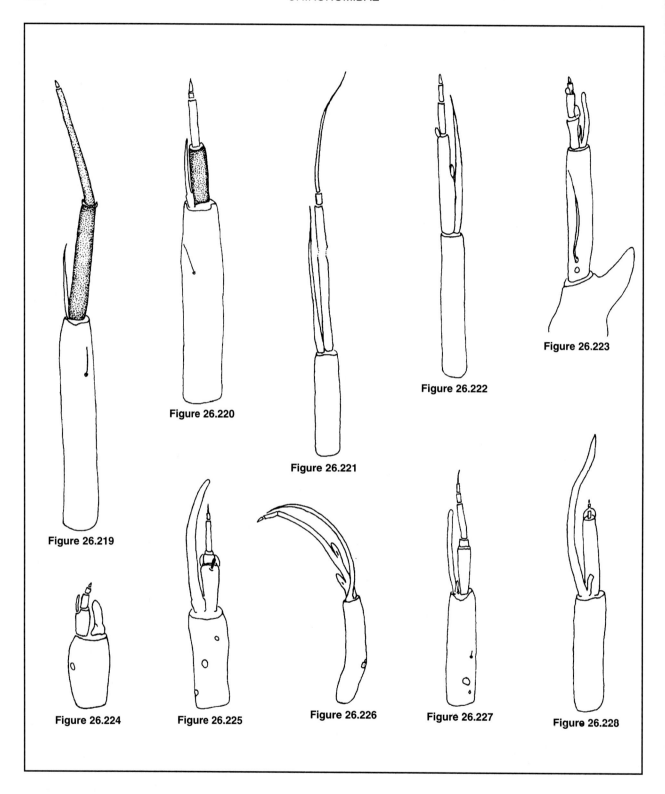

Figure 26.219. Antenna of *Corynoneura* sp.
Figure 26.220. Antenna of *Thienemanniella* sp.
Figure 26.221. Antenna of *Lopescladius* sp.
Figure 26.222. Antenna of *Rheosmittia* sp.
Figure 26.223. Antenna of *Abiskomyia* sp. (redrawn with modification from Pankratova [1970]).
Figure 26.224. Antenna of *Baeoctenus bicolor* Saether (redrawn with modification from Saether [1977]).
Figure 26.225. Antenna of *Georthocladius* sp.
Figure 26.226. Antenna of *Heterotanytarsus perennis* Saether (redrawn with modification from Saether [1975]).
Figure 26.227. Antenna of *Heterotrissocladius* sp.
Figure 26.228. Antenna of *Heleniella thienemanni* Gowin.

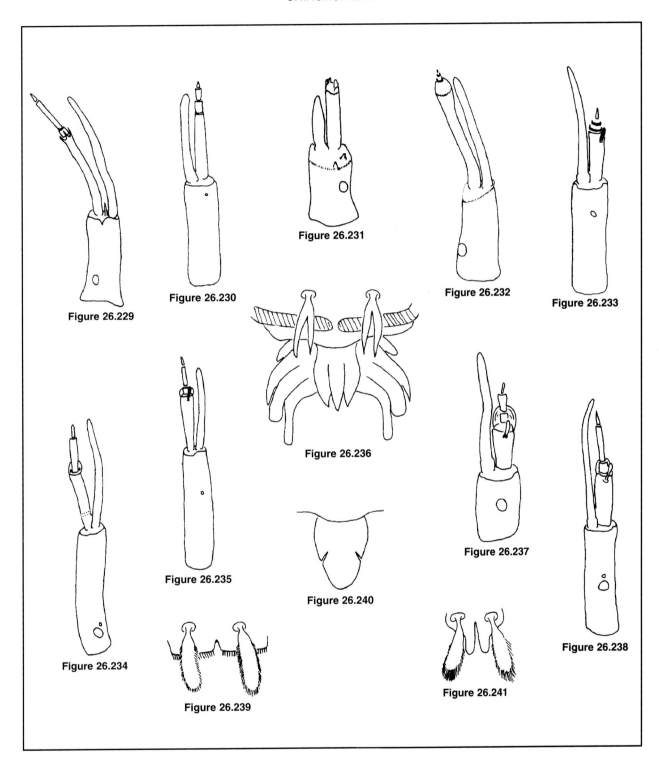

Figure 26.229. Antenna of *Gymnometriocnemus* sp.
Fiogure 26.230. Antenna of *Bryophaenocladius* sp.
Figure 26.231. Antenna of *Smittia* sp.
Figure 26.232. Antenna of *Smittia aquatilis* Goetghebuer (redrawn with modification from Thienemann and Strenzke [1941]).
Figure 26.233. Antenna of *Epoicocladius* sp.
Figure 26.234. Antenna of *Brillia* sp.
Figure 26.235. Antenna of *Parametriocnemus* sp.
Figure 26.236. Pecten epipharyngis and associated labral setae and other armature of *Cricotopus* (*Cricotopus*) sp.
Figure 26.237. Antenna of *Paraphaenocladius* sp.
Figure 26.238. Antenna of *Psilometriocnemus* sp.
Figure 26.239. SI and labral lamellae of *Brillia* sp.
Figure 26.240. Pecten epipharyngis of *Cricotopus* (*Isocladius*) sp.
Figure 26.241. SI and labral lamellae of *Heterotrissocladius* sp.

31(29').	Second antennal segment elongate, more than two-thirds the length of the 1st segment; antennal segments 3-5 very reduced, their combined lengths equal to about one-fifth the length of segment 2; antennal blade curved and extending well beyond 5th antennal segment (Fig. 26.228); mentum with 2 broad median teeth and 5 smaller lateral teeth; outermost lateral tooth more elongate than preceding lateral tooth (Fig. 26.247) ....*Heleniella* Gowin
31'.	Antenna and mentum not as in Figures 26.228 and 26.247 32
32(31').	Antenna 6-segmented; mentum broadly convex, similar to Figure 26.257, 26.258, or 26.259; SI plumose or with multiple strong or weak apical dissections 33
32'.	If mentum similar to Figure 26.257, 26.258, or 26.259, then antenna 4- or 5-segmented 35
33(32).	Median tooth with distinct median cusp (Fig. 26.258) ....*Parakiefferiella* Brundin
33'.	Median tooth without medium cusp...34
34(33').	Median tooth truncated (Fig. 26.259); SI plumose..*Oliveridia* Saether
34'.	Mentum with paired median teeth (Fig. 26.257); SI with multiple apical dissections...*Hydrobaenus* Fries
35(32').	SI plumose, labral lamella well developed (Fig. 26.239); 1st antennal segment slightly or strongly bent (Fig. 26.234, 26.310); mentum similar to Figure 26.251 or 26.253 61
35'.	Not with the above combination of characters 36
36(35').	Mentum with strongly arcuate median tooth and 2 pairs of small lateral teeth (Fig. 26.255); larvae living in submerged decomposing wood ....*Orthocladius (Symposiocladius)*
36'.	Mentum not as in Figure 26.255; larvae occurring in a variety of habitats 37
37(36').	Apical tooth of mandible clearly longer than the combined widths of the lateral teeth (Fig. 26.301) 38
37'.	Apical tooth of mandible not longer than the combined widths of the lateral teeth.....................................39
38(37).	Mentum consisting of a broad median tooth with a single median cusp, and 10 pairs of sharp pointed, distinct, lateral teeth (Fig. 26.271); larvae apparently restricted to high alpine or arctic regions ....*Lapposmittia* Thienemann
38.	Mentum consisting of a broad median tooth with 2 median cusps, and 6 or fewer pairs of lateral teeth, some of which are often small or indistinct; ventromental plates well developed, usually extending beyond posterolateral corner of mentum (Fig. 26.270) ....*Nanocladius* Kieffer
39(37').	SI seta simple...40
39'.	SI seta bifid, plumose, palmate, or with weak (Fig. 26.218) or strong apical (Fig. 26.217) dissections 42
40(39).	SI seta thick and heavily sclerotized; median tooth of mentum broad, heavily sclerotized (Fig. 26.260); procercus reduced; 2 procercal setae distinctly larger in diameter and longer in length than the remaining setae ....*Cardiocladius* Kieffer
40'.	If SI seta thick and heavily sclerotized then median tooth of mentum not broad and heavily sclerotized; procerus not reduced; all procercal setae similar in diameter and length....................41
41(40').	Antenna reduced, less than one-half the length of the mandible, or antenna about the length of the mandible and larvae occurring in the leaves of pitcher plants 51
41'.	Antenna not reduced, subequal to or longer than the length of the mandible; larvae occurring in a wide variety of habitats but never in the leaves of pitcher plants; a very species rich and commonly encountered genus, mentum of some common species illustrated in Figures 26.263-26.265 ....*Eukiefferiella* Thienemann
42(39').	SI seta bifid 43
42'.	SI seta plumose, palmate, or with weak or strong apical dissections 52
43(42).	Mentum with broad median tooth and 9 pairs of lateral teeth (Fig. 26.269) ....*Orthocladius (Euorthocladius)* Thienemann (in part)
43'.	Mentum with fewer than 9 pairs of lateral teeth, median tooth variable 44
44(43').	Posterolateral corners of at least abdominal segments 1 through 6 with a group of setae, all of which arise from a single point (Fig. 26.292)..45
44'.	Posterolateral corners of abdominal segments without a group of setae arising from a single point...46
45(44).	Pecten epipharyngis composed of 3 nearly equal sized plates (Fig. 26.236) 68
45'.	Pecten epipharyngis composed of 1 large conical plate, vestigial lateral plates only indicated by lateral notches in median plate (Fig. 26.240).............................*Cricotopus (Isocladius)* Kieffer (in part)
46(44').	Mandible with spines along the inner margin (Fig. 26.302) ....*Cricotopus (Cricotopus)* Hirvenoja (in part)
46'.	Mandible smooth along inner margin 47
47(46').	Antenna 4-segmented; lauterborn organs poorly developed; eyespots dark and tripartite with 2 small anterior and 1 larger posterior pigment patch; body green ....*Orthocladius (Pogonocladius)* Brundin
47'.	Antenna 5-segmented, or if 4-segmented then not with the combination of poorly developed lauterborn organs, tripartite eyespots, and green body 48
48(47').	Head capsule dark brown to black 49

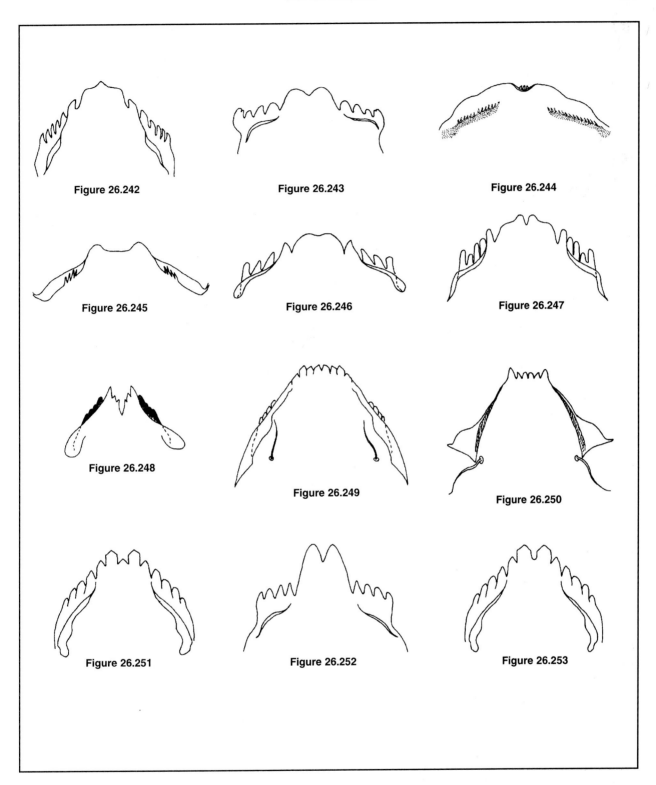

Figure 26.242. Mentum of *Krenosmittia* sp.
Figure 26.243. Mentum of *Parachaetocladius* sp.
Figure 26.244. Mentum of *Acamptocladius* sp.
Figure 26.245. Mentum of *Symbiocladius* sp.
Figure 26.246. Mentum of *Georthocladius* sp.
Figure 26.247. Mentum of *Heleniella thienemanni* Gowin.
Figure 26.248. Mentum of *Heterotanytarsus perennis* Saether (redrawn with modification from Seather [1975]).
Figure 26.249. Mentum of *Baeoctenus bicolor* Saether (redrawn with modification from Saether [1977]).
Figure 26.250. Mentum of *Epoicocladius* sp.
Figure 26.251. Mentum of *Brillia* sp.
Figure 26.252. Mentum of *Xylotopus* sp.
Figure 26.253. Mentum of *Brillia* sp.

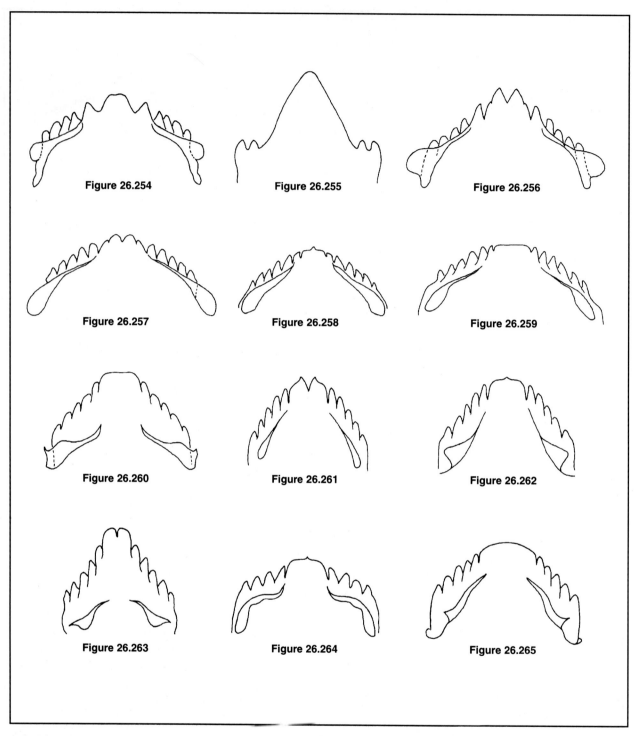

Figure 26.254. Mentum of *Heterotrissocladius maeaeri* group species.

Figure 26.255. Mentum of *Orthocladius (Symposiocladius)* sp.

Figure 26.256. Mentum of *Heterotrissocladius marcidus* group species.

Figure 26.257. Mentum of *Hydrobaenus* sp.

Figure 26.258. Mentum of *Parakiefferiella* sp.

Figure 26.259. Mentum of *Oliveridea tricornis* (Oliver) (redrawn with modification from Saether [1976]).

Figure 26.260. Mentum of *Cardiocladius* sp.

Figure 26.261. Mentum of *Tvetenia bavarica* group species.

Figure 26.262. Mentum of *Tvetenia discoloripes* group species.

Figure 26.263. Mentum of *Eukiefferiella pseudomontana* group species.

Figure 26.264. Mentum of *Eukiefferiella devonica* group species.

Figure 26.265. Mentum of *Eukiefferiella potthasti* group species.

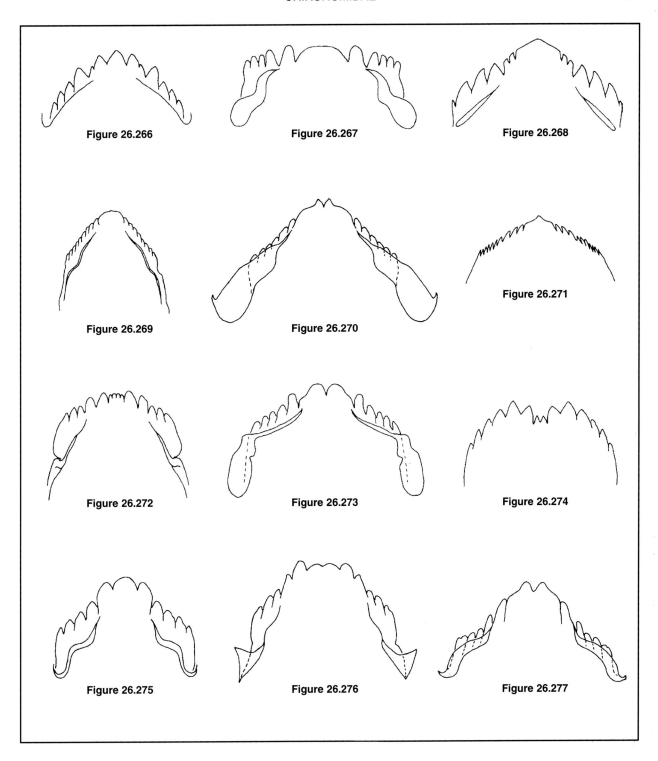

Figure 26.266. Mentum of *Cricotopus* (*Isocladius*) *elegans* Johannsen (redrawn with modification from Hirvenoja [1973]).

Figure 26.267. Mentum of *Paraphaenocladius* sp.

Figure 26.268. Mentum of *Cricotopus* (*Cricotopus*) *trifascia* Edwards.

Figure 26.269. Mentum of *Orthocladius* (*Euorthocladius*) sp.

Figure 26.270. Mentum of *Nanocladius* sp.

Figure 26.271. Mentum of *Lapposmittia* sp. (redrawn with modification from Thienemann [1944]).

Figure 26.272. Mentum of *Metriocnemus knabi* Coquillett.

Figure 26.273. Mentum of *Parametriocnemus* sp.

Figure 26.274. Mentum of *Metriocnemus* cf. *fuscipes* (Meigen).

Figure 26.275. Mentum of *Paracricotopus* sp.

Figure 26.276. Mentum of *Chaetocladius* sp.

Figure 26.277. Mentum of *Psilometriocnemus* sp.

48'.	Head capsule yellow to light brown	50
49(48).	Head capsule reddish brown to dark brown	*Orthocladius (Eudactylocladius)* Thienemann
49'.	Head capsule light brown	*Orthocladius (Euorthocladius)* Thienemann (in part)
50(48').	Median tooth of mentum broad, 1st and 2nd lateral teeth appearing partially fused with median tooth and smaller than remaining lateral teeth (Fig. 26.268)	*Cricotopus (Cricotopus) trifascia* Edwards
50'.	Mentum not as in Figure 26.268	*Cricotopus (Cricotopus)* (in part), *Orthocladius (Orthocladius)*
51(41).	SI seta short and peglike; mentum with an even number of teeth (Figs. 26.272, 26.274)	*Metriocnemus* Van der Wulp
51'.	SI seta more elongate, mentum with an odd number of teeth (Fig. 26.266)	*Cricotopus (Isocladius)* Kieffer (in part)
52(42').	SI seta plumose; mentum with 2 large, rounded, median teeth and 5 lateral teeth that form a gentle arch (Fig. 26.273); ventromental plates with a distinct curve such that they are anteriorly oriented parallel to the arch of the lateral mental teeth, and posteriorly to the basal margin of the mentum	53
52'.	SI seta not plumose or if plumose then mentum and ventromental plates not as in Figure 26.273	55
53(52).	Antenna shorter than mandible (Fig. 26.237)	*Paraphaenocladius* Thienemann (in part)
53'.	Antenna longer than mandible	54
54(53').	Antennal segments 2 through 5 reduced and antennal blade much longer than segments 2 through 5	*Paraphaenocladius* Thienemann (in part)
54'.	Antennal segments 2 through 5 not abnormally reduced, antennal blade equal to or shorter than segments 2 through 5 (Fig. 26.235)	*Parametriocnemus* Thienemann
55(52').	SI seta plumose, mentum with single, broad median tooth and ventromental plates as in Figure 26.267; antenna shorter than mandible	*Paraphaenocladius* Thienemann (in part)
55'.	Not with the above combination of characters	56
56(55').	SI seta weak, with several fine apical dissections (Fig. 26.218)	57
56'.	SI seta well developed, with several strong apical dissections	58
57(56).	Mentum with a single round median tooth and 1st lateral teeth that are set off from the remaining lateral teeth (Fig. 26.275); procercus with a preapical spine and a basal spur; inner margin of mandible never with spines	*Paracricotopus* Thienemann and Harnish
57'.	Mentum with a single or double median tooth Figs. 26.261, 26.262; procercus lacking preapical spine and basal spur; however, a well-developed preapical seta usually present; inner margin of mandible with spines (Fig. 26.297) which are easily worn away or are often difficult to distinguish; a very species rich and commonly encountered genus	*Tvetenia* Kieffer
58(56').	Median pair of mental teeth extending distinctly anterior to 1st lateral teeth	59
58'.	Median pair of mental teeth shorter than 1st lateral teeth or subequal in length to 1st laterals (Fig. 26.276) or mentum with single median tooth	60
59(58).	Antenna with 3rd segment much shorter than 4th segment, and blade extending distinctly beyond apex of 5th segment; length of antennal segment 1 about equal to the combined lengths of segments 2 through 5 (Fig. 26.238); mentum as in Figure 26.277 or 26.313; mature larvae 5-7 mm	66
59'.	If 3rd antennal segment much shorter than 4th segment then blade not extending beyond apex of 5th antennal segment, or 1st antennal segment distinctly larger than the combined lengths of segments 2 through 5; mentum variable, but occasionally with reduced or vestigial ventromental plates; length of mature larvae variable, but many species very small, less than 4 mm	*Limnophyes* Eaton
60(58').	Mentum with odd number of teeth, median tooth shorter than 2nd lateral pair of teeth and 1st lateral pair partially fused to 2nd lateral pair (Fig. 26.304)	*Unniella* Saether
60'.	Mentum with even number of teeth and median pair shorter than or subequal in length to 1st laterals (Fig. 26.276)	*Chaetocladius* Kieffer
61(35).	Second antennal segment divided in the basal 1/3 by an area of reduced sclerotization (Fig. 26.234) and antennal blade shorter than or subequal in length to segments 2-4 combined (Fig. 26.234)	*Brillia* Kieffer
61'.	Second antennal segment not divided, blade conspicuously longer than the combined lengths of segments 2-4 (Fig. 26.305)	*Euryhapsis* Oliver
62(14).	Mentum with single median tooth (Fig. 26.306)	*Stilocladius* Saether
62'.	Mentum with paired median teeth (Fig. 26.307)	*Doncricotopus* Saether

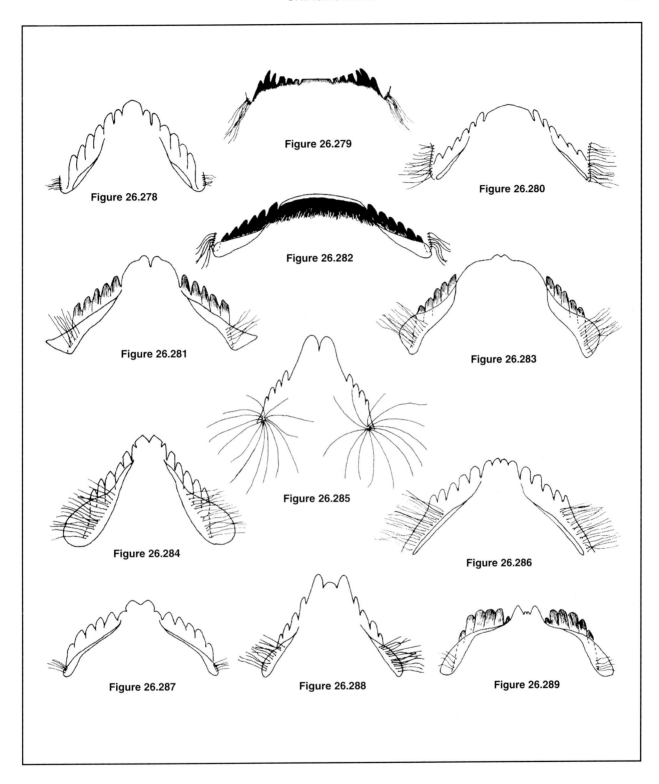

Figure 26.278. Mentum of *Halocladius* (*Halocladius*) sp. (redrawn with modification from Hirvenoja [1973]).

Figure 26.279. Mentum of *Halocladius* (*Psammocladius*) sp. (redrawn with modification from Hirvenoja [1973]).

Figure 26.280. Mentum of *Acricotopus* sp.

Figure 26.281. Mentum of *Psectrocladius* sp.

Figure 26.282. Mentum of *Paracladius* sp.

Figure 26.283. Mentum of *Psectrocladius* sp.

Figure 26.284. Mentum of *Rheocricotopus* sp.

Figure 26.285. Mentum of *Synorthocladius* sp.

Figure 26.286. Mentum of *Diplocladius* sp.

Figure 26.287. Mentum of *Zalutschia* sp.

Figure 26.288. Mentum of *Parorthocladius* sp. (redrawn with modification from Thienemann [1944]).

Figure 26.289. Mentum of *Zalutschia zalutschicola* Lipina (redrawn with modification from Saether [1976]).

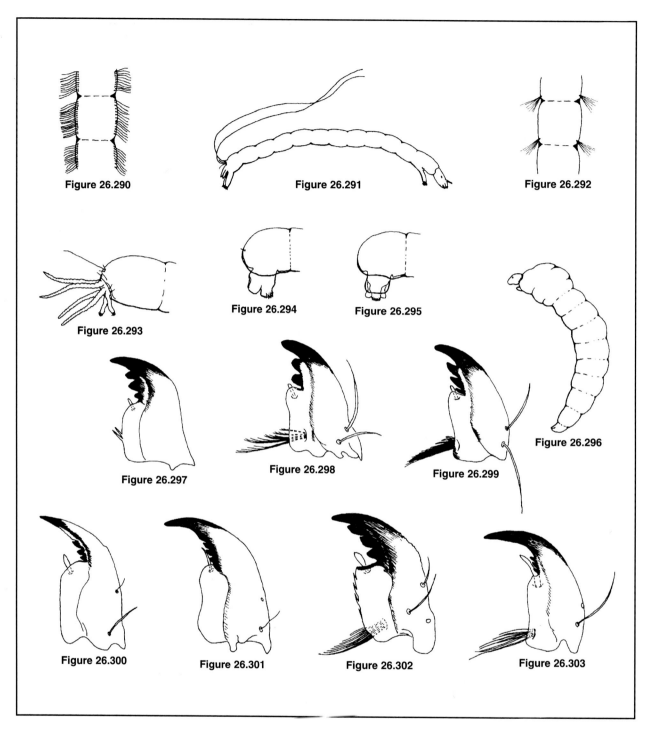

Figure 26.290. Abdominal segments of *Xylotopus* sp. showing lateral hair fringe setae.

Figure 26.291. Larva of *Krenosmittia* sp.

Figure 26.292. Abdominal segments of *Cricotopus* sp. showing setal groups arising from single points.

Figure 26.293. Posterior abdominal segment, anal tubules and posterior prolegs of *Georthocladius* sp.

Figure 26.294. Posterior abdominal segment and posterior prolegs of *Gymnometriocnemus* sp. (redrawn from Thienemann and Kruger [1941]).

Figure 26.295. Posterior abdominal segment and posterior prolegs of *Bryophaenocladius* sp.

Figure 26.296. Larva of *Symbiocladius* sp.

Figure 26.297. Mandible of *Eukiefferiella* sp.

Figure 26.298. Mandible of *Parachaetocladius* sp.

Figure 26.299. Mandible of *Pseudorthocladius* sp.

Figure 26.300. Mandible of *Psectrocladius* sp.

Figure 26.301. Mandible of *Nanocladius* sp.

Figure 26.302. Mandible of *Cricotopus (Cricotopus)* cf. *bicinctus* (Meigen).

Figure 26.303. Mandible of *Krenosmittia* sp.

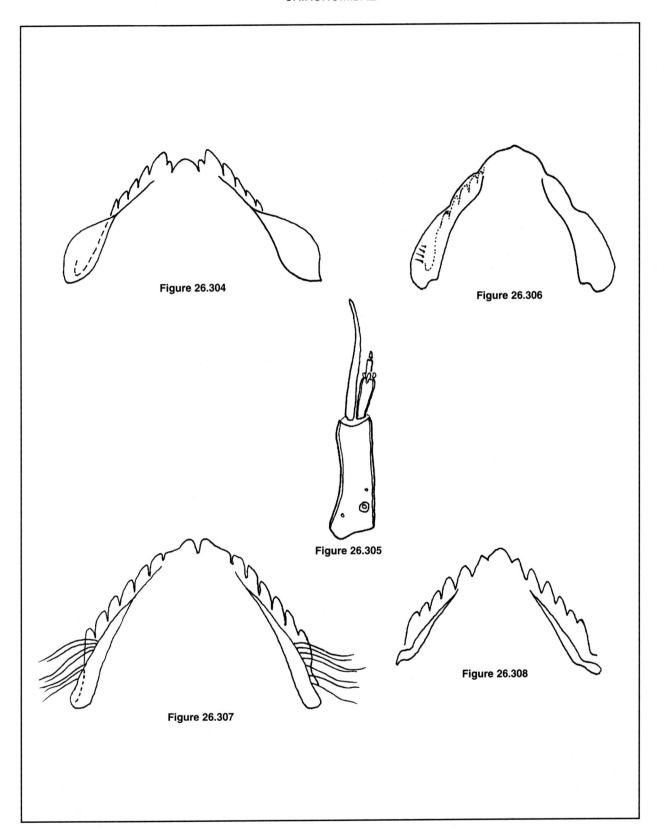

Figure 26.304. Mentum and ventromental plates of *Unniella* sp.

Figure 26.305. Antenna of *Euryhapsis* sp.

Figure 26.306. Mentum and ventromental plates of *Stilocladius* sp.

Figure 26.307. Mentum and ventromental plates of *Doncricotopus* sp.

Figure 26.308. Mentum and ventromental plates of *Stackelbergina* sp.

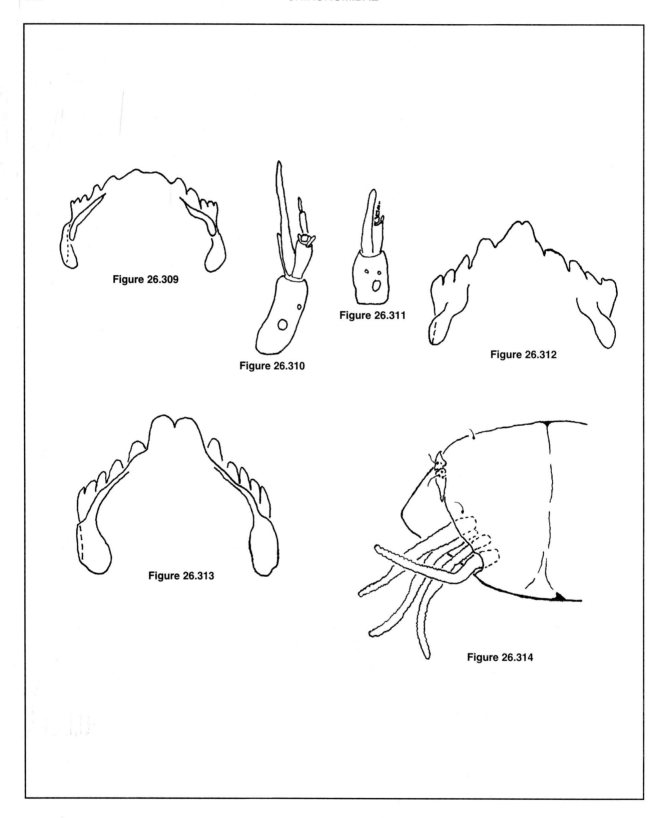

Figure 26.309. Mentum and ventromental plates of *Doithrix* sp.

Figure 26.310. Antenna of *Doithrix* sp.

Figure 26.311. Antenna of *Antillocladius* sp.

Figure 26.312. Mentum and ventromental plates of *Antillocladius* sp.

Figure 26.313. Mentum and ventromental plates of *Georthocladius* (*Atelopodella*) sp.

Figure 26.314. Apex of abdomen in *Georthocladius* (*Atelopodella*) sp.

63(14').	Cardinal beard consisting of bristlelike, often branched, stellate setae; mentum with an even number of teeth (Fig. 26.285); mature larvae very small, less than 4 mm total length ....***Synorthocladius*** Thienemann
63'.	Cardinal beard consisting of numerous simple setae; mentum with an odd number of teeth (Fig. 26.288); mature larvae 3.5-7.0 mm total length...***Parorthocladius*** Thienemann
64(28).	Mentum as in Figure 26.252 ....***Xylotopus*** Oliver
64'.	Mentum as in Figure 26.308 ....***Stackelbergina*** Shilova and Zelentsov
65(26)	Mandible with 3 distinct inner teeth; 5 procercal setae of various lengths present; mentum and antenna as in Figures 26.309 and 26.310 ....***Doithrix*** Saether and Sublette
65'.	Mandible with 2 inner teeth (however, in some species the mola may be darkly pigmented and incorrectly considered to be a third inner tooth); only one well developed procercal seta..***Georthocladius*** (*Georthocladius*)
66(59).	Posterior prolegs present; procerci normal and with 6 setae; mentum as in Figure 26.277 ....***Psilometriocnemus*** Saether
66'.	Posterior prolegs absent; procerci represented as small triangular shaped lobes and with 2-3 setae (Fig. 26.314); mentum as in Fig. 26.313 ....***Georthocladius*** (*Atelopodella*)
67(26').	Antenna and mentum as in Figure 26.311 and 26.312; pecten epipharyngis consisting of 3 scales that are each divided into 2-3 teeth apically ....***Antillocladius*** Saether
67'	Antenna similar to Figures 26.231 or 26.232; mentum not as in Figure 26.312 ....***Smittia*** Holmgren
68(45).	Median tooth of mentum narrow and extending well beyond apices of first lateral teeth; first lateral teeth closely appressed to median tooth, almost giving the impression of a trifid median tooth (Fig. 26.332); lauterborn organs well developed..***Orthocladius annectens***
68'.	Median tooth of mentum variable in width and length, rarely extending beyond apices of first lateral teeth; first lateral teeth usually not closely appressed to median tooth; lauterborn organs variable, often small.......................................***Cricotopus*** (*Cricotopus*) Hirvenoja (in part)

Diamesinae

1.	Head capsule with conspicuous dorsal tubercles (Fig. 26.320); abdominal segments with numerous small, closely set, stellate setae ....Boreoheptagyini—***Boreoheptagyia*** Brundin
1'.	Head capsule lacking dorsal tubercles; abdominal segments either lacking setae or if present then not stellate 2
2(1').	Labrum with large scales either forming a contiguous row across the anterior portion (Fig. 26.322), or when not contiguous then not separated by more than the width of the largest scale; mentum as in Figure 26.321..Protanypini—***Protanypus*** Kieffer
2'.	Labrum lacking large scales, or if scales are present then not forming a contiguous row; mentum not as in Figure 26.321.. 3
3(2').	Three large, sharp pointed teeth clearly visible in central area of mentum (Fig. 26.326); lateral teeth covered by very darkly pigmented ventromental plates (Fig. 26.325) ....***Pseudodiamesa*** Goetghebuer
3'.	Central area of mentum either truncate and lacking teeth (Fig. 26.328), with one broad tooth, or with numerous small teeth; when numerous teeth are present, ventromental plates reduced or vestigial, rarely obscuring lateral teeth 4
4(3').	Mentum lacking teeth entirely; mandible with hook-shaped lateral tooth ....***Potthastia*** Kieffer (in part)
4'.	Mentum with at least lateral teeth, though often covered by ventromental plates (Fig. 26.327); mandible without hook-shaped lateral tooth 5
5(4').	Ventromental plates large and entirely covering at least some lateral teeth of the mentum...7
5'.	Ventromental plates not conspicuous, or if discernible then not entirely covering any lateral teeth of the mentum 6
6(5').	Abdomen with numerous elongate and erect setae that are subequal to or longer in length than their corresponding abdominal segment ....***Pseudokiefferiella*** Zavrel
6'.	Abdominal segments either lacking elongate and erect setae, or if occasional setae are present, then distinctly shorter in length than corresponding abdominal segment 10
7(5).	Median portion of mentum truncate; ventromental plates pigmented golden brown to dark brown and entirely covering all lateral teeth (Figs. 26.327-26.328) ....***Pagastia*** Oliver

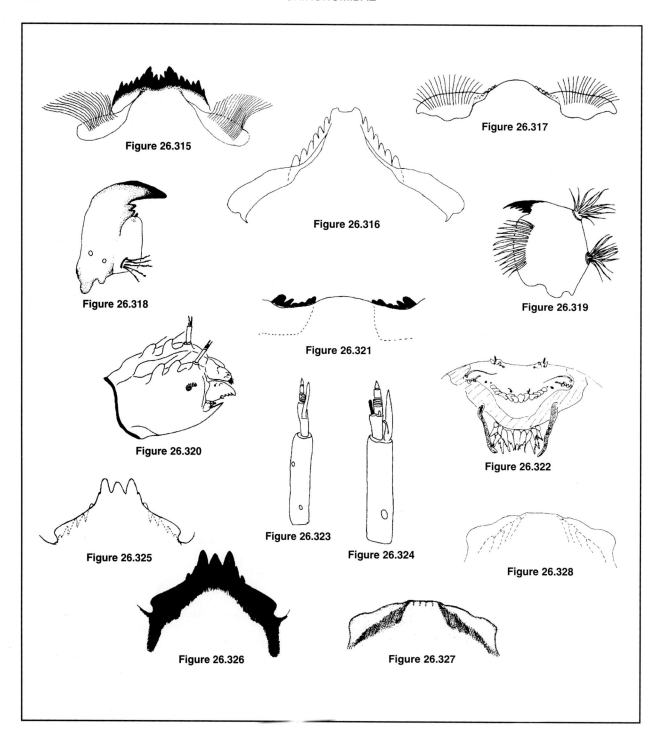

Figure 26.315. Mentum of *Prodiamesa olivacea* (Meigen).
Figure 26.316. Mentum of *Monodiamesa* sp.
Figure 26.317. Mentum of *Odontomesa* sp.
Figure 26.318. Mandible of *Monodiamesa* sp.
Figure 26.319. Mandible of *Odontomesa* sp.
Figure 26.320. Head capsule of *Boreoheptagyia* sp.
Figure 26.321. Mentum of *Protanypus* sp.
Figure 26.322. Anterior view of labral region of *Protanypus ramosus* Saether (redrawn with modification from Saether [1975]).
Figure 26.323. Antenna of *Potthastia gaedii* (Meigen).
Figure 26.324. Antenna of *Diamesa* sp.
Figure 26.325. Schematic drawing of mentum of *Pseudodiamesa* sp. showing lateral teeth covered by ventromental plates.
Figure 26.326. Mentum of *Pseudodiamesa* sp.
Figure 26.327. Mentum of *Pagastia* sp.
Figure 26.328. Schematic drawing of mentum of *Pagastia* sp. showing lateral teeth covered by ventromental plates.

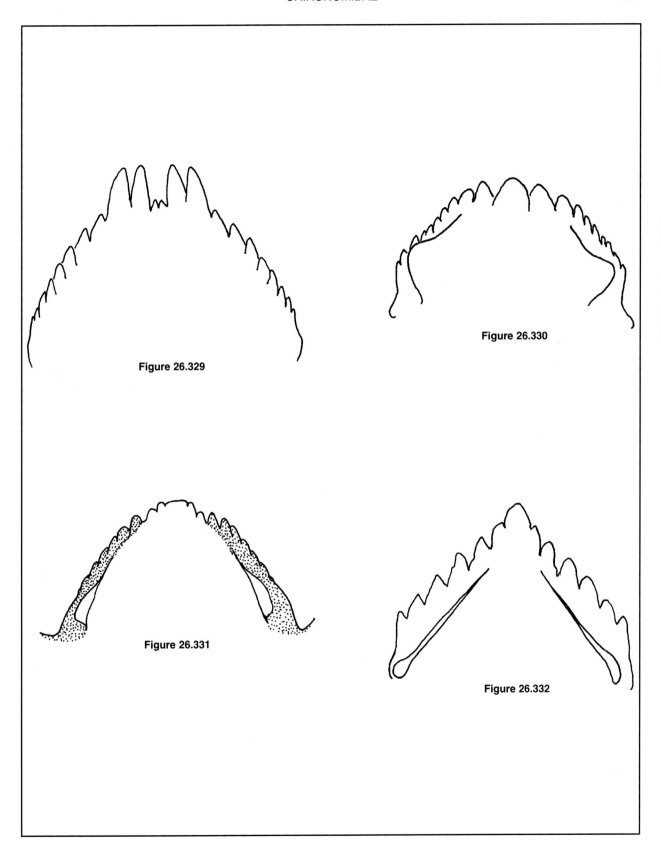

Figure 26.329. Mentum of *Arctodiamesa appendiculata* (Lundstrom).

Figure 26.330. Mentum and ventromental plates of *Syndiamesa* sp.

Figure 26.331. Mentum and ventromental plates of *Lappodiamesa* sp.

Figure 26.332. Mentum and ventromental plates of *Orthocladius annectens*.

7'.	Median portion of mentum with one broad and occasionally somewhat rounded tooth; ventromental plates not darkly pigmented, rarely entirely covering all the lateral teeth of the mentum	8
8(7').	Second antennal segment with blade arising near middle of segment (Fig. 26.323)	***Potthastia*** Kieffer (in part)
8'.	Second antennal segment with blade arising from near base (Fig. 26.324), or blade absent or not conspicuous	9
9(8').	Premandible with 2-4 teeth	***Sympotthastia*** Pagast
9'.	Premandible with more than 4 teeth	***Diamesa*** Meigen (in part)
10(6').	Mentum as in Figure 26.329	***Arctodiamesa appendiculata*** (Lundstrom)
10'.	Mentum not as in Figure 26.329	11
11(10').	Third antennal segment longer than second segment; procercus well developed and with 2 subapical and 5-6 apical setae; mentum as in Figure 26.330	***Syndiamesa*** Kieffer
11'.	Third antennal segment shorter than second segment; procercus vestigial, lacking subapical setae and with 4 apical setae or procercus well developed, with 2 subapical setae and 8-9 apical setae; mentum not as in Figure 26.330	12
12(11').	Procercus well developed, with 2 subapical setae and 8-9 apical setae; mentum as in Figure 26.331	***Lappodiamesa*** Serra-Tosio
12'.	Procercus vestigial, lacking subapical setae and with 4 apical setae; mentum not as in Figure 26.331	***Diamesa*** Meigen (in part)

Podonominae

1.	Median tooth of mentum subdivided into 3-4 dents and deeply recessed from remaining lateral teeth (Fig. 26.338)	Boreochlini (in part)—***Trichotanypus*** Kieffer
1'.	Median tooth of mentum not subdivided, or if appearing somewhat divided then not deeply recessed	2
2(1').	Procercus unicolorous	Podonomini—***Parochlus*** Edwards
2'.	Procercus hyaline anteriorly and darkly pigmented posteriorly (Fig. 26.337)	Boreochlini (in part) 3
3(2').	Eighth abdominal segment with paired spiracular rings; supraanal seta, if present, not longer than posterior prolegs (Fig. 26.340)	4
3'.	Eighth abdominal segment lacking paired spiracular rings; supraanal seta conspicuous and longer than posterior prolegs (Fig. 26.339)	***Paraboreochlus*** Thienemann
4(3).	Mentum with less than 9 pairs of lateral teeth (usually 7 pairs present); procercus with 5 apical setae; larvae more commonly encountered in small springs, seeps, and 1st or 2nd order streams	***Boreochlus*** Edwards
4'.	Mentum with 10 or more pairs of lateral teeth (Fig. 26.336); procercus with more than 10 apical setae (usually 11, 13, or 15 setae present); larvae more commonly encountered in acid bogs or littoral regions of lakes	***Lasiodiamesa*** Kieffer

Prodiamesinae

1.	Mentum with an odd number of teeth, median tooth dome-shaped (Fig. 26.317); mandible with 2 setal bunches on outer margin and inner margin greatly expanded (Fig. 26.319)	***Odontomesa*** Pagast
1'.	Mentum with an even number of teeth (Figs. 26.315-26.316); mandible not as in Figure 26.319	2
2(1').	Mentum strongly arched as in Figure 26.316; ventromental plates with either fine setae that do not extend beyond the anterior edge of the plate or lacking easily discernible setae; mandible as in Figure 26.318	***Monodiamesa*** Kieffer
2'.	Mentum with median teeth recessed (Fig. 26.315); ventromental plates with well-developed cardinal beard; mandible not as in Figure 26.318	***Prodiamesa*** Kieffer

KEY TO SUBFAMILIES OF NORTH AMERICAN FRESHWATER CHIRONOMIDAE* PUPAE

1.	Thoracic horn with a conspicuous plastron plate (e.g., Figs. 26.348, 26.350, 26.359, 26.364, 26.366, 26.378, 26.388)	2
1'.	Thoracic horn, when present, without a conspicuous plastron plate (e.g., Figs. 26.396, 26.407, 26.434, 26.636, 26.642, 26.713)	3

*Genera of Diamesinae, Prodiamesinae, and Orthocladiinae are keyed together since these subfamilies are not easily separated without reference to difficult characters (see couplets 6 and 7).

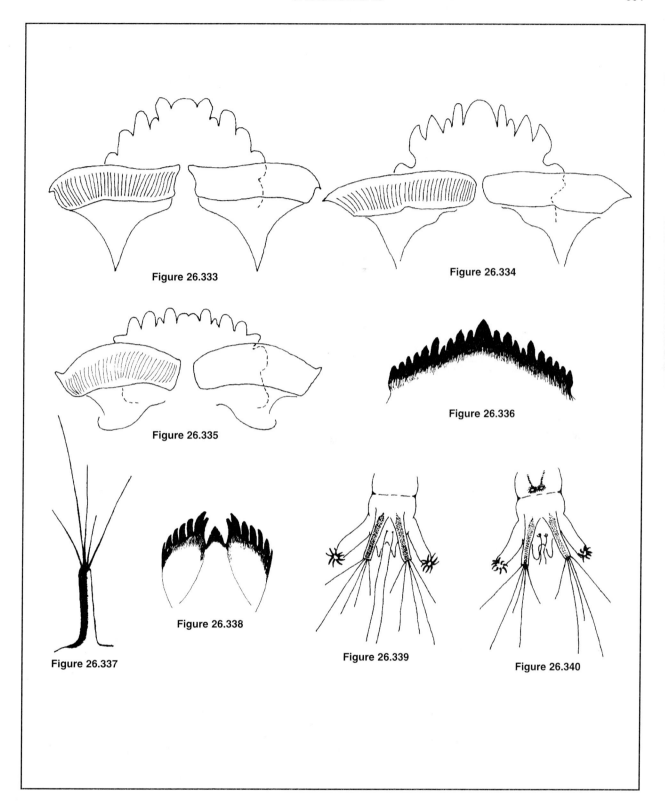

Figure 26.333. Mentum of *Pseudochironomus* cf. *articaudatus* Saether.
Figure 26.334. Mentum of *Pseudochironomus* sp.
Figure 26.335. Mentum of *Pseudochironomus fulviventris* (Johannsen).
Figure 26.336. Mentum of *Lasiodiamesa* sp.
Figure 26.337. Procercus of *Boreochlus* sp.
Figure 26.338. Mentum of *Trichotanypus* sp.
Figure 26.339. Posterior abdominal segments of *Paraboreochlus* sp.
Figure 26.340. Posterior abdominal segments of *Boreochlus* sp.

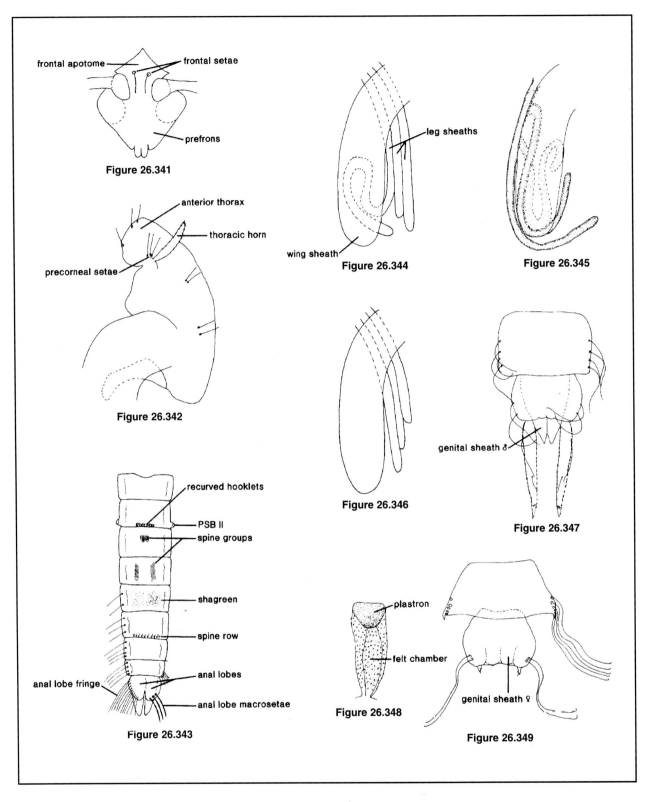

Figure 26.341. Frontal apotome and ocular field of chironomid pupa.

Figure 26.342. Thorax (left side) of chironomid pupa.

Figure 26.343. Generalized abdomen of chironomid pupa (dorsal view).

Figure 26.344. Leg sheath pattern of Diamesinae.

Figure 26.345. Leg sheath pattern of Chironomidae (except Diamesinae and a few Orthocladiinae).

Figure 26.346. Leg sheath pattern of *Lopescladius*.

Figure 26.347. Anal lobes and segment 8 of *Lasiodiamesa* sp.

Figure 26.348. Thoracic horn of *Lasiodiamesa* sp.

Figure 26.349. Anal lobes and segment 8 of *Parochlus*

2(1).	Anal lobes with 2 large macrosetae which insert approximately in middle third of lobe margin (Figs. 26.357, 26.360, 26.369) or anal lobe with a fringe of setae or spines along the margin (Figs. 26.377, 26.381, 26.383, 26.393)	*Tanypodinae* (in part) (p. 695)
2'.	Anal lobe macrosetae often more distal, smaller or spinelike and frequently more than 2, (Figs. 26.347, 26.349, 26.351, 26.353); anal lobe never with a fringe	***Podonominae*** (p. 693)
3(1').	Thoracic horn large with a surface meshwork (Figs. 26.396, 26.401-26.402, 26.405) or more slender without a meshwork (Figs. 26.406-26.410, 26.412); when slender and without meshwork, anal lobes with 2 large lateral macrosetae (Figs. 26.369, 26.400)	***Tanypodinae*** (in part)(p. 695)
3'.	Thoracic horn not large with a meshwork, frequently absent; anal lobe often with less than or more than 2 macrosetae, when 2 are present they are not lateral	4
4(3').	Thoracic horn, when present, unbranched, usually distinct and at least somewhat erect (Figs. 26.515, 26.555, 26.585, 26.606, 26.624, 26.635)	5
4'.	Thoracic horn with 2 or more (often many more) branches (Figs. 26.667, 26.696, 26.699, 26.713), often collapsed in slide mounts	***Chironominae*** (in part)(p. 720)
5(4).	Caudolateral angles of segment 8 almost always with a spine or a group of spines (Figs. 26.623, 26.630, 26.641, 26.646, 26.652, 26.655); wing sheaths almost always with a subterminal "Nase" (Fig. 26.656); anal lobe fringe usually present (Fig. 26.655) but may be reduced (Figs. 26.630, 26.643) or absent, never with terminal macrosetae	***Chironominae*** (in part)(p. 720)
5'.	Caudolateral angles of segment 8 rarely with a spine; when present, anal lobes with macrosetae (Fig. 26.571); wing sheaths without "Nase"	6
		(***Orthocladiinae, Prodiamesinae, Diamesinae***) (p. 700)
6(5').	Posterior pair of leg sheaths curved under the wing sheaths, the other 2 pair distally straight (Fig. 26.344)	***Diamesinae*** (p. 700)
6'.	All leg sheaths curved under wing sheaths (Fig. 26.345) or, rarely, all leg sheaths distally straight (Fig. 26.346)	7
7(6').	Large species (often 8 mm or more in length) with large thoracic horns (Figs. 26.464-26.465); anal lobes with more than 3 terminal (subterminal) macrosetae (Figs. 26.467-26.468) or with a few weak setae on lobe surface similar to Figure 26.462; when more than 3 macrosetae, terga are without median spine groups as in Figures 26.441-26.442; when a few weak setae on the lobe surface, posterior margins of terga without a row(s) of strong, blunt spines as in Figure 26.462	***Prodiamesinae*** (p. 700)
7'.	Often smaller species (frequently less than 5 mm in length); thoracic horns usually smaller (e.g., Figs. 26.492, 26.500, 26.502, 26.508, 26.529, 26.562) often absent; anal lobes usually with 3 or fewer marginal macrosetae (e.g., Figs. 26.425, 26.460, 26.488, 26.501, 26.548, 26.569); when 4 or more macrosetae, then terga with median spine groups (Figs. 26.441-26.442); when a few weak setae on anal lobe surface, then terga with strong posterior spines (Fig. 26.462)	***Orthocladiinae*** (p. 700)

KEYS TO THE GENERA OF NORTH AMERICAN FRESHWATER CHIRONOMIDAE PUPAE

Podonominae

1.	Segment 8 with 4-5 long, filamentous or lamellar lateral setae (Figs. 26.347, 26.349, 26.351)	2
1'.	Segment 8 with 2-3 short, spinelike lateral setae (Figs. 26.353, 26.356)	4
2(1).	Anal lobes with very long slender posterior extensions (Fig. 26.347); thoracic horn as in Figure 26.348	***Lasiodiamesa***
2'.	Anal lobes with, at most, short posterior extensions (Figs. 26.349, 26.351); thoracic horn as in Figure 26.350 or 26.352	3
3(2').	Anal lobes with a short posterior extension (Fig. 26.349); segment 8 with small caudolateral spines (Fig. 26.349); thoracic horn as in Figure 26.350	***Parochlus***
3'.	Anal lobes without posterior extension (Fig. 26.351); segment 8 with large lobelike caudolateral processes (Fig. 26.351); thoracic horn as in Figure 26.352	***Trichotanypus***
4(1').	Lateral margins of segment 8 with two spinelike setae (Fig. 26.353); thoracic horn as in Figure 26.354 or 26.355	***Boreochlus***
4'.	Lateral margins of segment 8 with three spinelike setae (Fig. 26.356); thoracic horn as in Figure 26.354	***Paraboreochlus***

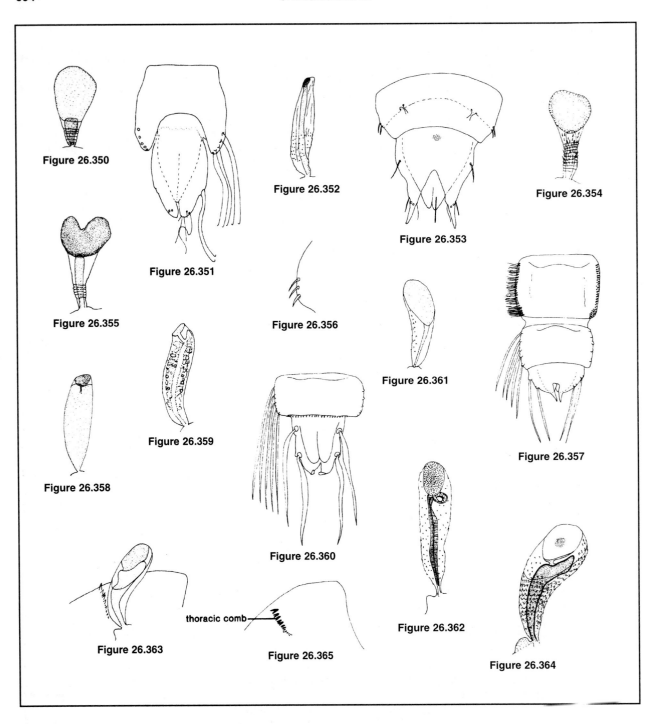

Figure 26.350. Thoracic horn of *Parochlus* sp.
Figure 26.351. Anal lobes and segment 8 of *Trichotanypus* sp.
Figure 26.352. Thoracic horn of *Trichotanypus* sp.
Figure 26.353. Anal lobes and segment 8 of *Boreochlus* sp.
Figure 26.354. Thoracic horn of *Boreochlus* sp. 1.
Figure 26.355. Thoracic horn of *Boreochlus* sp. 2.
Figure 26.356. Lateral spines on segment 8 of *Paraboreochlus* sp.
Figure 26.357. Anal lobes and segments 7 and 8 of *Djalmabatista* sp.
Figure 26.358. Thoracic horn of *Djalmabatista* sp.
Figure 26.359. Thoracic horn of *Larsia* sp.
Figure 26.360. Anal lobes and segment 8 (ventral view) of *Nilotanypus* sp.
Figure 26.361. Thoracic horn of *Nilotanypus* sp.
Figure 26.362. Thoracic horn of *Xenopelopia* sp. (after Fittkau 1962).
Figure 26.363. Thoracic horn of *Pentaneura* sp.
Figure 26.364. Thoracic horn of *Reomyia* sp.
Figure 26.365. Thoracic comb on anterior thorax of *Pentaneura* sp.

Tanypodinae

1.	Thoracic horn with a distinct plastron plate (e.g., Figs. 26.359, 26.361, 26.364, 26.388 26.392), but it may be small (e.g., Fig. 26. 366)	2
1'.	Thoracic horn without a distinct plastron plate, thoracic horn either large with a reticulate meshwork (e.g., Figs. 26.396, 26.398-26.399, 26.401-26.402, 26.405) or more slender (e.g., Figs. 26.407-26.410)	34
2(1).	Anal lobes without a fringe of setae or setalike spines (e.g., Figs. 26.357, 26.360, 26.367, 26.374)	3
2'.	Anal lobes with a fringe of setae (e.g., Figs. 26.377, 26.379, 26.393) or setalike spines (Fig. 26.383) in addition to the 2 macrosetae	21
3(2).	Anal lobes with terminal processes (Fig. 26.357); abdominal segments 4-7 with a dense fringe of setae (Fig. 26.357)	*Djalmabatista*
3'.	Anal lobes without terminal processes, although the lobes may be long and pointed (e.g., Figs. 26.360, 26.367, 26.369, 26.374); abdominal segments with, at most, 5-6 lateral setae	4
4(3').	Thoracic horn with numerous convolutions of the felt chamber and less than 10X as long as wide (Fig. 26.359)	*Larsia*
4'.	Thoracic horn with few, if any, convolutions of the felt chamber (e.g., Figs. 26.362, 26.364, 26.371), when some convolutions are present, the thoracic horn is at least 10X as long as wide (Fig. 26.727)	5
5(4').	Posterior margin of sternum 8 with a row of spines (Fig. 26.360); thoracic horn as in Figure 26.361	*Nilotanypus*
5'.	Posterior margin of sternum 8 without a row of spines; thoracic horn not as above	6
6(5').	Thoracic comb present and usually well developed (Figs. 26.363, 26.365)	7
6'.	Thoracic comb absent or weakly developed (e.g., Fig. 26.370)	14
7(6).	Tergites and sternites densely covered with single, bifid or multibranched shagreen spinules (e.g, Fig. 26.725)	*Telopelopia*
7'.	Tergites and sternites with, at most, arching shagreen (e.g, Fig. 26.726); otherwise shagreen never dense on the tergites and sternites	8
8(7').	Thoracic horn with felt chamber narrow with a loop at base of plastron plate (Fig. 26.362)	*Xenopelopia*
8'.	Thoracic horn with felt chamber broader or not looped (e.g, Figs. 26.363, 26.366, 26.368, 26.371, 26.376)	9
9(8').	Plastron plate broadly joined to the felt chamber (Fig. 26.363), tergal shagreen sparse and simple	*Pentaneura*
9'.	Plastron plate usually narrowly joined to the felt chamber by 1 or 2 narrow connections (Figs. 26.364, 26.366, 26.368), when broadly joined, tergites with dense arching shagreen as in Figure 26.726)	10
10(9').	Tergites with strong arching shagreen, especially posteriorly (Fig. 26.726); plastron plate may join broadly with the felt chamber but 2 short narrow connections can sometimes be seen	*Trissopelopia*
10'.	Tergites with small, simple, spinule shagreen, or very weak arching shagreen; plastron plate clearly joined to the felt chamber by a single narrow connection (e.g., Figs. 26.366, 26.368, 26.727)	11
11(10').	Plastron plate large and connected to the felt chamber by a narrow short stalk (Fig. 26.364); exuviae usually pale yellow	*Reomyia*
11'.	Plastron plate usually smaller with connecting stalk often longer and/or thicker (Figs. 26.366, 26.368, 26.727); exuviae often with distinct color patterns	12
12(11').	Thoracic horn at least 10X as long as wide, with a long narrow connection between the plastron plate and felt chamber; felt chamber with distinct convolutions (folds) and occupies nearly the entire horn lumen (Fig. 26.727); exuviae brown	*Denopelopia*
12'.	Thoracic horn much shorter with the connection between the plastron plate and felt chamber much shorter and/or thicker (e.g., Figs. 26.366, 26.368); exuviae often with a distinct color pattern	13
13(12').	Genital sheaths of male usually longer than anal lobes (Fig. 26.367)	*Paramerina*
13'.	Genital sheaths of male usually not longer than anal lobes	*Zavrelimyia*
14(6').	Tergites and sternites densely covered with single, bifid or multibranched shagreen spinules (e.g, Fig. 26.725)	15
14'.	Tergites and sternites with, at most, small points or arching shagreen similar to Figure 26.726	17

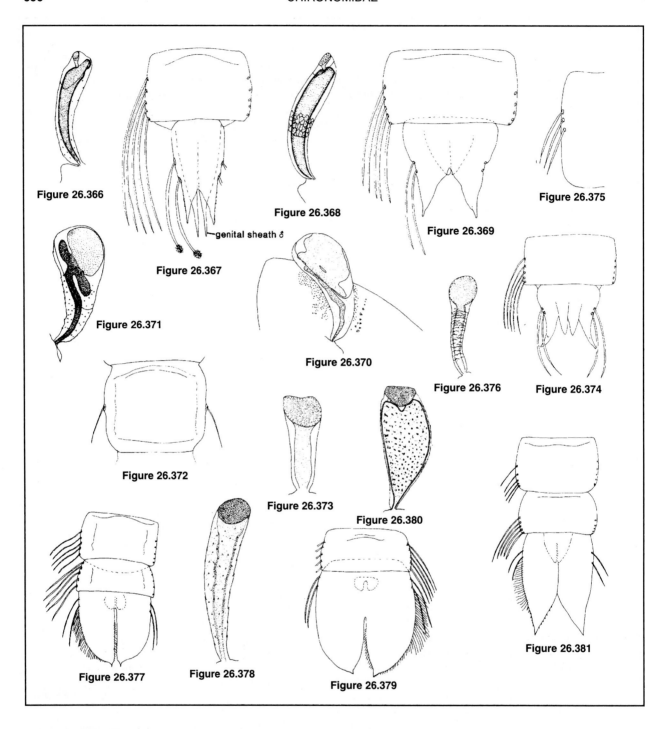

Figure 26.366. Thoracic horn of *Paramerina* sp.
Figure 26.367. Anal lobes and segment 8 of *Paramerina* sp.
Figure 26.368. Thoracic horn of *Zavrelimyia* sp.
Figure 26.369. Anal lobes and segment 8 of *Conchapelopia* sp. 1.
Figure 26.370. Thoracic horn of *Conchapelopia* sp. 1.
Figure 26.371. Thoracic horn of *Conchapelopia* sp. 2.
Figure 26.372. Segment 7 of *Natarsia* sp.
Figure 26.373. Thoracic horn of *Natarsia* sp.
Figure 26.374. Anal lobes and segment 8 of *Krenopelopia* sp.
Figure 26.375. Lateral margin of segment 7 of *Krenopelopia* sp.
Figure 26.376. Thoracic horn of *Krenopelopia* sp.
Figure 26.377. Anal lobes and segments 7 and 8 of *Alotanypus* sp.
Figure 26.378. Thoracic horn of *Alotanypus* sp.
Figure 26.379. Anal lobes and segment 8 of *Apsectrotanypus* sp.
Figure 26.380. Thoracic horn of *Apsectrotanypus* sp.
Figure 26.381. Anal lobes and segments 7 and 8 of *Macropelopia* sp. 1.

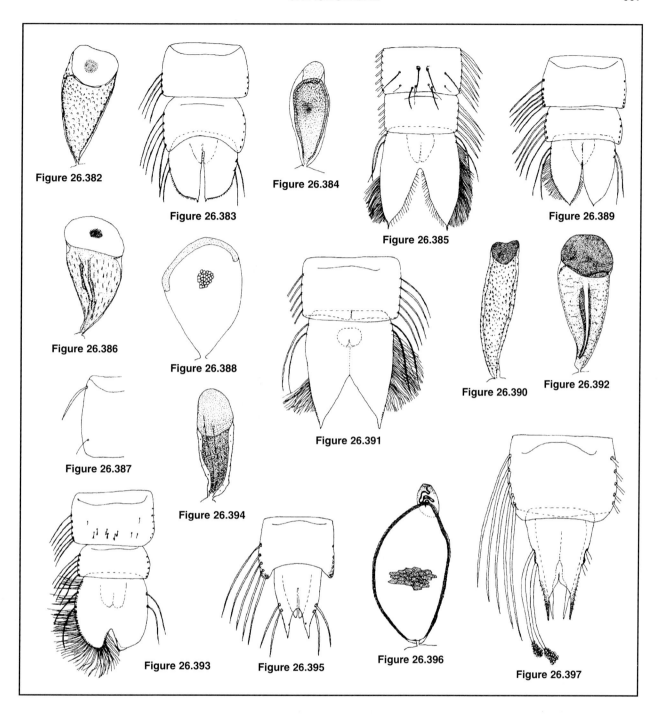

Figure 26.382. Thoracic horn of *Macropelopia* sp. 1.

Figure 26.383. Anal lobes and segments 7 and 8 of *Procladius (Holotanypus)* sp.

Figure 26.384. Thoracic horn of *Procladius (Holotanypus)* sp.

Figure 26.385. Anal lobes and segments 7 and 89 of *Brundiniella* sp. 1.

Figure 26.386. Thoracic horn of *Brundiniella* sp. 1.

Figure 26.387. Lateral margin of segment 6 of *Brundiniella* sp. 1.

Figure 26.388. Thoracic horn of *Radotanypus* sp.

Figure 26.389. Anal lobes and segments 7 and 8 of *Psectrotanypus* sp.

Figure 26.390. Thoracic horn of *Psectrotanypus* sp.

Figure 26.391. Anal lobes and segment 8 of *Macropelopia* sp. 2.

Figure 26.392. Thoracic horn of *Macropelopia* sp. 2.

Figure 26.393. Anal lobes and segment 7 and 8 of *Clinotanypus* sp.

Figure 26.394. Thoracic horn of *Clinotanypus* sp.

Figure 26.395. Anal lobes and segment 8 of *Guttipelopia* sp.

Figure 26.396. Thoracic horn of *Guttipelopia* sp.

Figure 26.397. Anal lobes and segment 8 of *Labrundinia* sp.

15(14).	Thoracic horn with a distinct clear area surrounding the plastron area(e.g, Figs. 26.370-26.371)	16
15'.	Thoracic horn without such a clear area (Fig. 26.728)	*Meropelopia*
16(15).	Thoracic horn long and relatively narrow, plastron plate usually small (Fig. 26.729)	*Helopelopia*
16'.	Thoracic horn generally shorter and/or broader; plastron plate usually larger (e.g., Fig. 26.370, 26.371)	*Conchapelopia*
17(14').	Abdominal segment 7 with one strong lateral seta (Fig. 26.372)	*Natarsia*
17'.	Abdominal segment 7 with 3-4 strong lateral setae (Fig. 26.375)	18
18(17').	Anal lobes directed posteriorly as in Figure 26.369; segment 7 usually with 4 lamellar lateral setae spread along much of the segment margin	19
18'.	Anal lobes directed laterally (e.g., Fig. 26.374); segment 7 with 3-4 lateral setae in a group (Fig. 26.375)	20
19(18).	Many of the dorsal setae of most tergites branched (Fig. 26.730)	*Monopelopia*
19'.	No dorsal setae of the tergites branced	*Telmatopelopia*
20(18').	Anal lobes relatively short, i.e., the lateral macrosetae insert approximately midway along margin (Fig. 26.374)	*Krenopelopia*
20'.	Anal lobes much longer, i.e., the lateral macrosetae insert at a point about one-third along margin from base	*Hudsonimyia*
21(2').	Segment 7 with a lateral fringe of many more than 12 hairlike lateral setae (Fig. 26. 731)	*Fittkauimyia*
21'.	Segment 7 with, at most, 12 rather stiff setae	22
22(21').	Segment 7 with 5-6 very thin, short lateral setae (Fig. 26.732)	*Bethbilbeckia*
22'.	Segment 7 often with more than 6 or fewer than 4 lateral setae, when 5-6 the setae are not very thin and short	23
23(22').	Thoracic horn with a pair of longitudinal bands of stacked "cells" within the felt chamber (Fig. 26.733)	*Derotanypus*
23'.	Thoracic horn not as above	24
24(23').	Segment 7 with 5 lateral setae (Fig. 26.377), or, when with 6 setae, thoracic horn with numerous dark "pegs" in the surface meshwork (Fig. 26.380)	25
24'.	Segment 7 with less than or more than 5 lateral setae (Figs. 26.383, 26.385)	27
25(24).	Felt chamber of thoracic horn with distinct undulations (Fig. 26.378)	*Alotanypus*
25'.	Felt chamber without undulations (Figs. 26.380, 26.382)	26
26(25').	Thoracic horns with numerous dark "pegs" in the surface structure (Fig. 26.380)	*Apsectrotanypus*
26'.	Thoracic horns without dark "pegs" (Fig. 26.301)	*Macropelopia* (in part)
27(24').	Segment 7 with 4 lateral setae (Fig. 26.383); anal lobes with spines along margins (Fig. 26.383)	28
27'.	Segment 7 with more than 5 lateral setae (Fig. 26.389); anal lobes with fringe of setae (Figs. 26.385, 26.389, 26.391, 26.393)	29
28(27).	Anal lobes with small medial projections (Fig. 26.383); exuviae often more than 5 mm in length	*Procladius* (*Holotanypus*) (in part)
28'.	Anal lobes without small medial projections; exuviae often less than 5 mm in length	*Procladius* (*Psilotanypus*)
29(27').	Some tergites with at least 2 enlarged dorsal setae (lamellar or setiform), often on distinct tubercles, never branched (e.g., Fig. 26.385)	30
29'.	Tergites with all dorsal setae small (Fig. 26.393) or, when enlarged, branched	33
30(29).	Segment 6 with a large anterior lateral seta and a smaller posterior seta (Fig. 26.387)	*Brundiniella*
30'.	Segment 6 without a large anterior lateral seta	31
31(30').	Thoracic horn with a narrow marginal plastron plate (Fig. 26.388); segment 7 with about 11 strong lateral setae	*Radotanypus* (in part)
31'.	Thoracic horn with plastron plate variable, but never as above; segment 7 with 8 or fewer strong lateral setae.	32
32(31').	Inner margins of anal lobes with setae or setalike spines (Fig. 26.389)	*Psectrotanypus*
32'.	Inner margins of anal lobes with, at most, short spines	*Macropelopia* (in part)
33(29').	Anal lobes with medial setiferous projections (Fig. 26.393); usually 6-7 lateral setae on segment 7	*Clinotanypus*
33'.	Anal lobes rounded uniformly; usually 8 lateral setae on segment 7	*Coelotanypus*
34(1').	Thoracic horns large, usually dark and with a conspicuous reticulate meshwork on the surface (Figs. 26.396, 26.398-26.399, 26.401-26.402, 26.405)	35
34'.	Thoracic horns smaller (sometimes slender) lighter and without meshwork (Figs. 26.406-26.412)	40
35(34).	Segments 7 and 8 each with no more than 5 lateral setae	36

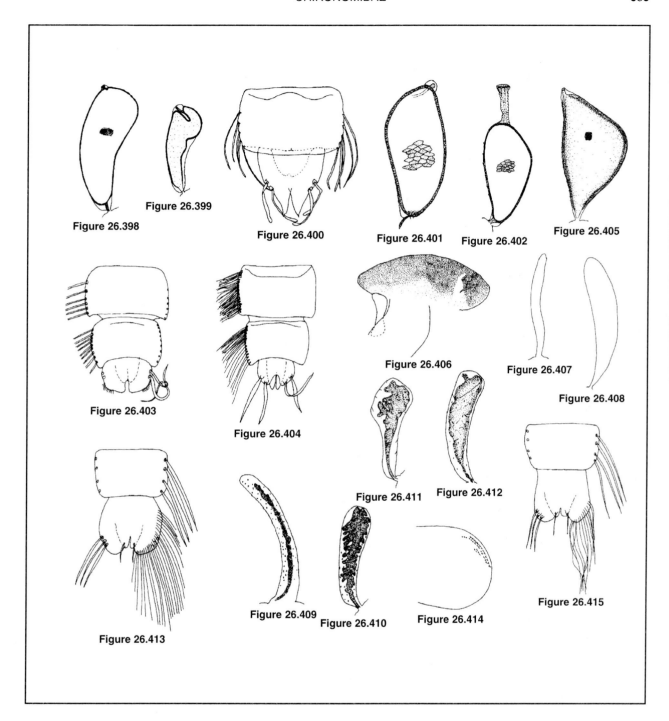

Figure 26.398. Thoracic horn of *Labrundinia* sp. 1.
Figure 26.399. Thoracic horn of *Labrundinia* sp. 2.
Figure 26.400. Anal lobes and segment 8 of *Ablabesmyia* sp. 1.
Figure 26.401. Thoracic horn of *Ablabesmyia* sp. 1.
Figure 26.402. Thoracic horn of *Ablabesymia* sp. 2.
Figure 26.403. Anal lobes and segments 7 and 8 of *Tanypus* sp. 1.
Figure 26.404. Anal lobes and segments 7 and 8 of *Tanypus* sp. 2.
Figure 26.405. Thoracic horn of *Tanypus* sp. 2.
Figure 26.406. Thorax (one-half) of *Rheopelopia* sp. 1.
Figure 26.407. Thoracic horn of *Rheopelopia* sp. 1.
Figure 26.408. Thoracic horn of *Rheopelopia* sp. 2.
Figure 26.409. Thoracic horn of *Arctopelopia* sp.
Figure 26.410. Thoracic horn of *Thienemannimyia* sp. 1.
Figure 26.411. Thoracic horn of *Thienemannimyia* sp. 2.
Figure 26.412. Thoracic horn of *Hayesomyia* sp.
Figure 26.413. Anal lobes and segment 8 of *Cornynoneura* sp. 1.
Figure 26.414. "Pearl rows" on wing sheath of *Corynoneura* sp. 1.
Figure 26.415. Anal lobes and segment 8 of *Corynoneura* sp. 2.

35'.	Segments 7 or 8, or both, with more than 5 lateral setae	39
36(35).	Posterior margins of segment 8 produced laterally as lobelike projections (Fig. 26.395)	*Guttipelopia*
36'.	Posterior margin of segment 8 not so produced (e.g, Figs. 26.397, 26.400)	37
37(36').	Anal lobes truncated or rounded with spines along the margin as in Figure 26.383	*Procladius* (*Holotanypus*) (in part)
37'.	Anal lobes triangular (Figs. 26.397, 26.400)	38
38(37').	Meshwork of thoracic horn usually fine (Figs. 26.398-26.399); anal lobes long and narrow (Fig. 26.397)	*Labrundinia*
38'.	Meshwork of thoracic horn coarse (Fig. 26.401-26.402); anal lobes relatively short and broad (Fig. 26.400)	*Ablabesmyia*
39(35').	Anal lobes short and rounded (Figs. 26.404-26.404); dorsal setae of terga small; thoracic horn (Fig. 26.405)	*Tanypus*
39'.	Anal lobes long and pointed as in Figure 26.385; some dorsal setae large as in Figure 26.385	*Radotanypus* (in part)
40(34').	Segment 7 with, at most, 2 small lateral setae; segment 8 with 3-5 small setae; thoracic horn pale and slender (Figs. 26.406-26.408)	*Rheopelopia*
40'.	Segments 7 and 8 with larger, lamellar setae (Fig. 26.369); thoracic horn not so pale and/or more stout (Figs. 26.409-26.412)	41
41(40').	Thoracic horn curved and of nearly uniform width along its length (Fig. 26.409)	*Arctopelopia*
41'.	Thoracic horn not as above, particularly not of uniform width along its length (Figs. 26.410-26.412)	42
42(41').	Thoracic horn as in Figures 26.410-26.411; segments 2-6 with 3-6 lateral lamellar setae	*Thienemannimyia*
42'.	Thoracic horn as in Figure 26.412; segments 2-6 without lateral lamellar setae	*Hayesomyia*

Orthocladiinae, Diamesinae, and Prodiamesinae

1.	Anal lobes with a fringe that consists of at least a partial row(s) of long (e.g., Figs. 26.413, 26.415-26.416, 26.429, 26.444, 26.446, 26.450, 26.452, 26.455, 26.458, 26.460, 26.462, 26.466) or short setae (Figs. 26.417-26.418, 26.421, 26.424-26.426)	2
1'.	Anal lobe without a fringe of setae (e.g., Figs. 26.471, 26.474, 26.481, 26.485, 26.497, 26.501, 26.504, 26.507, 26.525, 26.572)	51
2(1).	Thoracic horns absent (examine thorax carefully for small, hyaline, or broken horns)	3
2'.	Thoracic horns present but sometimes small and/or hyaline (e.g., Figs. 26.420, 26.427, 26.430, 26.432-26.435, 26.438, 26.448-26.449, 26.464-26.465)	7
3(2).	Anal lobe fringe setae long (Figs. 26.413, 26.415-26.416); lateral abdominal setae not branched (Figs. 26.413, 26.415-26.416); exuviae usually less than 3mm in length	4
3'.	Anal lobe fringe setae short (Fig. 26.418) or very short (Fig. 26.417); some lateral setae of abdomen may be branched; exuviae usually greater than 3mm in length	6
4(3).	All lateral setae of abdomen hairlike; exuviae greater than 3mm in length	*Nanocladius* (*Plecopteracoluthus*) (in part)
4'.	At least 1 lamellar lateral seta on abdominal segments 3-8; exuviae usually less than 3mm in length	5
5(4').	One or more pearl rows at tip of wing sheaths (Fig. 26.414)	*Corynoneura*
5'.	No pearl rows on wing sheaths	*Thienemanniella*
6(3').	Anal lobe fringe setae very short (Fig. 26.417); some lateral setae of abdomen branched (Fig. 26.417)	*Sympotthastia*
6'.	Anal lobe fringe setae short (Fig. 26.418); all lateral setae simple; Note: thoracic horns appear to be absent but short, brown stalks indicate the attachment points of these horns that are usually lost (Figs. 26.419-26.420)	*Orthocladius* (*Orthocladius*) (in part)
7(2').	Anal lobes with 3 macrosetae that are evenly spaced and confined to the distal halves of the lobes (Figs. 26.421, 26.425, 26.429, 26.444, 26.446, 26.450, 26.452, 26.455); Note: In *Brillia* there are usually 0-1 macrosetae but "scars" indicate the lost setae (Figs. 26.446, 26.450)	8
7'.	Anal lobes with less than or more than 3 macrosetae (Figs. 26.446, 26.458, 26.460, 26.466-26.467) or when 3 macrosetae are present they are unevenly or widely spaced (Fig. 26.459); in a few forms the macrosetae are on the lobe surface (Fig. 26.462)	41

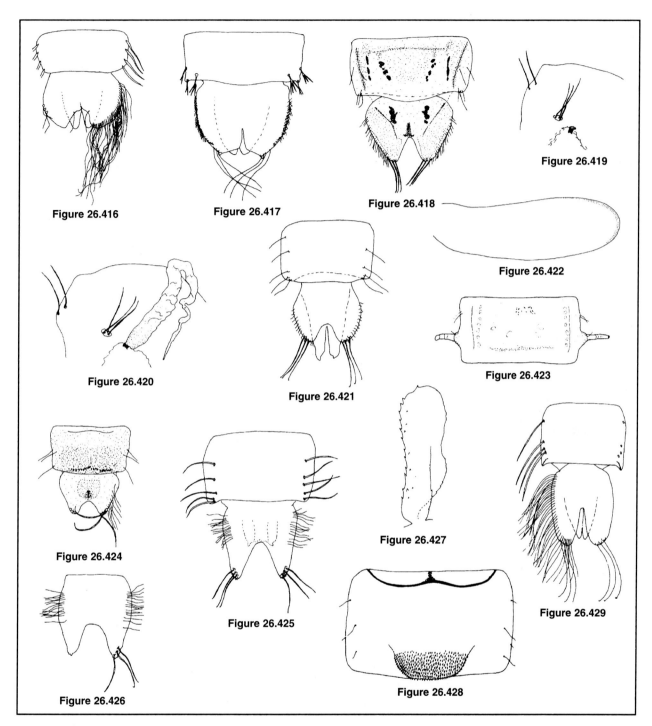

Figure 26.416. Anal lobes and segment 8 of *Thienemaniella* sp.

Figure 26.417. Anal lobes and segment 8 of *Sympotthastia* sp.

Figure 26.418. Anal lobes and segment 8 of *Orthocladius (Orthocladius)* sp. 1

Figure 26.419. Anterior thorax with stalk of broken thoracic horn of *Orthocladius (Orthocladius)* sp. 1

Figure 26.420. Anterior thorax with complete thoracic horn of *Orthocladius (Orthocladius)* sp. 1

Figure 26.421. Anal lobes and segment 8 of *Orthocladius (Orthocladius)* sp. 2.

Figure 26.422. Wing sheath with "pearl rows" of *Parametriocnemus* sp.

Figure 26.423. Segment 2 of *Parametriocnemus* sp.

Figure 26.424. Anal lobes and segment 8 of *Parametrionemus* sp.

Figure 26.425. Anal lobes and segment 8 of *Oliveridia* sp. (after Saether 1976).

Figure 26.426. Anal lobes of *Hydrobaenus* sp.

Figure 26.427. Thoracic horn of *Hydrobaenus* sp.

Figure 26.428. Segment 2 of *Zalutschia* sp.

Figure 26.429. Anal lobes and segment 8 of *Zalutschia* sp.

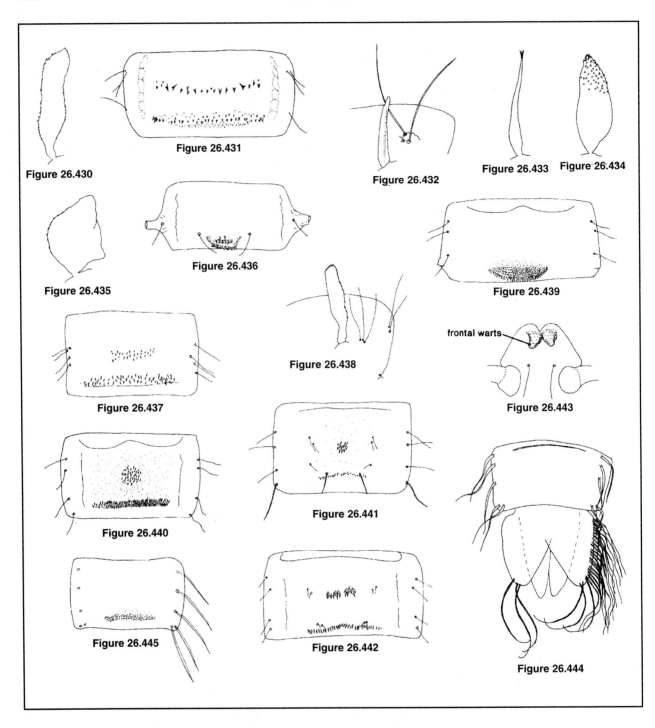

Figure 26.430. Thoracic horn of *Zalutschia* sp.
Figure 26.431. Tergum 5 of *Paracricotopus* sp.
Figure 26.432. Anterior thorax and thoracic horn of *Nanocladius* sp. 1.
Figure 26.433. Thoracic horn of *Nanocladius* sp. 2.
Figure 26.434. Thoracic horn of *Nanocladius* sp. 3.
Figure 26.435. Thoracic horn of *Nanocladius* sp. 4.
Figure 26.436. Segment 2 of *Nanocladius* sp. 1.
Figure 26.437. Segment 5 of *Nanocladius* sp. 1.
Figure 26.438. Anterior thorax and thoracic horn of *Rheocricotopus* sp.
Figure 26.439. Segment 2 of *Rheocricotopus* sp.
Figure 26.440. Segment 5 of *Rheocricotopus* sp.
Figure 26.441. Segment 5 of *Psectrocladius* (*Monopsectrocladius*) sp.
Figure 26.442. Segment 5 of *Psectrocladius* (*Psectrocladius*) sp. 1.
Figure 26.443. Frontal apotome with frontal setae and frontal warts of *Psectrocladius* (*Psectrocladius*) sp. 1.
Figure 26.444. Anal lobes and segment 8 of *Psectrocladius* (*Psectrocladius*) sp. 1.
Figure 26.445. Segment 8 (ventral view) of male *Brillia* sp. 1.

8(7).	Fringe setae of anal lobe short, i.e., no more than one-half the length of the lobe (Figs. 26.421, 26.424-26.426); usually no more than 30 setae; Note: this must be examined carefully since long transparent setae may appear to be short and there is a tendency for the setae to break close to the lobes	9
8'.	Most fringe setae of anal lobe more than one-half the length of the lobe (Figs. 26.429, 26.444, 26.446, 26.450, 26.452, 26.455); often more than 30 fringe setae	17
9(8).	Thoracic horn large, flat and usually with a deep distal notch (Fig. 26.734)	*Euryhapsis*
9'.	Thoracic horn not as above	10
10(9').	Fringe setae of anal lobe very short, thin, weakly sclerotized, and generally concentrated in the distal half (Fig. 26.421)	*Orthocladius (Orthocladius)* (in part)
10'.	Fringe setae of anal lobes longer, i.e., closer to one-half length of lobe (Figs. 26.424-26.426)	11
11(10').	Wing sheaths with pearl rows (Fig. 26.422); PSB II elongated and pointed (Fig. 26.423); anal lobes with a few fringe setae (Fig. 26.424)	*Parametriocnemus* (in part)
11'.	Wing sheaths without pearl rows; PSB if present, not elongated; anal lobes often with numerous setae	12
12(11').	Fringe setae of anal lobes concentrated in basal half; distal ends of lobes with groups of short terminal spines (Fig. 26.425) in addition to the 3 macrosetae	*Oliveridia*
12'.	Fringe setae of anal lobes variable (Figs. 26.426, 26.429); anal lobes without terminal spines, but sometimes with terminal rugulosity	13
13(12').	Fringe setae of anal lobes mostly concentrated in basal half (e.g., Fig. 26.426)	14
13'.	Fringe setae not concentrated in basal half (e.g., Fig. 26.429); thoracic horn with (Figs. 26.427, 26.430) or without (Fig. 26.420) serrations	15
14(13).	Segments 7 and 8 with 2 very broad setae on the caudolateral margins, the other lateral setae of these segments are much thinner (Fig. 26.735)	*Vivacricotopus*
14'.	Segments 7 and 8 with all lateral setae of approximately the same size	*Hydrobaenus* (in part)
15(13').	Thoracic horn weakly sclerotized, elongated, and tapering (Fig. 26.420)	*Orthocladius (Orthocladius)* (in part)
15'.	Thoracic horn much shorter and serrate (Figs. 26.427, 26.430)	16
16(15').	Posterior margins of terga 3-5(6) with raised areas of coarse spines (Fig. 26.428); embedded spines often present at posterior angles of segments 6-7, 6-8, or 7-8 (Fig. 26.429)	*Zalutschia* (in part)
16'.	Posterior margins of terga 3-5(6) without raised areas; embedded spines absent	*Hydrobaenus* (in part)
17(8').	Tergum 5 (and usually some others) with sharply demarcated groups of spines in or near the center (Figs. 26.431, 26.437, 26.440-26.442), rarely are there central spine groups on other terga when not on 5	18
17'.	Tergum 5, and others, without central groups of spines, at most, the terga may have shagreen of slightly larger size in the center (Fig. 26.451)	25
18(17).	Tergum 5, and others, with 1-2 transverse bands of spines that extend almost completely across the segment (Fig. 26.431)	*Paracricotopus* (in part)
18'.	Tergum 5 with 1-2 groups of central spines, but never extending across more than one-half the segment (Figs. 26.437, 26.440, 26.442)	19
19(18').	Two precorneal setae large, 1 small, all insert on a distinct process (Fig. 26.432); PSB II often very large (Fig. 26.436); spines on tergum 5 variable (e.g., Fig. 26.437)	20
19'.	Precorneal setae usually smaller (Fig. 26.438), when 2 are larger than the third they are not on a distinct process and/or are shorter than or only slightly longer than the thoracic horn; PSB II absent or weak (Fig. 26.439)	21
20(19).	Anal lobe macrosetae longer than the lobes and generally thick (Fig. 26.736); tergites 4-6 with nearly circular fields of spines in middle of tergites (Fig. 26.736); thoracic horn elongate and serrate	*Doncricotopus*
20'.	Anal lobe macrosetae usually shorter and/or thinner; armature of tergites 4-6 variable but spine fields rarely nearly circular; thoracic horn shorter or thinner and not serrate (e.g., Figs. 26.432-26.435)	*Nanocladius (Nanocladius)* (in part)
21(19').	Central spine groups on terga 4-6(7) consist of weak to moderately developed spines that are in one group per tergum (Figs. 26.437, 26.440); frontal warts absent	22
21'.	Central spine groups on terga 3(4)-6(7) consist of strong spines in 1-2 groups (Figs. 26.441-26.442); frontal warts often present (Fig. 26.443)	24
22(21).	Two precorneal setae distinctly larger than the third	*Nanocladius (Plecopteracoluthus)* (in part)
22'.	All precorneal setae approximately the same size	23

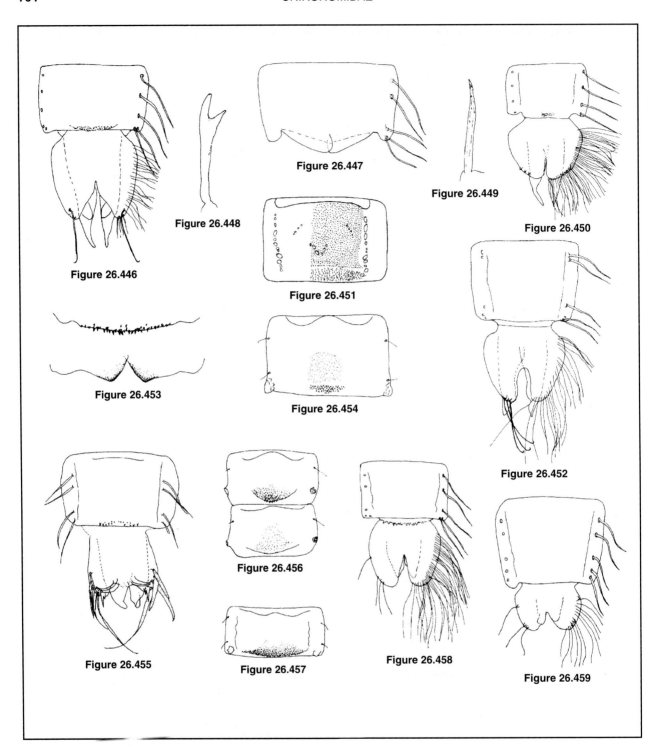

Figure 26.446. Anal lobes and segment 8 of *Brillia* sp. 1.
Figure 26.447. Segment 8 (ventral view) of female *Brillia* sp. 1.
Figure 26.448. Thoracic horn of *Brillia* sp. 1.
Figure 26.449. Thoracic horn of *Brillia* sp. 2
Figure 26.450. Anal lobes and segment 8 of *Brillia* sp. 2.
Figure 26.451. Shagreen (partially indicated) on segment 6 of *Brillia* sp. 2.
Figure 26.452. Anal lobes and segment 8 of *Heterotrissocladius* sp.
Figure 26.453. Posterior margin of sternum 8 of male (upper) and female (lower) of *Heterotrissocladius* sp.
Figure 26.454. Segment 2 of *Heterotrissocladius* sp.
Figure 26.455. Anal lobes and segment 8 of Genus 2.
Figure 26.456. Segments 2 and 3 of *Mesocricotopus* sp.
Figure 26.457. Segment 2 of Genus 4.
Figure 26.458. Anal lobes and segment 8 of Genus 4.
Figure 26.459. Anal lobes and segment 8 of *Mesocricotopus* sp.

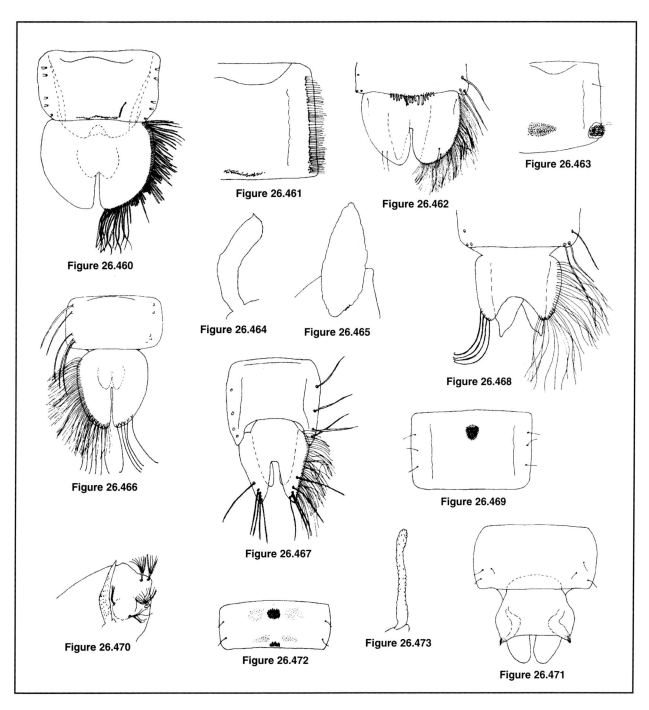

Figure 26.460. Anal lobes and segment 8 of *Xylotopus*.
Figure 26.461. Lateral margin of segment 5 of *Xylotopus*.
Figure 26.462. Anal lobes and posterior margin of segment 8 of *Psectrocladius* (*Allopsectrocladius*) sp.
Figure 26.463. Right half of segment 2 of *Monodiamesa* sp. 1.
Figure 26.464. Thoracic horn of *Monodiamesa* sp. 1.
Figure 26.465. Thoracic horn of *Monodiamesa* sp. 2.
Figure 26.466. Anal lobes and segment 8 of *Psectrocladius* (Psectrocladius) sp. 2.
Figure 26.467. Anal lobes and segment 8 of *Odontomesa* sp.
Figure 26.468. Anal lobes and segment 8 of *Prodiamesa* sp.
Figure 26.469. Segment 6 of *Abiskomyia* sp.
Figure 26.470. Anterior thorax and thoracic horn of *Abiskomyia* sp.
Figure 26.471. Anal lobes and segment 8 of *Abiskomyia* sp.
Figure 26.472. Segment 5 of *Orthocladius* (*Euorthocladius*) sp. 1.
Figure 26.473. Thoracic horn of *Orthocladius* (*Euorthocladius*) sp. 1.

23(22').	Posterior margin of tergum 2 with a dense group of small hooklets (Fig. 26.439); central spine groups of terga nearly circular (Fig. 26.440); moderately heavy shagreen often present on terga (Fig. 26.440); thoracic horn as in Figure 26.438 ....**Rheocricotopus** *(Rheocricotopus)* (in part)
23'.	Posterior margin of tergum 2 with a less dense group of larger hooklets (Fig. 26.436); central spine groups on terga often in short transverse rows (Fig. 26.437); shagreen on terga very weak (Fig. 26.437); thoracic horn variable in shape (Figs. 26.432-26.433, 26.435) ....**Nanocladius** *(Nanocladius)* (in part)
24(21').	Posterior dorsal setae of most terga almost spinelike (Fig. 26.441) ....**Psectrocladius** *(Monopsectrocladius)*
24'.	Posterior dorsal setae not spinelike (Fig. 26.442) ....**Psectrocladius** *(Psectrocladius)* (in part)
25(17').	Thoracic horn long and slightly bifurcate at tip (Fig. 26.448) or short, dark, and slender (Fig. 26.449); anal lobe almost always with no more than 1 macroseta, but "scars" indicate the presence of 3 (Figs. 26.446, 26.450); posterior margin of sternum of male often with a strong row of spinules (Fig. 26.446); sternum 8 of female without posterior spines (Fig. 26.447) ....**Brillia** (in part)
25'.	Not with above combination of characters 26
26(25').	Male with row(s) of sharp spinules on posterior margin of sternum 8 (Fig. 26.453); female with 2 triangular "lobes" (Fig. 26.453); PSB II moderately developed (Fig. 26.454); 8-40 setae in anal lobe fringe (e.g., Fig. 26.452); wing sheaths with or without pearl rows; weak hooklets on posterior margin of tergum 2 (Fig. 26.454) ....**Heterotrissocladius**
26'.	Not with above combination of characters 27
27(26').	PSB on segments 2 or 2 and 3 (Fig. 26.456); no lateral setae of abdomen short and spinelike 28
27'.	PSB usually absent, if present, then some lateral setae of the abdomen short and spinelike 36
28(27).	PSB present on segment 2 only 29
28'.	PSB present on segments 2 and 3 (Fig. 26.456) 35
29(28).	Anal lobes with a "fringe" consisting of a few very broad lamellar setae grading to spinelike setae; anal lobe macrosetae subterminal and inconspicuous (Fig. 26.455); wing sheaths with pearl rows ....**Genus 2**
29'.	Not with above combination of characters 30
30(29').	Posterior margin of tergum 2 with a distinct group of hooklets near the midline (Fig. 26.436) 31
30'.	Posterior margin of tergum 2 with, at most, a row(s) of spines across the margin (Fig. 26.423) 34
31(30).	Group of hooklets on tergum 2 weakly developed; exuviae nearly transparent ....**"Paratrissocladius"**
31'.	Group of hooklets on tergum 2 more developed (Figs. 26.436, 26.439); exuviae almost always with pigmentation 32
32(31').	Hooklets on tergum 2 as in Figure 26.439; a trace of a spine group often present in the center of tergum 5 ....**Rheocricotopus** *(Rheocricotopus)* (in part)
32'.	Hooklets on tergum 2 as in Figure 26.436; usually no trace of a central spine group on tergum 5 33
33(32').	Precorneal setae not on a distinct tubercle ....**Unniella**
33'.	At 2 large precorneal setaea on a distinct tuberecle (Fig. 26.432) ....**Nanocladius** *(Nanocladius)* (in part)
34(30').	PSB II well developed, extended to a point (Fig. 26.423); wing sheaths with pearl rows (Fig. 26.341) ....**Parametriocnemus** (in part)
34'.	PSB II developed, but not pointed (Fig. 26.457); wing sheaths without pearl rows ....**Genus 4**
35(28').	Anal lobe macrosetae all terminal or slightly subterminal ....**Rheocricotopus** *(Rheocricotopus)* (in part)
35'.	Two anal lobe macrosetae distal and close together, the 3rd much more basal, all more or less inconspicuous (Fig. 26.459) ....**Mesocricotopus** (in part)
36(27').	Posterior margins of terga 3-6 without distinct rows of spines or swollen areas with spines, but may have shagreen; frontal apotome with warts as in Figure 26.443; some lateral setae of abdomen short and spinelike; caudolateral margins of terga 6-8 without embedded spines ....**Heterotanytarsus**
36'.	Posterior margins of terga 3-6(7) with rows of spines or swollen areas with spines (Fig. 26.428, 26.437, 26.440); frontal apotome with or without warts; embedded spines sometimes present on terga (Fig. 26.429) 37
37(36').	Posterior margins of terga 3-6 with swollen areas that have groups of triangular spines (Fig. 26.428); thoracic horn weakly serrate (Fig. 26.430); terga often with embedded spines (Fig. 26.429); frontal apotome with or without warts ....**Zalutschia** (in part)

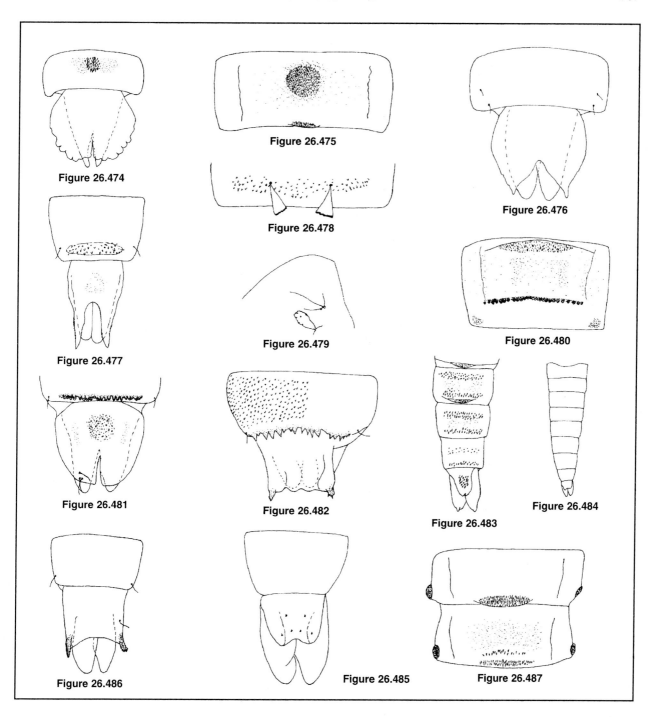

Figure 26.474. Anal lobes and segment 8 of *Orthocladius (Euorthocladius)* sp. 1.

Figure 26.475. Segment 5 of *Orthocladius (Euorthocladius)* sp. 2.

Figure 26.476. Anal lobes and segment 8 of *Orthocladius (Euorthocladius)* sp. 2.

Figure 26.477. Anal lobes and segment 8 of *Rheosmittia* sp. 1.

Figure 26.478. Posterior margin of segment 3 of *Rheosmittia* sp. 1.

Figure 26.479. Anterior thorax and thoracic horn of Genus 5.

Figure 26.480. Segment 4 of *Metriocnemus* sp.

Figure 26.481. Anal lobes and segment 8 of *Metriocnemus* sp.

Figure 26.482. Anal lobes and segment 8 of *Georthocladius* sp.

Figure 26.483. Anal lobes and segments 5-8 of *Pseudosmittia* sp.

Figure 26.484. Abdomen of *Symbiocladius* sp.

Figure 26.485. Anal lobes and segment 8 of *Symbiocladius* sp.

Figure 26.486. Anal lobes and segment 8 of *Cricotopus (Nostococladius)* sp.

Figure 26.487. Segments 2 and 3 of *Orthocladius (Euorthocladius)* sp. 3.

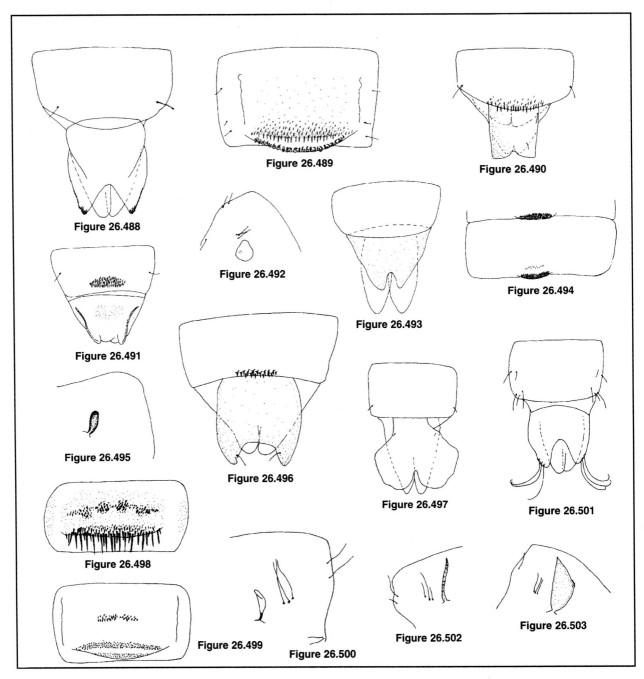

Figure 26.488. Anal lobes and segment 8 of *Orthocladius (Euorthocladius)* sp. 3.

Figure 26.489. Segment 4 of *Tokunagaia* sp. 1.

Figure 26.490. Anal lobes and segment 8 of *Tokunagaia* sp. 1.

Figure 26.491. Anal lobes and segment 8 of *Orthocladius (Euorthocladius)* sp. 4.

Figure 26.492. Anterior thorax and thoracic horn of *Orthocladius (Euorthocladius)* sp. 4.

Figure 26.493. Anal lobes and segment 8 of *Orthocladius (Euorthocladius)* sp. 5.

Figure 26.494. Segments 4 and 5 of *Orthocladius (Euorthocladius)* sp. 5.

Figure 26.495. Anterior thorax and thoracic horn of *Orthocladius (Euorthocladius)* sp. 5.

Figure 26.496. Anal lobes and segment 8 of *Orthocladius (Euorthocladius)* sp. 6

Figure 26.497. Anal lobes and segment 8 of *Bryophaenocladius* sp.

Figure 26.498. Segment of *Cardiocladius* sp.

Figure 26.499. Segment 4 of *Orthocladius (Eudactylocladius)* sp.

Figure 26.500. Anterior thorax and thoracic horn of *Orthocladius (Eudactylocladius)* sp.

Figure 26.501. Anal lobes and segment 8 of *Orthocladius (Eudactylocladius)* sp.

Figure 26.502. Anterior thorax and thoracic horn of *Acricotopus* sp.

Figure 26.503. Anterior thorax and thoracic horn of *Orthocladius (Pogonocladius)* sp.

37'.	Posterior margins of terga 3-6(7) with distinct rows of spines (Fig. 26.437, 26.440); thoracic horn with or without serrations; frontal lobe with or without warts; no embedded spines on caudolateral angles of terga 6-8	38
38(37').	Anal lobe fringe with a few, very broad setae (Fig. 26.455); wing sheaths with pearl rows	**Genus 2**
38'.	Anal lobe fringe setae more numerous or not so broad; wing sheaths without pearl rows	39
39(38').	Frontal setae on prefrons, similar to Figure 26.602)	**Rheocricotopus** (*Psilocricotopus*)
39'.	Frontal setae on apotome, similar to Figure 26.603)	40
40(39').	Posterior margins of tergites 3-8 with transverse rows of spines, these are relatively short on tergites 3-5 but much longer on tergites 6-8 (Fig. 26. 737)	***Psectrocladius*** (*Mesopsectrocladius*)
40'.	Posterior margins of tergites 3-7 with small spines, 8 without	***Nanocladius*** (*Plecopteracoluthus*) (in part)
41(7').	Anal lobes with no true macrosetae (Figs. 26.450, 26.460)	42
41'.	Anal lobes with at least one macroseta (Figs. 26.446, 26.458-26.459, 26.462, 26.467)	44
42(41).	Thoracic horn short and conical (Fig. 26. 738)	***Nanocladius*** (*Plecopteracoluthus*) (in part)
42'.	Thoracic horn long and/or slender (Fig. 26. 449), or broader (Fig. 26.448), often with a distal notch	43
43(42').	Abdominal segments with a lateral fringe of setae (Fig. 26.461)	***Xylotopus***
43'.	Abdominal segments without a fringe	***Brillia*** (in part)
44(41').	Anal lobes with one macroseta and 2 insertion scars (Fig. 26.446)	***Brillia*** (in part)
44'.	Anal lobes usually with 3 or more macrosetae (Figs. 26.458-26.459, 26.466-26.468); when there is only one it is on the surface of the lobe (Fig. 26.462)	45
45(44').	Two marginal anal macrosetae located distally of the 3rd (Fig. 26.459)	***Mesocricotopus*** (in part)
45'.	Anal lobe macrosetae not as above	46
46(45').	Anal lobe with 3 macrosetae located on the surface (Fig. 26.462); long blunt spines on the posterior margins of most terga (Fig. 26.462)	***Psectrocladius*** (*Allopsectrocladius*)
46'.	Not as above	47
47(46').	Anal lobe with 1-3 macrosetae on the surface, when 3, 2 are in the distal half similar to Figure 26.462; PSB 11 well developed and with spinules (Fig. 26.463)	***Monodiamesa***
47'.	Anal lobe with more than 3 macrosetae (Figs. 26.466-26.468); PSB II, when present, without spinules	48
48(47').	Terga 4 and 5, at least, with central patches of spines (Fig. 26.442)	***Psectrocladius*** (*Psectrocladius*) (in part)
48'.	Only shagreen present on terga	49
49(48').	Segment 8 extended as lobes parallel to the anal lobes (Fig. 26.467)	***Odontomesa***
49'.	Segment 8 without caudolateral lobes	50
50(49').	Anal lobes with 4 macrosetae, all of which are terminal (Fig. 26.468)	***Prodiamesa***
50'.	Anal lobes with 3 terminal and 1 more basal macroseta (Fig. 26.458)	**Genus 4**
51(1').	Anal lobes without macrosetae (Figs. 26.471, 26.474, 26.476-26.477, 26.482-26.483), although short spines other than reduced macrosetae may be present (Figs. 26.471, 26.476, 26.482, 26.488); Note: some taxa with very small "macrosetae," which may be overlooked, may be keyed either way from this couplet - in doubtful cases specimens should be run in both directions from this couplet	52
51'.	Anal lobes with macrosetae that may be hairlike (Figs. 26.481, 26.485, 26.496), spinelike (Figs. 26.516, 26.519, 26.524), or long and slender (Figs. 26.501, 26.509, 26.548)	66
52(51).	Distinct, nearly circular groups of spines on the central surface of at least some terga (Figs. 26.472, 26.474-26.475)	53
52'.	Terga without such central groups of spines	54
53(52).	Spine groups on terga 4-6 (Fig. 26.472); thoracic horn weakly serrate (Fig. 26.470); some thoracic setae branched (Fig. 26.470)	***Abiskomyia***
53'.	Spine groups on terga as in Figures 26.474-26.475; thoracic horn long and slender without serrations (Fig. 26.473); all thoracic setae simple	***Orthocladius*** (*Euorthocladius*) (in part)
54(52').	Anal lobes elongate, narrow and tapering to a point (Fig. 26.477)	55
54'.	Anal lobes, if pointed, not shaped as above (Figs. 26.476, 26.482, 26.488)	56
55(54).	Posterior margins of terga 2-4 with a pair of broad scalelike setae (Figs. 26.397, 26.739); thoracic horn absent	***Rheosmittia***

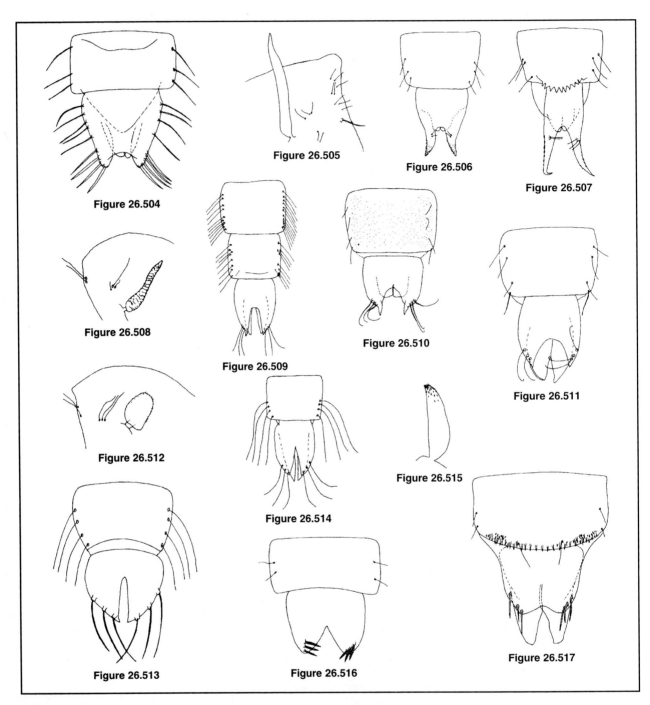

Figure 26.504. Anal lobes and segment 8 of *Protanypus* sp.

Figure 26.505. Anterior thorax and thoracic horn of *Protanypus* sp.

Figure 26.506. Anal lobes and segment 8 of Genus 5 (see Fig. 479).

Figure 26.507. Anal lobes and segment 8 of *Krenosmittia* sp.

Figure 26.508. Anterior thorax and thoracic horn of *Krenosmittia* sp.

Figure 26.509. Anal lobes and segments 7 and 8 of *Epoicocladius* sp. 1

Figure 26.510. Anal lobes and segment 8 of *Parakiefferiella* sp. 1.

Figure 26.511. Anal lobes and segment 8 of *Parakiefferiella* sp. 2.

Figure 26.512. Anterior thorax and thoracic horn of *Parakiefferiella* sp. 1.

Figure 26.513. Anal lobes and segment 8 of *Acamptocladius* sp. (after Saether 1971).

Figure 26.514. Anal lobes and segment 8 of *Lapposmittia* sp. (after Thienemann 1944).

Figure 26.515. Anal lobes and segment 8 of *Lapposmittia* sp. (after Thienemann 1944).

Figure 26.516. Anal lobes and segment 8 of *Baeoctenus* sp. (after Saether 1977).

Figure 26.517. Anal lobes and segment 8 of *Eukiefferiella* sp. 1.

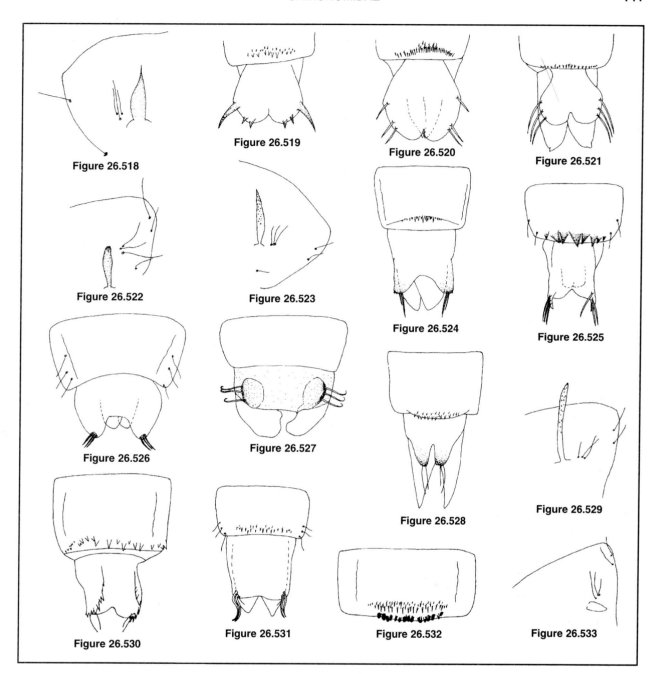

Figure 26.518. Anterior thorax and thoracic horn of *Eukiefferriella* sp. 1.

Figure 26.519. Anal lobes and segment 8 of *Chaetocladius* sp. 1.

Figure 26.520. Anal lobes and segment 8 of *Chaetocladius* sp. 2.

Figure 26.521. Anal lobes and segment 8 of *Chaetocladius* sp. 3.

Figure 26.522. Anterior thorax and thoracic horn of *Chaetocladius* sp. 2.

Figure 26.523. Anterior thorax and thoracic horn of *Chaetocladius* sp. 1.

Figure 26.524. Anal lobes and segment 8 of *Synorthocladius* sp.

Figure 26.525. Anal lobes and segment 8 of *Parachaetocladius* sp.

Figure 26.526. Anal lobes and segment 8 of *Halocladius* sp.

Figure 26.527. Anal lobes and segment 8 of *Boreoheptagyia* sp.

Figure 26.528. Anal lobes and segment 8 of *Paraphaenocladius* sp.

Figure 26.529. Anterior thorax and thoracic horn of *Paraphaenocladius* sp.

Figure 26.530. Anal lobes and segment 8 of *Pseudorthocladius* sp.

Figure 26.531. Anal lobes and segment 8 of *Eukiefferiella* sp. 2.

Figure 26.532. Segment 5 of *Eukiefferiella* sp. 2.

Figure 26.533. Anterior thorax and thoracic horn of *Eukiefferiella* sp. 2.

55'.	Posterior margins of terga with simple setae; thoracic horn a short, stout sac (Fig. 26.479)	**Genus 5**
56(54').	Posterior margins of terga 4-6, at least, with a single row of distinct spines that may be blunt or sharp (Figs. 26.480-26.482)	57
56'.	Posterior margins of terga 4-6 without a posterior row of spines or with more than 1 row (Figs. 26.483, 26.485-26.486, 26.488-26.489, 26.491)	59
57(56).	Spines on posterior margins of terga forming a continuous row (Fig. 26.480); the anterior third of most terga with fine needlelike spinules (Fig. 26.480); anal lobes with or without macrosetae (Fig. 26.481)	***Metriocnemus*** (in part)
57'.	Spines on posterior margins in a less continuous row (Fig. 26.482);	58
58(57').	Anal lobes short and somewhat triangular, often with a few short spines at the apex (Fig. 26.482)	***Georthocladius***
58'.	Anal lobes of male somewhat crescent shaped, lobes with a few terminal spines or more numerous spines which extend anteriorly along the dorsal ridge of each lobe (Figs. 26. 740, 26.741)	***Doithrix***
59(56').	Small (2 mm), nearly transparent exuviae; terga with coarse shagreen (Fig. 26.483); conjunctiva 2/3-5/6, (7/8) with groups of weak spinules in middle third (Fig. 26.483)	***Pseudosmittia***
59'.	Not with above combination of characters	60
60(59').	Anterior abdominal segments wide, diminishing to a narrow 8th (Fig. 26.484); especially in female	61
60'.	Reduction in width of posterior segments much less	62
61(60).	Anal lobes terminating in spined processes (Fig. 26.486); terga covered with fine dense shagreen; a weak anal macroseta sometimes present (Fig. 26.486)	***Cricotopus*** (*Nostococladius*)
61'.	Anal lobes without terminal processes (Fig. 26.485); terga with posterior bands of shagreen; tiny "macrosetae" present (Fig. 26.485)	***Symbiocladius***
62(60').	PSB well developed on segments 2 and 3 (Fig. 26.487); anal lobes with many terminal spines (Fig. 26.488); thoracic horn as in Figure 26.473	***Orthocladius*** (*Euorthocladius*) (in part)
62'.	PSB, at most, developed on segment 2; anal lobes without spines; thoracic horn, when present, not as above	63
63(62').	Posterior margins (or conjunctiva) of terga 3-5 with a row of large recurved hooks (Fig. 26.489)	***Eukiefferiella*** or ***Tokunagaia*** (in part)
63'.	No such large recurved hooks on conjunctiva	64
64(63').	Posterior margins of, at least, some terga with groups of several rows of weak to strong spines (Figs. 26.491, 26.494, 26.496); thoracic horn a small yellow (Fig. 26.492) or brown sac (Fig. 26.495)	***Orthocladius*** (*Euorthocladius*) (in part)
64'.	Not with above combination of characters	65
65(64').	Anal lobes as in Figure 26.497; terga with fine dense shagreen that is slightly larger in posterior bands	***Bryophaenocladius*** (in part)
65'.	Anal lobes not shaped as above; terga with coarse shagreen; (Note: difficult to separate semiaquatic genera)	***Bryophaenocladius*** (in part), ***Gymnometriocnemus*** (*Gymnometriocnemus*), ***Smittia***
66(51').	Terga 4 and 5, at least, with central groups or rows of stronger spines or spinules (coarse shagreen) in addition to smaller shagreen that may be present (Figs. 26.498-26.499); thoracic horn usually present (e.g., Figs. 26.500, 26.502-26.503) but not as in Figure 26.512, when absent, spines on tergum 4 as in Figure 26.498, no hooklets on conjunctiva as in Figures 26.532, 26.561, 26.563)	67
66'.	Terga 4 and 5 usually without central groups or rows of spines or coarse spinules, but anterior and posterior bands or rows of strong spines may be present; if such spines are present, then the thoracic horn is either absent or present; when thoracic horn is present it is not as in Figures 26.500, 26.502-26.503; if thoracic horn is absent, the tergal armature is not as in Figure 26.498; hooklets may be present on some conjunctiva (Figs. 26.532, 26.561, 26.563)	71
67(66).	Tergites 3-6 with a median transverse band of strong spines (e.g., Fig. 26.431)	***Paracricotopus*** (in part)
67'.	Tergites not armed as above.	68
68(67).	Median spine fields on terga 2-7 often in 4 somewhat vaguely to sharply defined groups Fig. 26.498); posterior margins of terga with very long spines (Fig. 26.498); thoracic horn absent or present, when present it is a very small nearly spherical, smooth, colorless sac	***Cardiocladius***

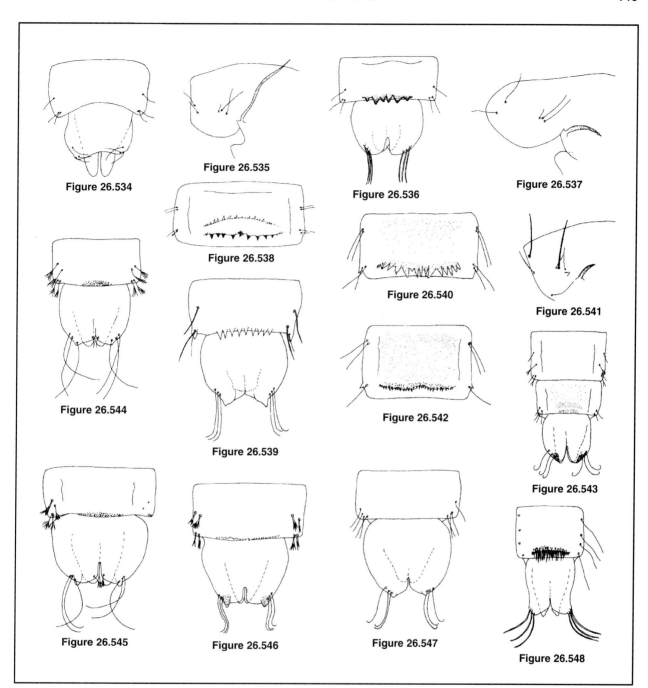

Figure 26.534. Anal lobes and segment 8 of *Cricotopus* sp. 1.

Figure 26.535. Anterior thorax and thoracic horn of *Diamesa* sp. 1.

Figure 26.536. Anal lobes and segment 8 of *Diamesa* sp. 1.

Figure 26.537. Anterior thorax and thoracic horn of *Diamesa* sp. 2.

Figure 26.538. Segment 5 of *Diamesa* sp. 2.

Figure 26.539. Anal lobes and segment 8 of *Pseudokiefferiella* sp. 1.

Figure 26.540. Segment 5 of *Pseudokiefferiella* sp. 1.

Figure 26.541. Anterior thorax and thoracic horn of *Pseudokiefferiella* sp. 1.

Figure 26.542. Segment 5 of *Pseudokiefferiella* sp. 2.

Figure 26.543. Anal lobes and segments 7 and 8 of *Potthastia* sp. 1.

Figure 26.544. Anal lobes and segment 8 of *Pagastia* sp. 1.

Figure 26.545. Anal lobes and segment 8 of *Pagastia* sp. 2.

Figure 26.546. Anal lobes and segment 8 of *Potthastia* sp. 2.

Figure 26.547. Anal lobes and segment 8 of *Pseudodiamesa* sp.

Figure 26.548. Anal lobes and segment 8 of *Limnophyes* sp.

68'.	Median spine fields on terga 2-7 in 2 groups (Fig. 26.499); posterior margins of terga without long spines; thoracic horn not as above (Figs. 26.500, 26.502-26.503) 69
69(68').	Thoracic horn an elongated, stalked sac without spinules (Fig. 26.500) ....***Orthocladius* (*Eudactylocladius*)**
69'.	Thoracic horn not as above (Figs. 26.502-26.503) 70
70(69').	Thoracic horn slender and tapering (Fig. 26.502) ....***Acricotopus***
70'.	Thoracic horn broad (Fig. 26.503) ....***Orthocladius* (*Pogonocladius*)**
71(66').	Anal lobes with 8-12 spinelike macrosetae (often lost) (Fig. 26.504); thoracic horn large (Fig. 26.505) ....***Protanypus***
71'.	Not with above characters 72
72(71').	Anal lobes extremely long, narrow and pointed with 2-3 small macrosetae inserted subterminally, near the midpoint of the lateral margin (Figs. 26.506-26.507) 73
72'.	Anal lobes not as above, when pointed, not as long and tapering (Figs. 26.510, 26.525); anal macrosetae usually larger and inserted closer to the apex (Figs. 26.510, 26.517, 26.528, 26.546, 26.548) 74
73(72).	Thoracic horn a more or less spherical sac similar to Figure 26.512 ....***Genus 5***
73'.	Thoracic horn long and covered with scale-like ridges (Fig. 26.508) ....***Krenosmittia***
74(72').	Anal lobes terminating in sharp points (Figs. 26.509-26.511, 26.513-26.514); thoracic horn usually a more or less spherical sac (Fig. 26.512) but may be more elongate (Fig. 26.515) 75
74'.	Anal lobes not as above; thoracic horn very rarely as above, frequently absent 78
75(74).	Tergites and sternites with anterior transverse dark lines (apophyses) (Fig. 26.742) 76
75'.	Tergites and sternites without such dark lines ....***Parakiefferiella***
76(75).	Tergite 2 with a very small but strongly elevated lobe-like pad of hooks on the posterior margin (Fig. 26.743); anal lobe macrosetae about as long as the anal lobes ....***Acamptocladius***
76'.	Tergite 2 without an elevated pad of hooks although a broad less elevated group of hooks may be present; anal lobe macrosetae as long as above, or shorter 77
77(76').	Abdominal segments sometimes with more than 4 lateral setae (e.g., Fig. 26.509); when 4 or fewer, the macrosetae of the anal lobes are not situated on tubercles and are usually much shorter than the anal lobes ....***Epoicocladius***
77'.	Abdominal segments never with more than 4 lateral setae; anal lobe macrosetae situated on large tubercles (Fig. 26.513) and nearly as long as the anal lobes. Note: not yet known from the Nearctic ....***Lapposmittia***
78(74').	One-3 short, spinelike anal macrosetae, none hairlike (Figs. 26.516-26.517, 26.519-26.521, 26.524-26.527); Note: variation between species of a genus and among specimens of a species, as well as, problems in the interpretation of spinelike and hairlike, make it necessary to key in both directions from this couplet in doubtful cases 79
78'.	One or more long (Figs. 26.536, 26.548, 26.560, 26.568, 26.582) or short (Figs. 26.528, 26.530, 26.534) and hairlike anal macrosetae 86
79(78).	Thoracic horn present 80
79'.	Thoracic horn absent; Note: the thoracic horn-like structure on the thorax of *Boreoheptagyia* is not considered to be a thoracic horn in this key 82
80(79).	Posterior margin of tergum 8 with a row(s) of spines (Figs. 26.517, 26.519-26.521) 81
80'.	Posterior margin of tergum 8 without a row of spines (Fig. 26.516) ....***Baeoctenus***
81(80).	Thoracic horn as in Figure 26.518 ....***Euklefferiella* (in part)**
81'.	Thoracic horn as in Figures 26.522 or 26.523 ....***Chaetocladius* (in part)**
82(79').	Anal lobes with 2-3 approximately equal macrosetae (e.g., Figs. 26.525, 26.527) 83
82'.	Anal lobes with 2-3 unequal macrosetae (Fig. 26.524) ....***Synorthocladius***
83(82).	Anal lobe macrosetae insert on a narrow, pointed, terminal process (Fig. 26.525); posterior margin of tergum 8 with large spines (Fig. 26.525) ....***Parachaetocladius***
83'.	Anal lobe macrosetae not so inserted on a process (Figs. 26.526-26.527) 84
84(83').	Anal lobe macrosetae insert terminally (Fig. 26.526) ....***Halocladius***
84'.	Anal lobe macrosetae insert laterally (Fig. 26.527) 85
85(84').	A conspicuous, dark, triangular thoracic horn-like structure which is covered by small "hairs" is located on the thorax close to the position where the thoracic horn would be found; anal lobe macrosetae more than half the lobe length and somewhat sinuate (Fig. 26.527) ....***Boreoheptagyia***
85'.	No such structure on the thorax; macrosetae of anal lobes very short and thorn-like, much less than half the length of the lobes (Fig. 26.744) ....***Antillocladius***

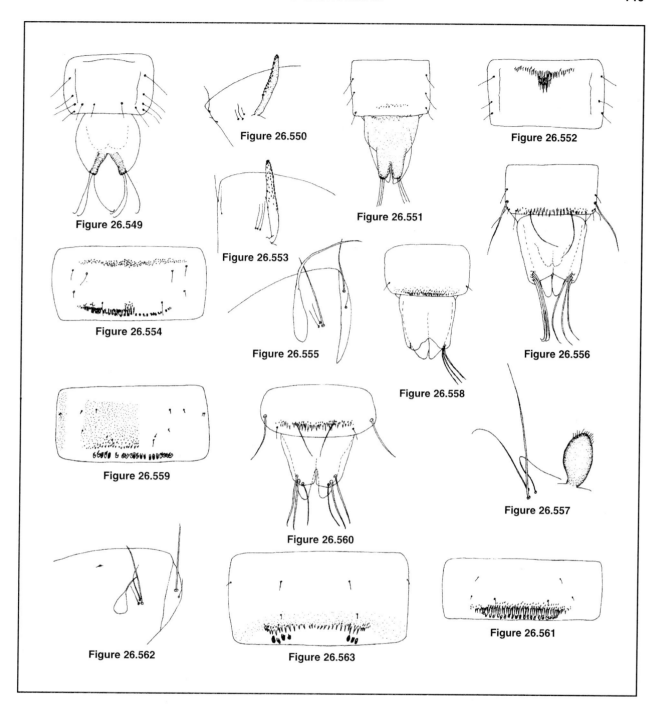

Figure 26.549. Anal lobes and segment 8 of *Diplocladius* sp.

Figure 26.550. Anterior thorax and thoracic horn of *Diplocladius* sp.

Figure 26.551. Anal lobes and segment 8 of *Heleniella* sp.

Figure 26.552. Segment 2 (ventral view) of *Heleniella* sp.

Figure 26.553. Anterior thorax and thoracic horn of *Heleniella* sp.

Figure 26.554. Segment 5 of *Eukiefferiella* sp. 3.

Figure 26.555. Anterior thorax and thoracic horn of *Eukeifferiella* sp. 3.

Figure 26.556. Anal lobes and segment 8 of *Eukiefferiella* sp. 4.

Figure 26.557. Thoracic horn and precorneal setae of *Eukiefferiella* sp. 4.

Figure 26.558. Anal lobes and segment 8 of *Tokunagaia* sp. 2.

Figure 26.559. Segment 5 of *Tokunagaia* sp. 2.

Figure 26.560. Anal lobes and segment 8 of *Eukiefferiella* sp. 5.

Figure 26.561. Segment 5 of *Eukiefferiella* sp. 6.

Figure 26.562. Anterior thorax and thoracic horn of *Eukiefferiella* sp. 6.

Figure 26.563. Segment 5 of *Eukiefferiella* sp. 7.

86(78').	Anal lobe macrosetae all short and hairlike (Figs. 26.528, 26.530, 26.534, but not Fig. 26.531), sometimes inconspicuous, and sometimes with more than or less than 3 (Fig. 26.496)	87
86'.	Anal lobe macrosetae longer and usually strong (Figs. 26.536, 26.548, 26.560, 26.568, 26.582), but not spinelike; almost always with 3 macrosetae	95
87(86).	Anal lobes with 4 weak macrosetae (Fig. 26. 745); tergites 2-9 with transverse anterior and posterior bands of strong spines (Fig. 26. 745)	*Camptocladius*
87'.	Anal lobes with no more than 3 macrosetae	88
88(87').	Anal lobes with 2 widely spaced and hairlike macrosetae (Fig. 26.746); anal lobes distally truncate (Fig. 26. 746)	*Gymnometriocnemus* (*Raphidocladius*)
88'.	Macrosetae and anal lobe shape not as above	89
89(88').	Posterior margins of terga 3(4)-8 with distinct groups of spinules (Figs. 26.494, 26.496); thoracic horn as in Figure 26.495	*Orthocladius* (*Euorthocladius*) (in part)
89'.	Posterior margins of terga 3(4)-8 without such spines; thoracic horn, when present, not as above	90
90(89') .	Thoracic horn present	91
90'.	Thoracic horn absent	92
91(90).	Wing sheaths with pearl rows similar to Figure 26.422; PSB developed, at most, on segment 2; anal lobe with 2 weak macrosetae (Fig. 26.528); thoracic horn with a few spinules (Fig. 26.529)	*Paraphaenocladius*
91'.	Wing sheaths without pearl rows; PSB developed on segments 2 and 3 (Fig. 26.487); anal lobe with 2-3 short macrosetae (Fig. 26.488); thoracic horn without spinules (Fig. 26.473)	*Orthocladius* (*Euorthocladius*) (in part)
92(90').	Anal lobes without spines (Figs. 26.481, 26.490, 26.531)	93
92'.	Anal lobes with numerous spines (Fig. 26.530)	*Pseudorthocladius*
93(92).	Some dorsal abdominal conjunctiva usually with a few large hooklets (Fig. 26.489)	*Eukiefferiella* (in part)
93'.	Dorsal conjunctiva without hooklets	94
94(93').	Terga with posterior rows of spines (Fig. 26.481)	*Metriocnemus* (in part)
94'.	Terga without posterior rows of spines (Fig. 26.534)	*Cricotopus* (*Cricotopus*) (in part)
95(86').	Posterior margins of most terga and usually sterna with heavy spines that are mostly in a single row (Figs. 26.536, 26.538-26.540, 26.542); thoracic horn a thin filament (Fig. 26.535) or a short spur (Fig. 26.537, 26.541) or absent	96
95'.	Not with above combination of characters	98
96(95).	Posterior margins of terga and sterna (3)4-7(8) with heavy spines (Figs. 26.536, 26.538); thoracic horn usually a filament (Fig. 26.535) but may be spurlike (Fig. 26.537)	97
96'.	Posterior margins of only the terga with heavy spines (Figs. 26.539-26.540, 26.542); thoracic horn spurlike (Fig. 26.541) or absent	*Pseudokiefferiella*
97(96).	A strong tubercle on the thorax between the thoracic horn and the anterior central thoracic seta (Fig. 26. 747); many thoracic and abdominal setae long, dark and thick; thoracic horn a long filament.	*Syndiamesa*
97'.	No such tubercle on the thorax (Figs. 26. 535, 26.537); thoracic and abdominal setae usually shorter and paler (Figs. 26.535-26.537); thoracic horn a thin filament (Fig. 26. 535) or spurlike (Fig. 26. 537)	*Diamesa*
98(95').	Anal lobes broad, usually with short, terminal (subterminal) pointed projections (Figs. 26.543-26.547); no thoracic horn; lateral abdominal setae often branched (Figs. 26.543-26.546)	99
98'.	Not with above combination of characters	103
99(98).	Some lateral abdominal setae branched (Figs. 26.543-26.546)	100
99'.	No branched abdominal setae (Fig. 26.547)	101
100(99).	Anal lobe with a 4th seta along inner margin (Figs. 26.544-26.545)	*Pagastia*
100'.	Anal lobes without a 4th seta (Figs. 26.543, 26.546)	*Potthastia* (in part)
101(99').	Anal lobes distally rounded or truncate	*Lappodiamesa*
101'.	Anal lobe with inset, distally pointed processes (Figs. 26. 543, 26.546-26.547)	102
102(101').	Terminal anal lobe projections with scalelike spines (Fig. 26.546)	*Potthastia* (in part)
102'.	Terminal projections smooth, without scalelike spines (Fig. 26.547)	*Pseudodiamesa*
103(98').	Posterior margins of terga 2-8 with rows of very long, needlelike spines (Fig. 26.548); thoracic horn absent	*Limnophyes*
103'.	Posterior margins of terga with or without long spines, when present then thoracic horn is present (Fig. 26.555)	104

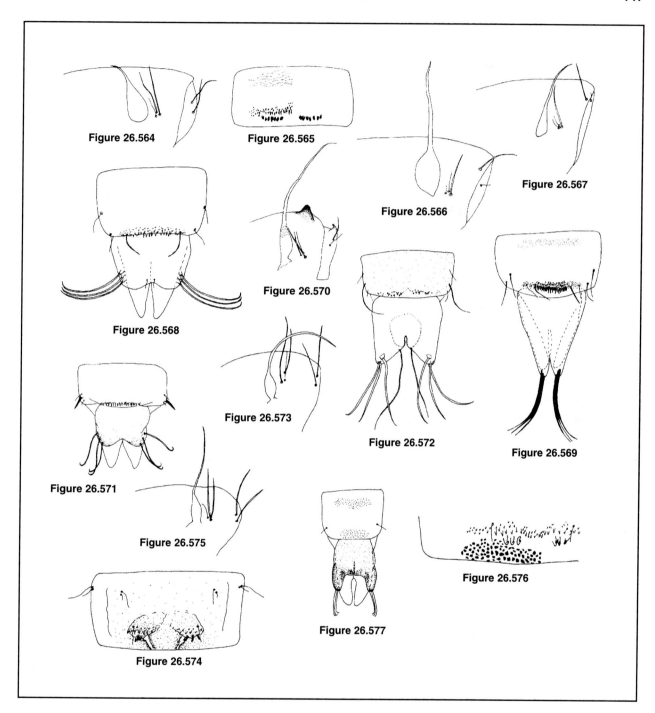

Figure 26.564. Anterior thorax and thoracic horn of *Eukiefferiella* sp. 8.

Figure 26.565. Segment 5 of *Eukiefferiella* sp. 8.

Figure 26.566. Anterior thorax and thoracic horn of *Eukiefferiella* sp. 9.

Figure 26.567. Anterior thorax and thoracic horn of *Eukiefferiella* sp. 10.

Figure 26.568. Anal lobes and segment 8 of *Tokunagaia* sp. 3.

Figure 26.569. Anal lobes and segment 8 of "*Cardiocladius*" sp.

Figure 26.570. Anterior thorax and thoracic horn of "*Cardiocladius*" sp.

Figure 26.571. Anal lobes and segment 8 of *Tokunagaia* sp. 4.

Figure 26.572. Anal lobes and segment 8 of *Tvetenia* sp. 1.

Figure 26.573. Anterior thorax and thoracic horn of *Tvetenia* sp. 1.

Figure 26.574. Segment 5 of *Tvetenia* sp. 1.

Figure 26.575. Anterior thorax and thoracic horn of *Tvetenia* sp. 2.

Figure 26.576. Posterior margin of segment 5 of *Tvetenia* sp. 2.

Figure 26.577. Anal lobes and segment 8 of *Parorthocladius* sp.

104(103').	Anal lobes subcylindrical (e.g., Fig. 26. 549); posterior margins of terga without a row(s) of larger spines	105
104'.	If anal lobes are subcylindrical then posterior margins of most terga with a row(s) of strong spines	107
105(104).	Tergites and sternites demarcated by dark transverse lines (apophyses); anal lobes and thoracic horn as in figsures 26.468, 26.469; frontal setae present	*Diplocladius*
105'.	Tergites and sternites without dark lines; frontal setae present or absent	106
106(105').	Frontal setae present; thoracic horn a weak, unarmed, elongated sac	*Parorthocladius* (in part)
106'.	Frontal setae absent; thoracic horn elongate with small spines	*Plhudsonia*
107(104').	Sterna 2 and 3 with anterior groups of needlelike spines (Fig. 26.552); thoracic horn present (Fig. 26.553)	*Heleniella*
107'.	Sterna 2 and 3 without needlelike spines, or, if present, thoracic horn absent	108
108(107').	Terga 3-4, 3-5, or 4-5 with rows of large recurved hooklets on conjunctiva (Figs. 26.554, 26.559, 26.561, 24.563, 26.565); thoracic horn often "onion" shaped (Figs. 26.555, 26.562, 26.564, 26.566-26.567), sometimes absent; wing sheaths without pearl rows	109
108'.	Terga 3-5 without such hooklets, or, when present thoracic horn as in Figures 26.573 and 26.575 and wing sheaths with pearl rows similar to Figure 26.422	112
109(108).	Terga 3-5 with recurved hooklets on conjunctiva (Figs. 26.554, 26.559, 26.561, 26.563, 26.565); thoracic horn usually present (Figs. 26.555, 26.562, 26.564, 26.566-26.567)	110
109'.	Terga 3-4 or 4-5 with recurved hooklets on conjunctiva; thoracic horn "onion" shaped (Fig. 26.570) or absent	111
110(109).	Thoracic horn absent and tergites and dorsal pleurites nearly covered with fields of shagreen	*Tokunagaia* (in part)
110'.	Thoracic horn usually present (e.g., Figs., 26.557, 26.562, 26.564, 26.566); tergites and dorsal pleurites less extensively and less uniformly covered with shagreen fields	*Eukiefferiella* (in part)
111(109').	Terga 3-4 with recurved hooklets on conjunctiva; thoracic horn present (Fig. 26.570); anterior areas of thorax with dark tubercles (Fig. 26.570)	*"Cardiocladius"*
111'.	Terga 4-5 with recurved hooklets on conjunctiva; thoracic horn absent; anterior areas of thorax without tubercles (also see Fig. 26.571)	*Tokunagaia* (in part)
112(108).	Thoracic horn "onion" shaped (Figs. 26.573, 26.575); a few recurved hooklets usually present on conjunctiva 3/4-5/6 (Figs. 26.574, 26.576); wing sheaths with pearl rows (similar to Fig. 26.422)	*Tvetenia*
112'.	Not with above combination of characters	113
113(112').	Anal lobes cylindrical with 3 strong macrosetae inserted at tips (Figs. 26.577, 26.579, 26.582); thoracic horn present or absent	114
113'.	Anal lobes broader with macrosetae inserted terminally or somewhat subterminally	117
114(113).	Thoracic horn present (Figs. 26.578, 26.580-26.581)	115
114'.	Thoracic horn absent	116
115(114).	Thoracic horn a pale, spineless sac (Fig. 26.578); abdominal segments without long setae (Fig. 26.577)	*Parorthocladius* (in part)
115'.	Thoracic horn yellowish, long and slender with some spines (Figs. 26.580); abdomen with long setae (Fig. 26.579)	Genus 8
116(114').	Posterior margins of terga 2-8 with single rows of spines which are large on 2-5 and small on 6-8 (Fig. 26.583); sterna 5-7 with posterior rows of sharp spines	Genus 9
116'.	Posterior margins of terga 2-8 with rows of spines which are approximately the same size on all terga (Fig. 26.582); sterna 2-8 with posterior spines; all leg sheaths straight (Fig. 26. 346)	*Lopescladius*
117(113').	Thoracic horn brown, contrasting strongly with the thorax (Figs. 26.585-26.586); exuviae small (3 mm or less)	*Stilocladius*
117'.	Thoracic horn present or absent, when present and brown, exuviae larger than 3 mm	118
118(117')	PSB 11 elongate and pointed (Fig. 26.423)	*Parametriocnemus* (in part)
118'.	PSB 11, when present, not as above (Fig. 26.592)	119
119(118').	Most terga with a posterior row of blunt spines similar to Figure 26.480; thoracic horn similar to Figure 26.473; tergites 6-8 with median fields of shagreen	*Thienemannia*
119'.	Terga without posterior rows of blunt spines, but sharp spines may be present; when present, thoracic horn not as above	120

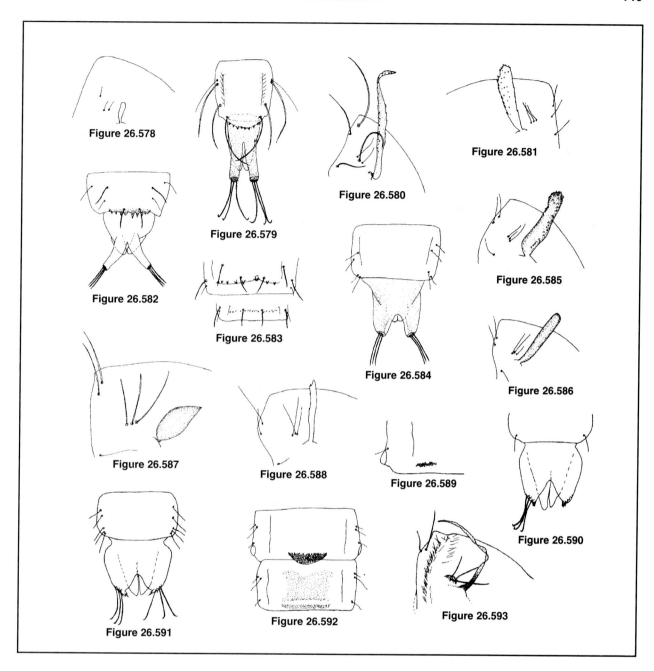

Figure 26.578. Anterior thorax and thoracic horn of *Parorthocladius* sp.

Figure 26.579. Anal lobes and segment 8 of Genus 8.

Figure 26.580. Anterior thorax and thoracic horn of Genus 8.

Figure 26.581. Anterior thorax and thoracic horn of *Psilometriocnemus* sp.

Figure 26.582. Anal lobes and segment 8 of *Lopescladius*.

Figure 26.583. Posterior margins of segments 5 and 7 of Genus 9.

Figure 26.584. Anal lobes and segment 8 of *Stilocladius* sp. 1.

Figure 26.585. Anterior thorax and thoracic horn of *Stilocladius* sp. 1.

Figure 26.586. Anterior thorax and thoracic horn of *Stilocladius* sp. 2.

Figure 26.587. Anterior thorax and thoracic horn of *Paracladius* sp.

Figure 26.588. Anterior thorax and thoracic horn of *Orthocladius (Orthocladius)* sp. 3.

Figure 26.589. Lateral margin of segment 2 of *Orthocladius (Orthocladius)* sp. 3.

Figure 26.590. Anal lobes and posterior margin of segment 8 of *Orthocladius (Orthocladius)* sp. 3.

Figure 26.591. Anal lobes and segment 8 of *Orthocladius (Orthocladius)* sp. 4.

Figure 26.592. Segments 2 and 3 of *Orthocladius (Orthocladius)* sp. 4.

Figure 26.593. Anterior thorax and thoracic horn of *Orthocladius (Orthocladius)* sp. 5.

120(119'). Frontal setae and frontal warts present (Fig. 26.748), and, abdominal tergites
2-8 with transverse posterior rows of sharp teeth ....**Psilometriocnemus**
120'. Frontal setae and frontal warts present or absent, but, when present
tergites 2-8 without posterior rows of sharp teeth 121
121(121'). Posterior margins of tergum 8 with rows of sharp spines (Fig. 26.521); thoracic
horn as in Figures 26.522 and 26.523 ....**Chaetocladius** (in part)
121'. Only shagreen on posterior margins of tergum 8 122
122(121'). Thoracic horn short and broad (Fig. 26.587); frontal setae on large tubercles ....**Paracladius**
122'. Thoracic horn not as above (e.g., Figs. 26.588, 26.593-26.594, 26.598-26.600);
frontal setae, when present, not on tubercles although warts may be
present (Fig. 26.595) 123
123(122'). Abdominal segments 2-6 with 2-3 multibranched lateral setae ....**Stackelbergina**
123'. Lateral setae of abdominal segments not branched 124
124(123') Frontal setae usually large (Fig. 26.595) thoracic horn usually large and well pigmented
(Figs. 26.588, 26.593-26.594), never ovoid as in Figure 26.599; recurved hooklets
on tergum 2 almost always in more than 2 rows (Fig. 26.592); anal lobes often with
terminal spines (Figs. 26.590-26.591); exuviae frequently yellow-golden-brown;
middle abdominal segments sometimes with chitinous rings (Fig. 26.589); conjunctiva
sometimes with a reticulate pigmentation similar to Figure 26.576; frontal setae,
when present, always on frontal apotome (Fig. 26.595); frontal warts sometimes
present (Fig. 26.595) ....**Orthocladius** (*Orthocladius* and *Symposiocladius*) (in part)
124'. Frontal setae usually small (Figs. 26.602-26.603); thoracic horn usually small and
weakly pigmented (Figs. 26.598-26.600), sometimes absent; recurved hooklets on
tergum 2 almost always in 2 rows (Fig. 26.596); anal lobes, at most, with tiny terminal
spines (Fig. 26.601); exuviae often with little pigment, but may be yellow or brown;
terga never with chitinous rings; conjunctiva rarely with reticulate pigmentation; frontal
setae on frontal apotome (Fig. 26.603) or prefrons (Fig. 26.602) or absent; frontal
warts absent; anal lobe macrosetae sometimes unequal
(Fig. 26.597) ....***Cricotopus*** (*Cricotopus*) (in part) and **Paratrichocladius**

Chironominae

1. Thoracic horn always unbranched (e.g., Figs. 26.606, 26.611, 26.618, 26.621,
26.624); wing sheaths almost always with a subterminal tubercle ("Nase") (Fig.
26.656); if subterminal tubercle of wing sheath is absent, then at least some
terga with conspicuous groups of spines (e.g., Figs. 26.604, 26.615, 26.645) 2
1'. Thoracic horn almost always with 2 or more branches (Figs. 26.696, 26.699,
26.713); when unbranched, most abdominal terga have large circular areas of
fine spinules (Fig. 26.658) and anal lobes without fringe (Fig. 26.659); wing
sheaths almost never with a "Nase" 19
2(1). Tergum 4 (and some others) with one or more distinct groups of short and/or long
spines (e.g., Figs. 26.604, 26.607, 26.613, 26.615, 26.622, 26.645) 3
2'. Tergum 4 (and others) with a more or less uniform field of shagreen, although this
may be loosely divided into 2-4 subfields (Figs. 26.651, 26.654) 18
3(2). Tergum 4, and sometimes 3 and 5, with conspicuous groups of long needlelike
spines (e.g., Figs. 26.604-26.605, 26.607, 26.609, 26.612) 4
3'. Terga 4 and 5 without groups of needlelike spines (e.g., Figs. 26.619, 26.622,
26.629, 26.633, 26.645) although some long spines may be present on tergum 3
and tergum 4 and 5 may have groups of short spines 8
4(3). Tergum 4 with 2 longitudinal rows of needlelike spines that are angled medially at
their anterior ends and usually meet (Figs. 26.604-26.605) ....**Micropsectra** (in part)
4'. Tergum 4 not as above, i.e., the 2 spine rows are not joined anteriorly (e.g., Figs.
26.610, 26.612, 26.614), sometimes additional spine groups are present (e.g.,
Figs. 26.607, 26.609) or the rows are transverse (Fig. 26.615) 5
5(4'). Tergum 4 with 2 longitudinal rows of spines (although sometimes very weakly
developed) and 1-2 anterior median groups of spines (Figs. 26.607, 26.609);
wing sheaths with or without a pearl row similar to Figure 26.422 ....**Paratanytarsus** (in part)
5'. Tergum 4 with 2 longitudinal rows of spines but without median groups (Figs.
26.610, 26.612-26.614) or, tergum 4 with spine rows transverse (Fig. 26.615);
wing sheath without pearl rows 6

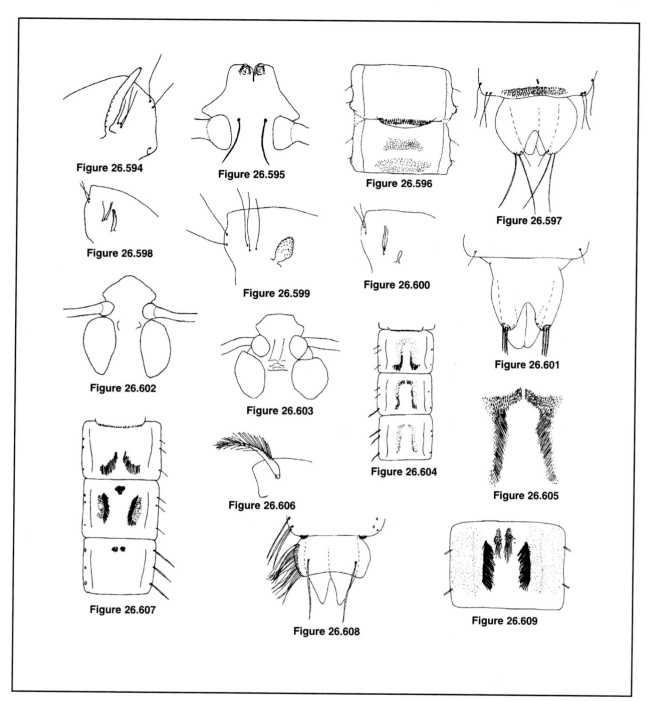

Figure 26.594. Anterior thorax and thoracic horn of *Orthocladius (Orthocladius)* sp. 6.

Figure 26.595. Frontal apotome of *Orthocladius (Orthocladius)* sp. 7.

Figure 26.596. Segments 2 and 3 of *Cricotopus* sp. 2.

Figure 26.597. Anal lobes and posterior margin of segment 8 of *Cricotopus* sp. 3.

Figure 26.598. Anterior thorax and thoracic horn of *Cricotopus* sp. 4.

Figure 26.599. Anterior thorax and thoracic horn of *Cricotopos* sp. 5.

Figure 26.600. Anterior thorax and thoracic horn of *Cricotopus* sp. 6.

Figure 26.601. Anal lobes of *Cricotopus* sp. 7.

Figure 26.602. Frontal apotome of some *Cricotopus* spp.

Figure 26.603. Frontal apotome of some *Cricotopus* spp.

Figure 26.604. Segments 3-5 of *Micropsectra* sp.1.

Figure 26.605. Groups of spines on segment 4 of *Micropsectra* sp. 1.

Figure 26.606. Thoracic horn of *Micropsectra* sp. 1.

Figure 26.607. Segments 3-5 of *Paratanytarsus* sp. 1.

Figure 26.608. Anal lobes and posterior margin of segment 8 of *Paratanytarsus* sp. 1.

Figure 26.609. Segment 4 of *Paratanytarsus* sp. 2.

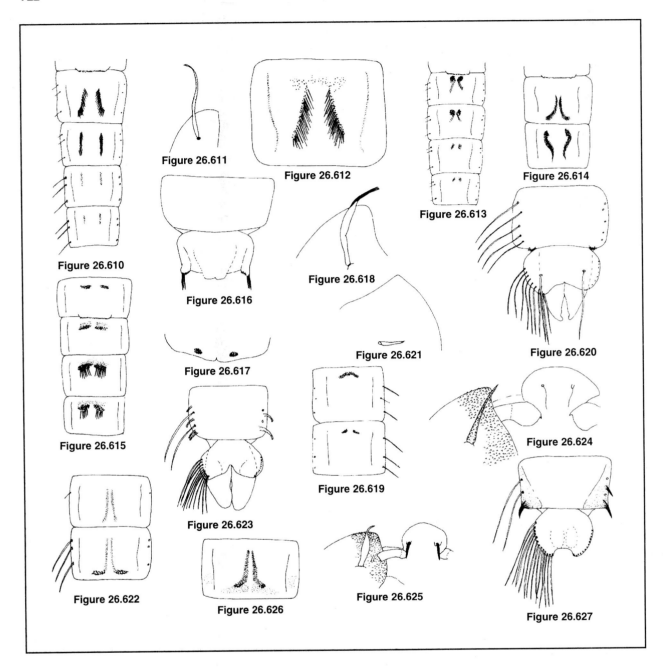

Figure 26.610. Segments 3-6 of *Tanytarsus (Tanytarsus)* sp. 1.

Figure 26.611. Thoracic horn of *Tanytarsus (Tanytarsus)* sp. 1.

Figure 26.612. Segment 3 of *Tanytarsus (Tanytarsus)* sp. 1.

Figure 26.613. Segments 3-6 of *Tanytarsus (Tanytarsus)* sp. 2.

Figure 26.614. Segments 3 and 4 of *Tanytarsus (Tanytarsus)* sp. 3.

Figure 26.615. Segments 2-5 of *Sublettea* sp.

Figure 26.616. Anal lobes and segment 8 of *Sublettea* sp.

Figure 26.617. Posterior margin of segment 8 (ventral view) of *Sublettea* sp.

Figure 26.618. Thoracic horn of *Sublettea* sp.

Figure 26.619. Segments 4 and 5 of *Paratanytarsus* sp. 3.

Figure 26.620. Anal lobes and segment 8 of *Paratanytarsus* sp. 3.

Figure 26.621. Thoracic horn of *Paratanytarsus* sp. 3.

Figure 26.622. Segments 4 and 5 of *Stempellina* sp.

Figure 26.623. Anal lobes and segment 8 of *Stempellina* sp.

Figure 26.624. Frontal apotome and anterior thorax with thoracic horn of *Stempellina* sp.

Figure 26.625. Frontal apotome and anterior thorax with thoracic horn of *Constempellina* sp. 1.

Figure 26.626. Segment 5 of *Constempellina* sp. 1.

Figure 26.627. Anal lobes and segment 8 of *Constempellina* sp. 1.

6(5').	Tergum 4 with transverse rows of spines (Fig. 26.615)	**Sublettea**
6'.	Tergum 4 with 2 longitudinal rows of spines (Figs. 26.610, 26.612-26.614)	7
7(6').	Longitudinal spine rows of tergite 4 sinuate, with anterior spines much shorter than posterior spines and directed postero-medially (Fig. 26.749); anal comb very broad (Fig. 26.750)	**Nimbocera**
7'.	If longitudinal spine rows of tergite 4 are sinuate, the anterior spines are about the same length as the posterior spines (Fig. 26.614); anal comb much less broad (e.g., Fig. 26.647)	**Tanytarsus** (in part)
8(3').	Tergum 4 with a single median group of short spines (e.g., Fig. 26.619)	**Paratanytarsus** (in part)
8'.	Tergum 4 with paired groups of short spines (e.g., Figs. 26.622, 26.631, 26.633, 26.638, 26.648)	9
9(8').	Strong lateral spines on segment 8 (Figs. 26.623 and 26.627); shagreen and spine groups on terga as in Figures 26.622, 26.626, and 26.629	10
9'.	Spines on segment 8 usually in the form of a caudolateral comb (e.g., Figs. 26.630, 26.639, 26.641, 26.643); however, when single (e.g., Fig. 26.634), tergal spine groups never as above	11
10(9).	Frontal setae not spinelike (Fig. 26.624); spine and shagreen groups on terga 4 and 5 as in Figure 26.622	**Stempellina**
10'.	Frontal setae spinelike (Figs. 26.625, 26.628) and the spine and shagreen groups on tergum 5 as in Figure 26.626 or 26.629	**Constempellina**
11(9').	Terga 5 and 6 with large oval groups of small spines (Fig. 26.631); anal lobes with fringe setae limited to distal ends (Fig. 26.630); generally lotic; 2-3 mm in length	**Neozavrelia**
11'.	Terga 5 and 6 without such large areas of spinules but smaller groups may be present (Figs. 26.633, 26.638, 26.645, 26.648); anal lobes usually with more complete fringe (Figs. 26.634, 26.639, 26.641, 26.643, 26.647)	12
12(11').	Caudolateral spine on segment 8 usually a simple spur (Fig. 26.634) less often with a small accessory spine	**Rheotanytarsus**
12'.	Caudolateral spines on segment 8 in the form of a comb (Figs. 26.639, 26.641, 26.643, 26.646-26.647)	13
13(12').	Each anal lobe with 1 or 2 dorsal seta (e.g., Fig. 26.639), if 2, then paired patches of short spines are present on terga 3-6 and most tergites and pleurites are strongly shagreened	14
13'.	Each anal lobe with 2 dorsal setae (Figs. 26.641, 26.643, 26.647); when paired patches of short spines are present on terga 3-6 the tergites and pleurites are not strongly shagreened	15
14(13).	Paired groups of short spines on terga (2)3-6 (Fig. 26.637)	**Parapsectra**
14'.	Paired groups of short spines on terga 4-5 or 4-6 (Fig. 26.638)	**Micropsectra** (in part)
15(13').	Combs on caudolateral corners of segment 8 wide (Fig. 26.641); precorneal setae lamellar and inserting on a mound (Figs. 26.640, 26.642)	**Cladotanytarsus**
15'.	Combs on 8 usually less wide (Figs. 26.643, 26.646); precorneal setae not as above	16
16(15').	Fringe of anal lobe limited to no more than the distal half of the margin (Fig. 26.643)	17
16'.	Fringe of anal lobes usually extending along entire margin (e.g., Fig. 26.647),	**Tanytarsus** (in part)
17(16).	Thoracic horn with long hairs (Fig. 26.751)	**Krenopsectra**
17'.	Thoracic horn without setae (Fig. 26.644)	**Corynocera**
18(2').	Terga 2-6 with shagreen occupying about one-third to one-half of the surface; fields of pleura with, at most, weak shagreen (Fig. 26.651)	**Stempellinella**
18'.	Terga 2-6 with larger fields of shagreen, those on tergite 2 as large as those on tergite 3 (Fig. 26.752); Note: a few unplaced species of Tanytarsisni, possibly belonging to the genus *Neostempellina*, will also key here (Figs. 26.654-26.655)	**Zavrelia**
19(1').	Thoracic horn unbranched (Fig. 26.657); terga with large circular areas of fine spinules (Fig. 26.658); anal lobes without fringe (Fig. 26.659)	**Pseudochironomus** (in part)
19'.	Thoracic horn with 2 or more branches (e.g., Figs. 26.667, 26.696, 26.699, 26.713); anal lobes with, at least, a partial fringe (e.g., Figs. 26.669, 26.686, 26.704, 26.709)	20
20(19').	Row of hooklets on posterior margin of segment 2 distinctly interrupted (Figs. 26.662, 26.664, 26.672) or, rarely, absent	21
20'.	Row of hooklets on segment 2, at most, very narrowly interrupted (Figs. 26.684, 26.702-26.703, 26.710, 26.721), always present	31
21(20).	Caudolateral margins of segment 8 with a spine or group of spines (e.g., Figs. 26.685-26.687, 26.690, 26.697)	22

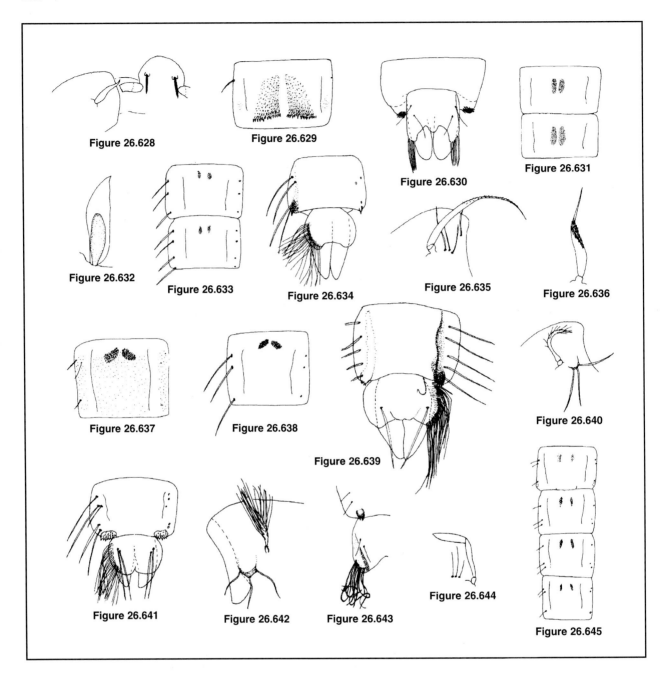

Figure 26.628. Frontal apotome and anterior thorax with thoracic horn of *Constempellina* sp. 2.

Figure 26.629. Segment 5 of *Constempellina* sp. 2.

Figure 26.630. Anal lobes and segment 8 of *Neozavrelia* sp.

Figure 26.631. Segments 5 and 6 of *Neozavrelia* sp.

Figure 26.632. Thoracic horn of *Neozavrelia* sp.

Figure 26.633. Segments 5 and 6 of *Rheotanytarsus* sp. 1.

Figure 26.634. Anal lobes and segment 8 of *Rheotanytarsus* sp. 1.

Figure 26.635. Anterior thorax and thoracic horn of *Rheotanytarsus* sp. 1.

Figure 26.636. Thoracic horn of *Rheotanytarsus* sp. 2.

Figure 26.637. Segment 4 of *Parapsectra* sp.

Figure 26.638. Segment 5 of *Micropsectra* sp. 2.

Figure 26.639. Anal lobes and segment 8 of *Micropsectra* sp. 2.

Figure 26.640. Anterior thorax and thoracic horn of *Cladotanytarsus* sp. 1

Figure 26.641. Anal lobes and segment 8 of *Cladotanytarsus* sp. 1

Figure 26.642. Anterior thorax and thoracic horn of *Cladotanytarsus* sp. 2

Figure 26.643. Anal lobe and segment 8 (left side) of *Corynocera* sp.

Figure 26.644. Thoracic horn and precorneal setae of *Corynocera* sp.

Figure 26.645. Segment 2-5 of *Tanytarsus (Tanytarsus)* sp. 4.

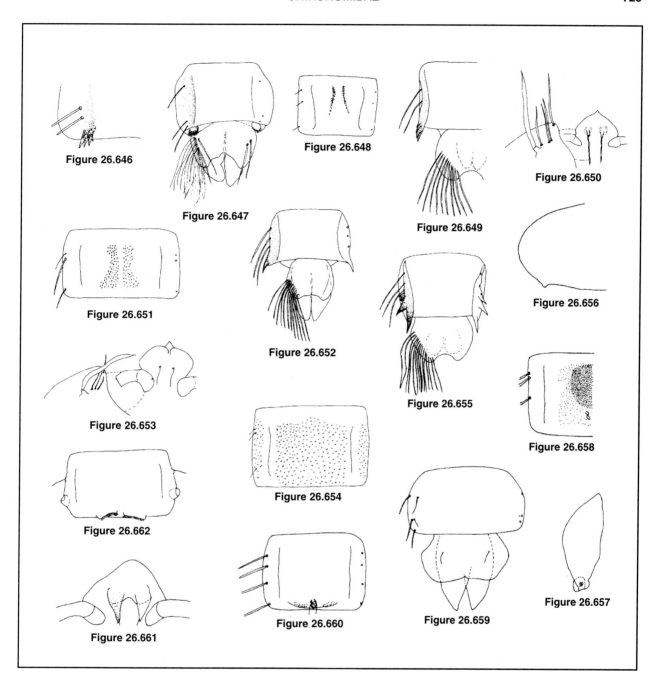

Figure 26.646. Spines on caudolateral margin of segment 8 of *Tanytarsus (Tanytarsus)* sp. 4.

Figure 26.647. Anal lobes and segment 8 of *Tanytarsus (Tanytarsus)* sp. 5.

Figure 26.648. Segment 4 of *Tanytarsus (Tanytarsus)* sp. 5.

Figure 26.649. Anal lobe and segment 8 (left half) of "*Stempellinella* " sp.

Figure 26.650. Frontal apotome and anterior thorax with thoracic horn of "*Stempellinella* " sp. 1.

Figure 26.651. Segment 4 of *Stempellinella* sp. 1.

Figure 26.652. Anal lobes and segment 8 of *Stempellinella* sp. 2.

Figure 26.653. Frontal apotome and anterior thorax with thoracic horn of "*Zavrelia* "sp.

Figure 26.654. Segment 4 of "*Zavrelia* "sp.

Figure 26.655. Anal lobes and segment 8 of "*Zavrelia* "sp.

Figure 26.656. Tip of wing sheath with "Nase" of "*Zavrelia* "sp.

Figure 26.657. Thoracic horn of *Pseudochironomus* sp. 1.

Figure 26.658. Segment 6 (left half) of *Pseudochironomus* sp. 1.

Figure 26.659. Anal lobes and segment 8 of *Pseudochironomus* sp. 1

Figure 26.660. Segment 6 of *Cladopelma* sp.

Figure 26.661. Frontal apotome of *Microchironomus* sp.

Figure 26.662. Segment 2 of *Microchironomus* sp.

21'.	Caudolateral margins of segment 8 without spines	25
22(21).	Thoracic horn exceptionally long (Fig. 26.667)	*Cryptotendipes* (in part)
22'.	Thoracic horn never as long as above (Figs. 26.696, 26.699, 26.713)	23
23(22').	Tergum 6 with a posterior, median spiniferous process (Fig. 26.660)	*Cladopelma*
23'.	Tergum 6 with, at most, rows of spines along the posterior margin	24
24(23').	Cephalic tubercles long (Fig. 26.661); PSB II developed (Fig. 26.662)	*Microchironomus*
24'.	Cephalic tubercles short (Fig. 26.663); PSB II absent	Genus 12
25(21').	Hooklets on posterior margin of tergum 2 absent; large tubercles on terga (Fig. 26.666); thoracic horn exceptionally long (Fig. 26.667)	*Cryptotendipes* (in part)
25'.	Not with above characters	26
26(25').	Caudal region with a forked posterior extension (Fig. 26.669); frontal apotome often with ornate cephalic tubercles (e.g., Figs. 26.670 and 26.671)	*Cryptochironomus*
26'.	Caudal region without a forked process; cephalic tubercles, when present, never ornate	27
27(26').	PSB II with spinules (Fig. 26.672)	*Beckidia*
27'.	PSB II, when present, without spinules	28
28(27').	Posterior margins of tergites 2-5(6) with a row(s) of long needle-line spines	*Kloosia*
28'.	Tergites without posterior needle-like spines	29
29(28').	Cephalic tubercles absent	*Chernovskiia*
29'.	Cephalic tubercles present	30
30(29').	Each caudal lobe with more than 100 fringe setae	Genus 13
30'.	Each caudal lobe with no more than about 50-60 setae	*Harnischia*
31(20').	Caudolateral margins of segment 8 with a spine or a group of spines (e.g., Figs. 26.690-26.692, 26.697, 26.705)	32
31'.	Caudolateral margins of segment 8 without spines	89
32(31).	Thoracic horn with only 2 branches, one of which has numerous short spines distally (Fig. 26.753); terga 2-5(6) with paired transverse patches of spines (Fig. 26.754)	*Lauterborniella*
32'.	If the thoracic horn has only 2 branches there are never distal spines (e.g, Fig. 26.696); terga usually without paired spine patches	33
33(32').	Cephalic tubercles present (e.g., Figs. 26.677-26.680)	34
33'.	Cephalic tubercles absent	78
34(33).	Thoracic horn exceptionally long (Fig. 26.667)	*Cryptotendipes* (in part)
34'.	Thoracic horn shorter (e.g., Figs. 26.696, 26.699, 26.713)	35
35(34').	Terga 2-6, 3-6, or 4-5 with small to large unpaired groups of spines or spiniferous processes (Figs. 26.673, 26.675-26.676)	36
35'.	Terga without such groups of spines or processes	39
36(35).	Terga 2-6 with large spiniferous processes (Fig. 26.673)	*Glyptotendipes* (*Glyptotendipes*) (in part)
36'.	Terga 3-6 or 4-5(6) with smaller spine groups (Figs. 26.675-26.676)	37
37(36').	Epaulettes on terga 3-6	38
37'.	Small epaulettes on terga 4-5(6) (Fig. 26.676)	*Demijerea*
38(37).	Epaulettes on terga 3-6 all generally of about the same size (Fig. 26.675)	*Glyptotendipes* (*Trichotendipes*)
38'.	Epaulettes distinctly increasing in size from tergite 3 to tergite 6	*Glyptotendipes* (*Caulochironomus*)
39(35').	Cephalic tubercles truncate and with a cluster of spinules (Figs. 26.677, 26.757-26.758)	40
39'.	Cephalic tubercles not truncate and without spinules	43
40(39).	Conjunctives 3-4 and/or 4-5 with a few large dark spines on each side of the midline (Fig. 26.755)	*Hyporhygma*
40'.	Conjunctives with, at most, a field(s) of small spinules	41
41(40').	Anterior conjunctives conspicuously darkened laterally (Fig. 26.756)	*Phaenopsectra*
41'.	Anterior conjuctives without lateral darkening	42
42(41').	Frontal setae long, about twice the diameter of the truncated apexes of the cepahlic tubercles (Fig. 26.757)	*Endotribelos*
42'.	Frontal setae at most slightly longer than the truncated apexex of the cepahlic tubercles (Fig. 26.758)	*Sergentia*
43(39').	Cephalic tubercles extremely long and tapering; frontal setae attached near the bases of the tubercles (Fig. 26. 678)	*Polypedilum* (in part)
43'.	If cephalic tubercles very long, not as above (e.g., Figs. 26.679-26.680)	44

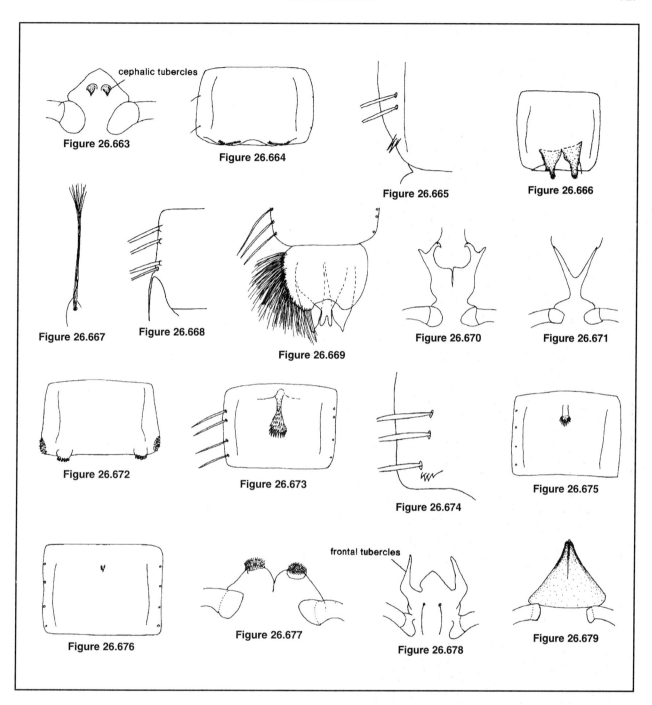

Figure 26.663. Frontal apotome of Genus 12.
Figure 26.664. Segment 2 of Genus 12.
Figure 26.665. Caudolateral margin of segment 8 of Genus 12.
Figure 26.666. Segment 5 of *Cryptotendipes* sp.
Figure 26.667. Thoracic horn of *Cryptotendipes* sp.
Figure 26.668. Caudolateral margin of segment 8 (left half) of *Cryptotendipes* sp.
Figure 26.669. Anal lobes and segment 8 of *Cryptochironomus* sp. 1.
Figure 26.670. Frontal apotome of *Cryptochironomus* sp. 2.
Figure 26.671. Frontal apotome of *Cryptochironomus* sp. 3.
Figure 26.672. Segment 2 of *Beckidia* sp.
Figure 26.673. Segment 6 of *Glyptotendipes (Glyptotendipes)* sp.
Figure 26.674. Caudolateral margin of segment 8 (left half) of *Glyptotendipes (Glyptotendipes)* sp.
Figure 26.675. Segment 6 of *Glyptotendipes (Trichotendipes)* sp.
Figure 26.676. Segment 5 of *Demeijerea* sp.
Figure 26.677. Frontal apotome of *Phaenopsectra* sp.
Figure 26.678. Frontal apotome of *Polypedilum* sp. 1.
Figure 26.679. Frontal apotome of *Polypedilum* sp. 2.

44(43').	Cephalic tubercles large, heavily sclerotized, and fused (Fig. 26.679)	*Polypedilum* (in part)
44'.	Cephalic tubercles not as above (e.g., Figs. 26.661, 26.663, 26.680)	45
45(44').	Cephalic tubercles long to extremely long, with frontal setae attached near the tips; and, caudolateral armature of segment 8 with blunt teeth which are weakly serrate at the tips (Fig. 26.759)	*Lipiniella*
45'.	Cephalic tubercles not so long, or, armature of segment 8 consisting of sharp spines (teeth)	46
46(45').	Frontal apotome with large frontal warts and cephalic tubercles (Fig. 26.680), and, caudolateral armature of segment 8 not a compound spur as in Figures 26.697-26.698	*Einfeldia* (in part)
46'.	Only cephalic tubercles present, or, caudolateral armature of segment 8 a compound spur as in Figures 26.697-26.698	47
47(46').	Sternum 2 and sometimes 1 and 3 with rows of needle-like spines (Figs. 26.681-26.682), and, caudolateral armature of segment 8 always strong (e.g., Figs. 26.689-26.690, 26.760)	48
47'.	Sternum 1-3 without such spines, or, when present, caudolateral armature of segment 8 weak	50
48(47).	Needle-like spines on sternum 2 in transverse rows only (e.g., Fig. 26.600)	49
48'.	Needle-like spines on sternum 2 in transverse and longitudinal rows (Fig. 26.682)	*Kiefferulus* (*Wirthiella*)
49(48).	Armature of caudolateral margins of segment 8 usually a simple spine - sometimes with 1 or 2 accessory spines (e.g., Figs. 26.689-26.690)	*Dicrotendipes* (in part)
49'.	Armature of segment consisting of 4-6 parallel spines with a common base (e.g., Fig. 26.760)	*Goeldichironomus* (in part)
50(47').	Caudolateral armature of segment 8 usually a single (or double), often sinuate, spine (e.g., Figs. 26.689-26.690); rarely the armature of 8 includes a few spines along the entire margin; gonopod sheaths of male never with distal spines; tergites 2-8 without paired spine patches	51
50'.	Caudolateral armature of segment 8 a comb of weak to strong spines or a compound spur; if the armature is single, the gonopod sheath of the male has distal spines, or, tergites 2-6 have paired spine patches	52
51(50).	Caudolateral armature of segment 8 usually a single straight or sinuate spine (e.g., Figs. 26.689-26.690); the largest shagreen spinules on tergites 3-6 tend to be in the middle of the tergites or along the posterior margin, or, both, but not with distinct anterior bands on tergites 2-6; conjunctives 3/4-5/6 usually with very small shagreen	*Dicrotendipes* (in part)
51'.	Caudolateral armature of segment 8 simple, but sometimes there are additional spines spread along the entire margin; shagreen of tergites 2-6 strongest in anterior transverse bands; only conjunctive 4/5 with shagreen	*Beardius*
52(50').	Comb on segment 8 composed of 4-5 or more strong, nearly parallel spines (Fig. 26.760), and, segment 8 has 5 lateral lamellar setae	53
52'.	If the comb on segment 8 is as above, segment 8 has 4 lateral lamellar setae	55
53(52).	Conjunctiva 3/4-5/6 without fields of shagreen	*Goeldichironomus* (in part)
53'.	Conjunctiva 3/4-5/6 with fields of shagreen	54
54(53').	Most of tergites 2-5 covered with shagreen of nearly uniform size	*Kiefferulus* (*Kiefferulus*)
54'.	Most of tergites 2-5 free of shagreen although shagreen is present it may be confined to relatively small patches	*Einfeldia* (in part)
55(52').	Tergite 6 with the posterior armature stronger than on other tergites	*Parachironomus* (in part)
55'.	Not as above	56
56(55').	Segment 5 with 1-3 lamellar lateral setae (Figs. 26.683, 26.688)	57
56'.	Segment 5 with 4-5 lamellar lateral setae (Fig. 26.700)	66
57(56).	Segment 5 with 1-2 lamellar lateral setae (Fig. 26.683)	*"Paratendipes"* (in part)
57'.	Segment 5 with 3 lamellar lateral setae (Fig. 26.688)	58
58(57').	Segment 6 with 3 lamellar lateral setae (Fig. 26.714)	59
58'.	Segment 6 with 4 lamellar lateral setae (Fig. 26.688)	62
59(58).	Terga 2-6 with paired groups of spines (Fig. 26.684); spine(s) on caudolateral margin of segment 8 as in Figure 26.685	*Zavreliella*
59'.	Terga 2-6 without paired groups of spines; spines on segment 8 as in Figure 26.686	60
60(59').	Fringe of setae on caudal lobes completely uniserial (Fig. 26.686)	*Polypedilum* (in part)
60'.	Fringe of setae on caudal lobes at least partially multiserial	61
61(60').	Segment 8 with 3-4 lateral lamellar setae	*Stictochironomus*

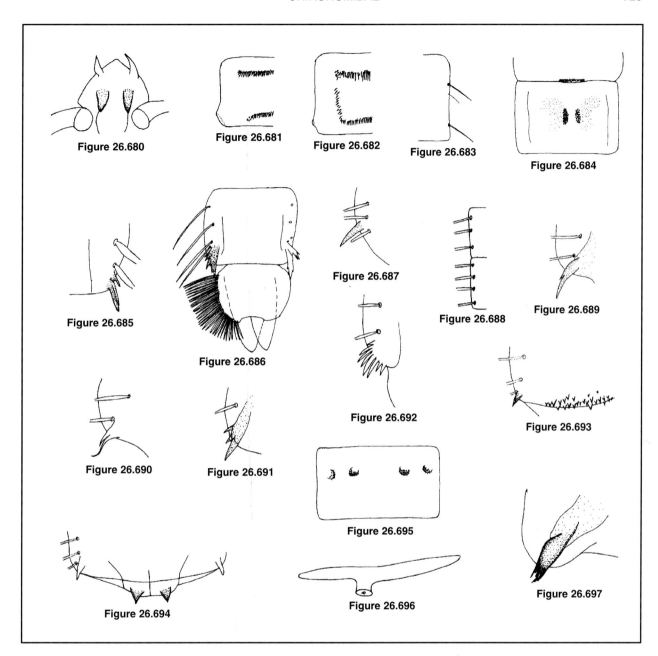

Figure 26.680. Frontal apotome of *Einfeldia* sp. 1.

Figure 26.681. Segment 2 (ventral view, left half) of *Dicrotendipes* sp. 1

Figure 26.682. Segment 2 (ventral view, left half) of *Kiefferulus (Wirthiella)* sp.

Figure 26.683. Segment 5 (right half) of Genus 15.

Figure 26.684. Posterior margin of segment 2 and segment 3 of *Zavreliella* sp.

Figure 26.685. Caudolateral margin of segment 8 (right half) of *Zavreliella* sp.

Figure 26.686. Anal lobes and segment 8 of *Polypedilum* sp. 3

Figure 26.687. Caudolateral margin of segment 8 (left half) of *Pagastiella* sp.

Figure 26.688. Segments 5 and 6 (left half) of *Paratendipes* sp.

Figure 26.689. Caudolateral margin of segment 8 (left half) of *Dictrotendipes sp. 2*.

Figure 26.690. Caudolateral margin of segment 8 (left half) of *Dicrotendipes sp. 3*.

Figure 26.691. Caudolateral margin of segment 8 (left half) of *Paralauterborniella* sp.

Figure 26.692. Caudolateral margin of segment 8 (left half) of *Cyphomella* sp.

Figure 26.693. Segment 8 (ventral view, left side) of male of *Demicryptochironomus* sp.

Figure 26.694. Segment 8 (ventral view of posterior margin) of female of *Demicryptochironomus* sp.

Figure 26.695. Segment 1 (ventral view) of *Pseudochironomus* sp. 2.

Figure 26.696. Thoracic horn of *Pseudochironomus* sp. 2.

Figure 26.697. Caudolateral margin of segment 8 (left half) of *Chironomus* sp. 1.

61'.	Segment 8 with 5 lateral lamellar setae	*Tribelos*
62(58').	Segment 8 with 3-4 lamellar lateral setae, frontal setae present	*Paratendipes* (in part)
62'.	Segment 8 with 5 lamellar lateral setae, if 4, then frontal setae absent	63
63(62').	Armature on caudolateral margins of segment 8 in the form of a compound spur (e.g., Fig. 26.697)	*Chironomus* (in part)
63'.	Armature on margins of segment 8 consisting of a single spine (e.g, Fig. 26.687) or groups of spines	64
64(63').	Armature on margins of segment 8 often consisting of single spines (Fig. 26.687); anal lobes with 15 or fewer fringe seate	*Pagastiella*
64'.	Armature on margins of segment 8 usually consisting of multiple spines; anal lobes usually with many more than 15 finge setae	65
65(64').	Terga 2-5 or 3-5 with paired groups of dark spines similar to Figure 26.754; frontal setae present	*Omisus*
65'.	Terga without such paired groups, although isolated fields of shagreen may be present; frontal setae absent, but short to rather long cephalic tubercles are present	*Microtendipes*
66(56').	Segment 8 with 4-5 lamellar lateral setae, when 5, caudolateral armature on segment 8 similar to Figures 26.689-26.690	67
66'.	Segment 8 with 5 lateral setae (5 lamellar or 4 lamellar and 1 hairlike); armature on segment 8 usually not like Figs. 26.689-26.690	69
67(66).	Spines on caudolateral margins of segment 8 usually simple as in Figures 26.689-26.690)	*Dicrotendipes* (in part)
67'.	Spines on caudolateral margins of segment 8 multiple (e.g., Figs. 26.691, 26.692)	68
68(67').	Caudolateral margins of segment 8 with 3 or more short straight spines (Fig. 26.691)	*Paralauterborniella*
68'.	Caudolateral margins of segment 8 with a series of closely set curved spines (Fig. 26.692)	*Cyphomella*
69(66').	Posterior margin of sternum 8 of female with 2 sclerotized spines (e.g, Fig. 26.694)	70
69'.	No such spines present	71
70(69).	Posterior shagreen of tergites 2-3(4) very strong and dark; posterior margin of sternite 8 of male with a row of spines (Fig. 26.693)	*Demicryptochironomus* (Irmakia)
70'.	Posterior shagreen of tergites 2-3(4) not very strong and dark; posterior margin of sternite 8 of male without a row of spines	*Demicryptochironomus* (Demicryptochironomus)
71(69').	Sternum I with 2 pairs of spiniferous processes (Fig. 26.695)	*Pseudochironomus* (in part)
71'.	Sternum I without such processes	72
72(71').	Spines on caudolateral corners of segment 8 in the form of a compound spur (Figs. 26.697-26.698)	*Chironomus* (in part)
72'.	Spines on segment 8 not as above (e.g., Figs. 26.701, 26.705, 26.711-26.712)	73
73(72').	Spines on caudolateral margins of segment 8 single, and, gonopod sheaths of male with distal spines	*Gillotia*
73'.	Spines on segment 8 multiple (e.g., Figs. 26.701, 26.705, 26.711)	74
74(73').	Spines on caudolateral margins of segment 8 long, yellow, and slightly curved (Fig. 26.701)	*Einfeldia* (in part)
74'.	Spines on segment 8 not as above	75
75(74').	PSB 11 present (Fig. 26.702)	76
75'.	PSB 11 absent	77
76(75).	Segment 4 with 1-2 lateral lamellar setae	*Apedilum*
76'.	Segment 4 with no lateral lamellar setae	*Parachironomus* (in part)
77(75").	Fringe of setae on anal lobes uniserial (Fig. 26.704); armature on segment 8 may be as in Figure 26.705	*Paracladopelma* (in part)
77'.	Fringe of setae on anal lobes multiserial	*Genus 17*
78(33').	Segment 5 with no lamellar lateral setae (Fig. 26.708)	79
78'.	Segment 5 with 3-4 lamellar lateral setae (Figs. 26.700, 26.714)	80
79(78).	Anterior regions of terga 2-5 with single or multiple rows of strong spines (Fig. 26.707)	*Asheum*
79'.	Anterior regions of terga 2-6 with multiple rows of smaller spines (Fig. 26.708), a tuft of fringe setae on the apex of each anal lobe	*Endochironomus*
80(78').	Segment 5 with 3 lateral lamellar setae (Fig. 26.714)	81
80'.	Segment 5 with 4 lateral lamellar setae (Fig. 26.700)	86
81(80).	Segment 6 with 3 lateral lamellar setae (Fig. 26.714)	82
81'.	Segment 6 with 4 lateral lamellar setae (Fig. 26.700)	85

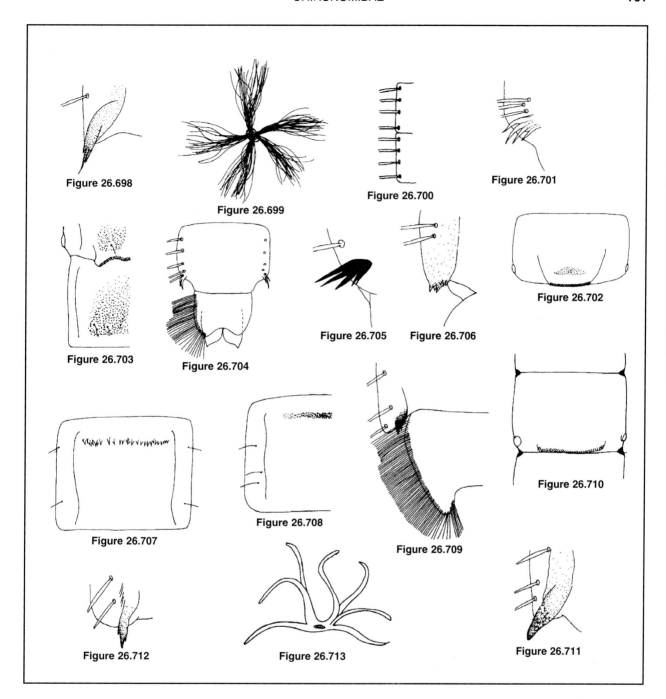

Figure 26.698. Caudolateral margin of segment 8 (left side) of *Chironomus* sp. 2.

Figure 26.699. Thoracic horn of *Chironomus* sp. 2.

Figure 26.700. Segments 5 and 6 (left side) of *Chironomus* sp. 2.

Figure 26.701. Caudolateral margin of segment 8 (left side) of *Einfeldia* sp. 2.

Figure 26.702. Segment 2 of *Parachironomus* sp.

Figure 26.703. Segments 2 and 3 (left side) of *Kiefferulus* sp.

Figure 26.704. Anal lobes and segment 8 of *Paracladopelma* sp. 1.

Figure 26.705. Caudolateral margin of segment 8 (left side) of "*Paracladopelma*" sp.

Figure 26.706. Caudolateral margin of segment 8 (left side) of Genus 17.

Figure 26.707. Segment 5 of *Asheum* sp.

Figure 26.708. Segment 5 (left side) of *Endochironomus* sp.

Figure 26.709. Caudolateral margin of segment 8 and left anal lobe of Genus 18.

Figure 26.710. Conjunctiva 1/2 and 2/3 of *Polypedilum* sp. 4.

Figure 26.711. Caudolateral margin of segment 8 (left side) of *Polypedilum* sp. 5.

Figure 26.712. Caudolateral margin of segment 8 (left side) of *Polypedilum* sp. 6.

Figure 26.713. Thoracic horn of *Polypedilum* sp. 6.

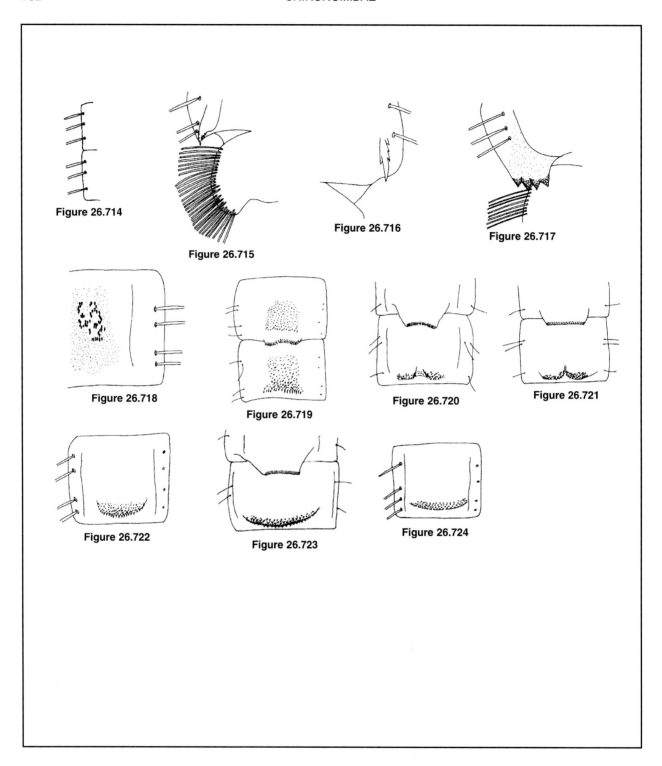

Figure 26.714. Segments 5 and 6 (left side) of *Polypedilum* sp. 6.

Figure 26.715. Anal lobe and caudolateral margin of segment 8 (left side) of *Graceus* sp.

Figure 26.716. Caudolateral margin of segment 8 (right side) of *Nilothauma sp.*

Figure 26.717. Caudolateral margin of segment 8 (left side) of *Stenochironomus* sp.

Figure 26.718. Segment 5 (right half) of *Xenochironomus* sp.

Figure 26.719. Segments 2 and 3 of *Saetheria* sp. 1

Figure 26.720. Posterior margin of segment 2 and segment 3 of *Paracladopelma* sp. 2.

Figure 26.721. Posterior margin of segment 2 and segment 3 of *Paracladopelma* sp. 3.

Figure 26.722. Segment 6 of *Paracladopelma* sp. 3.

Figure 26.723. Posterior margin of segment 2 and segment 3 of *Saetheria* sp. 2

Figure 26.724. Segment 6 of *Saetheria* sp. 2

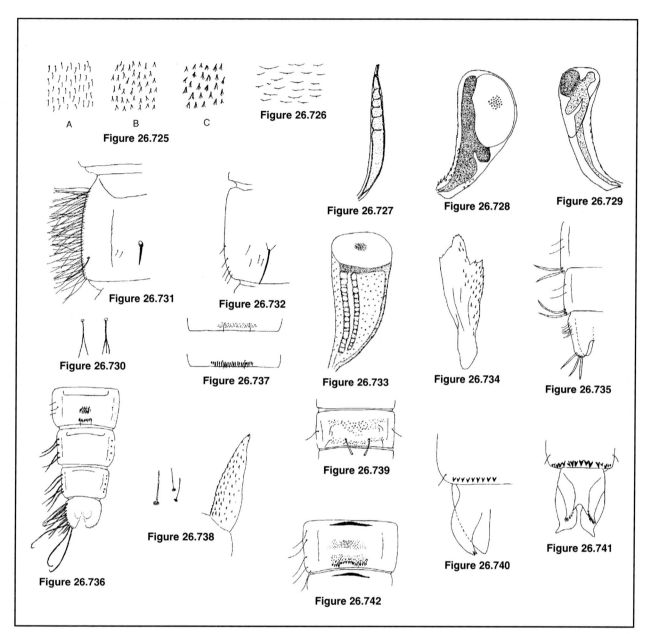

Figure 26.725. Dense a. simple, b. bifid or c. multi-branched shagreen of abdomen found in many Tanypodinae genera. (after Roback 1981).

Figure 26.726. Arching shagreen of abdominal tergites of *Trissopelopia* sp.

Figure 26.727. Thoracic horn of *Denopelopia* sp. (after Roback and Rutter 1988)

Figure 26.728. Thoracic horn of *Meropelopia* sp.

Figure 26.729. Thoracic horn of *Helopelopia* sp.

Figure 26.730. Branched dorsal setae of abdominal tergites of *Monopelopia* sp.

Figure 26.731. Segment 7 of *Fittkauimyia* sp.

Figure 26.732. Segment 7 of *Bethbilbeckia* sp. (after Fittkau and Murrary 1986)

Figure 26.733. Thoracic horn of *Derotanypus* sp. (after Fittkau and Murrary 1986)

Figure 26.734. Thoracic horn of *Euryhapsis* sp.

Figure 26.735. Segments 7-8 and anal lobes of *Vivacricotopus* sp. (after Saether 1988).

Figure 26.736. Segments 6-8 and anal lobes of *Doncricotopus* sp.

Figure 26.737. Posterior margins of tergites 3 and 7 of *Psectrocladius (Mesopsectrocladius)* sp. (after Coffman et al. 1986).

Figure 26.738. Thoracic horn of *Nanocladius (Plecopteracoluthus)* sp.

Figure 26.739. Posteriomedial dorsal setae of middle abdominal segments of *Rheosmittia* sp. 2.

Figure 26.740. Anal lobes of *Doithrix* sp. 1 (after Coffman et al. 1986).

Figure 26.741. Anal lobes of *Doithrix* sp. 2.

Figure 26.742. Apophyses (dark transverse markings) on abdominal segments of *Epoicocladius* sp. 2

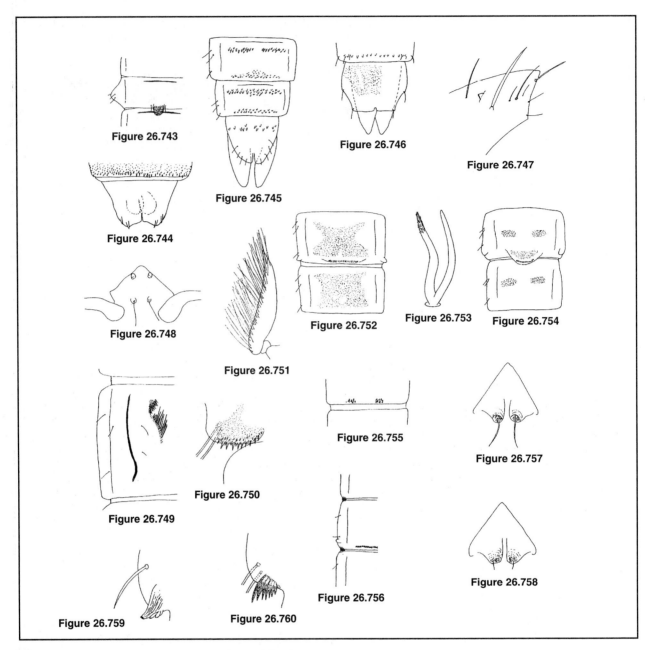

Figure 26.743. Pad of hooklets on posterior margin of tergite 2 of *Acamptocladius* sp. (after Coffman et al. 1986)

Figure 26.744. Anal lobes of *Antillocladius* sp.

Figure 26.745. Segments 8-9 and anal lobes of *Camptocladius* sp. (after Coffman et al. 1986).

Figure 26.746. Anal lobes of *Gymnometriocnemus (Raphidocladius)* sp. (after Coffman et al. 1986).

Figure 26.747. Anterior thorax and spur of *Syndiamesa* sp. (after Oliver 1986).

Figure 26.748. Frontal apotome with cephalic tubercles and frontal warts of *Psilometriocnemus* sp.

Figure 26.749. Armature of tergite 4 of *Nimbocera* sp.

Figure 26.750. Anal comb of segment 8 of *Nimbocera* sp.

Figure 26.751. Thoracic horn of *Krenopsectra* sp. (after Pinder and Reiss 1986).

Figure 26.752. Shagreen of tergites 2 and 3 of *Zavrelia* sp. (after Pinder and Reiss 1986).

Figure 26.753. Thoracic horn of *Lauterborniella* sp.

Figure 26.754. Armature of middle tergites of *Lauterborniella* sp.

Figure 26.755. Large spines of conjunctiva behind tergites 3 and 4 of *Hyporhygma* sp. (after Pinder and Reiss 1986).

Figure 26.756. Darkened lateral areas of conjunctiva of *Phaenopsectra* sp.

Figure 26.757. Cephalic tubercles and frontal setae of *Endotribelos* sp. (after Grodhaus 1987).

Figure 26.758. Cephalic tubercles and frontal setae of *Sergentia* sp. (after Pinder and Reiss 1986).

Figure 26.759. Caudolateral armature of segment 8 of *Lipiniella* sp. (after Pinder and Reiss 1986).

Figure 26.760. Anal comb of segment of *Goeldichironomus* sp. (after Pinder and Reiss 1986).

82(81).	Fringe of setae on anal lobe extends along inner margin (Fig. 26.709)	***Genus 18***
82'.	Fringe of setae does not extend along inner margin of anal lobe	83
83(82').	Thoracic horn rarely with more than 16 branches; fringe of setae on anal lobes usually uniserial as in Figure 26.686, but may be completely multiserial in some species	***Polypedilum*** (in part)
83'.	Thoracic horn plumose; fringe of setae on anal lobes distally multiserial (Fig. 26.715)	84
84(83').	Anal lobes with 2 rows of fringe setae along the distal margins (e.g., Fig. 26.715)	***"Graceus"***
84'.	Anal lobes with 2-4 rows of fringe setae along most of the margins	***Synendotendipes***
85(81').	Sternum 1 with 2 pairs of spiniferous processes (Fig. 26.695)	***Pseudochironomus*** (in part)
85'.	Sternum 1 without such processes	***Axarus***
86(80').	Armature on caudolateral margins of segment 8 as in Figure 26.716, median shagreen patches on terga 7 and 8	***Nilothauma***
86'.	Armature on caudolateral margins of segment 8 not as above	87
87(86').	Armature on caudolateral margin of segment 8 similar to Figure 26.717	***Stenochironomus***
87'.	Armature on segment 8 not as above	88
88(87').	Row of hooklets on segment 2 narrowly interrupted; frontal apotome with a pair of swollen mounds	***Xestochironomus***
88'.	Hook row on segment 2 entire; frontal apotome flat	***Stelechomyia***
89(31').	Cephalic tubercles present (e.g., Figs. 26.661, 26.663, 26.680)	90
89'.	Cephalic tubercles absent	98
90(89).	Terga 2-6 with median spiniferous processes (Fig. 26.592)	***Glyptotendipes*** (*Glyptotendipes*) (in part)
90'.	Terga 2-6 without such processes	91
91(90').	Terga 2-5 with coarse shagreen arranged in a fenestrated pattern with many spinules located on pigmented spots of cuticle (Fig. 26.718)	***Xenochironomus*** (in part)
91'.	Shagreen not as above	92
92(91').	Frontal tubercles and cephalic tubercles present (Fig. 26.680)	***Einfeldia*** (in part)
92'.	Only cephalic tubercles present	93
93(92').	Segment 5 with 0 or 3 lateral lamellar setae (e.g., Fig. 26.688)	94
93'.	Segment 5 with 4 lateral lamellar setae (e.g., Fig. 26.700)	96
94(93).	Segment 5 with no lateral lamellar setae	***Polypedilum*** (in part)
94'.	Segment 5 with 3 lateral lamellar seteae	95
95(94').	Segment 6 with 3 lamellar lateral setae (e.g., Fig. 26.714)	***Parachironomus*** (in part)
95'.	Segment 6 with 4 lamellar lateral setae (Fig. 26.688)	***Pseudochironomus*** (in part)
96(93').	Tergum 2 with shagreen area about the same size as that of tergum 3 (Fig. 26.719)	***Saetheria*** (in part)
96'.	Tergum 2 without shagreen	97
97(96').	Rows of spines on posterior margins of terga 3-6 usually with some spines distinctly larger and darker than the others (Fig. 26.720), or with spines on tergum 6 as large or larger than those on tergum 3 (Fig. 26.722)	***Paracladopelma*** (in part)
97'.	Rows of spines on posterior margins of terga 3-6 all of about the same size on any one tergum (Fig. 26.723) but those on tergum 6 distinctly smaller than those on tergum 3 (Fig. 26.724)	***Saetheria*** (in part)
98(89').	Shagreen on terga as in Figure 26.718	***Xenochironomus*** (in part)
98'.	Shagreen on terga not as above	99
99(98').	Sternum 2, at least, with rows of needle-like spines similar to Figure 26.681	***Robackia***
99'.	Sterna without needle-like spines	***Parachironomus*** (in part)

KEY TO SUBFAMILIES AND TRIBES OF FRESHWATER CHIRONOMIDAE ADULTS*

(Males Only)

1.	Wing with crossvein M—Cu present (Figs. 26.762, 26.764, 26.771-26.772)	2
1'.	Wing with crossvein M—Cu absent (Figs. 26.761, 26.763, 26.782-26.785)	12
2(1).	Wing vein R_{2+3} present (Figs. 26.762-26.763, 26.765, 26.771-26.772)	3
2'.	Wing vein R_{2+3} absent (Fig. 26.764)	Podonominae ...11
3(2).	Vein R_{2+3} forked (Figs. 26.765, 26.771-26.772)	Tanypodinae ...4
3'.	Vein R_{2+3} simple (Fig. 26.762)	Diamesinae ...8
4(3).	A double comb present on distal end of tibia 3 (3rd, or hind leg) (Fig. 26.766); 4th tarsal segment cordiform (heart-shaped) (Fig. 26.767)	**Coelotanypodini**

* Couplets 4-7 modified from Roback (1971); couplet 14 modified from Pinder (1978) and Cranston *et al.* (1989); couplets 15 and 16 modified from Brundin (1956).

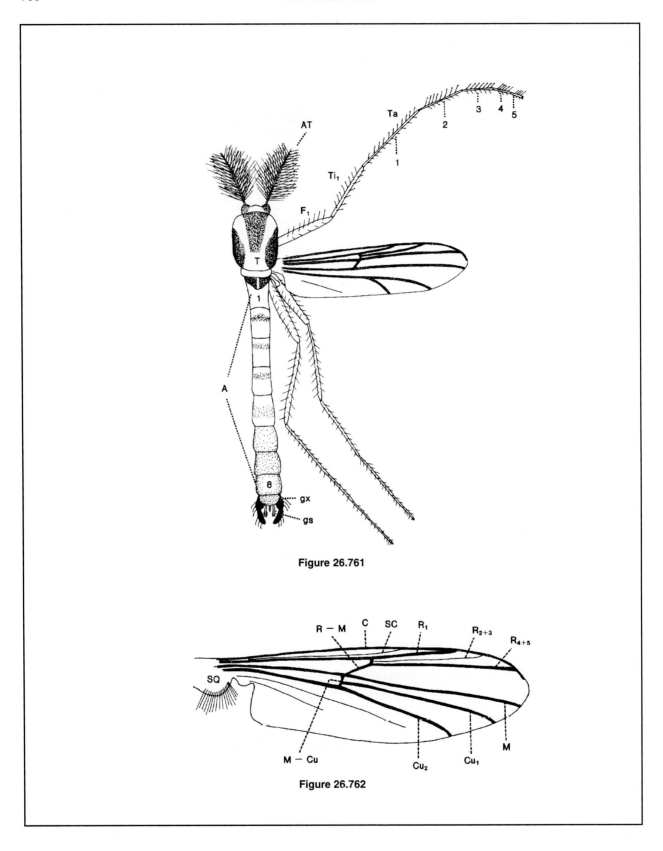

Figure 26.761. Adult chironomid male (Chironomini) in dorsal view: AT, antenna; T, thorax; A, abdominal tergites 1-8; gx, gonocoxite of male genitalia; gs, gonostylus of male genitalia; F_1, femur of foreleg; Ti_1, tibia of foreleg; Ta_{1-5}, tarsal segments of foreleg.

Figure 26.762. Diamesini wing: SQ, squama; R-M, radial-medial crossvein; C, costa; SC, subcosta: R_1, first radial; R_{2+3}, second and third radials; R_{4+5}, fourth and fifth radials; M, medius; Cu_1, first cubital; Cu_2, second cubital; M-Cu, medial-cubital crossvein.

CHIRONOMIDAE

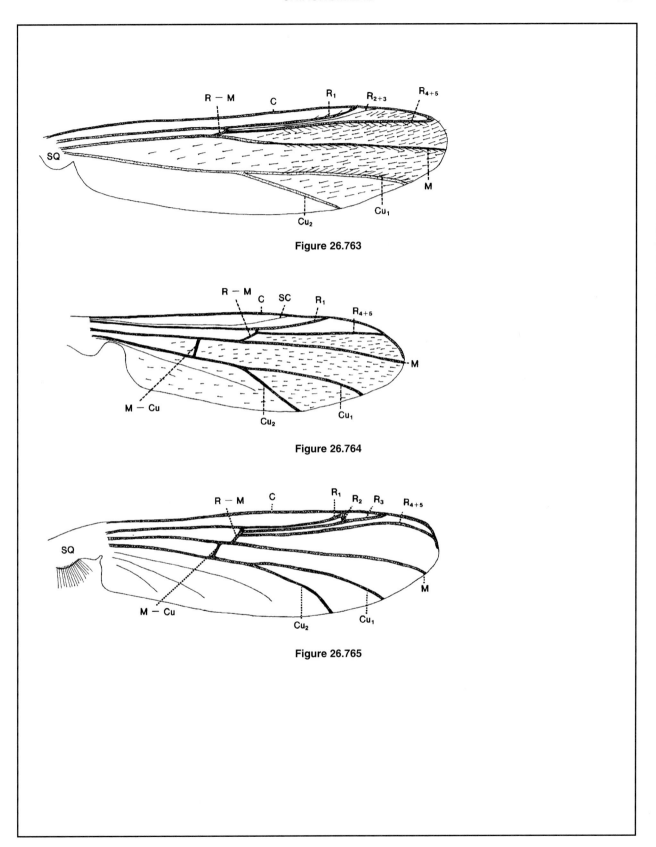

Figure 26.763. *Tanytarsini* wing (abbreviations as in Fig. 26.762).

Figure 26.764. *Boreochlini* wing (abbreviations as in Fig. 26.762; after Goetghebuer 1939).

Figure 26.765. *Tanypodini* wing (abbreviations as in Fig. 26.762).

4'.	A single comb (or none) on distal end of tibia 3 (Fig. 26.768); 4th tarsal segment cylindrical (Fig. 26.769)	5
5(4').	Fork of Cu (fCu) sessile (Figs. 26.770-26.771)	6
5'.	Fork of Cu (fCu) petiolate (stalked) (Figs. 26.765, 26.772)	7
6(5).	Costa (anterior marginal wing vein) produced beyond R_{4+5} by about twice the length of R—M (Fig. 26.771)	*Macropelopiini* (in part)
6'.	Costa not or only slightly produced beyond R_{4+5} (Fig. 26.765)	*Pentaneurini*
7(5').	Petiole of cubital fork short, one-third or less the length of Cu_2 (Fig. 26.765)	*Tanypodini*
7'.	Petiole of cubital fork long, one-half or more the length of Cu_2 (Fig. 26.772)	*Macropelopiini* (in part)
8(3').	Gonostylus of genitalia narrow, simple, and attached mesally at the base of the distal third of gonocoxite (Fig. 26.773)	*Protanypini*
8'.	Gonostylus attached at the distal end of the gonocoxite (Figs. 26.774-26.776)	9
9(8').	Gonostylus very short and robust (Fig. 26.774)	*Boreoheptagyini*
9'.	Gonostylus usually longer and more narrow (Figs. 26.775-26.776)	10
10(9').	Gonostylus with a large dorsobasal appendage (Fig. 26.775)	*Prodiamesini* (in part)
10'.	Gonostylus simple (Fig. 26.776) ....*Diamesini*............*Prodiamesini* (in part)	
11(2').	Gonostylus with 2 lobes (Fig. 26.777)	*Podonomini*
11'.	Gonostylus simple (Figs. 26.778-26.779)	*Boreochlini*
12(1').	Gonostylus rigidly directed backwards (somewhat flexible in one genus) (Fig. 26.780)	Chironominae ... 13
12'.	Gonostylus movable, often flexed inwardly toward the midline (Fig. 26.781)	Orthocladiinae ... 14
13(12).	Wing surface with macrotrichia (small hairs), especially near the apex; squama without a fringe of hairs; crossvein R—M almost parallel to the long axis of the wing (Fig. 26.763)	*Tanytarsini*
13'.	Wing surface with or without macrotrichia; when macrotrichia are present, squama with a fringe of hairs; crossvein R—M sharply oblique to the long axis of the wing (Fig. 26.782)	*Chironomini*
14(12').	Cu_2 straight or weakly curved; squama usually with a well-developed hair fringe (Fig. 26.783)	*"Orthocladiini"*
14'.	Cu_2 usually somewhat strongly curved; hair fringe on squama usually reduced, often absent (Fig. 26.784); if Cu_2 is straight, then wing as in Fig. 26.785	15
15(14').	Vein R_{4+5} visible (Fig. 26.784)	*"Metriocnemini"*
15'.	Vein R_{4+5} completely fused with the thickened costa (Fig. 26.785)	*"Corynoneurini"*

KEYS TO TRIBES OF MARINE CHIRONOMIDAE*

Larvae (Marine)

1.	Larval head with a ventral median mentum (labial plate), and striate ventromental plates (Figs. 26.1, 26.3)	Chironominae (go to freshwater key)
1'.	Larval head with a distinct mentum, but never with striate ventromental plates	2
2(1').	Larval antenna with 5 segments (Figs. 26.669-26.670)	3
2'.	Larval antenna with 4 segments (Fig. 26.671)	Telmatogetoninae
3(2).	Preanal papillae present, but short, with 5-6 terminal setae and 1-2 lateral setae (Fig. 26.672)	Orthocladiinae—*"Orthocladiini"* (in part) (*Halocladius*)
3'.	Preanal papillae absent, replaced by 1-3 short setae or a single long seta (Fig. 26.673)	4
4(3').	Preanal papillae replaced by 1-3 short setae ....Orthocladiinae (in part) ......(*Thalassosmittia*, tribe uncertain)	
4'.	Preanal papillae replaced by a single (may be branched) stout, long seta (Fig. 26.673)	Orthocladiinae—*Clunionini*

Pupae (Marine)

1.	Anal lobes with at least a partial fringe of long hairs along edge (e.g., Fig. 26.639); thoracic horn multibranched (Figs. 26.699, 26.713) or simple (Figs. 26.606, 26.611); a sclerotized spine or comb of spines usually present on caudolateral edges of segment 8 (Figs. 26.641, 26.646, 26.697); surface of anal lobes in same plane as abdomen	Chironominae (go to freshwater key)
1'.	Anal lobes with or without a fringe of hairs, when hairs present, they are short and the surface of the anal lobe is in a plane oblique to that of the abdomen (Fig. 26.791); thoracic horn, when present, always simple; no spines on caudolateral edges of segment 8	2
2(1').	Thorax with a thoracic horn; anal lobes with or without a fringe of hairs, in either case, the surface is oblique to the plane of the abdomen (Fig. 26.791)	Telmatogetoninae

*The first couplet in each of the life stage keys separates the Chironominae, which has many brackish water species, from the truly marine forms. The genus *Halocladius* (Orthocladiinae: "Orthocladiini") is also known to occur in saline inland waters.

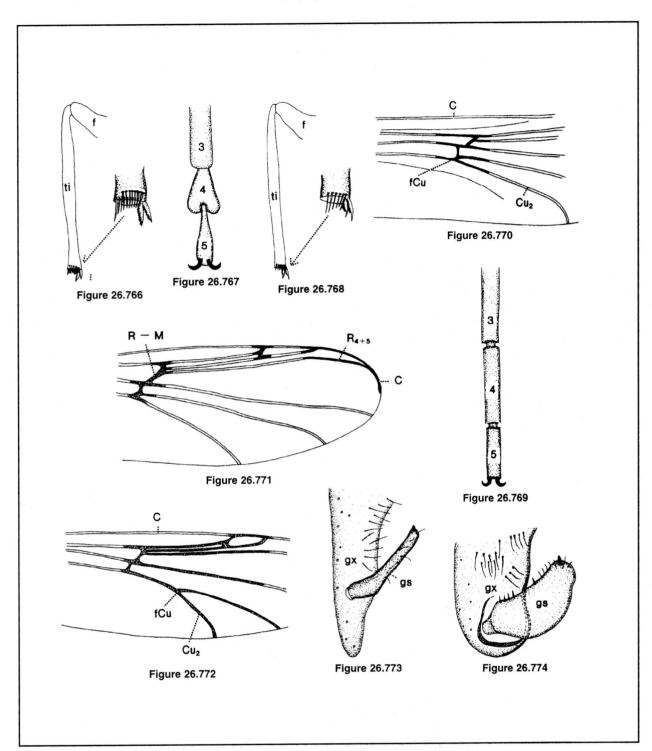

Figure 26.766. Third tibia of a *Coelotanypodini* adult (after Roback 1971). ti, tibia; f, femur.

Figure 26.767. Terminal tarsal segments of a *Coelotanypodini* adult (tarsal segments numbered).

Figure 26.768. Third tibia of a *Pentaneurini* adult (after Roback 1971): ti, tibia; f, femur.

Figure 26.769. Terminal tarsal segments of a *Pentaneurini* adult (tarsal segments numbered).

Figure 26.770. Central portion of a *Pentaneurini* wing: fCu, fork of cubitus; other abbreviations as in Fig. 26.762.

Figure 26.771. *Macropelopiini* (*Natarsia*) wing (abbreviations as in Fig. 26.762.).

Figure 26.772. Central portion of a *Macropelopiini* (*Procladius*) wing (abbreviations as in Figs. 26.762, 26.770).

Figure 26.773. Dorsal view of left gonocoxite and gonostylus of a male *Protanypini* (after Saether 1975); gx, gonocoxite; gs, gonostylus.

Figure 26.774. Dorsal view of left gonocoxite and gonostylus of a male *Boreoheptagyini* (after Brundin 1966).

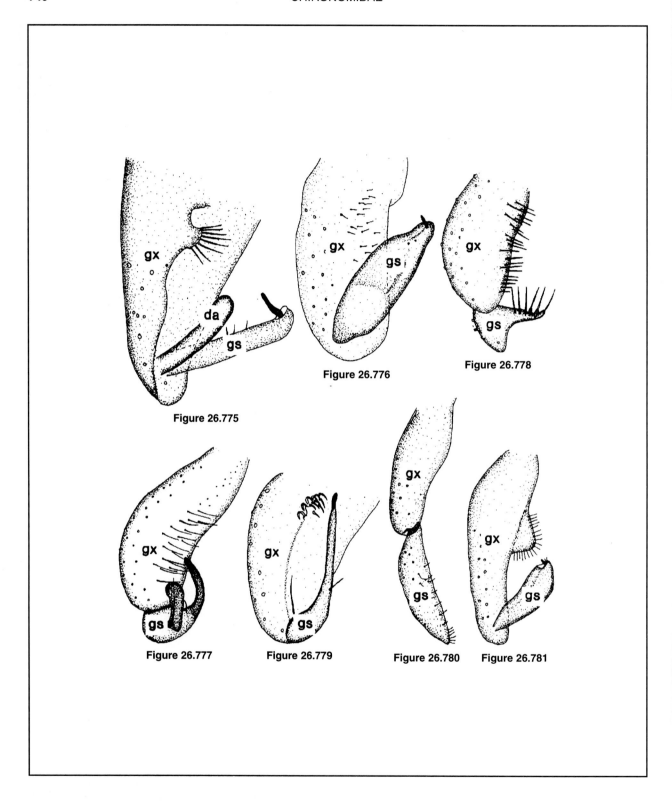

Figure 26.775. Dorsal view of left gonocoxite and gonostylus of a male *Prodiamesini* (*Prodiamesa* only).

Figure 26.776. Dorsal view of left gonocoxite and gonostylus of a male *Diamesini*.

Figure 26.777. Dorsal view of left gonocoxite and gonostylus of a male *Podonomini* (*Parochlus*, the only genus known from North America) (after Brundin 1966).

Figure 26.778. Dorsal view of left gonocoxite and gonostylus of a male *Boreochlini*.

Figure 26.779. Dorsal view of left gonocoxite and gonostylus of a male *Boreochlini* (after Brundin 1966).

Figure 26.780. Dorsal view of left gonocoxite and gonostylus of a male *Chironomini*.

Figure 26.781. Dorsal view of left gonocoxite and gonostylus of a male "*Orthocladiini*."

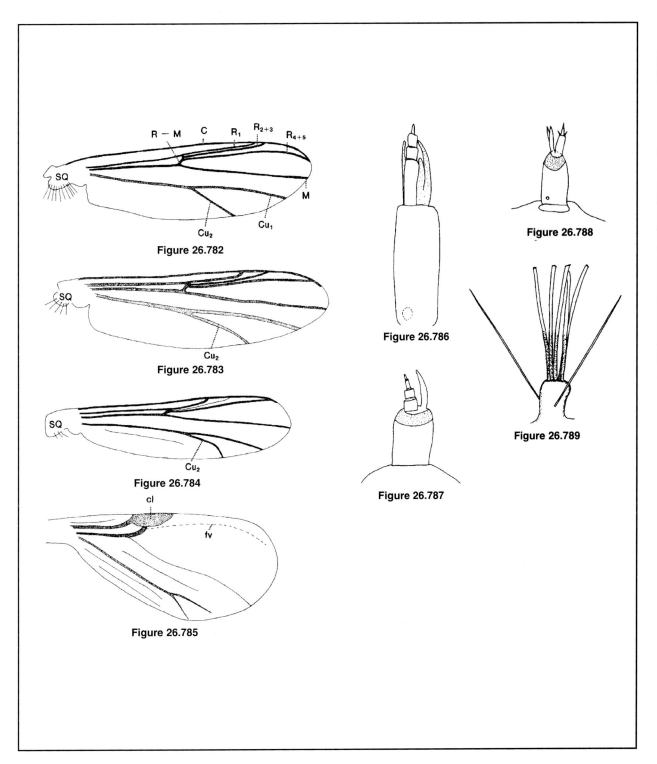

Figure 26.782. *Chironomini* wing (abbreviations as in Fig. 26.762).

Figure 26.783. "*Orthocladiini*" wing (abbreviations as in Fig. 26.762).

Figure 26.784. "*Metriocnemini*" wing (abbreviations as in Fig. 26.762).

Figure 26.785. "*Corynoneurini*" wing. Cl, clavus; fv, false vein.

Figure 26.786. Five-segmented antenna of larva of *Halocladius* sp. ("Orthocaldiini") (redrawn from Hirvenoja [1973]).

Figure 26.787. Five-segmented antenna of larva of *Tethymyia* sp. (Clunionini) (redrawn from Wirth [1949]).

Figure 26.788. Four-segmented antenna of larva of *Thalassomya* sp. (Telmatogetoninae) (redrawn from Wirth [1947]).

Figure 26.789. Preanal papilla of larva of *Halocladius* sp. ("Orthocladiini") (redrawn from Hirvenoja [1973]).

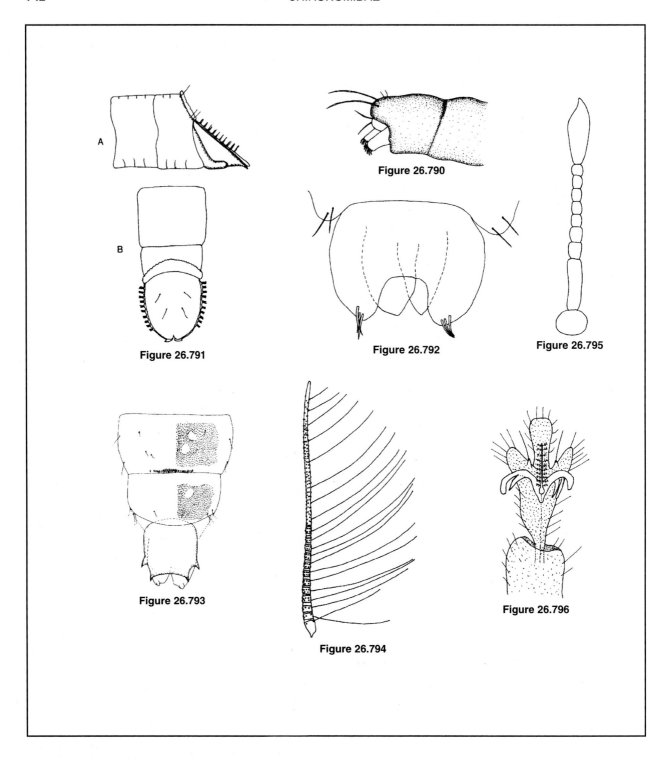

Figure 26.790. Caudal segments of larva of *Clunio* sp. (Clunionini) (redrawn from Lenz [1950]).

Figure 26.791. *A.* Lateral view of terminal abdominal segments of pupa of *Thalassomya* sp. (Telmatogetoninae). *B.* Dorsal view of same. Both redrawn from Wirth (1947).

Figure 26.792. Dorsal view of caudal lobes of pupa of *Halocladius* sp. ("Orthocladiini").

Figure 26.793. Dorsal view of segments 7 and 8 and caudal lobes of pupa of *Thalassosmittia* sp. (Orthocladiinae).

Figure 26.794. Adult male antenna of *Cricotopus* sp. (very similar to *Halocladius*); basal segment not included; only a few of the plume hairs drawn.

Figure 26.795. Adult male antenna of *Clunio* sp. (Clunionini); basal segment not included (redrawn from Goetghebuer [1950]).

Figure 26.796. Ventral view of trilobed fifth tarsal segment of *Telmatogeton* sp. (Telmatogetoninae) (redrawn from Wirth [1949]).

2'.	Thorax without a thoracic horn; anal lobes never with a fringe of hairs; surface of lobes in same plane as abdomen	3
3(2').	Anal lobes each with 3 terminal (or subterminal) setae (Fig. 26.792) ....Orthocladiinae—*"Orthocladiini"* (in part) *(Halocladius)*	
3'.	Anal lobes without terminal setae, but there may be 1-2 spinelike projections from the margin (Fig. 26.793)	4
4(3').	Dorsal surfaces of segments 2-7 (at least) heavily and somewhat uniformly shagreened (rough-surfaced) (Fig. 26.793)	5
4'.	Dorsal surfaces of segments 2-7 not heavily and uniformly shagreened, at most with anterior and posterior transverse rows of spines or hooks on some segments ....Orthocladiinae—***Clunionini*** (in part) *(Clunio)*	
5.	Segment 8 with only very fine dorsal shagreen; caudolateral corners of segment 9 (anal lobe) with 2 small, closely set spines ....Orthocladiinae—***Clunionini*** (in part) *(Tethymyia)*	
5'.	Segment 8 with heavy dorsal shagreen (as heavy as on segments 2-7) (Fig. 26.793); caudolateral corners of anal lobes with an elongated spinelike projection, another located at base of distal third of the anal lobe ....Orthocladiinae (in part) *(Thalassosmittia,* tribe uncertain)	

Adults (Marine)

1.	Antenna of male with 11-13 flagellomeres, the terminal one longer, or nearly as long, as all other segments combined (e.g., Fig. 26.794); all tarsal segments cylindrical (Fig. 26.769); gonostylus of male directed rigidly backward from gonocoxite (Fig. 26.780; compare this to Fig. 26.781) ....Chironominae (go to freshwater key)	
1'.	Not with above combination of characters	2
2(1').	Antenna of male with 14 flagellomeres, the terminal one nearly as long as segments 1-13 combined (Fig. 26.794); all tarsal segments cylindrical (Fig. 26.769); gonostylus of male directed medially and anteriorly (Fig. 26.781) ....Orthocladiinae—*"Orthocladiini"* (in part) *(Halocladius)*	
2'.	Antenna of male with 6-13 flagellomeres, the terminal one never longer than combined lengths of the preceding 4 (Fig. 26.795); 4th tarsal segment often cordiform (heart-shaped) (Fig. 26.767); 5th tarsal segment often trilobed at apex (Fig. 26.796)	3
3(2').	Antenna of male with 8, 12, or 13 flagellomeres; all tarsal segments cylindrical (Fig. 26.769) ....Orthocladiinae (in part) *(Thalassosmittia,* tribe uncertain)	
3'.	Antenna of male with 6, 7, or 11 flagellomeres (e.g., Fig. 26.795); 4th tarsal segment often cordiform (Fig. 26.767); 5th tarsal segment often trilobed at apex (Fig. 26.796)	4
4(3').	Fourth and 5th tarsal segments cylindrical (Fig. 26.769) ....Orthocladiinae-Clunionini	
4'.	Either 4th tarsal segment cordiform (Fig. 26.767), or 5th tarsal segment trilobed at apex (Fig. 26.796) ....Telmatogetoninae	

ADDITIONAL TAXONOMIC REFERENCES*

General

Johannsen (1905, 1937); Goetghebuer and Lenz (1936-1950); Townes (1945); Sublette and Sublette (1965); Frommer (1967); Hamilton et al. (1969); Bryce and Hobart (1972); Mason (1973); Fittkau et al. (1976); Hashimoto (1976); Pinder (1978); Saether (1979, 1980); Murray (1980); Hoffrichter and Reiss (1981); Oliver (1981); Simpson (1982); Wilson and McGill (1982); Ashe (1983); Wiederholm (1983, 1986, 1989); Oliver et al. (1990); Armitage et al. (1995).

Regional faunas

California: Sublette (1960, 1964b).
Canada: Webb (1969); Oliver et al. (1978); Oliver and Roussel (1983).
Carolinas: Webb and Brigham (1982).
Connecticut: Johannsen and Townes (1952).
Eastern United States: Roback (1976, 1977, 1978, 1980, 1981, 1985, 1986a, 1986b, 1987).
Florida: Beck and Beck (1966, 1969); Epler (1992).
Kansas: Ferrington (1981, 1982, 1983a,b); Goldhammer et al. (1992).
Louisiana: Sublette (1964a).
New York: Johannsen (1905); Simpson and Bode (1980).
Northeastern United States: Bode (1990).
Pennsylvania (Philadelphia area): Roback (1957).
Puerto Rico: Ferrington et al. (1992).
Southeastern United States: Beck (1968, 1975); Hudson et al. (1990).

Taxonomic treatments at the subfamily, tribe, and generic levels

(L=larvae; P=pupae; A=adults)

Chironomini: Roback (1964)-L, P; Maschwitz (1976)-L, P; Saether (1977, 1983a)-L, P; Borkent (1984)-L, P, A; Bosel (1985)-L, P, A; Reiss and Sublette (1985)-L, P; Epler (1987, 1988a,b)-L, P,A; Grodhaus (1987)-L, P,A.
Clunionini: Wirth (1949)-A; Goetghebuer and Lenz (1950)-A; Lenz (1950)-L, A.
Diamesinae: Saether (1969)-L, P, A; Hansen and Cook (1976)-A; Oliver and Roussel (1982)-L.
Orthocladiinae: Brundin (1956)-L, P, A; Cranston (1982)-L; Saether (1969, 1975, 1976, 1977, 1981, 1982, 1983b, 1985b)-L, P, A; Saether and Sublette (1983)-L, P; Cranston and Saether (1986)-L,P, A; Hirvenoja (1973)-L, P, A; Jackson (1977)-L,P,A; Soponis (1977)-L, P,A; LeSage and Harrison (1980)-L,P; Oliver (1982, 1985,)-L,P; Bode (1983)-L; Boesel (1983)-A;Simpson et al. 1982-L, P; Oliver and Roussel (1983b)-L, P; Saether and Ferrington (1993)-L, P.
Podonominae: Brundin (1966)-L, P, A; Saether (1969)-L, P, A.
Prodiamesinae: Saether (1985a)-L, P.
Pseudochironomini: Saether (1977)-L, P.
Tanypodinae: Goetghebuer and Lenz (1939)-A; Fittkau (1962)-P, A; Roback (1971)-A.; Roback (1974, 1985, 1987)-L, P; Murray and Fittkau (1985)-L, P; Epler (1986)-L; Roback and Rutter (1988)-L, P.
Tanytarsini: Bilyj and Davies (1989)-L, P.
Telmatogetoninae: Wirth (1947)-A.

*This is a very limited list and includes only publications of general interest and those specifically treating the Nearctic fauna.

Table 26A. Summary of ecological and distributional data for *Chironomidae (Diptera)*. (For definition of terms see Tables 6A-6C; table prepared by M. B. Berg, R. W. Merritt, K. W. Cummins, W. P. Coffman, and L. Ferrington.)

Taxa* (number of species in parentheses)†	Habitat	Habit	Trophic Relationships	North American Distribution	Ecological References‡
Chironomidae (=Tendipedidae) (midges)	Essentially all types of aquatic habitats, including marine, springs, tree holes	Generally burrowers (most are tube builders)	Generally of two types: (1) collectors—gatherers and filterers (2) Predators (engulfers and piercers)	Widespread	1985, 1986, 2459, 2970, 3076, 3373, 2213, 4303, 4511, 135, 534, 583, 2441, 3143, 3144, 1282, 2959, 3909, 119, 2316
Telmatogetoninae	Generally marine (tidal pools, salt marshes, estuaries)			East and West Coasts	
Telmatogeton(A)	Beach zone—marine (rocks of intertidal)	Clingers (tube builders)	Scrapers, shredders—herbivores (chewers—macroalgae), collectors—gatherers	West Coast, Southeast Coast	613, 2884, 3545, 4511
Thalassomya(A)	Marine		Collectors—gatherers, scrapers	Florida Coast	2884, 2762, 2884, 4511, 3368, 3409
Tanypodinae (= Pelopiinae)	All types of lentic and lotic habitats	Generally sprawlers—swimmers (very active predators, not tube builders)	Generally predators (engulfers and piercers)	Widespread	1985, 2970, 3076, 4511, 275, 534, 2959
Coelotanypodini	Generally lentic—littoral	Generally burrowers	Generally predators (engulfers)	Eastern United States, Canada south to California	
Clinotanypus(A)	Lentic—littoral, lotic—depositional	Burrowers	Predators (engulfers; Oligochaeta, Ostracoda, Chironomidae)	Eastern United States, Canada south to California	1266, 2775, 3376
Coelotanypus(A)	Lentic—littoral	Burrowers	Predators, (engulfers; Oligochaeta, Cladocera, Chironomidae)	East	258, 2775, 3376
Macropelopiini			Generally predators (engulfers and piercers)	Widespread	3371
Alotanypus(A)	Lotic	Burrowers—sprawlers		Widespread	
Apsectrotanypus(A)	Lotic	Burrowers—sprawlers	Predators (engulfers; Chironomidae)	Widespread	3371
Bethbilbeckia (A)	Lotic			Florida (Ohio)	
Brundiniella(A)	Lotic	Burrowers—sprawlers	Predators, (engulfers; Protozoa, Cladocera, Ostracoda, Tardigrada, Hydracarina, Chironomidae)	Widespread	3371
Derotanypus(A)	Lotic and lentic			North and west	
Fittkauimyia(A)	Lotic and lentic			Florida	
Macropelopia (A)	Lotic—erosional, lentic—littoral	Sprawlers	Predators, (engulfers); Protozoa, Cladocera, Ostracoda, Crustacea, Ceratopogonidae, Chironomidae)	North	1715, 3371
Psectrotanypus(B)	Lotic—depositional	Sprawlers	Predators, (engulfers); Protozoa, Cladocera, Ostracoda, Crustacea, Trichoptera, Chironomidae)	Widespread	865, 1715, 3371

* The North American chironomid fauna includes many undescribed species and genera, especially in the subfamilies Orthocladiinae and Chironominae. New taxa are to be expected from unusual habitats and from areas of North America that have been poorly collected.

† The number of species per genus has been estimated in terms of probable range A = 1-5 species; B = 6-19 species; C = 20 or more species. The range for any genus was derived by a subjective process based on known North American and European diversities and material in the collection of W. P. Coffman.

‡ Emphasis on trophic relationships.

Table 26A. Continued

Taxa* (number of species in parentheses)†	Habitat	Habit	Trophic Relationships	North American Distribution	Ecological References‡
Radotanypus(A)	Lotic			West	
Procladiini					
Djalmabatista(A)	Lotic	Sprawlers	Predators (engulfers)	Eastern United States	3371
Procladius(C)	Lentic—profundal (some littoral), lotic—depositional	Sprawlers	Predators (engulfers; Protozoa, microcrustacea, Ephemeroptera, Ceratopogonidae, Gastrotricha), collectors—gatherers (winter and early instars)	Widespread (especially North)	115, 150, 211, 313, 609, 669, 671, 811, 840, 865, 1154, 1266, 2038, 2039, 2043, 2347, 2425, 2730, 2747, 2755, 2775, 3187, 3376, 4005, 4013, 258, 1695, 4028, 4511, 3563, 3627, 4136
Natarsiini					
Natarsia(A)	Lotic—erosional	Sprawlers	Predators (engulfers; Cladocera, Ostracoda, Copepoda, Ceratopogonidae)	Widespread	669
Pentaneurini	Generally lotic—erosional and lentic—littoral	Generally sprawlers	Generally predators (engulfers and piercers)	Widespread	
Ablabesmyia(C)	Lentic—littoral, lotic—erosional and depositional	Sprawlers	Predators (engulfers and piercers) (Rotifera, microcrustacea, Chironomidae), collectors—gatherers (early instars)	Widespread	115, 609, 840, 2425, 258, 781, 2730, 3376, 3563, 3707
Arctopelopia(A)	Lotic—erosional, lentic—littoral	Sprawlers	Predators (engulfers)	Rocky Mountains	781
Cantopelopia(A)	Lentic			North (East only?)	
Conchapelopia(B)	Lotic—erosional, lentic—littoral	Sprawlers	Predators (engulfers and piercers) (Chironomidae, Trichoptera, Ephemeroptera)	Widespread	
Denopelopia(A)	Lentic—drainage ditch			Florida	
Guttipelopia(A)	Lentic—littoral	Sprawlers	Predators (engulfers)	Widespread	
Hayesomyia(A)	Lotic			Widespread	
Helopelopia(A)	Lotic			Widespread (especially North and East)	
Hudsonimyia(A)	Lotic—erosional (hygropetric)	Sprawlers		East	
Krenopelopia(A)	Lotic—erosional	Sprawlers	Predators (engulfers)	Primarily North	
Labrundinia(B)	Lotic—erosional, lentic—littoral	Sprawlers	Predators (engulfers and piercers, Oligochaeta, Cladocera, Ostracoda)	Widespread	3376
Larsia(B)	Lotic—erosional, lentic—littoral	Sprawlers	Predators (engulfers)	Widespread	
Meropelopia(A)	Lotic			Widespread	
Monopelopia(A)	Lentic—littoral, lotic?	Sprawlers	Predators (engulfers)	Widespread	
Nilotanypus(A)	Lotic—erosional	Sprawlers		Widespread	

* The North American chironomid fauna includes many undescribed species and genera, especially in the subfamilies Orthocladiinae and Chironominae. New taxa are to be expected from unusual habitats and from areas of North America that have been poorly collected.
† The number of species per genus has been estimated in terms of probable range A = 1-5 species; B = 6-19 species; C = 20 or more species. The range for any genus was derived by a subjective process based on known North American and European diversities and material in the collection of W. P. Coffman.
‡ Emphasis on trophic relationships.

Table 26A. Continued

Taxa* (number of species in parentheses)[†]	Habitat	Habit	Trophic Relationships	North American Distribution	Ecological References[‡]
Paramerina(B)	Lotic—erosional	Sprawlers		Widespread	
Pentaneura(A)	Lentic—littoral (vascular hydrophytes and detritus), lotic—erosional and depositional	Sprawlers	Predators (engulfers and piercers, Chironomidae), collectors—gatherers (diatoms and detritus)	Eastern North America (especially Southeast)	115, 609, 669, 840, 1266, 1696, 2755, 3376, 3078
Reomyia(A)	Lotic—including springs			North	
Rheopelopia(A)	Lotic—erosional	Sprawlers	Predators (engulfers and piercers)	Widespread	
Telopelopia(A)	Lotic-erosional (sandy bottom rivers)		Predators (engulfers and piercers)	Midwest	
Thienemannimyia(A)	Lotic—erosional, lentic—littoral	Sprawlers	Predators (engulfers and piercers, Protozoa, Cladocera, Ostracoda, Chironomidae)	Widespread	669, 671, 2060, 3376, 258, 3707, 4018, 3388, 3563
Trissopelopia(A)	Lotic—erosional, lentic—littoral	Sprawlers	Predators (engulfers)	Primarily North	1715
Xenopelopia(A)	Lentic			California	
Zavrelimyia(B)	Lotic—erosional, lentic—littoral	Sprawlers	Predators (engulfers and piercers, Oligochaeta, Ostracoda, Chironomidae)	Widespread	1715, 3376, 3707
Tanypodini			Generally predators (engulfers and piercers)	Widespread	
Tanypus(B) (=*Pelopia*)	Lentic littoral	Sprawlers	Predators (engulfers and piercers), collectors—gatherers (diatoms, filamentous green algae, detritus)	Widespread	115, 151, 670, 865, 1154, 2038, 2039, 2043, 2290, 2347, 3375, 3953, 4013
Podonominae	Generally lotic—erosional and depositional and lentic (at high altitudes and latitudes)	Generally burrowers (tube builders)		Primarily North	534, 4511
Boreochlini	Generally lotic—erosional, lentic—littoral	Generally sprawlers	Generally collectors—gatherers, scrapers		4511
Boreochlus(A)	Lotic—erosional	Sprawlers	Collectors—gatherers, scrapers	North	
Lasiodiamesa(A)	Lentic—littoral (including bogs)	Sprawlers	Collectors—gatherers, scrapers		
Paraboreochlus(A)	Lotic—erosional	Sprawlers	Collectors—gatherers, scrapers	Appalachians	672
Trichotanypus(A)	Lotic—erosional	Sprawlers	Collectors—gatherers, scrapers	North	
Podonomini *Parochlus*(A)	Lotic—erosional	Sprawlers	Collectors—gatherers, scrapers	Primarily North	
Diamesinae	Generally lotic erosional and depositional (cold streams), lentic—erosional (oligotrophic lakes)	Generally—clingers and some burrowers (tube builders)	Generally collectors—gatherers, scrapers	Primarily North	1985, 2970, 3076, 534, 4511

* The North American chironomid fauna includes many undescribed species and genera, especially in the subfamilies Orthocladiinae and Chironominae. New taxa are to be expected from unusual habitats and from areas of North America that have been poorly collected.
[†] The number of species per genus has been estimated in terms of probable range A = 1-5 species; B = 6-19 species; C = 20 or more species. The range for any genus was derived by a subjective process based on known North American and European diversities and material in the collection of W. P. Coffman.
[‡] Emphasis on trophic relationships.

Table 26A. Continued

Taxa* (number of species in parentheses)[†]	Habitat	Habit	Trophic Relationships	North American Distribution	Ecological References[‡]
Boreoheptagyini *Boreoheptagyia*(A)	Lotic—erosional (cold, fast streams)	Sprawlers—clingers	Collectors—gatherers, scrapers?	North and mountains	
Diamesini					1470, 2060, 2061, 2627, 4511
Arctodiamesa(A)	Lotic			Alaska	
Diamesa(C)	Lotic—erosional	Sprawlers	Collectors—gatherers, scrapers?	Widespread (in uplands and mountains)	277, 1537, 4236, 276, 3078
Lappodiamesa(A)	Lotic?			Ohio	
Pagastia(A)	Lotic			North	277,. 276
Potthastia(B)	Lotic—erosional	Sprawlers	Collectors—gatherers, scrapers	Widespread	4302
Pseudodiamesa(A)	Lentic—littoral (erosional)	Sprawlers	Collectors—gatherers	Probably primarily North	858, 1470
Pseudokiefferiella(A)	Lotic—erosional (small mountain springs)	Sprawlers	Collectors—gatherers?	West, North	2061, 4236
Sympotthastia(A)	Lotic—erosional	Sprawlers	Collectors—gatherers, scrapers?	Widespread	222
Syndiamesa(A)	Lotic—erosional	Sprawlers		North	
Protanypini *Protanypus*(A)	Lentic—profundal	Burrowers?	Collectors—gatherers	Widespread (primarily North)	
Prodiamesinae	Generally lotic erosional	Generally sprawlers	Generally collectors—gatherers		534
Compteromesa(A)	Lotic—seeps			Southeast	
Monodiamesa(A)	Lotic—erosional	Sprawlers	Collectors—gatherers?	Widespread	
Odontomesa(A)	Lotic—erosional	Sprawlers	Collectors—gatherers?	Widespread	2730
Prodiamesa(A)	Lotic—erosional and depositional (detritus)	Burrowers—sprawlers	Collectors—gatherers	Widespread	1266, 1962, 1963, 2290, 2425, 3581
Orthocladiinae (=Hydrobaeninae)	Primarily lotic, but with many lentic representatives (especially oligotrophic lakes)	Generally burrowers (tube builders)	Generally collectors—gatherers, scrapers	Widespread, particularly North	603, 669, 671, 745, 1985, 2970, 3076, 4511, 275, 534
Clunionini					
Clunio (A)	Beach zone—marine (rocky shore)	Clingers (sand tube builders)	Collectors—gatherers, scrapers	East and West Coasts	613, 2884, 2967, 4511
Eretmoptera(A)	Beach zone—marine (rocks of intertidal)	Clingers (tube builders)	Scrapers, shredders—herbivores (chewers—macroalgae)	West Coast	4511
Tethymyia(A)	Beach zone—marine (rocks of intertidal)	Clingers (tube builders)	Scrapers, shredders—herbivores (chewers—macroalgae)	West Coast	4511
"Corynoneurini"	Generally lotic—erosional and depositional, lentic—littoral	Generally sprawlers	Generally collectors—gatherers		
Corynoneura(B)	Lotic—depositional (aquatic hydrophytes), lentic—littoral (some instar 1 planktonic)	Sprawlers	Collectors—gatherers	Widespread	669, 671, 840, 851, 258, 2425, 3372, 4302, 781, 3078

* The North American chironomid fauna includes many undescribed species and genera, especially in the subfamilies Orthocladiinae and Chironominae. New taxa are to be expected from unusual habitats and from areas of North America that have been poorly collected.
† The number of species per genus has been estimated in terms of probable range A = 1-5 species; B = 6-19 species; C = 20 or more species. The range for any genus was derived by a subjective process based on known North American and European diversities and material in the collection of W. P. Coffman.
‡ Emphasis on trophic relationships.

Table 26A. Continued

Taxa* (number of species in parentheses)†	Habitat	Habit	Trophic Relationships	North American Distribution	Ecological References‡
Thienemanniella(B)	Lotic—erosional and depositional, lentic—littoral	Sprawlers	Collectors—gatherers	Widespread	258, 669, 671, 4302, 374, 3078
"Orthocladiini" and "Metriocnemini"‡‡	Wide range of lotic and lentic habitats	Wide range of habits	Generally collectors, but other functional groups also		
Abiskomyia(A)	Lentic			Far North	
Acamptocladius(A)	Lentic—littoral	In colonial algae	Collectors—gatherers (chewers—macroalgae)	North	3505
Acricotopus(A)	Lotic—erosional, lentic—littoral	Sprawlers		North	
Antillocladius(A)	Flood plain soils—seeps			East	
Apometriocnemus(A)				Tennessee	
Baeoctenus(A)	Lentic—littoral			Manitoba, New Brunswick	
Brillia(A)	Lotic—erosional and depositional (detritus)	Burrowers (miners) in rotted wood, sprawlers (in detritus)	Shredders—detritivores (chewers and miners), collectors—gatherers	Widespread	669, 671, 980, 3082, 90, 2974, 3874, 3335, 4302, §
Bryophaenocladius(B)	Lotic—erosional	Sprawlers		Widespread	
Camptocladius(A)	Terrestrial (dung)			Widespread?	
Cardiocladius(A)	Lotic—erosional	Burrowers (loose tube construction), clingers (tube makers)	Engulfers (predators of black fly larvae)	Widespread	810, 2975, 4511
Chaetocladius(B)	Lotic—erosional	Sprawlers	Collectors—gatherers	Widespread	90, 746, 2061
Chasmatonotus(B)					
Compterosmittia(A)	Lotic—seeps			South Carolina	
Cricotopus(C)	Lentic—vascular hydrophytes (and algal mats, sediments, detritus), lotic—erosional and depositional (some instar 1 planktonic)	Clingers (tube builders), burrowers (miners and tube builders)	Shredders—herbivores (miners and chewers), collectors—gatherers (detritus and algae)	Widespread	151, 265, 432, 811, 840, 851, 1985, 2060, 2425, 2627, 2794, 3757, 4499, 258, 277, 4236, 4511, 3078, 3522, 4242, 781, 926, 1696, 2047, 2432, 2433, 2931, 4227, 4228, 1173, 3563, 4416, 276
Diplocladius(A)	Lotic—erosional	Sprawlers	Collectors—gatherers?	Widespread	
Diplosmittia(A)	Lotic?			Michigan	
Doithrix(B)	Lotic—springs and seeps			Widespread	
Doncricotopus(A)				Northwest Territories	
Epoicocladius(B)	Lotic—erosional and depositional		Collectors—gatherers commensals—parasites on lotic insects	Widespread	
Eukiefferiella(C)	Lotic—erosional, lentic—littoral	Sprawlers	Collectors—gatherers, scrapers, predators (engulfers of chironomid eggs and larvae)	Widespread	669, 671, 891, 1470, 277, 2060, 3867, 4005, 2213, 4227, 4302, 276, 3078, 3563
Euryhapsis(B)	Lotic			Western	

* The North American chironomid fauna includes many undescribed species and genera, especially in the subfamilies Orthocladiinae and Chironominae. New taxa are to be expected from unusual habitats and from areas of North America that have been poorly collected.
† The number of species per genus has been estimated in terms of probable range A = 1-5 species; B = 6-19 species; C = 20 or more species. The range for any genus was derived by a subjective process based on known North American and European diversities and material in the collection of W. P. Coffman.
‡ Emphasis on trophic relationships.
‡‡ Since the relationships among genera are poorly known in these tentative tribes, they have not been separated.
§ Unpublished data, R. H. King, G. M. Ward and K. W. Cummins, Kellogg Biological Station.

Table 26A. Continued

Taxa* (number of species in parentheses)†	Habitat	Habit	Trophic Relationships	North American Distribution	Ecological References‡
Georthocladius(B)	Lotic—erosional			North and East	
Gymnometriocnemus(A)	Semiaquatic (lentic—margins)	Sprawlers		North	
Halocladius(A)	Beach zone—marine (in *Fucus*)	Clingers (sand tube builders)	Collectors—gatherers, scrapers	Northeast Coast	2884
Heleniella(A)	Lotic—erosional	Sprawlers		North	1169, 2060
Heterotanytarsus(A)	Lotic—erosional and lentic-littoral			North	
Heterotrissocladius(B)	Lotic—erosional, lentic—littoral and profundal	Sprawlers, burrowers	Collectors, gatherers, scrapers?	Widespread	858, 2604, 2747
Hydrobaenus(B)	Lotic—erosional, lentic—littoral	Sprawlers	Scrapers, collectors—gatherers	Widespread	603, 671, 858, 3078
Krenosmittia(B)	Lotic—erosional (hyporheic)	Sprawlers	Collectors—gatherers	Widespread	350, 3078, 4241
Lapposmittia(A)	High latitude pools			Not yet known from North America	
Limnophyes(B)	Lentic—littoral (macroalgae), lotic—depositional	Sprawlers	Collectors—gatherers?	Widespread	90, 2794, 4236
Lipurometriocnemus(A)	Semiaquatic?			South Carolina and Yukon Territory	
Lopescladius(B)	Lotic (often hyporheic)	Sprawlers	Collectors—gatherers	Widespread	1436, 3078, 4241
Mesocricotopus(A)	Lotic—erosional	Sprawlers?	Collectors—gatherers	North	
Mesosmittia(B)	Lotic			Widespread	
Metriocnemus(B)	Lotic—erosional and depositional (detritus), lentic—littoral (oligotrophic) (2 pitcher plant species)	Burrowers, sprawlers	Collectors—gatherers, predators (engulfers)	Widespread	669, 811, 2019, 2114, 2290, 4371
Nanocladius(B)	Lotic—erosional, lentic—littoral	Sprawlers	Collectors—gatherers	Widespread	669, 671, 4302
Oliveridia(A)	Lentic			Arctic	1176
Oreadomyia(A)				Alberta	
Orthocladius(C)	Lentic—littoral (erosional) and profundal, lotic—erosional	Sprawlers, burrowers (detritus, diatoms, filamentous algae)	Collectors—gatherers	Widespread	669, 858, 1470, 2290, 276, 2627, 2794, 4005, 90, 277, 746, 4302, 1537, 1701, 4227, 3078, 4234, 4242
Parachaetocladius(A)	Lotic—erosional	Sprawlers	Collectors—gatherers	North	1169
Paracladius(A)	Lentic—littoral	Sprawlers	Collectors—gatherers	Wyoming, Montana	
Paracricotopus(A)	Lotic—erosional	Sprawlers	Collectors—gatherers	Widespread	
Parakiefferiella(B)	Lotic—erosional, lentic—littoral	Sprawlers	Collectors—gatherers	Widespread	258, 3078, 4353, 3563, 4241
Paralimnophyes(A)	Lentic—pools			Alaska and Arctic Canada	
Parametriocnemus(B)	Lotic—erosional and depositional	Sprawlers	Collectors—gatherers	Widespread	276, 671, 3563 3078

* The North American chironomid fauna includes many undescribed species and genera, especially in the subfamilies Orthocladiinae and Chironominae. New taxa are to be expected from unusual habitats and from areas of North America that have been poorly collected.
† The number of species per genus has been estimated in terms of probable range A = 1-5 species; B = 6-19 species; C = 20 or more species. The range for any genus was derived by a subjective process based on known North American and European diversities and material in the collection of W. P. Coffman.
‡ Emphasis on trophic relationships.

Table 26A. Continued

Taxa* (number of species in parentheses)†	Habitat	Habit	Trophic Relationships	North American Distribution	Ecological References‡
Paraphaeno-cladius(B)	Lotic—erosional and depositional	Sprawlers?	Collectors—gatherers	Primarily North	
Parasmittia(A)	Lotic?			Nova Scotia and Pennsylvania	
Paratrichocladius(A)	Lotic—erosional	Sprawlers	Collectors—gatherers	Widespread	
Paratrissocladius(A)	Lotic			Northeast	
Parorthocladius(A)	Lotic—erosional	Sprawlers	Collectors—gatherers	West	4234
Platysmittia(A)	Lotic—springs?			Tennessee and Manitoba	
Plhudsonia(A)	Lotic—spring seeps and pools			East	
Psectrocladius(C)	Lentic—littoral, lotic—depositional (macroalgae)	Sprawlers, burrowers	Collectors—gatherers, shredders—herbivores	Widespread	669, 840, 1266, 2794, 374, 1537, 4227, 1173, 3563
Pseudortho-cladius(B)	Lotic—erosional	Sprawlers	Collectors—gatherers	East	
Pseudosmittia(B)	Lotic			Widespread	1175
Psilometrio-cnemus(A)	Lotic—erosional	Sprawlers	Collectors—gatherers	Primarily North	1169
Rheocricotopus(B)	Lotic—erosional	Sprawlers	Collectors—gatherers, shredders—herbivores (chewers—macro algae), predators (engulfers)	Widespread	669, 671, 2290, 4302
Rheosmittia(A)	Lotic—sandy substrates			Widespread	3749
Saetheriella(A)	Lotic—springs?			Tennessee	
Smittia(B)	Semiaquatic—lentic margins	Burrowers?	Collectors—gatherers	Widespread	2794
Stackelbergina(A)	Lentic—ponds			Northern Quebec	
Stilocladius(A)	Lotic			Widespread	
Sublettiella(A)	Lotic?			South Carolina	
Symbiocladius(A)	Lotic—erosional	On mayfly and stonefly nymphs	Parasites	Widespread	643, 3076, 3372, 4511
Synorthocladius(A)			Collectors—gatherers, scrapers?		2425, 4302
Thalassosmittia(A) (=*Saunderia*)	Beach zone—marine (rocks of intertidal)	Clingers (tube builders)	Scrapers, shredders—herbivores (chewers—macroalgae), collectors—gatherers	Pacific Northwest coast	2762, 2884, 3409, 3368
Thienemannia(A)	Lotic—springs in moss			Tennessee	
Tokunagaia(A)	Lotic—erosional	Sprawlers	Collectors—gatherers	Widespread	
Trichochilus(A)	Lotic?			Pennsylvania	
Tvetenia(B)	Lotic	Sprawlers	Collectors—gatherers	Widespread	277, 276, 4236, 3078, 4234
Unniella(A)	Lotic			East (especially Southeast)	
Vivacricotopus(A)	Lotic			Northwest	
Xylotopus(A)	Lotic depositional	Burrowers in wood		East	2059, 2976
Zalutschia(B)	Lentic—littoral, lotic			Widespread	781

* The North American chironomid fauna includes many undescribed species and genera, especially in the subfamilies Orthocladiinae and Chironominae. New taxa are to be expected from unusual habitats and from areas of North America that have been poorly collected.

† The number of species per genus has been estimated in terms of probable range A = 1-5 species; B = 6-19 species; C = 20 or more species. The range for any genus was derived by a subjective process based on known North American and European diversities and material in the collection of W. P. Coffman.

‡ Emphasis on trophic relationships.

Table 26A. Continued

Taxa* (number of species in parentheses)†	Habitat	Habit	Trophic Relationships	North American Distribution	Ecological References‡
Chironominae	Lentic—littoral and profundal, lotic depositional and erosional	Generally burrowers and clingers	Generally collectors—gatherers and collectors—filterers	Widespread	274, 1986, 2970, 3076, 275, 534, 4511
Chironomini	Generally lentic littoral and profundal, lotic depositional (some instar 1 planktonic)	Generally burrowers	Generally collectors—gatherers	Widespread	2970, 3076, 4511
Acalcarella(A)	Lotic—sandy areas of rivers			North (perhaps upper Midwest)	
Apedilum(A)	Lentic-lotic, submerged vegetation			Widespread	
Asheum(A)	Lentic			Southeast	
Axarus(A)	Lotic depositional, lentic	Sprawlers, burrowers	Collectors gatherers	Widespread	1170
Beardius(A)	Lentic			Texas-Florida	
Beckidia(A)	Lotic			West	
Chernovskiia(A)	Lotic—sandy areas of rivers			Widespread	
Chironomus(C) (=*Tendipes*)	Lentic—littoral and profundal, lotic depositional	Burrowers (tube builders)	Collectors—gatherers (a few filterers), shredders—herbivores (miners)	Widespread	265, 396, 670, 811, 838, 840, 851, 1154, 1266, 1725, 1791, 1986, 2017, 2044, 2159, 2290, 2346, 2347, 2618, 2626, 2794, 3187, 3216, 3282, 3342, 3413, 4005, 4206, 4210, 349, 1087, 1988, 4511, 2010, 2011, 2014, 2331, 3251, 3253, 276, 2398, 4267, 1173, 3536, 4302, 2966, 4242
Cladopelma(B)	Lentic—littoral	Burrowers	Collectors—gatherers	Widespread	2730, 2747, 4013
Cryptochironomus(C)	Lentic—littoral and profundal, lotic depositional	Sprawlers, burrowers	Predators (engulfers of Protozoa, microcrustacea, Chironomidae and piercers of Oligochaeta)	Widespread	115, 811, 840, 1043, 2038, 2039, 2347, 2425, 2730, 2775, 2794, 3982, 4005, 258, 3581, 4013, 4511, 2966, 3563, 4227
Cryptotendipes(B)	Lentic littoral, lotic depositional	Sprawlers		Widespread	
Cyphomella(A)	Lentic and lotic (sandy rivers)	Burrowers	Collectors—gatherers	Widespread	
Demijerea(A)	Lentic			Widespread	
Demicryptochironomus(A)	Lotic depositions	Burrowers	Collectors—gatherers	Widespread	258
Dicrotendipes(C) (=*Limnochironomus*)	Lentic—littoral (wide range of microhabitats)	Burrowers	Collectors—gatherers and filterers, scrapers?	Widespread	115, 116, 669, 671, 2425, 2747, 2794, 2948, 3187, 3563
Einfeldia(B)	Lentic—littoral and profundal	Burrowers	Collectors—gatherers	Widespread	837

* The North American chironomid fauna includes many undescribed species and genera, especially in the subfamilies Orthocladiinae and Chironominae. New taxa are to be expected from unusual habitats and from areas of North America that have been poorly collected.

† The number of species per genus has been estimated in terms of probable range A = 1-5 species; B = 6-19 species; C = 20 or more species. The range for any genus was derived by a subjective process based on known North American and European diversities and material in the collection of W. P. Coffman.

‡ Emphasis on trophic relationships.

Table 26A. Continued

Taxa* (number of species in parentheses)†	Habitat	Habit	Trophic Relationships	North American Distribution	Ecological References‡
Endochironomus(A)	Lentic—littoral (algal mats) and profundal (instar 1 planktonic)	Clingers (tube builders)	Shredders—herbivores (miners and chewers—macroalgae), collectors—filterers and gatherers	Widespread	838, 851, 1926, 2730, 2047, 2165, 2794, 4210, 2966, 4099
Endotribelos(A)	Lentic			California and Florida	
Gillotia(A)	Lotic (sandy rivers)	Burrowers	Collectors—gatherers	Midwest	
Glyptotendipes(B)	Lentic—littoral and profundal, lotic—depositional (rarely) (some instar 1 planktonic)	Burrowers (miners and tube builders), clingers (net spinners)	Shredders—herbivores (miners and chewers—filamentous algae), collectors—filterers and gatherers	Widespread	265, 494, 811, 840, 851, 1154, 2044, 2092, 2425, 2730, 2747, 2794, 3187, 2966, 3216, 4005, 4210, 349, 3251, 3253, 3563, 4099
Goeldichironomus(B)	Lentic (small stagnant ponds)	Burrowers	Collectors—gatherers	Widespread	313, 1173
Graceus(A)	Lentic—littoral, lotic	Sprawlers		Ontario	
Harnischia(A)	Lentic—vascular hydrophytes (submerged zone)	Climbers—clingers	Collectors—gatherers, scrapers?	Widespread	670, 1154, 2618, 2730, 258, 2775, 3563, 3581
Hyporhygma(A)	Lentic (vascular hydrophytes)	Burrowers	Shredders (miners in stems and petioles)		
Kiefferulus(A)	Lentic	Burrowers	Collectors—gatherers	Widespread	2826, 3009, 4210, 4511, 1491
Kloosia(A)	Lotic?			South Carolina	
Lauterborniella(A)	Lentic—littoral and profundal	Climbers—sprawlers—clingers, burrowers (portable sand tube builders)	Collectors—gatherers	Widespread	858
Lipiniella(A)	Lentic			Northwest Territories	
Microchironomus(A)	Lentic (sandy littoral)	Burrowers	Collector—gatherers	Widespread	
Microtendipes(B)	Lentic—littoral, lotic—depositional	Clingers (net spinners)	Collectors—filterers and gatherers	Widespread	265, 669, 671, 2794, 2966, 3187, 4210, 4227
Nilothauma(A)	Lotic—depositional			Widespread	
Omisus(A)				Widespread	
Pagastiella(A)	Lentic—littoral and profundal			Widespread	
Parachironomus(C)	Lentic—littoral	Sprawlers (some parasites in Mollusca)	Predators (engulfers), collectors—gatherers, parasites	Widespread	2019, 2320, 2425, 2730, 2794, 2966, 3187
Paracladopelma(B)	Lentic—littoral, lotic—depositional	Sprawlers		Widespread	
Paralauterborniella(A)	Lentic—vascular hydrophytes	Clingers (tube builders on plants)	Collectors—gatherers	Widespread	
Paratendipes(B)	Lotic—depositional, lentic—littoral	Burrowers (tube builders)	Collectors—gatherers	Widespread	1168, 1175, 4229, 4230, 1436
Phaenopsectra(B)	Lentic—littoral	Clingers (tube builders)	Scrapers, collectors—gatherers (and filterers?)	Widespread	669, 671, 840, 2425, 3563

* The North American chironomid fauna includes many undescribed species and genera, especially in the subfamilies Orthocladiinae and Chironominae. New taxa are to be expected from unusual habitats and from areas of North America that have been poorly collected.

† The number of species per genus has been estimated in terms of probable range A = 1-5 species; B = 6-19 species; C = 20 or more species. The range for any genus was derived by a subjective process based on known North American and European diversities and material in the collection of W. P. Coffman.

‡ Emphasis on trophic relationships.

Table 26A. Continued

Taxa* (number of species in parentheses)†	Habitat	Habit	Trophic Relationships	North American Distribution	Ecological References‡
Polypedilum(C)	Lentic—vascular hydrophytes (floating zone)	Climbers, clingers	Shredders—herbivores (miners), collectors—gatherers (and filterers?), predators (engulfers)	Widespread	265, 669, 671, 811, 2042, 2044, 2618, 2747, 2748, 2775, 2794, 4005, 90, 4013, 4210, 4511, 258, 277, 2151, 3581, 4099, 4227, 4353, 276, 3563, 4234
Robackia(A)	Lentic and lotic (sandy bottom)	Burrowers	Collectors—gatherers	Widespread	258, 3749, 4353, 3078
Saetheria(A)	Lentic and lotic (sandy bottom)	Burrowers	Collectors—gatherers	Widespread	2776
Sergentia(A)	Lentic			North	
Stelechomyia(A)	Lentic and lotic?	Wood		East and South	
Stenochironomus(B)	Lentic—vascular hydrophytes, lotic—wood	Burrowers (miners)	Collectors—gatherers shredders (wood gougers)	Widespread	90, 980, 3082, 4511, 258, 360, 1174, 2948
Stictochironomus(B)	Lotic—depositional (organic sediments)	Burrowers (tube makers)	Collectors—gatherers, shredders—herbivores (miners)	Widespread	811, 1778, 2093, 2730, 276, 3563
Synendotendipes(A)	Lentic and slow lotic			Western	
Tribelos(A)	Lentic and lotic (depositional)	Burrowers (wood miners)	Collectors—gatherers	Widespread	258, 4099
Xenochironomus(A)	Lotic (and lentic) in sponges	Burrowers (in sponges)	Predators (engulfers of sponges)	Widespread	258
Xestochironomus(A)	Lotic and lentic?			South	
Zavreliella(A)	Lentic			Widespread	
Pseudochironomini	Lentic and lotic—erosional (in algae) and depositional	Burrowers	Collectors—gatherers	Widespread	811, 2794
Pseudochironomus(B)	Lentic and lotic—erosional (in algae) and depositional	Burrowers	Collectors—gatherers	Widespread	811, 2794, 3190, 4236
Tanytarsini (=Calopsectrini)	Generally lotic—erosional and depositional, lentic—littoral	Generally burrowers or clingers (tube builders)	Generally collectors—filterers and gatherers	Widespread	2970, 3076, 4160, 4511
Cladotanytarsus(B)	Lentic—vascular hydrophytes, lotic—depositional		Collectors—gatherers and filterers	Widespread	115, 840, 2794, 3187, 258, 277, 276, 1168
Constempellina(A)	Lotic—erosional			Widespread	1171
Corynocera(A)	Lentic—littoral			Wyoming (Rocky Mountains), Northwest Canada	
Krenopsectra(A)	Lotic			North	
Micropsectra(C)	Lentic—littoral (including brackish), lotic—depositional	Climbers, sprawlers	Collectors—gatherers	Widespread	669, 671, 840, 2060, 2627, 2794, 2971, 4210, 3078, 4302
Neozavrelia(A)	Lotic—erosional			Eastern United States	
Nimbocera(A)	Lentic			South	
Parapsectra(A)	Lotic			North and East	

* The North American chironomid fauna includes many undescribed species and genera, especially in the subfamilies Orthocladiinae and Chironominae. New taxa are to be expected from unusual habitats and from areas of North America that have been poorly collected.

† The number of species per genus has been estimated in terms of probable range A = 1-5 species; B = 6-19 species; C = 20 or more species. The range for any genus was derived by a subjective process based on known North American and European diversities and material in the collection of W. P. Coffman.

‡ Emphasis on trophic relationships.

Table 26A. Continued

Taxa* (number of species in parentheses)†	Habitat	Habit	Trophic Relationships	North American Distribution	Ecological References‡
Paratanytarsus(B)	Lotic—erosional, lentic—littoral	Sprawlers		Widespread	781, 1476, 1696, 4302
Rheotanytarsus(B)	Lotic—erosional	Clingers (tube and net builders)	Collectors—filterers	Widespread	258, 669, 671, 4210, 277, 276, 2205
Skutzia(A)	Lotic?			Western Canada	
Stempellina(B)	Lotic—erosional, lentic—littoral	Climbers—sprawlers—clingers (portable, mineral tube builders)	Collectors—gatherers (detritus, algae)	Widespread	115, 669, 671, 4210
Stempellinella(A)	Lotic—erosional, lentic—littoral	Sprawlers		Widespread	
Sublettea(A)	Lotic			Widespread	
Tanytarsus(C) (= *Calopsectra*)	Lentic—vascular hydrophytes (floating zone) and profundal, lotic—erosional (some instar 1 planktonic)	Climbers, clingers (net spinners)	Collectors—filterers and gatherers, a few scrapers	Widespread	115, 151, 265, 586, 669, 671, 811, 840, 851, 1266, 2044, 2290, 2346, 2618, 2627, 2747, 2794, 3187, 258, 277, 276, 4210, 1778, 4227, 4267, 4302
Zavrelia(A)	Lotic	Climbers—sprawlers—clingers (portable, mineral tube builders)	Collectors—gatherers	Widespread	603, 669, 671, 4210

* The North American chironomid fauna includes many undescribed species and genera, especially in the subfamilies Orthocladiinae and Chironominae. New taxa are to be expected from unusual habitats and from areas of North America that have been poorly collected.

† The number of species per genus has been estimated in terms of probable range A = 1-5 species; B = 6-19 species; C = 20 or more species. The range for any genus was derived by a subjective process based on known North American and European diversities and material in the collection of W. P. Coffman.

‡ Emphasis on trophic relationships.

BIBLIOGRAPHY

1. Abbott, J. C., and K. W. Stewart. 1993. Male search behavior of the stonefly *Pteronarcella badia* (Hagen) (Plecoptera: Pteronarcyidae). J. Insect Beh. 6:467-481.
2. Abdelnur, O. M. 1968. The biology of some black flies (Diptera: Simuliidae) of Alberta. Quaest. Ent. 4:113-174.
3. Acorn, J. H. 1983. New distribution records of Odonata from Alberta, Canada. Not. Odonatol. 2:17-19.
4. Addicott, J. F. 1974. Predation and prey community structure: an experimental study of the effect of mosquito larvae on the protozoan communities of pitcher plants. Ecology 55:475-92
5. Adler, P. H., and K. C. Kim. 1986. The black flies of Pennsylvania (Simuliidae, Diptera). Bionomics, taxonomy, and distribution. Penn. State Univ. Agr. Exp. Sta. Bull. 856:1-88.
6. Adler, P. H., R. W. Light, and E. A. Cameron. 1985. Habitat characteristics of *Palaeodipteron walkeri* (Diptera: Nymphomyiidae). Ent. News 96:211-213.
7. Aiken, R. B. 1979. A size selective underwater light trap. Hydrobiologia 65:65-68.
8. Aiken, R. B., and C. W. Wilkinson. 1985. Bionomics of *Dytiscus alaskanus* (Coleoptera: Dytiscidae) in a central Alberta lake. Can. J. Zool. 63:1316-1323.
9. Ainslie, G. G., and W. B. Cartwright. 1922. Biology of the lotus borer *Pyrausta penitalis* Grote. U.S. Dept. Agric. Tech. Bull. 1076:1-14.
10. Akov, S. 1961. A qualitative and quantitative study of the nutritional requirements of *Aedes aegyptii* L. larvae. J. Insect Physiol. 8:319-335.
11. Akre, B. G., and D. M. Johnson. 1979. Switching and sigmoid functional response curves by damselfly naiads with alternative prey available. J. Anim. Ecol. 48:703-720.
12. Alarie, Y., P. P. Harper, and A. Maire. 1989. Rearing dytiscid beetles (Coleoptera: Dytiscidae). Ent. Bais. 13:147-149.
13. Albrecht, M. L. 1959. Die quantitative Untersuchung der Bodenfauna fliessender Gewässer, (Untersuchchungsmethoden und Arbeitsergebnisse). Z. Fisch. 8:481-550.
14. Aldrich, J. M. 1912. The biology of some western species of the dipterous genus *Ephydra*. J. N.Y. Ent. Soc. 20:77-99.
15. Aldrich, J. M. 1916. *Sarcophaga* and allies in North America. Thomas Say Found. Ent. Soc. Am. 1:1-302.
16. Alekseev, N. K. 1965. Plankton feeding of Chironomidae during larval stage. Nauch. Dok. Vysshei Shkolz. Biol. Nauk. 1:19-21.
17. Alexander, C. P. 1919. The crane flies of New York. Part I. Distribution and taxonomy of the adult flies. Mem. Cornell Univ. Agric. Exp. Sta. 25:765-993.
18. Alexander, C. P. 1920. The crane flies of New York. Part II. Biology and plylogeny. Mem. Cornell Univ. Agric. Exp. Sta. 38:691-1133.
19. Alexander, C. P. 1927. The interpretation of the radial field of the wing in the nematocerous Diptera, with special reference to the Tipulidae. Proc. Linn. Soc. N.S.W. 52:42-72.
20. Alexander, C. P. 1930. Observations on the dipterous family Tanyderidae. Proc. Linn. Soc. N.S.W. 44:221-230.
21. Alexander, C. P. 1931. Deutsche limnologische Sunda-Expedition. The crane flies (Tipulidae, Diptera). Arch. Hydrobiol. Suppl. 9, Tropische Binnengewasser 2:135-191.
22. Alexander, C. P. 1934. Family Tipulidae—the crane flies, pp. 33-58. In C. H. Curran. The families and genera of North American Diptera. Ballou, N.Y. 512 pp.
23. Alexander, C. P. 1942. Family Tipulidae. Guide to the insects of Connecticut. VI. The Diptera or true flies of Connecticut. Fasc. 1. Bull. Conn. State Geol. Nat. Hist. Surv. 64:196-485.
24. Alexander, C. P. 1949. Records and descriptions of North American craneflies (Diptera). Part VIII. The Tipuloidea of Washington, I. Am. Midl. Nat. 42:257-333.
25. Alexander, C. P. 1954. Records and descriptions of North American craneflies (Diptera). Part IX. The Tipuloidea of Oregon, I. Am. Midl. Nat. 51:1-86.
26. Alexander, C. P. 1958. Geographical distribution of the netwinged midges. Proc. 10th Int. Congr. Ent. 1:813-828.
27. Alexander, C. P. 1963. Family Deuterophlebiidae. Guide to the insects of Connecticut. VI. The Diptera or true flies of Connecticut. Fasc. 8. Bull. Conn. State Geol. Nat. Hist. Surv. 93:73-83.
28. Alexander, C. P. 1965. Family Tipulidae, pp. 16-90. In A. Stone, C. W. Sabrosky, W. W. Wirth, R. H. Foote, and J. R. Coulson (eds.). A Catalog of the Diptera of America north of Mexico. U.S. Dept. Agric. Handbk. 276. Washington, D.C. 1696 pp.
29. Alexander, C. P. 1967. The crane flies of California. Bull. Calif. Insect Surv. 8:1-269.
30. Alexander, C. P. 1981a. Chap. 22. Ptychopteridae, pp. 325-328. In J. F. McAlpine, B. V. Peterson, G. E. Shewell, H. J. Teskey, J. R. Vockeroth, and D. M. Wood (coords.). Manual of Nearctic Diptera, Vol. 1. Res. Branch, Agric. Can. Monogr. 27. Ottawa. 674 pp.
31. Alexander, C. P. 1981b. Chap. 6. Tanyderidae, pp. 149-152. In J. F. McAlpine, B. V. Peterson, G. E. Shewell, H. J. Teskey, J. R. Vockeroth, and D. M. Wood (coords.). Manual of Nearctic Diptera, Vol. 1. Res. Branch Agric. Can. Monogr. 27. Ottawa. 674 pp.
32. Alexander, C. P., and G. W. Byers. 1981. Tipulidae, pp.153-190. In J. F. McAlpine, B. V. Peterson, G. E. Shewell, H. J. Teskey, J. R. Vockeroth, and D. M. Wood (coords.). Manual of Nearctic Diptera, Vol. 1. Res. Branch, Agric. Can. Monogr. 27. Ottawa. 674 pp.
33. Ali, A. 1984. A simple and efficient sediment corer for shallow lakes. J. Environ. Qual. 13:63-66.
34. Ali, A. 1989. Chironomids, algae, and chemical nutrients in a lentic system in central Florida, USA. Acta Biol. Debr. Oecol. Hungary 3:15-27.
35. Ali, A. 1990. Seasonal changes of larval food and feeding of *Chironomus crassicaudatus* (Diptera: Chironomidae) in a subtropical lake. J. Am. Mosq. Control Ass. 6:84-88.
36. Ali, A., P. K. Chaudhuri, and D. K. Guha. 1987. Description of *Stictochironomus affinis* (Johannsen) (Diptera: Chironomidae), with notes on its behavior. Fla. Ent. 70:259-67.
37. Allan, I.R.H. 1951. A hand-operated quantitative grab for sampling river beds. J. Anim. Ecol. 21:159-160.
38. Allan, J. D. 1982. Feeding habits and prey consumption of three setipalpian stoneflies (Plecoptera) in a mountain stream. Ecology 63:26-34.
39. Allan, J. D. 1995. Stream ecology. Structure and function of running waters. Chapman and Hall, London. 388p.

40. Allan, J. D., A. S. Flecker, and N. L. McClintock. 1986. Diel epibenthic activity of mayfly nymphs, and its nonconcordance with behavioral drift. Limnol. Oceanogr. 31:1057-1065.
41. Allan, J. D., A. S. Flecker, and N. L. McClintock. 1987a. Prey preference of stoneflies: sedentary vs mobile prey. Oikos 49:323-331.
42. Allan, J. D., A. S. Flecker, and N. L. McClintock. 1987b. Prey size selection by carnivorous stoneflies. Limnol. Oceanogr. 32:864-872.
43. Allan, J. D., and A. S. Flecker. 1988. Prey preference in stoneflies: A comparative analysis of prey vulnerability. Oecologia 76:496-503.
44. Allan, J., and E. Russek. 1985. The quantification of stream drift. Can. J. Fish. Aquat. Sci. 42:210-215.
45. Allard, M., and G. Moreau. 1986. Leaf decomposition in an experimentally acidified stream channel. Hydrobiologia 139:109-117.
46. Allen, B. L., and D. C. Tarter. 1985. Life history and ecology of *Eccoptura xanthenes* (Plecoptera: Perlidae) from a small Kentucky (USA) stream. Trans. Ky. Acad. Sci. 46:87-91.
47. Allen, R. K. 1965. A review of the subfamilies of Ephemerellidae (Ephemeroptera). J. Kansas Ent. Soc. 38:262-266.
48. Allen, R. K. 1967. New species of new world Leptohyphinae (Ephemeroptera: Tricorythidae). Can. Ent. 99:350-375.
49. Allen, R. K. 1973. New species of *Leptohyphes* Eaton. Pan-Pac. Ent. 49:363-372.
50. Allen, R. K. 1974. *Neochoroterpes*, a new subgenus of *Choroterpes* Eaton from North America (Ephemeroptera: Leptophlebiidae). Can. Ent. 106:161-68.
51. Allen, R. K. 1980. Geographic distribution and reclassification of the subfamily Ephemerellinae (Ephemeroptera: Ephemerellidae), pp. 71-91. In J. F. Flannagan and K. E. Marshall (eds.). Advances in Ephemeroptera biology. Plenum, N.Y. 552 pp.
52. Allen, R. K. 1984. A new classification of the subfamily Ephemerellinae and the description of a new genus. Pan-Pac. Ent. 60:245-247.
53. Allen, R. K., and G. F. Edmunds, Jr. 1962. A revision of the genus *Ephemerella* (Ephemeroptera: Ephemerellidae). IV. The subgenus *Dannella*. J. Kansas Ent. Soc. 35:333-338.
54. Allen, R. K., and G. F. Edmunds, Jr. 1963a. A revision of the genus *Ephemerella* (Ephemeroptera: Ephemerellidae). VI. The subgenus *Serratella* in North America. Ann. Ent. Soc. Am. 56:583-600.
55. Allen, R. K., and G. F. Edmunds, Jr. 1963b. A revision of the genus *Ephemerella* (Ephemeroptera: Ephemerellidae). VII. The subgenus *Eurylophella*. Can. Ent. 95:597-623.
56. Allen, R. K., and G. F. Edmunds, Jr. 1965. A revision of the genus *Ephemerella* (Ephemeroptera: Ephemerellidae). VIII. The subgenus *Ephemerella* in North America. Ent. Soc. Am. Misc. Publ. 4:243-282.
57. Allen, R. K., and G. F. Edmunds, Jr. 1976. A revision of the genus *Ametropus* in North America (Ephemeroptera: Ametropodidae). J. Kans. Ent. Soc. 49:625-635.
58. Almazan, M.L.P., and K. L. Heong. 1992. Fall-off rate of *Nilaparvata lugens* (Stål) and efficiency of the predator *Limnogonus fossarum* (F.). Int. Rice Res. Newsltr. 17:17.
59. Alrutz, R. W. 1992. Additional records of dragonflies (Odonata) from Ohio. Ohio J. Sci. 92:119-120.
60. Alstad, D. N. 1979. Comparative biology of the common Utah Hydropsychidae (Trichoptera). Am. Midl. Nat. 103:167-174.
61. Aly, C. 1985. Feeding rate of larval *Aedes vexans* stimulated by food substances. J. Am. Mosq. Contr. Assoc. 1:506-510.
62. Aly, C., and R. H. Dadd. 1989. Drinking rate regulation in some freshwater mosquito larvae. Physiol. Ent. 14:241-256.
63. Aly, C., and M. S. Mulla. 1986. Orientation and ingestion rates of larval *Anopheles albimanus* in response to floating particles. Ent. Exp. Appl. 42:83-90.
64. Ambühl, H. 1959. Die Bedeutung der Strömung als ökologischer Faktor. Schweiz. Z. Hydrol. 21:133-264.
65. Ameen, M., and T. M. Iversen. 1978. Food of *Aedes* larvae (Diptera: Culicidae) in a temporary forest pool. Arch. Hydrobiol. 83:552-64.
66. American Public Health Association. 1971. Standard methods for the examination of water and waste-water (13th ed.). Am. Public Health Assoc., N.Y. 874 pp.
67. Amrine, J. W., and L. Butler. 1978. An annotated list of the mosquitoes of West Virginia. Mosquito News 38:101-104.
68. Andersen, N. M. 1973. Seasonal polymorphism and developmental changes in organs of flight and reproduction in bivoltine pondskaters. Ent. Scand. 4:1-20.
69. Andersen, N. M. 1975. The *Limnogonus* and *Neogerris* of the old World, with character analysis and a reclassification of the Gerrinae (Hemiptera: Gerridae). Ent. Scand. (Suppl.) 7:1-96.
70. Andersen, N. M. 1981a. Adaptations, ecological diversifications, and the origin of higher taxa of semiaquatic bugs (Gerromorpha). Rostria 33 (Suppl.):3-16.
71. Andersen, N. M. 1981b. Semiaquatic bugs: phylogeny and classification of the Hebridae (Heteroptera: Gerromorpha) with revisions of *Timasius, Neotimasius* and *Hyrcanus*. Syst. Ent. 6:377-412.
72. Andersen, N. M. 1982. The semiaquatic bugs (Hemiptera, Gerromorpha). Phylogeny, adaptations, biogeography and classification. Entomonograph, Vol. 3. Scandinavian Sci. Press, Klampenborg, Denmark. 455 pp.
73. Andersen, N. M. 1990. Phylogeny and taxonomy of water striders, genus *Aquarius* Schellenberg (Insecta, Hemiptera, Gerridae), with a new species from Australia. Steenstrupia 16:37-81.
74. Andersen, N. M. 1991. Marine insects: genital morphology, phylogeny and evolution of sea skaters, genus *Halobates* (Hemiptera: Gerridae). Zool. J. Linn. Soc. 103:21-60.
75. Andersen, N. M., and J. R. Spence. 1992. Classification and phylogeny of the holarctic water strider genus *Limnoporus* Stål (Hemiptera, Gerridae). Can. J. Zool. 70:753-785.
76. Andersen, N. M., and J. T. Polhemus. 1976. Water-striders (Hemiptera: Gerridae, Veliidae, etc.), pp. 187-224. In L. Cheng (ed.). Marine Insects. North Holland, Amsterdam. 581 pp.
77. Andersen, N. M., and J. T. Polhemus. 1980. Four new genera of Mesoveliidae (Hemiptera, Gerromorpha) and the phylogeny and classification of the family. Ent. Scand. 11:369-392.
78. Anderson, J. B., and W. T. Mason, Jr. 1968. A comparison of benthic macroinvertebrates collected by dredge and basket sampler. J. Wat. Poll. Contr. Fed. 40:252-259.
79. Anderson, J. F., and S. W. Hitchcock. 1968. Biology of *Chironomus atrella* in a tidal cove. Ann. Ent. Soc. Am. 61:1597-1603.
80. Anderson, J.M.E. 1976. Aquatic Hydrophilidae (Coleoptera). The biology of some Australian species with descriptions of immature stages reared in the laboratory. J. Austr. Ent. Soc. 15:219-228.
81. Anderson, J. R. 1960. The biology and taxonomy of Wisconsin black flies (Diptera: Simuliidae). Ph.D. diss., University of Wisconsin, Madison. 185 pp.
82. Anderson, J. R., and R. J. Dicke. 1960. Ecology of the immature stages of some Wisconsin black flies (Simuliidae: Diptera). Ann. Ent. Soc. Am. 53:386-404.
83. Anderson, L. D. 1932. A monograph of the genus *Metrobates*. Univ. Kans. Sci. Bull. 20:297-311.
84. Anderson, N. H. 1967. Life cycle of a terrestrial caddisfly, *Philocasca demita* (Trichoptera: Limnephilidae), in North America. Ann. Ent. Soc. Am. 60:320-323.
85. Anderson, N. H. 1974a. Observations on the biology and laboratory rearing of *Pseudostenophylax edwardsi* (Trichoptera: Limnephilidae). Can. J. Zool. 52:7-13.
86. Anderson, N. H. 1974b. The eggs and oviposition behaviour of *Agapetus fuscipes* Curtis (Trich., Glossosomatidae). Entomol. Mon. Mag. 109:129-131.
87. Anderson, N. H. 1976a. Carnivory by an aquatic detritivore, *Clistoronia magnifica* (Trichoptera: Limnephilidae). Ecology 57:1081-1085.
88. Anderson, N. H. 1976b. The distribution and biology of the Oregon Trichoptera. Ore. Agric. Exp. Sta. Tech. Bull. 134:1-152.
89. Anderson, N. H. 1978. Continuous rearing of the limnephilid caddisfly, *Clistoronia magnifica* (Banks), pp. 317-329. In M. I. Crichton (ed.). Proc. 2nd Int. Symp. Trichoptera. Junk, The Hague, Netherlands. 359 pp.

90. Anderson, N. H. 1989. Xylophagous Chironomidae from Oregon streams. Aquat. Insects 11:33-45.
91. Anderson, N. H., and M. J. Anderson. 1974. Making a case for the caddisfly. Insect World Digest 1:1-6.
92. Anderson, N. H., and J. R. Bourne. 1974. Bionomics of three species of glossosomatid caddisflies (Trichoptera: Glossosomatidae) in Oregon. Can. J. Zool. 52:405-411.
93. Anderson, N. H., and K. W. Cummins. 1979. The influence of diet on the life histories of aquatic insects. J. Fish. Res. Bd. Can. 36:335-342.
94. Anderson, N. H., and E. Grafius. 1975. Utilization and processing of allochthonous material by stream Trichoptera. Verh. Int. Verein. Limnol. 19:3083-3088.
95. Anderson, N. H., J. R. Sedell, L. M. Roberts, and F. J. Triska. 1978. The role of aquatic invertebrates in processing wood debris in coniferous forest streams. Am. Midl. Nat. 100:64-82.
96. Anderson, N. H., R. J. Steedman, and T. Dudley. 1984. Patterns of exploitation by stream invertebrates of wood debris (xylophagy). Verh. Int. Verein. Limnol. 22:1847-1852.
97. Anderson, R. D. 1962. The Dytiscidae (Coleoptera) of Utah: keys, original citation, types, and Utah distribution. Great Basin Nat. 22:54-75.
98. Anderson, R. D. 1971. A revision of the Nearctic representatives of *Hygrotus* (Coleoptera: Dytiscidae). Ann. Ent. Soc. Am. 64:503-512.
99. Anderson, R. D. 1976. A revision of the Nearctic species of *Hycrotus* groups II and III (Coleoptera: Dytiscidae). Ann. Ent. Soc. Am. 69:577-584.
100. Anderson, R. D. 1983. Revision of the Nearctic species of *Hygrotus* groups IV, V and VI (Coleoptera: Dytiscidae). Ann. Ent. Soc. Am. 76:173-296.
101. Anderson, R. I. 1980. Chironomidae toxicity tests—biological background and procedures, pp. 70-80. *In* A. L. Buikema, Jr., and J. Cairns, Jr. (eds.). Aquatic invertebrate bioassays. Am. Soc. Test. Mater. Philadelphia. 209 pp.
102. Andre, P., P. Legerdre, and P. P. Harper. 1981. La selectivite de trois engins d'echantillonnage du benthos lacustre. Annls. Limnol. 17:25-40.
103. Andreeva, R. V. 1989. The morphological adaptations of horse fly larvae (Diptera: Tabanidae) to developmental sites in the Palearctic region and their relationship to the evolution and distribution of the family. Can. J. Zool. 67:2286-2293.
104. Andrewartha, H. G., and L. C. Birch. 1954. The distribution and abundance of animals. Univ. of Chicago, IL.
105. Anholt, B. R. 1990. An experimental separation of interference and exploitative competition in a larval damselfly. Ecology 71:1483-1493.
106. Annecke, D. P., and R. L. Doutt. 1961. The genera of the Mymaridae (Hym.: Chalciodoidea). Rep. S. Afr. Dept. Agric. Tech. Serv. Ent. Mem. 5:1-71.
107. APHA (American Public Health Association, American Water Works Association, Water Pollution Control Federation). 1989. Standard methods for the examination of water and wastewater. (17th ed.) Am. Publ. Hlth. Assoc., Washington, DC. 1525 pp.
108. Apperson, C. S., and D. G. Yows. 1976. A light trap for collecting aquatic organisms. Mosquito News 36:205-206.
109. Applegate, R. L. 1973. Corixidae (water boatmen) of the South Dakota glacial lake district. Ent. News 84:163-170.
110. Arbona, F. L., Jr. 1989. Mayflies, the angler, and the trout. Lyons and Burford, New York.
111. Arens, W. 1989. Comparative functional morphology of the mouthparts of stream animals feeding on epilithic algae. Arch. Hydrobiol./Suppl. 83:253-354.
112. Arens, W. 1990. Wear and tear on mouthparts: a critical problem in stream animals feeding on epilithic algae. Can. J. Zool. 68:1896-1914.
113. Armitage, B. J. 1991. Diagnostic atlas of the North American caddisfly adults. I. Philopotamidae (2nd ed.). The Caddis Press, Athens, AL.
114. Armitage, B. J., and S. W. Hamilton. 1990. Diagnostic atlas of the North American caddisfly adults. II. Ecnomidae, Polycentropodidae, Psychomyiidae, and Xiphocentronidae. The Caddis Press, Athens, AL.
115. Armitage, P. D. 1968. Some notes on the food of chironomid larvae of a shallow woodland lake in South Finland. Ann. Zool. Fenn. 5:6-13.
116. Armitage, P. D. 1974. Some aspects of the ecology of the Tanypodinae and other less common species of Chironomidae in Lake Kuusijarvi, South Finland. Ent. Tidskr. Suppl. 95:13-17.
117. Armitage, P. D. 1978. Catches of invertebrate drift by pump and net. Hydrobiologia 60:229-233.
118. Armitage, P. D. 1979. Folding artificial substratum sampler for use in standing water. Hydrobiologia 66:245-249.
119. Armitage, P., P. S. Cranston, and L.C.V. Pinder (eds.). 1995. Chironomidae: The biology and ecology of non-biting midges. Chapman and Hall, London. 572p.
120. Arnaud, P. H., and G. W. Byers. 1990. *Holorusia hespera*, a new name for *Holorusia grandis* (Bergroth) (*Holorusia rubiginosa* Loew) (Diptera: Tipulidae). Myia 5:1-9.
121. Arnaud, P. H., Jr., and I. A. Boussy. 1994. The adult Thaumaleidae (Diptera: Culicomorpha) of western North America. Myia 5:141-152.
122. Arnett, R. H. 1960. The beetles of the United States. Catholic University America Press, Washington, D.C. 1112 pp.
123. Arnold, D. C. 1974. Black flies (Diptera: Simuliidae) of the costal plains and sandhills of South Carolina. M.Sc. Thesis, Clemson Univ., Clemson, SC.
124. Arnold, D. C., and W. A. Drew. 1987. A preliminary survey of the Megaloptera of Oklahoma. Proc. Oklahoma Acad. Sci. 67:23-26.
125. Arrow, G. J. 1924. Vocal organs in the coleopterous families Dytiscidae, Erotylidae, and Endomychidae. Trans. Ent. Soc. Lond. 72:134-143.
126. Arsuffi, T. L., and K. Suberkropp. 1984. Leaf processing capabilities of aquatic hyphomycetes: interspecific differences and influence on shredder feeding preferences. Oikos 42:144-154.
127. Arsuffi, T. L., and K. Suberkropp. 1989. Selective feeding by shredders on leaf-colonizing stream fungi: comparison of macroinvertebrate taxa. Oecologia 79:30-37.
128. Arts, M. T., E. J. Maly, and M. Pasitschniak. 1981. The influence of *Acilius* (Dytiscidae) predation on *Daphina* in a small pond. Limnol. Oceanogr. 26:1172-1175.
129. Ashe, P. 1983. A catalogue of chironomid genera and subgenera of the world including synonyms (Diptera: Chironomidae). Ent. Scand. Suppl. 17:1-68.
130. Ashley, D. L., D. C. Tarter, and W. D. Watkins. 1976. Life history and ecology of *Diploperla robusta* Stark and Gaufin (Plecoptera: Perlodidae). Psyche 83:310-318.
131. Askevold, I. S. 1987. The genus *Neohaemonia* Szekessy in North America (Coleoptera: Chrysomelidae). Donaciinae systematics reconstructed phylogeny and geographic history. Trans. Am. Ent. Soc. 113:361-430.
132. Asmus, B. S. 1973. The use of the ATP assay in terrestrial decomposition studies. Bull. Ecol. Res. Comm. 17:223-234.
133. ASTM. 1987. Annual book of ASTM standards. Pesticides; resource recovery; hazardous substances and oil spill response; waste disposal; biological effects. Vol. 11:4. ASTM, Philadelphia, PA.
134. Aussel, J.-P., and J. R. Linley. 1994. Natural food and feeding behavior of *Culicoides furens* larvae (Diptera: Ceratopogonidae). J. Med. Ent. 31:99-104.
135. Austin, D. A., and J. H. Baker. 1988. Fate of bacteria ingested by larvae of the freshwater mayfly, *Ephemera danica*. Microb. Ecol. 15:323-332.
136. Axtell, R. C. 1976. Coastal horseflies and deerflies (Diptera: Tabanidae), pp. 415-445. *In* L. Cheng (ed.). Marine insects. North Holland, Amsterdam. 581 pp.
137. Azam, K. M. 1969. Life history and production studies of *Sialis californica* Banks and *Sialis rotunda* Banks (Megaloptera: Sialidae), Ph.D. diss., Oregon State University, Corvallis. 111 pp.
138. Azam, K. M., and N. H. Anderson. 1969. Life history and habits of *Sialis californica* Banks and *Sialis rotunda* Banks in western Oregon. Ann. Ent. Soc. Am. 62:549-558.
139. Azevedo-Ramos, C., M. van Sluys, J. M. Hero, and W. E. Magnusson. 1992. Influence of tadpole movement on predation by odonate naiads. J. Herpetol. 26:335-338.

140. Back, C., and P. P. Harper. 1978. Les mouches noires (Diptera: Simuliidae) de deux ruisseaux des Laurentides, Québec. Ann. Soc. Ent. Québec 23:55-66.
141. Back, C., and D. M. Wood. 1979. *Paleodipteron walkeri* (Diptera: Nymphomyiidae) in northern Quebec. Can. Ent. 111:1287-1291.
142. Bacon, J. A. 1956. A taxonomic study of the genus *Rhagovelia* of the Western Hemisphere. Univ. Kans. Sci. Bull. 38:695-913.
143. Bacon, M., and R. H. McCauley, Jr. 1959. Black flies (Diptera:Simuliidae) in a newly developed irrigation district (Columbia Basin, Washington). Northwest Sci. 33:103-110.
144. Badcock, R. M. 1949. Studies on stream life in tributaries of a Welsh Dee. J. Anim. Ecol. 18:193-208.
145. Badcock, R. M. 1953. Observation of oviposition under water of the aerial insect *Hydropsyche angustipennis* (Curtis) (Trichoptera). Hydrobiologia 5:222-225.
146. Bae, Y. J., and W. P. McCafferty. 1991. Phylogenetic systematics of the Potamanthidae (Ephemeroptera). Trans. Am. Ent. Soc. 117:1-143.
147. Baekken, T. 1981. Growth patterns and food habits of *Baetis rhodani, Capnia pygmaea* and *Diura nanseni* in a west Norwegian river. Holarct. Ecol. 4:139-144.
148. Bahr, A., and G. Schulte. 1976. Distribution of shore bugs (Heteroptera: Saldidae) in the brackish and marine littoral of the North American Pacific Coast. Mar. Biol. 36:37-46.
149. Bailey, P.C.E. 1986. The feeding behaviour of a sit-and wait-predator, *Ranatra dispar* (Heteroptera: Nepidae): optimal foraging and feeding dynamics. Oecologia (Berlin) 68:291-297.
150. Baker, A. S., and A. J. McLachlan. 1979. Food preferences of Tanypodinae larvae (Diptera: Chironomidae) . Hydrobiologia 62:283-288.
151. Baker, F. C. 1918. The productivity of invertebrate fish food on the bottom of Oneida Lake, with special reference to mollusks. Tech. Publ. N.Y. State Coll. For. 18:1-265.
152. Baker, J. H., and L. A. Bradnum. 1976. The role of bacteria in the nutrition of aquatic detritivores. Oecologia 24:95-104.
153. Baker, J. R., and H. H. Neunzig. 1968. The egg masses, eggs, and first instar larvae of eastern North American Corydalidae. Ann. Ent. Soc. Am. 61:1181-1187.
154. Baker, R. L. 1981a. Behavioral interactions and use of feeding areas by nymphs of *Coenagrion resolutum* (Coenagrionidae: Odonata). Oecologia 49:353-358.
155. Baker, R. L. 1981b. Life cycles and food of *Coenagrion resolutum* (Coenagrionidae: Odonata) and *Lestes disjunctus disjunctus* (Lestidae: Odonata) populations from the boreal forest of Alberta, Canada. Aquat. Insects 3:179-191.
156. Baker, R. L. 1982. Effects of food abundance of growth, survival, and use of the space by nymphs of *Coenagrion resolutum* (Zygoptera). Oikos 38:47-51.
157. Baker, R. L. 1986. Food limitation of larval dragonflies: a field test of spacing behaviour. Can. J. Fish. Aquat. Sci. 43:1720-1725.
158. Baker, R. L. 1987. Dispersal of larval damselflies: do larvae exhibit spacing behaviour in the field? J. N. Am. Benthol. Soc. 6:35-45.
159. Baker, R. L. 1988. Effects of previous diet and frequency of feeding on development of larval damselflies. Freshwat. Biol. 19:191-195.
160. Baker, R. L. 1989. Condition and size of damselflies: a field study of food limitation. Oecologia 81:111-119.
161. Baker, R. L., and B. W. Feltmate. 1987. Development of *Ischnura verticalis* (Coenagrionidae: Odonata): effects of temperature and prey abundance. Can. J. Fish. Aquat. Sci. 44:1658-1661.
162. Balagi, A., M. Vaitheiswaran, and K. Venkataraman. 1990. Laboratory observations on the life cycle patterns of two *Baetis* species. Geobios 17:15-17.
163. Balciunas, J. K., and T. D. Center. 1981. Preliminary host specificity tests of a Panamanian *Parapoynx rugosalis* as a potential biological control agent for *Hydrilla verticillata*. Environ. Ent. 10:462-467.
164. Balciunas, J. K., and M. C. Minno. 1985. Insects damaging *Hydrilla* in the USA. J. Aquat. Plant Manage. 23:77-83.
165. Balduf, W. V. 1935. The bionomics of entomophagous Coleoptera. John S. Swift, N.Y. 220 pp.
166. Balduf, W. V. 1939. The bionomics of entomophagous insects. Part II. John S. Swift. St. Louis. 384 pp.
167. Balfour Browne, F. 1910. On the life history of *Hydrobius fuscipes* L. Trans. R. ent. Soc. Edinb. 47:310-340.
168. Balfour Browne, F. 1947 . On the false chelate leg of the water beetle. Proc. R. ent. Soc. Lond (A) 22:38-41.
169. Balfour Browne, F., and J. Balfour Browne. 1940. An outline of the habits of a water beetle: *Noterus capricornis* Herbst (Coleopt.). Proc. Zool. Soc. Lond. 15:10-12.
170. Ball, R. C., N. R. Kevern, and K. J. Linton. 1969. Red Cedar River report. II. Bioecology. Publ. Mich. State Univ. Mus. Biol. Ser. 4:105-160.
171. Balling, S. S., and V. H. Resh. 1984. Life history variability in the water boatman, *Trichocorixa reticulata* (Hemiptera: Corixidae), in San Francisco Bay salt marsh ponds. Ann. Ent. Soc. Am. 77:14-19.
172. Balling, S. S., and V. H. Resh. 1985. Seasonal patterns of pondweed standing crop and *Anopheles occidentalis* densities in Coyote Hills marsh. Proc. Calif. Mosq. Vector Contr. Assoc. 52:122-125.
173. Balsbaugh. E. U., Jr., and K. L. Hays. 1972. The leaf beetles of Alabama (Coleoptera: Chrysomelidae). Auburn Univ. Agric. Exp. Sta. Bull. 441:1-223.
174. Bane, C. T., and O. T. Lind. 1978. The benthic invertebrate standing crop and diversity of a small desert stream in the Big Bend National Park, Texas. Southwest. Nat. 23:215-226.
175. Banks, B., and T.J.C. Beebee. 1988. Reproductive success of Natterjack Toads *Bufo calamita* in two contrasting habitats. J. Anim. Ecol. 57:475-492.
176. Barber, W. E., and N. R. Kevern. 1974. Seasonal variation of sieving efficiency in a lotic habitat. Freshwat. Biol. 4:293-300.
177. Barbier, R., and G. Chauvin. 1974. The aquatic egg of *Nymphula nympheata* (Lepidoptera: Pyralidae). Cell Tiss. Res. 149:473-479.
178. Bare, C. O. 1926. Life histories of some Kansas "backswimmers." Ann. Ent. Soc. Am. 19:93-101.
179. Bärlocher, F., R. J. Mackay, and G. B. Wiggins. 1978. Detritus processing in a temporary vernal pool in southern Ontario. Arch. Hydrobiol. 81:269-295.
180. Barlow, A. E. 1991. New observations on the distribution and behaviour of *Tachopteryx thoreyi* (Hag.) (Anisoptera: Petaluridae). Not. Odonatol. 3:131-132.
181. Barmuta, L. A. 1984. A method for separating benthic arthropods from detritus. Hydrobiologia 112:105-108.
182. Barmuta, L. A., S. D. Cooper, S. K. Hamilton, K. W. Kratz, and J. M. Melack. 1990. Responses of zooplankton and zoobenthos to experimental acidification in a high-elevation lake (Sierra Nevada, California, U.S.A.). Freshwat. Biol. 23:571-586.
183. Barnard, P. C. 1971. The larva of *Agraylea sexmaculata* Curtis (Trichoptera: Hydroptilidae). Ent. Gaz. 22:253-257.
184. Barnby, M. A., and V. H. Resh. 1988. Factors affecting the distribution of an endemic and a widespread species of brine fly (Diptera: Ephydridae) in a northern California thermal saline spring. Ann. Ent. Soc. Am. 81:437-446.
185. Barnes, H. F. 1937. Methods of investigating the bionomics of the common crane fly, *Tipula paludosa* Meigen, together with some results. Ann. Appl. Biol. 24:356-368.
186. Barnes, J. R., and G. W. Minshall. 1983. Stream ecology. Application and testing of general ecological theory. Plenum Press, New York. 399p.
187. Barr, A. R. 1958. The mosquitoes of Minnesota. Univ. Minn. Agric. Exp. Sta. Tech. Bull. 228:1-154.
188. Barr, A. R., and P. Guptavanij. 1989. *Anopheles hermsi* n. sp., an unrecognized American species of the *Anopheles maculipennis* group (Diptera: Culicidae). Mosq. Syst. 20:352-356.
189. Barr, C. B., and J. B. Chapin. 1988. The aquatic Dryopoidea of Louisiana, USA (Coleoptera: Psephenidae: Dryopidae: Elmidae). Tulane Stud. Zool. Bot. 26:89-164.
190. Barr, C. B., and P. J. Spangler. 1992. A new genus and species of stygobiontic dryopid beetle, *Stygoparnus comalensis* (Coleoptera: Dryopidae), from Comal Springs, Texas. Proc. Biol. Soc. Wash. 105:40-54.

191. Barr, C. B., and P. J. Spangler. 1994. Two new synonymies: *Alabameubria* Brown, a junior synonym of *Dicranopselaphus* Guerin-Meneville; and *Alabameubria starki* Brown, a synonym of *Dicranopselaphus variegatus* Horn (Coleoptera: Psephenidae: Eubriinae). Ent. News 105:299-302.
192. Bartholomae, P. G., and P. G. Meier. 1977. Notes on the life history of *Potamanthus myops* in southeastern Michigan (Ephemeroptera: Potamanthidae). Great Lakes Ent. 10:227-232.
193. Barton, D. R. 1980. Observations on the life histories and biology of Ephemeroptera and Plecoptera in northeastern Alberta. Aquat. Insects 2:97-111.
194. Barton, D. R., C. W. Pugsley, and H.B.N. Hynes. 1987. The life history and occurrence of *Parachaetocladius abnobaeus* (Diptera: Chironomidae). Aquat. Insects 9:189-194.
195. Bartsch, A. F., and W. M. Ingram. 1966. Biological analysis of water pollution in North America. Verh. Int. Verein. Limnol. 16:786-800.
196. Bates, M. 1949. The natural history of mosquitoes. Harper and Row, N.Y. 378 pp.
197. Batra, S.W.T. 1977. Bionomics of the aquatic moth *Acentropus niveus* (Olivier) a potential biological control agent for Eurasian watermilfoil and *Hydrilla*. J. N.Y. Ent. Soc. 85:143-152.
198. Batta Spinelli, G., and G. P. Moretti. 1992. Morphology, ecology and dietary regime of *Tinodes antonioi* Bot. Tat. (Trichoptera: Psychomyiidae), pp. 39-42. *In* C. Otto (ed.). Proc. VII Int. Symp. Trichoptera. Umea, Sweden, Backhuys Publ., Leiden, Netherlands.
199. Battle, F. V., and E. C. Turner. 1971. The Insects of Virginia. III. A systematic review of the genus *Culicoides* (Diptera: Ceratopogonidae) in Virginia with a geographic catalog of the species occurring in the eastern United States north of Florida. Bull. Res. Div. Va. Poly. Inst. State Univ. 44:1-129.
200. Batzer, D. P., and V. H. Resh. 1991. Trophic interactions among a beetle predator, a chironomid grazer, and periphyton in a seasonal wetland. Oikos 60:251-257.
201. Bauer, L. S., and J. Granett. 1979. The black flies of Maine. Maine Life Sci. Agr. Exp. Sta. Tech. Bull. 95:1-18.
202. Baumann, R. W. 1975. Revision of the stonefly family Nemouridae (Plecoptera): a study of the world fauna at the generic level. Smithson. Contr. Zool. 211:1-74.
203. Baumann, R. W. 1976. An annotated review of the systematics of North American stoneflies (Plecoptera). Perla 2:21-23.
204. Baumann, R. W., and A. R. Gaufin. 1970. The *Capnia projecta* complex of Western North America (Plecoptera: Capniidae). Trans. Am. Ent. Soc. 96:435-468.
205. Baumann, R. W., and K. W. Stewart. 1980. The nymph of *Lednia tumana* (Ricker) (Plecoptera: Nemouridae). Proc. Ent. Soc. Wash. 82:655-659.
206. Baumann, R. W., A. R. Gaufin, and R. F. Surdick. 1977. The stoneflies (Plecoptera) of the Rocky Mountains. Mem. Am. Ent. Soc. 31:1-208.
207. Baumann, R. W., and D. R. Lauck. 1987. *Salmoperla*, a new stonefly genus from northern California (Plecoptera: Perlodidae). Proc. Ent. Soc. Wash. 89:825-830.
208. Baumann, R. W., and J. D. Unzicker. 1981. Preliminary checklist of Utah caddisflies (Trichoptera). Encyclia 58:25-29.
209. Baumgartner-Gamauf, M. 1959. Einige ufer-und wasserbewohnende Collembolen des Seewinkels. Ost. Akad. Wiss. Math. Nat. Kl. 168:363-369.
210. Bay, E. C. 1967. An inexpensive filter-aquarium for rearing and experimenting with aquatic invertebrates. Turtox News 45:146-148.
211. Bay, E. C. 1972. An observatory built in a pond provides a good view of aquatic animals and plants. Sci. Am. 227:114-118.
212. Bay, E. C. 1974. Predator-prey relationships among aquatic insects. Ann. Rev. Ent. 19:441-453.
213. Bay, E. C., and J. R. Caton. 1969. A benthos core sampler for wading depths. Calif. Vector Views 16:88-89.
214. Bayer, L. J., and H. J. Brockmann. 1975. Curculionidae and Chrysomelidae found in aquatic habitats in Wisconsin. Great Lakes Ent. 8:219-226.
215. Bayne, B. L. 1989. Measuring the biological effects of pollution: the Mussel Watch approach. Wat. Sci. Technol. 21:1089-1100.
216. Beak, T. W., T. C. Griffing, and A. G. Appleby. 1973. Use of artificial substrate samplers to assess water pollution, pp. 227-241. *In* Biological methods for the assessment of water quality. Am. Soc. Test. Mater. 528 pp.
217. Beam, B. D., and G. B. Wiggins. 1987. A comparative study of the biology of five species of *Neophylax* (Trichoptera: Limnephilidae) in southern Ontario. Can. J. Zool. 65:1741-1754.
218. Beardsley, J. W. 1992. Review of the genus *Aspidogyrus* Yoshimoto, with descriptions of three new species (Hymenoptera: Cynipoidea: Eucoilidae). Proc. Hawaii. Ent. Soc. 31:139-150.
219. Beatty, A. F., G. H. Beatty, and H. B. White, III. 1969. Seasonal distribution of Pennsylvania Odonata. Proc. Penn. Acad. Sci. 43:119-126.
220. Beatty, G. H., and A. F. Beatty. 1968. Checklist and bibliography of Pennsylvania Odonata. Penn. Acad. Sci. 42:120-129.
221. Beck, H. 1960. Die Larvalsystematik der Eulen (Noctuidae). Abhandlungen zur Larvalsystematik der Insekten. Nr. 4. Akademie, Berlin. 406 pp.
222. Beck, W. M., Jr. 1968. Chironomidae, pp. V.l-V.22. *In* F. K. Parish (ed.). Keys to water quality indicative organisms of the southeastern United States. Fed. Wat. Poll. Contr. Adm., Atlanta. 195 pp.
223. Beck, W. M., Jr. 1975. Chironomidae, pp. 159-180. *In* F. K. Parish (ed.). Keys to water quality indicative organisms of the southeastern United States (2nd ed.). EMSL/EPA, Cincinnati. 195 pp.
224. Beck, W. M., Jr., and E. C. Beck. 1966. Chironomidae (Diptera) of Florida. I. Pentaneurini (Tanypodinae). Bull. Fla. State Mus. 10:305-379.
225. Beck, W. M., Jr., and E. C. Beck. 1969. Chironomidae (Diptera) of Florida. III. The *Harnischia* complex (Chironominae). Bull. Fla. State Mus. 13:227-313.
226. Becker, C. D. 1973. Development of *Simulium* (*Psilozia*) *vittatum* Zett. (Diptera: Simuliidae) from larvae to adults at thermal increments from 17.0 to 27.0°C. Am. Midl. Nat. 89:246-251.
227. Becker, G. 1990. Comparison of the dietary composition of epilithic trichopteran species in a first order stream. Arch. Hydrobiol. 120:13-40.
228. Becker, G. 1993. Age structure and colonization of natural substrate by the epilithic caddisfly *Tinodes rostoki* (Trichoptera: Psychomyiidae). Arch. Hydrobiol. 127:423-436.
229. Becker, P. 1958. The behavior of larvae of *Culicoides circumscriptus* Kieff. (Dipt., Ceratopogonidae) towards light stimuli as influenced by feeding with observations on the feeding habits. Bull. Ent. Res. 49:785-802.
230. Becker, T. 1926. Ephydridae. Fam. 56, pp. 1-115. *In* E. Lindner (ed.). Die Fliegen der palaearktischen Region 6, Part I. Stuttgart. 280 pp.
231. Bednarik, A. F., and G. F. Edmunds. 1980. Descriptions of larval *Heptagenia* from the Rocky Mountain region. Pan-Pacific. Ent. 56:51-62.
232. Bednarik, A. F., and W. P. McCafferty. 1977. A checklist of the stoneflies or Plecoptera of Indiana. Great Lakes Ent. 10:223-226.
233. Bednarik, A. F., and W. P. McCafferty. 1979. Biosystematic revision of the genus *Stenonema* (Ephemeroptera: Heptageniidae). Can. Bull. Fish. Aquat. Sci. 201:1-73.
234. Beerbower, F. V. 1944. Life history of *Scirtes orbiculatus* Fabricius (Coleoptera: Helodidae). Ann. Ent. Soc. Am. 36:672-680.
235. Beisser, M. C., S. Testa, III, and N. G. Aumen. 1991. Macroinvertebrate trophic composition and processing of four leaf species in a Mississippi stream. J. Freshwat. Ecol. 6:23-33.
236. Bekker, E. 1938. On the mechanism of feeding in larvae of *Anopheles*. Zool. Zh. 17:741-762.
237. Belkin, J. N., and W. A. McDonald. 1955. A population of *Corethrella laneana* from Death Valley, with descriptions of all stages and discussion of the Corethrellini (Diptera: Culicidae). Bull. Soc. Calif. Acad. Sci. 54:82-96.
238. Belle, J. 1973. A revision of the New World Genus *Progomphus* Sélys, 1854 (Anisoptera: Gomphidae). Odonatologica 2:191-348.
239. Belluck, D., and J. Furnish. 1981. *Psephidonus brunneus* (Say): a stream inhabiting staphylinid beetle (Coleoptera: Staphylinidae) associated with blackfly larvae (Diptera: Simuliidae). Coleopt. Bull. 35:204.

240. Bendell, B. E., and D. K. McNicol. 1987. Estimation of nektonic insect populations. Freshwat. Biol. 18:105-108.
241. Benech, V. 1972. La fécondité de *Baetis rhodani* Pictet. Freshwat. Biol. 2:337-354.
242. Benedetta, C. 1970. Observations on the oxygen needs of some European Plecoptera. Int. Revue ges. Hydrobiol. 55:505-510.
243. Benedetto, L. A. 1970. Tagesperiodik der Flugaktivität von vier *Leuctra*-Arten am Polarkreis. Oikos Suppl. 13:87-90.
244. Benfield, E. F. 1972. A defensive secretion of *Dineutes discolor* (Coleoptera: Gyrinidae). Ann. Ent. Soc. Am. 65:1324-1327.
245. Benfield, E. F. 1974. Autohemorrhage in two stoneflies and its effectiveness as a defensive secretion. Ann. Ent. Soc. Am. 67:739-742.
246. Benfield, E. F., D. S. Jones, and M. F. Patterson. 1977. Leaf pack processing in a pastureland stream. Oikos 29:99-103.
247. Bengtsson, J. 1977. Food preference experiments with nymphs of *Nemoura cinerea* (Retz.) (Plecoptera). Flora Fauna 83:36-39.
248. Bengtsson, J. 1986. Growth and life cycle of *Leptophlebia vespertina*. Flora-Fauna 92:94-96.
249. Benke, A. C. 1970. A method for comparing individual growth rates of aquatic insects with special reference to the Odonata. Ecology 51:328-331.
250. Benke, A. C. 1976. Dragonfly production and prey turnover. Ecology 57:915-927.
251. Benke, A. C. 1978. Interactions among coexisting predators-a field experiment with dragonfly larvae. J. Anim. Ecol. 47:335-350.
252. Benke, A. C. 1979. A modification of the Hynes method for estimating secondary production with particular significance for multivoltine populations. Limnol. Oceanogr. 24:168-171.
253. Benke, A. C. 1984. Secondary production of aquatic insects, pp. 289-322. *In* V. H. Resh, and D. M. Rosenberg (eds.). The ecology of aquat. insects. Praeger Publs., NY. 625p.
254. Benke, A. C., and S. S. Benke. 1975. Comparative analysis and life histories of coexisting dragonfly populations. Ecology 56:302-317.
255. Benke, A. C., D. M. Gillespie, F. K. Parrish, T. C. Van Arsdall, R. J. Hunter, and R. L. Henry. 1979. Biological basis for assessing impacts of channel modifications: Invertebrate production, drift and fish feeding in a Southeastern blackwater river. Environ. Res. Center, Ga. Inst. Tech. Atlanta Rept. No. 06-79. 187 pp.
256. Benke, A. C., and K. A. Parsons. 1990. Modeling blackfly production dynamics in blackwater streams. Freshwat. Biol. 24:167-180.
257. Benke, A. C., and J. B. Wallace. 1980. Trophic basis of production among netspinning caddisflies in a southern Appalachian stream. Ecology 61:108-118.
258. Benke, A. C., T. Van Arsdall, Jr., D. M. Gillespie, and F. K. Parrish. 1984. Invertebrate productivity in a subtropical blackwater river: the importance of habitat and life history. Ecol. Monogr. 54:25-63.
259. Bennefield, B. L. 1965. A taxonomic study of the subgenus *Ladona* (Odonata: Libellulidae). Univ. Kans. Sci. Bull. 45:361-396.
260. Bennett, D. V., and E. F. Cook. 1981. The semi-aquatic Hemiptera of Minnesota (Hemiptera: Heteroptera). Minn. Agric. Exp. Sta. Tech. Bull. 332:1-59.
261. Bentinck, W. C. 1956. Structure and classification, pp. 68-73. *In* R. L. Usinger (ed.). Aquatic insects of California. Univ. Calif. Press, Berkeley. 508 pp.
262. Bentley, M. D., and J. F. Day. 1989. Chemical ecology and behavioral aspects of mosquito oviposition. Ann. Rev. Ent. 34:401-421.
263. Berg, C. O. 1949. Limnological relations of insects to plants of the genus *Potamogeton*. Trans. Am. Microsc. Soc. 68:279-291.
264. Berg, C. O. 1950a. Biology of certain aquatic caterpillars (Pyralididae. *Nymphula* spp.) which feed on *Potamogeton*. Trans. Am. Microsc. Soc. 69:254-266.
265. Berg, C. O. 1950b. Biology and certain Chironomidae reared from *Potamogeton*. Ecol. Monogr. 20:83-101.
266. Berg, C. O. 1950c. *Hydrellia* (Ephydridae) and some other acalyptrate Diptera reared from *Potamogeton*. Ann. Ent. Soc. Am. 43:374-398.
267. Berg, C. O. 1953. Sciomyzid larvae (Diptera) that feed on snails. J. Parasit. 39:630-636.
268. Berg, C. O. 1961. Biology of snail-killing Sciomyzidae (Diptera) of North America and Europe. Proc. 10th Int. Congr. Ent. 1:197-202.
269. Berg, C. O. 1964. Snail-killing sciomyzid flies: Biology of the aquatic species. Verh. Int. Verein. Limnol. 15:926-932.
270. Berg, C. O., B. A. Foote, L. V. Knutson, J. K. Barnes, S. L. Arnold, and K. Valley. 1982. Adaptive differences in phenology in sciomyzid flies. Mem. Ent. Soc. Wash. 10:15-36.
271. Berg, C. O., and L. Knutson. 1978. Biology and systematics of the Sciomyzidae. Ann. Rev. Ent. 23:239-258.
272. Berg, K. 1937. Contributions to the biology of *Corethra* Meigen (*Chaoborus* Lichtenstein). Biol. Med. 13:1-101.
273. Berg, K. 1942. Contributions to the biology of the aquatic moth *Acentropus niveus* (Oliv.). Vidensk. Medd. Dansk Naturhist. Foren. 105:59-139.
274. Berg, K. 1948. Biological studies of the River Susaa. Folia Limnol. Scand. 4:1-318.
275. Berg, M. B. 1995. Chapter 7. Larval food and feeding behaviour, pp. 136-168. *In* P. Armitage, P.S. Cranston, and L.C.V. Pinder (eds.). Chironomidae: The biology and ecology of non-biting midges. Chapman and Hall, London.
276. Berg, M. B., and R. A. Hellenthal. 1991. Secondary production of Chironomidae (Diptera) in a north temperate stream. Freshwat. Biol.25: 497-505.
277. Berg, M. B., and R. A. Hellenthal. 1992. Life histories and growth of lotic chironomids (Diptera: Chironomidae). Ann. Ent. Soc. Am. 85:578-589.
278. Bergelson, J. M. 1985. A mechanistic interpretation of prey selection by *Anax junius* larvae (Odonata: Aeschnidae). Ecology 66:1699-1705.
279. Bergey, E. A., and V. H. Resh. 1994a. Effects of burrowing by a stream caddisfly on case-associated algae. J. N. Am. Benthol. Soc. 13:379-390.
280. Bergey, E. A., and V. H. Resh. 1994b. Interactions between a stream caddisfly and the algae on its case: factors affecting algal quantity. Freshwat. Biol. 31:153-163.
281. Bernardo, M. J., E. W. Cupp, and A. E. Kiszewski. 1986. Rearing black flies (Diptera: Simuliidae) in the laboratory: Colonization and life table statistics for *Simulium vittatum*. Ann. Ent. Soc. Am. 79:610-621.
282. Berner, L. 1955. The southeastern species of *Baetisca* (Ephemeroptera: Baetiscidae). Quart. J. Fla. Acad. Sci. 18:1-19.
283. Berner, L. 1956. The genus *Neoephemera* in North America (Ephemeroptera: Neoephemeridae). Ann. Ent. Soc. Am. 49:33-42.
284. Berner, L. 1959. A tabular summary of the biology of North American mayfly nymphs (Ephemeroptera). Bull. Fla. State Mus. 4:1-58.
285. Berner, L. 1975. The mayfly family Leptophlebiidae in the southeastern United States. Fla. Ent. 58:137-156.
286. Berner, L. 1978. A review of the mayfly family Metretopodidae. Trans. Am. Ent. Soc. 104:91-137.
287. Berner, L., and M. L. Pescador. 1980. The mayfly family Baetiscidae (Ephemeroptera). Part I., pp. 511-524. *In* J. F. Flannagan and K. E. Marshall (eds.). Advances in Ephemeroptera biology. Plenum, N.Y. 552 pp.
288. Berner, L., and M. L. Pescador. 1988. The mayflies of Florida (rev. ed.). Univ. Presses of Florida, Gainesville. 415 p.
289. Berté, S. B., and G. Pritchard. 1986. The life histories of *Limnephilus externus* (Hagen), *Anabolia bimaculata* (Walker), and *Nemotaulius hostilis* (Hagen) (Trichoptera: Limnephilidae) in a pond in southern Alberta, Canada. Can. J. Zool. 64:2348-2356.
290. Bertrand, H. 1954. Les insects aquatiques d'Europe. Encyclopédie Entomologique. Ser. A. 31. 2:1-547.
291. Bertrand, H.P.I. 1972. Larves et nymphes des Coléopteres aquatiques du globe. Centre National de la Recherche Scientifique, Paris. 804 pp.
292. Betsch, J. M. 1980. Eléments pour une monographie des Collemboles Symphypléones (Hexapodes, Aptérygotes). Mem. Mus. Nat. Hist. Natur. (Paris), Ser. A. 116:1-227.
293. Betten, C. 1934. The caddis flies or Trichoptera of New York State. Bull. N.Y. State Mus. 292:1-576.

294. Betten, C. 1950. The genus *Pycnopsyche* (Trichoptera). Ann. Ent. Soc. Am. 43:508-522.
295. Beyer, A. 1939. Morphologische, ökologische und physiologische Studien an den Larven der Fliegen: *Ephydria riparia* Fallen, *E. micans* Haliday und *Cania fumosa* Stenhammar. Kieler Meereforsch. 3:265-320.
296. Bick, G. H. 1941. Life-history of the dragonfly, *Erythemis simplicollis* (Say). Ann. Ent. Soc. Am. 34:215-230.
297. Bick, G. H. 1950. The dragonflies of Mississippi. Am. Midl. Nat. 43:66-78.
298. Bick, G. H. 1951. The early nymphal stages of *Tramea lacerata* (Odonata:Libellulidae). Ent. News 62:293-303.
299. Bick, G. H. 1957a. The dragonflies of Louisiana. Tulane Studies Zool. 5:1-135.
300. Bick, G. H. 1957b. The Odonata of Oklahoma. Southwest. Nat. 2:1-18.
301. Bick, G. H. 1959. Additional dragonflies (Odonata) from Arkansas. Southwest. Nat. 4:131-133.
302. Bick, G. H. 1984a. Endangered U. S. species. Selysia 13:21.
303. Bick, G. H. 1984b. Odonata at risk in conterminous United States and Canada. Odonatologica 12:209-226.
304. Bick, G. H., and J. F. Aycock. 1950. The life history of *Aphylla williamsoni* Gloyd. Proc. Ent. Soc. Wash. 52:26-32.
305. Bick, G. H., and J. C. Bick. 1958. The ecology of the Odonata at a small creek in southern Oklahoma. J. Tenn. Acad. Sci. 33:240-251.
306. Bick, G. H., and J. C. Bick. 1970. Oviposition in *Archilestes grandis* (Rambur) (Odonata: Lestidae). Ent. News 81:157-163.
307. Bick, G. H., and J. C. Bick. 1983. New records of adult Odonata from Alabama, United States, in the Florida State Collection of Arthropods. Not. Odonatol. 2:19-21.
308. Bick, G. H., J. C. Bick, and L. E. Hornuff. 1977. An annotated list of the Odonata of the Dakotas Fla. Ent. 60:149-165.
309. Bick, G. H., and L. E. Hornuff. 1972. Odonata collected in Wyoming, South Dakota, and Nebraska. Proc. Ent. Soc. Wash. 74:1-8.
310. Bick, G. H., and L. E. Hornuff. 1974. New records of Odonata from Montana and Colorado. Proc. Ent. Soc. Wash. 76:90-93.
311. Bickley, W. E., S. R. Joseph, J. Mallack, and R. A. Berry. 1971. An annotated checklist of the mosquitoes of Maryland. Mosquito News 31:186-190.
312. Biever, K. D. 1965. A rearing technique for the colonization of chironomid midges. Ann. Ent. Soc. Am. 58:135-136.
313. Biever, K. D. 1971. Effect on diet and competition in laboratory rearing of chironomid midges. Ann. Ent. Soc. Am. 64:1166-1169.
314. Bilyj, B., and I. J. Davies. 1989. Descriptions and ecological notes on seven new species of *Cladotanytarsus* (Chironomidae: Diptera) collected from an experimentally acidified lake. Can. J. Zool. 67:948-962.
315. Bird, G. A., and N. K. Kausik. 1984. Survival and growth of early-instar nymphs of *Ephemerella subvaria* fed various diets. Hydrobiologia 119:227-233.
316. Bird, G. A., and N. K. Kaushik. 1985. Processing of decaying maple leaf, *Potamogeton* and *Cladophora* packs by invertebrates in an artificial stream. Arch. Hydrobiol. 105:93-104.
317. Bird, G. A., and N. K. Kaushik. 1985. Processing of elm and maple leaf discs by collectors and shredders in laboratory feeding studies. Hydrobiologia 126:109-120.
318. Bird, G. A., and N. K. Kaushik. 1987. Processing of maple leaf, grass and fern packs and their colonization by invertebrates in a stream. J. Freshwat. Ecol. 4:177-190.
319. Bird, R. D. 1932. Dragonflies of Oklahoma. Publ. Univ. Okla. Biol. Surv. 4:51-57.
320. Bishop, J. E. 1973. Observations on the vertical distribution of benthos in a Malaysian stream. Freshwat. Biol. 3:147-156.
321. Bishop, J. E., and H.B.N. Hynes. 1969. Downstream drift of the invertebrate fauna in a stream ecosystem. Arch. Hydrobiol. 66:56-90.
322. Bistrom, O. 1978. Dytiscidae from Newfoundland and adjacent areas (Coleoptera). Ann. Ent. Fenn. 44:65-71.
323. Bjarnov, N., and J. Thorup. 1970. A simple method for rearing running water insects, with some preliminary results. Arch. Hydrobiol. 67:201-209.
324. Blackwelder, R. E. 1932. The genus *Endeodes* LeConte (Coleoptera, Melyridae). Pan-Pacif. Ent. 8:128-136.
325. Blanckenhorn, W. U. 1991. Life-history differences in adjacent water strider populations: phenotypic plasticity or heritable responses to stream temperature? Evolution 45:1520-1525.
326. Blanckenhorn, W. U. 1991a. Foraging in groups of water striders (*Gerris remigis*): effects of variability in prey arrivals and handling times. Behav. Ecol. Sociobiol. 28:221-226.
327. Blanckenhorn, W. U. 1991b. Fitness consequences of foraging success in water striders (*Gerris remigis*; Heteroptera: Gerridae). Behav. Ecol. 2:46-55.
328. Blanton, F. S., and W. W. Wirth. 1979. The sand flies (*Culicoides*) of Florida (Diptera: Ceratopogonidae). Arthropods of Florida. Fla. Dep. Agric., Gainesville. 10:1-204.
329. Blatchley, W. S. 1910. An illustrated descriptive catalogue of the Coleoptera beetles (exclusive of the Rhyncophora) known to occur in Indiana. Bull. Ind. Dept. Geol. Nat. Res. 1:1-1386.
330. Blatchley, W. S. 1920. The Orthoptera of northeastern America. Nature, Indianapolis. 784 pp.
331. Blatchley, W. S. 1926. Heteroptera or true bugs of eastern North America. Nature, Indianapolis. 1116 pp.
332. Blickle, R. L. 1962. Hydroptilidae (Trichoptera) of Florida. Fla. Ent. 45:153-155.
333. Blickle, R. L. 1964. Hydroptilidae (Trichoptera) of Maine. Ent. News 75:159-162.
334. Blickle, R. L. 1979. Hydroptilidae (Trichoptera) of America north of Mexico. Bull. Univ. N.H. Agric. Exp. Sta. 509:1-97.
335. Blickle, R. L., and D. G. Denning. 1977. New species and a new genus of Hydroptilidae (Trichoptera). J. Kans. Ent. Soc. 50:287-300.
336. Blickle, R. L., and W. J. Morse. 1966. The caddis flies (Trichoptera) of Maine, excepting the family Hydroptilidae. Maine Agric. Exp. Sta. Tech. Bull. 24:1-12.
337. Blinn, D. W., C. Pinney, and M. W. Sanderson. 1982. Nocturnal planktonic behavior of *Ranatra montezuma* Polhemus (Nepidae: Hemiptera) in Montezuma Well, Arizona. J. Kans. Ent. Soc. 55:481-484.
338. Blinn, D. W., C. Runck, and R. W. Davies. 1993. The impact of prey behaviour and prey density on the foraging ecology of *Ranatra montezuma* (Heteroptera): a serological examination. Can. J. Zool. 71:387-391.
339. Blinn, D. W., and M. W. Sanderson. 1989. Aquatic insects in Montezuma Well, Arizona, USA: A travertine spring mound with high alkalinity and dissolved carbon dioxide. Great Basin Nat. 49:85-88.
340. Blois, C. 1985. Diets and resource partitioning between larvae of three anisopteran species. Hydrobiologia 126:221-227.
341. Blois-Heulin, C. 1990. Influence of prey densities on prey selection in *Anax imperator* larvae (Aeshnidae: Odonata). Aquat. Insects 12:209-217.
342. Blois-Heulin, C., P. H. Crowley, M. Arrington, and D. M. Johnson. 1990. Direct and indirect effects of predators on the dominant invertebrates of two freshwater littoral communities. Oecologia 84:295-306.
343. Bloomfield, E. N. 1897. Habits of *Sericomyia borealis* Fln. Entomol. mon. Mag. 33:222-223.
344. Blum, M. S. 1981. Chemical defenses of arthropods. Academic Press, NY.
345. Bobb, M. L. 1951a. Life history of *Ochterus banksi* Barber. Bull. Brooklyn Ent. Soc. 46:92-100.
346. Bobb, M. L. 1951b. The life history of *Gerris canaliculatus* Say in Virginia (Hemiptera: Gerridae). Va. J. Sci. 2:102-108.
347. Bobb, M. L. 1953. Observations on the life history of *Hesperocorixa interrupta* (Say) (Hemiptera: Corixidae). Va. J. Sci. 4:111-115.
348. Bobb, M. L. 1974. The aquatic and semiaquatic Hemiptera of Virginia. The Insects of Virginia: No. 7. Bull. Res. Div. Va. Poly. Inst. State Univ. 87:1-195.
349. Bode, R. W. 1983. Larvae of North American *Eukiefferiella* and *Tvetenia* (Diptera: Chironomidae). Bull. N. York St. Mus. 452:1-40.
350. Bode, R. W. 1990. Chironomidae. Chap. 14. *In* B.L. Peckarsky, P.R. Fraissinet, M.A. Penton, and D.J. Conklin, Jr. (eds.). Freshwater macroinvertebrates of northeastern North America. Cornell Univ. Press, Ithaca.

351. Boesel, M. W. 1983. A review of the genus *Cricotopus* in Ohio, with a key to adults of species of the northeastern United States (Diptera: Chironomidae). Ohio J. Sci. 83:74-90.
352. Boesel, M. W. 1985. A brief review of the genus *Polypedilum* in Ohio, with keys to known stages of species occurring in northeastern United States (Diptera: Chironomidae). Ohio J. Sci. 5:245-262.
353. Boesel, M. W., and E. G. Snyder. 1944. Observations on the early stages of the grass punky, *Atrichopogon levis* (Coquillett) (Diptera: Heleidae). Ann. Ent. Soc. Am. 37:37-46.
354. Bohart, B. E., and J. E. Gressitt. 1951. Filth-inhabiting flies of Guam. Bull. Bishop Mus. 204:1-152.
355. Bohart, R. M., and R. K. Washino. 1978. Mosquitoes of California (3rd ed.). Univ. Calif. Div. Agric. Sci. Printed Publ. No. 4084. 154 pp.
356. Bokerman, W.C.A. 1957. Frog eggs parasitized by dipterous larvae. Herpetologica 13:231-232.
357. Boling, R. H., E. D. Goodman, J. A. VanSickle, J. O. Zimmer, K. W. Cummins, R. C. Petersen, and S. R. Reice. 1975b. Toward a model of detritus processing in a woodland stream. Ecology 56:141-151.
358. Boling, R. H., Jr., R. C. Petersen, and K. W. Cummins. 1975a. Ecosystem modeling for small woodland streams, pp. 183-204. *In* B. C. Patten (ed.). Systems analysis and simulation in ecology, Vol. 3. Academic, N.Y. 601 pp.
359. Borger, G. A. 1980. Naturals. Stackpole Bks., Harrisburg, PA. 223p.
360. Borkent, A. 1984. The systematics and phylogeny of the *Stenochironomus* complex (*Xestochironomus*, *Harrisius*, and *Stenochironomus*) (Diptera: Chironomidae). Mem. Ent. Soc. Can. 128:269.
361. Borkent, A., and B. Bissett. 1990. A revision of the Holarctic genus of *Serromyia* Meigen (Diptera: Ceratopogonidae). Sys. Ent. 15:153-217.
362. Borkert, A. 1979. Systematics and bionomics of the species of the subgenus *Schadonophasma* Dyar and Shannon (*Chaoborus*, Chaoboridae. Diptera). Quaest. Ent. 15:122-255.
363. Borror, D. J. 1934. Ecological studies of *Argia moesta* Hagen (Odonata: Coenagrionidae) by means of marking. Ohio J. Sci. 34:97-108.
364. Borror, D. J. 1937. An annotated list of the dragonflies (Odonata) of Ohio. Ohio J. Sci. 37:185-196.
365. Borror, D. J. 1942. A revision of the libelluline Genus *Erythrodiplax* (Odonata). Ohio State Univ. Grad. School Stud. Contr. Zool. Ent. 4:1-286.
366. Borror, D. J. 1944. An Annotated List of the Odonata of Maine. Can. Ent. 76:134-150.
367. Borror, D. J. 1945. A key to the new world genera of Libellulidae (Odonata). Ann. Ent. Soc. Am. 38:168-194.
368. Borror, D. J., C. A. Triplehorn, and N. F. Johnson. 1989. An introduction to the study of insects (6th ed.). Saunders College Publ., Philadelphia. 875 pp.
369. Borror, D. J., and R. E. White. 1970. A field guide to the insects of America north of Mexico. Houghton Mifflin, Boston. 404 pp.
370. Bose, C., and N. S. Sen. 1985. Studies on the preferential feeding habits of the common water beetle, *Cybister tripunctatus asiaticus* (Sharp) (Dytiscidae: Coleoptera). Bangladesh J. Zool. 13:61-62.
371. Bosquet, Y. (ed.). 1991. Checklist of beetles of Canada and Alaska. Res. Br. Agric. Can. Publ. 1861/E. 430pp.
372. Bottorff, R. L., and A. W. Knight. 1989. Stonefly (Plecoptera) feeding modes: variation along a California river continuum. USDA Forest Service Gen. Tech. Rep. PSW. 110:235-241.
373. Bottorff, R. L., K. W. Stewart, and A. W. Knight. 1989. Description and drumming of *Susulus*, a new genus of stonefly (Plecoptera: Perlodidae) from California. Ann. Ent. Soc. Am. 82:545-554.
374. Botts, P. S., and B. C. Cowell. 1992. Feeding electivity of two epiphytic chironomids in a subtropical lake. Oecologica 89:331-337.
375. Boucek, Z., M.W.R. de V. Graham, and G. J. Kerrich. 1963. A revision of the European species of the genus *Mestocharis* Forster (Hym.: Chalcidoidea: Eulophidae). Entomologist 96:4-9.
376. Boulton, A. J. 1985. A sampling device that quantitatively collects benthos in flowing or standing waters. Hydrobiologia 127:31-39.
377. Boulton, A. J., and P. I. Boon. 1991. A review of methodology used to measure leaf litter decomposition in lotic environments: time to turn over an old leaf? Aust. J. Mar. Freshwat. Res. 42:1-43.
378. Boulton, A. J., H. M. Valett, and S. G. Fisher. 1992. Spatial distribution and taxonomic composition of the hyporheos of several Sonoran Desert streams. Arch. Hydrobiol. 125:37-61.
379. Bourassa, J. P., Y. Alarie, and R. LeClair, Jr. 1986. Distribution and habitat selection of dytiscid beetles in characteristic vegetal units of southern Quebec. Ent. Basiliensia 11:289-295.
380. Bournaud, M. 1963. Le courant, facteur écologique et éthologique de la vie aquatique. Hydrobiologia 22:125-165.
381. Boving, A. G. 1910. Natural history of the larvae of Donaciinae. Int. Revue ges. Hydrobiol. Biol. Suppl. 1:1-108.
382. Boving, A. G., and F. C. Craighead. 1930. An illustrated synopsis of the principal larval forms of the order Coleoptera. J. Ent. Soc. Am. 11:1-351.
383. Bowen, T. W. 1983. Production of the predaceous midge tribes Sphaeromiini and Palpomyiini (Diptera: Ceratopogonidae) in Lake Norman, N. C. Hydrobiologia 99:81-87.
384. Bowles, D. E. 1989. New records of *Sialis* (Megaloptera: Sialidae) from Arkansas and Oklahoma. Ent. News. 100:27-28.
385. Bowles, D. E. 1990. Life history and variability of secondary production estimates for *Corydalus cornutus* (Megaloptera: Corydalidae) in an Ozark stream. J. Agric. Ent. 7:61-70.
386. Bowles, D. E., and R. T. Allen. 1992. Life histories of six species of caddisflies (Trichoptera) in an Ozark Stream, U.S.A. J. Kans. Ent. Soc. 65:174-184.
387. Bowles, M. L., and M. L. Mathis. 1989. Caddisflies (Insecta: Trichoptera) of mountainous regions in Arkansas, with new state records for the order. J. Kansas Ent. Soc. 62:234-244.
388. Bowles, D. E., and M. L. Mathis. 1992a. A preliminary checklist of the caddisflies (Insecta: Trichoptera) of Oklahoma. Insecta Mundi 6:29-35.
389. Bowles, D. E., and M. L. Mathis. 1992b. Variation in the terminalia of *Neohermes concolor* with a key to males of *Neohermes* in eastern North America (Megaloptera: Corydalidae: Chauliodinae). Insecta Mundi 6:145-150.
390. Bradshaw, W. E. 1970. Interaction of food and photoperiod in the termination of larval diapause in *Chaeoborus americanus* (Diptera: Culicidae). Biol. Bull. 139:476-484.
391. Bradshaw, W. E. 1980. Thermoperiodism and the thermal environment of the pitcher-plant mosquito, *Wyeomyia smithii*. Oecologia 46:13-17.
392. Braimah, S. A. 1985. Mechanisms of filter-feeding in aquatic insects. Ph.D. thesis, Univer. of Alberta, Edmonton, Alberta.
393. Braimah, S. A. 1987a. Mechanisms of filter feeding in immature *Simulium bivittatum* Malloch (Diptera: Simuliidae) and *Isonychia campestris* McDunnough (Ephemeroptera: Oligoneuriidae). Can. J. Zool. 65:504-513.
394. Braimah, S. A. 1987b. Pattern of flow around filter-feeding structures of immature *Simulium bivittatum* Malloch (Diptera: Simuliidae) and *Isonychia campestris* McDunnough (Ephemeroptera: Oligoneuriidae). Can. J. Zool. 65:514-521.
395. Braimah, S. A. 1987c. The influence of water velocity on particle capture by the labral fans of larvae of *Simulium bivittatum* Malloch (Diptera: Simuliidae). Can. J. Zool. 65:2395-2399.
396. Branch, H. E. 1923. The life history of *Chironomus cristatus* Fabr. with descriptions of the species. J. N.Y. Ent. Soc. 1:15-30.
397. Branham, J. M., and R. R. Hathaway. 1975. Sexual differences in the growth of *Pteronarcys californica* Newport and *Pteronarcella badia* (Hagen) (Plecoptera). Can. J. Zool. 53:501-506.
398. Bratt, A. D., L. V. Knutson, B. A. Foote, and C. O. Berg. 1969. Biology of *Pherbellia* (Diptera:Sciomyzidae). Mem. Cornell Univ. Exp. Sta. 404:1-247.
399. Braun, A. F. 1917. The Nepticulidae of North America. Trans. Am. Ent. Soc. 43:155-201.
400. Breeland, S. G., and T. M. Loyless. 1982. Illustrated keys to the mosquitoes of Florida. Adult females and fourth stage larvae. J. Fla. Anti-mosq. Assoc. 53:63-84.

401. Breland, O. P. 1958. A report on *Haemogus* mosquitoes in the United States with notes on identification. Ann. Ent. Soc. Am. 51:217-221.
402. Brenner, R. J., and E. W. Cupp. 1980. Rearing black flies (Diptera: Simuliidae) in a closed system of water circulation. Tropenmed. Parasit. 31:247-258.
403. Bretschko, G., and W. Klemens. 1985. Quantitative sampling of the fauna of gravel streams (Project RITRODAT-LUNZ). Verh. Int. Verein. Limnol. 22:2049-2052.
404. Bretschko, G., and W. Klemens. 1986. Quantitative methods and aspects in the study of the interstitial fauna of running waters. Stygologia 2:297-316.
405. Bridges, C. A. 1993. Catalogue of the family-group, genus-group and species-group names of the Odonata of the World. Charles A. Bridges, Urbana, IL. 774pp.
406. Brigham, A. R., W. U. Brigham, and A. Gnilka (eds.). 1982. The aquatic insects and oligochaetes of North and South Carolina. Midwest Aquatic Enterprises, Mahomet, Ill. 837 pp.
407. Brigham, A. R., and D. D. Herlong. 1982. Aquatic and semi-aquatic Lepidoptera, p.12.1-12.36. *In* A. R. Brigham, W. U. Brigham, and A. Gnilka, (eds.). Aquatic insects and oligochaetes of North and South Carolina, Midwest Aquat. Enterprises, Mahomet, IL.
408. Brigham, W. U. 1981. *Ectopria leechi*, a new false water penny from the United States (Coleoptera: Eubriidae). Pan-Pacif. Ent. 57:313-320.
409. Brigham, W. U. 1982. Aquatic Coleoptera, pp. 10.1-10.136. *In* A. R. Brigham, W. U. Brigham, and A. Gnilka (eds.). Aquatic insects and oligochaetes of North and South Carolina. Midwest Enterprises, Mahomet, Ill. 837 pp.
410. Brimley, C. S. 1938. The Insects of North Carolina. Carolina Dep. Agric. Div. Ent. pp. 36-42.
411. Brinck, P. 1949. Studies on Swedish stoneflies. Opusc. Ent. Suppl. 11:1-250.
412. Brinck, P. 1956. Reproductive system and mating in Plecoptera. Opusc. Ent. 21:57-127.
413. Brinck, P. 1958. On a collection of stoneflies (Plecoptera) from Newfoundland and Labrador. Opusc. Ent. 23:47-58.
414. Brindle, A. 1973. Taxonomic notes on the larvae of British Diptera, 28. The larvae and pupae of *Hydrodromia stagnalis* (Holiday). Entomologist 106:249-252.
415. Brinkhurst, R. O. 1959. Alary polymorphism in the Gerroidea. J. Anim. Ecol. 28:211-230.
416. Brinkhurst, R. O. 1960. Studies on the functional morphology of *Gerris najas* DeGeer. Proc. Zool. Soc. Lond. 133:531-559.
417. Brinkhurst, R. O., P. M. Chapman, and M. A. Farrell. 1983. A comparative study of respiration rates of some aquatic oligochaetes in relation to sublethal stress. Int. Revue ges. Hydrobiol. 68:683-699.
418. Brinkhurst, R. O., K. E. Chua, and E. Batoosingh. 1969. Modifications in sampling procedures as applied to studies on bacteria and tubificid oligochaetes inhabiting aquatic sediments. J. Fish. Res. Bd. Can. 26:2581-2593.
419. Britt, N. W. 1967. Biology of two species of Lake Erie mayflies, *Ephoron album* (Say) and *Ephemera simulans* Walker. Bull. Ohio Biol. Surv. 1:1-70.
420. Brittain, J. E. 1973. The biology and life cycle of *Nemoura avicularis* Morton. Freshwat. Biol. 3:199-210.
421. Brittain, J. E. 1974. Studies on the lentic Ephemeroptera and Plecoptera in southern Norway. Nor. Ent. Tidsskr. 21:135-154.
422. Brittain, J. E. 1976. Experimental studies on nymphal growth in *Leptophlebia vespertina* (L.) (Ephemeroptera). Freshwat. Biol. 6:445-449.
423. Brittain, J. E. 1980. Mayfly strategies in a Norwegian subalpine lake, pp. 179-186. *In* J. F. Flannagan, and K. E. Marshall (eds.). Advances in Ephemeroptera biology. Plenum, N.Y. 552 pp.
424. Brittain, J. E. 1982. Biology of mayflies. Ann. Rev. Ent. 27:119-147.
425. Brittain, J. E. 1990. Life history strategies in Ephemeroptera and Plecoptera, pp. 1-12. *In* I. C. Campbell (ed.). Mayflies and stoneflies: life histories and biology. Kluwer Acad. Publ., Dordrecht, The Netherlands. 366pp.
426. Brittain, J. E., and T. J. Eikeland. 1988. Invertebrate drift: a review. Hydrobiologia 166:77-93.
427. Brittain, J. E., A. Lillehammer, and S. J. Saltveit. 1986. Intraspecific variation in the nymphal growth rate of the stonefly, *Capnia atra* (Plecoptera). J. Anim. Ecol. 55:1001-1006.
428. Brittain, J. E., and R. A. Mutch. 1984. The effect of water temperature on the egg incubation period of *Mesocapnia oenone* (Plecoptera) from the Canadian Rocky Mountains. Can. Ent. 116:549-554.
429. Britton, E. B. 1966. On the larva of *Sphaerius* and the systematic position of the Sphaeriidae (Coleoptera). Austr. J. Zool. 14:1193-1198.
430. Britton, L. J., and P. E. Greeson (eds.). 1987. Methods for collection and analysis of aquatic biological and microbiological samples. Tech. Water-Resources Inv., Bk. 5, Chap. A4. U. S. Geological Survey, Denver, CO.
431. Britton, W. E. (ed.) 1923. Guide to the insects of Connecticut. Part IV. The Hemiptera or sucking insects of Connecticut. Conn. State Geol. Nat. Hist. Surv. Bull. 34:1-807.
432. Brock, E. M. 1960. Mutualism between the midge *Cricotopus* and the alga *Nostoc*. Ecology 41:474-483.
433. Brock, M. L., R. G. Wiegert, and T. D. Brock. 1969. Feeding of *Paracoenia* and *Ephydra* (Diptera: Ephydridae) on the microorganisms of hot springs. Ecology 50:192-200.
434. Brock, T. D., and M. L. Brock. 1968. Life in a hot water basin. Nat. Hist. 77:47-53.
435. Brock, T.C.M., and G. Van Der Velde. 1983. An autecological study on *Hydromyza livens* (Fabricius) (Diptera, Scatomyzidae), a fly associated with nymphaeid vegetation dominated by *Nuphar*. Tijdschr. Ent. 126:59-90.
436. Brodo, F. 1967. A review of the subfamily Cylindrotominae in North America (Diptera: Tipulidae). Univ. Kans. Sci. Bull. 47:71-115.
437. Brodo, F. 1987. A revision of the genus *Prionocera* (Diptera: Tipulidae). Evol. Monogr. 8:1-93.
438. Brodsky, K. A. 1980. Mountain torrent of the Tien Shan. A faunistic-ecology essay. (Translated from Russian by V. V. Golosov). Junk, The Hague. 311 pp.
439. Brömark, C. B. Malmqvist, and C. Otto. 1984. Anti-predator adaptations in a neustonic insect (*Velia caprai*). Oecologia 61:189-191.
440. Brooker, M. P. 1979. The life cycle and growth of *Sialis lutaria* L. (Megaloptera) in a drainage channel under different methods of plant management. Ecol. Ent. 4:111-117.
441. Brooks, A. R., and L. A. Kelton. 1967. Aquatic and semi-aquatic Heteroptera of Alberta, Saskatchewan and Manitoba. Mem. Ent. Soc. Can. 51:1-92.
442. Brothers, D. B. 1971. A checklist of the mosquitoes of Idaho. Tebiwa, Idaho State Univ. Mus. 14:72-73.
443. Brown, A. V., and L. C. Fitzpatrick. 1978. Life history and population energetics of the dobson fly, *Corydalus cornutus*. Ecology 59:1091-1108.
444. Brown, A. V., M. D. Schram, and P. P. Brussock. 1987. A vacuum benthos sampler suitable for diverse habitats. Hydrobiologia 153:241-247.
445. Brown, C. L. 1984. Improved above-substrate sampler for macrophytes and phytomacrofauna. Progr. Fish. Cult. 20:142-144.
446. Brown, C.J.D. 1934. A Preliminary List of Utah Odonata. Occ. Pap. Univ. Mich. Mus. Zool. 291:1-17.
447. Brown, D. S. 1960. The ingestion and digestion of algae by *Cloeon dipterum* L. (Ephemeroptera). Hydrobiologia 16:81-96.
448. Brown, D. S. 1961. The food of the larvae of *Cloeon dipterum* L. and *Baetis rhodani* (Pictet) (Ephemeroptera). J. Anim. Ecol. 30:55-75.
449. Brown, H. P. 1951. *Climacia areolaris* (Hagen) parasitized by a new pteromalid (Hym.: Chalcidoidea). Ann. Ent. Soc. Am. 44:103-110.
450. Brown, H. P. 1952. The life history of *Climacia areolaris* (Hagen), a neuropterous "parasite" of freshwater sponges. Am. Midl. Nat. 47:130-160.
451. Brown, H. P. 1968. *Psephenus* (Coleoptera: Psephenidae) parasitized by a new chalcidoid (Hymenoptera: Eulophidae). II. Biology of the parasite. Ann. Ent. Soc. Am. 61:452-456.
452. Brown, H. P. 1970a. A key to the dryopoid genera of the new world. Ent. News 81:171-175.

453. Brown, H. P. 1970b. *Neocyllopus*, a new genus from Texas and Central America (Coleoptera: Elmidae). Coleopt. Bull. 24:1-28.
454. Brown, H. P. 1972a. Synopsis of the genus *Heterelmis* Sharp in the United States, with description of a new species from Arizona (Coleoptera, Dryopoidea, Elmidae). Ent. News 83:229-238.
455. Brown, H. P. 1972b. Aquatic dryopoid beetles (Coleoptera) of the United States. Biota of freshwater ecosystems identification manual no. 6. Wat. Poll. Conf. Res. Ser., E.P.A., Washington, D.C. 82 pp.
456. Brown, H. P. 1973. Survival records for elmid beetles with notes on laboratory rearing of various dryopoids. Ent. News 84:278-284.
457. Brown, H. P. 1981a. A distribution survey of the world genera of aquatic dryopoid beetles (Coleoptera: Dryopoidea: Elmidae and Psephenidae sens. lat.). water quality indicators. Pan-Pacif. Ent. 57:133-148.
458. Brown, H. P. 1981b. *Huleechius*, a new genus of riffle beetles from Mexico and Arizona (Coleoptera: Drypoidea: Elmidae). Pan-Pacif. Ent. 57:228-244.
459. Brown, H. P. 1981c. Key to the world genera of Larinae (Coleoptera: Dryopoidea: Elmidae) with descriptions of new genera from Hispaniola, Columbia, Australia, and New Guinea. Pan-Pacif. Ent. 57:76-104.
460. Brown, H. P. 1985. *Xenelmis sandersoni* new species of riffle beetle from Arizona and northern Mexico (Coleoptera: Dryopoidea: Elmidae). Southwest. Nat. 30:53-58.
461. Brown, H. P. 1987. Biology of riffle beetles. Ann. Rev. Ent. 38:253-274.
462. Brown, H. P., and C. M. Murvosh. 1970. *Lutrochus arizonicus* new species, with notes on ecology and behavior (Coleoptera, Dryopoidea, Limnichidae). Ann. Ent. Soc. Am. 63:1030-1035.
463. Brown, H. P., and C. M. Murvosh. 1974. A revision of the genus *Psephenus* (waterpenny beetles) on the United States and Canada (Coleoptera, Dryopoidea, Psephenidae). Trans. Ent. Soc. Am. 100:289-340.
464. Brown, H. P., and C. M. Shoemake. 1964. Oklahoma riffle beetles (Coleoptera: Dryopoidea). III. Additional state and county records. Proc. Okla. Acad. Sci. 44:42-43.
465. Brown, H. P., and D. S. White. 1978. Notes on separation and identification of North American riffle beetles (Coleoptera: Dryopoidea: Elmidae). Ent. News 89:1-13.
466. Brown, J. M. 1929. Freshwater Collembola. Naturalist 2:111-113.
467. Brues, C. T. 1928. Studies on the fauna of hot springs in the Western United States and the biology of thermophilous animals. Proc. Am. Acad. Arts Sci. 63:139-228.
468. Brues, C. T., A. L. Melander, and F. M. Carpenter. 1954. Classification of insects. Bull. Harvard Mus. Comp. Zool. 108:1-917.
469. Brundin, L. 1949. Chironomiden und andere Bodentiere der südschwedischen Urgebirgsseen. Rep. Inst. Freshwat. Res. Drottningholm 30:1-94.
470. Brundin, L. 1956. Zur Systematik der Orthocladiinae (Diptera: Chironomidae). Rep. Inst. Freshwat. Res. Drottningholm 37:5-185.
471. Brundin, L. 1965. On the real nature of transantarctic relationships. Evolution 19:496-505.
472. Brundin, L. 1966. Transantarctic relationships and their significance, as evidenced by chironomid midges, with a monograph of the subfamilies Podonominae and Aphroteniinae and the Austral Heptagyinae. Kungl. Svenska Vetenskapsakad. Handl., Fjorde Sev. (4)11:1-472.
473. Brusven, M. A. 1970. Drift periodicity of some riffle beetles (Coleoptera: Elmidae). J. Kans. Ent. Soc. 43:364-371.
474. Brusven, M. A., and A. C. Scoggan. 1969. Sarcophagus habits of Trichoptera larvae on dead fish. Ent. News 80:103-105.
475. Bryce, D., and A. Hobart. 1972. The biology and identification of the larvae of the Chironomidae (Diptera). Ent. Gaz. 23:175-217.
476. Buckingham, G. R., and B. M. Ross. 1981. Notes on the biology and host specificity of *Acentria nivea*. J. Aquat. Plant Manage. 19:32-36.
477. Buckingham, G. R., and C. A. Bennett. 1989. Laboratory host range of *Parapoynx diminutalis* (Lepidoptera: Pyralidae), an Asian aquatic moth adventive in Florida and Panama on *Hydrilla vericillata* (Hydrocharitaceae). Environ. Ent. 18:526-530.
478. Bueler, C. M. 1985. Feeding preference of *Pteronarcys pictetii* (Plecoptera: Insecta) from a small, acidic, woodland stream. Fla. Ent. 67:383-401.
479. Buikema, A. L., Jr., and J. Cairns, Jr. 1980. Aquatic invertebrate bioassays. Am. Soc. Test. Mater, Philadelphia. 209 pp.
480. Buikema, A. L., Jr., and J. R. Voshell, Jr. 1993. Toxicity studies using freshwater benthic macroinvertebrates, pp. 344-398. *In* D. M. Rosenberg and V. H. Resh (eds.). Freshwater biomonitoring and benthic macroinvertebrates. Chapman and Hall, NY. 488 pp.
481. Burger, J. F. 1977. The biosystematics of immature Arizona Tabanidae (Diptera). Trans. Am. Ent. Soc. 103:145-258.
482. Burger, J. F., J. R. Anderson, and M. F. Knudsen. 1980. The habits and life history of *Oedoparena glauca* (Diptera: Dryomyzidae), a predator of barnacles. Proc. Ent. Soc. Wash. 82:360-377.
483. Burger, J. F., and R. V. Andreeva. 1988. Rheophilic larvae of horse flies (Diptera: Tabanidae): ecology, morphology, distribution. Vestn. Zool. 1988 2:17-23.
484. Burger, J. F., D. J. Lake, and M. L. McKay. 1981. The larval habitats and rearing of some common *Chrysops* species (Diptera: Tabanidae) in New Hampshire. Proc. Ent. Soc. Wash. 83:373-389.
485. Burger, J. F., L. A. Martinez, L. L. Pechuman, and L. V. Bermudez. 1990. A revision of the horse fly genus *Agkistrocerus* Philip (Diptera: Tabanidae). Pan-Pac. Ent. 66:181-194.
486. Burger, J. F., and L. L. Pechuman. 1986. A review of the genus Haematopota (Diptera: Tabanidae): in North America. J. Med. Ent. 23:345-352.
487. Burghele, A. 1959. New Rumanian species of Dachnusini (Hym.: Braconidae) and some ecological observations upon them. Entomol. mon. Mag. 95:121-126.
488. Burke, H. R. 1963. Notes on Texas riffle beetles (Coleoptera: Elmidae). Southwest. Nat. 8:111-114.
489. Burks, B. D. 1953. The mayflies, or Ephemeroptera, of Illinois. Bull. Ill. Nat. Hist. Surv. 26:1-216.
490. Burks, B. D. 1967. The North American species of *Aprostocetus* Westwood (Hymenoptera: Eulophidae). Ann. Ent. Soc. Am. 60:756-760.
491. Burks, B. D. 1968. *Psephenus* (Coleoptera: Psephenidae) parasitized by a new chalcidoid (Hymenoptera: Eulophidae). I. Description of the parasite. Ann. Ent. Soc. Am. 61:450-452.
492. Burton, G. J. 1973. Feeding of *Simulium hargreavesi* Gibbons larvae on *Oedogonium* algal filaments in Ghana. J. Med. Ent. 10:101-106.
493. Burton, G. J., and T. M. McRae. 1972a. Observations on trichopteran predators on aquatic stages of *Simulium damnosum* and other *Simulium* species in Ghana. J. Med. Ent. 9:289-294.
494. Burtt, E. T. 1940. A filter-feeding mechanism in a larva of the Chironomidae (Diptera:Nematocera). Proc. R. ent. Soc. Lond. (A) 15:113-121.
495. Busacca, J. D., and B. A. Foote. 1978. Biology and immature stages of two species of *Notiphila*, with notes on other shore flies occurring in cattail marshes. Ann. Ent. Soc. Am. 71:457-466.
496. Buschman. L. L. 1984. Biology of the firefly *Pyractomena lucifera* (Coleoptera: Lampyridae). Fla. Ent. 67:529-542.
497. Butcher, F. G. 1930. Notes on the cocooning habits of *Gyrinus*. J. Kans. Ent. Soc. 3:64-66.
498. Butcher, F. G. 1933. Hymenopterous parasites of Gyrinidae with descriptions of new species of *Hemiteles*. Ann. Ent. Soc. Am. 26:76-85.
499. Butler, E. A. 1923. A biology of the British Hemiptera-Heteroptera. H. F. Witherby, Lond. 696 pp.
500. Butler, M. C. 1984. Life histories of aquatic insects, pp. 24-55. *In* V. H. Resh and D. M. Rosenberg (eds.). The ecology of aquatic insects. Praeger, Publs., NY. 622 pp.
501. Butler, M. G. 1982. A 7-year life cycle for two *Chironomus* species in arctic Alaskan tundra ponds (Diptera: Chironomidae). Can. J. Zool. 60:58-70.
502. Butler, M. G., and D. H. Anderson. 1990. Cohort structure, biomass, and production of a merovoltine *Chironomus* population in a Wisconsin bog lake. J. N. Am. Benthol. Soc. 9:180-192.
503. Byers, C. F. 1927. An annotated list of the Odonata of Michigan. Occ. Pap. Univ. Mich. Mus. Zool. 183:1-16.

504. Byers, C. F. 1930. A contribution to the knowledge of Florida Odonata. Univ. Fla. Publ. 1:1-327.
505. Byers, C. F. 1937. A review of the dragonflies of the genera *Neurocordulia* and *Platycordulia*. Misc. Publ. Univ. Mich. Mus. Zool. 36:1-36.
506. Byers, C. F. 1939. A study of the dragonflies of the genus *Progomphus* (Gomphoides), with a description of a new species. Proc. Fla. Acad. Sci. 4:19-85.
507. Byers, C. F. 1941. Notes of the emergence and life history of the dragonfly *Pantala flavescens*. Proc. Fla. Acad. Sci. 6:14-25.
508. Byers, C. F. 1951. Some notes on the Odonata fauna of Mountain Lake, Virginia. Ent. News 62:164-167.
509. Bystrak, P. G., and W. W. Wirth. 1978. The North American species of *Forcipomyia*, subgenus *Euprojoannisia* (Diptera: Ceratopogonidae). U.S. Dep. Agric. Tech. Bull. 1591:1-51.
510. Caillere, L. 1972. Dynamics of the strike of *Agrion* (Syn. *Calopteryx*) *splendens* Harris 1782 larvae (Odonata: Calopterygidae). Odonatologica 11-19.
511. Cain, D. J., S. N. Luoma, J. L. Carter, and S. V. Fend. 1992. Aquatic insects as bioindicators of trace element contamination in cobble-bottom rivers and streams. Can. J. Fish. Aquat. Sci. 49:2141-2154.
512. Cairns, J., Jr., D. W. Albaugh, F. Busey, and M. D. Chanay. 1968. The Sequential Comparison Index - a simplified method for nonbiologists to estimate relative differences in biological diversity in stream pollution studies. J. Wat. Poll. Contr. Fed. 40:1607-1613.
513. Cairns, J., Jr., and J. R. Pratt. 1993. A history of biological monitoring using benthic macroinvertebrates, pp. 10-27. *In* D. M. Rosenberg and V. H. Resh (eds.). Freshwater biomonitoring and benthic macroinvertebrates. Chapman and Hall, NY. 488 pp.
514. Calabrese, D. M. 1974. Keys to the adults and nymphs of the species of *Gerris* Fabricius occurring in Connecticut. Mem. Conn. Ent. Soc. 1974:228-266.
515. Calabrese, D. M. 1977. The habitats of *Gerris* F. (Hemiptera: Heteroptera: Gerridae) in Connecticut. Ann. Ent. Soc. Am. 70: 977-983.
516. Calabrese, D. M. 1978. Life history data for ten species of waterstriders (Hemiptera: Heteroptera: Gerridae) in Connecticut. Trans. Kans. Acad. Sci. 61:257-264.
517. Calabrese, E. J., C. C. Chamberlain, R. Coler, and M. Young. 1987. The effects of trichloroacetic acid, a widespread product of chlorine disinfection, on the dragonfly nymph respiration. J. Environ. Sci. Health A22:343-355.
518. Caldwell, B. A. 1976. The distribution of *Nigronia serricornis* and *Nigronia fasciatus* in Georgia and water quality parameters associated with the larvae (Megaloptera: Corydalidae). Bull. Ga. Acad. Sci. 34:24-31.
519. Call, S. M. 1982. Distributional records and observations of *Nigronia serricornis* (Megaloptera: Corydalidae) in eastern Kentucky. Trans. Ky. Acad. Sci. 43:146-149.
520. Callahan, J. R. 1974. Observations on *Gerris incognitus* and *Gerris gillettei* (Heteroptera: Gerridae). Proc. Ent. Soc. Wash. 76:15-21.
521. Calow, P., and G. E. Petts (eds.). 1993. The rivers handbook, Vol I. Blackwell Sci. Publ. Oxford, U.K. 526p.
522. Calvert, P. P. 1893. Catalogue of the Odonata (dragonflies) of the vicinity of Philadelphia. Trans. Am. Ent. Soc. 20:152-272.
523. Calvert, P. P. 1934. The rates of growth, larval development and seasonal distribution of the dragonflies of the genus *Anax* (Aeshnidae). Proc. Am. Phil. Soc. 73:1-70.
524. Calvert, P. P. 1937. Methods of rearing Odonata, pp. 270-273. *In* J. G. Needham (ed.). Culture methods for invertebrate animals. Comstock, Ithaca, N.Y. 590 pp.
525. Camargo, J. A. 1991. Toxic effects of residual chlorine on larvae of *Hydropsyche pellucidula* (Trichoptera, Hydropsychidae): a proposal of biological indicator. Bull. Environ. Contam. Toxicol. 47:261-265.
526. Camargo, J. A. 1992a. Changes in a hydropsychid guild downstream from a eutrophic impoundment. Hydrobiologia 239:25-32.
527. Camargo, J. A. 1992b. Structural and trophic alterations in macrobenthic communities downstream from a fish farm outlet. Hydrobiologia 242:41-49.
528. Cameron, A. E. 1926. Bionomics of the Tabanidae (Diptera) of the Canadian prairie. Bull. Ent. Res. 17:1-42.
529. Cameron, G. N. 1976. Do tides affect coastal insect communities? Am. Midl. Nat. 95:279-287.
530. Camilo, G. R., and M. R. Willig. 1995. Dynamics of a food chain model from an arthropod-dominated lotic community. Ecol. Model. (in press).
531. Camilo, R. C., and M. R. Willig. 1993. Diet of some common insects in the South Llano River. Tex. J. Sci. 45:100-104.
532. Campbell, B. C. 1979. The spatial and seasonal abundance of *Trichocorixa verticalis* (Hemiptera: Corixidae) in salt marsh intertidal pools. Can. Ent. 111:1005-1011.
533. Campbell, I. C. (ed.). 1990. Mayflies and stoneflies: life history and biology. Kluwer Acad. Publ., Dordrecht, The Netherlands.
534. Campeau, S., H. R. Murkin, and R. D. Titman. 1994. Relative importance of algae and emergent plant litter to freshwater marsh invertebrates. Can. J. Fish. Aquat. Sci. 51:681-692.
535. Cannings, R. A. 1978. The distribution of *Tanypteryx hageni* (Odonata:Petaluridae) in British Columbia. J. Ent. Soc. Brit. Columbia 75:18-19.
536. Cannings, R. A. 1982a. Notes on the biology of *Aeshna sitchensis* Hagen (Anisoptera: Aeshnidae). Odonatologica 11: 219-223.
537. Cannings, R. A. 1982b. Dragonfly days. Nature, Can. 11:12-17.
538. Cannings, R. A., and G. P. Doerksen. 1979. Description of the larva of *Ischnura erratica* (Odonata: Coenagrionidae) with notes on the species of British Columbia. Can. Ent. 111:327-331.
539. Cannings, R. A., S. G. Cannings, and R. J. Cannings. 1980. The distribution of the genus *Lestes* in a saline lake series in central British Columbia, Canada (Zygoptera: Lestidae). Odonatologica 9:19-28.
540. Cannings, R. A., and K. M. Stuart. 1977. The dragonflies of British Columbia. Brit. Columbia Provin. Mus. Handbk. No. 35, 256 pp.
541. Cannings, S. G. 1981. New distributional records of Odonata from northwestern British Columbia. Syesis 13:13-15.
542. Cannings, S. G., and R. A. Cannings. 1980. The larva of *Coenagrion interrogatum* (Odonata: Coenagrionidae). with notes on the species in the Yukon. Can. Ent. 112:437-441.
543. Cannings, S. G., R. A. Cannings, and R. J. Cannings. 1991. Distribution of the dragonflies (Insecta: Odonata) of the Yukon Territory, Canada with notes on ecology and behaviour. Contrib. Nat. Sci. 13:1-27.
544. Canterbury, L. E. 1978. Studies of the genus *Sialis* (Sialidae: Megaloptera) in eastern North America. Ph.D. diss., University of Louisville. 93 pp.
545. Canterbury, L. E., and S. E. Neff. 1980. Eggs of *Sialis* (Sialidae: Megaloptera) in eastern North America. Can. Ent. 112:409-419.
546. Canton, S. P., and J. W. Chadwick. 1984. A new modified Hess sampler. Progr. Fish Cult. 46:57-59.
547. Capek, M. 1970. A new classification of the Braconidae (Hymenoptera) based on the cephalic structures of the final instar larva and biological evidence. Can. Ent. 102:846-875.
548. Capps, H. W. 1956. Keys for the identification of some lepidopterous larvae frequently intercepted at quarantine. U.S. Dep. Agric. E-475. 37 pp.
549. Carey, W. E., and F. W. Fisk. 1965. The effect of water temperature and dissolved oxygen content on the rate of gill movement of the hellgrammite *Corydalus cornutus*. Ohio J. Sci. 65:137-141.
550. Cargill, A. S., II, K. W. Cummins, B. J. Hanson, and R. R. Lowry. 1985. The role of lipids as feeding stimulants for shredding aquatic insects. Freshwat. Biol. 15:455-464.
551. Cargill, A. S., II, K. W. Cummins, B. J. Hanson, and R. R. Lowry. 1985. The role of lipids, fungi, and temperature in the nutrition of a shredder caddisfly, *Clistoronia magnifica*. Freshwat. Invertebr. Biol. 4:64-78.
552. Carle, F. L. 1979a. Two new *Gomphus* (Odonata: Gomphidae) from eastern North America with adult keys to the subgenus *Hylogomphus*. Ann. Ent. Soc. Am. 72:418-426.
553. Carle, F. L. 1979b. Environmental monitoring potential of the Odonata, with a list of rare and endangered Anisoptera of Virginia, United States. Odonatologica 8:319-323.
554. Carle, F. L. 1980. A new *Lanthus* (Odonata: Gomphidae) from eastern North America with adult and nymphal keys to American Octogomphines. Ann. Ent. Soc. Am. 73:172-179.

555. Carle, F. L. 1981. A new species of *Ophiogomphus* from eastern North America, with a key to the regional species (Anisoptera: Gomphidae). Odonatologica 10:271-278.
556. Carle, F. L. 1982. *Ophiogomphus incurvatus.* a new name for *Ophiogomphus carolinus* Hagen (Odonata: Gomphidae). Ann. Ent. Soc. Am. 75:335-339.
557. Carle, F. L. 1983. A new *Zoraena* (Odonata: Cordulegastridae) from eastern North America, with a key to the adult Cordulegastridae of America. Ann. Ent. Soc. Am. 76:61-68.
558. Carle, F. L. 1986. The classification, phylogeny and biogeography of the Gomphidae (Anisoptera). I. Classification. Odonatologica 15:275-326.
559. Carle, F. L. 1992. *Ophiogomphus* (Ophionurus) *australis* spec. nov. from the Gulf coast of Louisiana, with larval and adult keys to American *Ophiogomphus* (Anisoptera: Gomphidae). Odonatologica 21:141-152.
560. Carlson, D. 1971. A method for sampling larval and emerging insects using an aquatic black light trap. Can. Ent. 103:1365-1369.
561. Carlson, D. 1972. Comparative value of black light and cool white lamps in attracting insects to aquatic traps. J. Kans. Ent. Soc. 45:194-199.
562. Carlsson, G. 1962. Studies on Scandinavian black flies. Opusc. Ent. 21:1-280.
563. Carlsson, M., L. M. Nilsson, Bj. Svensson, S. Ulfstrand, and R. S. Wotton. 1977. Lacustrine seston and other factors influencing the blackflies (Diptera: Simuliidae) inhabiting lake outlets in Swedish Lapland. Oikos 29:229-238.
564. Carpenter, F. M. 1992. Treatise on invertebrate paleontology, part R: Arthropoda (4), Vol. III: Hexapoda. Geol. Soc. Am. Univ. Kansas, Boulder. 656pp. (2 parts, Odonata and Protodonata: pp. 59-89, 524-525).
565. Carpenter, S. J. 1941. The mosquitoes of Arkansas. Ark. State Bd. Health, Little Rock. 87 pp.
566. Carpenter, S. J. 1968. Review of recent literature on mosquitoes of North America. Calif. Vector Views 15:71-98.
567. Carpenter, S. J. 1970. Review of recent literature on mosquitoes of North America. Suppl. Calif. Vector Views 17:39-65.
568. Carpenter, S. J. 1974. Review of recent literature on mosquitoes of North America. Suppl. II. Calif. Vector Views 21:73-99.
569. Carpenter, S. J., and W. J. LaCasse. 1955. Mosquitoes of North America. Univ. Calif. Press, Berkeley. 360 pp.
570. Carpenter, S. J., W. W. Middlekauff, and R. W. Chamberlain. 1946. The mosquitoes of the southern United States east of Oklahoma and Texas. Am. Midl. Nat. Monogr. 3292 pp.
571. Carpenter, S. R. 1982. Stemflow chemistry: Effects on population dynamics of detritivorous mosquitoes in tree-hole ecosystems. Oecologia 53:1-6.
572. Carpenter, V. 1991. Dragonflies and damselflies of Cape Cod. The Cape Cod Mus. Nat. Hist. 80pp.
573. Carr, R. I., and M. S. Topping. 1983. Aspects of the life history of the hellgrammite in southwestern Missouri, *Corydalus cornutus.* Trans. Missouri Acad. Sci. 17:196.
574. Cashatt, E. D. 1987. New state records, range extensions, and confirmed records of Odonata from Illinois, United States. Not. Odonatol. 2:152-153.
575. Caspers, H. 1951. Rhythmische Erscheinungen in der Fortpflanzung von *Clunio marinus* (Dipt. Chiron.) und das Problem der lunaren Periodizität bein Organismen. Arch. Hydrobiol. Suppl. 18:415-594.
576. Caspers, N. 1980. Zur Larvalentwicklung und Produktionsökologie von *Tipula maxima* Poda (Diptera, Nematocera, Tipulidae). Arch. Hydrobiol. 58:273-309.
577. Caspers, V. N., and R. Wagner. 1982. Investigations on the insect emergence from a woodland brook near Bonn. VII. Emergence of Empididae and Dolichopodidae in 1976 (Insecta, Diptera, Brachycera). Arch. Hydrobiol. 93:209-237.
578. Cassani, J. R. 1985. Biology of *Simyra henrici* (Lepidoptera: Noctuidae). Fla. Ent. 68:645-652.
579. Cassani, J. R., and H. D. Newson. 1980. An annotated list of mosquitoes reported from Michigan. Mosquito News 40:356-368.
580. Cather, M. R., and A. R. Gaufin. 1975. Life history of *Megarcys signata* (Plecoptera: Perlodidae), Mill Creek, Wasatch Mountains, Utah. Great Basin Nat. 35:39-48.
581. Cather, M. R., and A. R. Gaufin. 1976. Comparative ecology of three *Zapada* species of Mill Creek, Wasatch Mountains, Utah (Plecoptera: Nemouridae). Am. Midl. Nat. 95:464-471.
582. Cather, M. R., B. P. Stark, and A. R. Gaufin. 1975. Records of stoneflies (Plecoptera) from Nevada. Great Basin Nat. 35:49-50.
583. Cattaneo, A. 1983. Grazing on epiphytes. Limnol. Oceanogr. 28:124-32.
584. Caucci, A., and R. Nastasi. 1975. Hatches. Comparahatch, N.Y. 320 pp.
585. Caudell, A. N. 1922. A diving wasp. Proc. Ent. Soc. Wash. 24:125-126.
586. Cavenaugh, W. J., and J. E. Tilden. 1930. Algal food, feeding, and case building habits of the larva of the midge fly *Tanytarsus dissimilis* Johannsen. Ecology 5:105-115.
587. Center, T. D. 1984. Dispersal and variation in infestation intensities of water hyacinth moth, *Sameodes albiguttalis* (Lepidoptera: Pyralidae), populations in peninsular Florida. Environ. Ent. 13:482-491.
588. Chadwick, J. W., and S. P. Canton. 1992. Comparison of multiplate and surber samplers in a Colorado mountain stream. J. Freshwat. Ecol. 2:287-292.
589. Chaffee, D. L., and D. C. Tarter. 1979. Life history and ecology of *Baetisca bajkovi* Neave, in Beech Fork of Twelvepole Creek, Wayne County, West Virginia (Ephemeroptera: Baetiscidae). Psyche 86:53-61.
590. Chagnon, G. 1922. A hymenopteran of aquatic habits. Can. Ent. 65:24.
591. Chagnon, G., and O. Fournier. 1948. Contribution a l'etude des Hempiteres aquatiques du Québec. Contr. L'Inst. Biol. Univ. Montréal, Quebec. 21:1-66.
592. Chamberlin, J. C., and G. F. Ferris. 1929. On *Liparocephalus* and allied genera (Coleoptera: Staphylinidae). Pan-Pacif. Ent. 5:137-143, 5:153-162.
593. Chan, K. L., and J. R. Linley. 1990. Distribution of immature *Atrichopogon wirthi* (Diptera: Ceratopogonidae) on leaves of the water lettuce, *Pistia stratiotes.* Environ. Ent. 19:286-292.
594. Chan, K. L., and J. R. Linley. 1991. Distribution of immature *Dasyhelea chani* (Diptera: Ceratopogonidae) on leaves of *Pistia stratiotes.* Ann. Ent. Soc. Am. 84:61-66.
595. Chance, M. M. 1970. The functional morphology of the mouthparts of black fly larvae (Diptera: Simuliidae). Quaest. Ent. 6:245-284.
596. Chance, M. M., and D. A. Craig. 1986. Hydrodynamics and behaviour of Simuliidae larvae (Diptera). Can. J. Zool. 64:1295-1309.
597. Chandler, H. P. 1954. Four new species of dobsonflies from California. Pan Pacif. Ent. 30:105-111.
598. Chandler, H. P. 1956a. Aquatic Neuroptera, pp. 234-236. *In* R. L. Usinger (ed.). Aquatic insects of California. Univ. Calif. Press. Berkeley. 508 pp.
599. Chandler, H. P. 1956b. Megaloptera, pp. 229-233. *In* R. L. Usinger (ed.). Aquatic insects of California. Univ. Calif. Press, Berkeley. 508 pp.
600. Chang, S. L. 1966. Some physiological observations on two aquatic Collembola. Trans. Am. Microsc. Soc. 85:359-371.
601. Chao, H-fu, and Y. Zhang. 1981. Two new species of *Agriotypus* from Jilin province (Hymenoptera: Agriotypidae). Entomotaxonomia 3:79-86.
602. Chapin. J. W. 1978. Systematics of Nearctic *Micrasema* (Trichoptera: Brachycentridae). Ph.D. diss., Clemson University, Clemson, S.C. 136 pp.
603. Chapman, D. W., and R. Demory. 1963. Seasonal changes in the food ingested by aquatic insect larvae and nymphs in two Oregon streams. Ecology 44:140-146.
604. Chapman, H. C. 1958. Notes on the identity, habitat and distribution of some semi-aquatic Hemiptera of Florida. Fla. Ent. 41:117-124.
605. Chapman, H. C. 1959. Distributional and ecological records for some aquatic and semi-aquatic Heteroptera of New Jersey. Bull. Brooklyn Ent. Soc. 54:8-12.

606. Chapman, H. C. 1962. The Saldidae of Nevada (Hemiptera). Pan-Pacif. Ent. 38:147-159.
607. Chapman, J. A., and J. M. Kinghorn. 1955. Window-trap for flying insects. Can. Ent. 82:46-47.
608. Chapman, R. F. 1982. The insects: structure and function (3rd ed.). Harvard Univ. Press, Cambridge. 919 pp.
609. Charles, W. N., K. East, and T. D. Murray. 1976. Production of larval Tanypodinae (Insecta: Chironomidae) in the mud at Loch Leven, Kinross. Proc. R. Soc. Edinb. 75:157-169.
610. Cheary, B. S. 1971. The biology, ecology and systematics of the genus *Laccobius (Laccobius)* of the new world. Ph.D. diss., University of California, Riverside. 178 pp.
611. Chen, C. S., Y. N. Lin, C. L. Chung, and H. Hung. 1979. Preliminary observations on the larval breeding sites and adult resting places of a bloodsucking midges, *Forcipomyia* (*Lasiohelea*) *taiwana* (Shiraki) (Diptera: Ceratopogonidae). Bull. Soc. Ent. 14:51-59.
612. Cheng, L. 1967. Studies on the biology of the Gerridae (Hem., Heteroptera). I: Observations on the feeding of *Limnogonus fossarum* (F.) Entomol mon. Mag. 102:121-129.
613. Cheng, L., (ed.). 1976. Marine insects. North Holland, Amsterdam. 581 pp.
614. Cheng, L. 1985. Biology of *Halobates* (Heteroptera: Gerridae). Ann. Rev. Ent. 30:111-135.
615. Cheng, L., and C. H. Fernando. 1970. The waterstriders of Ontario. Misc. Life Sci. Publ. Roy. Ont. Mus. 23 pp.
616. Cheng, L., and C. H. Fernando. 1971. Life history and biology of the riffle bug, *Rhagovelia obesa* Uhler (Heteroptera: Veliidae) in Southern Ontario. Can. J. Zool. 49:435-442.
617. Cheng, L., and J. H. Frank. 1993. Marine insects and their reproduction, pp 479-506. In A. D. Ansell, R. N. Gibson, and M. Barnes (eds.). Oceanogr. Mar. Biol. Annu. Rev. UCLA Press.
618. Cheng, L., and C. L. Hogue. 1974. New distribution and habitat records of biting midges and mangrove flies from the coasts of southern Baja California, Mexico (Diptera: Ceratopogonidae, Culicidae, Chironomidae, and Phoridae). Ent. News 85:211-218.
619. Chesson, J. 1984. Effect of notonectids (Hemiptera: Notonectidae) on mosquitoes (Diptera: Culicidae): Predation or selective oviposition? Environ. Ent. 13:531-538.
620. Chesson, J. 1989. The effect of alternative prey on the functional response of *Notonecta hoffmanni*. Ecology 70:1227-1235.
621. Chevenet, F., S. Dolédec, and D. Chessel. 1994. A fuzzy coding approach for the analysis of long-term ecological data. Freshwat. Biol. 31:295-309.
622. Chillcott, J. G. 1961. A revision of the genus *Roederioides* Coquillett (Diptera, Empididae). Can. Ent. 93:419-428.
623. China, W. E. 1955. The evolution of the water bugs. Bull. Nat. Inst. Sci. India 7:91-103.
624. China, W. E., and N.C.E. Miller. 1959. Checklist and keys to the families and subfamilies of the Hemiptera—Heteroptera. Bull. Brit. Mus. Nat. Hist. Ent. 8:1-45.
625. Chow, V. T. 1959. Open channel hydraulics. McGraw-Hill Book Co., New York, NY. 680p.
626. Chowdhury, S. H., and P. S. Corbet. 1988. Feeding rate of larvae of *Enallagma cyathigerum* in the presence of conspecifics and predators (Zygoptera: Coenagrionidae). Odonatologica 17:115-119.
627. Chowdhury, S. H., and P. S. Corbet. 1989. Feeding-related behaviour in larvae of *Enallagma cyathigerum* (Zygoptera: Coenagrionidae). Odonatologica 18:285-288.
628. Chowdhury, S. H., P. S. Corbet, and I. F. Harvey. 1989. Feeding and prey selection by larvae of *Enallagma cyathigerum* in relation to size and density of prey (Zygoptera: Coenagrionidae). Odonatologica 18:1-11.
629. Christiansen, K. 1964. Bionomics of the Collembola. Ann. Rev. Ent. 9:147-178.
630. Christiansen, K. A., and B. F. Bellinger. 1988. Marine littoral Collembola of North and Central America. Bull. Marine Sci. 42:215-245.
631. Christiansen, K., and P. Bellinger. 1980-81. The Collembola of North America north of the Rio Grande. Grinnell College. Grinnell, Iowa. 1322 pp.
632. Christman, V. D., and J. R. Voshell, Jr. 1992. Life history, growth and production of Ephemeroptera in experimental ponds. Ann. Ent. Soc. Am. 85:705-712.
633. Christophers, S. R. 1960. *Aedes aegypti* (L.) the yellow fever mosquito. Its life history, bionomics and structure. Univ. Press, Cambridge. 739pp.
634. Chu, H. F. 1949. How to know the immature insects. Wm. C. Brown, Dubuque. 234 pp.
635. Chu, H. F. 1956. The nomenclature of the chaetotaxy of lepidopterous larvae and its application. Acta Ent. Sinica 6:323-333.
636. Chutter, F. M. 1971. A reappraisal of Needham and Usinger's data on the variability of a stream fauna when sampled with a Surber sampler. Limnol. Oceanogr. 17:139-141.
637. Chutter, F. M. 1972. An empirical biotic index of the quality of water in South African streams and rivers. Wat. Res. 6:19-30.
638. Cianciara, S. 1979. Life cycles of *Cloeon dipterum* (L.) in natural environment. Polish Arch. Hydrobiol. 26:501-513.
639. Cianciara, S. 1980. Food preference of *Cloeon dipterum* (L.) larvae and dependence of their development and growth on the type of food. Polish Arch. Hydrobiol. 27:143-160.
640. Ciborowski, J. H., and D. A. Craig. 1991. Factors influencing dispersion of larval black flies (Diptera: Simuliidae): Effects of the presence of an invertebrate predator. Can. J. Zool. 69:1120-1123.
641. Ciborowski, J.J.H. 1991. Estimating processing time of stream benthic samples. Hydrobiologia 222:101-107.
642. Claassen, P. W. 1921. *Typha* insects: their ecological relationships. Mem. Cornell Univ. Agric. Exp. Sta. 57:459-531.
643. Claassen, P. W. 1922. The larva of a chironomid (*Trissolcadius equitans* n. sp.) which is parasitic upon a mayfly nymph (*Rhithrogena sp.*). Univ. Kans. Sci. Bull. 14:395-405.
644. Claassen, P. W. 1931. Plecoptera nymphs of America (north of Mexico). Thomas Say Found. Ent. Soc. Am. 3:1-199.
645. Clark, W. H. 1985. First record of *Climacia californica*, (Neuroptera:Sisyridae) and its host sponge *Ephydatia muelleri* (Porifera: Spongillidae) from Idaho with water quality relationships. Great Basin Nat. 45:391-394.
646. Clark, W. H., and G. L. Ralston. 1974. *Eubrianax edwardsi* in Nevada with notes on larval pupation and emergence of adults (Coleoptera: Psephenidae). Coleopt. Bull. 28:217-218.
647. Clarke, G. M. 1993. Fluctuating asymmetry of invertebrate populations as a biological indicator of environmental quality. Environ. Poll. 82:207-211.
648. Clastrier, J., and W. W. Wirth. 1978. The *Leptoconops kerteszi* complex in North America. U.S. Dep. Agric. Tech. Bull. 1573:1-58.
649. Clausen, C. P. 1931. Biological observations on *Agriotypus* (Hymenoptera). Proc. Ent. Soc. Wash. 33:29-37.
650. Clausen, C. P. 1940. Entomophagous insects. McGraw-Hill, N.Y. 688 pp.
651. Clausen, C. P. 1950. Respiratory adaptations in the immature stages of parasitic insects. Arthropoda 1:198-224.
652. Clay, M. E., and C. E. Venard. 1972. Larval diapause in the mosquito *Aedes triseriatus*. Effects of diet and temperature on photoperiod induction. J. Insect Physiol. 18:1441-1446.
653. Clemens, W. A. 1917. An ecological study of the mayfly *Chirotenetes*. Univ. Toronto Stud. Biol. Ser. 17:5-43.
654. Clements, A. N. 1963. The physiology of mosquitoes. Pergamon, N.Y. 393 pp.
655. Clements, A. N. 1992. The biology of mosquitoes, Vol. I. Development, nutrition, and reproduction. Chapman and Hall, NY. 509pp.
656. Clements, W. H. 1991. Characterization of stream benthic communities using substrate-filled trays: colonization, variability, and sampling selectivity. J. Freshwat. Ecol. 6:209-221.
657. Clifford, H. F. 1976. Observations on the life cycle of *Siphloplecton basale* (Walker) (Ephemeroptera: Metretopodidae). Pan-Pacif. Ent. 52:265-271.
658. Clifford, H. F. 1982. Life cycles of mayflies (Ephemeroptera), with special reference to voltinism. Quaest. Ent. 18:15-90.
659. Clifford, H. F. 1991. Aquatic invertebrates of Alberta. Univ. Alberta Press, Edmonton, Alberta, Canada. 538p.

660. Clifford, H. F., and D. R. Barton. 1979. Observations on the biology of *Ametropus neavei* (Ephemeroptera: Ametropidae) from a large river in northern Alberta, Canada. Can. Ent. 111:855-858.
661. Clifford, H. F., and H. Boerger. 1974. Fecundity of mayflies (Ephemeroptera), with special reference to mayflies of a brown-water stream of Alberta, Canada. Can. Ent. 106:1111-1119.
662. Clifford, H. F., and R. J. Casey. 1992. Differences between operators in collecting quantitative samples of stream macroinvertebrates. J. Freshwat. Ecol. 7:271-276.
663. Clifford, H. F., H. Hamilton, and B. A. Killins. 1979. Biology of the mayfly *Leptophlebia cupida* (Say) (Ephemeroptera: Leptophlebiidae). Can. J. Zool 57:1026-1045.
664. Clifford, H. F., and H. R. Hamilton. 1987. Volume of material ingested by mayfly nymphs of various sizes from some Canadian streams. J. Freshwat. Ecol. 4:259-261.
665. Clifford, H. F., and K. A. Zelt. 1972. Assessment of two mesh sizes for interpreting life cycles, standing crop, and percentage composition of stream insects. Freshwat. Biol. 2:259-269.
666. Cloarec, A. 1988. Behavioral adaptations to aquatic life in insects: An example. Adv. in the Study of Behavior 18:99-151.
667. Cobben, R. H. 1968. Evolutionary trends in Heteroptera. Part I. Eggs, architecture of the shell, gross embryology, and eclosion. Centre Agric. Publ. Documentation, Wageningen. 475 pp.
668. Cobben, R. H. 1978. Evolutionary trends in Heteroptera. Part II. Mouthpart structures and feeding strategies. Meded. Lab. Ent. 289:1-407.
669. Coffman, W. P. 1967. Community structure and trophic relations in a small woodland stream, Linesville Creek, Crawford County, Pennsylvania. Ph.D. diss., University of Pittsburgh, Pittsburgh.
670. Coffman, W. P. 1973. Energy flow in a woodland stream ecosystem: II. The taxonomic composition and phenology of the Chironomidae as determined by the collection of pupal exuviae. Arch. Hydrobiol. 71:281-322.
671. Coffman, W. P., K. W. Cummins, and J. C. Wuycheck. 1971. Energy flow in a woodland stream ecosystem. I. Tissue support trophic structure of the autumnal community. Arch. Hydrobiol. 68:232-276.
672. Coffman, W. P., L. C. Ferrington, Jr., and R. W. Seward. 1988. *Paraboreochlus stahli*, n. sp. A new species of Podonominae (Diptera: Chironomidae from North America. Aquat. Insects 10:189-200.
673. Cofrancesco, A. F., Jr., and F. G. Howell. 1982. Influence of temperature and time of day on ventilatory activities of *Erythemis simplicicollis* Say (Odonata) naiads. Environ. Ent. 11:313-317.
674. Colbo, M. H., and G. N. Porter. 1979. Effects of the food supply on the life history of Simuliidae (Diptera). Can. J. Zool. 57:301-306.
675. Colbo, M. H., and G. N. Porter. 1981. The interaction of rearing, temperature and food supply on the life history of two species of Simuliidae (Diptera). Can. J. Zool. 59:158-163.
676. Cole, F. R. (with collaboration of E. I. Schlinger). 1969. The flies of western North America. Univ. Calif. Press, Berkeley. 693 pp.
677. Cole, G. A. 1983. Textbook of limnology (3rd ed.). C. V. Mosby Co., St. Louis. 401 pp.
678. Coleman, M. J., and H.B.N. Hynes. 1970a. The vertical distribution of the invertebrate fauna in the bed of a stream. Limnol. Oceanogr. 15:31-40.
679. Coleman, M. J., and H.B.N. Hynes. 1970b. The life-histories of some Plecoptera and Ephemeroptera in a Southern Ontario stream. Can. J. Zool. 48:1333-1339.
680. Coler, R. A., and R. C. Haynes. 1966. A practical benthos sampler. Progr. Fish Cult. 28:95.
681. Colletti, P. J., D. W. Blinn, A. Pickart, and V. T. Wagner. 1987. Influence of different densities of the mayfly grazer *Heptagenia criddlei* on lotic diatom communities. J. N. Am. Benthol. Soc. 6:270-280.
682. Collier, J. E. 1970. A taxonomic revision of the genus *Optioservus* (Coleoptera: Elmidae) in the Nearctic region. Diss. Abstr. Int. 30(B):46-48.
683. Collins, F. H., and R. K. Washino. 1985. Insect predators, pp. 25-41. *In* H. C. Chapman (ed.). Biological control of mosquitoes. Bull. Am. Mosq. Contr. Assoc. 6. 218 pp.
684. Collins, N. C. 1975. Population biology of the brine fly (Diptera: Ephydridae) in the presence of abundant algal food. Ecology 56:1139-1148.
685. Common, I.F.B. 1970. Lepidoptera, pp. 765-866. *In* The insects of Australia. CSIRO, Melbourne Univ. Press Melbourne. 1029 pp.
686. Connell, T. D., and J. F. Scheiring. 1981. The feeding ecology of the larvae of the shore fly *Scatella picea* (Walker) (Diptera: Ephydridae). Can. J. Zool. 59:1831-1835.
687. Conroy, J. C., and J. L. Kuhn. 1977. New annotated records of Odonata from the province of Manitoba with notes on their parasitism by larvae of water mites. Manitoba Ent. 11:27-40.
688. Contreras-Ramos, A. 1990. The immature stages of *Platyneuromus* (Corydalidae) with a key to the genera of larval Megaloptera of Mexico. M.S. Thesis. Univ. of Alabama, Tuscaloosa. 109 pp.
689. Contreras-Ramos, A. 1995. A remarkable range extension for the fishfly genus Dysmicohermes (Megaloptera: Corydalidae). Ent. News 106:123-126.
690. Cook, E. F. 1956. The Nearctic Chaoborinae (Diptera: Culicidae). Univ. Minn. Agric. Exp. Sta. Tech. Bull. 218:1-102.
691. Cook, E. F. 1981. Chap. 24. Chaoboridae, pp. 335-340. *In* J. F. McAlpine, B. V. Peterson, G. E. Shewell, H. J. Teskey, J. R. Vockeroth, and D. M. Wood (coords.). Manual of Nearctic Diptera, Vol. 1. Res. Branch, Agric. Can. Monogr 27. Ottawa 674 pp.
692. Cooper, C. M. 1987. Benthos in Bear Creek, Mississippi: Effects of habitat variation and agricultural sediments. J. Freshwat. Ecol. 4:101-114.
693. Cooper, S. D. 1983. Selective predation on cladocerans by common pond insects. Can. J. Zool. 61:879-886.
694. Cooper, S. D., and L. A. Barmuta. 1993. Field experiments in biomonitoring, pp. 399-441. *In* D. M. Rosenberg and V. H. Resh (eds.). Freshwater biomonitoring and benthic macroinvertebrates. Chapman and Hall, NY. 488 pp.
695. Cooper, S. D., D. W. Smith, and J. R. Bence. 1985. Prey selection by freshwater predators with different foraging strategies. Can. J. Fish. Aquat. Sci. 42:1720-1732.
696. Cooper, S. D., S. J. Walde, and B. L. Peckarsky. 1990. Prey exchange rates and the impact of predators on prey populations in streams. Ecology 71:1503-1514.
697. Coquillett, D. W. 1898. The buffalo-gnats, or black-flies,of the United States. [A synopsis of the dipterous family Simuliidae.]. U.S. Dept. Agr., Div. Ent. Bull. 10:66-69.
698. Corbet, P. S. 1955. The immature stages of the emperor dragonfly, *Anax imperator* (Leach) (Odonata: Aeshnidae). Ent. Gaz. 6:189-204.
699. Corbet, P. S. 1956. The life-histories of *Lestes sponsa* (Hansemann) and *Sympetrum striolatum* (Charpentier) (Odonata). Tijdschr. Ent. 99:217-229.
700. Corbet, P. S. 1957a. The life histories of two spring species of dragonfly (Odonata: Zygoptera). Ent. Gaz. 8:79-89.
701. Corbet, P. S. 1957b. The life-history of the emperor dragonfly *Anax imperator* (Leach) (Odonata: Aeshnidae). J. Anim. Ecol. 26:1-69.
702. Corbet, P. S. 1960. Fossil history, pp. 149-163. *In* P. S. Corbet, C. Longfield, and N. W. Moore (eds.). Dragonflies. Collins, London. 260 pp.
703. Corbet, P. S. 1963. A biology of dragonflies. Quandrangle, Chicago. 247 pp.
704. Corbet, P. S. 1964. Temporal patterns of emergence in aquatic insects. Can. Ent. 96:264-279.
705. Corbet, P. S. 1965. An insect emergence trap for quantitative studies in shallow ponds. Can. Ent. 97:845-848.
706. Corbet, P. S. 1978. Concluding remarks. Symposium: Seasonality in New Zealand insects. N. Z. Ent. Soc. 6:367.
707. Corbet, P. S. 1979. Odonata. *In* H. V. Danks (ed.). Canada and its insect fauna. Mem. Ent. Soc. Can. 108:308-311.
708. Corbet, P. S. 1980. Biology of Odonata. Ann. Rev. Ent. 25:189-217.
709. Corbet, P. S., C. Longfield, and N. W. Moore. 1960. Dragonflies. Collins, London. 260 pp.
710. Corkum, L. D. 1978a. The influence of density and behavioral type on the active entry of two mayfly species (Ephemeroptera) into the water column. Can. J. Zool. 56:1201-1206.

711. Corkum, L. D. 1978b. The nymphal development of *Paraleptophlebia adoptiva* (McDunnough) and *Paraleptophlebia mollis* (Eaton) (Ephemeroptera: Leptophlebiidae) and the possible influence of temperature. Can. J. Zool. 56:1842-1846.
712. Corkum, L. D. 1989. Habitat characterization of the morphologically similar mayfly larvae, *Caenis* and *Tricorythodes* (Ephemeroptera). Hydrobiologia 179:103-110.
713. Corkum, L. D., and P. J. Pointing. 1979. Nymphal development of *Baetis vagans* McDunnough (Ephemeroptera: Baetidae) and drift habits of large nymphs. Can. J. Zool. 57:2348-2354.
714. Corpus, L. D. 1981a. A brief survey of the Dolichopodidae. Proc. Wash. State Ent. Soc. 43:617-618.
715. Corpus, L. D. 1981b. Preliminary data on the immature stages of *Pelastoneurus vagans* Loew (Diptera: Dolichopodidae). Proc. Wash. State Ent. Soc. 43:610-612.
716. Corpus, L. D. 1986a. Immature stages of *Liancalus similis* (Diptera: Dolichopodidae). J. Kans. Ent. Soc. 59:635-640.
717. Corpus, L. D. 1986b. Biological notes and descriptions of the immature stages of *Pelastoneurus vagans* Loew (Diptera: Dolichopodidae). Proc. Ent. Soc. Wash. 88:673-679.
718. Corpus, L. D. 1988. Immature stages of *Tachytrechus auratus* (Diptera: Dolichopodidae). Pan-Pac. Ent. 64:9-15.
719. Correa, M., E. J. Calabrese, and R. A. Coler. 1985a. Effects of trichloroacetic acid, a new contaminant found from chlorinating water with organic material, on dragonfly nymphs. Bull. Environ. Contam. Toxicol. 34:271-274.
720. Correa, M., R. A. Coler, and C.-M. Yin. 1985b. Changes in oxygen consumption and nitrogen metabolism in the dragonfly *Somatochlora cingulata* exposed to aluminum in acid waters. Hydrobiologia 121:151-156.
721. Correa, M., R. A. Coler, C.-M. Yin, and E. Kaufman. 1986. Oxygen consumption and ammonia excretion in the detritivore caddisfly *Limnephilus* sp. exposed to low pH and aluminum. Hydrobiologia 140:237-241.
722. Coulson, J. C. 1962. The biology of *Tipula subnodicornis* Zetterstedt with comparative observations on *Tipula paludosa* Meigen. J. Anim. Ecol. 31:1-21.
723. Courtemanch, D. L., and S. P. Davies. 1987. A coefficient of community loss to assess detrimental change in aquatic communities. Wat. Res. 21:217-222.
724. Courtney, G. W. 1986. Discovery of the immature stages of *Parasimulium crosskeyi* Peterson (Diptera: Simuliidae), with a discussion of a unique black fly habitat. Proc. Ent. Soc. Wash. 88:280-286.
725. Courtney, G. W. 1989. Morphology, systematics and ecology of mountain midges (Diptera: Deuterophlebiidae). Ph.D. Diss., Univ. of Alberta, Edmonton.
726. Courtney, G. W. 1990a. Cuticular morphology of larval mountain midges (Diptera: Deuterophlebiidae): implications for the phylogenetic relationships of Nematocera. Can. J. Zool. 68:556-578.
727. Courtney, G. W. 1990b. Revision of Nearctic mountain midges (Diptera: Deuterophlebiidae). J. Nat. Hist. 2:81-118.
728. Courtney, G. W. 1991a. Life history patterns of Nearctic mountain midges (Diptera: Deuterophlebiidae). J. N. Am. Benthol. Soc. 10:177-197.
729. Courtney, G. W. 1991b. Phylogenetic analysis of the Blephariceromorpha, with special reference to mountain midges (Diptera: Deuterophlebiidae). Syst. Ent. 16:137-172.
730. Courtney, G. W. 1993. Archaic black flies and ancient forests: conservation of Parasimulium habitats in the Pacific Northwest. Aquat. Conserv. Mar. Freshwat. Ecosys. 3:361-373.
731. Courtney, G. W. 1994a. Biosystematics of the Nymphomyiidae (Insecta: Diptera): life history, morphology and phylogenetic relationships. Smithson. Contr. Zool. 550:1-41.
732. Courtney, G. W. 1994b. Revision of Palaearctic mountain midges (Diptera: Deuterophlebiidae), with phylogenetic and biogeographic analyses of world species. Sys. Ent. 19:1-24.
733. Cowan, C. A., and B. L. Peckarsky. 1990. Feeding by a lotic grazer as quantified by gut fluorescence. J. N. Am. Benthol. Soc. 9:368-378.
734. Cowan, C. A., and B. L. Peckarsky. 1993. Diel feeding and positioning periodicity of a grazing mayfly in a trout stream and a fishless stream. Can. J. Aquat. Sci. 51:450-459.
735. Cowardin, L. M., V. Carter, F. C. Golet, and E. T. LaRoe. 1979. Classification of wetlands and deepwater habitats of the United States. U. S. Fish Wildlife Serv. Pub. FWS/OBS-79/31, Washington, D. C., 103 p.
736. Craig, D. A. 1966. Techniques for rearing stream-dwelling organisms in the laboratory. Tuatara 14:65-72.
737. Craig, D. A. 1977a. Mouthparts and feeding behaviour of Tahitian larval Simuliidae (Diptera: Nematocera). Quaest. Ent. 13:195-218.
738. Craig, D. A. 1977b. A reliable chilled water stream for rheophilic insects. Mosquito News 37:773-774.
739. Craig, D. A. 1990. Behavioural hydrodynamics of *Cloeon dipterum* larvae (Ephermeroptera: Baetidae). J. N. Am. Benthol. Soc. 9:346-357.
740. Craig, D. A., and M. M. Galloway. 1986. Hydrodynamics of larval back flies, pp. 170-185. *In* K. C. Kim and R. W. Merritt (eds.). Black flies: ecology, population management, and annotated world list. Penn. State Univ. Press, Univ. Park, PA.
741. Craig, P. 1970. The behavior and distribution of the intertidal sand beetle, *Thinopinus pictus*. Ecology 51:1012-1017.
742. Crampton, G. C. 1930. A comparison of the more important structural details of the larva of the archaic tanyderid dipteran *Protoplasa fitchii*, with other Holometabola, from the standpoint of phylogeny. Bull. Brooklyn Ent. Soc. 35:235-258.
743. Crampton, G. C. 1942. Guide to the insects of Connecticut. Diptera or true flies of Connecticut. Part VI, Fasc. I. External Morphology. Bull. Conn. State Geol. Nat. Hist. Surv. 64:10-165.
744. Cranston, P. S. 1982a. The metamorphosis of *Symposiocladius lignicola* (Kieffer) n. gen., n. comb., a wood-mining Chironomidae (Diptera). Ent. Scand. 13:419-429.
745. Cranston, P. S. 1982b. A key to the larvae of the British Orthocladiinae (Chironomidae). Sci. Publ. Freshwat. Biol. Assoc. 45:1-152.
746. Cranston, P. S., and D. R. Oliver. 1988. Aquatic xylophagous Orthocladiinae systematics and ecology (Diptera: Chironomidae). Spixiana 14:143-154.
747. Cranston, P. S., and O. A. Saether. 1986. Rheosmittia (Diptera: Chironomidae): a generic validation and revision of the western Palaearctic species. J. Nat. Hist. 20:31-51.
748. Crawford, D. O. 1912. The petroleum fly of California, *Psilopa petrolei* Coq. Pomona Coll. J. Ent. 4:687-697.
749. Credland, P. F. 1973. A new method for establishing a permanent laboratory culture of *Chironomus riparius* Meigen (Diptera: Chironomidae). Freshwat. Biol. 3:45-51.
750. Creed, R. P., Jr., and S. P. Sheldon. 1993. The effect of feeding by a North American weevil, *Euhrychiopsis lecontei*, on Eurasian watermilfoil (Myriophyllum: Spicatum). Aquat. Bot. 45:245-256.
751. Creed, R. P., Jr., and S. P. Sheldon. 1994b. Aquatic weevils (Coleoptera: Curculionidae) associated with northern watermilfoil *Myriophyllum sibiricum* in Alberta, Canada. Ent. News 105:98-102.
752. Creed, R. P., Jr., and S. P. Sheldon. 1994a. The effect of two herbivorous insect larvae on Eurasian watermilfoil. J. Aquat. Plant Manage. 32:21-26.
753. Creed, R. P., Jr., S. P Sheldon, and D. M. Cheek. 1992. The effect of herbivore feeding on the buoyancy of Eurasian watermilfoil. J. Aquat. Plant Manage 30:75-76.
754. Crichton, M. I. 1957. The structure and function of the mouth parts of adult caddis flies (Trichoptera). Phil. Trans. Roy. Soc. Lond. (B) 241:45-91.
755. Crisp, D. T. 1962. Observations on the biology of *Corixa germari* (Fieb.) (Hemiptera: Heteroptera) in an upland reservoir. Arch. Hydrobiol. 58:261-280.
756. Cromar, G. L., and D. D. Williams. 1991. Centrifugal flotation as an aid to separating invertebrates from detritus in benthic samples. Hydrobiologia 209:67-70.
757. Cronin, J. T., and J. Travis. 1986. Size-limited predation on larval *Rana areolata* (Anura: Ranidae) by two species of backswimmer (Insecta: Hemiptera: Notonectidae). Herpetologica 42:171-174.
758. Crosby, T. K. 1975. Food of the New Zealand trichopterans *Hydrobiosis parumbripennis* McFarlane and *Hydropsyche colonica* McLachlan. Freshwat. Biol. 5:105-114.

759. Cross, W. H. 1955. Anisopteran Odonata of the Savannah river plant, South Carolina. J. Elisha Mitchell Sci. Soc. 71:9-17.
760. Crosskey, R. W. 1973. Simuliidae (Black-flies), pp. 109-153. *In* K.G.V. Smith (ed.). Insects and other arthropods of medical importance. Bull. Ent. Br. Mus. Nat. Hist., London. 561 pp.
761. Crosskey, R. W. 1987. Part VIII. Black fly species of the world. Chap. 32. An annotated checklist of the world black flies (Diptera: Simuliidae), pp. 425-520. *In* K. C. Kim and R. W. Merritt, (eds.). Black flies. Ecology, population management, and annotated world list. Penn. State Univ. Press, Univ. Park, PA.
762. Crosskey, R. W. 1990. The natural history of blackflies. John Wiley and Sons, Chichester, England. 711pp.
763. Crossman, J. S., and J. Cairns. 1974. A comparative study between two different artificial substrate samplers and regular sampling techniques. Hydrobiologia 44:517-522.
764. Crowell, R. M. 1967. The immature stages of *Albia caerulea* (Acarina: Axonopsidae), a water mite parasitizing caddisflies (Trichoptera). Can. Ent. 99:730-734.
765. Crowell. R. M. 1968. Supplementary observations of the early developmental stages of *Albia caerulea* (Acarina: Axonopsidae). Can. Ent. 100:178-180.
766. Crowl, T. A., and J. E. Alexander. 1988. Parental care and foraging ability in male water bugs (*Belostoma flumineum*). Can. J. Zool. 67:513-515.
767. Crowley, P. H. 1979. Behavior of zygopteran nymphs in a simulated weed bed. Odonatologica 8:91-101.
768. Crowley, P. H., P. M. Dillon, D. M. Johnson, and C. N. Watson. 1987. Intraspecific interference among larvae in a semivoltine dragonfly population. Oecologia 71:447-456.
769. Crowley, P. H., and K. R. Hopper. 1994. How to behave around cannibals: a density-dependent dynamic game. Am. Nat. 143:117-154.
770. Crowley, P. H., and E. K. Martin. 1989. Functional responses and interference within and between year classes of a dragonfly population. J. N. Am. Benthol. Soc. 8:211-221.
771. Crowson, R. A. 1981. The biology of the Coleoptera. Academic Press, N.Y. 802 pp.
772. Cruden, R. W. 1962. A preliminary survey of West Virginia dragonflies (Odonata). Ent. News 73:156-160.
773. Crumb, S. E. 1929. Tobacco cutworms. U.S. Dep. Agric. Tech. Bull. 88:1-180.
774. Crumb, S. E. 1956. The larvae of the Phalaenidae. U.S. Dep. Agric. Tech. Bull. 1135:1-356.
775. Cudney, M. D., and J. B. Wallace. 1980. Life cycles, microdistribution and production dynamics of six species of net-spinning caddisflies in a large Southeastern (U.S.A.) river. Holarct. Ecol. 3:169-182.
776. Cuffney, T. F., M. E. Gurtz, and M. R. Meador. 1993a. Methods for collecting benthic invertebrate samples as part of the National Water-Quality Assessment Program. U. S. Geol. Surv. Open-File Rept. 93-406. 66p.
777. Cuffney, T. F., M. E. Gurtz, and M. R. Meador. 1993b. Guidelines for the processing and quality assurance of benthic invertebrate samples collected as part of the National Water-Quality Assessment Program: U. S. Geol. Surv. Open-File Rept. 93-407, 80p.
778. Cuffney, T. F., and G. W. Minshall. 1981. Life history and bionomics of *Arctopsyche grandis* (Trichoptera) in a central Idaho stream. Holarct. Ecol. 4:252-262.
779. Cuffney, T. F., J. B. Wallace, and G. J. Lugthart. 1990. Experimental evidence quantifying the role of benthic invertebrates in organic matter dynamics of headwater streams. Freshwat. Biol. 23:281-299.
780. Cuffney, T. F., J. B. Wallace, and J. R. Webster. 1984. Pesticide manipulation of a headwater stream: invertebrate responses and their significance for ecosystem processes. Freshwat. Invert. Biol. 3:153-171.
781. Cuker, B. E. 1983. Competition and coexistence among the grazing snail *Lymnaea*, Chironomidae, and microcrustacea in an arctic epilithic lacustrine community. Ecology 64:10-15.
782. Cullen, M. J. 1969. The biology of giant water bugs (Hemiptera: Belostomatidae) in Trinidad. Proc. R. ent. Soc. Lond. (A) 44:123-136.
783. Culley, C. E. 1967. Field notes on the aquatic moth, *Parargyractis truckeealis* (Dyar) (Lepidoptera: Pyralidae). Pan-Pac. Ent. 43:94-95.
784. Culp, J. M., and G. J. Scrimgeour. 1993. Size-dependent diel foraging periodicity of a mayfly grazer in streams with and without fish. Oikos 68:242-250.
785. Cumming, J. M., and B. E. Cooper. 1992. A revision of the Nearctic species of the tachydromiine fly genus *Stilpon* Loew (Diptera: Empidoidea). Can. Ent. 124:951-998.
786. Cummings, C. 1933. The giant water bugs. Univ. Kans. Sci. Bull. 21:197-219.
787. Cummins, K. W. 1962. An evaluation of some techniques for the collection and analysis of benthic samples with special emphasis on lotic waters. Am. Midl. Nat. 67:477-504.
788. Cummins, K. W. 1964. Factors limiting the microdistribution of larvae of the caddisflies *Pycnopsyche lepida* (Hagen) and *Pycnopsyche guttifer* (Walker) in a Michigan stream. Ecol. Monogr. 34:271-295.
789. Cummins, K. W. 1972. What is a river?-zoological description, pp. 33-52. *In* R. T. Oglesby, C. A. Carlson, and J. A. McCann (eds.). River ecology and man. Academic, N.Y. 465 pp.
790. Cummins, K. W. 1973. Trophic relations of aquatic insects. Ann. Rev. Ent. 18:183-206.
791. Cummins, K. W. 1974. Structure and function of stream ecosystems. BioScience 24:631-641.
792. Cummins, K. W. 1975. Macroinvertebrates, pp. 170-198. *In* B. A. Whitton (ed.). River ecology. Blackwell, England. 725 pp.
793. Cummins, K. W. 1980a. The multiple linkages of forests to streams, pp. 191-198. *In* R. H. Waring (ed.). Forests: Fresh perspectives from ecosystem analysis. Proc. 40th Ann. Biol. Colloq., Oregon State Univ., Corvallis. 198 pp.
794. Cummins, K. W. 1980b. The natural stream ecosystem, pp. 7-24. *In* J. V. Ward and J. A. Stanford (eds.). The ecology of regulated streams. Plenum, N.Y. 398 pp.
795. Cummins, K. W. 1988a. Rapid bioassessment using functional analysis of running water invertebrates, pp. 49-54. *In* T. P. Simon, L. L. Holst, and L. J. Shepard (eds.). Proceedings of the First National Workshop on Biological Criteria. EPA-905/9-89/003. U.S. Environmental Protection Agency, Chicago. 129 pp.
796. Cummins, K. W. 1988b. The study of stream ecosystems: A functional view, pp. 247-262. *In* L. R. Pomeroy and J. J. Alberts (eds.). Concepts of ecosystem ecology: A comparative view. Ecol. Stud. 67. Springer-Verlag Inc., NY.
797. Cummins, K. W. 1993. Invertebrates, pp. 234-250. *In* P. Calow, and G. E. Petts (eds.). Rivers handbook. Blackwell Sci. Publ. Oxford, U.K. 526p.
798. Cummins, K. W. 1996. The ecology of streams and rivers. Chapman and Hall, London. (in press.)
799. Cummins, K. W., and M. J. Klug. 1979. Feeding ecology of stream invertebrates. Ann. Rev. Ecol. Syst. 10:147-172.
800. Cummins, K. W., M. J. Klug, G. M. Ward, G. L. Spengler, R. W. Speaker, R. W. Ovink, D. C. Mahan, and R. C. Petersen. 1981. Trends in particulate organic matter fluxes, community processes, and macroinvertebrate functional groups along a Great Lakes drainage basin river continuum. Verh. Int. Verein. Limnol. 21:841-849.
801. Cummins, K. W., L. D. Miller, N. A. Smith, and R. M. Fox. 1965. Experimental entomology. Reinhold, N.Y. 160 pp.
802. Cummins, K. W., G. W. Minshall, J. R. Sedell, C. E. Cushing, and R. C. Petersen. 1984. Stream ecosystem theory. Verh. Int. Verein. Limnol. 22:1818-1827.
803. Cummins, K. W., R. C. Petersen, F. O. Howard, J. C. Wuycheck, and V. I. Holt. 1973. The utilization of leaf litter by stream detritivores. Ecology 54:336-345.
804. Cummins, K. W., and M. A. Wilzbach. 1985. Field procedures for analysis of functional feeding groups of stream macroinvertebrates. Contrib. 1611, Appalachian Environ. Lab., Univ. Maryland, Frostburg. 18 pp.
805. Cummins, K. W., and M. A. Wilzbach. 1988. Do pathogens regulate stream invertebrate populations? Verh. Int. Verein. Limnol. 23:1232-1243.
806. Cummins, K. W., M. A. Wilzbach, D. M. Gates, J. B. Perry, and W. B. Taliaferro. 1989. Shredders and riparian vegetation. BioScience 39:24-30.

807. Cupp, E. W., and A. E. Gordon (eds.). 1983. Notes on the systematics, distribution, and bionomics of black flies (Diptera: Simuliidae) in the Northeastern United States. Search: Agriculture. Ithaca, N.Y.: Cornell Univ. Agric. Exp. Sta. No. 25:1-76.
808. Currie, D. C. 1986. An annotated list of and keys to the immature black flies of Alberta (Diptera: Simuliidae). Mem. Ent. Soc. Can. 134:1-90.
809. Currie, D. C., and D. A. Craig. 1986. Feeding strategies of larval black flies, pp 155-170. In K. C. Kim and R. W. Merritt (eds.). Black flies: ecology, population management, and annotated world list. Penn. State Univ. Press, Univ. Park, PA.
810. Curry, L. L. 1954. Notes on the ecology of the midge fauna of Hunt Creek, Montmorency County, Michigan. Ecology 35:541-550.
811. Curry, L. L. 1966. Freshwater invertebrate food preferences. Unpubl. Rept. Dept. Biol., Central Michigan Univ., Mt. Pleasant.
812. Curtis, L. C. 1967. The mosquitos of British Columbia. Occ. Pap. Brit. Columbia Prov. Mus. 15:1-90.
813. Cushing, C. E. 1964. An apparatus for sampling drifting organisms in streams. J. Wildl. Manage. 28:592-594.
814. Cushing, C. E., and R. T. Rader. 1983. A note on the food of *Callibaetis* (Ephemeroptera: Baetidae). Great Basin Nat. 41:431-432.
815. Cushman, R. M. 1983. An inexpensive, floating, insect-emergence trap. Bull. Environ. Contam. Toxicol. 31:547-550.
816. Cushman, R. M., J. W. Elwood, and S. G. Hildebrand. 1975. Production dynamics of *Alloperla mediana* Banks (Plecoptera: Chloroperlidae) and *Diplectrona modesta* Banks (Trichoptera: Hydropsychidae) in Walker Branch, Tennessee. Publ. Env. Sci. Div. Oak Ridge Nat. Lab. Tenn. 785:1-66.
817. Cushman, R. M., and J. C. Goyert. 1984. Effects of a synthetic crude oil on pond benthic insects. Environ. Poll. (Ser. A) 33:163-186.
818. Cutten-Ali-Khan, E. A., and D. K. McE. Kevan. 1970. The Nymphomyiidae (Diptera), with special reference to *Palaeodipteron walkeri* Ide and to larva in Quebec, and a description of a new genus and species from India. Can. J. Zool. 48:1-24.
819. Cuyler, R. D. 1956. Taxonomy and ecology of larvae of sialoid Megaloptera of east-central North Carolina with a key to and description of larvae of genera known to occur in the United States. M.S. thesis. North Carolina State Univ., Raleigh. 150 pp.
820. Cuyler, R. D. 1958. The larvae of *Chauliodes* Latrielle (Megaloptera: Corydalidae). Ann. Ent. Soc. Am. 51:582-586.
821. Cuyler, R. D. 1965. The larva of *Nigronia fasciatus* Walker (Megaloptera: Corydalidae). Ent. News 76:192-194.
822. Cuyler, R. D. 1968. Range extensions of Odonata in southeastern states. Ent. News 79:29-34.
823. Cuyler, R. D. 1984. Range extensions of Odonata in North Carolina, United States. Not. Odonatol. 2:55-57.
824. Dadd, R. H. 1970a. Relationship between filtering activity and ingestion of solids by larvae of the mosquito *Culex pipiens*: A method for assessing phagostimulant factors. J. Med. Ent. 7:708-712.
825. Dadd, R. H. 1970b. Comparison of rates of ingestion of particulate solids by *Culex pipiens* larvae: Phagostimulant effect of water-soluble yeast extract. Ent. Exp. Appl. 13:407-419.
826. Dadd, R. H. 1971. Effects of size and concentration of particles on rates of ingestion of latex particulates by mosquito larvae. Ann. Ent. Soc. Am. 64:687-692.
827. Dadd, R. H. 1973. Autophagostimulation by mosquito larvae. Ent. exp. appl. 16:295-300.
828. Dadd, R. H. 1975. Ingestion of colloid solutions by filter-feeding mosquito larvae: Relationship to viscosity. J. exp. Zool. 191:395-406.
829. Dahl, C., D. A. Craig, and R. W. Merritt. 1990. Sites of possible mucus-producing glands in the feeding system of mosquito larvae (Diptera: Culicidae). Ann. Ent. Soc. Am. 83:827-833.
830. Dahl, C., L. E. Widahl, and C. Nilsson. 1988. Functional analysis of the suspension feeding system in mosquitos (Culicidae: Diptera). Ann. Ent. Soc. Am. 81:105-27
831. Dahm, E. 1972. Zur Biologie von *Notonecta glauca* (Insecta, Hemiptera) unter besonderer Berücksichtigung der fischereilichen Schadwirkung. Int. Revue ges. Hydrobiol. 57:429-461.
832. Daigle, J. J. 1991. Florida damselflies (Zygoptera): A species key to the aquatic larval stages. Fla. Dept. Environ. Reg. Tech. Series 11:1-12.
833. Daigle, J. J. 1992. Florida dragonflies (Anisoptera): A species key to the aquatic larval stages. Fla. Dept. Environ. Reg. Tech. Series 12:1-29.
834. Daly, H. V., J. T. Doyen, and P. R. Ehrlich. 1978. Introduction to insect biology and diversity. McGraw-Hill N.Y. 564 pp.
835. Danecker, E. 1961. Studien zur hygropetrischen Fauna. Biologie and Okologie von *Stactobia* und *Tinodes* (Insect., Trichopt.). Int. Revue ges. Hydrobiol. 46:214-254.
836. Danks, H. V. 1971a. Overwintering of some north temperate and arctic Chironomidae. II. Chironomid biology. Can. Ent. 103:1875-1910.
837. Danks, H. V. 1971b. Life history and biology of *Elinfeldia synchrona* (Diptera: Chironomidae). Can. Ent. 103:1597-1606.
838. Danks, H. V. 1978. Some effects of photoperiod, temperature and food on emergence in three species of Chironomidae (Diptera). Can. Ent. 110:289-300.
839. Danks, H. V., and D. R. Oliver. 1972. Seasonal emergence of some high arctic Chironomidae (Diptera). Can. Ent. 104:661-686.
840. Darby, R. E. 1962. Midges associated with California rice fields, with special reference to their ecology (Diptera: Chironomidae). Hilgardia 32:1-206.
841. Darsie, R. F., D. MacCreary, and L. A. Stearns. 1951. An annotated list of the mosquitoes of Delaware. N.J. Mosquito Exterm. Assoc. Proc. 38:137-146.
842. Darsie, R. F., and R. A. Ward. 1981. Identification and geographical distribution of the mosquitoes of North America, north of Mexico. Mosquito Sys. Suppl. No. 1. Amer. Mosquito Contr. Assoc., Fresno, Calif. 313 pp.
843. Darsie, R. F., Jr. 1992. Key characters for identifying *Aedes bahamensis* and *Aedes albopictus* in North America, north of Mexico. J. Am. Mosq. Cont. Assoc. 8:323-324.
844. Darsie, R. F., Jr., and A. W. Anderson. 1985. A revised list of the mosquitoes of North Dakota including new additions to the fauna. J. Am. Mosq. Contr. Assoc. 1:76-79.
845. Darsie, R. F., Jr., and R. A. Ward. 1989. Review of new nearctic mosquito distributional records north of Mexico, with notes on additions and taxonomic changes of the fauna, 1982-1989. J. Am. Mosq. Contr. Assoc. 5:552-557.
846. Darville, R. G., and J. L. Wilhm. 1984. The effect of naphthalene on oxygen consumption and hemoglobin concentration in *Chironomus attenuatus* and on oxygen consumption and life cycle of *Tanytarsus dissimilis*. Environ. Toxicol. Chem. 3:135-141.
847. Dasch, C. E. 1979. Ichneumon - Flies of America north of Mexico: 8. Subfamily Cremastinae. Mem. Am. Ent. Instit. 29:702.
848. Dauble, D. D., W. E. Fallon, H. Gray, and R. M. Bean. 1982. Effects of coal liquid water-soluble fractions on growth and survival of four aquatic organisms. Arch. Environ. Contam. Toxicol. 11:553-560.
849. Daussin, G. L. 1979. The aquatic Coleoptera of Minnesota: A species list with references on taxonomy, biology and ecology (Coleoptera: Haliplidae, Gyrinidae, Dytiscidae, Hydrophilidae, Hydraenidae, Elmidae, Dryopidae, Psephenidae, Helodidae, Chrysomelidae, Curculionidae). Newsltr. Ass. Minn. Ent. 12:6-18.
850. Davenport, C. B. 1903. The Collembola of Cold Spring Beach, with special reference to the movements of the Poduridae. Cold Spring Harbor Monogr. 2:1-32.
851. Davies, B. R. 1976. The dispersal of Chironomidae: a review. J. Ent. Soc. S. Afr. 39:39-62.
852. Davies, D. M. 1991. Additional records of predators upon black flies (Simuliidae: Diptera). Bull. Soc. Vect. Ecol. 16:256-268.
853. Davies, D. M., and B. V. Peterson. 1956. Observations on the mating, feeding, ovarian development, and oviposition of adult black flies (Simuliidae, Diptera). Can. J. Zool. 34:615-655.
854. Davies, D. M., B. V. Peterson, and D. M. Wood. 1962. The black flies (Diptera: Simuliidae) of Ontario. Part I. Adult identification and distribution with descriptions of six new species. Proc. Ent. Soc. Ont. 92:70-154.
855. Davies, D.A.L. 1981. A synopsis of the extant genera of the Odonata. Soc. Int. Odonatol. Rapid Comm. 3:1-60.

856. Davies, D.A.L., and P. Tobin. 1984. The dragonflies of the world: A systematic list of the extant species of Odonata. Vol. 1. Zygoptera, Anisozygoptera. SIO Rapid Comm. (Suppl.) No. 3. Utrecht, The Netherlands. 127pp.

857. Davies, D.A.L., and P. Tobin. 1985. The dragonflies of the world: A systematic list of the extant species of Odonata. Vol. 2. Anisoptera. SIO Rapid Comm. (Suppl.) No. 5. Utrecht, The Netherlands. 151pp.

858. Davies, I. J. 1975. Selective feeding in some arctic Chironomidae. Verh. Int. Verein. Limnol. 19:3149-3154.

859. Davies, I. J. 1980. Relationships between dipteran emergence and phytoplankton production in the Experimental Lakes Area, northwestern Ontario. Can. J. Fish. Aquat. Sci. 37:523-533.

860. Davies, I. J. 1984. Sampling aquatic insect emergence, pp. 161-227. In J. A. Downing and F. H. Rigler (eds.) (2nd ed.) A manual on methods for the assessment of secondary productivity in freshwaters. IBP Handbook No. 17, Blackwell Scien. Publ., Oxford, England.

861. Davies, I. J. 1991. Canadian freshwater biomonitoring: the program of the Department of Fisheries and Oceans, pp. 75-88. In Y. A. Izrael, S. M. Semenov, G. E. Insarov, V. A. Abakumov, and T. A. Golovina (eds.). Problems of ecological monitoring and ecosystem modelling. Vol. XIII. USSR Academy of Sciences, Leningrad. 313 pp. (In Russian; English version issued as Can. Transl. Fish. Aquat. Sci. 5551, 1992, 24 p. Freshwater Institute Reprint 1087a).

862. Davies, L., and C. D. Smith. 1958. The distribution and growth of *Prosimulium* larvae (Diptera: Simuliidae) in hill streams in Northern England. J. Anim. Ecol. 27:335-348.

863. Davies, L. 1965a. On spermatophores in Simuliidae (Diptera). Proc. R. ent. Soc. Lond., Ser. A. 40:30-34.

864. Davies, L. 1965b. The structure of certain atypical Simuliidae (Diptera) in relation to evolution within the family, and the erection of a new genus for the Crozet Island black-fly. Proc. Linn. Soc. London 176:159-180.

865. Davies, R. W., and V. J. McCauley. 1970. The effects of preservatives on the regurgitation of gut contents by Chironomidae (Diptera) larvae. Can. J. Zool. 48:519-522.

866. Davis, C. C. 1961a. A study of the hatching process in aquatic invertebrates. I. The hatching process in *Amnicola limosa*. II. Hatching in *Ranatra fusca* P. Beauvois. Trans. Am. Microsc. Soc. 80:227-234.

867. Davis, C. C. 1961b. Hatching in *Dineutes assimilis* Aube (Coleoptera, Gyrinidae). Int. Revue ges. Hydrobiol. 46:429-433.

868. Davis, C. C. 1962. Ecological and morphological notes on *Hydrophylita aquivalans* (Math. and Crosby). Limnol. Oceanogr. 7:390-392.

869. Davis, C. C. 1964. A study of the hatching process in aquatic invertebrates. VII. Observations on hatching in *Notonecta melaena* Kirkaldy (Hemiptera, Notonectidae) and on *Ranatra absona* D. and DeC. (Hemiptera, Nepidae). Hydrobiologia 23:253-259.

870. Davis, C. C. 1965a. A study of the hatching process in aquatic invertebrates. XIX. Hatching in *Psephenus herricki* (DeKay) (Coleoptera, Psephenidae). Am. Midl. Nat. 74:443-450.

871. Davis, C. C. 1965b. A study of the hatching process in aquatic invertebrates. XII. The eclosion process in *Trichocorixa naias* (Kirkaldy). Trans. Am. Microsc. Soc. 84:60-65.

872. Davis, C. C. 1966. Notes on the ecology and reproduction of *Trichocorixa reticulata* in a Jamaican salt-water pool. Ecology 47:850-852.

873. Davis, J. R. 1987. A new species of *Farrodes* (Ephemeroptera: Leptophlebiddae), Atalophlebiinae) from southern Texas. Proc. Ent. Soc. Wash. 89:407-416.

874. Davis, K. C. 1903. Sialididae of North America. Bull. N.Y. State Mus. 18:442-486.

875. Davis, L. V., and I. E. Gray. 1966. Zonal and seasonal distribution of insects in North Carolina salt marshes. Ecol. Monogr. 36:275-295.

876. Davis, M. M., and G. R. Finni. 1974. Habitat preference of *Macronychus glabratus* Say (Coleoptera: Elmidae). Proc. Penn. Acad. Sci. 48:95-97.

877. Davis, W. S. 1995. Biological assessment and criteria: building on the past, pp. 15-29. In W. S. Davis and T. P. Simon (eds.). Biological assessment and criteria. Tools for water resource planning and decision making. Lewis Publs., Boca Raton, Fla.

878. Davis, W. T. 1933. Dragonflies of the Genus *Tetragoneuria*. Bull. Brooklyn Ent. Soc. 28:87-104.

879. Dawson, C. L., and R. A. Hellenthal. 1986. A computerized system for the evaluation of aquatic habitats based on environmental requirements and pollution tolerance associations of resident organisms. Project summary. EPA/600/S3-86/019. Environmental Research Laboratory, U.S. Environmental Protection Agency, Corvallis. 5 pp.

880. Day, W. C. 1956. Ephemeroptera, pp. 79-105. In R. L. Usinger (ed.). Aquatic insects of California. Univ. Calif. Press, Berkeley. 508 pp.

881. De Marmels, J., and J. Racenis. 1982. An analysis of the *cophysa*- group of *Tramea* Hagen, with descriptions of two new species (Anisoptera: Libellulidae). Odonatologica 11:109-128.

882. De Pauw, N., and G. Vanhooren. 1983. Method for biological quality assessment of watercourses in Belgium. Hydrobiologia 100:153-168.

883. De Ruiter, L., H. P. Wolvekamp, A. J. van Tooren, and A. Vlasblom. 1952. Experiments on the efficiency of the "physical gill" (*Hydrous piceus* L., *Naucoris cimicoides* L., and *Notonecta glauca* L.). Acta Physiol. Pharmacol. Neerl. 2:180-213.

884. Deay, H. O., and G. E. Gould. 1936. The Hemiptera of Indiana. I. Family Gerridae. Am. Midl. Nat. 17:753-769.

885. DeCoursey, R. M. 1971. Keys to the families and subfamilies of the nymphs of North American Hemiptera-Heteroptera. Proc. Ent. Soc. Wash. 73:413-428.

886. Deevy, E. S. 1941. Limnological studies in Connecticut. VI. The quantity and composition of the bottom fauna of thirty-six Connecticut and New York lakes. Ecol. Monogr. 11:413-455.

887. Delamare-Deboutteville, C. 1953. Collemboles marins de la zone souterraine humide de sables littoraux. Vie Milieu 4:290-319.

888. Delamare-Deboutteville, C. 1960. Biologie des eaux souterraines littorales et continentales. Vlg. Hermann, Paris. 740 pp.

889. DeLong, M. D., J. H. Thorp, and K. H. Haag. 1993. A new device for sampling macroinvertebrates from woody debris (snags) in nearshore areas of aquatic systems. Am. Midl. Nat. 130:413-417.

890. Delucchi, C. M., and B. L. Peckarsky. 1989. Life history patterns of insects in an intermittent and a permanent stream. J. N. Am. Benthol. Soc. 8:308-321.

891. Dendy, J. S. 1973. Predation on chironomid eggs and larvae by *Nanocladius alternantherae* Dendy and Sublette (Diptera: Chironomidae, Orthocladiinae). Ent. News 84:91-95.

892. DeNicola, D. M., C. D. McIntire, G. A. Lamberti, S. V. Gregory, and L. R. Ashkenas. 1990. Temporal patterns of grazer-periphyton interactions in laboratory streams. Freshwat. Biol. 23:475-489.

893. Denning, D. G. 1943. The Hydropsychidae of Minnesota (Trichoptera). Entomologica am. 23:101-171.

894. Denning, D. G. 1950. Order Trichoptera, the caddisflies, pp. 12-23. In D. L. Wray (ed.). The insects of North Carolina (2nd suppl.). N. C. Dep. Agric. 59 pp.

895. Denning, D. G. 1956. Trichoptera, pp. 237-270. In R. L. Usinger (ed.). Aquatic insects of California. Univ. Calif. Press, Berkeley. 508 pp.

896. Denning, D. G. 1958. The genus *Farula* (Trichoptera: Limnephilidae). Ann. Ent. Soc. Am. 51:531-535.

897. Denning, D. G. 1964. The genus *Homophylax* (Trichoptera: Limnephilidae). Ann. Ent. Soc. Am. 57:253-260.

898. Denning, D. G. 1970. The genus *Psychoglypha* (Trichoptera: Limnephilidae). Can. Ent. 102:15-30.

899. Denning, D. G. 1975. New species of Trichoptera from western North America. Pan-Pacif. Ent. 51:318-326.

900. Denning, D. G., and R. L. Blickle. 1972. A review of the genus *Ochrotrichia* (Trichoptera:Hydroptilidae). Ann. Ent. Soc. Am. 65:141-151.

901. Denno, R. F. 1976. Ecological significance of wing-polymorphism in Fulgoridae which inhabit tidal salt marshes. Ecol. Ent. 1:257-266.

902. Deonier, D. L. 1964. Ecological observations on Iowa shore flies (Diptera, Ephydridae). Proc. Iowa Acad. Sci. 71:496-510.

903. Deonier, D. L. 1971. A systematic and ecological study of Nearctic *Hydrellia* (Diptera: Ephydridae). Smithson. Contr. Zool. 68:1-147.
904. Deonier, D. L. 1972. Observations on mating and food habits of certain shore flies (Diptera: Ephydridae). Ohio J. Sci. 72:22-29.
905. Deonier, D. L. 1974. Biology and descriptions of immature stages of the shore fly, *Scatophila iowana* (Diptera: Ephydridae). Iowa State J. Res. 49:17-21.
906. Deonier, D. L. (ed.). 1979. First symposium on the systematics and ecology of Ephydridae (Diptera). N. Am. Benthol. Soc. 147 pp.
907. Deonier, D. L., S. Kincaid, and J. Scheiring. 1976. Substrate and moisture preference in the common toad bug, *Gelastocoris oculatus*. Ent. News 87:257-264.
908. Deonier, D. L., W. N. Mathis, and J. T. Regensburg. 1978a. Natural history and life-cycle stages of *Notiphila carinata* (Diptera: Ephydridae). Proc. Biol. Soc. Wash. 91:798-814.
909. Deonier, D. L., and J. T. Regensburg. 1978b. Biology and immature stages of *Parydra quadrituberculata*. Ann. Ent. Soc. Am. 71:341-353.
910. Deshefy, G. S. 1980. Anti-predator behavior in swarms of *Rhagovelia obesa* (Hemiptera: Veliidae). Pan-Pacif. Ent. 56:111-112.
911. Dettner, K. 1987. Chemosystematics and evolution of beetle chemical defenses. Ann. Rev. Ent. 32:17-48.
912. Deutsch, W. G. 1980. Macroinvertebrate colonization of acrylic plates in a large river. Hydrobiologia 75:65-72.
913. Dickinson, J. C., Jr. 1949. An ecological reconnaissance of the biota of some ponds and ditches in Northern Florida. Quart. J. Fla. Acad. Sci. 11:1-28.
914. Dickinson, W. E. 1944. The mosquitoes of Wisconsin. Milwaukee Pub. Mus. Bull. 8:269-385.
915. Diggins, T. P., and K. M. Stewart. 1993. Deformities of aquatic larval midges (Chironomidae: Diptera) in the sediments of the Buffalo River, New York. J. Great Lakes Res. 19:648-659.
916. Dimond, J. B., and W. G. Hart. 1953. Notes on the blackflies (Simuliidae) of Rhode Island. Mosq. News 13:238-242.
917. Disney, R. H. L. 1975. A key to the British Dixidae. Sci. Pub. Freshwat. Biol. Assoc. 31:4-78.
918. Disney, R. H. L. 1991. The aquatic Phoridae (Diptera). Ent. Scand. 22:171-191.
919. Dixon, R. D., and R. A. Burst. 1971. Mosquitoes of Manitoba. III. Ecology of larvae in the Winnipeg area. Can. Ent. 104:961-968.
920. Dixon, R. W., and F. J. Wrona. 1992. Life history and production of the predatory caddisfly *Rhyacophila vao* Milne in a spring-fed stream. Freshwat. Biol. 27: 1-11.
921. Dobson, M. 1991. An assessment of mesh bags and plastic leaf traps as tools for studying macroinvertebrate assemblages in natural leaf packs. Hydrobiologia 222:19-28.
922. Dodds, G. S., and F. L. Hisaw. 1924a. Ecological studies of aquatic insects. I. Adaptations of mayfly nymphs to swift streams. Ecology 5:137-148.
923. Dodds, G. S., and F. L. Hisaw. 1924b. Ecological studies of aquatic insects. II. Size of respiratory organs in relation to environmental conditions. Ecology 5:262-271.
924. Dodds, W. K. 1990. Hydrodynamic constraints on evolution of chemically mediated interactions between aquatic organisms in unidirectional flows. J. Chem. Ecol. 16:1417-1430.
925. Dodds, W. K. 1991. Community interaction between the filamentous alga *Cladophora glomerata* Kuetzing, its epiphytes, and epiphyte grazers. Oecologia 85:572-580.
926. Dodds, W. K., and J. L. Marra. 1989. Behaviors of the midge, Cricotopus (Diptera: Chironomidae) related to the mutualism with *Nostoc parmelioides* (Cyanobacteria). Aquat. Insects 11:201-208.
927. Dodson, S. I. 1970. Complementary feeding niches sustained by size-selective predation. Limnol. Oceanogr. 15:131-137.
928. Dodson, S. I., and J. E. Havel. 1988. Indirect prey effects: some morphological and life history responses of *Daphnia pulex* exposed to *Notonecta undulata*. Limnol. Oceanogr. 33:1274-1285.
929. Doeg, T., and P. S. Lake. 1981. A technique for assessing the composition and density of the macroinvertebrate fauna of large stones in streams. Hydrobiologia 80:3-6.
930. Doherty, F. G., and W. D. Hummon. 1980. Respiration of aquatic insect larvae (Ephemeroptera, Plecoptera) in acid mine water. Bull. Environ. Contam. Toxicol. 25:358-363.
931. Dolin, P. S., and D. C. Tarter. 1982. Life history and ecology of *Chauliodes rastricornis* and *Chauliodes pectinicornis* (Megaloptera: Corydalidae) in Greenbottom Swamp, Cabell County, West Virginia. Brimleyana 7:111-120.
932. Dominguez, T. M., E. J. Calabrese, P. T. Kostecki, and R. A. Coler. 1988. The effects of tri- and dichloroacetic acids on the oxygen consumption of the dragonfly nymph *Aeschna umbrosa*. J. Environ. Sci. Health A23:251-271.
933. Domizi, E. A., A. L. Estevez, J. A. Schnack, and G. R. Spinelli. 1978. Ecologia y estrategia de una poblacíon de *Belostoma oxyurum* (Dufour) (Hemiptera, Belostomatidae). Ecosur 5:157-168.
934. Don, A. W. 1967. Aspects of the biology of *Microvelia macgregori* Kirkaldy (Heteroptera: Veliidae). Proc. R. ent. Soc. Lond. (A). 42:171-179.
935. Donald, D. B., and R. S. Anderson. 1977. Distribution of the stoneflies (Plecoptera) of the Waterton River drainage, Alberta, Canada. Can. Wild. Serv. Syesis 10:111-120.
936. Donald, G. L., and C. G. Paterson. 1977. Effect of preservation on wet weight biomass of chironomid larvae. Hydrobiologia 53:75-80.
937. Donnelly, T. W. 1961a. The Odonata of Washington, D.C. and vicinity. Proc. Ent. Soc. Wash. 63:1-13.
938. Donnelly, T. W. 1961b. *Aeshna persephone*, a new species of dragonfly from Arizona, with notes on *Aeshna arida* Kennedy (Odonata: Aeshnidae). Proc. Ent. Soc. Wash. 63:193-202.
939. Donnelly, T. W. 1962. *Somatochlora margarita*, a new species of dragonfly from eastern Texas. Proc. Ent. Soc. Wash. 64:235-240.
940. Donnelly, T. W. 1965. A new species of *Ischnura* from Guatemala, with revisionary notes on related North and Central American damselflies (Odonata: Coenagrionidae). Fla. Ent. 48:57-63.
941. Donnelly, T. W. 1966. A new gomphine dragonfly from eastern Texas. Proc. Ent. Soc. Wash. 68:102-105.
942. Donnelly, T. W. 1970. The Odonata of Dominican British West Indies. Smithson. Contr. Zool. 37:1-20.
943. Donnelly, T. W. 1978. Odonata of the Sam Houston National Forest and vicinity, east Texas, United States, 1960-1966. Notul. Odonatol. 1:67.
944. Donnelly, T. W. 1992. The Odonata of New York. Bull. Am. Odonatol. 1:1-27.
945. Dorer, R. E., W. E. Bickley, and H. P. Nicholson. 1944. An annotated list of the mosquitoes of Virginia. Mosquito News 4:48-50.
946. Dosdall, L., and D. M. Lehmkuhl. 1979. Stoneflies (Plecoptera) of Saskatchewan. Quaest. Ent. 15:3-116.
947. Douglas, B. 1958. The ecology of the attached diatoms and other algae in a small stony stream. Ecology 46:295-322.
948. Doutt, R. L. 1949. A synopsis of North American Anaphoidea. Pan-Pacif. Ent. 25:155-160.
949. Doutt, R. L., and G. Viggiani. 1968. The classification of the Trichogrammatidae (Hym.: Chalciodoidea). Proc. Calif. Acad. Sci. 35:477-586.
950. Dowling, J. A., and H. Cyr. 1985. Quantitative estimation of epiphytic invertebrate populations. Can. J. Fish Aquat. Sci. 42:1570-1579.
951. Downe, A.E.R., and V. G. Caspary. 1973. The swarming behavior of *Chironomus riparius* (Diptera: Chironomidae) in the laboratory. Can. Ent. 105:165-171.
952. Downes, J. A. 1974. The feeding habits of adult Chironomidae. Tijdschr. Ent. Suppl. 95:84-90.
953. Downes, J. A., and W. W. Wirth. 1981. Chap. 28. Ceratopogonidae, pp. 393-422. In J. P. McAlpine, B. V. Peterson, G. E. Shewell, H. J. Teskey, J. R. Vockeroth, and D. M. Wood (coords.). Manual of Nearctic Diptera, Vol. 1. Res. Branch, Agric. Can. Monogr. 27. 674 pp.
954. Downing, J. A. 1984. Sampling the benthos of standing waters, pp. 87-130. In J.A. Downing and F.H. Rigler (eds.) (2nd ed.). A manual on methods for the assessment of secondary productivity in fresh waters. IBP Handbook No. 17, Blackwell Scien. Publ., Oxford, England 501 pp.

955. Downing, J. A., and F. H. Rigler (eds.). 1984. A manual on methods for the assessment of secondary productivity in fresh waters. (2nd ed.). I.B.P. Handbook 17. Blackwell Sci. Pubs., Oxford, England.

956. Doyen, J. T. 1975. Intertidal insects: order Coleoptera, pp. 446-452. *In* R. I. Smith (ed.). Intertidal invertebrates of the central California coast. Univ. Calif. Press, Berkeley. 716 pp.

957. Doyen, J. T. 1976. Marine beetles (Coleoptera excluding Staphylinidae), pp. 497-519. *In* L. Cheng (ed.). Marine insects. North Holland, Amsterdam. 581 pp.

958. Drake, C. J. 1917. A survey of the North American species of *Merragata*. Ohio J. Sci. 17:101-105.

959. Drake, C. J. 1950. Concerning North American Saldidae (Hemiptera). Bull. Brooklyn Ent. Soc. 45:1-7.

960. Drake, C. J. 1952. Alaskan Saldidae (Hemiptera). Proc. Ent. Soc. Wash. 54:145-148.

961. Drake, C. J., and H. C. Chapman. 1953. A preliminary report on the Pleidae of the Americas. Proc. Biol. Soc. Wash. 66:53-60.

962. Drake, C. J., and H. C. Chapman. 1954. New American water-striders. Fla. Ent. 37:151-155.

963. Drake, C. J., and H. C. Chapman. 1958a. New Neotropical Hebridae, including a catalog of the American species (Hemiptera). J. Wash. Acad. Sci. 48:317-326.

964. Drake, C. J., and H. C. Chapman. 1958b. The subfamily Saldoidinae (Hemiptera: Saldidae). Ann. Ent. Soc. Am. 51:480-485.

965. Drake, C. J., and H. M. Harris. 1932. A survey of the species of *Trepobates* Uhler. Bull. Brooklyn Ent. Soc. 27:113-123.

966. Drake, C. J., and H. M. Harris. 1934. The Gerrinae of the Western Hemisphere. Ann. Carnegie Mus. 23:179-240.

967. Drake, C. J., and L. Hoberlandt. 1950. Catalog of genera and species of Saldidae. Acta Ent. Mus. Nat. Prague 26(376):1-12.

968. Drake, C. J., and F. C. Hottes. 1950. Saldidae of the Americas (Hemiptera). Great Basin Nat. 10:51-61.

969. Drake, C. J., and R. F. Hussey. 1955. Concerning the genus *Microvelia* Westwood with descriptions of two new species and checklist of the American forms (Hemiptera: Veliidae). Fla. Ent. 38:95-115.

970. Drake, C. J., and D. R. Lauck. 1959. Description, synonomy, and checklist of American Hydrometridae (Hemiptera-Heteroptera). Great Basin Nat. 19:43-52.

971. Drake, C. J., and J. Maldonado-Capriles. 1956. Some pleids and water-striders from the Dominican Republic (Hemiptera). Bull. Brooklyn Ent. Soc. 51:53-56.

972. Drake, C. M., and J. M. Elliott. 1982. A comparative study of three air-lift samplers used for sampling benthic macro-invertebrates in rivers. Freshwat. Biol. 12:511-533.

973. Drake, C. M., and J. M. Elliott. 1983. A new quantitative air-lift sampler for collecting macroinvertebrates on stony bottoms in deep rivers. Freshwat. Biol. 13:545-559.

974. Dray, F. A., Jr., T. D. Center, and D. H. Habeck. 1989. Immature stage of the aquatic moth *Petrophila drumalis* (Lepidoptera: Pyralidae, Nymphulinae) from *Pistia stratiotes* (water-lettuce). Fla. Ent. 72:711-714.

975. Dray, F. A., Jr., T. D. Center, and D. H. Habeck. 1993. Phytophagous insects associated with *Pistia stratiotes* in Florida. Environ. Ent. 22:1146-1155.

976. Drees, B. M., L. Butler, and L. L. Pechuman. 1980. Horse flies and deer flies of West Virginia: An illustrated key (Diptera, Tabanidae). Bull. W.Va. Agric. For. Exp. Sta. 674:1-67.

977. Dudgeon, D. 1990. Feeding by the aquatic heteropteran, *Diplonychus rusticum* (Belostomatidae): an effect of prey density on meal size. Hyrobiologia 190:93-96.

978. Dudgeon, D., and J. S. Richardson. 1988. Dietary variations of predaceous caddisfly larvae (Trichoptera: Rhyacophilidae, Polycentropodidae and Arctopsychidae) from British Columbian streams. Hydrobiologia 160:33-44.

979. Dudley, T. L. 1982. Population and production ecology of *Lipsothrix* spp. (Diptera: Tipulidae). M.S. thesis. Oregon State University, Corvallis. 161 pp.

980. Dudley, T. L., and N. H. Anderson. 1982. A survey of invertebrates associated with wood debris in aquatic habitats. Melanderia 39:1-21.

981. Dudley, T. L., and N. H. Anderson. 1987. The biology and life cycles of *Lipsothrix spp*. (Diptera: Tipulidae) inhabiting wood in Western Oregon streams. Freshwat. Biol. 17:437-451.

982. Dudley, T. L., S. D. Cooper, and N. Hemphill. 1986. Effects of macroalgae on a stream invertebrate community. J. N. Am. Benthol. Soc. 5:93-106.

983. Dudley, T. L., C. M. D'Antonio, and S. D. Cooper. 1990. Mechanisms and consequences of interspecific competition between two stream insects. J. Anim. Ecol. 59:849-866.

984. Dudley, T. L., and C. M. D'Antonio. 1991. The effects of substrate texture, grazing and disturbances on macroalgae establishment in streams. Ecology 72:297-309.

985. Dumont, B., and J. Verneaux. 1976. Edifice trophique partiel du cours supérieur d'un ruisseau forestier. Ann. Limnol. 12:239-252.

986. Dunham, M. 1994. The effect of physical characters on foraging in *Pachydiplax longipennis* (Burmeister) (Anisoptera: Libellulidae). Odonatologica 23:55-62.

987. Dunkle, S. W. 1976. Larva of the dragonfly *Ophiogomphus arizonicus* (Odonata: Gomphidae). Fla. Ent. 59:317-320.

988. Dunkle, S. W. 1977a. The larva of *Somatochlora filosa* (Odonata: Corduliidae). Fla. Ent. 60:187-191.

989. Dunkle, S. W. 1977b. Larvae of the genus *Gomphaeschna* (Odonata: Aeshnidae). Fla. Ent. 60:223-225.

990. Dunkle, S. W. 1978. Notes on adult behavior and emergence of *Paltothemis lineatipes* Karsch, 1890 (Anisoptera: Libellulidae). Odonatologica 7:277-279.

991. Dunkle, S. W. 1981. The ecology and behavior of *Tachopteryx thoreyi* (Hagen) (Anisoptera: Petaluridae). Odonatologica 10:189-199.

992. Dunkle, S. W. 1985. New records of Bahamian Odonata. Not. Odonatol. 2:99-100.

993. Dunkle, S. W. 1988. Odonata of Bermuda. Not. Odonatol. 3:13-14.

994. Dunkle, S. W. 1989. Dragonflies of the Florida peninsula, Bermuda and the Bahamas. Scien. Publ. Nat. Guide No. 1. Sci. Publ., Gainesville, Florida. 154pp.

995. Dunkle, S. W. 1990. Damselflies of the Florida, Bermuda and the Bahamas. Scien. Publ. Nat. Guide No. 3., Sci. Publ., Gainesville, FL. 148pp.

996. Dunkle, S. W. 1992. Distribution of dragonflies and damselflies (Odonata) in Florida. Bull. Am. Odonatol. 1:29-50.

997. Dunkle, S. W., and M. J. Westfall, Jr. 1982. Order Odonata, pp. 32-45. *In* R. Franz (ed.). Invertebrates, Vol. 6. Rare and Endangered Biota of Florida. Univ. Presses of Florida.

998. Dunn, C. E. 1979. A revision and phylogenetic study of the genus *Hesperocorixa* Kirkaldy (Hemiptera: Corixidae). Proc. Acad. Nat. Sci. Philadelphia 131:158-190.

999. Dunson, W. A. 1980. Adaptations of nymphs of a marine dragonfly, *Erythrodiplax berenice*, to wide variations of salinity. Physiol. Zool. 53:445-452.

1000. DuPorte, E. M. 1959. Manual of insect morphology. Reinhold, N.Y. 224 pp.

1001. Dyar, H. G., and R. C. Shannon. 1927. The North American two-winged flies of the family Simuliidae. Proc. U.S. Nat. Mus. 69:1-54.

1002. Dyte, C. E. 1959. Some interesting habitats of larval Dolichopodidae (Diptera). Ent. Mon. Mag. 95:139-143.

1003. Dyte, C. E. 1967. Some distinctions between the larvae and pupae of the Empididae and Dolichopodidae (Diptera). Proc. R. Ent. Soc. Lond., Ser. A 42:119-128.

1004. Eastham, L.E.S. 1934. Metachronal rhythms and gill movements of the nymph of *Cunis horaria* (Ephemeroptera) in relation to water flow. Proc. R. Soc. Lond., Ser. B. 115:30-48.

1005. Eastham, L.E.S. 1936. The rhythmical movements of the gills of nymphal *Leptophlebia marginata* (Ephemeroptera) and the currents produced by them in water. J. exp. Biol. 13:443-453.

1006. Eastham, L.E.S. 1937. The gill movements of nymphal *Ecdyonurus venosus* (Ephemeroptera) and the currents produced by them in water. J. exp. Biol. 14:219-228.

1007. Eastham, L.E.S. 1939. Gill movements of nymphal *Ephemera vulgata* (Ephemeroptera) and the water currents caused by them. J. exp. Biol. 16:18-33.

1008. Eastin, W. C. 1970. The biology and immature stages of certain shoreflies (Diptera: Ephydridae). Unpubl. M. S. Thesis, Kent State Univ. 112p.
1009. Eastin, W. C., and B. A. Foote. 1971. Biology and immature stages of *Dichaeta caudata* (Diptera: Ephydridae). Ann. Ent. Soc. Am. 64:271-279.
1010. Eaton, K. A. 1983. The life history and production of *Chaoborus punctipennis* (Diptera: Chaoboridae) in Lake Norman, North Carolina, U.S.A. Hydrobiologia 106:247-252.
1011. Eckblad, J. W. 1973. Experimental predation studies of malacophagous larvae of *Sepedon fuscipennis* (Diptera: Sciomyzidae). Exp. Parasitol. 33:331-342.
1012. Eckhart, P., and R. Snetsinger. 1969. Black flies (Diptera: Simuliidae) of northeastern Pennsylvania. Melsheimer Ent. Ser. 4:1-7.
1013. Eddington, J. M. 1968. Habitat preferences in net-spinning caddis larvae with special reference to the influence of water velocity. J. Anim. Ecol. 37:675-692.
1014. Eddington, J. M., and A. H. Hildrew. 1973. Experimental observations relating to the distribution of net-spinning Trichoptera in streams. Verh. Int. Verein. Limnol. 18:1549-1558.
1015. Eddy, S., and A. C. Hodson. 1961. Taxonomic keys to the common animals of the north central states (3rd ed.). Burgess, Minneapolis 162 pp.
1016. Edman, J. D., and K. R. Simmons. 1985. Rearing and colonization of black flies (Diptera: Simuliidae). J. Med. Ent. 22:1-17.
1017. Edmondson, W. T. (ed.). 1959. Freshwater Biology (2nd ed.). John Wiley & Sons, N.Y. 1248 pp.
1018. Edmondson, W. T., and G. G. Winberg (eds.). 1971. A manual on methods for the assessment of secondary productivity in freshwaters. IBP Handbook 17. Blackwell, Oxford. 358 pp.
1019. Edmunds, G. F., Jr. 1957. The predaceous mayfly nymphs of North America. Proc. Utah Acad. Sci. Arts Lett. 34:23-24.
1020. Edmunds, G. F., Jr. 1959. Ephemeroptera, pp. 908-916. *In* W. T. Edmondson (ed.). Freshwater biology (2nd ed.). John Wiley & Sons. 1248 pp.
1021. Edmunds, G. F., Jr. 1960. The food habits of the nymph of the mayfly *Siphlonurus occidentalis*. Proc. Utah Acad. Sci. Arts Lett. 37:73-74.
1022. Edmunds, G. F., Jr. 1961. A key to the genera of known nymphs of the Oligoneuriidae (Ephemeroptera). Proc. Ent. Soc. Wash. 63:255-256.
1023. Edmunds, G. F. 1972. Biogeography and evolution of Ephemeroptera. Ann. Rev. Ent. 17:21-43.
1024. Edmunds, G. F. 1975. Phylogenetic biogeography of mayflies. Ann. Mo. Bot. Gard. 62:251-263.
1025. Edmunds, G. F., Jr., and R. K. Allen. 1964. The Rocky Mountain species of *Epeorus (Iron)* Eaton (Ephemeroptera: Heptageniidae). J. Kans. Ent. Soc. 37:275-288.
1026. Edmunds, G. F., Jr., L. Berner, and J. R. Traver. 1958. North American mayflies of the family Oligoneuridae. Ann. Ent. Soc. Am. 51:375-382.
1027. Edmunds, G. F., and S. L. Jensen. 1975. A new genus and subfamily of North American Heptageniidae (Ephemeroptera). Proc. Ent. Soc. Wash. 76:495-497.
1028. Edmunds, G. F., Jr., S. L. Jensen, and L. Berner. 1976. The mayflies of North and Central America. Univ. Minnesota Press, Minneapolis. 330 pp.
1029. Edmunds, G. F., Jr., and R. W. Koss. 1972. A review of the Acanthametropodinae with the description of new genus. Pan-Pac Ent. 48:136-143.
1030. Edmunds, G. F., Jr., and W. P. McCafferty. 1988. The mayfly subimago. Ann. Rev. Ent. 33:509-529.
1031. Edmunds, G. F., Jr., and J. R. Traver. 1954. The flight mechanics and evolution of the wings of Ephemeroptera, with notes on the archetype insect wing. J. Wash. Acad. Sci. 44:390-400.
1032. Edmunds, G. F., Jr., and J. R. Traver. 1959. The classification of the Ephemeroptera. I. Ephemeroidea: Behningiidae. Ann. Ent. Soc. Am. 52:43-51.
1033. Edward, D.H.D. 1963. The biology of a parthenogenetic species of *Lundstroemia* (Diptera: Chironomidae), with descriptions of immature stages. Proc. R. ent. Soc. Lond.(A) 38:165-170.
1034. Edwards, J. G. 1951. Amphizoidae (Coleoptera) of the world. Wasmann J. Biol. 8:303-332.
1035. Edwards, J. G. 1954. Observations on the biology of Amphizoidae. Coleopt. Bull. 8:19-24.
1036. Edwards, R. T., and J. L. Meyer. 1990. Bacterivory by deposit-feeding mayfly larvae (*Stenonema* spp.). Freshwat. Biol. 24:453-462.
1037. Edwards, S. W. 1961. The immature stages of *Xiphocentron mexico* (Trichoptera). Tex. J. Sci. 13:51-56.
1038. Edwards, S. W. 1966. An annotated list of the Trichoptera of middle and west Tennessee. J. Tenn. Acad. Sci. 41:116-128.
1039. Edwards, S. W. 1973. Texas caddis flies. Tex. J. Sci. 24:491-516.
1040. Efford, I. E. 1960. A method of studying the vertical distribution of the bottom fauna in shallow waters. Hydrobiologia 16:288-292.
1041. Ege, R. 1915. On the respiratory function of the air stores carried by some aquatic insects (Corixidae, Dytiscidae and Notonectidae). Z. Allg. Physiol. 17:81-124.
1042. Eggert, S. L., and T. M. Burton. 1994. A comparison of *Acroneuria lycorias* (Plecoptera) production and growth in northern Michigan hard- and soft-water streams. Freshwat. Biol. 32:21-31.
1043. Eggleton, F. E. 1931. A limnological study of the profundal bottom fauna of certain freshwater lakes. Ecol. Monogr. 1:231-332.
1044. Egglishaw, H. J., and D. W. Mackay. 1967. A survey of the bottom fauna and plant detritus in streams. J. Anim. Ecol. 33:463-476.
1045. Ekman, S. 1911. Neue Apparate zur qualitativen und quantitativen Untersuchung der Bodenfauna der Binnenseen. Int. Revue ges. Hydrobiol. 3:553-561.
1046. Elliott, J. M. 1965. Daily fluctuations of drift invertebrates in a Dartmoor stream. Nature 205:1127-1129.
1047. Elliott, J. M. 1970a. Life history and biology of *Sericostoma personatum* Spence (Trichoptera). Oikos 20:110-118.
1048. Elliott, J. M. 1970b. Methods of sampling invertebrate drift in running water. Ann. Limnol. 6:133-159.
1049. Elliott, J. M. 1971. The life history and biology of *Apatania muliebris* McLachlan (Trichoptera). Ent. Gaz. 22:245-251.
1050. Elliott, J. M. 1972. Effect of temperature on the time of hatching in *Baetis rhodani* (Ephemeroptera: Baetidae). Oecologia 9:47-51.
1051. Elliott, J. M. 1977. Some methods for the statistical analysis of samples of benthic invertebrates (2nd ed.). Sci. Publ. Freshwat. Biol. Assoc. 25. 160 pp.
1052. Elliott, J. M. 1978. Effect of temperature on the hatching time of eggs of *Ephemerella ignita* (Poda) (Ephemerellidae). Freshwat. Biol. 8:51-58.
1053. Elliott, J. M. 1982. The life cycle and spatial distribution of the aquatic parasitoid *Agriotypus armatus* (Hymenoptera: Agriotypidae and its caddis host *Silo pallipes* (Trichoptera: Goeridae). J. Anim. Ecol. 51:923-941.
1054. Elliott, J. M. 1983a. Life cycle and growth of *Dicranota maculata* (Schummel) (Diptera: Tipulidae). Ent. Gaz. 34:291-296.
1055. Elliott, J. M. 1983b. The responses of aquatic parasitoid *Agriotypus armatus* (Hymenoptera: Agriotypidae) to the spatial distribution and density of its caddis host *Silo pallipes* (Trichoptera: Goeridae). J. Anim. Ecol. 52:315-330.
1056. Elliott, J. M. 1987a. Egg hatching and resource partitioning in stoneflies: the six British *Leuctra* spp. (Plecoptera: Leuctridae). J. Anim. Ecol. 56:415-426.
1057. Elliott, J. M. 1987b. Temperature-induced changes in the life cycle of *Leuctra nigra* (Plecoptera: Leuctridae) from a Lake District stream. Freshwat. Biol. 18:177-184.
1058. Elliott, J. M. 1987c. Life cycle and growth of *Leuctra geniculata* (Stephens) (Plecoptera: Leuctridae) in the River Leven. Ent. Gaz. 38:129-136.
1059. Elliott, J. M. 1988a. Interspecific and intraspecific variation in egg hatching for British populations of the stoneflies *Siphonoperla torrentium* and *Chloroperla tripunctata* (Plecoptera: Chloroperlidae). Freshwat. Biol. 20:11-18.

1060. Elliott, J. M. 1988b. Egg hatching and resource partitioning in stoneflies (Plecoptera): ten British species in the family Nemouridae. J. Anim. Ecol. 57:201-215.
1061. Elliott, J. M. 1991. The effect of temperature on egg hatching for three populations of *Isoperla grammatica* and one population of *Isogenus nubecula* (Plecoptera: Perlodidae). Ent. Gaz. 42:61-65.
1062. Elliott, J. M. 1992. The effect of temperature on hatching for three populations of *Perlodes microcephala* (Pictet) and three populations of *Diura bicaudata* (Linnaeus) (Plecoptera: Perlodidae). Ent. Gaz. 43:115-123.
1063. Elliott, J. M., and C. M. Drake. 1981a. A comparative study of seven grabs used for sampling benthic macroinvertebrates in rivers. Freshwat. Biol. 11:99-120.
1064. Elliott, J. M., and C. M. Drake. 1981b. A comparative study of four dredges used for sampling benthic macroinvertebrates in rivers. Freshwat. Biol. 11:245-261.
1065. Elliott, J. M., C. M. Drake, and P. A. Tullett. 1980. The choice of a suitable sampler for benthic macroinvertebrates in deep rivers. Poll. Rep. Dep. Environ. 8:36-44.
1066. Elliott, J. M., and U. H. Humpesch. 1980. Eggs of Ephemeroptera. Freshwat. Biol. Assoc. Ann. Rept. 48:41-52.
1067. Elliott, J. M., P. A. Tullett, and J. A. Elliott. 1993. A new bibliography of samplers for freshwater benthic invertebrates. Occ. Publ. Freshwat. Biol. Assn. No. 30. 92pp.
1068. Ellis, L. L. 1952. The aquatic Hemiptera of southeastern Louisiana (exclusive of Corixidae). Am. Midl. Nat. 48:302-329.
1069. Ellis, L. L. 1965. An unusual habitat for *Plea striola* (Hemiptera: Pleidae). Fla. Ent. 48:77.
1070. Ellis, R. A., and D. M. Wood. 1974. First Canadian record of *Corethrella brakeleyi* (Diptera: Chaoboridae). Can. Ent. 106:221-222.
1071. Ellis, R. A., and J. H. Borden. 1969. Laboratory rearing of *Notonecta undulata* Say (Hemiptera, Heteroptera, Notonectidae). J. Ent. Soc. Brit. Columbia 66:51-53.
1072. Ellis, R. A., and J. H. Borden. 1970. Predation by *Notonecta undulata* on larvae of the yellow fever mosquito. Ann. Ent. Soc. Am. 63:963-973.
1073. Ellis, R. J. 1962. Adult caddisflies (Trichoptera) from Houghton Creek, Ogemaw County, Michigan. Occ. Pap. Univ. Mich. Mus. Zool. 624:1-15.
1074. Ellis, R. J. 1970. *Alloperla* stonefly nymphs: predators or scavengers on salmon eggs and alevins. Trans. Am. Fish. Soc. 99:677-683.
1075. Ellis, R. J. 1975. Seasonal abundance and distribution of adult stoneflies of Sashin Creek, Baranof Island, southern Alaska (Plecoptera). Pan-Pacif. Ent. 51:23-30.
1076. Ellis, W. N., and P. F. Bellinger. 1973. An annotated list of the generic names of Collembola (insects) and their type species. Monogr. med. ent. Veren. 71:1-74.
1077. Ellison, A. M. 1991. Ecology of case-bearing moths (Lepidoptera: Coleophoridae) in a New England salt marsh. Ann. Ent. Soc. Am. 20:857-864.
1078. Elpers, C., and I. Tomka. 1994. Structure of mouthparts and feeding habits of *Potamanthus luteus* (Linne) (Ephemeroptera: Potamanthidae). Arch. Hydrobiol./Suppl. 99:73-96.
1079. Elton, C. S. 1956. Stoneflies (Plecoptera, Nemouridae), a component of the aquatic leaf-litter fauna in Wytham Woods, Berkshire. Entomol. mon. Mag. 91:231-236.
1080. Elwood, J. W., and R. M. Cushman. 1975. The life history and ecology of *Peltoperla maria* (Plecoptera: Peltoperlidae) in a small spring-fed stream. Verh. Int. Verein. Limnol. 19: 3050-3056.
1081. Emsley, M G. 1969. The Schizopteridae (Hemiptera: Heteroptera) with the description of new species from Trinidad. Mem. Am. Ent. Soc. 25:1-154.
1082. Engblom, E., P. E. Lingdell, A. N. Nilsson, and E. Savolainen. 1993. The genus *Metretopus* (Ephemeroptera: Siphlonuridae) in Fennoscandia - identification, faunistics, and natural history. Ent. Fennica 4:213-222.
1083. Engelke, M. J., Jr. 1980. Aestivation of a water scavenger beetle *Helophorus* sp. in southwestern Wyoming (Coleoptera: Hydrophilidae). Coleopt. Bull. 34:176.
1084. English, W. 1987. Three inexpensive aquatic invertebrate samplers for the benthos, drift and emergent fauna. Ent. News. 98:171-179.
1085. Englund, G. 1993. Effects of density and food availability on habitat selection in a net-spinning caddis larva, *Hydropsyche siltalai*. Oikos 68:473-480.
1086. Epler, J. H. 1986. The larva of *Radotanypus submarginella* (Sublette) (Diptera: Chironomidae). Spixiana 9:285-287.
1087. Epler, J. H. 1987. Revision of the Nearctic *Dicrotendipes* Kieffer,1913 (Diptera: Chironomidae). Evol. Monogr. 9. 102p.
1088. Epler, J. H. 1988a. A reconsideration of the genus *Apedilum* Townes, 1945 (Diptera: Chironomidae: Chironominae). Spixiana Suppl. 14:105-116.
1089. Epler, J. H. 1988b. Biosystematics of the genus *Dicrotendipes* Kieffer, 1913 (Diptera: Chironomidae: Chironominae) of the world. Mem. Am. Ent. Soc. 36:1-214.
1090. Epler, J. H. 1992. Identification manual for the larval Chironomidae (Diptera) of Florida. Fla. Dept. Environ. Reg., Tallahassee, FL.
1091. Epperson, C. R., and R. A. Short. 1987. Annual production of *Corydalus cornutus* (Megaloptera) in the Guadalupe River, Texas. Am. Mid. Nat. 118:433-438.
1092. Eriksen, C. H. 1963a. Respiratory regulation in *Ephemera simulans* (Walker) and *Hexagenia limbata* (Serville) (Ephemeroptera). J. exp. Biol. 40:455-468.
1093. Eriksen, C. H. 1963b. The relation of oxygen consumption to substrate particle size in two burrowing mayflies. J. exp. Biol. 40:447-453.
1094. Eriksen, C. H. 1968. Ecological significance of respiration and substrate for burrowing Ephemeroptera. Can. J. Zool. 46:93-103.
1095. Eriksen, C. H. 1984. The physiological ecology of larval *Lestes disjunctus* Selys (Zygoptera, Odonata). Freshwat. Invertebr. Biol. 3:105-117.
1096. Eriksen, C. H. 1986. Respiratory roles of caudal lamellae (gills) in a lestid damselfly. J. N. Am. Benthol. Soc. 5:16-27.
1097. Eriksen, C. H., and C. R. Feldmeth. 1967. A water-current respirometer. Hydrobiologia 29:495-504.
1098. Eriksen, C. H., and J. E. Moeur. 1990. Respiratory functions of motile tracheal lamellae (gills) in Ephemeroptera, as exemplified by *Siphlonurus occidentalis* Eaton, pp. 109-118. In: I. C. Campbell (ed.). Mayflies and stoneflies: life history and biology. Kluwer Acad. Publ., Dorcrecht. The Netherlands. 366pp.
1099. Eriksson, M.O.G., L. Henrikson, B.-I. Nilsson, G. Nyman, H. G. Oscarson, and A. E. Stenson. 1980. Predator-prey relationships important for the biotic changes in acidified lakes. Ambio 9:248-249.
1100. Erman, D. C., and W. C. Chouteau. 1979. Fine particulate organic carbon output from fens and its effect on benthic macroinvertebrates. Oikos 32:409-415.
1101. Erman, N. A. 1981. Terrestrial feeding migration and life history of the stream-dwelling caddisfly, *Desmona bethula* (Trichoptera: Limnephilidae). Can. J. Zool. 59:1658-1665.
1102. Ernst, M. R., B. C. Poulton, and K. W. Stewart. 1986. *Neoperla* (Plecoptera: Perlidae) of the southern Ozark and Ouachita mountain region, and two new species of *Neoperla*. Ann. Ent. Soc. Am. 79:645-661.
1103. Ernst, M. R., and K. W. Stewart. 1985. Growth and drift of nine stonefly species (Plecoptera) in an Oklahoma Ozark foothills stream, and conformation to regression models. Ann. Ent. Soc. Am. 78:635-646.
1104. Ernst, M. R., and K. W. Stewart. 1986. Microdistribution of stoneflies (Plecoptera) in relation to organic matter in an Ozark foothills stream. Aquat. Insects 8:237-254.
1105. Esaki, T., and W. E. China. 1927. A new family of aquatic Hemiptera. Trans. Ent. Soc. Lond. 1927(II):279-295.
1106. Eskafi, F. M., and E. F. Legner. 1974. Descriptions of immature stages of the cynipid *Hexacola* sp. near *websteri* (Eucoilinae: Hymenoptera), a larval-pupal parasite of *Hippelates* eye gnats (Diptera: Chloropidae). Can. Ent. 106:1043-1048.
1107. Espinosa, L. R., and W. E. Clark. 1972. A polyprophylene light trap for aquatic invertebrates. Calif. Fish Game 58:149-152.
1108. Essenberg, C. 1915. The habits and natural history of the backswimmers Notonectidae. J. Anim. Behav. 5:381-390.

1109. Etnier, D. A. 1965. An annotated list of the Trichoptera of Minnesota with a description of a new species. Ent. News 76:141-152.

1110. Etnier, D. A., and G. A. Schuster. 1979. An annotated list of Trichoptera (caddisflies) of Tennessee. J. Tenn. Acad. Sci. 54:15-22.

1111. Ettinger, W. S. 1979. A collapsible insect emergence trap for use in shallow standing water. Ent. News 90:114-117.

1112. Ettinger, W. 1984. Variation between technicians sorting benthic macroinvertebrate samples. Freshwat. Invertebr. Biol. 3:147-149.

1113. Euliss, N. H., Jr., G. A. Swanson, and J. Mackay. 1992. Multiple tube sampler for benthic and pelagic invertebrates in shallow wetlands. J. Wildl. Manage. 56:186-191.

1114. Evans, E. D. 1972. A study of the Megaloptera of the Pacific coastal region of the United States. Ph.D. diss., Oregon State University, Corvallis. 210 pp.

1115. Evans, E. D. 1984. A new genus and a new species of a dobson fly from the far western United States (Megaloptera: Corydalidae). Pan-Pacif. Ent. 60:1-3.

1116. Evans, E. D., and H. H. Neunzig. 1984. Megaloptera and aquatic Neuroptera, p. 261-270. In R. W. Merritt and K. W. Cummins (eds.). An introduction to aquatic insects of North America (2nd ed.) Kendall/Hunt Publ. Co., Dubuque, IA. 722 pp.

1117. Evans, H. E. 1949. The strange habits of Anoplius depressipes Banks: A mystery solved. Proc. Ent. Soc. Wash. 51:206-208.

1118. Evans, H. E. 1950-51. A taxonomic study of the Nearctic spider wasp belonging to tribe Pompilini (Hym.: Pompilidae). Trans. Am. Ent. Soc. I, 75:133 270;II, 76:207-361;III, 77:203-330.

1119. Evans, H. E. 1959. The larvae of Pompilidae. Ann. Ent. Soc. Am. 52:430-444.

1120. Evans, H. E. 1987a. Cynipoidea, pp. 665-666. In F. W. Stehr (ed.). Immature Insects. Vol. 1. Kendall/Hunt Publ. Co., Dubuque, IA. 754pp.

1121. Evans, H. E. 1987b. Proctotrupoidea, pp. 666-667. In F. W. Stehr (ed.). Immature Insects. Vol. 1. Kendall/Hunt Publ. Co., Dubuque, IA. 754pp.

1122. Evans, H. E. 1987c. Pompilidae (Pompiloidea), pp. 680-681. In F. W. Stehr (ed.). Immature Insects. Vol. 1. Kendall/Hunt Publ. Co., Dubuque, IA. 754pp.

1123. Evans, H. E., T. Finlayson, R. J. McGinley, W. W. Middlekauff and D. R. Smith. 1987. Key to families of Hymenoptera (mature larvae), pp. 602-617. In F. W. Stehr (ed.). Immature insects. Vol. 1. Kendall/Hurt Publ. Co., Dubuque, IA. 754pp.

1124. Evans, H. E., and C. M. Yoshimoto. 1962. The ecology and nesting behavior of the Pompilidae (Hymenoptera) of the northeastern United States. Misc. Publ. Ent. Soc. Am. 3:67-119.

1125. Evans, M. A. 1988. Checklist of the Odonata of Colorado. Great Basin Nat. 48:96-101.

1126. Exner, K., and D. A. Craig. 1976. Larvae of Alberta Tanyderidae (Diptera: Nematocera). Quaest. Ent. 12:219-237.

1127. Exner, K. K., and R. W. Davies. 1979. Comments on the use of a standpipe corer in fluvial gravels. Freshwat. Biol. 9:77-78.

1128. Eymann, M. 1991. Flow patterns around cocoons and pupae of black flies of the genus *Simulium* (Diptera: Simuliidae). Hydrobiologia 215:223-229.

1129. Fager, E. W., A. O. Flechsig, R. F. Ford, R. I. Clutter, and R. J. Ghelardi. 1966. Equipment for use in ecological studies using SCUBA. Limnol. Oceanogr. 11:503-509.

1130. Fahy, E. 1972a. An automatic separator for the removal of aquatic insects from detritus. J. Appl. Ecol. 6:655-658.

1131. Fahy, E. 1972b. The feeding behavior of some common lotic insect species in two streams of differing detrital content. J. Zool. Lond. 167:337-350.

1132. Fairbairn, D. J. 1984. Microgeographic variation in body size and development time in the waterstrider, *Limnoporus notabilis*. Oecologia (Berlin) 61:126-133.

1133. Fairchild, W. L., M.C.A. O'Neill, and D. M. Rosenberg. 1987. Quantitative evaluation of the behavioral extraction of aquatic invertebrates from samples of spagnum moss. J. N. Am. Benthol. Soc. 6:281-287.

1134. Fairchild, W. L., and G. B. Wiggins. 1989. Immature stages and biology of the North American caddisfly genus *Phanocelia* (Trichoptera: Limnephilidae). Can. Ent. 121:515-519.

1135. Falkenhan, H. H. 1932. Biologische Beobachtungen an *Sminthurides aquaticus* (Collembola). Z. Wiss. Zool. 141:525-580.

1136. Fall, H. C. 1901. List of the Coleoptera of Southern California with notes on habits and distribution and descriptions of new species. Occ. Pap. Calif. Acad. Sci. 8:1-282.

1137. Fall, H. C. 1919. The North American species of *Coelambus*. John D. Sherman, Jr., Mt. Vernon, N.Y. 20 pp.

1138. Fall, H. C. 1922a. A revision of the North American species of *Agabus* together with a description of a new genus and species of the tribe Agabini. John D. Sherman, Jr., Mt. Vernon, N.Y. 36 pp.

1139. Fall, H. C. 1922b. The North American species of *Gyrinus*. Trans. Ent. Soc. Am. 47:269-306.

1140. Fall, H. C. 1923. A revision of the North American species of *Hydroporus* and *Agaporus*. Privately printed by H. C. Fall. 129 pp.

1141. Farmer, R. G. 1979. The life history, population dynamics and energetics of the riffle beetle *Stenelmis exigua* (Elmidae) in Barebone Creek, Trimble County, Kentucky. Diss. Abstr. Int. (B)39:5212.

1142. Farmer, R. G., and D. C. Tarter. 1976. Distribution of the superfamily Nemouroidea in West Virginia (Insecta: Plecoptera). Ent. News 87:17-24.

1143. Farris, J. L., and G. L. Harp. 1982. Aquatic macroinvertebrates of three acid bogs on Crowley's Ridge in northeast Arkansas. Ark. Acad. Sci. Proc. 36:23-27.

1144. Fast, A. W. 1972. A new aquatic insect trap. Pap. Mich. Acad. Sci. Arts Lett. 5:115-124.

1145. Faucheux, M. J. 1978. Contribution a l'etude de la biologie de la larvae de *Statiomyia longicornis* (Diptera, Stratiomylidae): prise de nourriture et respiration. Ann. Soc. Ent. Fr. (N.S.). 14:49-72.

1146. Federley, H. 1908. Einige Libelluliden wanderungen uber die zoologische Station bei Tvärminne. Acta Soc. Fauna Flora Fenn. 31:1-38.

1147. Fedorenko, A. Y., and M. C. Swift. 1972. Comparative biology of *Chaoborus americanus* and *Chaoborus trivittatus* in Eunice lake, British Columbia. Limnol. Oceanogr. 17:721-730.

1148. Fee, E. J., and R. E. Hecky. 1992. Introduction to the Northwest Ontario Lake Size Series (NOLSS). Can. J. Fish. Aquat. Sci. 49:2434-2444.

1149. Fee, E. J., J. A. Shearer, E. R. DeBruyn, and E. U. Schindler. 1992. Effects of lake size on phytoplankton photosynthesis. Can. J. Fish. Aquat. Sci. 49:2445-2459.

1150. Feldmann, A. M., W. Brouwer, J. Meeussen, and J. Oude Vashaar. 1989. Rearing of larvae of *Anopheles stephensi*, using water replacement, purification and automated feeding. Ent. Exp. Appl. 52:57-68.

1151. Feldmeth, C. R. 1968. The respiratory energetics of two species of stream caddisflies in relation to water flow. Ph.D. diss. University of Toronto, Canada.

1152. Feldmeth, C. R. 1970. The respiratory energetics of two species of stream caddisfly larvae in relation to water flow. Comp. Biochem. Physiol. 32:193-202.

1153. Feltmate, B. W., R. L. Baker, and P. J. Pointing. 1986. Distribution of the stonefly nymph *Paragnetina media* (Plecoptera: Perlidae): influence of prey, predators, current speed, and substrate compostion. Can. J. Fish. Aquat. Sci. 43:1582-1587.

1154. Felton, H. L. 1940. Control of aquatic midges with notes on the biology of certain species. J. Econ. Ent. 33:252-264.

1155. Feminella, J. W., M. E. Power, and V. H. Resh. 1989. Periphyton responses to invertebrate grazing and riparian canopy in three northern California coastal streams. Freshwat. Biol. 22:445-457.

1156. Feminella, J. W., and V. H. Resh. 1990. Hydrologic influences, disturbance, and intraspecific competition in a stream caddisfly population. Ecology 71:2083-2094.

1157. Feminella, J. W., and V. H. Resh. 1991. Herbivorous caddisflies, macroalgae, and epilithic microalgae: dynamic interactions in a stream grazing ecosystem. Oecologia 87:247-256.

1158. Feminella, J. W., and K. W. Stewart. 1986. Diet and predation by three leaf-associated stoneflies (Plecoptera) in an Arkansas mountain stream. Freshwat. Biol. 16:521-538.
1159. Ferguson, A. 1940. A preliminary list of the Odonata of Dallas County, Texas. Field Lab. 8:1-10.
1160. Ferkinhoff, W. D., and R. W. Gundersen. 1983. A key to the whirligig beetles of Minnesota and adjacent states and Canadian provinces (Coleoptera: Gyrinidae). Sci. Publ. Sci. Mus. Minn.(N. S.) 5:1-53.
1161. Ferrar, P. 1987. A guide to the breeding habits and immature stages of Diptera Cyclorrhapha. E. J. Brill/Scandinavian Science Press, Leiden. 907pp.
1162. Ferrier, M. D., and T. E. Wissing. 1979. Larval retreat and food habits of the netspinning trichopteran *Cheumatopsyche analis* (Trichoptera: Hydropsychidae). Great Lakes Ent. 12:157-164.
1163. Ferrington, L. C., Jr. 1981. Kansas Chironomidae, Part I pp. 45-51. *In* R. Brooks (ed.). New records of the fauna and flora of Kansas for 1980. St. Biol. Surv. Kansas Tech. Pub. 10.
1164. Ferrington, L. C., Jr. 1982. Kansas Chironomidae, Part II: The Tanypodinae, pp. 49-60. *In* R. Brooks (ed.). New records of the fauna and flora of Kansas for 1981. St. Biol. Surv. Kansas Tech. Publ. 12.
1165. Ferrington, L. C., Jr. 1983a. Kansas Chironomidae, Part III: The *Harnischia* complex, pp. 48-62. *In* R. Brooks (ed.). New records of the fauna and flora of Kansas for 1982. St. Biol. Surv. Kansas Tech. Publ. 13.
1166. Ferrington, L. C., Jr. 1983b. Interdigitating broadscale distributional patterns in some Kansas Chironomidae. Mem. Ent. Soc. Am. 34:101-113.
1167. Ferrington, L. C., Jr. 1984. Evidence for the hyporheic zone as a microhabitat of *Krenosmittia* spp. larvae (Diptera: C hironomidae). J. Freshwat. Ecol. 2:353-358.
1168. Ferrington, L. C., Jr. 1987b. Chironomidae in the capillary fringe of the Cimarron River. J. Kans. Ent. Soc. 60:153-156.
1169. Ferrington, L. C., Jr. 1987a. Microhabitat preferences of larvae of three orthocladiinae species (Diptera: Chironomidae) in big springs, a sand bottom spring in the high plains of western Kansas. Ent. Scand. (suppl.) 29:361-368.
1170. Ferrington, L. C., Jr. 1992. Habitat and sediment preferences of *Axarus festivus* larvae. Neth. J. Aquat. Ecol. 26:347-354.
1171. Ferrington, L. C., Jr. 1995. Utilization of anterior headcapsule structures in locomotion by larvae of *Constempellina* sp. (Diptera: Chironomidae). Proc. XII. Int. Conf. on Chironomidae, CSIRO, Canberra, Australia.
1172. Ferrington, L. C., Jr., K. M. Buzby, and E. C. Masteller. 1993. Composition and temporal abundance of Chironomidae emergence from a tropical rainforest stream at El Verde, Puerto Rico. J. Kans. Ent. Soc. 66:167-180.
1173. Ferrington, L. C., Jr., and N. H. Crisp. 1989. Water chemistry characteristics of receiving streams and the occurrence of *Chironomus riparius* and other Chironomidae in Kansas. Acta Biol. Debrecina Oecol. Hungary 3:115-126.
1174. Ferrington, L. C., Jr., and C. Christianssen. 1985. Biological statistical significance of *Stenochironomus* larvae on artificial substrate samplers with masonite discs. J. Kans. Ent. Soc. 58:724-726.
1175. Ferrington, L. C., Jr., and D. S. Goldhammer. 1992. The biological significance of capillary fringe habitat in large sandy-bottomed rivers of the Central Plains, pp. 203-225. *In* Bohart and Bothwell (eds.). Proc. Symp Aquat Res Semi-Arid Environ. N. H. R. I. Symp. VII. 375pp.
1176. Ferrington, L. C., Jr., and O. A. Saether. 1987. The male, female pupa and biology of *Oliveridia hugginsi* n. sp. (Diptera: Chironomidae, Orthocladiinae) from Kansas. J. Kans. Ent. Soc. 60:451-461.
1177. Fiance, S. B. 1977. The genera of eastern North American Chloroperlidae (Plecoptera): Key to larval stages. Psyche 83:308-316.
1178. Fiance, S. B., and R. E. Moeller. 1974. Immature stages and ecological observations on *Eoparargyractis plevie* (Pyralidae: Nymphulinae). J. Lepidopterist's Soc. 31:81-88.
1179. Field, G., R. J. Duplessis, and A. P. Breton. 1967. Progress report on laboratory rearing of black flies (Diptera: Simuliidae). J. med. Ent. 4:304-305.
1180. Fink, T. J. 1986. The reproductive life history of the predaceous, sand-burrowing mayfly *Dolania americana* (Ephemeroptera: Behningiidae). Ph.D. diss., Florida State Univ., Tallahassee, FL. 160pp.
1181. Fink, T. J., T. Soldan, J. G. Peters, and W. L. Peters. 1991. The reproductive life history of the predaceous, sand-burrowing mayfly *Dolania americana* (Ephemeroptera: Behningiidae) and comparisons with other mayflies. Can. J. Zool. 69:1083-1103.
1182. Finlayson, T. 1987a. Ichneumonidae (Ichneumoidea), pp. 650-657. *In* F. W. Stehr (ed.). Immature Insects. Vol. 1. Kendall/Hunt Publ. Co., Dubuque, IA. 754pp.
1183. Finlayson, T. 1987b. Braconidae (Ichneumonoidea), pp. 657-663. *In* F. W. Stehr (ed.). Immature Insects. Vol. 1. Kendall/Hunt Publ. Co., Dubuque, IA. 754pp.
1184. Finlayson, T., and K. S. Hagen. 1979. Final instar larvae of parasitic Hymenoptera. Pest Manage. Pap. No. 10, Simon Fraser Univ., Brit. Columbia. 111 pp.
1185. Finni, G. R. 1973. Biology of winter stoneflies in a central Indiana stream (Plecoptera). Ann. Ent. Soc. Am. 66:1243-1248.
1186. Finni, G. R. 1975. Feeding and longevity of the winter stonefly, *Allocapnia granulata* (Claassen) (Plecoptera: Capniidae). Ann. Ent. Soc. Am. 68:207-208.
1187. Finni, G. R., and B. A. Skinner. 1975. The Elmidae and Dryopidae (Coleoptera: Dryopoidea) of Indiana. J. Kans. Ent. Soc. 48:388-395.
1188. Finni, G. R., and L. Chandler. 1977. Post diapause instar discrimination and life history of the capniid stonefly, *Allocapnia granulata* (Claassen) (Insecta: Plecoptera). Am. Midl. Nat. 98:243-250.
1189. Finni, G. R., and L. Chandler. 1979. The micro-distribution of *Allocapnia* naiads (Plecoptera: Capniidae). J. Kans. Ent. Soc. 52:93-102.
1190. Finnish IBP/PM Group. 1969. Quantitative sampling equipment for the littoral benthos. Int. Revue ges. Hydrobiol. 54:185-193.
1191. Fischer, M. 1964. Die Opiinae der Nearctic Region (Hymenoptera: Braconidae) Pol. Pismo Ent. Bull. Ent. Pol. I, 44:197-530; II, 35:1-212.
1192. Fischer, M. 1971. World Opiinae (Hymenoptera), pp. 1-189. *In* V. Delucchi and G. Remaudiere (eds.). Index of entomophagous insects, 5. LeFrancois, Paris.
1193. Fischer, T. W., and R. E. Orth. 1969. A new *Dictya* from California, with biological notes. Pan-Pac. Ent. 45:222-228.
1194. Fischer, Z. 1966. Food selection and energy transformation in larvae of *Lestes sponsa* (Odonata) in Asiatic waters. Verh. Int. Verein. Limnol. 16:600-603.
1195. Fish, D. 1976. Insect-plant relationships of the insectivorous pitcher plant *Sarracenia minor*. Fla. Ent. 59:199-203.
1196. Fish, D., and D. W. Hall. 1978. Succession and stratification of aquatic insects inhabiting leaves of the insectivorous pitcher plant *Saracenia purpurea*. Am. Midl. Nat. 99:172-183.
1197. Fish, D., and S. R. Carpenter. 1982. Leaf litter and larval mosquito dynamics in treehole ecosystems. Ecology 63:283-88.
1198. Fisher, E. G. 1940. A list of Maryland Odonata. Ent. News 51:37-42, 51:67-72.
1199. Fisher, S. G., and L. J. Gray. 1983. Secondary production and organic matter processing by collector macroinvertebrates in a desert stream. Ecology 64:1217-1224.
1200. Fisher, T. W., and R. W. Orth. 1964. Biology and immature stages of *Antichaeta testacea* Melander (Diptera: Sciomyzidae). Hilgardia 36:1-29.
1201. Fittkau, E. J. 1962. Die Tanypodinae (Diptera: Chironomidae). Abh. Laralsyst. Insekt. 6:1-453.
1202. Fittkau, E. J., and D. A. Murray. 1988. *Bethbilbeckia floridensis*: a new genus and species of Macropelopiini from the South Eastern Nearctic. Spixiana Suppl. 14:253-259.
1203. Fittkau, E. J., F. Reiss, and O. Hoffrichter. 1976. A bibliography of the Chironomidae. Gunneria 26:1-177.
1204. Fitzgerald, V. J. 1986. Social behavior of adult whirligig beetles (*Dineutus nigrior* and *D. discolor*) (Coleoptera: Gyrinidae). Am. Midl. Nat.118:439-448.
1205. Fjellberg, A. 1983. Collembola from the Colorado front Range, USA. Arctic and Alpine Res. 16:193-208.

1206. Fjellberg, A. 1985. Arctic Collembola I- Alaskan Collembola of the families Poduridae, Hypogastruridae, Odontellidae, Brachystomellidae and Neanuridae. Ent. Scand. (Suppl.) 21:1-125.

1207. Fjellheim, A., and G. G. Raddum. 1990. Acid precipitation: biological monitoring of streams and lakes. Sci. Tot. Environ. 96:57-66.

1208. Flannagan, J. F. 1970. Efficiencies of various grabs and corers in sampling freshwater benthos. J. Fish. Res. Bd. Can. 27:1691-1700.

1209. Flannagan, P. M., and J. F. Flannagan. 1982. Present distribution and the post-glacial origin of the Ephemeroptera, Plecoptera and Trichoptera of Manitoba. Manitoba Dept. Nat. Res. Tech. Rep. 82-1. 79 pp.

1210. Flannagan, J. F., W. L. Lockhart, D. G. Cobb, and D. Metner. 1978. Stonefly (Plecoptera) head cholinesterase as an indicator of exposure to fenitrothion. Man. Ent. 12:42-48.

1211. Flannagan, J. F., and D. M. Rosenberg. 1982. Types of artificial substrates used for sampling freshwater benthic macroinvertebrates, pp. 237-266. In J. Cairns, Jr. (ed.). Artificial substrates. Ann Arbor Sci. Publ., Mich. 279 pp.

1212. Flint, O. S. 1956. The life history of the genus *Frenesia* (Trichoptera: Limnephilidae). Bull. Brooklyn Ent. Soc. 51:93-108.

1213. Flint, O. S. 1958. The larva and terrestrial pupa of *Ironoquia parvula* (Trichoptera, Limnephilidae). J. N.Y. Ent. Soc. 66:59-62.

1214. Flint, O. S. 1960. Taxonomy and biology of Nearctic limnephilid larvae (Trichoptera), with special reference to species in eastern United States. Entomologica Am. 40:1-117.

1215. Flint, O. S. 1961. The immature stages of the Arctopsychinae occurring in eastern North America (Trichoptera: Hydropsychidae). Ann. Ent. Soc. Am. 54:5-11.

1216. Flint, O. S. 1962. Larvae of the caddis fly genus *Rhyacophila* in eastern North America (Trichoptera: Rhyacophilidae). Proc. U.S. Nat. Mus. 113:465-493.

1217. Flint, O. S. 1964a. The caddisflies (Trichoptera) of Puerto Rico. Univ. Puerto Rico Agric. Exp. Sta. Tech. Pap. 40.

1218. Flint, O. S. 1964b. Notes on some Nearctic Psychomyiidae with special reference to their larvae (Trichoptera). Proc. U.S. Nat. Mus. 115:467-481.

1219. Flint, O. S. 1965. The genus *Neohermes*. Psyche 72:255-263.

1220. Flint, O. S. 1968. Bredin-Archbold-Smithsonian biological survey of Dominica. 9. The Trichoptera (caddisflies) of the Lesser Antilles. Proc. U.S. Nat. Mus. 125:1-86.

1221. Flint, O. S. 1970. Studies of Neotropical caddisflies, X: *Leucotrichia* and related genera from North and Central America (Trichoptera: Hydroptilidae). Smithson. Contr. Zool. 60:1-64.

1222. Flint, O. S. 1973. Studies of Neotropical caddisflies, XVI.: The genus *Austrotinodes* (Trichoptera: Psychomyiidae). Proc. Biol. Soc. Wash. 86:127-142.

1223. Flint, O. S. 1974. Studies of Neotropical caddisflies, XVII: The genus *Smicridea* from North and Central America (Trichoptera: Hydropsychidae). Smithson. Contr. Zool. 167:1-65.

1224. Flint, O. S. 1984. The genus *Brachycentrus* in North America, with a proposed phylogeny of the genera of Brachycentridae (Trichoptera). Smithsonian Contr. Zool. 398.

1225. Flint, O. S., Jr., and J. Bueno-Soria. 1982. Studies of Neotropical caddisflies, XXXII: The immature stages of *Macronema variipenne* Flint and Bueno, with the division of *Macronema* by the resurrection of *Macrostemum* (Trichoptera: Hydropsychidae). Proc. Biol. Soc. Wash. 95:358-370.

1226. Flint, O. S., J. R. Voshell, and C. R. Parker. 1979. The *Hydropsyche scalaris* group in Virginia, with the description of two new species (Trichoptera: Hydropsychidae). Proc. Biol. Soc. Wash. 92:837-862.

1227. Flint, O. S., and G. B. Wiggins. 1961. Records and descriptions of North American species in the genus *Lepidostoma*, with a revision of the *vernalis* group (Trichoptera: Lepidostomatidae). Can. Ent. 93:279-297.

1228. Flowers, R. W. 1980. Two new genera of Nearctic Heptageniidae (Ephemeroptera). Fla. Ent. 63:296-307.

1228'. Flowers, R. W. 1982. Review of the genus *Macdunnoa* (Ephemeroptera: Heptageniidae) with description of a new species from Florida. Gr. Lakes Ent. 15:25-30.

1229. Flowers, R. W., and W. L. Hilsenhoff. 1975. Heptageniidae (Ephemeroptera) of Wisconsin. Great Lakes Ent. 8:201-218.

1230. Floyd, M. A. 1993. The biology and distribution of *Oecetis* larvae in North America (Trichoptera: Leptoceridae), pp. 87-90. In C. Otto (ed.). Proc. VII. Int. Symp. Trichoptera, Umea, Sweden. Backhuys Publ., Leiden, Netherlands.

1231. Folkerts, G. W. 1967. Mutualistic cleaning behavior in an aquatic beetle (Coleoptera). Coleopt. Bull. 21:27-28.

1232. Folkerts, G. W. 1979. *Spanglerogyrus albiventris,* a primitive new genus and species of Gyrinidae (Coleoptera) from Alabama. Coleopt. Bull. 33:1-8.

1233. Folkerts, G. W., and L. A. Donavan. 1973. Resting sites of stream-dwelling gyrinids. Ent. News 84:198-201.

1234. Folkerts, G. W., and L. A. Donavan. 1974. Notes on the ranges and habitats of some little-known aquatic beetles of the southeastern U.S. (Coleoptera: Gyrinidae, Dytiscidae). Coleopt. Bull. 28:203-208.

1235. Folsom, F. C., and K. C. Manuel. 1983. The life cycle of *Pteronarcys scotti* (Plecoptera: Pteronarcyidae) from the southern Appalachians, U.S.A. Aquat. Insects 5:227-232.

1236. Folsom, J. W. 1916. North American collembolous insects of the subfamilies Achorutinae, Neanurinae, and Poduridae. Proc. U.S. Nat. Mus. 50:477-525.

1237. Folsom, J. W. 1917. North American collembolous insects of the subfamily Onychiuridae. Proc. U.S. Nat. Mus. 53:637-659.

1238. Folsom, J. W. 1937. Nearctic Collembola or springtails of the family Isotomidae. Bull. U.S. Nat. Mus. 168:1-138.

1239. Folsom, J. W., and H. B. Mills. 1938. Contribution to the knowledge of the genus *Sminthurides* Boerner. Bull. Mus. Comp. Zool. 82:231-274.

1240. Folsom, T. C., and N. C. Collins. 1982a. Food availability in nature for the larval dragonfly *Anax junius* (Odonata: Aeshnidae). Freshwat. Invert. Biol. 1:33-40.

1241. Folsom, T. C., and N. C. Collins. 1982b. An index of food limitation in the field for the larval dragonfly *Anax junius* (Odonata: Aeshnidae). Freshwat. Invert. Biol. 1:25-32.

1242. Folsom, T. C., and N. C. Collins. 1984. The diet and foraging behavior of the larval dragonfly *Anax junius* (Aeshnidae), with an assessment of the role of refuges and prey activity. Oikos 42:105-113.

1243. Foote, B. A. 1959a. Biology and life history of the snail-killing flies belonging to the genus *Sciomyza* Fallen (Diptera: Sciomyzidae). Ann. Ent. Soc. Am. 52:31-43.

1244. Foote, B. A. 1959b. Biology of *Pherbellia prefixa* (Diptera: Sciomyzidae), a parasitoid-predator of the operculate snail *Valvata sincera* (Gastropoda: Valatidae). Proc. Ent. Soc. Wash. 75:141-149.

1245. Foote, B. A. 1971. Biology of *Hedria mixta* (Diptera: Sciomyzidae). Ann. Ent. Soc. Am. 64:931-941.

1246. Foote, B. A. 1976. Biology and larval feeding habits of three species of *Renocera* (Diptera: Sciomyzidae) that prey on fingernail clams (Mollusca: Sphaeriidae). Ann. Ent. Soc. Am. 68:121-133.

1247. Foote, B. A. 1977. Utilization of blue-green algae by larvae of shore flies. Environ. Ent. 6:812-814.

1248. Foote, B. A. 1979. Utilization of algae by larvae of shore flies, pp. 61-71. In D. L. Deonier (ed.). First symposium on the systematics and ecology of Ephydridae (Diptera). N. Am. Benthol. Soc. 147 pp.

1249. Foote, B. A. 1981a. Biology and immature stages of *Lytogaster excavata*, a grazer of blue-green algae (Diptera: Ephydridae). Proc. Ent. Soc. Wash. 83:304-315.

1250. Foote, B. A. 1981b. Biology and immature stages of *Pelina truncatula*, a consumer of blue-green algae (Diptera: Ephydridae). Proc. Ent. Soc. Wash. 83:607-619.

1251. Foote, B. A. 1982. Biology and immature stages of *Setacera atrovirens*, a grazer of floating algal mats (Diptera: Ephydridae). Proc. Ent. Soc. Wash. 84:828-844.

1252. Foote, B. A. 1983. Biology and immature stages of *Nostima approximata* (Diptera: Ephydridae), a grazer of the blue-green algal genus *Oscillatoria*. Proc. Ent. Soc. Wash. 85:472-484.

1253. Foote, B. A. (coord.). 1987. Chap. 37. Order Diptera, pp. 690-915. In. F.W. Stehr (ed.). Immature insects. Vol. 2. Kendall/Hunt Publ. Co., Dubuque, IA. 975pp.

1254. Foote, B. A. 1990. Biology and immature stages of *Coenia curicauda* (Diptera: Ephydridae). J. N. York Ent. Soc. 98:93-102.

1255. Foote, B. A. 1995. Biology of shore flies. Ann. Rev. Ent. 40:417-442.

1256. Foote, B. A., and W. C. Eastin. 1974. Biology and immature stages of *Discocerina obscurella* (Diptera: Ephydridae). Proc. Ent. Soc. Wash. 76:401-408.

1257. Foote, B. A., and L. V. Knutson. 1970. Clam-killing fly larvae. Nature, Lond. 226:466.

1258. Foote, B. A., S. E. Neff, and C. O. Berg. 1960. Biology and immature stages of *Atrichomelina pubera* (Diptera: Sciomyzidae). Ann. Ent. Soc. Am. 53:192-199.

1259. Forbes, S. A. 1887. The lake as a microcosm. Bull. Ill. Nat. Hist. Surv. 15:537-550.

1260. Forbes, W.T.M. 1910. The aquatic caterpillars of Lake Quinsigamond. Psyche 17:219-227.

1261. Forbes, W.T.M. 1911. Another aquatic caterpillar (*Elophila*). Psyche 18:120-121.

1262. Forbes, W.T.M. 1923. The Lepidoptera of New York and neighboring states. Mem. Cornell Univ. Agric. Exp. Sta. 68:574-581.

1263. Forbes, W.T.M. 1938. *Acentropus* in America (Lepidoptera, Pyralidae). J. N.Y. Ent. Soc. 46:338.

1264. Forbes, W.T.M. 1954. Lepidoptera of New York and neighboring states. III. Noctuidae. Mem. Cornell Univ. Agric. Exp. Sta. 329:1-433.

1265. Forbes, W.T.M. 1960. Lepidoptera of New York and neighboring states. IV. Agaristidae through Nymphalidae, including butterflies. Mem. Cornell Univ. Agric. Exp. Sta. 371:1-188.

1266. Ford, J. B. 1962. The vertical distribution of larval Chironomidae (Dipt.) in the mud of a stream. Hydrobiologia 19:262-272.

1267. Forey, P. L., C. J. Humphries, I. J. Kitching, R. W. Scotland, D. J. Siebert, and D. M. Williams. 1992. Cladistics: a practical course in systematics. Clarendon Press, Oxford. 191pp.

1268. Formanowicz, D. R., Jr. 1982. Foraging tactics of larvae of *Dytiscus verticalis* (Coleoptera: Dytiscidae) the assessment of prey density. J. Anim. Ecol. 51:757-768.

1269. Formanowicz, D. R., Jr. 1987. Foraging tactics of *Dytiscus verticalis* larvae (Coleoptera: Dytiscidae) prey detection reactive distance and predator size. J. Kans. Ent. Soc. 60:92-99.

1270. Formanowicz, D. R., Jr., and E. D. Brodie, Jr. 1981. Prepupation behavior and pupation of the predaceous diving beetle *Dytiscus verticolis* (Coleoptera: Dytiscidae). J. N.Y. Ent. Soc. 89:152-157.

1271. Forno, I. W. 1983. Life history and biology of a water hyacinth moth, *Argyractis subornata* (Lepidoptera: Pyralidae, Nymphulinae). Ann. Ent. Soc. Am. 76:624-627.

1272. Forsyth, A. B., and R. J. Robertson. 1975. K reproductive strategy and larval behavior of the pitcher plant sarcophagid fly, *Baesoxipha fletcheri*. Can. J. Zool. 53:174-179.

1273. Foster, S. A., V. B. Garcia, and M. Y. Town. 1988. Cannibalism as the cause of an ontogenetic shift in habitat use by fry of the threespine stickleback. Oecologia 74:577-585.

1274. Foster, W. A. 1975. The life history and population biology of an intertidal aphid, *Pemphigus trehernei* Foster. Trans. Ent. Soc. Lond. 127:193-207.

1275. Foster, W. A., and J. E. Treherne. 1976. Insects of marine salt marshes: Problems and Adaptations, pp. 5-42. *In* L. Cheng (ed.). Marine Insects. Elsevier, N.Y. 581 pp.

1276. Fox, H. M. 1920. Methods of studying the respiratory exchange in smaller aquatic organisms, with particular reference to the use of flagellates as an indicator for oxygen consumption. J. Gen. Physiol. 3:565-573.

1277. Fox, H. M., and J. Sidney. 1953. The influence of dissolved oxygen on the respiratory movements of caddis larvae. J. exp. Biol. 30:235-237.

1278. Fox, L. R. 1975a. Some demographic consequences of food shortage for the predator *Notonecta hoffmanni*. Ecology 56:868-880.

1279. Fox, L. R. 1975b. Factors influencing cannibalism, a mechanism of population limitation in the predator *Notonecta hoffmanni*. Ecology 56:933-941.

1280. Fracker, S. B. 1915. The classification of lepidopterous larvae. Univ. Ill. Biol. Monogr. 2:1-165.

1281. France, R. L., and B. D. LaZerte. 1987. Empirical hypothesis to explain the restricted distribution of *Hyalella azteca* (Amphipoda) in anthropogenically acidified lakes. Can. J. Fish. Aquat. Sci. 44:1112-1121.

1282. France, R. L., E. T. Howell, M. J. Paterson, and P. M. Welbourn. 1991. Relationship between littoral grazers and metaphytic algae in fine soft water lakes. Hydrobiologia 220:9-28.

1283. Frank, J. H., and G. A. Curtis. 1981. Bionomics of the bromeliad-inhabiting mosquito *Wyeomyia vanduzeei* and its nursery plant *Tillandsia utriculata*. Fla. Ent. 64:491-506.

1284. Frank, J. H., and L. P. Lounibos (eds.). 1983. Phytotelmata: terrestrial plants as hosts for aquatic insect communities. Plexus, Medford, NJ. 293pp.

1285. Fraser, F. C. 1929. A revision of the Fissilabioidea (Cordulegasteridae, Petaliidae, and Petaluridae) (Order Odonata). Part I. Cordulegasteridae. Mem. India Mus. Zool. Surv. 9:69-167.

1286. Fraser, F. C. 1933. A revision of the Fissilabioidea (Cordulegasteridae, Petaliidae, and Petaluridae) (Order Odonata). Part II. Petaliidae and Petaluridae, and Appendix to Part I. Mem. India Mus. Zool. Surv. 9:205-260.

1287. Fraser, F. C. 1957. A reclassification of the order Odonata. R. Zool. Soc. New Wales 12:1-133.

1288. Fredeen, F.J.H. 1958. Black flies (Diptera: Simuliidae) of the agricultural areas of Manitoba, Saskatchewan, and Alberta. Proc. X. Int. Congr. Ent. 3:819-823.

1289. Fredeen, F.J.H. 1959. Rearing black flies in the laboratory (Diptera: Simuliidae). Can. Ent. 91:73-83.

1290. Fredeen, F.J.H. 1960. Bacteria as a source of food for black fly larvae. Nature 187:963.

1291. Fredeen, F.J.H. 1964. Bacteria as food for black fly larvae in laboratory cultures and in natural streams. Can. J. Zool. 42:527-548.

1292. Fredeen, F.J.H. 1969. *Culicoides (Selfia) denningi,* a unique riverbreeding species. Can. Ent. 101:539-544.

1293. Fredeen, F.J.H. 1977. A review of the economic importance of black flies (Simuliidae) in Canada. Quaest. Ent. 13:219-229.

1294. Fredeen, F.J.H. 1981. Keys to the black flies (Simuliidae) of the Saskatchewan River in Saskatchewan. Quaest. Ent. 17:189-210.

1295. Fredeen, F.J.H. 1985. The black flies (Diptera: Simuliidae) of Saskatchewan. Sask. Culture and Recreation Mus. Nat. Hist., Nat. Hist. Contr. 8:1-41.

1296. Freeman, B. E. 1967. Studies on the ecology of larval Tipulinae (Diptera, Tipulidae). J. Anim. Ecol. 36:123-146.

1297. Freeman, B. J., H. S. Greening, and J. D. Oliver. 1984. Comparison of three methods for sampling fishes and macroinvertebrates in a vegetated freshwater wetland. J. Freshwat. Ecol. 2:603-609.

1298. Freeman, J. V. 1987. Immature stages of *Tabanus conterminus* and keys to larvae and pupae of common Tabanidae from United States east coast salt marshes. Ann. Ent. Soc. Am. 80:613-623.

1299. Freeman, M. C., and J. B. Wallace. 1984. Production of net-spinning caddisflies (Hydropsychidae) and black flies (Simuliidae) on rock outcrop substrates in a small southeastern Piedmont stream. Hydrobiologia 112:3-15.

1300. Freilich, J. E. 1986. Contact behavior of the whirligig beetles *Dineutus assimilis* (Coleoptera: Gyrinidae). Ent. News. 97:215-221.

1301. Freilich, J. E. 1991. Movement pattern and ecology of *Pteronarcys* nymphs (Plecoptera): observations of marked individuals in a Rocky Mountain stream. Freshwat. Biol. 25:379-394.

1302. Fremling, C. R. 1960a. Biology and possible control of nuisance caddisflies of the upper Mississippi River. Tech. Res. Bull. Iowa State Univ. 483:856-879.

1302'. Fremling, C. R. 1960b. Biology of a large mayfly, *Hexagenia bilineata* (Say), of the upper Mississippi River. Res. Bull. Iowa Agric. Exp. Sta. 482:842-852.

1303. Fremling, C. R. 1967. Methods for mass-rearing *Hexagenia* mayflies (Ephemeroptera: Ephemeridae). Trans. Am. Fish. Soc. 96:407-410.

1304. Fremling, C. R., and W. L. Mauck. 1980. Methods for using nymphs of burrowing mayflies (Ephemeroptera, *Hexagenia*) as toxcity test organisms, pp. 81-97. *In* A. L. Buikema, Jr., and J. Cairns, Jr. (eds.). Aquatic invertebrate bioassays. Am. Soc. Test. Mater., Philadelphia. 209 pp.

1305. Friberg, N., and D. Jacobsen. 1994. Feeding plasticity of two detritivore-shredders. Freshwat. Biol. 32:133-142.
1306. Frick, K. E. 1949. Biology of *Microvelia capitata* Guerin, 1957, in the Panama Canal Zone and its role as a predator on anopheline larvae. Ann. Ent. Soc. Am. 42:77-100.
1307. Frison, T. 1929. Fall and winter stoneflies, or Plecoptera, of Illlinois. Bull. 181. Nat. Hist. Surv. 18:343-409.
1308. Frison, T. H. 1935. The stoneflies, or Plecoptera, of Illinois. Bull. Ill. Nat. Hist. Surv. 20:281-471.
1309. Fritz, G. N., D. L. Kline, and E. Daniels. 1989. Improved techniques for rearing *Anopheles freeborni*. J. Am. Mosq. Contr. Assoc. 5:201-207.
1310. Froelich, C. G. 1964. The feeding apparatus of the nymph of *Arthroplea congener* Bengtsson (Ephemeroptera). Opusc. Ent. 29:188-208.
1311. Froeschner, R. C. 1949. Contribution to a synopsis of the Hemiptera of Missouri. Part IV. Hebridae, Mesoveliidae, Cimicidae, Anthocoridae, Cryptostemmatidae, Isometopidae, Miridae. Am. Midl. Nat. 42:123-188.
1312. Froeschner, R. C. 1962. Contribution to a synopsis of the Hemiptera of Missouri. Part V. Hydrometridae, Gerridae, Veliidae, Saldidae, Ochteridae, Gelastocoridae, Naucoridae, Belostomatidae, Nepidae, Notonectidae, Pleidae, Corixidae. Am. Midl. Nat. 67:208-240.
1313. Frohne, W. C. 1938. Contribution to knowledge of the limnological role of the higher aquatic plants. Trans. Am. Microsc. Soc. 57:256-268.
1314. Frohne, W. C. 1939a. Biology of certain subaquatic flies reared from emergent water plants. Pap. Mich. Acad. Sci. Arts Lett. 24:139-147.
1315. Frohne, W. C. 1939b. Biology of *Chiloforbesellus* Fernald, an hygrophilous crambine moth. Trans. Am. Microsc. Soc. 58:304-326.
1316. Frohne, W. C. 1939c. Observations on the biology of three semiaquatic lacustrine moths. Trans. Am. Microsc. Soc. 58:327-348.
1317. Frohne, W. C. 1939d. Semiaquatic Hymenoptera in north Michigan lakes. Trans. Am. Microsc. Soc. 58:228-240.
1318. Frommer, S. I. 1963. Gross morphological studies of the reproductive system in representative North American crane flies (Diptera:Tipulidae). Univ. Kans. Sci. Bull. 44:535-626.
1319. Frommer, S. I. 1967. Review of the anatomy of adult Chironomidae. Bull. Calif. Mosquito Cont. Assoc. Tech. Ser. 1:1-39.
1320. Frost, S. W. 1949. The Simuliidae of Pennsylvania (Diptera). Ent. News 60:129-131.
1321. Frost, S., A. Huni, and W. E. Kershaw. 1970. Evaluation of a kicking technique for sampling stream bottom fauna. Can. J. Zool. 49:167-173.
1322. Fuller, R. L., and P. A. Destaffan. 1988. A laboratory study of the vulnerability of prey to predation by three aquatic insects. Can. J. Zool. 66:875-878.
1323. Fuller, R. L., and T. J. Fry. 1991. The influence of temperature and food quality on the growth of *Hydropsyche betteni* (Trichoptera) and *Simulium vittatum* (Diptera). J. Fresh. Ecol. 6:75-86.
1324. Fuller, R. L., T. J. Fry, and J. A. Roelofs. 1988. Influence of different food types on the growth of *Simulium vittatum* (Diptera) and *Hydropsyche betteni* (Trichoptera). J. N. Am. Benthol. Soc. 7:197-204.
1325. Fuller, R. L., and H.B.N. Hynes. 1987a. Feeding ecology of three predaceous aquatic insects and two fish in a riffle of the Speed River, Ontario. Hydrobiologia 150:243-256.
1326. Fuller, R. L., and H.B.N. Hynes. 1987b. The life cycle, food habits and production of *Antocha saxicola* O.S.(Diptera: Tipulidae). Aquat. Insects 9:129-135.
1327. Fuller, R. L., and R. J. Mackay. 1980a. Field and laboratory studies of net-spinning activity by *Hydropsyche* larvae (Trichoptera: Hydropsychidae). Can. J. Zool. 58:2006-2014.
1328. Fuller, R. L., and R. J. Mackay. 1980b. Feeding ecology of three species of *Hydropsyche* (Trichoptera:Hydropsychidae) in southern Ontario. Can. J. Zool. 58:2239-2251.
1329. Fuller, R. L., and R. J. Mackay. 1981. Effects of food quality on the growth of three *Hydropsyche* species (Trichoptera: Hydropsychidae). Can. J. Zool. 59:1133-1140.
1330. Fuller, R. L., J. L. Roelofs, and T. J. Fry. 1986. The importance of algae to stream invertebrates. J. N. Am. Benthol. Soc. 5:290-296.
1331. Fuller, R. L., and K. W. Stewart. 1977. The food habits of stoneflies (Plecoptera) in the Upper Gunnison River, Colorado. Environ. Ent. 6:293-302.
1332. Fuller, R. L., and P. S. Rand. 1990. Influence of substrate type on vulnerability of prey to predacious aquatic insects. J. N. Am. Benthol. Soc. 9:1-8.
1333. Fuller, R. L., and K. W. Stewart. 1979. Stonefly (Plecoptera) food habits and prey preference in the Dolores River, Colorado. Am. Midl. Nat. 101:170-181.
1334. Fullington, K. E., and K. W. Stewart. 1980. Nymphs of the stonefly genus *Taeniopteryx* (Plecoptera: Taeniopterygidae) of North America. J. Kans. Ent. Soc. 53:237-259.
1335. Funk, D. H., and B. W. Sweeney. 1994. The larvae of eastern North American *Eurylophella* Tiensuu (Ephemeroptera: Ephemerellidae). Trans. Am. Ent. Soc. 120:209-286.
1336. Funk, D. H., B. W. Sweeney, and R. L. Vannote. 1988. Electrophoretic study of eastern North American *Eurylophella* (Ephemeroptera: Ephemerellidae) with the discovery of morphologically cryptic species. Ann. Ent. Soc. Am. 81:174-186.
1337. Furse, M. T., D. Moss, J. F. Wright, and P. D. Armitage. 1984. The influence of seasonal and taxonomic factors on the ordination and classification of running-water sites in Great Britain and on the prediction of their macro-invertebrate communities. Freshwat. Biol. 14:257-280.
1338. Galbraith, D. F., and C. H. Fernando. 1977. The life history of *Gerris remigis* (Heteroptera: Gerridae) in a small stream in southern Ontario. Can. Ent. 109:221-228.
1339. Galbreath, G. H., and A. C. Hendricks. 1992. Life history characteristics and prey selection of larval *Boyeria vinosa* (Odonata: Aeshnidae). J. Freshwat. Ecol. 7:201-207.
1340. Gale, W. F. 1971. Shallow-water core sampler. Progr. Fish Cult. 33:238-239.
1341. Gale, W. F. 1981. A floatable, benthic corer for use with SCUBA. Hydrobiologia 77:273-275.
1342. Gale, W. F., and D. Thompson. 1974a. Placement and retrieval of artificial substrate samplers by SCUBA. Progr. Fish Cult. 36:231-233.
1343. Gale, W. F., and D. Thompson. 1974b. Aids to benthic sampling by SCUBA divers in rivers. Limnol. Oceanogr. 19:1004-1007.
1344. Gale, W. F., and D. Thompson. 1975. A suction sampler for quantitatively sampling benthos on rocky substrates in rivers. Trans. Am. Fish. Soc. 104:398-405.
1345. Gallepp, G. W. 1974a. Behavioral ecology of *Brachycentrus occidentalis* Banks during the pupation period. Ecology 55:1283-1294.
1346. Gallepp, G. W. 1974b. Diel periodicity in the behavior of the caddisfly *Brachycentrus americanus* (Banks). Freshwat. Biol. 4:193-204.
1347. Gallepp, G. W. 1977. Responses of caddisfly larvae (*Brachycentrus* spp.) to temperature, food availability and current velocity. Am. Midl. Nat. 98:59-84.
1348. Ganguly, D. N., and B. Mitra. 1961. Observations on the fish-fry destroying capacity of certain aquatic insects and the suggestion for their eradication. Indian Agric. 5:184-188.
1349. Garcia, R., and E. I. Schlinger. 1972. Studies of spider predation on *Aedes dorsalis* (Meigen) in a salt marsh. Proc. Calif. Mosquito Contr. Assoc. 40:117-118.
1350. Garcia-Diaz, J. 1938. An ecological survey of the fresh water insects of Puerto Rico, I. The Odonata, with new life histories. J. Agric. Univ. Puerto Rico 22:43-97.
1351. Gardner, A. E. 1952. The life history of *Lestes dryas* Kirby. Ent. Gaz. 3:4-26.
1352. Garman, H. 1924. Odonata from Kentucky. Ent. News 25:285-288.
1353. Garman, P. 1917. The Zygoptera, or damselflies, of Illinois. Bull. Ill. State Lab. Nat. Hist. 12:411-587.
1354. Garman, P. 1927. The Odonata or dragonflies of Connecticut. Bull. Conn. State Geol. Nat. Hist. Surv. 39:1-331.

1355. Garrison, R. W. 1981. Description of the larva of *Ischnura gemina* with a key and new characters for the separation of sympatric *Ischnura* larvae. Ann. Ent. Soc. Am. 74:525-530.

1356. Garrison, R. W. 1984. Revision of the genus *Enallagma* of the United States west of the Rocky Mountains and identification of certain larvae by discriminant analysis (Odonata: Coenagrionidae). Univ. Calif. Publ. Ent. 105-130.

1357. Garrison, R. W. 1990. A synopsis of the genus *Hetaerina* with descriptions of four new species (Odonata: Calopterygidae). Trans. Am. Ent. Soc. 116:175-259.

1358. Garrison, R. W. 1991. A synonymic list of the New World Odonata. Argia 3:1-30.

1359. Garrison, R. W., and J. E. Hafernik. 1981. The distribution of *Ischnura gemina* (Kennedy) and a description of the andromorph female (Zygoptera: Coenagrionidae). Odonatologica 10:85-91.

1360. Gates, T. E., D. J. Baird, F. J. Wrona, and R. W. Davies. 1987. A device for sampling macroinvertebrates in weedy ponds. J. N. Am. Benthol. Soc. 6:133-139.

1361. Gatjen, L. 1926. Nahrungsuntersuchung bei Phryganidenlarven (*Phryganea* and *Neuronia*). Arch. Hydrobiol. 16:649-667.

1362. Gaufin, A. R. 1956. Annotated list of stoneflies of Ohio. Ohio J. Sci. 56:321-324.

1363. Gaufin, A. R. 1964a. Systematic list of Plecoptera of the Intermountain region. Proc. Utal Acad. Sci. 41:221-227.

1364. Gaufin, A. R. 1964b. The Chloroperlidae of North America. Gewasser und Abwasser 34/35:37-49.

1365. Gaufin, A. R., A. V. Nebeker, and J. Sessions. 1966. The stoneflies (Plecoptera) of Utah. Univ. Utah Biol. Ser. 14:4-93.

1366. Gaufin, A. R., and W. E. Ricker. 1974. Additions and corrections to a list of Montana stoneflies. Ent. News 85:285-288.

1367. Gaufin, A. R., W. E. Ricker, M. Miner, P. Milam, and R. A. Hays. 1972. The stoneflies (Plecoptera) of Montana. Trans. Am. Ent. Soc. 98:1-161.

1368. Gaufin, R. F., and A. R. Gaufin. 1961. The efllect of low oxygen concentrations on stoneflies. Proc. Utah Acad. Sci. Arts Lett. 38:57-64.

1369. Geigy, R., and G. Sarasin. 1958. Die ökologische Abhanggigkeit des Metamorphose Geschehens bei *Sialis lutaria* L. Rev. Suisse Zool. 65:323-329.

1370. Geijskes, D. C. 1943. Notes on Odonata of Surinam, IV. Nine new or little known zygopterous nymphs from the inland waters. Ann. Ent. Soc. Am. 36:165-184.

1371. Gelhaus, J. K. 1986. Larvae of the crane fly genus *Tipula* in North America (Diptera: Tipulidae). Univ. Kansas Sci. Bull. 53:121-182.

1372. Gelhaus, J. K., and C. W. Young. 1991. The immature instars and biology of the crane fly genus *Brachypremna* Osten Sacken (Diptera: Tipulidae). Proc. Ent. Soc. Wash. 93:613-621.

1373. Georgian, T., and J. H. Thorp. 1992. Effects of microhabitat selection on feeding rates of net-spinning caddisfly larvae. Ecology 73:229-240.

1374. Georgian, T. J., and J. B. Wallace. 1981. A model of seston capture by net-spinning caddisflies. Oikos 36:147-157.

1375. Georgian, T. J., and J. B. Wallace. 1983. Seasonal production dynamics in a guild of periphyton-grazing insects in a southern Appalachian stream. Ecology. 64:1236-1248.

1376. Gerasimov, A. 1937. Bestimmungstabelle der Familien von Schmetterlingsrapuen (Lep.) Stett. Ent. Zeitung. 98:281-300.

1377. Gerberg, E. J. 1970. Manual for mosquito rearing and experimental techniques. Bull. Am. Mosquito Contr. Assoc. 5:1-109.

1378. Gerhardt, A. 1992. Effects of subacute doses of iron (Fe) on *Leptophlebia marginata* (Insecta: Ephemeroptera). Freshwat. Biol. 27:79-84.

1379. Gerking, S. D. 1957. A method of sampling the littoral macrofauna and its application. Ecology 38:219-226.

1380. Gersabeck, E. F., and R. W. Merritt. 1979. The effect of physical factors on the colonization and relocation behavior of immature black flies. Environ. Ent. 8:34-39.

1381. Gessner, M. O., and M. Dobson. 1993. Colonization of fresh and dried leaf litter by macroinvertebrates. Arch. Hydrobiol. 127:141-149.

1382. Giani, N., and H. Laville. 1973. Cycle biologique et production de *Sialis lutaria* L. (Megaloptera) dans le lac de Port-Bielh (Pyrenees Centrales). Ann. Limnol. 9:45-61.

1383. Gibbs, K. E. 1980. The occurrence and biology of *Siphlonisca aerodromia* Needham (Ephemeroptera: Siphlonuridae) in Maine, USA, pp. 167-168. *In* J. F. Flannigan and K. E. Marshall (eds.). Advances in Ephemeroptera biology. Plenum, N.Y. 552 pp.

1384. Gibbs, K. E., and T. M. Mingo. 1985. The life history, nymphal growth rates, and feeding habitats of *Siphlonisca aerodromia* Needham (Ephemeroptera: Siphlonuridae) in Maine. Can. J. Zool. 64:427-430.

1385. Gibbs, R. H., Jr., and S. P. Gibbs. 1954. The Odonata of Cape Cod, Massachusetts. J. N.Y. Ent. Soc. 62:167-184.

1386. Giberson, D. J., and T. D. Galloway. 1985. Life history and production of *Ephoron album* (Ephemeroptera: Polymitarcyidae) in the Valley River, Manitoba. Can. J. Zool. 63:1668-1674.

1387. Giberson, D. J., and R. J. Mackay. 1991. Life history and distribution of mayflies (Ephemeroptera) in some acid streams in south central Ontario, Canada. Can. J. Zool. 69:899-910.

1388. Giberson, D. J., and D. M. Rosenberg. 1992. Effects of temperature, food quality, and nymphal rearing density on life history traits of a northern population of *Hexagenia* (Ephemeroptera: Ephemeridae). J. N. Am. Benthol. Soc. 11:181-193.

1389. Gill, G. D., and L. S. West. 1955. Notes on the ecology of certain species of Simuliidae in the Upper Peninsula of Michigan. Paps. Mich. Acad. Sci., Arts, Lett. 40:119-124.

1390. Giller, P. S. 1986. The natural diet of the Notonectidae: field trials using electrophoresis. Ecol. Ent. 11:163-172.

1391. Giller, P. S., and N. Sangpradub. 1993. Predatory foraging behaviour and activity patterns of larvae of two species of limnephilid - cased Caddis. Oikos 67:351-357.

1392. Giller, P. S., and S. McNeill. 1981. Predation strategies, resource partitioning and habitat selection in *Notonecta* (Hemiptera/Heteroptera). J. Anim. Ecol. 50:789-808.

1393. Gillespie, D. M., and C.J.D. Brown. 1966. A quantitative sampler for macroinvertebrates associated with aquatic macrophytes. Limnol. Oceanogr. 11:404-406.

1394. Gillespie, D. M., D. L. Stites, and A. C. Benke. 1985. An inexpensive core sampler for use in sandy substrata. Freshwat. Biol. 4:147.

1395. Gillespie, J. 1941. Some unusual dragonfly records from New Jersey (Odonata). Ent. News 52:225-226.

1396. Gillespie, J. M., W. F. Barr, and S. T. Elliott. 1994. Taxonomy and biology of the immature stages of species of *Thaumalea* occurring in Idaho and California (Diptera: Thaumaleidae). Myia 5:153-193.

1397. Gillett, J. D. 1972. The mosquito, its life, activities, and impact on human affairs. Doubleday, N.Y. 358 pp.

1398. Gillott, C. 1980. Entomology. Plenum, N.Y. 729 pp.

1399. Gilpin, B. R., and M. A. Brusven. 1970. Food habits and ecology of mayflies of the St. Mary's River in Idaho. Melanderia 4:19-40.

1400. Gilpin, B. R., and M. A. Brusven. 1976. Subsurface sampler for determining vertical distribution of stream-bed benthos. Progr. Fish Cult. 38:192-194.

1401. Gilson, W. E. 1963. Differential respirometer of simplified and improved design. Science 141:531-532.

1402. Girault, A. A. 1911a. Descriptions of North American Mymaridae with synonymic and other notes on described genera and species. Trans. Am. Ent. Soc. 37:253-324.

1403. Girault, A. A. 1911b. Synonymic and descriptive notes on the chalcidoid family Trichogrammatidae with descriptions of new species. Trans. Am. Ent. Soc. 37:43-83.

1404. Gisin, H. 1960. Collembolenfauna Europas. Geneve Educ. Mus. Nat. Hist. 312 pp.

1405. Gíslason, G. M. 1985. The biology of the blackfly *Simulium vittatum* Zett. (Diptera: Simuliidae) in the river Laxa, northern Iceland. Natturufraedingurinn 55:175-194.

1406. Gíslason, G. M. 1993. The life cycle of *Limnephilus griseus* (L.) (Trichoptera, Limnephilidae) in temporary rock pools in nothern England, pp. 171-175. *In* C. Otto (ed.). Proc. VII. Int. Symp. Trichoptera. Umea, Sweden, Backhuys Publ., Leiden, Netherlands.

1407. Gíslason, G. M., and V. Johannsson. 1991. Effects of food and temperature on the life cycle of *Simulium vittatum* Zett. (Diptera: Simuliidae) in the river Laxa, North Iceland. Verh. Int. Verein. Limnol. 24:2912-2916.

1408. Gíslason, G. M., T. Hrafnsdottir, and A. Gardarsson. 1994. Long-term monitoring of numbers of Chironomidae and Simuliidae in the River Laxa, North Iceland. Verh. Int. Verein. Limnol. 25:1492-1495.

1409. Gittelman, S. H. 1974a. Locomotion and predatory strategy in backswimmers. Am. Midl. Nat. 92:496-500.

1410. Gittelman, S. H. 1974b. The habitat preference and immature stages of *Neoplea striola* (Hemiptera: Pleidae). J. Kans. Ent. Soc. 47:491-503.

1411. Gittelman, S. H. 1975. Physical gill efficiency and winter dormancy in the pigmy backswimmer, *Neoplea striola* (Hemiptera: Pleidae). Ann. Ent. Soc. Am. 68:1011-1017.

1412. Gittelman, S. H. 1978. Optimum diet and body size in backswimmers (Heteroptera: Notonectidae, Pleidae). Ann. Ent. Soc. Am. 71:737-747.

1413. Gittelman, S. H., and P. W. Severance. 1975. The habitat preference and immature stages of *Buenoa confusa* and *B. margaritacea* (Hemiptera: Notonectidae). J. Kans. Ent. Soc. 48:507-518.

1414. Givens, D. R., and S. D. Smith. 1980. A synopsis of the western Arctopsychinae (Trichoptera: Hydropsychidae). Melanderia 35:1-24.

1415. Gjullin, C. M., and G. W. Eddy. 1972. The mosquitos of the northwestern United States. U.S. Dept. Agric. Tech. Bull. 1447. 111 pp.

1416. Gjullin, C. M., R. I. Sailer, A. Stone, and B. V. Travis. 1961. The mosquitoes of Alaska. U.S. Dept. Agric., Agric. Res. Serv. Agric. Handbk. 182:1-98.

1417. Gladney, W. J., and E. C. Turner, Jr. 1969. Insects of Virginia No. 2. The mosquitoes of Virginia. Bull. Res. Div. Va. Poly. Inst. State Univ. 49. 24 pp.

1418. Glasgow, J. P. 1936. The bionomics of *Hydropsyche colonica* McLach. and *H. philopotti* Till. Proc. R. ent. Soc. Lond. (A) 11:122-128.

1419. Glorioso, M. J. 1981. Systematics of the dobsonfly subfamily Corydalinae (Megaloptera: Corydalidae). Syst. Ent. 6:253-290.

1420. Glotzel, V. R. 1973. Populationsdynamik und Ernahrungsbiologie von Simuliidenlarven in einem mit organischen Abwassern verunreinigten Gebirgsbach. Arch. Hydrobiol. Suppl. 42: 406-451.

1421. Glover, J. B. 1993. The taxonomy and biology of the larvae of the North American caddisflies in the genera *Triaenodes* and *Ylodes* (Trichoptera: Leptoceridae). Ph.D. diss., Univ. Louisville, Louisville, KY.

1422. Gloyd, L. K. 1944. A new species of *Stylurus* from Mexico (Odonata: Gomphinae). Occ. Pap. Univ. Mich. Mus. Zool. 482:1-4.

1423. Gloyd, L. K. 1951. Records of some Virginia Odonata. Ent. News 62:109-114.

1424. Gloyd, L. K. 1958. The dragonfly fauna of the Big Bend region of Trans-Pecos Texas. Occ. Pap. Univ. Mich. Mus. Zool. 593:1-23.

1425. Gloyd, L. K. 1959. Elevation of the *Macromia* group to family status (Odonata). Ent. News 70:197-205.

1426. Gloyd, L. K. 1968a. The union of *Argia fumipennis* (Burmeister 1839) with *Argia violacea* (Hagen, 1861), and the recognition of three subspecies (Odonata). Occ. Pap. Univ. Mich. Mus. Zool. 658:1-6.

1427. Gloyd, L. K. 1968b. The synonymy of *Diargia* and *Hyponeura* with the genus *Argia* (Odonata: Coenagrionidae: Argiinae). Mich. Entomol. 1:271-274.

1428. Gloyd, L. K. 1973. The status of the generic names *Gomphoides, Negomphoides, Progomphus,* and *Ammogomphus* (Odonata: Gomphidae). Occ. Pap. Univ. Mich. Mus. Zool. 668:1-7.

1429. Gloyd, L. K., and M. Wright. 1959. Odonata, pp. 917-940. *In* W. T. Edmondson (ed.). Freshwater biology (2nd ed.). John Wiley & Sons, N.Y. 1248 pp.

1430. Glozier, N. E., and J. M. Culp. 1989. Experimental investigations of diel vertical movements by lotic mayflies over substrate surfaces. Freshwat. Biol. 21:253-260.

1431. Glukhova, V. M. 1977. Midges of the family Certatopogonidae (synonym Heleidae), pp. 431-457. *In* L. A. Kutikova and Y. I. Starobogatov (eds.). Determination of the freshwater invertebrates of the European regions of the USSR (plankton and benthos). Zool. Inst. USSR Acad. Sci. Nauka "Hydrometeo" Pub., Lennigrad. 510p. (in Russian).

1432. Glukhova, V. M. 1979. Midge larvae of the subfamilies Palpomyiinae and Ceratopogoninae in the USSR (Diptera: Ceratopogonidae=Heleidae). Identification of the fauna of the USSR. Zool. Inst. USSR Acad. Sci. Nauka Pub., Lennigrad. 121:1-231.

1433. Goetghebuer, M., and F. Lenz. 1936-1950. Tendipedidae. A. Die Imagines. Pelopiinae; Tendipedinae; Diamesinae; Orthocladiinae, pp. 13b, 97:1-48, 100:49-81; 13c, 1937:1-72, 1938:73-128; 1954:129-168; 13d:1-30; 13f:1-19; 13g, 1940:7-24, 1942:25-64, 1943:113-114, 1950:145-208. *In* E. Lindner (ed.). Fliegen Palaerktischen Region. 280 pp.

1434. Goetghebuer, M., and F. Lenz. 1939. Tendipedidae-Podonominae. A. Die Imagines, pp. 1-29. *In* E. Lindner (ed.). Fliegen Palaerktischen Region 13e. 280 pp.

1435. Goetghebuer, M., and F. Lenz. 1950. Tendipedidae-Clunioninae. A. Die Imagines, pp. 1-23. *In* Lindner (ed.). Fliegen Palaerktischen Region 13h. 280 pp.

1436. Goldhammer, D. S., and L. C. Ferrington, Jr. 1992. Emergence of aquatic insects from epirhreic zones of capillary fringe habitats in the Cimarron River, Kansas, pp. 155-164. *In* J. A. Stanford and J. J. Simons (eds.). Proc. 1st. Internat. Conf. Ground Wat. Ecol. Am. Wat. Res. Assoc., Bethesda, MD. 420p.

1437. Goldhammer, D. S., C. A. Wright, M. A. Blackwood, and L. C. Ferrington, Jr. 1992. Composition and phenology of Chironomidae of the Nelson Environmental Studies Area, Univ. Kans. Neth. J. Aquat. Ecol. 26:281-291.

1438. Goldie-Smith, E. K. 1989. Laboratory rearing and egg development in two species of *Dixella* (Diptera: Dixidae). Ent. Gaz. 40:53-65.

1439. Goldman, C. R., and A. J. Horne. 1983. Limnology. McGraw-Hill, NY. 464pp.

1440. Goldspink, C. R., and D.B.C. Scott. 1971. Vertical migration of *Chaoborus flavicans* in a Scottish loch. Freshwat. Biol. 1:411-421.

1441. Golubkov, S. M., T. M. Tiunova, and S. L. Kocharina. 1992. Dependence of the respiration rate of aquatic insects upon the oxygen concentration in running and still water. Aquat. Insects 14:137-144.

1442. Gonsoulin, G. J. 1973a. Seven families of aquatic and semi-aquatic Hemiptera in Louisiana. I. Hydrometridae. Ent. News 84:9-16.

1443. Gonsoulin, G. J. 1973b. Seven families of aquatic and semi-aquatic Hemiptera in Louisiana. Part II. Family Naucoridae Fallen, 1814, "Creeping water bugs." Ent. News 84:83-88.

1444. Gonsoulin, G. J. 1973c. Seven families of aquatic and semi-aquatic Hemiptera in Louisiana. Part III. Family Belostomatidae Leach, 1815. Ent. News 84:173-189.

1445. Gonsoulin, G. J. 1974. Seven families of aquatic and semi-aquatic Hemiptera in Louisiana. IV. Family Gerridae. Trans. Am. Ent. Soc. 100:513-546.

1446. Gonsoulin, G. J. 1975. Seven families of aquatic and semi-aquatic Hemiptera in Louisiana. Part V. Family Nepidae Latreille, 1802. Ent. News 86:23-32.

1447. Good, H. G. 1924. Notes on the life history of *Prionocyphon limbatus* Lec. (Helodidae, Coleoptera). J. N.Y. Ent. Soc. 32:79-84.

1448. Good, N. E. 1945. A list of the mosquitoes of the District of Columbia. Proc. Ent. Soc. Wash. 47:168-179.

1449. Goodwin, J. T. 1973a. Immature stages of some eastern Nearctic Tabanidae (Diptera) II. Genera of the tribe Diachlorini. J. Ga. Ent. Soc. 8:5-11.

1450. Goodwin, J. T. 1973b. Immature stages of some eastern Nearctic Tabanidae (Diptera). IV. The genus *Merycomyia*. J. Tenn. Acad. Sci. 48:115-118.

1451. Goodwin, J. T. 1974. Immature stages of some eastern Nearctic Tabanidae (Diptera). V. *Stenotabanus (Aegialomyia) magnicallus* (Stone). J. Tenn. Acad. Sci. 49:14-15.
1452. Goodwin, J. T. 1987. Immature stages of some eastern Nearctic Tabanidae (Diptera): VIII. Additional species of *Tabanus* Linnaeus and *Chrysops* Meigen. Fla. Ent. 70:268-277.
1453. Gordon, A. E. 1974. A synopsis and phylogenetic outline of the Nearctic members of *Cheumatopsyche*. Proc. Acad. Nat. Sci. Philadelphia 126:117-160.
1454. Gordon, N. D., T. A. McMahon, and B. L. Finlayson. 1992. Stream hydrology: an introduction for ecologists. J. Wiley and Sons, Chichester, England. 526p.
1455. Gordon, R. D. 1981. New species of North American *Hydroporus, niger-tenebrosus* group (Coleoptera: Dytiscidae). Pan Pacif. Ent. 57:105-123.
1456. Gordon, R. D., and R. L. Post. 1965. North Dakota water beetles. North Dakota Insects. Publ. N. D. State Univ. Agric. Exp. Sta. 5:1-53.
1457. Gotceitas, V. 1985. Formation of aggregations by overwintering fifth instar *Dicosmoecus atripes* larvae (Trichoptera). Oikos 44:313-318.
1458. Gotceitas, V., and H. F. Clifford. 1983. The life history of *Dicosmoecus atripes* (Hagen) (Limnephilidae: Trichoptera) in a Rocky Mountain stream of Alberta, Canada. Can. J. Zool. 61:586-596.
1459. Gotte, S. W. 1992. Testudines. Herpetol. Rev. 23:80.
1460. Gouteux, J.-P. 1977. Observations morphologiques et écologiques sur les stades préimaginaux de *Xenomyia oxycera* Emden, 1951 (Muscidae Limnophorinae), diptère prédateur de *Simulium damnosum* s.l. (Diptera Simuliidae). Revue Zool. Afr. 91:618-622.
1461. Gower, A. M. 1967. A study of *Limnephilus lunatus* Curtis (Trichoptera: Limnephilidae) with reference to its life cycle in watercress beds. Trans. R. ent. Soc. Lond. 119:283-302.
1462. Grafius, E. J., and N. H. Anderson. 1973. Literature review of foods of aquatic insects. Coniferous Forest Biome Ecosyst. Anal. Stud., Oregon State Univ., Corvallis, Internal Rept. 129. 52 pp.
1463. Grafius, E. J., and N. H. Anderson. 1980. Population dynamics and role of two species of *Lepidostoma* (Trichoptera: Lepidostomatidae) in an Oregon coniferous forest stream. Ecology 61:808-816.
1464. Graham, A. C., J. F. Turmel, and R. F. Darsie, Jr. 1991. New state mosquito records for Vermont including checklist of the mosquito fauna. J. Am. Mosq. Contr. Assoc. 7:502-503.
1465. Graham, M.W.R. de V. 1959. Keys to the British genera and species of Elachertinae, Eulophinae, Entedontrinae and Euderinae (Hym.: Chalcidoidea). Trans. Soc. Brit. Ent. 13:169-204.
1466. Graham, M.W.R. de V. 1969. The Pteromalidae of northwestern Europe (Hym.: Chalcidoidea). Bull. Brit. Mus. Nat. Hist. Ent. 16:1-908.
1467. Grant, P. M., and K. W. Stewart. 1980. The life history of *Isonychia sicca* (Ephemeroptera: Oligoneuriidae) in an intermittent stream in North Central Texas. Ann. Ent. Soc. Am. 73:747-755.
1468. Grant, P. R., and R. J. Mackay. 1969. Ecological segregation of systematically related stream insects. Can. J. Zool. 47:691-694.
1469. Gray, L. J. 1981. Species composition and life histories of aquatic insects in a lowland Sonoran desert stream. Am. Midl. Nat. 106:229-242.
1470. Gray, L. J., and J. V. Ward. 1979. Food habits of stream benthos at sites of differing food availability. Am. Midl. Nat. 102:157-167.
1471. Green, R. H. 1979. Sampling design and statistical methods for environmental biologists. Wiley-Interscience, N.Y. 257 pp.
1472. Greene, C. T. 1923. A contribution to the biology of N. A. Diptera. Proc. Ent. Soc. Wash. 25:82-89.
1473. Greenstone, M. H. 1979. A sampling device for aquatic arthropods active at the water surface. Ecology 60:642-644.
1474. Grenier, P. 1949. Contribution a l'etude biologique des Simuliides de France. Physiologia Comp. Oecol. 1:165-330.
1475. Grenier, S. 1970. Biologie d'*Agriotypus armatus* Curtis (Hymenoptera: Agriotypidae), parasite de nymphs de Trichopteres. Ann. Limnol. 6:317-361.
1476. Gresens, S. E., and R. L. Lowe. 1994. Periphyton patch preference in grazing chironomid larvae. J. N. Am. Benthol. Soc. 13:89-99.
1477. Gribbin, S. D., and D. J. Thompson. 1990. Asymmetric intraspecific competition among larvae of the damselfly *Ischnura elegans* (Zygoptera: Coenagrionidae). Ecol. Ent. 15:37-42.
1478. Griesinger, P., and T. Bauer 1990. Predation and visual space in *Saldula saltatoria* (Linnaeus, 1758) (Heteroptera: Saldidae). Zool. Anz. 225:20-36.
1479. Grieve, E. G. 1937a. Culture methods for the damselfly, *Ischnura verticalis*, pp. 268-270. *In* J. G. Needham (ed.). Culture methods for invertebrate animals. Comstock, Ithaca, N.Y. 590 pp.
1480. Grieve, E. G. 1937b. Studies on the biology of the damselfly (*Ischnura verticalis* Say), with notes on certain parasites. Am. Ent. 17:121-153.
1481. Griffith, M. B., and S. A. Perry. 1991. Leaf pack processing in two Appalachian mountain streams draining catchment with different management histories. Hydrobiologia 220:247-254.
1482. Griffith, M. B., and S. A. Perry. 1993. Colonization and processing of leaf litter by macro-invertebrate shredders in streams of contrasting pH. Freshwat. Biol. 30:93-103.
1483. Griffith, M. B., S. A. Perry, and W. B. Perry. 1993. Growth and secondary production of *Paracapnia angulata* Hanson (Plecoptera, Capniidae) in an Appalachian stream affected by acid precipitation. Can. J. Zool. 71:735-743.
1484. Griffith, M. E. 1945. The environment, life history and structure of the water boatman, *Ramphocorixa acuminata* (Uhler). Univ. Kans. Sci. Bull. 30:241-365.
1485. Griffiths, G.C.D. 1972. Studies on the phylogenetic classification of Diptera Cyclorrhapha, with special reference to the structure of the male post abdomen. Series Entomologica, 8. The Hague, Netherlands. 340 pp.
1486. Grigarick, A. A. 1959a. A floating pantrap for insects associated with the water surface. J. Econ. Ent. 52:348-349.
1487. Grigarick, A. A. 1959b. Bionomics of the rice leaf miner, *Hydrellia griseola* (Fallen), in California (Diptera: Ephydridae). Hilgardia 29:1-80.
1488. Grigarick, A. A. 1975. The occurrence of a second genus of spongilla-fly (*Sisyra vicaria*, Walker.) at Clear Lake, Lake County, California. Pan-Pacif. Ent. 51:296-297.
1489. Grimaldi, D., and J. Jaenike. 1983. The Diptera on skunk cabbage, *Symplocarpus foetidus* (Araceae). J. N. York Ent. Soc. 91:83-89.
1490. Grissell, E. E., and M. E. Schauff. 1990. A handbook of the families of Nearctic Chalcidoidea (Hymenoptera). Ent. Soc. Wash. 85pp.
1491. Grodhaus, G. G. 1980. Aestivating chironomid larvae associated with vernal pools, pp. 315-322. *In* D. A. Murray (ed.). Chironomidae ecology, systematics, cytology and physiology. Pergammon Press, Oxford, England. 354 p.
1492. Grodhaus, G. 1987. Endochironomus Kieffer, Tribelos Townes, Synendotendipes, n. gen., and Endotribelos, n. gen. (Diptera: Chironomidae) of the Nearctic region. J. Kans. Ent. Soc. 60:167-247.
1493. Grogan, W. L., and P. G. Bystrak. 1976. Description of the pupa, larva, and larval habitat of *Alluaudomyia parva* Wirth (Diptera: Ceratopogonidae). Proc. Ent. Soc. Wash. 78:314-316.
1494. Grogan, W. L., and D. H. Messersmith. 1976. The immature stages of *Alluaudomyia paraspina* (Diptera: Ceratopogonidae). Ann. Ent. Soc. Am. 69:687-690.
1495. Grogan, W. L., and W. W. Wirth. 1975. A revision of the genus *Palpomyia* Meigen of northeastern North America (Diptera: Ceratopogonidae). Contr. Md. Agric. Exp. Sta. 5076:1-49.
1496. Grogan, W. L., Jr., and W. W. Wirth. 1979. The North American predaceous midges of the genus *Palpomyia* Meiger (Diptera: Ceratopogonidae). Mem. Ent. Soc. Wash. 8:1-125.
1497. Growns, I. O. 1990. Efficiency estimates of a stream benthos suction sampler. Aust. J. Mar. Freshwat. Res. 41:621-626.
1498. Grunewald, J. 1973. Die hydrochemischen Lebensbeingungen der praimaginalen Stadien von *Boophthora erythrocephala* De Geer (Diptera: Simuliidae). 2. Die Entwicklung einer Zucht unter experimentellen Bedingungen. Z. Tropenmed. Parasit. 24:232-249.

1499. Gulliksen, B., and K. M. Deras. 1975. A diver-operated suction sampler for fauna on rocky bottoms. Oikos 26:246-249.
1500. Gundersen, R. W. 1978. Nearctic *Enochrus* biology, keys, description (Coleoptera: Hydrophilidae). Dept. Biol. Sci., St. Cloud State Univ., St. Cloud, Minn. 54 pp.
1501. Gundersen, R. W., and C. Otremba. 1988. Haliplidae of Minnesota, USA. Sci. Publ. Sci. Mus. Minn. 6:3-43.
1502. Günter, G., and J. Y. Christmas. 1959. Corixid insects as part of the off-shore fauna of the sea. Ecology 40:724-725.
1503. Günther, K. K. 1975. Das genus *Neotridactylus* Günther, 1972 (Tridactylidae: Saltatoria:Insecta). Mitt. Zool. Mus. Berlin 51:305-365.
1504. Gupta, A. 1993. Life histories of two species of *Baetis* (Ephemeroptera: Baetidae) in a small Northeast Indian stream. Arch. Hydrobiol. 127:105-114.
1505. Gupta, S., R. G. Michael, and A. Gupta. 1993. Laboratory studies of the life cycle and growth of *Cloeon* (Ephemeroptera: Baetidae) in Meghalaya State, India. Aquat. Insects 15:49-55.
1506. Gurney, A. B., and S. Parfin. 1959. Neuroptera, pp. 973-980. In W. T. Edmondson (ed.). Freshwater biology (2nd ed.). John Wiley & Sons, N.Y. 1248 pp.
1507. Guthrie, J. E. 1903. The Collembola of Minnesota. Rep. Geol. Nat. Hist. Surv., Minn. Zool. Ser. 4:1-110.
1508. Guthrie, M. 1989. Animals of the surface film. Nat. Hndbk. No. 12. Richmond Publ. Co., Ltd., Slough, England. 87p.
1509. Guyer, G., and R. Hutson. 1955. A comparison of sampling techniques utilized in an ecological study of aquatic insects. J. Econ. Ent. 48:662-665.
1510. Haag, K. H., V. H. Resh, and S. E. Neff. 1984. Changes in the adult caddisfly (Trichoptera) community of the Salt River, Kentucky. Trans. Kentucky Acad. Sci. 45:101-108.
1511. Haage, P. 1970. On the feeding habits of two Baltic species of caddis larvae (Trichoptera). Ent. Scand. 1:282-290.
1512. Habeck, D. H. 1974a. *Arzama densa* as a pest of Dasheen. Fla. Ent. 57:409-410.
1513. Habeck, D. H. 1974b. Caterpillars of *Parapoynx* in relation to aquatic plants in Florida. Hyacinth Contr. J. 12:15-18.
1514. Habeck, D. H. 1976. The life cycle of *Neoerastria caduca* (Grote) (Lepidoptera: Noctuidae). Fla. Ent. 59:101-102.
1515. Habeck, D. H. 1979. Host plant of *Pseudolampsis guttata* (LeConte) (Coleoptera: Chrysomelidae). Coleop. Bull. 33:150.
1516. Habeck, D. H. 1988. *Neargyractis slossonalis* (Lepidoptera: Pyralidae, Nymphulinae): larval description and biological notes. Fla. Ent. 71:588-592.
1517. Habeck, D. H. 1991. *Synclita obliteralis* (Walker), the water lily leafcutter. (Lepidoptera: Pyralidae: Nymphulinae). Fla. Dept. Agric. and Consumer Services, Ent. Circular No. 345. 2p.
1518. Habeck, D. H., K. Haag, and G. Buckingham. 1986. Native insect enemies of aquatic macrophytes—moths. Aquatics 8:17-22.
1519. Habeck, D. H., and R. Wilkerson. 1980. The life cycle of *Lysathia ludoviciana* (Fall) (Coleoptera: Chrysomelidae) on parrot feather, *Myriophyllum aquaticum* (Velloso) Verde. Coleop. Bull. 34:167-170.
1520. Haddock, J. D. 1977. The biosystematics of the caddis fly genus *Nectopsyche* in North America with emphasis on the aquatic stages. Am. Midl. Nat. 98:382-421.
1521. Hadley, M. 1971. Aspects of the larval ecology and population dynamics of *Molophilus ater* Meigen (Diptera: Tipulidae) on Pennine moorland. J. Anim. Ecol. 40:445-456.
1522. Hagen, H. A. 1880. On an aquatic sphinx larva on *Nymphaea*. Psyche 3:113.
1523. Hagen, K. S. 1956. Aquatic Hymenoptera, pp. 289-292. In R. L. Usinger (ed.). Aquatic insects of California. Univ. Calif. Press, Berkeley. 508 pp.
1524. Hagen, K. S. 1964. Developmental stages of parasites, pp. 168-246. In P. Debach (ed.). Biological control of insects pests and weeds. Chapman and Hall, London. 844 pp.
1525. Haggett, G. 1955-1961. Larvae of the British Lepidoptera not figured by Buckler. Soc. Lond. Ent. Nat. Hist. Proc. Trans. Part I, 1955:152-163; Part II, 1957:94-99; Part IV, 1959:207-214; Part V, 1960:136-137.
1526. Hagstrum, D. W., and E. B. Workman. 1971. Interaction of temperature and feeding rate in determining the rate of development of larval *Culex tarsalis* (Diptera: Culicidae). Ann. Ent. Soc. Am. 64:668-671.
1527. Hair, J. A., and E. C. Turner, Jr. 1966. Laboratory colonization and mass-production procedures for *Culicoides guttipennis*. Mosquito News 26:429-433.
1528. Hall, F. 1972. Observations on black flies of the genus Simulium in Los Angeles County, California. Calif. Vect. Views 19:54-58.
1529. Hall, F. 1974. A key to the *Simulium* larvae of southern California (Diptera: Simuliidae). Calif. Vector Views 21:65-71.
1530. Hall, H. A., and G. Pritchard. 1975. The food of larvae of *Tipula sacra* Alexander in a series of abandoned beaver ponds (Diptera: Tipulidae). J. Anim. Ecol. 44:55-66.
1531. Hall, R. E., and J. J. Harrod. 1963. A method of rearing *Simulium ornatum* var. *nitidifrons* (Diptera: Simuliidae) in the laboratory. Hydrobiologia 22:450-453.
1532. Hall, R. J. 1975. Life history, drift and production of the stream mayfly *Tricorythodes atratus* McDunnough in the headwaters of the Mississippi River. Ph.D. diss., University of Minnesota. 228 pp.
1533. Hall, R. J., and G. E. Likens. 1981. Chemical flux in an acid-stressed stream. Nature 292:329-331.
1534. Hall, R. J., G. E. Likens, S. B. Fiance, and G. R. Hendrey. 1980 Experimental acidification of a stream in the Hubbard Brook Experimental Forest, New Hampshire. Ecology 61:976-989.
1535. Hall, R. J., T. F. Waters, and E. F. Cook. 1980. The role of drift dispersal in production ecology of a stream mayfly. Ecology 61:37-43.
1536. Hall, T. J. 1982. Colonizing macroinvertebrates in the upper Mississippi River with a comparison of basket and multiplate samplers. Freshwat. Biol. 12:211-215.
1537. Hambrook, J. A., and R. G. Sheath. 1987. Grazing of freshwater Rhodophyta. J. Phycol. 23:656-62.
1538. Hamilton, A. L. 1969. A new type of emergence trap for collecting stream insects. J. Fish. Res. Bd. Can. 26:1685-1689.
1539. Hamilton, A. L., W. Burton, and J. P. Flannagan. 1970. A multiple corer for sampling profundal benthos. J. Fish. Res. Bd. Can. 27:1867-1869.
1540. Hamilton, A. L., O. Saether, and D. Oliver. 1969. A classification of the Nearctic Chironomidae. Fish. Res. Bd. Can. Tech. Rept. 124:1-42.
1541. Hamilton, D. A., and D. C. Tarter. 1977. Life history and ecology of *Ephemerella funeralis* McDunnough (Ephemeroptera: Ephemerellidae) in a small West Virginia stream. Am. Midl. Nat. 98:458-462.
1542. Hamilton, J. D., and J. Timmons. 1980. Effects of mild tannery pollution on growth and emergence of two aquatic insects *Rhithrogena semicolorata* and *Ephemerella ignita*. Wat. Res. 14:723-727.
1543. Hamilton, S. W. 1985. The larva and pupa of *Beraea gorteba* Ross (Trichoptera: Beraeidae). Proc. Ent. Soc. Wash. 87:783-789.
1544. Hamilton, S. W., and G. A. Schuster. 1978. Hydroptilidae from Kansas (Trichoptera). Ent. News 89:201-205.
1545. Hamilton, S. W., and G. A. Schuster. 1979. Records of Trichoptera from Kansas, II: The families Glossosomatidae, Helicopsychidae, Hydropsychidae and Rhyacophilidae. State Biol. Surv. Kans. Tech. Publ. 8:15-22.
1546. Hamilton, S. W., and G. A. Schuster. 1980. Records of Trichoptera from Kansas, III: The families Limnephilidae, Phryganeidae, Polycentropodidae, and Sericostomatidae. State Biol. Surv. Kans. Tech. Publ. 9:20-29.
1547. Hamilton, S. W., G. A. Schuster, and M. B. DuBois. 1983. Checklist of the Trichoptera of Kansas. Trans. Kans. Acad. Sci. 86:10-23.
1548. Hammer, M. 1953. Investigations on the microfauna of northern Canada. Part II. Collembola. Acta Arct. 6:1-108.
1549. Hamrum, C. L., M. A. Anderson, and M. Boole. 1971. Distribution and habitat preference of Minnesota dragonfly species (Odonata, Anisoptera). II. J. Minn. Acad. Sci. 37:93-96.

1550. Hanes, E. C., and J.J.H. Ciborowski. 1992. Effects of density and food limitations on size variation and mortality of larval *Hexagenia rigida* (Ephemeroptera: Ephemeridae). Can. J. Zool. 70:1824-1832.

1551. Hanna, H. M. 1957. A study of the growth and feeding habits of the larvae of four species of caddis flies. Proc. R. ent. Soc. Lond. (A) 32:139-146.

1552. Hanna, H. M. 1959. The growth of larvae and their cases and the life cycles of five species of caddis flies (Trichoptera). Proc. R. ent. Soc. Lond. (A) 34:121-129.

1553. Hansell, M. H. 1972. Case building behavior of the caddis fly larva, *Lepidostoma hirtum*. J. Zool. Lond. 167:179-192.

1554. Hansen, D. C., and E. F. Cook. 1976. The systematics and morphology of the Nearctic species of *Diamesa* Meigen, 1835 (Diptera: Chironomidae). Mem. Am. Ent. Soc. 30:1-203.

1555. Hansen, M. 1991. The hydrophiloid beetles. Phylogeny, classification and a revision of the genera (Coleoptera, Hydrophiloidea). Royal Danish Acad. Sci. Let., Biologiske Skrifter 40:1-367.

1556. Hansen, S. R., and R. R. Garton. 1982. The effects of diflubenzuron on a complex laboratory stream community. Arch. Environ. Contam. Toxicol. 11:1-10.

1557. Hansford, R. G. 1978. Life-history and distribution of *Simulium austeni* (Diptera: Simuliidae) in relation to phytoplankton in some southern English rivers. Freshwat. Biol. 8:521-531.

1558. Hanson, B. J., K. W. Cummins, A. S. Cargill, II, and R. R. Lowry. 1983. Dietary effects on lipid and fatty acid composition of *Clistoronia magnifica* (Trichoptera; Limnephilidae). Freshwat. Invert. Biol. 2:2-15.

1559. Hanson, B. J., K. W. Cummins, A. S. Cargill, and R. R. Lowry. 1985. Lipid content, fatty acid composition, and the effect of diet on fats of aquatic insects. Comp. Biochem. Physiol. 80:257-276.

1560. Hanson, M. 1991. A review of the genera of the beetle family Hydraenidae (Coleoptera). Steenstrupia 17:1-52.

1561. Harbach, R. E. 1977. Comparative and functional morphology of the mandibles of some fourth stage mosquito larvae. Zoomorphologie 87:217-236.

1562. Harbach, R. E., and K. L. Knight. 1980. Taxonomists' glossary of mosquito anatomy. Plexus, NJ. 415pp.

1563. Harden, P., and C. Mickel. 1952. The stoneflies of Minnesota (Plecoptera). Univ. Minn. Agric. Exp. Sta. Tech. Bull. 201:1-84.

1564. Hardin, F. W., H. R. Hepburn, and B. J. Ethridge. 1967. A history of mosquitoes and mosquito-borne diseases in Mississippi. 1699-1965. Mosquito News 27:60-66.

1565. Hardy, D. E., and M. D. Delfinado. 1980. Insects of Hawaii. Vol. 13. Diptera: Cyclorrhapha III. Univ. Press of Hawaii. 451pp.

1566. Hare, L. 1992. Aquatic insects and trace metals: bioavailability, bioaccumulation, and toxicity. Crit. Rev. Toxicol. 22:327-369.

1567. Hare, L., P. G. C. Campbell, A. Tessier, and N. Belzile. 1989. Gut sediments in a burrowing mayfly (Ephemeroptera: Hexagenia). Their contribution to animal trace element burdens, their removal, and the efficacy of a correction for their presence. Can. J. Fish. Aquat. Sci. 46:451-456.

1568. Hargeby, A. 1986. A simple trickle chamber for rearing aquatic invertebrates. Hydrobiologia 133:271-274.

1569. Harmston, F. C. 1949. An annotated list of the mosquito records from Colorado. Great Basin Nat. 9:65-75.

1570. Harmston, F. C., and F. A. Lawson. 1967. Mosquitoes of Colorado. U.S. Dept. Health Educ. Welfare. 140 pp.

1571. Harp, G. L., and J. D. Rickett. 1977. The dragonflies of Arkansas. Proc. Ark. Acad. Sci. 31:50-54.

1572. Harp, G. L., and P. A. Harp. 1980. Aquatic macroinvertebrates of Wapanocca National Wildlife Refuge. Proc. Ark. Acad. Sci. 34:115-117.

1573. Harper, P. P. 1973a. Emergence, reproduction and growth of setipalpian Plecoptera in southern Ontario. Oikos 24:94-107.

1574. Harper, P. P. 1973b. Life histories of Nemouridae and Leuctridae in southern Ontario (Plecoptera). Hydrobiologia 41:309-356.

1575. Harper, P. P. 1980. Phenology and distribution of aquatic dance flies (Diptera: Empididae) in a Laurentian watershed. Am. Midl. Nat. 104:110-117.

1576. Harper, P. P. 1990. Life cycles of *Leuctra duplicata* and *Ostrocerca prolongata* in an intermittent streamlet in Quebec (Plecoptera: Leuctridae and Nemouridae). Great Lakes Ent. 23:211-216.

1577. Harper, P. P., and F. Harper. 1983. Biogéographie et associations des Plécoptères d'hiver du Québec méridional (Plecoptera: Capniidae et Taeniopterygidae). Can. Ent. 115:1465-1476.

1578. Harper, P. P., and H.B.N. Hynes. 1970. Diapause in the nymphs of Canadian winter stoneflies. Ecology 51:925-927.

1579. Harper, P. P., and H.B.N. Hynes. 1971a. The Leuctridae of eastern Canada (Insecta: Plecoptera). Can. J. Zool. 49:915-920.

1580. Harper, P. P., and H.B.N. Hynes. 1971b. The Capniidae of eastern Canada (Insecta: Plecoptera). Can. J. Zool. 49:921-940.

1581. Harper, P. P., and H.B.N. Hynes. 1971c. The nymphs of the Taeniopterygidae of eastern Canada (Insecta: Plecoptera). Can. J. Zool. 49:941-947.

1582. Harper, P. P., and H.B.N. Hynes. 1971d. The nymphs of the Nemouridae of eastern Canada (Insecta: Plecoptera). Can. J. Zool. 49: 1129-1142.

1583. Harper, P. P., and H.B.N. Hynes. 1972. Life histories of Capniidae and Taeniopterygidae in southern Ontario (Plecoptera). Arch. Hydrobiol. Suppl. 40:274-314.

1584. Harper, P. P., and M. Lauzon. 1985. The crane fly fauna of a Laurentian woodland, with special reference to the aquatic species (Diptera: Tipulidae). Revue D'Entomologie Du Quebec 3:3-22.

1585. Harper, P. P., and M. Lauzon. 1989. Life cycle of the nymph fly *Palaeodipteron walkeri* Ide 1965 (Diptera: Nymphomyiidae) in the White Mountains of southern Québec. Can. Ent. 121:603-607.

1586. Harper, P. P., M. Lauzon, and F. Harper. 1991. Life cycles of 12 species of winter stoneflies from Quebec (Plecoptera: Capniidae and Taeniopterygidae). Can. J. Zool. 69:787-796.

1587. Harper, P. P., M. Lauzon, and F. Harper. 1994. Life cycles of sundry stoneflies (Plecoptera) from Québec. Rev. Ent. Qué. 36: 28-42.

1588. Harper, P. P., L. LeSage, and M. Lauzon. 1993. The life cycle of *Podmosta macdunnoughi* (Ricker) in the Lower Laurentians, Quebec (Plecoptera: Nemouridae), with a discussion on embryonic diapause. Can. J. Zool. 71:2136-2139.

1589. Harper, P. P., and E. Magnin. 1969. Cycles vitaux de quelques Plecopteres des Laurentides (Insecta). Can. J. Zool. 47:483-494.

1590. Harpster, H. T. 1941. An investigation of the gaseous plastron as a respiratory mechanism in *Heliehus striatus* LeConte (Dryopidae). Trans. Am. Ent. Soc. 60:329-358.

1591. Harpster, H. T. 1944. The gaseous plastron as a respiratory mechanism in *Stenelmis quadrimaculata* Horn (Dryopidae). Trans. Am. Microsc. Soc. 63:1-26.

1592. Harris, H. M. 1937. Contribution to the South Dakota list of the Hemiptera. Iowa State Coll. J. Sci. 11:169-176.

1593. Harris, H. M., and W. E. Shull. 1944. A preliminary list of Hemiptera of Idaho. Iowa State Coll. J. Sci. 18:199-208.

1594. Harris, S. C., and R. B. Carlson. 1978. Distribution of *Bittacomorpha clavipes* (Fabricius) and *Ptychoptera quadrlasciata* Say (Diptera: Ptychopteridae) in a sandhill springbrook of southeastern North Dakota. Ann. Proc. N.D. Acad. Sci. 29:59-66.

1595. Harris, S. C., P. K. Lago, and R. B. Carlson. 1980. Preliminary survey of the Trichoptera of North Dakota. Proc. Ent. Soc. Wash. 82:39-43.

1596. Harris, S. C., P. K. Lago, and R. W Holzenthal. 1982a. An annotated checklist of the caddisflies (Trichoptera) of Mississippi and southeastern Louisiana. Part II: Rhyacophiloidea Proc. Ent. Soc. Wash. 84:509-512.

1597. Harris, S. C., P. K. Lago, and J. F. Scheiring. 1982b. An annotated list of Trichoptera of several streams on Eglin Air Force Base, Florida. Ent. News 93(3):79-84.

1598. Harris, S. C., P. E. O'Neil, and P. K. Lago. 1991. Caddisflies of Alabama. Geol. Surv. Alabama Bull. 142. 442 pp.

1599. Harrold, J. F. 1978. Relation of sample variations to plate orientation in the Hester-Dendy plate sampler. Progr. Fish Cult. 40:24-25.

1600. Hart, C. A. 1895. On the entomology of the Illinois River and adjacent waters. Bull. Ill. State Lab. Nat. Hist. 4:1-273.

1601. Hart, C. W., and S.L.H. Fuller (eds.). 1974. Pollution ecology of freshwater invertebrates. Academic, N.Y. 389 pp.
1602. Hart, D. D. 1985a. Causes and consequences of territoriality in a grazing stream insect. Ecology 66:404-414.
1603. Hart, D. D. 1985b. Grazing insects mediate algal interactions in a stream benthic community. Oikos 44:40-46.
1604. Hart, D. D., S. L. Kohler, and R. G. Carlton. 1991. Harvesting of benthic algae by territorial grazers: the potential for prudent predation. Oikos 60:329-335.
1605. Hart, D. D., and S. C. Latta. 1986. Determinants of ingestion rates in filter-feeding larval blackflies (Diptera: Simuliidae). Freshwat. Biol. 16:1-14.
1606. Hart, D. D., and V. H. Resh. 1980. Movement patterns and foraging ecology of a stream caddisfly larva. Can. J. Zool. 58:1174-1185.
1607. Hart, D. D., and C. T. Robinson. 1990. Resource limitation in a stream community: phosphorus enrichment effects on periphyton and grazers. Ecology 71:1494-1502.
1608. Hart, J. W. 1970. A checklist of Indiana Collembola. Proc. Ind. Acad. Sci. 79:249-252.
1609. Hart, J. W. 1971. New records of Indiana Collembola. Proc. Ind. Acad. Sci. 80:246.
1610. Hart, J. W. 1973. New records of Indiana Collembola. Proc. Ind. Acad. Sci. 82:231.
1611. Hart, J. W. 1974. Preliminary studies of Collembola at the Brookville Ecological Research Center, including new records of Indiana Collembola. Proc. Ind. Acad. Sci. 83:334-339.
1612. Hart, R. C. 1985. Seasonality of aquatic invertebrates in low latitude and Southern Hemisphere inland waters. Hydrobiologia 125:151-178.
1613. Hartland-Rowe, R. 1964. Factors influencing the life-histories of some stream insects (Ephemeroptera, Plecoptera) in Alberta. Verh. Int. Verein, Limnol. 15:17-925.
1614. Hartley, C. F. 1955. Rearing simuliids in the laboratory from eggs to adults. Proc. Helminth. Soc. Wash. 22:93-95.
1615. Hartley, J. C. 1961. A taxonomic account of the larvae of some British Syrphidae. Proc. Zool. Soc. Lond. 136:505-573.
1616. Harvey, R. S., R. L. Vannote, and B. W. Sweeney. 1979. Life history, developmental processes, and energetics of the burrowing mayfly *Dolania americana*. pp. 211-229. *In* J. F. Flannagan and K. E. Marshall (eds.). Advances in Ephemeroptera biology. Plenum, N.Y. 552 pp.
1617. Harwood, P. D. 1971. Synopsis of James G. Needham's (Cornell University) unpublished manuscript. "The dragonflies of West Virginia." Proc. W. Va. Acad. Sci. 43:72-74.
1618. Harwood, R. F., and M. T. James. 1979. Entomology in human and animal health (7th ed.). Macmillan, N.Y. 548 pp.
1619. Hase, A. 1926. Zur Kenntniss der Lebensgewohneiten und der Umwelt des marinen Kafer *Ochthebius quadricollis* Mulsant. Int. Revue ges. Hydrobiol. 16:141-179.
1620. Hasenfuss, I. 1960. Die Larvalsystematik der Zunsler (Pyralidae). Abhandlungen zur Larvalsystematik der Insekten. 5:1-263.
1621. Hashimoto, H. 1976. Non-biting insects of marine habitats, pp. 377-414. *In* L. Cheng (ed.). Marine insects. North Holland, Amsterdam. 581 pp.
1622. Hassage, R. L., and K. W. Stewart. 1990. Growth and voltinism of five stonefly species in a New Mexico mountain stream. Southwest. Nat. 35:130-134.
1623. Hatakeyama, S. 1987. Chronic effects of Cd on reproduction of *Polypedilum nubifer* (Chironomidae) through water and food. Environ. Poll. 48:249-261.
1624. Hatch, M. C. 1925. An outline of the ecology of Gyrinidae. Bull. Brooklyn Ent. Soc. 20:101-104.
1625. Hatch, M. H. 1930. Records and new species of Coleoptera from Oklahoma and western Arkansas, with subsidiary studies. Publ. Okla. Biol. Surv. 2:15-26.
1626. Hatch, M. H. 1962, 1965. The beetles of the Pacific Northwest. Part III: Pselaphidae and Diversicornia (in collaboration with O. Part, J. A. Wagner, K. M. Fender, W. F. Barr, G. E. Woodraffe, and C. W. Coombs). Part IV: *Marcodactyles, Palpicornes,* and *Heteromera* (in collaboration with D. V. Miller, D. V. McCorkle, F. Werner, and D. W. Boddy). Univ. Wash. Publ. Biol. 16:1-503; 19:1-268.
1627. Hauer, F. R., and A. C. Benke. 1991. Rapid growth rates of snag-dwelling chironomids in a black water river: the influence of temperature and discharge. J. N. Am. Benthol. Soc. 10:154-164.
1628. Hauer, F. R., and J. A. Stanford. 1981. Larval specialization and phenotypic variation in *Arctopsyche grandis* (Trichoptera: Hydropsychidae). Ecology 62:645-653.
1629. Hauer, F. R., and J. A. Stanford. 1982. Bionomics of *Dicosmoecus gilvipes* (Trichoptera: Limnephilidae) in a large western montane river. Am. Midl. Nat. 108:81-87.
1630. Hauer, F. R., and J. A. Stanford. 1986. Ecology and co-existence of two species of *Brachycentrus* (Trichoptera) in a Rocky Mountain river. Can. J. Zool. 64:1469-1474.
1631. Havas, M., and T. C. Hutchinson. 1983. Effect of low pH on the chemical composition of aquatic invertebrates from tundra ponds at the Smoking Hills, N.W.T., Canada. Can. J. Zool. 61:241-249.
1632. Havel, J. E., J. Link, and J. Niedzwiecki. 1993. Selective predation by *Lestes* (Odonata: Lestidae) on littoral microcrustacea. Freshwat. Biol. 29:47-58.
1633. Hawkes, H. A. 1979. Invertebrates as indicators of river water quality, pp. 2-1 - 2-45. *In* A. James and L. Evison (eds.). Biological indicators of water quality. J. Wiley and Sons, Chichester, England. 569 pp.
1634. Hawkins, C. P. 1982. Ecological relationships among western Ephemerellidae: Growth, life cycles, food habits and habitat relationships. Ph.D. diss., Oregon State University, Corvallis. 213 pp.
1635. Hawkins, C. P. 1984. Substrate associations and longitudinal distributions in species of Ephemerellidae (Ephemeroptera: Insecta) from western Oregon. Freshwat. Invertebr. Biol. 3:181-188.
1636. Hawkins, C. P. 1985. Food habits of species of ephemerellid mayflies (Ephemeroptera: Insecta) in streams in Oregon. Am. Midl. Nat.. 113:343-352.
1637. Hawkins, C. P. 1986. Variation in individual growth rates and population densities of ephemerellid mayflies. Ecology 67:1384-1395.
1638. Hawkins, C. P., and J. A. MacMahon. 1989. Guilds: the multiple meanings of a concept. Ann. Rev. Ent. 34:423-451.
1639. Hawkins, C. P., and J. R. Sedell. 1981. Longitudinal and seasonal changes in functional organization of macroinvertebrate communities in four Oregon streams. Ecology 62:387-397.
1640. Hayden, W., and H. T. Clifford. 1974. Seasonal movements of the mayfly *Leptophlebia cupida* (Say) in a brown water stream in Alberta, Canada. Am. Midl. Nat. 91:90-102.
1641. Hazard, E. E. 1960. A revision of the genera *Chauliodes* and *Nigronia* (Megaloptera: Corydalidae). M.S. thesis, Ohio State University, Columbus. 53 pp.
1642. Headlee, T. J. 1945. The mosquitoes of New Jersey and their control. Rutgers Univ. Press, New Brunswick. 326 pp.
1643. Heads, P. A. 1985. The effect of invertebrate and vertebrate predators on the foraging movements of *Ischnura elegans* larvae (Odonata: Zygoptera). Freshwat. Biol. 15:559-572.
1644. Heads, P. A. 1986. The costs of reduced feeding due to predator avoidance: potential effects on growth and fitness in *Ischnura elegans* larvae (Odonata: Zygoptera). Ecol. Ent. 11:369-377.
1645. Hearle, E. 1932. The blackflies of British Columbia (Simuliidae, Diptera). Proc. Brit. Colum. Ent. Soc. 1932:5-19.
1646. Heath, B. L., and W. P. McCafferty. 1975. Aquatic and semiaquatic Diptera of Indiana. Purdue Univ. Res. Bull. 930:1-17.
1647. Hedqvist, K. J. 1967. Hymenoptera, pp. 242-244. *In* J. Illies (ed.). Limnofauna Europaea. Gustav Fischer Verlag, Stuttgart. 417 pp.
1648. Hefti, D., and I. Tomka. 1990. Abundance, growth and production of three mayfly species (Ephemeroptera: Insecta) from the Swiss Prealps. Arch. Hydrobiol. 120:211-228.
1649. Heiman, D. R., and A. W. Knight. 1970. Studies on growth and development of the stonefly *Paragnetina media* Walker (Plecoptera: Perlidae). Am. Midl. Nat. 84:274-278.
1650. Heiman, D. R., and A. W. Knight. 1975. The influence of temperature on the bioenergetics of the carnivorous stonefly nymph, *Acroneuria californica* Banks (Plecoptera: Perlidae). Ecology 56:105-116.

1651. Heinis, F., and T. Crommentuijn. 1989. The natural habitat of the deposit feeding chironomid larvae *Stictochironomus histrio* (Fabricius) and *Chironomus anthracinus* Zett, in relation to their responses to changing oxygen concentrations. Acta Biol. Debr. Oecol. Hungary 3:135-140.

1652. Heinis, F., K. R. Timmermans, and W. R. Swain. 1990. Short-term sublethal effects of cadmium on the filter feeding chironomid larva *Glytotendipes pallens* (Meigen) (Diptera). Aquat. Toxicol. 16:73-86.

1653. Heinrich, B., and D. Vogt. 1980. Aggregation and foraging behavior of whirligig beetles (Gyrinidae). Behav. Ecol. Sociobiol. 7:179-186.

1654. Heinrich, C. 1916. On the taxonomic value of some larval characters in the Lepidoptera. Proc. Ent. Soc. Wash. 18:154-164.

1655. Heinrich, C. 1940. Some new American pyralidoid moths. Proc. Ent. Soc. Wash. 42:31-44.

1656. Heise, B. A., J. J. Flannagan, and T. D. Galloway. 1987. Life histories of *Hexagenia limbata* and *Ephemera simulans* (Ephemeroptera) in Dauphin Lake, Manitoba. J. N. Am. Benthol. Soc. 6:230-240.

1657. Heise, B. A., J. F. Flannagan, and T. D. Galloway. 1988. Production of *Hexagenia limbata* (Serville) and *Ephemera simulans* Walker (Ephemeroptera) in Dauphin Lake, Manitoba, with a note on weight loss due to preservatives. Can. J. Fish. Aquat. Sci. 45:774-781.

1658. Heitzman, R. L., and D. H. Habeck. 1979. (1981). Taxonomic and biological notes of *Bellura gortynoides* Walker (Noctuidae). J. Res. Lepidoptera 18:228-231.

1659. Helfer, J. R. 1953. The grasshoppers, cockroaches and their allies. Wm. C. Brown Publ. Co., Dubuque, IA. 353 p.

1660. Hellawell, J. M. 1978a. Biological surveillance of rivers. Nat. Environ. Res. Council and Water Res. Centre, Stevenage, England. 332 pp.

1661. Hellawell, J. M. 1978b. Chap. 4. Macroinvertebrate methods, pp. 35-90. *In* Biological surveillance of rivers. A biological monitoring handbook. Dorset Press, Dorchester, England. 332 pp.

1662. Hellawell, J. M. 1986. Biological indicators of freshwater pollution and environmental management. Elsevier, London. 546 pp.

1663. Hemmingsen, A. M. 1965. The lotic cranefly, *Tipula saginata* Bergroth, and the adaptive radiation of the Tipulinae, with a test of Dyar's law. Vidensk. Medd. Dansk Naturhist. Foren 18:93-150.

1664. Hemphill, N., and S. D. Cooper. 1983. The effect of physical disturbance on the relative abundances of two filter-feeding insects in a small stream. Oecologia 58:378-383.

1665. Henderson, J., A. G. Hildrew, and C. R. Townsend. 1990. Detritivorous stoneflies of an iron-rich stream: Food and feeding, pp. 249-254. *In* I. C. Campbell (ed.). Mayflies and stoneflies: life histories and biology. Kluwer Academic Publishers, Dordrecht, The Netherlands. 366 pp.

1666. Hennig, W. 1943. Ubersicht über die bisher bekannten Metamorphosesstadien der Ephydriden. Neubeschreibungen nach dem Material der Deutschen Limnologischen Sundaexpedition (Diptera: Ephydridae). Arb. Morph. Tax. Ent. Berlin 10:105-138.

1667. Hennig, W. 1948, 1950, 1952. Die Larvenformen der Dipteren. Akademie-Verlag, Berlin. Pt. 1 185 pp.; Pt. 2 458 pp.; Pt. 3 628 pp.

1668. Hennig, W. 1966. Phylogenetic systematics. Translated by D. D. Davis and R. Zangerl, Univ. Ill. Press, Urbana. 263 pp.

1669. Hennig, W. 1967. Diptera:Muscidae, pp. 423-424. *In* J. Illies (ed.). Limnofauna Europaea. Gustav Fischer Verlag, Stuttgart. 417 pp.

1670. Hennig, W. 1973. Diptera. *In* M. Beier and W. de Gruyter (eds.). Handbuch der Zool. 4. Spezielles 31:1-337.

1671. Henriksen, K. L. 1922. Notes on some aquatic Hymenoptera. Ann. Biol. Lacustre 11:19-37.

1672. Henrikson, B. I. 1988. The absence of antipredator behaviour in the larvae of *Leucorrhinia dubia* (Odonata) and the consequences for their distribution. Oikos 51:179-183.

1673. Henrikson, B. I. 1990. Predation on amphibian eggs and tadpoles by common predators in acidified lakes. Holarctic Ecol. 13:201-206.

1674. Henrikson, L., and H. Oscarson. 1978. A quantitative sampler for air-breathing aquatic insects. Freshwat. Biol. 8:73-77.

1675. Henriquez, N. P., and J. R. Spence. 1993. Studies of *Lathromeroidea* sp. nov. (Hymenoptera: Trichogrammatidae) a parasitoid of gerrid eggs. Can. Ent. 125:693-702.

1676. Henry, B. C. 1993. A revision of *Neochoroterpes* (Ephemeroptera: Leptophlebiidae) new status. Trans. Am. Ent. Soc. 119:317-333.

1677. Henry, T. J., and R. C. Froeschner (eds.). 1988 Catalog of the Heteroptera, or true bugs, of Canada and the continental United States. E. J. Brill, Leiden, NY. 958 pp.

1678. Heong, K. L., and G. B. Aquino. 1990. Arthropod diversity in tropical rice ecosystems. Int. Rice Res. Newsltr. 15: 31-32.

1679. Heong, K. L., G. B. Aquino, and A. T. Barrion. 1991. Arthropod community stuctures of rice ecosystems in the Philippines. Bull. Ent. Res. 81:407-416.

1680. Heong, K. L., G. B. Aquino, and A. T. Barrion. 1992. Population dynamics of plant- and leafhoppers and their natural enemies in rice ecosytems in the Philippines. Crop Protect. 11:371-379.

1681. Hepburn, H. R., and J. P. Woodring. 1963. Checklist of the Collembola (Insects) of Louisiana. Proc. La. Acad. Sci. 26:5-9.

1682. Heppner, J. B., and D. H. Habeck. 1976. Insects associated with *Polygonum* (Polygonaceae) in north central Florida. I. Introduction and Lepidoptera. Fla. Ent. 59:231-239.

1683. Herlong, D. D. 1979. Aquatic Pyralidae (Lepidoptera: *Nymphulinae*) in South Carolina. Fla. Ent. 62:188-193.

1684. Herman, L. 1972. Revision of *Bledius* and related genera. Part I. The *aequatorialis, mandibularis* and semiferrigueous groups and two new genera (Coleoptera, Staphylinidae, Oxytelinae). Bull. Am. Mus. Nat. Hist. 149:111-254.

1685. Hermann, S. J. 1988. New record and range extension for *Ceraclea resurgens* (Trichoptera: Leptoceridae) from Colorado, with notes on ecological conditions. Ent. News 99:253-259.

1686. Hermann, S. J., D. E. Ruiter, and J. D. Unzicker. 1986. Distribution and records of Colorado Trichoptera. Southwest. Nat. 31:421-457.

1687. Herring, J. L. 1950. The aquatic and semiaquatic Hemiptera of northern Florida. Part II: Veliidae and Mesoveliidae. Fla. Ent. 33:145-150.

1688. Herring, J. L. 1951a. The aquatic and semiaquatic Hemiptera of northern Florida. Part III. Nepidae, Belostomatidae, Notonectidae, Pleidae and Corixidae. Fla. Ent. 34:17-29.

1689. Herring, J. L. 1951b. The aquatic and semiaquatic Hemiptera of northern Florida. Part IV. Classification of habitats and keys to the species. Fla. Ent. 34:146-161.

1690. Herring, J. L. 1958. Evidence for hurricane transport and dispersal of aquatic Hemiptera. Pan-Pacif. Ent. 34:174-175.

1691. Herring, J. L. 1961. The genus *Halobates* (Hemiptera: Gerridae). Pacif. Insects 3:223-305.

1692. Herring, J. L., and P. D. Ashlock. 1971. A key to the nymphs of the families of Hemiptera of America north of Mexico. Fla. Ent. 54:207-212.

1693. Herrmann, J., and K. G. Andersson. 1986. Aluminum impact on respiration of lotic mayflies at low pH. Water, Air and Soil Pollut. 30:703-709.

1694. Herrmann, S. J., and H. L. Davis. 1991. Distribution records of *Corydalus cornutus* (Megaloptera: Corydalidae) in Colorado. Ent. News. 102:25 30.

1695. Hershey, A. E. 1986. Selective predation by *Procladius* in an arctic Alaskan lake. Can. J. Fish. Aquat. Sci. 43:2523-28.

1696. Hershey, A. E. 1987. Tubes and foraging behavior in larval Chironomidae: implications for predator avoidance. Oecologica 73:236-241.

1697. Hershey, A. E., W. B. Bowden, L. A. Deegan, J. E. Hobbie, B. J. Peterson, G. W. Kipphut, G. W. Kling, M. A. Lock, R. W. Merritt, M. C. Miller, J. R. Vestal, and J. A. Schuldt. 1995b. The Kuparuk River: A long-term study of biological and chemical processes in an arctic river, pp. 1-44. *In* A. Milner and M. Oswood (eds.). Alaskan freshwaters. Springer-Verlag, Berlin.

1698. Hershey, A. E., and S. I. Dodson. 1985. Selective predation by a sculpin and a stonefly on two chironomids in laboratory feeding trials. Hydrobiologia 124:269-273.

1699. Hershey, A. E., and S. I. Dodson. 1987. Predator avoidance by *Cricotopus*: Cyclomorphosis and the importance of being big and hairy. Ecology 68:913-920.

1700. Hershey, A. E., and A. L. Hiltner. 1988. Effect of a caddisfly on black fly density: interspecific interactions limit black flies in an arctic river. J. N. Am. Benthol. Soc. 7:188-196.

1701. Hershey, A. E., A. L. Hiltner, M.A J. Hullar, M. C. Miller, J. R. Vestal, M. A. Lock, S. Rundle, and B. J. Peterson. 1988. Nutrient influence on a stream grazer: *Orthocladius* microcommunities respond to nutrient input. Ecology 69:1383-1392.

1702. Hershey, A. E., R. W. Merritt, and M. C. Miller. 1995a. Insect diversity, life history, and trophic dynamics in arctic streams, with particular emphasis on black flies (Diptera: Simuliidae), pp. 281-293. *In* F. S. Chapin, III. and C. Korner (eds.). Arctic and alpine biodiversity: Patterns, causes, and ecosystem consequences. Springer-Verlag, Berlin.

1703. Hershey, A. E., J. Pastor, B. J. Petersen, and G. W. Kling. 1993. Stable isotopes resolve the the drift paradox for *Baetis* mayflies in an arctic river. Ecology 74:2315-2325.

1704. Hess, A. D. 1941. New limnological sampling equipment. Limnol. Soc. Am. Spec. Publ. 6:1-5.

1705. Hester, F. E., and J. S. Dendy. 1962. A multiple-plate sampler for aquatic macroinvertebrates. Trans. Am. Fish. Soc. 91: 420-421.

1706. Heymons, R., and H. Heymons. 1909. Collembola: Die Süsswasserfauna Deutschlands 7:1-16.

1707. Hickin, N. E. 1967. Caddis larvae. Hutchinson, London. 480 pp.

1708. Hickman, J. R. 1930a. Life-histories of Michigan Haliplidae (Coleoptera). Pap. Mich. Acad. Sci. Arts Lett. 11:399-424.

1709. Hickman, J. R. 1930b. Respiration of the Haliplidae (Coleoptera). Pap. Mich. Acad. Sci. Arts Lett. 13:277-289.

1710. Hickman, J. R. 1931. Contribution to the biology of the Haliplidae (Coleoptera). Ann. Ent. Soc. Am. 24:129-142.

1711. Hildrew, A. G. 1977. Ecological aspects of life history in some net-spinning Trichoptera. pp. 269-281. *In* M. I. Crichton, (ed.). Proc. 2nd Internat. Symp. Trichop., Dr. W. Junk, Publ., The Hague. 359 p.

1712. Hildrew, A. G., and J. M. Edington. 1979. Factors facilitating the coexistence of hydropsychid caddis larvae (Trichoptera) in the same river system. J. Anim. Ecol. 48:557-576.

1713. Hildrew, A. G., and C. R. Townsend. 1976. The distribution of two predators and their prey in an iron-rich stream. J. Anim. Ecol. 45:41-57.

1714. Hildrew, A. G., C. R. Townsend, J. Francis, and K. Finch. 1984. Cellulolytic decomposition in streams of contrasting pH and its relationship with invertebrate community structure. Freshwat. Biol. 14:323-328.

1715. Hildrew, A. G., C. R. Townsend, and A. Hasham. 1985. The predatory Chironomidae of an iron-rich stream: feeding ecology and food web structure. Ecol. Ent. 10:403-413.

1716. Hiley, P. D. 1969. A method of rearing Trichoptera larvae for taxonomic purposes. Entomol. mon. Mag. 105:278-279.

1717. Hiley, P. D., J. F. Wright, and A. D. Berrie. 1981. A new sampler for stream benthos, epiphytic macrofauna and aquatic macrophytes. Freshwat. Biol. 11:79-85.

1718. Hill, P., and D. Tarter. 1978. A taxonomic and distributional study of adult limnephilid caddisflies of West Virginia (Trichoptera: Limnephilidae). Ent. News 89:214-216.

1719. Hill, S. B., F. W. Stehr, and W. R. Enns. 1987. Key to orders of immature insects and selected arthropods, pp. 19-46. *In* F. W. Stehr (ed.). Immature Insects. Vol. 1. Kendall/Hunt Publ. Co., Dubuque, IA. 754pp.

1720. Hill, W. R. 1992. Food limitation and interspecific competition in snail-dominated steams. Can. J. Fish. Aquat. Sci. 49:1257-1267.

1721. Hill, W. R., and A. W. Knight. 1988. Concurrent grazing effects of two stream insects on periphyton. Limnol. Oceanogr. 33:15-26.

1722. Hill, W. R., S. C. Weber, and A. J. Stewart. 1992. Food limitation of two lotic grazers: quantity, quality, and size-specificity. J. N. Am. Benthol. Soc. 11:420-432.

1723. Hill-Griffin, A. L. 1912. New Oregon Trichoptera. Ent. News 23:17-21.

1724. Hillis, D. M., and C. Moritz. 1990. Molecular systematics. Sinauer Associates. Sunderland, MA. 588pp.

1725. Hilsenhoff, W. L. 1966. The biology of *Chironomus plumosus* (Diptera: Chironomidae) in Lake Winnebago, Wisconsin. Ann. Ent. Soc. Am. 59:465-473.

1726. Hilsenhoff, W. L. 1969. An artificial substrate sampler for stream insects. Limnol. Oceanogr. 14:465-471.

1727. Hilsenhoff, W. L. 1970. Corixidae of Wisconsin. Proc. Wisc. Acad. Sci. Arts Lett. 58:203-235.

1728. Hilsenhoff, W. L. 1973. Notes on *Dubiraphia* (Coleoptera: Elmidae) with descriptions of five new species. Ann. Ent. Soc. Am. 66:55-61.

1729. Hilsenhoff, W. L. 1974. The unusual larva and habitat of *Agabus confusus* (Dytiscidae). Ann. Ent. Soc. Am. 67: 703-705.

1730. Hilsenhoff, W. L. 1975. Notes on Nearctic *Acilius* (Dytiscidae), with the description of a new species. Ann. Ent. Soc. Am. 68:271-274.

1731. Hilsenhoff, W. L. 1977. Use of arthropods to evaluate water quality of streams. Wisc. Dept. Nat. Res. Tech. Bull. 100:1-15.

1732. Hilsenhoff, W. L. 1980. *Coptotomus* (Coleoptera: Dytiscidae) in Eastern North America with descriptions of 2 new species. Trans. Am. Ent. Soc. 105:461-472.

1733. Hilsenhoff, W. L. 1981. Aquatic insects of Wisconsin. Keys to Wisconsin genera and notes on biology, distribution and species. Natural Hist. Council, Univ. Wisconsin-Madison, Publ. No. 2.

1734. Hilsenhoff, W. L. 1982. Using a biotic index to evaluate water quality in streams. Wisc. Dept. Nat. Res. Tech. Bull. 132:1-22.

1735. Hilsenhoff, W. L. 1984. Aquatic Hemiptera of Wisconsin. Great Lakes Ent. 17:29-50.

1736. Hilsenhoff, W. L. 1986a. Life history strategies of some nearctic Agabini (Coleoptera: Dytiscidae). Ent. Basiliensia 11:385-390.

1737. Hilsenhoff, W. L. 1986b. Semiaquatic Hemiptera of Wisconsin. Great Lakes Ent. 19:7-19.

1738. Hilsenhoff, W. L. 1987a. An improved biotic index of organic stream pollution. Great Lakes Ent. 20:31-39.

1739. Hilsenhoff, W. L. 1987b. Effectiveness of bottle traps for collecting Dytiscidae (Coleoptera). Coleopt. Bull. 4:377-380.

1740. Hilsenhoff, W. L. 1988. Rapid field assessment of organic pollution with a family-level biotic index. J. N. Am. Benthol. Soc. 7:65-68.

1741. Hilsenhoff, W. L. 1990. Gyrinidae of Wisconsin, USA with a key to adults of both sexes and notes on distribution and habitat. Great Lakes Ent. 23:77-92.

1742. Hilsenhoff, W. L. 1991. Diversity and classification of insects and Collembola, pp. 593-663. *In* J. H. Thorp and A. P. Covich (eds.). Ecology and classification of North American freshwater invertebrates. Academic Press, NY. 911p.

1743. Hilsenhoff, W. L. 1992. Dytiscidae and Noteridae of Wisconsin (Coleoptera). I. Introduction, key to genera of adults, and distribution, habitat, life cycle, and identification of Agabetinae, Laccophilinae and Noteridae. Great Lakes Ent. 25:57-69.

1744. Hilsenhoff, W. L. 1993a. Dytiscidae and Noteridae of Wisconsin (Coleoptera). II. Distribution, habitat, life cycle, and identification of species of Dytiscidae. Great Lakes Ent. 26: 35-53.

1745. Hilsenhoff, W. L. 1993b. Dytiscidae and Noteridae of Wisconsin (Coleoptera). III. Distribution, habitat, life cycle, and identification of species of Colymbetinae, except Agabini. Great Lakes Ent. 26:121-136.

1746. Hilsenhoff, W. L. 1993c. Dytiscidae and Noteridae of Wisconsin (Coleoptera). IV. Distribution, habitat, life cycle, and identification of species of Agabini (Colymbetinae). Great Lakes Ent. 26:173-197.

1747. Hilsenhoff, W. L. 1994. Dytiscidae and Noteridae of Wisconsin (Coleoptera): V. Distribution, habitat, life cycle, and identification of species of Hydroporinae, except *Hydroporus* Clairville Sensu Lato. Great Lakes Ent. 26:275-295.

1748. Hilsenhoff, W. L., and S. J. Billmyer. 1973. Perlodidae (Plecoptera) of Wisconsin. Great Lakes Ent. 6:1-14.

1749. Hilsenhoff, W. L., and W. U. Brigham. 1978. Crawling water beetles of Wisconsin (Coleoptera: Haliplidae). Great Lakes Ent. 11:11-22.

1750. Hilsenhoff, W. L., J. L. Longridge, R. P. Narf, K. T. Tennessen, and C. P. Walton. 1972. Aquatic insects of the Pine-Popple River, Wisconsin. Wisc. Dept. Nat. Res. Tech. Bull. 54:1-44.

1751. Hilsenhoff, W. L., and R. F. Narf. 1968. Ecology of Chironomidae, Chaoboridae, and other benthos in fourteen Wisconsin lakes. Ann. Ent. Soc. Am. 61:1173-1181.

1752. Hilsenhoff, W. L., and K. L. Schmude. 1992. Riffle beetles of Wisconsin (Coleoptera: Dryopidae, Elmidae, Lutrochidae, Psephenidae) with notes on distribution, habitat, and identification. Great Lakes Ent. 25:191-213.

1753. Hilsenhoff, W. L., and B. H. Tracy. 1985. Techniques for collecting water beetles from lentic habitats. Proc. Acad. Nat. Sci. Philadelphia 137:8-11.

1754. Hiltner, A. L., and A. E. Hershey. 1992. Black fly (Diptera: Simuliidae) response to phosphorus enrichment of an arctic tundra stream. Hydrobiologia 240:259-265.

1755. Hilton, D.F.J. 1990. The Odonata of the province of Prince Edward Island, Canada. Not. Odonatol. 3:68-73.

1756. Hilton, D.F.J. 1992. Odonata from the Magdalen Islands, Quebec, Canada. Entomologist 111:102-108.

1757. Hinman, E. H. 1934. Predators of the Culicidae. I. The predators of larvae and pupae exclusive of fish. J. Trop. Med. Hyg. 37:129-134.

1758. Hinshaw, S. H., and B. K. Sullivan. 1990. Predation on *Hyla versicolor* and *Pseudacaris crucifer* during reproduction. J. Herpetol. 24:196-197.

1759. Hinton, H. E. 1936. Notes on the biology of *Dryops luridus* Erichs. (Coleoptera: Dryopidae). Trans. Soc. Brit. Ent. 3:76-78.

1760. Hinton, H. E. 1946. On the homology and nomenclature of the setae of lepidopterous larvae, with some notes on the phylogeny of the Lepidoptera. Trans. R. ent. Soc. Lond. 97:1-37.

1761. Hinton, H. E. 1948. The dorsal cranial areas of caterpillars. Ann. Mag. Nat. Hist. 14:843-852.

1762. Hinton, H. E. 1955. On the respiratory adaptations, biology and taxonomy of the Psephenidae with notes on some related families (Coleoptera). Proc. Zool. Soc. Lond. 130:543-568.

1763. Hinton, H. E. 1956. The larvae of the Tineidae of economic importance. Bull. Ent. Res. 47:251-346.

1764. Hinton, H. E. 1958a. The pupa of the fly *Simulium* feeds, and spins its own cocoon. Entomol. mon. Mag. 94:14-16.

1765. Hinton, H. E. 1958b. The phylogeny of the panorpoid orders. Ann. Rev. Ent. 3:181-206.

1766. Hinton, H. E. 1960. Cryptobiosis in the larva of *Polypedilum vanderplanki* Hint. (Chironomidae). J. Insect Physiol. 5:286-300.

1767. Hinton, H. E. 1963. The origin and function of the pupal stage. Proc. R. ent. Soc. Lond. (A) 38:77-85.

1768. Hinton, H. E. 1966. Respiratory adaptations of the pupae of beetles of the family Psephenidae. Phil. Trans. R. Soc. (B) 251:211-245.

1769. Hinton, H. E. 1967. On the spiracles of the larvae of the suborder Myxophaga (Coleoptera). Aust. J. Zool. 15:955-959.

1770. Hinton, H. E. 1968. Spiracular gills. Adv. Insect Physiol. 5:65-162.

1771. Hinton, H. E. 1969. Plastron respiration in adult beetles of the suborder Myxophaga. J. Zool. Lond. 159:131-137.

1772. Hinton, H. E. 1971a. Some neglected phases in metamorphosis. Proc. R. ent. Soc. Lond. (C). 35:55-64.

1773. Hinton, H. E. 1971b. A revision of the genus *Hintonelmis* Spangler (Coleoptera: Elmidae). Trans. ent. Soc. Lond. 123:189-208.

1774. Hinton, H. E. 1976a. Plastron respiration in bugs and beetles. J. Insect Physiol. 22:1529-1550.

1775. Hinton, H. E. 1976b. Respiratory adaptations of marine insects, pp. 43-78. *In* L. Cheng (ed.). Marine Insects. North Holland, Amsterdam. 581 pp.

1776. Hinton, H. E. 1977. Enabling mechanisms. Proc. XV Int. Congr. Ent. (Washington), Ent. Soc. Am. 15:71-83.

1777. Hinton, H. E. 1981. Biology of insect eggs. Vols. I-III. Pergamon, Oxford. 1125 pp.

1778. Hiroaki, N., and M. Miyazaki. 1986. Injury to rice leaves by chironomid larvae (Diptera: Chironomidae). Jap. J. Appl. Ent. Zool. 30:66-68.

1779. Hirvenoja, M. 1973. Revision der Gattung *Cricotopus* van der Wulp und ihrer Verwandten (Diptera: Chironomidae). Ann. Zool. Fenn. 10:1-363.

1780. Hirvonen, H. 1992. Effects of backswimmer (*Notonecta*) predation on crayfish (*Pacifastacus*) young: autotomy and behavioural responses. Ann. Zool. Fenn. 29:261-271.

1781. Hissom, F. K., and D. C. Tarter. 1976. Taxonomy and distribution of nymphal Perlodidae of West Virginia (Insecta: Plecoptera). J. Ga. Ent. Soc. 11:317-323.

1782. Hitchcock, S. W. 1968. *Alloperla* (Chloroperlidae: Plecoptera) of the Northeast with a key to species. J. N.Y. Ent. Soc. 76:39-46.

1783. Hitchcock, S. W. 1974. Guide to the insects of Connecticut. Part VII. The Plecoptera or stoneflies of Connecticut. Bull. Conn. State Geol. Nat. Hist. Surv. 107:1-262.

1784. Hocking, B., and L. R. Pickering. 1954. Observations on the bionomics of some northern species of Simuliidae (Diptera). Can. J. Zool. 32:99-119.

1785. Hocking, B., W. R. Richards, and C. R. Twinn. 1950. Observations on the bionomics of some northern mosquito species (Culicidae: Diptera). Can. J. Res. 28:58-80.

1786. Hodgden, B. B. 1949a. A monograph of the Saldidae of North and Central America and the West Indies. Ph.D. diss., University of Kansas, Lawrence. 511 pp.

1787. Hodgden, B. B. 1949b. New Saldidae from the western hemisphere. J. Kans. Ent. Soc. 22:149-165.

1788. Hodges, R. W. 1962. A revision of the Cosmopterygidae of America north of Mexico, with a definition of the Momphidae and Walshiidae (Lepidoptera: Gelechioidea). Entomologica 42:1-171.

1789. Hodgson, C. E. 1940. Collection and laboratory maintenance of Dytiscidae (Coleoptera). Ent. News 64:36-37.

1790. Hodkinson, I. D. 1973. The immature stages of *Ptychoptera lenis lenis* (Diptera: Ptychopteridae) with notes on their biology. Can. Ent. 105:1091-1099.

1791. Hodkinson, I. D., and K. A. Williams. 1980. Tube formation and distribution of *Chironomus plumosus* L. (Diptera: Chironomidae) in a eutrophic woodland pond, pp. 331-337. *In* D. A. Murray, (ed.). Chironomidae: ecology, systematics, cytology and physiology. Pergamon, N.Y.

1792. Hoffman, C. H. 1932a. Hymenopterous parasites from the eggs of aquatic and semi-aquatic insects. J. Kans. Ent. Soc. 5:33-37.

1793. Hoffman, C. H. 1932b. The biology of three North American species of *Mesovelia*. Can. Ent. 64:88 94, 113-120, 126-133.

1794. Hoffman, C. H. 1937. How to rear *Mesovelia*, pp. 305-306. *In* J. G. Needham (ed.). Culture methods for invertebrate animals. Comstock, Ithaca. 590 pp.

1795. Hoffman, C. H. 1940a. Limnological relationships of some northern Michigan Donaciini (Chrysomelidae: Coleoptera). Trans. Am. Microsc. Soc. 59:259-274.

1796. Hoffman, C. H. 1940b. The relation of *Donacia* larvae (Chrysommelidae, Coleoptera) to dissolved oxygen. Ecology 20:176-183.

1797. Hoffmann, W. E. 1924. The life history of three species of gerrids (Heteroptera: Gerridae). Ann. Ent. Soc. Am. 17:419-430.

1798. Hoffmann, W. E. 1925. The life history of *Velia watsoni* Drake (Heteroptera, Veliidae). Can. Ent. 57:107-112.

1799. Hoffrichter, O., and F. Reiss. 1981. Supplement 1 to "A bibliography of the Chironomidae." Gunneria 37:1-68.

1800. Hofsvang, T. 1972. *Tipula excisa* Schum. (Diptera, Tipulidae), life cycle and population dynamics. Nord. Ent. Tidsskr. 19:43-48.

1801. Hogue, C. L. 1973a. A taxonomic review of the genus *Maruina* (Diptera: Psychodidae). Los Angeles Co. Nat. Hist. Mus. Sci. Bull. 17:1-69.

1802. Hogue, C. L. 1973b. The net-winged midges or Blephariceridae of California. Bull. Calif. Insect Surv. 15:1-83.

1803. Hogue, C. L. 1981. Chap. 8. Blephariceridae, pp. 191-198. *In* J. F. McAlpine, B. V. Peterson, G. E. Shewell, H. J. Teskey, J. R. Vockeroth, and D. M. Wood (coords.). Manual of Nearctic Diptera, Vol. 1. Res. Branch, Agric. Can. Monogr. 27. Ottawa. 674 pp.

1804. Hogue, C. L. 1987. Blephariceridae. In G.C.D. Griffiths (ed.). Flies of the Nearctic region, 2:1-172.

1805. Hogue, C. L., and T. Georgian. 1986. Recent discoveries in the *Blepharicera tenuipes* group, including descriptions of two new species from Appalachia (Diptera: Blephariceridae). Contr. Sci., Nat. Hist. Mus. Los Angeles Co. 377:1-20.

1806. Hokari, N. 1992. Comparison of the habitats and feeding habits of two species of *Epeorus* (Ephemeroptera: Insecta). Biol. Inland Waters 7:10-19.
1807. Holdsworth, R. 1941a. The life history and growth of *Pteronarcys proteus* Newman. Ann. Ent. Soc. Am. 34:495-502.
1808. Holdsworth, R. 1941b. Additional information and a correction concerning the growth of *Pteronarcys proteus* Newman. Ann. Ent. Soc. Am. 34:714-715.
1809. Holomuzki, J. R. 1985. Life history aspects of the predaceous diving beetle, *Dytiscus dauricus* (Gebler), in Arizona. Southwest. Nat. 30:485-490.
1810. Holomuzki, J. R., and S. H. Messier. 1993. Habitat selection by the stream mayfly *Paraleptophlebia guttata*. J. N. Am. Benthol. Soc. 12:126-135.
1811. Holopainen, I. J., and J. Sarvala. 1975. Efficiencies of two corers in sampling soft-bottom invertebrates. Ann. Zool. Fenn. 12:280-284.
1812. Holzenthal, R. W. 1982. The caddisfly genus *Setodes* in North America (Trichoptera: Leptoceridae). J. Kans. Ent. Soc. 55:253-271.
1813. Holzenthal, R. W., S. C. Harris, and P. K. Lago. 1982. An annotated checklist of the caddisflies (Trichoptera) of Mississippi and southeastern Louisiana. Part III: Limnephiloidea and conclusions. Proc. Ent. Soc. Wash. 84:513-520.
1814. Hopkins. P. S., K. W. Kratz, and S. D. Cooper. 1989. Effects of an experimental acid pulse on invertebrates in a high altitude Sierra Nevada stream. Hydrobiologia 171:45-58.
1815. Hora, S. L. 1930. Ecology, bionomics and evolution of the torrential fauna, with special reference to the organs of attachment. Phil. Trans. R. Soc. (B) 218:171-282.
1816. Horn, G. H. 1873. Revision of the genera and species of the tribe Hydrobiini. Proc. Am. Phil. Soc. 13:118-137.
1817. Horsfall, W. R. 1955. Mosquitoes: Their behavior and relation to disease. Ronald, N.Y. 723 pp.
1818. Horst, T. J. 1976. Population dynamics of the burrowing mayfly *Hexagenia limbata*. Ecology 57:199-204.
1819. Horst, T. J., and G. R. Marzolf. 1975. Production ecology of burrowing mayflies in a Kansas reservoir. Verh. Int. Verein. Limnol. 19:3029-3038.
1820. Hosseinig, S. O. 1966. Studies on the biology and life histories of aquatic beetles of the genus *Tropisternus* (Coleoptera: Hydrophilidae). Diss. Abstr. 261:4903.
1821. Houlihan, D. F. 1969a. The structure and behavior of *Notiphila riparia* and *Erioptera squalida* (Dipt.). J. Zool. Lond. 159:249-267.
1822. Houlihan, D. F. 1969b. Respiratory physiology of the larva of *Donacia simplex*, a root-piercing beetle. J. Insect Physiol. 15:1517-1536.
1823. Houlihan, D. F. 1970. Respiration in low oxygen partial pressure: The adults of *Donacia simplex* that respire from the roots of aquatic plants. J. Insect Physiol. 16:1607-1622.
1824. Howard, F. O. 1974. Natural history and ecology of *Pycnopsyche lepida, P. guttifer* and *P. scabripennis* (Trichoptera: Limnephilidae) in a woodland stream. Ph.D. diss., Michigan State University, East Lansing. 115 pp.
1825. Howarth, F. G., and D. A. Polhemus. 1991. A review of the Hawaiian stream insect fauna, pp. 40-50. *In* W. Devick (ed.). New directions in research management and conservation of Hawaiian freshwater stream ecosystems. Dept. Land Nat. Res. Honolulu, HA. 318pp.
1826. Howe, R. H. 1917-1923. Manual of the Odonata of New England. Mem. Thoreau Mus. Nat. Hist. (Parts 1-6). Vol. II:1-149.
1827. Howland, L. J. 1930. The nutrition of mosquito larvae, with special reference to their algal food. Bull. Ent. Res. 21:431-439.
1828. Howmiller, R. P. 1971. A comparison of the effectiveness of Ekman and Ponar grabs. Trans. Am. Fish. Soc. 100:560-564.
1829. Hrbacek, J. 1950. On the morphology and function of the antennae of the central European Hydrophilidae (Coleoptera). Trans. R. ent. Soc. Lond. 101:239-256.
1830. Hrbacek, J. (ed.) 1962. Hydrobiologicke Metody. Praha. 130 pp.
1831. Hribar, L. J. 1990. A review of methods for recovering biting midge larvae (Diptera: Ceratopogonidae) from substrate samples. J. Agric. Ent. 5:71-77.
1832. Hribar, L. J. 1993. Mouthpart morphology and feeding behavior of biting midge larvae (Diptera: Ceratopogonidae), pp. 43-58. *In* C. W. Schaefer and R. A. B. Leschen (eds.). Functional morphology of insect feeding. Thomas Say Publ. Ent: Proc. Ent. Soc. Am., Lanham, MD. 162 pp.
1833. Hribar, L. J., and C. S. Murphree. 1987. *Heleidomermis* sp. (Nematoda: Mermithidae) infecting *Culicoides variipennis* (Diptera: Certatopogonidae) in Alabama. J. Am. Mosq. Contr. Assoc. 3:332.
1834. Hribar, L. J., and G. R. Mullen. 1991a. Comparative morphology of mouthparts of biting midge (Diptera: Ceratopogonidae) larvae. Contrib. Am. Ent. Inst. 26:1-71.
1835. Hribar, L. J., and G. R. Mullen. 1991b. Predation by *Bezzia* larva (Diptera: Culicidae). Ent. News 102:183-186.
1836. Hribar, L. J., and G. R. Mullen. 1991c. Alimentary tract contents of some biting midge larvae (Diptera: Ceratopogonidae). J. Ent. Sci. 26:430-435.
1837. Hubbard, M. D. 1994. The mayfly family Behningiidae (Ephemeroptera: Ephemeroidea): Keys to the recent species with a catalog of the family. Great Lakes Ent. 27:161-168.
1838. Huckett, H. C., and J. R. Vockeroth. 1987. Chap. 105. Muscidae, pp. 1115-1131. *In* J. F. McAlpine, B. V. Peterson, G. E. Shewell, H. J. Teskey, J. R. Vockeroth, and D. M. Wood (coords.). Manual Nearctic Diptera, Vol. 2. Res. Branch, Agr. Can. Mon. 28. 1332p.
1839. Hudson, P. L., D. R. Lenat, B. A. Caldwell, and D. Smith. 1990. Chironomidae of the Southeastern United States: A checklist of species and notes on biology distribution, and habitat. Fish and Wildlife Res. Rept. 7. USDI, Fish and Wildlife Service, Wash. D.C.
1840. Hudson, P. L., and J. L. Oliver. 1983. Portable suction sampler for quantitatively sampling macroinvertebrates and periphyton on bedrock. Progr. Fish Cult. 45:123-124
1841. Hudson, P. L., J. C. Morse, and J. R. Voshell. 1981. Larva and pupa of *Cernotina spicata*. Ann. Ent. Soc. Am. 74:516-519.
1842. Huffman, I. E. W. 1955. The biology of an aquatic pyralid, *Cataclysta fulicalis* Clem. (Lepidoptera), and its tachinid parasitoid, *Ginglymia acricalis* Towns. (Diptera). Ph. D. diss., Ohio State Univ., Columbus, OH. 55p.
1843. Huggins, D. G. 1978a. Description of the nymph of *Enalagma divagans* Selys (Odonata: Coenagrionidae). J. Kans. Ent. Soc. 51:140-143.
1844. Huggins, D. G. 1978b. Redescription of the nymph of *Enalagma basidens* Calvert (Odonata: Coenagrionidae). J. Kans. Ent. Soc. 51:222-227.
1845. Huggins, D. G. 1980. The spongillaflies (Neuroptera: Sisyridae) of Kansas. Tech. Publ. State Biol. Surv. Kans. 9:67-70.
1846. Huggins, D. G. 1983. New Kansas records of Odonata. Tech. Publ. Biol. Surv. Kansas 13:24-25.
1847. Huggins, D. G., and W. U. Brigham. 1982. Chap. 4, Odonata, pp. 4.1-4.100. *In* A. R. Brigham, W. U. Brigham, and A. Gnilka (eds.). Aquatic insects and oligochaetes of North and South Carolina. Midwest Aquatic Enterprises, Mahomet, Ill. 837 pp.
1848. Huggins, D. G., P. Liechti, and D. W. Roubik. 1976. New records of the fauna and flora for 1975. Tech. Publ. Kans. State Biol. Surv. 1:13-44.
1849. Hughes, B. D. 1975. A comparison of four samplers for benthic macroinvertebrates inhabiting coarse river deposits. Wat. Res. 9:61-69.
1850. Hughes, R. M., D. P. Larsen, and J. Omernik. 1986. Regional reference sites: a method for assessing stream potentials. Environ. Manage. 10:629-635.
1851. Hummel, S., and A. C. Haman. 1975. Notes on the Odonata of Black Hawk County, Iowa. Ent. News 86:63-64.
1852. Humpesch, U. H. 1979. Life cycles and growth rates of *Baetis* spp. (Ephemeroptera: Baetidae) in the laboratory and in two stony streams in Austria. Freshwat. Biol. 9:467-479.
1853. Humphrey, C. L., and P. L. Dostine. 1994. Development of biological monitoring programs to detect mining-waste impacts upon aquatic ecosystems of the Alligator Rivers Region, Northern Territory, Australia. Mitt. Int. Verein. Limnol. 24:293-314.
1854. Hungerford, H. B. 1917a. Food habits of corixids. J. N.Y. Ent. Soc. 25:1-5.

1855. Hungerford, H. B. 1917b. Life history of a boatman. J. N.Y. Ent. Soc. 25:112-122.
1856. Hungerford, H. B. 1917c. The life history of *Mesovelia mulsanti* White. Psyche 24:73-84.
1857. Hungerford, H. B. 1920. The biology and ecology of aquatic and semi-aquatic Hemiptera. Univ. Kans. Sci. Bull. 21:1-341.
1858. Hungerford, H. B. 1922a. Oxyhaemoglobin present in backswimmer, *Buenoa margaritacea*. Can. Ent. 54:262-263.
1859. Hungerford, H. B. 1922b. The life history of the toad bug *Gelastocoris oculatus* Fabr. Univ. Kans. Sci. Bull. 14:145-171.
1860. Hungerford, H. B. 1922c. The Nepidae of North America north of Mexico. Univ. Kans. Sci. Bull. 14:423-469.
1861. Hungerford, H. B. 1924. A new *Mesovelia* with some biological notes regarding it, *Mesovelia douglasensis*. Can. Ent. 56:142-144.
1862. Hungerford, H. B. 1927. The life history of the creeping water bug *Pelocoris carolinensis* Bueno (Naucoridae). Univ. Kans. Sci. Bull. 22:77-82.
1863. Hungerford, H. B. 1933. The genus *Notonecta* of the world. Univ. Kans. Sci. Bull. 21:5-195.
1864. Hungerford, H. B. 1948. The Corixidae of the Western Hemisphere (Hemiptera). Univ. Kans. Sci. Bull. 32:1-827.
1865. Hungerford, H. B. 1954. The genus *Rheumatobates* Bergroth. Univ. Kans. Sci. Bull. 36:529-588.
1866. Hungerford, H. B. 1958. Some interesting aspects of the World distribution and classification of aquatic and semiaquatic Hemiptera. Proc. 10th Int. Congr. Ent. 1:337-348.
1867. Hungerford, H. B. 1959. Hemiptera, pp. 958-972. *In* W. T. Edmondson (ed.). Freshwater Biology (2nd ed.). John Wiley & Sons, N.Y. 1248 pp.
1868. Hungerford, H. B., and N. E. Evans. 1934. The Hydrometridae of the Hungarian National Museum and other studies in the family. Ann. Mus. Nat. Hungar. 28:31-112.
1869. Hungerford, H. B., and R. Matsuda. 1960. Keys to the subfamilies, tribes, genera and subgenera of the Gerridae of the world. Univ. Kans. Sci. Bull. 41:3-23.
1870. Hungerford, H. B., P. J. Spangler, and N. A. Walker. 1955. Subaquatic light traps for insects and other animal organisms. Trans. Kans. Acad. Sci. 58:387-407.
1871. Hunt, B. P. 1953. The life history and economic importance of a burrowing mayfly, *Hexagenia limbata*, in southern Michigan lakes. Bull. Inst. Fish. Res. Ann Arbor, Mich. 4:1-151.
1872. Hunter, F. F., and A. K. Maier. 1994. Feeding behavior of predatory larvae of *Atherix lantha* (Diptera: Athericidae). Can. J. Zool. 72:1695-1699.
1873. Huryn, A. D. 1990. Growth and voltinism of lotic midge larvae: patterns across an Appalachian Mountain basin. Limnol. Oceanogr. 35:339-351.
1874. Huryn, A. D., and B. A. Foote. 1983. An annotated list of the caddisflies (Trichoptera) of Ohio. Proc. Ent. Soc. Wash. 85:783-796.
1875. Huryn, A. D., and J. B. Wallace. 1984. New eastern Nearctic limnephilid (Trichoptera) with unusual zoogeographical affinities. Ann. Ent. Soc. Am. 77:284-292.
1876. Huryn, A. D., and J. B. Wallace. 1985. Life history and production of *Goerita semata* Ross (Trichoptera: Limnephilidae) in the southern Appalachian Mountains. Can. J. Zool. 63:2604-2611.
1877. Huryn, A. D., and J. B. Wallace. 1987. The exopterygote insect community of a mountain stream in North Carolina, U.S.A.: life histories, production, and functional structure. Aquat. Insects 9:229-251.
1878. Huryn, A. D., and J. B. Wallace. 1988. Community structure of Trichoptera in a mountain stream: spatial patterns of production and functional organization. Freshwat. Biol. 20:141-155.
1879. Husbands, R. C. 1967. A subsurface light trap for sampling aquatic insect populations. Calif. Vector Views 14:81-82.
1880. Hussey, R. F., and J. L. Herring. 1949. Notes on the variation of the *Metrobates* of Florida (Hemiptera, Gerridae). Fla. Ent. 32:166-170.
1881. Hussey, R. F., and J. L. Herring. 1950. A remarkable new belostomatid from Florida and Georgia. Fla. Ent. 33:84-89.
1882. Hutchinson, G. E. 1931. On the occurrence of *Trichocorixa* Kirkaldy in salt water and its zoogeographical significance. Am. Nat. 65:573-574.
1883. Hutchinson, G. E. 1945. On the species of *Notonecta* (Hemiptera-Heteroptera) inhabiting New England. Trans. Conn. Acad. Arts Sci. 36:599-605.
1884. Hutchinson, G. E. 1957. A treatise on limnology. Vol. I. John Wiley & Sons, Inc., N.Y. 1015 pp.
1885. Hutchinson, G. E. 1981. Thoughts on aquatic insects. BioScience 31:495-500.
1886. Hutchinson, G. E. 1993. A treatise on limnology. Vol. IV. J. Wiley and Sons, Inc. N.Y. 944pp.
1887. Hutchinson, R., and A. Larochelle. 1977. Catalogue des Libellules du Quebec. Cordulia (Suppl.) 3:1-45.
1888. Hyland, K., Jr. 1948. New records of Pennsylvania caddis flies (Trichoptera). Ent. News 59:38-40.
1889. Hynes, C. D. 1969a. The immature stages of *Conomyodes tacoma* Alex. Pan-Pacif. Ent. 45:116-119.
1890. Hynes, C. D. 1969b. The immature stages of the genus *Rhabdomastix* (Diptera: Tipulidae). Pan-Pacif. Ent. 45:229-237.
1891. Hynes, H.B.N. 1941. The taxonomy and ecology of the nymphs of British Plecoptera with notes on the adults and eggs. Trans. R. ent. Soc. Lond. 91:459-557.
1892. Hynes, H.B.N. 1948. Notes on the aquatic Hemiptera-Heteroptera of Trinidad and Tobago, B.W.I., with a description of a new species of *Martarega* B. White (Notonectidae). Trans. R. ent. Soc. Lond. 99:341-360.
1893. Hynes, H.B.N. 1961. The invertebrate fauna of a Welsh mountain stream. Arch. Hydrobiol. 57:344-388.
1894. Hynes, H.B.N. 1963. Imported organic matter and secondary productivity of streams. Int. Congr. Zool. 4:324-329.
1895. Hynes, H.B.N. 1970a. The ecology of running waters. Univ. Toronto Press, Toronto. 555 pp.
1896. Hynes, H.B.N. 1970b. The ecology of stream insects. Ann. Rev. Ent. 15:25-42.
1897. Hynes, H.B.N. 1971. Benthos of flowing water, pp. 66-80. *In* W. T. Edmondson and G. G. Winberg (eds.). A manual on methods for the assessment of secondary productivity in freshwaters. IBP Handbook 17, Blackwell, Oxford. 358 pp.
1898. Hynes, H.B.N. 1974. Further studies on the distribution of animals within the substratum. Limnol. Oceanogr. 19:92-99.
1899. Hynes, H.B.N. 1976. The biology of Plecoptera. Ann. Rev. Ent. 21:135-153.
1900. Hynes, H.B.N. 1984. The relationships between the taxonomy and ecology of aquatic insects, pp. 9-23. *In* V. H. Resh and D. M. Rosenberg (eds.). The ecology of aquatic insects. Praeger, New York, NY. 625p.
1901. Hynes, H.B.N. 1988. Biogeography and origins of the North American stoneflies (Plecoptera). Mem. Ent. Soc. Can. 144:31-37.
1902. Hynes, H.B.N., and M. E. Hynes. 1975. The life histories of many of the stoneflies (Plecoptera) of southeastern mainland Australia. Australian J. Mar. Freshwat. Res. 26:113-153.
1903. Hynes, H.B.N., and N. K. Kaushik. 1968. Experimental study of the role of autumn shed leaves in aquatic environments. J. Ecol. 56:229-243.
1904. Ide, F. P. 1930. Contribution to the biology of Ontario mayflies with descriptions of new species. Can. Ent. 62:204-213, 62:218-231.
1905. Ide, F. P. 1935. Life history notes on *Ephoron*, *Potamanthus*, *Lepthophlebia* and *Blasturus* with descriptions (Ephemeroptera). Can. Ent. 67:113-125.
1906. Ide, F. P. 1940. Quantitative determination of the insect fauna of rapid water. Publ. Ontario Fish. Res. Lab. 47:1-24.
1907. Ide, F. P. 1965. A fly of the archaic family Nymphomyiidae (Diptera) from North America. Can. Ent. 97:496-507.
1908. Illies, J. 1965. Phylogeny and zoogeography of the Plecoptera. Ann. Rev. Ent. 10:117-141.
1909. Illies, J. 1966. Katalog der rezenten Plecoptera. Das Tierreich, 82. Walter de Gruyter, Berlin. 623 pp.
1910. Illies, J. 1969. Retardierte Schlupfzeit von *Baetis*-Gelegen (Ins., Ephem.). Naturwissenschaften 46:119-120.
1911. Imms, A. D. 1906. *Anurida*. Marine Biol. Mem. Liverpool 13:1-99.
1912. Imms, A. D. 1948. A general textbook of entomology (7th ed.).Dutton, New York. 727 pp.

1913. Ingram, B. R. 1976. Life histories of three species of Lestidae in North Carolina, United States (Zygoptera). Odonatologica 5:231-244.
1914. Ingram, B. R., and C. E. Jenner. 1976. Life histories of *Enallagma hageni* (Walsh) and *E. aspersum* (Hagen) (Zygoptera: Coenagrionidae). Odonatologica 5:331-345.
1915. Irons, J. G., L. K. Miller, and M. W. Oswood. 1992. Ecological adaptations of aquatic macroinvertebrates to overwintering in interior Alaska (U.S.A.) subarctic streams. Can. J. Zool. 71:98-108.
1916. Isely, D., and H. H. Schwardt. 1934. The rice water weevil, *Lissoroptrus simplex* (Say) (Curculionidae). Ark. Agr. Exp. Sta. Bull. 299. 44p.
1917. Istock, C. A. 1966. Distribution, coexistence, and competition of whirligig beetles. Evolution 20:211-239.
1918. Istock, C. 1967. Transient competitive displacement in natural populations of whirligig beetles. Ecology 48:929-937.
1919. Istock, C. A. 1972. Population characteristics of a species ensemble of waterboatmen (Corixidae). Ecology 54:535-544.
1920. Istock, C. A., S. E. Wasserman, and H. Zimmer. 1975. Ecology and evolution of the pitcher-plant mosquito: I. Population dynamics and laboratory responses to food and population density. Evolution 29:296-312.
1921. Ivanova, S. S. 1958. Nutrition of some mayfly larvae. Proc. Mikoyan Moscow Tech. Inst. Fish. Indust. 9:102-109.
1922. Iversen, T. M. 1973. Life cycle and growth of *Sericostoma personatum* Spence (Trichoptera: Sericostomatidae) in a Danish spring. Ent. Scand. 41:323-327.
1923. Iversen, T. M. 1974. Ingestion and growth in *Sericostoma personatum* (Trichoptera) in relation to the nitrogen content of ingested leaves. Oikos 25:278-282.
1924. Iversen, T. M. 1980. Densities and energetics of two stream living larval populations of *Sericostoma personatum* (Trichoptera). Holarct. Ecol. 3:65-73.
1925. Izvekova, E. I. 1971. On the feeding habits of chironomid larvae. Limnologica 8:201-202.
1926. Izvekova, E. I., and A. A. Lvova-katchanova. 1972. Sedimentation of suspended matter by *Dreissena polymorpha* Pallas and its subsequent utilization by chironomid larvae. Pol. Arch. Hydrobiol. 19:203-210.
1927. Jaag, O., and H. Ambuhl. 1964. The effect of the current on the composition of biocoenoses in flowing water streams, pp. 31-44. In B. A. Southgate (ed.). Advances in water pollution research; proceedings of the international conference. London, 1962. Pergamon, Oxford.
1928. Jackman, R., S. Nowicki, S. Aneshansley, and T. Eisner. 1983. Predatory capture of toads by fly larvae. Science 222:515-516.
1929. Jackson, D. J. 1956a. Dimorphism of the metasternal wings in *Agabus raffrayi* Sharp and *A. labiatus* Brahn. (Coleoptera, Dytiscidae) and its relation to capacity of flight. Proc. R. ent. Soc. Lond. (A) 131:1-11.
1930. Jackson, D. J. 1956b. Notes on hymenopterous parasitoids bred from eggs of Dytiscidae in Fife. J. Soc. Brit. Ent. 5:144-149.
1931. Jackson, D. J. 1958a. A further note on a *Chrysocharis* (Hym.: Eulophidae) parasitizing the eggs of *Dytiscus marginalis* L., and comparison of its larva with that of *Caraphractus cinctus* Walker (Hym.: Mymaridae). J. Soc. Brit. Ent. 6:15-22.
1932. Jackson, D. J. 1958b. Egg-laying and egg-hatching in *Agabus bipustulatus* L., with notes on oviposition in other species of *Agabus* (Coleoptera: Dytiscidae). Trans. Ent. Soc. Lond. I 10:53-80.
1933. Jackson, D. J. 1958c. Observations on the biology of *Caraphractus cinctus* Walker (Hym.: Mymaridae), a parasitoid of the eggs of Dytiscidae. I—Methods of rearing and numbers bred on different host eggs. Trans. R. ent. Soc. Lond. I 10:533-554.
1934. Jackson, D. J. 1961a. Diapause in an aquatic mymarid. Nature 192:823-824.
1935. Jackson, D. J. 1961b. Observations on the biology of *Caraphractus cinctus* Walker (Hym.: Mymaridae), a parasitoid of the eggs of Dytiscidae (Coleoptera). 2. Immature stages and seasonal history with a review of mymarid larvae. Parasitology 51:269-294.
1936. Jackson, D. J. 1964. Observations on the life-history of *Mestocharis bimacularis* (Dalman) (Hym.: Eulophidae), a parasitoid of eggs of Dytiscidae. Opusc. Ent. 29:81-97.
1937. Jackson, D. J. 1966. Observations on the biology of *Caraphractus cinctus* Walker (Hym. Mymaridae), a parasitoid of the eggs of Dytiscidae. III. The adult life and sex ratio. Trans. R. ent. Soc. Lond. 118:2349.
1938. Jackson, G. A. 1977. Nearctic and Palaearctic Paracladopelma Harnisch and Saetheria n. gen. (Diptera: Chironomidae). J. Fish. Res. Bd. Can. 34:1321-1359.
1939. Jackson, J. K., and V. H. Resh. 1989. Distribution and abundance of adult aquatic insects in the forest adjacent to a northern California stream. Ann. Ent. Soc. Am. 18:278-283.
1940. Jackson, J. K., and V. H. Resh. 1991. Periodicity in mate attraction and flight activity of three species of Trichoptera. J. N. Am. Benthol. Soc. 10:198-209.
1941. Jackson, J. K., and V. H. Resh. 1992. Variation in genetic structure among populations of the caddisfly *Helicopsyche borealis* from three streams in northern California, U.S.A. Freshwat. Biol. 27:29-42.
1942. Jacobi, D. I., and A. C. Benke. 1991. Life histories and abundance patterns of snag-dwelling mayflies in a blackwater Coastal Plain river. J. N. Am. Benthol. Soc. 10:372-387.
1943. Jacobi, G. Z. 1971. A quantitative artificial substrate sampler for benthic macroinvertebrates. Trans. Am. Fish. Soc. 100:136-138.
1944. Jacobi, G. Z. 1978. An inexpensive circular sampler for collecting benthic macroinvertebrates in streams. Arch. Hydrobiol. 83:126-131.
1945. Jacobsen, D., and K. Sand-Jensen. 1994. Growth and energetics of a trichopteran larva feeding on fresh submerged and terrestrial plants. Oecologia 97:412-418.
1946. Jacoby, J. M. 1987. Alterations in periphyton characteristics due to grazing in a Cascade foothill stream. Freshwat. Biol. 18:495-508.
1947. Jaczewski, T. 1930. Notes on the American species of the genus *Mesovelia* Muls. Ann. Mus. Zool. Polon. 9:1-12.
1948. Jaczewski, T., and A. S. Kostrowicki. 1969. Number of species of aquatic and semi-aquatic Heteroptera in the fauna of various parts of the Holarctic in relation to the world fauna. Mem. Soc. Ent. Ital. 48:153-156.
1949. Jagg, O., and H. Ambühl. 1964. The effect of the current on the composition of biocoenoses in flowing water streams. Internat. Conf. Water Pollut. Res., London, Sept. 1962. pp. 31-49.
1950. James, H. G. 1933. Collembola of the Toronto region, with notes on the biology of *Isotoma palustris* Muller. Trans. Can. Inst. 19:77-116.
1951. James, H. G. 1957. *Mochlonyx velutinus* (Ruthe) (Diptera: Culicidae), an occasional predator of mosquito larvae. Can. Ent. 89:470-480.
1952. James, H. G. 1961. Some predators of *Aedes stimulans* (Walk.) and *Aedes trichurus* (Dyar) in woodland pools. Can. J. Zool. 39:533-540.
1953. James, H. G. 1964a. Insect and other fauna associated with the rock pool mosquito *Aedes atropalpus* (Coq.). Mosquito News 23:325-329.
1954. James, H. G. 1964b. The role of Coleoptera in the natural control of mosquitoes in Canada. Proc. 12th Int. Congr. Ent. 12:357-358.
1955. James, H. G. 1969. Immature stages of five diving beetles (Coleoptera: Dytiscidae), notes on their habits and life history, and a key to aquatic beetles of vernal woodland pools in southern Ontario. Proc. Ent. Soc. Ont. 100:52-97.
1956. James, H. G., G. Wishart, R. E. Bellamy, M. Maw, and P. Belton. 1969. An annotated list of mosquitoes of southeastern Ontario. Proc. Ent. Soc. Ont. 100:200-230.
1957. James, M. T. 1959. Diptera, pp. 1057-1079. In W. T. Edmondson (ed.). Freshwater biology (2nd ed.). John Wiley & Sons, N.Y. 1248 pp.
1958. James, M. T. 1981. Chap. 36. Stratiomyidae, pp. 497-512. In J. F. McAlpine, B. V. Peterson, G. E. Shewell, H. J. Teskey, J. R. Vockeroth, and D. M. Wood (coords.). Manual of Nearctic Diptera, Vol. 1. Res. Branch, Agric. Can. Monogr. 27. Ottawa. 674 pp.

1959. Jamieson, G. S., and G.G.E. Scudder. 1977. Food consumption in *Gerris* (Hemiptera). Oecologia 30:23-41.
1960. Jamieson, G. S., and G.G.E. Scudder. 1979. Predation in *Gerris* (Hemiptera): Reactive distances and locomotion rates. Oecologia 44:13-20.
1961. Jamnback, H. 1969. Bloodsucking flies and other outdoor nuisance arthropods of New York state. Mem. N.Y. State Mus. Sci. Serv. 19. 90 pp.
1962. Jankovic, M. 1974. Feeding and food assimilation in larvae of *Prodiamesa olivacea*. Ent. Tidskr. Suppl. 95:116-119.
1963. Jankovic, M. 1978. Role of plant debris in the feeding of *Prodiamesa olivacea* larvae. Acta Universitatis Carolinae, Biologica 1978:77-82.
1964. Jansson, A. 1976. Audiospectrographic analysis of stridulatory signals of some North American Corixidae (Hemiptera). Ann. Zool. Fenn. 13:48-62.
1965. Jansson, A. 1977. Micronectae (Heteroptera, Corixidae) as indicators of water quality in two lakes in southern Finland. Ann. Zool. Fenn. 14:118-124.
1966. Jansson, A. 1979. A new species of *Callicorixa* from Northwestern North America. Pan-Pacif. Ent. 54:261-266.
1967. Jansson, A. 1981. Generic name *Ahuautlea* De La Llave, 1832 (Insecta, Heteroptera, Corixidae): Proposed suppression under the plenary powers Z. N. (S.) 2299. Bull. Zool. Nom. 38:197-200.
1968. Jansson, A., and G.G.E. Scudder. 1972. Corixidae (Hemiptera) as predators: rearing on frozen brine shrimp. J. Ent. Soc. Brit. Columbia 69:44-45.
1969. Jansson, A., and G.G.E. Scudder. 1974. The life cycle and sexual development of *Cenocorixa* species (Hemiptera: Corixidae) in the Pacific Northwest. Freshwat. Biol. 4:73-92.
1970. Jansson, A., and J. T. Polhemus. 1987. Revision of the genus *Morphocorixa* Jaczewski (Heteroptera, Corixidae). Ann. Ent. Fenn. 53:105-118.
1971. Jansson, A., and T. Vuoristo. 1979. Significance of stridulation in larval Hydropsychidae (Trichoptera). Behaviour 71:167-186.
1972. Jeffrey, R. W. 1877. *Hydrocampa stagnalis* Bred. Entomol. mon. Mag. 14:116.
1973. Jeffries, M. 1988. Individual vulnerability to predation: the effect of alternative prey types. Freshwat. Biol. 19:49-56.
1974. Jenkins, D. W. 1964. Pathogens, parasites and predators of medically important arthropods, annotated list and bibliography. Bull. World Health Org. Suppl. 30:1-150.
1975. Jenkins, M. F. 1960. On the method by which *Stenus* and *Dianous* (Coleoptera: Staphylinidae) return to the banks of a pool. Trans. R. ent. Soc. Lond. 112:1-14.
1976. Jewett, S. G., Jr. 1956. Plecoptera, pp. 155-181. *In* R. L. Usinger (ed.). Aquatic insects of California. Univ. Calif. Press, Berkeley. 508 pp.
1977. Jewett, S. G., Jr. 1959. The stoneflies of the Pacific Northwest. Ore. State Monogr. Stud. Ent. 3:1-95.
1978. Jewett, S. G., Jr. 1960. The stoneflies (Plecoptera) of California. Bull. Calif. Insect Surv. 6:122-177.
1979. Jewett, S. G., Jr. 1963. A stonefly aquatic in the adult stage. Science 139:484-485.
1980. Jewett, S. G., Jr. 1971. Some Alaskan stoneflies (Plecoptera). Pan-Pacif. Ent. 47:189-192.
1981. Johannsen, O. A. 1903. Part 6. Aquatic nematocerous Diptera, pp. 328-441. *In* J. G. Needham, A. D. MacGillivray, O. A. Johannsen, and K. C. Davis (eds.). Aquat. Insects of New York State. N.York State Mus. Bull. 68 (Ent. 18). 197-517. (NY. State Univ. Bull. 295).
1982. Johannsen, O. A. 1905. Aquatic nematocerous Diptera. II-Chironomidae, pp. 76-331. *In* J. Needham, K. Morton, and O. Johannsen (eds.). Mayflies and midges of New York. Bull. New York State Mus. 86:1-352.
1983. Johannsen, O. A. 1922. Stratiomyiid larvae and puparia of the northeastern states. J. N.Y. Ent. Soc. 30:141-153.
1984. Johannsen, O. A. 1934, 1935. Aquatic Diptera. Part I. Nematocera, exclusive of Chironomidae and Ceratopogonidae. Part II. Orthorrhapha-Brachycera and Cyclorrhapha. Mem. Cornell Univ. Agric. Exp. Sta. 164:1-71; 171:1-62.
1985. Johannsen, O. A. 1937. Aquatic Diptera. III. Chironomidae: subfamilies Tanypodinae, Diamesinae, and Orthocladiinae. Mem. Cornell Univ. Agric. Exp. Sta. 205:3-84.
1986. Johannsen, O. A. 1938. Aquatic Diptera. IV. Chironomidae: subfamily Chironominae. Mem. Cornell Univ. Agric. Exp. Sta. 210:3-80.
1987. Johannsen, O. A., and H. K. Townes. 1952. Tendipedidae (Chironomidae) except Tendipedini, pp. 31-47. *In* Guide to the insects of Connecticut. Part VI. The Diptera or true flies. Fasc. 5. Midges and gnats. Bull. Conn. State Geol. Nat. Hist. Surv. 80:1-254.
1988. Johannsson, O. E., and J. L. Beaver. 1983. Role of algae in the diet of *Chironomus plumosus* f. *semireductus* from the Bay of Quinte, Lake Ontario. Hydrobiologia. 107:237-247.
1989. Johansson, A., and A. N. Nilsson. 1992. *Dytiscus latissimus* and *Dytiscus circumcinctus* (Coleoptera: Dytiscidae) larvae as predators on three case-making caddis larvae. Hydrobiologia 248:201-213.
1990. Johansson, F. 1991. Foraging modes in an assemblage of odonate larvae-effects of prey and interference. Hydrobiologia 209:79-87.
1991. Johansson, F. 1992. Effects of zooplankton availability and foraging mode on cannibalism in three dragonfly larvae. Oecologia 91:179-183.
1992. Johansson, F. 1993a. Diel feeding behavior in larvae of four odonate species. J. Insect Behav. 6:253-264.
1993. Johansson, F. 1993b. Intraguild predation and cannibalism in odonate larvae: effects of foraging behaviour and zooplankton availability. Oikos 66:80-87.
1994. Johansson, F. 1994. Effects of prey type, prey density and predator presence on behavior and predation risk in a larval damselfly. Oikos 68:481-489
1995. Johnson, C. 1968. Seasonal ecology of the dragonfly *Oplonaeschna armata* Hagen (Odonata: Aeshnidae). Am. Midl. Nat. 80:449-457.
1996. Johnson, C. 1972. The damselflies (Zygoptera) of Texas. Bull. Fla. State Mus. 16:55-128.
1997. Johnson, C. 1974. Taxonomic keys and distributional patterns for Nearctic species of *Calopteryx* damselflies. Fla. Ent. 57:231-248.
1998. Johnson, C., and M. J. Westfall. 1970. Diagnostic keys and notes on the damselflies (Zygoptera) of Florida. Bull. Fla. State Mus. 15:1-89.
1999. Johnson, D. M. 1973. Predation by damselfly naiads on cladoceran populations: fluctuating intensity. Ecology 54: 251-268.
2000. Johnson, D. M. 1986. The life history of *Tetragoneuria cynosura* in Bays Mountain Lake, Tennessee United States (Anisoptera: Corduliidae). Odonatologica 15:81-90.
2001. Johnson, D. M., B. G. Akre, and P. H. Crowley. 1975. Modeling arthropod predation: wasteful killing by damselfly naiads. Ecology 56:1081-1093.
2002. Johnson, D. M., and P. H. Crowley. 1980. Habitat and seasonal segregation among coexisting odonate larvae. Odonatologica 9:297-308.
2003. Johnson, D. M., P. H. Crowley, R. E. Bohanan, C. N. Watson, and T. H. Martin. 1985. Competition among larval dragonflies: a field enclosure experiment. Ecology 66:119-128.
2004. Johnson, D. M., C. L. Pierce, T. H. Martin, C. N. Watson, R. E. Bohanon, and P. H. Crowley. 1987. Prey depletion by odonate larvae: combining evidence from multiple field experiments. Ecology 68:1459-1465.
2005. Johnson, G. H. 1972. Flight behavior of the predaceous diving beetle, *Cybister fimbriolatus* (Say) (Coleoptera:Dytiscidae). Coleopt. Bull. 26:23-24.
2006. Johnson, G. H., and W. Jakinovich, Jr. 1970. Feeding behavior of the predaceous diving beetle, *Cybister fimbriolatus* (Say). BioScience 20:1111.
2007. Johnson, J. H. 1981. Food habits and dietary overlap of perlid stoneflies (Plecoptera) in a tributary of Lake Ontario. Can. J. Zool. 59:2030-2037.
2008. Johnson, J. H. 1983. Diel food habits of two species of setipalpian stoneflies (Plecoptera) in tributaries of the Clearwater Reiver, Idaho. Freshwat. Biol. 13: 105-111.
2009. Johnson, J. H. 1985. Diel feeding ecology of the nymphs of *Aeshna multicolor* and *Lestes unguiculatus* (Odonata). Freshwat. Biol. 15:749-755.

2010. Johnson, R. K. 1985. Feeding efficiencies of *Chironomus plumosus* (L.) and *C. anthracinus* Zett. (Diptera: Chironomidae) in mesotrophic Lake Erken. Freshwat. Biol. 15:605-612.

2011. Johnson, R. K. 1987. Seasonal variation in diet of *Chironomus plumosus* (L.) and *C. anthracinus* Zett. (Diptera: Chironomidae) in mesotrophic Lake Erken. Freshwat. Biol. 17:525-532.

2012. Johnson, R. K. 1989. Classification of profundal chironomid communities in oligotrophic/ humic lakes of Sweden using environmental data. Acta Biol. Debr. Oecol. Hung. 3:167-175.

2013. Johnson, R. K., and T. Wiederholm. 1989. Classification and ordination of profundal macroinvertebrate communities in nutrient poor, oligo-mesohumic lakes in relation to environmental data. Freshwat. Biol. 21:375-386.

2014. Johnson, R. K., B. Bostrom, and W. van de Bund. 1989. Interactions between *Chironomus plumosus* (L.) and the microbial community in surficial sediments of a shallow, eutrophic lake. Limnol. Oceanogr. 34:992-1003.

2015. Johnson, R. K., T. Wiederholm, and D. M. Rosenberg. 1993. Freshwater biomonitoring using individual organisms, populations, and species assemblages of benthic macroinvertebrates, pp. 40-158. *In*: D. M. Rosenberg and V.H. Resh (eds.). Freshwater biomonitoring and benthic macroinvertebrates. Chapman and Hall, NY. 488 pp.

2016. Jonasson, P. M. 1954. An improved funnel trap for capturing emerging aquatic insects, with some preliminary results. Oikos 5:179-188.

2017. Jonasson, P. M., and J. Kristiansen. 1967. Primary and secondary production in Lake Esrom. Growth of *Chironomus anthracinus* in relation to seasonal cycles of phytoplankton and dissolved oxygen. Int. Revue ges. Hydrobiol. 52:163-217.

2018. Jones, C. M., and D. W. Anthony. 1964. The Tabanidae of Florida. U.S.D.A. Tech. Bull. 1295. 85 pp.

2019. Jones, F. M. 1916. Two insect associates of the California pitcher plant, *Darlingtonia californica* (Dipt.). Ent. News 27:385-392.

2020. Jones, F. M. 1918. *Dohrniphora venusta* Coquillett (Dipt.) in *Sarracenia flava*. Ent. News 29:299-302.

2021. Jones, J.R.E. 1950. A further ecological study of the River Rheidol: the food of the common insects of the main stream. J. Anim. Ecol. 19:159-174.

2022. Jónsson, E., A. Gardarsson, and G. Gíslason. 1986. A new window trap used in the assessment of the flight periods of Chironomidae and Simuliidae (Diptera). Freshwat. Biol. 16:711-719.

2023. Joosse, E.N.G. 1966. Some observations on the biology of *Anurida maritima* (Guerin), (Collembola). S. Morph. Okol. Tiere 57:320-328.

2024. Joosse, E.N.G. 1976. Littoral Apterygotes (Collembola and Thysanura) pp. 151-186. *In* L. Cheng (ed.) Marine Insects, North Holland Publ. Co. Amsterdam.

2025. Jop, K. M., and K. W. Stewart. 1987. Annual stonefly (Plecoptera) production in a second order Oklahoma Ozark stream. J. N. Am. Benthol. Soc. 6: 26-34.

2026. Jop, K. M., and S.W. Szczytko. 1984. Life cycle and production of *Isoperla signata* (Banks) in a central Wisconsin trout stream. Aquat. Insects 6:81-100.

2027. Joy, J. E., C. A. Allman, and B. T. Dowell. 1994. Mosquitoes of West Virginia: an update. J. Am. Mosq. Contr. Assoc. 10:115-118.

2028. Judd, W. W. 1950. *Acentropus niveus* (Pyralid.) on the north shore of Lake Erie with a consideration of its distribution in North America. Can. Ent. 82:250-252.

2029. Judd, W. W. 1953. A study of the population of insects emerging as adults from the Dundas Marsh, Hamilton, Ontario, during 1948. Am. Midl. Nat. 49:801-824.

2030. Juliano, S. A. 1981. *Trichogramma spp.* (Hymenoptera: Trichogrammatidae) as egg parasitoids of *Sepedon fuscipennis* (Diptera:Sciomyzidae) and other aquatic Diptera. Can. Ent. 113:271-279.

2031. Juliano, S. A. 1982. Influence of host age on host aceptability and suitability for a species of *Trichogramma* (Hymenoptera: Trichogrammatidae) attacking aquatic Diptera. Can. Ent. 114:713-720.

2032. Julin, A. M., and H. O. Sanders. 1978. Toxicity of the IGR, diflubenzuron, to freshwater invertebrates and fishes. Mosquito News 38:256-259.

2033. Kaiser, E. W. 1961. On the biology of *Sialis fuliginosa* Pict. and *S. nigripes* Ed. Pict. Flora Fauna, Silkeborg. 67:74-96.

2034. Kaisin, F. J., and A. S. Bosnia. 1987. Annual production of *Caenis* sp. in the E. Ramos Mexia reservoir. Physis Sec. B. Aguas Cont. Org. 45:53-63.

2035. Kaitala, A. 1987. Dynamic life-history strategy of the wasterstrider *Gerris thoracicus* as an adaptation to food and habitat variation. Oikos 48:125-131.

2036. Kajak, Z. 1963. Analysis of quantitative benthic methods. Ekol. Polska (A) 11:1-56.

2037. Kajak, Z. 1971. Benthos of standing water, pp. 25-65. *In* W. T. Edmondson and G. G. Winberg (eds.). A manual on methods for the assessment of secondary productivity in fresh waters. IBP Handbook 17. Blackwell, Oxford. 358 pp.

2038. Kajak, Z. 1980. Role of invertebrate predators (mainly *Procladius* sp.) in benthos, pp. 339-348. *In* D. Murray (ed.). Chironomidae: ecology, systematics, cytology, and physiology. Pergamon, N.Y.

2039. Kajak, Z., and R. Dusage. 1970. Production efficiency of *Procladius choreus* Mg. (Chironomidae, Diptera) and its dependence on the trophic conditions. Pol. Arch. Hydrobiol. 17:217-224.

2040. Kajak, Z., K. Kacprzak, and R. Polkowski. 1965. Chwytacz rurowy do pobierania prob mikro-i makrobentosu, orax prob o niezaburzonej strukturze mulu dia celow ekspery mentalnych. Ekol. Polska (B) 11:159-165 (English Summary).

2041. Kajak, Z., and B. Ranke-Rybica. 1970. Feeding and production efficiency of *Chaoborus flavicans* Meigen (Diptera, Culicidae) larvae in eutrophic and dystrophic lakes. Pol. Arch. Hydrobiol. 17:225-232.

2042. Kajak, Z., and J. Rybak. 1970. Food conditions for larvae of Chironomidae in various layers of bottom sediments. Bull. Pol. Acad. Sci. Biol. 18:193-196.

2043. Kajak, Z., and A. Stanczykowska. 1968. Influence of mutual relations of organisms, especially Chironomidae in natural benthic communities, on their abundance. Ann. Zool. Fenn. 5:49-56.

2044. Kajak, Z., and J. Warda. 1968. Feeding of benthic non-predatory Chironomidae in lakes. Ann. Zool. Fenn. 5:57-64.

2045. Kalmus, H. 1963. 101 simple experiments with insects. Doubleday & Co., Inc., Garden City, N.Y. 194 pp.

2046. Kaminski, R. M., and H. R. Murkin. 1981. Evaluation of two devices for sampling nektonic invertebrates. J. Wildl. Manage. 45:493-496.

2047. Kangasniemi, B. J., and D. R. Oliver. 1983. Chironomidae (Diptera) associated with *Myriophyllum spicatum* in Okanagan Valley Lakes, British Columbia. Can. Ent. 115:1545-1546.

2048. Kansanen, P. H. 1985. Assessment of pollution history from recent sediments in Lake Vanajavesi, southern Finland. II. Changes in the Chironomidae, Chaoboridae and Ceratopogonidae (Diptera) fauna. Ann. Zool. Fenn. 22:57-90.

2049. Kansanen, P. H. 1986. Information value of chironomid remains in the uppermost sediment layers of a complex lake basin. Hydrobiologia 143:159-165.

2050. Kansanen, P. H., and T. Jaakkola. 1985. Assessment of pollution history from recent sediments in Lake Vanajavesi, southern Finland. I. Selection of representative profiles, their dating and chemostratigraphy. Ann. Zool. Fenn. 22:13-55.

2051. Kapoor, N. N. 1972a. A recording device for measuring respiratory movements of aquatic insects. Proc. Ent. Soc. Ont. 102:71-78.

2052. Kapoor, N. N. 1972b. Rearing and maintenance of Plecoptera nymphs. Hydrobiologia 40:51-53.

2053. Kapoor, N. N. 1976. The effect of copper on the oxygen consumption rates of the stonefly nymph, *Pasganophora capitata* (Pictet) (Plecoptera). Zool. J. Linnean Soc. 59:209-215.

2054. Kardatzke, J. T., and W. A. Rowley. 1971. Comparison of *Culicoides* larval habitats and populations in central Iowa. Ann. Ent. Soc. Am. 64:215-218.

2055. Karlsson, M., T. Bohlin, and J. Stenson. 1976. Core sampling and flotation: two methods to reduce costs of a chironomid population study. Oikos 27:336-338.
2056. Karny, H. H. 1934. Biologie der Wasserinsekten. F. Wagner, Vienna. 311 pp.
2057. Karouna, N. K., and R. L. Fuller. 1992. Influence of four grazers on periphyton communities associated with clay tiles and leaves. Hydrobiologia 245:53-64.
2058. Karr, J. R. 1991. Biological integrity: a long-neglected aspect of water resource management. Ecol. Appl. 1:66-84.
2059. Kaufman, M. G., and R. H. King. 1987. Colonization of wood substrates by the aquatic xylophage *Xylotopus par* (Diptera: Chironomidae) and a description of its life history. Can. J. Zool. 65:2280-2286.
2060. Kawecka, B., and A. Kownacki. 1974. Food conditions of Chironomidae in the River Raba. Ent. Tidskr. Suppl. 95:120-128.
2061. Kawecka, B., A. Kownacki, and M. Kownacka. 1978. Food relations between algae and bottom-fauna communities in glacial streams. Verh. Int. Verein. Limnol. 20:1527-1530.
2062. Kehr, A. I., and J. A. Schnack. 1991. Predator prey relationship between giant water bugs (*Belostoma oxyurum*) and larval anurans (*Bufo arenarum*). Alytes (Paris) 9:61-69.
2063. Keilin, D. 1944. Respiratory systems and respiratory adaptations in larvae and pupae of Diptera. Parasitology 36:1-66.
2064. Kellen, W. R. 1954. A new bottom sampler. Limnol. Oceanogr. Soc. Am. Spec. Publ. 22:1-3.
2065. Kelley, R. 1982. The micro-caddisfly genus *Oxyethira* (Trichoptera: Hydroptilidae): Morphology, biogeography, evolution and classification. Ph.D. diss., Clemson University, Clemson, S.C. 432 pp.
2066. Kelley, R. W. 1984. Phylogeny, morphology and classification of the micro-caddisfly genus *Oxyethira* Eaton (Trichoptera: Hydroptilidae). Trans. Am. Ent. Inst. 110:435-463.
2067. Kelley, R. W., and J. C. Morse. 1982. A key to the females of the genus *Oxyethira* (Trichoptera: Hydroptilidae) from the southern United States. Proc. Ent. Soc. Wash. 84:256-269.
2068. Kellicott, D. S. 1899. The Odonata of Ohio. Spec. Pap. Ohio Acad. Sci. 2:1-114.
2069. Keltner, J., and W. P. McCafferty. 1986. Functional morphology of burrowing in the mayflies *Hexagenia limbata* and *Pentagenia vittigera*. Zool. J. Linn. Soc. London 87:139-162.
2070. Kelts, L. J. 1979. Ecology of a tidal marsh corixid, *Trichocorixa verticalis* (Insecta, Hemiptera). Hydrobiologia 64:37-57.
2071. Kennedy, C. H. 1915. Notes on the life history and ecology of the dragonflies (Odonata) of Washington and Oregon. Proc. U.S. Nat. Mus. 49:259-345.
2072. Kennedy, C. H. 1917. Notes on the life history and ecology of the dragonflies (Odonata) of central California and Nevada. Proc. U.S. Nat. Mus. 52:483-635.
2073. Kennedy, H. D. 1958. Biology and life history of a new species of mountain midge, *Deuterophlebia nielsoni*, from eastern California (Diptera: Deuterophlebiidae). Trans. Am. Microsc. Soc. 89:201-228.
2074. Kennedy, H. D. 1960. *Deuterophlebia inyoensis*, a new species of mountain midge from the alpine zone of the Sierra Nevada Range, California (Diptera: Deuterophlebiidae). Trans. Am. Microsc. Soc. 79:191-210.
2075. Kennedy, H. D. 1981. Chap. 9. Druterophlebiidae, pp. 199 202. In J. F. McAlpine, B. V. Peterson, G. E. Shewell, H. J. Teskey, J. R. Vockeroth, and D. M. Wood (coords.). Manual of Nearctic Diptera, Vol. 1. Res. Branch Agric. Can. Monogr. 27. Ottawa. 674 pp.
2076. Kennedy, J. H., and H. B. White, III. 1979. Description of the nymph of *Ophiogomphus howei* (Odonata: Gomphidae). Proc. Ent. Soc. Wash. 81:64-69.
2077. Kerr, J. D., and G. B. Wiggins. 1995. A comparative morphological study of the lateral line systems in larvae and pupae of Trichoptera. Zool. J. Linnean Soc. (in press).
2078. Kershaw, W. E., T. R. Williams, S. Frost, R. E. Matchett, M. L. Mills, and R. D. Johnson. 1968. The selective control of *Simulium* larvae by particulate insecticides and its significance in river management. Trans. Roy. Soc. Trop. Med. Hyg. 62:35-40.

2079. Kerst, C. D., and N. H. Anderson. 1975. The Plecoptera community of a small stream in Oregon, U.S.A. Freshwat. Biol. 5:189-203.
2080. Kesler, D. H., and W. R. Munns, Jr. 1989. Predation by *Belostoma flumineum* (Hemiptera): an important cause of mortality in freshwater snails. J. N. Am. Benthol. Soc. 8:342-350.
2081. Kettle, D. S. 1977. Biology and bionomics of bloodsucking ceratopogonids. Ann. Rev. Ent. 22:33-51.
2082. Kettle, D. S., and J.W.H. Lawson. 1952. The early stages of the British biting midges, *Culicoides* Latreille (Diptera: Ceratopogonidae) and allied genera. Bull. Ent. Res. 43:421-467.
2083. Kettle, D. S., C. H. Wild, and M. M. Elson. 1975. A new technique for rearing individual *Culicoides* larvae (Diptera:Ceratopogonidae). J. Med. Ent. 12:263-264.
2084. Kevan, D. K. 1979. Megaloptera, pp. 351-352. In H. V. Danks (ed.). Canada and its insect fauna. Mem. Ent. Soc. Can. 108. 573 pp.
2085. Kevan, D. K., and F. E. A. Cutten. 1981. Chap. 10. Nymphomyiidae, pp. 203-208. In J. F. McAlpine, B. V. Peterson, G. E. Shewell, H. J. Teskey, J. R. Vockeroth, and D. M. Wood (coords.). Manual of Nearctic Diptera, Vol. 1. Res. Branch, Agric. Can. Monogr. 27. Ottawa. 674 pp.
2086. Kevan, D. K., and F.E.A. Cutten-Ali-Khan. 1975. Canadian Nymphomyiidae (Diptera). Can. J. Zool. 53:853-866.
2087. Khalaf, K. T. 1969. Distribution and phenology of *Culicoides* (Diptera: Ceratopogonidae) along the Gulf of Mexico. Ann. Ent. Soc. Am. 62:1153-1161.
2088. Khoo, S. G. 1964. Studies on the biology of *Capnia bifrons* (Newman) and notes on the diapause in the nymphs of this species. Gewasser und Abwasser 34/35:23-30.
2089. Kieffer, J. J. 1916. Diapriidae. Das Tierreich, 44. Friedlander und Sonn, Berlin. 627 pp.
2090. Kim, K. C., and R. W. Merritt (eds.). 1987. Black flies: Ecology, population management, and annotated world list. Penn. State Univ. Press, Univ. Park, PA. 528 pp.
2091. Kimerle, R. A., and N. H. Anderson. 1967. Evaluation of aquatic insect emergence traps. J. Econ. Ent. 60:1255-1259.
2092. Kimerle, R. A., and N. H. Anderson. 1971. Production and bioenergetic role of the midge *Glyptotendipes barbipes* (Staeger) in a waste stabilization lagoon. Limnol. Oceanogr. 16:646-659.
2093. King, R. H. 1977. Natural history and ecology of *Stictochironomus annulicrus* (Townes) (Diptera: Chironomidae), Augusta Creek, Michigan. Ph.D. diss., Michigan State University, East Lansing. 160 pp.
2094. King, W. V., G. H. Bradley, C. N. Smith, and W. C. McDuffy. 1960. A handbook of the mosquitoes of the southeastern United States. U.S. Dep. Agric. Handbk. 173. 188 pp.
2095. Kingsley, K. J. 1985. *Eretes sticticus* (Coleoptera: Dytiscidae) life history observations and an account of a remarkable event of synchronous emigration from a temporary desert pond. Coleop. Bull. 39:7-10.
2096. Kinser, P. D., and H. H. Neunzig. 1981. Description of the immature stages and biology of *Synclita tinealis* Munroe (Lepidoptera: Pyralidae: Nymphulinae). J. Lep. Soc. 35:137-146.
2097. Kirk, E. J., and S. A. Perry. 1994. A comparison of three artificial substrate samplers: Macroinvertebrate densities, taxa richness, and ease of use. Water Environ. Res. 66:193-198.
2098. Kirkaldy, G. W., and J. R. de la Torre-Bueno. 1909. A catalogue of American aquatic and semi-aquatic Hemiptera, Proc. Ent. Soc. Wash. 10:173-215.
2099. Kissinger. D. G. 1964. Curculionidae of America north of Mexico. A key to genera. Taxonomic Publ., South Lancaster, Mass. 143 pp.
2100. Kitakami, S. 1931. The Blephariceridae of Japan. Mem. Coll. Sci., Kyoto Imp. Univ., Ser. B. 6:53-108.
2101. Kitakami, S. 1950. The revision of the Blephariceridae of Japan and adjacent territories. J. Kumamoto Womens Univ. 2:15-80.
2102. Kittle, P. D. 1977a. A revision of the genus *Trepobates* Uhler (Hemiptera: Gerridae). Ph.D. diss., University of Arkansas, Fayetteville. 255 pp.

2103. Kittle, P. D. 1977b. Notes on the distribution and morphology of the water strider *Metrobates alacris* Drake (Hemiptera: Gerridae). Ent. News 88:67-68.

2104. Kittle, P. D. 1977c. The biology of water striders (Hemiptera: Gerridae) in northwest Arkansas. Am. Midl. Nat. 97:400-410.

2105. Kittle, P. D. 1980. The water striders (Hemiptera: Gerridae) of Arkansas. Proc. Ark. Acad. Sci. 34:68-71.

2106. Kittle, P. D. 1982. Two new species of water striders of the genus *Trepobates* (Hemiptera: Gerridae). Proc. Ent. Soc. Wash. 84:157-164.

2107. Klemm, D. J., P. A. Lewis, F. Fulk, and J. M. Lazorchak. 1990. Macroinvertebrate field and laboratory methods for evaluating the biological integrity of surface waters. EPA/600/4-90/030, Environ. Monitoring Systems Lab., U.S.E.P.A., Cincinnati, OH. 256 pp.

2108. Kline, D. L., and R. C. Axtell. 1975 *Culicoides melleus* (Coq.) (Diptera: Ceratopogonidae): seasonal abundance and emergence from sandy intertidal habitats. Mosq. News 35:328-334.

2109. Kline, D. L., and R. C. Axtell. 1976. Salt marsh *Culicoides* (Diptera: Ceratopogonidae): species, seasonal abundance and comparison of trapping methods. Mosq. News 36:1-10.

2110. Klink, A. 1989. The lower Rhine: palaeoecological analysis, pp. 183-201. *In* G. E. Petts (ed.). Historical change of large alluvial rivers: Western Europe. John Wiley, Chichester, England. 355 pp.

2111. Kloter, K. O., L. R. Penner, and W. J. Widmer. 1977. Interactions between the larvae of *Psychoda alternara* and *Dohrniphora cornuta* in a trickling filter sewage bed, with descriptions of the immature stages of the latter. Ann. Ent. Soc. Am. 70:775-781.

2112. Klots, E. B. 1966. The new field book of freshwater life. G. P. Putnam's Sons, N.Y. 398 pp.

2113. Klug, M. J., and S. Kotarski. 1980. Bacteria associated with the gut tract of larval stages of the aquatic cranefly *Tipula abdominalis* (Diptera; Tipulidae). Appl. Environ. Microbiol. 40:408-416.

2114. Knab, F. 1905. A chironomid inhabitant of *Sarracenia purpurea*, *Metriocnemus knabi* Coq. J. N.Y. Ent. Soc. 13:69-73.

2115. Knabke, J. J. 1976. Diapause in the rice water weevil, *Lissorhoptrus oryzophilus* Kuschel (Coleoptera: Curculionidae) in California. Ph.D diss., University of California, Davis.

2116. Knausenberger, W. I. 1987. Contributions to the autecology and ecosystematics of immature Ceratopogonidae (Diptera), with emphasis on the tribes Heteromyiini and Sphaeromini in the middle Atlantic United States. Ph. D. diss., V. P. I. and State Univ., Blacksburg, VA. 729p.

2117. Knight, A. W. 1963. Description of the tanyderid larva *Protanvderus marginata* Alexander from Colorado. Bull. Brooklyn Ent. Soc. 58:99-102.

2118. Knight, A. W., and A. R. Gaufin. 1963. The effect of water flow, temperature, and oxygen concentration on the Plecoptera nymph, *Acroneuria pacifica* Banks. Proc. Utah Acad. Sci. Arts Lett. 40:175-184.

2119. Knight, A. W., and A. R. Gaufin. 1964. Relative importance of varying oxygen concentration, temperature and water flow on the mechanical activity and survival of the Plecoptera nymph, *Pteronarcys californica* Newport. Proc. Utah Acad. Sci. Arts Lett. 41:14-28.

2120. Knight, A. W., and M. A. Simmons. 1975a. Factors influencing the oxygen consumption of the hellgrammite, *Corydalus cornutus* (L.) (Megaloptera: Corydalidae). Comp. Biochem. Physiol. 50A:827-833.

2121. Knight, A. W., and M. A. Simmons. 1975b. Factors influencing the oxygen consumption of larval *Nigronia serricornis* (Say) (Megaloptera: Corydalidae). Comp. Biochem. Physiol. 51A:177-183.

2122. Knight, A. W., M. A. Simmons, and C. S. Simmons. 1976. A phenomenological approach to the growth of the winter stonefly, *Taeniopteryx nivalis* (Fitch) (Plecoptera: Taeniopterygidae). Growth 40:343-367.

2123. Knight, K. L., and M. Wonio. 1969. Mosquitoes of Iowa. Spec. Rept. Iowa State Univ. 79 pp.

2124. Knight, S. S., and C. M. Cooper. 1989. New records for *Tortopus incertus* (Ephemeroptera) in Mississippi and notes on microhabitat. Ent. News 100:21-26.

2125. Knoph, K. W., and D. H. Habeck. 1976. Life history and biology of *Samea multiplicalis*. Environ. Ent. 5:539-542.

2126. Knopf, K. W., and K. J. Tennessen. 1980. A new species of *Progomphus* Selys, 1854 from North America (Anisoptera: Gomphidae). Odonatologica 9:247-252.

2127. Knowlton, G. F., and F. C. Harmston. 1938. Notes on Utah Plecoptera and Trichoptera. Ent. News 49:284-286.

2128. Knudsen, M. 1968. The biology and life history of *Oedoparena glauca* (Diptera: Dryomyzidae), a predator of barnacles. M.S. thesis, University of California, Berkeley.

2129. Knutson, L .V., J. W. Stephenson, and C. O. Berg. 1965. Biology of a slug-killing fly, *Tetanocera elata* (Diptera: Sciomyzidae). Proc. Malac. Soc. Lond. 36:213-220.

2130. Knutson, L. V. 1970. Biology and immature stages of malacophagous flies: *Antichaeta analis*, *A. brevipennis* and *A. obliviosa* (Diptera: Sciomyzidae). Trans. Am. Ent. Soc. 92:67-101.

2131. Knutson, L. V. 1972. Description of the female of *Pherbecta limenitis* Steyskal (Diptera: Sciomyzidae), with notes on biology, immature stages, and distribution. Ent. News 83:15-21.

2132. Knutson, L. V. 1987. Chap. 84. Sciomyzidae, pp. 927-940. *In* J. F. McAlpine, B. V. Peterson, G. E. Shewell, H. J. Teskey, J. R. Vockeroth, and D. M. Wood. (coords.). Manual of Nearctic Diptera, Vol. 2. Res. Branch, Agr. Can. Monogr. 28. 1332pp.

2133. Knutson, L. V. 1988. Life cycles of snail-killing flies: *Pherbellia griseicollis*, *Sciomyza dryomyzina*, *S. simplex*, and *S. testacea* (Diptera: Sciomyzidae). Ent. scand. 18:383-391.

2134. Knutson, L. V., and J. Abercrombie. 1977. Biology of *Antichaeta melansoma* (Diptera: Sciomyzidae), with notes on parasitoid Braconidae and Ichneumonidae (Hymenoptera). Proc. Ent. Soc. Wash. 79:111-125.

2135. Knutson, L. V., and C. O. Berg. 1964. Biology and immature stages of snail-killing flies: the genus *Elgiva* (Diptera: Sciomyzidae). Ann. Ent. Soc. Am. 57:173-192.

2136. Knutson, L. V., and O. S. Flint, Jr. 1971. Pupae of Empididae in pupal cocoons of Rhyacophilidae and Glossosomatidae. Proc. Ent. Soc. Wash. 13:314-320.

2137. Knutson, L. V., and O. S. Flint, Jr. 1979. Do dance flies feed on caddisflies?—further evidence (Diptera: Empididae; Trichoptera). Proc. Ent. Soc. Wash. 81:32-33.

2138. Koehler, O. 1924. Sinnesphysiologische Untersuchungen an Libellenlaren. Verh. Dtsch. Zool. Ges. 29:83-91.

2139. Koenekoop, R., and T. Livdahl. 1986. Cannibalism among *Aedes triseriatus* larvae. Ecol. Ent. 11:111-114.

2140. Kohler, S. L. 1984. Search mechanism of a stream grazer in patchy environments: the role of food abundance. Oecologia 62:209-218.

2141. Kohler, S. L. 1985. Identification of stream drift mechanisms: an experimental and observational approach. Ecology 66:1749-1761.

2142. Kohler, S. L. 1992. Competition and the structure of a benthic stream community. Ecol. Monogr. 62:165-188.

2143. Kohler, S. L., and M. A. McPeek. 1989. Predation risk and the foraging behavior of competing stream insects. Ecology 70:1811-1825.

2144. Kohler, S. L., and M. J. Wiley. 1992. Parasite-induced collapse of a population of a dominant grazer in Michigan streams. Oikos 65: 443-449.

2145. Kolenkina, L. V. 1951. Nutrition of the larvae of some caddisflies (Trichoptera). Trudy Vsesoyuznogo Gidrobiologicheskogo Obshchest-va. 3:44-57. *In* S. G. Lepneva (ed.). Fauna of the U.S.S.R. Trichoptera. (Russian, translated by Israel Program for Scientific Translation, 1970.) U.S. Dep. Comm., Springfield, Va. 638 pp.

2146. Kolkwitz, R., and M. Marsson. 1908. Ökologie der pflanzlichen Saprobien. Berichte der Deutschen Botanischen Gesellschaft 26A:505-519.

2147. Kolkwitz, R., and M. Marsson. 1909. Ökologie der tierischen Saprobien. Beiträge zur Lehre von der biologischen Gewässerbeurteilung. Int. Revue ges. Hydrobiol. Hydrogr. 2:126-152.

2148. Kolmes, S. A. 1983. Ecological and sensory aspects of prey capture by the whirligig beetle *Dineutes discolor* (Coleoptera: Gyrinidae). J. N. York Ent. Soc. 91:405-412.

2149. Kolpelke, J.-P., and I. Müller-Liebenau. 1981. Eistrukturen bei Ephemeroptera und deren Bedeutung für die Aufstellung von Artengruppen am Beispiel der europäischen Arten der Gattung *Baetis* Leach, 1815. Teil III: *buceratus-, atrebatinus-, niger-, gracilis-,* und *muticus-*Gruppe (Ephemeroptera: Baetidae). Deutsche Ent. Zeitschrift 28:1-6.

2150. Komnick, H. 1977. Chloride cells and chloride epithelia of aquatic insects. Int. Rev. Cytol. 49:285-329.

2151. Kondo, S., S. Hamashima, and H. Hashimoto. 1989. Life history and seasonal occurrence of *Pentapedilum tigrinum* Hashimoto associated with *Nymphoides indica* O. Kuntze in an irrigation reservoir. Acta Biol. Debr. Oecol. Hungary 2:237-245.

2152. Kondratieff, B. C., and R. F. Kirchner. 1987. Additions, taxonomic corrections and faunal affinities to the list of stoneflies (Plecoptera) of Virginia. Proc. Ent. Soc. Wash. 89:24-30.

2153. Kondratieff, B. C., R. F. Kirchner, and K. W. Stewart. 1988. A review of *Perlinella* Banks (Plecoptera: Perlidae). Ann. Ent. Soc. Am. 81:19-27.

2154. Kondratieff, B. C., and J. R. Voshell, Jr. 1979. A checklist of the stoneflies (Plecoptera) of Virginia. Ent. News 90:241-246.

2155. Kondratieff, B. C., and J. R. Voshell, Jr. 1980. Life history and ecology of *Stenonema modestum* (Banks) (Ephemeroptera: Heptageniidae) in Virginia, U.S.A. Aquat. Insects 2:177-189.

2156. Kondratieff, B. C., and J. R. Voshell, Jr. 1981. Influence of a reservoir with surface release on life history of the mayfly *Heterocloeon curiosum* (McDunnough) (Ephemeroptera: Baetidae). Can. J. Zool. 59:305-314.

2157. Kondratieff, B. C., and J. R. Voshell, Jr. 1983. Subgeneric and species group classification of the mayfly genus *Isonychia* in North America (Ephemeroptera: Oligoneuriidae). Proc. Ent. Wash. 85:128-138.

2158. Kondratieff, B. C., and J. R. Voshell, Jr. 1984. The North and Central American species of *Isonychia* (Ephemeroptera: Oligoneuriidae). Trans. Am. Ent. Soc. 110:129-244.

2159. Konstantinov, A. S. 1971. Feeding habits of the chironomid larvae and certain ways to increase the food content of the water basins. Fish. Res. Bd. Can. Translation Ser. 1853.

2160. Kopelke, J. P. 1981. Morphologische und biologische studien an Belostomatiden am Beispiel der mittelamerikanischen arten *Belostoma ellipticum* und *B. thomasi* (Heteroptera). Ent. Abh. 44:59-80.

2161. Korch, P. P., and J. E. McPherson, III. 1987. Life history and laboratory rearing of *Gerris argenticollis* (Hemiptera: Gerridae) with descriptions of immature stages. Great Lakes Ent. 20:193-204.

2162. Kormondy, E. J. 1958. Catalogue of the Odonata of Michigan. Misc. Publ. Univ. Mich. Mus. Zool. 104:1-43.

2163. Kormondy, E. J. 1959. The systematics of *Tetragoneuria*, based on ecological, life history, and morphological evidence (Odonata: Corduliidae). Misc. Publ. Univ. Mich. Mus. Zool. 107:1-79.

2164. Kormondy, E. J., and J. L. Gower. 1965. Life history variations in an association of Odonata. Ecology 46:882-886.

2165. Kornijow, R. 1992. Seasonal migration by larvae of an epiphytic chironomid. Freshwat. Biol. 27:85-89.

2166. Kosalwat, P., and A. W. Knight. 1987. Acute toxicity of aqueous and substrate-bound copper to the midge, *Chironomus decorus*. Arch. Environ. Contam. Toxicol. 16:275-282.

2167. Koshima, S. 1984. A novel cold-tolerant insect found in a Himalayan glacier. Nature 310:225-227.

2168. Koslucher, D. G., and G. W. Minshall. 1973. Food habits of some benthic invertebrates in a northern cool-desert stream (Deep Creek, Curlew Valley, Idaho-Utah). Trans. Am. Microsc. Soc. 92:441-452.

2169. Koss, R. W. 1968. Morphology and taxonomic use of Ephemeroptera eggs. Ann. Ent. Soc. Am. 61:696-721.

2170. Koss, R. W. 1970. The significance of the egg stage to taxonomic and phylogenetic studies of the Ephemeroptera. Proc. 1st. Int. Conf. Ephem. 1:73-78.

2171. Koss, R. W., and G. F. Edmunds, Jr. 1974. Ephemeroptera eggs and their contribution to phylogenetic studies of the order. J. Zool. Linn. Soc. Lond. 58:61-120.

2172. Koster, K. C. 1934. A study of the general biology, morphology of the respiratory system and respiration of certain aquatic *Stratiomyia* and *Odontomyia* larvae (Diptera). Pap. Mich. Acad. Sci. Arts Lett. 19:605-658.

2173. Kovalak, W. P. 1976. Seasonal and diel changes in the positioning of *Glossosoma nigrior* Banks (Trichoptera: Glossosomatidae) on artificial substrates. Can. J. Zool. 54:1585-1594.

2174. Kovalak, W. P. 1978. On the feeding habits of *Phasganophora capitata* (Plecoptera: Perlidae). Great Lakes Ent. 11:45-50.

2175. Kraatz, W. C. 1918. *Scirtes tibialis* Guer. (Coleoptera, Dascyllidae), with observations on its life history. Ann. Ent. Soc. Am. 11:393-400.

2176. Kraft, K. J., and B. M. Sabol. 1980. The stonefly *Isoperla bilineata* in Lake Superior. J. Great Lakes Res. 6:258-259.

2177. Krantzberg, G. 1992. Ecosystem health as measured from the molecular to the community level of organization, with reference to sediment bioassessment. J. Aquat. Ecosyst. Health 1:319-328.

2178. Krawany, H. 1930. Trichopterenstudien im Gebiete der Lunzer Seen. Int. Revue ges. Hydrobiol. 23:417-427.

2179. Krecker, F. H. 1919. The fauna of rock bottom ponds. Ohio J. Sci. 19:427-474.

2180. Kristensen, N. P. 1975. The phylogeny of hexapod "orders." A critical review of recent accounts. Z. Zool. Syst. Evolutions Forsh 13:1-44.

2181. Kristensen, N. P. 1981a. Amphiesmenoptera. Trichoptera. Lepidoptera (Revisionary notes), pp. 325-330, 412-415. *In* W. Hennig, A. C. Pont, and D. Schlee (eds.). Insect Phylogeny. J. Wiley and Sons, Chichester, England.

2182. Kristensen, N. P. 1981b. Phylogeny of insect orders. Ann. Rev. Ent. 26:135-157.

2183. Krivda, W. V. 1961. Notes on the distribution and habitat of *Chilostigma areolatum* (Walker) in Manitoba (Trichoptera: Limnephilidae). J. N.Y. Ent. Soc. 69:68-70.

2184. Krivosheina, M. G., and A. L. Oserov. 1989. Ecology and morphology of the larvae of *Athyroglossa glabra* Meigen (Diptera, Ephydridae) living in the corpses of vertebrate animals. Biol. Nauki 5:47-50 (in Russian).

2185. Krivosheina, N. P. 1971. The family Glutopidae, fam. n. and its position in the system of Diptera Brachycera Orthorrhapha (in Russian). Ent. Obozr. 50:681-694. (transl. in Ent. Rev., Wash. 50:387-395).

2186. Krogh, A. 1920. Studien uber Tracheenrespiration. II. Uber Gasdiffusion in den Tracheen. Pflugers Arch. ges. Physiol. 179:95-112.

2187. Krogh, A. 1943. Some experiments on the osmoregulation and respiration of *Eristalis* larvae. Entomol. Medd. 23:49-65.

2188. Krogstad, B. O. 1974. Aquatic stages of *Stratiomys normula unilimbata* Loew (Diptera: Stratiomyiidae). J. Minn. Acad. Sci. 38:86-88.

2189. Krombein, K. V., and B. D. Burks. 1967. Hymenoptera of America north of Mexico synoptic catalog. U.S. Dep. Agric. Monogr. 2. 305 pp.

2190. Krombein, K. V., P. D. Hurd, Jr., D. R. Smith, and B. D. Burks. 1979. Catalog of Hymenoptera in America north of Mexico (3 volumes). Smithson. Instit. Press, Washington, D.C. 2735 pp.

2191. Krotzer, S., and M. J. Krotzer. 1992. Significant new records of Odonata from the southeastern United States. Not. Odonatol. 3:168-169.

2192. Krueger, C. C., and E. F. Cook. 1981. Life cycles, drift and standing crops of some stoneflies (Insects: Plecoptera) from streams in Minnesota, U.S.A. Hydrobiologia 83:85-92.

2193. Krueger, C. C., and E. F. Cook. 1984a. Life cycles, standing stocks and drift of some Megaloptera, Ephemeroptera and Diptera from streams in Minnesota. Aquat. Insects 6:101-108.

2194. Krueger, C. C., and E. F. Cook. 1984b. Life cycles and drift of Trichoptera from a woodland stream in Minnesota. Can. J. Zool. 62:1479-1484.

2195. Krueger, C. C., and F. B. Martin. 1980. Computation of confidence intervals for the size-frequency (Hynes) method of estimating secondary production. Limnol. Oceanogr. 25:773-777.

2196. Krull, W. H. 1929. The rearing of dragonflies from eggs. Ann. Ent. Soc. Am. 22:651-658.

2197. Kruse, K. C. 1983. Optimal foraging by predaceous diving beetle larvae on toad tadpoles. Oecologia 58:383-388.

2198. Kuehne, R. A. 1962. A classification of streams, illustrated by fish distribution in an eastern Kentucky creek. Ecology 43:608-614.

2199. Kuenzel, W. J., and R. G. Wiegert. 1977. Energetics of an insect predator, *Tachytrechus augustipennis* (Diptera). Oikos 28:201-209.

2200. Kuhlhorn, F. 1961. Investigations on the importance of various representatives of the hydrofauna and flora as natural limiting factors for *Anopheles* larvae. Z. Angew. Zool. 48:129-161.

2201. Kuitert, L. C. 1942. Gerrinae in the University of Kansas collections. Univ. Kans. Sci. Bull. 28:113-143.

2202. Kukalova-Peck, J. 1978. Origin and evolution of insect wings and their relation to metamorphosis, as documented by the fossil record. J. Morphol. 156:53-126.

2203. Kukalova-Peck, J. 1985. Ephemeroid wing venation based upon new gigantic Carboniferous mayflies and basic morphology, phylogeny, and metamorphosis of pterygote insects. Can. J. Zool. 63:933-955.

2204. Kukalova-Peck, J. 1991. Chap. 6. Fossil history and the evolution of hexapod structure, pp. 141-179. *In* I. D. Naumann, P. B. Carne, J. F. Lawrence, E. S. Nielsen, J. P. Spradbery, R. W. Taylor, M. J. Whitten, and M. J. Littlejohn (eds.). The insects of Australia: a textbook for students and research workers (2nd ed.). Div. Ent., CSIRO, Australia. Melbourne Univ. Press, Melbourne, Australia.

2205. Kullberg, A. 1988. The case, mouthparts, silk and silk formation of *Rheotanytarsus muscicola* Kieffer (Chironomidae: Tanytarsini). Aquat. Insects 10:249-255.

2206. Kurtak, D. C. 1978. Efficiency of filter feeding of blackfly larvae. Can. J. Zool. 56:1608-1623.

2207. Kurtak, D. C. 1979. Food of black fly larvae (Diptera: Simuliidae): Seasonal changes in gut contents and suspended material at several sites in a single watershed. Quaest. Ent. 15:357-374.

2208. Kuusela, K., and H. Pulkkinen. 1978. A simple trap for collecting newly emerged stoneflies (Plecoptera). Oikos 31:323-325.

2209. Lacey, L. A., and M. S. Mulla. 1979. Factors affecting feeding rates of black fly larvae. Mosquito News 39:315-319.

2210. Lacoursiere, J. O., and D. A. Craig. 1993. Fluid transmission and filtration efficiency of the labral fans of black fly larvae (Diptera: Simuliidae): hydrodynamic, morphological, and behavioural aspects. Can. J. Zool. 71:148-162.

2211. Ladle, M. 1972. Larval Simuliidae as detritus feeders in chalk streams. Mem. Inst. Ital. Idrobiol. Suppl. 29:429-439.

2212. Ladle, M. 1982. Organic detritus and its role as a food source in chalk streams. Ann. Rept. Freshwat. Biol. Assoc. 50:30-37.

2213. Ladle, M. 1990. Long-term investigations of trophic relationships in southern chalk streams. Freshwat. Biol. 23:113-118.

2214. Ladle, M., J. A. Bass, and W. R. Jenkins. 1972. Studies on production and food consumption by the larval Simuliidae (Diptera) of a chalk stream. Hydrobiologia 39:429-448.

2215. Ladle, M., and A. Esmat. 1973. Records of Simuliidae (Diptera) from the Bere Stream, Dorset with details of the life-history and larval growth of *Simulium (Eusimulium) latipes* Meigen. Entomol. mon. Mag. 108:167-172.

2216. Ladle, M., and R. Radke. 1990. Burrowing and feeding behavior of the larva of *Ephemerea danica* Mueller (Ephemeroptera: Ephemeridae). Ent. Gaz. 41:113-118.

2217. LaFontaine, G. 1981. Caddisflies. Winchester Press, New Jersey. 336p.

2218. Lager, T. M. 1985. Range extensions and ecological data for southern United States Ephemeroptera. Proc. Ent. Soc. Wash. 87:255-256.

2219. Lager, T. M., M. D. Johnson, S. N. Williams, and J. L. McCulloch. 1979. A preliminary report on the Plecoptera and Trichoptera of northeastern Minnesota. Great Lakes Ent. 12:109-114.

2220. Lago, P. K., R. W. Holzenthal, and S. C. Harris. 1982. An annotated checklist of the caddisflies (Trichoptera) of Mississippi and southeastern Louisiana. Part I: Introduction and Hydropsychoidea. Proc. Ent. Soc. Wash. 84:495-508.

2221. Lago, P. K., D. F. Stanford, and P. D. Hartfield. 1979. A preliminary list of Mississippi damselflies (Insects: Odonata: Zygoptera). J. Miss. Acad. Sci. 24:72-76.

2222. Laird, M. 1956. Studies of mosquitoes and freshwater ecology in the South Pacific. Bull. R. Soc. N. Z. 6:1-213.

2223. Laird, M. 1988. The natural history of larval mosquito habitats. Academic Press, Ltd., London. 555pp.

2224. Lake, R. W. 1980. Distribution of the stoneflies (Plecoptera) of Delaware. Ent. News 9:43-48.

2225. Lake, R. W. 1984. Distribution of caddisflies (Trichoptera) in Delaware. Ent. News 95:215-224.

2226. Lamberti, G. A., L. R. Ashkenas, S. V. Gregory, and A. D. Steinman. 1987. Effects of three herbivores on periphyton communities in laboratory streams. J. N. Am. Benthol. Soc. 6:92-104.

2227. Lamberti, G. A., J. W. Feminella, and V. H. Resh. 1987. Herbivory and intraspecific competition in a stream caddisfly population. Oecologia 73:75-81.

2228. Lamberti, G. A., and V H. Resh. 1978. Substrate relationships, spatial distribution patterns, and sampling variability in a stream caddisfly population. Environ. Ent. 8:561-567.

2229. Lamberti, G. A., and V. H. Resh. 1983. Stream periphyton and insect herbivores: an experimental study of grazing by a caddisfly population. Ecology 64:1124-1135.

2230. Lamberti, G. A., and V. H. Resh. 1985. Comparability of introduced tiles and natural substrates for sampling lotic bacteria, algae and macroinvertebrates. Freshwat. Biol. 15:21-30.

2231. Lammers, R. 1977. Sampling insects with a wetland emergence trap: Design and evaluation of the trap and preliminary results. Am. Midl. Nat. 97:381-389.

2232. Lamp, W. O., and N. W. Britt. 1981. Resource partitioning by two species of stream mayflies (Ephemeroptera: Heptageniidae). Great Lakes Ent. 14:151-157.

2233. Lanciani, C. A. 1987. Rearing immature *Mesovelia mulsanti* (Hemiptera: Mesoveliidae) on a substratum of duckweed. Fla. Ent. 70:286-288.

2234. Lanciani, C. A. 1991. Laboratory rearing of *Hydrometra australis* (Hemiptera: Hydrometridae). Florida Ent. 74:356-357.

2235. Landa, V., and T. Soldan. 1985. The phylogeny and higher classification of the order Ephemeroptera: A discussion from the comparative anatomical point of view. Ceskoslovenska Akad. Ved., Praha. 121pp.

2236. Landin, J. 1976a. Methods of sampling aquatic beetles in the transitional habitats at water margins. Freshwat. Biol. 6:81-87.

2237. Landin, J. 1976b. Seasonal patterns in abundance of water beetles belonging to the Hydrophiloidea (Coleoptera). Freshwat. Biol. 6:89-108.

2238. Landry, B., and P. P. Harper. 1985. The aquatic dance fly fauna of a subarctic river system in Québec, with the description of a new species of *Hemerodromia* (Diptera: Empididae). Can. Ent. 117:1379-1386.

2239. Lane, R. S. 1975. Immatures of some Tabanidae (Diptera) from Mendocino County, Calif. Ann. Ent. Soc. Am. 68: 803-819.

2240. Lane, R. S. 1976. Density and diversity of immature Tabanidae (Diptera) in relation to habitat type in Mendicino County, California. J. Med. Ent. 12:683-691.

2241. Lane, R. S., and J. R. Anderson. 1976. Extracting larvae of *Chrysops hirsuticallus* (Diptera: Tabanidae) from soil: efficiency of two methods. Ann. Ent. Soc. Am. 69:854-856.

2242. Lange, W. H. 1956a. Aquatic Lepidoptera, pp. 271-288. *In* R. L. Usinger (ed.). Aquatic insects of California. Univ. Calif. Press, Berkeley. 508 pp.

2243. Lange, W. H. 1956b. A generic revision of the aquatic moths of North America: (Lepidoptera: Pyralidae, Nymphulinae). Wasmann J. Biol. 14:59-144.

2244. Langford, T. E., and J. R. Daffern. 1975. The emergence of insects from a British river warmed by power station coolingwater. Part I. The use and performance of insect emergence traps in a large, spate-river and the effects of various factors on total catches, upstream and downstream of the coolingwater outfalls. Hydrobiologia 46:71-114.

2245. Lansbury, I. 1954. Some notes on the ecology of *Corixa (Halicorixa) stagnalis* Leach with some information on the measurement of salinity of brackish habitats. Entomol. mon. Mag. 90:139-140.

2246. Lansbury, I. 1956. Further observations on the ecology of *Cymatia coleoptrata* (Fabr.), (Hemiptera-Heteroptera, Corixidae) in southern England. Entomologist 89:188-195.

2247. Lansbury, I. 1957. Observations on the ecology of water-bugs (Hemiptera-Heteroptera) and their associated fauna and flora in southeast Kent. Entomologist 90:167-177.

2248. Lansbury, I. 1960. The Corixidae (Hemiptera-Heteroptera) of British Columbia. Proc. Ent. Soc. Brit. Col. 57:34-43.
2249. LaRivers, I. 1940. A preliminary synopsis of the dragonflies of Nevada. Pan-Pacif. Ent. 16:111-123.
2250. LaRivers, I. 1948. A new species of *Pelocoris* from Nevada, with notes on the genus in the United States. Ann. Ent. Soc. Am. 41:371-376.
2251. LaRivers, I. 1950a. The Dryopoidea known or expected to occur in the Nevada area (Coleoptera). Wasmann J. Biol. 8:97-111.
2252. LaRivers, I. 1950b. The staphylinoid and dascilloid aquatic Coleoptera of the Nevada area. Great Basin Nat. 10:66-70.
2253. LaRivers, I. 1951. A revision of the genus *Ambrysus* in the United States. Univ. Calif. Publ. Ent. 8:277-338.
2254. LaRivers, I. 1954. Nevada Hydrophilidae (Coleoptera). Am. Midl. Nat. 52:164-175.
2255. LaRivers, I. 1971. Catalogue of taxa described in the family Naucoridae (Hemiptera). Mem. Biol. Soc. Nev. 2:65-120.
2256. LaRivers, I. 1974. Catalogue of taxa described in the family Naucoridae (Hemiptera). Supplement No. 1: Corrections, emendations and additions, with descriptions of new species. Occ. Pap. Nev. Biol. Soc. 38:1-17.
2257. LaRivers, I. 1976. Supplement No. 2 to the catalogue of taxa described in the family Naucoridae (Hemiptera), with descriptions of new species. Occ. Pap. Nev. Biol. Soc. 41:1-17.
2258. LaRow, E. J. 1970. The effect of oxygen tension on the vertical migration of *Chaoborus* larvae. Limnol. Oceanogr. 15:375-362.
2259. Larson, D. J. 1975. The predaceous water beetles (Coleoptera: Dytiscidae) of Alberta: systematics, natural history and distribution. Quaest. Ent. 11:245-498.
2260. Larson, D. J. 1987. Revision of North American species of *Ilybius* Erichson (Coleoptera: Dytiscidae) with systematic notes on Palearctic species. J. New York Ent. Soc. 95:341-413.
2261. Larson, D. J. 1989. Revision of North American *Agabus* Leach (Coleoptera: Dytiscidae): Introduction, key to species groups, and classification of the *ambiguus tristis* and *arcticus* groups. Can. Ent. 121:861-919.
2262. Larson, D. J., and J. R. LaBonte. 1994. *Stygoporus oregonensis*, a new genus and species of subterranean water beetle (Coleoptera: Dytiscidae: Hydroporini) from the United States. Coleopt. Bull. 48:371-379.
2263. Larson, D. J., and G. Pritchard. 1974. Organs of possible stridulatory function in waterbeetles (Coleoptera: Dytiscidae). Coleopt. Bull. 28:53-63.
2264. Larson, D. J., and R. E. Roughley. 1990. A review of the species of *Liodessus* Guignot of North America north of Mexico, with the description of a new species (Coleoptera: Dytiscidae). J. N. York Ent. Soc. 98:233-245.
2265. Lattin, J. D. 1956. Equipment and techniques, pp. 50-67. In R. L. Usinger (ed.). Aquatic insects of California. Univ. Calif. Press, Berkeley. 508 pp.
2266. Lauck, D. R. 1959. The taxonomy and bionomics of the aquatic Hemiptera of Illinois. M.S. thesis, University of Illinois, Urbana. 353 pp.
2267. Lauck, D. R. 1962. A monograph of the genus *Belostoma* (Hemiptera). Part I. Introduction and *B. dentatum* and *subspinosum* groups. Bull. Chi. Acad. Sci. B. 11:34-81.
2268. Lauck, D. R. 1964. A monograph of the genus *Belostoma* (Hemiptera). Part III. *B. triangulum, bergi, minor, bifoveolatum* and *flumineum* groups. Bull. Chi. Acad. Sci. B. 11:102-154.
2269. Lauck, D. R., and A. S. Menke. 1961. The higher classification of the Belostomatidae (Hemiptera). Ann. Ent. Soc. Am. 54:644-657.
2270. Lauff, G. H., and K. W. Cummins. 1964. A model stream for studies in lotic ecology. Ecology 45:188-190.
2271. Lauff, G. H., K. W. Cummins, C. H. Eriksen, and M. Parker. 1961. A method of sorting bottom fauna samples by elutriation. Limnol. Oceanogr. 6:462-466.
2272. Laughlin, R. 1960. Biology of *Tipula oleracea* L.: growth of the larva. Ent. exp. Appl. 3:185-197.
2273. Laughlin, R. 1967. Biology of *Tipula paludosa*: growth of the larva in the field. Ent. exp. Appl. 10:52-68.
2274. Lauzon, M., and P. P. Harper. 1986. Life history and production of the stream-dwelling mayfly *Habrophlebia vibrans* Needham (Ephemeroptera: Leptophlebiidae). Can. J. Zool. 64:2038-2045.
2275. Lauzon, M., and P. P. Harper. 1988. Seasonal dynamics of a mayfly (Insecta: Ephemeroptera) community in a Laurentian stream. Holarct. Ecol. 11:220-234.
2276. Lauzon, M., and P. P. Harper. 1993. The life cycle of the aquatic snipe fly *Atherix lantha* Webb (Diptera: Brachycera; Athericidae) in Quebec. Can. J. Zool. 71:1530-1533.
2277. Lavallee, A. G., and J. B. Wallace. 1974. Immature stages of Milesiinae (Syrphidae) II. *Sphegina keeniana* and *Chrysogaster nitida*. J. Ga. Ent. Soc. 9:8-15.
2278. Lavandier, P. 1979. Cycle biologique, regime alimentaire et production *d'Arcynopteryx compacta* (Plecoptera, Perlodidae) dans un torrent de haute altitude. Bull. Soc. Hist. Nat. Toulouse. 115:140-150.
2279. Lavery, M. A., and R. R. Costa. 1972. Reliability of the Surber sampler in estimating *Parargyractic fulicalis* (Clemens) (Lepidoptera: Pyralidae) populations. Can. J. Zool. 50:1335-1336.
2280. Lavery, M. A., and R. R. Costa. 1973. Geographic distribution of the genus *Parargyractis* Lange (Lepidoptera: Pyralidae) throughout the Lake Erie and Lake Ontario watersheds (U.S.A.). J. N.Y. Ent. Soc. 81:42-49.
2281. Lavery, M. A., and R. R. Costa. 1976. Life history of *Parargyractis canadensis* Munroe (Lepidoptera: Pyralidae). Am. Midl. Nat. 96:407-417.
2282. Lawrence, S. G. (ed.). 1981. Manual for the culture of selected freshwater invertebrates. Can. Spec. Publ. Fish. Aquat. Sci. 54:1-169.
2283. Lawson, D. L., and R. W. Merritt. 1979. A modified Ladell apparatus for the extraction of wetland macroinvertebrates. Can. Ent. 111:1389-1393.
2284. Lawson, F. A. 1959. Identification of the nymphs of common families of Hemiptera. J. Kans. Ent. Soc. 32:88-92.
2285. Lawson, H. R., and W. P. McCafferty. 1984. A checklist of Megaloptera and Neuroptera (Planipennia) of Indiana. Great Lakes Ent. 17:129-131.
2286. Lawton, J. H. 1970. Feeding and food energy assimilation in larvae of the damselfly *Pyrrhosoma nymphula (Sulz.)* (Odonata: Zygoptera). J. Anim. Ecol. 39:669-689.
2287. Lawton, J. H. 1971. Maximum and actual field feeding-rates in larvae of the damselfly *Pyrrhosoma nymphula* (Sulzer) (Odonata: Zygoptera). Freshwat. Biol. 1:99-111.
2288. Leader, J. P. 1971. Effect of temperature, salinity, and dissolved oxygen concentration upon respiratory activity of the larva of *Philanisus plebeius* (Trichoptera). J. Insect Physiol. 17:1917-1924.
2289. Leathers, A. L. 1922a. Ecological study of aquatic midges and some related insects with special reference to feeding habits. U.S. Bur. Fish. 38:1-61.
2290. Leathers, A. L. 1922b. Ecological study of aquatic midges and some related insects with special reference to feeding habits. Bull. U.S. Bur. Fish. 38:1-61.
2291. Lechleitner, R. A., D. S. Cherry, J. Cairns, Jr., and D. A. Stetler. 1985. Ionoregulatory and toxicological responses of stonefly nymphs (Plecoptera) to acidic and alkaline pH. Arch. Environ. Contam. Toxicol. 14:179-185.
2292. Lechleitner, R. A., and B. Kondratieff. 1983. The life history of *Pteronarcys dorsata* (Say) in southwestern Virginia. Can. J. Zool. 61:1981-1985.
2293. LeClair, R., Jr., Y. Alarie, and J. P. Bourassa. 1986. Prey choice in larval *Dytiscus harrisii* Kirby and *D. verticalis* Say (Coleoptera: Dytiscidae). Ent. Basiliensia 11:337-342.
2294. Lee, F. C. 1967. Laboratory observations on certain mosquito predators. Mosquito News 27:332-338.
2295. Lee, V. F., R. S. Lane, and C. B. Philip. 1976. Confirmation of the beach habitation of *Apatolestes actites* Philip and Steffan (Diptera: Tabanidae) on the California coast. Pan-Pacif. Ent. 52:212.
2296. Leech, H. B. 1938. A study of the Pacific Coast species of *Agabus* Leach, with a key to the nearctic species (Coleoptera: Dytiscidae). M.S. thesis, University of California, Berkeley. 106 pp.

2297. Leech, H. B. 1940. Description of a new species of *Laccornis*, with a key to the Nearctic species. Can. Ent. 72:122-128.
2298. Leech, H. B. 1942a. Key to the Nearctic genera of water beetles of the tribe Agabini, with some generic synonymy. Ann. Ent. Soc. Am. 35:76-80.
2299. Leech, H. B. 1942b. Dimorphism in the flying wings of a species of water beetle, *Agabus bifarius* (Kirby) (Coleoptera: Dytiscidae). Ann. Ent. Soc. Am. 35:355-362.
2300. Leech, H. B., and H. P. Chandler. 1956. Aquatic Coleoptera, pp. 293-371. In R. L. Usinger (ed.). Aquatic insects of California. Univ. Calif. Press, Berkeley. 508 pp.
2301. Leech, H. B., and M. W. Sanderson. 1959. Coleoptera, pp. 981-1023. In W. T. Edmondson (ed.). Freshwater biology (2nd ed.). John Wiley & Sons, N.Y. 1248 pp.
2302. Legner, E. F., R. A. Medved, and R. D. Sjogren. 1975. Quantitative water column sampler for insects in shallow aquatic habitats. Proc. Calif. Mosquito Contr. Assoc. 43:110-115.
2303. Lehmann, J. 1971. Die Chironomiden der Fulda. Arch. Hydrobiol. Suppl. 37:466-555.
2304. Lehmkuhl, D. M. 1970. A North American trichopteran larva which feeds on freshwater sponges (Trichoptera: Leptoceridae, Porifera: Spongillidae). Am. Midl. Nat. 84:278-280.
2305. Lehmkuhl, D. M. 1971. Stoneflies (Plecoptera: Nemouridae) from temporary lentic habitats in Oregon. Am. Midl. Nat. 85:514-515.
2306. Lehmkuhl, D. M. 1972a. Changes in thermal regime as a cause of reduction of benthic fauna downstream of a reservoir. J. Fish. Res. Bd. Can. 29:1329-1332.
2307. Lehmkuhl, D. M. 1972b. *Baetisca* (Ephemeroptera: Baetiscidae) from the western interior of Canada with notes on the life cycle. Can. J. Zool. 50:1017-1019.
2308. Lehmkuhl, D. M. 1976a. Mayflies. Blue Jay 34:70-81.
2309. Lehmkuhl, D. M. 1976b. Additions to the taxonomy, zoogeography, and biology of *Analetris eximia* (Acanthametropodinae: Siphlonuridae: Ephemeroptera). Can. Ent. 108:199-207.
2310. Lehmkuhl, D. M. 1979a. How to know the aquatic insects. Wm. C. Brown, Dubuque, Iowa. 168 pp.
2311. Lehmkuhl, D. M. 1979b. A new genus and species of Heptageniidae (Ephemeroptera) from western Canada. Can. Ent. 111:859-862.
2312. Lehmkuhl, D. M., and N. H. Anderson. 1970. Observations on the biology of *Cinygmula reticulata* McDunnough in Oregon. Pan. Pac. Ent. 46:268-274.
2313. Lehmkuhl, D. M., and N. H. Anderson. 1971. Contributions to the biology and taxonomy of the *Paraleptophlebia* of Oregon. Pan. Pac. Ent. 47:85-93.
2314. Leischner, T. G., and G. Pritchard. 1973. The immature stages of the alderfly, *Sialis cornuta* (Megaloptera: Sialidae). Can. Ent. 105:411-418.
2315. Leiser, E., and R. H. Boyle. 1982. Stoneflies for the angler. Stackpole Bks., Harrisburg, Penn. 174p.
2316. Leland, H. V., S. V. Fend, J. L. Carter, and A. D. Mahood. 1986. Composition and abundance of periphyton and aquatic insects in a Sierra Nevada, California stream. Great Basin Nat. 46:595-611.
2317. Lenat, D. R. 1988. Water quality assessment of streams using a qualitative collection method for benthic macroinvertebrates. J. N. Am. Benthol. Soc. 7:222-233.
2318. Lenat, D. R. 1993. A biotic index for the southeastern United States: derivation and list of tolerance values, with criteria for assigning water-quality ratings. J. N. Am. Benthol. Soc. 12:279-290.
2319. Lenz, F. 1950. Tendipedidae (Chironomidae). H. Clunioninae. B. Die Metamorphose den Clunioninae, pp. 8-25. In E. Lindner (ed.). Die Fliegen der Palaearktischen Region. 280 pp.
2320. Lenz, F. 1951. Neuue Beobachtungen zur Biologie der Jungendstadien der Tendipedidengattung *Parachironomus* Lenz. Zool. Anz. 147:95-111.
2321. Lenz, F. 1955. Der Wert der Exuviensammlung fur die Beurteilung der Tendipedidenbesielung eines Sees. Arch. Hydrobiol. Suppl. 22:415-421.
2322. Leonard, J. W., and F. A. Leonard. 1949. Noteworthy records of caddisflies from Michigan with descriptions of new species. Occ. Pap. Univ. Mich. Mus. Zool. 520:1-8.
2323. Leonard, J. W., and F. A. Leonard. 1962. Mayflies of Michigan trout streams. Cranbrook Inst. Sci., Bloomfield Hills, Mich. 139 pp.
2324. Lepneva, S. G. 1964. Larvae and pupae of Annulipalpia, Trichoptera. Fauna of the U.S.S.R. Zool. Inst. Akad. Nauk. S.S.S.R., New Ser. 88:1-638. (Transl. Israel Program Sci. Transl. 1971.)
2325. Lepneva, S. G. 1966. Larvae and pupae of Integripalpia, Trichoptera. Fauna of the U.S.S.R. Zool. Inst. Akad. Nauk. S.S.S.R., New Ser. 95:1-560. (Transl. Israel Program Sci. Transl. 1971.
2326. LeSage, L., and A. D. Harrison. 1979. Improved traps and techniques for the study of emerging aquatic insects. Ent. News 90:65-78.
2327. LeSage, L., and A. D. Harrison. 1980. Taxonomy of species (Diptera: Chironomidae) from the Salem Creek, Ontario. Proc. Ent. Soc. Ont. 111:57-114.
2328. LeSage, L., and P. P. Harper. 1976a. Cycles biologiques Elmidae (Coleopteres) de Ruisseaux des Laurentides, Quebec. Ann. Limnol. 12:139-174.
2329. Lesage, L., and P. P. Harper. 1976b. Notes on the life history of the toe-winged beetle *Anchytarsus bicolor* (Melsheimer) (Coleoptera: Ptilodactylidae). Coleopt. Bull. 30:233-238.
2330. Leston, D., and J. W. Pringle. 1963. Acoustic behaviour of Hemiptera, pp. 391-411. In R. G. Busnel (ed.). Acoustic behavior of animals. Elsevier, Amsterdam.
2331. Leuchs, H., and D. Neumann. 1990. Tube texture, spinning and feeding behaviour of *Chironomus* larvae. Zool. Jb Abt. Syst. Okol. Geograph. Tiere 117:31-40.
2332. Leuven, R., T. Brock, and H. Van Druten. 1985. Effects of preservation on dry- and ash-free dry weight biomass of some common aquatic macroinvertebrates. Hydrobiologia 127:151-160.
2333. Levine, E. 1974. Biology of *Bellura gortynoides* Walker (= *vulnifera* Grote), the yellow water lily borer (Lepidoptera: Noctuidae). Ind. Acad. Sci. 83:214-215.
2334. Levine, E., and L. Chandler. 1976. Biology of *Bellura gortynoides* (Lepidoptera: Noctuidae), a yellow water lily borer, in Indiana. Ann. Ent. Soc. Am. 69:405-414.
2335. Lewis, D. J., and G. F. Bennett. 1973. The blackflies (Diptera: Simuliidae) of insular Newfoundland. I. Distribution and bionomics. Can. J. Zool. 51:1181-1187.
2336. Lewis, D. J., and G. F. Bennett. 1979. An annotated list of the black flies (Diptera: Simuliidae) of the Maritime Provinces of Canada. Can. Ent. 111:1227-1230.
2337. Lewis, P. A. 1974. Taxonomy and ecology of *Stenonema* mayflies (Heptageniidae: Ephemeroptera). U.S. EPA, Environmental Monitoring Ser. Rept. EPA-670/4-74-006. 81 pp.
2338. Lewis, P. A., W. T. Mason, and C. I. Weber. 1982. Evaluation of 3 bottom grab samplers for collecting river benthos. Ohio J. Sci. 82:107-113.
2339. Lillehammer, A. 1985a. Temperature influence on egg incubation and nymphal growth of the stoneflies *Leuctra digitata* and *L. fusca* (Plecoptera: Leuctridae). Ent. General. 11:59-67.
2340. Lillehammer, A. 1985b. The coexistence of stoneflies in a mountain lake outlet biotope. Aquat. Insects 7:173-187.
2341. Lillehammer, A. 1986. The effect of temperature on egg incubation period and nymphal growth in two *Nemoura* species (Plecoptera) from subarctic Fennoscandia. Aquat. Insects 8:223-235.
2342. Lillehammer, A. 1987. Intraspecific variation in the biology of eggs and nymphs of Norwegian populations of *Leuctra hippopus* (Plecoptera). J. Nat. Hist. 21:29-41.
2343. Lillehammer, A., J. E. Brittain, S. J. Saltreit, and P. S. Nielsen. 1989. Egg development, nymphal growth and life cycle strategies in Plecoptera. Holarct. Ecol. 12:173-186.
2344. Lilly, C. K., D. L. Ashley, and D. C. Tarter. 1978. Observations on a population of *Sialis itasca* Ross in West Virginia (Megaloptera: Sialidae). Psyche 85:209-217.
2345. Lindeberg, B. 1958. A new trap for collecting emerging insects from small rock pools, with some examples of results obtained. Ann. Ent. Fenn. 24:186-191.
2346. Lindegaard, C., and P. Jonasson. 1975. Life cycles of *Chironomus hyperboreus* Staeger and *Tanytarsus gracilentus* (Holmgren) (Chironomidae, Diptera) in Lake Myvatn, Northern Iceland. Verh. Int. Verein. Limnol. 19:3155-3163.

2347. Lindeman, R. L. 1941. Seasonal food cycle dynamics in a senescent lake. Am. Midl. Nat. 27:428-444.
2348. Lindeman, R. L. 1942. The trophic-dynamic aspect of ecology. Ecology 23:399-418.
2349. Lindskog, P. 1968. The relations between transpiration, humidity reaction, thirst and water content in the shorebug *Saldula saltatoria* L. (Heteroptera: Saldidae). Arkiv. Zool. Ser. 2, 20:465-493.
2350. Lindskog, P., and J. T. Polhemus. 1992. Taxonomy of *Saldula*: Revised genus and species group definitions, and a new species of the *pallipes* group from Tunisia (Heteroptera: Saldidae). Ent. scand. 23:63-88.
2351. Linduska, J. P. 1942. Bottom type as a factor influencing the local distribution of mayfly nymphs. Can. Ent. 74:26-30.
2352. Linley, J. R. 1968. Studies on the larval biology of *Leptoconops becquaerti* (Kieff). (Diptera: Ceratopogonidae). Bull. Ent. Res. 58:1-24.
2353. Linley, J. R. 1969a. Seasonal changes in larval populations of *Leptoconops beaquaerti* (Kieff) (Diptera: Ceratopogonidae) in Jamaica, with observations on the ecology. Bull. Ent. Res. 59:47-64.
2354. Linley, J. R. 1969b. Studies on larval development in *Culicoides furens* (Poey) (Diptera: Ceratopogonidae). I.-Establishment of a standard rearing technique. Ann. Ent. Soc. Am. 62:702-711.
2355. Linley, J. R. 1976. Biting midges of mangrove swamps and salt marshes (Diptera: Ceratopogonidae), pp. 335-376. *In* L. Cheng (ed.). Marine insects. North Holland, Amsterdam. 581 pp.
2356. Linley, J. R. 1985. Growth and survival of *Culicoides melleus* larvae (Diptera: Ceratopogonidae) on four prey organisms. J. Med. Ent. 22:178-189.
2357. Linsenmair, K. E., and R. Jander. 1963. Das "Entspannungsschwimmen" von *Velia* und *Stenus*. Naturwissenschaften 50:231.
2358. Lippert, G., and L. Butler. 1976. Taxonomic study of Collembola of West Virginia. Bull. W. Va. Univ. Agric. Exp. Sta. 643:1-27.
2359. Lloyd, E. C., and S. J. Ormerod. 1992. Further studies on the larvae of the golden-ringed dragonfly, *Cordulegaster boltoni* (Donovan) (Odonata: Cordulegasteridae), in upland streams. Ent. Gaz. 43:275-281.
2360. Lloyd, J. T. 1914. Lepidopterous larvae from rapid streams. J. N.Y. Ent. Soc. 22:145-152.
2361. Lloyd, J. T. 1915. Note on *Ithytrichia confusa* Morton. Can. Ent. 47:117-121.
2362. Lloyd, J. T. 1919. An aquatic dipterous parasite and additional notes on its lepidopterous host. J. N. Y. Ent. Soc. 27:262-265.
2363. Lloyd, J. T. 1921. The biology of North America caddis fly larvae. Bull. Lloyd Lib. 21:1-124.
2364. Lock, M. A., and D. D. Williams. 1981. Perspectives in running water ecology. Plenum Press, New York. 430p.
2365. Lodge, D. M. 1991. Herbivory on freshwater macrophytes. Aquat. Bot. 41:195-224.
2366. Logan, E. R., and S. D. Smith. 1966. New distributional records of Intermountain stoneflies (Plecoptera). Occ. Pap. Biol. Soc. Nevada 9:1-3.
2367. Lohmann, H. 1992. Revision der Cordulegastridae. 1. Entwurf einer neuen Klassifizierung der Familie (Odonata: Anisoptera). Opusc. Zool. Flumin. 96:1-18.
2368. Lohmann, H. 1993. Revision der Cordulegastridae. 2. Besehreibung neuen Arten in den Gattungen *Cordulegaster*, *Anotogaster*, *Neallogaster* und *Sonjagaster* (Anisoptera). Odonatologica 22:273-294.
2369. Longfield, C. 1948. A vast immigration of dragonflies into the South Coast of Co. Cork. Irish. Nat. J. 9:133-141.
2370. Longley, G., and P. J. Spangler. 1977. The larva of a new subterranean water beetle: *Haideoporus texanus* (Coleoptera: Dytiscidae: Hydroporinae). Proc. Biol. Soc. Wash. 90:532-535.
2371. Longridge, J. W., and W. L. Hilsenhoff. 1973. Annotated list of Trichoptera (Caddisflies) in Wisconsin. Wisconsin Acad. Sci., Arts and Letters, 61:173-183.
2372. Lotspeich, F. B., and B. H. Reid. 1980. Tri-tube freeze-core procedure for sampling stream gravels. Progr. Fish Cult. 42:96-99.
2373. Lounibos, L. P., J. R. Rey, and J. H. Frank (eds.). 1985. Ecology of mosquitoes: Proceedings of a workshop. Vero Beach, Fla., Fla. Med. Ent. Lab. 579pp.
2374. Louton, J. A. 1982. Lotic dragonfly (Anisoptera: Odonata) nymphs of the southeastern United States: identification, distribution and historical biogeography. Ph.D. diss., University of Tennessee, Knoxville. 357 pp.
2375. Lowen, R. G., and J. F. Flannagan. 1990a. The nymph and male of *Centroptilum infrequens* McD. (Baetidae), pp. 311-321. *In* I. C. Campbell (ed.). Mayflies and stoneflies: life history and biology. Kluwer Academic Publishers, Dordrecht, The Netherlands.
2376. Lowen, R. G., and J. F. Flannagan. 1990b. *Centroptilum infrequens* McDunnough (Ephemeroptera: Baetidae), a junior synonym of *Pseudocentroptilum pennulatum* (Eaton). Can Ent. 122:173-174.
2377. Lowen, R. G., and J. F. Flannagan. 1991. Four Manitoba species of *Centroptilum* Eaton (Ephemeroptera: Baetidae) with remarks on the genus, pp. 189-206. *In* J. Alba-Tercedor and A. Sanchez-Ortega (eds.). Overview and Strategies of Ephemeroptera and Plecoptera, Sandhill Crane Press, Gainesville, FA. 588pp.
2378. Lubbock, J. 1863. On 2 aquatic Hymenoptera, 1 of which uses its wings in swimming. Trans. Linn. Soc. Lond. 24:135-141.
2379. Lugthart G. J., J. B. Wallace, and A. D. Huryn. 1991. Secondary production of chironomid communities in insecticide-treated and untreated headwater streams. Freshwat. Biol. 23:143-156.
2380. Lundbock, W. 1916. Diptera Danica. Genera and species of flies hitherto found in Denmark. Vol. 5. Lonchopteridae-Syrphidae. Copenhagen. 603 pp.
2381. Lutz, P. E. 1964. Life-history and photoperiodic responses of nymphs of *Tetragoneuria cynosura* (Say). Biol. Bull. 127:304-316.
2382. Lutz, P. E. 1968. Effects of temperature and photoperiod on larval development in *Lestes eurinus* (Odonata: Lestidae). Ecology 49:637-644.
2383. Lutz, P. E. 1974. Effects of temperature and photoperiod on larval development in *Tetragoneuria cynosura* (Odonata: Libellulidae). Ecology 55:370-377.
2384. Lyman, F. E. 1956. Environmental factors affecting distribution of mayfly nymphs in Douglas Lake, Michigan. Ecology 37:568-576.
2385. Macan, T. T. 1938. Evolution of aquatic habitats with special reference to the distribution of Corixidae. J. Anim. Ecol. 7:1-19.
2386. Macan, T. T. 1949. Survey of a moorland fishpond. J. Anim. Ecol. 18:160-186.
2387. Macan, T. T. 1950. Descriptions of some nymphs of the British species of *Baetis* (Ephem.). Trans. Soc. Brit. Ent. 10:143-166.
2388. Macan, T. T. 1958. Methods of sampling the bottom fauna in stony streams. Mitt. Int. Verh. Limnol. 8:1-21.
2389. Macan, T. T. 1962. The ecology of aquatic insects. Ann. Rev. Ent. 7:261-288.
2390. Macan, T. T. 1964. Emergence traps and the investigation of stream faunas. Rev. Idrobiol. 3:75-92.
2391. Macan, T. T. 1965. Predation as a factor in the ecology of water bugs. J. Anim. Ecol. 34:691-698.
2392. Macan, T. T. 1973. Ponds and lakes. Allen & Unwin, London. 148 pp.
2393. Macan, T. T. 1974. Freshwater ecology (2nd ed.). John Wiley & Sons, N.Y. 343 pp.
2394. Macan, T. T. 1976. A twenty-one-year study of the water-bugs in a moorland fishpond. J. Anim. Ecol. 45:913-922.
2395. Macan, T. T. 1977. The influence of predation on the composition of freshwater animal communities. Biol. Rev. Cambridge Philos. Soc. 52:45-70.
2396. Macan, T. T., and A. Kitching. 1976. The colonization of squares of plastic suspended in midwater. Freshwat. Biol. 6:33-40.
2397. Macan, T. T., and E. B. Worthington. 1968. Life in lakes and rivers. Collins, Lond. 272 pp.
2398. Macchiusi, F., and R. L. Baker. 1992. Effects of predators and food availability on activity and growth of *Chironomus tentans* (Chironomidae: Diptera). Freshwat. Biol. 28:207-216.

2399. MacDonald, J. F. 1988. New synonyms pertaining to *Chelifera* and generic key for North American Hemerodromiinae (Diptera: Empididae). Proc. Ent. Soc. Wash. 90:98-100.

2400. MacDonald, J. F. 1989. Review of Nearctic *Metachela* Coquillett, with description of a new species (Diptera: Empididae; Hemerodromiinae). Proc. Ent. Soc. Wash. 91:513-522.

2401. MacDonald, J. F. 1993. Review of the genus *Chelipoda* Macquart of America north of Mexico (Diptera: Empididae; Hemerodromiinae). Proc. Ent. Soc. Wash. 95:327-350.

2402. MacDonald, J. F. 1994. Review of the Nearctic species of the genus *Chelifera* Macquart (Diptera: Empididae; Hemerodromiinae). Proc. Ent. Soc. Wash. 96:236-275.

2403. MacDonald, J. F. 1995. Review of the genus *Hemerodromia* Meigen of America north of Mexico (Diptera: Empididae; Hemerodromiinae). Proc. Ent. Soc. Wash.

2404. MacDonald, J. F., and W. J. Turner. 1993. Review of the genus *Neoplasta* Coquillett of America north of Mexico (Diptera: Empididae; Hemerodromiinae). Proc. Ent. Soc. Wash. 95:351-376.

2405. MacKay, M. R. 1959. Larvae of North American Olethreutinae (Lepidoptera). Can. Ent. 91:3-338.

2406. MacKay, M. R. 1962a. Additional larvae of the North American Olethreutinae (Lepidoptera: Tortricidae). Can. Ent. 94:626-643.

2407. MacKay, M. R. 1962b. Larvae of North American Tortricinae (Lepidoptera: Tortricidae). Can. Ent. Suppl. 28:1-182.

2408. MacKay, M. R. 1963a. Problems in naming the setae of lepidopterous larvae. Can. Ent. 95:996-999.

2409. MacKay, M. R. 1963b. Evolutional adaptation of larval characters in the Tortricidae. Can. Ent. 95:1321-1344.

2410. MacKay, M. R. 1964. The relationship of form and function of minute characters of lepidopterous larvae, and its importance in life history studies. Can. Ent. 96:991-1004.

2411. MacKay, M. R. 1972. Larval sketches of some Microlepidoptera, chiefly North American. Mem. Ent. Soc. Can. 88:1-83.

2412. MacKay, M. R., and E. W. Rockburne. 1958. Notes on life-history and larval description of *Apamea apamiformis* (Guenee), a pest of wild rice (Lepidoptera: Noctuidae). Can. Ent. 90:579-582.

2413. Mackay, A. P., D. A. Cooling, and A. D. Berrie. 1984. An evaluation of sampling strategies for qualitative surveys of macroinvertebrates in rivers, using pond nets. J. Appl. Ecol. 21:515-534.

2414. Mackay, R. J. 1969. Aquatic insect communities of a small stream on Mont St. Hilaire, Quebec. J. Fish. Res. Bd. Can. 26:1157-1183.

2415. Mackay, R. J. 1972. Temporal patterns in life history and flight behavior of *Pycnopsyche gentilis*, *P. luculenta*, and *P. scabripennis* (Trichoptera: Limnephilidae). Can. Ent. 104:1819-1835.

2416. Mackay, R. J. 1977. Behavior of *Pycnopsyche* on mineral substrates in laboratory streams. Ecology 58:191-195.

2417. Mackay, R. J. 1978. Larval identification and instar association in some species of *Hydropsyche* and *Cheumatopsyche* (Trichoptera: Hydropsychidae). Ann. Ent. Soc. Am. 71:499-509.

2418. Mackay, R. J. 1979. Life history patterns of some species of *Hydropsyche* (Trichoptera: Hydropsychidae) in southern Ontario. Can. J. Zool. 57:963-975.

2419. Mackay, R. J. 1984. Life history patterns of *Hydropsyche bronta* and *H. morosa* (Trichoptera: Hydropsychidae) in summer-warm rivers of southern Ontario. Can. J. Zool. 62:271-275.

2420. Mackay, R. J. 1986. Life cycles of *Hydropsyche riola*, *H. slossonae*, and *Cheumatopsyche pettiti* (Trichoptera: Hydropsychidae) in a spring-fed stream in Minnesota. Am. Midl. Nat. 115:19-24.

2421. Mackay, R. J., and G. B. Wiggins. 1979. Ecological diversity in Trichoptera. Ann. Rev. Ent. 24:185-208.

2422. Mackay, R. J., and J. Kalff. 1973. Ecology of two related species of caddisfly larvae in the organic substrates of a woodland stream. Ecology 54:499-511.

2423. Mackay, R. J., and K. E. Kersey. 1985. A preliminary study of aquatic insect communities and leaf decomposition in acid streams near Dorset, Ontario. Hydrobiologia 122:3-11.

2424. Mackey, A. P. 1972. An air-lift sampler for sampling freshwater benthos. Oikos 23:413-415.

2425. Mackey, A. P. 1979. Trophic dependencies of some larval Chironomidae (Diptera) and fish species in the River Thames. Hydrobiologia 62:241-247.

2426. Mackey, H. E., Jr. 1972. A life history survey of *Gelastocoris oculatus* in eastern Tennessee. J. Tenn. Acad. Sci. 47:153-155.

2427. Mackie, G. L., and R. C. Bailey. 1981. An inexpensive stream bottom sampler. J. Freshwat. Ecol. 1:61-69.

2428. Macklin, J. M. 1960. Techniques for rearing Odonata. Proc. N. Central Br. Ent. Soc. Am. 15:67-71.

2429. Macklin, J. M. 1963. Notes on the life history of *Anax junius* (Drury) (Odonata: Aeshnidae). Proc. Ind. Acad. Sci. 73:154-163.

2430. MacLean, S. F., Jr. 1973. Life cycle and growth energetics of the Arctic crane fly *Pedicia hannai antennata*. Oikos 24:436-443.

2431. MacNeill, N. 1960. A study of the caudal gills of dragonfly larvae of the suborder Zygoptera. Proc. Roy. Irish Acad. 61:115-140.

2432. Macrae, I. V., N. N. Winchester, and R. A. Ring. 1989. An evaluation of *Cricotopus myriophylli* as a potential biocontrol for Eurasian Watermilfoil (*Myriophyllum spicatum*). Acta Biol. Debr. Oecol. Hungary 3:241-248.

2433. Macrae, I. V., N. N. Winchester, and R. A. Ring. 1990. Feeding activity and host preference of the milfoil midge, *Cricotopus myriophylli* Oliver (Diptera: Chironomidae). J. Aquat. Pl. Manag. 28:89-92.

2434. Maddux, D. E. 1954. A new species of dobsonfly from California (Megaloptera: Corydalidae). Pan-Pacif. Ent. 30:70-71.

2435. Madsen, B. L. 1972. Detritus on stones in small streams. Mem. Inst. Ital. Idrobiol. Suppl. 29:385-403.

2436. Madsen, B. L. 1974. A note on the food of *Amphinemura sulcicollis*. Hydrobiologia 45:169-175.

2437. Maeda, M., and K. Yano. 1988. Biology of *Chironomus kiiensis* (Diptera: Chironomidae). Bull. Fac. Agric. Yamaguchi Univ., No. 36:31-45.

2438. Magan, B. P. 1992. Oviposition of the dobsonfly (*Corydalus cornutus*, Megaloptera) on a large river. Am. Midl. Nat. 127:348-354.

2439. Magdych, W. P. 1979. The microdistribution of mayflies (Ephemeroptera) in *Myriophyllum* beds in Pennington Creek, Johnston County, Oklahoma. Hydrobiologia 66:161-175.

2440. Magnarelli, L. A., and J. F. Anderson. 1978. Distribution and development of immature salt marsh Tabanidae (Diptera). J. Med. Ent. 14:573-578.

2441. Mahato, M., and D. M. Johnson. 1991. Invasion of the Bays Mountain Lake dragonfly assemblage by *Dromogomphus spinosus* (Odonata: Gomphidae). J. N. Am. Benthol. Soc. 10:165-176.

2442. Maibach, A., and P. G. de Tiefenau. 1993. Description et clé de détermination des stades immatures de plusieurs espèces du genre *Neoascia* Williston de la région paléarctique occidentale (Diptera, Syrphidae). Mitt. Schw. Ent. Ges. 66:337-357.

2443. Maier, C. T. 1977. The behavior of *Hydrometra championiana* (Hemiptera: Hydrometridae) and resource partitioning with *Tenagogonus quadrilineatus* (Hemiptera: Gerridae). J. Kans. Ent. Soc. 50:263-271.

2444. Maier, C. T. 1978. The immature stages and biology of *Mallota posticata* (Fabricius) (Diptera: Syrphidae). Proc. Ent. Soc. Wash. 80:424-440.

2445. Maier, C. T. 1982. Larval habitats and mate-seeking sites of flower flies (Diptera: Syrphidae, Eristalinae). Proc. Ent. Soc. Wash. 84:603-609.

2446. Maitland, P. S. 1979. The habitats of British Ephemeroptera, pp. 123-139. In J. F. Flannagan and K. E. Marshall, (eds). Proc. 3rd. Internat. Conf. Ephemeroptera, Plenum Press, N.Y. 552 p.

2447. Maitland, P. S., and M. M. Penny. 1967. The ecology of the Simuliidae in a Scottish river. J. Anim. Ecol. 35:179-206.

2448. Makarchenko, E. A., L. A. Chubareva, and M. A. Makarchenko. 1989. New data on the distribution, karyology and biology of the archaic insect (Diptera, Nymphomyiidae) from the Soviet Far East, pp 15-19. In Systematics and ecology of river organisms. (in Russian).

2449. Makarchenko, E. A., and M. A. Makarchenko. 1983. The archaic flies Nymphomyiidae (Diptera) from the far east of the USSR, pp. 92-95. In O. A. Skarlato (ed.). Diptera (Insecta), their systematics, geographic distribution and ecology. Zool. Inst. Lennigrad, Acad. Sci. USSR. (in Russian.).

2450. Maki, A. W., K. W. Stewart, and J. K. G. Silvey. 1973. The effects of dibrom on respiratory activity of the stonefly, *Hydroperla crosbyi*, hellgrammite, *Corydalus cornutus* and the golden shiner, *Notemigonus crysoleucas*. Trans. Am. Fish. Soc. 102:806-815.

2451. Malaise, R. 1937. A new insect trap. Ent. Tidskr. 58:148-160.

2452. Malas, D., and J. B. Wallace. 1977. Strategies for coexistence in three net-spinning caddisflies (Trichoptera) in second-order southern Appalachian streams. Can. J. Zool. 55:1829-1840.

2453. Malcolm, S. E. 1971. The water beetles of Maine: including the families Gyrinidae, Haliplidae, Dytiscidae, Noteridae, and Hydrophilidae. Univ. Maine Tech. Bull. 48:1-49.

2454. Malcolm, S. E. 1979. Two new species of *Laccobius spangleri*, *Laccobius reftexipenis* from Eastern North America (Coleoptera: Hydrophilidae). J. N.Y. Ent. Soc. 87:59-65.

2455. Malcolm, S. E. 1980. *Oreomicrus explanatus* new genus, and elaboration of the tribe Omicrini (Coleoptera: Hydrophilidae: Sphaeridiinae). Ann. Ent. Soc. Am. 73:185-188.

2456. Malicky, H. 1973. Trichoptera (Köcherfliegen). Handb. Zool. 4:1-114.

2457. Malicky, H. 1980. Evidence for seasonal migration of the larvae of two species of philopotamid caddisflies (Trichoptera) in a mountain stream in lower Austria. Aquat. Insects 2:153-160.

2458. Malloch, J. R. 1914. American black flies or buffalo gnats. U.S. Dept. Agr. Bur. Ent. Tech. Ser. 26:1-83.

2459. Malloch, J. R. 1915. The Chironomidae, or midges, of Illinois, with particular reference to the species occurring in the Illinois River. Bull. Ill. State Lab. Nat. Hist. 10:275-543.

2460. Malmqvist, B. 1994. Preimaginal blackflies (Diptera: Simuliidae) and their predators in a central Scandinavian lake outlet stream. Ann. Zool. Fennici 31:245-255.

2461. Malmqvist, B., and P. Sjöström. 1984. The microdistribution of some lotic insect predators in relation to their prey and to abiotic factors. Freshwat. Biol. 14:649-656.

2462. Malmqvist, B., P. Sjöström, and K. Frick. 1991. The diet of two species of *Isoperla* (Plecoptera: Perlodidae) in relation to season, site, and sympatry. Hydrobiologia 213:191-204.

2463. Mangum, F. A., and R. N. Winget. 1991. Environmental profile of *Drunella* (*Eatonella*) *doddsi* (Needham) (Ephemeroptera: Ephemerellidae). J. Freshwat. Ecol. 6:11-22.

2464. Mangum, F. A., and R. N. Winget. 1993. Environmental profile of *Drunella grandis* Eaton (Ephemeroptera: Ephemerellidae) in the Western United States. J. Freshwat. Ecol. 8:133-140.

2465. Manton, S. M., and D. T. Anderson. 1979. Polyphyly and the evolution of arthropods, pp. 269-321. In M. R. House (ed.). The origin of major invertebrate groups. Academic, London. 515 pp.

2466. Mantula, J. 1911. Untersuchungen über die Funktionen der Zentralnervensystems bei Insekten. Pfluegers Arch. Ges. Physiol. 138:388-456.

2467. Manuel, K. L., and T. C. Folsom. 1982. Instar sizes, life cycles, and food habits of five *Rhyacophila* (Trichoptera: Rhyacophilidae) species from the Appalachian Mountains of South Carolina, U.S.A. Hydrobiologia 97:281-285.

2468. Manuel, K. L., and A. P. Nimmo. 1984. The caddisfly genus *Ylodes* in North America (Trichoptera: Leptoceridae), pp. 219-224. In J. C. Morse (ed.). Proc. IV. Int. Symp. Trichoptera, Clemson Univ., SC. Dr. W. Junk Publ., The Hague, Netherlands.

2469. Maquet, B., and D. Rjosillon. 1985. The life cycle of the mayfly *Baetis rhodani* Pictet in two Belgian salmonid rivers. Verh. Int. Ver. Theor. Angew. Limnol. 22:3244-3249.

2470. Marchal, P. 1900. Sur un nouvel hymenoptere aquatique, le *Limnodytes gerriphagus* n. gen. n. sp. Ann. Soc. Ent. France 69:171-176.

2471. Marchand, W. 1917. An improved method of rearing tabanid larvae. J. Econ. Ent. 10:469-472.

2472. Marchand, W. 1923. The larval stages of *Limnophora discreta* Stein (Diptera, Anthomyiidae). Bull. Brooklyn Ent. Soc. 18:58-62.

2473. Marks, E. P. 1957. The food pump of *Pelocoris* and comparative studies on other aquatic Hemiptera. Psyche 64:123-134.

2474. Marsh, P. M., S. R. Shaw and R. A. Wharton. 1987. An identification manual for the North America genera of the family Braconidae (Hymenoptera). Mem. Ent. Soc. Wash. No. 13. 98pp.

2475. Marshall, J. E. 1979. A review of the genera of the Hydroptilidae (Trichoptera). Bull. Brit. Mus. Nat. Hist. 39:135-239.

2476. Marshall, J. S., and D. J. Larson. 1982. The adult caddisflies (Trichoptera: Insecta) of insular Newfoundland. Memorial Univ. of Newfoundland Occ. Pap. in Biology 6. 85 pp.

2477. Marten, M., and P. Zwick. 1989. The temperature dependence of embryonic and larval development in *Protonemura intricata* (Plecoptera: Nemouridae). Freshwat. Biol. 22:1-14.

2478. Martin, C. H. 1927. Biological studies of 2 hymenopterous parasites of aquatic insect eggs. Entomologica Am. 8:105-151.

2479. Martin, C. H. 1928. An exploratory survey of characters of specific value in the genus *Gelastocoris* Kirkaldy, and some new species. Univ. Kans. Sci. Bull. 18:351-369.

2480. Martin, I. D. 1985. Microhabitat selection and life cycle patterns of two *Rhyacophila* species (Trichoptera: Rhyacophilidae) in southern Ontario streams. Freshwat. Biol. 15:1-14.

2481. Martin, I. D., and R. J. Mackay. 1983. Growth rates and prey selection of two congeneric predatory caddisflies (Trichoptera: Rhyacophilidae). Can. J. Zool. 61:895-900.

2482. Martin, I. D., W. D. Taylor, and D. R. Barton. 1991. Experimental analysis of density dependent effects on two caddisflies and their algal food. J. N. Am. Benthol. Soc. 10:404-418.

2483. Martin, J. O. 1900. A study of *Hydrometra lineata*. Can. Ent. 32:70-76.

2484. Martin, M. M., J. J. Kukor, J. S. Martin, and R. W. Merritt. 1981a. Digestive enzymes of larvae of three species of caddisflies (Trichoptera). Insect Biochem. 11:501-505.

2485. Martin, M. M., J. J. Kukor, J. S. Martin, and R. W. Merritt. 1985. The digestive enzymes of larvae of the black fly, *Prosimulium fuscum* (Diptera, Simuliidae). Comp. Biochem. Physiol. 82B:37-39.

2486. Martin, M. M., J. S. Martin, J. J. Kukor, and R. W. Merritt. 1981b. The digestive enzymes of detritus-feeding stonefly nymphs (Plecoptera: Pteronarcyidae). Can. J. Zool. 59: 1947-1951.

2487. Martin, M. M., J. S. Martin, J. J. Kukor, and R. W. Merritt. 1980. The digestion of protein and carbohydrate by the stream detritivore, *Tipula abdominalis* (Diptera, Tipulidae). Oecologia 46:360-364.

2488. Martin, P. J. S., and J. D. Edman. 1993. Assimilation rates of different particulate foods for *Simulium verecundum* (Diptera: Simuliidae). Ann. Ent. Soc. Am. 30:805-809.

2489. Martin, R. 1907. Cordulines. Collections zoologiques du Baron Edm. de Selys Longchamps, Catalogue systematique et descriptif. Coll. Selys Longchamps 17:1-94.

2490. Martin, R. 1908. Aeschines. Collections zoologiques du Baron Edm. de Selys Longchamps, Catalogue systématique at descriptif. Coll. Selys. Longchamps 18:1-84.

2491. Martin, R. 1909a. Aeschnines. Collections zoologiques du Baron Edm. de Selys Longchamps, Catalogue systematique et descriptif. Coll. Selys Longchamps 19:85-156.

2492. Martin, R. 1909b. Aeschnines. Collections zoologiques du Baron Edm. de Selys Longchamps, Catalogue systematique et descriptif. Coll. Selys Longchamps 20:157-223.

2493. Martin, R.D.C. 1939. Life histories of *Agrion aequabile* and *Agrion maculatum* (Agriidae: Odonata). Ann. Ent. Soc. Am. 35:601-619.

2494. Martin, T. H., D. M. Johnson, and R. D. Moore. 1991. Fish-mediated alternative life-history strategies in *Epitheca cynosura*. J. N. Am. Benthol. Soc. 10:271-279.

2495. Martinson, R. J., and J. D. Ward. 1982. Life history and ecology of *Hesperophylax occidentalis* (Banks) (Trichoptera: Limnephilidae) from three springs in the Piceance Basin, Colorado. Freshwat. Invert. Biol. 1:41-47.

2496. Martynova, E. F. 1972. Springtails (Collembola) inhabiting the outlets of subterranean water in the Kirghis and Uzbek SSR. Trudy Zool. Inst., Leningrad. 51:147-150.

2497. Marx, E.J.F. 1957. A review of the subgenus *Donacia* in the Western Hemisphere (Coleoptera: Donaciidae). Bull. Am. Mus. Nat. Hist. 112:191-278.

2498. Maschwitz, D. E. 1976. Revision of the Nearctic species of the subgenus *Polypedilum*, (Chironomidae: Diptera). Ph. D. diss., Univ. Minn., St. Paul, MN. 325pp.

2499. Masner, L. 1972. The classification and interrelationships of Thoronini (Hym., Proctotrupoidea, Scelionidae). Can. Ent. 104:833-849.

2500. Masner, L. 1976. Revisionary notes and keys to world genera of Scelionidae (Hym.: Proctotrupoidea). Mem. Ent. Soc. Can. 97:1-87.

2501. Masner, L. 1980. Key to genera of Scelionidae of the Holarctic region, with descriptions of new genera and species (Hymenoptera: Proctotrupoidea). Mem. Ent. Soc. Can. 113:1-54.

2502. Masner, L., and M. A. Kozlov. 1965. Four remarkable egg parasites in Europe (Hym., Scelionidae, Telenominae). Acta Ent. Bohem. 62:287-293.

2503. Mason, J. C. 1976. Evaluating a substrate tray for sampling the invertebrate fauna of small streams, with comments on general sampling problems. Arch. Hydrobiol. 78:51-70.

2504. Mason, W. R. M. 1968. New Canadian Cryptinae (Ichneumonidae; Hymenoptera). Can. Ent. 100:17-23.

2505. Mason, W. T., J. B. Anderson, and G. E. Morrison. 1967. A limestone-filled, artificial substrate sampler-float unit for collecting macroinvertebrates in large streams. Progr. Fish Cult. 29:1-74.

2506. Mason, W. T., Jr. 1973. An introduction to the identification of chironomid larvae. MERC/EPA, Cincinnati. 90 pp.

2507. Mason, W. T., Jr. 1991. Sieve sample splitter for benthic invertebrates. J. Freshwat. Ecol. 6:445-449.

2508. Mason, W. T., Jr., and P. A. Lewis. 1970. Rearing devices for stream insect larvae. Progr. Fish Cult. 32:61-62.

2509. Mason, W. T., Jr., C. I. Weber, P. A. Lewis, and E. C. Julian. 1973. Factors affecting the performance of basket and multiplate macroinvertebrate samplers. Freshwat. Biol. 3:409-436.

2510. Mason, W.R.M. 1971. An Indian *Agriotypus* (Hym.: Agriotypidae). Can. Ent. 103:1521-1524.

2511. Masteller, E. C. 1977. An aquatic emergence trap on a shale stream of western Pennsylvania. Melsheimer Ent. Ser. 23:10-15.

2512. Masteller, E. C., and O. S. Flint. 1979. Light trap and emergence trap records of caddisflies (Trichoptera) of the Lake Erie region of Pennsylvania and adjacent Ohio. Great Lakes Ent. 12:165-177.

2513. Masteller, E. C., and O. S. Flint Jr. 1992. The Trichoptera (caddisflies) of Pennsylvania: An annotated checklist. J. Penn. Acad. Sci. 66: 68-78.

2514. Mather, T. N. 1981. Larvae of alderfly (Megaloptera: Sialidae) from pitcher plant. Ent. News. 92:32.

2515. Matheson, R. 1912. The Haliplidae of America north of Mexico. J. N.Y. Ent. Soc. 21:91-123.

2516. Matheson, R. 1914. Life-history of a dytiscid beetle *Hydroporus septentrionalis* (Gyll.). Can. Ent. 46:37-50.

2517. Matheson, R., and C. R. Crosby. 1912. Aquatic Hymenoptera in America. Ann. Ent. Soc. Am. 5:65-71.

2518. Mathis, M. L., and D. E. Bowles. 1989. A new microcaddisfly genus (Trichoptera: Hydroptilidae) from the interior Highlands of Arkansas. J. N. Y. Ent. Soc. 97:187-191.

2519. Mathis, M. L., and D. E. Bowles. 1992. A preliminary survey of the Trichoptera of the Ozark Mountains, Missouri, U. S. A. Ent. News 103:19-29.

2520. Mathis, W. N. 1975. A systematic study of *Coenia* and *Paracoenia* (Diptera: Ephydridae). Great Basin Nat. 35:65-85.

2521. Mathis, W. N. 1977. A revision of the genus *Rhysophora* Cresson with a key to related genera (Diptera: Ephydridae). Proc. Biol. Soc. Wash. 90:921-945.

2522. Mathis, W. N. 1978. A revision of the Nearctic species of *Limnellia* Malloch (Diptera: Ephydridae). Proc. Biol. Soc. Wash. 91:250-293.

2523. Mathis, W. N. 1979a. Studies of Notiphilinae (Diptera: Ephydridae), I: revision of the Nearctic species of *Notiphila* Fallen, excluding the *caudata* group. Smithson. Contr. Zool. 287:1-111.

2524. Mathis, W. N. 1979b. Studies of Ephydrinae (Diptera: Ephydridae), II: phylogeny, classification, and zoogeography of Nearctic *Lamproscatella* Hendel. Smithson. Contr. Zool. 295:1-41.

2525. Mathis, W. N. 1982. Studies of Ephydrinae (Diptera: Ephydridae), VII: revision of the genus *Setacera* Cresson. Smithson. Cont. Zool. 350:1-57.

2526. Mathis, W. N. 1983. A revision of the genus *Asmeringa* Becker (Diptera: Ephydridae). Israel J. Ent. 17:67-79.

2527. Mathis, W. N. 1986. Studies of Psilopinae (Diptera: Ephydridae). I: a review of the shore fly genus *Placopsidella* Kertész. Smithson. Cont. Zool. 430:1-30.

2528. Mathis, W. N. 1988a. First record of the genus *Procanace* Hendel from North America, with the description of a new species (Diptera: Canacidae). Proc. Ent. Soc. Wash. 90:329-333.

2529. Mathis, W. N. 1988b. First record of the shore-fly genus *Placopsidella* Kertész from North America (Diptera: Ephydridae). Proc. Ent. Soc. Wash. 90:334-337.

2530. Mathis, W. N. 1990. A revision of the shore-fly genus *Diphuia* Cresson (Diptera: Ephydridae). Proc. Ent. Soc. Wash. 92:746-756.

2531. Mathis, W. N. 1992. World catalog of the beach-fly family Canacidae (Diptera). Smithson. Cont. Zool. 536:1-18.

2532. Mathis, W. N. 1993a. A revision of the shore-fly genera *Hostis* Cresson and *Paratissa* Coquillett (Diptera: Ephydridae). Proc. Ent. Soc. Wash. 95:21-47.

2533. Mathis, W. N. 1993b. Studies of Gymnomyzinae (Diptera: Ephydridae), IV: a revision of the shore-fly genus *Hecamede* Haliday. Smithson. Cont. Zool. 541:1-46.

2534. Mathis, W. N. 1994. A revision of the shore-fly genus *Clasiopella* (Diptera: Ephydridae). Proc. Ent. Soc. Wash. 96 (in press).

2535. Mathis, W. N., and K. Ghorpade. 1985. Studies of Parydrinae (Diptera: Ephydridae), I: a review of the genus *Brachydeutera* Loew from the Oriental, Australian, and oceanian regions. Smithson. Cont. Zool. 406:1-25.

2536. Mathis, W. N., and G. E. Shewell. 1978. Studies of Ephydrinae (Diptera: Ephydridae), I: Revisions of *Parascatella* Cresson and the *triseta* group of *Scatella* Robineau-Desvoidy. Smithson, Contr. Zool. 285:1-44.

2537. Mathis, W. N., and K. W. Simpson. 1981. Studies of Ephydrinae (Diptera: Ephydridae), V. The genera *Cirrula* Cresson and *Dimecoenia* Cresson in North America. Smithson. Contr. Zool. 329:1-51.

2538. Mathis, W. N., and W. E. Steiner, Jr. 1986. An adventive species of Brachydeutera Loew in North America (Diptera: Ephydridae). J. New York Ent. Soc. 94:56-61.

2539. Mathis, W. N., T. Zatwarnicki, and M. G. Krivosheina. 1993. Studies of Gymnomyzinae (Diptera: Ephydridae), V: a revision of the shore-fly genus *Mosillus* Latreille. Smithson. Cont. Zool. 548:1-38.

2540. Matsuda, R. 1960. Morphology, evolution and a classification of the Gerridae. Univ. Kans. Sci. Bull. 41:25-632.

2541. Matsuda, R. 1965. Morphology and evolution of the insect head. Mem. Am. Ent. Inst. 4:1-334.

2542. Matsuda, R. 1970. Morphology and evolution of the insect thorax. Mem. Ent. Soc. Can. 76:1:431.

2543. Matsuda, R. 1976. Morphology and evolution of the insect abdomen. Pergamon, Oxford. 534 pp.

2544. Matta, J. F. 1974. The insects of Virginia: No. 8. The aquatic Hydrophilidae of Virginia (Coleoptera: Polyphaga). Bull. Res. Div. Va. Poly. Inst. State Univ. 94:1-144.

2545. Matta, J. F. 1976. Haliplidae of Virginia (Coleoptera: Adephaga). The insects of Virginia. No. 10. Va. Poly. Inst. State Univ., Res. Div. Bull. 109:1-26.

2546. Matta, J. F. 1978. An annotated list of the Odonata of southeastern Virginia. Va. J. Sci. 29:180-182.

2547. Matta, J. F. 1979. New species of Nearctic *Hydroporus* (Coleoptera: Dytiscidae) *Hydroporus sulphurius, Hydroporus ouachitus, Hydroporus allegenianus.* Proc. Biol. Soc. Wash. 92:287-293.

2548. Matta, J. F. 1982. The bionomics of two species of *Hydrochara* (Coleoptera: Hydrophilidae) with descriptions of their larvae. Proc. Ent. Soc. Washington 84:461-467.

2549. Matta, J. F. 1983. Description of the larva of *Uvarus granarius* (Coleoptera: Dytiscidae), with a key to the Nearctic Hydroporinae larvae. Coleop. Bull. 37:203-207.

2550. Matta, J. F. 1986. *Agabus* (Coleoptera: Dytiscidae) larvae of southeastern United States. Proc. Ent. Soc. Washington 88:515-520.

2551. Matta, J. F. and G. W. Wolf. 1981. A revision of the subgenus *Heterosternuta* Strand of *Hydroporus* clairville (Coleoptera: Dytiscidae). Pan-Pacif. Ent. 57:176-219.

2552. Matteson, J. D., and G. Z. Jacobi. 1980. Benthic macroinvertebrates found on the freshwater sponge *Spongilla lacustris*. Great Lakes Ent. 13:169-172.

2553. Matthews, K. A., and D. C. Tarter. 1989. Ecological life history, including laboratory respiratory investigations of the mayfly, *Ameletus tarteri* (Ephemeroptera: Siphlonuridae). Psyche 96:21-38.

2554. Mattingly, P. F. 1973. Culicidae (Mosquitoes), pp. 37-107. *In* K.G.V. Smith (ed.). Insects and other arthropods of medical importance. Bull. Brit. Mus. Nat. Hist. 561 pp.

2555. Mattingly, R. L. 1986. Trophic and habitat requirements of two deposit-feeding stream invertebrates, *Ptychoptera townesi* (Diptera) and *Paraleptophlebia* spp. (Ephemeroptera). Ph.D. diss., Oregon State Univ., Corvallis, OR. 119pp.

2556. Mattingly, R. L. 1987a. Handling of coarse and fine particulate organic matter by the aquatic insects *Paraleptophlebia gregalis* and *P. temporalis* (Ephemeroptera: Leptophlebiidae). Freshwat. Biol. 18:255-266.

2557. Mattingly, R. L. 1987b. Resource utilization by the freshwater deposit feeder *Ptychoptera townesi* (Diptera: Ptychopteridae). Freshwat. Biol. 18:241-253.

2558. Mattingly, R. L., K. W. Cummins, and R. H. King. 1981. The influence of substrate organic content on the growth of a stream chironomid. Hydrobiologia 77:161-165.

2559. May, M. L. 1992. Morphological and ecological differences among species of *Ladona* (Anisoptera: Libellulidae). Bull. Am. Odonatol. 1:51-56.

2560. Mayer, M. S., and G. E. Likens. 1987. The importance of algae in a shaded headwater stream as food for an abundant caddisfly (Trichoptera). J. N. Am. Benthol. Soc. 6:262-269.

2561. Maynard, E. A. 1951. A monograph of the Collembola or springtail insects of New York State. Comstock, Ithaca. 339 pp.

2562. Mayr, E. 1969. Principles of systematic zoology. McGraw-Hill, N.Y. 428 pp.

2563. Mayr, E., and P. D. Ashlock. 1991. Principles of systematic zoology (2nd ed.). McGraw-Hill, Inc., NY. 475pp.

2564. Mc Auliffe, J. R., and N. E. Williams. 1983. Univoltine life cycle of *Parargyractis confusalis* (Walker) (Lepidoptera: Pyralidae) in the northern part of its range. Am. Midl. Nat. 110:440-443.

2565. McAlpine, J. F., B. V. Peterson, G. E. Shewell, H. J. Teskey, J. R. Vockeroth, and D. M. Wood (coords.). 1981. Manual of Nearctic Diptera, Vol. 1. Res. Branch, Agric. Can. Monogr. 27. 674 pp.

2566. McAlpine, J. F., B. V. Peterson, G. E. Shewell, H. J. Teskey, J. R. Vockeroth, and D. M. Wood (coords.). 1987. Manual of Nearctic Diptera, Vol. 2. Res. Branch, Agr. Can. Monogr. 28. 1332pp.

2567. McAlpine, J. F., and D. M. Wood (coords.). 1989. Manual of Nearctic Diptera, Vol.3. Res. Branch Agric. Can. Monogr. 32. 1581pp.

2568. McAtee, W. L., and J. R. Malloch. 1925. Revision of bugs of the family Cryptostemmatidae in the collection of the United States National Museum. Proc. U.S. Nat. Mus. 67:1-42.

2569. McAuliffe, J. R. 1982. Behavior and life history of *Leucotrichia pictipes* (Banks) (Trichoptera: Hydroptilidae) with special emphasis on case reoccupancy. Can. J. Zool. 60:1557-1561.

2570. McAuliffe, J. R. 1983. Resource depression by a stream herbivore: effects on distributions and abundances of other grazers. Oikos 42:327-333.

2571. McAuliffe, J. R. 1984. Competition for space, disturbance, and the structure of a benthic stream community. Ecology 65:894-908.

2572. McAuliffe, J. R., and N. E. Williams. 1983. Univoltine life cycle of *Parargyractis confusalis* (Walker) (Lepidoptera: Pyralidae) in the northern part of its range. Am. Midl. Nat. 110:440-444.

2573. McCafferty, W. P. 1975. The burrowing mayflies (Ephemeroptera: Ephemeroidea) of the United States. Trans. Am. Ent. Soc. 101:447-504.

2574. McCafferty, W. P. 1981. Aquatic entomology. Science Books Internat., Boston. 448 pp.

2575. McCafferty, W. P. 1988. Neotype designation for *Raptoheptagenia cruentata* (Walsh) (Ephemeroptera: Heptageniidae). Proc. Ent. Soc. Wash. 90:97.

2576. McCafferty, W. P. 1989. Characterization and relationships of the subgenera of *Isonychia* (Ephemeroptera: Oligoneuriidae). Ent. News 100:72-78.

2577. McCafferty, W. P. 1990. Chapter 2. Ephemeroptera, pp. 20-50. *In* D.A. Grimaldi (ed.). Insects from the Santana Formation, Lower Cretaceous, of Brazil. Bull. Am. Mus. Nat. Hist., 195.

2578. McCafferty, W. P. 1991a. Toward a phylogenetic classification of the Ephemeroptera (Insecta): A commentary on systematics. Ann. Ent. Soc. Am. 84:343-360.

2579. McCafferty, W. P. 1991b. The cladistics, classification, and evolution of the Heptagenioidea (Ephemeroptera), pp. 87-102. *In* J. Alba-Tercedor and A. Sanchez-Ortega (eds.). Overview and Strategies of Ephemeroptera and Plecoptera. Sandhill Crane Press, Gainesville, FA. 588pp.

2580. McCafferty, W. P. 1991c. Comparison of old and new world *Acanthametropus* (Ephemeroptera: Acanthametropodidae) and other psammophilous mayflies. Ent. News 102:205-214.

2581. McCafferty, W. P. 1992a. *Ephemerella apopsis*, a new species from rocky mountain high (Ephemeroptera: Ephemerellidae). Ent. News 103:135-138.

2582. McCafferty, W. P. 1992b. New larval descriptions and comparisons of North American *Choroterpes* (Ephemeroptera: Leptophlebiidae). Great Lakes Ent.. 25:71-78.

2583. McCafferty, W. P. 1992c. The sand-dwelling predatory mayfly *Pseudiron centralis* in Michigan (Ephemeroptera: Pseudironidae). Great Lakes Ent. 25:133-134.

2584. McCafferty, W. P. 1994. Distributional and classificatory supplement to the burrowing mayflies (Ephemeroptera: Ephemeroidea) of the United States. Ent. News 105:1-13.

2585. McCafferty, W. P., and Y. J. Bae. 1990. *Anthopotamus*, a new genus for North American species previously known as *Potamanthus* (Ephemeroptera: Potamanthidae). Ent. News 101:200-202.

2586. McCafferty, W. P., and Y. J. Bae. 1992. Filter-feeding habits of the larvae of *Anthopotamus* (Ephemeroptera: Potamanthidae). Ann. Limnol. 28:27-34.

2587. McCafferty, W. P., and Y. J. Bae. 1994. Life history aspects of *Anthopotamus verticis* (Ephemeroptera: Potomanthidae). Great Lakes Ent. 27:57-67.

2588. McCafferty, W. P., R. S. Durfee, and B. C. Kondratieff. 1993. Colorado mayflies (Ephemeroptera): an annotated inventory. Southwest. Nat. 38:252-274.

2589. McCafferty, W. P., and G. F. Edmunds, Jr. 1976. Redefinition of the family Palingeniidae and its implications for the higher classification of Ephemeroptera. Ann. Ent. Soc. Am. 69:486-490.

2590. McCafferty, W. P., and G. F. Edmunds, Jr. 1979. The higher classification of the Ephemeroptera and its evolutionary basis. Ann. Ent. Soc. Am. 72:5-12.

2591. McCafferty, W. P., and B. L. Hull, Jr. 1978. The life cycle of the mayfly *Stenacron interpunctatum* (Ephemeroptera: Heptageniidae). Great Lakes Ent. 11:209-216.

2592. McCafferty, W. P., and M. C. Minno. 1979. The aquatic and semiaquatic Lepidoptera of Indiana and adjacent areas. Great Lakes Ent.12:179-187.

2593. McCafferty, W. P., and A. V. Provonsha. 1984. The first adult of *Spinadis* (Ephemeroptera: Heptageniidae). Ent. News 95:173-179.

2594. McCafferty, W. P., and A. V. Provonsha. 1985. Systematics of *Anepeorus* (Ephemeroptera: Heptageniidae). Great Lakes. Ent.18:1-6.

2595. McCafferty, W. P., and A. V. Provonsha. 1986. Comparative mouthpart morphology and evolution of the carnivorous Heptageniidae (Ephemeroptera). Aquat. Insects 8:83-89.

2596. McCafferty, W. P., and A. V. Provonsha. 1988. Revisionary notes on predaceous Heptageniidae based on larval and adult associations (Ephemeroptera). Great Lakes Ent. 21:15-17.

2597. McCafferty, W. P., B. Stark, and A. V. Provonsha. 1990. Ephemeroptera, Plecoptera, and Odonata, pp. 43-58. *In* M. Kosztarab and C.W. Schaefer (eds.). Systematics of the North

American insects and arachnids: status and needs. Virginia Agr. Exp. Sta. Informat. Ser. 90-1, Virginia Polytech. Inst. St. Univ., Blacksburg.

2598. McCafferty, W. P., and R. D. Waltz. 1990. Revisionary synopsis of the Baetidae (Ephemeroptera) of North and Middle America. Trans. Am. Ent. Soc. 116:769-799.

2599. McCafferty, W. P., and R. D. Waltz. 1995. *Labiobaetis* (Ephemeroptera: Baetidae): New status, new North American species, and related new genus. Ent. News 27:209-216.

2600. McCafferty, W. P., and T. Wang. 1994a. Relationships of the genera *Acanthametropus*, *Analetris*, and *Siphluriscus*, with a re-evaluation of the higher classification (Ephemeroptera: Pisciforma). Great Lakes Ent. 27:209-216.

2601. McCafferty, W. P., and T. Wang. 1994b. Phylogenetics and the classification of the *Timpanoga* complex (Ephemeroptera: Ephemerellidae). J. N. Am. Benthol. Soc. 13:569-579.

2602. McCafferty, W. P., M. J. Wigle, and R. D. Waltz. 1994. Contribution to the taxonomy and biology of *Acentrella turbida* (McDunnough) (Ephemeroptera: Baetidae). Pan-Pac. Insects 70:301-308.

2603. McCaskill, V. H., and R. Prins. 1968. Stoneflies (Plecoptera) of northwestern South Carolina. J. Elisha Mitchell Sci. Soc. 84:448-453.

2604. McCauley, V. J. E. 1976. Efficiency of a trap for catching and retaining insects emerging from standing water. Oikos 27:339-346.

2605. McClean, E. B. 1990. Sexual dimorphism and predaceous feeding habits of the waterstrider *Gerris remigis* Say (Heteroptera: Gerridae). Can. J. Zool. 68:2688-2691.

2606. McClure, R. G., and K. W. Stewart. 1976. Life cycle and production of the mayfly *Choroterpes* (*Neochoroterpes*) *mexicanus* Allen (Ephemeroptera: Leptophebiidae). Ann. Ent. Soc. Am. 69:134-148.

2607. McCorkle, D. V. 1967. A revision of the species of *Elophorus* Fabricius in America north of Mexico. Ph.D. diss., University of Washington, Seattle.

2608. McCreadie, J. W., and M. H. Colbo. 1991. A critical examination of four methods of estimating the surface area of stone substrate from streams in relation to sampling Simuliidae (Diptera). Hydrobiologia 220:205-210.

2609. McCullough, D. A., G. W. Minshall, and C. E. Cushing. 1979a. Bioenergetics of lotic filter-feeding insects *Simulium* spp. (Diptera) and *Hydropsyche occidentalis* (Trichoptera) and their function in controlling organic content in streams. Ecology 60:585-596.

2610. McCullough, D. A., G. W. Minshall, and C. E. Cushing. 1979b. Bioenergetics of a stream "collector" organism *Tricorythodes minutus* (Insecta: Ephemeroptera). Limnol. Oceanogr. 24:45-58.

2611. McDiffett, W. F. 1970. The transformation of energy by a stream detritivore *Pteronarcys scotti* (Plecoptera). Ecology 51:975-988.

2612. McDonald, J. L., T. P. Sluss, J. D. Lang, and C. C. Roan. 1973. The mosquitoes of Arizona. Univ. Arizona Agric. Exp. Sta. Tech. Bull. 205. 21 pp.

2613. McDunnough, J. 1933. Notes on the biology of certain tortricid species with structural details of the larvae and pupae. Can. J. Res. 9:502-517.

2614. McElravy, E. P., T. L. Arsuffi, and B. A. Foote. 1977. New records of caddisflies (Trichoptera) for Ohio. Proc. Ent. Soc. Wash. 79:599-604.

2615. McElravy, E. P., H. Wolda, and V. H. Resh. 1982. Seasonality and annual variability of caddisfly adults (Trichoptera) in a "nonseasonal" tropical environment. Arch. Hydrobiol. 94:302-317.

2616. McFadden, M. W. 1967. Soldier fly larvae in America north of Mexico. Proc. U.S. Nat. Mus. 121:1-72.

2617. McFadden, M. W. 1972. The soldier flies of Canada and Alaska (Diptera: Stratiomyidae) I. Beridinae, Sarginae, and Clitellariinae. Can. Ent. 104:531-562.

2618. McGaha, Y. J. 1952. The limnological relations of insects to certain aquatic flowering plants. Trans. Am. Microsc. Soc. 71:355-381.

2619. McGaha, Y. J. 1954. Contribution to the biology of some Lepidoptera which feed on certain aquatic flowering plants. Trans. Am. Microsc. Soc. 73:167-177.

2620. McIntire, C. D. 1983. A conceptual framework for process studies in lotic ecosystems, pp 43-67. *In* T. D. Fontaine, III., and S. M. Bartell (eds.). Dynamics of lotic ecosystems. Ann Arbor Sci. Publ. 494p.

2621. McKeever, S., and F. E. French. 1991a. *Corethrella* (Diptera: Corethrellidae) of eastern North America: Laboratory life history and field responses to anuran calls. Ann. Ent. Soc. Am. 84:493-497.

2622. McKeever, S., and F. E. French. 1991b. *Corethrella* (Diptera: Corethrellidae) of North America north of Mexico: Distribution and morphology of immature stages. Ann. Ent. Soc. Am. 84:522-530.

2623. McKeever, S., and F. E. French. 1992. Observations on the laboratory life history of *Chrysops atlanticus*, *C. univittatus* and *C. vittatus* (Diptera: Tabanidae). J. Ent. Sci. 24:458-460.

2624. McKinnon, C. N., and J. T. Polhemus. 1986. Notes on the genus *Ioscytus* with the description of a new species and a key to species (Hemiptera: Heteroptera: Saldidae). J. New York Ent. Soc. 94:434-441.

2625. McKinstry, A. P. 1942. A new family of Hemiptera-Heteroptera proposed for *Macrovelia hornii* Uhler. Pan-Pacif. Ent. 18:90-96.

2626. McLachlan, A. J. 1977. Some effects of tube shape on the feeding of *Chironomus plumosus* L. J. Anim. Ecol. 46:139-146.

2627. McLachlan, A. J., A. Brennan, and R. S. Wotton. 1978. Particle size and chironomid (Diptera) food in an upland river. Oikos 31:247-252.

2628. McLachlan, A. J., and M. A. Cantrell. 1980. Survival strategies in tropical rainpools. Oecologia 47:344-351.

2629. McLaughlin, R. E. 1990. Predation rate of larval *Corethrella brakeleyi* (Diptera: Chaoboridae) on mosquito larvae. Florida Ent. 73:143-146.

2630. McMahon, J. A., D. J. Schimph, D. C. Anderson, K. G. Smith, and R. L. Bayr, Jr. 1981. An organism-centered approach to some community and ecosystem concepts. J. Theor. Biol. 88:287-307.

2631. McPeek, M. A. 1990. Determination of species composition in the *Enallagma* damselfly assemblages of permanent lakes. Ecology 71:83-98.

2632. McPeek, M. A., and P. H. Crowley. 1987. The effects of density and relative size on the aggressive behaviour, movement and feeding of damselfly larvae (Odonata: Coenagrionidae). Anim. Behav. 35:1051-1061.

2633. McPherson, J. E. 1965. Notes on the life history of *Notonecta hoffmanni* (Hemiptera: Notonectidae). Pan-Pacif. Ent. 41:86-89.

2634. McPherson, J. E. 1966. Notes on the laboratory rearing of *Notonecta hoffmanni* (Hemiptera: Notonectidae). Pan-Pacif. Ent. 42:54-56.

2635. McPherson, J. E. 1986. Life history of *Neoplea striola* (Hemiptera: Pleidae). Great Lakes Ent. 19:217-220.

2636. McPherson, J. E., and R. J. Packauskas. 1986. Life history and laboratory rearing of *Belostoma lutarium* (Heteroptera: Belstomatidae) with descriptions of immature stages. J. New York Ent. Soc. 94:154-162.

2637. McPherson, J. E., and R. J. Packauskas. 1987. Life history and laboratory rearing of *Nepa apiculata* (Heteroptera: Nepidae) with descriptions of immature stages. Ann. Ent. Soc. Am. 80:680-685.

2638. McShaffrey, D. 1988. Behavior, functional morphology, and ecology related to feeding in aquatic insects with particular reference to *Stenacron interpunctatum*, *Rhithrogena pellucida* (Ephemeroptera: Heptageniidae), and *Ephemerella needhami* (Ephemeroptera: Ephemerellidae). Ph.D. diss., Purdue Univ., Lafayette, IN. 247pp.

2639. McShaffrey, D., and W. P. McCafferty. 1986. Feeding behavior of *Stenacron interpunctatum* (Ephemeroptera: Heptageniidae). J. N. Am. Benthol. Soc. 5:200-210.

2640. McShaffrey, D., and W. P. McCafferty. 1987. The behavior and form of *Psephenus herricki* Dekay (Coleoptera: Psephenidae) in relation to water flow. Freshwat. Biol. 18:319-324.

2641. McShaffrey, D., and W. P. McCafferty. 1988. Feeding behavior of *Rhithrogena pellucida* (Ephemeroptera: Heptageniidae). J. N. Am. Benthol. Soc. 7:87-99.

2642. McShaffrey, D., and W. P. McCafferty. 1990. Feeding behavior and related functional morphology of the mayfly *Ephemerella needhami* (Ephemeroptera: Ephemerellidae). J. Insect Behav. 3:673-688.
2643. McShaffrey, D., and W. P. McCafferty. 1991. Ecological association of the mayfly *Ephemerella needhami* (Ephemeroptera: Ephemerellidae) with the green alga *Cladophora* (Chlorophyta: Cladophoraceae). J. Freshwat. Ecol. 6:383-394.
2644. McWilliams, K. L. 1969. A taxonomic revision of the North American species of the genus *Thermonectus* Dejean (Coleoptera: Dytiscidae). Diss. Abstr. 29(B):3781.
2645. Mead, A. R. 1938. New subspecies and notes on *Donacia* with key to the species of the Pacific States (Coleoptera, Chrysomelidae). Pan-Pacif. Ent. 14:113-120.
2646. Means, R. G. 1979. Mosquitoes of New York, Part I. The genus *Aedes* Meigen with identification keys to genera of Culicidae. N.Y. State Mus., Albany. 221 pp.
2647. Means, R. G. 1987. Mosquitoes of New York, part II. Genera of Culicidae other than *Aedes* occurring in New York. Bull. 430b, New York State Museum, Albany. 180pp.
2648. Mecom, J. O. 1970. Evidence of diurnal feeding activity in Trichoptera larvae. J. Grad. Res. Centr. S. Methodist Univ. 38:44-57.
2649. Mecom, J. O. 1972. Feeding habits of Trichoptera in a mountain stream. Oikos 23:401-407.
2650. Mecom, J. O., and K. W. Cummins. 1964. A preliminary study of the trophic relationships of the larvae of *Brachycentrus americanus* (Banks) (Trichoptera: Brachycentridae). Trans. Am. Microsc. Soc. 83:233-243.
2651. Meier, P. G., and H. C. Torres. 1978. A modified method for rearing midges (Diptera: Chironomidae). Great Lakes Ent. 11:89-91.
2652. Meier, P. G., and P. G. Bartholomae. 1980. Diel periodicity in the feeding activity of *Potamanthus myops* (Ephemeroptera). Arch. Hydrobiol. 88:1-8.
2653. Melin, B. E., and R. C. Graves. 1971. The water beetles of Miller Blue Hole, Sandusky County, Ohio (Insecta: Coleoptera). Ohio J. Sci. 71:73-77.
2654. Menke, A. E. (ed.). 1979. The semiaquatic and aquatic Hemiptera of California (Heteroptera: Hemiptera). Bull. Calif. Insect Surv. 21:1-166.
2655. Menke, A. S. 1958. A synopsis of the genus *Belostoma* Latreille, of America north of Mexico, with the description of a new species. Bull. S. Calif. Acad. Sci. 57:154-174.
2656. Menke, A. S. 1960. A taxonomic study of the genus *Abedus* Stål (Hemiptera: Belostomatidae). Univ. Calif. Publ. Ent. 16:393-440.
2657. Menke, A. S. 1963. A review of the genus *Lethocerus* in North and Central America, including the West Indies. Ann. Ent. Soc. Am. 56:261-267.
2658. Merlassino, M. B., and J. A. Schnack. 1978. Estructura communitaria y variation estacional de la mesofauna de artropodos en el pleuston de dos afluentes de la Laguna de Chascomus. Rev. Soc. Ent. Argentina 37:1-8.
2659. Merrill, D., and G. B. Wiggins. 1971. The larva and pupa of the caddisfly genus *Setodes* in North America (Trichoptera: Leptoceridae). Occ. Pap. Life Sci. Roy. Ont. Mus. 19:1-12.
2660. Merrill, R. J., and D. M. Johnson. 1984. Dietary niche overlap and mutual predation among coexisting larval *Anisoptera*. Odonatologica 13:387-406.
2661. Merritt, R. W. 1976. A review of the food habits of the insect fauna inhabiting cattle droppings in north central California. Pan-Pacif. Ent. 52:13-22.
2662. Merritt, R. W. 1987. Do different instars of *Aedes triseriatus* feed on particles of the same size? J. Am. Mosq. Contr. Assoc. 3:94-96
2663. Merritt, R. W., and D. A. Craig. 1987. Larval mosquito (Diptera: Culicidae) feeding mechanisms: Mucosubstance production for capture of fine particles. J. Med. Ent. 24:275-78.
2664. Merritt, R. W., D. A. Craig, E. D. Walker, H. A. Vanderploeg, and R. S. Wotton. 1992. Interfacial feeding behavior and particle flow patterns of *Anopheles quadrimaculatus* (Diptera: Culicidae). J. Insect Behav. 5:741-761.

2665. Merritt, R. W., and K. W. Cummins (eds.). 1978. An introduction to the aquatic insects of North America. Kendall/Hunt, Dubuque, Iowa. 441 pp.
2666. Merritt, R. W., and K. W. Cummins (eds.). 1984. An introduction to the aquatic insects of North America (2nd ed.). Kendall/Hunt Publ., Dubuque, IA. 722p.
2667. Merritt, R. W., and K. W. Cummins. 1996. Trophic relations of macroinvertebrates. *In* F. R. Hauer, and G. A. Lamberti (eds.). Methods in stream ecology. Academic Press, Orlando, FLA (in press).
2668. Merritt, R. W., K. W. Cummins, and J. R. Barnes. 1979. Demonstration of stream watershed community processes with some simple bioassay techniques, pp. 101-113. *In* V. H. Resh and D. M. Rosenberg (eds.). Innovative teaching in aquatic entomology. Can. Spec. Publ. Fish. Aquat. Sci. 43:1-118.
2669. Merritt, R. W., K. W. Cummins, and T. M. Burton. 1984. The role of aquatic insects in the processing and cycling of nutrients, pp. 134-163. *In* V. H. Resh and D. M. Rosenberg (eds.). The ecology of aquatic insects. Praeger Publishers, N.Y. 638 p.
2670. Merritt, R. W., R. H. Dadd, and E. D. Walker. 1992. Feeding behavior, natural food, and nutritional relationships of larval mosquitoes. Ann. Rev. Ent. 37:349-376.
2671. Merritt, R. W., and D. L. Lawson. 1979. Leaf litter processing in floodplain and stream communities, pp. 93-105. *In* R. R. Johnson and J. F. McCormick (coords.). Strategies for protection and management of floodplain wetlands and other riparian ecosystems. For. Serv./U.S.D.A. Gen. Tech. Rept. WO-12, Washington, D.C. 410 pp.
2672. Merritt, R. W., and D. L. Lawson. 1981. Adult emergence patterns and species distribution and abundance of Tipulidae in three woodland floodplains. Environ. Ent. 10:915-921.
2673. Merritt, R. W., and D. L. Lawson. 1992. The role of macroinvertebrates in stream-floodplain dynamics. Hydrobiologia 248:65-77.
2674. Merritt, R. W., M. M. Mortland, E. F. Gersabeck and D. H. Ross. 1978. X-ray diffraction analysis of particles ingested by filter-feeding animals. Ent. exp. appl. 24:27-34.
2675. Merritt, R. W., and H. D. Newson. 1978. Chap. 6. Ecology and management of arthropod populations in recreational lands, pp. 125-162. *In* G. W. Frankie and C. S. Koehler (eds.). Perspectives in urban entomology. Academic Press, New York. 417 pp.
2676. Merritt, R. W., E. J. Olds, and E. D. Walker. 1990. Natural food and feeding ecology of larval *Coquillettidia perturbans*. J. Am. Mosq. Contr. Assoc. 6:35-42
2677. Merritt, R. W., D. H. Ross, and B. V. Peterson. 1978. Larval ecology of some lower Michigan black flies with keys to the immature stages. Great Lakes Ent. 11:177-208.
2678. Merritt, R. W., D. H. Ross, and G. J. Larson. 1982. Influence of stream temperature and seston on the growth and production of overwintering larval black flies (Diptera: Simuliidae). Ecology 63:1322-1331.
2679. Merritt, R. W., and J. B. Wallace. 1981. Filter-feeding insects. Sci. Am. 244:132-144.
2680. Merritt, R. W., M. S. Wipfli, and R. S. Wotton. 1991. Changes in feeding habits of selected nontarget aquatic insects in response to live and *Bacillus thuringiensis* var. *israelensis* De Barjac-killed black fly larvae (Diptera: Simuliidae). Can. Ent. 123:179-186.
2681. Merritt, R. W., and R. S. Wotton. 1988. The life history and behavior of *Limnophora riparia* (Diptera: Muscidae), a predator of larval black flies. J. N. Am. Benthol. Soc. 7:1-12.
2682. Metcalfe, J. L. 1989. Biological water quality assessment of running waters based on macroinvertebrate communities: history and present status in Europe. Environ. Poll. 60:101-139.
2683. Metcalfe, J. L., and M. N. Charlton. 1990. Freshwater mussels as biomonitors for organic industrial contaminants and pesticides in the St. Lawrence River. Sci. Tot. Environ. 97/98:595-615.
2684. Meyer, J. L., and J. O'Hop. 1983. Leaf-shredder insects as a source of dissolved organic carbon in headwater streams. Am. Midl. Nat. 109:175-183.
2685. Meyer, R. P., and S. L. Durso. 1993. Identification of the mosquitoes of California. Calif. Mosq. Vector Contr. Assoc. Sacramento. 80pp.

2686. Miall, L. C. 1893. *Dicranota*; a carnivorous tipulid larva. Trans. R. ent. Soc. Lond. 1893:235-253.
2687. Miall, L.C. 1895. The natural history of aquatic insects. MacMillan, London. 395 pp.
2688. Miall, L. C., and R. Shelford. 1897. The structure and life history of *Phalacrocera replicata*. Trans. R. ent. Soc. Lond. 1897:343-361.
2689. Michael, A. G., and J. F. Matta. 1977. The Dytiscidae of Virginia (Coleptera: Adephaga) (subfamilies: Laccophininae, Colymbetinae, Dytiscinae, Hydaticinae and Cybestrinae). The insects of Virgina. No. 12. Virginia Polytech. Inst. State Univ., Res. Div. Bull. 124:1-53.
2690. Michener, C. D. 1953. Comparative morphological and systematic studies of bee larvae with a key to the families of hymenopterous larvae. Univ. Kans. Sci. Bull. 35:987-1102.
2691. Mickel, C. E., and H. E. Milliron. 1939. Rearing the caddice fly, *Limnephilus indivisus* Walker and its hymenopterous parasite *Hemiteles biannulatus* Grav. Ann. Ent. Soc. Am. 32:575-580.
2692. Middlekauff, W. W., and R. S. Lane. 1980. Adult and immature Tabanidae (Diptera) of California. Bull. Calif. Ins. Surv. 22:1-99.
2693. Milbrink, G. 1983. Characteristic deformities in tubificid oligochaetes inhabiting polluted bays of Lake Vänern, southern Sweden. Hydrobiologia 106:169-184.
2694. Milbrink, G., and T. Wiederholm. 1973. Sampling efficiency of four types of mud bottom samplers. Oikos 24:479-482.
2695. Mill, P. J. 1973. Respiration: Aquatic insects, pp. 403-467. *In* M. Rockstein (ed.). The physiology of Insecta, Vol. 6. 2nd ed. Academic, N.Y. 548 pp.
2696. Mill, P. J., and G. M. Hughes. 1966. The nervous control of ventilation in dragonfly larvae. J. exp. Biol. 44:297-316.
2697. Mill, P. J., and R. S. Pickard. 1972. Anal valve movement and normal ventilation in aeshnid dragonfly larvae. J. exp. Biol. 56:537-543.
2698. Miller, A. 1939. The egg and early development of the stonefly *Pteronarcys proteus* Newman. J. Morph. 64:555-609.
2699. Miller, D. C. 1964. Notes on *Enochrus* and *Cymbiodyta* from the Pacific Northwest (Coleoptera: Hydrophilidae). Coleopt. Bull. 18:69-78.
2700. Miller, D. C. 1974. Revision of the New World *Chaetarthria* (Coleoptera: Hydrophilidae). Entomologica am. 49:1-123.
2701. Miller, D. E., and W. P. Kovalak. 1979. Distribution of *Peltoperla arcuata* Needham (Ins., Plecoptera) in a small woodlawn stream. Int. Revue ges. Hydrobiol. 64:795-800.
2702. Miller, N. 1906. Some notes on the dragonflies of Waterloo Iowa. Ent. News 17:375-361.
2703. Miller, N.C.E. 1971. The biology of the Heteroptera (2nd ed.). E. Classey, Hampton. 206 pp.
2704. Miller, P. L. 1961. Some features of the respiratory system of *Hydrocyrius columbiae* (Belostomatidae, Hemiptera). J. Insect Physiol. 6:243-271.
2705. Miller, P. L. 1964a. Respiration - aerial gas transport, pp. 557-615. *In* M. Rockstein (ed.). The physiology of Insecta, Vol. 3. Academic, N.Y. 692 pp.
2706. Miller, P. L. 1964b. The possible role of haemoglobin in *Anisops* and *Buenoa* (Hemiptera: Notonectidae). Proc. R. ent. Soc. Lond. (A). 39:166-175.
2707. Miller, P. L. 1966. The function of haemoglobin in relation to the maintenance of neutral buoyancy in *Anisops pellucens* (Notonectidae, Hemiptera). J. exp. Biol. 44:529-543.
2708. Mills, H. B. 1934. A monograph of the Collembola of Iowa. Collegiate Press Inc., Ames. 143 pp.
2709. Millspaugh, D. D. 1939. Bionomics of the aquatic and semi-aquatic Hemiptera of Dallas County, Texas. Field and Lab. 7:67-87.
2710. Milne, L. J., and M. Milne. 1978. Insects of the water surface. Sci. Am. 238(4):134-142.
2711. Milne, M. J. 1938. The "metamorphotype method" in Trichoptera. J. N.Y. Ent. Soc. 46:435-437.
2712. Minckley, W. L. 1963. The ecology of a spring stream, Doe Run, Meade County, Kentucky. Wildl. Monogr. 11:1-124.
2713. Minckley, W. L. 1964. Upstream movements of *Gammarus* (Amphipoda) in Doe Run, Meade County, Kentucky. Ecology 45:195-197.
2714. Mingo, T. M. 1979. Distribution of aquatic Dryopoidea (Coleoptera) in Maine. Ent. News 90:177-185.
2715. Minno, M. C. 1992. Lepidoptera of the Archibold Biological Station, Highlands County, Florida. Fla. Ent. 75:297-329.
2716. Minshall, J. N. 1964. An ecological life history of *Epeorus pleuralis* (Banks) in Morgan's Creek, Meade County, Kentucky. M.S. thesis, University of Louisville, Ky. 79 pp.
2717. Minshall, J. N. 1967a. Life history and ecology of *Eperous pleuralis* (Banks) (Ephemeroptera: Heptageniidae). Am. Midl. Nat. 78:369-388.
2718. Minshall, G. W. 1967b. Role of allochthonous detritus in the trophic structure of a woodland spring brook community. Ecology 48:139-144.
2719. Minshall, G. W., and J. N. Minshall. 1966. Notes on the life history and ecology of *Isoperla clio* (Newman) and *Isogenus decisus* Walker (Plecoptera: Perlodidae). Am. Midl. Nat. 76:340-350.
2720. Minshall, G. W., and J. N. Minshall. 1977. Microdistribution of benthic invertebrates in a Rocky Mountain (USA) stream. Hydrobiologia 55:231-249.
2721. Minshall, G. W., R. C. Petersen, K. W. Cummins, T. L. Bott, J. R. Sedell, C. E. Cushing, and R. L. Vannote. 1983. Interbiome comparison of stream ecosystem dynamics. Ecol. Monogr. 53:1-25.
2722. Minto, M. L. 1977. A sampling device for invertebrate fauna of aquatic vegetation. Freshwat. Biol. 7:425-430.
2723. Mittelbach, G. G. 1981. Patterns of invertebrate size and abundance in aquatic habitats. Can. J. Fish. Aquat. Sci. 38:896-904.
2724. Miura, T., and R. M. Takahashi. 1987. Augmentation of *Notonecta unifasciata* eggs for suppressing *Culex tsarsalis* larval population densities in rice fields. Proc. Calif. Mosq. Vect. Contr. Assoc. 55:45-49.
2725. Miura, T., and R. Takahashi. 1988. Predation of Micro*velia pulchella* (Hemiptera: Veliidae) on mosquito larvae. J. Am. Mosq. Contr. Assoc. 4:91-93.
2726. Moffet, J. W. 1936. A quantitative study of the bottom fauna in some Utah streams variously affected by erosion. Univ. Utah Biol. Ser. 3:1-33.
2727. Molles, M. C., Jr., and R. D. Pietruszka. 1983. Mechanisms of prey selection by predaceous stoneflies: roles of prey morphology, behavior and predator hunger. Oecologia 57:25-31.
2728. Molles, M. C., and R. D. Pietruszka. 1987. Prey selection by a stonefly: the influence of hunger and prey size. Oecologia 72:473-478.
2729. Molnar, D. R., and R. J. Lavigne. 1979. The Odonata of Wyoming (dragonflies and damselflies). Sci. Monogr. Agric. Exp. Sta. Univ. Wyoming 37:1-142.
2730. Monakov, A. V. 1972. Review of studies on feeding of aquatic invertebrates conducted at the Institute of Biology of Inland Waters. Academy of Science, USSR. J. Fish. Res. Bd. Can. 29:363-383.
2731. Montgomery, B. E. 1940. The Odonata of South Carolina. J. Elisha Mitchell Sci. Soc. 56:283-301.
2732. Montgomery, B. E. 1941. Records of Indiana dragonflies, X,1937-1940. Proc. Ind. Acad. Sci. 50:229-241.
2733. Montgomery, B. E. 1947. The distribution and relative seasonal abundance of Indiana species of five families of dragonflies (Odonata, Calopterygidae, Petaluridae, Cordulegasteridae, Gomphidae and Aeshnidae). Proc. Ind. Acad. Sci. 56:163-169.
2734. Montgomery, B. E. 1948. The distribution and relative seasonal abundance of the Indiana species of Lestidae (Odonata: Zygoptera). Proc. Ind. Acad. Sci. 57:113-115.
2735. Montgomery, B. E. 1967. Geographical distribution of the Odonata of the North Central States. Proc. N. Cent. Br. Ent. Soc. Am. 22:121-129.
2736. Montgomery, B. E. 1968. The distribution of western Odonata. Proc. N. Cent. Br. Ent. Soc. Am. 23:126-136.
2737. Montgomery, B. E., and J. M. Macklin. 1962. Rates of development in the later instars of *Neotetrum pulchellum* (Drury) (Odonata, Libellulidae). Proc. N. Cent. Br. Ent. Soc. Am. 17:21-23.
2738. Moon, H. P. 1935. Methods and apparatus suitable for an investigation of the littoral region of oligotrophic lakes. Int. Revue ges. Hydrobiol. 32:319-333.
2739. Moon, H. P. 1938. The growth of *Caenis horaria, Leptophlebia marginata* and *L. vespertina*. Proc. Zool. Soc. Lond. 108:507-512.

2740. Moon, H. P. 1939. Aspects of the ecology of aquatic insects. Trans. Brit. Ent. Soc. 6:39-49.
2741. Moore, C. G., D. B. Francy, D. A. Eliason, and T. P. Monath. 1988. *Aedes albopictus* in the United States: rapid spread of a potential disease vector. J. Am. Mosq. Cont. Assoc. 4:356-361.
2742. Moore, I. 1956. A revision of the Pacific coast Phytosini with a review of the foreign genera (Coleoptera: Staphylinidae). Trans. San Diego Soc. Nat. Hist. 12:103-152.
2743. Moore, I., and E. F. Legner. 1974. Keys to the genera of Staphylinidae of America north of Mexico exclusive of the Aleocharinae (Coleoptera: Staphylinidae). Hilgardia 42:548-563.
2744. Moore, I., and E. F. Legner. 1975. Revision of the genus *Endeodes* LeConte with a tabular key to the species (Coleoptera: Melyridae). J. N.Y. Ent. Soc. 85:70-81.
2745. Moore, I., and E. F. Legner. 1976. Intertidal rove beetles (Coleoptera: Staphylinidae), pp. 521-551. *In* L. Cheng (ed.). Marine insects. North Holland, Amsterdam. 581 pp.
2746. Moore, J. W. 1977. Some factors affecting algal consumption in subarctic Ephemeroptera, Plecoptera and Simuliidae. Oecologia 27:261-273.
2747. Moore, J. W. 1979a. Factors influencing algal consumption and feeding rate in *Heterotrissocladius changi* Saether and *Polypedilum nubeculosum* (Meigen). Oecologia 40:219-227.
2748. Moore, J. W. 1979b. Some factors affecting the distribution, seasonal abundance and feeding of subarctic Chironomidae (Diptera). Arch, Hydrobiol. 85:203-225.
2749. Moore, K. A., and D. D. Williams. 1990. Novel strategies in the complex defense of a stonefly (*Pteronarcys dorsata*) nymph. Oikos 57:49-56.
2750. Moore, M. V. 1988. Differential food resources by the instars of *Chaoborus punctipennis*. Freshwat. Biol. 19:249-268.
2751. Moreira, G. R. P., and B. L. Peckarsky. 1994. Multiple developmental pathways of *Agnetina capitata* (Plecoptera: Perlidae) in a temperate forest stream. J. N. Am. Benthol. Soc. 13:19-29.
2752. Morgan, A. H. 1911. Mayflies of Fall Creek. Ann. Ent. Soc. Am. 4:93-119.
2753. Morgan, A. H. 1913. A contribution to the biology of mayflies. Ann. Ent. Soc. Am. 6:371-413.
2754. Morgan, A. H., and H. D. O'Neil. 1931. The function of the tracheal gills in larvae of the caddisfly, *Macronema zebratum* Hagen. Physiol. Zool. 4:361-379.
2755. Morgan, J. J. 1949. The metamorphosis and ecology of some species of Tanypodinae (Dipt., Chironomidae). Entomol. mon. Mag. 85:119-126.
2756. Morgan, N. C. 1971. Factors in the design and selection of insect emergence traps, pp. 93-108. *In* W. T. Edmondson and G. G. Winberg (eds.). A manual on methods for the assessment of secondary productivity in fresh waters. IBP Handbook 17. Blackwell, Oxford. 358 pp.
2757. Morgan, N. C., A. B. Waddell, and W. B. Hall. 1963. A comparison of the catches of emerging aquatic insects in floating box and submerged funnel traps. J. Anim. Ecol. 32:203-219.
2758. Morihara, D. K., and W. P. McCafferty. 1979. The *Baetis* larvae of North America (Ephemeroptera: Baetidae). Trans. Am. Ent. Soc. 105:139-221.
2759. Morin, A., and P. P. Harper. 1986. Phénologie et microdistribution des adultes et des larvaes de Trichoptères filtreurs dans un ruisseau des basses Laurentides (Québec). Arch. Hydrobiol. 108:167-183.
2760. Morin, P. J. 1984. Odonate guild composition: experiments with colonization history and fish predation. Ecology 65:1866-1873.
2761. Moring, J. B., R. L. Hassage, and K. W. Stewart. 1991. Behavioral interactions among male and female nymphs of the predatory stonefly *Hydroperla crosbyi* (Plecoptera: Perlodidae). Ann. Ent. Soc. Am. 84:207-211.
2762. Morley, R. L., and R. A. Ring. 1972. The intertidal Chironomidae (Diptera) of British Columbia II. Life history and population dynamics. Can. Ent. 104:1099-1121.
2763. Morofsky, W. F. 1939. Survey of insect fauna of some Michigan trout streams in connection with improved and unimproved streams. J. Econ. Ent. 29:749-754.
2764. Morrigetta, R. 1979. Les Coleopteres Gyrinidae du Quebec. Cordulia Supp. 8:1-43.
2765. Morris, D. L., and M. P. Brooker. 1979. The vertical distribution of macroinvertebrates in the substratum of the upper reaches of the River Wye, Wales. Freshwat. Biol. 9:573-583.
2766. Morris, R. W. 1963. A modified Barcroft respirometer for study of aquatic animals. Turtox News 41:22-23.
2767. Morse, J. C. 1972. The genus *Nyctiophylax* in North America. J. Kans. Ent. Soc. 45:172-181.
2768. Morse, J. C. 1975. A phylogeny and revision of the caddisfly genus *Ceraclea* (Trichoptera, Leptoceridae). Contr. Am. Ent. Inst. 11:1-97.
2769. Morse, J. C. 1981. A phylogeny and classification of family-group taxa of Leptoceridae (Trichoptera), pp. 257-264. *In* G. P. Moretti (ed.). Proc. 3rd Internat. Symp. Trichoptera. Ser. Entomologica 20. Junk, The Hague, Netherlands. 472 pp.
2770. Morse, J. C., J. W. Chapin, D. D. Herlong, and R. S. Harvey. 1980. Aquatic insects of Upper Three Runs Creek, Savannah River Plant, South Carolina. Part I: Orders other than Diptera. J. Ga. Ent. Soc. 15:73-101.
2771. Morse, W. J., and R. L. Blickle. 1953. A check list of the Trichoptera (caddisflies) of New Hampshire. Ent. News 64:68-73, 97-102.
2772. Morse, W. J., and R. L. Blickle. 1957. Additions and corrections to the list of New Hampshire Trichoptera. Ent. News 68:127-131.
2773. Mosher, E. 1916. A classification of the Lepidoptera based on characters of the pupae. Bull. Ill. State Lab. Hist. 12:14-159.
2774. Motyka, G. L., R. W. Merritt, M. J. Klug, and J. R. Miller. 1985. Food-finding behavior of selected aquatic detritivores: Direct or indirect behavioral mechanism? Can. J. Zool. 63:1388-1394.
2775. Mozley, S. C. 1968. The integrative roles of the chironomid (Diptera: Chironomidae) larvae in the trophic web of a shallow, five hectare lake in the Piedmont region of Georgia. Ph.D. diss., Emory University, Atlanta. 109 pp.
2776. Mozley, S. C., and L. C. Garcia. 1972. Benthic macrofauna in a coastal zone of southeastern Lake Michigan. Proc. 15th Conf. Great Lakes, pp. 102-116.
2777. Mueller, U. G., and D. Dearing. 1994. Predation and avoidance of tough leaves by aquatic larvae of the moth *Parapoynx rugosalis* (Lepidoptera: Pyralidae). Ecol. Ent. 19:155-158.
2778. Muesebeck, C.F.W. 1949. A new flightless *Phaenopria* (Hym.: Diapriidae). Can. Ent. 81:234-235.
2779. Muesebeck, C.F.W. 1950. Two new genera and three new species of Braconidae. Proc. Ent. Soc. Wash. 52:77-81.
2780. Muesebeck, C.F.W. 1972. A new reared species of *Trichopria* (Proctotrueoidea: Diapriidae). Ent. News 83:141-143.
2781. Muesebeck, C.F.W., K. V. Krombein, and H. K. Townes. 1951. Hymenoptera of America north of Mexico synoptic catalog. U.S. Dep. Agric. Monogr. 21. 420p.
2782. Muirhead-Thompson, R. C. 1966. Blackflies, pp. 127-144. *In* C. N. Smith (ed.). Insect colonization and mass production. Academic, N.Y. 618 pp.
2783. Muirhead-Thompson, R. C. 1969. A laboratory technique for establishing *Simulium* larvae in an experimental channel. Bull. Ent. Res. 59:533-536.
2784. Mulhern, T. D. 1942. New Jersey mechanical trap for mosquito surveys. New Jersey Agric. Exp. Sta. Circ. 421:1-8.
2785. Mulla, M. S., and L. A. Lacey. 1976. Feeding rates of *Simulium* larvae on particulates in natural streams (Diptera: Simuliidae). Environ. Ent. 5:283-287.
2786. Mullen, G. R., and L. J. Hribar. 1988. Biology and feeding behavior of ceratopogonid larvae (Diptera: Ceratopogonidae) in North America. Bull. Soc. Vect. Ecol. 13:60-81.
2787. Mullens, B. A., and D. A. Rutz. 1983. Voltinism and seasonal abundance of *Culicoides variipennis* (Diptera: Ceratopogonidae) in central New York State. Ann. Ent. Soc. Am. 76:913-917
2788. Müller, G. W. 1892. Beobachtungen an im Wasserlebenden Schmetterlingsraupen. Zool. Jb. Syst. 6:617-630.
2789. Müller, K. 1954. Investigation on the organic drift in North Swedish streams. Rep. Inst. Freshwat. Res. Drottningholm 35:133-148.
2790. Müller, K. 1974. Stream drift as a chronobiological phenomenon in running water ecosystems. Ann. Rev. Ecol. Syst. 5:309-323.

2791. Müller-Liebenau, I. 1974. *Rheobaetis*: a new genus from Georgia (Ephemeroptera: Baetidae). Ann. Ent. Soc. Am. 67:555-567.
2792. Mundahl, N. D., and K. J. Kraft. 1988. Abundance and growth of three species of aquatic insects exposed to surface-release hydropower flows. J. N. Am. Benthol. Soc. 7:100-108.
2793. Mundie, J. H. 1956. Emergence traps for aquatic insects. Mitt. Int. Verh. Limnol. 7:1-13.
2794. Mundie, J. H. 1957. The ecology of Chironomidae in storage reservoirs. Trans. Ent. Soc. Lond. 109:149-232.
2795. Mundie, J. H. 1964. A sampler for catching emerging insects and drifting materials in streams. Limnol. Oceanogr. 9:456-459.
2796. Mundie, J. H. 1966. Sampling emerging insects and drifting materials in deep flowing water. Gewasser und Abwasser. 41/42:159-162.
2797. Mundie, J. H. 1971a. Sampling benthos and substrate materials, down to 50 microns in size, in shallow streams. J. Fish. Res. Bd. Can. 28:849-860.
2798. Mundie, J. H. 1971b. Insect emergence traps. pp. 80-93. *In* W. T. Edmondson and G. G. Winberg (eds.). A manual on methods for the assessment of secondary productivity in fresh waters. IBP Handbook 17. Blackwell, Oxford 358 pp.
2799. Munn, M. D., and R. H. King. 1987. Ecology of *Potamanthus myops* (Walsh) (Ephemeroptera: Potamanthidae) in a Michigan stream (USA). Hydrobiologia 117: 119-125.
2800. Munroe, E. G. 1947. Further North American records of *Acentropus niveus* (Lepidoptera, Pyralidae). Can. Ent. 79:120.
2801. Munroe, E. G. 1951a. A previously unrecognized species of *Nymphula* (Lepidoptera: Pyralidae). Can. Ent. 83:20-23.
2802. Munroe, E. G. 1951b. The identity and generic position of *Chauliodes disjunctus* Walker (Megaloptera: Corydalidae). Can. Ent. 83:33-35.
2803. Munroe, E. G. 1958. *Chauliodes disjunctus* Walker: a correction with the descriptions of a new species of a new genus (Megaloptera: Corydalidae). Can. Ent. 85:190-192.
2804. Munroe, E. G. 1972-1973. Pyraloidea. Pyralidae (in part), pp. 1-304, Fasc. 13.1, A-C. *In* R. B. Dominick (ed.). The moths of America north of Mexico. E. W Classey Ltd., London. (ongoing series).
2805. Munz, P. A. 1919. A venational study of the suborder Zygoptera. Mem. Ent. Soc. Am. 3:1-307.
2806. Muotka, T. 1990. Coexistence in a guild of filter-feeding caddis larvae: do different instars act as different species? Oecologia 85:281-292.
2807. Murdoch, W. W., and J. Bence. 1987. General predators and unstable prey populations, pp. 17-30. *In* W. C. Kerfoot and A. Sih (eds.). Predation: Direct and indirect impacts on aquatic communities. Univ. Press New England, Hanover, NH. 386 pp.
2808. Murdoch, W. W., J. Chesson, and P. L. Chesson. 1985. Biological control in theory and practice. Am. Nat. 125:344-366.
2809. Murdoch, W. W., M. A. Scott, and P. Ebsworth. 1984. Effects of the general predator, *Notonecta* (Hemiptera) upon a freshwater community. J. Anim. Ecol. 53:791-808.
2810. Murphee, C. S., and G. R. Mullen. 1991. Comparative larval morphology of the genus *Culicoides* Latreille (Diptera: Ceratopogonidae) in North America with a key to species. Bull. Soc. Vect. Ecol. 16:269-399.
2811. Murphey, R. K. 1971. Sensory aspects of the control of orientation to prey by the water strider, *Gerris remigis*. Z. Vergl. Physiol 72:168-185.
2812. Murphy, H. 1919. Observations on the egg laying of the caddisfly *Brachycentrus nigrosoma* Banks, and the habits of the young larvae. J. N.Y. Ent. Soc. 27:154-158.
2813. Murphy, H. E. 1937. Rearing mayflies from egg to adult, pp. 266-267. *In* J. G. Needham (ed.). Culture methods for invertebrate animals. Comstock, Ithaca. 590 pp.
2814. Murray, D. A. (ed.) 1980. Chironomidae: ecology, systematics, cytology and physiology. Pergamon, N.Y.
2815. Murray, D. A., and E. J. Fittkau. 1985. *Hayesomyia*, a new genus of Tanypodinae from the Holarctic (Diptera: Chironomidae). Spixiana Suppl. 11:195-207.
2816. Murray, T. D., and W. N. Charles. 1975. A pneumatic grab for obtaining large, undisturbed mud samples: its construction and some applications for measuring the growth of larvae and emergence of adult Chironomidae. Freshwat. Biol. 5:205-210.

2817. Murvosh, C. M. 1971. Ecology of the water penny beetle *Psephenus herricki* DeKay. Ecol. Monogr. 41:79-96.
2818. Murvosh, C. M. 1992. On the occurrence of the water penny beetle *Eubrianax edwardsii* in lentic ecosystems (Coleoptera: Psephenidae). Coleop. Bull. 46:43-51.
2819. Murvosh, C. M. 1993. Microdistribution of the water penny *Psephenus montanus* (Coleoptera: Psephenidae), with notes on life history and zoology. Southwest. Nat. 38:119-126.
2820. Murvosh, C. M., and H. P. Brown. 1976. Mating behavior of water penny beetles (Coleoptera: Psephenidae): a hypothesis. Coleopt. Bull. 30:57-59.
2821. Murvosh, C. M., and B. W. Miller. 1974. Life history of the western water penny beetle, *Psephenus falli* (Coleoptera: Psephenidae). Coleopt. Bull. 28:85-92.
2822. Musgrave, P. N. 1935. A synopsis of the genus *Helichus* Erichson in the United States and Canada, with descriptions of new species. Proc. Ent. Soc. Wash. 37:137-145.
2823. Musser, R. J. 1962. Dragonfly nymphs of Utah. (Odonata: Anisoptera). Univ. Utah Biol. Ser. 12:1-71.
2824. Mutch, R. A., and G. Pritchard. 1982. The importance of sampling and sorting techniques in the elucidation of the life cycle of *Zapada columbiana* (Nemouridae: Plecoptera). Can. J. Zool. 60:3394-3399.
2825. Mutch, R. A., and G. Pritchard. 1984. The life history of *Zapada columbiana* (Plecoptera: Nemouridae) in a Rocky Mountain stream. Can. J. Zool. 62:1273-1281.
2826. Muthukrishnan, J., and A. Palavesam. 1992. Secondary production and energy flow through *Kiefferulus barbitarsis* (Diptera:Chironomidae) in tropical ponds. Arch. Hydrobiol 125:207-226.
2827. Muttkowski, R. A. 1908. Review of the dragonflies of Wisconsin. Bull. Wisc. Nat. Hist. Soc. 6:57-127.
2828. Muttkowski, R. A. 1910. Catalogue of the Odonata of North America. Bull. Mus. Milwaukee 207 pp.
2829. Muttkowski, R. A. 1918. The fauna of Lake Mendota – a qualitative and quantitative survey with special reference to insects. Trans. Wisc. Acad. Sci. Arts Lett. 19:374-482.
2830. Muttkowski, R. A., and G. M. Smith. 1929. The food of trout stream insects in Yellowstone National Park. Ann. Roosevelt Wildl. 2:241-263.
2831. Nagagawa, A. 1952. Food habits of hydropsychid larvae. Jap. J. Limnol. 16:130-138.
2832. Nagaraja, H., and S. Nagarkatti. 1973. A key to some New World species of *Trichogramma* (Hymenoptera; Trichogrammatidae), with descriptions of four new species. Proc. Ent. Soc. Wash. 75:288-297.
2833. Nagatomi, A. 1961a. Studies in the aquatic snipe flies of Japan. Part III. Descriptions of the larvae (Diptera: Rhagionidae). Mushi 35:11-27.
2834. Nagatomi, A. 1961b. Studies in the aquatic snipe flies of Japan. Part. IV. Descriptions of the pupae (Diptera, Rhagionidae). Mushi 35:29-38.
2835. Nagatomi, A. 1962. Studies in the aquatic snipe flies of Japan. Part V. Biological notes (Diptera, Rhagionidae). Mushi 36:103-149.
2836. Nagell, B. 1973. The oxygen consumption of mayfly (Ephemeroptera) and stonefly (Plecoptera) larvae at different oxygen concentration. Hydrobiologia 42:461-489.
2837. Nagell, B. 1974. The basic picture and the ecological implications of the oxygen consumption/oxygen concentration curve of some aquatic insect larvae. Ent. Tidskr. (Suppl.) 95:182-187.
2838. Nakasuji, F., and V. A. Dyck. 1984. Evaluation of the role of *Microvelia douglasi atrolineata* (Bergroth) (Heteroptera: Veliidae) as a predator of the brown planthopper *Nilaparvata lugens* (Stål) (Homoptera: Delphacidae). Res. Pop. Ecol. 26:134-149.
2839. National Transportation Safety Board. 1989. Collapse of the S.R. 675 bridge spans over the Pocomoke River near Pocomoke City, Maryland. PB89-916205 Washington, DC: National Transportation Safety Board/HAR-89/04. 80pp.
2840. Nayar, J. K. (ed.). 1982. Bionomics and physiology of *Culex nigripalpus* (Diptera: Culicidae) of Florida: an important vector of diseases. Fla. Agr. Exp. Sta. Tech. Bull. 852. 73p.

2841. Nayar, J. K. (ed.). 1985. Bionomics and physiology of *Aedes taeniorhynchus* and *Aedes sollicitans*, the salt marsh mosquitoes of Florida. Fla. Agr. Exp. Sta. Tech. Bull. 827. 148p.

2842. Naylor, C., L. Maltby, and P. Calow. 1989. Scope for growth in *Gammarus pulex*, a freshwater benthic detritivore. Hydrobiologia 188/189:517-523.

2843. Neave, F. 1930. Migratory habits of the mayfly, *Blasturus cupidus* Say. Ecology 11:568-576.

2844. Nebeker, A. V. 1971. Effect of water temperature on nymphal feeding rate, emergence, and adult longevity of the stonefly, *Pteronarcys dorsata*. J. Kans. Ent. Soc. 44:21-26.

2845. Nebeker, A. V., and A. R. Gaufin. 1965. The *Capnia columbiana* complex of North America (Capniidae, Plecoptera). Trans. Am. Ent. Soc. 91:467-487.

2846. Needham, J. G. 1899. Directions for collecting and rearing dragonflies, stoneflies, and mayflies. Bull. U.S. Nat. Mus. 39:1-9.

2847. Needham, J. G. 1900. The fruiting of the Blue Flag (*Iris versicolor* L.). Am. Nat. 34:361-386.

2848. Needham, J. G. 1901a. Aquatic insects of the Adirondacks. Ephemeridae. Bull. N.Y. State Mus. 47:418-429.

2849. Needham, J. G. 1901b. Aquatic insects of the Adirondacks. Odonata. Bull. N.Y. State Mus. 47:429-540.

2850. Needham, J. G. 1903a. Aquatic insects in New York state. Part 3. Life histories of Odonata suborder Zygoptera. Bull. N.Y. State Mus. 68:218-279.

2851. Needham, J. G. 1903b. A genealogic study of dragonfly wing venation. Proc. U. S. Nat. Mus. 26:703–764.

2852. Needham, J. G. 1907. Notes on the aquatic insects of Walnut Lake, pp. 252-271. *In* T. C. Hankinson (ed.). A biological survey of Walnut Lake, Michigan. Rept. Mich. Geol. Surv. 1907. 288 pp.

2853. Needham, J. G. 1928. A list of the insects of New York. Mem. Cornell Univ. Agric. Exp. Sta. 101:45-56.

2854. Needham, J. G. 1941. Life history studies on Progomphus and its nearest allies (Odonata: Aeschnidae). Trans. Am. Ent. Soc. 67:221-245.

2855. Needham, J. G. 1951. Prodrome for a manual of the dragonflies of North America, with extended comments on wing venation systems. Trans. Am. Ent. Soc. 77:21-62.

2856. Needham, J. G., and C. A. Hart. 1901. The Dragonflies (Odonata) of Illinois Part 1. Petaluridae, Aeschnidae, and Gomphidae. Bull. Ill. State Lab. Nat. Hist. Surv. 6:1-94.

2857. Needham, J. G., and C. Betten. 1901a. Family Sialidae. Aquatic insects of the Adirondacks. Bull. N.Y. State Mus. 47:541-544.

2858. Needham, J. G., and C. Betten. 1901b. Aquatic insects in the Adirondacks. Diptera. Bull. N.Y. State Mus. 47:545-612.

2859. Needham, J. G., and P. W. Claassen. 1925. A monograph on the Plecoptera or stoneflies of America north of Mexico. Thomas Say Found. Ent. Soc. Am. 2:1-397.

2860. Needham, J. G., and E. Fisher. 1936. The nymphs of North American libelluline dragonflies (Odonata). Trans. Am. Ent. Soc. 62:107-116.

2861. Needham, J. G., and H. B. Heywood. 1929. A handbook of the dragonflies of North America. C. C. Thomas, Springfield. Ill. 378 pp.

2862. Needham, J. G., and J. T. Lloyd. 1928. The life of inland waters. The Amer. Viewpoint Soc., N.Y. 438 pp.

2863. Needham, J. G., and P. R. Needham. 1962. A guide to the study of fresh-water biology (5th ed.). Holden-Day, San Francisco. 108 pp.

2864. Needham, J. G., and L. W. Smith. 1916. The stoneflies of the genus *Peltoperla*. Can. Ent. 48:80-88.

2865. Needham, J. G., and M. J. Westfall, Jr. 1955. A manual of the dragonflies of North America (Anisoptera) including the Greater Antilles and the provinces of the Mexican border. Univ. Calif. Press, Berkeley. 615 pp.

2866. Needham, J. G., J. R. Traver, and Y. C. Hsu. 1935. The biology of mayflies with a systematic account of North American species. Comstock, Ithaca. 759 pp.

2867. Needham, P. R. 1934. Quantitative studies of stream bottom foods. Trans. Am. Fish. Soc. 64:238-247.

2868. Needham, P. R., and R. L. Usinger. 1956. Variability in the macrofauna of a single riffle in Prosser Creek, California, as indicated by the Surber sampler. Hilgardia 24:383-409.

2869. Neff, S. E. 1955. Studies on a Kentucky knobs lake. II. Some aquatic Nematocera (Diptera) from Tom Wallace Lake. Trans. Ky. Acad. Sci. 16:1-13.

2870. Neff, S. E., and C. O. Berg. 1962. Biology and immature stages of *Hoplodictya spinicornis* and *H. setosa* (Diptera: Sciomyzidae). Trans. Am. Ent. Soc. 88:77-93.

2871. Neff, S. E., and C. O. Berg. 1966. Biology and immature stages of malacophagous Diptera of the genus *Sepedon* (Sciomyzidae). Bull. Va. Polytech. Inst. Agric. Exp. Sta. 566. 113 pp.

2872. Neff, S. E., and J . B. Wallace. 1969a. Observations on the immature stages of *Cordilura* (*Achaetella*) *deceptiva* and *C.* (*A.*) *varipes*. Ann. Ent. Soc. Am. 62:775-785.

2873. Neff, S. E., and J. B. Wallace. 1969b. Biology and description of immature stages of *Orthacheta hirtipes*, a predator of *Cordilura* spp. Ann. Ent. Soc. Am. 62:785-790.

2874. Nel, A., X. J. C. Martinez-Delclos, J. C. Paicheler, and M. Henortay. 1993. Les "Anisozygoptera" fossiles: phylogenie et classification (Odonata). Martinia (hors ser.). 3:1-311.

2875. Nelson, C. H. 1982. Notes on the life histories of *Strophopteryx limata* (Frison) and *Oemopteryx contorta* (Needham and Claassen) (Plecoptera: Taeniopterygidae) in Tennessee. J. Tenn. Acad. Sci. 57:9-15.

2876. Nelson, C. H., and J. F. Hanson. 1971. Contribution to the anatomy and phylogeny of the family Pteronarcidae (Plecoptera). Trans. Am. Ent. Soc. 97:123-200.

2877. Nelson, C. H., and J. F. Hanson. 1973. The genus *Perlomyia*. J. Kans. Ent. Soc. 46:187-199.

2878. Nelson, C. H., D. C. Tarter, and M. L. Little. 1977. Description of the adult male of *Allonarcys comstocki* (Smith) (Plecoptera: Pteronarcidae). Ent. News 88:33-36.

2879. Nelson, C. R., and R. W. Baumann. 1987a. New winter stoneflies of the genus *Capnia* with notes and an annotated checklist of the Capniidae of California. Entomography 5:485-521.

2880. Nelson, C. R., and R. W. Baumann. 1987b. The winter stonefly genus *Capnura* (Plecoptera: Capniidae) in North America: systematics, phylogeny, and zoogeography. Trans. Am. Ent. Soc. 113:1-28.

2881. Nelson, C. R., and R. W. Baumann. 1989. Systematics and distribution of the winter stonefly genus *Capnia* (Plecoptera: Capniidae) in North America. Great Basin Nat. 49:289-363.

2882. Nelson, D. J., and D. C. Scott. 1962. Role of detritus in the productivity of a rock-outcrop community in a Piedmont Stream. Limnol. Oceanogr. 7:396-413.

2883. Nelson, H. G. 1989. *Postelichus*, a new genus of nearctic Dryopidae (Coleoptera). Coleop. Bull. 43:19-24.

2884. Neumann, D. 1976. Adaptations of chironomids to intertidal environments. Ann. Rev. Ent. 21:387-414.

2885. Neunzig, H. H. 1966. Larvae of the genus *Nigronia* Banks. Proc. Ent. Soc. Wash. 68:11-16.

2886. Neunzig, H. H., and J. R. Baker. 1991. Megaloptera, p. 112-122. *In* Stehr, F.W. (ed.). Immature insects Vol. II, Kendall/Hunt Dubuque. 722 p.

2887. Neves, R. J. 1979. A checklist of caddisflies (Trichoptera) from Massachusetts. Ent. News 90:167-175.

2888. Neveu, A. 1972. Introduction a l'étude de la faune des dipteres a larves aquatiques d'un ruisseau des Pyrénées—Atlantiques, le Lissuraga. Ann. Hydrobiol. 3:173-196.

2889. Neveu, A. 1973a. Le cycle de développement des Simuliidae (Diptera, Nematocera) d'un ruisseau des Pyrenees-Atlantiques, le Lissuraga. Ann. Hydrobiol. 4:51-73.

2890. Neveu, A. 1973b. Estimation de la production de populations larvaires du genre *Simulium* (Diptera, Nematocera). Ann. Hydrobiol. 4:183-189.

2891. Neveu, A. 1976. Ecologie des larves d'Athericidae (Diptera, Brachycera) dans un ruisseau des Pyrénées-atlantiques. I. Structure et dynamique des populations. Ann. Hydrobiol. 7:73-90.

2892. Neveu, A. 1977. Ecologie des larves d'Athericidae (Diptera, Brachycera) dans un ruisseau Pyrénées-atlantiques. II. Production comparison de diffrentes methodes de calcul. Ann. Hydrobiol. 8:45-66.

2893. Nevin, F. R. 1929. Larval development of *Sympetrum vicinum* Trans. Am. Ent. Soc. 55:79-102.

2894. Newbold, J. D., B. W. Sweeney, and R. L. Vannote. 1994. A model for seasonal synchrony in stream mayflies. J. N. Am. Benthol. 13:3-18.
2895. Newbury, R. W. 1984. Hydrologic determinants of aquatic insect habitats, pp 323-357. In V.H. Resh and D.M. Rosenberg (eds.). The ecology of aquatic insects. Praeger Publishers, New York, NY. 625p.
2896. Newell, R. L., and G. W. Minshall. 1976. An annotated list of the aquatic insects of southeastern Idaho. Part 1. Plecoptera. Great Basin Nat. 36:501-504.
2897. Newell, R. L., and G. W. Minshall. 1978. Life history of a multivoltine mayfly, *Tricorythodes minutus*: an example of the effect of temperature on life cycle. Ann. Ent. Soc. Am. 71:876-881.
2898. Newell, R. L., and D. S. Potter. 1973. Distribution of some Montana caddisflies (Trichoptera). Proc. Mont. Acad. Sci. 33:12-21.
2899. Newell, R.L. 1970. Checklist of some aquatic insects from Montana. Proc. Montana Acad. Sci. 30:45-56.
2900. Newkirk, M. 1981. The biology of *Cloeon cognatum* (Ephemeroptera: Baetidae). Ann. Ent. Soc. Am. 74:204-208.
2901. Newman, D. L., and R. C. Funk. 1984. Drift of riffle beetles (Coleoptera: Elmidae) in a small Illinois stream. Great Lakes Ent. 17:211-214.
2902. Newman, R. M. 1991. Herbivory and detritivory on freshwater macrophytes by invertebtates: a review. J. N. Am. Benthol. Soc. 10:89-114.
2903. Newman, R. M., and J. A. Perry. 1989. The combined effects of chlorine and ammonia on litter breakdown in outdoor experimental streams. Hydrobiologia 184:69-78.
2904. Newman, R. M., J. A. Perry, E. Tam, and R. L. Crawford. 1987. Effects of chronic chlorine exposure on litter processing in outdoor experimental streams. Freshwat. Biol. 18:415-428.
2905. Nicholson, H. P., and C. E. Mickel. 1950. The black flies of Minnesota (Simuliidae). Univ. Minn. Agric. Exp. Sta. Tech. Bull. 192:1-64.
2906. Nicola, S. 1968. Scavenging by *Alloperla* (Plecoptera: Chloroperlidae) nymphs on dead pink *(Oncorhynchus gorbuscha)* and chum *(O. keta)* salmon. Can. J. Zool. 46:787-796.
2907. Nielson, A. 1942. Uber die Entwicklung und Biologie der Trichopteren mit besonderer Berücksichtigung der Quelltrichopteren Himmerlands. Arch. Hydrobiol. Suppl. 17:255-631.
2908. Nielsen, A. 1948. Postembryonic development and biology of the Hydroptilidae. Biol. Skr. Dan. Vid. Selsk. 5:1-200.
2909. Nielsen, A. 1950. The torrential invertebrate fauna. Oikos 2:176-196.
2910. Nielsen, A. 1951a. Contributions to the metamorphosis and biology of the genus *Atrichopogon* Kieffer (Diptera, Ceratopogonidae), with remarks on the evolution and taxonomy of the genus. Kongr. Dan. Vidensk. Selsk. Biol. Skr. 6:1-95.
2911. Nielsen, A. 1951b. Is dorsoventral flattening of the body an adaptation to torrential life? Verh. Int. Verein. Limnol. 11:264-267.
2912. Nielsen, A. 1980. A comparative study of the genital segments and the genital chamber in female Trichoptera. Biol. Skr. (Kon. Danske Vid. Selsk.) 23:1-200.
2913. Nielsen, A. 1981. On the evolution of the phallus and other male terminalia in the Hydropsychidae with a proposal for a new generic name, pp. 273-278. In G. P. Moretti (ed.). Proc. 3rd Internat. Symp. on Trichoptera. Ser. Entomologica 20. Junk, The Hague, Netherlands. 472 pp.
2914. Nielsen, L. T. 1968. A current list of mosquitoes known to occur in Utah with a report of new records. Proc. Utah Mosquito Abate. Assoc. 21:34-37.
2915. Nielsen, L. T., and D. M. Rees. 1961. An identification guide to the mosquitoes of Utah. Univ. Utah Biol. Ser. 12:1-63.
2916. Nielsen, P., O. Ringdahl, and S. L. Tuxen. 1954. The zoology of Iceland. III. part 48a, Diptera. 1. Ejnar Munksgaard, Copenhagen, and Reykjavik. 189 pp.
2917. Nieser, N. 1977. A revision of the genus *Tenagobia* Bergroth (Heteroptera: Corixidae). Stud. Neotrop. Fauna Environ. 12:1-56.
2918. Nigmann, M. 1908. Anatomie und Biologie von *Acentropus niveus* Oliv. Zool. Jb. Syst. 26:489-560.
2919. Nilsson, A. N., and R. B. Angus. 1992. A reclassification of the Deronectes-group of genera (Coleoptera; Dytiscidae) based on a phylogenetic study. Ent. scand. 23:275-288.
2920. Nimmo, A. P. 1971. The adult Rhyacophilidae and Limnephilidae (Trichoptera) of Alberta and eastern British Columbia and their post-glacial origin. Quaest. Ent. 7:3-234.
2921. Nimmo, A. P. 1974. The adult Trichoptera (Insecta) of Alberta and eastern British Columbia, and their post-glacial origins. II. The families Glossosomatidae and Philopotamidae. Quaest. Ent. 10:315-349.
2922. Nimmo, A. P. 1977a. The adult Trichoptera (Insecta) of Alberta and eastern British Columbia, and their post-glacial origins. I. The families Rhyacophilidae and Limnephilidae. Suppl. I Quaest. Ent. 13:25-67.
2923. Nimmo, A. P. 1977b. The adult Trichoptera (Insecta) of Alberta and eastern British Columbia, and their post-glacial origins. II. The families Glossosomatidae and Philopotamidae. Suppl. 1. Quaest. Ent. 13:69-71.
2924. Nimmo, A. P. 1986a. The adult Polycentropodidae of Canada and adjacent United States. Quaest. Ent. 22:143-252.
2925. Nimmo, A. P. 1986b. Preliminary annotated checklist of the Trichoptera (Insecta) of Alaska. B.C. Prov. Mus. Contr. Nat. Sci. No. 5. 7 pp.
2926. Nimmo, A. P. 1987. The adult Arctopsychidae and Hydropsychidae of Canada. Quaest. Ent. 23:1-189.
2927. Nimmo, A. P., and G.G.E. Scudder. 1978. An annotated checklist of the Trichoptera (Insecta) of British Columbia. Syesis 11:117-134.
2928. Nimmo, A. P., and G.G.E. Scudder. 1983. Supplement to an annotated checklist of the Trichoptera (Insecta) of British Columbia. Syesis 16:71-83.
2929. Nimmo, A. P., and R. Wickstrom. 1984. Preliminary annotated checklist of the Trichoptera (Insecta) of the Yukon. Syesis 17:3-9.
2930. Noblet, R., W. A. Gardner, D. C. Arnold, R. E. Moore, Jr., E. L. Snoddy, and T. R. Adkins, Jr. 1979. An annotated list of the Simuliidae (Diptera) of South Carolina. J. Ga. Ent. Soc. 13:333-338.
2931. Noda, H., M. Miyazaki, and H. Hashimoto. 1986. Injury to rice leaves by chironomid larvae (Diptera: Chironomidae). Jap. J. Appl. Ent. Zool. 30:66-68.
2932. Nolte, U., and T. Hoffman. 1992. Fast life in cold water: *Diamesia incallida* (Chironomidae). Ecography 15:25-30.
2933. Noonan, G. R. 1986. Distribution of insects in the Northern Hemisphere: continental drift and epicontinental seas. Bull. Ent. Soc. Am. 32:80-84.
2934. Norling, U. 1982. Structure and ontogeny of the lateral abdominal gills and the caudal gills in Euphaeidae (Odonata: Zygoptera) larvae. Zool. Jb. Anat. 107:343-389.
2935. Norris, R. H. 1980. An appraisal of an air-lift sampler for sampling stream macroinvertebrates. Bull. Austral. Soc. Limnol. 7:9-15.
2936. Norris, R. H., and A. Georges. 1986. Design and analysis for assessment of water quality, pp. 555-572. In P. De Deckker and W. D. Williams (eds.). Limnology in Australia. Junk Publs., Dordrecht, The Netherlands. 671 pp.
2937. Norris, R. H., and A. Georges. 1993. Analysis and interpretation of benthic macroinvertebrate surveys, pp. 234-286. In D. M. Rosenberg and V. H. Resh (eds.). Freshwater biomonitoring and benthic macroinvertebrates. Chapman and Hall, New York. 488 pp.
2938. Norris, R. H., E. P. McElravy, and V. H. Resh. 1992. Chap. 13. The sampling problem, pp. 282-306. In P. Calow and G. E. Petts (eds.). The rivers handbook. Vol. 1. Hydrological and ecological principles. Blackwell Sci. Publ. Ltd., London, England. 526 pp.
2939. Nost, T. 1985. Distribution and food habits of mayflies (Ephemeroptera) in streams in the Dovrefijell Mountains, central Norway. Fauna Norv. Ser. B. 32:100-105.
2940. Notestine, M. K. 1993. Function of gills and mesonotal shield of *Baetisca rogersi* nymphs (Ephemeroptera: Baetiscidae). Fla. Ent. 76:423-427.
2941. Novak, K., and F. Sehnal. 1963. The development cycle of some species of the genus *Limnephilus* (Trichoptera). Cas. Cs. Spol. Ent. 60:68-80.
2942. Nowell, W. R. 1951. The dipterous family Dixidae in western North America (Insecta: Diptera). Microentomology 16:187-270.
2943. Nowell, W. R. 1963. Guide to the insects of Connecticut Part VI, Fas. 3. Dixidae. Conn. Bull. State Geol. Nat. Hist. Surv. 93:85-102.

2944. Noyes, A. A. 1914. The biology of the net-spinning Trichoptera of Cascadilla Creek. Ann. Ent. Soc. Am. 7:251-272.
2945. Nummelin, M., and K. Vepsäläinen. 1988. Does size and abundance distribution of food explain habitat segregation among developmental stages of waterstriders (Heteroptera: Gerridae)? Ann. Ent. Fenn. 54:69-71.
2946. O'Brien, C. W. 1975. A taxonomic revision of the new world subaquatic genus *Neochetina* (Coleoptera: Curculionidae: Bagoini). Ann. Ent. Soc. Am. 69:165-174.
2947. O'Brien, C. W., and G. B. Marshall. 1979. U. S. Bagous, bionomic notes, a new species, and a new name (Bagoini, Erirhininae, Curculionidae, Coleoptera). Southwest. Ent. 4:141-149.
2948. O'Connor, N. A. 1992. Quantification of submerged wood in a lowland Australian stream system. Freshwat. Biol. 27:387-395.
2949. O'Hop, J., J. B. Wallace, J. D. Haefner. 1984. Production of a stream shredder, *Peltoperla maria* (Plecoptera: Peltoperlidae) in disturbed and undisturbed hardwood catchments. Freshwat. Biol. 14:13-21.
2950. O'Neill, W. L. 1973. Biology of *Trichopria popei* and *T atrichomelinae* (Hym., Diapriidae) parasitoids of the Sciomyzidae (Diptera). Ann. Ent. Soc. Am. 66:1043-1054.
2951. Oakley, B., and J. M. Palka. 1967. Prey capture by dragonfly larvae. Am. Zool. 7:727-728.
2952. Oberndorfer, R. Y., J. V. McArthur, and J. R. Barnes. 1984. The effect of invertebrate predators on leaf litter processing in an alpine stream. Ecology 65:1325-1331.
2953. Oberndorfer, R. Y., and K. W. Stewart. 1977. The life cycle of *Hydroperla crosbyi* (Plecoptera: Perlodidae). Great Basin Nat. 37:260-273.
2954. Ode, P. R., and S. A. Wissinger. 1993. Interaction between chemical and tactile cues in mayfly detection of stoneflies. Freshwat. Biol. 30:351-357.
2955. Oehme, G. 1972. Zur maximalen Lebensdaur von *Cloëon dipterum L.*, (Eph. Baëtidae). Ent. Nachr. 1972/10:131-133.
2956. Oemke, M. P. 1984a. Diatom feeding preferences of *Glossosoma nigrior* (Banks) larvae, stream grazers in two southern Michigan streams, pp. 285-290. *In* J.C. Morse (ed.). Proc. IV. Int. Symp. Trichoptera, Clemson Univ., SC. Dr. W. Junk Publ., The Hague, Netherlands.
2957. Oemke, M. P. 1984b. Interactions between a stream grazer and the diatom flora, pp. 291-299. *In* J. C. Morse (ed.). Proc. IV. Int. Symp. Trichoptera, Clemson Univ., SC. Dr. W. Junk Publ., The Hague, Netherlands.
2958. Oemke, M. P. 1987. The effect of temperature and diet on the larval growth of *Glossosoma nigrior*, pp. 257-262. *In* M. Bourmaud, and H. Tachet (eds.). Proc. V. Int. Symp. Trichoptera. Lyon, France. Dr. W. Junk Publ., The Hague, Netherlands.
2959. Oertli, B. 1993. Leaf litter processing and energy flow through macroinvertebrates in a woodland pond (Switzerland). Oecologia 96:466-477.
2960. Oertli, B., and P. C. Dall. 1993. Population dynamics and energy budget of *Nemoura avicularis* (Plecoptera) in Lake Esrom, Denmark. Limnology 23:115-122.
2961. Ogilvie, G. A., and H. F. Clifford. 1986. Life histories, production, and microdistribution of two caddisflies (Trichoptera) in a Rocky Mountain stream. Can. J. Zool. 64:2706-2717.
2962. Ohio EPA (Environmental Protection Agency). 1987. Biological criteria for the protection of aquatic life. Vols. I-III. Surface Waters Section, Division of Water Quality Monitoring and Assessment, Ohio EPA, Columbus.
2963. Okland, B. 1991. Laboratory studies of egg development and diapause in *Isoperla obscura* (Plecoptera) from a mountain stream in Norway. Freshwat. Biol. 25:485-495.
2964. Okland, J. 1962. Litt om teknikk ved insammling og konservering av freshkvannsdyr. Fauna 15:69-92.
2965. Okumura, G. T. 1961. Identification of lepidopterous larvae attacking cotton-with illustrated key. Calif. Dept. Agric. Spec. Publ. 282. 50 pp.
2966. Olafsson, J. S. 1992. A comparative study on mouthpart morphology of certain larvae of Chironomini (Diptera: Chironomidae) with reference to larval feeding habits. J. Zool. Lond. 228:183-204.
2967. Olander, R., and E. Palmen. 1968. Taxonomy, ecology and behaviour of the northern Baltic *Clunio marinus* Halid. Ann. Zool. Fenn. 5:97-110.
2968. Old, M. C. 1933. Observations on the Sisyridae (Neuroptera). Pap. Mich. Acad. Sci. Arts Lett. 17:681-684.
2969. Olds, E. J., R. W. Merritt, and E. D. Walker. 1989. Sampling, seasonal abundance, and mermithid parasitism of larval *Coquillettidia perturbans* in south-central Michigan. J. Am. Mosq. Contr. Assoc. 5:586-592
2970. Oliver, D. R. 1971. Life history of the Chironomidae. Ann. Rev. Ent. 16:211-230.
2971. Oliver, D. R. 1976. Chironomidae (Diptera) of Char Lake, Cornwallis Island, N.W.T., with descriptions of two new species. Can. Ent. 108:1053-1064.
2972. Oliver, D. R. 1981a. Chap. 29. Chironomidae, pp. 423-458. *In* J. F. McAlpine, B. V. Peterson, G. E. Shewell, H. J. Teskey, J. R. Vockeroth, and D. M. Wood (coords.). Manual of Nearctic Diptera, Vol. 1. Res. Branch Agric. Can. Monogr. 27. Ottawa. 674 pp.
2973. Oliver, D. R. 1981b. Description of *Euryhapsis* new genus including three new species (Diptera: Chironomidae). Can. Ent. 113:711-722.
2974. Oliver, D. R. 1982. *Xylotopus*, a new genus of Orthocladiinae (Diptera: Chironomidae). Can. Ent. 114:167-168.
2975. Oliver, D. R. 1985. Review of *Xylotopus* Oliver and description of *Isobrillia* n. gen. (Diptera: Chironomidae). Can. Ent. 117:1093-1110.
2976. Oliver, D. R., and R. W. Bode. 1985. Description of the larva and pupa of *Cardiocladius albiplumus* Saether (Diptera:Chironomidae). Can. Ent. 117:803-809.
2977. Oliver, D. R., M. E. Dillon, and P. S. Cranston. 1990. A catalog of Nearctic Chironomidae. Publication 1857/B, Res. Br. Agric. Can. 89pp.
2978. Oliver, D. R., D. McClymont, and M. E. Roussel. 1978. A key to some larvae of Chironomidae (Diptera) from the Mackenzie and Porcupine River watersheds. Fish. Environ. Can. Fish. Mar. Serv. Tech. Rept. No. 791:1-73.
2979. Oliver, D. R., and M. E. Roussel. 1982. The larvae of *Pagastia* Oliver (Diptera: Chironomidae) with descriptions of three Nearctic species. Can. Ent. 114:849-854.
2980. Oliver, D. R., and M. E. Roussel. 1983a. The genera of larval midges of Canada (Diptera: Chironomidae). Part 11. The insects and arachnids of Canada. Biosys. Res. Inst., Ottawa. 263pp.
2981. Oliver, D. R., and M. E. Roussel. 1983b. Redescription of *Brillia* Kieffer (Diptera: Chironomidae) with descriptions of Nearctic species. Can. Ent. 115:257-279.
2982. Omerod, E. A. 1889. *Ranatra linearis* attacking small fish. Entomologist 11:119-120.
2983. Oosterbroek, P., and Br. Theowald. 1991. Phylogeny of the Tipuloidea based on characters of larvae and pupae (Diptera, Nematocera). Tijds. voor Ent. 134:211-267.
2984. Opler, P. A. (ed.) 1985. Species of special concern in Pennsylvania. Chapter 2. Invertebrates. Spec. Publ. Carnegie Mus. Nat. Hist. 11:81-168.
2985. Orr, B. K., W. W. Murdoch, and J. R. Bence. 1990. Population regulation, convergence, and cannibalism in *Notonecta* (Hemiptera). Ecology 71:68-82.
2986. Ortal, R., and S. Ritman. 1985. Israel inland water ecology data base, pp. 73-76. *In* P.S. Glaeser (ed.). The role of data in scientific progress. Elsevier, New York. 549 pp.
2987. Orth, D. J., and O.E . Maughan. 1983. Microhabitat preferences of benthic fauna in a woodland stream. Hydrobiologia. 106:157-168.
2988. Orth, R. E., and L. K. Knutson. 1987. Systematics of snail-killing flies of the genus Elgiva in North America and biology of *E. divisa* (Diptera: Sciomyzidae). Ann. Ent. Soc. Am. 80:829-840.
2989. Oscarson, H. G. 1987. Habitat segregation in a water boatman (Corixidae) assemblage - the role of predation. Oikos 49:133-140.
2990. Osten-Sacken, C. R. 1869. The North American Tipulidae. Monogr. N. Am. Diptera IV. Smithson. Misc. Coll. 8:1-345.

2991. Oswood, M. W. 1976. Comparative life histories of the Hydropsychidae (Trichoptera) in a Montana lake outlet. Am. Midl. Nat. 96:493-497.
2992. Oswood, M. W. 1979. Abundance patterns of filter-feeding caddisflies (Trichoptera: Hydropsychidae) and seston in a Montana (U.S.A.) lake outlet. Hydrobiologia 63:177-183.
2993. Otte, D. 1981. The North American grasshoppers, Acrididae. Vol. 1. Harvard Univ. Press, Cambridge. 275 p.
2994. Otto, C. 1985. Effects of temporal and spatial variations in food availability on life cycle and palatability of a chrysomelid beetle (*Donacia cinerea*). Aquat. Insects 7:19-28.
2995. Otto, C., and B. S. Svensson. 1981a. A comparison between food, feeding and growth of two mayflies, *Ephemera danica* and *Siphlonurus aestivalis* (Ephemeroptera) in a South Swedish stream. Arch. Hydrobiol. 91:341-350.
2996. Otto, C., and B. S. Svensson. 1981b. How do macrophytes growing in or close to water reduce their consumption by aquatic herbivores? Hydrobiologia 78:107-112.
2997. Otto, C., and J. B. Wallace. 1989. Life cycle variation and habitat longevity in waterlily leaf beetles. Holarct. Ecol. 12:144-151.
2998. Packard, A. S. 1984. Habits of an aquatic pyralid caterpillar. Am. Nat. 18:824-826.
2999. Packauskas, R. J., and J. E. McPherson. 1986. Life history and laboratory rearing of *Ranatra fusca* (Hemiptera: Nepidae) with descriptions of immature stages. Ann. Ent. Soc. Am. 79:566-571.
3000. Paelt, J. 1956. Biologie der primar flugellosen Insekten. GusLav Fischer Verlag, Jena. 258 p.
3001. Page, T. L., and D. A. Neitzel. 1979. A device to hold and identify rock-filled baskets for benthic sampling in large rivers. Limnol. Oceanogr. 24:988-990.
3002. Pajunen, V. I. 1970. Adaptations of *Arctocorisa carinata* (Sahlb.) and *Callocorixa producta* (Reut.) populations to a rock pool environment. Proc. Adv. Stud. Dynam. Pop. 1970:148-158.
3003. Pajunen, V. I. 1977. Population structure in rock-pool corixids (Hemiptera: Corixidae) during the reproductive season. Ann. Zool. Fenn. 14:26-47.
3004. Pajunen, V. I. 1979a. Competition between rock pool corixids. Ann. Zool. Fenn. 16:138-143.
3005. Pajunen, V. I. 1979b. Quantitative analysis of competition between *Arctocorisa carinata* (Sahlb.) and *Callicorixa producta* (Reut.) (Hemiptera, Corixidae). Ann. Zool. Fenn. 16:195-200.
3006. Pajunen , V. I., and I. Pajunen. 1991. Oviposition and egg cannibalism in rock pool corixids (Hemiptera: Corixidae). Oikos 60:83-90.
3007. Pajunen, V. I., and J. Salmi. 1991. The influence of corixids on the bottom fauna of rock-pools. Hydrobiologia 222:77-84.
3008. Pajunen, V. I., and M. Ukkonen. 1987. Intra- and interspecific predation in rock pool corixids (Hemiptera, Corixidae). Ann. Zool. Fenn. 24:295-304.
3009. Palavesam, A., and J. Muthukrishnan. 1992. Influence of food quality and temperature on fecundity of *Kiefferulus barbitarsis* (Kieffer) (Diptera: Chironomidae). Aquat. Insects 14:145-152.
3010. Palmer, C. G., and J. H. O'Keeffe. 1992. Feeding patterns of four macroinvertebrate taxa in the headwaters of the Buffalo River, eastern Cape. Hydrobiologia 228:157-173.
3011. Palmer, C., J. O'Keefe, and A. Palmer. 1993a. Macroinvertebrate functional feeding groups in the middle and lower reaches of the Buffalo River, eastern Cape, South Africa. II. Functional morphology and behaviour. Freshwat. Biol. 29:455-462.
3012. Palmer, C., J. O'Keefe, A. Palmer, T. Dunne, and S. Radloff. 1993b. Macroinvertebrate functional feeding groups in the middle and lower reaches of the Buffalo River, eastern Cape, South Africa. I. Dietary variability. Freshwat. Biol. 29:441-453.
3013. Pardue, W. J., and D. H. Webb. 1985. A comparison of aquatic macroinvertebrates occurring in association with Eurasian watermilfoil *Myriophyllum spicatum* with those found in the open littoral zone. J. Freshwat. Ecol. 3:69-80.
3014. Parfin, S. I. 1952. The Megaloptera and Neuroptera of Minnesota. Am. Midl. Nat. 47:421-434.
3015. Parfin, S. I., and A. B. Gurney. 1956. The spongilla-flies, with special reference to those of the Western Hemisphere (Sisyridae, Neuroptera). Proc. U.S. Nat. Mus. 105:421-529.
3016. Parker, C. R., and G. B. Wiggins. 1985. The Nearctic caddisfly genus *Hesperophylax* (Trichoptera: Limnephilidae). Can. J. Zool. 61:2443-2472.
3017. Parker, C. R., and G. B. Wiggins. 1987. Revision of the caddisfly genus *Psilotreta* (Trichoptera: Odontoceridae). Life Sci. Contr., Roy. Ont. Mus. 144.
3018. Parker, C. R., and J. R. Voshell. 1981. A preliminary checklist of the caddisflies (Trichoptera) of Virginia. J. Ga. Ent. Soc 16:l-7.
3019. Parker, C. R., and J. R. Voshell, Jr. 1982. Life histories of some filter-feeding Trichoptera in Virginia. Can. J. Zool. 60:1732-1742.
3020. Parker, C. R., and R. Voshell, Jr. 1983. Production of filter-feeding Trichoptera in an impounded and a free flowing river. Can. J. Zool. 61:70-87.
3021. Parker, H. L. 1924. Recherches sur les formes post-embryonnaires des chalcidiens. Ann. Soc. Ent. France 93:261-379.
3022. Parkinson, A., and R. A. Ring. 1983. Osmoregulation and respiration in a marine chironomid larva, *Paraclunio alaskensis* Coquillett (Diptera, Chironomidae). Can. J. Zool. 61:1937-1943.
3023. Parr, M. J. 1970. The life histories of *Ischnura elegans* (van der Linden) and *Coenagrion puella* (L.) (Odonata) in south Lancashire. Proc. R. ent. Soc. Lond.(A) 45:172-181.
3024. Parrish, F. K. 1975. Keys to water quality indicative organisms of the southeastern United States (2nd ed.). EMSL/EPA, Cincinnati. 195 pp.
3025. Parshley, H. M. 1925. A bibliography of the North American Hemiptera-Heteroptera. Smith College, Northampton. 252 pp.
3026. Parsons, M. A., R. L. Berry, M. Jalil, and R. A. Masterson. 1972. A revised list of the mosquitoes of Ohio with some new distribution and species records. Mosquito News 32:223-226.
3027. Parsons, M. C. 1970. Respiratory significance of the thoracic and abdominal morphology of the three aquatic bugs *Ambrysus*, *Notonecta* and *Hesperocorixa* (Insecta. Heteroptera). Z. Morph. Tiere 66:242-298.
3028. Parsons, M. C. 1972. Respiratory significance of the thoracic and abdominal morphology of *Belostoma* and *Ranatra* (Insecta, Heteroptera). Z. Morph. Tiere. 73:163-194.
3029. Parsons, M. C. 1974. Anterior displacements of the metathoracic spiracle and lateral intersegmental boundary in the pterothorax of Hydrocorisae. Z. Morphol. Tiere. 79:165-198.
3030. Parsons, M. C., and R. J. Hewson. 1974. Plastral respiratory devices in adult *Cryphocricos* (Naucoridae: Heteroptera). Psyche 81:510-527.
3031. Parsons, R. E., and D. E. Howell. 1971. A list of Oklahoma mosquitoes. Mosquito News 31:168-169.
3032. Patocka, J. 1955. Die Puppen der Schmetterlinge Schaedlinge der Eichen, die in oder an der Erde ruhen. Lesn. Sborn. 2:20-101.
3033. Patrick, R., J. Cairns, Jr., and S. S. Roback. 1967. An ecosystematic study of the fauna and flora of the Savannah River. Proc. Acad. Nat. Sci. Phila. 118:109-407.
3034. Patterson, J. W., and R. L. Vannote. 1979. Life history and population dynamics of *Heteroplectron americanum*. Environ. Ent. 8:665-669.
3035. Paulson, D. R. 1966a. New records of Bahamian Odonata. Quart. J. Fla. Acad. Sci. 29:97-110.
3036. Paulson, D. R. 1966b. The dragonflies (Odonata: Anisoptera) of southern Florida. Ph.D. Diss., University of Miami, Miami, Fla. 603 pp.
3037. Paulson, D. R. 1970. A list of the Odonata of Washington with additions to and deletions from the state list. Pan-Pacif. Ent. 46:194-198.
3038. Paulson, D. R., and R. A. Cannings. 1980. Distribution, natural history and relationship of *Ischnura erratica* Calvert (Zygoptera: Coenagrionidae). Odonatologica 9:147-153.
3039. Paulson, D. R., and R. W. Garrison. 1977. A list and new distributional records of Pacific coast Odonata. Pan-Pacif Ent. 53:147-160.
3040. Paulson, D. R., and C. E. Jenner. 1971. Population structure in overwintering larval Odonata in North Carolina in relation to adult flight season. Ecology 52:96-107.
3041. Peach, W. J., and J. A. Fowler. 1986. Life cycle and laboratory culture of *Dixella autumnalis* Meigen (Dipt., Dixidae). Ent. Mon. Mag. 122:59-62.

3042. Pearlstone, P.S.M. 1973. The food of damselfly larvae in Marion Lake, British Columbia. Syesis 6:33-39.

3043. Pearse, A. S. 1932. Animals in brackish water ponds and pools at Dry Tortugas. Papers Tortugas Lab. Carnegie Inst. Washington 28:125-142.

3044. Pearson, R. G., M. R. Litterick, and N. V. Jones. 1973. An airlift for quantitative sampling of the benthos. Freshwat. Biol. 3:309-315.

3045. Pearson, W. D., and R. H. Kramer. 1972. Drift and production of two aquatic insects in a mountain stream. Ecol. Monogr. 42:365-385.

3046. Pechuman, L. L., and H. J. Teskey. 1981. Chap. 31. Tabanidae, pp. 463-478. In J. F. McAlpine, B. V. Peterson, G. E. Shewell, J. J. Teskey, J. R. Vockeroth, and D. M. Wood (coords.). Manual of Nearctic Diptera, Vol. I. Res. Branch, Agric. Can. Monogr. 27. Ottawa. 674 pp.

3047. Pechuman, L. L., D. W. Webb, and H. J. Teskey. 1983. The Diptera, or true flies, of Illinois. I. Tabanidae. Ill. Nat. Hist. Surv. Bull. 33:1-122.

3048. Peck, D. L., and S. D. Smith. 1978. A revision of the *Rhyacophila coloradensis* complex (Trichoptera: Rhyacophilidae). Melanderia 27:1-24.

3049. Peckarsky, B. L. 1979. A review of the distribution, ecology and evolution of the North American species of *Acroneuria* and six related genera (Plecoptera: Perlidae). J. Kans. Ent. Soc. 52:787-809.

3050. Peckarsky, B. L. 1980. Predator-prey interactions between stoneflies and mayflies: behavioral observations. Ecology 61:932-943.

3051. Peckarsky, B. L. 1984a. Predator-prey interactions among aquatic insects, pp. 196-254. In V. H. Resh, and D. M. Rosenberg (eds.). The ecology of aquatic insects. Praeger, New York. 625 pp.

3052. Peckarsky, B. L. 1984b. Sampling the stream benthos, pp. 131-160. In J. A. Downing and F. H. Rigler (eds.). (2nd ed.). A manual on methods for the assessment of secondary productivity in fresh waters. IBP Handbook No. 17, Blackwell Scien. Publ., Oxford, England. 501 pp.

3053. Peckarsky, B. L. 1985. Do predaceous stoneflies and siltation affect the structure of stream insect communities colonizing enclosures? Can. J. Zool. 63:1519-1530.

3054. Peckarsky, B. L. 1987. Mayfly cerci as defense against stonefly predation: deflection and detection. Oikos 48:161-170.

3055. Peckarsky, B. L. 1988. Why predaceous stoneflies do not aggregate with their prey. Verh. Int. Ver. Limnol. 23:2135-2140.

3056. Peckarsky, B. L. 1991a. A field test of resource depression by predatory stonefly larvae. Oikos 61:3-10.

3057. Peckarsky, B. L. 1991b. Habitat selection by stream-dwelling predatory stoneflies. Can. J. Fish. Aquat. Sci. 48:1069-1076.

3058. Peckarsky, B. L. 1991c. Is there a coevolutionary arms race between predators and prey? A case study with stoneflies and mayflies. Adv. Ecol. 1:167-180.

3059. Peckarsky, B. L. 1991d. Mechanisms of intra- and interspecific interference between larval stoneflies. Oecologia 85:521-529.

3060. Peckarsky, B. L., and C. A. Cowan. 1991. Consequences of larval interspecific competition to stonefly growth and fecundity. Oecologia 88:277-288.

3061. Peckarsky, B. L., C. A. Cowan, M. A. Penton, and C. Anderson. 1993. Sublethal consequences of stream-dwelling predatory stoneflies on mayfly growth and fecundity. Ecology 74:1836-1846.

3062. Peckarsky, B. L., C. A. Cowan, and C. R. Anderson. 1994. Consequences and plasticity of the specialized predatory behavior of stream-dwelling stonefly larvae. Ecology 75:166-181.

3063. Peckarsky, B. L., and S. I. Dodson. 1980. Do stonefly predators influence benthic distributions in streams? Ecology 61:1275-1282.

3064. Peckarsky, B. L., S. C. Horn, and B. Statzner. 1990. Stonefly predation along a hydraulic gradient: a field test of the harsh-benign hypothesis. Freshwat. Biol. 24:181-191.

3065. Peckarsky, B. L., and M. A. Penton. 1985. Is predaceous stonefly behavior affected by competition? Ecology 66:1718-1728.

3066. Peckarsky, B. L., and M. A. Penton. 1988. Why do *Ephemerella* nymphs scorpion posture: a "ghost of predation past"? Oikos 53:185-193.

3067. Peckarsky, B. L., and M. A. Penton. 1989a. Early warning lowers risk of stonefly predation for a vulnerable mayfly. Oikos 54:301-309.

3068. Peckarsky, B. L., and M. A. Penton. 1989b. Mechanisms of prey selection by stream-dwelling stoneflies. Ecology 70(5):1203-1218.

3069. Peckarsky, B. L., and M. A. Penton. 1990. Effects of enclosures on stream microhabitat and invertebrate community structure. J. N. Am. Benthol. Soc. 9:249-261

3069'. Peckarsky, B. L., P. R. Fraissinet, M. A. Penton and D. J. Conklin, Jr. 1990. Freshwater macroinvertebrates of Northeastern United States. Cornell Univ. Press. 442 pp.

3070. Peckarsky, B. L., and R. S. Wilcox. 1989. Stonefly nymphs use hydrodynamic cues to discriminate between prey. Oecologia 79:265-270.

3071. Pedigo, L. P. 1968. Pond shore Collembola: a redescription of *Salina banksi* MacGillivray (Entomobryidae) and new Sminthuridae. J. Kans. Ent. Soc. 41:548-556.

3072. Pedigo, L. P. 1970. Activity and local distribution of surface active Collembola II: pond-shore populations. Ann. Ent. Soc. Am. 63:753-760.

3073. Pellerin, P., and J. Pilon. 1975. Cycle biologique de *Lestes eurinus* Say (Odonata: Lestidae), methode d'elevage en milieu conditionne. Odonatologica 6:83-96.

3074. Penland, D. R. 1953. A detailed study of the life cycle and respiratory system of a new species of western dobsonfly, *Neohermes aridus*. M. S. thesis, Chico State College, Chico, Calif. 34 pp

3075. Pennak, R. W. 1945. Notes on mountain midges (Deuterophlebiidae) with a description of the immature stages of a new species from Colorado. Am. Mus. Nov. 1276:1-10.

3076. Pennak, R. W. 1978. Freshwater invertebrates of the United States (2nd ed.). J. Wiley & Sons, N.Y. 803 pp.

3077. Pennak, R. W. 1989. Fresh-water invertebrates of the United States (3rd ed.). J. Wiley & Sons, Inc., New York. 628p.

3078. Pennak, R. W., and J. V. Ward. 1986. Interstitial faunal communities of the hyporheic and adjacent groundwater biotopes of a Colorado mountain stream. Arch. Hydrobiol./Suppl. 74: 356-396.

3079. Percival, E., and H. Whitehead. 1928. Observations on the ova and oviposition of certain Ephemeroptera and Plecoptera. Proc. Leeds Phil. Lit. Soc. 1:271-288.

3080. Percival, E., and H. Whitehead. 1929. A quantitative study of the fauna of some types of streambeds. J. Ecol. 17:282-314.

3081. Pereira, C.R.D., and N. H. Anderson. 1982. Observations on the life histories and feeding of *Cinygma integrum* Eaton and *Ironodes nitidus* (Eaton) (Ephemeroptera: Heptageniidae). Melanderia 39:35-45.

3082. Pereira, C.R.D., N. H. Anderson, and T. Dudley. 1982. Gut content analyses of aquatic insects from wood substrates. Melanderia 39:23-33.

3083. Pericart, J., and J. T. Polhemus. 1990. Un appareil stridulatoire chez les Leptopodidae de l'Ancien Monde (Heteroptera). Ann. Soc. Ent. Fr. (N. S.) 26:9-17.

3084. Perkins, P. D. 1975. Biosystematics of Western Hemisphere aquatic beetles (Coleoptera: Hydraenidae). Ph.D. diss., University of Maryland, College Park. 765 pp.

3085. Perkins, P. D. 1976. Psammophilous aquatic beetles in southern California: A study of microhabitat preferences with notes on response to stream alteration (Coleoptera: Hydraenidae and Hydrophilidae). Coleopt. Bull. 30:309-324.

3086. Perkins, P. D. 1980. Aquatic beetles of the family Hydraenidae in the Western Hemisphere: classification, biogeography and inferred phylogeny (Insecta: Coleoptera). Quaest. Ent. 16:3-554.

3087. Perry, J. A., and N. H. Troelstrup, Jr. 1988. Whole ecosystem manipulation: a productive avenue for test system research? Environ. Toxicol. Chem. 7:941-951.

3088. Perry, S. A., W. B. Perry, and J. A. Stanford. 1985. Effects of stream regulation on density, growth, and emergence of two mayflies (Ephemeroptera: Ephemerellidae) and a caddisfly (Trichoptera: Hydropsychidae) in two Rocky Mountain rivers (USA). Can. J. Zool. 64:656-666.

3089. Perry, S. A., W. B. Perry, and J. A. Stanford. 1987. Effects of thermal regime on size, growth rates and emergence of two species of stoneflies (Plecoptera: Taeniopterygidae, Pteronarcyidae) in the Flathead River, Montana. Am. Midl. Nat. 117:83-93.

3090. Perry, T. E. 1983. Additions to state and local lists of dragonflies and damselflies (Odonata). Ohio J. Sci. 83:141.

3091. Perry, W. B., E. F. Benfield, S. A. Perry, and J. R. Webster. 1987. Energetics, growth, and production of a leaf-shredding stonefly in a Appalachian Mountain stream. J. N. Am. Benthol. Soc. 6:12-25.

3092. Pescador, M. L. 1985. Systematics of the Nearctic genus *Pseudiron* (Ephemeroptera: Heptageniidae: Pseudironidae). Florida Ent. 68:432-444.

3093. Pescador, M. L., and L. Berner. 1981. The mayfly family Baetiscidae (Ephemeroptera). Part II. Biosystematics of the genus *Baetisca*. Trans. Ann. Ent. Soc. 107:163-228.

3094. Pescador, M. L., and W. L. Peters. 1974. The life history of *Baetisca rogersi* Berner (Ephemeroptera: Baetiscidae). Bull. Fla. State Mus. Biol. Sci. 17:151-209.

3095. Pescador, M. L., and W. L. Peters. 1980. A revision of the genus *Homoeoneuria* (Ephemeroptera: Oligoneuriidae). Trans. Am. Ent. Soc. 106:357-393.

3096. Peters, T. M. 1981. Chap. 23. Dixidae, pp. 329-334. *In* J. F. McAlpine, B. V. Peterson, G. E. Shewell, H. J. Teskey, J. R. Vockeroth, and D. M. Wood (coords.). Manual of Nearctic Diptera, Vol. I. Res. Branch, Agric. Can. Monogr. 27. Ottawa. 674 pp.

3097. Peters, T. M., and E. F. Cook. 1966. The Nearctic Dixidae (Diptera). Misc. Publ. Ent. Soc. Am. 5:233-278.

3098. Peters, W. L. 1971. A revision of the Leptophlebiidae of the West Indies (Ephemeroptera). Smithsonian Contr. Zool. No. 62. 48pp.

3099. Peters, W. L. 1980. Phylogeny of the Leptophlebiidae (Ephemeroptera): An introduction, pp. 33-42. *In* J. F. Flannagan and K. E. Marshall (eds.). Advances in Ephemeroptera biology. Plenum, N.Y. 552 pp.

3100. Peters, W. L. 1988. Origins of the North American Ephemeroptera fauna, especially the Leptophlebiidae. Mem. Ent. Soc. Can. 144:13-24.

3101. Peters, W. L., and J. G. Peters. 1977. Adult life and emergence of *Dolania americana* in northwestern Florida (Ephemeroptera: Behningiidae). Int. Revue ges. Hydrobiol. 62:409-438.

3102. Peters, W., and J. Spurgeon. 1971. Biology of the waterboatmen *Kirzousacorixa femorata* (Heteroptera: Corixidae). Am. Midl. Nat. 86:197-207.

3103. Peters, W., and R. Ulbrich. 1973. The life history of the water boatman, *Trichocorisella mexicana* (Heteroptera: Corixidae) (Hem.). Can. Ent. 105:277-282.

3104. Petersen, C. E., and R. G. Wiegert. 1982. Coprophagous nutrition in a population of *Paracoenia bisetosa* (Ephydridae) from Yellowstone National Park, USA. Oikos 39:251-255.

3105. Petersen, L. B. M. 1987. Direct observations of *Hydropsyche* prey selection, pp. 293-297. *In* M. Bournaud and H. Tachet (eds.). Proc. V. Int. Symp. Trichoptera. Lyon, France. Dr. W. Junk Publ., The Hague, Netherlands.

3106. Petersen, L. B.-M., and R. C. Petersen, Jr. 1984. Effect of kraft pulp mill effluent and 4,5,6 trichloroguaiacol on the net spinning behavior of *Hydropsyche angustipennis* (Trichoptera). Ecol. Bulls. 36:68-74.

3107. Petersen, R. C. 1974. Life history and bionomics of *Nigronia serricornis* (Say) (Megaloptera: Corydalidae). Ph.D. diss., Michigan State Univ., East Lansing. 210 pp.

3108. Petersen, R. C. 1987. Seston quality as a factor influencing trichopteran populations, pp. 287-292. *In* M.Bournaud and H. Tachet (eds.). Proc. V. Int. Symp. Trichoptera. Lyon, France. Dr. W. Junk Publ., The Hague, Netherlands.

3109. Petersen, R. C., and K. W. Cummins. 1974. Leaf processing in a woodland stream ecosystem. Freshwat. Biol. 4:343-368.

3110. Petersen, R. C., K. W. Cummins, and G. M. Ward. 1989. Microbial and animal processing of detritus in a woodland stream. Ecol. Monogr. 59:21-39.

3111. Peterson, A. 1934. Entomological techniques. Edwards Bros., Ann Arbor. 435 pp.

3112. Peterson, A. 1948. Larvae of insects. (Lepidoptera and Plant Infesting Hymenoptera). Part I. Edwards Bros., Ann Arbor. 315 pp.

3113. Peterson, A. 1951. Larvae of insects. Part II. Coleoptera, Diptera, Neuroptera, Siphonaptera, Mecoptera, Trichoptera. Edwards Bros., Ann Arbor 416 pp.

3114. Peterson, B., B. Fry, L. Deegan, and A. Hershey. 1993. The trophic significance of epilithic algal production in a fertilized tundra river ecosystem. Limnol. Oceanogr. 38:872-878.

3115. Peterson, B. J., J. E. Hobbie, A. E. Hershey, M. A. Lock, T. E. Ford, J. R. Vestal, V. L. McKinley, M. A. J. Hullar, M. C. Miller, R. M. Ventullo, and G. S. Volk. 1985. Transformation of a tundra river from heterotrophy to autotrophy by addition of phosphorus. Science 229:1383-1386.

3116. Peterson, B.V. 1955. A preliminary list of the black flies (Diptera: Simuliidae) of Utah. Proc. Utah Acad. Sci., Arts and Letters 32:113-115.

3117. Peterson, B. V. 1956. Observations on the biology of Utah black flies (Diptera: Simuliidae). Can Ent. 88:496-507.

3118. Peterson, B. V. 1960. The Simuliidae (Diptera) of Utah, Part I. Keys, original citations, types and distribution. Great Basin Nat. 20:81-104.

3119. Peterson, B. V. 1970. The *Prosimulium* of Canada and Alaska (Diptera: Simuliidae). Mem. Ent. Soc. Can. 69:1-216.

3120. Peterson, B. V. 1977. A synopsis of the genus *Parasimulium* Malloch (Diptera: Simuliidae), with descriptions of one new subgenus and two new species. Proc. Ent. Soc. Wash. 79:96-106.

3121. Peterson, B. V. 1981. Simuliidae, pp. 355-391. *In* J. F. McAlpine, B. V. Peterson, G. E. Shewell, H. J. Teskey, J. R. Vockeroth, and D. M. Wood (coords.). Manual of Nearctic Diptera, Vol. I. Res. Branch, Agric. Can. Monogr. 27. Ottawa. 674 pp.

3122. Peterson, B.V. 1989. An unusual black fly (Diptera: Simuliidae), representing a new genus and new species. J. N.Y. Ent. Soc. 97:317-331.

3123. Peterson, B.V. 1993. The black flies of the genus *Simulium*, subgenus *Psilopelmia* (Diptera: Simuliidae), in the contiguous United States. J. N. York. Ent. Soc. 101:301-390.

3124. Peterson, B.V., and B.C. Kondratieff. 1995. The black flies (Diptera: Simuliidae) of Colorado: An annotated list with keys, illustrations and descriptions of three new species. Mem. Am. Ent. Soc. 42:1-121.

3125. Peterson, C.G.J., and P. Boysen Tensen. 1911. Valuation of the sea. I. Animal life of the sea bottom: its food and quantity. Rept. Dan. Biol. Sta. 20:1-76.

3126. Peterson, D. G., and L. S. Wolfe. 1958. The biology and control of black flies (Diptera: Simuliidae) in Canada. Proc. 10th Int. Congr. Ent. 3:551-564.

3127. Pettry, D. K., and D. C. Tarter. 1985. Ecological life history of *Baetisca carolina* in Panther Creek, Nicholas County, West Virginia (Ephemeroptera: Baetiscidae). Psyche 92:355-368.

3128. Pfau, H. K. 1991. Contributions of functional morphology to the phylogenetic systematics of Odonata. Adv. Odonatology 5:109-141.

3129. Philip, C. B. 1931. The Tabanidae (horseflies) of Minnesota, with special reference to their biologies and taxonomy. Univ. Minn. Agric. Exp. Sta. Tech. Bull. 80:1-128.

3130. Philipson, G. N. 1953. A method of rearing Trichoptera larvae collected from swift flowing waters. Proc. R. ent. Soc. Lond. (A) 28:15-16.

3131. Philipson, G. N. 1954. The effect of water flow and oxygen concentration on six species of caddisfly (Trichoptera) larvae. Proc. Zool. Soc. Lond. 124:547-564.

3132. Philipson, G. N. 1977. The undulatory behavior of larvae of *Hydropsyche pellucidula* Curtis and *Hydropsyche siltalai* Dohler, pp. 241-247. *In* M. I. Crichton (ed.). Proc. 2nd Int. Symp. Trichoptera. Junk, The Hague. 359 pp.

3133. Philipson, G. N., and B.H.S. Moorhouse. 1974. Observations on ventilatory and net-spinning activities of larvae of the genus *Hydropsyche* Pictet (Trichoptera, Hydropsychidae) under experimental conditions. Freshwat. Biol. 4:525-533.

3134. Philipson, G. N., and B.H.S. Moorhouse. 1976. Respiratory behaviour of larvae of four species of the family Polycentropodidae (Trichoptera). Freshwat. Biol. 6:347-353.

3135. Phillips, D. J. H., and P. S. Rainbow. 1993. Biomonitoring of trace aquatic contaminants. Elsevier, London. 371 pp.
3136. Phillips, E. C., and R. V. Kilambi. 1994. Utilization of coarse woody debris by Ephemeroptera in three Ozark streams of Arkansas. Southwest. Nat. 39:58-62.
3137. Phillips, R. S. 1955. Seeking the secrets of the water springtail. Nature 48:241-243.
3138. Pierce, C. L. 1988. Predator avoidance, microhabitat shift, and risk-sensitive foraging in larval dragonflies. Oecologia 77:81-90.
3139. Pierce, C. L., P. H. Crowley, and D. M. Johnson. 1985. Behavior and ecological interactions of larval Odonata. Ecology 66:1504-1512.
3140. Pilon, J.-G., L. Pilon, and D. Lagace. 1990. La faune odonatologique de la zone boreale du Quebec. Soc. Int. Odonatol. Rapid Comm. (Suppl.) 11:1-42.
3141. Pilon, J.-G., L. Pilon, and D. Lagace. 1991. Les odonates de la zone temperee du Quebec: zygoptères. Soc. Int. Odonatol. Rapid Comm. (Suppl.) 13:1-37.
3142. Pinder, L.C.V. 1978. A key to the adult males of the British Chironomidae (Diptera). Sci. Publ. Freshwat. Biol. Assoc. 37:1-169.
3143. Pinder, L. C. V. 1986. Biology of freshwater Chironomidae. Ann. Rev. Ent. 31:1-23.
3144. Pinder, L. C. V. 1992. Biology of epiphytic Chironomidae (Diptera: Nematocera) in chalk streams. Hydrobiologia 248:39-51.
3145. Ping, C. 1921. The biology of *Ephydra subopaca* Loew. Mem. Cornell Univ. Agric. Exp. Sta. 49:557-616.
3146. Pinkham, C. F. A., and J. G. Pearson. 1976. Applications of a new coefficient of similarity to pollution surveys. J. Wat. Poll. Contr. Fed. 48:717-723.
3147. Pinkovsky, D. D., and J. F. Butler. 1978. Black flies of Florida I. Geographic and seasonal distribution. Florida Ent. 61:257-267.
3148. Plafkin, J. L., M. T. Barbour, K. D. Porter, S. K. Gross, and R. M. Hughes. 1989. Rapid bioassessment protocols for use in streams and rivers. Benthic macroinvertebrates and fish. EPA/444/4-89/001. Office of Water Regulations and Standards, U.S. Environmental Protection Agency, Washington, DC. 162 pp.
3149. Pobst, D. 1990. Trout stream insects. An Orvis streamside guide. Lyons and Burford, New York. 81p.
3150. Poff, N. L., and J. V. Ward. 1992. Heterogeneous currents and algal resources mediate in situ foraging activity of a mobile stream grazer. Oikos 65: 465-478.
3151. Poirrier, M. A. 1969. Some freshwater sponge hosts of Louisiana and Texas spongillaflies, with new locality records. Am. Midl. Nat. 81:573-575.
3152. Poirrier, M. A., and Y. M. Arceneaux. 1972. Studies on southern Sisyridae (spongillaflies) with a key to the third-instar larvae and additional sponge-host records. Am. Midl. Nat. 88:455-458.
3153. Poirrier, M. A., and R. W. Holzenthal. 1980. Records of Spongillaflies (Neuroptera: Sisyridae) from Mississippi. J. Miss. Acad. Sci. 15:1-2.
3154. Polhemus, J.T. 1967. Notes on North American Saldidae (Hemiptera). Proc. Ent. Soc. Wash. 69:24-30.
3155. Polhemus, J. T. 1966. Some Hemiptera new to the United States (Notonectidae, Saldidae). Proc. Ent. Soc. Wash. 68:57.
3156. Polhemus, J. T. 1973. Notes on aquatic and semi-aquatic Hemiptera from the southwestern United States (Insecta: Hemiptera). Great Basin Nat. 33:113-119.
3157. Polhemus, J. T. 1974. The *austrina* group of the genus Microvelia (Hemiptera: Veliidae). Great Basin Nat. 34:207-217.
3158. Polhemus, J. T. 1976a. A reconsideration of the status of the genus *Paravelia* Breddin, with other notes and a checklist of species. J. Kans. Ent. Soc. 49:509-513.
3159. Polhemus, J. T. 1976b. Notes on North American Nepidae (Hemiptera: Heteroptera). Pan-Pacif. Ent. 52:204-208.
3160. Polhemus, J. T. 1976c. Shore bugs (Hemiptera: Saldidae, etc.), pp. 225-261. *In* Cheng, L. (ed.). Marine insects. North-Holland, Amsterdam. 581 pp.
3161. Polhemus, J. T. 1985. Shore bugs (Heteroptera, Hemiptera: Saldidae). A world overview and taxonomy of Middle America forms. The Different Drummer, Englewood, Colorado. 252 p.
3162. Polhemus, J. T. 1994a. A new species of *Pentacora* (Heteroptera: Saldidae) from the Ouachita Mountains of Arkansas. J. Kansas Ent. Soc. 66:455-457, 1993.
3163. Polhemus, J. T. 1994b. Stridulatory mechanisms in aquatic and semiaquatic Heteroptera. J. New York Ent. Soc. 102:270-274.
3164. Polhemus, J. T., and H. C. Chapman. 1966. Notes on some Hebridae from the United States with the description of a new species (Hemiptera). Proc. Ent. Soc. Wash. 68:209-211.
3165. Polhemus, J. T., and P. Linskog. 1994. The stridulatory mechanism of *Nerthra* Say, a new species, and synonymy (Gelastocoridae: Heteroptera). J. New York Ent. Soc. 102:242-248..
3166. Polhemus, J. T., and C. N. McKinnon. 1983. Notes on the Hebridae of the Western Hemisphere with a description of two new species (Heteroptera: Hemiptera). Proc. Ent. Soc. Wash. 85:110-115.
3167. Polhemus, J. T., and M. S. Polhemus. 1976. Aquatic and semiaquatic Heteroptera of the Grand Canyon (Insecta: Hemiptera). Great Basin Nat. 36:221-226.
3168. Polhemus, J. T., and D. A. Polhemus. 1993a. Two new genera for New World Veliinae (Heteroptera: Veliidae). J. New York Ent. Soc. 101:391-398.
3169. Polhemus, J. T., and D. A. Polhemus. 1993b. The Trepobatinae (Gerridae) of New Guinea and surrounding regions, with a review of the world fauna. Part 1. Tribe Metrobatini. Ent. scand. 24:241-284.
3170. Polhemus, J. T., and D. A. Polhemus. 1994. A new species of *Ambrysus* from Ash Meadows, Nevada (Naucoridae: Heteroptera). J. New York Ent. Soc. (in press).
3171. Polhemus, J. T., and D. A. Polhemus. 1995. The Trepobatinae (Gerridae) of New Guinea and surrounding regions, with a review of the world fauna. Part 3. Tribe Trepobatini. Ent. scand. (in press).
3172. Polhemus, J. T., and P. J. Spangler. 1989. A new species of *Rheumatobates* from Ecuador and distribution of the genus (Heteroptera: Gerridae). Proc. Ent. Soc. Wash. 91:421-428.
3173. Pollard, J. E. 1981. Investigator differences associated with a kicking method for sampling macroinvertebrates. J. Freshwat. Ecol. 1:215-224.
3174. Pollard, J. E., and S. M. Melancon. 1984. Field washing efficiency in removal of macroinvertebrates from aquatic vegetation mats. J. Freshwat. Ecol. 2:383-292.
3175. Pommen, G. D., and D. A. Craig. 1995. Flow patterns around gills of pupal net-winged midges (Diptera: Blephariceridae): possible implications for respiration. Can. J. Zool. 73 (in press).
3176. Pontasch, K. W. 1988. Predation by *Paragnetina fumosa* (Plecoptera: Perlidae) on mayflies- the influence of substrate complexity. Am. Midl. Nat. 119:441-443.
3177. Poole, W. C., and K. W. Stewart. 1976. The vertical distribution of macrobenthos within the substratum of the Brazos River, Texas. Hydrobiologia 50:151-160.
3178. Pope, R. D. 1975. Nomenclatural notes on the British Scirtidae (= Helodidae) Entomol. mon. Mag. 111:186-187.
3179. Popham, E. J. 1960. On the respiration of aquatic Hemiptera Heteroptera with special reference to the Corixidae. Proc. Zool. Soc. Lond. 135:209-242.
3180. Popham, E. J. 1962. A repetition of Ege's experiments and a note on the efficiency of the physical gill of *Notonecta* (Hemiptera-Heteroptera). Proc. R. ent. Soc. Lond. (A) 37:154-160.
3181. Popham, E. J. 1964. The migration of aquatic bugs with special reference to the Corixidae (Hemiptera, Heteroptera). Arch. Hydrobiol. 60:450-496.
3182. Popowa, A. N. 1927. Uber die Ernahrung der Trichopterenlarven *(Neureclipsis bimaculata* L. und *Hydropsyche ornatula* McLach). Arch. Wiss. Insektenbiol. 22:147-159.
3183. Porch, S. S., W. L. Grogan, Jr., and L. J. Hribar. 1992. Biting midges (Diptera: Certatopogonidae) captured by sundews. Proc. Ent. Soc. Wash. 94:172-173.
3184. Porter, T. W. 1950. Taxonomy of the American Hebridae and the natural history of selected species. Ph.D. diss., University of Kansas, Lawrence. 165 pp.
3185. Portier, P. 1911. Recherches physiologiques sur les insectes aquatiques. Arch. Zool. Exp. Gen. 8:89-379.
3186. Portier, P. 1949. La biologie des Lepidopteres. Paul Lechevalier, Paris. 643 pp.

3187. Potter, D.W.B., and M. A. Learner. 1974. A study of the benthic macroinvertebrates of a shallow eutrophic reservoir in South Wales with emphasis on the Chironomidae (Diptera); their life histories and production. Arch. Hydrobiol. 74:186-226.

3188. Poulton, B. C., and K. W. Stewart. 1991. The stoneflies of the Ozark and Ouachita Mountains (Plecoptera). Mem. Am. Ent. Soc. 38:1-116.

3189. Powell, J. A. 1964. Biological and taxonomic studies on tortricine moths, with reference to the species in California. Univ. Calif. Publ. Ent. 32:1-317.

3190. Power, M. E. 1991. Shifts in the effects of tuft-weaving midges on filamentous algae. Am. Midl. Nat 125:275-285.

3191. Power, M. E., A. J. Stewart, and W. J. Matthews. 1988. Grazer control of algae in an Ozark mountain stream: effects of a short-term exclusion. Ecology 69:1894-1898.

3192. Power, M. E., R. J. Stout, C. E. Cushing, P. P. Harper, F. R. Hauer, W. J. Mathews, P. B. Moyle, B. Statzner, and I. R. Wais De Badgen. 1988. Biotic and abiotic controls in river and stream communities. J. N. Am. Benthol. Soc. 7:456-479.

3193. Powers, C. F., and A. Robertson. 1967. Design and evaluation of an all-purpose benthos sampler, pp. 126 131. In J. C. Ayers and D. C. Chandler (eds.). Studies on the environment and eutrophication of Lake Michigan. Univ. Mich. Great Lakes Res. Div. Spec. Rept. 30. 163 pp.

3194. Prepas, E. E. 1984. Some statistical methods for the design of experiments and analysis of samples, pp. 266-335. In J. A. Downing and F. H. Rigler (eds.) (2nd ed.). A manual on methods for the assessment of secondary productivity in fresh waters. IBP Handbook No. 17. Blackwell Scien. Publ., Oxford, England. 501 pp.

3195. Pringle, C. M. 1985. Effects of chironomid (Insecta: Diptera) tube-building activities on stream diatom communities. J. Phycol. 21:185-194.

3196. Pringle, C. M., G. A. Blake, A. P. Covich, K. M. Buzby, and A. Finley. 1993. Effects of omnivorous shrimp in a montane tropical stream: sediment removal, disturbance of sessile invertebrates and enhancement of understory algal biomass. Oecologia 93:1-11.

3197. Pringle, C. M., R. J. Naiman, G. Bretschko, J. R. Karr, M. W. Oswood, J. R. Webster, R. L. Welcomme, and M. J. Winterbourn. 1988. Patch dynamics in lotic systems: the stream as a mosaic. J. N. Am. Benthol. Soc. 7:503-524.

3198. Pritchard, G. 1964. The prey of dragonfly larvae (Odonata: Anisoptera) in ponds in northern Alberta. Can. J. Zool. 42:785-795.

3199. Pritchard, G. 1976. Growth and development of larvae and adults of *Tipula sacra* Alexander (Insecta: Diptera) in a series of abandoned beaver ponds. Can. J. Zool. 54:266-284.

3200. Pritchard, G. 1978. Study of dynamics of populations of aquatic insects: the problem of variability in life history exemplified by *Tipula sacra* Alexander (Diptera; Tipulidae). Verh. Int. Verein. Limnol. 20:2634-2640.

3201. Pritchard, G. 1980a. Life budgets for a population of *Tipula sacra* (Diptera: Tipulidae). Ecol. Ent. 5:165-173.

3202. Pritchard, G. 1980b. The life cycle of *Argia vivida* Hagen in the northern part of its range (Zygoptera: Coenagrionidae). Odonatologica 9:101-106.

3203. Pritchard, G. 1983. Biology of Tipulidae. Ann. Rev. Ent. 28:1-22.

3204. Pritchard, G. 1988. Dragonflies of the Cave and Basin Hot Springs, Banff National Park, Alberta, Canada. Not. Odonatol. 3:8-9.

3205. Pritchard, G. 1991. Insects in thermal springs. Mem. Ent. Soc. Can. 155:89-106.

3206. Pritchard, G., and H. A. Hall. 1971. An introduction to the biology of craneflies in a series of abandoned beaver ponds, with an account of the life cycle of *Tipula sacra* Alexander (Diptera: Tipulidae). Can. J. Zool. 49:467-482.

3207. Pritchard, G., and T. G. Leischner. 1973. The life history and feeding habits of *Sialis cornuta* Ross in a series of abandoned beaver ponds (Insecta: Megaloptera). Can. J. Zool. 51:121-131.

3208. Pritchard, G., M. H. McKee, E. M. Pike, G. J. Scrimgeour, and J. Zloty. 1993. Did the first insects live in water or in air? Biol. J. Linnean Soc. 49:31-44.

3209. Pritchard, G., and M. Stewart. 1982. How cranefly larvae breathe. Can. J. Zool. 60:310-317.

3210. Pritchard, G., and P. J. Scholefield. 1980. An adult emergence trap for use in small shallow ponds. Mosquito News 40:294-296.

3211. Proctor, D.L.C. 1973. The effect of temperature and photoperiod in Odonata. Can J. Zool. 51:1165-1170.

3212. Provonsha, A. V. 1990. A revision of the genus *Caenis* in North America (Ephemeroptera: Caenidae). Trans. Am. Ent. Soc. 116:801-884.

3213. Provonsha, A. V., and W. P. McCafferty. 1975. New techniques for associating the stages of aquatic insects. Great Lakes Ent. 8:105-109.

3214. Provonsha, A. V., and W. P. McCafferty. 1982. New species and previously undescribed larvae of North American Ephemeroptera. J. Kansas Ent. Soc. 55:23-33.

3215. Provonsha, A. V., and W. P. McCafferty. 1985. *Amercaenis*: new Nearctic genus of Caenidae (Ephemeroptera). Int. Quart. Ent. 1:1-7.

3216. Provost, M. W., and N . Branch. 1959. Food of chironomid larvae in Polk County Lakes. Fla. Ent. 42:49-62.

3217. Pruess, K. C. 1968. Checklist of Nebraska Odonata. Proc. N. Cent. Br. Ecol. Soc. Am. 22:112.

3218. Pruess, K. C., and B. V. Peterson 1987. The black flies (Diptera: Simuliidae) of Nebraska: An annotated list. J. Kansas Ent.Soc. 60:528-534.

3219. Pucat, A. M. 1965. The functional morphology of the mouthparts of some mosquito larvae. Quaest. Ent. 1:41-86.

3220. Puchkova, L. V. 1969. On the trophic relationships of water crickets (Corixidae). Zool. Zh. 48:1581-1583.

3221. Pugsley, C. W., and H. B. N. Hynes. 1985. Summer diapause and nymphal development in *Allocapnia pygmaea* (Burmeister), (Plecoptera: Capniidae), in the Speed River, southern Ontario. Aquat. Insects 7:53-63.

3222. Pugsley, C. W., and H. B. N. Hynes. 1986a. Three-dimensional distribution of winter stonefly nymphs, *Allocapnia pygmaea*, within the substrate of a southern Ontario river. Can. J. Fish. Aquat. Sci. 43:1812-1817.

3223. Pugsley, C., and H. B. N. Hynes. 1986b. A modified freeze-core technique to quantify the depth distribution of fauna in stony streambeds. Can. J. Fish. Aquat. Sci. 40:637-643.

3224. Pupedis, R. J. 1978. Tube feeding by *Sisyridivora cavigena* (Hymenoptera: Pteromalidae) on *Climacia areolaris* (Neuroptera: Sisyridae). Ann. Ent. Soc. Am. 71:773-775.

3225. Pupedis, R. J. 1980. Generic differences among new world Spongilla-fly larvae and a description of the female of *Climacia striata* (Neuroptera: Sisyridae). Psyche. 87:305-314.

3226. Pupedis, R. J. 1987. Foraging behavior and food of adult spongilla-flies (Neuroptera: Sisyridae). Ann. Ent. Soc. Am. 80:758-760.

3227. Puri, I. M. 1925. On the life history and structure of the early stages of Simuliidae (Diptera: Nematocera). Parasitology 17:295-334.

3228. Quail, A. 1904. On the tubercles of thorax and abdomen in first larval stage of Lepidoptera. Entomologist 37:269.

3229. Quate, L. W. 1955. A revision of the Psychodidae (Diptera) in American north of Mexico. Univ. Calif. Publ. Ent. 10: 103-273.

3230. Quate, L. W., and J. R. Vockeroth. 1981. Chap. 17. Psychodidae, pp. 293-300. In J. F. McAlpine, B. V. Peterson, G. E. Shewell, H. J. Teskey, J. R. Vockeroth, and D. M. Wood (coords.). Manual of Nearctic Diptera, Vol. 1. Res. Branch, Agric. Can. Monogr. 27. Ottawa. 674 p.

3231. Quate, L. W., and W. W. Wirth. 1951. A taxonomic revision of the genus *Maruina* (Diptera: Psychodidae). Wasmann J. Biol. 9:151-166.

3232. Quickenden, K. L., and V. C. Jamison. 1979. Montana mosquitoes part I: Identification and biology. Montana Dept. Hlth. Vector Cont. Bull. No. 1 (Rev.). 35pp.

3233. Quinby, G. E., R. E. Serfling, and J. K. Neel. 1944. Distribution and prevalence of the mosquitoes of Kentucky. J. Econ. Ent. 37:547-550.

3234. Quinlan, J. 1967. The brachypterous genera and species of Eucoilidae (Hymenoptera), with descriptions and figures of some type species. Proc. R. ent. Soc. Lond. (B) 36:1-10.

3235. Rabeni, C. F., and G. W. Minshall. 1977. Factors affecting microdistribution of stream benthic insects. Oikos 29:33-43.
3236. Rabeni, C. F., and K. E. Gibbs. 1978. Comparison of two methods used by divers for sampling benthic invertebrates in deep rivers. J. Fish. Res. Bd. Can. 35:332-336.
3237. Raddum, G. G., A. Fjellheim, and T. Hesthagen. 1988. Monitoring of acidification by the use of aquatic organisms. Verh. Int. Verein. Limnol. 23:2291-2297.
3238. Raddum, G. G., and O. A. Sæther. 1981. Chironomid communities in Norwegian lakes with different degrees of acidification. Verh. Int. Verein. Limnol. 21:399-405.
3239. Rader, R. B., and J. V. Ward. 1989. Influence of impoundments on mayfly diets, life histories, and production. J. N. Am. Benthol. Soc. 8: 64-73.
3240. Rader, R. B., and J. V. Ward. 1990. Mayfly growth and population density in constant and variable temperature regimes. Great Basin Nat. 50:97-106.
3241. Radford, D. S., and R. Hartland-Rowe. 1971. The life-cycles of some stream insects (Ephemeroptera, Plecoptera) in Alberta. Can. Ent. 103:609-617.
3242. Radinovsky, S. 1964. Cannibal of the pond. Nat. Hist. 73:16-25.
3243. Rahn, H., and C. V. Paganelli. 1968. Gas exchange in gas gills of diving insects. Resp. Physiol. 5:145-164.
3244. Rai, K. S. 1991. *Aedes albopictus* in the Americas. Ann. Rev. Ent. 36:459-484.
3245. Randolph, N. M., and K. O'Neill. 1944. The mosquitoes of Texas. Texas State Hlth. Dept. 100 pp.
3246. Rankin, K. P. 1935. Life history of *Lethocerus americanus* Leidy (Belostomatidae, Hemiptera). Univ. Kans. Sci. Bull. 22:479-491.
3247. Ranta, E., and J. Espo. 1989. Predation by the rock-pool insects *Arctocorisa carinata*, *Callicorixa producta* (Het. Corixidae) and *Potamonectes griseostriatus* (Col. Dytiscidae). Ann. Zool. Fenn. 26:53-60.
3248. Rao, T. K. R. 1962. On the biology of *Ranatra elongata* Fabr. (Heteroptera: Nepidae) and *Sphaerodema annulatum* Fabr. (Heteroptera: Belostomatidae). Proc. R. Ent. Soc. Lond. (A) 37:61-64.
3249. Rapoport, E. H., and L. Sanchez. 1963. On the epineuston or superaquatic fauna. Oikos 14:96-109.
3250. Rashed, S. S., and M. S. Mulla. 1990. Comparative functional morphology of the mouth brushes of mosquito larvae (Diptera: Culicidae). J. Med. Ent. 27:429-439.
3251. Rasmussen, J. B. 1984a. Comparison of gut contents and assimilation efficiency of fourth instar larvae of two coexisting chironomids, Chironomus *riparius* Meigen and *Glyptotendipes paripes* (Edwards). Can. J. Zool. 62:1022-1026.
3252. Rasmussen, J. B. 1984b. The life-history, distribution, and production of *Chironomus riparius* and *Glyptotendipes paripes* in a prairie pond. Hydrobiologia 119:65-72.
3253. Rasmussen, J. B. 1985. Effects of density and microdetritus enrichment on the growth of chironomid larvae in a small pond. Can. J. Fish. Aquat. Sci. 42:1418-1422.
3254. Rasmussen, J. B. 1988. Habitat requirements of burrowing mayflies (Ephemeridae: *Hexagenia*) in lakes, with special reference to the effects of eutrophication. J. N. Am. Benthol. Soc. 7:51-64.
3255. Rau, G. H., and N. H. Anderson. 1981. Use of $^{13}C/^{12}C$ to trace dissolved and particulate organic matter utilization by populations of an aquatic invertebrate. Oecologia 48:19-21.
3256. Rawat, B. L. 1939. On the habits, metamorphosis and reproductive organs of *Naucoris cimicoides* L. (Hemiptera-Heteroptera). Trans. R. ent. Soc. Lond. 88:119-138.
3257. Raybould, J. N. 1967. A method of rearing *Simulium damnosum* Theobald (Diptera: Simuliidae) under artificial conditions. Bull. World Health Org. 37:447-453.
3258. Raybould, J. N., and J. Grunewald. 1975. Present progress towards the laboratory colonization of African Simuliidae (Diptera). Tropenmed. Parasit. 26:155-168.
3259. Raybould, J. N., and H. K. Mhiddin. 1974. A simple technique for maintaining *Simulium* adults, including onchocerciasis vectors, in the laboratory. Bull. World Health Org. 51:309-310.
3260. Rees, B. E., and G. F. Ferris. 1939. The morphology of *Tipula reesi* Alexander (Diptera: Tipulidae). Microentomology 4:143-178.
3261. Rees, D. M. 1943. The mosquitoes of Utah. Bull. Univ. Utah 33:1-99.
3262. Reger, S. J., C. F. Brothersen, T. G. Osborn, and W. T. Helm. 1982. Rapid and effective processing of macroinvertebrate samples. J. Freshwat. Ecol. 1:451-465.
3263. Rehn, J.A.G., and D. C. Eades. 1961. The tribe Leptysmini (Orthoptera: Acrididae: Cyrtacanthacridinae) as found in North America and Mexico. Proc. Acad. Nat. Sci. Phila. 113:81-134.
3264. Rehn, J.A.G., and H. J. Grant. 1961. A monograph of the Orthoptera of North America (north of Mexico). Monogr. Acad. Nat. Sci. Phila. 12:1-257.
3265. Rehn, J.A.G., and M. Hebard. 1915a. Studies in American Tettigoniidae (Orthoptera). IV. A synopsis of the species of the genus *Orchelimum*. Trans. Am. Ent. Soc. 41:11-83.
3266. Rehn, J.A.G., and M. Hebard. 1915b. Studies in American Tettigoniidae (Orthoptera); V. A synopsis of the species of the genus *Conocephalus* found in North America north of Mexico. Trans. Am. Ent. Soc. 41:155-224.
3267. Reice, S. R. 1980. The role of substratum in benthic macroinvertebrate microdistribution and litter decomposition in a woodland stream. Ecology 61:580-590.
3268. Reice, S. R., and M. Wohlenberg. 1993. Monitoring freshwater benthic macroinvertebrates and benthic processes: measures for assessment of ecosystem health, pp. 287-305. *In* D. M. Rosenberg and V. H. Resh (eds.). Freshwater biomonitoring and benthic macroinvertebrates. Chapman and Hall, New York. 488 pp.
3269. Reichart, C. V. 1949. An ecological and taxonomic study of the aquatic insects of Blacklick Creek (Franklin County, Ohio). Abstr. Doct. Dissert., Ohio State Univ. 56:105-111.
3270. Reichart, C. V. 1971. A new *Buenoa* from Florida. Fla. Ent. 54:311-313.
3271. Reichart, C. V. 1976. Aquatic Hemiptera of Rhode Island. Part I. Notonectidae. Biol. Notes, Providence College 1:1-15.
3272. Reichart, C. V. 1977. Aquatic Hemiptera of Rhode Island. Part II. Naucoridae. Biol. Notes, Providence College 2:1-4.
3273. Reichart, C. V. 1978. Aquatic Hemiptera of Rhode Island. Part III. Belostomatidae. Biol. Notes Providence College 3:1-10.
3274. Reid, G. K. 1965. Ecology of inland waters and estuaries. Reinhold Publ. Co., N.Y. 375 pp.
3275. Reilly, P., and T. K. McCarthy. 1990. Observations on the natural diet of *Cymatia bonsdorfi* (C. Sahlb.). (Heteroptera: Corixidae): an immunological analysis. Hydrobiologia 196:159-166.
3276. Reisen, W. K. 1973. Invertebrate and chemical serial progression in temporary pool communities at Turner's Falls, Murray County, Oklahoma. J. Kans. Ent. Soc. 46:294-301.
3277. Reisen, W. K. 1975a. Quantitative aspects of *Simulium virgatum* Coq. and S. species life history in a southern Oklahoma stream. Ann. Ent. Soc. Am. 68:949-954.
3278. Reisen, W. K. 1975b. The ecology of Honey Creek, Oklahoma: spatial and temporal distributions of the macroinvertebrates. Proc. Okla. Acad. Sci. 55:25-31.
3279. Reiss, F. 1968. Okologische und systematische Untersuchungen und Chironomiden (Diptera) des Bodensees. Ein Beitrag zur lacustrischen Chironomidenfauna des nördlichen Alpenvorlandes. Arch. Hydrobiol. 64:176-323.
3280. Reiss, F., and J. E. Sublette. 1985. *Beardius* new genus with notes on additional Pan-American taxa. Spixiana Suppl. 11:179-193.
3281. Remmert, H. 1962. Der Schlupfrhythmus der Insekten. Franz Steiner, Wiesbaden. 73 pp.
3282. Rempel, J. G. 1936. The life history and morphology of *Chironomus hyperboreus*. J. Biol. Bd. Can. 2:209-220.
3283. Rempel, J. G. 1953. The mosquitoes of Saskatchewan. Can. J. Zool. 31:433-509.
3284. Rempel, J. G. 1975. The evolution of the insect head: The endless dispute. Quaest. Ent. 11:7-25.
3285. Renn, C. E. 1941. The food economy of *Anopheles quadrimaculatus* and *A. crucians* larvae. pp. 329-342. *In* A symposium on hydrobiology. Univ. Wisc. Press, Madison. 405 pp.
3286. Rennie, J. 1917. On the biology and economic significance of *Tipula paludosa*. Ann. Appl. Biol. 3:116-137.

3287. Rensing, L. 1962. Bietrage zur vergleichenden Morphologie, Physiologie and Ethologie der Wasserlaufer (Gerroidea). Zool. Beitr. 7:447-485.
3288. Resh, V. H. 1972. A technique for rearing caddisflies (Trichoptera). Can. Ent. 104:1959-1961.
3289. Resh, V. H. 1975. A distributional study of the caddisflies of Kentucky. Trans. Ky. Acad. Sci. 36:6-16.
3290. Resh, V. H. 1976a. Life cycles of invertebrate predators of freshwater sponge, pp. 299-314. In F. W. Harrison and R. R. Cowden (eds.). Aspects of sponge biology. Academic, N.Y. 354 pp.
3291. Resh, V. H. 1976b. The biology and immature stages of the caddisfly genus *Ceraclea* in eastern North America (Trichoptera: Leptoceridae). Ann. Ent. Soc. Am. 69:1039-1061.
3292. Resh, V. H. 1976c. Life histories of coexisting species of *Ceraclea* caddisflies (Trichoptera: Leptoceridae): The operation of independent functional units in a stream ecosystem. Can. Ent. 108:1303-1318.
3293. Resh, V. H. 1977. Habitat and substrate influences on population and production dynamics of a stream caddisfly, *Ceraclea ancylus* (Leptoceridae). Freshwat. Biol. 7:261-277.
3294. Resh, V. H. 1979a. Sampling variability and life history features: basic considerations in the design of aquatic insect studies. J. Fish. Res. Bd. Can. 36:290-311.
3295. Resh, V. H. 1979b. Biomonitoring, species diversity indices, and taxonomy, pp. 241-253. In J. F. Grassle, G. P. Patil, W. K. Smith, and C. Taillie (eds.). Ecological diversity in theory and practice. Internat. Coop. Publ. House, Fairland, Md. 365 pp.
3296. Resh, V. H. 1983. Spatial differences in the distribution of benthic macroinvertebrates along a springbrook. Aquat. Insects 5:193-200.
3297. Resh, V. H. 1992. Year-to-year changes in the age structure of a caddisfly population following loss and recovery of a springbrook habitat. Ecography 15:314-317.
3298. Resh, V. H. 1993. Recent trends in the use of Trichoptera in water quality monitoring, pp. 285-291. In C. Otto (ed.). Proc. VII Int. Symp. Trichoptera, Umeå, Sweden. Backhuys Publ., Leiden, The Netherlands.
3299. Resh, V. H. 1995. Freshwater benthic macroinvertebrates and rapid assessment procedures for water quality monitoring in developing and newly industrialized countries, pp. 167-177. In W. S. Davis and T. P. Simon (eds.). Biological assessment and criteria. Tools for water resource planning and decision making. Lewis Publs., Boca Raton, Fla.
3300. Resh, V. H., A. V. Brown, A. P. Covich, M. E. Gurtz, H. W. Li, G. W. Minshall, S. R. Reice, A. L. Sheldon, J. B. Wallace, and R. Wissmar. 1988. The role of disturbance in stream ecology. J. N. Am. Benthol. Soc. 7:433-455.
3301. Resh, V. H., J. W. Feminella, and E. P. McElravy. 1990. Sampling aquatic insects. Videotape. Office of Media Services, Univ. Calif, Berkeley.
3302. Resh, V. H., T. S. Flynn, G. A. Lamberti, E. P. McElravy, K. L. Sorg, and J. R. Wood. 1981. Responses of the sericostomatid caddisfly *Gumaga nigricula* (McL.) to environmental disruption, pp.311-318. In G.P. Moretti (ed.). Proc. III. Int. Symp. Trichoptera, Perugia. Dr. W. Junk Publ., The Hague, Netherlands.
3303. Resh, V. H., and G. Grodhaus. 1983. Aquatic insects in urban environments, pp. 247-276. In G. W. Frankie and C. S. Koehler (eds.). Urban entomology: interdisciplinary perspectives. Praeger, New York. 493 pp.
3304. Resh, V. H., and K. H. Haag. 1974. New records of parasitism of caddisflies by erythraeid mites. J. Parasit. 60:382-383.
3305. Resh, V. H., and R. E. Houp. 1986. Life history of the caddisfly *Dibusa angata* and its association with the red alga *Lemanea australis*. J. N. Am. Benthol. Soc. 5:28-40.
3306. Resh, V. H., and J. K. Jackson. 1993. Rapid assessment approaches to biomonitoring using benthic macroinvertebrates, pp. 195-233. In D. M. Rosenberg and V. H. Resh (eds.). Freshwater biomonitoring and benthic macroinvertebrates. Chapman and Hall, New York. 488 pp.
3307. Resh, V. H., and W. Jamieson. 1988. Parasitism of the aquatic moth *Petrophila confusalis* (Lepidoptera: Pyralidae) by the aquatic wasp *Tanychela pilosa* (Hymenoptera: Ichneumonidae). Ent. News 99:185-188.
3308. Resh, V. H., G. A. Lamberti, and J. R. Wood. 1984a. Biological studies of *Helicopsyche borealis* (Hagen) in a coastal California stream, pp. 315-319. In J.C. Morse (ed.). Proc. IV. Int. Symp. Trichoptera, Clemson, Univ., SC. Dr. W. Junk Publ., The Hague, Netherlands.
3309. Resh, V. H., G. A. Lamberti, and J. R. Wood. 1984b. Biology of the caddisfly *Helicopsyche borealis*: a comparison of North American populations. Freshwat. Invertebr. Biol. 3:172-180.
3310. Resh, V. H., and E. P. McElravy. 1993. Contemporary quantitative approaches to biomonitoring using benthic macroinvertebrates, pp. 159-194. In D. M. Rosenberg and V. H. Resh (eds.). Freshwater biomonitoring and benthic macroinvertebrates. Chapman and Hall, NY. 488 pp.
3311. Resh, V. H., J. C. Morse, and J. D. Wallace. 1976. The evolution of the sponge feeding habit in the caddisfly genus *Ceraclea* (Trichoptera: Leptoceridae). Ann. Ent. Soc. Am. 69:937-941.
3312. Resh, V. H., M. J. Meyers, and M.J. Hannaford. 1996. Macroinvertebrates as biotic indicators of environmental quality, pp. 647-667. In F.R. Hauer and G.A. Lamberti (eds). Methods in stream ecology. Academic Press. San Diego, CA. 674 pp.
3313. Resh, V. H., R. H. Norris, and M. T. Barbour. 1995. Design and implementation of rapid assessment approaches for water resource monitoring using benthic macroinvertebrates. Austr. J. Ecol. 20:108-121.
3314. Resh, V. H., and D. M. Rosenberg (eds.). 1979. Innovative teaching in aquatic entomology. Can. Spec. Publ. Fish. Aquat. Sci. 43:1-118.
3315. Resh, V. H., and D. M. Rosenberg (eds.). 1984. The ecology of aquatic insects. Praeger Publishers, N.Y. 625 p.
3316. Resh, V. H., and D. M. Rosenberg. 1989. Spatial-temporal variability and the study of aquatic insects. Can. Ent. 121:941-963.
3317. Resh, V. H., and F. de Szalay. 1994. Streams and rivers of oceania. In C. E. Cushing, K. W. Cummins, and G. W. Minshell (eds.). River Ecosystems of the World. Elsevier Publishers, Amsterdam. (in press.)
3318. Resh, V. H., and K. L. Sorg. 1983. Distribution of the Wilbur Springs shore bug (Hemiptera: Saldidae): Predicting occurrence using water chemistry parameters. Environ. Ent. 12:1628-1635.
3319. Resh, V. H., and J. D. Unzicker. 1975. Water quality monitoring and aquatic organisms: the importance of species identification. J. Wat. Poll. Contr. Fed. 47:9-19.
3320. Reynolds, J. D., and G. G. E. Scudder. 1987a. Experimental evidence of the fundamental feeding niche in *Cenocorixa* (Hemiptera: Corixidae). Can. J. Zool. 65:967-973.
3321. Reynolds, J. D., and G. G. E. Scudder. 1987b. Serological evidence of the realized feeding niche in *Cenocorixa* species (Hemiptera: Corixidae). Can. J. Zool. 65:974-980.
3322. Reynoldson, T. B, K. E. Day, C. Clarke, and D. Milani. 1994. Effect of indigenous animals on chronic end points in freshwater sediment toxicity tests. Environ. Toxicol. Chem. 13:973-977.
3323. Reynoldson, T. B., and J. L. Metcalfe-Smith. 1992. An overview of the assessment of aquatic ecosystem health using benthic invertebrates. J. Aquat. Ecosyst. Health 1:295-308.
3324. Rhame, R. E., and K. W. Stewart. 1976. Life cycles and food habits of three Hydropsychidae (Trichoptera) species in the Brazos River, Texas. Trans. Am. Ent. Soc. 102:65-99.
3325. Rice, L. A. 1954. Observations on the biology of ten notonectoid species found in the Douglas Lake, Michigan region. Am. Midl. Nat. 51:105-132.
3326. Richards, C. 1986. Distribution and foraging behavior of a grazing stream mayfly *Baetis bicaudatus*. Ph.D. diss., Idaho State Univ., Pocatello, ID. 99pp.
3327. Richards, C., G. E. Host, and J. W. Arthur. 1993. Identification of predominant environmental factors structuring stream macroinvertebrate communities within a large agricultural catchment. Freshwat. Biol. 29:285-294.
3328. Richards, C., and G. W. Minshall. 1988. The influence of periphyton abundance on *Baetis bicaudatus* distribution and colonization in a small stream. J. N. Am. Benthol. Soc. 7:77-86.
3329. Richards, L. J. 1983. Feeding and activity patterns of an intertidal beetle. J. Exp. Mar. Biol. Ecol. 73:213-224.

3330. Richards, L. J. 1984. Field studies of foraging behavior of an intertidal beetle. Ecol. Ent. 9:189-194.
3331. Richards, O. W. 1956. Hymenoptera, introduction and keys to families. Handbooks for identification of British insects 6 (pt. I) 94 pp.
3332. Richardson, J. S. 1984a. Effects of seston quality on the growth of a lake-outlet filter feeder. Oikos 43:386-390.
3333. Richardson, J. S. 1984b. Prey selection and distribution of a predacious, net-spinning caddisfly, *Neureclipsis bimaculata* (Polycentropodidae). Can. J. Zool. 62:1561-1565.
3334. Richardson, J. S. 1991. Seasonal food limitation of detritivores in a montane stream: an experimental test. Ecology 72:873-887.
3335. Richardson, J. S. 1992. Food, microhabitat, or both? Macroinvertebrate use of leaf accumulations in a montaine stream. Freshwat. Biol. 27:169-176.
3336. Richardson, J. S., and H. F. Clifford. 1983. Life history and microdistribution of *Neureclipsis bimaculata* (Trichoptera: Polycentropodidae) in a lake outflow stream of Alberta, Canada. Can. J. Zool. 61:2434-2445.
3337. Richardson, J. S., and H. F. Clifford. 1986. Phenology and ecology of some Trichoptera in a low-gradient boreal stream. J. N. Am. Benthol. Soc. 5:191-199.
3338. Richardson, J. S., and R. J. Mackay. 1984. A comparison of the life history and growth of *Limnephilus indivisus* (Trichoptera: Limnephilidae) in three temporary pools. Arch. Hydrobiol. 99:515-528.
3339. Richardson, J. S., and R. J. Mackay. 1991. Lake outlets and the distribution of filter feeders: an assessment of hypotheses. Oikos 62:370-380.
3340. Richardson, J. W., and A. R. Gaufin. 1971. Food habits of some Western stonefly nymphs. Trans. Am. Ent. Soc. 97:91-121.
3341. Richardson, M. Y., and D. C. Tarter. 1976. Life histories of *Stenonema vicarium* (Walker) and *S. tripunctatum* (Banks) in a West Virginia stream (Ephemeroptera:Heptageniidae). Am. Midl. Nat. 95:1-9.
3342. Richardson, R. E. 1928. The bottom fauna of the middle Illinois River,1913-1925. Bull. Ill. Nat. Hist. Surv. 8:363-522.
3343. Richmond, E. A. 1920. Studies on the biology of the aquatic Hydrophilidae. Bull. Am. Mus. Nat. Hist. 42:1-94.
3344. Ricker, W. E. 1943. Stoneflies of southwestern British Columbia. Ind. Univ. Publ. Sci. Ser. 12:1-145.
3345. Ricker, W. E. 1944. Some Plecoptera from the far North. Can. Ent. 76:174-185.
3346. Ricker, W. E. 1945. A first list of Indiana stoneflies (Plecoptera). Proc. Ind. Acad. Sci. 54:225-230.
3347. Ricker, W. E. 1946. Some prairie stoneflies (Plecoptera). Trans. Roy. Can. Inst. 26:3-8.
3348. Ricker, W. E. 1947. Stoneflies of the Maritime Provinces and Newfoundland. Trans. Roy. Can. Inst. 26:401-411.
3349. Ricker, W. E. 1949. The North American species of *Paragnetina* (Plecoptera, Perlidae). Ann. Ent. Soc. Am. 42:279-288.
3350. Ricker, W. E. 1952. Systematic studies in Plecoptera. Ind. Univ. Publ. Sci. Ser. 18:1-200.
3351. Ricker, W. E. 1959a. The species of *Isocapnia* Banks (Insecta, Plecoptera, Nemouridae). Can. J. Zool. 37:639-653.
3352. Ricker, W. E. 1959b. Plecoptera, pp. 941-957. *In* W. T. Edmondson (ed.). Freshwater biology. John Wiley & Sons, N.Y. 1248 pp.
3353. Ricker, W. E., R. Malouin, P. Harper, and H. H. Ross. 1968. Distribution of Quebec stoneflies (Plecoptera). Nat. Can. 95:1085-1123.
3354. Ricker, W. E., and H. H. Ross. 1968. North American species of *Taeniopteryx* (Plecoptera, Insecta). J. Fish. Res. Bd. Can. 25:1423-1439.
3355. Ricker, W. E., and H. H. Ross. 1969. The genus *Zealeuctra* and its position in the family Leuctridae. Can. J. Zool. 47:1113-1127.
3356. Ricker, W. E., and H. H. Ross. 1975. Synopsis of the Brachypterinae (Insecta: Plecoptera: Taeniopterygidae). Can. J. Zool. 53:132-153.
3357. Ricker, W. E., and G.G.E. Scudder. 1975. An annotated checklist of the Plecoptera (Insecta) of British Columbia. Syesis 8:333-348.
3358. Riek, E. F. 1970. Hymenoptera, pp. 867-983. *In* Insects of Australia. Melbourne Univ. Press, Melbourne. 1029 pp.
3359. Riek, E. F. 1971. The origin of insects. Proc. 13th Int. Congr. Ent. 13:292-293.
3360. Ries, M. D. 1967. Present state of knowledge of the distribution of Odonata in Wisconsin. Proc. N. Cent. Br. Ent. Soc. Am. 22:113-115.
3361. Ries, M. D. 1969. Odonata new to the Wisconsin list. Mich. Ent. 2:22-27.
3362. Rigler, F. H., and J. A. Downing. 1984. The calculation of secondary productivity, pp. 19-58. *In* J. A. Downing and F. H. Rigler (eds.). (2nd ed.). A manual on methods for the assessment of secondary productivity in fresh waters. IBP Handbook No. 17. Blackwell Scien. Publ., Oxford, England.
3363. Riley, C. V. 1874. Descriptions and natural history of two insects which brave the dangers of *Sarracenia variolaris*. Trans. St. Louis Acad. Sci. 3:235-240.
3364. Riley, C. V. 1879. On the larval characteristics of *Corydalus* and *Chauliodes* and on the development of *Corydalus cornutus*. Can. Ent. 11:96-98.
3365. Riley, C.F.C. 1918. Food of aquatic Hemiptera. Science 48:545-547.
3366. Rimski-Korsakov, M. N. 1917. Observations biologiques sur les Hymenopteres aquatiques. Rev. Russe Ent. 16:209-225.
3367. Rimski-Korsakov, M. N. 1940. Key to the freshwater Collembola of U.S.S.R. with descriptive notes. Freshwat. Life, U.S.S.R. 1:108-110.
3368. Ring, R. A. 1989. Intertidal Chironomidae of B.C., Canada. Acta Biol. Debr. Oecol. Hungary 3:275-288.
3369. Ris, F. 1909-1916. Libellulinen. Collections zoologiques du Baron Edm. de Selys Longchamps. Impr. Acad. Hayez, Brussels 9-13:1-1245.
3370. Ris, F. 1930. A revision of the libelluline genus *Perithemis* (Odonata). Misc. Publ. Univ. Mich. Mus. Zool. 21:1-50.
3371. Roback, S. 1978. The immature chironomids of the eastern United States III. Tanypodinae-Anatopyniini, Macropelopiini and Natarsiini. Proc. Acad. Nat. Sci. Philadelphia 129:151-202.
3372. Roback, S. S. 1953. Savannah River tendipedid larvae (Diptera: Tendipedidae-Chironomidae). Proc. Acad. Nat. Sci. Philadelphia 105:91-132.
3373. Roback, S. S. 1957. The immature tendipedids of the Philadelphia area. Monogr. Acad. Nat. Sci. Philadelphia 9:1-152.
3374. Roback, S. S. 1963. The genus *Xenochironomus* (Diptera: Tendipedidae) Kieffer, taxonomy and immature stages. Trans. Am. Ent. Soc. 88:235-245.
3375. Roback, S. S. 1968. The immature stages of the genus *Tanypus* Meigen (Diptera: Chironomidae: Tanypodinae). Trans. Am. Ent. Soc. 94:407-428.
3376. Roback, S. S. 1969. Notes on the food of Tanypodinae larvae. Ent. News 80:13-18.
3377. Roback, S. S. 1971. The adults of the subfamily Tanypodinae in North America (Diptera: Chironomidae). Monogr. Acad. Nat. Sci. Philadelphia 17:1-410.
3378. Roback, S. S. 1974. The immature stages of the genus *Coelotanypus* (Chironomidae: Tanypodinae: Coelotanypodinae) in North America. Proc. Acad. Nat. Sci. Phil. 126:9-19.
3379. Roback, S. S. 1976. The immature chironomids of the eastern United States. I. Introduction and Tanypodinae—Coelotanypodini. Proc. Acad. Nat. Sci. Philadelphia 127:147-201.
3380. Roback, S. S. 1977. The immature chironomids of the eastern United States II. Tanypodinae-Tanypodini. Proc. Acad. Nat. Sci. Philadelphia 128:55-87.
3381. Roback, S. S. 1978. The immature chironomids of the eastern United States III. Tanypodinae-Anatopyniini, Macropelopiini and Natarsiini. Proc. Acad. Nat. Sci. Philadelphia 129:151-202.
3382. Roback, S. S. 1980. The immature chironomids of the eastern United States IV. Tanypodinae-Procladiini. Proc. Acad. Nat. Sci. Philadelphia 132:1-63.
3383. Roback, S. S. 1981. The immature chironomids of the eastern United States V. Pentaneurini-*Thienemannimyia* group. Proc. Acad. Nat. Sci. Philadelphia 133:73-129.
3384. Roback, S. S. 1985. The immature chironomids of the eastern United States. VI. Pentaneurini- Genus *Ablabesmyia*. Proc. Acad. Nat. Sci. Phila. 137:153-212.

3385. Roback, S. S. 1986a. The immature chironomids of the eastern United States VII. Pentaneurini-Genus *Monopelopia*, with redescription of the male adults and description of some Neotropical material. Proc. Acad. Nat. Sci. Phil. 138:350-365.

3386. Roback, S. S. 1986b. The immature chironomids of the eastern United States VIII. Pentaneurini-Genus *Nilotanypus*, with the description of a new species from Kansas. Proc. Acad. Nat. Sci. Phil. 138:443-465.

3387. Roback, S. S. 1987. The immature chironomids of the eastern United States IX. Pentaneurini- Genus *Labrundinia* with the description of some Neotropical material. Proc. Acad. Nat. Sci. Phila. 139:159-209.

3388. Roback, S. S., and L. C. Ferrington, Jr. 1983. The immature stages of *Thienemannimvia barberi* (Coquillett) (Diptera: Chironomidae: Tanypodinae). Freshwat. Invertebr. Biol. 2:107-111.

3389. Roback, S. S., and J. W. Richardson. 1969. The effects of acid mine drainage on aquatic insects. Proc. Acad. Nat. Sci. Philadelphia 21:81-107.

3390. Roback, S. S., and R. P. Rutter. 1988. *Denopelopia atria*, new genus and species of Pentaneurini (Diptera: Chironomidae: Tanypodinae) from Florida. Spixiana Suppl. 14:117-127.

3391. Roback, S. S., and M. J. Westfall, Jr. 1967. New records of Odonata nymphs from the United States and Canada with water quality data. Trans. Am. Ent. Soc. 93:101-124.

3392. Robert, A. 1963. Les libellules du Quebec. Bull. Sta. Biol. Mt. Tremblant Quebec 1:1-223.

3393. Roberts, C. H. 1895. The species of *Dineutes* of America north of Mexico. Trans. Ent. Soc. Am. 22:279-288.

3394. Roberts, C. H. 1913. Critical notes on the Haliplidae. J. N. Y. Ent. Soc. 21:91-123.

3395. Roberts, D. R., L. W. Smith, and W. R. Enns. 1967. Laboratory observations on predation activities of *Laccophilus* beetles on the immature stages of some dipterous pests found in Missouri oxidation lagoons. Ann. Ent. Soc. Am. 60:908-910.

3396. Roberts, M. J. 1970. The structure of the mouthparts of syrphid larvae (Diptera) in relation to feeding habits. Acta Zool. (Stockholm) 51:43-65.

3397. Roberts, M. J. 1971. The structure and mouthparts of some calypterate dipteran larvae in relation to their feeding habits. Acta Zool. (Stockholm) 52:171-188.

3398. Roberts, R. H. 1962. Notes on the biology of *Tabanus dorsifer* (Tabanidae: Diptera). Ann. Ent. Soc. Am. 55:436-438.

3399. Roberts, R. H. 1966. A technique for rearing the immature stage of Tabanidae (Diptera). Ent. News 77:79-82.

3400. Roberts, R. H., and R. J. Dicke. 1964. The biology and taxonomy of some immature Nearctic Tabanidae (Diptera). Ann. Ent. Soc. Am. 57:31-40.

3401. Robinson, C. T., L. M. Reed, and G. W. Minshall. 1992. Influence of flow regime on life history, production, and genetic structure of *Baetis tricaudatus* (Ephemeroptera) and *Hesperoperla pacifica* (Plecoptera). J. N. Am. Benthol. Soc. 11:278-289.

3402. Robinson, H., and J. R. Vockeroth. 1981. Chap. 48. Dolichopodidae, pp. 625-639. In J. F. McAlpine, B. V. Peterson, G. E. Shewell, H. J. Teskey, J. R. Vockeroth, and D. M. Wood (coords.). Manual of Nearctic Diptera, Vol. 1. Res. Branch Agric. Can. Monogr. 27. Ottawa. 647 p.

3403. Robinson, J. V., L. R. Shaffer, D. D. Hagemeier, and N. J. Smatresk. 1991. The ecological role of caudal lamellae loss in the larval damselfly *Ischnura posita* (Odonata: Zygoptera). Oecologia 87:1-7.

3404. Robinson, J. V., and G. A. Wellborn. 1987. Mutual predation in assembled communities of odonate species. Ecology 68:921-927.

3405. Robinson, W. H., and B. A. Foote. 1978. Biology and immature stages of *Antichaeta borealis* (Diptera: Sciomyzidae), a predator of snail eggs. Proc. Ent. Soc. Wash. 80:388-396.

3406. Roble, S. M. 1984. First records of *Limnephilus submonilifer*, (Trichoptera: Limnephilidae) and *Neohermes concolor*, (Megaloptera: Corydalidae) for Kansas. Trans. Kansas Acad. Sci. 87:69-70.

3407. Roble, S. M. 1985. Submergent capture of *Dolomedes triton* (Araneae, Pisauridae) by *Anophius depressipes* (Hymenoptera, Pompilidae). J. Arachnology 13:391-392.

3408. Robles, C. 1982. Disturbance and predation in an assemblage of herbivorous Diptera and algae on rocky shores. Oecologia 54:23-31.

3409. Robles, C. D., and J. Cubit. 1981. Influence of biotic factors in an intertidal community: Dipteran larvae grazing on algae. Ecology 62:1536-1547.

3410. Roby, K. B., J. D. Newbold, and D. C. Erman. 1978. Effectiveness of an artificial substrate for sampling macroinvertebrates in small streams. Freshwat. Biol. 8:1-8.

3411. Rockel, E. G., and E. J. Hansens. 1970. Distribution of larval horse flies and deer flies (Tabanidae: Diptera) of a New Jersey salt marsh. Ann. Ent. Soc. Am. 63:681-684.

3412. Rockwood, J. P., D. S. Jones, and R. A. Coler. 1990. The effect of aluminum in soft water at low pH on oxygen consumption by the dragonfly *Libellula julia* Uhler. Hydrobiologia 190:55-59.

3413. Rodina, A. G. 1971. The role of bacteria in the feeding of the tendipedid larvae. Trans. Ser., Fish. Res. Bd. Can. 1848.

3414. Rodionov, Z. 1928. *Helophorus micans*, un ennemi des Graminees. Def. Veget. (Leningrad) 4:951-954.

3415. Roeding, C. E., and L. A. Smock. 1989. Ecology of macroinvertebrate shredders in a low-gradient sandy-bottomed stream. J. N. Am. Benthol. Soc. 8:149-161.

3416. Roell, M. J. and D. J. Orth. 1991. Production of dobsonfly (*Corydalus cornutus*) larvae in the New River of West Virginia. J. Freshwat. Ecol. 6:1-9.

3417. Roemhild, G. 1976. Aquatic Heteroptera (true bugs) of Montana. Montana Agr. Exp. Sta. Res. Rpt. 102. 70 pp.

3418. Roemhild, G. 1982. Trichoptera of Montana with distributional and ecological notes. Northwest Sci. 56:8-13.

3419. Rogers, J. S. 1926. On the biology and immature stages of *Gonomyia pleuralis* Williston. Fla. Ent. 10:33-38.

3420. Rogers, J. S. 1927. On the biology and immature stages of *Geranomyia: I. Geranomyia rostrata* (Say). Fla. Ent. 11:17-26.

3421. Rogers, J. S. 1930. The summer crane-fly fauna of the Cumberland Plateau in Tennessee. Occ. Pap. Univ. Mich. Mus. Zool. 215:1-50.

3422. Rogers, J. S. 1933. The ecological distribution of the craneflies of northern Florida. Ecol. Monogr. 3:2-74.

3423. Rogers, J. S. 1937. Craneflies, pp. 368-376. In J. G. Needham (ed.). Culture methods for invertebrate animals. Comstock, Ithaca. 590 pp.

3424. Rogers, J. S. 1942. The crane-flies (Tipulidae) of the George Reserve, Michigan. Misc. Publ. Univ. Mich. Mus. Zool. 53:1-128.

3425. Rogers, J. S. 1949. The life history of *Megistocera longipennis* (Macquart) (Tipulidae, Diptera), a member of the neuston fauna. Occ. Pap. Mus. Zool. Univ. Mich. 52:1-17.

3426. Rogers, J. S., and G. W. Byers. 1956. The ecological distribution, life history, and immature stages of *Lipsothrix sylvia* (Diptera: Tipulidae). Occ. Pap. Mus. Zool. Univ. Mich. 572:1-14.

3427. Rohdendorf, B. 1974. The historical development of Diptera. Transl. from Russian by J. E. Moore, I. Thiele, B. Hocking, H. Oldroyd, and G. Ball. Univ. Alberta Press, Edmonton. 360 pp.

3428. Root, R. B. 1973. Organization of a plant-arthropod association in simple and diverse habitats: The fauna of collards (*Brassica oleracea*). Ecol. Monogr. 67:95-124.

3429. Rose, J. H. 1963. Supposed larva of *Protanyderus vipio* (Osten Sacken) discovered in California (Diptera: Tanyderidae). Pan Pacif. Ent. 39:272-275.

3430. Rosenberg, D. M. 1978. Practical sampling of freshwater macrozoobenthos: A bibliography of useful texts, reviews, and recent papers. Can. Fish. Mar. Serv. Tech. Rept. 790:1-15.

3431. Rosenberg, D. M. 1992. Freshwater biomonitoring and Chironomidae. Neth. J. Aquat. Ecol. 26:101-122.

3432. Rosenberg, D. M., and V. H. Resh. 1982. The use of artificial substrates in the study of freshwater macroinvertebrates, pp. 175-235. In J. Cairns, Jr. (ed.). Artificial substrates. Ann Arbor Sci. Publ., Mich. 279 pp.

3433. Rosenberg, D. M., and V. H. Resh (eds.). 1993a. Freshwater biomonitoring and benthic macroinvertebrates. Chapman and Hall, New York. 488 pp.

3434. Rosenberg, D. M., and V. H. Resh. 1993b. Introduction to freshwater biomonitoring and benthic macroinvertebrates, pp. 1-9. In D. M. Rosenberg and V. H. Resh (eds.). Freshwater biomonitoring and benthic macroinvertebrates. Chapman and Hall, New York. 488pp.
3435. Rosenberg, D. M., and A. P. Wiens. 1978. Effect of sediment addition on macrobenthic invertebrates in a northern Canadian river. Wat. Res. 12:753-763.
3436. Rosenberg, D. M., and A. P. Wiens. 1983. Efficiency of modifications in the design and use of submerged funnel traps for sampling Chironomidae (Diptera). Hydrobiologia 98:113-118.
3437. Rosenberg, D. M., A. P. Wiens, and B. Bilyj. 1980. Sampling emerging Chironomidae (Diptera) with submerged funnel traps in a new northern Canadian reservoir, Southern Indian Lake, Manitoba. Can. J. Fish. Aquat. Sci. 37:927-936.
3438. Rosenberg, D. M., A. P. Wiens, and O. A. Saether. 1977. Life histories of Cricotopus (Cricotopus) bicinctus and C. (C.) machenziensis (Diptera: Chironomidae) in the Fort Simpson area, Northwest Territories. J. Fish. Res. Bd. Can. 34:247-253.
3439. Rosillon, D. 1986. Life cycle, growth, mortality and production of Ephemerella major Klapalek (Ephemeroptera) in a trout stream in Belgium. Freshwat. Biol. 16:269-277.
3440. Rosillon, D. 1988. Food preference and relative influence of temperature and food quality on life history characteristics of a grazing mayfly, Ephemerella ignita. Can. J. Zool. 66:1474-1481.
3441. Ross, D. H., and D. A. Craig. 1980. Mechanisms of fine particle capture by larval black flies (Diptera: Simuliidae). Can. J. Zool. 58:1186-1192.
3442. Ross, D. H., and R. W. Merritt. 1978. The larval instars and population dynamics of five species of blackflies (Diptera: Simuliidae) and their responses to selected environmental factors. Can. J. Zool. 56:163-164.
3443. Ross, D. H., and R. W. Merritt. 1987. Factors affecting larval black fly distributions and population dynamics, pp. 90-108. In Black flies: Ecology, population management, and annotated world list. Penn. State Univ. Press, University Park, PA.
3444. Ross, D. H., and J. B. Wallace. 1981. Production of Brachycentrus spinae (Trichoptera: Brachycentridae) and its role in seston dynamics of a southern Appalachian stream. Environ. Ent. 10:240-246.
3445. Ross, D. H., and J. B. Wallace. 1983. Longitudinal patterns of production, food consumption, and seston utilization by net-spinning caddisflies (Trichoptera) in a southern Appalachian stream (U.S.A.). Holarct. Ecol. 6:270-284.
3446. Ross, H. H. 1937. Nearctic alder flies of the genus Sialis (Megaloptera, Sialidae). Bull. Ill. Nat. Hist. Surv. 21:57-78.
3447. Ross, H. H. 1944. The caddis flies, or Trichoptera, of Illinois. Bull. Ill. Nat. Hist. Surv. 23:1-326.
3448. Ross, H. H. 1946. A review of the Nearctic Lepidostomatidae (Trichoptera). Ann. Ent. Soc. Am. 39:265-291.
3449. Ross, H. H. 1947. The mosquitoes of Illinois. Bull. Ill. Nat. Hist. Surv. 24:1-96.
3450. Ross, H. H. 1948. New species of sericostomatoid Trichoptera. Proc. Ent. Soc. Wash. 50:151-157.
3451. Ross, H. H. 1949. Xiphocentronidae, a new family of Trichoptera. Ent. News 60:1-7.
3452. Ross, H. H. 1950. Synoptic notes on some Nearctic limnephilid caddisflies (Trichoptera, Limnephilidae). Am. Midl. Nat. 43:410-429.
3453. Ross, H. H. 1956. Evolution and classification of mountain caddisflies. Univ. Ill. Press, Urbana. 213 pp.
3454. Ross, H. H. 1959. Trichoptera, pp. 1024-1049. In W. T. Edmondson (ed.). Freshwater biology (2nd ed.). John Wiley & Sons, N.Y. 1248 pp.
3455. Ross, H. H. 1963. Stream communities and terrestrial biomes. Arch. Hydrobiol. 59:235-242.
3456. Ross, H. H. 1965a. A textbook of Entomology (3rd ed.). John Wiley & Sons, N.Y. 539 pp.
3457. Ross, H. H. 1965b. The evolutionary history of Phylocentropus (Trichoptera: Psychomyiidae). J. Kans. Ent. Soc. 38:398-400.
3458. Ross, H. H. 1967a. Aquatic insects and ecological problems. Bull. Ent. Soc. Am. 13:112-113.
3459. Ross, H. H. 1967b. The evolution and past dispersal of the Trichoptera. Ann. Rev. Ent. 12:169-207.
3460. Ross, H. H. 1974. Biological systematics. Addison-Wesley, Mass. 345 pp.
3461. Ross, H. H., and W. R. Horsfall. 1965. A synopsis of the mosquitoes of Illinois. Biol. Notes Nat. Hist. Surv. Div. St. Ill. 52:1-50.
3462. Ross, H. H., and E. W. King. 1952. Biogeographic and taxonomic studies in Atopsyche (Trichoptera, Rhyacophilidae). Ann. Ent. Soc. Am. 45:177-204.
3463. Ross, H. H., and D. R. Merkley. 1950. The genus Tinodes in North America. J. Kans. Ent. Soc. 23:64-67.
3464. Ross, H. H., and D. R. Merkley. 1952. An annotated key to the Nearctic males of Limnephilus (Trichoptera: Limnephilidae). Am. Midl. Nat. 47:435-455.
3465. Ross, H. H., and W. E. Ricker. 1971. The classification, evolution, and dispersal of the winter stonefly genus Allocapnia. Univ. Ill. Biol. Monogr. 45:1-166.
3466. Ross, H. H., and D. C. Scott. 1974. A review of the caddisfly genus Agarodes, with descriptions of new species (Trichoptera: Sericostomatidae). J. Ga. Ent. Soc. 9:147-155.
3467. Ross, H. H., and G. J. Spencer. 1952. A preliminary list of the Trichoptera of British Columbia. Proc. Ent. Soc. Brit. Columbia 48:43-51.
3468. Ross, H. H., and J. D. Unzicker. 1977. The relationships of the genera of American Hydropsychinae as indicated by phallic structures (Trichoptera, Hydropsychidae). J. Ga. Ent. Soc. 12:298-312.
3469. Ross, H. H., and J. B. Wallace. 1974. The North American genera of the family Sericostomatidae (Trichoptera). J. Ga. Ent. Soc. 9:42-48.
3470. Ross, L. C. M., and H. R. Murkin. 1993. The effect of above-normal flooding of a northern prairie marsh on Agraylea multipunctata Curtis (Trichoptera: Hydroptilidae). J. Freshwat. Ecol. 8:27-35.
3471. Roth, J. C., and S. Parma. 1970. A Chaohorus bibliography. Bull. Ent. Soc. Am. 16:100-110.
3472. Rotheray, G. E. 1991. Larval stages of 17 rare and poorly known British hoverflies (Diptera: Syrphidae). J. Nat. Hist. 25:945-969.
3473. Roughley, R. E. 1990. A systematic revision of species of Dytiscus Linnaeus (Coleoptera: Dytiscidae) Part 1. Classification based on adult stage. Quaest. Ent. 26:383-557.
3474. Roughley, R. E., and D. H. Pengelly. 1981. Classification phylogeny and zoogeography of Hydaticus (Coleoptera: Dytiscidae) of North America. Quaest. Ent. 17:249-312.
3475. Roy, D., and P. P. Harper. 1975. Nouvelles mentions de trichopteres du Québec et description de Limnephilus nimmoi sp. nov. (Limnephilidae). Can. J. Zool. 53:1080-1088.
3476. Roy, D., and P. P. Harper. 1979. Liste préliminaire des Trichopteres (insectes) du Québec. Ann. Ent. Soc. Quebec 24: 148-172.
3477. Roy, D., and P. P. Harper. 1980. Females of the Nearctic Molanna (Trichoptera: Molannidae). Proc. Ent. Soc. Wash 82:229-236.
3478. Rozeboom, L. E. 1942. The mosquitoes of Oklahoma. Okla. Agric. Exp. Sta. Tech. Bull. 16:1-56.
3479. Rozkosny, R. 1965. Neue Metamorphosestadien mancher Tetanocera—Arten (Diptera, Sciomyzidae). Zool. Listy 14:367-371.
3480. Rozkosny, R. 1982. A biosystematic study of the European Stratiomyidae (Diptera). Vol. 1. Introduction, Beridinae, Sarginae and Stratiomyinae. Dr. W. Junk Publishers, The Hague. 401pp.
3481. Rozkosny, R. 1983. A biosystematic study of the European Stratiomyidae (Diptera). Vol. 2. Clitellarinae, Hermetiinae, Pachgasterinae and bibliography. Dr. W. Junk Publ., The Hague. 431pp.
3482. Rozkosny, R. 1987. A review of the Palaearctic Sciomyzidae / Diptera /. Univerzita J. E. Purkyne v Brne. 97pp.
3483. Rozkosny, R., and L. V. Knutson. 1970. Taxonomy, biology, and immature stages of Palearctic Pteromicra, snail-killing Diptera (Sciomyzidae). Ann. Ent. Soc. Am. 63:1434-1459.

3484. Ruggles, K. K., and D. C. Tarter. 1991. Ecological life history of *Peltoperla tarteri* (Plecoptera: Peltoperlidae) from Big Hollow of Paint Creek, Fayette County, West Virginia. Psyche 98:33-46.

3485. Ruhoff, E. A. 1968. Bibliography and index to scientific contributions of Carl J. Drake for the years 1914-1967. Bull. U.S. Nat. Mus. 267:1-81.

3486. Ruiter, D. E., and R. J. Lavigne. 1985. Distribution of Wyoming Trichoptera. Univ. of Wyoming, Agric. Exp. Stn., Publ. 5 M 47, 102pp.

3487. Runck, C., and D. W. Blinn. 1990. Population dynamics and secondary production by *Ranatra montezuma* (Heteroptera: Nepidae). J. N. Am. Benthol. Soc. 9:262-270.

3488. Runck, C., and D. W. Blinn. 1992. The foraging ecology of *Ranatra montezuma* (Heteroptera): an optimal forager in Montzuma Well, Arizona. J. Arizona-Nevada Acad. Sci. 26:111-122.

3489. Runck, C., and D. W. Blinn. 1993. Secondary production by *Telebasis salva* (Odonata) in a thermally constant aquatic ecosystem. J. N. Am. Benthol. Soc. 12:136-147.

3490. Runyan, J. T., and D. L. Deonier. 1979. A comparative study of *Pseudohecamede* and *Allotrichoma* (Diptera: Ephydridae). pp. 123-137. *In* D. L. Deonier (ed.). First symposium on the systematics and Ecology of Ephydridae (Diptera). N. Am. Benthol. Soc. 147 pp.

3491. Rupprecht, R. 1967. Das Trommeln der Plecopteren. Z. Vergl. Physiol. 59:38-71.

3492. Rupprecht, R. 1972. Dialektbildung bei den Trommelsignalen von *Diura* (Plecoptera). Oikos 23:410-412.

3493. Rupprecht, R. 1977. Nachweis von Trommelsignalen bei einem europaischen Vertreter der Steinfliegen-familie Leuctridae (Plecoptera). Ent. Germ. 3:333-336.

3494. Ruschka, F., and A. Thienemann. 1913. Zur Kenntnis der Wasser-Hymenopteren. A Wiss. Insektenbiol. 9:48-52, 82-87.

3495. Ruttner, F. 1953. Fundamentals of limnology (Translated by D. G. Frey and F.E.J. Fry). Univ. of Toronto Press, Toronto, Canada. 242 pp.

3496. Ryker, L. C. 1972. Acoustic behavior of four sympatric species of water scavenger beetles (Coleoptera, Hydrophilidae, *Tropisternus*). Occ. Pap. Mus. Zool. Univ. Mich. 666:1-19.

3497. Ryker, L. C. 1975a. Calling chirps in *Tropisternus natator* (D'Orchymont) and *T. lateralis nimbatus* (Say) (Coleoptera: Hydrophilidae). Ent. News 86:179-186.

3498. Ryker, L. C. 1975b. Observations on the life cycle and flight dispersal of a water beetle, *Tropisternus ellipticus* LeConte, in western Oregon (Coleoptera: Hydrophilidae). Pan-Pacif. Ent. 51:184-194.

3499. Ryker, L. C. 1976. Acoustic behavior of *Tropisternus ellipticus*, *T. columbianus*, and *T. lateralis limbalis* in western Oregon (Coleoptera: Hydrophilidae). Coleopt. Bull. 30:147-156.

3500. Saber Hussain, M., and K. Jamil. 1989. Bioaccumulation of heavy metal ions and their effect on certain biochemical parameters of water hyacinth weevil *Neochtina eichhorniae* (Warner). J. Environ. Sci. Health B24:251-264.

3501. Saber Hussain, M., and K. Jamil. 1990. Bioaccumulation of mercury and its effect on protein metabolism of the water hyacinth weevil *Neochetina eichhornae* [sic] (Warner). Bull. Environ. Contam. Toxicol. 45:294-298.

3502. Saether, O. A. 1969. Some Nearctic Podonominae, Diamesinae, and Orthocladiinae (Diptera: Chironomidae). Bull. Fish. Res. Bd. Can. 170:1-154.

3503. Saether, O. A. 1970. Nearctic and Palaearctic *Chaoborus* (Diptera: Chaoboridae). Bull. Fish. Res. Bd. Can. 174:1-57.

3504. Saether, O. A. 1971a. Notes on general morphology and terminology of the Chironomidae (Diptera). Can. Ent. 103:1237-1260.

3505. Saether, O. A. 1971b. Four new and unusual Chironomidae (Diptera). Can. Ent. 103:1799-1827.

3506. Saether, O. A. 1972. Chaoboridae, pp. 257-280. *In* H. J. Elster and W. Ohle (eds.). Das Zooplankton der Binnengewasser. Die Binnengewasser 26 Stuttgart, E. Schweizebart'sche Verlagsbuchhandlung. 294 pp.

3507. Saether, O. A. 1975a. Nearctic and Palaearctic *Heterotrissocladius* (Diptera: Chironomidae). Bull. Fish. Res. Bd. Can. 193:1-67.

3508. Saether, O. A. 1975b. Two new species of *Protanypus* Kieffer, with keys to Nearctic and Palaearctic species of the genus (Diptera: Chironomidae). J. Fish. Res. Bd. Can. 32:367-388.

3509. Saether, O. A. 1976. Revision of *Hydrobaenus*,*Trissocladius*, *Zalutschia*, *Paratrissocladius*, and some related genera. Bull. Fish. Res. Bd. Can. 195:1-287.

3510. Saether, O. A. 1977a. *Habrobaenus hudsoni*, n. gen., n. sp. and the immatures of *Baeoctenus bicolor* Saether (Diptera: Chironomidae). J: Fish. Res. Bd. Can. 34:2354-2361.

3511. Saether, O. A. 1977b. Taxonomic studies on Chironomidae: *Nanocladius*, *Pseudochironomus* and the *Harnischia* complex. Bull. Fish. Res. Bd. Can. 196:1-143.

3512. Saether, O. A. (ed.). 1979a. Recent developments in chironomid studies (Diptera: Chironomidae). Ent. Scand. 10:1-150.

3513. Saether, O. A. 1979b. Chironomid communities as water quality indicators. Holarct. Ecol. 2:65-74.

3514. Saether, O. A. 1980a. Glossary of chironomid morphology terminology (Diptera: Chironomidae). Ent. Scand., Suppl. 14:1-51.

3515. Saether, O. A. 1980b. The influence of eutrophication on deep lake benthic invertebrate communities. Prog. Wat. Tech. 12:161-180.

3516. Saether, O. A. 1981a. *Doncricotopus bicaudatus* n. gen., n. sp. (Diptera: Chironomidae, Orthocladiinae) from the Northwest Territories, Canada. Ent. Scand. 12:223-229.

3517. Saether, O. A. 1981b. Orthocladiinae (Chironomidae: Diptera) from the British West Indies with descriptions of *Antillocladius* n. gen., *Lipurometriocnemus* n. gen., and *Diplosmittia* n. gen. Ent. Scand. Suppl. 16:1-46.

3518. Saether, O. A. 1982. Orthocladiinae (Diptera: Chironomidae) from SE U. S. A., with descriptions of *Plhudsonia*, *Unniella* and *Platysmittia* n. genera and *Atelopodella* n. subgen. Ent. Scand. 13:465-510.

3519. Saether, O. A. 1983a. *Oschia dorsenna* n. gen. n. sp. and *Saetheria hirta* n. sp., two members of the *Harnischia* complex (Diptera: Chironomidae). Ent. Scand. 14:395-404.

3520. Saether, O. A. 1983b. Three new species of *Lopescladius* Oliveira, 1967 (syn. "*Cordites*" Brudin, 1966, n. syn.), with a phylogeny of the *Parakiefferiella* group. Mem. Am. Ent. Soc. 34:279-298.

3521. Saether, O. A. 1985a. A review of *Odontomesa* Pagast, 1947 (Diptera, Chironomidae, Prodiamesinae). Spixiana Suppl. 11:15-29.

3522. Saether, O. A. 1985b. A review of the genus *Rheocricotopus* Thienemann and Harnisch, 1932, with the description of three new species (Diptera: Chironomidae). Spixiana Suppl. (Muench) 11:59-108.

3523. Saether, O. A., and L. C. Ferrington, Jr. 1993. Redescription of *Prosmittia jemdandica* (Brundin, 1947), with a review of the genus. J. Kans. Ent. Soc. 66:257-262.

3524. Saether, O. A., and J. E. Sublette. 1983. A review of the genera *Doithrix* n. gen., *Georthocladius* Strenzke, *Parachaetocladius* Wulker, and *Pseudorthocladius* Goetghebuer (Diptera: Chironomidae, Orthocladiinae). Ent. Scand. Suppl. 20:1-100.

3525. Saha, T. C., and S. K. Raut. 1992. Bioecology of the water-bug *Sphaerodema annulatum* Fabricius (Heteroptera: Belostomatidae). Arch. Hydrobiol. 124:239-253.

3526. Sailer, R. I. 1948. The genus *Trichocorixa*. Univ. Kans. Sci. Bull. 32:289-407.

3527. Sailer, R. I. 1972. Biological control of aquatic weeds, recent progress. Proc. N.E. Weed Sci. Soc. 26:180-182.

3528. Sailer, R. I., and S. E. Lienk. 1954. Insect predators of mosquito larvae and pupae in Alaska. Mosquito News 14:14-16.

3529. Salmon, J. T. 1964. An index to the Collembola. Roy. Soc. New Zealand 7:1-644.

3530. Salt, G. 1937. The egg-parasite of *Sialis lutaria*:. a study of the influence of the host upon a dimorphic parasite. Parasitology 29:539-558.

3531. Sanderson, M. W. 1938. A monographic revision of the North American species of *Stenelmis* (Dryopidae: Coleoptera). Univ. Kans. Sci. Bull. 25:635-717.

3532. Sanderson, M. W. 1953. A revision of the Nearctic genera of Elmidae (Coleoptera). I. J. Kans. Ent. Soc. 26:148-163.

3533. Sanderson, M. W. 1954. A revision of the Nearctic Elmidae (Coleoptera). II. J. Kans. Ent. Soc. 27:1-13.

3534. Sanderson, M. W. 1982a. Aquatic and semiaquatic Heteroptera. pp. 6.1-6.94. In A. R. Brigham, W. V. Brigham, and A. Gnilka (eds.). Aquatic insects and oligochaetes of North and South Carolina. Midwest Aquatic Enterprises, Mahomet, Ill. 837 pp.

3535. Sanderson, M. W. 1982b. Gyrinidae, pp. 10.29-10.38. In A. R. Brigham, W. U. Brigham, and A. Gnilka (eds.). Aquatic insects and oligochaetes of North and South Carolina. Midwest Aquatic Enterprises, Mahomet, Ill. 837 pp.

3536. Sankarperumal, G., and T. J. Pandian. 1991. Effect of temperature and *Chlorella* density on growth and metamorphosis of *Chironomus circumdatus* (Kieffer) (Diptera). Aquat. Insects 13:167-177.

3537. Santamarina Mijares, A., and R. Gonzáles Broche. 1985. Biologia y capacidad depredadora del *Belostoma boscii* en el laboratorio. Rev. Cubana Higiene Epidem. 23:140-146.

3538. Saouter, E., R. LeMenn, A. Boudou, and F. Ribeyre. 1991. Structural and ultrastructural analysis of gills and gut of *Hexagenia rigida* nymphs (Ephemeroptera) in relation to contamination mechanisms. Tissue & Cell 23:929-938.

3539. Satchell, G. H. 1947. The ecology of the British species of *Psychoda* (Diptera: Psychodidae). Ann. Appl. Biol. 34:611-621.

3540. Satchell, G. H. 1949. The early stages of the British species of *Pericoma* Walker (Diptera: Psychodidae). Trans. R. Ent. Soc. Lond. 100:411-447.

3541. Satija, G. R. 1959. Food in relation to mouth parts of *Wormaldia occipitalis* Pictet. Res. Bull. Punjab Univ. N.S. 10:169-178.

3542. Satija, R. C., and G. R. Satija. 1959. Food, mouth parts and alimentary canal of *Limnophilus stigma* Curtis. Res. Bull. Punjab Univ. N.S. 11:11-24.

3543. Sattler, W. 1963. Uber den Korperbau, die Okologie und Ethologie der Larvae und Puppe von *Macronema* Pict. (Hydropsychidae). Arch. Hydrobiol. 59:26-60.

3544. Saunders, L. G. 1924. On the life history and the anatomy of the early stage of *Forcipomyia* (Diptera, Nemat., Ceratopogonidae). Parasitology 16:164-213.

3545. Saunders, L. G. 1928. Some marine insects of the Pacific coast of Canada. Ann. Ent. Soc. Am. 21:521-545.

3546. Savage, A. A. 1986. The distribution, life cycle and production of *Leptophlebia vespertina* (Ephemeroptera) in a lowland lake. Hydrobiologia 133:3-19.

3547. Savan, B. I., and D. L. Gibo. 1974. A mass culture technique ensuring synchronous emergence for *Leucorrhinia intacta* (Hagen) (Anisoptera: Libellulidae). Odonatologica 3:269-272.

3548. Sawchyn, W. E., and N. S. Church. 1974. The life histories of three species of *Lestes* (Odonata: Zygoptera) in Saskatchewan. Can. Ent. 106:1283-1293.

3549. Sawchyn, W. W., and C. Gillott. 1974. The life history of *Lestes congener* (Odonata: Zygoptera) on the Canadian prairies. Can. Ent. 106:367-376.

3550. Schaefer, K. F. 1966. The aquatic and semi-aquatic Hemiptera of Oklahoma. Ph.D. diss., Oklahoma State University, Stillwater. 102 pp.

3551. Schaefer, K. F., and W. A. Drew. 1964. Checklist of aquatic and semiaquatic Hemiptera (Insecta) of Oklahoma. Southwest. Nat. 9:99-101.

3552. Schaefer, K. F., and W. A. Drew. 1968. The aquatic and semi-aquatic Hemiptera of Oklahoma. Proc. Okla. Acad. Sci. 47:125-134.

3553. Schaeffer, C. 1925. Revision of the New World species of the tribe Donaciini of the coleopterous family Chrysomelidae. Bull. Brooklyn Mus. Sci. 3:45-165.

3554. Schaeffer, C. 1928. The North American species of *Hydrothassa* with notes on other Chrysomelidae and a description of new species and a variety (Col.). J. N.Y. Ent. Soc. 36:287-291.

3555. Schafer, D. A. 1950. Life history studies of *Psephenus lecontei* LeC. and *Ectopria nervosa* Melsh. (Coleoptera: Psephenidae: Dascillidae). M. S. thesis, Ohio State University, Columbus. 51 pp.

3556. Schaller, F. 1970. 1. Uberordnung und 1. Ordnung Collembola (Springschwanze). Handb. Zool. 4:1-72.

3557. Schauff, M. E. 1984. The holarctic genera of Mymaridae (Hymenoptera: Chalcidoidea) Mem. Ent. Soc. Wash. No. 12. 65pp.

3558. Schefter, P. W., and J. D. Unzicker. 1984. A review of the *Hydropsyche morosa-bifida*—complex in North America (Trichoptera: Hydropsychidae). In J. C. Morse (ed.). Proc. 4th Int. Symp. Trichoptera, Dr. W. Junk, Publ., The Hague (In press).

3559. Schefter, P. W., and G. B. Wiggins. 1986. A systematic study of the Nearctic larvae of the *Hydropsyche morosa* group (Trichoptera: Hydropsychidae). Roy. Ont. Mus. Misc. Publ. 100 pp.

3560. Schefter, P. W., G. B. Wiggins, and J. D. Unzicker. 1986. A proposal for assignment of *Ceratopsyche* as a subgenus of *Hydropsyche*, with new synonyms and a new species (Trichoptera: Hydropsychidae). J. N. Am. Benthol. Soc. 5:67-84.

3561. Scheiring, J. F. 1974. Diversity of shore flies (Diptera: Ephydridae) in inland freshwater habitats. J. Kans. Ent. Soc. 47:485-491.

3562. Scheiring, J. F. 1976. Ecological notes on the brine flies of northwestern Oklahoma (Diptera: Ephydridae). J. Kans. Ent. Soc. 49:450-452.

3563. Scheiring, J. F. 1985. Longitudinal and seasonal patterns of insect trophic structure in a Florida sand-hill stream. J. Kans. Ent. Soc. 58:207-219.

3564. Scheiring, J. F., and B. A. Foote. 1973. Habitat distribution of the shore flies of northeastern Ohio (Diptera: Ephydridae). Ohio J. Sci. 73:152-166.

3565. Schell, D. V. 1943. The Ochteridae of the Western Hemisphere. J. Kans. Ent. Soc. 16:29-47.

3566. Schildknecht, H., R. Sieverdt, and U. Maschivitz. 1966. A vertebrate hormone (cortisone) as defensive substance of the water beetle *Dytiscus marginalis*. Agnew. Chem. Int. Engl. 5:421-422.

3567. Schindler, D. W. 1974. Eutrophication and recovery in experimental lakes: implications for lake management. Science 184:897-899.

3568. Schindler, D. W., and E. J. Fee. 1974. Experimental Lakes Area: whole-lake experiments in eutrophication. J. Fish. Res. Bd. Can. 31:937-953.

3569. Schindler, D. W., E. J. Fee, and T. Ruszczynski. 1978. Phosphorus input and its consequences for phytoplankton standing crop and production in the Experimental Lakes Area and in similar lakes. J. Fish. Res. Bd. Can. 35:190-196.

3570. Schindler, D. W., K. H. Mills, D. F. Malley, D. L. Findlay, J. A. Shearer, I. J. Davies, M. A. Turner, G. A. Linsey, and D. R. Cruikshank. 1985. Long-term ecosystem stress: the effects of years of experimental acidification on a small lake. Science 228:1395-1401.

3571. Schlinger, E. I. 1975. Diptera, pp. 436-446. In R. I. Smith and J. T. Carlton (eds.). Light's manual: Intertidal invertebrates of the central California coast (3rd ed.). Univ. Calif. Press. Berkeley. 716 pp.

3572. Schliwa, W., and F. Schaller. 1963. Die Paarbildung des Springschwanzes *Podura aquatica*. Naturwissenschaften 50:698.

3573. Schmid, F. 1955. Contribution a l'etude des Limnophilidae (Trichoptera). Mitt. Schweiz. Ges. Ent. 28:1-245.

3574. Schmid, F. 1968. La famille des Arctopsychides (Trichoptera). Mem. Ent. Soc. Quebec 1:1-84.

3575. Schmid, F. 1970. Le genre *Rhyacophila* et la famille des Rhyacophilidae (Trichoptera). Mem. Ent. Soc. Can. 66:1-230.

3576. Schmid, F. 1979. On some new trends in trichopterology. Bull. Ent. Soc. Can. 11:48-57.

3577. Schmid, F. 1980. Genera des Trichopteres du Canada et des Etats adjacents. Les insectes et arachnides du Canada, Part. 7. Agric. Can. Publ. 1692. 296 pp.

3578. Schmid, F. 1981. Révision des Trichopteres Canadiens. I. La famille des Rhyacophilidae (Annulipalpia). Mem. Soc. Ent. Can. 116:1-83.

3579. Schmid, F. 1982. La famille des Xiphocentronides (Trichoptera: Annulipalpia). Mem. Ent. Soc. Can. 121.

3580. Schmid, F., and R. Guppy. 1952. An annotated list of Trichoptera collected on southern Vancouver Island. Proc. Ent. Soc. Brit. Columbia 48:41-42.

3581. Schmid, P. E. 1992. Population dynamics and resource utilization by larval Chironomidae (Diptera) in a backwater area of the River Danube. Freshwat. Biol. 28:111-127.

3582. Schmidt, D. A., and D. C. Tarter. 1985. Life history and ecology of *Acroneuria carolinensis* in Panther Creek, Nicholas County, West Virginia (Plecoptera: Perlidae). Psyche 92:393-406.

3583. Schneider, R. F. 1967. An aquatic rearing apparatus for insects. Turtox News 44:90.

3584. Schoonbee, H. J., and J. H. Swanepoel. 1979. Observations on the use of the radio-isotope 32p in the study of food uptake, pp. 343-352. In J. F. Flannagan and K. E. Marshall (eds.). Advances in Ephemeroptera biology. Plenum, N.Y. 552 pp.

3585. Schott, R. J., and M. A. Brusven. 1980. The ecology and electrophoretic analysis of the damselfly *Argia vivida* Hagen, living in a geothermal gradient. Hydrobiologia 69:261-265.

3586. Schremmer, F. 1950. Zue Morphologie und funktionellen Anatomie des Larvenkopfes von *Chaoborus (Corethra* auct.) *obscuripes* v.d. Wulp (Diptera, Chaoboridae). Ost. Zool. Z. 2:471-516.

3587. Schremmer, F. 1951. Die Mundteil der Brachycerenlarven und der Kopfbau der Larve von *Stratiomys chamaeleon* L. Ost. Zool. Z. 3:326-397.

3588. Schröder, P. 1984. Day-night observations of filter feeding in blackfly larvae (Diptera: Simuliidae). Arch. Hydrobiol./Supppl. 2:215-222.

3589. Schröder, P. 1986. Resource partitioning of food particles between associated larvae of *Prosimulium rufipes* and *Eusimulium cryophilum* (Diptera: Simuliidae) in Austrian mountain brooks. Arch. Hydrobiol. 107:497-509.

3590. Schuh, R. T. 1967. The shore bugs (Hemiptera: Saldidae) of the Great Lakes region. Contr. Am. Ent. Inst. 2:1-35.

3591. Schuh, R. T., B. Galil, and J. T. Polhemus. 1987. Catalog of Leptopodomorpha (Heteroptera). Bull. Am. Mus. Nat. Hist. 185:243-406.

3592. Schuster, G. A. 1984. *Hydropsyche?-Symphitopsyche?-Certopsyche?*. A taxonomic enigma, pp. 339-345. In J. C. Morse (ed.). Proc. 4th Int. Symp. Trichoptera, Dr. W. Junk, Publ., The Hague, Netherlands.

3593. Schuster, G. A., and D. A. Etnier. 1978. A manual for the identification of the larvae of the caddisfly genera *Hydropsyche* Pictet and *Symphitopsyche* Ulmer in eastern and central North America (Trichoptera: Hydropsychidae). U.S. Environ. Prot. Agency 600/4-78 060, Cincinnati. 129 pp.

3594. Schuster, G. A., and S. W. Hamilton. 1984. The genus *Phylocentropus* in North America (Trichoptera: Polycentropodiidae), pp. 347-362. In J.C. Morse (ed.). Proc. IV. Int. Symp. Trichoptera, Clemson Univ., SC. Dr. W. Junk Publ., The Hague, Netherlands.

3595. Schuster, R. 1965. Die Okologie der terrestrischen Kleinfauna des Meeresstrandes. Zool. Anz. Suppl. 28:492-521.

3596. Schwardt, H. H. 1937. Methods for collecting and rearing horseflies, pp. 405-409. In J. G. Needham (ed.). Culture methods for invertebrate animals. Comstock, Ithaca. 590 pp.

3597. Schwartz, E.A. 1914. Aquatic beetles, especially *Hydrosapha*, in hot springs in Arizona. Proc. Ent. Soc. Wash. 16:163-168.

3598. Schwarz, P. 1970. Autokologische Untersuchungen zum Lebenszyklus von Setipalpia-Arten (Plecoptera). Arch. Hydrobiol. 67:103-140.

3599. Schwiebert, E. 1973. Nymphs. Winchester, N.Y. 339 pp.

3600. Scoble, N. J. 1992. The Lepidoptera. Form, function and diversity. Oxford University Press, Natural History Museum Publications. 404 p.

3601. Scotland, M. B. 1940. Review and summary of studies of insects associated with *Lemna minor*. J. N.Y. Ent. Soc. 48:319-333.

3602. Scott H. M. 1924. Observations on the habits and life history of *Galerucella nymphaea* (Coleoptera). Trans. Am. Microsc. Soc. 43:11-16.

3603. Scott, D. 1958. Ecological studies on the Trichoptera of the River Dean, Cheshire. Arch. Hydrobiol. 54:340-392.

3604. Scott, D. B. Jr. 1956. Aquatic Collembola, pp. 74-78. In R. L. Usinger (ed.). Aquatic insects of California. Univ. Calif. Press, Berkeley. 508 pp.

3605. Scott, D. B., and R. Yosii. 1972. Notes on some Collembola of the Pacific coast of North America. Contr. Biol. Lab. Kyoto Univ., Kyoto 23:101-114.

3606. Scott, D. C., L. Berner, and A. Hirsch. 1959. The nymph of the mayfly genus *Tortopus* (Ephemeroptera: Polymitarcidae). Ann. Ent. Soc. Am. 52:205-213.

3607. Scrimgeour, G. J., and J. M. Culp. 1994. Foraging and evading predators: The effect of predator species on a behavioural trade-off by a lotic mayfly. Oikos 69:71-79.

3608. Scrimgeour, G. J., J. M. Culp, M. L. Bothwell, F. J. Wrona, and M. H. McKee. 1991. Mechanisms of algal patch depletion - importance of consumptive and non-consumptive losses in mayfly-diatom systems. Oecologia 85:343-348.

3609. Scrimgeour, G. J., J. M. Culp, and N. E. Glozier. 1993. An improved technique for sampling lotic invertebrates. Hydrobiologia 254:65-71.

3610. Scrimshaw, S., and W. C. Kerfoot. 1987. Chemical defenses of freshwater organisms: beetles and bugs, pp. 240-262. In W. C. Kerfott and A. Sih (eds.). Predation. Direct and indirect impacts on aquatic communities. Univ. Press New England, Hanover, NH.

3611. Scudder, G.G.E. 1965. The Notonectidae (Hemiptera) of British Columbia. Proc. Ent. Soc. Brit. Columbia 62:38-41.

3612. Scudder, G.G.E. 1971a. Comparative morphology of insect genitalia. Ann. Rev. Ent. 16:379-406.

3613. Scudder, G.G.E. 1971b. The Gerridae (Hemiptera) of British Columbia. J. Ent. Soc. Brit. Columbia. 68:3-10.

3614. Scudder, G.G.E. 1976. Water-Boatmen of saline waters (Hemiptera: Corixidae), pp. 263-289. In L. Cheng (ed.). Marine Insects. North Holland, Amsterdam. 581 pp.

3615. Scudder, G.G.E. 1977. An annotated checklist of the aquatic and semiaquatic Hemiptera (Insecta) of British Columbia. Syesis 10:31-38.

3616. Scudder, G.G.E., R. A. Cannings, and K. M. Stuart. 1976. An annotated checklist of the Odonata (Insecta) of British Columbia. Syesis 9:143-162.

3617. Seagle, H. H., Jr. 1980a. Flight periodicity and emergence patterns in the Elmidae (Coleoptera: Dryopoidea). Ann. Ent. Soc. Am. 73:300-306.

3618. Seagle, H. H., Jr. 1980b. Resource partitioning in three species of elmid beetles, *Stenelmis crenata* (Say), *Stenelmis mera* Sanderson and *Optioservus trivittatus* (Brown) (Coleoptera: Elmidae). Diss. Abstr. Internat. (B)40:4698.

3619. Seagle, H. H., Jr. 1982. Comparison of the food habits of three species of riffle beetles *Stenelmis crenata, Stenelmis mena*, and *Optioservus trivittatus* (Coleoptera: Dryopoidea: Elmidae). Freshwat. Invert. Biol. 1:33-38.

3620. Seawright, J. A., P. E. Kaiser, S. K. Narang, K. J. Tennessen, and S. E. Mitchell. 1992. Distribution of sibling species A, B, C, and D of the *Anopheles quadrimaculatus* complex. J. Agric. Ent. 9:289-300.

3621. Sebastien, R. J., D. M. Rosenberg, and A. P. Wiens. 1988. A method for subsampling unsorted benthic macroinvertebrates by weight. Hydrobiologia 157:69-75.

3622. Sedell, J. R. 1971. Trophic ecology and natural history of *Neophylax concinnus* and *N. oligius*. Ph.D. diss., University of Pittsburgh, Pittsburgh. 154 pp.

3623. Sedell, J. R., F. J. Triska, and N. M. Triska. 1975. The processing of conifer and hardwood leaves in two coniferous forest streams. I. Weight loss and associated invertebrates. Verh. Int. Verein. Limnol. 19:1617-1627.

3624. Seifert, R. P. 1980. Mosquito fauna of *Heliconia aurea*. J. Anim. Ecol. 49:687-697.

3625. Selby, D. A., J. M. Ihnat, and J. J. Messer. 1985. Effects of sub-acute cadmium exposure on a hardwater mountain stream microcosm. Wat. Res. 19:645-655.

3626. Sephton, D. H., and H. B. N. Hynes. 1984. The ecology of *Taeniopteryx nivalis* (Fitch) (Taeniopterygidae: Plecoptera) in a small stream in southern Ontario. Can. J. Zool. 62:637-642.

3627. Sephton, T. W. 1987. Some observations on the food of larvae of *Procladius bellus* (Diptera: Chironomidae). Aquat. Insects 9:195-202.

3628. Service, M. W. 1993. Mosquito ecology: Field sampling methods (2nd ed.). Chapman and Hall, London, England. 954 pp.

3629. Shapas, T. J., and W. L. Hilsenhoff. 1976. Feeding habits of Wisconsin's predominant lotic Plecoptera, Ephemeroptera and Trichoptera. Great Lakes Ent. 9:175-188.

3630. Sheldon, A. L. 1969. Size relationship of *Acroneuria californica* (Perlidae: Plecoptera) and its prey. Hydrobiologia 34:85-94.

3631. Sheldon, A. L. 1972. Comparative ecology of *Arcynopteryx* and *Diura* in a California stream. Arch. Hydrobiol. 69:521-546.

3632. Sheldon, A. L. 1980. Resource division by perlid stoneflies (Plecoptera) in a lake outlet ecosystem. Hydrobiologia 71:155-161.

3633. Sheldon, A. L. 1984. Cost and precision in stream sampling program. Hydrobiologia 111:147-152.

3634. Sheldon, A. L., and S. G. Jewett, Jr. 1967. Stonefly emergence in a Sierra Nevada stream. Pan-Pacif. Ent. 43:1-8.

3635. Sheldon, A. L., and M. W. Oswood. 1977. Blackfly (Diptera: Simuliidae) abundance in a lake outlet: test of a predictive model. Hydrobiologia 56:113-120.

3636. Shepard, W. D. 1979. Co-occurrence of a marine and freshwater species of Limnichidae (Coleoptera) in Aransas County, Texas. Ent. News 90:88.

3637. Shepard, W. D. 1990. *Microcylloepus formicoideus* new species (Coleoptera: Elmidae) a new riffle beetle from Death Valley National Monument California USA. Ent. News 101:147-153.

3638. Shepard, W. D., and C. B. Barr. 1991. Description of the larva of *Atractelmis* (Coleoptera: Elmidae) and new information on the morphology distribution and habitat of *Atractelmis wawona* Chandler. Pan-Pac. Ent. 67:195-199.

3639. Shepard, W. D., and K. W. Stewart. 1983. A comparative study of nymphal gills in North American stonefly (Plecoptera) genera and a new, proposed paradigm of Plecoptera gill evolution. Misc. Pubs. Ent. Soc. Am. 55:1-58.

3640. Sherberger, F. F., and J. B. Wallace. 1971. Larvae of the southeastern species of *Molanna*. J. Kans. Ent. Soc. 44:217-224.

3641. Sherman, J. D. 1913. Some habits of the Dytiscidae. J. N.Y. Ent. Soc. 21:43-54.

3642. Shewell, G. E. 1958. Classification and distribution of artic and subarctic Simuliidae. Proc. 10th Int. Contr. Ent. 1:635-643.

3643. Shewell, G. E. 1987. Chap. 108. Sarcophagidae, pp. 1159-1186. In J. F. McAlpine, B. V. Peterson, G. E. Shewell, H. J. Teskey, J. R. Vockeroth, and D. M. Wood (coords.). Manual Nearctic Diptera, Vol. 2. Res. Branch, Agr. Can. Mon. 28. 1332p.

3644. Short, J.R.T. 1952. The morphology of the head of larval Hymenoptera with special reference to the head of Ichneumonoidea, including a classification of the final instar larvae of Braconidae. Trans. Ent. Soc. Lond. 103:27-84.

3645. Short, J.R.T. 1959. A description and classification of the final instar larvae of the Ichneumonidae (Insecta, Hymenoptera). Proc. U.S. Nat. Mus. 110:391-511.

3646. Short, J.R.T. 1970. On the classification of the final instar larvae of the Ichneumonidae (Hymenoptera). Trans. Ent. Soc. Lond. Suppl. 122:185-210.

3647. Short. R. A., S. P. Canton, and J. V. Ward. 1980. Detrital processing and associated macroinvertebrates in a Colorado mountain stream. Ecology 61:727-732.

3648. Short, R. A., E. H. Stanely, J. W. Harrison, and C. R. Epperson. 1987. Production of *Corydalus cornutus* (Megaloptera) in four streams differing in size, flow and temperature. J. N. Am. Benthol. Soc. 6:105-114.

3649. Short, R. A., and J. V. Ward. 1980. Life cycle and production of *Skwala parallela* (Frison) (Plecoptera: Perlodidae) in a Colorado montane stream. Hydrobiologia 69:273-275.

3650. Short, R. A., and J. V. Ward. 1981. Trophic ecology of three winter stoneflies (Plecoptera). Am. Midl. Nat. 105:341-347.

3651. Siegfried, C. A., and A. W. Knight. 1976. Trophic relations of *Acroneuria* (*Calineuria*) *california* (Plecoptera: Perlidae) in a Sierra foothill stream. Environ. Ent. 5:575-581.

3652. Siegfried, C. A., and A. W. Knight. 1978. Aspects of the life history and growth of *Acroneuria* (*Calineuria*) *californica* in a Sierra foothill stream. Ann. Ent. Soc. Am. 71:149-154.

3653. Sih, A. 1981. Stability, prey density and age/dependent interference in an aquatic insect predator, *Notonecta hoffmanni*. J. Anim. Ecol. 50:625-636.

3654. Sih, A. 1987. Predators and prey lifestyles: An evolutionary and ecological overview, pp. 203-224. In W. C. Kerfoot and A. Sih. Predation: Direct and indirect impacts on aquatic communities, Univ. Press of New England, Hanover and London. 386 pp.

3655. Sih, A., and D. E. Wooster. 1994. Prey behavior, prey dispersal, and predator impacts on stream prey. Ecology 75:1199-1207.

3656. Siltala, A. J. 1907. Uber die Nahrung der Trichopteren. Acta. Soc. Flora Fauna Fenn. 29:1-34.

3657. Silvey, J.K.G. 1931. Observations on the life history of *Rheumatobates rileyi* (Berg.) (Hemiptera: Gerridae). Pap. Mich. Acad. Sci. Arts Lett. 13:433-446.

3658. Simberloff, D., and T. Dayan. 1991. The guild concept and the structure of ecological communities. Ann. Rev. Ecol. Syst. 22:115-143.

3659. Simmons, K. R., and J. D. Edman. 1981. Sustained colonization of the black fly *Simulium decorum* Walker (Diptera: Simuliidae). Can. J. Zool. 59:1-7.

3660. Simmons, K. R., and J. D. Edman. 1982. Laboratory colonization of the human onchocerciasis vector *Simulium damnosum* complex (Diptera: Simuliidae), using an enclosed, gravity-trough rearing system. J. Med. Ent. 19:117-126.

3661. Simmons, P., D. F. Barnes, C. K. Fisher, and G. F. Kaloostian. 1942. Caddisfly larvae fouling a water tunnel. J. Econ. Ent. 35:77-79.

3662. Simpson, K. W. 1975. Biology and immature stages of three species of Nearctic *Ochthera* (Diptera: Ephydridae). Proc. Ent. Soc. Wash. 77:129-155.

3663. Simpson, K. W. 1976. Shore flies and brine flies (Diptera: Ephydridae), pp. 465-495. In L. Cheng (ed.). Marine insects. North Holland, Amsterdam. 581 pp.

3664. Simpson, K. W. 1980. Abnormalities in the tracheal gills of aquatic insects collected from streams receiving chlorinated or crude oil wastes. Freshwat. Biol. 10:581-583.

3665. Simpson, K. W. 1982. A guide to basic chironomid literature for the genera of North American Chironomidae (Diptera)-adults, pupae, and larvae. Bull. N. Y. St. Mus. No. 447.

3666. Simpson, K., and R. Bode. 1980. Common larvae of Chironomidae (Diptera) from New York state streams and rivers, with particular reference to the fauna of artificial substrates. Bull. New York State Mus. 439:1-105.

3667. Simpson, K. W., R. W. Bode, and P. Albu. 1982. Keys for the genus *Cricotopus* adapted from "Revision der Gattung *Cricotopus* van der Wulp and Ihrer Verwandten (Diptera: Chironomidae)" by M. Hirvenoja. Bull. N. Y. St. Mus. 450:1-133.

3668. Sinclair, B. J. 1988. The madicolous Tipulidae (Diptera) of eastern North America, with descriptions of the biology and immature stages of *Dactylolabis montana* (O. S.) and *D. hudsonica* Alexander (Diptera: Tipulidae). Can. Ent. 120:569-573.

3669. Sinclair, B. J. 1989. The biology of *Euparyphus* Gerstaecker and *Caloparyphus* James occurring in madicolous habitats of eastern North America, with descriptions of adult and immature stages (Diptera: Stratiomyiidae). Can. J. Zool. 67:33-41.

3670. Sinclair, B. J. 1992. A new species of *Trichothaumalea* (Diptera: Thaumaleidae) from eastern North America and a discussion of male genitalic homologies. Can. Ent. 124:491-499.

3671. Sinclair, B. J. 1994. Revision of the Nearctic species of *Trichoclinocera* Collin (Diptera: Empididae; Clinocerinae). Can. Ent. 126:1007-1059.

3672. Sinclair, B. J., and S. A. Marshall. 1986. The madicolous fauna in southern Ontario. Proc. Ent. Soc. Ontario. 117:9-14.

3673. Sinclair, R. N. 1964. Water quality requirements for elmid beetles with larval and adult keys for the eastern genera. Tenn. Stream Poll. Contr. 15 pp.

3674. Singh, M. P., S. M. Smith, and A. D. Harrison. 1984. Life cycles, microdistribution, and food of two species of caddisflies (Trichoptera) in a wooded stream in Southern Ontario. Can. J. Zool. 62:2582-2588.

3675. Sinsabaugh, R. L., A. E. Linkins, and E. F. Benfield. 1985. Cellulose digestion and assimilation by three leaf-shredding aquatic insects. Ecology 66:1464-1471.

3676. Sioli, H. 1975. Tropical river: The Amazon, pp. 461-488. In B. A. Whitton (ed.). River Ecology. Univ. Calif. Press, Berkeley. 725 pp.

3677. Sison, P. 1938. Some observations on the life history, habits, and control of the rice caseworm, *Nymphula depunctalis* Guen. Philippine J. Agr. 9:273-299.

3678. Sites, R. W., and B. J. Nichols. 1990. Life history and descriptions of immature stages of *Ambrysus lunatus lunatus* (Hemiptera: Naucoridae). Ann. Ent. Soc. Am. 83:800-808.

3679. Sites, R. W., and B. J. Nichols. 1993. Voltinism, egg structure, and descriptions of immature stages of *Cryphocricos hungerfordi* (Hemiptera: Naucoridae). Ann. Ent. Soc. Am. 86:80-90.

3680. Sites, R. W., and J. T. Polhemus. 1994. Nepidae (Hemiptera) of the United States and Canada. Ann. Ent. Soc. Am. 87:27-42.

3681. Sites, R. W., and M. R. Willig. 1991. Microhabitat associations of three sympatric species of Naucoridae (Insecta: Hemiptera). Ann. Ent. Soc. Am. 20:127-134.

3682. Siverly, R. E. 1972. Mosquitoes of Indiana. Publ. Indiana State Bd. Hlth. 126 pp.

3683. Sjogren, R. D., and E. F. Legner. 1989. Survival of the mosquito predator, *Notonecta unifasciata* (Hemiptera: Notonectidae) embryos at low thermal gradients. Entomophaga 34:201-208.

3684. Skidmore, P. 1973. Notes on the biology of Palaearctic muscids (1); (2). Entomologist 106:25-48; 49-59.

3685. Skidmore, P. 1985. The biology of the Muscidae of the world. Series Ent. 29:1-550.

3686. Slack, H. D. 1936. The food of caddis fly (Trichoptera) larvae. J. Anim. Ecol. 5:105-115.

3687. Slack, K. V., L. J. Tilly, and S. S. Kennelly. 1991. Mesh-size effects on drift sample composition as determined with a triple net sampler. Hydrobiologia 209:215-226.

3688. Sladeckova, A. 1962. Limnological investigation methods for the periphyton ("aufwuchs") community. Bot. Rev. 28:286-350.

3689. Sladeckova, A., and E. Pieczynska. 1971. Periphyton, pp. 109-122. *In* W. T. Edmondson and G. G. Winberg (eds.). A manual of methods for the assessment of secondary productivity in fresh waters. IBP Handbook 17. Blackwell, Oxford. 358 pp.

3690. Slaff, M., and C. S. Apperson. 1989. A key to the mosquitoes of North Carolina and the mid-Atlantic states. North Carolina St. Univ., Agric. Exten. Serv., Publ. AG-412. 38pp.

3691. Slater, Alex. 1981. Aquatic and semiaquatic Heteroptera in the collection of the State Biological Survey of Kansas. State Biol. Surv. Kans. Tech. Publ. 10:71-88.

3692. Slater, J. A. 1974. A preliminary analysis of the derivation of the Heteroptera fauna of the northeastern United States with special reference to the fauna of Connecticut, 25th Anniv. Mem. Conn. Ent. Soc. pp. 145-213.

3693. Slater, J. A., and R. M. Baranowski. 1978. How to know the true-bugs (Hemiptera-Heteroptera). Wm. C. Brown, Dubuque. 256 pp.

3694. Smetana, A. 1974. Revision of the genus *Cymbiodyta* Bed. (Coleoptera: Hydrophilidae). Mem. Ent. Soc. Can. 93:1-113.

3695. Smetana, A. 1980. Revision of the genus *Hydrochara* Berth. (Coleoptera: Hydrophilidae). Mem. Ent. Soc. Can. 111:1-100.

3696. Smetana, A. 1985a. Synonymical notes on Hydrophilidae (Coleoptera). Coleop. Bull. 39:328.

3697. Smetana, A. 1985b. Revision of the subfamily Helophorinae of the Nearctic region (Coleoptera: Hydrophilidae). Mem. Ent. Soc. Can. 131:1-154.

3698. Smetana, A. 1988. Review of the family Hydrophilidae of Canada and Alaska, USA (Coleoptera). Mem. Ent.. Soc. Can. 142:3-316.

3699. Smirnov, N. N. 1962. On nutrition of caddis worms *Phryganea grandis* L. Hydrobiologia 19:252-261.

3700. Smit, H., E. Dudok van Heel, and S. Wiersma. 1993. Biovolume as a tool in biomass determination of Oligochaeta and Chironomidae. Freshwat. Biol. 29:37-46.

3701. Smith, C. L. 1980. A taxonomic revision of the genus *Microvelia* Westwood (Heteroptera: Veliidae) of North American including Mexico. Ph.D. diss., University of Georgia, Athens. 372 pp.

3702. Smith, C. L., and J. T. Polhemus. 1978. The Veliidae of America north of Mexico – Keys and checklist. Proc. Ent. Soc. Wash. 80:56-68.

3703. Smith, C. N. (ed.). 1966. Insect colonization and mass production. Academic, N.Y. 618 pp.

3704. Smith, E. L. 1970. Biology and structure of the dobsonfly, *Neohermes californicus* (Walker). Pan-Pacif. Ent. 46:142-150.

3705. Smith, G. E., S. G. Breeland, and E. Pickard. 1965. The Malaise trap – a survey tool in medical entomology. Mosquito News 25:398-400.

3706. Smith, J. A., and A. J. Dartnall. 1980. Boundary layer control by water pennies (Coleoptera: Psephenidae). Aquat. Insects 2:65-72.

3707. Smith, L. C., and L. A. Smock. 1992. Ecology of invertebrate predators in a Coastal Plain stream. Freshwat. Biol. 28:319-329.

3708. Smith, L. M., and H. Lowe. 1948. The black gnats of California. Hilgardia 18:157-183.

3709. Smith, L. W., Jr., and W. R. Enns. 1968. A list of Missouri mosquitoes. Mosquito News 28:50-51.

3710. Smith, M. E. 1952. Immature stages of the marine fly *Hypocharassus pruinosus* Wh., with a review of the biology of immature Dolichopodidae. Am. Midl. Nat. 48:421-432.

3711. Smith, M. E., B. J. Wyskowski, C. T. Driscoll, C. M. Brooks, and C. C. Cosentini. 1989. *Palaeodipteron walkeri* (Diptera: Nymphomyiidae) in the Adirondack Mountains, New York. Ent. News 100:122-124.

3712. Smith, M. H. 1986. A method for rearing caddisflies (Trichoptera) from the egg stage, or early instar larvae. Bull. R. Ent. Soc. London 10:171-172.

3713. Smith, R. F., and A. E. Pritchard. 1956. Odonata, pp. 106-153. *In* R. L. Usinger (ed.). Aquatic insects of California. Univ. Calif. Press, Berkeley. 508 pp.

3714. Smith, R. L. 1973. Aspects of the biology of three species of the genus *Rhantus* (Coleoptera: Dytiscidae) with special reference to the acoustical behavior of two. Can. Ent. 105:909-919.

3715. Smith, R. L. 1974. Life history of *Abedus herberti* in central Arizona (Hemiptera: Belostomatidae). Psyche 81:272-283.

3716. Smith, R. L. 1976a. Brooding behavior of a male water bug *Belostoma flumineum* (Hemiptera: Belostomatidae). J. Kans. Ent. Soc. 49:333-343.

3717. Smith, R. L. 1976b. Male brooding behavior of the water bug *Abedus herberti* (Hemiptera: Belostomatidae). Ann. Ent. Soc. Am. 69:740-747.

3718. Smith, R. I., and J. T. Carlton (eds.). 1975. Light's manual: Intertidal invertebrates of the central California coast (3rd ed.). Univ. of Calif. Press, Berkeley. 716 pp.

3719. Smith, R. L., and E. Larsen. 1993. Egg attendance and brooding by males of the giant water bug *Lethocerus medius* (Guerin) in the field (Heteroptera: Belostomatidae). J. Ins. Behav. 6:93-106.

3720. Smith, S. D. 1965. Distributional and biological records of Idaho caddisflies (Trichoptera). Ent. News 76:242-245.

3721. Smith, S. D. 1968a. The Arctopsychinae of Idaho (Trichoptera: Hydropsychidae). Pan-Pacif. Ent. 44:102-112.

3722. Smith, S. D. 1968b. The *Rhyacophila* of the Salmon River drainage of Idaho with special reference to larvae. Ann. Ent. Soc. Am. 61:655-674.

3723. Smith, W. A., T. E. Vogt, and K. H. Gaines. 1993. Checklist of Wisconsin dragonflies. Wisconsin Ent. Soc. Misc. Publ. No. 2:1-7.

3724. Smith, W., and A. D. McIntyre. 1954. A spring-loaded bottom sampler. J. Mar. Biol. Ass. U.K. 33:257-264.

3725. Smith-Cuffney, F. L., and J. B. Wallace. 1987. The influence of microhabitat on availability of drifting invertebrate prey to a net-spinning caddisfly. J. N. Am. Benthol. Soc. 17:91-98.

3726. Smock, L. A., E. Gilinsky, and D. L. Stonebrunner. 1985. Macroinvertebrate production in a southeastern United States blackwater stream. Ecology 66:1491-1503.

3727. Smock, L. A. 1980. Relationships between body size and biomass of aquatic insects. Freshwat. Biol. 10:375-383.

3728. Smock, L. A. 1988. Life histories, abundance, and distribution of some macroinvertebrates from a South Carolina, USA coastal plain stream. Hydrobiologia 157:193-208.

3729. Smock, L. A., and C. E. Roeding. 1986. The trophic basis of production of the macroinvertebrate community of a southeastern U.S.A. blackwater stream. Holarct. Ecol. 9:165-174.

3730. Sneath, P.H.A. and R. R. Sokal. 1973. Numerical taxonomy. W. H. Freeman, San Francisco. 573 pp.

3731. Snellen, R. K., and K. W. Stewart. 1979a. The life cycle and drumming behavior of *Zealeuctra claasseni* (Frison) and *Zealeuctra hitei* Ricker and Ross (Plecoptera: Leuctridae) in Texas, USA. Aquat. Insects 1:65-89.

3732. Snellen, R. K., and K. W. Stewart. 1979b. The life cycle of *Perlesta placida* (Plecoptera: Perlidae) in an intermittent stream in Northern Texas. Ann. Ent. Soc. Am. 72:659-666.

3733. Snider, R. J. 1967. An annotated list of the Collembola (springtails) of Michigan. Mich. Ent. 1:179-243.

3734. Snider, R. J., and S. J. Loring. 1982. *Sminthurus incognitus*, new species from Florida (Collembola: Sminthuridae). Fla. Ent. 65:216-221.

3735. Snoddy, E. L., and R. Noblet. 1976. Identification of the immature black flies (Diptera: Simuliidae) of the southeastern U.S. with some aspects of the adult role in transmission of *Leucocytozoon smithi* to turkeys. South Carolina Agr. Exp. Sta. Tech. Bull. 1057:1-58.

3736. Snodgrass, R. E. 1935. Principles of insect morphology. McGraw-Hill, N.Y. 667 pp.

3737. Snodgrass, R. E. 1954. The dragonfly larva. Smithson. Misc. Coll. 123:4175.

3738. Snyder, C. D., L. D. Willis, and A. C. Hendricks. 1991. Spatial and temporal variation in the growth and production of *Ephoron leukon* (Ephemeroptera: Polymitarcyidae). J. N. Am. Benthol. Soc. 10:57-67.

3739. Snyder, T. P., and D. G. Huggins. 1980. Kansas black flies (Diptera: Simuliidae) with notes on distribution and ecology. Tech. Publ. State Biol. Surv. Kansas 9:30-34.

3740. Soderstrom, O. 1988. Effects of temperature and food quality on life-history parameters in *Parameletus chelifer* and *P. minor* (Ephemeroptera): a laboratory study. Freshwat. Biol. 20:295-304.

3741. Soeffing, K. 1988. The importance of mycobacteria for the nutrition of larvae of *Leucorrhinia rubicunda* (L.) in bog water (Anisoptera: Libellulidae). Odonatologica 17:227-233.

3742. Sokal, R. R., and F. J. Rohlf. 1969. Biometry. W. H. Freeman, San Francisco. 776 pp.

3743. Sokal, R. R., and F. J. Rohlf. 1981. Biometry: the principles and practice of statistics in biological research (2nd ed.). W. H. Freeman, San Francisco. 859 pp.

3744. Soldán, T. 1979. The structure and function of the maxillary palps of *Arthroplea* congener (Ephemeroptera: Heptageniidae). Acta ent. Bohemosl. 66:300-307.

3745. Soldán, T. 1986. A revision of Caenidae with ocular tubercles in the nymphal stage (Ephemeroptera). Acta Univ. Carolinae-Biologica 1982-1984:289-362.

3746. Solem, J. O. 1973. Diel rhythmic pattern of *Leptophlebia marginata* L. and *L. vespertina* L. (Ephemeroptera). Aquilo Ser. Zool. 14:80-83.

3747. Solem, J. O., and V. H. Resh. 1981. Larval and pupal description, life cycle, and adult flight behaviour of the sponge-feeding caddisfly, *Ceraclea nigronervosa* (Retzius), in central Norway. Ent. Scand. 12:311-319.

3748. Soltesz, K. 1991. A survey of the damselflies and dragonflies of Cape May County, New Jersey. Cape May Bird Observatory, Cape May Pt., NJ. 54pp.

3749. Soluk, D. A. 1985. Macroinvertebrate abundance and production of psammophilous Chironomidae in shifting sand areas of a lowland river. Can. J. Fish. Aquat. Sci. 42:1296-1302.

3750. Soluk, D. A., and H. F. Clifford. 1984. Life history and abundance of the predaceous psammophilous mayfly *Pseudiron centralis* McDunnough (Ephemeroptera: Heptageniidae). Can. J. Zool. 62:1534-1539.

3751. Soluk, D. A., and H. F. Clifford. 1985. Microhabitat shifts and substrate selection by the psammophilous predator *Pseudiron centralis* McDunnough (Ephemeroptera: Heptageniidae). Can. J. Zool. 63:1539-1543.

3752. Soluk, D. A., and D. A. Craig. 1988. Vortex feeding from pits in the sand: a unique method of suspension feeding used by a stream invertebrate. Limnol. Oceangr. 33:638-645.

3753. Soluk, D. A., and D. A. Craig. 1990. Digging with a vortex: flow manipulation facilitates prey capture by a predatory stream mayfly. Limnol. Oceangr. 35:1201-1206.

3754. Sommerman, K. M. 1953. Identification of Alaskan black fly larvae (Diptera, Simuliidae). Proc. Ent. Soc. Wash. 55:258-273.

3755. Sommerman, K. M. 1962. Notes on two species of *Oreogeton*, predaceous on black fly larvae (Diptera: Empididae and Simuliidae). Proc. Ent. Soc. Wash. 64:123-129.

3756. Soponis, A. R. 1977. A revision of the Nearctic species of *Orthocladius (Orthocladius)* van der Wulp (Diptera: Chironomidae). Mem. Ent. Soc. Can. 102:1-187.

3757. Sorokin, Y. I. 1966. Use of radioactive carbon for the study of the nutrition and food relationship of aquatic animals. Inst. Biol. Vnutr. Vod. Trudy 12:83-132.

3758. Southwood, T.R.E. 1978. Ecological methods with particular reference to the study of insect populations (2nd ed.). Methuen, London. 391 pp.

3759. Spangler, P. J. 1960. A revision of the genus *Tropisternus* (Coleoptera: Hydrophilidae). Ph.D. diss., University of Missouri, Columbia. 365 pp.

3760. Spangler, P. J. 1961. Notes on the biology and distribution of *Sperchopsis tesselatus* (Ziegler), (Coleoptera, Hydrophilidae). Coleopt. Bull. 15:105-112.

3761. Spangler, P. J. 1962. A new species of the genus *Oosternum* and a key to the U.S. species (Coleoptera: Hydrophilidae). Proc. Biol. Soc. Wash. 75:97-100.

3762. Spangler, P. J. 1973. The bionomics, immature stages, and distribution of the rare predaceous water beetle, *Hoperius planatus* (Coleoptera: Dytiscidae). Proc. Biol. Soc. Wash. 86:423-434.

3763. Spangler, P. J., and J. L. Cross. 1972. A description of the egg case and larva of the water scavenger beetle, *Helobata striata* (Coleoptera, Hydrophilidae). Proc. Biol. Soc. Wash. 85:413-418.

3764. Spangler, P. J., and G. W. Folkerts. 1973. Reassignment of *Colpius inflatus* and a description of its larva (Coleoptera: Noteridae). Proc. Biol. Soc. Wash. 86:261-277.

3765. Spangler, P. J., R. C. Froeschner, and J. T. Polhemus. 1985. Comments on a water strider *Rheumatobates meinerti* from the Antilles and a checklist of the species of the genus (Hemiptera: Gerridae). Ent. News 96:196-200.

3766. Spence, J. R. 1980. Density estimation for water-striders (Heteroptera: Gerridae). Freshwat. Biol. 10:563-570.

3767. Spence, J. R. 1986a. Interactions between the scelionid egg parasitoid *Tiphodytes gerriphagus* (Hymenoptera) and its gerrid hosts (Heteroptera). Can. J. Zool. 64:2728-2738.

3768. Spence, J. R. 1986b. Relative impacts of mortality factors in field populations of the waterstrider *Gerris buenoi* Kirkaldy (Heteroptera: Gerridae). Oecologia 70:68-76.

3769. Spence, J. R., and N. M. Andersen. 1994. Biology of water striders: Interactions between systematics and ecology. Ann. Rev. Ent. 39:101-128.

3770. Spence, J. R., and H. A. Cárcamo. 1991. Effects of cannibalism and intraguild predation on pondskaters (Gerridae). Oikos 62:33-341.

3771. Spence, J. R., and G.G.E. Scudder. 1980. Habitats, life cycles, and guild structure among water striders (Heteroptera: Gerridae) on the Fraser Plateau of British Columbia. Can. Ent. 112:779-792.

3772. Spence, J. R., D. H. Spence, and G.G.E. Scudder. 1980. The effects of temperature on growth and development of water strider species (Heteroptera: Gerridae) of Central British Columbia and implications for species packing. Can. J. Zool. 58:1813-1820.

3773. Spencer, G. J. 1937. Rearing of Collembola, pp. 263. *In* J. G. Needham (ed.). Culture methods for invertebrate animals. Comstock, Ithaca. 590 pp.

3774. Speyer, W. 1958. Lepidopteren-Puppen an Obstgewaechsen und in ihrer naehern Umgebung: Versuch einer Bestimmungsuebersicht. Mitt. Biol. Reichsanst. Ld-U. Forstw. 93:1-40.

3775. Spieth, H. T. 1938. A method of rearing *Hexagenia* nymphs (Ephemerida). Ent. News 49:29-32.

3776. Spieth, H. T. 1941. Taxonomic studies on the Ephemeroptera. II. The genus *Hexagenia*. Am. Midl. Nat. 26:233-280.

3777. Spieth, H. T. 1947. Taxonomic studies on the Ephemeroptera: IV. The genus *Stenonema*. Ann. Ent. Soc. Am. 40:87-122.

3778. Spilman, R. L. 1961. On the immature stages of the Ptilodactylidae (Col.). Ent. News 72:105-107.

3779. Spilman, T. J. 1967. The heteromerous intertidal beetles. Pacif. Insects 9:1-21.

3780. Sprague, I. B. 1956. The biology and morphology of *Hydrometra martini* Kirkaldy. Univ. Kans. Sci. Bull. 38:579-693.

3781. Stach, J. 1947. The Apterygoten fauna of Poland in relation to the world fauna of this group of insects. Family: Isotomidae. Acta. Mon. Mus. Nat. Hist. Poland. 482 pp.

3782. Stach, J. 1949a. The Apterygoten fauna of Poland in relation to the world fauna of this group of insects. Families: Neogastruridae and Brachystomellidae. Acta. Mon. Mus. Hist. Nat. Poland. 341 pp.

3783. Stach, J. 1949b. The Apterygoten fauna of Poland in relation to the world fauna of this group of insects. Families: Anuridae and Pseudachorutidae. Acta. Mon. Mus. Nat. Hist. Poland. 122 pp.

3784. Stach, J. 1951. The Apterygoten fauna of Poland in relation to the world fauna of this group of insects. Family: Bilobidae. Acta. Mon. Mus. Hist. Nat. Poland. 97 pp.

3785. Stach, J. 1954. The Apterygoten fauna of this group of insects. Family: Onychiuridae. Acta. Mon. Mus. Nat. Hist. Poland. 219 pp.

3786. Stach, J. 1956. The Apterygoten fauna of Poland in relation to the world fauna of this group of insects. Family: Sminthuridae. Acta. Mon. Mus. Nat. Hist. Poland. 287 pp.

3787. Stach, J. 1957. The Apterygoten fauna of Poland in relation to the world fauna of this group of insects. Families: Neelidae and Dicyrtomidae. Acta. Mon. Mus. Nat. Hist. Poland. 113 pp.

3788. Stach, J. 1960. The Apterygoten fauna of Poland in relation to the world fauna of this group of insects. Tribe: Orchesellini. Acta. Mon. Mus. Nat. Hist. Poland. 151 pp.

3789. Stach, J. 1963. The Apterygoten fauna of Poland in relation to the world fauna of this group of insects. Tribe: Entomobryini. Acta. Mon. Mus. Nat. Hist. Poland. 126 pp.

3790. Staines, Jr., C. L. 1986a. A preliminary checklist of the Hydradephaga (Coleoptera) of Maryland. Insecta Mundi 1:118,155.

3791. Staines, Jr., C. L. 1986b. A preliminary checklist of the Hydrophiloidea (Coleoptera) of Maryland. Insecta Mundi 1:259-260.

3792. Stains, G. S., and G. F. Knowlton. 1943. A taxonomic and distributional study of Simuliidae of western United States. Ann. Ent. Soc. Am. 36:259-280.

3793. Stanford, J. A., and A. R. Gaufin. 1974. Hyporheic communities of two Montana rivers. Science 185:700-702.

3794. Stanford, J. A., and J. V. Ward. 1988. The hyporheic habitat of river ecosystems. Nature 335: 64-66.

3795. Stanger, J. A., and R. W. Baumann. 1993. A revision of the stonefly genus *Taenionema* (Plecoptera: Taeniopterygidae). Trans. Am. Ent. Soc. 119:171-229.

3796. Stark, B. P. 1979. The stoneflies (Plecoptera) of Mississippi. J. Miss. Acad. Sci. 24:109-122.

3797. Stark, B. P. 1986. The Nearctic species of *Agnetina* (Plecoptera: Perlidae). J. Kans. Ent. Soc. 59:437-445.

3798. Stark, B. P. 1989. *Perlesta placida* (Hagen), an eastern Nearctic species complex (Plecoptera: Perlidae). Ent. scand. 20:263-286.

3799. Stark, B. P., and A. R. Gaufin. 1974a. The species of *Calineuria* and *Doroneuria* (Plecoptera: Perlidae). Great Basin Nat. 34:83-94.

3800. Stark, B. P., and A. R. Gaufin. 1974b. The genus *Diploperla* (Plecoptera: Peroldidae). J. Kans. Ent. Soc. 47:433-436.

3801. Stark, B. P., and A. R. Gaufin. 1976a. The Nearctic genera of Perlidae (Plecoptera). Misc. Publ. Ent. Soc. Am. 10:1-80.

3802. Stark, B. P., and A. R. Gaufin. 1976b. The nearctic species of *Acroneuria* (Plecoptera: Perlidae). J. Kans. Ent. Soc. 49:221-253.

3803. Stark, B. P., and A. R. Gaufin. 1979. The stoneflies (Plecoptera) of Florida. Trans. Am. Ent. Soc. 104:391-433.

3804. Stark, B. P., and R. W. Baumann. 1978. New species of Nearctic *Neoperla* (Plecoptera: Perlidae), with notes on the genus. Great Basin Nat. 38:97-114.

3805. Stark, B. P., and S. C. Harris. 1986. Records of stoneflies (Plecoptera) in Alabama. Ent. News 97:177-182.

3806. Stark, B. P., and P. K. Lago. 1980. New records of Nearctic *Sialis* (Megaloptera: Sialidae), with emphasis on Mississippi fauna. Ent. News 91:117-121.

3807. Stark, B. P., and P. K. Lago. 1983. Studies of Mississippi fishflies (Megaloptera: Corydalidae) Chauliodinae. J. Kans. Ent. Soc. 56:356-364.

3808. Stark, B., and D. L. Lentz. 1988. New species of Nearctic *Neoperla* (Plecoptera: Perlidae). Ann. Ent. Soc. Am. 81:371-376.

3809. Stark, B. P., and C. R. Nelson. 1994. Systematics, phylogeny and zoogeography of genus *Yoroperla* (Plecoptera: Peltoperlidae). Ent. scand. (in press).

3810. Stark, B. P., B. R. Oblad, and A. R. Gaufin. 1973. An annotated list of the stoneflies (Plecoptera) of Colorado. Ent. News 84:269-277.

3811. Stark, B. P., and D. H. Ray. 1983. A revision of the genus *Helopicus* (Plecoptera: Perlodidae). Freshwat. Invert. Biol. 2:16-27.

3812. Stark, B. P., and K. W. Stewart. 1973. Distribution of stoneflies (Plecoptera) in Oklahoma. J. Kans. Ent. Soc. 46:563-577.

3813. Stark, B. P., and K. W. Stewart. 1981. The Nearctic genera of Peltoperlidae (Plecoptera). J. Kans. Ent. Soc. 54:285-311.

3814. Stark, B. P., and K. W. Stewart. 1982. *Oconoperla*, a new genus of North American Perlodinae (Plecoptera: Perlodidae). Proc. Ent. Soc. Wash. 84:746-752.

3815. Stark, B. P., and K. W. Stewart. 1985. *Baumannella*, a new perlodine genus from California (Plecoptera: Perlodidae). Proc. Ent. Soc. Wash. 87:740-745.

3816. Stark, B. P., and S. W. Szczytko. 1976. The genus *Beloneuria* (Plecoptera: Perlidae). Ann. Ent. Soc. Am. 69:1120-1124.

3817. Stark, B. P., S .W. Szczytko, and B. C. Kondratieff. 1988. The *Cultus decisus* complex of eastern North America (Plecoptera: Perlodidae). Proc. Ent. Soc. Wash. 90:91-96.

3818. Stark, B. P., T. A. Wolff, and A. R. Gaufin. 1975. New records of stoneflies (Plecoptera) from New Mexico. Great Basin Nat. 35:97-99.

3819. Statzner, B. 1981. A method to estimate the population size of benthic macroinvertebrates in streams. Oecologia 51:157-161.

3820. Statzner, B., J. A. Gore, and V. H. Resh. 1988. Hydraulic stream ecology: observed patterns and potential applications. J. N. Am. Benthol. Soc. 7:307-360.

3821. Statzner, B., and B. Higler. 1985. Questions and comments on the river continuum concept. Can. J. Fish. Aquat. Sci. 42:1038-1044.

3822. Statzner, B., and T. F. Holm. 1982. Morphological adaptations of benthic invertebrates to stream flow an old question studied by means of a new technique (laser doppler anemometry). Oecologia 53:290-292.

3823. Statzner, B., and T. F. Holm. 1989. Morphological adaptation of shape to flow: microcurrents around lotic macroinvertebrates with known Reynolds numbers at quasi-natural flow conditions. Oecologia 78:145-157.

3824. Steedman, R. J., and N. H. Anderson. 1985. Life history and ecological role of the xylophagous aquatic beetle, *Lara avara* LeConte (Dryopoidea: Elmidae). Freshwat. Biol. 15:535-546.

3825. Steele, B. D., and D. C. Tarter. 1977. Distribution of the family Perlidae in West Virginia. Ent. News 88:18-22.

3826. Steelman, C. D. 1976. Effects of external and internal arthropod parasites on domestic livestock production. Ann. Rev. Ent. 21:155-178.

3827. Steelman, C. D., and A. R. Colmer. 1970. Some elffects of organic wastes on aquatic insects in impounded habitats. Ann. Ent. Soc. Am. 63:397-400.

3828. Stehr, F. W. (ed.). 1987. Immature Insects, Vol. 1. Kendall/Hunt Publ. Co. Dubuque, IA. 754p.

3829. Stehr, F. W. (ed.). 1991. Immature Insects, Vol. 2. Kendall/Hunt Publ. Co. Dubuque, IA. 974p.

3830. Steiner, W. E., Jr., and J. J. Anderson. 1981. Notes on the natural history of *Spanglerogyrus albiventris* Folkerts, with a new distribution record (Coleoptera: Gyrinidae). Pan-Pacif. Ent. 57: 124-132.

3831. Steinley, B. A., and J. T. Runyan. 1979. The life history of *Leptopsilopa atrimana* (Diptera: Ephydridae). pp. 139-147. *In* D. L. Deonier (ed.). First symposium on the systematics and ecology of Ephydridae (Diptera). N. Am. Benthol. Soc. 147 pp.

3832. Stewart, K. W., G. L. Atmar, and B. M. Solon. 1969. Reproductive morphology and mating behavior of *Perlesta placida* (Plecoptera: Perlidae). Ann. Ent. Soc. Am. 62:1433-1438.

3833. Stewart, K. W., R. W. Baumann, and B. P. Stark. 1974. The distribution and past dispersal of southwestern United States Plecoptera. Trans. Am. Ent. Soc. 99:507-546.

3834. Stewart, K. W., R. E. DeWalt and M. W. Oswood. 1991. *Alaskaperla*, a new stonefly genus (Plecoptera: Chloroperlidae), and further descriptions of related Chloroperlidae. Ann. Ent. Soc. Am. 84:239-247.

3835. Stewart, K. W., G. P. Friday, and R. E. Rhame. 1973. Food habits of hellgrammite larvae, *Corydalus cornutus* (Megaloptera: Corydalidae), in the Brazos River, Texas. Ann. Ent. Soc. Am. 66:959-963.

3836. Stewart, K. W., R. L. Hassage, S. J. Holder, and M. W. Oswood. 1990. Life cycles of six stonefly species (Plecoptera) in subarctic and arctic Alaska streams. Ann. Ent. Soc. Am. 83:207-214.

3837. Stewart, K. W., and D. W. Huggins. 1977. The stoneflies of Kansas. Tech. Publ. Kans. State Biol. Surv. 4:31-40.

3838. Stewart, K. W., and M. Maketon. 1991. Structures used by Nearctic stoneflies (Plecoptera) for drumming, and their relationship to behavioral pattern diversity. Aquat. Insects 13:33-53.

3839. Stewart, K. W., L. E. Milliger, and B. M. Solon. 1970. Dispersal of algae, protozoans, and fungi by aquatic Hemiptera, Trichoptera, and other aquatic insects. Ann. Ent. Soc. Am. 63:139-144.

3840. Stewart, K. W., and B. P. Stark. 1977. Reproductive system and mating of *Hydroperla crosbyi*: a newly discovered method of sperm transfer in Insecta. Oikos 28:84-89.

3841. Stewart, K. W., and B. P. Stark. 1988. Nymphs of North American stonefly genera (Plecoptera). Thomas Say Found. Ser., Ent. Soc. Am. 12:1-460.

3842. Stewart, K. W., B. P. Stark, and T. G. Huggins. 1976. The stoneflies (Plecoptera) of Louisiana. Great Basin Nat. 36:366-384.

3843. Stewart, K. W., S. W. Szczytko, and B. P. Stark. 1982. Drumming behavior of four species of North American Pteronarcyidae (Plecoptera): dialects in Colorado and Alaska *Pteronarcella badia*. Ann. Ent. Soc. Am. 75:530-533.

3844. Stewart, M., and G. Pritchard. 1982. Pharate phases in *Tipula paludosa* (Diptera: Tipulidae). Can. Ent. 114:275-278.

3845. Steyskal, G. C. 1957. A revision of the family Dryomyzidae. Pap. Mich. Acad. Sci. Arts Lett. 42:55-68.

3846. Steyskal, G. C. 1987. Chap. 83. Dryomyzidae, pp. 923-926. *In* J. F. McAlpine, B. V. Peterson, G. E. Shewell, H. J. Teskey, J. R. Vockeroth, and D. M. Wood (coords.). Manual Nearctic Diptera, Vol. 2. Res. Branch, Agr. Can. Mon. 28. 1332p.

3847. Steyskal, G. C., T. W. Fisher, L. Knutson, and R. E. Orth. 1978. Taxonomy of North American flies of the genus *Limnia* (Diptera: Sciomyzidae). Univ. Calif. Publ. Ent. 83:1-48.

3848. Steyskal, G. C., and L. V. Knutson. 1981. Chap. 47. Empididae, pp. 607-624. *In* J. F. McAlpine, B. V. Peterson, G. E. Shewell, H. J. Teskey, J. R. Vockeroth, and D. M. Wood (coords.). Manual of Nearctic Diptera, Vol. 1. Res. Branch, Agr. Can. Mon. 27. 674 pp.

3849. Stoakes, R. D., J. D. Neel, and R. L. Post. 1983. Observations on North Dakota sponges (Haplosclerina: Spongillidae) and Sisyrids (Neuroptera: Sisyridae). Great Lakes Ent. 16:171-176.

3850. Stobbart, R. H., and J. Shaw. 1974. Salt and water balance; excretions, pp. 361-446. *In* Rockstein (ed.). The physiology of Insecta. (2nd ed.) Vol. V. Academic Press, New York. 648 pp.

3851. Stock, M. W., and J. D. Lattin. 1976. Biology of intertidal *Saldula palustris* (Douglas) on the Oregon coast (Heteroptera: Saldidae). J. Kans. Ent. Soc. 49:313-326.

3852. Stocker, Z.S.J., and D. D. Williams. 1972. A freezing core method for describing the vertical distribution of sediments in a streambed. Limnol. Oceanogr. 17:136-139.

3853. Stockner, J. G. 1971. Ecological energetics and natural history of *Hedriodiscus truquii* (Diptera) in two thermal spring communities. J. Fish. Res. Bd. Can. 28:73-94.

3854. Stojanovich, C. J. 1961. Illustrated key to common mosquitoes of northeastern North America. C. J. Stojanovich, Campbell, Calif. 49 pp.

3855. Stone, A. 1952. The Simuliidae of Alaska (Diptera). Proc. Ent. Soc. Wash. 54:69-96.

3856. Stone, A. 1963. A new *Parasimulium* and further records for the type species (Diptera: Simuliidae). Bull. Brooklyn Ent. Soc. 58:127-129.

3857. Stone, A. 1964a. Guide to the insects of Connecticut. Part VI. The Diptera or true flies of Connecticut. Fasc. 9. Family Simuliidae. Bull. Conn. State Geol. Nat. Hist. Surv. 97:1-117.

3858. Stone, A. 1964b. Guide to the insects of Connecticut. Part VI. The Diptera or true flies of Connecticut. Fasc. 9. Family Thaumaleidae. Bull. Conn. State Geol. Nat. Hist. Surv. 97:119-122.

3859. Stone, A. 1965. Family Simuliidae, pp. 181-189. *In* A.Stone, C. W. Sabrosky, W. W. Wirth, R. H. Foote, and J. R. Coulson (eds.). A catalog of the Diptera of America north of Mexico. U.S. Dept. Agric. Handbk. 276. 1696 pp.

3860. Stone, A. 1981. Culicidae, pp. 341-350. *In* J. F. McAlpine, B. V. Peterson, G. E. Shewell, H. J. Teskey, J. R. Vockeroth, and D. M. Wood. (coords) Manual of Nearctic Diptera. Vol. I. Monogr. 27, Res. Branch, Agric. Can. Monogr 27 Ottawa, 674 pp.

3861. Stone, A., and H. A. Jamnback. 1955. The black flies of New York State (Diptera: Simuliidae). Bull. N.Y. State Mus. 349:1-144.

3862. Stone, A., and B. V. Peterson. 1981. Chap. 26. Thaumaleidae, pp. 351-353. *In* J. F. McAlpine, B. V. Peterson, G. E. Shewell, H. J. Teskey, J. R. Vockeroth, and D. M. Wood (coords.). Manual of Nearctic Diptera, Vol. I. Res. Branch, Agric. Can. Monogr 27,. Ottawa. 674 pp.

3863. Stone, A., C. W. Sabrosky, W. W. Wirth, R. H. Foote, and J. R. Coulson. 1965. A catalog of the Diptera of America north of Mexico. U.S. Dep. Agric. Handbk. 276. 1696 pp.

3864. Stone, A., and E. L. Snoddy. 1969. The black flies of Alabama (Diptera: Simuliidae). Bull. Alabama Agric. Exp. Sta. Auburn Univ. 390:1-93.

3865. Stonedahl, G. M., and J. D. Lattin. 1982. The Gerridae or water striders of Oregon and Washington (Hemiptera: Heteroptera). Oregon State Univ. Agric. Exp. Sta. Tech. Bull. 144:1-36.

3866. Stoner, A. W., H. S. Greening, J. D. Ryan, and R. J. Livingston. 1983. Comparison of macrobenthos collected with cores and suction sampler in vegetated and unvegetated marine habitats. Estuaries 6:76-82.

3867. Storey, A. W. 1987. Influence of temperature and food quality on the life history of an epiphytic chironomid. *In* O. A. Saether (ed). A conspectus of contemporary studies in Chironomidae (Diptera). Ent. Scand. Suppl. 29:339-347.

3868. Storey, A. W., D. H. Edward, and P. Gazey. 1991. Surber and kick sampling: a comparison for the assessment of macroinvertebrate community structure in streams of south-western Australia. Hydrobiologia 211:111-121.

3869. Storey, A., and L. Pinder. 1985. Mesh-size and efficiency of sampling larval Chironomidae. Hydrobiologia 124:194-198.

3870. Stout, B. M., E. F. Benfield, and J. R. Webster. 1992a. Effects of a forest disturbance on shredder production in southern Appalachian headwater streams. Freshwat. Biol. 29:59-69.

3871. Stout, B. M., III, K. K. Stout, and C. W. Stihler. 1992b. Predation by the caddisfly *Banksiola dossuaria* on egg masses of the Spotted Salamander *Amblystoma maculatum*. Am. Midl. Nat. 127:368-372.

3872. Stout, R. J. 1981. How abiotic factors affect the distribution of two species of tropical predaceous aquatic bugs (Family: Naucoridae). Ecology 62:1170-1178.

3873. Stout, R. J. 1992. Responses to extremely low frequency electromagnetic fields by a dragonfly naiad (*Ophiogomphus colubrinus*) in a northern Michigan stream: a five year study. J. Freshwat. Ecol. 7:33-35.

3874. Stout, R. J., and W. H. Taft. 1985. Growth patterns of a chironomid shredder on fresh and senescent tag alder leaves in two Michigan streams. J. Freshwat. Ecol. 3:147-153.

3875. Strand, R. M., and P. J. Spangler. 1994. The natural history, distribution, and larval description of *Brychius hungerfordi* Spangler (Coleoptera: Haliplidae). Proc. Ent. Soc. Wash. 96:208-213.

3876. Streams, F. A. 1974. Size and competition in Connecticut *Notonecta*. 25th Anniv. Mem. Connecticut Ent. Soc. pp. 215-225.

3877. Streams, F. A. 1987a. Foraging behavior in a notonectid assemblage. Am. Midl. Nat. 117:353-361.

3878. Streams, F. A. 1987b. Within habitat spatial separation of two *Notonecta* species: interactive vs. noninteractive resource partitioning. Ecology 68:935-945.

3879. Streams, F. A. 1992. Intrageneric predation by *Notonecta* (Hemiptera: Notonectidae) in the laboratory and in nature. Ann. Ent Soc. Am. 85:265-273.

3880. Streams, F. A. 1994. Effect of prey size on attack components of the functional response by *Notonecta undulata*. Oecologia 98:57-63.

3881. Streams, F. A., and S. Newfield. 1972. Spatial and temporal overlap among breeding populations of New England *Notonecta*. Univ. Conn. Occ. Pap. (Biol. Sci. Ser.) 2:139-157.
3882. Strenzke, K. 1955. Thalassabionte und thalassophile Collembola (Lief. 36), pp. 1-52. In A. Remane (ed.). Die Tierwelt der Nordund Ostee. Akad., Leipzig.
3883. Stribling, J. B. 1986. Revision of *Anchytarsus* (Coleoptera: Dryopoidea) and a key to the New World genera of Ptilodactylidae. Ann. Ent. Soc. Am. 79:219-234.
3884. Strickland, E. H. 1953. An annotated list of the Hemiptera (S.L.) of Alberta. Can. Ent. 85:193-214.
3885. Stubbs, A., and P. Chandler (eds.). 1978. A dipterist's handbook. Amat. Ent. 15:1-255.
3886. Stuckenberg, B. R. 1973. A new family in the lower Brachycera (Diptera). Ann. Nat. Mus. 21:646-673.
3887. Sturm, H. 1960. Die terrestrischen Puppengehäuse von *Xiphocentron sturmi* Ross (Xiphocentronidae, Trichoptera). Zool. Jb. (Syst.) 87:387-394.
3888. Sturtevent, A. H., and M. R. Wheeler. 1954. Synopsis of Nearctic Ephydridae (Diptera). Trans. Am. Ent. Soc. 79:151-261.
3889. Stys, P. 1970. On the morphology and classification of the family Dipsocoridae s. lat., with particular reference to the genus *Hypsipteryx* Drake (Heteroptera). Acta. Ent. Bohem. 67:21-46.
3890. Stys, P., and A. Jansson. 1988. Check-list of recent family-group and genus-group names of Nepomorpha (Heteroptera) of the world. Acta. Ent. Fenn. 50:1-44.
3891. Stys, P., and I. M. Kerzhner. 1975. The rank and nomenclature of higher taxa in recent Heteroptera. Acta. Ent. bohemoslav. 72:65-79.
3892. Suberkropp, K., T. L. Arsuffi, and J. P. Anderson. 1983. Comparison of the degradative ability, enzymatic activity and palatability of aquatic hyphomycetes grown on leaf litter. J. Appl. Environ. Microbiol. 46:237-244.
3893. Sublette, J. E. 1960. Chironomid midges of California I. Chironominae exclusive of Tanytarsini. Proc. U.S. Nat. Mus. 112:197-226.
3894. Sublette, J. E. 1964a. Chironomidae (Diptera) of Louisiana. I. Systematics and immature stages of some benthic chironomids of west-central Louisiana. Tulane Stud. Zool. 11:109-150.
3895. Sublette, J. E. 1964b. Chironomid midges of California II. Tanypodinae, Podonominae and Diamesinae. Proc. U.S. Nat. Mus. 115:85-136.
3896. Sublette, J. E., and M. S. Sublette. 1965. Family Chironomidae, pp. 142-181. In A. Stone, C. W. Sabrosky, W. W. Wirth, R. H. Foote, and J. R. Coulson (eds.). A catalog of the Diptera of America north of Mexico. U.S. Dept. Agric. Handbk. 276. 1,696 pp.
3897. Sublette, J. E., and M. S. Sublette. 1967. The limnology of playa lakes on the Llano Estacado, New Mexico and Texas. Southwest. Nat. 12:369-406.
3898. Sublette, M. S., and J. E. Sublette. 1970. Distributional records of mosquitoes in the southern high plains with a checklist of species from New Mexico and Texas. Mosquito News 30:533-538.
3899. Sudia, W. D., and R. W. Chamberlain. 1967. Collection and processing of medically important arthropods for arbovirus isolation. Nat. Commun. Disease Centr., Atlanta. 29 pp.
3900. Sughara, Y. 1938. An observation on the intertidal rock-dwelling beetle *Aegialites stejnegeri sugiharae* Kono in the Kuriles. Ent. World 6:6-12.
3901. Sulaiman, S. 1983. The difference in the instar composition of immature mosquitoes sampled by three sampling techniques. Japan. J. Med. Sci. Biol. 37:281-288.
3902. Sundholm, A. 1960. On *Diapria* Latreille and allied genera (Hym.: Diapriidae). Opusc. Ent. 25:215-223.
3903. Surber, E. W. 1937. Rainbow trout and bottom fauna production in one mile of stream. Trans. Am. Fish. Soc. 66:193-202.
3904. Surdick, R. F. 1981. Nearctic genera of Chloroperlinae (Plecoptera: Chloroperlidae). Ph.D. Thesis, Univ. of Utah. Univ. Microfilms #DA 8208937.
3905. Surdick, R. F. 1981. Nearctic genera of Chloroperlinae (Plecoptera: Chloroperlidae). Ph.D. Thesis, Univ. of Utah. Univ. Microfilms #DA 8208937.
3906. Surdick, R. F. 1985. Nearctic genera of Chloroperlinae (Plecoptera: Chloroperlidae). Illinois, Biol. Monogr. 54:1-146.
3907. Surdick, R. F., and K. C. Kim. 1976. Stoneflies (Plecoptera) of Pennsylvania, a synopsis. Bull. Penn. State Univ. Agric. Exp. Sta. 808:1073.
3908. Suren, A. M., and P. S. Lake. 1989. Edibility of fresh and decomposing macrophytes to three species of freshwater invertebrate herbivores. Hydrobiologia 178:165-178.
3909. Suren, A. M., and M. J. Winterboourn. 1991. Consumption of aquatic bryophytes by alpine stream invertebrates in New Zealand. N. Z. J. Mar. Freshwat. Res. 25:331-343.
3910. Surtees, G. 1959. Functional and morphological adaptations of the larval mouthparts in the subfamily Culicinae (Diptera) with a review of some related studies by Montchadsky. Proc. R. ent. Soc. Lond. (A) 34:7-16.
3911. Sutherland, D. W. S. and R. F. Darsie, Jr. 1960. A report on the blackflies (Simuliidae) of Delaware. Part I. Record of Delaware species and an introduction to a survey of the western branches of the Christiana River, New Castle County. Bull. Brooklyn Ent. Soc. 55:53-61.
3912. Sutton, M. 1951. On the food, feeding mechanism and alimentary canal of Corixidae (Hemiptera, Heteroptera). Proc. Zool. Soc. Lond. 121:465-499.
3913. Svensson, B. W. 1974. Population movements of adult Trichoptera at a south Swedish stream. Oikos 25:157-175.
3914. Svihla, A. 1959. The life history of *Tanypteryx hageni* Selys (Odonata). Trans. Am. Ent. Soc. 85:219-232.
3915. Swanson, G. A. 1978. A water column sampler for invertebrates in shallow wetlands. J. Wildl. Manage. 42:670-672.
3916. Swatchek, B. 1958. Die Larvalsystematik der Winkler (Tortricidae und Carposinidae). Abh. Larvalsyst. Insekt. 3. Akad., Berlin 269 pp.
3917. Sweeney, B. W. 1978. Bioenergetic and developmental response of a mayfly to thermal variation. Limnol. Oceanogr. 23:461-477.
3918. Sweeney, B. W. 1984. Factors influencing life-history patterns of aquatic insects, pp. 56-100. In V. H. Resh and D. M. Rosenberg (eds.). The ecology of aquatic insects. Praeger Publishers, NY. 625pp.
3919. Sweeney, B. W. 1993. Effects of streamside vegetation on macroinvertebrate communities of White Clay Creek in Eastern North America. Proc. Acad. Nat. Sci. Philadelphia 144:291-340.
3920. Sweeney, B. W., and J. A. Schnack. 1977. Egg development, growth and metabolism of *Sigara alternata* (Say) (Hemiptera: Corixidae) in fluctuating thermal environments. Ecology 58:265-277.
3921. Sweeney, B. W., and R. L. Vannote. 1978. Size variation and the distribution of hemimetabolous aquatic insects: two thermal equilibrium hypotheses. Science 200:444-446.
3922. Sweeney, B. W., and R. L. Vannote. 1981. *Ephemerella* mayflies of While Clay Creek: Bioenergetic and ecological relationships among six coexisting species. Ecology 62:1353-1369.
3923. Sweeney, B. W., and R. L. Vannote. 1984. Influence of food quality and temperature on life history characteristics of the parthenogenetic mayfly, *Cloeon triangulifer*. Freshwat. Biol. 14:621-630.
3924. Sweeney, B. W., and R. L. Vannote. 1986. Growth and production of a stream stonefly (*Soyedina carolinensis*): influences of diet and temperature. Ecology 67:1396-1410.
3925. Sweeney, B. W., R. L. Vannote, and P. J. Dodds. 1986a. Effects of temperature and food quality on growth and development of a mayfly, *Leptophlebia intermedia*. Can. J. Fish. Aquat. Sci. 43:12-18.
3926. Sweeney, B. W., R. L. Vannote, and P. J. Dodds. 1986b. The relative importance of temperature and diet to larval development and adult size of the winter stonefly *Soyedina carolinensis* (Plecoptera: Nemouridae). Freshwat. Biol. 16:39-48.
3927. Swisher, D., and C. Richards. 1971. Selective trout. Crown, N.Y. 184 pp.
3928. Swisher, D., and C. Richards. 1991. Emergers. Lyons and Burford, New York. 120p.
3929. Syms, E. E. 1934. Biological notes on British Megaloptera. Proc. Lond. Ent. Nat. Hist. Soc. 35:121-124.
3930. Szczytko, S. W., and R. L. Bottorff. 1987. *Cosumnoperla hypocrena*, a new genus and species of western Nearctic Isoperlinae (Plecoptera: Perlodidae). Pan-Pac. Ent. 63: 65-74.

3931. Szczytko, S. W., and K. W. Stewart. 1977. The stoneflies (Plecoptera) of Texas. Trans. Am. Ent. Soc. 103:327-378.

3932. Szczytko, S. W., and K. W. Stewart. 1979. The genus *Isoperla* of western North America; holomorphology and systematics, and a new stonefly genus *Cascadoperla*. Mem. Am. Ent. Soc. 32:1-120.

3933. Tachet, H. 1965a. Recherches sur l'alimentation des larves de *Polycentropus* (Trichoptera) dans leur milieu naturel. Ann. Soc. Ent. France 1:627-633.

3934. Tachet, H. 1965b. Influence du stade larvaire et de la saison sur l'alimentation des larves de *Polycentropus* (Trichoptera) dan des conditions naturelles. Ann. Soc. Ent. France 1:635-640.

3935. Takahashi, R. M., R. J. Stewart, C. H. Schaefer, and R. D. Sjogren. 1979. An assessment of *Plea striola* (Hemiptera: Pleidae) as a mosquito control agent in California. Mosq. News 39:514-519.

3936. Tanner, V. M. 1943. Study of subtribe Hydronomi with description of new species. Great Basin Nat. 4:1-38.

3937. Tarshis, I. B. 1968. Collecting and rearing black flies. Ann. Ent. Soc. Am. 61:1072-1083.

3938. Tarshis, I. B. 1971. Individual black fly rearing cylinders (Diptera: Simuliidae). Ann. Ent. Soc. Am. 64:1192-1193.

3939. Tarshis, I. B. 1973. Studies on the collection, rearing, and biology of the black fly *Cnephia ornithophilia*. U.S. Dep. Int., Fish Widl. Serv. Spec. Sci. Rept. 165. 16 pp.

3940. Tarshis, I. B., and W. Neil. 1970. Mass movement of black fly larvae on silken threads (Diptera: Simuliidae). Ann. Ent. Soc. Am. 63:607-610.

3941. Tarter, D. C. 1976. Limnology in West Virginia: A lecture and laboratory manual. Marshall Univ. Book Store, Huntington, W. Va. 249 pp.

3942. Tarter, D. C. 1988. New record of the alderfly *Sialis vagans* for West Virginia (Megaloptera: Sialidae). Ent. News 99:63-64.

3943. Tarter, D.C. 1990. A checklist of the caddisflies (Trichoptera) from West Virginia. Ent. News 101:236-245.

3944. Tarter, D. C., D. A. Atkins, and C. V. Covell, Jr. 1984. A checklist of the stoneflies (Plecoptera) of Kentucky. Ent. News 91:113-116.

3945. Tarter, D. C., and R. F. Kirchner. 1980. List of the stoneflies (Plecoptera) of West Virginia. Ent. News 91:49-53.

3946. Tarter, D. C., and L. A. Krumholz. 1971. Life history and ecology of *Paragnetina media* (Walker) in Doe Run, Meade County, Kentucky. Am. Midl. Nat. 86:169-180.

3947. Tarter, D. C., and W. D. Watkins. 1974. Distribution of the fishfly genera *Chauliodes* Latreille and *Nigronia* Banks in West Virginia. Proc. W. Va. Acad. Sci. 46:146-150.

3948. Tarter, D. C., W. D. Watkins, D. L. Ashley, and J. T. Goodwin. 1978. New state records and seasonal emergence patterns of alderflies east of the Rocky Mountains (Megaloptera: Sialidae). Ent. News 89:231-239.

3949. Tarter, D. C., W. D. Watkins, and D. A. Etnier. 1979. Larval description and habitat notes of the fishfly *Neohermes concolor* (Davis) (Megaloptera: Corydalidae). Ent. News 90:29-32.

3950. Tarter, D. C., W. D. Watkins, and M. L. Little. 1975. Life history of the fishfly *Nigronia fasciatus* (Megaloptera: Corydalidae). Psyche 82:81-88.

3951. Tarter, D. C., W. D. Watkins, and M. L. Little. 1976. Distribution, including new state records, of fishflies in Kentucky (Megaloptera: Corydalidae). Trans. Ky. Acad. Sci. 37:26-28.

3952. Tarter, D. C., W. D. Watkins, M. L. Little, and D. L. Ashley. 1977. Seasonal emergence patterns of fishflies east of the Rocky Mountains (Megaloptera: Corydalidae). Ent. News 88:69-76.

3953. Tarwid, M. 1969. Analysis of the contents of the alimentary tract of predatory Pelopiinae larvae (Chironomidae). Ekol. Polska (A) 17:125-131.

3954. Tate, P. 1935. The larvae of *Phaonia mirabilis* Ringdahl, predatory on mosquito larvae (Diptera, Anthomyiidae). Parasitology 27:556-560.

3955. Tauber, C. A. 1991. Neuroptera, p. 126-143. *In* Stehr, F.W. (ed.). Immature Insects Vol. II, Kendall/Hunt Dubuque. 975 pp.

3956. Tavares, A. F., and D. D. Williams. 1990. Life histories, diet, and niche overlap of three sympatric species of Elmidae (Coleoptera) in a temperate stream. Can. Ent. 122:563-577.

3957. Tawfik, M. F. S., M. M. El-Husseini, and H. Abou Bakr. 1986a. Ecological observations on aquatic insects attacking mosquitoes in Egypt. Bull. Soc. Ent. Egypte 66:117-126.

3958. Tawfik, M. F. S., M. M. El-Husseini, and H. Abou Bakr. 1986b. The biology of the notonectid *Anisops sardea* H. S.: An active mosquito predator in Egypt. Bull. Soc. Ent. Egypte 66:127-144.

3959. Tawfik, M. F. S., S. I. El-Sherif, and A. F. Lutfallah. 1978. Life history of the giant water-bug *Limnogeton fieberi* Mayr (Heteroptera: Belostomatidae), predator on some harmful snails. Z. ang. Ent. 86:138-145.

3960. Taylor, O. R., Jr. 1968. Coexistence and competitive interactions in fall and winter populations of six sympatric *Notonecla* (Hemiptera: Notonectidae) in New England: Univ. Conn. Occ. Pap. Biol. Sci. Ser. 1:109-139.

3961. Tennessen, K. J. 1975. Description of the nymph of *Somatochlora provocans* Calvert (Odonata: Corduliidae). Fla. Ent. 58:105-110.

3962. Tennessen, K. J. 1984. The nymphs of *Calopteryx amata* and *C. angustipennis* (Odonata: Calopterygidae). Proc. Ent. Soc. Wash. 86:602-607.

3963. Tennessen, K. J., and K. W. Knopf. 1975. Description of the nymph of *Enallagma minusculum* (Odonata: Coenagrionidae). Fla. Ent. 58:199-201.

3964. ter Braak, C. J. F., and I. C. Prentice. 1988. A theory of gradient analysis. Adv. Ecol. Res. 18:271-317.

3965. Terteryan, A. E., and V. S. Oganesyan. 1986. Morphology of the larva and pupa of *Silvius latifrons* Ols. (Diptera, Tabanidae). Ent. Rev. 65:148-153.

3966. Terwilliger, R. C. 1980. Structures of invertebrate hemoglobins. Am. Zool. 20:53-67.

3967. Teskey, H. J. 1962. A method and apparatus for collecting larvae of Tabanidae (Diptera) and other invertebrate inhabitants of wetlands. Proc. Ent. Soc. Ont. 92:204-206.

3968. Teskey, H. J. 1969. Larvae and pupae of some eastern North American Tabanidae (Diptera). Mem. Ent. Soc. Can. 63:1-147.

3969. Teskey, H. J. 1970. The immature stages and phyletic position of *Glutops rossi* (Diptera: Pelecorhynchidae). Can. Ent. 102:1130-1135.

3970. Teskey, H. J. 1981a. Chap. 3. Morphology and Terminology Larvae, pp. 65-88. *In* J. F. McAlpine, B. V, Peterson, G. E. Shewell, H. J. Teskey, J. R. Vockeroth, and D. M. Wood (coords). Manual of Nearctic Diptera, Vol. 1. Res. Branch, Agric. Can. Monogr. 27, Ottawa. 674 pp.

3971. Teskey, H. J. 1981b. Chap. 30. Pelecorhynchidae, pp. 459-462. *In* J. F. McAlpine, B. V. Peterson, G. E. Shewell, H. J. Teskey, J. R. Vockeroth, and D. M. Wood (coords). Manual of Nearctic Diptera, Vol. 1. Res. Branch, Agric. Can. Monogr. 27, Ottawa. 674 pp.

3972. Teskey, H. J. 1983a. A review of the *Atylotus insuetus* group from western North America including description of a new species and immature stages (Diptera: Tabanidae). Can. Ent. 115:693-702.

3973. Teskey, H. J. 1983b. A revision of the North American species of *Atylotus* (Diptera: Tabanidae) with keys to adult and immature stages. Proc. Ent. Soc. Ontario. 114:21-43.

3974. Teskey, H. J. 1984. Chap. 21. Aquatic Diptera. Part one. larvae of aquatic Diptera, pp. 448-466. *In*. R.W. Merritt and K.W. Cummins (eds.). An introduction to the aquatic insects of North America. Second edition. Kendall/Hunt Publishing Co. 722pp.

3975. Teskey, H, J. 1990. The insects and arachnids of Canada. Part 16. The horse flies and deer flies of Canada and Alaska, Diptera: Tabanidae. Canadian Government Publishing Centre, Ottawa

3976. Teskey, H. J. 1991. Diptera. Key to families of larvae, pp. 708-730. *In* F. W. Stehr (ed.). Immature Insects. Vol. 2. Kendall/Hunt Publ. Co., Dubuque, IA. 975pp.

3977. Teskey, H. J., and I. Valiela. 1977. The mature larva and puparium of *Canace macaleei* Malloch (Diptera: Canaceidae). Can. Ent. 109:545-547.

3978. Testa, S., III, and P. K. Lago. 1994. The aquatic Hydrophilidae (Coleoptera) of Mississippi. Miss. Agri. Forest. Expt. Sta.,Univ., MS., Tech. Bull. 193. 71 pp.

3979. Tetrault, R. C. 1967. A revision of the family Helodidae (Coleoptera) for America north of Mexico. Ph.D. diss., University of Wisconsin, Madison. 160 pp.
3980. Thienemann, A. 1910. Das Sammeln von Puppenhauten der Chironomiden. Arch. Hydrobiol. 6:213-214.
3981. Thienemann, A. 1944. Bestimmungstabellen fur die bis jetzt bekannten larven and puppen der Orthocladiinen (Diptera: Chironomidae). Arch. Hydrobiol. 39:551-664.
3982. Thienemann, A. 1954. *Chironomus*. Die Binnengewasser 20:1-834.
3983. Thier, R. W., and B. A. Foote. 1980. Biology of mud-shore Ephydridae (Diptera). Proc. Ent. Soc. Wash. 82:517-535.
3984. Thomas, A.G.B. 1974. Dipteres torrenticoles peu connus. Parts I, II. Athericidae (larves et imagos) du Sud de la France (Brachycera, Orthorrhapa). Ann. Limnol. 10:55-84; 121-130.
3985. Thomas, A. 1985. Diptères torrenticoles peu connus: les Athericidae et Rhagionidae européens et circum-méditerranéens. Mitt. Schweiz. Ent. Ges. 58:449-460.
3986. Thomas, E. 1966. Orientierung der Imagines von *Capnia atra* Morton (Plecoptera). Oikos 17:278-280.
3987. Thomas, E. S., and R. D. Alexander. 1962. Systematic and behavioral studies on the meadow grasshoppers of the *Orchelimum concinnum* group (Orthoptera: Tettigoniidae). Occ. Pap. Univ. Mich. Mus. Zool. 626:1-31.
3988. Thomas, L. J. 1946. Black fly incubator-aerator cabinet. Science 103:21-23.
3989. Thompson, C. R., and D. H. Habeck. 1989. Host specificity and biology of the weevil *Neohydronomus affinis* [Coleoptera: Curculionidae] a biological control agent of *Pistia stratiotes*. Entomophaga 34:299-396.
3990. Thompson, D. J. 1978a. The natural prey of larvae of the damselfly *Ischnura elegans* (Odonata: Zygoptera). Freshwat. Biol. 8:377-384.
3991. Thompson, D. J. 1978b. Prey size selection by larvae of the damselfly *Ischnura elegans* (Odonata). J. Anim. Ecol. 47:769-785.
3992. Thompson, D. J. 1987. Regulation of damselfly populations: the effects of weed density on larval mortality due to predation. Freshwat. Biol. 17:367-371.
3993. Thompson, P. L., and F. Bärlocher. 1989. Effect of pH on leaf breakdown in streams and in the laboratory. J. N. Am. Benthol. Soc. 8:203-210.
3994. Thomsen, L. 1937. Aquatic Diptera. Part V. Ceratopogonidae. Mem. Cornell Univ. Agric. Exp. Sta. 210:57-80.
3995. Thornton, K., and J. Wilhm. 1974. The effects of pH, phenol, and sodium chloride on survival and caloric, lipid, and nitrogen content of a laboratory population of *Chironomus attenuatus* (Walk.). Hydrobiologia 45:261-280.
3996. Thorp, J. H., and M. L. Cothran. 1984. Regulation of freshwater community structure at multiple intensities of dragonfly predation. Ecology 65:1546-1555.
3997. Thorp, J. H., and A. P. Covich. 1991. Ecology and classification of North American freshwater invertebrates. Academic Press, San Diego, CA. 911p.
3998. Thorpe, W. H. 1931. The biology of the petroleum fly. Science 73:101-103.
3999. Thorpe, W. H. 1950. Plastron respiration in aquatic insects. Biol. Rev. 25:344-390.
4000. Thorpe, W. H., and D. J. Crisp. 1947a. Studies on plastron respiration. I. The biology of *Aphelocheirus*, [Hemiptera, Aphelocheiridae (Naucoridae)] and the mechanism of plastron retention. J. exp. Biol. 24:227-269.
4001. Thorpe, W. H., and D. J. Crisp. 1947b. Studies on plastron respiration. II. The respiratory efficiency of the plastron of *Aphelocheirus*. J. exp. Biol. 24:270-303.
4002. Thorpe, W. H., and D. J. Crisp. 1947c. Studies on plastron respiration. III. The orientation responses of *Aphelocheirus*, [Hemiptera, Aphelocheiridae (Naucoridae)] in relation to plastron respiration; together with an account of specialized pressure receptors in aquatic insects. J. exp. Biol. 24:310-328.
4003. Thorpe, W. H., and D. J. Crisp. 1949. Studies on plastron respiration. IV. Plastron respiration in the Coleoptera. J. exp. Biol. 26:219-260.
4004. Thorup, J., and T. M. Iversen. 1974. Ingestion by *Sericostoma personatum* Spence (Trichoptera: Sericotomatidae). Arch. Hydrobiol. 74:39-47.
4005. Thut, R. N. 1969a. A study of the profundal bottom fauna of Lake Washington. Ecol. Monogr. 39:77-100.
4006. Thut, R. N. 1969b. Feeding habits of larvae of seven *Rhyacophila* species with notes on other life-history features. Ann. Ent. Soc. Am. 62:894-898.
4007. Thut, R. N. 1969c. Feeding habits of stonefly nymphs of the suborder Setipalpia. Rept. Weyerhaeuser Co., Longview, Wash. 4 pp.
4008. Tietze, N. S., and M. S. Mulla. 1989. Species composition and distribution of blackflies (Diptera: Simuliidae) in the Santa Monica Mountains, California. Bull. Soc. Vect. Ecol. 14:253-261.
4009. Tillyard, R. J. 1917. The biology of dragonflies (Odonata or Paraneuroptera). Cambridge Univ. Press. Cambridge. 396 pp.
4010. Tillyard, R. J. 1932. Kansas Permian insects 15. The order Plectoptera. Am. J. Sci. 23:97-134, 237-242.
4011. Tindall, A. R. 1960. The larval case of *Triaenodes bicolor* Curtis (Trichoptera: Leptoceridae). Proc. R. ent. Soc. Lond. (A) 35:93-96.
4012. Tindall, J. J., and W. P. Kovalak. 1979. Food particle sizes consumed by larval *Glossosoma nigrior* (Trichoptera: Glossosomatidae). Great Lakes Ent. 12:105-108.
4013. Titmus, G., and R. M. Badcock. 1981. Distribution and feeding of larval Chironomidae in a gravel-pit lake. Freshwat. Biol. 11:263-271.
4014. Tjonneland, A. 1960. The flight activity of mayflies as expressed in some East African species. Univ. Gergen Arbok Met. Naturv. Serv. 1:1-88.
4015. Tkac, M. A., and B. A. Foote. 1978. Annotated list of stoneflies (Plecoptera) from Stebbins Gulch in northeastern Ohio. Great Lakes Ent. 11:139-142.
4016. Todd, E. L. 1955. A taxonomic revision of the family Gelastocoridae. Univ. Kans. Sci. Bull. 37:277-475.
4017. Todd, E. L. 1961. A checklist of the Gelastocoridae (Hemiptera). Proc. Hawaii. Ent. Soc. 17:461-467.
4018. Tokeshi, M. 1991. On the feeding habits of *Thienemannimyia festiva* (Diptera: Chironomidae). Aquat. Insects 13:9-16.
4019. Torre-Bueno, J. R. de la. 1903. Brief notes toward the life history of *Pelecoris femorata Pal. B.* with a few remarks on habits. J. N.Y. Ent. Soc. 11:166-173.
4020. Torre-Bueno, J. R. de la. 1906. Life history of *Ranatra quadridentata*. Can. Ent. 38:242-252.
4021. Torre-Bueno, J. R. de la. 1910. Life histories of North American water bugs. III. *Microvelia americana* Uhler. Can. Ent. 42:176-186.
4022. Torre-Bueno, J. R. de la. 1917a. Aquatic Hemiptera. A study in the relation of structure to environment. Ann. Ent. Soc. Am. 9:353-365.
4023. Torre-Bueno, J. R. de la. 1917b. Life history of the northern *Microvelia—Microvelia borealis* Bueno. Ent. News 28:354-359.
4024. Torre-Bueno, J. R. de la. 1926. The family Hydrometridae in the western hemisphere. Entomologica Am. 7:83-128.
4025. Torre-Bueno, J. R. de la. 1937. A glossary of entomology. Science, Lancaster, Penn. 336 pp.
4026. Toth, R. S., and R. M. Chew. 1972a. Development and energetics of *Notonecta undulata* during predation on *Culex tarsalis*. Ann. Ent. Soc. Am. 65:1270-1279.
4027. Toth, R. S., and R. N. Chew. 1972b. Notes on behavior and colonization of *Buenoa scimitra*, a predator of mosquito larvae. Environ. Ent. 1:534-545.
4028. Townes, H. K. 1937. Studies on the food organisms of fish. Ann. Rept. N.Y. State Cons. Dep. 27 Suppl. 222:162-175.
4029. Townes, H. K. 1945. The Nearctic species of Tendipedini (Diptera: Tendipedidae). Am. Midl. Nat. 34:1-206.
4030. Townes, H. K. 1957. Nearctic wasps of the subfamilies Pepsinae and Ceropalinae. Bull. U.S. Nat. Mus. 209. 286 pp.
4031. Townes, H. K. 1962. Design for a malaise trap. Proc. Ent. Soc. Wash. 64:253-262.
4032. Townes, H. K. 1969, 1970, 1971. The genera Ichneumonidae. Parts 1, 2, 4. Mem. Am. Ent. Inst. 11:300 pp.; 12:537 pp.; 17:372 pp.

4033. Townes, H. K. 1972. A light-weight malaise trap. Ent. News 83:239-247.
4034. Towns, D. R. 1981. Life histories of benthic invertebrates in a kauri forest stream in northern New Zealand. Aust. J. Mar. Freshwat. Res. 32:191-211.
4035. Townsend, L. H. 1939. A new species of *Sialis* (Megaloptera: Sialidae) from Kentucky. Proc. Ent. Soc. Wash. 4:224-226.
4036. Tozer, W. 1979. Underwater behavioral thermoregulation in the adult stonefly, *Zapada cinctipes*. Nature 281:566-567.
4037. Tozer, W. E., V. H. Resh, and J. O. Solem. 1981. Bionomics and adult behavior of a lentic caddisfly, *Nectopsyche albida* (Walker). Am. Midl. Nat. 106:133-144.
4038. Tracy, B. H., and D. H. Hazelwood. 1983. The phoretic association of *Urnatella gracilis* (Entoprocta: Urnatellidae) and *Nanocladius downesi* (Diptera: Chironomidae) on *Corydalus cornutus* (Megaloptera: Corydalidae). Freshwat. Invertebr. Biol. 2:186-191.
4039. Trama, F. B. 1972. The transformation of energy by an aquatic herbivore *Stenonema pulchellum* (Ephemeroptera). Pol. Arch. Hydrobiol. 19:113-121.
4040. Traver, J. R., and G. F. Edmunds, Jr. 1967. A revision of the genus *Thraulodes* (Ephemeroptera: Leptophlebiidae). Misc. Publ. Ent. Soc. Am. 5:349-395.
4041. Travers, S. E. 1993. Group foraging facilitates food finding in a semiaquatic hemipteran, *Microvelia austrina* Bueno (Hemiptera: Veliidae). Pan-Pac. Ent. 69:117-121.
4042. Treat, A. E. 1954. *Acentropus niveus* in Massachusetts, remote from water. Lep. News 8:23-25.
4043. Treat, A. E. 1955. Flightless females of *Acentropus niveus* reared from Massachusetts progenitors. Lep. News 9:69-73.
4044. Trehen, P. 1969. Description des stades preimaginaux et donnees sur la biologie de *Phyllodromia melanocephala* Fabricius,1794 (Dipteres-Empididae). Rev. Ecol. Biol. Sol. 6:41-52.
4045. Treherne, J. R. 1951. The respiration of the larva of *Helodes minuta* (Col.). Proc. 9th Int. Congr. Ent. 1:311-314.
4046. Trelka, D. G., and B. A. Foote. 1970. Biology of slugkilling *Tetanocera* (Diptera: Sciomyzidae). Ann. Ent. Soc. Am. 63:877-895.
4047. Trelka, D. G., and C. O. Berg. 1977. Behavioural studies of the slug-killing larvae of two species of *Tetanocera* (Diptera: Sciomyzidae). Proc. Ent. Soc. Wash. 79:475-486.
4048. Trimble, R. M. 1972. Occurrence of *Culiseta minnesotae* and *Aedes trivittatus* in Manitoba, including a list of mosquitoes from Manitoba. Can. Ent. 104:1535-1537.
4049. Trost, L.M.W., and L. Berner. 1963. The biology of *Callibaetis floridanus* Banks (Ephemeroptera: Baetidae). Fla. Ent. 46:285-299.
4050. Trottier, R. 1971. Effect of temperature on the life-cycle of *Anax junius* (Odonata: Aeshnidae) in Canada. Can. Ent. 103:1671-1683.
4051. Truxal, F. S. 1949. A study of the genus *Martarega* (Hemiptera: Notonectidae). J. Kans. Ent. Soc. 22:1-24.
4052. Truxal, F. S. 1953. A revision of the genus *Buenoa*. Univ. Kans. Sci. Bull. 35:1351-1523.
4053. Tsou, Y. H. 1914. The body setae of lepidopterous larvae. Trans. Am. Miscrosc. Soc. 33:223-260.
4054. Tsuda, S. 1991. A distributional list of world Odonata. Osaka, Japan. 362pp.
4055. Tsui, P.T.P., and M. D. Hubbard. 1979. Feeding habits of the predaceous nymphs of *Dolania americana* in northwestern Florida (Ephemeroptera: Behningiidae). Hydrobiologia 67:119-123.
4056. Tubb, R. A., and T. C. Dorris. 1965. Herbivorous insect populations in oil refinery effluent holding pond series. Limnol. Oceanogr. 10:121-134.
4057. Tuchman, N. C., and R. H. King. 1993. Changes in mechanisms of summer detritus processing between wooded and agricultural sites in a Michigan headwater stream. Hydrobiologia 268:115-127.
4058. Tucker, E. S. 1912. The rice water-weevil and methods of its control. U.S. Dept. Agric. Circ. 152. 20 pp.
4059. Turnbow, R., Jr., and C. L. Smith. 1983. An annotated checklist of the Hydradephaga (Coleoptera) of Georgia. J. Georgia Ent. Soc. 18:429-443.
4060. Tuskes, P. M. 1977. Observations on the biology of *Paragyractis confusalis*, an aquatic pyralid (Lepidoptera: Pyrelidae). Can. Ent. 109:695-699.
4061. Tuskes, P. M. 1981. Factors influencing the abundance and distribution of two aquatic moths of the genus *Parargyractis* (Pyralidae). J. Lepid. Soc. 35:161-168.
4062. Tuxen, S. L. 1944. The hot springs, their animal communities and their zoogeographical significance. Zool. Iceland 1:1-206.
4063. Tuxen, S. L. 1970. Taxonomist's glossary of genitalia in insects. Munksgaard, Copenhagen. 359 pp.
4064. Twinn, C. R. 1936. The blackflies of eastern Canada (Simuliidae, Diptera). Can. J. Res. 14:97-150.
4065. Twinn, C. R. 1938. Blackflies from Utah and Idaho, with descriptions of new species (Simuliidae, Diptera). Can. Ent. 70:48-55.
4066. Ubukata, H. 1981. Survivorship curve and annual fluctuation in the size of emerging population of *Cordulia aenea amurensis* Selys (Odonata: Corduliidae). Jap. J. Ecol. 31:335-346.
4067. Uchida, H. 1971. Tentative key to the Japanese genera of Collembola, in relation to the world general of this order. I. Sci. Rep. Hirosaki Univ. 18:64-76.
4068. Uchida, H. 1972a. Tentative key to the Japanese genera of Collembola, in relation to the world genera of this Order. II. Sci. Rep. Hirosaki Univ. 19:19-42.
4069. Uchida, H. 1972b. Tentative key to the Japanese genera of Collembola, in relation to the world genera of this order. III. Sci. Rep. Hirosaki Univ. 19:79-114.
4070. Ulfstrand, S. 1968. Life cycles of benthic insects in Lapland streams (Ephemeroptera, Plecoptera, Trichoptera, Diptera: Simuliidae). Oikos 19:167-190.
4071. Ulrich, G. W. 1986. The larvae and pupae of *Helichus suturalis* Leconte and *Helichus productus* Leconte (Coleoptera: Dryopidae). Coleop. Bull. 40:325-334.
4072. Unzicker, J. D., L. Aggus, and L. O. Warren. 1970. A preliminary list of the Arkansas Trichoptera. J. Ga. Ent. Soc. 5:167-174.
4073. Unzicker, J. D., V. H. Resh, and J. C. Morse. 1982. Trichoptera, pp. 9.24-9.124. In A. R. Brigham, W. U. Brigham, and A. Gnilka (eds.). Aquatic insects and oligochaetes of North Carolina. Midwest Aquatic Enterprises, Mahomet, Ill. 837 pp.
4074. Usinger, R. L. (ed.). 1956a. Aquatic insects of California. Univ. Calif. Press, Berkeley. 508 pp.
4075. Usinger, R. L. 1956b. Aquatic Hemiptera, pp. 182-228. In R. L. Usinger (ed.). Aquatic insects of California. Univ. Calif. Press, Berkeley. 508 pp.
4076. Usinger, R. L. 1941. Key to the subfamilies of Naucoridae with a generic synopsis of the new subfamily Ambrysinae. Ann. Ent. Soc. Am. 34:5-16.
4077. Usinger, R. L. 1945. Notes on the genus *Cryptostemma* with a new record for Georgia and a new species from Puerto Rico. Ent. News 56:238-241.
4078. Usinger, R. L. 1946. Notes and descriptions of *Ambrysus* Stål, with an account of the life history of *Ambrysus mormon* Montd. Univ. Kans. Sci. Bull. 31:185-210.
4079. Usinger, R. L., and P. R. Needham. 1956. A drag-type rifflebottom sampler. Progr. Fish Cult. 18:42-44.
4080. Usseglio-Polatera, P. 1994. Theoretical habitat templets, species traits, and species richness: aquatic insects in the Upper Rhône and its floodplain. Freshwat. Biol. (in press.)
4081. Uutala, A. J. 1986. Paleolimnological assessment of the effects of lake acidification on Chironomidae (Diptera) assemblages in the Adirondack region of New York. Ph.D. diss., State Univ. of New York College of Environmental Science and Forestry, Syracuse. 156 pp.
4082. Uutala, A. J. 1990. *Chaoborus* (Diptera: Chaoboridae) mandibles-paleolimnological indicators of the historical status of fish populations in acid-sensitive lakes. J. Paleolimnol. 4:139-151.
4083. Vaillant, F. 1951. Un empidide destructeur de simulies. Bull. Soc. Zool. France 76:371-379.
4084. Vaillant, F. 1952. *Kowarzia barbatula* Mik et *Dolichocephala ocellata* Costa, deux empidides a larves hygropétriques (Dipteres). Bull. Soc. Zool. France 77:286-291.
4085. Vaillant, F. 1953. *Hemerodromia seguyi*, nouvel empidide d'Algerie destructeur de simulies. Hydrobiologia 5:180-188.

4086. Vaillant, F. 1967. Diptera: Dolichopodidae, Empididae, pp. 401-409. *In* J. Illies (ed.). Limnofauna Europaea. Gustav Fischer Verlag, Stuttgart. 474 pp.

4087. Vaillant, F. 1971. 9d. Psychodidae—Psychodinae, pp. 1-48. *In* E. Lindner (ed.). Die Fliegen der Palaearktischen Region. E. Schweizerbart'sche Verlagsbuchhandlung, Stuttgart.

4088. Vaillant, F. 1989. Les Psychodinae dendrolimnophiles et dendrolimnobiontes paléarctiques et néarctiques (Insecta, Diptera, Nematocera, Psychodidae). Spixiana 12:193-208.

4089. Valley, K., and C. O. Berg. 1977. Biology, immature stages, and new species of snail-killing Diptera of the genus *Dictya* (Sciomyzidae). Search Agric., Cornell Agric. Exp. Sta. 7:1-44.

4090. Van Buskirk, J. 1987. Density-dependent population dynamics in larvae of the dragonfly *Pachydiplax longipennis*: a field experiment. Oecologia 72:221-225.

4091. Van Buskirk, J. 1988. Interactive effects of dragonfly predation in experimental pond communities. Ecology 69:857-867.

4092. Van Buskirk, J. 1989. Density-dependent cannibalism in larval dragonflies. Ecology 70:1442-1449.

4093. Van Buskirk, J. 1992. Competition, cannibalism, and size class dominance in a dragonfly. Oikos 65:455-464.

4094. Van Buskirk, J. 1993. Population consequences of larval crowding in the dragonfly *Aeshna juncea*. Ecology 74:1950-1958.

4095. Van Dam, L. 1938. On the utilization of oxygen and regulation of breathing in some aquatic animals. Volharding, Groningen.

4096. Van den Berg, M. A. 1993. *Enallagma glaucum* (Burmeister), a newly recorded predator of the citrus psylla, *Triosa erytreae* (Del Guercio) (Zygoptera: Coenagrionidae; Hemiptera: Triozidae). Not. Odonatol. 4:29-31.

4097. Van Der Valde, G., and T. C. M. Brock. 1980. The life history and habits of *Notiphila brunnipes* Robineau-Desvoidy (Diptera, Ephydridae), an autecological study on a fly associated with nymphaeid vegetations. Tijdschr. Ent. 123:105-127.

4098. Van Der Valde, G., H. J. G. Meuffels, M. Heine, and P. M. Peeters. 1985. Dolichopodidae (Diptera) of a nymphaeid-dominated system in The Netherlands: species composition, diversity, spatial and temporal distribution. Aquat. Insects 7:189-207.

4099. Van Der Velde, G., and R. Hiddink. 1987. Chironomidae mining in *Nuphar lutea* (L.) Sm. (Nymphaeaceae), in a conspectus of contemporary studies in Chironomidae (Diptera). Ent. Scand. 29:255-264.

4100. Van Duzee, E. P. 1917. Catalogue of the Hemiptera of America north of Mexico. Univ. Calif. Publ. Ent. 2:1-902.

4101. Van Dyke, E. C. 1918. New intertidal rock dwelling Coleoptera from California. Ent. News 29:303-308.

4102. Van Emden, F. 1956. The *Georyssus* larva, a hydrophilid. Proc. R. ent. Soc. Lond. (A) 31:20-24.

4103. van Frankenhuyzen, K., and G. H. Geen. 1987. Effects of low pH and nickel on growth and survival of the shredding caddisfly *Clistoronia magnifica* (Limnephilidae). Can. J. Zool. 65:1729-1732.

4104. Van Tassel, E. R. 1963. A new *Berosus* from Arizona, with a key to the Arizona species (Coleoptera, Hydrophilidae). Coleopt. Bull. 17:1-5.

4105. Van Tassell, E. R. 1966. Taxonomy and biology of the subfamily Berosinae of North and Central America and the West Indies (Coleoptera: Hydrophilidae). Ph.D. diss., Catholic University of Amer., Washington, D.C. 329 pp.

4106. Van Urk, G., F. Kerkum, and C. J. Van Leeuwen. 1993. Insects and insecticides in the Lower Rhine. Wat. Res. 27:205-213.

4107. Vandel, A. 1964. Biospéologie. La biologie des animaux cavernicoles. Gauthier-Villars, Paris, 619 pp.

4108. Vannote, R. L., G. W. Minshall, K. W. Cummins, J. R. Sedell, and C. E. Cushing. 1980. The river continuum concept. Can. J. Fish. Aquat. Sci. 37:130-137.

4109. Vannote, R. L., and B. W. Sweeney. 1980. Geographic analysis of thermal equilibria: a conceptual model for evaluating the effect of natural and modified thermal regimes on aquatic insect communities. Am. Nat. 115:667-695.

4110. Vannote, R. L., and B. W. Sweeney. 1985. Larval feeding and growth rate of the stream cranefly *Tipula abdominalis* in gradients of temperature and nutrition. Acad. Nat. Sci. Phil. 137:119-128.

4111. Vargas, L. 1945. Simulidos del Nuevo Mundo. Monogr. Inst. Salub. Enferm. Trop. 1:1-241.

4112. Varley, G. C. 1937. Aquatic insect larvae which obtain oxygen from the roots of plants. Proc. R. ent. Soc. Lond. (A) 12:55-60.

4113. Varvio-Aho, S. 1981. The effects of ecological differences on the amount of enzyme gene variation in Finnish water-strider *(Gerris)* species. Hereditas 94:35-39.

4114. Vaughn, C. C. 1984. Ecology of *Helicopsyche borealis* (Trichoptera: Helicopsychidae) in Oklahoma. Am. Midl. Nat. 113:76-83.

4115. Vaughn, C. C. 1985a. Evolutionary ecology of case architecture in the snailcase caddisfly, *Helicopsyche borealis*. Freshwat. Invertebr. Biol. 4:178-186.

4116. Vaughn, C. C. 1985b. Life history of *Helicopsyche borealis* (Trichoptera: Helicopsychidae) in Oklahoma. Am. Midl. Nat. 113:76-83.

4117. Vaughn, C. C. 1986. The role of periphyton abundance and quality in the microdistribution of a stream grazer, *Helicopsyche borealis* (Trichoptera: Helicopsychidae). Freshwat. Biol. 16:485-493.

4118. Vaughn, C. C. 1987. Substratum preference of the caddisfly *Helicopsyche borealis* (Trichoptera: Helicopsychidae). Hydrobiologia 154:201-205.

4119. Vaught, G. L., and K. W. Stewart. 1974. The life-history and ecology of the stonefly *Neoperla clymene* (Newman) (Plecoptera: Perlidae). Ann. Ent. Soc. Am. 67:167-178.

4120. Veneski, R., and R. K. Washino. 1970. Ecological studies of *Hydrophilus triangularis* in the laboratory and in a rice field habitat, a preliminary report. Proc. Calif. Mosquito Contr. Assoc. 38:95-97.

4121. Venkatesan, P., and S. Sivaraman. 1984. Changes in the functional response of instars of *Diplonychus indicus* Venk. & Rao (Hemiptera: Belostomatidae) in its predation of two species of mosquito larvae of varied size. Entomonograph 9:191-196.

4122. Venkatesh, M. G. 1976. Some observations on the biology of *Brachydeutera longipes* Hendel (Insecta: Diptera: Ephydridae). Sci. Cult. 42:175-176.

4123. Vepsäläinen, K. 1971a. The role of gradually changing daylength in determination of wing length, alary polymorphism and diapause in a *Cerris odontogaster* (Zett.) population in South Finland. Ann. Acad. Sci. Fenn. (A) 183:1-25.

4124. Vepsäläinen, K. 1971b. The roles of photoperiodism and genetie switch in alary polymorphism in *Gerris*. Acta Ent. Fenn. 28:101-102.

4125. Vepsäläinen, K. 1974. Determination of wing length and diapause in waterstriders. Hereditas 77:163-176.

4126. Vepsäläinen, K., and M. Nummelin. 1986. Habitat selection by waterstrider larvae (Heteroptera: Gerridae) in relation to food and imagoes. Oikos 47:374-381.

4127. Verollet, G., and H. Taehet. 1978. A suction sampler for sampling benthic macroinvertebrates in large rivers. Arch. Hydrobiol. 84:55-64.

4128. Vickery, V. R., and D. E. Johnstone. 1970. Generic status of some Nemobiinae (Orthoptera: Gryllidae) in northern North America. Ann. Ent. Soc. Am. 63:1740-1749.

4129. Victor, R., and J. C. Wigwe. 1989. Hoarding - a predatory behaviour of *Sphaerodema nepoides* Fabricius (Heteroptera: Belostomatidae). Arch. Hydrobiol. 116:107-111.

4130. Villa, J. 1980. "Frogflies" from Central and South America with notes on other organisms of the amphibian egg microhabitat. Brenesia 17:49-68.

4131. Vineyard, R. N., and G. B. Wiggins. 1987. Seven new species from North America in the caddisfly genus *Neophylax* (Trichoptera: Limnephilidae). Ann. Ent. Soc. Am. 80:62-73.

4132. Vineyard, R. N., and G. B. Wiggins. 1988. Further revision of the caddisfly family Uenoidae (Trichoptera): evidence for inclusion of Neophylacinae and Thremmatidae. Syst. Ent. 13:361-372.

4133. Vockeroth, J. R. 1967. Diptera Scatophagidae, p. 422. *In* J. Illies (ed.). Liomnofauna Europaea. Gustav Fischer Verlag, Stuttgart. 417 pp.

4134. Vockeroth, J. R. 1987. Chap. 103. Scathophagidae, pp. 1085-1097. *In* J. F. McAlpine, B. V. Peterson, G. E. Shewell, H. J. Teskey, J. R. Vockeroth, and D. M. Wood (coords.). Manual Nearctic Diptera, Vol. 2. Res. Branch, Agr. Can. Mon. 28. 1332p.

4135. Vockeroth, J. R., and F. C. Thompson. 1987. Chap. 52. Syrphidae, pp. 713-743. *In* J. F. McAlpine, B. V. Peterson, G. E. Shewell, H. J. Teskey, J. R. Vockeroth, and D. M. Wood. (coords.). Manual of Nearctic Diptera, Vol. 2. Res. Branch, Agr. Can. Monogr. 28. 1332pp.

4136. Vodopich, D. S., and B. C. Cowell. 1984. Interaction of factors governing the distribution of a predatory aquatic insect. Ecology 65:39-52.

4137. Voelz, N. J., and J. V. Ward. 1992. Feeding habits and food resources of filter feeding Trichoptera in a regulated mountain stream. Hydrobiologia 23:187-196.

4138. Vogel, E. and A. D. Oliver, Jr. 1969. Life history and some factors affecting the population of *Arzama densa* in Louisiana. Ann. Ent. Soc. Am. 16:374-383.

4139. Vogel, S. 1981. Life in moving fluids: the physical biology of flow. Willard Grant Press, Boston, MA. 352p.

4140. Vogel, S. 1988. Life's Devices: The physical world of animals and plants. Princeton Univ. Press. 367pp.

4141. Vogel, S., and M. LaBarbera. 1978. Simple flow tanks for teaching and research. BioScience 28:638-643.

4142. Vogt, G. G., J. U. McGuire, Jr., and A. D. Cushman. 1979. Probable evolution and morphological variation in South American disonychine flea beetles (Coleoptera: Chrysomelidae) and their amaranthaceous hosts. U.S. Dept. Agric. Tech. Bull. 1593:1-148.

4143. Voigt, W. G., and R. Garcia. 1976. Keys to the *Notonecta* nymphs of the West Coast United States (Hemiptera: Notonectidae). Pan-Pacif. Ent. 52:172-176.

4144. Voigts, D. K. 1976. Aquatic invertebrate abundance in relation to changing marsh vegetation. Am. Midl. Nat. 95:313-322.

4145. Vorhies, C. 1909. Studies on the Trichoptera of Wisconsin. Trans. Wisc. Acad. Sci. Arts Lett. 16:647-738.

4146. Voshell, J. R. 1982. Life history and ecology of *Siphlonurus mirus* Eaton (Ephemeroptera: Siphlonuridae) in an intermittant pond. Freshwat. Invert. Biol. 1:17-26.

4147. Voshell, J. R., Jr. 1991. Life cycle of *Simulium jenningsi* (Dipterar: Simuliidae) in Southern West Virginia. Ann. Ent. Soc. Am. 84:1220-1226.

4148. Voshell, J. R., S. W. Hiner and R. J. Layton. 1992. Evaluation of a benthic macroinvertebrate sampler for rock outcrops in rivers. J. Freshwat. Ecol. 7:1-6.

4149. Voshell, J. R., Jr., R. J. Layton, and S. W. Hiner. 1989. Field techniques for determining the effects of toxic substances on benthic macroinvertebrates in rocky-bottomed streams, pp. 134-155. *In* U. M. Cowgill and L. R. Williams (eds.). Aquatic toxicology and hazard assessment: 12th Vol. Am. Soc. Test. Mater. Spec. Tech. Pub. 1027. Am. Soc. Test. Mater., Philadelphia. 444 pp.

4150. Voshell, J. R., and G. M. Simmons, Jr. 1977. An evaluation of artificial substrates for sampling macrobenthos in reservoirs. Hydrobiologia 53:257-269.

4151. Voshell, J. R., Jr., and G. M. Simmons, Jr. 1978. The Odonata of a new reservoir in the southeastern United States. Odonatologica 7:67-76.

4152. Vulinec, K. 1987. Swimming in whirligig beetles (Coleoptera: Gyrinidae) a possible role of the pygidial gland secretion. Coleop. Bull. 41:151-153.

4153. Vulinec, K., and S. A. Kolmes. 1987. Temperature contact rates and interindividual distance in whirligig beetles (Coleoptera: Gyrinidae). J. N. York Ent. Soc. 95:481-496.

4154. Vuori, K.-M. 1994. Rapid behavioural and morphological responses of hydropsychid larvae (Trichoptera, Hydropsychidae) to sublethal cadmium exposure. Environ. Poll. 84:291-299.

4155. Waldbauer, G. 1968. The consumption and utilization of food by insects. Adv. Insect Physiol. 5:229-288.

4156. Walde, S. J., and R. W. Davies. 1984a. Invertebrate predation and lotic prey communities: evaluation of in situ enclosure/exclosure experiments. Ecology 65:1206-1213.

4157. Walde, S. J., and R. W. Davies. 1984b. The effect of intraspecific interference on *Kogotus nonus* (Plecoptera) foraging behavior. Can. J. Zool. 62:2221-2226.

4158. Walde, S. J., and R. W. Davies. 1985. Diel feeding periodicity of two predatory stoneflies (Plecoptera). Can. J. Zool. 63:883-887.

4159. Walde, S. J., and R. W. Davies. 1987. Spatial and temporal variation in the diet of a predaceous stonefly (Plecoptera: Perlodidae). Freshwat. Biol. 17:109-116.

4160. Walentowicz, A. T., and A. J. McLachlan. 1980. Chironomids and particles: A field experiment with peat in an upland stream, pp. 179-185. *In* D. A. Murray (ed.). Chironomidae: Ecology, systematics, cytology and physiology. Pergamon Press, N.Y.

4161. Walker, C. R. 1955. A core sampler for obtaining samples of bottom muds. Progr. Fish Cult. 17:140.

4162. Walker, E. D., D. L. Lawson, R. W. Merritt, W. T. Morgan, and M. J. Klug. 1991. Nutrient dynamics, bacterial populations, and mosquito productivity in tree hole ecosystems and microcosms. Ecology 72:1529-1546.

4163. Walker, E. D., and R. W. Merritt. 1988. The significance of leaf detritus to mosquito (Diptera: Culicidae) productivity from treeholes. Environ. Ent. 17:199-206.

4164. Walker, E. D., and R. W. Merritt. 1991. Behavior of larval *Aedes triseriatus* (Diptera: Culicidae) larvae. J. Med. Ent. 28:581-589.

4165. Walker, E. D., E. J. Olds, and R. W. Merritt. 1988. Gut content analysis of mosquito larvae (Diptera: Culicidae) using DAPI stain and epifluorescence microscopy. J. Med. Ent. 25:551-54.

4166. Walker, E. M. 1912. The North American dragonflies of the genus *Aeshna*. Toronto Stud. Biol. Ser. 11:1-213.

4167. Walker, E. M. 1925. The North American dragonflies of the genus *Somatochlora*. Univ. Toronto Stud. Biol. Ser. 26:1-202.

4168. Walker, E. M. 1928. The nymphs of the *Stylurus* group of the genus *Gomphus* with notes on the distribution of the group in Canada (Odonata). Can. Ent. 60:79-88.

4169. Walker, E. M. 1933. The nymphs of the Canadian species of *Ophiogomphus* Odonata, Gomphidae. Can. Ent. 65:217-229.

4170. Walker, E. M. 1953. The Odonata of Canada and Alaska. Part I, General, Part II. The Zygoptera-damselflies. Vol. I. Univ. Toronto Press, Toronto. 292 pp.

4171. Walker, E. M. 1958. The Odonata of Canada and Alaska. Anisoptera. Vol. 2. Univ. Toronto Press, Toronto. 318 pp.

4172. Walker, E. M., and P. S. Corbet. 1975. The Odonata of Canada and Alaska. Anisoptera, Macromiidae, Corduliidae, Libellulidae. Vol. 3. Univ. Toronto Press, Toronto. 307 pp.

4173. Walker, I. R. 1993. Paleolimnological biomonitoring using freshwater benthic macroinvertebrates, pp. 306-343. *In* D.M. Rosenberg and V.H. Resh (eds.). Freshwater biomonitoring and benthic macroinvertebrates. Chapman and Hall, New York. 488 pp.

4174. Walker, I. R., R. J. Mott, and J. P. Smol. 1991a. Allerød-Younger Dryas lake temperatures from midge fossils in Atlantic Canada. Science 253:1010-1012.

4175. Walker, I. R., J. P. Smol, D. R. Engstrom, and H. J. B. Birks. 1991b. An assessment of Chironomidae as quantitative indicators of past climatic change. Can. J. Fish. Aquat. Sci. 48:975-987.

4176. Walker, T. J. 1971. *Orchelimum carinatum*, a new meadow katydid from the southeastern United States (Orthoptera: Tettigoniidae). Fla. Ent. 54:277-281.

4177. Walkotten, W. J. 1976. An improved technique for freeze sampling streambed sediments. U.S. Dep. Agric. For. Serv. Res. Note PNW-281:1-9.

4178. Wallace, J. B. 1975a. The larval retreat and food of *Arctopsyche*, with phylogenetic notes on feeding adaptations in Hydropsychidae larvae (Trichoptera). Ann. Ent. Soc. Am. 68:167-173.

4179. Wallace, J. B. 1975b. Food partitioning in net-spinning Trichoptera larvae: *Hydropsyche venularis*, *Cheumatopsyche etrona* and *Macronema zebratum* (Hydropsychidae). Ann. Ent. Soc. Am. 68:463-472.

4180. Wallace, J. B., T. F. Cuffney, C. C. Lay, and D. Vogel. 1987. The influence of an ecosystem-level manipulation on prey consumption by a lotic dragonfly. Can. J. Zool. 65:35-40.

4181. Wallace, J. B., T. F. Cuffney, B. S. Goldowitz, K. Chung, and G. J. Lugthart. 1991a. Long-term studies of the influence of invertebrate manipulations and drought on particulate organic matter export from headwater streams. Verh. Int. Verein. Limnol. 24:1676-1680.

4182. Wallace, J. B., T. F. Cuffney, J. R. Webster, G. J. Lugthart, K. Chung, and B. S. Goldowitz. 1991b. Export of fine organic particles from headwater streams: effects of season, extreme discharges, and invertebrate manipulation. Limnol. Oceanogr. 36:670-682.

4183. Wallace, J. B., A. D. Huryn, and G. J. Lugthart. 1991. Colonization of a headwater stream during three years of seasonal insecticidal applications. Hydrobiologia 211:65-76.

4184. Wallace, J. B., G. J. Lugthart, T. F. Cuffney, and G. A. Schurr. 1989. The impact of repeated insecticidal treatments on drift and benthos of a headwater stream. Hydrobiologia 179:135-147.

4185. Wallace, J. B., and D. Malas. 1976a. The fine structure of capture nets of larval Philopotamidae (Trichoptera), with special emphasis on *Dolophilodes distinctus*. Can. J. Zool. 54:1788-1802.

4186. Wallace, J. B., and D. Malas. 1976b. The significance of the elongate, rectangular mesh found in capture nets of fine particle filter feeding Trichoptera larvae. Arch. Hydrobiol. 77:205-212.

4187. Wallace, J. B., and R. W. Merritt. 1980. Filter-feeding ecology of aquatic insects. Ann. Rev. Ent. 25:103-132.

4188. Wallace, J. B., and S. E. Neff. 1971. Biology and immature stages of the genus *Cordilura* (Diptera: Scatophagidae) in the eastern United States. Ann. Ent. Soc. Am. 64:1310-1330.

4189. Wallace, J. B., and J. O'hop. 1979. Fine particle suspension-feeding capabilities of *Isonychia* spp. (Ephemeroptera: Siphlonuridae). Ann. Ent. Soc. Am. 72:353-357.

4190. Wallace, J. B., and J. O'Hop. 1985. Life on a fast pad: waterlily leaf beetle impact on water lilies. Ecology 66:1534-1544.

4191. Wallace, J. B., and H. H. Ross. 1971. Pseudogoerinae: a new subfamily of Odontoceridae (Trichoptera). Ann. Ent. Soc. Am. 64(4):890-894.

4192. Wallace, J. B., and F. F. Sherberger. 1970. The immature stages of *Anisocentropus pyraloides* (Trichoptera: Calamoceratidae). J. Ga. Ent. Soc. 5(4):217-224.

4193. Wallace, J. B., and F. F. Sherberger. 1972. New Nearctic species of *Lepidostoma* in the *vernalis* group from the southern Appalachians (Trichoptera: Lepidostomatidae). Ent. News 83:222-228.

4194. Wallace, J. B., and F. F. Sherberger. 1974. The larval retreat and feeding net of *Macronema carolina* Banks (Trichoptera: Hydropsychidae). Hydrobiologia 45:177-184.

4195. Wallace, J. B., and F. F. Sherberger. 1975. The larval retreat and feeding net of *Macronema transversum* Hagen (Trichoptera: Hydropsychidae). Anim. Behav. 23:592-596.

4196. Wallace, J. B., J. R. Webster, and T. F. Cuffney. 1982. Stream detritus dynamics: regulation by invertebrate consumers. Oecologia 53:197-200.

4197. Wallace, J. B., J. R. Webster, and W. R. Woodall. 1977. The role of filter feeders in flowing waters. Arch. Hydrobiol. 79:506-532.

4198. Wallace, J. B., W. R. Woodall, and F. F. Sherberger. 1970. Breakdown of leaves by feeding of *Peltoperla maria* nymphs (Plecoptera: Peltoperlidae). Ann. Ent. Soc. Am. 63:563-567.

4199. Wallace, J. B., W. R. Woodall, and A. A. Staats. 1976. The larval dwelling-tube, capture net and food of *Phylocentropus placidus* (Trichoptera: Polycentropodidae). Ann. Ent. Soc. Am. 69:149-154.

4200. Wallace, J. R., F. D. Howard, H. E. Hays, and K. W. Cummins. 1992. The growth and natural history of the caddisfly *Pycnopsyche luculenta* (Trichoptera: Limnephilidae). J. Freshwat. Ecol. 7:399-405.

4201. Wallis, J. B. 1933. Revision of the North American species (north of Mexico) of the genus *Haliplus* Latreille. Trans. Roy. Can. Inst. 19:1-76.

4202. Wallis, J. B. 1939a. The genus *Graphoderus* Aube in North America (north of Mexico). Can. Ent. 71:128-130.

4203. Wallis, J. B. 1939b. The genus *Hybius* Er. in North America. Can. Ent. 82:50-52.

4204. Walsh, B. D. 1963. Notes on the Neuroptera. Proc. Ent. Soc. Phila. 15:182-272.

4205. Walsh, J. F. 1985. The feeding behaviour of *Simulium* larvae, and the development, testing and monitoring of the use of larvicides, with special reference to the control of *Simulium damnosum* Theobald s. l. (Diptera: Simuliidae): a review. Bull. Ent. Res. 75:549-594.

4206. Walshe, B. M. 1947a. Feeding mechanisms of *Chironomus* larvae. Nature 160:474.

4207. Walshe, B. M. 1947b. On the function of haemoglobin in *Chironomus* after oxygen lack. J. exp. Biol. 24:329-342.

4208. Walshe, B. M. 1947c. The function of haemoglobin in *Tanytarsus* (Chironomidae). J. exp. Biol. 24:343-351.

4209. Walshe, B. M. 1950. The function of haemoglobin in *Chironomus plumosus* under natural conditions. J. exp. Biol. 27:73-95.

4210. Walshe, B. M. 1951. The feeding habits of certain chironomid larvae (subfamily Tendipedinae). Proc. Zool. Soc. Lond. 121:63-79.

4211. Walsingham, L. 1907. Microlepidoptera. Fauna Hawaiiensis 1:549-640.

4212. Walton, E., Jr. 1980. Invertebrate drift from predator-prey associations. Ecology 61:1486-1497.

4213. Walton, O. E., Jr., S. R. Reice, and R. W. Andrews. 1977. The effects of density, sediment, particle size and velocity on drift of *Acroneuria abnormis* (Plecoptera). Oikos 28:291-298.

4214. Waltz, R. D. 1994. Field recognition of adult *Acentrella* and *Heterocloeon* (Ephemeroptera: Baetidae). Great Lakes Ent. 26:321-323.

4215. Waltz, R. D., and W. P. McCafferty. 1979. Freshwater springtails (Hexapoda: Collembola) of North America. Purdue Univ. Agric. Exp. Sta. Res. Bull. 960. Lafayette, Ind.

4216. Waltz, R. D., and W. P. McCafferty. 1983. *Austrotinodes* Schmid (Trichoptera: Psychomyiidae), a first U.S. record from Texas. Proc. Ent. Soc. Wash. 85:181-182.

4217. Waltz, R. D., and W. P. McCafferty. 1983. The caddisflies of Indiana (Insecta: Trichoptera). Purdue Univ. Agr. Exp. Stn. Bull. 978. 25 pp.

4218. Waltz, R. D., and W. P. McCafferty. 1985. Redescription and new lectotype designation for the type species of *Pseudocloeon*, *P. kraepelini* Klapálek (Ephemeroptera: Baetidae). Proc. Ent. Soc. Wash. 87:800-804.

4219. Waltz, R. D., and W. P. McCafferty. 1987a. Revision of the genus *Cloeodes* Traver (Ephemeroptera: Baetidae). Ann. Ent. Soc. Am. 80:191-207.

4220. Waltz, R. D., and W. P. McCafferty. 1987b. Generic revision of *Cloeodes* and description of two new genera (Ephemeroptera: Baetidae). Proc. Ent. Soc. Wash. 89:177-184.

4221. Waltz, R. D., and W. P. McCafferty. 1987c. New genera of Baetidae for some Nearctic species previously included in *Baetis* Leach (Ephemeroptera: Baetidae). Ann. Ent. Soc. Am. 80:667-670.

4222. Waltz, R. D., and W. P. McCafferty. 1987d. Systematics of *Pseudocloeon*, *Acentrella*, *Baetiella*, and *Liebebiella*, new genus (Ephemeroptera: Baetidae). J. N. York Ent. Soc. 95:553-568.

4223. Waltz, R. D., and W. P. McCafferty. 1989. New species, redescriptions, and cladistics of the genus *Pseudocentroptiloides* (Ephemeroptera: Baetidae). J. N. York Ent. Soc. 97:151-158.

4224. Waltz, R. D., and W. P. McCafferty. 1996. Keys to the Holarctic and selected Oriental genera of larval Baetidae (Ephemeroptera). J. N. Am. Benthol. Soc.

4225. Waltz, R. D., W. P. McCafferty, and J. H. Kennedy. 1985. *Barbaetis*: A new genus of eastern Nearctic mayflies (Ephemeroptera: Baetidae). Great Lakes Ent. 18:161-165.

4226. Waltz, R. D., W. P. McCafferty, and A.G.B. Thomas. 1994. Systematics of the genera *Alainites* n. gen., *Diphetor*, *Indobaetis*, *Nigrobaetis* n. stat., and *Takobia* n. stat. (Ephemeroptera: Baetidae). Bull. Soc. Hist. Nat., Toulouse 130:33-36.

4227. Ward, A. F., and D. D Williams. 1986. Longitudinal zonation and food of larval chironomids (Insecta:Diptera) along the course of a river in temperate Canada. Holarct. Ecol. 9:48-57.

4228. Ward, A. K., C. N. Dahm, and K. W. Cummins. 1985. *Nostoc* (Cyanophyta) productivity in Oregon stream ecosystems: invertebrate influences and differences between morphological types. J. Phycol. 21:223-227.

4229. Ward, G. M., and K. W. Cummins. 1978. Life history and growth pattern of *Paratendipes albimanus* in a Michigan headwater stream. Ann. Ent. Soc. Am. 71:272-284.

4230. Ward, G. M., and K. W. Cummins. 1979. Effects of food quality on growth of a stream detritivore, *Paratendipes albimanus* (Meigen) (Diptera: Chironomidae). Ecology 60:57-64.
4231. Ward, H. B., and G. C. Whipple (eds.). 1918. Fresh-water biology. John Wiley & Sons, N.Y. 1111 pp.
4232. Ward, J. V. 1976. Comparative limnology of differentially regulated sections of a Colorado mountain river. Arch. Hydrobiol. 78:319-342.
4233. Ward, J. V. 1984. Stream regulation of the upper Colorado River: Channel configuration and thermal heterogeneity. Verh. Internat. Verein. Limnol. 22: 1862-1866.
4234. Ward, J. V. 1986. Altitudinal zonation in a Rocky Mountain stream. Arch. Hydrobiol./Suppl. 74: 133-199.
4235. Ward, J. V. 1992. Aquatic insect ecology. 1. Biology and habitat. J. Wiley & Sons, Inc. NY. 438pp.
4236. Ward, J. V. 1994a. Ecology of alpine streams. Freshwat. Biol. 32:277-294.
4237. Ward, J. V. 1994b. The structure and dynamics of lotic ecosystems, pp. 195-218. *In* R. Margalef (ed.). Limnology now: A paradigm of planetary problems. Elsevier Sci., B.V.
4238. Ward, J. V., and B. C. Kondratieff. 1992. An illustrated guide to the mountain stream insects of Colorado. 191pp. Univ. Press of Colorado, Niwot, Colorado.
4239. Ward, J. V., and J. A. Stanford (eds.). 1979. The ecology of regulated streams. Plenum, N.Y. 398 pp.
4240. Ward, J. V., and J. A. Stanford. 1982. Thermal responses in the evolutionary ecology of aquatic insects. Ann. Rev. Ent. 27:97-117.
4241. Ward, J. V., and N. J. Voelz. 1990. Gradient analysis of interstitial meiofauna along a longitudinal stream profile. Stygologia 5: 93-99.
4242. Ward, J. V., H. J. Zimmermann, and L. D. Cline. 1986. Lotic zoobenthos of the Colorado system, pp. 403-423. *In* B. R. Davies and K. F. Walker (eds.). The ecology of river systems. Dr. W. Junk Publs., Dordrecht, Netherlands.
4243. Ward, R. A., and R. F. Darsie, Jr. 1982. Corrections and additions to the publication, Identification and Geographical Distribution of the Mosquitoes of North America, North of Mexico. Mosq. Syst. 14:209-219.
4244. Warren, C. E. 1971. Biology and water pollution control. W. B. Saunders, Philadelphia. 434 pp.
4245. Wartinbee, D. C., and W. P. Coffman. 1976. Quantitative determination of chironomid emergence from enclosed channels in a small lotic ecosystem. Am. Midl. Nat. 95:479-484.
4246. Warwick, W. F. 1980a. Chironomidae (Diptera) responses to 2800 years of cultural influence: a palaeolimnological study with special reference to sedimentation, eutrophication, and contamination processes. Can. Ent. 112:1193-1238.
4247. Warwick, W. F. 1980b. Palaeolimnology of the Bay of Quinte, Lake Ontario: 2800 years of cultural influence. Can. Bull. Fish. Aquat. Sci. 206:1-117.
4248. Warwick, W. F. 1989. Morphological deformities in larvae of *Procladius* Skuse (Diptera: Chironomidae) and their biomonitoring potential. Can. J. Fish. Aquat. Sci. 46:1255-1271.
4249. Warwick, W. F., and N. A. Tisdale. 1988. Morphological deformities in *Chironomus*, *Cryptochironomus*, and *Procladius* larvae (Diptera: Chironomidae) from two differentially stressed sites in Tobin Lake, Saskatchewan. Can. J. Fish. Aquat. Sci. 45:1123-1144.
4250. Washington, H. G. 1984. Diversity, biotic and similarity indices. A review with special relevance to aquatic ecosystems. Wat. Res. 18:653-694.
4251. Washino, R. K. 1969. Progress in biological control of mosquitoes—invertebrate and vertebrate predators. Proc. Calif. Mosquito Contr. Assoc. 27:16-19.
4252. Washino, R. K., and Y. Hokama. 1968. Quantitive sampling of aquatic insects in a shallow-water habitat. Ann. Ent. Soc. Am. 61:785-786.
4253. Wasscher, M. T. 1991. A list of recorded Odonata of Bays Mountain Park, Sullivan County, Tennessee, United States. Not. Odonatol. 3:108-110.
4254. Waters, T. F. 1965. Interpretation of invertebrate drift in streams. Ecology 46:327-334.
4255. Waters, T. F. 1969a. The turnover ratio in production ecology of freshwater invertebrates. Am. Nat. 103:173-185.
4256. Waters, T. F. 1969b. Subsampler for dividing large samples of stream invertebrate drift. Limnol. Oceanogr. 14:813-815.
4257. Waters, T. F. 1972. The drift of stream insects. Ann. Rev. Ent. 17:253-272.
4258. Waters, T. F. 1977. Secondary production in inland waters. Adv. Ecol. Res. 10:91-164.
4259. Waters, T. F. 1979a. Influence of benthos life history upon the estimation of secondary production. J. Fish. Res. Bd. Can. 36:1425-1430.
4260. Waters, T. F. 1979b. Benthic life histories: summary and future needs. J. Fish. Res. Bd. Can. 36:342-45.
4261. Waters, T. F. 1988. Fish production - benthos production relationships in trout streams. Polskie Arch. Hydrobiol. 35:548-561.
4262. Waters, T. F., and G. W. Crawford. 1973. Annual production of a stream mayfly population: a comparison of methods. Limnol. Oceanogr. 18:286-296.
4263. Waters, T. F., and J. C. Hokenstrom. 1980. Annual production and drift of the stream amphipod *Gammarus pseudolimnaeus* in Valley Creek, Minnesota. Limnol. Oceanogr. 25:700-710.
4264. Waters, T. F., and R. J. Knapp. 1961. An improved stream bottom fauna sampler. Trans. Am. Fish. Soc. 90:225-226.
4265. Waters, W. E., and V. H. Resh. 1979. Ecological and statistical features of sampling insect populations in forest and aquatic environments, pp. 569-617. *In* G. P. Patil and M. Rosenzweig (eds.). Contemporary quantitative ecology and related ecometrics. Internat. Coop. Publ. House, Fairland, Md. 695 pp.
4266. Watkins, W. D., D. C. Tarter, M. L. Little, and S. D. Hopkin. 1975. New records of fishflies for West Virginia (Megaloptera: Corydalidae). Proc. W. Va Acad. Sci. 47:1-5.
4267. Way, M. O., and R. G. Wallace. 1989. First record of midge damage to rice in Texas. Southwest. Ent. 14:27-33.
4268. Weatherley, N. S., E. C. Lloyd, S. D. Rundle, and S. J. Omerod. 1993. Management of conifer plantations for the conservation of stream macroinvertebrates. Biol. Conserv. 63:171-176.
4269. Weaver, J. S., III. 1983. The evolution and classification of Trichoptera, with a revision of the Lepidostomatidae and a North American synopsis of this family. Ph.D. diss., Clemson University, Clemson, SC. 411pp.
4270. Weaver, J. S., III. 1984. The evolution and classification of Trichoptera, (part 1): the ground plan of Trichoptera, pp. 413-419. *In* J. C. Morse (ed.). Proc. IV. Int. Symp. Trichoptera, Clemson, Univ., S. C. Dr. W. Junk Publ., The Hague, Netherlands. 486pp.
4271. Weaver, J. S., III. 1988. A synopsis of the North American Lepidostomatidae (Trichoptera). Contrib. Am. Ent. Inst. 24. 141 pp.
4272. Weaver, J. S., III. 1992a. Remarks on the evolution of Trichoptera: a critique of Wiggins and Wichard's classification. Cladistics 8:171-180.
4273. Weaver, J. S., III. 1992b. Further remarks on the evolution of Trichoptera: a reply to Wiggins. Cladistics 8:187-190.
4274. Weaver, J. S., III, and J. C. Morse. 1986. Evolution of feeding and case-making behavior in Trichoptera. J. N. Am. Benthol. Soc. 5:150-158.
4275. Weaver, J. S. and J. L. Sykora. 1979. The *Rhyacophila* of Pennsylvania, with larval descriptions of *R. banksi* and *R. carpenteri* (Trichoptera: Rhyacophilidae). Ann. Carnegie Mus. 48:403-423.
4276. Weaver, J. S., III, J. A. Wojtowicz, and D. A. Etnier. 1981. Larval and pupal descriptions of *Dolophilodes (Fumonta) major* (Banks) (Trichoptera: Philopotomidae). Ent. News 92:85-90.
4277. Webb, D. W. 1969. New species of chironomids from Costello Lake, Ontario (Diptera: Chironomidae). J. Kans. Ent. Soc. 42:91-108.
4278. Webb, D. W. 1977. The Nearctic Athericidae. J. Kans. Ent. Soc. 50:473-495.
4279. Webb, D. W. 1981. Chap. 32. Athericidae, pp. 479-482. *In* J. F. McAlpine, B. V. Peterson, G. E. Shewell, H. J. Teskey, J. R. Vockeroth, and D. M. Wood (coords). Manual of Nearctic Diptera, Vol. I. Res. Branch, Agric. Can. Monogr. 27. Ottawa. 674 pp.

4280. Webb, D. W. 1995. The immature stages of *Suragina concinna* (Williston) (Diptera: Athericidae). J. Kans. Ent. Soc. 67:421-425.

4281. Webb, D. W., and W. U. Brigham. 1982. Aquatic Diptera, pp. 11.1-11.11. *In* A. R. Brigham, W. U. Brigham, and A. Gnilka, eds. Aquatic insects and oligochaetes of the Carolina Piedmont. Duke Power Training Manual. Duke Power, Charlotte, N.C. 837 pp.

4282. Webb, J. L., and H. W. Wells. 1924. Horseflies: biologies and relation to western agriculture. Bull. U.S. Dept. Agric. 1218. 35 pp.

4283. Webb, K. M., and R. W. Merritt. 1987. The influence of diet on the growth of *Stenonema vicarium* (Ephemeroptera: Heptageniidae). Hydrobiologia 153:253-260.

4284. Weber, C. I. (ed.). 1973. Biological field and laboratory methods for measuring the quality of surface waters and effluents. NERC/EPA, Cincinnati. 176 pp.

4285. Webster, J. R., and E. F. Benfield. 1986. Vascular plant breakdown in freshwater ecosystems. Ann. Rev. Ecol. System. 17:567-594.

4286. Wefring, D. R., and J. C. Teed. 1980. Device for collecting replicate artificial substrate samples of benthic invertebrates in large rivers. Progr. Fish-Cult. 42:26-29.

4287. Weiss, H. B., and E. West. 1920. Notes on *Galerucella nymphaea* L. the pond-lily leaf-beetle (Coleoptera). Can. Ent. 52:237-239.

4288. Welch, P. S. 1914a. Habits of the larva of *Bellura melanopyga* Grote (Lepidoptera). Biol. Bull. 27:97-114.

4289. Welch, P. S. 1914b. Observations on the life history and habits of *Hydromyza confluens* Loew (Diptera). Ann. Ent. Soc. Am. 7:135-147.

4290. Welch, P. S. 1915. The Lepidoptera of the Douglas Lake region, northern Michigan. Ent. News 26:115-119.

4291. Welch, P. S. 1916. Contributions to the biology of certain aquatic Lepidoptera. Ann. Ent. Soc. Am. 9:159-187.

4292. Welch, P. S. 1917. Further studies on *Hydromyza confluens* Loew (Diptera). Ann. Ent. Soc. Am. 10:35-45.

4293. Welch, P. S. 1919. The aquatic adaptations of *Pyrausta penitalis* Grt. Ann. Ent. Soc. Am. 12:213-226.

4294. Welch, P. S. 1922. The respiratory mechanisms in certain aquatic Lepidoptera. Trans. Am. Microsc. Soc. 41:29-50.

4295. Welch, P. S. 1924. Observations on the early larval activities of *Nymphula maculalis* Clemens (Lepidoptera). Ann. Ent. Soc. Am. 17:395-402.

4296. Welch, P. S. 1948. Limnological methods. McGraw-Hill, N.Y. 382 pp.

4297. Welch, P. S. 1959. Lepidoptera, pp. 1050-1056. *In* W. T. Edmondson (ed.). Freshwater biology (2nd ed.). John Wiley & Sons. N.Y. 1,248 pp.

4298. Welch, P. S., and G. L. Sehon. 1928. The periodic vibratory movements of the larva of *Nymphula maculalis* Clemens (Lepidoptera) and their respiratory significance. Ann. Ent. Soc. Am. 21:243-258.

4299. Weld, L. H. 1952. Cynipoidea (Hym.) 1905-1950. Privately printed, Ann Arbor, Mich. 351 pp.

4300. Wells, R. M. G., M. J. Hudson, and T. Brittain. 1981. Function of the hemoglobin and the gas bubble in the backswimmer *Anisops assimilis* (Hemiptera: Notonectidae). J. Comp. Physiol. 142:515-522.

4301. Welton, J. S., D. A. Cooling, and M. Ladle. 1982. A comparison of two colonisation samplers with a conventional technique for quantitative sampling of benthic macroinvertebrates in the gravel substratum of an experimental recirculating stream. Int. Rev. ges. Hydrobiol. 67:901-906.

4302. Welton, J. S., M. Ladle, J. A. B. Bass, and R. T. Clarke. 1991. Grazing of epilithic chironomid larvae at two different water velocities in recirculating streams. Arch. Hydrobiol. 121:405-418.

4303. Wene, G. 1940. The soil as an ecological factor in the abundances of aquatic chironomid larvae. Ohio J. Sci. 40:193-199.

4304. Wentworth, C. K. 1922. A scale of grade and class terms for clastic sediments. J. Geol. 30:377-392.

4305. Wesenberg-Lund, C. 1911. Biologische Studien uber netzspinnende Trichopteren-Larven. Int. Revue ges. Hydrobiol. Suppl. 3:1-64.

4306. Wesenberg-Lund, C. 1943. Biglogie der Süsswasserinsekten. Springer, Berlin. 682 pp.

4307. West, L. S. 1929. Life history notes on *Psephenus lecontei* (Coleoptera: Dryopoidea; Psephenidae). Bull. Battle Cr. Coll. 3:3-20.

4308. Westfall, M. J. 1942. A list of the dragonflies (Odonata) taken near Brevard, North Carolina. Ent. News 53:94-100,127-132.

4309. Westfall, M. J. 1952. Additions to the list of dragonflies of Mississippi (Odonata: Anisoptera). Ent. News 63:200-203.

4310. Westfall, M. J. 1953. Notes on Florida Odonata, including additions to the state list. Fla. Ent. 36:165-173.

4311. Westfall, M. J., and K. J. Tennessen. 1979. Taxonomic clarification within the genus *Dromogomphus* Selys (Odonata: Gomphidae). Fla. Ent. 62:266-273.

4312. Westfall, M. J., Jr. 1956. A new species of *Gomphus* from Alabama (Odonata). Quart. J. Fla. Acad. Sci. 19:251-258.

4313. Westfall, M. J., Jr. 1957. A new species of *Telebasis* from Florida (Odonata: Zygoptera). Fla. Ent. 40:19-27.

4314. Westfall, M. J., Jr. 1965. Confusion among species of *Gomphus*. Quart. J. Fla. Acad. Sci. 28:245-254.

4315. Westfall, M. J., Jr. 1974. A critical study of *Gomphus modestus* Needham, 1942, with notes on related species. Odonatologica 3:63-73.

4316. Westfall, M. J., Jr. 1975. A new species of *Gomphus* from Arkansas (Odonata: Gomphidae). Fla. Ent. 58:91-95.

4317. Westfall, M. J., Jr., and K. J. Tennessen. 1973. Description of the nymph of *Lestes inaequalis* (Odonata: Lestidae). Fla. Ent. 56:291-293.

4318. Westfall, M. J., Jr., and R. P. Trogdon. 1962. The true *Gomphus consanguis* Selys (Odonata: Gomphidae). Fla. Ent. 45:29-41.

4319. Westlake, D. F. 1969. Macrophytes, pp. 32-41. *In* R. A. Vollenweider (ed.) A manual on methods for measuring primary production in aquatic environments. IBP Handbook 12. Blackwell Scientific, Oxford. 225 pp.

4320. Wetmore, S. H., R. J. Mackay, and R. W. Newbury. 1990. Characterization of the hydraulic habitat of *Brachycentrus occidentalis*, a filter-feeding caddisfly. J. N. Am. Benthol. Soc. 9:157-169.

4321. Wetzel, R. G. 1983. Limnology (2nd ed.). W. B. Saunders, Philadelphia.

4322. Wevers, M. J., and R. W. Wisseman. 1987. Larval development, substrate preference, and feeding habits of *Polycentropus variegatus* Milne in model stream channels, pp. 263-268. *In* M. Bourmaud and H. Tachet (eds.). Proc. V. Int. Symp. on Trichoptera, Lyon, France. Dr. W. Junk Publ., The Hague, Netherlands.

4323. Whedon, A. D. 1914. Preliminary notes on the Odonata of southern Minnesota. Rept. Minn. State Ent. 77-103.

4324. Whedon, A. D. 1942. Some observations on rearing Odonata in the laboratory. Ann. Ent. Soc. Am. 35:339-342.

4325. Wheeler, A. G. 1973. Studies on the arthropod fauna of alfalfa. IV. Species associated with the crown. Can. Ent. 105:353-366.

4326. Whiles, M. R., J. B. Wallace, and K. Chung. 1993. The influence of *Lepidostoma* (Trichoptera: Lepidostomatidae) on recovery of leaf-litter processing in disturbed headwater streams. Am. Midl. Nat. 130:356-363.

4327. White, D. S. 1976. *Climacia areolaris* (Neuroptera: Sisyridae) in Lake Texoma, Texas and Oklahoma. Ent. News 87:287-291.

4328. White, D. S. 1978a. Life cycle of the riffle beetle, *Stenelmis sexlineata* (Elmidae). Ann. Ent. Soc. Am. 71:121-125.

4329. White, D. S. 1978b. A revision of the Nearctic *Optioservus* (Coleoptera: Elmidae) with descriptions of new species. System. Ent. 3:59-74.

4330. White, D. S. 1978c. Coleoptera (Dryopoidea), pp. 94-99. *In* J. C. Morse, J. W. Chapin, D. D. Herlong, and R. S. Harvey (eds.). Aquatic insects of Upper Three Runs Creek, Savanah River Plant, South Carolina. Part 1: Orders other than Diptera. J. Ga. Ent. Soc. 15:73-101.

4331. White, D. S. 1982. Elmidae pp. 10.99-10.110. *In* A. R. Brigham, W. U. Brigham, and A. Gnilka (eds.). Aquatic insects and oligochaetes of North and South Carolina. Midwest Aquatic Enterprises, Mahomet, Ill. 837 pp.

4332. White, D. S. 1989. Defense mechanisms in riffle beetles (Coleoptera: Dryopoidea). Ann. Ent. Soc. Am. 82:237-241.

4333. White, D. S. 1993. Perspectives on defining and delineating hyporheic zones. J. N. Am. Benthol. Soc. 12:61-69.

4334. White, D. S., and D. E. Jennings. 1973. A rearing technique for various aquatic Coleoptera. Ann. Ent. Soc. Am. 66:1174-1176.

4335. White, H. B. 1989. Dragonflies and damselflies (Odonata) of Acadia National Park and vicinity, Maine. Ent. News 100:89-103.

4336. White, H. B., and W. J. Morse. 1973. Odonata (Dragonflies) of New Hampshire: an annotated list. N.H. Agric. Exp. Sta. 30:1-46.

4337. White, H. B., III, and R. A. Raff. 1970. The nymph of *Williamsonia lintneri* (Hagen) (Odonata: Corduliidae). Psyche 77:252-257.

4338. White, J. H. 1951. Observations on the life history and biology of *Tipula lateralis* Meig. Ann. Appl. Biol. 38:847-858.

4339. White, J. S., and M. A. Brusven. 1994. Terrestrial behavior of a larval caddisfly (*Eocosmoecus schmidi*) in a north Idaho stream. Bull. N. Am.. Benthol. Soc. 11:191.

4340. White, M. M. 1989. Age class and population genetic differentiation in *Pteronarcys proteus* (Plecoptera: Pteronarcyidae). Am. Midl. Nat. 122:242-248.

4341. White, T. R., P. H. Carlson, and R. C. Fox. 1979. Emergence patterns of fall and winter stoneflies (Plecoptera: Filipalpia) in northwestern South Carolina. Proc. Ent. Soc. Wash. 81:379-390.

4342. White, T. R., and R. C. Fox. 1980. Recolonization of streams by aquatic insects following channelization. Rept. Water Resour. Res. Inst. Clemson Univ. 87, pt. 1:120 pp., pt. 2:57 pp.

4343. White, T. R., K. J. Tennessen, R. C. Fox, and P. H. Carlson. 1980. The aquatic insects of South Carolina, Part 1: Anisoptera (Odonata). Bull. S.C. Agric. Exp. Sta. 632:1-153.

4344. White, T. R., K. J. Tennessen, R. C. Fox, and P. H. Carlson. 1983. The aquatic insects of South Carolina, Part II: Zygoptera (Odonata). S. C. Agric. Exp. Sta. Bull. 648:1-72.

4345. Whitehouse, F. C. 1941. British Columbia dragonflies (Odonata) with notes on distribution and habits. Am. Midl. Nat. 26:488-557.

4346. Whiteside, M. C., and C. Lindegaard. 1980. Complementary procedures for sampling small benthic invertebrates. Oikos 35:317-320.

4347. Whiting, E. R., and D. M. Lehmkuhl. 1987a. *Raptoheptagenia cruentata*, gen. nov. (Ephemeroptera: Heptageniidae), new association of the larva previously thought to be *Anepeorus* with the adult of *Heptagenia cruentata* Walsh. Can. Ent. 119:405-407.

4348. Whiting, E. R., and D. M. Lehmkuhl. 1987b. *Acanthomola pubescens,* a new genus and species of Heptageniidae (Ephemeroptera) from western Canada. Can. Ent. 119:409-417.

4349. Whiting, M. F. 1991a. A distributional study of *Sialis* (Megaloptera: Sialidae) in North America. Ent. News. 102:50-56.

4350. Whiting, M. F. 1991b. Scanning electron microscopic study of the male genitalia of the North American genus, *Sialis* (Megaloptera: Sialidae). Great Basin Nat.. 51:404-410.

4351. Whiting, M. F. l991c. New species of *Sialis* from southern California (Megaloptera: Sialidae). Great Basin Nat. 51:411-413.

4352. Whitlock, D. 1982. Dave Whitlock's guide to aquatic trout foods. N. Lyons Books, N.Y. 224 pp.

4353. Whitman, R. L., and W. J. Clark. 1984. Ecological studies of the sand-dwelling community of an East Texas stream. Freshwat. Invertebr. Biol. 3:59-79.

4354. Whitman, R. L., J. M. Inglis, W. J. Clark, and R. W. Clary. 1983. An inexpensive and simple elutriation device for separation of invertebrates from sand and gravel. Freshwat. Invertebr. Biol. 2:159-163.

4355. Wiberg-Larsen, P. 1993. Notes on the feeding biology of *Ecnomus tenellus* (Rambur,1842) (Trichoptera: Ecnomidae). Ent Meddr. 61:29-38.

4356. Wichard, W. 1976. Morphologische Komponenten bei der Osmoregulation von Trichopterenlarven, pp. 171-177. *In* H. Malicky, (ed.). Proc. 1st Int. Symp. on Trichoptera, Lunz Am See (Austria), 1974. Junk, The Hague, Netherlands. 213 pp.

4357. Wichard, W. 1978. Structure and function of the tracheal gills of *Molanna angustata* Curt., pp. 293-296. *In* M. I. Crichton (ed.), Proc. 2nd Internat. Symp. Trichoptera. Junk, The Hague. 359 pp.

4358. Wichard, W., and K. Hauss. 1975. Der Chloridzellenfehlbetrag als Okomorphologischer zeigerwert fur die salinitat von Binnengewassern. Acta Hydrochim. Hydrobiol. 3:347-356.

4359. Wichard, W., and H. Komnick. 1973. Fine structure and function of the abdominal chloride epithelia in caddisfly larvae. Z. Zellforsch. Mikrosk. Anat. 136:579-590.

4360. Wichard, W., and H. Komnick. 1974. Structure and function of the respiratory epithelium in the tracheal gills of stonefly larvae. J. Ins. Physiol. 20:2397-2406.

4361. Wichard, W., P.T.P. Tsui, and A. Dewall. 1975. Chloridzellen der larven von *Caenis diminuta* Walker (Ephemeroptera, Caenidae) bei unterschiedlicher Salinität. Int. Revue ges. Hydrobiol. 60:705-709.

4362. Wickham, H. F. 1894. On some aquatic larvae, with notice of their parasites. Can. Ent. 26:39-41.

4363. Widahl, L. E. 1988. Some morphometric differences between container and pool breeding Culicidae. J. Am. Mosq. Contr. Assoc. 4:76-81

4364. Widahl, L. E. 1992. Flow patterns of suspension feeding mosquito larvae (Diptera: Culicidae). Ann. Ent. Soc. Am. 85:91-95.

4365. Wiederholm, T. 1976. Chironomids as indicators of water quality in Swedish lakes. Naturvårdsverkets Limnologiska Undersökningar 10:1-17.

4366. Wiederholm, T. 1980. Use of benthos in lake monitoring. J. Wat. Poll. Contr. Fed. 52:537-547.

4367. Wiederholm, T. (ed.). 1983. Chironomidae of the holarctic region. Keys and diagnoses. Part 1. Larvae. Ent. Scand. Suppl. 19:1-457.

4368. Wiederholm, T. (ed.). 1986. Chironomidae of the holarctic region. Keys and diagnoses. Part 2. Pupae. Ent. Scand. Suppl. 28:1-482.

4369. Wiederholm, T. (ed.). 1989. Chironomidae of the holarctic region. Keys and diagnoses. Part 3. Adult Males. Ent. Scand. Suppl. 34:1-524.

4370. Wiederholm, T., and L. Eriksson. 1977. Effects of alcohol preservation on the weight of some benthic invertebrates. Zoon 5:29-31.

4371. Wiens, A. P. 1972. Bionomics of the pitcher plant midge *Metriocnemus knabi* Coquillett (Diptera: Chironomidae). Ph.D. diss., University of Manitoba, Winnipeg.

4372. Wiggins, G. B. 1954. The caddisfly genus *Beraea* in North America. Life Sci. Contr. Roy. Ont. Mus. 39:1-18.

4373. Wiggins, G. B. 1956. A revision of the North American caddisfly genus *Banksiola* (Trichoptera: Phryganeidae). Life Sci. Contr. Roy. Ont. Mus. 43:1-12.

4374. Wiggins, G. B. 1959. A method of rearing caddisflies (Trichoptera). Can. Ent. 91:402-405.

4375. Wiggins, G. B. 1960a. A preliminary systematic study of the North American larvae of the caddisfly family Phryganeidae (Trichoptera). Can. J. Zool. 38:1153-1170.

4376. Wiggins, G. B. 1960b. The unusual pupal mandibles in the caddisfly famlly Phryganeidae (Trichoptera). Can. Ent. 92:449-457.

4377. Wiggins, G. B. 1961. The rediscovery of an unusual North American phryganeid, with some additional records of caddisflies from Newfoundland (Trichoptera) Can. Ent. 93:695-702.

4378. Wiggins, G. B. 1962. A new subfamily of phryganeid caddisflies from western North America (Trichoptera: Phryganeidae). Can. J. Zool. 40:879-891.

4379. Wiggins, G. B. 1965. Additions and revisions to the genera of North American caddisflies of the family Brachycentridae with special reference to the larval stages (Trichoptera). Can. Ent. 97:1089-1106.

4380. Wiggins, G. B. 1966. The critical problem of systematics in stream ecology, pp. 55-58. *In* K. W. Cummins, C. A. Tryon, Jr., and R. T. Hartman (eds.). Organism-substrate relationships in streams. Spec. Publ. Pymatuning Lab. Ecol. 4:1-145.

4381. Wiggins, G. B. 1973a. A contribution to the biology of caddisflies (Trichoptera) in temporary pools. Life Sci. Contr. Roy. Ont. Mus. 88:1-28.

4382. Wiggins, G. B. 1973b. New systematic data for the North American caddisfly genera *Lepania, Goeracea* and *Goerita* (Trichoptera: Limnephilidae). Life Sci. Contr. Roy. Ont. Mus. 91:1-33.

4383. Wiggins, G. B. 1973c. Contributions to the systematics of the caddisfly family Limnephilidae (Trichoptera). I. Life Sci. Contr. Roy. Ont. Mus. 94:1-32.

4384. Wiggins, G. B. 1975. Contributions to the systematics of the caddisfly family Limnephilidae (Trichoptera). II. Can. Ent. 107:325-336.

4385. Wiggins, G. B. 1976. Contributions to the systematics of the caddisfly family Limnephilidae (Trichoptera). III, pp. 7-19. *In* H. Malicky (ed.). Proc. 1st Int. Symp. Trichoptera, Lunz am See (Austria), 1974. Junk, The Hague, Netherlands.

4386. Wiggins, G. B. 1977. Larvae of the North American caddisfly genera. Univ. Toronto Press, Toronto. 401 pp.

4387. Wiggins, G. B. 1982. Trichoptera, pp. 599-612. *In* S. P. Parker (ed.). Synopsis and classification of living organisms. McGraw-Hill, N.Y.

4388. Wiggins, G. B. 1992. Comments on the phylogeny of pupation behavior in Trichoptera: a response to Weaver. Cladistics 8:181-185.

4389. Wiggins, G. B. 1995. Larvae of the North American caddisfly genera (2nd ed.). Univ. Toronto Press, Toronto (in press).

4390. Wiggins, G. B., and N. H. Anderson. 1968. Contributions to the systematics of the caddisfly genera *Pseudostenophylax* and *Philocasca* with special reference to the immature stages (Trichoptera: Limnephilidae). Can. J. Zool. 46:61-75.

4391. Wiggins, G. B., and N. A. Erman. 1987. Additions to the systematics and biology of the caddisfly family Uenoidae (Trichoptera). Can. Ent. 119:867-872.

4392. Wiggins, G. B., and D. J. Larson. 1989. Systematics and biology of a new Nearctic genus in the caddisfly family Phryganeidae (Trichoptera). Can. J. Zool. 67:1550-1556.

4393. Wiggins, G. B., and R. J. Mackay. 1978. Some relationships between systematics and trophic ecology in Nearctic aquatic insects, with special reference to Trichoptera. Ecology 59:1211-1220.

4394. Wiggins, G. B., and R. J. Mackay. 1979. Some relationships between systematics and trophic ecology in nearctic aquatic insects, with special reference to Trichoptera. Ecology 59:1211-1220.

4395. Wiggins, G. B., R. J. Mackay, and I. M. Smith. 1980. Evolutionary and ecological strategies of animals in annual temporary pools. Arch. Hydrobiol./Suppl. 58:97-206.

4396. Wiggins, G. B., and J. S. Richardson. 1982. Revision and synopsis of the caddisfly genus *Dicosmoecus* (Trichoptera: Limnephilidae, Dicosmoecinae). Aquat. Ins. 4:181-217.

4397. Wiggins, G. B., and J. S. Richardson. 1987. Revision of the *Onocosmoecus unicolor* group (Trichoptera: Limnephilidae: Dicosmoecinae). Psyche 93:187-217.

4398. Wiggins, G. B., and J. S. Richardson. 1989. Biosystematics of *Eocosmoecus*, a new Nearctic caddisfly genus (Trichoptera: Limnephilidae, Dicosmoecinae). J. N. Am. Benthol. Soc. 8:355-369.

4399. Wiggins, G. B., J. S. Weaver, and J. D. Unzicker. 1985. Revision of the caddisfly family Uenoidae (Trichoptera). Can. Ent. 117:763-800.

4400. Wiggins, G. B., and W. Wichard. 1989. Phylogeny of pupation in Trichoptera, with proposals on the origin and higher classification of the order. J. N. Am. Benthol. Soc. 8:260-276.

4401. Wiggins, G. B., and N. N. Winchester. 1984. A remarkable new caddisfly genus from northwestern North America (Trichoptera, Limnephilidae, Limnephilinae). Can. J. Zool. 62:1853-1858.

4402. Wiggins, G. B., and R. W. Wisseman. 1990. Revision of the North American caddisfly genus *Desmona* (Trichoptera: Limnephilidae). Ann. Ent. Soc. Am. 82:155-161.

4403. Wiggins, G. B., and R. W. Wisseman. 1992. New North American species in the genera *Neothremma* and *Farula*, with hypotheses on phylogeny and biogeography (Trichoptera: Uenoidae). Can. Ent. 124:1063-1074.

4404. Wigglesworth, V. B. 1938. The regulation of osmotic pressure and chloride concentration in the haemolymph of mosquito larvae. J. exp. Biol. 15:235-247.

4405. Wigglesworth, V. B. 1972. The principles of insect physiology. Chapman and Hall, London. 827 pp.

4406. Wilcox, J. A. 1965. A synopsis of the North American Galerucinae (Coleoptera: Chrysomelidae). N.Y. State Mus. Bull. 400:1-226.

4407. Wilcox, R. S. 1979. Sex discrimination in *Gerris remigis*:. Role of a surface wave signal. Science 206:1325-1327.

4408. Wilder, D. D. 1981a. A revision of the genus *Niphogenia* Melander (Diptera: Empididae). Pan-Pacif. Ent. 57:422-428.

4409. Wilder, D. D. 1981b. A revision of the genus *Oreothalia* Melander (Diptera: Empididae). Proc. Ent. Soc. Wash. 83:461-471.

4410. Wilder, D. D. 1981c. A review of the genus *Roederoides* Coquillett with the description of a new species (Diptera: Empididae). Pan-Pacif. Ent. 57:415-421.

4411. Wilding, J. L. 1940. A new square-foot aquatic sampler. Limnol. Soc. Am. Spec. Publ. 4:1-4.

4412. Wiley, E. O. 1981. Phylogenetics: The theory and practice of phylogenetic systematics. J. Wiley and Sons, NY. 439pp.

4413. Wiley, E. O., D. Siegel-Causey, D. R. Brooks, and V. A. Funk. 1991. The compleat cladist. Univ. Kansas. Mus. Nat. Hist. No. 19. 158 pp.

4414. Wiley, G. O. 1922. Life history notes on two species of Saldidae. Univ. Kans. Sci. Bull. 14:301-311.

4415. Wiley, G. O. 1924. On the biology of *Curicta drakei* Hungerford. Ent. News 35:324-331.

4416. Wiley, M. J., and G. L. Warren. 1992. Territory abandonment, theft, and recycling by a lotic grazer: a foraging strategy for hard times. Oikos 63:495- 505.

4417. Wiley, M. J., and S. L. Kohler. 1993. Community responses to epizootic induced collapse of *Glossoma nigrior* populations in Michigan trout streams. Bull. Ecol. Soc. Am. 74:486-487.

4418. Wilhm, J. L., and T. C. Dorris. 1968. Biological parameters of water quality. BioScience 18:447-481.

4419. Wilkey, R. F. 1959. Preliminary list of the Collembola of California. Bull. Dep. Agric. Calif. 48:222-224.

4420. Willem, M. 1907. *Nymphula stratiotata*. Ann. Soc. Ent. Belgique 51:289-290.

4421. Willey, R. L., and H. O. Eiler. 1972. Drought resistance in subalpine nymphs of *Somatochlora semicircularis* Sélys. Am. Midl. Nat. 87:215-221.

4422. Williams, C. E. 1976. *Neurocordulia (Platycordulia) xanthosoma* (Williamson) in Texas (Odonata: Libellulidae: Corduliinae). Great Lakes Ent. 9:63-73.

4423. Williams, C. E. 1977. Courtship display in *Belonia croceipennis* (Selys), with notes on copulation and oviposition (Anisoptera: Libellulidae). Odonatologica 6:283-287.

4424. Williams, C. E. 1979. Observations on the behavior of the nymph of *Neurocordulia xanthosoma* (Williamson) under laboratory conditions (Anisoptera: Corduliidae). Notul. Odonatol. 1:44-46.

4425. Williams, C. E. 1982. The dragonflies of McLennan County, central Texas, United States. Notul. Odonatol. 1:160-161.

4426. Williams, C. E., and S. W. Dunkle. 1976. The larva of *Neurocordulia xanthosoma* (Odonata: Corduliidae). Fla. Ent. 59:429-433.

4427. Williams, D. C., A. T. Read, and K. A. Moore. 1983. The biology and zoogeography of *Helicopsyche borealis* (Trichoptera: Helicopsychidae): a Nearctic representative of a tropical genus. Can. J. Zool. 61:2288-2299.

4428. Williams, D. D. 1981a. Migrations and distributions of stream benthos, pp. 155-207. *In* M. A. Lock and D. D. Williams (eds.). Perspectives in running water ecology. Plenum, N.Y. 430 pp.

4429. Williams. D. D. 1981b. Evaluation of a standpipe corer for sampling aquatic interstitial biotopes. Hydrobiologia 83:257-260.

4430. Williams, D. D. 1984. The hyporheic zone as a habitat for aquatic insects and associated arthropods, pp. 430-455. *In* V. H. Resh and D. M. Rosenberg (eds.). The ecology of aquatic insects. Praeger, New York, NY. 625p.

4431. Williams, D. D. 1987a. A laboratory study of predator-prey interactions of stoneflies and mayflies. Freshwat. Biol. 17:471-490.

4432. Williams, D. D. 1987b. The ecology of temporary waters. Croom Helm Ltd., London. 205p.

4433. Williams, D. D., J. A. Barnes, and P. C. Beach. 1993. The effects of prey profitability and habitat complexity on the foraging success and growth of stonefly (Plecoptera) nymphs. Freshwat. Biol. 29:107-117.

4434. Williams, D. D., and B. W. Feltmate. 1992. Aquatic insects. C.A.B. Int., Wallingford, Oxon, U.K.

4435. Williams, D. D., and I. D. Hogg. 1988. Ecology and production of invertebrates in a Canadian coldwater spring-springbrook system. Holarct. Ecol. 11:41-54.

4436. Williams, D. D. and H.B.N. Hynes. 1974. The occurrence of benthos deep in the substratum of a stream. Freshwat. Biol. 4:233-256.

4437. Williams. D. D., and H.B.N. Hynes. 1976. The recolonization mechanisms of stream benthos. Oikos 27:265-272.

4438. Williams, D. D., and H.B.N. Hynes. 1979. Reply to comments by Exner and Davies on the use of a standpipe corer. Freshwat. Biol. 9:79-80.

4439. Williams, D. D., and G. P. Levens. 1988. Evidence that hunger and limb loss can contribute to stream invertebrate drift. J. N. Am. Benthol. Soc. 7:180-187.

4440. Williams, D. D., A. T. Read, and K. A. Moore. 1983. The biology and zoogeography of *Helicopsyche borealis* (Trichoptera: Helicopsychidae): a Nearctic representative of a tropical genus. Can. J. Zool. 61:2288-2299.

4441. Williams, D. D., A. F. Tavares and E. Bryant. 1987. Respiratory device or camouflage? - A case for the caddisfly. Oikos 50:42-52.

4442. Williams, D. D., A. Tavares-Cromar, D. J. Kushner, and J. R. Coleman. 1993. Colonization patterns and life-history dynamics of *Culex* mosquitoes in artificial ponds of different character. Can. J. Zool. 71:568-578.

4443. Williams, D. D., and N. E. Williams. 1975. A contribution to the biology of *Ironoquia punctatissima* (Trichoptera: Limnephilidae). Can. Ent. 107:829-832.

4444. Williams, D. D., and N. E. Williams. 1981. Some aspects of the life history and feeding ecology of *Dolophilodes distinctus* (Walker) in two Ontario streams. Ser. Ent. 20:433-442.

4445. Williams, D. D., and N. E. Williams. 1982. Morphological and dietary variations in a riverine population of *Pycnopsyche guttifer* (Trichoptera: Limnephilidae). Aquat. Insects 4:21-27.

4446. Williams, F. X. 1938. Biological studies in Hawaiian waterloving insects. Part III. Diptera or flies. A. Ephydridae and Anthomyiidae. Proc. Hawaii Ent. Soc. 10:85-119.

4447. Williams, F. X. 1939. Biological studies in Hawaiian waterloving insects. Part III. Diptera or flies. B. Asteiidae, Syrphidae, and Dolichopodidae. Proc. Hawaii Ent. Soc. 10:281-315.

4448. Williams, F. X. 1944. Biological studies in Hawaiian waterloving insects. Part IV. Lepidoptera or moths and butterflies. Proc. Hawaii Ent. Soc. 12:180-185.

4449. Williams, N. E., and H.B.N. Hynes. 1973. Microdistribution and feeding of the net-spinning caddisflies (Trichoptera) of a Canadian stream. Oikos 24:73-84.

4450. Williams, N. E., and D. D. Williams. 1980. Distribution and feeding records of the caddisflies (Trichoptera) of the Matamek River region Quebec. Can. J. Zool. 57:2402-2412.

4451. Williams, R. W. 1951. Observations on the bionomics of *Culicoides tristratulus* Hoffman with notes on *C. alaskensis* Wirth and other species at Valdez, Alaska, summer 1949 (Diptera, Heleidae). Ann. Ent. Soc. Am. 44:173-440.

4452. Williams, T. R., R. Connolly, H.B.N. Hynes, and W. E. Kershaw. 1961a. Size of particles ingested by *Simulium* larvae. Nature 189:78.

4453. Williams, T. R., R. C. Connolly, H.B.N. Hynes, and W. E. Kershaw. 1961b. The size of particulate material ingested by *Simulium* larvae. Ann. Trop. Med. Parasit. 55:125-127.

4454. Williamson, E. B. 1900. Dragonflies of Indiana. Dept. Geol. Ann. Rept. Nat. Res. Ind. 24:229-333, 1003-1011.

4455. Williamson, E. B. 1903. The dragonflies (Odonata) of Tennessee, with a few records for Virginia and Alabama. Ent. News 14:221-229.

4456. Williamson, E. B. 1917. An annotated list of the Odonata of Indiana. Misc. Publ. Univ. Mich. Mus. Zool. 2:1-13.

4457. Williamson, E. B. 1932. Dragonflies collected in Missouri. Occ. Pap. Univ. Mich. Mus. Zool. 240:1-40.

4458. Willis, L. D., and A. C. Hendricks. 1992. Life history, growth, survivorship, and production of *Hydropsyche slossonae* in Mill Creek, Virginia. J. N. Am. Benthol. Soc. 11:290-303.

4459. Willoughby, L. G., and R. G. Mappin. 1988. The distribution of *Ephemerella ignita* (Ephemeroptera) in streams: The role of pH and food resources. Freshwat. Biol. 19:145-156.

4460. Wilmot, T. R., D. S. Zeller, and R. W. Merritt. 1992. A key to container-breeding mosquitoes of Michigan (Diptera: Culicidae), with notes on their biology. Great Lakes Ent. 25:137-148.

4461. Wilson, C. A. 1958. Aquatic and semiaquatic Hemiptera of Mississippi. Tulane Stud. Zool. 6:115-170.

4462. Wilson, C. A., R. C. Barnes, and H. L. Fulton. 1946. A list of the mosquitoes of Pennsylvania with notes on their distribution and aundance. Mosquito News 6:78-84.

4463. Wilson, C. B. 1923a. Life history of the scavenger water-beetle *Hydrous (Hydrophilus) triangularis,* and its economic importance to fish breeding. Bull. U.S. Bur. Fish. 39:9-38.

4464. Wilson, C. B. 1923b. Water beetles in relation to pond-fish culture, with life histories of those found in fish ponds at Fairport, Iowa. Bull. U.S. Bur. Fish. 39:231-345.

4465. Wilson, D. S., M. Leighton, and D. R. Leighton. 1978. Interference competition in a tropical ripple bug (Hemiptera: Veliidae). Biotropica 10:302-306.

4466. Wilson, R. B. 1967. The Hydrophilidae of Michigan with keys to species of the Great Lakes region. M.S. thesis, Michigan State University. 100 pp.

4467. Wilson, R. S. 1989. The modification of chironomid pupal exuvial assemblages by sewage effluent in rivers within the Bristol Avon catchment, England. Acta Biol. Debr. Oecol. Hung. 3:367-376.

4468. Wilson, R. S., and J. D. McGill. 1982. A practical key to the genera of pupal exuviae of the British Chironomidae. Univ. Bristol. 62 pp.

4469. Wilson, R. S., and P. L. Bright. 1973. The use of chironomid pupal exuviae for characterizing streams. Freshwat. Biol. 3:283-302.

4470. Wilson, S. W., and J. E. McPherson. 1981. Life history of *Megamelus davisi* with description of immature stages. Ann. Ent. Soc. Am. 74:345-350.

4471. Wilton, D. P., L. E. Fetzer, Jr., and R. W. Fay. 1972. Quantitative determination of feeding rates of *Anopheles albimanus* larvae. Mosquito News 32:23-27.

4472. Wilzbach, M. A. 1990. Non-concordance of drift and benthic activity in *Baetis*. Limnol. Oceanogr. 35:945-952.

4473. Wilzbach, M. A., and K. W. Cummins. 1989. An assessment of short-term depletion of stream macrobenthos by drift. Hydrobiologia 185:29-39.

4474. Wilzbach, M. A., K. W. Cummins, and J. D. Hall. 1986. Influence of habitat manipulations on interactions between cutthroat trout and invertebrate drift. Ecology 67:898-911.

4475. Wilzbach, M. A., K. W. Cummins, and R. A. Knapp. 1988. Toward a functional classification of stream invertebrate drift. Verh. Int. Verein. Limnol. 23:1244-1254.

4476. Winchester, N. N. 1993. Feature species: the caddisfly *Sphagnophylax meips*. Arctic Insect News 4:14-17.

4477. Winchester, N. N., G. B. Wiggins, and R. Ring. 1993. The immature stages and biology of the unusual North American caddisfly *Sphagnophylax meips*, with consideration of the phyletic relationships of the genus (Trichoptera: Limnephilidae). Can. J. Zool. 71:1212-1220.

4478. Winget, R. N. 1993. Habitat partitioning among three species of Ephemerelloidea. J. Freshwat. Ecol. 8: 227-233.

4479. Winget, R. N., and F. A. Mangum. 1979. Biotic Condition Index: integrated biological, physical, and chemical stream parameters for management. U.S. Forest Service Intermountain Region, U.S. Department of Agriculture, Ogden, UT. 67 pp.

4480. Winget, R. N., and F. A. Mangum. 1991. Environmental profile of *Tricorythodes minutus* Traver (Ephemeroptera: Tricorythidae) in western United States. J. Freshwat. Ecol. 6:335-344.

4481. Wingfield, C. A. 1939. The function of the gills of mayfly nymphs from different habitats. J. exp. Biol. 16:363-373.

4482. Winterbourn, M. J. 1971a. The life histories and trophic relationships of the Trichoptera of Marion Lake, British Columbia. Can. J. Zool. 49:623-635.

4483. Winterbourn, M. J. 1971b. An ecological study of *Banksiola crotchi* Banks (Trichoptera, Phryganeidae) in Marion Lake, British Columbia. Can. J. Zool. 49:636-645.

4484. Winterbourn, M. J. 1974. The life histories, trophic relations and production of *Stenoperla prasina* (Plecoptera) and *Deleatidium sp.* (Ephemeroptera) in a New Zealand river. Freshwat. Biol. 4:507-524.

4485. Winterbourn, M. J. 1978. The macroinvertebrate fauna of a New Zealand forest stream. N.Z.J. Zool. 5:157-169.

4486. Winterbourn, M. J. 1985. Sampling stream invertebrates, pp. 241-258. *In* R. D. Pridmore and A. B. Cooper (eds.). Biological monitoring in freshwaters, Part. 2. Proc. of a Seminar, Hamilton, Nov 21-23, 1984. Water and Soil Misc. Publ. No. 83, Nat. Water and Soil Conserv. Authority, Wellington, NZ. 191 pp.

4487. Winterbourn, M. J., and N. H. Anderson. 1980. The life history of *Philanisus plebeius* Walker (Trichoptera: Chathamiidae), a caddisfly whose eggs were found in a starfish. Ecol. Ent. 5:293-303.

4488. Winterbourn, M. J., A. G. Hildrew, and A. Box. 1985. Structure and grazing of stone surface organic layers in some acid streams of southern England. Freshwat. Biol. 15:363-374.

4489. Winters, F. E. 1927. Key to the subtribe Helocharae Orchym. (Coleoptera-Hydrophilidae) of Boreal America. Pan-Pacif. Ent. 4:19-29.

4490. Wipfli, M. S., and R. W. Merritt. 1994a. Disturbance to a stream food web by a bacterial larvicide specific to black flies: feeding responses of predatory macroinvertebrates. Freshwat. Biol. 32:91-103.

4491. Wipfli, M. S., and R. W. Merritt. 1994b. Effects of *Bacillus thuringiensis* var. *israelensis* on nontarget benthic insects through direct and indirect exposure. J. N. Am. Benthol. Soc. 13:190-205.

4492. Wirth, E. 1947. Notes on the genus *Thalassomya* Schiner. with descriptions of two new species (Diptera: Tendipedidae). Proc. Hawaii Ent. Soc. 13:117-139.

4493. Wirth, W. 1949. A revision of the clunionine midges with descriptions of a new genus and four new species (Diptera: Tendipedidae). Univ. Calif. Publ. Ent 8:151-182.

4494. Wirth, W. W. 1951a. A new mountain midge from California (Diptera: Deuterophlebiidae). Pan-Pacif. Ent. 27:49-57.

4495. Wirth, W. W. 1951b. A revision of the dipterous family Canaceidae. Occ. Pap. Bishop Mus. Honolulu 20:245-275.

4496. Wirth, W. W. 1952a. The Heleidae of California. Univ. Calif. Publ. Ent. 9:95-266.

4497. Wirth, W. W. 1952b. Three new species of *Systenus* (Diptera, Dolichopodidae), with a description of the immature stages from tree cavities. Proc. Ent. Soc. Wash. 54:236-244.

4498. Wirth, W. W. 1954. A new genus and species of Ephydridae (Diptera) from a California sulfur spring. Wasmann J. Biol. 12:195-202.

4499. Wirth, W. W. 1957. The species of *Cricoptopus* midges living in the blue-green alga *Nostoc* in California (Diptera: Tendipedidae). Pan-Pacif. Ent. 33:121-126.

4500. Wirth, W. W. 1964. A revision of the shore flies of the genus *Brachydeutera* Loew (Diptera: Ephydridae). Ann. Ent. Soc. Am. 57:3-12.

4501. Wirth, W. W. 1971a. *Platygymnopa,* a new genus of Ephydridae reared from decaying snails in North America (Diptera). Can. Ent. 103:266-270.

4502. Wirth, W. W. 1971b. The brine flies of the genus *Ephydra* in North America (Diptera: Ephydridae). Ann. Ent. Soc. Am. 64:357-377.

4503. Wirth, W. W. 1978. New species and records of intertidal biting midges of the genus *Dasyhelea* Kieffer from the Gulf of California (Diptera: Ceratopogonidae). Pac. Insects 18:191-198.

4504. Wirth, W. W. 1987. Chap. 102. Canacidae, pp. 1079-1083. *In* J. F. McAlpine, B. V. Peterson, G. E. Shewell, H. J. Teskey, J. R. Vockeroth, and D. M. Wood. (coords.). Manual of Nearctic Diptera, Vol. 2. Res. Branch, Agr. Can. Monogr. 28. 1332pp.

4505. Wirth, W. W., and W. R. Atchley. 1973. A review of the North American *Leptoconops* (Diptera: Ceratopogonidae). Texas Tech. Univ. Grad. Stud. 5. 57 pp.

4506. Wirth, W. W., and W. L. Grogan, Jr. 1977. Notes on the systematics and biology of the biting midge *Forcipomyia elegantula* Malloch (Diptera: Ceratopogonidae). Proc. Ent. Soc. Wash. 80:94-102.

4507. Wirth, W. W., and W. L. Grogan, Jr. 1988. The predaceous midges of the world (Diptera: Ceratopogonidae; tribe Ceratopogonini). Flora and Fauna Hdbk. No. 4. E.J. Brill. Leiden.

4508. Wirth, W. W., and F. G. Howarth. 1982. The "*Forcipomyia ingrami*" complex in Hawaii (Diptera: Ceratopogonidae). Proc. Hawaiian Ent. Soc. 24:127-151.

4509. Wirth, W. W., W. N. Mathis, and J. R. Vockeroth. 1987. Chap. 98. Ephydridae, pp. 1027-1047. *In* J. F. McAlpine, B. V. Peterson, G. E. Shewell, H. J. Teskey, J. R. Vockeroth, and D. M. Wood. (coords.). Manual of Nearctic Diptera, Vol. 2. Res. Branch, Agr. Can. Monogr. 28. 1332pp.

4510. Wirth, W. W., N. C. Ratanaworabhan, and D. H. Messersmith. 1977. Natural history of Plumbers Island, Maryland. XXII. Biting midges (Diptera: Ceratopogonidae). I. Introduction and key to genera. Proc. Biol. Soc. Wash. 90:615-647.

4511. Wirth, W. W., and A. Stone. 1956. Aquatic Diptera, pp. 372-482. *In* R. L. Usinger (ed.). Aquatic insects of California. Univ. Calif. Press, Berkeley. 508 pp.

4512. Wiseman, S. W., and S. D. Cooper. 1988. The predatory effects of water striders on emerging insects. Verh. Int. Verein. Limnol. 23:2141-2144.

4513. Wisseman, R. W., and N. H. Anderson. 1987. The life history of *Cryptochia pilosa* (Trichoptera: Limnephilidae) in an Oregon coast range watershed, pp. 243-246. *In* M. Bournaud and J. Tachet (eds.). Proc. V. Int. Symp. Trichoptera, Lyon, France. Dr. W. Junk Publ., The Hague, Netherlands.

4514. Wisseman, R. W., and N. H. Anderson. 1991. The life history of *Onocosmoecus unicolor* (Limnephilidae: Dicosmoecinae) in an Oregon coast watershed, pp.159-163. *In* C. Tomaszewski (ed.). Proc. VI. Int. Symp. Trichoptera, Lodz-Zakopane, Poland. Adam Mickiewicz Univ. Press, Poznan, Poland.

4515. Wissinger, S. A. 1988a. Effects of food availability on larval development and inter-instar predation among larvae of *Libellula lydia* and *Libellula luctuosa* (Odonata: Anisoptera). Can. J. Zool. 66:543-549.

4516. Wissinger, S. A. 1988b. Life history and size structure of larval dragonfly populations. J. N. Am. Benthol. Soc. 7:13-28.

4517. Wissinger, S. A. 1989. Seasonal variation in the intensity of competition and predation among dragonfly larvae. Ecology 70:1017-1027.

4518. Wissinger, S. A. 1992. Niche overlap and the potential for competition and intraguild predation between size-structured populations. Ecology 73:1431-1444.

4519. Wissinger, S. A., and J. McGrady. 1993. Intraguild predation and competition between larval dragonflies: direct and indirect effects on shared prey. Ecology 74:207-218.

4520. Wissinger, S. A., and G. Sparks. 1992. Salamander predation counteracts the effects of intraguild predation between two species of pond-dwelling caddisflies. Am. Zool. 32:98.

4521. Wodsedalek, J. E. 1912. Natural history and general behavior of the Ephemeridae nymphs *Heptagenia interpunctata* (Say). Ann. Ent. Soc. Am. 5:31-40.

4522. Wohlschlag, D. E. 1950. Vegetation and invertebrate life in a marl lake. Invest. Ind. Lakes 3:321-372.

4523. Wolcott, L. T., M. B. Griffith, S. A. Perry, and W. B. Perry. 1992. Evaluation of a pump/core sampler in second order streams. J. Freshwat. Ecol. 2:419-424.

4524. Wolf, G. W., and J. F. Matta. 1981. Notes on nomenclature and classification of *Hydroporus* subgenera with the description of a new genus of Hydroporini (Coleoptera: Dytiscidae). Pan-Pacif. Ent. 57:149-175.

4525. Wolfe, G. W., and R. E. Roughley. 1990. A taxonomic phylogenetic and zoogeographic analysis of *Laccornis* Gozis (Coleoptera: Dytiscidae) with the description of Laccornini, a new tribe of Hydroporinae. Quaest. Ent. 26:273-354.

4526. Wolfe, G. W., and P. J. Spangler. 1985. A synopsis of the *Laccornis difformis* species group with a revised key to North American species of *Laccornis* des Gozis (Coleoptera: Dytiscidae). Proc. Biol. Soc. Wash. 98:61-71.

4527. Wood, D. M. 1978. Taxonomy of the Nearctic species of *Twinnia* and *Gymnopais* (Diptera: Simuliidae) and a discussion of the ancestry of the Simuliidae. Can. Ent. 110:1297-1337.

4528. Wood, D. M., and A. Borkent. 1989. Chap. 114. Phylogeny and classification of the Nematocera, pp. 1333-1370. *In* J.F. McAlpine and D.M. Wood (coords.). Manual of Nearctic Diptera, Vol. 3. Res. Branch, Agric. Can. Monogr. 32, Ottawa.

4529. Wood, D. M., P. T. Dang, and R. A. Ellis. 1979. The insects and arachnids of Canada. Part 6. The mosquitoes of Canada. Agric. Can. Publ. 1686:1-390.

4530. Wood, D. M., and D. M. Davies. 1964. The rearing of simuliids. Proc. 12th Int. Congr. Ent. 12:821-823.

4531. Wood, D. M., and D. M. Davies. 1966. Some methods of rearing and collecting blackflies (Diptera: Simuliidae). Proc. Ent. Soc. Ont. 96:81-90.

4532. Wood, D. M., B. V. Peterson, D. M. Davies, and H. Gyorkos. 1963. The black flies (Diptera: Simuliidae) of Ontario. Part II. Larval identification, with descriptions and illustrations. Proc. Ent. Soc. Ont. 93:99-129.

4533. Wood, F. E. 1962. A synopsis of the genus *Dineutus* (Coleoptera: Gyrinidae) in the Western Hemisphere. M.S. thesis, University of Missouri, Columbia. 99 pp.

4534. Wood, J. R. 1988. The bionomics and secondary production of caddisflies in the genus *Gumaga* (Trichoptera: Sericostomatidae) in California. Ph.D. diss., Univ. California, Berkeley.

4535. Wood, J. R., and V. H. Resh. 1991. Morphological and ecological variation in stream and spring populations of *Gumaga nigricula* (McLachlan) in the California (USA) coast ranges, pp. 15-20. *In* C. Tomaszewski (ed.). Proc. VI. Int. Symp. Trichoptera, Lodz-Zakopane, Poland. Adam Mickiewicz Univ. Press, Poznan, Poland.

4536. Wood, K. G. 1956. Ecology of *Chaoborus* (Diptera: Culicidae) in an Ontario lake. Ecology 37:639-643.

4537. Woodall, W. R., Jr., and J. B. Wallace. 1972. The benthic fauna in four small southern Appalachian streams. Am. Midl. Nat. 88:393-407.

4538. Woodiwiss, F. S. 1964. The biological system of stream classification used by the Trent River Board. Chemy. Indy. 83:443-447.

4539. Woodley, N. E. 1989. Chap. 115. Phylogeny and classification of the "Orthorrhaphous" Brachycera, pp. 1371-1395. *In* J. F. McAlpine and D. M. Wood (coords.). Manual of Nearctic Diptera, Vol.3. Res. Branch Agric. Can. Monogr. 32.

4540. Woodrum, J. E., and D. C. Tarter. 1972. The life history of the alderfly *Sialis aequalis* Banks, in an acid mine stream. Am. Midl. Nat. 89:360-368.

4541. Wooldridge, D. P. 1965. A preliminary checklist of the aquatic Hydrophilidae of Illinois. Trans. Ill. State Acad. Sci. 58:205-206.

4542. Wooldridge, D. P. 1966. Notes on nearctic *Paracymus* with descriptions of new species (Coleoptera: Hydrophilidae). J. Kans. Ent. Soc. 39:712-725.

4543. Wooldridge, D. P. 1967. The aquatic Hydrophilidae of Illinois. Trans. Ill. State Acad. Sci. 60:422-431.

4544. Wootton, R. J. 1972. The evolution of insects in freshwater ecosystems, pp. 69-82. *In* R. B. Clark and R. J. Wootton (eds.). Essays in hydrobiology. Univ. of Exeter, Exeter. 136 pp.

4545. Wootton, R. J. 1988. The historical ecology of aquatic insects: an overview. Paleogeography, Paleoclimatology, Paleoecology 62:477-492.

4546. Worswick, J. M., Jr., and M. T. Barbour. 1974. An elutriation apparatus for macroinvertebrates. Limnol. Oceanogr. 19:538-540.

4547. Worthington, C. E. 1878. Miscellaneous memoranda. Can. Ent. 10:15-17.

4548. Worthington, E. B. 1931. Vertical movements of freshwater macroplankton. Int. Revue ges. Hydrobiol. 25:394-436.

4549. Wotton, R. S. 1976. Evidence that blackfly larvae can feed on particles of colloidal size. Nature 261:697.

4550. Wotton, R. S. 1977. The size of particles ingested by moorland stream blackfly larvae (Simuliiudae). Oikos 29:332-335.

4551. Wotton, R. S. 1978a. Life-histories and production of blackflies (Diptera: Simuliidae) in moorland streams in Upper Teesdale, Northern England. Arch. Hydrobiol. 83:232-250.

4552. Wotton, R. S. 1978b. Growth, respiration, and assimilation of blackfly larvae (Diptera: Simuliidae) in a lake-outlet in Finland. Oecologia 33:279-290.

4553. Wotton, R. S. 1978c. The feeding-rate of *Metacnephia tredecimatum* larvae (Diptera: Simuliidae) in a Swedish lake outlet. Oikos 30:121-125.

4554. Wotton, R. S. 1979. The influence of a lake on the distribution of blackfly species (Diptera: Simuliidae) along a river. Oikos 32:368-372.

4555. Wotton, R. S. 1980a. Bacteria as food for blackfly larvae (Diptera: Simuliidae) in a lake - outlet in Finland. Ann. Zool. Fennici. 17:127-130.

4556. Wotton, R. S. 1980b. Coprophagy as an economic feeding tactic in black fly larvae. Oikos 34:282-286.

4557. Wotton, R. S. 1984a. The relationship between food particle size and larval size in *Simulium noelleri* Friederichs. Freshwat. Biol. 14:547-550.

4558. Wotton, R. S. 1984b. The importance of identifying the origins of microfine particles in aquatic systems. Oikos 43:217-221.

4559. Wotton, R. S. 1987. Lake outlet blackflies - the dynamics of filter feeders at very high population densities. Hol. Ecol. 10:65-72.

4560. Wotton, R. S. 1988. Dissolved organic material and trophic dynamics. BioScience 38:172-177.

4561. Wotton, R. S. 1992. Feeding by blackfly larvae (Diptera: Simuliidae) forming dense aggregations at lake outlets. Freshwat. Biol. 27:139-149.

4562. Wotton, R. S. 1994a. The biology of particles in aquatic systems (2nd ed.). Lewis Publ., Boca Raton, FL. 325p.

4563. Wotton, R. S. 1994b. Particulate and dissolved organic matter as food, pp. 235-288. *In* R. S. Wotton (ed.). The biology of particles in aquatic systems (2nd ed.). Lewis Publ., Boca Raton, FL. 325p.

4564. Wotton, R. S., and R. W. Merritt. 1988. Experiments on predation and substratum choice by larvae of the muscid fly, *Limnophora riparia*. Holarc. Ecol. 11:151-159.

4565. Wotton, R. S., M. S. Wipfli, L. Watson, and R. W. Merritt. 1993. Feeding variability among individual aquatic predators in experimental channels. Can. J. Zool. 71:2033-2037.

4566. Wray, D. L. 1967. Insects of North Carolina. Third Suppl. N.C. Dep. Agric. Div. Ent. 4:158-159.

4567. Wright, F. N. 1957. Rearing of *Simulium damnosum* Theobald (Diptera, Simuliidae) in the laboratory. Nature 180:1059.

4568. Wright, J. F., P. D. Armitage, M. T. Furse, and D. Moss. 1988. A new approach to the biological surveillance of river quality using macroinvertebrates. Verh. Int. Verein. Limnol. 23:1548-1552.

4569. Wright, J. F., J. H. Blackburn, D. F. Westlake, and P. D. Armitage. 1991. Anticipating the consequences of river management for the conservation of macroinvertebrates, pp. 137-149. *In* P. J. Boon, P. Calow, and G. E. Petts (eds.). River conservation and management. J. Wiley & Sons, New York. 470p.

4570. Wright, J. F., D. Moss, P. D. Armitage, and M. T. Furse. 1984. A preliminary classification of running-water sites in Great Britain based on macro-invertebrate species and the prediction of community type using environmental data. Freshwat. Biol. 14:221-256.

4571. Wright, M. 1938. A review of the literature on the Odonata of Tennessee. Tenn. Acad. Sci. 13:26-33.

4572. Wright, M. 1946a. A description of the nymph of *Sympetrum ambiguum* (Rambur) with habitat notes. J. Tenn. Acad. Sci. 21:135-138.

4573. Wright, M. 1946b. A description of the nymph of *Agrion dimidiatum* (Burmeister). J. Tenn. Acad. Sci. 21:336-338.

4574. Wright, M., and A. Peterson. 1944. A key to the genera of anisopterous dragonfly nymphs of the United States and Canada (Odonata, suborder Anisopteral). Ohio J. Sci. 44:151-166.

4575. Wrona, F. J. P. Calow, I. Ford, D. J. Baird, and L. Maltby. 1986. Estimating the abundance of stone-dwelling organisms: a new method. Can. J. Fish. Aquat. Sci. 43:2025-2035.

4576. Wrona, F. J., J. M. Culp, and R. W. Davies. 1982. Macroinvertebrate subsampling: A simplified apparatus and approach. Can. J. Fish. Aquat. Sci. 39:1051-1054.
4577. Wrubleski, D. A., and D. M. Rosenberg. 1984 Overestimates of Chironmidae (Diptera) abundance from emergence traps with polystyrene floats. Am. Midl. Nat. 111:145.
4578. Wu, C. 1923. Morphology, anatomy and ethology of *Nemoura*. Bull. Lloyd Lib. 3:1-81.
4579. Wu, Y. F. 1931. A contribution to the biology of *Simulium* (Diptera). Pap. Mich. Acad. Sci. Arts Lett. 13:543-599.
4580. Wurtzbaugh, W. A. 1992. Food web modification by an invertebrate predator in the Great Salt Lake (Utah). Oecologia (Berlin) 89:168-175.
4581. Xiang, V. J., P. Schroder, and J. Schwoerbel. 1984. Phanologie und Nahrung der Larven von *Hydropsyche angustipennis* und *H. siltalai* (Trichoptera: Hydropsychidae) in einem Seeabflur. Arch. Hydrobiol. Suppl. 66:255-292.
4582. Yamamoto, T., and G. B. Wiggins. 1964. A comparative study of the North American species in the caddisfly genus *Mystacides* (Trichoptera: Leptoceridae). Can. J. Zool. 42:1105-1126.
4583. Yamamoto, T., and H. H. Ross. 1966. A phylogenetic outline of the caddisfly genus *Mystacides* (Trichoptera: Leptoceridae). Can. Ent. 98:627-632.
4584. Yonke, T. R. 1991. Chap. 29. Order Hemiptera, pp. 22-65. *In* F. W. Stehr (ed.). Immature insects. Vol. 2. Kendall Hunt, Dubuque, Iowa. 975 pp.
4585. Yoshimoto, C. M. 1976. Synopsis of the genus *Mestocharis* Forster in America north of Mexico (Chalcidoidea: Eulophidae) Can. Ent. 108:755-758.
4586. Yoshimoto, C. M. 1990. A review of the genera of New World Mymoridae (Hymenoptera: Chalcidoidea). Flora and Fauna Hdbk. No. 7. Sandhill Crane Press, Inc., FL. 166pp.
4587. Yoshitake, S. 1974. Studies on the contents of the digestive tract of *Chironomus sp.* on the bottom of Lake Yunoko. Jap. J. Limnol. 35:25-31.
4588. Yoshitake, S., and H. Fukushima. 1984. Interrelation between epilithic or drifting algae and algae contained in the digestive tracts of some aquatic insects. Verh. Int. Verein. Limnol. 22:2838-2844.
4589. Yosii, R. 1960. Studies on the Collembola genus *Hypogastrura*. Am. Midl. Nat. 64:257-281.
4590. Young, A. M. 1966. The culturing of the diving beetle, *Dytiscus verticalis*, in the laboratory for observation of holometabolic development in aquatic insects. Turtox News 44:224-228.
4591. Young, A. M. 1967. Predation in the larvae of *Dytiscus marginalis* Linnaeus. Pan-Pacif. Ent. 43:113-117.
4592. Young, C. W. 1978. Comparison of the crane flies (Diptera: Tipulidae) of two woodlands in eastern Kansas, with a key to the adult crane flies of eastern Kansas. Univ. Kans. Sci. Bull. 51:407-440.
4593. Young, D. K., and J. B. Stribling. 1990. Systematics of the North American *Cyphon collaris* species complex with the description of a new species (Coleoptera: Scirtidae). Proc. Ent. Soc. Wash. 92:194-204.
4594. Young, E. C. 1965. Flight muscle polymorphism in British Corixidae: Ecological observations. J. Anim. Ecol. 34:353-390.
4595. Young, F. N. 1953. Two new species of *Matus*, with a key to the known species and subspecies of the genus (Coleoptera: Dytiscidae). Ann. Ent. Soc. Am. 46:49-55.
4596. Young, F. N. 1954. The water beetles of Florida. Univ. Fla. Stud. Biol. Ser. 5:1-238.
4597. Young, F. N. 1956. A preliminary key to the species of *Hydrovatus* of the Eastern United States (Coleoptera: Dytiscidae). Coleopt. Bull. 10:53-54.
4598. Young, F. N. 1958. Notes on the care and rearing of *Tropisternus* in the laboratory (Coleoptera: Hydrophilidae). Ecology 39:166-167.
4599. Young, F. N. 1960. Notes on the water beetles of Southampton Island in the Canadian Arctic (Coleoptera: Dytiscidae and Haliplidae). Can. Ent. 92:275-278.
4600. Young, F. N. 1961a. Effects of pollution on natural associations of water beetles. Eng. Bull. Purdue Univ. 45:373-380.
4601. Young, F. N. 1961b. Geographical variation in the *Tropisternus mexicanus* (Castelnau) complex (Coleoptera, Hydrophilidae). Proc. 11th Int. Congr. Ent. 11:112-116.
4602. Young, F. N. 1961c. Pseudosibling species in the genus *Peltodytes* (Coleoptera: Haliplidae). Ann. Ent. Soc. Am. 54:214-222.
4603. Young, F. N. 1963. The Nearctic species of *Copelatus* Erickson (Coleoptera: Dytiscidae). Quart. J. Fla. Acad. Sci. 26:56-77.
4604. Young, F. N. 1967. A key to the genera of American bidessine water beetles, with descriptions of three new genera (Coleoptera: Dytiscidae, Hydroporinae). Coleopt. Bull. 21:75-84.
4605. Young, F. N. 1974. Review of the predaceous water beetles of the genus *Anodocheilus* (Coleoptera: Dytiscidae, Hydroporinae). Occ. Pap. Univ. Mich. Mus. Zool. 670:1-128.
4606. Young, F. N. 1978. The new world species of the water beetle genus *Notomicrus* (Noteridae). Syst. Ent. 3:285-293.
4607. Young, F. N. 1979a. A key to Nearctic species of *Celina* with descriptions of new species (Coleoptera: Dytiscidae) *Celina imitatrix, Celina hubbelli, Celina occidentalis, Celina palustris*, new taxa, from North America. J. Kans. Ent. Soc. 52:820-830.
4608. Young, F. N. 1979b. Water beetles of the genus *Suphisellus* Crotch in the Americas north of Colombia (Coleoptera: Noteridae). Southwest. Nat. 24:409-429.
4609. Young, F. N. 1981a. Predaceous water beetles of the genus *Desmopachria*: The *convexa-grana* group (Coleoptera: Dytiscidae). Occ. Pap. Fla. Coll. Arthropods 2:1-11.
4610. Young, F. N. 1981b. Predaceous water beetles of the genus *Desmopachria* the subgenera with descriptions of new taxa (Coleoptera: Dytiscidae). Rev. Biol. Trop. 28:305-321.
4611. Young, F. N. 1985. A key to the American species of *Hydrocanthus* Say, with descriptions of new taxa (Coleoptera: Noteridae). Proc. Acad. Nat. Sci. Philadelphia 137:90-98.
4612. Young, F. N. 1989. A new species of *Agaporomorphus* from Florida, a new Nearctic record for the genus (Coleoptera: Dytiscidae). Fla. Ent. 72:263-264.
4613. Young, F. N., and G. Longley. 1976. A new subterranean aquatic beetle from Texas (Coleoptera: Dytiscidae: Hydroporinae). Ann. Ent. Soc. Am. 69:787-792.
4614. Young, W. C., and C. W. Bayer. 1979. The dragonfly nymphs (Odonata: Anisoptera) of the Guadelupe River Basin, Texas. Texas J. Sci. 31:85-98.
4615. Yount, J. 1966. A method for rearing large numbers of pond midge larvae, with estimates of productivity and standing crop. Am. Midl. Nat. 76:230-238.
4616. Yount, J. D., and G. J. Niemi (eds.). 1990. Recovery of lotic communities following disturbance: theroy and applications. Environ. Managt. 4:515-762.
4617. Zack, R. S. 1983a. Biology and immature stages of *Paracoenia bisetosa* (Coquillett) (Diptera: Ephydridae). Ann. Ent. Soc. Am. 76:487-497.
4618. Zack, R. S. 1983b. Biology and immature stages of *Setacera needhami* Johannsen (Diptera: Ephydridae). Proc. Ent. Soc. Wash. 85:10-25.
4619. Zack, R. S., Jr., and B. A. Foote. 1978. Utilization of algal monocultures by larvae of *Scatella stagnalis*. Environ. Ent. 7:509-511.
4620. Zahar, A. R. 1951. The ecology and distribution of black flies in southeast Scotland. J. Anim. Ecol. 20:33-36.
4621. Zahner, R. 1959. Uber die Bindung der Mitteleuropäichen *Calopteryx*-arten (Odonata) an den Lebensraum des strömenden Wassers. I. Der Anteil der Larven an der Biotopbindung. Int. Revue ges. Hydrobiol. 44:51-130.
4622. Zalom, F. G. 1977. The Notonectidae (Hemiptera) of Arizona. Southwest Natur. 22:327-336.
4623. Zalom, F. G. 1978. A comparison of predation rates and prey handling times of adult *Notonecta* and *Buenoa* (Hemiptera: Notonectidae). Ann. Ent. Soc. Am. 71:143-148.
4624. Zalom, F. G. 1980. Diel flight periodicities of some Dytiscidae (Coleoptera) associated with California rice paddies. Ecol. Ent. 5:183-187.
4625. Zalom, F. G., and A. A. Grigarick. 1980. Predation by *Hydrophilus triangularis* and *Tropisternus lateralis* in California rice fields. Ann. Ent. Soc. Am. 73:167-171.

4626. Zar, J. H. 1984. Biostatistical analysis (2nd ed.). Prentice Hall, Englewood Cliffs, NJ.
4627. Zatwarnicki, T. 1992. A new classification of Ephydridae based on phylogenetic reconstruction (Diptera: Cyclorrhapha). Genus 3:65-119.
4628. Zeigler, D. D., and K. W. Stewart. 1977. Drumming behavior of eleven Nearctic stonefly (Plecoptera) species. Ann. Ent. Soc. Am. 70:495-505.
4629. Zelinka, M., and P. Marvan. 1961. Zur Präzisierung der biologischen Klassifikation der Reinheit fliessender Gewässer. Arch. Hydrobiol. 57:389-407.
4630. Zenger, J. T., and R. W. Baumann. 1995. Revision of the winter stonefly genus *Isocapnia* (Plecoptera: Capniidae) in North America: systematics, phylogeny and zoogeography. Great Basin Nat. (in press).
4631. Zimmerman, E. C. 1978. Insects of Hawaii (Microlepidoptera, Part II). Univ. Press, Hawaii, Honolulu. 9:1-1903.
4632. Zimmerman, J. R. 1960. Seasonal population changes and habitat preferences of the genus *Laccophilus* (Coleoptera: Dytiscidae). Ecology 41:141-152.
4633. Zimmerman, J. R. 1970. A taxonomic revision of the aquatic beetle genus *Laccophilus* (Dytiscidae) of North America. Mem. Ent. Soc. Am. 16:1-275.
4634. Zimmerman, J. R. 1980. Use of multivariate procedures in studies of species problems in the *sculptilis* group of North American *Colymbetes* (Coleoptera: Dytiscidae). Coleopt. Bull. 34:213-226.
4635. Zimmerman, J. R. 1980b. Reticulate evolution in *Colymbetes* (Coleoptera: Dytiscidae). Univ. British Columbia: Vancouver, B.C., Canada. 406 pp.
4636. Zimmerman, J. R. 1981. A revision of the *Colymbetes* of North America (Dytiscidae). Coleopt. Bull 35:1-52.
4637. Zimmerman, J. R. 1982. The *Deronectes* (Coleoptera: Dytiscidae) of the southwestern U.S.A., Mexico and Guatemala. Coleopt. Bull. 36:412-438.
4638. Zimmerman, J. R. 1985. A revision of the genus *Oreodytes* in North America (Coleoptera: Dytiscidae). Proc. Acad. Nat. Sci. Philadelphia 137:99-127.
4639. Zimmerman, J. R., and R. L. Smith. 1975a. The genus *Rhampus* (Coleoptera: Dytiscidae) in North America. Part I. General account of the species. Trans. Am. Ent. Soc. 101:33-123.
4640. Zimmerman, J. R., and R. L. Smith. 1975b. A survey of the *Deronectes* (Coleoptera: Dytiscidae) of Canada, United States and northern Mexico. Trans. Am. Ent. Soc. 101:651-722.
4641. Zimmerman, M. C., and T. E. Wissing. 1978. Effects of temperature on gut-loading and gut-clearing times of the burrowing mayfly *Hexagenia limbata*. Freshwat. Biol. 8:269-277.
4642. Zimmerman, M. C., and T. E. Wissing. 1979. The nutritional dynamics of the burrowing mayfly, *Hexagenia limbata*. pp. 231-257. *In* J. F. Flannagan, and K. E. Marshall (eds.). Advances in Ephemeroptera biology. Plenum, N.Y. 552 pp.
4643. Zimmerman, M. C., T. E. Wissing, and R. P. Rutter. 1975. Bioenergetics of the burrowing mayfly, *Hexagenia limbata* in a pond ecosystem. Verh Int. Verein. Limnol. 19:3039-3049.
4644. Zimmerman, R. H., and E. C. Turner, Jr. 1982. *Cordilura varipes* (Scatophagidae), a predator of *Culicoides variipennis* (Ceratopogonidae). Mosquito News 42:279.
4645. Zischke, J. A., J. W. Arthur, K. J. Nordlie, R. O. Hermanutz, D. A. Standen, and T. P. Henry. 1983. Acidification effects on macroinvertebrates and fathead minnows (*Pimephales promelas*) in outdoor experimental channels. Wat. Res. 17:47-63.
4646. Zismann, L. 1969. A light trap for sampling aquatic organisms. Israel J. Zool. 18:343-348.
4647. Zloty, J., G. Pritchard, and R. Krishnaraj. 1993. Larval insect identification by cellulose acetate gel electrophoresis and its application to life history evaluation and cohort analysis. J. N. Am. Benthol. Soc. 12:270-278.
4648. Zwick, P. 1973. Insecta: Plecoptera. Phylogenetisches System und Katalog. Das Tierreich 94. Walter de Gruyter, Berlin. 465 pp.
4649. Zwick, P. 1977. Australian Blephariceridae (Diptera). Aust. J. Zool. Suppl. Ser. 46:1-121.
4650. Zwick, P. 1981. Diapause development of *Protonemura intricata* (Plecoptera: Nemouridae). Verh. Int. Ver. Limol. 21:1607-611.
4651. Zwick, P. 1991. Biometric studies of the growth and development of two species of *Leuctra* and of *Nemurella picteti* (Plecoptera: Leuctridae and Nemouridae), pp. 515-526. *In* J. Alba-Tercedor and A. Sanchez-Ortega (eds.). Overview and strategies of Ephemeroptera and Plecoptera. Sandhill Crane Press, Inc., Gainesville, FL. 588 pp.
4652. Zwick, P. 1992. Growth and emergence of *Leuctra prima* (Plecoptera): habitat-species interactions. Verh. Int. Verein. Limnol. 24:2886-2890.

INDEX

A

Abdita, 597, 601, 611
abdomen, 572
abdominalis, 549
Abedus, 271, 293
Abedus herberti, 52
Abedus herbti, 62
Abiskomyia, 639, 673, 709, 748
Ablabesmyia, 646, 700, 745
Acalcarella, 663, 751
Acamptocladius, 638, 673, 714, 748
Acanthagrion, 179, 183, 210
Acanthametropodidae, 129, 132, 150, 158-159
Acanthametropus, 129, 132, 159
Acanthamola, 129, 136, 160
Acanthamola pubescens, 129
Acanthocnema, 546
accessory blade, 638
Acentrella, 135, 152, 159
Acentria, 392, 397
Acentria nivea, 387, 390
Acentropus, 33, 397
Acerpenna, 136, 152, 159
Achaetella, 68
Acherontiella, 115
Acigona, 397
Acilius, 65, 415, 418, 423, 463
Acneus, 400, 440, 468
Acricotopus, 673, 714, 748
Acrididae, 212, 214, 216
Acroneuria, 231, 258, 264
Acroneuria abnormis, 90
Acroneuria evoluta, 62
Acroneuria lycorias, 62, 90
Acroneuriinae, 264
Aculeata, 474, 477
Ademon, 474, 482
Adephaga, 65, 399, 401-402, 462
adfrontal sclerites, 389
Adicrophleps, 343, 351
aedeagus, 11, 390
Aedes, 49, 69, 101, 571, 577, 583, 590
Aedes aegypti, 58
Aedes albopictus, 101
Aedes taeniorhynchus, 66
Aegialites, 401, 468
Aegialomyia, 498
aerenchyma, 32
Aeschnidae, 207
Aeshna, 165, 185, 188, 207
Aeshna cyanea, 166
Aeshna grandis, 166
Aeshna umbrosa, 90
Aeshnidae, 47, 61, 75, 165-167, 169, 174, 185, 205, 207
Agabetes, 418, 423, 463
Agabetini, 418, 423
Agabini, 418, 422
Agabinus, 418, 422, 463
Agabus, 65, 415, 418, 423, 463, 474, 482-483
Agabus bipustulatus, 476
Agabus erichsoni, 56

Agapetinae, 341
Agapetus, 64, 341, 351
Agapetus bifidus, 48, 54
Agapetus fuscipes, 54
Agapetus illini, 64
Agapetus occidentis, 64
Agaporomorphus, 415, 423, 463
Agarodes, 347, 383
Agarodes libalis, 64
Agasicles, 455, 471
Agathon, 487, 537
Agkistrocerus, 542
Agnetina, 258, 264
Agnetina capitata, 62
Agraylea, 342, 359
Agrenia, 121, 125
Agriidae, 209
Agrion, 209-210
Agrionidae, 209-210
Agriotypidae, 56, 474, 476, 481
Agriotypus, 56, 68
Agrypnia, 104, 343, 380
Ahautlea, 295
Aino, 209
Alabameubria, 440
Alaskaperla, 235, 255, 266
Albistylus, 206
Aleocharinae, 436
Alisotrichia, 342, 350
Allacma, 123
Allan grab, 15, 18
Allocapnia, 62, 101, 106, 228, 246, 263
Allocapnia nivicola, 62
Allocapnia rickeri, 62
Allocosmoecus, 344, 371
Allognosta, 496
Allomyia, 347, 365
Alloperla, 231, 255, 266
Allotrichoma, 503, 544
Alluaudomyia, 493, 537
Alotanypus, 646, 698, 744
Alternanthera, 471
Alternanthera philoxeroides, 452
Alticinae, 452, 455
Ambiini, 396
Amblopusa, 436, 467
Ambrysinae, 278
Ambrysus, 278, 294
Ameletidae, 60, 129, 132, 150, 159
Ameletus, 44, 60, 129, 132, 150, 159
Amercaenis, 147, 156, 162
Americabrya, 117
ametabolous, 109, 115
Ametor, 429, 432, 466
Ametropodidae, 132, 148, 158, 160
Ametropus, 132, 148, 160
Amiocentrus, 343, 351
Amphiagrion, 179, 181, 210
Amphicosmoecus, 344, 371
Amphinemura, 61, 224, 242, 263
Amphinemura nigritta, 52
Amphinemurinae, 263
amphipneustic, 486
Amphizoa, 403, 408, 411, 462

Amphizoidae, 32, 401, 403, 408, 411, 461-462
Ampumixis, 448, 450, 469
Anabolia, 345, 374
Anacaena, 429, 434, 466
Anacroneuria, 231, 258, 264
Anagapetus, 341, 351
Anagapetus bernea, 64
Anagrus, 477, 483
anal angle, 167
anal claw, 313
anal combs, 389
anal crossing, 169
anal gills, 550
anal lobe, 641
anal lobe fringe, 641
anal lobe macrosetae, 641
anal margin, 167
anal papillae, 592
anal processes, 315
anal prolegs, 313
anal sclerite, 592
anal triangle, 169
anal tubules, 638
anal vein, 169
Analetris, 45, 129, 132, 150, 159
Anatopyniini, 644
anautogenous, 572
Anax, 70, 185, 188, 207
Anax imperator, 50, 71
Anax junius, 61
Anchycteis, 441, 471
Anchytarsus, 441, 471
Anchytarsus bicolor, 66
Ancyronyx, 66, 446, 450, 470
Anepeorus, 129, 136, 152, 161
Anepeorus rusticus, 129
Anepeorus simplex, 129
Angarotipula, 551
Anisocentropus, 348, 351
Anisops, 37
Anisoptera, 36, 40, 61, 164, 166-167, 169, 171, 185, 205-206
annulations, 638
Anodocheilus, 415, 422, 463
Anomalagrion, 210
Anomalura, 210
Anopheles, 46, 571-573, 577, 583, 590
Anoplius, 482
Anoplius depressipes, 474
Anoplodonta, 541
Anreris, 482
antehumeral stripe, 167
antennae, 389, 516, 638
antennal blade, 639
antennal sheath, 515, 592
antennal tubercle, 639-640
antenodal crossveins, 169
antepenultimate, 638
anterior labral region, 639
anterior lamina, 169
anterior prolegs, 639
Anthicidae, 405, 410
Anthomyiidae, 536, 546
Anthopotamus, 129-130, 148, 163

849

Antichaeta, 482-483, 503, 547
Antillocladius, 687, 714, 748
Antocha, 484, 549, 551, 561, 568
Antocha saxicola, 66
Anurida, 117, 125
Anurida ashbyae, 113
Anurida maritima, 113
Anuridella, 117, 125
Anurophorus, 119
anus, 11
Apamea, 389
Apanisagrion, 179, 181, 210
Apatania, 344, 365
Apataniinae, 310, 344, 365
Apatolestes, 498, 542
Apedilum, 668, 730, 751
Aphanta, 482
Aphelocheiridae, 267
Aphelocheirus, 33, 35
Aphrosylus, 540
Aphylla, 45, 165, 192, 194, 206
Aphylla williamsoni, 164
apical dissections, 639
apical planate, 169
apneustic, 30, 35, 486
Apobaetis, 135, 152, 159
Apocrita, 474, 476
Apometriocnemus, 748
apomorphic, 98
Aprostocetus, 483
Apsectrotanypus, 646, 698, 744
Apsilops, 474, 482
Apteraliplus, 413, 462
apterous, 11
apterygota, 108
Aquarius, 268, 275, 293
Araeopidius, 402, 441
Archanara, 389-390, 398, 482
Archaphorura, 115
Archilestes, 176, 210
Archilestes grandis, 50, 61
Archips, 389-390, 398
Archisotoma, 121, 125
Archisotoma besselsi, 113
Arctiidae, 387, 389-390, 392
Arctoconopa, 553, 565, 568
Arctocorisa, 275, 294
Arctocorisa chancea, 275
Arctodiamesa, 747
Arctodiamesa appendiculata, 690
Arctopelopia, 653, 700, 745
Arctopora, 345, 374
Arctopsyche, 340, 356
Arctopsyche grandis, 63
Arctopsyche irrorata, 63
Arctopsychinae, 310, 340, 356
arculus, 167
Arcynopteryx, 237, 249, 265
Areopidius, 441
Argia, 176, 210
Argia vivida, 90
Argiallagma, 211
Argyactini, 396
Argyra, 540
Argyractini, 387
Argyractis subornata, 387
Arigomphus, 194, 196, 206
arista, 517
armata, 207
Arrhopalites, 113, 123
Arthroplea, 136, 152, 161
Arthropleidae, 161
artificial substrates, 19
Arzama, 398
Asheum, 666, 730, 751
Asmeringa, 545
Asobara, 482
Aspidogyrus, 476
Astenophylax, 346
Astoneomyia, 606
Asynarchus, 345, 374
Asyndetus, 540
Atarba, 561

Athericidae, 67, 484-486, 498, 509, 515-516, 524, 530, 540
Atherix, 106, 498, 540
Atherix lantha, 67
Athripsodes, 349
Athyroglossa, 544
Atissa, 545
atmospheric breathers, 30, 32
Atopsyche, 319, 341, 350
Atractelmis, 448, 450, 470
Atrichomelina, 482, 547
Atrichopogon, 493, 537
Attaneuria, 231, 258, 261, 264
Attenella, 140, 156, 161
Atylotus, 500, 542
Auchenhorrhyncha, 269
Auleutes, 458, 472
Austrolimnophila, 563
Austrotinodes, 339, 350
autogenous, 572
Axarus, 668, 735, 751
Axelsonia, 121, 125
Axysta, 545
Aylmeri, 487
Azola caroliniana, 452

B

Bacopa, 396
Bactra, 392
Badiusalis, 57
Baeoctenus, 674, 714, 748
Baetidae, 48, 50, 60, 126-127, 129, 132, 148, 158-159
Baetis, 44, 48, 50, 60, 69, 106, 126-127, 135, 152, 159
Baetis diablis, 152
Baetis magnus, 152
Baetis pygmaeus, 60
Baetis quilleri, 60
Baetis rhodani, 51, 68, 104
Baetis tricaudatus, 126
Baetisca, 130, 148, 162
Baetisca rogersi, 60, 126
Baetiscidae, 60, 130, 148, 158, 162
Baetodes, 135, 152, 159
bag sampler, 13-14
Bagous, 455, 458, 472
Bagous americanus, 455
Balteata, 209
Banksiola, 343, 380
Banksiola crotchi, 64
Barbaetis, 135, 152, 159
Basiaeschna, 185, 188, 207
basistyles, 550
basitarsi, 593
Bathythrix, 482
Baumannella, 239, 252, 265
beach zone, 75
Beardius, 668, 728, 751
Beckidia, 636, 668, 726, 751
behavioral drift, 104
behavioral extraction procedure, 17
Behningiidae, 60, 127, 130, 147, 158, 163
Belgian Biotic Index, 91
Bella, 209
Bellura, 389-390, 398
Bellura gortynoides, 65
Beloneuria, 62, 231, 258, 261, 264
Belonia, 208
Belostoma, 33, 267, 271, 293
Belostoma malkini, 62
Belostomatidae, 32-33, 37, 52, 62, 101, 267, 269, 271, 288, 293-294
Belostomatinae, 271
Beothukus, 343, 380
Beraea, 347, 350
Beraeidae, 310, 313, 315, 322, 326, 333, 338, 347, 350
Berlaise funnels, 17
Berosini, 432
Berosus, 401, 427, 432, 466

Bethbilbeckia, 653, 698, 744
Bezzia, 493, 537
biordinal, 389
Bibiocephala, 487, 537
Bidessini, 419
Bidessonotus, 415, 419, 422, 463
bifid, 639
Biological Condition Index, 91
biomonitoring, 87-88, 93, 96
biomonitoring survey, 95
biotic indices, 88, 92
Bisancora, 235, 255, 266
Bittacomorpha, 487, 539
Bittacomorphella, 487, 539
Blaesoxipha, 503, 536, 546
Bledius, 436, 467
Blepharicera, 487, 537
Blepharicera williamsae, 66
Blephariceridae, 32, 36, 40, 43, 66, 71, 75, 106, 484-487, 509, 515, 519, 524, 537
Blera, 547
Blissia, 119
blood gills, 389
Bohr effect, 37
Bolbodimyia, 542
Bolitocharini, 436
Bolotoperla, 224, 242, 262
Bolshecapnia, 228, 246, 263
Bonetograstrura, 117
Bonetrura, 121
Boreochlini, 690, 738, 746
Boreochlus, 638, 690, 693, 746
Boreoheptagyia, 687, 714, 747
Boreoheptagyini, 687, 738, 747
Bothriovulsus, 121
boundary layer, 30
Bourletiella, 113, 123, 125
box sampler, 13
Boyeria, 185, 188, 207
Boyeria vinosa, 61
Brachinus, 403
Brachybamus, 461, 472
Brachycentridae, 64, 310, 315, 319, 326, 333, 338, 343, 350
Brachycentrus, 44, 64, 69, 72, 310, 343, 350
Brachycentrus occidentalis, 43
Brachycentrus spinae, 64
Brachycera, 485-486, 496, 515, 517, 524
Brachycera-Cyclorrhapha, 542
Brachycera-Orthorrhapha, 540
Brachycercus, 147, 156, 162
Brachydeutera, 508, 543
Brachymesia, 200, 208
Brachypogon, 537
Brachypremna, 551, 558, 565, 568
Brachypterinae, 262
Brachypterous, 390
Brachystomella, 117, 125
Brachyvatus, 415, 419, 463
Bracon, 482
Braconidae, 477, 481-482
branchial chamber, 36
Brasenia, 396
Brechmorhoga, 200, 202, 204, 208
Brennania, 542
Brevistylus, 206
Brillia, 682, 706, 709, 748
Broweri, 387
Brundiniella, 653, 698, 744
Brychius, 413, 462
Bryophaenocladius, 674, 712, 748
Bryothinusa, 436, 467
Buenoa, 37, 278, 295
burrowers, 75
Byssodon, 627

C

Caborius, 344
Caenidae, 60, 75, 126-127, 130, 150, 158, 162
Caenis, 35, 60, 126, 147, 156, 162

Caenis amica, 60
Calacanthia, 284, 296
Calamoceratidae, 54, 64, 310, 322, 329, 333, 348, 351
calcipala, 593
Calineuria, 231, 258, 264
Calineuria californica, 62
Callibaetis, 42, 69, 127, 135, 150, 160
Callibaetis floridanus, 50, 60
Callicera, 547
Callicorixa, 275, 294
Callinapaea, 543
Calliperla, 237, 252, 255, 266
Calocoenia, 543
Caloparyphus, 496, 541
Calopsectra, 754
Calopsectrini, 753
Calopterygidae, 106, 164, 171, 205, 483
Calopteryx, 176, 209
Calopteryx aequabilis, 165
calypteres, 517
Camelobaetidius, 127, 132, 150, 160
campaniform sensillum, 639
Campsicnemus, 540
Campsurus, 127, 147, 158, 163
Camptocladius, 716, 748
Canace, 500, 542
Canaceoides, 500, 542
Canacidae, 32, 79, 485-486, 500, 536, 542
Canada balsam, 643
Cannacria, 208
Cannaphila, 208
Cantharidae, 400
Cantopelopia, 745
Capnia, 218, 228, 246, 264
Capnia vernalis, 62
Capniidae, 62, 69, 217-219, 239, 261, 263
Capnura, 218, 228, 246, 264
Cara, 210
Carabidae, 400-401, 403, 408, 462
Caraphractus, 477, 483
Caraphractus cinctus, 474
cardinal beard, 639
Cardiocladius, 678, 712, 718, 748
cardo, 639
Carex, 397, 546
Caricea, 546
Carmine, 40
Carpelimus, 436, 467
Carrhydrus, 415, 423, 463
Cascadoperla, 235, 252, 255, 266
catastrophic, 44
Catatasina, 496
caudal filaments, 129
caudal gills, 36
caudal horn, 389
Caudatella, 140, 156, 161
caudolateral armature, 8, 641
Caulochironomus, 726
Caurinella, 129-130, 140, 161
Celaenura, 210
Celina, 408, 415, 419, 463
Celinini, 415, 419
Celithemis, 61, 200, 204, 208
Celithemis fasciata, 61
Cenocorixa, 275, 294
Centrarchidae, 166
Centrobiopsis, 483
Centrocorisa, 272, 294
Centroptilum, 106, 135, 150, 160
cephalic rods, 486
cephalic tubercles, 641
Cephalobyrrhinae, 444
cephalopharyngeal skeleton, 486
cephalothoracic suture, 515
cephalothorax, 484, 572
Cephalozia, 387
Cephalozia fruitans, 396
Ceraclea, 54, 105, 310, 349, 362
Ceraclea ancylus, 106
Ceraclea excisa, 65
Ceraclea transversa, 65
Ceratocombidae, 267, 292

Ceratocombus, 292
Ceratophyllum, 392, 397, 413
Ceratophysella, 115
Ceratopogon, 493, 537
Ceratopogonidae, 23, 32, 66, 166, 485, 491, 493, 509, 515, 519, 530, 537, 546, 744-745
Ceratopogoninae, 537
Ceratopsyche, 340, 356
cerci, 11, 129, 167, 169
Cercobrachys, 147, 156, 162
Ceriana, 547
Cernotina, 333, 339, 380
Ceropsilopa, 543
cervical shield, 389
cervix, 5
Chaenusa, 474, 483
Chaetarthria, 429, 432, 466
Chaetocladius, 682, 714, 720, 748
chaetosemata, 389
chaetotaxy, 573
chalaza, 389
Chalcidoidea, 474, 476-477, 483
Chalcosyrphus, 548
Chaoboridae, 32, 36, 40, 66, 71, 75, 94, 484-486, 491, 509, 515, 519, 530, 538
Chaoborus, 46-47, 104, 491, 538
Chaoborus americanus, 66
Chaoborus punctipennis, 66
Chaoborus trivittatus, 66
Chasmatonotus, 748
Chathamiidae, 54
Chauliodes, 300, 304, 308, 483
Chauliodinae, 308
Cheilotrichia, 565
Chelifera, 500, 541
Chelipoda, 541
Chernokrilus, 239, 249, 252, 265
Chernovskiia, 661, 726, 751
Cheumatopsyche, 54, 340, 356
Cheumatopsyche pasella, 63
Chilo, 389, 397, 482
Chilostigma, 345, 350, 365
Chilostigmini, 368, 374
Chilostigmodes, 345, 350, 365
Chiloxanthinae, 284
Chiloxanthus, 284, 296
Chimarra, 63, 339, 377
Chimarra aterrima, 63
Chionea, 563
Chironomidae, 1, 21-23, 32, 36-37, 40, 45, 47-48, 58, 67, 69, 71, 90, 94, 101, 105, 110, 163, 209-210, 264-266, 484-486, 491, 509, 515, 517, 519, 530, 538, 635, 643, 744, 746, 751
Chironomidae Pupae, 690
Chironominae, 75, 101, 641, 643, 693, 720, 738, 743-754
Chironomini, 636, 638-640, 644, 661, 738, 743, 751
Chironomus, 37, 47, 67, 668, 730, 751
Chironomus attenuatus, 38
Chironomus plumosus, 58, 67
Chironomus riparius, 67, 90
Chironomus tenuistylus, 67
Chitonophora, 130
Chloroperlidae, 62, 217-218, 223, 239, 261, 266
Chloroperlinae, 266
Chloropidae, 483
Chlorotabanus, 498, 542
Chorebidea, 474, 483
Chorebidella, 474, 483
Chorebus, 474, 477, 483
Choristoneura, 398
Choroterpes, 139, 156, 162
Christobella, 117
Chromagrion, 167, 179, 183, 210
Chrysendeton, 396
Chrysogaster, 33, 47, 500, 548
Chrysomelidae, 32, 47, 66, 104, 399-401, 403, 410, 452, 455, 461, 471
Chrysomelinae, 452, 455
Chrysops, 67, 482, 498, 542
Chyranda, 345, 368
Cicadellidae, 269

Cinygma, 45, 139, 152, 161
Cinygmula, 139, 154, 161
Cinygmula ramaleyi, 60
Cinygmula reticulata, 60
Cirrula, 508, 543
Claassenia, 228, 258, 264
Claassenia sabulosa, 62
Cladocera, 210
Cladopelma, 663, 726, 751
Cladotantarsus mancus, 67
Cladotanytarsus, 658, 723, 753
Cladura, 563
Clanoneurum, 543
Clasiopella, 543
claspers, 11, 315
clavus, 268
Cleptelmis, 448, 450, 470
Climacia, 304, 308, 483
Climacia areolaris, 63
Climacia californica, 304
climbers, 75
clingers, 75
Clinocera, 500, 541
Clinohelea, 537
Clinotanypus, 644, 698, 744
Clioperla, 237, 255, 266
Clioperla clio, 62
Clistoronia, 345, 371
Clistoronia magnifica, 54, 64, 90
Cloeodes, 136, 152, 160
Cloeon, 127, 135, 160
Cloeon dipterum, 35, 43
Cloeon triangulifer, 60
Closed Respiratory System, 40
Clostoeca, 345, 371
Clunio, 747
Clunio marinus, 104
Clunionini, 738, 743, 747
clypeus, 5, 389
Cnephia, 597, 601, 611, 618, 634
Cnophomyia, 565
Coelopidae, 485
Coelotanypodini, 639, 644, 735, 744
Coelotanypus, 644, 698, 744
Coenagrion, 181, 183, 210
Coenagrion interrogatum, 183
Coenagrionidae, 36, 61, 99, 166-167, 171, 176, 205, 210, 483
Coenia, 508, 543
Coleophora, 389
Coleophoridae, 387, 398
Coleoptera, 1-3, 11, 23, 32, 35, 41, 43, 46, 48, 56, 65, 68-69, 71, 75, 80, 101, 104, 106, 109-112, 399
Colephora, 398
collectors, 76, 79
Collembola, 1, 11, 23, 46, 108-109, 112-113
Collophora, 123
collophore, 113
Colobaea, 547
Coloburella, 119
colonization cages, 13
Colymbetes, 65, 415, 418, 423, 463
Colymbetes sculptilis, 56
Colymbetinae, 418-419
Colymbetini, 418, 423
combination indices, 93
complete metamorphosis, 5
compound eyes, 389
compressible gill, 33
Compteromesa, 747
Compterosmittia, 748
Conchapelopia, 653, 698, 745
Conditum, 210
conjunctiva, 641
Conocephalinae, 216
Conocephalus, 212, 216
Constempellina, 658, 723, 753
Consumnoperla, 252
Contiger, 396
Copelatini, 418, 423
Copelatus, 423, 463
Copelatus glyphicus, 65

Coptotomini, 418, 423
Coptotomus, 65, 415, 418, 423, 463
Coquillettidia, 33, 485, 573, 577, 583
Coquillettidia perturbans, 573
Cordilura, 68, 546
Cordiluridae, 546
Cordulegaster, 45, 174, 207
Cordulegastridae, 165, 174, 205, 207
Cordulia, 197-198, 208
Corduliidae, 61, 165-167, 174, 176, 205, 207
Corduliinae, 196-197, 208
core sampler, 15-16
corers, 18
Corethra, 538
Corethrella, 491, 538
Corethrellidae, 486, 491, 515, 519, 530, 538
Corisella, 272, 294
corium, 268
Corixidae, 32-33, 36, 40, 52, 63, 75, 267-269, 271, 288, 294
Corixidea, 292
Corixiinae, 271
Corixini, 272
Corydalidae, 36, 40, 42, 52, 63, 69, 106, 298-300, 304, 308, 483
Corydalinae, 308
Corydalus, 37, 299-300, 308
Corydalus cornutus, 63
Corynocera, 658, 723, 753
Corynoneura, 67, 673, 700, 747
Corynoneurini, 738, 747
Corynothrix, 119
Coryphaeschna, 185, 188, 207
Cosmopterigidae, 387, 390, 392, 397
Cosmopteriginae, 390, 397
Cosmopteryx, 389, 397
Cossidae, 387, 392
costa, 167
costal vein, 11
Cosumnoperla, 237, 255, 266
counting, 22
coxa, 5, 167
CPOM, 76, 81, 84
CPOM/FPOM, 85
Crambinae, 397
Crambus, 389, 392
creeping welts, 486, 550
cremaster, 389
Cremastus, 474, 482
Crenitis, 427, 434, 466
Crenitulus, 427, 432
crenulate, 639-640
Cressonomyia, 543
Cricotopus, 67, 678, 682, 687, 712, 716, 720, 748
Cricotopus bicinctus, 67
Cricotopus triannulatus, 67
Cricotopus trifascia, 67
critical point, 38
crochets, 389, 484
Crocothemis, 202
Crocothemis servilia, 202
Cryphocricinae, 278
Cryphocricos, 278, 294
Cryphocricos hungerfordi, 63
Cryptochia, 344, 368
Cryptochironomus, 661, 663, 726, 751
Cryptolabis, 555, 563, 568
Cryptopygus, 121
Cryptostemma, 292
Cryptostemmatidae, 292
Cryptotendipes, 663, 726, 751
cubensis, 208
cubital vein, 11
cubito-anal crossvein, 169
cubitus, 167
Culex, 46, 101, 571-572, 577, 583, 590
Culex nigripalpus, 66
Culex pipiens, 42, 58
Culex resturans, 66
Culicidae, 1, 23, 32, 40, 46, 48, 58, 66, 71, 75, 110, 484-486, 491, 509, 515, 519, 530, 538, 573, 590
Culicoides, 493, 537
Culicoides variipennis, 66

Culiseta, 571, 577, 583, 590
Culoptila, 341, 351
Cultellatum, 211
Cultus, 252, 265
Curculionidae, 32, 47, 66, 104, 112, 400-401, 403, 405, 455, 458, 461, 472
Curicta, 278, 293
Curictini, 278
Cushing-Mundie drift sampler, 13
cutaneous, 32, 36
cutaneous respiration, 35
Cybiste, 463
Cybister, 415, 419, 423
Cybister fimbriolatus, 65
Cybistrinae, 419
Cyclorrhaphous, 496, 524
Cyclorrhaphous-Brachycera, 524
Cylindrotoma, 558
Cylindrotominae, 549, 558, 568
Cylloepus, 446, 452, 470
Cymatia, 271, 294
Cymatiinae, 271
Cymbiodyta, 429, 434, 466
Cynipoidea, 474, 477, 483
Cyperaceae, 396
Cyphoderinae, 117
Cyphoderus, 117
Cyphomella, 663, 730, 751
Cyphon, 438, 469
Cyphonidae, 402
Cyretes, 462
Cyrnellus, 339, 380

D

D-Vac vacuum sampler, 16
Dacnusa, 474, 483
Dactylobaetis, 160
Dactylolabis, 555, 563, 569
Dagamaea, 121
Dannella, 130, 140, 162
darners, 207
Dascillidae, 440-441
Dasycorixa, 272, 294
Dasyhelea, 493, 537
Dasyheleinae, 537
Deinocerites, 573, 577, 583, 590
Delphacidae, 269
Demicryptochironomus, 661, 730, 751
Demijerea, 726, 751
Denaria, 597, 601, 611
Denisiella, 121
Denopelopia, 653, 695, 745
Dentatella, 130, 140, 162
Derallus, 429, 432, 466
Deronectes, 464-465
Derotanypus, 646, 698, 744
Derovatellus, 415, 419, 463
Desmona, 345, 368
Desmona bethula, 64
Desmopachria, 415, 419, 463
Desoria, 121
Despaxia, 219, 228, 249, 263
detergent, 40
Deuterophlebia, 66, 487, 538
Deuterophlebiidae, 32, 44, 66, 71, 484-487, 509, 515-516, 519, 524, 538
Deutonura, 117
Diachlorus, 498
Diamesa, 47, 690, 716, 747
Diamesa incallida, 67
Diamesa nivoriunda, 67
Diamesinae, 638, 641, 644, 687, 693, 700, 735, 743, 746
Diamesini, 738, 747
Diapriidae, 474, 477, 481-482
Diaspididae, 269
Diaulota, 436, 467
Dibolocelus, 427, 432, 466
Dibusa, 310, 342, 359
Dibusa angata, 64
Dicercomyzon, 106

Dichaeta, 58, 508
Dichaeta caudata, 67
Diclasiopa, 544
Dicosmoecinae, 310, 344, 368, 371, 374
Dicosmoecus, 69, 344, 371
Dicosmoecus gilvipes, 64
Dicranopselaphus, 440, 469
Dicranoptycha, 561
Dicranota, 66, 549, 553, 561, 569
Dicranota bimaculata, 66
Dicrotendipes, 67, 668, 728, 730, 751
Dictya, 482, 503, 547
Dictyacium, 547
Dicyrtoma, 121
Dicyrtomina, 121
Didymops, 196-197, 207, 208
Diglotta, 436, 467
Dimecoenia, 503, 508, 543
Dineutes, 483
Dineutus, 56, 65, 411, 413, 462
Diotrepha, 555
Diphetor, 136, 152, 160
Diphuia, 544
Diplax, 209
Diplectrona, 340, 356
Diplectrona modesta, 63
Diplectroninae, 340, 356
Diplocladius, 674, 718, 748
Diploperla, 239, 249, 265
Diplosmittia, 748
Dipseudopsinae, 310, 340, 380
Dipsocoridae, 267, 288, 292
Dipsocoromorpha, 267
Diptera, 1, 5, 11, 23, 32, 35-36, 45, 47, 58, 66, 69, 71, 75, 82, 101, 104, 106, 108-110, 112, 484
Discocerina, 503, 544
Discocerina obscurella, 67
Discomyza, 543
Discomyzinae, 543
Disonycha, 452, 455, 471
dissolved oxygen, 29
distal, 639
dististyles, 550
Distosimulium, 610
Ditrichophora, 544
Diura, 217, 237, 239, 252, 265
Diura bicaudata, 104
divers, 75
diversity indices, 92
diving tube, 40
Dixa, 491, 539
Dixella, 491, 539
Dixidae, 23, 71, 484-486, 491, 509, 515, 519, 530, 539
Djalmabatista, 646, 695, 745
Doddsia, 224, 242, 262
Dohrniphora, 546
Doithrix, 687, 712, 748
Dolania, 45, 68, 127, 130, 147, 163
Dolania americana, 60
Dolichocephala, 500, 541
Dolichopeza, 558
Dolichopodidae, 32, 485-486, 500, 515, 524, 536, 540
Dolichopus, 540
Dolomedes, 474, 482
Dolophilodes, 339, 377
Dolophilodes distinctus, 63
Dolophiloides, 106
Dolophilus, 339
DOM, 81
Donacia, 33, 47, 104, 425, 452, 455, 471
Donacia cinerea, 66
Donaciinae, 452, 455
Doncricotopus, 682, 703, 748
Dorocordulia, 197-198, 208
Doroneuria, 231, 258, 264
dorsal central setae, 641
dorsal hooks, 166
dorsal setae, 641
dorsomental plate, 638-639
dorsomental teeth, 638-639
dorsomentum, 638

Douglas method, 15
Draeculacephala, 269
drift, 44
drift net, 13
drift samplers, 18
Dromogomphus, 192, 194, 206
Drosophila, 28
Drunella, 60, 140, 156, 162
Drunella doddsi, 44, 60, 106
Drusinus, 345
Dryomyzidae, 485, 503, 509, 536, 542
Dryopidae, 32, 36, 101, 106, 110, 399-401, 405, 410, 441, 444, 461, 469
Dryopoidea, 399
Dryops, 444, 469
Dubiraphia, 446, 450, 470
Dubiraphia quadrinotata, 66
ductus bursae, 390
duplex gill, 36
Dysmicohermes, 300, 308
Dythemis, 200, 204, 208
Dytiscidae, 32-33, 37, 40, 56, 65, 69, 75, 106, 399, 401, 403, 408, 415, 427, 461, 463, 482-483
Dytiscinae, 418-419
Dytiscini, 419, 423
Dytiscus, 33, 37, 65, 415, 419, 423, 464, 482-483

E

Ecclisocosmoecus, 344, 371
Ecclisomyia, 344, 368
Eccoptura, 231, 258, 264
Eccoptura xanthenes, 62
Ecdyonurus, 68
ecdysial lines, 313
ecdysis, 109
ECI, 81, 85
eclosion, 71
Ecnomidae, 310, 319, 333, 339, 350
Ectemnaspis, 604
Ectemnia, 597, 604, 611, 634
Ectopria, 440, 469
Edmundsius, 132, 150, 159
Eichhornia, 458, 472
Eichhornia crassipes, 387
Eichornia, 398
Einfeldia, 668, 728, 730, 735, 751
Einfeldia synchrona, 67
Ekman grab, 14-17
Ekman pole, 16
Elatine, 397
electroshocking, 22
Eleocharis, 397
Elephantomyia, 561
Elgiva, 482-483, 503, 547
Elgiva rufa, 68
Ellipes, 214, 216
Elliptera, 555, 561, 569
Elliptio complanata, 90
Elmidae, 32-33, 48, 56, 66, 101, 106, 399-401, 405, 410, 441, 444, 446, 461, 469
Elodea, 75, 397
Elodes, 438, 469
elutriation, 17
emergence traps, 16-17
Empedomorpha, 565
Emphyastes, 458, 472
Empididae, 32, 67, 484, 486, 500, 509, 515, 519, 524, 536, 540
empodium, 517
Enallagma, 179, 181, 183, 210
Enallagma pollutum, 164
Enallagma aspersum, 165
Enallagma hageni, 61
Enallagma vesperum, 164
enclosed channels, 14
end hook, 166
Endeodes, 468
Endochironomus, 668, 730, 752
Endonura, 117
Endotribelos, 668, 726, 752
Eneopteri, 216
Enochrus, 401, 429, 434, 466

Entomobrya, 119, 125
Entomobryidae, 117, 119, 123, 125
Entomobryinae, 117, 119
Entomobryomorpha, 115
enumerations, 92
Eobrachycentrus, 343, 351
Eocosmoecus, 64, 344, 374
Eoparargyractis, 387, 396
Eoparargyractis plevie, 65
Epeorus, 44, 106, 129, 136, 154, 161
Epeorus pleuralis, 50
Ephemera, 47, 147, 158, 163
Ephemera simulans, 61
Ephemerella, 60, 140, 156, 162
Ephemerella aurivillii, 130
Ephemerella ignita, 50, 60
Ephemerellida, 127
Ephemerellidae, 50, 60, 106, 129-130, 148, 158, 161
Ephemeridae, 11, 50, 61, 75, 108, 129-130, 148, 158, 163
Ephemeroptera, 1, 11, 23, 32, 35-36, 38, 40, 42, 46-48, 50, 60, 68-69, 75, 91-92, 94, 98-99, 101, 104-106, 108-111, 126, 745
Ephidatia, 208
Ephoron, 35, 126-127, 147, 158, 163
Ephoron album, 61
Ephoron leukon, 61
Ephydatia, 308
Ephydra, 508, 543
Ephydra cinerea, 42
Ephydridae, 32, 58, 67, 482-486, 503, 509, 515, 536, 543
Ephydrinae, 543
Epiaeschna, 185, 188, 207
Epicordulia, 197, 208
epicranial suture, 5, 389, 515
Epimetopidae, 402, 427, 432, 465
Epimetopus, 427, 432, 465
epiphragma, 563
epiproct, 11, 167, 169
Epitheca, 61, 164, 197-198, 208
Epitheca cynosura, 61
Epoicocladius, 674, 714, 748
EPT richness, 92
Eretes, 418, 423, 464
Eretini, 418, 423
Eretmoptera, 747
Eriocera, 569
Eriococcidae, 269
Eriococcus, 269
Erioptera, 553, 555, 565, 569
Eriopterini, 549, 556, 561
Eriplanus, 482
Eristalinus, 548
Eristalis, 32, 37, 101, 500, 548
Erpetogomphus, 192, 194, 206
Erythemis, 198, 205, 208
Erythrodiplax, 200, 202, 205, 208
Estigmene, 390
Eubrianax, 440-441, 468
Eubriidae, 402, 440
Eubriinae, 440, 468
Eucapnopsis, 228, 246, 264
Eucoilidae, 474, 476-477, 481, 483
Eucorethra, 491, 538
Eudactylocladius, 682, 714
Eugnophomyia, 565
Euhrychiopsis, 472
Eukiefferiella, 638, 678, 712, 714, 716, 718, 748
Eukiefferiella brevinervis, 67
Eukiefferiella devonica, 67
Eulachadidae, 402
Eulalia, 541
Eulophidae, 476-477, 481, 483
Euorthocladius, 678, 682, 709, 712, 716
Euparal, 643
Euparyphus, 496, 541
Eupatorium, 397
Eupteromalus, 483
Euryhapsis, 682, 703, 748
Eurylophella, 130, 140, 156, 162
Eurylophella doris, 60
Eurylophella funeralis, 60
Eurylophella temporalis, 60
Eurystethidae, 468

Eusimulium, 618
Eutaenionotum, 544
Euterophlebiidae, 484
Euthycera, 547
exclamationis, 211
exuviae, 109
eye-cap, 389
eyespots, 636, 639

F

Fabria, 343, 380
Fallceon, 136, 152, 160
family biotic index, 95
Farrodes, 129, 140, 156, 162
Farula, 347, 383
Fattigia, 347, 383
Fattigia pele, 64
felt chamber, 641
femur, 5, 167
Ferruginea, 209
Fick's law, 30
field washing, 17
filaments, 572
filtering, 79
Fittkauimyia, 646, 698, 744
flagellomeres, 593
flagellum, 639
Flavohelodes, 469
Fletcherimyia, 546
floating drift trap, 14
floating emergence traps, 14
flotation, 22
Folsomia, 119
Folsomides, 121
Folsomina, 119
Fonscolombia, 207
Fontinalis, 75
forceps, 129
Forcipomyia, 493, 537
Forcipomyiinae, 537
forks, 389
FPOM, 76, 81, 84
Frenesia, 345, 374
Friesea, 117, 125
Frisonia, 237, 249
Frisonla, 265
frontal apotome, 641
frontal setae, 641
frontal suture, 517
frontal tubercles, 641
frontal warts, 641
frontoclypeal apotome, 313, 636, 639
frontoclypeal sutures, 313
frontoclypeus, 5
Froude number, 43
Fucus, 472
functional feeding-group measures, 93
funnel-trap, 16
furcula, 115

G

Galerucella, 47, 452, 472
Galerucella nymphaeae, 66, 72
Galerucellinae, 452
Galerucinae, 455
Gammarus, 44
Gammarus pulex, 90
gas concentrations, 40
Gasterophilidae, 37
Gastrops, 544
gathering, 79
GCOL, 85
Gelastocoridae, 268-269, 275, 288, 296
Gelastocorinae, 275
Gelastocoris, 275, 296
Gelechiidae, 387, 392
Gelechoidea, 392
genae, 5, 636
genital pocket, 169
genital sheaths, 641

Georthocladius, 687, 712, 749
Georyssidae, 400, 403, 410
Geranomyia, 569
Gerking sampler, 15
Gerridae, 41, 46, 52, 62, 75, 101, 112, 267-268, 271, 275, 288, 293, 482-483
Gerrinae, 275, 293
Gerris, 52, 268, 275, 293, 482-483
Gerris agenticollis, 62
Gerris canaliculatus, 62
Gerris remigis, 62
Gerromorpha, 267-268
Gillespie and Brown sampler, 17
Gillotia, 661, 730, 752
gills, 30, 36, 167
Glaenocorisa, 272, 294
Glaenocorisini, 272
Glenanthe, 544
Glossosoma, 69, 72, 106, 341, 351
Glossosoma intermedium, 64
Glossosoma nigrior, 20, 64, 85
Glossosoma pentium, 64
Glossosomatidae, 20, 43, 54, 64, 309, 315, 319, 326, 329, 338, 341, 351
Glossosomatinae, 341
Glutops, 498, 540
Glyphopsyche, 345, 374
Glyphotaelius, 346
Glyptotendipes, 668, 726, 735, 752
Glyptotendipes pallens, 90
Glyptotendipes paripes, 67
Goeldichironomus, 638, 666, 728, 752
Goera, 68, 106, 346, 365
Goeracea, 346, 365
Goereilla, 329, 346, 365
Goeridae, 68
Goerinae, 310, 322, 329, 346, 365
Goerini, 333
Goerita, 346, 365
Goerita semata, 64
Gomphaeschna, 185, 188, 207
Gomphidae, 11, 45, 61, 106, 164-166, 171, 174, 192, 205-206
Gomphurus, 194, 196, 206
Gomphus, 194, 196, 206
Gonempeda, 565
Gonielmis, 448, 452, 470
gonocoxite, 642
Gonomyia, 555, 563, 565, 569
Gonomyodes, 553, 565, 569
gonostylus, 642
grab samplers, 14, 18
Graceus, 735, 752
Graminaceae, 398
Grammotaulius, 49, 345, 374
Granisotoma, 121
Graphoderus, 65, 419, 425, 464
Graphomya, 546
Graptocorixa, 272, 295
Graptocorixa serrulata, 63
Graptocorixini, 271
Greniera, 597, 601, 611, 618, 634
Grensia, 345, 371
Griseosilvius, 498
Gross production rate, 85
Gryllidae, 212, 214, 216
Gryllotalpa, 214
Gryllotalpidae, 212, 214, 216
gula, 268
Gumaga, 347, 383
Guttipelopia, 649, 700, 745
Guttipsilopa, 543
Gymnometriocnemus, 674, 712, 716, 749
Gymnomyzinae, 544
Gymnopais, 485, 593, 601, 606, 633-634
Gynacanrha, 207
Gynacantha, 185, 188
Gynacantha nervosa, 164
Gyretes, 411, 413
Gyrinidae, 32, 36, 46, 56, 65, 69, 106, 399-402, 405, 411, 461-462, 482-483
Gyrinophagus, 483
Gyrinus, 400, 411, 413, 462, 482-483

H

habitat, 74
habitat selection, 48
Habrophlebia, 35, 139, 156, 163
Habrophlebia vibrans, 60
Habrophlebiodes, 139, 156, 163
Haemagogus, 573, 577, 583, 590
Haematopota, 498, 542
Hagenella, 343, 380
Hageni, 206
Hagenius, 45, 166, 169, 192, 194, 206
Hagenius brevistylus, 164-165
Haideoporus, 399, 418, 422, 464
Halesochila, 345, 371
Haliaspis spartinae, 269
Haliplidae, 32, 56, 65, 399, 401-403, 408, 413, 461-462
Haliplus, 56-57, 413, 462
Haliplus immaculicollis, 65
Halisotoma, 121
Halobates, 41, 101, 268, 275, 293
Halobatinae, 275, 293
Halocladius, 673-674, 714, 738, 743, 749
halteres, 11, 517
Halticoptera, 483
hamules, 169
Hansonoperla, 231, 258, 264
Haploperla, 235, 255, 266
Haploperla brevis, 62
Hardy plankton sampler, 13-14
Harlomillsia, 117
Harnischia, 636, 663, 726, 752
harpagones, 11
Hayesomyia, 653, 700, 745
head, 572
head capsule, 639
head ratio, 639
Hearlea, 627
Hebridae, 271, 278, 288, 295
Hebrus, 278, 296
Hecamede, 544
Hecamedoides, 544
Hedria, 503, 547
Hedriodiscus, 496, 541
Hedriodiscus truquii, 42
Helaeomyia, 508, 543
Heleidae, 537
Heleniella, 67, 678, 718, 749
Heleodromia, 541
Helianthus, 392
Helichus, 400, 444, 446, 469
Helicopsyche, 315, 348, 350
Helicopsyche borealis, 64, 76
Helicopsyche limnella, 64
Helicopsyche mexicana, 64
Helicopsychidae, 64, 76, 310, 313, 315, 326, 333, 348, 350
Helius, 555, 561, 569
Hellichiella, 618
Helobata, 429, 434, 466
Helochara communis, 269
Helochares, 429, 434, 466
Helocombus, 429, 434, 466
Helocordulia, 197-198, 208
Helodidae, 32, 65, 402, 438, 469
Helodon, 611
Helopelopia, 653, 698, 745
Helophilus, 548
Helophoidae, 427
Helophorinae, 402, 427, 432, 466
Helophorus, 427, 432, 466
Helopicus, 62, 237, 249, 265
Helotrephidae, 35, 267-268
hemelytra, 268
Hemerodromia, 500, 541
Hemerodromia empiformis, 67
Hemicnetha, 618
hemimetabolous, 109
Hemiptera, 2-3, 5, 11, 23, 32-33, 35, 41, 46, 48, 52, 62, 68-70, 75, 101, 104, 109-112, 267-268, 292
Hemiteles, 482

hemoglobin, 32, 37, 636
Heptagenia, 139, 154, 161
Heptagenia cruentata, 129
Heptageniidae, 36, 50, 60, 75, 77, 129, 136, 148, 158, 160
Hercostomus, 540
Hermione, 541
heros, 207
Hesperagrion, 176, 181, 183, 210
Hesperoconopa, 549, 551, 565, 569
Hesperocorixa, 272, 295
Hesperocorixa interrupta, 63
Hesperoperla, 228, 258, 264
Hesperophylax, 49, 345, 377
Hess sampler, 13, 18
Hetaerina, 176, 209
Heterelmis, 446, 452, 470
Heterlimnius, 448, 452, 470
Heteroceridae, 399-400, 405, 410
Heterocloeon, 135, 152, 160
Heterocloeon curiosum, 60
Heterocloeon frivolum, 152
Heterodoxum, 210
Heteromurus, 117
Heteroplectron, 54, 310, 348, 351
Heteroplectron californicum, 45, 64
Heteroptera, 94, 267-269
Heterotanytarsus, 673, 706, 749
Heterotrissocladius, 47, 674, 706, 749
Heumatopsyche pettiti, 63
Hexacola, 474, 483
Hexacylloepus, 448, 452, 470
Hexagenia, 47, 50, 106, 108, 147, 158, 163
Hexagenia bilineata, 61
Hexagenia limbata, 61, 108, 126
Hexagenia munda, 61
Hexagenia rigida, 39
Hexatoma, 549, 556, 561, 569
Hexatomini, 549-550, 561
Himalopsyche, 341, 383
Hippelates, 483
Histeridae, 408
histoblasts, 592
holometabolous, 109
Holorusia, 551, 558, 565, 568
Holotanypus, 698
Homoeoneuria, 136, 152, 160
Homophylax, 345, 368
Homoplectra, 340, 356
Homoptera, 268-269
hookplates, 315
hooks, 592
Hoperius, 418, 423, 464
Hoplitimyia, 541
Hoplodictya, 547
Hudsonimyia, 649, 698, 745
Hudsonimyia karelena, 638
Huleechius, 448, 452, 470
humeral crossvein, 11
humeral suture, 167
humps, 313
Hungaroscelio, 482
Husseyella, 271, 288, 292
Hyadina, 545
Hybomitra, 498, 542
Hydaticini, 419, 423
Hydaticus, 65, 419, 423, 464
Hydatophylax, 346, 368
Hydra, 70
Hydracarina, 166
Hydraena, 434, 468
Hydraenidae, 32, 399-401, 405, 408, 434, 461, 468
Hydraeninae, 434
Hydraporus, 415
Hydraticus, 415
Hydrellia, 474, 482-483, 508, 545
Hydrelliinae, 545
Hydrilla, 397
Hydrilla vericillata, 387
Hydrobaeninae, 747
Hydrobaenus, 678, 703, 749
Hydrobiomorpha, 429, 432, 466
Hydrobiosidae, 309, 319, 326, 329, 338, 341, 350

Hydrobius, 427, 429, 432, 466
Hydrocampa, 396
Hydrocampinae, 396
Hydrocanthus, 425, 465
Hydrochara, 429, 432, 466
Hydrochasma, 544
Hydrochidae, 402, 427, 432, 466
Hydrochus, 427, 432, 466
Hydrocyrius, 37
Hydrodromia, 500
hydrofuge, 32
Hydroisotoma, 121, 125
Hydroisotoma schaefferi, 113
Hydrometra, 292
Hydrometra martini, 62
Hydrometridae, 62, 271, 288, 292
Hydromyza, 546
Hydroperla, 237, 249, 265
Hydroperla crosbyi, 52, 62
Hydrophilidae, 32, 56, 65, 399-403, 427, 432, 461, 465-466
Hydrophilinae, 432
Hydrophiloidea, 400-403, 410, 427
Hydrophilus, 70, 427, 429, 432, 467
Hydrophilus triangularis, 56
Hydrophorus, 540
Hydrophylita, 483
Hydroporinae, 415, 419
Hydroporini, 415, 419
Hydroporus, 415, 418, 422, 464
Hydropsyche, 54-55, 63, 310, 340, 356
Hydropsyche angustipennis, 90
Hydropsyche bifida, 356
Hydropsyche contubernalis, 90
Hydropsyche morosa, 356
Hydropsyche slossonae, 63, 72
Hydropsychidae, 44, 54, 63, 75, 106, 264, 310, 315, 326, 333, 338, 343, 350, 356
Hydropsychinae, 340, 356
Hydropsychoidea, 310, 313, 315, 339
Hydroptila, 342, 359
Hydroptilidae, 64, 309, 315, 319, 326, 329, 338, 341-342, 350, 359
Hydroptilinae, 342, 359
Hydroptilini, 342
Hydropyrus, 543
Hydroscapha, 399, 403, 405, 427, 465
Hydroscaphidae, 32, 399-400, 403, 405, 427, 465
Hydrothassa, 452, 455, 471
Hydrotrupes, 418, 422, 464
Hydrous, 33
Hydrovatini, 418-419
Hydrovatus, 418-419, 464
Hygrobiidae, 401, 403, 408
Hygrotus, 418, 422, 464
Hylogomphus, 194, 196
Hymenoptera, 1, 5, 23, 56, 109-112, 474
Hyphydrini, 415, 419
Hypocharassus, 540
Hypogastrura, 115
hypogastruridae, 115, 123, 125
hypognathous, 5, 389
Hyponeura, 210
hypopharynx, 5
hypopygium, 642
hyporheic, 44
Hyporhygma, 663, 726, 752
Hyposmocoma, 387, 390, 397
hypostoma, 592
hypostomal bridge, 550
hypostomal cleft, 592

I

Ichneumonidae, 477, 481-482
Ichneumonoidea, 474, 476, 482
Idiataphe, 200, 204, 208
Idiocoris, 35
Idiognophomyia, 565
Ilybius, 65, 415, 418, 423, 464
Ilytheinae, 545
imago, 109

incompressible gill, 33
indicator communities, 87
indicator organisms, 87
inferior appendages, 167, 315
insect emergence traps, 14
instar, 109
insularis, 208
intersternum, 167
invertebrate community index, 91
Ioscytus, 284
Iris versicolor, 165
Irmakia, 730
Iron, 35, 129, 136, 154, 161
Ironodes, 45, 136, 154, 161
Ironopsis, 129, 136, 154, 161
Ironoquia, 344, 374
Ischnura, 181, 183, 210, 483
Ischnuridia, 210
Isocapnia, 217, 224, 246, 264
Isocladius, 678, 682
Isocytus, 296
Isogenoides, 237, 249, 265
Isogenoides olivaceus, 62
Isonychia, 44, 60, 69, 77, 127, 129, 136, 148, 160
Isonychiidae, 60, 129, 136, 148, 158, 160
Isoperla, 62, 235, 237, 239, 255, 266
Isoperla fulva, 62
Isoperla signata, 62
Isoperlinae, 249, 266
Isotoma, 121, 125
Isotomidae, 117, 119, 123, 125
Isotomiella, 121
Isotomodes, 121
Isotomurus, 121
Issidae, 269
Ithytrichia, 342, 362

J

Janata, 207
Jenkinshelea, 537
Johannsenomyia, 538
Jordan's organs, 389
Juncus, 33, 397-398, 546
Juxta, 390

K

Kathroperla, 231, 255, 266
Kellen grab, 15-16
kick net, 12
kick samplers, 18
Kiefferulus, 668, 728, 752
Kloosia, 726, 752
Knowltonella, 115
Kogotus, 235, 252, 265
KOH, 643
Krenopelopia, 649, 653, 698, 745
Krenopsectra, 661, 723, 753
Krenosmittia, 673, 714, 749
Krizousacorixa, 295

L

labial palpi, 389
Labiobaetis, 135, 152, 160
labiohypopharyngeal apparatus, 592
labium, 5, 166
Labostigmina, 541
labral fans, 592
labral lamella, 639
labral sensilla, 639
labrum, 5, 389, 572, 592, 636
Labrundinia, 649, 700, 745
Laccobius, 429, 432
Laccodytes, 415, 419, 464
Laccophilinae, 418-419
Laccophilini, 418-419

Laccophilus, 415, 418-419, 464
Laccophilus maculosus, 65
Laccornis, 415, 422, 464
Lachlania, 136, 152, 160
Lacobius, 467
Ladona, 198, 204, 208
Ladona deplanata, 61
Lais, 210
Lambourn sampler, 13, 15
lamellar lateral setae, 641
Lampracanthia, 284, 296
Lamproscatella, 544
Lampsilis radiata, 90
Lampyridae, 400, 403, 410
Langessa, 396
Lanthus, 192, 196, 206
Lanthus vernalis, 61
Lappodiamesa, 690, 716, 747
Lapposmittia, 678, 714, 749
Lara, 446, 450, 470
Lara avara, 45, 66
Larsia, 649, 695, 745
larvae, 109
Laser Doppler Anemometry, 43
Lasiodiamesa, 690, 693, 746
lateral abdominal fringe, 641
lateral appendages, 167
lateral hair fringe, 638-639
lateral lobes, 166
lateral palatal brushes, 572
lateral sclerite, 592
lateral setae, 166
lateral spine, 166
lateral tubercles, 313
Lathromeroidea, 476, 483
Lauterborn organ, 639
Lauterborn organ stalk, 639
Lauterborniella, 668, 726, 752
Lavernidae, 397
Lednia, 224, 246, 263
leg sheaths, 642
Leiodidae, 400
Lemanea, 310
Lemna, 75, 125, 396, 473, 508
Lemnaphila, 508, 545
Lenarchulus, 345
Lenarchus, 346, 374
lentic-limnetic, 75
lentic-littoral, 75
lentic-profundal, 75
Lepania, 329, 346, 365
Lepiceridae, 402, 408
Lepidium, 473
Lepidocyrtus, 117
Lepidoptera, 5, 23, 32, 36, 40, 47, 56, 65, 101, 104, 106, 109-112, 309, 387, 389
Lepidostoma, 64, 344, 362
Lepidostoma bryanti, 64
Lepidostomatidae, 64, 310, 315, 322, 329, 333, 338, 344, 362
Leprysma, 216
Lepthemis, 208
Leptocella, 349
Leptoceridae, 54, 65, 106, 310, 319, 329, 333, 338, 349-350, 362
Leptocerus, 349, 362
Leptoconopinae, 538
Leptoconops, 493, 538
Leptohyphes, 147, 156, 162
Leptonannus, 292
Leptonema, 340, 356
Leptophlebia, 85, 107, 139, 154, 163
Leptophlebia cupida, 50
Leptophlebiidae, 50, 60, 75, 129, 132, 148, 158, 162
Leptophylax, 346, 350, 365
Leptopodidae, 267-268, 288
Leptopodomorpha, 267-268
Leptopsilopa, 543
Leptotarsus, 551, 558, 568
Leptysma, 214
Lestes, 61, 165, 167, 176, 210, 483
Lestes congener, 50
Lestes disjunctus, 35

Lestes sponsa, 104
Lestidae, 36, 40, 61, 165, 171, 209, 483
Lethemurus, 117
Lethocerinae, 271
Lethocerus, 267, 271, 293
Lethocerus americanus, 62
Lethocerus maximus, 62
Leucocytozoon, 591
Leucorrhinia, 200, 202, 204, 208
Leucotabanus, 498, 542
Leucotrichia, 342, 359
Leucotrichiinae, 359
Leucotrichiini, 310, 342
Leucrocuta, 60, 139, 154, 161
Leuctra, 62, 217, 228, 249, 263
Leuctridae, 62, 217-219, 239, 261, 263
Leuctrinae, 263
Liancalus, 540
Libellula, 61, 198, 204, 208
Libellula quadrimaculata, 164
Libellula semifasciata, 202
Libellulidae, 45, 49, 61, 75, 106, 164-165, 167, 169, 174, 198, 205, 208
light traps, 17
ligula, 166, 638-639
limbata, 108
Limnebiinae, 434
Limnebius, 434, 468
Limnellia, 544
Limnephilidae, 47, 54, 64, 82, 310, 313, 315, 322, 326, 329, 333, 338, 344, 347, 350, 365, 482
Limnephilinae, 310, 345, 368, 371, 374
Limnephilini, 371, 374
Limnephiloidea, 310, 313, 342
Limnephilus, 346, 374
Limnephilus indivisus, 49, 54, 64
Limnephilus lunatus, 47, 54
Limnephilus pallens, 90
limnetic zone, 46
Limnia, 547
Limnichidae, 400, 402, 405, 410, 441, 444, 461, 471
Limnichinae, 444
Limnichus, 441, 444
Limnochironomus, 751
Limnocorinae, 278
Limnocoris, 278, 294
Limnodytes, 482
Limnogonus, 275, 293
Limnophila, 549, 556, 563, 569
Limnophora, 484, 546
Limnophora riparia, 68
Limnophyes, 682, 716, 749
Limnoporus, 268, 275, 293, 482-483
Limnoporus notabilis, 62
Limonia, 46, 555-556, 558, 561, 569
Limoniinae, 549-551, 568
Limoniini, 549, 558
Lineatipes, 209
Liodessus, 415, 418, 422, 464
Liogma, 558
Lioporeus, 415, 422
Lioporius, 464
Liparocephalus, 436, 467
Lipiniella, 668, 728, 752
Lipochaeta, 544
Lipogomphus, 278, 296
Lipsothrix, 45, 66, 484, 555, 565, 569
Lipsothrix migrilinea, 58
Lipsothrix sylvia, 66
Lipurometriocnemus, 749
Liriope, 539
Liriopeidae, 539
Lispe, 546
Lispocephala, 546
Lispoides, 546
Lissorhopterus simplex, 66
Lissorhaptrus, 33
Lissorhoptrus, 104, 461, 472
Lissorhoptrus oryzophilus, 47
Lissorhoptrus simplex, 455
Listronotus, 458, 472
Litobrancha, 147, 158, 163
Littoral Zone, 46

Littorina, 547
Lixellus, 461, 472
Lixus, 458, 472
Llythea, 545
lobes, 550, 592
lobules, 592
longipennis, 209
longitudinal incisions, 486
longitudinal seam, 517
Longurio, 551, 558, 568
Lopescladius, 673, 718, 749
Lophognathella, 115
lotic habitats, 42
lotic-depositional, 75
lotic-erosional, 75
Ludwigia, 396
Lutrochidae, 101, 399, 402, 405, 410, 441, 444, 461, 471
Lutrochus, 402, 441, 444, 471
Lutzomiops, 491, 538
Lymnaecia, 389, 397
Lype, 339, 383
Lype diversa, 63
Lysathia ludoviciana, 452
Lytogaster, 508, 545

M

M-Appendage, 639
Macan sampler, 15, 17
Macdunnoa, 139, 154, 161
Mackenziella, 121
Mackenziellidae, 121
Macrelmi, 448
Macrelmis, 450, 470
Macrodiplax, 200, 204, 209
Macromia, 165, 196-197, 208
Macromiidae, 45
Macromiinae, 196-197, 207
Macronema, 35, 340
Macronematinae, 310, 340, 356
Macronemum, 340
Macronychus, 448, 450, 470
Macronychus glabratus, 66
Macropelopia, 646, 698, 744
Macropelopiini, 639, 646, 738, 744
macrophyte piercers, 76
Macroplea, 104
Macrostemum, 340, 356
Macrostemum carolina, 63
Macrostemum zebratum, 63
Macrothemis, 200, 202, 204, 209
Macrovelia, 278, 292
Macroveliidae, 268, 271, 278, 288, 292
Madeophylax, 347, 365
Maenas, 387
Malaise traps, 17
Malenka, 224, 242, 263
Malirekus, 237, 252, 265
Mallochohelea, 493, 538
Mallota, 548
mandibles, 5, 592, 639
Manophylax, 347, 365
Mansonia, 33, 47, 571, 573, 577, 583, 590
Marcella, 209
Marilia, 348, 377
Martarega, 284, 295
Maruina, 485, 487, 493, 539
Matini, 418, 423
Matrioptila, 341, 351
Matus, 418, 423, 464
Matus bicarinatus, 65
maxilla, 640
maxillae, 5
maxillary palp, 315, 636, 640
maxillary palpi, 389
maxillary palpus, 573
maxillary plates, 550
Mayacnephia, 597, 601, 611, 634
Mayatrichia, 342, 362
media, 167
medial vein, 11

medial-cubital crossvein, 11
median planate, 169
Medophron, 474, 482
Megadytes, 415, 423, 464
Megaleuctra, 228, 246, 263
Megaleuctrinae, 263
Megaloptera, 11, 23, 32, 35, 37, 47-48, 52, 63, 68-69, 71, 104, 106, 108-112, 298-299, 308
Megalopteran, 298
Megamelus davisi, 269
Megarcys, 235, 249, 265
Megaselia, 546
Megistocera, 551, 558, 568
Melanderia, 540
Melyridae, 400-401, 405, 436, 461, 468
Membranule, 167
Mendax, 208
mental plates, 550
mental setae, 166
mentum, 550, 636, 638, 640
Meringodixa, 491, 539
Meropelopia, 653, 698, 745
Merragata, 278, 296
Merycomyia, 498, 542
Mesachorutes, 115
Mesocapnia, 218, 228, 246, 264
mesocosms, 94-96
Mesocricotopus, 706, 709, 749
Mesoleptus, 482
mesonotum, 313, 517
Mesopsectrocladius, 709
Mesoseries, 389
Mesosmittia, 749
Mesothemis, 208
mesothorax, 5, 389, 573
Mesovelia, 267-268, 271, 295
Mesoveliidae, 268, 271, 288, 295
Mestocharis, 477, 483
Metachela, 541
Metacnephia, 597, 604, 611, 634
Metaleptea, 214, 216
metamorphosis, 5, 70, 109
metanotum, 313
Metaphorura, 115
metathorax, 5, 390, 573
Methlinae, 415, 419
Metisotoma, 119
Metretopodidae, 132, 148, 158-159
Metretopus, 139, 154, 159
Metriocnemini, 738, 748
Metriocnemus, 682, 712, 716, 749
Metrobates, 278, 293
Miathyria, 200, 209
Micracanthia, 288, 296
Micralymma, 436, 467
Micranurida, 117
Micranurophorus, 119
Micrasema, 343, 351
Micrasema kluane, 64
Micrathyria, 204, 209
Micrathyria hageni, 202
Micrisotoma, 121
Microbledius, 436
Microbracon, 482
Microcara, 438, 469
Microchironomus, 663, 726, 752
microcosms, 94-95
Microcylloepus, 448, 452, 470
Microgastrura, 117
Microlytogaster, 545
Micronectinae, 271
Micropsectra, 67, 661, 720, 723, 753
Microsporidae, 400-403, 408, 461, 465
Microtendipes, 730, 752
Microvelia, 267, 288, 292
Microveliinae, 288
midbasal space, 169
midventral bulge, 592
Minto sampler, 15
Mitchellania, 115
Mochlonyx, 491, 538
modified Gerking sampler, 15
Mogoplistinae, 216
Molanna, 106, 348, 377

Molanna angustata, 36
Molanna flavicornis, 64
Molannidae, 64, 310, 315, 322, 329, 333, 338, 348, 377
Molannodes, 348, 377
molecular-systematic methods, 107
Molophilus, 553, 563, 569
molting, 109
Monodiamesa, 690, 709, 747
Monohelea, 493, 538
Mononyx), 296
Monopelopia, 649, 653, 698, 745
Monopsectrocladius, 706
Morphocorixa, 272, 295
Morulina, 117
Morulodes, 117
Moselia, 228, 249, 263
Moselyana, 347, 365
Mosillus, 544
movable hook, 166
multiple core sampler, 17
multiserial, 389
multiserial fringe, 642
multivariate analysis, 88
multivariate approaches, 93
Mundie pyramid trap, 14
Munroessa, 387, 390, 396
Muscidae, 68, 484, 508, 536, 546
Mymaridae, 477, 481, 483
Myolepta, 548
Myriophyllum, 396-397, 472
Myriophyllum aquaticum, 452
Mystacides, 349, 362
Mystacides interjecta, 65
Myxophaga, 399, 402, 465
Myxosargus, 496

N

naiads, 109
Namamyia, 348, 377
Nannocoris, 292
Nannothemis, 200, 202, 209
Nanocladius, 678, 700, 703, 706, 709, 749
Narpus, 446, 450, 470
nase, 642
Nasiaeschna, 185, 188, 207
Nasiternella, 551, 561
Natarsia, 638, 644, 698, 745
Natarsiini, 639, 644, 745
Naucoridae, 32-33, 36, 40, 63, 267-268, 271, 278, 288, 294
Naucorinae, 278
Neanura, 117
Neargyractis, 396
Neargyractis slossonalis, 387
Neaviperla, 231, 255, 266
Nebria, 400
Nebrioparus, 415
Nebrioporus, 415, 419, 422, 464
Nectopsyche, 349, 362
Nectopsyche albida, 65
Neelidae, 121
Neelus, 121
Nehalennia, 171, 179, 181, 211
nektonic, 46
Nelumbo, 398
Nelumbo lutea, 389
Nematocera, 486, 515, 517, 524
Nemobiinae, 216
Nemocapnia, 228, 246, 264
Nemotaulius, 346, 371
Nemotelus, 496, 541
Nemoura, 224, 242, 246, 263
Nemoura trispinosa, 52, 61
Nemouridae, 37, 52, 61, 82, 104, 111, 217-219, 239, 261-262
Nemourinae, 263
Neoascia, 548
Neobidessus, 415, 422, 464
Neocataclysta, 387, 390, 396
Neochetina, 458, 472

Neochetina eichhorniae, 90
Neochoroterpes, 129, 139, 156, 163
Neochthebius, 468
Neoclatura, 563
Neoclypeodytes, 415, 418, 422, 464
Neocorixa, 272, 295
Neocrastria, 389
Neocurtilla, 212, 214, 216
Neocylloepus, 448, 452, 470
Neocysta, 209
Neoelmis, 448, 452, 470
Neoephemera, 130, 147, 162
Neoephemeridae, 130, 147, 158, 162
Neoerastria, 398
Neoerythromma, 181, 183, 211
Neogerris, 275, 293
Neohaemonia, 104, 452, 455, 471
Neohermes, 300, 308
Neohydronomus affinis, 458
Neolimnophila, 563
Neonaphorura, 115
Neoneura, 185, 210
Neoperla, 228, 255, 264
Neoperla clymene, 62
Neophylax, 64, 69, 347, 383
Neophylax consimilis, 64
Neophylax mitchelli, 64
Neoplasta, 541
Neoplea, 284, 294
Neoplea striola, 62
Neoptera, 109
Neoscutopterus, 418, 423, 464
Neosminthurus, 123
Neostricklandia, 482
Neothremma, 347, 383
Neothremma alicia, 64
Neotrichia, 342, 362
Neotrichiini, 342
Neotridactylus, 214, 216
Neozavrelia, 658, 723, 753
Nepa, 267, 278, 294, 482-483
Nepa apiculata, 62
Nephrotoma, 558
Nepidae, 32-33, 62, 101, 267-268, 278, 288, 293, 482-483
Nepinae, 278
Nepini, 278
Nepomorpha, 267-268
Nepticula, 389, 397
Nepticulidae, 387, 389-390, 392, 397
Nerophilus, 348, 377
Nerthra, 275, 296
Nerthra rugosa, 269
Nerthrinae, 275
Nervosa, 207
net production rate, 85
netted samplers, 18
Neureclipsis, 310, 339, 380
Neureclipsis bimaculatus, 63
Neureclipsis crepuscularis, 63
Neurocordulia, 164, 166, 197, 208
Neurocordulia molesta, 196
Neurocordulia molesta, 61
Neuroptera, 23, 32, 35, 63, 71, 104, 109-112, 298-299, 308
Nevermannia, 618
Nigronia, 80, 299-300, 304, 308
Nigronia serricornis, 63
Nilobezzia , 538
Nilotanypus, 649, 695, 745
Nilothauma, 663, 735, 752
Nimbocera, 658, 723, 753
Nippotipula, 549
Nitella, 413
Nixe, 60, 139, 154, 161
Nocticanace, 542
Noctuidae, 65, 387, 389-390, 392, 398, 482
nodus, 167
Nostima, 508, 545
Nostocladius, 712
nota, 167
notal setae, 313
Noteridae, 65, 104, 400-401, 403, 408, 425, 461, 465

Noterus, 104
Noterus capricornis, 425
Nothifixis, 209
Notiodes, 461, 472
Notiphila, 33, 58, 503, 508, 545
Notomicrus, 425, 465
Notonecta, 267, 284, 295, 483
Notonecta hoffmanni, 63
Notonecta undulata, 52, 63, 267
Notonectidae, 32, 36-37, 40, 52, 63, 267-268, 271, 278, 288, 295, 483
notum, 5
Nuphar, 47, 269, 390, 396, 398, 452, 546
Nyctiophylax, 310
Nymphaceae, 397
Nymphaea, 33, 75, 387, 390, 396, 452, 455, 472
Nymphoides, 396
Nymphomyia, 66, 487, 539
Nymphomyiidae, 66, 484-487, 509, 515, 517, 524, 539
nymphs, 109
Nymphula, 56, 396, 482
Nymphuliella, 396
Nymphuliella daeckealis, 387
Nymphulinae, 387, 389-390, 396
Nymphulini, 387, 396

O

oblique vein, 167
obtect, 389
Occidentalia, 397, 482
occiput, 5, 167
ocelli, 167, 315, 389
Ochrotrichia, 342, 359
Ochteridae, 268-269, 288, 297
Ochterus, 297
Ochthebius, 434, 468
Ochthera, 508, 544
Oconoperla, 237, 252, 265
Octogomphus, 192, 194, 206
Odonata, 1, 11, 23, 32, 48, 61, 68-69, 71, 75, 98-99, 101, 104-106, 108-109, 111, 164
Odontella, 117
Odontoceridae, 310, 322, 329, 333, 338, 348, 377
Odontocerinae, 377
Odontomesa, 690, 709, 747
Odontomyia, 496, 541
Odontomyiina, 496
Odontonyx, 441
Oecetis, 349, 362
Oecetis immobilis, 65
Oecetis inconspicua, 65
Oecotelma, 482
Oedematops, 547
Oedenops, 545
Oedoparena, 485, 503, 542
Oemopteryx, 224, 242, 262
Oemopteryx contorta, 61
Olethreutidae, 387, 389, 392
Oleuthreutinae, 392
Oligia, 389, 398
Oligoneuria, 160
Oligoneuriidae, 129, 136, 147, 158, 160
Oligophlebodes, 347, 383
Oligophlebodes zelti, 64
Oligoplectrum, 343
oligopneustic tracheal system, 30
Oligostigmoides, 396
Oligostomis, 343, 380
Oligotricha, 343, 380
Oliveridia, 678, 703, 749
Omaliinae, 436
Omisus, 661, 730, 752
Onchocerca volvulus, 591
Oncopodura, 117
Oncopodurinae, 117
Onocosmoecus, 344, 374
Onocosmoecus unicolor, 64
Onychiuridae, 115, 123, 125
Onychiurus, 115, 125
Onychylis, 461, 472

Onyciurus dentatus, 113
open respiratory system, 40
Ophiogomphus, 192, 194, 206
Opisthognathous, 5
Opius, 474, 477, 483
Oplonaeschna, 185, 188, 207
Optioservus, 446, 448, 452, 470
Optioservus fastiditus, 66
Ora, 438, 469
Oravelia, 268, 278, 292
Orchelimum, 212, 216
Orchelimum bradleyi, 212
Orchesella, 119
Orcus, 206
order, 108
Ordobrevia, 448, 450, 470
Oreadomyia, 749
Oreodytes, 415, 422, 465
Oreogeton, 500, 541
Oreothalia, 541
Orimarga, 555, 561, 569
Ormosia, 553, 555, 565, 569
Ornithodes, 551, 561
Orohermes, 300, 308
Orohermes crepusculus, 52
Oroperla, 235, 249, 265
Oropsyche, 340, 350, 356
Orortium, 396
Orthacheta, 546
Orthacheta hirtipes, 68
Orthemis, 198, 204, 209
Orthocladiinae, 106, 638-644, 673, 693, 700, 738, 743-754
Orthocladiini, 738, 743, 748
Orthocladius, 678, 682, 700, 703, 709, 712, 714, 716, 720, 749
Orthocladius annectens, 687
Orthocladius consobrinus, 90
Orthocladius obumbratus, 67
Orthonevra, 548
Orthopodomyia, 577, 583, 590
Orthoptera, 1, 11, 108-111, 212
Orthorrhaphous, 496
Orthotrichia, 342, 362
Orthotrichiini, 342
Oryza, 397
Oryzophilus, 66
Osmobranchiae, 36
osmoregulation, 41
Osobenus, 235, 249, 265
Ostium bursae, 390
Ostrinia, 397
Ostrocerca, 224, 242, 246, 263
Ostrocerca albidipennis, 61
Oudemansia, 117, 125
Oulimnius, 448, 452, 470
Oxycera, 496, 541
Oxyelophila, 397
Oxyethira, 342, 359

P

Pachydiplax, 202, 204, 209
Pachydrus, 415, 419, 465
paddles, 516, 572
Paduniella, 339, 350, 383
Paduniellinae, 339
Pagastia, 67, 687, 716, 747
Pagastiella, 661, 730, 752
Palaeagapetus, 341, 359
Palaeodipteron, 66, 539
paleolimnological techniques, 94
Paleoptera, 109
Palmacorixa, 272, 295
palmate, 640
palmate process, 640
Palpada, 548
palpal lobes, 166
palpal setae, 166
Palpomyia, 493, 538
Palpomyiini, 66
palps, 5

palpus, 592
Paltothemis, 200, 204, 209
Palustra, 387
Panchaetoma, 121
Pantala, 202, 204, 209
Pantala flavescens, 48
Paraboreochlus, 638, 690, 693, 746
Parabyssodon, 627
Parac, 432
Paracanace, 542
Paracapnia, 228, 246, 264
Paracapnia angulata, 62
Paracentrobia, 483
Parachaetocladius, 673, 714, 749
Parachaetocladius abnobaeus, 67
Parachironomus, 666, 728, 730, 735, 752
Paracladius, 673, 720, 749
Paracladopelma, 663, 730, 735, 752
Paracloeodes, 135, 152, 160
Paracoenia, 508, 544
Paracricotopus, 682, 703, 712, 749
Paracymus, 429, 467
Paradelphomyia, 555, 563, 569
Paraglenanthe, 544
paraglossa, 640
Paragnetina, 228, 258, 264
Paragnetina media, 52, 62
Paragyractis, 106, 387
Parahelodon, 610
Parakiefferiella, 678, 714, 749
Paralauterborniella, 668, 730, 752
Paraleptophlebia, 35, 60, 84, 140, 154, 163
Paraleptophlebia volitans, 60
Paraleuctra, 219, 228, 249, 263
Paralimna, 545
Paralimnophyes, 749
Parameletus, 126, 132, 150, 159
Paramerina, 646, 649, 695, 746
Parametriocnemus, 682, 703, 706, 718, 749
paranal cells, 169
Paranemoura, 224, 242, 263
Paranura, 117
Paranurophorus, 119
Paranyctiophylax, 339, 380
Paraperla, 217, 231, 255, 266
Paraperlinae, 266
Paraphaenocladius, 682, 716, 750
Paraplea, 284, 294
Paraponyx, 387, 396
Paraponyx diminutalis, 387
Parapoynx, 387, 390, 396
Parapoynx maculalis, 387
paraprocts, 11, 167, 169
Parapsectra, 661, 723, 753
Parapsyche, 340, 356
Parapsyche apicalis, 63
Parapsyche cardis, 63
Parargyractis, 396
Parasimuliinae, 604
Parasimulium, 593, 601, 604, 606, 633-634
parasites, 76
Parasitica, 474, 476
Parasmittia, 750
Paratanytarsus, 658, 720, 723, 754
Paratendipes, 661, 728, 730, 752
Paratendipes albimanus, 67
Paratissa, 543
Paratrichocladius, 720, 750
Paratrissocladius, 706, 750
Paravelia, 268
Parochlus, 690, 693, 746
Paronellinae, 119
Parorthocladius, 687, 718, 750
Parthina, 348, 377
Parydra, 482-483, 485, 508, 544
Paskia, 35
Patagia, 389
Patasson, 483
Paucicalcaria, 342, 350, 359
pearl rows, 642
pecten epipharyngis, 636, 640
pecten hypopharyngis, 640
pectinate, 640
Pedicia, 549, 553, 561, 570

Pediciini, 549-550, 561
pedisulcus, 593
Pedomoecus, 347, 365
Pelastoneurus, 540
Pelecorhynchidae, 498, 509, 524, 530, 540
Pelenomus, 458, 472
Pelina, 508, 545
Pelocoris, 278, 294
Pelonomus, 444, 469
Pelopia, 746
Pelopiinae, 744
Peltodytes, 56-57, 65, 413, 463
Peltoperla, 219, 242, 262
Peltoperlidae, 61, 217-219, 239, 261-262
Pemphigus trehernei, 269
penellipse, 389
penes, 129
penis, 11
Pentacantha, 207
Pentacanthella, 119
Pentacora, 284, 296
Pentagenia, 129, 147, 158, 163
Pentaneura, 638, 649, 695, 746
Pentaneurella, 653
Pentaneurini, 638-639, 644, 738, 745
Percidae, 166
Perenthis, 458, 472
Pericoma, 493, 539
Perithemis, 198, 204, 209
Perithemis tenera, 61
Perlesta, 231, 258, 265
Perlesta placida, 62
Perlidae, 52, 62, 217-219, 239, 261, 264
Perlinae, 264
Perlinella, 231, 258, 265
Perlinodes, 235, 249, 265
Perlodidae, 36, 52, 62, 217-219, 239, 261, 265
Perlodinae, 249, 265
Perlomyia, 219, 228, 249, 263
permanent air store, 32, 33, 36
Petaluridae, 165, 171, 174, 188, 205-206
Peterson-type grab, 15, 17
petite Ponar grab, 16
Petrophila, 387, 390, 396, 482
Petrophila canadensis, 65
Petrophila confusalis, 65, 104, 387
Petrophila jaliscalis, 387
Phaenocarpa, 483
Phaenopria, 482
Phaenopsectra, 668, 726, 752
Phalacrocera, 550, 558, 561, 565, 568
Phalaenidae, 398
phallobase, 11
phallus, 11
Phanocelia, 346, 371
Phanocerus, 446, 450, 470
Phaonia, 546
pharate, 71
Phasganophora, 228, 264
Pherbecta, 547
Pherbellia, 68, 547
Philanisus plebeius, 54
Philarctus, 346, 374
Philocasca, 346, 368
Philocasca demita, 64
Philopotamidae, 44, 63, 101, 104, 310, 319, 326, 329, 338-339, 377
Philopotamus ludificatus, 104
Philorus, 487, 537
Philosepedon, 539
Philotelma, 544
Philygria, 546
Phoridae, 500, 509, 536, 546
Phryganea, 104, 343, 380
Phryganeidae, 64, 310, 315, 319, 329, 338, 342, 377
Phryganeinae, 343, 377
Phygadeuon, 482
Phyllogomphoides, 192, 194, 206
Phylloicus, 349, 351
Phyllolabis, 563
Phylocentropus, 310, 340, 380
physical gill, 33
physiological, 88

Phytobius, 33, 458, 473
Phytosus, 436, 467
phytotelmata, 572
Pictetiella, 237, 252, 265
piercers, 79
Piezosimulium, 601, 606, 633
Piezosimulium jeanninae, 591
Pilaria, 556, 563, 570
pinaculum, 389
Pinkham-Pearson Index, 91
Pisauridae, 482
Pistia, 387
Pistia stratiotes, 458
Placopsidella, 545
plankton net, 13
planktonic, 46, 75
plant breathers, 30, 32
planta, 389
plastic window screen, 40
plastron, 33
Plateumaris, 452
Plathemis, 198, 204, 209
Platycentropus, 346, 374
Platygymnopa, 545
Platyphylax, 345
Platysmittia, 750
Platyvelia, 268, 288, 292
Plecoptera, 1, 5, 11, 23, 32, 35-38, 40, 42, 48, 52, 61, 68-69, 82, 91-92, 94, 98-99, 101, 104-106, 109, 111, 217, 239
Plecopteracoluthus, 700, 703, 709
Pleidae, 32, 62, 267-268, 271, 284, 288, 294
Pleronarcys, 262
plesiomorphic, 98
pleura, 5, 167, 642
pleurites, 516
Pleurogyrus, 482
pleuston, 46
plexiglass, 40
plexiglass tubes, 15
Plhudsonia, 718, 750
Plumiperla, 235, 255, 266
plumose, 640
Plutomurus, 117
Pnigodes, 458, 473
Podmosta, 224, 246, 263
Podonominae, 636, 638, 640-641, 643, 690, 693, 735, 743, 746
Podonomini, 690, 738, 746
Podura, 125
Podura aquatica, 113
Poduridae, 115, 125
Poduromorpha, 115
Poecilographa, 547
Pogonocladius, 678, 714
Pogonognathellus, 117
Polycentropodidae, 63, 310, 313, 315, 319, 326, 333, 338-339, 380
Polycentropodinae, 339, 380
Polycentropus, 339, 380
Polycentropus centralis, 63
Polycentropus maculatus, 63
Polygonum, 397-398
Polymera, 556, 563, 570
Polymitarcyidae, 61, 127, 130, 148, 158, 163
Polynema, 483
Polypedilum, 666, 726, 728, 735, 753
Polypedilum nubifer, 90
Polypedilum vanderplanki, 69
Polyphaga, 65, 399, 401-402, 465
Polyplectropus, 339, 380
Polytrichophora, 545
POM, 85
Pompilidae, 477, 481-482
Ponar grab, 15-17
Pontamalota, 436
Pontederia, 390, 398
Pontomalota, 467
Pontomyia, 658
Populus, 392
Portable invertebrate box sampler, 13
Postelichus, 400, 444, 446, 469
posterior spine row(s), 642
postnodal crossveins, 169

postnotum, 517
postocular spots, 167
Potamanthidae, 129-130, 148, 158, 163
Potamanthus, 106, 129
Potamogeton, 75, 396-397, 546
Potamogeton pectinatus, 75
Potamogeton penitalis, 397
Potamyia, 340, 356
Potthastia, 687, 690, 716, 747
Prasocurini, 455
Prasocuris, 452, 455, 471
Pratanurida, 117
preanal segment, 638, 640
precorneal setae, 642
predators, 76, 79
prefrons, 642
prehalter, 517
premandible, 640
premandibles, 636
premental setae, 166
prementohypopharyngeal complex, 636, 640
prescutum, 517
preservation, 22
Prestwichia, 483
primary aquatic associates, 113
primary setae, 389
Priocnemus, 482
Prionocera, 551, 558, 565, 568
Prionocyphon, 438, 469
Prionoxystus, 392
Probezzia, 493, 538
proboscis, 517, 573
Procanace, 542
procerci, 638
procercus, 640
Procladiini, 639, 646, 745
Procladius, 646, 698, 700, 745
Proclinopyga, 541
Procloeon, 135, 150, 160
Proctotrupoidea, 474, 476-477, 482
Prodiamesa, 690, 709, 747
Prodiamesinae, 639, 641, 644, 690, 693, 700, 743, 747
Prodiamesini, 738
profundal zone, 47
prognathous, 5, 389, 517
Progomphus, 166, 192, 194, 206
Proisotoma, 121
projections, 516
Prokelisia marginata, 269
prolegs, 486, 592, 638, 640, 642
Prolimnophila, 563
Promoresia, 448, 452, 470
Promoresia elegans, 66
Pronoterus, 425, 465
pronotum, 167, 313
pronymph, 165
Prosimulium, 593, 604, 610, 633-634
Prosimulium mixtum/fuscum, 67
Prosimulium ursinum, 591
prosternal horn, 313
Prostoia, 224, 246, 263
Prostoia besametsa, 61
Prostoia similis, 61
Protachoutes, 117
Protanyderus, 539
Protanypini, 687, 738, 747
Protanypus, 638, 687, 714, 747
Protaphorura, 115
prothoracic, 389
prothorax, 5, 389, 517
Protochauliodes, 300, 304, 308
Protoneura, 183, 185, 210
Protoneuridae, 166, 171, 183, 210
Protoplasa, 539
Protoptila, 341, 351
Protoptilinae, 341
Psamathobledius, 436
Psammisotoma, 121
Psectrocladius, 674, 706, 709, 750
Psectrotanypus, 646, 698
Psectrotanypus(B), 744
Pselaphidae, 400
Psephenidae, 32, 36, 43, 56, 65, 77, 101, 104, 106, 399-402, 405, 410, 440, 461, 468, 482-483

Psephenivorus, 483
Psephenoides, 104
Psephenoididae, 402
Psephenus, 106, 440-441, 468, 482-483
Psephenus herricki, 43, 440
Psephenus montanus, 65
Psephenus falli, 56
Psephidonus, 436, 467
Pseudachorutes, 117
Pseudanteris, 482
Pseudanurida, 113, 117, 125
Pseudanurophorus, 119
Pseudiron, 45, 129, 136, 150, 160
Pseudiron centralis, 60
Pseudironidae, 60, 129, 136, 150, 158, 160
Pseudobourletiella, 125
Pseudocentroptiloides, 135, 150, 160
Pseudochironomini, 638, 640-641, 644, 743, 753
Pseudochironomus, 723, 730, 735, 753
Pseudocorixa, 295
Pseudodiamesa, 687, 716, 747
Pseudogoera, 348, 377
Pseudogoerinae, 377
Pseudohecamede, 545
Pseudohyadina, 546
Pseudokiefferiella, 687, 716, 747
Pseudolampsis guttata, 452
Pseudoleon, 202, 209
Pseudolimnophila, 549, 553, 555, 563, 570
pseudoradula, 640
Pseudorthocladius, 67, 673, 716, 750
Pseudosinella, 117, 125
Pseudosmittia, 674, 712, 750
Pseudostenophylacinae, 310, 345, 371
Pseudostenophylax, 345, 371
Pseudostenophylax edwardsi, 368
Psilocricotopus, 709
Psilometriocnemus, 687, 720, 750
Psilopa, 543
Psilopelmia, 618
Psilotanypus, 698
Psilotreta, 310, 348, 377
Psilozia, 627
Psorophora, 571-573, 577, 583, 590
Psychoda, 493, 538
Psychodidae, 32, 485-487, 493, 509, 515-517, 519, 530, 539
Psychoglypha, 346, 368
Psychomyia, 339, 383
Psychomyiidae, 44, 63, 106, 310, 315, 319, 326, 333, 338-339, 350, 383
Psychomyiinae, 339
Psychoronia, 346, 374
Ptenothrix, 121
Pteromalidae, 476-477, 481, 483
Pteromicra, 547
Pteronarcella, 217, 219, 239, 262
Pteronarcyidae, 5, 61, 217-219, 239, 261-262
Pteronarcys, 219, 239
Pteronarcys dorsata, 39, 61
Pteronarcys proteus, 61
Pteronarcys scotti, 61
pterostigma, 167
pterothorax, 167
Pterygota, 108
Ptiliidae, 400, 405, 408, 410
Ptilimnium, 473
ptilinal suture, 517
Ptilocolepinae, 329, 341, 359
Ptilodactylidae, 66, 400, 402, 405, 410, 441, 461, 471
Ptilomyia, 545
Ptilostomis, 343, 380
Ptychoptera, 487, 539
Ptychoptera lenis, 58
Ptychoptera lenis lenis, 66
Ptychopteridae, 32, 58, 66, 484-487, 509, 517, 519, 524, 539
pulvilli, 517
pupa, 109
pupal exuviae, 636, 641, 643
pupal mandibles, 313
puparium, 516
Pycnopsyche, 64, 69, 72, 82, 346, 368

Pycnopsyche guttifer, 38, 72, 82
Pycnopsyche lepida, 38
Pyralidae, 32, 56, 65, 387, 390, 392, 396, 482
Pyrausta, 397
Pyrausta penitalis, 387
Pyraustidae, 396
Pyraustinae, 397
Pyroderces, 392
Pyrrhalta, 452, 455, 472
Pyrrhalta nymphaeae, 455
Pyrrhosoma, 70

Q

quadrangle, 169
quadrat clipping, 15
quadratum, 210

R

radial crossvein, 11
radial planate, 169
radial sector, 167
radial sector vein, 11
radial vein, 11
radial-medial crossvein, 11
radius, 167
Radotanypus, 646, 698, 700, 745
Ramphocorixa, 272, 295
Ramphocorixa acuminata, 52
Ranatra, 33, 278, 294, 482-483
Ranatra montezuma, 62
Ranatrinae, 278
Ranunculus, 75, 471
Raphidocladius, 716
Raphidolabina, 553
Rapoportella, 117
Raptoheptagenia, 129, 136, 154, 161
Rasvena, 235, 255, 266
rearing methods, 21
recurved hooks, 642
regulation, 37
Remenus, 235, 252, 265
Removal substrates, 15, 17
Renocera, 503, 547
Reomyia, 695, 746
respiratory conformer, 38
respiratory filaments, 592
respiratory mechanisms, 39
respiratory pigments, 37
respiratory processes, 39
respiratory regulator, 38
respiratory system, 29
reverse vein, 169
Reynolds number, 43
Rhabdomastix, 556, 565, 570
Rhagadotarsinae, 293
Rhagadotarsus, 53
Rhagionidae, 530, 540
Rhagovelia, 271, 288, 292
Rhagoveliinae, 288
Rhamphomyia, 541
Rhantus, 418, 423, 465
Rhantus binotatus, 65
Rhatus, 415
Rheocricotopus, 673, 706, 709, 750
Rheopelopia, 653, 700, 746
Rheosmittia, 67, 673, 709, 750
Rheotanytarsus, 106, 658, 723, 754
Rheotanytarsus exiquus, 67
Rheumatobates, 271, 275, 293
Rhinonapaea, 544
Rhinoncus, 458, 473
Rhithrogena, 44, 106, 136, 154, 161
Rhithrogena semicolorata, 60
Rhizelmis, 446, 450, 470
Rhyacophila, 64, 309, 341, 383
Rhyacophila vao, 64
Rhyacophilidae, 64, 106, 309, 315, 319, 326, 329, 338, 341, 383

Rhyacophiloidea, 309, 313, 341
Rhysophora, 543
Rickera, 235, 252, 255, 266
Rioptila, 342
Robackia, 636, 663, 735, 753
Robackia demeijerei, 67
rock-outcrop sampler, 13
Roederiodes, 500, 541
Rossiana, 347, 365
rostrum, 268
Rupisalda, 284, 296
Rusekella, 117

S

Saetheria, 663, 735, 753
Saetheriella, 750
Sagittaria, 398, 472
Salda, 284, 296
Saldidae, 267-268, 271, 284, 288, 296
Saldinae, 284
Saldoida, 284, 296
Saldula, 288, 296
Salicornia, 436
Salina, 119, 125
Salix, 392, 398
Salmonidae, 166
Salmoperla, 235, 249, 265
Salpingidae, 400-401, 405, 410, 461, 468
Salticornia, 398
Salvalina, 458
sample preservation, 21
sampling design, 12, 19
sampling processing time, 22
Saprobic Index, 91
Saprobity, 87
Sarabandus, 438, 469
Sarcophagidae, 503, 536, 546
Sargus, 542
Saunderia, 750
scale, 640
scalloped, 639-640
scape, 389
Scapteriscus, 214
Scatella, 508, 544
Scatella thermarum, 42
Scathophagidae, 68, 484, 508, 536, 546
Scatophagidae, 546
Scatophila, 508, 544
Scelionidae, 474, 477, 481-482
Schaefferia, 117
Schema, 545
Schizoptera, 292
Schizopteridae, 267, 288, 292
Schloenobius, 482
Schoenobiinae, 397
Schoenobius, 397
Schoetella, 115
scientific name, 108
Sciomyzidae, 23, 58, 68, 482-483, 485-486, 503, 509, 515, 536, 547
Scirpus, 389, 397-398, 546
Scirtes, 438, 469
Scirtes tibialis, 65
Scirtidae, 63, 400, 402, 405, 410, 438, 461, 469
sclerite, 516
Scleroprocta, 553
sclerotized teeth, 592
Scopeumatidae, 546
scrapers, 76, 79
SCUBA diving, 14, 16-17
scutellum, 517
scutum, 517
Scydmaenidae, 400
secondary setae, 389
sectors of the arculus, 167
Seira, 117
Semeodes albiguttalis, 387
Semicerura, 117
Sensillanura, 117
Sensiphorura, 115
sentinel organisms, 88

Sepedomerus, 503, 547
Sepedon, 58, 68, 482-483, 503, 547
Sequential Comparison Index, 96
Sergentia, 668, 726, 753
Sericomyia, 548
Sericostomatidae, 64, 310, 315, 322, 329, 333, 338, 347, 383
Sericostriata, 347, 383
Serratella, 60, 127, 140, 156, 162
Serromyia, 493, 538
Setacera, 508, 544
setae, 572
setal brushes, 592
setal warts, 315
Setodes, 349, 362
Setvena, 235, 249, 265
shagreen, 642
Shannon's Index, 91
Shannonomyia, 563
Shewellomyia, 618
Shipsa, 224, 246, 263
Shipsa rotunda, 61
shovel sampler, 18
shredders, 76, 79
shurtleffi, 208
SI seta, 640
Sialidae, 52, 63, 298-299, 304, 308, 483
Sialis, 63, 299, 308, 483
Sialis aequalis, 63
Sialis dreisbachi, 63
Sialis rotunda, 52
Sierraperla, 219, 239, 262
sieves, 22
sieving, 22
Sigara, 275, 295
SII seta, 640
Silo, 68
Silvius, 498, 542
similarity indices, 92
simple, 5
simplex gill, 36
Simuliidae, 1, 23, 32, 35-36, 40, 44, 58, 67, 69, 71, 264-266, 484-486, 491, 509, 517, 530, 539, 541, 546, 591, 634
Simuliinae, 604
Simulium, 48, 58, 69, 106, 597, 604, 611, 627, 633-634
Simulium damnosum, 591
Simulium jenningsi, 67
Simyra, 389-390, 398
single core sampler, 17
single corer, 14
Siphlonisca, 132, 150, 159
Siphlonuridae, 75, 129, 132, 150, 159
Siphlonurus, 132, 150, 159
Siphlonurus occidentalis, 35
Siphloplecton, 139, 154, 159
Siphon, 572
Sisridivora, 477, 483
Sisyra, 304, 308
Sisyridae, 32, 63, 110, 298-299, 304, 308, 483
skaters, 75
Skutzia, 754
Skwala, 237, 249, 265
Slylogomphus, 206
Smicridea, 340, 356
Sminthuridae, 121, 123, 125
Sminthurides, 113, 121, 123, 125
Sminthurinus, 123
Sminthurus, 123, 125
Smittia, 687, 712, 750
Soliperla, 219, 242, 262
Somatochlora, 197-198, 208
somites, 5
Sortosa, 339
Soyedina, 224, 242, 263
Soyedina carolinesis, 61
Soyedina vallicularia, 61
Spanganium, 398
Spanglerogyrus, 411, 462
Sparganium, 211, 398
Spartina, 269
Spazipbora, 547
Specularis, 206

Sperchiidae, 402, 427
Sperchius, 427
Sperchopsis, 429, 432, 467
Sphaeridiinae, 427, 432
Sphaeriidae, 32, 402, 465, 503
Sphaerius, 403, 408, 465
Sphaeroceridae, 485
Sphaeromias, 493, 538
Sphaeromiini, 66
Sphagnophylax, 346, 374
Sphagnophylax meiops, 64
Sphagnum, 371, 539
Sphingidae, 387, 389-390, 392
Sphyrotheca, 123
Spilogona, 546
Spilomyia, 548
Spinadis, 129, 161
spine groups, 642
spinneret, 389
spiracles, 30, 516, 572
spiracular disk, 550
spiracular gills, 32, 35
spiracular ring, 640
Spodoptera, 389
Spongilla, 308
sprawlers, 75
squama, 517
Stachanorema, 119
Stackelbergina, 638-639, 687, 720, 750
Stactobiella, 342, 359
Stactobiini, 342
standpipe corer, 13
Staphylinidae, 400-401, 403, 408, 436, 461, 467
stationary screen trap, 14
Stegopterna, 597, 601, 611, 634
Steinovelia, 268, 288, 292
Stelechomyia, 668, 735, 753
Stemmata, 313, 389
Stempellina, 658, 723, 754
Stempellinella, 658, 723, 754
Stenacris, 214, 216
Stenacron, 139, 154, 161
Stenacron interpunctatum, 60, 126
Stenelmis, 33, 446, 448, 450, 471
Stenelmis crenata, 66
Stenelmis nr. bicarinata, 66
Stenelmis sexlineata, 56, 66
Steninae, 436
Stenochironomus, 643, 668, 735, 753
Stenocolus, 402, 441, 471
Stenogastrura, 115
Stenonema, 139, 154, 161
Stenonema modestum, 60
Stenonema vicarium, 60
Stenopelmus, 458, 473
Stenophylacini, 368, 371
Stenophylax, 106
Stenotabanus, 498, 542
Stenus, 436, 468
Steremnius, 458, 473
Sterna, 167, 642
Sternorrhyncha, 269
sternum, 5
Stethophyma, 214, 216
Stictochironomus, 84, 661, 728, 753
Stictotarsus, 415, 419, 422, 465
Stictotarusus, 415
Stigmella, 397
Stigmellidae, 397
Stilobezzia, 493, 538
Stilocladius, 682, 718, 750
Stilpon, 541
stock biomass, 86
stovepipe sampler, 13
Stratiomyia, 542
Stratiomyidae, 32, 484-486, 496, 509, 515, 524, 530, 541
Stratiomys, 46, 496, 542
striations, 640
Strophopteryx, 224, 242, 262
Strophopteryx limata, 61
Stygoparnus, 399, 444, 469
Stygoporus, 399
Stygoporus oregonensis, 415

style, 516
styli, 11
styliger plate, 129
Stylogomphus, 192, 196
Stylurus, 169, 194, 196, 207
subanal plate, 129
subcosta, 167
subcostal vein, 11
subfamilies, 108
subgenital plate, 129
subimago, 99, 109, 126
Sublettea, 658, 723, 754
Sublettiella, 750
subprimary setae, 389
subsampling, 22
subtriangle, 169
suction sampler, 15-16
suction samplers, 13-14, 16
suctorial discs, 486
Sulcarius, 482
superbus, 209
superfamilies, 108
superior appendage, 167
supersaturation, 29
supertriangle, 169
Suphis, 425, 465
Suphisellus, 425, 465
Suphisellus puncticollis, 65
supraanal setae, 638, 640
supracoxal processes, 166
Suragina, 498, 540
Surber sampler, 13, 18, 19
surface film, 46
surface film sampler, 14
suspension, 40
Susulus, 235, 249, 265
suture, 5
Suwal, 266
Suwallia, 231, 255
Sweltsa, 62, 231, 255, 266
Sweltsa onkos, 62
swimmers, 75
Symbiocladius, 674, 712, 750
Sympetrum, 164-165, 200, 202, 204, 209
Symphitopsyche, 340, 356
Symplocarpus, 398
Symposiocladius, 484, 678, 720
Sympotthastia, 690, 700, 747
Sympycnus, 540
Synclita, 387, 390, 396
Syndiamesa, 690, 716, 747
Synendotendipes, 668, 735, 753
Synorthocladius, 687, 714, 750
synthorax, 166
Syrphidae, 32, 37, 40, 485-486, 500, 509, 515, 530, 547
Systenus, 540

T

T-sampler, 13
Tabanidae, 23, 32, 58, 67, 481-482, 484-486, 496, 509, 515-516, 524, 530, 542
Tabanus, 482, 500, 542
Tabanus atratus, 58, 67
Tabanus fumipennis, 67
Tachopteryx, 188, 206
Tachytrechus, 540
Taeniogaster, 207
Taenionema, 224, 242, 262
Taenionema atlanticum, 61
Taeniopterygidae, 36, 61, 69, 217-219, 239, 261, 262
Taeniopteryginae, 262
Taeniopteryx, 36, 106, 219, 242, 262
Taeniopteryx nivalis, 61
Taeniorhynchus, 33
Tafallia, 115
Tallaperla, 72, 219, 242, 262
Tallaperla maria, 61
Tanychela, 474, 476, 482
Tanyderidae, 32, 485-486, 491, 509, 515, 519, 524, 539

Tanypodinae, 636, 638-643, 693, 695, 735, 743-744
Tanypodini, 639, 644, 738, 746
Tanypteryx, 188, 206
Tanypus, 644, 700, 746
Tanysphyrus, 461, 473
Tanytarsini, 638-640, 642, 644, 653, 658, 738, 743, 753
Tanytarsus, 47, 159, 661, 723, 754
Tanytarsus balteatus, 104
Tanytarsus dissimilis, 38
Tarnetrum, 209
tarsal claws, 5, 167
tarsus, 5, 167
Tascobia, 342
Tasiocera, 553, 563
Tauriphila, 200, 205, 209
Tauriphila argo, 164
taxon, 108
tegmina, 212
tegulae, 390
tegumen, 390
Teleallagma, 210
Teleas, 482
Telebasis, 179, 183, 211
Telebasis salva, 61
Telenomus, 482
Telmatogeton, 744
Telmatogetoninae, 738, 743-744
Telmatopelopia, 653, 698
Telmatoscopus, 493, 539
Telmaturgus, 540
Teloleuca, 284, 296
Telopelopia, 653, 695, 746
temperature, 40, 42
temporary, 36
temporary air store, 32
temporary air stores, 33
tenaculum, 115
Tenagobia, 271, 295
Tendipedidae, 744
Tendipes, 751
Tenebrionidae, 405, 410
tentorial arms, 486
terga, 642
tergite, 516
tergum, 5
terminal filament, 129
tertiary aquatic associates, 113
Tetanocera, 68, 483, 503, 547
Tetanocera loewi, 68
Tetanoceratidae, 547
Tetanoceridae, 547
Tethymyia, 743, 747
Tetracanthella, 119
Tetragoneuria, 50, 197-198, 208
Tetrastichus, 483
Tetrigidae, 212, 214, 216
Tettigoniidae, 212, 214, 216
Teucholabis, 563
Teuchophorus, 540
Thalasselephas, 458, 473
Thalassomya, 744
Thalassosmittia, 738, 743, 750
Thalassotrechus, 462
Thaumalea, 540
Thaumaleidae, 484-486, 491, 509, 519, 530, 540
Thaumastoptera, 556, 561, 570
Thecaphora, 207
Theliopsyche, 344, 362
Theobaldia, 590
Therascopus, 482
thermal equilibrium hypothesis, 85
Thermonectini, 418, 423
Thermonectus, 415, 419, 425, 465
Thermonectus ornaticollis, 65
Thienemannia, 718, 750
Thienemanniella, 673, 700, 748
Thienemannimyia, 649, 653, 700, 746
Thienemanniola, 658
Thinobius, 436, 467
Thinophilus, 540
Thinopinus, 436, 467
Thinusa, 436, 468

thoracic comb, 642
thoracic horn, 642
thoracic spiracle, 516
thorax, 5, 572, 640
Thoreyi, 206
Thoron, 482
Thraulodes, 140, 156, 163
Threticus, 539
Throscinus, 441, 444, 471
tibia, 5, 167
tibial keels, 167
tibial spurs, 315
Timpanoga, 130, 140, 156, 162
Tinodes, 339, 383
Tiphodytes, 482
Tipula, 11, 78, 82, 549, 551, 558, 565, 568
Tipula abdominalis, 66
Tipula sacra, 58, 66, 549
Tipulidae, 11, 23, 32, 40, 58, 66, 71, 82, 484-487, 509, 515, 517, 519, 524, 540, 546, 549, 568
Tipulinae, 549-551, 558, 568
Tipulini, 549
Tokunagaia, 712, 718, 750
Tomocerina, 117
Tomocerinae, 117
Tomocerus, 117
Tomolonus, 117
Torridincolidae, 32, 403, 408
Tortopus, 126-127, 147, 158, 163
Tortricidae, 387, 389-390, 392, 398
toxicants, 38-39
toxicity tests, 94-95
Toxorhina, 563
Toxorhynchites, 572-573, 577, 583, 590
tracheae, 30
tracheal, 29
tracheal gills, 32, 35-36, 313, 389
tracheal system, 30
tracheoles, 30
Tramea, 202, 205, 209
transitory, 113
transitory neuston, 113
Trapezostigma, 209
Traverella, 139, 156, 163
Trentonius, 339
Trepobates, 278, 293, 482
Trepobatinae, 275, 293
Triacanthagyna, 185, 188, 207
Triacanthella, 115
Triaenodes, 104, 349, 362
Triaenodes injusta, 65
triangle, 169
tribe, 108
Tribelos, 666, 730, 753
Trichestema, 482
Trichobothria, 268
Trichochilus, 750
Trichoclinocera, 500, 541
Trichocorixa, 272, 295
Trichocorixa reticulata, 63
Trichocryptus, 482
Trichogramma, 476, 483
Trichogrammatidae, 476-477, 481, 483
trichomes, 592
Trichopria, 474, 477, 482
Trichoptera, 1, 11, 20, 22-23, 32, 35-37, 40, 42, 47-48, 54, 63, 69, 75-76, 80-82, 91-92, 94, 101, 104-106, 108-112, 309, 350, 390, 744-745
Trichotanypus, 690, 693, 746
Trichotendipes, 726
Trichothaumalea, 540
Tricorythidae, 48, 60, 126-127, 130, 150, 158, 162
Tricorythodes, 35, 126, 140, 156, 162
Tricorythodes allectus, 60
Tricorythodes atratus, 60
Tricorythodes dimorphus, 60
Tricorythodes minutus, 60
Tridactylidae, 212, 214, 216
trifida, 207
trigonal planate, 169
Trigonidiinae, 216
Trimerina, 543
Trimerinoides, 543
Trimicra, 553, 569

Triogma, 550, 558, 568
Trissopelopia, 649, 695, 746
Triznaka, 231, 255, 266
Triznaka signata, 62
trochanter, 5, 167
trochantin, 313
Trochopus, 271, 288, 292
Tropisternus, 70, 427, 429, 432, 467
Tropisternus ellipticus, 65
truckeealis, 387
trumpets, 572
Trypanosoma, 591
tubercles, 516, 592
Tullbergia, 115
Tvetenia, 682, 718, 750
Twinnia, 593, 601, 606, 633-634
tympanal organs, 390
Typha, 33, 75, 389-390, 397-398
Typhlogastrura, 117
Typopsilopa, 545

U

u-shaped glass burrows, 40
Uenoidae, 64, 310, 313, 322, 326, 329, 338, 347, 383
Ula, 561
Ulomorpha, 556, 563, 570
uncus, 390
Undulambia, 396
unguiculus, 115
unguis, 115
uniordinal, 389
uniserial, 389
uniserial fringe, 642
Unniella, 682, 706, 750
unstable substrates, 44
Uranotaenia, 571, 573, 577, 583, 590
urogomphi, 402
Usingeriessa, 397
Usingerina, 294
Utacapnia, 217-218, 228, 246, 264
Utaperla, 231, 255, 266
Uvarus, 418-419, 465
Uzelia, 119

V

Vallisneria, 396
valvae, 11, 390
Vatellini, 415, 419
veins, 11
Veliidae, 46, 268, 271, 288, 292
Veliinae, 288
ventilation, 37-38
ventral apotome, 313
ventral tubules, 638, 640
ventromental plate, 640
ventromental plate, 638, 640
ventromentum, 638
verruca, 389
vertex, 5, 167, 516
Vesicephalus, 121
Vespoidea, 477, 482
Viehoperla, 219, 242, 262
Visoka, 224, 242, 263
Vivacricotopus, 703, 750

W

water column sampler, 15, 16
water current generation, 40
Weberacantha, 121
Whitneyomyia, 500
Wiedemannia, 500, 541
Wilding sampler, 13-17

Willemia, 115
Williamsonia, 197, 208
Willowsia, 119
window traps, 17
wing sheaths, 516, 642
wings, 5
Wirthiella, 668
wood-associated insects, 45
Wormaldia, 339, 377
Wormaldia copiosa, 104
Wormaldia moestus, 63
Wyeomyia, 577, 583, 590
Wyeomyia smithii, 42

X

Xenelmis, 446, 471
Xenochironomus, 666, 735, 753
Xenopelopia, 653, 695, 746
Xenylla, 115
Xenyllodes, 117
Xestochironomus, 643, 668, 735, 753
Xiphocentron, 319, 339, 350
Xiphocentronidae, 310, 319, 326, 333, 338-339, 350
Xylotopus, 484, 638-639, 687, 709, 750
Xylotopus par, 67

Y

Yamatotipula, 549
Ylodes, 349, 362
Ylodes frontalis, 65
Yoraperla, 219, 239
Yoroperla, 262
Yphria, 342, 377
Yphriinae, 342, 377
Yugus, 237, 252, 266

Z

Zaitzevia, 448, 450, 471
Zalutschia, 674, 703, 706, 750
Zapada, 82, 224, 242, 263
Zapada cinctipes, 37, 61
Zapada columbiana, 61
Zavrelia, 658, 723, 754
Zavreliella, 668, 728, 753
Zavrelimyia, 649, 695, 746
Zealeuctra, 62, 218-219, 228, 249, 263
Zeros, 546
Zeugloptera, 101
Zeuximyia, 498
Zon, 183
Zoniagrion, 179, 183, 211
Zoraena, 207
Zumatrichia, 342, 359
Zygoptera, 35-36, 61, 164-167, 169, 205, 209